BERGEY'S MANUAL® OF
Systematic Bacteriology
Volume 2

BERGEY'S MANUAL® OF
Systematic Bacteriology
Volume 2

PETER H. A. SNEATH
EDITOR, VOLUME 2

NICHOLAS S. MAIR
ASSOCIATE EDITOR, VOLUME 2

M. ELISABETH SHARPE
ASSOCIATE EDITOR, VOLUME 2

JOHN G. HOLT
EDITOR-IN-CHIEF

WITH CONTRIBUTIONS FROM
61 COLLEAGUES

WILLIAMS & WILKINS
Baltimore • Hong Kong • London • Sydney

Editor: John P. Butler
Associate Editor: Victoria M. Vaughn
Copy Editor: Caral Shields Nolley
Design: JoAnne Janowiak
Illustration Planning: Lorraine Wrzosek
Production: Raymond E. Reter

Copyright © 1986
Williams & Wilkins
428 East Preston Street
Baltimore, MD 21202, U.S.A.

Printed in the United States of America

Library of Congress Cataloging in Publication Data
(Revised for volume 2)

Main entry under title:

Bergey's manual of systematic bacteriology.

   Based on: Bergey's manual of determinative bacteriology.
   Vol. 2: Peter H. A. Sneath, Editor; N. S. Mair, E. Sharpe, associate editors.
   Includes bibliographies and indexes.
   1. Bacteriology—Classification—Collected works. I. Bergey, D. H. (David Hen-
dricks), 1860–1937. II. Krieg, Noel R. III. Holt, John G. IV. Bergey's manual of
determinative bacteriology. [DNLM: 1. Bacteriology—Terminology. 2. Bacteria—
Classification. QW 4 B832m]
QR81.B46 1984   589.9′0012   82-21760
ISBN 0-683-04108-8 (v. 1)
ISBN 0-683-07893-3 (v. 2)

Contributors

N. S. Agre
Institute of Biochemistry and Physiology of Microorganisms, USSR Academy of Sciences, Poustchino-on-the-Oka, Moscow Region 142292, USSR

Tom Bergan
University of Oslo, Department of Microbiology, PO Box 1108, BLINDERN, Oslo, 3, Norway

R. C. W. Berkeley
Department of Microbiology, Medical School, University Walk, Bristol BS8 1TD, England

George H. Bowden
Department of Oral Biology, University of Manitoba, Winnipeg, Manitoba, Canada R3E 0W3

J. F. Bradbury
Commonwealth Mycological Institute, Ferry Lane, Kew, Richmond, Surrey TW9 3AF, England

B. W. Brooks
Department of Veterinary Microbiology, University of Guelph, Guelph, Ontario, Canada

Marvin P. Bryant
Department of Dairy Science, University of Illinois, Urbana, Illinois 61801 USA

L. Leon Campbell
Provost Vice-President for Academic Affairs, 104 Hullihen Hall, University of Delaware, Newark, Delaware 19711 USA

Ercole Canale-Parola
Department of Microbiology, University of Massachusetts, Amherst, Massachusetts 01002 USA

Lester E. Casida, Jr.
Pennsylvania State University, 101 S. Frear Laboratory, Microbiology, University Park, Pennsylvania 16802 USA

Elizabeth P. Cato
Department of Anaerobic Microbiology, Virginia Polytechnic Institute and State University, Blacksburg, Virginia 24061 USA

D. Claus
Deutsche Sammlung von Mikroorganismen, D-3400 Göttingen, Grisebachstrasse 8, Federal Republic of Germany

M. D. Collins
NCDO, National Institute for Research in Dairying, Shinfield, Reading RG2 9AT, England

W. H. J. Crombach
Laboratory of Bacteriology, St. Maartens Gasthuis, Tegelseweg 210, 5912 BL VENLO, The Netherlands

C. S. Cummins
Department of Anaerobic Microbiology, Virginia Polytechnic Institute and State University, Blacksburg, Virginia 24061 USA

James B. Evans
Department of Microbiology, North Carolina State University, Box 5476, Raleigh, North Carolina 27650 USA

L. I. Evtushenko
Institute of Biochemistry and Physiology of Microorganisms, USSR Academy of Sciences, Poustchino-on-the-Oka, Moscow Region 142292 USSR

Fatma Fahmy
Faculty of Science, Botany Department, Moharrem Bey, Alexandria, Egypt

Sydney M. Finegold
Chief, Infectious Disease Section 691/111F, Veterans Administration, Wadsworth Medical Center, Wilshire and Sawtelle Blvds., Los Angeles, California 90073 USA

J. L. Fryer
Department of Microbiology, Oregon State University, Corvallis, Oregon 97331 USA

Ellen I. Garvie
Mullach A'Chnuic, 1, Conordon, Braes Portree, Isle of Skye IV51 9LH, Scotland

W. Lance George
Veterans Administration, Wadsworth Medical Center, Wilshire and Sawtelle Blvds., Los Angeles, California 90073 USA

Mary Ann Gerencser
Department of Microbiology, West Virginia University Medical Center, Morgantown, West Virginia USA

Norman E. Gibbons (Deceased)
64 Fuller St., Ottawa, Ontario, Canada

Thomas Gibson (Deceased)
The Edinburgh School of Agriculture, West Mains Road, Edinburgh, EH9 3JG, Scotland

Michael Goodfellow
Department of Microbiology, The Medical School, University of Newcastle-upon-Tyne, Newcastle-upon-Tyne NE1 7RU, England

J. R. Greenwood
Orange County Public Health Laboratory, 1729 W. 17th St., Santa Ana, California 92706 USA

Jeremy M. Hardie
Department of Oral Microbiology, Turner Building, The London Hospital Medical College, Turner St., London E1 2AD, England

Aino Henssen
Fachbereich Biologie der Philipps-Universität Marburg, Botanik, 3550 Marburg/Lahn, Lahnberge, Federal Republic of Germany

John G. Holt
Department of Microbiology, 205 Sciences I, Iowa State University, Ames, Iowa 50011 USA

John L. Johnson
Department of Anaerobic Microbiology, Virginia Polytechnic Institute and State University, Blacksburg, Virginia 24061 USA

Dorothy Jones
Department of Microbiology, The Medical School, University of Leicester, Leicester LE1 7RH, England

L. V. Kalakoutskii
Institute of Biochemistry and Physiology of Microorganisms, USSR Academy of Sciences, Poustchino-on-the-Oka, Moscow Region 142292 USSR

Otto Kandler
Botanisches Institüt der Universität München, Menzinger Str. 67, D-8000 München 19, Federal Republic of Germany

Ronald M. Keddie
Department of Microbiology, The University London Road, Reading RG1 5BB, England

Wesley E. Kloos
Department of Genetics, North Carolina State University, Box 5487, Raleigh, North Carolina 27650 USA

Miloslav Kocur
Czechoslovak Collection of Microorganisms, J. E. Purkyne University, 662 43 Brno, Czechoslovakia

Kazuo Komagata
Institute of Applied Microbiology, The University of Tokyo, Bunkyo-ku, Tokyo, Japan

Noel R. Krieg
Department of Biology, Virginia Polytechnic Institute and State University, Blacksburg, Virginia 24061 USA

George P. Kubica
Laboratory Training and Consultation Division, Centers for Disease Control, Atlanta, Georgia 30333 USA

J. Lacey
Plant Pathology Department, Rothamstead Experimental Station, Harpenden, Herts. AL5 2JQ, England

Hubert A. Lechevalier
Waksman Institute of Microbiology, State University of New Jersey, Rutgers, PO Box 759, Piscataway, New Jersey 08854-0759 USA

Mary P. Lechevalier
Waksman Institute of Microbiology, State University of New Jersey, Rutgers, PO Box 759, Piscataway, New Jersey 08854-0759 USA

Nicholas S. Mair
Department of Microbiology, The Medical School, University of Leicester, Leicester LE1 7RH, England

Lillian V. Holdeman Moore
Department of Anaerobic Microbiology, Virginia Polytechnic Institute and State University, Blacksburg, Virginia 24061 USA

W. E. C. Moore
Department of Anaerobic Microbiology, Virginia Polytechnic Institute and State University, Blacksburg, Virginia 24061 USA

J. Orvin Mundt (Deceased)
Department of Microbiology, University of Tennessee, Knoxville, Tennessee 37996 USA

R. G. E. Murray
Department of Bacteriology and Immunology, University of Western Ontario, London, Ontario N6A 5C1, Canada

M. J. Pickett
Department of Microbiology, University of California, Los Angeles, California 90024 USA

Helmut Prauser
Institute of Microbiology and Experimental Therapy, Beutenbergstrasse 11, 69 - Jena, German Democratic Republic

Alice Reyn
Vesterled 12, 2100 København 0, Denmark

Jiri Rotta
Department of Epidemiology and Microbiology, Institute of Hygiene and Epidemiology, Prague, Czechoslovakia

J. E. Sanders
Department of Microbiology, Oregon State University, Corvallis, Oregon 97331 USA

Vittorio Scardovi
Universita Degli Studi di Bologna, Facolta di Scienze Agrarie, Istituto di Microbiologia Agraria, Via Filippo Re, 6, Bologna, Italy

Klaus P. Schaal
Institut für Med. Mikrobiologie, u. Immunologie der Universität Bonn, Sigmund-Freud-Strasse 25, 5300 Bonn 1, Federal Republic of Germany

Karl Heinz Schleifer
Lehrstuhl für Mikrobiologie, Institut Botanik und Mikrobiologie, Technische Universität München, Arcisstrasse 21, D-8000 München 2, Federal Republic of Germany

H. P. R. Seeliger
Institut für Hygiene und Mikrobiologie der Universität Wurzburg, Josef-Schneider-Strasse 2, Bau 17, 8700-Wurzburg, Federal Republic of Germany

M. Elisabeth Sharpe
National Institute for Research in Dairying, University of Reading, Shinfield, Reading RG2 9AT, England

S. Shaw
Central Laboratory, Watney Mann and Truman Brewers Ltd., 14 Mortlake High Street, London SW14 8JD, England

Rivers Singleton, Jr.
School of Life and Health Sciences, University of Delaware, Newark, Delaware 19711 USA

Peter H. A. Sneath
Department of Microbiology, The Medical School, University of Leicester, Leicester LE1 7RH, England

Erko Stackebrandt
Lehrstuhl für Allgemeine Mikrobiologie, Christian-Albrechts-Universität, Olshausenstr. 40, Biologiezentrum, D-2300 Kiel, Federal Republic of Germany

James T. Staley
Department of Microbiology, University of Washington, Seattle, Washington 98195 USA

Ken-Ichiro Suzuki
Japan Collection of Microorganisms, RIKEN, Wako-shi, Saitama 351, Japan

Ralph S. Tanner
Rohm and Haas Company, Research Laboratories, 727 Norristown Road, Spring House, Pennsylvania 19477 USA

William C. Trentini
Box 984, Sackville, New Brunswick E0A 3C0, Canada

Lawrence G. Wayne
Tuberculosis Research Laboratory, Veterans Administration Hospital, 5901 East Seventh Street, Long Beach, California 90822 USA

Norbert Weiss
Deutsche Sammlung von Mikroorganismen, Menzinger Strasse 67, D-8000 München 19, Federal Republic of Germany

Jürgen K. W. Wiegel
Department of Microbiology, The University of Georgia, Athens, Georgia 30602 USA

Advisory Committee Members

The Board of Trustees is grateful to all who served on the Advisory Committees and assisted materially in the preparation of this edition of the *Manual.* Chairpersons of committees are indicated by an asterisk.

1. *Micrococci and allies:* Wesley E. Kloos, M. Kocur, R. G. E. Murray,* G. Pulverer, K. H. Schleifer

2. *Streptococci and allies:* J. M. Hardie, L. V. H. Moore, the late J. O. Mundt,* M. T. Parker

3. *Endospore forming bacteria:* D. Claus, L. V. H. Moore,* J. R. Norris, T. Willis

4. *Regular nonsporing rods:* F. Gasser,* Dorothy Jones, O. Kandler, W. E. C. Moore, M. Rogosa, M. Elizabeth Sharpe

5. *Irregular nonsporing rods:* C. S. Cummins, D. Jones,* R. M. Keddie, K. Komagata, K. P. Schaal

6. *Mycobacteria and nocardioform bacteria:* M. Goodfellow, M. P. Lechevalier, H. Prauser, L. G. Wayne*

In addition, the Board extends special thanks to S. T. Williams for his invaluable help.

Preface to First Edition of Bergey's Manual® of Systematic Bacteriology, Volume 2

Many microbiologists advised the Trust that a new edition of the *Manual* was urgently needed. Of great concern to us was the steadily increasing time interval between editions; this interval reached a maximum of 17 years between the seventh and eighth editions. To be useful the *Manual* must reflect relatively recent information; a new edition is soon dated or obsolete in parts because of the nearly exponential rate at which new information accumulates. A new approach to publication was needed, and from this conviction came our plan to publish the *Manual* as a sequence of four subvolumes concerned with systematic bacteriology as it applies to taxonomy. The four subvolumes are divided roughly as follows: (a) the Gram-negatives of general, medical or industrial importance; (b) the Gram-positives other than actinomycetes; (c) the archaeobacteria, cyanobacteria and remaining Gram-negatives; and (d) the actinomycetes. The Trust believed that more attention and care could be given to preparation of the various descriptions within each subvolume, and also that each subvolume could be prepared, published, and revised as the area demanded, more rapidly than could be the case if the *Manual* were to remain as a single, comprehensive volume as in the past. Moreover, microbiologists would have the option of purchasing only that particular subvolume containing the organisms in which they were interested.

The Trust also believed that the scope of the *Manual* needed to be expanded to include more information of importance for systematic bacteriology and bring together information dealing with ecology, enrichment and isolation, descriptions of species and their determinative characters, maintenance and preservation, all focused on the illumination of bacterial taxonomy. To reflect this change in scope, the title of the *Manual* was changed and the primary publication becomes *Bergey's Manual® of Systematic Bacteriology*. This contains not only determinative material such as diagnostic keys and tables useful for identification, but also all of the detailed descriptive information and taxonomic comments. Upon completion of each subvolume, the purely determinative information will be assembled for eventual incorporation into a much smaller publication which will continue the original name of the *Manual*, *Bergey's Manual® of Determinative Bacteriology*, which will be a similar but improved version of the present *Shorter Bergey's Manual®*

So, in the end there will be two publications, one systematic and one determinative in character.

An important task of the Trust was to decide which genera should be covered in the first and subsequent subvolumes. We were assisted in this decision by the recommendations of our Advisory Committees, composed of prominent taxonomic authorities to whom we are most grateful. Authors were chosen on the basis of constant surveillance of the literature of bacterial systematics and by recommendations from our Advisory Committees.

The activation of the 1976 Code had introduced some novel problems. We decided to include not only those genera that had been published in the Approved Lists of Bacterial Names in January 1980 or that had been subsequently validly published, but also certain genera whose names had no current standing in nomenclature. We also decided to include descriptions of certain organisms which had no formal taxonomic nomenclature, such as the endosymbionts of insects. Our goal was to omit no important group of cultivated bacteria and also to stimulate taxonomic research on "neglected" groups and on some groups of undoubted bacteria that have not yet been cultivated and subjected to conventional studies.

Some readers will note the consistent use of the stem -var instead of -type in words such as biovar, serovar and pathovar. This is in keeping with the recommendations of the Bacteriological Code and was done against the wishes of some of the authors.

We have deleted much of the synonymy of scientific names which was contained in past editions. The adoption of the new starting date of January 1, 1980 and publication of the Approved Lists of Bacterial Names has made mention of past synonymy obsolete. We have included synonyms of a name only if they have been published since the new starting date, or if they were also on the Approved Lists and, in rare cases, if the mention of an old name would help readers associate the organism with a clinical problem. If the reader is interested in tracing the history of a name we suggest he or she consult past editions of the *Manual* or the *Index Bergeyana* and its *Supplement*. In citations of names we have used the abbreviation *AL* to denote the inclusion of the name on the Approved Lists of Bacterial Names and *VP* to show the name has been validly published.

In the matter of citation of the *Manual* in the scientific literature, we again stress the fact that the *Manual* is a collection of authored chapters and the citation should refer to the author, the chapter title and its inclusive pages, not the Editors.

To all contributors, the sincere thanks of the Trust is due; the Editors are especially grateful for the good grace with which the authors accepted comments, criticisms and editing of their manuscripts. It is only because of the voluntary and dedicated efforts of these authors that the *Manual* can continue to serve the science of bacteriology on an international basis.

A number of institutions and individuals deserve special acknowledgment from the Trust for their help in bringing about the publication of this volume. We are grateful to the University of Leicester for providing space, facilities and, above all, tolerance for the diverted time taken by the Editor during the preparation of the book. The Department of Microbiology at Iowa State University of Science and Technology continues to provide a welcome home for the main editorial offices and archives of the Trust and we acknowledge their contin-

ued support. A grant (LM-03707) from the National Library of Medicine, National Institutes of Health to assist in the preparation of this and the first volume of the *Manual* is gratefully acknowledged.

A number of individuals deserve special mention and thanks for their help. Professor Thomas O. MacAdoo of the Department of Foreign Languages and Literatures at the Virginia Polytechnic Institute and State University has given invaluable advice on the etymology and correctness of scientific names. Those assisting the Editors in the Leicester office were Dorothy Jones, Grace Redfern, Wynn Rutt, Brenda Jones and Pauline Carr, and their help is sincerely appreciated. In the Ames office, we were ably assisted by Cynthia Pease who had the major responsibility for keying and sorting the list of references and index and Susan Blakely who assisted in keying the list of references.

Comments on this edition of the *Manual* will be welcomed and should be addressed to the Bergey's Manual® Trust, c/o Williams & Wilkins, 428 E. Preston St., Baltimore, Md. 21202, U.S.A.

Preface to First Edition of Bergey's Manual® of Determinative Bacteriology

The elaborate system of classification of the bacteria into families, tribes and genera by a Committee on Characterization and Classification of the Society of American Bacteriologists (1917, 1920) has made it very desirable to be able to place in the hands of students a more detailed key for the identification of species than any that is available at present. The valuable book on "Determinative Bacteriology" by Professor F. D. Chester, published in 1901, is now of very little assistance to the students, and all previous classifications are of still less value, especially as earlier systems of classification were based entirely on morphologic characters.

It is hoped that this manual will serve to stimulate efforts to perfect the classification of bacteria, especially by emphasizing the valuable features as well as the weaker points in the new system which the Committee of the Society of American Bacteriologists has promulgated. The Committee does not regard the classification of species offered here as in any sense final, but merely a progress report leading to more satisfactory classification in the future.

The Committee desires to express its appreciation and thanks to those members of the society who gave valuable aid in the compilation of material and the classification of certain species . . .

The assistance of all bacteriologists is earnestly solicited in the correction of possible errors in the text; in the collection of descriptions of all bacteria that may have been omitted from the text; in supplying more detailed decriptions of such organisms as are described incompletely; and in furnishing complete descriptions of new organisms that may be discovered, or in directing the attention of the Committee to publications of such newly described bacteria.

DAVID H. BERGEY, *Chairman*
FRANCIS C. HARRISON
ROBERT S. BREED
BERNARD W. HAMMER
FRANK M. HUNTOON
Committee on Manual

August, 1923.

DAVID HENDRICKS BERGEY
1860–1937
Bergey set up the Trust on January 2, 1936

History of the Manual

The first edition of *Bergey's Manual® of Determinative Bacteriology* was initiated by action of the Society of American Bacteriologists (now called the American Society of Microbiology) by appointment of an Editorial Board consisting of David H. Bergey, Chairman, Francis C. Harrison, Robert S. Breed, Bernard W. Hammer, and Frank M. Huntoon. This Board, under auspices of the Society of American Bacteriologists who, then as now, published the *Journal of Bacteriology* as a service to science, brought the first edition of the *Manual* into print in 1923. The Board, with some changes in membership and Dr. David Bergey as Chairman, published a second edition of the *Manual* in 1925 and a third edition in 1930.

In 1934, during preparation of the fourth edition, Dr. Bergey requested that the Society of American Bacteriologists make available the royalties paid to the Treasurer of the Society from the sale of the earlier editions to defray the expense of preparing the fourth edition for publication. The Society made such provision, but the use of the Society's fiscal machinery proved cumbersome, both to the Society and the Editorial Board. Subsequently, it was agreed by the Society and Dr. Bergey that the Society would transfer to Dr. Bergey all of its rights, title, and interest in the *Manual* and that Dr. Bergey would, in turn, create an educational trust to which all rights would be transferred.

Dr. Bergey was then the nominal owner of the *Manual* and he executed a Trust Indenture on January 2, 1936 designating David H. Bergey, Robert S. Breed, and E. G. D. Murray as the initial trustees, and transferring to the Trustees and their successors the ownership of the *Manual*, its copyrights, and the right to receive the income arising from its publication. The Trust is a nonprofit organization and its income is used solely for the purpose of preparing, editing, and publishing revisions and successive editions of the *Manual* and any supplementary publications, as well as providing for any research that may be necessary or desirable in such activities.

Since the creation of the Trust, the Trustees have published, successively, the fourth, fifth, sixth, seventh, and eighth editions of the *Manual* (dated 1934, 1939, 1948, 1957, and 1974, respectively). In 1977 the Trust published an abbreviated version of the eighth edition, called *The Shorter Bergey's Manual® of Determinative Bacteriology*; this contained the outline classification of the bacteria, the descriptions of all genera and higher taxa, all of the keys and tables for the diagnosis of species, all of the illustrations, and two of the introductory chapters; however, it did not contain the detailed species descriptions, most of the taxonomic comments, the etymology of names, and references to authors.

Other ventures in producing books to assist those engaged in bacteriology and bacterial taxonomy in particular include the *Index Bergeyana* (1966), a *Supplement to Index Bergeyana* (1981), and a planned future volume bringing the lists of published names up to date. The Trust is presently publishing the first edition of *Bergey's Manual® of Systematic Bacteriology*, which has a much broader scope than the previous publications and is intended to act as the amplified source for revision of the determinative *Manual*.

Through the years the *Manual* has become a widely used international reference work for bacterial taxonomy. Similarly, the Bergey's Manual Trust has become international in its composition, in the location of its meetings and in the breadth of its consultations. In addition to its publication activities, the Trust attempts to foster and support various aspects of taxonomic research. One of the ways in which it does this is by recognizing those individuals who have made outstanding contributions to bacterial taxonomy, through its periodic presentation of the Bergey Award, an effort jointly supported by funds from the Trust and Williams & Wilkins who have been involved in the production of the *Manual* from its beginning.

On Using the Manual

Peter H. A. Sneath

ARRANGEMENT OF THE MANUAL

The present section is largely based on that by N. R. Krieg prepared for Volume 1. One important goal of the *Manual* is to assist in the identification of bacteria, but another goal, equally important, is to indicate the relationships that exist between the various kinds of bacteria. The methods of molecular biology have now made it possible to envision the eventual development of a comprehensive classification of bacteria based on their phylogenetic relatedness to one another. Such a general classification scheme would lead, hopefully, to more unifying concepts of bacterial taxa, to greater stability and predictability, to the development of more reliable identification schemes, and to an understanding of how bacteria have evolved.

Such a general scheme, however, cannot yet be perceived fully. The relatedness within and between some bacterial groups has been intensively studied, but for other groups very little work has been done. Moreover, the relatedness studies that have been done often have involved the use of one or another method without confirmation by other methods. Studies have been done at differing levels of resolution, and the interpretation of the data may not yet be entirely clear. Still another major difficulty is the conflict between "practical" classification vs. strange groupings that may be indicated by molecular biology methods. This is because some of the phenotypic characteristics traditionally used in bacterial classification (e.g. cell shape, flagellar arrangement, fermentative vs. respiratory types of metabolism, etc.) do not always correlate well with groups established on the basis of relatedness. This conflict will, one hopes, eventually be relieved by the finding of nontraditional, easily determined, phenotypic characteristics that *do* correlate well with relatedness groups, but much work needs to be done in this regard.

Such considerations have forced the present edition of the *Manual* to adhere largely to traditional characteristics in arranging bacterial taxa. It should be understood, however, that reassessments of these groupings will soon need to be made on a broad, comprehensive scale. The present classification, although of considerable practical value, must be regarded as only an interim arrangement.

THE SECTIONS

The *Manual* is presented as various "sections" based on a few readily determined criteria. Each section bears a vernacular name. All accepted genera have been placed in what seems the most appropriate section, although allocation of certain genera has presented difficulties, as indicated by the following examples:

(a) The genus *Oscillospira* remains poorly studied, and the account given is that from the 8th edition of *Bergey's Manual® of Determinative Bacteriology*. It is only tentatively placed in Section 13, because it is still uncertain whether it produces true endospores.

(b) The genus *Gardnerella*. It is not entirely clear whether this genus should be placed with Gram-negative or with Gram-positive bacteria. It has been reprinted from Volume 1 and is placed for convenience in Section 15.

(c) The genera *Lachnospira* and *Butyrivibrio* stain Gram-negative under usual conditions, but whether or not they possess cell walls of the Gram-negative type is not clear. They are, therefore, reprinted from Volume 1 and are placed in Section 15.

(d) The practical distinction between organisms of Sections 14 and 15 rests largely on the more regular shape of the cells of those in Section 14. The reader is advised, therefore, to consult both sections when considering an organism that may belong to either of them. The placement of some of these genera is in considerable doubt.

(c) The organisms of Sections 16 and 17 share a tendency to produce branching filaments, particularly in early phases of growth. This is less marked in the mycobacteria (Section 16) than the others, but the mycobacteria are acid-fast (as are some organisms in Section 17).

(f) The genera *Corynebacterium* (Section 15), *Mycobacterium* (Section 16), *Nocardia* and *Rhodococcus* (Section 17) share many chemotaxonomic properties, which could justify their placement in one grouping. However, for determinative purposes they have been arranged as noted above.

As an interim solution to some of these problems, some taxa are described not only in Volume 2 but in an appropriate subsequent volume as well.

SECTIONS VS. TAXONOMIC NAMES

Each section bears a vernacular name, but may also bear the name of a taxon. Thus, Section 16 (The Mycobacteria) is the family *Mycobacteriaceae*. As indicated previously, no attempt has been made to provide a complete formal hierarchy of higher taxa throughout the *Manual*, and the vernacular names of the sections form the primary basis for the organization of the *Manual*; however, a suggested hierarchy for higher taxa has been proposed in one of the introductory articles (see The Higher Taxa, or a Place for Everything).

ARTICLES

Each article dealing with a bacterial genus is presented wherever possible in a definite sequence as follows.

(a) *Name of the Genus.* Accepted names are in **boldface,** followed by the authority for the name, the year of the original description, and the page on which the taxon was named and described. The superscript *AL* indicates that the name was included on the Approved Lists of Bacterial Names, published in January 1980. The superscript *VP* indicates that the name, although not on the Approved Lists of Bacterial Names, was subsequently validly published in the International Journal of Systematic Bacteriology. Names given within quotation marks have no standing in nomenclature; as of the date of preparation of the *Manual* they had not been validly published in the International Journal of Systematic Bacteriology, although they had been "effectively published" elsewhere. Names followed by the term "gen. nov." are newly proposed but will not be validly published until they appear in the International Journal of Systematic Bacteriology; their proposal in the *Manual* constitutes only "effective publication," not valid publication.

(b) *Name of Author(s).* The person or persons who prepared the Bergey article are indicated. The address of each author can be found in the list of Contributors at the beginning of the *Manual.*

(c) *Synonyms.* In some instances a list is given of synonyms which have been used in the past for the same genus. The synonymy may not always be complete, and usually is not given at all, as the Editorial Board believes that the earlier synonyms have been covered adequately in the *Index Bergeyana* or the *Supplement to the Index Bergeyana.*

(d) *Etymology of the Genus Name.* Etymologies are provided as in previous editions, and many (but undoubtedly not all) errors have been corrected. It is often difficult, however, to determine why a particular name was chosen, or the nuance intended, if the details were not provided in the original publication. Those authors who propose new names are urged to consult a Greek and Latin authority before publishing, in order to ensure grammatical correctness and also to ensure that the name means what it is intended to mean. An excellent authority to communicate with in this regard is Dr. Thomas O. MacAdoo, Department of Foreign Languages, Virginia Polytechnic Institute and State University, Blacksburg, Virginia U.S.A. 24061.

(e) *Capsule Description.* This is a brief resume of the salient features of the genus. The most important characteristics are given in **boldface.** The name of the type species of the genus is also indicated.

(f) *Further Descriptive Information.* This portion elaborates on the various features of the genus, particularly those features having significance for systematic bacteriology. The treatment serves to acquaint the reader with the overall biology of the organisms but is not meant to be a comprehensive review. The information is represented in sequence, as follows:

Morphological characteristics
Colonial morphology and pigmentation
Growth conditions and nutrition
Physiology and metabolism
Genetics, plasmids and bacteriophages
Antigenic structure
Pathogenicity
Ecology

(g) *Enrichment and Isolation.* A few selected methods are presented, together with the pertinent media formulations.

(h) *Maintenance Procedures.* Methods used for maintenance of stock cultures and preservation of strains are given.

(i) *Procedures for Testing Special Characters.* This portion provides methodology for testing for unusual characteristics or performing tests of special importance.

(j) *Differentiation of the Genus from Other Genera.* Those characteristics that are especially useful for distinguishing the genus from similar or treated organisms are indicated here, usually in a tabular form.

(k) *Taxonomic Comment.* This summarizes the available information about the taxonomic placement of the genus and indicates the justification for considering genus to be a distinct taxon. Particular emphasis is given to the methods of molecular biology for estimating the relatedness to other taxa, where such information is available. Taxonomic information regarding the arrangement and status of the various species within the genus follows. Where taxonomic controversy exists, the problems are delineated and the various alternative viewpoints are discussed.

(l) *Further Reading.* A list of selected references, usually of a general nature, is given to enable the reader to gain access to additional sources of information about the genus.

(m) *Differentiation of the Species of the Genus.* Those characteristics that are important for distinguishing from one another the various species within the genus are presented, usually with reference to a table summarizing the information.

(n) *List of the Species of the Genus.* The citation of each species is given, followed in some instances by a brief list of objective synonyms. The etymology of the specific epithet is indicated. Descriptive information for the species is usually presented in tabular form, but special information may be given in the text. Because of the emphasis on tabular data the species descriptions are usually brief. The type strain of each species is indicated, together with the collection in which it can be found. (Addresses of the various culture collections are given in the chapter List of Culture Collections.)

(o) *Species Incertae Sedis.* The List of Species may be followed in some instances by a listing of additional species under the heading "Species Incertae Sedis." The taxonomic placement or status of such species is questionable and the reasons for the uncertainty are presented.

(p) *Literature Cited.* All references given in the article are listed alphabetically at the end of the volume rather than at the end of each article.

TABLES

In each article dealing with a genus, there are generally three kinds of tables: (a) those that differentiate the genus from similar or related genera, (b) those that differentiate the species within the genus, and (c) those that provide additional information about the species, such information not being particularly useful for differentiation. Unless otherwise indicated, the meanings of symbols are as follows:

+ 90% or more of the strains are positive.

d 11–89% of the strains are positive.

− 90% or more of the strains are negative.

D different reactions occur in different taxa (species of a genus or genera of a family).

v strain instability (NOT equivalent to "d").

Exceptions to use of these symbols, as well as the meaning of additional symbols, are clearly indicated in footnotes to the tables.

USE OF THE MANUAL FOR DETERMINATIVE PURPOSES

Entry into the *Manual* is best achieved by studying the titles of the various sections, as listed in the Contents. These titles provide an elementary, but by no means perfect, key to the various kinds of bacteria. Each section has keys or tables for differentiation of the various taxa contained therein. Suggestions on identification may be found in the article Identification of Bacteria. For identification of species, it is important to read both the generic and species descriptions because characteristics listed in the generic descriptions are not usually repeated in the species descriptions.

The index is useful in locating the names of unfamiliar taxa or in discovering what has been done with a particular taxon. Every bacterial name mentioned in the *Manual* is listed in the index.

ERRORS, COMMENTS, SUGGESTIONS

As indicated in the Preface to the first edition of *Bergey's Manual®* *of Determinative Bacteriology,* the assistance of all bacteriologists is earnestly solicited in the correction of possible errors in the text. Comments on the presentation will also be welcomed, as well as suggestions for future editions. Correspondence should be addressed to the Bergey's Manual® Board of Trustees c/o Williams & Wilkins, 428 East Preston St., Baltimore, Md. 21202, U.S.A.

Contents

SECTION 17
Nocardioforms **1458**

Classification of Procaryotic Organisms: An Overview

James T. Staley and Noel R. Krieg

CLASSIFICATION, NOMENCLATURE AND IDENTIFICATION

Classification, nomenclature, and identification are the three separate, but interrelated, areas of taxonomy. **Classification** is the arranging of organisms into taxonomic groups (taxa) on the basis of similarities or relationships. **Nomenclature** is the assignment of names to the taxonomic groups according to international rules. **Identification** is the process of determining that a new isolate belongs to one of the established, named taxa.

There are numerous procaryotic organisms and great diversity in their types. In any endeavor aimed at an understanding of large numbers of entities it is convenient to arrange, or classify, the objects into groups based upon their similarities. Thus, classification has been used to organize the bewildering and seemingly chaotic array of individual bacteria into an orderly framework.

Classification of organisms requires knowledge of their characteristics. For procaryotes, this knowledge is obtained by experimental as well as observational techniques, because biochemical, physiological and genetic characteristics are often necessary, in addition to morphological features, for an adequate description of a taxon.

The process of classification may be applied to existing, named, taxa or to newly described organisms. If the taxa have already been described, named, and classified, either new characteristics about the organisms or a reinterpretation of existing knowledge of characteristics is used to formulate a new classification. However, if the organisms are new, i.e. cannot be identified as existing taxa, they are named according to the rules of nomenclature and placed in an appropriate position in an existing classification.

Taxonomic Ranks

Several levels or ranks are used in bacterial classification. All procaryotic organisms are placed in the Kingdom *Procaryotae*. Divisions, classes, orders, families, genera and species are successively smaller, nonoverlapping subsets of the Kingdom, and the names of these subsets are given formal recognition (have "standing in nomenclature"). An example is given in Table I.1.

In addition to these formal, hierarchical taxonomic categories, informal or vernacular groups that are defined by common descriptive names are often used; the names of such groups have no official standing in nomenclature. Examples of such groups are: the procaryotes, the spirochetes, dissimilatory sulfate- and sulfur-reducing bacteria, the methane-oxidizing bacteria, etc.

Species

The basic taxonomic group in bacterial systematics is the species. The concept of a bacterial species is less definitive than for higher organisms. This difference should not seem surprising, because bacteria, being procaryotic organisms, differ markedly from higher organisms. Sexuality, for example, is not used in bacterial species definitions because relatively few bacteria undergo conjugation. Likewise, morphological features alone are usually of little classificatory significance; this is because most procaryotic organisms are too simple morphologically to provide much useful taxonomic information. Consequently, morphological features are relegated to a less important role in bacterial taxonomy in comparison with the taxonomy of higher organisms.

A bacterial species may be regarded as a collection of strains that share many features in common and differ considerably from other strains. (A strain is made up of the descendants of a single isolation in pure culture, and usually is made up of a succession of cultures ultimately derived from an initial single colony.) One strain of a species is designated as the **type strain**; this strain serves as the name-bearer strain of the species and is the permanent example of the species, i.e. the *reference specimen for the name*. (See the chapter on *Nomenclature* for more detailed information about nomenclatural types.) The type strain has great importance for classification at the species level, because *a species consists of the type strain and all other strains that are considered to be sufficiently similar to it as to warrant inclusion with it in the species.* This concept of a species obviously involves making subjective judgments, and it is not surprising that some bacterial species have greater phenotypic and genetic diversity than others. A more uniform and rigorous species definition would be desirable. For example, the level of DNA homology exhibited among a group of strains might be used as a basis for defining a species, i.e. definition on the basis of a particular degree of genetic relatedness. The advantage of adopting this or a similarly restrictive species definition must be weighed against its potential impact on well established and accepted bacterial groups. For practical reasons, classifications and nomenclature should remain stable because changes create confusion, particularly at the genus and species levels and result in costly modifications of identification schemes and texts. However, classifications have *never* remained static and probably never will, because new information bearing on the taxonomy of bacteria is continually being generated by researchers.

Though classification schemes based on genetic relatedness are rather recent, they promise to be quite reliable and stable. This view may have to be reassessed, however, when we more fully understand the impact that transposable elements might have upon the stability of the procaryotic genome. Genetic studies have already resolved many instances of confusion concerning which strains belong to a given

Table I.1.

Taxonomic ranks

Formal Rank	Example
Kingdom	*Procaryotae*
Division	*Gracilicutes*
Class	*Scotobacteria*
Order	*Spirochaetales*
Family	*Leptospiraceae*
Genus	*Leptospira*
Species	*Leptospira interrogans*

Table I.2.

Infrasubspecific ranks

Preferred Name	Synonym	Applied to Strains Having:
Biovar	Biotype	Special biochemical or physiological properties
Serovar	Serotype	Distinctive antigenic properties
Pathovar	Pathotype	Pathogenic properties for certain hosts
Phagovar	Phagotype	Ability to be lysed by certain bacteriophages
Morphovar	Morphotype	Special morphological features

species, and DNA homology is increasingly being used for establishing new species and for resolving taxonomic problems at the species level.

Subspecies

A species may be divided into two or more subspecies based on minor but consistent phenotypic variations within the species or on genetically determined clusters of strains within the species. It is the lowest taxonomic rank that has official standing in nomenclature.

Infrasubspecific Ranks

Ranks below subspecies, such as biovars, serovars, phagovars, are often used to indicate groups of strains that can be distinguished by some special character, such as antigenic makeup, reactions to bacteriophage, or the like. Such ranks have no official standing in nomenclature but often have great practical usefulness. A list of some common infrasubspecific categories is given in Table I.2.

Genus

All species are assigned to a genus (although not always with a high degree of certainty as to which genus is the best choice). In this regard, bacteriologists conform to the binomial system of nomenclature of Linnaeus in which the organism is designated by its combined genus and species name. The bacterial genus is usually a well-defined group that is clearly separated from other genera, and the thorough descriptions of genera in this edition of *Bergey's Manual* exemplify the depth to which this taxonomic group is usually known. However, there is so far no general agreement on the *definition* of a genus in bacterial taxonomy, and considerable subjectivity is involved at the genus level. Indeed, what is perceived to be a genus by one person may be perceived as being merely a species by another systematist. The use of genetic relatedness (e.g. ribosomal RNA (rRNA) homology or rRNA oligonucleotide cataloging) offers hope for greater objectivity and has already been useful in several instances.

Higher Taxa

Classificatory relationships at the familial and ordinal levels are even less certain than those at the genus and species levels. Frequently there is little basis for ascription of taxa at these higher levels, except in a few cases (e.g. the family *Enterobacteriaceae*) where there is evidence for genetic relatedness. Thus, rather than formalize families and orders upon uncertain relationships, many systematists frequently adopt a provisional, ad hoc ranking in which purely descriptive and vernacular names for groups are applied (e.g. in this edition of *Bergey's Manual* see Section 7 on "Dissimilatory Sulfate- or Sulfur-Reducing Bacteria"). As more is learned about the similarities among these bacteria, familial and ordinal placements will likely ensue. A recent example that illustrates the effect that increased knowledge has on the taxonomy of groups concerns the methane-producing bacteria. In the eighth edition of the *Manual,* the methanogens were treated as a single family of bacteria, with three genera. Authorities for this group now propose that three *orders* are required for the circumscription of these organisms (Balch et al., 1979).

In this edition of the *Manual* the procaryotes have been classified into four divisions, these being subdivided into classes (see the chapter by Murray on "The Higher Taxa"). There is no general agreement about this or any other arrangement of divisions and classes, however, and even at the kingdom level of classification controversy exists. Recent information based on rRNA oligonucleotide catalogues and biochemical features has led some authorities to propose that not all bacteria are procaryotes, and that some represent a kingdom of life distinct from both procaryotes and eucaryotes (i.e. the so-called "archaebacteria") (see Fox et al., 1980, and Woese, 1981, for summaries). That this group possesses a number of unique features is beyond question, and there is strong evidence that it has taken an evolutionary path distinct from that of other bacteria, but so far there is no general agreement as to what level of classification is applicable to the group.

MAJOR DEVELOPMENTS IN BACTERIAL CLASSIFICATION

A century elapsed between Antony van Leeuwenhoek's discovery of bacteria and Müller's initial acknowledgement of bacteria in a classification scheme (Müller, 1773). Another century passed before techniques and procedures had advanced sufficiently to permit a fairly inclusive and meaningful classification of these organisms. For a comprehensive review of the early development of bacterial classification, readers should consult the introductory sections of the first, second, and third editions of *Bergey's Manual*. A less detailed treatment of early classifications can be found in the sixth edition of the *Manual* in which post-1923 developments were emphasized.

Two primary difficulties beset early bacterial classification systems. First, they relied heavily upon morphological criteria. For example, cell shape was often considered to be an extremely important feature. Thus, the cocci were often classified together in one group (family or order). In contrast, contemporary schemes rely much more strongly on physi-

ological characteristics. For example, the fermentative cocci are now separated from the photosynthetic cocci, which are separated from the methanogenic cocci, which are in turn separated from the nitrifying cocci, and so forth. Secondly, the pure culture technique which revolutionized microbiology was not developed until the latter half of the 19th century. In addition to dispelling the concept of "polymorphism," this technical development of Robert Koch's laboratory had great impact on the development of modern procedures in bacterial systematics. Pure cultures are analogous to herbarium specimens in botany. However, pure cultures are much more useful because they can be (a) maintained in a viable state, (b) subcultured, (c) subjected indefinitely to experimental tests, and (d) shipped from one laboratory to another. A natural outgrowth of the pure culture technique was the establishment of *type strains* of species which are deposited in repositories referred to as "culture collections" (a more suitable term would be "strain collections"). These type strains can be obtained from culture collections and used as reference strains for direct comparison with new isolates.

Before the development of computer-assisted numerical taxonomy and subsequent taxonomic methods based on molecular biology, the traditional method of classifying bacteria was to characterize them as thoroughly as possible and then to arrange them according to the intuitive judgment of the systematist. Although the subjective aspects of this method resulted in classifications that were often drastically revised by other systematists who were likely to make different intuitive judgments, many of the arrangements have survived to the present day, even under scrutiny by modern methods. One explanation for this is that the systematists usually *knew their organisms thoroughly*, and their intuitive judgments were based on a wealth of information. Their data, while not computer processed, were at least processed by an active mind to give fairly accurate impressions of the relationships existing between organisms. Moreover, some of the characteristics that were given great weight in classification were, in fact, highly correlated with many other characteristics. This principle of *correlation of characteristics* appears to have started with the Winslows (1908), who noted that parasitic cocci tended to grow poorly on ordinary nutrient media, were strongly Gram-positive, and formed acid from sugars, in contrast to saprophytic cocci which grew abundantly on ordinary media, were generally only weakly Gram-positive and formed no acid. This division of the cocci that were studied by the Winslows (equivalent to the present genus *Micrococcus* (the saprophytes) and the genera *Staphylococcus* and *Streptococcus* (the parasites)) has held up reasonably well even to the present day.

Other classifications have not been so fortunate. A classic example of one which was not is that of the genus "*Paracolobactrum*." This genus was proposed in 1944 and is described in the seventh edition of *Bergey's Manual* in 1957. It was created to contain certain lactose-negative members of the family *Enterobacteriaceae*. Because of the importance of a lactose-negative reaction in *identification* of enteric pathogens (i.e. *Salmonella* and *Shigella*), the reaction was mistakenly given great taxonomic weight in *classification* as well. However, for the organisms placed in "*Paracolobactrum*," the lactose reaction was not highly correlated with other characteristics. In fact, the organisms were merely lactose-negative variants of other lactose-positive species; for example, "*Paracolobactrum coliforme*" resembled *Escherichia coli* in every way except in being lactose-negative. Absurd arrangements such as this eventually led to the development of more objective methods of classification, i.e. numerical taxonomy, in order to avoid giving great weight to any single characteristic.

Phylogenetic Classifications

Classification systems for many higher organisms are based to a large extent upon evolutionary evidence obtained from the fossil record and appropriate sedimentary dating procedures. Such classifications are termed "natural" or "phylogenetic," and are distinguished from "practical" or "artificial" classifications, which are based entirely on phenotypic characteristics. Until about 20 years ago, however, there was no convincing evidence of fossil microorganisms. Now, micropaleontological evidence indicates that microorganisms existed during the Precambrian period. Indeed, many scientists believe that bacteria existed at least 3.5 billion years ago on an earth that is 4.5 billion years old. Of course, the discovery of fossil microorganisms in early sedimentary rocks tells very little about the phylogeny of procaryotic groups. Micropaleontologists are far from reconstructing an evolutionary scheme based upon the presently available fossil record.

Despite the absence of a complete fossil record, proposals have been made since the early part of this century regarding the evolution of bacteria. Until recently, these proposals have been entirely speculative in nature. Orla-Jensen (1909) proposed that autotrophic bacteria were the most primitive group, and he devised an extensive phylogenetic scheme based on this premise. Today, most microbiologists would agree that the premise is probably incorrect, but Orla-Jensen's classification

did provide a coherent framework for thinking about the relationships among bacteria. Another notable phylogenetic scheme was that devised by Kluyver and van Niel (1936); in contrast to Orla-Jensen's scheme, which had been based almost entirely on physiological characteristics, Kluyver and van Niel's scheme was based on morphology. The basic premise was that the simplest morphological form, the coccus, was also the most primitive, and from this form developed more complex forms such as spirilla, rods, and branching filaments.

As recently as the seventh edition of *Bergey's Manual* (i.e. 1957), before convincing evidence of Precambrian microbes had been discovered, the view was expressed that bacteria were a primitive group of organisms, and the classification scheme presented in that edition of the *Manual* claimed to be a natural scheme in which the photosynthetic bacteria were treated first, because they were regarded as the most primitive bacterial group. However, because of the lack of objective evidence for this (or any other) phylogenetic scheme, the eighth edition of the *Manual* abandoned all attempts at a phylogenetic approach to bacterial classification and concentrated instead on providing groupings of organisms under vernacular headings for purposes of recognition and identification; i.e. it was a purely practical and admittedly artificial classification.

Phylogenetic information has increased since the eighth edition, however, largely through the increasing use of methods for measuring genetic relatedness (i.e. DNA/DNA hybridization, DNA/rRNA hybridization, rRNA oligonucleotide cataloging, and protein sequencing). A record of bacterial evolution appears to exist in the amino acid sequences of bacterial proteins and in the nucleotide sequences of bacterial DNA and RNA. Unfortunately, the phylogenetic information is still in a fragmentary form, and it seems probable that the interpretation of the data is still not entirely clear. Not all of the bacterial groups have been surveyed, and it is likely that surprises and strange associations will continue to come from further work. Available phylogenetic information is presented throughout this *Manual* in the *Taxonomic Comments* sections of the various chapters, and some preliminary rearrangements of taxa have already been made based upon phylogenetic information.

Official Classifications

Some microbiologists seem to have the impression that the classification presented in *Bergey's Manual* is the "official classification" to be used in microbiology. It seems important to correct that impression. **There is no "official" classification of bacteria.** (This is in contrast to bacterial *nomenclature,* where each taxon has one and only one valid name, according to internationally agreed-upon rules, and judicial decisions are rendered in instances of controversy about the validity of a name.) The closest approximation to an "official" classification of bacteria would be one that is widely accepted by the community of microbiologists. A classification that is of little use to microbiologists, no matter how fine a scheme or who devised it, will soon be ignored or significantly modified.

It also seems worthwhile to emphasize something that has often been said before, viz. **bacterial classifications are devised for microbiologists, not for the entities being classified.** Bacteria show little interest in the matter of their classification. For the systematist, this is sometimes a very sobering thought!

Further Reading

Cowan, S.T. 1971. Sense and nonsense in bacterial taxonomy. J. Gen. Microbiol. 67: 1–8.
   An incisive, personal view of bacterial taxonomy, with some "heretical" suggestions.
Cowan, S.T. 1974. *Cowan and Steel's manual for the identification of medical bacteria.* Cambridge University Press, Cambridge, England.
   Chapters 1 and 9 of this work provide a concise statement of many principles of bacterial taxonomy.
Gerhardt, P., R.G.E. Murray, R.N. Costilow, E.W. Nester, W.A. Wood, N.R. Krieg and G.B. Phillips (Editors). 1981. Manual of methods for general bacteriology. American Society for Microbiology, Washington, D. C.
   Section V of this book gives a brief introduction to phenotypic characteriza-

tion, numerical taxonomy, genetic characterization, and classification of bacteria.

Johnson, J.L. 1973. Use of nucleic acid homologies in the taxonomy of anaerobic bacteria. Int. J. Syst. Bacteriol. *23:* 308–315.

This paper proposes a unifying concept of a bacterial species and stresses the importance of correlating nucleic acid homology with phenotypic tests to allow differentiation among species.

Margulis, L. 1968. Evolutionary criteria in Thallophytes: a radical alternative. Science (Washington) *161:* 1020–1022.

This paper presents the hypothesis that eucaryotic organisms evolved from procaryotic organisms through endosymbioses.

Schopf, J.W. 1978. The evolution of the earliest cells. Sci. Amer. *239:* 110–138.

A micropaleontologist's view of microbial evolution.

Schwartz, R.M. and M.O. Dayhoff. 1978. Origins of procaryotes, eucaryotes, mitochondria, and chloroplasts. Science (Washington) *199:* 395–403.

A discussion of results obtained from the analysis of protein and nucleic acid sequence data as they pertain to the phylogeny of organisms.

Sneath, P.H.A. 1978. Classification of microorganisms. *In* Norris and Richmond (Editors), *Essays in Microbiology.* John Wiley , Chichester, UK, pp. 9/1–9/31.

An excellent general introduction to bacterial classification.

Trüper, H.G. and J. Krämer. 1981. Principles of characterization and identification of prokaryotes. *In* Starr, Stolp, Trüper, Balows and Schlegel (Editors), *The Prokaryotes: a Handbook on Habitats, Isolation and Identification of Bacteria.* Springer-Verlag, Berlin, pp. 176–193.

A brief overview of systematic bacteriology, including developments and trends in taxonomy.

Woese, C.R. and G.E. Fox. 1977. Phylogenetic structure of the procaryotic domain: the primary kingdoms. Proc. Nat. Acad. Sci. USA *74:* 5088–5090.

The authors recognize three distinct groups: the eubacteria, the archaebacteria, and the urcaryotes (cytoplasmic components of eucaryotes).

BACTERIAL CLASSIFICATION II

Numerical Taxonomy

Peter H. A. Sneath

Numerical taxonomy (sometimes called **taxometrics**) developed in the late 1950s as part of multivariate analyses and in parallel with the development of computers. Its aim was to devise a consistent set of methods for classification of organisms. Much of the impetus in bacteriology came from the problem of handling the tables of data that result from examination of their physiological, biochemical and other properties. Such tables of results are not readily analyzed by eye, in contrast to the elaborate morphological detail that is usually available from examination of higher plants and animals. There was thus a need for an objective method of taxonomic analyses, whose first aim was to sort individual strains of bacteria into homogeneous groups (conventionally species), and which would also assist in the arrangement of species into genera and higher groupings. Such numerical methods also promised to improve the exactitude in measuring taxonomic, phylogenetic, serological, and other forms of relationship, together with other benefits that can accrue from quantitation (such as improved methods for bacterial identification; see the discussion by Sneath of numerical identification on p. 990 of this *Manual*).

Numerical taxonomy has been broadly successful in most of these aims, particularly in defining homogeneous **clusters** of strains, and in integrating data of different kinds (morphological, physiological, antigenic). There are still problems in constructing satisfactory groups at high taxonomic levels, e.g. families and orders, although this may be due to inadequacies in the available data rather than any fundamental weakness in the numerical methods themselves.

The application of the concepts of numerical taxonomy was made possible only through the use of computers, because of the heavy load of routine calculations. However, the principles can be easily illustrated in hand-worked examples. In addition, two problems had to be solved: the first was to decide how to weight different variables or characters; the second was to analyze similarities so as to reveal the **taxonomic structure** of groups, species, or clusters. A full description of numerical taxonomic methods may be found in Sneath (1972) and Sneath and Sokal (1973). Briefer descriptions and illustrations in bacteriology are given by Skerman (1967), Lockhart and Liston (1970), and Sneath (1978a). A thorough review of applications to bacteria is that of Colwell (1973).

It is important to bear in mind certain definitions. Relationships between organisms can be of several kinds. Two broad classes are as follows.

Similarity on Observed Properties. Similarity, or **resemblance,** refers to the attributes that an organism possesses today, without reference to how those attributes arose. It is expressed as proportions of similarities and differences, for example, in existing attributes, and is called **phenetic relationship.** This includes both similarities in phenotype (e.g. motility) and in genotype (e.g. DNA pairing).

Relationship by Ancestry, or Evolutionary Relationship. This refers to the **phylogeny** of organisms, and not necessarily to their present attributes. It is expressed as the time to a common ancestor, or the amount of change that has occurred in an evolutionary lineage. It is not expressed as a proportion of similar attributes, or as the amount of DNA pairing and the like, although evolutionary relationship may sometimes be *deduced* from phenetics *on the assumption* that evolution has indeed proceeded in some orderly and defined way. To give an analogy, individuals from different nations may occasionally look more similar than brothers or sisters of one family: their phenetic resemblance (in the properties observed) may be high though their evolutionary relationship is distant.

Numerical taxonomy is concerned primarily with phenetic relationships. It has in recent years been extended to phylogenetic work, by using rather different techniques: these seek to build up on the assumed regularities of evolution so as to give, from *phenetic data,* the *most probable phylogenetic reconstructions.* Relatively little has been done so far in bacteriology, but a review of the area is given by Sneath (1974).

The basic taxonomic category is the species. It is noted in the chapter on "Nomenclature" that it is useful to distinguish a **taxospecies** (a cluster of strains of high mutual phenetic similarity) from a **genospecies** (a group of strains capable of gene exchange), and both of these from a **nomenspecies** (a group bearing a binominal name whatever its status in other respects). Numerical taxonomy attempts to define taxospecies. Whether these are justified as genospecies or nomenspecies turns on other criteria. It should be emphasized that groups with high genomic similarity are not necessarily genospecies: genomic resemblance is included in phenetic resemblance; genospecies are defined by gene exchange.

Groups can be of two important types. In the first, the possession of certain invariant properties defines the group without permitting any exception. All triangles, for example, have three sides, not four. Such groupings are termed **monothetic.** Taxonomic groups are, however, not of this kind. Exceptions to the most invariant characters are always possible. Instead, taxa are **polythetic,** that is they consist of assemblages whose members share a high proportion of common attributes, but not necessary any invariable set. Numerical taxonomy produces polythetic groups and thus permits the occasional exception on any character.

LOGICAL STEPS IN CLASSIFICATION

The steps in the process of classification are as follows:

1. Collection of data. The **bacterial strains** that are to be classified have to be chosen, and they must be examined for a number of relevant properties (**taxonomic characters**).
2. The data must be coded and scaled in an appropriate fashion.
3. The **similarity** or **resemblance** between the strains is calculated. This yields a table of similarities (**similarity matrix**) based on the chosen set of characters.
4. The similarities are analyzed for **taxonomic structure**, to yield the groups or clusters that are present, and the strains are arranged into **phenons** (phenetic groups), which are broadly equated with taxonomic groups (**taxa**).

5. The properties of the phenons can be tabulated for publication or further study, and the most appropriate characters (**diagnostic characters**) can be chosen on which to set up **identification systems** that will allow the best identification of additional strains.

It may be noted that those steps must be carried out in the above order. One cannot, for example, find diagnostic characters before finding the groups of which they are diagnostic. Furthermore, it is important to obtain complete data, determined under well standardized conditions.

Data for numerical taxonomy

The data needed for numerical taxonomy must be adequate in quantity and quality. It is a common experience that data from the literature are inadequate on both counts: most often it is necessary to examine bacterial strains afresh by an appropriate set of tests.

Organisms

Most taxonomic work with bacteria consists of examining individual strains of bacteria. However, the entities that can be classified may be of various forms,—strains, species, genera,— for which no common term is available. These entities, t in number, are therefore called **operational taxonomic units (OTUs)**. In most studies OTUs will be strains. A numerical taxonomic study, therefore should contain a good selection of strains of the groups under study, together with type strains of the taxa and of related taxa. Where possible, recently isolated strains, and strains from different parts of the world, should be included.

Characters

A **character** is defined as any property that can vary between OTUs. The values it can assume are **character states**. Thus, "length of spore" is a character and "1.5 μm" is one of its states. It is obviously important to compare the same character in different organisms, and the recognition that characters are the same is called the **determination of homology**. This may sometimes pose problems, but in bacteriology these are seldom serious. A single character treated as independent of others is called a **unit character**. Sets of characters that are related in some way are called **character complexes**.

There are many kinds of characters that can be used in taxonomy. The descriptions in the *Manual* give many examples. For numerical taxonomy, the characters should cover a broad range of properties: morphological, physiological, biochemical. It should be noted that certain data are not characters in the above sense. Thus the degree of serological cross-reaction or the percent pairing of DNA are analogous, not to character states, but to similarity measures.

Numbers of Characters

Although it is well to include a number of strains of each known species, numerical taxonomies are not greatly affected by having only a few strains of a species. This is not so, however, for characters. The similarity values should be thought of as estimates of values that would be obtained if one could include a very large number of phenotypic features. The accuracy of such estimates depends critically on having a reasonably large number of characters. The number, n, should be 50 or more. Several hundred are desirable, though the taxonomic gain falls off with very large numbers.

Quality of Data

The quality of the characters is also important. Microbiological data are prone to more experimental error than is commonly realized. The average difference in replicate tests on the same strain is commonly about 5%. Efforts should be made to keep this figure low, particularly by rigorous standardization of test methods. It is very difficult to obtain reasonably reproducible results with some tests, and they should be excluded from the analysis. As a check on the quality of the data, it is useful to reduplicate a few of the strains and carry them through as separate OTUs: the average test error is about half the percentage discrepancy in similarity of such replicates (e.g. 90% similarity implies about 5% of experimental variation).

Coding of the Results

The test reactions and character states now need coding for numerical analysis. There are several satisfactory ways of doing this, but for the present purposes of illustration only one common scheme will be described. This is the familiar process of coding the reactions or states into positive and negative form. The resulting table, therefore, contains entries + and − (or 1 and 0, which are more convenient for computation), for t OTUs scored for n characters. Naturally, there should be as few gaps as possible.

The question arises as to what weight should be given to each character relative to the rest. The usual practice in numerical taxonomy is to give each character equal weight. More specifically, it may be argued that unit characters should have unit weight, and if character complexes are broken into a number of unit characters (each carrying one unit of taxonomic information) it is logical to accord unit weight to each unit character. The difficulties of deciding what weight should be given *before* making a classification (and hence in a fashion that does not prejudge the taxonomy) are considerable. This philosophy derives from the opinions of the 18th century botanist Adanson, and therefore numerical taxonomies are sometimes referred to as Adansonian.

Similarity

The $n \times t$ table can then be analyzed to yield similarities between OTUs. The simplest way is to count, for any pair of OTUs, the number of characters in which they are identical (i.e. both are positive or both are negative). These **matches** can be expressed as a percentage or a proportion, symbolized as S_{SM} (for simple matching coefficient). This is the commonest coefficient in bacteriology. Other coefficients are sometimes used because of particular advantages. Thus the Gower coefficient S_G accommodates both presence-absence characters and quantitative ones, the Jacquard coefficient S_J discounts matches between two negative results, and the Pattern coefficient S_P corrects for apparent differences that are caused solely by differences between strains in growth rate and hence metabolic vigor. These coefficients emphasize different aspects of the phenotype (as is quite legitimate in taxonomy) so one cannot regard one or other as necessarily the correct coefficient, but fortunately this makes little practical difference in most studies.

The similarity values between all pairs of OTUs yields a checkerboard of entries, a square table of similarities known as a **similarity matrix** or *S* **matrix**. The entries are percentages, with 100% indicating identity and 0% indicating complete dissimilarity between OTUs. Such a table is symmetrical (the similarity of *a* to *b* is the same as that of *b* to *a*), so that usually only one half, the left lower triangle, is filled in.

These similarities can also be expressed in a complementary form, as *dissimilarities*. Dissimilarities can be treated as analogues of distances when preparing "taxonomic maps" of the OTUs, and it is a convenient property that the quantity $d = \sqrt{(1 - S_{SM})}$ is equivalent geometrically to a *distance* between points representing the OTUs in a space of many dimensions (a **phenetic hyperspace**).

Taxonomic structure

A table of similarities does not of itself make evident the **taxonomic structure** of the OTUs. The strains will be in an arbitrary order which will not reflect the species or other groups. These similarities therefore require further manipulation. It will be seen that a table of serological cross-reactions, if complete and expressed in quantitative terms, is analogous to a table of percentage similarities, and the same is true of a table of DNA pairing values. Such tables can be analyzed by the methods described below, though in serological and nucleic studies there are some particular difficulties on which further work is needed.

There are two main types of analyses to reveal the taxonomic structure, **cluster analysis**, and **ordination**. The result of the former is a tree-like diagram or **dendrogram** (more precisely a **phenogram**, because it expresses phenetic relationships), in which the tightest bunches of twigs represent clusters of very similar OTUs. The result of the latter is an **ordination diagram** or **taxonomic map**, in which closely similar OTUs are placed close together. The mathematical methods can be elaborate, so only a nontechnical account is given here.

In cluster analysis, the principle is to search the table of similarities for high values that indicate the most similar pairs of OTUs. These form the nuclei of the clusters and the computer searches for the next highest similarity values and adds the corresponding OTUs onto these cluster nuclei. Ultimately all OTUs fuse into one group, represented by the basal stem of the dendrogram. Lines drawn across the dendrogram at descending similarity levels define, in turn, phenons that correspond to a reasonable approximation to species, genera, etc. The commonest cluster methods are the **unweighted pair group method with averages (UPGMA)** and **single linkage.**

In ordination, the similarities (or their mathematical equivalents) are analyzed so that the phenetic hyperspace is summarized in a space of only a few dimensions. In two dimensions this is a scattergram of the positions of OTUs from which one can recognize clusters by eye. Three-dimensional perspective drawings can also be made. The commonest ordination methods are **principal components analysis** and **principal coordinates analysis.**

A number of other representations are also used. One example is a similarity matrix in which the OTUs have first been rearranged into the order given by a clustering method and then the cells of the matrix have been shaded, with the highest similarities shown in the darkest tone. In these "shaded *S* matrices," clusters are shown by dark triangles. Another representation is a table of the mean similarities between OTUs of the same cluster and of different clusters (**inter-** and **intra-group similarity table**): if based on S_{SM} with UPGMA clustering, this table expresses the positions and radii of clusters (Sneath, 1979a) and consequently the distance between them and their probable overlap—properties of importance in numerical identification as discussed later.

For general purposes a dendrogram is the most useful representation, but the others can be very instructive, since each method emphasizes somewhat different aspects of the taxonomy.

The analysis for taxonomic structure should lead logically to the establishment or revision of taxonomic groups. We lack, at present, objective criteria for different taxonomic ranks, that is, one cannot automatically equate a phenon with a taxon. It is however commonly found that phenetic groups formed at about 80% *S* are equivalent to bacterial species. Similarly, we lack good tests for the statistical significance of clusters and for determining how much they overlap, though some progress is being made here (Sneath, 1977, 1979b). The fidelity with which the dendrogram summarizes the *S* matrix can be assessed by the **cophenetic correlation coefficient**, and similar statistics can be used to compare the **congruence** between two taxonomies if they are in quantitative form (e.g. phenetic and serological taxonomies). Good scientific judgement in the light of other knowledge is indispensible for interpreting the results of numerical taxonomy.

Descriptions of the groups can now be made by referring back to the original table of strain data. The better diagnostic characters can be chosen—those whose states are very constant within groups but vary between groups. It is better to give percentages or proportions than to use symbols such as +, (+), *v*, *d*, or − for varying percentages, because significant loss of statistical information can occur with these simplified schemes. It would, however, be superfluous to list percentages based on very few strains. As systematic bacteriology advances, it will be increasingly important to publish the actual data on individual strains or deposit it in archives; such data will show its full value when test methods become very highly standardized.

It is evident that numerical taxonomy (and also numerical identification; see the chapter on "Identification of Bacteria" in this *Manual*) place considerable demands on laboratory expertise. New test methods are continually being devised. New information is continually being accumulated. It is important that progress should be made toward agreed data bases (Krichevsky and Norton, 1974), as well as toward improvements in standardization of test methods in determinative bacteriology, if the full potential of numerical methods is to be achieved.

Nucleic Acids in Bacterial Classification

John L. Johnson

Historically, classification of bacteria has been based on similarities in phenotypic characteristics. Although this method has been quite successful, it has not been precise enough for distinguishing superficially similar organisms or for determining phylogenetic relationships among the bacterial groups. Nucleic acid studies were first applied to such problems in bacterial classification more than 20 years ago and have since become of major importance. There are several advantages to be gained by basing classification on genomic relatedness:

1. A more unifying concept of a bacterial species is possible,
2. Classifications based on genomic relatedness tend not to be subject to frequent or radical changes,
3. Reliable identification schemes can be prepared after organisms have been classified on the basis of genomic relatedness,
4. Information can be obtained that is useful for understanding how the various bacterial groups have evolved and how they can be arranged according to their ancestral relationships.

The purpose of this chapter is to provide an overview of the principles involved in nucleic acid methodology, to give a brief description of the procedures being used, to compare the results obtained by one procedure with those obtained by another, and to indicate how the results are being used in bacterial classification.

PROPERTIES OF NUCLEIC ACIDS

DNA Base Composition

The first unique feature of DNA that was recognized as having taxonomic importance was its mole percent guanine plus cytosine content (mol% G + C). Among the bacteria, the mol% G + C values range from ca. 25–75 and the value is constant for a given organism. Closely related bacteria have similar mol% G + C values. However, it is important to recognize that two organisms that have similar mol% G + C values are not necessarily closely related; this is because the mol% G + C values *do not take into account the linear arrangement of the nucleotides in the DNA.*

Mol% G + C values were initially determined by acid-hydrolyzing the DNA, separating the nucleotide bases by paper chromatography, and then eluting and quantifying the individual bases. Other methods have since become more popular.

Thermal Denaturation Method. During the controlled heating of a preparation of double-stranded DNA in an ultraviolet spectrophotometer, the absorbance increases by $\sim 40\%$. This is due to the disruption of the hydrogen bonds between the base pairs that link the two DNA strands. *The temperature at the midpoint of the curve obtained by plotting temperature versus absorbance is called the "melting temperature," or T_m.* The T_m is correlated in a linear manner with the mol% G + C content of the DNA (Marmur and Doty, 1962). The higher the T_m, the higher the mol% G + C of the DNA (see Johnson, J.L. (1981) for further details).

Buoyant Density Method. When DNA is subjected to centrifugation in a cesium chloride density gradient (isopycnic centrifugation), it will become located in the form of a band at a position where its density exactly matches that of the cesium chloride solution. The higher the density of cesium chloride where the DNA forms a band, the higher is the mol% G + C value of the DNA (Schildkraut et al. (1962), also see Mandel et al. (1968) for further details).

Although these methods are widely used for estimating DNA base composition, technical problems occasionally do arise because of contamination of the DNA preparation by polysaccharides or pigments, or because of excessive fragmentation of the DNA during its purification. Recent developments in high pressure liquid chromatography have resulted in methods that will accurately and rapidly quantify the free bases, nucleosides, or nucleotides of DNA (e.g. see Ko et al. 1977).

DNA Denaturation and Renaturation

A unique physical property of double-stranded (native) DNA is that under certain conditions (high temperature or high pH) the complementary strands will dissociate (denature). When the resulting single-stranded DNA is then subjected to a somewhat lower temperature and a rather high salt concentration, the complementary strands will reassociate (renature) to form double-stranded DNA/DNA structures (duplexes) that are very similar if not identical to the native DNA (Marmur and Doty, 1961). The renaturation rate is inversely proportional to the genome size (see the following references for further details: Wetmer and Davidson, 1968; Wetmer, 1976).

RNA/DNA Hybrids

Since only one strand of DNA is used by a cell as a template for RNA synthesis, RNA is complementary only to that strand. Since RNA is single-stranded, RNA molecules do not associate with other RNA molecules; however, when mixed with denatured DNA they can pair with a complementary DNA strand (hybridization) (see Galau et al., 1977 for further details).

Heterologous DNA Duplexes or RNA Hybrids

If denatured DNA from one organism is mixed with denatured DNA from a second organism, heterologous duplexes may form (i.e. duplexes consisting of one strand from the first DNA hybridized with one strand from the second DNA). Similarly, heterologous RNA duplexes may be formed when RNA from one organism is mixed with denatured DNA from a second organism. However, in order for heterologous DNA duplexes or RNA hybrids to occur, the two strands must be complementary in their nucleotide base sequence. A perfect match is not required, and estimates of the amount of base pair mismatch that is tolerated range from $\sim$ 8–10% (Ullman and McCarthy, 1973). The thermal stability is usually determined by measuring strand separation during stepwise increases in temperature, and the results mimic the optical melting profile previously discussed under DNA Base Composition. The thermal stability is usually represented by the term "$T_{m(e)}$," which is the midpoint of the thermal stability profile (i.e. analogous to T_m of native DNA). The difference between the $T_{m(e)}$ of a heterologous duplex and that of a homologous duplex is referred to as the $\Delta T_{m(e)}$ and is used as a measure of the degree of base-pair mismatching in the heterologous duplex. The $\Delta T_{m(e)}$ values for heterologous duplexes range from 0 (no mismatching) to 18°C (considerable mismatching). In general, as the fractions of the genomes which can form heterologous duplexes decrease, the thermal stabilities of the duplexes that do form also decrease.

DNA AND RNA HOMOLOGY EXPERIMENTS

Such experiments attempt to answer one question: does DNA or RNA from organism A have a base sequence that is sufficiently similar to that from organism B to allow the formation of DNA heteroduplexes or heterologous RNA hybrids?

DNA Homology Values

These are average measurements of similarity in which the *entire genome of one organism is compared with that of another.*

RNA Homology Values

These values are specific for each type of RNA:

Messenger RNA (mRNA) Homology Values. These are similar to those obtained by DNA homology (at least for bacteria) because a large portion of the genome is used for transcribing the mRNA molecules. For this reason, and because mRNA is difficult to label, mRNA homology has not been widely used in bacterial taxonomy.

Ribosomal RNA (rRNA) and Transfer RNA (tRNA) Homology Values. In contrast to mRNA, rRNA and tRNA are coded for by *only a small fraction of the bacterial genome*; therefore, in homology experiments using either of these two types of RNA, only those fractions of the genome are being compared, not the entire genomes. In all groups of bacteria so far studied, the arrangement of nucleotides in the rRNA and tRNA cistrons of the DNA appears to have evolved less rapidly than the bulk of the cistrons in the DNA. This is probably due to their role in determining the structural and functional aspects of the ribosome (Woese et al., 1975).

Therefore, DNA homology experiments are used to detect similarities between *closely related* organisms, whereas RNA homology experiments are used to detect similarities between *more distantly related* organisms.

METHODS FOR HOMOLOGY EXPERIMENTS

Many procedures have been developed for detecting heterologous DNA duplexes or RNA hybrids. A brief description of some of these follows.

Heavy Isotopes

The earliest efforts to quantify the formation of heteroduplexes were made by incorporating a heavy base (5-bromouracil) or a heavy isotope (^{15}N) into one of the DNA preparations. After the labeled and unlabeled DNA preparations were mixed and allowed to reassociate, the mixture was subjected to ultracentrifugation with cesium chloride. This allowed the separation of heteroduplexes (which had an intermediate buoyant density) from the homologous duplexes (which had either a light or heavy density). These experiments were time-consuming and worked best only for small genomes such as those of viruses.

Agarose Gels

In 1963, McCarthy and Bolton immobilized high molecular weight denatured DNA in an agarose gel. The gel was then cut into small particles by forcing the agar through a small mesh screen. The agar particles were then incubated with radioactive-labeled RNA or fragmented DNA. The smaller RNA molecules or DNA fragments could diffuse through the agar and form hybrids or duplexes with complementary immobilized DNA. The immobilization of the high molecular weight DNA prevented it from reassociating with other high molecular weight DNA and also provided a means for washing unreacted labeled nucleic acid fragments away from those that had formed hybrids or duplexes with the immobilized DNA. The results from such experiments were quantitative and could be readily applied to broad taxonomic studies (Hoyer et al., 1964).

Binding to Nitrocellulose

In 1963, Nygaard and Hall found that native DNA, denatured DNA, and RNA/DNA hybrids would bind to nitrocellulose whereas RNA would not. This provided another means for immobilizing denatured DNA for use in RNA/DNA hybridization experiments and also for separating RNA/DNA hybrids from free RNA. The parameters for these experiments were worked out in detail by Gillespie and Spiegelman (1965).

In 1966, Denhardt described a procedure for covering the DNA binding sites on nitrocellulose membranes. This made it possible to first immobilize a given amount of denatured DNA on the membrane and then treat the membrane with a mixture that prevented additional DNA from binding to the membrane (unless it was complementary to the immobilized DNA on the membrane). Thus the membrane procedure became readily applicable to DNA homology experiments and has completely replaced the agarose gel method.

By the use of nitrocellulose membranes, DNA or RNA homology values can be determined by either *direct binding* or by *competition* experiments.

Direct Binding Method. In the direct binding method, a given amount of denatured labeled DNA or RNA is incubated under standardized conditions with various single-stranded DNA preparations that have been immobilized on nitrocellulose membranes. After incubation the unbound labeled nucleic acid is washed away and the radioactivity remaining on the membrane (due to duplex or hybrid formation) is measured. The *percent homology* is expressed as the *amount of heterologous binding divided by the amount of homologous binding × 100*. The results are somewhat variable because it is difficult to consistently get the same amounts of DNA on the membranes. This problem is circumvented with the competition method.

Competition Method. In the competition method, unlabeled denatured reference DNA is fixed onto nitrocellulose membranes. A direct binding reaction, used for a reference point, is performed between the homologous denatured labeled DNA in solution and membrane-bound reference DNA. The competitive reactions have the same components as the direct binding reaction but additionally contain high concentrations of unlabeled denatured DNA fragments in solution. If the competitor DNA is homologous to the labeled DNA in solution and to the unlabeled DNA bound to the membrane, the competitor DNA will form duplexes with both the labeled DNA and the immobilized DNA: consequently, the amount of labeled DNA that forms duplexes with the immobilized DNA will be much lower than that occurring in the direct binding reaction. The homologous competition will be ~90% effective. On the other hand, if the competitor DNA is not related, it will not form duplexes with the labeled DNA and immobilized DNA and there will be no competition. The percent homology is the ratio of the heterologous competition to the homologous competition × 100. Such competition experiments give very reproducible results but do require relatively large quantities of DNA (Johnson, J.L., 1981).

Free Solution Reassociation

In this method all of the component nucleic acids are in solution rather than being immobilized in some manner. Reassociation of DNA may be monitored optically by ultraviolet spectrophotometry or by means of a labeled probe.

Optical Procedure. In the optical procedure, the rates of reassociation are determined. Since DNA reassociation is a 2nd-order reaction, the rate will be proportional to the square of the concentration. The general procedure for comparing the DNAs from two organisms is to measure the reassociation rates of equivalent concentrations from each of the organisms separately and compare those rates with that of an equal mixture of the two DNA preparations. If the two organisms are identical, the reassociation rates in the three cuvettes will be the same. If the two organisms are unrelated, then each kind of DNA in the mixture will reassociate independently of the other and, since they are each at half the concentration as that used in the cuvettes with a single DNA component, the overall rate will be one-half. De Ley et al. (1970) have studied the parameters of the method in detail and have derived equations for calculating the homology values.

Labeled DNA Probe. The most popular procedure for free solution reassociation involves the use of a labeled DNA probe. As discussed above, the rate of DNA reassociation is a function of DNA concentration and, because the labeled probe DNA is used at a very low concentration, very little of it will reassociate. The unlabeled test DNA with which the probe DNA is incubated is at a much higher concentration and most of it will reassociate. Therefore, if the probe DNA is identical to the unlabeled test DNA, it will reassociate with the unlabeled DNA at the rate at which the unlabeled DNA is reassociating. On the other hand, if the two DNAs are unrelated the unlabeled DNA will reassociate but most of the probe DNA will remain single-stranded. To determine the amount of probe DNA that has duplexed with the unlabeled DNA, either *hydroxylapatite* or *S1 nuclease* is usually used.

Hydroxylapatite is used to separate single-stranded (denatured) DNA from double-stranded DNA. At a phosphate concentration of 0.14 M, only double-stranded DNA will adsorb to hydroxylapatite and single-stranded DNA can be washed away. The double-stranded DNA can then be desorbed by increasing the phosphate concentration. Although originally used as a column chromatography procedure (Bernardi, 1969a, b; Miyazawa and Thomas, 1965), the batch procedure described by Brenner et al. (1969) has been widely used.

Under suitable conditions, S1 nuclease will have little effect on double-stranded DNA but will hydrolyze single-stranded DNA. Consequently, the extent of duplex formation by the probe DNA can be determined by the amount of S1 nuclease-resistant (i.e. acid-precipitable) radioactivity (Crosa et al., 1973).

Comparison of the Various Homology Methods

In spite of the diversity of the DNA homology methods, they are all used to measure the same phenomenon and so it is comforting to find that, for the most part, they all give similar results. The major experimental parameters that affect homology results are the sodium ion concentration and the reassociation temperatures. The most commonly used sodium ion concentration is about 0.4 M although concentrations up to 1 M do not alter the results significantly. The reassociation temperature can have a profound effect on the homology values and therefore a standardized temperature of about 25°C below the T_m (T_m − 25 °C) is most commonly used (Marmur and Doty, 1961). The reassociation temperature effect is approximately linear for the membrane competition and the hydroxylapatite procedures: for organisms having less than 50% homology, the homology values will increase by about 20% at 10 C below the T_m − 25°C temperature and decrease by about 20% at 10°C above the T_m − 25°C temperature. Reassociation temperature differences do not have as great an effect on the optical (De Ley et al., 1970) or the S1 nuclease methods (Grimont et al., 1980).

Under similar conditions of reassociation, the hydroxylapatite, membrane competition and spectrophotometric methods give very similar results (Kurtzman et al., 1980). The S1 nuclease procedure results in somewhat lower (15–20%) homology values, particularly between organisms having less than 50% homology.

The rRNA cistrons have been found to be very conserved in all groups of organisms that have been investigated. The nitrocellulose membrane procedures, such as competition, direct binding and thermal stability of hybrids have been used for most of the rRNA homology studies. Results from these experiments appear to reflect nucleotide sequence differences that are similar to those found in the DNA homology experiments discussed above.

rRNA OLIGONUCLEOTIDE CATALOGUES

Besides the use of RNA/DNA homology experiments for comparison of the rRNA cistrons from various bacteria, rRNA molecules have been compared directly by determining the nucleotide sequences in oligonucleotides. The rRNA preparation is first digested with T1 ribonuclease which cleaves between the 3′-guanylic acid and the 5′hydroxyl group of the adjacent nucleotide. This results in a guanine residue at the 3′ end of each oligonucleotide. The oligonucleotides are then separated by two-dimensional electrophoresis (Sanger et al., 1965; Uchida et al., 1974). The first dimension is on cellulose acetate at pH 3.5. The oligonucleotides are then transferred from the cellulose acetate strip onto DEAE cellulose and electrophoresed in the second dimension in 6.5% formic acid. The oligonucleotide spots form three-to-four series of wedge-shaped patterns (Sanger et al., 1965). Within each pattern the oligonucleotides contain a constant number of uracil residues and the locations of the spots within a pattern indicate the number of adenine and cytosine residues. Therefore, by inspecting the pattern one can predict the nucleotide sequence of the shorter oligonucleotides and the base compositions for the longer ones. The spots containing the longer nucleotides are then cut out for secondary analysis. After digestion with other ribonucleases they are again electrophoresed on

DEAE cellulose. If the nucleotide sequence still is not clear, a tertiary analysis is required. The unique oligonucleotides (usually only one per rRNA molecule) of each organism are entered (cataloged) into computer storage. The oligonucleotide catalog from one organism can then be compared with that of another. The similarity values between two organisms is the number of unique oligonucleotides (in each of their rRNA molecules) that they both share divided by the average total number of unique oligonucleotides. This procedure compares the sequence for a rather large portion of the rRNA molecules.

Most recently, procedures have been developed for rapidly sequencing long segments of DNA and RNA (Maxam and Gilbert, 1977; Peattie, 1979; Sanger et al., 1977). DNA from several viruses have been sequenced. Sequencing all of the DNA of a bacterium would generate a rather formidable amount of data; however, specific cistrons have been compared by sequence analysis, such as the genes of the tryptophan operon of *Escherichia coli* and *Salmonella typhimurium* (Crawford et al., 1980).

CONTRIBUTIONS OF NUCLEIC ACID STUDIES TO BACTERIAL TAXONOMY

Concept of a Bacterial Species

A major contribution of DNA homology studies has been to provide a more unifying concept of a bacterial species. Although the exact level of DNA homology above which one considers organisms as belonging to the same species is arbitrary, similar homology clusters have been found in all bacterial groups that have been investigated. I have previously suggested what seemed to be reasonable cut-off points for delineating subspecies, species, and closely related species (Johnson, 1973). These are illustrated in Figure III.1. DNA heterogeneity in the species range (*A*) has been found for many bacterial groups that are phenotypically very similar. In some instances the homology values will tend to cluster in the 80–90% homology range (*B*). Examples of this are the clustering of *Propionibacterium acnes* (Johnson and Cummins, 1972) and *Bacteroides uniformis* (Johnson, 1978). In other instances there may also be clustering at the lower end of the species range (*C*). *Bacteroides fragilis*, for example, clusters into two groups where the intergroup homology values are in the range of 60–70% and the intragroup homology values are in the 80–90% range (Johnson, 1978). It is important to note that the thermal stabilities of heteroduplexes between organisms in the 80–90% DNA homology range will be very similar to those of homoduplexes ($\Delta T_{m(e)}$ values of 0–3°C), whereas with heteroduplexes between 60–70% homology they will be substantially lower ($\Delta T_{m(e)}$ values of 6–9°C). Therefore, it appears that 60–70% homology is a transitional point between genetic events that may be largely cistron-rearranging in nature and genetic events where there are also many changes in the base sequences (Johnson, 1973). In other instances, e.g. *Bacteroides ovatus* (Johnson, 1978), multiple groups within the 60–70% homology range make subgrouping at this level rather complicated so that, unless there are other important considerations, such as pathogenicity (Krych et al., 1980), it may not be justified.

The DNA homology groups in the lower homology range (*D* in Fig. III.1) often are quite distinct phenotypically from the species with which they are being compared, although in some instances they may differ only in a few characters (Johnson and Ault, 1978; Johnson, 1981; Mays et al., 1982).

It is important to remember that few bacteria have read Fig. III.1; therefore, the exact limits chosen for a given group of organisms will have to remain at the discretion of the individual investigator.

Identification Schemes

A major practical use of DNA homology data is for correlation with individual phenotypic tests. It is common to find variability for a trait

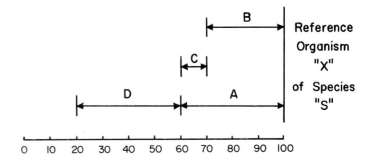

PERCENT DNA HOMOLOGY

Figure III.1. Proposed taxonomic groupings based upon DNA homology data. *A*, organisms belonging to Species "*S*"; *B*, varieties within subspecies to which "*X*" belongs; *C*, other subspecies that belong to "*S*"; *D*, species that are closely related to Species "*S*".

among strains within a DNA homology group as well as distinct DNA homology groups that differ from each other by only a few traits (Johnson and Ault, 1978; Johnson, 1980; Mays et al., 1982; Holdeman et al., 1982). Correlating phenotypic test results with DNA homology groups enables investigators to select phenotypic tests that are required for the accurate identification of organisms belonging to these groups.

Concept of a Bacterial Genus and Higher Taxa

Comparisons of rRNA cistrons by rRNA homology experiments and by 16S oligonucleotide catalog similarities are providing data from which a more unifying phylogenetic concept for higher bacterial taxa is possible. De Ley and his associates (De Ley et al., 1978; De Smedt et al., 1980) have proposed the establishment of several genera on the basis of rRNA homology results. On the basis of 16S rRNA oligonucleotide similarity, Woese (in Fox et al., 1980) has proposed the reestablishing of the higher bacterial taxa which were dropped from the eighth edition of *Bergey's Manual* because it was thought that the higher taxa listed in the seventh edition did not represent phylogenetic relationships. As examples, the 16S rRNA oligonucleotide similarity values have contributed greatly to the present taxonomic scheme of the methanogenic bacteria (Balch et al., 1979) and to the establishment of Division IV *Mendosicutes* in the Kingdom *Procaryotae* (see the chapter on The Higher Taxa by Murray in this *Manual*).

BACTERIAL CLASSIFICATION IV

Genetic Methods

Dorothy Jones

The use of genetic characteristics in bacterial classification is comparatively recent. It dates from the mid 1950s when bacterial gene transfer was discovered and Watson and Crick demonstrated the molecular basis of genetic information in the sequence of bases on the deoxyribonucleic acid (DNA) molecule. Since that time the development of physicochemical techniques for the analysis of the genetic material, together with the exploitation of bacteria as genetic tools, has resulted in the accumulation of material which has proved significant for bacterial systematics.

In the last 2 decades it has become clear that the genetic complement of a bacterial cell lies not only in the main chromosome but, in many cases, also in extrachromosomal elements such as plasmids, transposons and lysogenic or temperate phages. All these elements carry genetic material capable of phenotypic expression. What contribution such extrachromosomal entities make to a particular bacterial phenotype, either by direct expression or interaction with the chromosomal DNA of the cell, is only just beginning to be understood (see Broda, 1979; Harwood, 1980; Hardy, 1981).

For the bacterial taxonomist the genetic approach to systematics has great appeal both for its potential to reveal biologically significant, stable groupings (taxa) and for the elucidation of bacterial evolutionary relationships (phylogeny). Consequently several of the newer taxonomic methods have been and are being directed towards the characterization of the genetic complement of bacteria.

Physicochemical methods for the analysis of bacterial genomes have been discussed in the previous chapter. The present chapter is concerned with genetic methods used in bacterial classification, i.e. methods based on the transfer of genes between bacteria.

CHROMOSOMAL GENE EXCHANGE

The three main classes of chromosomal gene exchange are: (a) those in which genes are transferred as soluble DNA molecules, i.e. **transformation**; (b) those involving transfer by bacteriophage i.e. **transduction**; and (c) those involving cell contact followed by transfer of whole or part of the bacterial chromosome, i.e. **conjugation**. Of these classes, transformation studies have so far proved the most useful for determining relationships between bacteria.

Transformation

Transformation has been demonstrated usually between different taxospecies and only rarely between taxa presently recognized as different genera. Interspecific transformation has revealed three distinct homology groups amongst neisseriae and moraxellae. Transformation studies have indicated a close relationship between *Rhizobium leguminosarum* and *Agrobacterium tumefaciens*. Studies with the micrococci have shown a close relationship between *Micrococcus luteus* and *Micrococcus lylae* and, in this case, was confirmed by DNA reassociation studies in vitro. Similar studies have shown a low rate of transformation between *Pasteurella multocida*, *P. haemolytica*, *P. ureae* and *P. pneumotropica*, taxa which are also closely related on phenetic and DNA reassociation criteria. A great deal of transformation work has been done on the genus *Haemophilis*. *Haemophilus influenzae*, *H. aegyptius* and *H. parainfluenzae* all appear to be closely related.

Transformation of chromosomal DNA has been demonstrated also among other taxa and there is no doubt that it is a good indication of the degree of relatedness between different taxospecies and can highlight areas of taxonomic homogeneity and heterogeneity (Jones and Sneath, 1970; Bøvre, 1980).

Transduction

In transduction, host chromosomal material is incorporated into a bacteriophage by several mechanisms and transmitted from one bacteriophage host to another by phage-mediated transduction. Only a small range of bacterial groups are presently known to be susceptible to transduction, e.g. the *Enterobacteriaceae*, the genus *Bacillus*, pseudomonads and some streptococci. Not much is known about how readily strains of the same species can be transduced, but the host range pattern of the transducing bacteriophages is probably a major limiting factor. It has been suggested also that the greater difficulties associated with transduction are due to the larger sizes of the DNA fragments involved in transduction as compared with transformation, the larger fragments being less easily integrated into the recipient chromosome. Again this mechanism of genetic transfer appears to have significance for bacterial classification only at the taxospecies level, and its usefulness is further restricted by the host range of bacteriophages (see Jones and Sneath, 1970).

It is appropriate here to mention the other roles of bacteriophages in bacterial classification. As noted earlier, a temperate bacteriophage can lysogenize in a host bacterium and express its genetic information as phenotypic characters different from those typical of the bacterium devoid of phage. The consequences of this for bacterial classification will be dealt with later (see Extrachromosomal Elements). Additionally, and this is perhaps their best known feature, virulent bacteriophages infect and lyse bacteria. The process is referred to as **phage lysis**.

The inclusion of phage lysis in a section dealing with genetic methods may cause the reader some surprise. Phage lysis of the bacterial cell (as distinct from the much less specific phage adsorption, or killing of the cell followed by lysis from without) involves phage infection with phage multiplication but without lysogenization. In bacteriophage infection, the genes of the virulent phage are transferred and expressed even though they are not integrated into the host chromosome nor of course in the lineage of the recipient. Specific phage receptors are necessary for the adsorption of virulent phage on to the recipient bacterial cell; once in the cell the phage may be repressed if the bacterium is carrying a homologous prophage or it may be restricted enzymically. The ability of two bacterial strains to support the growth of a given virulent phage may reflect similarity in only one or two host genes. Therefore, the technique has little value for *bacterial classification*. However, the value of phage lysis cross-reactions for *bacterial identification* is high. The reported host range of bacteriophages extends from those specific for very few strains of one taxospecies, to those that can lyse bacteria which are currently placed in different bacterial genera, families and even orders. However, most reports in the literature show that most phages lyse a significant proportion of strains belonging to the same taxospecies as the propagating strain. Phage-typing schemes are playing increasingly important epidemiological and identification roles among a number of bacterial groups, e.g. some pyogenic streptococci, staphylococci and enterobacteria.

Conjugation

This method of gene exchange refers to the transfer of the whole or a portion of the bacterial chromosome following cell-to-cell contact. The conjugation system is best understood among the coliforms (Curtiss, 1969). Similar systems have been noted amongst other genera such as *Pseudomonas*, *Vibrio*, *Pasteurella* and *Rhizobium* and are known to occur among other groups, but the mechanism is less well understood. In the streptococci there is evidence that in some cases the bacteria make use of sex pheromones to generate cell-to-cell contact (Clewell, 1981). Transfer of bacterial chromosomal material by conjugation has not been reported so frequently as by transformation or transduction. However, evidence suggests that it takes place only between closely related taxa.

Bacterial taxonomy has not, to date, benefited greatly from studies involving genetic exchange of chromosomal material and the concept of a bacterial genospecies is far from being realised. However, bacteriologists no longer believe that gene transfer is so rare among bacteria that it is of no consequence for natural bacterial populations. In the past 2 decades it has been recognized that gene transfer particularly involving phages, plasmids and transposons, their interaction with each other, and the bacterial chromosome together with the gene transfer mediated by insertion sequences, can be a significant factor in bacterial variation. This variation has obvious consequences for bacterial systematics.

EXTRACHROMOSOMAL ELEMENTS

Plasmids, transposons and phages are collectively referred to as extrachromosomal elements (Novick, 1969; Broda, 1979; Hardy, 1981). Their transfer between bacteria is essentially by the same mechanisms as those described under Genetic Methods. Phages play a role in the transduction of all genetic material between bacteria and it is now recognized that the F′ factor is a plasmid. It is therefore probably artificial to make too clear a distinction between the transfer of chromosomal DNA and that of extrachromosomal elements between bacteria. Transformation by chromosomal DNA may or may not be accompanied by plasmid DNA. In transduction phages can carry a portion of the chromosomal or plasmid DNA, and in conjugation plasmid and chromosomal DNA can be transferred at the same time. The situation is far more complex than was previously realized (Novick, 1969; Hardy, 1981; Clewell, 1981).

A range of methods now exists for the isolation of extrachromosomal genetic elements from bacteria and for their analyses by physicochemical methods. A good account is given by Hardy (1981).

The two aspects of extrachromosomal elements which are of prime interest to the bacterial taxonomist are their ability to code for phenotypic traits in a range of bacteria, and their significance in evolution.

Phenotypic Traits

Plasmids have been observed in virtually every bacterial genus examined. Many plasmids detected by physical screening methods are not known to code for any phenotypic trait in the host bacterium. They are called **cryptic plasmids**. The fact that their presence has not been correlated with a phenotypic characteristic does not mean that they do not code for such a trait. It may be that their particular phenotypic traits have not been identified.

Phenotypic traits known to be coded for by plasmids include resistance to a variety of antibiotics, heavy metal ions and ultraviolet light; production of enterotoxin, exfoliate toxin, the surface antigens K88 and K89, hemolysins, proteases, bacteriocins, urease and H_2S; metabolism of lactose, sucrose, raffinose and citrate; degradation of a variety of organic compounds such as camphor, octanol and toluene (at least part of the remarkable diversity shown by pseudomonads in the degradation of organic compounds is due to the presence of degradative

plasmids); and nitrogen fixation. Preliminary evidence suggests that the production of gas vacuoles in *Halobacterium* is controlled by a plasmid. There is also evidence that pigment, coagulase and fibrolysin production in staphylococci are plasmid determined. It also seems highly probable that among the streptococci the production of serum opacity factor, M protein production, nisin production and the ability to ferment galactose and xylose are plasmid-coded.

Transposons found on the plasmids of Gram-negative bacteria have been shown to code for resistance to a number of antibiotics, lactose fermentation (in *Yersinia enterocolitica*), heat stable toxin (in *E. coli*) and doubtless others coding for other phenotypic traits will be found. Full accounts of the phenotypic traits conferred on bacteria by plasmids and transposons are given by Harwood (1980), Clewell (1981) and Hardy (1981).

The classic example of a phage-encoded phenotypic trait is the diphtheria toxin which was shown by Freeman in 1950 to be produced only when *Corynebacterium diphtheriae* is lysogenized by a particular phage. The structural gene for the protein toxin is on the phage chromosome. This phage can lysogenize and synthesize toxin in a number of closely related corynebacteria viz. *C. diphtheriae*, *C. ulcerans* and *C. ovis* (Barksdale, 1970).

Effect of Extrachromosomal Elements on Classification and Identification

Since these elements confer extra phenotypic traits on their hosts they could have a marked effect on bacterial classification if those characters were ones on which the classification was based. Two examples of the presence of plasmids which relate to species nomenclature are the plasmid-coded hemolysin of *Streptococcus faecalis* which resulted in the naming of such plasmid-bearing strains as *Streptococcus faecalis* var. *zymogenes*, and the plasmid-determining citrate utilization in *Streptococcus lactis* which appears to be responsible for the name *Streptococcus lactis* subsp. *diacetylactis*. However, the effect of an extra chromosomal-coded trait on a classification based on a large number of characters would normally be expected to be small and this has proved to be the case in the few preliminary studies so far conducted.

Such characters can, however, affect the identification of bacteria when the identification is based on a small number of characters and considerable weight is placed on individual features, e.g. lactose fermentation in the identification of enterobacteria. It is best therefore if identification schemes are based on stable features chosen as a result of a taxonomic study where a large number of characters have been employed, e.g. computer-assisted classifications (numerical taxonomy). Ideally, computer-based identification matrices derived from such studies should be employed. The risk of a misidentification due to the loss or gain of one or two phenotypic characters is thereby reduced to a minimum.

It has been suggested that strains known to carry extrachromosomal elements should be excluded from taxonomic studies. Such a policy is not practical because present methods do not always detect such strains; further, it is believed that many bacterial populations depend on the presence of these elements for their survival. It has also been suggested that known or suspected extrachromosomal coded characters should be excluded when classifications are constructed. Again, present methods do not allow all such characters to be determined, besides, such characters may have taxonomic relevance.

Bacteriologists should accept that extrachromosomal elements do contribute to bacterial variation. This variation should therefore be recognized, and due allowance made, when bacterial taxa are described and when identification schemes are constructed.

Extrachromosomal Elements and Evolution

At the present time the relative contributions of mutation and recombination to bacterial evolution is difficult to assess. Mutation results in changes in the protein structure of the organism. Recombination leads to the rearrangement of existing genes. Until recently little attention was paid to the possible involvement of gene rearrangement in evolution. The recognition that gene transfer involving extrachromosomal elements can be a significant factor in bacterial variation has led to a view that these elements have played an important role in bacterial evolution. Whether or not the role which these elements play in contemporary bacterial variation and adaptation is one of the major ways in which bacteria have evolved from the earliest times is still not resolved (Cullum and Saedler, 1981; Hardy, 1981; Koch, 1981; Reanney, 1976).

Serology and Chemotaxonomy

Dorothy Jones and Noel R. Krieg

Serology and chemotaxonomy are both methods for investigating the molecular architecture of the bacterial cell although the methodologies used in the two techniques are quite different.

SEROLOGY

Serological techniques depend on the ability of the chemical constituents of bacterial cells to behave as antigens, i.e. to elicit the production of antibodies in vertebrate animals. The antibodies used in serological studies are the humoral antibodies found in the blood serum and referred to as antiserum.

Serological techniques used include agglutination, precipitation (including many refinements, e.g. use of gels and electrophoretic techniques), complement fixation and immunofluorescence. Details of the techniques may be found in a number of immunological or microbiological text books.

Serological studies of value in bacterial taxonomy can be divided into two broad classes: (a) those concerned with detecting differences or similarities between bacteria **on the basis of their cell surface and associated antigenic complement** (e.g. flagella, pili, cell walls, cytoplasmic membranes, capsules and slime layers) and (b) the use of antisera raised against purified enzymes to assess **structural similarities between homologous proteins from different bacteria.**

Cell Surface and Associated Antigens

On the basis of the antigenic complexity of their surface antigens (cell wall lipopolysaccharide, flagella and capsule constituents), the genera of the family *Enterobacteriaceae* can be divided into many serovars; e.g. more than 1000 serovars have been detected within the genus *Salmonella* (Kauffmann, 1966). Contrary to the view of Kauffmann (1966) these serovars do not represent separate taxospecies. The information derived from serological studies of a group such as the enterobacteria is now so large, and so many cross-reactions occur that, in the absence of any methods (e.g. computer programs) for analyzing the plethora of data in an objective fashion, it is generally accepted that these techniques are of little value in classification but are valuable in epidemiological studies.

Serological studies of the streptococci based on the use of acid-extracted polysaccharide antigens (Lancefield, 1933, 1934) have resulted in the division of the genus *Streptococcus* into a number (now approaching 30) of serological groups labeled A, B, C, etc. Until fairly recently, very great emphasis was placed on the serological grouping of streptococci for purposes of both classification and identification.

Although some serological groups correspond to distinct taxospecies (e.g. serological group A (*S. pyogenes*) and serological group B (*S. agalactiae*)) other serolological groups comprise more than one taxo-

species (e.g. serological groups C, D and N), while serological groups G, H and K do not serve to define any good taxa (see Jones, 1978).

Other serological studies of this kind have been based on the use of different classes of antigenic material, e.g. cell walls, spore suspensions, etc. A review of the serochemical specificity and location of antigens in the bacterial cell together with observations on the significance of such serological studies for bacterial classification has been provided by Cummins (1962).

Use of Antisera Raised Against Purified Proteins

The basis of this approach is that one antiserum raised against a purified enzyme can be used to detect the serological cross-reactions of homologous proteins in crude extracts of other bacteria if the bacteria possess the same enzyme. The use of microcomplement fixation techniques makes this approach a very sensitive one. Comparative studies on purified proteins of known primary structure have indicated that there is a very high correlation between the amino acid sequence of the proteins and the degree of serological similarity (see later section on Amino Acid Sequences). Examples of this approach include studies of the muconate-lactonizing enzymes of the *Pseudomonadaceae* (Stanier et al., 1970), the fructose diphosphate aldolases of the lactic acid bacteria (London and Kline, 1973) and the catalases of staphylococci and micrococci. In the instance of the staphylococci, a very high correlation has been shown to exist between the serological relationships of their catalases and genetic relatedness based on DNA/DNA homology data (see Kandler and Schleifer, 1980).

Similar studies on the transaldolases of several species of bifidobacteria (Sgorbati and Scardovi, 1979; Sgorbati, 1979) indicate that the genus *Bifidobacterium* contains several distinct clusters based on the index of dissimilarity of their respective aldolases. In some instances there was good correlation between the clusters so obtained and clusters formed on the basis of other criteria, but in others the correlation was not so high.

Baumann et al. (1980) and Bang et al. (1981) found the immunological relationships among glutamine synthetases and superoxide dismutases of *Vibrio* and *Photobacterium* species were in good agreement with relationships based on rRNA/DNA homology experiments. The amino acid sequence of the glutamine synthetases was conserved to a greater extent than that of the superoxide dismutases, and this supports the idea that the study of proteins having different evolutionary rates can permit the resolution of close, intermediate, and distant relationships among organisms.

It should be noted that serological homology studies of proteins, like many other techniques, have their limitations. There is evidence that the approach is useful only for the study of proteins with relatively high (70% or greater) sequence homologies. Further, serological techniques measure similarities only at the surface of proteins and it is at the protein surface that the greatest number of amino acid changes occur. The results can be influenced also by the number of antigenic sites per protein molecule. Nevertheless, serological techniques of this kind provide a rapid and convenient method for assessing structural similarities between homologous proteins and are useful in the classification of bacteria and can also cast some light on possible phylogenetic relationships.

CHEMOTAXONOMY

During the past 20 years or so, the application of chemical and physical techniques to elucidate the chemical composition of whole bacterial cells or parts of cells has produced information of great value in the classification and identification of bacteria. Indeed, so useful have some of the data generated proved to be, that the word "chemotaxonomy" used to describe the classification of bacteria on the basis of their chemical composition, is now firmly entrenched in the literature.

In addition, techniques such as gas chromatography have allowed the more precise analysis of the products of fermentation and there is a growing awareness of the taxonomic significance of enzyme systems and their regulation as opposed to the detection of individual enzymes.

Cell Wall Composition

The characteristic cell wall polymer of many procaryotes, present in Gram-negative and Gram-positive bacteria and in the cyanobacteria, is peptidoglycan. Peptidoglycan is not found in the mycoplasmas nor in the archaebacteria. The chemical structure of the peptidoglycan of Gram-negative bacteria is, with few exceptions, reasonably uniform. However, the variation in qualitative amino acid and/or sugar composition, especially the variation in the primary structure of the peptidoglycans of various Gram-positive bacteria has provided information of enormous taxonomic value.

The cell wall composition of Gram-positive bacteria was one of the earliest useful chemotaxonomic characters. On the basis of the analysis of the purified cell walls of Gram-positive bacteria, Cummins and Harris (1956) suggested that the cell wall amino acid composition might prove to be an important taxonomic criterion at the generic level and that the sugar composition might help to distinguish between species. Subsequent studies have indicated this to be the case. The amino acids present in the cell wall are now an accepted important part of the generic description. Information from cell wall analysis of the type done by Cummins and Harris (1956) has proved of especial value among the coryneform group of bacteria (see Keddie and Bousfield, 1980, for a comprehensive review).

Information of even greater taxonomic value has resulted from the methods devised by Schleifer and Kandler (1967) and used by them and their associates to determine the peptidoglycan types of a wide range of Gram-positive bacteria (see Schleifer and Kandler, 1972; Kandler and Schleifer, 1980). This approach has revealed differences between bacteria which could not possibly have been detected by qualitative cell wall analysis. The methods for determining differences in peptidoglycan types are, however, quite specialized and cannot be used routinely to screen large numbers of bacteria.

More recently a number of "rapid" methods have been developed for the routine screening of bacteria to determine those cell wall components which have been shown to be of the greatest discriminatory value in bacterial classification and identification. The review of Keddie and Bousfield (1980) contains references to the pertinent literature as does that of Kandler and Schleifer (1980).

A novel peptidoglycan, the so-called pseudomurein, characterized by the replacement of muramic acid by talosaminouronic acid, has been found to be the typical cell wall constituent of the genus *Methanobacterium*, which taxon is now recognized as a member of the archaeobacteria (see Kandler and Schleifer, 1980).

Lipid Composition

Among the procaryotes there are two quite distinct lipid categories. The eubacteria possess acyl lipids (ester-linked) while the archaeobacteria possess ether-linked lipids. Thus, the presence of ether-linked lipids serves to distinguish the archaeobacteria. Further details of the archaeobacterial lipids are given by Kates (1978).

Lipids occur in the cytoplasmic membranes of all eubacteria and in the cell wall complex of Gram-negative bacteria and certain Gram-positive bacteria such as the genera *Corynebacterium* and *Mycobacterium*. The eubacterial lipids comprise a number of different classes and, in the last decade, it has become increasingly clear that at least some of these lipids have chemotaxonomic potential (see Lechevalier, 1977).

The fatty acid composition of the bacterial cell has proved useful in the classification of certain bacteria and in some cases the fatty acid pattern may be characteristic for a particular taxon (see Lechevalier, 1977). However, it should be noted that the fatty acid patterns obtained may be influenced by a number of factors: composition of growth medium, temperature of incubation, age of culture and by the techniques employed to analyze the sample.

A special category of fatty acids free from the aforementioned limitations are the mycolic acids. These long-chain 3-hydroxy 2-branched acids have been found, so far, only in the taxa *Bacterionema, Corynebacterium, Micropolyspora, Mycobacterium, Nocardia* and *Rhodococcus*. Differences in the structure of their component mycolic acids have proved to be a valuable criterion in the classification and identification of members of these taxa (see Minnikin and Goodfellow, 1980; Collins et al., 1982).

Another class of lipids of recognized chemotaxonomic potential are the polar lipids which occur in all bacteria. The most common polar lipid types are the phospholipids and the glycolipids. Phospholipids occur in many bacteria but certain actinomycetes and coryneform bacteria contain very characteristic phospholipids, the phosphatidylinositol mannosides. Other highly characteristic phospholipids include phosphosphingolipids found in certain Gram-negative taxa, e.g. *Bacteroides* (see Lechevalier, 1977). Glycolipids (glycosyl diacylglycerols) are widely distributed amongst Gram-positive bacteria and can also be used as chemotaxonomic markers (see Shaw, 1975).

Other lipids with chemotaxonomic potential include hopanoids, hydrocarbons, and carotenoids.

Isoprenoid Quinones

Isoprenoid quinones are a class of terpenoid lipids located in the cytoplasmic membranes of many bacteria. They play important roles in electron transport, oxidative phosphorylation and possibly active transport. Their potential as an aid to the classification of bacteria was recognized by Jeffries et al. (1969), Yamada et al. (1976) and others (see Collins and Jones, 1981). Representatives of one, or more than one, of the three main types, ubiquinones, menaquinones and demethylmenaquinones are present in the majority of procaryotes so far examined. The cyanobacteria contain neither ubiquinones nor menaquinones. However, they do contain phylloquinones and plastoquinones which are indigenous to the plant kingdom but not normally found in bacteria. All the mycoplasmas so far examined contain menaquinones only. Among the archaeobacteria no isoprenoid quinones have been detected in the fastidious anaerobic species *Methanobacterium thermoautotrophicum*, a situation in keeping with that most commonly, but not invariably, found among the strictly anaerobic eubacteria. An unusual terpenoid, caldariellaquinone, has been detected in the extreme acidophile "*Caldariella acidophila.*" The other archaeobacteria examined possess menaquinones.

The majority of the strictly aerobic, Gram-negative bacteria produce only ubiquinones, with the exception of cytophagas and myxobacters which produce only menaquinones. Facultatively anaerobic, Gram-negative bacteria contain ubiquinones, menaquinones or demethylmenaquinones or a combination of the three. Strictly anaerobic, Gram-negative bacteria (e.g. the genus *Bacteroides*) produce only menaquinones.

The majority of the aerobic and facultatively anaerobic Gram-positive bacteria produce only menaquinones. Most streptococci do not contain any isoprenoid quinones but demethylmenaquinones are present in *Streptococcus faecalis* and menaquinones have been detected in "*S. faecium* subsp. *casseliflavus*" and *S. lactis*. Similarly members of the genus *Lactobacillus* generally lack isoprenoid quinones but recently low levels of an uncharacterized menaquinone have been detected in one strain of *L. brevis*. Uncharacterized menaquinones have also been reported in some strains of the strictly anaerobic genus *Clostridium* although, in general, this genus lacks quinones.

The current data on isoprenoid quinone structural types in bacteria and their implications for taxonomy are reviewed by Collins and Jones (1981). From the available data on procaryotes in general, it appears that menaquinones have far greater discriminatory value that ubiquinones. Menaquinones possess not only a greater range of isoprenologues but additional modifications such as ring demethylation and partial hydrogenation of the polyprenyl side chain occur. The available data strongly suggest that these compounds will be of considerable value in the classification of micrococci, staphylococci, coryneform bacteria and certain actinomycetes (see Collins and Jones, 1981).

Cytochrome Composition

Cytochromes are specialized forms of hemoproteins which are involved in a variety of redox processes in the procaryote cell. They can be assigned to four main classes, *a*, *b*, *c* and *d* according to the structures of their heme prosthetic groups. Cytochrome *o* is an autoxidizable *b*-type cytochrome.

Two basic methods are available which use cytochromes as an aid to classification and identification of bacteria; the "pattern" and the "structure" approach. The former compares the cytochrome patterns of different bacterial species as compared by conventional difference spectrophotometry (see Meyer and Jones, 1973); the latter compares primary structures and, where possible, the tertiary structures of easily purified cytochrome *c* as determined by amino acid sequence and x-ray diffraction (see Ambler, 1976, and see later section on Amino Acid Sequences).

Cytochrome patterns show greater variation among the procaryotes than among eucaryotes and can therefore be a useful aid in bacterial classification. Qualitative analyses of the cytochrome composition of over 200 species of bacteria have now been done. The results indicate that the heterotrophic Gram-positive bacteria comprise a rather homogeneous grouping with cytochromes $bcaa_3o$ forming the predominant pattern. There are some variations however and cytochrome *c* is often absent from facultatively anaerobic Gram-positive bacteria. Some lactic acid bacteria contain only cytochrome *b* when grown on a heme-containing medium. Propionibacteria exhibit a cytochrome bda_1 pattern. The genus *Clostridium* lacks cytochromes. It is of interest that the cytochrome $bcaa_3o$ pattern of logarithmic growth-phase cells of the aerobe *Arthrobacter globiformis* changes to $bcaa_3od$ when the cells become oxygen-limited and lose their ability to retain the crystal violet-iodine complex in the Gram stain. Cytochrome *d* is characteristic of many Gram-negative bacteria.

In contrast, the Gram-negative heterotrophic bacteria form a much less homogeneous group on the basis of cytochrome composition. The majority have the basic pattern $bcdoa_1$ from which *c* may often be absent. Cytochrome c_{co} appears to be characteristic of methylotrophs; however, it is also present in the nonmethylotrohpic genus *Chromobacterium*. The phototrophic bacteria contain cytochromes *b* and *c* when grown photosynthetically but there are differences between the taxa when grown aerobically in the absence of light. The obligately aerobic chemolithotrophs exhibit the cytochrome pattern $bcaa_3oa$, with some occasional omissions. Neither the phototrophs nor the chemolithotrophs have been shown to produce cytochrome *d*.

There is now sufficient evidence available to indicate that cytochrome patterns, in conjunction with other evidence, are useful guides in bacterial classification. There is, however, little evidence that cytochrome patterns will be useful for purposes of identification mainly because bacteria contain relatively few types of spectrally distinct cytochromes. A comprehensive review of the use of cytochrome patterns in bacterial classification is that of Jones (1980).

It should be stressed that when cytochrome patterns are used for taxonomic purposes, the influence of the growth environment should be taken into account. Growth conditions can influence the quantitative and, to a lesser extent, the qualitative cytochrome content of bacteria.

Amino Acid Sequences of Various Proteins

Comparison of the amino acid sequence of specific kinds of proteins, or of properties such as antigenic reactivity which reflect the amino acid sequence of these proteins, have been used as measures of phylogenetic relationships among organisms. The fundamental concept involved is that most extant proteins are likely to have evolved from a very small number of archetypal proteins by the processes of genetic duplication and modification. In comparing the proteins of any particular group (such as cytochrome *c*, superoxide dismutase, ferredoxin or other enzymes), the greater the difference in amino acid sequence in the protein of one organism from the corresponding protein of another organism, the greater is believed to be the evolutionary divergence between the two organisms. Conversely, if the amino acid sequence of corresponding proteins from two organisms is very similar, the two organisms are believed to be closely related phylogenetically. Even distant relationships between organisms can be deduced by this approach, and various phylogenetic schemes have been constructed to reflect the perceived evolutionary development of a great variety of organisms, both procaryotic and eucaryotic (see the review by Schwartz and Dayhoff, 1978). Among the proteins that have been used for such studies are ferredoxins, flavodoxins, azurins, plastocyanins, and cytochrome *c*. For example, a remarkable similarity in the structure of cytochrome *c* exists between certain nonsulfur purple photosynthetic bacteria (i.e. *Rhodopseudomonas capsulatus* and *R. sphaeroides*), the nonphotosynthetic respiring bacterium *Paracoccus denitrificans*, and the mitochondria of eucaryotic organisms; this and other kinds of congruent data have led to the view that *P. denitrificans* descended from nonsulfur purple bacteria by loss of photosynthetic properties and that this species is the procaryote that most closely resembles the putative procaryotic ancestor of mitochondria (see the exposition by Dickerson (1980) as well as a critical review of the various theories for the endosymbiont origin of mitochondria and chloroplasts by Gray and Doolittle (1982). Reservations about many of the conclusions based on cytochrome *c* sequences have been expressed by Ambler et al. (1979 a, b), and an analysis of some of the limitations involved in comparing amino acid sequences of proteins has been given by Doolittle (1981).

Protein Profiles

The basic premise here is that closely related organisms should have similar or identical kinds of cellular proteins. Two-dimensional electrophoretic and isoelectric focusing procedures (O'Farrell, 1975) have made it possible to resolve several hundred proteins from a cell extract. The protein "fingerprint" so obtained for one bacterial strain is a reflection of the genetic background of that strain and can be compared with the "fingerprints" from other strains as a measure of relatedness.

For examples of the application of this method, see the comparison of *Rhizobium* strains made by Roberts et al. (1980), and the comparison of *Spiroplasma* strains made by Mouches et al. (1979). The method requires a considerable degree of standardization in order to yield optimum results.

One-dimensional polyacrylamide gel electrophoresis (PAGE) of cellular proteins can yield patterns of up to ca. 30 bands and, although it is not comparable in resolving power to the two-dimensional separation method, it can distinguish related organisms from unrelated organisms. In general, whole cells or cellular membrane fractions are used, and the proteins are solubilized by means of a detergent such as sodium dodecyl sulfate (SDS); however, many studies have employed merely the water-soluble proteins ("soluble" fraction) from disintegrated cells. A few examples of the application of PAGE are: identification of mycoplasmas (Razin and Rottem, 1967), taxonomy of *Haemophilus* strains (Nicolet et al., 1980), comparison of isolates from gingival crevice floras (Moore et al., 1980), and differentiation of isolates of indigenous *Rhizobium* populations (Noel and Brill, 1980). By use of rigorously standardized conditions, extremely reproducible protein patterns can be obtained which are amenable to rapid, computerized, numerical analysis (Kersters and De Ley, 1975).

In an analysis of the patterns of soluble cellular proteins from strains of 70 *Clostridium* species, Cato et al. (1982) found that strains having greater than 80% DNA/DNA homology usually produced identical patterns, strains related by ca. 70% homology showed overall similarity of the total patterns but also showed minor differences, and strains unrelated by DNA homology showed major differences. In many instances, the patterns obtained within 24 h of isolating an organism were sufficiently distinctive so that the identity of the organism could be strongly suspected.

Enzyme Characterization

It is now recognized that the functional and structural patterns displayed by certain bacterial enzymes provide data of use in classification. Good examples are the diverse regulatory and molecular size patterns exhibited by bacterial citrate synthases and succinate thiokinases. Both these are enzymes of the citric acid (Krebs) cycle and the near universal occurrence of this cycle in living cells makes it a very suitable pathway for comparative studies between different organisms (see Weitzman, 1980).

In general, the citrate synthases of Gram-negative bacteria are inhibited by reduced nicotinamide adenine dinucleotide (NADH) while those of Gram-positive bacteria are not. The citrate synthases of Gram-negative bacteria can be further divided into two classes on the basis of whether or not their NADH sensitivity is overcome by adenosine monophosphate (AMP). Citrate synthases from the majority of strictly aerobic Gram-negative bacteria are reactivated by AMP while those of the facultatively anaerobic Gram-negative bacteria are not. Citrate synthases of the Gram-negative facultative anaerobes are also inhibited by α-oxoglutarate but the enzymes from the aerobic Gram-negative bacteria and from Gram-positive bacteria are not. Citrate synthases of the cyanobacteria are not inhibited by NADH but they are inhibited by α-oxoglutarate and by succinyl-CoA.

Bacterial citrate synthases fall into two groups, "large" and "small," on the basis of molecular size. The majority of Gram-negative bacteria possess large citrate synthases (mol. wt. = ~250,000) while the majority of Gram-positive bacteria produce citrate synthases of the small type (mol. wt. = ~100,000).

Exceptions to the broad general pattern occur. The citrate synthases of the Gram-negative genus *Acetobacter* do not appear to be inhibited by NADH although the enzyme is of the large type. On the other hand, the citrate synthase of *Thermus aquaticus* is both insensitive to NADH and is of the small type. In both respects it resembles the citrate synthases of the archaebacterial genus *Halobacterium* and the majority of the citrate synthases of Gram-positive bacteria.

Similar molecular size patterns occur among bacterial succinate thiokinases. All succinate thiokinases from the Gram-positive bacteria so far studied are of the small type (mol. wt. = 70,000–75,000) whereas those of Gram-negative bacteria, cyanobacteria and *Halobacterium* species are of the large type (mol. wt. = 140,000–150,000). Bacterial succinate thiokinases can be further subdivided on the basis of their specificity for nucleotide substrates (guanosine diphosphate or inosine diphosphate) and preliminary results point to interesting patterns of enzyme diversity of possible potential in bacterial classification.

Rapid methods are now available for the routine laboratory screening of bacterial citrate synthases and as further such methods are developed it is likely that the regulatory and molecular properties of these and other enzymes will prove useful in the classification of bacteria (see Weitzman, 1980).

Fermentation Product Profiles

The use of gas liquid chromatographic methods to analyze the fatty acids formed as end products of protein or carbohydrate metabolism is particularly useful in the classification and identification of the anaerobic genera *Clostridium*, *Bacteroides*, *Eubacterium*, etc. (see Holdeman et al., 1977).

Bacterial Nomenclature

Peter H. A. Sneath

THE SCOPE OF NOMENCLATURE

Nomenclature has been called the handmaid of taxonomy. The need for a stable set of names for living organisms, and rules to regulate them, has been recognized for over a century. The rules are embodied in *international codes of nomenclature*. There are separate codes for animals, noncultivated plants, cultivated plants, bacteria and viruses. But partly because the rules are framed in legalistic language (so as to avoid imprecision) they are often difficult to understand. Useful commentaries are found in Ainsworth and Sneath (1962), Cowan (1978) and Jeffrey (1977).

The nomenclature of the different kinds of living creatures falls into two parts: (a) informal or vernacular names, or very specialized and restricted names, and (b) scientific names of taxonomic groups (taxon, plural taxa).

Examples of the first are vernacular names from a disease, strain numbers, the symbols for antigenic variants, and the symbols for genetic variants. Thus one can have a vernacular name like the tubercle bacillus, a strain with the designation K12, a serological form with the antigenic formula Ia, and a genetic mutant requiring valine for growth labeled *val*. These names are usually not controlled by the codes of nomenclature, although the codes may recommend good practice for them.

Examples of scientific names are the names of species, genera, and higher ranks. Thus *Mycobacterium tuberculosis* is the scientific name of the tubercle bacillus, a species of bacterium.

These scientific names are regulated by the codes (with few exceptions) and have two things in common: (a) they are all Latinized in form so as to be easily recognized as scientific names, and (b) they possess definite positions in the taxonomic hierarchy. These names are international: thus microbiologists of all nations know what is meant by *Bacillus anthracis*, but few would know it under vernacular names like Milzbrandbacillus or Bactéridie de charbon.

The scientific names of bacteria are regulated by the *International Code of Nomenclature of Bacteria* (Lapage et al., 1975). This is the most recent edition, and is also referred to as the *Revised Code*. This edition authorized a new starting date for names of bacteria on January 1, 1980, and the starting document is the *Approved Lists of Bacterial Names* (Skerman et al., 1980), which contains all the scientific names of bacteria that retain their nomenclatural validity from the past. The operation of these Lists will be referred to later. The Code and the Lists are under the aegis of the International Committee on Systematic Bacteriology, which is a constituent part of the International Union of Microbiological Associations. The Committee is assisted by a number of Taxonomic Subcommittees on different groups of bacteria, and by the Judicial Commission which considers amendments to the Code and any exceptions that may be needed to specific Rules.

LATINIZATION

Since scientific names are in latinized form, they obey the grammar of classic or medieval Latin. Fortunately the necessary grammar is not very difficult, and the commonest point to watch is that adjectives agree in gender with the substantives they qualify. Some examples are given later. The names of genera and species are normally printed in italics (or underlined in manuscripts to indicate italic font). For higher categories conventions vary: in Britain they are often in ordinary roman type, but in America they are usually in italics, which is preferable because this reminds the reader they are latinized scientific names.

THE TAXONOMIC HIERARCHY

The taxonomic hierarchy is a conventional arrangement. Each level above the basic level of species is increasingly inclusive. The names belong to successive **categories**, each of which possesses a position in the hierarchy called its **rank**. The lowest category ordinarily employed is that of species, though sometimes these are subdivided into subspe-

cies. The main categories in decreasing rank, with their vernacular and latin forms, and examples, are shown in Table VI.1.

Additional categories may sometimes be intercalated (e.g. subclass below class, and tribe below family).

Table VI.1.

The ranking of taxonomic categories

Category	Example[a]
Kingdom (*Regnum*)	*Procaryotae*
Phylum (*Phylum*) in zoology or Division (*Divisio*) in botany and bacteriology	*Gracilicutes*
Class (*Classis*)	*Scotobacteria*
Order (*Ordo*)	*Rickettsiales*
Family (*Familia*)	*Rickettsiaceae*
Genus (*Genus*)	*Coxiella*
Species (*Species*)	*Coxiella burnetii*

[a] Based on the classification given by Murray in the chapter on The Higher Taxa, in this *Manual*.

FORM OF NAMES

The form of latinized names differs with the category. The species name consists of two parts. The first is the **genus name**. This is spelled with an initial capital letter, and is a latinized substantive. The second is the **specific epithet**, and is spelled with a lower case initial letter. The epithet is a latinized adjective in agreement with the gender of the genus name, or a Latin word in the genitive case, or occasionally a noun in apposition. Thus in *Mycobacterium tuberculosis*, the epithet *tuberculosis* means "of tubercle," so the species name means the mycobacterium of tuberculosis. The species name is called a **binominal name**, or **binomen**, because it has two parts. When subspecies names are used, a trinominal name results, with the addition of an extra **subspecific epithet**. An example is the subspecies of *Lactobacillus casei* that is called *Lactobaccillus casei* subsp. *rhamnosus*. In this name, *casei* is the specific epithet and *rhamnosus* is the subspecific epithet. The existence of a subspecies such as *rhamnosus* implies the existence of another subspecies, in which the subspecific and specific epithets are identical, i.e. *Lactobacillus casei* subsp. *casei*.

One problem that frequently arises is the scientific status of a species. It may be difficult to know whether an entity differs from its neighbors in certain specified ways. A useful terminology was introduced by Ravin (1963). It may be believed, for example, that the entity can undergo genetic exchange with a nearby species, in which event they could be considered to belong to the same **genospecies**. It may be believed that the entity is not phenotypically distinct from its neighbors, in which event they could be considered to belong to the same **taxospecies**. Yet the conditions for genetic exchange may vary greatly with experimental conditions and the criteria of distinctness depend on what properties are considered, so that it may not be possible to make clear-cut decisions on these matters. Nevertheless, it may be convenient to give the entity a species name and to treat it in nomenclature as a separate species, a **nomenspecies**. It follows that all species in nomenclature should strictly be regarded as nomenspecies.

Genus names, as mentioned above, are latinized nouns, and so are subgenus names (now rarely used) which are conventionally written in parentheses after the genus name, e.g. *Bacillus* (*Aerobacillus*) indicates the subgenus *Aerobacillus* of the genus *Bacillus*. As in the case of subspecies, this implies the existence of a subgenus *Bacillus* (*Bacillus*).

Above the genus level most names are plural adjectives in the feminine gender, agreeing with the word *procaryotae*, so that *Brucellaceae* means *procaryotae brucellaceae*, for example.

PURPOSES OF THE CODES OF NOMENCLATURE

The codes have three main aims:

1. Names should be stable,
2. Names should be unambiguous,
3. Names should be necessary.

These three aims are sometimes contradictory, and the rules of nomenclature have to make provision for exceptions where they clash. The principles are implemented by three main devices: (a) priority of publication to assist stability, (b) establishment of nomenclatural types to ensure the names are not ambiguous, and (c) publication of descriptions to indicate that different names do refer to different entities. These are supported by subsidiary devices such as the latinized forms of names, and the avoidance of synonyms for the same taxon.

PRIORITY OF PUBLICATION

In order to achieve stability the first name given to a taxon (provided the other rules are obeyed), is taken as the correct name. This is the **principle of priority**. But to be safeguarded in this way a name obviously has to be made known to the scientific community: one cannot use a name that has been kept secret. Therefore names have to be published in the scientific literature, together with sufficient indication of what they refer to. This is called **valid publication**. If a name is merely published in the scientific literature it is called **effective publication**: to be valid it also has to satisfy additional requirements, which are summarized later.

The earliest names that must be considered are those published after an official starting date. For many groups of organisms this is Linnaeus' *Species Plantarum* of 1753, but the difficulties of knowing to what the early descriptions refer, and of searching of voluminous and growing literature, have made the principle of priority increasingly hard to obey.

The Code of nomenclature for bacteria, therefore, has established a new starting date of 1980, with a new starting document, the *Approved Lists of Bacterial Names* (Skerman et al., 1980). This list contains names of bacterial taxa that are recognizable and in current use. Names not on the lists lost standing in nomenclature on January 1, 1980, although there are provisions for reviving them if the taxa are subsequently rediscovered or need to be reestablished. In order to prevent the need to search the voluminous scientific literature, the new provisions for bacterial nomenclature require that for valid publication new names (including new names in patents) must be published in certain official publications. Alternatively, if the new names were effectively published in other scientific publications they must be announced in the official publications to become validly published. Priority dates from the official publication concerned. At present the only official publication is the *International Journal of Systematic Bacteriology*.

NOMENCLATURAL TYPES

In order to make clear what names refer to, the taxa must be recognizable by other workers. In the past it was thought sufficient to publish a description of a taxon. This has been found over the years to be inadequate. Advances in techniques and in knowledge of the many undescribed species in nature have shown that old descriptions are usually insufficient. Therefore an additional principle is employed, that of **nomenclatural types**. These are actual specimens (or names of subordinate taxa that ultimately relate to actual specimens). These type specimens are deposited in museums and other institutions. For bacteria (like some other microorganisms that are classified according to their properties in artificial culture) instead of type specimens, **type strains** are employed. The type specimens or strains are intended to be typical specimens or strains which can be compared with other material when classification or identification is undertaken, hence the word "type." However, a moment's thought will show that if a type specimen has to be designated when a taxon is *first* described and named, this will be done at a time when little has yet been found out about the new group. Therefore it is impossible to be sure that it is indeed a typical specimen. By the time a completely typical specimen can be chosen the taxon may be so well known that a type specimen is unnecessary: no one would now bother to designate a type specimen of a bird so well known as the common house sparrow.

The word "type" thus does *not* mean it is typical, but simply that it is a **reference specimen for the name**. This use of the word, type, is a very understandable cause for confusion that may well repay attention by the taxonomists of the future.

In recent years other type concepts have been suggested. Numerical taxonomists have proposed the hypothetical median organism (Liston et al., 1963), or the centroid: these are mathematical abstractions, not actual organisms. The most typical strain in a collection is commonly taken to be the **centrotype** (Silvestri et al., 1962), which is broadly equivalent to the strain closest to the center (centroid) of a species cluster. Some workers have suggested that several type strains should be designated. Gordon (1967) refers to this as the "population concept." One strain, however, must be the official nomenclatural type in case the species must later be divided. Gibbons (1974) proposed that the official type strain should be supplemented by reference strains that indicated the range of variation in the species, and that these strains could be termed the "type constellation." It may be noted that some of these concepts are intended to define not merely the center but, in some fashion, the limits of a species. Since these limits may well vary in different ways for different characters, or classes of characters, it will be appreciated that there may be difficulties in extending the type concept in this way. The centrotype, being a very typical strain, has often been chosen as the type strain, but otherwise these new ideas have not had much application to bacterial nomenclature.

Type strains are of the greatest importance for work on both classification and identification. These strains are preserved (by methods to minimize change to their properties) in culture collections from which they are available for study. They are obviously required for new classificatory work, so that the worker can determine if he has new species among his material. They are also needed in diagnostic microbiology, because one of the most important principles in attempting to identify a microorganism that presents difficulties is to compare it with authentic strains of known species. The drawback that the type strain may not be entirely typical is outweighed by the fact that the type strain is by definition authentic.

Not all microorganisms can be cultured, and for some the function of a type can be served by a preserved specimen, a photograph, or some other device. In such instances, these are the nomenclatural types, though it is commonly considered wise to replace them by type strains when this becomes possible.

Sometimes types become lost, and new ones (**neotypes**) have to be set up to replace them: the procedure for this is described in the Code. In the past it was necessary to define certain special classes of types, but most of these are now not needed.

Types of species and subspecies are type specimens or type strains. For categories above the species the function of the type—to serve as a point of reference—is assumed by a *name*, e.g. that of a species or subspecies. The species or subspecies is, of course, tied to its type specimen or type strain.

Types of genera are **type species** (one of the included species) and types of higher names are usually **type genera** (one of the included genera). This principle applies up to and including the category of Order. This can be illustrated by the types of an example of a taxonomic hierarchy shown in Table VI.2.

Just as the type specimen, or type strain, must be considered a member of the species whatever other specimens or strains are excluded, so the **type species of a genus must be retained in the genus even if all other species are removed from it**. A type, therefore, is sometimes called a **nominifer** or **name bearer**: it is the reference point for the name in question.

Table VI.2.

An example of taxonomic types.

Category	Taxon	Type
Family	*Pseudomonadaceae*	*Pseudomonas*
Genus	*Pseudomonas*	*Pseudomonas aeruginosa*
Species	*Pseudomonas aeruginosa*	American Type Culture Collection strain number 10145

DESCRIPTIONS

The publication of a name, with a designated type, does in a technical sense create a new taxon—insofar as it indicates that the author believes he has observations to support the recognition of a new taxonomic group. But this does not afford evidence that can be readily assessed from the bald facts of a name and designation of a type. From the earliest days of systematic biology it was thought important to describe the new taxon for two reasons: (a) to show the evidence in support of a new taxon, and (b) to permit others to identify their own material with it—indeed this antedated the type concept (which was introduced later to resolve difficulties with descriptions alone).

It is, therefore, a requirement for valid publication that a description of a new taxon is needed. However, just how full the description should be, and what properties must be listed, is difficult to prescribe.

The codes of nomenclature recognize that the most important aspect of a description is to provide a list of properties that distinguish the new taxon from others that are very similar to it, and that consequently fulfill the two purposes of adducing evidence for a new group and allowing another worker to recognize it. Such a brief differential description is called a **diagnosis**, by analogy with the characteristics of diseases that are associated with the same word. Although it is difficult to legislate for adequate diagnoses, it is usually easy to provide an acceptable one: inability to do so is often because insufficient evidence has been obtained to support the establishment of the new taxon.

The Code provides guidance on descriptions, in the form of recommendations. Failure to follow the recommendations does not of itself invalidate a name, though it may well lead later workers to dismiss the taxon as unrecognizable or trivial. The code for bacteria recommends that as soon as minimum standards of description are prepared for various groups, workers should thereafter provide that minimum information; this is intended as a guide to good practice, and should do much to raise the quality of systematic bacteriology. For an example of minimum standards, see the report of the International Committee on Systematic Bacteriology Subcommittee on the Taxonomy of *Mollicutes* (1979).

CLASSIFICATION DETERMINES NOMENCLATURE

The student often asks how an organism can have two different names. The reason lies in the fact that a name implies acceptance of some taxonomy, and on occasion no taxonomy is generally agreed. Scientists are entitled to their own opinions on taxonomies: there are no rules to force the acceptance of a single classification.

Thus opinions may be divided on whether the bacterial genus *Pectobacterium* is sufficiently separate from the genus *Erwinia*. The soft-rot bacterium was originally called *Bacterium carotovorum* in the days when most bacteria were placed in a few large genera such as *Bacillus* and *Bacterium*. As it became clear that these unwieldy genera had to be divided into a number of smaller genera, which were more homogeneous and convenient, this bacterium was placed in the genus *Erwinia* (established for the bacterium of fireblight, *Erwinia amylovora*) as *Erwinia carotovora*. When further knowledge accumulated, it was considered by some workers that the soft-rot bacterium was sufficiently

distinct to merit a new genus, *Pectobacterium*. The same organism, therefore, is also known as *Pectobacterium carotovorum*. Both names are correct in their respective positions. If one believes that two separate genera are justified, then the correct name for the soft-rot bacterium is *Pectobacterium carotovorum*. If one considers that *Pectobacterium* is not justified as a separate genus, the correct name is *Erwinia carotovora*.

Classification, therefore, determines nomenclature, not nomenclature classification. Although unprofitable or frivolous changes of name should be avoided, the freezing of classification in the form it had centuries ago is too high a price to pay for stability of names. Progress in classification must reflect progress in knowledge (for example, no one now wants to classify all rod-shaped bacteria in *Bacillus* as was popular a century ago). Changes in name must reflect progress in classification: some changes in name are thus inevitable.

CHANGES OF NAME

Most changes in name are due to moving species from one genus to another or dividing up older genera. Another cause, however, is the rejection of a commonly-used name because it is incorrect under one or more of the Rules. A much-used name, for example, may not be the earliest, because the earliest name was published in some obscure journal and had been overlooked. Or there may already be another identical name for a different microorganism in the literature. Changes can be very inconvenient if a well-established name is found to be **illegitimate** (contrary to a Rule) because of a technicality. The codes of nomenclature therefore make provision to allow the organizations that are responsible for the codes to make exceptions if this seems necessary. A name thus retained by international agreement is called

a **conserved name**, and when a name is conserved the type may be changed to a more suitable one.

When a species is moved from one genus into another, the specific epithet is retained (unless there is by chance an earlier name which forms the same combination, when some other epithet must be chosen), and this is done in the interests of stability. The new name is called a **new combination**. An example has been given above. When the original *Bacterium carotovorum* was moved to *Erwinia*, the species name became *Erwinia carotovora*. The gender of the species epithet becomes the same as that of the genus *Erwinia*, which is feminine, so the feminine ending, -*a*, is substituted for the neuter ending, -*um*.

NAMES SHOULD BE NECESSARY

The codes require that names should be necessary, that is **there is only one correct name for a taxon** in a given or implied taxonomy. This is sometimes expressed by the statement that an organism with a

given position, rank and circumscription can have only one correct name.

NAMES ARE LABELS, NOT DESCRIPTIONS

In the early days of biology there was no regular system of names, and organisms were referred to by long Latin phrases which described them briefly, such as *Tulipa minor lutea italica folio latiore*, "the little yellow Italian tulip with broader leaves." The Swedish naturalist Linnaeus tried to reduce these to just two words for species, and in doing

so he founded the present *binominal system* for species. This tulip might then become *Tulipa lutea*, just "the yellow tulip." Very soon it would be noted that a white variant sometimes occurred. Should it then still be named "the yellow tulip"? Why not change it to "the Italian tulip"? Then someone would find it in Greece and point out that the record from Italy was a mistake anyway. Twenty years later an orange form would be found in Italy after all. Soon the nomenclature would be confused again.

After a time it was realized that the original name had to be kept, even if it was not descriptive, just as a man keeps his name of Fairchild Goldsmith as he grows older, and even if he becomes a farmer. The scientific names or oganisms are today only **labels**, to provide a means of referring to taxa, just like personal names.

A change of name is therefore only rarely justified, even if it sometimes seems inappropriate. Provisions exist for replacement when the name causes great confusion.

CITATION OF NAMES

A scientific name is sometimes amplified by a *citation*, i.e., by adding after it the author who proposed it. Thus the bacterium that causes crown-galls is *Agrobacterium tumefaciens* (Smith and Townsend) Conn. This indicates that the name refers to the organism first named by Smith and Townsend (as *Bacterium tumefaciens*, in fact, though this is not evident in the citation) and later moved to the genus *Agrobacterium* by Conn, who therefore created a **new combination**. Sometimes the citation is expanded to include the date (e.g. *Rhizobium* Frank 1889), and more rarely to include also the publication, e.g. *Proteus morganii* Rauss 1936 *Journal of Pathology and Bacteriology* Vol. 42, p. 183.

It will be noted that citation is only necessary to provide a suitable reference to the literature or to distinguish between inadvertent duplication of names by different authors. A citation is *not* a means of giving credit to the author who described a taxon: the main functions of citation would be served by the bibliographic reference without mentioning the author's name. Citation of a name is to provide a **means of referring** to a name, just as a name is a means of referring to a taxon.

SYNONYMS AND HOMONYMS

A homonym is a name identical in spelling to another name but based on a different type, so they refer to different taxa under the same name. They are obviously a source of confusion, and the one that was published later is suppressed. The first published name is known as the **senior homonym**, and later published names are **junior homonyms**. Names of higher animals and plants that are the same as bacterial names are not treated as homonyms of names of bacteria, but to reduce confusion among microorganisms, bacterial names are suppressed if they are junior homonyms of names of fungi, algae, protozoa or viruses.

A synonym is a name that refers to the same taxon under another scientific name. Synonyms thus come in pairs or even swarms. They are of two kinds:

1. **Objective synonyms** are names with the same nomenclatural type, so that there is no doubt that they refer to the same taxon. These are often called nomenclatural synonyms. An example is *Erwinia carotovora* and *Pectobacterium carotovorum*: they have the same type strain, American Type Culture Collection strain 15713.

2. **Subjective synonyms** are names that are believed to refer to the same taxon but which do not have the same type. They are matters of taxonomic opinion. Thus *Pseudomonas geniculata* is a subjective synonym of *Pseudomonas fluorescens* for a worker who believes that these taxa are sufficiently similar to be included in one species, *P. fluorescens*. They have different types, however, (American Type Culture Collection strains #19374 and 13525, respectively) and another worker is entitled to treat them as separate species if he so wishes.

There are senior and junior synonyms, as for homonyms. The synonym that was first published is known as the **senior synonym**, and those published later are **junior synonyms.** Junior synonyms are normally suppressed.

PROPOSAL OF NEW NAMES

The valid publication of a new taxon requires that it be named. The Code insists that an author should make up his mind about the new taxon: if he feels certain enough to propose a new taxon with a new name then he should say he does so propose; if he is not sure enough to make a definite proposal then his name will not be afforded the protection of the Code. He cannot expect to suggest provisional names—or possible names, or names that one day might be justified—and then expect others to treat them as definite proposals at some unspecified future date: how can a reader possibly know when such vague conditions have been fulfilled?

If a taxon is too uncertain to receive a new name it should remain with a vernacular designation (e.g. the marine form, group 12A). If it is already named, but its affinities are too uncertain to move it to another genus or family, it should be left where it is. There is one exception, and that is that a new species should be put into some genus even if it is not very certain which is the most appropriate, or if necessary a new genus should be created for it. Otherwise, it will not be validly published, it will be in limbo, and it will be generally overlooked, because no one else will know how to index it or whether they should consider it seriously. If it is misplaced it can later be moved to a better genus.

The basic needs for publication of a new taxon are four: (a) the publication should contain a new name in proper form that is not a homonym of an earlier name of bacteria, fungi, algae, protozoa or viruses; (b) the taxon should not be a synonym of an earlier taxon; (c) a description or at least a diagnosis should be given; (d) the type should be designated. A new species is indicated by adding the Latin abbreviation *sp. nov.*, a new genus by *gen. nov.*, and a new combination by *comb. nov.* The most troublesome part is the search of the literature to cover the first two points. This is now greatly simplified for bacteria, because the new Starting Date means that one need only search the *Approved Lists of Bacterial Names* and the issues of the *International Journal of Systematic Bacteriology* from January 1980 onwards for all validly published names that have to be considered. However, the new name has to be published in that journal, with its description and designation of type, or—if published elsewhere—the name must be announced in that journal to render it validly published.

Identification of Bacteria

Noel R. Krieg

THE NATURE OF IDENTIFICATION SCHEMES

Identification schemes are not classification schemes, although there may be a superficial similarity. An identification scheme for a group of organisms can be devised only **after** that group has first been classified (i.e. recognized as being different from other organisms); it is based on one or more characters, or on a pattern of characters, which all the members of the group have and which other groups do not have. The characters used are often not those that were involved in classification of the group; for example, classification might be based on a DNA/DNA hybridization study, whereas identification might be based on a phenotypic character that is found to correlate well with the genetic information. In general, the characters chosen for an identification scheme should be **easily determinable**, whereas those used for classification may be quite difficult to determine (such as DNA homology values). The characters should also be **few in number**, whereas classification may involve large numbers of characters, such as in a numerical taxonomy study. These ideal features of an identification scheme may not always be possible, particularly with genera or species that are not susceptible to being characterized by traditional biochemical or physiological tests. In such cases, one may need to resort to relatively difficult procedures in order to achieve an accurate identification—procedures such as polyacrylamide gel electrophoresis (PAGE) of cellular proteins, cellular lipid patterns, genetic transformation, or even nucleic acid hybridization.

Serological reactions, which generally have only limited value for classification, often have enormous value for identification. Slide agglutination tests, fluorescent antibody techniques, and other serological methods can be performed simply and rapidly and are usually highly specific; therefore, they offer a means for achieving quick, presumptive identification of bacteria. Their specificity is frequently not absolute, however, and confirmation of the identification by additional physiological or biochemical tests is usually required.

With many genera and species, identification may not be based on only a few tests, but rather on the pattern given by applying a whole battery of tests. The members of the family *Enterobacteriaceae* represent one example of this. To alleviate the need for inoculating large numbers of tubed media, a variety of convenient and rapid multitest systems have been devised and are commercially available for use in identifying various taxa, particularly those of medical importance. A summary of some of these systems has been given by Smibert and Krieg (1981), but new systems are being developed continually. Each manufacturer provides charts, tables, coding systems, and characterization profiles for use with the particular multitest system being offered.

NEED FOR STANDARDIZED TEST METHODS

One difficulty in devising identification schemes is that the results of characterization tests may vary depending on the size of the inoculum, incubation temperature, length of the incubation period, composition of the medium, the surface-to-volume ratio of the medium, and the criteria used to define a "positive" or "negative" reaction. Therefore, the results of characterization tests obtained by one laboratory often do not match exactly those obtained by another laboratory, although the results within each laboratory may be quite consistent. The blind acceptance of an identification scheme without reference to the particular conditions employed by those who devised the scheme can lead to error (and, unfortunately, such conditions are not always specified). Ideally, it would be desirable to standardize the conditions used for

testing various characteristics, but this is easier said than done, especially on an international basis. The use of commercial multitest systems offers some hope of increasing the standardization among various laboratories because of the high degree of quality control exercised over the media and reagents, but no one system has yet been agreed on for universal use for any given taxon. **It is therefore always advisable to include strains whose identity has been firmly established** (type or reference strains, available from national culture collections) **for comparative purposes when making use of an identification scheme**, to make sure that the scheme is valid for the conditions employed in one's own laboratory.

NEED FOR DEFINITIONS OF "POSITIVE" AND "NEGATIVE" REACTIONS

Some tests may be found to be based on plasmid- or phage-mediated characteristics; such characteristics may be highly mutable and therefore unreliable for identification purposes. Even with immutable characteristics, certain tests may not be well suited for use in identification schemes because they may not give highly reproducible results (e.g. the catalase test, oxidase test, Voges-Proskauer test, and gelatin liquefaction are notorious in this regard). Ideally, a test should give reproducible results that are clearly either positive or negative, without equivocal reactions. In fact, no such test may exist. The Gram reaction of an organism may be "Gram-variable," the presence of endospores in a strain that makes only a few may be very difficult to determine by staining or by heat-resistance tests, acid production from sugars may be difficult to distinguish from no acid production if only small amounts of acid are produced, and a weak growth response may not be clearly distinguishable from "no growth." A precise (although arbitrary) definition of what constitutes a "positive" and "negative" reaction is often important in order for a test to be useful for an identification scheme.

PURE CULTURES

Although a few bacteria are so morphologically remarkable as to make them identifiable without isolation, pure cultures are nearly always a necessity before one can attempt identification of an organism. **It is important to realize that the single selection of a colony from a plate does not assure purity.** This is especially true if selective media are used; live but non-growing contaminants may often be present in or near a colony and can be subcultured along with the chosen organism. It is for this reason that **non-selective media are preferred for final isolation**, because they allow such contaminants to develop into visible colonies. Even with non-selective media, apparently well-isolated colonies should not be isolated too soon; some contaminants may be slow growing and may appear on the plate only after a longer incubation. Another difficulty occurs with bacteria that form extracellular slime or that grow as a network of chains or filaments; contaminants often become firmly embedded or entrapped and are difficult to penetrate. In the instance of cyanobacteria, contaminants frequently penetrate and live in the gelatinous sheaths that surround the cells, making pure cultures difficult to obtain.

In general, colonies from a pure culture that has been streaked on a solid medium are similar to one another, providing evidence of purity. Although this is generally true, there are exceptions, as in the case of S → R variation, capsular variants, pigmented or nonpigmented variants, etc., which may be selected by certain media, temperatures, or other growth conditions. Another criterion of purity is morphology: organisms from a pure culture generally exhibit a high degree of morphological similarity in stains or wet mounts. Again, there are exceptions, depending on the age of the culture, the medium used, and other growth conditions: coccoid body formation, cyst formation, spore formation, pleomorphism, etc. For example, examination of a broth culture of a marine spirillum after 2 or 3 days may lead one to believe the culture is highly contaminated with cocci unless one is previously aware that such spirilla generally develop into thin-walled coccoid forms following active growth.

APPROACHES TO IDENTIFICATION OF AN ISOLATE

The vernacular headings of the various sections of *Bergey's Manual* indicate major categories of the procaryotes and are a good starting point for identification. The categories are concerned with such phenotypic characteristics as the Gram staining reactions, morphology, and general type of metabolism. It is therefore important to establish whether the new isolate is a chemolithotrophic autotroph, a photosynthetic organism or a chemoheterotrophic organism. Living cells should be examined by phase-contrast microscopy and Gram-stained cells by light microscopy; other stains can be applied if this seems appropriate. If some outstanding morphological property, such as endospore production, sheaths, holdfasts, acidfastness, cysts, stalks, fruiting bodies, budding division, or true branching, is obvious, then further efforts in identification can be confined to those groups having such a property. Whether or not the organisms are motile, and the type of motility (swimming, gliding) may be very helpful in restricting the range of possibilities. Gross growth characteristics, such as pigmentation, mucoid colonies, swarming, or a minute size, may also provide valuable clues to identification. For example, a motile, Gram-negative rod that produces a water-soluble fluorescent pigment is likely to be a *Pseudomonas* species, whereas one that forms bioluminescent colonies is likely to belong to the *Vibrionaceae*.

The source of the isolate can also help to narrow the field of possibilities. For example, a spirillum isolated from coastal sea water is likely to be an *Oceanospirillum*, whereas Gram-positive cocci occurring in grape-like clusters and isolated from the human nasopharynx are likely to belong to the genus *Staphylococcus*.

The relation of the isolate to oxygen (i.e. whether it is aerobic, anaerobic, facultatively anaerobic, or microaerophilic) is often of fundamental importance in identification. For example, a small, microaerophilic vibrio isolated from a case of diarrhea is likely to be a *Campy-*lobacter, whereas an anaerobic, Gram-negative rod isolated from a wound infection is probably a member of the *Bacteroidaceae*. Similarly, it is important to test the isolate for its ability to dissimilate glucose (or other simple sugar) to determine if the type of metabolism is oxidative or fermentative, or whether sugars are catabolized at all.

Above all, common sense should be used at each stage where the possibilities are narrowed in deciding what additional tests should be performed. There should be a reason for the selection of each test, in contrast to a "shotgun" type of approach where many tests are used but most provide little pertinent information for the particular isolate under investigation. As the category to which the isolate belongs becomes increasingly delineated, one should follow the specific tests indicated in the particular diagnostic tables or keys that apply to that category.

The following summary is taken from "The Mechanism of Identification" by S. T. Cowan and J. Liston in the eighth edition of the *Manual*, with some modifications:

1. Make sure that you have a pure culture.
2. Work from broad categories down to a smaller, specific category of organism.
3. Use all the information available to you in order to narrow the range of possibilities.
4. Apply common sense at each step.
5. Use the minimum number of tests to make the identification.
6. Compare your isolate to type or reference strains of the pertinent taxon to make sure the identification scheme being used actually is valid for the conditions in your particular laboratory.

If, as may well happen, you cannot identify your isolate from the information contained in the *Manual*, neither despair nor immediately assume that you have isolated a new genus or species; many of the problems of microbial classification are the result of people jumping to this conclusion prematurely. When you fail to identify your isolate, check (a) its **purity**, (b) that you have carried out the **appropriate tests**, (c) that your **methods are reliable**, and (d) that you have used correctly the various keys and tables of the *Manual*. It has been said that the most frequent cause of mistaken identity of bacteria is error in the determination of shape, Gram-staining reaction and motility. In most cases, you should have little difficulty in placing your isolate into a genus; allocation to a species or subspecies may need the help of a specialized reference laboratory.

On the other hand, it is always possible that you have actually isolated a new genus or species. A comparison of the present edition of the *Manual* with the previous edition indicates that a number of new genera and species have been added. Some prime examples can be found in the family *Legionellaceae, Other Genera of the Family Enterobacteriaceae*, the genus *Azospirillum, Dissimilatory Sulfate- or Sulfur-Reducing Bacteria*, the genus *Meniscus*, etc. Undoubtedly, there exist in nature a great number of bacteria that have not yet been classified, and therefore cannot yet be identified by existing schemes. Yet, before describing and naming a new taxon, one must be **very sure that it is really a new taxon** and not merely the result of an inadequate identification.

Further Readings for Bacterial Identification

Goodfellow, M. and R.G. Board (Editors). 1980. Microbiological classification and identification. *Society for Applied Bacteriology Symposium Series No. 8.* Academic Press, London.

Hedén, C. and T. Illéni (Editors). 1975. *New Approaches to the Identification of Microorganisms.* John Wiley & Sons, New York.

Holding, J.A. and J.G. Colee. 1971. Routine biochemical tests. *In* Norris and Ribbons (Editors), *Methods in Microbiology*, Vol. 6A, Academic Press, New York, pp. 1–32.

Mitruka, B.J. 1976. *Methods of Detection and Identification of Bacteria.* CRC Press, Cleveland, Ohio.

Skerman, V.B.D. 1967. *A Guide to the Identification of the Genera of Bacteria*, 2nd Ed., Williams & Wilkins, Baltimore.

Skerman, V.B.D. 1969. *Abstracts of Microbiological Methods.* Wiley-Interscience, New York.

Skerman, V.B.D. 1974. A key for the determination of the generic position of organisms listed in the *Manual*. *In* Buchanan and Gibbons (Editors), *Bergey's Manual of Determinative Bacteriology*, 8th Ed., Williams & Wilkins, Baltimore, pp. 1098–1146.

Skinner, F.A. and D.W. Lovelock. 1979. Identification methods for microbiologists. *Society for Applied Bacteriology Technical Series No. 14.* Academic Press, New York.

Smibert, R.M. and N.R. Krieg. 1981. General characterization. *In* Gerhardt, Murray, Costilow, Nester, Wood, Krieg and Phillips (Editors), *Manual of Methods for General Bacteriology*, American Society for Microbiology, Washington, D.C, pp. 409–443.

NUMERICAL IDENTIFICATION

Peter H. A. Sneath

The success of numerical taxonomy has in recent years led to the development of a new diagnostic method based upon it, called **numerical identification**. The rapidly growing field is well reviewed by Lapage et al. (1973), and Willcox et al. (1980). The essential principles can be illustrated geometrically (Sneath, 1978) by considering the columns of percent positive test reactions in a new table, a table of q taxa for m diagnostic characters. If an object is scored for two variables, its position can be represented by a point on a scatter diagram. Three variables determines a position in a three-dimensional model. Objects that are very similar on the variables will be represented by clusters of points in the diagram or the model, and a circle or sphere can be drawn round each cluster so as to define its position and radius. The same principles can be extended to many variables or tests, which then represent a multidimensional space or "hyperspace." A column representing a species defines, in effect, a region in hyperspace, and it is useful to think of a species as being represented by a hypersphere in that space, whose position and radius are specified by the numerical values of these percentages. The tables form a reference library, or data base, of properties of the taxa.

The operation of numerical identification is to compare an unknown strain with each column of the table in turn, and to calculate a distance (or its analogue) to the center of each taxon hypersphere. If the unknown lies well within a hypersphere, this will identify it with that taxon. Further, such systems have important advantages over most other diagnostic systems. The numerical process allows a likelihood to be attached to an identification, so that one can know to some order of magnitude the certainty that the identity is correct. The results are not greatly affected by an occasional aberrant property of the unknown, or an occasional experimental mistake in performing the tests. Furthermore, the system is robust toward missing information, and quite good identifications can be obtained if only a moderate proportion of the tests have been performed.

Numerous applications of numerical identification are now being made. Most commercial testing kits or automatic instruments for microbial identification are based on these concepts, and they require the comparison of results on an unknown strain with a data base using computer software or with printed material prepared by such means. Research sponsored by the Bergey's Manual® Trust (Feltham et al., 1984) shows that these concepts can be extended to a very wide range of genera.

Further Readings for Numerical Identification

Feltham, R.K.A., P.A. Wood and P.H.A. Sneath. 1984. A general-purpose system for characterizing medically important bacteria to genus level. J. Appl. Bacteriol. *57:*279–290

Lapage, S.P., S. Bascomb, W.R. Willcox and M.A. Curtis. 1973. Identification of bacteria by computer. I. General aspects and perspectives. J. Gen. Microbiol. *77:* 273–290.

Sneath, P.H.A. 1978. Identification of microorganisms. *In* Norris and Richmond (Editors), *Essays in Microbiology*. John Wiley, Chichester, England, pp. 10/1–10/32.

Willcox, W.R., S.P. Lapage and B. Holmes. 1980. A review of numerical methods in bacterial identification. Antonie van Leeuwenhoek J. Microbiol. Serol. *46:* 233–299.

Reference Collections of Bacteria—
The Need and Requirements
for Type Strains

The Late Norman E. Gibbons
Revised by Peter H. A. Sneath and Stephen P. Lapage

As it became possible to grow bacteria in liquid and solid media, microbiologists began to exchange cultures with their colleagues for information and comparison. Each investigator kept his own isolates, added those received from others and in this way built up his own reference and working collection.

About the turn of the century, Professor František Král of Prague realized the value of a central collection and began to collect cultures which he made available for a fee to other workers. After Král's death in 1911, the collection was acquired by Professor Ernst Pribram and transferred to the University of Vienna in 1915. Pribram brought part of the collection to Loyola University in Chicago some years before the Second World War. He was killed in a car accident in 1940, but the fate of his collection is not known. The cultures left in Vienna were destroyed during World War II.

The next oldest collection—Centraalbureau voor Schimmelcultures—was founded in 1906 by the Association internationale des Botanistes. Although the founding association did not survive the First World War, the collection is still in existence at Baarn under the auspices of The Royal Netherlands Academy of Sciences. This collection provides a holding and distribution center for fungi and an identification service.

Since then many other collections have developed, some general, some specialized, some oriented to service. A full account of the history of culture collections is given by Porter (1976). Some salient developments may be mentioned briefly. About 1946, Professor P. Hauduroy established a centralized information facility at Lausanne, the "Centre de Collections de Types Microbiens" which provided information on which collections held cultures of various bacterial species. In 1947, the Lausanne Centre became associated with the International Association of Microbiological Societies (IAMS, now the International Union of Microbiological Societies, IUMS) and, in cooperation with it, an International Federation of Type Culture Collections was formed. This Federation had ambitious plans which were never realized, and the Federation went out of existence within a few years.

In 1962, therefore, a Conference on Culture Collections (Martin, 1963), held after the VIIth International Congress for Microbiology, asked IAMS to form a Section on Culture Collections. The Section was set up in 1963 and, on the reorganization of IAMS in 1970, became the World Federation of Culture Collections (WFCC). The WFCC is also a multidisciplinary Commission of the International Union of Biological Sciences in the Divisions of Botany and Zoology, linking it with other organizations concerned with problems of biological preservation, such as herbaria, zoological gardens and museums. It has collected information on several hundred collections throughout the world and the *World Directory of Collections of Cultures of Microorganisms* (Martin and Skerman, 1972) has been published. This has recently been updated (V. F. McGowan and V. B. D. Skerman (Eds.), 1982, *World Directory of Collections of Cultures of Microorganisms*, 2nd Ed., World Data Centre on Microorganisms, University of Queensland, Brisbane, Australia). Pridham (1974) has also compiled a useful list of the acronyms and abbreviations for numerous culture collections. A number of national Federations of Culture Collections have also been formed which are affiliated to the WFCC. The aims of the WFCC include the collection of information on strains held by the collections and more detailed information on the strains themselves.

In the preservation of cultures, satisfactory methods of maintenance, with minimal change in the cultures, are essential, and an accepted system of taxonomy should be used. Particularly stringent standardization is required in the case of cooperative and comparative studies. In pursuit of these and similar aims, the WFCC has held training courses for curators and workers in culture collections at which other important functions of culture collections are also discussed. The WFCC also works in close cooperation with the International Committee on Systematic Bacteriology (ICSB) and its Judicial Commission, and other related national or international bodies dealing with all aspects of the preservation of various groups of microorganisms. It has also sponsored a number of international conferences on culture collections. A list of these, together with a summary of other WFCC activities is given by Lapage (1975). The WFCC supports the development of the World Data Centre at the University of Queensland, Brisbane, Australia, which is collecting cultural, physiological and other data on strains of microorganisms, and is exploring methods of recording such information in a standard format.

THE NEED FOR CULTURE COLLECTIONS

It is essential for the orderly development of bacteriology that cultures of organisms described or mentioned in publications be available for independent study. Because microbiologists are mortal and their interests vary during their working life, collections are necessary to provide an element of stability and continuity.

While some microbiologists spend a lifetime on one or two groups of organisms and build up large specialized collections, others move from one organism to another, abandoning old favorites. Both approaches generate problems in the preservation of organisms. The specialized collection may become so large and so specialized that it is hard to find a willing successor to the original enthusiastic curator. The worker whose interests are more fickle seldom worries about the systematic aspects, which make the preservation of cultures so desirable to the taxonomist.

Until the 1920s, the main reason for the existence of collections was their value for taxonomic and epidemiological studies. In the 1930s, the burgeoning interest in microbial physiology and biochemistry gave rise to a need for preserving organisms that produced or gave better yields of specific compounds. This greatly increased the value of culture collections.

More recently, studies on bacterial genetics have resulted in the isolation of numerous mutants which have, in turn, necessitated specialized collections. Some of these mutants are concerned with genetic loci useful in studies of nutrient and of biochemical pathways. The 1972 Stockholm Conference on the Environment recognized the importance of genetic pools and of collections of microorganisms.

Current developments in culture collections are diverse. Reviews of these can be found in Lapage (1971), the volume edited by Colwell (1976) and Kirsop (Crit. Rev. Biotechnol. 2:287–314. 1985). Methods of preservation are undergoing change, with increasing use of storage at very low temperatures to reduce the risk of genetic change. Loss of plasmids is a problem with some methods. Preservation methods are described in Kirsop and Snell (1984). Recent legislation on patents has led to the need for deposition of strains used in industry. A review of requirements for patents is given by Crespi (*Patenting in the Biological Sciences.* John Wiley & Sons, New York, 1982). Cultures are also needed for teaching of microbiology and for quality control in many fields. The growth of numerous new diagnostic aids requires that large sets of strains from numerous species shall be available for establishing the data bases that are needed (Sneath, 1977). Culture collections may also in the future expand associated activities, such as storage and supply of dried material of microbial origin, standard antisera, nucleic acid preparations, and the like.

TYPE STRAINS

A particularly important function of culture collections is to preserve type strains and make them available to microbiologists who are undertaking taxonomic revisions. The nomenclatural aspects of type strains are discussed in the article on Bacterial Nomenclature, but some related points are briefly summarized here.

Type strains of bacterial species and subspecies are essential for the advance of taxonomy. They are required for comparison with strains that an author may believe belong to a new species or subspecies. Descriptions have never proved to be sufficient, because new techniques in systematics are continually being devised, and there is no substitute for an authentic strain when one wishes to make a critical comparison.

Type strains are of such taxonomic and nomenclatural importance that in this edition as many of them as possible are listed by their designation and catalog number in the main collections; a list of collections mentioned is given in the next article.

The new *International Code of Bacteriological Nomenclature* (Lapage et al., 1975; *Revised Code*) has made several special provisions for types. It is now a requirement for valid publication of a cultivable new species or subspecies of bacteria that a type strain shall be designated (alternative provisions exist for noncultivable bacteria). The Code urges that a type strain should be deposited in one or more of the permanently established culture collections. The numerous problems caused in the past by taxa for which there were no type strains should thus be largely overcome. Type cultures should, in the future, be available for all cultivable species of bacteria.

In the past it was frequently necessary to distinguish between different classes of type culture, in particular between types and neotypes, but the *Revised Code* has made most of these distinctions unnecessary. The new starting document for bacterial nomenclature, which came into force on January 1, 1980 (*Approved Lists of Bacterial Names*; Skerman, McGowan and Sneath, 1980) lists the type strains for the names of bacterial species that are currently recognized and given in the *Lists*. In the past, when many species had no type strains, it was necessary to establish **neotypes** for taxa where no type existed or had been lost. A neotype was thus a replacement for a type, and there should rarely be a need in the future for neotypes, except in the case of loss of the types. The procedure for establishing a neotype is given in the *Revised Code* and, needless to say, it should be deposited in one, or preferably several of the main culture collections.

While culture collections maintain type strains and neotypes as described above, they also keep typical and atypical strains, reference strains, and strains with particular properties of interest to biochemistry, genetics, serology, bacteriophage studies and the like; they also carry out many other functions, of which a general account can be found in Lapage (1971). Culture collections are therefore of great value not only to systematists but to all bacteriologists, and are essential to the development of the subject.

List of Culture Collections

There are several hundred culture collections in the world, the majority being small specialized collections, often collected by one individual. Details of most of these may be found in the *World Directory of Collections of Cultures of Microorganisms* (edited S. M. Martin and V. B. D. Skerman, 1972). This has recently been updated (V. F. McGowan and V. B. D. Skerman (Eds.), 1982, *World Directory of Collections of Cultures of Microorganisms*, 2nd Ed., World Data Centre on Microorganisms, University of Queensland, Brisbane, Australia). A smaller number of collections are frequently referred to in bacteriological work, and a selection of these is given below with commonly used abbreviations.

AMRC	FAO-WHO International Reference Centre for Animal Mycoplasmas, Institute for Medical Microbiology, University of Aarhus, Aarhus, Denmark.
ATCC	American Type Culture Collection, 12301 Parklawn Drive, Rockville, Maryland 20852, U.S.A.
BKM	See VKM.
BKMW	See VKM.
CBS	Centraalbureau voor Schimmelcultures, Oosterstraat 1, Baarn, The Netherlands.
CCEB	Culture Collection of Entomophagous Bacteria, Institute of Entomology, Czechoslovak Academy of Sciences, Flemingovo N2, Prague 6, Czechoslovakia.
CCM	Czechoslovak Collection of Microorganisms, J. E. Purkyne University, Tr. Obr. Miru 10, Brno, Czechoslovakia.
CDC	Centers for Disease Control, Atlanta, Georgia, 30333 U.S.A.
CIP	Collection of the Institut Pasteur, Rue du Dr. Roux, Paris 15, France.
CNC	Czechoslovak National Collection of Type Cultures, Institute of Epidemiology and Microbiology, Srobarova 48, Prague 10, Czechoslovakia.
DSM	Deutsche Sammlung von Mikroorganismen, Grisebachstrasse 8, Gottingen, Federal Republic of Germany.
IAM	Institute of Applied Microbiology, University of Tokyo, Bunkyo-ku, Tokyo, Japan.
ICPB	International Collection of Phytopathogenic Bacteria, University of California, Davis, California 95616, U.S.A.
IFO	Institute for Fermentation, 4-54 Jusonishinocho, Osaka, Japan.
IMET	Institutes für Mikrobiologie und Experimentelle Therapie, Deutsche Akademie der Wissenschaften zu Berlin, Beuthenbergstrasse 11, Jena 69, German Democratic Republic.
IMRU	Institute of Microbiology, Rutgers—The State University, New Brunswick, New Jersey 08903, U.S.A.
IMV	Institute of Microbiology and Virology, Academy of Sciences of the Ukrainian S.S.R., Kiev, U.S.S.R.
INA	Institute for New Antibiotics, Bolshaya Pirogovskaya II, Moscow, U.S.S.R.
INMI	Institute for Microbiology, U.S.S.R. Academy of Sciences, Moscow, U.S.S.R.
IPV	Istituto di Patologia Vegetale, Milan, Italy.
KCC	Kaken Chemical Company Ltd., 6-42 Jujodai-1-Chome, Tokyo 114, Japan.
LIA	Museum of Cultures, Leningrad Research Institute of Antibiotics, 23 Ogorodnikov Prospect, Leningrad L-20, U.S.S.R.
LSU	Louisiana State University, Baton Rouge, Louisiana 70803, U.S.A.
LMD	Laboratorium voor Microbiologie, Technische Hogeschool, Julianalaan 67a, 2623 BC Delft, The Netherlands.
NCDO	National Collection of Dairy Organisms, National Institute for Research in Dairying, University of Reading, Shinfield, Reading, England, U.K.
NCIB	National Collection of Industrial Bacteria, Torry Research Station, Aberdeen AB9 8DG, Scotland, U.K.
NCPPB	National Collection of Plant Pathogenic Bacteria, Plant Pathology Laboratory, Hatching Green, Harpenden, England, U.K.
NCTC	National Collection of Type Cultures, Central Public Health Laboratory, Colindale, London NW9 5HT, England, U.K.
NIAID	National Institute of Allergy and Infectious Diseases, Hamilton, Montana 59840, U.S.A.
NIHJ	National Institute of Health, Tokyo, Japan.
NRC	National Research Council, Sussex Drive, Ottawa 2, Canada.
NRL	Neisseria Reference Laboratory, U.S. Public Health Service Hospital, Seattle, Washington 98114, U.S.A.
NRRL	Northern Utilization Research and Development Division, U.S. Department of Agriculture, Peoria, Illinois 61604, U.S.A.
PDDCC	Culture Collection of Plant Diseases Division, New Zealand Department of Scientific and Industrial Research, Auckland, New Zealand
TC	Thaxter Collection, Farlow Herbarium, Harvard University, Cambridge, Massachusetts 02138, U.S.A.
TPH	Microbiological Culture Collection, Public Health Laboratory, Ontario Department of Health, Toronto 116, Canada.
UMH	University of Missouri Herbarium, Columbia, Missouri 65201, U.S.A.

UQM   Culture Collection, Department of Microbiology, University of Queensland, Herston, Brisbane 4006, Australia.

VKM   Department of Culture Collection, Institute of Biochemistry and Physiology of Microorganisms, U.S.S.R. Academy of Sciences, Pushchino, Moscow region, 142292, U.S.S.R.

VPI   Anaerobe Laboratory, Virginia Polytechnic Institute and State University, Blackburg, Virginia 24061, U.S.A.

WINDSOR   Culture Collection, University of Windsor, Windsor, Ontario, Canada

WVU   West Virginia University, Department of Microbiology, Medical Center, Morgantown, W. Virginia, 26506, USA.

The Higher Taxa, or,
A Place for Everything . . . ?

R. G. E. Murray

"Quot homines tot sententiae; suo quoque mos." (so many men, so many opinions; each to his own taste).

Terence, Phormio.

When the eighth edition of *Bergey's Manual* was in preparation, a major taxonomic concern was the provision of a clear statement of where the bacteria fitted among living things. This was set out in *A Place for Bacteria in the Living World* (Murray, 1974), which summarized the reasons for recognizing the Kingdom *Procaryotae*, inclusive of the bacteria and the "blue-green algae" (Cyanobacteria). This concept, based on cellular organization, is now a part of the fundamental training of all biologists and a formal repetition is no longer a necessity. The student who wishes to relive that era should consult the major essays for details and references (Stanier, 1961; Stanier and van Niel, 1962; Murray, 1962; Allsopp, 1969; Stanier, 1970). Taxonomy is not static, however, and new horizons are being explored that give perspective and greater definition to higher taxa as well as the lower categories of genus and species.

The prefatory chapter mentioned above included a tentative proposal of appropriate higher taxa. The arguments and proposals that have arisen since then concern the levels of dissection of the kingdoms of the living world (Whittaker and Margulis, 1978; Woese and Fox, 1977a), the definition and levels of dissection of the major procaryotic groups (Gibbons and Murray, 1978; Whittaker and Margulis, 1978) and the integration of evolutionary information (Stackebrandt and Woese, 1981). There is a renewal of interest in bacterial taxonomy stimulated by the recognition of novel groups of bacteria that do not fit comfortably into current systematic schemes and by new understanding of the taxonomic utility and phylogenetic significance of molecular and genetic data. The system of "superphyla" and phyla proposed by Whittaker and Margulis (1978) is not sensitive to current interpretations of biochemical relatedness based on wall chemistry or other unique features of well-established groups of procaryotes. There is no advantage, at this stage of our understanding, in debating the relative value of recognizing "super kingdoms" (Whittaker and Margulis, 1978) or "primary kingdoms" (the Urkingdoms of Woese and Fox, 1977a) to accomodate views of cellular organization in protists, plants and animals as well as speculations about the nature of putative progenitors. We should be content for now to deal with the *Procaryotae* and the systematic problems that arise within that circumscription; there must be sufficient time for assimilation and consolidation of the burgeoning data. There are attractive features in the dendrograms generated by C. R. Woese and his colleagues; their time will come when the patterns of associations are less fragmentary. For now it would appear sensible to look to the Gibbons and Murray (1978) proposal as capable of modifi-

cation as an interim broad classification with a few areas of taxonomic validity.

The conceptual changes deriving from genetic and molecular studies are making inroads into the cherished beliefs of taxonomists and bacteriologists. We have to agree with the moderate statement taken from Stackebrandt and Woese (1981): ". . . what bacterial classification we have (say up through the eighth edition of *Bergey's Manual*, 1923–1974) is probably not in very good accord with the natural relationships that exist among organisms." This is true enough because it is only in the past decade that sequencing of biopolymers and molecular genetics has provided convincing data on relatedness and because the intent and role of *Bergey's Manual* has always been to provide a basis for the determination of the identity of a pure culture. It is unfortunate that the expression of nomenclatural decisions, hierarchical arrangements (even if all but abandoned in the eighth edition), and the mask of authority has tended to induce undue confidence in the relationships implied in earlier editions.

Even if our perception of "natural relationships" is flawed by ignorance as well as inadequate information, the practical bacteriologist needs a simple scheme of classification as a framework for recognition. At this stage, for example, the possibility that some micrococci are more closely related to *Arthrobacter* than to other spherical Gram-positive cocci and the implication for a further splitting of the genus *Micrococcus* (Stackebrandt and Woese 1979, 1981) would be confusing to the practical bench worker or the physician and is unhelpful. At the higher taxonomic levels some sort of serviceable scheme that can recognize and accomodate to a degree the possibilities will be of service, will cushion the shocks to come, and will stimulate appropriate research. This sort of practicality is almost realized in the proposal by Gibbons and Murray (1978). But any scheme will require future modification because it will take time to attain a complete reassessment of the taxonomic significance and validity of those characters that are reasonably easy to determine and apply effectively to each level of identification. The alternative possibility is that we should maintain two entirely independent schemes: a practical taxonomy and an academic (phylogenetic) taxonomy. The dichotomy of these phenotypic and genotypic approaches is with us because people of such persuasions work in semi-isolation and because, as argued by Stackebrandt and Woese (1981), ". . . the classically defined taxonomic categories, the genera and families, do not correspond to fixed (minimal) S_{AB} values." Furthermore, the genetics of today is beginning to clarify the mechanisms operating in the grand evolutionary experiment, blessed with minimal constraints of time and circumstance, conducted in nature's laboratory. It is clear that point mutations are less important than effective reassociations of determinants with their modifying segments and the mechanisms allowing the exchange, chromosomal integration and amplification of operative sets of determinants (Campbell, 1981; Cullum and Saedler, 1981). It is conceivable, nowadays, that major complex characters of physiological and taxonomic significance (involving a considerable number of genes) could be transferred between organisms both closely and distantly related in the clonal arborizations of the phylogenetic tree. All that is required is an occasional evolutionarily successful experiment in a time frame measured in thousands, millions or billions of years (or cell divisions, for that matter).

An overall taxonomic scheme that is capable of incorporating phylogenetic data (as well as providing a primary key) would be helpful in minimizing the dichotomy of interests and understanding among bacteriologists. It is desirable to bridge the growing gap between the practical applied fields and the academic substratum with something more than the perfidy of plasmids and technological legerdemain. The perpetual quandary is how and when to incorporate into systematic bacteriology the generalizations derived by intensive and expensive study of "model" organisms or of a limited set.

The most exciting and evocative of recent explorations of the possibilities of extracting phylogenetic information from highly conserved biopolymers (the "semantides" of Zuckerkandl and Pauling, 1965) are the comparisons of 16S ribosomal RNA catalogs undertaken by Carl Woese and his colleagues (Woese and Fox 1977b). This approach together with that involving the functional structural homologies and interchangeability of whole ribosome parts and some protein components shows promise of providing comparative data with evolutionary significance for the whole living world (Brimacombe et al., 1978; Kandler, 1981). Of most interest to us is the capability of the technique of RNA nucleotide analysis, expensive and slow though it may be, in spanning the widest range of procaryotic clones. The stability of most of this fairly large (1540 residues) molecule is the basis of the application to assessment of taxa at the level of family and higher. The fact that a few variable domains exist in the molecule allows for the partially realized possibility of contributing a degree of resolution at the level of genus. This could add to the data on relations within genus and species generated by utilizing DNA/DNA (Johnson, 1973) and RNA/DNA hybridization (De Smedt and DeLey, 1977). The figures that are generated (either the number of shared oligonucleotides or an association coefficient, the S_{AB} value) are based on a computer comparison of the catalogs of oligonucleotides (liberated from the 16S RNA by ribonuclease T_1) large enough to show individuality (larger than pentamers). This means that a considerable portion of the molecule yielding oligonucleotides smaller than hexamers is not taken into account and, therefore, the sequence data cannot be directly related to all other hybridization data. Nevertheless the results of comparing some 200 representative bacteria, surprises and all, support the directing thesis that most of the 16S rRNA sequence has drifted only slowly with time. We surmise (Stackebrandt and Woese, 1981) that the comparison provides a measure of the "depth" of the separation between the phylogenetic units or branches of the phylogenetic tree. It is an article of faith that the degree of cleavage (a low S_{AB} value) is proportional to time and is reasonable for initial purposes, although the time scale may be different for different major taxa.

Clones giving rise to unique and now recognizable groups of bacteria must have separated at various stages of procaryotic evolution (Fox et al., 1980). Earliest among these departures from the main stem so far detected by this oligonucleotide cataloging are the Archaeobacteria*, which comprise the methanogens, halobacteria and thermoacidophiles; these, it is now known (Kandler 1981), possess peculiar lipids and either no murein or a pseudomurein in their cell walls. The eight or so major groups of photosynthetic and chemosynthetic bacteria (all designated "Eubacteria" in the papers cited) arose somewhat later in this imprecise evolutionary time scale. The data suggest that some genera are truly ancient (e.g. Clostridium, Spirochaeta) and older, in fact, than some very complex associations of genera (e.g. the actinomycetes).

Those involved in the comparative studies of 16 S rRNA, ribosomes and ribosomal proteins have come to the enthusiastic conclusion that the procaryotes are made up of two kingdoms, the Eubacteria and the Archaeobacteria. The molecular and biochemical bases for the separation have been summarized by Woese (1981) and Kandler (1981). There is no doubt whatever that the Archaeobacteria are distinguished by a number of specialized characters from the rest of the procaryotes ("Eubacteria" has had too many meanings in the past to be a useful term). They include the number of ribosomal proteins, the size and shape of the ribosomal S unit, the proportion of acidic ribosomal proteins, the constitution of tRNA initiator, the presence of ether-linked rather than ester-linked lipids, and the absence of muramic acid or the normal form of peptidoglycan from cell walls. These and some other intimate features make for interesting thoughts about eucaryotes, mitochondria and chloroplasts, as well as the procaryotes. But these distinctions are not suitable for kingdom status. Stackebrandt and Woese (1981) sum up the situation as follows: ". . . the general conclusion that seems to be emerging with regard to the differences among archaebacteria, true bacteria and eucaryotes is that all are identical in the basic aspects of their basic processes, yet all differ from one another in the details of these processes." An examination of any of the methanogens, strict halophiles or thermoacidophiles would place them in the kingdom Procaryotae as presently defined. It is not appropriate to separate kingdoms on any basis but a major, reasonably easily determined, difference in organization. Therefore, it seems sensible to treat the Archaeobacteria as a major taxon within the Procaryotae and, if necessary amend its status at some later date when more of the evidence is collected and digested. Perhaps we will soon recognize other equally distinctive clones that diverged very early from the stem clones of primitive microbes.

There is a comforting sequel to pondering compilations of articles regarding biochemical evolution (Wilson et al., 1977; Carlile et al., 1981). Although these studies suggest that "strange bedfellows" may be assigned to some of the more complex groups or point to unexpected separations (e.g. among the photosynthetic bacteria, Gibson et al., 1979; among the micrococci, Stackebrandt and Woese, 1979) there are concordant features. Gram-positiveness and Gram-negativeness are still unassailable characters except in what are now known to be phylogenetically and biochemically separate groups, the Archaeobacteria (Balch et al., 1979), the radiation-resistant cocci (Brooks et al., 1980) and, of course, the wall-less Mollicutes. Among the Gram-positives, it is encouraging to see that the many peptidoglycan types form consistent patterns in the branches of the dendrograms generated by comparison of the oligonucleotide catalogs (Schleifer and Kandler, 1972; Kandler and Schleifer, 1980; Kandler, 1981).

The infinitely diverse groupings of Gram-negative bacteria have yet to be surveyed to an extent that allows of any decisive taxonomic proposals and many of those included up to now exhibit phylogenetic and phenotypic incoherence. A major surprise arising from the analysis of rRNA oligonucleotides has been that each of the three coherent phylogenetic groupings of anoxygenic photosynthesizers contain a variety of seemingly related, diverse and well-known nonphotosynthetic genera showing subordinate S_{AB} values (Gibson et al., 1979; Stackebrandt and Woese, 1981). The implication is that photosynthetic clones may spawn apochlorotic derivatives; a thesis often directed to the cyanobacteria in the past (Pringsheim, 1967) but not yet subjected to this sort of phylogenetic analysis. It would seem wise not to make phototrophism an overriding taxonomic unit until the situation clarifies. This makes for difficulties in more conventional schemes such as that of Gibbons and Murray (1978), which separated in simplistic fashion the photosynthetic (Photobacteria) and the nonphotosynthetic (Scotobacteria) included in the Gram-negative bacteria (Gracilicutes). But it may be too early to be either discouraged or encouraged and there is still room for the exercise of one's prejudices.

Classifications cannot be final and there are many ways in which

---

* Equivalent to the term Archaebacteria. Because the word is formed by a combination of two Greek words (archaios, ancient, and bakterion, a small rod) the letter o should be used as the combining vowel, hence, Archaeobacteria.

bacteria can be classified (Cowan, 1968); this one has no more permanence than those that went before it. But the doubts and criticisms that come to the mind of the reader are the stimuli to further work and a deeper consideration of the taxonomic implications of the new mix of biochemical and phylogenetic data. Changes from the earlier versions (Murray, 1974; Gibbons and Murray, 1978) were inevitable. The hierarchical levels needed to be raised to give greater scope for classifying the range of organisms included at each major level. For example, students of the cyanobacteria, even those most sympathetic to their incorporation into bacterial taxonomy, despair of being able to accommodate their charges within the single order assigned by Gibbons and Murray (1978). As already indicated, it would take more than ordinary taxonomic agility to make a phylogenetically sensitive classification of the phototrophic bacteria and their derivatives in our present state of understanding.

The following is proposed as an arrangement of higher taxa which can serve during this time of taxonomic transition. It involves some amendments of rank and new names.

Kingdom *Procaryotae* Murray 1968, 252.
- Division I.   *Gracilicutes* Gibbons and Murray 1978, 3.
- Class I.   *Scotobacteria* Gibbons and Murray 1978, 4.
- Class II.   *Anoxyphotobacteria* (Gibbons and Murray) classis nov. (Subclassis *Anoxyphotobacteria* Gibbons and Murray 1978, 4.)
- Class III.   *Oxyphotobacteria* (Gibbons and Murray) classis nov. (Subclassis *Oxyphotobacteria*, Gibbons and Murray 1978, 3.)
- Division II.   *Firmicutes* Gibbons and Murray 1978, 5. (*Firmacutes* (sic) Gibbons and Murray 1978, 5.)
- Class I.   *Firmibacteria* classis nov.; L. adj. *firmus* strong; Gr. dim.n. *bakterion* a small rod; M.L. fem.pl.n. *Firmibacteria* strong bacteria, indicative of simple Gram-positive bacteria.
- Class II.   *Thallobacteria* classis nov.; Gr. n. *thallos* branch; Gr. dim.n. *bakterion* a small rod; M.L. fem.pl.n. *Thallobacteria* branching bacteria.

   (These new names are proposed to express the general basis of splitting the division into the simple Gram-positive bacilli and those Gram-positive bacteria showing a branching habit, the actinomycetes and related organisms.)
- Division III.   *Tenericutes* div. nov.; L. adj. *tener* soft, tender; L. fem.n. *cutis* skin; M.L. fem.n. *Tenericutes* procaryotes of pliable, soft nature, indicative of lack of a rigid cell wall.
- Class I.   *Mollicutes* Edward and Freundt 1967, 267.

   (The *Mollicutes* are a distinctive group of wall-less procaryotes of sufficiently diverse phylogenies that separate classes may well be required in the future.)
- Division IV.   *Mendosicutes* Gibbons and Murray 1978, 2. (*Mendocutes* (sic) Gibbons and Murray 1978, 2.)
- Class I.   *Archaeobacteria* (Woese and Fox) classis nov. (Kingdom *Archaebacteria* (sic) Woese and Fox 1977a, 5089.)

   (The *Archaeobacteria* are defined in terms of being procaryotes with unusual walls, membrane lipids, ribosomes and RNA sequences (Kandler, 1981). The future may bring further classes into the *Mendosicutes* when (and if) truly primitive organisms as envisioned by Woese and Fox (1977b) are isolated and recognized as related to a "universal ancestor" or progenote.)

This arrangement of the procaryotes continues to recognize the absence or presence and nature of cell walls as determinative at the highest level. The omission of Photobacteria (Gibbons and Murray, 1978) as a class and the elevation of *Oxyphotobacteria* and *Anoxyphotobacteria* to class status is intended to provide more scope than offered by Gibbons and Murray (1978) for the inevitable arrangement and rearrangement of the lower taxa within these categories as new under-

standing of lineage and relationships is brought to bear. For instance, the groups of phototrophic bacteria are formed of several major subgroups; separation of these from the level of class would be appropriate to the deep phylogenetic clefts that may be established in both the *Oxyphotobacteria* (the Cyanobacteria and the *Prochlorales*) and the *Anoxyphotobacteria* as pointed out by Stackebrandt and Woese (1981). Furthermore, it may not be appropriate to place all the nonphotosynthetic, Gram-negative bacteria in the *Scotobacteria* if the molecular evidence points clearly to derivation from photosynthetic ancestors; undoubtedly this would require separation at a high level within the class. The same need for broad scope is apparent in the *Firmicutes* with its two major divisions, the simple Gram-positive bacteria (*Clostridium* and relatives) and the actinomycetes.

The *Tenericutes* would have less support from the molecular phylogenetic evidence as a taxon at the highest level because of their probable origin from Gram-positive bacteria and the possibility that they may not have a single common ancestry (Woese et al., 1980). However, they form a stable and distinctive group; they are not obviously a subset of the *Firmicutes*, and their wallless state puts them clearly in a division by themselves as long as we base our classification on the presence or absence and character of the cell wall. On the other hand, we must recognize that an organism may lose a component of a very complex wall and still merit consideration as a member of that class; e.g. the members of the genus *Chlamydia* have no muramic acid but other characters, including some concerned with the relict wall, suggest a relationship to organisms that are definitive members of the *Gracilicutes*.

The *Archaeobacteria*, for their part, are a very diverse group in terms of cell-wall attributes (all the way from a complex wall including pseudomurein to wall-less) and there are at least five groupings with S_{AB} values of less than 0.3. There is sufficient scope within a division for the apparent complexity laid out by Balch et al., (1979) and any extraordinary "primitive" organisms that may be isolated. The sensible approach would seem to be maintenance of the consistency of the higher taxa by including in the class *Mendosicutes* all those procaryotes with a cell-wall composition inconsistent with that defined for *Gracilicutes* and *Firmicutes* (e.g. in simplest terms, not possessing muramic acid). This view is supported by Starr and Schmidt (1981). The inclusion of wall-less thermoacidophiles among the *Archaeobacteria* will be necessary even if seemingly inconsistent. There will come a time, without doubt, when we can set up taxa that are precisely defined in terms of molecular genetics; but that time has not yet arrived.

We cannot assume that all possible procaryotic organisms have been observed and isolated for study. The possibility exists that organisms will be found that are even more "primitive" (i.e. separated from the main stem even earlier) than the *Archaeobacteria* and have a constitution revealing more of the nature of the "universal common ancestor" (Woese and Fox, 1977; Stackebrandt and Woese, 1981). A class can be formed in the future as a suitable home for any organism whose proteins and genetic translation apparatus do not fit into the line of evolution represented by the procaryotes and eucaryotes studied up to now. They may have characteristics that foreshadow the fundamental eucaryotic cell.

No place is provided in our scheme for the fossil microbes being described from specimens of precambrian and possibly archean cherts (see Walter, 1977). This is because they can only be described in terms of size, shape and associations. The oldest are in stratified structures closely resembling the stromatolites and "algal mats" that can be found today, which are complex consortia of cyanobacteria, algae and bacteria with equally complex layering of metabolic and physiological characteristics. There are several attractive morphological resemblances. Size is about all that distinguishes the interpretation of forms as bacterial or algal. A further interpretation that a particular form is photosynthetic is entirely circumstantial and assumptive. Names have been assigned in binominal form (usually based on the Botanical Code) as a means of classifying the varied types being observed. This is a legitimate and stimulating activity but it is not yet helpful in terms of procaryotic classification or evolutionary taxonomy. Because of the uncertainties of alignment, they should be classified for determinative purposes in a separate group of microbiota.

Our view of classification and the taxonomic edifice is based on a century of experience supporting the contention that there is a reasonable degree of fixity in the characters describing a species. Such variation as there is within the clusters (as we now see in computer-assisted studies) can be included in the circumscription offered in describing the species or other categories. Certainly the species is a concept and not an entity (Cowan, 1968); but phenetic studies utilizing large numbers of strains and the widest possible range of characters has, if anything, clarified our concept of taxonomic groups (Sneath, 1978). Despite a growing appreciation of the Adansonian approach to species, the attitude to the definition of higher taxa involves the selection of seemingly single, but very complex characters as exemplified by this essay.

The alternative and extreme view that the bacteria are so pleomorphic that the species concept has no reality, is misleading, and draws the attention away from the important facts of life in clonal populations. Views of this character have been put forward by Sonea (1971) and Sonea and Panisset (1976, 1980). They believe, along with the rest of us, that the procaryotic clones are united by their lineage but differ in their interpretations of the stability of taxonomic units. They argue that genetic exchange (and "communications" of all sorts) between diverse clones make nonsense of the species concept. The extension of their analysis into a concept of unity for the entire world population of bacteria and their interactions with the environment (a sort of global organism as real as a horse or an elephant) is an interesting but philosophical curiosity. The approach stimulates thought but it seems evident that modern pleomorphists will have to modify and adapt their views to practical necessities dictated by new knowledge as much as our views, expressed in this chapter, will have to bend with the winds of change. A century of bacteriological analysis and studies of cultures can convince one that there is sufficient stability to allow of the recognition of most taxonomic clusters.

No doubt, there are many and variant attitudes to the details of bacterial taxonomy. But the substratum of fact is now beginning to be revealed beneath the veneer of fancy or prejudice and to allow judgement to operate. Perhaps bacteriology is maturing after about 150 years of seeking a stable basis for classification.

"For now we see through a glass, darkly; but then face to face: now I know in part . . . "

The first Epistle of Paul
to the Corinthians (XIII:12)

That we can now perceive, albeit dimly, aspects of classification and phylogeny in the macromolecules of the *Procaryotae* is the legacy of many more scientists than have been cited so far in the prefatory chapters. The views that we espouse today still reflect the prejudices and enthusiams of our teachers, our teachers' teachers and the influential observers and exponents of each stage in the maturation of bacteriology. The praise and the blame cannot easily be apportioned but undoubtedly we can recognize the pervasive influence of strong minded people of the modern era who enjoyed trying to make order out of chaos. These included D. H. Bergey, R. S. Breed, R. E. Buchanan and N. E. Gibbons whose efforts have brought *Bergey's Manual* into print in its various editions. But they, like others before and after them and despite their special interests, not to say prejudices, were collectors of the intellections and arrangements of "authorities." Despite the arguments concerning validity that undoubtedly surfaced at the time, their efforts could not prevent the perpetuation of numerous unstable taxa that we still struggle with today: form genera, color genera, physiological genera, etc., encompassing diverse and probably unrelated species, as genetical and biochemical criteria now force us to realize. But it is more important, perhaps, to realize that the most pervasive influences on all manner of approaches have been the writings of the "Delft school" (M. W. Beijerinck and A. J. Kluyver) and the product of that school, C. B. van Niel. The discussions accompanying "the van Niel course" (attended by an equally remarkable collection of microbiologists) started many thinking about microbes in new ways and set them and their students on productive lines of work. The comparative studies that resulted from these stimuli were of great significance in microbial biochemistry, physiology and ecology; a high proportion of the studies made significant contributions to systematic bacteriology. Not least among those influenced by van Niel was R. Y. Stanier, whose death came just as this volume was being readied for the press. It is obvious that we owe a particular debt to this lineage of bacteriologists. Stanier's arguments in our meetings, when he was a member of the Board of Trustees, were largely responsible for major changes in attitude and format expressed in the eighth and in this edition of the *Manual*; we sharpened our judgements with the help of other well established heretics such as S. T. Cowan (1970). And now, as strongly expressed and supported in this essay, a new breed of heretic is influencing bacterial systematics, as we had been warned would be the case by another former trustee, A. W. Ravin.

In the end, a reassessment of the diverse characters used in classifications will have as important consequences as the major generalization based on cellular organization used in defining the kingdom. We must identify characters of proven reliability and validity encompassing more of the genome than seems to be the case today. The techniques must allow the comparison of groups whatever their ecological niche or the professional proclivities of those that study them. Happily, we can echo the Rabbi ben Ezra and proclaim: "The best is yet to be."

Editorial Note

A publication presenting much of the data and examining the perspectives revealed in research on the *Archaeobacteria* appeared after the completion of this essay. It includes contributions from most if not all of the laboratories engaged in this thrilling task and starts with "an overview" by C. R. Woese. This comprehensive collection of papers is in Zentralblatt für Bakteriologie, Mikrobiologie und Hygiene, I Abt. Orig. C.*3*(1/2): 1–345, March/May 1982.

SECTION 12

Gram-Positive Cocci

Karl Heinz Schleifer

Table 12.1.

Differential properties of genera of gram-positive cocci[a]

Characteristics	Aerobic genera		
	Micrococcus	*Planococcus[b]*	*Deinococcus*
Predominant arrangement of cells (other than single cells), commonest appearance listed first	Clusters, tetrads	Pairs, tetrads	Pairs, tetrads
Strict aerobes	+	+	+
Facultative anaerobes or microaerophils	−[c]	−	−
Strict anaerobes	−	−	−
Catalase reaction	+	+	+
Cytochromes present	+	+	+
Major fermentation products from carbohydrates anaerobically (if not fermentative, acidity recorded)	NA (often acid)	NA (no acid)	NA (weak or no acid)
Peptidoglycan			
Position 1	Ala	Ala	Ala
Position 3	Lys	Lys	Orn
Main interpeptide bridge	Peptide subunit Ala$_{1-4}$, Asp, Thr-Ala$_3$, or Ser$_2$-D-Glu	D-Glu	Gly$_2$
Teichoic acid in cell wall	−	−	−
Major menaquinones[d]	MK-7–MK-9, often hydrogenated	MK-8	MK-8
Mol% G + C of DNA	64–75	39–52	62–70

[a] Symbols: +, 90% or more of strains positive; −, 10% or less of strains positive; d, 11–89% of strains positive; D, substantial proportion of species differ; NA, not applicable; and ND, not determined.

[b] Most strains are motile.

[c] One species sometimes facultatively anaerobic.

[d] See Collins and Jones (1981): most streptococci lack menaquinones but some strains of serological groups D and N contain MK-8, MK-9 or demethylmenaquinone containing nine isoprene units.

[e] Most strains are capsulated.

[f] Rarely strictly aerobic.

[g] Grows best at reduced oxygen concentration, sometimes does not grow anaerobically.

[h] Weak anaerobic growth.

[i] Rarely strictly anaerobic.

[j] Rarely negative.

[k] Sometimes weak catalase or pseudocatalase reaction.

[l] A few strains synthesize cytochrome on aerobic media supplemented with hemin.

[m] Present in strains of serological groups D and N.

[n] Sometimes aerotolerant.

[o] *P. productus* ferments carbohydrates to acetate, formate, succinate and lactate. *P. tetradius* produces acid from glucose.

[p] A$_2$pm, diaminopimelic acid (Symbolism of Schleifer and Kandler, 1972).

Table 12.1.—*continued*

Characteristics	Facultatively anaerobic genera						
	Staphylococcus	*Stomatococcus*[e]	*Streptococcus*	*Leuconostoc*	*Pediococcus*	*Aerococcus*	*Gemella*
Predominant arrangement of cells (other than single cells), commonest appearance listed first	Clusters, pairs	Clusters, some pairs	Chains, pairs	Pairs, chains	Tetrads, some pairs	Tetrads. pairs	Pairs, short chains
Strict aerobes	—[f]	—	—	—	—	—	—
Facultative anaerobes or microaerophils	+	+	+	+	+	+[g]	+[h]
Strict anaerobes	—	—	—[i]	—	—	—	—
Catalase reaction	+[j]	+ (weak)	—[k]	—	—	—[k]	—
Cytochrome present	+	+	—[l]	—	—	—	—[l]
Major fermentation products from carbohydrates anaerobically (if not fermentative, acidity recorded)	Lactate	ND (acid)	Lactate	Lactate, ethanol	Lactate	ND (acid)	ND
Peptidoglycan							
Position 1	Ala	Ala	Ala	Ala	**Ala**	Ala	Ala
Position 3	Lys	Lys	Lys	Lys	**Lys**	Lys	Lys
Main interpeptide bridge	Gly$_{5-6}$, Ala-Gly$_4$ or Gly$_{3-5}$, Ser$_{1-2}$	Ala, Gly or Ser	Ala$_{1-4}$, Thr-Ala, Thr-Gly, Ala-Gly, D-Asp, or none	Ala$_2$, Ser-Ala$_2$ or Ser$_2$	**D-Asp or none**	None	Ala$_{2-3}$
Teichoic acid in cell wall	+	—	D[m]	ND	ND	+	ND
Major menaquinones[d]	MK-6–MK-8	ND	D[d]	None	None	ND	ND
Mol% G + C of DNA	30–39	56–61	34–46	38–44	34–42	35–40	33–35

[a] Symbols: +, 90% or more of strains positive; −, 10% or less of strains positive; d, 11–89% of strains positive; D, substantial proportion of species differ; NA, not applicable; and ND, not determined.

[b] Most strains are motile.

[c] One species sometimes facultatively anaerobic.

[d] See Collins and Jones (1981): most streptococci lack menaquinones but some strains of serological groups D and N contain MK-8, MK-9 or demethylmenaquinone containing nine isoprene units.

[e] Most strains are capsulated.

[f] Rarely strictly aerobic.

[g] Grows best at reduced oxygen concentration, sometimes does not grow anaerobically.

[h] Weak anaerobic growth.

[i] Rarely strictly anaerobic.

[j] Rarely negative.

[k] Sometimes weak catalase or pseudocatalase reaction.

[l] A few strains synthesize cytochrome on aerobic media supplemented with hemin.

[m] Present in strains of serological groups D and N.

[n] Sometimes aerotolerant.

[o] *P. productus* ferments carbohydrates to acetate, formate, succinate and lactate. *P. tetradius* produces acid from glucose.

[p] A$_2$pm, diaminopimelic acid (Symbolism of Schleifer and Kandler, 1972).

This section consists of 15 different genera which are phylogenetically and phenotypically quite diverse (Table 12.1). They have only a few morphological (Gram-positive cocci) and physiological (chemoorganotrophic, mesophilic, nonsporing) characters in common. Table 12.1 shows differential features of these genera. Arrangement of the cells, oxygen requirement and catalase are the most convenient features for initial separation of possible identities.

The presence or absence of catalases and cytochromes separates the Gram-positive cocci into two groups. The families *Micrococcaceae* (*Staphylococcus*, *Micrococcus*, *Planococcus*), *Deinococcaceae* and members of the genus *Stomatococcus* possess catalase or cytochromes or both, whereas all the other genera lack these compounds or exhibit only a weak catalase activity but contain no cytochromes. In a few species (*Streptococcus faecalis*, *Streptococcus lactis*, *Gemella haemolysans*) cytochromes can be synthesized when cells are grown aerobically in the presence of hemin. The family *Micrococcaceae* is not a phylogenetic coherent group (Stackebrandt and Woese, 1979). Comparative sequence analysis of 16S-rRNA indicates that species of the genus *Micrococcus* are phylogenetically related to the so-called actinomycete-coryneform subbranch (Stackebrandt and Woese, 1981) that comprises all Gram-positive bacteria possessing a high DNA base composition (>55 mol%). Members of the genus *Micrococcus* are closely related to certain coryneforms, particularly arthrobacters, and in some cases are even inseparable from them (Stackebrandt and Woese, 1979; Schleifer and Lang, 1980). They can be regarded as degenerate forms of arthrobacters. Micrococci can be clearly distinguished from all other Gram-positive cocci on the basis of their cell wall compositions, G + C content of DNA and menaquinone pattern.

Staphylococci are phylogenetically a coherent group of organisms. Even the least related species (*Staphylococcus sciuri*, *Staphylococcus caseolyticus*) still exhibit significant DNA homology values to the other species. This was supported by oligonucleotide cataloging in 16S-rRNA. The latter studies also demonstrated that staphylococci are phylogenetically not closely related to micrococci but belong to the so-called clostridium-bacillus-streptococcus subbranch that comprises all the Gram-positive bacteria possessing a DNA base composition lower than 55 mol% (Stackebrandt and Woese, 1981). The cell wall composition and the low G + C content of DNA are valuable chemotaxonomic markers to separate staphylococci from the other catalase-positive cocci. There are some strains of staphylococci such as *Staphylococcus saccharolyticus* and *Staphylococcus aureus* subsp. *anaerobius* (De la Fuente et al., 1985) that do not produce catalase and grow predominantly under anaerobic conditions. However, they possess all the other characteristic features mentioned above. In particular, their cell wall

Table 12.1.—*continued*

Characteristics	Anaerobic genera				
	Peptococcus	*Peptostreptococcus*	*Ruminococcus*	*Coprococcus*	*Sarcina*
Predominant arrangement of cells (other than single cells), commonest appearance listed first	Pairs, tetrads	Pairs, chains, tetrads	Pairs, chains	Pairs, chains of pairs	Cuboidal packets
Strict aerobes	−	−	−	−	−
Facultative anaerobes or microaerophils	−	−	−	−	−
Strict anaerobes	+	+	+	+	$+^n$
Catalase reaction	$−^k$	$−^k$	−	−	−
Cytochromes present	−	−	ND	ND	ND
Major fermentation products from carbohydrates anaerobically (if not fermentative, acidity recorded)	NA (no acid)	NA^o	Acetate, lactate, succinate, formate, ethanol, CO_2, H_2	Butyrate, acetate, lactate, formate or proprionate, H_2	Acetate, ethanol, H_2, CO_2 or butyrate
Peptidoglycan					
Position 1	Ala	Ala or GLy	Ala	Ala	Ala
Position 3	Lys	Lys or Orn	m-A_2pm^p	m-A_2pm^m	LL-A_2pm^p
Main interpeptide bridge	D-Asp	D-Asp, D-Glu, Gly, Lys-D-Glu, or Gly-D-Asp	None	None	Gly
Teichoic acid in cell wall	ND	ND	ND	ND	ND
Major menaquinones^d	ND	ND	ND	ND	ND
Mol% G + C of DNA	50–51	27–45	39–46	39–42	28–31

^{a} Symbols: +, 90% or more of strains positive; −, 10% or less of strains positive; d, 11–89% of strains positive; D, substantial proportion of species differ; NA, not applicable; and ND, not determined.
^{b} Most strains are motile.
^{c} One species sometimes facultatively anaerobic.
^{d} See Collins and Jones (1981): most streptococci lack menaquinones but some strains of serological groups D and N contain MK-8, MK-9 or demethylmenaquinone containing nine isoprene units.
^{e} Most strains are capsulated.
^{f} Rarely strictly aerobic.
^{g} Grows best at reduced oxygen concentration, sometimes does not grow anaerobically.

^{h} Weak anaerobic growth.
^{i} Rarely strictly anaerobic.
^{j} Rarely negative.
^{k} Sometimes weak catalase or pseudocatalase reaction.
^{l} A few strains synthesize cytochrome on aerobic media supplemented with hemin.
^{m} Present in strains of serological groups D and N.
^{n} Sometimes aerotolerant.
^{o} *P. productus* ferments carbohydrates to acetate, formate, succinate and lactate. *P. tetradius* produces acid from glucose.
^{p} A_2pm, diaminopimelic acid (Symbolism of Schleifer and Kandler, 1972).

composition characterizes them as typical staphylococci. Recently, three new species, namely *Staphylococcus arlettae*, *Staphylococcus equorum* and *Staphylococcus kloosii* have been described and validly published (Schleifer et al., 1984a, 1985b). They were isolated mainly from farm animals.

Members of the genus *Planococcus* are phylogenetically closely related to those of the genera *Bacillus* and *Sporosarcina* and should therefore be placed in the family *Bacillaceae*, despite the fact that they are nonsporing organisms. They exhibit characteristics which readily differentiate them from other morphologically and physiologically similar taxa (Table 12.1).

The monospecific genus *Stomatococcus* is phylogenetically related to the actinomycete-coryneform group. However, it represents a separate line of descent within this group (Stackebrandt et al., 1983b). Differences in the peptidoglycan type and the cytochrome pattern between *Stomatococcus* and members of the other genera as well as the results of the DNA-rRNA hybridization studies confirm the conclusions drawn from the comparative 16S-rRNA sequence analysis. Several phenotypic characters, in particular the presence of a thick capsule, the strong adherence to the agar surface and the poor growth on nutrient agar, together with the unusual peptidoglycan type, differentiate *Stomatococcus* from micrococci and other cocci.

The radioresistant cocci assigned to the genus *Deinococcus* are a distinct phylogenetic group of eubacteria which, on the basis of 16S-rRNA sequence data, cannot be aligned with any other hierarchical group. They are also different with regard to other features. They possess cell walls that are atypical for Gram-positive bacteria; the profile of the cell wall includes an outer membrane; the polar lipid composition is unusual and the fatty acid pattern resembles that of Gram-negative bacteria.

The catalase-negative cocci can be divided into two groups: group 1 consists of facultatively anaerobic or microaerophilic organisms and group 2 consists of strictly anaerobic ones. *Streptococcus* is the largest genus of group 1. Streptococci are distinguished by homolactic fermentation and the arrangements of cells in pairs or, more characteristically, in chains. A few strictly anaerobic cocci are also included in the genus description. However, their allocation should be considered as temporary as long as nucleic acid hybridization studies have not proved their relatedness to other streptococci.

Serology has been of major importance in the classification and identification of streptococci. However, the serological criteria cannot be used as the sole or main basis for establishing a valid taxonomic differentiation of the genus *Streptococcus* since some of the serological groups are not in accordance with the biochemical data or the nucleic acid hybridization studies (Jones, 1978; Kilpper-Bälz and Schleifer, 1984). After the description of the genus *Streptococcus* had been completed, several studies on the systematics of streptococci were published which will certainly lead to a major revision of the genus over the next few years. Comparative analysis of ribonuclease T$_1$-resistant oligonucleotides of 16S-rRNA has shown that species of the genus *Streptococcus* form a loosely related group that is phylogenetically equivalent to those of lactobacilli, bacilli and other related taxa within the clostridium subbranch (Ludwig et al., 1985). Moreover, the streptococci studied can be subdivided into three moderately related clusters. This finding has been confirmed and extended by DNA-rRNA hybridization studies (Kilpper-Bälz et al., 1982; Kilpper-Bälz and Schleifer, 1984) which have led to the proposal to divide the genus *Streptococcus* into three genera: 1. The genus *Streptococcus sensu stricto* comprising the majority of known species, in particular the pyogenic and oral streptococci including the pneumococci. 2. The genus *Enterococcus* encompassing the enterococcal group (Schleifer and Kilpper-Bälz, 1984; Collins et al., 1984a). 3. The genus "*Lactococcus*" to which all lactic streptococci (group N) are transferred (Schleifer et al., 1985a). These findings are in reasonable agreement with those from numerical taxonomy (Bridge and Sneath, 1983).

Subdivisions within the genus and relationships at the species level have been determined using numerical taxonomic, chemotaxonomic methods and DNA-DNA hybridization studies. The term pyogenic streptococci might well be confined to the serological groups A, A-variant, C, G (type II) and L (Kilpper-Bälz and Schleifer, 1984). There are some indications that "*Streptococcus equisimilis*" and strains of serological groups G and L may belong to *Streptococcus dysgalactiae*, whereas "*Streptococcus zooepidemicus*" may belong to *Streptococcus equi* (Kilpper-Bälz and Schleifer, 1984; Farrow and Collins, 1984c) but this is not supported by the results of Bridge and Sneath (1983). A common feature of the pyogenic streptococci as thus restricted may be their capability to bind to mammalian fibrinogens and human fibronectin (Switalski et al., 1982; Myhre and Kuusela, 1983; Lämmler et al., 1983). *Streptococcus agalactiae* (serological group B) and streptococci of serological groups R and S are not closely related to the other pyogenic streptococci and form separate lines of descent within the genus *Streptococcus*. Serological groups E, P, U and V are also not closely related to *Str. pyogenes* and have been designated as a new species, *Streptococcus porcinus* (Collins et al., 1984b). Strains labeled *Streptococcus bovis* and *Streptococcus equinus* were found to be very heterogeneous (Farrow et al., 1984). Strains of *Str. equinus* from pigs and chickens were found to be rather different from the type strain of *Str. equinus* and were considered to represent a new species *Streptococcus alactolyticus*. Some atypical strains of *Str. bovis* were also transferred to a new species, for which the name *Streptococcus saccharolyticus* has been proposed.

The "*Streptococcus mutans* group" can be subdivided into five distinct species (Coykendall, 1977; Schleifer et al., 1984b). Strains of "*Streptococcus milleri*" could be allocated to either *Streptococcus anginosus* or *Streptococcus constellatus* (Kilpper-Bälz et al., 1984). Despite the progress made in the last few years, a number of taxonomic and nomenclatural problems remain in particular within the oral (viridans) and anaerobic groups but they may be resolved by applying nucleic acid hybridization, numerical taxonomic and chemotaxonomic methods.

The genera *Leuconostoc*, *Pediococcus*, *Aerococcus* and *Gemella* complete group 1 of catalase-negative cocci. Oligonucleotide cataloging of 16S-rRNA has clearly shown that they all belong to the clostridium-bacillus-streptococcus subbranch of Gram-positive bacteria. *Leuconostoc* and *Pediococcus*, together with *Streptococcus* and *Lactobacillus*, form a coherent supercluster within the clostridium subbranch (Stackebrandt et al., 1983a). *Aerococcus* and *Gemella* represent independent lines of descent within this subbranch (Stackebrandt et al., 1982; Ludwig et al., 1985). Gas production from glucose under anaerobic growth conditions separates leuconostocs from the other catalase-negative cocci. However, the separation of leuconostocs from hetero-

fermentative lactobacilli is not easy since there are morphological overlaps and *Leuconostoc* is phenetically close to a number of streptococci (Bridge and Sneath, 1983). The characteristic occurrence in tetrads is very helpful in separating pediococci from other catalase-negative cocci. Sometimes they may be confused with aerococci because the latter also exhibit a strong tendency to tetrad formation. However, most aerococci show weak growth under anaerobic conditions and produce a greening reaction on blood-agar.

Gemella, like *Aerococcus*, is a monospecific genus. Originally, *Gemella* was classified as belonging to the genus *Neisseria* but later studies indicated that it is a Gram-positive bacterium. It grows under aerobic as well as under anaerobic conditions to only a low cell density. It exhibits superoxide dismutase activity and can synthesize cytochromes during aerobic growth in the presence of hemin (Stackebrandt et al., 1982).

The genera *Peptococcus*, *Peptostreptococcus*, *Ruminococcus*, *Coprococcus* and *Sarcina* are strictly anaerobic and form group 2 of the catalase-negative cocci. The phylogenetic analysis of these organisms by 16S-rRNA oligonucleotide cataloging is still incomplete. Only a few strains "*Peptococcus aerogenes*," *Peptococcus glycinophilus*, *Peptostreptococcus anaerobius*, *R. bromii* and *Sarcina ventriculi*) have been analyzed so far (Stackebrandt and Woese, 1981). All of them appear to be off-shoots of the clostridia. Nothing is known about the phylogenetic position of *Peptococcus* and *Coprococcus*.

Peptococcus is now considered to be a monospecific genus. It has a significantly higher DNA base composition (50–51 mol% G + C) than all the other strictly anaerobic cocci. A distinct separation of peptococci and peptostreptococci can be obtained only by determining the G + C content of the DNA. Extensive DNA-rRNA and DNA-DNA hybridization studies indicate that the peptostreptococci are genetically rather heterogeneous (Huss et al., 1984). The strains studied could be divided into seven groups. The first group comprises strains by *Peptostreptococcus magnus*, the second group consists of *Peptostreptococcus prevotii* and strains of Hare groups I, III and VIII. The type strain of *Peptostreptococcus asaccharolyticus*, together with strains of *Peptostreptococcus indolicus*, form the fourth group. The fifth group comprises another strain of *P. asaccharolyticus* and a strain of Hare group VIII. Strains of *Peptostreptococcus micros* and of Hare group IX form the sixth group. The type strain of *Streptococcus parvulus* is related to this group. The last group consists of *Peptostreptococcus anaerobius*, *Eubacterium tenue* and *Clostridium lituseburense*. The specific relatedness of these organisms has also been demonstrated by 16S-rRNA oligonucleotide cataloging (Stackebrandt and Woese, 1981). The heterogeneity of the strains designated *Peptostreptococcus* indicates that their allocation within one genus should be considered as temporary. Further extensive studies are necessary to clarify the taxonomic position of these organisms.

Ruminococci can be separated from the other strictly anaerobic cocci by their ability to ferment sugars and their inability to ferment peptone or to produce butyrate or longer carbon-chained fatty acids. The phylogenetic position of only one of the eight species has been studied. *Ruminococcus bromii* is distantly related to certain clostridia. Very few taxonomic details are known about members of the genus *Coprococcus*. They are physiologically similar to *Ruminococcus*. However, the formation of butyrate as a characteristic fermentation product facilitates the separation of coprococci from ruminococci.

The packet-forming *Sarcina* can be readily distinguished morphologically from other strictly anaerobic cocci. Moreover, they produce both acid and gas as major fermentation products. According to 16S-rRNA oligonucleotide cataloging, *Sarcina ventriculi* is phylogenetically related to clostridia.

In conclusion, the 15 different genera listed in this section certainly do not belong to the same family or order. The majority (11 genera) cluster within the clostridium-bacillus-streptococcus subbranch of Gram-positive bacteria. Two genera (*Micrococcus*, *Stomatococcus*) can be allotted to the actinomycete-coryneform subbranch and the genus *Deinococcus* represents a unique line of descent within the eubacteria.

FAMILY I. **MICROCOCCACEAE** PREVOT 1961, 31[AL*]

KARL HEINZ SCHLEIFER

Mi.cro.coc.ca′ce.ae. M.L. masc. n. *Micrococcus* type genus of the family; *-aceae* ending to denote family; M.L. fem. pl. n. *Micrococcaceae* the *Micrococcus* family.

Cells spherical, 0.5–2.5 μm in diameter, characteristically dividing in more than one plane to form regular or irregular clusters or packets. Gram-positive. Nonmotile or less frequently motile. Resting stages and endospores are not produced. The **diamino acid** present in the **peptidoglycan is L-lysine.**

Aerobic or facultatively anaerobic. **Chemoorganotrophic,** metabolism respiratory and/or fermentative. Utilize carbohydrates and/or amino acids as carbon and energy sources.

Nutritional requirements variable. All strains grow in presence of 5% NaCl, many grow in 10–15%.

Usually catalase-positive.

Some species are opportunistic pathogens of animals and/or man.

Type genus: *Micrococcus* Cohn 1872, 151.

Further Comments

The family consists of four genera, *Micrococcus, Staphylococcus, Stomatococcus* and *Planococcus* and their differential characteristics are listed in Table 12.2. It has to be stressed that these **genera are not closely related** according to the phylogenic concepts deduced from the wide-ranging G + C content of the DNA (30–75 mol%), their different cell wall composition (Schleifer, 1973), nucleic acid hybridization studies (Schleifer, Heise and Myer, 1979; Schleifer, Meyer and Rupprecht, 1979; Kilpper-Bälz and Schleifer, 1981) and comparative analysis of 16S rRNA sequences (Stackebrandt and Woese, 1979; Ludwig et al., 1981; Stackebrandt, Scheuerlein and Schleifer, 1983b). These studies clearly indicate that the four genera should not be combined in one family. The 16S rRNA oligonucleotide catalog data suggest that the micrococci are closely related to arthrobacters (Stackebrandt et al., 1980), planococci to certain bacilli and the genus *Staphylococcus* belongs phylogenetically to the broad *Bacillus-Lactobacillus-Streptococcus* cluster. The genus *Stomatococcus* represents an independent line of descent within the Gram-positive eubacteria with a high G + C content of DNA (> 55 mol %). The close relatedness of morphologically quite different bacteria demonstrate that shape may be a treacherous criterion for defining phylogenetic relationships. The genera *Micrococcus, Stomatococcus, Staphylococcus* and *Planococcus* should probably be allocated to different families when the phylogenetic studies of actually and potentially related groups are completed.

* *AL*, denotes the inclusion of this name on the Approved Lists of Bacterial Names (1980).

Table 12.2.
Differential characteristics of the genus **Micrococcus, Stomatococcus, Planococcus** *and* **Staphylococcus**[a]

Characteristics	Micrococcus	Stomatococcus	Planococcus	Staphylococcus
Irregular clusters	+	+	+	+
Tetrads	+	−	−	−
Capsule	−	+	−	−
Motility	−[b]	−	+	−
Growth on (FTO) furazolidone agar	+	−	−	−
Anaerobic fermentation of glucose	−[c]	+	−	+
Oxidase and benzidine tests[d]	+	−	ND	−[e]
FDP-aldolase (class)[f]	II	ND	ND	I
Resistance to lysostaphin[g]	R	R	R	S
Menaquinones[h]	Hydrogenated	ND	Normal	Normal
Glycine present in peptidoglycan[i]	−	−	−	+
Teichoic acid present in cell wall[i]	−	−	−	+
Mol% G+C of DNA[j]	65–75	56–60	39–52	30–39

[a] Symbols: +, 90% or more of strains are positive; −, 90% or more of strains are negative; d, 11–89% of strains are positive; ND, not determined; R, resistant; and S, susceptible.
[b] *M. agilis* is motile.
[c] *M. kristinae* produces acid from glucose anaerobically.
[d] Tested by method of Faller and Schleifer (1981).
[e] Strains of *S. sciuri* and *S. caseolyticus* possess cytochrome *c* and are positive in the modified oxidase and benzidine tests.
[f] Data from Götz et al. (1979); Fischer et al. (1982).
[g] Data from Schleifer and Kloos (1975b).
[h] Data from Jeffries (1969); Yamada et al. (1976).
[i] Data from Schleifer and Kandler (1972); Schleifer and Kloos. (1976).
[j] Data from Kocur et al. (1971); Holländer and Pohl (1980).

Genus I. *Micrococcus* Cohn 1872, 151[AL]

MILOSLAV KOCUR

Mi.cro.coc'cus. Gr. adj. *micrus* small; Gr. n. *coccus* a grain or berry; M.L. masc. n. *Micrococcus* small coccus.

Cells spherical 0.5–2.0 μm in diameter, occurring mostly in pairs, tetrads or irregular clusters. Gram-positive. **Usually nonmotile.** Spores are not formed. Chemoorganotrophic, metabolism strictly respiratory. **Aerobic;** one species may grow facultatively anaerobically. Catalase- and oxidase-positive. Acid without gas produced from glucose, when attacked. Acid production from other carbohydrates varies with species. Most species produce carotenoid pigments. All species grow in the presence of up to 5% NaCl. Optimum temperature 25–37°C. The primary natural habitat is mammalian skin; the secondary habitat is meat and dairy products, soil and water. Nonpathogenic; some strains may be opportunist pathogens. The mol% of the DNA is 64.0–75.0 (T_m, Bd).

Type species: *Micrococcus luteus* (Schroeter) Cohn 1872, 153.

Further Descriptive Comments

In usual complex media such as nutrient agar, micrococci appear as typical Gram-positive or Gram-variable spheres arranged predominantly in tetrads or diplococci. In some species tetrads may form large adherent clusters. The cell morphology of micrococci does not change with the kind of medium or with culture age.

Micrococci have the ultrastructure of the cell similar to that of other Gram-positive cocci (Chapman, 1960; Hunter et al., 1973; Kocur unpublished results).

The cell wall of micrococci consists of thick, rigid layers of peptidoglycan. Six different types of peptidoglycan have been found in their cell walls. Teichoic acid is absent (Schleifer and Kandler, 1972). *M. luteus* contains teichuronic acid (Perkins, 1963; Yamada et al., 1975). *M. varians*, containing *N*-acetylglocosamine 1-phosphate polymers (Partridge et al., 1973) has been reclassified as *Staphylococcus caseolyticus* (Schleifer et al., 1982).

Colonies of micrococci are usually circular, entire, convex and smooth. Some strains may produce matted colonies. Colonies may have a yellow, orange, orange-red, pink-red or red pigment. The differences in colony morphology or pigment are usually associated with differences in other characteristics of the strains. Colonies of some species, e.g. *M. varians* and *M. nishinomiyaensis* may be confused with those of certain coryneform bacteria (Kloos et al., 1974).

Nutritional requirements of micrococci are variable. Strains of some species (e.g. *M. roseus*, *M. luteus*), can grow with glutamic acid as carbon, nitrogen and energy source and thiamine and/or biotin as growth factors (Eisenberg and Evans, 1963; Cooney and Thierry, 1966; Perry and Evans, 1966). Carbon-containing compounds, oxidized to carbon dioxide and water, include acetate, lactate, pyruvate, succinate, fructose, galactose, glucose, glycerol, maltose and sucrose. Variable oxidation of mannitol, sorbitol, arabinose, rhamnose, ribose, xylose and starch. Dulcitol is not oxidized (Saz and Krampitz, 1954; Rosypal and Kocur, 1963; Perry and Evans, 1960, 1966). Glucose is metabolized by hexose monophosphate pathway and citric acid enzymes; the glycolysis cycle functions only under aerobic conditions (Dawes and Holmes, 1958; Perry and Evans, 1966; Blevins et al., 1969).

Some *Micrococcus* species can grow with ammonium phosphate as sole nitrogen source; *M. luteus* strains grow on inorganic nitrogen agar; *M. varians* strains grow on Simmons citrate agar (Kloos et al., 1974).

Chemically defined media for the growth of *M. luteus* and for the growth and the pigmentation of *M. roseus* have been devised (Wolin and Naylor, 1957; Grula et al., 1961; Cooney and Thierry, 1966). Strains of *M. luteus* may grow in a defined medium containing pyruvic acid as carbon and energy source, glutamic acid, biotin and mineral salts (Perry and Evans, 1966). Some strains require a more complex medium containing a number of amino acids (Grula et al., 1961). Micrococci require one or more amino acids and vitamins for growth (Aaronson, 1955; Wolin and Naylor, 1957; Eisenberg and Evans, 1963; Farrior and Kloos, 1975).

The major fatty acids component of *Micrococcus* strains is a methyl-branched C_{15}-saturated acid (Girard, 1971; Onishi and Kamekura, 1972; Thirkell and Gray, 1974; Janzen et al., 1974; Brooks et al., 1980). Micrococci contain relatively high amounts of long chain aliphatic hydrocarbons in the range C_{22} to C_{33} (Tornabene et al., 1970; Morrison et al., 1971; Kloos et al., 1974). Cardiolipin and phosphatidylglycerol are the major phospholipids. Unidentified phospholipids a and b were described by Komura et al. (1975).

Micrococcus species contain hydrogenated menaquinones of the type MK-7, MK-8 and MK-9 (Jeffries, 1969; Jeffries et al., 1969; Yamada et al., 1976). They possess, like most bacteria, a class II D-fructose 1,6-biphosphate aldolase (Götz et al., 1979) and contain cytochromes of *a*-, *b*-, *c*-, and *d*-types (Faller et al., 1980). Oxidase activity was demonstrated in *M. luteus*, *M. varians*, *M. kristinae* and *M. nishinomiyaensis* only when conventional methods were used. The remaining *Micrococcus* species were oxidase-negative (Steel, 1962; Boswell et al., 1972; Kloos et al., 1974). A modified oxidase test has been devised allowing all micrococci and *Staphylococcus sciuri* to develop blue color, whereas all of the other staphylococci exhibited no coloration (Faller and Schleifer, 1981).

The chemical nature of pigment has been determined in two *Micrococcus* species only. The pigment of *M. luteus* is a carotenoid (dihydroxy C_{50} carotenoid) (Littwack and Carlucci, 1958; Thirkell and Strang, 1967; Thirkell and Hunter, 1969a, b); the main pigment of *M. roseus* is canthaxanthin; further pigments are α or β carotene derivatives (Cooney et al., 1966; Ungers and Cooney, 1968; Schwartzel and Cooney, 1970).

Most *Micrococcus* species grow on nutrient agar containing 0–12% NaCl. The only exception is *M. agilis* which grows solely in 0–5% NaCl in a medium and *M. halobius* which does not grow on nutrient agar without salt added. *M. halobius* is a moderate halophilic organism capable of growth in the presence of 1–4 M NaCl in a medium. It produces a considerable amount of halophilic amylase of unique salt response (Onishi, 1972a, b; Onishi and Sonoda, 1979).

DNA/DNA hybridization studies among *M. luteus*, *M. lylae*, *M. varians* and *M. kristinae* revealed a discrete genetic relationship between *M. luteus* and *M. lylae* but only low homology values among the other species (Schleifer et al., 1979). The lack of close genetic relatedness between *M. luteus* and *M. varians* was also demonstrated by DNA homology studies of Ogasawara-Fujita and Sakaguchi (1976). Double immunodiffusion tests with cell free extracts from various *Micrococcus* species and antisera against catalase of *M. luteus* confirmed the above results. Good cross-reaction between *M. luteus* and *M. lylae* was found, *M. varians* reacted very weakly and no reaction was recorded with the preparation of *M. kristinae*, *M. nishinomiyaensis*, *M. roseus* and *M. sedentarius* (Rupprecht and Schleifer, 1977).

Micrococci are resistant to a wide spectrum of staphylococcal phages (Peters et al., 1976; Bauske et al., 1978) and do not adsorb polyvalent phages isolated from coagulase-negative staphylococci (Schumacher-Perdreau et al., 1979). Several bacteriophages have been isolated from micrococci particularly from *M. luteus* and *M. varians*, which lyse micrococci only (Naylor and Burgi, 1956; Peters and Pulverer, 1975).

No serological relationship has been found among species of the genera *Micrococcus*, *Stomatococcus* and *Staphylococcus*. *M. luteus*, *M. varians* and *M. roseus* have a number of agglutinogens without any systematic pattern. No correlation between their biochemical features and specific agglutinogens was demonstrated (Hasselgren and Oeding, 1972; Oeding and Hasselgren, 1972). Simple serological tests for the separation of micrococci from staphylococci based on the presence or absence of cell wall teichoic acid and peptidoglycan types have been devised (Hamada et al., 1978; Seidl and Schleifer, 1978).

Most micrococci are susceptible to erythromycin, streptomycin, penicillin, methicillin, novobiocin, tetracycline, chloramphenicol, neomycin, vancomycin, kanamycin and polymyxin B. *M. luteus* and *M. sedentarius* strains are susceptible to lysozyme; strains of other species are slightly resistant or resistant to lysozyme (Jeffries, 1968; Kloos et al., 1974); *M. sedentarius* strains are penicillin and methicillin resistant (Kloos et al., 1974). Micrococci are resistant to lysostaphin (Klesius and Schuhardt, 1968; Schleifer and Kloos, 1975a).

Up to now there is very little information regarding the pathogenicity of micrococci. The *Micrococcus* strains isolated from various infections were most probably misclassified staphylococci. This was due to updated classification and lack of tests for the clear-cut separation of micrococci from staphylococci (Roberts, 1967; Person et al., 1969). Recent reports, however, confirm that micrococci may be associated with human infections, particularly in immunosuppressed patients (Peters and Pulverer, 1978; Souhami et al., 1979; Schleifer et al., 1981).

Mammalian skin is now considered to be a primary natural habitat of micrococci. Strains isolated from various other sources may have been, initially, contaminants disseminated by host animal carriers (Marples, 1965; Glass, 1973; Noble and Sommerville, 1974; Kloos et al., 1976; Carr and Kloos, 1977; Schleifer et al., 1981).

Enrichment and Isolation Procedures

For the isolation of micrococci from human and other mammalian skin the procedures described by Kloos and Musselwhite (1975) and Kloos et al., (1976) are recommended. In this case, the nonselective medium (P agar*) can be used which may be supplemented with the mold inhibitor cycloheximide (50 μg/ml) (Kloos et al., 1974, 1976) or if

Bacillus species are numerous on skin the P medium should be supplemented with 7% NaCl to inhibit their large or spreading colonies (Schleifer et al., 1981).

A selective medium (FTO agar†) for direct, selective isolation of micrococci from skin was described by Curry and Borovian (1976) and modified by von Rheinbaben and Hadlok (1981). The medium allows the growth of micrococci only, whereas the growth of staphylococci is strongly repressed. For the isolation of micrococci from marine environment sea water agar§ is recommended.

Maintenance Procedures

Micrococci are initially cultured on nutrient agar or on yeast agar; *M. halobius* is cultured on nutrient agar with 5% NaCl. After incubation at 30–37°C (*M. agilis* at 22–25°C) to allow abundant growth, the cultures may be stored in a refrigerator (5°C) for 3–5 months if they are in perfectly sealed tubes. They may also be stored on the above media under liquid paraffin in a refrigerator (5°C) for 1–2 years and in lyophilized state for 5 or more years.

Procedures for Testing Special Characters

The basic methods for the separation of micrococci from other Gram-positive cocci are fully described in papers by Baird-Parker (1979) and Schleifer et al. (1981). For the examination of oxidase and benzidine tests the modifications described by Faller and Schleifer (1981) are recommended. For testing the ability of micrococci to grow on inorganic nitrogen agar the procedure of Kloos et al. (1974) should be used.

Differentiation of the genus **Micrococcus** from other genera

Table 12.2 indicates the characteristics of *Micrococcus* that distinguish it from other species or genera of morphologically or physiologically similar genera.

Taxonomic Comments

The genus *Micrococcus* is not considered to be a phylogenetically acceptable one (Stackebrandt and Woese, 1979). The data on G + C content in the DNA, chemical cell wall analysis and a comparative analysis of 16S rRNA sequences suggest that the genus *Micrococcus* is more closely related to the genus *Arthrobacter* than it is to the other coccoid genera *Staphylococcus* and *Planococcus* (Kocur et al., 1971; Kloos et al., 1974; Stackebrandt and Woese, 1979). In the opinion of

Keddie (1974) *Micrococcus* species should be regarded as degenerate forms of arthrobacters, locked into the coccoid stage of the *Arthrobacter* life cycle. These data suggest that micrococci should be accommodated in one family with the genus *Arthrobacter*.

Further Reading

Baird-Parker, A.T. 1979. Methods for identifying staphylococci and micrococci. *In* Skinner and Lovelock (Editors), Identification Methods for Microbiologists, 2nd Ed, Academic Press, London, pp. 201–210.

Schleifer, K.H., W.E. Kloos and M. Kocur. 1981. The genus *Micrococcus. In* Starr, Stolp, Trüper, Balows and Schlegel (Editors), The Prokaryotes. A Handbook on Habitats, Isolation, and Identification of Bacteria, Springer-Verlag, New York, pp. 1539–1547.

Differentiation and characteristics of the species of the genus **Micrococcus**

The differential characteristics of the species of *Micrococcus* are indicated in Table 12.3. Other characteristics of the species are presented in Table 12.4.

List of species of the genus **Micrococcus**

1. **Micrococcus luteus** (Schroeter) Cohn 1872, 153.[AL] (*Bacteridium luteus* Schroeter 1872, 126.)

Lu'te.us. L. adj. *luteus* golden-yellow.

Spheres 0.9–1.8 μm in diameter occurring in tetrads and in irregular clusters of tetrads. Strains growing as tetrads or irregular cell clusters form smooth, convex colonies with regular edge. Colonies of strains forming cubical packets usually have a granular surface and a matt appearance. Colonies are yellow, yellowish green or orange pigmented. Some strains form a violet pigment which diffuses into the medium.

Physiological and biochemical characteristics are presented in Tables 12.3 and 12.4. The optimal growth temperature range is 25–37°C.

The primary habitat is mammalian skin.

The mol% G + C of the DNA is 70–75.5 (T_m, Bd) (Kocur et al., 1971).

Type strain: ATCC 4698 (CCM 169, NCTC 2665).

2. **Micrococcus lylae** Kloos, Tornabene and Schleifer 1974, 83.[AL]

ly'lae. M.L. gen. n. *lylae* of Lyla.

---

* P agar: peptone (Difco), 10 g; yeast extract (Difco), 5 g; sodium chloride, 5 g; glucose, 1 g; agar (Difco), 15 g; distilled water, 1000 ml (Naylor and Burgi, 1956).

† Furazolidone (FTO) agar: peptone, 10 g; yeast extract, 5 g; sodium chloride, 5 g; glucose, 1 g; agar, 12 g; distilled water, 1000 ml; pH 7.0. After autoclaving and cooling to 48°C, 100 ml of a 0.02% acetone solution of furazolidone are mixed under slow stirring with the basal medium. Before pouring the plates the flasks are left open or loosely covered in a water bath for 3–5 min. to allow the evaporation of acetone (von Rheinbaben and Hadlok, 1981).

§ Sea water agar: beef extract, 10 g; peptone, 10 g; agar, 20 g; tap water, 250 ml; sea water, 750 ml; pH 7.2.

Table 12.3.
Characteristics differentiating the species of the genus **Micrococcus**[a,b]

Characteristics	1. *M. luteus*	2. *M. lylae*	3. *M. varians*	4. *M. roseus*	5. *M. agilis*	6. *M. kristinae*	7. *M. nishinomiyaensis*	8. *M. sedentarius*	9. *M. halobius*
Major pigment	Y	CW	Y	PR, OR	R	PO	O	CW, BY	N
Aerobic acid from									
Glucose	−	−	+	+	−	+	d	−	+
Glycerol	−	−	−	−	−	+	−	−	+
Esculin hydrolysis	−	−	−	−	+	+	−	−	ND
Arginine dihydrolase	−	−	−	−	−	−	−	+	−
Nitrate to nitrite	−	−	+	+	−	−	d	−	−
Growth at 37°C	+	+	+	+	−	+	+	+	+
Growth on nutrient agar with 7.5% NaCl	+	+	+	+	−	+	−	+	+
Growth on inorganic nitrogen agar	+	−	−	−	−	−	−	−	ND
Growth on Simmons citrate agar	−	−	+	−	−	−	−	−	−
Motility	−	−	−	−	+	−	−	−	−

[a] Data from Kloos et al. (1974); Schleifer et al. (1981).

[b] Symbols: see Table 12.2; also Y, yellow; CW, cream-white; PR, pastel red; OR, orange red; R, red; PO, pale orange; O, orange; BY, buttercup yellow; N, unpigmented.

Spheres 0.8–1.6 μm in diameter occurring predominantly in tetrads. Colonies are circular, convex, entire, smooth, usually nonpigmented or cream-white. Physiological and biochemical characteristics are presented in Table 12.3 and 12.4. The optimal growth temperature range is 25–37°C.

The mol% G + C of the DNA is 68.6 ± 0.9 (T_m) (Kloos et al., 1974). The primary habitat is mammalian skin.

Type strain: ATCC 27566 (CCM 2693, NCTC 11037).

3. **Micrococcus varians** Migula 1900, 135.[AL]

va′ri.ans. L. part. adj. *varians* varying.

Spheres 0.9–1.5 μm in diameter occurring in tetrads and in irregular clusters. Colonies are circular, slightly convex, smooth and glistening. Some strains may form rough, wrinkled, matted and dry colonies. Produce a light dark yellow pigment. Physiological and biochemical characteristics are presented in Table 12.3 and 12.4. The optimal growth temperature range is 25–37°C.

Distributed on mammalian skin, beach sand and water.

The mol% G + C of the DNA is 66–72 (T_m) (Kocur et al., 1971).

Type strain: ATCC 15306 (CCM 884, NCTC 7564).

4. **Micrococcus roseus** Flügge 1886, 183.[AL]

ro.se.us. L. adj. *roseus* rose colored.

Spheres 1.0–1.5 μg in diameter occurring in pairs or tetrads. Colonies on agar are circular, slightly convex, smooth, occasionally rough, pink or red. Physiological and biochemical characteristics are presented in Table 12.3 and 12.4. The optimal growth temperature range is 25–35°C.

Distributed in soil and water.

The mol% G + C of the DNA is 66–75 (T_m) (Boháček et al., 1969).

Type strain: ATCC 186 (CCM 679, NCTC 7523).

5. **Micrococcus agilis** Ali-Cohen 1889, 36.[AL]

a.gi′lis. L. adj. *agilis* agile.

Spheres 0.8–1.2 μm in diameter, occurring in pairs and tetrads. Motile by means of one or three flagella. Nonmotile strains may occur. Colonies on agar are circular, entire, slightly convex, smooth and matted. Produce a dark rose-red, water-insoluble pigment. Physiological and biochemical characteristics are presented in Table 12.3 and 12.4. The optimal growth temperature range is 20–30°C.

Isolated from water, soil and human skin. Not common.

The mol% G + C of the DNA is 67.0–69.0 (T_m) (Boháček et al., 1969).

Type strain: ATCC 966 (CCM 2390, NCTC 7509).

6. **Micrococcus kristinae** Kloos, Tornabene and Schleifer 1974, 87.[AL]

kris.ti′nae. M.L. gen. n. *kristinae* of Kristin.

Spheres 0.7–1.1 μm in diameter occurring in tetrads which may form large adherent clusters. Nonmotile. Colonies on agar are circular, entire or crenate, convex, smooth or rough. Produce a pale cream to pale orange pigment. Slightly facultative anaerobic. Physiological and biochemical characteristics are presented in Table 12.3 and 12.4. Optimal growth temperature range is 25–37°C.

The primary habitat is human skin.

The mol% G + C of the DNA is 66.8 ± 0.2 (T_m) (Kloos et al., 1974).

Type strain: ATCC 27570 (CCM 2690, NCTC 2665).

7. **Micrococcus nishinomiyaensis** Oda 1935, 1202.[AL]

nisih.no′miya.ensis. M.L. adj. pertaining to *Nishinomiya* a city in Japan.

Spheres 0.9–1.5 μm in diameter, occurring in pairs, tetrads and irregular clusters. Colonies on agar are circular, entire, slightly convex, smooth with matted or glistening surface. Produce a bright orange pigment. Some strains may produce orange exopigment. Physiological and biochemical characteristics are presented in Table 12.3 and 12.4. The optimal growth temperature range is 25–37°C.

Distributed on mammalian skin and in water.

The mol% G + C of the DNA is 66.4–71.1 (T_m) (Kocur et al., 1975).

Type strain: ATCC 29093 (CCM 2140, NCTC 11039).

8. **Micrococcus sedentarius** ZoBell and Upham 1944.[AL]

se.den.ta′rius. L. adj. *sedere* to sit.

Spheres 0.8–1.1 μm in diameter occurring predominantly in tetrads or tetrads in cubical packets. Colonies on agar are circular, entire, convex to pulvinate, smooth, cream-white or deep buttercup yellow. Some strains produce brownish exopigment. Physiological and biochemical characteristics are presented in Table 12.3 and 12.4. The optimal growth temperature range is 28–37°C.

The primary habitat is human skin.

The mol% G + C of the DNA is 67.7 ± 0.6 (T_m) (Kloos et al., 1974).

Type strain: ATCC 14392 (CCM 314).

9. **Micrococcus halobius** Onishi and Kamekura 1972, 234.[AL]

ha′lo.bius. Gr. n. *hals* salt; Gr. n. *bius* life; M.L. adj. living on salt.

Spheres 0.8–1.5 μm in diameter occurring singly, in pairs and occasionally in tetrads or clusters. Colonies on nutrient agar containing 5% NaCl are circular, smooth, opaque and nonpigmented. Moderate halo-

Table 12.4.

Other characteristics of the species of the genus **Micrococcus**[a]

Characteristics	1. M. luteus	2. M. lylae	3. M. varians	4. M. roseus	5. M. agilis	6. M. kristinae	7. M. nishinomiyaensis	8. M. sedentarius	9. M. halobius
Water soluble exopigment	−	−	−	−	−	−	−/+	+	−
Aerobic acid from									
Mannose	−	−	−	−	−	+	−	−	−[b]
Lactose	−	−	−	−	−	−	−	−	+
Galactose	−	−	−	−	−	−	−	−	+
β-Galactosidase (ONPG test)	−	−	−	−	−	−/+	−	−	+
Tween 80 hydrolysis	−	D	−	−	−	−	D	−	−
Starch hydrolysis	−	−	D	D	+	−	D	−	+
Gelatin hydrolysis	+	+	+	−	+	−	+	+	−
Acetoin	−	−	weak or −	weak or −	−	+	−	−	+
Oxidase	+	+	+	−	−	+	+	−	+
Urease	D	−	+	−	−	D	+	−	−
Phosphatase	−	−	−	−	−	−	−	−	−
Hemolysis	−	−	−	−	−	−	−	−	ND
Egg-yolk reaction	−	−	−	−	−	+	−	−	+
Growth on nutrient agar with 10% NaCl	D	D	−	−	−	+	−	−	+
Lysozyme susceptibility[c]	S	SR	R	SR-R	R	R	SR-R	SR-R	R
Peptidoglycan type[c]	L-Lys-peptide subunit	L-Lys-Asp	L-Lys-L-Ala$_{3-4}$	L-Lys-L-Ala$_{3-4}$	L-Lys-Thr-L-Ala$_3$	L-Lys-L-Ala$_3$	L-Lys-L-Ser$_2$-D-Glu	Uncertain	ND
Aminosugar in cell wall[c] polysaccharide	Mannosamin-uronic acid	Galactosamine	Galactosamine	Galactosamine	Glucosamine	Glucosamine	Galactosamine	ND	ND
Major aliphatic hydrocarbons (br-Δ-C)[d]	C27, C28, C29	C27, C28, C29	C25, C26, C27	C24, C25	ND	C28, C29	C22, C23, C25, C26, C27	C30, C31, C32, C33	ND

[a] Symbols: see Table 12.2; also D, different reactions in different taxa (species of a genus or genera of a family); S, susceptible; SR, slightly resistant; and R, resistant.
[b] Tested in MOF medium of Leifson (1963).
[c] Data from Schleifer et al. (1981).
[d] Data from Kloos et al. (1974).

philic. Best growth on media containing 1–2 M NaCl. Physiological and biochemical characteristics are presented in Table 12.3 and 12.4. The optimal growth temperature range is 20–40°C.

Isolated from unrefined salt. Presumably distributed in saline environment.

The mol% G + C of the DNA is 70.0–71.5 (T_m) (Onishi and Kamekura, 1972).

Type strain: ATCC 21727 (CCM 2591).

Genus II. **Stomatococcus** Bergan and Kocur, 1982, 375[VP*]

TOM BERGAN AND MILOSLAV KOCUR

Sto.ma.to.coc′cus. Gr. n. stoma mouth; Gr. n. coccus a grain or berry; M.L. masc. n. *Stomatococcus* coccus pertaining to the mouth.

Cells spherical, 0.9–1.3 μm in diameter, arranged mostly in clusters, occasionally in pairs and tetrads. Gram-positive. **Nonmotile. Encapsulated.** Spores are not formed. Chemoorganotrophic, metabolism facultatively fermentative. **Facultatively anaerobic. The catalase reaction is either weakly positive or negative.** Acid, but not gas, is produced from glucose and a variety of sugars. Optimum growth temperature, 30–37°C. Colonies are usually mucoid, transparent or whitish, and adherent to agar surface. Probably a normal inhabitant of the mouth and the upper respiratory tract of humans. The mol% G + C of the DNA is 56–60.4 (T_m, Bd).

Type species: *Stomatococcus mucilaginosus* Bergan and Kocur, 1982, 375.

Further Descriptive Information

In usual complex media such as blood agar or tryptose agar *S. mucilaginosus* grows as cocci arranged in large or irregular clusters (Fig. 12.1). The cells form voluminous capsules (Fig. 12.2) containing polysaccharide (Silva et al., 1977). The peptidoglycan interpeptide bridge of *Stomatococcus* consists of a single monocarboxylic amino acid, either L-alanine, L-serine or glycine (Schleifer and Kandler, 1972). The cell wall does not contain teichoic acids (Bowden, 1969). Fatty acids predominant in the cell wall are 12-methyltridecanoic (3–12 relative percent), 12-methyltetradecanoic (27–43%), *n*-pentadecanoic (trace, 12%), 14-methylpentadecanoic (12–20%), *n*-hexadecanoic (8–17%). and 14-methyldecanoic (2–12%). The relative proportion of 14-methylpentadecanoic acid in stomatococci is above 12% in contrast to less than 5% in micrococci (Jantzen et al., 1979).

After growth overnight, colonies on blood agar are 1 mm in diameter, round, entire, convex, mucoid, transparent or white and nonhemolytic. They adhere to the agar surface; this property increases upon further incubation. Under anaerobic conditions, colonies are smaller (0.5 mm) and firmly adherent to the agar surface (Gordon, 1967).

S. mucilaginosus fails to grow at 10 or 45°C and does not initiate growth at pH 9.6, in the presence of 40% bile, or in medium containing 5% NaCl (Gordon, 1967; Bergan et al., 1970a, b; Rubin et al., 1978). Growth of *S. mucilaginosus* is stimulated when nitrate is added to the medium (Gordon, 1967; Rubin et al., 1978).

Nearly one-half of the *S. mucilaginosus* strains so far isolated have been without catalase activity. Catalase-negative strains remained negative also after growth on a medium without carbohydrate, indicating that acid production probably did not depress catalase production per milligram of cell free protein. Two catalase-positive strains, respectively, utilized 147 and 40 μmol of H_2O_2 per min. None of six similarly grown cultures from catalase-negative strains exhibited catalase activity (Gordon, 1967).

Most of *S. mucilaginosus* strains are susceptible to ampicillin, bacitracin, benzylpenicillin, chloramphenicol, erythromycin, fusidic acid, lincomycin, neomycin, novobiocin, oleandomycin, and oxytetracycline. They are resistant to lysostaphin and lysozyme (Gordon, 1967; Bergan et al., 1970a).

Stomatococci are resistant to staphylococcal phages but susceptible to micrococcal phages; a single strain has been found susceptible to 12 of 14 micrococcal phages (Bauske et al., 1978). There is no serological relationship between stomatococci and either micrococci or staphylococci. Strains examined so far have similar capsular antigen (P. Oeding, personal communication).

Enrichment and Isolation Procedures

S. mucilaginosus appears to be an important component of the normal oral flora, comprising about 3.5% of the predominant cultivable aerobic microorganisms of the human oral cavity. It may be isolated from tongue, throat, nasopharynx, bronchial secretions and from blood cultures (Gordon, 1967; Bergan et al., 1970a; Rubin et al., 1978).

Swab samples from the human tongue are inoculated onto blood agar

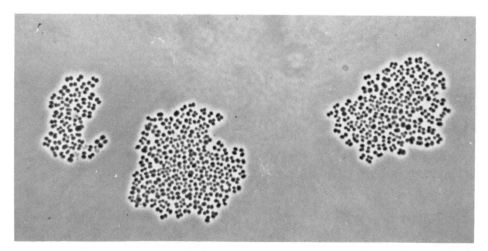

Figure 12.1.   Phase contrast photomicrograph of *S. mucilaginosus* grown in nutrient broth (× 1000).

---

* *VP*, denotes that this name has been validly published in the official publication, International Journal of Systematic Bacteriology.

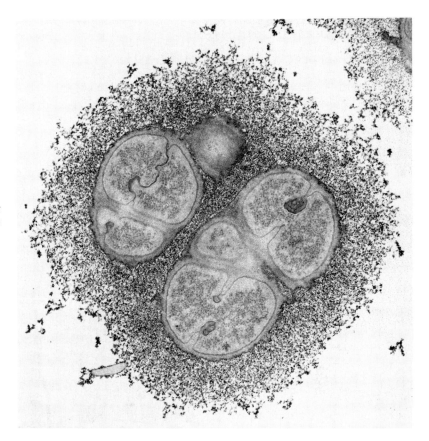

Figure 12.2. Thin section of *S. mucilaginosus* showing rather copious capsular substance. Fixation by the complete R-K procedure (× 40,000).

or trypticase soy agar. After 2–5 days, the colonies, which are adherent to agar surface, are transferred to fresh blood agar and rechecked after 1–2 days.

Maintenance Procedures

Strains of *S. mucilaginosus* are initially cultured on blood agar or on trypticase yeast extract agar. After incubation at 37°C to allow abundant growth, the cultures may be maintained at 4°C for 1 month. The cultures easily loose viability upon desiccation, consequently, sufficient moisture should be present and cultures should be stored in tightly sealed or screw-capped tubes. For long term preservation lyophilization is recommended.

Procedures for Testing Special Characters

No special tests are required for the identification of *Stomatococcus*. The same morphological and biochemical tests as used for staphylococci and micrococci are recommended.

Differentiation of the genus **Stomatococcus** from other genera

Table 12.5 provides the key characteristics for differentiation between the genus *Stomatococcus* and biochemically similar members of *Micrococcus*. Several phenotypic characters, such as the presence of a thick capsule, poor growth on nutrient agar, weak or no catalase reaction, and lack of growth on agar with 5% NaCl differentiate between *S. mucilaginosus* and *Micrococcus* species. Isolates of *S. mucilaginosus* which are catalase-negative may be misidentified as *Strep-*tococcus spp. However its strong adherence to the agar surface is a simple indicator of its identity.

Taxonomic Comments

The genus *Stomatococcus* is a newly described genus (Bergan and Kocur, 1982), created because, although the stomatococci resemble *Micrococcaceae* in part, they differ from both the genera *Micrococcus*

Table 12.5.
Differential characteristics of the genera **Stomatococcus** *and* **Micrococcus**[a]

Characteristics	*Stomatococcus mucilaginosus*	*Micrococcus kristinae*	*Micrococcus* spp.
Cell arrangement	Clusters, diplococci	Tetrads	Tetrads, clusters
Capsule	+	−	−
Catalase	w/−	+	+
Growth on nutrient agar containing 5% NaCl	−	+	+
Glycerol (acid)	+	+	−
Trehalose (acid)	+	−	−
Esculin hydrolysis	+	+	−
G + C content of DNA (mol%)	56–60.4	66.8	64–75

[a] Symbols: +, positive reaction; −, negative reaction; and w, weak reaction.

and *Staphylococcus* in key characteristics by (a) having an intermediate mol% G + C content of DNA and (b) grow slowly under anaerobic conditions. The whole cell composition of fatty acids distinguishes stomatococci from both micrococci and staphylococci. This applies to (c) fatty acid composition; the relative proportion of 14-methylpenta-decanoic acid in stomatococci is above 12% in contrast to less than 5% in micrococci. The *n*-hexadecanoic acid is also higher than in micrococci (Jantzen et al., 1974). There is (d) no serological relationship between stomatococci and micrococci or staphylococci (P. Oeding, personal communication). The (e) comparative oligonucleotide cataloging of 16S rRNA of *S. mucilaginosus* supports the above distinction

and recognition of *Stomatococcus* as a new genus (Ludwig et al., 1981). Phylogenetically, *S. mucilaginosus* represents an independent line of descent within a broad group of Gram-positive bacteria that contains arthrobacters, micrococci, cellulomonads, brevibacteria and microbacteria (Stackebrandt et al., 1983).

Further Reading

Bergan, T. and M. Kocur. 1982. *Stomatococcus mucilaginosus* gen. nov., sp. nov., ep. rev., a member of the family *Micrococcaceae*. Int. J. Syst. Bacteriol. 32: 374–377.

List of species of the genus **Stomatoccccus**

1. **Stomatococcus mucilaginosus** Bergan and Kocur 1982, 375.[VP]
mu.ci.la.gi.no′sus. L. masc. adj. *mucilaginosus* slimy.

The cell morphology and colonial morphology are as given for the genus.

Physiological and nutritional characteristics are presented in Table 12.6.

The species may be pathogenic for humans (Rubin et al., 1978) and produce local abscesses in mice after subcutaneous injection. Large doses may produce fatal disease in mice. Not pathogenic for guinea pigs (Bergan et al., 1970a).

The mol% G + C of DNA is 56–60.4 (T_m, Bd).

Type strain: ATCC 25296 (CCM 2417, NCTC 10663).

Table 12.6.
Other characteristics of **Stomatococcus mucilaginosus**[a]

Test	Stomatococcus mucilaginosus
Acetoin production	+
Esculin hydrolysis	+
Gelatin hydrolysis	+
Nitrate reduction	+
Benzidine test	+
Catalase	d
Growth in 5% NaCl	−
Acid production from:	
Glucose	+
Fructose	+
Sucrose	+
Trehalose	+
Mannose	+
Salicin	+
Maltose	d
Glycerol	d
Mannitol	−
Adonitol	−
Raffinose	−
Sorbitol	−
Arabinose	−
Sorbose	−
Xylose	−
Coagulase	−
Oxidase	−
Phosphatase	−
Tween 80 hydrolysis	−
Starch hydrolysis	−
Urease	−
Citrate (Simmons)	−
H_2S production	−
Indole production	−
Arginine dihydrolase	−
Phenylalanine deaminase	−

[a] Symbols: +, positive reaction; −, negative reaction; d, different reactions

Genus III. *Planococcus* Migula 1894, 236[AL]

MILOSLAV KOCUR

Plan.o.coc′cus. Gr. comb. form *planos* wandering; Gr. n. coccus a grain, berry; *Planococcus* motile coccus.

Cells spherical, 1.0–1.2 μm in diameter, occurring singly, in pairs, in groups of three cells, occasionally in tetrads. Gram-positive to Gram-variable. **Motile.** The motile cells usually possess one or two flagella. Spores are not formed. Chemoorganotrophic, metabolism respiratory. Aerobic. The catalase reaction is positive. Colonies are circular, slightly convex, smooth, glistening and **yellow-orange in color**. Good growth in sea water media as well as on nutrient agar with 1–12% NaCl. Most strains hydrolyze gelatin, **do not attack carbohydrates** in normal media, do not reduce nitrate, do not hydrolyze esculin, starch and Tween 80. Growth factors are usually not required. Optimum temperature, 27–37°C. Distributed in sea water, marine clams, shrimps and prawns. The mol% G + C of the DNA is either 39–42 or 48–52 (T_m, Bd).

Type species: *Planococcus citreus* Migula 1894, 236.

Further Descriptive Information

In media such as sea water agar or on nutrient agar with 1–5% NaCl planococci form spherical cells. Motile cells usually possess one or two flagella (Fig. 12.3); occasionally cells with three or four flagella are found. The flagella are often irregular but some show a regular sine curve with a wave length of 3.0 μm and an amplitude of 0.5–1.0 μm (Fig. 12.3). Motile cells occur in both liquid and solid media. Their cell wall peptidoglycan is of the L-Lys-D-Glu type (Schleifer and Kandler, 1970). No teichoic acid was found in the cell wall of *P. citreus* strains (Endresen and Oeding, 1973). The poly-γ-D-glutamyl polymer is found in the cell walls of *P. halophilus* (Kandler et al., 1983).

The fine structure of the cells of planococci is similar to that of other Gram-positive, catalase-positive cocci (Novitsky and Kushner, 1976; Kocur, unpublished). The cell wall of *P. citreus* is double layered and its thickness varies with the age of culture from 25–35 nm (Fig. 12.4).

The analysis of fatty acids has shown that in common with other Gram-positive bacteria *P. citreus* has a branched saturated C_{15} acid as the major component of free fatty acids, mono- and diglycerides, most

of the glycolipids and the phospholipids (Thirkell and Summerfield, 1977b).

P. citreus produces a yellow-orange, water-insoluble carotenoid pigment. The biosynthesis of carotenoids and the type of carotenoids appear to be influenced by the concentration of sea salt in the medium and by the age of the culture (Thirkell and Summerfield, 1980). The

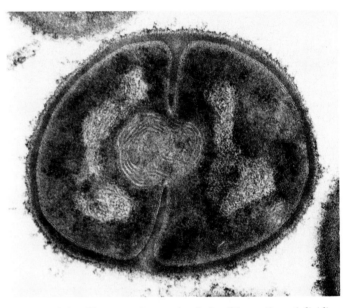

Figure 12.4. Electron micrograph of ultrathin section of dividing cell of *Planococcus citreus* (× 65,000).

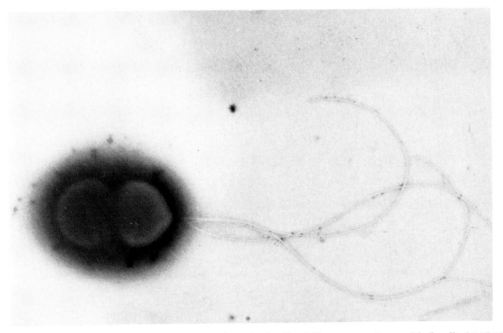

Figure 12.3. Electron micrograph of negatively stained cells of *Planococcus citreus* with flagella (× 17,100).

hydrostatic pressure of 40 MPa has no influence on the pigment production of *P. citreus* (Courington and Goodwin, 1955).

The concentration of salt in the medium affects the amount of membrane in the cell of *P. citreus*. Salt concentrations above and below the normal 3% of sea water, apparently, reduce the membrane material present (Thirkell and Summerfield, 1977a). *P. citreus* was found to be one of the few marine bacteria that reproduced at the hydrostatic pressure from 20–40 MPa (Oppenheimer and ZoBell, 1952).

P. citreus contains normal menaquinones with the MK-8 as the major component and MK-7 as the minor component (Jeffries, 1969; Yamada et al., 1976). The phospholipid pattern of planococci is similar to that of *Sporosarcina* and contains a large amount of cardiolipins (Yamada et al., 1976; Thirkell and Summerfield, 1977b). *P. halophilus* is capable of growth in medium containing no added NaCl and in medium containing 5.5 M NaCl (Novitsky and Kushner, 1975).

A serological examination of *P. citreus* has shown no antigenic relationship to staphylococci and micrococci (Oeding, 1971). Planococci are sensitive to lysozyme, chloramphenicol, erythromycin, novobiocin, oleandomycin, penicillin, tetracycline (Jeffries, 1969; Kocur et al., 1970; Novitsky and Kushner, 1976). *P. halophilus* is resistant to streptomycin (5 µg) and to sulfisoxazole (150 µg) (Novitsky and Kushner, 1976).

Planococci have been isolated from marine environments, sea water

(ZoBell and Upham, 1944), fish brining tanks (Georgala, 1967), marine clams (Leifson, 1964), salted mackerel (Novitsky and Kushner, 1976) and shrimps (Alvarez, 1982).

Enrichment and Isolation Procedures

For the isolation of planococci, sea water agar* or nutrient agar containing 10% NaCl may be used.

Maintenance Procedures

For short term preservation (several months) planococci can be stored in screw-capped tubes containing semisolid sea water agar. The tubes are stab inoculated and after overnight growth at 30°C they are tightly closed and kept at 4°C. For closing the tubes, rubber stoppers can be used as well. The tubes should be perfectly protected against dessication. For long term preservation, freeze-drying is recommended.

Procedures for Testing Special Characters

Marine oxidation-fermentation medium (MOF) for the determination of carbohydrate utilization: Casitone (Difco), 0.1%; yeast extract, 0.01%; ammonium sulfate, 0.05%; tris buffer, 0.05%; agar, 0.3%; phenol red, 0.001% (1.0 ml of 0.1% aqueous solution/100 ml of medium); HCl to pH 7.5 (about 0.3 ml of 1 N HCl/100 ml of medium); artificial sea water, half-strength; carbohydrate, 0.5–1.0% (Leifson, 1963).

Differentiation of the genus **Planococcus** from other genera

Table 12.7 indicates the characteristics of *Planococcus* that distinguish it from other genera of morphologically or physiologically similar taxa.

Taxonomic Comments

Until now the genus *Planococcus* has been classified into the family *Micrococcaceae*; however, at present it is considered to be a phylogenetically invalid family. A recent comparative analysis of 16S rRNA sequences has shown that planococci bear no specific relationship to the genera *Micrococcus* and *Staphylococcus* (Stackebrandt and Woese, 1979). Surprisingly they show a specific relationship to the genus *Bacillus*, particularly to *Bacillus pasteurii* and the genus *Sporosarcina* (Boháček et al., 1968b; Pechman et al., 1976; Stackebrandt and Woese, 1979). The data on G + C content in the DNA and chemical cell wall composition support these observations (Boháček et al., 1968b; Kocur et al., 1971; Schleifer and Kandler, 1970). Therefore, in the light of above facts, the genus *Planococcus* should be placed into the family *Bacillaceae* as a separate genus. The only objection against this transfer could be that a nonsporeforming genus will be accommodated among sporeforming bacteria. But from the point of view of present knowledge the morphology of cell is a less valuable criterion than it used to be. There is an opinion that the spherical shape of cell is phylogenetically a useful character at the lowest taxonomic level only (Stackebrandt and Woese, 1979).

There are some taxonomic problems of the genus *Planococcus* which have not been solved up to now. The first is the existence of a group of strains with 39.5–42.2% G + C in the DNA which bears no species name and has been tentatively designated *Planococcus* sp.

The second problem is the placement of *P. halophilus* in the genus *Planococcus*. At present there is only one single strain available of this species which differs significantly in the cell wall composition from *P. citreus* having *meso*-diaminopimelic acid (m-DAP) in the peptidoglycan. Its accommodation in the genus *Planococcus* should be considered as temporary. The definite solution of its appropriate classification among Gram-positive cocci will be possible when more data on its genetic relatedness are made available.

Table 12.7.

Differential characteristics of the genus **Planococcus** *and other morphologically or biochemically similar taxa*[a]

Characteristics	Planococcus	Sporosarcina	Micrococcus
Spores	−	+	−
Motility	+	+	−
Mol% G + C of the DNA[b]	39–52	40–44	65–75
Peptidoglycan type[c]	L-Lys-D-Glu or m-DAP	L-Lys-Gly-D-Glu	Mostly L-Lys-peptide subunit, or L-Lys-L-Ala$_{3-4}$
Menaquinone pattern[d]	Normal	Normal	Hydrogenated
Phosphatidyl-ethanolamine[e]	+	+	
Aliphatic hydrocarbons[f]	Absent	ND	Present
Yellow-orange pigment	+	−	−
Growth on nutrient agar containing 12% NaCl	+	−	−
Gelatin hydrolysis	+	−	D
Urease	−	+	D
NO$_3$–NO$_2$	−	+	D

[a] Symbols: see Table 12.2; also D, different reactions in different taxa (species of a genus or genera of a family).
[b] Data from Boháček et al. (1968a, b); Kocur et al. (1970).
[c] Data from Schleifer and Kandler (1970); Novitsky and Kushner (1976).
[d] Data from Jeffries (1969); Yamada et al. (1976).
[e] Data from Komura et al. (1975).
[f] Data from Morrison et al. (1971).

---

* Sea water agar: beef extract, 10 g; peptone, 10 g; agar, 20 g; tap water 250 ml; sea water 750 ml; pH 7.2.

Further Reading

Kocur, M. and K. H. Schleifer. 1981. The genus *Planococcus. In* Starr, Stolp, Trüper, Ballows and Schlegel (Editors), The Prokaryotes. A Handbook on Habitats, Isolation, and Identification of Bacteria. Springer-Verlag, New York, pp. 1570–1571.

Differentiation and characteristics of the species of the genus **Planococcus**

The differential characteristics of the species of *Planococcus* are indicated in Table 12.8. Other characteristics of the species are presented in Table 12.9.

Table 12.8.

Characteristics differentiating the species of the genus **Planococcus**[a, b]

Characteristics	1. *P. citreus*	2. *P. halophilus*
Growth on nutrient agar containing 20% NaCl	−	+
Susceptibility to streptomycin (5 µg)	S	R
Peptidoglycan type	L-Lys-D-Glu	mDpm

[a] Data from Kocur et al. (1970); Schleifer and Kandler (1970); Novitsky and Kushner (1976).

[b] Symbols: see Table 12.2; also S, susceptible; R, resistant.

Table 12.9.

Other characteristics of the species of the genus **Planococcus**[a]

Test	1. *P. citreus*	2. *P. halophilus*
Catalase	+	+
Growth in nutrient agar containing 12% NaCl	+	+
Yellow-orange pigment	+	+
Acid from glucose[b]	+	+
Hydrolysis of		
Gelatin	+	+
Starch	−	−
Esculin	−	−
Tween 80	−	−
Oxidase	−	−
NO_3–NO_2	−	−
Phosphatase	−	−
Urease	−	−
Indole	−	−
H_2S	−	−
Voges-Proskauer reaction	−	−
Growth on citrate agar (Simmons)	−	−
Arginine dihydrolase	−	−
Ornithine and lysine decarboxylase	−	−
Hemolysis	−	−
Phenylalanine deaminase	−	−

[a] Symbols: see Table 12.2.

[b] Positive reaction if tested in MOF medium of Leifson (1963).

List of species of the genus **Planococcus**

1. **Planococcus citreus** Migula 1894, 236.[AL]

ci′tre.us. L. masc. adj. *citreus,* lemon yellow.

The cell morphology and colonial morphology are as given for the genus. Physiological and biochemical characteristics are listed in Table 12.8.

The mol% G + C of the DNA is 48–52 (T_m, Bd).

Type strain: ATCC 14404 (CCM 316).

2. **Planococcus halophilus** Novitsky and Kushner 1976, 53.[AL]

hal.o.phi′lus. Gr. n. *hals, halos* the sea, salt; Gr. adj. loving; M.L. adj. *halophilus* salt loving.

The cell morphology and colonial morphology are as given for the genus. Physiological and biochemical characteristics are listed in Table 12.8.

The mol% G + C of the DNA is 50–51 (T_m).

Type strain: ATCC 27964 (CCM 2706, NRCC 14033).

Genus IV. **Staphylococcus** *Rosenbach 1884, 18*[AL], (Nom. Cons. Opin. 17 Jud. Comm. 1958, 153)

WESLEY E. KLOOS AND Karl Heinz SCHLEIFER

Staph. y. lo. coc′ cus. Gr. n. *staphyle* bunch of grapes; Gr. n. *coccus* a grain or berry; M.L. masc. n. *Staphylococcus* the grape-like coccus.

Cells spherical, 0.5–1.5 µm in diameter, occurring singly, in pairs, and tetrads and characteristically dividing in more than one plane to form irregular clusters. **Gram-positive.** Nonmotile. Resting stages not produced. **Cell wall contains peptidoglycan and teichoic acid. The diamino acid present in the peptidoglycan is L-lysine.**

Facultative anaerobes. With the exception of the anaerobic species *S. saccharolyticus,* growth is more rapid and abundant under aerobic conditions. **Usually catalase-positive.** Most strains grow in the presence of 10% NaCl and between 18 and 40°C.

Chemoorganotrophs: **metabolism respiratory and fermentative.**

Some species are mainly respiratory or mainly fermentative. Unsaturated menaquinones and cytochromes *a* and *b* (and *c* in the species *S. caseolyticus* and *S. sciuri*) form the electron transport system. Carotenoid pigments may be present in most species. D- and/or L-lactic acid may be produced from glucose under anaerobic conditions. Lactose or D-galactose is metabolized via the D-tagatose-6-phosphate or Leloir pathways, depending upon the particular species. Carbohydrates and/or amino acids are utilized as carbon and energy sources. A variety of carbohydrates may be utilized aerobically with the production of acid. For most species, the main product of glucose fermentation is lactic acid; in the presence of air the main products are acetic acid and CO_2. Most species possess the fructose-1, 6-biphosphate (FDP) aldolase of class I.

Nutritional requirements are variable. Most of the species identified require an organic source of nitrogen, i.e. certain amino acids, and B group vitamins. Some others can grow with $(NH_4)_2SO_4$ as a sole source of substrate nitrogen. Uracil and/or a fermentable carbon source may be required by certain species for anaerobic growth.

Susceptible to lysis by lysostaphin, but resistant to lysis by lysozyme. Species or strains which have significant amounts of L-serine or L-alanine replacing glycine in the cell wall peptidoglycan are generally less susceptible to lysostaphin than those with an interpeptide bridge consisting solely of glycine residues. Some species are resistant to novobiocin; whereas, others are typically susceptible. Usually susceptible to antibacterials such as phenols and their derivatives, salicylanilides, carbanilides and halogens (chlorine and iodine) and their derivatives.

Host for a variety of bacteriophages which may have a narrow or wide host range. Transfer of characters by transduction, transformation and cell to cell contact has been demonstrated in some species.

Natural populations are mainly associated with skin, skin glands and mucous membranes of warm-blooded animals. Host or niche range may be narrow or wide, depending upon the particular species or subspecies. Some organisms may be isolated from a variety of animal products (e.g. meat, milk, cheese) and environmental sources (e.g. fomites, soil, sand, dust, air or natural waters). **Some species are opportunistic pathogens of humans and/or animals.**

The **G + C content of DNA** ranges from **30–39 mol%** (T_m, Bd). Type species: *Staphylococcus aureus* Rosenbach 1884, 18.

Further Comments

Nineteen species are recognized in the genus *Staphylococcus* and several others are currently under investigation.

The genus *Staphylococcus* can be subdivided into at least four species groups on the basis of DNA/DNA relationships and phenotypic characterization (Kloos et al., 1976; Schleifer et al., 1979; Kloos, 1980; Kloos and Schleifer, 1981). The *S. epidermidis* species group is composed of the species *S. epidermidis*, *S. capitis*, *S. warneri*, *S. haemolyticus*, *S. hominis* and *S. saccharolyticus*. The *S. saprophyticus* species group is composed of the species *S. saprophyticus*, *S. cohnii* and *S. xylosus*. The *S. simulans* species group is composed of the species *S. simulans* and *S. carnosus*. The *S. sciuri* species group is composed of the species *S. sciuri* and *S. lentus* (proposed to be elevated in taxonomic rank from *S. sciuri* subsp. *lentus*). The species *S. aureus*, *S. auricularis*, *S. intermedius*, *S. hyicus* and *S. caseolyticus* cannot be easily accommodated in any of the above species groups and are too distantly related from each other for the formation of another group(s).

Species included in the genus *Staphylococcus* are distinctly and unequivocally related on the basis of DNA/rRNA binding and the thermal stability of DNA/rRNA hybrids (Kilpper-Bälz and Schleifer, 1981b; Schleifer et al., 1982), DNA/DNA binding and the thermal stability of DNA/DNA hybrids (Meyer and Schleifer, 1978; Kloos and Wolfshohl, 1979, 1982, 1983; Schleifer et al., 1979; Kloos, 1980; Schleifer et al., 1982), comparative sequence analysis (cataloging) of 16S rRNA (Stackebrandt and Woese, 1979; Ludwig et al., 1981), and the immunological relationship (distance) of their catalases (Rupprecht and Schleifer, 1979; Schleifer et al., 1979) and their fructose-1,6-biphosphate aldolases (Fischer et al., 1983). On the other hand, these species have diverged from one another to an extent expected for that of a separate species status. Under optimal DNA/DNA reassociation conditions, the different species demonstrate 60% or less DNA homology and under restrictive conditions demonstrate 20% or less homology. The heterologous DNA/DNA hybrids formed between various staphylococcal species also exhibit low thermal stabilities, with $\Delta T_{m(e)}$ values of greater than 13°C. Species identity is also made on the basis of a variety of phenotypic characters, including cell wall composition, colony morphology, activities and molecular properties of various en-

zymes, production of acid from various carbohydrates, products of glucose metabolism, resistance to certain antibiotics, nutritional requirements, cellular fatty acid composition, oxygen requirements, electron transport systems and susceptibility to certain phages (see individual species descriptions below).

The various staphylococcal species can be identified by conventional methods (e.g. those recommended in the simplified schemes of Kloos and Schleifer (1975b) and Varaldo and Satta (1978) for the identification of human-adapted species and schemes for the identification of animal-adapted species proposed by Hajek (1976), Kloos et al. (1976), Devriese et al. (1978), Varaldo et al. (1978), Schleifer and Fischer (1982) and Schleifer et al. (1982)) or by commercially prepared micromethods utilizing miniaturized conventional and chromogenic tests (e.g. in the form of indicator-substrate impregnated strips and microcupules) for rapid identification. The commercially available systems reduce the need for preparing a variety of test media and reagents and the time required for interpretation of results, thereby making the identification of various staphylococcal species more plausible in the routine laboratory.

Enrichment and Isolation Procedures

Staphylococcus aureus has been confirmed to be a major causative agent of food poisoning. Its identity in foods, therefore, is of major concern. Furthermore, when it or its enterotoxins are found in processed foods poor sanitation is usually indicated. Procedures to be used during outbreaks of bacterial food poisoning have been reviewed by Bryan (1980). The enrichment procedures of Heidelbaugh et al (1973) may be used for processed foods containing a small number of cells which may have been injured as a result of heating, freezing, desiccation or storage. Following incubation of the food slurry in trypticase soy broth (TSB) supplemented with 10% NaCl, cultures are examined for the presence of *S. aureus* on Vogel-Johnson agar (Leininger, 1976) or Baird-Parker agar (Holbrook et al., 1969; Food and Drug Administration, 1976). For raw food ingredients and unprocessed foods expected to contain small numbers of *S. aureus* cells and a large population of competing species, the most probable number (MPN) enrichment procedure recommended by Baer et al. (1976), as outlined by the Association of Official Analytical Chemists (AOAC) (1975), is often the method of choice. Brewer et al. (1977) have recommended the addition of filter-sterilized catalase or pyruvate to TSB containing 10% NaCl in an MPN procedure for good recovery of heat-stressed *S. aureus* cells from foods. For the detection of *S. aureus* in raw or processed foods where there is minimal interference by competing species, direct surface plating procedures may be used (AOAC, 1975; Baer et al., 1976). Direct plating is sometimes preferred by laboratories; it is generally regarded as more accurate for the enumeration of staphylococci than MPN procedures. Baird-Parker agar is widely used for this purpose; however, tellurite polymyxin egg yolk agar (Crisley et al., 1964), KRANEP agar (Sinell and Baumgart, 1966), and the selective medium of Schleifer and Krämer (1980) also may be used with success.

Procedures used in the isolation of staphylococci from clinical specimens including blood, pus, purulent fluids, sputum, urine and feces have been reviewed adequately by Isenberg et al. (1980). The primary plate culture is usually made on blood agar containing 5% sterile defibrinated sheep (preferable), rabbit or bovine blood. Blood agar base is available commercially. Inoculated plates should be incubated at 34–37°C for at least 18–24 h under aerobic conditions. Well isolated colonies of staphylococci will generally be 1–3 mm in diameter within 24 h; they are usually circular, smooth, raised, butyrous and opaque at this time. For the detection of *S. saccharolyticus*, incubation under anaerobic conditions is also recommended. Specimens from heavily contaminated sources (e.g. feces) should also be inoculated on a selective medium such as mannitol-salt agar, phenylethyl alcohol agar, Columbia CNA agar, or the selective medium of Schleifer and Krämer (1980). A fluid medium such as thioglycolate broth may also be inoculated.

Several basic methods are available for isolating natural populations of staphylococci from skin, including washing or swabbing, impression and biopsy (reviewed by Noble and Somerville, 1974). Most sampling of aerobic bacteria has been performed using swabbing methods. The swab technique described by Kloos and Musselwhite (1975) is satisfactory for the isolation of staphylococci from human as well as other mammalian skin. The primary isolation plate may be made of P agar (recommended by Kloos et al., 1974), blood agar, trypticase soy agar, brain heart infusion agar, or other suitable nonselective medium which will permit the growth and distinction of staphylococcal colonies.

Colonies should be allowed to grow at least 3 days at 34–35°C for the development of distinctive genus and species characteristics.

Maintenance Procedures

Stocks may be stored for up to one to several years in broth (e.g. CY (Novick, 1963) or 50% glycerol) at −75°C or by desiccation on ceramic beads (Norris, 1963). To ensure viability for longer periods of storage, cultures should be lyophilized.

Differentiation of the genus *Staphylococcus* from other genera

The genus *Staphylococcus* can be differentiated from the genus *Micrococcus*, which often shares its habitat, and *Planococcus* by characters listed above in the description of the family *Micrococcaceae*. It can be distinguished from members of the family *Streptococcaceae* on the basis of cell wall composition (see review by Schleifer and Kandler, 1972) and occurrence of cytochromes. *Staphylococcus* contains cytochromes and so will be benzidine-positive. *Streptococcaceae* are benzidine-negative. Differences in catalase activity, colony and cell morphology, products of glucose metabolism and serological properties are also noteworthy.

Further Reading

Cohen, J.O. (Editor). 1972. The Staphylococci, Wiley-Interscience New York.
Easmon, C.S.F. and C. Adlam (Editors). 1983. Staphylococci and Staphylococcal Infections (2 vols.) Academic Press, New York.

Jeljaszewicz, J. (Editor). 1976. Staphylococci and Staphylococcal Diseases, Fischer-Verlag, Stuttgart.
Jeljaszewicz, J. (Editor). 1981. Staphylococci and Staphylococcal Infections. Fischer-Verlag, Stuttgart.
Kloos, W.E. 1980. Natural populations of the genus *Staphylococcus*. Annu. Rev. Microbiol. *34:* 559–592.
Kloos, W.E. and K.H. Schleifer. 1981. The genus *Staphylococcus*. *In* Starr, Stolp, Trüper, Balows and Schlegel (Editors), The Prokaryotes. A Handbook on Habitats Isolation, and Identification of Bacteria. Springer-Verlag, Berlin, pp. 1548–1569.
Kloos, W.E. and P.B. Smith. 1980. Staphylococci. *In* Lennette, Balows, Hausler and Truant (Editors), Manual of Clinical Microbiology, 3rd Ed., American Society for Microbiology, Washington, pp. 83–87.

Differentiation and characteristics of the species of the genus *Staphylococcus*

Some key differential characteristics of the species of *Staphylococcus* are indicated in Table 12.10. Those shown can be tested in most laboratories.

List of species of the genus *Staphylococcus*

1. **Staphylococcus aureus** Rosenbach 1884, 18.[AL]

au′ re.us. L. adj. aureus golden.

Spheres, 0.5–1.0 μm in diameter. Cells occur singly and in pairs. Division may be in more than one plane, giving rise to irregular clusters. Some uncommon strains produce cells with a capsule or pseudocapsule (reviewed by Wiley, 1972; Smith et al., 1977; Wilkinson et al., 1981) and these often demonstrate increased virulence over unencapsulated strains. There are a variety of capsular types. Some capsular polysaccharides have been shown to contain N-acetyl-D-aminogalacturonic acid, N-acetyl-D-fucosamine and taurine (Liau and Hash, 1977). Some strains produce cell wall-defective (e.g. L-form) cells which may pass through filters with a pore size as small as 50–450 nm (reviewed by Kagan, 1972). They may be induced (e.g. by benzylpenicillin, methicillin or lysostaphin) or isolated naturally from a variety of disease processes, and are of concern in chro. ′c or recurrent infections. Since L-forms are osmotically fragile, they may missed during isolation procedures, unless a hypertonic environment is supplied.

Peptidoglycan is of the type L-Lys-Gly$_{5-6}$. Cell wall teichoic acid is composed of ribitol with either α- or β-glycosidically linked N-acetylglucosamine residues. Some uncommon *S. aureus* strains (in particular phage type 187) contain a cell wall teichoic acid composed of glycerol and α-glycosidically linked N-acetyl-galactosamine residues (Endl *et al.*, 1983). (For details on the biochemistry of the cell wall see: Baddiley et al., 1961; Oeding, 1965; Tipper and Strominger, 1966; Archibald et al. 1968a; Ghuysen, 1968; Archibald, 1972; Schleifer and Kandler, 1972; Labischinski et al., 1981). Most strains possess the species-specific precipitinogen, protein A (Forsgren, 1969; Hjelm et al., 1972; Movitz, 1976; Masuda and Kondo, 1981).

Cell membranes contain the glycolipids, mono- and diglucosyldiglyceride and the phospholipids, lysyl-phosphatidylglycerol, phosphatidylglycerol, and cardiolipin (White and Frerman, 1967). The membranes

of stable L-forms may contain large quantities of cholesterol (Hayami et al., 1979). The major cellular fatty acids include C_{15Br}, C_{20}, C_{18}, and C_{17Br} components (White and Frerman, 1968; Durham and Kloos, 1978).

Colonies are smooth, raised, glistening, circular, entire and translucent, and single colonies may obtain a size of 6–8 mm in diameter on nonselective media used for the propagation of staphylococci (Kloos and Schleifer, 1975a). With increasing age, colonies become nearly transparent on the above media. Under conditions inhibitory to normal growth, rough (R) or dwarf (G) colony variants may be produced. Encapsulated strains produce colonies which are usually smaller and more convex in profile than unencapsulated strains, and have a glistening wet appearance. Upon storage of these, growth may become slimy and run down the surface of agar under normal gravitational force. Many L-forms produce slow growing, small, characteristic "fried-egg" colonies on salt serum agar. Colonial pigment is variable; however, most strains demonstrate some degree of colony or cell pigmentation, ranging from gray or gray-white with a yellowish tint, through yellow-orange to orange. The pigments are triterpenoid carotenoids or derivatives of them and are located in the cell membrane (Hammond and White, 1970; Marshall and Wilmoth, 1981); unpigmented cells contain very low quantities of carotenoid pigments. The production of pigment may be influenced by growth conditions; it can be intensified by growth on a medium containing 10% bovine cream (O'Conner et al., 1966). Growth on agar slants is abundant, slightly sticky in consistency, and translucent to nearly opaque. In broth, growth changes from a uniform turbidity to a fine, easily suspended deposit, often in the form of a ring pellicle.

Facultative anaerobes: growth is best under aerobic conditions. Growth in the anaerobic portion of a semisolid thioglycolate medium (Evans and Kloos, 1972) is rapid and uniformly dense. Terminal pH in

Table 12.10.

Characteristics differentiating the species of the genus **Staphylococcus**[a]

Characteristics	1. S. aureus	2. S. epidermidis	3. S. capitis	4. S. warneri	5. S. haemolyticus	6. S. hominis	7. S. saccharolyticus	8. S. auricularis	9. S. saprophyticus	10a. S. cohnii subsp. 1	10b. S. cohnii subsp. 2	11. S. xylosus	12. S. simulans	13. S. carnosus	14. S. intermedius	15a. S. hyicus subsp. hyicus	15b. S. hyicus subsp. chromogenes	16. S. caseolyticus	17a. S. sciuri	17b. S. lentus	18. S. gallinarum	19. S. caprae
Colony diameter >5 mm[b]	+	−	−	−	+	−	−	−	+	+	+	+	+	+	+	+	+	−	+	−	+	d
Colony pigment (carotenoid)	+w	−	−	d	d	d	−	−	d	−	d	d	+	+	−	−	+	d	d	d	d	+
Aerobic growth	+	+	+	+	+	+	−w	+	+	+	+	+	+	+	+	+	+	+	+	+	+	+
Anaerobic growth (thioglycolate)	+	+	(+)	+	(+)	−w	+	−w	(+)	d	(+)	d	+	+	(+)	+	+	−w	(+)	−w	+	+
Growth on NaCl agar																						
10% (w/v)	+	w	+	+	+	w	ND	w	+	+	+	+	+	+	+	+	+	ND	+	+	ND	ND
15% (w/v)	w	−	−w	w	d	−	ND	−	d	−	d	d	w	+	d	−w	−w	ND	d	−w	ND	ND
Growth at																						
15°C	+	−w	−	d	−w	−w	−w	−	d	d	d	−w	+	+	−	−w	−w	ND	+	−	ND	ND
45°C	+	+	−	+	+	+	+	+	−	+	−	−	+	+	+	+	+	+	+	+	+w	+
Cytochrome c (oxidase test)	−	−	−	−	−	−	−	−	−	−	−	−	−	−	−	−	−	+	+	+	−	−
Lactic acid																						
L isomer	+	+	+	+	−	d	w	w	w	w	w	w	+	+	+	+	+	w	+	+	+	+
D isomer	+	+	−w	+	+	+	−	−	+	−	w	−w	d	+	d	+	+	−	−	−	−	+
Acetoin production	+	+	d	+	d	+	ND	d	d	d	d	d	−w	+	+	+	+	ND	−w	−w	+	+
FDP-aldolase																						
Class I	+	+	+	+	+	+	+	+	+	+	ND	−	+	+	+	+	ND	−	+	+	+	+
Class II	−	−	−	−	−	−	−	−	−	+	ND	−	+	+	+	+	ND	+	+	+	−	−
Acid (aerobically) from																						
D-Xylose	−	−	−	−	−	−	−	−	−	−	−	+	−	−	−	−	−	−	−	−w	+	−
L-Arabinose	−	−	−	d	+	−	ND	−	−	−	−	+	−	−	−	−	−	ND	d	d	+	−
D-Cellobiose	−	+	+	d	+	(+)	ND	(+)	+	(d)	d	d	d	−	−	−	d	ND	+	+	w	−
D-Fucose	+	(+)	+	d	d	d	ND	(+)	−	(d)	d	d	d	d	d	+	+	ND	(d)	d	+	+
Raffinose	−	d	−	d	d	d	ND	−	−	+	+	d	d	d	d	+	+	ND	+	+	w	−
Salicin	−	d	−	ds	d	d	ND	−	+	+	+	d	−w	d	d	+	d	ND	+	+	d	−
Sucrose	+	+	(+)	d	d	(+)	−	d	+	+	+	+	−w	+	+	+	+	d	(d)	d	+	−
Maltose	+	+	+	d	d	d	−	(+)	d	(+)	d	+	−w	−	(w)	−	d	+	d	d	+	d
D-Mannitol	+	(+)	+	d	+	−	(+)	(+)	−	(d)	d	d	d	d	(d)	+	d	ND	(d)	d	w	d
D-Mannose	+	d	d	d	d	d	(+)	−	−	+	+	d	d	d	d	+	+	+	+	+	+	+
D-Trehalose	+	d	d	d	d	d	ND	−	+	+	+	d	d	d	d	+	+	+	+	+	+	+
α-Lactose	+	d	d	ds	d	d	ND	−	−	+	+	d	−w	d	d	+	+	ND	−w	d	d	+
D-Galactose	+	d	d	d	d	d	(+)	+	d	+	+	d	d	d	d	+	+	ND	(+)	(+)	+	+
β-D-Fructose	+	+	+	d	+	+	ND	+	+	+	+	+	+	+	+	+	+	+	+	+	+	+
D-Melezitose	−	(d)	−	ds	−	−	ND	−	d	d	−	d	−w	d	d	−	d	−	d	d	−	−
D-Turanose	+w	d	−	d	d	−	ND	(d)	+	+	−	+	−w	+	+	−	+	ND	+	+	−	−
D-Ribose	−	d	−	d	d	d	ND	−	d	d	−	d	−	ND	d	−	d	−	d	−	d	−
Xylitol	−	−	−	−w	−	−	ND	−	−	(d)	(d)	−w	ND	ND	−	+	−	−	−	−	−	−
Hyaluronidase	+	d	ND	ND	ND	ND	ND	ND	ND	ND	ND	ND	ND	ND	+	+	−	−	ND	ND	ND	ND
Growth on (NH₄)₂SO₄ (nitrogen source)	−	−	ND	−	ND	ND	ND	ND	d	d	d	+	−	ND	−	ND	ND	ND	ND	ND	−	−
Nitrate reduction	+	+w	d	−w	−	d	+	(d)	−	−	d	+	+	ND	+	+	−	+	+	+	−	−
Alkaline phosphatase	+	+	d	d	−	d	ND	(d)	−	−	+w	d	w	ND	+	+	+	ND	w	w	ND	+
Arginine dihydrolase	+w	+w	d	d	d	d	+	d	−w	−w	−w	−	+	ds	+	+	+	ND	−	w	−	+

1016

Urease	+w	+	−	+	−	+	ND	−	+	−	+	+	+	−	+	d	d	ND	−	−	+	+
Coagulase (rabbit plasma)	+	−	−	−	−	−	−	−	−	−	−	−	−	−	+	d	−	−	−	−	−	−
Clumping factor	+	−	−	−	−	−	−	−	−	−	−	−	−	−	d	−	−	−	−	−	−	−
Fibrinolysin	de	d	ND	ND	ND	ND	ND	ND	ND	ND	ND	ND	ND	ND	−	d	−	ND	ND	ND	−	−
Hemolysis	+	−w	−w	(ds)	(+)	−w	−	−	−	(d)	(d)	−w	−w	−	d	−	−	−	−	−	w	(+)
Deoxyribonuclease (DNase agar)	+	−w	w	ds	ds	−w	ND	−w	−	−w	−w	−w	w	w	+	+	w	ND	+w	+w	ND	+
Heat-stable nuclease	+	−w	−	−	−	−	ND	−	−	−	−	−	−w	d	+	+	−w	ND	ND	ND	−	−
β-glucosidase	+	(d)	−	+	d	−	ND	−	d	−	−	+	−	−	d	d	d	ND	+	+	+	−
β-glucuronidase	−	−	−	d	d	−	ND	−	−	−	+	d	d	−	−	d	−	ND	−	−	−	−
β-galactosidase	−	−	−	−	−	−	ND	(d)	d	−	+	+	+	+	ds	−	−	ND	−	−	−w	−
Novobiocin resistance (MIC ⩾ 1.6 μg/ml)	−	−	−	−	−	−	−	−	+	+	+	+	−	−	−	−	−	−	+	+	+	−

[a] Symbols: +, 90% or more strains positive; −, 90% or more strains negative; d, 11–89% strains positive; ( ) delayed reaction; w, weak reaction; −w, negative to weak reaction; +w, positive to weak reaction; ds, test differentiates subspecies (not separated out above in heading); de, test differentiates ecotypes; ND, test not determined.

[b] Colony diameter is determined after incubation on P agar (Kloos et al., 1974; Kloos and Schleifer, 1975) at 34–35°C for 3 days and room temperature (≃25°C) for an additional 2 days.

glucose broth, under anaerobic conditions, is 4.3–4.6 (Schleifer and Kocur, 1973). Catalase is produced by cells growing aerobically. It is immunologically distinct from the catalases of other species, as determined by double immunodiffusion and microcomplement fixation analyses (Rupprecht and Schleifer, 1979). Catalase may be absent in respiratory deficient mutants (Jensen, 1963). Several catalase-negative strains have been isolated from disease processes (Tu and Palutke, 1976; Schumacher-Perdreau et al., 1981).

Growth is good at NaCl concentrations up to 10% and is relatively poor at 15%. Most strains grow between 10 and 45°C (optimum: 30–37°C) and at pH values between 4.2 and 9.3 (optimum pH 7.0–7.5).

Chemoorganotrophs: metabolism respiratory and fermentative. Unsaturated menaquinones (mainly Mk-8; Mk-7 and Mk-9 as minor isoprenologues) (Jeffries et al., 1969; Collins, 1981) and cytochromes *a* and *b* (Taber and Morrison, 1964; Faller et al., 1980) form the membrane-bound electron transport system. Glucose is metabolized to pyruvate via the Embden-Meyerhof (EM) glycolytic pathway and/or hexose monophosphate (HMP) pathway (see review by Blumenthal, 1972). Aerobically, pyruvate is oxidized to acetate which through acetyl-CoA, can be further oxidized via the tricarboxylic acid (TCA, citric acid) cycle. Addition of glucose to a growth medium suppresses operation of the TCA cycle, resulting in the accumulation of acetate (glucose effect) (Collins and Lascelles, 1962; Montiel and Blumenthal, 1965). Acetate and CO_2 are the major end-products of aerobic glucose metabolism; whereas, lactate is the major end product of anaerobic glucose metabolism (Strasters and Winkler, 1963; Theodore and Schade, 1965). Both D-and L-lactate isomers are produced by glucose fermentation (Schleifer and Kocur, 1973). Rare strains produce only L-lactate. Typical strains of *S. aureus* possess two NAD-dependent lactate dehydrogenases: an L- and a D-lactate dehydrogenase (Götz and Schleifer, 1976; Schleifer et al., 1976). Acetyl methyl carbinol (acetoin) is usually a minor end point product of glucose metabolism; production may be enhanced in an iron-deficient medium at low pH (Morse, 1969) and in heat-injured cells (Bluhm and Ordal, 1969). Mannitol can be an intracellular product of glucose metabolism, probably via mannitol-1-phosphate (Edwards et al., 1981). D-galactose and lactose are metabolized via the D-tagatose-6-phosphate pathway (Bissett and Anderson, 1973; Schleifer et al., 1978). *S. aureus* has an unusually heat-stable and acid/base-stable class I fructose-1,6-biphosphate (FDP) aldolase (Götz et al., 1980). A phosphoenol pyruvate (PEP)-dependent phosphotransferase system (PTS) is present for the active transport and phosphorylation of a broad range of carbohydrates (e.g., lactose, galactose, glucose, fructose, mannitol, mannose and sucrose)(Hengstenberg et al., 1969; Friedman and Hays, 1977).

Acid is produced aerobically and anaerobically from glucose, lactose, maltose and mannitol. Acid is produced aerobically from fructose, galactose, mannose, ribose, sucrose, trehalose, turanose and glycerol. A few strains do not produce detectable acid from lactose, galactose, turanose or mannitol. Acid is not produced from arabinose, xylose, xylitol, melezitose, cellobiose, gentiobiose, sorbitol, inositol, salicin, adonitol, dulcitol, arabitol, erythritol, erythrose, raffinose, melibiose, fucose, rhamnose, lyxose, sorbose or dextrin (Baird-Parker, 1963, 1965; Kloos and Schleifer, 1975a).

Esculin and starch are not usually hydrolyzed. Hyaluronic acid may be hydrolyzed by hyaluronic acid lyase (staphylococcal hyaluronidase) (Abramson, 1972). Suitable cell wall glycans (e.g. *Micrococcus luteus* cell walls) may be degraded by an endo-β-*N*-acetylglucosaminidase (staphylococcal lysozyme) (Hawiger, 1968; Arvidson et al., 1970; Valisena et al., 1982).

An organic nitrogen source (from 5–12 different amino acids) and B group vitamins (from 2–3, including thiamine and nicotinic acid, or nicotinamide) are required for growth (Fildes et al., 1936; Gladstone, 1937; Mah et al., 1967; Yamaguchi and Kurokawa, 1972; Emmett and Kloos, 1975, 1979). For anaerobic growth, uracil and a fermentative carbon source (e.g., pyruvate) are also required (Richardson, 1936; Evans, 1976).

Nitrates are reduced by nitratases (nitrate reductase) and nitritases with the formation of ammonia. Most strains accumulate moderate quantities of nitrite after incubation in standard nitrate broth containing 0.1% potassium nitrate. Respiratory nitrate reductase in the cytoplasmic membrane can couple with membrane dehydrogenases via cytochrome *b* (Burke and Lascelles, 1975; Burke et al., 1981). This enzyme can also occur in the cytoplasm. Acid and alkaline phosphatases are produced (Malveaux and San Clemente, 1967; Tirunarayanan, 1968).

Ammonia is produced from arginine by arginine dihydrolase and from urea by urease (Krasuski, 1981). Trace amounts of hydrogen sulfide may be formed from cysteine. Lysine and ornithine are not decarboxylated. Most strains will hydrolyze native animal proteins (e.g. hemoglobin, fibrin, egg white and casein, and polypeptides such as gelatin). Proteases (Drapeau, 1978a, b), lipases (Brunner et al., 1981; Tyski, 1981) and esterases (Saggers and Stewart, 1968; Schleifer et al., 1976; Zimmerman and Kloos, 1976) are produced. Various lipids, Tweens, Spans and phospholipoproteins, are hydrolyzed with the release of fatty acids. Some strains may produce lecithinase A (Nygren et al., 1966). Baird-Parker agar, which utilizes the ability of *S. aureus* to clear egg yolk, has gained wide acceptance as a diagnostic medium for this species (see review by Baird-Parker (1969) on diagnostic and selective media). Egg yolk is cleared by a phospholipoprotein lipase which splits the lipid moiety from lipovitellin (Tirunarayanan and Lundbeck, 1967).

Coagulases are produced by nearly all strains; several antigenically distinct and substrate-specific coagulases are produced. Detection of coagulases is very important in the routine identification of *S. aureus*; however, additional characters (Table 12.10) need to be considered when it is necessary to identify *S. aureus* apart from the other coagulase-positive (animal) species *S. intermedius* (Hajek, 1976) and the coagulase-variable (animal) species *S. hyicus* (Devriese et al., 1978). Coagulases should be tested under carefully standardized conditions. (See standard methods proposed by the ICSB Subcommittee on Taxonomy of Staphylococci and Micrococci (1965) and recommendations made by Kloos and Smith (1980) in the *Manual of Clinical Microbiology*.) A tube test is recommended for the detection of free coagulase (staphylocoagulase) and a slide test for clumping factor (bound coagulase). The tube test is usually regarded as the more definitive one; the slide test has been recommended as a screening technique. A variety of plasmas may be used for either test, but rabbit plasma is readily available and satisfactory for human and most animal strains. It should be noted that some poultry (Evans et al., 1983) and goat (Gray, 1980; Adegoke, 1981) strains may fail to clot rabbit plasma in the tube test. Some staphylocoagulase-negative strains may clot plasmas as a result of producing very active proteases (pseudocoagulase) (Bulanda et al., 1981; Heczko et al., 1981). Some staphylocoagulase-positive strains may also produce pseudocoagulase. Rare coagulase-negative strains should be examined for other differentiating characteristics outlined in Table 12.10 for appropriate identity. Plasma containing ethylenediaminetetraacetate (EDTA) is superior to citrated plasmas; citrate-utilizing bacteria (e.g. fecal streptococci and some members of the *Enterobacteriaceae*) may cause clotting of citrated plasma, but have no effect on EDTA and in its presence are prevented from giving false-positive reactions (Evans et al., 1952; Baer, 1968). Clumping factor (CF)-positive reactions may be false when undiluted plasma is used in the slide-testing of strong protein A-producing strains. Protein A reacts with the immunoglobulin G in plasma (Grov et al., 1970). For the precise determination of CF itself, it is better to use a 2% fibrinogen solution, instead of plasma (Blobel et al., 1981). Recently, a latex slide agglutination test has been described which may possibly substitute for the slide and tube coagulase tests (Essers and Radebold, 1980; Myrick and Ellner, 1982). (See reviews by Abramson (1972) and Zajdel et al. (1976) for the purification, properties, and mechanism of action of staphylocoagulase and properties of CF.) Many strains produce staphylokinase (Muller factor, SAK, or fibrinolysin), especially those of human origin. Strains producing large amounts of SAK may be incorrectly classified as coagulase-negative if they are grown in vitro for

extended periods (Stankiewicz, 1962). SAK may be present in multiple molecular forms (Vesterberg and Vesterberg, 1972; Makino, 1981). The production of SAK in many strains is mediated by a lysogenic (phage) conversion (Kondo et al., 1981).

At least four different hemolysins (exotoxins) are produced, including *α*-, *β*-, *γ*-, and *δ*-hemolysins. Nearly all strains produce one or a combination of several hemolysins. *β*-Hemolysin is produced more frequently by strains of animal origin; it is a phospholipase C, specific for sphingomyelin, and may be referred to as sphingomyelinase C. (For reviews of the chemistry, properties and cellular effects of the hemolysins, see Jeljaszewicz, 1972, Wadstrom et al., 1974; and Thalestam, 1976).

Some strains produce an epidermolytic toxin (exfoliatin, ET, exfoliative toxin) which has been implicated in the staphylococcal scalded skin syndrome (Ritter's disease, toxic epidermal necrolysis) (Melish et al., 1976, 1981; Kondo et al., 1976; Wiley et al., 1976). At least two types of ET may occur; one is under plasmid control, while the other is under chromosomal control (Rogolsky et al., 1976; Arbuthnott, 1981; Johnson, 1981). Most ET-producing strains belong to phage group II.

Many strains produce enterotoxins, which if ingested (e.g. via contaminated foods) may produce symptoms of staphylococcal food poisoning. At least five different types of enterotoxin have been identified (A, B, C, D, and E) (see reviews by Bergdoll, 1972 and Bergdoll et al., 1974). At least three different forms of enterotoxin C (C_1, C_2, and C_3) have been identified, which differ in their isoelectric points and antigenicity (Avena and Bergdoll, 1967; Borja and Bergdoll, 1967; M. S. Bergdoll, personal communication). Most strains produce enterotoxin A. The various types of enterotoxin may be produced singly or in combinations of two or more. Detection of low levels of enterotoxins in food may be a problem and require some purification and concentration of the food extract. A microslide test has been developed for the detection of enterotoxin in foods involved in food poisoning outbreaks (Bergdoll, 1972). More recently, Miller et al. (1978) have developed a very sensitive radioimmunoassay (RIA) which can detect as little as 1 ng of enterotoxin/ml of extract, eliminating the need for extract concentration. An enterotoxin-like protein (enterotoxin F, pyrogenic exotoxin C) is produced by many of the strains associated with toxic shock syndrome (TSS) (Bergdoll et al., 1982; Altemeier et al., 1982). It appears that a higher percentage of these strains belong to phage group I than to other phage groups.

Heat-resistant staphylococcal nuclease (thermonuclease, TNase, DNase, phosphodiesterase) having endo- and exonucleolytic properties, which can cleave either DNA or RNA, is produced by most strains (Cunningham et al., 1956; Wadstrom, 1967; Cuatrecasas et al., 1969). This nuclease may be used as an indicator for the detection of *S. aureus* in foods (Chesbro and Auburn, 1967; Lachica and Deibel, 1969). Rapid screening methods are available for the detection of TNase (e.g. using metachromatic-agar diffusion procedures and DNA-toluidine blue agar (Lachica et al., 1971, 1972) or buffered peptone-DNA agar (Gemmel et al., 1981)). Several other species (e.g. *S. intermedius*, *S. hyicus*, and *S. simulans*) may also produce detectable heat-stable nucleases (Hajek, 1976; Devriese et al., 1978; Gramoli and Wilkinson, 1978; Gemmell et al., 1981). Lachica et al. (1979) have developed a TNase seroinhibition test to distinguish *S. aureus* TNase from those of other species.

Some strains produce antibiotic-like substances described as staphylococcins (Bacteriocins, sta-cins) (Lachowicz and Sienienska, 1976; Brandis, 1981; Smarda and Obdrzalek, 1981) and micrococcins (Pulverer and Jeljaszewicz, 1976) which may be bacteriostatic and/or bactericidal to other staphylococci and certain other bacteria.

Differentiation of strains of *S. aureus* by phage typing is widely used; the system is well developed for typing human strains and systems for typing various animal strains are in different stages of development (see reviews by Smith, 1972, and Live, 1972). The methods of Blair and Williams (1961), or modifications of them, are commonly used for the propagation of phages and for the typing of cultures. Phages are normally applied to cultures at RTD (routine test dilution) as well as at $100 \times$ RTD. *S. aureus* typing phages in the international basic set, used routinely for the typing of human strains, include group I: 29, 52, 52A, 79, 80; group II: 3A, 3C, 55, 71; group III: 6, 42E, 47, 53, 54, 75, 77, 83A, 84, 85; and not allocated: 81, 94, 95, 96. Some strains are untypable by these phages.

Serological typing of *S. aureus* strains may have limited value in epidemiological studies (see review by Cohen, 1972). Progress in using this form of typing has been hampered by the complex nature of *S. aureus* antigens; the identification of a particular antigen may be difficult and it is a problem to produce large quantities of specific sera (Fleurette and Modjadedy, 1976). At present there is no international standardization of reference strains or methods. The systems of *S. aureus* serotyping in current use include the Cowan-Mercier-Pillet system, which recognizes thermolabile structures on the cell surface, and the Oeding-Haukenes system, which combines the recognition of thermolabile and thermostable surface antigens by using specific antisera for given antigens (i.e. factor sera) (Haukenes, 1967; Pillet et al., 1967; Oeding, 1974; Fleurette and Modjadedy, 1976; Flandrois et al., 1981). Continued cultivation in the laboratory may lead to antigenic changes (Torres Pereira, 1961).

Chromosomal and plasmid DNA may be transferred to appropriate recipient cells via transduction (Ritz and Baldwin, 1958; Morse, 1959; Pattee and Baldwin, 1961; Novick, 1963; Asheshov and Porthouse, 1976), transformation (Lindberg et al., 1972; Pattee and Neveln, 1975; Pattee et al., 1977), or protoplast fusion (Götz et al., 1981; Lindberg, 1981), resulting in the exchange of genetic information. Conjugative transfer of certain plasmids between *S. aureus* strains and *S. aureus* and *S. epidermidis* strains (Forbes and Schaberg, 1982; Cohen et al., 1982) and between *Streptococcus faecalis* and *S. aureus* strains (Schaberg et al., 1982) has been suggested, although the mechanisms of such a conjugative system remain to be determined. Pattee et al. (1982) have published an updated version of the chromosome map of *S. aureus*.

Most strains are susceptible to novobiocin (MIC < 0.4 μg/ml). Novobiocin resistance when it occurs, is usually chromosomally determined (Pattee and Baldwin, 1961). Susceptibility to benzylpenicillin (penicillin G), erythromycin, tetracycline, neomycin, kanamycin, gentamicin, chloramphenicol and streptomycin is variable. Resistance to one or more of these antibiotics and/or heavy metals (e.g. mercury, lead, cadmium, arsenate, arsenite, bismuth, antimony) is often plasmid-determined (see reviews by Novick and Bouanchaud, 1971; Richmond, 1972; Lacey and Richmond, 1974; Lacey, 1975; and Novick et al., 1977). Resistance to penicillin G is quite common and is due usually to the production of β-lactamases (penicillinase) (Richmond, 1963; Schaeg and Blobel, 1970, reviewed by Gale et al., 1981). Methicillin- and gentamicin-resistant strains appear to be increasing in numbers in many localities (Perceval et al, 1976; Shanson et al., 1976; Buckwold et al., 1979; Rosendal et al., 1981; Schaefler et al., 1981).

DNA/DNA hybridization studies have indicated that strains of *S. aureus* isolated from humans and a variety of animal sources (e.g., cattle, sheep, pigs, hares) are closely related, demonstrating over 75% relative DNA binding (DNA homology) under restrictive or optimal reassociation conditions (Meyer and Schleifer, 1978; Rosenblum and Tyrone, 1979; Kloos and Wolfshohl, 1982). DNA heteroduplexes, formed by the reassociation of DNA strands from different strains, are relatively stable to heat denaturation as indicated by $\Delta T_{m(e)}$ values of 0–3°C (Rosenblum and Tyrone, 1979; Mordarski et al., 1981). Based on these values, it would appear that the DNA of *S. aureus* strains has not diverged more than 3%. DNA relationships between *S. aureus* and other species are shown in Figure 12.5. *S. aureus* is not closely related to other species which have been examined to date.

On the basis of phenotypic characterization, *S. aureus* strains living on certain different host species (e.g. cattle, pigs, hares, poultry, humans) can be distinguished from one another and may be regarded as representing different ecovars (or biovars) (Meyer, 1967; Hajek and Marsalek, 1971, 1976; Oeding et al., 1971; Live, 1972; Rische et al., 1973). The distinction of ecovars is made on the basis of susceptibility to phages, antigenic components of the cell wall, nutritional requirements, coagulation of different plasmas, hemolysins, fibrinolysin activ-

ity, serological differences of nucleases and crystal violet type. Cross-contamination may give rise to the isolation of more than one ecovar from an individual host, such as that made possible by frequent contact between certain host species (e.g. different species of animals on the same farm, humans and their pets or farm animals).

The major habitats of *S. aureus* include the nasal membranes (anterior nares, nasopharynx) and skin and to a somewhat lesser extent the perineum, gastrointestinal tract and genital tract of warm-blooded animals (see reviews by Marples, 1965; Noble and Somerville, 1974; Kloos, 1980; Kloos and Schleifer, 1981). This species may be isolated from the nares (headquarter region) in less than 10% to somewhat more than 40% of nonhospitalized, human adults, depending upon the population sampled. Nasal carrier rates may increase considerably with prolonged hospitalization (Noble et al., 1964; Zeirdt, 1982). In preadolescent children, the carriage rate is often higher than in adults and populations are often more dispersed over the body (Williams, 1963; Noble and Somerville, 1974; Kloos and Musselwhite, 1975). Nonhuman primates carry frequently large populations of *S. aureus* in the nares and on skin (Kloos and Wolfshohl, 1979; Kloos, 1980).

S. aureus is a potential pathogen causing a wide range of infections (see reviews by Elek, 1959; Shulman and Nahmias, 1972; and Noble and Somerville, 1974; and reports of international symposia edited by Jeljaszewicz, 1973, 1976, 1981). Some of the major infections include: furuncles (boils), carbuncles, impetigo, toxic epidermal necrolysis, pneumonia, osteomyelitis, meningitis, endocarditis, mastitis, bacteremia, various abscesses, food poisoning (via enterotoxin), enterocolitis, urogenital infections and toxic shock syndrome.

The mol% G + C of the DNA is 32–36 (T_m, Bd).

Type strain: ATCC 12600 (NCTC 8352; CCM885) (Opin. 17, Jud. Comm. 1958, 153).

2. **Staphylococcus epidermidis** (Winslow and Winslow) Evans 1916, 449, *emend. mut. char.* Schleifer and Kloos 1975, 52.[AL] (*Albococcus epidermidis* Winslow and Winslow 1908, 201.)

e. pi. der′ mi. dis. Gr. n. *epidermidis* the outer skin; M.L. gen. n. *epidermidis* of the epidermis

Spheres, 0.5–1.5 μm in diameter. Cells occur predominantly in pairs and tetrads; occasionally single cells are observed. Some strains produce a slime (probably a polysaccharide, as indicated by staining with alcian blue) which encases cells and aids in their adherence and accumulation on the smooth surfaces of medical devices (e.g. prostheses, catheters, shunts, etc.) (Christensen et al., 1982; Peters et al., 1982). Amount of slime produced is variable among strains and may be influenced by growth medium composition. Some strains produce cell wall-deficient cells, and these may produce human chorionic gonadotropin (HCG)-like immunoreactive material (IRM) (Acevedo et al., 1980; Backus and Affronti, 1981). Most of the IRM-producing strains have been isolated from various malignant tumors or body fluids of compromised patients with cancer.

Peptidoglycan is of the type L-Lys-Gly$_{4-5}$, L-Ser$_{0.7-1.5}$. Cell wall teichoic acid is composed of glycerol, glucose and N-acetylglucosamine. Glucose and N-acetylglucosamine are either α- or β-glycosidically linked to glycerol. The teichoic acids are serologically distinct. The α-linked glucosyl-glycerol teichoic acid is called Bα and the β-linked is called Bβ (Oeding et al., 1967). The cell wall teichoic acid is necessary for specific phage adsorption (Schleifer and Steber, 1974).

The major cellular fatty acids include C_{20}, C_{15Br}, and C_{18} components (Komaratat and Kates, 1975; Durham and Kloos, 1978). The fatty acid compositions of *S. epidermidis* and *S. hominis* are nearly identical, but different from those of other species.

Colonies are smooth to mucoid, raised, glistening, circular, entire and translucent to nearly opaque, and single colonies may obtain a size of 2.5–6 mm in diameter on nonselective media used for the propagation of staphylococci (Kloos and Schleifer, 1975a; Schleifer and Kloos, 1976). With increasing age, elevated temperature (above 35°C), or crowding, colonies develop depressed dark centers and become more sticky in consistency. Slime-producing strains become very sticky in

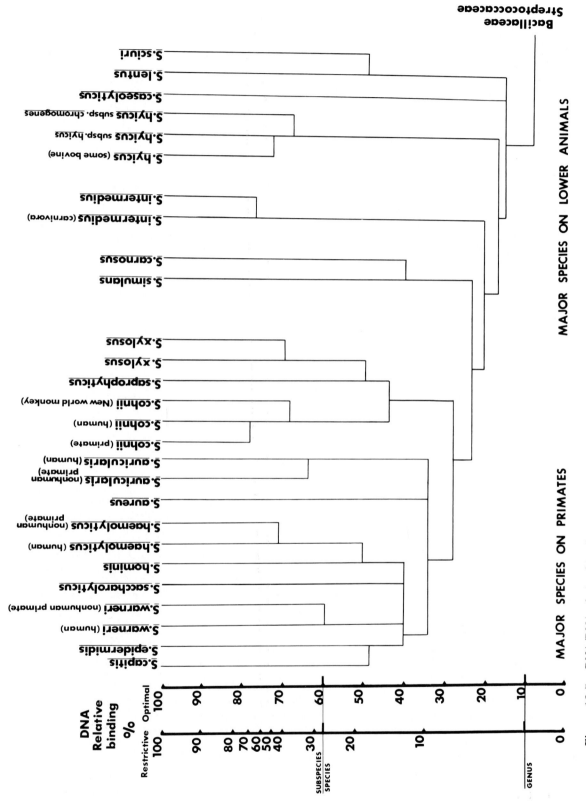

Figure 12.5. DNA/DNA relationships (homologies) of *Staphylococcus* species and subspecies. Optimum DNA reassociation conditions were performed at 29–31°C below the T_m of the DNA. Restrictive DNA reassociation conditions were performed at 16–18°C below the T_m of the DNA.

consistency. Cell wall-deficient (including partially revertant) strains produce slow growing, small colonies. Colonial pigment is not usually detected. Most strains produce colonies that are gray or grayish white. Rare strains may produce colonies that are yellowish, brownish or violet; the pigment being more intense in the center of the colony. Growth on agar slants is moderate, sticky in consistency, and translucent (becoming nearly transparent with storage). In broth, growth is first turbid, later becoming somewhat clear with a mucoid deposit. Slime-producing strains produce large aggregates of cells, many of which adhere to the culture tube or flask.

Facultative anaerobes: growth is best under aerobic conditions. Growth in the anaerobic portion of a semisolid thioglycolate medium (Evans and Kloos, 1972) is rapid and uniformly dense. Terminal pH in glucose broth, under anaerobic conditions, is 4.3–4.6 (Schleifer and Kocur, 1973; Schleifer and Kloos, 1975b). Catalase is produced; it is distinct, but closely related immunologically to those of S. capitis and S. warneri (Schleifer et al., 1979). The electrophoretic mobility of catalase from S. epidermidis is different from that of catalase from S. capitis and S. warneri (Zimmerman, 1976).

Growth is good at NaCl concentrations up to 7.5% and is relatively poor at 10%. Most strains grow between 15 and 45°C (optimum: 30–37°C).

Chemoorganotrophs: metabolism respiratory and fermentative. Unsaturated menaquinones (mainly Mk-7) (Collins, 1981) and cytochromes a and b (Faller et al., 1980) form the electron transport system. Glucose is metabolized to pyruvate via the EM glycolytic pathway and/or HMP pathway (Pan and Blumenthal, 1962). Aerobically, pyruvate is oxidized to acetate which, through acetyl-CoA, can be further oxidized via the TCA cycle. Addition of glucose to a growth medium suppresses operation of the TCA cycle (Ivler, 1965). However, at a low level of 0.5% glucose, there is a stimulation in the amount of aconitase and isocitrate dehydrogenase in extracts of S. epidermidis; whereas, with the same concentration of glucose, these enzymes are completely repressed in S. aureus. Acetate and CO_2 are the major end products of aerobic glucose metabolism; whereas, lactate is the major end product of anaerobic glucose metabolism. S. epidermidis produces predominantly L-lactic acid from glucose under anaerobic conditions (Schleifer and Kocur, 1973; Schleifer and Kloos, 1975b). Typical strains possess a FDP-activated NAD-dependent L-lactate dehydrogenase (Schleifer and Kloos, 1975b; Götz and Schleifer, 1975). Acetyl methyl carbinol is an end product of glucose metabolism. Mannitol can be an intracellular product of glucose metabolism, probably via mannitol-1-phosphate (Edwards et al., 1981). D-Galactose and lactose are metabolized via the D-tagatose-6-phosphate pathway (Schleifer et al., 1978). FDP-aldolase belongs to the class I type (Götz et al., 1979).

Acid is produced aerobically and anaerobically from glucose. Acid is produced aerobically from fructose, maltose, sucrose and glycerol; 70–90% of the strains also produce acid from lactose, galactose, mannose and/or turanose. Less than 20% of strains produce acid from ribose or melezitose. Rare strains produce acid from trehalose. No acid is produced from mannitol, rhamnose, xylose, xylitol, arabinose, gentiobiose, cellobiose, sorbitol, inositol, salicin, adonitol, dulcitol, arabitol, erythritol, erythrose, raffinose, melibiose, fucose, tagatose, lyxose or sorbose (Schleifer and Kloos, 1975b; Kloos and Wolfshohl, 1982).

Esculin and starch are not usually hydrolyzed. Hyaluronic acid may be hydrolyzed by hyaluronidases (Abramson and Friedman, 1967). Suitable cell wall glycans (e.g. Micrococcus luteus cell walls) may be degraded by a bacteriolytic enzyme demonstrating glycosidase activity (Valisena et al., 1981; 1982).

An organic nitrogen source (from 5–12 different amino acids) and B group vitamins (from 2–4, including thiamine and nicotinic acid (or nicotinamide) and often biotin and/or pantothenic acid) are required for growth (Gretler et al., 1955; Emmett and Kloos, 1975). For good anaerobic growth, most strains also require the addition of uracil and/or a fermentable carbon source (e.g. pyruvate) (Jones et al., 1963; Evans, 1976).

Most strains reduce nitrates to nitrites and/or ammonia, as a result of nitratase and nitritase activities, respectively (Baird-Parker, 1963; 1965; Jones et al., 1963; Schleifer and Kloos, 1975b). Alkaline phosphatase is produced by most strains (Baird-Parker, 1965; Pennock and Huddy, 1967; Schleifer and Kloos, 1975b; Cove et al., 1981; Pal and Ray, 1982). Acid phosphatase may be produced. Usually less than 10% of strains isolated from various sources are phosphatase-negative. Most strains demonstrate arginine dihydrolase and urease activities resulting in the production of ammonia from arginine and urea, respectively. Traces of H_2S may be produced from cysteine. Glutamic acid, lysine and ornithine are not decarboxylated (Hugh and Ellis, 1968; Nord et al., 1976). Proteases, lipases, phospholipases, lipoprotein lipases, esterases and deoxyribonucleases (DNase) may be produced (Baird-Parker, 1963, 1965; Schleifer and Kloos, 1975b). Esterases of S. epidermidis are distinctly different from those of other species (Zimmerman and Kloos, 1976). DNase activity is usually weak, when detected. Hemolysins (exotoxins) may be produced. These are rather similar to the α- and δ-hemolysins of S. aureus, but are probably not identical (Gemmell et al., 1976; Wadstrom et al., 1976). Hemolysin activity, when present, is usually weak. Coagulase is not produced.

Some strains produce antibiotic-like substances (e.g. bacteriocins, staphylococcins or micrococcins) which may be bacteriostatic and/or bactericidal to other staphylococci and certain other bacteria (Hsu and Wiseman, 1971; Lachowicz and Sienienska, 1976; Pulverer and Jeljaszewicz, 1976; Eady et al., 1981).

Differentiation of strains of S. epidermidis by phage typing has been accomplished with limited success. Problems still exist with the discrimination, reproducibility, and typeability of strains (DeSaxe et al., 1981). Standardization of phage-typing methods and the development of an international set of phages are still under investigation. Some of the more commonly used phage systems are those of Verhoef et al. (1971, 1972), Dean et al. (1973), Pulverer et al. (1973, 1975, 1976), Blouse et al. (1975), and Parisi et al. (1978).

Some progress has been made toward the use of serotyping in identifying coagulase-negative Staphylococcus species, but, as yet, only a few different sera have been evaluated for the typing of strains within a species, such as S. epidermidis (Tierno and Stotzky, 1978; Pillet and Orta, 1981).

Chromosomal and plasmid DNA may be transferred to appropriate recipient cells via transduction, resulting in the exchange of genetic information (Minshew and Rosenblum, 1970; Yu and Baldwin, 1971; Rosendorf and Kayer, 1974; Posten and Palmer, 1977; Olsen et al., 1979; Totten et al., 1981). Conjugative transfer of certain plasmids between S. aureus and S. epidermidis strains has been suggested (Cohen et al., 1982; Forbes and Schaberg, 1982), although the mechanisms of such a conjugative system remain to be determined. Naidoo and Noble (1981) have reported on the transmission of gentamicin resistance from a strain of S. epidermidis to S. aureus on human and mouse skin; however, the mechanism of transfer was not determined.

Most strains are susceptible to novobiocin (MIC $\leq$ 0.2 μg/ml). Susceptibility to benzylpenicillin (penicillin G), tetracycline, chloramphenicol, neomycin, erythromycin and gentamicin is variable, and resistance to these antibiotics may be often plasmid-determined (Minshew and Rosenblum, 1973; Rosendorf and Kayer, 1974; Poston and Palmer, 1977; Groves, 1979; Olson et al., 1979; Totten et al., 1981; Archer et al., 1982; Cohen et al., 1982; Forbes and Schaberg, 1982). Susceptibility to kanamycin, streptomycin, tobramycin, lividomycin, amikacin, lincomycin, clindamycin, nafcillin, methicillin, oxacillin, doxycycline and cephalosporins is also variable; the usual location of genes determining resistance for these antibiotics remains to be determined (Sabath et al., 1976; Archer, 1978; John et al., 1978; Archer and Tenenbaum, 1980; Wilkinson et al., 1980; Brun et al., 1981; Marples and Richardson, 1981). Multiple antibiotic resistance is common in this species; it often accumulates following antibiotic therapy (e.g. penicillin, tetracycline and erythromycin treatment).

DNA/DNA hybridization studies have indicated that strains of S. epidermidis are closely related, demonstrating over 85% relative DNA binding (DNA homology) under restrictive or optimal reassociation conditions (Kloos and Wolfshohl, 1982). The $\Delta T_{m(e)}$ values of inter-strain DNA heteroduplexes ranged from 0-2°C (W. E. Kloos, unpublished data). DNA relationships between *S epidermidis* and other species are shown in Figure 12.5. *S. epidermidis* is more closely related to *S. capitis*, *S. warneri*, *S. hominis*, *S. haemolyticus* and *S. saccharolyticus* (members of an *S. epidermidis* species group) (30-54% DNA homology, optimal; 11-19% DNA homology, restrictive) than to other species (15-33% DNA homology, optimal; 5-10% DNA homology, restrictive) (Kloos and Wolfshohl, 1979; Schleifer et al., 1979; Kilpper et al., 1980; Kloos, 1980; Kloos and Schleifer, 1981; Schleifer and Fischer, 1982; Schleifer et al., 1982; Kloos and Schleifer, 1983).

The major habitat of *S. epidermidis* is human skin; it is occasionally found on the skin of other mammals, particularly those living in association with man, (e.g. domestic animals and their products) (Kloos and Musselwhite, 1975; reviewed by Kloos, 1980 and Kloos and Schleifer, 1981). Populations found on mammals other than humans are usually small and transient. On humans, this species is considered a resident and may inhabit many of the habitats found in the cutaneous ecosystem, including also the mucous membranes of the nasopharynx and other areas adjoining the various body openings. It is usually the most predominant *Staphylococcus* species found living on human skin.

S. epidermidis is now recognized as an opportunistic pathogen which may colonize various indwelling medical devices, such as prosthetic heart valves (Dismukes et al., 1973; Speller and Mitchell, 1973; Johnson, 1976; Gardner et al., 1977; Archer, 1978; Marples et al., 1978; Archer et al., 1982), cerebrospinal fluid shunts (Holt, 1969, 1971; Schoenbaum et al., 1975), orthopaedic prostheses or appliances (Patterson and Brown, 1972; Wilson et al., 1972, 1973; Wilson, 1977), and intravascular catheters (Liekweg and Greenfield, 1977; Archer, 1978; Bender and Hughes, 1980; Peters et al., 1982). This species may be also responsible for postoperative infections in patients undergoing cardiac surgery (Archer and Tenenbaum, 1980), subacute bacterial endocarditis (perhaps as many as 1-10% of all cases) (Speller and Mitchell, 1973; Kaye, 1976; Pulverer and Halswick, 1977; Gemmell and Dawson, 1982), peritonitis in patients receiving continuous peritoneal dialysis (Rubin et al., 1980), urinary tract infections (Nord et al., 1976; John et al., 1978; Akatov et al., 1981b; Gemmell and Dawson, 1982), endophthalmitis (Valenton et al., 1973), otitis media (Feigin et al., 1973), and a variety of wound infections (Nord et al., 1976). Compromised patients undergoing immunosuppressive therapy are particularly susceptible to infection by this species.

The mol% G + C of the DNA is 30-37 (T_m, Bd).

Type strain: ATCC 14990 (Hugh and Ellis 1968, 237).

3. **Staphylococcus capitis** Kloos and Schleifer 1975a, 64.[AL]

ca′pi.tis. L. n. *caput* head; L. gen. n. *capitis* of the head; pertaining to that part of the human body where cutaneous populations of this species are usually the largest and most frequent.

Spheres, 0.8-1.2 µm in diameter. Cells occur predominantly in pairs and in tetrads.

Peptidoglycan is of the type L-Lys-Gly$_{3.5-4.3}$, L-Ser$_{0.8-1.2}$. Cell wall teichoic acid is composed of glycerol and *N*-acetylglucosamine. *N*-acetylglucosamine is either α- or β-glycosidically linked to glycerol (Endl et al., 1983). Cell walls also contain additional glutamic acid probably in the form of a glutamic acid polymer.

The major cellular fatty acids include C$_{20}$, C$_{18}$, and C$_{15Br}$ components (Durham and Kloos, 1978). The fatty acid profile of *S. capitis* is distinct from other related species of the *S. epidermidis* species group.

Colonies are smooth, slightly convex, glistening, entire and opaque, and single colonies may obtain a size of 2-3.5 mm in diameter on nonselective media used for the propagation of staphylococci. The colonies of *S. capitis* are typically smaller than those of most other species (with the notable exceptions of *S. auricularis* and *S. saccharolyticus*). Colonial pigment is absent at room or incubation temperatures (25-37°C), but in some strains may become yellowish or yellow-orange with prolonged storage at refrigeration temperatures (4-10°C). Most strains produce chalk white colonies; a few strains produce gray-white

colonies. Growth on agar slants is moderate to abundant, smooth in consistency and opaque.

Facultative anaerobes: growth is best under aerobic conditions. Growth in the anaerobic portion of a semisolid thioglycolate medium is often moderate to dense; some strains produce light growth as indicated by the appearance of individual colonies in the anaerobic portion of this medium. Terminal pH in glucose, under anaerobic conditions, is 4.2-4.7. Catalase is produced; it is distinct, but closely related immunologically to those of *S. epidermidis* and *S. warneri* (Schleifer et al., 1979). The electrophoretic mobility of catalase from *S. capitis* is different from that of catalases of *S. epidermidis* and *S. warneri* (Zimmerman, 1976).

Growth is good at NaCl concentrations up to 10%. Most strains grow between 18 and 45°C (optimum 30-40°C).

Chemoorganotrophs: metabolism respiratory and fermentative. Unsaturated menaquinones (mainly Mk-7) (Collins, 1981) and cytochromes *a* and *b* (Faller et al., 1980) form the electron transport system. Predominantly L-lactate is produced by glucose fermentation. D-lactate is produced in less amounts (5-25%). Acetyl methyl carbinol (acetoin) production is variable. FDP-aldolase is the class I type.

Acid is produced aerobically and anaerobically from glucose. Acid is produced aerobically from fructose, glycerol, mannose and mannitol. Most strains produce acid slowly from sucrose. Rare strains may produce acid slowly from maltose or lactose or fail to produce acid from mannitol. Acid is not produced from trehalose, galactose, rhamnose, xylose, xylitol, arabinose, ribose, turanose, gentiobiose, cellobiose, melezitose, sortibol, arabitol, inositol, salicin, adonitol, dulcitol, erythritol, erythrose, raffinose, melibiose, fucose, tagatose, lyxose or sorbose.

Suitable cell wall glycans (e.g. *Micrococcus luteus* cell walls) may be degraded by a bacteriolytic enzyme(s) (Kloos and Schleifer, 1975a; Varaldo and Satta, 1978).

An organic nitrogen source (from 9-13 different amino acids) and B group vitamins are required for growth (Emmett and Kloos, 1975, 1979). For anaerobic growth, most strains also require uracil and a fermentative carbon source (e.g. pyruvate) (Evans, 1976).

Most strains reduce nitrates. Alkaline or acid phosphatases are not usually produced (Kloos and Schleifer, 1975a; Hirunmitnakorn and Blumenthal, 1978).

Many strains demonstrate arginine dihydrolase activity, producing ammonia from arginine; most strains do not produce ammonia from urea and are urease-negative (Nord et al., 1976; Brun et al., 1978; API System S.A., 1979; Analytab Products, 1981; Kloos and Wolfshohl, 1982). Ornithine and lysine are not decarboxylated (Nord et al., 1976).

Lipases are produced. Hemolysins may be produced by some strains. Hemolysin activity on human blood, when present, is usually moderate; on bovine or sheep blood it is usually weak or not detectable. Coagulase is not produced. DNase activity is weak and not usually thermostable. Casein and gelatin are not usually hydrolyzed. Some of the esterases of *S. capitis* are distinctly different from those of other species (Zimmerman and Kloos, 1976).

Nearly all strains are susceptible to novobiocin (MIC $\leq$ 0.1 µg/ml), chloramphenicol, methicillin, cephalosporins, clindamycin, kanamycin, neomycin and streptomycin. Susceptibility to benzylpenicillin (penicillin G), erythromycin and/or tetracycline is somewhat more variable, although in most populations sampled the frequency of resistant strains seldom exceeded 15% (Kloos and Schleifer, 1975a; John et al., 1978; Kloos et al., 1983). *S. capitis* strains usually carry plasmids, but their relationships to antibiotic resistance have not been determined (Kloos et al., 1981).

DNA/DNA hybridization studies have indicated that strains of *S. capitis* are closely related, demonstrating over 90% relative DNA binding (DNA homology) under restrictive or optimal reassociation conditions (Kloos, 1980; Kloos and Wolfshohl, 1982). DNA relationships between S. *capitis* and other species are shown in Figure 12.5. *S. capitis* is more closely related to species of the *S. epidermidis* species group than to other species (see comments and references in above description of *S. epidermidis*).

The major habitats of *S. capitis* include the skin of the human scalp, forehead, eyebrows, face, neck and ears (including the external auditory meatus), regions where sebaceous glands are numerous and well developed or close by (Kloos and Musselwhite, 1975; Kloos, 1980, 1982). Much smaller, often transient, populations of this species may be found on the skin of other areas of the human body (e.g. arms, legs, chest). This species exhibits a rather strong niche and host preference.

S. capitis appears to be seldom associated with infections (Nord et al., 1976; Oeding and Digranes, 1977; John et al., 1978; Akatov et al., 1981a; Gemmel and Dawson, 1982). This species has been isolated occasionally from urinary tract and wound infections, but accounted for less than 4% of the total coagulase-negative strains obtained from these sources. Even fewer isolations have been made from the blood of septicemic patients.

The mol% G + C of the DNA is 31–36 (T_m).

Type strain: ATCC 27840 (Kloos and Schleifer 1975a, 66).

4. **Staphylococcus warneri** Kloos and Schleifer 1975a, 63.[AL]

war.ner'i. M.L. gen. n. *warneri* of Warner; named for Arthur Warner, Jr., from whom this organism was originally isolated.

Spheres, 0.5–1.2 μm in diameter. Cells occur singly and in pairs, and occasionally in tetrads. Some strains produce cell wall-defective cells, which may produce HCG-like IRM (see comments and references above in the description of *S. epidermidis*).

Peptidoglycan is of the type L-Lys-Gly$_{3.3-4.5}$, L-Ser$_{0.6-1.4}$. Cell wall teichoic acid is composed of glycerol, glucose and *N*-acetylglucosamine (Endl et al., 1983).

The major cellular fatty acids include C_{20}, C_{15Br}, and C_{18} components (Durham and Kloos, 1978). The C_{20} component constitutes an unusually high percentage (48%) of the total fatty acids. This species also produces a unique C_{22} fatty acid fraction.

Colonies are smooth, usually raised with a slightly elevated center, glistening, entire to undulate and translucent to opaque, and single colonies may obtain a size of 3–6 mm in diameter on nonselective media used in the propagation of staphylococci. Colonies become sticky in consistency with age. Cell wall-defective (including partial revertant) strains produce small, slow-growing colonies. Colonial pigment is variable; however, most strains produce yellowish gray, yellow, or yellow-orange pigment. In some human strains, colonies are gray-white with a yellow or yellow-orange edge. This pigment is enhanced greatly at room or refrigeration temperatures, but is often genetically unstable and lost with subculture. Growth on agar slants is moderate to abundant, sticky in consistency and opaque.

Facultative anaerobes: growth is best under aerobic conditions. Growth in the anaerobic portion of a semisolid thioglycolate medium is rapid and uniformly dense. Terminal pH in glucose broth, under anaerobic conditions, is 4.7–5.0. Catalase is produced; it is distinct but closely related immunologically to the catalases of *S. epidermidis* and *S. capitis* (Schleifer et al., 1979). The electrophoretic mobility of *S. warneri* catalase is different from that of *S. capitis* and *S. epidermidis* catalases (Zimmerman, 1976).

Growth is good at NaCl concentrations up to 10% and is usually poor at 15%. Most strains grow between 15 and 45°C (optimum: 30–40°C).

Chemoorganotrophs: metabolism respiratory and fermentative. Unsaturated menaquinones (mainly Mk-7) (Collins, 1981) and cytochromes *a* and *b* (Faller et al., 1980) form the electron transport system. Both D- and L-lactate isomers, in nearly equal amounts, are produced by glucose fermentation. The D-lactate dehydrogenase (D-LDH) of *S. warneri* is stabilized against heat inactivation by NADH; whereas, the D-LDH of *S. hominis* is heat resistant in the presense of NADH and pyruvate and the D-LDH of *S. haemolyticus* is stabilized in the presence of D(−)-lactate and NAD (Götz and Schleifer, 1976). Acetyl methyl carbinol (acetoin) is produced. The FDP-aldolase is of the class I type.

Acid is produced aerobically and anaerobically from glucose. Acid is produced aerobically from fructose, sucrose, trehalose and glycerol. Most strains produce acid from mannitol and acid slowly from maltose.

Some strains produce acid from ribose, galactose, melezitose, lactose, turanose or mannose. Human strains seldom produce acid from lactose or melezitose; nonhuman primate strains produce acid from these carbohydrates more frequently (Kloos and Wolfshohl, 1979.) Rare strains fail to produce acid from trehalose. No acid is produced from rhamnose, xylose, xylitol, arabinose, gentiobiose, cellobiose, sorbitol, inositol, salicin, adonitol, dulcitol, arabitol, erythritol, erythrose, raffinose, melibiose, fucose, tagatose, lyxose or sorbose.

Suitable cell wall glycans (e.g. *Micrococcus luteus* cell walls) may be degraded by a bacteriolytic enzyme(s) (Kloos and Schleifer, 1975a; Varaldo and Satta, 1978).

An organic nitrogen source (from 6–12 different amino acids) and B group vitamins are required for growth (Emmett and Kloos, 1975, 1979). For anaerobic growth, most strains also require uracil and/or a fermentative carbon source (e.g. pyruvate) (Evans, 1976).

Most strains do not reduce nitrates or produce phosphatases. Some strains produce ammonia from arginine by the action of arginine dihydrolase; nearly all strains produce ammonia from urea by the action of urease (see references above in the description of *S. capitis*). Nearly all *S. warneri* strains demonstrate strong β-glucosidase activity and many strains demonstrate strong β-glucuronidase activity; whereas, by contrast, *S. hominis* and *S. capitis* do not produce detectable levels of these enzymes (Analytab Products, 1981; Kloos and Wolfshohl, 1982). Lysine and ornithine are not decarboxylated (Nord et al., 1976). Lipases are produced. The esterase of *S. warneri* is distinctly different from those of *S. hominis*, *S. capitis* and *S epidermidis*, but has a similar electrophoretic mobility as the esterase of *S. haemolyticus* (Zimmerman and Kloos, 1976). Gelatin and casein are not usually hydroyzed.

Coagulases are not produced. Hemolysins may be produced, but activity is usually weak. Some strains, especially those of nonhuman primates, demonstrate moderate hemolytic activity. Human strains may demonstrate weak to moderate deoxyribonuclease (DNase) activity. Nonhuman primate strains produce often stronger DNase activity than human strains. The DNase of *S. warneri* is not usually thermostable.

Nearly all strains are susceptible to novobiocin (MIC ≤ 0.1 μg/ml). Most strains are susceptible to erythromycin, tetracycline, streptomycin, chloramphenicol, neomycin, kanamycin, benzylpenicillin (penicillin G), methicillin, cephalosporins and gentamicin; however, following prolonged penicillin, tetracycline and erythromycin treatment of the host, strains resistant to these and other antibiotics may be found (Kloos et al., 1983). Multiple resistance in this species is often genetically unstable. *S. warneri* may carry plasmids, but their possible relationships to antibiotic resistance have not been determined (Kloos et al., 1981). Nonhuman primate *S. warneri* strains often carry a large variety of different plasmids (usually 3–10 per strain).

DNA/DNA hybridization studies have indicated that human and nonhuman primate *S. warneri* strains represent different DNA homology or subspecies groups (Kloos and Wolfshohl, 1979). The relative binding (DNA homology) of human to nonhuman primate *S. warneri* DNA in reciprocal reactions at optimal reassociation conditions is 50–67% and at restrictive reassociation conditions is 19–36%; the $\Delta T_{m(e)}$ of intersubspecies heterologous DNA duplexes is 10–12°C. The extent of divergence suggests a near-separate species status. Different strains of human *S. warneri* demonstrate over 80% DNA homology under optimal or restrictive conditions. Different strains of nonhuman primate *S. warneri* isolated from major lines of divergence from *Prosimii* to *Pongidae* (great apes) also demonstrate over 80% DNA homology. DNA relationships between *S. warneri* and other species are shown in Figure 12.5. *S. warneri* is more closely related to members of the *S. epidermidis* species group than to other species (see comments and references above in the description of *S. epidermidis*).

S. warneri is usually found on human skin in very small populations; though some rare individuals may carry large population of this species (Kloos and Musselwhite, 1975; Kloos et al., 1983). It may occupy a variety of cutaneous niches and, as yet, a predominant niche has not been determined.

The nonhuman primate *S. warneri* is one of the major *Staphylococcus* species found living on the skin and nasal membranes of various prosimians and monkeys (Kloos and Wolfshohl, 1979). It and *S. aureus* are often the predominant species of these animals. *S. warneri* is seldom found living on other mammalian hosts; a few isolations have been made from domestic animals (Kloos et al., 1976; Devriese, 1979).

S. warneri may be associated with a variety of human infections, such as septicemia, endocarditis, conjunctivitis and urinary tract and wound infections, but it usually accounts for less than 6% of the total coagulase-negative strains isolated from these sources (see references on surveys of infections above in the description of *S. capitis*). At present it remains a questionable pathogen. In the neonatal mouse weight gain test, *S. warneri* demonstrates moderate virulence (Jonsson et al., 1981).

Occurrence of the nonhuman primate *S. warneri* in animal infections has not been reported.

The mol% G + C of the DNA is 34–35 (T_m).

Type strain: ATCC 27836 (Kloos and Schleifer 1975a, 64).

5. **Staphylococcus haemolyticus** Schleifer and Kloos 1975, 56.[AL]
hae.mo.ly'ti.cus. M.L. adj. *haemolyticus* blood-dissolving.

Spheres, 0.8–1.3 μm in diameter. Cells occur predominantly in pairs and tetrads. Some strains produce cell wall-defective cells, which may produce (HCG)-like IRM (see comments and references above in the description of *S. epidermidis*).

Peptidoglycan is of the type L-Lys-Gly$_{3.3-4.0}$, L-Ser$_{0.9-1.5}$. Cell wall teichoic acid is composed of glycerol and *N*-acetylglucosamine.

The major cellular fatty acids include C_{15Br}, C_{18}, C_{20} and C_{17Br} components (Durham and Kloos, 1978). The fatty acid profile is similar to that of *S. aureus*, with the minor exception of differences in the C_{19Br} fraction. It is also rather similar to the fatty acid profile of *S. hominis*.

Colonies are smooth, raised to slightly convex, glistening, usually entire, circular and opaque, and single colonies may obtain a large size of 5–9 mm in diameter on nonselective media used in the propagation of staphylococci. Cell wall-defective (including partial revertant) strains produce smaller, slow-growing colonies. Colonial pigment is variable; however, most strains produce colonies that are unpigmented (gray-white or white) or have a slight yellow tint which increases slightly in intensity with age. Pigmented strains produce yellowish colonies. Growth on agar slants is abundant and opaque.

Facultative anaerobes: growth is best under aerobic conditions. Growth in the anaerobic portion of a semisolid thioglycolate medium is variable; however, most strains produce light growth or individual colonies in the deeper, more anaerobic, portion of the medium. Terminal pH in glucose broth, under anaerobic conditions is 4.9–5.5. Catalase is produced; it is distinct from those of other species (Zimmerman, 1976; Schleifer et al., 1979).

Growth is good at NaCl concentrations up to 10% and is poor or absent at 15%. Most strains grow between 18 and 45°C (optimum: 30–40°C).

Chemoorganotrophs: metabolism respiratory and fermentative. Unsaturated menaquinones (mainly Mk-7) (Collins, 1981) and cytochromes *a* and *b* (Faller et al., 1980) form the electron transport system. Most strains produce only or predominantly the D-lactate isomer by glucose fermentation. The D-lactate dehydrogenase (D-LDH) of *S. haemolyticus* appears to be distinct from those of other species with respect to heat inactivation properties, but demonstrates a rather similar electrophoretic mobility as the D-LDH of *S. hominis* (Götz and Schleifer, 1976). Most strains produce acetyl methyl carbinol (acetoin). The FDP-aldolase is of the class I type.

Acid is produced aerobically and anaerobically from glucose. Acid is produced aerobically from maltose, sucrose, trehalose and glycerol. Nearly 50% of strains produce acid from lactose, galactose, fructose, turanose or mannitol. Some strains produce acid from ribose; rare stains may produce acid from mannose or melezitose. No acid is produced from rhamnose, xylose, xylitol, arabinose, gentiobiose, cello-biose, sorbitol, inositol, salicin, adonitol, dulcitol, arabitol, erythritol, erythrose, raffinose, melibiose, fucose, tagatose, lyxose or sorbose.

Suitable cell wall glycans (e.g. *Micrococcus luteus* cell walls) may be degraded by a bacteriolytic enzyme(s) (Schleifer and Kloos, 1975b; Varaldo and Satta, 1978).

An organic nitrogen source (from 5–8 different amino acids) and B group vitamins are required for growth (Emmett and Kloos, 1975; 1979). For anaerobic growth, most strains also require uracil and a fermentative carbon source (e.g. pyruvate) (Evans, 1976).

Most strains reduce nitrates, but do not exhibit phosphatase or deoxyribonuclease (DNase) activity. Nonhuman primate strains produce weak to moderate DNase activity more frequently than human strains. The DNase of these strains is not usually thermostable. Most strains produce ammonia from arginine (arginine dihydrolase-positive); nearly all strains do not produce ammonia from urea (urease-negative) (see references above in the description for *S. capitis*). Many strains (>40%) of *S. haemolyticus* demonstrate strong β-glucosidase and/or β-glucuronidase activity (Analytab Products, 1981; Kloos and Wolfshohl, 1982). By contrast, the related species *S. hominis* is usually urease-positive and arginine dihydrolase-negative and is β-glucosidase- and β-glucuronidase-negative. Lysine and ornithine are rarely decarboxylated (Nord et al., 1976). Lipases are produced. The esterase of *S. haemolyticus* is distinctly different from those of *S. hominis*, *S. capitis*, and *S. epidermidis*, but has a similar electrophoretic mobility as the esterase of *S. warneri* (Zimmerman and Kloos, 1976). Gelatin and casein are not usually hydrolyzed.

Coagulases are not produced. Hemolysins are produced. They are probably not identical with the classical α-, β-, γ-, or δ-hemolysins of *S. aureus* (Wadstrom et al., 1976). Hemolytic activity is usually moderate to strong, detectable within 48–72 h on blood agar. Hemolysis is not as rapid or as strong as that of *S. aureus*. Uncommon strains may produce only weak hemolytic activity.

Nearly all strains are susceptible to novobiocin (MIC $\leq$ 0.2 μg/ml), chloramphenicol, streptomycin, neomycin, kanamycin, methicillin, cephalosporins and gentamicin. Susceptibility to benzylpenicillin (penicillin G), tetracycline, and erythromycin is variable. Multiple resistance to these antibiotics is not uncommon. *S. haemolyticus* strains usually carry plasmids, but to date few reports have established their relationships to antibiotic resistance (Schaefler, 1971; Kloos et al., 1981).

DNA/DNA hybridization studies have indicated that human and nonhuman primate *S. haemolyticus* strains represent different DNA homology or subspecies groups (Kloos and Wolfshohl, 1979). The relative binding of human to nonhuman primate *S. haemolyticus* DNA in reciprocal reactions at optimal reassociation conditions is 65–77% and at restrictive reassociation conditions is 34–48%; the $\Delta T_{m(e)}$ of intersubspecies heterologous DNA duplexes is 8–10°C. Different strains of human *S. haemolyticus* demonstrate over 75% (usually over 80%) relative DNA binding (DNA homology) under optimal or restrictive conditions. Different strains of nonhuman primate *S. haemolyticus* isolated from major lines of divergence from *Prosimii* to *Pongidae* (great apes) also demonstrate over 75% (usually over 80%) DNA homology. DNA relationships between *S. haemolyticus* and other species are shown in Figure 12.5. *S. haemolyticus* is more closely related to members of the *S. epidermidis* species group than to other species; it is most closely related to *S. hominis*, but the extent of divergence ($\Delta T_{m(e)}$ of 14–14.5 °C) is sufficient to warrant a separate species status (see comments and references above in the description of *S. epidermidis*).

S. haemolyticus is usually found on human skin in small to medium-size populations (Kloos and Musselwhite, 1975; Kloos, 1980). It may occupy a variety of cutaneous niches; it is isolated less frequently from the nasal membranes and most frequently from the axillae, perineum and inguinal area, arms and legs. Most individuals carry only one or two detectable strains (or clonal populations) of this species (Kloos and Musselwhite, 1975; Kloos et al., 1983). *S. haemolyticus* is also a relatively minor species living on the skin of various prosimians and monkeys (Kloos and Wolfshohl, 1979). It is seldom found living on other mammalian hosts; a few isolations have been made from domestic animals.

S. haemolyticus may be associated with a variety of human infections, such as septicemia, conjunctivitis, urinary tract and wound infections; it usually accounts for less than 15% of the total coagulase-negative strains isolated from these sources (see references on surveys of infections above in the description of *S. capitis*). At present it remains a questionable pathogen; although it probably has some potential pathogenic properties. Occurrence of the nonhuman primate *S. haemolyticus* in animal infections has not been reported.

The mol% G + C of the DNA is 34–36 (T_m).

Type strain: ATCC 29970 (DSM 20263) (Schleifer and Kloos 1975b, 57).

6. Staphylococcus hominis Kloos and Schleifer 1975a, 68.[AL]

ho'mi.nis. L. n. *homo* man; L. gen. n. *hominis* of man; named for the host on whose skin this species is commonly found.

Spheres, 1.0–1.5 μm in diameter. In most strains, cells occur predominantly in tetrads and occasionally in pairs. Some strains have approximately equal numbers of tetrads and pairs. Some strains produce cell wall-defective cells, which may produce HCG-like IRM (see comments and references above in the description of *S. epidermidis*).

Peptidoglycan is of the type L-Lys-Gly$_{3.5-4.5}$, L-Ser$_{0.6-1.3}$. Cell wall teichoic acid is composed of glycerol and *N*-acetylglucosamine.

The major fatty acids include C_{15Br}, C_{20} and C_{18} components (Durham and Kloos, 1978). The fatty acid profile is rather similar to that of *S. haemolyticus*.

Colonies are smooth and somewhat butyrous in texture, raised to slightly umbonate, entire, circular, slightly glistening to dull and opaque, and single colonies may obtain a size of 3–5 mm in diameter on most nonselective media used in the propagation of staphylococci. On blood (5%)-supplemented P agar, colonial growth is reduced significantly (colony diameter ~2/3–3/4 of that obtained on unsupplemented P agar), a feature which distinguishes this species from other members of the *S. epidermidis* species group, which are usually stimulated by the addition of blood to growth media. Cell wall-deficient (including partial revertant) strains produce smaller, slow growing colonies. Colonial pigment is variable; many strains produce colonies with a wide yellowish or yellow-orange center and two or more alternating light and dark (e.g. whitish and grayish) concentric rings proceeding out from near the center to the edge. Unpigmented strains are relatively common; their colonies exhibit often a pattern of alternating whitish and grayish concentric rings. Some strains produce colonies which are uniformly pigmented or unpigmented. Growth on agar slants is moderate to abundant and opaque. In general, *S. hominis* cannot be stored on ordinary agar media used in the propagation of staphylococci for extended periods of time, even at refrigeration temperatures, compared with most other species. Some strains will not remain viable for more than 1–2 months under these conditions.

Facultative anaerobes: growth is much better under aerobic conditions. Growth in the anaerobic portion of a semisolid thioglycolate medium is usually very slow and slight. Except for the development of a few tiny individual colonies in the deeper, more anaerobic, portion of the medium, many strains would appear to be strictly aerobic by this test. A few strains produce a gradient of growth down the anaerobic portion of the medium, from moderately dense to slight. Even though anaerobic growth is slow, most strains will ferment glucose and produce a terminal pH in glucose broth of 4.4–4.9. Catalase is produced; it is distinct from those of other species (Zimmerman, 1976; Schleifer et al., 1979).

Growth is good at NaCl concentrations up to 7.5% and poor at 10%. Most strains grow between 20 and 45°C (optimum: 30–40°C).

Chemoorganotrophs: metabolism respiratory and fermentative. Unsaturated menaquinones (mainly Mk-7) (Collins, 1981) and cyto-chromes *a* and *b* (Faller et al., 1980) form the electron transport system. Some strains produce only the D-lactate isomer; whereas, others produce both L- and D-lactate isomers, in nearly equal amounts, by glucose fermentation. The D-lactate dehydrogenase (D-LDH) of *S. hominis* appears to be distinct from those of other species with respect to heat inactivation properties, but demonstrates a rather similar electrophoretic mobility as the D-LDH of *S. haemolyticus* (Götz and Schleifer, 1976). Acetyl methyl carbinol (acetoin) production is variable. Galactose and lactose are metabolized via the D-tagatose-6-phosphate pathway (Schleifer et al., 1978). The FDP-aldolase is of the class I type.

Acid is produced aerobically and anaerobically from glucose. Acid is produced aerobically from fructose, maltose, sucrose (often slowly) and glycerol. Most strains produce acid from trehalose, melezitose and turanose; nearly 50% of strains produce acid from lactose or galactose. Some strains produce acid from mannitol, mannose, or more rarely, ribose. No acid is produced from rhamnose, xylose, xylitol, arabinose, gentibiose, cellobiose, sorbitol, inositol, salicin, adonitol, dulcitol, arabitol, erythritol, erythrose, raffinose, melibiose, fucose, tagatose, lyxose or sorbose.

Suitable cell wall glycans (e.g. *Micrococcus luteus* cell walls) may be degraded by a bacteriolytic enzyme(s); activity is usually weak (Kloos and Schleifer, 1975a; Varaldo and Satta, 1978).

An organic nitrogen source (6–10 different amino acids) and B group vitamins are required for growth (Emmett and Kloos, 1975; 1979). For anaerobic growth, most strains also require a fermentative carbon source (e.g. pyruvate); some strains require both uracil and a fermentative carbon source (Evans, 1976). Most strains reduce nitrates, although some only weakly, and do not demonstrate phosphatase or significant deoxyribonuclease (DNase) activity. Most strains do not produce ammonia from arginine (arginine dihydrolase-negative), but do produce ammonia from urease (urease-positive); these activities are the reverse of those found in the related species *S. haemolyticus* (see references above in the description for *S. capitis*). Lysine and ornithine are rarely decarboxylated (Nord et al., 1976). Lipases are produced. Esterases of *S. hominis* are distinctly different from those of *S. haemolyticus* and other species (Zimmerman and Kloos, 1976). Gelatin and casein are not usually hydrolyzed.

Coagulases are not produced. Hemolysins demonstrating weak activity may be produced; however, most strains do not produce detectable hemolysis on blood (5%)-supplemented agars.

Nearly all strains are susceptible to novobiocin (MIC < 0.2 μg/ml), methicillin, cephalosporins, neomycin, chloramphenicol and gentamicin. Susceptibility to benzylpenicillin (penicillin G), tetracycline, erythromycin, kanamycin and streptomycin is variable. Multiple resistance to these antibiotics is not uncommon. *S. hominis* strains usually carry plasmids, but their possible relationships to antibiotic resistance have not been determined (Kloos et al., 1981).

DNA/DNA hybridization studies have indicated that strains of *S. hominis* are closely related demonstrating over 85% relative DNA binding (DNA homology) under restrictive or optimal reassociation conditions (Kloos, 1980; Kloos and Wolfshohl, 1982). DNA relationships between *S. hominis* and other species are shown in Figure 12.5. *S. hominis* is more closely related to members of the *S. epidermidis* species group than to other species; it is most closely related to *S. haemolyticus*, but the extent of divergence ($\Delta T_{m(e)}$ of 14–14.5°C) is sufficient to warrant a separate species status (see comments and references above in the description of *S. epidermidis*).

S. hominis is one of the major *Staphylococcus* species found living on human skin (Kloos and Musselwhite, 1975; Kloos, 1980). It may be distributed over a variety of habitats in the cutaneous ecosystem, but demonstrates some preference for the axillae, arms, pubic and inguinal (including perineum) regions and legs, the largest populations being produced in areas richly supplied with both apocrine and eccrine glands. *S. hominis* appears to be quite host (human) specific.

S. hominis may be associated with a variety of human infections, such as septicemia, conjunctivitis, urinary tract and wound infections;

it usually accounts for less than 10% of the total coagulase-negative strains isolated from these sources (see references on surveys of infections above in the description of *S. capitis*). At present it remains a questionable pathogen. In the neonatal mouse weight gain test, *S. hominis* demonstrates moderate virulence (Jonsson et al., 1981).

The mol% G + C of the DNA is 30–36 (T_m).

Type strain: ATCC 27844 (Kloos and Schleifer 1975a, 69).

7. Staphylococcus saccharolyticus (Foubert and Douglas) Kilpper-Bälz and Schleifer 1984, 91.[VP] (Effective publication: Kilpper-Bälz and Schleifer 1981a, 324.) (*Micrococcus saccharolyticus* Foubert and Douglas 1948, 31.)

sac.cha.ro.ly'ti.cus. Gr. n. *sacchar* sugar; Gr. adj. *lyticus* able to loose; M.L. adj. *saccharolyticus* sugar-digesting.

Spheres, 0.6–1.0 μm in diameter. Cells occur singly, in pairs, tetrads or irregular masses.

Peptidoglycan is of the type L-Lys-Gly$_4$, L-Ser$_{0.7-1.2}$. Cell wall teichoic acid is composed of glycerol and *N*-acetylglucosamine.

Colonies are smooth, slightly convex, entire, circular, glistening and opaque, and single colonies may obtain a size of 0.5–2.0 mm in diameter under anaerobic conditions or <0.5 mm in diameter under aerobic conditions. They are usually grayish white (unpigmented), but on hemin-supplemented agar may become slightly yellow.

Anaerobes: growth is good under anaerobic conditions, but poor or absent under aerobic conditions. Catalase reaction is positive after growth on hemin-supplemented medium, but weak or absent after growth on medium without hemin supplementation.

Optimal growth temperature is between 30 and 37°C.

Chemoorganotrophs: metabolism is mainly fermentative. Cytochromes *a* and *b* are present; therefore the benzidine test is positive, although in some strains it is weak. *S. saccharolyticus* differs from all other described staphylococci with respect to products of glucose metabolism. Glucose is fermented to CO_2, ethanol, acetic acid and small amounts of formic and lactic acid. Both D- and L-lactate isomers are produced by glucose fermentation, though only in low amounts. FDP-aldolase is of the class I type.

Acid is produced slowly to moderately from glucose, fructose, mannose and glycerol. Acid is not produced from xylose, xylitol, arabinose, sucrose, maltose, lactose, trehalose, mannitol or cellobiose.

Nutritional requirements have not been determined. Nitrates are reduced. Ammonia is produced from arginine. Coagulases are not produced. Hemolysin activity is not detected.

Strains are susceptible to novobiocin. Susceptibilities to other antibiotics have not been adequately assessed.

DNA/DNA and DNA/rRNA hybridization studies (Kilpper et al., 1980; Kilpper-Bälz and Schleifer, 1981b) and comparative 16S rRNA analysis (Ludwig et al., 1981) demonstrate clearly a separate species status and definite relationship to the genus *Staphylococcus*. DNA relationships between *S. saccharolyticus* and other species are shown in Figure 12.5. *S. saccharolyticus* appears to be more closely related to members of the *S. epidermidis* species group than to other species.

S. saccharolyticus is found frequently on the skin of the human forehead; some less extensive studies have also demonstrated its occurrence on the back and arms (Evans and Mattern, 1978; Evans et al., 1978). Due to its slow growth and frequent outnumbering by propionibacteria, it may have been missed in some ecological studies. A comprehensive study of the habitat range of this species has not been reported. It is present occasionally in clinical material (Werner and Rintelen, 1973).

The mol% G + C of the DNA is 33–34 (T_m).

Type strain: ATCC 14953 (Kilpper-Bälz and Schleiffer 1981a, 329).

8. Staphylococcus auricularis Kloos and Schleifer 1983, 9.[VP]

au.ri.cu.la'ris. N.L. adj. *auricularis* pertaining to the ear; named for the region of the body (external auditory meatus, external ear) from which this species is commonly found.

Spheres, 0.8–1.2 μm in diameter. Cells occur predominantly in tetrads and pairs.

Peptidoglycan is of the type L-Lys-Gly$_{4.3-4.8}$, L-Ser$_{0-0.6}$. Cell wall teichoic acid is of the poly (*N*-acetylglucosamine 1-phosphate) type.

Colonies are usually smooth and butyrous in texture, convex, opaque, circular and slightly glistening, and single colonies may obtain a size of 1.4–2.8 mm in diameter on nonselective media used in the propagation of staphylococci. Compared with most other species, growth is slow and colony size is quite small. Some uncommon strains produce colonies with a granular texture and ornate, wrinkled or rough surface. Colony edge is entire to slight undulate and becomes usually crimpled after 4–5 days of incubation. Young colonies exhibit an unusually high profile for staphylococci. Nearly all strains are unpigmented and produce white colonies; with extended (> 5 day) incubation or storage at refrigeration temperatures some may develop a pale cream color.

Facultative anaerobes: growth is much better under aerobic conditions. Growth in the anaerobic portion of a semisolid thioglycolate medium is slow and slight, much like that found with *S. hominis*. Catalase is produced.

Growth is good at NaCl concentrations up to 10%, and rather poor to moderate at 15%. Most strains grow between 20 and 45°C (optimum: 30–40°C).

Chemoorganotrophs: metabolism respiratory and fermentative. Cytochromes *a* and *b* are present. The L-lactate isomer is produced by glucose fermentation. Acetyl methyl carbinol (acetoin) production is variable; when it is detected it is usually weak. The FDP-aldolase is of the class I type.

Acid is produced aerobically and anaerobically from glucose. Acid is produced aerobically from glycerol. Most strains produce acid from fructose, trehalose and maltose. Some strains produce acid from sucrose and turanose (weak). Uncommon strains produce acid from lactose, mannose and ribose. Acid is not produced from mannitol, xylose, xylitol, arabinose, rhamnose, gentiobiose, cellobiose, raffinose, sorbitol, sorbose, salicin, fucose, inositol or melezitose.

Nutritional requirements have not been determined. Some strains reduce nitrates, and then usually slowly. Alkaline phosphatases are not produced. Most strains demonstrate only weak deoxyribonuclease (DNase) activity. Some strains produce ammonia from arginine (arginine dihydrolase-positive). Ammonia is not produced from urea (urease-negative). Some strains exhibit weak to moderate β-galactosidase activity. Lipases are produced. Coagulases are not produced. Hemolysis of blood agar is very weak and only partial, requiring prolonged incubation for detection.

Susceptible to novobiocin (MIC 0.1–0.8 μg/ml), benzylpenicillin (penicillin G), methicillin, erythromycin, tetracycline, chloramphenicol, streptomycin, kanamycin, gentamicin, lincomycin, neomycin, vancomycin and bacitracin. Compared to other species *S. auricularis* is unusually susceptible to a variety of antibiotics; resistance has not yet been reported. Furthermore, this species appears not to be affected by penicillin treatment (via oral or intramuscular routes) of the host, perhaps due to the inadequate penetration of this antibiotic into their natural habitat (i.e. the external auditory canal bathed with ceruminous gland secretions: ear wax) (Kloos et al., 1983). The majority of strains apparently do not carry plasmids (Kloos, 1982).

DNA/DNA hybridization studies have indicated that strains of *S. auricularis* are closely related demonstrating over 90% relative DNA binding (DNA homology) under restrictive or optimal reassociation conditions. DNA relationships between *S. auricularis* and other species are shown in Figure 12.5. This species is not closely related to other staphylococci.

The major habitat of *S. auricularis* is the external auditory meatus of the ear. *S. auricularis* and *S. capitis* are often the predominant species occupying this niche. At present, only human and nonhuman primate populations of this species have been observed. Its occurrence in other mammals and in infections remains to be determined.

The mol% G + C of the DNA is 38–39 (T_m).

Type strain: ATCC 33753 (Kloos and Schleifer 1983, 12).

9. Staphylococcus saprophyticus (Fairbrother) *emend. mut. char.* Shaw, Stitt and Cowan 1951, 1021 and Schleifer and Kloos 1975b, 53.[AL] sa.pro.phy'tic.us. Gr. adj. *sapros* putrid; Gr. n. *phyton* plant; M.L. adj. *saprophyticus* saprophytic, growing on dead tissues.

Spheres, 0.8–1.5 μm in diameter, occurring singly, in pairs, and much less frequently as tetrads.

Peptidoglycan is of the type L-Lys-Gly$_{4-5}$, L-Ser$_{0.6-0.8}$. Cell wall teichoic acid is composed of glycerol, ribitol and *N*-acetylglucosamine. *N*-acetylglucosamine residues are usually β-linked to ribitol and α-linked to glycerol (Endl et al., 1983). The *N*-acetylglucosamine-glycerol and -ribitol teichoic acids are characteristic antigens (poly AβC) (Oeding and Hasselgren, 1972).

The major cellular fatty acids include C$_{15Br}$, C$_{20}$, C$_{18}$, and C$_{16}$ components (Durham and Kloos, 1978). The fatty acid profile is distinct from those of other species.

Colonies are raised to slightly convex, circular, usually entire, smooth, glistening and usually opaque, and single colonies may obtain a size of 5–9 mm in diameter on nonselective media used for the propagation of staphylococci. Colonial pigment is variable; however, most strains produce colonies which are unpigmented or have a slight yellow tint which increases in intensity with age. The colonies of pigmented strains are yellow to yellow-orange. Growth on agar slants is abundant, smooth, glistening and opaque.

Facultative anaerobes: growth is best under aerobic conditions. Growth in the anaerobic portion of a semisolid thioglycolate medium is usually dense or in a gradient from dense to light growth down the tube to the more anaerobic portion of this medium. Some uncommon strains may produce only slight growth or individual colonies in the anaerobic portion. Many strains only ferment glucose weakly and would be misclassified as micrococci on the basis of the O/F-test. Terminal pH in glucose broth, under anaerobic conditions, is 5.0–5.5. Catalase is produced; it is distinct, but very closely related immunologically to the catalase of *S. xylosus* (Schleifer et al., 1979).

Growth is good at NaCl concentrations up to 10% and is usually poor to moderate at 15%. Most strains grow between 10 and 40°C (optimum: 28–35°C).

Chemoorganotrophs: Metabolism mainly respiratory. Unsaturated menaquinones (mainly Mk-7) (Collins, 1981) and cytochromes *a* and *b* (Faller et al., 1980) form the electron transport system. Both D- and L-lactate isomers (predominantly the L-lactate isomer) are produced only in low amounts. Acetyl methyl carbinol (acetoin) is produced. Mannitol can be an intracellular product of glucose metabolism, probably via mannitol-1-phosphate (Edwards et al., 1981). Lactose is metabolized via the Leloir pathway (Schleifer et al., 1978). FDP-aldolase is of the class I type. The phosphoenolpyruvate (PEP)-phosphotransferase (PTS) system is not involved in the uptake of xylitol (Lehmer and Schleifer, 1980). The dehydration of xylitol is inducible and phosphorylation probably occurs on ketopentoses.

Acid is produced aerobically from glucose, maltose, sucrose, turanose and glycerol. Most strains produce acid from mannitol, fructose, trehalose, lactose and xylitol (often slowly). Uncommon strains produce acid from galactose, ribose or gentiobiose. No acid is produced from mannose, rhamnose, xylose, arabinose, ribose, cellobiose, melezitose, sorbitol, inositol, salicin, adonitol, dulcitol, arabitol, erythritol, erythrose, raffinose, melibiose, fucose, tagatose, lyxose or sorbose.

Esculin and starch are not usually hydrolyzed (Nord et al., 1976). Under certain conditions suitable cell wall glycans (e.g. *Micrococcus luteus* cell walls) may be degraded by an endo-β-*N*-acetylglucosaminidase (Valisena et al., 1981).

An organic nitrogen source (from 1–3 different amino acids) and B group vitamins are required for the growth of most strains (Emmett and Kloos, 1975). Some strains are capable of growing on a medium containing (NH$_4$)$_2$SO$_4$ as a sole source of substrate nitrogen. These minimal requirements are in marked contrast to the relatively large number of amino acids required by strains of *S. aureus* or members of the *S. epidermidis* species group. For anaerobic growth, uracil and a fermentative carbon source (e.g. pyruvate) or uracil alone are also required (Evans, 1976).

Most strains do not reduce nitrate. Acid and/or alkaline phosphatases are not usually produced. Lysine and ornithine are not decarboxylated. Some strains may hydrolyze casein or gelatin. Lipases and esterases may be produced (Zimmerman and Kloos, 1976). Coagulase and extracellular deoxyribonuclease (DNase) are not produced. Hemolysis is only rarely detected. Ammonia is usually not produced or is produced only weakly from arginine (no or weak arginine dihydrolase activity), but is produced very strongly from urea (urease-positive) (see references above in the description of *S. capitis*). Many strains demonstrate strong β-galactosidase activity; a property which distinguishes this species from human *S. cohnii* strains and certain other species (Analytab Products, 1981; Kloos and Wolfshohl, 1982).

Resistant to novobiocin (MIC 3.1–12.5 μg/ml; most strains >6.2 μg/ml). Susceptibility to benzylpenicillin (penicillin G) is partial (MIC 0.2–0.8 μg/ml), characteristic of other novobiocin-resistant species. Susceptibility to erythromycin, tetracycline, chloramphenicol or streptomycin is variable. Susceptible to neomycin, clindamycin, kanamycin, gentamicin, vancomycin and lincomycin. *S. saprophyticus* strains usually carry plasmids, but their possible relationships to antibiotic resistance have not been determined (Kloos et al., 1981).

DNA/DNA hybridization studies have indicated that strains of *S. saprophyticus* isolated from humans and variety of other primate sources (i.e. the major lines of divergence from *Prosimii* to *Pongidae* (great apes)) are closely related, demonstrating over 85% relative DNA binding (DNA homology) under restrictive or optimal reassociation conditions (Kloos and Wolfshohl, 1982; unpublished data). DNA relationships between *S. saprophyticus* and other species are shown in Figure 12.5. *S. saprophyticus* is more closely related to *S. cohnii* and *S. xylosus*, members of an *S. saprophyticus* species group, than to other species (Schleifer et al., 1979; Kloos and Wolfshohl, 1982).

S. saprophyticus is isolated occasionally from the skin of humans and other mammals and their products (Baird-Parker, 1963; Schleifer and Kloos, 1975b; Kloos and Musselwhite, 1975; Kloos, 1980). Populations on human skin are usually small and transient.

S. saprophyticus and *S. epidermidis* are the predominant coagulase-negative species isolated from human urinary tract infections (Torres-Pereira, 1962; Maskell, 1974; Telander and Wallmark, 1975; Nord et al., 1976; Oeding and Digranes, 1976; John et al., 1978; Hovelius et al., 1979; Jordan et al., 1980; Akatov et al., 1981b; Bollgren et al., 1981; Marrie et al., 1982). Some of the more common types of urinary tract infections include cystitis, urethritis and pyelonephritis, usually accompanied with a significant bacteriuria. *S. saprophyticus* appears to be the predominant staphylococcal species in acute urinary tract infections of young adult women (Mabeck, 1969; Maskell, 1974). This species appears to have a higher capacity to adhere to uroepithelial cells than to buccal or skin cells and does so better than other staphylococcal species (Colleen et al., 1979; Mardh et al., 1979). It also may be implicated in prostatitis (Akatov et al., 1981).

The mol% G + C of the DNA is 31–36 (T_m).

Type strain: ATCC 15305 (NCTC 7292 CCM 883) (Shaw, Stitt and Cowan 1951, 1021).

10. **Staphylococcus cohnii** Schleifer and Kloos 1975b, 54.[AL]

coh'ni.i. M.L. gen. n. of Cohn; named for Ferdinand Cohn, a German botanist and bacteriologist.

Spheres, 0.5–1.2 μm in diameter, occurring singly, in pairs, and less frequently as tetrads.

Peptidoglycan is of the type L-Lys-Gly$_{5-6}$. Cell wall teichoic acid is composed of glycerol, usually glucose, and low amounts of *N*-acetylglucosamine. Some strains also contain *N*-acetylgalactosamine or, more rarely, ribitol in the teichoic acid.

The major cellular fatty acids include C$_{15Br}$, C$_{20}$, C$_{18}$ and C$_{17}$ components (Durham and Kloos, 1978). The fatty acid composition is similar to that found in the related species *S. xylosus*, but different from that of the related species *S. saprophyticus*.

Human-specific *S. cohnii* (subsp. 1) strains produce colonies which are convex to slightly umbonate (peaked), entire, circular, smooth, glistening or very glossy and opaque, and single colonies may obtain a size of 4–7 mm in diameter on nonselective media used for the propagation of staphylococci; colonies are usually white. By contrast, primate (hosts include major lines of divergence from Tree shrews and *Prosimii* to Man) *S. cohnii* (subsp. 2) strains produce colonies which are slightly larger (5.5–8.0 mm) and less opaque, lower in profile, and have a concentric pigment or gray and gray-white unpigmented ring pattern (Kloos and Wolfshohl, 1983). This difference in morphology and pigment pattern is even more profound with growth on 1% maltose (purple agar base) agar. Many *S. cohnii* (subsp. 2) strains isolated from nonhuman primates have brilliantly pigmented colonies with alternating yellow-orange, gray, gray-white, orange, and gray bands or rings. Some strains produce colonies with only a subtle pigment ring pattern or a yellow-green hue. Human strains of this group (subsp. 2) are usually unpigmented. Growth on agar slants is abundant, smooth, glistening to glossy and slightly translucent to opaque.

Facultative anaerobes: growth is best under aerobic conditions. Growth in the anaerobic portion of a semisolid thioglycolate medium is usually dense or in a gradient from dense to light growth down the tube to the more anaerobic portion of this medium. Some strains may produce only slight growth or individual colonies in the anaerobic portion. Many strains ony ferment glucose weakly and would be misclassified as micrococci on the basis of the O/F-test. Terminal pH in glucose broth, under anaerobic conditions, is 5.3–6.0. Catalase is produced; it is distinctly different immunologically from those of the related species *S. xylosus* and *S. saprophyticus* (Schleifer et al., 1979).

Growth is good at NaCl concentrations up to 10% and is usually poor at 15%. Most strains grow between 15 and 45°C (optimum: 30–40°C).

Chemoorganotrophs: metabolism mainly respiratory. Unsaturated menaquinones (mainly Mk-7) (Collins, 1981) and cytochromes *a* and *b* (Faller et al., 1980) form the electron transport system. Human (subsp. 1) strains produce predominantly the L-lactate isomer; whereas, primate (subsp. 2) strains produce both D- and L-lactate isomers in near equal proportion, by glucose fermentation. In both groups lactic acid is produced only in low amounts. Acetyl methyl carbinol (acetoin) production is variable, most strains produce weak to moderate amounts. Mannitol can be an intracellular product of glucose metabolism, probably via mannitol-1-phosphate (Edwards et al., 1981). FDP-aldolase is of the class I type.

Acid is produced aerobically from glucose, fructose, trehalose and glycerol. Most strains produce acid from maltose and mannose, often slowly, and mannitol. Some strains produce acid from xylitol (usually slowly). Human (subsp. 1) strains rarely produce acid from lactose or galactose, whereas, most primate (subsp. 2) strains produce acid from these carbohydrates (Kloos and Wolfshohl, 1983). Acid is only rarely produced from sucrose. Acid is not produced from rhamnose, turanose, ribose, gentiobiose, cellobiose, melezitose, sorbitol, inositol, salicin, adonitol, dulcitol, arabitol, erythritol, erythrose, raffinose, melibiose, fucose, tagatose, lyxose or sorbose. Human (subsp. 1) and primate (subsp.2) strains do not produce acid from xylose or arabinose; however, a small group of New World monkey *S. cohnii* (subsp. 3) strains, which have been examined recently, do produce weak to moderate amounts of acid from these carbohydrates (Kloos and Wolfshohl, 1983). In other features, the New World monkey (subsp. 3) strains are rather similar to the primate (subsp. 2) strains.

Suitable cell wall glycans (e.g. *Micrococcus luteus* cell walls) may be degraded by a bacteriolytic enzyme(s) (Schleifer and Kloos, 1975b; Varaldo and Satta, 1978).

An organic nitrogen source (from 1–3 different amino acids) and B group vitamins are required for the growth of most human (subsp. 1) strains (Emmett and Kloos, 1975). Most primate (subsp. 2) strains and some of the human group (above) are capable of growing on a medium containing $(NH_4)_2SO_4$ as a sole source of substrate nitrogen. For anaerobic growth, uracil and a fermentative carbon source (e.g., pyruvate) or uracil alone are also required (Evans, 1976). Most primate (subsp. 2) strains were stimulated by the addition of uracil alone.

Most strains do not reduce nitrate. Human (subsp. 1) strains do not usually demonstrate acid and/or alkaline phosphatase activity;

whereas, primate (subsp. 2) strains usually have weak to moderate alkaline phosphatase activity. Some strains may hydrolyze casein or gelatin. Lipases and esterases may be produced. Most of the esterases of human (subsp. 1) strains are different from those of primate (subsp. 2) strains (Zimmerman Kloos, 1976). Coagulase is not produced. Most strains do not demonstrate deoxyribonuclease (DNase) or hemolysin activity. Ammonia is usually not produced or is produced only weakly from arginine (no or weak arginine dihydrolase activity); in primate (subsp. 2) strains it is produced from urea (urease-positive), but is usually not or only weakly produced from urea in human (subsp. 1) strains (see references above in the description of *S. capitis*). Nearly all primate (subsp. 2) strains demonstrate strong β-glucuronidase and β-galactosidase activity; whereas, human (subsp. 1) strains do not exhibit activity for these enzymes (Analytab Products, 1981; Kloos and Wolfshohl, 1982).

Resistant to novobiocin (MIC 3.1–12.5 µg/ml; most strains >6.2 µg/ml) and usually lincomycin (MIC > 1.6 µg/ml). Susceptibility to benzylpenicillin (penicillin G) is partial (MIC 0.1–0.8 µg/ml), characteristic of other novobiocin-resistant species. Susceptibility to erythromycin, tetracycline, chloramphenicol or streptomycin is variable. Susceptible to neomycin, clindamycin, kanamycin, gentamicin and vancomycin. *S. cohnii* strains usually carry plasmids, but their possible relationships to antibiotic resistance have not been determined (Kloos et al., 1981).

DNA/DNA hybridization studies have indicated that human (subsp. 1), primate (subsp. 2), and New World monkey (subsp. 3) *S. cohnii* strains represent different DNA homology or subspecies groups (Kloos and Wolfshohl, 1983). The relative binding (DNA homology) of human (subsp. 1) to primate (subsp. 2) *S. cohnii* DNA in reciprocal reactions at optimal reassociation conditions is 71–88% and at restrictive reassociation conditions is 59–76%; the $\Delta T_{m(e)}$ of intersubspecies heterologous DNA duplexes is 6–7°C. The relative binding of human (subsp. 1) to New World monkey (subsp. 3) *S. cohnii* DNA is 61–63% (optimal) and 22–38% (restrictive), respectively; whereas, the relative binding of primate (subsp. 2) to New World monkey (subsp. 3) *S. cohnii* DNA is 64–71% (optimal) and 26–36% (restrictive), respectively. The $\Delta T_{m(e)}$ of intersubspecies heterologous DNA duplexes including DNA from New World monkey (subsp. 3) strains and the other subspecies is 10–13°C. This extent of divergence suggests a near separate species status for the New World monkey group. Different strains of human *S. cohnii* (subsp. 1) demonstrate over 89% DNA homology under optimal or restrictive conditions. Different strains of primate *S. cohnii* (subsp. 2) demonstrate over 82% DNA homology. DNA relationships between *S. cohnii* and other species are shown in Figure 12.5. *S. cohnii* is more closely related to *S. saprophyticus* and *S. xylosus* than to other species (see comment and references above in the description of *S. saprophyticus*).

Human *S. cohnii* (subsp. 1) is isolated occasionally from the skin of humans; there it produces usually small and transient populations (Kloos and Musselwhite, 1975). The primate *S. cohnii* (subsp. 2) is isolated from humans less frequently, but represents the major *S. cohnii* subspecies group found living on nonhuman primates (Kloos and Wolfshohl, 1983). This subspecies often accounts for a rather large proportion of the staphylococci isolated from lower primates (e.g. Tree shrews, prosimii, and certain New World monkeys). The New World monkey *S. cohnii* (subsp. 3) has been isolated sporadically from the New World Howler (*Alouatta*) and Spider (*Ateles*) monkey. *S. cohnii* is seldom isolated from the skin or products of nonprimate mammals.

S. cohnii (subsp 1) has been isolated from human urinary tract and wound infections, and less frequently from endocarditis or septicemia; it usually accounts for less than 5% of the total coagulase-negative strains isolated from these sources (see references on surveys of infections above in the description of *S. capitis* and *S. saprophyticus*). At present it remains a questionable pathogen. Occurrence of the other *S. cohnii* subspecies in human or animal infections has not been reported.

The mol % G + C of the DNA is 36–38 (T_m).

Type strain: ATCC 29974 (DSM 20260) (Schleifer and Kloos 1975b,

55). (This strain is a member of the human *S. cohnii* (subsp. 1) group. A formal proposal has not been made yet to name the above subspecies.)

11. Staphylococcus xylosus Schleifer and Kloos 1975, 57.[AL]

xy. lo'sus. M.L. adj. *xylosus* xylose.

Spheres, 0.8–1.2 μm in diameter. Cells are arranged predominantly in pairs or as single cells, occasionally as tetrads.

Peptidoglycan is of the type L-Lys-Gly$_{5-6}$. Cell wall teichoic acid contains, like that of *S. saprophyticus*, two polyols: glycerol and ribitol. Ribitol residues are usually substituted by β-linked and glycerol by α-linked *N*-acetylglucosamine (Endl et al., 1983).

The major cellular fatty acids include C_{15Br}, C_{18} C_{20}, and C_{17Br} components; the fatty acid profile is similar to the related species *S. cohnii*, but is different from the related species *S. saprophyticus* (Durham and Kloos, 1978).

Colony morphology is quite variable and includes some forms not observed in other species. Colonies are raised to slightly convex, circular, smooth to rough, dull to glistening and usually opaque, and single colonies may obtain a size of 4–10 mm in diameter on nonselective media used for the propagation of staphylococci. The colony edge may be entire, undulate or crenate. On a soft-agar medium (0.4–0.75% agar) surface, the growth of most strains will fan or spread widely from the site of inoculation, a feature which has been observed only with this species (Kloos and Schleifer, 1981) and *S. gallinarum*. Colony pigment is also variable. Most strains produce colonies which are orange-yellow, yellowish or gray to gray-white with a yellowish tint. Some strains are gray or gray-white (unpigmented). Growth on agar slants is moderate to abundant, variable in consistency, and usually opaque.

Facultative anaerobes: growth is best under aerobic conditions. Growth in the anaerobic portion of a semisolid thioglycolate medium is usually dense or in a gradient from dense to light growth down the tube to the more anaerobic portion of this medium. Some strains may produce only slight growth or individual colonies in the anaerobic portion. Some strains only ferment glucose weakly and would be misclassified as micrococci on the basis of the O/F-test. Terminal pH in glucose broth, under anaerobic conditions, is 4.9–5.6. Catalase is produced; it is distinct, but very closely related immunologically to the catalase of *S. saprophyticus* (Schleifer et al., 1979).

Growth is good at NaCl concentrations up to 10%. All strains grow between 10 and 40°C (optimum: 25–35°C).

Chemoorganotrophs: metabolism mainly respiratory. Unsaturated menaquinones (mainly Mk-7) (Collins, 1981) and cytochromes *a* and *b* (Faller et al., 1980) form the electron transport system. Predominantly the L-lactate isomer is produced by glucose fermentation; however, it is produced only in low amounts. Acetyl methyl carbinol (acetoin) is usually produced in low quantity or is not produced. Mannitol can be an intracellular product of glucose metabolism, probably via mannitol-1-phosphate (Edwards et al., 1981). Some *S. xylosus* strains produce very high levels of intracellular mannitol, compared to other species. Galactose is metabolized via the Leloir pathway (Schleifer et al., 1978). FDP-aldolase is of the class I type. The PEP-PTS system is not involved in the uptake of pentoses and xylitol (Lehmer and Schleifer, 1980). Isomerization of pentoses and dehydration of xylitol is inducible and phosphorylation probably occurs on ketopentoses.

Acid may be produced from a wide variety of carbohydrates. It is produced aerobically from glucose, fructose, mannose, maltose, xylose, mannitol, sucrose and glycerol. Rare strains may not produce acid from sucrose, maltose or mannitol. About 80% of strains produce acid from galactose, arabinose, lactose, trehalose or turanose. Less than 30% of strains produce acid from ribose, rhamnose, gentiobiose, xylitol, sorbitol or inositol. Rare strains may produce acid from salicin. Acid is not produced from cellobiose, melezitose, adonitol, dulcitol, arabitol, erythritol, erythrose, raffinose, melibiose, fucose, tagatose, lyxose or sorbose. Some strains isolated from soybean oil that are rather similar to *S. xylosus* can produce acid from cellobiose (Bucher et al., 1980).

Suitable cell wall glycans (e.g. *Micrococcus luteus* cell walls) may be

degraded by a bacteriolytic enzyme(s); although activity is usually weak (Schleifer and Kloos, 1975b; Varaldo and Satta, 1978).

Most strains do not require an organic nitrogen source, but usually require one or more B group vitamins for growth (Emmett and Kloos, 1975; Emmett, 1976). Most strains are stimulated by the addition of uracil to the basal medium for anaerobic growth (Evans, 1976).

About 80% of strains reduce nitrates and/or show alkaline phosphatase activity. Acid phosphatase may be produced. About 40–70% of strains show lipase activity. Several different esterases are produced; patterns often showed considerable polymorphism (Zimmerman and Kloos, 1976). Most strains do not show protease activity. Coagulase is not produced. Most strains do not exhibit extracellular deoxyribonuclease (DNase) or hemolysin activity. Ammonia is usually not produced from arginine (arginine dihydrolase-negative), but is produced from urea (urease-positive) (see references above in the description of *S. capitis*). Most strains produce strong β-glucosidase, β-glucuronidase, and β-galactosidase activity, features which taken together can separate *S. xylosus* from the related species *S. saprophyticus* and *S. cohnii* (Analytab Products, 1981; Kloos and Wolfshohl, 1982).

Resistant to novobiocin (MIC 1.6-12.5 μg/ml; most strains >3.1 μg/ml) and lincomycin (MIC > 1.6 μg/ml). Susceptibility to benzylpenicillin (penicillin G) is usually partial (MIC 0.1–0.8 μg/ml), characteristic of other novobiocin-resistant species. Susceptibility to erythromycin, tetracycline, chloramphenicol or streptomycin is variable. Multiple resistance is relatively common. Susceptible to neomycin, kanamycin, gentamicin and vancomycin. *S. xylosus* strains usually carry plasmids, but their possible relationships to antibiotic resistance have not been determined (Kloos et al., 1981). An arsenate resistance plasmid was isolated and characterized (Götz et al., 1983b).

DNA/DNA hybridization studies have indicated that selected strains of *S. xylosus* isolated from humans and a variety of animal sources are closely related, demonstrating over 70% relative DNA binding (DNA homology) under restrictive or optimal reassociation conditions (Kloos and Wolfshohl, 1982; unpublished data). However, one group of *S. xylosus* strains isolated from New World (*Alouatta*) monkeys appears to represent a distinctly separate subspecies population, demonstrating 38–39% relative DNA binding in reassociation reactions with DNA from the major group (above), at restrictive conditions. DNA relationships between *S. xylosus* and other species are shown in Figure 12.5. *S. xylosus* is more closely related to *S. saprophyticus* and *S. cohnii* than to other species (Schleifer et al., 1979; Kloos and Wolfshohl, 1982).

S. xylosus is found only occasionally on the skin of humans and other higher primates (e.g. Old World monkeys to Great Apes) (Kloos and Musselwhite, 1975; Kloos, 1980). This species is commonly found on the skin of lower primates and a variety of other mammals. It has been also isolated from their products and some environmental sources (e.g. soil, beach sand, natural waters). On many lower mammals, it may be one of the predominant species present.

S. xylosus has been rarely associated with human or animal infections (see references above on surveys of infections in the descriptions of *S. capitis* and *S. saprophyticus*). Recently, Tselenis-Kotsowilis et al. (1982) described a rather convincing case of acute pyelonephritis in a young woman caused by *S. xylosus*.

The mol% G + C of the DNA is 30–36 (T_m, Bd).

Type strain: ATCC 29971 (DSM 20266) (Schleifer and Kloos, 1975, 59).

12. Staphylococcus simulans Kloos and Schleifer 1975a, 69.[AL]

sim'u.lans. L. part. adj. *simulans* imitating; named for having similarities to certain coagulase-positive staphylococci, including *S. aureus*.

Spheres, 0.8–1.5 μm in diameter. Cells are arranged singly, in pairs, and in tetrads.

Peptidoglycan is of the type L-Lys-Gly$_{5-6}$. Cell wall teichoic acid consists of glycerol *N*-acetylgalactosamine, and *N*-acetylglucosamine. In all strains studied *N*-acetylgalactosamine is α-glycosidically linked to glycerol, whereas *N*-acetylgucosamine is α- or β-linked The lysostaphin-producing strain (NRRL B-2628) designated *S. simulans* biovar *staphylolyticus* (Sloan et al., 1982) has a peptidoglycan of the type L-Lys-Gly$_{2.3}$, L-Ser$_{1.3}$. The difference in cell wall composition is not surprising considering that this biovar produces an enzyme that lyses most other staphylococcal cells, including other strains of *S. simulans* (Robinson et al., 1979).

The major cellular fatty acids include C$_{15Br}$, C$_{18}$, and C$_{20}$ components; the fatty acid profile is distinctly different from those of other species examined (Durham and Kloos, 1978).

Colonies are raised, circular, entire, smooth, slightly glistening, translucent to nearly-opaque and usually gray-white (unpigmented). Single colonies may obtain a size of 5–7.5 mm in diameter on nonselective media used for the propagation of staphylococci. Growth on agar slants is abundant, smooth and usually opaque.

Facultative anaerobes: growth is best under aerobic conditions. Growth in the anaerobic portion of a semisolid thioglycolate medium is rapid and uniformly dense. Terminal pH in glucose broth, under anaerobic conditions, is 4.7–5.0. Catalase is produced; it is distinctly different from those other species, immunologically (Schleifer et al., 1979) and on the basis of electrophoretic mobility (comprising two molecular forms (Zimmerman, 1976).

Growth is good at NaCl concentrations up to 10% and usually poor at 15%. Most strains grow between 15 and 45°C (optimum: 25–40°C).

Chemoorganotrophs: metabolism is respiratory and fermentative. Unsaturated menaquinones (mainly Mk-7) (Collins, 1981) and cytochromes a and b (Faller et al., 1980) form the electron transport system. Some strains produce only the L-lactate isomer; whereas, others produce both D- and L-lactate isomers by glucose fementation. Acetyl methyl carbinol (acetoin) is usually not produced or produced only in small amounts. Mannitol can be an intracellular product of glucose metabolism, probably via mannitol-1-phosphate (Edwards et al., 1981). *S. simulans* produces relatively high levels of intracellular mannitol, compared with other species (with the exception of some strains of *S. xylosus*). FDP-aldolase is of the class I type.

Acid is produced aerobically and anaerobically from glucose and usually mannitol. Acid is produced aerobically from fructose, sucrose and glycerol. Most strains produce acid from lactose, mannose and trehalose. About 25% of strains produce acid from ribose. Most strains do not produce acid from maltose or galactose and those which do, produce acid only weakly. Acid is not produced from rhamnose, xylose, xylitol, arabinose, turanose, gentiobiose, cellobiose, melezitose, sorbitol, inositol, salicin, adonitol, dulcitol, arabitol, erythritol, erythrose, raffinose, melibiose, fucose, tagatose, lyxose, or sorbose.

Suitable cell wall glycans (e.g. *Micrococcus luteus* cell walls) may be degraded by an endo-β-N-acetylglucosaminidase (Wadstrom and Vesterberg, 1971; Iverson and Grov, 1973; Valisena et al., 1982). *S. simulans* biovar *staphylolyticus* also produces a staphylolytic lysostaphin endopeptidase and an acetylmuramic acid-L-alanine amidase, both of which act on cell wall peptidoglycan.

An organic nitrogen source (from 4–7 different amino acids) and B group vitamins are required for growth (Emmett and Kloos, 1975; 1979). *S. simulans*, in contrast to other species examined (with the exception of some strains of *S. epidermidis*), does not require uracil and/or pyruvate for vigorous anaerobic growth (Evans, 1976).

Most strains reduce nitrates and have weak to moderate deoxyribonuclease (DNase) activity. Some strains produce a heat-stable DNase (TNase) (Gramoli and Wilkinson, 1978; Sloan et al., 1982). Most strains produce weak or no alkaline phosphatase activity. About 10–20% of strains produce moderate alkaline phosphatase activity. Lipases and esterases are produced. The esterases are distinctly different from those of other species (Zimmerman and Kloos, 1976). Most strains do not hydrolyze casein or gelatin. Coagulase is not produced. Hemolysins may be produced; they are probably not identical with the classical α-, β-, γ-, or δ-hemolysins of *S. aureus* (Wadstrom et al., 1976). Hemolysin activity on bovine or sheep blood is usually weak; on human blood activity it is somewhat stronger. Ammonia is usually produced from arginine (arginine dihydrolase-positive) and is produced from urea

(urease-positive) (see references above in the description of *S. capitis*). Strong β-galactosidase activity is exhibited and many strains also exhibit β-glucuronidase activity (Analytab Products, 1981; Kloos and Wolfshohl, 1982).

Susceptible to novobiocin (MIC 0.1–0.2 μg/ml). Susceptibility to benzylpenicillin (penicillin G), methicillin, erythromycin, tetracycline, kanamycin, vancomycin and streptomycin is variable. Susceptible to neomycin, gentamicin, chloramphenicol and lincomycin. Some strains carry plasmids, but their possible relationships to antibiotic resistance have not been determined (Kloos et al., 1981).

DNA/DNA hybridization studies have indicated that strains of *S. simulans* from human and other primate sources are closely related, demonstrating over 80% relative DNA binding (DNA homology) under restrictive or optimal reassociation conditions (Kloos and Wolfshohl, 1982; unpublished data). The *S. simulans* biovar *staphylolyticus* demonstrated 100% DNA homology with the *S. simulans* type strain at optimal reassociation conditions and 82% relative DNA binding at restrictive reassociation conditions (Sloan et al., 1982). DNA relationships between *S. simulans* and other species are shown in Figure 12.5. *S. simulans* is more closely related to *S. carnosus* than to other species (Schleifer and Fischer, 1982).

S. simulans is found occasionally on the skin of humans and other primates (Kloos and Musselwhite, 1975; unpublished data). A few individuals carry relatively persistent populations of this species. Its possible presence on other mammals has not been adequately assessed.

S. simulans may be associated with a variety of human infections including urinary tract and wound infections, and less frequently with endocarditis and septicemia (see references above on surveys of infections in the descriptions of *S. capitis* and *S. saprophyticus*). This species accounts usually for less than 8% of the total coagulase-negative staphylococcal strains isolated from these sources. It remains a questionable pathogen.

The mol% G + C of the DNA is 34–38% (T_m).

Type strain: ATCC 27848 (Kloos and Schleifer 1975a, 73).

13. **Staphylococcus carnosus** Schleifer and Fischer 1982, 153.[VP]

car.no′sus. L. adj. *carnosus* pertaining to flesh.

Spheres, 0.5–1.5 μm in diameter, occurring predominantly in pairs and singly.

Peptidoglycan is of the type L-Lys-Gly$_{5-6}$. The cell wall teichoic acid is composed of glycerol, glucose, N-acetylgalactosamine, and small amounts of N-acetylglucosamine (Fischer and Schleifer, 1980).

Colonies are raised, smooth, circular, entire, glistening to glossy and usually gray-white (unpigmented). Single colonies may obtain a size of 5–7 mm in diameter on nonselective media used for the propagation of staphylococci. Some strains produce a brownish or violet pigment which is intensified with age or storage at refrigeration temperatures. Growth on agar slants is abundant, smooth, glistening and opaque.

Facultative anaerobes: growth is best under aerobic conditions. Growth in the anaerobic portion of a semisolid thioglycolate medium is usually moderate to dense. Terminal pH in glucose broth, under anaerobic conditions, is 4.2–4.8. Catalase is produced.

Growth is good at NaCl concentrations up to 15%. All strains grow well at 15–45°C (optimum: 30–40°C).

Chemoorganotrophs: metabolism respiratory and fermentative. Cytochromes a and b are present. Both D- and L-lactate isomers are produced from glucose under anaerobic growth conditions. Acetyl methyl carbinol (acetoin) is produced. FDP-aldolase is of the class I type.

Acid is produced aerobically and anaerobically from glucose. Acid is produced aerobically from fructose, mannose, mannitol and glycerol. Many strains produce acid from trehalose and some produce acid from lactose or galactose. Acid is not produced from sucrose, maltose, xylose, xylitol, arabinose, rhamnose, melezitose, turanose, raffinose, fucose, cellobiose, melibiose or salicin.

Nutritional requirements have not been reported.

Nitrates are reduced: nitrites may be reduced further to ammonia.

Alkaline phosphatase activity is usually moderate. None of the strains produce coagulase or hemolysin activity. A heat-stable deoxyribonuclease (DNase, TNase) may be produced by some strains (M. S. Bergdoll, personal communication). Ammonia is usually produced from arginine (arginine dihydrolase-positive), but not usually from urea (urease-negative). Many strains demonstrate strong β-galactosidase activity.

Susceptible to novobiocin (MIC 0.1–0.2 μg/ml). Susceptible to benzylpenicillin (penicillin G), tetracycline, erythromycin, kanamycin, chloramphenicol and streptomycin. Susceptibility patterns with other antibiotics remain to be determined. Some strains carry plasmids. Several plasmids isolated from various staphylococci and *Bacillus subtilis* can be transferred into *S. carnosus* by protoplast fusion (Götz et al., 1983a).

DNA/DNA hybridization studies have indicated that strains of *S. carnosus* are closely related, demonstrating nearly 100% relative DNA binding (DNA homology) under restrictive or optimal reassociation conditions. DNA relationships between *S. carnosus* and other species are shown in Figure 12.5. *S. carnosus* is more closely related to *S. simulans* than to other species. DNA homology between *S. carnosus* and *S. simulans* at optimal reassociation conditions is 32–40% and at stringent reassociation conditions is 20–28%, which is sufficiently low enough to justify their separate species status.

S. carnosus may be isolated from dry sausage and is used as a starter culture for the production of dry sausage (Fischer and Schleifer, 1980).

The natural habitat of this species has not been adequately determined although it might be expected to be in relation with animals and their products. *S. carnosus* has not been reported as causing infections of humans or animals.

The mol% G + C of the DNA is 35–36% (T_m).

Type strain: DSM20501 (Schleifer and Fischer 1982, 155).

14. Staphylococcus intermedius Hajek 1976, 401.[AL]

in.ter.me′di.us. L. prep. *inter* between, among; L. adj. *medius* middle; M.L. adj. *intermedius* in between, intermediate; intended to indicate that this species possesses some properties of *S. aureus* and *S. epidermidis.* Many strains of this species were designated previously as *S. aureus* biotype E or F (Hajek and Marsalek, 1971).

Spheres, 0.5–1.5 μm in diameter, occurring singly, in pairs, and irregular clusters.

Peptidoglycan is of the type L-Lys-Gly$_{4-5}$, L-Ser$_{0.2-1.0}$. Cell wall teichoic acid is composed of glycerol, *N*-acetylglucosamine and/or glucose (Schleifer et al., 1976; Endl et al., 1983). Strains of *S. intermedius* possess at least two different cell wall teichoic acid antigens. Strains isolated from pigeons contain poly(C) (Endresen et al., 1974) and strains isolated from dogs contain poly(P) (Endresen and Grov, 1976). No protein A is present.

Colonies are slightly convex, circular, entire, smooth, butyrous, glistening, translucent and gray-white (unpigmented). Single colonies may obtain a size of 5–8 mm in diameter on nonselective media used for the propagation of staphylococci. Some strains produce a faint violet pigment which intensifies with age or storage at refrigeration temperatures. On crystal violet agar, growth occurs as white colonies of the positive type E, rather than as yellow or violet colonies typical for *S. aureus* of positive type A/B or C/D (Marsalek and Hajek. 1973). Growth on agar slants is abundant, smooth, glistening and nearly opaque. In broth, growth is diffuse with sediment. Some strains form fine ring pellicles.

Facultative anaerobes: growth is best under aerobic conditions. Growth in the anaerobic portion of a semisolid thioglycolate medium is usually weak to moderate; individual colonies are often visible in the deeper, more anaerobic, portion of the medium. Catalase is produced; it is distinct and not closely related immunologically to the catalases of other species (Rupprecht and Schleifer, 1979).

Growth is good at NaCl concentrations up to 12.5%; it is somewhat poorer at 15%. Most strains grow between 15 and 45°C (optimum: 30–40°C).

Chemoorganotrophs: metabolism respiratory and fermentative. Un-

saturated menaquinones (mainly Mk-7) (Collins, 1981) and cytochromes *a* and *b* (Faller et al., 1980) form the electron transport system. Predominantly the L-lactate isomer is produced by glucose fermentation (Schleifer et al., 1976). *S. intermedius* contains an NAD-dependent L-lactate dehydrogenase (L-LDH) which is specifically activated by fructose-1,6-diphosphate (FDP). The FDP-dependent L-LDH of this species migrates electrophoretically faster than the FDP-dependent L-LDH of *S. epidermidis.* Acetyl methyl carbinol (acetoin) is not usually detected. Galactose and lactose are metabolized via the Leloir pathway and not via the D-tagatose-6-phosphate pathway as is the case for *S. aureus* (Schleifer et al., 1978). Strains tested possess both class I and II FDP-aldolase (Fischer et al., 1982).

Acid is produced aerobically and anaerobically from glucose. Acid is produced aerobically from fructose, trehalose, galactose, mannose, sucrose and glycerol. Most strains produce acid from lactose. Most strains produce weak to moderate acid from mannitol and very slow, weak acid from maltose (Phillips and Kloos, 1981). Rare strains fail to produce acid from trehalose. Acid is not produced from xylose, xylitol, arabinose, melezitose, salicin, cellobiose, raffinose, fucose, rhamnose or sorbitol.

Esculin is not hydrolyzed.

Strains require an organic nitrogen source (3–6 different amino acids) and B group vitamins for growth (Tschäpe, 1973).

Nitrates are reduced and alkaline and acid phosphatases are produced. Casein and gelatin are usually hydrolyzed. Lipases may be produced. Esterases are produced; their patterns are complex (polymorphic) and their electrophoretic migrations are slower than the major esterase of *S. aureus* (Schleifer et al., 1976; Zimmerman and Kloos, 1976). Most carnivora and equine strains produce ammonia from arginine (arginine dihydrolase-positive). Pigeon strains do not usually show arginine dihydrolase activity. Ammonia is usually produced from urea (urease-positive).

Coagulases are produced. Rabbit and bovine plasmas are coagulated; most strains isolated from pigeons also coagulate human plasma. Clumping factor (CF, bound coagulase) may be produced by some strains. See discussion and references above in the description of *S. aureus* for coagulase testing procedures. Fibrinolysin is not produced.

Beta- and δ-hemolysins are produced; most strains isolated from pigeons also produce an α-hemolysin. β-D-Galactosidase activity is demonstrated and is usually strong. Heat-stable (TNase) and heat-labile nucleases are produced. TNase from *S. intermedius* can be distinguished from the TNases of *S. aureus* and *S. hyicus* by a seroinhibition test (Lachica et al., 1979).

S. intermedius strains are resistant to the basic sets of human and bovine phages of *S. aureus,* but may be susceptible to canine phages.

Susceptible to novobiocin (MIC < 0.6 μg/ml), cephaloridine, lincomycin, rifampin, spiramycin, chloramphenicol and vancomycin. Susceptibility to benzylpenicillin (penicillin G), tetracycline, kanamycin and erythromycin is variable. *S. intermedius* strains may carry plasmids, but their possible relationships to antibiotic resistance have not been adequately determined (Kloos et al., 1979; 1981).

DNA/DNA hybridization studies have indicated that carnivora (together with equine) strains and pigeon *S. intermedius* strains represent different DNA homology or subspecies groups (Meyer and Schleifer, 1978). The relative binding (DNA homology) of carnivora to pigeon *S. intermedius* DNA in reciprocal reactions at optimal reassociation conditions is 75–82% and at restrictive reassociation conditions is 50–65%. DNA relationships between *S. intermedius* and other species are shown in Figure 12.5. *S. intermedius* is not closely related to other staphylococci.

S. intermedius is a common, and often predominant, staphylococcal species inhabiting the nasal membranes (anterior nares, nasopharynx) and skin of Carnivora (e.g. dogs, mink, raccoons, foxes). It has also been isolated from the anterior nares of horses and pigeons. It is rarely isolated from humans or other primates. People keeping pet Carnivora (e.g. dogs) may, on occasion, carry small transient populations of this species.

This species has been implicated in a variety of infections in dogs, such as otitis externa (Devriese and Oeding, 1976), wounds, pyoderma and mastitis (Phillips and Kloos, 1981). It is a potential pathogen for animals.

The mol% G + C of the DNA is 31–36 (T_m).

Type strain: ATCC 29663 (CCM 5739, NCTC 11048) (Hajek 1976, 406). (This strain is a member of the pigeon subspecies group. A formal proposal has not been made yet to name the above subspecies.)

15. **Staphylococcus hyicus** (Sompolinsky 1953) Devriese, Hajek, Oeding, Meyer and Schleifer 1978, 482.[AL] (*Micrococcus hyicus* Sompolinsky 1953, 307.)

hy′i.cus. Gr. noun *hyos* hog, pig. M. L. adj. *hyicus.*

Currently, two subspecies are recognized: *S. hyicus* subsp. *hyicus* and *S. hyicus* subsp. *chromogenes* (chro.mo′ge.nes. Gr. n. *chroma* color; Gr. v. *gennaio* to produce, M.L. adj. *chromogenes* producing color).

Spheres, 0.6–1.3 μm in diameter. Cells are arranged in pairs, tetrads, and clusters; occasionally they occur singly.

Peptidoglycan is of the type L-Lys-Gly$_{4-5}$, L-Ser$_{0-0.3}$. The cell wall teichoic acid is composed of glycerol and *N*-acetylglucosamine. The structure of the teichoic acid was determined for the type strain of *S. hyicus* subsp. *hyicus* and is rather complex, like the teichoic acid found in *S. sciuri* (Endl et al., 1983). It is a poly (-glycerolphosphate-glycosylphosphate) teichoic acid. The glycosyl residues consist of mono-, di- or trisaccharides of *N*-acetylglucosamine. The cell wall teichoic acid may act as a species-specific antigen designated A1 (Devriese and Oeding, 1975). Some strains (especially from porcine sources) of *S. hyicus* subsp. *hyicus* produce protein A-like material, although it has a lower molecular weight and more acid isoelectric point than the protein A of *S. aureus* (Muller et al., 1980).

Colonies are low convex, glistening, circular, entire and opaque and single colonies obtain a size of 4–7 mm in diameter on nonselective media used for the propagation of staphylococci. Pigments are not produced in strains of *S. hyicus* subsp. *hyicus.* Strains of *S. hyicus* subsp. *chromogenes* produce usually yellow, orange or cream-colored pigment. Rare strains of this subspecies may be unpigmented. Growth on agar slants is abundant, smooth, glistening and opaque. In broth, growth is uniformly turbid with a sediment.

Facultative anaerobes: growth is best under aerobic conditions. Growth in the anaerobic portion of a semisolid thioglycolate medium is usually moderate. Terminal pH in glucose broth, under anaerobic conditions, is 4.8–5.3. Catalase is produced; it is distinct and not closely related immunologically to the catalases of other staphylococci (Rupprecht and Schleifer, 1979).

Growth is good at NaCl concentrations up to 10%. Most strains grow between 15 and 40°C (optimum 30–35°C).

Chemoorganotrophs: metabolism respiratory and fermentative; some strains mainly respiratory. Cytochromes *a* and *b* are present (Faller et al., 1980). Predominantly, the L-lactate isomer is produced by glucose fermentation. Acetyl methyl carbinol (acetoin) is not produced. Galactose is metabolized via the D-tagatose-6-phosphate pathway (Schleifer, 1981). Strains tested possess both class I and II FDP-aldolases (Fischer et al., 1982).

Acid is produced aerobically and anaerobically from glucose. Acid is usually produced aerobically from sucrose, mannose, ribose, trehalose, glycerol, fructose, lactose and galactose. Most strains of *S. hyicus* subsp. *hyicus* do not produce acid from mannitol; whereas, about 58% of *S. hyicus* subsp. *chromogenes* strains produce weak to moderate acid from this carbohydrate. *S. hyicus* subsp. *hyicus* does not produce acid from turanose or maltose; whereas, about half of *S. hyicus* subsp. *chromogenes* strains produce weak acid from one or both of these carbohydrates. Acid is not produced from xylose, xylitol, arabinose, rhamnose, melezitose, salicin, cellobiose, tagatose, sorbitol, sorbose, raffinose, melibiose, lyxose, gentiobiose, fucose, dulcitol, arabitol, adonitol or amygdalin.

Esculin is not hydrolyzed. Hippurate is hydrolyzed. Hyaluronidase is usually produced by *S. hyicus* subsp. *hyicus*, but not by *S. hyicus* subsp. *chromogenes.*

Nutritional requirements have not been determined.

Nitrates are reduced beyond nitrite. Phosphatases, proteases, and lipases are produced. Tween 80 and 85 are hydrolyzed by *S. hyicus* subsp. *hyicus*, but not by *S. hyicus* subsp. *chromogenes.*

Ammonia is usually produced from arginine (arginine dihydrolase-positive) and urea (urease-positive).

Coagulase is produced by some (24–56%) strains of *S. hyicus* subsp. *hyicus* although it is often weak in activity and may require 18–24 h for detection in the tube test (with rabbit plasma). Less than 10% of strains produce a positive coagulase reaction within 4 h. *S. hyicus* subsp. *chromogenes* does not demonstrate coagulase activity in the above test. Clumping factor (CF, bound coagulase) is not produced. See discussion and references above in the description of *S. aureus* for coagulase testing procedures. Fibrinolysin may be produced by some strains of *S. hyicus* subsp. *hyicus.*

Hemolysins (e.g., α, β, and δ) are not produced. Deoxyribonuclease (DNase, including heat-stable TNase) activity is moderate to strong in *S. hyicus* subsp. *hyicus*, but weak or negative in *S. hyicus* subsp. *chromogenes.* Many strains of *S. hyicus* subsp. *hyicus* produce β-glucosidase and/or β-glucuronidase activity; β-glucuronidase is not produced by the coagulase-positive species *S. aureus* or *S. intermedius* (Analytab Products, 1981; Kloos and Wolfshohl, 1982). Using similar testing procedures, α-galactosidase activity cannot be detected in *S. hyicus* or *S. aureus*, but is detected in *S. intermedius.*

S. hyicus strains are not susceptible to phages of the international *S. aureus* typing set or those used for typing bovine *S. aureus* strains.

Susceptible to novobiocin (MIC 0.05–0.1 μg/ml), neomycin, cloxacillin, rifamycin, trimethoprim and bacitracin. Susceptibility to benzylpenicillin (penicillin G), erythromycin, tetracycline, chloramphenicol, lincomycin and streptomycin is variable. *S. hyicus* strains may carry plasmids, but their possible relationships to antibiotic resistance have not been determined (Kloos et al., 1981).

DNA/DNA hybridization studies have indicated that *S. hyicus* subsp. *hyicus, S. hyicus* subsp. *chromogenes* and some bovine *S. hyicus* strains (the nucleus of possibly another subspecies as yet unnamed) represent different DNA homology groups (Devriese et al., 1978; Phillips and Kloos, 1981). The relative binding (DNA homology) of *S. hyicus* subsp. *hyicus* to *S. hyicus* subsp. *chromogenes* DNA in reciprocal reactions at restrictive reassociation conditions is 32–55%; whereas, within each of these subspecies, strains demonstrate over 90% DNA homology. The DNA homology between some bovine *S. hyicus* strains (A$_4$ and 451) and *S. hyicus* subsp. *hyicus*, at restrictive reassociation conditions, is 30–52% and with *S. hyicus* subsp. *chromogenes* is 30–47%, indicating a separate subspecies status. DNA relationships between *S. hyicus* and other species are shown in Figure 12.5. This species is not closely related to other staphylococci.

S. hyicus subsp. *hyicus* occurs commonly on the skin of pigs (Devriese, 1977) and less frequently on the skin or in milk of cattle (Brown et al., 1967). It may also be isolated from poultry (Sato et al., 1972). Porcine strains are opportunistic pathogens and may be implicated in infectious exudative epidermitis (EE, impetigo contagiosa suis) (Sompolinsky, 1950, 1953; Underdahl et al., 1965; Devriese and Oeding, 1975). They also may be implicated in septic polyarthritis of pigs (Phillips et al., 1980). This subspecies is only rarely isolated from the milk of cows suffering from mastitis (Brown et al., 1967; Phillips and Kloos, 1981). *S. hyicus* subsp. *chromogenes* may be isolated from the skin of pigs and cows. It is commonly isolated from the milk of cows suffering from mastitis, although its role as an etiologic agent is not yet clear.

The mol% G + C of the DNA is 33–34 (T_m).

Type strain of *S. hyicus* subsp. *hyicus:* ATCC 1124 (NCTC 10350, CCM 2368) (Devriese, Hajek, Oeding, Meyer and Schleifer 1978, 488).

Type strain of *S. hyicus* subps. *chromogenes:* NCTC 10530 (Devriese, Hajek, Oeding, Meyer and Schleifer 1978, 488).

16. **Staphylococcus caseolyticus** Schleifer, Kilpper-Bälz, Fischer, Faller and Endl 1982, 19.[VP]

ca.se.o.ly'ti.cus. L. n. *caseus* cheese; Gr. adj. *lyticus* able to loose; M.L. adj. *caseolyticus* casein-dissolving.

Spheres, 0.8–1.2 µm in diameter, occurring predominantly in pairs and clusters, occasionally singly.

Peptidoglycan is the type L-Lys-Gly$_4$, L-Ser$_{0.8-1.3}$. Cell wall teichoic acid is of the poly (glycosylphosphate) type (Endl et al., 1983). It is composed of *N*-acetylglucosamine or *N*-acetylgalactosamine plus glucose.

Colonies are slightly convex, circular, usually entire, smooth, glistening and opaque, and single colonies may obtain a size of 2–5 mm in diameter on nonselective media used for the propagation of staphylococci. Colonial pigment may be somewhat variable. Colonies are white or pale cream with a yellow ring around the edge.

Facultative anaerobes: growth is much better under aerobic conditions. Growth in the anaerobic portion of a semisolid thioglycolate medium is usually very weak. Strains of this species may be misclassified as micrococci on the basis of the O/F-test. The two strains studied to date were originally classified as *Micrococcus caseolyticus* (ATCC 13548) and *M. varians* (ATCC 29750), respectively, (Evans, 1916; Hucker, 1948; Archibald et al., 1968b). Catalase is produced.

Chemoorganotrophs: metabolism mainly respiratory. Cytochromes aa_3, *b*, and *c* are present. Therefore, modified benzidine and oxidase tests are positive. Acetyl methyl carbinol (acetoin) is not produced. A class II FDP-aldolase is produced. A small amount of the L-lactate isomer is produced from glucose under anaerobic growth conditions.

Acid is produced aerobically from glucose, trehalose, fructose, galactose, lactose, ribose and maltose. One of the two strains studied also produces acid from sucrose. Acid is not produced from mannose, mannitol, xylose, xylitol or turanose.

Nutritional requirements have not been determined.

Nitrates are reduced and casein is hydrolyzed. Coagulases and hemolysin activity are not produced.

Susceptible to novobiocin. Susceptibilities to other antibiotics have not been determined.

DNA-DNA hybridization studies have indicated that the two strains of *S. caseolyticus* are closely related, demonstrating over 80% relative DNA binding (DNA homology) under optimal reassociation conditions. DNA relationships between *S. caseolyticus* and other species are shown in Figure 12.5. This species is not closely related to other staphylococci. DNA/rRNA studies have indicated that *S. caseolyticus* is related to various staphylococci on the basis of conserved cistrons coding for RNA.

S. caseolyticus may be found in milk and dairy products. Its natural habitat has not been determined clearly.

It has not been found associated with human or animal infections.

The mol% G + C of the DNA is 38–39 (T_m).

Type strain: ATCC 13548 (Schleifer, Kilpper-Bälz, Fischer, Faller and Endl 1982, 19).

17. **Staphylococcus sciuri** Kloos, Schleifer and Smith 1976, 23.[AL]

sci'ur.i. L. masc. n. *Sciurus* generic name of a squirrel on whose skin this species is commonly found in large populations; L. gen. n. *sciuri* of the squirrel.

Spheres, 0.7–1.2 µm in diameter, occurring singly, in pairs and tetrads.

Peptidoglycan is of the type L-Lys-L-Ala-Gly$_4$. In some strains, small amounts of glycine are replaced by L-serine. Cell wall teichoic acid is composed of glycerol and *N*-acetylglucosamine. It is a poly (-glycerol-phosphate-glycosylphosphate) teichoic acid much like that found in *S. hyicus* (Endl et al., 1983). Some unusual strains may contain ribitol instead of glycerol or glucose in place of *N*-acetylglucosamine.

Colonies are raised with a slight elevated center, smooth, glistening, circular, usually slight to moderate undulate and opaque, and if isolated may obtain a large size of 7–11 mm in diameter on nonselective media used for the propagation of staphylococci. Colonies often develop numerous pin-point depressions near the edge within 5–7 days. Most strains produce colonies that are gray-white with a yellowish or cream-colored tint toward the center; some strains produce yellowish or yellow-gray colonies; fewer produce gray-white (unpigmented) colonies. Pigment is intensified considerably during growth at low temperatures (15–20°C). Compared with most other species, colonies of *S. sciuri* grow very large (8–12 mm in diameter) on 1% maltose agar (purple agar base). Growth on agar slants is moderate to abundant, smooth, glistening and opaque.

Facultative anaerobes: growth is much better under aerobic conditions. Growth in the anaerobic portion of a semisolid thioglycolate medium is variable, but is usually in a gradient from moderate to light growth down the tube to the more anaerobic portion of this medium. Some strains may produce moderate to dense growth; whereas, others may even fail to produce detectable growth in the anaerobic portion. Many strains only ferment glucose weakly and would be misclassified as micrococci on the basis of the O/F-test. Terminal pH in glucose broth, under anaerobic conditions, is 5.7–6.0. Catalase is produced; it is distinct and not closely related immunologically to catalases of other species tested (Schleifer et al., 1979).

Growth is good at NaCl concentrations up to 10% and usually poor at 15%. Most strains grow between 15 and 40°C (optimum: 25–35°C).

Chemoorganotrophs: metabolism mainly respiratory. Unsaturated menaquinones (mainly Mk-6) and cytochromes aa_3, *b* and *c* form the electron transport system. Therefore, modified benzidine and oxidase tests are positive. Low amounts of the L-lactate isomer are produced by glucose fermentation. Acetyl methyl carbinol (acetoin) is not produced or is rarely produced in small amounts. FDP-aldolase is of the class I type.

Acid is produced aerobically from a wide variety of carbohydrates, including glucose, fructose, ribose, fucose, cellobiose, mannitol, galactose, sucrose and glycerol. Most strains also produce acid from trehalose, β-gentiobiose, sorbitol and salicin, and very small to moderate amounts of acid from maltose. Some strains produce acid from melezitose, arabinose, and/or rhamnose and small amounts of acid from mannose. Less than 5% of strains produce small amounts of acid from xylose and less than 10% of strains produce acid from lactose. No acid is produced from xylitol, arabitol, dulcitol, erythritol, erythrose, tagatose, melibiose, lyxose or sorbose.

Esculin is hydrolyzed.

An organic nitrogen source is not usually required for growth (Emmett, 1976). With rare exceptions strains can grow with $(NH_4)_2SO_4$ as a sole source of substrate nitrogen. One or more B group vitamins may be required by some strains; for most strains vitamin supplementation has a slight stimulatory effect.

Nitrates are reduced and nearly all strains demonstrate weak to moderate deoxyribonuclease (DNase) and alkaline phosphatase activity. Casein and gelatin are hydrolyzed. Lipase activity is very weak or not detected. Elastase is not produced. Most strains do not hydrolyze Tween 80. An esterase is produced which is electrophoretically highly mobile (Zimmerman and Kloos, 1976). Coagulases and hemolysin activity are not produced. Ammonia is not produced from arginine (arginine dihydrolase-negative) or urea (urease-negative) (see references above in the description of *S. capitis*). This species produces β-glucosidase activity, but not β-glucuronidase or β-galactosidase activity (Analytab Products, 1981; Kloos and Wolfshohl, 1982).

Resistant to novobiocin (MIC 1.6–12.5 µg/ml) and lincomycin (MIC > 1.6 µg/ml). Susceptibility to benzylpenicillin (penicillin G) is partial (MIC 0.4–6.2 µg/ml), characteristic of other novobiocin-resistant species. Susceptibility to tetracycline and erythromycin is variable. Susceptible to vancomycin, gentamicin, kanamycin, chloramphenicol, neomycin and bacitracin. *S. sciuri* strains may occasionally carry plasmids, but their possible relationships to antibiotic resistance have not been determined (Kloos et al., 1981).

Currently, two subspecies are recognized: *S. sciuri* subsp. *sciuri* and *S. sciuri* subsp. *lentus* (len'tus. L. adj. *lentus* slow; pertaining to slow growth). However, the latter subspecies would be more appropriately elevated to a separate species rank, based on recent information of its DNA relatedness to *S. sciuri* subsp. *sciuri* (Meyer, 1979; Kloos, 1980).

A proposal is being made to change the name and rank of S. sciuri subsp. lentus to S. lentus as recommended previously (Schleifer et al., 1983). The relative binding (DNA homology) of S. sciuri subsp. sciuri (S. sciuri) DNA to S. sciuri subsp. lentus (S. lentus) DNA is less than 50% at optimal reassociation conditions and 20% or less at restrictive reassociation conditions. DNA relationships between S. sciuri and S. lentus and other species are shown in Figure 12.5. These two species are more closely related to one another than to other species. Although they are not closely related to other staphylococci, DNA/rRNA hybridization studies have indicated a definite relationship to the genus Staphylococcus (Kilpper-Bälz and Schleifer, 1981b). This has been confirmed by comparative sequence analysis of 16S rRNA (Ludwig et al., 1981).

S. lentus shares many phenotypic characters with S. sciuri, but is different in the following respects. Colonies of S. lentus are small (2.2–3.5 mm in diameter), slow growing, and have a glistening to wet-glossy appearance. They are more convex and are usually white, gray-white, cream or yellow. Most strains produce acid aerobically from lactose, arabinose, raffinose and mannose; some strains produce acid from xylose and/or melibiose. Phosphatase, caseinolytic and gelatinase activities are usually weaker than those found in S. sciuri strains. Most strains produce very weak or no detectable anaerobic growth in a thioglycolate medium.

S. sciuri is commonly isolated from the skin of rodents and somewhat less frequently from the skin of ungulates, carnivora and marsupials; it also may be isolated occasionally from other mammals and environmental sources (e.g. soil, sand and natural waters) (Kloos, 1980; Kloos and Schleifer, 1981). This species has been very rarely isolated from humans and occasionally isolated from nonhuman primates. S. lentus is commonly isolated from the skin and udders of goats and sheep.

Both species are apparently only rarely pathogenic.

The mol% G + C of the DNA is 30–36 (T_m).

Type strain of S. sciuri subsp. sciuri: ATCC 29062 (Kloos, Schleifer and Smith 1976, 29).

Type strain of S. sciuri subsp. lentus: ATCC 29070 (Kloos, Schleifer and Smith 1976, 30).

18. Staphylococcus gallinarum Devriese, Poutrel, Kilpper-Bälz and Schleifer 1983, 481.[VP]

gal.li.na′rum. L. fem. n. gallina hen; L. fem. gen. pl. n. gallinarum of hens.

Spheres, 0.5–1.8 μm in diameter. Cells occur singly, in pairs, or in small groups or short chains.

Peptidoglycan is of the type L-Lys-Gly$_{4-5}$, L-Ser$_{0.7-1.0}$. Cell wall teichoic acid is composed of glycerol, glucose and small amounts of N-acetylglucosamine.

Colonies are 10–15 mm in diameter on nonselective media used for the propagation of staphylacocci. However, growth is highly variable and may demonstrate outgrowths measuring up to 40 mm at times. Colony edge is lobate or crenate, only small colony types have entire edges. The colonies are opaque and flat and have dry surfaces. Small colony types have smooth glistening surfaces. Most strains produce yellow or yellowish colonies, a few strains are unpigmented.

Facultative anaerobes: growth is better under aerobic conditions. Catalase is produced.

All strains grow well between 25 and 42°C, but colonies are smaller after 24 or 48 h at 25°C than at higher temperatures.

Chemoorganotrophs: metabolism respiratory and fermentative. Predominantly L-lactic acid is produced by glucose fermentation. Acetyl methyl carbinol (acetoin) is not produced. The FDP-aldolase is of the class I type.

Acid is produced aerobically and anerobically from glucose. Most strains produce acid aerobically from amygdalin, arabinose, arbutin, cellobiose, fructose, fucose (weak reaction), galactose, gentiobiose, glucitol, glycerol, maltose, mannitol, mannose, melezitose, melibiose, ribose, salicin, sucrose, trehalose, turanose and xylose. About 70% of the strains produce acid from lactose and xylitol. Acid is usually not produced from rhamnose.

Nutritional requirements have not been determined. All strains reduce nitrates and produce alkaline phosphatase. Heat-stable deoxyribonuclease is not produced. Esculin and urea are hydrolyzed. Proteases are produced. Hyaluronidase, staphylokinase and lipases are not produced. Coagulase and clumping factor are not produced. Some strains show weak hemolysis on sheep blood agar.

All strains are resistant to novobiocin (MIC, 4–32 μg/ml) and lysozyme. The lysostaphin MICs are 34–64 μg/ml. The levels of susceptibility to the following antibiotics are higher than the levels normally found for S. aureus: penicillin G (MIC, 0.1–0.5 μg/ml), cloxacillin (MIC, 1–2 μg/ml), erythromycin (MIC for most strains, 0.5–1 μg/ml) and lincomycin (MIC for all strains except one, 2–8 μg/ml). All strains are susceptible to neomycin (MIC, 0.06–0.25 μg/ml), sulfonamides (sodium sulfamethoxazole MIC, 4–32 μg/ml) and chloramphenicol (MIC 4–16 μg/ml). Most strains show resistance to tetracyclines (tetracycline hydrochloride MIC, 32–128 μg/ml) and most are inhibited at streptomycin concentrations of 2–4 μg/ml.

DNA/DNA hydridization studies indicated that the strains of S. gallinarum are closely related, whereas the levels of DNA homology with other staphylococci are rather low (10–25%).

S. gallinarum is easily differentiated from the novobiocin-resistant organisms Staphylococcus saprophyticus, Staphylococcus cohnii, and Staphylococcus xylosus by its wide range of positive carbohydrate reactions. None of these other species produces acid from fucose, cellobiose, melezitose and raffinose, but S. gallinarum does. Also, the very irregular colony morphology with strongly crenate edges is a useful diagnostic trait; this characteristic is found only in S. gallinarum and certain S. xylosus strains. In particular, strains which produce only colonies with entire edges can be confused with Staphylococcus lentus. The ability of these organisms to grow anaerobically in thioglycolate cultures, their positive urease reactions and their negative oxidase reactions are useful distinguishing characteristics. However, it is important to use the highly sensitive oxidase test described by Faller and Schleifer (1981).

The major habitat of S. gallinarum is the skin of poultry. The mol% G + C of the DNA is 34–35 (T_m).

Type strain: CCM 3572 (Devriese, Poutrel, Kilpper-Bälz and Schleifer, 1983, 482).

19. Staphylococcus caprae Devriese, Poutrel, Kilpper-Bälz and Schleifer 1983, 483.[VP]

ca′prae. L. fem. n. capra goat, L. fem. gen. n. caprae of a goat.

Spheres, 0.8–1.2 μm in diameter. Cells occur singly, in pairs, in short chains, or in small groups.

Peptidoglycan is of the type L-Lys-Gly$_{4-5}$, L-Ser$_{0.8-1.2}$. Cell wall teichoic acid is composed of glycerol, N-acetylglucosamine and small amounts of glucose.

Colonies are circular with entire edges, low convex, opaque, nonpigmented and glistening. Colony diameters reach 5–8 mm on nonselective media used in the propagation of staphylococci.

Facultative anaerobes. Catalase is produced.

All strains grow between 25 and 45°C, but growth is much slower at 25°C.

Chemoorganotrophs: metabolism respiratory and fermentative. Predominantly L-lactic acid is produced by glucose fermentation. Acetyl methyl carbinol (acetoin) is produced from glucose and pyruvate. The FDP-aldolase is of the class I type.

Acid is produced aerobically and anaerobically from glucose. All strains produce acid aerobically from galactose, glycerol, lactose, mannose and trehalose. Most strains produce acid from maltose. Acid is not produced from amygdalin, arabinose, arbutin, cellobiose, fructose, fucose, β-gentiobiose, mannitol, melezitose, melibiose, raffinose, rhamnose, ribose, salicin, glucitol, sucrose, turanose, xylitol, or xylose.

Nutritional requirements have not been determined. All strains reduce nitrates, produce alkaline phosphatase and hydrolyze arginine. Most strains are urease-positive. All strains are negative for the following reactions: coagulase with rabbit plasma, clumping factor with rabbit and human plasma, oxidase, protease on casein and gelatin, hyaluronidase, staphylokinase, hydrolysis of polysorbate (Tween 80) and esculin hydrolysis. All strains are negative in the heat-stable deoxyribonuclease (DNase) test but produce heat-labile DNase.

Weak hemolysis is observed irregularly after overnight incubation on sheep blood agar. After 2 days at 37°C all strains show a narrow clear hemolysis zone surrounded by a broad zone of faint discoloration. These broad zones become more visible after the plates are kept at 4°C in a refrigerator and are reminiscent of the hemolytic effects of β-hemolysin (sphingomyelinase C), which is produced by many animal *S. aureus* strains and most *Staphylococcus intermedius* strains. *S. caprae* zones differ from zones produced by *S. aureus* and *S. intermedius* in their delayed appearance and in their diffuse edges. On human blood all *S. caprae* strains produce only a narrow zone of clear hemolysis, which is visible after incubation for 1 day at 37°C.

All strains are susceptible to novobiocin (MIC, 0.1 μg/ml), penicillin G (MIC, ≤ 0.06 μg/ml), tetracycline hydrochloride (MIC, ≤ 0.5 μg/ml),

erythromycin (MIC, ≤ 0.5 μg/ml), lincomycin (MIC, ≤ 1 μg/ml), streptomycin (MIC, ≤ 16 μg/ml), neomycin (MIC, ≤ 1 μg/ml), furazolidone (MIC, ≤ 8 μg/ml), and sulfamethoxazole (MIC, 4–8 μg/ml).

The strains are resistant to lysozyme (MIC, > 800 μg/ml). The lysostaphin MIC is 128–256 μg/ml.

The results of DNA/DNA hybridization studies indicate a phylogenetic relationship to other staphylococci, in particular a relationship to *S. capitis* and *S. epidermidis*.

S. caprae is differentiated from other novobiocin-sensitive, coagulase-negative staphylococci by its unique fermentation pattern. In particular, the negative mannitol, fructose and sucrose reaction is useful in the identification of *S. caprae*.

S. caprae has been isolated from goat milk.

The mol% G + C of the DNA is 35–37 (T_m).

Type strain: CCM 3573 (Devriese, Poutrel, Kilpper-Bälz and Schleifer 1983, 484).

FAMILY II. **DEINOCOCCACEAE** BROOKS AND MURRAY 1981, 356[VP]

R. G. E. MURRAY

Dei.no.coc.ca′ce.ae. M.L. masc. n. *Deinococcus* type genus of the family; *-aceae* ending to denote family; M.L. fem. pl. n. *Deinococcaceae* the *Deinococcus* family.

Cells usually **spherical** (showing dividing **pairs or tetrads**) but may be elongate (showing dividing pairs). Nonmotile. Resting stages are not produced. **Gram-positive. Cell walls show a complex electronmicroscopic profile** with several layers of distinct components.

Aerobic. Chemoorganotrophic. Metabolism is respiratory. Generally **inactive towards sugars.** Nutritional requirements are variable. **Catalase-positive.**

Lipid composition includes **palmitoleate** (16:1), absence or slight amounts of branched chain fatty acids and an **unusual spectrum of polar lipids** without representation of phosphatidylglycerol.

The mol % G + C of the DNA is in the range of 60–70 (T_m). Type genus: *Deinococcus* Brooks and Murray, 1981, 354.

Further Descriptive Information

The Family contains only a single genus and the descriptive details are given under *Deinococcus*. Despite a superficial morphological and physiological resemblance to *Micrococcus roseus*, in particular, it was found (Stackebrandt and Woese, 1979; Brooks et al., 1980; Brooks and Murray, 1981) that the included species could not be related to the *Micrococcaceae*. The strongest evidence was provided by the oligonucleotide catalogs derived from the 16S rRNA, which was distinctive and the S_{AB} values (0.17–0.29) indicated that these few species form a unique phylogenetic group of bacteria (Stackebrandt and Woese, 1981). These organisms could not be aligned with any one of the accepted hierarchical groups and so they were included in a new family (Brooks and Murray, 1981).

Taxonomic Comments

The main reason for maintaining the family *Deinococcaceae* is to draw attention to a distinctive genus and to the need to search for relatives and the ecological niches that they occupy. It is maintained among the Gram-positive bacteria and near to *Micrococcaceae* for

determinative reasons: the known members are Gram-positive and are cocci.

It is probable that the species included in the family are, in reality, more closely allied to Gram-negative bacteria despite the fact that they are indubitably Gram-positive. The profile of the cell wall of *Deinococcus* sp. includes an outer membrane, which contains lipids and proteins (Thompson et al., 1980; Thompson and Murray, 1981), and this must be considered uncharacteristic of any veritable Gram-positive species. Furthermore, Brooks et al., (1980) confirm the observation of Girard (1971) that palmitoleate (16:1) is predominant and extend it to all the species (particularly *D. radiodurans*, *D. radiophilus* and *D. proteolyticus* in which it forms more than 60% of the fatty acids), while no branched chain fatty acids are present, except as a minor component in *D. radiopugnans*. This fatty acid profile is unlike that of most Gram-positive bacteria, including the *Micrococcaceae*, and resembles that of Gram-negative bacteria in showing a high proportion of even numbered, straight chain, saturated and unsaturated acids (Lechevalier 1977; Shaw, 1974; Brooks et al., 1980).

So one could doubt if all the relatives that might be found to add to *Deinococcaceae* would have to be Gram-positive. They might exhibit shapes other than spherical (see *Neisseria* for a similar problem). They might show considerable variation from the extraordinary pattern of polar lipids and the lack of phosphatidylglycerol that characterizes the species of *Deinococcus*. They might not all be resistant to ionizing or UV-radiation because sensitive mutants of *D. radiodurans* have been isolated (Sweet and Moseley, 1974).

It is most unlikely that a family of great phylogenetic age would have no relatives and contain only a single genus. There may be unrecognized relatives among described bacteria as well as undescribed and at large in nature. The characters available for recognition are limited and are not routine procedures; also it is not yet certain how many of the peculiarities of *Deinococcus* would be likely to be descriptive of the broader related group.

Genus I. **Deinococcus** Brooks and Murray, 1981, 354[VP]

R. G. E. MURRAY AND B. W. BROOKS

Dei.no.coc′ cus. Gr. adj. *deinos* strange or unusual; Gr. n. *coccus* a grain or berry: M.L. masc. n. *Deinococcus* unusual coccus.

Cells are **spherical**, 0.5–3.5 μm in diameter in **pairs or tetrads**, and **appear large** in relation to other cocci. **Several distinct cell**

wall layers are visible by electron microscopy of sections **and an outer membrane** is included. **Gram-positive**. No resting stages known. Nonmotile. **Aerobic**. Optimum growth temperature is 25–35°C. **The peptidoglycan contains L-ornithine. Palmitoleate** accounts for at least 25% of total fatty acid composition. **The phospholipids do not include phosphatidylglycerol, diphosphatidylglycerol or derivatives.** Chemoorganotrophic; metabolism is respiratory. May be proteolytic. Glucose may be metabolized but **acid is produced from**

only a limited number of carbohydrate substrates, if at all. **Catalase is produced.** Nearly all strains are **radiation resistant** and **desiccation resistant.** The mol% G + C of the DNA is 62–70 (T_m).

Type species: *Deinococcus radiodurans* Brooks and Murray 1981, 357.

Further Descriptive Information

Cells occur singly, in pairs and in tetrads (Fig. 12.6); they divide in two planes, alternately. Some mutant strains divide with such synchrony, notably in *D. radiodurans*, that they form square tablets of cells which are clearly visible at the margin of colonies (Fig. 12.7).

The cell walls are structurally complex and layered (Thornley et al., 1965) (Figs. 12.8 and 12.9), which is reflected in comparable chemical complexity (Work and Griffiths 1968; Brooks et al., 1980). The wall of *D. radiodurans* includes an outer membrane which, for many strains, has a hexagonal protein array on its outer aspect (Kubler, et al., 1980; Thompson et al., 1982); we now know that all four of the recognized species in the genus have an outer membrane in the cell wall (T. Counsell and R. G. E. Murray, unpublished data). No lipopolysaccharide based on heptoses, hydroxy fatty acids and lipid A has been found. The other species have not been subjected to the same level of analysis but the complex wall profiles suggest that a similar chemical complexity would be found.

The possibility exists that the Gram-positive reaction conceals an underlying relationship to Gram-negative bacteria (see *Deinococcaceae*, p. 1035) and the reaction is likely to be due to the thickness of the peptidoglycan component.

The peptidoglycan type for the veritable species in the genus is L-Orn-Gly$_2$ (Schleifer and Kandler, 1972; Sleytr et al., 1976; Brooks et al., 1980) and this is a clear distinction from the *Micrococcaceae* and the red-pigmented species, *M. roseus* and *M. agilis*, with which they are easily confused. There are strains with some resemblances to *Deinococcus* but they have other diamino acids in the peptide (see Brooks et al., 1980, and Brooks and Murray, 1981), and require further study or should be *incertae sedis*. The peptidoglycan does not seem to be associated with another polymer, even in the case of the fenestrated layer mentioned above (Thompson and Murray, 1982), and teichoic acids have not been detected.

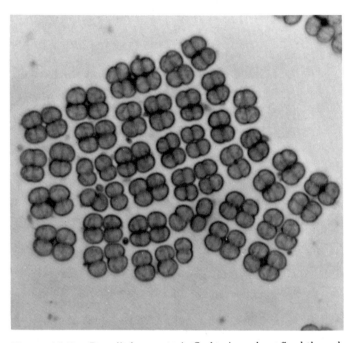

Figure 12.6. *D. radiodurans* strain Sark microcolony fixed through the agar with Bouin's fixative and the cell wall stained with methyl green after treatment with phosphomolybdic acid (Hale, 1953) to show tetrad formation and the regularity of division. Photomicrograph by C. F. Robinow. Average cell diameter 2.5 µm.

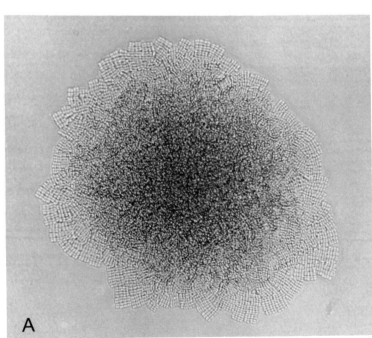

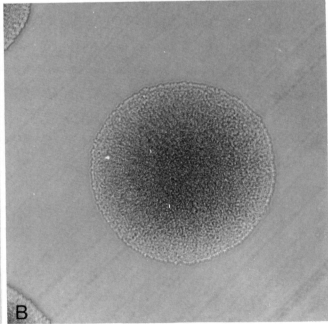

Figure 12.7. Low power photomicrographs of living microcolonies of *M. radiodurans* R$_1$ variants: *A*, a colony showing coherence of cells following division, and *B*, a smooth, noncoherent colony.

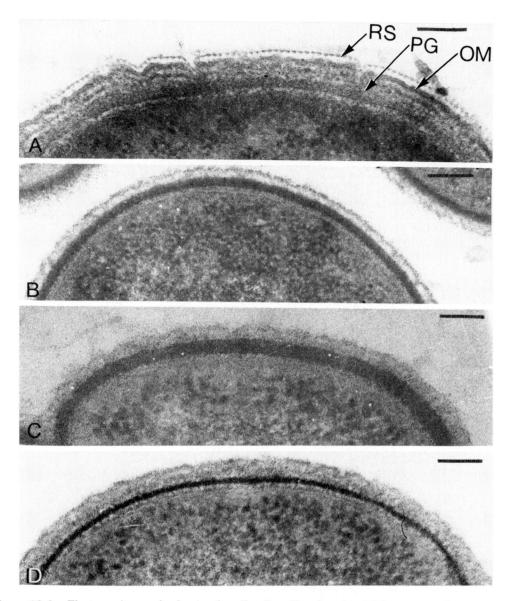

Figure 12.8. Electron micrographs showing the cell wall profiles of species of *Deinococcus*: *A, D. radiodurans*; *B, D. radiophilus*; *C, D. proteolyticus*; *D, D. radiopugnans*. The sections are stained with uranyl acetate and lead citrate. Note: a fenestrated PG layer (a character for *A* and *D*) is not visible when sections are stained with uranyl acetate but is shown by lead staining alone (Thompson and Murray 1982, and see Fig. 12.9). *RS*, regular surface protein array; *PG*, peptidoglycan layer; *OM*, outer membrane. *Bar*, 100 nm.

An unusual feature of fine structure of division in *D. radiodurans* lies in the asymmetrical septa, which originate from finite opposing points at the cell periphery and traverse the cell as two curtains (Thornley et al., 1965; Murray et al., 1983). The new polarity is established before the septal curtains meet; so rapidly dividing cells may have four communicating compartments (Fig. 12.10). The cytoplasm in the compartments appears to be free of mesosomes but may contain rounded inclusions that appear to be polysaccharide in nature (B. G. Thompson and R. G. E. Murray, unpublished data). The nucleoplasms have no unusual structural features but Moseley and Evans (1981) have concluded from studying unlinked chromosomal markers that there are a number of genome copies (~.5) in each nucleoid. Both the communicating compartments and polyploidy may contribute to the resistance to radiation killing.

The predominant fatty acid component is palmitoleate in *D. radiodurans*, *D. radiophilus* and *D. proteolyticus*. No branched chain fatty acids are present except for minor amounts in strains of *D. radiopugnans*. This fatty acid profile also suggests some alliance to Gramnegative bacteria.

A variety of carotenoids are present in all species (Brooks et al., 1980); colonies are usually pink to brick red in color and isolated plasma membrane is bright red. The MK-8 menaquinone system is present in *D. radiodurans*, *D. radiophilus* and *D. proteolyticus* (Yamada et al., 1977).

An unusual feature of membrane composition of *D. radiodurans* (Rebeyrotte et al., 1979; Thompson et al., 1980; Thompson and Murray, 1981) is the absence of phosphatidylglycerol and the derivative phospholipids (di-phosphatidyl glycerol, phosphatidyl-ethanolamine, -serine, -choline). The phospholipids that are present give an unusual pattern on thin layer chromatography (Fig. 12.11), which is repeated in principle for all the named species (T. Counsell, R. Anderson and R. G. E. Murray, unpublished data). *D. radiodurans* has been studied

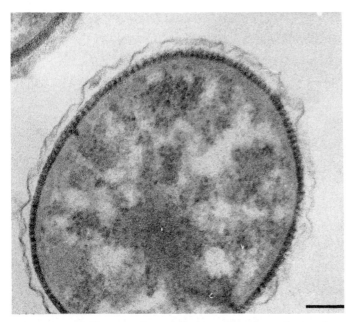

Figure 12.9. Electron micrograph of a section of a stationary phase cell of *D. radiodurans* Sark showing the appearance of fenestrated peptidoglycan (section stained with lead citrate, only). Bar, 100 nm.

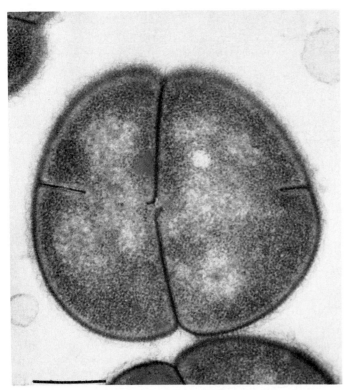

Figure 12.10. Electron micrograph of a section of *D. radiodurans* Sark showing a dividing cell and the beginning aversion of the septal advancing edge towards the polarity for the second division plane. Section stained with uranyl acetate and lead citrate. Bar, 100 nm.

in the most detail (Thompson et al., 1980) and there are some 12 polar lipids including a sulfolipid as well as several phosphoglycolipids and glycolipids. These have not been chemically characterized.

Many strains are resistant to 1.5 Mrad of gamma radiation and to 500 Jm^{-2} of ultraviolet radiation.

Cultural Characteristics

The strains of *Deinococcus* species have been isolated and maintained on rich media, usually variants of that of Work and Griffith (1968), which consisted of 5 g tryptone, 3 g yeast extract, 1 g glucose and 1 g DL-methionine/liter, and 1.5% w/v agar if solid medium (TGYM) is required. For most strains there is no evidence that either the glucose or the methionine is an absolute requirement, and adequate growth is attained on tryptone-yeast extract medium, usually at 30°C. Growth is not particularly rapid and stationary phase is reached in shaken (aerated) cultures in 48–96 h. The lag period is long and an adequate growth in fluid media requires a heavy inoculum. This suggests that nutritional requirements are not fully satisfied and that supplements (Raj et al., 1960; Shapiro et al., 1977) such as vitamins (biotin, cobalamin, nicotonic acid, pyridoxine and thiamin), metals (especially iron) and others unidentified might play a role in stimulating growth. They are mesophiles and the majority of strains will grow at 37°C, with thermal limitation in the range of 41–44°C. They are not heat resistant.

Colonies are matt to smooth with entire edges for most of the species but the tablet-forming strains of *D. radiodurans* show projections of sheets of cells at the margins (Fig. 12.7). Colonies are very small after 24 h of incubation, and require 2–7 d for adequate development. The colonies are red pigmented (most deeply on media with increased content of yeast extract). Some isolates form coherent colonies that are not easily broken-up with a loop; most are pasty and emulsify easily.

Defined media have been developed for *D. radiodurans* but have not

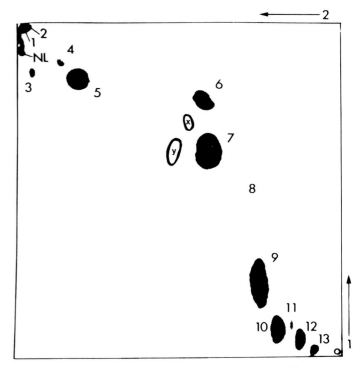

Figure 12.11. Two-dimensional autoradiogram of [1-^{14}C] acetate-labeled whole cell lipids extracted from *D. radiodurans* Sark. *Arrows* indicate direction of movement of solvent 1 (chloroform-methanol-28% ammonium hydroxide (65:35:5, v/v) and solvent 2 (chloroform-acetone-methanol-acetic acid-water (10:4:2:2:1, v/v)). *NL*, neutral lipids; *O*, origin, *x*, site where phosphatidylglycerol would have been found if it were present; *y*, site where phosphatidylethanolamine would have been found if it were present. Polar lipid spots *3, 10, 11* and *13* are glycolipids; spots *4, 5, 6, 7, 9* and *12* are phosphoglycolipids, and spot *8* is a sulfophospholipid. (Reproduced with permission from Thompson et al., Canadian Journal of Microbiology, *26:* 1408–1411, 1980, National Research Council of Canada.)

been refined to the point of providing a minimal medium (Raj et al. 1960; Little and Hanawalt, 1973; Shapiro et al. 1977). Growth in these media is slow or uncertain, and varies with the strain.

Cultural conditions may be crucial to the determination of acid production from sugar substrates by strains of *Deinococcus* because there are conflicting results in the literature. Standardization is advised to avoid further confusion. Acid was produced from only a limited number of substrates by the strains tested by Brooks and Murray (1980) using bromocresol purple as the pH indicator and either one of two basal media (including that recommended for diagnostic purposes in Table 12.12, below). However, acid production has been observed (Ito et al., 1983) in media containing a number of sugars (including glucose and sucrose) by some of the strains previously reported as not producing acid reactions. In our experience (Brooks and Murray, unpublished data) this has been observed when using other basal media or using pH indicators with an active range closer to neutrality. The precise reasons for the anomalous results are not known to us.

Isolation Procedures

The natural habitat of the *Deinococcus* species is not known. Resistance to radiation has been used as a selective factor for isolation from ground meat (Anderson et al., 1956), fish (Davis et al., 1963; Lewis, 1973), fecal samples (Kobatake et al., 1973) and sawdust (Ito, 1977). Radiation-resistant strains have also been isolated without exposure to radiation as an air contaminant (R.G.E. Murray and C.F. Robinow, Abstr. 7th Int. Congr. Microbiol., p. 427, 1958) and from creek water (Krabbenhoft et al., 1965). Moseley (1967) has shown that radiation resistance is mutable and, therefore, some relatives may be expected to be found that are not radiation resistant.

The radiation resistance of the vegetative cells of the species of *Deinococcus* is remarkable and is a major feature that may be utilized for selective isolation; some endospores are the only cells that may be expected to survive similar doses. All the veritable species can provide some survivors after exposure to a gamma-ray dose of 1.0–1.5 Mrad. It is more useful to know that the dose/response curve for *D. radiodurans* shows a large shoulder (i.e. no loss of viability up to a certain dose) before viability drops as an exponential (Moseley and Mattingly 1971). Extrapolation of the exponential curve gives an intercept value for the shoulder of about 500 Jm^{-2} for UV-radiation and about 700 Krad for gamma radiation. The D_{37} values (37% survival dose, which approximates the dose to kill one viable unit) are of the same order, presumably because most plating units are tetrads or pairs. In practical terms, these doses are so large in relation to killing doses for other vegetative cells that the major differences (up to 100%) in killing doses according to conditions of humidity or desiccation, in air or anaerobic, etc., are of little consequence in using radiations for selective isolation. However, there may well be radiation-sensitive relatives of *Deinococcus* because radiation-sensitive mutants of *D. radiodurans* have been derived (Moseley, 1967).

It has also been observed that *D. radiodurans* is unusually resistant to desiccation, and this seems to apply to other radiation-resistant bacteria. Desiccation has been used by Sanders and Maxcy (1979b) for the isolation of radiation-resistant bacteria without using radiation as a selective mechanism. Interestingly, two of the strains isolated in this way (M-4 and M-7 in Sanders and Maxcy, 1979a) proved to have thick and complex cell wall profiles, are Gram-positive and their colonies are pink to red in color (R. G. E. Murray, unpublished data). They appear to be members of the *Deinococcaceae* but they have not been sufficiently studied to form conclusions. The desiccation resistance of representative strains of the species in this genus is unusual and extreme (T. Counsell and R. G. E. Murray, unpublished data); they survive when dried on glass and maintained in air in a desiccator for 8 months and long after the demise of other vegetative bacteria.

Christensen and Kristensen (1981) have found that highly radiation-resistant, Gram-positive cocci resembling *D. radiodurans* can be isolated (about 0.2% of colonies isolated) from the air of clean rooms and laboratories of an institute involved in public health microbiology and biological products, and also from the clothing of people who work there. But similar organisms were recovered from nonlaboratory (home and office) environments using towels and underwear to get recovery (Kristensen and Christensen, 1981). These observations suggest that *Deinococcus* species may be widely distributed in nature, but in modest numbers, from many sites including skin.

Maintenance Procedures

For short term preservation stock cultures are maintained on TGYM medium (Work and Griffiths, 1968) with the addition of 1.5% agar; incubation being at 30°C.

Slant cultures remain viable for long periods at room temperature but occasional failures after 1–6 months storage argue for some caution. For long term preservation, lyophilization is effective and so is storage in liquid nitrogen.

Differentiation of the genus **Deinococcus** from other taxa

It is evident from the rRNA catalog data (Stackebrandt and Woese, 1981) that closely related taxa have been identified up to now. But there is no doubt that *M. roseus* and *M. agilis* are very similar in superficial phenotype. This includes the red pigmentation, positive catalase test and respiratory and nonfermentative metabolism as well as a long list of negative characters. All of these characters have been assessed (Brooks et al., 1980; Brooks and Murray, 1981) and the conclusion is that differentiation requires the determination of structural, biochemical and biophysical parameters (Table 12.11). None of these are simple determinations but at this time there is no small group of alternative, reliable and simple bench tests allowing a quick diagnosis. However, a numerical approach (Feltham, 1979) using a large number of unweighted characters indicated that the radiation-resistant cocci formed a cluster separable from the rest of the micrococci.

Given red-pigmented, Gram-positive, catalase-positive coccus-forming tetrads, a rough and unproven test to give slight substance to a suspicion would be: spread TGYM plates with the test isolate and other plates with a veritable *Micrococcus* (preferable *M. roseus*); place these open for exposure under a 25-W UV germicidal lamp at 150-mm distance and mask appropriately so that regions of the plates are exposed for 0.5, 1, 2, 4 and 8 min, and then incubate. This is likely to provide in 1–2 min a dose in the range of 100–300 Jm^{-2}/min. *D. radiodurans* will survive 5–10 times more exposure than will *M. roseus*.

If the culture is radiation resistant the suspicion could be supported by determining the peptidoglycan type, and/or by examining the cell wall profile in sections by electron microscopy, and/or by determining the polar lipid using thin layer chromatography compared to appropriate standard phospholipid preparations.

It should be noted the one or more of the unusual polar lipids may cofractionate on one-dimensional thin layer chromatography plates with a phosphatidylglycerol standard. However, we note that these spots generally react with reagents for amino (ninhydrin) and/or carbohydrate (α-naphthol) as well as phosphorus components, thus providing a distinction (T. Counsell, R. Anderson and R. G. E. Murray, unpublished results).

Taxonomic Comments

Anderson et al., (1956) isolated a radiation-resistant organism (strain R_1) which was later given the name "*Micrococcus radiodurans*" (Raj et al., 1960) (names in quotation marks not included on the Approved Lists of Bacterial Names (Skerman et al., 1980)). The organism was aligned with the genus *Micrococcus* on the basis of morphological, cultural and physiological characteristics but was designated as a new species on the basis of resistance to radiation. Another strain, isolated by Murray and Robinow (Abstr. 7th Int. Congr. Microbiol., p. 427, 1958) and designated the Sark strain, was later shown to be closely

Table 12.11.

*Differentiation of the genus **Deinococcus** and red-pigmented members of the genus **Micrococcus***

Characteristics	Deinococcus	Red-Pigmented Species of *Micrococcus* (*M. roseus*, *M. agilis*)
Number of cell wall layers external to cytoplasmic membrane, (Brooks et al., 1980)	Three or more, with an outer membrane	Single layered, homogeneous wall
Diamino acid in peptide subunit of peptidoglycan	Ornithine	Lysine
Interpeptide bridge (Schleifer & Kandler, 1972)	Glycine	Alanine, alanine and Threonine[a]
Presence of the fatty acid palmitoleate (Brooks et al., 1980)	Yes	Not detected
Menaquinone system (Collins & Jones, 1981)	MK-8[b]	MK-8(H_2),[c] MK-9(H_2)[d]
Resistance to gamma or UV radiation (Moseley, 1982)	Yes	No
Presence of phosphatidylglycerol in membranes (Thompson et al., 1980)	No[e]	Yes

[a] In *M. agilis* (Kocur and Schleifer, 1975).

[b] In *D. radiodurans*, *D. radiophilus* and *D. proteolyticus* (Yamada et al., 1977).

[c] In *M. roseus* (see Yamada et al., 1977).

[d] In *M. agilis* (see Yamada et al., 1977).

[e] Observations extended to all included species (T. Counsell, R. Anderson and R. G. E. Murray, unpublished data).

related to the Anderson R_1 strain (Brooks et al., 1980). Subsequent taxonomic studies by Baird-Parker (1965, 1970) demonstrated distinct differences between the cell wall composition of the Anderson R_1 strain and other red-pigmented micrococci and suggested that "*M. radiodurans*" should be excluded from the genus *Micrococcus*. This suggestion was supported by further studies on cell wall composition (Work and Griffiths, 1968; Schleifer and Kandler, 1972; Sleytr et al., 1973) and on fatty acid components of the cell (Knivett et al., 1965; Work, 1970; Girard, 1971; Jantzen et al., 1974).

In addition to "*M. radiodurans*," other red-pigmented and radiation-resistant strains have been isolated (Davis et al., 1963; Kobatake et al., 1973; Lewis, 1973). Some were described as new species: "*M. radioproteolyticus*" and "*M. radiophilus*." All the radiation-resistant strains were included as species incertae sedis in the *Micrococcaceae* in the eighth edition of Bergey's Manual (Baird-Parker, 1974) under the common name of "*M. radiodurans*."

Although it was understood that these radiation-resistant organisms were distinct from other micrococci, effective evidence for a taxonomic separation was slow to develop. Observations on cell structure (Sleytr et al., 1976; Lancy and Murray, 1978) comparative studies of menaquinone systems (Yamada et al., 1977), and the recent computer-assisted taxonomic study by Feltham (1979) all pointed the way. Brooks et al., (1980) confirmed many earlier findings, indicating little or no relationship to *M. roseus* or to other micrococci, and provided the strongest evidence for taxonomic and phylogenetic separation by the comparison of 16S rRNA oligonucleotide sequence catalogs. Other substantial characters have been identified that now add strength to the description

of a unique taxon, e.g. the strange phospholipid complement (Rebeyrotte et al., 1979; Thompson et al., 1980). We can expect that more such characteristics will be identified. Brooks and Murray (1981) proposed the present nomenclature for these organisms. No disrespect was intended toward the original authors of the specific epithets; provisions of the Bacteriological Code (1976 revision) did not allow their inclusion as a contribution to the "new" names.

There are four named species, which are properly assignable to *Deinococcus* (compare Tables 12.12 and 12.13, below) and identifiable using appropriate tests if the generic determination has been made accurately.

There is good reason to examine with care the miscellany of new and old strains that are hard to assign to species. As pointed out by Brooks et al. (1980) and Brooks and Murray (1981) there are not only strains of doubtful or uncertain alignment but also there are variations that demand assessment of relatedness among the strains presently assigned to *D. radiodurans*. It should be noted that the DNA/DNA homology between the named *Deinococcus* species is not significant (see Table 12.14) and that between strains of *D. radiodurans* is somewhat low. Other methods of inter- and intrageneric comparison should be attempted. Fortunately, it is clear from the 16S rRNA catalog comparisons that three of the species are certainly related to each other (Fig. 12.12).

Further Comments

The lack of information on the ecological distribution and natural associations of the *Deinococcus* species adds to the difficulty of a significant taxonomic study. Despite a considerable interest in radiation resistance only a small number of species has been identified and limited comparative studies have been performed. Serological techniques have not been applied in search of intraspecific or interspecific common antigens or as a means of typing within a species having a number of independently isolated strains, i.e. *D. radiodurans*.

There may be limited opportunity for genetic exchange in nature if the populations are small and dispersed, as would appear to be the case because of the relatively infrequent isolations. Furthermore, no bacteriophages have been found; so that infective transfer is not a likely feature. Plasmids exist in all four species (B. E. B. Moseley, University of Edinburgh, personal communication, 1982) but they have yet to be characterized.

There are genetic peculiarities in *D. radiodurans* (see Moseley, 1982) and among them, as might be expected, is a general resistance to the mutagenic effects of radiations and chemicals. There are effective repair mechanisms for breaks in single- and double-stranded DNA as well as an excision repair mechanism for thymine dimers. Photoreactivation and error-prone (SOS) repair capabilities are not present. The error-free repair is mediated by some aspect of nuclear structure that allows juxtaposition of homologous regions of multiple genome copies so that recombinational exchange can take place. Such a mechanism would obviate the need for error-prone (SOS) repair. The effectiveness of the available mechanisms means that expressed mutations are of low frequency, and experience shows that the only effective chemical mutagens are nitrosoguanidine and nitrosourea derivatives. The multiplicity of genome copies (about five per nucleoid) means that considerable time must be allowed for segregation following exposure to a mutagen so that mutations can be isolated. Among the results of mutagenesis are radiation sensitive strains (Sweet and Moseley, 1974) and, usefully, a strain that is more sensitive to the nitrosoguanidine mutagens.

It is difficult to predict what circumstances dictate the maintenance, let alone the origin, of an extreme radiation resistance, which is of an order that must involve structural repair at levels other than genetic. It is possible that the content and construction of membranes is critical and that the peculiar polar lipids, as well as the broad selection of carotenoids, play their part in minimizing phase effects and the consequences of free radical formation. There are few places on this earth where the radiation flux might be expected to be an effective selective mechanism and, naturally, they do provide types of resistant bacteria

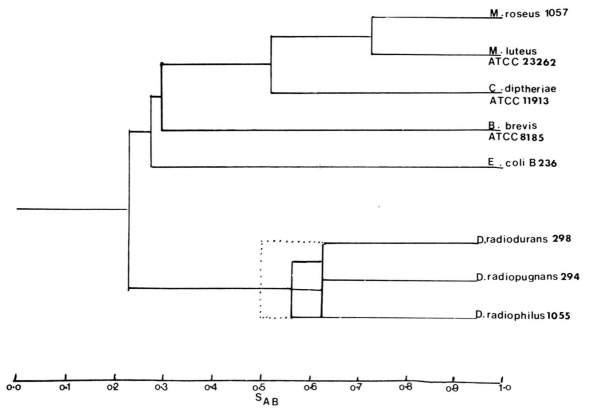

Figure 12.12. Dendrogram to show the relationship, in terms of 16S, rRNA association coefficients (S_{AB}), of *D. radiodurans*, *D. radiophilus* and *D. radiopugnans* to each other and to species of *Micrococcus* and diverse other bacteria. The relative positions of the three radiation-resistant species to each other is uncertain and is in the range shown by the *dotted line* (S_{AB} 0.5–0.6). (Reproduced with permission from Brooks et al., International Journal of Systematic Bacteriology, *30:* 627–646, 1980, American Society for Microbiology. For a more complete demonstration of relationships see Stackebrandt and Woese (1981).)

for study (Yoshinaka et al., 1973). These properties must have been maintained and selected in the *Deinococcus* sp. over the eons as indicated by phylogenetic data. No consistent site or ecological niche has yet been identified although, as might be expected, many strains are isolated in association with atomic reactors or radiation sources, which are recent artifacts.

Acknowledgements

The authors are grateful to B. E. B. Moseley (University of Edinburgh) for exchanges of information by correspondence. We are indebted to C. R. Woese (University of Illinois), J. L. Johnson (Virginia Polytechnic Institute and State University) and to E. Stackebrandt (University of Kiel, Germany) for undertaking special analyses and for their advice. We are appreciative of the technical help provided by M. Hall and D. Moyles, the provision of data by R. Anderson and T. Counsell (Department of Microbiology, University of Western Ontario) and photomicrographs by C. F. Robinow. The support of our research by the Medical Research Council of Canada is gratefully acknowledged.

Further Reading

Moseley, B.E.B. 1983. Photobiology and Radiobiology of *Micrococcus* (*Deinococcus*) *radiodurans*. Photochem. Photobiol. Rev. *7:* 223–274.
This is the only detailed review of the considerable range of work done on the radiation resistance of these organisms and it describes much of what is known about the general biology and genetics of the type species.

List of species of the genus **Deinococcus**

The diagnostic and descriptive features of the species listed below are set out in Tables 12.12–12.14 which provide a selection and rearrangement of the data from Brooks et al., (1980) and Brooks and Murray (1981).

1. **Deinococcus radiodurans** Brooks and Murray 1981, 354.[VP]
ra.di.o.du′rans. M.L. n. *radiatio* radiation; L. part. adj. *durans* enduring; *radiodurans* resisting radiation.
Variation in the superficial cell wall structure, reduction of nitrates, tolerance to sodium chloride, proportion of fatty acid components, radiation sensitivity, mutagenic response and pigmentation have been observed at some time in special strains or mutants.
Growth is enhanced by trace levels of copper, cobalt and iron (R. G. E. Murray, unpublished data) and by many other supplements (Shapiro et al., 1977).
Type strain: ATCC 13939 (UWO 288).

2. **Deinococcus radiophilus** Brooks and Murray 1981, 357.[VP]
ra.di.o′phil.us. M.L. n. *radiatio* radiations; Gr. adj. *philos* loving; M.L. adj. *radiophilus* radiation loving.
This species, represented by derivatives of a single isolate studied by Lewis (1973), is even more resistant to UV and ionizing radiation than *D. radiodurans* (Moseley, 1982).
Type strain: ATCC 27603 (UWO 1055).

3. **Deinococcus proteolyticus** Brooks and Murray 1981, 357.[VP]
pro.te.o.ly′ti.cus. M.L. n. adj. *proteolyticus* proteolytic.
Type strain: CCM 2703 (UWO 1056; ATCC 35074).
The specific epithet implies that this species peptonizes proteins (milk, soy and gelatin) as was observed by the original isolators (Kobatake et al. 1973). The other species also show some activity but this species is the most active.

Table 12.12.
*Differential characteristics for species of the genus **Deinococcus**[a]*

Characteristics	1. *D. radiodurans*	2. *D. radiophilus*	3. *D. proteolyticus*	4. *D. radiopugnans*
Predominant fatty acid	16:1	16:1	16:1	16:1, 17:0
Branched chain fatty acid	ND	ND	ND	15:0
Fenestrated peptidoglycan layer	+	−	−	+
Growth in presence of 5% NaCl	18/24[b]	+	−	−
Nitrate reduction	18/24[b]	−	−	+
Acid from glucose in standard medium[c]	−	−	+	−
Mol% G + C of the DNA (T_m)	67(70)[d]	62	65	70

[a] Symbols: See Table 12.2.
[b] The fraction of 24 strains negative for the character.
[c] Medium recommended by ICBN Subcommittee on Taxonomy of Staphylococci and Micrococci, 1965.
[d] (70) is the extreme figure for one included strain (see Brooks and Murray, 1981).

Table 12.13.
*Other characteristics of species of the genus **Deinococcus**[a]*

Characteristics	1. *D. radiodurans*	2. *D. radiophilus*	3. *D. proteolyticus*	4. *D. radiopugnans*
Cell size (μm)	1.5–3.0	1.0–2.0	1.0–2.0	1.0–2.0
Catalase	+	+	+	+
Esculin hydrolysis	− ✓	− ✓	+	− ✓
Acid from fructose	−	−	+	−
ONPG	−	−	−	+
Growth at 37°C	+	+	+	+
Resistance to 1 Mrad gamma radiation	+	+	+	+
Multiple carotenoids	Present	Present	Present	Present
Colony color	Red	Orange-red	Orange-red	Orange-red
Peptidoglycan type	L-Orn-Gly₂ or Gly₃	L-Orn-Gly₂	L-Orn-Gly₂	Orn-Gly₂
Gram reaction	Positive	Positive	Positive	Positive
Colonial morphology	Smooth and convex with regular edge	Smooth and convex with regular edge	Smooth and convex with regular edge	Smooth and convex with regular edge

[a] Symbols: see Table 12.2.

Table 12.14.
*DNA homology among species of **Deinococcus**[a–c]*

Species	*D. radiodurans* UWO 298	*D. proteolyticus* UWO 1056	*D. radiopugnans* UWO 294
1. *D. radiodurans* UWO 288	33	6	17
D. radiodurans UWO 298	(100)	8	17
2. *D. radiophilus* UWO 1055	4	18	4
3. *D. proteolyticus* UWO 1056	10	(100)	8
4. *D. radiopugnans* UWO 294	13	6	(100)
5. *D. erythromyxa* UWO 1045[d]	1	3	3

[a] Data abstracted from Brooks et al., 1980.
[b] Percentage of homology with DNA from strain.
[c] UWO, collection in the Department of Microbiology and Immunology, University of Western Ontario.
[d] Species incertae sedis.

4. **Deinococcus radiopugnans** Brooks and Murray 1981, 358.[VP]

ra.di.o.pug'nans. M.L., n. *radiatio* radiation; L. part. adj. *pugnans* fighting or resisting; M.L. adj. *radiopugnans* radiation resisting.

Smooth and rough variants as well as variants with less pigment may occur. The 15-carbon, saturated, branched chain fatty acid component may be absent.

Type strain: ATCC 19172 (UWO 293).

Species Incertae Sedis

Brooks and Murray (1981) included the organism formerly known as "Sarcina erythromyxa" as a species incertae sedis, *Deinococcus erythromyxa*. Strains of this species had been retrieved from the Kral Collection and maintained as CCM706 (UWO 1045 and ATCC 187). The general phenotype and the fatty acid profile bore some resemblance to *Deinococcus* and this included considerable resistance to radiation. However, the peptidoglycan type was L-Lys-L-Ala₃₋₄, which was more appropriate to *Micrococcus* than to *Deinococcus*. When this was the only basis of difference, it could be argued that there might be variation within the taxon and retention as a species incertae sedis would serve the purpose of stimulating research on the group and the organisms that might be related to it.

Another criterion is added now that we have recognized a unique and distinctive pattern of polar lipids and the lack of phosphatidylglycerol or its derivatives as characteristic of *Deinococcus*. Recent comparative studies using one- and two-dimensional thin layer chromatography (T. Counsell, R. Anderson and R. G. E. Murray, unpublished data) have shown that the strains of *D. erythromyxa* possess phosphatidyl- and diphosphatidylglycerol. The determination was not made easy by the presence of a number (~10) of unconventional polar lipids, two of which comigrated with the phosphatidylglycerol of *Eschericia coli*. The reaction of the comigrating phospholipid with ninhydrin added to the confusion on one-dimensional chromatography plates.

The conclusion must be, on the basis of both peptidoglycan and phospholipid typing, that *D. erythromyxa* is not a *Deinococcus*. Perhaps it should be reconsidered as a member of the genus *Micrococcus*, but that will require further taxonomic study.

We should add, as a lesson in humility for taxonomists, that some confusion was created by the unwitting submission to two laboratories of tubes of lyophilized cultures mislabeled as UWO 1045. The mistake was recognized when it was apparent that the cells of this red-pigmented culture were coccobacillary, divided in a single plane, were not radiation resistant (B. E. B. Moseley, University of Edinburgh, personal communication), and showed a 16S rRNA oligonucleotide catalog resembling some *Actinomycetales* and quite unlike *Deinococcus* species (C. R. Woese, University of Illinois, personal communicatin). It serves as a warning that some bacteria aligned with the actinomycetes and coryneform groups may show only phosphatidylglycerol and/or diphosphatidylglycerol of the conventional polar lipids (Minnikin and Goodfellow, 1980).

Genus *Streptococcus* Rosenbach 1884, 22[AL]

JEREMY M. HARDIE

Strep.to.coc′cus. Gr. adj. *streptus* pliant; Gr. n. *coccus* a grain, berry; M.L. masc. n. *Streptococcus* pliant coccus.

Cells normally spherical or ovoid, less than 2 μm in diameter, **occurring in pairs or chains** when grown in liquid media. Usually non-motile. **Endospores are not formed. Gram-positive.** Most are **facultatively anaerobic,** but some require additional CO_2 for growth and some may be strictly anaerobic. **Chemoorganotrophs, with fermentative metabolism.** Carbohydrates are fermented with the **production of mainly lactic acid,** but no gas. Some species ferment organic acids (malic and citric) and amino acids (serine and arginine). **Catalase-negative. Nutritional requirements are complex and variable.**

Temperature optimum usually about 37°C but maximum and minimum temperatures vary among species. Many species exist as **commensals or parasites on man or animals,** some are highly pathogenic. A few are saprophytes and occur in the natural environment.

The mol% G + C of the DNA is 34–46 (T_m, Bd).

Type species: *Streptococcus pyogenes* Rosenbach, 1884, 23.

Further Descriptive Information

Streptococcus cells are usually spherical or ovoid in shape and are arranged in chains or pairs. Chain formation is best seen in liquid cultures. Some species, notably *S. mutans*, grow as short rods under certain cultural conditions, and several of the oral streptococci appear to be highly pleomorphic on primary isolation. Growth is by elongation in the axis of the chain with cell division in one plane, at right angles to the long axis. Chain length varies considerably among different species and strains, and is also dependent on medium composition; long chains of over 50 cells may sometimes be produced. Cells of *S. pneumoniae* characteristically form pairs (diplococci).

Not all species form capsules. Some may form hyaluronic acid capsules during the early phase of growth, while pneumococci possess different polysaccharide capsules, giving rise to many type-specific antigenic varieties. Several streptococcal species produce both soluble and insoluble extracellular polysaccharides when grown in the presence of sucrose, but these do not necessarily form a morphologically distinctive capsule.

Several surface structures and appendages have been described. Fimbriae have been observed in *S. sanguis* (Henriksen and Henrichsen, 1975), and in *S. salivarius* (Handley et al., 1985); and three morphological types of filamentous appendages occur in "*S. mitior*" (Handley and Carter, 1979). Short, fine filaments have been seen to cover the entire surface of *S. pyogenes* (Swanson and McCarty, 1969). Although usually nonmotile, 1–5 flagella have been described in some motile, saprophytic strains of enterococci (Graudal, 1955).

The cell wall composition is characteristic of Gram-positive bacteria, consisting mainly of peptidoglycan to which are attached a variety of carbohydrate, teichoic acid and surface protein antigens. Several peptidoglycan chemotypes have been recognized (Schleifer and Kandler, 1972). Cell walls always contain the amino sugars glucosamine and muramic acid, while galactosamine is a variable component. The commonly found reducing sugars are glucose, galactose and rhamnose, which occur in various combinations in different species (Slade and Slamp, 1962; Colman and Williams, 1965). Some streptococci contain cell-wall teichoic acids in addition to the membrane-associated lipoteichoic acid which is common to many Gram-positive bacteria (Wicken and Knox, 1975).

Growth on solid media is enhanced by the addition of blood, serum or glucose. Colonies on blood agar are usually 0.5–1.00 mm in diameter after 24 h at 37°C and show little or no increase after prolonged incubation. Most species are nonpigmented, except for some strains of *S. agalactiae* and some enterococci, which may have yellow-, orange- or red-colored colonies. Growth in liquid media is increased greatly by the addition of glucose, but the rapid fall in pH quickly inhibits growth unless the medium is highly buffered, as in Todd-Hewitt broth (Todd and Hewitt, 1932), or the medium is pH-controlled by continuous addition of alkali. In batch culture, some strains produce a granular type of growth in broth with a clear supernatant fluid; these are usually found to have formed long chains. Other strains tend to produce a more diffuse, even turbidity. In continuous culture, both the macroscopic and microscopic appearance of the growth may vary according to the dilution rate and the limiting nutrient (Ellwood et al., 1974).

Several types of changes in blood agar are produced by different streptococci. The original descriptions of these types of hemolysis were based on the appearance of deep colonies in pour plates (Brown, 1919) but reasonably reliable assessments can be made from surface growth on carefully prepared, layered blood agar plates (Parker, 1983). Blood-containing media should not include reducing sugars which may inhibit hemolysis of *S. pyogenes* (Facklam and Wilkinson, 1981). Either sheep or horse blood are generally used.

β-Hemolysis is characterized by a sharply defined zone of clearing around the colonies; the size of the zone varies from strain to strain. With α-hemolysis, a zone of greenish discoloration occurs around the colony, usually 1–3 mm in width, and the margin is indistinct. A third type of hemolysis, α-prime, has also been described and resembles α-hemolysis with an obvious outer ring of clearing around the zone of discolored (green) erythrocytes (Parker 1983). Both the type and extent of hemolysis are influenced by the composition of the basal medium, the type and concentration of blood used and the cultural conditions.

In the United States, 5% sheep blood agar is generally used for recognition of hemolytic streptococci, especially as this medium does not support the growth of hemolytic *Haemophilus* species. In the United Kingdom most laboratories prefer 5% horse blood agar, on which good hemolytic zones may be seen around surface colonies.

Most species of *Streptococcus* are facultatively anaerobic. Some strains of *S. mutans*, "*S. milleri*" and *S. pneumoniae* require the addition of 5% CO_2 to the atmosphere for growth, and many other streptococci grow better under such conditions. Some obligately anaerobic cocci are also included in the genus description in this edition of the *Manual* for the first time.

All streptococci ferment carbohydrates, producing predominantly lactic acid; minor amounts of acetic and formic acids, ethanol and CO_2 may also be produced. Although all species ferment glucose, a wide range of other carbohydrates is utilized and variations in fermentation patterns between species can be useful for identification purposes. Streptococci cannot synthesize heme compounds. Although normally described as catalase-negative, some species, such as *S. faecalis*, are able to produce catalase or pseudocatalase activities when grown aerobically in the presence of hemin (Whittenbury, 1978). Since *S. faecalis* appears to have many properties that enable the species to grow well and survive in aerobic conditions, including several enzymes for coping with H_2O_2, the fundamental anaerobic nature of the streptococci has been challenged (Whittenbury, 1978).

The complex nutritional requirements, generally including amino acids, peptides, purines, pyrimidines and vitamins, are normally provided by using complex media, often containing fresh meat extract. For optimal growth in liquid media, addition of a fermentable carbohydrate is also necessary. Some strains will not grow unless media are supplemented with pyridoxine. The least nutritionally demanding of the streptococci appears to be *S. bovis*, many strains being able to use ammonium salts as a source of nitrogen (Wolin et al., 1959).

The pyogenic streptococci generally grow within the temperature range 20–40°C, with an optimum of about 37°C. Most of the group D streptococci (enterococci) grow at 45°C; enterococci, lactic streptococci and *S. uberis* also grow at 10°C. The enterococci can also tolerate high salt concentrations (6.5% NaCl) and will grow at pH 9.6. Some streptococci, including most of the group D streptococci and some strains of *S. agalactiae* and *S. mutans*, will grow on 40% bile agar, but representatives of other species may also share this property.

Active virulent and temperate phages have been described. Some virulent phages are able to produce a phage-associated lysin which attacks the cell walls of groups A, C and E streptococci (Maxted, 1957); this enzyme has been used for cell wall studies (Krause, 1958) and the induction of L forms (Gooder and Maxted, 1961). Production of erythrogenic toxin by some strains of *S. pyogenes* is due to lysogenization with a temperate phage, which can be transferred to other strains (Zabriskie, 1964), and many other streptococci are lysogenic. A phage typing method has been described for *S. agalactiae* (Group B streptococci) which allows subdivisions to be made within serotypes (Stringer, 1980).

Bacteriocins are produced by many streptococci (Tagg et al., 1976). Some of these inhibit a wide range of Gram-positive organisms while the activities of others are sufficiently restricted and specific to be used as a means of streptococcal typing (Tagg and Bannister, 1979). Bacteriocin typing of pyogenic streptococci has shown some interesting correlations with antigenic structure (Tagg and Bannister, 1979; Johnson et al., 1979). Bacteriocin typing systems for *S. mutans* have been used to study the distribution of particular types within individuals and among family groups, and also to investigate transmission between mothers and their children (Kelstrup et al., 1970; Berkovitz and Jordan, 1975; Rogers, 1976).

Serology has been of major importance in the classification and identification of streptococci since the grouping scheme of Lancefield (1933) was first proposed. The antigens known as *group-specific antigens*, also called C-substances, are either polysaccharides (as in groups A, B, C, E, F and G), which are associated with the cell wall, or teichoic acids (groups D and N) situated in the region between the cell membrane and the inner surface of the cell wall. The Lancefield grouping system, in which organisms are designated by letters of the alphabet, is mainly useful for distinguishing between pathogenic β-hemolytic streptococci from human and animal infections. It is not an all-embracing system which can be used to classify all streptococci and several species, particularly nonhemolytic and α-hemolytic varieties, do not possess any of the recognized Lancefield group antigens.

In some cases the chemical nature of the group antigenic determinants is known. For example, in group A, the group polysaccharide is composed of *N*-acetylglucosamine and rhamnose, while in group C it is *N*-acetylgalactosamine linked to rhamnose (Parker, 1983). Further details of some of the other antigens are given in the species descriptions in following sections. In addition to the major group antigens, a number of type-specific surface antigens (such as M, T and R antigens of *S. pyogenes*) are also recognized. Such antigens are commonly proteins, although polysaccharide type antigens also occur, as in group F (Ottens and Winkler, 1962). Although much is known about the antigenic structure of some streptococci, especially the major human pathogens, there is far less information about many of the other species.

Most streptococci are sensitive to a wide range of antimicrobial chemotherapeutic agents, including the penicillins. The enterococci are the exception to this rule, being relatively resistant to benzylpenicillin, but synergistic combinations of penicillin and an aminoglycoside are effective against these organisms. The pyogenic streptococci and most pneumococci are highly sensitive to benzylpenicillin. Use of antibiotics in humans commonly results in the emergence of resistant strains in the normal flora, such as in the mouth, and such strains may still be detectable for several weeks after the antibiotic has been withdrawn (Phillips et al., 1976; Sukchotiratana et al., 1975).

Extensive use of sulfonamides during the Second World War led to the development of resistance among the pyogenic streptococci, but this is less common today. Tetracycline resistance is relatively common in several countries, including Japan where resistance to erythromycin and chloramphenicol has also been noted (Miyamoto et al., 1978). Plasmid-determined antibiotic resistance has been reported (Parker, 1983).

The sensitivity of *S. pyogenes* to bacitracin is utilized in the disk test which allows rapid presumptive identification of this species when isolated from human clinical specimens (Maxted, 1953). It is recommended that the disks should contain 0.04 units of bacitracin (Coleman et al., 1977) and only β-hemolytic strains should be tested. Since bacitracin sensitivity is not confined to group A streptococci, additional confirmatory identification tests are normally required (Facklam and Wilkinson, 1981).

Streptococci are responsible for a large number of important diseases of man (Parker, 1978) and animals (Wilson and Salt, 1978). The pyogenic group are the major pathogens, but many of the other species are commonly involved in disease processes, often as opportunistic pathogens. Much of the available evidence about toxins and pathogenic mechanisms is related to *S. pyogenes*, which produces hemolysins (streptolysins O and S), erythrogenic toxin, pyrogenic exotoxin, streptokinase, nucleases, proteinase, hyaluronidase and NADase, in addition to other extracellular products (Parker, 1983; Maxted, 1978). Group A streptococci can cause various types of localized infection in man, including tonsillitis and impetigo, as well as scarlet fever, erysipelas and other spreading infections, and two serious poststreptococcal sequelae, rheumatic fever and glomerulonephritis (Read and Zabriskie, 1980). There is an association between some of these pathological conditions and particular M-protein antigen types of *S. pyogenes* (Fox, 1974; Maxted, 1978). Further information about the pathogenicity of other species is given with the species descriptions in following sections.

Streptococci are found on the mucous membranes of the mouth, respiratory, alimentary and genitourinary tracts, and the skin of man and animals (including insects). They are also present in milk and dairy products, in some food and plant material, soil and fecally-contaminated water (Mundt, 1982). Pathogenic species, such as *S. pyogenes* and *S. agalactiae*, may form part of the normal commensal flora in asymptomatic carriers (Facklam and Wilkinson, 1981). There is a considerable degree of animal species and site specificity; even within a particular habitat, such as the mouth, there are preferred niches for particular species (Hardie and Marsh, 1978).

Enrichment and Isolation Procedures

Since streptococci are isolated from a wide range of habitats and types of sample, and since the genus includes species with a considerable variety of characters, it is not surprising that numerous isolation and enrichment media have been described for different purposes. Colonies of hemolytic streptococci can often be picked from nonselective, blood-containing agar media, although their presence may be masked if large numbers of other organisms are present. Three chemical selective agents in common use are crystal violet, sodium azide and thallous acetate (Barnes et al., 1978; Parker, 1983). Antimicrobial compounds are also used in selective media, including fusidic acid, nalidixic acid, polymyxin and the aminoglycosides gentamycin, neomycin and kanamycin. For selective agar media, several different bases can be used; Columbia blood agar, trypticase soy blood agar, Hartley's digest blood agar, Todd-Hewitt blood agar and other blood agar bases are among the many that have been recommended (Facklam and Wilkinson, 1981). The recognition of particular properties, such as bile tolerance and esculin hydrolysis in group D streptococci (Sabbaj et al., 1971, Facklam, 1973) or extracellular polysaccharide production from sucrose by *S. bovis* and some oral streptococci (Chapman, 1944; Hardie and Marsh, 1978), is often useful for isolation. Specific recommendations for appropriate selective isolation procedures have been published for human clinical isolates, animal disease isolates, oral streptococci, normal intestinal flora, dairy streptococci, food isolates and water (Barnes et al., 1978, Facklam and Wilkinson, 1981). Several enrichment broth media also have been advocated for streptococci, with particular reference to clinical isolates of groups A, B and D, containing similar selective/inhibitory agents to those employed in solid media, but often in lower concentrations (Facklam and Wilkinson, 1981).

Maintenance Procedures

Most strains can be maintained for short periods by weekly subculture on appropriate media, preferably at 4°C on slopes, in agar stabs or, with strains able to grow at 10°C, in litmus milk + 1% chalk, supplemented with 0.3% yeast extract with 1% glucose if required (Garvie et al., 1981). For long term maintenance, strains can be preserved at −80°C, either deep frozen or in liquid nitrogen. The technique of freezing cultures on glass beads is particularly useful (Feltham et al., 1978), since one or two beads can be removed whenever the culture needs to be regenerated. Cells for freezing are harvested from blood agar and suspended in a diluent containing 1% tryptone, 0.5% yeast extract, 0.1% glucose, 0.1% cysteine HCl and 2% bovine serum (Bowden and Hardie, 1971). The most reliable method for long term preservation is lyophilization, using standard methods (Kirsop and Snell, 1984).

Procedures for Testing Special Characters

Hemolysis

Using a suitable commercially available blood agar base such as trypticase soy, proteose peptone, brain heart infusion, neopeptone infusion or Todd-Hewitt agars plus 5% defibrinated sheep or horse blood, streak stab or pour plates are used. For streak plates, layered plates are used, with a thin layer of blood agar poured on top of a nutrient agar base. When inoculating such plates it is recommended that, in addition to normal streaking to produce separate surface colonies, the loop should also be used to stab an unused portion of the agar surface, thus allowing deep growth to occur (Facklam and Wilkinson, 1981). Incubate anaerobically in an atmosphere of N_2 (90–95%) + CO_2 (5–10%) or N_2 + H_2 + CO_2, but not in a candle jar or CO_2 incubator which may encourage H_2O_2 formation and inhibit β-hemolysis (Facklam and Wilkinson, 1981).

Pour plates are prepared by adding the inoculum to 15 ml of melted agar base cooled to 50–55°C, and then adding 0.6 ml defibrinated blood, mixing and pouring into a Petri dish. If desired, one-half of the plate can then be streaked conventionally to produce surface colonies (pour-streak plate). Anaerobic incubation is preferable but not essential (Facklam and Wilkinson, 1981).

It is recommended that plates be examined microscopically (magnification approximately × 60) so that unlysed erythrocytes can be recognized within zones. This is especially important for recognition of α-prime hemolysis (Parker, 1981; Facklam and Wilkinson, 1981).

Extraction and Serological Grouping

Grouping of β-hemolytic streptococci is achieved conventionally by extracting the group antigen from harvested cells (grown either in liquid media or as surface colonies on plates) and testing the extract against group-specific precipitating antisera (which are available commercially). The original technique (Lancefield, 1933) involved extraction with boiling HCl (0.2 M HCl in 0.85% NaCl), although less concentrated acid is sometimes preferable. Numerous other extraction methods have been described, including hot formamide (Fuller, 1938), autoclaving (Rantz and Randall, 1955), nitrous acid (El Kholy et al., 1974), *Streptomyces albus* enzyme (Maxted, 1948), Pronase B enzyme (Ederer et al., 1972) and *Streptomyces albus*—lysozyme enzyme (Watson et al., 1975). Ring precipitin tests can be performed in small capillary tubes, by layering the streptococcal extract on top of the antiserum. Alternative precipitation methods are double diffusion in gels (Rotta et al., 1971) and countercurrent electrophoresis (Dajani, 1973).

Coagglutination methods for grouping streptococci are also available. In these, the streptococcal suspension is mixed with a series of suspensions of *Staphylococcus aureus* (a protein A forming strain) which are coated with group-specific antibodies (Christensen et al., 1973). For such tests the streptococcal cells must normally be trypsinized, although formamide extracts have been used (Efstratiou and Maxted, 1979). Other agglutination methods involving antibody-coated erythrocytes or latex particles have also been developed.

Immunofluorescent staining methods are available for identifying some groups (Cars et al., 1975), particularly for recognition of Group A (Moody et al., 1958) and Group B (Romero and Wilkinson, 1974) streptococci. The main application of immunofluorescent reagents is for rapid identification of pathogenic streptococci in clinical specimens, although they can also be used for grouping isolated strains.

The methods result in the allocation of certain streptococci to one of the several Lancefield groups, being based on the presence and detection of a group antigen. Within several of the groups, notably group A (*S. pyogenes*), group B (*S. agalactiae*) and group D, a number of type antigens are also demonstrable by serological methods and are of considerable epidemiological significance (Parker, 1983).

Serological typing of *S. pneumoniae* depends on recognition of different capsular polysaccharide antigens by means of the Quellung reaction. A small drop of culture suspension is placed on a microscope slide, a loopful of appropriate antiserum is added and mixed with a loopful of saturated aqueous methylene blue. The mixture is covered with a cover slip and, after 10 min, examined microscopically (Austrian, 1976). A positive result is recognized by swelling and increased visibility of the capsule around the cocci.

Biochemical and Physiological Tests

A wide range of tests has been used for characterizing streptococci, including carbohydrate fermentations, production of acetyl methyl carbinol from glucose, production of ammonia from arginine, hydrolysis of esculin, hippurate and starch, reduction of litmus milk, tellurite and tetrazolium, production of H_2O_2, NaCl and bile tolerance, and formation of extracellular polysaccharides from sucrose (Facklam and Wilkinson, 1981; Parker, 1983). Such tests may be carried out conventionally in meat extract or serum water broths with appropriate additives (Cowan and Steel, 1974), in miniaturized systems with laboratory prepared ingredients (Jayne-Williams, 1976) or in commercially available systems such as the API-20 STREP system (API Systems S.A., Montalieu-Vercieu; also known in the United States as the RAPID STREP SYSTEM, supplied by DMS Laboratories, Inc., Flemington, N.J.) and a system based on substrate-impregnated disks (MINITEK, Becton-Dickinson, BBL Microbiology Systems, Cockeysville, MD.) (Holloway et al., 1979). The API-20 STREP system incorporates both traditional types of test, such as sugar fermentations, and some preformed enzyme tests (pyrrolidonylarylamidase, α-galactosidase, β-glucuronidase, β-galactosidase, phosphatase and leucine aminopeptidase). Such tests are useful for generating data for taxonomic studies (Bridge and Sneath, 1983) and for identification purposes (Appelbaum et al., 1984; Facklam et al., 1984; Colman and Ball, 1984).

A few tests are especially useful for presumptive identification of particular streptococci. Sensitivity to 0.04 unit bacitracin disks is a valuable screening test for *S. pyogenes* (Maxted, 1953), while SXT disks, containing 1.25 mg trimethoprim and 23.75 mg sulfamethoxazole can be used to distinguish the resistant *S. pyogenes* (group A) and *S. agalactiae* (group B), from other streptococci which are sensitive (Gunn, 1976; Facklam and Wilkinson, 1981). Group B strains can be also distinguished by means of the CAMP test, in which a single streak of the streptococcus is made perpendicular to (but not touching) a streak of β-lysin-producing *Staphylococcus aureus*, either on sheep or bovine blood agar. After incubation aerobically or in a candle jar, CAMP-positive strains produce an arrowhead-shaped enlargement of the zone of lysis around the *Staphylococcus* at the junction between the two streaks of growth (Darling, 1975).

Hydrolysis of sodium hippurate is a character shared by *S. agalactiae* (group B), *S. uberis*, *S. acidominimus* and some group D streptococci (enterococci) (for technique see Facklam et al., 1974). Tolerance to 6.5% NaCl, and also to 10% or 40% bile, can be tested either in liquid or solid media, but results from different laboratories are sometimes contradictory, indicating the necessity for improved standardization of the methods used. A combined bile-esculin test for presumptive iden-tification of group D streptococci is recommended (Facklam and Moody, 1970; Facklam et al., 1974).

Streptococcus pneumoniae can be distinguished from other α-hemo-lytic streptococci on the basis of being sensitive to a 5 μg Optochin (ethylhydrocupreine) disk on a blood agar plate (Bowers and Jeffries, 1955). The bile solubility test also differentiates this species, but is more complicated and less reliable than the Optochin test (Lund, 1959).

Production of extracellular polysaccharides (glucans (dextrans) and fructans (levans)) from sucrose is an important characteristic of *S. bovis* and several of the oral streptococci (including *S. mutans*, *S. salivarius*, *S. sanguis* and "*S. mitior*"). Such polysaccharides can be detected in the supernatant of cultures in buffered 5% sucrose broth, either by observing gel formation (only in some cases) or by differential precipitation with ethanol. A precipitate following addition of 1.2 vol ethanol indicates glucan formation, whereas fructans are precipitated only by 2.5 vol (Hehre and Neill, 1946). Cross-reactivity of sucrose culture supernatants with type 2 pneumococcal antiserum is also indic-ative of glucan production (Hehre and Neill, 1946).

Standardization of the tests for polysaccharide production can prove to be difficult in practice. A useful indication of the presence of these products can be obtained more simply by observation of colonial morphology on 5% sucrose agar plates (Parker, 1983; Colman and Ball, 1984).

Differentiation of the genus **Streptococcus** from other taxa

Streptococci are traditionally separated from other "related" genera on the basis of their mode of cell division and the major products of glucose fermentation (Deibel and Seeley, 1974), although the validity of these criteria is perhaps debatable (Jones, 1978). A recent numerical taxonomic study has shown that species of *Aerococcus*, *Gemella*, *Leu-conostoc* and *Pediococcus* are apparently closely related to *Streptococcus* (Bridge and Sneath, 1983).

Streptococci have in the past been separated from the obligately anaerobic Gram-positive cocci partly on the basis of degree of anaero-biosis and partly because the latter group of organisms produce gas and a variety of end products from glucose, as opposed to the major product being lactic acid. With the broadening of the genus description to include some anaerobic streptococci, the relationships between various Gram-positive cocci require further examination, using molecular and genetic approaches in addition to phenotypic characters.

Taxonomic Comments

A number of changes have been made to the genus *Streptococcus* since the 8th edition of the *Manual*, including the addition of several new species and the inclusion of some obligate anaerobes within the genus description. During the preparation of this section, after most of the species descriptions had been completed, several significant tax-onomic papers were published which may lead to further changes in subsequent editions. One particularly important recent proposal is for the creation of a new genus, *Enterococcus*, to encompass the enterococ-cal group, such as *S. faecalis* and *S. faecium* (Schleifer and Kilpper-Bälz, 1984). The formation of this genus was originally suggested some years earlier (Kalina, 1970), but it is now fully supported by data from DNA/DNA and DNA/rRNA hybridization studies (Kilpper-Bälz et al., 1982; Schleifer and Kilpper-Bälz, 1984). Further work has led to the suggestion that other streptococcal species be transferred to the genus *Enterococcus* as *E. avium*, *E. casseliflavus*, *E. durans*, *E. gallinarum* and *E. malodoratus* (Collins et al., 1984a).

Because of the diversity of the streptococci, several attempts have been made over the years to organize the organisms into broad groups. The designation of four main divisions (pyogenic, viridans, lactic and enterococcus) by Sherman (1937) proved to be useful, at least for descriptive purposes. More recently, Jones (1978) divided the genus into 7 groups designated pyogenic, pneumococci, oral, fecal, lactic, anaerobic and other streptococci. Although such groupings are admit-tedly artificial and, in most cases, have no strict taxonomic validity, they provide a convenient framework within which to describe the various aspects. A similar arrangement to that of Jones (1978) has been adopted here, except that *S. pneumoniae* has been included in the pyogenic group. A recent numerical study of *Streptococcus* revealed 28 reasonably distinct phenons, which could be grouped into seven main areas named enterococcal species group, paraviridans species group, lactic species group, thermophilic species group, viridans species group, pyogenic species group and a parapyogenic species group (Bridge and Sneath, 1983). *S. pneumoniae* was included in the viridans species group in this study.

A recent trend in streptococcal taxonomy has been towards the more extensive application of numerical, chemotaxonomic and genetic tech-niques, with less emphasis on serological critiera for classification. Although serology remains important for identification and typing of some major pathogens, it cannot be used as the main basis for estab-lishing fundamental taxonomic relationships between streptococci. Subdivisions within the genus are indicated by DNA/rRNA hybridi-zation studies (Garvie and Farrow, 1981; Schleifer and Kilpper-Bälz, 1984). Genetic relationships at the species level have been investigated by DNA/DNA hybridization, particularly among the "*S. mutans*-group" (Coykendall, 1971, 1974, 1977; Schleifer et al., 1984), with the result that five distinct species are now recognized (*S. mutans*, *S. rattus*, *S. sobrinus*, *S. cricetus*, *S. ferus*).

As indicated in some of the later species descriptions, a number of taxonomic and nomenclatural uncertainties remain to be resolved within the genus *Streptococcus*. Further work is required to clarify the species designation of some of the streptococci presently described as Lancefield groups C, E, G, L, M, P, U and V. Recently, it has been proposed the strains belonging to groups E, P, U and V, mainly isolated from pigs, represent a new species designated *S. porcinus* (Collins et al., 1984b). Problems still exist among the oral streptococci, notably those named *S. mitis* and "*S. mitior*," and also with the "*S. milleri*-group" (including *S. anginosus*, *S. constellatus*, *S. intermedius* and "*S. milleri*").

The inclusion of some obligately anaerobic species (*S. morbillorum*, *S. hansenii*, *S. pleomorphus* and *S. parvulus*) is a new, possibly contro-versial, departure, although their fermentative activity, with the pro-duction of predominantly lactic acid from glucose, would appear to support this move (Holdeman and Moore, 1974; Barnes et al., 1977; Cato, 1983).

In view of the intensive research activities currently in progress in several laboratories around the world, it is inevitable that further changes in the taxonomy and nomenclature of streptococci will occur over the next few years. Almost certainly a major revision of the genus *Streptococcus* will be necessary by the time of the next edition of this *Manual*.

Further Reading

Facklam, R. and Wilkinson, H.W. 1981. The family *Streptococcaceae* (medical aspects). *In* Starr, Stolp, Trüper, Balows and Schlegel (Editors), The Prokaryotes. A Handbook on Habitats, Isolation and Identification of Bacteria, Springer-Verlag, New York.

Holm, S.E. and Christensen, P. (Editors). 1982. Basic Concepts of Streptococci and Streptococcal Diseases, Reedbooks, Chertsey, Surrey, Great Britain.

Jones, D. 1978. Composition and differentiation of the genus *Streptococcus*. *In*

Skinner and Quesnel (Editors), Streptococci, Society for Applied Bacteriology Symposium Series No. 7, Academic Press, London.

Parker, M.T. (Editor). 1979. Pathogenic Streptococci, Reedbooks, Chertsey, Surrey, Great Britain.

Parker, M.T. 1983. *Streptococcus* and *Lactobacillus*. *In* Wilson, Miles and Parker (Editors), Topley and Wilson's Principles of Bacteriology, Virology and Immunity, 7th Ed., Vol. II, Edward Arnold, London.

Read, S.E. and Zabriskie, J.B. (Editors). 1980. Streptococcal Disease and the Immune Response, Academic Press, London.

Sharpe, M.E. 1979. Identification of the lactic acid bacteria. *In* Skinner and Lovelock (Editors), Identification Methods for Microbiologists, Society for Applied Bacteriology Technical Series No. 14, 2nd Ed., Academic Press, London.

Wannamaker, L.W. and Matsen, J.M. (Editors). 1972. Streptococci and Streptococcal Diseases, Academic Press, New York.

Differentiation of the species of the genus **Streptococcus**

Characteristics useful for differentiating the species of *Streptococcus* are listed in Table 12.15.

List of species of the genus **Streptococcus**

The following list of species and their descriptions is broken down into the broad categories of pyogenic, oral, enterococci, lactic, anaerobic and other streptococci (see Table 12.15).

Pyogenic Hemolytic Streptococci

JIRI ROTTA

The term "hemolytic streptococci" is currently used to designate streptococci producing β- (complete) hemolysis on blood agar plates and usually containing a polysaccharide antigen in the cell wall. This polysaccharide antigen was used by Lancefield (1933) for the classification of these organisms into serological groups (so-called Lancefield's scheme).

Hemolytic streptococci are pathogenic for man or animals, or for both. A medically outstanding biological property is to produce various diseases with symptoms of pyogenic infection. However, a few strains of pyogenic streptococci produce α- or no hemolysis, and occasional strains of species other than pyogenic streptococci produce β, α or no hemolysis and can provoke pyogenic infection in animals or even in man.

In general, pyogenic hemolytic streptococci have the morphological appearance characteristic of the genus *Streptococcus* growing in short, medium or long chains. An exception is *S. pneumoniae*, which grows in pairs of cocci rather than chains. For good growth a nutritionally complex medium is required, e.g. one containing defibrinated animal blood, or serum. The optimal temperature for growth is 37°C. Can grow under aerobic conditions. Killed by heating to 56°C. Metabolic pathways are generally comparable with those of the other members of the genus *Streptococcus*. Particular metabolic products and identification of characteristic cellular or extracellular products for classification and diagnostic purposes are indicated in the descriptions of the particular species. Strains of pyogenic streptococci can be preserved by a number of methods: for short periods it is convenient to keep them frozen in a broth culture, for a long term period lyophilization is used.

Biochemical, antigenic and other characteristics of the pyogenic streptococci are shown in Tables 12.16 and 12.17.

1. **Streptococcus pyogenes** Rosenbach 1884, 23.[AL]

py.og'en.es. Gr. n. *pyum* pus; Gr. v. *gennaio* beget; M.L. adj. *pyogenes* pus-producing.

Constitutes Lancefield's (1933) group A streptococci.

Spherical cells; ovoid forms may occur, but usually in older cultures; diameter of cells 0.5–1.0 μm. Growth in short or moderately long chains, in clinical material also occur in pairs, long chains frequent in broth cultures.

Chemically the cell wall is made up of four constituents: protein, polysaccharide and peptidoglycan are interwoven rather than arranged in separate concentrical layers as assumed previously; the fourth component is teichoic acid. Cell wall thickness is about 20 nm. On the outer layer of the cell wall is a fringe of fimbriae, earlier referred to as a microcapsule. In some strains the cell wall with fimbriae can be enveloped by a capsule of hyaluronic acid.

The rigidity of the cell wall structure is provided by the peptidoglycan. The peptidoglycan molecule is of the A3 α-type, (Schleifer and Kandler, 1972), containing a polysaccharide polymer with tetrapeptide subunits that consist of L-Ala, D-*iso*Gln, L-Lys, and D-Ala cross-linked through an interpeptide bridge L-Ala-L-Ala located between the terminal D-Ala and subterminal L-Lys. Some of the peptide subunits are not cross-linked, possess one more D-Ala and thus form pentapeptides. The cell wall polysaccharide ("C" polysaccharide), which is specific to the species (group A-specific), is attached to the peptidoglycan by phosphate-containing bridges composed of one or more units having glycerol or glyceryl-rhamnoside as an organic moiety (Barkulis, 1968). The antigens of the protein layer are linked to the peptidoglycan by covalent bonds.

After overnight growth on blood agar three major visual colony types exhibiting β- (complete) hemolysis may form: mucoid, matt (dehydrated mucoid) or glossy. The colony type depends largely on production of hyaluronic acid and growth conditions.

The optimal temperature for growth is 37°C. Growth is enhanced by supplementation of broth with blood or serum. Growth requirements of 22 amino acids, 6 vitamins and some other substances have been established on a chemically defined medium (Ginsburg and Grossowicz, 1957).

Energy yielding metabolism is fermentative and the final pH in glucose broth is 4.8–6.0. Acid is produced from glucose, maltose, sucrose and salicin. No acid from inulin, raffinose, arabinose or common polyols.

Substances with a bactericidal effect are produced either as a typical bacteriocin (Tagg et al., 1973) or as a low weight and dialyzable bacteriocidal substance (Wolff and Duncan, 1974).

Virulent bacteriophages for group A strains (e.g. for serovars 1, 6, 12 and 25) exist, but their lytic properties are not strictly group specific. Temperate bacteriophages are very common in group A strains and may function as vectors of genetic information. For example, lysogenicity is closely associated with erythrogenic toxin production (Zabriskie, 1964a, b) and temperate bacteriophage may transduce resistance to erythromycin, lincomycin and streptogramin B (Malke, 1975).

Phage-typing has not been established, except for phage types re-

Table 12.15.
Differential reactions of species of **Streptococcus**[a]

Characteristics	Pyogenic streptococci					Oral streptococci									Enterococci				Lactic streptococci		Anaerobic streptococci				Other streptococci				
	1. *S. pyogenes*	2. *S. agalactiae*	3. *S. equi*	4. *S. iniae*	5. *S. pneumoniae*	6. *S. salivarius*	7. *S. sanguis*	8. *"S. mitior"*	9. *"S. milleri"*	10. *S. mutans*	11. *S. rattus*	12. *S. cricetus*	13. *S. sobrinus*	14. *S. ferus*	15. *S. faecalis*	16. *S. faecium*	17. *S. avium*	18. *S. gallinarum*	19. *S. lactis*	20. *S. raffinolactis*	21. *S. morbillorum*	22. *S. hansenii*	23. *S. pleomorphus*	24. *S. parvulus*	25. *S. acidominimus*	26. *S. uberis*	27. *S. bovis*	28. *S. equinus*	29. *S. thermophilus*
Growth at 10°C	−	d	−	+	−	−	−	−	−	−	−	−	−	−	+	+	+	+	+	+	NT	−	−	−	−	+	−	−	−
Growth at 45°C	−	−	−	−	−	d	d	d	d	d	d	d	d	−	+	+	+	+	−	−	NT	+	+	+	−	−	d	+	+
Growth at 6.5% NaCl	−	d	−	−	−	−	−	−	−	−	−	d	d	−	+	+	+	+	−	−	NT	NT	NT	−	NT	−	−	−	−
Growth at pH 9.6	−	−	−	−	−	d	d	−	d	d	d	d	d	NT	+	+	+	+	+	−	NT	NT	NT	NT	−	−	−	−	−
Growth with 40% bile	−	d	−	−	−	d	+	+	−	−	−	−	−	−	+	d	d	+	+	−	d	NT	NT	−	NT	d	+	+	−
α-Hemolysis	−	−	−	+[b]	+	−	−	+	−[d]	−	−	−	−	−	+	d	d	d	d	d	−	−	−	−	d	d	wk[c]	wk[c]	d
β-Hemolysis	+	d	+	+	−	−	−	−	d	−	+	−	−	−	+	+	−	d	d	−	−	−	NT	−	−	+	−	−	−
Arginine hydrolysis	+	+	+	NT	+	−	+	−	d	−	+	−	−	−	+	+	−	d	d	d	−	−	−	−	+[e]	+	−	−	d
Hippurate hydrolysis	−	+	−	+	−	−	−	−	−	−	+	−	−	+	+	+	+	+	d	d	−	d	−	−	−	+	−	−	−
Esculin hydrolysis	d	−	d	−	d	+	d	−	d	+	+	d	d	+	−	−	−	−	−	+	−	−	NT	+	−	+	+	+	−
Obligate anaerobe	−	−	−	−	−	−	−	−	−[f]	−[f]	−[f]	−[f]	−[f]	−[f]	−	−	−	−	−	−	+	+	+	+	−	−	−	−	−

[a] Symbols: see Table 12.2; also NT, not tested.
[b] Outer ring of α-hemolysis.
[c] Usually weak reaction.
[d] Strains called *S. anginosus* may be positive.
[e] Often slow.
[f] Some strains are microaerophilic or require added CO_2.

Table 12.16.

Biochemical characteristics of the pyogenic streptococci[a,b]

Characteristics	S. pyo-genes	S. agalac-tiae	S. equi	S. dysga-lactiae	Groups C, G and L	S. iniae	S. pneumo-niae	Groups E, P, U, V (S. porcinus)
Acid from								
Inulin	−	−	−	−	−	−	d	−
Lactose	+	d	−	+	+	−	+	d
Mannitol	−	−	−	−	−	+	−	+
Raffinose	−	−	−	−	−	−	+	−
Ribose	−	+	−	+	+	NT	−	+
Salicin	+	d	+	d	NT	+	NT	+
Sorbitol	−	−	−	d	−	−	−	+
Trehalose	+	+	−	+	+	+	+	+
Hydrolysis of								
Arginine	+	+	+	+	+	NT	+	+
Esculin	d	−	d	−	−	+	d	+
Hippurate	−	+	−	−	d	−	−	−
Voges-Proskauer	−	+	−	−	−	NT	−	+
Production of								
Alkaline phosphatase	+	+	+	+	+	NT	−	+
α-Galactosidase	−	d	−	−	−	NT	+	−
β-Glucuronidase	d	d	+	+	+	NT	−	+
Pyrrolidonylarylamidase	+	−	−	−	−	NT	d	−
β-hemolysis	+	+	+	−	d	+	−	+
Sensitive to optochin	−	−	−	−	−	−	+	−
Lancefield group	A	B	C	C	C, G, L	−	−	E, P. U, V
Resistant to 40% bile	−	d	−	−	NT	−	NT	NT

[a] Symbols: see Table 12.2; also NT, not tested.

[b] No strains produce polysaccharides from sucrose or acid from arabinose.

Table 12.17.

Some antigenic, chemical and pathogenic properties of pyogenic hemolytic streptococci[a]

Species	Group antigen (polysaccharide)		Type antigen(s)			Other major cell antigens	Pathogenicity for
	Symbol	Major chemical constituents	Symbol	Chemical nature	Location		
S. pyogenes	A	N-Acetylglucosamine, rhamnose	M	Protein	Fimbriae	LTA	Man
			SOF	Protein	CW	MAP	Monkey, mouse
			T	Protein	CW	R protein	
S. agalactiae	B	Rhamnose, galactose, N-acetylglucosamine	I–IV	Polysaccharides	Capsule		Man
			Ib, c, R, X	Proteins	CW		Animals
S. equi	C	N-Acetylgalactosamine, rhamnose		Proteins	CW	R protein	Horse
S. sp. group C	C	N-Acetylgalactosamine, rhamnose		Proteins	CW	R protein	Man Animals
S. sp. group G	G	Rhamnose, galactosamine		Proteins	CW	R protein	Man Dog
S. sp. group L	L	N-Acetylglucosamine, N-acetylgalactosamine, galactose, rhamnose		Proteins	CW	R protein	Dog Man
S. sp. group M	M	Not known		Proteins			Dog, man
S. sp. group P	P	Not known					Pig
S. sp. group U	U	Not known					Pig
S. sp. group V	V	Not known					Pig
S. iniae		Not known					Dolphin
S pneumoniae		Ribitol teichoic acid, choline phosphate	SSS	Polysaccharides	Capsule	M-like proteins	Man

[a] Symbols: LTA, lipoteichoic acid; SOF, serum opacity factor; CW, cell wall; MAP, M-associated protein; SSS, specific soluble substance.

ported within one serological type, the M type 49 (Skjold and Wanna-maker, 1976). A plasmid involved in resistance to erythromycin, lincomycin and streptogramin B has been isolated from *S. pyogenes* and its molecular weight has been found to be 17×10^6 (Clewell and Franke, 1974).

Cell wall antigenic components. Antigenic components essential in classification and diagnosis are located mainly in the cell wall.

The group A specificity of *S. pyogenes* is, due to the presence of a polysaccharide, a multibranch structure and composed of N-acetylglucosamine and rhamnose in a molar ratio of 1:2.47, molecular weight

8000. The immunodominant component of the polysaccharide is *N*-acetylglucosamine, located in the terminal positions of the molecule. Enzymatic or chemical degradation of this compound unmasks the subterminal rhamnose-rhamnose structure, which is also antigenic and represents the immunodeterminant structure of the "group A-variant" polysaccharide.

M protein type antigen. This heat stable, trypsin-sensitive protein is situated in the fimbriae on the surface of the cell wall. It is the principal factor of virulence, rendering the organism resistant to phagocytosis, and is, by definition, the type-specific substance. Presence of peptides in the medium and absence of proteinase formation are essential for its production (Fox, 1974). M protein can be isolated by a number of procedures such as HCl extraction at pH 2.0, treatment with phage-associated lysin, mild pepsin digestion, treatment with detergents or other techniques, and can be further purified. Structural studies indicate that the molecule of M protein is composed of subunits made of repeating covalently bound polypeptides, some of them having identical amino acid sequence in certain portions. For example, the M24 protein is composed of seven polypeptides with molecular weights of 4000–5000 or 9000 (Beachey et al., 1978). The M protein contains a number of immunodominant groups, with some degree of heterogeneity, located within the polypeptides. Partially related polypeptides occur in different M types, which provide some degree of biological relationship between the types concerned. Serovars M1–M71 have been identified and a number of serovars above M71 have been described but taxonomically are not yet accepted. Numbers M7 and M21 are not used because the corresponding reference strains have been shown to belong to group C. Similarly M16 belongs to group G. Serovars 35 and 49 are the same, and serovars 64 and 52; only the numbers 49 and 52 are used. Antigenic M protein relationships have been found between serovars 5, 6 and 24; 33, 41, 43 and 52; 3 and 31; 46 and 51; 55 and 60; and they may possibly exist between other serovars as well. The term "nephritogenic type" is applied to serovars 12, 49, above 52 and some others which cause infections followed with some frequency by acute glomerulonephritis. The term "rheumatogenic" strain or serovar is of rather limited value since a strain of any serovar may cause infection followed by acute rheumatic fever. In routine M typing of group A streptococci, an HCl extract of harvested cells is used as antigen against absorbed rabbit hyperimmune sera in the capillary test (Swift et al., 1943) or against unabsorbed sera in the agar double-diffusion test (Rotta et al., 1971). The M-associated protein (MAP) is an antigen which is difficult to separate from M protein. It is antigenically related to sarcolemmal components in the myocardium and is considered responsible for the delayed hypersensitivity skin reactions to most M protein preparations.

T protein type antigens which are trypsin resistant occur in several antigenic patterns, most of them shared with a number of M types. T antigen has also been found in some group C and G strains. Typing of group A streptococci by the agglutination reaction is based on the demonstration of T antigens by absorbed rabbit hyperimmune sera (Griffith, 1934).

R protein antigens are also trypsin resistant and occur in some strains of some types only. They have also been found in strains of groups other than A and have been identified in four antigenic forms, the most common being R28 and R3 (Wilkinson, 1972). They have no relation to virulence.

Serum opacity factor (SOF) is a trypsin-sensitive protein antigen causing opacity of serum. It is produced by all strains belonging to some M serovars, e.g. M 2, 4, 9, 11, 13, 22 and about ten others, while it is absent from all strains belonging to other M serovars (see Maxted 1978). Being M-type specific, it is used for typing group A streptococci that are nontypable by M antigen (Widdowson et al., 1971).

The Fc reactive factor (Kronvall, 1973) is a protein localized on the surface of some strains. It binds the Fc fragment of the heavy chain of IgG. This factor, however, is also present in some strains of groups C and G.

Lipoteichoic acid is a polymer of glycerol attached by 1,3-phospho-diester bonds with ester-linked substitutes of D-Ala and various glycosyl and fatty acid residues (Beachey, 1975). It is a constituent of fimbriae on the surface of the cell wall and is responsible for the adherence of streptococci to epithelial cells of the mucosa.

Cytoplasm. In the cytoplasm the following substances have been studied: a nucleoprotein antigenically related to the nucleoproteins of staphylococci and pneumococci, an intracellular hemolysin, a β-glucuronidase, an immunosuppressive substance and endostreptosin.

Extracellular products. These include substances of diagnostic and biological importance. Titration of antibodies to several of these products (e.g. antistreptolysin O, anti-DNase B) is used in diagnostic work. Erythrogenic toxin (scarlet fever toxin, streptococcal pyrogenic exotoxin) elicits the rash in scarlet fever. It exists in at least three antigenically distinct types, A, B and C (Watson, 1979). For its production, the streptococcus must be in a lysogenic state (Zabriskie, 1964a, b). The majority of strains produce toxins of types B and C simultaneously.

Streptolysin O is an oxygen-labile hemolysin which is reversibly activated from the reduced form. Biologically it is closely related to pneumolysin, tetanolysin, cereolysin and the hemolysins of some other bacteria. It is antigenic and the antigen and its antibody can be quantified in a standard hemolytic system using a reference serum.

Streptolysin S is an oxygen-stable, heat-labile, nonantigenic hemolysin producing β-hemolysis around colonies on blood agar. Ribonucleic acid, serum albumin, detergents and other substances can act as carriers for the hemolytic component.

Streptokinase (once termed fibrinolysin) is produced by strains of group A and some other groups. It exists in at least two antigenic types (Dillon and Wannamaker, 1965). The enzyme activates the conversion of plasminogen to plasmin, and this principle is used in the titration of the enzyme and its antibody.

DNase exists in four antigenic variants known as types A, B, C and D (Wannamaker, 1962). Most strains produce two or more types of the enzyme, type B being the most common. DNase is only active in the presence of bivalent cations. For the detection of the antigen and its antibody several methods have been elaborated, the most common one being based on colorimetry of the DNA-methyl green complex after hydrolysis by the enzyme. To titrate antibody in human sera, a type-B enzyme is used.

NADase is produced by strains of groups A, C, and G (Carlson et al., 1957). There is some relationship between production of the enzyme and the M type of the strain. One antigenic type only is known.

Hyaluronidase production is common in group A streptococci, since corresponding antibody develops very frequently after infection. In contrast, production of the enzyme in vitro is rare. The enzyme is also produced by strains of groups B, C and G. The hyaluronidases associated with temperate bacteriophages, e.g. derived from M types 49 and 4 (Benchetrit et al., 1979), are antigenically different from the nonphage-associated enzyme.

Proteinase is frequently produced by group A strains. It is responsible for the degradation of M protein in cultures, rendering typing impossible. The enzyme is produced as an inactive precursor; it has been prepared in crystalline form, molecular weight 44,000 (Elliot, 1950).

Antibiotic sensitivity: All group A strains are highly sensitive to penicillin the minimal inhibitory concentration (MIC) being 0.016 μg/ml, but in many strains as low as 0.004 μg/ml. Strains are also sensitive to a number of other antibiotics; bacitracin sensitivity of group A streptococci is used to differentiate this group from strains of other serological groups (Maxted, 1953).

Pathogenicity: Under natural conditions, man: possibly but (if so) rarely, some animals, e.g. monkeys.

Sources: Human upper respiratory tract, skin lesions, inflammatory exudates, blood, contaminated dust in the environment.

The mol% G + C of the DNA ranges from 34.5 and 38.5 (T_m) to 38–40 (method not stated).

Type strain: ATCC 12344.

2. **Streptococcus agalactiae** Lehmann and Neumann 1896, 126[AL]; nom. cons. Opin. 8 Jud. Comm. 1954, 152.

a.ga.lac′ti.ae. Gr. n. *Agalactia* want of milk, agalactia; M.L. gen. n. *agalactiae* of agalactia.

Spherical or ovoid cells 0.6–1.2 µm in diameter, occurring in chains of seldom less than four cells and frequently very long. Chains may appear to be composed of paired cocci.

The taxonomic classification of *S. agalactiae* is conveniently based on a specific cell wall polysaccharide, which is characteristic for Lancefield's group B, and on some distinct physiological properties. Attempts have been made to differentiate, taxonomically, human and animal strains of this species into clinical group B streptococci and *S. agalactiae*, respectively. Since no clear differences have yet been established, both human and animal pathogens will be considered in this section as one species.

The group B-specific cell wall polysaccharide antigen is composed of rhamnose, N-acetylglucosamine and galactose (Curtis and Krause, 1964). As rhamnose is expressed more than the other two sugars in the immunodominant structure of the polysaccharide, serological cross-reactivity between groups B and G is sometimes encountered because, in the latter group, rhamnose also has a dominant position in the group-specific polysaccharide.

Energy yielding metabolism is fermentative with lactic acid constituting the chief end product. Final pH in glucose broth is 4.2–4.8. Acid produced from glucose, maltose, sucrose and trehalose. Glycerol fermented only aerobically. Lactose usually fermented by strains from bovine sources but this characteristic may be variable in strains isolated from other animals and humans. Xylose, arabinose, raffinose, inulin, mannitol and sorbitol are not fermented.

Many strains can grow in media containing 40% bile. All strains hydrolyze sodium hippurate and this test can be conveniently used to differentiate them from enterococci, which do not hydrolyze sodium hippurate.

Group B streptococci grow readily on blood agar, and exhibit various types of hemolysis viz. typical β- but with a narrow zone, α- double-zone, or no hemolysis. Some strains produce a characteristic opaque β-hemolysis, probably due to a soluble hemolysin with a low hemolytic activity. This hemolysin is different from both hemolysin O and S. The CAMP factor (Christie et al., 1944) produced by group B streptococci binds to erythrocyte membrane altered by *Staphylococcus* sphingomyelinase C and this results in profound lysis of the erythrocytes. This CAMP test is not specific to group B streptococci, however, since it can also be positive with strains of some other groups, e.g. C, F or G.

Some strains produce a yellow, orange or brick-red pigment and production may be enhanced by addition of starch to the medium or by anaerobic incubation.

Bacteriocin production has been encountered although infrequently (Tagg et al., 1975), and so has sensitivity to bacteriocins in other strains; both properties may coexist in one strain. Group B streptococcus phages have been identified and recently used for differentiation into phage types (Stringer and Maxted, 1979).

Eight serological types have been established so far: Ia, Ib, Ic, II, III, IV, R and X. They are based on the capsular polysaccharide antigens Ia, Ib, II, III and IV, and on the protein antigens Ic, R and X (Lancefield, 1934, 1938; Wilkinson and Eagon, 1971; Pattison et al., 1955; Jelínková, 1982). The capsular polysaccharide antigens are virulence factors and the respective antibodies confer immunity, which is type specific.

Little basic information is available on hyaluronidase, although frequently produced by many strains, or on DNase, three serological types different from those of group A streptococci being known (Ferrieri et al., 1980). Neuraminidase, has been identified in the culture filtrate of some strains.

Antibiotic sensitivity: The MIC for penicillin is about 0.064 µg/ml and that for ampicillin much higher than in group A streptococci. Erythromycin-resistant strains have appeared recently.

Pathogenicity: In man, group B streptococci can produce a variety of clinical conditions, some of them very serious, others less so. In neonates and children may cause sepsis, pneumonia, meningitis, conjunctivitis and otitis media. In adults may cause meningitis, pneumonia, pyelonephritis and (in women) infections of the urogenital tract. In animals these organisms are one of the main causes of bovine mastitis. Sources of infection in man are the vaginal mucosa, the upper respiratory tract, urine, stool and, in animals, milk and udder tissues.

The mol% G + C of the DNA is 34.0 (T_m; Garvie and Farrow, 1981)

Type strain: NCTC 8181.

3. **Streptococcus equi** Sand and Jensen 1888, 436.[AL]

e′qui. L. n. *equus* horse; L. gen. n. *equi* of a horse.

Ovoid or spherical cells 0.6–1.0 µm in diameter; sometimes in pus the long axis of the cells is transverse to the long axis of the chain and at other times parallel with the long axis of the chain, in the latter case resembling streptobacilli; bacillary forms may occur; grow in pairs, short or long chains, very long chains being common in broth. Capsules are demonstrable in some strains when young cultures are examined or when serum is added to the growth medium.

Belong to Lancefield's group C (see *Streptococcus* sp. (group C) for description of group antigen). Only one serotype; the type-specific antigen is a protein, but apparently not associated with immunity to infection. A protective antigen distinct from the envelope type antigen can be demonstrated in capsular material.

Final pH in glucose broth is 4.8–5.5; a restricted fermentative pattern: acid from glucose, maltose, sucrose and salicin, but no acid from arabinose, lactose, trehalose, raffinose, inulin, glycerol, mannitol or sorbitol.

Wide zones of β-hemolysis observed on blood agar where the small, watery colonies dry out rapidly and ultimately leave flat, glistening colonies. A soluble hemolysin, distinct from streptolysins O and S, is produced in serum-fortified broth culture.

Growth in common laboratory media is poor unless fortified with serum. The minimum nutritional requirements of some strains have been partially identified; in a casein hydrolysate medium, at least two B vitamins and uracil are required. The amino acid requirements have not been determined.

Minimum temperature for growth is approximately 20°C.

Pathogenicity: Causes equine strangles. Isolated from abscesses in submaxillary glands and mucopurulent discharges of the upper respiratory system of horses and their immediate environment. Rarely isolated from other animals.

Type strain: NCTC 9682.

4. **Streptococcus iniae** Pier and Madin 1976, 552.[AL]

in′i.ae. M.L. of the dolphin, *Inia*.

Spherical cells, encapsulated, up to 1.5 µm diameter, in broth culture cocci arranged in long chains. On blood agar small colonies up to 1 mm diameter, with opaque center and translucent border, colonies surrounded by a small to moderate area of β-hemolysis passing to α-hemolysis.

Good growth in Todd Hewitt broth when incubated overnight at 37°C under aerobic conditions. Acid is produced from dextran, fructose, galactose, glucose and some other sugars; starch and esculin are hydrolyzed, sodium hippurate and gelatin are not; no growth in bile-esculin media or at 45°C.

Lancefield's serological group not yet established. However, this streptococcus contains a specific antigen extractable by HCl or formamide, so that it can be compared with the group antigens of other Lancefield groups. The antigen does not react with group-specific antisera A-V, but strongly reacts by precipitation reaction with rabbit hyperimmune antisera to *S. iniae*.

Antibiotic sensitivity: This closely resembles that of groups A, C and G streptococci, particularly with regard to the aminoglycoside antibiotics. Strains of *S. iniae* are also susceptible to gentamicin and to bacitracin, but less susceptible than group A streptococci.

Pathogenicity: Not pathogenic for guinea pig, mouse and rabbit in model infections using an infective dose of 10^7–10^8 cfu.

Source: A freshwater dolphin, *Inia geoffrensis*, living in the Amazon river. *Streptococcus iniae* produces subcutaneous abscess foci on the thorax and abdomen.

The mol% G + C of the DNA is 32.9.

Type strain: ATCC 29178.

5. **Streptococcus pneumoniae** (Klein) Chester 1901, 63.[AL] (*Micrococcus pneumoniae* Klein 1884, 329.)

pneu.mo′ni.ae. Gr. n. *pneumon* the lungs; M.L. fem. n. *pneumonia* pneumonia. M.L. gen. n. *pneumoniae* of pneumonia.

Oval or spherical, coccal-like forms 0.5–1.25 μm typically in pairs, occasionally singly or in short chains. The distal ends of each pair of organisms tend to be pointed or lance shaped. On primary isolation, generally heavily encapsulated with polysaccharide (termed SSS, specific soluble substance). Continued growth in laboratory media promotes chain formation. Gram-positive reaction of young cells may be lost as culture ages and subsequently stains Gram-negative.

A tentative structure of the peptide subunit of the peptidoglycan (Mosser and Tomasz, 1970) indicates identity with that of *S. pyogenes*. No bridge amino acid or peptide described. The major polymeric components of the cell wall are the peptidoglycan and the choline-ribitol teichoic acid complex (Mosser and Tomasz, 1970) which constitutes the species-specific substance.

Mucoid colonies result from copious capsular polysaccharide synthesis. Smooth colonies are glistening and dome shaped, and reflect decreased capsular polysaccharide. Rough colonies occur rarely and have a wrinkled, mycelium-like appearance. "Phantom" colonies reflect early and rapid partial autolysis of a mucoid colony which is suppressed by incubation under increased CO_2 tension.

Strong α-hemolysis on blood agar when cultures incubated aerobically. Anaerobic incubation results in β-hemolysis due to pneumolysin O (identical to streptolysin O) activity.

The addition of blood, serum or ascitic fluid to media enhances growth especially on primary isolation.

In contrast to other streptococci, the pneumococci require choline for growth in defined media. Ethanolamine replaces choline but not on a molar basis (Badger, 1944). Reducing agents are almost essential. Most strains require for growth at least 4 of the B vitamins, also adenine, guanine and uracil and 7–10 amino acids.

The addition of bile or bile salts to a neutralized culture activates an autolytic amidase which cleaves the bond between alanine and muramic acid in the peptidoglycan (Mosser and Tomasz, 1970); thus the organisms are bile soluble.

The energy-yielding metabolism is fermentative, yielding primarily low levels of lactic acid. Optimum pH is 7.8 with a range of 6.5–8.3. Final pH in glucose broth approximately 5.0. Aerobically, a significant quantity of H_2O_2 accumulates as well as acetic and formic acids. Glucose, galactose, fructose, sucrose, lactose, maltose, raffinose, glycogen and inulin are fermented. Slow acid production from glycerol (aerobic incubation), xylose, arabinose and erythritol. Some strains may ferment mannitol. No acid from dulcitol or sorbitol.

Facultative anaerobe with marked tendency to accumulate H_2O_2 on aerobic culture. Temperature range 25–42°C.

Many genetic markers have been transformed among various strains of pneumococci. Reciprocal intraspecies and intrageneric transformations have been demonstrated (Bracco et al., 1957; Pakula et al., 1958; Ravin and De Sa, 1964; Chen and Ravin, 1966).

A species-specific, somatic carbohydrate distinct from that of other species has been demonstrated (Tillett and Francis, 1930) and consists of a ribitol teichoic acid with choline phosphate (Brundish and Baddiley, 1968). The antigenic determinant of this polysaccharide is the *N*-acetylgalactosamine phosphate residue (Gotschlich and Liu, 1967). The choline phosphate component is responsible for the immunological reactivity of pneumococci with C-reactive protein (Kaplan and Volanakis, 1974).

The capsular polysaccharide, SSS, is the virulence factor and type-specific antigen. The antibody to this polysaccharide provides immunity to pneumococcal infection, i.e. immunity is type specific. The polysaccharide can be prepared from cells grown in overnight culture in a medium with 5% serum, using the following procedure: lysis of cells with bile, separation of constituents other than polysaccharide from the lysate by precipitation with acetic acid and precipitation of the polysaccharide from the supernatant fluid with ethanol. Presence of the polysaccharide in the capsule can be demonstrated by the Neufeld reaction; polyvalent antipneumococcal serum binds with the capsular polysaccharide and this results in clear visibility of the capsule and in agglutination of the cells. For rapid diagnosis of pneumococcal meningitis or pneumonia, the capsular polysaccharide can be demonstrated in cerebrospinal fluid or blood by counter immunoelectrophoresis (Artman et al., 1980) or indirect enzyme immunoassay (Drow and Manning, 1980).

Two typing systems have been elaborated for pneumococci. The Danish scheme after Kauffmann et al. (1960) is the one most frequently used. It is based on the principle that types sharing common antigens are combined in groups, which are designated by figures, and within the groups types are differentiated. At present, a scheme of 46 groups marked 1–48 (the figures 26 and 30 are not used) is employed. The types within the groups are marked with letters in alphabetical order, except for the first type identified in most groups, which is labeled F (for "first").

Until the 1960s, pneumococci were very sensitive to penicillin. Later, the appearance of penicillin-resistant strains was repeatedly reported (Tarpay, 1978). The resistance of some pneumococcal strains to β-lactam antibiotics is of an intrinsic nature, since these strains do not produce β-lactamases (Zighelboim and Tomasz, 1981). The trend of increasing antibiotic resistance has prompted the development of vaccine against pneumococcal infection, which is now available for prophylactic purposes.

Pneumococci are inhibited by approximately 1:400 ethylhydrocupreine HCl (Optochin); this sensitivity test and the bile-solubility test can be used to differentiate pneumococci from other streptococci (Bowers and Jeffries, 1955).

Pathogenicity: The clinical patterns of pneumococcal infection in man are numerous. They include pneumonia, meningitis, otitis media and some less frequent conditions such as abscesses, conjunctivitis, pericarditis and arthritis. In animals, pneumococci may occasionally cause mastitis and septicaemia in cows, sheep and goats, and respiratory tract infections in monkeys.

Source: Upper respiratory tract, inflammatory exudates and various body fluids of diseased humans and, rarely, domestic animals. Upper respiratory tract of normal humans and domestic animals.

The mol% G + C of the DNA ranges from 38.5 (chemical analysis) to 30 (T_m) and 42 (Bd).

Type strain: NCTC 7465.

Species Incertae Sedis

Several biological taxa of streptococci have been identified for which species names have not yet been proposed or the species have been poorly defined and the names are, therefore, not generally used. The identification of these taxa is based on particular characteristics, of which the presence of an antigenically group-specific polysaccharide is a major criterion. These taxa are referred to as serological groups. The taxonomic relationships among them have not been satisfactorily explored so far.

a. Streptococcus sp. (group C)

This taxon includes streptococci pathogenic for man or animals and which carry the group C-specific polysaccharide but which do not belong to *S. equi*. In the past, most of these streptococci were divided into three species on the basis of some biochemical properties. These "species" and their markers were as follows: *S. equisimilis*" Frost and Engelbrecht 1936, 3 (type B, Ogura 1929, 174; Human C, Sherman 1937, 35), acid from trehalose and glycerol; *S. dysgalactiae* Diernhofer

1932, 369 (Group II, Minett 1934, 511; "*S. pseudoagalactiae*" Plastridge and Hartsell 1937, 110.), acid from trehalose but not glycerol; and "*S. zooepidemicus*" Frost and Engelbrecht 1936, 3. (Animal pyogenes, type A Edwards 1934, 527; "*S. pyogenes animalis*" Seelemann 1942, 8), no acid from trehalose or glycerol but acid from sorbitol. While the strains classified as "*S. equisimilis*" were predominantly pathogens of man and rarely of animals, those of the other two "species" were animal pathogens only. The strains classified as *S. dysgalactiae* were α-hemolytic; strains belonging to "*S. equisimilis*" and "*S. zooepidemicus*" were typically β-hemolytic.

Since the biological differences between these three organisms are at present not considered to substantiate sufficiently the establishment of separate species, the strains are taxonomically referred to one taxon, namely group C.

Their cellular and colonial morphology is similar to *S. pyogenes*. Some animal strains produce a large quantity of hyaluronic acid and grow on blood agar in mucoid colonies.

The major known constituents of the cell wall in these streptococci are protein, polysaccharide and peptidoglycan.

Final pH in glucose broth 4.4–5.4. Acid is produced by all strains from glucose, maltose and sucrose, and in some strains from salicin, lactose, trehalose, glycerin and sorbitol; it is not produced from raffinose, inulin and mannitol.

Grow only on nutritionally complex media; the minimal nutritional requirements are unknown. Some strains (human) produce streptolysin O; the production of a hemolysin (S) unrelated to streptolysin O or streptolysin S is frequent.

The polysaccharide of the cell wall is the group-specific antigen. It is composed of rhamnose and *N*-acetylgalactosamine, the latter sugar being the immunodominant component of the antigen. The structure of the molecule is very similar to that of the group A polysaccharide. Only 50% of *N*-acetylgalactosamine residues are localized in the terminal nonreducing structures, while the rest are bound as 1,3-units inside the molecule (Coligan et al., 1975). The immunodominant unit is present in the molecule as a disaccharide 3-O-α-*N*-acetylgalactosamine (Coligan et al., 1977).

Protein antigens present in the cell wall have not yet been satisfactorily identified and characterized. Some mimic the protein antigens of group A streptococci, e.g. the organisms originally erroneously proposed in Griffith's (1934) scheme as M serovars 7, 20 and 21 of group A were later recognized to be group C strains. Some strains may carry proteins closely related to or identical with T proteins, 2, 4, 8 or 25, or the R 28 antigen, of group A streptococci. A typing scheme for group C streptococci based on protein antigens has been proposed, but it is not used because of inconclusive results.

The extracellular products are similar to those of group A streptococci, e.g. streptokinase, NADase, esterase and DNase (Smyth and Fehrenbach, 1974).

Antibiotic sensitivity: The MIC for penicillin varies between 0.004 and 0.064 μg/ml. In some strains tolerance to penicillin may show a 32- to 512-fold difference between MIC and the minimal bactericidal concentration. This may be a reason for the poor effect of penicillin in the therapy of some serious infections in man (Portnoy et al., 1981). Erythromycin-resistant strains have also been reported.

Pathogenicity: In man, group C streptococci can be associated with respiratory tract, skin and wound infections. It has repeatedly been demonstrated as an etiological agent in endocarditis, meningitis, urinary tract infection and other clinical conditions (Stamm and Cobbs, 1980). In animals, group C strains produce septicemia in cows, rabbits and swine. Frequently isolated from wound infections of horses (Stableforth, 1959). Sometimes associated with various avian diseases (Peckham, 1966). Isolated from milk and udders of cows with acute or sometimes mild mastitis. Also from blood and tissues of lambs suffering from polyarthritis (joint-ill).

Sources: Upper respiratory tract of normal and diseased humans and animals; blood, inflammatory exudates and lesions of diseased humans and animals; bovine milk.

Further Taxonomic Comments. Since the preceding section was written, the name *S. dysgalactiae* has been revived and formally proposed as a species (Garvie et al., 1983). The following description has been given.

Streptococcus dysgalactiae (ex Diernhofer 1932, 368) Garvie et al. 1983, 404.[VP]

dys.ga.lac'ti.ae. Gr. pref. *dys* ill, hard; Gr. n. *galactia* pertaining to milk; N.L. n. *dysgalactia* loss or impairment of milk secretion; N.L. gen. n. *dysgalactiae* of dysgalactia.

Gram-positive cocci or oval cells in short to medium length chains. Growth in glucose broth is poor with a final pH of 4.7–4.9. A wide zone of α-hemolysis on blood agar plates.

Optimum temperature for growth is 37°C; no growth at 10° or 45°C, in 6.5% NaCl, in 10% bile, or at pH 9.6. Cells do not survive 60°C for 30 min. Complex media needed for growth, growth requirements not known. Litmus milk is usually reduced, acidified and clotted. Acid is formed from glucose, lactose, maltose, sucrose and trehalose; no acid from raffinose, inulin, glycerol or mannitol. Variable reaction in sorbitol. Esculin is not hydrolyzed but some strains hydrolyze salicin. Ammonia is produced from arginine; hippurate is not attacked. Hyaluronidase is usually produced.

Reacts with Lancefield group C antiserum. The cell wall peptidoglycan contains L-Lys-L-Ala linkage (Schleifer and Kandler, 1972). A fibrinolysin for bovine fibrin but not for human fibrin may be produced. Low pathogenicity for mice.

The lactate dehydrogenase is activated by fructose 1,6-diphosphate, but inhibition by phosphate is slight, particularly at pH 5.5.

Commonly isolated from mastitic bovine udders (McDonald and McDonald, 1976), and produces mastitis experimentally (Higgs et al., 1980).

The mol% G + C of the DNA is 39–40 (T_m, Garvie and Farrow, 1981).

DNA-DNA hybridization shows that *S. dysgalactiae* is a distinct cluster with a low relationship to *S. agalactiae*, *S. acidominimus*, *S. uberis* or *S. bovis* (Garvie and Bramley, 1979a, b); each of these species belongs to a single cluster as determined by DNA/RNA hybridization (Garvie and Farrow, 1981).

Type strain: NCDO 2023.

b. *Streptococcus sp. (group G)* Lancefield and Hare, 1935, 346.

Spherical or ovoid cells 0.6–1.0 μm in diameter in medium to long chains.

Characteristic, "matt"-type colonies (indistinguishable from *S. pyogenes*).

Final pH in glucose broth 4.8–5.2. Acid produced from glucose, lactose, sucrose, trehalose and glycerol (aerobically only). Salicin fermentation is variable and occasional strains ferment inulin. No acid from mannitol, sorbitol or raffinose.

The broad zones of β-hemolysis on blood agar may be larger than those of *S. pyogenes*. Streptolysin O is produced.

Nutritional requirements are not known.

The cell wall polysaccharide is the group G-specific antigen, rhamnose being its immunodominant component.

Six serological types have been described, but the formerly widely employed typing by slide agglutination is no longer used. Some strains contain proteins immunologically related to T proteins 2 and 4 of group A streptococci. Human as well as animal strains have a surface component analogous to the Fc factor of group A streptococci that combines with IgG. While human strains bind both human and animal IgG, bovine strains bind human IgG only (Myhre et al., 1979).

Antibiotic sensitivity: The MIC for penicillin is 0.004–0.064 μg/ml. Although sensitivity to erythromycin is generally good, resistance to this antibiotic has already been reported.

Pathogenicity: In man group G streptococci can provoke various clinical conditions, such as endocarditis (Blair and Martin, 1978), occasionally pharyngitis, gastrointestinal tract infections, cutaneous infection or neonatal infections. Cases of infectious arthritis have also been reported (Lin et al., 1982; Fujita et al., 1982). In animals, pathogenic for dogs (Biberstein et al., 1980) and cats (Swindle et al., 1980), possibly also for other animal species.

Source: Human upper respiratory tract, skin lesions, inflammatory exudates, contaminated foods, and skin lesions and exudates of animals.

The mol% G + C of the DNA is 41 (Bd).

c. *Streptococcus sp. (group L)* Hare and Fry 1938, 1537. (See Laughton (1948) for additional information.)

Spherical or ovoid cells in long chains. Grow as β-hemolytic glossy and intermediate type colonies on blood agar.

Acid from maltose, lactose, sucrose, trehalose and salicin. Glycerol and sorbitol may or may not be fermented. Final pH in glucose broth is 4.7–5.2

The cell wall group-specific antigen is a polysaccharide, immunologically partly related to the group A polysaccharide. It is composed of *N*-acetylglucosamine, *N*-acetylgalactosamine, galactose and rhamnose. The immunodominant structure of the polysaccharide molecule is *N*-acetylglucosamine (Karakawa et al., 1971).. A protein antigen identical with the R 28 protein of group A streptococcus is present in the cell wall of some strains. Other protein antigens, which are trypsin labile, also occur in the cell wall. They can be used for type differentiation, but typing is not employed.

Antibiotic sensitivity: Highly sensitive to penicillin, erythromycin, chloramphenicol and tetracycline.

Pathogenicity: In dogs, give rise to infections of the urogenital tract and miscellaneous infections. In man group L strains rarely occur. Their pathogenic role has been documented in cases of thrombophlebitis and endocarditis (Bevanger and Stamnes, 1979).

Source: Inflammatory exudates and skin lesions in dogs.

d. *Streptococcus sp. (group M)*, Fry 1964, 721; Skadhauge and Perch, 1959.

Spherical or ovoid cells in long chains.

Acid from glucose, maltose, lactose, sucrose and no acid from melibiose, melezitose, inositol, rhamnose, dulcitol, xylose or adonitol. Generally no growth in 40% bile-blood agar, in media containing either 6.5% NaCl or 0.04% tellurite.

Three biovars are distinguished: Biovar I consists of α-hemolytic human strains that fail to hydrolyze arginine and have a final pH in glucose broth of 4.6–5.2. Biovar II strains are of animal origin, β-hemolytic, hydrolyze arginine and attain a final pH of 6.3–7.2. Biovar III strains are also of animal origin, β-hemolytic, hydrolyze arginine but produce more acid from glucose (final pH 5.9–6.7). Only three strains of this last biovar have been isolated.

Belongs to serological group M. The three biovars are also distinguished serologically in that biotype I contains a "group" antigen that is heat stable (127°C for 2 h) and resistant to pepsin. Biovar III contains a "group" antigen that is heat labile and sensitive to pepsin. Biovar II has both antigens. Some biovar I strains cross-react with group K antiserum, probably reflecting common type-specific antigens.

Antibiotic sensitivity: human strains are sensitive to penicillin, erythromycin and chloramphenicol.

Pathogenicity: isolated from cases of human subacute endocarditis

(Broome et al., 1976). Other human sources include abscesses, nasopharynx and vagina (Rifkind and Cole, 1962). Isolated from urethra, vagina and tonsillar area of dogs.

e. *Streptococcus sp. (group P)* Moberg and Thall 1954, 69.

Strongly β-hemolytic. Most strains grow on blood agar with 40% bile, and in broth containing 6.5% NaCl.

Acid produced from glucose, sucrose, maltose, trehalose, mannitol, sorbitol and salicin; acid production varies from lactose, starch, glycerol and melibiose. CAMP test frequently positive. Esculin hydrolyzed, sodium hippurate not hydrolyzed; ammonia produced from arginine (de Moor and Thal, 1968).

The antigen responsible for group P specificity can be extracted from cells with formamide; this extract, however, cross-reacts with group U antiserum. The cross-reactivity can be eliminated by absorption with group B or group U cells, while the homologous group P-specific reaction is retained (Shuman and Nord, 1974).

A common antigen is shared by groups P and U strains, but antibody to it is not demonstrable in group P sera.

Biologically the group is closely related to group U streptococci.

Pathogenicity: pathogenic for swine, producing abscesses of the pharyngeal region and sepsis.

Further taxonomic comments. Streptococci of serological groups E, P, U and V have recently been shown to share a number of biochemical and chemotaxonomic characters and have been proposed as a new species, *S. porcinus* (Collins et al., 1984b, 1985).

Type strain: NCTC 10999.

f. *Streptococcus sp. (group U)* Thal and Grabell 1964, 223.

β-Hemolytic strains usually grow on blood agar plates with 40% bile and in broth with 6.5% NaCl. Pattern of acid production from sugars and some other biochemical tests closely resemble those of group P streptococci.

Group-specific antigen is extractable with formamide. The extract cross-reacts with group P sera, but this cross-reactivity can be eliminated by absorption, which renders the reaction group U specific (Shuman and Nord, 1974).

Biologically the organism is closely related to group P streptococci.

Pathogenicity: Produces lymphadenitis and sepsis in swine.

Further taxonomic comments. See comments about group P streptococci and *S. porcinus* (Collins et al., 1984b).

g. *Streptococcus sp. (group V)* Jelíková and Kubín 1974, 434.

β-Hemolytic on blood agar after some time. Grow in broth with 4% NaCl and in methylene blue milk, produce ammonia from arginine and acid from trehalose, lactose, glucose, sucrose, sorbitol, mannitol, esculin, arabinose, inulin, dulcitol, melezitose, melibiose, glycerol and salicin. CAMP test positive. No growth in presence of 40% or 10% bile. Biochemical characteristics resemble those of groups E, P and U, being differentiated by growth in broth with 4% NaCl, and not hydrolyzing esculin or sodium hippurate.

The group-specific substance is extractable with formamide, precipitable with acetone and resistant to digestion by trypsin and pepsin.

Pathogenic for swine, provoking lymphadenitis.

Further taxonomic comments. See comments about group P streptococci and *S. porcinus* (Collins et al., 1984b).

Oral Streptococci

JEREMY M. HARDIE

The streptococci grouped together here for convenience as "Oral Streptococci" are commonly found in the oral cavity and upper respiratory tract of man and other animals (Hardie and Marsh, 1978). Although the mouth appears to be their main habitat, many of the species have also been isolated, on occasions, from other sites and from a variety of clinical infections (Parker and Ball, 1976; Facklam, 1977; Ruoff and Kinz, 1983). As pointed out by Jones (1978), grouping these species together in such a way for descriptive purposes does not

necessarily imply that they are more closely related to one another taxonomically than to other streptococci.

Several of the species considered here would have fallen into the earlier "Viridans Group" of Sherman (1937). However, while most of the species include strains which produce α-hemolysis (greening) on blood-containing media, the type of hemolysis shown is not a constant or reliable taxonomic feature of the group as a whole.

Interest in the oral streptococci has increased significantly in recent years, partly because some of the species are known to be important opportunist pathogens in other parts of the body, and partly because of the association of certain species (notably *Streptococcus mutans*) with dental caries. Early attempts to classify the entire Viridans Group by serological methods resulted in much confusion, but the more recent application of numerical and other taxonomic approaches has led to considerable clarification. It is now possible to distinguish several well defined species although, as indicated below, some taxonomic and nomenclatural problems remain to be resolved. Although serology does not appear to offer the best approach to species recognition, serological subdivision within some of the species is both possible and useful.

A guide to the biochemical characteristics of this group of streptococci is given in Table 12.18.

6. Streptococcus salivarius Andrewes and Horder 1906, 712.[AL]

sa.li.va′ri.us. L. adj. *salivarius* salivary, slimy.

Spherical or ovoid cells 0.8–1.0 µm in diameter. Chain length may vary from short to very long. Grows readily on suitable media in the presence of O_2. Most strains are nonhemolytic on blood agar (Sherman et al., 1943), although occasional α- and β-hemolytic strains are found. The hemolysis shown by some strains is dependent upon the use of horse blood in a base containing starch (Saunders and Ball, 1980). Smooth and rough variants occur, with the rough variant often reverting after subculture in broth. On sucrose agar most isolates produce soluble fructan (levan) which results in the development of large mucoid colonies (Niven et al., 1941a, b). Some strains also produce insoluble glucans (dextrans). Colonies on sucrose agar vary from smooth to rough depending upon the relative proportions of the different extracellular polysaccharides synthesized.

The minimal nutritional requirements were determined by Smiley et al. (1943) and it was reported that nine amino acids, five vitamins and uracil were required for growth. More recently, Carlsson (1971) showed that ammonia could serve as the major N source in a medium containing glucose, cysteine, nicotinic acid, biotin, thiamin, riboflavin, pantothenic acid and inorganic salts; some strains also required glutamic acid. The cysteine requirement could be replaced by cystine, homocysteine, homocystine or thiosulfate. Some strains could use urea as the source of N.

The final pH in glucose broth is 4.0–4.4. Acid is produced from glucose, sucrose, maltose, raffinose, inulin, salicin and, usually, trehalose and lactose. No acid from glycerol, mannitol, sorbitol, xylose or arabinose. Some strains ferment only the terminal fructofuranose portion of raffinose (forming polysaccharide) and melibiose accumulates. Most strains hydrolyze esculin and urea (Colman, 1976), but not arginine, and most also produce acetoin from glucose. Growth usually occurs at 45°C and not at 10°C.

The serology of this species has still not been completely resolved. Strains designated serovar I (Sherman et al., 1943) usually react with group K antiserum (Williams, 1956; Montague and Knox, 1968) and cross-react with antiserum against *Streptococcus MG* (*S. anginosus* or "*S. milleri*") according to Mirick et al. (1944a). *S. salivarius* type II strains do not react with either of these antisera (Montague and Knox, 1968). Approximately 50% of *S. salivarius* strains studied in two reports on large series of clinical isolates were found to be group K positive (Parker and Ball, 1976; Facklam, 1977).

S. salivarius is found in the mouths of man and other animals, being associated particularly with the tongue and saliva, and in feces. It is occasionally isolated from the blood in cases of infective endocarditis. Some strains have been shown to be cariogenic in gnotobiotic rats.

The mol% G + C of the DNA is 39–42 (T_m).

Type strain: ATCC 7073.

Further taxonomic comments. Several numerical taxonomic studies have shown that *S. salivarius* forms a distinct and homogeneous cluster, apart from other oral streptococci (Colman, 1968; Carlsson, 1968; Hardie et al., 1982; Bridge and Sneath, 1983). Recent nucleic acid

Table 12.18.
Some biochemical characteristics of the oral streptococci[a]

Characteristics	S. salivarius	S. sanguis	"S. mitior"	"S. milleri"	S. mutans	S. rattus	S. cricetus	S. sobrinus	S. ferus	S. oralis	S. mitis
Fermentation of											
Mannitol	−	−	−	−	+	+	+	+	+	d	−
Sorbitol	−	−	−	−	+	+	+	d	+	d	−
Inulin	d[b]	d[b]	d[b]	+	+	+	d	d	NT	d	d
Raffinose	d	d	d	d[b]	+	+	+	d	−	d	−
Lactose	+	+	d[b]	d	+	+	+	+	NT	+	+
Trehalose	d[b]	d[b]	d	+	+	+	+	d	NT	d	+
Hydrolysis of											
Arginine	−	+	−	d[b]	−	+	−	−	−	d	+
Esculin	+	d	−	d[b]	+	+	d	d	+	d	+
Starch	d	d	d	d[b]	d	d	d	d	NT	d	+
Urea	d	d	−	−	−	−	−	−	NT	d	NT
Production of											
Acetoin	d	−	d	d[b]	+	+	+	+	d	d	−
H_2O_2	−	+	+	−	−	−	−	+	−	+	+
Glucan	−	d[b]	d	−	+	+	+	+	+	d	−
Fructan	+	−	−	−	−	−	−	−	−	−	−
Growth in/at											
4% NaCl	d	d	−	d	+	d	+	+	NT	−	−
6.5% NaCl	−	−	−	−	−	−	d	d	−	−	−
10% Bile	+	d	−	d[b]	d	+	d	d	NT	d	+
40% Bile	d	d	−	d[b]	d	d	d	d	NT	−	d[b]
45°C	d	d	d	d	d	d	d	d	−	d	d[b]
Lancefield antigens	K or −	H	− (Some H, O or K)	F, G, C or A (or −)	− (Some E)	−	−	−	−	−	− (Some H)
Mol% G + C	39–42	40–46	39–43	34–39	36–38	41–43	42–44	44–46	43–45	39–40	38–40

[a] Symbols: see Table 12.2; also NT, not tested.
[b] Reported results vary from different laboratories.

hybridization studies have indicated a close genetic relationship between *S. salivarius* and *S. thermophilus*. This has been demonstrated by DNA/RNA studies, where the two species were shown to be closely related at the generic level, not only to each other but also to *S. bovis* and *S. equinus* (Garvie and Farrow, 1981; Kilpper-Bälz et al., 1982). DNA/DNA homology between four strains of *S. salivarius* and *S. thermophilus* was found to be in the range 75–97% (Kilpper-Bälz et al., 1982). This close relationship has been confirmed by further DNA/DNA hybridization experiments and by examination of long chain fatty acid profiles, using larger numbers of strains (Farrow and Collins, 1984) and the results indicate that *S. thermophilus* should be reclassified as a subspecies of *S. salivarius* (*S. salivarius* subsp. *thermophilus* Farrow and Collins, 1984a, 1984b). Although the genetic data supporting this relationship between *S. salivarius* and *S. thermophilus* appear clear-cut, they have not so far been supported by numerical taxonomic studies using phenotypic characters (Carlsson, 1968; Bridge and Sneath, 1983).

7. **Streptococcus sanguis** White and Niven 1946, 722.[AL]
san'guis. L. n. *sanguis* blood.

Spherical or ovoid cells 0.8–1.2 μm in diameter occurring in medium or long chains. Rod-shaped and pleomorphic cells are occasionally seen, especially when cultures are grown aerobically.

Colonies on blood agar are 0.7–1.00 mm in diameter and may be smooth or matt surfaced. Most strains are α-hemolytic (best seen on aerobic incubation), but both β- and nonhemolytic isolates are found. On sucrose agar plates, extracellular polysaccharide-producing strains form hard, rubbery or glassy adherent colonies often rough and heaped up, which may distort the surrounding agar and are difficult to remove from the surface. Watery polysaccharide material may be seen on or around such colonies. A minority of strains are nonpolysaccharide producers; these form round, soft, smooth colonies on sucrose agar.

Some of the biochemical reactions are shown in Table 12.19. The final pH in glucose broth is 4.6–5.2. Acid is produced from glucose, maltose, sucrose, salicin and usually trehalose. Variable reactions occur with inulin and raffinose. Arabinose, xylose, glycerol and mannitol are usually not fermented; occasional sorbitol-positive strains have been reported although this substrate is not usually metabolized. Most strains produce ammonia from arginine and form H$_2$O$_2$; over 50% of isolates also hydrolyze esculin. Extracellular glucan production from sucrose is a common feature of most strains; this polysaccharide cross-reacts with type 2 pneumococcus serum and can also be demonstrated in sucrose broth supernatants by precipitation with 1.2 vol ethanol (Hehre and Neill, 1946; Niven et al., 1946a).

May or may not grow at 45°C.

The minimal nutritional requirements of *S. sanguis* include the organic compounds arginine, cysteine, glutamic acid, leucine, methionine, valine, biotin, thiamin, riboflavin, pyridoxine, pantothenic acid, nicotinic acid and glucose (Carlsson, 1970). Some strains (including the type strain ATCC 10556) may require additional amino acids and vitamins for transferable growth in chemically defined media.

Most strains of *S. sanguis* appear to possess the Lancefield group H antigen, although there has been some controversy over the nature of this antigen and different immunizing strains are used for the production of commercial grouping antisera in different parts of the world. In the earliest studies on the serology of *S. sanguis*, three serovars were described (designated I, II and I/II) but no common group antigen was identified (Washburn et al., 1946). Strains at that time called type II have subsequently been shown to belong to a separate species, "*S. mitior*" (Colman and Williams, 1972; Coykendall and Specht, 1975; Cole et al., 1976). Rosan (1973, 1976) has shown that isolates of *S. sanguis* contain up to five antigens in autoclaved extracts, one of which is common to most strains and represents the H antigen. This group-specific antigen is a glycerol teichoic acid, containing glycerol, phosphate and glucose in a molar ratio of 1:0.9:0.3, (Rosan and Argenbright, 1982). The major antigenic determinant appears to be an α-glucose linked to the glycerol phosphate backbone. This antigen is considered

Table 12.19.
Some biochemical characteristics of **Streptococcus sanguis,** **Streptococcus mitis** *and* **"Streptococcus mitior"**[a, b]

Characteristics	S. sanguis		S. mitis	"S. mitior"	
	Pooled data	ATCC 10556	NCTC 3165	Pooled data	NCTC 10712
Fermentation of					
Lactose	+	+	+	d[c]	+
Trehalose	d[c]	+	+	d	−
Raffinose	d	−	−	d	+
Inulin	d	+	d[c]	d[c]	−
Cellobiose	d	+	+	d	−
Hydrolysis of					
Arginine	d[c]	+[c]	+	d[c]	−
Esculin	d	+[d]	+	d[c]	−
Starch	d	+	+	d	−
Growth in/at					
10% Bile	d	+	+	d[c]	−
40% Bile	d	−	d[c]	d[c]	−
45°C	d	−	d[c]	d	−
4% NaCl	d	−	−	d[c]	−
6.5% NaCl	−	−	−	−	−
Production of					
Acetoin	−	−	−	d	
H$_2$O$_2$	+	+	+	+	+
Glucan	d	+	−	d	
α-Hemolysis	d[c]	+	+	+	+
Cell walls contain					
Rhamnose	+	+	+	d[c]	−
Ribitol	−	−	−	+	+

[a] All strains positive in glucose, sucrose, maltose; all strains negative in mannitol, sorbitol, glycerol, arabinose, melezitose, xylose, hippurate and urea, and survival at 60°C for 30 min.
[b] Symbols: see Table 12.2.
[c] Reported results vary from different laboratories.

to be analogous to the membrane lipoteichoic acids found in other streptococci and lactobacilli (Wicken and Knox, 1975).

Streptococcus sanguis strains have been used extensively for intraspecies, intrageneric and intergeneric transformation studies (Pakula, 1963; Perry and Slade, 1964; Willers et al., 1968; Dobrzanski et al., 1968). Two genetic groups within *S. sanguis* have been revealed by DNA/DNA homology, showing 85–100% intragroup and 40–60% intergroup similarity (Coykendall and Specht, 1975). The suggestion that these groups be designated as named subspecies of *S. sanguis* (subsp. *sanguis* and subsp. *carlssonii*) has not so far found general acceptance and further confirmatory studies are required before such subdivisions can be recommended.

S. sanguis is consistently isolated from dental plaque, where it constitutes a significant part of the flora (Carlsson, 1965, 1967a) and, in lower numbers, from other parts of the mouth. Only becomes established in the mouths of infants after eruption of the first primary teeth (Carlsson et al., 1970). Isolated from blood and heart valves in some cases of bacterial endocarditis (Parker and Ball, 1976; Facklam, 1977). Low levels may be detected in human feces (van Houte et al., 1971) and the organism has also been isolated from soil (Gledhill and Casida, 1969). L-Forms may be associated with recurrent aphthous stomatitis (Barile et al., 1968).

The mol% G + C of the DNA is 40–46.4% (T_m). The two genetically distinct subgroups have slightly different values, 40.8–42.8% and 43.8–46.4% (Coykendall and Specht, 1975).

Type strain: ATCC 10556 (NCTC 7863).

Further taxonomic comments. The numerical taxonomic study of

Carlsson (1968) revealed two clusters (I:A and I:B), at that time considered to correspond to *S. sanguis*. The strains grouped in cluster I:B are equivalent to *S. sanguis* as recognized by other workers (e.g. Colman and Williams, 1972; Hardie and Bowden, 1976; Feltham, 1979; Hardie et al., 1982; Bridge and Sneath, 1983), whereas those in cluster I:A, which includes strains ATCC 10057 (NCTC 7864, listed as "*S. sanguis* type II") and ATCC 9811 (listed as *S. mitis*), are now considered to represent "*S. mitior*." The clear distinction between these two species has been confirmed by physiological and biochemical tests (Colman and Williams, 1972; Cole et al., 1976), nutritional requirements (Carlsson, 1970), serology (Rosan, 1973, 1976), protein profiles (Whiley et al., 1982) and DNA/DNA hybridization studies (Coykendall and Specht, 1975; Coykendall and Munzenmaier, 1978; Welborn et al., 1983).

Unfortunately, some confusion persists because certain reference strains continue to be listed in culture collections under their original names although they have been shown to belong to other species. Further difficulty arises when following the identification scheme proposed by Facklam (1977). According to this system, *S. sanguis* I, *S. sanguis* II and *S. mitis* are differentiated, the latter two being separated from one another on the basis of raffinose fermentation. This scheme is used in the Streptococcal Reference Laboratory at the Centers for Disease Control, Atlanta, Ga., and has been adopted by at least one producer of rapid identification kits (API Systems S.A., Montalien-Vercieu, France) (Ruoff and Kunz, 1983). According to the classification recommended in this chapter, Facklam's *S. sanguis* II and *S. mitis* should be combined into one species ("*S. mitior*") while his *S. sanguis* I is analogous to *S. sanguis* as described above.

Some of the biochemical characteristics of *S. sanguis*, *S. mitis* and "*S. mitior*" are listed in Table 12.19.

8. **"Streptococcus mitior"** Schottmüller 1903, 850.

mi.ti′or. M.L. adj. (from *mitis*, mild) milder.

This description is based on published reports of the properties of streptococci variously called "*S. mitior*" (Colman and Williams, 1972; Mejare and Edwardsson, 1975; Parker and Ball, 1976; Hardie et al., 1982), *S. sanguis* group I:A (Carlsson, 1968) and *S. sanguis* type II (Facklam, 1977).

Spherical or oval cocci, 0.6–1.0 μm in diameter, arranged singly, in pairs or chains; long chains, up to 100 cells, may occur in broth cultures. Most strains are α-hemolytic on blood agar, while a few produce clear zones of β-hemolysis. Colonies on sucrose agar may either be soft or hard and rubbery and adherent, due to the production of the extracellular polysaccharide (glucan). Smooth and rough colony variants may occur.

Some of the biochemical characteristics are summarized in Table 12.19. Good growth occurs aerobically and anaerobically. Most strains do not produce NH_3 from arginine or hydrolyze esculin and these tests are useful for distinguishing "*S. mitior*" from *S. sanguis*.

All strains produce acid from glucose, sucrose and maltose, while variable results are obtained with raffinose (which has been used by some authors to distinguish *S. mitis* from *S. sanguis* type II; Facklam, 1977), lactose, trehalose, cellobiose, inulin (usually negative), production of acetoin, and various tolerance tests. Hydrogen peroxide is produced by most strains, while glucan formation from sucrose is a variable character.

One of the main distinguishing features of "*S. mitior*" is the absence of significant amounts of rhamnose in the cell wall and the presence of ribitol teichoic acid (Colman and Williams, 1965, 1972; Cole et al., 1976). Strains of "*S. mitior*" (described previously as *S. sanguis* type II) have been shown to possess the L-Lys- direct type of peptidoglycan, in contrast to the L-Lys-L-Ala$_{2-3}$ type found in type 1 *S. sanguis* strains (Schleifer and Kandler, 1972).

The serology of "*S. mitior*" is complex and no common group antigen has been demonstrated. Some strains cross-react with Lancefield Group O or Group K antisera (Colman and Williams, 1972), and with certain Group H antisera (Cole et al., 1976). There does not appear at present

to be any clear serological reaction of value in recognizing and identifying this species.

DNA/DNA hybridization studies have shown, firstly, that "*S. mitior*" strains are genetically distinct from *S. sanguis* (Coykendall and Specht, 1975) and, secondly, that two homology groups can be distinguished within "*S. mitior*" (Coykendall and Munzenmaier, 1978). The second of these homology groups contains two of the strains (OS51, NS51) included in the recently proposed new species, *S. oralis* (Bridge and Sneath, 1982, 1983). Several numerical taxonomic studies have revealed distinct clusters of strains which correspond to "*S. mitior*" (Colman, 1968; Carlsson, 1968; Drucker and Melville, 1971; Feltham, 1979; Hardie et al., 1982; Bridge and Sneath, 1983).

This species has been isolated from human saliva, dental plaque, sputum, feces and from clinical infections, particularly bacterial endocarditis. It has also been found in the mouths of animals (Dent et al., 1978).

The mol% G + C of the DNA is 39–43 (T_m). Two genetic groups with slightly different G + C mol% values have been described, group 1 is 41.3–42.6 and group 2 is 39.9–41.0 (Coykendall and Munzenmaier, 1978).

Type strain: No formally proposed type strain of "*S. mitior*" exists at present, but a suitable candidate would be NCTC 10712 (FW 75; G. Colman, personal communication). As mentioned elsewhere, the type strain of *S. mitis* (NCTC 3165) differs in several important respects from that of "*S. mitior*" (Table 12.19).

Further taxonomic comments. There is some confusion surrounding the correct nomenclature of streptococci variously referred to in the past as "*S. mitior*," "*S. mitis*," "*S. viridans*" and "*S. sanguis* type 2." The name currently included on the Approved Lists of Bacterial Names (1980) is *S. mitis* (Andrewes and Horder, 1906), the properties of which were described in greater detail by Sherman et al. (1943). However, the latter authors regarded *S. mitis* as an "... ill-defined and heterogeneous group." The description of *S. mitis* given in the eighth edition of *Bergey's Manual* and reproduced in this chapter also referred to a "... heterogeneous group of α-hemolytic streptococci associated with the human respiratory tract. It has no unique identifiable characteristics and some nonhemolytic varieties of typically hemolytic species may be confused with it." To further confuse the situation, the designated type strain of *S. mitis* (NCTC 3165) has been shown to be more characteristic of the species *S. sanguis* (Colman and Williams, 1972; Cole et al., 1976; Welborn et al., 1983) (Table 12.19).

The name "*S. mitior seu viridans*" was introduced by Schottmüller (1903) for certain α-hemolytic streptococci isolated from the human respiratory tract and from various clinical conditions, including infective endocarditis. From the very brief original description of this species, it is difficult to relate presently available strains to those examined by Schottmüller, with any degree of certainty. However, exactly the same problem arises with the early description of *S. mitis* given by Andrewes and Horder (1906). In both cases the original descriptions were based largely on colonial and morphological features, with few biochemical characters recorded, and no original strains are available for examination.

Colman and Williams (1972) have argued cogently in favor of retaining the name "*S. mitior*," which has priority over *S. mitis*, for a group of α-hemolytic streptococci that can be defined with a reasonable degree of precision.

The existing descriptions of *S. mitis* indicate a heterogeneous group of streptococci that is difficult to define precisely and, as mentioned above, the type strain appears to be different from "*S. mitior*" as described here.

Since *S. mitis* was not very clearly defined in the original descriptions and as there may well be more than one taxon within the confines of its "heterogeneous group," the use of the name "*S. mitior*" for a particular cluster of recognizably distinct streptococci within this area is preferred. This conclusion, in line with the proposal by Colman and Williams (1972), does not preclude the possibility that other taxa may exist among the former "*S. mitis* group" of streptococci. The use of the name "*S. mitior*" has also been recommended by Coykendall and Munzenmaier (1978). Clusters of strains corresponding to the species "*S. mitior*" as described above have been recognized by numerical taxonomic analysis of conventional biochemical and physiological test data (Hardie et al., 1982) and by analysis of sodium dodecyl sulfate-polyacrylamide gel electrophoresis (SDS-PAGE) protein patterns (Whiley et al., 1982). In each case, the "*S. mitior*" clusters included some reference strains previously named *S. mitis* and *S. sanguis* type 2 (or group 1:A).

A recently described property which may be useful to distinguish "*S. mitior*" from *S. sanguis* is the possession of a cell-associated neuraminidase (Murray et al., 1984).

It has been suggested that strains designated *S. sanguis* type 2, which were shown to have a different peptidoglycan pattern from *S. sanguis* type 1, should be re-classified as "*S. pseudosanguis*" (Hladny, Ph.D. thesis, Technical University, Munich 1971; quoted by Schleifer and Kandler, 1972). This proposal has not apparently found general acceptance and it is probable that the organisms studied correspond to the species "*S. mitior*" described in this section.

In view of the current doubts about the taxonomic position of *S. mitis* and its relationship to "*S. mitior*," as described here, and to *S. sanguis*, the name is listed below under the heading Species Incertae Sedis.

"*Streptococcus anginosus-milleri* Group"

This group of streptococci includes strains which have been referred to variously as *S. anginosus* (Andrewes and Horder, 1906), *Streptococcus MG* (Mirick et al., 1944a–c), *S. constellatus* (Prevot, 1924; Holdeman and Moore, 1974) *S. intermedius* (Prevot, 1925; Holdeman et al., 1977), "*S. milleri*" (Guthof, 1956) and minute β-hemolytic streptococci of Lancefield groups F and G (Long and Bliss, 1934; Bliss, 1937). It seems likely that most, if not all, of these streptococci are sufficiently similar to be regarded as members of the same species. However, at the time of writing, genetic information on the relationships between all of these named varieties is incomplete.

For the purposes of description, on the basis of available data, it seems sensible to group these streptococci together under a single heading. The name *S. anginosus* (Andrewes and Horder, 1906) has priority, is listed in the Approved Lists of Bacterial Names (Skerman et al., 1980), and is represented by a designated type strain. However, the name which is preferred by many workers, and which is in common use in numerous laboratories throughout the world, is "*S. milleri*" (Guthof, 1956; Colman and Williams, 1972; Mejare and Edwardsson, 1975; Hardie and Bowden, 1976; Ball and Parker, 1979; Bridge and Sneath, 1983; Drucker and Lee, 1983; Welborn et al., 1983). Unfortunately, this name does not appear on the Approved Lists, but it is likely that approval will be sought for the revival of "*S. milleri*" in the near future. Current differences in nomenclature and taxonomy between American and British workers are discussed by Facklam (1984). Some of the biochemical characteristics of these streptococci are shown in Table 12.20.

9. "**Streptococcus milleri**" Guthof 1956, 558.

mil.le.ri. M.L. gen. n. *milleri* named after W. D. Miller, an American oral microbiologist.

This species description includes organisms referred to as *S. anginosus* (Andrewes and Horder, 1906), *Streptococcus MG* (Mirick et al., 1944), *S. constellatus*, (Prevot, 1924; Holdeman and Moore, 1974), *S. intermedius* (Prevot, 1925; Holdeman, et al., 1977), "*S. milleri*" (Guthof, 1956) and minute β-hemolytic streptococci of groups F and G (Long and Bliss, 1934; Bliss, 1937).

Spherical or ovoid cells arranged in pairs or chains which may be long or short. Growth of many strains on solid media either requires, or is enhanced by, the addition of CO_2 to the atmosphere. In some cases, the inability of strains to grow in air alone has led to isolates being described initially as obligate anaerobes. Hemolysis on blood agar is a variable characteristic. In one study on 346 strains, 56% were

Table 12.20.
Some biochemical characteristics of the "**Streptococcus milleri**" *group*[a,b]

Characteristics	S. anginosus	NCTC 10713	S. constellatus	S. intermedius	S. MG	S. milleri	S. parvulus
Fermentation of							
Inulin	−	−	−	−	−	−	+
Lactose	d[c]	+	−	+	+	d	+
Raffinose	d	+	−	d	−	−	−
Salicin	+	+	+	+	+	+	+
Sucrose	+	+	+	+	+	+	+
Trehalose	+	+	d[c]	+	d	+	+
Hydrolysis of							
Esculin	d	+	+	+	+	d[c]	+
Arginine	d	+	+	d[c]	+	+	−
Starch	d	+	−	d	−	−	−
Production of							
Acetoin	NT	+	−	−	NT	d	−
H_2O_2	NT	−	NT	NT	NT	−	NT
Growth in/at							
4% NaCl	d	−	d	d	−	d	NT
10% Bile	d	+	d	d	d[c]	d	NT
40% Bile	d	+	d	d	d	d	NT
45°C	d	+	d	d	−	d	NT

[a] All strains positive in glucose and maltose; all strains negative in arabinose, glycerol, mannitol (a few mannitol-positive strains have been reported (Ball and Parker, 1979; Ruoff and Kunz, 1982), melezitose, sorbitol, xylose and hydrolysis of hippurate.
[b] Symbols: see Table 12.2; also NT, not tested.
[c] Results vary in different published reports.

nonhemolytic, 19% α-hemolytic and 25% β-hemolytic (Ball and Parker, 1979), but the proportions of strains displaying each type of hemolysis vary in different studies. The majority of group F and group G-type I strains form minute colonies on blood agar with relatively large hemolytic zones after incubation for 48–96 h. The hemolytic zones may appear before the colonies are visible to the naked eye. Incubation under 10% CO_2 stimulates hemolysis and growth significantly (Deibel and Niven, 1955b). Strains may be isolated on a selective medium containing sucrose and sulfonamide (MC agar; Carlsson, 1967), on which both smooth and rough colonies may be found (Mejare and Edwardsson, 1975).

For growth of some strains in synthetic medium it has been shown that cultures must contain oleic acid or be incubated under increased CO_2 tension (Deibel and Niven, 1955a). Folic acid and four other B vitamins are required. Other vitamins, a reducing substance and a peptide factor are stimulatory (Niven et al., 1946b; Deibel and Niven, 1955a).

Acid is produced from glucose, sucrose, maltose and usually from salicin, lactose and trehalose. Mannitol, sorbitol, glycerol, inulin, xylose, melezitose and arabinose are not usually fermented, although some strains allotted to "*S. milleri*;" have been described as fermenting mannitol, among other carbohydrates (Ball and Parker, 1979; Ruoff and Kunz, 1982); raffinose, although usually negative, shows some variation between published results. Acetoin is produced from glucose. Most strains hydrolyze esculin and arginine, but not hippurate. Hydrogen peroxide is not usually produced and extracellular polysaccharides are not formed from sucrose. Most strains are resistant to 10% bile and many also to 40% bile. Growth in 4% NaCl is variable and very few strains can tolerate 6.5% NaCl. No growth occurs at 10°C while growth at 45°C is variable (Table 12.20).

The antigenic composition of strains within this species is complex and heterogeneous. Some strains react with Lancefield grouping serum, most commonly group F, but also occasionally groups A, C, G or K. The small colony of hemolytic streptococci belonging to groups F and

G (Long and Bliss, 1934; Bliss, 1937) are included among these "groupable" strains. Both hemolytic and nonhemolytic strains possess a number of type antigens. Among group F strains, five serovars have been described, of which I–III are relatively common while serovars IV and V are less frequently isolated (Ottens and Winkler, 1962). Organisms described originally as *Streptococcus* MG (Mirick et al., 1944a–c) were shown to possess group F, serovar III antigen (Willers et al., 1964).

The predominant determinant of the group F antigen is glucosyl-*N*-acetyl-D-galactosamine (Michel and Willers, 1964). The serovar I antigen contains *N*-acetylgalactosamine; serovar II, rhamnose, glucose, galactose, galactosamine (ratio 2:2:2:1); (Michel and Krause, 1967) and serovar III, glucose, galactose and rhamnose (5:3:1); Willers and Alderkamp, 1967).

Strains are generally sensitive to penicillin, erythromycin and trimethoprim but resistant to sulfamethoxazole and nitrofurazone. Most are resistant to bacitracin, and some to tetracycline (Poole and Wilson, 1976).

Strains of "*S. milleri*" have been isolated from a variety of infectious conditions in man, particularly from abscesses in various parts of the body. Sites that may be infected include the mouth, liver, brain, female genital tract, appendix and blood stream (Poole and Wilson, 1976, 1979; Ball and Parker, 1979). Some strains have been shown to be capable of inducing dental caries in gnotobiotic rats (Drucker and Green, 1978).

"*S. milleri*" is found in healthy individuals in the mouth, predominantly on the surface of teeth and in the gingival crevice (Mejare and Edwardsson, 1975), nasopharynx (Hare, 1935), throat (Long et al., 1934), vagina (Lancefield and Hare, 1935; Wort, 1975) and in feces (Poole and Wilson, 1979).

The mol% G + C of the DNA is 34–39% (T_m).

Type strain: There is no designated type strain of "*S. milleri*." NCTC 10708 and NCTC 10709 are representative strains of "*S. milleri*" from Guthof (1956). The type strain of *S. anginosus* is NCTC 10713. Type strains for the other named species now grouped together are as follows: *S. intermedius*, NCTC 27335; *S. constellatus*, ATCC 27823.

Further taxonomic comments. The species variously named *S. anginosus*, *S. intermedius*, *S. constellatus* and *Streptococcus* MG appear to share a number of similarities in their biochemical properties. A summary of the known reactions of these streptococci is shown in Table 12.20 alongside the comparable results for the anaerobic or microaerophilic species *S. parvulus*. The data given in this table have been compiled from the published work of Deibel and Seely, 1974; Holdeman and Moore, 1974; Mejare and Edwardsson, 1975; Mirick et al., 1944a; Facklam, 1977; Colman and Williams, 1972; Hardie and Bowden, 1976a; Ball and Parker, 1979; Bridge and Sneath, 1983; and Cato, 1983.

Although these organisms display a considerable degree of heterogeneity with respect to oxygen sensitivity, hemolysis, antigenic composition and cellular fatty acid profile (Drucker and Lee, 1981), their overall similarity in conventional biochemical reactions means that in practice different "species" may be separated on the basis of one or two test differences. For example, *S. constellatus/S. anginosus* have been distinguished from *S. intermedius* on the basis of lactose fermentation (Facklam, 1977). Several independent numerical taxonomic studies have revealed clusters corresponding to "*S. milleri*" (Colman, 1968; Colman and Williams, 1972; Drucker and Melville, 1971; Feltham, 1979; Hardie et al., 1982). Unfortunately, none of these studies has included representative strains of all the possible candidates for inclusion in the species "*S. milleri*" (i.e. *S. anginosus*, *S. intermedius*, *S. constellatus*, *S. parvulus*, *Streptotoccus* MG, minute group F streptococci and group G streptococci). A close genetic relationship between strains representing some of these named varieties, including the type strains of *S. intermedius* and *S. constellatus*, *Streptococcus* MG and two group F strains, has been demonstrated by DNA/DNA hybridization studies (Welborn et al., 1983). The relative binding ratios were reported to be within the range of 60–99%, with most values greater than 85%. These results support the evidence from other sources that these

strains, currently known by several different names, should be grouped together in a single species. However, some heterogeneity within the species, possibly indicative of different biotypes or subspecies, has been reported (Drucker and Lee, 1983).

Although the name *S. anginosus* (Andrewes and Horder, 1906) would appear to have priority over other suggested epithets, it is not at all clear whether the organisms described above correspond to the original species description.

As pointed out by Colman (1976), Andrewes and Horder's early, rather incomplete, description of *S. anginosus* referred to the variety of *Steptococcus* associated with scarlet fever (presumably *S. pyogenes*). No original isolates are available for comparative studies.

In view of the difficulty in relating presently available strains to earlier published descriptions, it would seem appropriate to allocate this group of streptococci to the more recently described and better defined species "*S. milleri*" (Guthof, 1956), as recommended by several authors (Mejare and Edwardsson, 1975; Colman and Williams, 1972; Hardie and Bowden, 1976; Parker and Ball, 1976; Ball and Parker, 1979; Bridge and Sneath, 1983; Welborn et al., 1983).

"*Streptococcus mutans* Group"

The organism designated *Streptococcus mutans* was first isolated from carious human teeth by Clarke (1924). It was not included in the eighth edition of the *Manual* but has subsequently been added to the Approved Lists of Bacterial Names.

Although *S. mutans* was largely overlooked for many years after its discovery, a vast amount of published work on this species has appeared since the early 1960s (for review, see Hamada and Slade, 1980a), mainly prompted by the observation that many strains are highly cariogenic in experimental animals. Detailed descriptions of collections of *S. mutans* strains have been provided by several authors (e.g. Edwardsson, 1968; Colman and Williams, 1972; Facklam, 1974; Perch et al., 1974; Hardie and Bowden, 1976; Parker and Ball, 1976) and have shown that there is a recognizable set of characteristics which distinguishes *S. mutans* from other oral streptococci (Table 12.18).

Numerical taxonomic studies on streptococci have usually included only small numbers of *S. mutans* strains, often limited to the same few reference strains. Isolates corresponding to *S. mutans* have been found to occupy a single cluster in some studies (Carlsson, 1968; Colman and Williams, 1972; Drucker and Melville, 1971), although strains have been observed to fall into more than one cluster in more recent investigations (Feltham, 1979; Hardie et al., 1982; Bridge and Sneath, 1983). The earlier numerical studies did not include representatives of all the biovars, serovars and genovars now known to exist within the "*S. mutans* group".

Serological examination of *S. mutans* revealed the existence of several serovars, now designated a–h (Zinner et al., 1965; Bratthall, 1970; Perch et al., 1974; Beighton et al., 1981). Prior to detailed immunochemical analysis of the antigenic determinants of the type antigens (Hamada and Slade, 1980a), it was found that the carbohydrate cell wall patterns of *S. mutans* correlated well with serovar (Hardie and Bowden, 1974).

Studies on the DNA base composition and DNA/DNA homology led Coykendall (1974) to propose, firstly, that four subspecies (*mutans*, *rattus*, *cricetus*, *sobrinus*) be recognized within *S. mutans* and, subsequently, that these subspecies be elevated to separate species status, together with a new, fifth named variety, *ferus*, (Coykendall, 1977). Three of the proposed names, *S. mutans*, *S. rattus* and *S. cricetus*, appeared on the Approved Lists of Bacterial Names, while *S. sobrinus* and *S. ferus* were omitted (possibly because the designated type strains did not, at the time of publication, have official type collection numbers). The latter species names have subsequently been validly published (Coykendall, 1983).

For practical purposes, it is arguable that the creation of five species names from a group of phenotypically similar streptococci which have not, until quite recently, been "officially" recognized, may cause considerable confusion. The name *S. mutans* is widely used for all members of this group at present, and some authors have recommended that the species should remain as a single entity, within which there exist several serovars, biovars and genovars (Hamada and Slade, 1980a; Jones, 1978). The genetic subdivisions of Coykendall (1974, 1977) appear to correspond closely to the patterns of whole cell proteins obtained by SDS-PAGE (Russell, 1976; Whiley et al., 1982).

The combined weight of evidence from numerical taxonomic, chemotaxonomic and genetic studies would seem to indicate that there is a case for dividing *S. mutans* into more than one species, although, for practical convenience, one species name would be easier to manage. Since the species names proposed by Coykendall (1977) have been validly published, descriptions of them as separate entities are given below. A summary of some of the properties of these species is shown in Table 12.21. The properties of the recently described *S. mutans* serovar h (Beighton et al., 1981) are also listed in this Table for comparison.

10. **Streptococcus mutans** Clarke 1924, 144.[AL]

mu'tans. L. part. adj. *mutans*, changing.

Gram-positive cocci, about 0.5–0.75 μm in diameter, occurring in pairs, short or medium length chains, without capsules. Under acid conditions in broth, and on some solid media, may form short rods from 1.5–3.0 μm in length. Rod-shaped morphology is often evident on primary isolation from oral specimens. Nonmotile and catalase-negative. On blood agar, incubated anaerobically for 2 days, colonies are white or gray, circular or irregular, 0.5–1.0 mm in diameter, sometimes rather hard and coherent and tending to adhere to the surface of the agar. Usually either α- or nonhemolytic, but occasional β-hemolytic strains are found (Perch et al., 1974; Wolff and Liljemark, 1978). On sucrose-containing agar (such as MSA or TYC) most strains produce rough, heaped colonies, about 1 mm in diameter, often with beads, droplets or puddles of liquid (containing soluble extracellular polysaccharide) on or around the colonies, while some may form smooth or mucoid colonies (Edwardsson, 1968, 1970). Most strains will grow to some extent in air but growth is enhanced under anaerobic conditions; most grow well in air or N₂ + CO₂ and a few are CO₂-dependent. Optimum growth occurs at 37°C; some strains grow at 45°C (although there is considerable interlaboratory variation in published reports) but none grows at 10°C.

Glucose is fermented by homolactic fermentation, normally to L-lactic acid, without gas formation. Under glucose-limited conditions in a chemostat, significant amounts of formate, acetate and ethanol are produced (Carlsson and Griffith, 1974; Ellwood et al., 1974). Glucose

is transported into *S. mutans* cells via a phosphoenolpyruvate-dependent phosphotransferase system (Schachtele 1975; Ellwood et al., 1979). The terminal pH in batch-grown glucose broth cultures is 4.0–4.3.

Some of the biochemical characteristics of *S. mutans* are shown in Table 12.21. Most strains produce acid from mannitol, sorbitol, raffinose, lactose, inulin, salicin, mannose and trehalose, but not from arabinose, xylose, glycerol or melezitose. With the exception of strains of serovar b, now designated *S. rattus* (Coykendall, 1977), ammonia is not produced from arginine. Esculin is hydrolyzed, but not hippurate or gelatin. Hydrogen peroxide is not produced, except by strains of serovars d and g (now called *S. sobrinus*).

S. mutans synthesizes several types of extracellular polysaccharides from sucrose which are considered to be important in the colonization of hard tissue surfaces in the mouth (Gibbons and Van Houte, 1973, 1975; Hamada and Slade, 1980b). Both glucans, formed by the enzymatic action of glucosyltransferase (EC 2.4.1.5), and fructans, due to the activity of fructosyltransferase (EC 2.4.1.10) are produced (Hamada and Slade, 1980a). The glucans include a water-soluble, α(1→6)-linked linear glucose polymer with α(1→3) glucosidic branch linkages (Long and Edwards, 1972), and essentially water-insoluble, cell-associated polymers. The latter possess a higher proportion of α(1→3) glucosidic linkages than the water-soluble glucans, and are more resistant to the enzymatic action of dextranase (α(1→6) glucanase; Guggenheim, 1970; Guggenheim and Schroeder, 1967; Hamada and Slade, 1976). The amount and chemical structure of the extracellular glucans produced from sucrose vary among different serovars, especially with respect to water solubility and proportions of α(1→3) linkages. Both water-soluble and water-insoluble fructans are also produced by strains of *S. mutans*; the predominant linkage in these polymers appears to be an inulin-type α(2→1) fructofuranoside rather than the β(2→6) levan-type previously suspected (Baird et al., 1973; Ebisu et al., 1975; Birkhed et al., 1979).

Most strains of *S. mutans* also produce intracellular iodine-staining polysaccharides (IPS) from sucrose, which may contribute to their caries-inducing potential. The IPS is a glycogen-like glucan with α(1→4) and β(1→6) linkages which are susceptible to α-amylase (van Houte et al., 1970; Critchley et al., 1976).

Mutant strains of *S. mutans* with decreased cariogenicity have been described; these have either been associated with reduced ability to produce extracellular polysaccharides (De Stoppelaar et al., 1971) or with loss of production of IPS (Tanzer et al., 1976). Some strains have been shown to carry lysogenic phages (Greer et al., 1971; Klein and Frank, 1973), and plasmids have been demonstrated in several strains

Table 12.21.
Characteristics differentiating species within the **Streptococcus mutans** *group[a]*

Characteristics	S. mutans	S. ferus	S. cricetus	S. sobrinus	S. rattus	Serotype "h"
Fermentation of						
Mannitol	+	+	+	+	+	+
Sorbitol	+	+	+	d	+	−
Melibiose	+	NT	d	d	+	+
Raffinose	+	−	+	d	+	−
Hydrolysis of						
Arginine	−	−	−	−	+	−
Esculin	+	+	d	d	+	−
Voges-Proskauer	+	d	+	+	+	+
H₂O₂ produced	−	−	−	+	−	−
Resistant to						
Bacitracin (2 units/ml)	+	−	−	+	+	−
Serovar	c, e, f	c	a	d, g	b	h
Mol% G + C	36–38	43–45	42–44	44–46	41–43	NT
Source	Human	Wild rats	Hamster, human, wild rats	Human	Rat, human	Monkey

[a] Symbols: see Table 12.2; also NT, not tested.

(Dunny et al., 1973; Higuchi et al., 1973; Clewell et al., 1976; Macrina et al., 1977; Katayama et al., 1978).

Bacteriocins which inhibit the growth of a variety of Gram-positive bacteria have been described and are sometimes referred to as mutacins (Kelstrup and Gibbons, 1969; Hamada and Ooshima, 1975). Mutacin-typing schemes have developed for use in epidemiological studies on *S. mutans* (Kelstrup et al., 1970; Berkowitz and Jordan, 1975; Rogers, 1975, 1976).

The serological heterogeneity of strains of *S. mutans*, first demonstrated by Zinner et al., (1965), was used to develop a serotyping scheme by Bratthall (1969, 1970). Initially, five serovars, a–e, were described, to which serovars f and g were subsequently added by Perch et al. (1974). More recently, a seventh serovar, h, has been isolated from the mouths of macaque monkeys (Beighton et al., 1981). As discussed elsewhere, some of these serovars of *S. mutans* have now been proposed as separate species (Coykendall, 1977).

Purified type-specific cell wall polysaccharide antigens from representative strains of each serovar of *S. mutans* have been prepared, using a variety of extraction procedures, and have been shown to be composed mainly of combinations of glucose, galactose and rhamnose. Significant quantities of *N*-acetylglucosamine and *N*-acetylgalactosamine are normally absent, although small amounts of these amino sugars are present in the antigens of serovars a and b. A summary of the known chemical composition and antigenic determinants of the type-specific antigens of serovar a–g is shown in Table 12.22, which is based on publications by Hamada et al. (1976), Mukasa and Slade (1973a, 1973b), Wetherell and Bleiweiss (1975, 1978), Iacono et al. (1975) and Linzer et al. (1975, 1976).

The peptidoglycan of *S. mutans* cell walls usually contains glutamic acid, alanine, lysine, glucosamine and muramic acid in the approximate molar ratio of 2:2–4:1:1:1. In serovars a, d and g, the presence of threonine has also been reported; the molar ratio of this amino acid to glutamic acid, when it occurs, is approximately 0.7–1:1 (Hamada and Slade, 1980a). Membrane-associated lipoteichoic acid is present in all serotypes of *S. mutans* (Hamada et al., 1976).

S mutans strains are normally sensitive to penicillin, ampicillin, erythromycin, cephalothin, methicillin and other antimicrobial agents. For short term suppression of *S. mutans* in the mouth, topically applied agents such as vancomycin, kanamycin and iodine have been used. Fluoride, bis-biguanidines and surfactants have also been reported to inhibit *S. mutans* in the oral cavity. Most strains are lysed by an enzyme, "mutanolysin," which has been derived from the soil bacterium, *Streptomyces globisporus* (Yokagawa et al., 1974, 1975).

Many strains of *S. mutans* have been shown to be cariogenic in experimental animals, including rats, hamsters, gerbils, mice and monkeys, and there is evidence for an association between this species (or

group of species) and human dental caries (Hardie, 1981; Tanzer, 1981). *S. mutans* is also one of the oral streptococci which may be isolated from cases of infective endocarditis.

The primary habitat of *S. mutans* is the tooth surface of man, but it can also be isolated from feces. Colonization of the tooth is favored by high levels of dietary sucrose.

The mol% G + C of the DNA is 36–38 (by T_m, Bd).

Type strain: ATCC 25175; (NCTC 10449).

11. **Streptococcus rattus** Coykendall 1977, 28.[AL]

rat'tus. M.L. n. *rattus* generic name of the rat (from which the original isolate was derived).

Gram-positive cocci, 0.5 μm in diameter, in pairs and chains. Mannitol, sorbitol, raffinose, sucrose, lactose, maltose and inulin are fermented, but not glycerol, melezitose, rhamnose or xylose. The terminal pH in glucose broth is 4.2–4.4. Arginine and esculin are hydrolyzed, but not starch. Hydrogen peroxide is not produced. Growth occurs in air but is generally improved by an atmosphere of reduced O_2 content, with added CO_2. An adhesive extracellular glucan is produced from sucrose. Firm, "rubbery" colonies are formed on sucrose agar by some strains, often rough, heaped and with beads or puddles of liquid (containing glucan).

S. rattus strains were originally described as a distinct serovar of *S. mutans*, designated *type b* (Bratthall, 1970). Two polysaccharide antigens, apparently with identical immunodeterminants, have been purified from this serovar (Mukasa and Slade, 1973b), and a glycerol teichoic acid substituted with a galactosyl moiety has also been described (Vaught and Bleiweiss, 1974).

S. rattus was first obtained from a laboratory rat, but has also been isolated from the human mouth.

The mol% G + C of the DNA is 41–43% (T_m, Bd).

Type strain: ATCC 19645 (FAI).

Further taxonomic comments. This species (or serovar) is less commonly isolated from humans than the other *S. mutans* serovars. Several authors have recognized *S. rattus* as a separate biovar of *S. mutans* (Shklair and Keene, 1974; Perch et al., 1974) and the production of ammonia from arginine is one simple phenotypic character which allows it to be distinguished from the other biovars. Strains of *S. rattus* can also be separated on the basis of SDS-PAGE protein patterns and comprised a distinct cluster in one numerical taxonomic study (Whiley et al., 1982; Hardie et al., 1982). Taxonomic studies in which *S. rattus* strains have been included have invariably been confined to a small number of representative strains (maximum four).

12. **Streptococcus cricetus** Coykendall 1977, 28.[AL]

cri'ce.tus. M.L. n. *cricetus* hamster (from which the original isolate was derived).

Gram-positive cocci, 0.5 μm in diameter, occurring in pairs or chains. Colonies on sucrose agar are about 1 mm in diameter, rough, heaped, often glossy and may be surrounded by liquid containing soluble extracellular glucan. Colonies on blood agar are 2–3 mm in diameter, smooth and round; some strains produce a zone of α-hemolysis, but most are nonhemolytic. Facultatively anaerobic, growing best in an atmosphere with reduced O_2 and added CO_2. Mannitol, sorbitol, raffinose, mannose, inulin, salicin, sucrose and lactose are fermented, but not arabinose, glycerol, melezitose, rhamnose or xylose. The final pH in glucose broth is 4.1–4.2. Ammonia is not producued from arginine. Esculin is usually hydrolyzed but not starch or hippurate. Human isolates are susceptible to 2 unit/ml bacitracin.

Most strains react with *S. mutans* serovar a antiserum, but some strains isolated from wild rats lack the a antigen. A high degree of DNA homology has been demonstrated between *S. cricetus* strains obtained from different sources (Coykendall et al., 1976).

S. cricetus has been isolated from the mouth of hamsters, wild rats and, occasionally, man.

The mol% G + C of the DNA is 42–44% (T_m).

Type strain: ATCC 19642 (HS6).

Table 12.22.
Proposed antigenic determinants of serovars within the
Streptococcus mutans *group*

Serovar	Name	Proposed antigenic determinant
a	*S. cricetus*	Glucose-β(1,6)-glucose[a]
b	*S. rattus*	α-Galactose[b]
c	*S. mutans*	Glucose-α(1,4)-glucose[c,d]
d	*S. sobrinus*	Galactose-β(1,6)-glucose[e]
e	*S. mutans*	Glucose-β(1,6)-glucose[f]
		or
		Glucose-β(1,4)-glucose[g]
f	*S. mutans*	Glucose-α(1,6)-glucose[h]
g	*S. sobrinus*	β-Galactose[i]

[a] Mukasa and Slade, 1973a
[b] Mukasa and Slade, 1973b
[c] Linzer et al., 1976
[d] Wetherell and Bleiweis, 1975
[e] Linzer et al., 1975.
[f] Hamada and Slade, 1976
[g] Wetherell and Bleiweis, 1978
[h] Hamada et al., 1976
[i] Iacono et al., 1975

Further taxonomic comments. S. cricetus, also referred to in the literature as *S. mutans* serotype a, was included on the Approved Lists of Bacterial Names. Although shown to comprise a distinct genetic group on the basis of DNA/DNA homology studies, it is not easy to distinguish *S. cricetus* from other members of the "*S. mutans* group" by means of phenotypic characters (Table 12.21).

Numerical taxonomic studies and SDS-PAGE protein profiles indicate that these strains are very similar to *S. mutans* serovars d and g (otherwise known as *S. sobrinus*) (Hardie et al., 1982; Whiley et al., 1982). It is possible that further work may lead to the regrouping of serovars a, d and g into a single species or subspecies.

13. Streptococcus sobrinus Coykendall 1983, 883.[VP]

so.bri′nus. L. masc. n. *sobrinus* male cousin on mother's side (referring to "distant relationship" between this species and *S. mutans*).

Gram-positive cocci, 0.5 μm in diameter, occurring in pairs and chains, often long chains. Colonies on sucrose agar are about 1 mm in diameter, rough, heaped, often showing drops of glucan-containing liquid on or around the colony. Some strains are α-hemolytic on blood agar, others nonhemolytic. Acid is produced from mannitol, inulin and lactose, but strains vary in their ability to ferment sorbitol, melibiose and raffinose. Ammonia is not produced from arginine. Most strains produce H_2O_2 and do not hydrolyze esculin. Significant amount of IPS are not synthesized.

Strains described as *S. sobrinus* usually react with *S. mutans* serovar d or g antisera and many cross-reactions occur between these serovars (Perch et al., 1974). However, the type strain (SL1) designated by Coykendall (1977) does not apparently possess either the d or g type antigens. The strains grouped together in this species share extensive common DNA base sequences (Coykendall, 1974; Coykendall et al., 1976).

The habitat of *S. sobrinus* is the human tooth surface. Strains have been shown to be cariogenic in experimental animals and may be associated with human dental caries.

The mol% G + C of the DNA is 44–46% (T_m).

Type strain: SL1 (ATCC 33478).

Further taxonomic comments. Although *S. sobrinus* has been proposed as a separate species, some investigators have preferred to regard these strains as a biovar or serovar within *S. mutans* (Perch et al., 1974; Hardie and Marsh, 1978; Hamada and Slade, 1980a). As mentioned previously, numerical taxonomic studies and analysis of SDS-PAGE protein patterns indicate a close relationship between strains of serovars a (*S. cricetus*), d and g (*S. sobrinus*) (Hardie et al., 1982; Whiley et al., 1982).

14. Streptococcus ferus Coykendall 1983, 883.[VP]

fe′rus L. adj. *ferus* wild (referring to wild rats from which the organism was isolated).

Gram-positive cocci, 0.5 μm in diameter, in pairs or chains. Colonies on sucrose agar are about 1 mm in diameter, raised, somewhat adherent but not showing drops of glucan-containing liquid on or around the colony. Both extracellular and intracellular glucans are produced from sucrose. Mannitol and sorbitol are fermented, but not raffinose. The final pH in glucose broth is 4.2–4.5. Ammonia is not produced from arginine. No growth at 45°C or in 6.5% NaCl.

Strains are inhibited by bacitracin (2 unit/ml).

S. ferus strains react with *S. mutans* type c antiserum, but DNA homology studies indicate that they form a distinct genetic group, not closely related to *S. mutans* (Coykendall et al., 1976).

The organism has been isolated only from the mouths of wild rats (Coykendall et al., 1974); it has not been found in humans.

The mol% G + C of the DNA is 43–45% (T_m).

Type strain: 8S1 (ATCC 33477).

Further taxonomic comments. This proposed species was not included in the Approved Lists of Bacterial Names and has not been widely studied. Further data are required before the relationship of *S. ferus* to other members of the "*S. mutans* group" can be fully assessed.

Species Incertae Sedis

h. *Streptococcus oralis* Bridge and Sneath, 1982, 414.[VP]

o.ra′lis. M.L. adj. *oralis*, of the mouth.

Gram-positive cocci in short chains, without capsules. They are nonmotile, nonsporing, facultatively anaerobic, fermentative and catalase-negative. Produce α-hemolysis around colonies on blood agar. Some of the characteristic biochemical reactions are shown in Table 12.18. Strains of *S. oralis* reduce tetrazolium, grow in the presence of 0.0004% crystal violet, and reduce tetrathionate (Bridge and Sneath, 1982).

All strains described to date were originally isolated by Carlsson (1968) from human mouths.

The mol% G + C of the DNA is 39.9 (T_m).

Type strain: LVG1 (NCTC 11427; PB 182).

Further taxonomic comments. This recently described species formed a distinct cluster of 13 strains in the numerical taxonomic study of Bridge and Sneath (1983). The nearest clusters in the phenogram were *S. milleri, S. sanguis* and *S. mitis*, which joined *S. oralis* (or oral group II) at a similarity level (SG) between 0.75 and 0.8. In the earlier numerical study, from which all the strains were obtained, the isolates fell into several different clusters, mostly cluster I (*S. sanguis*) and cluster V (*S. mitis*) (Carlsson, 1968). It may also be relevant to note that these strains were first isolated from the mouths of only four individuals (Carlsson, 1967a).

As mentioned by Bridge and Sneath (1982), the properties of *S. oralis* do resemble *S. sanguis* in some respects and although the proportions of strains which hydrolyze esculin and ferment raffinose differ, both of these tests are variable characters in the two species (Table 12.18). Some strains designated *S. oralis* also give biochemical reactions very similar to those of "*S. mitior.*"

The validity and usefulness of this species needs to be tested by further chemotaxonomic and genetic studies, particularly DNA/DNA hybridization. On the basis of phenotypic characters it is not easy to define clear-cut criteria for differentiating *S. oralis* from other oral streptococci.

i. *Streptococcus mitis* Andrewes and Horder 1906, 712.[AL]

mi′tis. L. adj. *mitis* mild.

Description based on Sherman et al. (1943).

Spherical or ellipsoidal cells 0.6–0.8 μm in diameter. Long chains in broth cultures.

Smooth and rough colony variants occur with frequent reversion of rough to smooth upon subculture in broth. No group antigen has been shown but many serological types are demonstrable by the precipitin test. Using the cell wall agglutination procedure two "groups" have been defined (Kalonaros and Bahn, 1965). One group was agglutinated by group O antiserum while the other was agglutinated by group O and N antisera. Only slight reactions occurred in the precipitin test. Apparently, serological reactions are of little value in identifying this species.

Final pH in glucose broth is 4.2–5.8, averaging about 4.5. Acid produced from glucose, maltose, sucrose and usually from lactose and salicin. Occasional strains ferment raffinose and trehalose. No acid from inulin, mannitol, sorbitol, glycerol, arabinose or xylose. Rarely, a strain produces a typical mucoid colony on sucrose agar or a colony resembling small bits of broken glass. Some strains oxidize butyric acid (Wolin et al., 1952) with the accumulation of H_2O_2 in the medium.

On blood agar incubated aerobically a pronounced α-reaction occurs. Growth may or may not occur at 45°C.

Source: Human saliva, sputum and feces.

This species comprises a heterogeneous group of α-hemolytic streptococci associated with the human respiratory tract. It has no unique identifiable characteristic and some nonhemolytic varieties of typically hemolytic species may be confused with it. (Tables 12.18; 12.19).

Type strain: NCTC 3165.

Further taxonomic comments. The designated type strain of *S. mitis* is NCTC 3165 (ATCC 33399 and NCDO 2495). The mol% G + C of the DNA of this strain has been estimated by T_m as 39.5% (Farrow and Collins, 1984) or as 38.3% (R. A. Whiley, personal communication).

DNA/DNA hybridization studies with this strain have shown fairly low relative binding ratios (36–40%) to another "more typical" *S. mitis* strain and to ATCC 10557 (*S. sanguis* type II or "*S. mitior*") (Welborn et al., 1983). A relative binding ratio of 47% was found with the type strain of *S. sanguis* (ATCC 10556).

The physiological, biochemical and serological characteristics of strain NCTC 3165 indicate that it should probably be included within *S. sanguis* (Colman and Williams, 1972; Cole et al., 1976; Rosan, 1973) (Table 12.15). However, the DNA/DNA hybridization data referred to above show that this strain is not entirely homologous with *S. sanguis* and may, in fact, represent a separate species distinct from both *S. sanguis* and "*S. mitior*" as described in this chapter. For this reason, the three named species are listed separately, pending the availability of further genetic data.

Nutritionally-variant streptococci

The existence of nutritionally-variant α-hemolytic streptococci and their involvement in bacterial endocarditis and septicemia has been known for many years. Such isolates have been variously described as "satelliting streptococci," "thiol-dependent streptococci," "vitamin B6-dependent streptococci" and "symbiotic streptococci," but the term "nutritionally variant streptococci" is perhaps the most appropriate (Cooksey et al., 1979).

It is not entirely clear whether these streptococci represent a distinct taxonomic entity or, alternatively, such variants occur within several different species. According to one report, vitamin B6-dependent streptococci, identified as "*Streptococcus mitior*" were found to account for 5–6% of streptococcal isolates from bacterial endocarditis (Roberts et al., 1979). Other authors have also described such isolates as belonging to either "*S. mitior*" (Schiller and Roberts, 1982) or *S. mitis* (Bouvet et al., 1981; van de Rijn and Bouvet, 1984). In contrast to these reports, however, Cooksey et al. (1979) found that nutritionally variant streptococci could be allocated to one of five different named species on the basis of biochemical tests (*S. salivarius, S. sanguis, S. mitis, S. morbillorum* and *S. anginosus*).

Nutritionally variant streptococci do not grow on regular blood agar plates and are one of the recognized causes of "culture-negative" endocarditis. Supplementation of media with pyridoxal appears to enable the growth of all strains (Reimer and Reller, 1981) which may also grow as satellite colonies around *Staphylococcus aureus*. An interesting recent observation which may be of some taxonomic value is the production of a red chromophore when strains are boiled in 2M HCl (Bouvet et al., 1981). This property appears to be unique to nutritionally variant streptococci and strains of *S. mitis* (Van de Rijn and Bouvet, 1984), and might be useful as an identification test.

Enterococci*

The Late J. Orvin Mundt

Streptococcus faecalis, S. faecium, "*S. avium*" and *S. gallinarum* comprise the group of enterococci. *S. faecalis* and *S. faecium* are residents of the intestinal tracts of humans and most animals, and the latter two species occur in poultry. *S. bovis* and *S. equinus,* residents of the bovine and equine intestinal tracts and possessing the group D antigen, differ physiologically and in the properties of their fructose-diphosphate aldolases (London and Kline, 1973). They also have a high minimum temperature for growth, 20–22°C, and they do not survive apart from the animal host. Typical and atypical *S. faecalis* and *S. faecium* and streptococci which bear similarities to them occur commonly on plants and in insects and wild animals (Geldreich et al., 1964; Mundt, 1963a, b; Martin and Mundt, 1972). The majority of *S. faecalis* from plants and insects digest starch (Geldreich et al., 1964) and produce stratiform digestion of a soft, rennet-like curd in litmus milk (Mundt, 1973). However, these properties have not been observed among human isolates: Bridge and Sneath (1983), employing a sensitive method, detected acid production from starch by nearly all strains of enterococci. Interspecific DNA duplexes between *S. faecalis* and *S. faecium* exhibited high levels of formation at 60° and 70°C, and low levels between them and strains of streptococci from nonhuman sources (Roop et al., 1974).

The ability to initiate growth in 6.5% NaCl broth and in broth at pH 9.6 are commonly used to segregate the enterococci from other streptococci. However, strains of anaerobic *S. bovis* with these properties have been isolated from the bovine alimentary canal (Latham et al., 1979), and approximately 80% of strains of group B streptococci and a low percent of other groups have been found to grow in 6.5% NaCl broth (Facklam et al., 1974).

The enterococci utilize many carbohydrates through a fermentative pathway. The final pH in glucose broth is usually 4.2–4.6, with occasional lower values. Discrepancies in sugars fermented have been noted in reports of various authors. These may result from the extent to which the basal medium is buffered, the sensitivity of the indicator, the time of incubation and strain differences. Dependence in identification should be placed on a spectrum of characteristics possessed by the strain in question (Deibel, 1964), obtained with culture media which duplicate as nearly as possible those used by other investigators.

The enterococci usually do not reduce nitrate, and do not digest cellulose or pectin or hydrolyze triglycerides, although lipases have been detected in some strains of each species (Bridge and Sneath, 1983). Only the proteolytic biovar of *S. faecalis* is known to digest gelatin and casein.

Enrichment and Isolation

Rothe's azide-glucose broth (Mallmann and Seligmann, 1950) is selective for the enrichment and recovery of *S. faecalis* and *S. faecium.* Solid media which may be surface-plated directly or from enrichment tubes are the medium of De Man et al. (1960) as modified by Mundt et al. (1967), KF medium (Kenner et al., 1961) and thallous acetate-citrate agar (Lachica and Hartman, 1968). Streptococci resembling *S. faecalis* and *S. faecium,* the lactic acid streptococci, *Leuconostoc* spp. and *Pediococcus* spp. may also appear on the plates. Therefore, it is essential that outgrowths are identified by criteria customarily used for the species. "*S. avium*" may be isolated in media containing sorbose as the energy source and 0.01% sodium azide, and adjusted to pH 10.

Procedures for Testing Special Characters

Several of the criteria which characterize the enterococci represent critical conditions which must be determined with accuracy. Inocula should be young, vigorously growing cultures. Incubation at 10°C for 7–10 days, and at 45°C for 2 days, must be done in accurately controlled water baths. The NaCl content of formulated media should be taken into consideration in the preparation of 6.5% NaCl broth. Media at pH 9.6 may be prepared according to the method of Chesbro and Evans (1959) or, alternatively, sterilized basal broth may be adjusted aseptically with 20% tripotassium (not trisodium) phosphate and promptly distributed into screw-capped tubes which are sealed to prevent absorption of CO_2 from the atmosphere. The pH of uninoculated control

---

* **Editorial note:** As mentioned previously, a new genus *Enterococcus* has recently been proposed to include the species described in this section (Schleifer and Kilpper-Bälz, 1984; Collins, et al., 1984). This proposal was not published until after this section was completed.

tubes should be checked at the time of inoculation and after incubation. Detection of ammonia is normally done with L-arginine with medium adjusted to pH 7.0 at the time of preparation (Niven et al., 1942).

Some distinguishing features of the different species within the enterococci are shown in Table 12.23.

15. **Streptococcus faecalis** Andrewes and Horder 1906,713[AL]; nom. cons. Opin. 30, Jud. Comm. 1963, 167. (*Enterococcus faecalis* (Andrewes and Horder) Schleifer and Kilpper-Bälz 1984, 33.)

fae.cal'is. L. n. *faex, faecis*, dregs. M.L. adj. *faecalis* relating to feces.

Ovoid cells elongated in direction of the chain; 0.5-1.0 μm in diameter, mostly in pairs and short chains. Generally nonmotile. Rarely pigmented (false pigmentation may be due to precipitation of metal ions, Jones et al., 1963). Colonies on solid media smooth, cream or white and entire.

Lancefield's group D. The group-specific antigenic determinant is an intracellular (Jones and Shattock, 1960) glycerol teichoic acid associated with the cytoplasmic membrane (Slade and Shockman, 1963). The antigen is released in the conversion of cells to protoplasts (Shattock and Smith, 1963). Eleven type-specific antigens are polysaccharides which contain *N*-acetylglucosamine (Willers and Michel, 1966) and are located in the cell wall (Elliott, 1960; Sharpe, 1964). Cells contain dimethyl quinones with nine isoprenes as the major unit, and with lesser amounts of 7–8 isoprene units (Collins and Jones, 1979). The peptide subunit of the peptidoglycan consists of L-Ala, D-Glu, L-Lys and D-Ala. A bridge tripeptide of L-Ala joins peptide subunits through L-Lys and D-Ala (Kandler et al., 1968).

Fermentation of glucose normally yields primarily lactic acid. Increased amounts of formic and acetic acids and ethanol are formed if neutrality of the medium is maintained (Gunsalus and Niven, 1942). Glycerol is phosphorylated and oxidized aerobically to lactate (Jacobs and Van Demark, 1960). With vigorous aeration (not simply aerobic incubation) end products are changed to a mixture of acetic acid, acetyl methyl carbinol and CO_2 (London and Appleman, 1962). Malate in the presence of glucose is metabolized without gas formation (Whittenbury, 1965).

Final pH in glucose broth is 4.1–4.6. Fermentation of melezitose and inability to ferment melibiose help to distinguish *S. faecalis* from *S. faecium*. Arabinose and raffinose are rarely fermented.

Grows on 0.1% thallous acetate agar. Rapid reduction of tetrazolium.

Table 12.23.
Distinguishing features of the enterococci[a,b]

Characteristics	S. faecalis	S. faecium	"S. avium"	S. gallinarum
Growth in 0.1% methylene blue milk	+	+	d	— (Delayed)[c]
Ammonia from arginine	+	+	—	d
Reduction of tetrazolium	+	—	—	+ (Delayed)[c]
Reduction of potassium tellurite	+	d	d	—
Tyrosine decarboxylated	+	—	ND	ND
Acid from L-arabinose	—	d	+	+
Acid from arbutin	—	+	+	+
Acid from melezitose	+	—	+	d
Acid from melibiose	—	+	d	+
Acid from sorbitol	+	d	+	+
Acid from sorbose	—	+	+	—

[a] All strains produced acid from amygdalin, cellobiose, fructose, galactose, glucose, glycerol, lactose, maltose, mannose, mannitol, ribose, salicin and sucrose, but variable results are reported with *S. faecium* in media with arabinose, sorbitol and sucrose. All strains hydrolyze esculin. None ferment erythritol or reduce nitrate.
[b] Symbols: see Table 12.2.
[c] Data of Barnes et al., 1978.

Survives heating to 60°C 30 min in neutral media. Resistance to heat is markedly affected by the pH of the medium (White, 1963) and the age of the cells (Beauchat and Leschwich, 1968).

Reduction of litmus precedes formation of curd in milk, Casein and gelatin may be digested.

Tyrosine is decarboxylated to tyramine.

Transmissible plasmids regulate resistance to antibiotics (Jacob and Hobbs, 1974) and the production of β-hemolysin and bacteriocin (Jacob et al., 1975; Dunny and Clewell, 1975).

Sources: Feces of humans and homeothermic and poikilothermic animals; insects; and plants. Common in many nonsterile foods, with presence often not related to fecal contamination. Pathogenic agent in urinary tract infections and subacute endocarditis.

The mol% G + C of the DNA ranges from 33.5 (chemical analysis) to 38.

Type strain: NCTC 775 (ATCC 19433; NCDO 5681).

Recent numerical taxonomic studies (Bridge and Sneath, 1983; Jones et al., 1972) group the subspecies formerly known as "*S. faecalis* var. *liquefaciens*" and "*zymogenes*" into one cluster with *S. faecalis* with little evidence to justify retention of the subspecies. These subspecies and the subspecies "*S. faecalis* subsp. *faecalis*" which were included in the previous edition of the *Manual* (Deibel and Seeley, 1974) are no longer recognized.

16. **Streptococcus faecium** Orla-Jensen 1919,139.[AL] (*Enterococcus faecium* (Orla-Jensen) Schleifer and Kilpper-Bälz 1984, 33.)

fae'ci.um. L. n. *faex, faecis*, dregs; L. gen. pl. n. *faecium* of the dregs, of feces.

Spherical to ovoid cells, chiefly in pairs and short chains. Elongated cells may be formed. Nonmotile. Motile strains differ in esterase and protease patterns on polyacrylamide gel (Lund, 1967). Colonies on solid media smooth, white, entire.

Lancefield's group D. Chemistry and location of group antigen the same as that of *S. faecalis*. Many type-specific antigens (Barnes, 1965). No isoprenoid quinones (Collins and Jones, 1979).

Final pH in glucose broth 4.0–4.4. Alteration of glucose metabolism with vigorous aeration is identical to that of *S. faecalis*. Anaerobic growth in glycerol medium is restricted by peroxide formation (London and Appleman, 1962). Arabinose and melibiose fermented. Melezitose, sorbitol and inulin not fermented. Malate in the presence of glucose is metabolized with vigorous gas formation.

No growth or sparse gray rather than black colonies on 0.04% potassium tellurite. Growth in 0.1% thallous acetate. Tetrazolium reduction to pale pink or none.

Survives heating to 60°C 30 min in neutral media.

Reduction of litmus precedes curd formation in milk. Curd formation may be delayed to 5–7 days.

Gelatin and casein not digested.

Tyrosine not decarboxylated.

Peroxidase produced in heated blood medium containing dianisidine. Catalase not produced.

α-Hemolysis on blood agar.

Sources: Feces of humans and homeothermic and poikilothermic animals; insects; and plants. Common in many nonsterile foods, usually not related to fecal contamination. Strictly anaerobic strains which can adapt to grow aerobically have been isolated from the bovine alimentary tract (Latham et al., 1979). Strains containing a plasmid-coded urease have been isolated from the rumens of sheep (Cook, 1976).

The mol% G + C of the DNA is 38.3–39.0 (T_m).

Type strain: NCTC 7171 (ATCC 19434; NCDO 942).

"*Streptococcus faecium* subsp. *casseliflavus*" Mundt and Graham 1968,2007. (*Streptococcus casseliflavus* Vaughan, Riggsby and Mundt 1979, 212; *Enterococcus casseliflavus* (Vaughan, Riggsby and Mundt 1979), Collins, Jones, Farrow, Kilpper-Bälz, and Schleifer 1984, 221)

Bluntly pointed cells in pairs and pairs within chains. Both motile and nonmotile strains common. Grows at 10°C and in broth containing 6.5% NaCl or broth adjusted to pH 9.6; many strains fail to grow at

45°C. Ammonia usually not produced from arginine. Growth on light or contrasting agar, or of cells precipitated from broth, slight to deeply pigmented yellow resembling Cassel's yellow.

Final pH in glucose broth 4.5–4.7. Has the same sugar fermentation pattern as *S. faecium*. Glycerol may be fermented anaerobically. Inulin and dulcitol may be fermented. Melezitose usually not fermented.

Gray, punctate colonies on agar with 0.04% potassium tellurite. Tetrazolium reduced to intermediate pink pigment. Peroxidase produced on heated blood agar; catalase is not produced. Hippurate usually hydrolyzed.

Malate dissimilated vigorously in the presence of glucose with production of gas.

Isoprenoid quinones contain menaquinones with 7–8 isoprene units (Collins and Jones, 1979). Differs from *S. faecium* in having higher cellular levels of $C_{16:0}$ fatty acids and lower levels of cyclopropane $C_{19:0}$ fatty acids (Amstein and Hartman, 1973).

None of 28 strains hybridized at a level of more than 21% with strains of either *S. faecalis* or *S. faecium* (Vaughan et al., 1979). Two groups were recognized, one hybridizing at a level of at least 75% with the DNA of the reference strain and the other which hybridized at about 10%.

The mol% G + C of the reference strain is 42.

Reference strain: ATCC 25789 (The type strain of *Enterococcus casseliflavus* Collins et al. 1984, 221 is ATCC 25788).

"*Streptococcus faecium* subsp. *mobilis*" Langston, Guiterrez and Bouma 1960, 716.

Morphology and properties similar to those of *S. faecium*. Motile. Ammonia produced from arginine.

Final pH in glucose broth 4.2–4.6. Acid from same sugars fermented by *S. faecium*. Variable in raffinose, starch and salicin. Inositol not fermented. Esculin hydrolyzed.

Moderate gray growth on 0.04% potassium tellurite.

Methylene blue in milk rapidly reduced. Tetrazolium weakly reduced. Litmus reduced, with delayed curd formation.

Acetyl methyl carbinol produced. Nitrate may be reduced in neutral to alkaline media (Langston and Williams, 1962).

Isoprenoid quinones contain menaquinones with 7–8 isoprene units (Collins and Jones, 1979).

Source: From grass silage.

Reference strains: ATCC 14432, 14436; type strain not designated.

17. "**Streptococcus avium**" Nowlan and Deibel 1967,295. (*Enterococcus avium* (ex Nowlan and Deibel 1967) Collins, Jones, Farrow, Kilpper-Bälz and Schleifer 1984,220.)

av'i.um. L. n. *avis* bird; L. gen. pl. *avium* of birds.

Ovoid cells in pairs and short chains. Nonmotile. Most strains produce α-hemolysis on blood. Ammonia usually not produced from arginine. Growth may be slow or absent in broth with 6.5% NaCl.

Lancefield's group Q. The group antigen, which is associated with the cell wall, is not found in all strains. Most of the strains also possess the group D antigen, which is located between cell wall and membrane (Smith and Shattock, 1964).

Average final pH in glucose broth 4.2. Reported to ferment many more sugars than are fermented by *S. faecium* (Bridge and Sneath, 1983). Erythritol, inositol and raffinose not fermented. Starch may be hydrolyzed. Esculin hydrolyzed. Pyruvate, malate, citrate and arginine and serine not utilized as energy sources.

Nitrate not reduced.

No reduction of methylene blue in milk and rarely reduction of potassium tellurite.

Sources: Characteristically from chickens, also feces of humans, dogs and pigs, appendicitis, otitis and abscesses of the brain.

The species can be separated from *S. gallinarum* by its ability to grow in broth containing 0.01% sodium azide at pH 10 with sorbose as an energy source.

The mol% G + C of the DNA is 39.9 (T_m).

Type strain: Collins et al. (1984) list NCDO 2369 (ATCC 14025) as the type strain of *Enterococcus avium*.

18. **Streptococcus gallinarum** Bridge and Sneath 1982, 414.[VP] (*Enterococcus gallinarum* (Bridge and Sneath) Collins, Jones, Farrow, Kilpper-Bälz and Schleifer 1984, 222.)

gal.lin.ar'um. L. gen. pl. *gallinarum* of hens.

Cells in pairs and short chains, somewhat pleomorphic. Nonmotile. Catalase-negative. Grows slowly on thallous acetate tetrazolium agar at room temperature, producing pink colonies. Hemolytic on horse blood agar, producing greening or complete lysis.

Lancefield's group D.

Ammonia from L-arginine; no ammonia from D-arginine.

Hippurate usually hydrolyzed. Esculin hydrolyzed. Gelatin not digested.

Final pH in glucose broth below 4.25. No acid from D-arabinose, adonitol, sorbose or dulcitol. Malonate may be fermented. No growth on Christensen citrate. Acid from starch.

Source: Intestines of domestic fowls.

The mol% G + C of the DNA is 37.4 (T_m).

Type strain: F87/276 (NCTC 11428; PB21)

Differs from "*S. avium*" in being adonitol negative, L-sorbose negative, raffinose positive, often malonate positive, and reacts only with group D antigen. The description by Barnes et al. (1978), who isolated 37 strains, differs from that of *S. avium* in the following: methylene blue in milk is reduced only on 10 days incubation; all strains hydrolyze hippurate; and not all strains produced acid from starch.

Addendum. Collins et al. (1984) also describe two other species that belong in this group: *Enterococcus durans* (ex Sherman and Wing 1937) and *Enterococcus malodoratus*. The reader is referred to this paper for further information on the classification of the enterococci.

Lactic Acid Streptococci

The Late J. Orvin MUNDT

The lactic acid streptococci consist of two species which grow at 10°C or less but not at 45°C. The optimum is near 30°C. They do not grow in broth adjusted to pH 9.6 or in broth containing 6.5% NaCl. They exhibit α-hemolysis or none on blood agar.

"*Streptococcus lactis* subsp. *diacetylactis*" and *S. cremoris* are here combined with *S. lactis* to form a single species. The subspecies *diacetylactis* was first described by Matuszewski et al. (1936) as *S. diacetilactis* because of its ability to produce diacetyl. It differs from *S. lactis* only in plasmid-regulated ability (Møller-Madsen and Jensen, 1962) to bring about a reaction between acetyl-CoA and hydroxyethylthiamin pyrophosphate with the production of diacetyl and CO_2 from citrate (Cogan, 1976).

S. cremoris was proposed by Orla-Jensen (1919) to accommodate streptococci which differed from *S. lactis* in having a lower temperature for growth, producing detectable quantity of CO_2 and in morphological differences. These differences were confirmed in later descriptions by easily performed tests, i.e. inability to produce ammonia from arginine, failure to grow in broth adjusted to pH 9.2 or in broth containing 4% NaCl and inability to grow in milk containing 0.3% methylene blue (Deibel and Seeley, 1974). Strains recognized as *S. cremoris* have a high level of β-galactosidase activity and exhibit slow fermentation of lactose (Farrow, 1980). They are unable to transport large peptide molecules into the cell (Law et al., 1976). All strains lack arginine deiminase and many lack ornithine transcarbamylase (Crow and Thomas, 1982). However, recent taxonomic studies strongly indicate that this organism is very similar to *S. lactis*. *S. lactis*, *S. cremoris* and "*S. lactis* subsp. *diacetylactis*" all possess identical isoprenoid quinones (Collins and Jones, 1979), indistinguishable lactic dehydrogenases

(Garvie, 1978a), possess in common with the nondairy strains of *S. lactis* the enzyme β-phosphogalactase (Farrow, 1980) and are all identical in mol% G + C (Garvie et al., 1981). They have in common at least 50% of the base sequences (Jarvis and Jarvis, 1981).

The distinguishing features of *S. lactis* and *S. raffinolactis* are given in Table 12.24.

Enrichment and Isolation

Stark and Sherman (1935) isolated *S. lactis* from plants by introducing plant material into litmus milk and incubating at 30°C. Mundt et al. (1967) isolated the organism from plants and from frozen and dried foods (Mundt, 1976) by surface plating serially diluted homogenates on the medium of De Man et al. (1960) modified to include 0.02% sodium azide, 0.01% tetrazolium and 1.5% agar, and replacement of 40% of the required glucose with mannitol.

Maintenance Procedures

Cultures can be maintained under refrigeration in litmus milk containing 0.1–0.5% CaCO₃, as frozen or lyophilized cultures. Survival of cultures in broth or on agar slants is poor.

Further Taxonomic Comments

The division of *S. lactis* into three subspecies has recently been proposed (Garvie and Farrow, 1982). These are: *S. lactis* subsp. *lactis*,[VP] type strain NCDO 604; *S. lactis* subsp. *diacetilactis*,[VP] type strain NCDO 176; and *S. lactis* subsp. *cremoris*,[VP] type strain NCDO 607.

As noted earlier, there is evidence that some of the properties which distinguish them are plasmid controlled, and that they show high DNA homology, but these subspecies are of particular importance in the dairy industry. The second subspecies produces diacetyl, which is important in giving flavor to cheese and the third is favored by cheese manufacturers also because it gives flavors to cheese.

19. **Streptococcus lactis** (Lister) Lohnis 1909, 554.[AL] (*Bacterium lactis* Lister 1873, 408).

lac′tis. L. n. *lac* milk; L. gen. n. *lactis* of milk.

Ovoid cells elongated in the direction of the chain; 0.5–1.0 m in

Table 12.24.
Distinguishing features of the lactic acid streptococci[a]

Characteristics	*S. lactis*	*S. raffinolactis*
Growth at 40°C	+	−
Growth in broth at pH 9.2	+	−
Growth in broth with 4% NaCl	+	−
Growth in 0.3% methylene blue in milk	+	−
Growth on 40% bile agar	+	−
Hydrolysis of arginine	+	−
Isoprenoid quinones	+	−
Acid from		
Dextrin	−	+
Raffinose	−	+
Rhamnose	−	+
Ribose	+	−
Sorbitol	−	+

[a] Symbols: see Table 12.2.

diameter. Mostly in pairs or short chains. Motile forms rarely observed. Microaerophilic. In broth culture growth frequently absent in uppermost several millimeters. Aerobic growth on agar media containing fermentable carbohydrate and supplemented with yeast extract. Surface colonies very small and discrete.

The peptidoglycan in the cell wall is similar to that of *S. pyogenes* except that the cross-bridge consists of D-isoasparagine (Schleifer and Kandler, 1967).

Belongs to Lancefield's group N. The group antigenic determinant is glycerol teichoic acid containing galactose phosphate (Elliott, 1963) which is located within the cell wall (Smith and Shattock, 1964). The teichoic acid cross reacts with certain type specific antipneumococcal sera (Heidelberger and Elliott, 1966). Proteinase(s) when present are located near the cell wall (Thomas et al., 1974).

When cultivated on artificial substrates many strains lose the ability to ferment lactose rapidly and to proteolyze milk. These lost characteristics may be restored through regular cultivation in milk. The same observation has been made with arabinose and xylose (Orla-Jensen, 1919). The ability to ferment lactose and to digest casein is regulated by two plasmids (Efstathiou and McKay, 1976) which may be transferred to nonfermenting and nonproteolyzing strains by conjugation or transformation (McKay et al., 1971, 1980). Phosphoenolpyruvate-dependent phosphotransferase and β-phosphogalactosidase may be prerequisite for rapid growth in milk (Thomas, 1976) and activity of cheese starter strains (Farrow, 1980). The less active nondairy strains also contain β-galactosidase.

Final pH in glucose broth 4.0–4.5. Acid from galactose, glucose, maltose and lactose. Arabinose, xylose, sucrose, trehalose, mannitol and salicin may not be fermented. No acid from raffinose, inulin, glycerol or sorbitol.

Tyrosine is not decarboxylated. No growth on 0.04% potassium tellurite.

Some strains produce an antibiotic, nisin (Hurst, 1978) that inhibits many Gram-positive organisms. Production of nisin may be plasmid mediated (Fuchs et al., 1975).

Sources: Raw milk and milk products, plants and nonsterile frozen and dry foods.

The mol% G + C of the DNA is 38.6 (T_m Deibel and Seeley, 1974).

Type strain: NCDO 604 (ATCC 19435; NCTC 6681).

20. **Streptococcus raffinolactis** Orla-Jensen and Hansen 1932, 152.[AL] Description from Garvie (1979).

raf′fin.o.lac.tis. L. n. *raffinosum*, raffinose; L. gen. n. *lactis* of milk.

Ovoid cells in pairs and short chains. Lactic dehydrogenases are in two bands, with the slower moving band traveling the same distance as the single band detected in *S. lactis*. Both components are activated by fructose diphosphate and inhibited by low concentration of phosphate, in contrast to the dehydrogenase of *S. lactis* (Garvie, 1978b). Hybridization with the DNA of *S. lactis* is less than 50%.

The mol% G + C of the DNA is 40.3–41.5 (T_m).

Type strain: NCDO 617.

Lancefield's group N.

Acid from glucose, fructose, galactose, mannose, lactose, maltose, sucrose, trehalose, glycerol and salicin. Arabinose, inulin, mannitol and starch are not fermented.

Ammonia is not produced from L-arginine. Acetoin is not produced. Gelatin is not digested. Acid in litmus milk with slow coagulation.

Grows in broth with 2%, but not 4%, NaCl. Does not grow in broth at pH 9.2. No isoprenoid quinones.

Source: Isolated from souring milk.

Anaerobic Streptococci

Jeremy M. Hardie

The streptococci grouped together here consist of three strictly anaerobic species (*S. hansenii*, *S. pleomorphus* and *S. parvulus*) and one which is aerotolerant (*S. morbillorum*). Two other aerotolerant species. *S. intermedius* and *S. constellatus*, have previously been included among the anaerobic streptococci (Holdeman and Moore, 1974; Jones, 1978) but, as described earlier, there is a strong case for considering these as a new grouping with "*S. milleri*." These anaerobic or microaerophilic cocci are now considered to belong to the genus *Strep-*

tococcus because of their lactic fermentation (Rogosa, 1974; Holdeman and Moore, 1974; Barnes et al., 1977; Jones, 1978). One of the species described in this section, *S. parvulus*, was until recently classified as *Peptostreptococcus parvulus* (Weinberg et al., 1937; Smith, 1957) but has been transferred back to *Streptococcus* following the discovery of increased biochemical activity under appropriate growth conditions (Cato, 1983).

Although it is convenient, at present, to consider these species together as a group, further work is required to establish their relationships to one another and to other streptococci.

21. **Streptococcus morbillorum** (Prévot 1933) Holdeman and Moore 1974, 269.[AL] (*Diplococcus morbillorum* Prévot 1933, 148.)

mor.bil'lor.um. M.L. g. n., *morbillorum* of measles; once considered to be associated with measles.

Anaerobic to aerotolerant Gram-positive cocci, nonmotile and nonsporing, in pairs or short chains. The cocci often elongate and cells occurring in pairs may be of unequal size, ranging from $0.3–0.8 \times 0.5–1.4$ μm. Colonies on blood agar plates, after 2 days, are pinpoint to 0.5 mm in diameter, circular, entire convex, translucent, shiny and smooth. Some strains produce slight greening (α-hemolysis), others are nonhemolytic. Most strains do not grow in air alone or air + CO_2, at least on first isolation, and the neotype strain remains obligately anaerobic (Holdeman and Moore, 1974).

Growth is enhanced by fermentable carbohydrates and by the addition of Tween 80 to fluid media. Optimum growth is at 35–37°C. Some of the biochemical reactions are shown in Table 12.25. H_2S, acetyl methyl carbinol, catalase, lecithinase and lipase are not produced, and hippurate and esculin are not hydrolyzed. Ammonia is not produced from arginine (Facklam, 1977).

The major fermentation product from glucose is lactic acid, with smaller amounts of acetic; sometimes formic and trace amounts of succinic and pyruvic acids, or ethanol, are detected. Lactate and gluconate are not utilized and threonine is not converted to propionate.

S. morbillorum has been isolated from human clinical specimens and intestinal contents; the neotype strain came from a lung abscess.

Type strain: ATCC 27824 (VPI 5424; Prévot 2917B).

Further taxonomic comments. Strains of *S. morbillorum* have not generally been included in numerical taxonomic, chemotaxonomic or genetic studies on the streptococci. Further work is required to establish the relationship of this species to other streptococci.

Table 12.25.
Some biochemical characteristics of the anaerobic streptococci[a]

Characteristics	S. morbillorum	S. hansenii	S. pleomorphus	S. parvulus
Aerotolerance	+	—	—	—
Acid from				
Cellobiose	—	—	—	+
Fructose	—	—	+	+
Galactose	—	+	—	+
Inulin	—	—	NT	+
Lactose	—	+	—	+
Maltose	W	+	—	+
Mannose	W	—	d	+
Salicin	—	—	—	+
Sucrose	W	—	—	+
Raffinose	—	+	NT	—
Production of				
Ammonia from arginine	—	—	NT	—
Acetyl methyl carbinol	—	d	NT	—
H_2S	—	+	W	—
Hydrolysis of esculin	—	d	NT	+
Mol% G + C	NT	37–38	39.4	46.0

[a] All species fail to ferment mannitol, sorbitol and starch.
[b] Symbols: see Table 12.2; also NT, not tested; W, weak reaction.

22. **Streptococcus hansenii** Holdeman and Moore 1974,266.[AL]

han.sen'i.i. M.L. gen. n. *hansenii*, named after P. Arne Hansen, a Danish-American bacteriologist.

Obligately anaerobic, nonmotile, Gram-positive cocci, with rounded or slightly tapered ends, in pairs or chains of pairs. Cells of the type strain in glucose broth cultures 1.6–2.3 μm in diameter. Colonies after 5 days of incubation in rumen fluid-glucose-cellobiose-agar (RGCA, Holdeman et al. 1977) are 0.5–2 mm in diameter, lenticular and translucent, or white with translucent edges. On anaerobic blood agar plates incubated for 2 days, surface colonies are 2 mm in diameter, circular, entire, convex, opaque, shiny, smooth and nonhemolytic. No growth occurs aerobically or in a candle jar (air + CO_2 atmosphere). Cultures in peptone-yeast extract-glucose (PYG) broth usually turbid with a smooth or stringy sediment; terminal pH 4.9–5.2 after 5 days. Moderate to poor growth in PY broth (without glucose), unaffected by addition of 0.1% Tween 80, 10% rumen fluid, or 20% bile. Good growth in PYG broth at 37° and 45°C, poor to moderate growth at 30°C and usually no growth at 25°C.

The biochemical characteristics are shown in Table 12.25. Some strains, including the type strain, ferment galactose and inulin, and reduce neutral red. No acid is produced from adonitol, dextrin, dulcitol, glycerol or sorbose and hippurate is not hydrolyzed. Strains do not grow in 6.5% NaCl-glucose broth and little or no ammonia is produced in chopped meat cultures. D(−)-lactic, acetic and succinic acids are produced in PYG broth; lactate and gluconate not utilized and no propionate produced from threonine.

The species has been isolated from human feces.

The mol% G + C of the DNA is 37–38% (by T_m).

Type strain: ATCC 27752 (VPI C7-24).

Further taxonomic comments. *S. hansenii* has not been studied by many investigators and appears not to have been included in any numerical taxonomic or DNA/DNA hybridization studies. According to the properties described by Holdeman and Moore (1974), this species differs only by a few biochemical reactions from *S. constellatus* and *S. morbillorum* (from which it is distinguished by fermentation of lactose) and *S. intermedius* (which ferments cellobiose). It is possible that further comparative studies might indicate a close relationship between strains currently designated *S. hansenii* and other streptococci here described collectively as the "*Streptococcus anginosus-milleri group*".

23. **Streptococcus pleomorphus** Barnes, Impey, Stevens and Peel 1977,52.[AL]

ple. o.mor'phus. Gr. adj. *pleomorphus*, many forms.

Obligately anaerobic, pleomorphic cocci, occuring singly, in pairs or short chains; variations are observed in size and shape according to the media and conditions of growth. Cells are Gram-positive in young cultures, often becoming Gram-negative within 24 h. Colonies after 3 days of incubation on growth media such as RCM (Hirsch and Grinsted, 1954; Barnes and Impey, 1968), supplemented BGP agar (Barnes et al., 1978) or VL agar (Beerens et al., 1963; Barnes and Impey, 1968, 1970) are 2–3 mm in diameter, circular, convex, with irregular edge. Some strains are weakly β-hemolytic on VL blood agar. Growth in broth is flocculent.

No growth occurs in air or air plus 10% CO_2. Good growth at 37°C and 45°C, but none at 20°C. Catalase-negative.

Carbohydrates are required for growth. Glucose is fermented to L-lactic acid; no gas produced. Traces of butyric, formic and sometimes acetic and succinic acids are also found. The terminal pH is in the range 4.4–5.0.

Acid is produced from glucose, fructose and, usually, mannose. No acid from arbinose, cellobiose, dextrin, galactose, inositol, lactose, maltose, mannitol salicin, starch, sucrose or xylose. No change in cysteine milk, gelatin is not liquefied, indole is not produced and nitrates are not reduced. Small amounts of H_2S are detected in media containing ferrous sulfate and sodium thiosulfate.

Growth occurs in the presence of polymyxin B (10 μg ml⁻¹), neomycin (100 μg ml⁻¹), kanamycin (100 μg ml⁻¹) and most strains are resistant to brilliant green (1/100.000).

A comparison of some of the properties of *S. pleomorphus* with other anaerobic streptococci is given in Table 12.25.

The species has been isolated from the intestines of chickens and, occasionally, from human feces.

The mol% G + C of the DNA (type strain) is 39% (by T_m).

Type strain: NCTC 11087 (EBF 61/60B).

24. **Streptococcus parvulus** (*ex* Weinberg, Nativelle and Prévot 1937) Cato 1983, 83.[VP] (*Streptococcus parvulus* Weinberg, Nativelle and Prévot 1937, 1011; *Peptostreptococcus parvulus* (Weinberg, Nativelle and Prévot) Smith 1957, 538.)

par′vu.lus. L. dim. adj. *parvulus*, somewhat small.

Obligately anaerobic, nonmotile, nonspore forming, Gram-positive cocci; 0.3–0.6 µm in diameter; in short chains or, occasionally, in pairs. Colonies after 2 days of incubation on anaerobic supplemented brain heart infusion blood agar plates are minute to 1.0 mm in diameter, circular, entire, transparent, grayish, slightly peaked; no hemolysis. No visible growth on plates incubated aerobically or in a candle jar.

Growth in liquid media is markedly stimulated by the addition of 0.02% Tween 80. This discovery enabled Cato (1983) to demonstrate a greater range of biochemical activity than previously recognized. Optimum temperature for growth is 37°C, but equally good growth occurs at 45°C; barely visible growth at 25°C. In carbohydrate broth media, slight turbidity is produced with a smooth to flocculent sediment. Growth completely inhibited by 20% bile or 6.5% NaCl. Final pH in peptone-pepticase-yeast extract-glucose-Tween 80 broth after 5 days of incubation is 4.0–4.2; when Tween 80 is omitted, the pH achieved is 5.8.

The biochemical reactions, based on two strains, are shown in Table 12.25. Acid is produced from cellobiose, esculin, fructose, galactose, glucose, inulin, lactose, maltose, mannose, salicin, sucrose and trehalose; erythritol and xylose weakly fermented; no acid from amygdalin, arabinose, glycerol, inositol, mannitol, melezitose, melibiose, pectin, raffinose, rhamnose, ribose, sorbitol or starch.

Esculin is hydrolyzed, but not starch or hippurate. Nitrate not reduced and indole not formed.

A solid acid curd is formed in milk; neither milk, gelatin nor meat is digested. Neither catalase, urease, DNAse, lecithinase nor lipase activity has been detected.

The fermentation products of the type strain after 5 days of incubation in glucose broth were found to be (meq/100 ml): lactic acid, 7.3–8.4; acetic acid, 0.3; and succinic acid, 0.03.

Pyruvate is converted to acetate, but neither lactate nor threonine are utilized.

No gas is produced in glucose agar deeps and H_2, ammonia and H_2S are not produced.

Both strains examined are susceptible to chloramphenicol (12 µg/ml) clindamycin (1.6 µg/ml), erythromycin (3 µg/ml), penicillin G (2 unit/ml), and tetracycline (6 µg/ml).

The source of the two described strains of this species is not known, but it has been reported that their principal habitat was the respiratory tract (Weinberg et al., 1937).

The mol% G + C of the DNA of the type strain is 46% (by T_m).

Type strain: ATCC 33793 (VPI 0546).

Further taxonomic comments. These strains were, until recently, classified as *Peptostreptococcus parvulus*, until the increased biochemical activity stimulated by addition of Tween 80 was observed (Cato, 1983) and led to their transfer back to *Streptococcus*. Apart from the DNA base ratio, which is high, there is little or no chemotaxonomic or genetic information available. Phenotypically, this species resembles most closely *S. anginosus*, *S. intermedius* and *S. morbillorum* (Table 12.25). Further studies are needed to determine the relationship, if any, of *S. parvulus* to these representatives of the "*S. anginosus-milleri* group.″

Other Streptococci

Jeremy M. Hardie

25. **Streptococcus acidominimus** Ayers and Mudge 1922, 49.[AL]

a.ci.do.mi′ni.mus. L. adj. acidus sour, acid; L. sup. adj. minimus very least, probably intended to mean that this organism produces the least amount of acid.

Spherical cells occurring in short chains.

No group-specific antigen has been demonstrated.

Weakly fermentative; most strains fail to decrease pH value of carbohydrate-containing media below 6.0. Acid produced from glucose, lactose and sucrose and, generally, maltose and trehalose. Glycerol, raffinose, inulin, arabinose and xylose are not fermented. The high limiting pH value offers difficulty in determining fermentation reactions.

α-Hemolysis on blood agar.

Growth only on complex media; minimal nutritional requirements unknown.

Source: Common in bovine vagina, occasionally found on skin of calves and in raw milk.

This species may be confused with *S. agalactiae* because of its slow hydrolysis of hippurate. Differentiation is effected by serological reactions and the inability of *S. acidominimus* to ferment glycerol or to hydrolyze arginine.

The mol% G + C of the DNA is 39.7 (T_m). (Garvie and Farrow, 1981).

Type strain: NCDO 2025.

Further taxonomic comments. The above description is taken from the eighth edition of the *Manual*. This species is listed in the Approved Lists (Skerman et al., 1980), but does not appear to have been included in recent numerical and chemotaxonomic studies on the genus.

26. **Streptococcus uberis** Diernhofer 1932, 370.[AL]

u′ber.is. L. n. uber udder, teat; L. gen. n. uberis of an udder.

Spheres occurring in pairs to chains of moderate length. Nonmotile. Microaerophilic.

The species is serologically heterogeneous. Nearly one-half the strains studied by Roguinsky (1971) were ungroupable, one-third reacted with group E antiserum, and small numbers reacted with C, D, P and U group antisera.

Deibel and Seeley (1974) noted a superficial similarity between *S. uberis* and group D streptococci, because of the common properties of growth at 10° and 45°C and survived heating of to 60°C for 30 min. These properties seem to apply to American strains, but not to European strains. Very few of the latter survived heating (Roguinsky, 1971) and neither he nor Garvie and Bramley (1979a) obtained growth at 45°C. Many strains failing to grow at 10°C resemble *S. agalactiae* in the ability to hydrolyze hippurate, to produce ammonia from arginine and the positive CAMP test. *S. uberis* does not, however, react with group B antiserum and does not hybridize with *S. agalactiae* in hybridization studies. Two distinct genotypes exist, one growing at 10°C with mol% G + C of the DNA of 34.8–35.9 and the other not growing at 10°C with mol% G + C of 36.3–37.0.

Final pH in glucose broth 4.6–4.9. Acid is produced from cellobiose, esculin, glucose, fructose, galactose, inulin, maltose, mannitol, mannose, ribose, salicin, sorbitol, starch, sucrose and trehalose. Arabinose, adonitol, erythritol, glycerol, sorbose and xylose are not fermented. A few strains ferment dulcitol, melezitose or raffinose.

Reaction on blood agar may be a weak α-hemolysis or none.

Ammonia is produced from arginine. Sodium hippurate is hydrolyzed. Tyrosine is not decarboxylated. DNase is not produced. Tetrazolium is reduced.

Grows in broth with 4%, but not 6.5% NaCl. Does not grow in broth at pH 9.6. Slow growth or none at 10°C No growth at 45°C. Does not survive heating to 60°C for 30 min.

Litmus milk reaction varies with the ability of the strain to ferment lactose.

Found on lips and skin of cows, in raw milk and on udder tissue.

Enrichment and isolation. Cullen (1966) used milk containing 0.02% potassium tellurite and 0.5% boric acid, followed by streaking on blood or other agar.

Differentiation from other species. *S. uberis* is differentiated from other species of streptococci by a combination of positive and negative properties which have been recorded with the descriptive information.

The mol% G + C of the DNA is 34.8–37.0 (T_m).

Type strain: NCTC 3858 (ATCC 19436).

27. **Streptococcus bovis** Orla-Jensen 1919, 137; *emend mut. char.* Sherman and Wing 1937, 57.[AL]

bo'vis. L. n. *bos* cow; L. gen. n. *bovis* of a cow.

Spherical to ovoid cells 0.8–1.0 μm in diameter in pairs and moderate to long chains. Grows readily on suitable media in the presence of oxygen. Latham et al. (1979) isolated anaerobic strains capable of growth in broth at pH 9.6 or broth containing 6.5% NaCl. Nonmotile. Large amounts of polysaccharide are produced in sucrose broth (Niven et al., 1948) and copious quantities of glucan are produced on sucrose agar (Dain et al., 1956). Starch is hydrolyzed to maltose and glucose by an enzyme resembling α-amylase (Dunican and Seeley, 1962). The species comprises strains with heterogeneous properties. Jones et al. (1972) and Garvie and Bramley (1979b) were unable to identify subgroups proposed by Medrek and Barnes (1962) and Iverson and Millis (1976). Niven et al. (1948) found no differences between human and bovine strains. Strains producing urease have been isolated from hogs (Raibaud et al., 1961). Garvie and Bramley (1979b) isolated strains which failed to grow at 45°C and which showed a relationship with *S. mutans* through DNA/DNA hybridization.

Lancefield's group D. Many serotypes have been isolated with specificity associated with large capsules surrounding the cells (Medrek and Barnes, 1962). Cross-reactions occur with groups E and N streptococci (Perry et al., 1958).

Final pH in glucose broth 4.0–4.5. Acid produced from glucose, fructose, galactose, mannose, lactose, maltose, sucrose and salicin. Variations occur in the fermentation of arabinose, xylose, mannitol, sorbitol, trehalose and inulin. No acid from glycerol. Esculin is hydrolyzed.

No growth in the presence of 0.04% potassium tellurite or 0.1% methylene blue in milk. Tyrosine is not decarboxylated. Most strains survive heating to 60°C for 30 min. Most strains produce an α-reaction on blood agar.

Minimum temperature for growth 22°C.

Acid production in litmus milk may or may not be followed by curd formation which precedes reduction.

This is the least nutritionally fastidious species among the streptococci. Many strains can utilize ammonium salts as a source of nitrogen (Wolin et al., 1959).

Source: Alimentary tract of cow, sheep and other ruminants; feces of pigs; occasionally occurring in large numbers in human feces; occasionally encountered in cases of human endocarditis; and from raw and pasteurized milk, cream and cheese.

The mol% G + C of the DNA is 36.6–39.1 (T_m).

Type strain: NCDO 597.

28. **Streptococcus equinus** Andrewes and Horder 1906, 712.[AL]

e.qui'nus. L. adj. *equinus* pertaining to a horse.

Spheres occurring in moderately long chains. Chaining pronounced in broth cultures. Grows on suitable media in the presence of oxygen. Nonmotile. Glucan production in sucrose-containing media has not been reported. Differing from *S. bovis*, starch is hydrolyzed only to oligosaccharides in the presence of small concentrations of a fermentable carbohydrate (Dunican and Seeley, 1962). Anaerobic conditions

inhibit hydrolysis. *S. equinus* resembles *S. bovis* in morphology and in fermentative properties. Mieth (1962) suggested that it is a lactose-negative variant of *S. bovis*. The aldolases of the two species fall into the same antigenic group (London and Kline, 1973).

Lancefield's group D. Cellular disruption of the antigenic preparation is required (Smith and Shattock, 1962). A weak antigenic reaction is obtained after many injections, and reaction of the antiserum with *S. faecalis* and *S. faecium* is obtained if the antiserum is concentrated by precipitation with ethanol (Fuller and Newland, 1963).

Final pH range in glucose broth 4.0–4.5. Acid from glucose, fructose, galactose, maltose and usually from sucrose and salicin. Raffinose and inulin are seldom fermented. No acid from arabinose, xylose, lactose, mannitol or glycerol.

No growth on agar containing 0.04% potassium tellurite or in broth containing 4% NaCl. No reaction in litmus milk. Weak α-hemolysis on blood agar. Does not survive heating to 60°C for 30 min.

Source: The predominant streptococcus in the alimentary tract of the horse.

Further taxonomic comments. Although thought to be an ill-defined species, *S. equinus* was considered to be sufficiently distinctive physiologically to merit species status in the eighth edition of the *Manual* and is listed in the Approved Lists (Skerman et al., 1980). Strains of *S. equinus* formed a distinct phenon, close to but distinct from *S. bovis*, in a numerical taxonomic study (Bridge and Sneath, 1983).

The mol% of the DNA is 35.6 (T_m).

Type strain: ATCC 9812.

29. **Streptococcus thermophilus** Orla-Jensen 1919, 136.[AL]

ther.mo'phil.us. Gr. n. *therme* heat; Gr. adj. *philus* loving. M.L. adj. *thermophilus* heat loving.

Spherical to ovoid cells 0.7–0.9 m in diameter, in pairs to long chains. Irregular segments and cells occur at 45°C.

No group-specific antigen has been demonstrated. The structure of the peptidoglycan is identical to that of *S. faecalis* (Schleifer and Kandler, 1967).

Habitat unknown. Jones (1978) suggests that it is a heat-resistant variant of another organism which currently cannot be recognized. Feltham (1978) observed a close cultural similarity to *S. pneumoniae*. Its aldolase relates it to *S. lactis* (London and Kline, 1973). A close relationship to *S. agalactiae*, *S. dysgalactiae* and *S. acidominimus*, as shown by DNA/rRNA hybridization, has also been observed (Garvie and Farrow, 1981). It has two lactic dehydrogenases which form L-lactate, differing from other lactic acid bacteria with more than one dehydrogenase which form DL-lactate (Garvie, 1978b).

Final pH in glucose broth is 4.0–4.5. A marked preferential fermentation of sucrose and lactose on primary isolation may disappear with continued cultivation. Acid is produced from fructose, glucose, mannose and lactose. Many strains ferment sucrose and some ferment maltose weakly. The glucose moiety of lactose is fermented to lactic acid and the galactose either is excreted into the medium (O'Leary and Woychick, 1976) or it is weakly fermented (Orla-Jensen, 1919). Arabinose, dextrin, glycerol, inulin, mannitol, rhamose, salicin, sorbitol, starch and xylose are not fermented.

α-Hemolysis or none on blood agar. Freshly isolated strains may not grow on blood.

Ammonia is not produced from arginine. No growth on agar containing 40% bile or 0.04% potassium tellurite. Tetrazolium is not reduced.

Grows in broth with 2.5% but not 4% NaCl. No growth at pH 9.6 or in milk with 0.1% methylene blue. Survives heating to 60°C for 30 min. Minimum temperature for growth 19–21°C, maximum 52°C. Excellent growth at 37°C.

Gelatin and casein are not digested. Usually poor growth on meat peptone media. Good growth on casein digest media.

Source: heated and pasteurized milks.

The mol% G + C of the DNA is 40.0 (T_m).

Type strain: NCDO 573 (ATCC 19258).

Enrichment and Isolation. Enrichment is obtained in milk incubated at 45–50°C. Colonies can be selected after growth on casein digest agar containing a fermentable carbohydrate.

Differentiation from other species. S. thermophilus may be recognized by rapid growth in litmus milk incubated at 45°C. It does not grow on bile agar, thereby separating it from *S. bovis* and *S. equinus*. *S. lactis* and *S. raffinolactis* grow at 10°C.

Further taxonomic comments. As discussed in the section on *S. salivarius*, it has recently been proposed that *S. thermophilus* be reclassified as *S. salivarius* subsp. *thermophilus* (Farrow and Collins, 1984). This recommendation is based on nucleic acid hybridization studies and long chain fatty acid profiles (Garvie and Farrow, 1981; Kilpper-Bälz et al., 1982; Farrow and Collins, 1984).

Species Incertae Sedis

j. "*Streptococcus suis*" Elliott 1966, 211 (Group S *Streptococcus* de Moor 1963, 272).

Cocci occuring singly, in pairs and rarely short chains as examined in clinical material (Field et al., 1954).

Reacts with Lancefield group D antiserum, possessing a lipid-bound teichoic acid antigen closely related to that of other group D streptococci (Elliot et al., 1977). Originally two serovars were recognized, based on capsular polysaccharide antigens: six new serovars have more recently been described (Perch et al., 1983).

Acid from trehalose, salicin, inulin, lactose, glucose, maltose, fructose, sucrose and galactose; no acid from sorbitol, mannitol, raffinose, arabinose, dulcitol, glycerol, inositol, rhamnose or xylose. Glucans not produced. Slow growth in media containing 40% bile. α-Hemolytic on calf blood agar, variable types of hemolysis on 5% horse blood agar. Most strains of serovars 7 and 8 produce hyaluronidase, other serovars negative. The majority of strains do not hydrolyze hippurate with the exception of some belonging to serovars 3 and 4.

Isolated from bacteremia in young pigs, frequently with involvement of the brain and joints; also from throat cultures of diseased pigs and their healthy littermates.

No information available on the mol% G + C of the DNA.

Type strain: not designated.

Further taxonomic comments. In the eighth edition of the *Manual* it was considered that this species more closely resembled *S. bovis* than other group D streptococci in view of its hydrolysis of starch and esculin, intolerance to heat and the capsular location of the type-specific antigen. Although the species has not been formally proposed, it appears to be identical to the streptococci previously described as groups R, S and RS (Perch et al., 1983). Strains of "*S. suis*" did not group with the enterococci in numerical taxonomic study of the genus, and were also distinct from *S. bovis* and *S. equinus* (Bridge and Sneath, 1983).

Recently Published Streptococcal Species

Since writing the species descriptions included in this Chapter, some new species names have recently been proposed and validly published.

1. **Streptococcus alactolyticus** Farrow, Kruze, Phillips, Bramley and Collins 1985b, 224.[VP] (Effective publication: Farrow et al. 1985a.)
Type strain: NCDO 1091.

2. **Streptococcus cecorum** Devriese, Dutta, Farrow, van de Kerckhove and Phillips 1983, 774.[VP]
Carboxyphilic strains, with some similarity to *S. bovis* but lacking Lancefield Group D reaction, isolated from the ceca of chickens.
Type strain: NCDO 2674 (A60).

3. **Streptococcus equi** subsp. **zooepidemicus** Farrow and Collins 1985b, 224.[VP] (Effective publication: Farrow and Collins 1985a.)
Type strain: NCDO 1358.

4. **Streptococcus garvieae** Collins, Farrow, Phillips and Kandler 1983h, 270.[VP] (Effective publication: Collins et al., 1983g.)

Strains isolated from bovine mastitis. Capable of growing at 10°C and 40°C.
Type strain: NCDO 2155.

5. **Streptococcus macacae** Beighton, Hayday, Russell and Whiley 1984, 333.[VP]
Oral streptococci isolated from dental plaque of monkeys. Differs from other oral strains in sugar reactions.
Type strain: 25-1 (NCTC 11558).

6. **Streptococcus plantarum** Collins, Farrow, Phillips and Kandler 1983h.[VP] (Effective publication: Collins et al., 1983g.)
Strains isolated from frozen peas. Grow at 10°C but not at 40°C. Contain Lancefield Group N antigen.
Type strain: NCDO 1869.

7. **Streptococcus porcinus** Collins, Farrow, Katic and Kandler 1985, 224.[VP] (Effective publication: Collins, Farrow, Katic and Kandler 1984.)
Type strain: NCTC 10999.

8. **Streptococcus saccharolyticus** Farrow, Kruze, Phillips, Bramley and Collins 1985b, 224.[VP] (Effective publication: Farrow et al. 1985a.)
Type strain: NCDO 2594.

Genus Incertae Sedis

Melissococcus Bailey and Collins 1983,672.[VP] (Effective publication: Bailey and Collins 1982b, 216.)

me.lis'so.coc'cus. Gr. n. *melissa* bee; Gr. n. *coccus* berry; M.L. masc. n. *Melissococcus* the coccus of the (honey) bee.

The Gram-positive coccus which causes European foulbrood of honey bees (*Apis* spp.) has been variously named "*Bacillus pluton*" and "*Streptococcus pluton*" (White, 1912, 1920; Bailey, 1957; Bailey and Gibbs, 1962; Glinski, 1972, Bailey and Collins, 1982a). The species was not described in the eighth edition of the *Manual* and was not included in the Approved Lists of Bacterial Names (Skerman et al., 1980).

Recent taxonomic studies (Bailey and Collins, 1982a) have further defined the characteristics of this organism and led to the proposal that it be allocated to a new genus, *Melissococcus* (Bailey and Collins, 1982b, 1983).

Description of the genus Melissococcus. Chains of lanceolate cocci, sometimes pleomorphic and rod shaped. Gram-positive (but easily destained), nonsporeforming, nonacid fast and nonmotile. Anaerobic to microaerophilic; no growth in air or anaerobically without CO_2. Usually ferments glucose and fructose only. No growth with citrate. Slight acid production. Requires a Na:K ratio of 1. Requires cysteine or cystine for growth in addition to complex sources of organic nitrogen and other nutrients.

The cell wall peptidoglycan is based on Lys. Respiratory quinones are absent. The long chain fatty composition is primarily straight chain, monounsaturated (cis-, vaccenic acid series) and cyclopropane ring acids.

The mol% G + C of the DNA is 29–30% (T_m).

Type species: Melissococcus pluton Bailey and Collins 1983,672.

List of the species of the genus Melissococcus.

1. *Melissococcus pluton* Bailey and Collins 1983,672.[VP] (Effective publication: Bailey and Collins 1982b,216.)
plu'ton. L. n. God of the lower world.

Based partly on previous descriptions of "*Streptococcus pluton*", White 1912, 1920; Bailey 1962a,b, 1974; Bailey and Gibbs 1962, Glinski 1972a,b; Bailey and Collins 1982a).

Cellular morphology as given in genus description. Colonies up to 1 mm in diameter, ranging from dense white or granular, entire, smooth and dome-shaped to relatively transparent umbonate or annular forms with clear centers and granular peripheries. All strains require some CO_2, but some are inhibited by concentrations over 5% (v/v). Glucose and fructose are usually the only sugars to support growth; some strains dissimilate sucrose, melezitose and salicin. Major amounts of lactic acid are produced; small amounts of acetic, isobutyric and succinic acids also produced. Final pH, 5.3. Optimum growth at 35°C; some growth between 20° and 45°C. Optimum pH, 6.5–6.6. Requires a Na:K ratio of 1. Requires free cysteine or cystine in addition to peptones or most yeast extracts; little or no growth on most ordinary media. Reacts with Lancefield Group D antiserum.

The cell wall peptidoglycan type is Lys-Ala. The major long chain fatty acids are hexadecanoic and lactobacillic acids.

Causative agent of European foulbrood of the honeybee. Isolated from larvae of *A. millifera* and *A. cerana* with European foulbrood.

The mol% G + C of the DNA is 29–30% (T_m).

Type strain: NCDO 2443

Further taxonomic comments. The descriptions given above are based on those published by Bailey and Collins (1982a, b). Comparative studies of *M. pluton* with other Gram-positive cocci, using modern numerical, chemical or genetic taxonomic techniques are not available.

Genus **Leuconostoc** *van Tieghem 1878, 198^AL emend mut. char. Hucker and Pederson 1930, 66^AL*

ELLEN I. GARVIE

Leu.co.nos′toc. Gr. Adj. *leucus* clear, light; M.L. neut. n. *Nostoc* algal generic name; M.L. neut. n. *Leuconostoc* colorless nostoc.

Cells may be **spherical but often lenticular** particularly when growing on agar, cells usually occur in **pairs and chains**. Gram-positive, nonmotile, spores not formed. Facultative anaerobes.

Colonies are small usually less than 1.00 mm in diameter, smooth, round, grayish white. In stab cultures growth occurs along the stab with little surface growth. Broth cultures often have uniform turbidity but strains forming long chains tend to sediment. **Optimum temperature 20–30°C** and growth occurs between 5°C and 30°C. Chemoorganotrophs, requiring a rich medium **often having complex growth factors and amino acid requirements.** All species require nicotinic acid + thiamine + biotin and either pantothenic acid or a pantothenic acid derivative. No strains require cobalamin, or *p*-aminobenzoic acid.

Growth is dependent on the presence of a fermentable carbohydrate and glucose is fermented by a combination of the hexose-monophosphate and phosphoketolase pathways. However, the pathway of glucose fermentation in *Leuconostoc oenos* has not been fully confirmed. **Fructose 1,6-diphosphate aldolase is absent,** and an active **glucose-6-phosphate dehydrogenase is present. CO_2** and D-ribulose-5-P are formed from glucose. Xylulose 5-P phosphoketolase is present and the resulting end products are **ethanol and D-(−)-lactic acid.** Some strains have an oxidative mechanism and **acetic acid** is formed in place of ethanol. Polysaccharides and alcohols (except mannitol) are usually not fermented. **Malate can be utilized and converted to L-(+)-lactate.**

Catalase-negative. Cytochromes are absent. Arginine is not hydrolyzed and **milk is usually not acidified and curdled.** Nonproteolytic. Indole is not formed. Nitrates not reduced. Nonhemolytic.

Nonpathogenic to plants and animals (including humans). Properties separating the species are given in Table 12.26 and further information is given in Table 12.27.

The amino acid composition of the cross-linking peptide of the cell wall peptidoglycan is of the alanine, serine, lysine type (Table 12.28). The mol% G + C in the DNA is 38–44 (T_m and Bd) (Table 12.29). The type species is *Leuconostoc mesenteroides* (Tsenkovskii) van Tieghem 1878, 191.

Further Descriptive Information

Growth conditions may affect cell morphology, and not all strains will be influenced in the same way. Cultured in milk (or supplemented milk), most strains form coccoid cells in chains. Chain length varies with the strain. Cultured in broth, cells are elongated and can be mistaken for rods, appearing morphologically closer to the lactobacilli than to the streptococci. Cultured on agar, spherical cells are seldom formed.

The cell wall of dextran-forming strains contains dextran-sucrase and the cell wall structure is affected by growth in sucrose broth, to which strains differ in their response (Brooker, 1976, 1977). Although capsular material is apparent in some strains, a true bacterial capsule is not formed.

The composition of the cross-linked peptide in the cell wall peptidoglycan is given in Table 12.28.

Growth is never rapid, the active strains of *L. mesenteroides* subsp. *mesenteroides* have the shortest generation time and good growth can be obtained in 24 h incubation at 30°C. On the other hand, *L. mesen-*

Table 12.26.
Diagnostic characteristics of the species of the genus **Leuconostoc**[a]

| Characteristics | 1. *L. mesenteroides*, subsp.[b] | | | 2. *L. paramesenteroides* | 3. *L. lactis* | 4. *L. oenos* |
	1a. *mesenteroides*	1b. *dextranicum*	1c. *cremoris*			
Acid from						
Arabinose	+	−	−	d	−	d
Cellulose	d	d	−	(d)	−	d
Fructose	+	+	−	+	+	+
Sucrose	+	+	−	+	+	−
Trehalose	+	+	−	+	−	+
Hydrolysis of esculin	d	d	−	d	−	+
Dextran formation	+	+	−	−	−	−
Growth at pH 4.8	−	−	−	d	−	+
Requirement for TJF	−	−	−	−	−	d
Growth in 10% ethanol	−	−	−	−	−	+
NAD-dependent G-6-PDH present	+	+	+	+	+	−

[a] Symbols: see Table 12.2; also (d) delayed reaction; and TJF, glucopantothenate (tomato juice factor).

Table 12.27.

Differential characteristics of the species of the genus **Leuconostoc**[a, b]

Characteristics	1. *L. mesenteroides*, subsp.			2. *L. paramesenteroides*	3. *L. lactis*	4. *L. oenos*
	1a. *mesenteroides*	1b. *dextranicum*	1c. *cremoris*			
Acid from						
Amygdalin	d	d	−	(d)	−	ND
Arabinose	+	−	−	d	−	d
Arbutin	d	−	−	−	−	ND
Cellobiose	d	d	−	(d)	−	d
Fructose	+	+	−	+	+	+
Galactose	+	d	d	+	+	d
Glucose	+	+	+	+	+	+
Lactose	(d)	+	+	(d)	+	−
Maltose	+	+	d	+	+	−
Mannitol	d	d	−	(d)	−	−
Mannose	+	d	−	+	d	d
Melibiose	d	d	−	+	d	d
Raffinose	d	d	−	d	d	−
Ribose	+	ND	ND	ND	ND	ND
Salicin	d	d	−	−	d	d
Sucrose	+	+	−	+	+	−
Trehalose	+	+	−	+	−	+
Xylose	d	d	−	d	−	d
Hydrolysis of esculin	d	d	−	d	−	+
Required for growth						
Uracil	−	−	+	−	−	−
Guanine + adenine + xanthine + uracil	−	d	+	d	−	+
Riboflavin	d	d	+	+	+	+
Pyridoxal	d	d	+	+	−	+
Folic acid	d	d	+	+	−	+
Tomato juice factor	−	−	−	−	−	d
Destruction of tomato juice factor	−	−	−	−	−	d
Dextran formation	+	+	−	−	−	−
Dissimilate citrate (carbohydrate present)	d	d	+	d	d	d
Dissimilate malate						
No carbohydrate present	d	−	−	d	−	ND
Carbohydrate	d	−	−	d	−	+
Yeast glucose litmus milk	+	+	+	+	+	d
Acid clot	d	d	d	d	d	−
Reduction	d	d	−	d	(d)	d
Gas	d	d	−	−	−	−
Growth in						
3.0% NaCl	+	d	−	d	d	ND
6.5% NaCl	d	−	−	d	−	ND
Growth at pH						
4.8 (initial)	−	−	−	d	−	+
6.5 (initial)	+	+	+	+	+	d
Growth at 37°C	d	+	−	d	+	d
Final pH in glucose broth	4.5	4.5	5.0	4.4	4.7	ND

[a] Symbols: see Table 12.2; (d), delayed reaction; TJF, glucopantothenate (tomato juice factor); ND, not determined.

[b] DNA/DNA hybridization shows that strains previously classified as *L. dextranicum* belong to the same genotype as *L. mesenteroides* NCDO 523. This may explain the failure to find satisfactory properties for separating these groups.

teroides subsp. *cremoris* may require 48 h incubation and prefers 22°C to 30°C. The slower growing strains prefer reducing conditions and 0.05% cysteine HCl added to broth media encourages growth. *L. oenos* has many differences from other species and grows best in acid media (initial pH 4.2–4.8) containing tomato juice. Growth is slow and 5–7 days incubation at 22°C may be needed. Other species of leuconostoc will not grow in the acid media preferred by *L. oenos*.

Milk is a poor medium for leuconostocs, although most strains will grow in milk supplemented with yeast extract and glucose. *L. mesenteroides* subsp. *mesenteroides* usually acidifies and clots milk media with gas formation, while other species are less active and *L. paramesenteroides* and other species with a high requirement for amino acids fail to clot milk. Nutritional requirements vary (Garvie, 1967b) Table 12.30. *L. oenos* will grow with a very high level of pantothenic acid but prefers a gluco-derivative of the vitamin, probably 4′o-(β-glucopyranosyl)D-pantothenic acid (Amachi et al. 1970). The degree of depend-

Table 12.28.

Amino acid sequence of the interpeptide bridge of cell wall peptidoglycan of the species of the genus **Leuconostoc**[a]

Species	Peptidoglycan
1a. *L. mesenteroides* subsp. *mesenteroides*	L-Lys-L-Ser-L-Ala$_2$; L-Lys-L-Ala$_2$
1b. *L. mesenteroides* subsp. *dextranicum*	L-Lys-L-Ser-L-Ala$_2$
1c. *L. mesenteroides* subsp. *cremoris*	L-Lys-L-Ser-L-Ala$_2$
2. *L. paramesenteroides*	L-Lys-L-Ser-L-Ala$_2$; L-Lys-L-Ala$_2$
3. *L. lactis*	L-Lys-L-Ser-L-Ala$_2$; L-Lys-L-Ala$_2$
4. *L. oenos*	L-Lys-L-Ala-L-Ser; L-Lys-L-Ser-L-Ser

[a] Adapted from W. H. Holzapfel, Inaugural dissertation der Technischen Hochschule, München, 1969.

Table 12.29.

Mol% G + C of the DNA of the **Leuconostoc** *species*[a]

Species	T_m and buoyant density
1. *L. mesenteroides*	
a. Subsp. *mesenteroides*	37–39, with some strains 40–41
b. Subsp. *dextranicum*	37–40
c. Subsp. *cremoris*	38–40
2. *L. lactis*	43–45
3. *L. paramesenteroides*	37–38
4. *L. oenos*	37–39

[a] Compiled from Garvie et al. (1974) and Hontebeyrie and Gasser (1977).

ence on this growth factor varies with different strains, but most strains will destroy it when present in media. Other species of *Leuconostoc* do not attack glucopantothenic acid (Garvie and Mabbitt, 1967).

L. mesenteroides subsp. *mesenteroides* requires only glutamic acid and valine while other subspecies and species require a variety of amino acids, the requirements varying with the strain (Garvie, 1967b).

Leuconostocs are dependent on the presence of a fermentable carbohydrate, and fermentation ability varies in different species (Table 12.27). Glucose is used by all species but fructose is prefered by all except *L. mesenteroides* subsp. *cremoris*. Glucose is phosphorylated and all species have an active glucose-6-phosphate dehydrogenase (G-6-PDH). In species 1–3, NAD or NADP will serve as coenzyme with a preference for the former but in *L. oenos* the G-6-PDH is active only with NADP (Garvie, 1975). Gluconate is decarboxylated and pentose converted to D-(−)-lactate and ethanol by the phosphoketolase pathway. Acetate as well as ethanol may be formed by some strains. The D-(−)-lactate dehydrogenase (LDH) of *L. oenos* migrates slowly on electrophoresis and there is evidence of differences between strains. The LDH of other species has a fast migration which is the same for them all (Garvie, 1969). Immunological studies have separated the LDHs and G-6-PDHs of different species. More than one type of these enzymes has been found in different strains of *L. mesenteroides* subsp. *mesenteroides*, indicating that it is a heterologous species (Hontebeyrie and Gasser, 1975).

Early work indicated that most leuconostocs dissimilate citrate (Hucker and Pederson, 1930). This property appears to be lost in strains kept in the laboratory but is important in strains which are components of cheese and butter starters (species 1c and 3). *L. oenos* can also dissimilate citrate. Malate is attacked by *L. oenos* and also by *L. mesenteroides* subsp. *mesenteroides*. Information on other species is lacking. Malate is converted to L-(+)-lactate, and an LDH is not involved (Alizade and Simon, 1973; Radler, 1975). It is important to exclude malate from media used for cultures when the type of lactic acid produced from glucose is to be determined. Acetate and tartrate are not utilized.

The mol% G + C in the DNA of *L. lactis* is 43–45 while that of other species is 37–40. There are indications from the values obtained that *L. mesenteroides* subsp. *mesenteroides* is a heterologous subspecies (Garvie et al. 1974; Hontebeyrie and Gasser, 1977). This situation is confirmed by DNA/DNA hybridization studies (Garvie, 1976; Hontebeyrie and Gasser, 1977). In addition high hybridization occurs between the DNA of the three species, *L. mesenteroides*, *L. dextranicum* and *L. cremoris*, showing that they belong to a single genospecies, as put forward by Garvie (1983). The other species are clearly identified (Table 12.31). RNA/DNA hybridization separates *L. oenos* from the other leuconostocs which all belong to a single RNA group (Garvie, 1981).

Phage attack on *L. oenos* may occur in wine making, and bacteriophages have been described (Sozzi et al. 1982). Bacteriophage for *L. mesenteroides* is reported (Sozzi et al. 1978), but very little is known about bacteriophages attacking leuconostocs.

Sensitivity to antibiotics and drugs is unknown as no species are pathogenic.

Leuconostocs are found on plants and to a lesser extent in milk and milk products. *L. mesenteroides* subsp. *cremoris* and *L. lactis* may be components of cheese and butter starters. Dextran-forming species occur on sugar cane and sugar beet where they may cause widespread spoilage. *L. oenos* is known only in wine and related habitats; no other leuconostoc has been isolated from these sources.

Enrichment and Isolation Procedures

Leuconostocs on plants can be isolated on media containing thallous acetate and crystal violet (Cavett et al. 1965). Enrichment in broth may be necessary before plating on agar. Citrate-utilizing strains in dairy starters can be isolated on whey agar (Galesloot et al. 1961) and *L. oenos* can be isolated on tomato juice agar with initial pH below 4.5 with cyclohexamide to inhibit yeasts (Kunkee, 1967). Growth of *L. oenos* on agar may be poor and growth in broth often slow.

Maintenance Procedures

All species can be preserved by lyophilization in horse serum + 7.5% glucose, but cells of an actively growing culture in late logarithmic or early stationary phase should be used. Care is needed with *L. mesenteroides* subsp. *cremoris* and *L. oenos*. It is important to wait for high turbidity in a culture before lyophilization. Once dried, cultures can be kept under vacuum at 10°C. Cultures should be revived in media giving optimum growth conditions.

Nonacidophilic species can be kept for 3–4 months in litmus milk + 0.3% yeast + 1% glucose and 1% calcium carbonate. Preliminary incubation for 18–24 or 48 h (depending on the strain) at 30°C is necessary before storage. *L. oenos* can be kept in tomato juice agar stabs.

Taxonomic Comments

Leuconostocs occur in the same habitats as lactobacilli and lactic streptococci. Gas production from glucose will separate the leuconos-

Table 12.30.
Growth factor requirements of the **Leuconostoc** *species[a]*

Characteristics	1. *L. mesenteroides*, subsp.			2. *L. paramesenteroides*	3. *L. lactis*	4. *L. oenos*
	1a. *mesenteroides*	1b. *dextranicum*	1c. *cremoris*			
Uracil	−	−	+	−	−	−
Guanine + adenine + xanthine + uracil	−	−	+	d	−	+
Riboflavin	d	d	+	+	+	+
Pyridoxal	d	d	+	d	−	−
Folic acid	d	d	+	+	−	+
Tween 80	−	d	−	d	−	d

[a] Symbols: see Table 12.2.

Table 12.31.
DNA/DNA homology between the species of **Leuconostoc**[a, b]

Species		Labeled DNA from			
		L. mesenteroides		*L. lactis*	*L. paramesenteroides*
L. mesenteroides	A	85–100	35–50	20–50	8–22
	B	46–78	90–100	20–50	8–22
L. dextranicum		85–100	38–50	19–30	7–19
L. cremoris		78–100	46	24–35	5
L. lactis		32–47	40–60	70–100	9–46
L. paramesenteroides		11–16	17–21	6–14	60–100
L. oenos		11–15	10	2–11	5

[a] There are indications that the *L. mesenteroides/L. dextranicum* group contains more than two genotypes.
[b] Compiled from Garvie (1976) and Hontebeyrie and Gasser (1977).

tocs from streptococci but this property should be tested only with actively growing strains, otherwise gas production in the former may not be evident. Normal streptococcal media are unsuitable for leuconostocs and if used can result in misidentification owing to poor growth. Type of lactic acid produced also separates the D-(−)-forming leuconostocs from L-(+)-forming streptococci.

Separation of leuconostocs from gas-forming lactobacilli is not easy (Sharpe et al. 1972). Morphology can overlap. *Lactobacillus viridescens* does not hydrolyze arginine, forms predominantly D-(−)-lactate but some L-(+)-is usually found, and has a mol% G + C in its DNA between that of *L. mesenteroides* and *L. lactis*. Generally cells of *Lactobacillus viridescens* are more elongated than those of any leuconostoc. *Lactobacillus confusus* usually hydrolyzes arginine and has a mol% G + C in its DNA similar to that of *L. lactis*, and forms DL-lactate.

Early classification relied heavily on morphology and the leuconostocs, being more coccoid than rod-like were placed with the streptococci, while the heterofermentative species with cells more rod-like than coccoid (heterofermentative lactobacilli) were placed with the homofermentative lactobacilli. The significance of the physiological similar-

ities between the leuconostocs and heterofermentative lactobacilli, in particular *Lactobacillus confusus* and *Lactobacillus viridescens*, require reassessing. These latter organisms have similar LDHs and cell wall peptides to leuconostocs but belong to different DNA and RNA homology groups. However, the nonacidophilic leuconostocs *mesenteroides*, *paramesenteroides* and *lactis* appear to have more in common with *Lactobacillus confusus* and *Lactobacillus viridescens* than with *L. oenos*.

The nomenclature of the genus, discussed in detail in the last edition has been modified because of the results of enzyme studies and DNA/DNA homology. These studies have shown that *L. mesenteroides* contains three subspecies viz *mesenteroides*, *dextranicum* and *cremoris*. The change in status of *L. dextranicum* and *L. cremoris* is discussed elsewhere (Garvie, 1984).

Further Reading

Garvie, E.I. 1984. Separation of species of the genus *Leuconostoc* and the differentiation of the leuconostocs from other lactic acid bacteria. Methods Microbiol *16:* 147–178.

List of species of the genus **Leuconostoc**

1. **Leuconostoc mesenteroides** (Tsenkovskii) van Tieghem 1879, 198.[AL] (*Ascococcus mesenteroides* Tsenkovskii 1878, 159.)

me.sen.ter.oi'des. Gr. n. *mesenterium* the mesentery; Gr. n. *oides* form, shape; M.L. adj. *mesenteroides* mesentery-like.

Morphology as in general description.

1a. *Leuconostoc mesenteroides* subsp. *mesenteroides* (Tsenkovskii) van Tieghem 1879 198.[AL] A characteristic slime of dextran is formed from sucrose, the production being favored by growing at 20–25°C. Different colonial types are formed on sucrose agar depending on the characteristic chemical structure of the dominant type of dextran formed. These differences have not proved to be of taxonomic value.

Some strains produce a heme-requiring catalase (Whittenbury, 1964). In glucose broth cells do not survive heating to 55°C for 30 min but in slimy sugar solutions they may withstand heating to 80–85°C.

Temperature range 10–37°C, optimum 20–30°C.

The mol% G + C of the DNA: see Table 12.29.

Type strain: ATCC 8293 (NCDO 523).

1b. *Leuconostoc mesenteroides* subsp. *dextranicum* (Beijerinck) Garvie 1983, 118.[VP] (*Leuconostoc dextranicum* (Beijerinck) Hucker and Pederson 1930, 67; *Lactococcus dextranicus* Beijerinck 1912, 27.)

dex.tra'ni.cum. M.L. n. *dextranum* dextran; M.L. neut. adj. *dextranicum* relating to dextran.

Morphology as in general description.

Dextran is formed but less actively than with *L. mesenteroides* subsp. *mesenteroides*.

This subspecies ferments fewer substrates than subspecies *mesenteroides* and requires a few more amino acids and vitamins for growth.

Optimum growth temperatures and range are the same as for 1a.

Differentiation between *L. mesenteroides* subsp. *mesenteroides* and subsp. *dextranicum* has always been blurred and unsatisfactory. This is probably because they are a single genospecies.

Type strain: NCDO 529 (ATCC 19255).

1c. **Leuconostoc mesenteroides** subsp. *cremoris* (Knudsen and Sorensen) Garvie 1983, 118.[VP] (*Leuconostoc cremoris* (Knudsen and Sorensen) Garvie 1960, 288; *Betacoccus cremoris* Knudsen and Sorensen 1929, 81.)

cre.mor'is. L. n. *cremor* cream; L. gen. no. *cremoris* of cream.

Morphology as general description, but cultures often form long chains with resultant flocculent growth in broth.

Citrate is normally dissimilated and under certain conditions acetoin and diacetyl are formed (Speckman and Collins, 1968). These end products are not always detected because the pyruvate formed from citrate is probably used for the regeneration of NAD, and D-(−)-lactate results.

Most strains do not attack sucrose but mutant colonies in soft agar cultures have been reported (Whittenbury, 1966).

This subspecies is the least active and requires a large number of vitamins and amino acids. It prefers reducing conditions and a temperature of 18–25°C for growth.

It appears to be an adaption of *L. mesenteroides* subsp. *mesenteroides* to the dairy environment. All known strains have come from milk, dairy starter or related habitats. Truly wild sources are unknown.

Difficulties in separating some strains of *L. mesenteroides* subsp. *cremoris* from subsp. *dextranicum* are probably due to the fact that they belong to the same genospecies.

Type strain: NCDO 543 (ATCC 19254).

2. **Leuconostoc paramesenteroides** Garvie 1967b, 446.[AL]

pa.ra.me.sen.ter.oi'des. Gr. prep. *para* resembling; M.L. *mesenteroides* a specific epithet; M.L. adj. *paramesenteroides* resembling *L. mesenteroides*.

Morphology as general description.

Dextran is not formed from sucrose and amino acid requirements are complex and variable.

Many strains grow well at 30°C but some prefer reducing conditions and a temperature of 18–24°C (Garvie, 1967). **Pseudocatalase** may be present if organisms are grown in a medium with a low glucose content (Whittenbury, 1964).

Tolerance of NaCl is higher than for other species, particularly those strains isolated from foods containing high levels of salt.

More tolerant of acid pH than species 1 or 3 and may grow in media with an initial pH below 5.0.

At one time strains were considered to be nondextran-forming variants of *L. mesenteroides* but genetic studies have shown this to be incorrect. However, it would be difficult to distinguish *L. paramesenteroides* from nondextran-forming strains of *L. mesenteroides* by phenotypic tests.

Type strain: NCDO 803.

3. **Leuconostoc lactis** Garvie 1960, 290.[AL]

lac'tis. L. n. *lac* milk; L. gen. n. *lactis* of milk.

Morphology as general description.

The amino acid requirements are complex. However, lactose is fermented more readily than by other species and strains may acidify and even clot unsupplemented milk.

Citrate may be dissimilated and acetoin and diacetyl formed (Cogan et al., 1981).

Heat resistance is higher than in other species and cells may survive 60°C for 30 min.

The species may not be widely distributed as recorded isolations are few and are mostly from dairy sources.

Type strain: NCDO 533 (ATCC 19256).

4. **Leuconostoc oenos** Garvie 1967a. 431.[AL]

oe.nos.' Gr. n. *oinos* wine; Gr. gen. n. *oenos* of wine.

Morphologically resembles the other species but is different in many other respects.

Growth is slow and variations in properties between strains may be due in part to unsuitable growth conditions. Division of wine leuconostocs into separate species has been proposed but a variety of strains from these proposed species have been found to belong to a single genospecies (Garvie and Farrow, 1981). Differences in LDHs in different strains have been found and the species may not be homologous.

Isolated only from wine and related habitats.

Type strain: NCDO 1674 (ATCC 23279).

Genus *Pediococcus* Claussen 1903, 68[AL*]

Ellen I. Garvie

Pe.di.o.coc'cus. Gr. n. *pedium* a plane surface; Gr. n, *coccus* a berry; M.L. masc. n, *Pediococcus*, coccus growing in one plane.

Cells **spherical, never elongated, division occurs alternately in two planes at right angles** (Gunther, 1959) to form tetrads, however, these may not always be present and only pairs of cells occur. **Single cells are rare and chains of cells are not formed.** Grampositive, nonmotile. Spores not formed. Facultative anaerobes, but tolerance to oxygen varies in different species.

Colonies vary in size from 1.0–2.5 mm in diameter, smooth, round, grayish white. In stab culture, growth is along the stab with little surface growth. Broth cultures usually have uniform turbidity. All species will grow at 30°C but optimum temperatures range from 25–40°C.

Chemoorganotrophs **requiring a rich medium; having complex growth factors and amino acid requirements.** All species require nicotinic acid, pantothenic acid and biotin, while none requires thiamine, *p*-aminobenzoic acid or cobalamin.

Growth is dependent on the presence of a fermentable carbohydrate and glucose is fermented, probably by the Embden-Meyerhof pathway, to DL or L-(+) lactate. Gas is not formed. Under certain growth conditions end products in addition to lactate can be formed.

Catalase-negative. Cytochromes are absent. Milk usually not acidified or curdled. Nonproteolytic. Indole not formed. Nitrates not reduced. Sodium hippurate not hydrolyzed. Nonpathogenic to plants and animals.

The mol% G + C in the DNA is in the range 34–42. (T_m).

The type species is *Pediococcus damnosus* Claussen, 1903, 68.

Further Descriptive Information

Pediococci are the only lactic acid bacteria dividing in two planes. Short chains may be seen but these are formed by pairs of cells and

---

* **Editorial note:** The citation of this name in the Approved Lists to Balcke is incorrect (See Judicial Commission, 1976, Opinion 52).

not by division in a single plane. Some strains form small cells, 0.6 μm while others are larger, 1.0 μm. Within one culture, cells are of uniform size.

Cells are noncapsulated. The cross-linked peptide of the cell wall peptidoglycan is L-Lys-L-Ala-D-Asp except in *P. urinae-equi* where D-Asp is missing.

Species can be separated by tolerance to temperature, pH and NaCl (Table 12.32). The acid tolerant, low temperature species, *P. damnosus* and *P. parvulus*, require the most anaerobic conditions. Cysteine hydrochloride (0.05–0.1%) added to broth improves growth while poor growth on an agar surface is improved by incubation in an atmosphere of H_2 + 10% CO_2. These species are slow to grow and 2–5 days incubation may be required. *P. pentosaceus* and *P. acidilactici* grow rapidly in suitable broth media. They grow equally well on an agar surface aerobically and in an atmosphere of H_2 + 10% CO_2. *P. halophilus* and *P. urinaeequi* are more aerobic than other species and no special incubation conditions are required. *P. halophilus* fails to grow in the absence of 5% NaCl (approx) but other species have no requirement for salt. MRS broth and agar can be used for growing all pediococci. Modifications are needed for species associated with beer and for salt tolerant species. For the former, MRS base is mixed 1:1 with beer, and for the latter, the pH is raised to 7.0 and 4% NaCl added (Back 1978a). *P. urinaeequi* requires an initial pH of 7.0–8.0 and no added NaCl.

Milk is a poor medium for pediococci, as lactose is not a readily available carbohydrate and other required growth factors are unavail-

able. Amino acid requirements are not known for all species. *P. pentosaceus* and *P. acidilactici* require most amino acids for growth (Jensen and Seeley, 1954). Vitamin requirements are similar for most species (Table 12.33). Some strains of *P. pentosaceus* have a requirement for folinic acid which can be partially replaced by thymidine (Nakagawa and Kitahara, 1959; Jensen and Seeley, 1954). This requirement for folinic acid is not shared by other species (Deibel and Niven, 1960; Gunther and White, 1961). However, other workers have shown folinic acid dependence in *P. urinaeequi* and *P. halophilus* (Sakaguchi and Mori, 1969) and some strains of *P. dextrinicus* (Garvie et al. 1961). Some strains of *P. damnosus* require mevalonic acid.

Growth of strains is dependent on the presence of a fermentable carbohydrate; glucose and most other monosaccharides are fermented, while the ability to use pentoses is limited to *P. pentosaceus*, *P. acidilactici*, *P. urinae-equi* and some strains of *P. halophilus* (Table 12.34). Polysaccharides are generally not fermented except dextrin by *P. dextrinicus*. Little information is available about the pathway of glucose fermentation but some species are known to contain aldolase (London et al. 1976) and all contain lactate dehydrogenase (LDH). Back (1978a) found that electrophoresis of the LDHs could be used to separate species. *P. halophilus* forms L-(+)-lactate with only a trace of D-(−), but so far, only an L-(+)-LDH has been found. *P. dextrinicus* also forms L-(+)-lactate with an LDH activated by fructose-1,6-diphosphate and does not have a D-(−)-LDH (Back 1978a). Other species form DL-lactate and posses both an L-(+)- and a D-(−)-LDH. The ratio of D/L lactate formed may vary with growth conditions. It is probable

Table 12.32.

*Differential conditions of growth of species of the genus **Pediococcus**[a]*

Characteristics	1. *P. damnosus*	2. *P. parvulus*	3. *P. inopinatus*	4. *P. dextrinicus*	5. *P. pentosaceus*	6. *P. acidilactici*	7. *P. halophilus*	8. *P. urinaeequi*
Growth at								
35°C	−	+	+	+	+	+	+	+
40°C	−	−	−	+	+	+	−	+
50°C	−	−	−	−	−	+	−	−
Growth at								
pH 4.2	+	+	−	−	+	+	−	−
pH 7.5	−	+	d	+	+	+	+	+
pH 8.5	−	−	−	−	d	d	+	+
Growth in								
4% NaCl	−	+	+	+	+	+	d	+
6.5% NaCl	−	+	d	−	+	+	+	+
18% NaCl	−	−	−	−	−	−	+	−

[a] Symbols: see Table 12.2.

Table 12.33.

*Vitamin requirement of species of the genus **Pediococcus**[a]*

Characteristics	1. *P. damnosus*	2. *P. parvulus*	3. *P. inopinatus*	4. *P. dextrinicus*	5. *P. pentosaceus*	6. *P. acidilactici*	7. *P. halophilus*	8. *P. urinaeequi*
Riboflavin	s	ND	ND	ND	−	s	−	−
Pyridoxal	s	ND	ND	ND	−	−	±	±
Tween 80	−	+	ND	+	ND	−	−	−
Folinic acid	−	±	ND	±	±	−	−	−

[a] Symbols: s, stimulatory; ND, not determined; −, not required; +, required; ±, variable.

Table 12.34.
*Properties of the species of the genus **Pediococcus**[a,b]*

Characteristics	1. P. damnosus	2. P. parvulus	3. P. inopinatus	4. P. dextrinicus	5. P. pentosaceus	6. P. acidilactici	7. P. halophilus	8. P. urinaeequi
Growth at 45°C	−	−	−	−	d	+	−	−
Growth at pH 7.0	−	+	+	+	+	+	+	+
Growth at pH 4.5	+	+	ND	−	+	ND	−	−
Growth in 10% NaCl	−	−	−	−	d	−	+	−
Production of "catalase"	−	−	ND	−	+	+	−	−
Hydrolysis of arginine	−	−	−	−	+	+	−	−
Acid produced from								
Arabinose	−	−	−	−	+	d	+	d
Ribose	−	−	−	−	+	+	+	ND
Xylose	−	−	−	−	d	+	−	d
Rhamnose	−	−	−	−	d	d	−	ND
Lactose	−	−	+	d	d	d	−	d
Maltose	d	+	+	+	+	−	+	+
Melezitose	d	−	−	−	−	−	+	−
Sucrose	d	−	d	d	−	−	+	+
Trehalose	+	d	+	−	+	d	+	+
Maltotriose	d	d	d	+	−	−	+	ND
Dextrin	−	−	d	+	−	−	−	+
Starch	−	−	−	+	−	−	−	−
Glycerol	−	−	−	−	−	−	+	−
Mannitol	−	−	−	−	−	−	−	d
Sorbitol	−	−	−	−	−	−	−	−
Arbutin	d	ND	ND	ND	+	ND	ND	ND
α-Methylglucoside	d	d	d	+	−	−	+	ND
Lactate formed	DL	DL	DL	L-(+)	DL	DL	L-(+) D-(−)	L-(+)

[a] Most strains ferment fructose, galactose, glucose, mannose, cellobiose, amygdalin, esculin and salicin. Most strains do not ferment sorbose, melibiose, inulin, dulcitol, and inositol. Raffinose may be fermented by strains of *P. pentosaceus*, *P. acidilactici* and *P. urinaeequi*.
[b] Symbols: see Table 12.2.

Table 12.35.
*DNA/DNA homology among species of the genus **Pediococcus**[a]*

Characteristics	P. pentosaceus	P. acidilactici	P. parvulus	P. inopinatus	P. damnosus	P. dextrinicus	P. halophilus	P. urinaeequi
P. pentosaceus	100							
P. acidilactici	20–30	100						
P. parvulus	0–10	0–10	100					
P. inopinatus	<10	<10	30–40	100				
P. damnosus	8–18	0–10	0–40	40–50	100			
P. dextrinicus	<10	<10	<10	<10	<10	100		
P. halophilus	<10	<10	<10	<10	<10	<10	100	
P. urinaeequi	<10	<10	<10		<10		<10	100

[a] Compiled from Dellaglio et al. (1981); Back (1978); and Back and Stackebrandt (1978).

that fermentation of glucose follows the Embden-Meyerhof pathway, and under near optimal conditions lactate is the major end product. The pyruvate formed, however, can be diverted to other end products and diacetyl/acetoin production often occurs with *P. damnosus* although seldom with other species. In many lactic acid bacteria, reduction of pyruvate to lactate is controlled by the requirement of the strain for NAD. Growth with adequate glucose present but restricted by other nutrients can cause pyruvate conversion to acetoin/diacetyl. Production of diacetyl has been used as a diagnostic character for *P. damnosus* but this is unreliable for the reason given above.

Citrate utilization has not been studied. *P. pentosaceus*, *P. acidilactici* and *P. damnosus* can attack malate.

The relationships between species as shown by DNA/DNA hybridization are given in Table 12.35.

Cytochromes are absent (Deibel and Evans, 1960), but *P. pentosaceus* possesses a pseudocatalase which can give a false-positive when colonies are tested with H_2O_2 (Whittenbury, 1964). This activity is most readily shown in cultures grown on a medium with a low glucose content (Delwiche, 1961). Some *P. acidilactici* strains may possess a heme-requiring catalase (Whittenbury, 1964).

Bacteriophages attacking pediococci have not been reported. Sensitivity to drugs and antibiotics is unknown. Tolerance to hop antiseptic is important in *P. damnosus* which grows in beer, while other species are inhibited (Nakagawa and Kitahara, 1959) by a hop concentration equivalent to that in beer.

No strains are pathogenic to plants or animals.

P. damnosus occurs in beer and brewery habitats almost to the exclusion of other species. Growth in beer causes cloudiness, and other faults. This species also occurs in wine and cider. *P. dextrinicus* has been found in beer but more often occurs in silage. *P. parvulus* and *P. inopinatus* occur in sauerkraut, the former also in silage and the latter in beer. *P. pentosaceus* is widespread on vegetable material and often occurs along with *P. acidilactici*. These two species may occur in milk and dairy products but are less important than other lactic acid bacteria. *P. pentosaceus* is used as a starter in fermented sausages. The salt tolerant *P. halophilus* occurs in soy sauce and picking brines. *P. urinaeequi* has been seldom isolated. The original strains came from horse urine but the general distribution is unknown.

Enrichment and Isolation

Variation in cultural characters means that there is no single isolation procedure suitable for all species. Tolerance to hop antiseptic and NaCl can be used for isolating *P. damnosus* and *P. halophilus*, respectively. *P. pentosaceus* and *P. acidilactici* will grow on the SL acetate agar designed as a selective medium for lactobacilli while other species have been found only occasionally and fortuitously on various media. Addition of thallium acetate and crystal violet to media inhibits Gram-negative bacteria and spore formers where these occur along with pediococci. Actidione can be used to suppress yeast growth.

Maintenance Procedures

Pediococci will survive lyophilization when cells are suspended in horse serum ± 7.5% glucose. Cells from a culture in late logarithmic or early stationary phase should be used. It is important to grow cultures for lyophilization and to revive dried cultures under optimum conditions. Dried cultures will survive for long periods at 10°C under vacuum. Cultures can be preserved for 3–4 months at 4°C in skim milk supplemented with glucose 1.0%, yeast extract 0.3% and calcium carbonate 1.0%. Cultures for storage should not be allowed to overgrow. Species which will not grow in the above medium can be stored in agar stabs of the appropriate medium.

Taxonomic Comments

Pediococci are unlikely to be confused with other lactic acid bacteria because they are morphologically distinct. They are more likely to be confused with micrococci as they are morphologically similar and as they can sometimes apparently be weakly catalase-positive. Micrococci are tolerant of NaCl and so may be confused with *P. halophilus*. However, micrococci normally grow well in the absence of sugar, are frequently pigmented and do not produce large amounts of lactic acid.

The separation of pediococci from aerococci requires further clarification. Dellaglio et al. (1981) showed that some strains named as *P. homari* had high DNA homology with *P. urinaeequi*, while others belonged to a different DNA homology group. The latter strains were aerococci (Schultes and Evans, 1971).

Pediococci occur along with lactobacilli and leuconostocs in plant environments and physiologically have more in common with these organisms than with the streptococci, which are more closely associated with animal habitats. The pediococci are undoubtedly lactic acid bacteria despite their division in two planes.

The nomenclature of the species became confused when the folinic acid requirement of a strain of *Leuconostoc citrovorum* ATCC 8081 was studied. This strain was finally identified as a pediococcus and renamed *P. cerevisiae* (Felton and Niven, 1953). In addition, the pediococci of plants and beer were not separated and both were called *P. cerevisiae*.

The work of Nakagawa and Kitahara (1959) in Japan and of Gunther and White (1961) and Coster and White (1964) in London showed that the species of pediococci found on plants and those found in beer were distinct. Gunther and White (1962a) proposed that ATCC 8081 should be the type strain of *P. cerevisiae* and this strain belonged to the plant pediococci. Studies of the situation were undertaken by several groups of microbiologists and eventually passed to a small subcommittee of the Subcommittee on Lactobacilli and Related Organisms. As the result of a detailed study of the nomenclature from 1884 onward, this subcommittee proposed that because *P. cerevisiae* Balcke 1884 was not validly published and had become a source of confusion, the name should be rejected and the genus name *Pediococcus* should be conserved with *P. damnosus* Claussen as the type species (Garvie, 1974). Thus the genus name *Pediococcus* became ascribed to Claussen (1903). This proposal was accepted (Judicial Commission, Opinion 52, 1976).

Since then this nomenclature has become generally accepted and the names given in the Approved List of Bacterial Names (Skerman et al. 1980) are now used.

Differentiation of the species of the genus **Pediococcus**

Table 12.32 gives the conditions under which the different species will grow, and which can be used for a preliminary separation.

1. **Pediococcus damnosus** Claussen 1903, 68.[AL]

dam.no'sus. L. adj. *damnosus*—destructive.

Morphology as general description.

Growth is often slow and 2–3 days incubation at 22°C may be required for good growth. Broth cultures will develop without anaerobic conditions but cysteine added to media improves growth. On an agar surface colony development is poor unless cultures are incubated anaerobically. The final pH in broth is about 4.0. Optimum pH for growth is 5.5.

Acetoin or diacetyl are readily produced and cause "sarcina" (buttery) odor in spoiled beer. Slime-forming strains are reported (Shimwell, 1948, Carr, 1970). The nature of the slime has not been fully investigated but it may be due to capsular material.

Hop humalone may cause the formation of giant cells 5.15 μm (Nakagawa and Kitahara, 1962).

Cells are readily destroyed by heat, i.e. 60°C for 10 min.

The mol% G + C of the DNA is 37–42 (T_m) (Kocur et al., 1971).

Type strain: NCDO 1832 (ATCC 29358; DSM 20331).

2. **Pediococcus parvulus** Gunther and White 1961, 195.[AL]

par'vu.lus. L. dim. adj. *parvulus* very small.

Morphology as general description.

Colonies on tomato juice agar were originally reported to be very small but improved growth can be obtained particularly when anaerobic incubation is used and larger colonies result—usually 48 h at 30°C is required.

Broth cultures are improved by the addition of cysteine and Tween 80 but 48 h incubation may be needed (Garvie and Gregory, 1961). Some strains require asparagine (Garvie et al. 1961) but not folinic acid. The final pH in broth is about 4.0 in well grown strains. Optimum pH 6.5. Optimum temperature 30°C.

Tolerance to hop antiseptic is unknown.

The mol% G + C of the DNA is 40.5–41.6 (T_m) (W. Back, unpublished).

The distribution of the species is unknown as reported isolations are few.

Habitat: fermenting vegetables.

Type strain: NCDO 1634 (ATCC 19371; DSM 20332).

3. **Pediococcus inopinatus** Back 1978a, 245.[VP]

in.o.pin.a'tus. L. adj. *inopinatus* unexpected.

Morphology as general description.

On agar, growth is slow and colonies may take 3–5 days to develop. The final pH in MRS broth is about 4.0.

There is a close similarity between *P. inopinatus* and *P. parvulus* which both occur in the same habitat. DNA/DNA hybridization showed some relationship between the two species and also to *P. damnosus* (Back and Stackebrandt 1978).

The mol% G + C of the DNA is 39–40 (T_m) which is close to that of the other pediococcus species (W. Back, unpublished).

Further separation of *P. parvulus* and *P. inopinatus* was obtained by the electrophoresis of the L-(+)- and D-(−)-LDHs (Back, 1978a).

Habitat: fermenting vegetables and beverages (beer and wine).

Type strain: DSM 20285.

4. **Pediococcus dextrinicus** (Coster and White) Back 1978b, 523.[AL]

(*Pediococcus cerevisiae* subsp. *dextrinicus* Coster and White 1964, 29.)

dex.trin'i.cus. M.L. n. *dextrinosum*—dextrin. M.L. neut. adj. *dextrinicus*—relating to dextrin.

Morphology as general description. Less anaerobic than previously described species. Colonies will develop on agar aerobically but growth is improved in an atmosphere of H_2 + 10% CO_2.

In MRS broth the final pH is about 4.4

Optimum pH for growth 6.5. Optimum temperature 30–35°C.

Growth requirements have not been studied. Growth occurs in weakly hopped beer.

Habitat: fermenting vegetables and beer.

The mol% G + C of the DNA is 40–41 (T_m) (Back, 1978b).

Type strain: DSM 20335 (NCDO 1561; ATCC 33087).

5. **Pediococcus pentosaceus** Mees 1934, 96.[AL]

pen.to.sa'ce.us. M.L. neut. n. *pentosum*—a pentose sugar. M.L. adj. *pentosaceus*—relating to a pentose.

Morphology and colonial appearance as general description.

Anaerobic incubation is not necessary and colonies should be visible on agar after incubating aerobically for 24 h at 30°C.

Litmus milk reactions are variable and may be related to growth requirements. The requirement for folinic acid varies between strains.

A limited study of the aldolases found in pediococci have shown that *P. pentosaceus* and *P. acidilactici* are closely related species which are separate from *P. parvulus*. There is some evidence that not all strains of *P. pentosaceus* have the same aldolase (London and Chance, 1976).

In broth growth can be very rapid and the final pH in MRS broth is usually below 4.0. Optimum pH 6.0–6.5. Optimum temperature 28–32°C. Low heat resistance, cells being destroyed at 65°C in 8 min.

The rapid growth, low final pH and absence of cytochromes distinguish *P. pentosaceus* from micrococci. *P. pentosaceus* could be confused with micrococci as it can form small colonies on sugar-free agar and can grow at pH 9.0. It may also be weakly catalase-positive when grown in a medium with low glucose content (Whittenbury 1964).

The mol% G + C of the DNA is 35–39 (T_m) (various authors).

Type strain: NCDO 990 (ATCC 33161; DSM 20336).

6. **Pediococcus acidilactici** Lindner 1887, 440.[AL]

a.ci.di.lac.ti'ci. M.L. n. *acidium lacticum* lactic acid. M.L. gen. n. *acidilactici* of lactic acid.

Morphological, cultural and physiological properties do not readily separate *P. acidilactici* from *P. pentosaceus*.

Optimum temperature of growth 40°C.

Heat tolerant, destroyed at 70°C in 10 min while some strains may be even more heat tolerant particularly when freshly isolated.

The mol% G + C of the DNA is 38–44 (T_m) (various authors). DNA/DNA hybridization shows *P. acidilactici* as a distinct species (Back and Stackebrandt 1978; Dellaglio et al., 1981), while studies of aldolases suggest that *P. pentosaceus* and *P. acidilactici* are closely related.

Type strain: NCDO 1859. The reference strain used by Dellaglio et al., 1981) was ATCC 25742. The work on DNA/DNA hybridization suggests that NCDO 1859, the type strain of *P. acidilactici* is a strain of *P. pentosaceus*. The Judicial Commission should be asked to replace NCDO 1859 with a more suitable strain.

In early work *P. pentosaceus* and *P. acidilactici* were not separated and the properties given may be a combined study of both species. DNA and enzyme studies clearly separate the two species, but when these characteristics cannot be determined difficulties could still arise.

7. **Pediococcus halophilus** Mees 1934, 96.[AL]

hal.o.phi'lus. Gr. n. *halos*. salt. Gr. Adj. *philus*. loving. M.L. adj. *halophilus*, salt loving.

Morphology as general description.

Growth on agar is slow and colonies develop aerobically.

Growth in broth is also slow and 4–5 days incubation may be required. The final pH is about 5.0 and turbidity is less than more acid-tolerant species. Media suitable for the acid-tolerant species do not support good growth of *P. halophilus*, which has an optimum pH for growth between 7.0 and 8.0 and optimum temperature 30–35°C.

Growth will take place in 18% NaCl, and 20–26% may be tolerated (Sakaguchi 1958).

The ratio of D-(−) : L-(+) lactate formed by cultures growing on glucose is about 3:97.

Growth takes place in hopped wort at pH above 5.5 (Sakaguchi, 1958).

The mol% G + C of the DNA is 34–36% (T_m) (various authors).

Type strain: NCDO 1635 (ATCC 33315; DSM 20339).

Comment: In a comparative study of the salt-tolerant pediococci and some aerococci, Deibel and Niven (1960) found that the tetracocci from brine were the same as *P. halophilus* but considered the strains to be *Aerococcus homari* (*Pediococcus homari*, *Aerococcus viridans*). Clearly the relationship between these species requries clarification, as does their relationship to *P. urinaeequi* (see comment after *P. urinaeequi*).

8. **Pediococcus urinaeequi** (*ex* Mees) nom. rev.

u.ri'nae-e.qui. L. fem. n. *urina*—urine. L. mas. n. *equs*—horse. M.L. gen. n. *urinae-equi*, horse urine.

Morphology as general description.

Growth is generally improved if the initial pH of the medium is alkaline. Optimum pH is between 8.5 and 9.0 (Nakagawa and Kitahara, 1959) although growth will take place in media with an initial pH of 6.5–7.0. The final pH is about 5.0 (Gunther and White, 1961). Optimum temperature 25–30°C. Growth can occur in media which do not contain added carbohydrate (Sakaguchi and Mori, 1969).

L-(+)-lactate is formed from glucose. The LDHs have not been studied. It is not known whether a trace of D-(−)-lactate is formed, as with *P. halophilus*.

The mol% G + C of the DNA is 39.5% (T_m) (Sakaguchi and Mori, 1969; Dellaglio et al. 1974).

The species does not appear to be widely distributed and reported isolations are few.

Type strain: NCDO 1636 (ATCC 29723; DSM 20341).

Comment: The taxonomic position of *P. urinaeequi* requires clarification. There are similarities with *P. halophilus* but the composition of the cell wall murein and salt tolerance are different. Gunther and White (1961) placed *P. urinaeequi* as a variant of their *P. cerevisiae* but Whittenbury (1965) grouped it with *Aerococcus viridans*. It is now known that these species have the same cross-linkage in their cell wall peptide. There is no DNA/DNA homology between *P. urinaeequi* and either *P. halophilus* or *A. viridans* (Dellaglio et al. 1981). However, two strains identified as *Pediococcus homari* (*A. viridans*) had high DNA homology with *P. urinaeequi*.

More work is necessary to clarify the taxonomy of the alkaline-tolerant tetrad forming cocci. Meantime the existing nomenclature is used.

Genus *Aerococcus* Williams, Hirch and Cowan 1953, 475[AL]

JAMES B. EVANS

A.ë.ro.coc′cus. Gr. masc. noun *aër* air, gas; Gr. n. *coccus* a grain, berry; M.L. masc. n. *Aerococcus* air coccus.

Spheres, 1.0–2.0 μm in diameter with a **strong tendency toward tetrad formation** when grown in suitable liquid media. Gram-positive. Nonmotile. **Microaerophilic**. In shake cultures or soft sugar agar a heavy band of discrete colonies is produced just beneath the surface. Anaerobic growth frequently is absent and when it does occur is delayed and often consists of only a few discrete colonies. **Catalase activity is absent or weak** and when present is a nonheme pseudocatalase. Porphyrin respiratory enzymes are absent. H_2O_2 is produced during aerobic growth. Growth on solid media is generally sparse and beaded (small discrete colonies). **Greening reaction produced on blood agar.** The mol% G + C of the DNA is 35–40 (T_m).

Type species: *Aerococcus viridans* Williams, Hirch and Cowan 1953, 477.

Further Descriptive Information

Spherical cells occur singly, in pairs, in tetrads and in larger clusters. Distinct tetrads may be observed frequently in young cultures in rich broth media. The cell wall murein contains no interpeptide bridge. The carboxyl group of D-alanine is bound to the ϵ-amino group of L-lysine of an adjacent peptide subunit. The cell wall contains glucose, galactose and galactosamine but no rhamnose, glycerol or ribitol.

Agar colonies are round, 0.5–1.0 mm in diameter, semitransparent, white or gray. On blood agar colonies are larger and are surrounded by a zone of greening, presumably the result of H_2O_2 production. In broth there is moderate uniform turbidity that tends to settle rather quickly into a packed sediment. The vitamins pantothenic acid, nicotinic acid and biotin are either required or markedly stimulatory. Guanine or another purine base is required. Amino acids are required but the requirement is not specific or sharply defined.

Chemoorganotrophic: acid but no gas is produced from glucose, fructose, galactose, mannose, maltose and sucrose, and usually is pro-duced from lactose, trehalose and mannitol. Final pH in glucose broth is 5.0–5.5.

Acetyl methyl carbinol is not produced. Gelatin is not liquefied. Ammonia is not produced from arginine. Nitrates are not reduced to nitrites. Growth is not prevented by 40% bile, 10% NaCl, 0.01% potassium tellurite or a pH of 9.6. Most strains will grow at 10 but not at 45°C.

Generally saprophytic but has been associated with human endocarditis and is the primary cause of a fatal disease of lobsters called gaffkemia.

Frequently isolated from two widely divergent sources: as a common airborne organism in hospital environments and as a marine organism causing a fatal disease of lobsters. The taxonomic kinship of strains from these two widely divergent sources has been confirmed by numerical taxonomy, serological and DNA homology studies.

Enrichment and Isolation

Aerococcus strains will grow on a wide range of selective and nonselective media that are generally employed for other Gram-positive cocci (e.g. staphylococci and fecal streptococci) but because of their comparatively slower and weaker growth are generally overlooked in most isolation procedures. Blood agar containing 0.001% potassium tellurite and 0.00025% crystal violet is a useful selective medium. APT broth or agar is a very satisfactory nonselective medium.

Maintenance Procedures

Aerococcus cultures can be maintained on APT agar slants, by the usual lyophilization procedures, or as frozen cultures (preferably at ultralow temperature).

Procedures for Testing Special Characters

Procedures are described in the papers cited for further reading.

Differentiation of the genus *Aerococcus* from other genera

This genus is differentiated from other genera of Gram-positive cocci primarily by its rather wide range of negative characteristics that are positive for other genera. These include their rather sparse growth on and in agar media, particularly anaerobically, their inability to produce acetyl methyl carbinol from glucose or ammonia from arginine and their failure to reduce nitrate. Among their positive features are the greening of blood agar, tetrad formation, and tolerance to several commonly used inhibitory growth conditions.

Further Reading

Colman, G. 1967. Aerococcus-like organisms isolated from human infections. J. Clin. Pathol. *20:* 294–297.
Deibel, R.H. and C.F. Niven, Jr. 1960. Comparative study of *Gaffkya homari*, *Aerococcus viridans*, tetrad-forming cocci from meat curing brines, and the genus *Pediococcus*. J. Bacteriol. *79:* 175–180.
Hanna, B.A. and W.J. Untereker. 1975. Endocarditis and osteomyelitis caused by *Aerococcus viridans*. Abst. Annu. Meet. Am. Soc. Microbiol. p. 49.

Hitchner, E.R. and S.F. Snieszko. 1947. A study of a microorganism causing a bacterial disease of lobsters. J. Bacteriol. *54:* 48.
Kelly, K.F. and J.B. Evans. 1974. Deoxyribonucleic acid homology among strains of the lobster pathogen "*Gaffkya homari*" and *Aerococcus viridans*. J. Gen. Microbiol. *81:* 257–260.
Kerbaugh, M.A. and J.B. Evans. 1968. *Aerococcus viridans* in the hospital environment. Appl. Microbiol. *16:* 519–523.
Miller, T.L., and J.B. Evans. 1970. Nutritional requirements for growth of *Aerococcus viridans*. J. Gen. Microbiol. *61:* 131–135.
Nakel, M., J.M. Ghuysen and O. Kandler. 1971. Wall peptidoglycan in *Aerococcus viridans* strains 201 Evans and ATCC 11563 and in *Gaffkya homari* strain 10400. Biochemistry *10:* 2170–2175.
Steenbergen, J.F., H.S. Kimball, D.A. Low, H.C. Shapiro and L.N. Phelps. 1977. Serological grouping of virulent and avirulent strains of the lobster pathogen *Aerococcus viridans*. J. Gen. Microbiol. *99:* 425–430.
Williams, R.E.O., A. Hirch and S.T. Cowan. 1953. *Aerococcus*, a new bacterial genus. J. Gen. Microbiol. *8:* 475–480.

List of species of the genus *Aerococcus*

1. **Aerococcus viridans** Williams, Hirch and Cowan 1963, 477.[AL]
vi. ri′ dans. L. part adj. *viridans* producing a green color.

This is the only recognized species, so the preceeding description of the genus applies to the species.

Type strain: NCTC 8251 (ATCC 11563).

Genus **Gemella** Berger 1960, 253[AL]

ALICE REYN

Ge.mel′la. M.L. dim. n. *gemellus* a twin; M.L. fem. n. *Gemella* a little twin.

Cocci singly or in pairs with adjacent sides flattened; short chains occur. Gram-indeterminate but cell wall is of the Gram-positive type (Fig. 12.13).

Endospores not formed; nonmotile. On blood agar small smooth colonies resemble those of β-hemolytic streptococci; hemolysin is produced on rabbit or horse blood agar. Pigment is not produced. Chemoorganotroph: fermentative, acid is produced from several carbohydrates. The products of fermentation have not been analyzed. Growth requirements complex.

Nitrates not reduced; nitrites reduced by some strains. Aerobic or facultatively aerobic. Respiratory system resistant to KCN. Optimal temperature 37°C, grows at 22°C but not at 10° or 45°C. Parasites of man. Mol% G + C of the DNA is 33.5 ± 1.6 (Bd).

Type species: *Gemella haemolysans* (Thjøtta and Bøe) Berger 1960, 253.

Further Descriptive Information

Gram-stained smears contain distinct Gram-positive, Gram-negative and Gram-variable cells. Originally *G. haemolysans* was classified as belonging to the genus *Neisseria* under the name *N. haemolysans* (Thjøtta and Bøe, 1938). By gas chromatography Yamakawa and Ueta (1964) demonstrated that the fatty acid content and sugars in whole cells of *N. haemolysans* differed from those of other *Neisseria* species studied. Similar results were obtained by Lambert et al. (1971); Brooks et al. (1971, 1972); Morse et al. (1977). The fine structure resembles that of many Gram-positive organisms, although the cell wall is comparatively thin (about 10 nm) and of varying thickness; this probably accounts for the variability in decolorization after Gram-staining. Outside the cell wall but in close connection with it a slimy layer is found, presumably produced by the organism (Reyn et al., 1966, 1970) (Fig. 12.13).

Hemolysis occurs on rabbit or horse blood agar. Poor growth on media without protein. Growth on medium with tellurite very weak or absent. Pigment not formed.

Acid produced fermentatively from glucose, maltose, fructose, sucrose, glycogen, starch and dextrin, and occasionally from mannitol, arabinose, sorbitol and inulin. Polysaccharide not produced on 5% sucrose. Very feeble growth in Hugh and Leifson's (1953) medium under anaerobic conditions. Esculin not hydrolyzed. Arginine not hydrolyzed. Indole, H$_2$S, gelatinase, urease, cytochrome oxidase, catalase and peroxidase not produced. H$_2$O$_2$ formed under aerobic conditions. Resistant to optochin. Nitrates not reduced; nitrites reduced by some strains.

Aerobic, facultatively anaerobic. Weak growth under completely anaerobic conditions. Serologically different from *N. perflava*, *N. pharyngis* and *N. sicca*. Further serological analysis is lacking.

Sensitive to penicillin, streptomycin, tetracycline, sulfathiazole, chloramphenicol, macrolides, vancomycin and related drugs. Found in bronchial secretions and slime from the human gingiva. No known pathogenicity.

Maintenance Procedures

G. haemolysans can be lyophilized by the common procedures used for aerobic organisms.

Taxonomic Comments

Berger (1960, 1961) proposed *Gemella* as an aerobic, oxidase-negative, catalase-negative genus within the *Neisseriaceae*, but later studies have indicated that it does not belong to this family of Gram-negative cocci (Reyn et al., 1966; Reyn, 1970; Reyn et al., 1970).

The mol% G + C of the DNA is 33.5 ± 1.6 (Bd, Reyn et al., 1970) compared to 38–55 in *Neisseriaceae* (*Manual*, volume 1, p. 289) and 40–44 (Bd) in *Veillonella* (*Manual*, volume 1, p. 681). This corresponds to the peroxidase-negative, catalase-negative family *Streptococcaceae*, but although *Gemella* has many physiological features in common with *Streptococcaceae*, as for example, resistance to H$_2$O$_2$ and KCN (Berger 1970), it does not fit completely into any of the genera described within this family.

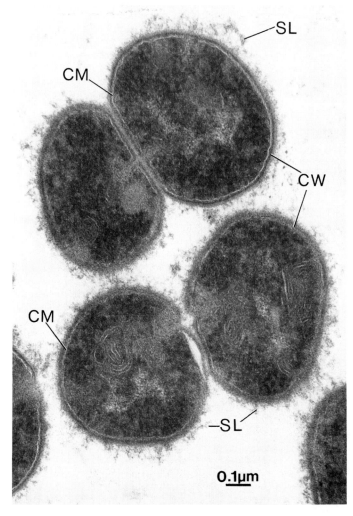

Figure 12.13. Section of dividing cells of *Gemella haemolysans*, (ATCC 10379). Note slime layer (*SL*), cell wall (*CW*) and cytoplasmic membrane (*CM*). Stained with magnesium uranyl acetate and lead citrate (× 90,000). From unpublished work of A. Reyn and A. Birch-Andersen

List of species of the genus **Gemella**

Gemella haemolysans (Thjøtta and Bøe) Berger 1960, 253.[AL] (*Neisseria haemolysans* Thøtta and Bøe 1938, 531.)

hae.mo.ly′sans. Gr. n. *haema* blood; Gr. v. *lyo* loosen; M.L. part. adj. *haemolysans* dissolving blood.

Cocci 0.5 × 0.5–0.6 μm singly or in pairs. Considerable variation in size; division in several planes.

Type strain: ATCC 10379.

Genus **Peptococcus** Kluyver and van Niel 1936, 400[AL]

LILLIAN V. HOLDEMAN MOORE, JOHN L. JOHNSON, AND W. E. C. MOORE

Pep.to.coc′cus. Gr. v. *pepto* cook, digest; Gr. n. *coccus* a grain, berry; M.L. masc. n. *Peptococcus* the digesting-coccus.

Nonsporing cocci that are Gram-positive and anaerobic. Chemoorganotrophs. Metabolize peptone and amino acids to combinations of C1 through C6 normal and branched chain fatty acids.

The mol% G + C of the DNA is 50 for the type species.

Type species: *Peptococcus niger* (Hall) Kluyver and van Niel 1936, 400.

Further Descriptive Information

Neither catalase production nor cellular arrangement (e.g. chains vs. diplococci or tetrads) is reliable for differentiation between species in the genera *Peptococcus* and *Peptostreptococcus*. The mol% G + C of species of *Peptococcus* is 50–51; that of most species of *Peptostreptococcus* is 28–34.

Information regarding *isolation, maintenance and procedures and methods for characterization tests* are given in the description of the genus *Peptostreptococcus*.

Taxonomic Comments

Five species of *Peptococcus* were listed in the eighth edition of the *Manual*. Only *P. niger*, the type species of the genus, is retained in this edition.

Ezaki et al. (1983) have shown that mol% G + C of the DNA of *Peptococcus asaccharolyticus, Peptococcus prevotii, Peptococcus magnus*, and *Peptoccus indolicus* is 29–34, more like the 33 mol% G + C of *Peptostreptococcus anaerobius*, the type species of the genus *Peptostreptococcus*, than like the 50 mol% of *Peptococcus niger*, the type species of *Peptococcus*. Ezaki et al., therefore, have proposed the transfer of these species to the genus *Peptostreptococcus*.

Other species listed in the eighth edition of the *Manual* and not included in the genus in this edition are "*Peptococcus aerogenes*", *Peptococcus anaerobius*", and "*Peptococcus constellatus*". Strain ATCC 14963, deposited in ATCC to represent "*Peptococcus aerogenes*", does not ferment carbohydrates and therefore was considered to be a strain of *P. asaccharolyticus*. When the 1980 Approved Lists of Names were compiled, ATCC 14963 was given as a reference strain to represent *P.*

asaccharolyticus and "*P. aereogenes*" was not included in the lists. The proposed reference strains were designated type strains of the species (Skerman et al., 1980). Therefore, the name "*Peptococcus aerogenes*" has no taxonomic standing and the strain that originally represented this species is now the type strain of *P. asaccharolyticus*.

Because there was no available strain to represent "*P. activus*," it was not included on the 1980 Approved Lists of Names (Skerman et al., 1980).

The name "*Peptococcus anaerobius*" (Hamm) Douglas has been placed on the list of *nomina rejicienda* (West and Holdeman, 1973; Opinion 56, 1982. Int. J. Syst. Bacteriol. *32:* 468). Most strains previously associated with the name "*Peptococcus anaerobius*" are strains of *P. magnus*.

"*P. constellatus*" was transferred to the genus *Streptococcus* as *S. constellatus* (Holdeman and Moore, 1974). Lactic acid is the sole major fermentation acid produced by *S. constellatus* and the mol% G + C content of the DNA of *S. constellatus* is 37–38, similar to that of other species of streptococci. Subsequently, Welborn et al. (1983) showed that the type strains of *S. constellatus* and *S. intermedius* have 76–89% reciprocal DNA homology. The name *S. constellatus* (Prévot 1924) Holdeman and Moore 1974 has priority. However, Moore et al. (1982) observed that the type strain of *S. constellatus* and the type strain of *S. anginosus* have extremely similar phenotypic reactions and polyacrylamide gel electrophoresis patterns and suggest that they are the same species, in which case *S. anginosus* Andrewes and Horder 1906 would have priority.

Peptococcus saccharolyticus was not included among the peptococci in the eighth edition of the *Manual* because its characteristics did not conform to those of the genus (Rogosa, 1974). However, the species was recognized in other manuals (e.g. Holdeman et al., 1977). Using nucleic acid studies, Kilpper et al. (1980) and Ludwig et al. (1981) have shown that *Peptococcus saccharolyticus* is more closely related to species of *Staphylococcus* than to species of *Peptococcus*. The species therefore has been transferred to *Staphylococcus* (Kilpper-Bälz and Schleifer, 1981), where it is described in this edition of the *Manual*.

List of species of the genus **Peptococcus**

1. **Peptococcus niger** (Hall 1930) Kluyver and van Niel 1936, 400.[AL] (*Micrococcus niger* Hall 1930, 409)

ni′ger. L. adj. *niger* black.

From glucose broth cultures, cells are 0.3–1.3 μm in diameter and occur singly and in pairs, tetrads, and irregular masses (Fig. 12.14).

Surface colonies on blood agar are black, minute to 0.5 mm, circular with entire margins, convex, shiny, smooth and nonhemolytic. The black colonies become light gray when exposed to air. After several transfers or lyophilization, cultures may not produce black colonies on

the surface of blood agar but will continue to produce black colonies in (but not on the surface of) chopped meat agar (Wilkins et al., 1975). (We have encountered one strain that did not produce black colonies in chopped meat agar but had all of the other characteristics described for the species.) No lecithinase or lipase activity is detected on McClung-Toabe egg yolk agar plates (Holdeman et al., 1977).

Broth cultures have a smooth white or grayish white sediment and are not turbid.

The optimum temperature for growth is 37°C; growth may occur at 25 and 45°C. Oleate (Tween 80) does not stimulate growth and inhibits growth of an occasional strain.

Carbohydrates are not fermented. Carbohydrates tested were amygdalin, arabinose, dulcitol, erythritol, esculin, fructose, glucose, glycerol, glycogen, inositiol, lactose, maltose, mannitol, mannose, melezitose, melibiose, raffinose, rhamnose, ribose, salicin, sorbitol, sorbose, starch, sucrose and xylose.

10 μm   PY   PYG

Figure 12.14. *Peptococcus niger.*

Hydrogen sulfide (in sulfide-indole-motility (SIM) medium, BBL) and ammonia are produced. A small amount of catalase activity may be detected. Indole, urease, and coagulase are not produced. Nitrate is not reduced. Neither esculin nor starch is hydrolyzed.

Pyruvate is metabolized. Products in peptone-yeast extract (PY)-pyruvate broth culture (in average milliequivalents per 100 ml of culture) are acetic (3.5), isobutyric (0.3), butyric (1.2), isovaleric (0.6) and caproic (0.4) acids. Occasionally trace amounts of propionic, valeric, lactic and succinic acids are produced. Products in 1% PY broth are similar to those from pyruvate, but in lesser amounts, and caproic acid usually is not detected. Neither lactate nor fumarate is utilized. Copious H_2 is produced from peptone. Small amounts of gas may be produced in glucose agar deep cultures.

Strains are susceptible to 2 units of penicillin/ml; 3 μg erythromycin/ml; 1.6 μg clindamycin/ml; 12 μg chloramphenicol/ml; and 6 μg tetracycline/ml.

Isolated from the human umbilicus, vaginal area and occasionally from clinical specimens.

The mol% G + C of the DNA of the type strain is 50–51 (T_m) (Wilkins et al., 1975; Ezaki et al., 1983).

Type strain: ATCC 27731 (DSM 20475).

Further comments. I. C. Hall (1930) described as "*Micrococcus niger*" a species of obligately anaerobic Gram-positive cocci producing black colonies on blood agar. The species was placed in *Peptococcus* by Kluyver and van Niel (1936). Although Prévot et al. (1959) reported the isolation of "*M. niger*" from various human clinical specimens (Prévot et al., 1967), it was not isolated commonly by other workers and no representative or type strain was extant when Rogosa (1974) concluded that Hall's isolate "unfortunately was most probably . . . a mutant strain of vague origin." Wilkins et al. (1975) reported isolation and study of strains of anaerobic Gram-positive cocci that had characteristics originally described by Hall for "*M. niger*" and that were different from other described species of *Peptococcus*. Although several species of Gram-positive anaerobic cocci produce gray-to-black colonies on media containing cysteine, these other species do not form black colonies on blood agar or in chopped meat agar, and do not produce caproic acid from pyruvate (Wilkins et al., 1975). In all probability, not all of the strains designated "*M. niger*" by Prévot et al. (1959) belonged to the species as presently defined, since they reported indole production by some of their strains.

Genus **Peptostreptococcus** *Kluyver and van Niel 1936, 401*[AL]

LILLIAN V. HOLDEMAN MOORE, JOHN L. JOHNSON AND W. E. C. MOORE

Pep.to.strep.to.coc′cus. Gr. v. *pepto* cook, digest; N.L. masc. n. *Streptococcus* a generic name; N.L. masc. n. *Peptostreptococcus* the digesting streptococcus.

Nonsporeforming cocci that are Gram-positive and anaerobic. Cells may occur in pairs, tetrads, irregular masses or chains. Chemoorganotrophs. Metabolize peptone and amino acids to acetic acid, often with isobutyric, butyric, isovaleric or isocaproic acids; the one highly saccharoclastic species (*P. productus*) metabolizes carbohydrates to combinations of acetic, formic, succinic and lactic acids.

The mol% G + C of the DNA of the type species is 33; that of *P. productus* is 44–45; that of other species in the genus is 27–35.

Type species: *Peptostreptococcus anaerobius* (Natvig 1905) Kluyver and van Niel 1936, 401.

Further Descriptive Information

Differentiation of Peptostreptococcus and Peptococcus. In this edition of the *Manual*, the anaerobic Gram-positive cocci that metabolize peptones and amino acids and that have DNA with a G + C content of 28–35 mol% (except for *Peptostreptococcus productus*, q.v.) have been placed in the genus *Peptostreptococcus*, following the recommendation of Ezaki et al. (1983), and *Peptococcus niger*, with a mol% G + C of 50, is the sole species of *Peptococcus*. As presently defined, the major difference between the genera *Peptostreptococcus* and *Peptococcus* is the G + C content of the DNA. *Neither arrangement of the cells (tetrads, masses or chains) nor production of catalase is useful to differentiate between the two genera.*

Ecology. Most described species are part of the normal flora of man; some also have been detected in other animals.

Isolation Procedures

In general, strains of anaerobic Gram-positive cocci found in clinical infections can be isolated on usual complex media appropriate for culture of anaerobes. Such media include blood agar plates or prereduced media in roll tubes with a peptone base including or supplemented with 0.5% (w/v) yeast extract. Additional supplementation with vitamin K and hemin (required by some other anaerobes) does not affect growth. Although Tween 80 enhances growth of many strains, it usually is not required for their isolation.

The general rule that anaerobes are most sensitive to the toxic effects of O_2 when the cells are in the logarithmic phase of growth has been confirmed for *Peptostreptococcus anaerobius* (Nyberg and Carlsson, 1981) in an investigation of the basis of O_2 toxicity in this species by Carlsson and colleagues (Frölander and Carlsson, 1977; Carlsson et al.,

1978; and Carlsson et al., 1979). The rate of killing is high in media that auto-oxidize rapidly because hydrogen peroxide accumulates. Both metal ions and active cellular metabolism are implicated in the bactericidal effect of hydrogen peroxide.

Marshall et al. (1981) report that media preparation, age and storage conditions are critical for isolation of the anaerobic Gram-positive cocci of clinical importance. The organisms tested required freshly prepared media for optimal recovery. Recovery of reference strains was poor on agar media (thioglycolate-yeast extract-hemin-menadione-sodium bicarbonate) stored at 4°C in GasPak jars in the refrigerator for more than 8 h and, regardless of storage time, no growth was obtained if the aerobically stored agar medium was remelted before inoculation. Rapid growth of test strains of peptostreptococci (and strains then identified as peptococci, see Taxonomic Comments) occurred in freshly prepared thioglycolate-yeast extract broth. Although Tween 80 (0.1%) did not affect efficiency of recovery, the colony size of some species was markedly increased in media containing Tween 80. Of the strains used in this study, growth of only *Peptostreptococcus productus* (rarely, if ever, isolated from properly collected clinical specimens) was stimulated by the addition of sodium bicarbonate.

A modified stainless steel chamber for homogenizing tissue samples anaerobically for recovery of O_2-sensitive species is described by Spengler et al. (1979).

Maintenance Procedures

Lyophilization of cultures in the early stationary phase of growth in a medium containing no more than 0.1–0.2% fermentable carbohydrate is recommended for long term storage of most strains. Strains also survive well when frozen in blood and stored at −80°C (a heavy suspension of cells from surface growth on solid media is dispensed in tubes of small diameter or coated onto glass beads, frozen quickly in dry ice and alcohol, and placed in a freezer). For short term storage, cultures grown in a nutritionally adequate medium containing no, or minimal, fermentable carbohydrate can be stored at room temperatures (~22°C). Strains should be transferred every 2–3 weeks.

Procedures and Method for Characterization Tests

A heavy inoculum (~5% v/v of broth culture) of a young culture (late logarithmic or early stationary phase of growth) gives the most reliable results for biochemical reactions based upon growth of the organism in the substrate. Results of tests for constitutive enzymes with heavy inocula on dehydrated substrates may differ from results based upon growth of the organism in the substrate where adaptive enzymes also are detected. Unless otherwise cited, the characteristics listed here for the species were determined with prereduced anaerobically sterilized media by the methods described in Holdeman et al. (1977) and susceptibility to antimicrobial agents was tested by the broth-disk method of Wilkins and Thiel (1973). The basal prereduced peptone-yeast extract (PY) medium contained (per 100 ml): 0.5 g trypticase, 0.5 g peptone, 1 g yeast extract, 0.1 μl vitamin K_1, 0.5 g hemin, 0.05 g cysteine hydrochloride, and salts solution (Holdeman et al., 1977).

Morphology and Gram-reaction. For determination of Gram-reaction, a buffered Gram's stain (such as Kopeloff's modification) of cells from young cultures (logarithmic phase of growth) is recommended. Cells of many Gram-positive species stain Gram-negative in older cultures.

Hydrogen production. Hydrogen production was determined by gas chromatography of the headspace gas of cultures grown in tubes closed with rubber stoppers (Holdeman et al., 1977). Hydrogen production may not disrupt the agar in PY-glucose agar deep cultures.

Gas production. Gas production was determined by disruption of agar in loosely covered PY-glucose deep agar cultures. Accumulation of small (lenticular) bubbles near the colonies in the deep agar was recorded as "1" gas, a small split of the agar column across the tube as "2" gas, agar separated and displaced in one or more places as "3" gas and agar forced to the top of the tube as "4" gas.

Esculin hydrolysis and esculin fermentation. These are independent reactions. Hydrolytic products are detected by the appearance of a black color on addition of ferric ammonium citrate reagent to PY-esculin broth cultures. Acid production is determined by pH measurement.

Indole. Cultures in media containing sufficient tryptophan (e.g. chopped meat medium, tryptone-yeast extract medium, etc.) and no fermentable carbohydrate are required. The culture medium should be extracted with xylene and the reagent (either Ehrlich's or Kovac's) poured down the side of the tube to layer next to the xylene. The tube should not be shaken after the reagent has been added. Alternatively, a loopful of growth from a *pure culture* can be measured on filter paper saturated with 1% paradimethylaminocinnamaldehyde in 10% (v/v) hydrochloric acid (Sutter et al., 1980). Development of a blue color indicates indole. This method may not detect weak indole production.

Gelatin digestion. Incubated cultures of PY medium containing 10% gelatin and 1% glucose and uninoculated tubes of the same medium are chilled until the originally liquid controls are solid (about 15 min at 4°C). Failure of the gelatin cultures to solidify at 4°C indicates complete digestion (+). Liquefaction of cultures at room temperature within 30 min, or in less than half the time of the control tube, indicates partial digestion (w).

Milk culture reactions. Acid production in milk often causes protein coagulation (c). The curd may or may not show evidence of streaks caused by evolution of gas. Shrinking of the curd may leave a clear whey. This liquid is sometimes mistaken for digestion (proteolysis) of the milk. Digestion of curd is first evidenced by dissolution of the curd with increasing turbidity of the whey, and often takes many days. Digestion may occur without previous curd formation and results (usually slowly) in clear liquid after extended incubation (up to 3 weeks). *Clostridium sporogenes* is a good positive control species for milk proteolysis.

Acid production. Positive (+) reactions listed for carbohydrate fermentation represent a pH below 5.5 and a decrease in pH at least 0.5 pH units below the control PY-basal medium cultures. Weak (w) acid production represents a pH of 5.5–5.9 and at least 0.3 pH units below the PY culture control. A pH of 5.9 or above, or within 0.3 pH units of the PY control culture pH, is considered negative (−). When in doubt, the amount of growth in the sugar-containing medium compared with that in medium without sugar can help in interpretation of weak reactions; i.e. with slight pH decrease, if growth is much better in the sugar-containing medium, the carbohydrate probably was fermented.

Uninoculated xylose and arabinose, under CO_2, are often pH 5.9. For cultures in these media, a final pH of 5.4 or below is considered strong acid production.

Effect of Tween 80 on growth. Growth in media with Tween is compared with growth in the same medium (e.g. PY-glucose) without Tween.

Determination that a fermentable carbohydrate is required for growth. Growth in media to which no sugar has been added (e.g. PY) is compared with growth in sugar-containing media. If there is good growth in a medium that contains a carbohydrate and no growth in the medium without sugar, a fermentable carbohydrate is required for growth. If the inoculum was from a medium containing carbohydrate and slight growth occurs in the sugar-free medium, the culture in the sugar-free medium is transferred to another tube of sugar-free medium and the second transfer is checked for growth.

Differentiation of the genus **Peptostreptococcus** from other genera

Characteristics useful for differentiating the genus from the other closely related genera are given in Table 12.37, next chapter. For strains that do not belong to described species, it will be difficult, if not impossible, to determine whether a strain is a peptococcus or a peptostreptococcus unless the mol% G + C of the DNA is determined.

Although Wren et al. (1977) report that common human clinical isolates of peptococci are resistant to 5 μg of novobiocin/ml and common clinical isolates of peptostreptococci are sensitive (as assayed by antibiotic disk diffusion on a seeded blood agar plate), the peptococci that are resistant have been transferred now to the genus *Peptostreptococcus* (see Taxonomic Comments) and this test is not useful to separate the two genera.

Taxonomic Comments

Five species of *Peptostreptococcus* were listed in the eighth edition of the *Manual.* Three of these species (*P. anaerobius, P. micros* and *P. productus*) are retained in this edition. *Peptostreptococcus parvulus* was transferred to the genus *Streptococcus* (Cato, 1983). "*Peptostreptococcus lanceolatus*" was not included in the 1980 Approved Lists of Bacterial Names (Skerman et al., 1980) because no strain with characteristics described for the species was extant.

Six additional species are included in *Peptostreptococcus* in this edition of the *Manual.* Ezaki et al. (1983) proposed the transfer of *Peptococcus asaccharolyticus, P. magnus, P. indolicus* and *P. prevotii* to the genus *Peptostreptococcus* because the mol% G + C of the DNA of these species is 28–34, similar to the 33 mol% G + C of the DNA of *Peptostreptococcus anaerobius,* the type species of *Peptostreptococcus,* and unlike the 50 mol% G + C of the DNA of *Peptococcus niger,* the type species of *Peptococcus.*

We also have included in *Peptostreptococcus* the species described by Lanigan (1976) as *Peptococcus heliotrinreducens* (corrig.) because the mol% G + C of the DNA of this species is 35–37 (Lanigan, 1976).

Ezaki et al. (1983) proposed the name *Peptostreptococcus tetradius* for those strains with characteristics originally described for "*Gaffkya anaerobia*" (Choukévitch 1911) Prévot 1933.

Acknowledgements

We gratefully acknowledge the microbiological or technical assistance of Loretta P. Albert, Pauletta C. Atkins, Ella W. Beaver, Ruth Z. Beyer, Maeve N. Crowgey, Ann P. Donnelly, Jackie C. Eudaly, Luba S. Fabrycky, Barbara A. Harich, Donald E. Hash, Linda R. Hoffman, Carolyn L. Hubbard, Jane L. Hungate, Daniel M. Linn, June M. McElwee, Ann C. Ridpath, Carolyn W. Salmon, Gail S. Selph, Debra B. Sinsabaugh, Sue C. Smith, Christina J. Spittle, Susan E. Stevens, Barbara C. Thompson, Dianne M. Wall, Catherine A. Waters and Bonnie M. Williams; the support service assistance of Eleanor A.

Johnston, Julia J. Hylton, Ruth E. McCoy, Claudine P. Saville, Phyllis V. Sparks and Margaret L. Vaught; the secretarial assistance of Donya B. Stephens and Mary P. Harvey, whose work has contributed to our taxonomic studies since 1974.

We are especially grateful to Elizabeth P. Cato for collaborative assistance in the taxonomic studies, and to Thomas O. MacAdoo, Department of Foreign Languages and Literature, for advice concerning the etymology of specific epithets.

For support of research for studies of many of these species, we appreciate the assistance of project 2022820 from the Commonwealth of Virginia, contracts NO1-CP-33334 from the National Cancer Institute and 9-12601 from the National Aeronautics and Space Administration, and Public Health Service Grants DE-05054 and DE-05139 from the National Institute of Dental Research, and AI-15244 from the Institute of Allergy and Infectious Disease.

Further Reading

Incidence in infections

Finegold, S.M. 1977. Anaerobic Bacteria in Human Disease. Academic Press, New York.

Chapel, T., W.J. Brown, C. Jeffries and J.A. Steward. 1978. The microbiological flora of penile ulcerations. J. Infect. Dis. *137*: 50–56.
Petrini, B., T. Welin-Berger and C.E. Nord. 1979. Anaerobic bacteria in late infections following orthopedic surgery. Med. Microbiol. Immunol. *167*: 155–159.

Other

Weiss, N. 1981. Cell wall structure of anaerobic cocci. Rev. Inst. Pasteur Lyon *14*: 53–59.
Wong, M., A. Catena and W.K. Hadley. 1980. Antigenic relationships and rapid identification of *Peptostreptococcus* species. J. Clin. Microbiol. *11*: 515–521.

General

McClung, L.S. 1982. The Anaerobic Bacteria, Their Activities in Nature and Disease. Marcel Dekker, New York. Part I, Vols. 1–4; Part II, Vols. 1 and 2.
Mitsuoka, T. 1980. The World of Intestinal Bacteria—the Isolation and Identification of Anaerobic Bacteria; A Color Atlas of Anaerobic Bacteria (English title given). Sobunsha (Sobun Press) (Hisamatsu Building, 1-4-5 Enraku-cho, Kanada, Chiyoda-ku), Tokyo.

Differentiation of the species of the genus **Peptostreptococcus**

Characteristics by which species in the genus can be differentiated are given in the key to the species in the genus and in Table 12.36. Additional characteristics are given in the text concerning each species.

Unless otherwise specified, colony descriptions are from blood agar plates or streak-tube cultures incubated anaerobically for 2 days. The drawings of each species, given in the text, are composites from different cultures of the type and other strains of the species.

Lecithinase and lipase are not produced on McClung-Toabe egg yolk agar (Holdeman et al., 1977).

Nonsaccharoclastic species do not ferment amygdalin, arabinose, cellobiose, erythritol, esculin, fructose, glucose, glycogen, inositol, lactose, maltose, mannitol, mannose, melezitose, melibiose, raffinose, rhamnose, ribose, salicin, sorbitol, starch, sucrose, trehalose or xylose.

Unless otherwise specified, gelatin is not liquefied and milk and meat are not digested.

Table 12.36.

Some differential characteristics of **Peptostreptococcus** *species*[a]

Characteristics	1. P. anaerobius	2. P. asaccharolyticus	3. P. heliotrinreducens	4. P. indolicus	5. P. magnus	6. P. micros	7. P. prevotii	8. P. productus	9. P. tetradius
Butyrate produced	−, tr	+	d	+	−	−	+	−	+
Succinate from fumarate	NT	−	+	−	−	−	−	NT	NT
Propionate from lactate	−	−	−	+	−	−	−	−	−
Indole	−	+	−	+	−	−	−	−	−
Nitrate	−	−	−	+[b]	−	−	−	−	−
Coagulase	−	−	NT	+[b]	−	−	−	NT	−
Urease	−	−	−	−	−	−	−, +	d	+
Catalase	−	v	−	−	+, −	−	v	−	v
Esculin hydrolysis	−	−	−	−	−	−	−, w	+	−, w
Acid from									
Glucose	w, −	−	−	−	−, w	−	w	+	+
Lactose	−	−	−	−	−	−	−	+	−
Maltose	w, −	−	−	−	−	−	w, −	+	+
Mannose	−	−	−	−	−	−	v	+	+
Sucrose	−, w	−	−	−	−	−	−, w	+	+
Ammonia from									
Glutamate	−	+	NT	+	−	−	−	NT	−
Glycine	−	−	NT	−	+	w	−	NT	−
Production of									
Alkaline phosphatase	−	−	NT	+	− (d)	+	+	NT	−
β-Glucuronidase	−	−	NT	−	−	−	−	NT	+
α-Glucosidase	+	−	NT	−	−	−	−	NT	w
Mol% G + C	33–34	31–32	35–37	32–34	32–34	27–28	29–33	44–45	30–32

[a] −, negative reaction; tr, trace; +, positive reaction; d, different results in different laboratories; NT, not tested; v, variable among strains; and w, weak reaction. When two reactions are given, the first is the usual reaction.
[b] An occasional strain may be negative.

Key for differentiating species in the genus **Peptostreptococcus**

I. Branched chain fatty acids are produced.
                    1. *P. anaerobius*
II. Branched chain fatty acids are not produced.
  A. Fumarate is reduced to succinate.
                    3. *P. heliotrinreducens*
  B. Fumarate is not reduced to succinate.
    1. Butyrate is produced.
        a. Indole is produced.
            b. Lactate is converted to propionate.
                  4. *P. indolicus*
           bb. Lactate is not converted to propionate.
                  2. *P. asaccharolyticus*
        aa. Indole is not produced.
            b. Urease is produced, carbodydrates are fermented.
                  9. *P. tetradius*
           bb. Urease is not produced, carbohydrates usually not fermented strongly.
                  7. *P. prevotii*
    2. Butyrate is not produced.
        a. Carbohydrates are fermented.
                  8. *P. productus*
        aa. Carbohydrates are not fermented.
            b. Alkaline phosphatase is produced.
                  6. *P. micros*
           bb. Alkaline phosphatase is not produced.
                  5. *P. magnus*

List of species of the genus **Peptostreptococcus**

1. **Peptostreptococcus anaerobius** (Natvig 1905) Kluyver and van Niel 1936, 401.[AL] (*Streptococcus anaerobius* Natvig 1905, 724.)

an.a.e.ro′bi.us. Gr. pref. *an* not; Gr. n. *aër* air; Gr. n. *bius* life; N.L. adj. *anaerobius* not living in air, anaerobic.

Types species of the genus *Peptostreptococcus*.

Unless otherwise cited, this description is from the study of the type strain and over 100 similar strains.

Cells are 0.5–0.6 μm in diameter and occur in pairs and chains. Cells of young cultures in particular may be somewhat elongate.

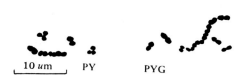

Figure 12.15. *P. anaerobius*

Surface colonies on blood agar plates are minute to 1.0 mm, circular, entire, convex, opaque, white, shiny, smooth and nonhemolytic on rabbit blood. Colonies are gray on supplemented brain heart infusion agar (Holdeman et al., 1977) without blood.

PY-glucose broth cultures have a granular to viscous sediment and usually are not turbid. The sediment often is gray in media-containing cysteine hydrochloride.

Growth often is stimulated by oleate (0.02% Tween 80). Growth is inhibited by 0.1% sodium polyanetholsulfonate (Graves et al., 1974; Kocka et al., 1974; Wideman et al., 1976); 1.2% gelatin overcomes the inhibitory effect (Wilkins and West, 1976). Growth is not inhibited by sodium amylosulfate at concentrations of 0.05 or 0.5% (Kocka et al., 1974).

The optimum temperature for growth is 37°C. Cultures grow less well at 25°C and 30°C and rarely grow at 45°C.

Moderate to copious gas is produced in glucose agar deep cultures. Ammonia is produced from peptone. Hydrogen sulfide is produced in SIM (sulfide-indole-motility (BBL)) medium.

Tyrosine is degraded (Babcock, 1979). Lactulose is utilized (Sahota et al., 1982). Extracellular DNase is produced (Porschen and Sonntag, 1974); production of DNase and RNase is variable among strains (Marshall and Kaufman, 1981). Reaction for alkaline phosphatase is negative or weak (Porschen and Spaulding, 1974). No superoxide dismutase, catalase or peroxidase activities have been detected in strain VPI 4330-1 (Frölander and Carlsson, 1977; Rolfe et al., 1978).

A few carbohydrates may be fermented weakly (Table 12.36).

Acid products (milliequivalents (meq) per 100 ml of culture) in PY-glucose cultures of the type strain are: acetic (1.0–1.5), isobutyric (0.05), butyric (0.05–0.15), isovaleric (0.05) and isocaproic (0.4–0.5). Similar products are produced by the other strains tested. Isovaleric acid is not detected in occasional strains and only about two-thirds of the strains produce isocaproic acid. More of the branched chain acids (isobutyric, isovaleric and isocaproic) may be present in cultures grown in media containing more than 1% peptone. Although only trace amounts of isobutyric, butyric, isovaleric and isocaproic acids may be detected, their presence is significant and a presumptive identification of *P. anaerobius* can be made when these products are detected in a culture of an anaerobic Gram-positive coccus isolated from a human clinical specimen. Abundant H_2 is produced. Lactate is not utilized. Threonine is converted to moderate amounts of propionate.

Pyruvate is metabolized principally to acetate and ethanol; leucine is metabolized to 4-methylvalerate under anaerobic conditions and to α-ketoisocaproate under aerobic conditions. Under anaerobic conditions, CO_2 and H_2 are produced from pyruvate and CO_2 is produced from glucose (Hoshino et al., 1978).

Resting cell suspensions convert leucine to isovaleric and isocaproic acids, in the ratio of 1 isovaleric to 2 isocaproic, via the Stickland reaction. The reaction is stimulted by glucose (Britz and Wilkinson, 1982).

In addition to these products, Lambert and Moss (1980) report that benzoic, hydrocinnamic and p-hydroxyhydrocinnamic acids are produced from growth in Schaedler broth. If 0.05% tyrosine is added to Schaedler broth, an increased amount of p-hydroxyhydrocinnamic acid is produced.

The type strain produces butyrate (C4) esterase, some (weak) caprylate (C8) esterase lipase, and no leucine arylamidase or β-galactosidase in the API ZYM tests (Ezaki et al., 1983).

Crude cell-free extracts have NADH-oxidase and NADPH-oxidase activities. The specific activity of NADH-oxidase, which reduces O_2 to water, is much higher than the specific activity of NADPH-oxidase, which reduces O_2 to superoxide radicals and hydrogen peroxide (Hoshino et al., 1978).

Major cellular fatty acids are n-C8:0, i-C10:0, n-C10:0, i-C12, n-C12:0, i-C13, i-C14, n-C14:0, i-C15, i-C16, and n-C16:0, usually with n-C18:0 and i-C18:1 (Wells and Field, 1976; Moss et al., 1977; Lambert and Armfield, 1979, Figure 4; Ezaki et al., 1983).

Soluble antigens are species-specific (Graham and Falkler, 1979).

All of 35 strains tested were susceptible to 1.6 μg of clindamycin/ml; one was resistant to 12 μg of chloramphenicol/ml; two were resistant to 3 μg of erythromycin/ml; four were resistant to 2 units of penicillin/ml and a different four were resistant to 6 μg of tetracycline/ml. Chow et al. (1975) report that 88% of 51 strains were susceptible to 2.5 μg of minocycline/ml, whereas only 69% of these strains were susceptible to 6.25 μg of tetracycline/ml.

Other characteristics of the species are given in Table 12.36.

Isolated from a wide variety of human clinical specimens including abscesses of the brain, jaw, pleural cavity, ear, pelvic region, urogenital area and abdominal region, as well as from blood, spinal fluid, joint cultures and specimens from osteomyelitis. The species often is part of the flora of the gingival crevice of persons with gingivitis or periodontal diseases and also has been reported from clinical specimens from animals. The species is part of the normal flora of the vagina. Although it is not part of the normal flora of the healthy gingival crevice and is not predominant in fecal flora, the mouth, intestinal tract and genital tract represent probable sources of infecting organisms in many of the clinical specimens.

Also isolated from clinical specimens from dogs and cats (Berg et al., 1979), usually in association with other bacteria. One strain was isolated in pure culture from a brain abscess in a dog. Isolated from subcutaneous abscesses in cats following cat fights (Love et al., 1979), empyema in cats (Love et al., 1982) and subgingival plaque from *Macaca arctoides* monkeys (Mashimo et al., 1979).

The mol% G + C of the DNA is 33–34 (Ezaki and Yabuuchi, 1983).

Type strain: ATCC 27337 (CIPP (Prévot) 4372).

Further reading: Thadepalli, H., M.D. Appleman, J.E. Maidman, J.J. Arce and E.C. Davidson, Jr. 1981. Antimicrobial effect of amniotic fluid against anaerobic bacteria. Am. J. Obstet. Gynecol. *127*: 250–254.

2. **Peptostreptococcus asaccharolyticus** (Distaso 1912) Ezaki, Yamamoto, Ninomiya, Suzuki and Yabuuchi 1983, 690.[VP] (*Peptococcus asaccharolyticus* (Distaso 1912) Douglas 1957.[AL])

a.sac.cha.ro.ly'.ti.cus. Gr. pref. *a* not; Gr. n. *sacchar* sugar; Gr. adj. *lyticus* able to loosen; N.L. adj. *asaccharolyticus* not digesting sugar.

Unless otherwise specified, the description is based upon that of Ezaki et al. (1983), who studied seven strains with 97–107% DNA homology with the type strain, and our study of the type strain.

Cells from 24-h-old PY-glucose broth cultures are 0.5 to 1.6 μm in diameter and are arranged in pairs, tetrads, or in irregular clumps. Cells in other cultures may stain Gram-negative.

Figure 12.16. *P. asaccharolyticus*

The peptidoglycan type of the type strain is L-ornithine-D-glutamic acid. In addition, the cell walls contain alanine, muramic acid and glucosamine (Schleifer and Nimmermann, 1973; Huss et al., 1982). After incubation for 48–72 h, surface colonies are minute to 2.0 mm, circular with entire margins, low convex, translucent to opaque, white to buff, slightly shiny and smooth. There is no action on sheep or rabbit blood.

Glucose broth cultures are turbid with a smooth sediment. Growth is stimulated slightly by oleate (0.02% Tween 80). Bile (2% oxgall) and 6.5% NaCl completely inhibit growth.

The optimum temperature for growth is 37°C. Slight growth may occur at 25°C and 30°C.

Nitrate is not reduced (Ezaki et al., 1983); nitrate may be reduced (our laboratory). A small amount of catalase activity may be detected from some strains. Most strains produce ammonia from glutamate, threonine and serine, but not from glycine. Tests for C4 esterase, C8 esterase lipase, leucine arylamidase and β-galactosidase are negative (API ZYM tests).

Ten human clinical strains identified as *P. asaccharolyticus* did not produce coagulase or hemolyze sheep erythrocytes; reactions for DNase and RNase were variable among strains (Marshall and Kaufman, 1981).

Carbohydrates are not fermented (Table 12.36).

Acid products of the type strain detected in PY or PY-glucose cultures (in meq/100 ml of culture) are acetic (2.6), butyric (1.6) and propionic (0.3). Copious H_2 is produced. Pyruvate is metabolized to acetate and butyrate. Lactate is not utilized. A moderate (0.3–0.5 meq/100 ml of culture) amount of propionate may be formed from threonine. Ezaki et al. (1983) report that major amounts of butyrate are produced by all eight strains tested, seven strains produced major amounts of acetate, and an occasional strain produced small amounts of propionate or lactate from PY-glucose broth cultures.

Cell suspensions of the type strain metabolize purines and pyrimidines primarily to acetic and lactic acids, CO_2, and ammonia (Whiteley, 1952; Vogels and Van der Drift, 1976).

C18:1 fatty acid and C18:1 aldehyde are major cellular components (Ezaki et al., 1983).

Oligonucleotide sequence of the 16S ribosomal RNA of the type strain (referred to *Peptococcus aerogenes*) is given in Tanner et al. (1982).

The type strain is susceptible to 2 units of penicillin/ml, 12 μg of chloramphenicol/ml, 1.6 μg of clindamycin/ml, 6 μg of tetracycline/ml, and 3 μg of erythromycin/ml.

Other characteristics of the species are given in Table 12.36.

The source of the type strain is unknown. Strains homologous with the type strain have been isolated from vaginal discharge, skin abscess and peritoneal abscess. Strains phenotypically similar to the type strain have been isolated from various types of clinical specimens. Russell (1979) reported that they constitute 8.3% of the microbial flora of the hog intestinal tract.

The mol% G + C of the DNA of the type strain is 31–32 (Huss et al., 1982; Ezaki et al., 1983).

Type strain: ATCC 14963 (NCIB 10074; DSM 20463).

Further comments. ATCC 14963, the type strain of *P. asaccharolyticus* (Skerman et al., 1980) originally was deposited in ATCC as a representative of *Peptococcus aerogenes*.

There is considerable genetic diversity among strains that generally are recognized (phenotypically) as *P. asaccharolyticus*. Huss et al. (1982) report that four clinical isolates identified as *P. asaccharolyticus* have no DNA homology with ATCC 14963 (type strain) but have 70–78% homology with DSM 20364, which was labeled *P. asaccharolyticus*. There was no DNA homology between DSM 20364 and the type strain of *P. asaccharolyticus*. Also, 14 strains (mostly from human clinical specimens) that we had identified as *P. asaccharolyticus* had negligible DNA homology with the type strain (K. Ninomiya and J. L. Johnson, unpublished data).

Ezaki et al. (1983) studied seven strains phenotypically similar to *P.*

asaccharolyticus that had only moderate (32–67%) DNA homology with the type strain. These strains also were mostly from vaginal discharge. Because no tested phenotypic reaction clearly differentiated these strains (their homology group 2) from those with high DNA homology with the type strain, these authors did not propose a new species for the strains in their homology group 2.

Because of the reported genetic diversity among phenotypically similar strains, reports of isolation and incidence of *P. asaccharolyticus* in present literature should be interpreted cautiously.

3. **Peptostreptococcus heliotrinreducens** (corrig.) (Lanigan 1976) comb. nov. (*Peptococcus helotrinreducans* (sic) Lanigan 1976, 7*)

he.li.o.trin.re.du′cens. Chemical term *heliotrine*, a pyrrolizidine alkaloid; L. part. adj. *reducens* reducing; *heliotrinreducens* reducing heliotrine, referring to the organism's ability to bring about reductive cleavage of the heliotrine molecule.

The description is principally from Lanigan (1976), supplemented with results from study of the type strain.

Cells are not encapsulated and are 0.5–0.6 μm in diameter in stained smears and up to 0.7 μm in diameter in wet preparations. They occur in pairs, in small clusters, and occasionally in chains of 3–6 cells.

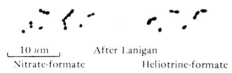

10 μm   After Lanigan
Nitrate-formate   Heliotrine-formate

Figure 12.17. *P. heliotrinreducens*

Subsurface colonies in tryptone-yeast-mineral salts (TYM) agar are lenticular with an entire margin and smooth surface. After incubation for 7–10 days, they are 0.6–0.8 mm in diameter, grayish-white, and translucent. Surface colonies are 1–2 mm in diameter, effuse, colorless, transparent, and have an an entire edge. A few large colonies show varying amounts of radial striation and crenation of the edges. Confluent growth on agar is glistening, beaded, colorless, transparent and has a slightly raised confluent margin. No growth occurs on blood agar.

Cultures in TYM broth are lightly turbid initially, after which a highly viscous sediment that is difficult to disperse is formed. Growth is maximal in 24–30 h.

Cultures do not survive in media in which resazurin is oxidized.

The temperature for optimal growth is between 38°C and 42°C. All strains grow at temperatures between 30°C and 46°C. Maximal growth occurs between pH 6.5 and 7.0 and cultures grow reasonably well at pH 6.2 and 7.2. No growth occurs below pH 5.4 or above pH 8.2. Strains grow poorly in TYM medium with 1% NaCl and not at all in TYM with 2% NaCl. Growth is not affected by the addition of 1% (v/v) serum or 1% (v/v) Tween 80. Arginine (10–25 mM) stimulates growth in TYM medium; glycine (40 mM) is inhibitory. Addition of alanine, glutamate, histidine, ornithine, proline, serine, threonine and combinations of alanine plus ornithine or alanine plus proline do not affect growth.

Ammonia is formed from tryptone, yeast extract, adenine, uracil and arginine. Nitrates are completely reduced to ammonia if sufficient electron donor (H_2 or formate) is present.

Sulfates are not reduced. Creatinine is not hydrolyzed. No gas is formed in TYM-agar medium.

Very small amounts of acetate, propionate and butyrate, sometimes with isovalerate, are produced in TYM cultures. Greater amounts of isovalerate are produced by cultures in TYM-arginine medium (Lanigan, 1976). Only a trace of acetate is detected in PY-glucose culture with the chromatographic procedures recommended by Holdeman et

al. (1977) but a small amount of butyrate can be detected if 4 ml of culture is extracted and 28 μl of this ether extract is chromatographed.

Small amounts of H_2 and CO_2 are produced. Fumarate is reduced to succinate if formate or H_2 is present. Lactate, malate and pyruvate are not metabolized.

Pyrrolizidine alkaloids (heliotrine, europine, heleurine, supinine and lasiocarpine) are reduced to 1-methylene derivatives in the presence of H_2 or formate (electron donors). Some macrocyclic di-esters are metabolized, but more slowly than the mono-esters. Heliotridine, anacrotine, retrorsine, cynaustraline and sarracine are not metabolized.

A c-type cytochrome is present in ultrasonic extracts of cells.

Growth is inhibited in media containing 10 units/ml penicillin.

Other characteristics of the species are given in Table 12.36.

Isolated from the rumen of sheep maintained on various dry rations (Australia) and from a rumen-fistulated sheep fed alfalfa pellets (Davis, CA). Procedures for isolation are given in Further Comments.

The mol% G + C of the DNA is 35–37 (Bd).

Type strain: ATCC 29202 (NCTC 11029; DSM 20476).

Further comments. To enrich for the organism, heliotrine (2 mg/ml^{-1}) and chloral hydrate (0.2 mg/ml^{-1}) were added to sheep rumen contents that had been strained through gauze. The flask was gassed with H_2-CO_2 (4:1), sealed with a rubber stopper and incubated at 39°C with constant shaking until almost all of the heliotrine had been metabolized (24–30 h). Successive subcultures were made in a medium containing 30% rumen fluid, 0.2% heliotrine and salts (Lanigan, 1976) until, with 5% inoculum, the time for utilization of the heliotrine became minimal (about 16 h). Dilutions of this last enrichment were cultured in roll tubes in medium containing 30% rumen fluid, 0.1% heliotrine, salts and 1% (w/v) ionagar (Oxoid no. 2). After incubation for 7–8 days at 38°C, colonies resembling those of *P. heliotrinreducens* were selected and subcultured (Lanigan, 1976).

Methods for assay of the alkaloids and their metabolites are given in Lanigan and Smith (1970).

4. **Peptostreptococcus indolicus** (Christiansen 1934) Ezaki, Yamamoto, Ninomiya, Suzuki and Yabuuchi 1983, 692.[VP] (*Micrococcus indolicus* Christiansen 1934a, 366; *Peptococcus indolicus* (Christiansen 1934) Høi Sorensen 1975.)

in.do′li.cus. Chemical term *indole*; L. suffix *-icus* related to; *indolicus* related to indole, referring to the ability of the organism to produce indole.

The description is from Høi Sorensen (1973, 1975), Schwan (1979), Ezaki et al. (1983), and our study of the type and four other strains.

Cells are 0.5–0.6 μm in diameter and occur singly or in pairs, short chains, tetrads, or small clusters.

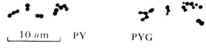

10 μm   PY   PYG

Figure 12.18. *P. indolicus*

The peptidoglycan type of most strains is L-Orn-D-Glu (see discussion of serovars below) (Huss et al., 1982).

Surface colonies on blood agar plates are 0.5–1.0 mm in diameter, grayish to yellow with glistening surface, circular with an entire margin, and convex to slightly pulvinate. On freshly prepared plating medium, colonies are butyrous or viscous; on stored media they may be friable (Høi Sorensen, 1973). Although most strains are nonhemolytic on calf's blood, colonies of some strains are surrounded by small zones of almost complete hemolysis (α-hemolysis) and colonies of other strains are surrounded by broad zones of incomplete hemolysis (Høi Sorensen, 1975). Colonies in agar are 1–2 mm, lenticular, and grayish-yellow after

---

* **Editorial note:** the species name *Peptococcus heliotrinreducans* Lanigan 1976 was validated and corrected in Validation List No. 11, Int. J. Syst. Bacteriol. *33:* 673, as an erratum to a previous validation list. The species is here being transferred to the genus *Peptostreptococcus*.

incubation for 4–5 days. Gas that disrupts the agar is apparent in 1–3 days.

Broth cultures are turbid with a viscous, stringy or granular sediment.

The optimum temperature for growth is 37°C; little if any growth occurs at 22°C or 45°C. Growth is stimulated by 10% CO_2 in the gaseous atmosphere. Oleate (0.02% Tween 80) has little effect on growth. Growth is slightly inhibited by 0.02% ferrous sulfate and is markedly inhibited by 0.03% sodium thiosulfate. There is no growth in citrate medium (Høi Sorensen, 1973).

Coagulase (peptocoagulase) for rabbit plasma is produced by most strains (267 of 274; Høi Sorensen, 1973); 263 of these 267 strains also coagulated calf plasma. The peptocoagulase is cell-associated in all 75 strains studied and present in culture filtrates of 93% of these strains (Switalski et al., 1978).

Hydrogen sulfide is produced in SIM and lead acetate strips suspended over chopped meat cultures are darkened. Most strains reduce nitrate to nitrite and will reduce the nitrite upon continued incubation. All 79 strains studied by Schwan (1979) reduce nitrate.

Alkaline and acid phosphatases are produced (API-ZYM tests), 88% of the strains weakly hydrolyze ribonucleic acid and 6 of 79 strains produce DNase. Glucosidases and trypsin, chymotrypsin, lipase and valine amino peptidases are not produced. Schwan (1979) reported that butyrate esterase is produced (API-ZYM tests) and that weak reactions are obtained for leucine and cystine aminopeptidases and caprylate esterase-lipase. Ezaki et al. (1983) reported that the type and three other strains do not produce butyrate (C4) esterase, caprylate (C8) esterase lipase or β-galactosidase (API-ZYM tests).

Carbohydrates are not fermented (Table 12.36).

Products in PY-glucose cultures (meq/100 ml of culture) are acetic (1–2), butyric (1) and propionic (0.3–0.5) acids. Abundant H_2 is produced. Products in PY-lactate cultures are acetate (4–5), propionate (5–6), and butyrate (1); products in PY-pyruvate cultures are acetate (3–5), butyrate (1.5–2) and propionate (0.3–0.5).

The major cellular fatty acid is C18:1 (Ezaki et al., 1983).

Høi Sorensen (1973) differentiated 217 bovine strains into 6 serovars (A–F) by double-diffusion in agar; most strains belonged to serovars B, C, D or E. Three of nine porcine strains failed to react with any of the six antisera available. Schwan (1979) also found that the majority (72 of 79) of the strains tested belong to serovars B, C, D or E. Huss et al. (1982) report that the peptidoglycan type of 22 of 26 strains (of serovars A–F) tested is L-Orn-D-Glu. The peptidoglycan type of two of seven serovar C strains is L-Lys-L-Thr-Gly; that of one of four serovar E strains and one of two serovar F strains is L-Lys-L-Ala. These strains with the unusual peptidoglycan type have not been included in DNA homology studies (Huss et al., 1982).

All of 20 strains tested (Rosco Sensitabs on seeded blood agar plates incubated anaerobically for 48 h) were susceptible to penicillin, bacitracin, tetracycline and chloramphenicol and relatively resistant to streptomycin; 15 were relatively resistant to neomycin and polymyxin B (Høi Sorensen, 1973).

There is little, if any, pathogenicity for mice or guinea pigs inoculated intraperitoneally (Høi Sorensen, 1973).

Other characteristics of the species are given in Table 12.36.

Isolated from summer mastitis secretions of cattle, tonsils and various mucous membranes of apparently healthy cattle, clinically healthy swine, insects (*Hydrotaea irritans* (Fallen) (sheep head fly), and species of *Simulian* (black fly) and *Culicoides*) (Høi Sorensen, 1973, 1974, 1976) and skin lesion of a sheepherder (Bourgault and Rosenblatt, 1979).

The mol% G + C of the DNA of the type strain is 32–34 by T_m (Huss et al., 1982; Ezaki et al., 1983).

Type strain: ATCC 29427 (DSM 20464; CCM 5987).

Further comments. In older literature, the species was recognized as *Micrococcus indolicus* (Christiansen, 1934a) and *Staphylococcus asaccharolyticus* var. *indolicus* (Prévot et al., 1967).

P. indolicus is most easily differentiated from *P. asaccharolyticus*, which it most closely resembles, by the ability of *P. indolicus* to convert lactate to large amounts of propionate and acetate and by the coagulase produced by most strains of *P. indolicus*. Schwan (1979) reports that

reactions for alkaline phosphatase (+), nitrate reduction (+), and ribonuclease (usually +) also helps to differentiate *P. indolicus* from strains with biochemical characteristics of *P. asaccharolyticus*.

5. **Peptostreptococcus magnus** (*ex* Smith 1957) (Prévot 1933) Ezaki, Yamamoto, Ninomiya, Suzuki and Yobuuchi 1983, 696.[VP] (*Diplococcus magnus* Prévot 1933, 140; *Peptococcus magnus* (Prévot 1933) Holdeman and Moore 1972.)

mag'nus. L. adj. *magnus* large.

This description is based on study of the type and approximately 100 similar strains.

Cells of the type strain grown in PY-glucose + 0.02% Tween 80 are 0.7–1.2 μm in diameter and occur singly and in pairs, tetrads and clusters.

Figure 12.19. *P. magnus*

Surface colonies on blood agar plates incubated anaerobically for 48 h are minute to 0.5 mm, circular with an entire margin, raised, dull and smooth. There is no action on rabbit blood.

In the cell walls, the glycine cross-linkage is between the ϵ-amino group of lysine and the carboxyl group of D-alanine (Schleifer and Nimmerman, 1973).

Glucose broth cultures are turbid with a smooth to granular sediment.

Growth of most strains, especially original isolates, is stimulated by oleate (0.02% Tween 80). The optimum temperature for growth is 37°C. Many strains produce limited growth in broth at temperatures ranging from 25–45°C.

Growth occurs in anaerobic media at an Eh of −50 to +350 mv but not in aerated broth at an Eh of approximately −50 mv (Walden and Hentges, 1975).

Trace to moderate amounts of acid phosphatase but no alkaline phosphatase are detected (API-ZYM tests) (Porschen and Spaulding, 1974; Cato et al., 1983). Butyrate (C4) esterase and leucine arylamidase are produced (API-ZYM tests); most strains produce caprylate (C8) esterase lipase; no strain produces β-galactosidase (Ezaki et al., 1983). Catalase is produced by most strains. Ammonia is produced from glycine but not from glutamine; urease is not produced (Ezaki et al., 1983). Hippurate is hydrolyzed by the type and many other strains. Ammonia is produced from peptone.

None of 21 strains tested produced DNase, RNase, coagulase, or hemolysins for sheep erythrocytes (Marshall and Kaufman, 1981).

Fructose or glucose may be weakly fermented (Ezaki et al., 1983). The pH is seldom lower than pH 6.0 in PY-carbohydrate cultures (our laboratories), weak acid production probably is neutralized by ammonia produced from peptone. About two-thirds of the strains hydrolyze gelatin, particularly in media containing Tween 80.

Gas bubbles may accumulate in glucose deep agar cultures.

The major product in PY-glucose cultures is acetic acid (1.5–2.5 meq/ 100 ml of culture); trace to moderate amounts of lactic and/or succinic acids also may be present. Small amounts of H_2 are detected in headspace gas of cultures of the type strain and about one-half of the other strains. The type strain converts glycine to acetate, ammonia and CO_2 (Cato et al., 1983). The type strain does not metabolize pyruvate, lactate or fumarate.

The major cellular fatty acid is C18:1; trace to moderate amounts of C16:0 acid and C18:1 aldehyde also are present.

All strains tested are susceptible to 2 units of penicillin/ml and 12 μg chloramphenicol/ml. Many strains are resistant to 6 μg tetracycline/ ml and 3 μg erythromycin/ml. Only one of more than 100 strains tested was resistant to 1.6 μg of clindamycin/ml (unpublished data).

Isolated from human clinical specimens, principally wounds and abscesses of abdominal, peritoneal, appendiceal and urogenital sites.

Less than 20% of the strains are from anatomical sites above the diaphragm. The principal natural habitat probably is the urogenital tract. We have one isolate (among 8,000 randomly picked colonies) from human feces, two (among 12,000 randomly picked colonies) from human gingival crevice and one from bovine rumen. Russell (1979) reported that *P. magnus* is the ninth most numerous organism in hog intestinal flora (2.6% of the flora).

The mol% G + C of the DNA is 32 by T_m for the type strain and 32–34 for 22 other strains examined (Ezaki et al., 1983).

Type strain: ATTC 15794 (DSM 20470).

Further comments. The inability of *P. magnus* to produce alkaline phosphatase is helpful to differentiate *P. magnus* from *Peptostreptococcus micros*, the species it most closely resembles phenotypically. Cato et al. (1983) also have demonstrated that the polyacrylamide gel electrophoresis (PAGE) patterns of the soluble proteins of the two species are different. (Also see Further Comments, *P. micros*.)

Strains of "*Peptococcus variabilis*" and *P. magnus* are indistinguishable by usual phenotypic tests. Therefore, West and Holdeman (1973) proposed that "*Peptococcus variabilis*" (Foubert and Douglas) Douglas 1957 was a later synonym of *P. magnus*, and "*P. variabilis*" was not included in the Approved Lists of Bacterial Names (Skerman et al., 1980). Two strains (Foubert and Douglas BU and U3) on which the original description of "*P. variabilis*" was based are available from the American Type Culture Collection. Cato et al. (1983) report that PAGE patterns of soluble proteins of strain U3, ATCC 14956 (labeled "*P. variabilis*"), and ATCC 15794 (type strain of *P. magnus*) are identical. However, the PAGE pattern of the designated type strain of "*P. variabilis*" (strain BU, ATCC 14955) is different from the pattern of the type strain of *P. magnus*. Pending additional studies, these authors made no recommendation concerning the taxonomic status of strain BU or of "*P. variabilis*." The cell walls of ATCC 14955 contain glucose, galactose, D-glutamic acid, L-lysine, glycine, D-alanine, muramic acid and glucosamine (Schleifer and Nimmermann, 1973).

In the eighth edition of the *Manual*, *P. magnus* was given as a synonym of "*Peptococcus anaerobius*" (Rogosa, 1974). West and Holdeman (1973) requested that the name "*Peptococcus anaerobius*" be placed on the list of nomina rejicienda because (a) the original description very probably represented several species of anaerobic cocci and (b) the confusion caused by two species of anaerobic cocci with names "*Peptococcus anaerobius*" and *Peptostreptococcus anaerobius* is directly contrary to Principle I of the International Code of Nomenclature and could be a potential medical hazard. The Judicial Commission (1979, 1982) subsequently rejected the name "*Peptococcus anaerobius*."

6. **Peptostreptococcus micros** (Prévot 1933) Smith 1957, 537.[AL]
(*Streptococcus micros* Prévot 1933, 195.)

mi′cros. Gr. adj. *mikros* small.

This description is based on Cato et al. (1983) and study of the type strain and about 100 strains with similar phenotypic characteristics and PAGE patterns of the soluble cellular proteins.

Cells are 0.3–0.7 μm in diameter and typically occur in pairs and chains of 6–20 elements.

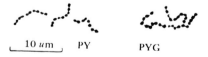

10 μm   PY   PYG

Figure 12.20. *P. micros*

Surface colonies on anaerobic rabbit blood agar plates are nonhemolytic, minute to 1.0 mm in diameter, circular, entire, convex, smooth, shiny and opaque white to translucent gray. Colonies in enriched brain heart infusion agar (Moore et al., 1982) are 0.5–2.0 mm in diameter, lenticular to dish shaped, and translucent.

Peptone-yeast extract and PY-glucose broth cultures have light turbidity and small amounts of smooth, ropy or flaky sediment. Broth growth seldom exceeds 2+ density on a scale of 0–4+. Carbohydrates and sugar alcohols are not fermented.

Growth is stimulated by 0.02% Tween 80 and by 5–10% serum. Bile (2% oxgall) usually inhibits growth. No growth occurs in PY-glucose media with 6.5% NaCl.

The optimum temperature for growth is 37°C. Less growth is produced at 30°C, and growth is minimal at 25°C or 45°C.

Acid and alkaline phosphatases are produced (Porschen and Spaulding, 1974; Cato et al., 1983). The type strain produces butyrate (C4) esterase (weak), caprylate (C8) esterase lipase, and leucine arylamidase but not β-galactosidase (API-ZYM test, Ezaki et al., 1983).

No gas is produced in glucose agar deep cultures, and little or no H_2 is detected in the headspace gas of stoppered broth cultures. Neutral red is reduced by the type and most other strains. Ammonia is produced from peptone in well-growing cultures (in media supplemented with Tween 80). Hydrogen sulfide production in SIM medium is variable among strains. Extracellular DNase is not produced (Porschen and Sonntag, 1974).

Carbohydrates are not fermented (Table 12.36).

Acid products (meq/100 ml of culture) in PY-glucose broth cultures with Tween 80 are acetic (0.3–1.8), frequently with small amounts of succinic (0.01–0.3), and occasionally with small amounts of lactic (0.01–0.3). Some strains metabolize pyruvate, principally to acetic acid; formic acid also may be detected. Lactate is not utilized. Acetic acid, ammonia and CO_2 are produced from glycine via the glycine synthase system (Vogels and Van der Drift, 1976).

The major cellular fatty acid of the type strain is C18:1, C18:1 aldehyde also is present (Ezaki et al., 1983).

Of 20 representative clinical and dental strains tested, all were susceptible to 1.6 μg of clindamycin/ml, 12 μg of chloramphenicol/ml, and 3 μg of erythromycin/ml. One strain, from a brain abscess, was resistant to 2 units of penicillin G/ml and to 6 μg of tetracycline/ml. The other 19 strains were susceptible to these two antibiotics. Chow et al. (1975) report that 67% of 27 strains tested were susceptible to 6.25 μg/ml of either minocycline or tetracycline.

The oligonucleotide catalog of the 16S rRNA of ATCC 23195, labeled *Peptococcus glycinophilus*, a later synonym of *P. micros* (Cato et al., 1983), is given by Tanner et al. (1982).

Other characteristics of the species are given in Table 12.36.

P. micros may be synergistic with other organisms to produce experimental otitis media in guinea pigs (Thore et al., 1982).

Frequently isolated from human clinical specimens including brain, lung, jaw, head, neck and bite abscesses, spinal fluid, blood and abscesses of other body sites. *Peptostreptococcus micros* is often a major component of the flora of the gingival sulcus in periodontal disease. It occurs infrequently in the gingival crevice of healthy gingiva and in the intestinal tract. It also is present in the flora of the reproductive tract. Its more frequent occurrence, alone or with other species, in infections above the diaphragm suggests that the gingival sulcus in periodontitis is its primary reservoir. Some persons have humoral antibody, detected by indirect fluorescent antibody, to *P. micros* (Graham et al., 1982).

The mol% G + C of the DNA of the type strain is 27–28 (Cato et al., 1983; Ezaki and Yabuuchi, 1983).

Type strain: VPI 5464 (ATCC 33270; Prévot 3119B; DSM 20468).

Further comments. Cato et al. (1983) report that *Peptococcus glycinophilus* (Cardon and Barker) Douglas 1957 is a later synonym of *P. micros.* DNA of the type strain of *P. glycinophilus* (ATCC 23195) is 84% homologous with DNA of the type strain of *P. micros* (ATCC 33270). PAGE patterns of soluble proteins of the two strains are identical.

Differentiation of *P. micros* from *Peptostreptococcus magnus* has been difficult. Strains of these two species are genetically unrelated and often are presumptively differentiated by cell size and arrangement (chains vs. tetrads). However, this procedure is not reliable because both size and arrangement can be similar in some strains and under some growth conditions. Other known phenotypic properties, except phosphatase and electrophoretic patterns of soluble proteins, are similar. As reported by Porschen and Spaulding (1974) and Cato et al. (1983), *P. micros* produces strong acid phosphatase and moderate alkaline phosphatase reactions, whereas *P. magnus* produces only weak or moderate acid phosphatase and no alkaline phosphatase reactions. Ezaki et al. (1983), however, report that 23 strains of *P. magnus* gave weak to strong reactions for alkaline phosphatase (API-ZYM tests) and that the type strain of *P. micros* gave a weak reaction. These authors suggest that liquefaction of gelatin and production of catalase by those strains of *P. magnus* that are positive for either of these characteristics can help differentiate between the species. Only one of their strains of *P. magnus* was negative for both gelatin and catalase. The PAGE patterns are exceedingly helpful and reliable to differentiate between the species (Cato et al., 1983). Graham et al. (1982) report that the two species are distinct serologically.

The oral flora usually is the origin of clinical isolates of *P. micros*, whereas the vagina appears to be the usual origin of clinical isolates of *P. magnus*.

Further reading. Sundqvist, G.K., M.I. Eckerbom, Å.P. Larsson, and U.T. Sjögren. 1979. Capacity of anaerobic bacteria from necrotic dental pulps to induce purulent infections. Infect. Immun. *25:* 686–693.

7. **Peptostreptococcus prevotii** (Foubert and Douglas 1948) Ezaki, Yamamoto, Ninomiya, Suzuki and Yabuuchi, 1983, 693.[VP] [*Micrococcus prevotii* Foubert and Douglas 1948, 31; *Peptococcus prevotii* (Foubert and Douglas 1948) Douglas 1957, 475.)

pre.vo′ti.i. N.L. gen. n. *prevotii* of Prévot, named after A.R. Prévot, a French microbiologist.

Unless otherwise specified, the description is based on Ezaki et al. (1983) and our study of the type strain.

Cells of the type strain from PY-glucose broth with 0.02% Tween 80 are 0.6–0.9 μm in diameter and are arranged principally in tetrads and irregular groups, although short chains of 6–8 cells occasionally are seen.

Figure 12.21. *P. prevotii*

Cell walls contain glucose, galactose, glucosamine, an unidentified sugar, D-glutamic acid, L-lysine, glycine, D-alanine, and muramic acid. The interpeptide bridge is glutamic, which is the ε-amino group of L-lysine and the carboxyl group of D-alanine (Schleifer and Nimmermann, 1973).

Surface colonies on blood agar plates incubated anaerobically for 48–72 h are minute, circular with entire margins, raised to low convex, semiopaque, white, and smooth. There is indefinite clearing under the heavy growth on sheep blood agar.

Growth is stimulated by oleate (0.02% Tween 80). The optimum temperature for growth is 37°C. Good growth occurs between 30°C and 45°C but growth usually is poor at 25°C. Growth in PY + 6.5% NaCl broth is equal to that in PY broth. Bile (2% oxgall) partially inhibits growth. Abundant gas is produced in PY-glucose agar deep cultures.

Carbohydrates are fermented weakly (Table 12.36).

Reactions for butyrate (C4) esterase, caprylate (C8) esterase lipase, leucine arylamidase, and β-galactosidase are variable among homologous strains (Ezaki et al., 1983).

Sixteen human clinical isolates identified as *P. prevotii* did not produce coagulase or hemolyze sheep erythrocytes; five produced DNase and one produced RNase (Marshall and Kaufman, 1981).

Acids detected in PY-glucose Tween 80 cultures (meq/100 ml of culture) are butyric (2–3), often with lactic (1.5–2.5); trace to moderate amounts of acetic, propionic and succinic acids also may be detected. Pyruvate is converted to butyrate (5–8 meq/100 ml of culture). Lactate is not utilized.

Cellular fatty acids (as percentage of total acids) of the type strain are *n*-C14:0 (9), *n*-C16:1 (38); *n*-C18:0 (26), *i*-C15 (20), *n*-C17:1 (5); and *n*-C19:0 (2) (Lambert and Armfield, 1979). Ezaki et al. (1983) report that 60–61% of the total cellular fatty acid is C18:1 and that C18:1 aldehyde is 35–39%.

Other characteristics of the species are given in Table 12.36.

Isolated from human plasma, skin, vagina and tonsils.

The mol% G + C of the DNA is 29–33 (T_m) (Ezaki et al., 1983).

Type strain: ATCC 9321 (DSM 20548).

Further comments. ATCC strains 9321 and 14952 represent two of the six strains upon which Foubert and Douglas (1948) based the description of the species. The type strain (ATCC 9231) ferments several sugars (fructose, glucose, mannose, melibiose, raffinose, sucrose and ribose) weakly and ferments galactose strongly (pH ~5.2). In complex media ATCC 14952 does not ferment any sugars tested; the pH is not lowered in PY media containing these sugars and the acid products of PY cultures with and without glucose are the same. This type of variation, described by Foubert and Douglas (1948), led Rogosa to conclude in the eighth edition of the *Manual* that the original description was a "composite description of two organisms," one of which "is indistinguishable from the organism with the rejected name *Gaffkya anaerobia*" and the other which, "except for a negative indole reaction . . . is indistinguishable from *P. asaccharolyticus.*"

Although the type strain of the species is saccharoclastic, the species generally is considered nonsaccharoclastic, probably because descriptions of *P. prevotii* published in recent manuals (Holdeman et al., 1977 and Sutter et al., 1980) were based on reactions obtained with ATCC 14952 (the nonsaccharoclastic strain deposited by Foubert and Douglas). The relatedness of the DNA of these two strains has not been published. In this edition of the *Manual*, the characteristics given for *P. prevotii* are those of the type strain and strains that have >70% DNA homology with the type strain (Ezaki et al., 1983).

Serine and threonine, the only amino acids metabolized by the nonsaccharoclastic strain ATCC 14952, are deaminated to pyruvate and α-ketoglutarate, which are converted via the thioclastic reaction to acetyl-CoA and propionyl-CoA, respectively, with the production of one ATP (Bentley and Dawes, 1974). Washed cells, harvested just before the stationary phase of growth, ferment (in order of decreasing rate) adenosine, inosine, glucose, xanthine, ribose, hypoxanthine and thymidine (Reece et al., 1976).

8. **Peptostreptococcus productus** (Prévot 1941) Smith 1957, 536[AL] (*Streptococcus productus* Prévot 1941, 105.)

pro.duc′tus. L. adj. *productus* produced.

Unless otherwise noted, this description is based on study of the type strain and several hundred similar isolates from human fecal samples.

Cells are elliptical, 0.6–0.9 × 0.8–2.0 μm, occurring singly, in pairs, and in chains of up to 10 or more pairs. Pairs of dividing cells often remain joined to give the appearance of boat-shaped cells up to 5.0 μm long.

Figure 12.22. *P. productus*

Surface colonies on anaerobic rabbit blood agar are 0.5–1 mm in diameter, circular, entire, convex, translucent, gray-white, shiny and smooth. Colonies in roll tube cultures in rumen fluid-glucose-cellobiose agar are 0.5–2.0 mm, white and lenticular.

PY-glucose broth cultures have slight turbidity with abundant powdery or flocculent sediment and have a terminal pH of 4.4–5.4.

Growth is greatly stimulated by fermentable carbohydrate. Acid production, but not growth, often is stimulated by 0.02% Tween 80.

Bile (2% oxgall) inhibits acid production but usually does not visibly affect growth yield. No growth is produced in PY-glucose broth containing 6.5% NaCl.

Optimum temperature for growth is 37°C. The type strain grows well at 30°C and moderately well at 25°C or 45°C.

Gas production in glucose deep agar cultures is variable, from none to abundant.

Amygdalin, arabinose, cellobiose, fructose, glucose, lactose, maltose, mannitol, mannose, melibiose, raffinose, salicin, sorbitol, sucrose and xylose are fermented to a pH of 4.9–5.3. Esculin is fermented weakly (pH 5.4–5.6). Most strains produce acid from melezitose and trehalose; fermentation of starch and ribose is variable among strains. Most strains do not ferment inositol and rhamnose; glycogen and dulcitol are not fermented. Growth but no lipase or lecithinase is produced on egg yolk agar.

The type strain ferments dextrin and galactose, weakly ferments sorbose, and does not ferment adonitol, dulcitol, glycerol or inulin.

Strains can use urea as sole nitrogen source but are negative for urease production in usual urease tests in complex media (Varel et al., 1974).

The type strain produces butyrate (C4) esterase, caprylate (C-8) esterase lipase, β-galactosidase, and α-glucosidase but not alkaline phosphatase, leucine arylamidase, or β-glucuronidase (API-ZYM tests) (Ezaki et al., 1983).

Products (meq/100 ml of culture) in PY-glucose broth cultures with Tween 80 are acetic (3.0–7.0), succinic (0.3–2.0), and lactic (0.02–1.0) acids and ethanol. Variable amounts of H_2 are produced in the headspace gas above prereduced carbohydrate broth cultures in stoppered culture tubes. Acetate is produced from gluconate and from pyruvate. Propionate is not produced from threonine. Lactate is not utilized. The type strain produces about 2% H_2 in the head space gas.

The major cellular fatty acids are C18:1 (29% of total) and C17:1 (28% of total) (Ezaki et al., 1983).

The antibiotic susceptibility patterns of strains in the species have not been tested; the species is seldom if ever detected in properly collected clinical specimens.

Hirano and Masuda strain b-52 contains an NADP-dependent 7-β-hydrosteroid dehydrogenase that is active on both conjugated and unconjugaged bile acids (Hirano and Masuda, 1982).

Other characteristics of the species are given in Table 12.36.

Isolated from the human colon and feces, where it is one of the predominant members of the flora (Moore and Holdeman, 1974; Holdeman et al, 1976; Finegold et al, 1974) and from swine intestinal flora (Russell, 1979). Reported from the intestinal tract of the American cockroach *Periplaneta americana* L. (Bracke et al, 1978).

The mol% G + C of the DNA of the type strain is 44–45 (Romond et al, 1966; Ezaki et al., 1983).

Type strain: ATCC 27340 (CIPP (Prévot) 2396; VPI 4299).

Further comments. In their studies of the human intestinal flora, Holdeman, Cato and Moore (unpublished data) differentiated subgroups of *P. productus* on the basis of several phenotypic characteristics that seemed to indicate the occurrence of distinct subspecies. All of these groups ferment many carbohydrates. *P. productus* I strains uniformly produce major amounts of formic acid in addition to acetic, succinic and lactic acids. Many of these strains fail to ferment sucrose. The strains that do not produce major amounts of formic acid have been referred to as *P. productus* II strains (Moore and Holdeman, 1974; Holdeman et al, 1976), but there is considerable heterogeneity within this subgroup. *P. productus* IIa strains ferment inositol but not rhamnose. *P. productus* IIb strains (including the type strain of *P. productus*) ferment inositol and rhamnose. The fermentation of rhamnose is unusual in that propionic acid is produced in addition to acetic, succinic

and sometimes lactic acid. *P. productus* IIc, IId and IIe strains do not ferment inositol. IIc strains do not produce H_2 and fermentation of rhamnose (with production of propionic acid) is variable. IId strains do not ferment rhamnose, uniformly ferment ribose, usually ferment trehalose and uniformly produce abundant H_2. IIe strains uniformly ferment rhamnose with production of propionic acid, uniformly do not ferment trehalose and uniformly produce abundant H_2. The genetic relatonships of these groups have not been studied; however, it appears that the species is heterogeneous or possibly contains several related entities.

The *P. productus* groups have many phenotypic and morphologic similarities to strains classified as *Ruminococcus callidus, Ruminococcus gnavus, Ruminococcus lactaris, Ruminococcus obeum,* and *Ruminococcus torques* (Holdeman and Moore, 1974; Moore et al., 1976). It is likely that these ovoid, highly fermentative intestinal species represent a genus that is distinct from the spherical *Ruminococcus* species (*R. albus, R. flavefaciens* and *R. bromii*). *P. productus* ferments raffinose, arabinose and cellobiose and can, therefore, be distinguished from members of the genus *Ruminococcus* (see p. 1093).

9. **Peptostreptococcus tetradius** Ezaki et al., 1983, 696.[VP] ("*Gaffkya anaerobia*" (Choukévitch) Prévot 1933, 203.)

te.tra′di.us. Gr. adj. *tetradios* by fours; N.L. adj. *tetradius*, occurring in groups of four.

The description is from Ezaki et al. (1983), who studied five strains, and from our study of the type strain.

Cells are 0.8–1.8 μm in diameter and often occur as pairs or in tetrads; short chains or masses also are present.

Figure 12.23. *P. tetradius*

Surface colonies on sheep blood agar are minute to 1 mm in diameter and are not hemolytic. Colonies do not have black pigment on either rabbit or sheep blood agar after incubation for 7 days.

Glucose broth cultures are turbid with a smooth sediment and pH of 5.0.

Strains grow at 35–37°C. The temperature for optimum growth has not been determined.

Gas is produced in PY-glucose broth.

Fermentation of carbohydrates is stimulated by 0.02% Tween 80.

Fructose, glucose, maltose, mannose and sucrose are fermented. No acid is produced from arabinose, esculin, lactose, mannitol, raffinose, rhamnose or sorbitol.

Ammonia is produced from threonine and serine. Starch is not hydrolyzed. Butyrate (C4) esterase and β-galactosidase are not produced. Of five strains tested, two produced a small amount of catalase and a different two strains weakly hydrolyzed esculin. Of the five strains, only the type gave weak reactions for caprylate (C8) esterase lipase and leucine arylamidase (Ezaki et al., 1983).

Products (meq/100 ml of PY-glucose culture) of the type strain are butyrate (2.0), lactate (2.0) and acetate (0.5). Propionate is not produced from lactate.

Isolated from human vaginal discharge and various purulent secretions.

The mol% G + C of the DNA is 30 (T_m) for the type strain and 32 for four other strains tested (Ezaki et al., 1983).

Type strain: GIFU 7672 (ATCC 35098; JCM 1964; CCM 3634).

Genus **Ruminococcus** *Sijpesteijn 1948, 152^{AL}*

Marvin P. Bryant

Ru.min.o.coc'cus. L. adj. *ruminalis* of the rumen; Gr. n. *coccus* a grain, berry; M.L. masc. n. *Ruminococcus* coccus of the rumen.

Coccoid cells, usually 0.3–1.5×0.7–1.8 μm, when elongated, ends are flat, rounded or pointed and lancet, and **cells are arranged as diplococci and chains.** A few are motile with one to three flagella. Resting stages are not known. **Gram-positive cell wall structure but many stain Gram-negative.** Some produce amylopectin type reserves. **Optimum temperature, 37–42°C;** maximum, about 45°C; minimum, about 20–30°C. **Chemoorganotrophic and anaerobic having a fermentative type of catabolism. Carbohydrates are used as fermentable substrates.** Fermentation of carbohydrate yields various proportions of **acetate,** formate, **succinate, lactate, ethanol, CO$_2$ and H$_2$. CO$_2$ may show net uptake in succinate producers. Amino acids and peptides are not fermented. Catalase and indole are not produced, nitrate is not reduced and ammonia is not produced from amino acids or peptides.** Isolated from rumen, large bowel or cecum of many animals. The mol% G + C of the DNA is 39–46 (T_m).

Type species: *Ruminococcus flavefaciens* Sijpesteijn 1948, 152.

Further Descriptive Information

R. flavefaciens was shown to possess a prominent glycoprotein coat which contained rhamnose, glucose and galactose and adhered strongly by means of this coat to cotton cellulose and to damaged cell walls of forage grass (Latham et al., 1978). *R. albus* also has a "polysaccharide" coat layer and the coat material mediates the adhesion to cellulose fibers (Patterson et al., 1975). Nothing is known about cell walls in other species except that they generally stain at least weakly Gram-positive.

In fine structure *R. albus* and *R. flavefaciens* have the outer coat indicated above, a thick peptidoglycan layer, Gram-positive structure and glucan ("amylopectin") reserve materials (Patterson et al. 1975; Roger Cole quoted by Rogosa, 1971; Latham et al., 1978). Patterson et al. (1975) noted an outer membrane difficult to resolve but others have not, and both species are highly sensitive to monensin (Chen and Wolin, 1979) and have an easily detected extracellular cellulase and are, therefore, physiologically Gram-positive.

Nutritionally, only *R. flavefaciens, R. albus* and *R. bromii* have been studied (Allison et al., 1958; Bryant and Robinson, 1961a, b; Dehority et al., 1967; Scott and Dehority, 1965; Bryant, 1973, 1974; Hungate and Stack, 1982). All strains require one or more B vitamins and biotin, pyridoxine, *p*-aminobenzoate, folate, riboflavin, thiamin, cobalamin (sometimes replaced by methionine), pantothenate, pantethine and tetrahydrofolate may be required, depending on the strain. All strains so far adequately studied **require** ammonia as the main source of nitrogen and have a very limited ability to utilize amino acids or peptides as nitrogen source (Bryant and Robinson, 1963). A few strains contain a repressible urease and, thus, do not require ammonia (Wozny et al., 1977). Very few strains require nucleic acid hydrolysis products. All strains have a requirement for CO$_2$ or bicarbonate, and high levels are required by strains producing large amounts of succinate (Dehority, 1971). Most strains so far studied require isobutyrate and/or 2-methyl butyrate, and some strains require isovalerate and/or hydrocinnamate (phenylpropanoate). The amounts of acids required are less than 0.1 mM and the branch chained acids are usually required for long chain fatty acid and aldehyde (some species contain plasmalogens, Allison et al., 1962b) and sometimes for amino acid biosynthesis via reductive carboxylation reactions (Allison et al., 1962a; Allison, 1969). ^{14}C-Phenylacetate in certain *R. albus* strains is recovered in phenylalanine (Stack et al., 1983). Most strains of the three species grow well with sulfide and, sometimes, methionine, as sole S source. *R. bromii* utilizes sulfate as a sole S source and this could be used in a selective medium for the species in rumen contents (Herbeck and Bryant, 1974). Na$^+$ is required

by rumen strains of *R. albus* and *R. flavefaciens* (Caldwell and Hudson, 1974).

Pathways of metabolism have only been studied in rumen strains of *R. albus* and *R. bromii* (see Miller and Wolin, 1979 for much information and references). All exhibit the Embden-Meyerhoff-Parnas (EMP) pathway from hexose to pyruvate but *R. flavefaciens,* and some strains of *R. albus,* probably do not transport glucose and have a cellobiose phosphorylase to initiate the EMP pathway (Ayres, 1958). Ruminococci do not contain cytochromes but at least some contain the low potential electron carrier ferredoxin. Lactate is formed from pyruvate by oxidation of NADH formed in glycolysis. Carboxylation of pyruvate or phosphoenolypyruvate leads to oxalacetate and eventually to succinate via fumarate. Pyruvate also leads to acetyl CoA, CO$_2$ and reduced ferredoxin. Formate is formed when excess electrons (H$_2$) are present and appears to involve a formate-forming CO$_2$ reductase system. Acetyl CoA leads to acetaldehyde and CoA and, hence, to ethanol. H$_2$ is produced from NADH via ferredoxin or more directly from reduced ferredoxin. H$_2$ production is inhibited in *R. albus* by 10^{-5} M molybdate when sulfide is in the growth medium and formate then accumulates (Wolin and Miller, 1980). Acetate and ATP are produced from acetyl CoA via phosphotransacetylase and acetate kinase. Cellulases of ruminococci have received some attention (Pettipher and Latham, 1979a, b; Leatherwood, 1973; Smith et al., 1973; Yu and Hungate, 1979).

There are a multiplicity of antigenic types among the cellulolytic ruminococci (Jarvis, 1967).

Pathogenicity for humans or animals has not been detected.

Antibiotic sensitivity has been studied for a few strains of cellulolytic *R. albus* and *R. flavefaciens* (Fulghum et al., 1968). All Gram-positive fine-structured rumen bacteria studied, including the above species, are sensitive to monensin and lasalocid and do not rapidly adapt to become resistant, while essentially Gram-negative rumen bacteria are resistant or rapidly become resistant (Chen and Wolin, 1979; Dennis et al., 1981).

In general ecology of the genus, *Ruminococcus* species have so far been found only in the rumen, rumen-like forestomachs, the cecum and large bowel, including man. Since all produce H$_2$ from fermentable carbohydrate, they all will interact metabolically in interspecies hydrogen transfer with H$_2$-CO$_2$ utilizing methanogens such as *Methanobrevibacter ruminantium* or *Methanobrevibacter smithii*, allowing the *Ruminococcus* species to produce more acetate, CO$_2$ and H$_2$ and little or no ethanol, formate, succinate or lactate in the natural ecosystem (Wolin, 1981). In the absence of methanogens, they can interact with H$_2$-CO$_2$ utilizing acetate and butyrate- or acetate-producing bacteria such as *Eubacterium limosum* (Rode et al., 1981) or *Peptostreptococcus productus* (Lorowitz and Bryant, 1983) allowing the *Ruminococcus* species again to produce more acetate, CO$_2$ and H$_2$ and less reduced products such as formate, ethanol and lactate (see also Prins and Lankhorst, 1977; Hungate, 1966). However, the methanogens have much greater affinity for H$_2$-CO$_2$ than do the acetogens.

Cellulolytic (and probably hemicellulolytic) species (not identified) were found in the forestomach of leaf-eating langur monkeys (Bauchop and Martucci, 1968) and cecum and colon of swine and horses (Davies, 1965; swine, V. Varel, personal communication, 1983).

R. flavefaciens, an important cellulose and hemicellulose fermenter and pectin hydrolyzer, has been isolated from the rumen of many different species (Hungate, 1966; Sijpesteijn, 1951); the cecum of rabbits (Hall, 1952), guinea pigs (Dehority, 1977), and rats (Montgomery and Macy, 1982).

R. albus, with similar function but producing more ethanol, H$_2$ and CO$_2$ and less succinate than *R. flavefaciens* (Hungate, 1966), has been isolated from the rumen of many animals and from human feces; but the strains from human feces are not known to ferment cellulose (Holdeman et al., 1976) and some of these produce urease (Wozny et al., 1977). The human fecal strains need study with more specific tests to be certain that they are *R. albus* (Wozny et al., 1977).

Ruminococcus bromii is sometimes a major starch fermenter in the rumen and human colon, producing fermentation products identical with *R. albus* and sometimes urease. It has been isolated as a predominant organism from the cow rumen (+C-S group, Bryant and Burkey, 1953; Wozny et al., 1977) and from human and swine feces (group 10, Eller et al., 1971; Moore et al., 1972).

The other species of *Ruminococcus* have been isolated only from human fecal or colon samples (Holdeman and Moore, 1974; Moore et al., 1976).

Isolation and Enrichment Procedures

Nonselective enumerations and isolation procedures using various modifications of the anaerobic Hungate technique (Hungate, 1950, 1966) have often been used with roll-tube media based on 30–40% (v/v) rumen fluid, minerals, glucose, cellobiose, starch (or maltose), HCO_3^-, CO_2 gas phase, cysteine or cysteine-Na_2S reducing agents, 1.5 or 2% agar and pH 6.6–6.8. This technique required the picking of large numbers of colonies and checking isolates for morphology, motility, hydrolysis of amorphous cellulose, starch and/or fermentation of certain sugars and hemicellulose, for presumptive identification. These techniques have been used by Bryant and Burkey (1953), Bryant and Robinson (1961c), Moore (1966), Latham and Sharpe (1971), Holdeman et al. (1977), Grubb and Dehority (1976) and many others. Similar techniques and media have been used but with rumen fluid replaced with about 30 mM acetate, 0.1 mM each of valerate, isovalerate, 2-methyl butyrate, isobutyrate, (hydrocinnamate, i.e. phenylpropanoate, should also be added, Hungate and Stack (1982), for some cellulolytic ruminococci), 1 μM hemin, 0.2% trypticase and 0.05% yeast extract (for nonselectivity, about 50 μM each of 1-4-naphthoquinone, tetrahydrofolate and pantethine might also be added (Herbeck and Bryant, 1974; Gomez-Alarcon et al., 1982)). This type of medium (medium 10 and modification) was used by Caldwell and Bryant (1966), Thorley et al. (1968) and Latham and Sharpe (1971) for bacteria from rumen contents and by Eller et al. (1971) for human fecal flora.

Very good, more selective methods based on rumen fluid-amorphous cellulose-agar roll tubes have been used by Hungate (1950), Sijpesteijn (1948), Kistner (1960) and others for isolations of cellulolytic bacteria from rumen contents. With these techniques one picks colonies associated with zones of cellulose digestion, and colony type often allows presumptive identification of the cellulose digesters. A similar technique was used for isolation of *R. flavefaciens* and other cellulolytic bacteria, e.g. *Bacteroides succinogenes*, from the cecum of rabbits (Hall, 1952) and rats (Montgomery and Macy, 1982).

A very effective method for enumeration, isolation and presumptive identification using specially prepared rumen fluid-agar plates and anaerobic chamber has been developed by Leedle and Hespell (1980) and used for large scale sampling of rumen contents (Leedle et al., 1982), but could be used for other anaerobic ecosystems where the organism to be enumerated and isolated is in high enough numbers to form isolated colonies in nonselective medium. It involves growth of isolated colonies on plates inoculated with high dilutions of samples, and then replicate plating the colonies to plates which are selective for organisms using only one energy source such as various sugars, starch, cellulose, xylan, pectin, rutin, etc. One can also detect protease- or urease-producing colonies, etc., by flooding appropriate plates with

appropriate staining or precipitating reagents. One can also make the method more selective by plating diluted samples directly on medium with only one energy source, e.g. insoluble cellulose, hemicellulose such as xylan, rutin or starch, or any soluble energy source. Zones of hydrolysis of energy sources under and, usually, around colonies, and/or size of colonies allow the detection of bacteria using the energy source.

For those without resources for, or who dislike, anaerobic chambers, those who detest cellulose agar roll tubes and those who do not wish to pick 20–500 colonies from relatively nonselective media to find the organism(s) they seek, somewhat selective methods are available or can be developed. These methods can involve three-tube MPN with 0.2% amorphous cellulose-30% rumen fluid-liquid medium with CO_2 gas phase. With the highest tubes showing cellulose disappearance after 1- or 2-week incubation at 37–39°C, transfers can be made to the same medium at 2- to 3-day intervals for one or two transfers. The latter is diluted and inoculated into rumen fluid-cellobiose agar roll tubes and incubated 3–5 days. Colonies can then be picked to cellobiose agar slants and studied for morphology and inoculated back into liquid cellulose medium to presumptively identify species. *R. albus* and/or *R. flavefaciens* often outnumber other cellulolytics (Hall, 1952: Bryant and Burkey, 1953) and can be easily isolated. *Bacteroides succinogenes* can be easily isolated with this protocol but using more crytalline cellulose such as dewaxed cotton fibers or Whatman cellulose powder (for chromatography, Bryant, 1973) and/or 2–5 μg of monensin/ml, either added to the MPN medium or to the ruminant diet (Chen and Wolin, 1979; Brulla and Bryant, 1980).

Chemically defined minimal media are already available for many *R. albus* and *R. flavefaciens* (Bryant, 1973; Hungate and Stack, 1982) and for *R. bromii* (Herbeck and Bryant, 1974). Thus, one can visualize the use of MPN with chemically defined media with appropriate energy sources and/or appropriate antibiotics to select for cellulolytic ruminococci, rather than *B. succinogenes* or less actively cellulolytic bacteria (most of these in ecosystems that support the ruminococci and *B. succinogenes* are less numerous or less actively cellulolytic). One can also visualize MPN medium containing rumen fluid treated by acid-ether extraction to remove lipids and free fatty acids (Bryant and Doetsch, 1955) and with only isobutyrate, isovalerate, 2-methyl butyrate, phenylacetate and hydrocinnamate added back to select for cellulolytic ruminococci. *B. succinogenes* requires both 2-methyl butyrate or isobutyrate and a straight chain saturated fatty acid, *n*-valerate or longer chain acid for growth (Bryant and Doetsch, 1955).

Even moderately selective methods have not yet been developed for *R. bromii*, *R. callidus*, *R. torques*, *R. gnavus*, *R. lactaris* or *R. obeum* (but see Bryant, 1974, for chemically defined roll tube media).

Maintenance Procedures

Ruminococcus strains can be lyophilized, stored under liquid N_2, stored at −60°C as stabbed slants or stored at −20°C in liquid media containing 20% glycerol using procedures commonly used for anaerobic bacteria (Teather, 1982; Phillips et al., 1975).

Procedures and Methods for Characterization Tests

See procedures under the genus *Eubacterium* Holdeman and Moore. See Hungate (1950, 1957), Bryant et al. (1958), Kistner and Gouws (1964), van Gylswyk and Roché (1970), Latham and Sharpe (1971), and Dehority (1963, especially for nutritional screening) and references therein for techniques for cellulolytic species.

Differentiation of the genus **Ruminococcus** from other genera

Table 12.37 shows a few features to be used in differentiation of the genus from other genera of anaerobic Gram-positive (staining or cell wall structure), anaerobic cocci. Some types of *Peptostreptococcus productus* are difficult to differentiate from some species of *Ruminococcus* (q.v. *Peptostreptococcus productus*, this volume p. 1091). Note that human fecal strains of *R. albus* and *R. flavefaciens* may be found to be different species because none of them are known to ferment cellulose.

Taxonomic Comments

The Woese type analysis of 16S rRNA sequence place *R. bromii* as well as several *Peptococcus* species in the *Clostridium* group (order?) of bacteria (Fox et al., 1980). It is probable that the other *Ruminococcus* species will also be placed here. It is possible that the *Ruminococcus* species with pointed ends and/or oval cells will be found to belong to a genus separate from *R. albus*, *R. flavefaciens* and *R. bromii* (see discussion by Moore and Holdeman Moore, this volume, on *Peptostreptococcus productus*). The latter *Ruminococcus* species have flattened to rounded ends.

Acknowledgements

The author (sine 1947) as well as many of the people in references cited are indebted to the fantastic creativity and devotion of Professor Robert E. Hungate who brought order out of chaos to our understanding of the taxonomy and biochemical ecology of anaerobic microorganisms.

He is a member of the first order of the Delft School of Microbiology (Beijerinck → Kluyver → van Niel → Hungate, H. A. Barker, R. Y. Stanier and many others) which believed in the comparative biochemistry and biochemical ecology of microorganisms for the advancement of biology.

Further Reading

Bryant, M.P., Nola Small, Cecelia Bouma and I.M. Robinson. 1958. The characteristics of ruminal anaerobic cellulolytic cocci and *Cillobacterium cellulosolvens* n. sp. J. Bacteriol. 76: 529–537.
Holdeman, L.V. and W.E.C. Moore. 1974. New Genus, *Coprococcus*, twelve new species, and emended descriptions of four previously described species of bacteria and human feces. Int. J. Syst. Bact. 24: 260–277.
Hungate, R.E. 1957. Microorganisms in the rumen of cattle fed a constant ration. Can. J. Microbiol. 3: 289–311.
Moore, W.E.C., J.L. Johnson and L.V. Holdeman. 1976. Emendation of *Bacteriodaceae* and *Butyrivibrio* and descriptions of *Desulfomonas* gen. nov. and ten new species in the genera *Desulfomonas*, *Butyrivibrio*, *Eubacterium*, *Clostridium*, and *Ruminococcus*. Int. J. Syst. Bacteriol. 25: 238–352.
Moore, W.E.C., E.P. Cato and L.V. Holdeman. 1972. *Ruminococcus bromii* sp. n. and emendation of the description of *Ruminococcus* Sijpesteijn Int. J. Syst. Bacteriol. 22: 78–80.
Sijpesteijn, A.K. 1951. On *Ruminococcus flavefaciens*, a cellulose-decomposing bacterium from the rumen of sheep and cattle. J. Gen Microbiol. 5: 869–879.
van Gylswyk, N.O. and C.E.G. Roché. 1970. Characteristics of *Ruminococcus* and cellulolytic *Butyrivibrio* species from the rumens of sheep fed differently supplemented teff (*Eragrostis tef*) hay diets. J. Gen. Microbiol. 64: 11–17.

Table 12.37.
Differential features of the genus **Ruminococcus** *and other genera of Gram-positive (staining or structure of cell wall), anaerobic, fermentative cocci arranged as diplococci or chains*[a]

Characteristics	Ruminococcus	Peptostreptococcus	Peptococcus	Coprococcus
Sugar fermented	+[b]	D[c]	D	+[d]
Peptone fermented	−[e]	D[c]	+	−
Butyrate or longer carbon-chained acids produced	−	D	D	+

[a] Symbols: see Table 12.2; also D, different reactions in different taxa (species of a genus or genera of a family).
[b] Required for growth.
[c] *Peptostreptococcus productus* strains may require sugars for growth and not ferment peptone (Bryant, 1974).
[d] Required or highly stimulatory for growth.
[e] Rumen strains of *R. flavefaciens* and *R. albus* and strains of *R. bromii* require ammonia as the main nitrogen source and do not grow with mixtures of amino acids or peptides as the main nitrogen source. These species almost always require one or more of the acids isobutyric, 2-methylbutyric and isovaleric.

Differentiation of the species of the genus **Ruminococcus**

See Table 12.38 for some features differentiating the species of *Ruminococcus*. Strains of cellulolytic *R. flavefaciens* that produce mainly lactic acid have been placed in *R. flavefaciens* subsp. "*lacticus*" (van Gylswyk and Roché, 1970), with which the author agrees, but the name is not yet validated.

Further features of the species of the genus are listed in the generic description and in Table 12.39.

List of species of the genus **Ruminococcus**

1. **Ruminococcus flavefaciens** Sijpesteijn 1948, 152.[AL]
fla.ve.fac'i.ens. L. adj. *flavus* yellow. L. pres. part. *faciens* producing; M.L. part. adj. *flavefaciens* yellow producing.
Further described by Sijpesteijn (1951).
Features listed in Tables 12.38 and 12.39 are mainly from Bryant, 1959; Bryant et al., 1958; Hall, 1952; Hungate, 1950, 1957, 1966; Kistner and Gouws, 1964; Montgomery and Macy, 1982; Rogosa, 1974; and van Gylswyk and Roché, 1970; all of which were cellulolytic. Human fecal strains may in the future be placed in different species (W. E. C. Moore, personal communication).
The mol% G + C of the DNA is 39–44 (T_m, Rogosa, 1974).

Type strain: ATCC 19208 (NCDO 2213, Bryant C94; Bryant et al., 1958). Perhaps a better type strain is NCDO 2215 (Bryant et al., 1958, strain FDI) because of its more typical size and excellent ability to degrade more resistant cellulose (Halliwell and Bryant, 1963; A. Kistner, personal communication, 1983).

2. **Ruminococcus albus** Hungate 1957, 307.[AL]
al'bus. L. adj. *albus* white.
Features listed in Tables 12.38 and 12.39 are mainly from Bryant, 1959; Bryant et al., 1958; Kistner and Gouws, 1964; Rogosa, 1974; van Gylswyk and Roché, 1970; and exclude human fecal strains which may

Table 12.38.
Characteristics differentiating species of the genus **Ruminococcus**[a]

Characteristics	1. R. flavefaciens	2. R. albus	3. R. bromii	4. R. callidus	5. R. torques	6. R. gnavus	7. R. lactaris	8. R. obeum
Oval to pointed ends	−	−	−	+	+	+	+	+
Major fermentation products								
Succinate	+	−	−	+	−	−	−	−
Lactate	d[b]	−	−	−	+	−	+	−
Ethanol	−	+	+	−	+	+	−	+
Hydrolyze								
Cellulose	d[c]	d	−		−	−	−	−
Starch	−	−	+	−	−	d	−	−
Ferment								
Arabinose, rhamnose and xylose	d	d	−	−	−	+	−	+
Cellobiose	+	+	−	+	−	−	−	−
Fructose	−	d	+	−	+	+	+	+
Lactose	d	d	−	d	+	d	+	+
Maltose	−	−	+	+	−	+	d	+
Mannitol	−	−	−	−	−	−	+	d
Raffinose	−	−	−	+	−	+	−	+
Sucrose	d	d	−	+	−	d	−	+

[a] Symbols: see Table 12.2.
[b] Strains producing mainly lactate have been placed in the subspecies *R. flavefaciens* subsp. "*lacticus*" (van Gylswyk and Roché, 1970).
[c] Many strains produce yellow to yellow-orange pigment especially when grown with cellulose as energy source.

Table 12.39.
Other features of the species of the genus **Ruminococcus**[a, b]

Characteristics	1. R. flavefaciens	2. R. albus	3. R. bromii	4. R. callidus	5. R. torques	6. R. gnavus	7. R. lactaris	8. R. obeum
Cell width								
0.3–0.9 μm					+		+	
0.7–1.6 μm	+	+	+	+		+		+
Require for growth								
Ammonia	+	+	+					
Isobutyrate and/or isoval-erate and/or 2-methyl butyrate	+	+	+					
High CO_2–HCO_3^-	+	−	−					
Na^+	d	d						
Produce								
Formate	d	d	d	d	d	+	+	−
Propionate	−	−	d	−	−	−	−	−
H_2S	−	−	−			d	−	d
Amylopectin-like reserve	d	d						
Xylanase	d	d						
Urease	−	d	d	−	−	−		−
Curdle milk	−	−	−	d	d	d	+	+
Ferment								
Amygdalin	−	−	−	d	−	+	−	d
Esculin	d	d	−	d	d	d	−	−
Gluconate				−	−	+		−
Glucose	d	d	d	d	+	+	+	+
Glycerol	−	−	−					
Glycogen	−	−	+	−	−	−	−	−
Inulin	−	−	−	d				
Mannose	d	d	d	d	d	−	d	+
Melezitose	−	−	−	d	−			
Melibiose	−	−	d	d	−	−	−	+
Salicin	−	d	−	d	d	+	−	d
Sorbitol	−	−	−	−	−	−	d	d
Trehalose	−	−	−	d	−	−	−	d

[a] Symbols: see Table 12.2.
[b] None of the species ferment inositol or lactate, produce ammonia from peptone (except a few *R. torques*), are motile (except some *R. obeum*), hemolyze blood, produce reactions on egg yolk agar, reduce nitrate, produce indole, liquify gelatin (except some weakly), or have a mol % G + C of the DNA below 39 or above 46 (T_m). All produce H_2 and acetate and produce some chains except many strains of *R. albus*. None have so far been isolated except from mammalian G.I. tract ecosystems. All have a lower limit for growth between 20 and 30°C, optimum at 37–42°C and upper limit between 44 and 46°C. None have as yet been shown to utilize H_2-CO_2, formate, methanol or CO as major energy sources, although we can look forward to this because strains of *Peptococcus* and *Peptostreptococcus productus* do use one or more of these energy sources and these plus *R. bromii* are related to the *Clostridium* group (Fox et al., 1980; Tanner et al., 1982; Lorowitz and Bryant, 1984). Additionally, those that produce acetate from one-carbon energy sources undoubtedly contain the nickel-containing CO dehydrogenase (Wood et al., 1982).

in the future be placed in a different species (W. E. C. Moore, personal communication; Wozny et al., 1977).

The mol% G + C of the DNA is 42.6–45.8 (T_m, Rogosa, 1974).

Type strain: ATCC 27210 (NCDO 2250, Bryant 7; Bryant et al., 1958 from cow rumen).

3. **Ruminococcus bromii** Moore et al., 1972, 80.[AL]

brom'i.i. M.L. gen. n. *bromii* god of alcohol.

Features listed in Tables 12.38 and 12.39 are mainly from Bryant and Burkey, 1953 (+C-S group); Eller et al., 1971 (group 10); and Moore et al., 1972).

The mol% G + C of the DNA is 39–40 (T_m, Moore et al., 1972).

Type strain: ATCC 27255 (from human feces, Moore et al., 1972).

4. **Ruminococcus callidus** Holdeman and Moore 1974, 264.[AL]

cal'li.dus. Gr. adj. *callidus* clever, expert.

The mol% G + C of the DNA is 43 (T_m, Holdeman and Moore, 1974).

Type strain: VPI 57–31.

5. **Ruminococcus torques** Holdeman and Moore 1974, 265.[AL]

tor'ques. L. n. *torques* twisted necklace.

The mol% G + C of the DNA is 40–42 (T_m, Holdeman and Moore, 1974).

Type strain: ATCC 27756, isolated from human feces, Holdeman and Moore, 1974.

6. **Ruminococcus gnavus** Moore et al., 1976, 243.[AL]

gna'vus. L. adj. *gnavus* busy, active.

The mol% G + C of the DNA is 43 (T_m, Moore et al., 1976).

Type strain: ATCC 29149 isolated from human feces, Moore et al., 1976.

7. **Ruminococcus lactaris** Moore et al. 1976, 244.[AL]

lac.ta'ris. L. adj. *lactaris* milk drinking.

The mol% G + C of the DNA is 45 (T_m, Moore et al., 1976).

Type strain: ATCC 29176 isolated from human feces (Moore et al., 1976).

8. **Ruminococcus obeum** Moore et al. 1976, 245.[AL]

o'be.um. Gr. n. *obeum* egg.

The mol% G + C of the DNA is 42 (T_m, Moore et al., 1976).

Type strain: ATCC 29174 isolated from human feces (Moore et al. 1976).

Genus **Coprococcus** *Holdeman and Moore 1974, 260*[AL]

LILLIAN V. HOLDEMAN MOORE AND W. E. C. MOORE

Co' pro. coc' cus. Gr. n. *copro* feces; Gr. n. *coccus* berry; M.L. masc. n. *Coprococcus* fecal coccus.

Cocci that are **Gram-positive**, nonmotile, and **obligately anaerobic chemoorganotrophs. Fermentable carbohydrates** are either **required** or are **highly stimulatory** for growth. **Fermentation products** include **butyrate** and acetate with formate or propionate and/or lactate.

The mol% G + C of the DNA is 39–42 T_m.

Type species: *Coprococcus eutactus* Holdeman and Moore, 1974, 261.

Further Descriptive Information

Morphology. In stains, cells occur as pairs or chains of pairs. Cells of some strains are slightly elongate, particularly when grown in medium containing a fermentable carbohydrate.

Ecology. All described species are from human intestinal contents or feces. The characteristics of unnamed species of the genus *Coprococcus* from the bovine rumen have been reported by Tsai and Jones (1975) and by Latham et al. (1979).

Isolation Procedures

Strains have been isolated on prereduced anaerobically sterilized rumen fluid-glucose-cellobiose agar (Holdeman et al., 1977) roll tubes. Pure cultures of strains of most species grow on anaerobically incubated blood agar plates (supplemented brain heart infusion agar with 5% sheep blood; Holdeman et al., 1977).

Maintenance Procedures

Lyophilization of early stationary phase cultures grown in complex medium containing yeast extract and 0.1–0.2% fermentable carbohydrate is recommended for long term storage. Most strains can be maintained in the laboratory by weekly transfer of cultures grown in a complex medium with ~0.2% fermentable carbohydrate.

Procedures for Testing Characteristics

Prereduced media and methods described in Holdeman et al. (1977) were used to determine characteristics of the species (also see the description of Procedures and Methods for Characterization Tests in the chapter on *Eubacterium.*)

Acknowledgements

The assistance of those persons and the support of those agencies listed in the chapter on *Eubacterium* also are gratefully acknowledged for the work on *Coprococcus*.

Further Reading

Holdeman, L.V. and W.E.C. Moore. 1974. New genus, *Coprococcus*, twelve new species, and emended descriptions of four previously described species of bacteria from human feces. Int. J. Syst. Bacteriol. *24:* 260–277.

Holdeman, L.V., I.J. Good and W.E.C. Moore. 1976. Human fecal flora: variation in bacterial composition within individuals and a possible effect of emotional stress. Appl. Environ. Microbiol. *31:* 359–375.

Latham, M.J., M.E. Sharpe and N. Weiss. 1979. Anaerobic cocci from the bovine alimentary tract, the amino acids of their cell wall peptidoglycans and those of various species of anaerobic streptococcus. J. Appl. Bacteriol. *47:* 209–221.

Moore, W.E.C., E.P. Cato, I.J. Good and L.V. Holdeman. 1981. The effect of diet on the human fecal flora. *In:* Banbury Report 7: Gastrointestinal Cancer: Endogenous Factors. Cold Spring Harbor Laboratory, Cold Spring Harbor, NY, pp. 11–24.

Moore, W.E.C. and L.V. Holdeman. 1974. Human fecal flora: the normal flora of 20 Japanese-Hawaiians. Appl. Microbiol. *27:* 961–979.

Tsai, C.-G. and G.A. Jones. 1975. Isolation and identification of rumen bacteria capable of anaerobic phloroglucinol degradation. Can. J. Microbiol. *21:* 794–801.

Differentiation of the species of the genus **Coprococcus**

Characteristics by which the described species in the genus can be differentiated are given in Tables 12.40 and 12.41. Additional characteristics are given in the text concerning each species.

Table 12.40.

Characteristics differentiating species of **Coprococcus**[a]

Characteristics	1. *C. eutactus*	2. *C. comes*	3. *C. catus*
Acid produced from			
Sucrose	+	+	−
Melezitose	+	−	−
Xylose	−	+	−
Fermentation	Fbla	Lba	BPa
products	(2, py, s)	(s, py)	(ls)

[a] Symbols: see Table 12.2; also f, formic acid; b, butyric acid; l, lactic acid; a, acetic acid; p, propionic acid; 2, ethanol; py, pyruvic acid; s, succinic acid. Capital letters indicate 1 meq (or more)/100 ml of culture; lower case letters indicate less than 1 meq/100 ml of culture. Products in parentheses are not detected uniformly.

Table 12.41.

Characteristics of species of the genus **Coprococcus**[a–c]

Characteristics	1. *C. eutactus*	2. *C. comes*	3. *C. catus*
Acid produced from			
Amygdalin	+	w−	−
Arabinose	−	+w	−
Cellobiose	+	−w	−
Esculin	d	−	−
Fructose	+	+	+
Glucose	+	+	−w
Lactose	+	+w	−
Maltose	+	+	w−
Mannitol	−	d	+−
Mannose	+	−w	−
Melezitose	+	−	−
Melibiose	+	d	−
Raffinose	+	+w	−
Ribose	−	−w	−
Salicin	+w	−+	−
Sorbitol	−	w−	−
Starch	+w	−w	−
Sucrose	+	+	−
Trehalose	−	−	−w
Xylose	−	+	−
Hydrolysis of			
Esculin	+	d	−
Starch	d	−	−
Gas (glucose agar)	3, 2	4, 2	3

[a] Reactions from Holdeman and Moore (1974) from study of 59 isolates of *C. eutactus*, 14 isolates of *C. comes*, and 12 isolates of *C. catus*.

[b] Symbols: +, pH below 5.5; w, pH 5.5–5.9; −, pH above 5.9; d, variable among strains; numbers (gas), amount estimated on "− to 4+" scale. Where two symbols are given, the first indicates the reaction of most strains.

[c] All strains were negative for fermentation of erythritol, glycogen, inositol, and rhamnose. No strain completely digested gelatin or digested milk or cooked meat. No strain produced indole or catalase or reduced nitrate.

List of species of the genus **Coprococcus**

1. **Coprococcus eutactus** Holdeman and Moore 1974, 261.[AL]

eu. tac′ tus. Gr. adj. *eutactus* orderly, well disciplined (referring to the uniform reactions of the different strains).

Cells usually are round and 0.7–1.3 μm in diameter; they often are slightly elongate in peptone-yeast extract-glucose (PYG) cultures (Fig. 12.24).

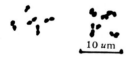

Figure 12.24. *Coprococcus eutactus.*

After incubation for 5 days in rumen fluid-glucose-cellobiose agar (RGCA) roll tubes, colonies are 0.5–2.0 mm in diameter, white or tan, lenticular and translucent; the centers of the colonies occasionally are granular or opaque. On anaerobic BAP (sheep blood) incubated for 2 days, surface colonies of the type strain are punctiform, circular, entire, convex, translucent, whitish, smooth, shiny and without hemolytic activity.

There is poor or no growth in PY broth without fermentable carbohydrate. PYG cultures have abundant growth and a smooth (occasion-ally ropy) sediment, usually with a slight to moderate turbidity. After incubation for 5 days the pH of PYG cultures is 4.7–5.0.

Growth occurs equally well at 37°C and 45°C; growth is poor to moderate at 25°C and 35°C.

In addition to the characteristics of 60 isolates from eight humans given in Table 12.41, the type strain and five other strains tested ferment dextrin, reduce neutral red and produce acetyl methyl carbinol. The six strains tested do not ferment adonitol, glycerol, inulin or sorbose and therefore grow poorly, if at all, in PY media containing these substrates. No reaction is produced on egg yolk agar (EYA) by five strains; one strain did not grow on the surface of EYA incubated anaerobically. No ammonia is detected from cultures in chopped meat, PY, PYG or arginine medium. Hippurate is not hydrolyzed, and H₂S is not detected in SIM (sulfide-indole-motility) medium (Baltimore Biological Laboratories).

Fermentation products (average meq/100 ml of cultures) are as follows.

From PY-glucose: formate (1.5–2.0), butyrate (0.9), lactate (0.7) and acetate (0.5), sometimes with ethanol and trace amounts of pyruvate and succinate. Abundant H₂ is produced.

From PY-pyruvate: formate (1.3) and acetate (0.8), sometimes with trace amounts of lactate, butyrate and succinate.

From PY: Occasionally trace amounts of acetate, succinate or butyrate.

Threonine is not converted to propionate. Lactate and gluconate are not used.

The mol% G + C of the DNA of the type strain is 41 (T_m).

Type strain: ATCC 27759 (Virginia Polytechnic Institute (VPI) C33-22); isolated from human feces.

Further comments. This species is easily recognized by its production of formic, lactic and butyric acids from glucose, by its poor growth in the absence of fermentable carbohydrates and by its rather uniform fermentation of the substrates indicated in Table 12.41.

2. Coprococcus comes Holdeman and Moore 1974, 263.[AL]

co' mes. L. n. *comes* companion, fellow traveler (referring to the presence of the species in human feces).

Cells are elongate cocci with tapered ends, occurring singly and in pairs and chains of 4–20 elements. Cell size ranges from 0.8–2.0 μm in diameter by 0.8–5.0 μm in length (Fig. 12.25).

Figure 12.25. *Coprococcus comes.*

After incubation for 5 days in RGCA roll tubes, the colonies are 0.5–1.0 mm in diameter, lenticular or bifoleate, and translucent. On blood agar plates incubated anaerobically for 2–3 days, surface colonies of the type strain are 0.5–1.0 mm in diameter, circular, entire, convex, opaque, smooth, shiny, and white with slight α-hemolysis. Other strains of the species are nonhemolytic on sheep blood agar.

There is only moderate growth in broth without fermentable carbohydrate. PYG cultures have a smooth (sometimes stringy) sediment with no, or only moderate, turbidity; after incubation for 1 day, the pH is 4.8–5.2.

No growth occurs in PY-6.5% NaCl-glucose broth. Growth usually is slightly inhibited by 20% bile. Growth in PYG is unaffected by 0.1% Tween 80 or 10% rumen fluid.

The temperature for optimal growth is 37°C; most strains grow well at 45°C but not 30°C.

In addition to the characteristics of 23 isolates from nine humans given in Table 12.41, the type strain and six other strains tested do not ferment adonitol, dextrin, glycerol or sorbose, and do not hydrolyze hippurate or produce acetyl methyl carbinol. Neutral red is reduced. Small to moderate amounts of ammonia are produced in chopped meat and PY cultures; no ammonia is detected from arginine or PYG cultures. The type strain and five of six other strains tested ferment galactose. The fermentation of inulin is variable among strains. Growth on anaerobically incubated EYA is variable; no lecithinase or lipase is produced by strains that grow on this medium.

Fermentation products (average meq/100 ml of culture) are as follows.

From PY-glucose: D(−)-lactate (3.0), butyrate (0.7) and acetate (0.5), sometimes with small amounts of succinate and pyruvate. No H_2 detected.

From PY-pyruvate: acetate (5) and butyrate (1–2), sometimes with formate.

From PY: acetate (1.5) and butyrate (0.2) acids, sometimes with trace amounts of pyruvate, lactate and succinate.

Threonine is not converted to propionate. Lactate and gluconate are not utilized.

The mol% G + C of the DNA is 40 (T_m) for the type strain.

Type strain: VPI C1-38 (ATCC 27758), isolated from human feces.

3. Coprococcus catus Holdeman and Moore 1974, 263.[AL]

ca' tus. L. adj. *catus* clever (referring to the unusual property of producing large quantities of both propionate and butyrate).

Cells usually occur in pairs that form long chains. Coccoid and oval cells of unequal size can occur in the pairs. Cells range in size from 0.8–1.4 μm in diameter by 1.6–1.9 μm in length (Fig. 12.26).

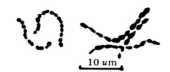

Figure 12.26. *Coprococcus catus.*

After incubation for 5 days in RGCA roll tubes, colonies are 0.5–1.0 mm in diameter, lenticular (occasionally trifoleate) and white to tan. On blood agar plates incubated anaerobically for 2 days, surface colonies of the type strain are 0.5–1.0 mm in diameter, circular, erose, umbonate, opaque, smooth, shiny and white, and produce slight α-hemolysis.

Slight to moderate growth occurs in PY broth. PY-fructose cultures have abundant growth with some turbidity and a smooth, crumb-like, or slightly ropy, sediment; after incubation for 5 days, the pH is 4.8–5.5. Although the pH in glucose broth does not go below 6.0, growth usually is stimulated by glucose, and propionate and butyrate production exceed that of PY cultures.

There is little growth in PY-6.5% NaCl-glucose broth. Growth in PYG is not affected by 20% bile, 0.1% Tween 80 or 10% rumen fluid.

There usually is no growth at 30°C; good growth occurs at 37°C and 45°C.

In addition to the characteristics of 23 isolates from nine humans given in Table 12.41, the type strain and four other strains tested do not ferment adonitol, dextrin, dulcitol, glycerol, inulin or sorbose; do not produce lecithinase or lipase (EYA), acetyl methyl carbinol, or urease; and do not hydrolyze hippurate. Neutral red is reduced. Galactose is very weakly fermented by two strains but not by the type strains and two other strains tested. Ammonia is not produced from cultures in chopped meat, PY, or arginine medium, but small amounts sometimes are detected in PYG and PY-gluconate cultures.

Fermentation products (average meq/100 ml of cultures) are as follows.

From PY-fructose: butyrate (5–7) and propionate (1–2) with trace amounts of acetate and sometimes lactate and succinate. Abundant H_2 is produced.

From PY-lactate: propionate (greater than 3.5), acetate (2) and butyrate (1.5), sometimes with very slight amounts of succinate, pyruvate, isobutyrate or isovalerate.

From PY-pyruvate: acetate (3.5), butyrate (2) and propionate (1), sometimes with trace amounts of lactate, succinate, isobutyrate or valerate.

From PY: butyrate (0.3), with trace amounts of propionate, succinate and acetate.

Threonine is not converted to propionate; gluconate is not used.

The cell wall of the type strain is reported to contain *meso*-diaminopimelic acid (*meso*-DAP) (Latham et al., 1979).

The mol% G + C of the DNA is 41 (T_m) for the type strain and 39 mol% for VPI C1-32.

Type strain: VPI C6-61 (ATCC 27761), isolated from human feces.

The production of large amounts of propionate with butyrate, and the limited number of carbohydrates fermented, are distinctive for this species.

Genus **Sarcina** Goodsir 1842, 434[AL]

ERCOLE CANALE-PAROLA

Sar.ci′na. L. fem. n. *Sarcina* a package, bundle.

Nearly spherical cells, 1.8–3 μm in diameter, **occurring in packets** of eight or more. Some of the cells in cultures may be present singly, or as groups of fewer than eight cells. Generally the cells are flattened in the areas of contact with adjacent cells. **Division occurs in three perpendicular planes.** Spore formation by these organisms has been reported (Knöll, 1965; Knöll and Horschak, 1971). **Grampositive. Nonmotile. Chemoorganotrophic anaerobes, having an exclusively fermentative metabolism.** Relatively aerotolerant. **Carbohydrates are the fermentable substrates.** The main products of glucose fermentation are CO_2, H_2, acetic acid as well as ethanol for *Sarcina ventriculi* and butyric acid for *Sarcina maxima*. Not pigmented. Catalase-negative. The minimal growth requirements include numerous amino acids and few vitamins, in addition to a fermentable substrate and inorganic salts. Grow at pH values near 1 and up to pH 9.8. The mol% G + C of the DNA ranges from 28–31 (Bd).

Type species: *Sarcina ventriculi* Goodsir 1842, 437.

Further Descriptive Information

Usually, packets of *S. ventriculi* consist of a greater number of cells than those of *S. maxima*. Many of the cells in large packets of *S. ventriculi* (e.g. packets comprising approximately 64 or more cells) have flattened shapes and are irregularly arranged (Canale-Parola et al., 1961; Holt and Canale-Parola, 1967; Smit, 1930). Thus, large packets tend to have a distorted appearance. Holt and Canale-Parola (1967) isolated strains of *S. ventriculi* that consistently form small packets. These strains are especially difficult to distinguish from *S. maxima* by means of light microscopy.

Cells of *S. ventriculi* are surrounded by a thick, fibrous layer of cellulose usually 150–200 nm in thickness (Canale-Parola et al., 1961; Canale-Parola and Wolfe, 1964). The cellulose layer around each cell is either continuous with, or attached to, the cellulose layer surrounding adjacent cells in the same packet (Canale-Parola et al., 1961; Holt and Canale-Parola, 1967). Apparently the cellulose layer functions as a matrix or cementing material that holds together cells of *S. ventriculi* into the large packets typical of this bacterium (Canale-Parola et al., 1961). In fact, strains of *S. ventriculi* that produce little or no cellulose form packets consisting of relatively few cells, loosely bound to one another (Canale-Parola et al., 1961; Holt and Canale-Parola, 1967). In contrast, cells of strains that form large packets are surrounded by a thick, fibrous layer of cellulose, with the result that the packets are crisscrossed by an intercellular network of this polymer. It has been suggested that nutrients diffuse from the external environment into the interior regions of the packets via this network of cellulose fibers (Canale-Parola, 1970). In this manner nutrients may reach cells that have no direct contact with the growth medium because they are located in the interior regions of the packets (Canale-Parola, 1970).

The cellulose layer is absent from cells of *S. maxima* (Canale-Parola, 1970).

The peptidoglycans of *S. ventriculi* and *S. maxima* contain LL-diaminopimelic acid. The interpeptide bridge consists of one glycine residue (Kandler et al., 1972). This type of peptidoglycan is not found in aerobic packet-forming cocci (Kandler et al., 1972.).

Subsurface colonies of *S. ventriculi* in agar media grow to several millimeters in diameter and are irregularly cubical or star-shaped. *S. maxima* forms smaller subsurface colonies, which may be cuboid with protuberances or shaped as uneven spheres. On the surface of agar media, in an anaerobic atmosphere, both organisms form roundish colonies which frequently have jagged edges.

Knöll (1965) and Knöll and Horschak (1971) reported that both *S. ventriculi* and *S. maxima* form endospores. The procedure they used to induce sporulation involved incubation of growing cells in CO_2 atmosphere, followed by addition of phosphate buffer and alkali to raise the pH of the cultures rapidly to 7.5 for *S. ventriculi* and to 9–10 for *S.*

maxima. Spore formation occurred during further incubation of the cultures in N_2. Spherical spores were formed by *S. ventriculi*, oval spores by *S. maxima*. The spores were heat resistant and stained green with Wirtz's (1908) spore stain (Knöll and Horschak,1971; Knöll, personal communication).

The optimum temperature range for growth is 30–37°C for *S. ventriculi* and 30–35°C for *S. maxima*. Growth of both species occurs between pH 1 and 9.8 (Smith, 1930, 1933), an extraordinarily wide pH range. *S. ventriculi* strains freshly isolated from natural environments can be subcultured indefinitely in media of very low pH, but when the strains are subcultured repeatedly at near neutral pH they tend to lose their ability to grow in media of pH 4.5 or lower (Canale-Parola, 1970).

S. ventriculi and *S. maxima* require for growth a fermentable carbohydrate (see Table 12.43), vitamins, amino acids, as well as inorganic salts. For example, *S. ventriculi* strain EC-1 requires biotin, nicotinic acid and 11 amino acids (serine, histidine, isoleucine, leucine, tyrosine, methionine, tryptophan, phenylalanine, arginine, valine and glutamic acid) (Canale-Parola and Wolfe, 1960b). A chemically defined medium containing glucose, inorganic salts and the above-mentioned vitamins and amino acids supports abundant growth of the organism (Canale-Parola and Wolfe, 1960b). A strain of *S. maxima* studied required thiamine, threonine, alanine and asparatic acid, in addition to the vitamins and amino acids required by *S. ventriculi* EC-1 (Knöll and Horschak, 1964).

In complex media (e.g. 2 g each of glucose and yeast extract/100 ml distilled water (DW) *S. ventriculi* EC-1 grew to cell yields of 8–12 g (wet weight)/liter (Canale-Parola et al., 1961). A complex medium for *S. maxima* consists of (g/100 ml DW): glucose and peptone, 1.0 each; yeast extract, 0.5; L-cysteine, 0.05; and $FeSO_4 \cdot 7H_2O$, 0.005 (Kupfer and Canale-Parola, 1967). One liter of this medium yielded approximately 2.4 g (wet weight) of *S. maxima* strain 11 cells. Cultivation procedures for both species of *Sarcina* have been described (Canale-Parola et al., 1961; Canale-Parola, 1970; Kupfer and Canale-Parola, 1967).

Carbohydrate fermentation by *S. ventriculi* yields mainly ethanol, acetate, CO_2 and H_2, whereas *S. maxima* ferments sugars primarily to butyrate, acetate, CO_2, and H_2 (see Table 12.44). Both species utilize the Embden-Meyerhof pathway for the fermentation of carbohydrates (Canale-Parola, 1970). Two enzymatic systems for pyruvate metabolism are present in *S. ventriculi*: a yeast-type decarboxylase that produces acetaldehyde and CO_2 from pyruvate, and a clostridial-type ferredoxin-dependent pyruvate clastic system that yields acetyl phosphate, CO_2 and H_2. Furthermore, CO_2 and H_2 are produced from formate by *S. ventriculi* via a reaction catalyzed by formate hydrogenlyase (Stephenson and Dawes, 1971). Like *S. ventriculi*, *S. maxima* possesses a clostridial-type pathway for pyruvate cleavage and metabolism. Cleavage of pyruvate by *S. maxima* results in the production of acetyl-CoA, CO_2 and electrons which are transferred to ferredoxin (Kupfer and Canale-Parola, 1967). Phosphotransacetylase (EC 2.3.1.8) catalyzes the conversion of acetyl-CoA to acetyl phosphate, which is metabolized to acetate in a reaction catalyzed by acetate kinase (EC 2.7.2.1). Furthermore, acetyl-CoA is metabolized to butyrate via a pathway similar to that present in saccharolytic clostridia and involving butyryl-CoA dehydrogenase (EC 1.3.99.2), phosphate butyryltransferase (EC 2.3.1.19), and butyrate kinase (EC 2.7.2.7) (Kupfer and Canale-Parola, 1968). *S. maxima* utilizes a hydrogenlyase system to convert formate to CO_2 and H_2 (Kupfer and Canale-Parola, 1967). The occurrence of the latter enzymatic activity, and the accumulation of small amounts of formate in cultures fermenting glucose suggest that, in addition to the clostridial-type system, *S. maxima* possesses a coliform-type pyruvate clastic system (Kupfer and Canale-Parola, 1967). The latter enzymatic system catabolizes pyruvate to acetyl phosphate and formate in coliform bacteria.

Growth of *S. ventriculi* takes place in the human stomach as a result of the development of certain pathological conditions (e.g. pyloric ulceration, stenosis) that retard the flow of food to the intestine. Under these abnormal circumstances, at the acid pH of the stomach and in the presence of carbohydrates and other growth nutrients contained in food, *S. ventriculi* thrives and multiplies rapidly (Smit, 1933). The sarcinae, which occur commonly in soil, are ingested with soil particles present in food.

S. ventriculi has been isolated from soil, mud, contents of diseased human stomach, rabbit and guinea pig stomach contents, elephant dung, human feces and the surface of cereal seeds. *S. maxima* has been isolated from the hull or outer coat of cereal grains, such as wheat, oat, rice and rye. It was also isolated from fresh wheat bran, horse manure and soil.

Isolation and Enrichment Procedures

Selective isolation procedures for *S. ventriculi* and *S. maxima* are based on the ability of these bacteria to grow anaerobically at very low pH (e.g. 2.0–2.5) in the presence of a fermentable carbohydrate.

The following is a procedure for the isolation of *S. ventriculi* (Canale-Parola, 1970). "The enrichment medium contains (g/100 ml of distilled or tap water): maltose, technical (Pfanstiehl Lab. Inc., Waukegan, Ill.), 2.0; malt extract broth (powder, BBL or Difco), 5.0; peptone (Difco), 0.5. The pH of the medium is adjusted to 2.2 ± 0.1 with diluted acid (e.g. 1 vol of H_2SO_4 (specific gravity 1.84) to 9 vol of H_2O). The medium is boiled for 2 or 3 min and, while still hot, it is poured into 60-ml glass-stopper bottles. The bottles, completely filled with medium, are cooled to 40°C in a cold-water bath. Garden soil (preferably saturated with water or growth medium to decrease the amount of air introduced in the bottles) is added to form a layer (2–4 mm) on the bottom of each bottle. The bottles are stoppered without trapping air bubbles and are incubated at 37°C.

After 16–48 h, successful enrichments exhibit vigorous gas production. Fine gas bubbles originating from the sarcinae in the sediment rise through the medium and form a layer of foam in the upper part of the bottles. Unless gas production is so active that it resuspends part of the sediment, the supernatant liquid is clear and essentially free of microbial growth since the large sarcina packets remain settled on the bottom. The supernatant liquid of enrichments containing a large number of contaminating organisms (frequently rod-shaped bacteria or yeasts) is turbid; the contamination develops as a result of a rise in pH, generally due to reactions between the acid in the culture and material present in the soil. When very alkaline soil is used, the initial pH of the medium should be lower than 2.2.

Second enrichment cultures, also in bottles, are prepared without delay by using the same procedure, except that 1–2 ml of sediment from the first enrichments is used as the inoculum. After 16–48 h of incubation, cells from the second enrichments may be used to inoculate identical third enrichments for the purpose of accomplishing further dilution of contaminating organisms. Serial dilutions of the growth in these cultures are plated to obtain isolated colonies.

Preparation of third enrichments may not be necessary since the sarcinae in the second enrichments are often so numerically predominant over other organisms that it is advantageous to plate serial dilutions directly from the second enrichments. The following medium (MYA) is used for plating (g/100 ml of distilled water): malt extract broth (powder, BBL or Difco), 2; maltose, technical (Pfanstiehl), 2; yeast extract (Difco), 0.1; agar, 2. The pH of the medium is adjusted to 6.0 ± 0.2 with 5% (w/v) KOH and the medium is sterilized. Serial dilutions are prepared in tubes of melted medium MYA at 45°C. These are poured in sterile petri dishes and, after solidification, are incubated anaerobically, or 15 ml of medium MYA is poured on the surface of the medium in each plate (double-layer plates) and the cultures are incubated in air.

Colonies of *S. ventriculi* appear after 10–32 h. Cells from the colonies are transferred to tubes containing medium MYA from which the agar

has been omitted. The medium in the tubes is heated for 5–10 min in a boiling-water bath, then cooled to 40°C before inoculation. Serial dilutions of the growth in these tubes are plated as described above to obtain pure cultures."*

A selective method for the isolation of *S. maxima* from soil has been described by Claus and Wilmanns (1974). In this method, D-xylose is used as the fermentable substrate in the enrichment medium to exclude growth of *S. ventriculi*, which does not ferment this pentose. The enrichment medium contains (g/100 ml DW): peptone (Difco), 0.5; yeast extract (Difco),0.5; D-xylose, 2.0. The medium is adjusted to pH 2 or 3 with HCl. The isolation procedure is similar to that described above for *S. ventriculi*.

Maintenance Procedures

Cells of *S. ventriculi* and *S. maxima* are viable only for 2–4 days in broth cultures in which nutrients are present at concentrations that do not limit growth. Thus, it is prudent to transfer the organisms on alternate days so that viable cells may be maintained in such cultures. The physiological bases for the rapid loss of viability have not been elucidated. Viability of *S. ventriculi* cells can be prolonged by using growth-limiting concentrations of fermentable carbohydrate or of other nutrients in the culture medium (Canale-Parola and Wolfe, 1960a). Furthermore, when K_2HPO_4 or $NaHCO_3$ is added to culture media containing a growth-limiting concentration of fermentable substrate, most strains of *S. ventriculi* survive for 30 days under anaerobic conditions (Claus et al., 1970).

A convenient method for maintenance of *S. ventriculi* or *S. maxima* involves heavy inoculation of the organism in a small well, melted through the surface of a relatively large volume of agar medium contained in an Erlenmeyer flask (Canale-Parola and Wolfe, 1960a). These flask stock cultures must be kept at a temperature (e.g. 30 or 37°C) that allows continuous growth of the cells. Cells grow only within or near the well and remain viable for approximately 2 months.

Cells of *S. ventriculi* and *S. maxima* retain viability for at least several years in liquid N_2 storage (Canale-Parola, 1970).

Procedures for Testing Special Characters

The fact that cells of *S. ventriculi* are surrounded by a layer of cellulose (see Further Descriptive Information), whereas those of *S. maxima* are not, is a useful characteristic for differentiation. The following staining procedure may be used to demonstrate the presence or absence of the cellulose layer in sarcina packets (E. Canale-Parola, unpublished data).

Approximately 5 g (wet weight) of *S. ventriculi* or *S. maxima* cells are refluxed in 100 ml boiling 2% (w/v) KOH for 1 h to remove most of the cytoplasmic material from the cells. Then the cells are harvested by centrifugation and washed four times (centrifugation) with DW. The white, washed pellet consists of almost "empty" cells still arranged in packets. Material from the pellet, from which as much water as possible has been drained, is smeared thickly on a glass microscope slide and the staining solution applied dropwise directly on the smear.

An Iodine-Zinc Chloride ($I-ZnCl_2$) solution used for staining is prepared as follows. Twenty grams $ZnCl_2$ are dissolved in 8.5 ml water and the solution is allowed to cool to room temperature. Then, iodine solution (DW, 60 ml; KI, 3 g; I, 1.5 g) is added to the $ZnCl_2$ solution dropwise until iodine begins to precipitate. Only a few ml of iodine solution are required. When the $I-ZnCl_2$ solution is added to the treated *S. ventriculi* cells smeared on a glass slide, the cellulose present in *S. ventriculi* packets stains blue and, consequently, the smear on the glass slide appears blue. In contrast, smears of *S. maxima* (refluxed with KOH solution and washed as described above) remain colorless when treated with $I-ZnCl_2$ solution. Cellulose powder (Nutritional Biochemicals), treated with the same procedure, and untreated filter paper stain blue with the $I-ZnCl_2$ solution.

Smit's cellulose-staining solution (Smit, 1930) may be used instead

---

*Reproduced with permission from E. Canale-Parola, *Bacteriological Reviews*, *34:* 82–97, 1970, American Society for Microbiology.

of the I-ZnCl₂ solution. Smit's solution consists of: DW, 100 ml; KI, 26 g; I, 0.4 g; and ZnCl₂, 85 g. Prior to staining, the cells are refluxed with KOH solution, washed and smeared on a glass slide as described above. Smears of *S. ventriculi* cells stained with Smit's solution become dark brown with a reddish-purplish tinge. The same color appears when treated cellulose powder or untreated filter paper is stained with Smit's solution. Smears of treated *S. maxima* cells remain colorless.

Ethanol, which is a major fermentation end product of *S. ventriculi* but is not formed by *S. maxima*, may be determined enzymatically (Ethyl Alcohol Reagent Set, Worthington Diagnostics, Freehold, N.J., USA) or by gas-liquid chromatography techniques. Butyrate, produced by *S. maxima* but not by *S. ventriculi*, and acetate (produced by both species) are assayed by gas-liquid chromatography.

Taxonomic Comments

John Goodsir discovered and named *Sarcina ventriculi* in 1842 (Goodsir, 1842). Subsequently, other investigators described and cultivated various packet-forming cocci, which were assigned to the genus *Sarcina*. Some of these packet formers were aerobes (e.g. *Sarcina lutea*), others anaerobes (e.g. *Sarcina methanica*). Some of the aerobic packet-forming cocci were flagellated and formed spores (e.g. *Sarcina ureae*).

In time it became evident that *S. ventriculi*, and the closely related *S. maxima*, are phylogenetically distant from other packet-forming cocci (Canale-Parola et al., 1967; Canale-Parola, 1970). Aerobic, nonspore-forming, nonmotile, packet-forming cocci were assigned to the genus *Micrococcus*; anaerobic, methanogenic, packet-forming cocci to the genus *Methanosarcina*; and aerobic, sporeforming sarcinae to the genus *Sporosarcina*. Only the anaerobic, sugar-fermenting species (*S. ventriculi* and *S. maxima*) were retained in the genus *Sarcina* (Canale-Parola, 1970).

According to 16S rRNA sequence characterization, *S. ventriculi* is closely related phylogenetically to *Clostridium butyricum* and other clostridia, but it is distant from packet-forming cocci in the genera *Micrococcus*, *Methanosarcina* and *Sporosarcina* (Fox et al., 1980). The 16S rRNA sequence analysis of *S. maxima* has not been reported.

Further Reading

Canale-Parola, E. 1970. Biology of the sugar-fermenting sarcinae. Bacteriol. Rev. *34:* 82–97.

Kandler, O., D. Claus and A. Moore. 1972. Die Aminosäuresequenz des Mureins von *Sarcina ventriculi* und *Sarcina maxima*. Arch. Mikrobiol. *82:* 140–146.

Claus, D. and H. Wilmanns. 1974. Enrichment and selective isolation of *Sarcina maxima* Lindner. Arch. Microbiol. *96:* 201–204.

Differentiation and characteristics of the species of the genus **Sarcina**

The differential characteristics of the species of *Sarcina* are listed in Table 12.42. Other characteristics of the species are indicated in Tables 12.43 and 12.44.

Table 12.42.

Differential characteristics of the species of the genus **Sarcina**[a]

Characteristics	1. *S. ventriculi*	2. *S. maxima*
Ethanol production	+	−
Cellulose formation	+	−
Butyrate production	−	+
Fermentation of D-xylose	−	+

[a] Symbols: see Table 12.2.

Table 12.43.

Characteristics of the species of the genus **Sarcina**[a, b]

Characteristics	1. *S. ventriculi*	2. *S. maxima*
Cell diameter (μm)	1.8–2.4	2–3
Packet formation	+	+
Anaerobic	+	+
Pigmentation	−	−
Catalase	−	−
Growth at pH 2	+	+
Cellulose formation	+	−
Mol% G + C of DNA (Bd)	30.6	28.6
Products from sugars		
CO₂, H₂ and acetate	+	+
Ethanol	+	−
Butyrate	−	+
Fermentable substrates[c]:		
D-Arabinose	−	d
L-Arabinose	d	+
D-Ribose	−	−
D-Xylose	−	+
D-Fructose	+	+
D-Galactose	+	+
D-Glucose	+	+
D-Mannose	+	+
Cellobiose	d	−
Lactose	+	−[d]
Maltose	+	+
Melibiose	+	−
Sucrose	+	+

[a] Symbols: see Table 12.2.

[b] The following substrates are not fermented by either species: trehalose, raffinose, dextrin, starch, glycogen, glycerol, dulcitol, mannitol and amino acids. Citrate, gluconate and succinate are not fermented by *S. ventriculi* (fermentation of these compounds by *S. maxima* was not tested).

[c] Data from Smit, 1933; Canale-Parola and Wolfe, 1960b; Claus and Wilmanns, 1974.

[d] Smit (1933) reported that *S. maxima* ferments lactose.

Table 12.44.
Glucose fermentation by **S. ventriculi** *and* **S. maxima**[a, b]

Product[c]	Amount of product[d]				
	S. ventriculi			S. maxima	
	1	2	3	4	5
Carbon dioxide	195	190	190	149	197
Hydrogen	41	170	140	230	223
Formic acid	3	Trace	NR	4	NP[e]
Acetic acid	20	90	60	30	40
Butyric acid	NR	NR	NP	76	77
Lactic acid	NR	NR	10	21	NP
Succinic acid	NR	NR	NR	5	NR
Ethanol	171	80	100	Trace	NP
Acetoin	4	NR	Trace	NR	NP
Neutral volatile (as butanol)	NR	NR	NR	NR	9
Percent carbon recovered	99.3	88.3	90.0	100.0	103.5
Oxidation-reduction balance	1.0	1.15	1.12	0.80	0.95

[a] Table adapted from Canale-Parola (1970), with permission of the publisher.
[b] Symbols: NR, not reported; and NP, not present in detectable amounts.
[c] Fermentation products of growing cells, except for the data in column 2 which were obtained with cell suspensions. Data in column 1 are from Kluyver (1931), in column 2 from Milhaud et al. (1956), in column 3 from Canale-Parola and Wolfe (1960a), in column 4 from Smit (1930) and in column 5 from Kupfer and Canale-Parola (1967). Data in columns 1 and 4 were originally reported as percentages of glucose fermented.
[d] Expressed as micromoles of product/100 μM of glucose fermented.
[e] Detected when cells were grown in media containing $CaCO_3$.

List of species of the genus **Sarcina**

1. **Sarcina ventriculi** Goodsir 1842, 437.[AL]
ven.tri'cu.li. L. n. *ventriculus* the stomach; L. gen. n. *ventriculi* of the stomach.

Nearly spherical cells, 1.8–2.4 μm in diameter, occurring in packets of eight to several hundred or more. Large packets (e.g. consisting of approximately 60 or more cells) tend to have an irregular or distorted appearance. Frequently, cells in these packets exhibit flattened shapes and are irregularly arranged. A fibrous layer 150–200 nm thick, composed either totally or in great part of cellulose, is present on the outer surface of the cell wall. This layer is absent in some strains. Reported to form spherical spores (Knöll, 1965). Subsurface colonies in agar media are star-shaped or irregularly cubical, and measure up to several mm in diameter. Surface colonies (anaerobic) are roundish, often with rugged edges.

Carbohydrates are fermented. Amino acids are not fermented. Carbohydrate fermentation patterns and other physiological characteristics are summarized in Table 12.43. The main products of glucose fermentation are ethanol, acetate, CO_2, and H_2 (Table 12.44). Two strains studied required for growth two vitamins (biotin, nicotinic acid and eleven amino acids (see Further Descriptive Information, above), in addition to a fermentable carbohydrate and inorganic salts.

Temperature optimum: 30–37°C.

Isolated from soil, mud, contents of diseased human stomach, rabbit and guinea pig stomach contents, elephant dung, human feces and the surface of cereal seeds.

The mol % G + C content of the DNA is 30.6 ± 1 (Bd).

Phase contrast photomicrographs and electron micrographs of this species have been published (Canale-Parola et al., 1961; Canale-Parola, 1970).

Type strain: DSM 286 (ATCC 19633).

2. **Sarcina maxima** Lindner 1888, 54.[AL]
max'i.ma. L. sup. adj. *maximus* greatest, largest.

Nearly spherical cells, 2–3 μm in diameter, occurring in packets of eight or more. The cells lack a cellulose outer layer. Reported to form oval spores (Knöll, 1965). Subsurface colonies in agar media are cuboid with protuberances or unevenly spherical. Surface colonies (anaerobic) are roundish, often with rugged edges.

Carbohydrates are fermented. Amino acids are not fermented. Carbohydrate fermentation patterns and other physiological characteristics are summarized in Table 12.43. The main products of glucose fermentation are butyrate, acetate, CO_2, and H_2 (Table 12.44). A strain studied required for growth three vitamins (thiamine, biotin, nicotinic acid) and 14 amino acids (see Further Descriptive Information, above), in addition to a fermentable carbohydrate and inorganic salts.

Temperature optimum: 30–35°C.

Isolated from the hull or outer coat of cereal grains, such as wheat, oat, rice and rye. Fresh wheat bran has been used as a source. Also isolated from horse manure, field soil, and garden soil.

The mol % G + C of the DNA is 28.6 ± 1 (Bd).

Phase contrast photomicrographs and electron micrographs of this species have been published (Smit, 1930; Canale-Parola, 1970).

Type strain: DSM 316.

SECTION 13

Endospore-forming Gram-Positive Rods and Cocci

Peter H. A. Sneath

The endospore-forming bacteria, most of which are Gram-positive motile rods, from a diverse assemblage that is a grouping of convenience. For determinative purposes most of these bacteria are retained in the present section, although some branching filamentous forms that are morphologically similar to streptomycetes are treated in the forthcoming Volume 4.

Distinguishing features of the genera in the present section, with some that may be confused if spore formation is not observed, are shown in Table 13.1. The two genera forming endospores that are treated in Volume 4 comprise *Thermoactinomyces* and *Pasteuria*. *Thermoactinomyces* forms abundant branched filaments. Its colonies produce aerial mycelium bearing endospores; these may be on short sporophores or may be sessile (Cross, 1981). The status of *Pasteuria* is at present uncertain. The organism listed in the Approved Lists of Bacterial Names (Skerman et al. 1980) has a type culture that is not in close agreement with the original description of Metchnikoff (1888). What appears to be Metchnikoff's organism has recently been shown to possess spores borne on the tips of short branching filaments, and although it has not yet been cultivated the spores have the appearance of true endospores on electron microscopy (for details see Sayre et al., 1983 and Starr et al., 1983).

A number of other organisms have been reported that have not been cultivated, but which appear on morphological grounds to possess endospores. The best known is *Oscillospira* which is included in the present section. Others include "*Sporospirillum*" (Delaporte, 1964a) and "*Fusosporus*" (Delaporte, 1964b, 1969) from the alimentary tract

Table 13.1.

Differential characters of endospore-forming bacteria and similar genera[a, b]

Characteristics	Genera with endospores						Genera without endospores		
	Bacillus	*Sporolactobacillus*	*Clostridium[c]*	*Desulfotomaculum*	*Sporosarcina*	*Oscillospira[d]*	*Planococcus*	*Lactobacillus*	*Kurthia*
Rod-shaped	+	+	+	+	−	+	−	+	+[e]
Diameter over 2.5 μm	−	−	−	−	−	+	−	−	−
Filaments	−	−	D	−	−	+[f]	−	−[g]	−
Rods or filaments curved	−	−	D	D	NA	+	NA	−	d
Cocci in tetrads or packets	−	−	−	−	+	−	d	−	−
Endospores produced	+	+	+	+	+	+	−	−	−
Motile	+	+	+	+	+	+	+	−	+
Stain Gram-positive at least in young cultures	+	+	+[h]	−[i]	+	−[j]	+	+	+
Strict aerobes	D	−	−	−	+	ND	+	−	+
Facultative anaerobes or microaerophils	D	+	−[k]	−	−	ND	−	+	−
Strict anaerobes	−	−	+	+	−	+[j]	−	−	−
Homolactic fermentation	D	+	−	−	−	ND	−	−	−
Sulfate actively reduced to sulfide	−	−	−	+	−	ND	−	−	−
Catalase	+	−	−	−[l]	+	ND	+	−	+
Oxidase	D	ND	−	ND	+	ND	−	−	−
Marked acidity from glucose	+	+	D	−	−	ND	+	+	−
Nitrate reduced to nitrite	D	−	D	ND	D	ND	−	−	−
Mol% G + C	32–69	38–40	24–54	37–50	40–42	ND	39–52	32–53	36–38

[a] Symbols: +, 90% or more of strains positive; −, 10% or less of strains positive; d, 11–89% of strains positive; D substantial proportion of species differ; NA not applicable; ND not determined;
[b] For *Thermoactinomyces* and *Pasteuria*, see text.
[c] Strains producing few spores may be confused with *Eubacterium*, *Bifidobacterium*, *Bacteroides*, *Fusobacterium* or *Propionibacterium*: see also *Sarcina*
[d] Morphologically similar to *Caryophanon*.

[e] Sometimes coccoid in old cultures.
[f] Broad discs of cells form filament.
[g] Chains common.
[h] Rarely Gram-negative.
[i] *D. nigrificans* reported Gram-positive (McClung and McCoy, 1957).
[j] Uncertain, see text.
[k] Rarely aerotolerant.
[l] Norris et al. (1981).

of amphibian larvae. These bacteria are large, (several micrometers in diameter and up to 150 μm long) and often contain two spores per cell. "*Sporospirillum*" is helical and "*Fusosporus*" is a rod with pointed ends. Where reported the Gram reaction was said to be negative. "*Metabacterium*" from the guinea pig caecum is another large rod-shaped organism containing up to eight spores per cell (Chatton and Perard, 1913). Brief notes on "*Sporospirillum*" are given in Volume 1, pp. 89–90.

Those unfamiliar with handling strict anaerobes should note that some strains of *Clostridium* may sporulate very poorly, and if spores are not observed such strains may appear to belong to *Eubacterium*, *Bifidobacterium* or *Propionibacterium*, or to the Gram-negative genera *Bacteroides* or *Fusobacterium* (see Volume 1). Similar confusion may sometimes occur between aerobic genera. Spores have been reported in *Sarcina* (see the contribution on that genus).

For determinative work, therefore, the observation of endospores is important. Bacterial endospores are round or oval (occasionally cylindrical) structures that form within the bacterial cell. They are highly refractile and contain dipicolinic acid. In electron microscope sections they show a thin outer spore coat, a thick spore cortex and an inner spore membrane surrounding the spore contents. They are highly resistant to lethal effects of heat, drying and many disinfectants; they survive in nature in a dormant state for long periods (up to centuries in some instances, Sneath, 1962; Seaward et al., 1976). This longevity is of great importance in their ecology (Slepecky and Leadbetter, 1983).

Endospores are recognized by their regular size and shape, their consistent position when seen within the cell, and by their highly refractile appearance. Because of their impermeability they stain poorly with the usual stains, but they may be stained by special methods. They may be acid-fast when stained by the stains used for mycobacteria. Endospores may occasionally be confused with lipid inclusions. On germination in suitable media they first lose their refractibility and then the new cell grows out. Endospores also show a characteristic reaction to N-HNO$_3$, whereby they "explode" (Robinow, 1951). It is wise to confirm that the structures seen are endospores. This is best done by testing that cultures survive heating: temperatures of 70–80°C for 10 min are usually recommended, followed by cultivation under suitable conditions. Alternatively, ability to survive in 95% ethanol for 45 min at 20°C may be tested. Technical suggestions may be found in the contributions on *Bacillus* and *Clostridium* and in Hobbs and Cross (1983), and Slepecky and Leadbetter (1983). Notes on optimal conditions for sporulation may also be found there. Endospores are best sought in old cultures, which should be kept for at least 10 days before discarding. Endospore formation may be suppressed by unfavorable conditions. Strains of *Bacillus* may require media with additional manganous ions. Clostridia often do not spore if the medium is not highly anaerobic, or if it becomes excessively acidic.

The detailed taxonomic relationships of the endospore-forming bacteria are currently uncertain. Evidence from rRNA shows that *Bacillus* and *Clostridium* belong to a major branch of procaryotes which also contains genera such as *Staphylococcus, Streptococcus, Lactobacillus, Eubacterium, Acetobacterium, Sarcina* and probably the mycoplasmas (Stackebrandt and Woese, 1981; Kandler, 1982). *Bacillus* and *Clostridium* are very diverse phenotypically as well as genomically, as shown, for example, by the wide range of G + C ratios and by the difficulty in finding constant characters for these two genera. *Clostridium* in particular is represented by a number of rRNA subbranches, one of which is allied to *Peptococcus, Peptostreptococcus* and *Ruminococcus*.

Sporosarcina and *Planococcus* are close to *Bacillus* on 16S-rRNA evidence (Pechman et al., 1976; Stackebrandt and Woese, 1979). *Thermoactinomyces vulgaris* is also closer to *Bacillus subtilis* than to *Streptomyces* and similar organisms (Stackebrandt et al., 1983); this is supported by menaquinone data of Collins et al. (1982).

There is, therefore, an emerging picture showing that the endospore formers are a distinctive branch of bacteria, but that this branch also contains genera that do not form endospores. But considerable rearrangement of these organisms is also implied: thus *Sporosarcina ureae* appears closer to *Bacillus pasteurii* than either is to *Bacillus subtilis* (Pechman et al., 1976). It is evident that current generic groupings place much emphasis on sporulation and cell shape; if these are discounted, the resemblances between some species in genera such as *Bacillus, Sporosarcina* and *Planococcus*, or *Bacillus, Sporolactobacillus* and *Lactobacillus* are very strong. Mention has been made of the risk of confusion between *Clostridium, Eubacterium, Propionibacterium, Bacteroides* and *Fusobacterium*. It is not clear how often strains of the last four are later found to produce spores and are then considered to be clostridia, but there is evidence that some *Bacteroides* and allied forms, and some *Clostridium* strains, are both phenetically and phenotypically very similar (Barnes and Goldberg, 1968). It is therefore very likely that some species in various genera will ultimately be recognized as species of *Clostridium* that are unable to form spores perhaps because of unsuitable conditions or limited genetic defects. Nevertheless, the present circumscription of *Clostridium* is evidently arbitrary, depending on spore formation, and it seems very likely that *Clostridium* itself, and probably *Bacillus*, will be divided into a number of separate genera in the future.

Further Reading

Hobbs, G. and T. Cross. 1983. Identification of endospore-forming bacteria. *In* Hurst and Gold (Editors), The Bacterial Spore, Vol. II, Academic Press, London, pp. 49–78.
Kandler, O. 1982. Cell wall structures and their phylogenetic implications. Zentralbl. Bakteriol. Mikrobiol. Hyg. I. Abt. Orig. C. *3:* 149–160.
Slepecky, R.E. and E.R. Leadbetter. 1983. On the prevalence and roles of spore-forming bacteria and their spores in nature. *In* Hurst and Gould (Editors), The Bacterial Spore Volume 2, Academic Press, London, pp. 79–99.
Stackebrandt, E. and C.R. Woese. 1981. The evolution of prokaryotes. *In* Molecular and Cellular Aspects of Microbial Evolution. *In* Carlile, Collins and Moseley (Editors), Symp. Soc. Gen. Microbiol. *32:* 1–31, Cambridge University Press, London.

Genus **Bacillus** Cohn 1872, 174[AL]*

D. CLAUS AND R. C. W. BERKELEY

Ba.cil′lus. L. dim. n. *bacillum* a small rod; M.L. n. *Bacillus* a rodlet.

Cells rod-shaped, straight or nearly so; **endospores, very resistant to many adverse conditions, formed;** not more than one per cell; **sporulation not repressed by exposure to air. Gram-positive,** or positive only in early stages of growth, or negative. **Flagella peritrichous** or degenerately peritrichous. **Aerobic or facultatively anaerobic.** Oxygen is the terminal electron acceptor replaceable in some species by alternatives. Colony morphology and size very variable; pigments may be produced on certain media. Exhibit a wide diversity of physiological ability; psychrophilic to thermophilic; acidophilic to alkaliphilic; some strains are salt tolerant, others have specific requirements for salts. **Catalase formed by most species.** Oxidase-positive or negative. **Chemoorganotrophs;** one species a facultative chemolithotroph: **prototrophs to auxotrophs requiring several growth factors. The cell wall peptidoglycan of most species belongs to the directly cross-linked *meso*-diaminopimelic acid type.** The main isoprenoid quinone is menaquinone with seven isoprene units (MK-7). Terminally methyl-branched *iso-* and *anteiso*-fatty acids having 12 to 17 carbons predominate. Phospholipids that occur most commonly are phosphatidylethanolamine and phosphatidylglycerol. Most species widely distributed in nature. Occurence not necessarily related to the natural habitat because of passive distribution and persistence of spores. *B. anthracis* is a pathogen of humans and some

---

* *AL* denotes the inclusion of this name on the Approved Lists of Bacterial Names (1980).

other animals; *B. thuringiensis*, *B. larvae*, *B. lentimorbus*, *B. popilliae* and certain strains of *B. sphaericus* are insect pathogens; certain strains of *B. cereus* play some role in food-transmitted gastroenteritis; other species may be opportunistic pathogens. The mol% G + C of the DNA is 32–69 (T_m, Bd).

Type species: *Bacillus subtilis* Cohn 1872, 174.

Further Descriptive Information

Cell Morphology. Cells of *Bacillus* may occur singly or in chains which may be of considerable length. The rods may have rounded or squared ends and may be quite small (0.5 × 1.2 μm) or rather large (2.5 × 10 μm). The cytoplasm may be vacuolate or may stain uniformly. Cells may contain parasporal bodies or protein crystals (Fig. 13.1). The form of the endospore and the shape of the spore-bearing mother cell, the sporangium, is a characteristic feature of *Bacillus* species. Endospores are usually cylindrical, or ellipsoidal, or oval or round, but some strains of certain species produce either kidney- or banana-shaped endospores. The spore may be located in a central, paracentral, subterminal, terminal or lateral position within the sporangium. During the formation of endospores the mother cell may not change its shape or may swell (Fig. 13.2). Gordon et al. (1973) arranged the strains they studied in three groups based on the shape of the spore and swelling of the sporangium. Group 1 has ellipsoidal spores which do not cause the sporangium to swell, group 2 has ellipsoidal spores swelling the sporangium and group 3, spherical spores with swollen sporangia. Correlating fairly well with this grouping are the nutritional requirements of these organisms. Mature endospores may be released from the sporangium as a result of lysis. In any one culture of some species, spherical and

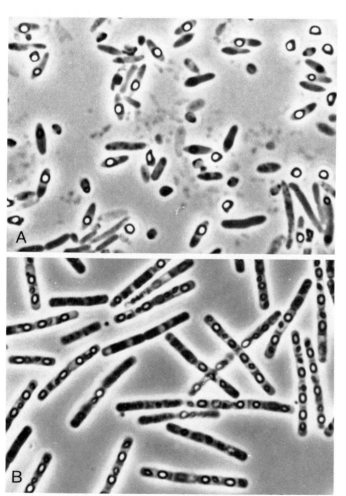

Figure 13.1 A and B.

ellipsoidal spores may be seen. The spore coat may have well defined markings and in some an exosporium may be present. Capsules are synthesized under some conditions by some strains. Black, yellow, orange/brown and red pigments may be produced under some conditions of culture by some species.

Cell Wall Composition. The predominant murein type of strains of the genus *Bacillus* is of the direct linked *meso*-diaminopimelic acid (*meso*-DAP or mA_2pm) type (Schleifer and Kandler, 1972). However, only 17 type strains of the 34 validly published species have been studied in respect to their peptidoglycan (murein) composition (see Table 13.3.)

According to Ranftl (1972) the murein type of five strains of *B. sphaericus* belongs to the L-Lys-D-Asp type confirming an earlier study of Hungerer and Tipper (1969). The type strain of *B. sphaericus* was not, however, included. Since within the species more than five DNA/DNA homology groups have been found (Krych et al., 1980) the murein type of *B. sphaericus sensu stricto* can not be deduced.

Whereas four strains identified as *B. pasteurii* have been found to belong to the murein type L-Lys-L-Ala-D-Asp (Ranftl and Kandler, 1973) the type strain of the species has been reported to belong to the L-Lys-D-Asp type as was found in *B. sphaericus* (Schleifer and Kandler, 1972). A reexamination of the *B. pasteurii* type strain (DSM 33, a subculture of ATCC 11859) by O. Kandler (personal communication) has shown that this strain, too, forms the L-Lys-L-Ala-D-Asp type of the murein.

In the peptidoglycan of spores of *B. pasteurii* and *B. sphaericus meso*-DAP has been found (Powell and Strange, 1957; Ranftl and Kandler, 1973).

Cell wall teichoic acids or teichuronic acids have been detected in several *Bacillus* species (Ranftl, 1972). Their taxonomic implications have not yet been established (Schleifer and Kandler, 1972), as is also the case for the cell wall polysaccharides found in the genus *Bacillus*.

Fine Structure. *Capsules.* Several *Bacillus* species produce carbohydrate capsules; levan and dextran structures being not uncommon. Chemically more complex polysaccharides are also produced. *B. circulans* elaborates a glucose and glucuronic acid extracellular polymer, for example (Cagle, 1974). The production of these structures and their chemical nature is not of any great taxonomic consequence. Indeed, several *Bacillus* strains produce polysaccharides which cross-react with antisera against those from other genera; *B. mycoides* with *Streptococcus pneumoniae* type III, *B. pumilus* with *Neisseria meningitidis* group A and *Haemophilus influenzae* type b, and *B. alvei* with *S. pneumoniae* Type III and *H. influenzae* type b (Myerowitz, et al., 1973). There was, however, no reaction with 15 strains of *B. circulans* or with representatives of several other species.

B. anthracis produces a poly-γ-D-glutamyl capsule in vivo or on serum plates in a 5% carbon dioxide atmosphere (Meynell and Meynell, 1964). This is the basis of the McFadyean reaction which involves staining with polychrome methylene blue to give blue rods surrounded by purple/pink-stained capsular material in a positive test. *B. subtilis* (Bovarnik, 1942), *B. licheniformis* (Gardner and Troy, 1979), *B. megaterium* (Guex-Holzer and Tomcsik, 1956), *Sporosarcina halophilia* and *Planococcus halophilus* (Kandler et al., 1983) are also known to produce this polymer. The species most closely related to *B. anthracis* do not, however, appear to synthesize this polymer.

B. megaterium can synthesize a capsule containing both polypeptide and polysaccharide; the former located laterally and the latter at the poles and equators of cells (Guex-Holzer and Tomcsik, 1956). When grown on a fructose mineral salts medium the structure is polysaccharide in nature and appears to be fibrillar in character when examined after thin-sectioning, freeze-etching and critical point drying by transmission electron microscopy and scanning electron microscopy (Cassity et al., 1978). The capsule in *B. circulans* also appears to be fibrillar when examined using a ruthenium red staining technique (Cagle, 1974).

Flagella. Most *Bacillus* species are motile by means of peritrichous flagella. In some instances only very few may be seen and they are never particularly numerous. Flagellation is not a character of tax-

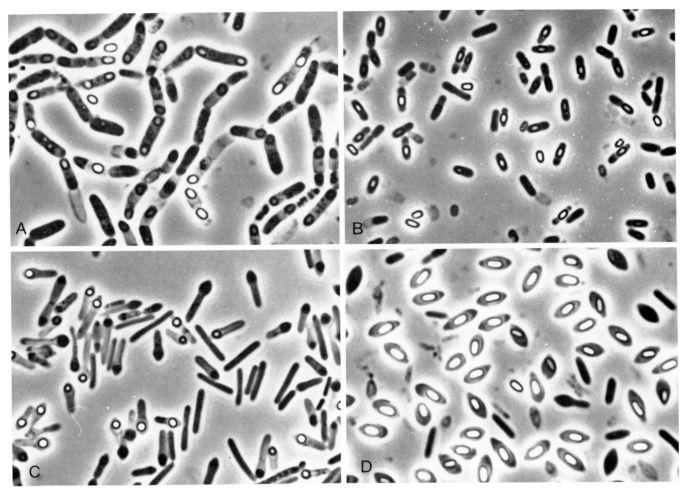

Figure 13.2 A–D.

onomic use in this genus although some use has been made of lack of motility. *B. anthracis* is normally nonmotile but one motile strain has been identified (Brown and Cherry, 1955); *B. cereus* is normally motile but nonmotile strains are sometimes encountered. For example, Logan and Berkeley (1984) found 5 of 149 strains to be nonmotile.

More use has been made of the H-serotyping schemes available for *B. cereus* (Le Mille *et al.*, 1969b; Gilbert and Parry, 1977), *B. thuringiensis* and *B. sphaericus* (de Barjac, 1981a; 1981b). The number of new H-serotypes of *B. thuringiensis* is increasing rapidly and a procedure for avoiding confusion in their numbering has been proposed (Burges et al., 1982). A considerable amount of information is available on the chemistry of the flagella of *B. subtilis*. For one strain, 168, the entire amino acid sequence is known (DeLange et al., 1976) and limited molecular interpretation of serological phenomena is possible in some instances.

Sixteen *Bacillus subtilis* strains were divided into five groups on the basis of the antigenicity of their flagellar filaments. These groups correlated well with the grouping established on the basis of phenylalanine and tyrosine content. In contrast, the antigenicity of flagellar hooks from *B. subtilis* strains and, indeed, from *Salmonella* was remarkably homogenous. It has been suggested that the structures of filaments are suited to the environment of the organism (Simon et al., 1977). In keeping with this idea is the low, basic amino acid content of the flagella of *B. firmus* RAB, an alkaliphile, probably rendering it more stable at environmental pH values up to 11.0 (Guffanti and Eisenstein, 1983).

Spores. The shape of the spore, its position in the sporangium and its size relative to the sporangium are of taxonomic use (Table 13.2

and see Table 13.4). In addition, the appearance of the spore surface as revealed by the carbon replica technique or scanning electron microscopy appears to be characteristic of some species (Bradley and Franklin, 1958), although Nadirova and Aleksandruskina (1979), having recognized three surface patterns in 10 strains of *B. megaterium*, concluded that surface appearance is not a sufficient basis for allocation to a particular taxon. In some species exosporia are produced but in *B. megaterium*, at least, this is a variable character. Two different types of exosporium can be recognized and some strains do not produce either type of exosporium (Koshikawa et al., 1984).

In cross-section, *Bacillus* spores show a more complex ultrastructure to that seen in vegetative cell sections. The spore protoplast (spore core) is surrounded by the germ cell wall, the cortex and then the spore coats. Depending on the species, an exosporium may be present. Whereas the interior of all *Bacillus* spores seems to be essentially the same, there are major differences between species in the organization and number of layers surrounding the protoplast (Fitz-James and Young, 1969; Russell, 1982). The taxonomic implications of these ultrastructure differences are, on the whole, not clear.

Cell walls. The vegetative *Bacillus* cells so far studied, like many other procaryotes, form paracrystalline cell wall surface layers (S-layers), which completely cover the cell surface. These layers, which can be detected by freeze-etching methods, consist of proteins or glycoproteins and often possess a high degree of structural regularity (Sleytr and Messner, 1983).

In *Bacillus* species, the type of lattice has been found to be hexagonal in *B. alvei*, *B. anthracis*, *B. brevis* and few strains of *B. stearothermophilus*. A square lattice has been observed with *B. cereus*, *B. fasti-*

Table 13.2.
Morphology of Bacillus species[a]

Characteristics	1. B. subtilis	2. B. acidocaldarius	3. B. alcalophilus	4. B. alvei	5. B. anthracis	6. B. azotoformans	7. B. badius	8. B. brevis	9. B. cereus	10. B. circulans	11. B. coagulans	12. B. fastidiosus	13. B. firmus	14. B. globisporus	15. B. insolitus	16. B. larvae	17. B. laterosporus
Width of rod (μm)	0.7-0.8	0.9-1.1	0.7-0.9	0.5-0.8	1.0-1.2	0.9-1.0	0.8-1.2	0.6-0.9	1.0-1.2	0.5-0.7	0.6-1.0	1.5-2.5	0.6-0.9	0.6-1.0	1.1-1.5	0.5-0.6	0.5-0.6
Length of rod (μm)	2-3	2-3	3-4	2-5	3-5	3-10	1.5-4	1.5-4	3-5	2-5	2.5-5	3-6	1.2-4	1.5-5	1.1-2.5	1.5-6	1.5-6
Sporangium swollen	-	+	-	+	-	+	-	+	-	+	d	-	-	+	-	+	+
Spore shape	E	E	E	E	E	E	E	E	E	E	E	E	E	S	S	E	E
Spore position	C	T	T	C/T	C	T	C/T	C/T	C	C/T	C/T	C/T	C	T	C/T	C/T	CL

Characteristics	18. B. lentimorbus	19. B. lentus	20. B. licheniformis	21. B. macerans	22. B. macquariensis	23. B. marinus	24. B. megaterium	25. B. mycoides	26. B. pantothenticus	27. B. pasteurii	28. B. polymyxa	29. B. popilliae	30. B. pumilus	31. B. schlegelii	32. B. sphaericus	33. B. stearothermophilus	34. B. thuringiensis
Width of rod (μm)	0.5-0.7	0.6-0.9	0.6-0.8	0.5-0.7	0.5-0.7	0.9-1.2	1.2-1.5	1.0-1.2	0.5-0.7	0.5-1.2	0.6-0.8	0.5-0.8	0.6-0.7	0.6-0.8	0.6-1	0.6-1	1.0-1.2
Length of rod (μm)	1.8-7	1.2-4	1.5-3	2.5-5	2-6	2-4	2-5	3-5	2-5	1.3-4	2-5	1.3-5.2	2-3	2.5-5.8	1.5-5	2-3.5	3-5
Sporangium swollen	+	-	-	+	+	+	-	-	+	+	+	+	-	+	-	d	-
Spore shape	E	E	E	E	E	S	E	E	E & S	S	E	E	E	S	S	E	E
Spore position	C/T	C	C	T	T	T	C/T	C	T	T	T	C	C	T	T	T	C

[a] Symbols: −, 90% or more of strains are negative; +, 90% or more of strains are positive; d, 11–89% of strains are positive; E, ellipsoidal; S, spherical; C, central; T, terminal; and CL, central and lateral.

diosus, "*B. macroides,*" *B. megaterium, B. polymyxa, B. psychrophilus, B. schlegelii,* and some strains of *B. stearothermophilus* (Beveridge, 1981; Burley and Murray, 1983; Sleytr and Messner, 1983). Other strains of *B. stearothermophilus* form an oblique lattice (Messner et al., 1984).

The taxonomic implications of the structure and assembly of S-layers in *Bacillus* species are not known since most studies have been done by using only single, selected strains from a given species.

Comparative studies on the existence of S-layers have been performed with 39 strains of *B. stearothermophilus* (Messner et al., 1984) and with 61 strains of *B. sphaericus* (Word et al., 1983). A crystalline S-layer was found in 30 of the 39 strains of *B. stearothermophilus*. A hexagonal lattice was observed in only one strain, a square lattice in 15 strains and an oblique lattice in 14 strains. It has been found that strains thought to be closely related as a result of the taxonomic study of Klaushofer and Hollaus (1970) show a remarkable heterogeneity in the molecular weights of the S-layer subunits and the geometry and constants of the S-layer lattices (Messner et al., 1984).

With *B. sphaericus* strains, regularly structured S-layers (RS) have been observed in only 30 of 61 strains studied. Since *B. sphaericus* is now considered to be a "species complex" rather than a genospecies (Krych et al., 1980), it is of some taxonomic interest that the presence of RS is partly related to the different DNA homology groups of this species (Word et al., 1983).

Colony Characteristics. The appearance of surface colonies of *Bacillus* strains generally vary a great deal with environmental factors. These include those that are imposed on the culture (composition of medium, temperature of incubation, humidity, etc.) as well as those resulting from changes in the environment caused by the development of the culture itself (Knaysi, 1951; Henneberg and Marwitz, 1963). With the exception of strains which consistently form only small colonies, colony diameter depends largely on the number of colonies developing on the plate as well as on the concentration of nutrients and the amount of nutrient agar poured into the Petri dish.

A description of the colonies of *Bacillus* species has been given by Smith et al. (1952). Although different strains of a particular species often form colonies which are similar enough to be characteristic, colonial variants of most species also have been observed. In certain cases such variants may not belong to the species *sensu stricto* but to a different group within the "species complex." A color atlas showing the characteristics of colonies of strains of most *Bacillus* species on blood agar plates has been published by Parry et al. (1983).

Extreme variations in colony morphology can be often observed with *B. subtilis* and related species. For example, with *B. pumilus* (ATCC 27142) smooth and rough colonies with 25 variants have been observed by Pokallus et al. (1976). Subcultures of these colonies reverted to any of the other colony types. Colony variation was found to be influenced by the type of peptone. Colony variants of *B. subtilis,* even though usually highly irregular, can show a remarkable degree of symmetry (see cover of *Science 162,* No. 3860, 1968).

A very peculiar colony structure is seen with strains of *Bacillus mycoides*. They show a rhizoid growth covering the plate within 48 h. The rhizoid growth usually turns to the left. Only few strains isolated turn to the right (Stapp and Zycha, 1931). With certain strains, however, the direction may be influenced by the pH of the medium (von Klopotek, 1969). The rhizoid growth may be lost spontaneously (Stapp and Zycha, 1931) or by cultivating small inocula in fairly large volumes of nutrient broth. The nonrhizoid variants are stable and have the characteristic appearance of *B. cereus* (Gordon et al., 1973).

Some *Bacillus* species tend to swarm on solid media, especially if the plates are not dried to remove surface moisture before inoculation. The migration of colonies is considered to be a special type of swarming (Henrichsen, 1972). Motile colonies have been observed with strains of *B. alvei, B. circulans* and *B. sphaericus* ("*B. rotans*") under conditions suboptimal for swarming, i.e. on dried plates (Smith and Clark, 1938; Shinn, 1938; Murray and Elder, 1949; Shukla and Lyer, 1961). Wandering and rotating colonies can be easily observed in transmitted light and at a magnification of about × 50. They migrate predominantly counterclockwise at a velocity of up to about 15 mm/h on a curved or spiral path (Shinn, 1938; Murray and Elder, 1949). As a result of this plates are often overgrown within 1 or 2 days. A film on the structure and behavior of motile colonies of *B. circulans* has been published by Gillert and Institut für den wissenschaftlichen Film (1961).

Colonies of pure cultures on a single plate sometimes show more or less pronounced differences in translucence or opaqueness or are more or less whitish or cream-colored. This may be due to different degrees of sporulation within the colonies.

Most *Bacillus* species are unpigmented. Strains of *B. megaterium sensu stricto* form a yellow pigment especially on casein agar (Hunger and Claus, 1981). A yellowish pigment is also found in colonies of *B. fastidiosus* upon prolonged incubation on allantoin agar (F. Fahmy, personal communication). According to Smith et al. (1952) some pigmented variants have been observed with *B. firmus* (pink), *B. licheniformis* (brownish red), *B. pumilus* (slightly yellowish), *B. pulvifaciens* (buff to red, orange), *B. sphaericus* (pink) and *B. subtilis* (pink, yellow, orange, brown). The pigments may not be produced on all media. "*B. subtilis* subsp. *aterrimus*" forms a blue-black pigment only in the presence of utilizable carbohydrates. The property may be lost during serial transfer (Gordon et al., 1973). On tyrosine agar, black pigments are formed by strains of *B. megaterium* belonging to the "*B. carotarum*" group (Hunger and Claus, 1981) and by "*B. subtilis* subsp. *niger*." With the latter organism this property is easily lost (Gordon et al., 1973). A few strains of *B. cereus* and *B. subtilis* have been found to excrete a yellowish green, fluorescent pigment (Smith et al., 1952) and a yellow diffusible pigment is produced by some strains of *B. mycoides* (Nowak, 1965).

In contrast to terrestrial isolates, a very high proportion of *Bacillus* colonies developing from pasteurized samples of salt marsh soils form yellow, orange/brown or pink pigments (Turner and Jervis, 1968; Slepecky and Leadbetter, 1977; Fahmy, 1978). Isolates studied belong to the *B. firmus* or *B. circulans* complex (Gordon et al., 1977; Fahmy, 1978).

The terms used in the description of *Bacillus* colonies are sometimes difficult to understand. It is therefore recommended that the terms proposed by Salle (1973) or by Smibert and Krieg (1981) should be used.

Life Cycle. Unlike most other bacteria, *Bacillus* species are potentially able to form resting cells after the end of exponential cell growth or if vegetative cells are transferred from a rich to a poor medium. Resting cells are formed intracellularly and are therefore designated as endospores. They differ from vegetative cells in many respects like optical refractility, ultrastructure, chemical composition and resistance to chemical and physical stresses which lead to vegetative cells being killed rapidly. The degree of resistance of endospores depends largely on the environmental conditions under which they are formed.

Endospores can be detected by phase contrast microscopy due to their optical brightness. To observe the shape of the sporangium and the position of the endospore within the sporangium it is often necessary to use cultures of different age. Microscopical preparations are best studied using an oil immersion objective and slides coated with a thin layer (about 0.5 mm) of a 2% water agar on which the cells are pressed against the cover glass and all cells present in the microscopic field are in focus. This method is especially useful in preparing photomicrographs.

Inexperienced workers may confuse lipid or other cell inclusions with endospores. Their presence may be confirmed by staining procedures (Conklin, 1934; Schaeffer and Fulton, 1933; Snyder, 1934), by the "acid popping test" described by Robinow (1951) and Warth (1979), where endospores will "explode" after treatment with certain strong acids, or by the microscopic observation of the germination and outgrowth of endospores.

Bacillus strains do not form endospores under all cultural conditions. This is most important from a taxonomic point of view and especially in the identification of bacterial isolates where nonsporulating *Bacillus* strains may be misidentified as members of other genera. Most *Bacillus* strains form endospores on nutrient agar supplemented with 10–50 mg

manganous salts/liter of medium, but may lose this property or may sporulate only with a very low frequency after repeated transfer. Such cultures are referred to as oligosporogenous (Osp). Other factors affecting endospore formation include the temperature of growth, the pH of the medium, aeration, presence of minerals, presence of certain carbon or nitrogen compounds and the concentration of the carbon or nitrogen source.

The formation of endospores is a multiphasic process which is similar in all *Bacillus* strains so far studied: stage 0—vegetative cell growth; stage I—preseptation, the DNA forms an axial filament; stage II—septation, separation of chromosomes resulting in asymmetric cell formation; stage III—engulfment of the forespore, the membrane of the developing spore becomes completely detached from that of the mother cell and surrounds the spore protoplast; stage IV—cortex formation starts; stage V—spore coats are synthesized; stage VI—development of refractility and heat resistance, spore maturation; and stage VII—lysis of the sporangium and liberation of the mature spore.

Endospores usually show the phenomenon of dormancy. The transformation of a dormant endospore into a vegetative cell usually involves three sequential processes which are known as activation, germination and outgrowth.

Activation is a treatment which results in an endospore which is poised for germination, and hence is the process responsible for breaking of dormancy. A generally effective procedure for the activation of endospores is not known. The method usually employed is heat treatment for 10–30 min at an appropriate temperature, depending upon the species or strain under study. It should be borne in mind, however, that application of heat in this way may kill or damage some of the endospores or may change their germination requirements.

Endospores of many strains may show germination and outgrowth without any activation step. This depends mainly on the conditions under which endospores have been formed. For example, endospores of *B. fastidiosus* need heat activation only if the medium used for their production includes manganous ions (Aoki and Slepecky, 1973).

Activation of endospores may also be achieved by aging even at low temperatures. Thus spores of *Bacillus* strains maintained under these conditions often become activated and will germinate at once when placed on a suitable medium. However, only a few endospores of a population may germinate, which may lead to strain selection.

Germination is the change of an activated endospore from the dormant to a metabolically active state. Under suitable conditions germination may take place within minutes. The main changes occurring during germination are the depolymerization and excretion of spore constituents accompanied by loss of optical refractility (Fig. 13.3), onset of stainability, and loss of resistance to heat and other physical or chemical stresses.

Germination can be initiated by certain amino acids, ribosides and sugars, but is often achieved most rapidly and completely by using combinations of germinants or complex culture media. Environmental factors necessary for optimal germination, like temperature, pH value, and ionic strength of the medium, vary widely and depend very much on the organism.

Outgrowth is defined as the development of a vegetative cell from a germinated endospore. It will take place only in a medium that can support cell growth. After germination is completed, the young vegetative cell emerges, elongates and divides (Fig. 13.4).

The manner in which the coat opens is often strain characteristic and is generally not strongly influenced by environmental factors. Opening may be polarly; equatorially, by splitting along the transverse cell axis; by comma-shaped expansion with two halves of the spore coat remaining at either end of the growing cell; or by lysis of the coat. Well defined coat residues are left behind by the small-celled organisms like *B. subtilis*, whereas the large-celled organisms like *B. cereus* or *B. megaterium* often lyse their coats (Lamanna, 1940; Gould, 1962). Whether the mode of coat opening is a character of taxonomic value is not fully established.

The complete development from endospore to endospore was followed firstly by Cohn (1876) for *B. subtilis* and by Koch (1876) for *B.*

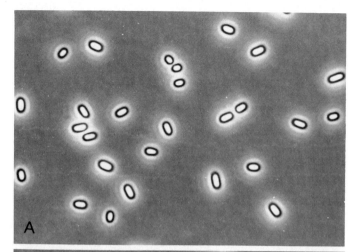

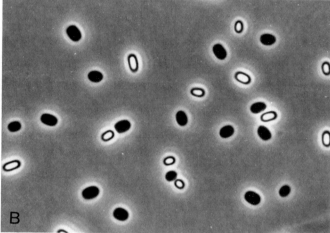

Figure 13.3 A and B.

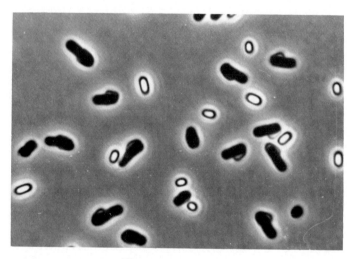

Figure 13.4.

anthracis. Since then much work has been done in respect to the sporulation, germination and outgrowth of *Bacillus* endospores as well as on their resistance and destruction. This work has been reviewed by Knaysi (1951), Robinow (1960), Gould and Hurst (1969), Russell

(1982), Hurst and Gould (1984) and Dring et al. (1985). Methods for studying bacterial endospores have been described by Gould (1971).

Nutrition and Growth Conditions. Nowhere is the diversity in the genus *Bacillus* more striking than in the wide range of nutritional requirements and growth conditions spanned by its members. Nevertheless, the majority of strains will grow satisfactorily on nutrient agar. Exceptions are *B. fastidiosus,* which requires an allantoin or uric acid-containing medium, *B. pasteurii,* which needs a medium at a high pH and containing a high concentration of NH_4Cl, and "*B. laevolacticus*" and "*B. racemilacticus,*" which fail to grow on media not containing carbohydrate. Details of suitable media for these organisms are given in Berkeley et al. (1984). *B. larvae, B. lentimorbus* and *B. popilliae* also fail to grow or grow poorly for a limited number of transfers in nutrient broth. These organisms grow well on J-medium.

For many species the requirements for chemically defined media have been determined although minimal needs have not always been established: *B. alvei* (Katznelson and Lochhead, 1947); *B. anthracis* (Ristroph and Ivins, 1983); *B. cereus* (Summers et al., 1979); *B. circulans* (Proom and Knight, 1955); *B. coagulans* (Marshall and Beers, 1967); *B. fastidiosus* (Kaltwasser, 1971); *B. firmus* and *B. lentus* (Proom and Knight, 1950); *B. licheniformis* and *B. macerans* (Knight and Proom, 1950); *B. megaterium* (White, 1972); *B. pumilus* and *B. polymyxa* (Knight and Proom, 1950); *B. sphaericus* (White and Lotay, 1980); *B. stearothermophilus* (Tandom and Gollakota, 1971); *B. subtilis* (Knight and Proom, 1950) and *B. thuringiensis* (Proom and Knight, 1955).

The nutritional requirements of the organisms studied by Gordon et al. (1973) correlate fairly well with the grouping they established on the basis of spore shape and sporangium swelling. Organisms in group 1 had simple or relatively simple needs, those in group 2 had more complex requirements and those in group 3 were the most fastidious.

A chemically defined medium enables *B. schlegelii* to grow autotrophically in an $O_2/CO_2/H_2$ or O_2/CO atmosphere; in air it grows heterotrophically given an appropriate carbon and energy source such as acetate or succinate (Schenk and Aragno, 1979). Such facultative chemolithoautotrophy is unique in the genus.

Establishment of minimal growth requirements is not a simple task and, although impurities in medium components are less likely now to lead to conflicting evidence than was once the case, the interplay of nutritional and environmental factors is an important consideration. Koser (1968), who has reviewed the literature on vitamin requirements of members of the genus, has pointed out that the needs of the thermophiles for both vitamins and amino acids may vary considerably with incubation temperature.

The need for nitrogen can be at least partially satisfied by the ability of some species to fix nitrogen: *B. polymyxa* (Hino and Wilson, 1958), *B. macerans* (Witz et al., 1967) as well as others (Rhodes-Roberts, 1981).

Nitrate reduction is a common property (see Table 13.4); *B. azotoformans, B. licheniformis* and some other species producing a considerable amount of N_2 or N_2O. This property may be lost after repeated transfer on nutrient agar.

B. pumilus has been reported to be resistant to and to utilize cyanide (Skowronski and Strobel, 1969).

That *Bacillus* species have the ability to metabolize aromatic and hydroxyaromatic compounds, has begun to emerge recently (Pichinoty and Mandel, 1978; Aftring and Taylor, 1981).

The ability of *Bacillus* species to grow at both low and high temperatures and at low and high pHs has been discussed by Norris et al. (1981). The thermophilic and alkaliphilic species are of considerable interest because of the character of commercially valuable enzymes which they produce and a large literature relating to them is emerging (e.g. Friedman, 1978; Horikoshi and Akiba, 1982).

The alkaliphilic strains share with the marine isolates an obligate requirement for Na^+ (Rüger and Hentzschel 1980; Horikoshi and Akiba, 1982).

A *Bacillus* strain of unknown taxonomic position, isolated from seeds of *Astralgalus crotolariae*, a selenium-accumulating plant, can grow in nutrient broth only in the presence of a rather high concentration of sodium selenite. The optimum concentration is 30–100 mM. Selenite can be replaced by selenate or tellurate. Elemental selenium (red) or tellurium (black) is produced during growth. The strain, however, needs no selenite if grown in brain heart infusion broth or trypticase soy broth (Lindblow-Kull et al., 1982).

Genetics. Following the demonstration that *B. subtilis* is transformable (Spizizen, 1958), the development of *Bacillus* genetics was rapid; by 1982 *B. subtilis* had a well defined map on which more than 355 genes had been located (Henner and Hoch, 1982) and more are continually being added (e.g. Lamont and Mandelstam, 1984). At least some of the impetus for this progress stems from the fact that this organism sporulates and has been an important model of procaryotic differentiation studies (Szulmajster, 1979). Mutants are available in which only sporulation is affected and which, when in the vegetative phase, are not distinguishable from wild type bacteria, thus enabling the analysis, mostly by transformation and transduction, of functions necessary for differentiation (Doi, 1977).

Since Spizizen's demonstration of transformation (1958), manipulation of *B. subtilis* by generalized and specialized transduction, and transformation of protoplasts by plasmid DNA has been achieved. Young et al. (1983) have reviewed the literature on the plasmid and bacteriophage vectors used for cloning in this species. There are also other recent and more extensive reviews relating to genetic engineering in *B. subtilis* and its relevance to commercial processes (Sakaguchi and Okanishi, 1980; Dubnau, 1982; Ganesan et al., 1982).

Transfer of DNA to and/or from several other *Bacillus* species has also been achieved with *B. brevis* (Takahashi et al., 1983); *B. cereus* (Gonzâlez and Carlton, 1982); *B. coagulans* (Cornelis et al., 1982); *B. licheniformis* (Imanaka et al., 1981; Fujii et al. 1982); *B. megaterium* (Vorobjeva et al., 1980); *B. sphaericus* (Ganesau et al., 1983); *B. stearothermophilus* (Fugii et al., 1982; Mielenz, 1983) and *B. thuringiensis* (Martin et al., 1981; Miteva et al., 1981).

The use of bacteriophage for typing *B. cereus, B. anthracis, B. thuringiensis, B. sphaericus, B. stearothermophilus* and alkaliphilic strains in the *B. firmus-B. lentus* spectrum has been reviewed recently (Berkeley et al., 1984).

DNA base composition. According to DeLey (1978) species belonging to a natural genus should not differ in their DNA base composition more than 10–15%. The mol% G + C of the DNA of strains of the genus *Bacillus* have been found to differ by more than 30%, indicating that the genus as presently defined is genetically heterogenous. The lowest reported value (31.7 mol% G + C) is for a strain of *B. cereus* (McDonald et al., 1963), whereas the highest value of 69 mol% G + C was found with "*B. thermocatenulatus*" (Golovacheva et al., 1975).

On the species level, strains belonging to a single species should not differ in their DNA base composition more than 2% (DeLey, 1978). Published data on the DNA base composition of *Bacillus* strains considered to belong to a single species according to phenotypic properties, however, often differ by more than this (Priest, 1981), indicating the genetic unrelatedness of such strains. Where detailed studies have been carried out such strains have often been found to belong to different DNA/DNA homology groups within the species (see below). A striking example is *B. circulans* where a study of 123 strains showed a continuous spectrum in the DNA base composition from 37–61 mol% G + C (Nakamura and Swezey, 1983a).

One of the main problems in evaluating data on the DNA base composition of *Bacillus* strains is the fact that figures published by different authors for the same strain, including many type strains, sometimes differ considerably (up to 14 mol% G + C). Such values may have been obtained by applying unsuitable methods or by studying mislabeled cultures (Fahmy et al., 1985).

A comparative study on the DNA base composition of most *Bacillus* type strains by thermal denaturation (T_m) and buoyant density (Bd) measurements has been done by Fahmy et al. (1985). The type strains were obtained directly from the collections listed in the Approved Lists

of Bacterial Names (Skerman et al., 1980) as keeping the type strains. To ensure their identity was correct the strains were tested for selected phenotypic properties.

Table 13.3 lists the DNA base composition of the *Bacillus* type strains in an increasing order (T_m) as determined by Fahmy et al. (1985) together with the corresponding Bd values. Few data have been included from other sources.

The DNA base composition (mol% G + C) has been studied also for some invalid species: "*B. agarexedens*," 44.4–52.7 (Hunger and Claus, 1981); "*B. amyloliquefaciens*", 43.4–44.9 (Welker and Campbell, 1967); "*B. amylosolvens,*" 46.9 (Baptist et al., 1978); "*B. aneurinolyticus*", 42.2 (Ikeda et al., 1965); "*B. caldolyticus,*" 52.3, "*B. caldotenax,*" 64.8, "*B. caldovelox,*" 65.1 (Sharp et al., 1980); "*B. macroides,*" 42.0 (Bennett and Canale-Parola, 1965); "*B. nitritollens,*" 37.8 and 41.8 (Pichinoty et al., 1978); "*B. psychrosaccharolyticus,*" 31.0 (Eiroma et al., 1971); *B. pulvifaciens,* 45.9–46.9 (Baptist et al., 1978); "*B. thermodenitrificans,*" 40.8 and 52.0 (Pichinoty et al., 1978); "*B. thermoflavus,*" 61.0 (Heinen et al., 1982); "*B. thermoruber,*" 67.0 (Guicciardi et al., 1968).

DNA reassociation. The use of conventional methods for the allocation of *Bacillus* strains to species has only been partly successful. This is evident from the broad range of the DNA base composition found within some species. The phenotypic properties of strains at the extremes of some species are such as to justify their allocation to separate species. The existence of intermediate strains, however, makes decisions, based on phenotypic characters, about the proper allocation to appropriate taxa difficult.

DNA homology studies have contributed to a more satisfying concept of what a bacterial species is and the technique of DNA reassociation may be a basis for the recognition of strains at the species level (DeLey, 1978; Johnson, 1984). By applying this technique some progress in the taxonomy of the genus *Bacillus* has been made (Priest, 1981). Further studies including as many strains as possible and especially those differing in specific phenotypic characters from the type strains are needed.

Results of species identification by DNA-DNA hybridization data showed good agreement with those obtained by conventional taxonomic methods for *B. pumilus* (8 strains, Lovett and Young, 1969; 16 strains, Seki et al., 1978; 8 strains, O'Donnell et al., 1980) and *B. licheniformis* (16 strains, Seki et al., 1975; 8 strains, O'Donnell et al., 1980). The genetic homogeneity of 11 strains of *B. coagulans* (Seki et al., 1978) probably reflects the fortuitous selection of strains which happened to be closely related, since the species shows a wide range of DNA base composition. Seventeen strains of *B. fastidiosus* showed a high homology of 80–100% (F. Fahmy, personal communication).

With 27 strains of *B. subtilis* two groups of high intra- but low interstrain homology have been observed by Seki et al. (1975), confirming an earlier study of Welker and Campbell (1967) which showed that strains described as "*B. amyloliquefaciens*" should have a separate species status. Strains belonging to the *B. subtilis* or "*B. amyloliquefaciens*" homology groups have also been found to differ in other properties (O'Donnell et al., 1980).

DNA reassociation measurements revealed that 50 of 111 *B. circulans* strains examined could be separated into 10 groups of genetically related organisms (Nakamura and Swezey, 1983a, b). These groups can also be differentiated by classical tests. Similarly, *B. sphaericus* has been found to be genetically rather heterogeneous and, although all of the 62 strains studied have a DNA base composition differing by not more than 2%, five homology groups were identified (Krych et al., 1980). With *B. firmus*, according to Seki et al. (1983), 10 strains showed a high homology with the type strain. However, with five strains of *B. lentus* low homologies with each other (and with *B. firmus*) were found. Studies by Hunger and Claus (1981) have shown that within *B. megaterium*, also, there is heterogeneity. Four DNA homology groups can be distinguished and strains of "*B. agarexedens*" can be separated into five groups of genetically related organisms. A limited amount of data on the genetic relatedness of strains of *B. stearothermophilus* and other thermophiles has been published by Sharp et al. (1980).

Table 13.3.

DNA base composition, murein types and main menaquinones of **Bacillus** *type strains[a]*

Type strains	Mol% G + C		Murein type[c]	Main menaquinone[d]
	$T_m{}^b$	BD[b]		
B. anthracis	33.2[e]	ND	(meso-DAP direct)[f]	ND
B. thuringiensis	33.8	34.3	meso-DAP direct	MK-7[g]
B. mycoides	34.2	34.1	meso-DAP direct	(MK-7)[f,h]
B. fastidiosus	35.1	35.1	meso-DAP direct[i]	MK-7[i]
B. circulans	35.5	35.4	(meso-DAP direct)[f]	(MK-7)[f]
B. cereus	35.7	36.2	meso-DAP direct	(MK-7)[f]
B. insolitus	35.9	36.1	ND	MK-7
B. lentus	36.3	36.4	(meso-DAP direct)[f]	MK-9?
B. pantothenticus	36.9	36.8	meso-DAP direct	MK-7[h]
B. alcalophilus	37.0	36.7	ND	MK-7
B. megaterium	37.3	37.6	(meso-DAP direct)[f]	(MK-7)[f]
B. sphaericus	37.3	37.1	(L-Lys-D-Asp)[f]	MK-7[k]
B. marinus	37.6	38.0	ND	ND
B. lentimorbus	(37.7)[f,j]	ND	meso-DAP direct	ND
B. pasteurii	38.5	38.4	L-Lys-L-Ala-D-Asp[k]	MK-7
B. azotoformans	ND	39.0[m]	ND	ND
B. macquariensis	39.3	41.6	ND	MK-7
B. globisporus	39.8	39.7	ND	MK-7
B. laterosporus	40.2	40.5	meso-DAP direct	(MK-7)[f]
B. popilliae	41.3[j]	ND	ND	MK-7
B. firmus	41.4	40.7	(meso-DAP direct)[f]	(MK-7)[f]
B. pumilus	41.9	40.7	meso-DAP direct	(MK-7)[f]
B. subtilis	42.9	43.1	meso-DAP direct	(MK-7)[f]
B. badius	43.8	43.5	meso-DAP direct	MK-7
B. polymyxa	44.3	45.6	(meso-DAP direct)[f]	(MK-7)[f]
B. alvei	44.6	46.2	meso-DAP direct	(MK-7)[f]
B. licheniformis	46.4	44.7	meso-DAP direct	(MK-7)[f]
B. coagulans	47.1	44.5	meso-DAP direct	MK-7
B. brevis	47.3	47.4	meso-DAP direct	(MK-7)[f]
B. stearothermophilus	51.9	51.5	(meso-DAP direct)[f]	(MK-7)[f]
B. macerans	52.2	53.2	meso-DAP direct	(MK-7)[f]
B. acidocaldarius	60.3	62.3	ND	(MK-9)[n]?
B. schlegelii	64.6	66.3	meso-DAP direct[o]	MK-7
B. larvae	ND	ND	ND	(MK-7)[f]

[a] Symbols: ND, no data available.
[b] Data from Schleifer and Kandler (1972) except [i,l,o].
[c] Data from F. Fahmy (personal communication).
[d] Data from Hess et al. (1979) except [g,h,i,n].
[e] Data from Candeli et al. (1978).
[f] Type strain not studied. Data included in parentheses are from other strains of the species.
[g] Data from Kroppenstedt (unpublished).
[h] Data from Watanuki and Aida (1972).
[i] Data from Fahmy et al. (1985) except [e,j,l].
[j] Data from Manachini et al. (1968).
[k] Data from Collins and Jones (1979).
[l] Data from O. Kandler, personal communication.
[m] Data from Pichinoty et al. (1978).
[n] Data from DeRosa et al. (1973).
[o] Data from Krüger and Meyer (1984).

Although DNA reassociation studies have shown the genetic unrelatedness between certain *Bacillus* species (Seki et al., 1978), no systematic study of DNA-DNA homology of those *Bacillus* species, which have similar DNA base composition, has been carried out. However, studies by Sommerville and Jones (1972) and by Kaneko et al. (1978) have revealed that *B. anthracis*, *B. cereus* and *B. thuringiensis* have such a high DNA homology with each other that they should be considered to belong to a single species.

DNA-rRNA hybridization; comparative oligonucleotide cataloging of 16S rRNA. The wide range of the DNA base composition from about 30–70 mol% G + C suggests that the genus *Bacillus* comprises species of more than one "natural" genus. Support for this suggestion may also be derived from the studies of Seki et al. (1975, 1978) who, on the basis of DNA-DNA homology studies, found a reasonable genetic relationship between only a few *Bacillus* species. However, Schleifer and Stackebrandt (1983) pointed out that separation of genera should not be based solely on DNA homology values. Representatives of other genera, whose DNA homology with *Bacillus* is small, were found to be closely related when they were included in phylogenetic studies based on conserved ribosomal RNA molecules (Fox et al., 1980).

Extensive studies on the rRNA cistron similarity which allow decisions about the homogeneity of taxa above the species level (DeLey and Smedt, 1975) have not been done with the genus *Bacillus*. Herndon and Bott (1969), however, found a high degree of DNA-rRNA homology between *Sporosarcina ureae* and several *Bacillus* species having a similar DNA base composition.

The phylogenetic relatedness of some *Bacillus* species and *Sporosarcina ureae* emerging from sequence analyses of 16S rRNA has been considered by Pechman et al. (1976) and Fox et al. (1977). They found no reason to split the genus *Bacillus* but proposed the inclusion of *S. ureae* in the genus since this species is more closely related to *B. pasteurii* than the latter is to *B. subtilis* or *B. megaterium*. Data on the 16S rRNA sequence analyses of additional *Bacillus* species (in total 16), covering nearly the whole spectrum of DNA base composition, have been obtained by Stackebrandt and Woese (1971). Although detailed analyses have not yet been published, there seems to be good reason to consider the genus *Bacillus* as a phylogenetically homogenous taxon. The group, however, includes additionally *Planococcus, Sporolactobacillus, Staphylococcus* and *Thermoactinomyces,* as well as *S. ureae.*

It has been pointed out that, although homogeneous, this group cannot be considered to be a single genus because the relatedness of *B. subtilis* and *B. pumilus* according to 16S rRNA sequence analyses is the same as that seen among genera within the *Enterobacteriaceae* (Woese et al., 1976). Furthermore, the relatedness of *B. subtilis* and *B. stearothermophilus* is comparable to that seen between the families *Enterobacteriaceae* and *Vibrionaceae.* Clearly, for reasonable equivalence to exist between the taxonomy of organisms in the genus *Bacillus* and other genera, division is required (see below). To establish a sound basis for this division further genetic and phylogenetic studies are required.

Antigenic Structure. The use of serological methods in the identification of *Bacillus* species has recently been reviewed (Berkeley et al., 1984).

Antibiotic Sensitivity. Although some *Bacillus* species are opportunistic pathogens, the genus *Bacillus,* with the exception of *B. anthracis,* has attracted little interest from medical microbiologists. Because of this, studies on the sensitivity of *Bacillus* species to antibiotics, even as they relate to the taxonomy of the genus, are few.

Kundrat (1963) found that *Bacillus* species except *B. alvei, B. laterosporus* and *B. stearothermophilus* are resistant to sulfonamide (Debenal, Bayer AG, 50 μg/disk). It has been observed that *B. coagulans* has a greater overall sensitivity to several antibiotics than has *B. stearothermophilus* (Sharp et al., 1980). Strains of *B. marinus* show differences in their sensitivity to penicillin, terramycin and pteridin 0/129 (Rüger and Richter, 1979). According to Bernhard et al. (1981), *B. cereus* generally shows high resistance to ampicillin, colistin and polymyxin. Few of the strains studied were resistant to bacitracin, cephaloridin, kanamycin or tetracycline. *B. anthracis* strains are generally sensitive to penicillin but one resistant strain is now known. Bernhard et al. (1981) found that some strains of *B. subtilis* were resistant to streptomycin but all those examined were sensitive to many other antibiotics.

Resistance to erythromycin has been observed with strains of *B. licheniformis* by Docherty et al. (1981) and Yoshimura et al. (1983), to streptomycin with *B. sphaericus* (Burke and McDonald, 1983; Yoshimura et al., 1983), to chloramphenicol with *B. licheniformis* and *B. sphaericus* (Burke and McDonald, 1983) and to tetracycline with some strains of *B. cereus, B. licheniformis, B. sphaericus, B. stearothermophilus* and *B. subtilis* (Polak and Novick, 1982). With the exception of tetracycline, the antibiotic resistance generally is not associated with plasmids (Bernhard et al., 1981; Polak and Novick, 1982; Yoshimura et al., 1983).

The type of antibiotic and the concentration affecting vegetative growth, sporulation, germination or outgrowth of spores of *Bacillus* strains may differ (Russell, 1982; Ochi and Freese, 1983).

Pathogenicity. *B. anthracis* is the best known pathogen in the genus; some people regard it as the only mammalian pathogen. It is an important pathogen of farm animals, although in developed countries anthrax is of small consequence. Vaccination of those sections of the population at risk and the existence of a generally effective antibiotic, penicillin, have reduced the importance of the disease in such countries substantially. Recent isolation of a penicillin-resistant strain of *B. anthracis* should, however, serve as a warning that this organism should not be treated lightly. The molecular basis of resistance to this antibiotic has not yet been established. Elsewhere, given adverse conditions, this disease can be a major problem. Recent reviews on this topic by Norris et al. (1981) and Parry et al. (1983) are available.

Some further understanding of the molecular basis of virulence has now been obtained. It is now known that factor I (the edema factor) is an adenylate cyclase (Leppla, 1982) and there is evidence that toxin production is plasmid mediated (Mikesell et al., 1983).

B. cereus, to which *B. anthracis* is very closely related, is probably the next best known mammalian pathogen from this genus.

B. cereus is undoubtedly the causative organism of a variety of more or less severe infections of both man and other mammals. The relevant literature has been reviewed by Norris et al. (1981) and Turnbull (1981). This organism is also the cause of two types of food poisoning—an emetic type and a diarrheal type (Gilbert et al., 1981).

Other species too, *B. brevis, B. licheniformis, B. subtilis* (Gilbert et al., 1981) and *B. sphaericus* (Parry et al., 1983), have been implicated in food poisoning.

B. thuringiensis, very closely related to *B. anthracis,* as well as to *B. cereus,* is a well known insect pathogen and, of the five species of *Bacillus* insect pathogens, is also by far the most important in the biological control of insects. The other four are *B. larvae, B. lentimorbus, B. popilliae* and *B. sphaericus* (de Barjac, 1981a).

B. sphaericus and *B. thuringiensis* have in common that not only are they insect pathogens but also that they have been implicated in mammalian infections. Gordon (1977) reported the isolation of *B. thuringiensis* from a fatal case of bovine mastitis and Parry et al. (1983) recorded the laboratory infection by *B. thuringiensis,* with another organism, of the finger web. In view of the large amounts of biological insecticide used, it is perhaps surprising that there have not been more instances of the kind reported by Samples and Buettner (1983) in which a *B. thuringiensis* eye infection in a healthy person developed as a result of splashing of a commercial preparation of such a control agent. A healthy adult has also been infected by *B. sphaericus* leading to fatal meningitis (Allen and Wilkinson, 1969). The other recorded instances of infection by this organism, which led to meningitis, bacteremia and endocarditis (Farrar, 1963) and a fatal pulmonary pseudotumor (Isaacson et al., 1976), were, in contrast, seen in a chronic alcoholic and a chronic asthmatic, respectively.

Host predisposition is undoubtedly an important factor in the establishment of infection by many *Bacillus* species; hosts compromised for one reason or another are more often than not involved.

Other *Bacillus* species thought to be involved in mammalian infection are: *B. alvei,* meningitis (Park et al., 1976) and, with no details given, (Gordon et al., 1973); *B. brevis* with another organism, corneal ulceration (van Bijsterveld and Richards, 1965); *B. circulans,* meningitis (Boyette and Rights, 1952); *B. coagulans,* corneal ulceration (van Bijsterveld and Richards, 1965); *B. licheniformis,* bacteremia (Amador et al., 1976; Peloux et al., 1976) and septicemia (Fauchere et al., 1977; Sugar and McCloskey, 1977); *B. macerans,* infected site of melanoma excision (Ihde and Armstrong, 1973); *B. megaterium,* pharyngitis; and, with another organism, psoriasis, (Gordon, R. E., personal communication); *B. pumilus,* meningitis (Weinstein and Colburn, 1950), rectal abcess (Melles et al., 1969); *B. subtilis,* postoperative cellulitis (Behrend and Krouse, 1952), septicemia (Sathmary, 1958; Cox et al., 1959), respiratory disease (Greenberg et al., 1970), endocarditis (Reller, 1973); pneumonia (Ihde and Armstrong, 1973).

As Gilbert et al. (1981) have pointed out, in many instances the association of *Bacillus* species with infections is often not sufficiently rigid for them to be regarded unequivocally as the causative agent. The same is true of the relationship between species other than *B. cereus* with food poisoning. With regard to both cases, though, there are sufficiently strong indications of the role in mammalian pathogenicity of *Bacillus* species other than *B. anthracis* and *B. cereus,* for there to be a more general acceptance of the pathogenic potential of these organisms.

Ecology—Habitats. Due to the resistance of their endospores to air-drying, to other stresses and to their long term survival under adverse conditions, most aerobic sporeformers are ubiquitous and can be isolated from a wide variety of sources.

In spite of the wide interest in aerobic sporeformers in general and applied microbiology, and the vast literature that exists in these fields, it is still impossible to consider the species of the genus *Bacillus* in terms of ecosystems. There are many reports of the presence and numbers of *Bacillus* strains or species in certain habitats. However, the habitat approach has its pitfalls in that most *Bacillus* isolates originate from air-dried or heated samples and hence from endospores which are metabolically inactive.

Isolation data, therefore, mostly throw little light on the ecology of sporeformers, as in the case of the finding of endospores of the obligate thermophile *B. stearothermophilus* in deep sea sediments (Bartholomew and Paik, 1966). On the other hand, it may be assumed that high numbers of endospores found in a given sample indicates a relationship with earlier activity of vegetative *Bacillus* cells in the environment from which the sample was taken.

It is generally accepted that the primary habitat of the majority of *Bacillus* species is the soil where they were once considered part of the zymogenous bacterial flora (Winogradsky, 1949). These organisms are now usually regarded as r-strategists (Campbell, 1984). They become metabolically active when a suitable substrate is made available. Since most *Bacillus* species effectively degrade a series of biopolymers they are assumed to play an important role in the biological cycling of carbon and nitrogen.

From soil, aerobic sporeformers can contaminate everything by dust or other means. They may play an important role in such secondary habitats in degrading polymers or other chemical compounds. As a result, *Bacillus* species are especially important as food spoilage organisms.

Some comments on the habitats of *Bacillus* species can be found in the section on Isolation and Enrichment. The literature on the ecology and habitats of aerobic sporeformers has been recently reviewed by Norris et al. (1981) and Slepecky (1972).

Miscellaneous Descriptive Information. *Fatty acids and polar lipids.* The importance of fatty acids in the chemotaxonomy of the genus *Bacillus* has been reviewed by Kaneda (1977) and by Minnikin and Goodfellow (1981). The latter also discussed data on the distribution of polar lipids. The formation of both classes of compounds may be strongly influenced by cultural conditions, so that for any studies to be useful highly standardized growth conditions must be employed (Goldfine, 1982; Kaneda, 1977). Due to the wide diversity of their nutritional and physicochemical requirements, however, such standardization cannot be achieved for all *Bacillus* species.

The fatty acid composition of 19 *Bacillus* species has been studied by Kaneda (1977) who allocated these organisms to six groups (Kaneda groups A–F). All groups except D contain major amounts of branched chain acids.

Within groups A–C only insignificant amounts (~3%) of unsaturated fatty acids are found. The three groups can be separated on the basis of their predominant fatty acids. Group A (*B. alvei*, *B. brevis*, *B. circulans*, *B. licheniformis*, *B. macerans*, *B. megaterium*, *B. pumilus* and *B. subtilis*) contains anteiso-C_{14} (26–60%) and iso-C_{15} (13–30%) acids. The range of chain length of the acids is 14–17.

In group B (*B. polymyxa*, *B. larvae*, *B. lentimorbus*, *B. popilliae*) anteiso-C_{15} acid predominates (39–62%). Again, the range of chain length is 14–17.

Group C (*B. stearothermophilus*, "*B. caldolyticus*," "*B. caldotenax*") differs from A and B in having iso-C_{15} acid as the predominant compound. The range of chain length is as in the two other groups.

A unique fatty acid pattern has been observed with *B. acidocaldarius* (group D). The main compounds (up to 70%) are cyclohexane fatty acids with a chain length of 17–19 (De Rosa et al., 1971).

Group E includes *B. anthracis*, *B. cereus* and *B. thuringiensis*. In contrast to groups A–D, small proportions (7–12%) of unsaturated fatty acids are always present. The predominant fatty acids (19–21%) are of the iso-C_{15} type with a chain length of 12–17.

The main fatty acids found in *B. azotoformans* by Pichinoty et al. (1983) are iso-C_{15} (28–52%), iso-C_{14} (15–30%) and anteiso-C_{15} (16–20%).

A large proportion of unsaturated fatty acids (17–28%) is only found in the psychrophilic organism *B. globisporus* and *B. insolitus* (group F). The predominant branched chain fatty acids are anteiso-C_{15} acids with a chain length of 14–17.

The taxonomic significance of polar lipids has not been established. Most *Bacillus* strains tested contain phosphatidylglycerol and phosphatidylethanolamine. The latter is absent from *B. acidocaldarius*

(Langworthy et al., 1976) and certain strains of *B. coagulans* (Frade and Guenan-Siberil, 1975) and *B. stearothermophilus* (Minnikin et al., 1974). *B. acidocaldarius* contains unusual glycolipids containing hopane (De Rosa et al., 1973; Langworthy, 1982). Hopanoids and cyclohexane fatty acids have been detected also in some thermoacidophilic *Bacillus* strains of uncertain taxonomic status, isolated from nonacidic soil samples (Hippchen et al., 1981). The hopane glycolipid apparently has some significance in the stability of cell membranes (Kannenberg and Poralla, 1980; Poralla et al., 1980).

Menaquinones. The isoprenoid quinones from all *Bacillus* strains that have been studied are exclusively menaquinones. The major isoprenoid is usually an unsaturated menaquinone with seven isoprenoid units (MK-7). Minor components include MK-2 to MK-11, depending on the species or strain (Collins and Jones, 1981).

Although strains of nearly all *Bacillus* species have been studied, the menaquinone composition of only 14 type strains of the 34 presently recognized species is known. The heterogeneity of some species does not allow, however, any conclusions to be drawn with respect to the menaquinone composition on a species from results obtained with other strains not shown to be genetically highly related to the type strain. In table 13.3 therefore, results based on the study of strains other than the type strain are shown in parentheses.

Hess et al. (1979) have found MK-9 as the major component for the type strain of *B. lentus*, MK-8 for the type strain of *B. pantothenticus* and MK-8 and MK-7 for the type strain of *B. thuringiensis*. This is in contrast to the studies of Watanuki and Aida (1972) who found MK-7 in the type strain of *B. pantothenticus* and of Kroppenstedt (unpublished) who found in the type strains of *B. pantothenticus* and *B. thuringiensis* only MK-7. Seven other strains of *B. thuringiensis* also contained mainly MK-7 (98–99%). With the type strain of *B. lentus* MK-6 (23%), MK-7 (34%), MK-8 (30%) and MK-9 (12%) has been found.

For the type strain of *B. schlegelii* MK-7 (51%), MK-6 (36%) and minor amounts of MK-5 and MK-4 have been observed. Other strains of the species contain mainly (90–94%) MK-7 (Krüger and Meyer, 1984).

With a strain of *B. acidocaldarius*, De Rosa et al. (1973) have found MK-9 as the main menaquinone. According to Kroppenstedt (unpublished) the type strain of the species shows the following composition: MK-6 (2%), MK-7 (54%), MK-8 (3%), MK-9 (25%), MK-10 (5%) and MK-11 (11%).

Electrophoresis of enzymes. The comparative zone electrophoresis of enzymes extracted from *Bacillus* strains has been described as a useful technique for the study of taxonomic relationships within the genus (Baptist et al., 1978). Strains representing *B. alvei*, "*B. amyloliquefaciens*," *B. cereus*, *B. firmus*, *B. laterosporus*, *B. macerans*, *B. megaterium*, *B. polymyxa* and *B. subtilis* exhibited unique patterns of the migration of 11 different enzymes. *B. licheniformis* was similar to *B. pulvifaciens* and to a strain of *B. coagulans*. A second strain of the latter species was quite different from all other *Bacillus* strains studied. Strains of *B. pumilus* fell into two distinct groups. *B. pantothenticus* and strains of *B. thuringiensis* were not distinguishable from *B. cereus*.

Esterase patterns of 217 thermophilic *Bacillus* strains fell into three groups (Baillie and Walker, 1968) similar to those based on physiological and serological reactions (Walker and Wolf, 1971). Ten strains including "*B. caldolyticus*," "*B. caldotenax*" and "*B. caldovelox*" were studied by Sharp et al. (1980).

Studies on the classification or identification of *Bacillus* strains by electrophoresis of total soluble or envelope proteins have not yet been performed (Kersters and DeLey, 1980).

Enrichment and Isolation Procedures

Aerobic endospore-forming bacteria of the genus *Bacillus* can be isolated from almost all natural habitats and from many other sources. They are most commonly found in soil and in plant litter where they play an important role in the biological cycling of carbon and nitrogen. Other habitats like fresh water, polluted sea water, deep sea sediments, foods, milk, pharmaceuticals, etc., may have acquired these organisms from soil, by runoff, from dust, from infected plant materials, etc. Such habitats may provide conditions suitable for the growth of *Bacillus* species or may only harbor spores which, due to their remarkable power of resistance and dormancy, may survive in any habitat for long periods. Thus it is generally not possible to draw any conclusions from the site of isolation of a *Bacillus* strain as to its real natural habitat, although there are a few exceptions to this generalization (e.g. *B. acidocaldarius, B. marinus*).

Methods for the detection and isolation of *Bacillus* strains are based mainly on the resistance of their spores towards elevated temperatures. Whereas heating above 70°C for 10 min or longer will destroy vegetative cells of almost all known bacteria, the spores of *Bacillus* strains together with nonsporeforming, extremely thermophilic bacteria usually survive this treatment. It should be kept in mind, however, that spores in natural environments may possess different heat sensitivities from those which have been cultured. They also may have different heat activation requirements for optimal germination and outgrowth. Because of this it is generally not possible to determine exactly the real number of spores present in a sample from nature.

In contrast to vegetative cells of most bacteria, *Bacillus* spores are also very resistant towards dehydration. Within soil samples, air-drying will destroy most bacteria while spore-forming bacteria, cyst-forming bacteria like *Azotobacter* species, and *Arthrobacter* species may survive drying for longer periods (Mulder and Antheunisse, 1963; Vela, 1974).

Spores of *Bacillus* species are also known to be far more resistant towards destruction by disinfecting chemicals than are the cells of most bacteria. Ethanol seems to be a useful disinfectant to use for isolation of *Bacillus* strains as it kills vegetative cells in a sample whereas the more resistant endospores survive (Koransky et al., 1978). A study including some *Bacillus* type strains has shown that spores survive exposure to 50% (v/v) ethanol for at least 1 h, whereas vegetative cells of *Bacillus* and other species are normally killed (Claus, unpublished). Up to now the use of ethanol in selecting *Bacillus* strains from natural samples has not been commonly used. It is strongly recommended, however, as heat treatment of spores may lead to mutations (Hayase et al., 1974; Nishida et al., 1969; Kadota et al., 1978; Zamenhof, 1960). Whether ethanol-treated spores need an additional treatment for activation has yet to be established. Since commercial ethanol may contain bacterial spores it should be sterilized by filtration before use.

The isolation of strains of the predominant *Bacillus* flora present as spores in a given sample is an easy task once the samples under study have been treated by heat or ethanol or have been air-dried. The successful detection or isolation of strains of individual *Bacillus* species, however, may be a more difficult problem. If the organism of interest outnumbers all others in a given sample, it can be easily obtained by direct isolation, that is by plating of serial dilutions on agar media and incubation at appropriate temperatures.

The direct isolation of strains of specific *Bacillus* species present at only relatively low numbers in a mixed population needs, however, the application of selective media or other selective culture conditions. At present, direct isolation methods are available for only a few *Bacillus* species. Strains of other *Bacillus* species can be obtained from natural habitats by applying enrichment techniques using specific liquid media incubated under defined physicochemical conditions, followed by plating on aerobically incubated specific agar media. Methods available for the detection and selective isolation of *Bacillus* species are described below. In many cases it has been found useful to transfer 4 g of an air-dried soil sample to a beaker without infecting the walls. Sterile water (20 ml) is added and the beaker is heated in a water bath for 10 minutes at 80°C while the content is carefully agitated. The soil sample is then ready for inoculation.

B. acidocaldarius. This species has been isolated from a variety of thermal acid environments (water, soil from fumaroles) by enrichment at elevated temperatures and low pH (Darland and Brock, 1971; Loginova et al., 1978). Other thermoacidophilic isolates of uncertain taxonomic position have been isolated by Uchino and Doi (1967) from

Japanese hot springs and by Hippchen et al. (1981) from nonacidic soils.

Darland and Brock (1971) used the following medium for enrichment and isolation of *B. acidocaldarius*. Part A: $(NH_4)_2SO_4$, 0.4 g; $MgSO_4$, 1.0; $CaCl_2 \cdot 2H_2O$, 0.5 g; KH_2PO_4, 6.0 g; distilled water to 1000 ml; pH adjusted to 4.0. Part B: glucose, 2.0 g; yeast extract, 2.0 g; distilled water to 1000 ml; pH not adjusted. Parts A and B are sterilized separately at 121°C for 15 min and then combined after sterilization. For the preparation of solidified media 40 g of agar is added to Part B of the medium before sterilization.

Liquid media are inoculated with appropriate samples and are incubated at about 60°C. Pure cultures are obtained by streaking material from enrichment cultures on agar media.

"*B. agarexedens.*" Wieringa (1941) has described agarolytic *Bacillus* strains as "*B. agarexedens.*" Strains of this species are inhibited by peptone in a medium at neutral pH but not at an alkaline pH. Similar strains have been isolated by Hunger and Claus (1978, 1981) using the following enrichment medium: KH_2PO, 1.0 g; NH_4Cl, 1.0 g; $MgSO_4 \cdot 7H_2O$, 0.5 g; $CaCl_2 \cdot 2H_2O$, 0.05 g; $FeSO_4 \cdot 7H_2O$, 0.01 g; $MnSO_4 \cdot H_2O$, 0.005 g; agar, 1.0 g; distilled water, 1000 ml. The medium is adjusted to pH values in the range 7.2–8.2 and sterilized for 15 min at 121°C.

Portions (30 ml) of the medium in 300-ml flasks are inoculated with pasteurized soil suspensions (10 min, 80°C) and incubated at 30°C on a shaker for about 5 days. At pH 7.2 only a few agarolytic sporeformers can usually be detected, whereas at pH 7.8–8.2 the concentration of agarolytic bacteria may reach 10^5–10^7 cells/ml. They can be readily isolated by streaking one loopful up to 0.1 ml on mineral-agar plates containing 1.5% (w/v) agar, supplemented with 0.005% (w/v) yeast extract and adjusted to pH 7.8. Agar decomposing colonies are easily detected due to the formation of depressions on the agar surface. They are purified by restreaking on nutrient agar plates at pH 8.0.

Strains enriched at pH 7.2–7.5 may differ from those isolated from more alkaline enrichment cultures in some phenotypical properties and in DNA base composition (Hunger and Claus, 1981). Agarolytic *Bacillus* strains not inhibited by peptone may be obtained by plating material from pH 7.2 enrichments on nutrient agar of pH 6.8.

B. alcalophilus and related organisms. Alkaliphilic organisms of the genus *Bacillus* are widely distributed in nature. They have been isolated from soil, water and feces of man and other animals. In soils, up to 10^5 or 10^6 spores/g have been found. The highest frequency is usually observed in alkaline soils, although high numbers have sometimes also been detected in acid soils (Aunstrup et al., 1972; Horikoshi and Akiba, 1982).

The taxonomic relationships of most of the isolated alkaliphilic *Bacillus* strains to the type strain of *B. alcalophilus* (Vedder, 1934) are not fully established. Most alkaliphilic *Bacillus* strains are phenotypically related to the *B. firmus-B. lentus* group (Gordon and Hyde, 1982).

B. alcalophilus and other alkaliphilic sporeformers can be isolated by direct plating of pasteurized samples on a variety of alkaline nutrient agar media. The pH of such media should be adjusted to pH values between 8.5 and 11.0 by adding sodium hydrogen carbonate, sodium carbonate, sodium sesquicarbonate, trisodium phosphate, or sodium perborate in appropriate concentrations (Aunstrup et al., 1971; Aunstrup et al., 1972; Grant and Tindall, 1980; Horikoshi and Akiba, 1982).

A useful isolation medium has been described by Horikoshi and Akiba (1982). The composition is: glucose, 10.0 g; peptone, 5.0 g; yeast extract, 5.0 g; KH_2PO_4, 10.0 g; $MgSO_4 \cdot 7H_2O$, 0.2 g; distilled water to 900 ml. For solid media, 15 g of agar is added. The medium is sterilized at 121°C for 15 min and then 100 ml of 20% (w/v) $Na_2CO_3 \cdot 10H_2O$ (sterilized separately by autoclaving) is added. The final pH of the medium is 10.5. Lower pH values may be obtained by lowering the sodium carbonate concentration to 0.5% (w/v). Agar containing media should be mixed with the sodium carbonate solution at 60°C and poured immediately.

The use of potassium salts for adjusting the pH cannot be recommended since most alkaliphilic *Bacillus* strains tested have a requirement for sodium ions. In fact, some isolates can also be grown at neutral pH in the presence of sodium chloride (Horikoshi and Akiba, 1982).

Methods available for the determination of pH optima of strains and for maintaining high pH values by buffer systems have been described by Grant and Tindall (1980).

B. alvei. Strains of this species were isolated from honeycombs of bees infected with European foulbrood. A few strains have also been isolated from soil. Most strains form motile colonies which spread over the whole plate within 2 days.

Emberger (1970) found a high frequency of *B. alvei* colonies after anaerobic enrichment of heat-treated soil samples in nutrient broth at pH 6.0 and 28–37°C, followed by plating on nutrient agar, adjusted to pH 6.0 and incubated aerobically. For preliminary differentiation of an isolate from other species forming motile colonies the characteristic side-by-side arrangement of the spores in long rows can be helpful (Gordon et al., 1973).

"*B. aminovorans.*" Using trimethylamine as the sole source of carbon and nitrogen, den Dooren de Jong (1927) isolated from an enrichment culture a single *Bacillus* strain which he named "*B. aminovorans.*" Two similar strains, *Bacillus* sp. PM6 and S2A1, have been described by Myers and Zatman (1971) and by Colby and Zatman (1975).

From about 50 soil samples collected from different parts of the world Blumenstock (1981) isolated 25 trimethylamine utilizing *Bacillus* strains most of which are identical with "*B. aminovorans.*"

For enrichment the following medium is suitable: Part A: KH_2PO_4, 0.8 g; KH_2PO_4, 0.2 g; $MgSO_4\cdot7H_2O$, 0.5 g; $CaCl_2\cdot2H_2O$, 0.05 g, $FeSO_4\cdot7H_2O$, 0.01 g; distilled water, 900 ml. The pH is adjusted to 7.0. Part B: trimethylammonium chloride, 2.0 g; distilled water, 100 ml. The two parts are sterilized separately at 121°C for 15 min and combined after cooling to room temperature.

Portions (30 ml) of the medium are placed in 300-ml flasks, inoculated with pasteurized soil suspensions (10 min, 80°C) and incubated at 30°C on a shaker.

"*B. aminovorans*" can be detected in the enrichment cultures after 2 days by the typical cell morphology (thick rods), the tumbling movement of cells and the formation of round endospores which do not swell the sporangium.

After one or two transfers from young, just turbid cultures into liquid media, pure cultures can be obtained by plating serial dilutions on the above medium solidified by 1.2% agar and by restreaking single colonies composed of typical cells described above. In older enrichment cultures other *Bacillus* species may overgrow "*B. aminovorans.*"

B. anthracis. Several nonselective and selective media have been described for the detection and isolation of *B. anthracis* from pathological material, from hairs, hides, feedstuffs, fertilizers, and soil (Norris et al., 1981). A rapid screening test for *B. anthracis,* based on the morphology of microcolonies, was described by Chadwick (1963).

A selective medium (PLET medium) developed by Knisely (1966) contains the following filter-sterilized substances added to Difco heart infusion agar: polymyxin, 30 U/ml; lysozyme, 40 µg/ml; disodium ethylenediaminetetraacetate (EDTA), 300 µg/ml; thallous acetate, 40 µg/ml. The unadjusted pH of the medium is 7.35.

Aliquots (0.1 ml) of appropriate dilutions are spread onto the plates which are incubated at 37°C for 24–48 h. *B. anthracis* forms small, rather smooth colonies. Strains of *B. cereus* and *B. mycoides* and of other *Bacillus* species are generally inhibited on this medium.

B. azotoformans. Selective isolation can be performed by incubating pasteurized soil samples (10 min, 80°C) in a peptone broth at 32°C and under an atmosphere of pure N_2O (Pichinoty et al., 1976; 1978). Gassing or foaming cultures are transferred two or three times into fresh medium incubated under the same conditions. Pure cultures can be obtained by streaking material from enrichment cultures onto the same medium solidified with 1.2% (w/v) agar. The plates are incubated at 32°C in air.

The enrichment medium is composed of peptone, 10.0 g; $Na_2HPO_4\cdot12H_2O$, 3.6 g; KH_2PO_4, 1.0 g; NH_4Cl, 0.5 g; $MgSO_4\cdot7H_2O$, 0.03 g; heavy metals solution (Pichinoty et al., 1977), 0.2 ml; distilled water, 1000

ml. The medium is adjusted to pH 7.0 and sterilized for 15 min at 121°C.

B. badius. No specific methods for the isolation of this species are known to the authors.

B. brevis. Heigener (1935) described 10 new *Bacillus* species which were transferred by Smith et al. (1952) to *B. brevis.* The strains were obtained from soils by the following method: small amounts of soil were suspended in 3 ml of sterile water and heated for 5 min at 80°C. Aliquots (0.5 ml) of the pasteurized suspensions were transferred to agar slants composed of K_2HPO_4, 0.2 g; $MgSO_4\cdot7H_2O$, 0.02 g; NaCl, 0.02 g; $FeSO_4\cdot7H_2O$, 0.01 g; $MnSO_4\cdot H_2O$, 0.01 g; agar, 16.0 g; betaine, betaine hydrochloride or valine, 0.05 M; distilled water to 1000 ml.

The cultures were incubated for 4 weeks at 29°C. If profuse growth was observed, cultures were streaked onto agar plates of the same composition. Colonies showing the best growth were selected and purified.

B. brevis has been reported by Knight and Proom (1950) to need amino acids for growth. It can be assumed that these requirements were satisfied by the crude agar preparations in use at the time of Heigener's studies.

The enrichment used by Heigener also led in a few cases to the isolation of strains belonging to the *B. circulans* complex.

"*B. carotarum.*" The species was described by Koch (1888). It was later transferred to *B. megaterium* (Smith et al., 1952). Hunger and Claus (1981), however, have shown that strains similar to "*B. carotarum*" form a distinct group genetically unrelated to *B. megaterium.*

According to Koch (1888), "*B. carotarum*" can be regularly obtained on slices of carrots which have been boiled for a short time and incubated at room temperature in a moderately humid atmosphere. After 1–2 days small white, gummy colonies develop which normally can be identified as "*B. carotarum*" or *B. megaterium sensu stricto* ("*B. tumescens*" according to Koch). The two species can be easily distinguished microscopically (Hunger and Claus, 1981).

Stapp (1920) has described "*B. capri*", "*B. cobayae*" and "*B. musculi*" which are now considered to be identical with "*B. carotarum*" (Hunger and Claus, 1981). Strains of these species have been isolated from animal feces (guinea pig, goat, canary, mouse) by the following method: small samples of feces were transferred to tubes containing 2 ml of sterile water. The tubes were heated for 3 min at 80°C and 0.1-ml aliquots of the suspensions transferred to a mineral base medium containing uric or hippuric acid as the sole source of carbon and nitrogen.

Pure cultures were isolated on appropriate media solidified with agar. On these media all strains showed only rather weak growth. They were finally transferred to nutrient agar.

Bongaerts (1978), in addition to *B. fastidiosus,* isolated from enrichment cultures containing uric acid as the sole source of carbon and nitrogen other strains which he identified as *B. megaterium.* These can now be assumed to belong to "*B. carotarum*" and not to *B. megaterium sensu stricto.* This is concluded because F. Fahmy (personal communication) noted that in enrichment cultures for *B. fastidiosus,* cells typical of *B. megaterium* were never observed. Cells of "*B. carotarum*" are microscopically similar to *B. fastidiosus.*

B. cereus. This species is widely distributed in nature and is commonly found in soil, milk, cereals and other dried foodstuffs, and plays some role in food-transmitted gastroenteritis. Various selective media have been developed for its detection and direct isolation especially from foods.

All of these rely upon the suppression of Gram-negative organisms by polymyxin and the presumptive identification of *B. cereus* by means of the egg-yolk reaction (Bouwer-Hertzberger and Mossel, 1982; Gilbert and Taylor, 1976).

One of the media, polymyxin pyruvate egg-yolk mannitol bromothymol blue agar (PEMBA), has the following composition (Holbrook and Anderson, 1980): peptone, 1.0 g; D-mannitol, 10.0 g; $MgSO_4\cdot7H_2O$, 0.1 g; NaCl, 2.0 g; Na_2HPO_4, 2.5 g; KH_2PO_4, 0.25 g; bromothymol blue (water soluble), 0.12 g; agar, 18.0 g; distilled water to 1000 ml. The medium is adjusted to pH 7.4, dispensed in 90-ml amounts into bottles and sterilized at 121°C for 15 min. The final pH is 7.2. This medium is available from Oxoid (CM 617).

Prior to use the following solutions are added to 90 ml of the molten and cooled (50°C) basal agar: 20% (w/v) sodium pyruvate, filter sterilized, 5 ml; polymyxin, filter sterilized, to a final concentration of 100 units/ml; egg-yolk emulsion (Oxoid SR 47), 5 ml. If samples suspected to contain large numbers of fungi are to be examined, 1 ml of a filter-sterilized solution of 0.4 % (w/v) actidione can also be added.

Poured plates are dried for a short time and surface inoculated with 0.1 ml of decimal dilutions, incubated at 37°C and examined after 24 and 48 h.

On PEMBA, B. cereus forms crenate or fimbriate to slightly rhizoid colonies. They have a distinct turquoise to peacock blue color and are usually surrounded by an egg yolk precipitate of similar color. The detection of B. cereus may be difficult in the presence of B. mycoides which usually grows over the whole agar surface as a rhizoid, turquoise blue colony.

A liquid enrichment method for the specific isolation of B. cereus (Claus, 1965) makes use of a medium containing peptone, 5.0 g; meat extract, 3.0 g; KNO₃, 10.0 g; distilled water to 1000 ml. The medium is adjusted to pH 6.8–7.0 and sterilized at 121°C for 15 min.

Pasteurized soil samples (10 min, 80°C) are transferred in 1-ml amounts into sterile, glass-stoppered bottles and are then filled with the enrichment medium. After inserting the stopper, no air bubbles should be left.

After 24 h at 30°C cultures will be more or less turbid. They do not produce gas and microscopical examination shows mainly cells typical of B. cereus. After longer incubation times other sporeformers may predominate.

One loopful of 24-h cultures is streaked onto nutrient agar plates, consisting of peptone, 5.0 g; meat extract, 3.0 g; agar, 12.0 g; distilled water to 1000 ml, pH 6.8–7.0. Plates are incubated at 37°C to suppress the growth of B. mycoides. Colonies typical of B. cereus are purified by restreaking on the same medium. Colonies of B. cereus also can be easily detected by their egg-yolk reaction if the agar is supplemented with 50 ml of an egg-yolk emulsion (Oxoid SR 47).

Morris (1955) developed a selective medium for the isolation of B. anthracis which also allows growth of some B. cereus strains. Other aerobic sporeformers were reported to be inhibited on this medium.

B. circulans. Gibson and Topping (1938) and others have described B. circulans as a complex rather than a species. This was recently confirmed by the wide range found in the DNA base composition of strains of B. circulans (mol% G + C 37–61) and by DNA/DNA homology studies (Nakamura and Swezey, 1983a, b).

Strains belonging to the B. circulans complex have been isolated by Knight and Proom (1950) from air-dried soil samples suspended in nutrient broth, heated for 10 min at 70°C, followed by plating on nutrient agar and incubation at 37°C. The other main species developing on the plates was B. subtilis.

Cellulose degrading strains of B. circulans have been isolated by Kellerman and McBeth (1912). Similar strains have been isolated by Emerson and Weiser (1963). Cellulose degrading Bacillus isolates have also been described by Harmsen (1946) and by Imschenetski (1959). It is not known whether these isolates belong to the same species.

Tesic and Todorovic (1952, 1953) have proposed that the so-called "silicate bacteria" belong to the B. circulans group. They can be isolated by placing small soil particles on nitrogen-free agar media. Before the development of Azotobacter can be observed, small slimy colonies can be found on the soil particles consisting of rod-shaped bacteria which, in india ink preparations, show an extremely large slime layer. The same organisms have probably been described by Winogradsky (1949) as "B. gommeux" and by Elwan (1964), who observed a rapid increase in the numbers of such organisms in soil treated with streptomycin. Pure cultures can be obtained by restreaking on nutrient agar supplemented with 0.5% (w/v) glucose.

B. coagulans. Allen (1953) successfully isolated this species from a number of different sources. The medium used was composed of yeast extract, 5.0 g; glucose; 10.0 g; distilled water to 1000 ml, adjusted to pH 5.5 with lactic acid, distributed in 200-ml amounts into flasks and sterilized at 121°C for 15 min.

Flasks are inoculated with 1 ml of pasteurized soil suspensions and incubated at 50°C for 24 h. Aliquots (0.1 ml) of each culture are then transferred to fresh medium followed by incubation at 50°C. After 24 h a loopful of culture material is streaked on nutrient agar and incubated at 45°C for 2 days. Colonies developing on the plates are small, round, whitish, opaque or opalescent and generally belong to B. coagulans.

Emberger (1970) isolated B. coagulans together with B. licheniformis by incubating heat-treated soil samples in nutrient broth containing 1% (w/v) glucose, adjusted to pH 6.0 and incubated anaerobically at 54°C. Plates of the same medium solidified by 1.5% (w/v) agar were then streaked and incubated aerobically at 45°C.

B. fastidiosus. This species and its properties have been described by den Dooren de Jong (1929). It has been found to grow only in the presence of uric acid. Later studies have shown that growth of most strains is supported by allantoin and allantoic acid (Bongaerts, 1976; F. Fahmy, personal communication).

For the isolation of strains of B. fastidiosus the following medium has been found useful (F. Fahmy, personal communication): K₂HPO₄, 0.8 g; KH₂PO₄, 0.2 g; MgSO₄·7H₂O, 0.5 g; CaCl₂·2H₂O, 0.05 g; FeSO₄·7H₂O, 0.015 g; MnSO₄·H₂O, 0.01 g; uric acid, 10.0 g; distilled water to 1000 ml. The pH of the medium is not adjusted.

For enrichment of strains of this species, 30 ml of the medium in 300-ml Erlenmeyer flasks are inoculated with 5 ml of a soil suspension in water, pasteurized at 80°C for 5 min. Cultures are shaken at 30°C for 24 h. Aliquots (1 ml) of this primary enrichment are transferred to a fresh medium of the same composition and incubated under the same conditions.

After 24 h, one loopful, or 0.1 ml, of serial dilutions of the enrichment culture are streaked or plated on a mineral base agar of the same composition as given above, but without uric acid, overlayered by agar containing 1% (w/v) uric acid. For the preparation of the latter medium, a 3% (w/v) solution of agar in water and a 2% (w/v) suspension of uric acid in water are sterilized separately at 121°C for 15 min. Equal volumes of the two parts are combined and poured to a depth of about 3 mm onto the mineral agar plates. Due to the low solubility of uric acid the plates are milky.

Uric acid-degrading organisms not only grow on this medium but produce a clear halo around their colonies. This is due to both the utilization of the acid and to its solubilization because of the alkaline pH resulting from the splitting of urea formed by B. fastidiosus from uric acid.

On the medium described, B. fastidiosus usually forms colonies showing rhizoid outgrowths. Some other strains form irregular colonies. Cell material from selected colonies is suspended in nutrient broth and is streaked both on uric acid agar and nutrient agar. Cultures which develop only on the uric acid agar generally belong to B. fastidiosus. Strains are purified on the mineral base agar containing 2% (w/v) allantoin. The substrate can be sterilized together with the agar.

By a similar method Bongaerts and Vogels (1976) isolated, in addition to B. fastidiosus, other strains differing from B. fastidiosus in cell morphology (cell width <1 μm) and DNA base composition (F. Fahmy, personal communication). They belong to hitherto unknown species.

B. firmus. Like many other Bacillus species, B. firmus is a complex rather than a genospecies (Gordon et al., 1977; Seki et al., 1983). Strains belonging to the complex have been isolated by Emberger (1970) by incubating soil and food samples aerobically in nutrient broth at pH 7.2 and 28°C, followed by plating on nutrient agar. However, other species may be more numerous on the isolation plates.

In contrast to terrestrial Bacillus isolates which are usually unpigmented, a very high proportion of those from salt marsh soils are pigmented (Fahmy, 1978; Slepecky and Leadbetter, 1977; Turner and Jervis, 1968). According to Fahmy (1978) 77% of her pigmented isolates from salt marsh soil were identified by classical methods as *B. firmus*. The isolates, however, belong to more than five different DNA/DNA homology groups. Pigmented strains of different *Bacillus* species have also been isolated from marine and related habitats by Boeyé and Aerts (1976) and Bonde (1975).

Strains have been isolated by Fahmy (1978) by plating decimal dilutions of pasteurized (30 min at 60°C) samples on sea water agar (Lyman and Fleming, 1940) and incubating at 22°C for 7 days. Alternatively, Difco Marine Agar 2214 can be used.

B. globisporus. Larkin and Stokes (1966) isolated 90 psychrophilic *Bacillus* strains from 60 out of 75 samples of soil, mud and water. The isolates were enriched at 0°C in trypticase soy broth (BBL). After 2 weeks of incubation, cultures showing turbidity were streaked on trypticase soy agar and incubated for 2 weeks at 0°C. Isolated colonies were restreaked several times and the pure cultures obtained transferred to nutrient agar. They were retested for their ability to grow at 0°C and for their maximal growth temperature (25–30°C).

From 20 strains selected for further studies, 15 formed four distinct groups described as *B. globisporus*, *B. insolitus*, "*B. psychrophilus*" and "*B. psychrosaccharolyticus*" (Larkin and Stokes, 1967). The five remaining isolates differed from these groups and from each other.

Additional psychrophilic strains of uncertain taxonomic position have been isolated from food and milk by Gyllenberg and Laine (1971), Laine (1971) and Shehate and Collins (1971).

B. insolitus. See *B. globisporus*.

"*B. laevolacticus*". Strains of this species, together with strains of "*B. racemilacticus*" have been isolated from the rhizosphere of several plants. Both species are of uncertain taxonomic position but have some relationship with *B. coagulans*.

According to Nakayama and Yanoshi (1967) these mesophilic, facultatively anaerobic bacteria can be obtained by transferring a small piece of root with root hairs and adherent soil into 100 ml of a medium containing yeast extract (10 g of dried brewers yeast suspended in 100 ml of distilled water, heated at 100°C for 5 min, and centrifuged); soil extract, 100 ml (50 g of garden soil suspended in 100 ml distilled water, heated for 20 min at 130°C, and centrifuged); polypeptone, 10.0 g; glucose, 10.0 g; KH_2PO_4, 0.5 g; K_2HPO_4, 0.5 g; $MgSO_4 \cdot 7H_2O$, 0.3 g; NaCl, 0.01 g; $MnSO_4 \cdot 5H_2O$, 0.01 g; $CuSO_4 \cdot 5H_2O$, 0.001 g; $CoCl_2 \cdot 6H_2O$, 0.001 g; $FeSO_4 \cdot 7H_2O$, 0.001 g; trisodium citrate, 0.027 g; tap water to 1000 ml. The medium is adjusted to pH 6.4 and is sterilized at 115°C for 15 min.

Enrichment cultures were pasteurized at 80°C for 20 min and incubated anaerobically at 30°C. Tubes usually show production of gas due to the growth of clostridia. After prolonged incubation, cultures showing a drop in pH to below 4.0 were selected. One loopful of such cultures was plated on the same medium (but containing only 1% glucose), solidified by 1.8% (w/v) agar and supplemented with a small amount of $CaCO_3$ to make the agar plates opaque. The plates were incubated aerobically at 30°C and pinpoint colonies surrounded by a clear halo further purified.

B. larvae, B. lentimorbus, B. popilliae. Strains of these insect-pathogenic species can be isolated only with difficulty. Useful methods have been described by Krieg (1981).

B. laterosporus. Strains of this species have been isolated from water and soil. No specific method to enrich the organism from these habitats is known. Isolates have also been obtained from the larvae of diseased bees (McCray, 1917). Emberger (1970) isolated strains of this species, as well as some of other species, from soil and food samples after anaerobic enrichment in nutrient broth, pH 6.0, and aerobic plating on nutrient agar, pH 6.0.

B. lentus. According to Gibson (1935b) a direct plating method is most useful in isolating strains of *B. lentus*. A medium consisting of peptone, 10.0 g; meat extract, 10.0 g; agar, 15.0 g; distilled water to 1000 ml and adjusted to pH 7.0–7.5 is sterilized at 121°C for 15 min. Immediately before use, the agar is melted, 100 g of crystalline urea is added and, after steaming for 10 min, the agar is cooled and poured into plates.

Suspensions of soil dilutions are spread on the surface of the agar and the plates incubated at 25°C. *B. lentus* has been selected on the basis of its colony form and cell morphology. The species apparently is not present in all soils.

B. licheniformis. Strains of this species have been selectively obtained from soil by anaerobic enrichment in peptone-KNO_3 media at 37–45°C (Verhoeven, 1952). Under similar conditions Beijerinck and Minkman (1910) isolated "*B. nitroxus*," a rather polymorphic species of uncertain taxonomic position.

For enrichment and isolation Claus (1965) has proposed the following method: 2 ml of a pasteurized (10 min, 80°C) soil suspension are inoculated into a medium containing peptone, 5.0 g; meat extract, 3.0 g; KNO_3, 80.0 g; distilled water to 1000 ml. The medium is adjusted to pH 7.0 and sterilized at 121°C for 15 min before inoculation. Anaerobic conditions are provided by using glass-stoppered bottles, filled totally without leaving gas bubbles.

After incubation at 40–45°C for 48 h most bottles will show a more or less heavy turbidity and gas production. Material from the enrichment cultures is streaked on glucose mineral base agar consisting of K_2HPO_4, 0.8 g; KH_2PO_4, 0.2 g; $CaSO_4 \cdot 2H_2O$, 0.05 g; $FeSO_4 \cdot 7H_2O$, 0.01 g; $MgSO_4 \cdot 7H_2O$, 0.5 g; $(NH_4)_2SO_4$, 1.0 g; glucose, 10.0 g; agar, 12.0 g; distilled water to 1000 ml. After adjusting the pH to 6.8 the medium is sterilized at 121°C for 15 min.

On this medium *B. licheniformis* develops much better than the few other *Bacillus* species present in the enrichment culture. Colonies are often reddish and have mounds and lobes consisting of slime. Further purification may be achieved by streaking on nutrient agar.

"*B. loehnisii.*" This species is of uncertain taxonomic position but shows some relationship to *B. pasteurii*. For its isolation Gibson (1935b) proposed the following method: a medium consisting of peptone 10.0 g; meat extract, 10.0 g; urea, 100 g; distilled water to 1000 ml is distributed in 5-ml amounts into tubes. These are boiled, cooled and inoculated with 0.5 ml of decimal dilution of soil samples. The most useful dilutions are 10^{-3}–10^{-6}. The tubes are incubated at 22–30°C for up to 4 weeks. Urea-decomposing bacteria developing from the highest dilutions usually belong to "*B. loehnisii*".

B. macerans. Strains have been isolated by Schardinger (1905) from spontaneously fermenting mashed potatoes, previously heated over boiling water for 1 hr on 3 successive days and incubated at 37°C. Others strains have been isolated from gas-producing enrichment cultures inoculated with soil, water, grain, etc. (Porter et al., 1937).

Woldendorp (1963) observed a high increase of *B. macerans* (up to 10^7 cells/g oven-dry soil) in permanent grassland soil as a result of adding 2.1 g of potassium nitrate to sandy soil samples which were 11 cm in diameter and 7 cm thick. These were incubated at 55% of their water-holding capacity at 28°C in a nitrogen atmosphere. The treatment was repeated twice at intervals of 3 days. After incubation for 10 days soil dilutions were spread on a medium containing glycerol, 1.0 g; peptone, 1.0 g; yeast extract, 0.1 g; KNO_3, 3.0 g; agar, 10.0 g; tap water, 1000 ml. The agar plates were incubated at 28°C.

B. macquariensis. From three out of four samples of sub-Antarctic soil, 12 strains of the species were isolated by direct plating of pasteurized (10 min, 80°C) soil dilutions on nutrient agar incubated at 1°C. The organism accounted for about 1% of the total viable bacteria present in the soils before heating (Marshall and Ohye, 1966).

B. marinus. Among 126 cultures of sporeforming bacteria isolated from sediment samples collected from different depths of the Josephine Bank (North-East Atlantic) and from one station in the Iberian Deep Sea, 18 obligately halophilic, psychrophilic *Bacillus* strains were found which all belonged to the same species (Rüger and Richter, 1979).

Isolates were obtained from seawater agar pour plates inoculated with 1-ml amounts of serial dilutions of the sediment and incubated at 13–15°C for 1–2 weeks.

The seawater agar was composed of Bacto Petone (Difco), 5.0 g; yeast extract, 1.0 g; $FePO_4 \cdot 4H_2O$, 0.01 g; agar, 15.0 g; aged seawater, 750 ml; distilled water, 250 ml. The pH of the medium was 7.6.

B. megaterium. This species, as defined by Gordon et al. (1973), includes four groups which can be distinguished clearly by morphological, biochemical or genetic properties (Hunger and Claus, 1981). *B. megaterium sensu stricto* has the largest cells within the genus *Bacillus*. In general, it can be easily recognized in microscopical preparations by its typical cell morphology.

Strains of *B. megaterium sensu stricto* can be easily isolated from most soils by plating 0.1 ml of pasteurized soil dilutions (10^{-1}–10^{-3}) on a glucose mineral base agar of the following composition (Claus, 1965): K_2HPO_4, 0.8 g; KH_2PO_4, 0.2 g; $CaSO_4 \cdot 2H_2O$, 0.05 g; $FeSO_4 \cdot 7H_2O$, 0.01 g; $MgSO_4 \cdot 7H_2O$, 0.5 g; KNO_3 or $(NH_4)_2SO_4$, 1.0 g; glucose, 10.0 g; agar, 12.0 g; distilled water to 1000 ml. The medium is adjusted to pH 7.0 and sterilized at 121°C for 15 min.

With nitrate as the source of nitrogen, within 36–48 h at 30°C, completely white, round, smooth and shiny colonies (diameter 1–3 mm) of *B. megaterium* will often develop. Since not all strains of *B. megaterium* can use nitrate as nitrogen source, isolation should also be done on media containing ammonium ions. After 24–36 h incubation on this medium, different organisms will develop but *B. megaterium* colonies can be easily detected by their appearance. For isolation, cell material is checked microscopically for the presence of typical cells of *B. megaterium* and purified on nutrient agar plates.

B. mycoides. This species can be isolated very easily from many different types of material. A small drop of a freshly prepared or pasteurized soil suspension or a small particle of soil or any other material is placed in the middle of a nutrient agar plate. During incubation at 28°C *B. mycoides* will grow from the point of inoculation as a rhizoid colony which after 2 or 3 days will cover the whole plate. For isolation of pure cultures, material taken from the outer part of the rhizoid colony is suspended in a few drops of nutrient broth and restreaked several times on nutrient agar (Claus, 1965).

The chance of obtaining a pure culture originating from a single cell is much higher if cell material from primary isolation plates is transferred to a flask of nutrient broth which is shaken vigorously during incubation. This gives a more or less homogeneous suspension which, after serial dilution, can be plated on nutrient agar.

The rhizoid growth of *B. mycoides* normally turns to the left. A few strains, however, may turn to the right. This growth behavior may be pH-dependent (von Klopotek, 1969).

B. pantothenticus. Strains of this species have been repeatedly obtained by enrichment of soil samples in nutrient broth supplemented with 4% (w/v) NaCl. After incubation at 37°C cultures were isolated by plating on nutrient agar (Proom and Knight, 1950).

B. pasteurii. A specific enrichment method for the species has been described by Beijerinck (1901) and was also successfully used by Gibson (1935b).

The enrichment broth consists of peptone, 5.0 g; meat extract, 3.0 g; distilled water to 500 ml. The medium is distributed in 25-ml amounts into 1-liter Erlenmeyer flasks and sterilized at 121°C for 15 min without pH adjustment. Urea (100 g) is dissolved in distilled water to a total volume of 500 ml. The solution is brought to boiling point and to each flask containing 25 ml of enrichment broth, the same volume of urea solution is added.

The flasks are inoculated with about 5 g of air-dried soil. Pasteurization of the samples is not needed. During incubation at 26–30°C, urea hydrolysis proceeds vigorously and ammonium fumes are liberated. A succession of different urea-splitting bacteria may develop. These usually will be overgrown after some days by *B. pasteurii* if the enrichment culture is incubated in a shallow layer giving good aeration.

After 6–7 days a loopful of an enrichment culture is streaked on the surface of a nutrient agar plate supplemented with 2% (w/v) urea, added before sterilization, and incubated at 30°C for 1–2 days. Single colonies are suspended in a drop of sterile nutrient broth and streaked on nutrient agar, pH 6.8–7.0, as well as on the same medium containing 2% urea. Cultures developing only on the urea-containing medium generally belong to *B. pasteurii*. They are restreaked for final purification.

B. polymyxa. Methods for the enrichment and isolation of the species have been described by Bredemann (1909), Beijerinck and den Dooren de Jong (1922) and Knight and Proom (1950).

Another useful method was reported by Ledingham et al. (1945). One milliliter of a pasteurized soil suspension is transferred to a tube containing a medium composed of peptone, 10.0 g; lactose or starch, 10.0 g; distilled water to 1000 ml. The medium is adjusted to pH 6.8–7.0, distributed in 8-ml amounts into test tubes containing inverted Durham vials and sterilized at 121°C for 15 min.

After incubation at 30°C for 2 or 3 days, tubes showing gas formation are selected. From these a loopful of suspension is streaked on neutral red-peptone-starch-agar plates (produced by adding 1.2% (w/v) agar and 0.005% (w/v) neutral red) to the broth described above.

Colonies of *B. polymyxa* accumulate the indicator and are red pigmented. They are usually large and slimy and produce a fruity odor. On plates not dried before inoculation *B. polymyxa* may spread rapidly over the plate.

Single colonies are tested for gas production in the medium given above. Gas-forming strains generally belong to *B. polymyxa*. After purification the absence of obligate anaerobic bacteria should be confirmed.

B. pumilus. According to Knight and Proom (1950), soil suspensions in distilled water incubated at 37°C for 3 days, followed by plating on nutrient agar and incubation at 37°C produced a very mixed collection of colonies. *B. pumilus* and *B. circulans* were observed together with fewer numbers of colonies belonging to *B. subtilis*, *B. cereus* and *B. megaterium*.

Emberger (1970) detected *B. pumilus* in heat-treated soil and food samples plated directly on nutrient agar of pH 7.2 and incubated at 28°C.

B. psychrophilus. See *B. globisporus.*

"*B. racemilacticus.*" See "*B. laevolacticus.*"

B. schlegelii. This hydrogen-oxidizing organism was isolated by Aragno (1978) from the superficial layer of the sediment of a small eutrophic lake in Switzerland and was described by Schenk and Aragno (1979). Samples (~0.5 g) were inoculated into 100-ml bottles containing 20 ml of the following medium: $Na_2HPO_4 \cdot 2H_2O$, 4.5 g; KH_2PO_4, 1.5 g; NH_4Cl, 1.0 g; $MgSO_4 \cdot 7H_2O$, 0.2 g; $CaCl_2 \cdot 2H_2O$), 0.01 g; ferric ammonium citrate, 0.005 g; trace element solution, 5 ml; $NaHCO_3$, 0.5 g; distilled water to 1000 ml. The pH was adjusted to 7.0. The trace element solution was composed of $ZnSO_4 \cdot 7H_2O$, 0.1 g; $MnCl_2 \cdot 4H_2O$, 0.03 g; H_3BO_3, 0.3 g; $CoCl_2 \cdot 6H_2O$, 0.02 g; $CuCl_2 \cdot 2H_2O$, 0.01 g; $NiCl_2 \cdot 6H_2O$, 0.02 g; $Na_2MoO_4 \cdot 2H_2O$, 0.03 g; distilled water to 1000 ml.

Cultures were incubated in desiccators at 65°C under an atmosphere of 0.05 atm O_2 + 0.1 atm CO_2 + 0.45 atm H_2 (partial pressure measured at room temperature). After 4 days the samples gave rise to dense cultures. They were transferred twice to fresh medium using a loopful of the previous enrichment culture as inoculum. Pure cultures were obtained by streaking cell suspensions on plates of the same medium solidified by 1% Oxoid Agar No. 1, and incubated under the atmosphere given above.

Additional strains have been obtained by Kruger and Meyer (1984) from sludge samples of a settling pond of a sugar factory using carbon monoxide as the sole source of carbon and energy for enrichment. The strains were purified on nutrient agar supplemented with 0.1% (w/v) sodium pyruvate where they form golden brown to dark brown colonies. The strains can also grow autotrophically with hydrogen, carbon dioxide and oxygen in liquid medium.

Kruger and Meyer (1984) isolated one additional strain which, according to its DNA base composition and low DNA/DNA hybridization with the type strain, apparently belongs to another species.

B. sphaericus. This species, as defined by Gordon et al. (1973), includes strains belonging to different DNA/DNA homology groups (Krych et al., 1980). So far these groups have not been clearly separated by phenotypical properties (de Barjac et al., 1980).

Strains belonging to the species complex can be easily obtained by an enrichment method described by Beijerinck and Minkman (1910) for the isolation of "*B. sphaerosporus.*"

To 1000 ml of tap water 10–20 g of casein or its sodium salt is added. The medium is dispensed in 30-ml amounts into 300-ml Erlenmeyer flasks. After inoculation with about 1 g of soil, the flasks are heated to boiling, cooled and incubated at 30–37°C. After 3 days most cultures will show microscopically different sporulating organisms. Generally, round sporing forms with swollen sporangia will predominate.

The flasks are heated at 80°C for 5 min. A loopful of material is then streaked on nutrient agar plates, pH 7.0, and incubated at 30°C. Colonies showing microscopically swollen sporangia with round spores are further purified.

Some strains of *B. sphaericus,* especially those known to be extremely pathogenic for mosquito larvae, were found to be naturally resistant to streptomycin and to chloramphenicol (Burke and McDonald, 1983). This observation may lead to the development of selective media for the isolation of such strains. The selective retrieval of *B. sphaericus* from soil in a mosquito habitat has already been described by Hertlein et al. (1979).

Thermophilic strains similar to *B. sphaericus* have been isolated by Allen (1953) from enrichment cultures containing carboxylic acids as the sole source of carbon and energy. Similar strains have been isolated by Klaushofer and Hollaus (1970) from extraction installations in Austrian sugar factories.

B. stearothermophilus. The taxonomic position of the thermophilic aerobic sporeforming bacteria has been discussed by Allen (1953) and by Wolf and Sharp (1981). The only fact which seems to be clear is that *Bacillus* strains capable of growing at temperatures of 65°C and above do not belong to a single species.

Allen (1953) has pointed out that fresh isolates of thermophilic *Bacillus* strains tend to display a bewildering diversity of characteristics, while strains which have been maintained in culture for some time are usually readily classifiable. This observation should be kept in mind before describing new isolates. Another problem in the taxonomy of thermophilic *Bacillus* strains is the inadequate description of newly named species. Neither have the methods proposed by Gordon et al. (1973) been applied nor have other strains been included for comparative studies. One example is the recently described "*B. flavothermus*" (Heinen et al., 1982) which may be identical with "*B. thermocatenulatus*" (Golovacheva et al., 1975).

Strains of the *B. stearothermophilus* complex can be easily obtained by incubating soil samples or compost in a variety of media at 65°C and above. Enrichment methods for strains belonging to a certain morphological or physiological group have been described by Allen (1953).

A method of obtaining "*B. thermodenitrificans*" ("*Denitrobacterium thermophilum*"), an organism producing gas from nitrate, has been described by Ambroz (1913): samples of soil or dust are inoculated into nutrient broth supplemented with 1% (w/v) KNO_3. Cultures are incubated at 65–70°C under aerobic conditions. Within 24 h cultures produce gas leading to heavy foaming. Pure cultures are isolated by plating suspensions on nutrient agar at 55°C. Most strains spread on this medium forming typical colonies resembling ice ferns or similar structures.

B. subtilis. According to Zopf (1885), *B. subtilis,* the "hay bacillus," can regularly be obtained in a pure form using the following method: hay is soaked in water for 4 h at 36°C using as small a volume of water as possible. The extract is decanted and diluted to a specific gravity of 1.004. The pH is adjusted to 7.0 and 500 ml of the extract transferred to a sterile 1-liter Erlenmeyer flask, which is plugged with cotton wool and boiled for 1 h.

After incubation for about 28 h at 36°C, a pellicle has usually developed on the medium from which often only *B. subtilis* can be isolated.

The method used by Knight and Proom (1950) allows prolonged heating to be avoided. Air dried soil is added to 10 ml of nutrient broth to give a total volume of 15 ml. The tube is well shaken, and a loopful of the suspension plated on nutrient agar. Incubation at 37°C for 2 days gave a flora predominantly containing *B. subtilis.* Only an occasional colony of *B. circulans, B. megaterium* or *B. cereus* was obtained.

B. thuringiensis. Strains of the species are known as specific pathogens of lepidopterous larvae with greater or lesser virulence for the different insect species. They can be obtained from diseased larvae by methods used for the detection of strains of *B. cereus* from which they differ mainly in the presence within the cell of parasporal bodies which appear microscopically as lozenge-shaped, square, round or indefinite in shape. One of the subspecies described is specifically pathogenic to mosquito and *Simulidae* larvae (de Barjac, 1981a; Krieg, 1981).

Maintenance Procedures

As with other microorganisms, in order to avoid the selection of mutants, strains of *Bacillus* species should not be maintained by subculturing at short intervals.

If *Bacillus* strains are grown under conditions where spore formation occurs, their long term maintenance does not cause any problems. Sporulated cultures, preferably on agar slants and protected from drying, can be kept at room temperature, or below, for at least 1 year. Many strains will survive for 10 years or more, under these conditions.

Bacillus strains do not form spores on all media. Liquid media, especially, often do not allow spore formation. Except with those strains with special nutrient requirements it is generally best to grow *Bacillus* stock cultures on nutrient agar composed of peptone, 5.0 g; meat extract, 3.0 g; yeast extract, 1.0 g; agar, 15.0 g; distilled water to 1000 ml. The pH is adjusted to 6.8 (for alkaliphilic strains to pH 8.5).

For sporulation, generally higher concentrations of certain trace elements, especially manganous ions, are needed than for vegetative cell growth (Charney et al., 1951). Since peptones and similar organic substrates often contain only a very low concentration of such elements (Bovallius and Zacharias, 1971) their addition to stock culture media may drastically improve spore formation. For strains not forming spores on the nutrient agar described above or forming only few spores the nutrient agar should be supplemented with 50 mg of $MnSO_4 \cdot H_2Ol^{-1}$. Sometimes an additional supply of $CaCl_2 \cdot 2H_2O$ (100 mg l^{-1}) and $MgSO_4 \cdot 7H_2O$ (500 mg l^{-1}) can be useful.

Sporulation of other strains may be induced by adding soil extract to nutrient agar (Gordon and Rynearson, 1963). The extract is prepared as follows: garden soil, rich in organic matter, is air dried, crushed and sifted through a coarse sieve. Soil (400 g) is suspended in 1000 ml of distilled water in a 2-liter Erlenmeyer flask and sterilized at 121°C for 1 h. The suspension is left overnight to allow soil particles to sediment. The supernatant is then carefully decanted, centrifuged for 30 min, and bottled in 50- to 100-ml amounts before being again sterilized for 30 min at 121°C. The soil extract is then ready for use.

Soil extract agar consists of peptone, 5.0 g; meat extract, 3.0 g; agar, 15.0 g; soil extract, 100–750 ml; tap water to 1000 ml. The medium is adjusted to pH 7.0 and sterilized at 121°C for 20 min. Depending on the soil sample used and the strain being studied, the amount of soil extract needed for good sporulation may vary. It has often been found that a concentration of 25% (v/v) is useful.

Strains of some species which originally have formed spores on a particular medium may lose this property after a number of transfers. With such strains, spore formation should be tested on a variety of media and at different incubation temperatures. Sometimes low nutrient concentrations like 0.1% (w/v) yeast extract agar supplemented with trace elements may be useful. Other strains may form spores more readily on slants than on plates, or vice versa, or prefer a higher agar concentration (3% w/v) for sporulation.

For some species special media are always required to obtain sporulation. *B. acidocaldarius* usually forms spores on mineral base agar supplemented with yeast extract and glucose or starch (Darland and Brock, 1971), *B. fastidiosus* on allantoin agar (F. Fahmy, personal communication). Methods for the maintenance and preservation of *B. larvae, B. lentimorbus* and *B. popilliae* have been described by Gordon et al. (1973). Factors important for the sporulation of *B. popilliae* have been summarized by Bulla and Sharpe (1978).

Both sporulated and nonsporulated *Bacillus* strains can be preserved for long times by freeze-drying and storage under vacuum. Like other Gram-positive bacteria, strains of *Bacillus* generally survive this treatment and resuscitation well. A suitable suspension medium is skim milk prepared from skim milk powder (20% w/v) and supplemented with 5% (w/v) *meso*-inositol. Other protective media as well as general aspects of freeze-drying of bacteria have been discussed by Gherna (1981), Heckley (1978), Hill (1981), Kirsop and Snell (1984) and Sleesman (1982).

There appear to be no published reports on the survival of *Bacillus* strains at temperatures of −20 to −90°C. However, nearly all strains can be readily preserved for long term maintenance in liquid nitrogen using nutrient broth supplemented with 10% (v/v) glycerol or 5% (v/v) dimethyl sulfoxide as a cryoprotective medium. Methods used in freezing of microorganisms have been published by Gherna (1981), Kirsop and Snell (1984) and Rinfret and LaSalle (1975).

The long term maintenance of *Bacillus* strains in a genetically stable state is often important and some additional information may be helpful. It is sometimes recommended that *Bacillus* cultures should be pasteurized before a transfer to a fresh medium in order to avoid the selection of asporogenous mutants. However, according to Hayase et al. (1974), Nishida et al. (1969), Kadota et al. (1978) and Zamenhof (1960) heat treatment of bacterial spores may induce mutation.

Several other authors have reported the induction of mutations in cells which undergo dehydration. This occurs not only in the freeze-drying of vegetative cells (Asada et al., 1980; Ashwood-Smith and Grant, 1976; Servin-Massieu, 1971; Tanaka et al., 1979) but also during spore formation itself (Servin-Massieu, 1971; Zamenhof and Rosebaum-Oliver, 1968) when cells (spores) are dehydrated by a natural process. It is therefore recommended that vegetative cells of *Bacillus* strains are stored in liquid nitrogen in the presence of cryoprotective substances. Under these conditions mutations have not been observed.

Testing for Special Characters

The importance for inter- and intra-laboratory reproducibility of rigorous standardization of diagnostic tests is now widely recognized. In the context of the genus *Bacillus*, a reproducibility trial organized in 1975 by the International Committee on Systematic Bacteriology (ICSB) Sub-Committee on the genus *Bacillus* (BSC) underlined the need for standardization. Even in the hands of those in five laboratories expert in the genus, there was substantial disagreement about the results of certain tests even though test details had been provided as the basis of the trial (see Logan and Berkeley, 1981). Thus, details of the necessary media for the performance of all the tests for the differential characters listed in table 13.4 are given below. They are based on the work of Smith et al. (1952) and Gordon et al. (1973). Use of other methods for certain tests may give false-negative or false-positive results and lead to misidentification.

Media. *Medium components.* Difficulty may be experienced in obtaining the precise medium components specified below. The following information will help in the selection of alternative materials that are as similar as possible to those given: peptone (Oxoid L37) is meat hydrolyzed with trypsin; tryptone and trypticase are trypsin hydrolysates of casein. Proteosepeptone is an enzyme digest of meat and neopeptone is an enzyme digest of casein and meat.

Nutrient broth. Beef extract, 3 g; peptone, 5 g; distilled water, 1000 ml; pH 6.8. Distribute in test tubes and sterilize by autoclaving at 121°C for 20 min.

Nutrient agar. Beef extract, 3 g; peptone, 5 g; agar, 15 g; distilled water, 1000 ml; pH 6.8. Dissolve by steaming and mix well before distributing into final containers. Sterilize by autoclaving at 121°C for 20 min. For sporulation 5 mg hydrous manganese sulfate should be added per 1000 ml of medium.

Glucose agar. Glucose, anhydrous D-(+), 10 g; nutrient agar, 1 liter. Add the glucose to the nutrient agar and, after thorough mixing, dispense into test tubes or flasks and sterilize by autoclaving at 115°C for 20 min.

J-agar. Tryptone, 5 g; yeast extract, 15 g; dipotassium hydrogen phosphate, 3 g; agar, 20 g; distilled water, 1 liter. Thoroughly mix the solids and water, steam until all are dissolved and filter through a clarifying grade of filter paper. Adjust the pH to 7.3–7.5, distribute as

required and sterilize by autoclaving at 121°C for 20 min. Add aseptically 2 g/liter of glucose sterilized separately by autoclaving a 10% solution in distilled water at 115°C for 20 min.

For semisolid J-agar use only 10 g agar, for J-broth omit agar altogether and for basal J-agar omit glucose.

Chocolate agar. Melt nutrient agar by autoclaving at 115°C for 10 min. Cool to 60°C and add aseptically 10% by volume of sterile horse blood. (It is important to use horse blood containing no added preservative). Immerse the vessel containing the medium in boiling water for 1 min with constant mixing and immediately dispense the medium as slants or into Petri dishes as required.

Anaerobic agar. Trypticase, 20 g; glucose, 10 g; sodium chloride, 5 g; agar, 15 g; sodium thioglycolate, 2 g; sodium formaldehyde sulfoxylate, 1 g; distilled water, 1 liter. Adjust the pH to 7.2. Distribute into 15-mm test tubes in amounts sufficient to give a 75-mm depth of medium and sterilize by autoclaving at 121°C for 20 min. For cultures of *B. larvae*, *B. popilliae* and *B. lentimorbus* the agar should be supplemented with 15 g of yeast extract per liter.

Voges-Proskauer broth. Proteose peptone, 7 g; glucose, 5 g; sodium chloride, 5 g; distilled water, 1000 ml; pH 6.5. Distribute in 5-ml amounts in 20-mm test tubes and sterilize by autoclaving at 115°C for 20 min.

J-broth for Voges-Proskauer reaction. Tryptone, 5 g; yeast extract, 15 g; distilled water, 1 liter. Mix thoroughly until components are dissolved and adjust the pH to 7.3–7.5. Distribute as required and sterilize by autoclaving at 121°C for 20 min. Add aseptically 5 g/liter of glucose sterilized separately by autoclaving a 10% solution in distilled water at 115°C for 20 min.

Egg-yolk reaction medium. Tryptone, 10 g; disodium hydrogen phosphate, 5 g; potassium dihydrogen phosphate, 1 g; sodium chloride, 2 g; magnesium sulfate · 7H$_2$O, 0.1 g; glucose, 2 g; distilled water, 1000 ml. The pH of the basal medium is adjusted to 7.6 and the medium sterilized by autoclaving at 121°C for 20 min. Egg-yolk (1.5 ml) aspirated aseptically or a sterile commercial preparation used at the concentration recommended by the manufacturer, is added to 100 ml of basal medium. The medium is allowed to stand in a refrigerator overnight and the supernatant broth dispensed in sterile tubes in 2.5-ml amounts. Basal medium without added egg yolk is similarly dispensed. For the fastidious insect pathogens the basal medium should be replaced by J-broth for Voges-Proskauer reaction.

Resistance to lysozyme medium. Prepare a solution of lysozyme containing 10,000 enzyme units/ml of distilled water in a volumetric flask and sterilize by filtration. Mix 1 ml of the lysozyme solution with 99 ml of sterile nutrient broth and dispense in 2.5-ml amounts in sterile plugged tubes. For the fastidious insect pathogens 1 ml of the lysozyme solution (0.1%) is added to 99 ml of sterile semisolid J-agar.

Sabouraud dextrose agar. Neopeptone, 10 g; dextrose, 40 g; agar, 15 g; distilled water, 1000 ml; pH 5.6. For the fastidious insect pathogens supplement the medium with 1.5% yeast extract. Dispense as slants and sterilize by autoclaving at 121°C for 20 min.

Sabouraud dextrose broth. Neopeptone, 10 g; dextrose, 20 g; distilled water, 1000 ml; pH 5.7. For the fastidious insect pathogens supplement the medium with 1.5% yeast extract. Dispense into test tubes and sterilize by autoclaving at 121°C for 20 min.

Medium for acid production from carbohydrates. Basal medium: diammonium hydrogen phosphate, 1 g; potassium chloride, 0.2 g; magnesium sulfate, 0.2 g; yeast extract, 0.2 g; agar, 15 g; distilled water 1000 ml. Adjust the pH of the medium to 7.0 before adding 15 ml of a 0.04% (w/v) solution of bromcresol purple. Sterilize by autoclaving at 121°C for 20 min. Carbohydrate solutions: sterilize 10% (w/v) aqueous solutions of each test substrate by autoclaving at 121°C for 20 min. Then add aseptically sufficient carbohydrate solution to tubes of sterile basal medium to give a final concentration of 0.5% (w/v). Use as slants. The carbohydrates used are D-(+)-glucose, L-(+)-arabinose, D-(+)-xylose and D-(−)-mannitol. For the fastidious insect pathogens use J-broth with the glucose replaced by 0.5% of the test carbohydrate.

Starch agar. 1 g of potato starch is suspended in 10 ml of cold distilled water and mixed with 100 ml of nutrient agar or basal J-agar, for the

Table 13.4.

Differential characteristics of the species of the genus **Bacillus**[a, b]

Characteristics	1. B. subtilis	2. B. acidocaldarius[c]	3. B. alcalophilus[d]	4. B. alvei	5. B. anthracis	6. B. azotoformans[e]	7. B. badius	8. B. brevis	9. B. cereus	10. B. circulans	11. B. coagulans	12. B. fastidiosus[f]	13. B. firmus[b, g]	14. B. globisporus[b, h]	15. B. insolitus[b, h]	16. B. larvae	17. B. laterosporus
Cell diameter >1.0 μm	−	ND	−	−	+	−	v		+	−	−	+	−	+	v	−	−
Spores round	−	−	−	−	−	−	−	−	−	−	−	−	−	+	+	−	+
Sporangium swollen	−	+	+	+	−	+	−	+	−	+	v	−	−	+	−	+	+
Parasporal crystals	−	ND	−	−	−	−	−	−	−	−	−	−	−	−	−	−	−
Catalase	+	ND	+[g]	+	+	−	+	+	+	+	+	+	+	+	+	−	+
Anaerobic growth	−	−	−	+	+	−	−	−	+	d	+	−	−	−	−	+	+
Voges-Proskauer test	+	−	−	+	+	−	−	−	+	−	+	NG	−	−	−	−	−
pH in V-P broth																	
<6	d		ND	+	+	ND	−	−	+	+	+	NG	−	−	−	d	d
>7	−		ND	−	−	ND	+	+	−	−	−	NG	−	−	−	−	−
Acid from																	
D-Glucose	+	ND	+	+	+	−	−	d	+	+	+	NG	+	+	−[?]	+	+
L-Arabinose	+	ND	+	−	−	ND	−	−	−	+	d	NG	−	−	−	−	−
D-Xylose	+	ND	+	−	−	−	−	−	−	+	d	NG	−	−	−	−	−
D-Mannitol	+	ND	+	−	−	−	−	d	−	+	d	NG	+	−	−	d	+
Gas from glucose	−	−	−	−	−	−	−	−	−	−	−	−	−	−	−	−	−
Hydrolysis of																	
Casein	+	ND	+	+	+	ND	+	d	+	d	d	−	+	d	−	+	+
Gelatin	+	ND	+	+	+	−	ND	d	+	d	−	−	+	+	−	+	d
Starch	+	+	+	+	+	−	−	d	+	+	+	−	+	d	−	+	+
Utilization of																	
Citrate	+	−	−	−	d	+	−	d	+	d	d	−	−	−	−	−	−
Propionate	−	ND	−[g]	ND	ND	−	ND	ND	ND	ND	−	−	−	ND	ND	ND	ND
Degradation of tyrosine	−	ND	−[g]	d	d	−	+	+	+	−	−	−	d	−	−	−	+
Deamination of phenylalanine	−[i]	ND	ND	−	−	−	−	−	−	−	−	−	d	+	d	−	−
Egg-yolk lecithinase	−	ND	−[j]	−	+	−	ND	−	+	−	−	−	−	ND	ND	ND	+
Nitrate reduced to nitrite	+	ND	−	−	+	ND	−	d	+	d	d	−	d	d	−	d	+
Formation of																	
Indole	−[i]	−	−	+	ND	−	−	−	−[k]	−	−	ND	−	−	−	−	d
Dihydroxyacetone	ND	ND	−[g]	+	ND	−	−	−	ND	−	d	ND	−	−	−	−	−
NaCl and KCl required	−	−	−	−	−	−	−	−	−	−	−	−	−	−	−	−	−
Allantoin or urate required	−	−	−	−	−	−	−	−	−	−	−	+	−	−	−	−	−
Growth at pH																	
6.8, nutrient broth	+	−	+	+	+	+	+	+	+	+	+	−	+	+	+	−	+
5.7	+	d	−	−	+	−	−	d	+	d	+	−	−	−	−	−	−
Growth in NaCl																	
2%	+	ND	ND	ND	+	ND	ND	ND	ND	ND	+	+	+	+	d	+	ND
5%	+	ND	−	d	+	−	+	−	ND	d	−	+	+	−	−	−	d
7%	+	ND	−	−	+	−	ND	−	d	d	−	−	−	−	−	−	−
10%	ND	ND	−	−	ND	−	−	−	ND	−	−	−	−	ND	−	−	−
Growth at																	
5°C	−	−	ND	−	−	ND	−	−	−	−	−	−	−[b]	+	+	−	−
10°C	d	−	ND	−	−	ND	−	−	d	d	−	+	d[b]	+	+	−	−
30°C	+	−	+	+	+	+	+	d	+	+	+	+	+	d	−	+	+
40°C	+	−	+	+	+	+	+	+	d	+	+	+	+[b]	−	−	+	d
50°C	d	+	−	−	−	+	d	−	−	−	+	−	−	−	−	−	−
55°C	−	+	−	−	−	−	d	−	−	−	+	−	−	−	−	−	−
65°C	−	+	−	−	−	−	d	−	−	−	−	−	−	−	−	−	−
Growth with lysozyme present	d	ND	−[g]	+	+	−	−	d	+	d	−	ND	−	−	−	+	+
Autotrophic with H₂+ CO₂ or CO	−	−	−	−	−	−	−	−	−	−	−	−	−	−	−	−	−

[a] Symbols: −, 90% or more of strains are negative; +, 90% or more of strains are positive; v, strain instability (not equivalent to "d"); d, 11–89% of strains are positive; ND, no data available; NG, no growth; [b] compiled from Smith et al. (1952), Gordon et al. (1973) and Knight and Proom (1950), except[c-p]; [c] data from Darland and Brock (1971); [d] data from Boyer et al. (1973); [e] data from Pichinoty et al. (1978; 1983); [f] data from F. Fahmy (personal communication); [g] data from Gordon et al. (1977) and Gordon and Hyde (1982); [h] data from Larkin and Stokes (1967); [i] data from Hanáková-Bauerová et al. (1965); [j] data from McGaughey and Chu (1948); [k] data from Hanáková-Bauerová et al. (1966); [l] data from Gordon et al. (1977) and Gordon and Hyde (1982); [m] data from Marshall and Ohye (1966); [n] data from Rüger (1983); [o] data from Schenk and Aragno (1979) and Krüger and Meyer (1984); and [p] (−), few gas bubbles may be formed.

Table 13.4—continued

Characteristics	18. B. lentimorbus	19. B. lentus[l]	20. B. licheniformis	21. B. macerans	22. B. macquariensis[b,m]	23. B. marinus[n]	24. B. megaterium	25. B. mycoides	26. B. pantothenticus	27. B. pasteurii	28. B. polymyxa	29. B. popilliae	30. B. pumilus	31. B. schlegelii[o]	32. B. sphaericus	33. B. stearothermophilus	34. B. thuringiensis
Cell diameter > 1.0 μm	−	−	−	−	−	v	+	+	−	−	−	−	−	−	−	−	+
Spores round	−	−	−	−	−	+	v	−	v	+	−	−	−	+	+	−	−
Sporangium swollen	+	−	−	+	+	−	−	−	+	+	+	+	−	+	+	+	−
Parasporal crystals	−	−	−	−	−	−	−	−	−	−	−	+	−	−	−	−	d
Catalase	−	+	+	+	+	+	+	+	+	ND	+	−	+	+	+	d	+
Anaerobic growth	+	−	+	+	+	−	−	+	+	+	+	+	−	−	−	−	+
Voges-Proskauer test	−	−	+	−	−	−	−	+	−	−	+	−	+	−	−	−	d
pH in V-P broth																	
<6	d	−	+	+	+	ND	d	+	+	ND	d	d	+	ND	−	+	+
>7	−	ND	−	−	−	ND	−	−	−	ND	−	−	−	ND	+	−	−
Acid from																	
D-Glucose	+	+	+	+	+	+	+	+	+	ND	+	+	−	−	−	+	+
L-Arabinose	−	+	+	+	−	−	d	−	−	ND	+	−	+	−	−	d	−
D-Xylose	−	+	+	+	+	d	d	−	−	ND	+	−	+	−	−	d	−
D-Mannitol	−	+	+	+	+	−	d	−	−	ND	+	−	+	−	−	d	−
Gas from glucose	−	−	(−)[p]	+	−	−	−	−	−	−	+	−	−	−	−	−	−
Hydrolysis of																	
Casein	−	d	+	−	−	d	+	+	d	d	+	−	+	−	d	d	+
Gelatin	−	d	+	+	−	+	+	+	+	+	+	−	+	−	d	+	+
Starch	−	+	+	+	+	−	+	+	+	−	+	−	−	−	−	+	+
Utilization of																	
Citrate	−	−	+	d	−	−	+	d	−	ND	−	−	+	−	d	d	+
Propionate	ND	−	+	ND	ND	ND	ND	ND	ND	ND	ND	ND	−	−	ND	ND	ND
Degradation of tyrosine	−	−	−	−	−	ND	d	d	−	ND	−	−	−	−	−	−	d
Deamination of phenylalanine	−	d	−[i]	−	−	ND	d	−	d	ND	−	−	−[i]	ND	+	−	−
Egg-yolk lecithinase	ND	−	−	−	ND	ND	−	d[j]	−	−	−	ND	−	−	−	ND	d
Nitrate reduced to nitrite	−	d	+	+	−	d	d	+	d	+	+	−	−	+	−	d	+
Formation of																	
Indole	−	−	−[k]	−	−	−	−[k]	−[k]	−	ND	−	−	−[i]	−	−	−	−[k]
Dihydroxyacetone	−	−	ND	−	−	ND	ND	ND	−	ND	+	−	ND	ND	−	−	ND
NaCl and KCl required	−	−	−	−	−	+	−	−	−	−	−	−	−	−	−	−	−
Allantoin or urate required	−	−	−	−	−	−	−	−	−	−	−	−	−	−	−	−	−
Growth at pH																	
6.8, nutrient broth	−	+	+	+	+	−	+	+	+	−	+	−	+	+	+	+	+
5.7	−	−	+	+	−	ND	d	+	−	−	+	−	+	−	d	−	+
Growth in NaCl																	
2%	−	ND	+	ND	+	+	ND	ND	+	+	ND	+	+	+	ND	ND	+
5%	−	ND	+	−	−	ND	ND	ND	+	+	−	−	+	−	d	d	+
7%	−	d	+	−	−	d	d	d	+	+	−	−	+	−	d	−	+
10%	−	ND	ND	−	−	−	ND	ND	+	+	−	−	−	ND	−	−	ND
Growth at																	
5°C	−	ND	−	−	+	+	d	−	−	ND	d	−	−	−	−	−	−
10°C	−	ND	−	d	+	+	+	d	−	ND	+	−	+	−	+	−	d
30°C	+	+	+	+	−	d	+	+	+	+	+	+	+	−	+	−	+
40°C	−	ND	+	+	−	−	d	d	+	d	+	−	+	−	d	+	+
50°C	−	−	+	d	−	−	−	−	d	−	−	−	d	+	−	+	−
55°C	−	−	+	−	−	−	−	−	−	−	−	−	−	+	−	+	−
65°C	−	−	−	−	−	−	−	−	−	−	−	−	−	+	−	+	−
Growth with lysozyme present	+	−	d	−	−	ND	−	+	−	ND	d	−	d	+	d	−	+
Autotrophic with H₂ + CO₂ or CO	−	−	−	−	−	−	−	−	−	−	−	−	−	+	−	−	−

[a] Symbols: −, 90% or more of strains are negative; +, 90% or more of strains are positive; v, strain instability (not equivalent to "d"); d, 11–89% of strains are positive; ND, no data available; NG, no growth; [b] compiled from Smith et al. (1952), Gordon et al. (1973) and Knight and Proom (1950), except[c-p]; [c] data from Darland and Brock (1971); [d] data from Boyer et al. (1973); [e] data from Pichinoty et al. (1978; 1983); [f] data from F. Fahmy (personal communication); [g] data from Gordon et al. (1977) and Gordon and Hyde (1982); [h] data from Larkin and Stokes (1967); [i] data from Hanáková-Bauerová et al. (1965); [j] data from McGaughey and Chu (1948); [k] data from Hanáková-Bauerová et al. (1966); [l] data from Gordon et al. (1977) and Gordon and Hyde (1982); [m] data from Marshall and Ohye (1966); [n] data from Rüger (1983); [o] data from Schenk and Aragno (1979) and Krüger and Meyer (1984); and [p] (−), few gas bubbles may be formed.

fastidious insect pathogens. Autoclave at 121°C for 20 min. Cool to 45°C, thoroughly mix and pour into Petri dishes.

Citrate and propionate utilization medium. Trisodium citrate·2H₂O, 1 g (or sodium propionate, 2 g); magnesium sulfate·7H₂O, 1.2 g; diammonium hydrogen phosphate, 0.5 g; potassium chloride, 1 g; trace element solution (see below), 40 ml; agar, 15 g; distilled water, 920 ml; 0.04% (w/v) solution of phenol red, 20 ml. Adjust the pH of the agar to 6.8 before sterilizing at 121°C for 20 min and prepare slants.

The trace element solution contains ethylenediaminetetraacetate, 500 mg; FeSO₄·7H₂O, 200 mg; ZnSO₄·7H₂O, 10 mg; MnCl₂·4H₂O, 3 mg; H₃BO₃, 30 mg; CoCl₂·6H₂O, 20 mg; CuCl₂·2H₂O, 1 mg; NiCl₂·6H₂O, 2 mg; Na₂MoO₄·2H₂O, 3 mg; distilled water, 1000 ml. For the fastidious insect pathogens use semisolid J-agar without glucose but supplemented with sodium citrate or sodium propionate. Do not include phenol red in this medium.

Nitrate broth. Peptone, 5 g; beef extract, 3 g; potassium nitrate, 1 g; distilled water, 1000 ml; pH 7.0. Distribute the medium into test tubes containing inverted Durham's tubes and sterilize by autoclaving at 121°C for 20 min. For the fastidious insect pathogens use J-broth supplemented with 0.1% (w/v) potassium nitrate and with the glucose omitted. A semisolid medium containing peptone 0.5%, sodium chloride 0.5%, potassium nitrate 0.02% (w/v) is useful for marine isolates.

Indole production medium. Use 1% (w/v) tryptone broth or 1% trypticase broth or J-broth with double the usual amount of tryptone and without glucose, for the fastidious insect pathogens. Distribute approximately 5 ml into test tubes and sterilize by autoclaving at 121°C for 20 min.

Glycerol agar. Nutrient agar, 100 ml; yeast extract, 1 g; glycerol, 2 ml. Sterilize by autoclaving at 121°C for 20 min and pour large (100 × 20 mm) plates. For the fastidious insect pathogens, use basal J-agar 100 ml; glycerol, 2 ml.

Phenylalanine agar. Yeast extract, 3 g; DL-phenylalanine, 2 g; disodium hydrogen phosphate, 1 g; sodium chloride, 5 g; agar, 12 g; distilled water, 1000 ml. Final pH 7.3. Prepared as slants after sterilizing by autoclaving at 121°C for 20 min. Use J-agar with 2 g of DL-phenylalanine substituted for the glucose for the fastidious insect pathogens. Experience has shown that a commercial preparation (BBL 11537, Becton-Dickinson) is, for unknown reasons, better than self-prepared versions of this medium.

Milk agar. Skim milk powder, 5 g in 50 ml of distilled water; agar, 1 g in 50 ml of distilled water. Autoclave separately at 121°C for 20 min, cool to 45°C, mix together and pour into Petri dishes. For the fastidious insect pathogens supplement the agar part of J-agar with double concentrations of the normal ingredients but omit the glucose. Then mix with skim milk suspension as above. Allow plates to stand at room temperature for 3 days to dry the surface of the agar.

Tyrosine agar. L-tyrosine, 0.5 g; distilled water, 10 ml. Autoclave at 121°C for 20 min and then mix aseptically with 100 ml of sterile nutrient agar and pour into Petri dishes after cooling the medium to about 50°C. For cultures of fastidious insect pathogens mix the tyrosine suspension with 100 ml of basal J-agar. Care must be taken to ensure an even distribution of tyrosine crystals throughout the solidified agar. Dry the plates thoroughly before use.

Nutrient gelatin. Use a commercially available nutrient gelatin medium distributed in test tubes, or tubes of plain gelatin (gelatin 120 g; distilled water 1000 ml; pH 7.0). Alternatively, nutrient agar supplemented with 0.4% gelatin may be used.

Methods. *Temperature of incubation.* Cultures should be incubated at temperatures approximately 10–15°C below their maximum growth temperatures. Thus, cultures of psychrophiles should be incubated at 20°C, mesophiles at 30°C and thermophiles whose maximum growth temperature is 55–60°C should be incubated at 45°C. Strains capable of growth at 65°C should be incubated at 45°C and at 55°C.

Microscopic appearance. General morphology should be observed from young (18–24 h) cultures grown in nutrient broth and shaken to obtain good aeration. Shapes of cells from nutrient agar cultures may be heterogeneous due to the different conditions of nutrition and oxygen supply within colonies. *B. larvae, B. popilliae* and *B. lentimorbus* should be observed from cultures grown on J-agar. Longer incubation, and use of nutrient agar, may be required for observation of spores. Parasporal crystals (bodies) may be seen in cells grown for 3–7 days on nutrient agar. *B. larvae, B. popilliae* and *B. lentimorbus* do not sporulate on nutrient agar and may not sporulate at all satisfactorily on media in vitro.

The microscopic appearance of vegetative or spore-bearing cells and parasporal crystals (bodies) is best observed on slides freshly coated with a thin (0.5 mm) layer of 2% water agar. Purified agar preparations should be used as other types are often heavily contaminated with bacteria. To coat the slides, place a slide in a 90-mm Petri dish and add not more than 5 ml of hot agar so that the whole of the bottom of the dish and the slide are covered. After solidification of the agar, cut out the coated slide. After adding a small drop of a distinctly turbid suspension of the bacterium to be examined to the surface of the agar, place a cover glass on the preparation and examine using a phase-contrast microscope with an oil immersion objective.

Production of catalase. Flood cultures grown for 1 or 2 days on slants of nutrient agar with 0.5 ml of 10% hydrogen peroxide. Replace the tube closure immediately to avoid dispersal of an aerosol and observe for gas production. If no gas bubbles form, repeat, using cultures grown on chocolate agar. Cultures of *B. larvae, B. popilliae* and *B. lentimorbus* are conveniently handled by growing colonies on plates of J-agar and flooding colonies or the edges of confluent growth with 10% hydrogen peroxide. Cultures should be tested as soon as clearly visible growth is present. Weak formation of gas bubbles is better observed using about 3 ml of hydrogen peroxide.

Anaerobic growth. Inoculate a tube of anaerobic agar with a small (outside diameter 1.5 mm) loopful of nutrient broth culture by stabbing to the bottom of the culture tube. Alternatively, molten medium cooled to about 40°C may be inoculated thoroughly using a Pasteur pipette and allowed to solidify before incubation. At incubation temperatures below 45°C the growth should be recorded at 3 and 7 days, at temperatures of 45°C or higher growth should be recorded at 1 and 3 days. Cultures of the three species of fastidious insect pathogens may require as long as 14 days incubation before growth becomes apparent.

Voges-Proskauer reaction. Acetyl methyl carbinol production: inoculate tubes of Voges-Proskauer broth in triplicate and test for acetyl methyl carbinol production after incubation for 3, 5 and 7 days by mixing 3 ml of 40% (w/v) sodium hydroxide with the culture and adding 0.5–1 mg of creatine. Observe for the production of a red color after 30–60 min at room temperature. The fastidious insect pathogens should be grown on J-broth medium for Voges-Proskauer reaction for this test. Final pH produced in Voges-Proskauer broth: before cultures incubated for 7 days are tested for acetyl methyl carbinol, the pH is measured, preferably using a pH meter.

Maximum and minimum growth temperatures. Prepare slants of nutrient agar or, for the fastidious insect pathogens, tubes of semi solid J-agar. Determine ability to grow at 5°C intervals. Immerse tubes containing the medium in water baths at the appropriate temperatures until equilibrated and then inoculate. Observe growth of cultures after 3 days at temperatures of 55°C or higher, after 5 days at 30–50°C, after 14 days at 20°C and 25°C, and after 21 days at temperatures below 20°C. Care should be taken to ensure that the water levels in the baths are carefully maintained and that the temperatures are stable, with a variation not greater than ± 0.5°C.

Egg-yolk reaction. Cultures are inoculated into an egg-yolk broth and a tube of control broth without egg yolk. Observe after incubation for 1, 3, 5 and 7 days and, for the fastidious insect pathogens, 14 days for the appearance of a heavy white precipitate in or on the surface of the egg-yolk containing medium.

Resistance to lysozyme. Inoculate a small loopful of a broth culture into a tube of resistance-to-lysozyme medium and into a control tube of nutrient broth. Growth or its absence after incubation for 7–14 days should be observed. Treat the fastidious insect pathogens in the same way using semisolid J-agar as the base medium.

Growth in sodium chloride. Inoculate tubes of nutrient broth (3 ml/tube) containing 0, 5, 7 and 10% (w/v) sodium chloride with a small loopful of a culture grown in nutrient broth and incubate in a slanted position to improve aeration. For the fastidious insect pathogens use semisolid J-agar as the basal medium. Observe for growth in the various concentrations of sodium chloride after 7 and 14 days incubation.

Growth at pH 5.7. Inoculate a slope of Sabouraud dextrose agar and a tube of Sabouraud dextrose broth with a loopful of a culture grown in nutrient broth and inoculate a tube of nutrient broth as a control. Observe for growth in either or both Sabouraud media for up to 2 weeks of incubation.

Acid from carbohydrates. Slants of the medium for acid production from carbohydrates are inoculated and incubated at appropriate temperatures. Observe for growth and for production of acid and gas (gas bubbles throughout the agar or in the water of condensation or both) at 7 and 14 days. The fastidious insect pathogens are tested by aseptically removing a drop of culture to a spot plate, mixing with a drop of 0.04% (w/v) alcoholic bromcresol purple, and observing the color of the indicator.

Hydrolysis of starch. Inoculate duplicate plates of starch agar with each culture to be tested and incubate at appropriate temperatures. At 3 and at 5 days flood one of the plates with 95% ethanol. After 15–30 min the unchanged starch will become white and opaque. Observe for a clear zone underneath (after the growth is scraped off) and around the growth as an indicator of hydrolysis of starch. For the fastidious insect pathogens the plates are flooded by Gram's iodine after 5 and 10 days of incubation. Some cultures spread rapidly on this medium and examination before 3 days may be necessary.

Utilization of citrate and propionate. Inoculate slants of citrate and propionate utilization medium and incubate for 14 days. Observe for production of a red (alkaline) color indicating utilization of organic acids. For the fastidious insect pathogens inoculate the supplemented semisolid J-agar in test tubes with 2 or 3 drops of young cultures and incubate at 30°C. Remove a small amount of the culture after 14 and 21 days incubation and mix on a spot plate with phenol red indicator. Observe for production of a red (alkaline) color signifying utilization of organic acids.

Reduction of nitrate to nitrite. Grow cultures in nitrate broth or nitrate-supplemented J-broth, for the fastidious insect pathogens. Test after 3 and 7 days incubation by moistening a strip of potassium iodide/starch paper with a few drops of 1 N hydrochloric acid and then touching the paper with a loopful of the culture. Observe for the production of a purple color indicating the presence of nitrite and for the accumulation of nitrogen gas in the Durham's tube.

Cultures negative at 7 days are tested after 14 days by mixing 1 ml of the culture with 3 drops of each of the following solutions: (a) sulfanilic acid, 8 g; 5 N acetic acid (glacial acetic acid and water 1:2.5), 1000 ml, and (b) dimethyl-α-naphthylamine, 6 ml; 5 N acetic acid, 1000 ml. Observe for the development of a red or yellow color indicating the presence of nitrite. If the culture is still negative after 14 days, add 4–5 mg of zinc dust to the tube previously tested for nitrite. Look for the development of a red color indicating the presence of nitrate, i.e. the absence of reduction. This is to ensure that very rapid reduction has not occurred reducing nitrate beyond nitrite in less than 3 days.

Production of indole. Incubate cultures for 14 days in indole production medium and add 2 ml of the following test solution: *p*-dimethylaminobenzaldehyde, 5 g; *iso*-amyl alcohol, 75 ml; concentrated hydrochloric acid, 25 ml. Shake vigorously. Observe for a pink to red color in the alcohol layer which separates on standing. This indicates the presence of indole.

Production of dihydroxyacetone. Inoculate plates of glycerol agar or glycerol J-agar, for the fastidious insect pathogens, by streaking once across with inoculum. Incubate for 10 days and flood with a mixture of the following solutions: (a) hydrous copper sulfate, 34.66 g; distilled water, 500 ml, and (b) potassium sodium tartrate, 173 g; sodium hydroxide, 50 g; distilled water, 500 ml. Solutions (a) and (b) are stored in a refrigerator and mixed in a 1:1 ratio immediately before use. Observe 2 h after flooding the plate for the appearance of a red halo around the growth.

Deamination of phenylalanine. Incubate cultures on duplicate slants of phenylalanine agar for 7 days, then pipette 4 or 5 drops of 10% (w/v) ferric chloride solution over the growth on one of duplicate slants. Observe for the production of a green color beneath the growth indicating formation of phenylpyruvic acid from phenylalanine. If the test is negative repeat on the second culture after 21 days of incubation.

Decomposition of casein. Inoculate plates of milk agar with one streak of inoculum and examine after incubation at 7 and 14 days. The more slowly growing insect pathogens should also be examined after 21 days. Observe for clearing of the casein around and underneath the growth.

Decomposition of tyrosine. Inoculate plates of tyrosine agar with one streak of inoculum and incubate, taking care to prevent excessive drying. Examine at 7, 14 and, for the slowly growing insect pathogens, 21 days. Observe for clearing of the tyrosine crystals around and below the growth.

Liquefaction of gelatin. Inoculate tubes of nutrient gelatin and incubate at 28°C. Examine for liquefaction at 3- to 4-day intervals for 4 weeks. Before examination hold the tubes at 20°C for about 4 h to allow unchanged gelatin to harden. Failure to harden at 20°C indicates liquefaction of the gelatin. Alternatively, inoculate a plate of gelatin-supplemented nutrient agar with a single streak. After incubation for 3–5 days at a suitable temperature, flood the plates with 10 ml of 1 N H_2SO_4 saturated with Na_2SO_4. Hydrolysis is indicated by a clear zone, under and around the growth, visible in contrast to the opaque precipitate of unchanged gelatin, formed within 1 h.

Even with the greatest care it is difficult for the great majority of media departments to produce highly standardized test materials; the staff and time to carry out the necessary check procedures just do not exist. It is for this reason that in characterizing an unknown organism it is essential to test in parallel suitably chosen positive and negative controls. Appropriate strains are listed in Table 13.5.

Recently, however, the availability of highly standardized test materials from a commercial source, which has the resources to set up and operate a rigorous quality control system, and the investigation of the applicability of these materials to members of the genus *Bacillus* has led to the establishment of an alternative system for the characterization of the majority of species. The tests used are largely in the API system, from the API 20E and API 50CHB strips, with some morphological and supplementary tests. This approach has been described by Logan and Berkeley (1984). It was applied neither to the nutritionally fastidious species *B. larvae, B. lentimorbus, B. popillae* nor to the species requiring for growth very high or very low pH, *B. alcalophilus* and *B. acidocaldarius,* respectively. Similarly, those three species validly described since the work was carried out, *B. azotoformans, B. schlegelii* and *B. marinus,* have not been examined by these methods.

In common with other types of bacteria, there has recently been considerable interest in the application of instrumented approaches to the identification *Bacillus* strains. Pyrolysis mass-spectrometry has been applied to a number of areas of the genus (see Berkeley et al., 1984) and studies on the differentiation of *B. anthracis* from *B. cereus* by cytofluography (Phillips et al., 1983) and laser microprobe mass analysis (Böhm, 1981) have been carried out.

Table 13.5.

Control strains for diagnostic tests

Tests	Positive strains	ATCC	DSM	NCIB	NCTC	Negative strains	ATCC	DSM	NCIB	NCTC
Cell diameter > 1.0 μm	*B. cereus*	14579	31	9373	2599	*B. subtilis*	6051	10	3610	3610
Spores round	*B. sphaericus*	14577	28	9370	10338	*B. licheniformis* (spores el-lipsoid)	14580	13	9375	10341
Sporangium swollen	*B. polymyxa*	842	36	8158	10343	*B. cereus*	14579	31	9373	2599
Parasporal crystals	*B. thuringiensis*	10792	2046	9134		*B. cereus*	14579	31	9373	2599
Catalase	*B. cereus*	14579	31	9373	2599					
Anaerobic growth	*B. cereus*	14579	31	9373	2599	*B. megaterium*	14581	32	9376	10342
Voges-Proskauer test	*B. cereus*	14579	31	9373	2599	*B. megaterium*	14581	32	9376	10342
pH in V-P broth:										
<6	*B. cereus*	14579	31	9373	2599					
>7	*B. sphaericus*	14577	28	9370	10338					
Acid from:										
D-Glucose	*B. cereus*	14579	31	9373	2599	*B. sphaericus*	14577	28	9370	10338
L-Arabinose	*B. polymyxa*	842	36	8158	10343	*B. sphaericus*	14577	28	9370	10338
D-Xylose	*B. polymyxa*	842	36	8158	10343	*B. sphaericus*	14577	28	9370	10338
D-Mannitol	*B. polymyxa*	842	36	8158	10343	*B. sphaericus*	14577	28	9370	10338
Gas from glucose	*B. polymyxa*	842	36	8158	10343	*B. cereus*	14579	31	9373	2599
Hydrolysis of										
Casein	*B. megaterium*	14581	32	9376	10342	*B. macerans*	8244	24	9368	6355
Gelatin	*B. cereus*	14579	31	9373	2599	*B. coagulans*	7050	1	9365	10344
Starch	*B. cereus*	14579	31	9373	2599	*B. sphaericus*	14577	28	9370	10338
Utilization of										
Citrate	*B. cereus*	14579	31	9373	2599	*B. macerans*	8244	24	9368	6355
Propionate	*B. licheniformis*	14580	13	9375	10341	*B. subtilis*	6051	10	3610	3610
Degradation of tyrosine	*B. cereus*	14579	31	9373	2599	*B. sphaericus*	14577	28	9370	10338
Deamination of phenylala-nine	*B. megaterium*	14581	32	9376	10342	*B. cereus*	14579	31	9373	2599
Egg-yolk lecithinase	*B. cereus*	14579	31	9373	2599	*B. megaterium*	14581	32	9376	10342
Nitrate reduced to nitrite	*B. cereus*	14579	31	9373	2599	*B. megaterium*	14581	32	9376	10342
Formation of										
Indole	*B. alvei*	6344	29	9371	6352	*B. cereus*	14579	31	9373	2599
Dihydroxyacetone	*B. polymyxa*	842	36	8158	10343	*B. macerans*	8244	24	9368	6355
NaCl and KCl required	*B. marinus*	29841	1297			*B. megaterium*	14581	32	9376	10342
Allantoin or urate re-quired	*B. fastidiosus*	29313	325	10017		*B. cereus*	14599	31	9373	2599
Growth at pH										
6.8, nutrient broth	*B. cereus*	14579	31	9373	2599	*B. alcalophilus*	27647	485	10436	4553
5.7	*B. cereus*	14579	31	9373	2599	*B. alvei*	6344	29	9371	6352
Growth in NaCl										
2%	*B. licheniformis*	14580	13	9375	10341	*B. lentimorbus*	14707	2049	11202	
5%	*B. licheniformis*	14580	13	9375	10341	*B. macerans*	8244	24	9368	6355
7%	*B. licheniformis*	14580	13	9375	10341	*B. macerans*	8244	24	9368	6355
10%	*B. pantothenti-cus*	14576	26	8775	8162	*B. macerans*	8244	24	9368	6355
Growth at										
5°C	*B. globisporus*	23301	4	11434		*B. alvei*	6344	29	9371	6352
10°C	*B. globisporus*	23301	4	11434		*B. alvei*	6344	29	9377	6352
30°C	*B. cereus*	14579	31	9373	2599	*B. stearother-mophilus*	12980	22	8923	10339
40°C	*B. licheniformis*	14580	13	9375	10341	*B. globisporus*	23301	4	11434	
50°C	*B. licheniformis*	14580	13	9375	10341	*B. cereus*	14579	31	9373	2599
55°C	*B. coagulans*	7050	1	9365	10344	*B. cereus*	14579	31	9373	2599
65°C	*B. stearother-mophilus*	12980	22	8923	10339	*B. cereus*	14579	31	9373	2599
Growth with lysozyme present (10 μg ml^{-1})	*B. cereus*	14579	31	9373	2599	*B. megaterium*	14581	32	9376	10342

Differentiation of genus *Bacillus* from closely related taxa

I. Cells rod shaped.
    A. Aerobic or facultative, catalase usually produced.
        Bacillus
    B. Microaerophilic, catalase not produced.
        Sporolactobacillus
    C. Anaerobic.
        1. Sulfate not reduced to sulfide.
            Clostridium
        2. Sulfate reduced to sulfide.
            Desulfotomaculum
II. Cells spherical, in packets.
    1. Aerobic.
        Sporosarcina
    2. Anaerobic.
        Sarcina
III. Mycelium formed.
        Thermoactinomyces

Taxonomic Comments

The family *Bacillaceae* is distinctive in being well circumscribed by the single taxonomic character of endospore formation. As pointed out in the eighth edition of the *Manual*, this character is unusually reliable as, although asporogenous mutants are found in the laboratory, they apparently do not survive long in nature.

The genus *Bacillus* can be distinguished from other endospore-forming bacteria because it is rod shaped and either aerobic or faculative, usually producing catalase.

Within the genus, containing as it does, only one species which, in the past, has been generally recognized as an important mammalian pathogen, there has been no need for diagnosticians to try to devise methods for the rapid distinction between closely related organisms. Consequently, apart from the case of *B. anthracis*, there has been little or no conflict or confusion between the needs of determinative bacteriologists and those primarily interested in classification. Thus it might reasonably be expected that the taxonomic arrangement of the organisms in the group is better than in many others. That this is not the case implies no criticism of those workers who have contributed so much to *Bacillus* systematics; it is just a reflection of the fact that circumscription based on endospore formation leads to the inclusion of organisms with a very wide diversity of properties, as has long been recognized (Gibson and Gordon, 1974). In the fifth edition of the *Manual* (1939) 146 species were listed and, in all, more than 200 species have been described.

The first comprehensive studies on the taxonomy of endospore-forming, aerobic, rod-shaped bacteria were performed by Smith et al. (1946, 1952) who tried to collect for their work strains of all named *Bacillus* species available at that time. By applying a restricted number of tests, Smith and his colleagues concluded that most of the species then recognized could be allocated to about 25 species which might be taken as indicating that the genus is reasonably coherent. Recent studies, reviewed above, on DNA base composition and DNA reassociation, however, reveal unmistakably the unusually broad nature of this genus and emphasize the need for its division. Attempts have been made to do this in the past and have been unsuccessful (see Gibson and Gordon, 1974 and Norris et al., 1981) and any attempt to do so in the immediate future is likely to meet a similar fate. In spite of the powerful molecular tools now available to help in such an exercise, the uneven availability of strains representing the various areas of the genus is a barrier to a sensible new taxonomy which must take into account not only DNA base composition and DNA homology information, allowing soundly based allocation of organisms within the genus, as it currently exists, to new genera and species, but also DNA-rRNA hybridization data indicating the phylogenetic relationships with other genera.

From numerical taxonomic studies (Bonde, 1981; Logan and Berkeley, 1981; Priest, 1981) there are pointers to suggest that the genus may be best divided into five or six genera. As Gordon (1981) rightly says in any such exercise the taxonomist must have regard for the needs of diagnosticians and those who use members of the genus *Bacillus* in a wide variety of tasks. Thus, in view of the fact that some areas of the genus are as yet inadequately studied, it would be premature for taxonomists to split the genus, and so cause difficulties for diagnosticians and others who work with *Bacillus* species.

At the species level, also, in some instances the evidence of taxonomic heterogeneity is unequivocal. The division of some of these into genospecies, based on DNA/DNA reassociation studies, has already been successfully achieved. The degree of homogeneity of many other species is still unknown.

The description of new species based on only very few strains, especially if they are only obtained from culture collections, is an undesirable practice. It is to be hoped that the detailed description above of methods for the isolation of strains representative of the whole genus will assist workers to obtain strains from nature which exhibit a sufficiently wide spectrum of properties, so that newly erected taxa are soundly based.

Acknowledgments

We thank all members, past and present, of the International Committee on Systematic Bacteriology Subcommittee on the Taxonomy of the Genus *Bacillus* for the important contributions they have made to the work on which a substantial part of this article is based. Dr. R. E. Gordon and Professor G. J. Bonde, in addition to their major personal contributions to the taxonomy of the genus, have been particularly helpful in the provision of strains for study. Other workers, too numerous to mention individually, have also given strains and we wish to thank them for their help.

In preparing the descriptive material in the list of species we have leaned heavily on the information compiled by the late Dr. T. Gibson and Dr. R. E. Gordon for the eighth edition of the *Manual*.

We also thank Miss L. A. Shute and Mr. A. G. Capey who have given invaluable help in the preparation of this article and many other colleagues, especially Dr. N. A. Logan and Dr. A. J. Hedges, with whom we have had valuable discussions.

Further Readings

Berkeley, R.C.W. and M. Goodfellow. 1981. The Aerobic Endospore-Forming Bacteria. Academic Press, London.

Gordon, R.E., W.C. Haynes and C. Hor-Nay Pang. 1973. The Genus *Bacillus*. United States Department of Agriculture, Washington, D.C., Handbook No. 427.

Berkeley, R.C.W., N.A. Logan, L.A. Shute and A.G. Capey. 1984. Identification of *Bacillus* species. *In* Bergan (Editor), Methods in Microbiology, Vol. 16. Academic Press, London, pp. 291–328.

Norris, J.R., R.C.W. Berkeley, N.A. Logan and A.G. O'Donnell. 1981. The Genera *Bacillus* and *Sporolactobacillus*. *In* Starr, Stolp, Trüper, Balows and Schlegel (Editors), The Prokaryotes. A Handbook on Habitats, Isolation and Identification of Bacteria. Springer-Verlag, Berlin, pp. 1711–1742.

Differentiation and characteristics of the species of the genus *Bacillus*

Differential characteristics of the species of the genus *Bacillus* are shown in Tables 13.2–13.4 and 13.6.

Table 13.6.
Other characteristics of the species of the genus **Bacillus**[a]

Characteristics	1. B. subtilis	2. B. acidocaldarius	3. B. alcalophilus	4. B. alvei	5. B. anthracis	6. B. azotoformans[b]	7. B. badius	8. B. brevis	9. B. cereus	10. B. circulans[c]	11. B. coagulans	12. B. fastidiosus[d]	13. B. firmus[e]	14. B. globisporus[f]	15. B. insolitus[f]	16. B. larvae	17. B. laterosporus
Oxidase	d[g]			+[h]		+			−[i]	−		+		+	+		d[h]
β-Galactosidase[w]	d					−		−	−		+						
Arginine dihydrolase[i,g]	−								d								
Lysine decarboxylase[i,g,w]	−						−		−		−						
Lipase (olive oil)[i,g]	d								+								
Hydrolysis of																	
Alginate																	
Chitin																	
Dextran																	
Esculin[i,g,w]	+					−			+		+						
Tween 80									−								
Urea[j]	−		−[k]	−		−		−	d[i]	d	−		−[e]	+	−		−
Gas from nitrate[j]	−	−[l]		−		+			d	−	−	−	−	−	−		d
Growth at pH 5[j]		+[l]	−	−		−			−		+			−	−		
Growth factors required[m]	−	−[l]		+		+		+	+	+	+		+				+
Crystalline dextrins[j]			−[k]						d								
Degradation of																	
Tyrosine[n]	−		d	d	−	+	+	+	+	−	−		d	−	−	−	+
Hippurate[n]	−					−							d	+	+		
Reduction of methylene blue[j]			+[k]	+					+				−	+	−		
Acid from																	
Adonitol																	
Cellobiose										+							
Dulcitol									−[o]								
Fructose						−				+							
Galactose						−				+							
Glycerol			+[k]						+[o]					+	−		
Inositol									−[o]								
Lactose			+[k]			−			−[o]	+				+			
Maltose			+[k]			−			+[o]	+							
Mannose						−				+			−				
Melibiose										+			−				
Raffinose										−							
Rhamnose										+							
Ribose						−				+							
Salicin			+[k]						d[o]	+							
Sorbose						−											
Sorbitol			+[k]						−[o]	d				−			
Sucrose			+[k]						+[o]	+			+	+	−		
Trehalose									+[o]	+							
Formation of poly-β-hydroxy-butyrate or other storage products[n]	−			+		+			+		−		−				
Utilization of																	
Acetate						+			d[i]	+							
Aspartate						+			d[i]								
Formate									d[i]								
Fumarate										d							
Glutamate						+											
Glutarate						−											
Glycerol																	
Glycine						−											
Glycolate						−											
Lactate						+			d[i]								
Malate						+				d							
Malonate						−											
Mannitol																	
Pyruvate						+											
Succinate						+			d[i]	d							
Tartrate						−			−[i]								

[a] Symbols: see Table 13.4; also [b] Pichinoty et al. (1978, 1983); [c] Nakamura and Swezey (1983b); [d] F. Fahmy (personal communication); [e] Gordon et al. (1977); Gordon and Hyde (1982); [f] Larkin and Stokes (1967); [g] Hanáková-Bauerová et al. (1965); [h] Jurtshuk and Liu (1983); [i] Hanáková-Bauerová et al. (1966); [j] Smith et al. (1952); [k] Boyer et al. (1973); [l] Darland and Brock (1971); [m] Knight and Proom (1950); [n] Gordon et al. (1973); [o] Gilbert et al. (1981); [p] Marshall and Ohye (1966); [q] Rüger (1983); [r] Schenk and Aragno (1979); Krüger and Meyer (1984); [s] Krych et al. (1980); [t] Hunger and Claus (1981); [u] Proom and Knight (1950); [v] de Barjac et al. (1980); [w] LeMille at al. (1969a); [x] Sharp et al. (1980); [y] White and Lotay (1980).

Table 13.6—continued

Characteristics	18. B. lentimorbus	19. B. lentus[e]	20. B. licheniformis	21. B. macerans	22. B. macquariensis[p]	23. B. marinus[q]	24. B. megaterium	25. B. mycoides	26. B. pantothenticus	27. B. pasteurii	28. B. polymyxa	29. B. popilliae	30. B. pumilus	31. B. schlegelii	32. B. sphaericus	33. B. stearothermophilus	34. B. thuringiensis
Oxidase			d[g]			v	d[h]	-[i]			j[h]		d[g]	+	+[s]	d[q]	d[h]
β-Galactosidase[w]			+			-	+						+				
Arginine dihydrolase[i,g]			+				-	d					+				d
Lysine decarboxylase[i,g,w]			-				-	-					-				-
Lipase (olive oil)[i,g]			d					+					+				+
Hydrolysis of																	
Alginate						-											
Chitin						-	-[t]										
Dextran							-[t]										
Esculin[i,g,w]			+			+	+	+					+				+
Tween 80																	
Urea[j]	-	d	-	-	-	-	d	d[i]	-	+	-		-[g]	-	d	-	d
Gas from nitrate[j]			d	-			-		-	d	-		-	-		d	
Growth at pH 5[j]	-	-	-	-			-	-	-	+	-	-		-		-	
Growth factors required[m]		+	-		+		+		+	+[u]	+	+	+	-	d		+
Crystalline dextrins[j]			d								-						
Degradation of																	
Tyrosine[n]	-	-				-	d	d	-				-		-		d
Hippurate[n]		d	-			-							+				
Reduction of methylene blue[j]					+			d			+				+		
Acid from																	
Adonitol						-											
Cellobiose						-											
Dulcitol						-											
Fructose						+											
Galactose					+	-											
Glycerol						d											
Inositol						-											
Lactose					+	+									-[s]		
Maltose					+	+											
Mannose		+				+											
Melibiose		+															
Raffinose		+				-											
Rhamnose						-											
Ribose					+												
Salicin					+	-											
Sorbose																	
Sorbitol		d				-											
Sucrose		+			+	+											
Trehalose						+											
Formation of poly-β-hydroxy-butyrate or other storage products[n]							d	+						-			+
Utilization of																	
Acetate							d[i]	d[i]						+	+[y]		d[i]
Aspartate														-			+[v]
Formate							d[i]	d[i]						-			d[i]
Fumarate														+	-[y]		d[v]
Glutamate														+	-[y]		+[v]
Glutarate																	
Glycerol														-	-[y]		
Glycine														-	-[y]		d[v]
Glycolate															-[y]		
Lactate							+[i]	d[i]						-	-[y]		d[i]
Malate														+			
Malonate														-			
Mannitol														-			
Pyruvate														+	-[y]		d[v]
Succinate							d[i]	-[i]						+	-[y]		-[i]
Tartrate							d[i]	-[i]						-			-[i]

[a] Symbols: see Table 13.4; also [b] Pichinoty et al. (1978, 1983); [c] Nakamura and Swezey (1983b); [d] F. Fahmy (personal communication); [e] Gordon et al. (1977); Gordon and Hyde (1982); [f] Larkin and Stokes (1967); [g] Hanáková-Bauerová et al. (1965); [h] Jurtshuk and Liu (1983); [i] Hanáková-Bauerová et al. (1966); [j] Smith et al. (1952); [k] Boyer et al. (1973); [l] Darland and Brock (1971); [m] Knight and Proom (1950); [n] Gordon et al. (1973); [o] Gilbert et al. (1981); [p] Marshall and Ohye (1966); [q] Rüger (1983); [r] Schenk and Aragno (1979); Krüger and Meyer (1984); [s] Krych et al. (1980); [t] Hunger and Claus (1981); [u] Proom and Knight (1950); [v] de Barjac et al. (1980); [w] LeMille at al. (1969a); [x] Sharp et al. (1980); [y] White and Lotay (1980).

List of species of the genus **Bacillus**

1. Bacillus subtilis (Ehrenberg, 1835) Cohn, 1872, 174.[AL] *Nom. cons.* Nomencl. Comm. Intern. Soc. Microbiol. 1937, 28; Opin. A. Jud. Comm. 1955, 39. (*Vibrio subtilis* Ehrenberg 1835, 279.)

sub'ti.lis. L. adj. *subtilis* slender.

Rods seldom in chains; stain uniformly.

Colonies on agar media round or irregular; surface dull; become opaque; may be wrinkled and may become cream colored or brown. Active spreading occurs on agar with a moist surface. Cell material grown on agar does not disperse readily in liquids. In 1% glucose nutrient, agar stab surface growth becomes thick, often rugose and brown. A disk of reddish pigment may form below the growth. Deep growth starts but soon comes to a standstill. In broth dull, wrinkled, coherent pellicle; little or no turbidity.

Pectin and polysaccharides of plant tissues are decomposed, and some strains produce a rot in live potato tubers.

Levan and dextran are formed extracellularly from sucrose.

Pigments, which in particular cases have been identified as pulcherrimin or melanins, may be produced in colonies or the adjacent medium. In many strains they are brown or red, in fewer orange or black. Occurrence of each pigment is dependent on composition of medium.

A minimal medium for vegetative growth has no vitamins and contains glucose and an ammonium salt as the sole sources of carbon and nitrogen.

Aerobic, excepting that glucose and, less effectively, nitrate permit a much restricted anaerobic growth in complex media. Growth active at pH 5.5–8.5; pH limits not recorded.

Endospores are widespread. Vegetative organisms take part in the early stages of breakdown of various materials of plant and animal origin. Growth in nonacid foods if oxygen available. Causative agent of ropy (slimy) bread.

The mol% G + C of the DNA is reported to be 41.5–47.5 (T_m) for 31 strains, 41.8–46.3 (Bd) for 34 strains and for the type strain 42.9 (T_m), 43.1 (Bd).

Type strain: ATCC 6051 (DSM 10, NCIB 3610, NCTC 3610).

Strains forming a black pigment exclusively on carbohydrate-containing media have been named "*B. subtilis* var. *aterrimus*" Smith et al. (1946). Other strains forming a black pigment only in the presence of tyrosine have been described as "*B. subtilis* var. *niger*" Smith et al. (1946). Pigment formation in both the varieties is not fully stable. Comparative zone electrophoresis of enzymes suggests that the two varieties are different from *B. subtilis sensu stricto* (Baptist et al., 1978).

The species includes at least two DNA homology groups with high intra- but low inter-group relatedness (Seki et al., 1975; Welker and Campbell, 1967). Group 1 includes the type strain and hence can be defined as *B. subtilis sensu stricto*. Strains of group 2 have been earlier described as "*B. amyloliquefaciens*" (see list of species *incertae sedis*).

2. Bacillus acidocaldarius Darland and Brock, 1971, 9.[AL]

a.ci.do.cal.dar'i.us. M.L. n. *acidum* acid; L. adj. *caldarius* pertaining to warm or hot; M.L. adj. *acidocaldarius* pertaining to acid thermal (habitats).

Carbon-energy sources include glucose, galactose, glycerol and casamino acids. No growth with ethanol, sorbitol, acetate, succinate or citrate as sole C source. Growth occurs with NH_4^+ but not with NO_3^- as the sole nitrogen source. Growth factors are not required.

Only aerobic growth in media containing nitrate.

Limits of pH for growth: 2 and 5–6, except in 2 of 15 strains which have a lower limit of pH 3.

Spores have relatively weak heat resistance, half-time death at 86° C being 10–12 min.

Sources: Thermal, markedly acid water and soil. Enrichment from more nearly neutral soils has failed.

The mol% G + C of the DNA is reported to be 61.2–62.2 (Bd) for three strains and for the type strain 60.3 (T_m), 62.3 (Bd).

Type strain: ATCC 27009 (DSM 446, NCIB 11725).

3. Bacillus alcalophilus Vedder, 1934, 141.[AL]

al.cal.o.phil'us. M.L. *alcali* Eng. alkali from the Arabic *al* the end; *qaliy* soda ash; Gr. adj. *philus* loving; M.L. adj. *alcalophilus* liking alkaline (media).

Originally characterized chiefly by marked tolerance to alkali and inability to grow on media at pH 7 but the type strain grows on blood agar.

Unlike *B. pasteurii*, lacks urease activity and has no requirement for ammonia in addition to alkalinity.

Isolated from various materials using preliminary enrichment in broth at pH 10.

The mol% G + C of the DNA of the type strain is 37.0 (T_m), 36.7 (Bd).

Type strain: ATCC 27647 (DSM 485, NCIB 10436, NCTC 4553).

"*B. alcalophilus* subsp. *halodurans*" (Boyer et al., 1973) differs from the type strain in several phenotypic properties. Its mol% G + C is 42.6 (Bd).

4. Bacillus alvei Cheshire and Cheyne, 1885, 581.[AL]

al've.i. L. n. *alveus* a beehive; L. gen. n. *alvei* of a beehive.

Typical strains spread vigorously on agar and may show motile colonies. Their free spores may lie side by side in long rows on the agar.

The separation of this species from *B. circulans* relies principally on the production of dihydroxyacetone and indole. Some organisms give various intermediate results in tests for acetoin formation and action on pentoses, mannitol and proteins.

Minimal nutritional requirements are several amino acids plus thiamine, or thiamine and biotin, or more complex unidentified requirements.

Anaerobic growth occurs in complex media containing glucose.

Isolated from soil and from honeybee larvae suffering from European foulbrood, but not an insect pathogen.

The mol% G + C of the DNA is reported to be 44.7–46.8 (T_m) for two strains, 45.9–47.4 (Bd) for four strains and for the type strain 44.6 (T_m), 46.2 (Bd).

Type strain: ATCC 6344 (DSM 29, NCIB 9371, NCTC 6352).

5. Bacillus anthracis Cohn, 1872, 177.[AL]

an'thra.cis. Gr. n. *anthrax* charcoal, a carbuncle; M.L. n. *anthrax* the disease anthrax; M.L. gen. n. *anthracis* of anthrax.

This species is similar to *B. cereus* but differs in the following characteristics and those shown in Table 13.7. Virulent and avirulent strains can be differentiated from *B. cereus* using the API system.

Thiamine and a greater number of amino acids are required in a minimal medium.

No hemolysis of sheep cells in 24 h; spreads little.

Virulent strains of *B. anthracis* form capsules of glutamyl polypeptide during in vivo multiplication; they also form capsules on agar plus bicarbonate under CO_2, and their colonies are then mucoid.

The causative agent of the disease anthrax in man and animals. Spores persist for long periods in contaminated materials.

The mol% G + C of the DNA is reported to be 32.2–33.9 (T_m) for five strains and for the type strain 33.2 (T_m).

Type strain: ATCC 14578 (NCIB 9388, NCTC 10340).

DNA/DNA hybridization studies suggest that the species is identical with *B. cereus*. This has to be confirmed.

6. Bacillus azotoformans Pichinoty, de Barjac, Mandel and Asselineau, 1983, 660.[VP*]

a.zo.to.for'mans. French n. *azote* nitrogen; L. part. adj. *formans* forming; *azotoformans* nitrogen forming.

Gram-reaction negative. Colonies round with entire margins; partially translucent on yeast extract agar.

---

* VP denotes that this name has been validly published in the official publication, International Journal of Systematic Bacteriology.

Table 13.7.
Differential characteristics of **B. anthracis** *and* **B. cereus**[a]

Characteristics	B. anthracis	B. cereus
Motility[b]	−	+[c]
Lysis by γ phage[b]	+	−
String-of-pearls test[d-f] (10 U ml⁻¹ penicillin G)	+[g]	−
Growth on		
Chloralhydrate agar[f]	−	+
2-Phenylethanol agar[f]	d	+
Polymyxin-lysozyme-EDTA-thallous acetate agar[h]	+	−
Phosphatase[i]	−	d
Degradation of tyrosine[j]	−	d

[a] Symbols: see Table 13.4.
[b] Data from Brown et al. (1958).
[c] Motility may depend on the growth medium. Very good results have been obtained on plates containing yeast extract, 0.1%; K_2HPO_4, 0.01% and agar, 0.2%.
[d] Data from Lennette et al. (1980).
[e] Data from Jensen and Kleemeyer (1953).
[f] Data from Knisely (1965).
[g] One penicillin-resistant strain has been identified.
[h] Data from Knisely (1966).
[i] Data from Seidel (1962).
[j] Data from Gordon et al. (1973).

In yeast extract-salts liquid medium uniform, dense turbidity. Growth factors are required. Anaerobically, NO_3^-, NO_2^-, N_2O, $S_4O_6^{2-}$ and fumarate act as terminal electron acceptors. Considerable amounts of N_2 are produced by reduction of NO_3^-, NO_2^- and N_2O.

This species can be differentiated from *B. brevis*, which has a very different DNA base composition (Table 13.3), by its ability to grow anaerobically in the presence of NO_3^-, NO_2^-, $S_4O_6^{2-}$ and fumarate as well as by a number of other characters (see Table 13.4).

Isolated from soils.

The mol% G + C of the DNA is reported to be 39.0–43.9 (Bd) for 17 strains and for the type strain 39.0 (Bd).

Type strain: ATCC 29788 (DSM 1046).

7. Bacillus badius Batchelor, 1919, 23,[AL]

ba.di′us. L. adj. *badius* chestnut brown.

The type strain of *B. badius* grows as chains of rods with blunt or flat ends, and its colony has a folded hair structure and rhizoid outgrowths. Other strains appear to differ from the type culture only in the absence of chains and the production of smooth colonies.

Has been isolated from feces, dust, marine sources, foods and antacids.

The mol% G + C of the DNA is reported to be 45.2 and 50.0 (T_m) for one strain and for the type strain 43.8 (T_m), 43.5 (Bd).

Type strain: ATCC 14574 (DSM 123, NCIB 9364, NCTC 10333).

8. Bacillus brevis Migula, 1900, 583.[AL]

bre′vis. L. adj. *brevis* short.

Minimal nutritional requirement of most strains is a mixture of amino acids without vitamins.

Has been isolated chiefly from soil and foods.

The mol% G + C of the DNA is reported to be 42.5–47.0 (T_m) for four strains, 45.0–45.2 (Bd) for two strains and for the type strain 47.3 (T_m), 47.4 (Bd).

Type strain: ATCC 8246 (DSM 30, NCIB 9372, NCTC 2611).

The inclusion of strains having quite different temperature growth ranges suggests that the species is heterogeneous in spite of the phenetic evidence indicating that it is homogeneous (Gordon et al., 1973; Logan and Berkeley, 1981).

9. Bacillus cereus Frankland and Frankland, 1887, 257.[AL]

ce′re.us. L. adj. *cereus* waxen, wax-colored.

The rods tend to occur in chains; the stability of the chains determines the form of the colony, which may vary in different strains. Colonies have a dull or frosted glass appearance and often an undulate margin from which extensive outgrowths do not develop.

Extracellular products include hemolysin, soluble toxin lethal for mice, enzymes lytic for bacterial cells, proteolytic enzymes and phospholipase C.

The red pigment pulcherrimin is produced by some strains in starch media containing sufficient iron. Some strains produce a yellowish-green fluorescent pigment in various media. On nutrient agar some strains darken the medium slightly, and some produce a pinkish brown diffusible pigment.

Has an absolute requirement for one or several amino acids; strains differ in those needed. Vitamins are not required.

Spores widespread. Multiplication has been observed chiefly in foods and may lead to food poisoning.

The mol% G + C of the DNA is reported to be 31.7–40.1 (T_m) for 11 strains, 34.7–38.0 (Bd) for 11 strains and for the type strain 35.7 (T_m), 36.2 (Bd).

Type strain: ATCC 14579 (DSM 31, NCIB 9373, NCTC 2599).

DNA/DNA hybridization studies have been performed with only a few strains. From these studies the species appears to be genetically homogeneous. Also, high homologies have been found with strains of *B. anthracis*, *B. mycoides* and *B. thuringiensis*. This has to be confirmed. On the basis of phenotypic tests Logan and Berkeley (1984) suggested that *B. mycoides* and *B. thuringiensis* should be considered subspecies of *B. cereus*.

10. Bacillus circulans Jordan, 1890, 821.[AL]

cir′cu.lans. L. Part. adj. *circulans* circling.

Growth on nutrient agar is generally thin but may be enhanced by addition of carbohydrate; in some strains it spreads actively and may give rise to motile colonies.

Cellulose is degraded weakly by some strains.

Strains placed in this species by Knight and Proom (1950) and Proom and Knight (1950) showed a wide spectrum in minimal requirements from those that grew with NH_4-N without growth factors at one extreme to those that required complex undefined media at the other.

Spores numerous in soil; no special conditions which promote multiplication have been recognized.

The mol% G + C of the DNA is reported to be 31.6–61.0 (Bd) for 124 strains and for the type strain 35.7 (T_m), 36.2 (Bd).

Type strain: ATCC 4513 (DSM 11, NCIB 9374, NCTC 2610).

The species is genetically very heterogeneous. Several DNA homology groups have been established (Table 13.8), suggesting that in future separation of *B. circulans sensu stricto* into more than 10 different species may be necessary. These data confirm the long standing phenetic evidence of Gibson and Topping (1938) and more recent work of Logan and Berkeley (1981).

11. Bacillus coagulans Hammer, 1915, 119.[AL]

co.a′gu.lans. L. Part. adj. *coagulans* curdling, coagulating.

Morphological variation is considerable; in a diagram of species of *Bacillus*, Smith et al. (1952) placed *B. coagulans* on the line dividing group 1 (sporangia not appreciably swollen by oval or cylindrical spores) from group 2 (sporangia swollen by oval spores).

Minimal nutritional requirements are diverse, varying with strains and incubation temperature. Several amino acids and several vitamins fall within the range of requirements. Aciduric; for initiation of growth optimum pH level is close to 6; minimum 4.0–5.0 in different strains.

Spores relatively scarce in soil. May multiply in acid foods such as canned tomato juice and silage. Found in medicated creams and antacids.

The mol% G + C of the DNA is reported to be 44.3–50.3 (T_m) for seven strains, 45.4–56.0 (Bd) for the three strains and for the type strain 47.1 (T_m), 44.5 (Bd).

Type strain: ATCC 7050 (DSM 1, NCIB 9365, NCTC 10334).

Table 13.8.

Differential characteristics of DNA homology groups of the **Bacillus circulans** *complex[a,b]*

	A[c]	B	C	D	E	F	G	H	I	K
Anaerobic growth	+	+	−	−	+	+	−	+	+	+
Growth in 5% NaCl	+	−	−	−	d	−	−	+	−	d
Growth at pH 5.6	+	+	+	+	+	d	d	−	+	+
Growth with lysozyme	−	−	+	−	−	−	+	+	−	+
Hydrolysis of										
Casein	−	+	−	d	d	−	−	−	−	−
Starch	+	−	+	+	+	+	−	+	+	+
Nitrate reduced to nitrite	−	−	−	−	−	−	−	+	−	+
pH in V-P broth <5.5	+	−	−	−	d	+	−	+	−	+
Utilization of										
Citrate	−	−	−	−	−	−	−	−	+	−
Fumarate	+	+	+	−	−	−	+	−	+	−
Acid from										
L-Arabinose	+	+	+	+	+	+	−	+	+	+
D-Mannose	+	+	+	+	+	+	−	+	+	+
Sorbitol	+	+	−	−	d	d	−	−	+	−
Mol% G + C of DNA	37–39	45–46	47–49	47–48	48–50	47–50	50–52	50–52	53–54	53

[a] Symbols: see Table 13.4.

[b] Data from Nakamura and Swezey (1983b).

[c] Group A includes the type strain of *Bacillus circulans*.

Phenetic evidence points to this species being heterogeneous (Logan and Berkeley, 1984) and DNA/DNA hybridization studies of 25 strains revealed two DNA homology groups (I. Blumenstock, personal communication).

12. **Bacillus fastidiosus** den Dooren de Jong, 1929, 344.[AL]

fas.tid′i.os.us. L. adj. *fastidiosus* disdainful, fastidious.

Of many other organic compounds tested generally only uric acid, allantoin and allantoic acid were utilized. Peptone and glucose do not repress growth on uric acid. Some strains may grow on certain peptones especially at higher concentrations. No growth factors needed.

Colonies on uric acid (1%) agar become opaque and are usually unpigmented but may develop a yellowish color; often have a ragged outline and hair-like outgrowths or show marked rhizoidal structure; a zone of clearing around colonies is formed. Reaction becomes strongly alkaline.

Originally isolated from soil. Subsequent isolations from soil and poultry litter.

The mol% G + C of the DNA is reported to be 34.3–35.1 (T_m) for 17 strains and for the type strain 35.1 (T_m), 35.1 (Bd).

Type strain: LMD 29-14 (ATCC 29604; DSM 91, NCIB 11326).

On the basis of their mol% G + C (41.5–47.0 (T_m)) and their phenotypic characters four strains have been separated from the species (F. Fahmy, personal communication).

13. **Bacillus firmus** Bredemann and Werner in Werner 1933, 446.[AL]

fir′mus. L. adj. *firmus* strong, firm.

Minimal nutritional requirements are a mixture of amino acids and biotin, or both biotin and thiamine in a mineral salts medium.

Has been isolated chiefly from soil; pigmented strains mostly from salt marshes.

The mol% G + C of the DNA is reported to be 36.1–47.4 (T_m) for 42 strains, 41.8–42.3 (Bd) for two strains and for the type strain 41.4 (T_m), 40.7 (Bd).

Type strain: ATCC 14575 (DSM 12, NCIB 9366, NCTC 10335).

The species is genetically heterogeneous. Several DNA homology groups have been described by Fahmy (1978).

14. **Bacillus globisporus** Larkin and Stokes, 1967, 892.[AL]

glo.bis′por.us. L.n. *globus* a sphere; M.L. n. *spora* a spore; M.L. adj. *globisporus* with spherical spores.

Shows affinities with *B. sphaericus* but has a lower maximum temperature for growth, 25–30° C; grows and sporulates at 0° C.

The mol% G + C of the DNA is reported to be 36.8 and 40.6 (T_m) for one strain and for the type strain 39.8 (T_m), 39.7 (Bd).

Sources: Soil, river water.

Type strain: ATCC 23301 (DSM 4, NCIB 11434).

15. **Bacillus insolitus** Larkin and Stokes, 1967, 889.[AL]

in.so.li′tus. L. adj. *insolitus* unusual.

Growth and sporulation occur at 0° C.

Spores vary in shape from round to cylindrical, and in size from 0.7–1.4 μm in diameter and up to 2.4 μm in length, depending on the medium on which they are produced.

Vegetative cells are stout and often short.

Source: Soil.

The mol% G + C of the DNA is reported to be 41.0 (T_m) for one strain and for the type strain 35.9 (T_m), 36.1 (Bd).

Type strain: ATCC 23299 (DSM 5; NCIB 11433).

16. **Bacillus larvae** White, 1906, 42.[AL]

lar′vae. L. n. *larva* ghost; M.L. n. *larva* a larva; M.L. gen. n. *larvae* of a larva.

Cultures require thiamine and certain amino acids for growth and will not survive serial transfer in nutrient broth. Growth and sporulation are satisfactory on a tryptone-glucose-yeast extract J-medium.

Cause of American foulbrood of honeybees (*Apis mellifera*).

The mol% G + C of the DNA has not been reported.

Type strain: ATCC 9545.

17. **Bacillus laterosporus** Laubach, 1916, 505.[AL]

la.te.ro.spor′us. L. n. *latus, lateris* the side; M.L. n. *spora* spore; M.L. adj. *laterosporus* with lateral spore.

An important diagnostic feature is the production of a canoe-shaped body attached to the side of the spore. As a consequence the spore occupies a lateral position in a spindle-shaped sporangium. The parasporal body, which is stainable with fuchsin, remains firmly adherent to the spore after lysis of the sporangium. This feature is of limited use in diagnosis as the proportion of sporangia which contain parasporal bodies varies with the strain and with the growth medium. Also, lateral spores may occur in other species of *Bacillus*.

Moderate growth on nutrient agar, becoming dull and opaque; may spread actively if surface is moist. Growth thicker on glucose nutrient agar; may become wrinkled.

Has been isolated, only rarely, from dead honeybee larvae, soil, water and antacids.

The mol% G + C of the DNA is reported to be 39.8 and 42.7 (T_m) for one strain and for the type strain 40.2 (T_m), 40.5 (Bd).

Type strain: ATCC 64 (DSM 25; NCIB 9367; NCTC 6357).

Indications from phenetic and genetic evidence are that this species is homogeneous (Berkeley and Logan, 1984; Priest et al., 1981).

18. **Bacillus lentimorbus** Dutky, 1940, 57.[AL]

len.ti.mor'bus. L. adj. *lentus* slow; L. n. *morbus* disease; M.L. n. *lentimorbus* the slow disease.

This species, which is more fastidious nutritionally and less widespread than *B. popilliae*, also infects the larvae of the Japanese beetle (*Popillia japonica* Newman) and the European chafer (*Amphimallon majalis* Razoumowsky).

Isolated from diseased larvae or infected honeycombs.

The mol% G + C of the DNA is reported to be 37.7 (T_m) for one strain. The value for the type strain has not been determined.

Type strain: ATCC 14707 (DSM 2049; NCIB 11202).

19. **Bacillus lentus** Gibson, 1935a, 364.[AL]

len'tus. L. Adj. *lentus* slow.

Can be differentiated from *B. firmus* with classical and API tests because it produces acid from a wider range of carbohydrates.

Has been isolated from soil, food and spices.

The mol% G + C of the DNA is reported to be 36.5–47.3 (T_m) for 12 strains and for the type strain 36.3 (T_m), 36.4 (Bd).

Type strain: ATCC 10840 (DSM 9, NCIB 8773, NCTC 4824).

This species is closely related to *B. firmus* but DNA homology data suggest the two are distinct species to which intermediate strains may be allocated. *B. lentus* is, however, genetically heterogeneous.

20. **Bacillus licheniformis** (Weigmann, 1898) Chester, 1901, 287.[AL]
(*Clostridium licheniforme* Weigmann 1898, 822.)

li.che.ni.for'mis. Gr. n. *lichen* lichen; L. n. *forma* shape; M.L. adj. *licheniformis* lichen-shaped.

Rods often in chains.

Colonies on agar become opaque with dull to rough surface; hair-like outgrowths common; usually attached strongly to agar; mounds and lobes consisting largely of slime often accumulate on colony, especially on glucose agar or glutamate-glycerol agar.

Glutamyl polypeptide formed as an extracellular amorphous slime. Levan produced extracellularly from sucrose and raffinose.

Red pigment (presumably pulcherrimin) formed by many strains on carbohydrate media containing sufficient iron. Aged cultures may become brown.

Freshly isolated strains grow with ammonia as the sole source of nitrogen in the absence of growth factors.

Spores occur in soil; may survive severe heat treatment. Vegetative growth in many foods, especially if held at 30–50° C.

The mol% G + C of the DNA is reported to be 42.9–49.9 (T_m) for 12 strains, 44.9–46.4 (Bd) for 19 strains and for the type strain 46.4 (T_m), 44.7 (Bd).

Type strain: ATCC 14580 (DSM 13; NCIB 9375; NCTC 10341).

DNA/DNA hybridization studies of 16 strains have shown that the species is genetically homogeneous.

21. **Bacillus macerans** Schardinger, 1905, 772.[AL]

ma'ce.rans. L. part. adj. *macerans* softening by steeping, retting.

Colonies on nutrient agar thin, round to spreading. On glucose agar may be more opaque; not mucoid.

Pectin and polysaccharides of plant tissues decomposed. Weak or no action on cellulose.

A minimal medium for typical strains contains a carbon-energy source, NH$_4$-N, biotin and thiamine. N$_2$ is fixed under anaerobic conditions by the majority of strains examined.

Spores relatively scarce in soil; isolation from that source usually requires enrichment culture. Multiplication in plant materials at elevated temperatures. Growth has been recorded in canned fruits initially at pH 3.8–4.

The mol% G + C of the DNA is reported to be 45.9–51.0 (T_m) for eight strains, 44.9–46.4 (Bd) for eight strains and for the type strain 52.2 (T_m), 53.2 (Bd).

Type strain: ATCC 8244 (DSM 24, NCIB 9368, NCTC 6355).

22. **Bacillus macquariensis** Marshall and Ohye, 1966, 41.[AL]

mac.qua'ri.en.sis. M.L. adj. *macquariensis* pertaining to Macquarie Island.

Grows and sporulates at 0° C. In other properties similar to *B. circulans*.

Isolated from soil from Macquarie Island (subantarctic).

The mol% G + C of the DNA of the type strain is 39.3 (T_m), 41.6 (Bd).

Type strain: ATCC 23464 (DSM 2, NCIB 9934, NCTC 10419).

23. **Bacillus marinus** (Rüger and Richter) Rüger, 1983, 157.[VP]
(*Bacillus globisporus* subsp. *marinus* Rüger and Richter 1979, 196.)

ma.ri'nus. L. n. *marinus* marine.

Grows at 5–30° C but not at 37° C.

Has obligate requirement for Na$^+$.

Isolated from marine sediments.

The mol% G + C of the DNA is reported to be 36.9–39.5 (T_m) for four strains and for the type strain 37.6 (T_m), 38.0 (Bd).

Type strain: ATCC 29841 (DSM 1297).

DNA homology with *B. globisporus*, *B. aminovorans* and *B. insolitus* less than 39%.

24. **Bacillus megaterium** de Bary, 1884, 499.[AL]

me.ga.te'ri.um. Gr. adj. *mega* large; Gr. n. *teras, teratis* monster, beast; M.L. n. *megaterium* big beast.

Spores vary from round to elongate.

On nutrient agar, growth heaped and nonspreading, glossy or moderately dull, sometimes slightly rugose; on aging, usually some shade of yellow; on long incubation, growth and medium may become brown or black.

Growth on glucose agar mucoid to various degrees.

Multiplies without growth factors on ammonium salts or with nitrate and glucose as sole sources of nitrogen and carbon.

Spores occur in soil.

The mol% G + C of the DNA is reported to be 36.5–47.4 (T_m) for 67 strains, 38.0–45.0 (Bd) for 10 strains and for the type strain 37.3 (T_m), 37.6 (Bd).

Type strain: ATCC 14581 (DSM 32, NCIB 9376; NCTC 10342).

Gordon et al. (1973) recognized that the species was heterogeneous with two separable groups of strains with some intermediate strains, DNA/DNA hybridization studies have, however, shown the existence of four homology groups (Table 13.9).

25. **Bacillus mycoides** Flügge, 1886, 324.[AL]

my.co.i'des. Gr. n. *myces* fungus; Gr. n. *eidus* form, shape; M.L. adj. *mycoides* fungus-like.

Similar to *B. cereus* but nonmotile and forms distinctive rhizoid colonies on agar. Ability to form rhizoid colonies may be lost.

The mol% G + C of the DNA is reported to be 32.5–38.4 (T_m) for nine strains, 35.2–39.0 (Bd) for four strains and for the type strain 34.2 (T_m), 34.1 (Bd).

Type strain: ATCC 6462 (DSM 2048).

DNA/DNA hybridization studies have been performed with only a few strains; no comment on the genetic homogeneity of this species can therefore be made. With some strains a high degree of relatedness has been found with *B. anthracis*, *B. cereus* and *B. thuringiensis*. This has to be confirmed.

26. **Bacillus pantothenticus** Proom and Knight 1950, 539.[AL]

Table 13.9.
Differential characteristics of DNA homology groups of **Bacillus megaterium**[a, b]

Characteristics	DNA homology group			
	A[c]	B	C	D
Cell diameter 1.2 μm	+	−	−	−
Formation of poly-β-hydroxybutyrate	+	−	−	−
Phenylalanine deaminase	+	−	−	−
Hydrolysis of				
Esculin	+	−	−	+
Urea	+	−	+	−
Chitin	−	−	+	+
Splitting of amylose azure	+	−	+	+
Nitrite from nitrate	−	+	−	+
Growth with lysozyme present	−	−	−	+
Colonies				
Yellowish on milk agar	+	−	−	d
Brownish on tyrosine agar	−	+	−	−
Mol% G + C of DNA	37–38	40–41	37–39	34

[a] Symbols: see Table 13.4.
[b] Data from Hunger and Claus (1981).
[c] DNA homology group A includes the type strain of *Bacillus megaterium*.

pan.to.then′tic.us. M.L. n. *acidum pantothenicum* pantothenic acid; M.L. adj. *pantothenticus* relating to pantothenic (acid).

This species has a nutritional requirement, apparently unique in the genus *Bacillus*, for pantothenic acid. Also required are amino acids, thiamine and biotin. Growth stimulated by 4% NaCl.

Isolated from soil and antacids.

The mol% G + C of the DNA is reported to be 44.1 and 44.8 (T_m) for one strain, 36.7 (Bd) for one strain and for the type strain 36.9 (T_m), 36.8 (Bd).

Type strain: ATCC 14576 (DSM 26; NCIB 8775; NCTC 8162).

27. **Bacillus pasteurii** (Miquel) Chester, 1898, 47.[AL] (*Urobacillus pasteurii* Miquel 1889, 519.)

pas.teur′i.i. M.L. gen. n. *pasteurii* of Pasteur; named for Louis Pasteur, French chemist and bacteriologist.

Colonies on agar not distinctive; usually circular and glossy; size and opacity vary on different media.

Liquid media turbid; deposit slimy; rarely a fragile pellicle.

Urea converted to ammonium carbonate more actively than by any other known bacterium. Ureaclastic ability commonly decreases during maintenance on artificial media.

Alkaline media (optimum about pH 9) containing NH₃ (optimum about 1% NH₄Cl) are required.

A medium which supports growth contains casein hydrolysate at pH 8.5–9.5, ammonia, thiamine and, for some strains, biotin and nicotinic acid.

Isolated from soil, water, sewage and incrustations on urinals.

The mol% G + C of the DNA is reported to be 37.0 and 41.7 (T_m) for one strain and for the type strain 38.5 (T_m), 38.4 (Bd).

Type strain: ATCC 11859 (DSM 33, NCIB 8841, NCTC 4822).

28. **Bacillus polymyxa** (Prazmowski, 1880) Macé, 1889, 588.[AL] (*Clostridium polymyxa* Prazmowski 1880, 37.)

po.ly.my′xa. Gr. pref. *poly-* much, many; Gr. n. *myxa* slime or mucus M.L. n. *polymyxa* much slime.

Spore has parallel, longitudinal surface ridges so that it is star-like in cross-section.

Colonies on nutrient agar thin; often with amoeboid spreading. On glucose agar usually heaped, mucoid with matt surface.

Pectin and polysaccharides of plant tissues decomposed. Action on cellulose is weak or lacking.

Levan synthesized from sucrose; forming large capsules, not diffusing into the surrounding medium.

Nitrogen is fixed under anaerobic conditions by the majority of strains tested. A minimal medium consists of a carbon-energy source, NH₄-N and biotin.

Spores widespread; multiplication chiefly in decomposing vegetation. Often isolated from foods. Found in medicated creams and antacids.

The mol% G + C of the DNA is reported to be 41.0–51.4 (T_m) for 12 strains, 45.9–48.0 (Bd) for five strains and for the type strain 44.3 (T_m), 45.6 (Bd).

Type strain: ATCC 842 (DSM 36, NCIB 8158, NCTC 10343).

Phenentic evidence indicates a close relationship between *B. polymyxa*, *B. macerans* and *B. circulans*.

29. **Bacillus popilliae** Dutky, 1940, 57.[AL]

po.pil′li.ae. M.L. n. *Popillia* generic name of the Japanese beetle; M.L. gen. n. *popilliae* of *Popillia*.

Tryptophan and thiamine are essential for growth. Biotin, *meso*-inositol and niacin are stimulatory. Cultures will grow indefinitely upon serial transfer in liquid (shaken) or semisolid J-medium but will not survive more than four serial transfers in nutrient broth.

This species is a pathogen of scarabid beetles and causes the more widespread of two milky diseases of the Japanese beetle (*Popillia japonica* Newman). Together with *B. lentimorbus* it is a biological control agent for the Japanese beetle and the European chafer (*Amphimallon majalis* Razoumowsky). The larvae become milky white because of the prolific production of spores in the hemolymph.

Most strains will sporulate readily when injected into the hemolymph or fed to susceptible insects but, in laboratory media, only certain strains have been induced to sporulate (Bulla et al, 1978).

The parasporal body that distinguishes strains of *B. popilliae* from strains of *B. lentimorbus* has been described variously as hemispherical or subconical, triangular, rhombohedral or indefinite in shape. Other distinguishing characters have been reviewed by Bulla et al. (1978) as have proposals for subspecific division.

Isolated from the hemolymph of Japanese beetle grubs.

The mol% G + C of the DNA of the type strain is 41.3 (T_m).

Type strain: ATCC 14706 (DSM 2047).

30. **Bacillus pumilus** Meyer and Gottheil in Gottheil, 1901, 680.[AL]

pu′mi.lus. L. adj. *pumilus* little.

Colonies very variable on some media.

Requires biotin; some strains also require amino acids.

Spores ubiquitous; occur in soil more frequently than those of *B. subtilis*.

The mol% G + C of the DNA is reported to be 39.0–45.1 (T_m) for 12 strains, 40.0–46.9 (Bd) for 25 strains and for the type strain 41.9 (T_m), 40.7 (Bd).

Type strain: ATCC 7061 (DSM 27, NCIB 9369, NCTC 10337).

DNA/DNA hybridization studies of 32 strains indicate that the species is homogeneous genetically.

31. **Bacillus schlegelii** Schenk and Aragno, 1981, 215.[VP] (Effective publication: Schenk and Aragno, 1979.)

schle.gel′i.i. M.L. gen. n. *schlegelii* of Schlegel; named after H. G. Schlegel, a German microbiologist.

Thermophilic organism which can be differentiated from the other morphologically similar species or those with a high G + C content, *B. sphaericus* and *B. acidocaldarius*, by characters given in Table 13.2 and because it is facultatively chemolithotrophic. When growing in this fashion it uses either H₂ as the electron donor and CO₂ as the carbon source or CO which satisfies both requirements.

Isolated from lake sediment and sugar factory sludge.

The mol% G + C of the DNA is reported to be 63.9–65.4 (T_m) for four strains, 67.1–67.7 (Bd) for two strains and for the type strain 64.6 (T_m), 66.3 (Bd).

Type species: MA 48 (DSM 2000).

32. **Bacillus sphaericus** Meyer and Neide in Neide, 1904, 337.[AL]

sphae'ri.cus. Gr. adj. *sphaericus* spherical.

Growth on nutrient agar varies in different strains from compact and heaped to a widespreading over the surface. Uncommon strains produce pink colonies.

Isolated from soil, marine and freshwater sediments, milk, foods and antacids.

The mol% G + C of the DNA is reported to be 34.0–40.0 (T_m) for 68 strains, 38.3 (Bd) for one strain and for the type strain 37.3 (T_m), 37.1 (Bd).

Type strain: ATCC 14577 (DSM 28, NCIB 9370, NCTC 10338).

According to DNA/DNA hybridization studies the species is genetically heterogeneous. In a study of 62 strains, six DNA homology groups were established but this still left many strains ungrouped (Table 13.10). Strains of DNA homology group B were shown to be pathogenic for mosquito larvae.

33. **Bacillus stearothermophilus** Donk, 1920, 373.[AL]

ste.a.ro.ther.mo'phi.lus. Gr. n. *stear* fat; Gr. n. *thermus* heat; Gr. adj. *philus* loving; M.L. adj. *stearothermophilus* (presumably intended to mean) fat- and heat-loving.

The most distinctive diagnostic characters are capacity to grow at 65° C and a limited tolerance to acid.

Occurs in soil, hot springs, desert sand, Arctic waters, ocean sediments, food and compost.

The mol% G + C of the DNA is reported to be 43.5–62.2 (T_m) for 19 strains, 46.0–52.0 (Bd) for seven strains and for the type strain 51.9 (T_m), 51.5 (Bd).

Type strain: ATCC 12980 (DSM 22, NCIB 8923, NCTC 10339).

The heterogeneity of this species is indicated by the wide range of DNA base composition as well as of the phenotypic properties of strains assigned to this species (Wolf and Sharpe, 1981; Logan and Berkeley, 1984).

34. **Bacillus thuringiensis** Berliner, 1915, 29.[AL]

thur.in.gi.en'sis. M.L. gen. n. *thuringiensis* of Thuringia; named for Thuringia, a German province.

This species is distinguished from *B. cereus* by pathogenicity for larvae of *Lepidoptera* and by the production in the cell of a protein parasporal crystal, or, rarely, two or three, in parallel with spore formation. This structure is formed outside the exosporium and separates readily from the liberated spore. In the larval gut, toxin is released from the crystal by enzymatic action. The capacity to form crystals may be lost by laboratory cultures.

The mol% G + C of the DNA is reported to be 33.5–40.1 (T_m) for two strains, 35.7–36.7 (Bd) for four strains and for the type species 33.8 (T_m), 34.3 (Bd).

Type strain: ATCC 10792 (DSM 2046; NCIB 9134).

B. thuringiensis has been divided on the basis of H-antigens into a rapidly growing number of serotypes and a sensible proposal to avoid confusion in their numbering has recently been made (Burges et al., 1982). Division of the species on the basis of flagellar antigens correlates quite well with that based on esterase pattern (Norris and Burges, 1965) and phenotypic characterization (de Barjac, 1981b).

Burges (1984) has argued persuasively that the designation for the subspecific division of this species, corresponding with serovars, should be "variety" rather than "subspecies." Such usage accords with common practice among *B. thuringiensis* workers. The International Code of Nomenclature of Bacteria, however, seems to leave no alternative but to call such taxa "subspecies."

DNA/DNA reassociation studies indicate that this species is genetically closely related to *B. anthracis*, *B. cereus* and *B. mycoides* but this has to be confirmed. The closeness of the relationship with *B. cereus* seems, however, to be in no doubt.

The ability to synthesize parasporal crystals is plasmid coded (Ward and Ellar, 1983). Authentic cultures of *B. cereus* can acquire the ability to produce crystals as a result of growth in mixed culture with *B. thuringiensis* (Gonzales et al., 1982). Considerations of this kind have led at least one insect pathologist to question the status of *B. thuringiensis* as a species apart from *B. cereus* (Lysenko, 1983).

Species Incertae Sedis

More than 200 different species have been described as belonging to the genus *Bacillus* as defined above. Most of them have been described or listed in the publications of Krasil'nikov (1959), Smith et al. (1952) and Gordon et al. (1973). Others are listed in *Index Bergeyana* (Buchanan et al., 1966; Gibbons et al., 1981.) Some of the species names are considered by Smith et al. (1952) and Gordon et al. (1973) to be synonymous with species listed in the Approved Lists of Bacterial Names (Skerman et al., 1980). Others have been inadequately characterized and for most of them cultures are not available.

The following species may be considered for revival after more detailed studies have been performed or after additional strains have been obtained and examined:

a. "*B. agarexedens*" Wieringa 1941, 121.

Rods, 0.5–1.5 × 2–10 μm. Motile. Gram-positive in young cultures. Endospores oval, sporangia swollen or non-swollen.

Strictly aerobic. Temperature range for growth 10–40°C. Requires carbohydrates for growth. Decomposes agar. Does not grow on nutrient agar (pH 6.8–7.0) but good growth observed on that medium at pH 8.0. In mineral salts-glucose medium growth is inhibited by peptones at pH 6.8, but not at pH 8.0.

Source: Soil, compost and plant litter.

Reference strain: DSM 1327.

Original strains have been lost (K. T. Wieringa, personal communication). Studies of new isolates revealed that the species is heterogeneous. The mol% G + C of the DNA is 44–53 (T_m). Strains studied belong to five different DNA homology groups which can be differentiated by hydrolysis of starch, pectin, DNA, dextran and chitin, and by deamination of phenylalanine (Hunger and Claus, 1981).

Agarolytic strains not inhibited by peptones belong to three other homology groups.

b. "*B. agrestis*" Werner 1933, 468.

The species was accepted as a synonym of *B. megaterium* by Gordon et al. (1973) but differs from typical members of that species in that: cells are smaller (mean cell width <1.0 μm); poly-β-hydroxybutyrate is not formed; esculin not hydrolyzed; phenylalanine not deaminated.

Source: Soil.

Table 13.10.
Differential characteristics of DNA homology groups of **Bacillus sphaericus**[a, b]

Characteristics	A[c]	B	C	D	E	F
Hydrolysis of						
Casein	d	+	d	d	d	+
Gelatin	+	+	+	d	+	+
Urea	−	+	+	d	−	d
Utilization of citrate	+	−	−	d	d	−
Phenylalanine deaminase	+	+	d	d	d	+
Growth with						
7% NaCl present	−	+	+	d	d	d
Lysozyme present	d	+	d	−	d	d
Mosquito pathogenicity	−	+	−	−	−	−

[a] Symbols: see Table 13.4.
[b] Data from Krych et al. (1980).
[c] DNA homology group A includes the type strain of *Bacillus sphaericus*.

The mol% G + C of the DNA of the single strain known is 37.4 (T_m).
Representative strain: NRS 602 (DSM 1316).
This strain is genetically not closely related to *B. megaterium sensu stricto* (DNA homology 30%; Hunger and Claus, 1981).

c. *"B. aminovorans"* den Dooren de Jong 1926, 157.
Rods, 0.8–1.5 × 1.5–5.0 µm. Motile. Gram-positive in young cultures. Endospores spherical, central to paracentral. Sporangia not swollen.
Strictly aerobic. Temperature maximum 37°C. Growth on nutrient agar feeble. Better growth on trypticase peptone. Utilizes mono-, di- and trimethylamine and glucose. Fructose and maltose used by some strains. No growth with other carbohydrates, organic acids, amino acids, except gluconate, glutamate, 3-hydroxybenzoate, citrate. Some strains use betaine.
Source: Soil.
The mol% G + C of the DNA (about 20 strains) is 40.4–41.8 (T_m) (I. Blumenstock, personal communication).
Original strain: ATCC 7046 (DSM 1314).

d. *"B. amyloliquefaciens"* Fukumoto 1943, 488.
Similar to *B. subtilis*. The species has been separated from *B. subtilis* by Welker and Campbell (1967) on the grounds that the mol% G + C of the DNA is slightly higher (43.5–44.9) than for *B. subtilis* (42–43). DNA hybridization shows only about 15% homology with *B. subtilis* (Seki et al., 1975; Welker and Campbell, 1967). Strains belonging to the two species have also been found to differ in other properties (O'Donnell et al., 1980).
Source: Soil.
The mol% G + C of the representative strain is 46.2 (T_m) and 44.5 (Bd) (Fahmy et al., 1985).
Representative strain: Strain F (ATCC 23350, DSM 7).

e. *"B. aneurinolyticus"* Kimura and Aoyama in Aoyama 1952, 127.
Typically *Bacillus brevis* hydrolyses casein whereas *B. aneurinolyticus* does not. The significance of this distinction might be questioned since strains that have been classified in *B. brevis* show a graduation in proteolytic activity from moderately active to none. The special property of *B. aneurinolyticus*, the decomposition of thiamine has not been reported in *B. brevis*.
Source: Human feces.
The mol% G + C of the DNA is reported to be 42 (analysis).
Representative strain: ATCC 12856, (IAM 1077).

f. *"B. apiarius"* Katznelson 1955, 636.
Vegetative cells 0.6–0.8 µm in width, often less at poles. A special feature is the nature of the spore coat, which is ridged, rectangular in outline and unusually thick. The coat remains covered by stainable remnants of the sporangium for a considerable time.
In other properties, similar to *B. laterosporus* except that a parasporal body has not been described. Growth occurs at pH 5.7, acid is not formed from mannitol, starch is hydrolyzed and phenylalanine is deaminated.
Source: Dead larvae of honeybees.
Representative strain: NRS 1438 (ATCC 29575).

g. *"B. caldolyticus"* Heinen and Heinen 1972, 17.
Cell diameter 0.7 µm. Spores cylindrical, terminal, swelling the sporangium.
Aerobic, weak growth in glucose broth under anaerobic conditions. Catalase-negative. Oxidase-positive. Hydrolyzes casein, gelatin and starch. Acid without gas from fructose, glucose, glycerol, maltose, mannitol, mannose, sucrose and trehalose. Grows in 2% NaCl broth. No growth at pH 5.7 or in the presence of sodium azide. Nitrate reduced to nitrite (Sharp et al., 1980).
Optimum pH range 6.0–8.0. Optimum (maximum) temperature for growth 72°C (82°C).
Source: Hot natural pool, United States.
The mol% G + C of the DNA is 52.3 (T_m).
Original strain: YT-P (DSM 405).
No close DNA homology with *"B. caldotenax"* (Sharp et al., 1980).

h. *"B. caldotenax"* Heinen and Heinen 1972, 17.
Cell diameter 0.5 µm. Spores oval, terminal, swelling the sporangium.
Aerobic, weak growth in glucose broth under anaerobic conditions. Acid without gas from fructose, glucose, glycerol, maltose, mannitol, mannose and sucrose. Growth in 3% NaCl broth. No growth at pH 5.7 or in the presence of sodium azide. Nitrite reduced to nitrite (Sharp et al., 1980). Catalase-negative. Oxidase-positive. Hydrolysis of casein, gelatin and starch positive.
Optimum pH range 7.5–8.5. Optimum (maximum) temperature for growth 80°C (85°C).
Source: Superheated pool water, United States.
The mol% G + C of the DNA is 64.8 (T_m) (Sharp et al., 1980).
Original strain: YT-G (DSM 406).

i. *"B. caldovelox"* Heinen and Heinen 1972, 17.
Similar to *"B. caldotenax"* but differs from that species in the following properties:
Cell diameter is 0.6 µm. Acid produced from trehalose. Grows in 2% NaCl broth. Optimum pH range 6.3–8.5. Optimum temperature for growth 60–70°C, maximum 76°C (Sharp et al., 1980).
Source: Superheated pool water, United States.
The mol% G + C of the DNA is 65.1 (T_m).
Original strain: YT-F (DSM 411).
DNA/DNA pairing studies revealed close DNA homology to *"B. caldotenax"* (Sharp et al., 1980).

j. *"B. carotarum"* Koch 1888, 279.
The species was accepted as a synonym of *B. megaterium* by Gordon et al. (1973) but differs from typical members of that species in properties listed in Table 13.9 for DNA homology group B.
Original strains are not available. Strains described as *"B. capri"*, *"B. cobayae"*, *"B. cohaerens"*, *"B. musculi"* and *"B. simplex"* are accepted as synonyms of *"B. carotarum."*
Source: Soil, animal feces.
The mol% G + C of the DNA of strains of these five species is 40–41 (T_m). They are genetically closely related (DNA homology 57–100%) but less related (22–34%) to *B. megaterium sensu stricto* (Hunger and Claus, 1981).
Representative strain: NRS 607 (*"B. capri"*) (DSM 1317).

k. *"B. flavothermus"* Heinen, Lauwers and Mulders 1982, 270.
Rods, 0.8 × 2–7 µm. Motile. Gram-positive. Spores terminal. Colonies round, smooth, yellow.
Facultatively anaerobic. Catalase-positive. Oxidase-positive. Starch but not gelatin hydrolyzed. Acetoin, arginine dihydrolase, lysine decarboxylase, trytophan deaminase and β-galactosidase-positive. Nitrate reduced to nitrite. Urease, ornithine decarboxylase, indole and H$_2$S not produced. Growth in 2.5% NaCl broth.
Optimum pH for growth 6–9. No growth at pH 5.0. Temperature range for growth 30–72°C; optimum growth at 60°C (aerobic) and 65°C (anaerobic).
Source: Hot spring, New Zealand.
The mol% G + C of the DNA is 61 (T_m).
Original strain: d.y. (DSM 2641).

l. *"B. flexus"* Batchelor 1919, 32.
The species was accepted as a synonym of *B. megaterium* by Gordon et al. (1973) but differs from typical members of that species in that:
Cells are smaller (mean cell width 0.9 µm). Poly-β-hydroxybutyrate is not formed. Phenylalanine not deaminated. No acid formed from pentoses. Esculin not hydrolyzed.
Source: Feces.
The mol% G + C of the DNA of the single strain known is 38.4 (T_m).
Representative strain: NRS 665 (DSM 1320).
This strain is neither genetically closely related to *B. megaterium* nor to *"B. carotarum"* with which the DNA homology is 35% (Hunger and Claus, 1981).

m. *"B. freudenreichii"* (Miquel) Chester 1898, 110.

This species is very similar to *B. brevis* in morphology and physiology. It differs in that it produces a considerable titratable alkalinity in urea broth and is less tolerant of acid. Additionally, growth occurs in 5% NaCl broth and phenylalanine is deaminated. In nutritional requirements it appears to be similar to or identical with *B. brevis* (Bornside and Kallio, 1956). It might be regarded as an intermediate between *B. brevis* and *B. pasteurii*.

Source: Soil, river water and sewage.

Representative strain: ATCC 7053.

n. *"B. laevolacticus"* Nakayama and Yanoshi 1967, 149.

Cells 0.4–1 *μ*m in diameter. Motile. Spores oval, terminal or nearly so; swelling the sporangium slightly.

Glucose fermented to D-(−)-lactic acid equivalent to 94–99% of sugar consumed; final pH 3.8–3.2. No growth in carbohydrate-free media. Glucose supports active anaerobic growth. Catalase formed. Inulin and starch fermented. Glucose-gelatin slowly liquefied. Nitrate not reduced. Growth in 2% NaCl broth; slight or no growth in 5% NaCl.

Maximum temperature 45–50°C.

Source: Rhizosphere of various plants.

The mol% G + C of the DNA is 41.5–43.2 (T_m) (I. Blumenstock, personal communication).

Representative strain: ATCC 23492 (DSM 442, NCIB 10269).

Two of three strains of *"B. racemilacticus"* (Nakayama and Yanoshi, 1967) have the same properties as *"B. laevolacticus"* and show a high DNA homology with this species. One other strain has been identified as *B. coagulans* (I. Blumenstock, personal communication).

o. *"B. longisporus"* Delaporte, 1972, 826.

Rods, 0.7–0.9 × 2.6–3.5 *μ*m, single. Motile, Gram-negative. Spores cylindrical, up to 2.2 *μ*m in length. Sporangia not swollen.

Facultatively anaerobic. Catalase-positive. Oxidase-negative. Hydrolyzes casein, gelatin, starch and urea. V-P reaction negative. Indole not formed. Nitrate not reduced to nitrite. Acid but no gas from glucose, arabinose, xylose and mannitol. Citrate not utilized. Egg-yolk reaction negative. Growth at pH 6.0. Growth in the presence of 7% NaCl; some strains can tolerate up to 28% NaCl.

Good growth at 29–37°C, no growth at 56°C.

Source: Soil.

Representative strain: ATCC 29490 (DSM 477).

p. *"B. macroides"* Bennett and Canale-Parola 1965, 204.

Originally described as *"Lineola longa"* (Pringsheim, 1950). Characters conform to those of *B. sphaericus* with one exception: the spore is frankly oval and scarcely distends the sporangium. The spherical to slightly oval spores formed by many strains of *B. sphaericus* have so far been distingusihable. The properties of *"B. macroides"* are also not greatly different from those of *B. badius*.

Minimal nutritional requirements: a carbon energy source, NH_4-N, thiamine, biotin and, in one strain, guanine. Carbon sources: various amino acids and C_2–C_5 *n*-fatty acids but not sugars. Proteolytic action is not detected within 3 weeks.

Source: Cow dung, plant material decaying in water.

The mol% G + C content of the DNA is reported to be 42 (T_m).

Representative strain: ATCC 12905 (DSM 54, NCIB 8796) (possibly different).

q. *"B. maroccanus"* Delaporte and Sasson 1967, 2346.

Relatively large organisms distinguished from *B. megaterium* chiefly by the production of lecithinase and a failure to store poly-β-hydroxy-butyrate.

Source: Desert soil.

Original strain: ATCC 25099 (DSM 607, NCIB 10500).

r. *"B. nitritollens"* Delaporte 1972, 828.

Rods, 0.7–0.8 × 2.8–4.0 *μ*m, singly or in pairs. Motile. Gram-negative. Spores oval, swelling the sporangia.

Facultatively anaerobic. Catalase-positive. Oxidase-negative. Hydrolyzes gelatin and starch. V-P reaction negative. Indole not formed.

Citrate utilized. Nitrate reduced to nitrite and/or to gas. Acid without gas from glucose, arabinose, xylose. Growth in 7% NaCl broth, some strains with 18% NaCl.

Growth at 19°C much better than at 37°C.

Source: Soil.

The mol% G + C of the DNA for two strains is 37.8 and 41.8, respectively (Pichinoty et al., 1978).

Representative strain: ATCC 29491 (DSM 474).

s. *"B. pacificus"* Delaporte 1967, 3071.

Cells oval, exceptionally large, measuring 1.5–2.1 × 2.7–3.4 *μ*m; have capsules and lipid inclusions; motile; one or two flagella inserted at or near one pole, or both poles. Spores ellipsoid, 1.3–1.5 × 2.7–3.4 *μ*m in size.

Best medium reported is 0.1% tryptone in sea water. No growth on ordinary nutrient agar. Glucose broth reaches pH 6 in 10 days; no acetoin formed. Gelatin slowly liquefied. Nitrate reduced to nitrite. Catalase formed. Grows in 10% NaCl.

Growth good at 28–40°C; none at 4°C.

Source: Shore sand, Pacific ocean, California.

Original strain: ATCC 25098 (NCMB 1862).

t. *"B. psychrosaccharolyticus"* Larkin and Stokes 1967, 890.

Distinctly pleomorphic; varies from coccal to elongate. On glucose media may contain globules unstainable with fuchsin. If sporulation does not occur, organisms may swell and become faintly stainable, often forming pear-shaped bodies up to 2 *μ*m in diameter. The spore frequently fills most of the sporangium; it may occur in a lateral position. On agar media relatively thick opaque growth without spreading or outgrowths. Overgrowth of laboratory cultures by asporogenous mutants appears to occur frequently.

Glucose promotes anaerobic growth only slightly. Proteolysis relatively weak. Growth and sporulation occur at 0°C.

Source: Soil and marshes.

Representative strain: ATCC 23296 (DSM 6).

Direct plating of soil frequently yields organisms which have the characteristics of *"B. psychrosaccharolyticus,"* except that some of them may diverge from that species in their action on nitrate (none or denitrification), proteins, starch, particular sugars, or in the utilization of glucose for anaerobic growth. These organisms, which do not appear to have been named, have not been subjected to comparative studies and their possible relationship to *"B. psychrosaccharolyticus"* remains to be examined.

u. *B. pulvifaciens* (ex Katznelson 1950) Nakamura 1984c, 412.[VP] (*Bacillus pulvifaciens* Katznelson 1950, 155.)

Growth on agar media unpigmented or buff to red; thin and non-spreading on nutrient agar; heavier if glucose is added. Anaerobic growth in glucose media is moderate.

Minimal nutritional requirements are a mixture of amino acids plus biotin.

Catalase-negative on nutrient agar. Production dependent on medium (positive on J-medium).

Closely resembles *B. larvae* from which it is distinguished by an ability to grow at 20°C and also in ordinary nutrient broth on serial transfer.

Source: Dead larvae of honeybee.

Type strain: NRRL B-3685.

v. *"B. similibadius"* Delaporte 1972. 824.

Rods 0.8–1.0 × 3–5 *μ*m, single or in short chains. Motile. Gram-positive. Spores oval. Sporangia not swollen.

Strictly aerobic. Catalase-positive. Oxidase-negative. Hydrolysis of gelatin positive. Hydrolysis of starch and V-P reaction negative. Indole not formed. Nitrate not reduced to nitrite. Citrate is utilized. Egg-yolk reaction negative. Most strains can grow in the presence of 15–20% NaCl broth. No growth at pH 5.7.

Good growth at 29–37°C.

Source: Soil, intestine of tadpoles, feces of bear.
Representative strain: DSM 480.

w. *"B. thermocatenulatus"* Golovacheva, Longinova, Salikhov, Koleshikov and Zaitseva 1975, 265.

Rods, 0.9×6–8 μm, in long chains. Motile. Gram-positive. Spores cylindrical, terminal. Sporangia slightly swollen. Colonies yellowish.

Facultatively anaerobic. Starch, gelatin and urea not hydrolyzed. Casein weakly hydrolyzed. Acid without gas from cellobiose, fructose, glucose, galactose, glycerol, mannitol and sucrose. No acid from arabinose, xylose, rhamnose, lactose and ethanol. Citrate utilized. Nitrate is reduced to gaseous nitrogen. Acetoin, indole and H_2S not formed. Egg-yolk reaction negative. Growth in 4% NaCl broth.

Temperature for growth 35–78°C, optimum on agar media at 65–70°C, in liquid media at 55–60°C.

Source: Hot gas well, coatings inside tube, Southern Ural, USSR.

The mol% G + C of the DNA is 69 (by chromatography).

Original strain: 178 (BKM-1259, DSM 730).

x. *"B. thermodenitrificans"* (Ambroz 1913) Mishustin 1950.
Original strains not available.
Source: Soil, sugar beet juice from extraction plants.
Representative strain: E32-66 (ATCC 29493, DSM 466).

The species has been originally described by Ambroz as *"Denitrobacterium thermophilum"*. It belongs to the *"B. stearothermophilus"* complex, but was separated by a numerical study from typical strains of *B. stearothermophilus* (Klaushofer and Hollaus, 1970).

y. *"B. thiaminolyticus"* Kuno 1951, 364.
Decomposes thiamine actively.
Source: Human feces.
Representative strain: ATCC 11377.

In his original description, Kuno recognized the similarity of this species to *B. alvei.* After a study of 44 strains of *"B. thiaminolyticus,"* Hayashi and Nakayama (1953) reported that it was variable in some of the characteristics by which it was distinguished from *B. alvei.* The need for further study of more strains of *B. alvei* and *"B. thiaminolyticus"* is thus indicated. Decomposition of thiamine by *B. alvei* has not been reported.

z. *"B. xerothermodurans"* Bond and Favero 1977, 159.
Rods, 0.7–1.2×1.6–2.8 μm. Pleomorphic, especially in older cultures. Spores spherical to oval. Sporangia swollen. Scanning electron microscopy of spores revealed a surface honeycomb pattern of polygonal depressions surrounded by straight ridges. Unusual ultrastructure with an irregular, thick outer spore coat composed of globular subunits and laminated inner spore coat containing up to nine distinct layers.

Strictly aerobic. Catalase-positive. Growth in 10% NaCl broth. Potato starch hydrolyzed. Reactions in other properties studied are negative.

Cleaned spore preparations show extreme resistance to dry heat ($D_{125°C}$ = 139 h, $D_{150°C}$ = 2.5 h).

Source: Sandy soil, Cape Kennedy, Florida. The strain was isolated after heating dry samples at 125°C for 48 h.

Original strain: (ATCC 27380 (DSM 520).

aa. Fifteen *Bacillus* species have been described as occuring in the intestine of tadpoles (Collin, 1913; Delaporte, 1964b; Hollande, 1934). They have not been obtained in pure culture and have been described solely on the basis of microscopical appearance. Germination and outgrowth of endospores have not been observed. They are assumed to be aerobic and differ from other *Bacillus* species mainly in the size of the vegetative cells (2–5×10–95 μm, often in long chains). In certain species two endospores may be formed within a single nonseptate cell. In two of these species one of the two endospores have been found to move rapidly through much of the interior of the long vegetative cell (Delaporte, 1963)

"Fusosporus gyrini" Delaporte 1964b, 857, belongs to this group. It was separated from *Bacillus* because the vegetative cells taper to pointed ends.

Other organisms

"B. penetrans," a parasite of soil nematodes, has been described by Mankau (1975). It resembles a cladoceran parasite observed by Sayre and Wergin (1977).

The taxonomic position of *"B. penetrans,"* which has not yet been obtained in pure culture, is still obscure. The ultrastructure of the endospores is similar to that of other *Bacillus* species. Vegetative cells form dichotomously branched filaments. Endospores develop at the apices of the filaments similar to *Actinobifida* (Imbrian and Mankau, 1977).

According to Sayre et al. (1983) *"B. penetrans"* conforms to the original description of Metchnikoff (1888) for *Pasteuria ramosa.*

The following two species probably belong to the genus *Bacillus.* They were apparently separated because of the chemolithotropic properties:

"Sulfobacillus thermosulfidooxidans" Golovacheva and Karavaik 1978, 821.

Rods, 0.6–0.8×1.0–6.0 μm, with rounded or tapered ends. Nonmotile. Occur singly, in pairs or short chains. Gram-positive. Spores round or slightly oval, paracentral to terminal. Sporangia swollen. Colonies on agar containing Fe^{++} are round, shining, initially yellowish, becoming reddish-brown in color.

Strictly aerobic. Facultatively chemolithotrophic. Sulfur, iron and sulfur-containing minerals like pyrite are oxidized. Can be adapted to grow heterotrophically in mineral base media containing 0.1% glucose or sucrose. Acidophilic, optimum pH for growth 1.9–2.4.

Temperature range 28–60°C, optimum about 50°C.

Source: Isolated from zones of spontaneous heating of ores of copper-zinc pyrite deposits, Nikolaev, USSR.

The mol% G + C of DNA is 53.6–53.9 (method not specified).
Original strain: 1A, (BKM-1269).

"Thiobacillus thermophilica imshenetskii" Egorova and Deryugina 1963, 445.

Rods, 0.4–0.6×1.0–4.5, forming short or long chains. Motile. Gram-negative. Spores spherical, terminal. Sporangia swollen. Resistant to wet heat at 121°C for at least 8 min. Colonies small, round and nonpigmented.

Strictly aerobic. Obligately chemolithotrophic. Oxidize various sulfur compounds. Organic substrates may be inhibitory.

Temperature range of growth 40–80°C, optimum temperature at 55–60°C.

The original description was confirmed by Hutchinson et al. (1967).
Source: Bragunsk thermal spring.
Original strain: ATCC 23841.

Genera Incertae Sedis

Several trichome-forming bacteria, occuring in the alimentary tract of animals, have been reported to form endospores but have not been obtained in pure culture. These include the genera *"Anisomitus"* (Grassé, 1925), *"Arthromitus"* (Leidy, 1850) (syn. *"Entomitus"* (Grassé, 1924)), *"Coleomitus"* (Duboscq and Grassé, 1930) and *"Metabacterium"* (Chatton and Pérard, 1913b) as well as the spiral forming organisms of the genus *"Sporospirillum"* (Delaporte, 1964a) (syn. *"Bacillospira"* (Hollande, 1933)).

The response of all these organisms to oxygen, their physiology and the fine structure of their endospores is still unknown. Only the endospores of *"Metabacterium polyspora"* have been studied (Robinow, 1951; 1960) and have been found to be similar to those of *Bacillus* in certain cytological features. Their resistance to environmental stresses has not, however, been investigated.

In the fifth edition of the *Manual*, "*Arthromitus*" and *Oscillospira* were placed in the order *Caryophanales*.

Note Added in Proof

After the manuscript was completed the following new *Bacillus* species have been validly published.

B. amylolyticus (ex Kellerman and McBeth 1912) Nakamura, 1984b, 224.[VP]

am.y.lo.ly'ti.cus. L. n. *amylum* starch; Gr. adj. *lyticus* dissolving; M.L. masc. adj. *amylolyticus* starch dissolving.

The species corresponds to the DNA homology group K of the *B. circulans* complex (Table 13.8).

Habitat: soil.

The mol% G + C is 53 (Bd).

Type strain: NRRL NRS-290 (DSM 3034).

B. lautus (ex Batchelor 1919) Nakamura, 1984b, 225.[VP]

lau'tus. L. part. adj. *lautus* washed, splendid.

The species corresponds to the DNA homology group H of the *B. circulans* complex (Table 13.8).

Habitat: human intestinal tract; soil.

The mol% G + C is 50–52 (Bd).

Type strain: NRRL NRS-666 (DSM 3035).

B. pabuli (ex Schieblich 1923) Nakamura, 1984b, 225.[VP]

pa'bu.li. L. gen. n. *pabuli* of fodder.

The species corresponds to the DNA homology group E of the *B. circulans* complex (Table 13.8).

Habitat: fodder and soil.

The mol% G + C is 48–50 (Bd).

Type strain: NRRL NRS-924 (DSM 3036).

B. psychrophilus (ex Larkin and Stokes 1967) Nakamura, 1984a, 122.[VP]

psy.chro'phil.us. Gr. adj. *psychros* cold; Gr. adj. *philus* liking, preferring; M.L. adj. *psychrophilus* perferring cold.

Similar to *B. globisporus* but differs from this species in the following characters (positive reactions): growth in 3% NaCl; nitrate reduced to nitrite; acid produced from D-mannitol, D-ribose, sucrose, trehalose and D-xylose.

Habitat: soil and river water.

The mol% G + C is 40.5 (Bd) and 39.7 (T_m) (Fahmy et al., 1985). Nakamura (1984a) gives a value of 44.1 (Bd).

Type strain: NRRL NRS-1530 (ATCC 23304, DSM 3).

B. thermoglucosidasius Suzuki, 1984, 270.[VP] (Effective Publication: Suzuki et al., 1983, 493.)

ther'mo.glu.co.si.da'si.us. Gr. n. *therme* heat; M.L. adj. *glucosidasius* of glucosidase. M.L. adj. *thermoglucosidasius* indicating the production of heat-stable glucosidase.

Rods, 0.5–1.2 × 3.0–7.0 µm. Endospores ellipsoidal, terminal, swelling the sporangium. Gram-positive.

Obligate thermophile with an optimum at 61–63°C. Strictly aerobic, neutrophilic. Differs from strains of the *B. stearothermophilus* complex in several respects. Produce exooligo-1,6-glucosidase in large amounts.

Habitat: soil.

The mol% G + C is 45–46 (T_m).

Type strain: KP1006 (DSM 2542).

(Note: This species was designated in earlier publications as "*B. thermoglucosidius*").

B. validus (ex Bredemann and Heigener 1935) Nakamura, 1984b, 225.[VP]

val'i.dus. L. adj. *validus* strong, vigorous.

Habitat: soil.

The species corresponds to the DNA homology group 1 of the *B. circulans* complex (Table 13.8).

The mol% G + C is 53–54 (Bd).

Type strain: NRRL NRS-1000 (DSM 3037).

Genus **Sporolactobacillus** *Kitahara and Suzuki 1963, 69*[AL]

OTTO KANDLER AND NORBERT WEISS

Spo.ro.lac.to.ba.cil'lus. Gr. n. *spora* seed; L. n. *lac, lactis* milk; L. dim. n. *bacillus* a small rod; M.L. masc. n. *Sporolactobacillus* sporing milk rodlet.

Cells **straight rods**, 0.7–0.8 × 3–5 µm, occurring singly, in pairs, rarely in short chains. **Endospores formed. Gram-positive. Motile** by means of a small number of long peritrichous flagella.

Microaerophilic; homolactic fermentation of hexoses; does not contain compounds such as **catalase and cytochromes; contains menaquinones** but no ubiquinone; **nitrate is not reduced** and **indole is not formed;** contains odd-numbered saturated *anteiso-* and *iso*-branched but no unsaturated even-numbered or cyclopropane fatty acids (Table 13.11).

The **mol% G + C of DNA is 38** (T_m) (Weiss, unpublished) to **39.3** (Chem) (Suzuki and Kitahara, 1964).

Type species: *Sporolactobacillus inulinus* (Kitahara and Suzuki) Kitahara and Lai 1967,197.

Further Descriptive Information

Morphology. Cells are rather slender with rounded ends, sometimes slightly bent. Often granular appearance under the phase-contrast microscope. Growing cultures are highly motile especially when grown on gluconate or mannose. Motility is preserved for several days when the medium is buffered with $CaCO_3$.

Tadpole-shaped cells are frequently observed, especially when $(NH_4)_2SO_4$ and $CaCO_3$ are omitted from the medium. In the presence of 5% NaCl, 50% of the cells show this swollen state, which was found to be the result of autospheroplastization (Kitahara and Toyota, 1972). Ultrathin sections exhibit the usual **cell wall profile** of a Gram-positive organism (Kitahara and Toyota, 1972). The isolated **cell wall**

of *S. inulinus* contains about 25% murein of the **meso-diaminopimelic acid (meso-DAP) direct type,** a large amount of polysaccharides (not further analyzed), but no teichoic acid (Weiss et al., 1967; Okada et al., 1976). Since the formation of teichoic acid is species- and even strain-specific in other genera, teichoic acid-containing species of *Sporolactobacillus* may be found in the future. Hence, the lack of teichoic acid may not be characteristic of the genus, but rather of distinct species.

Colonies on agar surface are small (about 1 mm in diameter), grayish white, glistening. On semisolid agar (0.25%) colonies of up to 3 mm in diameter are formed, indicating the motility of the organism. Growth on agar slant is almost invisible. Filiform growth occurs in agar stab cultures along the stab canal without growth on and immediately beneath the surface. In agar shake tubes, small colonies are formed uniformly in the agar tube except for a 5-mm zone near the surface. Young liquid cultures exhibit a silky luster when shaken while the cells precipitate in older cultures.

Spore formation is very rare in most media (10^{-4}–10^{-6} of the population). About 1% of the population formed spores in the following medium (w/v): yeast extract 0.1%; meat extract 0.5%; $(NH_4)_2SO_4$ 1.0%, α-methylglucoside 0.5%; tomato serum 20% (v/v); $CaCO_3$ in excess; incubation at 37°C (Kitahara and Lai 1967). In a recent study of the effect of various media on sporulation frequency of *Sporolactobacillus*, the Kitahara medium without tomato serum was found superior (Doores and Westhoff, 1981). The fine structure of the spores is the same as that of spores of aerobic bacilli. The **spores** also contain **dipicolinic**

acid. However, **heat resistance** of the spores is significantly lower. They are killed at temperatures below 100°C (Kitahara and Lai, 1967). Depending on growth conditions the following decimal reduction times have been reported: $D_{75} = 20$–90 min, $D_{85} = 3$–12 min and $D_{90} = 1$–7.5 min (Doores and Westhoff, 1981). Vegetative cells are very heat sensitive ($D_{50} = <1$ min).

Growth conditions. The **nutritional requirements of** *Sporolactobacillus* are not so complex as those of lactobacilli. They resemble those of bacilli. A medium containing (w/v) yeast extract (Difco), (0.5%); peptone (0.5%); and glucose (2.0%) supports good growth, although the **addition of CaCO₃** (growth stops at pH 4.0) and the replacement of glucose by **α-methylglucoside** is favorable. Surface growth is favored by anaerobic incubation. No growth occurs without fermentable carbohydrate. **Biotin and pantothenate** are the only essential vitamins, **leucine** and **valine** the only essential amino acids, whereas none of the nucleotides is required (Kitahara and Suzuki, 1963). NaCl is not necessary. Concentrations of NaCl up to 3% are slightly, those of > 9% are completely inhibitory.

Physiology. **Hexoses** are fermented exclusively to **D-(−)-lactic acid** with less than 1% of volatile acids or ethanol. Pentoses are not fermented. The small amounts of titrable acid (<10%) found in pentose-containing broth (Kitahara and Suzuki 1963) may have originated from thermal destruction of pentoses during autoclaving. Hence, *Sporolactobacillus* is a typical homolactic acid fermenter, although a limited capacity for an electron transfer to oxygen is indicated by the typical microaerophilic growth in agar stabs and by the presence of menaquinones (Collins and Jones, 1979). However, no data on any oxidative pathways are yet known.

Habitat. So far only isolated from **chicken feed** (Kitahara and Suzuki, 1963) and **soil** (Nakayama and Yanoshi, 1967a); may be widespread in the environment, although in small numbers (Doores et al., 1982).

Enrichment and Isolation

Samples may be **heat-shocked for 5 min at 80°C**, plated on the above described *Sporolactobacillus* medium and screened for catalase-negative colonies consisting of Gram-positive slender rods.

An enrichment in MRS medium acidified to pH 5.5 by the addition of acetic acid before plating the heat-shocked samples may be useful (Doores and Westhoff, 1983). The strains obtained must be studied further with respect to taxonomically important characteristics.

Maintenance Procedures

Strains of *Sporolactobacillus* can be lyophilized by common procedures for lactobacilli and bacilli. They may also be stored for several months in neutralized medium at −20°C.

Taxonomic Comment

Sporolactobacillus was originally described as a subgenus of the genus *Lactobacillus*. However, in subsequent papers Kitahara and his coworkers referred to the type species as *Sporolactobacillus inulinus* and the recognition of generic rank of *Sporolactobacillus* has been attributed by Kitahara (1974) and by the Approved Lists of Bacterial Names (Skerman et al., 1980) to Kitahara and Lai (1967).

Since *Sporolactobacillus* exhibits characteristics typical of both genera *Lactobacillus* and *Bacillus* (Table 13.11), its taxonomic position

Table 13.11.

Differential characteristics of the genera **Sporolactobacillus**, **Bacillus** *and* **Lactobacillus**[a, b]

Characteristics	*Lactobacillus plantarum*	*Sporolactobacillus*	*Bacillus*
Homolactic fermentation	+	+	−/(+)[c]
Catalase	−/(+)[d]	−	+
Cytochromes	−	−	+
Nitrate reduction	−/(+)[d]	−	+
Indole formation	−	−	+
Motility	−/(+)[d]	+	+
Endospore formation	−	+	+
Peptidoglycan of meso-DAP, direct type	+	+	+
Teichoic acid	+	−	+
Menaquinones (MK-7)	−	+	+
Fatty acids[e]			
a-15:0			
a-17:0	−	+	+
i-15:0			
i-17:0	−	+	+
n-14:0			
n-16:0	+	+	+
n-14:1			
n-16:1	+	−	−
n-18:1			
cy-17:0			
cy-19:0	+	−	−

[a] Symbols: see Table 13.4.
[b] The genus *Lactobacillus* is here represented by *L. plantarum* which has more characteristics in common with *Sporolactobacillus* than any other species of *Lactobacillus*.
[c] Only a few species, and mainly under anaerobic conditions.
[d] Only in a few strains, or under special conditions (Davis 1964).
[e] *i*-, *a*-, *n*- and *cy*- denote iso-, anteiso-, normal (straight chain) and cyclopropane-fatty acid, respectively. The number on the *left* is the number of carbon atoms and that on the *right* is the number of double bonds. (Data according to Uchida and Mogi 1973.)

between the families *Lactobacillaceae* and *Bacillaceae* has been often discussed (Kitahara and Toyota, 1972; Collins and Jones, 1979). The pattern of fatty acids (Uchida and Mogi, 1973) and the presence of menaquinone (Collins and Jones, 1979) indicated a closer relationship with the genus *Bacillus* than with *Lactobacillus*. The closer genealogical relationship of *Sporolactobacillus inulinus* with various species of the aerobic sporeformers than with species of *Lactobacillus* was recently demonstrated by cataloging the oligonucleotide pattern of the ribosomal 16S-RNA (Fox et al., 1977; Stackebrandt et al., 1981, 1983). Thus *Sporolactobacillus inulinus* might be considered even as a species of the genus *Bacillus*. However, the lack of heme proteins such as catalase and cytochromes may still justify the generic rank of *Sporolactobacillus* within the family *Bacillaceae*.

List of species of the genus **Sporolactobacillus**

1. **Sporolactobacillus inulinus** (Kitahara and Suzuki) Kitahara and Lai 1967, 197.[AL] (*Sporolactobacillus (Lactobacillus) inulinus* Kitahara and Suzuki 1963, 69*.)

i.nu.li′nus.M.L. n. *inulum* inulin; M.L. adj. *inulinus* pertaining to inulin.

Morphology as for genus. **Elliptical endospores**, 0.8×1.0 μm, are formed in some cells in a terminal position, the sporangium swelling with maturation of the endospore. Spores tolerate heating for 10 min at 80°C.

Good growth in **glucose-yeast extract-peptone** (GYP) media. On

---

*****Editorial note:** This original designation inadvertently reduced *Lactobacillus* to subgeneric rank and was presumably a *lapsus calami* for *Lactobacillus (Sporolactobacillus) inulinus*. In subsequent papers Kitahara and co-workers referred to the organism as *Sporolactobacillus inulinus*,

GYP-agar colonies pinpoint; somewhat larger on or in semisolid agar. In deep agar, small colonies distributed equally throughout the agar except on the surface. Broth becomes turbid and the growth gradually precipitates.

Produces D-(−)-lactic acid, but no gas from fructose, glucose, inulin, maltose, mannose, raffinose, sucrose, trehalose, mannitol, sorbitol and α-methylglucoside. No acid from arabinose, xylose, galactose, lactose, melibiose, cellobiose, melezitose, dextrin, starch, glycerol, erythritol, adonitol, rhamnose and salicin. CO_2 and D-(−)-lactic acid are formed from gluconate. Limiting pH is 4.0; however, when $CaCO_3$ is present, 20% or more glucose may be fermented completely.

Gelatin not liquefied. Litmus milk unchanged.

The **cell wall** contains peptidoglycan of the *meso-DAP-direct* **type** (Weiss et al., 1967).

The **mol% G + C** of the DNA is **38** (T_m, Weiss, unpublished) to **39.3** (Chem; Suzuki and Kitahara, 1964).

Isolated from chicken feed.

Type strain: ATCC 15538.

Comments: Several catalase-negative, spore-bearing, lactic acid producing, motile strains isolated from the **rhizosphere** of various wild plants (Nakayama and Yoshi, 1967b) have been found to exhibit the same menaquinone pattern as *Sporolactobacillus* (Collins and Jones 1979). Since they differ from *S. inulinus* in forming DL-**lactic acid** and in fermenting a somewhat different range of saccharides, two new species (*"Sporolactobacillus laevus"* and *"Sporolactobacillus racemicus"*) have been proposed by Nakayama (1970). However, these species have not been recognized by Kitahara (1974) in his description of *Sporolactobacillus* in the eighth edition of *Bergey's Manual* and they are not included in the Approved Lists of Bacterial Names (Skerman et al., 1980). Further studies on these strains are required to validate the proposed species.

The high value of 47.3 mol% G + C of DNA reported by Miller et al. (1970) could not be verified.

Genus Clostridium Prazmowski 1880, 23[AL]

ELIZABETH P. CATO, W. LANCE GEORGE AND SYDNEY M. FINEGOLD

Clos.tri′di.um. Gr. n. *closter* a spindle; N.L. neut. dim. n. *Clostridium* a small spindle.

Rods, usually stain Gram-positive at least in very early stages of growth, although in some species Gram-positive cells have not been seen. Motile or nonmotile. When motile, cells usually are peritrichous. Form oval or spherical **endospores** that usually distend the cell.

Usually chemoorganotrophic; some species are chemoautotrophic or chemolithotrophic as well. Usually produce mixtures of organic acids and alcohols from carbohydrates or peptones. **May be saccharolytic, proteolytic, neither, or both.** May metabolize carbohydrates, alcohols, amino acids, purines, steroids or other organic compounds. Some species fix atmospheric nitrogen. **Do not carry out a dissimilatory sulfate reduction.** Usually catalase-negative, although trace amounts of catalase may be detected in some strains.

The cell wall usually contains *meso*-diaminopimelic acid (*meso*-DAP).

Most species are obligately **anaerobic**, although tolerance to oxygen varies widely; some species will grow but not sporulate in the presence of air at atmospheric pressure.

For most species, growth is most rapid at pH 6.5–7 and at temperatures between 30 and 37°C; the range of temperature for optimum growth is from 15–69°C.

The mol% G + C of the DNA of the type species is 27–28 (T_m). The mol% G + C of the DNA of other species examined is 22–55 (T_m).

Type species: *Clostridium butyricum* Prazmowski 1880, 24.

Further Descriptive Information

Cells of most strains occur as straight or slightly curved rods of varying length and width, with rounded, blunt or tapering ends; long filaments are formed by some species. Cells may occur singly, in pairs, in short or long chains or, in a few cases, in tight spirals or coils.

Nutritional and growth requirements, metabolism and metabolic pathways are diverse.

Most of the species are able to reduce the medium in which they grow, thus making their immediate environment more suitable for continued growth. This is shown by their ability to reduce neutral red in peptone-yeast extract-glucose (PYG) broth culture or resazurin in PYG deep agar cultures or both.

Many species have been shown to be infected by bacteriophages (Ogata and Hongo, 1979).

Some species are pathogenic for man or animals or both. For some species, toxin neutralization testing is required for definitive identification.

Clostridia are ubiquitous. Commonly found in soil, sewage, marine sediments, decaying vegetation, animal and plant products; in the intestinal tract of man, other vertebrates, and insects; and in wounds or soft tissue infections of man and animals.

Isolation and Enrichment Procedures

Strains of most species of *Clostridium* can be isolated on usual complex media appropriate for culture of anaerobes as described in the section on *Eubacterium*. Even those species that grow on the surface of blood agar plates incubated aerobically will grow as well or better under anaerobic conditions. Trace mineral elements may be stimulatory for some species as indicated in the species description. For selective isolation of clostridia from mixed cultures, intestinal contents, soil or other source, heating of samples at 70°C for 10 min prior to culture will destroy most vegetative cells and allow spores to predominate. Treatment of specimens 1:1 (v/v) with 95% or absolute ethanol for 45 min also will kill vegetative cells of *Clostridium* and nonspore-forming organisms (Koransky et al., 1978). For species (e.g. *C. septicum*, *C. tetani*) that swarm and cover the surface of solid media, the isolation of single colonies can usually be achieved by increasing the concentration of agar to 4.5% (Hayward and Miles, 1943; Hayward et al., 1978). Special growth factors are required for good growth of some species and are given in the descriptions of species. Growth of some species (*C. polysaccharolyticum*, *C. thermautotrophicum*) is stimulated by the presence of 20% rumen fluid. Although most species can be isolated at 37°C, some (*C. arcticum*, *C. putrefaciens*) require an incubation temperature of 22–25°C and some (the thermophilic clostridia) a temperature of at least 55°C. Most species grow well in an atmosphere of 100% CO_2 but many grow as well or better in an atmosphere of 90% N_2-10% CO_2 and some (*C. aceticum*, *C. formicaceticum*, *C. quercicolum*) require an atmosphere of 100% N_2 for growth. This information is given in the descriptions of species.

Maintenance Procedures

Information concerning maintenance and storage of cultures is given in the description of the genus *Eubacterium*. In addition, many *Clostridium* species can be maintained in tightly stoppered tubes of chopped

elevating *Sporolactobacillus* to generic rank. The Approved Lists of Bacterial Names (Skerman et al., 1980) cites Kitahara and Suzuki 1963 as the authors of the generic name and Kitahara and Lai 1967 as authors of the type species name, and we will follow that designation here.

meat broth for extended periods (months or years) at room temperature.

Procedures for Testing for Special Characters

Unless otherwise cited, characteristics of strains were determined using prereduced, anaerobically sterilized media and anaerobic methods described by Holdeman et al. (1977). Basal peptone-yeast extract (PY) media contained (per 100 ml): 0.5 g peptone, 0.5 g trypticase, 1 g yeast extract, 0.1 µl vitamin K, 0.5 g hemin, cysteine hydrochloride, and salts solution.

Fermentation products were determined chromatographically using a thermal conductivity detector. Those laboratories using flame ionization detectors will not detect formic acid.

Morphology and Gram reaction. Gram's stains were prepared from young cultures and a buffered Gram's stain such as Kopeloff's modification used. Cell sizes given in the descriptions were measured from a Dormer's nigrosin stain (Holdeman et al., 1977): 1 drop of culture on a clean slide, air-dried, with a very small amount of 10% nigrosin added, and allowed to dry in a slanted position. The drawings accompanying the species descriptions are composites made from Gram stains of the type and similar strains, unless otherwise indicated. Usually there will not be as much variation seen in single strains. Cell sizes may vary depending on culture medium or age of culture. Unless otherwise indicated, measurements were made from a 24-h culture in PYG medium.

Tests for H_2 production, gas production, esculin hydrolysis and fermentation, indole, gelatin digestion, curdling or proteolysis of milk, and acid production are interpreted as described in the chapter on *Eubacterium*.

For those species that require a pH of at least 7.0 for initiation of growth, the gas phase during inoculation should be either 90% N_2-10% CO_2 or 100% N_2.

H_2S *production.* This is determined in SIM (BBL, Cockeysville, Md) medium. A loopful or 3 or 4 drops of culture in a Pasteur pipet is inserted nearly to the bottom of the tube. Development of a black color (darker than uninoculated medium) indicates H_2S production.

Bile inhibition is determined by comparison of growth in PYG medium with that in PYG-2% oxgall medium (equivalent to 20% bile).

Neutral red reduction. Reaction is positive if red color disappears after 5 days incubation in 3 ml of PY-fructose broth containing 0.1 ml of neutral red stock solution (0.1% w/v neutral red dissolved in 60% absolute ethanol).

Requirement for fermentable carbohydrate. Growth in carbohydrate-free medium (PY) is compared with that in media with added carbohydrate. If growth occurs only in carbohydrate media that are fermented, the organism is considered to require fermentable carbohydrate. If the inoculum is from a carbohydrate-containing medium, there may be enough carryover to give slight growth in PY medium; the requirement may then be confirmed by transfer from the PY medium to another tube of PY medium.

Spore production. No one medium is optimal for stimulation of sporulation of all species of *Clostridium*. The most reliable general medium for demonstration of spores is chopped meat agar (Holdeman et al., 1977). Inoculated chopped meat agar slants are incubated at 30°C or at a less-than-optimal temperature; usually spores can be detected within 10 days in Gram stains either from the slant or from the water of syneresis. Some strains sporulate best when glucose is present, some without glucose. Egg yolk agar may stimulate spore production. Other additives that may be helpful are given in the descriptions of species. Spores are rarely seen in some species (e.g. *C. clostridioforme, C. perfringens, C. ramosum*). Demonstration of survival of these strains after heating for 10 min at 70°C or 80°C or after 45 min treatment with an equal volume of 95% ethanol indicates that spores are present. In some species, spores have been found in lyophilized cultures of strains in which no spores could be demonstrated in the original fresh isolates (Cato and Salmon, 1976). Possibly the lyophilization process destroys some vegetative cells and increases the percentage of the more resistant spore-forming cells. Tabor et al. (1976) have described a gas-liquid chromatographic method to detect spore dipicolinic acid present in cultures even when no spores could be detected microscopically from Gram stains.

Antibiotic susceptibility testing. Where no citation is given, susceptibility to antibiotics was tested by the broth disk method of Wilkins and Thiel (1973). Antibiotics tested and concentrations used were chloramphenicol (12 µg/ml), clindamycin (1.6 µg/ml), erythromycin (3 µg/ml), penicillin G (2 U/ml), and tetracycline (6 µg/ml). Minimal inhibitory concentrations (MIC) of antibiotics reported by Finegold and George were determined by methods described by Sutter et al. (1980). The wide range reported for many drugs emphasizes the necessity for making a susceptibility determination for any isolate that may be clinically significant.

Toxin testing and pathogenicity. Toxicity is determined by inoculating 0.4–0.5 ml of supernatant fluid from chopped meat carbohydrate or chopped meat glucose cultures intraperitoneally into mice. Some clostridial toxins (nonproteolytic *C. botulinum* types E, B and F toxins, and *C. perfringens* epsilon toxin) are produced as nontoxic protoxins which must be activated by proteolytic enzymes. The culture supernatant fluid is incubated with trypsin for 30 min at 37°C before inoculation into mice. See Smith (1975a) for further details. For determinations of neutralization of toxin by a specific antitoxin, the toxin and antitoxin are mixed, incubated at room temperature for 30 min, then inoculated intraperitoneally into mice. At least 2 mice should be inoculated with both the supernatant culture fluid and the supernatant fluid plus antitoxin. Antitoxin to most of the toxic clostridia is available from Burroughs Wellcome Company. Antitoxin to *C. difficile* is available from Tracy Wilkins, Department of Anaerobic Microbiology, VPI & SU, Blacksburg, VA. Commercial antisera are prepared to neutralize several lethal toxins produced by a species or a toxin type and are not specific for any single component. Where two different species have one lethal toxin in common among several toxins produced, antitoxin to one may neutralize that toxin in the other (e.g. *C. difficile* toxin neutralized by *C. sordellii* antitoxin).

To determine pathogenicity, equal volumes of a chopped meat carbohydrate culture in the late-log phase of growth and 10% (w/v in water) $CaCl_2$ are mixed. One-half milliliter of the mixture is injected into the muscle of the hind leg of a guinea pig, which is monitored for tissue changes or death. For further information on toxicity and pathogenicity, see Smith (1975a) and Holdeman et al. (1977).

Differentiation of the genus **Clostridium** from other closely related taxa

Clostridium species can be distinguished from other sporeforming genera as follows:

1. Aerobic, spherical cells (see *Sporosarcina*);
2. Aerobic or facultative, usually produce catalase (see *Bacillus*);
3. Microaerophilic (see *Sporolactobacillus*);
4. Reduce sulfates and sulfites to H_2S (see *Desulfotomaculum*).

Species of *Clostridium* that rarely or never stain Gram-positive can be differentiated from *Bacteroides* by demonstration of spores or of heat resistance; those Gram-positive species in which spores are difficult to detect can be differentiated from *Eubacterium* by demonstration of heat or alcohol resistance. It is not necessary to demonstrate spores in strains of *C. perfringens, C. ramosum* or *C. clostridioforme* for identification; other criteria as presented in their descriptions can lead to proper placement of these species and heat resistance will confirm identification. The recently described species "*Butyribacterium methylotrophicum*" (Zeikus et al., 1980), has been reported to produce "atypical" heat-resistant spores in a medium containing 0.05% yeast extract, 100 mM methanol, 50 mM acetate, and soil extract. Although we found other reactions to be as described, we have been unable to demonstrate spores in the proposed type strain of this species. It differs from all described species of *Clostridium* in utilizing methanol for growth, converting it to butyrate.

Taxonomic Comments

The wide diversity in metabolic activity, nutritional requirements, and mol% G + C of species of *Clostridium* suggests that the genus may eventually be divided into at least two genera: those species with a mol% G + C of 22–34 in one genus; those with a mol% G + C of 40–55 in another. From *r*RNA homology experiments, subgroups within the genus, which correlate with some phenotypic traits, have been indicated (Johnson and Francis, 1975). The genetic relatedness between species also has been investigated by oligonucleotide cataloging of their *r*RNA (Tanner et al., 1981, 1982) and this work should provide further insight into a logical reclassification of the genus.

Several species described in the eighth edition of *Bergey's Manual* (Smith and Hobbs, 1974), have been omitted from this edition for the following reasons:

"*C. rubrum.*" The type strain is homologous with the type strain of *C. beijerinckii* (Cummins and Johnson, 1971).

"*C. plagarum.*" The type strain is homologous with the type strain of *C. perfringens* (Nakamura et al., 1976).

"*C. pseudotetanicum.*" Synonymous with *C. butyricum* (Johnson and Francis, 1975; Cato et al., 1982b).

C. lentoputrescens. The type strain is homologous with the type strain of *C. cochlearium* (Nakamura et al., 1979).

C. paraperfringens. Synonymous with *C. baratii* (Cato et al., 1982b).

C. perenne. Synonymous with *C. baratii* (Cato et al., 1982b).

C. cylindrosporum. See discussion in description of *C. acidurici.* Species not included in Approved Lists (Skerman et al., 1980), but has been validly published (Andreesen, et al., 1985, 207).

"*C. brevifaciens.*" Species not included in Approved Lists; no strain available for study.

"*C. malacosomae.*" Species not included in Approved Lists; no strain available for study.

"*C. tetanomorphum.*" Strains of labeled *C. tetanomorphum* have been studied in several laboratories. The species was omitted from the eighth edition of *Bergey's Manual* because neither ATCC 15920 (DSM 528) nor ATCC 8606 have the reactions originally described for the species. These strains are nonproteolytic and asaccharolytic, fermenting only inositol; esculin is hydrolyzed, milk is curdled; butyrate, acetate, propionate and abundant H_2 are produced. For more complete descriptions, see McClung and McCoy (1957) and Gottschalk et al. (1981).

Acknowledgements

We gratefully acknowledge the current or former microbiological or technical assistance of Pauletta C. Atkins, Ruth Z. Beyer, Carole A. Dellinger, Ann P. Donnelly, Luba S. Fabrycky, Barbara A. Harich, Donald E. Hash, Linda K. Hoffman, Jane L. Hungate, Linda L. Long, Leesa M. Miller, Ann C. Ridpath, Carolyn W. Salmon, Gail S. Selph, Debra B. Sinsabaugh, Sue C. Smith, Dianne M. Wall, and Bonnie M. Williams; the support assistance of Linda B. Cook, Julia J. Hylton, Ruth E. McCoy, Virginia P. Saville, Faye M. Scott, Faye D. Smith, Phyllis V. Sparks, Lila A. Turpin and Margaret L. Vaught; and the secretarial assistance of Donya B. Stephens, Mary P. Harvey and Kimi Ishii.

We are especially grateful to W. E. C. Moore for his line drawings of representative cells of the species; to Lillian V. Holdeman Moore and W. E. C. Moore for their critical reviews of the manuscript, their helpful suggestions and their earlier taxonomic studies; to Vera L. Sutter, Diane Citron, Rial Rolfe and Hannah Wexler for their assistance in compiling data; and to Thomas O. MacAdoo, Department of Foreign Languages and Literature, Virgina Polytechnic Institute and State University, for advice concerning the derivation of specific epithets.

For support of research for studies of these species, we appreciate the assistance of project 2022820 from the Commonwealth of Virginia, Public Health Service grant AI-15244 from the National Institute of Allergy and Infectious Diseases, and contracts N01-CP-33334 from the National Cancer Institute and 9-12601 from the National Aeronautics and Space Administration; support for the susceptibility testing was received from the V.A. Medical Research Funds.

Further Reading

Pathogenicity and Incidence in Infections:

Finegold, S.M. 1977. Anaerobic bacteria in human disease, Academic Press, N.Y.

Hentges, D.J. (Editor). 1983. Human Intestinal Microflora in Health and Disease. Academic Press, N.Y.

Smith, L.DS. and B.L. Williams. 1984. The pathogenic anaerobic bacteria, 3rd Ed, Charles C Thomas, Springfield, Ill.

General

Gottschalk, G., J.R. Andreesen and H. Hippe. 1981. The genus *Clostridium* (nonmedical aspects), *In* Starr, Stolp, Trüper, Balows, and Schlegel (Editors). The Prokaryotes: A Handbook on Habitat, Isolation and Identification of Bacteria. Springer-Verlag, Berlin, pp. 1767–1803.

Willis, A.T. 1977. Anaerobic Bacteriology: Clinical and Laboratory Practice, 3rd Ed, Butterworths, London.

The bibliography compiled by L. S. McClung, a subject index to the literature from 1940–1975, is an invaluable aid to the study of any species of *Clostridium* described up to that time. McClung, 1982. The Anaerobic Bacteria, their Activities in Nature and Disease. Marcel Dekker, New York.

For excellent color photographs of colonies and of cells in many species of *Clostridium*, see Mitsuoka, T. 1980. The World of Anaerobic Bacteria; a Color Atlas of Anaerobic Bacteria. Sobunsta (Sobun Press) (Hisamatsu Building, 1-4-5 Enraku-cho, Kanada, Chiyoda-ku), Tokyo.

Key to the presumptive identification of species in the genus **Clostridium**

I. Glucose acid.
  A. Gelatin not hydrolyzed.
    1. No growth on aerobic blood agar plate.
      a. Optimum temperature 37°C.
        b. Lecithinase not produced.
          c. Caproic acid not produced.
            d. Indole not produced.
              e. Butyric acid produced.
                f. Esculin hydrolyzed.
                  g. Sucrose acid.
                    h. Spores subterminal.
                      i. Starch acid.
                        j. Fermentable carbohydrate not required.
                          k. Biotin only vitamin required for growth.
                13. *C. butyricum*
                          kk. No growth without other vitamins.

    10. *C. beijerinckii*
        jj. Fermentable carbohydrate required for growth.
    3. *C. acetobutylicum*
        ii. Starch not acid.
        j. Urease-positive (formic acid produced).
    16. *C. celatum*
        jj. Urease-negative (no formic acid).
    59. *C. rectum*
      hh. Spores terminal.
        i. Mannitol acid.
        j. Lactose acid.
    63. *C. sartagoforme*
        jj. Lactose not acid.
    35. *C. innocuum*
        ii. Mannitol not acid.
    48. *C. paraputrificum*
      gg. Sucrose not acid.
        h. Mannitol acid
        i. Sorbitol acid
    9. *C. barkeri*
        ii. Sorbitol not acid
    35. *C. innocuum*
      hh. Mannitol not acid.
    26. *C. fallax*
    ff. Esculin not hydrolyzed.
      g. Maltose acid.
    49. *C. pasteurianum*
      gg. Maltose not acid.
      h. Galactose acid.
    73. *C. symbiosum*
      hh. Galactose not acid.
    82. *C. tyrobutyricum*
  ee. Butyric acid not produced.
    f. Isovaleric acid produced.
    30. *C. glycolicum*
    ff. Isovaleric acid not produced.
      g. Sucrose acid.
        h. Cells coiled.
        i. Galactose acid.
    22. *C. cocleatum*
        ii. Galactose not acid.
    68. *C. spiroforme*
      hh. Cells not coiled.
        i. Maltose acid.
        j. Glycogen acid.
    20. *C. coccoides*
        jj. Glycogen not acid.
        k. Xylose acid.
        l. Inulin acid.
    46. *C. oroticum*
        ll. Inulin not acid.
    19. *C. clostridioforme*
        kk. Xylose not acid.
        l. Lactose acid
    58. *C. ramosum*
        ll. Lactose not acid.
    23. *C. colinum*
        ii. Maltose not acid.
    43. *C. nexile*
      gg. Sucrose not acid.
      h. Maltose acid.
    17. *C. cellobioparum*
      hh. Maltose not acid.
    47. *C. papyrosolvens*
  dd. Indole produced.
    e. Fermentable carbohydrate required for growth.
    61. *C. saccharolyticum*

        ee. Fermentable carbohydrate not required.
           f. Spores terminal.
              34. *C. indolis*
           ff. Spores subterminal.
              g. Mannitol acid.
              67. *C. sphenoides*
              gg. Mannitol not acid.
              19. *C. clostridioforme*
      cc. Caproic acid produced.
           64. *C. scatologenes*
    bb. Lecithinase produced.
         8. *C. baratii*
  aa. Optimum temperature not 37°C.
    b. Sucrose acid.
      c. Growth at 70°C.
         79. *C. thermohydrosulfuricum*
      cc. No growth at 70°C.
         80. *C. thermosaccharolyticum*
    bb. Sucrose not acid.
      c. Optimum temperature 25°C.
        d. Propionate produced.
          6. *C. arcticum*
        dd. Propionate not produced.
          47. *C. papyrosolvens*
      cc. No growth at 25°C or 37°C.
        d. Utilizes methanol for growth.
          77. *C. thermautotrophicum*
        dd. Cannot utilize methanol for growth.
          76. *C. thermaceticum*
2. Grows on aerobic blood agar plate.
  a. Mannitol acid.
    b. Lactose acid.
        74. *C. tertium*
    bb. Lactose not acid.
         see *Sporolactobacillus inulinus*
  aa. Mannitol not acid.
    b. Butyric acid produced.
        15. *C. carnis*
    bb. Butyric acid not produced.
        25. *C. durum*
B. Gelatin hydrolyzed.
  1. Indole produced.
    a. Lipase produced.
        12. *C. botulinum* type C,D
    aa. Lipase not produced.
      b. Lecithinase produced.
        c. Urease produced.
          66. *C. sordellii*
        cc. Urease not produced.
          d. Formic acid produced.
            11. *C. bifermentans*
          dd. Formic acid not produced.
            e. Toxin neutralized by species-specific antitoxin.
              44. *C. novyi* type B
              31. *C. haemolyticum*
    bb. Lecithinase not produced.
        14. *C. cadaveris*
  2. Indole not produced.
    a. Lactose acid.
      b. Sucrose acid.
        c. Lecithinase produced.
          d. Motile.
            62. *C. sardiniense*
          dd. Nonmotile.
            e. Salicin fermented.
              1. *C. absonum*
            ee. Salicin not fermented.

           50. *C. perfringens*
      cc. Lecithinase not produced.
         d.   Esculin hydrolyzed.
            e.   Starch hydrolyzed.
               f.   Colonies darken when exposed to air.
               7. *C. aurantibutyricum*
               ff. Colony pigmentation stable.
            53. *C. puniceum*
            ee. Starch not hydrolyzed.
               f.   Colonies darken when exposed to air.
            60. *C. roseum*
               ff. Colony pigmentation stable.
            27. *C. felsineum*
         dd. Esculin not hydrolyzed.
            18. *C. chauvoei*
     bb. Sucrose not acid.
           65. *C. septicum*
  aa. Lactose not acid.
     b.   Sucrose acid.
        c.   Mannitol acid.
            81. *C. thermosulfurigenes*
        cc. Mannitol not acid.
         d.   Lecithinase produced.
            40. *C. lituseburense*
         dd. Lecithinase not produced.
            e.   Lipase produced.
            12. *C. botulinum* types B, E, F (non-proteolytic)
            ee. Lipase not produced.
            53. *C. puniceum*
     bb. Sucrose not acid.
        c.   Mannitol acid.
            24. *C. difficile*
        cc. Mannitol not acid.
         d.   Mannose acid.
            e.   Lipase produced.
            12. *C. botulinum* types C,D
            ee. Lipase not produced.
               f.   Lactic acid major product.
            45. *C. oceanicum*
               ff. Lactic acid not major product.
            44. *C. novyi* type B
        dd. Mannose not acid.
         e.   Lecithinase produced.
               f.   Isovaleric and/or isobutyric acid produced.
            69. *C. sporogenes*
               ff. *Iso*-acids not produced.
            44. *C. novyi* type A
         ee. Lecithinase not produced.
            f.   Lipase produced.
               g.   Toxic to mice.
            12. *C. botulinum* types A, B, F (proteolytic)
               gg. Nontoxic to mice.
            69. *C. sporogenes*
            ff. Lipase not produced.
            56. *C. putrificum*
II. Glucose not acid.
  A. Gelatin hydrolyzed.
    1. Indole produced.
      a.   Esculin hydrolyzed.
            29. *C. ghonii*
     aa. Esculin not hydrolyzed.
        b.   Toxic to mice.
            75. *C. tetani*
        bb. Nontoxic to mice.
         c.   Butyric acid major product.
            41. *C. malenominatum*
        cc. Butyric acid not major product.

           42. *C. mangenotii*
2. Indole not produced.
   a. Lipase produced.
      b. Toxic to mice.
         12. *C. botulinum* types A, B, F (proteolytic)
      bb. Nontoxic to mice.
         69. *C. sporogenes*
  aa. Lipase not produced.
     b. Butyric acid produced.
        c. Toxic to mice.
           d. Isovaleric and/or isobutyric acid produced.
              12. *C. botulinum* type G
          dd. *Iso*-acids not produced.
            75. *C. tetani*
        cc. Nontoxic to mice.
           d. Large amount of H_2 produced.
              72. *C. subterminale*
          dd. Little or no H_2 produced.
             e. Motile.
               32. *C. hastiforme*
             ee. Nonmotile.
               83. *C. villosum*
    bb. Butyric acid not produced.
        c. Lecithinase produced.
           39. *C. limosum*
         cc. Lecithinase not produced.
           d. Growth at 37°C.
             e. *Iso*-acids formed.
               36. *C. irregulare*
             ee. *Iso*-acids not formed.
               33. *C. histolyticum*
          dd. No growth at 37°C.
             e. Optimum temperature 60°C.
               78. *C. thermocellum*
             ee. Optimum temperature 22°C.
               55. *C. putrefaciens*
B. Gelatin not hydrolyzed.
  1. Indole produced.
         41. *C. malenominatum*
  2. Indole not produced.
   a. Maltose acid.
         38. *C. leptum*
  aa. Maltose not acid.
     b. Esculin hydrolyzed.
        c. Cellobiose acid.
           51. *C. polysaccharolyticum*
        cc. Cellobiose not acid.
           5. *C. aminovalericum*
    bb. Esculin not hydrolyzed.
        c. Requires ethanol for growth.
           37. *C. kluyveri*
        cc. Ethanol not required for growth.
           d. Propionic acid major product.
             e. Fructose acid.
               57. *C. quercicolum*
             ee. Fructose not acid.
               52. *C. propionicum*
          dd. Propionic acid not major product.
             e. Fructose acid.
               f. Produces acetate from H_2 and CO_2.
               2. *C. aceticum*
               ff. Acetate not produced from H_2 and CO_2.
               28. *C. formicaceticum*
             ee. Fructose not acid.
               f. Motile.
                g. Butyric acid major product.
                  h. Isovaleric acid produced.

71. *C. sticklandii*
    hh. Isovaleric not produced.
21. *C. cochlearium*
    gg. Butyric acid not produced.
      h. Formic acid produced.
54. *C. purinilyticum*
    hh. Formic acid not produced.
 4. *C. acidurici*
ff. Nonmotile.
70. *C. sporosphaeroides*

List of species of the genus **Clostridium**

1. **Clostridium absonum** Nakamura, Shimamura, Hayase and Nishida 1973, 426.[AL]

ab′so.num. L. adj. *absonus* inharmonious, not corresponding with; intended to mean "deviating from *C. perfringens*."

This description, unless otherwise indicated, is based on study of the type strain, two other strains having a high degree of DNA homology with the type (Nakamura et al., 1973), and other phenotypically similar strains, as well as on the description by Nakamura et al. (1973) and Hayase et al. (1974).

Cells in PY broth cultures are reported to be nonmotile (Nakamura et al., 1973), but we have found that cells of vigorously growing young cultures of some strains are motile with a single subpolar flagellum. This observation also was made by Stanley M. Harmon (personal communication). Cells in PYG broth cultures are Gram-positive and 0.9–1.7 × 1.7–11.8 μm.

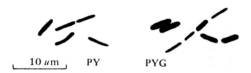

Figure 13.5. *Clostridium absonum.*

Spores are oval, subterminal, and do not swell the cell. Sporulation occurs most readily on egg-yolk agar incubated for 2 days. Spores are seldom seen in broth cultures although the organisms from such cultures survive heating at 80°C for 10 min.

Cell walls contain *meso*-DAP (Weiss et al., 1981).

Surface colonies on blood agar are 1–3 mm, circular to scalloped, slightly opaque, convex, grayish, shiny, smooth, and β-hemolytic.

Glucose broth cultures are turbid with a smooth to mucoid sediment and have a pH of 4.5–4.8 after incubation for 5 days.

Growth is equally abundant at 30, 37 and 45°C, and somewhat less at 25°C. Growth is inhibited by 6.5% NaCl or 20% bile.

Abundant gas is detected in PYG agar deep cultures.

Gelatin is slowly digested; there is no digestion of milk or chopped meat.

Some strains of the species produce hydroxysteroid dehydrogenases that are active in bile acid metabolism (Macdonald et al., 1981, 1983c; Macdonald and Roach, 1981; Sutherland and Macdonald, 1982; Macdonald and Hutchison, 1982).

Products in PYG broth cultures are butyric and acetic acids; lactic and formic acids may be present; ethanol and butanol may be detected. Abundant H_2 is produced. Pyruvate is converted to butyrate and acetate; lactate is not converted to propionate; threonine is not utilized. Ammonia is produced. Neutral red and resazurin are reduced.

All strains tested are susceptible to penicillin G, tetracycline, and chloramphenicol; the type strain is resistant to erythromycin, and four strains including the type are resistant to clindamycin.

Culture supernatants injected intraperitoneally are not toxic to mice; toxin is weakly produced by all strains (Nakamura et al., 1973). Intravenous injection of culture filtrates is lethal to mice (Nakamura et al., 1973; Hayase et al., 1974).

Other characteristics of the species are given in Table 13.12.

Isolated primarily from soil. Similar strains have been isolated from bear feces; one strain is from a case of gas gangrene in a human wound contaminated with soil (Nakamura et al., 1979).

The mol% G + C of the DNA of the type strain has not been reported.

Type strain: ATCC 27555 (DSM 599).

Further comments. C. absonum is most easily differentiated from *C. baratii* by its hydrolysis of gelatin, and from *C. perfringens* by not producing H_2S and by its relative lack of toxicity for mice. The lecithinase of *C. absonum* produced on half-antitoxin Nagler agar cannot be neutralized completely by *C. perfringens* type A antitoxin as can the lecithinases of *C. baratii* and *C. perfringens* (Nakamura et al., 1973).

2. **Clostridium aceticum** (*ex* Wieringa 1940) Gottschalk and Braun 1981, 476.[VP]

a.ce′ti.cum. L. n. *acetum* vinegar; L. adj. suff. *-icus* belonging to; N.L. neut. adj. *aceticum* related to acetic acid, which it produces.

This description is based on the descriptions by Adamse (1980), Braun et al., (1981), and Gottschalk and Braun (1981).

Cells in broth cultures in an atmosphere of 67% H_2 and 33% CO_2 as described by Braun et al. (1981), are Gram-negative rods, motile and peritrichous, and 0.3–1.0 × 4.0–8.0 μm. With fructose as a substrate, cells may be up to 40 μm in length.

Figure 13.6. *Clostridium aceticum.*

Spores are round, terminal and swell the cell. Sporulation occurs most readily on fructose agar medium incubated for 2 days.

Cell walls contain *meso*-DAP and are composed of at least two layers (Braun et al., 1981).

Colonies do not form readily. After prolonged incubation in roll tubes in an agar medium containing mud extract and a CO_2-H_2 (1:2) atmosphere (Adamse, 1980), "a barely visible, light-brown tuft of cellular material" can be seen.

The optimum temperature for growth is 30°C. Growth occurs between 25°C and 37°C, but is poor at 45°C. Optimum pH for autotrophic growth is 8.3; growth occurs between pH 7.5 and 9.5. Strains of this species grow chemolithotropically in an atmosphere of CO_2 and H_2, converting these substrates to acetate. They also utilize the organic

Table 13.12.
Characteristics of species in the genus **Clostridium.** *Acid from glucose, gelatin hydrolyzed.*[a]

Characteristics	1. C. absonum	7. C. auranti-butyricum	11. C. bifer-mentans	12. C. botulinum			14. C. cadav-eris	18. C. chauvoei	24. C. difficile	27. C. fel-sineum	31. C. haemo-lyticum	40. C. litus-eburense
				Types C, D	Types B, E, F (saccharo-lytic)	Types A, B, F (proteo-lytic)						
Products from PYG[b]	BAL (f2,4)	LBA	AF(iv icpibbls2)	BPA(vls)	BA(l)	ABiVib (icvp2,3,4)	BA2,4 (fpls)	ABF(ls)4	BAicivib (fvl2,4)	AB4 (lsf)	APB(s)	BAiV pfib (2,3,i4)
Motility	∓	+	+	±	+	±	±	d	±	d	+	+
H₂ produced	4	4	4	4	4	4	4	4	4	4	4	−
Indole produced	−	−	+	∓	−	−	+	−	−	−	+	−
Lecithinase produced	+	−	+	∓	−	−	−	−	−	−	+	+
Lipase produced	−	+	−	+	+	+	−	−	−	−	−	−
Esculin hydrolyzed	+	+	±	−	−	+	−	+	+	+	−	−
Starch hydrolyzed	∓	+	−	−	±	−	−	−	−	∓	−	−
Nitrate reduced	+	+	−	−	−	−	−	±	−	−	−	−
Acid produced from[c]												
Amygdalin	−w	−	−	−	−w	−	−	−	−	−	−	−
Arabinose	−	w	−	−	−	−	−	−	−	+w	−	−
Cellobiose	+	+	−	−	−	−	−	−	+w	+w	−	−
Fructose	+	+	d	w−	+w	−w	d	−w	+	+	d	+
Galactose	+	w	−	−w	∓	−	−	w+	−	+	−w	−
Glycogen	−	w	−	−	−w	−	−	−	−	−	−	−
Inositol	−	−	−	±	−w	−	−	−	−	−	d	−
Inulin	−	−	−	−	−w	−	−	−	−	d	−	−
Lactose	+	+	−	−	−	−	−	+w	−	+w	−	−
Maltose	+	+	w−	d	+w	−w	−	+w	−	∓	−w	+
Mannitol	−	−	−	−	−	−	−	−	±	−	−	−
Mannose	+	w	−w	d	+w	−	−w	+w	±	+	d	+w
Melezitose	−	−	−	−	d	−	−	−	d	−	−w	−
Melibiose	∓	w	−	−w	−	−	−	−	−	−	w−	−
Raffinose	−	+	−	−	−	−	−	−	−	∓	−w	−
Rhamnose	−	−	−	−	−	−	−	−	−	+w	−w	−
Ribose	+w	−	−	d	d	−	−	d	−	−	d	−w
Salicin	+w	w	−	−	−	−	−	−	−w	+	−	−
Sorbitol	−	−	−w	−	±	−w	−	−	−w	−	−	−
Starch	−w	w	−	−	d	−	−	−	−	∓	−	−
Sucrose	+	+	−	−	+w	−	−	+w	−	+	−	+w
Trehalose	w+	−	−	−	w+	−	−	−	−w	−	−w	−
Xylose	−	−	−	−	−	−	−	−	−w	+	−	−
Milk reaction	c	c	d	dc	c−	d	cd	c	−	c	cd	cd
Meat digested	−	−	+	±	−	+	+	−	−	−	∓	+

substrates fructose, ribose, glutamate, fumarate, malate, pyruvate, serine, formate, ethylene glycol, and ethanol, but in the presence of organic substrates CO_2 and H_2 are not converted to acetate (Braun and Gottschalk, 1981). Dulcitol, adonitol, citrate, succinate, glycine, threonine, lactate, methanol, isopropanol and glycerol are not utilized (Braun et al., 1981). Atmospheric N_2 is fixed (Rosenblum and Wilson, 1949).

H_2 is produced only in the stationary growth phase and inhibits growth in fructose medium at pH 8.5 if the bicarbonate concentration is very low (Braun and Gottschalk, 1981).

Gluconate is fermented to pyruvate and glyceraldehyde-3-phosphate by a modified Entner-Doudoroff pathway (Andreesen and Gottschalk, 1969).

Other characteristics of the species are given in Table 13.13.

Isolated from soil, lake sediment, and sewage sludge.

The mol% G + C of the DNA is 33 (T_m) (Braun et al., 1981).

Type strain: DSM 1496 (ATCC 35044).

Further comments. The species is most easily differentiated from *C.*

formicaceticum, which it most closely resembles phenotypically, by its ability to form acetate from CO_2 and H_2, and to utilize formate, serine or ethylene glycol, but not methanol as substrates. Also see Further Comments under *C. thermaceticum*.

3. **Clostridium acetobutylicum** McCoy, Fred, Peterson and Hastings 1926, 483.[AL]

a.ce.to.bu.ty′li.cum. English n. *acetone*; N.L. adj. *butylicum* butylic; N.L. neut. adj. *acetobutylicum* referring to production of acetone and butyl alcohol.

This description is based on the description by Smith and Hobbs (1974), Holdeman et al. (1977), and on study of the type and eight other strains.

Cells in PYG broth cultures are straight rods, motile and peritrichous, $0.5-0.9 \times 1.6-6.4$ μm. Granulose (a starch-like polymer) is often present (O'Brien and Morris, 1971). Gram-positive, becoming Gram-negative in older cultures.

Table 13.12.—*continued*

Characteristics	44. C. novyi types A	B	C	45. C. ocean-icum	50. C. per-fringens	53. C. puniceum	56. C. putri-ficum	60. C. roseum	62. C. sardin-iense	65. C. septicum	66. C. sordellii	69. C. sporo-genes	81. C. thermosul-furigenes
Products from PYG[b]	ABP	PBA	PBaf	ALb (fics pibiv2,3,4)	ABL (pfs)	AB4(lf)	ABibiv2 (fpicvls)	BAs4	BAL(Fp)	BA(Fpl2)	A(FiCp ibivl)	ABivib2 (picvls4)	2AL
Motility	±	±	+	±	−	+	+	∓	+	±	±	±	+
H₂ produced	4	d	1	4	4	4	4	4	4	4	4	4	4
Indole produced	−	±	+	−	−	−	−	−	−	−	+	−	−
Lecithinase produced	+	+	−	∓	+	−	−	−	+	−	+	−	NT
Lipase produced	+	−	−	−	−	−	−	−	−	−	−	+	NT
Esculin hydrolyzed	−	−	−	+	d	+	d	+	+	+	∓	+	NT
Starch hydrolyzed	−	−	−	d	±	+	−	−	d	−	−	−	+
Nitrate reduced	−	−	−	−	±	−	−	−	±	d	−	−	−
Acid produced from													
Amygdalin	−	−	−	−	−w	−w	−	−	−	−	−	−	+
Arabinose	−	−	−	−	−	−w	−	+	−	−	−	−	+
Cellobiose	−	−	−	d	∓	d	−	+	+w	+w	−	−	+
Fructose	−w	d	−	w+	+	w+	−w	+	+	+	d	−w	NT
Galactose	d	−	−	d	+w	+	−	+	+	+w	−	−	+
Glycogen	−	−	−	−	d	−	−	−	−	−	−	−	NT
Inositol	±	±	w	−	±	−	−	−	−	−	−	−	+
Inulin	−	−	−	−	−w	NT	−	−	−	−	−	−	NT
Lactose	−	−	−	−	+	∓	−	+	+w	+	−	−	−
Maltose	d	d	−	+w	+	+w	−w	−	+w	+	w+	−w	+
Mannitol	−	−	−	−	−	−	−	−	−	−	−	−	+
Mannose	−	w+	w	+w	+	d	−	+	+	+	−w	−	+
Melezitose	−	−	−	−	−	−	−	−	−	−	−	−	−
Melibiose	−	−	−	−	−w	−w	−	−	−	−	−	−	−
Raffinose	−	−	−	−	d	−w	−	−	−	−	−	−	−
Rhamnose	−	−	−	−	−	−	−	+	−	−	−	−	+
Ribose	d	−w	−	−	d	−w	−	−	d	d	−w	−	−
Salicin	∓	−	−	−w	−	±	−	+	+w	d	−	−	+
Sorbitol	−	−	−	−	∓	−	−	−	−	−	−	−	−
Starch	−	−	−	−	d	d	−	+	−w	−	−	−	+
Sucrose	−	−	−	−	+	+	−	+	+w	−	−	−	+
Trehalose	−	−	−	−w	d	∓	−	−	d	+w	−	−	+
Xylose	−	−	−	−	−	d	−	+	−	−	−	−	+
Milk reaction	c	d	−	d−	dc	c	d	cd	c	cd	d	d	NT
Meat digested	−	+	+	+	±	−	±	−	−	−	+	+	NT

[a] Symbols: +, reaction positive for 90–100% of strains (pH of sugars below 5.5); −, reaction negative for 90–100% of strains; ±, 61–89% of strains positive; ∓, 11–39% of strains positive; d, 40–60% of strains positive; w, weak reaction (pH of sugars 5.5–5.9); numbers (hydrogen) represent abundant (4) to negative on a "−" to "4+" scale; c (milk), curd; d (milk), digestion; NT, not tested. Where two reactions are listed, the first is the more usual and occurs in 60–90% of strains.

[b] Products (listed in the order of amounts usually detected): a, acetic acid; b, butyric acid; c, caproic acid; l, lactic acid; f, formic acid; p, propionic acid; s, succinic acid; v, valeric acid; ib, isobutyric acid; ic, isocaproic acid; iv, isovaleric acid; 2, ethanol; 3, propanol; 4, butanol; i4, isobutanol. Capital letters indicate at least 1 meq/100 ml of culture; small letters indicate less than 1 meq/100 ml. Products in parentheses are not detected uniformly.

[c] Ammonia produced from hydrolysis of peptone may mask acid production in this group.

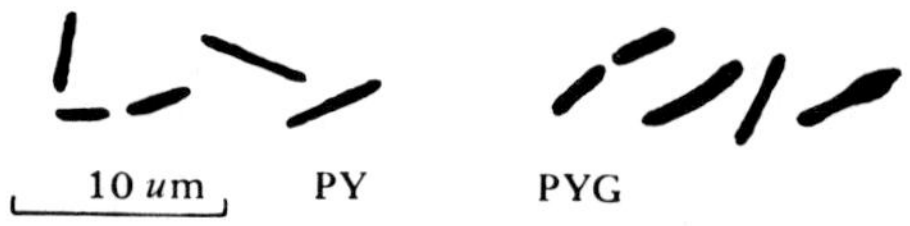

Figure 13.7. *Clostridium acetobutylicum.*

Spores are oval and subterminal, slightly swelling the cell.

Cell walls contain *meso*-DAP, and glucose, rhamnose, galactose and mannose (Cummins and Johnson, 1971). The wall is triple-layered (Cho and Doy, 1973).

Surface colonies on blood agar plates are 1–5 mm, flat to raised, granular, translucent to semiopaque with irregular margins and occasionally with a mosaic internal structure.

Cultures in PYG broth are turbid with a smooth sediment, and have a pH of 4.5–5.0 after incubation for 5 days.

The optimum temperature for growth is 37°C. A fermentable carbohydrate, biotin, and *p*-aminobenzoic acid are required. No growth in the presence of 6.5% NaCl or 20% bile.

Acetyl methyl carbinol is produced. Neutral red is reduced.

Abundant gas is produced in glucose agar deep cultures.

H₂S is produced by one of nine strains tested.

Fixes atmospheric N₂ (Rosenblum and Wilson, 1949).

Strains produce an inducible carboxymethyl cellulase and cellobiase (Allcock and Woods, 1981). NADH and NADPH-ferredoxin and rubredoxin oxidoreductases also are present (Petitdemange et al., 1977, 1981). Superoxide dismutase (Hewitt and Morris, 1975) and deoxyribonuclease (Johnson and Francis, 1975) are produced.

Table 13.13.
*Characteristics of species in the genus **Clostridium**. Gelatin not hydrolyzed, no acid from glucose.*[a,b]

Characteristics	2. C. aceticum	4. C. acidurici	5. C. aminovalericum	21. C. cochlearium	28. C. formicaceticum	37. C. kluyveri	38. C. leptum	41. C. malenominatum	51. C. polysaccharolyticum	52. C. propionicum	54. C. purinilyticum	57. C. quercicolum	70. C. sporosphaeroides	71. C. sticklandii
Products from PYG[c]	A	Af	Af	Bap(fls4)	A(Fs)	CBa	A(2)	BA(fpls)	fabp2	PiVibb as(1)	FA	APb3	ABp	Aivbpib
Motility	+	+	+	±	+	+	−	±	+	+	+	+	−	+
H₂ produced	−	−	4	4	−	2	4	4	3	4	−	4	4	1
Indole produced	NT	−	−	∓	−	−	−	+	−	−	−	−	−	−
Esculin hydrolyzed	−	−	+	−	−	−	±	−	+	−	−	−	−	−
Starch hydrolyzed	NT	−	+	−	−	−	−	−	+	−	−	−	−	−
Nitrate reduced	NT	−	−	−	−	−	−	∓	−	−	−	−	−	−
Acid produced from														
Amygdalin	−	−	−	−	−	−	−w	−	−	−	−	−	−	−
Cellobiose	NT	−	−	−	−	−	−	−	w	−	−	−	−	−
Fructose	+	−	−	−	w	−	−w	∓	−	−	−	+	−	−
Glycogen	NT	−	−	−	−	−	−w	−	w	−	−	−	−	−
Lactose	−	−	−	−	−	−	∓	−	−	−	−	−	−	−
Maltose	−	−	−	−	−	−	+	∓	−	−	−	−	−	−
Ribose	+	−	−	−	−	−	w−	−	−	−	−	−	−	−
Starch	−	−	−	−	−	−	−	−	w	−	−	−	−	−
Sucrose	−	−	−	−	−	−	±	−	−	−	−	−	−	−
Trehalose	NT	−	−	−	−	−	w−	−	−	−	−	−	−	−
Xylose	−	−	−	−	−	−	w−	−	−	−	−	−	−	−

[a] Strains in this group do not produce lecithinase or lipase; acid is not produced from arabinose, galactose, inositol, inulin, mannitol, mannose, melezitose, melibiose, raffinose, rhamnose, salicin or sorbitol; there is no reaction in milk; meat is not digested.
[b] Symbols: +, reaction positive for 90–100% of strains (pH of sugars below 5.5); −, reaction negative for 90–100% of strains; ±, 50–89% of strains positive; ∓, 11–50% of strains positive; w, weak reaction (pH of sugars 5.5–5.9); numbers (hydrogen) represent abundant (4) to negative on a "−" to "4+" scale; NT, not tested. Where two reactions are listed, the first is the more usual and occurs in 60–90% of strains.
[c] Products (listed in order of amounts usually detected): a, acetic acid; b, butyric acid; f, formic acid; c, caproic acid; l, lactic acid; p, propionic acid; s, succinic acid; iv, isovaleric acid; ib, isobutyric acid; 2, ethanol; 3, propanol; 4, butanol. Capital letters indicate at least 1 meq/100 ml of culture; small letters indicate less than 1 meq/100 ml. Products in parentheses are not detected uniformly.

Fermentation products include acetic, butyric and lactic acids, butanol, acetone, CO₂ and large amounts of H₂. Small amounts of succinic acid may be formed. Ethanol was detected with high pressure liquid chromatography (Ehrlich et al., 1981). During exponential growth, products are acetate and butyrate. Production of butanol and acetone is highest after 18 h when the organisms are in their stationary growth phase and is associated with morphological changes in the cells (Jones et al., 1982).

Pyruvate is converted to acetate, butyrate and butanol. Neither lactate nor threonine is utilized.

The type strain produces the amino acids lysine, arginine, aspartic acid, threonine, serine, glutamic acid, alanine, valine, isoleucine, leucine and tyrosine in broth (Matteuzzi et al., 1978).

For information on stimulation of solvent production by *C. acetobutylicum*, see reports of Bahl et al. (1982a, b), Andersch et al. (1982), Monot et al. (1982), Jones et al. (1982), George et al. (1983), and Lin and Blaschek (1983).

Strains are susceptible to chloramphenicol, clindamycin, erythromycin, penicillin G, and tetracycline.

Culture supernatants are nontoxic to mice.

Other characteristics of the species are listed in Table 13.14.

Isolated from soil. Also reported from lake sediment, well water, and clam gut (Ehrlich et al., 1981), from bovine feces (Princewill and Agba, 1982), canine feces (Balish et al., 1977), and human feces (Drasar et al., 1976; Finegold et al., 1983).

The mol% G + C of the DNA is 28–29 (T_m) (Cummins and Johnson, 1971).

Type strain: ATCC 824 (NCIB 8052, DSM 792).

Further comments. This species is most easily differentiated from *C. beijerinckii* by its absolute requirement for a fermentable carbohydrate for growth, and by requirement for biotin and *p*-aminobenzoic acid; *C. beijerinckii* requires other vitamins and amino acids as well. *C. acetobutylicum* differs from *C. aurantibutyricum* and *C. felsineum* by its failure to digest gelatin.

4. **Clostridium acidurici** (corrig.) (Liebert, 1909) Barker 1938, 323.[AL]

a.ci.du′ri.ci. N.L. n. *acidum uricum* uric acid; N.L.gen.n. *acidurici* of uric acid.

Based on the description by Smith and Hobbs (1974) and study of the type strain.

Cells in PY-0.3% uric acid broth are Gram-variable to Gram-negative, motile and peritrichous, 0.5–0.7 × 2.5–4.0 μm, occurring singly.

10 μm   PY   Uric acid

Figure 13.8. *Clostridium acidurici.*

Spores are oval, terminal and subterminal, and swell the cell. Sporulation occurs most reliably on chopped meat-uric acid agar slants incubated at 30°C for 1 week in an atmosphere of N₂.

Cell walls contain *meso*-DAP (Weiss et al., 1981).

Surface colonies on uric acid agar are spreading, rhizoid, transparent, colorless and flat, clearing the agar.

No growth in gelatin or milk, or on egg-yolk or blood agar.

Table 13.14.

Characteristics of species in the genus **Clostridium.** *Acid from glucose, gelatin not hydrolyzed, meat not digested.*[a]

Characteristics	3. C. aceto-butylicum	6. C. arcticum	8. C. baratii	9. C. barkeri	10. C. beijer-inckii	13. C. butyricum	15. C. carnis	16. C. celatum	17. C. cello-bioparum	19. C. clostrid-ioforme	20. C. coccoides	22. C. coclea-tum	23. C. colinum
Products from PYG[b]	BAl4(s)	PA(b)	BAL (fps)	BLpa	BA (Fpls,2,4)	BAF (ls2,4)	BALf(s)	AFb2(s)	Alf2	A(Fls2)	AS	AF(Ls)	FAp(l)
Motility	±	+	−	−	+	±	±	−	+	∓	−	−	+
H₂ produced	4	tr	4	4	4	4	4	4	4	4	4	4	4
Indole produced	−	+	−	−	−	−	−	−	−	∓	−	−	−
Lecithinase produced	−	NT	+	−	−	−	−	−	−	−	−	−	−
Lipase produced	−	NT	−	−	−	−	−	−	−	−	−	−	−
Esculin hydrolyzed	+	+	+	+	+	+	+	+	+	+	+	+	+
Starch hydrolyzed	+	−	d	−	±	+	−	−	−	−	−	−	−
Nitrate reduced	−	−	d	−	−	−	−	d	−	±	−	−	−
Acid produced from													
Amygdalin	−	−	−	−	∓	±	w	+	−	−w	+	±	w−
Arabinose	d	−	−	−	±	±	−	−	+	d	+	−	−
Cellobiose	+	w	+	−	+	+	w+	+	+	±	+	+	d
Fructose	+	+	+	+	+	+	d	+	+	+	+	+	+
Galactose	+	NT	+w	−	+	+	d	+	+	w+	+	+	w
Glycogen	d	−	∓	−	d	+	−	−	−	−	+	−	−
Inulin	−	NT	−	−	d	∓	−	−	−	−	−	+w	d
Lactose	d	−	+w	−	+	+	d	+	w	±	+	+	−
Maltose	+	−	+w	−	+	+	w+	+	+	+w	+	∓	+
Mannitol	∓	w	−	+	d	∓	−	−	−	−	+	−	d
Mannose	+	+	+	−	+	+	+w	+	+	+w	+	+	+
Melezitose	−	−	−	−	∓	∓	−	−	−	∓	−	−	−
Melibiose	−	−	−w	−	d	+	−	−	+	d	+	−	+w
Raffinose	−	−	−	−	±	+	−	−	−	±	+	∓	+
Rhamnose	−	−	−	−	∓	−	−	−	−	±	+	−	−
Ribose	−	−	−w	−	∓	+	−	d	+	d	+	−	w−
Salicin	+	w	+w	−	∓	+	w	+	w	±	+	d	w+
Sorbitol	−	−	−	+	∓	−	−	−	−	−	+	−	−
Starch	+	−	d	−	±	+	−w	−	−	−w	−	−	−w
Sucrose	+	−	+	−	+	+	+w	+	−	+	+	+	+
Trehalose	∓	−	−	−	±	+	−	+	−	±	+	+	+
Xylose	±	+	−	−	+	+	−	−	+	+	+	−	−
Milk reaction	c	a	c	−	c	c	−	c	−	c	c	c	−

Broth cultures supplemented with 0.3% uric acid have a smooth sediment with no turbidity and a pH of 7.4–7.7 after incubation under N₂ for 6 days. Most rapid growth occurs in media at an initial pH of 7.6–8.1; there is poor growth below pH 6.5 or above 9.0. Growth occurs between 19°C and 37°C (Barker and Beck, 1942). Uric acid, xanthine, guanine or hypoxanthine is required as a carbon and energy source (Barker and Beck, 1941). Selenite and tungstate stimulate xanthine dehydrogenase and formate dehydrogenase activity (Wagner and Andreesen, 1979).

Products in PY-urate broth are acetate, NH₃, and CO₂. No H₂ is produced. No carbohydrates are fermented.

The type strain is resistant to erythromycin, penicillin and tetracycline. It is moderately sensitive to chloramphenicol and clindamycin.

Culture supernatants of the type strain are nontoxic to mice.

Other characteristics of the species are given in Table 13.13.

Strains are widely distributed in soil. They have been isolated also from chicken droppings (Barker, 1978), and from wild birds (Barker and Beck, 1942).

The mol% of the G + C is 28 (T_m) (Dürre et al., 1981).

Type strain: ATCC 7906 (DSM 604).

Further comments. For reviews concerning the metabolic degradation of purines by this organism see Vogels and Van der Drift (1976) and Yoch and Carithers (1979). See also Champion and Rabinowitz (1977), Wagner and Andreesen (1977), Waber and Wood (1979), and Dürre and Andreesen (1983).

C. cylindrosporum was isolated at the same time and from the same sample as *C. acidurici*. *C. cylindrosporum* has been validated as a species (Andreesen et al., 1985, 207) and has been studied widely in conjunction with studies on *C. acidurici* (Barker and Beck, 1941, 1942; Champion and Rabinowitz, 1977; Wagner and Andreesen, 1977, 1979; Tanner et al., 1982). Metabolically the organisms are very similar; both ferment uric acid, neither ferments any carbohydrate. The mol% G + C of *C. cylindrosporum*, however, is 32 by T_m (Tonomura et al., 1965), and a partial catalog of the 16S-rRNA of *C. cylindrosporum* indicates that it is not closely related to *C. acidurici* (Tanner et al., 1982). Originally these organisms were differentiated on the basis of cell morphology and the size, shape and position of their spores (Barker and Beck, 1942). Champion and Rabinowitz (1977) have found that *C. cylindrosporum* forms formate from uric acid while formate is not produced by *C. acidurici*. Wagner and Andreesen (1977, 1979) have found that the two organisms have different metal ion requirements for formation of formate dehydrogenases and different requirements for hypoxanthine metabolism.

The mol% of the G + C is 28 (T_m) (Dürre et al., 1981).

C. cylindrosporum type strain: ATCC 7905 (DSM 605).

See Further Comments under *C. purinilyticum* for differentiation of these three species.

5. **Clostridium aminovalericum** Hardman and Stadtman 1960, 552.[AL]

Table 13.14—*continued*

Characteristics	25. C. durum	26. C. fallax	30. C. glycolicum	34. C. indolis	35. C. innocuum	43. C. nexile	46. C. oroticum	47. C. papyro-solvens	48. C. paraput-rificum	49. C. pasteur-ianum	58. C. ramosum	59. C. rectum
Products from PYG[b]	laf2	ABL(s)	AiViB2,3, i4i5(pfls)	AF2	BLa(fs)	AF2(ls)	AF2(ls)	AL2	BAL(sf)	ABf(ls)	FAl(s2)	Bapv
Motility	+	∓	±	+	−	−	−	+	±	∓	−	−
H₂ produced	4	4	4	4	4	4	4	4	4	4	d	4
Indole produced	−	−	−	+	−	−	−	−	−	−	−	−
Lecithinase produced	−	−	−	−	−	−	−	−	−	−	−	−
Lipase produced	−	−	−	−	−	−	−	−	−	−	−	−
Esculin hydrolyzed	+	+	∓	+	+	+	+	+	+	−	+	+
Starch hydrolyzed	−	−	−	±	−	−	−	−	+	−	−	−
Nitrate reduced	−	∓	−	±	−	−	±	−	∓	−	−	−
Acid produced from												
Amygdalin	−	−	−	−	−	−w	−	−	d	−	+	
Arabinose	−	−	−	−w	−	−	+	+	−	w	−	
Cellobiose	−	−w	−	+w	+	−	+	+	−	−	+	
Fructose	+	+	+w	+w	+	w+	+	±	+	+	+	−
Galactose	+	w+	−	w+	+	+w	+	+	+	w−	+	w
Glycogen	−	−	−	−	−	−	−	−	−	±	−	
Inulin	−	−	−	−	+	d	+w	−	−	−	−	
Lactose	−	w−	−	w+	−	+w	+	−	+	−w	+	w
Maltose	+	+	d	w+	−	−w	+	−	+	+	+	−
Mannitol	−	−	−	−w	+	−	±	−	−	+	±	−
Mannose	+	+w	−	d	+	−w	−	−	+	+	+	−
Melezitose	w	−	−	−	−	−	±	−	−	+	−	
Melibiose	−	−	−	−w	−	w−	−	−	−	+	±	
Raffinose	+	−	−	+w	−w	d	+	−	−	+	+	−
Rhamnose	−	−	−	∓	−	−	+	−	−	−	d	
Ribose	−	w+	−	−w	±	−	+	+	w−	−	d	
Salicin	−	−	−	w−	+	±	+	−	+	−	+	w
Sorbitol	−	−	±	−	−	−	−w	−	−	+	−	
Starch	−	w+	−w	−	−	−w	−	−	+	−w	∓	−
Sucrose	+	−	−	d	+	+w	+	−	+	+	+	w
Trehalose	+	−	−	w−	+	−w	±	−	∓	+	+	−
Xylose	−	−	±	d	−w	w+	+	+	−	w	−w	−
Milk reaction	−	c	−	c	−	−c	c	−	c	−	c	c

a.mi′no.va.ler′i.cum. L. adj. suff. *-icus* related to; N.L. neut. adj. *aminovalericum* referring to ability to ferment aminovaleric acid strongly.

Description based on those of Smith and Hobbs (1974) and Holdeman et al. (1977), and on study of the type strain (received from three different sources).

Cells in PYG broth cultures are straight rods, motile and peritrichous, 0.3–0.5 × 1.5–5.2 μm, occurring singly and in pairs. Cells stain Gram-positive but rapidly become Gram-negative as cultures reach maximum stationary phase.

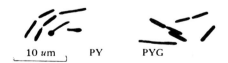

Figure 13.9. *Clostridium aminovalericum.*

Spores are small, spherical and terminal, swelling the cell. The sporulation occurs most reliably on chopped meat agar slants incubated at 30°C for 5 days.

Cell walls contain *meso*-DAP (Weiss et al., 1981).

Surface colonies on blood agar plates are 0.5–1 mm, circular, entire, flat to convex, translucent to opaque, granular, gray, dull, smooth and weakly hemolytic.

Cultures in PYG broth are turbid with a smooth sediment and have a pH of 5.9–6.2 after incubation for 1 week.

Optimum temperature for growth is 37°C; grows at 25°C and 30°C; poor growth at 45°C. Growth is inhibited by 6.5% NaCl and by 20% bile.

Hippurate is hydrolyzed by the type strain. Neutral red and resazurin are reduced.

Abundant gas is produced in PYG agar deep cultures.

Deoxyribonuclease is present (Smith 1975a).

Products in PYG broth at a pH of 6.1 include major amounts of acetic acid and abundant H₂. From aminovalerate as the sole energy source, at a pH of 7.4–7.7, acetate, ammonia, propionate and valerate are produced (Hardman and Stadtman, 1960). Phenylacetic acid has been detected in the one strain tested (Mayrand and Bourgeau, 1982).

Culture supernatants of the type strain are nontoxic to mice.

The type strain is sensitive to chloramphenicol, erythromycin, penicillin G and tetracycline. Resistance to clindamycin is variable.

Other characteristics of the species are given in Table 13.13.

Isolated from sewage sludge and from rumen contents of bloating calves (Jayne-Williams, 1979); also isolated from urine specimens from pregnant women with bacteriuria (Meijer-Severs et al., 1979), from hamster feces (Bartlett et al., 1978) and from human feces (Drasar et al., 1976; Finegold et al., 1983).

Table 13.14.—*continued*

Characteristics	61. C. saccharolyticum	63. C. sartagoforme	64. C. scatologenes	67. C. sphenoides	68. C. spiroforme	73. C. symbiosum	74. C. tertium	76. C. thermaceticum	77. C. thermautotrophicum[c]	79. C. thermohydrosulfuricum	80. C. thermosaccharolyticum	82. C. tyrobutyricum
Products from PYG[b]	Afl2	BAF(L)	AB (cfp ibivvls)	AF(ls2)	AFl(s2)	ABL(f2,4)	ABL(fs2)	A	A	AL2 (fBivic)	ABL	B(Asflp)
Motility	−	d	+	+	−	±	+	−	+	+	±	±
H$_2$ produced	4	4	4	4	4	4	4	2	4	+	4	4
Indole produced	+	−	∓	+	−	−	−	−	−	−	−	−
Lecithinase produced	−	−	−	−	−	−	−	NT	NT	NT	NT	−
Lipase produced	−	−	−	−	−	−	−	NT	NT	NT	NT	−
Esculin hydrolyzed	+	+	∓	+	∓	−	+	−	−	NT	+	−
Starch hydrolyzed	+	∓	−	d	−	−	d	−	−	NT	+	−
Nitrate reduced	+	∓	−	±	−	−	±	+	+	−	−	∓
Acid produced from												
Amygdalin	−	±	−	−	−w	−	d	−	−	d	+	−
Arabinose	+	−	−w	−w	−	d	−	−	+	±	+	−
Cellobiose	w	+	−w	+	∓	−	+	−	−	+	+	−
Fructose	+	+	+	+	+	+	+	+	+	±	+	+
Galactose	−	+	−	w+	−	+w	+	w	±	+	+	−
Glycogen	−	±	−	−w	−	−	+w	−	−	d	+	−
Inulin	−	−	−	−	+	−	−w	−	NT	−	−	−
Lactose	w	+	−	w+	+w	∓	+	−	−	d	+	−
Maltose	+w	±	−	+	−	−	+	−	−	+	+	−
Mannitol	w	+	−	w+	−	d	+w	−	−	d	−	−w
Mannose	+	+	+w	w+	+	d	+	−	−w	+	+	+w
Melezitose	w	∓	−	−w	−	−	d	−	−	NT	+	−
Melibiose	w	+	−	d	−	−	+w	−	−	NT	+	−
Raffinose	+	∓	−	+w	−	−	∓	−	−	d	+	−
Rhamnose	+	∓	d	+w	−	−	−w	−	+	∓	−	−
Ribose	−	d	−w	−w	−	−	+w	−	±	+	+	−
Salicin	w	+	−w	w+	∓	−	+	−	−	+	+	−
Sorbitol	−	−	−	−	−	−	−	−	−	d	−	−
Starch	−	±	−	w−	−	−	+w	−	−	+	+	−
Sucrose	w	+	−	w−	+	−	+	−	−	+	+	−
Trehalose	w	+	−	d	−	−	±	−	−	+	+	−
Xylose	w	∓	±	d	−	−	±	+	±	+	+	∓
Milk reaction	c	−c	−	c	c	−c	c	−	−	c−	c	−

[a] Symbols: +, reaction positive for 90–100% of strains (pH of sugars below 5.5); −, reaction negative for 90–100% of strains; ±, 61–89% of strains positive; ∓, 11–39% of strains positive; d, 40–60% of strains positive; w, weak reaction (pH of sugars 5.5–5.9); numbers (hydrogen) represent abundant (4) to negative on a "−" to "4+" scale; c (milk), curd; a (milk), acid; tr, trace; NT, not tested. Where two reactions are listed, the first is the more usual and occurs in 60–90% of strains.

[b] Products (listed in the order of amounts usually detected): a, acetic acid; b, butyric acid; l, lactic acid; s, succinic acid; p, propionic acid; f, formic acid; iv, isovaleric acid; ib, isobutyric acid; v, valeric acid; c, caproic acid; ic, isocaproic acid; 2, ethanol; 3, propanol; 4, butanol; i4, isobutanol; i5, isopentanol. Capital letters indicate at least 1 meq/100 ml of culture; small letters indicate less than 1 meq/100 ml. Products in parentheses are not detected uniformly.

[c] Gelatin is weakly digested by the type strain of *C. thermautotrophicum* after incubation for 3 weeks.

The mol% G + C of the DNA is 33 (T_m) (Johnson and Francis, 1975).

Type strain: ATCC 13725 (NCIB 10631, DSM 1283).

Further comments. C. aminovalericum is distinguished by its ability to grow with aminovalerate as its sole source of energy. It can be differentiated from other nonsaccharolytic species in the same group (Table 13.13) by its ability to hydrolyze esculin.

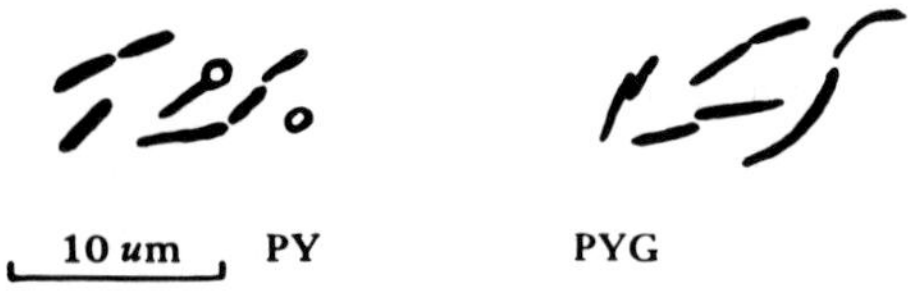

6. **Clostridium arcticum** (*ex* Jordan and McNicol 1979) nom. rev. arc'ti.cum. N.L. neut. adj. *arcticum* related to the Arctic.

This description is based on the description by Jordan and McNicol (1979), and on study of Jordan and McNicol strain III, one of the strains included in the original description.

Cells in PYG broth are straight or slightly curved motile rods, Gram-negative, and 0.5–0.7 × 3.2–4.6 μm. They occur singly or in pairs.

Figure 13.10. *Clostridium arcticum.*

Spores are round, terminal, and swell the cell.

Cell wall content has not been determined.

There is no growth on blood agar or egg-yolk agar plates. On trypticase soy agar, colonies are pinpoint, circular, convex and creamy or yellowish. On Jensen's N$_2$-free medium (Blasco and Jordan, 1976), colonies are yellow. In PYG deep agar cultures, colonies are white balls with slime that adheres to the sides of the tube.

Cultures grow slowly in PYG broth, which becomes turbid with little sediment; the pH after 7 days is 5.4.

The optimum temperature for growth is 22–25°C. Strains grow at 5°C and 37°C but more slowly and not as well. Growth is stimulated by fermentable carbohydrate.

Resazurin is reduced.

Atmospheric N_2 is fixed (Jordan and McNicol, 1979).

Products in PYG broth culture are propionic and acetic acids. Only traces of H_2 are detected. Lactate in chopped meat-carbohydrate is converted to propionate.

Culture supernatants are nontoxic for mice.

Other characteristics of the species are given in Table 13.14.

Isolated from Arctic soil where the species represented 19% of the anaerobic N_2-fixing strains isolated.

The mol% G + C of the DNA has not been determined.

Type strain: Jordan and McNicol No. III.

Further comments. Although this species has not been validated in any list, its properties indicate that it is distinct from any described species.

7. Clostridium aurantibutyricum Hellinger 1944, 46.[AL]

au.ran.ti.bu.ty'ri.cum. N.L. n. *aurantium* orange; N.L. n. *acidum butyricum* butyric acid; N.L. neut. adj. *aurantibutyricum* probably intended to mean the orange-colored organism producing butyric acid.

Based on descriptions by Smith and Hobbs (1974), Holdeman et al. (1977), and study of the type strain.

Cells in PYG broth cultures are motile and peritrichous, straight rods 0.5–0.8 μm × 2.8 to 6.3 μm occurring singly and in pairs, Gram-positive, rapidly becoming Gram-negative in older cultures, often granulose-positive.

Figure 13.11. *Clostridium aurantibutyricum.*

Spores are oval and subterminal, swelling the cell. Sporulation occurs most readily on chopped meat agar slants incubated at 30°C for 1 week.

Cell walls contain *meso*-DAP; cell wall sugars are rhamnose and traces of glucose, galactose and mannose (Cummins and Johnson, 1971).

Surface colonies on blood agar plates are 1–2 mm in diameter, circular to slightly irregular, entire, raised to low convex, translucent, gray to pink-orange, dull, smooth with a mosaic internal structure.

PYG broth cultures are turbid with a heavy ropy or viscous sediment and have a pH of 5.4 after incubation for 6 days.

The optimum temperature for growth is 37°C. Moderate growth occurs at 30°C but not at 25°C or 45°C. Growth is inhibited by 6.5% NaCl and by 20% bile.

Abundant gas is detected in PYG deep agar cultures.

Neutral red and resazurin are reduced.

On egg-yolk agar, lipase is produced with a zone of opacity extending beyond the area of lipase production.

There is stormy fermentation in milk; a solid curd is formed with 50% digestion in 3 weeks.

Products of fermentation in PYG broth are acetate and lactate with small amounts of butyrate, propionate and succinate. Large amounts of butyrate are produced in chopped meat carbohydrate broth. Large amounts of butanol as well as acetone and isopropanol are formed (George et al., 1983).

Pectic lyase enzymes and pectinesterase, but no pectic hydrolase, are formed. (Lund and Brocklehurst, 1978).

The type strain is susceptible to chloramphenicol, clindamycin, erythromycin, penicillin G and tetracycline.

Culture supernatants of the type strain are nontoxic to mice.

Other characteristics of the species are given in Table 13.12.

Isolated from rotting hibiscus stumps, flax, and rotting potatoes; soil (Smith, 1975), sewage sludge (Cox, 1978); also isolated from bovine, human infant and adult feces (Princewill and Agba, 1982; Borriello, 1980; Drasar et al., 1976; Finegold et al., 1983).

The mol% G + C of the DNA of the type strain is 27 (T_m) (Cummins and Johnson, 1971).

Type strain: ATCC 17777 (NCIB 10659, DSM 793).

Further comments. This species is most readily differentiated from *C. felsineum* by its hydrolysis of starch, and from *C. puniceum* by its reduction of nitrate.

8. Clostridium baratii (corrig.) (Prévot 1938) Holdeman and Moore 1970, 60.[AL] (*Inflabilis barati* Prévot 1938, 77; *Clostridium perenne* (Prévot 1940) McClung and McCoy, 1957, 673; *Clostridium paraperfringens* Nakamura, Tamai and Nishida 1970, 137.)

ba.ra'ti.i. N.L. gen. n. *baratii*, in honor of Barat, French bacteriologist.

Based on the descriptions by Holdeman et al. (1977), Cato et al. (1982b), and study of the type and 46 other strains.

Cells in PYG broth are Gram-positive, often granulose-positive, nonmotile, straight rods, 0.5–1.9 × 1.6–10.2 μm, and usually occur singly, occasionally in pairs.

Figure 13.12. *Clostridium baratii.*

Spores are round to oval, subterminal to terminal, and swell the cell. Strains sporulate poorly and spores may be found more readily in chopped meat-carbohydrate broth or PY broth cultures than in cultures on agar slants or plates.

Cell walls contain *meso*-DAP (C. S. Cummins, unpublished data).

Surface colonies on blood agar plates are 0.5–2 mm in diameter, circular to irregular, entire to lobate, flat to low convex, granular to mosaic, translucent to opaque, with a smooth, shiny surface. Most strains are β-hemolytic, but some show α- or no hemolysis.

Cultures in PYG broth are turbid with a heavy, sometimes ropy sediment, and have a pH of 4.5–4.8 after incubation for 5 days.

Growth is equally abundant at 30°C, 37°C, and 45°C, less at 25°C. Growth is inhibited by 6.5% NaCl and by 20% bile.

Abundant gas is produced in PYG deep agar cultures.

Ammonia is produced. Neutral red and resazurin are reduced. Acetyl methyl carbinol is formed by four of eight strains tested.

Products of fermentation in PYG broth are butyric, acetic and lactic acids; smaller amounts of formic, propionic and succinic acids are sometimes produced. Butanol is not detected. Abundant H_2 is produced. Pyruvate is converted to acetate and butyrate; most strains convert threonine to propionate. Lactate is not utilized.

All strains tested are sensitive to chloramphenicol, penicillin G and tetracycline; nearly all strains are resistant to clindamycin and erythromycin.

Culture supernatants are nontoxic to mice.

Other characteristics of the species are given in Table 13.14.

Isolated from normal human and rat feces, from war wounds, peritoneal fluid, infections of the eye, ear and prostate, and from soil; also isolated from sediments in Puget Sound (Matches and Liston, 1974) and from soil from Antarctica (Miwa, 1975).

The mol% of the DNA is 28 (T_m) (J. L. Johnson, unpublished data).

Type strain: ATCC 27638 (DSM 601).

Further comments. *C. baratii* is most readily differentiated from *C. perfringens* and *C. absonum*, which is resembles most closely phenotypically, by not producing butanol and not hydrolyzing gelatin. In addition, the patterns produced by its soluble cellular proteins using polyacrylamide gel electrophoresis are clearly distinct (Cato et al., 1982a). For a description of a synergistic hemolysis test for differentiation of these species, see Gubash (1980). For differentiation by lecithinase reactions on half-α-antitoxin Nagler agar plates, see Nakamura et al. (1973).

9. **Clostridium barkeri** Stadtman, Stadtman, Pastan and Smith 1972, 760.[AL]

bar'ker.i. N.L. gen. n. *barkeri* pertaining to Professor H. A. Barker, American biochemist.

Based on the description by Stadtman et al. (1972), Holdeman et al. (1977), and study of the type strain.

Cells in PYG broth are Gram-positive, nonmotile, and 0.3–0.5 × 1.6–9.7 µm, and occur singly or in pairs.

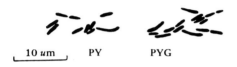

Figure 13.13. *Clostridium barkeri.*

Spores are oval, terminal and swell the cell. Sporulation occurs most reliably on chopped meat slants incubated at 30°C.

Cell walls do not contain DAP (C.S. Cummins, unpublished data). They do contain the B type of murein as described by Schleifer and Kandler (1972) and the cross-linking amino acids D-lysine and small amounts of D-ornithine (O. Kandler as cited in Tanner et al., 1981).

Surface colonies on blood agar are 0.5–1.0 mm, circular, entire, convex, translucent to slightly opaque, mosaic, white, shiny, smooth and nonhemolytic.

Broth cultures are turbid with a smooth sediment. After 5 days incubation, the pH of PYG cultures is 4.5.

Optimum temperature for growth is 37°C. Growth is good at 30°C, fair at 25°C, and poor at 45°C. Good growth at a pH of 8.5. No growth in 6.5% NaCl or 20% bile.

Ammonia is not produced; neutral red but not resazurin is reduced.

Fermentation products from glucose include butyric and lactic acids and large amounts of H_2; moderate amounts of acetic acid may be formed. Pyruvate is converted to acetate and butyrate; neither threonine nor lactate is converted to propionate.

Nicotinic acid is fermented to propionic and acetic acids, CO_2 and ammonia (Stadtman et al., 1972).

Vitamin B_{12} is formed by a pathway similar to that of *Eubacterium limosum*, with glycine and methionine as precursors (Höllriegel et al., 1982).

The type strain is sensitive to chloramphenicol, clindamycin, erythromycin and tetracycline, but resistant to penicillin G.

Culture supernatants are nontoxic to mice.

Other characteristics of the species are given in Table 13.14.

Isolated from Potomac River mud; also isolated from human feces (Finegold et al., 1983).

The mol% of the DNA is 45 (T_m) (Johnson and Francis, 1975).

Type strain: ATCC 25849 (NCIB 10623, DSM 1223).

Further comments. This species is most easily differentiated from other saccharolytic clostridia with terminal spores by its selective fermentation of commonly tested substrates, fermenting only glucose,

fructose, mannitol and sorbitol. Studies of 16S-rRNA by Tanner et al. (1981) have shown relationships between *C. barkeri*, *Eubacterium limosum*, and *Acetobacterium woodii*; although all three species have the same unusual type of murein cross-linkage in cell walls, the terminal spores of *C. barkeri* place it in the genus *Clostridium*.

10. **Clostridium beijerinckii** Donker 1926, 145.[AL]

bei.jer.inck'i.i. N.L. gen. n. *beijerinckii*, named for M. W. Beijerinck, Dutch bacteriologist.

This description is based on those by Smith and Hobbs (1974), Holdeman et al. (1977), and study of the type and 59 other strains including 20 strains found to have DNA homology by Cummins and Johnson (1971) and 4 DNA homologous strains studied by George et al. (1983).

Cells in PYG broth culture are straight rods with rounded ends, motile and peritrichous, 0.5–1.7 × 1.7–8.0 µm, occurring singly, in pairs, or in short chains. They are Gram-positive, becoming Gram-negative in older cultures.

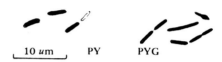

Figure 13.14. *Clostridium beijerinckii.*

Spores are oval, eccentric to subterminal, and swell the cell, with no exosporium or appendages. Sporulation occurs readily on chopped meat agar slants incubated at 30°C.

Cell walls contain *meso*-DAP; cell wall sugars are glucose and galactose (Cummins and Johnson, 1971).

Surface colonies on blood agar plates are 1–5 mm, circular to irregular, entire to scalloped, flat to raised, translucent, gray, shiny and smooth. Strains may be either β-, α- or nonhemolytic.

Cultures in PYG broth are turbid with a smooth to flocculent sediment and have a pH of 4.6–5.4 after incubation for 5 days.

Optimum temperature for growth is 37°C. Cultures grow well at 30 but poorly if at all at 25°C or 45°C. Growth is stimulated by a fermentable carbohydrate, inhibited by 6.5% NaCl or 20% bile. Strains are nutritionally fastidious, requiring a complex mixture of growth factors, such as are supplied by yeast extract (Cummins and Johnson, 1971).

Abundant gas is detected in deep cultures in PYG agar.

Ammonia is produced by 10 of 40 strains tested. Neutral red is reduced; most strains reduce resazurin.

Atmospheric N_2 is fixed (Rosenblum and Wilson, 1949).

Ammonium salts are utilized as the sole N_2 source to produce a wide variety of amino acids including alanine, valine, aspartic acid, and threonine (Matteuzzi et al., 1978).

A neuraminidase is produced by strains of this species (Müller, 1976).

Two of three strains tested produce an extracellular β-glucuronidase (Sakaguchi and Murata, 1983).

The activity of a ferro-flavoprotein hydrogenase isolated from one strain has been investigated by Peck and Gest (1957).

For a discussion of the regulation of tryptophane synthetic enzymes in *C. beijerinckii* (cited as *C. butyricum*), see Baskerville and Twarog (1972, 1974).

Products in PYG broth are butyric and acetic, moderate amounts of succinic, lactic and formic acids, and traces of propionic acid may also be detected. Although not a stable trait, most strains produce substantial amounts of *n*-butanol, and some produce moderate amounts of acetone or isopropanol (George et al., 1983). Pyruvate is converted to butyrate and acetate; neither threonine nor lactate is utilized.

All strains tested are sensitive to erythromycin and tetracycline. Of

21 strains, one is resistant to clindamycin, one to chloramphenicol, and one to penicillin G.

Culture supernatants are nontoxic to mice.

Other characteristics of the species are given in Table 13.14.

Isolated from soil, infected wounds, fermenting olives, spoiled candy; also isolated from human feces (Finegold et al., 1983).

The mol% G + C of the DNA is 26–28 (T_m) (Cummins and Johnson, 1971).

Type strain: ATCC 25752 (NCIB 9362, DSM 791).

Further comments. This species is most easily differentiated from *C. butyricum*, which it resembles most closely phenotypically, by its requirement for growth factors present in yeast extract (Cummins and Johnson, 1971). Patterns of soluble cellular proteins of these two species, as determined by polyacrylamide gel electrophoresis, are distinct (Cato et al., 1982a; Magot et al., 1983). Differential fermentation patterns reported by Magot et al. (1983) are helpful but not absolute: *C. butyricum* usually ferments ribose and glycerol but not inositol; *C. beijerinckii* usually ferments inositol but not ribose or glycerol.

11. **Clostridium bifermentans** (Weinberg and Séguin 1918) Bergey, Harrison, Breed, Hammer and Huntoon 1923, 323.[AL] (*Bacillus bifermentans* Weinberg and Séguin 1918, 128.)

bi.fer.men'tans. L. pref. *bis* twice; L. part. adj. *fermentans* leavening; N.L. adj. *bifermentans* fermenting both carbohydrates and amino acids.

This description is based on those by Smith and Hobbs (1974) and Holdeman et al. (1977), and on study of the type and 178 other strains.

Cells in PYG broth are Gram-positive straight rods, 0.6–1.9 × 1.6–11.0 μm, motile and peritrichous, occurring singly, in pairs or in short chains.

Figure 13.15. *Clostridium bifermentans.*

Spores are oval, central to subterminal, and usually do not swell the cell. Sporulation occurs readily both in PY broth and on chopped meat agar slants. The composition of sporulation medium can affect both the chemical content of spores, their germination rate, and their resistance to heat and to chemical agents (Waites et al., 1980). Spores have an exosporium and six different types of spores have been identified depending on the presence, type or absence of spore appendages (Pope et al., 1967; Rode and Smith, 1971; Samsonoff et al., 1970). The significance or functional role of these appendages is unknown.

Cell walls of most strains contain *meso*-DAP; in 9 of 32 strains tested, DAP was not detected (C. S. Cummins, unpublished data). Cell wall sugars of most strains are glucose, rhamnose and mannose; walls of some strains contain galactose rather than mannose; in some strains only glucose was detected (Rode and Smith, 1971).

Surface colonies on blood agar plates are 0.5–4 mm, circular with irregular margins, flat or raised, lobate or scalloped, translucent or opaque, granular or slightly mottled, gray, shiny and smooth. Individual colonies can often be seen best on 4% agar plates. Most strains are β-hemolytic. Addition of beef liver catalase to plating media increases the percentage recovery of the species (Harmon and Kautter, 1977).

Cultures in PYG broth are turbid with a heavy, often ropy, sediment.

The optimum temperature for growth is 30–37°C. Most strains grow nearly as well at 25°C and 45°C. Growth is inhibited by 6.5% NaCl and by 20% bile.

Abundant gas is detected in deep PYG agar cultures.

Ammonia is produced. Neutral red is reduced; reduction of resazurin is variable.

For discussions of the fatty acid composition of lipids and metabolism

of amino acids in this species, see Mead (1971), Elsden and Hilton (1978), Elsden et al. (1976, 1980) and Britz and Wilkinson (1982, 1983).

Reports of 7-α-dehydroxylation of cholic acid and hydrolysis of bile acid conjugates by *C. bifermentans* (Archer et al., 1981, 1982a, b) indicate that the studies were made on ATCC strain 9714, the type strain of *Clostridium sordellii.*

Fermentation products from PYG broth include large amounts of acetic and formic acids, smaller amounts of isobutyric, isovaleric, isocaproic, hydrocinnamic, benzoic and propionic acids, and ethyl alcohol. Trace amounts of butyric and phenylacetic acids and propyl and isobutyl alcohols are produced by some strains. In young cultures only acetic and formic acids may be detected. Abundant H_2 is produced. The glucose analog 1,2-0-*iso*-propylidene-β-glucofuranose ("monoacetone glucose") is utilized as a carbon source with different proportions of volatile fatty acids produced from those detected from glucose; a larger percentage of propionic, butyric, isovaleric and valeric acids, and a smaller percentage of isobutyric acid (Cmelik, 1980).

Valine is converted to isobutyrate, leucine to isovalerate and isocaproate, isoleucine to isovalerate, and threonine to propionate (Elsden and Hilton, 1978). Pyruvate is converted by most strains to acetate and formate; excess isobutyrate, isovalerate, and isocaproate may be produced. Lactate is not converted to propionate.

Proline, serine, threonine, arginine and aspartate are utilized; δ-aminovalerate and α-amino butyrate are produced (Mead, 1971).

DNAse is produced by 10 strains tested (Döll, 1973).

All strains tested are susceptible to chloramphenicol, erythromycin and penicillin G. One of 72 strains is resistant to clindamycin, one strain (a different one) is resistant to tetracycline. Three strains tested are sensitive to 8 μg of nalidixic acid/ml (Nadaud, 1977).

Culture supernatants are nontoxic to mice.

Other characteristics of the species are given in Table 13.12.

Strains have been isolated from soil; fresh water; marine sediments; human feces; normal cervical flora (Gorbach et al., 1973); snake venom; a goat stomach ulcer; wounds in horses and sheep; clinical specimens including wounds, abscesses and blood; clam gut; cheese fondue; canned tomatoes; and vacuum packed smoked fish.

The mol% G + C of the DNA is 27 (T_m) (Johnson and Francis, 1975).

Type strain: ATCC 638 (NCIB 10716).

Further comments. This species has been differentiated from *C. sordellii,* which it resembles closely phenotypically, by not producing urease. Differentiation by this characteristic has been shown to be not completely reliable, and Nakamura et al. (1975, 1976) have reported that growth of strains of *C. sordellii* is inhibited by 1% mannose while growth of strains of *C. bifermentans* is enhanced. The two species are reported to produce different patterns of synergistic hemolysis reactions on human blood agar plates and human blood agar plates supplemented with $CaCl_2$ (Gubash, 1980). The patterns of soluble cellular proteins, as determined by polyacrylamide gel electrophoresis, are distinct (Cato et al., 1982a). Tryptamine and β-phenylethylamine have been detected by flame ionization gas chromatography in growing cultures of *C. bifermentans* but not in cultures of *C. sordellii* (Brooks et al., 1969). *C. sordellii* produces isoamylamine and putrescine which are not detected in *C. bifermentans* (Brooks et al., 1973). Patterns of unidentified high boiling amines detected also differ (Brooks and Moore, 1969). Neuraminidase has been detected in strains of *C. sordellii* while none was detected in strains of *C. bifermentans* (Fraser, 1978). The type strain and five other strains of *C. bifermentans* produced hydrocinnamic acid in PYG broth while no hydrocinnamic acid was produced by the type and five other strains of *C. sordellii,* although Moss et al. (1970) reported that this reaction is variable.

12. **Clostridium botulinum** (van Ermengem 1896) Bergey, Harrison, Breed, Hammer, and Huntoon 1923, 328.[AL] (*Bacillus botulinus* van Ermengem 1896, 443.)

bo.tu.li'num. L. n. *botulus* sausage; N.L. adj. *botulinum* pertaining to sausage.

The species includes seven toxin types, A, B, C, D, E, F and G, differentiated by the antigenic specificity of their individual toxins. All strains of the species produce neurotoxins with similar effects on an affected host, but the toxins of the different types are serologically distinct. These distinctions do not necessarily correlate with observed phenotypic differences. The species was divided into three metabolic groups by Holdeman and Brooks (1970). Strains of *C. botulinum* type G, described by Giménez and Ciccarelli (1970) are metabolically distinct, and were placed in a separate group by Smith and Hobbs (1974). These groups and other species that are phenotypically similar are: (a) strains of type A, proteolytic strains of types B and F and *C. sporogenes*; (b) strains of type E and saccharolytic strains of types B and F; (c) strains of types C and D, and *C. novyi* type A; and (d) strains of type G and *C. subterminale*. The validity of this grouping has been confirmed by the data of Johnson and Francis (1975) in that the metabolic types correlate well with rRNA homology groups I-F, I-A, I-H, and I-K, respectively. The toxins of all types are pathogenic to laboratory animals through the action of a neurotoxin. Some toxins, particularly those of the nonproteolytic strains, require trypsin activation for effectiveness in laboratory toxin testing. Human disease (botulism) with similar symptoms is caused by toxins elaborated under anaerobic conditions usually by colonization of food, less often by colonization of wounds, or colonization of the intestinal tract as in infant botulism (Wilcke et al., 1980). For comprehensive reviews of the organism, its toxins, and the disease it causes, see Smith (1975a, 1977), Arnon et al. (1977), and Sugiyama (1980). For a review of amino acid degradation by members of the species, see Barker (1981).

This description is based on those by Smith and Hobbs (1974), and Holdeman et al. (1977), and on study of the type and 15 other strains of *C. botulinum* type A, 14 proteolytic strains of type B, 2 saccharolytic strains of type B, 24 strains of type C, 5 strains of type D, 9 strains of type E, 3 proteolytic strains of type F, 4 saccharolytic strains of type F, and 6 strains of type G.

(a) *Type A and proteolytic strains of types B and F.* Cells in PYG broth are usually motile and peritrichous, straight to slightly curved rods, 0.6–1.4 × 3.0–20.2 μm.

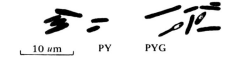

|_ 10 μm _| PY   PYG

Figure 13.16. *Clostridium botulinum*, type A.

Spores are oval and subterminal and swell the cell. Sporulation occurs most readily on egg-yolk agar plates incubated for 2 days or on chopped meat agar slants incubated at 30°C for 1 week.

Cell walls contain *meso*-DAP and glucose (Cummins and Johnson, 1971; and C. S. Cummins, unpublished data). A cell wall protein with a common antigenic specificity has been isolated from each of these types (Takumi and Kawata, 1974; Takumi et al., 1983).

Surface colonies on blood agar plates are 2–6 mm in diameter, circular to irregular with a scalloped or rhizoid margin, flat to raised, translucent to semiopaque, gray, often with a mottled or crystalline internal structure, and are β-hemolytic.

Cultures in PYG broth are turbid with a smooth or flocculent sediment and have a pH of 5.6–6.2 after incubation for 1 week. Ammonia produced from the deamination of amino acids often masks acid production from carbohydrates.

Optimum temperature for growth is 30–40°C. Some strains grow well at 25°C and a few at 45°C. Growth is inhibited by 6.5% NaCl, by 20% bile, and at a pH of 8.5. The bile acids lithocholic and chenodeoxycholic are the most inhibitory (Huhtanen, 1979). Strains of *C. perfringens* and *C. sporogenes* isolated from soil also can inhibit growth (Smith, 1975b).

Toxin production is delayed in an atmosphere of 100% CO_2 and pressurized CO_2 is lethal to strains, depending on the amount of pressure and the length of exposure (Doyle, 1983).

Gelatin, milk, and meat are digested. Ammonia and H_2S are produced.

Strains of *C. botulinum* type A and type B reduce proline to δ-aminovalerate. Arginine, glycine, phenylalanine, serine, tyrosine and tryptophan are also utilized for growth; valine, α-aminobutyrate, and γ-aminobutyrate are produced (Mead, 1971). The aromatic amino acids, phenylalanine, tyrosine and tryptophan are reduced to phenylpropionic, *p*-hydroxyphenylpropionic, and indolepropionic acids, respectively (Elsden et al., 1976). *C. botulinum* type A and proteolytic types B and F convert valine to isobutyrate, and isoleucine and leucine to isovalerate; types A and B but not type F produce isocaproate from leucine (Elsden and Hilton, 1978, 1979).

Fermentation products in PYG broth include large amounts of butyric and acetic acids with moderate amounts of isobutyric and isovaleric acids. Isocaproic, propionic and valeric acids, and ethanol, propanol and butanol may also be detected. Hydrocinnamic acid is produced in trypticase soy broth cultures by all strains of types A, and (presumably proteolytic) strains of types B, and F tested (Moss et al., 1970); abundant H_2 gas is produced. Pyruvate is converted to acetate, butyrate, ethanol and butanol; excess amounts of isovalerate from those produced in basal medium may also be detected. Most strains convert threonine to propionate; lactate is not utilized.

All strains are sensitive to chloramphenicol, penicillin G, and tetracycline. One strain of proteolytic type B is resistant to clindamycin and erythromycin; all other strains are sensitive to these antibiotics. Types A and B are resistant to cycloserine, sulfamethoxazole and trimethoprim (Dezfulian and Dowell, 1980). Swenson et al. (1980) have reported that of 177 strains in this group, all are sensitive in vitro to tetracycline (0.5 μg/ml), metronidazole (1 μg/ml), penicillin (4 μg/ml), rifampin (2 μg/ml), and erythromycin (4 μg/ml); more than 90% are sensitive to chloramphenicol (4 μg/ml), clindamycin (4 μg/ml), cefoxitin (1 μg/ml), and vancomycin (8 μg/ml); they are resistant to nalidixic acid and gentamycin.

Culture supernatants are toxic to mice (Smith, 1974, 1977). Culture supernatants of type A are toxic to chickens, turkeys, pheasants and peafowl (Gross and Smith, 1971). Although bacteriophages have been demonstrated in all types (Dolman and Chang, 1972) and two phages have been isolated from a strain of type A (Takumi et al., 1980; Kinouchi et al., 1981), there has been no report of the mediation of toxin production by phage in these strains.

Other characteristics are given in Table 13.12.

Commonly isolated from soil, marine and lake sediments. Found in animal, bird and fish intestines and in food (particularly improperly preserved vegetables, meat and fish). The types isolated reflect those present in the soil or sediment of the area. *C. botulinum* toxin type F has been implicated in one case of infant botulism (Hoffman et al., 1982), but otherwise only types A and B have been reported. These two types also are most frequently isolated in outbreaks of food poisoning and cases of wound botulism.

The mol% of the DNA is 26–28 (T_m) (Lee and Riemann, 1970a; Cummins and Johnson, 1971; Johnson and Francis, 1975).

Type strain: Type A, ATCC 25763 (NCIB 10640).

Reference strains: Proteolytic type B, ATCC 7949 (NCIB 10657); proteolytic type F, ATCC 25764 (NCIB 10658).

Further comments. *C. botulinum* type A and proteolytic strains of types B and F have high DNA-DNA homology with *C. sporogenes* (Lee and Riemann, 1970b) and cannot be distinguished metabolically or biochemically. Strains of these types are identified by toxin neutralization tests in mice. Although cellular protein electrophoretic patterns of the proteolytic *C. botulinum* strains cannot be used to identify the toxin type, they are distinct from patterns given by strains of nontoxic *C. sporogenes* (Cato et al., 1982a) as are gas chromatographic patterns of trimethylsilyl derivatives of whole cell hydrolysates of strains of the two species (Farshy and Moss, 1970).

(b) *Type E and saccharolytic strains of types B and F.* Cells in PYG broth cultures are motile and peritrichous, straight rods, 0.8–1.6 × 1.7–15.7 μm, occurring singly and in pairs. Spores are oval, eccentric to subterminal, and usually swell the cell. Sporulation occurs readily in broth and on solid media.

Cell walls contain *meso*-DAP acid (C. S. Cummins, unpublished data).

Surface colonies on blood agar plates are β-hemolytic, 1–5 mm in diameter, irregular with lobate or scalloped margin, raised, translucent to opaque, gray-white, with a mottled or mosaic internal structure.

Cultures in PYG broth are turbid with a smooth sediment, and have a pH of 5.2–5.5 after incubation for 1 to 2 days.

Optimum temperature for growth ranges from 25–37°C. Little or no growth occurs at 45°C. Growth is stimulated by a fermentable carbohydrate and is inhibited by 6.5% NaCl, 20% bile, or a pH of 8.5. Nonproteolytic strains of types B and F and toxic strains of type E are inhibited by soil strains of *C. perfringens*, while type E strains that have lost toxicity are unaffected (Smith, 1975b). A boticin isolated from a nontoxigenic strain of *C. botulinum* type E inhibited both vegetative growth and spore germination of 10 of 12 toxic and 3 of 6 nontoxic strains of *C. botulinum* E, as well as 2 of 2 strains of nonproteolytic strains of *C. botulinum* type B (Lau et al., 1974). Plasmids that may be related to production of "boticin E" have been demonstrated in both toxic type E and nontoxic type E-like strains (Scott and Duncan, 1978).

Gelatin is digested but not milk or meat.

Fermentation products in PYG broth cultures are butyric and acetic acids. Large amounts of H_2 are detected in headspace gas. Most strains convert pyruvate to acetate and butyrate. Neither lactate nor threonine is utilized.

All strains tested are susceptible to chloramphenicol, clindamycin, erythromycin, penicillin and tetracycline. They also are susceptible to metronidazole, rifampin, cefoxitin and vancomycin, but are resistant to nalidixic acid and gentamycin (Swenson et al., 1980).

Culture supernatants are toxic to mice. Supernatants of strains of type E are toxic to gallinaceous birds (Gross and Smith, 1971).

Other characteristics are given in Table 13.12.

Isolated from soil, marine and lake sediments, food, fish, birds, and mammals.

The mol% of the DNA is 27–29 (T_m) (Lee and Riemann, 1970b; Johnson and Francis, 1975).

Reference strains: Nonproteolytic type B, ATCC 25765 (NCIB 10642); nonproteolytic type F, ATCC 27321 (NCIB 10641); and type E, ATCC 9564 (NCIB 10660).

(c) *Type C and Type D.* Cells in PYG broth are straight rods, motile and peritrichous, 0.5–2.4 × 3.0–22.0 μm, and occur singly or in pairs. Spores are oval, subterminal and swell the cell. Sporulation of most strains occurs most readily on chopped meat agar slants incubated at 30°C for 1 week.

Cell walls contain *meso*-DAP (C. S. Cummins, unpublished data).

Surface colonies on blood agar plates are β-hemolytic, 1–5 mm in diameter, circular to slightly irregular, slightly scalloped or lobate, flat to raised, translucent, gray-white, with a mottled or mosaic internal structure.

Cultures in PYG broth are turbid with a smooth or flocculent sediment and have a pH of 5.2–5.7 after incubation for 1–2 days.

Optimum temperature for growth is 30–37°C; most strains grow well at 45°C, and poorly, if at all, at 25°C. Growth is stimulated by a fermentable carbohydrate but is inhibited by 6.5% NaCl, 20% bile or a pH of 8.5. There was no inhibition of growth by strains of *C. perfringens* isolated from United States soil samples (Smith, 1975), but several species of *Bacillus* isolated from samples of mud from England, France and Spain were inhibitory to growth of a strain of *C. botulinum* type C (Graham, 1978).

Gelatin is digested; milk is acidified, curdled, and digested by 20 of 29 strains tested; meat is digested by 20 of 28 strains tested. Production of ammonia and H_2S varies among strains.

Strains of type C utilize glutamic acid, serine, glycine, arginine and aspartic acid; δ-aminovalerate is not produced (Mead, 1971).

Products in PYG broth are butyric, propionic and acetic acids; traces of valeric, succinic and lactic acids may be detected. Abundant H_2 is formed. Lactate is converted to propionate. Pyruvate is converted to acetate and butyrate; propionate is formed from pyruvate by some strains. Five of 24 strains convert threonine to propionate.

All strains tested are susceptible to chloramphenicol, clindamycin, erythromycin, penicillin G and tetracycline. They also are susceptible to metronidazole, rifampin, cephalothin and cefoxitin, but resistant to nalidixic acid and gentamycin (Swenson et al., 1980).

Culture supernatants are toxic to mice; culture supernatants of strains of type C are toxic to gallinaceous birds (Gross and Smith, 1971); strains are pathogenic for laboratory animals. The toxin of *C. botulinum* type C is inactivated by bacteria in the rumen of cattle and sheep but the specific organisms involved are not known (Allison et al., 1976). Toxin production by *C. botulinum* type C and type D is phage-mediated, and the specific type of toxin produced is determined by the specific phage with which the culture is infected (Eklund et al., 1971; Iida et al., 1974; Eklund and Poysky, 1974; Hariharan and Mitchell, 1976; Oguma and Iida, 1979). *C. botulinum* type C can be cured of type C phage and of its toxin, then converted to *C. novyi* type A following infection by a *C. novyi* type A phage (Eklund et al., 1974).

Other characteristics of these types are given in Table 13.12.

Isolated from feces and carcasses of animals and birds and from soil (Smith, 1978; Serikawa et al., 1977), lake mud (Mason, 1968), rotting vegetation (Martinovich et al., 1972).

The mol% G + C of the DNA (T_m) is 26–28 (Lee and Riemann, 1970b; Johnson and Francis, 1975).

Reference strains: Type C, ATCC 25766 (NCIB 10618); Type D, ATCC 25767 (NCIB 10619).

Further comments. These species can be identified by toxin neutralization testing. They are most easily differentiated from *C. novyi* type A, which they resemble phenotypically, by distinct patterns of cellular proteins produced by polyacrylamide gel electrophoresis (Cato et al., 1982a).

(d) *Type G.* Cells in PYG broth cultures are straight rods, motile and peritrichous, 1.3–1.9 × 1.6–9.4 μm, and occur singly and in pairs. Spores, though rarely seen, are oval, subterminal and swell the cell (Giménez and Ciccarelli (1970). We have not detected spores from any medium.

Cell wall composition has not been reported.

Surface colonies on blood agar plates are β-hemolytic, 1–4 mm in diameter, circular to irregular, lobate to filamentous, raised, translucent, smooth and shiny. A spreading film may cover the entire plate. Large, rough, fried-egg colonies may be formed (Ciccarelli et al., 1977).

Cultures in PYG broth are turbid with a smooth white sediment and have a pH of 6.2–6.3 after 5 days incubation.

Optimum temperature for growth is 30–37°C. Cultures grow almost as well at 25°C and 45°C. Growth is inhibited by 6.5% NaCl and by 20% bile. The reference strain is inhibited by three strains of *C. perfringens* isolated from soil (Smith, 1975b).

Moderate gas is detected in PYG deep agar cultures.

Products in PY broth are acetic, butyric, isovaleric, isobutyric and phenylacetic acids and butyl and ethyl alcohols. When products are converted to butyl esters, hydroxyphenylacetic acid is also detected (Moss et al., 1980). Abundant H_2 is formed. Pyruvate is converted to acetate and ethyl alcohol; lactate and threonine are not utilized. Valine is converted to isobutyric acid and leucine to isovaleric acid (Elsden and Hilton, 1978). Indoleacetic acid is produced and lysine is utilized (Elsden and Hilton, 1979).

Gelatin and casein are digested rapidly; milk and meat are digested within 3 weeks. Ammonia and H_2S are produced.

The type strain is susceptible to chloramphenicol, clindamycin, erythromycin, penicillin G and tetracycline. It also is susceptible to metronidazole, rifampin, cephalothin and cefoxitin, but is resistant to vancomycin, nalidixic acid and gentamycin (Swenson et al., 1980).

Culture supernatants are toxic to mice. Monkeys, chickens, guinea pigs and mice are susceptible to the toxin; sheep and dogs are resistant (Ciccarelli et al., 1977).

Other characteristics are given in Table 13.15.

Isolated from soil (Giménez and Ciccarelli, 1970) and from human autopsy specimens (Sonnabend et al., 1981).

The mol% G + C of the DNA has not been reported.

Reference strain: ATCC 27322 (NCIB 10714).

Further comments. Unlike all other types of *C. botulinum*, strains of type G do not produce lipase on egg-yolk agar. They have been distinguished from strains of *C. subterminale*, which they resemble most closely phenotypically, by their toxicity for mice, but they can be differentiated readily by their distinct patterns of soluble cellular protein shown by polyacrylamide gel electrophoresis (Cato et al., 1982a).

In any other group of organisms, this species would have been divided into four separate species because of the distinct differences in metabolic activity exhibited by strains in the four groups and the lack of DNA homology among groups. However, because of the unique and similar action of the toxins produced by all strains and to facilitate communication between the microbiological and medical professions, they have been retained in one species.

13. **Clostridium butyricum** Prazmowski 1880, 24.[AL]

bu.ty′ri.cum. Gr. n. *boutyron* butter; N.L. neut. adj. *butyricum* related to butter, butyric.

Type species of the genus *Clostridium*.

Based on descriptions by Smith and Hobbs (1974) and Holdeman et al. (1977), and on study of the type strain and 76 similar strains, including 17 strains found to be homologous with the type strain by DNA/DNA homology determinations (Cummins and Johnson, 1971).

Cells in PYG broth are Gram-positive straight rods with rounded ends, motile and peritrichous, 0.5–1.7 × 2.4–7.6 µm, and occur singly, in pairs, or in short chains, occasionally as long filaments. Cells are often granulose-positive.

| 10 µm | PY | PYG |

Figure 13.17. *Clostridium butyricum.*

Spores are oval, central to subterminal, and usually do not swell the cell. No exosporium or appendages are present. Sporulation occurs readily both in broth and on solid media.

Cell walls contain *meso*-DAP and glucose (Cummins and Johnson, 1971). Glutamic acid and alanine are present (Schleifer and Kandler, 1972).

Surface colonies on blood agar plates are 1–6 mm in diameter, circular to irregular, lobate or slightly scalloped, raised to convex, translucent, gray-white, shiny or dull, smooth, with a granular or mottled internal structure.

Cultures in PYG broth are turbid with a smooth or flocculent sediment and have a pH of 4.6–5.0 after incubation for 5 days.

The optimum temperature for growth is 30–37°C; many strains grow equally well at 25°C and growth can occur at 10°C (Molongoski and Klug, 1976). Growth is stimulated by a fermentable carbohydrate and inhibited by 6.5% NaCl. Strains of *C. butyricum* grow readily in glucose-mineral salts medium with biotin as the only required vitamin (Cummins and Johnson, 1971).

Pectin is strongly fermented by all strains tested. Pectin degradation has been implicated in the formation of wetwood in living hardwood trees and in the anaerobic digestion of fruit and vegetable wastes, and strains of *C. butyricum* which carry out these processes have been isolated from these sources (Schink et al., 1981; Schink and Zeikus, 1982).

Chlorinated hydrocarbon pesticides can be degraded in the presence of glucose by strains of this species (Jagnow et al., 1977).

Atmospheric N_2 is fixed (Rosenblum and Wilson, 1949).

Strains contain an iron-sulfur-thiamin pyrophosphate enzyme involved in the reduction of ferredoxin by pyruvate (Yoch and Carithers, 1979).

DNase activity has been detected (Johnson and Francis, 1975).

Neutral red and resazurin are reduced.

Products in PYG broth are butyric, acetic and formic acids; lactic and succinic acids, and butanol and ethanol are sometimes produced. Pyruvate is converted to acetate, butyrate and sometimes formate. Neither lactate nor threonine is utilized. Products from the fermentation of pectin are large amounts of methanol, acetate, H_2, and CO_2, and moderate amounts of butyrate and ethanol (Schink et al., 1981; Schink and Zeikus, 1980, 1982).

All strains tested are susceptible to clindamycin, chloramphenicol, erythromycin, and tetracycline; 9 of 37 strains are resistant to penicillin. Three clinical isolates, resistant to penicillin, produce a β-lactamase

Table 13.15.

Characteristics of species in the genus **Clostridium**. *Gelatin hydrolyzed, no acid from carbohydrates, starch not hydrolyzed.*[a]

Characteristics	12. C. botulinum type G	29. C. ghonii	32. C. hastiforme	33. C. histolyticum	36. C. irregulare	39. C. limosum	41. C. malenominatum	42. C. mangenotii	55. C. putrefaciens	72. C. subterminale	75. C. tetani	78. C. thermocellum[b]	83. C. villosum
Products from PYG[c]	Abivib (l2,4)	Abivicib4i4 (fpls2,3)	ABivib (fpic)	A (fls)	Aiv(fpibl)	A(fls)	AB (fpls)	Afpibiv ic	afl(s)	ABiVib (fpicls2)	ABp4(2ls)	A2l	Baivfibl
Motility	+	+	+	±	+	d	±	−	−	±	∓	−	−
H_2 produced	4	1–3	−1	2−	1	−1	4	3	−	4	4	4	2
Indole produced	−	+	−	−	−	−	+	+	−	−	d	−	−
Lecithinase produced	−	+	−	−	−	+	−	−	−	∓	−	−	−
Lipase produced	−	+	−	−	−	−	−	−	−	−	−	−	−
Esculin hydrolyzed	−	+	−	−	−	−	−	−	−	−	−	+	−
Nitrate reduced	−	−	∓	−	−	−	∓	−	−	−	−	−	−
Milk reaction	d	d	−d	d	−	d	−	d	−	dc	d−	−	−
Meat digested	±	+	±	+	−	+	−	+	+	±	∓	−	−

[a] Symbols: +, reaction positive for 90–100% of strains (pH of sugars below 5.5); −, reaction negative for 90–100% of strains; ±, 61–89% of strains positive; ∓, 11–39% of strains positive; d, 40–60% of strains positive; numbers (hydrogen) represent abundant (4) to negative on a "−" to 4+ scale; c (milk), curd; d (milk) digestion. Where two reactions are listed, the first is the more usual and occurs in 60–90% of strains.

[b] *C. thermocellum* produces weak acid from cellobiose and cellulose.

[c] Products (listed in the order of amounts usually detected): a, acetic acid; b, butyric acid; f, formic acid; iv, isovaleric acid; ib, isobutyric acid; l, lactic acid; p, propionic acid; s, succinic acid; ic, isocaproic acid; 2, ethanol; 3, propanol; 4, butanol; i4, isobutanol. Capital letters indicate at least 1 meq/100 ml of culture; small letters less than 1 meq/100 ml. Products in parentheses are not detected uniformly.

(Carlson et al., 1981). Two clinical isolates tested are susceptible to cefoxitin (4 μg/ml), clindamycin (1 μg/ml), and metronidazole (1 μg/ml) (Finegold and George, unpublished data).

Culture supernatants are not toxic to mice.

A bacteriocin produced by the type strain is active against cell membrane functions of several species of clostridia, particularly those of *C. pasteurianum* (Clarke and Morris, 1976).

J. Möse strain M55 (ATCC 13732 (listed under *C. butyricum*); DSM 754 (labeled *Clostridium* sp.)) produces kininases that are active in selectively destroying tumor tissue in mice. This strain has been variously labeled *C. butyricum* (Möse and Möse, 1964; Möse et al., 1972a), *C. oncolyticum* s. *butyricum* (Gericke et al., 1979), and *C. oncolyticum* (Brantner and Schwager, 1979; Haller and Brantner, 1979). This strain, however, has the metabolic reactions and the electrophoretic protein pattern of *C. sporogenes*.

Other characteristics of the species are given in Table 13.14.

Isolated from soil (Smith, 1975b, Miwa, 1975a and b) freshwater and marine sediments (Matches and Liston, 1974; Molongoski and Klug, 1976), cheese (Matteuzzi et al., 1977), rumen of healthy calves (Jayne-Williams, 1979), animal and human feces, including feces of healthy infants (Stark and Lee, 1982; Finegold et al., 1983), snake venom, and although seldom in pure culture, from a wide variety of human and animal clinical specimens including those from blood, urine, lower respiratory tract, pleural cavity, abdomen, wounds and abscesses.

The mol% of the G + C is 27–28 (T_m) (Cummins and Johnson, 1971; Matteuzzi et al., 1977).

Type strain: ATCC 19398 (NCTC 7423; NCIB 7423; DSM 552; NCDO 1713; CCUG 4217).

Further comments. It has been established by DNA/DNA homology experiments that *C. butyricum* is a distinct species that includes strains previously identified as other species (Cummins and Johnson, 1971). In addition, many strains previously labeled *C. butyricum* are now recognized as *C. beijerinckii*. The two species are most easily differentiated by distinct electrophoretic patterns of cellular proteins (Cato et al., 1982a), and by nutritional requirements—*C. butyricum* will grow well even in a third serial transfer in a medium containing only glucose, mineral salts and biotin, while *C. beijerinckii* requires other vitamins and growth factors such as are provided by yeast extract (Cummins and Johnson, 1971). The two species also differ in phospholipid composition: *C. butyricum* has ethanolamine as its major nitrogenous phospholipid base, while *C. beijerinckii* has *N*-methylethanolamine and ethanolamine (Johnston and Goldfine, 1983).

Historically, "*C. pseudotetanicum*" has been differentiated from *C. butyricum* by spore location. However, both terminal and subterminal spores are seen in cultures of strains of the two species. The reference strain of "*C. pseudotetanicum*" (Prévot 1938) Smith and Hobbs 1974, 567 (ATCC 25779; NCIB 10630) is phenotypically indistinguishable from strains of *C. butyricum*. Bulk RNA preparations from ATCC 25779 are 100% homologous with the 23s RNA from ATCC 19398, the type strain of *C. butyricum* (Johnson and Francis, 1975). In addition, the electrophoretic patterns of soluble cellular proteins of the two strains are identical (Cato et al., 1982a). Thus "*C. pseudotetanicum*" appears to be a later synonym of *C. butyricum* pending confirmation by DNA/DNA homology determinations.

14. **Clostridium cadaveris** (Klein, 1899) McClung and McCoy 1957, 672.[AL] (*Bacillus cadaveris* Klein 1899, 280).

ca.dav′er.is. L. n. *cadaver* dead body; L. gen. n. *cadaveris* of a corpse.

Description based on those by Smith and Hobbs (1974), Holdeman et al. (1977), and study of the type and 63 other strains.

Cells in PYG broth are Gram-positive straight rods, usually motile and peritrichous, 0.5–1.3 × 1.4–9.4 μm, and occur singly or in pairs.

Spores are oval, terminal and swell the cell. Subterminal spores are seen occasionally. Sporulation of most strains occurs most readily on chopped meat agar slants incubated at 30°C for 1 week.

Cell walls of most strains contain *meso*-DAP; no DAP was detected in 5 of 12 strains (C. S. Cummins, unpublished data).

Surface colonies on blood agar plates are 0.5–3 mm, circular, entire to slightly scalloped, convex, translucent to opaque, smooth and shiny. Hemolysis is variable.

Cultures in PYG broth are turbid with a smooth sediment and have a pH of 4.9–5.4 after incubation for 5 days.

The optimum temperature for growth is between 30°C and 37°C. Most strains grow nearly as well at 25°C and 45°C. Growth is stimulated by fermentable carbohydrate and inhibited by 6.5% NaCl. Most strains are inhibited by 20% bile.

Ammonia and H_2S are produced. Neutral red is reduced; most strains reduce resazurin. Hippurate is hydrolyzed by 5 of 30 strains tested.

Abundant gas is detected in PYG deep agar cultures.

Deoxyribonuclease is formed (Smith, 1975a).

Fermentation products in PYG broth include large amounts of butyric and acetic acids, with ethyl and butyl alcohols. Traces of other alcohols and small amounts of lactic, formic, propionic and succinic acids may be detected. Abundant H_2 is formed. In chopped meat broth or PY broth cultures containing at least 2% peptone, large amounts of acetate and butyrate and moderate amounts of isobutyrate and isovalerate are produced. Lipid fatty acids of the type strain detected in a trypticase-yeast extract-thioglycolate medium without glucose have been identified by Elsden et al. (1980). Major fatty acids formed are normal saturated C_{16}, C_{14}, C_{18} and C_{12} acids with smaller amounts of unsaturated C_{16} acids with a double bond at the 7 or 9 position. Pyruvate is converted to acetate and butyrate; neither lactate nor threonine is converted to propionate. There is some increase in the amount of isobutyrate produced in valine broth without glucose but no utilization of leucine or isoleucine (Elsden and Hilton, 1978).

All strains tested are susceptible to penicillin G; of 37 strains, 4 are resistant to erythromycin and clindamycin, 5 are moderately resistant to clindamycin alone, 3 are resistant to tetracycline, and 2 are resistant to chloramphenicol. Three clinical isolates tested are susceptible to cefoxitin (1 μg/ml), moxalactam (4 μg/ml), cefoperazone (4 μg/ml), clindamycin (2 μg/ml), and metronidazole (0.25 μg/ml) (Finegold and George, unpublished data).

Culture supernatants are nontoxic to mice. Phage-like particles that inhibited growth of a strain of *C. septicum* have been demonstrated in one strain (Nieves et al., 1981).

Other characteristics of the species are given in Table 13.12.

Isolated from soil (Smith, 1975b, Miwa, 1975a and b), marine sediment (Matches and Liston, 1974), animal and human feces, snake venom and human clinical specimens from abscesses, wounds and blood.

The mol% of the G + C is 27 (T_m) (Johnson and Francis, 1975).

Type strain: ATCC 25783 (NCIB 10676; DSM 1284).

Further comments. Fermentation of glucose, production of indole, and lack of lecithinase activity differentiate *C. cadaveris* from other clostridial species. The formation of isobutyrate and isovalerate from PY broth but not from PYG broth is most helpful in identification.

15. **Clostridium carnis** (Klein, 1904) Spray 1939, 750.[AL] (*Bacillus carnis* Klein 1904, 459).

car′nis. L. gen. n. *carnis* of flesh.

This description is based on those of Smith and Hobbs (1974), Holdeman et al. (1977) and on study of the type and two other strains.

Cells in PYG broth cultures are Gram-positive, motile and peritrichous, straight to slightly curved rods, 0.5–1.1 × 1.6–9.9 μm, and occur singly or in pairs.

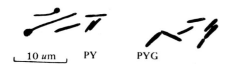

Figure 13.18. *Clostridium cadaveris.*

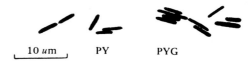

Figure 13.19. *Clostridium carnis.*

Spores are oval, terminal or subterminal, and swell the cell. Sporulation occurs most readily in chopped meat broth cultures incubated for 24 h.

Cell walls contain LL-DAP and glycine (Weiss et al., 1981).

Surface colonies on anaerobic blood agar plates are pinpoint to 2 mm, circular, entire, convex to slightly peaked, translucent, smooth, grayish white, with a mottled or mosaic internal structure. They may be slightly β-hemolytic. Colonies will grow on blood agar plates incubated aerobically, but spores are not formed under aerobic conditions.

Cultures in PYG broth are turbid with a smooth sediment and have a pH of 4.9–5.4 after incubation for 1 week. The optimum temperature for growth is 37°C. Growth is moderate at 30°C but little or no growth occurs at 25°C or 45°C. Growth is stimulated by fermentable carbohydrate, but inhibited by 6.5% NaCl, by 20% bile, or by a pH of 8.5.

Abundant gas is formed in deep agar cultures.

Neutral red and resazurin are reduced.

DNase activity has been demonstrated (Johnson and Francis, 1975).

Products in PYG broth are butyric, acetic and lactic acids, with small amounts of formic and usually succinic acids. Abundant H_2 is produced. Pyruvate is converted to acetate; neither lactate nor threonine is utilized.

Strains are susceptible to chloramphenicol (5 μg/ml), erythromycin (2 μg/ml), penicillin (2 U/ml), and tetracycline (5 μg/ml).

Culture supernatants are toxic to mice.

Other characteristics of the species are given in Table 13.14.

Isolated from soil (Prévot, 1948), from putrefying meat (Klein, 1904), from soft tissue infections and blood in humans (Gorbach and Thadepalli, 1975; Wort and Ozere, 1976), from human feces (Finegold et al., 1983), and from clinical specimens in animals.

The mol% G + C of the DNA is 28 (T_m) (Johnson and Francis 1975).

Type strain: ATCC 25777 (NCIB 10670; DSM 1293).

Further comments. This species is most easily differentiated from *C. tertium,* which it resembles most closely phenotypically, by its lack of fermentation of mannitol or melibiose.

16. Clostridium celatum Hauschild and Holdeman 1974, 479.[AL]

ce.la′tum. L. neut. part. adj. *celatum* hidden.

This description is based on that by Hauschild and Holdeman (1974) and on study of the type and 3 other strains received from A. H. W. Hauschild.

Cells in PYG broth are Gram-positive rods, straight or curved, nonmotile, 0.8–3.0 × 6.3 μm and occur singly, in pairs, or in long chains. Long filaments up to 200 μm are common.

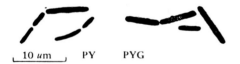

Figure 13.20.   *Clostridium celatum.*

Spores are oval, terminal, subterminal or central, and usually do not swell the cell. Sporulation occurs most readily on solid media (blood agar or egg-yolk agar plates or chopped meat slants).

Cell wall composition has not been reported.

Surface colonies on blood agar plates are 2–7 mm in diameter, circular, lobate to erose, flat to low convex, translucent to opaque, white, with a granular or mottled internal structure. Two of four strains tested were α-hemolytic, two were nonhemolytic on rabbit blood.

Cultures in PYG broth are turbid with a smooth or fluffy sediment and have a pH of 5.2–5.5 after incubation for 1 day.

The optimum temperature for growth is 37°C. Strains grow nearly as well at 30°C and poorly if at all at 25°C or 45°C. Growth is stimulated by fermentable carbohydrate, but inhibited by 6.5% NaCl or 20% bile.

Slight to moderate gas is detected in PYG deep agar cultures.

Milk is acidified and slightly curdled.

H_2S is produced from the reduction of bisulfite, but not from sulfate in SIM medium. Nitrate is reduced in Bacto-nitrate broth media supplemented with 0.3% agar, glycerol and galactose, but two of the four strains did not reduce nitrate in indole-nitrite medium (BBL).

Urease is produced.

Neutral red is reduced; resazurin is not reduced.

Fermentation products in PYG broth are acetic and formic acids and ethanol, with small amounts of butyric and pyruvic acids; succinic and fumaric acids are sometimes produced. Abundant H_2 is formed. Pyruvate is converted to acetate and formate; neither threonine, lactate nor gluconate is utilized.

All strains tested are susceptible to chloramphenicol, clindamycin, erythromycin, penicillin G and tetracycline.

Culture supernatants are nontoxic to mice.

Other characteristics of the species are given in Table 13.14.

Isolated from human feces.

The mol% G + C of the DNA has not been reported.

Type strain: ATCC 27791 (DSM 1785).

Further comments. Distinctive characteristics of this species are: production of acetic and formic acids and ethanol from glucose; production of urease and lack of production of lecithinase.

17. Clostridium cellobioparum Hungate 1944, 503.[AL]

cel.lo.bi.o′par.um. N.L. n. *cellobiosum* cellobiose; L. verb. adj. suff. *-parus* producing; N.L. neut. adj. *cellobioparum* (sic.) cellobiose-producing.

This description is based on that of Hungate (1944) and study of the type strain.

Cells in PYG broth culture stain Gram-negative and are straight or slightly curved rods, motile and peritrichous, 0.5–0.6 × 1.4–3.3 μm, and occur singly or in pairs.

Figure 13.21.   *Clostridium cellobioparum.*

Spores are spherical or oval, terminal and swell the cell. Sporulation occurs most readily on cellulose agar or in 3-week-old rumen fluid broth cultures.

Cell walls contain *meso*-DAP (Gottschalk et al., 1981).

Surface colonies on blood agar plates are 0.5–1.5 mm in diameter, circular, entire, convex to pulvinate, semiopaque, creamy white to yellowish, shiny and smooth. On rumen fluid cellobiose agar they may have a scalloped margin, raised elevation and translucent appearance.

Cultures in PYG broth supplemented with rumen fluid are turbid with a stringy sediment and have a pH of 5.3–5.5 after incubation for 5 days.

The optimum temperature for growth is 30–37°C. There is little or no growth at 25°C and none at 45°C. Growth is stimulated by rumen fluid, a fermentable carbohydrate, and a gaseous atmosphere of 90% N_2 and 10% CO_2 (rather than 100% CO_2 or 100% N_2). H_2 gas produced by this organism will limit growth unless removed either mechanically or by H_2-utilizing organisms growing concurrently (Chung, 1976).

Neither neutral red nor resazurin is reduced.

Cellulose is hydrolyzed to cellobiose; glucose is not formed (Hungate, 1944).

Fermentation products in PYG broth are acetic, lactic and formic acids, ethanol, CO_2 and large amounts of H_2. Pyruvate is converted to acetate; neither lactate nor threonine is utilized.

The type strain is susceptible to chloramphenicol, clindamycin, erythromycin, penicillin G and tetracycline.

Culture supernatants are nontoxic to mice.

Other characteristics of the species are given in Table 13.14.

Isolated from the bovine rumen (Hungate, 1944; Jayne-Williams, 1979), and from human feces (Finegold et al., 1983).

The mol% G + C of the DNA is 28 (T_m) (Johnson and Francis, 1975).

Type strain: ATCC 15832 (NCIB 10669; DSM 1351).

Further comments. This species is most easily differentiated from the cellulose-digesting *C. papyrosolvens*, which it resembles most closely phenotypically, by its fermentation of maltose and melibiose.

18. **Clostridium chauvoei** (Arloing, Cornevin, and Thomas 1887) Scott 1928, 260.[AL] (*Bacterium chauvoei* Arloing, Cornevin, and Thomas 1887, 82.)

chau'voe.i. N.L. gen. n. *chauvoei* pertaining to Professor J. A. B. Chauveau, French bacteriologist.

Description from those of Smith (1975a), Holdeman et al. (1977) and study of the type and 23 other strains.

Cells in PYG broth cultures are Gram-positive rods, usually motile and peritrichous. They may be quite pleomorphic with irregular staining, particularly in older cultures. Citron forms are seen often. Cells are 0.5–1.7 × 1.6–9.7 µm and occur singly or in pairs.

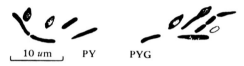

Figure 13.22. *Clostridium chauvoei.*

Spores are oval, central to subterminal and swell the cell. Sporulation occurs readily both in broth and on solid media.

Cell walls contain lysine (Weiss et al., 1981).

Surface colonies on blood agar plates are β-hemolytic, 0.5–3 mm in diameter, circular, low convex or raised, translucent to opaque, granular, shiny or dull, smooth, and have entire to erose margins.

Cultures in PYG broth are turbid with a smooth sediment and have a pH of 5.0–5.4 after incubation for 4 days.

The optimum temperature for growth is 37°C. There is poor growth at 25°C and 30°C, no growth at 45°C. Growth is stimulated by fermentable carbohydrate, inhibited by 6.5% NaCl, 20% bile, or a pH of 8.5.

Abundant gas is formed in PYG deep agar cultures.

Neutral red is reduced; reduction of resazurin is variable.

Products in PYG broth are acetic, butyric and formic acids, butanol, CO_2 and H_2. Small amounts of lactic, succinic and pyruvic acids may be detected. Pyruvate is converted to acetate and butyrate; neither lactate nor threonine is utilized.

All strains tested are susceptible to chloramphenicol, clindamycin, erythromycin, penicillin G, and tetracycline.

Culture supernatants are nontoxic to mice. Strains produce deoxyribonuclease (β-toxin) and hyaluronidase (γ-toxin) (Princewill and Oakley, 1976a). Neuraminidase is also produced (Müller, 1976; Fraser, 1978). Strains are pathogenic through tissue invasion for mice, guinea pigs and hamsters. Experimentally, $CaCl_2$ must be injected to provide some tissue destruction before infection can occur. Cattle, sheep, goats, swine, deer, mink (Langford, 1970) freshwater fish (Prévot et al., 1950), whales and frogs (Scott, 1928) are susceptible; humans, birds, cats, dogs and rabbits are resistant, although Berg and Fales (1977) have reported isolation of the organism from wounds in dogs and cats.

Other characteristics of the species are given in Table 13.12.

Isolated from infections in cattle, sheep, and other animals and from intestinal contents of cattle and dogs. No well-documented strain of *C. chauvoei* has been isolated from humans. The organism is best known as the cause of blackleg in cattle and sheep. Habitat is probably the soil, and outbreaks of the disease often follow soil excavation in areas where animals graze (Barnes et al., 1975).

The mol% G + C of the DNA is 27 (T_m) (Marmur and Doty, 1962).

Type strain: ATCC 10092 (NCIB 10665).

Further comments. This species is most easily differentiated from *C. septicum*, which it resembles most closely phenotypically, by its fermentation of sucrose and lack of fermentation of cellobiose or trehalose. Patterns of soluble cellular proteins are distinctive for each species (Cato et al., 1982a) and specific fluorescent antibody is available commercially for identification.

19. **Clostridium clostridioforme** (corrig.) (Burri and Ankersmit 1906) Kaneuchi, Watanabe, Terada, Benno and Mitsuoka 1976a, 202.[AL] (*Bacterium clostridiiforme* Burri and Ankersmit 1906, 115.)

clos.tri.di.o.for'me. Gr. n. *kloster* a spindle; Gr. diminutive suff. -idion; N.L. neut. n. *clostridium* a small spindle; L. suff. -*formis* in the form of; N.L. neut. adj. *clostridioforme* in the form of a small spindle, spindle-shaped.

This description is based on those of Holdeman and Moore (1974), Kaneuchi et al. (1976a), Cato and Salmon (1976), and on study of the type and 56 other strains.

Cells in PYG broth culture are Gram-negative straight rods with pointed ends, 0.3–0.9 × 1.4–9.0 µm. Cells usually occur in pairs but they also occur singly or in short chains.

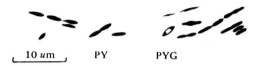

Figure 13.23. *Clostridium clostridioforme.*

Spores are oval, central to subterminal, and swell the cell. They are often difficult to demonstrate, particularly in young cultures or from freshly isolated strains. Sporulation occurs most reliably on chopped meat agar slants incubated at 30°C for 2–3 weeks or in 3-week-old chopped meat broth cultures. All strains resist heating at 70°C for 10 min, but many do not survive heating at 80°C (Kaneuchi et al., 1976a).

Motility is variable but sluggishly motile peritrichous cells, or cells with a subpolar tuft of flagella, can be detected in the majority of strains.

Cell walls contain *meso*-DAP (Weiss et al., 1981).

Surface colonies on blood agar plates are nonhemolytic, 0.5–2.0 mm in diameter, entire, slightly scalloped or erose, convex to slightly peaked, translucent to opaque, gray-white, usually with a mottled or mosaic internal structure.

Cultures in PYG broth are turbid with a heavy, sometimes viscous or ropy sediment and have a pH of 5.2–5.5 after incubation for 24 h.

Optimum temperature for growth is 37°C; many strains grow equally well at 30°C; growth is moderate at 25°C, usually poor at 45°C. Growth is inhibited by 6.5% NaCl or a pH of 8.5; strains vary in their reaction to 20% bile.

Abundant gas is detected in PYG deep agar cultures.

Ammonia is produced. H_2S is produced by 11 of 21 strains tested. Neutral red is reduced; reduction of resazurin is variable.

Low levels of superoxide dismutase have been reported from two clinical isolates (Tally et al., 1977). Deoxyribonuclease is formed by the one strain tested (Porschen and Sonntag, 1974).

Fermentation products in PYG broth cultures are acetic acid, usually formic and lactic acids and ethanol. Large amounts of H_2 are formed. Pyruvate is converted to acetate, usually with formate and ethanol. Some strains convert threonine to propionate; lactate is not utilized.

Susceptibility to antibiotics is variable. Of 42 strains tested, 28 are susceptible to chloramphenicol, 31 to clindamycin, 19 to erythromycin, 23 to penicillin G, and 27 to tetracycline. Edson et al. (1982) reported that their clinical isolates were most sensitive to metronidazole. An active β-lactamase has been demonstrated in one strain (Weinrich and Del Bene, 1976).

Culture supernatants are nontoxic for mice.

Other characteristics of the species are given in Table 13.14.

Isolated from intestinal contents of birds, humans and other animals, from calf rumen contents, from turkey liver lesions, from abdominal, cervical, scrotal, pleural and other infections, from septicemias, peritonitis, and appendicitis.

The mol% G + C of the DNA is 47–49 (T_m) (Kaneuchi et al., 1976a; Cato and Salmon, 1976).

Type strain: ATCC 25537 (DSM 933).

Further comments. Biochemical characteristics of strains in this species are quite variable, even those of a strain tested at different times. However, because of their morphological similarity and the close agreement in high G + C ratios of DNA found in all strains tested, they are retained in one species. When spores are not detected, they can be distinguished from other Gram-negative anaerobes by their fermentation products. When spores are detected, the spindle-shaped cells are distinctive. Fermentation of mannose distinguishes the species from *C. indolis*; failure to ferment mannitol distinguishes it from *C. sphenoides* and *C. ramosum*.

20. **Clostridium coccoides** Kaneuchi, Benno, and Mitsuoka 1976, 485.[AL]

coc.coi′des. Gr. n. *coccos* a berry; Gr. n. *eidos* shape; N.L. adj. *coccoides* berry-shaped.

This description is based on study of the type strain and on the description of Kaneuchi et al. (1976b).

Cells in PYG broth cultures are nonmotile, Gram-positive, coccoid short rods, 0.6–1.0 × 0.6–1.5 μm and occur singly, in pairs, and in short chains.

Figure 13.24.   *Clostridium coccoides.*

Spores are round, central to subterminal, and slightly swell the cell. Sporulation occurs most readily on chopped meat agar slants or on modified Eggerth-Gagnon agar plates (Mitsuoka et al., 1965) after incubation at 37°C for 2–5 days. All strains survive heating at 70°C for 10 min but resistance is variable to 80°C for 10 min.

Cell walls contain *meso*-DAP, lysine and alanine (Gottschalk et al., 1981; Weiss et al., 1981).

Surface colonies on blood agar plates are punctiform to 1 mm in diameter, circular, slightly irregular, slightly undulate, convex, gray-white, shiny, smooth and nonhemolytic. On modified Eggerth-Gagnon agar plates, colonies are 1.5–2.5 mm in diameter, circular, convex, entire, translucent, yellowish gray, smooth, shiny and nonhemolytic.

Cultures in PYG broth are turbid with a smooth sediment and have a pH of 4.4 after incubation for 24 h. Growth is stimulated by fermentable carbohydrate.

The optimum temperature for growth is 37°C. Strains grow well at 25°C and 45°C but poorly at 15°C.

Abundant gas is detected in PYG deep agar cultures.

Resazurin is reduced.

Acid is produced from inositol, dulcitol, α-methylglucoside, and α-methylmannoside. Orotic acid and tributyrin are not hydrolyzed. Small amounts of H_2S are produced.

Products in PYG broth culture after 5 days of incubation are acetic and succinic acids and abundant H_2. Neither lactate nor threonine is utilized.

The type strain is susceptible to chloramphenicol, clindamycin, erythromycin and tetracycline. It is resistant to penicillin G.

Culture supernatants are nontoxic to mice.

Other characteristics of the species are given in Table 13.14.

Isolated from the cecum of mice fed high lactose diets.

The mol% G + C of the DNA is 43–45 (T_m) (Kaneuchi et al., 1976b).

Type strain: ATCC 29236 (NCTC 11035; DSM 935).

Further comments. This species is readily differentiated from other clostridial species by its morphology, by its production of large amounts of succinic acid, by its fermentation of inositol and by its failure to produce indole.

21. **Clostridium cochlearium** (Douglas, Fleming, and Colebrook 1919) Bergey, Harrison, Breed, Hammer, and Huntoon 1923, 333.[AL] (*Bacillus cochlearius* Douglas, Fleming, and Colebrook *in* Bulloch et al. 1919, 40; *Clostridium lentoputrescens* Hartsell and Rettger 1934.)

coch.le.a′ri.um. L. n. *cochlear* spoon; N.L. neut. adj. *cochlearium* resembling a spoon.

This description is based on those of Holdeman et al. (1977), Nakamura et al. (1979), and on study of the type and 11 other strains.

Cells in PYG broth are motile and peritrichous, straight to slightly curved rods with rounded ends, 0.5–1.3 × 1.6–14.1 μm, and occur singly or in pairs.

Figure 13.25.   *Clostridium cochlearium.*

Spores are round to oval, swelling the cell, usually terminal although subterminal spores are seen occasionally. Sporulation occurs readily both in broth media or on solid media (blood agar plates or chopped meat slants). Three distinct spore coats and tubular appendages attached to one end of the spore have been demonstrated (Pope and Rode, 1969).

Cell walls contain *meso*-DAP (Weiss et al., 1981).

Surface colonies on blood agar plates are 0.5–2.0 mm, circular to irregular, slightly scalloped to lobate, flat to raised, translucent, gray-white, smooth, shiny, with a mottled or mosaic internal structure. Hemolysis is variable.

Cultures in PYG broth are turbid with a smooth sediment, and have a pH of 6.0–6.5 after 1 week of incubation.

Optimum temperature for growth is 37°C. Most strains grow well at 30°C and 45°C but poorly if at all at 25°C. Growth is inhibited by 6.5% NaCl, by 20% bile or at a pH of 8.5.

Gelatin may be weakly and slowly digested; neither meat, casein nor milk is digested. H_2S and ammonia are formed. Hippurate is hydrolyzed by three of eight strains tested. Neutral red is reduced.

One strain isolated from human subgingival plaque degrades fibrinogen, possibly contributing to periodontal disease (Wikström et al., 1983).

Products in PYG broth are butyrate, acetate and propionate; small amounts of lactate, formate, succinate and butanol are sometimes formed; abundant H_2 is produced. Pyruvate is converted to acetate and butyrate; threonine is converted to propionate; lactate is not utilized. Glutamate is fermented to butyrate, acetate, CO_2, and ammonia by the methylaspartate pathway (Buckel and Barker, 1974). Serine, aspartate, and histidine are utilized for growth (Mead, 1971). Phenylalanine is fermented to phenol (Elsden et al., 1976). Major cellular fatty acids produced in trypticase-yeast extract medium are principally of the straight-chain saturated series including C_{12}, C_{14}, C_{15}, C_{16}, and C_{18} acids; small amounts of straight-chain unsaturated acids, C_{15} and C_{18}, each with one double bond, are also formed (Elsden et al., 1980).

All strains tested are susceptible to clindamycin, erythromycin, penicillin G and tetracycline. One of six strains is resistant to chloramphenicol.

Culture supernatants are nontoxic to mice.

Other characteristics of the species are given in Table 13.13.

Isolated from soil, human oral cavity, human and horse feces, wounds and crabmeat.

The mol% G + C of the DNA is 27–28 (T_m) (Johnson and Francis, 1975).

Type strain: ATCC 17787 (NCIB 10633; DSM 1285).

Further comments. This species can be differentiated from *C. tetani*, which it resembles most closely phenotypically, by its lack of toxicity for mice, by its weak to negative digestion of gelatin, and by distinct patterns of soluble cellular proteins of the two species (Cato et al., 1982a).

22. **Clostridium cocleatum** Kaneuchi, Miyazato, Shinjo, and Mitsuoka 1979,10.[AL]

co.cle.a′tum. L. n. *coclea* a snail shell or whirlpool; N.L. neut. adj. *cocleatum* in the shape of a snail shell or whirlpool.

This description is based on that of Kaneuchi et al. (1979) and study of the type and two other strains.

Cells in PYG broth cultures are Gram-positive rods, nonmotile, 0.3–0.4 × 1.0–5 µm, semicircular, circular or spiral-shaped. They may grow in long chains, and be Gram-negative with filaments or swollen forms.

Figure 13.26. *Clostridium cocleatum.*

Spores are difficult to demonstrate but can be found most reliably on medium 10 agar (Caldwell and Bryant, 1966) or on chopped meat agar slants incubated at 30°C. Spores are round, terminal to subterminal and swell the cell.

Cell walls contain *meso*-DAP (Weiss et al., 1981).

Surface colonies on blood agar plates are pinpoint to 2 mm in diameter, circular, entire to erose, low convex to pulvinate, semiopaque to opaque, smooth, shiny, grayish white and nonhemolytic.

Cultures in PYG broth are turbid with a smooth to mucoid sediment and have a pH of 4.9–5.4 after incubation for 4 days.

The optimum temperature for growth is 37–45°C. Strains grow nearly as well at 30°C, poorly at 25°C and not at all at 15°C. Growth is stimulated by fermentable carbohydrate and by rumen fluid but inhibited by 6.5% NaCl or 20% bile.

Abundant gas is detected in PYG deep agar cultures.

Neutral red and resazurin are reduced.

Dextrin, dulcitol, α-methylglucoside, and α-methylmannoside are not fermented.

Products in PYG broth are acetic and formic acids; lactic acid usually is produced. Abundant H_2 is formed. Pyruvate is converted to acetate and formate; neither threonine nor lactate is utilized.

All strains tested are susceptible to chloramphenicol, erythromycin, penicillin G, and tetracycline but resistant to clindamycin.

Toxicity and pathogenicity have not been determined.

Other characteristics of the species are given in Table 13.14.

Isolated from feces of healthy humans and from cecal contents of mice, rats, and chickens.

The mol% G + C of the DNA is 28–29% (T_m) (Kaneuchi et al., 1979).

Type strain: ATCC 29902 (NCTC 11210; DSM 1551).

Further comments. This species is closely related to *C. spiroforme*, having 46–60% homology with strains of that species and being similar in morphology. *C. cocleatum* ferments galactose, while *C. spiroforme* does not. The patterns of soluble cellular protein as determined by polyacrylamide gel electrophoresis are also helpful in differentiating the two species (Cato et al., 1982a).

23. **Clostridium colinum** (*ex* Berkhoff, Campbell, Naylor and Smith 1974) Berkhoff 1985, 157.[VP] (*Clostridium colinum* Berkhoff et al. 1974, 203.)

co.li′num. L. n. *colinus* quail; N.L. neut. adj. *colinum* referring to the most susceptible host, the quail.

This description is based on that of Berkhoff et al. (1974), and on study of six strains included in that description.

Cells in PYG broth culture are Gram-positive rods that may rapidly become Gram-negative. They are motile, peritrichous, 1 × 3–4 µm, and occur singly or in pairs.

Figure 13.27. *Clostridium colinum.*

Spores are oval and subterminal. They are sparse in most media but transfers from chopped meat-glucose medium into PY-1% starch medium will survive heating at 70°C for 10 min.

Cell wall composition has not been reported.

Surface colonies on blood agar plates are pinpoint to 0.5 mm, circular to slightly irregular, low convex, transparent, grayish white to colorless, shiny and smooth. Most strains are α-hemolytic.

Cultures in PYG broth have a smooth white sediment without turbidity and a pH of 5.4 after incubation in an atmosphere of 100% CO_2 for 6 days. In brain-heart infusion broth with glucose and an atmosphere of 100% N_2, the final pH is 5.2–5.5.

The optimum temperature for growth is 37°C. Some strains grow equally well at 45°C. There is little growth at 30°C and none at 25°C. Growth is stimulated by fermentable carbohydrate. Tryptose-phosphate-glucose agar with 8% horse plasma has been recommended for isolation (Berkhoff et al., 1974; Smith, 1975a). Although isolation may be difficult on other media (e.g. supplemented brain-heart infusion agar or rumen fluid-glucose-cellobiose agar, Holdeman et al., 1977), stock cultures usually will grow on these media.

Moderate gas is detected in PYG deep agar cultures.

Products in PYG broth are formic, acetic and small amounts of lactic acids. Propionic acid usually is detected. Abundant H_2 is produced. Neither pyruvate, lactate nor threonine is utilized.

All strains tested are susceptible to chloramphenicol, clindamycin, erythromycin, penicillin G, and tetracycline.

Strains are pathogenic to quail and chickens, causing ulcerative enteritis or liver necrosis and death in less than 18 h. Guinea pigs are not susceptible (Smith, 1975).

Other characteristics of the species are given in Table 13.14.

Isolated from intestinal tracts of quail, pheasants, grouse, partridge, chickens, and turkeys.

The mol% of the DNA has not been reported.

Type strain: ATCC 27770.

24. **Clostridium difficile** (Hall and O'Toole, 1935) Prévot 1938, 84.[AL] (*Bacillus difficilis* Hall and O'Toole, 1935, 390.)

dif′fi.cile. L. neut. adj. *difficile* difficult (referring to "the unusual difficulty that was encountered in its isolation and study").

This description, unless otherwise indicated, is based on that of Holdeman et al. (1977) and on our study of the type and 32 other strains.

Cells are Gram-positive, usually motile in broth cultures, peritrichous and are 0.5–1.9 × 3.0–16.9 µm. Some strains produce chains consisting of two to six cells aligned end-to-end.

Figure 13.28. *Clostridium difficile.*

Spores are oval, subterminal (rarely terminal), and swell the cell. Sporulation by most strains occurs on brucella blood agar incubated for 2 days. Sporulation may be enhanced on solid media containing 0.1% sodium taurocholate (Wilson et al., 1982).

Cell walls contain *meso*-DAP (C. S. Cummins, unpublished data).

Surface colonies on blood agar are 2–5 mm, circular, occasionally rhizoid, flat or low convex, opaque, grayish or whitish, and have a matt to glossy surface. All strains produce an evanescent pale green fluorescence under long wave length ultraviolet light after 48 h incubation on brucella blood agar supplemented with hemin and vitamin K_1 (George et al., 1979).

Cultures in PYG broth are turbid with a smooth sediment and have a pH of 5.0–5.5 after incubation for 5 days.

The optimum temperature for growth is 30–37°C; growth also occurs at 25°C and 45°C. Proline, aspartic acid, serine, leucine, alanine, threonine, valine, phenylalanine, methionine, and isoleucine are utilized for growth; δ-aminovalerate and α-aminobutyrate are produced (Mead, 1971). A selective minimal medium described by Hubert et al. (1981), which contains selected amino acids as a source of carbon and energy, only a trace of fructose (0.1%), 2% bile as a growth stimulant, and 16 μg/ml cefoxitin and 500 μg/ml streptomycin for reduction of associated flora, has been found useful for isolation of these organisms from feces.

Abundant gas is produced in PYG deep agar cultures. Abundant H_2 is produced in PYG broth. Ammonia is produced. H_2S is produced by 8 of 17 strains tested.

The type strain metabolizes the bile acids cholic and chenodeoxycholic acids and splits the conjugate taurocholic acid (Midtvedt and Norman, 1967).

Hyaluronidase, chondroitin sulfatase, and collagenase are present in the one strain studied (Steffen and Hentges, 1981). One of two strains tested produces an extracellular β-glucuronidase (Sakaguchi and Murata, 1983).

Uracil is not utilized (Hilton et al., 1975).

Products in PYG broth include acetic, isobutyric, butyric, isovaleric, valeric, isocaproic, formic and lactic acids. Pyruvate is converted to acetate and butyrate; threonine is converted to propionate; lactate is not utilized. Phenylacetic, phenylpropionic, hydroxyphenylacetic, and indole acetic acids, and *p*-cresol are produced in trypticase medium supplemented with L-phenylalanine, L-tyrosine and L-tryptophan (Elsden et al., 1976; Mayrand and Bourgeau, 1982), *p*-hydroxyphenylacetic acid is converted to *p*-cresol (Phillips and Rogers, 1981). Isocaproate is produced from leucine, isovalerate from leucine and isoleucine, and isobutyrate from valine (Elsden and Hilton, 1978).

All strains are susceptible to 8 U of penicillin G/ml, 4 μg of ampicillin/ml, 4 μg of vancomycin/ml, < 1 μg of rifampin/ml, and 2 μg of metronidazole/ml; susceptibility to clindamycin, cephalosporins, cephamycins, tetracyclines, chloramphenicol and erythromycin is variable, whereas all strains are resistant to aminoglycosides (George et al., 1979; Nakamura et al., 1982). Tetracycline resistance has been demonstrated to be transferable from resistant to sensitive strains; the resistance determinant may be plasmid-mediated (Ionesco, 1980) or chromosomal (Smith et al., 1981).

Pseudomembranous colitis in humans is caused by overgrowth of the organism in the colon, usually after the flora has been disturbed by antimicrobial therapy. A very similar disease can be produced in hamsters and several other rodents by administration of antibiotics, but rats and mice are not affected. Although *C. difficile* has been reported to cause cecitis in rabbits (Rehg and Pakes, 1982) and hares (Debard et al., 1979), *C. spiroforme* appears to be a more common cause of the disease in these animals (Borriello and Carman, 1983).

C. difficile produces two large protein toxins (toxins A and B), and hamsters must be immunized against both toxins to survive *C. difficile* cecitis (Libby et al., 1982). Toxin A is lethal when given orally to hamsters but toxin B is not; either toxin is lethal when injected intraperitoneally into these rodents. Toxins A and B are lethal for mice (Taylor et al., 1981). Ruwart et al. (1983) have found that 16,16-

dimethyl-prostaglandin E_2 may inhibit production or release of cytopathic toxin(s) in vitro. Toxin A has been referred to as the enterotoxin because it causes fluid accumulation in the bowel, but the mechanism of action is not through stimulation of adenyl cyclase. Toxin B does not cause fluid accumulation, but is extremely cytopathic for all tissue-cultured cells tested. Exposure to less than a picogram of this toxin causes cells to become round, detach from supports and slowly die. The mechanism of action is unknown. A motility altering factor also has been described (Taylor et al., 1981; Banno et al., 1981; Libby et al., 1982; and Justus et al., 1982) that is different from the two known toxins, but the significance is unknown.

Other characteristics of the species are shown in Table 13.12.

Isolated from marine sediment (Matches and Liston, 1974); soil; sand (Hafiz and Oakley, 1976); the hospital environment (Mulligan et al., 1980); camel, horse and donkey dung (Hafiz and Oakley, 1976); feces of dogs, cats and domestic birds (Borriello et al., 1983); the human genital tract (Hafiz et al., 1975), feces of humans without diarrhea (Viscidi et al., 1981; Stark and Lee, 1982; Finegold et al., 1983) and rarely from blood and pyogenic infections in humans (Smith and King, 1962; Alpern and Dowell, 1971; Finegold et al., 1975; Gorbach and Thadepalli, 1975; and Lewis et al., 1980) and animals (Hirsh et al., 1979).

The mol% G + C of the DNA is 28 (Gottschalk et al., 1981).

Type strain: ATCC 9689 (NCIB 10666; DSM 1296; CCUG 4938).

Further comments. This species is most easily distinguished from *Clostridium sporogenes*, which it resembles most closely phenotypically, by its ability to ferment mannitol, and by its inability to digest meat or milk, or to produce lipase. *C. difficile* is one of the few species that produces isocaproic and valeric acids.

25. **Clostridium durum** Smith and Cato 1974, 1393.[AL]

dur'um. N.L. neut. adj. *durum* hard, tough, resistant, indicating ability to survive for long periods of time.

This description is based on that of Smith and Cato (1974).

Cells in PYG broth cultures incubated at 30°C in an atmosphere of 100% N_2 are Gram-positive rods that rapidly become Gram-negative. They are motile and peritrichous, 0.8–1.7 × 3.0–15 μm, and occur singly or in pairs.

10 μm   PY   PYG

Figure 13.29. *Clostridium durum.*

Spores are oval, subterminal to terminal, and are formed only when the organisms are grown anaerobically. Sporulation occurs most readily in chopped meat broth incubated at 30°C for 24 h.

Cell wall composition has not been determined.

Surface colonies on anaerobic blood agar plates are 0.5–1.5 mm in diameter, circular, entire, convex, translucent, grayish, shiny and may be slightly pebbled. Colonies of most strains are surrounded by a narrow zone of complete hemolysis. Growth is equally good on blood agar plates incubated in air, and similar colonies are produced. Although some strains grow to the surface in PYG deep agar cultures, they do not grow on the surface.

Cultures in PYG broth are turbid with a slightly flocculent sediment and have a pH of 5.3 after incubation for 24 h.

The optimum temperature for growth is 30°C. Strains grow nearly as well at 25°C and 37°C but not at all at 45°C. Growth is stimulated by fermentable carbohydrate and is inhibited by 20% bile, linoleic or linolenic acids. There is no growth on or in egg-yolk agar plates.

Abundant gas is detected in PYG deep agar cultures.

Acetyl methyl carbinol is produced by the type strain. Neutral red and resazurin are reduced.

Products in PYG broth are acetic, formic and lactic acids and ethanol. Abundant H_2 is formed. Pyruvate is converted to acetate and formate; gluconate is converted to acetate, formate and ethanol; neither lactate nor threonine is utilized.

The type strain is susceptible to clindamycin, chloramphenicol, erythromycin and tetracycline; it is resistant to penicillin G.

Toxicity and pathogenicity have not been reported.

Other characteristics of the species are given in Table 13.14.

Isolated from marine sediment.

The mol% G + C of the DNA is 50 (T_m) (Smith and Cato, 1974).

Type strain: ATCC 27763 (DSM 1735).

Further comments. This species can be differentiated from other aerotolerant clostridia by its inability to grow on egg-yolk agar; unlike *C. histolyticum*, it does not hydrolyze gelatin; unlike *C. carnis* and *C. tertium*, it does not produce butyric acid and does not ferment lactose or ribose.

26. **Clostridium fallax** (Weinberg and Séguin, 1915a) Bergey, Harrison, Breed, Hammer and Huntoon 1923, 325.[AL] (*Bacillus fallax* Weinberg and Séguin 1915a, 686.)

fal'lax. L. adj. *fallax* deceptive.

This description is based on those of Weinberg et al. (1937), Smith and Hobbs (1974), Holdeman et al. (1977), and study of the type and four other strains.

Cells in PYG broth cultures are Gram-positive rods with rounded ends, 0.5–1.4 × 1.6–15.4 µm that occur singly or in pairs. In freshly isolated strains and young cultures, cells are motile and peritrichous, but both motility and flagella may be lost on subsequent transfer.

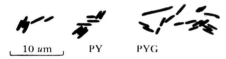

Figure 13.30. *Clostridium fallax.*

Spores are oval, central to subterminal, and swell the cell. Sporulation of most strains occurs most readily in chopped meat broth cultures.

Cell walls contain LL-DAP and the sugars glucose, galactose and rhamnose (Cummins and Johnson, 1971).

Surface colonies on blood agar plates are 1–5 mm in diameter, β-hemolytic, circular to slightly irregular, with entire to slightly erose margins, raised or convex, translucent, gray, shiny and smooth, often with a mottled or granular internal structure.

Cultures in PYG broth are turbid with a smooth sediment and have a pH of 4.8–5.3 after incubation for 1 week.

The optimum temperature for growth is 37°C. Strains grow nearly as well at 30°C and 45°C but poorly if at all at 25°C. Growth is inhibited by 20% bile, 6.5 NaCl, or a pH of 8.5.

Abundant gas is detected in PYG deep agar cultures.

Deoxyribonuclease is present (Johnson and Francis, 1975).

Products in PYG broth are acetic, butyric and lactic acids; pyruvic and succinic acids are usually detected. Abundant H_2 is produced. Pyruvate is converted to acetate; neither lacetate nor threonine is utilized.

Ammonia and H_2S are produced.

All strains tested are susceptible to chloramphenicol, clindamycin, erythromycin, penicillin G and tetracycline.

Culture supernatants are nontoxic to mice. The organism has been reported to be pathogenic for guinea pigs and mice but pathogenicity is quickly lost (Weinberg et al., 1937).

Other characteristics of the species are given in Table 13.14.

Isolated from soil, marine sediments, animal wounds and clinical specimens from soft tissue infections in humans (Gorbach and Thadepalli, 1975); also isolated from human feces (Finegold et al., 1983).

The mol% G + C of the DNA of the type strain is 26 (T_m) (Cummins and Johnson, 1971).

Type strain: ATCC 19400 (NCTC 8380; NCIB 10634; CCUG 4853).

27. **Clostridium felsineum** (Carbone and Tombolato, 1917) Spray 1939, 766.[AL] (*Bacillus felsineus* Carbone and Tombolato 1917, 563.)

fel.si.'ne.um. L. n. *Felsina,* an ancient Latin name for Bologna, Italy; N.L. neut. adj. *felsineum* pertaining to Bologna.

This description is based on those of Smith and Hobbs (1974) and of Holdeman et al. (1977) and on study of the type and 12 other strains.

Cells in PYG broth cultures are Gram-positive rods, Gram-negative in older cultures, and 0.5–1.3 × 3.1–25.7 µm. They are motile and peritrichous although motility may be lost in cultures that have been maintained in the laboratory for many years. They are granulose-positive in starch medium. Cells occur singly, in pairs, and in short or long chains.

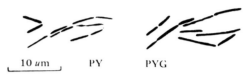

Figure 13.31. *Clostridium felsineum.*

Spores are oval, subterminal, and swell the cell. There are no appendages and no exposporium (Duda and Dobritsa, 1977). Spores are difficult to detect but usually can be found in PYG or PY-starch broth cultures incubated for 3–5 days.

Cell walls contain *meso*-DAP, major amounts of galactose and rhamnose, with lesser amounts of glucose and mannose (Cummins and Johnson, 1971).

These organisms grow very poorly if at all on anaerobic blood agar or egg-yolk agar plates. Surface colonies on brain-heart infusion agar roll streak tubes after incubation for 2–4 days are 1–4 mm in diameter, circular, flat to low convex, translucent to opaque, with a mottled or granular surface, a pebbled or mosaic internal structure and an entire to slightly scalloped or lobate margin. They may be white, yellow, orange, or brownish. When slight growth occurs on anaerobic blood agar plates (3 of 13 strains), colonies are β-hemolytic.

Cultures in PYG broth are turbid with a heavy, ropy or flocculent, often dark orange sediment, and have a pH of 4.7–5.4 after incubation for 24 h.

The optimum temperature for growth is 37°C. Most strains grow nearly as well at 30°C, but poorly if at all at 25°C or 45°C. Growth is markedly stimulated by fermentable carbohydrate but inhibited by 20% bile, 6.5% NaCl, or pH of 8.5.

Abundant gas is detected in PYG deep agar cultures.

Acetyl methyl carbinol is produced by 7 of 11 strains tested. Neutral red is reduced; resazurin is reduced by 4 of 13 strains tested.

Each of four strains tested, including the type strain, completely digests thin slices of carrot, turnip and radish in PY-broth cultures within 3 days. PY-pectin is strongly fermented (pH 4.8–5.2), and Lund and Brocklehurst (1978) have demonstrated pectic lyase, pectic hydrolase and polygalacturonase but no detectable pectinesterase activity in members of the species.

Spore suspensions of a strain of *C. felsineum* will germinate in and lyse tumor tissue but not healthy tissue in mice (Möse and Möse, 1964). Destruction of the tumor tissue, however, is neither complete nor permanent.

Atmospheric N_2 is fixed (Rosenblum and Wilson, 1949).

Products in PYG broth culture are butyric and acetic acids and butanol; lactic, formic and acetic acids may be detected. Pyruvate is converted to acetate, butyrate, and usually butanol. Neither lactate nor threonine is utilized. Abundant H_2 is detected. Acetone, CO_2, and ethanol may be produced also (Spray and McClung, 1957).

All strains tested are susceptible to clindamycin, erythromycin, penicillin G and tetracycline. One of five strains is resistant to chloramphenicol.

Culture supernatants are nontoxic to mice.

Other characteristics of the species are given in Table 13.12.

Isolated from retting flax, from soil in the United States (Smith, 1975) and in Antarctica (Miwa, 1975 a and b), and from human feces (Drasar et al., 1976; Finegold et al., 1983).

The mol% G + C of the DNA is 26 (T_m) (Cummins and Johnson, 1971).

Type strain: ATCC 17788 (NCIB 10690; DSM 794).

Further comments. Most of the strains available for study were isolated from enrichment cultures in 5% corn meal mash as described by McClung (1943). Pigmentation of colonies is most pronounced in this medium. Plating medium for purification of cultures was yeast infusion-starch agar and plates were incubated in an oat jar. Colonies of *C. felsineum* under these conditions are yellow and, unlike colonies of *C. roseum*, do not darken on exposure to air. *C. felsineum* may be distinguished from *C. aurantibutyricum* by failure to reduce nitrate and from *C. puniceum* by fermentation of rhamnose, by lack of pectinesterase and by colony pigmentation on corn meal agar.

28. **Clostridium formicaceticum** (corrig.) Andreesen, Gottschalk, and Schlegel 1970, 155.[AL]

for.mic.a.ce′ti.cum. L. n. *formica* an ant, N.L. adj. *formicus* pertaining to ants, to formic acid; L. n. *acetum* wine-vinegar, N.L. adj. *aceticus* pertaining to vinegar, to acetic acid; N.L. neut. adj. *formicaceticum* pertaining to formic and acetic acids.

This description is based on that of Andreesen et al. (1970) and on study of the type and three other strains.

Cells in PY-fructose broth cultures are Gram-negative, straight to slightly curved rods, motile and peritrichous, 1.2–2.0 × 5–12 μm, and occur singly or in pairs.

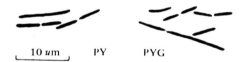

10 *um*   PY   PYG

Figure 13.32. *Clostridium formicaceticum.*

Spores are round, terminal or subterminal, and swell the cell. Sporulation occurs readily both in broth media and on chopped meat slants.

Cell walls contain *meso*-DAP (Weiss et al., 1981).

Surface colonies on brain-heart infusion agar streak tubes incubated in an atmosphere of 100% N_2 are 1–3 mm in diameter, circular to slightly irregular, entire, flat to low convex, semiopaque, white, shiny, smooth, with a mosaic internal structure.

Cultures in PY-fructose broth are slightly turbid with a smooth sediment and have a pH of 6.3–6.7 (compared with a pH of 7.2 in PY broth) after incubation for 4 days in an atmosphere of 100% N_2.

Ammonia is produced. Resazurin is not reduced.

The optimum temperature for growth is 37°C. Moderate growth occurs at 28°C, 32°C, and 44°C; there is no growth at 52°C. Bicarbonate or formate and fermentable carbohydrate are required for growth. Substrates utilized for growth include fructose, ribose, gluconate, glucuronate, galacturonate, 2-keto-3-deoxygluconate, mannonate, galacturonate, glutamate, malate, mannitol, glycerol, lactate, pyruvate, fumarate and pectin (Andreesen et al., 1970).

An α-isopropylmalate synthase has been identified in extracts of *C. formicaceticum* indicating that leucine is synthesized by the isopropylmalate pathway used by many aerobic organisms (Wiegel and Schlegel, 1977; Wiegel, 1981).

Enzymes utilized in the degradation of glutamate and those required for the assimilation of N_2 and NH_4^+ have been identified by Bogdahn et al. (1983).

Carbon monoxide is oxidized to CO_2 by cell suspensions from cultures grown in fructose broth and this reaction is coupled with the reduction of CO_2 to acetate (Diekert and Thauer, 1978).

During active growth in an atmosphere of either 100% N_2 or 90% N_2-10% CO_2, acetate is the major product detected in PY-fructose broth cultures; small amounts of succinate also may be formed. In the stationary phase of growth, both acetate and formate are produced. No H_2 is produced. L-malate is converted to acetate and CO_2; fumarate is reduced to succinate, acetate and CO_2 (Dorn et al., 1978). The electron carriers cytochrome *b* and menaquinone are present in cultures and possibly involved in this reaction (Gottwald et al., 1975). For further discussion of the fermentation of fructose and the synthesis of acetate from CO_2, see O'Brien and Ljungdahl (1972), Thauer et al. (1977), Ljungdahl and Wood (1982) and Zeikus (1983).

All strains tested are susceptible to chloramphenicol, clindamycin, erythromycin, penicillin G and tetracycline.

Culture supernatants are nontoxic to mice.

Other characteristics of the species are given in Table 13.13.

Isolated from sewage, pond and ditch mud, and stagnant river water.

The mol% G + C of the DNA is 34 (T_m) (Matteuzzi et al., 1978).

Type strain: ATCC 27076 (DSM 92).

Further comments. This species can be distinguished most readily from *C. aceticum* by its production of formate in later stages of growth, and by its inability to utilize H_2 as a reducing agent in the conversion of CO_2 to acetate (Andreesen et al., 1970). It differs from *C. thermaceticum* in its inability to grow at 55°C.

29. **Clostridium ghonii** (corrig.) Prévot 1938, 83.[AL]

gho′ni.i. N.L. gen. pertaining to Professor Ghon, German bacteriologist.

This description is based on that of Holdeman et al. (1977) and on study of the type and four other strains.

Cells in PYG broth culture are Gram-positive, straight rods, motile and peritrichous, and 0.5–1.4 × 1.6–6.3 μm. They usually occur singly, occasionally in pairs.

10 *u*m   PY   PYG

Figure 13.33. *Clostridium ghonii.*

Spores are central to subterminal, oval and swell the cell. Sporulation occurs most readily on egg-yolk agar plates or in chopped meat broth incubated for 2 days.

Cell walls contain mesoDAP (C. S. Cummins, unpublished data).

Surface colonies on blood agar plates are usually β-hemolytic, 0.5–2.0 mm in diameter, circular to slightly irregular, scalloped to lobate, translucent to semiopaque, flat to raised, white, shiny and with a granular or mosaic internal structure.

Cultures in PYG broth are turbid with a smooth to ropy sediment and a pH of 6.2–6.4 after incubation for 4 days. Growth is stimulated by carbohydrate even though no pH depression occurs. Good growth occurs at a pH of 8.5; growth is inhibited by 6.5% NaCl or 20% bile.

Abundant gas is detected in PYG deep agar cultures.

Gelatin, milk, meat and casein are rapidly digested. H_2S and ammonia are formed. Neutral red is reduced; two of the five strains reduce resazurin.

The organism contains a wide variety of cellular C_{12}–C_{18} fatty acids with unbranched, saturated fatty acids predominating, but with significant amounts of branched chain acids as well (Elsden et al., 1980).

Products in PYG broth after 4 days of incubation include acetate, isobutyrate, butyrate, isovalerate, isocaproate, and ethanol, butanol and isobutanol; formate, propionate and propanol also may be detected.

Moderate amounts of H_2 are formed. *C. ghonii* has been reported to form significant amounts of methane and dithia-2,3 butane, and moderate amounts of propanone and thiocyclopropane in sodium thioglycolate glucose broth cultures (Rimbault and Leluan, 1982), but this finding requires confirmation. Pyruvate is converted to acetate, butyrate and ethanol; threonine is converted to propionate; lactate is not utilized. Valine is converted to isobutyrate, leucine to isovalerate and isocaproate, isoleucine to isovalerate (Elsden and Hilton, 1978). Phenylpropionic and phenyllactic acids and indole are formed in trypticase-yeast extract-thioglycollate medium containing phenylalanine, tyrosine and tryptophan (Elsden et al., 1976).

All strains tested are susceptible to chloramphenicol, clindamycin, erythromycin, penicillin G and tetracycline.

Culture supernatants are nontoxic to mice.

Other characteristics of the species are given in Table 13.15.

Isolated from soil and marine sediment; also reported from soft tissue infections in humans (Gorbach and Thadepalli, 1975), and from human feces (Finegold et al., 1983).

The mol% G + C of the DNA is 27 (T_m) (Johnson and Francis, 1975).

Type strain: ATCC 25757 (NCIB 10636; CCUG 9282).

30. **Clostridium glycolicum** Gaston and Stadtman 1963, 356.[AL]
gly.co.li.cum. L. adj. suff. -*icus* related to, belonging to; N.L. neut. adj. *glycolicum* referring to ability to ferment ethylene glycol.

This description is based on that of Gaston and Stadtman (1963) and on study of the type and 40 other strains.

Cells in PYG broth cultures are Gram-positive, straight to slightly curved rods, motile and peritrichous, 0.3–1.3 × 1.8–15.4 μm, and occur singly or in pairs.

Figure 13.34. *Clostridium glycolicum.*

Spores are oval and usually subterminal, occasionally terminal, often occurring as free spores. Sporulation occurs most readily in chopped meat broth cultures or on chopped meat agar slants.

Cell walls do not contain DAP (C. S. Cummins, unpublished data). They do contain lysine and *iso*-D-asparagine (Weiss et al., 1981).

Surface colonies on blood agar plates are 1–4 mm, circular to irregular, raised to convex, translucent to semiopaque, grayish white, shiny and smooth with a granular, mottled or mosaic internal structure and entire, scalloped or erose margins.

Cultures in PYG broth are turbid with a smooth to stringy sediment and have a pH of 5.5–5.9 after incubation for 24 h. After 5 days of incubation the pH is 5.2–5.6.

The optimum temperature for growth is 30–37°C. Most strains grow moderately well at 25°C, poorly if at all at 45°C. Growth is inhibited by 6.5% NaCl, 20% bile and a pH of 8.5.

Abundant gas is detected in PYG deep agar cultures.

Deoxyribonuclease is present (Smith, 1975a). Ammonia is produced. H_2S is formed by 9 of 13 strains tested. Four of 13 strains hydrolyze hippurate.

This organism can ferment ethylene glycol to acetate and ethanol, and propylene glycol to propionate, propanol, and small amounts of acetate and ethanol (Gaston and Stadtman, 1963).

A wide variety of both normal straight chain saturated and *iso*-branched chain saturated and unsaturated fatty acids is present in strains of this species (Elsden et al., 1980).

C. glycolicum degrades uracil to β-alanine, ammonia and CO_2 (Mead et al., 1979).

Products in PYG broth are acetic, isovaleric, and isobutyric acids, and ethyl, propyl, isobutyl and isoamyl alcohols; propionic, formic, lactic and succinic acids also may be detected. Abundant H_2 is produced. Threonine is converted to propionate; lactate is not utilized.

All strains are susceptible to penicillin G; of 30 strains tested, five are resistant to erythromycin, two to clindamycin, one to chloramphenicol, and one to tetracycline. Susceptibility to metronidazole is variable (Chow et al., 1975).

Culture supernatants are notoxic to mice.

Other characteristics of the species are given in Table 13.14.

Isolated from soil, mud, snake venom, bovine intestine and from human clinical specimens including wounds, abscesses and peritoneal fluid; also isolated from human feces (Finegold et al., 1983; Drasar et al., 1976).

The mol% G + C of the DNA is 29 (T_m) (Johnson and Francis, 1975).

Type strain: ATCC 14880 (NCIB 10632; DSM 1288).

Further comments. The production of large amounts of isovaleric acid is most helpful in distinguishing this species from other saccharolytic, nonproteolytic clostridia.

31. **Clostridium haemolyticum** (Hall 1929) Scott, Turner, and Vawter 1935, 1972.[AL] (*Bacillus hemolyticus* Hall 1929, 156.)
hae.mo.ly'ti.cum. Gr. n. *haema* blood; Gr. adj. *lytikos* dissolving; N.L. neut. adj. *haemolyticum* blood-dissolving, hemolytic.

This description is based on those of Smith (1975a), Holdeman et al. (1977), and on study of the type and 15 other strains.

Cells in PYG broth cultures are motile and peritrichous, 0.6–1.6 × 1.9–17.3 μm, and occur singly or in pairs. In very young cultures, they are Gram-positive but they rapidly become Gram-negative.

Figure 13.35. *Clostridium haemolyticum.*

Spores are oval, subterminal and swell the cell. Sporulation occurs most readily in 2- to 3-week-old chopped meat broth cultures or on chopped meat slants incubated at 30°C.

Cell walls contain *meso*-DAP, alanine and glutamic acid (Cummins and Johnson, 1971; Schleifer and Kandler, 1972).

Surface colonies on blood agar plates are 1–3 mm in diameter, circular, raised to convex, translucent, gray, shiny, with a granular or mosaic surface and an erose on slightly scalloped margin.

Cultures in PYG broth are turbid, usually with a granular or flocculent sediment and have a pH of 5.0–5.5 after incubation for 24 h.

The optimum temperature for growth is 37°C; growth is slight at 30°C; there is little or no growth at 25°C or 45°C. The organisms are extremely sensitive to oxygen and require both prereduced media and stringent anaerobic conditions. Although they grow moderately well in carbohydrate-free media, growth is stimulated by fermentable carbohydrate. Growth is inhibited by 20% bile, 6.5% NaCl or a pH of 8.5.

Moderate gas is detected in PYG deep agar cultures.

Small amounts of acid phosphatase are produced.

Fermentation products in PYG broth are propionic, butyric and acetic acids. Lactate is converted to propionate; threonine is converted to propionate by 9 of 12 strains tested; pyruvate is converted to acetate, propionate and butyrate. Abundant H_2 is produced.

All strains tested are susceptible to chloramphenicol, clindamycin, erythromycin, penicillin G, and tetracycline; the one strain tested is resistant to >100 μg/ml of metronidazole (Chow et al., 1975).

Culture supernatants are toxic to mice and cultures are pathogenic for cattle, sheep and laboratory animals. The major lethal toxin is a phospholipase C (identical with *C. novyi* beta toxin) which hydrolyzes lecithin and sphingomyelin and hemolyzes red blood cells. The organism produces the beta, eta, and theta toxins of *C. novyi* B but not the alpha (necrotizing) toxin, the major lethal toxin of that species. In susceptible animals *C. haemolyticum* causes fatal bacillary hemoglobinuria. For a thorough review of the properties of *C. haemolyticum*, its toxins, the disease it causes, and its relationship to *C. novyi* B, see Smith (1975a).

Other characteristics of the species are given in Table 13.12.

Isolated from liver infections and muscle of cattle and sheep; also isolated from human feces (Finegold et al., 1983).

The mol% G + C of the DNA is 26–27 (T_m) (Johnson and Francis, 1975).

Type strain: ATCC 9650 (NCIB 10664).

Further comments. C. haemolyticum cannot be distinguished from *C. novyi* B by any phenotypic or morphological test yet devised except by toxin neutralization in laboratory animals. In addition, four strains of *C. haemolyticum* are 84–92% homologous with a reference strain of *C. novyi* B (Nakamura et al., 1983), and the patterns of soluble cellular proteins of the type strains as shown by polyacrylamide gel electrophoresis are nearly identical (Cato et al., 1982a). However, since the diseases they cause are quite different, the components of their lethal toxins are known to be different, and antisera to one does not protect against infection by the other, they will for practical purposes be retained as separate species. They can only be identified with certainty with toxin-specific antisera.

32. **Clostridium hastiforme** MacLennan 1939. 543.[AL]

has.ti.for′me. L. n. *hasta* a spear; L. n. *forma* shape; N.L. neut. adj. *hastiforme* spear-shaped.

This description is based on that of McClung and McCoy (1957) and on study of the type and 15 other strains.

Cells in PYG broth cultures are motile and peritrichous, straight to slightly curved rods with rounded to tapered ends, 0.6–1.6 × 1.9–9.3 μm. They are Gram-positive but may stain Gram-negative within 24 h.

L_10 μm_⌐   PY   Chopped meat agar

Figure 13.36. *Clostridium hastiforme.*

Spores are oval, terminal and swell the cell. Sporulation of most strains occurs readily in chopped meat broth or on chopped meat agar slants incubated at 30°C.

Cell walls contain *meso*-DAP (Weiss et al., 1981).

Surface colonies on blood agar plates are ponpoint to 3.0 mm, circular to slightly irregular, entire, convex to peaked, opaque, sometimes with a translucent edge, grayish white, granular to mottled, shiny and smooth. Some strains are β-hemolytic.

Cultures in PYG broth are slightly turbid with a smooth sediment and a pH of 6.0–6.4 after incubation for 4 days.

Strains grow equally well at 30°C and 37°C. There is little or no growth at 25°C or 45°C. Growth of some strains in an atmosphere of 90% N_2-10% CO_2 is superior to that in at atmosphere of 100% CO_2. Growth is inhibited by 20% bile or 6.5% NaCl.

Little or no gas is detected in PYG deep agar cultures.

Carbohydrates are not fermented; gelatin is rapidly digested; chopped meat is slowly digested; milk is slowly digested by 3 of 16 strains tested; casein is ot digested. H_2S and ammonia are produced.

Products in PYG broth are acetic, butyric and isovaleric acids, with smaller amounts of propionic and isobutyric acids. No H_2 is produced. Neither pyruvate, lactate nor threonine is utilized.

Of 11 strains tested, all are susceptible to chloramphenicol, penicillin G and tetracycline, 4 are resistant to erythromycin, and 2 are resistant to clindamycin.

Culture supernatants are not toxic to mice; cultures are not pathogenic to laboratory animals (Prévot et al., 1967).

Other characteristics of the species are given in Table 13.15.

Strains have been isolated from war wounds (MacLennan, 1943), from soil (Prévot et al., 1967), from bovine brain, and from human clinical specimens including blood, abdominal abscess and infected tissue. The principal habitat probably is soil.

The mol% G + C of the DNA has not been reported.

Type strain: VPI 12193 (Skerman et al., 1980, ATCC 33268); *not* ATCC 25772; *not* DSM 1786.

Further comments. Although there has been confusion in determining the characteristics most useful in separating *C. hastiforme* from *C. subterminale*, obvious difference in patterns of soluble cellular proteins between the types and similar strains of the two species, as shown on polyacrylamide gels, indicate that they are distinct species (Cato et al., 1982a). Characteristics listed in the 4th edition of the *Anaerobe Laboratory Manual* are those of *C. subterminale*, not those of *C. hastiforme*. Strains of *C. hastiforme* have terminal spores and do not produce H_2; strains of *C. subterminale* have subterminal spores and produce abundant H_2.

33. **Clostridium histolyticum** (Weinberg and Séguin, 1916) Bergey et al. 1923, 328.[AL] (*Bacillus histolyticus* Weinberg and Séguin 1916, 449.)

his.to.ly′ti.cum. Gr. n. *histos* tissue; Gr. adj. *lytikos* dissolving; N.L. neut. adj. *histolyticum* tissue-dissolving.

This description is based on those of Smith and Hobbs (1974) and Smith (1975a), and on study of the type and 18 other strains.

Cells in PYG broth culture usually are motile and peritrichous, straight rods, Gram-positive, 0.5–0.9 × 13.–9.2 μm, and occur slightly, in pairs, or in short chains.

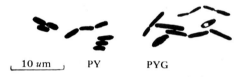

L_10 μm_⌐   PY   PYG

Figure 13.37. *Clostridium histolyticum.*

Spores are oval, central to subterminal and may slightly swell the cell. There are no exosporium or appendages. Sporulation of most strains occurs most readily in chopped meat broth cultures.

Cell walls contain *meso*-DAP, glutamic acid and alanine (Schleifer and Kandler, 1972).

Surface colonies on blood agar plates incubated anaerobically are β-hemolytic, 0.5–2 mm in diameter, circular to irregular, flat to low convex, translucent to semiopaque, gray-white, shiny, with a mosaic or granular surface and an entire to undulate margin. Similar colonies are formed on blood agar plates incubated aerobically but they are usually smaller and fewer in number. Spores are not formed during aerobic growth.

The optimum temperature for growth is 37°C; most strains grow nearly as well at 30°C; there is moderate growth at 25°C but little or none at 45°C. Although there is no pH depression, growth is stimulated by carbohydrate. Strains grow well at a pH of 8.5 but are inhibited by 6.5% NaCl or by 20% bile.

Moderate amounts of gas are detected in PYG deep agar cultures.

Strains are strongly proteolytic, digesting gelatin, meat, milk, casein, collagen, hemoglobin, fibrin, elastin, egg white, coagulated serum, muscle, liver, brain and Achilles tendon. For reviews concerning the proteolytic enzymes of *C. histolyticum* see Prévot et al. (1967), Schallehn and Müller (1973) and Smith (1975a).

Deoxyribonuclease is produced by the one strain tested (Döll, 1973).

The only major product detected in PYG broth is acetate; traces of formate, lactate or succinate are sometimes present. Small amounts of

H₂ are produced; ammonia is produced; H_2S is produced by 4 of 12 strains tested. Pyruvate is converted to acetate; neither lactate nor threonine is converted to propionate. Threonine is utilized but the product is acetate (Elsden and Hilton, 1978). Glycine, glutamate, serine, aspartate and arginine are utilized as sources of energy; valine, leucine isoleucine, lysine, proline and alanine are produced (Mead, 1971). Glycine is fermented to acetate, CO_2, and NH_3 (Guillaume et al., 1956).

Only straight chain fatty acids, principally saturated C_{14}, C_{12}, C_{16}, and C_{18} acids with small amounts of unsaturated acids of the same chain length are present in cells of this species (Elsden et al., 1980).

All strains tested are susceptible to chloramphenicol, clindamycin, erythromycin, penicillin G and tetracycline. Under anaerobic conditions, strains are susceptible to 0.5–1.0 μg/ml of metronidazole. As the O_2 level increases, they are increasingly resistant (Füzi, 1981).

Supernatant culture fluids are toxic to mice although the degree of toxigenicity varies among strains and may be lost on subculture. In addition, as the culture ages, the active proteases of the organism may destroy the toxins (Lettl, 1973). *C. histolyticum* is highly pathogenic to laboratory animals. Toxin found in fecal samples from guinea pigs with clindamycin-associated enterocolitis is neutralized by specific antitoxin to *C. histolyticum* (Knoop, 1979). Major lethal toxins are the α-toxin (necrotizing), and at least two collagenases (β-toxin).

Collagenase from atoxic *C. histolyticum* strains has been used to digest and remove necrotic tissue from burns (Lettl, 1973; Smith, 1975a). For a review of these and other toxins produced and the disease caused, see Smith (1975a).

Other characteristics of the species are given in Table 13.15.

Isolated from soil, war wounds, gas gangrene in humans and horses, and human intestinal contents; also isolated from gingival plaque of institutionalized and primitive populations (Leosche et al., 1974; Alfano et al., 1974). Principal habitat is probably the soil.

The mol% G + C of the DNA has not been reported.

Type strain: ATCC 19401 (NCIB 503; NCTC 503; DSM 2158; CCUG 4854).

Further comments. C. histolyticum can be differentiated readily from *C. limosum*, which it resembles most closely phenotypically, by its production of collagenase and lack of production of lecithinase. Unlike other aerotolerant clostridia, it is extremely proteolytic and does not ferment any carbohydrates tested. Unlike *C. irregulare* it produces only acetic acid in broth cultures.

34. **Clostridium indolis** McClung and McCoy 1957, 674.[AL]

in.do′lis. N.L. gen. n. *indolis* of indole.

This description is based on that of McClung and McCoy (1957) and study of the type and six other strains.

Cells in PYG broth cultures are motile and peritrichous, straight to slightly curved rods, 0.5–0.9 × 1.3–10.2 μm, and occur slightly and inpairs. They are Gram-negative in 24 h.

Figure 13.38. *Clostridium indolis.*

Spores are round to oval, swollen, and usually terminal, although subterminal spores are sometimes seen in the same preparation. Spores have no exosporium and no appendages (Smith and Hobbs, 1974). Sporulation occurs most readily on chopped meat slants incubated at 30°C.

Cell walls contain *meso*-DAP (C. S. Cumins, unpublished data).

Surface colonies on blood agar plates are nonhemolytic, 0.5–3.0 mm in diameter, circular to slightly irregular, convex, translucent to opaque, white, with a dull, granular surface and an entire to erose margin.

Cultures in PYG broth are turbid with a smooth sediment and have a pH of 5.3–5.7 after incubation for 3 days.

Optimum temperature for growth is 37°C; most strains grow nearly as well at 30°C; growth is poor at 25°C and there is little or no growth at 45°C. Growth is stimulated by fermentable carbohydrate but inhibited by 20% bile or 6.5% NaCl.

Abundant gas is detecrted in PYG deep agar cultures.

More than 30% of strains tested can dehydrogenate the steroid nucleus (Goddard et al., 1975; Hill, 1975).

Polypectate is liquefied; pectin, pectate and galacturonate are fermented (Ng and Vaughn, 1963).

Citrate is not utilized (Ehrlich et al., 1981).

Ammonia and H_2S are produced. Neutral red and resazurin are reduced. Of five strains tested, two produce acetyl methyl carbinol and hydrolyze hippurate.

Products in PYG broth include acetate, formate and ethanol. Abundant H₂ is detected. Pyruvate is converted to acetate, formate, ethanol and moderate amounts of butyrate; neither lactate nor threonine is utilized. Galacturonate is converted to acetate, butyrate, CO_2 and H_2 (Ng and Vaughn, 1963).

Of five strains tested, three are susceptible to chloramphenicol, four to clindamycin, four to erythromycin, four to tetracycline, and two to penicillin G.

Culture supernatant fluids are nontoxic to mice.

Other characteristics of the species are given in Table 13.14.

Isolated from soil (Smith, 1975b), from human feces (Finegold et al., 1983; Drasar et al., 1976), and from human clinical specimens from infections associated with the intestinal tract.

The mol% G + C of the DNA is 44 (T_m) (Johnson and Francis, 1975).

Type strain: ATCC 25771 (NCIB 9731; DSM 755).

Further comments. C. indolis can be differentiated from *C. sphenoides*, which it resembles most closely phenotypically, by the terminal spores that are produced by all strains; spores of *C. sphenoides* are subterminal. The electrophoretic patterns of soluble cellular proteins of the two species are distinctive (Cato et al., 1982). In addition, strains of *C. sphenoides* can ferment citrate with the production of acetate and ethanol, while strains of *C. indolis* do not utilize citrate (Ehrlich et al., 1981).

35. **Clostridium innocuum** Smith and King 1962, 939.[AL]

in. noc′u.um. L. neut. adj. *innocuum* harmless.

This description is based on that of Smith and Hobbs (1974) and on study of the type and 80 other strains.

Cells in PYG broth cultures are Gram-positive, nonmotile, straight rods with round or tapered ends, 0.4–1.6 × 1.6–9.4 μm. Sporulating cells tend to be among the larger ones. Cells occur singly or in pairs.

Figure 13.39. *Clostridium innocuum.*

Spores are oval, terminal, subterminal, or free and wider than the cell. Sporulation usually occurs readily in chopped meat broth cultures incubated for 24 h or on egg-yolk agar plates incubated for 48–72 h.

Cell walls contain glucose and galactose but no DAP (Cummins and Johnson, 1971). They contain a lysine-alanine type of peptidoglycan (Weiss et al., 1981).

Surface colonies on blood agar plates are 0.5–3 mm in diameter, circular, raised or convex, translucent, gray-white or yellowish, smooth, and shiny, with a mottled or mosaic internal structure and an entire or slightly scalloped margin. Hemolysis is variable between strains.

Cultures in PYG broth are turbid with a smooth sediment and a pH of 4.7–5.2 after incubation for 24 h.

Optimum temperature for growth is 37°C although many strains grow   equally well at 25°C, 30°C, and 45°C. Growth is stimulated by fermentable carbohydrate but inhibited by 6.5% NaCl.

Abundant gas is detected in PYG deep agar cultures.

Neither gelatin, milk, casein nor meat is digested. Milk may be weakly acidified.

Bacteriocin-like particles have been demonstrated in one strain of *C. innocuum* but no inhibitory activity was found against any strain tested (Nieves et al., 1981).

Products in PYG broth are butyric, lactic and acetic acids; small amounts of formic or succinic acids may be detected. Abundant H_2 is produced. Pyruvate is converted to acetate, butyrate and sometimes formate. Neither lactate nor threonine is utilized. Ethanol has been detected by high performance liquid chromatography PYG broth cultures (Ehrlich et al., 1981).

Susceptibility to chloramphenicol, clindamycin, erythromycin, penicillin G and tetracycline is variable among strains; two strains tested were susceptible to 1.6 μg/ml of metronidazole (Chow et al., 1977).

Supernatant culture fluids are not toxic to mice and cultures are not pathogenic in laboratory animals.

Other characteristics of the species are given in Table 13.14.

Strains are commonly isolated from human infections (Werner et al., 1973; Gorbach and Thadepalli, 1975; Finegold, 1977), particularly those associated with the intestinal tract. They have also been isolated from empyema fluids (Bartlett et al., 1974a). They are part of the normal intestinal flora of human infants and adults (Moore and Holdeman, 1974; Bartlett et al., 1974; Finegold et al., 1983; Drasar et al., 1976; Stark and Lee, 1982).

The mol% G + C of the DNA is 43–44 (T_m) (Johnson and Francis, 1975).

Type strain: ATCC 14051 (NCIB 10674; DSM 1286).

Further comments. This species is most easily differentiated from other saccharolytic species which it resembles, by its fermentation of mannitol, its failure to ferment lactose or sorbitol, and by the presence of terminal spores. Unlike *C. barkeri*, the occasional strain that does not ferment sucrose also does not ferment sorbitol.

36. **Clostridium irregulare** (corrig.) (Choukévitch 1911) Prévot 1938, 85.[AL] (*Bacillus irregularis* Choukévitch 1911, 348.)

ir.reg.u.lar′e. L. neut. adj. *irregulare* irregular, referring to pleomorphic, irregular cells.

This description is based on those of Prévot, Turpin, and Kaiser (1967), Holdeman et al. (1977), Mahony et al. (1977) and on study of the type and three other strains.

Cells in PYG broth culture are Gram-positive, straight to slightly curved rods, motile and peritrichous, and 0.8–1.6 × 3.5–12.6 μm. Cells in week-old cultures may be quite filamentous. They occur singly, in pairs or in short chains.

Figure 13.40. *Clostridium irregulare.*

Spores are oval, central or subterminal and swell the cell. Sporulation occurs most readily on chopped meat agar slants incubated at 30°C.

Surface colonies on blood agar plates are nonhemolytic, pinpoint to 0.5 mm, circular to irregular, convex, transparent to translucent, colorless, with a granular or mottled surface and an entire or scalloped margin.

Cultures in PY-glucose broth are turbid with a smooth or ropy sediment and have a pH of 6.0–6.3 after incubation for 4 days.

The optimum temperature for growth is between 30°C and 37°C.

There is little or no growth at 25°C or 45°C. Growth is inhibited by 6.5% NaCl and by 20% bile.

No gas is detected in PYG deep agar cultures. H_2S is formed in SIM medium. Ammonia is produced. Gelatin is digested but not milk, casein, or meat.

The type strain contains a 3-α-hydroxysteroid dehydrogenase (Mahony et al., 1977.)

Products detected in PYG broth are acetate and isovalerate; formate, isobutyrate and propionate also may be detected. H_2 is not formed. Pyruvate is converted to acetate and formate. Neither lactate nor threonine is utilized.

All strains are susceptible to chloramphenicol, clinidamycin, erythromycin, penicillin G, and tetracycline.

Supernatant culture fluids are nontoxic for mice and cultures are not pathogenic for laboratory animals.

Other characteristics of the species are given in Table 13.15.

Isolated from normal fecal flora of humans (Finegold et al., 1983) and horses (Choukévitch, 1911), from pond mud, from pharmaceutical products prepared in a laboratory adjacent to a stable (Prévot et al., 1967); from human penile lesions (Chapel et al., 1978) and from soil.

The mol% G + C of the DNA has not been reported.

Type strain: ATCC 25756.

Further comments. This species can be differentiated from *C. histolyticum* by its production of isovaleric and usually isobutyric acid, and by its failure to digest either casein, milk, or meat.

37. **Clostridium kluyveri** Barker and Taha 1942, 362.[AL]

kluy′ver.i. N.L. gen. n. *kluyveri* named for Professor A. J. Kluyver, Dutch microbiologist.

This description is based on those of Barker and Taha (1942), Bornstein and Barker (1948) and study of the type strain.

Cells in medium containing 20% yeast autolysate, 0.5% ethanol, and inorganic salts (Barker and Taha, 1942) are motile and peritrichous, 0.9–1.1 × 3–11 μm, and usually occur singly, occasionally in pairs, or chains. They are weakly Gram-positive but quickly become Gram-negative.

After Barker and Taha

Figure 13.41. *Clostridium kluyveri.*

Spores are oval, terminal or subterminal, and swell the cell.

Ultrastructural studies have demonstrated a five-layered cell wall and a three-layered plasma membrane (Cho and Doy, 1973). Cell walls contain *meso*-DAP (Gottschalk et al., 1981).

Surface colonies on blood agar plates are nonhemolytic, pinpoint to 0.5 mm in diameter, gray, with rhizoid margins. In laked blood streak tubes, colonies are low convex, gray-white, shiny, and smooth, with a scalloped margin and a mosaic internal structure. There is no growth on egg-yolk agar plates.

Cultures in PYG-1.5% ethanol broth have a stringy sediment and no turbidity after 5 days incubation in a 90% N_2-10% CO_2 atmosphere. The pH is 6.8.

The optimum temperature for growth is 35°C. Growth is slow and occurs between 19°C and 37°C. Strains require ethanol, CO_2 or sodium carbonate, and either a high concentration of yeast autolysate or acetate, propionate or butyrate for growth. A synthetic medium containing inorganic salts, acetate, ethanol, biotin and *p*-aminobenzoic acid will support growth (Bornstein and Barker, 1948).

Small amounts of gas are detected in PYG deep agar cultures.

Urease is produced; gelatin is not hydrolyzed; carbohydrates are not attacked.

For a review of references to the enzymes produced and the metabolic pathways of *C. kluyveri*, see Gottschalk et al. (1981). See also Thauer et al. (1977), Barker (1978, 1981), Yoch and Carithers (1979), Bader and Simon (1980), and Wiegel (1981).

In the presence of CO_2 or carbonate, and acetate or propionate, ethanol is converted to butyrate, caproate and H_2 (Gottschalk et al., 1981).

H_2 gas is formed and small amounts of butyrate may be detected. Neither pyruvate, lactate nor threonine is utilized. *C. kluyveri* can be adapted to utilize crotonate and produce acetate, butyrate and caproate (Bader et al., 1978; 1980).

The type strain is susceptible to chloramphenicol, clindamycin, erythromycin, penicillin G and tetracycline.

Other characteristics of the species are given in Table 13.13.

Isolated from fresh water and marine black mud. Also isolated from decaying plants and garden soil (Claus et al., 1983).

The mol% G + C of the DNA is 30 (T_m) (Bader et al., 1980).

Type strain: NCIB 10680 (ATCC 8527; ATCC 12489; DSM 555).

Further comments. This species is distinctive both in its nutritional requirements and in its ability to ferment ethanol to caproate.

38. Clostridium leptum Moore, Johnson, and Holdeman 1976, 250.[AL]

lep'tum. Gr. adj. *leptos* thin, delicate; N.L. neut. adj. *leptum* (referring to the morphological appearance of the cells).

This description is based on that of Moore et al. (1976) and their study of 27 strains.

Cells in PY-fructose broth cultures are nonmotile, slightly curved Gram-positive rods, 0.6–0.8 × 1.3–2.8 μm, that occur in pairs or short chains.

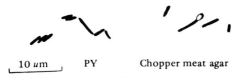

<pre>10 um   PY   Chopper meat agar</pre>

Figure 13.42.   *Clostridium leptum.*

Spores are oval, nearly terminal, and rarely seen. Heat-resistant cells can be demonstrated from chopped meat slants incubated at 30°C for 3 weeks, inoculated into PY-maltose broth, and heated at 80°C for 10 min.

Cell walls contain a lysine-serine-glycine type of peptidoglycan (Weiss et al., 1981).

Surface colonies on supplemented brain-heart infusion roll streak tubes (Holdeman et al., 1977) are pinpoint to 0.5 mm in diameter, circular, entire, low convex, tan and translucent. Neither of the 2 of eight strains tested that grew on the surface of anaerobic blood agar plates was hemolytic. Neither lecithinase, lipase nor urease was detected from the five of eight strains tested that grew on anaerobic egg-yolk agar plates.

Cultures in PY-maltose broth usually produce a smooth to stringy sediment with little or no turbidity and have a pH of 5.3–5.8 after incubation for 5 days.

The optimum temperature for growth is 37°C; some strains grow at 30°C and 45°C. Growth is markedly stimulated by some carbohydrates (glucose, cellobiose, mannitol, ribose) although there is little or no depression of pH of the medium. Growth is inhibited by 6.5% NaCl and by 20% bile.

Traces of gas may be detected in PYG deep agar cultures. Glucose may be fermented weakly.

A 7 α-dehydroxylase and a 12 α-hydroxysteroid dehydrogenase that may be significant in the breakdown of bile acids in the human colon have been isolated from a strain of this species (Stellwag and Hylemon, 1978, 1979; Harris and Hylemon, 1978).

Products in PY-maltose broth are acetic acid and abundant H_2; ethanol also may be detected. Neither lactate nor pyruvate is utilized.

Other characteristics of the species are given in Table 13.13.

Isolated from human feces and colonic contents.

The mol% G + C of the DNA is 51–52 (T_m) (Moore et al., 1976).

Type strain: ATCC 29065 (DSM 753).

Further comments. This species is most easily differentiated from *Eubacterium siraeum*, which it resembles most closely phenotypically, by demonstration of heat resistance and by lack of fermentation of cellobiose.

39. Clostridium limosum André in Prévot 1948, 165.[AL]

li.mo′sum. L. neut. adj. *limosum* muddy or slimy.

This description is based on that of Cato et al. (1970), Holdeman et al. (1977), and study of the type and 32 other strains.

Cells in PYG broth cultures are Gram-positive straigh rods, 0.6–1.6 × 1.7–16 μm, and occur singly, in pairs, or in short chains. Motility is variable; motile cells are peritrichous.

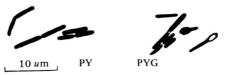

<pre>10 um   PY   PYG</pre>

Figure 13.43.   *Clostridium limosum.*

Spores are oval, central to subterminal and usually swell the cell. Sporulation occurs most reliably on chopped meat slants incubated at 30°C or on egg-yolk agar plates incubated for 48–72 h.

Cell walls contain *meso*-DAP, alanine and glutamic acid; cell wall sugars are glucose, galactose and small amounts of mannose and rhamnose (Cato et al., 1970).

Surface colonies on blood agar plates are β-hemolytic, 1–4 mm in diameter, circular to irregular, raised to convex, translucent, gray, shiny or dull, smooth, with a mosaic or granular internal structure and an entire, scalloped or undulate edge.

Cultures in PYG broth are turbid with a smooth to ropy sediment and a pH of 6.1–6.5 after incubation for 1 week.

The optimum temeprature for growth is 37°C. Some strains grow as well at 30°C or 45°C; growth is poor at 25°C. Growth is inhibited by 6.5% NaCl, 20% bile or a pH of 8.5

Trace to moderate gas is produced in PYG deep agar cultures.

Ammonia and H_2S are produced.

Lecithinase, collagenase, ribonuclease and deoxyribonuclease are produced. One of nine strains produces an extracellular β-glucuronidase.

Principal cellular lipid fatty acids are the saturated straight-chain C_{16}, C_{14}, and C_{12} acids (Elsden et al., 1980).

The principal product in PYG broth is acetate; small amounts of formate, succinate, or lactate may be detected. Little or no H_2 is produced. Pyruvate is converted to acetate; neither lactate nor threonine is utilized. Elsden and Hilton (1978) reported an increase in propionate in a threonine-trypticase-yeast extract-thioglycollate medium over that detected in the basal medium. β-Phenylethylamine, isoamylamine, and usually di-*n*-butylamine are produced (Brooks and Moore, 1969). Glutamate is fermented by the methylaspartate pathway (Buckel, 1980). Indole and phenol are produced from fermentation of tryptophane and tyrosine, respectively (Elsden et al., 1976), although indole formation is repressed in usual laboratory test media. Glutamic acid and histidine are fermented (Elsden et al., 1976).

All strains tested are susceptible to chloramphenicol, clindamycin, erythromycin, penicillin G, and tetracycline; the one strain tested by Phillips et al. (1981) was resistant to clindamycin.

Supernatant culture fluids of some strains are weakly toxic to mice;

some strains are pathogenic for guinea pigs. Pathogenicity appears to be related to the action of the collagenase and the lecithinase. Toxicity and pathogenicity are readily lost in laboratory cultures (Smith, 1975a).

Other characteristics of the species are given in Table 13.15.

Isolated from mud; infections in cattle, water buffalo, alligators and chickens (Cato et al., 1970); snake venom; home-preserved meat; from human feces (Finegold et al., 1983) and from human clinical specimens including blood, peritoneal fluid, pleural fluids (Gorbach and Thade-palli, 1973; Gorbach et al., 1974; Bartlett et al., 1974) and lung biopsy from pulmonary infections (Appelbaum et al., 1978.)

The mol% G + C of the DNA is 24 (T_m) (Johnson and Francis, 1975).

Type strain: ATCC 25620 (ATCC 25760; NCIB 10638; DSM 1400).

Further comments. C. limosum can be differentiated most easily from C. histolyticum, which it resembles phenotypically, by its production of lecithinase.

40. Clostridium lituseburense (Laplanche and Saissac in Prévot 1948a) McClung and McCoy 1957, 664.[AL] (*Inflabilis litus-eburense* Laplanche and Saissac in Prévot 1948a, 167.)

li.tus.e.bu.ren′se. L. n. *litus* coast; L. n. *ebur* ivory; N.L. adj. *lituseburense* pertaining to the Ivory Coast.

This description is based on that of Holdeman et al. (1977) and on study of the type strain.

Cells in PYG broth cultures are Gram-positive rods, straight or slightly curved, motile and peritrichous, 1.4–1.7 × 3.1–6.3 µm, and occur singly, in pairs or in short chains.

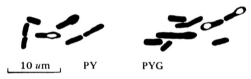

Figure 13.44. *Clostridium lituseburense.*

Spores are oval, central to subterminal, and may swell the cell or occur as free spores. Sporulation occurs readily in chopped meat broth cultures incubated for 24 h.

Cell walls contain L-lysine and aspartic acid (Tanner et al., 1981).

Surface colonies on blood agar plates are β-hemolytic, 1–3 mm in diameter, circular, low convex, opaque with translucent margins, white, shiny and smooth, with a coarse granular internal structure and an entire to scalloped margin.

Cultures in PYG broth are turbid with a ropy sediment and a pH of 5.3 after incubation for 5 days.

Growth is equally profuse at 35°C, 30°C, and 37°C, slightly less at 45°C. Growth is inhibited by 6.5% NaCl and 20% bile but unaffected by a pH of 8.5.

Traces of gas may be detected in PYG deep agar cultures.

Ammonia is produced. Neutral red and resazurin are reduced.

Gelatin, chopped meat, and casein are digested; milk is curdled and weakly digested in 3 weeks.

A complex mixture of straight and branched chain saturated and unsaturated cellular fatty acids is present with normal unsaturated C_{16} acid predominating (Elsden et al., 1980).

Products in PYG broth cultures include large amounts of butyric, acetic and isovaleric acids and small amounts of formic, propionic, and isobutyric acids. Small amounts of ethyl, propyl, and isobutyl alcohols may be detected. In chopped meat-carbohydrate broth, butyric and isovaleric acids are greatly increased, and moderate isobutyric acid is detected. Little or no H_2 gas is formed. Threonine is converted to propionate; an increased amount of acetate is detected from pyruvate; lactate is not utilized. Phenylalanine, tyrosine and tryptophane are converted to phenylacetic, hydroxyphenylacetic, and indole acetic acids

(Elsden et al., 1976). Valine is converted to isobutyric acid, leucine and isoleucine to isovaleric acid (Elsden and Hilton, 1978). C. lituseburense utilizes serine, threonine and arginine and produces 2-aminobutyric acid (Elsden and Hilton, 1979).

The type strain is susceptible to chloramphenicol, clindamycin, erythromycin, penicillin G and tetracycline.

Strains are not pathogenic for guinea pigs or mice.

Other characteristics of the species are given in Table 13.12.

Isolated from soil and humus from the Ivory Coast.

The mol% G + C of the DNA is 27 (T_m) (Johnson and Francis, 1975).

Type strain: ATCC 25759 (NCIB 10637; DSM 797).

Further comments. The oligonucleotide catalogs of the 16s-rRNAs of the type strains of C. lituseburense and Eubacterium tenue indicate that the two species are closely related (Tanner et al., 1981).

41. Clostridium malenominatum (Weinberg, Nativelle and Prévot 1937) Spray 1948, 786.[AL] (*Bacillus malenominatus* Weinberg, Nativelle and Prévot 1937, 763.)

ma.le.nom.i.na′tum. L. pref. *mal* ill; L. inf. *nominare* to name; N.L. past part. *malenominatum* poorly named.

This description is based on that of Holdeman et al. (1977) and on study of the type and 26 other strains.

Cells in PYG broth culture are Gram-positive, although usually only Gram-negative cells can be seen in 24-h cultures. They are straight rods, 0.3–0.9 × 1.4–10.8 µm, usually motile and peritrichous, and occur singly or in pairs.

Figure 13.45. *Clostridium malenominatum.*

Spores are oval or round, subterminal or terminal, and swell the cell. Sporulation occurs most readily on egg-yolk agar plates or on chopped meat slants incubated at 30°C for 1 week.

Cell walls contain L-lysine and aspartic acid (Gottschalk et al., 1981).

Surface colonies on blood agar plates are pinpoint to 2 mm, circular or slightly irregular, convex or raised, gray-white, and translucent, with a crystalline, mottled or granular internal structure, and an entire, erose or scalloped margin. The surface may be smooth or lumpy. Hemolysis is variable among strains.

Cultures in PYG broth are turbid with a smooth or ropy sediment and have a pH of 6.1–6.5 after incubation for 1 week.

The optimum temperature for growth is 37°C. Some strains grow equally well at 30°C. Growth is moderate at 25°C, poor at 45°C. Growth is inhibited by 20% bile, 6.5% NaCl, or a pH of 8.5.

Moderate amounts of gas are detected in PYG deep agar cultures.

Ammonia is produced.

Five of 27 strains tested weakly digest gelatin after incubation for 3 weeks. Neither casein, milk nor meat is digested. H_2S is produced in SIM by 22 of the 27 strains.

Uric acid but not urea is decomposed by one strain, an isolate from the chicken cecum (Barnes and Impey, 1974).

Cellular fatty acids detected are straight chain acids, principally saturated C_{14} and C_{16} and unsaturated C_{16} acids (Elsden et al., 1980).

Products in PYG broth include butyrate and acetate; propionate, lactate, and formate usually are detected; propyl and butyl alcohols may be present. Abundant H_2 is produced. Pyruvate is converted to acetate and butyrate; 19 strains, including the type strain, convert threonine to propionate, and 5 strains convert lactate to butyrate. Glutamic acid and tyrosine are utilized (Elsden and Hilton, 1979). Glutamate is converted by the type strain to butyrate and acetate by the methylaspartate pathway (Buckel, 1980). Traces of phenol and indole are formed from phenylalanine and tryptophan (Elsden et al., 1976).

All strains tested are susceptible to chloramphenicol and erythromycin; susceptibility to clindamycin or tetracycline is variable among strains; one of seven strains tested is resistant to penicillin G.

Supernatant fluids are not toxic to mice. Cultures have been reported pathogenic for the guinea pig and rabbit (Weinberg et al., 1937).

Other characteristics of the species are given in Tables 13.13 and 13.15.

Isolated from human feces and chicken cecal contents, soil and human and animal infections (Holdeman et al., 1977; Finegold et al., 1983; Barnes and Impey, 1974; Smith, 1975b; Chapel et al., 1978; Prescott, 1979).

The mol% G + C of the DNA is 28 (T_m) (Johnson and Francis, 1975).

Type strain: ATCC 25776 (NCIB 10667; DSM 1127).

Further comments. This species is most easily distinguished from other nonsaccharolytic, nonproteolytic *Clostridium* species by its production of indole. *C. malenominatum* can be distinguished from nonsaccharolytic species of *Fusobacterium* (*F. nucleatum*, *F. naviforme*, *F. russii*, *F. prausnitzii*, *F. gonidiaformans*) by detection of spores or demonstration of heat resistance, by its production of large amounts of H_2, and by the electrophoretic pattern of soluble cellular proteins (Cato et al., 1982a) which is quite unlike that of any of the fusobacteria.

42. Clostridium mangenotii (Prévot and Zimmès-Chaverou, 1947) McClung and McCoy 1957, 664.[AL] (*Inflabilis mangenoti* Prévot and Zimmès-Chaverou 1947, 603.)

man.ge.no'ti.i. N.L. gen. n. *mangenotii* pertaining to Professor Mangenot, Italian bacteriologist.

This description is based on that of Holdeman et al. (1977), and on study of the type strain.

Cells in PYG broth cultures are Gram-positive, nonmotile, 0.6–0.9 × 3.1–8.2 μm, and occur singly, in pairs and in short chains.

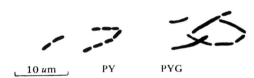

Figure 13.46. *Clostridium mangenotii.*

Spores are oval, subterminal and swell the cell. Sporulation occurs most readily on egg-yolk agar plates incubated for 72 h.

Cell walls contain *meso*-DAP (Weiss et al., 1981).

Surface colonies on blood agar plates are pinpoint to 0.5 mm, circular, low convex, translucent, granular, gray-white and dull, with an entire margin and a grainy surface.

Cultures in PYG broth are turbid, and have a stringy sediment and a pH of 6.2 after incubation for 1 week.

The optimum temperature for growth is 30–37°C. Strains grow moderately well at 25°C but not at all at 45°C. Growth is good at a pH of 8.5, but there is no growth with 6.5% NaCl or 20% bile.

Moderate gas is detected in PYG deep agar cultures.

Ammonia is produced in chopped meat broth and H_2S is produced in SIM.

A wide variety of straight and branched chain saturated and unsaturated cellular fatty acids is produced with the straight chain saturated C_{16} acid predominating (Elsden et al., 1980).

Products in PYG broth culture include acetate, formate, isocaproate, isovalerate, and isobutyrate; a small amount of propionate is present. A large amount of H_2 is formed. Pyruvate is converted to acetate; threonine is converted to propionate; lactate is not utilized. Valine is

converted to isobutyrate; leucine to isovalerate and isocaproate; isoleucine to isovalerate (Elsden and Hilton, 1978). Phenyl propionic and phenyl lactic acids are produced from phenylalanine, tyrosine, and tryptophane (Elsden et al., 1976). Proline is utilized and 5-aminovaleric acid and 2-aminobutyric acid are produced (Elsden and Hilton, 1979).

The type strain is susceptible to chloramphenicol, clindamycin, erythromycin, penicillin G and tetracyline.

Supernatant cultures are nontoxic to mice.

Other characteristics of the species are given in Table 13.15.

Isolated from soil (Prévot et al., 1967; Smith, 1975b), from marine sediments (Matches and Liston, 1974), and from human feces (Finegold et al., 1983).

The mol% G + C of the DNA has not been reported.

Type strain: ATCC 25761 (NCIB 10639; DSM 1289).

43. Clostridium nexile Holdeman and Moore 1974, 276.[AL]

nek'si.le. L. neut. adj. *nexile* tied together (referring to its chain formation).

This description is based on that of Holdeman and Moore (1974) and on their study of the type and 10 other strains.

Cells in PYG broth cultures are Gram-positive ovals or straight rods, nonmotile, 0.8–1.7 × 0.8–6.3 μm, and occur in pairs or chains.

Figure 13.47. *Clostridium nexile.*

Spores are round or oval, subterminal or nearly terminal, but are rarely seen, even in cultures that resist heating at 80°C for 10 min. Spores have been seen in 6-day-old cultures of PYG broth.

Surface colonies on blood agar plates are 0.5–1.0 mm in diameter, circular, convex to raised, semiopaque or with opaque centers and translucent edges, smooth, shiny, white to yellowish, nonhemolytic and with entire margins.

Cultures in PYG broth have a smooth or ropy sediment, often with no turbidity, and a pH of 4.9–5.3 after incubation for 5 days.

The optimum temperature for growth is 30–37°C. Most strains grow well at 45°C but poorly at 25°C. Growth is stimulated by fermentable carbohydrate, but is inhibited by 20% bile or 6.5% NaCl.

Abundant gas is produced in PYG deep agar cultures.

H_2S is produced by three of six strains tested. Neutral red and resazurin are reduced.

Products in PYG broth culture are acetic and formic acids and ethanol; moderate amounts of lactate and succinate are usually detected. Pyruvate is converted to acetate and ethanol; increased amounts of formate also are usually formed. Neither threonine nor lactate is utilized.

Orotic acid is not hydrolyzed.

Of seven strains tested, all are susceptible to chloramphenicol and clindamycin, five are susceptible to penicillin G, three to erythromycin, and two to tetracycline.

Other characteristics of the species are given in Table 13.14.

Isolated from human feces as part of the normal flora (Moore and Holdeman, 1974).

The mol% G + C of the DNA is 40–41 (T_m) (Holdeman and Moore, 1974).

Type strain: ATCC 27757 (DSM 1787).

Further comments. This species can be differentiated from *C. oroticum*, which it resembles most closely phenotypically, by not hydrolyzing orotic acid, and by lack of fermentation of arabinose, maltose and rhamnose. The electrophoretic patterns of soluble cellular proteins of the type strains of the two species are quite distinct (Cato et al., 1982a).

44. **Clostridium novyi** (Migula 1900, 872) Bergey et al., 1923, 236.[AL] (*Bacillus novyi* Migula 1900, 672.)

no'vy.i. N.L. gen. n. *novyi* pertaining to F. G. Novy, American bacteriologist.

This description is based on those of Smith (1975a) and of Holdeman et al. (1977) and on study of the type and 30 other strains of *C. novyi* type A, 7 strains of type B, and the strain of type C ("*C. bubalorum*") on which the description of this type was based.

Type A. Cells in PYG broth cultures are Gram-positive rods, 0.6–1.4 × 1.6–17 μm, and occur singly or in pairs. They are usually motile and peritrichous. Motility may be difficult to detect in wet mounts but peritrichous cells can usually be demonstrated.

Figure 13.48. *Clostridium novyi.*

Spores are oval, central or subterminal, and may swell the cell. There are no appendages or exosporium (Smith and Hobbs, 1974). Sporulation of most strains occurs most readily on blood agar plates incubated for 3 days.

Cell walls contain *meso*-DAP (Cummins and Johnson, 1971), glutamic acid and alanine (Schleifer and Kandler, 1972).

Surface colonies on blood agar plates are β-hemolytic, 1–5 mm in diameter, circular or irregular, flat or raised, translucent or opaque, gray, dull or glistening, with a crystalline or mosaic internal structure, and a scalloped, undulate, lobate or rhizoid margin. They may appear as a spreading film over the entire plate.

Cultures in PYG broth are turbid with a smooth or flocculent sediment and have a pH of 5.1–5.8 after incubation for 1 week.

The optimum temperature for growth is 45°C. Most strains grow nearly as well at 37°C and moderately at 30°C. There is little or no growth at 25°C. Strains require strictly anaerobic conditions and will not grow in the presence of even traces of O_2. Fermentable carbohydrate greatly stimulates growth. There is no growth in 20% bile, 6.5% NaCl, or at a pH of 8.5.

Moderate to abundant gas is produced in PYG deep agar cultures.

Both lecithinase and lipase are produced on egg-yolk agar plates. Gelatin is digested by all strains; there is weak and slow digestion of chopped meat by 6 of 31 strains; milk is curdled but neither milk nor casein is digested.

Deoxyribonuclease is produced by the type strain (Johnson and Francis, 1975) and by 16 of 25 other strains tested (Nord et al., 1975).

One strain, isolated from marine mud, causes the breakdown of chitin to *N*-acetylglucosamine (Timmis et al., 1974).

Of 25 strains tested, 20 produced protease on casein agar, 4 produced ribonuclease, 3 produced amylase, and 8 produced hyaluronatelyase (Nord et al., 1975).

Products in PYG broth culture include major amounts of butyric and propionic, and small amounts of acetic acids; small amounts of valeric acid and propanol may be detected. Abundant H_2 is produced. Pyruvate is converted to acetate, butyrate and sometimes propionate; lactate is converted to propionate; threonine is not utilized. Large amounts of *iso*-amylamine and phenethylamine are produced in chopped meat-glucose broth cultures (Holdeman and Brooks, 1970).

All strains tested are susceptible to chloramphenicol, clindamycin, erythromycin and penicillin G. One of 15 strains is resistant to tetracycline.

Cultures are pathogenic for guinea pigs, rabbits, mice, rats and pigeons, and supernatant culture fluids of about one-half of the pathogenic strains are toxic to mice. The principal lethal toxin of both types A and B is the necrotising alpha toxin. Type A strains also produce gamma (phospho-lipase C) and epsilon (lipase) toxins; some strains also produce delta (oxygen-labile hemolysin) toxin. The gamma toxin (lecithinase of type A) is active on horse red blood cells. The epsilon toxin (lipase) produces a pearly layer on colonies on egg-yolk agar plates and is useful in differentiating between type A and type B strains. Also, the type A lecithinase (gamma toxin) is antigenically distinct from the lecithinase (beta toxin) of *C. novyi* type B and *C. haemolyticum*. For further information on toxins produced by the different types, see Smith (1975a). Nontoxic strains can be converted to toxic strains through the mediation of specific bacteriophages (Schallehn et al., 1980; Imhoff and Schallehn, 1980). Plasmids have been found in both toxic and nontoxic variants of the same strains and may be involved in this transfer of toxigenicity (Schallehn and Krämer, 1981). For studies on phage-mediated conversion of *C. botulinum* type C to *C. novyi* type A, see Eklund et al. (1974, 1976). Phagelike particles that inhibit the growth of strains of *C. perfringens* and *C. tertium* have been detected in one strain of *C. novyi* type A; this strain of *C. novyi* type A was inhibited by a strain of *C. bifermentans* (Nieves et al., 1981).

High levels of ribosomal RNA homology exist between *C. botulinum* types C and D, *C. haemolyticum*, and *C. novyi* types A and B (Johnson and Francis, 1975). In addition, some strains of *C. novyi* type A have been shown to share antigens with strains of *C. botulinum* type C (Serikawa et al., 1977). For differentiation of these species, see Further Comments following their descriptions.

Other characteristics of the species are given in Table 13.12.

Isolated from soil (Nishida and Nakagawara, 1964; Smith, 1975b; Serikawa et al., 1977), from marine sediments (Finne and Matches, 1974; Matches and Liston, 1974), animal wounds (Berg and Fales, 1977; Berkhoff and Redenbarger, 1977; Smith 1975a), human wounds including gas gangrene (Smith, 1975a).

For further descriptions of the toxins produced and the diseases caused, see Prévot et al. (1967) and Smith (1975a).

The mol% G + C of the DNA is 29 (T_m) (Johnson and Francis, 1975).

Type strain: ATCC 17861 (NCIB 10661).

Type B. Phenotypic characteristics that help to distinguish strains of type B from those of type A are: Cells tend to be larger, 1.1–2.5 × 3.3–22.5 μm; mannose is fermented; milk and chopped meat are digested; lipase is not detected on egg-yolk agar; electrophoretic patterns of soluble cellular proteins are quite distinct (Cato et al., 1982a). Like type A strains, *C. novyi* type B strains produce the lethal, necrotizing *C. novyi* alpha toxin. However, the lecithinase of type B strains is the beta toxin, which also is produced by strains of *C. haemolyticum*. The beta toxin (lecithinase) can be identified by its hemolytic action on human red blood cells and can be neutralized by either *C. novyi* type B or *C. haemolyticum* antitoxin. Type B strains also produce zeta (hemolysin) and eta (tropomyosinase) toxins. Small amounts of theta toxin (lipase), commonly produced by strains of *C. haemolyticum*, may be produced by some strains of *C. novyi* type B. The lipase of the theta toxin does not produce a reaction on egg yolk agar.

Reference strain: ATCC 25758 (NCIB 10626).

Further comments. A nontoxigenic, nonpathogenic strain that is otherwise indistinguishable from *C. novyi* type A was isolated from water buffalo with osteomyelitis. It was designated *C. novyi* type C by Scott et al. (1935) and *C. bubalorum* by Prévot (1938). It produces only small amounts of lecithinase (gamma toxin) that are not detected on egg-yolk agar. Strains are quite commonly isolated that are phenotypically identical to *C. novyi* type A or to *C. botulinum* types C or D except that they are nontoxigenic, presumably through loss of their infecting phages. It has been shown that the heating process sometimes used in isolation of these strains can select for more actively sporulating strains and that there is an inverse relationship between sporulating potency and toxigenicity (Nishida and Nakagawara, 1965). For further discussion of the relationship between these species, see the description of *C. botulinum* types C and D.

Reference strain, *C. novyi* type C: ATCC 27323 (NCIB 9747).

45. Clostridium oceanicum Smith 1970, 811.[AL]

o.ce.an'i.cum. L. neut. adj. *oceanicum* belonging to the sea.

This description is based on that of Smith (1970), and on study of the type and 26 other strains.

Cells in PYG broth culture are Gram-positive rods, 0.3–1.6 × 1.7–25.7 μm and occur singly or in pairs. They are motile and peritrichous. Some strains are nonmotile when incubated at 37°C but motile at 25°C. Long filaments may be seen in older cultures.

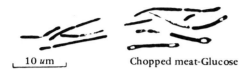

Figure 13.49. *Clostridium oceanicum.*

Spores are oval, terminal or subterminal, and usually do not swell the cell. There may be two spores, one at each end, in a single cell. Sporulation of most strains occurs most readily in chopped meat-glucose broth incubated at 30°C.

Cell walls contain *meso*-DAP (Weiss et al., 1981).

Surface colonies on blood agar plates are β-hemolytic, 1–6 mm in diameter, circular to irregular, flat or raised, translucent, gray and shiny, with an undulate or scalloped margin and a crystalline mosaic internal structure.

Cultures in PYG broth are slightly turbid with a smooth or flocculent sediment and a pH of 5.0–5.7 after incubation for 1 week.

The optimum temperature for growth is 30–37°C. Strains grow poorly at 25°C and not at all at 45°C. Growth is stimulated by carbohydrate, by Tween 80, and by a N_2 rather than a CO_2 atmosphere. Strains will grow in media with a pH from 6.5–8.5; there is little or no growth initiation at a pH of 6.0. Growth is inhibited by 20% bile or by 6.5% NaCl; 4% NaCl is not inhibitory.

Moderate to abundant gas is produced in PYG deep agar cultures.

Ammonia is produced; H_2S is produced in SIM medium by 20 of 26 strains tested.

Deoxyribonuclease and ribonuclease are produced (Smith, 1970).

Products in PYG broth include lactic, butyric and acetic acids; small amounts of formic, propionic, isobutyric, isovaleric, isocaproic and succinic acids, and ethanol, propanol and butanol may be detected. Abundant H_2 is produced. Acid production, both in amount and variety, is enhanced in PY broth without fermentable carbohydrate; production of acetic, propionic, isobutyric, isovaleric, and isocaproic acids is stimulated, and valeric acid usually is detected. Pyruvate is converted to acetate and butyrate. Threonine is converted to propionate by 19 of 26 strains. Utilization of lactate is variable.

All strains tested are susceptible to chloramphenicol, clindamycin, erythromycin, penicillin G, and tetracycline.

Culture supernatants are nontoxic to mice.

Other characteristics of the species are given in Table 13.12.

Isolated from marine sediments; also isolated from human feces (Finegold et al., 1983).

The mol% G + C of the DNA is 26–28 (T_m) (Smith, 1970).

Type strain: ATCC 25647 (NCIB 10625; DSM 1290).

46. Clostridium oroticum (Wachsman and Barker, 1954) Cato, Moore and Holdeman 1968, 9.[AL] (*Zymobacterium oroticum* Wachsman and Barker 1954, 400.)

o.ro'ti.cum. N.L. n. *acidum oroticum* orotic acid; *oroticum* pertaining to orotic acid.

This description is based on those by Wachsman and Barker (1954), Cato et al. (1968) and Holdeman et al. (1977), and on study of the type strain and one other strain.

Cells in PYG broth cultures are 0.6–1.6 × 1.3–3.9 μm, Gram-positive rods or ovals with tapering ends. They are nonmotile and occur in long tangled chains.

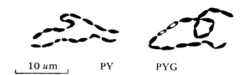

Figure 13.50. *Clostridium oroticum.*

Spores are round to ellipsoidal, central to subterminal, and do not swell the cell. Sporulation occurs most readily in PY broth cultures.

Cell walls contain *meso*-DAP (Gottschalk et al., 1981).

Surface colonies on blood agar plates are 1–2 mm in diameter, circular, convex, opaque, white or buff-colored, shiny, smooth and nonhemolytic. No internal structure is visible.

Cultures in PYG broth are not turbid and have a heavy, smooth to mucoid sediment and a pH of 5.1 after incubation for 4 days.

The optimum temperature for growth is 30–37°C. Growth is nearly as good at 45°C and moderate at 25°C. Growth is only slightly inhibited by 20% bile or a pH of 8.5 but completely inhibited by 6.5% NaCl.

Abundant gas is detected in PYG deep agar cultures.

Orotic acid is utilized with 90% of the substrate degraded in 4 days.

The organism contains high levels of the iron-sulfur flavoprotein dihydroorotate dehydrogenase (Aleman and Handler, 1967; Aleman et al., 1968). This enzyme catalyzes both the synthesis and degradation of pyrimidines (Yoch and Carithers, 1979). The organism also produces a zinc-containing metalloenzyme, dihydroorotase, which is active in pyrimidine degradation (Taylor et al., 1976). For the reactions involved, also see Vogels and Van der Drift (1976).

Products in PYG broth cultures are acetic and formic acids, ethanol, CO_2 and large amounts of H_2; trace amounts of lactic and succinic acids may be detected. Pyruvate is converted to acetate, formate and ethanol; neither lactate nor threonine is utilized. Ammonia is produced in orotic acid medium. H_2S is not produced in SIM.

The two strains tested are susceptible to chloramphenicol, erythromycin, penicillin G and tetracycline, but resistant to clindamycin. One clinical isolate of *C. oroticum* is susceptible to amoxicillin (0.25 μg/ml), carbenicillin (2 μg/ml), cephalexin (0.063 μg/ml), tiberal (0.063 μg/ml), clindamycin (2 μg/ml), metronidazole (0.063 μg/ml), chloramphenicol (2 μg/ml), LY 99638 (0.25 μg/ml), and Searle 28538 (0.125 μg/ml); this strain is moderately resistant to moxalactam, cefoperazone, cefoxitin and doxycycline, and resistant to cefamandole, erythromycin, rosaramicin, and tetracycline (Finegold and George, unpublished data).

Culture supernatants are nontoxic to mice.

Other characteristics of the species are given in Table 13.14.

Isolated from black mud from San Francisco Bay (Wachsman and Barker, 1954), human feces (Finegold et al., 1983), suprapubic bladder aspirates (Dankert et al., 1979) and a rectal abscess (Holdeman et al., 1977).

The mol% G + C of the DNA is 44 (T_m) (Johnson and Francis, 1975).

Type strain: ATCC 13619 (ATCC 25750; NCIB 10650; DSM 1287).

47. Clostridium papyrosolvens Madden, Bryder and Poole 1982, 90.[VP]

pa.py'ro.sol'vens. Gr. n. *papyros* paper; L. v. *solvere* to dissolve; N.L. part. adj. *papyrosolvens* paper-dissolving (intended to reflect the organisms' rapid fermentation of filter paper constituents).

This description is based on that of Madden et al. (1982) and on study of the type strain.

Cells in yeast extract-sea water-mineral solution-cellobiose broth are straight Gram-negative rods with a Gram-positive cell wall structure as shown by electron microscopy. They are motile and peritrichous, 0.5–0.8 × 2.0–5.0 μm, and occur singly and in pairs.

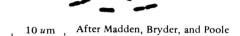

10 μm   After Madden, Bryder, and Poole

Figure 13.51. *Clostridium papyrosolvens.*

Spores are round, terminal, and swell the cell. Spores do not germinate unless they are heat shocked at 70°C for 15 min.

Cell wall composition has not been reported.

After incubation for 3 weeks, deep colonies in cellulose agar roll tubes are 1–2 mm in diameter, circular, translucent, colorless and granular, surrounded by clear zones of cellulose hydrolysis 1–2 cm in diameter. Surface colonies in rumen fluid agar roll tubes are circular to slightly irregular, entire to erose, convex, translucent, colorless, shiny and smooth.

Cultures in PYG broth are slightly turbid with a granular sediment and have a pH of 5.0 after incubation for 5 days.

Optimum temperature for growth is 25–30°C. The organism grows at 15°C and 37°C but only slightly at 45°C. Best growth occurs in an atmosphere of 90% N_2-10% CO_2. Fermentable carbohydrate is required for growth.

Abundant gas is detected in PYG deep agar cultures.

Cellulose, glycerol, and esculin are fermented.

Other characteristics of the species are given in Table 13.14.

Products in cellulose broth cultures are ethanol, acetate, lactate, H_2 and CO_2. Products in PYG broth are moderate amounts of lactic, formic, pyruvic, fumaric and lactic acids, and abundant H_2.

The type strain is susceptible to chloramphenicol, clindamycin, erythromycin, penicillin G and tetracycline.

Isolated from estuarine sediments.

The mol% G + C of the DNA is 30 (T_m) (Madden et al., 1982).

Type strain: NCIB 11394.

48. **Clostridium paraputrificum** (Bienstock 1906) Snyder 1936, 402.[AL] (*Bacillus paraputrificus* Bienstock 1906, 413.)

pa.ra.pu.tri′fi.cum. Gr. pref. *para* beside; N.L. n. *putrificum* a specific epithet; N.L. neut. adj. *paraputrificum* resembling (*Clostridium*) *putrificum.*

This description is based on that of Holdeman et al. (1977) and on study of the type and 50 other strains.

Cells in PYG broth cultures are straight or slightly curved rods, usually motile and peritrichous, 0.5–1.4 × 1.9–17.0 μm, and occur singly or in pairs. They are Gram-positive but rapidly become Gram-negative.

10 μm   PY   PYG

Figure 13.52. *Clostridium paraputrificum.*

Spores are oval, usually terminal and swell the cells; subterminal and free spores may be seen in the same preparations. Sporulation occurs readily in chopped-meat or PY broth after incubation for 24 h.

Cell walls do not contain DAP; the peptidoglycan bridge is composed of lysine, serine and glycine (Weiss et al., 1981). Glutamic acid and alanine also are present (Schleifer and Kandler, 1972). Traces of galactose are present in the wall of the type strain; other strains have been reported to contain glucose, rhamnose and mannose as well (Cummins and Johnson, 1971).

Surface colonies on blood agar plates are nonhemolytic, 1–5 mm in diameter, circular, low convex or flat, translucent or semiopaque, smooth, dull, with a mottled or mosaic internal structure and a scalloped, erose, or undulate margin.

Cultures in PYG broth are turbid with a smooth, ropy, or flocculent sediment and a pH of 4.5–5.0 after incubation for 5 days.

Optimum temperature for growth is 30–37°C. Most strains grow nearly as well at 45°C but poorly at 25°C. Growth is markedly stimulated by fermentable carbohydrate; 20% bile is not inhibitory; there is little or no growth with 6.5% NaCl.

Abundant gas is detected in PYG deep agar cultures.

Acetyl methyl carbinol is produced by 8 of 16 strains tested.

Strains of this species are active in the metabolism of bile acids and steroids and produce compounds that have been implicated in the incidence of colon and breast cancer (Goddard et al., 1975; Hill, 1975; Murray et al., 1980). For extensive descriptions of the metabolism of deoxycorticosterone and other steroids by *C. paraputrificum*, see Bokkenheuser et al. (1976), Winter and Bokkenheuser (1978), Winter et al. (1979, 1982), and Macdonald et al. (1983a and b).

One strain of this species is able to lyse Ehrlich ascites tumor tissue in mice (Möse and Möse, 1964). The lysis, however, is neither complete nor permanent. Other strains have been found to promote the formation of liver tumors in mice (Mizutani and Mitsuoka, 1979).

Products in PYG broth cultures include acetic, butyric and lactic acids; formic, pyruvic and succinic acids also may be detected. Abundant H_2 is produced.

Of 39 strains tested, one is resistant to chloramphenicol, 3 are resistant to penicillin G, 3 are resistant to tetracycline, 13 are resistant to erythromycin and 35 are resistant to clindamycin.

Culture supernatants are not toxic to mice and strains are not pathogenic for guinea pigs or rabbits.

Other characteristics of the species are given in Table 13.14.

Isolated from soil (Smith 1975b), marine sediments (Finne and Matches, 1974), avian feces (Lev and Briggs, 1956), human infant feces (Bullen et al., 1977; Rotimi and Duerden, 1981; Stark and Lee, 1982) human adult feces (Finegold et al., 1983; Drasar et al., 1976), porcine and bovine feces, and from human clinical specimens including blood, peritoneal fluid, wounds and appendicitis.

The mol% G + C of the DNA is 26–27 (T_m) (Cummins and Johnson, 1971).

Type strain: ATCC 25780 (NCIB 10671).

49. **Clostridium pasteurianum** Winogradsky 1895, 330.[AL]

pas.teu.ri.a′num. N.L. neut. adj. *pasteurianum* pertaining to Louis Pasteur, French microbiologist.

This description is based on that of Holdeman et al. (1977) and on study of the type and one other strain.

Cells in PYG broth culture are Gram-positive, becoming Gram-negative in old cultures, straight to slightly curved rods, 0.5–1.3 × 2.7–13.2 μm, and occur singly or in pairs. Motility is variable and may be lost on subculture; motile cells are peritrichous. They are granulose-positive (Darvill et al., 1977).

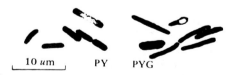

10 μm   PY   PYG

Figure 13.53. *Clostridium pasteurianum.*

Spores are oval, subterminal, and swell the cell. Sporulation occurs most readily on chopped meat slants incubated at 30°C for 1 week. There are no exosporium or appendages (Smith and Hobbs, 1974).

Cell walls contain *meso*-DAP; wall sugars are glucose, galactose, rhamnose and mannose (Cummins and Johnson, 1971). Glutamic acid and alanine also are present (Schleifer and Kandler, 1972). Squalene has been found in cell membranes, with greater amounts being present in sporulating than in vegetative cells (Mercer et al., 1979). Activity of magnesium-dependent membrane adenosine triphosphatase has been described (Clarke et al., 1979).

Surface colonies on blood agar plates are nonhemolytic, 1–3 mm in diameter, circular to irregular, low convex or flat, translucent to semiopaque, gray, shiny and smooth with an erose or rhizoid margin and a mosaic internal structure.

Cultures in PYG broth are turbid with a smooth sediment, and have a pH of 4.8–5.0 after incubation for 5 days. The optimum temperature for growth is 37°C. Strains grow nearly as well at 30°C, moderately well at 25°C and poorly if at all at 45°C. Growth is stimulated by fermentable carbohydrate but inhibited by 20% bile, 6.5% NaCl, or a pH of 8.5. Growth occurs in a synthetic medium described by Malette et al. (1974).

Abundant gas is detected in PYG deep agar cultures.

Deoxyribonuclease is produced.

Acetyl methyl carbinol is produced by the type strain.

Atmospheric N_2 is fixed; molybdenum is essential for biosynthesis and activity of the nitrogenase involved (Cardenas and Mortenson, 1975). There is an increase in cellular phospholipids with a high proportion of palmitic acid during N_2-fixing growth; during non-N_2-fixing growth, the proportion of palmitic acid decreases, accompanied by marked increases in shorter chain saturated fatty acids (Deo et al., 1979). Cellular fatty acids include mainly the C_{16} straight chain, monounsaturated straight chain C_{16}, C_{15} cyclopropane and unsaturated cyclopropane (Chan et al., 1971).

Leucine is synthesized by the α-iso-propylmalate pathway (Wiegel and Schlegel, 1977; Wiegel, 1981).

The type strain of *C. pasteurianum* can utilize crotonate slowly when the medium is supplemented with peptone and yeast extract (Bader et al., 1980).

Chlorinated hydrocarbon pesticides can be degraded by this species (Jagnow et al., 1977).

For extensive reviews of the properties and metabolic activity of the ferredoxin and hydrogenase of *C. pasteurianum*, see Gillum et al. (1977), Chen and Blanchard (1978), Yoch and Carithers (1979), Schönheit et al. (1979), and Adams et al. (1981). Sulfur metabolism by this species has been studied by McCready et al. (1976).

A nickel-requiring carbon monoxide dehydrogenase was reported by Diekert et al. (1979).

Products in PYG broth cultures include butyric and acetic acids and small amounts of formic acid; abundant CO_2 and H_2 are detected. Ethanol is produced (Ehrlich et al., 1981). Neither ammonia nor H_2S is detected. Pyruvate is converted principally to acetate, CO_2, H_2, and small amounts of butyrate (von Hugo et al., 1972); neither lactate nor threonine is utilized. Gluconate is fermented by way of 2-keto-3-deoxygluconate (Bender et al., 1971). The amino acids formed by resting cells in synthetic medium with ammonium salts as sole nitrogen source are alanine, threonine, aspartic acid, arginine, glutamic acid, lysine and valine, with traces of methionine, isoleucine and serine (Matteuzzi et al., 1978a).

Strains are susceptible to chloramphenicol, clindamycin, erythromycin, penicillin G and tetracycline.

Culture supernatants are not toxic to mice and strains are not pathogenic for laboratory animals. A bacteriocin produced by the type strain of *C. butyricum* is bactericidal against growing cultures of the type strain of *C. pasteurianum* (Clarke and Morris, 1976).

Other characteristics of the species are given in Table 13.14.

Isolated from soil.

The mol% G + C of the DNA is 26–28 (T_m) (Cummins and Johnson, 1971).

Type strain: ATCC 6013 (NCIB 9486; DSM 525).

Further comments. For more information on metabolic activities of this species, see Thauer et al. (1977), Kleiner (1979), and Gottschalk et al. (1981).

50. **Clostridium perfringens** (Veillon and Zuber, 1898) Hauduroy, Ehringer, Urbain, Guillot and Magrou 1937, 119.^{AL} (*Bacillus perfringens* Veillon and Zuber 1898, 539; "*Bacterium welchii*" Migula 1900, 392.)

per.frin′gens. L. part. adj. *perfringens* breaking through.

This species produces a number of soluble substances that cause a variety of toxic effects in in vitro, or in vivo conditions or both. *C. perfringens* has been divided into five types (A, B, C, D and E) on the basis of production of major lethal toxins as shown in Table 13.16 (Sterne and Warrack, 1964). A sixth type (type F) was proposed, but it is now considered type C, and type F designation has been abandoned (Sterne and Warrack, 1964; Smith, 1975a). The five types of *C. perfringens* cannot be differentiated reliably on the basis of cellular or colonial morphology, biochemical reactions, or gas-liquid-chromatographic analyses of fatty and organic acid end products of metabolism.

This description, unless otherwise indicated, is based on study of strains representing the five types (A, B, C, D and E) and 285 other strains, and on reviews of Willis (1969) and Smith (1972, 1975a).

Cells in PYG broth culture are Gram-positive, atrichous, nonmotile, straight rods with blunt ends, that occur singly or in pairs and are 0.6–2.4 × 1.3–19.0 µm.

<u>10 µm</u>   PY   PYG

Figure 13.54. *Clostridium perfringens,* type A.

Spores are rarely seen in vivo or in the usual in vitro conditions; when present they are large, oval, central or subterminal, and distend the cell (Rahman, 1978). There is no exosporium and spores lack appendages. Nakamura and Nishida (1974) found that a complex relationship existed between previous heat treatment, the ability to ferment certain sugars, and the ability to sporulate. Sacks and Thompson (1977) found that spore yield was markedly increased by addition of various methylxanthines to the medium. Spores from heat-resistant strains usually require heat activation to germinate, whereas those from heat-susceptible strains do not; spores from heat-resistant strains are also more resistant to the lethal effects of gamma radiation (Roberts, 1968).

The cell wall contains LL-DAP; cell wall sugars that may be present are galactose, glucose and rhamnose (Cummins and Johnson, 1971); however, cell walls of different strains or types may not possess all three sugars. Approximately three-fourths of strains possess a capsule that is composed largely of polysaccharides; the composition of the capsular polysaccharide may vary among strains (Lee and Cherniak, 1974; Cherniak and Frederick, 1977; and reviewed in Smith, 1975a).

Colonies on the surface of sheep blood agar are usually 2–5 mm in diameter, circular, entire, dome-shaped, gray to grayish yellow, and translucent with a glossy surface. Several other colonial morphologies (dwarfs, rough colonies with lobate margins and flat colonies with an irregular surface and filamentous margins) occur occasionally, even in the same culture (Wagner and Schallehn, 1982). The kind and extent

Table 13.16.

Classification of **C. perfringens** *by toxin type.*

C. perfringens type	Toxin produced			
	Alpha	Beta	Epsilon	Iota
A	+	−	−	−
B	+	+	+	−
C	+	+	−	−
D	+	−	+	−
E	+	−	−	+

of hemolysis present depends on both the species of blood and the type of *C. perfringens* being examined; the three types of hemolysins that may be produced in varying quantities are designated alpha, delta and theta (Table 13.17). On rabbit, sheep, cow, horse or human blood, most strains produce a narrow zone of complete hemolysis due to the theta toxin and a surrounding zone of incomplete hemolysis due to the alpha toxin. Some type B and C strains may produce a very wide zone of hemolysis on sheep or cow blood due to delta toxin.

Synergistic hemolysis (CAMP phenomenon) between this species and *Streptococcus agalactiae* has been described (Buchanan, 1982).

Colonies in agar are usually lenticular.

Cultures in PYG broth are turbid with a smooth, or occasionally ropy sediment and have a pH of 4.8–5.6 after incubation for 1 week.

The temperature for optimum growth of types A, D and E is 45°C; types B and C grow equally well at 37°C and 45°C. The range of temperatures that will support growth of most strains is 20–50°C; occasional strains will grow at 6°C for a limited number of passages, but they are not truly psychrotrophic (Beerens et al., 1965). Growth is stimulated by the presence of a fermentable carbohydrate and is not inhibited by 20% bile. Growth occurs readily from pH 5.5–8.0. Growth is not inhibited by NaCl concentrations up to 2%, but is markedly inhibited by 6.5% NaCl. Growth is not enhanced by addition of CO_2 to the atmosphere of incubation (Watt, 1973; Reilly, 1980). Hyperbaric oxygen (100% oxygen at 3 atm) is bactericidal for this species; partial protection is conferred by addition of whole blood, presumably because of enzymatic destruction of hydrogen peroxide by catalase (Hill and Osterhout, 1972).

Abundant gas is produced in PYG deep agar cultures.

Occasional strains weakly ferment glycerol, inulin, and sorbose; ammonia is produced; acetyl methyl carbinol and H_2S (Kawabata, 1980) are produced, and hippurate is hydrolyzed by some strains; neutral red is reduced; resazurin is reduced by most strains. Although more than 95% of strains ferment sucrose and produce lecithinase, an occasional isolate will be negative for one of these reactions.

Deoxyribonuclease (Döll, 1973), acid phosphatase (Ueno et al., 1970), ribonuclease, elastase, hyaluronidase and amylase (Nord et al., 1975), neuraminidase (Houdret et al., 1975; Fraser, 1978; Collee, 1965), a hemoagglutinin that is distinct from neuraminidase (Collee, 1965), exo-β-D-galactosidase (DiCioccio et al., 1980), ferredoxin-linked nitrate reductase (Chiba and Ishimoto, 1973; Seki et al., 1982), and superoxide dismutase (Hewitt and Morris, 1975; Tally et al., 1977) are produced by some or all strains.

Bile acid metabolism includes deconjugation of taurocholic and glycocholic acids by types A, B, C, D and E (Norman and Grubb, 1955), degradation of chenodeoxycholic acid and cholic acid (MacDonald et al., 1983b and c), and conversion of a number of 3-α-hydroxy bile acids to 3-β-hydroxy and 3-oxo-bile acids (Hirano et al., 1981). Occasional strains are reported to possess a Δ^4-steroid dehydrogenase and/or the ability to cause aromatisation of the Δ-ring of 4-androsten-3,17-dione (Goddard et al., 1975); these steps have been postulated to be important in converting steroids to colon carcinogens. Monoacetone glucose is fermented (Cmelik, 1980).

Krämer and Schallehn (1974) found that most strains were susceptible to bacteriocins produced by enterococci, particularly *Streptococcus faecium*. Smith (1975b) found that some soil isolates of *C. perfringens* produce inhibitors that are active against some or all of *C. botulinum* types A, B, E and F. *Clostridium perfringens* also produces bacteriocins that are active against other strains of *C. perfringens*. Wolff and Ionesco (1975) found that the bacteriocin produced by one strain was a single-chained polypeptide with a molecular weight of approximately 82,000. The mechanism of action of many bacteriocins involves inhibition of macromolecular synthesis (DNA, RNA, and proteins); an additional (but undefined) mechanism probably also exists (Ionesco and Wolff, 1975; Mahoney and Li, 1978). Bacteriocin production and resistance to that bacteriocin by several strains have been shown to be related to the presence of a plasmid (Ionesco et al., 1976; Mihelc et al., 1978; Li et al., 1980). Blascheck and Solberg (1981) reported that caseinase activity appears to be related to the presence of a plasmid in the one strain tested.

Bacteriophages of this species have been recognized since 1949 (see Smith, 1975a for a review). Virulent bacteriophage has been recovered from sewage or river water below sewage discharge points. Smith (1959) has reported that some strains of types A, B and C are lysogenic, whereas lysogeny has not been described for types D and E; in addition, smooth or rough strains usually are phage-susceptible whereas mucoid strains usually are resistant. Stewart and Johnson (1977) found that the rapidity of sporulation and the percentage of spores that were heat-resistant decreased when the one strain studied was cured of bacteriophage; these changes were reversed when the cured strain was reinfected with the temperate phage.

Cultures of all five types, when grown in PYG broth, produce large amounts of acetic, butyric and lactic acids; sometimes smaller amounts of propionic, formic, and succinic acids are detected. Abundant H_2 is produced. Lactate occasionally is converted to butyrate; pyruvate is

Table 13.17

Other possible virulence factors produced by **C. perfringens.**[a]

"Toxin"	Toxic activity	Comments
Gamma	Lethal toxin	Existence inferred from discrepancies in neutralization studies of toxic culture filtrates by various antitoxin preparations; probably not important in disease production
Delta	Hemolysin in vitro; lethal	Active only on erythrocytes of even-toed ungulates (deer, cattle, elk, sheep, goats, swine); produced by some type B and C strains
Eta	Lethal toxin	Same as comments for gamma toxin
Theta	Hemolysin; lethal and necrotizing in laboratory animals	Produced by most strains of all types; responsible for inner zone of complete hemolysis on blood agar
Kappa	Collagenase	Thought to contribute to softening of muscles in myonecrosis (gas gangrene); produced by all five types
Lambda	Protease	Produced by most strains of type B and E, and some of type D
Mu	Hyaluronidase	Probably not of major significance in disease; produced by most strains of type A and B, and some of types C and D
Nu	Deoxyribonuclease	Produced by strains of all types except type B from cases of necrotic enteritis (New Guinea); probably not important in disease production

[a] Modified from Sterne and Warrack (1964); Willis (1969); Smith (1975a); and Smith (1979).

converted to acetate and butyrate, and occasionally to formate; threonine is converted to propionate. Ishimoto et al. (1974) found that production of butyrate by one strain was abolished when nitrate was added to the medium. Cmelik (1980) found that one strain of type E produced isovaleric acid and larger amounts of propionic acid in a medium containing monoacetone-glucose (rather than glucose). Two of four strains tested by Mayrand and Bourgeau (1982) produced phenylacetic acid from a trypticase-yeast extract medium.

George et al. (1981) in a recent review of β-lactam antimicrobials noted that most agents in this class, particularly penicillin G, were quite active against *C. perfringens*. Marrie et al. (1981) subsequently reported relatively greater resistance of this species to penicillin G than had been noted previously. Chloramphenicol, clindamycin and metronidazole are active against most isolates, but less so on a weight basis than is penicillin G; erythromycin and tetracycline are generally less active than most of the penicillins, chloramphenicol, clindamycin and metronidazole (Marrie et al., 1981; Appelbaum and Chatterton, 1978; Martin et al., 1972; Staneck and Washington, 1974). Rood et al. (1978a) reported isolation from pig feces of a number of strains that were resistant to tetracycline, erythromycin, lincomycin and clindamycin; this degree of resistance appeared to correlate with the use of antimicrobial-containing animal feed. Aminoglycosides are inactive against *C. perfringens* (Martin, et al., 1972). Twelve of 25 strains were inhibited by 0.25 μg or less of nalidixic acid/ml whereas the remaining 13 strains required 4–64 μg/ml for inhibition (Nadaud, 1977). Plasmid-mediated resistance of *C. perfringens* to clindamycin-erythromycin, and to tetracycline-chloramphenicol (Brefort et al., 1977) and to tetracycline alone (Rood et al., 1978b) has been described.

C. perfringens produces a variety of substances (often referred to as "exotoxins") that have been suggested as possible virulence factors (see Tables 13.16 and 13.17). The four toxins shown in Table 13.16 often are referred to as the major lethal toxins.

The alpha toxin is a phospholipase C that hydrolyzes lecithin to phosphorylcholine and a diglyceride; it is produced by all five types of *C. perfringens*. Alpha toxin exerts its lethal effects by lysing cell membranes, presumably, as a consequence of hydrolysis of membrane lecithin. This toxin alone is the cause of muscle death in myonecrosis in cases of *C. perfringens* gas gangrene in humans and other animals following trauma or abortion. The alpha toxin may also cause intravascular hemolysis. Several other diseases in animals are known, or suspected to be caused by type A (presumably by the alpha toxin). These include a fatal enterotoxemia of lambs, newborn alpacas, captive wild goats, reindeer and possibly chickens (Kummeneje and Bakken, 1973; Al-Sheikhly and Truscott, 1977; Goldberg, 1980); and delayed hypersensitivity to alpha toxin in swine that results in arthritis with eventual joint deformity, parakeratosis and proliferative glomerulonephritis (Mansson et al., 1971). In addition to causing myonecrosis and/or intravascular hemolysis in humans via the alpha toxin, type A strains may also produce an enterotoxin that causes a food poisoning syndrome in humans (see below).

The beta toxin, produced by strains of types B and C, has not been completely characterized, but appears to be a highly trypsin-sensitive, single chain polypeptide with a molecular weight of approximately 30,000 (Sakurai and Duncan, 1978). It appears to exert its toxic effect by increasing capillary permeability (Sakurai and Duncan, 1978; Smith, 1979). Intravenously administered beta toxin appears to induce release of endogenous catecholamines (Sakurai et al., 1981). Beta toxin production is thought possibly to be plasmid-mediated (Duncan et al., 1978). The diseases produced by beta toxin involve the gastrointestinal tract. In humans a necrotic enteritis (pig-bel) develops soon after ingestion of roast pig that has been accidentally contaminated with feces containing *C. perfringens* type C; it is not known whether toxin is ingested or produced in the gut following ingestion of the organism. A necessary antecedent of pig-bel appears to be ingestion of large quantities of sweet potatoes which contain a trypsin inhibitor. A similar form of necrotic enteritis in humans (Darmbrand) was seen in Europe in the mid- to late 1940s; in this setting the presence of starvation,

which tends to produce low gastrointestinal levels of trypsin, was probably an important cofactor. Type C also causes enterotoxemia or necrotic enteritis in lambs, calves, piglets and sheep, whereas type B causes enterotoxemia or necrotic enteritis in foals, lambs, sheep and goats (Sterne and Warrack, 1964). An important cofactor for development of beta toxin-induced necrotic enteritis in newborn or very young animals may be the absence of trypsin in the gut (Smith, 1979).

Epsilon toxin is elaborated by types B and D in the form of a virtually nontoxic prototoxin that is converted to a potent heat-labile toxin by certain proteolytic enzymes such as trypsin (Smith, 1975a; Smith, 1979). This toxin appears to increase vascular permeability, possibly by adenyl cyclase activation (Buxton, 1978a), that ultimately leads to tissue necrosis (Smith, 1975a). Exposure of the gut to epsilon toxin (to which it is normally impervious) results in greater permeability to proteins, including epsilon toxin; this may be an important factor in pathogenesis of disease because oral administration of a single large dose of epsilon toxin may be without effect, whereas administration of the same total dose in several smaller portions may be lethal. The toxin probably produces an increase in vascular permeability of many organs, the most seriously affected of which is the brain; cerebral edema and necrosis of brain tissue is the most likely cause of death in afflicted animals (Gay et al., 1975; Smith, 1979). The toxin also is cytotoxic to guinea pig and rabbit peritoneal macrophages (Buxton, 1978b). Type D causes enterotoxemia of lambs, sheep, goats, cattle and possibly man (very rarely), chinchillas (Sterne and Warrack, 1964; Moore and Greenlee, 1975).

Iota toxin is produced only by type E (Sterne and Warrack, 1964). This toxin is elaborated as a prototoxin that is usually activated by proteolytic enzymes produced by the organism. Iota toxin markedly increases vascular permeability, produces necrosis on intradermal injection, and is lethal on intravenous injection. The organism is sometimes carried by normal sheep and cattle, and is purported to be a rare cause of enterotoxemia in calves (Hart and Hoper, 1967; Smith, 1979). Although LaMont et al. (1979) and Rehg and Pakes (1982) have implicated (by antitoxic neutralization) iota toxin as a cause of antimicrobial-induced colitis in rabbits, Borriello and Carman (1983) have shown that the causative organism probably is *C. spiroforme*; iota-like toxic activity is produced by strains tentatively identified as *C. spiroforme*.

Type A strains and some type C and D strains produce an enterotoxin (Uemura and Skjelkvalé, 1976; Skjelkvalé and Duncan, 1975); *C. perfringens* food poisoning in humans is produced by type A strains (Smith, 1979). Food poisoning can be caused by strains that produce only heat-sensitive spores, as well as by strains that produce heat-resistant spores (Hall et al., 1963). Spores that survive cooking may germinate and proliferate to high counts in food products (usually warm meats or meat products); when ingested, the vegetative cells sporulate in the gut and release enterotoxin (Finegold, 1977). Enterotoxin is a product of the sporulation process (Duncan et al., 1972; Duncan, 1973) and causes fluid accumulation in the small intestine of laboratory animals (Yamamoto et al., 1979). Feeding of purified enterotoxin to volunteers reproduces the food poisoning syndrome (Skjelkvalé and Uemura, 1977). Ochoa and de Velandia (1978) have presented preliminary data to implicate type A enterotoxin as the cause of a lethal enteritis in horses.

C. perfringens also produces a variety of other substances that have been referred to as toxins. The role of these factors as regards virulence and pathogenicity either is not significant or is not known (Table 13.17). Other possible "virulence factors" include sialidase and a non-alpha-delta-theta hemolysin (Smith, 1979).

Other characteristics of the species are given in Table 13.12. Serrano and Schneider (1978) recently have reported that some allegedly lecithinase-negative strains can be shown to produce lecithinase on a modified medium.

This species is more widely spread in nature than any other pathogenic microorganism. Although most investigators do not type their isolates, Smith (1975a) has stated that only type A strains are found as part of the microflora of both soil and intestinal tracts, and that types B, C, D and E seem to be obligate parasites of animals and occasionally are found in man. Sources yielding *C. perfringens* include soil and marine sediment samples worldwide (MacLennan, 1943; Smith and Gardner, 1949; Matches and Liston, 1974; Smith, 1975b; Miwa et al., 1976; Schrader and Schau, 1978); clothing (MacLennan, 1943), raw milk (Naguib and Shouman, 1972), cheese (Naguib and Shauman, 1973), semipreserved meat products (Skjelkvalé and Tjaberg, 1974), and venison (Sumner et al., 1977). According to Smith (1975a), *C. perfringens* has been isolated from the intestinal contents of virtually every animal that has been investigated. It has also been isolated from pheasant small intestine (Mead et al., 1973) and from rattlesnake venom (Goldstein et al., 1979).

In addition to causing specific toxin-induced diseases, *C. perfringens* has also been isolated from a variety of mixed anaerobic/aerobic pyogenic infections of a number of different species of domesticated animals (Ryff and Lee, 1946; Berkhoff and Redenbarger, 1977; Berg et al., 1979; Prescott, 1979; Hirsh et al., 1979).

Humans frequently carry *C. perfringens* as part of the normal endogenous flora. Although this species can be recovered in a small percentage of patients from the normal oral flora (Van Reenen and Coogan, 1970), the normal cervicovaginal flora (Gorbach et al., 1973; Goplerud et al., 1976; Bartlett et al., 1977; Hammerschlag et al., 1978), from urine (presumably reflecting the flora of the distal urethra) (Headington and Beyerlein, 1966), and from the skin of the antecubital fossae of approximately 20% of subjects (Ahmad and Darrell, 1976), the main site of carriage is the distal gastrointestinal tract.

Approximately 80% of cases of gas gangrene (clostridial myonecrosis) involve *C. perfringens* (MacLennan 1962). In addition, this species has been reported to cause bacteremia (with and without intravascular hemolysis) (Gorbach and Thadepalli, 1975; Stephenson and Wright, 1982).

C. perfringens is the species of *Clostridium* most commonly isolated from infections in humans (Lewis et al., 1980); such infections are often polymicrobial. Although virtually every type of infection in humans has yielded *C. perfringens* on one or more occasions, it is most commonly recovered from infections derived from the colonic flora (e.g. peritonitis, intra-abdominal abscess and soft tissue infections below the waist). For a review of infections yielding *C. perfringens* see Finegold (1977).

The mol% G + C of the DNA is 24–27 (T_m) (Cummins and Johnson, 1971; Johnson and Francis, 1975).

Type strain: ATCC 13124 (NCIB 6125; CCUG 1795; NCTC 8237; DSM 756).

Reference strains: type B: ATCC 3626, NCIB 10691; type C; ATCC 3628, NCIB 10662; type D: ATCC 3629, NCIB 10663; type E: ATCC 27324, NCIB 10748.

Further comments. This species closely resembles *C. absonum*; please see Further Comments on *C. absonum*. Isolates previously identified as "*Clostridium plagarum*" possess a high degree of DNA homology (Nakamura et al., 1976a) and 100% rRNA homology with *C. perfringens*; such isolates should be considered lecithinase-negative, theta toxin-negative variants of *C. perfringens* (Nakamura et al., 1976).

51. **Clostridium polysaccharolyticum** (van Gylswyk, 1981) van Gylswyk, Morris and Els 1983, 438.[VP] (*Fusobacterium polysaccharolyticum* van Gylswyk 1981, 382.)

po.ly.sac.ca.ro.ly'ti.cum. Gr. adj. *polys* many; Gr. n. *saccharon* sugar; Gr. adj. *lytikos* dissolving; N.L. neut. adj. *polysaccharolyticum* degrading polysaccharides.

This description is based on that of van Gylswyk (1980) and on study of the type strain.

Cells from cellobiose-rumen fluid agar medium are motile and peritrichous, Gram-negative straight rods with rounded ends, 0.6–1.1 × 2-15 μm. In broth cellobiose-rumen fluid cultures, aseptate filaments may be more than 50 μm in length. Cells occur singly, in pairs, or in short chains.

|_ 10 μm _| After Van Gylswyk

Figure 13.55.   *Clostridium polysaccharolyticum.*

Spores are usually oval, occasionally spherical and subterminal, although terminal spores may be seen. They swell the cell. Sporulation occurs most readily on cellulose-rumen fluid-agar slants (van Gylswyk et al., 1980).

Although the cells stain Gram-negative, cell walls have a peptidoglycan layer characteristic of Gram-positive cells (van Gylswyk et al., 1980).

Surface colonies on rumen fluid-glucose-cellobiose agar plates are 1 mm in diameter, circular, low convex, semiopaque, buff colored, shiny and smooth, with an entire edge and have no visible internal structure.

Cultures in PY-cellobiose broth with 30% rumen fluid are not turbid but have a smooth viscous sediment and a pH of 5.7 after incubation for 1 week.

Optimum temperature for growth is between 30°C and 38°C. There is no growth at 22°C and little or none at 45°C. Fermentable carbohydrate and CO_2 in the gas phase are required for growth. Acetate stimulates growth. Rumen fluid (30%) is greatly stimulatory.

Moderate gas is detected in cellobiose-rumen fluid deep agar cultures after incubation for 1 week.

Cellulose, xylan, starch and cellobiose are fermented consistently.

Products in PY-cellobiose broth are formate, butyrate, acetate, propionate, H_2 and small amounts of ethanol. In media containing acetate and propionate, these compounds are utilized to form more butyrate or formate (van Gylswyk, 1980).

Antibiotic susceptibility has not been determined.

Other characteristics of the species are given in Table 13.13.

Isolated from the sheep rumen.

The mol% G + C of the DNA is 42 (T_m) (van Gylswyk, 1980).

Type strain: strain B (ATCC 33142; DSM 1801).

52. **Clostridium propionicum** Cardon and Barker 1946, 631.[AL]

pro.pi.o'ni.cum. N.L. neut. adj. *propionicum* pertaining to propionic acid.

This description is based on that of Cardon and Barker (1946) and on study of the type strain.

Cells in PYG broth cultures are motile and peritrichous, Grampositive rods that rapidly become Gram-negative, 0.5–0.8 × 1.3–5.0 μm. They are straight or slightly curved with tapered or rounded ends, and usually occur in pairs, occasionally singly or in short chains.

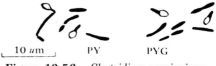

|_ 10 μm _| PY   PYG

Figure 13.56.   *Clostridium propionicum.*

Spores are oval, subterminal, and swell the cell. Sporulation occurs most readily on egg yolk agar plates or in PY broth.

Cell walls contain *meso*-DAP (Weiss et al., 1981).

Surface colonies on blood agar plates are pinpoint, circular, convex, translucent, gray, dull with a slightly shiny outer rim, smooth, with an entire to slightly scalloped margin and have no visible internal structure.

Cultures in PYG broth are turbid with a smooth sediment and have a pH of 6.0–6.2 after incubation for 1 week.

C. propionicum grows well between 25°C and 30°C. There is no growth at 45°C. Alanine, serine, threonine, lactate, pyruvate or acrylate is required for growth.

Moderate amounts of gas are formed in PYG deep agar cultures; resazurin is reduced.

Ammonia is produced; H_2S is formed in SIM medium.

A wide variety of lipid fatty acids, both straight and branched chain, saturated and unsaturated, is present in cells of the type strain of this species (Elsden et al., 1980).

Products in PYG broth are propionate, isovalerate, isobutyrate, butyrate, small amounts of acetate, succinate and sometimes lactate. Abundant H_2 is produced. Pyruvate is converted to propionate and acetate; threonine is converted to propionate and butyrate, CO_2 and ammonia. Lactate is converted to propionate by the acrylate pathway rather than by the more common fumarate pathway (Anderson and Wood, 1969). Acrylate can accumulate in resting cell solutions when alternative electron acceptors are provided (Akedo et al., 1978). Alanine is converted to propionate, acetate, NH_3, and CO_2 (Cardon and Barker, 1946). Valine is converted to isobutyrate; leucine and isoleucine are converted to isovalerate (Elsden and Hilton, 1978). Phenylalanine and tyrosine are oxidized to small amounts of phenylacetic and hydroxy-phenylacetic acids (Elsden et al., 1976).

Antibiotic susceptibility has not been determined.

Other characteristics of the species are given in Table 13.13.

Isolated from black mud in San Francisco Bay.

The mol% G + C of the DNA has not been reported.

Type strain: ATCC 25522 (NCIB 10656; DSM 1682; CCUG 9280).

53. **Clostridium puniceum** Lund, Brocklehurst and Wyatt 1981b, 216.[VP] (Effective publication: Lund, Brocklehurst and Wyatt 1981a, 17.)

pu.ni′ce.um. L. neut. adj. *puniceum* purplish (referring to pink color of colonies on potato infusion agar).

This description is based on that of Lund et al. (1981) and on study of the type and four other strains.

Cells in PYG broth culture are straight or curved rods, motile and peritrichous, 0.6 × 1.8–4.2 μm, granulose-positive, and occur singly, in pairs or in short chains. They are usually Gram-negative, but Gram-positive cells are sometimes present.

Figure 13.57. *Clostridium puniceum.*

Spores are oval, subterminal and do not swell the cell. Free spores are common and have an extensive exosporium (Lund et al., 1978).

Cell wall content has not been reported.

Surface colonies on blood agar plates are nonhemolytic, pinpoint to 2 mm, circular, raised to convex, opaque, white, slightly shiny, and smooth, with an entire margin and have a granular or mottled internal structure. On potato infusion agar, colonies are similar except that they are pale pink or deep pink and may have undulate or lobate margins (Lund et al., 1981).

Cultures in PYG broth are turbid with a viscous or flocculent sediment and have a pH of 5.2–5.5 after incubation for 6 days.

The optimum temperature for growth is 23–33°C. Growth range is 7–39°C. Strains grow moderately well in PY broth without fermentable carbohydrate; they do not grow in the defined medium of Lund et al. (1981) unless fermentable carbohydrate is present.

Abundant gas is detected in PYG deep agar cultures.

Slices of potato, carrot, radish and turnip are digested. Pectin is fermented. Pectate is hydrolyzed; tributyrin is not attacked. Cultures

grown in potato tissue form pectate lyase and pectinesterase but no pectic hydrolase (Lund and Brocklehurst, 1978).

Products in PYG broth include acetic, butyric and formic acids, and butanol. Abundant H_2 is produced.

All strains tested are susceptible to chloramphenicol, clindamycin, erythromycin, penicillin G and tetracycline.

Culture supernatants are not toxic for mice.

Other characteristics of the species are given in Table 13.12.

Isolated from rotting potatoes and from a cavity spot lesion in a carrot.

The mol% G + C of the DNA is 28–29 (T_m) (Lund et al., 1981).

Type strain: BL 70/20 (NCIB 11596).

54. **Clostridium purinilyticum** (corrig.) Dürre, Andersch and Andreesen 1981, 184.[VP]

pu.ri.ni.ly′ti.cum. N.L. n. *purum uricum*, condensed as "purin," a term proposed by E. Fisher for the basic ring system of uric acid; Gr. adj. *lytikos* dissolving; N.L. neut. adj. *purinilyticum* decomposing the purine ring.

This description is based on that of Dürre et al. (1981) and on study of the type strain.

Cells in PYG broth culture are Gram-positive straight rods, 1.1–1.6 × 2.7–9.6 μm, motile with lateral and subterminal flagella, and occur singly or in pairs.

Figure 13.58. *Clostridium purinilyticum.*

Spores are round, terminal and swell the cell. Sporulation occurs most readily on blood agar plates incubated anaerobically for 48 h or in chopped meat broth incubated in an atmosphere of 90% N_2-10% CO_2. Higher spore yields may be obtained in a medium with added hypoxanthine or guanine.

Cell walls contain *meso*-DAP (Dürre et al., 1981).

Surface colonies on blood agar plates prepared with rabbit blood are β-hemolytic, 2–3 mm in diameter, irregular, flat, transparent, slightly buff-colored, dull, and slightly pitted with an irregular or occasionally fringed margin, and have no visible internal structure.

Cultures in PYG broth incubated in an atmosphere of 90% N_2-10% CO_2 are turbid with a smooth sediment and have a pH of 7 after incubation for 4 days. Final pH of cultures after growth on purines or glycine is approximately 8.8.

The optimum temperature for growth is 36°C; growth occurs at 42°C. Selenite (0.1 μM) bicarbonate (0.2%), and thiamin are required for growth; molybdate, tungsten and yeast extract are stimulatory. Optimum pH for growth is 7.3–7.8; growth occurs at pH 6.5–9.0.

No gas is detected in PYG deep agar cultures.

Utilizes only purines (including adenine, guanine, xanthine, hypoxanthine, uric acid and others), and glycine and some glycine derivatives for growth (Dürre et al., 1981). Hippurate (benzoylglycine) is hydrolyzed.

Products from the fermentation of purines are acetate, formate, CO_2, and ammonia. The major product of PYG cultures is acetate; no H_2 is formed. Pyruvate and threonine are converted to acetate; lactate is not utilized.

The type strain is susceptible to chloramphenicol, clindamycin, penicillin G and tetracycline, but resistant to erythromycin.

Other characteristics of the species are given in Table 13.13.

Isolated from soils exposed to chicken manure and from sewage sludge enriched with adenine.

The mol% G + C of the DNA is 29 (T_m) (Dürre et al., 1981).

Type strain: WA-1 (DSM 1384).

Further comments. This species is very similar phenotypically to *C. aciduric i* and *C. cylindrosporum*. However, Dürre et al. (1981) report that there is very low DNA/DNA homology between the type strain of *C. purinilyticum* (DSM 1384), the type strain of *C. aciduric i* (ATCC 7906; DSM 604), and the type strain of *C. cylindrosporum* (ATCC 7905; DSM 605). They may be differentiated by the ability of *C. purinilyticum* to grow readily on adenine and hypoxanthine and to utilize glycine. The electrophoretic pattern of soluble cellular proteins of *C. purinilyticum* is distinct from that of *C. aciduric i* (Cato et al., 1982a). *C. sticklandii* can be distinguished easily from these species by its production of branched chain fatty acids in complex media (Holdeman et al., 1977).

55. Clostridium putrefaciens (McBryde, 1911) Sturges and Drake 1927, 125.[AL] (*Bacillus putrefaciens* McBryde 1911, 50.)

pu.tre.fa′ci.ens. L. adj. *putrefaciens* putrefying.

This description is based on those of Sturges and Drake (1927), Ross (1965), and Holdeman et al. (1977) and on study of the type strain.

Cells in PYG broth cultures are Gram-positive rods, nonmotile, 1.5–1.8 × 7.5–>15 µm. They often occur as long curving filaments. Cells occur singly, in pairs, in long chains or in tangled masses.

Figure 13.59. *Clostridium putrefaciens.*

Spores are round or oval, subterminal or terminal, and swell the cell. Sporulation occurs most readily in chopped meat cultures held at room temperature.

Cell walls contain LL-DAP and glycine (Weiss et al., 1981).

Surface colonies on blood agar plates are β-hemolytic, pinpoint to 0.5 mm, circular to irregular, flat to low convex, transparent to translucent, colorless, shiny and smooth, with a slightly scalloped or rhizoid margin and a crystalline or mosaic internal structure.

Cultures in PYG broth are only slightly turbid with a smooth sediment and have a pH of 7.0 after incubation under 90% N_2-10% CO_2 gas for 1 week.

The optimum temperature for growth is 15–22°C; growth is good at 25°C and 30°C, slow at 5°C; no growth at 37°C (Roberts and Hobbs, 1968). There is good growth between pH 6.2 and 7.4, moderate and slow growth at a pH of 5.8 and 8.5 (Barnes and Impey, 1968). Growth is inhibited by 6.5% NaCl or 20% bile.

No gas is detected in PYG deep agar cultures. Ammonia is formed slowly in chopped meat and gelatin cultures (Parsons and Sturges, 1927). Traces of H_2S are produced (Ross, 1965). Neutral red is reduced.

Cells contain both straight and branched chain fatty acids with saturated C_{16}, C_{12}, and unsaturated C_{18} acids predominating (Elsden et al., 1980).

Products in PYG broth are moderate amounts of acetate, formate, lactate and succinate; no H_2 is produced. Pyruvate is not utilized; propionate is not formed from lactate or threonine. The amino acids serine, threonine, glycine, and arginine, are utilized for growth, and alanine and valine are produced in 3% casein hydrolysate medium (Mead, 1971). Elsden and Hilton (1978) found no increase in products in threonine medium. Isobutyric acid is produced from valine; isovaleric acid is produced from leucine and *iso*-leucine (Elsden and Hilton, 1978). Phenylacetic, hydroxyphenylacetic, and indoleacetic acids are produced from phenylalanine, tyrosine and tryptophan, respectively (Elsden et al., 1976).

Other characteristics of the species are given in Table 13.15.

The type strain is nontoxic for mice.

Isolated from spoiled hams (McBryde, 1911); also found in hog muscle (Sturges and Drake, 1927), chicken carcasses (Barnes and Impey, 1968), human feces (Finegold et al., 1983) and in urine specimens from pregnant women with bacteriuria (Meijer-Severs et al., 1979).

The mol% G + C of the DNA is 22–25 (Gottschalk et al., 1981).

Type strain: ATCC 25786 (NCTC 9836; NCIB 11406; DSM 1291).

56. Clostridium putrificum (Trevisan 1889) Reddish and Rettger 1922, 9.[AL] (*Pacinia putrifica* Trevisan 1889, 23.)

pu.tri′fi.cum. L. neut. adj. *putrificum* making rotten.

This description is based on that of Holdeman et al. (1977) and on study of the type and 15 other strains.

Cells in PYG broth culture are Gram-positive rods, straight or slightly curved, motile and peritrichous, 0.3–1.3 × 1.3–11 µm, and occur singly or in pairs.

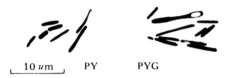

Figure 13.60. *Clostridium putrificum.*

Spores are oval or round, terminal or subterminal, and may swell the cell. Sporulation of most strains occurs readily in chopped meat or PYG broth after incubation for 24 h.

Cell walls contain *meso*-DAP; no wall sugars are detected (Cummins and Johnson, 1971); glutamic acid and alanine are present (Schleifer and Kandler, 1972).

Surface colonies on blood agar plates are usually β-hemolytic, 0.5–3 mm in diameter, circular or slightly irregular, flat or convex, translucent to semiopaque, white, shiny or dull, and smooth with entire, erose, or lobate margins and a mottled or mosaic internal pattern.

Cultures in PYG broth are turbid with a smooth or stringy sediment and have a pH of 5.4–6.0 after incubation for 1 week. The large amounts of ammonia produced may mask acid production by these strains.

The optimum temperature for growth is 37°C. Some strains grow as well at 30°C or 45°C but there is little or no growth at 25°C. Growth is inhibited by 6.5% NaCl or 20% bile. Growth is optimum at a pH of 7.0 or above.

Abundant gas is detected in PYG deep agar cultures.

Ammonia and H_2S are produced. Hippurate is hydrolyzed by 3 of 10 strains tested. Neutral red is reduced; reduction of resazurin is variable.

Extracellular DNase is produced by the one strain tested (Porschen and Sonntag, 1974).

The predominant fatty acids detected in cells of the species are saturated, straight chain C_{13}, C_{14}, C_{15} and C_{16} acids, although both unsaturated and branched chain acids are present (Elsden et al., 1980).

Products in PYG broth include acetic, butyric, isovaleric, isobutyric acids and ethanol; propionic, valeric, isocaproic, formic, lactic, and succinic acids may be detected as well. Abundant H_2 is produced. Pyruvate is converted to acetate and butyrate. Propionate is produced from threonine by four of nine strains tested; lactate is not converted to propionate and acetate. Elsden and Hilton (1978) reported that the type strain converts threonine to propionate, valine to isobutyrate, and isoleucine to isovalerate. Amino acids (particularly proline, lysine, alanine, serine, glutamic acid, and glycine) are utilized for growth in 6% casein hydrolysate medium; δ-amino valerate is produced (Mead, 1971).

Of 10 strains tested, all are susceptible to tetracycline, 9 are susceptible to chloramphenicol or penicillin G and 8 are susceptible to erythromycin or clindamycin.

Culture supernatants are not toxic to mice.

Other characteristics of the species are given in Table 13.12.

Isolated from the mouse cecum, from snake venom, from clinical specimens including abscesses, wounds and blood, from canned crabmeat, and from soil; also isolated from a sediment core from the Black Sea (Smith and Cato, 1974).

The mol% G + C of the DNA is 27 (T_m) (Johnson and Francis, 1975).

Type strain: ATCC 25784 (NCIB 10677; DSM 1734).

Further comments. RNA preparations of the type strain are 100% homologous with 23S-RNA from the type strain of *C. sporogenes* (Johnson and Francis, 1975). The two species can be differentiated by lipase production by *C. sporogenes* but not by *C. putrificum.*

57. **Clostridium quercicolum** Stankewich, Cosenza and Shigo 1971, 302.[AL]

quer.ci.co'lum. L. n. *quercus* oak; L. neut. adj. *quercicolum*, meaning associated with oak trees.

This description is based on the report by Stankewich et al. (1971) and on study of the type strain.

Cells grown on solid media are motile and peritrichous, 0.5 × 3.0 µm, and are uniformly Gram-negative. They occur singly or in pairs.

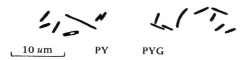

Figure 13.61. *Clostridium quercicolum.*

Spores are round and terminal or oval and central, and swell the cell. Sporulation occurs most reliably on chopped meat agar slants incubated at 30°C.

Surface colonies on blood agar plates are irregular, raised, gray and dull, with a lobate or rhizoid margin.

Cultures in PYG broth have a granular sediment with no turbidity and a pH of 7.0 after incubation for 7 days in an atmosphere of 100% N_2. Final pH in PY-fructose broth is 5.2.

The optimum temperature for growth is between 25°C and 30°C. Growth also occurs at 20°C and 45°C. Best growth occurs in an atmosphere of 100% N_2.

Moderate amounts of gas are detected in PYG deep agar cultures. Resazurin is reduced.

Weak fermentation of glycerol, inositol and ribose may occur in trypticase-yeast extract broth. Lipase and lecithinase are not produced.

Major products in PY-fructose broth are acetic and propionic acids, propanol and large amounts of H_2; a small amount of butyric acid is detected.

The type strain is susceptible to chloramphenicol, clindamycin, erythromycin and tetracycline but resistant to penicillin G.

Culture supernatants of the type strain are nontoxic to mice.

Other characteristics of the species are given in Table 13.13.

Isolated from discolored tissues of living oak trees.

The mol% G + C of the DNA is 52–54 (T_m) (Stankewich et al., 1971).

Type strain: ATCC 25974 (DSM 1736).

58. **Clostridium ramosum** (Vuillemin 1931) Holdeman, Cato and Moore 1971, 39.[AL] (*Nocardia ramosa* Vuillemin 1931, 32.)

ra.mo'sum. L. neut. adj. *ramosum* much-branched.

This description is based on those of Holdeman et al. (1971, 1977) and on study of the type and 133 other strains.

Cells in PYG broth cultures stain Gram-positive or Gram-negative and are nonmotile, straight rods, 0.5–0.9 × 2–12.8 µm, and occur singly, in pairs or in short chains often in "V" arrangements, with a "rail fence" appearance, or in irregular masses. Cells may have central or terminal swellings up to 1.6 µm in width.

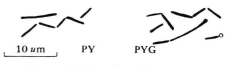

Figure 13.62. *Clostridium ramosum.*

Spores are round, thin-walled, usually terminal, and swell the cell, but are very rarely seen and often are difficult to detect by heat tests. They can be demonstrated most readily from 3-week-old chopped meat agar slants incubated at 30°C, or in old chopped meat or PYG broth cultures.

Cell walls contain *meso*-DAP (Holdeman et al., 1971); glutamic acid and alanine are present as well (Schleifer and Kandler, 1972).

Surface colonies on blood agar plates are nonhemolytic, 0.5–2 mm in diameter, circular to slightly irregular, convex or raised, colorless to gray-white, translucent or semiopaque and smooth, with an entire, scalloped or erose margin and a mottled, mosaic or granular internal structure.

Cultures in PYG broth are turbid with a smooth or ropy sediment and have a pH of 4.4–4.8 after incubation for 5 days.

The optimum temperature for growth is 37°C; most strains grow equally well at 30°C and 45°C and grow well at 25°C. Growth is stimulated by fermentable carbohydrate, inhibited by 6.5% NaCl and reduced in 20% bile.

Moderate gas is detected in glucose deep agar cultures. Production of ammonia and acetyl methyl carbinol is variable among strains.

One of four strains tested produces an extracellular β-glucuronidase (Sakaguchi and Murata, 1983).

Major products in PYG broth are acetic, formic and lactic acids, small amounts of pyruvic and succinic acids may be detected, and ethanol often is present. Pyruvate is converted to acetate and formate; neither threonine nor lactate is utilized. H_2 production is variable.

All strains tested are susceptible to chloramphenicol; of 61 strains tested, 4 are resistant to penicillin G, 7 to clindamycin, 13 to erythromycin and 31 to tetracycline. Strains are susceptible to achievable blood levels of carbenicillin and vancomycin but resistant to lincomycin, rifampin, and gentamicin; 37 of 48 strains tested are susceptible to metronidazole; the remaining strains are susceptible to high but achievable levels (25 µg/ml) of metronidazole (Tally et al., 1974). Niridazole, chemically similar to metronidazole, is approximately 15 times as effective as metronidazole (Hof et al., 1982).

Culture supernatants are not toxic to mice but strains are pathogenic for guinea pigs; pathogenicity may be lost in laboratory cultures (Prévot et al., 1967).

Other characteristics of the species are given in Table 13.14.

Isolated from infant (Stark and Lee, 1982) and adult feces (Moore and Holdeman, 1974; Finegold et al., 1983; Holdeman et al., 1976; Drasar et al., 1976); from the normal human cervix (Gorbach et al., 1973); frequently from human infections of the abdominal cavity, genital tract, lung (Bartlett and Finegold, 1972; Bartlett et al., 1974), biliary tract (Nielsen and Justesen, 1976) and from blood cultures (Gorbach and Thadepalli, 1975).

The mol% G + C of the DNA (T_m) is 26 (Johnson and Francis, 1975).

Type strain: ATCC 25582 (NCIB 10673; DSM 1402).

Further comments. This organism has been refered to in the past as "*Bacteroides terebrans,*" "*Bacteroides trichoides,*" "*Eubacterium filamentosum*" and "*Ramibacterium ramosum*" (which illustrates that cells often stain Gram-negative and spores are difficult to detect).

59. **Clostridium rectum** (Heller 1922) Holdeman and Moore 1972, 69.[AL] (*Hiblerillus rectus* Heller 1922, 17)

rec'tum. L. neut. adj. *rectum* straight.

This description is based on that of Holdeman et al. (1977) and on study of the type strain.

Cells in PYG broth cultures are nonmotile, Gram-positive straight rods, 0.5–1.1 × 1.6–3.1 µm, and occur singly or in pairs.

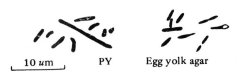

Figure 13.63. *Clostridium rectum.*

Spores are oval, subterminal and swell the cell. Sporulation occurs most readily in chopped meat broth cultures.

Cell walls contain *meso*-DAP (Weiss et al., 1981).

Surface colonies on blood agar plates are pinpoint to 1 mm in diameter, translucent to semiopaque, convex, grayish white, shiny and smooth, with entire margins and no visible internal structure. They may be slightly β-hemolytic.

Cultures in PYG broth are turbid with a smooth sediment and have a pH of 5.2–5.7 after incubation for 1 week.

The optimum temperature for growth is 37–45°C; moderate growth occurs at 25°C and 30°C. There is good growth at a pH of 8.5 or in 20% bile, no growth in 6.5% NaCl.

Abundant gas is detected in PYG deep agar cultures. Ammonia and H₂S are produced. Neutral red and resazurin are reduced.

The insecticide lindane can be degraded provided dithiothreitol or a leucine-proline mixture is present; isovaleric acid is formed (Ohisa et al., 1980).

The major product in PYG broth culture is butyric acid with moderate amounts of acetic and propionic, and a trace of valeric acid. Pyruvate is converted to acetate and butyrate; threonine is converted to propionate. Abundant H₂ is produced.

The type strain is susceptible to chloramphenicol, clindamycin, penicillin G, and tetracycline, but resistant to erythromycin.

Culture supernatants are nontoxic to mice.

Other characteristics of the species are given in Table 13.14.

Isolated from horse manure and beet rhizosphere (Prévot et al., 1967), and from rice paddy soil (Ohisa et al., 1980).

The mol% G + C of the DNA is 26 (T_m) (Johnson and Francis, 1975).

Type strain: ATCC 25751 (NCIB 10651; DSM 1295).

60. Clostridium roseum (*ex* McCoy and McClung 1935) nom. rev. (*Clostridium roseum* McCoy and McClung 1935, 237.)

ro′se.um. L. neut. adj. *roseum* rosy.

This description is based on those of McCoy and McClung (1935), Spray and McClung (1957) and on study of the type strain.

Cells in corn mash cultures (McClung, 1943) are 0.7–0.9 × 3.2–4.3 μm, granulose-positive in corn mash or glucose-tryptone broth, and occur singly, in pairs, or in short chains. They are Gram-positive but rapidly become Gram-negative. Vegetative cells are motile and peritrichous; sporulating cells are sluggish or nonmotile.

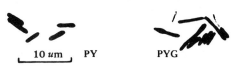

Figure 13.64. *Clostridium roseum.*

Spores are oval, subterminal and swell the cell. Sporulation occurs most rapidly in 5% corn meal mash medium.

Cell wall content has not been reported.

Surface colonies on blood agar plates are 4 mm in diameter, irregular, flat, grayish white with raised white centers, a dull rough surface, a rhizoid margin and are nonhemolytic. On beef PYG agar, surface colonies are raised and smooth, with irregular margins and a pink to orange pigmentation. Colonies become purplish black after exposure to

air. Pigmentation is most pronounced in 5% corn meal mash semisolid medium (McClung, 1943).

Cultures in PYG broth are only slightly turbid with a peach to orange ropy sediment and have a pH of 4.1 after incubation for 6 days.

The optimum temperature for growth is 37°C; growth occurs between 20°C and 47°C. Growth is greatly stimulated by fermentable carbohydrate and an atmosphere of 90% N₂-10% CO₂.

Abundant gas is formed in PYG deep agar cultures.

Gelatin is completely digested; there is a stormy fermentation in milk with an acid curd formed that is 50% digested in 3 weeks. H₂S is produced in 0.25% glucose-tryptone-sulfite or 0.25% glucose-tryptone-thiosulfate broth, but not in SIM which contains 0.02% thiosulfate and 0.02% sulfate.

Pectin is strongly fermented (final pH 4.8).

Gluconate is fermented to pyruvate and glyceraldehyde-3-phosphate by a modified Entner-Doudoroff pathway (Bender et al., 1971).

Products in PYG broth are butyric and acetic acids and butanol; a small amount of succinic acid also is detected. Neither pyruvate, lactate, nor threonine is utilized. In 5% corn mash medium, acetone and ethyl alcohol also are detected (McCoy and McClung, 1935).

Antibiotic susceptibility has not been determined.

Strains are not pathogenic for guinea pigs or rabbits.

Other characteristics of the species are given in Table 13.12.

Isolated from German maize. Probable habitat is the soil.

The mol% G + C of the DNA has not been reported.

Type strain: ATCC 17797 (DSM 51).

Further comments. For differentiation of this species from *C. felsineum* which it most closely resembles, see Further Comments following the description of that species. In addition, strains of *C. roseum* have been shown to be serologically distinct from strains of *C. felsineum* (McCoy and McClung, 1935; McClung and McCoy, 1935).

61. Clostridium saccharolyticum Murray, Khan and van den Berg 1982, 135.[VP]

sac′cha.ro.ly′ti.cum. Gr. n. *sacchar* sugar; Gr. adj. *lytikos* dissolving; N.L. neut. adj. *saccharolyticum* sugar-dissolving.

This description is based on that by Murray et al. (1982) and on study of the type strain.

Cells are Gram-negative, atrichous, nonmotile, spindle-shaped straight rods, 0.5–0.7 × 3.0 μm.

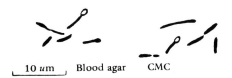

Figure 13.65. *Clostridium saccharolyticum.*

Spores are round, terminal or subterminal, and distend the cell. Sporulation occurs readily in chopped meat carbohydrate broth.

Cell wall composition has not been reported.

Surface colonies on cellobiose-yeast extract agar after 48 h incubation are 0.5–1.5 mm in diameter, circular with smooth margins, convex and white. On blood agar plates, colonies are pinpoint to 1 mm in diameter, circular or slightly irregular, low convex, with entire or slightly scalloped margins and a mottled or crystalline internal structure.

Cultures in PYG broth are turbid with a smooth sediment and have a pH of 5.1 after incubation for 4 days.

The temperature for optimum growth is 37°C; growth also occurs at 17°C and 43°C but not at 14°C or 45°C. Optimum pH for growth is 7.4; growth occurs between pH 6.0–8.8. Abundant growth occurs in PY media supplemented with vitamin K and heme (Holdeman et al., 1977), with or without carbohydrate. There is no growth in a defined carbohydrate-mineral salt-vitamin medium even when supplemented with synthetic mixtures of amino acids, purines and pyrimidines; good growth is obtained in this medium upon addition of yeast extract (Murray and Khan, 1983a) or in co-culture with *Acetivibrio cellulolyticus* (Khan and Murray, 1982). In minimal medium, 0.1% yeast extract, fermentable carbohydrate, iron and a reduced form of sulfur are required for growth; growth in this medium is enhanced by addition of B-vitamins and phosphate (Murray and Khan, 1983a).

Milk is curdled; acetyl methyl carbinol, ammonia, and H_2S are produced; and resazurin is reduced. Murray et al. (1982) report that starch is not hydrolyzed and nitrate not reduced; using media described by Holdeman et al. (1977), both reactions were repeatedly positive for the type strain.

Products of fermentation in cellobiose-yeast extract broth after 7 days incubation detected by a flame-ionization detector are acetic acid and ethanol, and traces of pyruvic and lactic acids (Murray et al., 1982). Products in PYG broth are acetic, formic and lactic acids. H_2 and CO_2 are produced. In naturally occurring habitats, the acetic acid and H_2 are utilized by methanogens and the ethanol and lactic acid are utilized by sulfate reducers to provide a favorable environment for cellulolytic anaerobes (Khan and Murray, 1982). The presence of calcium carbonate and of higher concentrations of yeast extract, or the presence of added H_2 in the headspace gas, produce a metabolic shift to less acetic acid production and greater ethanol production (Murray and Khan, 1983b). Pyruvate is utilized; neither lactate nor threonine is utilized.

The type strain is susceptible to chloramphenicol, clindamycin, erythromycin, penicillin G and tetracycline.

Other characteristics of the species are shown in Table 13.14.

Isolated from a methanogenic cellulose-enrichment culture from sewage sludge.

The mol% G + C of the DNA is 28 by ultraviolet spectroscopy (Murray et al., 1982).

Type strain: WM1 (NRC 2553).

62. Clostridium sardiniense (corrig.) Prévot 1938, 81.[AL]

sar.din.i.en′se. N.L. neut. adj. *sardiniense* from Sardinia.

This description is based on that of Holdeman et al. (1977) and on study of the type and 10 other strains.

Cells in PYG broth are Gram-positive, straight or slightly curved rods, motile and peritrichous. Motility may be lost on subculture. Cells occur singly, in pairs or in short chains, and are 0.5–1.7 × 1–10.0 μm.

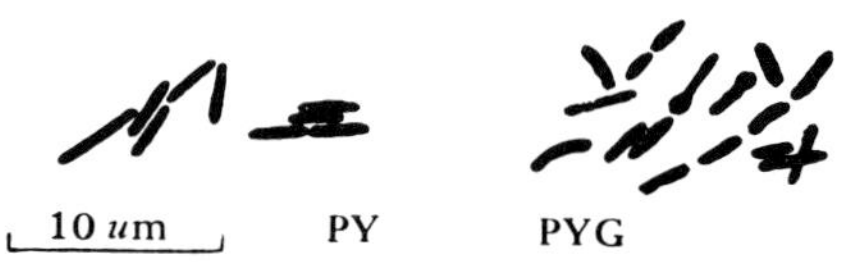

Figure 13.66. *Clostridium sardiniense.*

Spores are oval and subterminal, or occasionally terminal; very few spores are produced in usual media. Sporulation occurs most readily in 3-week-old chopped meat broth cultures.

Cell walls contain *meso*-DAP (C. S. Cummins, unpublished data).

Surface colonies on blood agar plates are β-hemolytic, 1–3 mm in diameter, circular to irregular, raised or low convex, translucent or semiopaque, gray-white, shiny, and smooth, with a lobate or erose margin and usually with a granular or mottled internal structure.

Cultures in PYG broth are turbid with a smooth or stringy sediment, and have a pH of 4.5–5.0 after incubation for 5 days.

The temperature for optimum growth is 25–37°C; growth is poor at 45°C. Growth is stimulated by fermentable carbohydrate but inhibited by 6.5% NaCl.

Abundant gas is produced in PYG deep agar cultures.

Ammonia is produced, neutral red and resazurin are reduced.

Products in PYG broth include large amounts of acetic, butyric and lactic acids, sometimes with small amounts of formic and propionic acids. Abundant H_2 is produced. Pyruvate is converted to acetate and butyrate. Neither lactate nor threonine is converted to propionate.

All strains tested are susceptible to chloramphenicol, penicillin G, and tetracycline; four of eight strains are resistant to clindamycin; two of eight strains are resistant to erythromycin.

Culture supernatants are nontoxic to mice. Prévot et al. (1967) reported that cultures injected subcutaneously or intramuscularly are pathogenic for sheep, guinea pigs, goats, dogs, rabbits, mice, rats and chickens.

Other characteristics of the species are given in Table 13.12.

Isolated from lesions of symptomatic anthrax in sheep (Prévot et al., 1967), from soil or water (Holdeman et al., 1977), and from feces of infants (Borriello, 1980).

The mol% G + C of the DNA has not been determined.

Type strain: VPI 2971 (ATCC 33455).

Further comments. This species is most easily differentiated from *C. perfringens*, which it closely resembles phenotypically, by its motility and its lack of toxicity for mice. At the present time the species is differentiated from *C. absonum* by motility but since some strains labeled *C. absonum* have been found to be motile (Stanley M. Harmon, personal communication), clear separation of the species must await DNA homology studies.

63. Clostridium sartagoforme (corrig.) Partansky and Henry 1935, 564.[AL]

sar.ta.go.for′me. L. n. *sartago* frying pan; L. adj. suffix *-formis* shaped like; N.L. neut. adj. *sartagoforme* intended to mean shaped like a frying pan (in reference to a sporulating cell).

This description, unless indicated otherwise, is based on that of Holdeman et al. (1977) and on study of the type and five other strains.

Cells in PYG broth culture are Gram-positive, straight to slightly curved rods that occur singly or in pairs, and are 0.3–0.9 × 2.2–8.0 μm. Motility is variable; cells of motile strains are peritrichous.

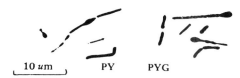

Figure 13.67. *Clostridium sartagoforme.*

Spores are oval (occasionally round), terminal, and swell the cell. Sporulation occurs most readily in chopped meat broth cultures.

Cell wall does not contain DAP (C. S. Cummins, unpublished data). The cell wall is susceptible to dissolution by lysozyme (Johnson and Francis, 1975).

Surface colonies on blood agar plates are 1–3 mm in diameter, circular, with entire or erose margins, flat or convex, gray, translucent, with a matt surface and are usually nonhemolytic. Colonies in agar are 1–1.5 mm in diameter and lenticular (Partansky and Henry, 1935).

Cultures in PYG broth are turbid with a smooth, or occasionally flocculent to ropy, sediment and have a pH of 4.8–5.1 after incubation for 5 days.

Temperature range for optimum growth is 30–37°C; growth also occurs at 25°C and 45°C. Growth is stimulated by a fermentable carbohydrate and is inhibited by 6.5% NaCl or 20% bile.

Abundant gas is produced in PYG deep agar cultures.

Adonitol, dulcitol, erythritol, glycerol, inositol and sorbose are not fermented. Ammonia is sometimes produced, hippurate is hydrolyzed by 1 of 3 strains tested, neutral red is reduced, and resazurin usually is reduced.

Deoxyribonuclease activity is not present (Johnson and Francis, 1975).

Products of fermentation in PYG broth are large amounts of acetic, butyric, and formic acids and sometimes large amounts of lactic acid. Abundant H_2 is produced. Neither lactate nor threonine is converted to propionate; pyruvate is converted to acetate and butyrate.

All strains tested are susceptible to chloramphenicol, erythromycin and tetracycline. Of four strains tested, one is resistant to clindamycin and a different one is resistant to penicillin G. Three strains studied by Perea et al. (1978) were inhibited by 0.1–0.5 μg of cefoxitin/ml.

Culture supernatants are nontoxic to mice.

Other characteristics of the species are given in Table 13.14.

This species has been isolated from soil and mud (Partansky and Henry, 1935; Smith and Hobbs, 1974), rumen fluid of healthy and bloating calves (Jayne-Williams, 1979), from the human gingival crevice (de Campos et al., 1981), feces of neonates and infants (Stark and Lee, 1982; Borriello, 1980) and from the feces of approximately 5% of adult subjects tested (Drasar et al., 1976; Finegold et al., 1983).

The mol% G + C of the DNA is 28 (T_m) (Johnson and Francis, 1975).

Type strain: ATCC 25778 (NCIB 10668; DSM 1292).

64. **Clostridium scatologenes** (Weinberg and Ginsbourg 1927) Prévot 1948a, 191.[AL] (*Clostridium scatol* Fellers and Clough 1925, 128; *Bacillus scatologenes* Weinberg and Ginsbourg 1927, 54.)

sca.to.lo′gen.es. Gr. n. *skatos* dung; Gr. v. *gennao* to produce; N.L. part. adj. *scatologenes* meaning either an organism that produces a dung-like odor or an organism that produces skatol.

This description, unless otherwise stated, is based on that of Holdeman et al. (1977) and on study of the type and three other strains.

Cells in PYG broth are Gram-positive straight rods that are motile and peritrichous, and are 0.5–1.6 × 3.1–21.2 μm.

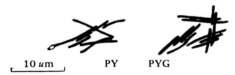

| 10 μm | PY | PYG |

Figure 13.68. *Clostridium scatologenes.*

Spores are oval, terminal or occasionally subterminal, and distend the cell slightly.

Cell walls contain *meso*-DAP (Weiss et al., 1981). The cell wall is susceptible to dissolution by lysozyme (Johnson and Francis, 1975).

Surface colonies on blood agar are α-hemolytic, 0.5–2.0 mm in diameter, circular, convex, translucent, gray, with an entire or scalloped margin, a matt surface and a granular or mottled internal structure.

Cultures in PYG broth are turbid with a smooth sediment and a pH of 5.2–5.4 after incubation for 5 days.

The temperature for optimum growth is 30–37°C; also grows at 25°C, but not at 45°C.

Grows well in absence of a fermentable carbohydrate. Growth inhibited by 20% bile, 6.5% NaCl or a pH of 8.5.

Abundant gas is produced in PYG deep agar cultures.

Produces ammonia and H_2S; two of three strains hydrolyze hippurate; reduces neutral red and usually reduces resazurin.

Uracil is not utilized (Hilton et al., 1975).

Products of fermentation in PYG broth are large amounts of acetic and butyric acids, and sometimes small amounts of formic, propionic, isobutyric, isovaleric, valeric, caproic, lactic and succinic acids. Abundant H_2 is produced. Addition of valine to the medium results in increased production of isobutyric acid, whereas addition of either leucine or isoleucine results in production of increaed amounts of isovaleric acid (Elsden and Hilton, 1978). Lactate is converted to acetate, butyrate and caproate and occasionally to valerate; pyruvate is converted to acetate and butyrate and occasionally caproate; and threonine is converted to propionate. Serine and arginine are utilized

and 2-aminobutyric acid and ornithine are produced in an amino acid-trypticase-yeast extract broth culture (Elsden and Hilton, 1979). Produces skatol (Fellers and Clough, 1925).

Elsden et al. (1980) have described the fatty acids present in this species. The major fatty acid is normal, saturated C_{16}; in addition, normal saturated C_{12}, C_{14}, C_{15}, C_{18} and normal unsaturated C_{16} acids represent appreciable proportions of the total fatty acids detected. Small amounts of branched chain fatty acids are also present.

Two strains tested are susceptible to chloramphenicol, erythromycin, clindamycin; two of three strains are susceptible to penicillin G. One strain tested is susceptible to the following antimicrobials in the concentrations indicated: penicillin V, 2 U/ml; cephalothin, 64 μg/ml; cephalexin, 32 μg/ml; cephradine, 16 μg/ml; cefoxitin, 128 μg/ml; tetracycline, 64 μg/ml; and doxycycline, 16 μg/ml.

Culture supernatants are nontoxic to mice. Reported to be nonpathogenic (Prévot et al., 1967).

Other characteristics of the species are given in Table 13.14.

Isolated from soil (Smith, 1975a), contaminated food (Smith and Hobbs, 1974), and feces of infants undersized at birth (Gothefors and Blenkharn, 1978).

The mol% G + C of the DNA is 27 (T_m) (Johnson and Francis, 1975).

Type strain: ATCC 25775 (NCIB 8855; DSM 757; CCUG 9283).

65. **Clostridium septicum** (Macé 1889) Ford 1927, 726.[AL] (*Bacillus septicus* Macé 1889, 445.)

sep′ti.cum. Gr. adj. *septicum* putrefactive.

This description, unless indicated otherwise, is based on that of Holdeman et al. (1977), Smith (1975a) and Willis (1969), and on study of the type and 46 other strains.

Cells in PYG broth culture are Gram-positive in young cultures but may be Gram-negative in older cultures; staining is often uneven, resulting in intensely Gram-positive bars or spots interspersed with decolorized areas. Cells in broth are straight or curved rods that occur singly or in pairs, are usually motile and peritrichous, and are 0.6–1.9 × 1.9–35.0 μm. Forms long filaments on the peritoneal surface of the liver of infected animals in contrast to other pathogenic species of *Clostridium*. Cells may be extremely pleomorphic under certain conditions (Willis, 1969; Sterne and Batty, 1975).

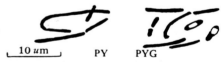

| 10 μm | PY | PYG |

Figure 13.69. *Clostridium septicum.*

Spores are oval, subterminal, and distend the cell. There is no exosporium and the spore lacks appendages (Smith and Hobbs, 1974).

Cell wall contains L-lysine in place of DAP (Weiss et al., 1981). Glutamic acid and alanine are present (Schleifer and Kandler, 1972); cell wall sugars are glucose, galactose, rhamnose and mannose (Cummins and Johnson, 1971). The cell wall is susceptible to dissolution by lysozyme (Johnson and Francis, 1975).

Colonies on the surface of blood agar are 1–5 mm in diameter, circular, with markedly irregular to rhizoid margins, slightly raised, translucent, gray, glossy and β-hemolytic. Surface growth of most strains on 1.5% agar is an invisible film over the entire agar surface; this swarming often may be prevented by shortening the incubation period or using 4 to 6% agar in the medium. Subsurface colonies in 1% agar are spherical or lenticular and transparent; in 2% agar, colonies are brownish yellow and heart shaped.

Cultures in PYG broth are turbid with a smooth sediment and have a pH of 4.7–5.3 after incubation for 5 days.

The temperature range for optimum growth is 37–40°C; most strains grow well at 44°C but not at all at 46°C. CO_2 is not required for growth

(Watt, 1973), but good growth occurs in atmospheres containing up to 100% CO$_2$. Growth is stimulated by the presence of a fermentable carbohydrate, serum, or peptic digest of blood. In chopped-meat medium, the meat particles commonly turn pink after 48 h incubation. Factors essential for growth have been summarized by Willis (1969) and Smith (1975a) and include biotin, nicotinic acid, pyridoxine, thiamin, cysteine, tryptophan and iron and, for some strains, pantothenate. Adenine, arginine, aspartic acid, histidine, isoleucine, phenylalanine, serine, threonine, tyrosine, and valine are also required (Hasen and Hall, 1976). Exposure of log phase cultures of this species to hyperbaric oxygen (100% O$_2$ at 3 atm) is relatively nonlethal compared with other species; approximately 50% of cells of *C. septicum* survive this treatment (Hill and Osterhout, 1972).

Abundant gas is produced in PYG deep agar cultures.

Adonitol, dulcitol, erythritol and glycerol are not fermented; sorbose occasionally is fermented weakly. Neutral red and resazurin are reduced and ammonia is produced by some strains; hippurate is hydrolyzed by 3 of 17 strains tested.

Phage-like particles produced by a strain of *C. septicum* inhibit the growth of a strain of *C. sporogenes*. This strain of *C. septicum* is in turn inhibited by strains of *C. bifermentans*, *C. butyricum*, *C. cadaveris*, *C. perfringens*, *C. tertium* and *C. tetani* (Nieves et al., 1981). Krämer and Schallehn (1974) found that all of 16 strains of *C. septicum* tested were susceptible to bacteriocins produced by 9 of 21 strains of *Streptococcus faecium* or *Streptococcus faecalis* studied. Schallehn and Krämer (1976) also studied bacteriocin production by *C. septicum*; the bacteriocin from one producer strain inhibited RNA and protein synthesis, and rapidly killed all of 16 strains of *C. septicum* and all of 10 strains of *C. chauvoei* tested, but none of a variety of other species of bacteria.

Norman and Grubb (1955) found that one strain of *C. septicum* tested could hydrolyze the conjugated bile acids taurocholate and glycocholate. All of 26 strains tested produce deoxyribonuclease (beta toxin) and hyaluronidase (gamma toxin) in both acidic and alkaline conditions (Princewill and Oakley, 1972a and b; 1976). An oxygen-stable hemolysin with necrotizing properties (alpha toxin), and an oxygen-labile hemolysin (delta toxin) are also produced. A neuraminidase and a hemagglutinin are produced (Gadalla and Collee, 1967, 1968; Müller, 1976; Fraser, 1978). Chitinase is produced (Clarke and Tracey, 1956). Uracil is not utilized (Hilton et al., 1975). Although lipase is not detected on egg-yolk agar plates, traces of lipidolytic activity, detected by a gas-liquid chromatographic method, have been reported in a strain of this species (Guillon and Chevrier, 1977).

Products in PYG broth include large amounts of acetic and butyric acids, and usually large amounts of formic acid. Abundant H$_2$ is produced. Brooks et al. (1976), using electron capture gas-liquid chromatography, have detected butanol, isoamyl alcohol, and propionic, isobutyric, isovaleric, and isocaproic as well as acetic and butyric acids in chopped meat-glucose cultures of *C. septicum*. Lactate usually is not utilized, but increased butyrate may be formed in PY-lactate broth. Pyruvate is converted to acetate and butyrate, and occasionally to formate. Threonine is not converted to propionate. Oleic acid is converted to hydroxystearic acid (Thomas, 1972).

Ethylamine was detected in the head-space gas above a chopped meat-glucose culture of one isolate (Larsson et al., 1978).

All strains tested are susceptible to chloramphenicol, clindamycin, erythromycin, penicillin G, and tetracycline.

This species is capable of causing rapidly fatal infections in humans and other animals by production of its lethal hemolytic and necrotizing alpha toxin. A lethal disseminated infection can be produced in mice and guinea pigs by intramuscular injection of fewer than 10 spores plus CaCl$_2$; postmortem examination reveals bloodstained edema fluid, a deep red color of the muscle and soft tissue, gas formation in the inoculated limb, and dissemination of the organism to all parts of the body. See Willis (1969) and Smith (1975a) for reviews of the possible mechanisms of action of the alpha toxin.

In cattle, wound infection produces "malignant edema," a disease similar to the experimental disease in laboratory animals, as described

above; and in sheep, braxy, a fatal bacteremic infection following penetration of the abomasal wall (Smith, 1975). Similar diseases have been noted in other species of domestic and wild animals. *C. septicum*, *C. perfringens* and *C. novyi* (MacLennan, 1962) are the three most common causes of clostridial myonecrosis (gas gangrene). Such infection may occur either by direct contamination of an infected traumatic wound (MacLennan 1962) or by metastatic infection following bacteremia from the gastrointestinal tract (Alpern and Dowell, 1969); in the latter case, a breech in the bowel mucosa (such as due to a malignancy) serves as the portal of entry. The mechanism of disease production apparently is similar to that of the animal model described above.

Other characteristics of the species are given in Table 13.12.

This species is present in soil (Miwa et al., 1976; Smith, 1975b); probably in the gastrointestinal tract contents of many domesticated animals; rattlesnake venom (Goldstein et al., 1979); feces of human infants and adults (Gothefors and Blankharn, 1978; Kahn, 1924; Finegold et al., 1983); animal infections including wound infections, myositis and enterotoxemia in ruminants and bacteremia (Smith, 1975; Nevig et al., 1981; Hirsh et al., 1979; Berkhoff and Redenbarger, 1977; Berg et al., 1979); and from human infections including bacteremia, suppurative infections, necrotizing enterocolitis, and myonecrosis or gas gangrene (Alpern and Dowell, 1969; Gorbach and Thadepalli, 1975; Lehman et al., 1977; Koransky et al., 1979; Ayyagari et al., 1981; Bignold and Harvey, 1979; Hopkins and Kushner, 1983).

The mol% G + C of the DNA is 24 (T_m) (Cummins and Johnson, 1971).

Type strain: ATCC 12464 (NCIB 547; NCTC 547; CCUG 4855).

66. **Clostridium sordellii** (Hall and Scott 1927) Prévot 1938, 83.^{AL} (*Bacillus sordellii* Hall and Scott 1927, 330.)

sor.del'li.i. N.L. gen. n. *sordellii* pertaining to Professor Sordelli, Argentinian bacteriologist.

This description is based on those of Smith (1975a), Holdeman et al. (1977), and on a study of the type and 80 other strains.

Cells in PYG broth cultures are Gram-positive straight rods, usually motile and peritrichous, 0.5–1.7 × 1.6–20.6 μm, and occur singly or in pairs.

$\underline{\quad 10\ \mu m \quad}$ CMC   PYG

Figure 13.70. *Clostridium sordellii.*

Spores are oval, and central or subterminal, and often occur as free spores. They swell the cell slightly. Spores have a thick exosporium; most have tubular appendages (Rode et al., 1971). Sporulation occurs readily in chopped meat broth cultures incubated for 24 h or on blood agar plates incubated 48h.

Cell walls contain *meso*-DAP, glucose and a trace of galactose (Cummins and Johnson, 1971). Glutamic acid and alanine also are present (Schleifer and Kandler, 1972). Urease-negative strains that are homologous by DNA/DNA homology determinations with authentic urease-positive strains of *C. sordellii* do not have glucose in their cell walls (Nakamura et al., 1976b).

Surface colonies on blood agar plates are 1–4 mm in diameter, circular to irregular, flat or raised, translucent or opaque, gray or chalk-white, with a dull or shiny surface, a granular or mottled internal structure, and a scalloped, lobate or entire margin. Hemolysis is variable, with most strains being slightly β-hemolytic on rabbit blood agar.

The optimum temperature for growth is 30–37°C. Strains grow moderately well at 25°C and 45°C. Growth is inhibited by 6.5% NaCl, by 20% bile, or by a pH of 8.5. Strains of *C. sordellii* are inhibited by phage-like particles found in strains of *C. bifermentans*; although similar particles have been found in these strains of *C. sordellii*, the inhibition is not reciprocal (Nieves et al., 1981). Strains of *C. sordellii* share a cross-reactive carbohydrate cell surface antigen with both *C. bifermentans* and *C. difficile* (Poxton and Byrne, 1981).

Abundant gas is detected in PYG deep agar cultures. Ammonia is produced; H_2S is produced in SIM by most strains; hippurate is hydrolyzed by most strains. Benzoic acid released from hydrolysis of hippurate can be detected by gas chromatography within 2–4 days (Kodaka et al., 1982).

Gelatin is digested by all strains; milk, casein and meat are digested by nearly all strains but more slowly.

Some strains produce DNase (Döll, 1973; Nord et al., 1975); five of eight strains tested produce hyaluronatelyase (Nord et al., 1975). Supernatant culture fluids hydrolyze bile acid conjugates (Masuda, 1981). Cholic acid is dehydroxylated at the 7α position to deoxycholic acid by the type strain (Archer et al., 1981; 1982a and b). Oleic acid is converted to hydroxystearic acid in vitro, a reaction that may be important in the pathogenesis of diarrhea in humans (Thomas, 1972). Supernatant culture fluids in protease peptone water broth contain substantial amounts of neuraminidase (Fraser, 1978). A wide variety of straight chain, *iso*-branched chain, and *anteiso*-branched chain fatty acids, both saturated and unsaturated, is present in the lipids of this species (Elsden et al., 1980).

Acetic acid is the major product in PYG broth cultures. Large amounts of formic and moderate amounts of isocaproic acid usually are detected, and trace amounts of propionic, isobutyric, butyric and isovaleric acids may be present. Ethanol and propanol also are detected. In the absence of glucose, the proportion of acids other than acetic is increased substantially. Abundant H_2 is produced. Pyruvate is converted to acetate and formate. Threonine is converted to propionate. Lactate is not utilized.

Phenylalanine and tryptophan are converted to phenylacetic, phenylpropionic, phenyllactic acids and indole (Elsden et al., 1976).

Leucine is converted to isovalerate, isocaproate, and CO_2 (Elsden and Hilton, 1978; Britz and Wilkinson, 1982). Isoleucine is converted to isovalerate and valine to isobutyrate (Elsden and Hilton, 1978). Phenylacetic acid is produced in trypticase-yeast extract medium (Mayrand and Bourgeau, 1982).

Proline, serine, threonine, alanine, aspartic acid, glycine, glutamic acid and methionine are amino acids utilized for growth in 3% casein hydrolysate medium; δ-aminovalerate; γ-aminobutyrate, and valine are produced (Mead, 1971).

Of 37 strains tested, all are susceptible to chloramphenicol, erythromycin and penicillin G; 3 are resistant to tetracycline, 1 is resistant to clindamycin.

Culture supernatants of only 9 of 71 strains tested were toxic to mice and this toxicity may be lost on subculture. *C. sordellii* is pathogenic for man, cattle, sheep, guinea pigs and mice. Pathogenicity also may be ephemeral, and nonpathogenic, urease-positive strains are isolated frequently. Although antitoxin to *C. sordellii* will neutralize the toxins of *C. difficile*, the toxins of the two species have been shown to have different properties (Aswell et al., 1979; Rehg, 1980). For a review of the toxins formed and diseases caused, see Smith (1975a).

Other characteristics of the species are given in Table 13.12.

Isolated from soil (Smith, 1975b; Miwa, 1975a and b); from normal human feces (Finegold et al., 1983); from human clinical specimens including wounds (Sanderson et al., 1979; Willis, 1969), penile lesions (Chapel et al., 1978), blood cultures (Lynch et al., 1980), abscesses and abdominal and vaginal drainage; from intestinal tracts of both normal and diseased pheasants (Mead et al., 1973), bony tissue from dogs with osteomyelitis (Walker et al., 1983), bovine intestinal inflammatory lesions (Al-Mashat and Taylor, 1983), and bovine uterus and muscle, alpaca and sheep infections, and chicken skin.

The mol% G + C of the DNA is 26 (T_m) (Johnson and Francis, 1975).

Type strain: ATCC 9714 (NCIB 10717; DSM 2141; CCUG 9284).

Further comments. For differentiation of this species from *C. bifer-*

mentans, which it resembles closely, see "Further Comments" following the description of that species.

67. **Clostridium sphenoides** (Douglas, Fleming and Colebrook 1919) Bergey, Harrison, Breed, Hammer and Huntoon 1923, 33.[AL] (*Bacillus sphenoides* Douglas, Fleming and Colebrook *in* Bulloch et al., 1919, 43.)

sphe.noi′des. Gr. adj. *sphenoides* wedge-shaped.

This description, unless otherwise indicated is based on that of Holdeman et al. (1977) and on study of the type and 22 other strains.

Cells usually stain Gram-negative, are straight rods with tapered or rounded ends, motile and peritrichous, occur singly, in pairs, or in short chains and are 0.3–1.1 × 1.3–8.6 μm.

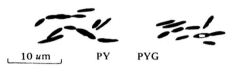

| 10 μm | PY | PYG |

Figure 13.71. *Clostridium sphenoides.*

Spores are oval and subterminal or occasionally terminal, and swell the cell. Sporulation occurs most readily on blood agar plates incubated 48 h or on chopped meat slants incubated at 30°C.

Cell wall contains *meso*-DAP (C. S. Cummins, unpublished data). The cell wall is susceptible to dissolution by lysozyme (Johnson and Francis, 1975).

Colonies on the surface of blood agar are 1–2 mm in diameter, nonhemolytic, circular with an entire or erose margin, low convex, translucent, gray, with a glossy surface, often with a mottled internal structure.

Cultures in PYG broth are turbid with a smooth, or occasionally ropy sediment and a pH of 4.9–5.4 after incubation for 5 days.

Optimum temperature for growth is 30–37°C. There is good growth at 25, little or none at 45°C. Growth is slightly stimulated by the presence of a fermentable carbohydrate and inhibited by 20% bile.

Abundant gas is produced in PYG deep agar cultures.

Adonitol, dulcitol, erythritol, glycerol and inositol are not fermented; sorbose is weakly fermented. Ammonia and H_2S are produced, acetyl methyl carbinol is produced occasionally, neutral red is reduced, and resazurin is reduced by most strains.

Citrate is used as both a carbon and an energy source (Walther et al., 1977) with production of acetic acid, ethanol, CO_2 and H_2. Citrate utilization by this species may be unique among clostridia. Taurocholic and glycocholic acids are hydrolyzed (Norman and Grubb, 1955); deoxyribonuclease is produced (Döll, 1973). The insecticide hexachlorocyclohexane is converted to tetrachlorocyclohexene (Heritage and MacRae, 1977).

Phage-like particles are produced by one strain tested (Nieves et al., 1981).

Products of metabolism in PYG broth are large amounts of acetic and formic acids, and occasionally, small amounts of lactic acid, succinic acid and ethanol. Lactate is not utilized, and threonine usually is not converted to propionate. Pyruvate is converted to acetate, occasionally with formate.

Of 20 strains tested, all are susceptible to erythromycin and tetracycline, 15 are susceptible to chloramphenicol, 12 to clindamycin, and 10 to penicillin G.

Culture supernatants are nontoxic to mice.

Although *C. sphenoides* occasionally is recovered from polymicrobial infections in humans, pathogenicity has not been reported.

Other characteristics of the species are given in Table 13.14.

Isolated from soil (Smith, 1975a), marine sediment (Finne and Matches, 1974), dog feces (Allo, 1980), normal human appendices (Löhr and Rabfeld, 1931), feces of from 4–6.4% of adult humans (Drasar et al., 1976; Legakis et al., 1981; Finegold et al., 1983), infections in range animals (Ryff and Lee, 1946), blood (Gorbach and Thadepalli, 1975), bone and soft tissue infections (Rustigan and Cipriani, 1947; Gorbach and Thadepalli, 1975; Isenberg et al., 1975), intraperitoneal infections (Thadepalli et al., 1973; Finegold et al., 1975), war wounds (Smith and George, 1946), visceral gas gangrene, and renal abscess (cited in Finegold, 1977).

The mol% G + C of the DNA is 41–42 (T_m) (Johnson and Francis, 1975).

Type strain: ATCC 19403 (NCIB 10627; NCTC 507; DSM 632).

Further comments. See comments under *C. indolis.*

68. Clostridium spiroforme Kaneuchi, Miyazato, Shinjo and Mitsuoka 1979, 10.[AL]

spi.ro.for'me. Gr. n. *spira* a coil; L. n. *forma* shape; N.L. neut. adj. *spiroforme* in the shape of a coil.

This description, unless otherwise indicated, is based on our study of the type and two other strains and on the description by Kaneuchi et al. (1979).

Cells on Eggerth-Gagnon agar after 2 days of incubation are Gram-positive, nonmotile, 0.3–0.5 × 2.0–10.0 μm and exhibit various degrees of coiling; long chains of organisms forming tight coils often are seen. After heating at 80°C for 10 min, cells may be nearly straight and subsequent subcultures do not have coiled cells (Kaneuchi et al., 1979).

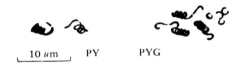

Figure 13.72. *Clostridium spiroforme.*

Most strains produce round, terminal or occasionally subterminal spores approximately 0.7 μm in diameter when incubated for 2 weeks at 30°C on medium 10 (Kaneuchi et al., 1979) or on chopped meat agar slants. Most strains survive heating to 70°C for 10 min; survival after heating to 80°C for 10 min is variable. Spore demonstration is sometimes difficult; fresh isolates from humans may not form spores or survive heating to 70°C for 10 min.

Colonies on the surface of Eggerth-Gagnon agar after 2 days incubation and on medium 10 agar after 3 days incubation are 0.7–1.5 mm in diameter, have an entire or slightly erose margin, are circular, convex to slightly pulvinate, smooth, shiny, semiopaque to opaque and are whitish to brownish gray. Colonies on blood agar plates are nonhemolytic.

PYG broth cultures are slightly turbid with a smooth to flocculent sediment and have a pH of 4.9–5.2 after incubation for 7 days.

The temperature for optimum growth is 30–37°C; most strains grow poorly, if at all, at 25°C or 45°C. There is no growth at 15°C. Growth is enhanced by the presence of a fermentable carbohydrate, or of 10% rumen fluid. Growth of most strains is inhibited by 20% bile or 6.5% NaCl.

Small to moderate amounts of gas are produced in PYG deep agar cultures.

Acetyl methyl carbinol is produced; neutral red and resazurin are reduced.

Esculin is fermented occasionally; dextrin, dulcitol, α-methylglucoside, and α-methylmannoside are not fermented. Urease is not produced.

Large amounts of acetic and formic acids, and small amounts of lactic acid are produced in PYG broth. Abundant H_2 is produced. Lactate and threonine are not converted to propionate.

The type strain is susceptible to chloramphenicol, clindamycin, erythromycin, penicillin G and tetracycline.

An organism that is identical to, or closely resembles *C. spiroforme*, has been implicated as the cause in rabbits of an apparently toxin-induced diarrhea that occurs spontaneously and in association with antimicrobial therapy. The toxic effect can be neutralized by antitoxin to *C. perfringens* type E iota toxin (Borriello and Carman, 1982); this toxic effect is not produced by the type strain of the species. A toxin produced in vitro is lethal to mice and causes dermonecrosis in guinea pigs, but toxin is not produced by all strains (Borriello and Carman, 1983).

Other characteristics of the species are given in Table 13.14.

Isolated from the feces of healthy humans. Reported to occur in the ceca of healthy chickens and rabbits, the ceca of rabbits with diarrhea (Borriello and Carman, 1983).

The mol% G + C of the DNA is 27 (T_m) (Kaneuchi et al., 1979).

Type strain: ATCC 29900 (NCTC 11211; DSM 1552).

Further comments. Clostridium spiroforme can be distinguished from most other species of *Clostridium* by its coiled morphology and its failure to produce butyric acid. It is most readily distinguished from *C. cocleatum*, which it most closely resembles, by its inability to ferment galactose. *Clostridium spiroforme* can be distinguished by biochemical testing from 2 other (unnamed) coiling clostridial species with which it is not homologous (Kaneuchi et al., 1979).

69. Clostridium sporogenes (Heller 1922) Bergey, Harrison, Breed, Hammer and Huntoon 1923, 329.[AL] (*Metchnikovillus sporogenes* Heller 1922, 29.)

spo.ro'ge.nes. Gr. n. *sporos* seed; Gr. suffix *-genes* born of; N.L. neut. adj. *sporogenes* spore-producing.

This description, unless otherwise indicated, is based on that of Holdeman et al. (1977), and on study of the type and 219 other strains.

Cells in PYG broth cultures are Gram-positive straight rods, motile and peritrichous, that occur singly and are 0.3–1.4 × 1.3–16.0 μm.

Figure 13.73. *Clostridium sporogenes.*

Spores are oval, subterminal and distend the cell; sporulation occurs readily on most media. Following sporulation, the vegetative material may disintegrate rapidly to leave only free spores. Outgrowth of spores in meat slurries is inhibited by nitrites and nisin (Rayman et al., 1981).

Cell wall contains *meso*-DAP; galactose is the only cell wall sugar (Cummins and Johnson, 1971). Glutamic acid and alanine are present (Schleifer and Kandler, 1972). The cell wall is susceptible to dissolution by lysozyme (Johnson and Francis, 1975).

Surface colonies on blood agar plates are 2–6 mm in diameter, irregularly circular, possess a coarse rhizoid edge, have a raised yellowish gray center and a flattened periphery composed of entangled filaments ("Medusa head" colony), are opaque, possess a matt surface, are usually β-hemolytic and are firmly adherent to the agar. Colonies on more moist agar are larger, flatter, gray and less adherent.

Colonies in agar are spherical, with an opaque center, and a woolly semitranslucent periphery; lenticular colonies with fine marginal outgrowths may be produced.

Cultures in PYG broth are turbid, with a smooth or occasionally ropy to flocculent sediment, and have a pH of 5.7–6.4 after incubation for 1 week. In chopped meat medium with iron filings, there is often marked blackening of the meat particles.

The temperature range for optimum growth is 30–40°C; will grow at 25°C and 45°C. Good growth occurs in an atmosphere containing up to 100% CO_2. Growth in 6.5% NaCl, 20% bile or at a pH of 8.5 is variable; there is no inhibition of some strains, others are completely inhibited.

Abundant gas is produced in PYG deep agar cultures.

Adonitol, dulcitol, erythritol, glycerol and sorbose are not fermented. Ammonia and H_2S are produced, neutral red is reduced and resazurin is reduced by some strains; a few strains produce urease and hydrolyze hippurate.

This species produces deoxyribonuclease (Döll, 1973; Nord et al., 1975), thiaminase (Princewill, 1980), chitinase (Timmis et al., 1974), kininase (Möse et al., 1972b), L-methioninase (Kreis and Hession, 1973), hyaluronatelyase (Nord et al., 1975), and superoxide dismutase (Hewitt and Morris, 1975). Uracil is reduced to dihydrouracil (Hilton et al., 1975). Cells contain ferredoxin (Valentine, 1964).

Princewill (1979) has described the existence of three flagellar antigens, four somatic antigens and four spore antigens.

Some strains of *C. sporogenes* are capable of inhibiting growth and toxin production by *C. botulinum* type A (Smith, 1975a).

Kiritani et al. (1973) found that only a few strains were lysed by mitomycin C, whereas all strains of *C. botulinum* tested were lysed. Nonlysogenic bacteriophages active against *C. sporogenes* have been isolated from soil, sewage and chicken feces (Betz and Anderson, 1964). Bacteriocin-like substances that are active against other strains of *C. sporogenes* are produced by *C. sporogenes* (Betz and Anderson, 1964).

The mutagen, 1-nitropyrene, is reduced to 1-aminopyrene with concomitant decrease in mutagenicity by a strain of this species (Kinouchi et al., 1982).

Products of metabolism in PYG broth include large amounts of acetic and butyric acids, and small amounts of isobutyric and isovaleric acids; propionic, valeric, isocaproic, lactic and succinic acids also may be produced. Ethanol and abundant H_2 are produced. Lactate is converted to butyrate; pyruvate is converted to acetate and butyrate; threonine usually is converted to propionate.

C. sporogenes is capable of carrying out the "Stickland reaction" in which energy is obtained by the coupled oxidation and reduction of various amino acid pairs (Stickland, 1934; Nisman, 1954; Costilow and Cooper, 1978). Betaine may also serve as an oxidant in this type of reaction (Naumann et al., 1983). When glycine is used as the oxidant, *C. sporogenes* has an absolute requirement for selenium (Costilow, 1977). Leucine and valine can be interconverted (Monticello and Costilow, 1982). In a thioglycollate-trypticase-yeast extract medium supplemented with casein hydrolysate, proline is converted to 5-aminovaleric acid, 2-aminobutyric acid is produced, phenylalanine is converted to phenylpropionic acid, tyrosine is converted to *p*-hydroxyphenylpropionic acid and tryptophan is converted to indole propionic acid (Elsden et al., 1976; Elsden and Hilton, 1979; Jellet et al., 1980). Arginine and serine also can be utilized (Mead, 1971). Addition of valine to a basal medium results in increased isobutyric acid production; addition of leucine results in increased isovaleric and isocaproic acid production and addition of isoleucine results in increased isovaleric acid production (Elsden and Hilton, 1978; Britz and Wilkinson, 1982). Cmelik (1980) has found that *C. sporogenes* can metabolize monoacetone glucose.

The major cellular fatty acids produced by this species are normal saturated C_{14}, C_{15}, C_{16}, C_{17} and C_{18} acids, and normal unsaturated C_{16} acid with a double bond at the 7 or 9 position (Elsden et al., 1980). Small amounts of other normal and branched chain fatty acids also are present.

Of 105 strains tested by the broth disk method of Wilkins and Thiel (1973), 1 is resistant to penicillin G, 1 to tetracycline, 4 to chloramphenicol, 5 to erythromycin, and 57 to clindamycin; all others are susceptible. Determinations of MICs of various antibiotics against 18 strains of *C. sporogenes* indicate that all are susceptible to penicillin G (2 U/ml), metronidazole (0.25 μg/ml), tinidazole (1 μg/ml), chloramphenicol (4 μg/ml), tetracycline (0.5 μg/ml), doxycycline (0.25 μg/ml), but resistant to streptomycin, neomycin, kanamycin, tobramycin and amikacin; susceptibility to cephalothin, erythromycin, clindamycin, lincomycin and gentamicin is variable between strains (Dornbusch et al., 1975).

Culture supernatants are not toxic to mice. Although *C. sporogenes* is isolated from infections, these infections are usually polymicrobial and the role, if any, of this species as a pathogen in such infections has not been established. Untoward effects are not seen when germ-free animals undergo gastrointestinal monoassociation (Yale, 1975; Yale and Balish, 1976). Self-limited, spontaneously healing abscesses are produced following intramuscular injection of a relatively large inoculum into guinea pigs (Smith, 1975a). A generalized lethal disease, possibly egg-borne, in newly hatched chicks has been attributed to *C. sporogenes* (Smith, 1975a; Peterson, 1967). Cerebrocortical necrosis in ruminant animals is thought to be due to thiamine deficiency; an association between this disease and colonization by thiaminase-producing strains of *C. sporogenes* has been postulated, but not proven (Shreeve and Edwin, 1974; Cushnie et al., 1979). The highly proteolytic nature of *C. sporogenes* is thought possibly to act as an adjuvant and promote invasiveness of other bacteria in various mixed infections of animals and humans (Katitch et al., 1964; Ryff and Lee, 1946; Smith and George, 1946; Smith, 1975a).

Other characteristics of the species are given in Table 13.12.

This species has been isolated from soil throughout the world (Smith, 1975b; Hamman and Ottow, 1976; Schrader and Schau, 1978; and Miwa et al., 1976), marine and fresh water lake sediment (Matches and Liston, 1974; Molongoski and Klug, 1976), preserved meat and dairy products (Skjelkvale and Tjaberg, 1974; Naguib and Shauman, 1972, 1973; Matteuzzi et al., 1977); snake venom (Holdeman et al., 1977); feces of sheep and dogs (Shreeve and Edwin, 1974; Balish et al., 1977); human infant and adult feces (Borriello, 1980; Rotimi and Duerden, 1982; Drasar et al., 1976; Marrie et al., 1978; Legakis et al., 1981); from infections in domestic animals (Katitch et al., 1964; Peterson, 1967; Hirsh et al., 1979; Walker et al., 1983; Berkhoff and Redenbarger, 1977); from infections in humans including bacteremia, infective endocarditis, central nervous sytem and pleuropulmonary infections, penile lesions, abscesses, war wounds and other pyogenic infections (Alpern and Dowell, 1971; Finegold, 1977; Malmborg et al., 1970; Chapel et al., 1978; Werner et al., 1973; Henry, 1917; Gorbach and Thadepalli, 1975; Lewis et al., 1980).

The mol% G + C of the DNA is 26 (T_m)(Johnson and Francis, 1975). *Type strain:* ATCC 3584 (NCIB 10696; DSM 795; IFO 13950).

Further comments. *C. sporogenes* can be differentiated from proteolytic strains of *C. botulinum* types A, B and F, which it closely resembles phenotypically, only by toxin neutralization in mice, by polyacrylamide gel electrophoretic examination of soluble cellular proteins, or by gas chromatography of trimethylsilyl derivatives of whole cell hydrolysates (see *C. botulinum*). *Clostridium sporogenes* and *C. difficile* also are morphologically similar and have similar fermentation products; they differ, however, in mannitol fermentation, proteolytic activity in milk and meat, and in lipase production.

70. **Clostridium sporosphaeroides** Soriano and Soriano 1948, 39.[AL]
spo.ro.sphae.roi′des. Gr. n. *sporos* seed; Gr. adj. *sphairoides* globular; N.L. neut. adj. *sporosphaeroides* having spherical spores.

This description, unless indicated otherwise, is based on that of Holdeman et al. (1977), and on study of the type strain.

Cells in PYG broth are Gram-positive but easily decolorized, nonmotile, straight rods, 0.5–0.6 × 1.8–5.5 μm.

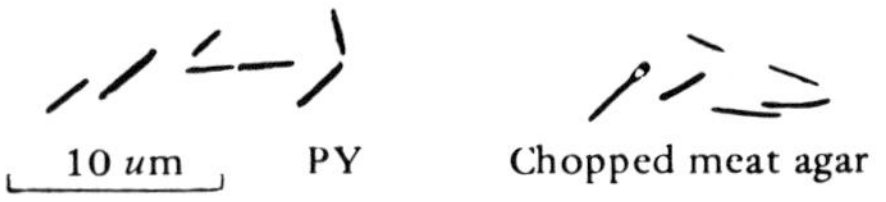

Figure 13.74. *Clostridium sporosphaeroides.*

Spores are oval or round, and terminal. Sporulation occurs most readily on chopped meat slants incubated at 30°C.

Cell wall contains *meso*-DAP (Weiss et al., 1981). The cell wall is susceptible to dissolution by lysozyme (Johnson and Francis, 1975).

Colonies on the surface of blood agar are 1–2 mm in diameter, circular, with a slightly lobate or scalloped margin, slightly raised, with a fried egg appearance, gray-white, have a glossy surface, a mosaic internal structure and are nonhemolytic.

Cultures in PYG broth are turbid with a smooth to ropy sediment and have a pH of 6.1 after incubation for 5 days.

The temperature range for optimum growth is 37–45°C. There is slight growth at 30°C, no growth at 25°C.

Growth is inhibited by 20% bile and by 6.5% NaCl.

A moderate amount of gas is produced in PYG deep agar cultures.

Adonitol, dulcitol, erythritol, glycerol and sorbose are not fermented. Ammonia and H_2S are produced, hippurate is hydrolyzed; neutral red and resazurin are reduced.

An NADP-dependent 7 α-hydroxysteroid dehydrogenase capable of dehydrogenation of bile salts is produced (Mahoney et al., 1977). Deoxyribonuclease activity is not present (Johnson and Francis, 1975).

Products of metabolism in PYG broth include large amounts of acetate and butyrate, and small amounts of propionate. Abundant H_2 is produced. Lactate is converted to propionate; pyruvate is converted to acetate; threonine is not utilized.

The type strain is susceptible to chloramphenicol, clindamycin, erythromycin, penicillin G and tetracycline.

C. sporosphaeroides does not metabolize phenylalanine, tyrosine or tryptophan (Elsden et al., 1976). Glutamate is metabolized to ammonia, CO_2, acetate and butyrate via a hydroxyglutarate pathway (Buckel, 1980).

Culture supernatants of the type strain are nontoxic to mice.

Other characteristics of the species are given in Table 13.13.

Isolated from canned food (Smith and Hobbs, 1974) and human feces (Finegold et al., 1983; Drasar et al., 1976; and Legakis et al., 1981).

The mol% G + C of the DNA is 27 (T_m) (Johnson and Francis, 1975).

Type strain: ATCC 25781 (NCIB 10672; DSM 1294).

71. Clostridium sticklandii Stadtman and McClung 1957, 218.[AL]

stick.lan'di.i. N.L. gen. n. sticklandii, pertaining to L. H. Stickland, British biochemist.

This description, unless stated otherwise, is based on that of Stadtman and McClung (1957) and on study of the type strain.

Cells in PYG broth are straight, slender Gram-positive rods that are motile and peritrichous, 0.3–0.5 × 1.3–3.8 μm, and occur singly, in pairs, and sometimes in short chains.

10 um | PY | PYG

Figure 13.75. *Clostridium sticklandii.*

Spores are oval, subterminal, distend the cell slightly, but are rarely seen, and then only in old cultures.

Cell wall composition has not been reported.

Surface colonies on blood agar are 1–2 mm in diameter, circular with an entire or slightly undulated margin, convex, grayish white, translucent or opaque, possess a glossy surface and a mottled internal structure, and are nonhemolytic.

Colonies in agar are 1–2 mm in size and lenticular, becoming lobate.

Cultures in PYG broth are turbid with a smooth sediment and have a pH of 6.0 after incubation for 6 days.

The temperature range for optimum growth is 30–37°C; moderate growth occurs at 25°C and 45°C.

Moderate gas is produced in PYG deep agar cultures.

Carbohydrates are not appreciably fermented. Ammonia is produced.

Energy for growth is obtained by coupled oxidation-reduction reactions between certain amino acid pairs ("Stickland reaction"). Uracil is not utilized (Hilton et al., 1975).

Products in PY broth are large amounts of acetate, butyrate, and isovalerate, small amounts of propionate and isobutyrate and a small amount of H_2. Lesser amounts of these products are detected when

glucose is added to the basal medium. Pyruvate is converted to acetate; lactate is not utilized; threonine is converted to acetate (Elsden and Hilton, 1978) and to propionate (Elsden and Hilton, 1979); threonine is reduced to α-aminobutyrate by dried cells and cell-free extracts of C. sticklandii (Stadtman, 1954). Ornithine is converted to δ-aminovalerate (Stadtman, 1954). In amino acid-supplemented broth cultures, 2-aminobutyric acid is produced; arginine, glycine, lysine, methionine and serine are utilized; proline is converted to 5-aminovaleric acid, phenylalanine to phenylacetic acid, tyrosine to p-hydroxyphenylacetic acid, valine to isobutyric acid, leucine and isoleucine to isovaleric acid, and tryptophan to indole acetic acid (Mead, 1971; Elsden et al., 1976; Elsden and Hilton, 1978, 1979). The purines adenine, hypoxanthine, xanthine, and uric acid can serve as hydrogen acceptors (Schäfer and Schwartz, 1976). Pathways for the conversion of lysine to acetic and butyric acids and ammonia have been described by Stadtman (1973). Glutamate is fermented slowly by way of the methylaspartate pathway (Buckel and Barker, 1974). For a review of amino acid metabolism of this species, see Barker (1981).

The major fatty acids of C. sticklandii are normal, saturated C_{14} and C_{16} acids; small amounts of other normal and branched chain fatty acids, both saturated and unsaturated, also are present (Elsden et al., 1980).

The type strain is susceptible to chloramphenicol, clindamycin, erythromycin, penicillin G and tetracycline.

Culture supernatants of the type strain are not toxic to mice; strains are not pathogenic for guinea pigs (Stadtman and McClung, 1957).

Other characteristics of the species are given in Table 13.13.

Isolated from soil (Smith, 1975b), San Francisco Bay black mud (Stadtman and McClung, 1957), and feces of one subject in Uganda (Drasar et al., 1976).

The mol% G + C of the DNA is 31 (T_m) (Johnson and Francis, 1975).

Type strain: ATCC 12662 (NCIB 10654; DSM 519; CCUG 9281).

72. Clostridium subterminale (Hall and Whitehead 1927) Spray 1948, 786.[AL] (*Bacillus subterminalis* Hall and Whitehead 1927, 67.)

sub.ter.mi.na'le. L. pref. sub under; L. adj. terminalis terminal; N.L. neut. adj. subterminale near the end, subterminal.

This description, unless indicated otherwise, is based on that of Holdeman et al. (1977) and on study of the type and 92 other strains.

Cells in PYG broth are Gram-positive, straight rods, usually motile and peritrichous, and are 0.5–1.9 × 1.6–11.0 μm.

10 um | PY | PYG

Figure 13.76. *Clostridium subterminale.*

Spores are without appendages and are oval and usually subterminal (occasionally central), and distend the cell. Sporulation occurs readily on blood agar, egg-yolk agar, and chopped meat agar.

Cell walls contain mesoDAP (Weiss et al., 1981). The cell wall is digested by lysozyme (Johnson and Francis, 1975).

Surface colonies on blood agar are 1–4 mm in diameter, raised or low convex, translucent, gray, irregularly circular with a lobate or scalloped margin and a matt surface, and usually are β-hemolytic. They often have a crystalline, mottled or mosaic internal structure.

Cultures in PYG broth are turbid with a smooth or ropy sediment and have a pH of 6.0–6.4 after incubation for 5 days.

The temperature for optimum growth is 37°C; most strains also grow at 25°C and 45°C. Growth is inhibited by 20% bile or 6.5% NaCl.

Moderate amounts of gas are produced in PYG deep agar cultures. Ammonia is produced, H_2S is produced by most strains, neutral red is reduced and resazurin occasionally is reduced.

An occasional strain may produce a slight amount of lecithinase (Smith, 1975a). Cells possess deoxyribonuclease (Johnson and Francis, 1975); neuraminidase is not produced (Fraser, 1978).

Metabolic products in PYG broth include large amounts of acetate, butyrate, and isovalerate and small amounts of isobutyrate; small amounts of formate, propionate, isocaproate, lactate and succinate may be produced. Ethanol and traces of other alcohols are usually detected. Phenylacetate is produced by the type strain. Abundant H_2 is produced. Threonine is converted to propionate by 6 of 22 strains tested; pyruvate is converted to acetate and usually to butyrate; lactate is not utilized. Valine is converted to isobutyrate; leucine and isoleucine to isovalerate (Elsden and Hilton, 1978). In amino acid-containing broth cultures, arginine, glycine, lysine and serine are utilized, and phenylalanine is converted to phenylacetic acid, tyrosine to p-hydroxyphenylacetic acid and tryptophan to indole acetic acid (Mead, 1971; Elsden et al., 1976; Elsden and Hilton, 1979).

The major fatty acids of *C. subterminale* are normal, saturated C_{12}, C_{14}, and C_{16} acids; lesser amounts of saturated *iso*-branched C_{15} and normal unsaturated C_{16} acids with a double bond at the 7 or 9 position; small amounts of other normal and branched chain fatty acids also are present (Elsden et al., 1980).

Of 31 strains tested, all are susceptible to erythromycin, penicillin G and tetracycline; 25 are susceptible to chloramphenicol, and 22 are susceptible to clindamycin. Werner et al. (1977) found that one strain tested was inhibited by 0.19 μg of erythromycin /ml.

Culture supernatants are nontoxic when injected into mice.

Other characteristics of the species are given in Table 13.15.

C. subterminale has been isolated occasionally from marine sediment (Matches and Liston, 1974), soil (Smith, 1975b), bovine feces (Princewill and Agba, 1982), from the small bowel contents of adult humans (Borriello, 1980), from feces of healthy humans (Finegold et al., 1983), of humans with antimicrobial-associated diarrhea (Bartlett et al., 1978), and from infection in animals (Berkhoff and Redenbarger, 1977). It has also been isolated rarely (and usually as part of a polymicrobial flora) from blood, biliary tract infections (England and Rosenblatt, 1977), empyema fluid and soft tissue or bone infections of humans (Finegold, 1977; Gorbach and Thadepalli, 1975; Werner et al., 1973).

The mol% G + C of the DNA is 28 (T_m) (Johnson and Francis, 1975).

Type strain: ATCC 25774 (NCIB 9384; DSM 758).

Further comments. *C. subterminale* is most easily distinguished from *C. hastiforme,* which it resembles most closely phenotypically, by the production of large amounts of H_2 and the subterminal location of spores. *C. subterminale* can be differentiated from *C. botulinum* G by its lack of toxicity for mice.

73. **Clostridium symbiosum** (Stevens 1956) Kaneuchi, Watanabe, Terada, Benno and Mitsuoka 1976, 202.[AL] (*Bacteroides symbiosus* Stevens 1956, 100.)

sym.bi.o′sum. Gr. n. *symbios* a companion; N.L. neut. adj. *symbiosum* living together with, symbiotic (refers to its use as a symbiote for cultivation of *Entamoeba histolytica*).

This description, unless stated otherwise, is based on study of the type and 21 other strains and on the description by Kaneuchi et al. (1976).

Cells on Eggerth-Gagnon agar are Gram-negative, are usually motile and peritrichous, straight rods, often with pointed ends, 0.5–0.6 × 1.5–2.0 μm, and occur singly, in pairs or in chains.

Figure 13.77. *Clostridium symbiosum.*

Spores are round or oval, and subterminal; spores survive heating to 70° C for 10 min but may not survive 80°C for 10 min.

Cell walls contain *meso*-DAP (Weiss et al., 1981).

Surface colonies on Eggerth-Gagnon agar after 3 days of incubation are minute to 1.0 mm in diameter, circular, entire, low convex, smooth (sometimes with a slightly irregular surface and margin), translucent, grayish to whitish, and have a whitish gray or reddish mottled center when viewed through a dissecting microscope. On blood agar, strains may be slightly β-hemolytic.

Cultures in PY-Fildes solution-glucose broth are turbid with a smooth sediment, abundant gas, and have a pH of 5.3–6.0 after 10 days incubation.

This species grows well at 37°C. Growth is nearly as good as 30°C but poor at 25°C or 45°C. Growth is stimulated by a fermentable carbohydrate and inhibited by 20% bile or 6.5% NaCl.

Dextrin, glycerol, inositol and sorbose are not fermented (Kaneuchi et al., 1976). Ammonia and H_2S are produced, acetyl methyl carbinol usually is produced, neutral red is reduced and resazurin usually is reduced.

Phage-like particles have been demonstrated from a strain of *C. symbiosum* (Foglesong and Markovetz, 1974).

Large amounts of acetic acid, butyric acid, and lactic acid are produced in PYG broth; ethanol and formic acid may be detected as well. Abundant H_2 is produced. Pyruvate is converted to acetate and butyrate; neither lactate nor threonine is utilized.

Glutamate is fermented to ammonia, CO_2, and acetic and butyric acids via a hydroxyglutarate pathway (Buckel, 1980; Buckel and Barker, 1974).

Other characteristics of the species are given in Table 13.14.

Isolated from the feces of healthy humans (Moore and Holdeman, 1974), from liver abscesses and blood infections in humans, and occasionally from human infections derived from the bowel flora.

The mol% G + C of the DNA is 46 (T_m) (Kaneuchi et al., 1976).

Type strain: ATCC 14940 (DSM 934).

Further comments. The taxonomy of this organism is extremely complicated; probable or definite synonyms of *C. symbiosum* include "*Fusobacterium symbiosum,*" "*Fusobacterium biacutus*" (Beerens, Type II), and "*Bacteroides symbiosus*" (Kaneuchi et al., 1976).

74. **Clostridium tertium** (Henry 1917) Bergey, Harrison, Breed, Hammer and Huntoon 1923, 332.[AL] (*Bacillus tertius* Henry 1917, 347.)

ter′ti.um. L. neut. adj. *tertium* third (refers to its being the anaerobe third most frequently isolated by Henry [1917] from open war wounds).

This description, unless otherwise stated, is based on that of Holdeman et al., 1977 and on study of the type and 34 other strains.

Cells in broth cultures are Gram-positive, motile and peritrichous, and are 0.5–1.4 × 1.5–10.2 μm. They occur singly or in pairs.

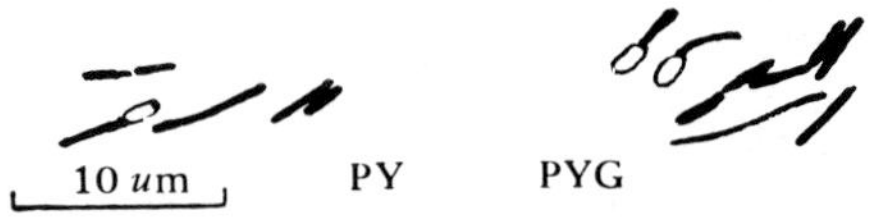

Figure 13.78. *Clostridium tertium.*

Spores are large, oval, terminal or occasionally subterminal, and markedly distend the cell. Although optimal growth is achieved by incubation under anaerobic conditions, the species is aerotolerant and will grow on the surface of freshly prepared blood agar incubated in air. Spores are formed readily in most media incubated anaerobically, but not under aerobic conditions.

The cell wall contains lysine rather than DAP; glucose or glucose and mannose are the cell wall sugars (Cummins and Johnson, 1971).

Colonies on the surface of blood agar incubated anaerobically are 2–4 mm in diameter, circular, low convex, have slightly irregular margins, are white to gray, have a matt surface and usually a mottled or granular internal structure. Hemolysis is variable and when present, colonies may be α- or β-hemolytic. Surface colonies after aerobic incubation are 1 mm, circular with entire edges, are dome shaped and have an opalescent appearance. Colonies in agar are small and lenticular.

Cultures in PYG broth grow well and are turbid with a smooth, ropy, or flocculent sediment. The optimum temperature for growth is 37°C; growth also occurs readily at 50°C (Willis, 1969). Growth is stimulated by the presence of a fermentable carbohydrate and inhibited by 20% bile or by 6.5% NaCl.

Abundant gas is produced in PYG deep agar cultures.

Adonitol, dulcitol, erythritol, glycerol, inositol and sorbose are not fermented. Ammonia is produced by some strains, and neutral red and resazurin are reduced.

Nitrate reduction is dissimilatory rather than assimilatory (Hasan and Hall, 1977). Enzymes produced include deoxyribonuclease (Döll, 1973), neuraminidase (Müller, 1976; Fraser, 1978), and an inducible chitinase (Timmis et al., 1974). Bile acid deconjugation (Norman and Grubb, 1955) and Δ^4-steroid dehydrogenase activity that may be related to the etiology of colon cancer have been reported (Hill, 1975; Goddard et al., 1975).

Products of fermentation of PYG broth include large amounts of acetic, butyric and lactic acids, and occasionally small amounts of formic and succinic acids and ethanol. Abundant H_2 is produced. Neither lactate nor threonine is utilized; pyruvate is converted to acetate and occasionally also to butyrate and formate.

Of 20 strains tested, all are susceptible to chloramphenicol and tetracycline, 19 to erythromycin, 10 to penicillin G and 7 to clindamycin.

C. tertium is relatively nonpathogenic. Culture supernatants are not toxic to mice. Debard et al. (1979) found that C. tertium increased the severity of a lethal enteritis induced by C. difficile in gnotobiotic newborn hares. Yale and Balish (1976) found C. tertium to be only weakly pathogenic when compared with several other species of Clostridium in a study of experimental intestinal strangulation in monoassociated gnotobiotic rats. Pneumatosis cystoides intestinalis can occur in these monoassociated animals (Yale, 1975).

Other characteristics of the species are shown in Table 13.14.

The organism has been isolated from soil (Smith, 1975b; Miwa et al., 1976); guano in Antarctica (Schrader and Schau, 1978), from the nares of a beagle dog (Balish et al., 1977), feces of healthy neonates and infants (Hall and Matsumura, 1924; Snyder, 1940; Borriello, 1980; Rotimi and Duerden, 1981; and Stark and Lee, 1982), appendices of healthy adults (Löhr and Rabfield, 1931), and the feces or colonic mucosa of healthy adults (Kahn, 1924; Borriello, 1980; Finegold et al., 1983). C. tertium also has been isolated from osteomyelitis in a dog (Walker et al., 1983), infection in a calf (Ryff and Lee, 1946), and from a variety of conditions in humans including brain abscess (Finegold, 1977), the gingival sulcus of two patients with periodontitis (De Campos et al., 1981), infections related to the intestinal tract (Cashore et al., 1981; Kahn, 1924; Cazzamali and Miglierina, 1933; Butler and Pitt, 1982); soft tissue infections (Gorbach and Thadepalli, 1975), war wounds (Henry, 1917; Smith and George, 1946; Rustigan and Cipriani, 1947) and blood (Thadepalli et al., 1973; Finegold, 1977).

The mol% G + C of the DNA is 24–26 (T_m) (Cummins and Johnson, 1971).

Type strain: ATCC 14573 (NCIB 10697).

75. Clostridium tetani (Flügge 1886) Bergey, Harrison, Breed, Hammer and Huntoon 1923, 330.[AL] (*Bacillus tetani* Flügge 1886, 274.)

te′ta.ni. Gr. n. *tetanos* tension; N.L. gen. n. *tetani* related to tension, tetanus.

This description, unless stated otherwise, is based on those of Smith (1975a), Holdeman et al. (1977) and on study of the type and 12 other strains.

Cells in young cultures are Gram-positive, but become Gram-negative

after approximately 24 h incubation. Most strains are motile and peritrichous. Cells are 0.5–1.7 × 2.1–18.1 µm and occur singly or in pairs.

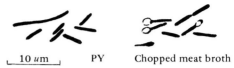

10 µm   PY   Chopped meat broth

Figure 13.79. *Clostridium tetani.*

Spores are usually round and terminal; occasionally they may be oval or subterminal, or both. Spores will germinate in liver broth at a starting O/R potential of up to +580 mV, but vegetative growth does not occur unless the culture is covered with liquid paraffin (Hachisuka et al., 1982). When spores of *C. tetani* were introduced orally or intrarectally into germ-free rats, the spores remained viable in the intestinal tract but were unable to germinate. Vegetative cells given by oral swabs did colonize the intestinal tract and stimulated the production of antitoxin but no toxin was detected (Wells and Balish, 1983).

Cell wall contains *meso*-DAP; cell wall sugars are primarily glucose and rhamnose with traces of galactose and mannose (Cummins and Johnson, 1971). Cell wall is susceptible to dissolution by lysozyme (Johnson and Francis, 1975).

Colonies on the surface of blood agar are 4–6 mm in diameter, flat, translucent, gray, with a matt surface, irregular and rhizoid margins, and usually cause a narrow zone of β-hemolysis. There is a tendency to swarm that is more marked on moist plates. Colonies in agar are transparent and very woolly (Smith, 1975a).

Cultures in PYG broth are turbid with a smooth or ropy sediment, and have a pH of 6.1–6.5 after incubation for 5 days.

The optimum temperature for growth is 37°C; there is moderate growth at 30°C, little or none at 25°C or 45°C. Growth is inhibited by 20% bile or 6.5% NaCl. Exposure of log phase cultures to hyperbaric oxygen (100% O_2 at 3 atm) resulted in death of more than 99.9% of cells (Hill and Osterhout, 1972).

Moderate to abundant gas is produced by most strains in PYG deep agar cultures.

Ammonia and H_2S are produced, neutral red is reduced, and resazurin is usually reduced.

Deoxyribonuclease is produced (Döll, 1973); neuraminidase is not produced (Fraser, 1978).

Products of metabolism in PYG broth are butanol, large amounts of acetate and butyrate, and small amounts of propionate; lactate and succinate may also be produced. Abundant H_2 is produced. Lactate and threonine are converted to propionate, pyruvate is converted to butyrate and acetate. Phenol is produced by all strains and indole by some strains in broth supplemented with L-phenylalanine, L-tryptophan and L-tyrosine (Elsden et al., 1976). All strains utilize aspartate, glutamate, histidine and serine for growth; some strains utilize methionine, threonine and tyrosine (Mead, 1971). Glutamate is converted to ammonia, CO_2, and acetate and butyrate via a methylaspartate pathway (Buckel, 1980; Buckel and Barker, 1974).

All of seven strains tested are susceptible to chloramphenicol, clindamycin, erythromycin, penicillin G and tetracycline.

Major lipid fatty acids formed in a trypticase-yeast extract-thioglycollate medium without glucose are normal saturated C_{12}, C_{14}, C_{16} and C_{18} acids and smaller amounts of unsaturated C_{18} and C_{16} acid with a double bond at the 7 or 9 position (Elsden et al., 1980). Branched chain fatty acids are not present.

Two pathogenic soluble antigens, tetanolysin and tetanospasmin, are produced. Injection of partially purified tetanolysin produces electrocardiographic changes in monkeys and mice and intravascular hemolysis in rabbits and monkeys (Hardegree et al., 1971); there is little evidence, however, that tetanolysin is normally involved in disease production. Tetanospasmin, the cause of clinical tetanus in man and other animals, is an extremely potent neurotoxin that probably spreads to the central nervous system both by passage up perineural tissue, and by lymphatic and hematogenous routes from a localized site of production; the toxin suppresses central inhibitory influences on motor neurons thereby leading to excessive muscle activity that is manifest by spasticity and tetanic contractions (Finegold, 1977). Localized forms of tetanus may also occur. A "nonspasmogenic toxin" has been studied (in partially purified form) and may interfere with function of motor nerve terminals (Feigen et al., 1963 and Parish et al., 1966). Mitsui et al. (1982) have reported detection of a high molecular weight hemolysin that is distinct from tetanolysin. Although bacteriophages have been recovered from some strains, they do not appear to be involved in toxigenicity (Prescott and Altenbern, 1967). Laird et al. (1980) have presented preliminary data that suggest toxigenicity is probably plasmid-related. For a review of the organism, its toxins and the disease caused, see Smith (1975a).

Other characteristics of the species are given in Table 13.15.

This is an ubiquitous organism. It has been isolated from soil throughout the world, human and other animal feces particularly horse feces, (Tenbroeck, 1922; Bauer and Meyer, 1926; Kerrin, 1928; Smith, 1975b), from the atmosphere in a hospital (Lowbury and Lilly, 1958) and occasionally from wounds in animals (Berkhoff and Redenbargen, 1977; Smith and George, 1946; Berg et al., 1979), and wounds in humans including infected gums and teeth, corneal ulcerations, mastoid and middle ear infections, intraperitoneal infection, omphalitis (tetanus neonatorum), postpartum uterine infections, and various soft tissue infections related to trauma, (including abrasions and lacerations) and use of contaminated needles and catgut (Finegold, 1977).

The mol% G + C of the DNA is 25–26 (T_m) (Cummins and Johnson, 1971).

Type strain: ATCC 19406 (NCTC 279; CCUG 4220).

76. Clostridium thermaceticum (corrig.) Fontaine, Peterson, McCoy and Johnson 1942, 707.[AL]

ther.ma.ce′ti.cum. Gr. adj. *thermos* hot; L. neut. adj. *aceticum* pertaining to vinegar, to acetic acid. N.L. neut. adj. *thermaceticum* producing acetic acid thermophilically.

This description, unless otherwise indicated, is based on that of Fontaine et al. (1942), and on study of the type strain.

Cells in PYG broth cultures are nonmotile although peritrichous cells have been seen in flagella stains (Fontaine et al., 1942). The cells are Gram-positive and are 0.4 × 2.8 μm. They may occur singly, in pairs, in short or long chains.

Figure 13.80. *Clostridium thermaceticum.*

Spores are round and terminal, and distend the cell slightly. They are rarely detected but cultures survive boiling for 10 min. Spores will survive heating to 100°C for 8 h and 120°C for 15 min, but not 100°C for 17 h or 120°C for 30 min.

Cell walls contain LL-DAP and glycine (Weiss et al., 1981).

Surface colonies on blood agar are 1 mm in diameter, circular, convex, translucent, shiny and smooth, with a scalloped margin and no visible internal structure. They are nonhemolytic.

Cultures in PYG broth have a crumby or flaky sediment, no turbidity, and a pH of 5.0 after incubation for 5 days. The optimum temperature for growth is 55–60°C; the minimum and maximum temperatures that will permit growth are 45°C and 65°C. Growth is stimulated by fer-

mentable carbohydrate. The organism will grow in a defined medium (Gottschalk et al., 1981). Trace elements such as tungsten, nickel and selenium are required (Andreesen et al., 1973).

α-Methylglucoside, dulcitol, erythritol, glycerol and inositol are not fermented. H_2S is not produced. Moderate gas is produced in PYG deep agar cultures.

C. thermaceticum grows chemolithotrophically in an atmosphere of either H_2 and CO_2 or carbon monoxide, and converts these substrates to acetate, or to acetate and CO_2, respectively (Diekert and Thauer, 1978; Kerby and Zeikus, 1983). There is good growth with pyruvate as a substrate but neither lactate nor threonine is utilized. Ragsdale et al. (1983) have purified a carbon monoxide dehydrogenase from *C. thermaceticum*; this enzyme is a major component of soluble cellular protein and may, therefore, play an important role in metabolism. *C. thermaceticum* grows chemoorganotrophically with glucose to form acetate as the sole fatty acid product of metabolism. Cytochrome *b* and menaquinone are present (Gottwald et al., 1975); a thermostable (to 80°C) ferredoxin with only one iron-sulfide cluster has been identified (Yang et al., 1977); a second ferredoxin isolated by Elliott and Ljungdahl (1982) contains two iron-sulfur clusters. Enzymes and metabolic pathways involved in the synthesis of acetate from CO_2 have been thoroughly investigated (Andreesen et al., 1973; Ljungdahl and Wood, 1982).

The metabolic pathway in the synthesis of leucine (Wiegel, 1981) and the pathway in synthesis of vitamin B_{12} (Höllriegel et al., 1982) utilized by *C. thermaceticum* have been studied extensively.

Other characteristics of the species are shown in Table 13.14.

The mol% G + C of the DNA is 54 (T_m) (Matteuzzi et al., 1978).

Isolated from horse manure.

Type strain:; DSM 521.

Further comments. Clostridium thermaceticum, C. aceticum and C. thermautotrophicum are chemolithotrophs that convert CO_2 and H_2 to acetate. *Clostridium thermaceticum* can be distinguished from *C. aceticum* by its thermophilic properties, and from *C. thermautotrophicum* by the inability of *C. thermaceticum* to utilize methanol for growth. For a review of the metabolism of substrates and stimulation of product formation by *C. thermaceticum*, see Ljungdahl and Wood (1969, 1982) and Zeikus (1980, 1983).

77. Clostridium thermautotrophicum (corrig.) Wiegel, Braun and Gottschalk 1982, 384.[VP] (Effective publication: Wiegel, Braun and Gottschalk 1981, 259.)

ther.mau.to.tro′phi.cum. Gr. adj. *thermos* hot; Gr. pron. *autos* self; Gr. n. *trophos* a feeder; *thermautotrophicum* self-feeding in heat, indicating that the organism grows at elevated temperatures and uses CO_2 as its principal carbon source for growth.

This description is based on that of Wiegel et al. (1981), and on study of the type strain.

Cells from the early logarithmic phase of growth have a Gram-positive reaction, whereas cells from older cultures have a Gram-negative reaction. Vegetative cells are slightly motile and peritrichous with three to eight flagella, and 0.8–1.0 × 3.0–6.0 μm. They occur singly or in pairs.

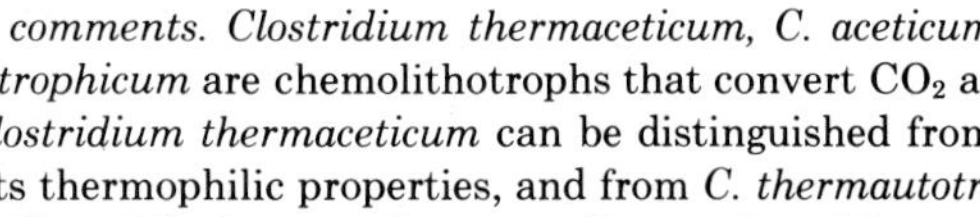

10 μm After Wiegel, Braun, and Gottschalk
Methanol   Salts-H_2-CO2
Figure 13.81. *Clostridium thermautotrophicum.*

Spores are round to slightly oval, terminal or subterminal, and distend the cell. Sporulation by most strains occurs most readily with formate or methanol as the carbon source either on agar or in broth.

Cell walls contain LL-DAP (Wiegel et al., 1981).

Colonies, after 10 days incubation on solid media containing glucose, are 3 mm, circular, flat to convex, smooth and "tannish white"; colonies may turn brown if incubated for longer than 10 days. There is no growth on egg-yolk agar plates.

Cultures in PYG broth are turbid with little or no sediment and have a pH of 5.5 after incubation for 3 days.

The optimum temperature range for growth is 55–60°C; minimum and maximum temperatures that support growth are 36°C and 70°C, respectively. Optimum pH for growth is pH 5.7; no growth at pH below 4.7 or above 7.5. Growth is stimulated by fermentable carbohydrate and by a final concentration of 20% rumen fluid in culture media. The organism grows chemolithotrophically in an atmosphere of CO_2 and H_2 (Wiegel et al., 1981) or of carbon monoxide (Wiegel, 1982) and converts these substrates to acetate, and grows chemoorganotrophically with glucose, fructose, galactose, methanol, glycerate, or formate to produce acetate. Some strains also can ferment arabinose, D-lactose, maltose, mannose, xylose or lactate.

Acetate is the only organic fermentation product formed. Abundant H_2 is detected in PY-fructose broth cultures grown in an atmosphere of 90% N_2–10% CO_2. The following cannot be used as carbon and energy sources: fucose, cellulose, dulcitol, erythritol, ethanol, ethylene glycol, glycerol, inositol, isopropanol, propanol, xylitol, citrate, hippurate, glutamate, malate, pyruvate, succinate, nicotinic acid or casamino acids.

C. thermautotrophicum contains high levels of corrinoids, formate dehydrogenase, tetrahydrofolate enzymes, carbon monoxide dehydrogenase and hydrogenase (Clark et al., 1982).

The type strain is susceptible to chloramphenicol, clindamycin, erythromycin, penicillin G and tetracycline.

Culture supernatants of the type strain are nontoxic to mice.

Other characteristics of the species are shown in Table 13.14.

Isolated from mud and wet soils in Africa, Europe, North America and Hawaii; not restricted to sites with elevated temperatures.

The mol% G + C is 54–55 (T_m) (Wiegel et al., 1981).

Type strain: JW 701/3 (DSM 1974).

Further comments: See Further comments under *Clostridium thermaceticum.*

78. **Clostridium thermocellum** Viljoen, Fred and Peterson 1926, 7.[AL]

ther.mo.cel′lum. Gr. adj. *thermos* hot; N.L. n. *cellulosum* cellulose; N.L. neut. adj. *thermocellum* a thermophile that digests cellulose.

This description is based on that by Ng et al. (1977) and on study of the type strain.

Cells in PY-cellobiose broth cultures are nonmotile, 0.5–0.7 × 2.5–5.0 µm and are Gram-negative. They are straight or slightly curved rods, often with tapered ends, and occur singly or in pairs.

Figure 13.82. *Clostridium thermocellum.*

Spores are oval, terminal, and swell the cell.

Surface colonies are watery, slightly convex, and frequently produce an insoluble yellow pigment. Deep colonies in cellulose-agar roll tubes (Weimer and Zeikus, 1977) are tannish yellow, round, and filamentous.

Cultures in PY-cellobiose broth are slightly turbid with a smooth sediment and have a pH of 5.6–5.8 after incubation for 4 days.

Gelatin is hydrolyzed when cellobiose is included in the medium.

Growth on most media requires the presence of cellulose, cellobiose, or one of the hemicelluloses. Medium 122, described in the 1983 DSM catalog of strains, has been recommended (Saddler and Chan, 1982). In a chemically defined medium, growth factors required are biotin,

pyridoxamine, vitamin B_{12}, and *p*-aminobenzoic acid (Johnson et al., 1981).

The optimum temperature for growth is 60–64°C; there is no growth at 37°C.

Acetyl methyl carbinol is produced and resazurin is reduced.

Although xylan does not serve as a carbon source, xylanase is produced when *C. thermocellum* is grown in cellobiose medium (Garcia-Martinez et al., 1980).

An extracellular, inducible cellulase that exists as a complex and is stable at 70°C, is produced (Ng et al., 1977; Aït et al., 1979). One component of this complex, an endo-β-1,4-glucanase, has been isolated by Petre et al. (1981). Approximately 5% of the total cellulase is an intracellular β-glucanase. A cell-bound β-glucosidase is present which increases the amount of glucose liberated during cellulose fermentation (Aït et al., 1982).

Products of metabolism in PY-cellobiose broth are acetic and lactic acids, ethanol, CO_2, and abundant H_2.

Other characteristics are given in Table 13.15.

Isolated from sewage digestor sludge (Ng et al., 1977) and is said to be present in "nearly all decaying organic material" (Wiegel, 1980). *C. thermocellum* is an important organism for its ability to convert cellulosic waste products to ethanol and H_2. Attempts to improve conversion of biomass to useful products have included induction of auxotrophic mutants (Mendez and Gomez, 1982) and co-culture or sequential culture with other organisms to remove inhibitors or to increase the rate of reaction (Weimer and Zeikus, 1977; Saddler and Chan, 1982; Ng et al., 1981).

The mol% G + C of the DNA is 38–39 (T_m) (Ng et al., 1977).

Type strain: ATCC 27405 (NCIB 10682; DSM 1237).

79. **Clostridium thermohydrosulfuricum** Klaushofer and Parkkinen 1965, 448.[AL]

ther.mo.hy.dro.sul.fur′i.cum. Gr. adj. *thermos* hot; N.L. neut. adj. *hydrosulfuricum* pertaining to hydrogen sulfide; N.L. neut. adj. *thermohydrosulfuricum* indicating that the organism grows at high temperatures and reduces sulfite to H_2S.

This description, unless otherwise indicated, is based on the descriptions by Hollaus and Sleytr (1972) and Wiegel et al. (1979).

Vegetative cells are single or in short chains, have a variable Gram reaction, are motile and peritrichous, and are 0.4–0.6 × 1.8–13.0 µm.

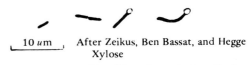

After Zeikus, Ben Bassat, and Hegge
Xylose

Figure 13.83. *Clostridium thermohydrosulfuricum.*

Spores are spherical, terminal and distend the cell. Cells in sporulating cultures are thinner and more elongated than in nonsporulating cultures. Most strains form spores on a glucose agar medium or when cultures grown at temperatures above 65°C are allowed to cool below 55°C (Wiegel et al., 1979).

The cell wall is composed of two layers. The outermost layer consists of hexagonally shaped particles. The center-to-center distance between adjacent particles is 13.5 nm; they are composed of glycoprotein containing glucose, galactose, mannose and rhamnose, which have a molecular weight of approximately 140,000 (Sleytr and Thorne, 1976). The outermost layer completely covers the cell and is resistant to digestion by proteolytic enzymes (Sleytr and Glauert, 1976). Cell walls contain *meso*-DAP (Weiss et al., 1981).

The optimum temperature for growth is 67–69°C; the maximum temperature at which growth occurs is 76–78°C; the minimum is 42°C. Growth occurs at pH 5.5–9.2; optimum growth from 6.9–7.5. Growth is inhibited by H_2 in the gas phase and by lactate (Wiegel et al., 1979).

H_2 and CO_2 are produced in media containing liver infusion. H_2S is produced from tryptone, peptone and yeast extract. Sulfite and thiosulfate are reduced to H_2S; sulfate is not reduced. Acetyl methyl carbinol is not produced.

Trehalose, pectin, esculin and salicin are fermented by all strains; fermentation of dextrin, potato starch, mannitol, dulcitol and sorbitol, and coagulation of litmus milk are variable; inositol, erythritol, glycerol, lactate, tartrate and cellulose are not fermented; nitrite, but not nitrate is reduced, coagulated albumin is not hydrolyzed and indole is not produced.

Products of metabolism in PYG broth are acetic and lactic acids, ethanol, CO_2 and H_2; formic, butyric, isovaleric and isocaproic acids, and propanol and isopropanol may be detected. Methanol is a major metabolic end product of growth in pectin (Schink and Zeikus, 1980). Ethanol yield from fermentation of cellulose by *C. thermocellum* can be doubled by co-culture with *C. thermohydrosulfuricum* (Ng et al., 1981). The metabolic basis for this reaction and enzyme activities involved have been clarified by Ng and Zeikus (1982). For reviews of the use of thermophilic clostridia in industrial fermentations, see Wiegel (1980) and Zeikus (1980).

Other characteristics of the species are shown in Table 13.14.

Isolated from juices extracted from beets during sugar manufacture, from mud and soil in several countries (Wiegel, 1980), from hot springs in Utah and Wyoming and from a sewage plant in Georgia (Wiegel et al., 1979).

The mol% of the DNA is 35–37 (T_m) (Wiegel et al., 1979).

Type strain: DSM 567 (NCIB 10956).

Further comments. Clostridium thermohydrosulfuricum is differentiated most easily from *C. thermosaccharolyticum*, which it resembles most closely, by the ability of the former to grow at 70°C.

80. **Clostridium thermosaccharolyticum** McClung 1935, 200.[AL]

ther.mo.sac.cha.ro.ly'ti.cum. Gr. adj. *thermos* hot; Gr. n. *sacchar* sugar; Gr. adj. *lytikos* dissolving; N.L. neut. adj. *thermosaccharolyticum* referring to thermophilic and sugar-dissolving.

This description, unless otherwise stated, is based on that of Mc-Clung (1935), Hollaus and Sleytr (1972), and on study of the type and one other strain.

Cells in PYG broth culture are Gram-negative, are usually motile and peritrichous, 0.4–0.7 × 2.4–16 μm, and occur singly or in pairs.

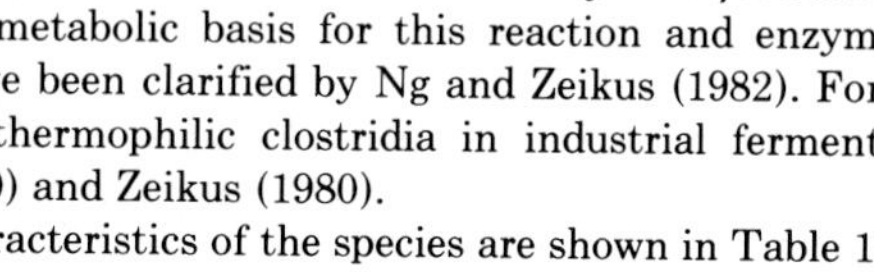

Figure 13.84. *Clostridium thermosaccharolyticum.*

Spores are 1.3–1.5 μm, round or oval, terminal, and distend the cell. Sporulation is enhanced in media containing any of several carbon sources that reduce the growth rate (α- or β-methylglucoside, cellobiose, galactose, salicin, starch) (Hsu and Ordal, 1969). In contrast, Pheil and Ordal (1967) achieved good sporulation in a basal medium supplemented with L-xylose or L-arabinose; the optimal pH for sporulation is 5.0–5.5.

Cell walls contain *meso*-DAP (Weiss et al., 1981). The cell wall possesses rectangular surface units whose centers are approximately 9.5 nm apart (Hollaus and Sleytr, 1972); these units are composed of glycoprotein with a molecular weight of 140,000 which contains glucose, galactose, mannose and rhamnose (Sleytr and Thorne, 1976). The surface units are resistant to proteolytic digestion (Sleytr and Glauert, 1976). The cell wall is susceptible to dissolution by lysozyme (Johnson and Francis, 1975).

Surface colonies on blood agar plates are nonhemolytic, pinpoint to 0.5 mm, circular to slightly irregular, scalloped, low convex, gray-white,

shiny and smooth, with a mottled internal structure. Colonies on the surface of pea infusion agar are 2–4 mm in diameter, flat with a raised center, have a rough indistinct "feather" edge, and are grayish white. Deep colonies in tryptone agar medium are small and lenticular.

Cultures in PYG broth are turbid with a smooth sediment, and have a pH of 4.6 after incubation for 5 days.

The optimum temperature for growth is 55–62°C; some strains will grow, but quite slowly, at 37°C; poorly, if at all, at 30°C; and not at all at 70°C. Good growth is produced in broth with or without a fermentable carbohydrate. Lee and Ordal (1976) found that the organism has a control mechanism whereby it can discriminate between a carbon source and an energy source; thus glucose is used in the synthetic pathway while energy is obtained from degradation of pyruvate.

Deoxyribonuclease is not produced (Johnson and Francis, 1975).

Abundant gas is produced in PYG deep agar cultures.

Dextrin and pectin are fermented (Hollaus and Sleytr, 1972); erythritol, inositol and glycerol are not fermented. Nitrite is reduced, and milk is coagulated. Sulfite and thiosulfate are reduced to H_2S in liver broth; small amounts of H_2S are produced in tryptone-glucose broth supplemented with phosphate (Hollaus and Sleytr, 1972). Ammonia is produced, neutral red and resazurin are reduced.

Products of metabolism in PYG broth include large amounts of acetic, butyric and lactic acids, ethanol, and H_2, and moderate amounts of succinic acid. Sporulating cells produce proportionally more ethanol than those in which sporulation is repressed (Hsu and Ordal, 1970). Pyruvate is converted to acetate and butyrate; neither lactate nor threonine is utilized.

Chan et al. (1971) studied the cellular fatty acids of one strain and found that normal saturated C_{14} and C_{16}, and *iso*-branched chain C_{15} and C_{17} acids predominate.

Culture supernatants are not toxic to mice. Broth cultures injected intravenously in rabbits or fed to rats are nonpathogenic.

Other characteristics of the species are given in Table 13.14.

The organism has been isolated from soil, "blown" canned foods, sugar, extraction juice in beet sugar factories (Hollaus and Sleytr, 1972, and Matteuzzi et al., 1978), and once from the gingival crevice of a human (De Campos et al., 1981).

The mol% G + C of the DNA is 29–32 (T_m) (Matteuzzi et al., 1978).

Type strain: ATCC 7956 (NCIB 9385; DSM 571).

Further comments. Hollaus and Sleytr (1972) proposed that "*Clostridium tartarivorum*" be considered a tartrate-decomposing biotype of *C. thermosaccharolyticum*. This proposal was confirmed by Matteuzzi et al. (1978) who found that two strains of "*Clostridium tartarivorum*" were 87% and 96% homologous with the type strain of *C. thermosaccharolyticum*. The name *C. thermosaccharolyticum* has priority.

81. **Clostridium thermosulfurigenes** (corrig.) Schink and Zeikus 1983c, 896.[VP] (Effective publication: Schink and Zeikus 1983a, 1156.)

ther.mo.sul.fur.i'ge.nes. Gr. adj. *thermos* hot. L. n. *sulfur* brimstone, Gr. suffix-*genes* born from; N.L. neut. adj. *thermosulfurigenes* releasing sulfur in heat.

This description is based on study of the type strain by Schink and Zeikus (1983a).

Cells in broth culture are Gram-negative, are straight rods, motile and peritrichous, 0.5 × 2 μm or more, and occur either singly or in filamentous chains.

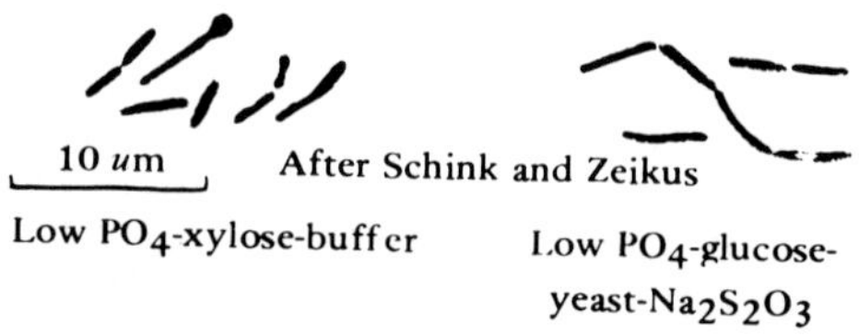

Figure 13.85. *Clostridium thermosulfurigenes.*

Spores are round, terminal and distend the cell. Sporulation occurs in xylose- or pectin-containing media, but not in glucose media; sporulation is particularly favored in a buffered medium containing yeast extract and xylose.

Cell wall composition has not been reported but electron microscopic analysis reveals numerous internal, and often vesicular, membrane-like structures and a double-layered cell wall.

Colonies in agar roll tubes containing low-phosphate buffered basal medium (Zeikus et al., 1980) and 0.1% yeast extract and 0.5% glucose are 0.5-1.5 mm in diameter, yellow and have a fluffy or brushlike appearance.

Addition of sodium thiosulfate to broth culture results in heavy turbidity and deposition of granules of elemental sulfur on cells and in the medium; neither sulfite nor sulfide is detected.

The optimum temperature for growth is 60°C; growth does not occur below 35°C or above 75°C. The optimum pH range for growth in glucose-containing media is 5.5-6.5; growth does not occur at pH below 4.0 or above 7.6. Growth is inhibited by 2% NaCl and by sulfite.

Esculin, pectin and polygalacturonic acid are fermented, whereas arabinogalactan, cellulose, citrate, galacturonate, glycerol, lactate, methanol, pyruvate and tartrate are not. Sulfate and sulfite are not reduced.

The principal products of carbohydrate fermentation are H_2, CO_2, ethanol, acetate and lactate. Both methanol and isopropanol are formed from pectin.

Pectin methylesterase and polygalacturonate hydrolase are produced (Schink and Zeikus, 1983b).

Growth is inhibited by 100 µg/ml of penicillin, streptomycin, cycloserine, tetracycline and chloramphenicol, and by 500 µg/ml of sodium azide.

Other characteristics of the species are given in Table 13.12.

Isolated from algal bacterial mat ecosystems associated with thermal volcanic springs.

The mol% G + C of the DNA is 33 (T_m) (Schink and Zeikus, 1983a).

Type strain: B4 (ATCC 33743; DSM 2229).

Further comments. C. thermosulfurigenes can be distinguished from other thermophilic clostridia by its ability to ferment glucose and liquefy gelatin. It differs from *C. thermohydrosulfuricum* by forming elemental sulfur instead of sulfide from thiosulfate.

82. **Clostridium tyrobutyricum** van Beynum and Pette 1935, 205.[AL]

ty.ro.bu.ty′ri.cum. Gr. n. *tyros* cheese; N.L. n. *acidum butyricum* butyric acid; N.L. neut. adj. *tyrobutyricum* the butyric acid-producing organism from cheese.

This description, unless otherwise indicated, is based on that by Roux and Bergère (1977) and on study of the type and three other strains.

Cells are Gram-positive, usually motile and peritrichous, 1.1-1.6 × 1.9-13.3 µm, and occur singly or in pairs.

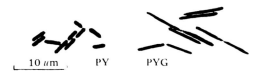

Figure 13.86. *Clostridium tyrobutyricum.*

Spores are oval, subterminal and swell the cell. Touraille and Bergère (1974) concluded that two enzymes (acetokinase and phosphotransacetylase) are involved in spore germination. Sporulation occurs most readily on chopped meat agar slants or in old PYG cultures.

Cell walls contain mesoDAP; glucose is the only cell wall sugar (Cummins and Johnson, 1971). Glutamic acid and alanine are present

(Schleifer and Kandler, 1972). The cell wall is susceptible to dissolution by lysozyme (Johnson and Francis, 1975).

Colonies on the surface of blood agar are circular, 0.5 mm in diameter, convex with an entire margin, gray, translucent, have a glossy surface and are often β-hemolytic.

Cultures in PYG broth are turbid with a smooth sediment and have a pH of 5.0-5.4 after incubation for 1 week.

Optimum temperature for growth is 30-37°C; growth is moderate at 25°C; poor or no growth at 45°C. Growth in broth is stimulated by a fermentable carbohydrate but inhibited by 6.5% NaCl or 20% bile.

Abundant gas is produced in PYG deep agar cultures.

Ammonia occasionally is produced; neutral red is reduced.

Roux and Bergère (1977) noted that of 77 strains studied, 82% fermented mannitol, 56% fermented mannose and 27% fermented xylose. Möse et al. (1972) reported kininase-like activity in one strain. Cells contain rubredoxin which is involved in NAD oxidation-reduction reactions (Petitdemange et al., 1981). Immunologically similar enoate reductases have been isolated from the type and three other strains that are 81-89% homologous with the type strain (Giesel and Simon, 1983). Deoxyribonuclease is not produced (Johnson and Francis, 1975).

Products of fermentation in PYG broth are large amounts of butyric and acetic acids, and small amounts of succinic acid; traces of formic acid may be detected. Abundant H_2 is produced. Pyruvate is converted to butyrate and acetate; threonine is not converted to propionate; in the presence of acetate, *C. tyrobutyricum* converts lactate to butyrate, CO_2 and H_2 (Bryant and Burkey, 1956).

One strain tested is susceptible to chloramphenicol, clindamycin, erythromycin, penicillin G and tetracycline.

Culture supernatants are not toxic to mice. Nonpathogenic for humans and other animals.

Other characteristics of the species are given in Table 13.14.

Isolated from gley soil (Hamman and Ottow, 1976), dairy products (Matteuzzi et al., 1977; Roux and Bergère, 1977; Naguib and Shauman, 1973), silage (Bryant and Burkey, 1956; Roux and Bergère, 1977), feces of beagle dogs (Balish et al., 1977), bovine feces (Princewill and Agba, 1982) and human adult and infant feces (Legakis et al., 1981; Smith et al., 1980).

The mol% G + C of the DNA is 28 (T_m) (Cummins and Johnson, 1971).

Type strain: ATCC 25755 (NCIB 10635).

83. **Clostridium villosum** Love, Jones and Bailey 1979, 241.[AL]

vil.lo′sum. L. neut. adj. *villosum* hairy, shaggy, rough haired (referring to the colonial morphology).

This description is based on that of Love et al. (1979) and on study of the type strain.

Cells are rods with parallel sides and rounded ends, nonmotile and atrichous, 0.6 × 4.0-6.0 µm. Filaments up to 24-30 µm in length are common in cultures more than 24-48 h old. Cells stain Gram-positive up to 18-24 h of incubation on agar or in broth; by 7 days, Gram reaction becomes variable with most cells exhibiting a negative reaction and with unstained gaps along the cells.

Figure 13.87. *Clostridium villosum.*

Spores are oval, subterminal and slightly swell the cell. Spores are not readily detected but chopped meat-carbohydrate slant cultures are heat-resistant.

Cell walls contain an ornithine-*iso*-D-asparagine peptidoglycan (Weiss et al., 1981).

Surface colonies on sheep blood agar and brain heart infusion agar are 0.5 mm in diameter at 24 h and approximately 5 mm at 3 days, irregular, rhizoid, convex, whitish to yellowish, rough, have a matt surface, adhere to the medium and are nonhemolytic.

Cultures in chopped meat-carbohydrate broth grow in a delicate membrane near the surface of the broth, with bacteria clumped as discrete colonies throughout the membrane; later the membrane collapses and settles to the bottom of the tube. Cultures in PYG broth have a ropy or flaky sediment and no turbidity.

Growth at 37°C is stimulated by 5% horse serum.

No gas is detected in PYG deep agar cultures. Ammonia is produced in chopped meat cultures.

Products of fermentation in chopped meat-carbohydrate and PY broth supplemented with 5% horse serum are acetate, butyrate, isobutyrate, formate and isovalerate. Trace amounts of lactate, succinate and methylmalonate may be detected (Love et al., 1979). Small amounts of H_2 are produced. Pyruvate is converted to butyrate; lactate is not utilized; increased amounts of acetate and butyrate are produced in threonine cultures but propionate is not produced.

Strains are susceptible to penicillin (2 U/ml), amoxycillin (2.5 µg/ml), carbenicillin (100 µg/ml), doxycycline (6 µg/ml), chloramphenicol (12 µg/ml) and erythromycin (3 µg/ml).

Culture supernatants are nontoxic to mice.

Other characteristics are given in Table 13.15.

Isolated as part of a mixed bacterial flora from "subcutaneous fight wound abscesses" of cats and as the predominant organism in some cases of empyema in cats (Love et al., 1979, 1982).

The mol% G + C of the DNA has not been reported.

Type strain: NCTC 11220 (DSM 1645).

Further comments. C. villosum may be confused with Gram-negative anaerobic nonsporing rods because of its negative Gram-reaction in older cultures and lack of readily detectable spores. The distinctive colonial morphology and heat resistance (80°C for 10 min) are helpful distinguishing characteristics. *C. villosum* can be distinguished from other gelatin-digesting species with similar products by its failure to digest or curdle milk. Unlike *C. hastiforme*, it is nonmotile.

Addendum

After this manuscript was completed, Madden (1983) published a description of a new species, *Clostridium stercorarium,*[VP] isolated from a compost heap. Strains of this species are obligate thermophiles that digest cellulose Cells are motile and peritrichous, and usually stain Gram-negative. Spores are oval, terminal and swell the cell. Optimum growth occurs at 65°C and a pH of 7.3; fermentable carbohydrate stimulates growth. Arabinose, cellobiose, cellulose, esculin, galactose, glucose, glycogen, lactose, maltose, mannose, melibiose, rhamnose, ribose, salicin, starch, xylan and xylose are fermented; not fermented are adonitol, amygdalin, dulcitol, erythritol, fructose, glycerol, inositol, inulin, mannitol, melezitose, raffinose, sorbitol, sorbose, sucrose or trehalose. Neither nitrate nor sulfate is reduced; gelatin is not hydrolyzed; lecithinase, lipase, urease, H_2S, and acetyl methyl carbinol are not produced. Products of fermentation in glucose broth are acetic and lactic acids, ethanol, H_2, and CO_2. *C. stercorarium* can be distinguished from the thermophilic cellulolytic *C. thermocellum* by its fermentation of pentoses, glycogen, maltose and starch. The mol% G + C of the DNA is 39 (T_m). The type strain is NCIB 11754.

Another species described after the manuscript was completed is *C. lortetii* Oren 1984[VP] (Effective publication: Oren 1983), an obligately anaerobic halophile isolated from Dead Sea sediment. Cells are Gram-negative, motile and peritrichous. Spores are round and terminal with gas vacuoles attached. Occasionally a spore can be seen at each end of the cell. Optimum growth occurs at 37–45°C, at a pH of 7.0. NaCl at a concentration of 1–2 M is required for growth. Glucose, fructose, maltose, sucrose and starch are highly stimulatory for growth, but they are poorly utilized. Neither lactose, cellobiose, galactose nor mannose stimulates growth. Indole is produced; gelatin is not hydrolyzed. Products detected include acetic acid, propionic acid, butyric acid, isobutyric acid, isovaleric acid and H_2. *C. lortetii* can be differentiated from other species of *Clostridium* by its absolute requirement for NaCl and by its production of gas vacuoles attached to its spores. The mol% G + C of the DNA is 31.5 by buoyant density. The type strain is strain MD-2 (ATCC 35059). For further description, see Oren (1983).

Editorial Note

The following is a list of species that have been validly published since the above was submitted:

1. *Clostridium bryantii* Stieb and Schink 1985b, 375.[VP] (Effective publication: Stieb and Schink 1985a, 390.)

    Type strain: CuCal (DSM 3014A.B).

2. *Clostridium cellulolyticum* Petitdemange, Caillet, Giallo and Gaudin 1984, 157.[VP]

    Type strain: H_{10} (ATCC 35319).

3. *Clostridium cellulovorans* Sleat, Mah and Robinson 1985, 223.[VP] (Effective publication: Sleat, Mah and Robinson 1984, 92.)

    Type strain: 743B (ATCC 35296).

4. *Clostridium magnum* Schink 1984b, 355.[VP] (Effective publication: Schink 1984a, 254.)

    Type strain: WoBdP1 (DSM 2767).

5. *Clostridium populeti* Sleat and May 1985, 160.[VP]

    Type strain: 743A (ATCC 35295).

Genus IV. **Desulfotomaculum** *Campbell and Postgate 1965, 361*[AL]

L. LEON CAMPBELL AND RIVERS SINGLETON, JR.

De.sul.fo.to.ma'cu.lum. L. pref. *de* from; L. n. *sulfur* sulfur; L. n. *tomaculum* sausage; M.L. neut. n. *Desulfotomaculum* a sausage (-shaped organism) that reduces sulfur compounds.

Straight or curved rods, 0.3–1.5 × 3–9 µm, with rounded or pointed ends, usually single but sometimes in chains. **Stain Gram-negative.** Electron microscopy has demonstrated a multilayered cell-wall structure (Sleytr et al., 1969). **Motile with peritrichous or polar flagella. Spores oval to round, terminal to subterminal, causing slight swelling of the cells.** Produce black colonies on a suitable carbon source and agar containing ferrous salts.

Strict anaerobes.

Chemoorganotrophs; **metabolism respiratory** (see Further Comments). **Sulfates, sulfites and reducible sulfur compounds act as electron acceptors and are reduced to H_2S.** In species using lactate and pyruvate (or homologue) as electron donors, oxidation is incomplete and leads to formation of acetate (or a homologue) and CO_2. Species using acetate as electron donor carry out a complete oxidation to CO_2. **Cells contain a cytochrome of the protoheme class** (cytochrome *b*) **and are characterized by the absence of cytochrome** c_3. Specialized media containing a reducible sulfur compound and organic growth factors are required for growth (Adams and Postgate, 1959; Campbell et al., 1957; Postgate and Campbell, 1963; Postgate, 1979).

Temperature optimum 35–55°C; maximum 70°C. Some strains grow at 20–30°C.

Common inhabitants of soil, fresh water, geothermal regions, certain spoiled foods, intestines of insects, rumen contents and stratal water of oil fields (Nazina and Rozanova, 1978). None has been reported to be pathogenic to man, guinea pig, mouse, rat or rabbit.

The G + C (buoyant density) content of the DNA ranges from **37–50 mol%** (Skyring and Jones, 1972; Widdel and Pfennig, 1977).

Type species: *Desulfotomaculum nigrificans* (Werkman and Weaver) Campbell and Postgate 1965, 360.

Isolation and Enrichment Procedures

An enrichment procedure and media for the isolation of lactate-oxidizing strains of *Desulfotomaculum* are described by Postgate (1979). Widdel and Pfennig (1981) describe methods for the enrichment of acetate-oxidizing strains.

Maintenance Procedures

There are little published data on the long term storage of desulfotomacula. Postgate (1979) describes a procedure for lyophilizing cultures. The organism also stores very well in liquid cultures (with FeS present) which have been sealed against air. Because the organism readily undergoes sporulation, spore cultures can be stored at liquid nitrogen temperatures.

Differentiation of genus **Desulfotomaculum** from other genera

Members of the genus *Desulfotomaculum* can be differentiated from members of the genus *Desulfovibrio*, the other major genus of sulfate-reducing bacteria, based on points enumerated in Table 13.18.

Further Comments

Recent work (Liu and Peck, 1981; Peck and LeGall, 1982) has demonstrated that lactate-oxidizing members of the genus *Desulfotomaculum* do not couple sulfate-reduction with electron transport-generated oxidative phosphorylation. Reducing equivalents from the oxidation of carbon compounds are disposed of by the reduction of sulfate to sulfide, without the concomitant generation of ATP via chemiosmosis. Since respiratory mechanisms are generally thought to involve the generation of ATP via oxidative phosphorylation, this observation raises questions regarding the respiratory nature of the genus *Desulfotomaculum*. However, a consideration of the potential biochemistry involved in acetate oxidation, suggests that either electron transport-coupled phosphorylation or some form of substrate-level phosphorylation must occur during acetate utilization.

Table 13.18.
Distinguishing characteristics of sulfate-reducing bacteria

Characteristics	*Desulfotomaculum*	*Desulfovibrio*
Spores	Present	Absent
Desulfoviridin[a]	Absent	Present
Cytochrome c_3	Absent	Present
Morphology	Generally rod-like	Generally vibrio-like
Flagella	Generally peritrichous	Generally single polar

[a] Serves as sulfite-reductase in the genus *Desulfovibrio*; it is replaced by P-582 in some species of the genus *Desulfotomaculum* (Trudinger 1970, Akagi and Adams, 1973).

List of species of the genus **Desulfotomaculum**

1. **Desulfotomaculum nigrificans** (Werkman and Weaver) Campbell and Postgate 1965, 361.[AL] (*Clostridium nigrificans* Werkman and Weaver 1927, 63.)

nig.ri'fi.cans. L. part. adj. *nigrificans* blackening.

Straight to curved rods 0.3–0.5 × 3–6 μm with rounded ends. Sometimes lenticulate to swollen. "Twisting and tumbling" motility.

Utilizes lactate, pyruvate, glucose and ethanol as hydrogen donors for sulfate reduction. Acetate and formate are not utilized. Pyruvate usually supports growth in presence or absence of sulfate. H_2S formed with cystine.

Thermophilic, temperature range 45–70°C. Optimal temperature 55°C. Can be adapted to grow slowly at 30–37°C.

Growth is inhibited by <0.1 μg/ml of hibitane.

The mol% G + C of the DNA is 48.5–49.9 (Bd).

Type strain: ATCC 19998 (Delft 74T; NCIB 8395).

The distinguishing characteristics of various species of the genus *Desulfotomaculum* are summarized in Table 13.19.

2. **Desulfotomaculum ruminis** Campbell and Postgate 1965, 361.[AL]

ru.min'is. L. n. *rumen* throat, adopted for first stomach (rumen) of a ruminant; L. gen. n. *ruminis* of a rumen.

Straight to curved rods, 0.5 × 3–6 μm, with rounded ends. Sometimes paired. Slight tumbling motility. Grows in specialized media containing

Table 13.19.
*Comparative characteristics of species of the genus **Desulfotomaculum**[a,b]*

Characteristics	*D. nigrificans*	*D. orientis*	*D. ruminis*	*D. antarcticum*	*D. acetoxidans*
Shape	Rod	Curved rod	Rod	Rod	Rod
Flagella	Peritrichous	Peritrichous	Peritrichous	Peritrichous	Polar (thick)
Mol% G + C[c]	48.5–49.9	44.7–45.9	48.5–49.9	ND	38
Growth with[d]					
Lac + SO₄	+	+	+	+	−
Pyr − SO₄	+	−	+	ND	−
Form + SO₄	−	−	+	−	ND
AC + SO₄	−	−	−	−	+
Glu + SO₄	−	−	−	+	−
Gelatinase	−	−	−	+	−
Thermophilic	+	−	−	−	−

[a] After Postgate 1979, 10.

[b] Symbols: see Table 13.4.

[c] Skyring and Jones, 1972; Widdel and Pfennig, 1977.

[d] Lac, lactate; Pyr, pyruvate; Form, formate; Ac, acetate; Glu, glucose.

lactate, pyruvate or formate (but not acetate or ethanol) plus sulfate and produces H_2S. Pyruvate supports growth with or without sulfate.

Mesophilic, temperature range 30–48°C; optimal temperature 37°C. Growth is inhibited by 1 μg/ml of hibitane.

The mol% G + C of the DNA is 48.5–49.9 (Bd).

Type strain: ATCC 23193 (Coleman DL; NCIB 8452).

3. **Desulfotomaculum orientis** (Adams and Postgate) Campbell and Postgate 1965, 361.[AL] (*Desulfovibrio orientis* Adams and Postgate 1959, 256.)

or.i.en′tis. L. part. adj. *oriens* rising (sun), hence the orient; L. gen. n. *orientis* of the orient.

Fat curved rods, 1.5 × 5 μm, sometimes paired. "Tumbling and twisting" motility. Spore round, central or paracentral (on rare occasions terminal), slightly swelling the cells.

Grows in specialized media containing lactate or pyruvate (but not formate, acetate or ethanol) plus sulfate and thioglycolate and produces H_2S. Does not grow without sulfate, even with pyruvate.

Mesophilic, temperature range 30–42°C. Optimal temperature between 30° and 37°C.

Growth is inhibited by <0.1 μg/ml of hibitane.

The mol% G + C of the DNA is 44.7–45.9 (Bd).

Type strain: ATCC 19365 (Singapore I; NCIB 8382).

4. **Desulfotomaculum acetoxidans** Widdel and Pfennig 1977, 121.[AL]

a.cet.o′xi.dans. L. n. *acetum* vinegar; M.L. n. *acidum aceticum* acetic acid; M.L. v. *oxido* make acid, oxidize; M.L. part. adj. *acetoxidans* oxidizing acetic acid.

Straight or slightly curved rods, 1–1.5 × 3.5–9 μm, with pointed ends. Motile by means of single, thick, polar flagellum. Spores spherical 1.5 μm in diameter, subterminal, causing swelling of the cells. Next to the spore is a refractile gas vacuole. Spore-forming cells typically spindle-shaped. Gram-negative.

Strictly anaerobic chemoorganotroph. Acetate, butanol, butyrate, ethanol and *iso*-butyrate are completely oxidized to CO_2; valerate is oxidized to CO_2 and propionate; formate is poorly utilized and requires the presence of some acetate as carbon source. Not utilized as electron

donors: H_2, propanol, higher fatty acids than valerate, lactate, pyruvate, succinate, fumarate, malate, benzoate and sugars. Not utilized as electron acceptors: sulfite, thiosulfate, elemental sulfur, fumarate, nitrate and oxygen. Sulfate is reduced to H_2S. Unable to ferment organic substances.

Growth requires mineral media with sulfide and dithionite as reductants. Biotin is needed for growth. Marine forms require 1–2% NaCl. Higher concentrations than 150 mmol NaCl/liter inhibit growth.

Mesophilic, temperature range 20–40°C. Optimal temperature 36°C. pH range 6.6–7.6. Optimal pH 7.1.

Pigments: cell membrane fractions contain *b*-type cytochromes (γ_{max} at 426, 530 and 559 nm). CO-difference spectrum of cytoplasmic fractions exhibits adsorption peaks at 412, 550 and 596 nm, indicating presence of the sulfite reductase P-582. No soluble *c*-type cytochromes are observed.

The mol% G + C of the DNA is 37.5 (T_m).

Habitats: manure and feces of higher animals, rumen content and anaerobic mud of fresh water especially if polluted with feces. Specific enrichment with acetate and sulfate, from feces and rumen also with butyrate.

Type strain: DSM 771 ("Göttingen," 5575).

5. **Desulfotomaculum antarcticum** (*ex* Iizuka, Okazaki and Seto) nom. rev.

M.L. adj. *antarticum* antarctic.

Fat rods, sometimes paired or in short chains; 1.0–1.2 × 4–6 μm with rounded ends. Motile with peritrichous flagella. Spores oval, central, or terminal, causing swelling of cells.

Gram-negative, obligate anaerobe.

Chemoorganotroph. Lactate and pyruvate serve as electron donors in the presence of sulfate. Cannot utilize acetate or formate. Acid from glucose, but not from fructose, sucrose, arabinose or xylose. Gelatin hydrolyzed. Sulfate reduced to sulfide, but nitrate not reduced.

Mesophilic, optimum temperature 20–30°C. Tolerates, but does not require, NaCl (2.5%) but higher concentrations inhibit growth.

Desulfovirdin and cytochrome c_3 absent, cytochrome *b* present.

Type strain: IAM 64.

Genus **Sporosarcina** *Kluyver and van Niel 1936. 401*[AL]

D. CLAUS AND FATMA FAHMY

Spo.ro.sar.ci′na. M.L. n. *spora* a spore; L. n. *sarcina* a package, bundle; M.L. fem. n. *Sporosarcina* a spore-forming package.

Single **cells spherical or oval.** Upon division, two hemispherical or nearly hemispherical cells are formed which then may develop into **diplococci or a pair of oval cells. Cells divide into two or three perpendicular planes** leading to the **formation of tetrads or packages of eight or more. Endospores are formed.** Gram-positive. Generally motile by one or more randomly spaced flagella on each cell. **Strictly aerobic.**

Colonies are round, smooth, semitransparent or opaque. Pigmentation is cream colored or pale yellow to bright orange. Almost all strains can grow between 15°C and 37°C. Catalase and oxidase reaction is positive.

Chemoorganotrophic, most strains require growth factors. The mol% G + C of the DNA is 40.0–41.5 (T_m).

Type species: *Sporosarcina ureae* (Beijerinck 1901) Kluyver and van Niel 1936, 401.

Further Descriptive Information

Although the typical cell shape is spherical (Fig. 13.88), some strains, especially in *S. halophila*, often form oval to egg- or pear-shaped cells (Fig. 13.89 and 13.90). Depending on strain and growth conditions, cultures of both species may be composed of different cell aggregates: single cells, pairs, threes, tetrads or packages may predominate. Within

pairs, threes and tetrads, the single cell may be spherical to oval. Interfacial flattening of cells often can be observed.

The diameter of spherical cells is 1.0–2.5 μm. Oval cells are about 1.0–2.0 × 2.0–3.0 μm. Abnormally large cells appearing singly or in packets often can be found. Due to asynchronous cell division, also short chains or irregular aggregates may be formed, especially in older cultures. At low magnification cells of *S. halophila* sometimes look like thick rods. At higher magnification it can be seen that these "rods" are composed of two cells.

Motility with strains of both species is tumbling. In case of *S. ureae* it is best observed with cells growing in nutrient broth containing 1% urea (Kocur and Martinec, 1963). Flagella have been described as exceedingly long (Hale and Bisset, 1958; Leifson, 1960; Claus et al., 1983).

Endospores are formed, often only under special conditions (see under maintenance procedures). They are highly refractile, round, 0.5–1.5 μm in diameter, located centrally or laterally (Fig. 13.89 and 13.91). They resist heating up to 70°C or more for 10 min and contain dipicolinic acid (Thompson and Leadbetter, 1963; Fahmy et al., 1985). Their ultrastructure is similar to that of *Bacillus* endospores (Silva et al., 1973; Fahmy et al., 1985).

Cell walls of vegetative cells of both species do not contain diami-

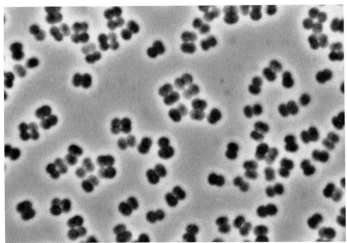

Figure 13.88. *S. ureae.* Vegetative cells forming tetrads. Phase contrast (× 2000).

The optimum temperature for *S. ureae* is about 26°C, for *S. halophila* 30°C.

S. halophila is an obligate, moderate halophilic bacterium and needs sodium, magnesium and chloride ions for growth. For good growth media have to be supplemented with 3% sodium chloride and 0.5% magnesium chloride. Growth of *S. ureae* may be enhanced by adding 0.5% ammonium chloride or 0.5–1.0% urea to complex media.

Strains of *S. ureae* usually can use acetate or glutamate as a sole source of carbon and energy. Most strains require biotin for growth either singly or in combination with niacin and/or thiamine. Other strains have additional requirements for aspartate or for unknown factors (Pregerson, 1973). Carbohydrates, in general, cannot be utilized with the exception of ribose and/or fructose which allows weak growth for certain strains (MacDonald and MacDonald, 1962; Pregerson, 1973). Ammonium salts as well as glutamate and peptones but not nitrate can be used as a source of nitrogen (Pregerson, 1973). A synthetic medium which supports excellent growth of *S. ureae* ATCC 13881 and some other strains has been described by Goldman and Wilson (1977).

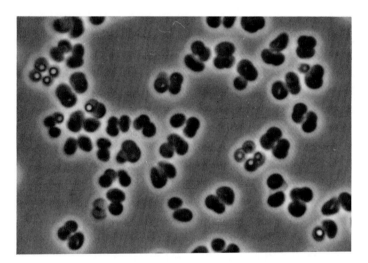

Figure 13.89. *S. halophila.* Vegetative cells forming tetrads and other aggregates. Endospores are formed in some cells. Phase contrast (× 2000).

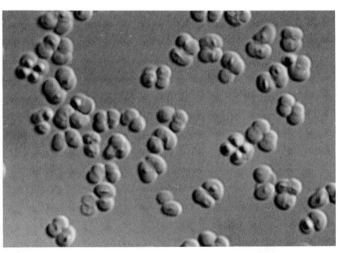

Figure 13.90. *S. halophila.* Same preparation as in Fig. 13.89. Nomarski differential interference contrast (× 2000).

nopimelic acid. This compound, however, is found in the walls of endospores (Ranftl, 1972; Claus et al., 1983). The two species differ in their murein types: In *S. ureae* the Lys-Gly-D-Glu-type (Linnett et al., 1974; Schleifer and Kandler, 1972), in *S. halophila* the Orn-D-Asp-type (Kandler et al., 1983) is found. Cell wall preparations of *S. halophila* also contain high amounts of a γ-D-glutamyl polymer, which is not found in *S. ureae* (Kandler et al., 1983).

Chemotaxonomical studies have revealed that cell membranes of *S. ureae* contain phosphatidylethanolamine (Komura et al., 1975). In both *Sporosarcina* species the menaquinone system MK-7 has been found (Yamada et al., 1976, Claus et al., 1983).

Studies on the fine structure of *S. ureae* have been published by several authors (Beveridge, 1979, 1980; Mazanec et al., 1965; Robinson and Spotts, 1983; Silva et al., 1973; Stewart and Beveridge, 1980).

Colonies of *S. halophila* constantly form orange pigments. With *S. ureae* colonies may be gray or cream becoming yellowish, orange or brownish in different strains and on different media. On certain media *S. ureae* may produce a yellowish diffusible pigment (Pregerson, 1973).

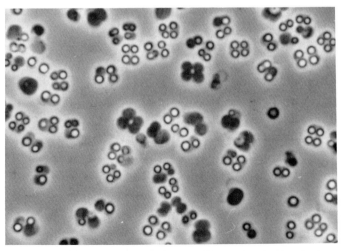

Figure 13.91. *S. ureae.* Endospores. Phase contrast (× 2000).

Minimal nutritional requirements of *S. halophila* have not been studied. Some strains grow in a mineral medium containing acetate or glutamate as carbon source and ammonium chloride as nitrogen source if small amounts of yeast extract are added.

Studies on metabolic pathways of *Sporosarcina* species have not been published except for the comparative allostery of 3-deoxy-D-arabino-heptulonosate-7-phosphate synthetase as a molecular basis for classification (Jensen and Stenmark, 1970).

S. ureae and *S. halophila* are considered to be not pathogenic for man, animals or plants. "*S. pulmonum*" has been isolated by several workers from the respiratory tract of man, chiefly in cases of phthisis. Although the species has been considered earlier to be responsible for severe infections of the lungs, its pathogenicity has been questioned (Lehmann and Neumann, 1927; Migula, 1900). Original isolates of this species are no longer available. According to Gibson (1935), "*S. pulmonum*" is probably identical with *S. ureae*.

S. ureae is widely distributed in garden and field soils (Gibson, 1935). Strains have been isolated from liquid manure (Sames, 1898) and from sea water (Wood, 1946). The primary habitat appears to be concentrated in certain urban soils closely associated with the activities of man and especially dogs. A good source for isolation apparently is soil from the base of trees where dogs have urinated (Pregerson, 1973). In nature, *S. ureae* possibly plays an active part in the decomposition of urea.

S. halophila, up to now has been isolated exclusively from salt marsh soils (North Sea, Germany). Its salt requirements and its temperature optimum suggest that such soils are the real habitat of the species.

Beijerinck (1901) has briefly described another spore-forming sarcina as "*Urosarcina dimorpha*" which developed only on media containing horse urine. Cultures of this organism, originally isolated from garden soils, are not available. Attempts to re-isolate the species have failed up to now.

Isolation and Enrichment Procedures

Strains of *S. ureae* and *S. halophila* are best isolated by plating soil dilutions (see above under habitats) on appropriate agar media. Selective enrichment methods in liquid media are not available. With both species the isolation often is not reproducible due to irregular background growth developing on dilution plates. In case of *S. ureae* this may be strongly depressed by adding at least 3% urea to the medium. Otherwise, most plates will be overgrown by *Bacillus mycoides*.

For the isolation of *S. ureae* two methods may be applied. In Gibson's method, modified by Claus (1981), about 5 g of air-dried garden or field soil is suspended in 20 ml of sterile tap water. Two ten-fold dilutions (10^{-1} and 10^{-2}) are prepared. Then 0.1 ml of the suspension and its dilutions are streaked onto plates containing nutrient agar (peptone, 5 g; meat extract, 3 g; agar, 15 g; distilled water, 1000 ml), supplemented with 30, 50 and 100 g urea/liter.

After incubation at about 25°C for 3 or more days colonies developing on the plates are examined under a dissecting microscope and transmitted light. At ten-fold magnification colonies of *S. ureae* appear round and black, but at a magnification of × 50 show a coarsely granulated structure, especially at the edges of the colonies. Similar colonies are formed mainly by *Bacillus megaterium*.

Material from appropriate colonies is checked microscopically for the presence of sarcinae and is restreaked for purification on nutrient agar containing 1% urea. The provisional identification of isolated strains as *S. ureae* should be confirmed by testing motility and the formation of endospores (see below under maintenance).

For the isolation of about 50 strains of *S. ureae*, Pregerson (1973) successfully has used an alkaline tryptic soy-yeast extract agar (Difco Tryptic Soy Broth, 27.5 g; Difco Yeast Extract, 5.0 g; Glucose, 5.0 g; Bacto Agar, 15.0 g; distilled water, 1000 ml). Before sterilization the pH is adjusted to 8.5 with NaOH. After sterilization the medium is supplemented with a filter-sterilized urea solution to give a final concentration of 1%. On this medium, colonies of *S. ureae* show a uniform surface granularity if viewed under a dissecting microscope,

smoothly opaque interiors and an orange or cream pigmentation (for details see also Claus, 1981). Selected colonies are checked microscopically for the presence of motile tetrads and/or packets and are restreaked on the same medium.

For the isolation of *S. halophila*, 5-g samples of salt marsh soils are suspended in 20 ml of sterile tap water. The suspensions are heated at 70°C for 10 min. Then 0.1 ml of each of four serial ten-fold dilutions are plated on Difco Marine Agar 2216. The plates are incubated at 30°C for 3 days.

On such plates, mainly spore-forming rods with pigmented colonies of different size and form will develop. On the more crowded plates usually some pinpoint colonies can be detected which, in transmitted light and at a magnification of about × 50, show a coarsely granulated structure at least at the edge of colonies whereas they appear black at a magnification of × 10. Material of such colonies is checked by phase-contrast microscopy for the presence of sarcina aggregates, which usually are motile or may have formed endospores.

For purification, cell material is restreaked on marine agar or on nutrient agar containing 3% NaCl and 0.5% MgCl$_2$. Colonies of *S. halophila* form an orange pigment which normally is not seen with the pinpoint colonies developing on the isolation plates.

Maintenance Procedures

Vegetative cultures of *S. ureae*, grown on nutrient agar slants, and of *S. halophila*, grown on nutrient agar slants supplemented with 3% NaCl and 0.5% MgCl$_2$, usually are viable for about 1 year if tightly closed to avoid drying and stored between 4°C and 10°C in the dark. Sporulated cultures of both species can be kept viable for several years in screw-capped tubes at 4°C up to room temperature.

Freshly isolated *Sporosarcina* strains may readily form spores on different media but may "lose" this property after a few transfers. Gibson (1935) observed rapid and extensive spore formation in *S. ureae* on an agar medium which contained 0.5% peptone, 0.5% meat extract and 0.5% ammonium chloride at a pH of 6.8–7.0 and an incubation temperature of 22°C. Most of the strains of *S. ureae* available at present from culture collections readily form spores on nutrient agar (peptone, 5.0 g; meat extract, 3.0 g; agar, 20 g; distilled water, 1000 ml; after sterilization, 20 ml of a filter-sterilized solution of urea (10% w/v) is added), if incubated below 25°C. The addition of 50 mg MnSO$_4$·H$_2$O/liter of medium may enhance spore formation considerably.

MacDonald and MacDonald (1962) have found good sporulation of *S. ureae* on the following medium: yeast extract, 2.0 g; peptone, 3.0 g; glucose, 4.0 g; malt extract, 3.0 g; K$_2$HPO$_4$, 1.0 g; (NH$_4$)$_2$SO$_4$, 4.0 g; CaCl$_2$, 0.1 g; MgSO$_4$, 0.8 g; MnSO$_4$ · H$_2$O, 0.1 g; FeSO$_4$ · 7 H$_2$O, 0.001 g; ZnSO$_4$, 0.01 g; CuSO$_4$ · 5H$_2$O, 0.01 g; agar, 30.0 g; distilled water, 1000 ml.

All isolates of *S. ureae* studied by Pregerson (1973) showed production of spores at 22°C on MacDonald's spore medium if the pH was adjusted to 8.8–9.0 before autoclaving. She observed, however, considerable variation in the onset and extent of sporulation. Crowded conditions on plates seemed to stimulate spore production.

Upon isolation, all strains of *S. halophila* showed good sporulation on Difco Marine Agar 2216. After purification and few transfers this property was lost by most strains. However, more or less numbers of spores can be obtained with most strains on the sea water agar of Lyman and Fleming (1940), if cultures are incubated at 22°C (Fahmy et al., 1985). The sea water agar has the following composition: Bacto Peptone, 5.0 g; yeast extract, 1.0 g; FePO$_4$ · 4H$_2$O, 0.1 g; MnCl$_2$, 10 mg; agar, 15 g; synthetic sea water, 1000 ml; pH adjusted to 7.1 before autoclaving. Synthetic sea water: distilled water, 1000 ml, NaCl, 24.32 g; MgCl$_2$ · 6H$_2$O, 10.99 g; Na$_2$SO$_4$, 4.06 g; CaCl$_2$ · 2H$_2$O, 1.51 g; KCl, 0.69 g; NaHCO$_3$, 0.20 g; KBr, 0.10 g; SrCl$_2$ · 6H$_2$O, 0.042 g; H$_3$BO$_3$, 0.027 g; Na$_2$SiO$_3$ · 9H$_2$O, 0.005 g; NaF, 0.003 g; NH$_4$NO$_3$, 0.002 g; FePO$_4$ · 4H$_2$O, 0.001 g.

For the long term preservation of *Sporosarcina* strains freeze-drying of vegetative cells or of spores is recommended. As a protective menstruum, skim-milk (20% w/v) or serum containing 5% meso-inositol is recommended. Both vegetative cells and spores can be successfully preserved for long periods in liquid nitrogen without severe loss in survival using glycerol (10%) or dimethylsulfoxide (5%) as cryoprotective agents. Data on the long term maintenance of strains frozen at −70°C to −90°C are not available.

Procedures for Testing Special Characters

The methods described by Gordon et al. (1973) are recommended for members of the genus *Bacillus* and also for the characterization of *Sporosarcina* isolates. The methods are described in the chapter on the genus *Bacillus*. For additional tests listed later in Tables 13.21 and 13.22 see Claus et al. (1983). For *S. halophila* all media have to be supplemented with 3% NaCl and 0.5% MgCl₂.

Morphological observations are best done using slides thinly coated with water agar (purified agar, 2% w/v) and observing preparations under differential-interference-contrast after Nomarski (Fig. 13.90). The formation of division walls at a right angle can also easily detected by using slides coated thinly with 20–30% (w/v) gelatin and phase-contrast microscopy (Fig. 13.92).

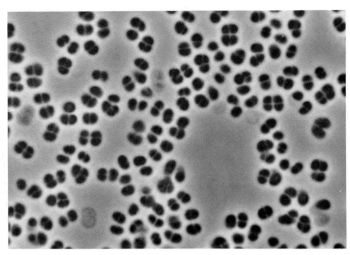

Figure 13.92. *S. halophila*. Cell division in two perpendicular planes. Phase contrast. Slides coated with a thin layer of 30% gelatin (× 2000).

Differentiation of genus **Sporosarcina** from other taxa

Table 13.20 provides the primary characteristics that can be used to differentiate the genus *Sporosarcina* from other endospore-forming or morphologically related taxa.

Taxonomic Comments

The taxonomic position of the genus *Sporosarcina* is still a matter of controversy. Beijerinck (1901) first described a packet and spore-forming, motile organism as "*Planosarcina ureae*." Orla-Jensen (1909) proposed the genus name *Sporosarcina* for the unusual coccus, but this name was equivocal and, therefore, illegitimate. The species was later transferred by Löhnis (1911) to the genus *Sarcina*. This genus, however, represented bacteria that, although morphologically similar, are very different physiologically and biochemically. Kluyver and van Niel (1936), therefore, accepted the proposal of Orla-Jensen (1909) and

circumscribed the genus *Sporosarcina*. This was supported by the studies of Kocur and Martinec (1963) and MacDonald and MacDonald (1962).

In the eighth edition of *Bergey's Manual of Determinative Bacteriology* (Gibson, 1974) the genus *Sporosarcina* has been given generic rank in the *Bacillaceae*. A close relationship of the spore-forming genera *Sporosarcina* and *Bacillus* was first expressed by Beijerinck (1901). His paper indicates the occurrence in soils of organisms being morphologically intermediate between *S. ureae* and *Bacillus megaterium*.

The genetic relationship between *S. ureae* and certain members of the genera *Bacillus* and *Micrococcus* has been studied by Herndon and Bott (1969) by DNA/RNA hybridizations. Although the homology values have been found higher between *S. ureae* and certain strains of *Bacillus* as between *S. ureae* and *Micrococcus*, the authors conclude

Table 13.20.
Differentiation of **Sporosarcina** *from related taxa*[a]

Characteristics	Sporosarcina	Sarcina	Bacillus	Micrococcus	Planococcus
Shape					
Spherical	+	+	−	+	+
Rods	−	−	+	−	−
Arrangement in tetrads or packages	+	+	−	d	d
Motility	+	−	D	−	+
Endospores formed	+	+	+	−	−
Obligate aerobic	+	−	D	+	+
Obligate anaerobic	−	+	−	−	−
Vegetative cell walls contain meso- or LL-diaminopimelate[b]	−	+	D	−	+
Menaquinone system[b]	MK-7	ND	MK-7	MK-8 (H₂) or MK-8 (MK-7), MK-9 (H₂)	MK-8 (MK-7)
DNA mol% G + C	40–41.5[d]	28–31[e]	32–68[e]	66–75[e]	39–52[e]

[a] Symbols: see Table 13.4.
[b] Data from Schleifer and Kandler (1972) and Kandler et al. (1983).
[c] Data from Yamada et al. (1976), Claus et al. (1983) and Collins and Jones (1981).
[d] Data from Bohácek et al. (1968) and Claus et al. (1983).
[e] Data from Buchanan and Gibbons (1974).

that the decision to include *S. ureae* in the family *Bacillaceae* should not be made until a thorough investigation of other representatives of this family has been performed. Further studies on this problem, however, have not been published. They would help to resolve the status of *Sporosarcina* within the *Bacillaceae*.

A close relationship of *S. ureae* to members of the genus *Bacillus* has been demonstrated by Pechman et al. (1976) and Fox et al. (1977) by comparative cataloguing 16S-rRNA. It was found that *S. ureae* is more closely related to *Bacillus pasteurii* than are several other species of *Bacillus*. The authors conclude that *S. ureae* is best classified as a member of the genus *Bacillus* and should be placed in the same subgroup

as is *B. pasteurii*. Phylogenetic relationships between the two *Sporosarcina* species have not been studied.

Studies on the DNA/DNA homology have shown that *S. halophila* is a homogeneous and discrete genospecies (Claus et al., 1983). Similar studies have not been done with strains of *S. ureae*. The appearance of groups differing clearly in growth factor requirements (Pregerson, 1973) may suggest that *S. ureae* is a heterogenous species.

Further Reading

Norris J.R. 1981. *Sporosarcina* and *Sporolactobacillus. In* Berkeley and Goodfellow (Editors), The Aerobic, Endospore-Forming Bacteria. Academic Press, London, pp. 337–357.

Differentiation and characterization of the species of the genus **Sporosarcina**

Differential features of the species *S. ureae* and *S. halophila* are indicated in Table 13.21. Other characteristics of the species are indicated in Table 13.22.

Table 13.21.
Differential characteristics of the species of the genus **Sporosarcina**[a]

	1. *S. ureae*	2. *S. halophila*
Growth in		
Nutrient broth	+	−
Nutrient broth plus 10% NaCl and 0.5% MgCl$_2$	−	+
Casein decomposed	−	+
Gelatin decomposed	−	+
Tyrosin decomposed	+	−
Starch hydrolyzed	−	+
Pullulan hydrolyzed	−	+
Nitrate reduced to nitrite	+	−
Urease	+	−

[a] Symbols: see Table 13.4.

Table 13.22.
Other characteristics of the species of the genus **Sporosarcina**[a]

Characteristics	1. *S. urea*	2. *S. halophila*
Growth in nutrient broth plus 3% NaCl and 0.5% MgCl$_2$	+	+
Growth at		
30°C	+	+
37°C	d	+
pH 5.7	−	−
Anaerobic growth	−	−
Catalase	+	+
Oxidase	+	+
Arginine dihydrolase	d	−
Phenylalanine deaminase	d	−
Phosphatase	−	d
Lecithinase	−	−
Indole formed	−	−
Acetoin formed from glucose	−	−
Citrate utilized	−	−
Cellulose hydrolyzed	−	−
Chitin hydrolyzed	−	−
DNA hydrolyzed	d (weak)	+
Dextran hydrolyzed	−	−
Tween 80 hydrolyzed	d	−
Acid or gas produced from D-glucose, D-mannose, L-arabinose or D-xylose	−	−
Gas produced from nitrate	−	−

[a] Symbols: see Table 13.4.

List of species of the genus **Sporosarcina**

1. **Sporosarcina ureae** (Beijerinck 1901) Kluyver and van Niel 1936, 401.[AL] (*Planosarcina ureae* Beijerinck 1901, 52.)

ure.ae. Gr. n. *urum* urine; M.L. n. *urea* urea; M.L. gen. n. *ureae* of urea.

The cell and colonial morphology are as given for the genus. On germination, the spore swells and acquires the properties of a vegetative cell (Thompson and Leadbetter, 1963). A germination system using calcium dipicolinate has been described by Iandolo and Ordal (1964).

Physiological and biochemical characteristics are presented in Tables 13.21 and 13.22. DNA/DNA hybridization studies have been done with two strains only. The binding between them is 93% (Claus et al., 1983).

The mol% G + C of the DNA of 11 strains is 40.0–41.5 (T_m) (Boháček et al., 1968).

Type strain: ATCC 6473.

2. **Sporosarcina halophila** Claus, Fahmy, Rolf and Tosunoglu, 1984, 270.[VP] (Effective Publication: Claus, et al., 1983, 503.)

ha.lo'phi.la. Gr. n. *halus* salt; Gr. adj. *philus* loving; M.L. fem. adj. *halophila* salt loving.

The cell and colony morphology are as given for the genus. Due to asynchronous or nonperpendicular cross-wall formation irregular cell aggregates may occur also in young cultures. Oval or similar cells more regularly found as in *S. ureae*. On germination the spore swells and acquires the properties of a vegetative cell. In some cases the ruptured spore coat is left visible after germination and outgrowth.

Physiological and biochemical characteristics are presented in Tables 13.21 and 13.22. The species is homogenous according to DNA/DNA hybridization studies. The binding degree of 22 strains is 72–100% (Claus et al., 1983).

The mol% G + C of the DNA of 22 strains is 40.1–40.9 (T$_m$) (Claus et al., 1983).

Type strain: DSM 2266.

GENUS OF UNCERTAIN AFFILIATION

Genus *Oscillospira* Chatton and Pérard 1913, 1159[AL]

THE LATE THOMAS GIBSON

Os.cil.lo.spi′ra. L. n. *oscillum* a swing; L. n. *spira* a spiral; M.L. fem. n. *Oscillospira* the oscillating spiral.

Large rods or filaments 3–6 μm in diameter, divided by closely spaced cross-walls into numerous disk-shaped cells. Reproduction by transverse fission. Motile by means of numerous lateral flagella. Endospores may be formed. Gram-negative.

Growth in pure culture has not been reported.

Exposure to air abolishes motility, thus suggesting the organisms are anaerobic.

Occur in the alimentary tract of herbivorous animals.

Type species (monotype): *Oscillospira guilliermondii* Chatton and Pérard 1913, 1159.

Further Comments

There has been little recent study of this genus and the description by the late Dr. Gibson in the eighth edition of *Bergey's Manual* is reprinted here with minor modifications.

Organisms of this genus have the multicellular structure shown by *Caryophanon* but have been distinguished from that genus by their usually larger size, the production of endospores (a property that is frequently not detected), the probability that they are anaerobes, their occurrence in the alimentary tract of animals and the fact that they have not been grown in pure culture. The reported difference in Gram-reaction may deserve further study. The endospores have not been seen to germinate, and it is not known if they possess all the properties that characterize endospores of other genera.

List of species of the genus *Oscillospira*

1. **Oscillospira guilliermondii** Chatton and Pérard 1913, 1159.[AL]

guil.lier.mon′dii M.L. gen. n. *guilliermondii* of Guilliermond; named for A. Guilliermond, a French biologist.

Large, often curved organism, 3–6 × 10–40 μm in size. Larger or smaller forms may be produced. The rod has rounded ends, and may taper to one pole. Closely spaced cross-walls, formed by diaphragm-like ingrowth from the outer wall, divide the rod into disk-shaped cells not more than 2 μm long.

An endospore, about 2.5 × 4 μm in size which lies longitudinally in the rod and occupies as much space as several disk-shaped cells may be formed. Rarely, there may be two spores in a single rod. The spore is refractile and resists cold stains; its other properties are unknown.

The occurrence of sporulation is variable; the host's diet is thought to be a controlling factor.

The cells frequently contain much polysaccharide which gives a reddish to mauve color with iodine.

Originally described in cecal contents of the guinea pig. Organisms that have the internal structure of *Oscillospira guilliermondii* have been found in the alimentary tract, chiefly in the rumen or caecum, of several species of herbivorous animals. Some differ from the original description of that species in certain morphological features. Moir and Masson (1952) identified as *Oscillospira guilliermondii* an organism in the rumen of sheep which had a spherical, not an elliptical, spore.

Illustrations: Chatton and Pérard, 1913, 1161.

Type strain: has not been cultivated.

SECTION 14

Regular, Nonsporing Gram-Positive Rods

Otto Kandler and Norbert Weiss

Table 14.1

Differential properties of the regular nonsporing rods[a]

Characteristics	Catalase-negative facultative anaerobes		Catalase-positive facultative anaerobes		Aerobes		
	Lactobacillus	*Erysipelothrix*	*Brochothrix*	*Listeria*	*Kurthia*	*Caryophanon*	*Renibacterium*
Cell morphology	Rods, usually straight, sometimes coccobacilli	Slender rods, often filaments	Slender rods, often filaments	Short rods, often short chains & filaments	Regular rods in chains, cocci in old cultures	Short rods in chains	Short rods, often in pairs
Multicellular rods (trichomes)	−	−	−	−	−	+	−
Diameter of rods, microns	0.5–1.1[b]	0.2–0.5	0.6–0.8	0.4–0.5	0.7–0.9	1.4–3.2	0.3–1.0
Motile (if motile, peritrichate flagella)	−[c]	−	−	+[d]	+[e]	+[e]	−
Strictly aerobic	−	−	−	−	+[f]	+	+
Facultatively anaerobic or microaerophilic	+	+	+	+	−	−	−
Catalase reaction	−[g]	−	+	+	+	+	+
Major fermentation products from carbohydrates anaerobically (if NA, acidity from glucose is noted)	Mainly lactate, but may give some acetate, ethanol, CO_2	Lactate	Mainly lactate	Lactate	NA (no acid)	NA (no acid)	NA (no acid)
Peptidoglycan group[h]	A	B	A	A	A	A	A[i]
Peptidoglycan: type of diamino acid[h]	Lys, mDAP, Orn	Lys	mDAP	mDAP	Lys	Lys	Lys[i]
Major fatty acids[j]	S, U, sometimes C	S, A, I, U	S, A, I	S, A, I	S, A, I	ND	A[k,l]
Major menaquinone[l]	None[m]	None	MK-7	MK-7	MK-7	MK-6	MK-9
Habitat	Widespread in fermentable materials, very rarely pathogenic	Widespread pathogen in vertebrates	Meat products, nonpathogenic	Widespread in decaying matter, may be vertebrate pathogen	Feces of farm animals, meat products, nonpathogenic	Cowdung, nonpathogenic	Pathogen in salmonid fish
Mol% G + C of DNA	32–53	36–40	36	36–38	36–38	41–46	53

[a] Symbols: +, 90% or more of strains are positive; −, 90% or more of strains are negative; NA, not applicable; and ND, not determined.

[b] Sometimes up to 1.6 μm.

[c] Rarely motile.

[d] At 20–25°C; poorly motile at 37°C.

[e] Numerous flagella.

[f] Rhizoid colonies.

[g] Some strains weak positive.

[h] Symbolism of Schleifer & Kandler (1972)

[i] Kusser and Fiedler (1983).

[j] S, straight-chain saturated; U, monounsaturated; A, *anteiso*-methyl-branched; I, *iso*-methyl-branched; C, cyclopropane ring fatty acids.

[k] Collins (1982).

[l] Collins and Jones (1981).

[m] *L. mali* contains MK-8 and MK-9; a menaquinone is also found in *L. casei* subsp. *rhamnosus*.

This section comprises a conglomerate of seven very different genera (Table 14.1) which have in common only a few morphological and physiological characteristics. They are all rod-shaped cells, (coccoid to elongated rods, filaments or trichomes) Gram-positive, nonsporing, nonpigmented (slight yellow pigmentation in *Caryophanon*), mesophilic, chemoorganotrophic, and grow only in complex media.

The largest genus, *Lactobacillus*, which is well characterized, with either homo- or heterolactic fermentation, comprises about 50 species, whereas each of the other six genera is monospecific or contains only a few (up to five) species. Most of the genera in the group exhibit unique characteristics which facilitate their differentiation and identification. The genus *Caryophanon* is easily recognized by the formation of trichomes consisting of disk-like cells, 1.5–2.0 μm wide and only 0.5–

1.0 μm long. *Caryophanon* is also well characterized by its habitat, cow feces, where it grows abundantly one to several days after the feces is voided.

The two species of the genus *Kurthia*, also commonly found in feces of farm animals, are recognizable by the characteristic "Medusa-head" appearance of their colonies on yeast extract nutrient agar, and "birds-feather" growth in nutrient gelatin.

Two monospecific genera cause unique diseases. *Erysipelothrix* is well known as the causative organism of swine erysipelas, and *Renibacterium* is an obligate pathogen of the subfamily Salmoninae of the salmon family, causing nephrotic syndromes. Species of the genus *Listeria* (e.g. *L. monocytogenes*) are characteristic pathogens involved in several inflammatory infections (listeriosis) in humans and animals.

Saprophytic species of *Listeria* are wide-spread in soil and decaying matter. They are often isolated from meat and meat products and may thus be confused with species of *Brochothrix* and *Kurthia*, nonpathogenic saprophytes also common in this habitat.

The differentiation of the facultatively anaerobic *Listeria*, containing meso-diaminopimelic acid in its peptidoglycan, from *Kurthia* is relatively easy, because *Kurthia* is strictly aerobic and possesses a lysine-containing peptidoglycan (Table 14.1). *Brochothrix*, however, shares numerous morphological and biochemical characteristics with *Listeria*. Therefore, the differentiation of these two genera is mainly based on differences in motility (Table 14.1) and minor physiological characteristics, e.g. inability of *Brochothrix* to grow at 37°C, pattern of fermented sugars, etc.

Metabolically, the seven genera may be divided into three groups: Group 1 consists of the two fermentative, saccharolytic, microaerophilic genera *Lactobacillus* and *Erysipelothrix*. They do not possess heme-containing catalase, cytochromes or menaquinones and they utilize oxygen only via flavin-containing oxidases and peroxidases.

Group 2 comprises the two aerobic, and facultatively anaerobic genera *Brochothrix* and *Listeria* which possess cofactors and enzymes for respiration. However, these organisms are also able to ferment sugars, mainly to lactic acid, under oxygen-limited or anaerobic conditions.

Group 3 contains the three strictly aerobic genera *Kurthia*, *Caryophanon* and *Renibacterium* which neither utilize glucose as carbon or energy source nor ferment sugars to organic acids.

These groupings have only limited taxonomic value as indicated by the low correlation with nonmetabolic characteristics. In fact, four genera, *Brochothrix* (formerly *Microbacterium thermosphactum*), *Listeria*, *Kurthia* and *Erysipelothrix*, have often been associated with the *Corynebacteriaceae* or at least with the coryneform group (*Bergey* 7). However, numerical taxonomic and chemotaxonomic studies have not supported this affiliation. Such studies rather suggest a remote relationship between coryneform organisms and the lactic acid bacteria (Wilkinson and Jones 1977). The presence of respiratory cofactors and enzymes in *Listeria*, *Brochothrix* and *Kurthia* is not in keeping with their inclusion within an enlarged family *Lactobacillaceae* (Collins et al., 1979). However the genera *Brochothrix*, *Listeria*, *Lactobacillus* and *Erysipelothrix* are close phenetically to each other and to *Streptococcus* and *Gemella* (Wilkinson and Jones, 1977).

The G + C ratios of the DNA of six of the seven genera fall within a range around 40 mol% (*Lactobacillus fermentum* is 50 mol%), whereas with *Renibacterium* 53 mol% is found. Comparative studies on the sequence homology of 16S-rRNA oligonucleotides in a large number of bacteria of different taxonomic affiliation showed that all the Gram-positive bacteria possessing a G + C content lower than about 55 mol% belong to the so-called *Clostridium-Lactobacillus-Bacillus* branch (Fox et al., 1980, Stackebrandt and Woese, 1981). In fact, detailed studies on the 16S-rRNA oligonucleotides of representatives of the six genera showed that they fit into this branch.

From the rRNA evidence the lactobacilli and streptococci together with the pediococci and leuconostocs are close to the genera *Bacillus*, *Brochothrix*, *Listeria*, *Staphylococcus*, *Gemella* and *Kurthia*. This position reflects the metabolism of the lactic acid bacteria, which is intermediate between aerobic and anaerobic metabolism. *Listeria* and *Brochothrix* are closely related to one another, and together with *Staphylococcus* are closer to *Bacillus* than are the lactic acid bacteria.

On the basis of 16S-rRNA cataloging, *Erysipelothrix* is related to the mycoplasmas, exhibiting nonisochronic evolution of their 16S-rRNA sequences (Ludwig et al., 1984). Thus, *Erysipelothrix* represents one of the many different lines of nonrespiratory organisms emerging from the *Clostridium* cluster. One of these lines comprises *Eubacterium limosum*, *Acetobacterium woodii* and *Clostridium barkeri* (Fox et al., 1980) which, in addition to other common properties, are characterized by the same very unusual peptidoglycan types of the cross-linking group B (Schleifer and Kandler, 1972), also found in *Erysipelothrix*. Position one of the peptide subunits of these group B peptidoglycan types is taken by a L-seryl residue, not by L-alanyl as in group A peptidoglycans, or by a glycyl residue as in those group B peptidoglycans occuring only in a certain section of the coryneform bacteria (*Arthrobacter*, *Clavibacter*, *Curtobacterium*, *Microbacterium*). Thus, comparative peptidoglycan chemistry corroborates the affiliation of *Erysipelothrix* with the *Clostridium* cluster which also harbors the *Eubacterium limosum* line, while the above mentioned group B peptidoglycan-containing coryneform genera belong to the *Actinomycetes* branch.

No data on 16S-rRNA cataloging are available for *Caryophanon* and *Renibacterium*. The low G + C content of the DNA of *Caryophanon* suggests an affiliation also with the *Clostridium-Lactobacillus-Bacillus* branch. However, the fairly high G + C content of 53 mol% in *Renibacterium* falls within the overlapping zone of the *Clostridium* and the *Actinomycetes* branch. Thus *Renibacterium* could be allotted to either branch. Phenetically, *Renibacterium* resembles the genus *Arthrobacter* in morphology, "Chinese letter" formation, slightly yellow pigmentation of colonies, aerobic metabolism, the presence of MK-9 instead of MK-7 menoquinone, and in its unusual peptidoglycan type, containing D-alaninamide (Kusser and Fiedler, 1983) found so far in only one other organism, *Arthrobacter* sp. NCIB 9423 (Fiedler et al., 1973). Therefore, *Renibacterium* is tentatively included in the Actinomycetes branch in Fig. 14.1 (see *Lactobacillus*). Final affiliation will only be possible on the basis of 16S-rRNA analysis or other sequence and chemotaxonomic or numerical taxonomic data.

In conclusion, the seven genera discussed in this section certainly do not belong to the same family. However, with the exception of *Renibacterium*, they may at present be allotted to the same order or superorder, in the event that the whole *Clostridium-Lactobacillus-Bacillus* branch may finally be recognized as a taxon at such a rank.

Genus *Lactobacillus* Beijerinck 1901, 212[AL]

OTTO KANDLER AND NORBERT WEISS

Lac.to.ba.cil′lus. L. n. *lac*, *lactis* milk; L. dim. n. *bacillus* a small rod; M.L. n. *Lactobacillus* milk rodlet.

Cells, varying from **long and slender, sometimes bent rods** to **short, often coryneform coccobacilli**; chain formation common. **Motility** uncommon; when present, by peritrichous flagella. **Nonsporing. Gram-positive.** Some strains exhibit bipolar bodies, internal granulations or a barred appearance with the Gram-reaction or methylene blue stain.

Metabolism fermentative; obligately saccharoclastic. At least half of **end product** carbon is **lactate**. Lactate is usually not fermented. Additional products may be acetate, ethanol, CO_2, formate or succinate. Volatile acids with more than two carbon atoms are not produced.

Microaerophilic; surface growth on solid media generally enhanced by anaerobiosis or reduced oxygen pressure and 5–10% CO_2; some are anaerobes on isolation.

Nitrate reduction highly unusual; if present, only when terminal pH is poised above 6.0. **Gelatin not liquefied.** Casein not digested but small amounts of soluble nitrogen produced by most strains. Indole and H_2S not produced.

Catalase and cytochrome negative (porphyrins absent); however, a few strains decompose peroxide by a pseudocatalase; benzidine reaction negative.

Pigment production rare; if present, yellow or orange-to-rust or brick red.

Complex nutritional requirements for amino acids, peptides, nucleic acid derivatives, vitamins, salts, fatty acids or fatty acid esters and fermentable carbohydrates. Nutritional requirements are generally characteristic for each species, often for particular strains only.

Growth temperature range 2–53°C; optimum generally 30–40°C.

Aciduric, optimal pH usually 5.5–6.2; growth generally occurs at 5.0 or less; the growth rate is often reduced at neutral or initially alkaline reactions.

Found in dairy products, grain products, meat and fish products, water, sewage, beer, wine, fruits and fruit juices, pickled vegetables, sauerkraut, silage, sour dough, and mash; they are a part of the normal flora in the mouth, intestinal tract and vagina of many homothermic animals including man. Pathogenicity is rare.

The **mol% G + C of the DNA** ranges from **32–53** (Bd, T_m).

Type species: *Lactobacillus delbrueckii* (Leichmann 1896) Beijerinck 1901, 229.

Further Descriptive Information

Cell morphology. The variability of lactobacilli from long, straight or slightly crescent rods to coryneform coccobacilli is depicted in Figure 14.1. The length of the rods and the degree of curvature is dependent on the age of the culture, the composition of the medium—e.g. availability of oleic acid esters (Jacques et al., 1980)—and the oxygen tension. However, the main morphological differences between the species usually remain clearly recognizable. Some species of the gas-producing lactobacilli (e.g. *L. fermentum*, *L. brevis*) always exhibit a mixture of long and short rods (Fig. 14.1E).

Coccobacilli may become so short that they may be confused with

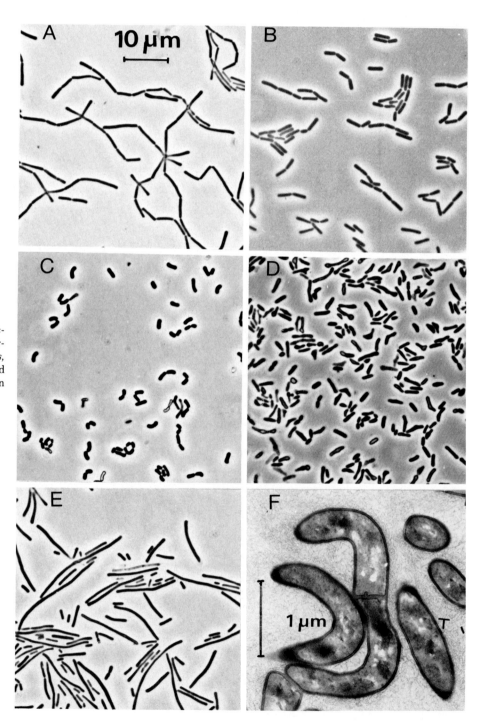

Figure 14.1. Phase contrast (*A–E*) and electron micrographs (*F*) showing different cell morphology of lactobacilli (*A, L. gasseri; B, L. agilis, C, L. curvatus; D, L. minor, E, L. fermentum;* and *F,* involution form of lactobacilli in a thin section of a kefir grain).

either *Leuconostoc* (e.g. *L. confusus*, originally considered as *Leuconostoc*) or streptococci. On the other hand, elongated streptococci have repeatedly been ascribed to the genus *Lactobacillus*, e.g. *L. xylosus* and "*L. hordniae*," recently found to belong to the genus *Streptococcus* (Garvie et al., 1981; Kilpper-Bälz et al., 1982). Cell division occurs only in one plane. The tendency towards chain formation varies between species and even strains. It depends on the growth phase and the pH of the medium (Rhee and Pack, 1980). The asymmetrical development of cells during cell division in coryneform lactobacilli (Fig. 14.2) leads to wrinkled chains or even ring formation. Irregular involution forms may be observed under symbiotic growth, e.g. in kefir grains (Fig. 14.1*F*) or under the influence of high concentrations of glycine, D-amino acids or cell wall-active antibiotics (Hammes et al., 1973; Schleifer et al., 1976). Motility by peritrichous flagellation is observed in only a few species. It is highly dependent on the medium and the age of the culture and is sometimes observed only during isolation, but lost after several transfers on artificial media.

All lactobacilli stain clearly Gram-positive. Only dying cells may give variable results. Internal granulation is often revealed by Gram or methylene blue stain especially in the homofermentative long rods. The large bipolar bodies probably contain polyphosphate and appear very electron-dense in electron microscopy.

Cell wall and fine structure. Electron micrographs of thin sections reveal a typical Gram-positive cell wall profile (Figs. 14.2 and 14.3). The cell wall contains peptidoglycan (murein) of various chemotypes of the cross-linkage group A. The Lys-D-Asp type is the most widespread peptidoglycan type (Schleifer and Kandler, 1972). The cell wall contains also polysaccharides attached to peptidoglycan by phosphodiester bonds (Knox and Hall, 1964). Membrane-bound teichoic acid is present in all species (Archibald and Baddiley, 1966), cell wall-bound teichoic acid only in some of the species (Knox and Wicken, 1973). Extracellular slime in large amounts is produced from sucrose by *L. confusus* and particular strains of some other heterofermentative species (Sharpe et al., 1972). Slime-forming strains of *L. delbrueckii* subsp. *bulgaricus* and *L. casei* are employed for the production of special sour milks.

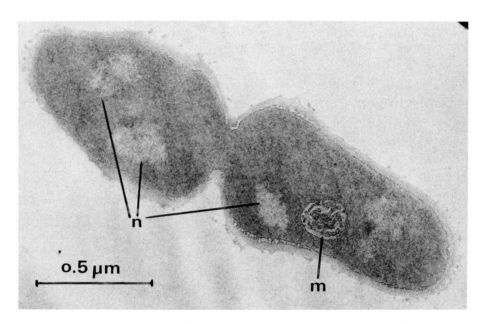

Figure 14.2. Electron micrograph of a dividing cell of *L. coryniformis* showing asymmetric growth (*m*, mesosome; *n*, nucleoid).

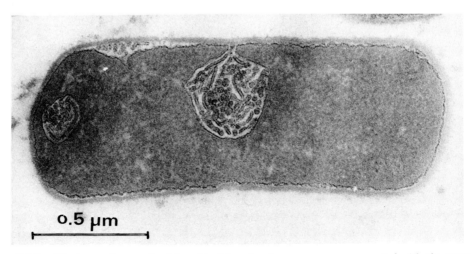

Figure 14.3. Electron micrograph of *L. acidophilus* showing a mesosome connected with the cytoplasmic membrane.

In addition to nucleoids and ribosomes typical of all procaryotes, electron micrographs of thin sections frequently show large mesosomes (Fig. 14.3). They are formed by invaginations of the cytoplasmic membrane and are filled with tubuli, probably derived from secondary membrane invaginations (Schötz et al., 1965; Sriranganathan et al., 1973).

Colony and cultural characteristics. Colonies on agar media are usually small (2–5 mm), with entire margins, convex, smooth, glistening, and opaque without pigment. In rare cases they are yellowish or reddish. Some species form rough colonies. Distinctly slimy colonies are only formed by *L. confusus.* Clearing zones caused by exoenzymes are usually not observed when grown on agar media containing dispersed protein or fat. However, most strains exhibit slight proteolytic activity due to cell wall-bound or cell wall-released proteases and peptidases (Law and Kolstad, 1983) and a weak lipolytic activity due to predominantly intracellular lipases. Distinct starch degradation leading to clearing zones on starch plates is only observed in a few species (e.g. *L. amylophilus, L. amylovorans*). Growth in liquid media generally occurs throughout the liquid, but the cells settle soon after growth ceases. The sediment is smooth and homogeneous, rarely granular or slimy. Pellicles are never formed.

Lactobacilli do not develop characteristic odors when grown in common media. However, they contribute to the flavor of fermented food by producing various volatile compounds, such as diacetyl and its derivatives, and even H_2S and amines in cheese (Sharpe and Franklin, 1962; Law and Kolstad, 1983).

Nutrition and growth conditions. Lactobacilli are extremely fastidious organisms, adapted to complex organic substrates. They require not only carbohydrates as energy and carbon source, but also nucleotides, amino acids and vitamins. While pantothenic acid and nicotinic acid are—with the exception of a few strains—required by all species, thiamine is only necessary for the growth of the heterofermentative lactobacilli. The requirement for folic acid, riboflavin, pyridoxal phosphate and *p*-aminobenzoic acid is scattered among the various species, riboflavin being the most frequently required compound. Biotin and B_{12} are required by only a few strains. Although the pattern of vitamin heterotrophy is considered to be characteristic of particular species (Rogosa et al., 1961), deviating strains are common (Abo-Elnaga and Kandler, 1965c; Ledesma et al., 1977). Vitamin-dependent strains are commonly in use for bioassays of vitamins and are listed in the catalogues of most culture collections. The pattern of the amino acid requirement also differs among species and even strains. By sequential mutagenesis, Morishita et al. (1974) were able to obtain quintuple mutants of *L. casei* which had lost their requirement for 5 amino acids. However, the mutants grew significantly slower and reverted frequently to their amino acid-dependent state when transferred back to the complete medium. Corresponding results were also obtained with four other species (Morishita et al., 1981).

These studies show that many—if not all—of the nutritional requirements of lactobacilli are the result of numerous minor defects within the genome, and that much of the information coding for the various biosynthetic pathways is still present in the chromosome. Thus, the multiple nutritional requirements of present-day lactobacilli reflect the stepwise natural selection of deficient mutants out of a chemoautotrophic population with a complement of biosynthetic pathways.

The various requirements for essential nutrients are normally met when the media contain fermentable carbohydrate, peptone, meat and yeast extract. Supplementations with tomato juice, manganese, acetate and oleic acid esters, especially Tween 80, are stimulatory or even essential for most species. Therefore, these compounds are included in the widely used MRS medium (De Man et al., 1960). Lactobacilli adapted to very particular substrates may require special growth factors. For instance D-mevalonic acid is necessary for rice wine (sake) spoilage organisms (Tamura, 1956) and a small peptide isolated from freshly prepared yeast extract was found to be required for luxurious growth of *L. sanfrancisco* (Berg et al., 1981), the sour dough organism. To meet the requirement of a still unknown growth factor, some of the original substrate must be added.

Lactobacilli grow best in slightly acidic media with an initial pH of 6.4–4.5. Growth ceases when pH 4.0–3.6 is reached, depending on the species and strain. Although most strains are fairly aerotolerant, optimal growth is achieved under microaerophilic or anaerobic conditions. Increased CO_2 concentration (~5%) may stimulate growth.

Most lactobacilli grow best at mesophilic temperatures with an upper limit of around 40°C. Some also grow below 15°C and some strains even below 5°C. The so-called "thermophilic" lactobacilli may have an upper limit of 55°C and do not grow below 15°C. Really thermophilic lactobacilli growing above 55°C are as yet unknown.

Metabolism. Metabolically, lactobacilli are at the threshold of anaerobic-to-aerobic life. They possess efficient carbohydrate fermentation pathways coupled to substrate level phosphorylation. A second substrate level phosphorylation site is the conversion of carbamyl phosphate to CO_2 and NH_3, the final step of arginine "fermentation," observed in most of the heterofermentative lactobacilli (cf. Abdelal, 1979). However, only some of the species forming NH_3 from arginine are able to grow on arginine as the only energy source. In addition to substrate-level phosphorylation, energy may be gained by the proton motive force generated by lactate efflux (Konings and Otto, 1983). Lactobacilli contain no isoprenoid quinones—except *L. yamanashiensis* and *L. casei* subsp. *rhamnosus* (Collins and Jones, 1981)—and no cytochrome systems to perform oxidative phosphorylation. However, they possess flavine-containing oxidases and peroxidases to carry out the oxidation of $NADH_2$ and O_2 as the final electron acceptor. They are also able to perform a manganese-catalyzed scavanging of superoxide (Götz et al., 1980; Archibald and Fridovich, 1981), although they do not possess superoxide dismutase and catalase.

The main fermentation pathways for hexoses are the Embden-Meyerhof pathway converting 1 mol of hexose to 2 mol of lactic acid (homolactic fermentation) and the 6-phosphogluconate pathway, resulting in 1 mol CO_2, 1 mol ethanol (or acetic acid) and 1 mol lactic acid (heterolactic fermentation). Under aerobic conditions, most strains are able to reoxidize $NADH_2$ with oxygen serving as the final electron acceptor, thus acetyl-CoA is not, or at least not completely, reduced to ethanol. Consequently, additional ATP is formed by substrate-level phosphorylation and varying ratios of acetic acid and ethanol are found, depending on the oxygen supply.

Pyruvate, intermediately formed in both pathways, may partly undergo several alternative conversions, yielding either the well-known aroma compound diacetyl and its derivatives, or acetic acid (ethanol); with hexose limitation, the latter pathway may become dominant and the homolactic fermentation may be changed to a heterofermentation with acetic acid, ethanol and formic acids as the main products. (DeVries et al., 1970; Thomas et al., 1979). Even lactate may partially be oxidized and broken down to acetic acid and formate or CO_2 by various little known mechanisms (cf. Kandler, 1983). The conversion of glycerol to 1,3-propanediol with glucose serving as electron donor is a peculiar metabolic activity observed in *L. brevis* isolated from wine (Schütz and Radler, 1984).

At the enzyme level, homo- and heterofermentative lactobacilli differ with respect to the presence or absence of FDP aldolase or phosphoketolase. Whereas the heterofermentative lactobacilli possess phosphoketolase but no aldolase, the obligate homofermentative ones possess FDP aldolase but no phosphoketolase. They are thus unable to ferment any of the pentoses, which are broken down by the heterofermenters via phosphoketolase, yielding equimolar amounts of lactic acid and acetic acid. However, one group of homofermentative lactobacilli, traditionally called "Streptobacteria" (Orla Jensen, 1919), possess an inducible phosphoketolase with pentoses acting as inducers. They are thus able to ferment pentoses upon adaptation to lactic acid and acetic acid, while hexoses are homofermentatively metabolized. Therefore, these lactobacilli must be called facultative heterofermenters (group II; see below). In rare cases, a homolactic fermentation of pentoses may be performed by lactic acid bacteria, as observed in some streptococci by Fukui et al. (1957) and in a so-far undescribed lactobacillus (Barre, 1978). Such fermentations may involve the transformation of pentoses to hexoses via transaldolase and transketolase reactions followed by glycolysis (Kandler, 1983) with lactic acid being the only fermentation product.

Carbohydrates may also contribute to other reactions: sucrose is not only a substrate for fermentation, but also for the formation of dextrans (slime) with the help of dextran sucrases, found in only a few species or strains. Fructose serves not only as a substrate for fermentation, but also as an electron acceptor and becomes reduced to mannitol by most heterofermentative lactobacilli. Correspondingly, glycerol is formed from triosephosphate and excreted into the medium by some heterofermentative strains.

The majority of saccharides and oligosaccharides are taken up with the help of specific permeases and are phosphorylated inside the cell. Oligosaccharides are split by the respective glycosidases prior to the phosphorylation of the resulting monosaccharides. However, at least lactose and galactose are taken up by some lactobacilli via the phosphoenolpyruvate-dependent phosphotransferase system (Chassy and Thompson, 1983). The lactose phosphate formed is split to glucose and D-galactose-6-phosphate. The latter is then metabolized via the D-tagatose-6-phosphate pathway (cf. Kandler, 1983). Little is known of the distribution of the various saccharide uptake mechanisms in the species of the genus *Lactobacillus*, although the presence or absence of such mechanisms determines the pattern of fermented sugars, an important characteristic for identification. Active transport of amino acids and peptides is also known (cf. Law and Kolstad, 1983). However, more information is available on streptococci than on lactobacilli.

Several organic acids, such as citric, tartaric and malic acids, are degraded via oxaloacetic acid and pyruvate to CO_2 and lactic or acetic acid (cf. Radler, 1975; Whiting, 1975). Detailed studies on the catalytic and regulatory properties of the DNA-dependent malic enzymes of lactic acid bacteria were performed by London et al. (1971). Alternatively, malic acid is split to CO_2 and L(+)-lactic acid in many lactobacilli by a multifunctional so-called "malolactic enzyme" with all intermediates remaining tightly bound to the enzyme complex (Radler, 1975).

Several amino acids are decarboxylated by lactobacilli, e.g. glutamic acid and tyrosine, but the decarboxylation product is not further metabolized (Blood, 1975).

Clorogenic acid is hydrolyzed and the resulting quinic acid is reduced to (−)-dehydroshikimic acid by heterofermentative lactobacilli. It is further reduced to dihydroxycyclohexane-1^c-carboxylic acid by homofermenters. Shikimic acid may be reduced to catechol by *L. plantarum* which also converts *p*-coumaric acid to *p*-ethylphenol. The electron source of these reactions is lactate which becomes oxidized to CO_2 and acetic acid (Cf. Whiting, 1975).

The lactic acid formed by the various fermentation pathways possesses either the L- or the D-configuration depending on the stereospecificity of the lactated dehydrogenase present in the cells. Racemate may be formed when both L- and D-lactate dehydrogenase are present in the same cell, or in rare cases, by the action of an inducible lactate racemase in combination with a constitutive L-lactate dehydrogenase (Stetter and Kandler, 1973). Lactate dehydrogenases of the various species often differ from each other considerably, e.g. with respect to their electrophoretic mobility and their kinetic properties. Most enzymes are nonallosteric but some species contain allosteric L-lactate dehydrogenases with FDP and Mn^{2+} acting as effectors (Hensel et al., 1977; cf. Garvie, 1980).

Mutagenesis. Spontaneous and induced mutants of lactobacilli are frequently selected to obtain strains exhibiting characters useful for biochemical studies or biotechnological application. The well-known mutagens *N*-methyl-*N'*-nitro-*N*-nitrosoguanidine, ethylmethane sulfonate and ultraviolet (UV) light have been applied successfully (Morishita et al., 1981).

Plasmids. No lactobacillus strain is known to be transformable or transducible and genetic engineering via recombinant DNA cannot be done at present in lactobacilli. However, plasmids are frequently found (Smiley and Fryder, 1978; Vescovo et al., 1981). They are often linked with drug resistance (Ishiwa and Iwata, 1980; Vescovo et al., 1982) or lactose metabolism (Chassy et al., 1976). The conjugal self-transmission of a plasmid that determines lactose metabolism in *L. casei* is the only

known naturally occurring genetic exchange in the genus (Chassy and Rokow, 1981). Extensive research, including cloning in *Escherichia coli*, is proceeding with the plasmids coding lactose metabolism (Chassy et al., 1983) in order to make the lactobacilli accessible to genetic engineering.

Phages. *Lactobacillus* phages causing slower acidification in food fermentation deserve much interest because of their commercial importance (cf. Sharpe, 1981). The morphology of numerous double-stranded DNA phages virulent to many species has been described. Physiochemical parameters of seven phages are known, the data being summarized by Sozzi et al. (1981) who grouped the lactobacillus phages in accordance with the system of Bradley (1967) and Ackermann (1974). With the exception of one tailless phage from *L. plantarum*, all phages belong to group A or B and possess hexagonal heads and long contractile or noncontractile tails. They are basically similar to phages against other groups of bacteria.

Lysogeny is widespread within the genus. Yokokura et al. (1974) found that 40 strains belonging to seven species, out of a total of 148 strains belonging to 15 different species were lysed with mitomycin C. Thirty-one of 40 lysates showed phage-like particles by electron microscopy. Some of these particles produced plaques while others were defective phages, unable to produce plaques. Stetter (1977) found that 17 out of 21 strains of streptobacteria were lysogenic when induced with mitomycin C. Two of these phages were homoimmune with the *L. casei* phage PL1, which showed a surprisingly narrow host range (Stetter et al., 1978). Thus, it is suggested that frequent lysogeny caused by homoimmune phages may be responsible for the very narrow host ranges of lactobacillus phages. It may also explain why attempts to initiate phage typing schemes were not successful (Coetzee et al., 1960).

Bacteriocins. Bacteriocinogenic strains have been found among homo- and heterofermentative species (cf. Tagg et al., 1976; cf. Konisky, 1978). Early papers on bacteriocins, especially those from *L. acidophilus*, reported a very broad activity spectrum. Thus it is questionable whether these substances represent true bacteriocins (Barefoot and Klaenhammer, 1983). Lactocin B, a well-defined bacteriocin recently isolated from *L. acidophilus*, has a very narrow activity spectrum, restricted to only a few homofermentative species related to *L. acidophilus* (Barefoot and Klaenhammer, 1983). Also, the bacteriocins isolated from *L. fermentum* (DeKlerk and Smit, 1967) and lactocin LP27 from *L. helveticus* (Upreti and Hinsdill, 1973, 1975) are only active against lactobacilli. Bacteriocin typing of a large number of strains (Filippov, 1976a, b; Filippov and Rubanenko, 1977) showed a fairly wide range of sensitive species on the one hand, but also led to a subdivision of many species into various types. This indicates that bacteriocin typing may be more useful to characterize specific strains rather than to identify species.

Antigenic structure. Many strains of lactobacilli can be assigned to seven serological groups based on specific antigenic determinants (cf. Sharpe, 1970, 1981, Table 14.2). Groups A, D, F and G are specific for *L. helveticus, L. plantarum, L. fermentum* and *L. salivarius*, respectively. A few strains belonging to *L. plantarum* according to phenotypical characteristics could not be assigned to group D. They do not contain ribitol teichoic acid, the typical D antigen, but an unusual glycerol teichoic acid (Adams et al., 1969; Archibald and Coapes, 1971). The chemical nature of the antigen of group G, an acid released polysaccharide with rhamnose as determinant, was recently studied by Knox et al. (1980).

Most strains of *L. casei* belong either to group B or C. However, strains of *L. casei* subsp. *rhamnosus* belong exclusively to group C. They possess a capsular, rhamnose-containing typing antigen. Its quantity is dependent on the cultural conditions (Wicken et al., 1983).

The homofermentative species *L. delbrueckii* and the two heterofermentative species *L. brevis* and *L. buchneri* belong to group E. The common antigen of these taxonomically distant species is a cell wall glycerol teichoic acid (Knox and Wicken, 1973, 1976).

An alternative serological nomenclature was proposed by Shimohashi and Mutai (1977). However, their scheme is based on a complex array of chemically undefined components and has thus no advantage over

Table 14.2.

Group antigens of lactobacilli[a]

Species	Group	Antigen	Location	Determinant
L. helveticus	A	GTA	Wall membrane	α-Glc
L. casei	B	Polysaccharide	Wall	α-Rha
L. casei	C	Polysaccharide	Wall	α-Glc
L. plantarum	D	RTA	Wall	α-Glc
L. delbrueckii subsp. *lactis* subsp. *bulgaricus*	E	GTA	Wall	
L. brevis	E	GTA	Wall	
L. buchneri	E	GTA	Wall	
L. fermentum	F	GTA	Membrane	α-Gal
L. salivarius	G	Polysaccharide	Wall	Rha

[a] Symbols: GTA, glycerol teichoic acid; RTA, ribitol teichoic acid; Glc, D-glucosyl; Rha, L-rhamnosyl; Gal, D-galactosyl. (After Sharpe, 1981.)

the nomenclature developed by Sharpe (1955) which is used in Table 14.2.

Antibiotic and drug sensitivity. Lactobacilli are sensitive toward most antibiotics active against Gram-positive bacteria (Sutter and Finegold, 1976). *L. delbrueckii* subsp. *bulgaricus* is often used to detect antibiotics in milk.

Studies on the sensitivity or resistance pattern of lactobacilli towards antibiotics originated mainly from problems created by the presence of antibiotics in milk derived from mastitis therapy (Marth and Ellickson, 1974; Sozzi and Smiley, 1980).

The sensitivity of intestinal lactobacilli toward antibiotics employed as feed additives has also been studied (Dutta and Devriese, 1981). Bile resistance was thought to be important for colonizing the intestine with lactobacilli. Therefore it was mainly studied in *L. acidophilus* (Klaenhammer and Kleeman, 1981).

Production of antibiotic substances by lactobacilli has repeatedly been claimed (Schrøder et al., 1980; Lindgren and Clevström, 1978a, b; DeKlerk and Coetzee, 1961). However, frequently there is no clear distinction between an antibiotic effect and the inhibition effects of lactic acid and/or H_2O_2 produced by the organism. No defined and commercially used antibiotic from lactobacilli is yet known.

Pathogenicity. Apart from dental caries (Rogosa et al., 1953), lactobacilli are generally considered to be apathogenic. However, there is an increasing number of reports that lactobacilli have been involved in human diseases (Sharpe et al., 1973a; Berger, 1974; Bayer et al., 1978; Bourne et al., 1978). Mainly *L. casei* subsp. *rhamnosus*, but also *L. acidophilus*, *L. plantarum* and occasionally *L. salivarius* have been found to be associated with subacute bacterial endocarditis, systemic septicemia and abscesses. In a recent study, a homofermentative lactobacillus was the only organism isolated in pure culture from a case of chorioamnionitis (Lorenz et al., 1982), and *L. gasseri* was found in a case of urosepsis (Dickgiesser et al., 1984). The many cases in which lactobacilli have been isolated from diseased tissue indicate their potential pathogenicity. However, the biochemical basis of such pathogenicity is as yet unknown. The finding that some rumen lactobacilli decarboxylate indoleacetic acid to skatol, a compound known to be responsible for acute bovine pulmonary emphysema, the naturally occurring form of the bovine respiratory disease (Yokoyama and Carlson, 1981), may be a first positive step in elucidating the pathogenicity of lactobacilli.

Ecology, habitats and biotechnology. Lactobacilli grow under anaerobic conditions or at least under reduced oxygen tension in all habitats providing ample carbohydrates, breakdown products of protein and nucleic acids, and vitamins. A mesophilic to slightly thermophilic temperature range is favorable. However, strains of some species (e.g.

L. viridescens, L. sake, L. curvatus, L. plantarum) grow—although slowly—even at low temperatures close to freezing point (e.g. refrigerated meat (Kitchell and Shaw, 1975), fish (Schröder et al., 1980). Lactobacilli are generally aciduric or acidophilic. They decrease the pH of their substrate by lactic acid formation to below 4.0, thus preventing, or at least severely delaying, growth of virtually all other competitors except other lactic acid bacteria and yeasts. These properties make lactobacilli valuable inhabitants of the intestinal tract of man and animals and important contributors to food technology.

Several individual species have adapted to specific ecological niches and are generally not found outside their specialized habitats. The relative ease with which such species can be reisolated from their respective sources since their first discovery, sometimes almost 100 years ago, indicates that these niches are, in fact, their natural habitats.

Plant sources. Lactobacilli occur in nature in low numbers at all plant surfaces (Keddie, 1959; Mundt and Hammer, 1968) and together with other lactic acid bacteria grow luxuriously in all decaying plant material, especially decaying fruits. Hence, lactobacilli are important for the production as well as the spoilage of fermented vegetable feed and food (e.g. silage, sauerkraut, mixed pickles) and beverages (e.g. beer, wine, juices). Species chiefly isolated have been: *L. plantarum, L. brevis, L. coryniformis, L. casei, L. curvatus, L. sake, L. fermentum* (cf. Carr et al., 1975; Sharpe, 1981; Steinkraus, 1983; Kandler, 1984).

Several species are typical of specific products. Thus, *L. delbrueckii* subsp. *delbrueckii* exhibiting a very narrow range of fermented carbohydrates is the characteristic thermophilic organism found in potato and grain mashes fermented at 40–55°C (Henneberg, 1903).

It is also employed in the fermentation of millet mash to produce Bantu beer (Novellie, 1968), and is used for industrial production of lactic acid from molasses (Buchta, 1983).

Another specifically adapted species is *L. sanfrancisco*, the dominant acid producer in Californian sour dough (Kline and Sugihara, 1977). The organism isolated from European sour dough, designated *L. brevis* var. "*lindneri*" by Spicher and Schroeder (1978) also proved to belong to the species *L. sanfrancisco* (Weiss, Schillinger and Spicher, personal communication). An organism specifically used for the production of mainly L(+)-lactic acid-containing sauerkraut is *L. bavaricus* (Stetter, 1974). *L. hilgardii* and *L. fructivorans* (Fornachon et al., 1949) are typical organisms of acidic and alcoholic beverages; *L. collinoides* (Carr and Davis, 1972) and *L. yamanashiensis* (Carr and Davis, 1970; Carr et al., 1977) are found in cider and other fruit juices.

Although many different species of lactobacilli have been found in spoiled beer (Rainbow, 1975; Kirsop and Dolezil, 1975), a very important lactobacillus in beer spoilage is probably "*L. lindneri*", a so far incompletely described species which requires the addition of beer to the medium for detection and isolation (Back, 1981). Among the slime-forming spoilage organisms in sugar factories (Tilbury, 1975), *L. confusus* is the most common species of lactobacilli (Sharpe et al., 1972), growing in sucrose concentrations up to 15%.

Milk and dairy products. Milk contains no lactobacilli when it leaves the udder, but becomes very easily contaminated with lactobacilli by dust, dairy utensils, etc. Since streptococci grow faster, the number of lactobacilli remains usually fairly low even in spontaneously soured milk. Only after prolonged incubation do lactobacilli take over, due to their higher acid tolerance. In sour whey, the most acid tolerant, and thus typical species, which produces as much as 3% lactic acid, is *L. helveticus*. It is traditionally used in starters for the production of Swiss cheese and other types of hard cheeses, e.g. Grana, Gorgonzola and Parmesan (Bottazzi et al., 1973). Nowadays *L. delbrueckii* subsp. *bulgaricus* or subsp. *lactis* are also used (Biede et al., 1976; Auclair and Accolas, 1983). In all types of cheese with ripening periods longer than about 14 days, several mesophilic lactobacilli (*L. plantarum, L. brevis, L. casei*, etc.) originating from the milk or the dairy environment, reach levels as high as 10^6–10^8/g cheese (Sharpe, 1962; Abo Elnaga and Kandler, 1965a; van Kerken and Kandler, 1966).

Very specifically adapted lactobacilli for the production of sour milks are *L. delbrueckii* subsp. *bulgaricus*, a component of the well-known yoghurt flora (Davis, 1975), and *L. kefir* (Kandler and Kunath, 1983), the heterofermentative component of the Caucasian sour milk kefir. These two sour milks are the only known habitats of these two lactobacilli.

Although several species of lactobacilli may contribute to spoilage of dairy products by slime or gas production, only two species cause specific spoilage. *L. maltaromicus* may be responsible for malty flavor in milk (Miller et al., 1974) and *L. bifermentans* has been found to cause the blowing of Edam cheese (Pette and Van Beynum, 1943).

Meat and meat products. Lactobacilli play an important role in the curing process of fermented sausages containing added sucrose. The most common naturally occurring species found in ripening raw sausages are *L. plantarum, L. brevis, L. farciminis, L. alimentarius* and "atypical" lactobacilli (Reuter, 1970, 1975) recently identified as *L. sake* and *L. curvatus* (Kagermeier, 1981; Kagermeier et al., 1985). In addition to streptococci, pediococci and micrococci/staphylococci, starters added to the meat mix often contain *L. plantarum* (Bacus and Brown, 1981; Robinson, 1983; Liepe, 1983).

Various species of lactobacilli multiply during cold storage of meat products. This delays spoilage by proteolytic bacteria, but may also lead to spoilage by producing off-flavor, acid taste, gas, slime or greening (Egan, 1983). While *L. viridescens* has been shown to cause greening (Niven and Evans, 1957), the role of the other species frequently isolated from stored meat—*L. plantarum, L. brevis*, and unidentified lactobacilli—is not clear. Some of the "atypical" homofermentative lactobacilli described by Hitchener et al. (1982) have been identified as *L. sake and L. curvatus* (Kandler, unpublished). The unidentified heterofermentative strains found by Hitchener et al. (1982), which are characterized by the production of L(+)-lactic acid, may be identical with *L. divergens,* the recently described new species isolated from vacuum-packaged raw minced meat in South Africa (Holzapfel and Gerber, 1983).

Fish and marinated fish. Although lactobacilli have not been considered to be indigenous to the marine environment, Kraus (1961) and Schrøder et al. (1980) have shown that herring caught far from populated areas and fish and krill from the arctic environment harbor cold-adapted lactobacilli resembling *L. plantarum.* However, one of these isolates, studied in more detail with respect to its ability to decarboxylate amino acids (Jonsson et al., 1983), forms exclusively L(+)-lactic acid, indicating that it represents a new, so far undescribed cryophilic species rather than *L. plantarum.* Homo- and heterofermentative lactobacilli play an important role in the spoilage of raw marinated herring (Blood, 1975). It is suggested that the acetic acid added to the herring provides the necessary acid environment for the action of proteinases present in the fish muscle (Meyer, 1962). The free amino acids thus liberated then provide the energy source for acetic acid-tolerant and salt-tolerant lactobacilli which are able to decarboxylate amino acids. The CO_2 formed is the first indication of spoilage. In carbohydrate-containing marinates, the carbohydrates may be the source of CO_2 liberated by heterofermentative lactobacilli. Therefore, Meyer (1956) distinguished between a "carbohydrate" swell and a "protein" swell. Lactobacilli isolated from marinated herring were mainly allotted to *L. plantarum, L. brevis* and *L. buchneri.* However, reinvestigation of such isolates employing modern biochemical and genomic characteristics is necessary to elucidate their true taxonomic position. Unidentified lactobacilli have also been isolated from fresh water salmonides (Evelyn and McDermott, 1961) and diseased rainbow trout (Cone, 1982).

Man and animals. The intestinal tract of man and animals harbors many species of lactobacilli (Lerche and Reuter, 1962; Mitsuoka, 1969) living as commensals intimately associated with the mucousous surface epithelium. This subject has been extensively reviewed (Savage, 1977; Sharpe, 1981). Only the few species found exclusively, or at least predominantly, in the intestinal tract will be discussed here.

L. salivarius may be the most typical species of the mouth flora, although it is also found in the intestinal tract (Rogosa et al., 1953). The other species found are much more universally distributed in nature.

The most prominent species, probably indigenous to the intestine, is *L. acidophilus,* which is believed to exert a beneficial effect on human and animal health. It is used on an industrial scale in preparing acidophilus sour milk and producing pharmaceutical preparations (Rehm, 1983) for restoring the normal intestinal flora after disturbance caused by diseases or treatment with antibiotics. Whether such preparations contain true *L. acidophilus* strains, and which strains, if any, have a beneficial influence in the particular individual remains a controversial topic (Lauer et al., 1980). The problem is further complicated by the finding that strains designated as *L. acidophilus* proved to belong to many different genotypes exhibiting only a low degree of DNA-DNA homology with each other (Johnson et al., 1980; Sarra et al., 1980; Lauer et al., 1980). While most genotypes cannot be distinguished on the basis of phenetic characteristics, two genotypes could be phenotypically separated, and one of them has been described as the new species *L. gasseri* (Lauer and Kandler, 1980). Another recently described homofermentative species, *L. animalis* (Dent and Williams, 1982) also phenotypically resembling *L. acidophilus,* was detected in dental plaques of primates and in the intestine of dog and mouse. It could be separated from *L. acidophilus* mainly on the basis of the protein pattern obtained in electrophoresis and the formation of exclusively L(+)-lactic acid.

Recently, strains belonging to an additional genotype of *L. acidophilus* or to the genotype IIB of *L. gasseri* were isolated from kefir. These strains represent the majority population of lactobacilli in the kefir grain, but only a minority in the final sour milk product, where the heterofermentative species *L. kefir* dominates (Kunath and Kandler, 1984). The distinct heterogeneity of the species *L. acidophilus* is a challenge to all intestinal microbiologists.

Among the heterofermentative intestinal lactobacilli, *L. fermentum* was considered to be the dominant species (Lerche and Reuter, 1962) in the intestine. A taxonomic study of several strains designated as *L. fermentum* based on the sugar fermentation pattern revealed that two groups of strains, representing two species exhibiting a G + C content of 53 mol% and 41 mol%, respectively, had been included together. Strains possessing the lower G + C value were described as the new species *L. reuteri* (Kandler et al., 1980), which includes most strains isolated from the intestine by Lerche and Reuter (1962). *L. reuteri* was also found to be the dominating heterofermentative species in the intestine of calves (Sarra et al., 1979). Thus *L. reuteri* may be the main heterofermentative lactobacillus species in the intestine, while *L. fermentum* seems to be more widespread in lactic acid fermented substrates. However, this suggestion needs further confirmation.

L. murinus, a recently described homofermentative species, has been isolated from the feces of mice and rats. It may be a typical species in the intestine of rodents (Hemme et al., 1980).

Lactobacilli are also found in the rumen of ruminants. However, they are rarely classified at the species level. Two anaerobic species, *L. ruminis* and *L. vitulinus,* have been described from the bovine rumen. *L. ruminis* has also been isolated from the human intestine (Sharpe et al., 1973b).

Sewage and manure. Sewage and manure are secondary habitats of all lactobacilli found in the intestine, but also of some other species not, or only rarely, found in the intestine. In manure, *L. coryniformis* and *L. curvatus,* neither recorded as intestinal, are frequently found (Abo Elnaga and Kandler, 1965a). *L. vaccinostercus* has only been found in cow dung as yet (Okada et al., 1979).

In municipal sewage, levels of 10^4–10^5 lactobacilli/ml have been found (Weiss et al., 1981). The heterofermentative strains (~25%) of the isolates have been classified as *L. fermentum, L. reuteri, L. brevis* and, to a lesser extent, as *L. confusus.* The homofermentative strains (~75%) of the isolates belonged to a larger number of different species. However, about 10% of the strains could not be allotted to any of the known species. They have been described as representatives of the two new species *L. sharpeae* and *L. agilis,* not as yet found in any other habitat (Weiss et al., 1981).

Enrichment and Isolation Procedure

Procedures for the isolation of lactobacilli must take into account their aciduric or acidophilic nature, their complex nutritional requirements and their preference for microaerophilic conditions. When lactobacilli are the predominant flora in the source material, the rather nonselective MRS* agar (de Man, Rogosa and Sharpe, 1960) or the somewhat similar APT agar (Evans and Niven, 1951) may be used for isolation. APT agar is commonly used for isolating *L. viridescens* and other lactobacilli from meat products. When lactobacilli occur only as part of a complex population, selective media are required. Most lactobacilli from many different sources have been successfully isolated on the widely used acetate medium† (SL) of Rogosa, Mitchell and Wiseman (1951). However, SL medium is not completely selective for lactobacilli as other lactic acid bacteria, e.g. leuconostocs, pediococci, enterococci, bifidobacteria (intestinal sources) and yeasts may also grow. Thus, colonies may have to be further examined. Yeasts may be eliminated by the addition of cycloheximide at a concentration of 100 mg/liter.

On the other hand, some lactobacilli, mainly from quite specialized environments, will not grow on SL medium. Depending on the source of isolation, minor modifications of SL medium, supplementing it with more or less specific growth factors such as meat extract, tomato juice, fresh yeast extract, malt extract, ethanol, mevalonic acid (sake) or even some of the natural substrate (beer, different juices) can improve the isolation of lactobacilli which are highly adapted to the conditions of their ecological niches. Replacement of glucose, either completely or partially, by other carbohydrates such as maltose, fructose, sucrose or arabinose is recommended in some cases, especially where heterofermentative lactobacilli play an important role. For the detection of beer-spoiling bacteria including nutritionally fastidious lactobacilli, a special medium (NBB medium) has been described by Back (1980). For further information reference is given to Sharpe (1981) where many media and methods of cultivating lactobacilli are compiled in detail.

For the isolation of anaerobic lactobacilli from intestinal sources 0.05% (w/v) cysteine should be added and it may be necessary to prereduce poured, dried plates by overnight incubation in an anaerobic jar.

Since most lactobacilli generally grow better either anaerobically or in the presence of increased CO_2 tension, agar plates should be incubated in jars evacuated and filled with 90% N_2 or H_2 + 10% CO_2 or in anaerobic jars (BBL, Oxoid) using H_2 + CO_2 generating kits.

Maintenance Procedures

For short-term preservation, cultures are preferably inoculated into MRS or optimal medium agar stabs, incubated until growth becomes visible, stored at 4–7°C and transferred monthly. Some species or strains, however, die out quite rapidly within a series of transfers. Alternatively, cultures grown to the early stationary growth phase may be deep frozen in the growth medium and stored at −20°C for several months.

The method of choice for long-term preservation is lyophilization. Cells grown to the late logarithmic growth phase are collected by centrifugation, resuspended in sterile skim milk or horse serum containing 7.5% (w/v) glucose and lyophilized. Ampules are sealed under vacuum and stored at 5–8°C. Most strains preserved by this method are still viable after 10–20 years, although some require more frequent relyophilization. Strains may also be kept for long periods (over 30 years) in liquid nitrogen.

Procedure for Testing Special Characters

Carbohydrate fermentation. MRS broth without meat extract and glucose with 0.05% (w/v) chlorophenol red is generally used as basal medium. Filter-sterilized solutions of the test carbohydrates are added to a final concentration of 1%. Tests are incubated at the optimum growth temperature and results recorded up to 7 days. In a few cases, e.g. some strains of *L. delbrueckii*, the addition of 0.2% meat extract broadens the pattern of fermented carbohydrates somewhat and the fermentation of glucose is distinctly improved. For strains which will not grow reasonably in MRS broth the optimal growth medium should be used as basal medium.

Lactic acid configuration. The amount of the isomers of lactic acid produced is best determined enzymatically using D-lactate (Gawehn and Bergmeyer, 1974) and L-lactate dehydrogenase (Gutmann and Wahlefeld, 1974).

Corrections must be made for the lactic acid content of the medium before inoculation. Care must be taken to analyze cultures after they have reached the stationary growth phase, since some DL-formers produce predominantly L(+)- or, in a few cases, D(−)-lactic acid during the early growth phase.

Cell wall analysis. The absence or presence of *meso*- or LL-diaminopimelic acid (*meso*-DAP; LL-DAP) in the cell wall may be tested by the following simple procedure: cells from about 1 ml of broth culture or a loopful of cell material taken from an agar plate or a slant are hydrolyzed with 0.5 ml 6 M HCl at 100°C, overnight, in a sealed ampule. HCl is removed by a gentle stream of air on the hydrolysate at about 50°C; the residue is taken up in a minimum of water, applied to a thin layer plate (precoated cellulose plastic sheets are recommended), developed in the solvent system: methanol:pyridine:water:10 M HCl (320:40:70:10 v/v/v/v) for 2–3 hours and sprayed with acidic ninhydrin; *meso*- and LL-DAP are well separated from all other amino acids due to their very low R_f value. They are further characterized by their olive green color which changes to yellow after several hours or days in the dark.

For details of the peptidoglycan composition, purified cell walls must be prepared. In most cases the rapid screening method, e.g. boiling the washed cells with trichloroacetic acid followed by digestion with trypsin (Schleifer and Kandler, 1972), is satisfactory. Lysine and ornithine can be distinguished by the chromatographic method described above. It is the least time-consuming test to differentiate *L. reuteri* from *L. fermentum*.

The peptidoglycan-type Lys-DAsp, most widely distributed within the genus *Lactobacillus*, is well characterized by the occurrence of N^6-(aminosuccinyl)-lysine, a derivative of N^6-(aspartyl)-lysine formed during acid hydrolysis of cell walls (4N HCl, 100°C, 16 hours). It can be easily detected by two-dimensional paper chromatography (first direction: isopropanol: acetic acid: water 75:10:15; second direction: α-picoline:25% NH_4OH:water 70:2:28). Other peptidoglycan types may be analyzed by the methods described in detail by Schleifer and Kandler (1972).

Teichoic acids may be extracted from cell walls with 70% hydrofluoric acid at 0°C and analyzed by liquid gas chromatography according to Fiedler et al. (1981).

Characterization of lactic acid dehydrogenases. The electrophoretic mobility of the lactic acid dehydrogenases (LDH) is determined by polyacrylamide gel electrophoresis at pH 7.5 using crude cell extracts according to Hensel et al. (1977). L-LDH rabbit Iso-I (Boehringer, Mannheim) serves as reference. Whether the L-LDH of an organism is allosteric or not is tested by spectrophotometric measurement of the rate of pyruvate reduction with and without the effectors fructose-1,6-diphosphate (FDP) and Mn^{2+} at pH 6.5 in dialyzed crude cell extracts (Hensel et al., 1977).

---

* MRS agar: casein peptone, 10.0 g; meat extract, 10.0 g; yeast extract, 5.0 g; glucose, 20.0 g; K_2HPO_4, 5.0 g; diammonium citrate, 2.0 g; Na acetate, 5.0 g; $MgSO_4 \cdot 7 H_2O$, 0.5 g; $MnSO_4 \cdot 4 H_2O$, 0.2 g; Tween 80, 1.0 g; agar, 15.0 g; distilled water 1000 ml; adjust pH to 6.2–6.4 and sterilize at 121°C for 15 min.

† Selective SL medium: casein peptone, 10.0 g; yeast extract, 5.0 g; KH_2PO_4, 6.0 g; diammonium citrate, 2.0 g; $MgSO_4 \cdot 7 H_2O$, 0.5 g; $MnSO_4 \cdot 4 H_2O$, 0.2 g; $FeSO_4 \cdot 7 H_2O$, 0.04 g; Tween 80, 1.0 g; glucose, 20.0 g; Na acetate·3 H_2O, 25.0 g; agar, 15.0 g; dissolve the agar separately by steaming in 500 ml distilled water; dissolve all the other ingredients without heating in 500 ml distilled water, adjust pH with glacial acetic acid to 5.4, then add this to the melted agar and boil for 5 min; no further sterilization is given.

Differentiation from Other Closely Related Taxa

Lactobacilli are metabolically very similar to the other genera of the so-called lactic acid bacteria. Only their rod shape readily distinguishes them from the coccal genera *Streptococcus, Leuconostoc* and *Pediococcus*. However, some species of the obligately heterofermentative lactobacilli form coccoid rods and may be confused with *Leuconostoc*. These species are differentiated from *Leuconostoc* by their formation of DL-lactic acid and not D(−)-lactic acid.

Strains of *Streptococcus* which form atypically elongated cells may also be confused with coccoid rods of lactobacilli. Here, differentiation may require nucleic acid hybridization as in the case of *L. xylosus* and "*L. hordniae*," both of which have been shown to belong to the genus *Streptococcus* (Garvie et al., 1981; Kilpper-Bälz et al., 1982).

The rod-shaped bifidobacteria, which until the eighth edition of *Bergey's Manual* had long been included in the genus *Lactobacillus* as "*Lactobacillus bifidus*," may be differentiated from lactobacilli on the basis of their characteristic hexose fermentation pathway which yields lactic acid and acetic acid at a molar ratio of 2:3, but no CO_2, instead of lactic acid, acetic acid (or ethanol) and CO_2 at a molar ratio of 1:1:1, the pattern of fermentation products typical of obligately heterofermentative lactobacilli.

Taxonomic Comments

The species of the genera *Lactobacillus, Leuconostoc, Pediococcus* and *Streptococcus* form a supercluster within the so-called clostridia subbranch of the Gram-positive bacteria, as shown by oligonucleotide cataloging of their 16S rRNA (Fig. 14.4; Stackebrandt et al., 1983). Bifidobacteria, already excluded from the family *Lactobacillaceae* in *Bergey's Manual*, eighth edition, have proved to be completely unrelated to lactobacilli. They belong to the so-called actinomycetales subbranch of the Gram-positive bacteria.

The neighborhood of the lactobacillus supercluster and the streptococcus cluster, and their position at the clostridia subbranch which also contains the aerobic bacilli (Fig. 14.4) is in accordance with Orla-Jensen's concept of "lactic acid bacteria" as a group of closely related microaerophilic genera. However, there is only limited agreement between the results obtained by oligonucleotide cataloging and the phylogenetic implications of serological studies involving antisera against malic enzymes (London, 1971), fructose-1,6-diphosphate aldolases (London and Kline, 1973; London and Chace, 1976) and glyceraldehyde-3-phosphate dehydrogenases (London and Chace, 1983) of various lactic acid bacteria and some anaerobic and aerobic bacteria. On the basis of the two techniques, only the very close interrelationship between the four genera of lactic acid bacteria and their origin from a common progenitor is certain. Different results were obtained not only regarding the relationship between the lactic acid bacteria and other phylogenetically more distant genera (*Eubacterium, Propionibacterium, Brochothrix, Acholeplasma, Aerococcus*) but also regarding the relationship within the lactic acid bacteria. The immunological grouping indicates a close relationship between streptococci and the *L. casei* group (London and Chace, 1983), whereas, on the basis of the 16S rRNA cataloging, only representatives of the genus *Streptococcus*, but not members of the genera *Pediococcus* and *Leuconostoc*, can be separated

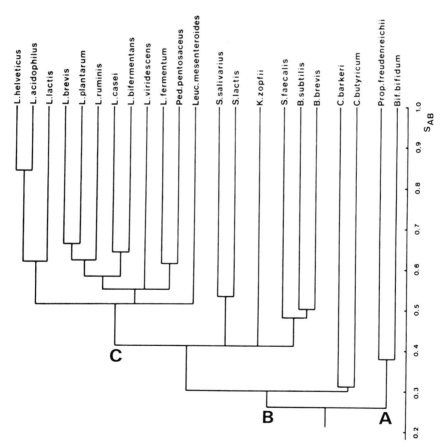

Figure 14.4. Dendrogram of relationship among representatives of the genera *Lactobacillus, Leuconostoc, Pediococcus, Streptococcus, Kurthia, Clostridium, Propionobacterium, Bifidobacterium* and *Bacillus* based on S_{AB} values (16S rRNA cataloging; Stackebrandt et al., 1983). *A*, actinomycetales subbranch; *B*, clostridia subbranch; and *C*, lactobacillus supercluster.

from the genus *Lactobacillus*. No subdivision of the genus *Lactobacillus* into three groups, corresponding to Orla-Jensen's genera "*Thermobacterium*," "*Streptobacterium*" and "*Betabacterium*," often referred to as subgenera (Sharpe, 1981), is indicated in the dendrogram based on S_{AB} values (Fig. 14.4). With the exception of the pair *L. helveticus* and *L. acidophilus*, which is related at the very high level of $S_{AB} = 0.83$, all investigated species exhibit low S_{AB} values between 0.47 and 0.65 indicating a considerable phylogenetic depth for each of the phenotypes. In addition, the small differences between the S_{AB} values suggest extensive speciation within a relatively short period of time, probably at the global "Pasteur point" when microaerophilic life became possible (Stackebrandt et al., 1983).

The high phylogenetic age of the genus *Lactobacillus* is also reflected by the wide range of the G + C content of DNA from 32–53 mol%—a span twice as large as is usually accepted for a single genus (cf. Schleifer and Stackebrandt, 1983), the lack of significant DNA/DNA homology between most of the species and the relatively high rate of amino acid exchange among pairs of lactobacillus species in the highly conserved substrate-binding region of L-lactic acid dehydrogenase (Hensel et al., 1981; Mayr et al., 1982).

More work is needed to elucidate the phylogenetic structure of the genus *Lactobacillus* and the other genera constituting the "lactic acid bacteria." Hence, we shall not at present follow the suggestion of Stackebrandt et al. (1983) to expand the description of the genus *Lactobacillus* so as to comprise also the genera *Leuconostoc* and *Pediococcus*.

We shall arrange the species of *Lactobacillus* into the traditional three groups resembling Orla-Jensen's three genera without designating them as formal subgeneric taxa since they do not represent phylogenetically defined clusters. Although the majority of strains of each of the new groups agree with the original definition of thermobacteria, streptobacteria and betabacteria, many of the recently described species do not fit these definitions. Hence, the following new definitions contain neither growth temperature nor morphology, the classical characteristics of Orla-Jensen's subgenera.

Group I, obligately homofermentative lactobacilli: hexoses are fermented almost exclusively to lactic acid by the Embden-Meyerhof pathway; pentoses or gluconate are not fermented. Rare reports on pentose fermentation by particular strains of members of group I should be reinvestigated. Fermentation balances should be determined, in order to get information on the possible fermentation mechanism of such atypical strains. In a few cases, we have obtained strains claimed to ferment pentoses. However, they either did not ferment pentoses in our hands or did not belong to a species of group I.

Group II, facultatively heterofermentative lactobacilli: hexoses are fermented almost exclusively to lactic acid by the Embden-Meyerhof pathway or, at least by some species, to lactic acid, acetic acid, ethanol and formic acid under glucose limitation; pentoses are fermented to lactic acid and acetic acid via an inducible phosphoketolase.

Group III, obligately heterofermentative lactobacilli: hexoses are fermented to lactic acid, acetic acid (ethanol) and CO_2; pentoses are fermented to lactic acid and acetic acid. In general, both pathways involve phosphoketolase. However, some species which probably possess other pathways for carbohydrate breakdown but performing also a heterofermentation including the production of gas from hexoses are tentatively also included in group III, e.g. *L. bifermentans*.

Group I harbors all the classical representatives of Orla-Jensen's thermobacteria and many recently described species. With regard to DNA/DNA homology, group I contains two complexes of related species or subspecies and many single species not related to any significant extent on the basis of present knowledge. One of the two complexes consists of the three subspecies of *L. delbrueckii*. The type strains of the four former species, *L. delbrueckii*, *L. bulgaricus*, *L. lactis* and *L. leichmannii*, were found to possess between each other more than 80% DNA/DNA homology (Weiss et al., 1983b) and the phenotypical differences are restricted to variations in the range of fermented carbo-

hydrates. Thus they have been considered to justify only the rank of subspecies. *L. delbrueckii* subsp. *lactis* exhibits the widest range of fermented carbohydrates and may be the common ancestor from which several variants, adapted to specialized niches (sour milk, grain mashes, etc.) have evolved by only minor changes of the phenotype and genotype.

The second complex is represented by *L. acidophilus* which was shown to exhibit a distinct genomic heterogeneity. A large number of strains designated originally *L. acidophilus* has been arranged in two main groups of genotypes each consisting of several subgroups based on DNA/DNA homology (Johnson et al., 1980; Lauer et al., 1980; Sarra et al., 1980). DNA/DNA homology is 75–100% between strains of the same subgroup, 25–50% between strains of different subgroups within each of the two main groups and below 25% between strains of the two main groups. The two main groups exhibit clear phenotypic differences and are thus considered to represent two different species. The main group containing the original type strain of the species retains the name *L. acidophilus*, while the other group has been described as the new species *L. gasseri* (Lauer and Kandler, 1980). Recently, the type strain of the earlier described species *L. crispatus* (Moore and Holdeman, 1970) was found to be 100% homologous with one of the subgroups of *L. acidophilus* (Cato et al., 1983).

L. helveticus may be considered a highly specialized derivative of the *L. acidophilus* complex, adapted to sour whey. It resembles *L. acidophilus* with respect to the G + C content of DNA and many biochemical characteristics and possesses DNA/DNA homology of 13–44% with representatives of the various genotypes of *L. acidophilus* (Johnson et al., 1980). It shares also a high S_{AB} value with the type strain of *L. acidophilus* (Fig. 14.4; Stackebrandt et al., 1983). Thus, *L. acidophilus*, *L. gasseri*, *L. crispatus* and *L. helveticus* form a cluster of closely related species within group I, which is only distantly related to the *L. delbrueckii* complex ($S_{AB} = 0.6$), represented by *L. delbrueckii* subsp. *lactis* in Figure 14.4.

Group II contains Orla-Jensen's streptobacteria and many newly described species. Three complexes of species or subspecies can be recognized, while the other species show no known phylogenetic relationship with each other.

One complex is formed by the strains designated *L. plantarum*. The phenotypical variation within this giant species have long been recognized. Strains exhibiting characteristics atypical for the genus *Lactobacillus*, e.g. motility, nitrate reduction, pseudocatalase, etc., have often been designated *L. plantarum*. A genomic heterogeneity of *L. plantarum* has been shown by DNA/DNA homology studies (Dellaglio et al., 1975). Although most of the strains investigated were related to the type strain at a homology level of 80–100%, a quarter of the strains was only related at a level of 30–70%. Three strains were highly related with a strain designated "*L. pentosus*" (Fred et al., 1921), a name considered to be synonymous with *L. plantarum* at present, but which may be revived in the future. Four other strains exhibited 57–70% DNA/DNA homology between each other and to the type strain of *L. plantarum*, thus indicating the existence of additional genotypes of *L. plantarum*.

A second complex of at least three genotypes is formed by the subspecies of *L. casei*. While the type strain and only two strains originally designated "*L. zeae*" (Kuznetzov, 1959) are related at a DNA/DNA homology level of 80–100%, the majority of the strains of *L. casei* subsp. *casei*, *L. casei* subsp. *pseudoplantarum* and *L. casei* subsp. *tolerans* form a second genotype at a homology level of 80–100% among each other, but with only 40% homology toward the genotype which contains the type strain. Strains of *L. casei* subsp. *rhamnosus* represent a third genotype which shares only 30–50% homology with strains of the other two genotypes. Because of the low DNA/DNA homology, and distinct phenetic differences to other subspecies (see Tables 14.7 and 14.8), *L. casei* subsp. *rhamnosus* is a candidate to be raised to the species status. The two other subspecies, although closely related to *L. casei* subsp. *casei* are phenotypically distinctly different by forming DL-lactic acid via a lactic acid racemase (*L. casei* subsp. *pseudoplantarum*; Stetter and Kandler, 1973) or by heat tolerance and an extremely sparse pattern of fermented carbohydrates (*L. casei* subsp. *tolerans*; see Tables 14.7 and 14.8), respectively. *L. casei* subsp. *tolerans* does not ferment ribose and gluconate and therefore does not fit the definition of group II. However, the high DNA/DNA homology with *L. casei* indicates that the lack of these characteristics is caused by minor genomic differences.

The third complex of species consists of *L. sake*, *L. curvatus* and *L. bavaricus*. The first two species are related at a DNA/DNA homology level of 50%. Both species are characterized by possessing inducible lactic acid racemase which converts the primarily formed L(+)-lactic acid to racemate (Stetter and Kandler, 1973). *L. bavaricus* (Stetter and Stetter, 1980) is phenotypically clearly different from the two species, by the lack of lactic acid racemase, but otherwise it is very similar. In fact, the type strain and most of the strains of *L. bavaricus* exhibit 100% DNA/DNA homology to *L. sake*, while a few strains are completely homologous with *L. curvatus* (Kagermeier, 1981). Thus *L. bavaricus* consists of two genotypes, one of which, including the type strain, may be considered as a subspecies of *L. sake*, the other as a subspecies of *L. curvatus*.

L. casei, *L. curvatus*, *L. sake* and *L. bavaricus* possess an allosteric L-lactic acid dehydrogenase with fructose-1,6-diphosphate and Mn^{2+} acting as effectors (Hensel et al., 1977). The properties of the L-lactic acid dehydrogenase enzymes of the various species are very similar. They show partial serological cross-reactions (Hensel, 1977) and their subunits form hybrids (Mayr et al., 1980). This could indicate a close phylogenetic relationship between these species. However, no significant DNA/DNA homology could be detected between *L. casei* and the other species. No S_{AB} values of these species are known as yet. An allosteric L-lactic acid dehydrogenase has also been found in *L. murinus*. However, this enzyme has not been studied in detail. None of the other species of group II show a significant DNA/DNA homology to any other species or possess phenotypic characters which would indicate a specific relationship between any pair of strains.

Group III contains all the obligately heterofermentative gas-forming lactobacilli of Orla-Jensen's genus "*Betabacterium*" and several more recently described species. Two species—*L. bifermentans* and *L. divergens*—which also form gas from glucose, probably do not possess the 6-phosphogluconate pathway. *L. bifermentans* ferments glucose homofermentatively to DL-lactic acid, but—depending on pH—the lactic acid formed is more or less completely split into acetic acid, CO_2 and H_2 (Pette and vanBeynum, 1943, Kandler et al., 1983). Although the formation of H_2 is a characteristic not included in the description of the genus *Lactobacillus*, the organism is kept in this genus because of its distinct relationship to lactobacilli as evidenced by the dendrogram based on 16S rRNA cataloging (Fig. 14.4) and, more recently, by rRNA/DNA hybridization (Schillinger, unpublished). Considering the S_{AB}-values and the rRNA/DNA hybridization data, *L. bifermentans* is closely related to *L. casei*, an organism able to form acetic acid, ethanol and formic acid instead of lactic acid when grown under glucose limitation. However, *L. casei* does not possess a dehydrogenase for H_2

evolution. It is tempting to suggest that *L. bifermentans* is derived from *L. casei* by evolving a formate hydrogen ligase.

The fermentation balance of *L. divergens* indicates that this organism also does not possess the 6-phosphogluconate pathway. On a molar basis, the proportion of the C_2 compounds (acetic acid and ethanol) and CO_2 formed from hexoses is too small compared to that of lactic acid. Also pentose fermentation does not yield the proper molar ratios of fermentation products. Here, the C_2 compounds are favored compared to lactic acid. Although the details of the fermentation mechanism remain to be elucidated, *L. divergens* is clearly heterofermentative and thus included in group III.

Most species of group III fall within a very narrow range of G + C content of DNA (40–46 mol%). However, they do not show significant DNA/DNA homology between each other (Vescovo et al., 1979). Except the pair *L. kefir* and *L. buchneri*, exhibiting a DNA/DNA homology of 40% (Kandler and Kunath, 1983), no reliable clustering of species within group III is possible. However, it is suggested that species in which the Lys-Asp type peptidoglycan—most common among lactobacilli—is replaced by the Lys-Ala-Ser or chemically similar types—typical of the genus *Leuconostoc*—may form a cluster related to *Leuconostoc*. Such a relationship has already been suggested in the case of *L. viridescens* and *L. confusus* (Garvie, 1975), species which contain Lys-Ser-Ala and Lys-Ala$_2$ type peptidoglycan, respectively. On the other hand, members of the two genera are sometimes confused. Thus, *L. confusus* was originally allotted to *Leuconostoc* because of its coccoid appearance and slime formation, whereas the heterofermentative, coccoid, "*L. batatas*" (ATCC 15520), described by Kitahara (1949), was found to form D(−)-lactic acid and belongs to *Leuconostoc* (Weiss, unpublished).

The phylogenetic structure of group III needs further elucidation with the help of DNA/RNA hybridization and the determination of S_{AB} values of a greater number of species.

Acknowledgment

We are indebted to the ATCC, DSM, NCIB, NCDO and VPI for supplying numerous strains. We owe many ideas and information to the stimulating discussions we have had over years with our friends from the subcommittee of lactobacilli and related organisms.

Further Reading

Carr, J.G., C.V. Cutting and G.C. Whiting (editors). 1975. Lactic acid bacteria in beverages and food. Academic Press, London.
Kandler, O. 1983. Carbohydrate metabolism in lactic acid bacteria. Antonie van Leeuwenhoek J. Microbiol. Serol. *49:* 209–224.
London, J. 1976. The ecology and taxonomic status of the lactobacilli. Annu. Rev. Microbiol. *30:* 279–301.
London, J. and N.M. Chace. 1983. Relationship among lactic acid bacteria demonstrated with glyceraldehyde-3-phosphate dehydrogenase as an evolutionary probe. Int. J. Syst. Bacteriol. *33:* 723–737.
Sharpe, M.E. 1981. The genus *Lactobacillus*. *In* Starr, Stolp, Trüper, Balows and Schlegel (Editors), The Prokaryotes. A Handbook on Habitats, Isolation, and Identification of Bacteria. Springer-Verlag, Berlin, pp. 1653–1679.
Stackebrandt, E., V.J. Fowler and C.R. Woese. 1983. A phylogenetic analysis of lactobacilli, *Pediococcus pentosaceus* and *Leuconostoc mesenteroides*. Syst. Appl. Microbiol. *4:* 326–337.

Differentiation and characteristics of the species of the genus Lactobacillus

The differential characteristics of the species of *Lactobacillus* are indicated in Tables 14.3 and 14.4. Other characteristics of the species are listed in Tables 14.5–14.10.

List of the species of the genus Lactobacillus

1. **Lactobacillus delbrueckii** (Leichmann 1896) Beijerinck 1901, 229.^{AL} (*Bacillus Delbrücki* (*sic*) Leichmann 1896, 284.)

Note. Because of the high phenotypic and genomic similarities between *L. delbrueckii*, *L. leichmannii*, *L. lactis* and *L. bulgaricus* only *L. delbrueckii* is here retained as a separate species, whereas both *L. lactis* and *L. leichmannii* are treated as *L. delbrueckii* subsp. *lactis* and *L.*

bulgaricus as *L. delbrueckii* subsp. *bulgaricus* (see Weiss et al., 1983b, 1984).

del.bruec'ki.i. M.L. gen. n. *delbrueckii* of Delbrück; named for M. Delbrück, a German bacteriologist.

Rods with rounded ends, 0.5–0.8 by about 2–9 μm, occurring singly and in short chains.

Table 14.3.

Differential characteristics of the obligately homofermentative and facultatively heterofermentative species of the genus **Lactobacillus**[a]

Species	Mol% G + C	Teichoic acid	Starch	Melibiose	Mannose	Mannitol	Maltose	Sucrose	D(−)-Lactic acid	L(+)-Lactic acid	DL-Lactic acid	Growth at 15°C	Ribose	mDpm in peptidoglycan
25. L. plantarum	45	+									+	+	+	+
23. L. maltaromicus	36	−								+				
16. L. agilis	44	−								+		−		
13. L. sharpeae	53	+				−	+				+		+	−
15. L. yamanashiensis	33	−			−	+								
11. L. ruminis	44	−								+		−		
14. L. vitulinus	35	−							+					
24. L. murinus	43	−				+				+		−	+	−
22. L. homohiochii	36	+				−		−			+	+		
19b. L. casei subsp. pseudoplantarum	46	−		−		+	+							
26. L. sake	43	−		+		−								
21. L. curvatus	43	−		−										
18. L. bavaricus	43	−				−				+				
19a. L. casei	46	−				+								
17. L. alimentarius	36	−				−								
20. L. coryniformis	45	−				+			+			+	−	
3. L. amylophilus	45	−	+			−			+					
7. L. farciminis	35	−	−											
5. L. animalis	42	−			+				+			−		
12. L. salivarius	35	−			−									
9. L. helveticus	39	+						−			+			
4. L. amylovorus	44	+	+						+					
2. L. acidophilus	36													
6. L. crispatus	36													
8. L. gasseri	34	−												
1. L. delbrueckii	50	+							+					
10. L. jensenii	35													

[a] Symbols: +, 90% or more of the strains are positive; −, 90% or more of the strains are negative.

Good growth at 45°C or even at 48–52°C.

Additional physiological and biochemical characteristics are presented in Tables 14.5 and 14.6.

Growth factor requirements: pantothenic acid and niacin generally essential; riboflavin, folic acid, vitamin B_{12} and thymidine are essential for particular strains; thiamine, pyridoxin, biotin and p-aminobenzoic acid are not required.

DNA/DNA homology: strains labeled L. delbrueckii, L. bulgaricus, L. lactis and L. leichmannii, including the respective type strains, were found highly homologous among each other (Weiss et al., 1983b); no genomic relationship could be detected to L. helveticus (Simonds et al., 1971; Dellaglio et al., 1973).

The mol% G + C of the DNA is 49–51 (Bd, T_m).

Three subspecies are presently recognized.

1a. Lactobacillus delbrueckii subsp. **delbrueckii** (Leichmann 1896) Weiss, Schillinger and Kandler 1984, 270.[VP*] (*Effective publication*: Weiss, Schillinger and Kandler 1983b, 556.)

Distinguishing characteristics are given in Tables 14.5 and 14.6.

Isolated mainly from plant material fermented at high temperatures (40–53°C).

Type strain: ATCC 9649.

1b. Lactobacillus delbrueckii subsp. **bulgaricus** (Orla-Jensen 1919) Weiss, Schillinger and Kandler 1984, 270.[VP] (*Effective publication*: Weiss, Schillinger and Kandler 1983b, 556). (*Thermobacterium bulgaricum* Orla-Jensen 1919, 164; *Lactobacillus bulgaricus* Rogosa and Hansen 1971, 181.)

bul.ga′ri.cus. M.L. adj. *bulgaricus* Bulgarian.

Ferments only a few carbohydrates as shown in Table 14.5. D-LDH migrates distinctly faster in electrophoresis than that of the other subspecies.

Isolated from yoghurt and cheese.

Type strain: ATCC 11842 (DSM 20081).

1c. Lactobacillus delbrueckii subsp. **lactis** (Orla-Jensen 1919) Weiss, Schillinger and Kandler 1984, 270.[VP] (*Effective publication*: Weiss, Schillinger and Kandler 1983b, 556.) (*Thermobacterium lactis* Orla-Jensen 1919, 164; *Lactobacillus leichmanni* (Henneberg) Bergey et al. 1923, 249.)

lac′tis. L. n. *lac* milk; L. gen. n. *lactis* of milk.

Distinguishing characteristics are given in Tables 14.5 and 14.6.

Isolated from milk, cheese, compressed yeast and grain mash.

Type strain: ATCC 12315 (DSM 20072).

2. Lactobacillus acidophilus (Moro 1900) Hansen and Mocquot 1970, 326.[AL†] (*Bacillus acidophilus* Moro 1900, 115.)

a.ci.do′ phi.lus. M.L. n. *acidum* acid; Gr. adj. *philus* loving; M.L. adj. *acidophilus* acid-loving.

Rods with rounded ends, generally 0.6–0.9 × 1.5–6 μm, occurring singly, in pairs and in short chains.

With rare exceptions good growth at 45°C. Starch is fermented by most strains.

---

* VP denotes that this name has been validly published in the official publication, International Journal of Systematic Bacteriology.

† AL denotes the inclusion of this name on the Approved Lists of Bacterial Names.

Table 14.4.
*Differential characteristics of the obligately heterofermentative species of the genus **Lactobacillus**[a]*

Species	Mol% G + C	Orn in peptidoglycan	Xylose	Melibiose	Mannose	Melezitose	Arabinose	Slime capsule	Cellobiose	Mannitol	Maltose	Sucrose	NH₃ from arginine	Growth at 15°C	DL-Lactic acid	L(+)-Lactic acid	Ribose	mDpm in peptidoglycan	CO₂ + H₂ from lactate
27. L. bifermentans	45												−		+		+	−	+
32. L. divergens	34												+	+			+	+	−
43. L. vaccinostercus	36												−	−		+			
44. L. viridescens	43												−	+	+			−	
42. L. sanfrancisco	37										+	+					−		
35. L. fructosus	47										−	−							
33. L. fermentum	53	+									+	+	+	−	+		+		
41. L. reuteri	41	−																	
31. L. confusus	46		+					+	+				+	+					
40. L. minor	44		−					−											
38. L. kandleri	39								−	+									
29. L. buchneri	45					+	+			−									
28. L. brevis	45																		
30. L. collinoides	45					−													
36. L. halotolerans	45				+		−												
39. L. kefir	41			+	−														
37. L. hilgardii	40		+	−															
34. L. fructivorans	40		−																

[a] Symbols: see Table 14.3.

1221

Table 14.5.

Pattern of fermented carbohydrates of the obligately homofermentative species of the genus **Lactobacillus** *(group 1)[a]*

Species	Amygdalin	Arabinose	Cellobiose	Esculin	Fructose	Galactose	Glucose	Gluconate	Lactose	Maltose	Mannitol	Mannose	Melezitose	Melibiose	Raffinose	Rhamnose	Ribose	Salicin	Sorbitol	Sucrose	Trehalose	Xylose
1a. *L. delbrueckii* subsp. *delbrueckii*	-	-	d	-	+	-	+	-	-	d	-	+	-	-	-	-	-	-	+	+	d	-
1b. *L. delbrueckii* subsp. *lactis*	+	-	d	+	+	d	+	-	+	+	-	+	-	-	-	-	-	+	+	+	+	-
1c. *L. delbrueckii* subsp. *bulgaricus*	-	-	-	-	+	-	+	-	+	-	-	+	-	-	-	-	-	-	+	-	-	-
2. *L. acidophilus*	+	-	+	+	+	+	+	-	+	+	-	+	-	d	d	-	-	+	+	+	d	-
3. *L. amylophilus*	-	-	-	-	+	+	+	-	-	+	-	+	-	-	-	-	-	-	+	-	+	-
4. *L. amylovorus*	+w	-	+	+w	+	+	+	-	-	+	-	+	-	+	+	-	-	+w	+	+	+	-
5. *L. animalis*	d	d	+	+	+	+	+	-	+	+	-	+	-	-	-	-	-	+	+	+	+	-
6. *L. crispatus*	+	-	+	+	+	+	+	-	+	+	-	+	-	-	+	-	-	+	+	+	+	-
7. *L. farciminis*	+	-	+	+	+	+	+	-	+	+	-	+	-	-	-	-	-	+	+	+	+	-
8. *L. gasseri*	+	-	+	+	d	+	+	-	d	d	-	d	-	d	d	-	-	+	+	+	d	-
9. *L. helveticus*	-	-	-	-	+	+	+	-	+	d	d	+	-	-	-	-	-	-	+	-	d	-
10. *L. jensenii*	+	-	+	+	+	+	+	-	-	d	-	+	-	-	-	-	-	+	+	-	d	-
11. *L. ruminis*	+	-	+	+	+	+	+	-	d	+	+	+	-	+	+	-	-	+	+	+	-	-
12. *L. salivarius*	-	-	-	d[b]	+	+	+	-	+	+	+	+	-	+	+	d[b]	-	d[b]	+	+	+	-
13. *L. sharpeae*	+	-	+	+	+	+	+	-	+	+	+	+	-	-	+	-	-	+	+	+	-	-
14. *L. vitulinus*	+	-	+	+	+	-	+	-	+	+	-	+	-	+	+	-	-	+	d	-	d	-
15. *L. yamanashiensis*	+	-	d	+	+	d	+	-	+	-	+	+	-	-	-	d	-	+	+	+	+	-

[a] Symbols: +, 90% or more strains positive; −, 90% or more strains negative; d, 11–89% strains positive; +w, positive to weak reaction.
[b] See text.

1222

Table 14.6.

Physiological and biochemical characteristics of the obligately homofermentative species of the genus **Lactobacillus** *(Group I)[a]*

Species	Peptidoglycan type[b]	Teichoic acid	Electrophoretic mobility[c]		Allosteric L-LDH	Mol% G + C	Lactic acid isomer(s)[d]	Growth at 15°C	NH₃ from arginine
			D-LDH	L-LDH					
1a. *L. delbrueckii* subsp. *delbrueckii*	Lys-DAsp	Glycerol	1.50	−	−	49–51	D	−	d
1b. *L. delbrueckii* subsp. *lactis*	Lys-DAsp	Glycerol	1.50	−	−	49–51	D	−	d
1c. *L. delbrueckii* subsp. *bulgaricus*	Lys-DAsp	Glycerol	1.70	−	−	49–51	D	−	−
2. *L. acidophilus*	Lys-DAsp	Glycerol	1.50[e]	1.30	−	34–37	DL	−	−
3. *L. amylophilus*	Lys-DAsp	None	1.60	1.40	−	44–46	L	+	ND
4. *L. amylovorus*	Lys-DAsp	Glycerol	1.15	1.20	−	40–41	DL	−	ND
5. *L. animalis*	Lys-DAsp	None	−	1.50	−	41–44	L	−	−
6. *L. crispatus*	Lys-DAsp	Glycerol	1.35	1.10	−	35–38	DL	−	−
7. *L. farciminis*	Lys-DAsp	None	1.15	1.20	−	34–36	L(D)	+	+
8. *L. gasseri*	Lys-DAsp	None	1.35[e]	0.95	−	33–35	DL	−	−
9. *L. helveticus*	Lys-DAsp	Glycerol	0.95	1.30	−	38–40	DL	−	−
10. *L. jensenii*	Lys-DAsp	Glycerol	1.50	−	−	35–37	D	−	+
11. *L. ruminis*	mDAP-Direct	None	ND	ND	−	44–47	L	−	−
12. *L. salivarius*	Lys-DAsp	None	−	1.35	−	34–36	L	−	−
13. *L. sharpeae*	mDAP-Direct	Glycerol	1.34	1.48	−	53	L	+	−
14. *L. vitulinus*	mDAP-Direct	None	ND	ND	−	34–37	D	−	−
15. *L. yamanashiensis*	mDAP-Direct	None	ND	ND	−	32–34	L	+	−

[a] Symbols: see Table 14.5; and ND, not determined.

[b] Abbreviations used by Schleifer and Kandler (1972).

[c] Determined in polyacrylamide disk gel electrophoresis pH 7.5; L-LDH rabbit Iso I served as reference.

[d] D or L, the isomer recorded makes up 90% or more of total lactic acid; DL, 25–75% of total lactic acid are of the L-configuration; and D(L) or L(D), the isomer given in brackets makes up 15–20% of total lactic acid.

[e] Strains of this species are known to give more than one band; the migration distance recorded is that obtained with the type strain.

Additional physiological and biochemical characteristics are presented in Tables 14.5 and 14.6.

Nutritional requirements: calcium pentothenate, folic acid, niacin and riboflavin are essential; pyridoxal, thiamine, thymidine and vitamin B₁₂ are not required.

DNA/DNA homology: the species comprises at least three homology groups which cannot be separated by simple phenotypical characteristics (groups A-1, A-3, A-4 of Johnson et al., 1980; groups Ia, Ib, Id, Ie of Lauer et al., 1980). Between individual strains of the different groups DNA/DNA homology values of about 20–50% are found. Group A-1 or group Ia, respectively, which include the type strain of *L. acidophilus*, can be differentiated from the other groups by studying the electrophoretical or immunological behavior of the L-LDH. Group A-2 of Johnson et al. (1980) and the corresponding group Ic of Lauer et al. (1980) were recently found to be homologous with *L. crispatus* (Cato et al., 1983). Among the lactobacilli species of group I (thermobacteria) having a similar mol% G + C as *L. acidophilus*, only very low DNA/DNA homology of *L. acidophilus* with *L. gasseri*, *L. helveticus* and *L. crispatus* but no homology with *L. salivarius* and *L. jensenii* could be detected.

Isolated from the intestinal tract of humans and animals, human mouth and vagina.

The mol% G + C of the DNA is 32–37 (Bd, T_m).

Type strain: ATCC 4356.

Further comments. *L. acidophilus* cannot be differentiated reliably from *L. gasseri*, *L. crispatus*, and *L. amylovorus* by any simple phenotypic test; electrophoretic analysis of soluble cellular proteins or lactate dehydrogenases, detailed cell wall studies or, in the case of *L. amylovorus*, determination of mol% G + C of the DNA are necessary.

3. **Lactobacillus amylophilus** Nakamura and Crowell 1981, 216.[VP]
(*Effective publication*: Nakamura and Crowell 1979, 539.)

a.my.lo′ phi.lus. Gr. n. *amylum* starch; Gr. adj. *philus* loving; M.L. adj. *amylophilus* starch loving.

Thin rods, 0.5–0.7 × 2–3 μm, occurring singly and in short chains.

No growth at 45°C. Actively ferments starch and displays extracellular amylolytic enzyme activity.

Additional physiological and biochemical characteristics are presented in Tables 14.5 and 14.6.

Growth factor requirements: riboflavin, pyridoxal, pantothenic acid, niacin, and folic acid are essential; thiamine is not required.

DNA/DNA homology: four strains form a narrow homology group not related to a number of homofermentative lactobacilli species studied (Nakamura, 1982).

Isolated from swine waste-corn fermentation.

The mol% G + C of the DNA is 44–46 (Bd).

Type strain: NRRL B-4437.

4. **Lactobacillus amylovorus** Nakamura 1981, 61.[VP]

a.my.lo.vo′rus. Gr. n. *amylum* starch; L. v. *vorare* to devour; M.L. adj. *amylovorus* starch destroying.

Rods, 1 × 3–5 μm, occurring singly and in short chains. Good growth at 45°C. Actively ferments starch and displays extracellular amylolytic enzyme activity.

Additional physiological and biochemical characteristics are presented in Tables 14.5 and 14.6.

Growth factor requirements: niacin, pantothenic acid, folic acid and riboflavin are essential; thiamine is not required.

DNA/DNA homology: three strains form a narrow homology group not related to the type strains of *L. acidophilus*, *L. leichmannii* and *L. amylophilus* (Nakamura, 1981).

Isolated from cattle waste-corn fermentation.

The mol% G + C of the DNA is 40.3 ± 0.1 (Bd).

Type strain: NRRL B-4540.

Further comments. Since many strains of *L. acidophilus*, *L. crispatus* and *L. gasseri* are able to ferment starch (Johnson et al., 1980), *L. amylovorus* cannot be reliably differentiated from these species by simple tests.

5. **Lactobacillus animalis** Dent and Williams 1983, 439.[VP] (*Effective publication*: Dent and Williams 1982, 384.)

a.ni.ma'lis. L. n. *animal* animal; L. gen. n. *animalis* of an animal.

Rods with rounded ends, generally 1.0–1.2 × 3–6 μm, occurring singly or in pairs.

Good growth at 45°C.

Additional physiological and biochemical characteristics are presented in Tables 14.5 and 14.6.

Isolated from dental plaques and alimentary canal of animals.

The mol% G + C of the DNA is 41–44 (T_m).

Type strain: NCDO 2425.

Further comments. Some of the strains on which the description of *L. animalis* was based originally ferment arabinose and also ribose weakly thus resembling *L. murinus*. DNA/DNA homology studies should be directed towards establishing the genomic relationship of the different strains of *L. animalis* among each other and with *L. murinus*.

6. **Lactobacillus crispatus** (Brygoo and Aladame 1953) Moore and Holdeman 1970, 15.[AL] (*Eubacterium crispatum* Brygoo and Aladame 1953, 641.)

Note. An emended description of *L. crispatus* is given by Cato et al. (1983).

cris.pa'tus. L. part. adj. *crispatus* curled, crisped, referring to morphology observed originally in broth media.

Straight to slightly curved rods with rounded ends, 0.8–1.6 × 2.3–11 μm, occurring singly and in short chains.

Generally good growth at 45°C.

Additional physiological and biochemical characteristics are presented in Tables 14.5 and 14.6.

DNA/DNA homology: *L. crispatus* was found highly homologous with "*L. acidophilus*" group A-2 of Johnson et al. 1980 (Cato et al. 1983).

Isolated from human feces, vagina and buccal cavities, crops and ceca of chicken; also found in patients with purulent pleurisy, leucorrhea and urinary tract infection.

The mol% G + C of the DNA is 35–38 (T_m).

Type strain: VPI 3199 (ATCC 33820).

Further comments. *L. crispatus* cannot reliably be differentiated from *L. acidophilus*, *L. gasseri* and *L. amylovorus* by any simple test: electrophoretic characterization of soluble cellular proteins or lactic acid dehydrogenases, detailed cell wall studies or, in the case of *L. amylovorus*, determination of mol% G + C of the DNA are necessary.

7. **Lactobacillus farciminis** Reuter 1983, 672.[VP] (*Effective publication*: Reuter 1983, 278)

far.ci'mi.nis. L. n. *farcimen* sausage; L. gen. n. *farciminis* of sausage.

Slender rods, 0.6–0.8 × 2–6 μm, occurring singly and in short chains.

No growth at 45°C. Grows in the presence of 10% NaCl and occasionally 12% NaCl.

Additional physiological and biochemical characteristics are presented in Tables 14.5 and 14.6.

DNA/DNA homology: no genomic relationship between *L. farciminis* and other species of group II (streptobacteria; Dellaglio et al., 1975). However, 26–28% DNA/DNA homology with *L. alimentarius* has been detected by Kagermeier et al. (1985).

Isolated from meat products (raw sausages) and sour dough.

The mol% G + C of the DNA is 34–36 (T_m).

Type strain: DSM 20184 (ATCC 29644).

8. **Lactobacillus gasseri** Lauer and Kandler 1980a, 601.[VP] (*Effective publication*: Lauer and Kandler 1980, 77.)

gas'se.ri. M.L. gen. n. *gasseri* of Gasser; named for F. Gasser, a French bacteriologist.

Rods with rounded ends, generally 0.6–0.8 × 3.0–5.0 μm, occurring singly and in chains. Formation of "mini cells" and snakes is frequently observed.

Generally good growth at 45°C. Starch is fermented by most strains.

Unlike all other lactobacilli, the D-alanyl-D-alanine termini of peptide subunits of peptidoglycan not involved in cross-linkage are preserved because of the lack of D,D-carboxypeptidase action.

Additional physiological and biochemical characteristics are presented in Tables 14.5 and 14.6.

DNA/DNA homology: the species comprises two DNA/DNA homology groups which cannot be separated by phenotypical characteristics (groups B-1 and B-2 of Johnson et al. 1980; groups IIa and IIb of Lauer et al. 1980). Between individual strains of the two groups, DNA/DNA homology values of about 30–60% are found. Among the species of group I (thermobacteria) having a similar mol% G + C to *L. gasseri*, only low DNA/DNA homology with *L. acidophilus* and *L. crispatus* but no homology with *L. helveticus*, *L. jensenii* and *L. salivarius* could be detected.

Isolated from the human mouth and vagina and from the intestinal tract of man and animals; also found in wounds, urine, blood and pus of patients suffering from septic infections.

The mol% G + C of the DNA is 33–35 (T_m).

Type strain: DSM 20243.

Further comments. *L. gasseri* cannot be differentiated reliably from *L. acidophilus*, *L. crispatus* and *L. amylovorus* by any simple phenotypic test; electrophoretic analysis of soluble cellular proteins or lactate dehydrogenases, detailed cell wall studies or, in the case of *L. amylovorus*, determination of mol% G + C of the DNA are required.

9. **Lactobacillus helveticus** (Orla-Jensen 1919) Bergey, Harrison, Breed, Hammer and Huntoon 1925 184.[AL] (*Thermobacterium helveticum* Orla-Jensen 1919, 164.)

hel.ve'ti.cus. L. adj. *helveticus* Swiss.

Good growth at 45°C; maximum growth temperature 50–52°C.

Additional physiological and biochemical characteristics are presented in Tables 14.5 and 14.6.

Growth factor requirements: Calcium pantothenate, niacin, riboflavin, pyridoxal or pyridoxamine are essential; thiamine, folic acid, vitamin B_{12} and thymidine are not required.

DNA/DNA homology: together with strains formerly labeled "*L. jugurti*," strains of *L. helveticus* form a narrow homology group genomically unrelated to *L. delbrueckii* subsp. *bulgaricus*, *L. delbrueckii* subsp. *lactis* (Simonds et al., 1971; Dellaglio et al., 1973), and *L. gasseri* (Johnson et al., 1980).

A closer phylogenetic relationship between *L. helveticus* and *L. acidophilus* is indicated by 13–44% DNA/DNA homology between the two species (Johnson et al., 1980) and by the relatively high S_{AB} value of 0.84 compared with the values of 0.47–0.59 found between other lactobacilli species (Stackebrandt et al., 1983).

Isolated from sour milk, cheese starter cultures and cheese, particularly Emmental and Gruyère cheese.

The mol% G + C of the DNA is 37–40 (Bd, T_m).

Type strain: ATCC 15009.

10. **Lactobacillus jensenii** Gasser, Mandel and Rogosa 1970, 221.[AL]

jen.se'ni.i. M.L. gen. n. *jensenii* of Jensen; named for S. Orla-Jensen, a Danish microbiologist.

Rods with rounded ends, 0.6–0.8 × 2.0–4.0 μm, occurring singly and in short chains.

Generally good growth at 45°C.

Additional physiological and biochemical characteristics are presented in Tables 14.5 and 14.6.

DNA/DNA homology: strains of *L. jensenii* form a narrow homology group genetically not related to *L. acidophilus*, *L. crispatus*, *L. delbrueckii* and *L. gasseri* (Gasser and Janvier, 1980; Johnson et al., 1980).

Isolated from human vaginal discharge and blood clot.

The mol% G + C of the DNA is 35–37 (Bd).

Type strain: ATCC 25258.

Further comments. *L. jensenii* is indistinguishable from *L. delbrueckii* by simple physiological tests. The slight difference in the migration distance of D-LDH of *L. jensenii* and *L. delbrueckii* in starch gel electrophoresis observed by Gasser (1970) could not be demonstrated by the polyacrylamide disk gel electrophoresis routinely used in our laboratory. Therefore, determination of mol% G + C of the DNA remains the most reliable characteristic to differentiate *L. jensenii* from *L. delbrueckii*.

11. **Lactobacillus ruminis** Sharpe, Latham, Garvie, Zirngibl and Kandler 1973, 47.[AL]

ru′mi.nis. L. n. *rumen* throat; M.L. n. *rumen* rumen; M.L. gen n. *ruminis* of rumen.

Rods, 0.6–0.8 × 3–5 μm, occurring singly, in pairs and in short chains.

Motile by peritrichous flagella; motility not always easy to demonstrate and often sluggish, best demonstrated as stab cultures in semisolid media containing low concentrations of glucose.

Surface growth is obtained only under reduced oxygen pressure; growth in liquid media is improved by the addition of cysteine-HCl.

Unlike the strains isolated from the rumen, many strains from sewage were nonmotile and failed to grow at 45°C.

Additional physiological and biochemical characteristics are presented in Tables 14.5 and 14.6

DNA/DNA homology: all strains studied form a narrow homology group not related to others, especially *meso*-DAP-containing species (Sharpe and Dellaglio, 1977; Weiss et al., 1981).

Isolated from rumen of cow and from sewage.

The mol% G + C of the DNA is 44–47 (T_m).

Type strain: ATCC 27780.

12. **Lactobacillus salivarius** Rogosa, Wiseman, Mitchell and Disraely 1953, 691.[AL]

sa.li.va′ri.us. L. adj. *salivarius* salivary.

Rods with rounded ends, 0.6–0.9 × 1.5–5 μm, occurring singly and in chains of varying length.

Generally growth at 45°C.

Additional physiological and biochemical characteristics are presented in Tables 14.5 and 14.6.

DNA/DNA homology: one strain tested showed no genomic relationship to *L. acidophilus* and *L. gasseri* (Lauer et al., 1980), *L. murinus* (E. Lauer, unpublished) or *L. sake, L. curvatus,* and *L. farciminis* (Kagermeier 1981).

Isolated from the mouth and intestinal tract of man and hamster; intestinal tract of chicken.

The mol% G + C of the DNA is 34–36 (Bd).

Two subspecies are recognized.

12a. **Lactobacillus salivarius** subsp. **salivarius** Rogosa, Wiseman, Mitchell and Disraely 1953, 691.[AL]

Description as for the species.

Ferments rhamnose but not salicin and esculin.

Type strain: ATCC 11741.

12b. **Lactobacillus salivarius** subsp. **salicinius** Rogosa, Wiseman, Mitchell and Disraely 1953, 691.[AL]

sa.li.ci′ni.us. M.L. adj. *salicinius* pertaining to salicin, a glycoside.

Description as for the species.

Ferments salicin and esculin but not rhamnose.

Type strain: ATCC 11742.

13. **Lactobacillus sharpeae** Weiss, Schillinger, Laternser and Kandler 1982, 266.[VP] (*Effective publication:* Weiss, Schillinger, Laternser and Kandler 1981, 251.)

shar′pe.ae. M.L. gen n. *sharpeae* of Sharpe; named for M. Elisabeth Sharpe, an English bacteriologist.

Rods with rounded ends, generally 0.6–0.8 × 3–8 μm, with a pronounced tendency to form "snakes" and, after prolonged incubation, long characteristically wrinkled chains. In broth cultures, a flocculent sediment is observed.

No growth at 45°C.

Additional physiological and biochemical characteristics are presented in Tables 14.5 and 14.6

DNA/DNA homology: three strains tested proved to be completely homologous to each other, whereas one single strain was more distantly related showing only 53% homology to the type strain. No genomic relationship could be detected to other *meso*-DAP-containing species of lactobacilli (Weiss et al., 1981).

Habitat unknown, isolated from municipal sewage.

The mol% G + C of the DNA is 53 (T_m).

Type strain: DSM 20505.

14. **Lactobacillus vitulinus** Sharpe, Latham, Garvie, Zirngibl and Kandler 1973, 47.[AL]

vi.tu.li′nus. L. adj. *vitulinus* of a calf.

Rods with rounded ends, 0.5–0.7 × 2–4 μm, occurring singly and in pairs.

Surface growth is only obtained under anaerobic conditions; grows in freshly boiled MRS broth with (w/v) 0.05% cysteine-HCl added.

Additional physiological and biochemical characteristics are presented in Tables 14.5 and 14.6

DNA/DNA homology: five strains belonged to three different homology groups completely unrelated to each other. No homology was detected to possibly related species (Sharpe and Dellaglio, 1977). Since no phenotypic characteristics are presently known to separate the different homology groups, *L. vitulinus* remains genotypically heterogenous.

Isolated from bovine rumen.

The mol% G + C of the DNA is 34–37 (T_m).

Type strain: ATCC 27783.

15. **Lactobacillus yamanashiensis** Nonomura 1983, 406.[VP] (*Lactobacillus mali* Carr and Davies 1970, 774.)

ya.ma.na.shi.en′sis. M.L. adj. *yamanashiensis* belonging to Yamanashi Prefecture, Japan, the source of wine must from which the organism was isolated.

Rods, 0.6–0.8 × 2–4 μm, occurring singly, in pairs, and in short chains.

Motile with a few peritrichous flagella; motility often sluggish, best demonstrated in semisolid MRS agar stab culture with only (w/v) 0.1% glucose. Most strains exhibit a weak pseudocatalase activity when grown on MRS agar containing (w/v) 0.1% glucose; benzidine test negative.

No growth at 45°C.

Additional physiological and biochemical characteristics are presented in Tables 14.5 and 14.6

DNA/DNA homology: strains of *L. mali* exhibited 70–95% homology between each other and with the type strain of *L. yamanashiensis* (Carr et al., 1977). No genomic relationship was detected to strains of group II (streptobacteria; Dellaglio et al., 1975) and to other *meso*-DAP-containing L(+)-lactic acid-producing lactobacilli (Weiss et al., 1981).

Isolated from cider and wine must.

The mol% G + C of the DNA is 32–34 (T_m).

Type strain: ATCC 27304.

Further comments. Nonomura (1983) mentioned two subspecies, namely *L. yamanashiensis* subsp. *yamanashiensis* and *L. yamanshiensis* subsp. *mali* in the title of the paper but, inconsequently, in the text only a description of the species, but not of the subspecies is given. The proposal of the subspecies *L. yamanashiensis* subsp. *mali* by Carr et al. (1977) is invalid since it is not mentioned in the Approved Lists of Bacterial Names (Skerman et al., 1980). ATCC 27502 is listed in the Approved Lists of Bacterial Names as type strain of *L. mali*.

Note. Significant amounts of menaquinones, predominantly with eight and nine isoprene units (MK-8, MK-9) have been found in *L. yamanashiensis* (Collins and Jones, 1981). All other lactobacilli studied so far lack both menaquinones and ubiquinones.

16. **Lactobacillus agilis** Weiss, Schillinger, Laternser and Kandler 1982, 266.[VP] (*Effective publication*: Weiss, Schillinger, Laternser and Kandler 1981, 252.)

a′gi.lis. L. adj. *agilis* agile, motile.

Rods with rounded ends, 0.7–1.0 × 3–6 μm, occurring singly, in pairs and in short chains.

Motile with peritrichous flagella; motility normally easy to demonstrate in MRS broth.

Good growth at 45°C.

Additional physiological and biochemical characteristics are presented in Tables 14.7 and 14.8.

DNA/DNA homology: five strains form a narrow homology group not related to representatives of any of the *meso*-DAP-containing and L(+)-lactic acid-forming species of lactobacilli (Weiss et al., 1981).

Habitat unknown, isolated from municipal sewage.

The mol% G + C is 43–44 (T_m).

Type strain: DSM 20509.

Further comment. "*Lactobacillus plantarum var. mobilis*" isolated from feces of turkey (Harrison and Hansen 1950) was only tentatively named and therefore omitted from the Approved Lists of Bacterial Names (Skerman et al., 1980). According to the original description and later investigations (Sharpe et al., 1973b) this organism may belong to *L. agilis*.

17. **Lactobacillus alimentarius** Reuter 1983, 672.[VP] (*Effective publication*: Reuter 1983, 278.)

a.li.men.ta′ri.us. L. adj. *alimentarius* pertaining to food.

Short, slender rods, generally 0.6–0.8 × 1.5–2.5 μm.

No growth at 45°C. Growth in the presence of 10% NaCl. Acetoin is produced from glucose.

Additional physiological and biochemical characteristics are presented in Tables 14.7 and 14.8.

DNA/DNA homology: no genomic relationship was found between *L. alimentarius* and other species of group II (streptobacteria, Dellaglio et al., 1975); however, 26–28% DNA/DNA homology with *L. farciminis* was detected by Kagermeier (1981).

Isolated from marinated fish products, meat products (raw sausages and sliced prepacked sausages) and sour dough.

The mol% G + C of the DNA is 36–37 (T_m).

Type strain: DSM 20249 (ATCC 29643).

18. **Lactobacillus bavaricus** Stetter and Stetter 1980, 601.[VP] (*Effective publication*: Stetter and Stetter 1980, 73.)

ba.va′ri.cus. M.L. adj. *bavaricus* Bavarian.

Rods with rounded ends, 0.8–1.0 × ~2–7 μm, occurring singly and in short chains; slightly curved, especially during stationary growth phase.

No growth at 45°C; growth from 2–37°C.

L-LDH is activated by FDP and Mn^{2+}. Does not contain lactic acid racemase.

Additional physiological and biochemical characteristics are presented in Tables 14.7 and 14.8.

Isolated from sauerkraut and fermented cabbage leaves.

The mol% G + C of the DNA is 41–43 (T_m).

Type strain: ATCC 31063.

Further comments. The type strain of *L. bavaricus* as well as five additional strains tested showed 80–95% DNA/DNA homology with *L. sake*, whereas one strain was completely homologous with *L. curvatus* (Kagermeier 1981). Therefore, organisms allocated to *L. bavaricus* may be regarded as racemase-free subspecies of *L. sake* or *L. curvatus*, respectively, rather than as members of a separate species. However, further studies are required before a formal description of two subspecies is possible.

19. **Lactobacillus casei** (Orla-Jensen 1916) Hansen and Lessel 1971, 71.[AL] (*Streptobacterium casei* Orla-Jensen 1919, 166.)

ca′se.i. L. n. *caseus* cheese; L. gen. n. *casei* of cheese.

Rods, generally 0.7–1.1 × 2.0–4.0 μm, often with square ends and tending to form chains.

No growth at 45°C (exception: *L. casei* subsp. *rhamnosus*).

L-LDH is activated by FDP and Mn^{2+}.

Growth factor requirements: riboflavin, folic acid, calcium pantoth-

enate and niacin are essential; pyridoxal or pyridoxamine is essential or stimulatory; thiamine, vitamin B_{12} and thymidine are not required.

The mol% G + C of the DNA is 45–47 (Bd).

Isolated from milk and cheese, dairy products and dairy environments, sour dough, cow dung, silage, human intestinal tract, mouth and vagina, sewage.

Four subspecies are recognized within this species:

19a. **Lactobacillus casei** subsp. **casei** (Orla-Jensen 1916) Hansen and Lessel 1971, 71.[AL]

Description as for the species.

Additional physiological and biochemical characteristics are presented in Tables 14.7 and 14.8.

Type strain: ATCC 393.

Note. The lactose-negative variant labeled *Lactobacillus casei* subsp. *alactosus* Mills and Lessel 1973, 67, should no longer be regarded as a separate taxon but included in *L. casei* subsp. *casei*.

19b. **Lactobacillus casei** subsp. **pseudoplantarum** Abo-Elnaga and Kandler 1965a, 26.[AL]

pseu′do.plan.ta′rum. Gr. adj. *pseudes* false; M.L. gen. n. *plantarum* a specific epithet; M.L. adj. *pseudoplantarum* not the true (*L.*) *plantarum*.

Inactive lactic acid is produced due to the activity of a L-lactic acid racemase (Stetter and Kandler 1973).

Additional physiological and biochemical characteristics are presented in Tables 14.7 and 14.8.

Type strain: ATCC 25598.

19c. **Lactobacillus casei** subsp. **rhamnosus** Hansen 1968, 76.[AL]

rham.no′sus. M.L. adj. *rhamnosus* pertaining to rhamnose.

These organisms are the only homofermentative lactobacilli which grow well at both 15°C and 45°C.

Additional physiological and biochemical characteristics are presented in Tables 14.7 and 14.8.

Type strain: ATCC 7469.

19d. **Lactobacillus casei** subsp. **tolerans** Abo-Elnaga and Kandler 1965a, 26.[AL]

to′le.rans. L. pres. part. *tolerans* tolerating, enduring; means survival during pasteurization of milk.

Survives heating at 72°C for 40 s and ferments a very narrow range of carbohydrates.

Additional physiological and biochemical characteristics are presented in Tables 14.7 and 14.8.

Type strain: ATCC 25599.

DNA/DNA homology: except the type strain of *L. casei* subsp. *casei* and the members of *L. casei* subsp. *rhamnosus*, all *L. casei* form a narrow homology group genomically not related to other species of group II (streptobacteria; Dellaglio et al., 1975). The type strain of *L. casei* subsp. *casei* is highly homologous only with "*Lactobacterium zeae*" whereas homology with other strains of *L. casei* is significantly lower than 50% indicating a heterogeneity of the species.

The strains of *L. casei* subsp. *rhamnosus* highly homologous among each other display only 30–50% homology with strains of other subspecies of *L. casei*. Thus, *L. casei* subsp. *rhamnosus* deserves the rank of a separate species rather than that of a subspecies of *L. casei*.

20. **Lactobacillus coryniformis** Abo-Elnaga and Kandler 1965a, 18.[AL]

co.ry′ni.for′mis. Gr. n. *coryne* a club; L. adj. *formis* shaped; M.L. adj. *coryniformis* club-shaped.

Short, often coccoid, rods; frequently somewhat pear-shaped, 0.8–1.1 × 1–3 μm, occurring singly, in pairs or short chains.

Generally no growth at 45°C.

Additional physiological and biochemical characteristics are presented in Tables 14.7 and 14.8.

Growth factor requirements: pantothenic acid, niacin, riboflavin,

Table 14.7.

Pattern of fermented carbohydrates of the facultatively heterofermentative species of the genus **Lactobacillus** *(group II)*[a]

Species	Amygdalin	Arabinose	Cellobiose	Esculin	Fructose	Galactose	Glucose	Gluconate	Lactose	Maltose	Mannitol	Mannose	Melezitose	Melibiose	Raffinose	Rhamnose	Ribose	Salicin	Sorbitol	Sucrose	Trehalose	Xylose
16. *L. agilis*	+	−	+	+	+	+	+	−	+	+	+	+	+	+	+	−	+	+	d	+	+	−
17. *L. alimentarius*	o	d	+	+	+	+	+	+	−	+	−	+	−	−	−	−	+	+	−	+	+	−
18. *L. bavaricus*	−	−	+	+	+	+	+	+	d	+	−	+	−	+	−	−	+	+	−	+	−	−
19a. *L. casei* subsp. *casei*	+	−	+	+	+	+	+	+	d	+	+	+	+	−	−	−	+	+	+	+	+	−
19b. *L. casei* subsp. *pseudoplantarum*	+	−	+	+	+	+	+	+	+	+	+	+	+	−	−	−	+	+	+	+	+	−
19c. *L. casei* subsp. *rhamnosus*	+	d	+	+	+	+	+	+	+	+	+	+	+	−	−	+	+	+	+	+	+	−
19d. *L. casei* subsp. *tolerans*	−	−	−	−	+	+	+	−	+	−	−	−	−	−	−	−	−	−	−	−	−	−
20a. *L. coryniformis* subsp. *coryniformis*	−	−	−	d	+	+	+	+	d	+	+	+	−	d	d	+	−	d	d	+	−	−
20b. *L. coryniformis* subsp. *torquens*	−	−	−	−	+	+	+	+	+	+	+	d	−	−	−	−	−	−	−	+	−	−
21. *L. curvatus*	−	−	+	+	+	+	+	+	d	+	−	+	−	−	−	−	+	+	−	−	−	−
22. *L. homohiochii*	−	−	d	o	+	−	+	−	−	+	−	+	−	−	−	−	+	d	−	−	d	−
23. *L. maltaromicus*	+	−	+	o	+	+	+	o	+	+	d	+	+	+	−	−	+	+	+	+	+	−
24. *L. murinus*	d	+	+	+	+	+	+	−	+	+	d	+	−	+	+	−	+	d	−	+	d	−
25. *L. plantarum*	+	d	+	+	+	+	+	+	+	+	+	+	d	+	+	−	+	+	+	+	+	d[b]
26. *L. sake*	+	+	+	+	+	+	+	+	+	+	−	+	−	+	−	−	+	+w	−	+	+	−

[a] Symbols: see Table 14.5; and o, reaction not determined.

[b] See text.

1227

Table 14.8.

Physiological and biochemical characteristics of the facultatively heterofermentative species of the genus **Lactobacillus** *(group II)[a]*

Species	Peptidoglycan type[b]	Teichoic acid	Electrophoretic mobility[c]		Allosteric L-LDH	Mol% G + C	Lactic acid isomer(s)[d]	Growth at 15°C	NH₃ from arginine
			D-LDH	L-LDH					
16. *L. agilis*	mDAP-Direct	None	1.40	1.20	−	43–44	L	−	−
17. *L. alimentarius*	Lys-DAsp	None	0.80	1.10	−	36–37	L(D)	+	−
18. *L. bavaricus*	Lys-DAsp	None	−	1.60	+	41–43	L	+	−
19a. *L. casei* subsp. *casei*	Lys-DAsp	None	1.22[e]	0.93	+	45–47	L	+	−
19b. *L. casei* subsp *pseudoplantarum*	Lys-DAsp	None	1.04	0.93	+	45–47	DL	+	−
19c. *L. casei* subsp *rhamnosus*	Lys-DAsp	None	0.75	0.93	+	45–47	L	+	−
19d. *L. casei* subsp *tolerans*	Lys-DAsp	None	−	0.93	+	45–47	L	+	−
20a. *L. coryniformis* subsp. *coryniformis*	Lys-DAsp	None	0.38	−	−	45	D(L)	+	−
20b. *L. coryniformis* subsp. *torquens*	Lys-DAsp	None	0.38	−	−	45	D	+	−
21. *L. curvatus*	Lys-DAsp	None	1.20	1.60	+	42–44	DL	+	−
22. *L. homohiochii*	Lys-DAsp	Glycerol	ND	ND	−	35–38	DL	+	−
23. *L. maltaromicus*	mDAP-Direct	None	ND	ND	−	36	L	+	ND
24. *L. murinus*	Lys-DAsp	None	−	0.92	+	43–44	L	−	−
25. *L. plantarum*	mDAP-Direct	Ribitol or glycerol	1.44	1.28	−	44–46	DL	+	−
26. *L. sake*	Lys-DAsp	None	1.20	1.60	+	42–44	DL	+	−

[a] Symbols: see Table 14.5; and ND, not determined.

[b–e] Footnotes: see Table 14.6.

biotin and *p*-aminobenzoic acid are essential for all or the majority of the strains tested; folic acid, pyridoxin, thiamine and vitamin B₁₂ are not required.

DNA/DNA homology: four strains representing both subspecies are highly homologous among each other, but no genomic relationship to other species of group II is found (Dellaglio et al., 1975).

Isolated from silage, cow dung, dairy barn air and sewage.

The mol% G + C of the DNA is close to 45 (T_m).

Two subspecies are recognized within *L. coryniformis*.

20a. Lactobacillus coryniformis subsp. **coryniformis** Abo-Elnaga and Kandler 1965a, 18.[AL]

The lactic acid produced from glucose contains appreciable amounts of the L-isomer (15–20% of total lactic acid).

Type strain: DSM 20001.

20b. Lactobacillus coryniformis subsp. **torquens** Abo-Elnaga and Kandler 1965a, 19.[AL]

tor′quens. L. pres. part. *torquens* twisting.

Exclusively D(−)-lactic acid is produced.

Type strain: ATCC 25600.

21. Lactobacillus curvatus (Troili-Petersson 1903) Abo-Elnaga and Kandler 1965a, 19.[AL] (*Bacterium curvatum* Troili-Petersson 1903, 137.)

cur.va′tus. L. v. *curvare* to curve; L. past. part. *curvatus* curved.

Curved, bean-shaped rods with rounded ends, 0.7–0.9 × 1–2 μm, occurring in pairs and short chains; closed rings of usually four cells or horseshoe forms frequently observed. Some strains at first motile; motility lost on subculture.

No growth at 45°C; some strains tested are able to grow even at 2–4°C.

L-LDH is activated by FDP and Mn²⁺. Possesses lactic acid racemase whose biosynthesis is induced by L(+)-lactic acid. Racemase induction generally not repressed by acetate.

Additional physiological and biochemical characteristics are presented in Tables 14.7 and 14.8.

DNA/DNA homology: strains of *L. curvatus* form a narrow homology group not related to other lactobacilli species except *L. sake; L. curvatus*

and *L. sake* have 40–50% homology with each other (Dellaglio et al., 1975; Kagermeier et al., 1985).

Isolated from cowdung, milk, silage, sauerkraut, prepacked finished dough and meat products.

The mol% G + C of the DNA is 42–44 (T_m).

Type strain: ATCC 25601.

Note. Some of the atypical streptobacteria from herbage, silage, fermented meat products and vacuum-packaged meat reported in the past belong to *L. curvatus*.

22. Lactobacillus homohiochii Kitahara, Kaneko and Goto 1957, 118.[AL]

ho′mo.hi.o′chi.i. Gr. adj. *homos* like, equal; Japanese n. *hiochi* spoiled sake; M.L. gen. n. *homohiochii* probably intended to mean homofermentative lactobacillus of hiochi.

Rods, with rounded ends, 0.7–0.8 × 2–4 μm or, occasionally, 6 μm in length.

Does not grow in MRS broth. In Rogosa SL broth supplemented with DL-mevalonic acid (30 mg/liter) and ethanol (40 ml/liter) copious growth is obtained at 30°C after a marked lag phase of 4–7 days.

No growth at 45°C and at an initial pH higher than 5.5. Resistant to 13–16% ethanol.

A redetermination of the lactic acid configuration produced by the type strain in our laboratory yielded equal amounts of D- and L(+)-lactic acid.

Additional physiological and biochemical characteristics are presented in Tables 14.7 and 14.8.

Growth factor requirements: D-mevalonic acid is essential or highly stimulatory; ethanol is promotive.

DNA/DNA homology: the type strain of *L. homohiochii* was found to be genetically highly related to the type strain of *L. sake* but unrelated to other streptobacteria (Dellaglio et al., 1975). In our laboratory, however, two strains of *L. homohiochii* were completely homologous with each other and had only 10% homology with *L. sake* (E. Lauer, unpublished results).

Isolated from spoiled sake.

The mol% G + C of the DNA is 35–38 (T_m). The values of 46% obtained by chemical analysis (Gasser and Sebald, 1966) and by T_m

(eighth edition Bergey's *Manual*) are in contrast with the values found by Momose et al., 1974 (34.6–36.8%) and our own unpublished results (38%).

Type strain: ATCC 15434.

23. **Lactobacillus maltaromicus** Miller, Morgan and Libbey 1974, 352.[AL]

malt.a.ro′mi.cus. M.E. n. *malte* ground dried sprouted barley; L. n. *aroma* pleasant flavor; M.L. adj. *maltaromicus*, producing a maltlike aroma.

Slender rods of varying length with a tendency to form filaments and long chains.

No growth at 45°C. Besides moderate amounts of L(+)-lactic acid (~1.5 g/liter) different aldehydes and alcohols such as 2-methylpropionaldehyde, 2-methylpropanol, 3-methylbutyraldehyde and 3-methylbutanol are produced in skim milk and trypticase soy broth.

Additional physiological and biochemical characteristics are presented in Tables 14.7 and 14.8.

Growth factor requirements: riboflavin and folic acid are essential, thiamine is not required.

DNA/DNA homology: no genetic relationship could be detected between *L. maltaromicus* and other *meso*-DAP-containing lactobacilli producing L(+)-lactic acid (Weiss et al., 1981).

Isolated from producers' milk samples possessing a malty flavor.

The mol% G + C content of the DNA is about 36.0 (T_m).

Type strain: ATCC 27865.

24. **Lactobacillus murinus** Hemme, Raibaud, Ducluzeau, Galpin, Sicard and van Heijenoort 1982, 384.[VP] (*Effective publication*: Hemme, Raibaud, Ducluzeau, Galpin, Sicard and van Heijenoort 1980, 306.)

mu.ri′nus. L. adj. *murinus* of mice.

Rods with rounded ends, 0.8–1.0 × 2.0–4.0 μm, frequently in chains.

Good growth at 45°C. Ribose and arabinose slowly fermented. L-LDH is activated by FDP and Mn^{2+}.

Additional physiological and biochemical characteristics are presented in Tables 14.7 and 14.8.

Growth factor requirement: riboflavin is essential, thiamine and vitamin B_{12} not required.

DNA/DNA homology: two strains tested were completely homologous to each other but unrelated to *L. alimentarius*, *L. casei*, *L. sake*, *L. curvatus* and *L. salivarius* (E. Lauer, unpublished results).

Isolated from the intestinal tract of mice and rats.

The mol% G + C of the DNA is 43.4–44.3 (T_m).

Type strain: CNRZ 220.

25. **Lactobacillus plantarum** (Orla-Jensen 1919) Bergey, Harrison, Breed, Hammer and Huntoon 1923, 250.[AL] (*Streptobacterium plantarum* Orla-Jensen 1919, 174.)

plan.ta′rum. L. fem. n. *planta* a sprout; M.L. n. *planta* a plant; M.L. gen. pl. n. *plantarum* of plants.

Rods with rounded ends, straight, generally 0.9–1.2 μm wide × 3–8 μm long, occurring singly, in pairs or in short chains.

No growth at 45°C. Some strains are able to reduce nitrate provided the concentration of glucose in the medium is limited and the pH thus poised at 6.0 or higher. Occasional strains exhibit pseudocatalase activity especially if grown under glucose limitation. Cell walls contain either ribitol or glycerol teichoic acid.

Additional physiological and biochemical characteristics are presented in Tables 14.7 and 14.8.

Growth factor requirements: calcium pantothenate and niacin required; thiamine, pyridoxal or pyridoxamine, folic acid, vitamin B_{12}, thymidine or deoxyribosides not required; riboflavin generally not required.

DNA/DNA homology: *L. plantarum* strains form two homology groups genomically related to each other at the level of 50–60%, but not related to other streptobacteria and other *meso*-DAP-containing lactobacilli species (Dellaglio et al., 1975; Weiss et al., 1981). "*L. pentosus*" Fred et al., 1921, and several other strains designated *L.*

plantarum form a third genotype only related at the 50% level to the two other genotypes. Therefore it should be regarded as a separate species (see Comments).

Isolated from dairy products and environments, silage, sauerkraut, pickled vegetables, sour dough, cow dung, and the human mouth, intestinal tract and stools, and from sewage.

The mol% G + C of the DNA is 44–46 (Bd, T_m).

Type strain: ATCC 14917.

Further comments. In the course of the last few years, a number of strains, mainly from sewage, have been isolated in the authors' laboratory. These strains shared high DNA/DNA homology with "*L. pentosus*" but only low homology with *L. plantarum*. Characteristically, these strains fermented glycerol whereas none of the strains within the *L. plantarum* homology group did so. A description of these organisms as *L. pentosus* nom. rev. is in preparation.

26. **Lactobacillus sake** Katagiri, Kitahara and Fukami 1934, 157.[AL]

sa′ke. Japanese n. *sake* rice wine; M.L. n. *sake* rice wine.

Rods with rounded ends, generally 0.6–0.8 × 2–3 μm, occurring singly and in short chains; frequently slightly curved and irregular, especially during stationary growth phase.

No growth at 45°C; many of the strains tested grow even at 2–4°C.

L-LDH is activated by FDP and Mn^{2+}. Possesses lactic acid racemase; induction of racemase in most strains is repressed by acetate. Therefore, the majority of strains produce L(+)-lactic acid in MRS broth whereas DL-lactic acid is produced in cabbage press juice. A few strains, whose identity with *L. sake* is confirmed by DNA/DNA homology, however, produce inactive lactic acid, also, in MRS broth.

Additional physiological and biochemical characteristics are presented in Tables 14.7 and 14.8.

DNA/DNA homology: the type strain and many other strains form a narrow homology group not significantly related to other lactobacilli except *L. bavaricus* (Kagermeier 1981) and *L. curvatus* (Dellaglio et al., 1975; Kagermeier 1981). While most strains of *L. bavaricus* exhibit complete DNA/DNA homology with *L. sake*, *L. curvatus* and *L. sake* are related to each other at a level of 40–50% homology. The high level of homology between *L. sake* and *L. homohiochii* reported by Dellaglio et al., (1975) could not be confirmed in our laboratory. Two strains of *L. homohiochii* completely homologous to each other show only about 10% homology to *L. sake*.

Originally isolated from sake starter; regularly found in sauerkraut and other fermented plant material, meat products and prepacked finished dough.

The mol% G + C of the DNA is 42–44 (T_m).

Type strain: ATCC 15521.

Note. Some of the atypical streptobacteria from herbage, silage, fermented meat products and vacuum packaged meat reported in the past probably belong to *L. sake*.

27. **Lactobacillus bifermentans** Kandler, Schillinger and Weiss 1983, 896.[VP] (*Effective publication*: Kandler, Schillinger and Weiss 1983, 409.)

bi.fer.men′tans. L. pref. *bis* twice; L. part. *fermentans* leavening; M.L. part. adj. *bifermentans* doubly fermenting.

Irregular rods with rounded or often tapered ends, 0.5–1.0 × 1.5–2.0 μm, occurring singly, in pairs or irregular short chains, often forming clumps.

No growth at 45°C.

Homofermentative with production of DL-lactic acid in media containing more than 1% fermentable hexoses. Lactic acid is fermented to acetic acid, ethanol, CO_2 and H_2 at pH >4.0.

Additional physiological and biochemical characteristics are presented in Tables 14.9 and 14.10.

DNA/DNA homology: no genomic relationship is detected between the type strain of *L. bifermentans* and heterofermentative lactobacilli (Vescovo et al., 1979).

In contrast to all other lactobacilli *L. bifermentans* ferments lactate

Table 14.9.

Pattern of fermented carbohydrates of the obligately heterofermentative species of the genus **Lactobacillus** *(group III)[a]*

Species	Amygdalin	Arabinose	Cellobiose	Esculin	Fructose	Galactose	Glucose	Gluconate	Lactose	Maltose	Mannitol	Mannose	Melezitose	Melibiose	Raffinose	Rhamnose	Ribose	Salicin	Sorbitol	Sucrose	Trehalose	Xylose
27. L. bifermentans	–	–	–	–	+	+	+	–	–	+	+	+	–	–	–	+	+	–	–	–	–	–
28. L. brevis	–	+	–	d	+	d	+	+	d	+	–	–	–	+	d	–	+	–	–	d	–	d
29. L. buchneri	–	+	–	d	+	d	+	+	d	+	–	–	+	+	d	–	+	–	–	d	–	d
30. L. collinoides	–	+	+	+	+	+	+	+	d	+	–	+	–	+	–	–	+	d	–	+	+	+
31. L. confusus	+	–	+	+	+	d	+	+	–	+	–	+	–	–	–	–	+	+	–	+	+	+
32. L. divergens	+	d	+	o	+	d	+	+	–	+	–	+	–	–	–	–	+	+	–	+	–	–
33. L. fermentum	–	d	d	–	+	+	+	d	+	+	–	+w	–	+	+	–	+	–	–	d	d	d
34. L. fructivorans	–	–	–	o	+	–	+	d	–	d	–	–	–	–	–	–	+w	–	–	–	–	–
35. L. fructosus	o	–	–	o	+	–	+	–	–	–	–	–	–	–	–	–	–	–	–	–	–	–
36. L. halotolerans	–	–	–	–	+	d	+	+	d	+	–	+	–	–	–	–	+	–	–	–	+	+
37. L. hilgardii	–	–	–	–	+	d	+	+	d	+	–	–	d	–	–	–	+	–	–	d	–	–
38. L. kandleri	–	–	–	–	+	+	+	+	–	–	+	–	–	–	–	o	+	–	–	–	+	–
39. L. kefir	–	d	–	–	+	–	+	+	+	+	–	–	–	+	–	–	+	–	–	–	–	–
40. L. minor	–	–	+	+	+	–	+	+	+	+	–	+	+	–	–	–	+	–	–	+	+	–
41. L. reuteri	o	+	–	o	+	+	+	+	+	+	–	–	–	+	+	–	+	–	–	+	–	–
42. L. sanfrancisco	o	–	–	o	–	+	+	+	–	+	–	–	–	–	–	–	–	–	–	–	–	–
43. L. vaccinostercus	–	+	+w	–	–	+w	+	–	–	+	–	–	–	–	–	–	+	–	–	–	–	+
44. L. viridescens	–	–	–	–	+	–	+	–	–	+	–	+	–	–	–	–	–	–	–	d	d	–

[a] Symbols: see Table 14.5; and o, reaction not determined.

1230

Table 14.10.
Physiological and biochemical characteristics of obligately heterofermentative species of the genus **Lactobacillus** *(group III)*[a]

Species	Peptidoglycan type[b]	Teichoic acid	Electrophoretic mobility[c]		Allosteric L-LDH	Mol% G + C	Lactic acid isomer(s)[d]	Growth at 15°C	NH₃ from arginine
			D-LDH	L-LDH					
27. *L. bifermentans*	Lys-DAsp	None	1.10	1.20	−	45	DL	+	−
28. *L. brevis*	Lys-DAsp	Glycerol	1.62	1.40	−	44–47	DL	+	+
29. *L. buchneri*	Lys-DAsp	Glycerol	1.33	1.26	−	44–46	DL	+	+
30. *L. collinoides*	Lys-DAsp	Glycerol	1.50	1.22	−	46	DL	+	+
31. *L. confusus*	Lys-Ala	None	2.08	1.82	−	45–47	DL	+	+
32. *L. divergens*	mDAP-Direct	None	−	1.30	−	33–35	L	+	+
33. *L. fermentum*	Orn-DAsp	None	1.85	−	−	52–54	DL	−	+
34. *L. fructivorans*	Lys-DAsp	None	ND	ND	−	38–41	DL	+	+
35. *L. fructosus*	Lys-Ala	None	1.32	1.14	−	47	D(L)	+	−
36. *L. halotolerans*	Lys-Ala-Ser	Glycerol	1.75	1.30	−	45	DL	+	+
37. *L. hilgardii*	Lys-DAsp	Glycerol	1.31	0.97	−	39–41	DL	+	+
38. *L. kandleri*	Lys-Ala-Gly-Ala₂	None	2.10	−	−	39	DL	+	+
39. *L. kefir*	Lys-DAsp	Glycerol	1.23	1.07	−	41–42	DL	+	+
40. *L. minor*	Lys-Ser-Ala₂	Glycerol	2.08	1.50	−	44	DL	+	+
41. *L. reuteri*	Lys-DAsp	None	1.74	0.88	−	40–42	DL	−	+
42. *L. sanfrancisco*	Lys-Ala	None	1.18	1.05	−	36–38	DL	+	−
43. *L. vaccinostercus*	mDAP-Direct	ND	1.32	1.18	−	36	DL	−	−
44. *L. viridescens*	Lys-Ala-Ser	Ribitol	2.03	−	−	41–44	DL	+	−

[a] Symbols: see Table 14.5; and ND, not determined.
[b–d] Footnotes: see Table 14.6.

and produces free H₂ and was therefore put on the list of species incertae sedis in the eighth edition of the *Manual*. 16S rRNA cataloging (Stackebrandt et al., 1983) and more recently DNA/rRNA hybridization studies (U. Schillinger, personal communication), however, gave strong evidence that *L. bifermentans* belongs to the genus *Lactobacillus* (see also under Taxonomic Comments).

Isolated from spoiled Edam and Gouda cheeses where it forms undesired small cracks ("Boekelscheuren" Pette and Beynum 1943).

The mol% G + C of the DNA is 45 (T_m).

Type strain: DSM 20003.

28. **Lactobacillus brevis** (Orla-Jensen 1919) Bergey, Breed, Hammer, Huntoon, Murray and Harrison 1934, 312.[AL] (*Betabacterium breve* Orla-Jensen 1919, 175.)

bre′vis. L. adj. *brevis* short.

Rods with rounded ends, generally 0.7–1.0 × 2–4 µm, occurring singly and in short chains.

No growth at 45°C.

Additional physiological and biochemical characteristics are presented in Tables 14.9 and 14.10.

Growth factor requirements: Calcium pantothenate, niacin, thiamine and folic acid are essential; riboflavin, pyridoxal and vitamin B₁₂ not required.

DNA/DNA homology: only 11 out of 24 strains originally labeled *L. brevis* form a narrow homology group including the type strain of *L. brevis*. The remaining strains were found homologous with *L. hilgardii*, *L. kefir*, *L. confusus* or *L. collinoides* or remained unassigned (Vescovo et al., 1979).

Isolated from milk, cheese, sauerkraut, sour dough, silage, cow manure, feces, mouth and intestinal tract of humans and rats.

The mol% G + C of the DNA is 44–47 (Bd, T_m).

Type strain: ATCC 14869.

Further comments. L. brevis is often difficult to distinguish clearly from *L. buchneri*, *L. hilgardii*, *L. collinoides* or *L. kefir* by simple physiological tests, especially carbohydrate fermentation reactions. In addition to DNA/DNA homology, characterization of the electrophoretic mobility of lactic acid dehydrogenases seems the most reliable procedure to separate these species.

29. **Lactobacillus buchneri** (Henneberg 1903) Bergey, Harrison, Breed, Hammer and Huntoon 1923, 251.[AL] (*Bacillus Buchneri* (*sic*) Henneberg 1903, 163.)

buch′ne.ri. M.L. gen. n. *buchneri* of Buchner; named for E. Buchner, a German bacteriologist.

Rods with rounded ends, generally 0.7–1.0 × 2–4 µm, occurring singly and in short chains.

No growth at 45°C.

L. buchneri is identical in almost all characteristics with *L. brevis*, except *L. buchneri* ferments melezitose and its L-LDH and D-LDH migrate distinctly slower in electrophoresis. However, at least one strain studied in detail in our laboratory did not ferment melezitose, although its LDH was electrophoretically identical with *L. buchneri*. Another strain was melezitose-positive but behaved like *L. brevis* in electrophoresis.

Additional physiological and biochemical characteristics are presented in Tables 14.9 and 14.10.

DNA/DNA homology: in spite of the high phenotypic similarities mentioned above, six strains of *L. buchneri* formed a narrow homology group completely unrelated to *L. brevis* and other heterofermentative lactobacilli (Vescovo et al., 1979).

Isolated from milk, cheese, fermenting plant material and human mouth.

The mol% G + C of the DNA is 44–46 (Bd, T_m).

Type strain: ATCC 4005.

30. **Lactobacillus collinoides** Carr and Davies 1972, 470.[AL]

col.li.no.i′des. L. adj. *collinus* hilly; Gr. n. *idus* form, shape; M.L. adj. *collinoides* hill-shaped, pertaining to colony form.

Rods with rounded ends, generally 0.6–0.8 × 3–5 µm; tendency to form long filaments, occurring singly, in palisades and irregular clumps.

No growth at 45°C. Growth in MRS broth is distinctly improved by the addition of 20% tomato juice and by replacement of glucose by maltose.

Additional physiological and biochemical characteristics are presented in Tables 14.9 and 14.10.

DNA/DNA homology: six strains tested form a narrow homology group not related to other heterofermentative lactobacilli (Vescovo et al., 1979).

Isolated from cider.

The mol% G + C of the DNA is 46 (T_m).
Type strain: ATCC 27612.

31. Lactobacillus confusus (Holzapfel and Kandler 1969) Sharpe, Garvie and Tilbury 1972, 396.[AL] (*Lactobacillus coprophilus* subsp. *confusus* Holzapfel and Kandler 1969, 665.)

con.fu′sus. L. v. *confundere* to confuse; L. past part. *confusus* confused, an allusion to its original confusion with *Leuconostoc*.

Short rods, 0.8–1.0 × 1.5–3 μm, with tendency to thicken at one end; occurring singly, rarely in short chains.

Growth at 45°C variable. Dextran is produced from sucrose.

Additional physiological and biochemical characteristics are presented in Tables 14.9 and 14.10.

DNA/DNA homology: four strains form a narrow homology group to which two strains are only distantly related (Vescovo et al., 1979). One of the deviating strains (DSM 20194) displayed 73% homology with the type strain when reinvestigated in our laboratory. No significant genomic relationship to other heterofermentative lactobacilli was detected.

Isolated from sugarcane and carrot juice, occasionally found in raw milk, saliva and sewage.

The mol% G + C of the DNA is 45–47 (Bd, T_m).
Type strain: ATCC 10881.

32. Lactobacillus divergens Holzapfel and Gerber 1984, 270.[VP] (*Effective publication:* Holzapfel and Gerber 1983, 530.)

di.ver′gens. L. part. *divergens* deviating, diverging.

Rods with rounded ends, 0.5–0.7 × 1.0–1.5 μm, occurring singly, in pairs and in short chains.

No growth at 45°C. Growth in MRS broth is relatively poor and visible gas is not, or only faintly, produced because of lack of an hitherto undetermined growth factor. This growth factor is contained in peptone from soybeans and certain yeast paste preparations ("Cenobis") and is produced by some molds which occur as laboratory infections. Pseudocatalase is produced on haem-containing media.

Additional physiological and biochemical characteristics are presented in Tables 14.9 and 14.10.

Isolated from vacuum-packaged, refrigerated meat.

The mol% G + C of the DNA is 33–35 (T_m).
Type strain: DSM 20623 (strain 66).

33. Lactobacillus fermentum Beijerinck 1901, 233.[AL] (*Lactobacillus cellobiosus* Rogosa, Wiseman, Mitchell and Disraely 1953, 693.[AL])

Note. Because of high phenotypic similarities and complete DNA/DNA homology, *L. cellobiosus* is here regarded as a biotype of *L. fermentum.*

fer.men′tum. L. n. *fermentum* ferment, yeast.

Rods, 0.5–0.9 μm thick and highly variable in length, mostly occurring singly or in pairs.

Generally good growth at 45°C.

Additional physiological and biochemical characteristics are presented in Tables 14.9 and 14.10.

Growth factor requirements: calcium pantothenate, niacin and thiamine are essential; riboflavin, pyridoxal and folic acid not required.

DNA/DNA homology: strains of *L. fermentum* and *L. cellobiosus* form a narrow homology group not related to other heterofermentative lactobacilli (Vescovo et al., 1979).

Isolated from yeast, milk products, sour dough, fermenting plant material, manure, sewage and mouth and feces of man.

The mol% G + C of the DNA is 52–54 (Bd, T_m).
Type strain: ATCC 14931.

Further comments. *L. fermentum* cannot be definitely distinguished from *L. reuteri* by simple physiological tests. Determinations of mol% G + C of the DNA, diamino acid of murein and electrophoretic mobility of LDH clearly separate the two species.

34. Lactobacillus fructivorans Charlton, Nelson and Werkman 1934, 1.[AL] (*Lactobacillus trichodes* Fornachon, Douglas and Vaughn 1949, 129[AL]; *Lactobacillus heterohiochii* Kitahara, Kaneko and Goto 1957, 117.[AL])

Note. Because of high phenotypic and genomic similarities found between *L. fructivorans, L. trichodes* and *L. heterohiochii, L. trichodes* and *L. heterohiochii* are here regarded as junior subjective synonyms of *L. fructivorans* (see Weiss et al., 1983a).

fruc.ti.vo′rans. L. n. *fructus* fruit; L. v. *vorare* to eat; M.L. pres. part. *fructivorans* fruit-eating, intended to mean fructose-devouring.

Rods with rounded ends, generally 0.5–0.8 × 1.5–4 μm, occurring singly, in pairs and in chains; very long, more or less curved or coiled filaments often observed.

No growth at 45°C.

Acidophilic; favorable pH is 5.0–5.5; no growth at an initial pH higher than 6.0.

Nutritionally very exacting, at least on primary isolation. Depending on the source of isolation, mevalonic acid, tomato juice and/or ethanol are required for growth. Some strains, especially those isolated from nonalcohol-containing sources, often become less fastidious during laboratory transfers and grow well in MRS broth.

Additional physiological and biochemical characteristics are presented in Tables 14.9 and 14.10.

DNA/DNA homology: the type strains of *L. fructivorans, L. trichodes* and *L. heterohiochii* and two additional strains are highly homologous among each other and genomically not related to other heterofermentative lactobacilli (Vescoco et al., 1979; Weiss et al., 1983a). The high homology values between *L. heterohiochii* and *L. buchneri* reported by Vescovo and co-workers could not be confirmed in our laboratory and may be caused by the use of an impure or mislabeled culture of *L. heterohiochii.*

Isolated from spoiled mayonnaise, salad dressings and vinegar preserves; from spoiled sake, dessert wines and aperitifs.

The mol% G + C of the DNA is 38–40 (T_m).
Type strain: ATCC 8288.

35. Lactobacillus fructosus Kodama 1956, 705.[AL]

fruc.to′sus. M.L. adj. *fructosus* of fructose, pertaining to fructose.

Rods, 0.5–0.8 × 2–4 μm, occurring singly, in pairs and in short chains.

No growth at 45°C. Growth in MRS broth is markedly improved if glucose is replaced by fructose.

Additional physiological and biochemical characteristics are presented in Tables 14.9 and 14.10.

Isolated from flowers.

The mol% G + C of the DNA is 47 (T_m).
Type strain: ATCC 13162.

36. Lactobacillus halotolerans Kandler, Schillinger and Weiss 1983, 672.[VP] (*Effective publication:* Kandler et al., 1983, 283.)

ha.lo.to′le.rans. Gr. n. *hals* salt; L. pres. part. *tolerans* tolerating, enduring; M.L. part. adj. *halotolerans* salt tolerating.

Irregular, short or even coccoid rods with rounded to tapered ends, generally 0.5–0.7 × 1–3 μm, sometimes longer, with tendency to form coiling chains, clumping together.

No growth at 45°C. Good growth in the presence of 12% NaCl and very weak growth in the presence of 14% NaCl.

Additional physiological and biochemical characteristics are presented in Tables 14.9 and 14.10.

DNA/DNA homology: no significant genomic relationship between *L. halotolerans* and other heterofermentative lactobacilli is detected (Vescovo et al., 1979).

Isolated from meat products.

The mol% G + C of the DNA is 45 (T_m).
Type strain: DSM 20190 (= strain R61).

37. Lactobacillus hilgardii Douglas and Cruess 1936, 115.[AL]

hil.gar′di.i. M.L. gen. n. *hilgardii* named for Hilgard.

Rods with rounded ends, generally 0.5–0.8 × 2–4 μm, occurring singly, in short chains and frequently in long filaments.

No growth at 45°C. Optimal initial pH for growth and carbohydrate fermentation reactions is in the range of 4.5–5.5. Grows in the presence of 15–18% ethanol.

Additional physiological and biochemical characteristics are presented in Tables 14.9 and 14.10.

DNA/DNA homology: genomically 13 strains, mostly isolated from wine and originally allocated to a variety of different species such as *L. brevis*, "*L. desidiosus*," *L. reuteri* and "*Betabacterium vermiforme*," are highly related to the type strain of *L. hilgardii*. No relationship to other heterofermentative lactobacilli is detected (Vescovo et al., 1979).

Originally isolated from California table wines but obviously widely distributed in wines of different origin.

The mol% G + C of the DNA is 39–41 (T_m).

Type strain: ATCC 8290.

38. Lactobacillus kandleri Holzapfel and van Wyk 1983, 439.[VP]

(*Effective publication*: Holzapfel and van Wyk 1982, 501.)

kand'le.ri. M.L. gen. n. *kandleri* of Kandler; named for O. Kandler, a German biologist.

Partly irregular rods, generally 0.7–0.8 × 1–5 μm, occurring singly or in pairs, seldom in short chains.

No growth at 45°C. Slime is produced from sucrose.

Additional physiological and biochemical characteristics are presented in Tables 14.9 and 14.10.

Isolated from a desert spring.

The mol% G + C of the DNA is 39 (T_m).

Type strain: DSM 20593.

39. Lactobacillus kefir Kandler and Kunath 1983, 672.[VP] (*Effective publication*: Kandler and Kunath 1983, 292.)

ke'fir. Turkish n. *kefir*, a Caucasian sour milk.

Rods with rounded ends, generally 0.6–0.8 × 3.0–15 μm, with tendency to form chains of short rods or long filaments.

No growth at 45°C.

Additional physiological and biochemical characteristics are presented in Tables 14.9 and 14.10.

DNA/DNA homology: strains isolated from kefir, together with two isolates from beer, form a narrow homology group genomically unrelated to other heterofermentative lactobacilli (Vescovo et al., 1979). *L. kefir* exhibits a DNA/DNA homology of about 40% to *L. buchneri* (Kandler and Kunath, 1983).

Isolated from kefir grains and drink kefir.

The mol% G + C of the DNA is 41–42 (T_m).

Type strain: DSM 20587 (strain A/K).

40. Lactobacillus minor Kandler, Schillinger and Weiss 1983, 672.[VP] (*Effective publication*: Kandler, Schillinger and Weiss 1983, 284.) (*Lactobacillus corynoides* subsp. *minor* Abo-Elnaga and Kandler 1965b, 128; *Lactobacillus viridescens* subsp. *minor* Kandler and Abo-Elnaga 1966, 754.)

mi'nor. L. comp. adj. *minor* smaller.

Irregular, short rods with rounded to tapered ends, generally 0.6–0.8 × 1.5–2.0 μm, sometimes longer, often bent with unilateral swellings, occurring in pairs or short chains with a tendency to form loose clusters.

No growth at 45°C.

Additional physiological and biochemical characteristics are presented in Tables 14.9 and 14.10.

DNA/DNA homology: no genetic relationship detected between *L. minor* and other heterofermentative lactobacilli (Vescovo et al., 1979).

Isolated from the sludge of milking machines.

The mol% G + C of the DNA is 44 (T_m).

Type strain: DSM 20014 (strain 3).

41. Lactobacillus reuteri Kandler, Stetter and Köhl 1982, 266.[VP] (*Effective publication*: Kandler, Stetter and Köhl 1980, 267.) (*Lactobacillus fermentum* Type II Lerche and Reuter 1962, 462.)

reu'te.ri. M.L. gen. n. *reuteri* of Reuter; named for G. Reuter, a German bacteriologist.

Slightly irregular, bent rods with rounded ends, generally 0.7–1.0 ×

2.0–5.0 μm, occurring singly, in pairs and in small clusters. Generally good growth at 45°C. In the original description, it was mistakenly stated that ammonia is not produced from arginine.

Additional physiological and biochemical characteristics are presented in Tables 14.9 and 14.10.

DNA/DNA homology: five strains tested form a narrow homology group not related to other heterofermentative lactobacilli (Vescovo et al., 1979; Dellaglio, personal communication). In addition, 218 strains isolated from feces of milking calves and indistinguishable from *L. fermentum* by physiological tests, displayed almost complete homology with *L. reuteri* but are genetically unrelated to *L. fermentum* (Sarra et al., 1979).

Isolated from feces of humans and animals and from meat products.

The mol% G + C of the DNA is 40–42.3 (Bd, T_m).

Type strain: DSM 20016.

Note. L. reuteri cannot be definitely distinguished from *L. fermentum* by simple physiological tests. Determination of mol% G + C, diamino acid of peptidoglycan or electrophoretic mobility of LDH clearly separates the two species.

42. Lactobacillus sanfrancisco Weiss and Schillinger 1984, 503.[VP]

(*Effective publication*: Weiss and Schillinger 1984, 231.)

Note. Kline and Sugihara (1971) proposed the name *L. sanfrancisco* with reservation as to results of pending DNA/DNA homology studies. Later on they confirmed briefly the proposal and designated a type strain (Sugihara and Kline, 1975). The name, however, was omitted from the Approved Lists of Bacterial Names and consequently has no standing in bacteriological nomenclature and was recently revived (Weiss and Schillinger, 1984).

san.fran.cis'co. M.L. n. *sanfrancisco* San Francisco, named after the city where the sour dough from which the organism was first isolated had been propagated for more than 100 years.

Rods with rounded ends, 0.6–0.8 × 2–4 μm, occurring singly and in pairs.

No growth at 45°C. Does not grow reasonably in MRS broth unless freshly prepared yeast extract is added and the initial pH is lowered to 5.6. A small peptide isolated from yeast extract was found responsible for the growth-promoting effect (Berg et al., 1981).

Additional physiological and biochemical characteristics are presented in Tables 14.9 and 14.10.

DNA/DNA homology: four strains tested were found to be highly homologous among each other but showed no significant genomic relationship with *L. acidophilus*, *L. helveticus* and *L. brevis* (Sriranganathan et al., 1973). No significant homology detected with other heterofermentative lactobacilli, especially *L. confusus* and *L. fructosus* containing the same peptidoglycan type (Weiss and Schillinger, 1984) as *L. sanfrancisco*.

Isolated from sour dough.

The mol% G + C of the DNA is 36–38 (T_m).

Type strain: NRRL B-3934.

Further comments. DNA-DNA hybridization studies have shown that other lactobacilli isolated from sour dough and labeled *L. brevis* var. *lindneri* (Spicher and Schröder, 1978) are identical with *L. sanfrancisco*.

43. Lactobacillus vaccinostercus Okada, Suzuki and Kozaki 1983, 439.[VP] (*Effective publication*: Okada, Suzuki and Kozaki, 1979, 217.)

vac.ci.no.ster'cus. L. adj. *vaccinus* from cows, L. n. *stercus* dung; M.L. adj. *vaccinostercus* from cow dung.

Rods with rounded ends, 0.9–1.0 × 1.5–2.5 μm, occurring mostly in pairs.

No growth at 45°C.

Additional physiological and biochemical characteristics are presented in Tables 14.9 and 14.10.

Growth factor requirements: thiamine, pantothenic acid, niacin, and biotin are essential; pyridoxal, *p*-aminobenzoic acid and folic acid not required.

Isolated from cow dung.

The mol% G + C of the DNA is 36 (T_m).

Type strain: ATCC 33310.

44. **Lactobacillus viridescens** Niven and Evans 1957, 758.[AL] (*Lactobacillus corynoides* subsp. *corynoides* Kandler and Abo-Elnaga 1966, 573.[AL])

Note. L. viridescens is incorrectly cited on the Approved Lists of Bacterial Names as *Lactobacillus viridescens* Kandler and Abo-Elnaga 1966, 573.

vi.ri.des'cens. M.L. pres. part. *viridescens* growing green, greening.

Small, often slightly irregular rods, generally 0.7–0.9 × 2.0–5.0 μm, with rounded to tapered ends, occurring singly or in pairs.

No growth at 45°C. In contrast to the data given in the eighth edition of the *Manual*, no fermentation of ribose and gluconate could be observed in our laboratory.

Additional physiological and biochemical characteristics are presented in Tables 14.9 and 14.10.

Growth factor requirements: pantothenate, niacin, thiamine, riboflavin, and biotin are essential; folic acid and pyridoxal may be stimulatory.

DNA/DNA homology: four strains tested form a narrow homology group genetically not related to other heterofermentative lactobacilli (Vescovo et al. 1979).

Isolated from discolored cured meat products and pasteurized milk.

The mol% G + C of the DNA is 41–44 (Bd, T_m).

Type strain: ATCC 12706.

Addendum I

Lactobacillus species included in the Approved Lists of Bacterial Names but, for reasons discussed below, not considered as belonging to the genus *Lactobacillus*.

Lactobacillus catenaforme (sic) (Eggerth 1935) Moore and Holdeman 1970, 15.[AL] (*Bacteroides catenaformis* Eggerth 1935, 286.)

ca.te.na.for'me. L. n. *catena* chain; L. n. *forma* form, shape; M.L. adj. *catenaforme* chainlike. *Note:* the correct specific epithet should read *catenaformis* because *Lactobacillus* is masculine in gender.

Small, slightly irregular rods, often in chains. Strictly anaerobic.

Good growth at 45°C. Acid is produced from amygdalin, cellobiose, esculin, fructose, glucose, glycogen, mannose, salicin, starch and sucrose, fermentation of lactose and maltose is recorded variable.

Main product from glucose fermentation is D(−)-lactic acid. No gas is produced from glucose.

Peptidoglycan of the type strain is of the Lys-Ala type (unpublished result); this peptidoglycan type is not found in other homofermentative lactobacilli.

Isolated from human feces, intestinal and pleural infections.

The mol% G + C of the DNA is 31–33 (T_m).

Type strain: ATCC 25536.

Further comments. 16S rRNA oligonucleotide sequence studies revealed no significant phylogenetic relationship to any of the lactobacilli investigated (S_{AB} 0.3; E. Stackebrandt, personal communication). The S_{AB} values for streptococci and *Clostridium innocuum* were 0.32 and 0.4, respectively. The taxonomic position of *L. catenaforme* therefore remains undetermined.

Lactobacillus minutus (Hauduroy, Ehringer, Urbain, Guillot and Magrou 1937) Moore and Holdeman 1972, 63.[AL] (*Bacteroides minutus* Hauduroy, Ehringer, Urbain, Guillot and Magrou 1937, 64.)

mi.nu'tus. L. adj. *minutus* minute, small.

Small, elliptical rods, occurring singly, in pairs and in short chains. Strictly anaerobic.

Generally no growth at 45°C. Acid is produced from glucose and variably or weakly from fructose, galactose and sucrose.

Main product from glucose fermentation is D(−)-lactic acid. No gas is produced from glucose.

Peptidoglycan of two strains studied including the type strain was of the Orn-Ser-DGlu type (unpublished result); this peptidoglycan type has not been found in other bacteria up to now.

The mol% G + C of the DNA is 45 (T_m).

Isolated from abscesses and wounds.

Type strain: VPI 9428 (ATCC 33267).

Further comments. 16S rRNA oligonucleotide sequence studies revealed no significant phylogenetic relationship to any of the lactobacilli investigated (S_{AB} 3.0; E. Stackebrandt, personal communication). The taxonomic position of *L. minutus* therefore remains undetermined.

Lactobacillus rogosae Holdeman and Moore 1974, 275.[AL]

ro.go'sae. M.L. gen. n. *rogosae* of Rogosa; named for M. Rogosa, an American bacteriologist.

Further comments. No strains which correspond to the original description are presently available. Two strains recently received from VPI, in our hands, were morphologically similar to propionibacteria; they produced mainly L(+)-lactic acid in PYG medium, but formed acetic acid and propionic acid in chopped meat media at the expense of the lactic acid naturally contained in these media. The peptidoglycan of these two strains were of the LL-DAP-Gly type, typical of propionibacteria. Moreover, the mol% G + C of 59 reported for one strain of *L. rogosae* is clearly outside the range determined in all other species of *Lactobacillus*. More investigations are needed to clarify the taxonomic position of *L. rogosae*.

Lactobacillus xylosus Kitahara 1938, 1449.[AL]

xy.lo'sus. M.L. adj. *xylosus* of xylose, pertaining to xylose.

Further comments. Both nucleic acid hybridization studies (Killper-Bälz et al., 1982) and immunological investigations of fructose-diphosphate aldolase and glyceraldehyde-3-phosphate dehydrogenase (London and Chace, 1983) have shown that *L. xylosus* has not been attributed to the appropriate genus. Because of the results of DNA/DNA homology and DNA/rRNA hybridization studies, Killper-Bälz et al. (1982) stated that *L. xylosus* "should be reclassified in the same genus or even the same species as *Streptococcus lactis*." A definite statement on the taxonomic position of *L. xylosus* is required.

Addendum II

The following *Lactobacillus* species are not included in the Approved Lists of Bacterial Names and have not been validly described since 1980. They have, therefore, no standing in bacteriological nomenclature.

"*Lactobacillus frigidus*" Bhandari and Walker 1953, 333.
Reference strain: ATCC 11307.

"*Lactobacillus malefermentans*" Russell and Walker 1953, 162.

"*Lactobacillus lindneri*" (Henneberg 1901) Bergey, Harrison, Breed, Hammer and Huntoon 1923, 245.

The three species are obligately heterofermentative lactobacilli and have been isolated from beer and brewery yeast. As shown by DNA/DNA homology studies, strains of the above-mentioned species are genomically not significantly related either among each other or to any other heterofermentative lactobacilli species (Vescovo et al., 1979; U. Schillinger, personal communication). They can, therefore, be regarded as additional separate species within the heterofermentative lactobacilli (group III). More comparative studies based on a greater number of strains are required for the revival of the presently invalid names.

Genus *Listeria* Pirie 1940, 383[AL]

H. P. R. SEELIGER AND DOROTHY JONES

Lis.te′ri.a. M.L. fem. n. *Listeria* named after Lord Lister, an English surgeon.

Regular, short rods, 0.4–0.5 μm in diameter and 0.5–2 μm in length with rounded ends. Some cells may be curved. Occur singly, in short chains or the cells may be arranged at an angle to each other to give V forms or in groups lying parallel along the long axis. **In older or rough cultures, filaments of 6–20 μm or more in length may develop. Gram-positive,** but some cells, especially in older cultures, lose the ability to retain the Gram stain. **Not acid fast. Capsules not formed. Do not form spores. Motile by a few peritrichous flagella when cultured at 20–25°C. Aerobic and facultatively anaerobic.** Colonies (24–48 h) on nutrient agar are 0.5–1.5 mm in diameter, round, translucent with dew-drop appearance, low convex with a finely textured surface and entire margin. **The colonies appear bluish gray by normal illumination and a characteristic blue-green sheen is produced by obliquely transmitted light.** May be sticky when removed from agar plate but emulsify easily. Colonies may leave an impression on the agar after removal. Older cultures (3–7 d) are larger, 3–5 mm in diameter, with a more opaque center and rough colonial forms may develop. In a 0.25% (w/v) agar, 8.0% (w/v) gelatin and 1.0% (w/v) glucose semisolid medium, growth along the stab in 24 h at 37°C is followed by irregular, cloudy extensions into the medium. Growth spreads slowly through the entire medium. An umbrella-like zone of maximal growth occurs 3–5 mm below the surface. Some species are β-hemolytic on blood agar. **Optimum growth temperature between 30 and 37°C. Temperature limits of growth 1–45°C. Does not survive heating at 60°C for 30 min.** Growth occurs between pH 6 and pH 9. Grow in nutrient broth supplemented with up to 10% (w/v) NaCl. **Catalase-positive, oxidase-negative. Cytochromes produced. Fermentative metabolism of glucose results in the production of mainly L(+)-lactic acid. Acid but no gas produced from a number of other sugars. Methyl red positive, Voges-Proskauer positive. Exogenous citrate is not utilized. Organic growth factors are required. Indole is not produced. Esculin and sodium hippurate are hydrolyzed. Urea is not hydrolyzed. Gelatin, casein and milk are not hydrolyzed.** The cell wall contains a directly cross-linked peptidoglycan based upon *meso*-diaminopimelic acid (*meso*-DAP) (variation A1γ of Schleifer and Kandler, 1972); the cell wall does not contain arabinose. Mycolic acids are not present. The long chain fatty acids consist predominantly of straight chain saturated, *anteiso*- and *iso*-methyl-branched chain types. The major fatty acids are 14-methylhexadecanoic (anteiso-$C_{17:0}$) and 12-methyltetradecanoic (anteiso-$C_{15:0}$). Menaquinones are the sole respiratory quinones; the major menaquinone contains seven isoprene units (MK-7). The mol% G + C content of the DNA is 36–38 (T_m). Widely distributed in nature, found in water, mud, sewage, vegetation and in the feces of animals and man. Some species are pathogenic for man and animals.

The species *Listeria grayi, L. murrayi* and *L. denitrificans* are not included in the genus description. The taxonomic position of these taxa is discussed under "Taxonomic Comment." Descriptions of all three are included at the end of the section.

Type species: *Listeria monocytogenes* (Murray, Webb and Swann) Pirie 1940, 383.

Further Descriptive Information

Coccoid forms 0.4–0.5 μm in diameter and 0.4–0.6 μm in length are frequently seen in smears from infected tissue or from liquid culture but are rarely seen in smears from colonies on solid media. These coccoid forms may be mistaken for streptococci; a test for the production of catalase is advisable. It should be done on cultures grown on a suitable medium.

Tests for motility should always be done at room temperature (20–25°C); at 37°C flagella development may be so poor that the organisms appear nonmotile. In hanging-drop preparations, tumbling and rotatory movements may alternate with periods of comparative rest. In doubtful cases motility can be demonstrated by inoculating strains into semisolid, nutrient agar in a modified Craigie tube (i.e. without added antiserum) or in a U tube. The medium at the top of the Craigie tube or at the top of one arm of the U tube is inoculated with the bacterium and incubated at room temperature. Progress of motile bacteria can be followed by a clouding of the medium either in the outer agar around the Craigie tube or just below the surface in the uninoculated arm of the U tube. Flagellation is poor, usually two or three peritrichous flagella and rarely more than five are seen either by the usual staining techniques or by electron microscopy.

On the basis of immunological and immune electron microscope studies, Smith and Metzger (1962) reported the presence of a mucopolysaccharide capsule in *L. monocytogenes.* However, there are no such other reports and the exhaustive studies of Seeliger and Bockemühl (1968) did not result in the detection of any capsule.

Care must be taken when examining cultures for the production of β-hemolysis. Strains of *L. ivanovii* (Fig. 14.5) tend to produce a very wide zone or multiple zones of hemolysis (Hunter, 1973; Seeliger et al.,

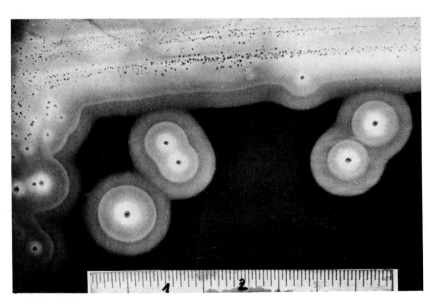

Figure 14.5. *Listeria ivanovii* strain SLCC 3765. Multiple zones of hemolysis on sheep blood (5% v/v) agar incubated for 3 days at 37°C. *Measurement* in centimeters.

1982). *L. monocytogenes* produces narrow zones of hemolysis which frequently do not extend beyond the edge of the colony and may be detected only by moving the colony. A weak or doubtful β-hemolytic reaction can be resolved by using the CAMP test (Christie et al., 1944) to potentiate the hemolytic activity (Brzin and Seeliger, 1975; Groves and Welshimer, 1977). The suspected hemolytic *Listeria* strain and β-toxin-producing *Staphylococcus aureus* strain are streaked at right angles to each other on washed sheep blood (5% v/v) agar. After incubation at 37°C for 24–48 h, hemolytic strains of *L. monocytogenes* or *L. seeligeri* exhibit enhanced hemolysis along the *Listeria* streak within the lytic zone of the β-toxin produced by the *Staphylococcus aureus*. A similar phenomenon is exhibited by strains of *L. ivanovii*, but with *Rhodococcus equi* and not with *Staphylococcus aureus* (Hunter, 1973; Rocourt et al., 1983b).

Hemolytic strains of *Listeria* exhibit β-hemolysis on sheep, cow, horse, rabbit and human blood. However, some blood, notably that of sheep, contains antibodies against *L. monocytogenes*. It is therefore preferable to use saline-washed red blood cells. *L. monocytogenes* produces a soluble hemolysin in liquid culture.

Listeria strains do not agglutinate washed erythrocytes of man, rabbit guinea pig or sheep (Seeliger, 1961).

Surface cultures of most strains may show different colonial types. Freshly isolated strains consist of circular, smooth colonies (S-form) which after 3 days incubation appear white with a glistening surface. After several days the smooth surface changes; the edges of the colony become raised and the center depressed. In addition to the S-form, more grayish, translucent colonies may be seen; the margins of these colonies are less entire and the central depression more marked. Laboratory strains often show this colonial type which is considered as a cultural variant of the S-form. It has a tendency to autoagglutinability. This variant of the S-form is not identical with the rough form (R-form) which occurs after varying periods of laboratory cultivation. R-forms have an undulating rough surface and a very uneven edge. They are difficult to emulsify. A number of other colonial variants have been described (see Seeliger, 1961 for full description).

L-forms have been reported to develop when *L. monocytogenes* is cultivated on media containing either penicillin or glycine, but whether or not these are true L-forms has not been resolved (see Seeliger, 1961; Gray and Killinger, 1966). It has been suggested that these forms and other granular forms play a role in the infectious process of *L. monocytogenes* and may be a reason for the difficulty experienced in isolating the organism, but there is little evidence to support this theory (see Seeliger, 1961; Gray and Killinger, 1966).

Listeria strains grow well on the usual bacteriological media, e.g. Blood Agar Base No. 2 (Difco) or Tryptose Agar (Difco). Growth is enhanced by the addition of glucose (0.2–1% w/v). Cultures grown on such media containing glucose have a characteristic sour, buttermilk-like odor which is especially noticeable on solid media.

All strains grow best at neutral to slightly alkaline pH. Some strains grow at pH 9.6 but all usually die at a pH lower than 5.5. Consequently, viable transfers can rarely be made from cultures used in fermentation studies where much acid is produced.

All strains grow in complex media containing 10% (w/v) NaCl. Some strains can tolerate 20% (w/v) NaCl and some strains have been reported to remain viable after 1 year in 16% (w/v) NaCl at pH 6.0 (see Seeliger, 1961). All strains grow on 10% (w/v) and 40% (w/v) bile agar. Wetzler et al. (1968) reported that generally growth on 40% (w/v) bile (Bacto ox gall was used) was better than the growth obtained on 10% (w/v) bile. Growth occurs on MacConkey agar. All strains grow in the presence of 0.025% (w/v) thallous acetate; 3.75% (w/v) potassium thiocyanate; 0.04% (w/v) potassium tellurite; 0.01% (w/v) 2,3,5-triphenyltetrazolium chloride; tellurite and tetrazolium are reduced. Strains do not grow in the presence of 0.02% (w/v) sodium azide. Wetzler et al. (1968) reported no growth in the presence of potassium cyanide. No growth occurs on the medium of Gardner (1966). Practically all strains are catalase-positive when grown on the usual laboratory media but may give a negative reaction if cultured on media containing low concentrations of meat and yeast extract. However, a few truly catalase-negative strains have been observed. Friedman and Alm (1962) reported that catalase activity is depressed in media containing higher (10% w/v) concentrations of glucose. Those strains of *L. monocytogenes, L. innocua* and *L. ivanovii* which have been examined contain cytochromes a_1bdo when cultured with shaking in Brain Heart Infusion Broth (Oxoid) (M. D. Collins, personal communication).

There is not a great deal of information on the nutritional requirements and metabolic pathways of *Listeria* strains, and such studies have probably been confined to strains of *L. monocytogenes*, although it is not possible to be certain from the strain designations cited in the literature. Biotin, riboflavin, thiamine, thioctic acid and several aminoacids including cysteine, glutamine, isoleucine, leucine and valine are required for growth. Arginine, histidine, methionine and tryptophan have a growth-stimulating effect (for reference, see Gray and Killinger, 1966). Carbohydrate is essential for growth of *Listeria* strains and glucose is the usual choice. In studies on *L. monocytogenes*, Miller and Silverman (1959) reported that glucose could not be replaced as an energy and carbon source by gluconate, xylose, arabinose or ribose. However, *L. innocua, L. ivanovii* and *L. seeligeri* produce acid from, and presumably utilize, xylose as a carbon source.

Catabolism of glucose apparently proceeds by the Embden-Meyerhof pathway both aerobically and anaerobically. Anaerobically the end product is mainly lactic acid; aerobically, pyruvate, acetoin, lactic acid and other end products are formed. There is no evidence for the Entner-Doudoroff pathway but glucose-6-phosphate dehydrogenase and 6-phosphogluconate dehydrogenase have been reported to be present (Miller and Silverman, 1959).

All strains are Methyl red (MR) positive and positive in the Voges-Proskauer test in 48 h when tested by the method of O'Meara or that of Barritt (see Cowan, 1974, for methods).

When testing for the production of acid from carbohydrates, the composition of the basal medium and the pH indicator used are important. *Listeria* spp. are actively saccharolytic and a basal medium which contains traces of fermentable carbohydrate and a pH indicator which changes color rather near neutrality can result in false-positive reactions. A variety of basal media and pH indicators have been used. A suitable, reliable basal medium for testing for acid production is Purple Broth Base (Difco) supplemented with 0.5 or 1% (w/v) of the carbohydrate. However, some workers prefer a peptone water medium with phenol red as indicator (Rocourt et al., 1983b) especially for the production of acid from L-rhamnose, D-xylose and α-methyl-D-mannoside. Where possible all carbohydrates should be sterilized by filtration and not by autoclaving.

All strains produce acid within 48 h from amygdalin, cellobiose, esculin, fructose, glucose, mannose and salicin. Acid is produced after longer periods of incubation from maltose (to 4 days), dextrin (to 8 days), α-methyl-D-glucoside (2–6 days) and glycerol (to 10 days).

Acid production from galactose, lactose, melezitose, α-methyl-D-mannoside, rhamnose, sorbitol, starch, sucrose, trehalose and xylose is variable, in some cases between species, but in others fermentation is variable between strains of the same species (Seeliger, 1961; Seeliger and Welshimer, 1974; Wetzler et al., 1968; Wilkinson and Jones, 1977; Rocourt et al., 1983b). Seeliger and Welshimer (1974) reported that all strains fermented starch. However, Wetzler et al. (1968) and Feresu and Jones (unpublished) found that only about 30% of the strains tested fermented this carbohydrate.

Acid is almost never produced from adonitol, arabinose, dulcitol, erythritol, glycogen, inositol, inulin, mannitol, melibiose, raffinose or sorbose. Wetzler et al. (1968) reported that one strain of *L. monocytogenes* serovar 4c produced acid from inositol and a freshly isolated human strain of serovar 4 produced acid from glycogen. Occasional strains produce acid from adonitol, arabinose, erythritol, glycogen or raffinose.

Exogenous citrate is not utilized. In a defined medium, neither pyruvate, acetate, citrate, isocitrate, α-ketoglutarate, succinate, fumarate nor malate support growth of *L. monocytogenes* in the absence of glucose, nor do they increase growth in the presence of glucose (Trivett and Meyer, 1971). Pyruvate, malate, succinate and α-ketoglutarate have been reported to be oxidized at low rates by *L. monocytogenes* (Friedman and Alm, 1962; Kolb and Seidel, 1960). In a complex medium pyruvate is utilized as a carbon source by some strains. *L. monocytogenes* appears to utilize a split noncyclic citrate pathway which has an oxidative portion and a reductive portion. The pathway is probably important in biosynthesis but not for a net gain in energy (Trivett and Meyer, 1971).

Iron is stimulatory for the growth of *L. monocytogenes* in stationary or aerated cultures (Sword, 1966; Trivett and Meyer, 1971), but aeration improves growth only in the presence of adequate iron (Trivett and Meyer, 1971).

All strains produce β-D-galactosidase. A few strains have been reported to hydrolyze starch (Wetzler et al., 1968; Feresu and Jones, unpublished). Cellulose, tyrosine and xanthine are not hydrolyzed. Casein is not hydrolyzed but care must be taken in performing the test. Traces of fermentable carbohydrate which may be present in yeast extract may result in slight acid production which gives false-positive reactions.

All strains produce alkaline phosphatases (Wetzler et al., 1968; Mrsvic et al., 1975); sulfatase is not produced. Phenylalanine-deaminase negative. Ornithine, lysine, glutamic acid and arginine decarboxylases are not produced, nor is arginine dihydrolase (Wetzler et al., 1968).

Weak deoxyribonuclease and ribonuclease activity is exhibited by most strains.

Tributyrinase-negative. Some strains produce lecithinase (egg-yolk medium). Tweens 20, 40, 60 and 80 are hydrolyzed to some degree by most strains but hydrolysis may require 7–14 days.

Although the literature contains conflicting reports (see Gray and Killinger, 1966), H$_2$S is not produced. Nitrates are not reduced to nitrites.

Lysozyme has been reported to be produced by some strains, especially those pathogenic for mice (Marinova and Manev, 1979).

Listeria strains are sensitive to a number of antibiotics. There are sometimes interesting, but as yet unexplained, differences between the sensitivity of new isolates and collection strains (see Gray and Killinger, 1966; Wetzler et al., 1968). When tested *in vitro* by the agar diffusion method, *Listeria* strains are reported to be sensitive to ampicillin, carbenicillin, cephaloridine, chloramphenicol, erythromycin, furazolidone, neomycin, novobiocin, oleandomycin, ticarcillin, azlocillin; less sensitive to chlortetracycline, oxytetracycline, tetracycline, gentamicin, kanamycin, nitrofurantoin, penicillin G, streptomycin. Most strains are sensitive to methicillin. Strains are resistant to colistin sulfate, nalidixic acid, polymyxin B and sulfonamides. It should be noted that these data are not necessarily relevant to the clinical effectiveness of the antibiotic in individual cases of listeriosis.

A cryptic plasmid has been demonstrated in *L. ivanovii* (Perez-Diaz et al., 1981; Yeasmin, unpublished).

Lysogeny is common in *Listeria* strains. Bacteriophages have been isolated from many strains designated *L. monocytogenes*, *L. innocua* and *L. ivanovii* (Schulz, 1945; Sword and Pickett, 1961; Guillot and McCleskey, 1963; Jasinska, 1964; Audurier et al., 1977; Audurier et al., 1979b; Ortel, 1981; Rocourt et al., 1982a). Phage typing schemes of practical value in epidemiological studies have been devised (Audurier et al., 1979a; Audurier et al., 1984). Strains of *L. monocytogenes* serovar 5 (*L. ivanovii*) appear to be particularly sensitive to phage lysis. There appears to be no correlation between the sugar fermentation patterns of the various *Listeria* species and the phage-typing pattern. Similarly, no correlation could be detected between the phage typing pattern and the source of the isolate (Rocourt et al., 1982a). Phages have been isolated from lysogenic strains with and without induction. Induction is used for research studies on *Listeria* phages but the process also frequently results in the production of bacteriocins which may interfere with the phage typing patterns (Rocourt et al., 1982a). Descriptions of the morphology of *Listeria* phages have been given by Sword and

Pickett (1961); Jasinska (1964); Chiron et al. (1977); Ackermann et al. (1981); Rocourt et al. (1983c). The phages so far isolated are members of the *Styloviridae* (noncontractile tails) and *Myoviridae* (contractile tails) families of phages. The latter have to date been found only from *L. innocua* strains (Rocourt et al., 1983c).

Bacteriocins are produced by a high proportion of strains examined (Sword and Pickett, 1961; Hamon and Péron, 1962, 1963; Bradley and Dewar, 1966; Ortel, 1978). The bacteriocins do not inhibit Gram-negative bacteria but are active against staphylococci and bacilli. Their general characteristics are given by Hamon and Péron (1962, 1963) and their morphological characteristics have been demonstrated by electron microscopy (Bradley and Dewar, 1966). A satisfactory bacteriocin typing scheme for epidemiological purposes has not been developed.

The antigenic composition of *Listeria* has been studied intensively (see Seeliger, 1961; Gray and Killinger, 1966 for references to early literature). The currently accepted scheme is based on that described by Paterson (1940) who recognized four types, this division being made primarily on the H and secondarily on the O antigens (for details see Seeliger and Höhne, 1979). This scheme was refined and expanded by Donker-Voet (1972) and Seeliger (1975, 1976). The Seeliger-Donker-Voet antigenic scheme presently consists of 16 serovars of the genus *Listeria* (see Table 14.13). The scheme does not include the taxa referred to as *Listeria denitrificans*, *Listeria grayi* and *Listeria murrayi*. These taxa are serologically different from members of the genus *Listeria*. *Listeria grayi* and *Listeria murrayi* are serologically related to each other (see Table 14.13) but *Listeria denitrificans* is serologically distinct.

Some minor shared antigens result in cross-reactions between *Listeria* strains and some Gram-positive bacteria, notably staphylococci and enterococci and also with *Escherichia coli*. It is therefore important that suitably absorbed antisera are used in serological tests (Seeliger, 1961). With the exception of serovar 5 strains, all of which examined to date are members of the species *L. ivanovii*, there is no strict correlation between serovar and species in the genus *Listeria*. According to Ullmann and Cameron (1969), the cell walls of serovars 1 and 2 (now designated serovar 1/2) contain predominantly glucosamine and rhamnose as the cell wall sugars; those of serovar 3 contain galactose, rhamnose and glucosamine and serovars 4a and 4b contain glucose and galactose. More recently, Fiedler et al. (1984) have more thoroughly investigated the nature and quantitative sugar substituents of the teichoic acids of all the serovars of the genus *Listeria*. The results are in general agreement with those of the more restricted study of Ullmann and Cameron (1969) and with the results of the studies of Kamisango et al. (1982). For a detailed discussion of the correlation between teichoic acid structure and the serological behavior of *Listeria* strains and the contribution of lipoteichoic acids (Hether and Jackson, 1983) to the biochemical basis of somatic antigenicity in *Listeria* strains, see Fiedler et al. (1984).

The cell wall peptidoglycan of all species investigated contains meso-diaminopimelic acid as the main diamino acid (A1γ variation of Schleifer and Kandler, 1972). Alanine and glutamic acid are also present (Robinson, 1968; Schleifer and Kandler, 1972; Srivastava and Siddique, 1973; Kamisango et al., 1982; Fiedler and Seger, 1983; Fiedler et al., 1984). In addition to N-acetylmuramic acid and N-acetylglucosamine of the peptidoglycan, glucosamine occurs as a component of the cell wall polysaccharide (Ullmann and Cameron, 1969; Hether et al., 1983; Fiedler and Seger, 1983). Furthermore, neutral sugars are present (Ullmann and Cameron, 1969; Keeler and Gray, 1960). All the strains contain teichoic acid (Fiedler et al., 1984).

The report of Mará and Michalec (1977) of mycolic acid-like substances in *L. monocytogenes* has not been confirmed by other workers. Jones et al. (1979) and M. D. Collins (personal communication) failed to detect any such substances in any of the *Listeria* strains examined.

All *Listeria* strains examined contain MK-7 as the major menaquinone component with MK-6 and MK-5 as minor components (Collins and Jones, 1981).

The polar lipid composition of *L. monocytogenes* has been shown to comprise phosphatidylglycerol, diphosphatidylglycerol, galactosylglucosyldiacylglycerol and an uncharacterized glycophospholipid (Kosaric and Carroll, 1971; Shaw, 1974). There is no information on the polar lipid composition of other species.

The fatty acid composition of *L. monocytogenes, L. innocua* and *L. ivanovii* is virtually identical. All consist of predominantly straight chain saturated, *anteiso-* and *iso*-methyl-branched chain types. The major fatty acids of all the strains examined are 14-methylhexadecanoic (*anteiso-*$C_{17:0}$) and 12-methyltetradecanoic (*anteiso-*$C_{15:0}$) (Carroll et al., 1968; Tadayon and Carroll, 1971; Collins, unpublished).

Jones et al. (1979) reported cytochromes abb_1 in *L. monocytogenes* NCTC 7973 but in a later study cytochromes a_1bdo were demonstrated in strains of *L. monocytogenes, L. innocua* and *L. ivanovii* (Feresu, Jones and Collins, unpublished).

Listeria monocytogenes is pathogenic for man and a large number of animal species. The occurrence of infections in man is sporadic although a few epidemics have been reported notably in Germany (Ortel, 1968). In man, meningeal involvement sometimes accompanied by septicemia is the usual clinical manifestation. The members of the population most at risk are neonates, the old and those compromised by pregnancy or an underlying illness such as malignancy or alcoholism or some condition which requires immunosuppressive procedures. Intrauterine infection of the fetus results in death, or an acutely ill infant with a septic disseminated form of listeriosis, *Granulomatosis infantiseptica* (Reiss et al., 1951a, b). Papular lesions of the skin may be found in listeriosis of the newborn. A similar cutaneous form has been reported in veterinarians working with infected animals. Other clinical forms of the disease occur more rarely in man (see Seeliger, 1961).

In animals, abortion, encephalitis and septicemia are the main clinical manifestations in cattle and sheep (see Seeliger, 1961). *L. ivanovii* especially seems to be very closely associated with abortion in sheep (Ivanov, 1975), whereas it has been isolated rarely from man.

The epidemiology and the basis of pathogenicity of *L. monocytogenes* and *L. ivanovii* is only just beginning to be understood (Seeliger, 1984).

Bacteria of the genus *Listeria* are widely distributed in nature and have been isolated from soil, vegetation, sewage, water, animal feed, fresh and frozen poultry, slaughter-house waste and healthy human and animal carriers (see Welshimer, 1981). Many of these isolates have in the past been designated *L. monocytogenes*. However, a number of these strains are nonhemolytic and nonpathogenic for mice and are now recognized as members of the species *L. innocua, L. seeligeri* and *L. welshimeri* (Rocourt and Grimont, 1983). Due to the high phenotypic similarity between strains of *L. monocytogenes, L. innocua, L. ivanovii, L. seeligeri* and *L. welshimeri*, it is particularly important that new isolates are examined for hemolytic activity and for experimental pathogenicity in mice.

Enrichment and Isolation Procedures

Although, once isolated, *Listeria* species grow well on the usual nutrient media, it is frequently necessary to employ either, or both, enrichment and selective procedures to isolate the bacteria from clinical and other material.

Difficulties in isolating *L. monocytogenes* were recognized when the organism was first isolated and described by Murray et al. (1926). Consequently, a great number of enrichment and isolation procedures have been developed (see Seeliger, 1961, 1972; Gray and Killinger, 1966; Kramer and Jones, 1969; Watkins and Sleath, 1981; Welshimer, 1981, for references to most procedures and media). In addition to the successful, but often very slow, cold enrichment technique (Gray et al., 1948) a variety of dyes, antibiotics and other chemicals such as potassium tellurite, potassium thiocyanate, thallous acetate and others are incorporated into enrichment and selective media used in different laboratories. Some of these procedures give excellent results in some laboratories but not in others. This is probably due to variations in the quality of the chemicals used, to slight differences in the composition of the suspending medium, and to variation in the techniques used.

Blood or spinal fluid samples from suspected clinical cases are placed in a flask containing 50 ml of Tryptose Broth (Difco) and mixed well. After 24 h, the sample incubated at 37°C is inoculated onto Tryptose Agar (Difco) and Blood Agar Base (Difco) containing 5% (v/v) washed sheep erythrocytes; both are incubated at 37°C for 24–48 h. The blood plate is examined for dew-drop-like colonies producing zones of β-hemolysis; the tryptose agar plate is examined by the oblique lighting technique of Henry (1933) as recommended by Gray (1957). The mirror is removed from a binocular scanning microscope and placed flat side up on the bench about 10 cm away from the microscope, which should have a glass stage. A white light source is so arranged that a beam of light at an angle of 45° is directed on to the mirror so that it is reflected at an angle of 40–45° through the glass microscope stage. The plate containing suspected colonies is placed on the stage and viewed (× 10 or × 15 magnification). The angle of light is important; direct transmitted light will not give satisfactory results. *Listeria* colonies appear as small, low convex, finely textured, bright blue green in color and with an entire edge. If no *Listeria*-like colonies are obtained, the sample is reincubated at 37°C and the procedure is repeated daily for up to 7 days.

Tissue specimens from suspected clinical cases should be macerated in tryptose broth and then treated as for blood samples.

Material which may be heavily contaminated with other bacteria, e.g. fecal material, silage, sewage, etc., should be selectively enriched and/or plated on to a selective medium.

Suitable enrichment media are thallous acetate-nalidixic acid (TN) medium* (Kramer and Jones, 1969), and potassium thiocyanate-nalidixic acid (PTN) medium† (Watkins and Sleath, 1981). Samples (1 to 2 g) of the infected material are added to two separate flasks containing 100 vol of enrichment broth. One flask is incubated at 37°C, the other at 4°C. The flask incubated at 37°C is sampled daily for 7 days by plating 0.1 ml amounts on to Tryptose Agar (Difco), incubating at 37°C for 24 h, and then examining the plate microscopically by the oblique lighting technique for the presence of distinctive blue green colonies. Flasks incubated at 4°C are sampled every 7 days for a period of up to 2 months if necessary. The samples are treated as described for the enrichment broth held at 37°C. Cold enrichment usually results in the detection of more positive samples but incubation at both 4°C and 37°C is recommended because positive results are obtained more quickly at 37°C.

A method for the isolation and enumeration of *Listeria* strains in sewage, sewage sludge and river water using PTN medium is described by Watkins and Sleath (1981).

A suitable selective medium is trypaflavin-nalidixic acid agar§ (Bockemühl et al., 1971). Samples such as macerated tissue, feces or vegeta-

---

* Thallous acetate-nalidixic acid (TN) medium: Nutrient Broth No. 2 (Oxoid) containing glucose 0.2% (w/v); thallous acetate 0.2% (w/v); nalidixic acid, 40 μg/ml. Crystalline nalidixic acid, 0.05 g, is dissolved in 0.5 ml 1N NaOH, when dissolved 4.5 ml of distilled water is added and the appropriate volume added to the basal medium before autoclaving at 121°C for 15 min.

† Potassium thiocyanate-nalidixic acid (PTN) medium: Nutrient Broth No. 2 (Oxoid) containing potassium thiocyanate, 3.75% (w/v) and nalidixic acid (Sigma Chemical Company) 100 μg/ml. The nutrient broth containing the potassium thiocyanate is sterilized at 121°C for 15 min. The nalidixic acid is then added to the sterilized medium as a solution.

§ Trypaflavin-nalidixic acid agar: Tryptose Agar (Difco), 41.0 g; nalidixic acid 0.04 g; distilled water 1000 ml. Crystalline nalidixic acid 0.8 g is dissolved in 10 ml 1 N NaOH and the volume then made up to 100 ml with distilled water; 5 ml of this nalidixic acid solution is added to the basal medium to give a final concentration of 40 μg/ml. After sterilization for 15 min at 120°C the medium is cooled to 70°C. Acriflavin neutral (Casella-Farbwerk) is dissolved in sterile distilled water to give a 0.5% (w/v) solution. Two milliliters of this solution are added to the tryptose-nalidixic acid medium to give a final concentration of 10 μg/ml. The medium is mixed thoroughly and dispensed in 13-ml amounts into dishes.

tion in a suitable suspending medium are spread on the surface of the medium and incubated for 24 h at 37°C and subsequently for 48 h at room temperature. The plates are then microscopically examined by oblique lighting for the presence of yellow-green colonies.

After obtaining pure cultures, all new isolates should be stained by the method of Gram, examined for hemolysis and motility and also by the appropriate biochemical and serological procedures.

All *Listeria* species are isolated by these methods.

As noted previously, there are a number of different published enrichment procedures and selective media. Recently, Skalka and Smola (1983) have devised a selective medium which they claim to be useful in the recognition of pathogenic *Listeria* spp. The medium is based on those described by Elischerová (1975) and Skalka et al. (1982). It consists of agar containing 5% (v/v) washed rabbit erythrocytes and supplemented with 10 μg acriflavin, 40 μg nalidixic acid and 7.5 activity units of "equi" factor per milliliter. The authors report that *L. innocua*

strains are nonhemolytic on this medium but *L. monocytogenes* and *L. ivanovii* exhibit easily detectable distinctive hemolytic zones, on the basis of which they can be differentiated (Skalka and Smola, 1983).

Maintenance Procedures

The organisms may be preserved for some months by stab inoculation into nutrient agar (Tryptose Agar (Difco) or other similar media), in screw-capped containers. After overnight incubation at 37°C the caps should be screwed tightly to prevent evaporation. To this end, tubes with paraffin waxed sealed stoppers may be preferred. The containers should be stored at room, or preferably refrigerator temperature in the dark.

Strains of *Listeria* and similar bacteria may be preserved for longer periods (at least 7 years) by freezing in glass beads at −70°C (Feltham et al., 1978) or by lyophilization.

Differentiation of the genus **Listeria** from other genera

Table 14.11 lists the features most useful in differentiating the genus *Listeria* from other Gram-positive, nonsporeforming, rod-shaped bacteria.

Identification of new isolates may be achieved by examination of cellular morphology, colonial morphology and hemolytic reaction on sheep or horse blood (5% v/v) agar and blue-green coloration of colonies when viewed microscopically with oblique lighting on Tryptose Agar (Difco); growth at 37°C, oxygen requirements, catalase production, hydrolysis of esculin and sodium hippurate, alkaline phosphatase production and production of acid from carbohydrates in a suitable medium. Serological identification may be necessary for the certain identification of *Listeria*. The composition of the cell wall and lipid composition are helpful but not necessary for routine identification.

Bacteria with which members of the genus *Listeria* are most likely to be confused are those of the genera *Brochothrix, Erysipelothrix, Lactobacillus* and *Kurthia*. As mentioned earlier, coccobacillary forms of the genus *Listeria* may be confused with streptococci but this problem should be resolved by the catalase test.

Listeria may be distinguished easily from the genus *Kurthia* by their

different oxygen requirements, *Kurthia* is a strictly aerobic genus; by fermentation of various sugars, *Kurthia* spp. produce little or no acid in sugar fermentation tests; and by the presence of *meso*-diaminopimelic acid in the cell wall, *Kurthia* spp. contain lysine (see Keddie, 1981).

Listeria spp. may be distinguished from *Erysipelothrix rhusiopathiae* by the catalase test, good growth on the usual nutrient media; producing colonies with a marked blue-green sheen on Tryptose Agar (Difco) when viewed microscopically by oblique light; exhibiting motility; possessing a more vigorous saccharolytic activity; hydrolyzing esculin and sodium hippurate, and not producing H_2S. *Listeria* spp. and *E. rhusiopathiae* are antigenically distinct. In addition, *Listeria* spp. contain cytochromes and menaquinones which are absent in *E. rhusiopathiae*. The two genera also differ in the chemical composition of the cell wall peptidoglycan, that of *Listeria* is based on *meso*-diaminopimelic acid while the cell wall of *E. rhusiopathiae* contains lysine.

Listeria spp. may be distinguished from most lactobacilli by the catalase test. There are, however, some catalase-positive lactobacilli and rare strains of *Listeria* that are catalase-negative. Members of the

Table 14.11.
Characters most useful in differentiating the genera **Brochothrix, Erysipelothrix, Listeria, Lactobacillus** *and* **Kurthia**[a,b]

Taxon	Motile	Oxygen requirements	Growth at 35°C	Cata-lase	H_2S pro-duction	Acid from glucose	Peptido-glycan group[c]	Major peptido-glycan diamino acid	Major mena-quinone	Fatty acid type[d,e]	Mol% G + C
Brochothrix	−	Facultative	−	+[f]	−	+	A	*meso*-DAP	MK-7	S, A, I	35.6–36.1
Erysipelothrix	−	Facultative	+	−	+	+	B	L-Lysine	−	S, A, I, U	36–40
Listeria	+[g]	Facultative	+	+	−	+	A	*meso*-DAP	MK-7	S, A, I	36–38
Lactobacillus	−[h]	Facultative	+	−[i]	−	+	A	Lysine or *meso*-DAP or ornithine	−[j]	S, U (C)	34–53
Kurthia	+[k]	Aerobic	+	+	+	−[l]	−	A	L-Lysine	MK-7	S, A, I	36.7–37.9

[a] Data from McLean and Sulzbacher, 1953; Davidson et al., 1968; Collins-Thompson et al., 1972; Tadayon and Carroll, 1971; Schleifer and Kandler, 1972; Stuart and Welshimer, 1973, 1974; Seeliger and Welshimer, 1974; Shaw, 1974; Jones, 1975a; Collins and Jones, 1981; Keddie, 1981; Sharpe, 1981; Rocourt et al., 1983b; Rocourt and Grimont, 1983; Seeliger et al., 1984; Flossmann and Erler, 1972; White and Mirikitani, 1976.

[b] Standard symbols: +, 90% or more of strains are postive; −, 90% or more of strains are negative.

[c] Group A: cross-linkage between positions 3 and 4 of two peptide subunits, group B cross-linkage between positions 2 and 4 of two peptide subunits (Schleifer and Kandler, 1972).

[d] S, straight-chain saturated; U, monounsaturated; A, *anteiso*-methyl-branched; I, *iso*-methyl branched; C, cyclopropane ring fatty acids.

[e] Those in parenthesis may be present.

[f] Catalase production dependent on medium and temperature of incubation (Davidson et al., 1968).

[g] All strains motile at 20–25°C, poorly or nonmotile at 37°C (Seeliger and Welshimer, 1974).

[h] Most strains nonmotile, but a few motile strains occur (Sharpe, 1981).

[i] Some strains give positive catalase reaction (Sharpe, 1981).

[j] *Lactobacillus mali* contains MK-8 and MK-9 as major menaquinones; a menaquinone has also been detected in *L. casei* subsp. *rhamnosus* (Collins and Jones, 1981).

[k] Majority of strains motile but nonmotile strains do occur (Keddie, 1981).

[l] Weak production of H_2S by some strains (Jones, 1975; Keddie, 1981).

genus *Listeria* grow very poorly or not at all on MRS medium (de Mann et al., 1960), a medium on which lactobacilli grow very well. When investigated at the correct incubation temperature, *Listeria* strains are motile while most lactobacilli are not. Lactobacilli do not exhibit a blue-green sheen when examined microscopically by oblique light on Tryptose Agar (Difco). There are also differences in the fermentation patterns of the two genera. All strains of *Listeria* examined possess *meso*-diaminopimelic acid in the cell wall. Some lactobacilli also possess this amino acid in the cell wall but in other lactobacilli it is replaced by lysine or ornithine (Schleifer and Kandler, 1972). Serologically, the genera *Listeria* and *Lactobacillus* are distinct.

In some respects, members of the genus *Listeria* can most easily be confused with *Brochothrix thermosphacta*. Both are frequently isolated from prepacked meats and poultry held at refrigerator temperatures. Both contain *meso*-diaminopimelic acid in the cell wall, MK-7 as the major isoprenoid quinone and have an identical fatty acid composition (Schleifer and Kandler, 1972; Feresu, Collins and Jones, unpublished). The inability of *B. thermosphacta* to grow at 37°C, and the inability of *Listeria* species to grow on the medium of Gardner (1966) serve to distinguish the two genera. Colonies of *B. thermosphacta* do not appear blue-green when grown on Tryptose Agar and viewed microscopically by oblique light. In addition, *B. thermosphacta* does not hydrolyze sodium hippurate but produces acid from a greater number of sugars. Serologically, the two taxa are distinct (Wilkinson and Jones, 1975).

Taxonomic Comments

A number of changes have been made since the eighth edition of the *Manual* in which the genus *Listeria* was listed as containing four species, *L. monocytogenes* (the type species), *L. denitrificans* (although the classification of this species in the genus *Listeria* was noted as requiring further consideration), *L. grayi* and *L. murrayi*. However, an editorial note drew attention to the, then, recent proposal of Stuart and Welshimer (1974) that the last two species should be transferred to a new monospecific genus "*Murraya*" as "*Murraya grayi* subspecies *grayi*," and "*M. grayi* subspecies *murrayi*."

It is now generally agreed that *L. denitrificans* is not a member of the genus *Listeria*. Details of the data on which this opinion is based are given in the "Taxonomic Comment" immediately following the description of the species *L. denitrificans*. It is retained in this section and listed as a species *incertae sedis* because, although it is not a member of the genus *Listeria*, its exact taxonomic affiliation has not been determined.

The taxonomic position of the species *L. grayi* and *L. murrayi* is controversial. Phenotypically they are, on the basis of tests carried out to date, very similar to each other and to *L. monocytogenes*, the type species of the genus *Listeria*. In numerical taxonomic studies, *L. grayi* and *L. murrayi* cluster as a distinct taxon related to the other listeriae examined at about 83%S (Wilkinson and Jones, 1977). On the basis of the results of cell wall, cytochrome, menaquinone and fatty acid studies, *L. grayi* and *L. murrayi* form a homogeneous group with *L. monocytogenes* (Fiedler and Seger, 1983; Fiedler et al., 1984; Feresu, Collins and Jones, unpublished). Consideration of data of this kind led Jones (1975b), Wilkinson and Jones (1977), and more recently, Fiedler and Seger (1983) and Fiedler et al. (1984), to conclude that *L. grayi* and *L. murrayi* were members of the genus *Listeria*. Wilkinson and Jones (1977) were also of the opinion that there are insufficient phenotypic differences between *L. grayi* and *L. murrayi* to warrant the retention of two species and suggested that they be reduced to synonymy as *L. grayi* (Errebo Larsen and Seeliger).

However, on the basis of DNA-DNA hybridization studies, Stuart and Welshimer (1973, 1974) concluded that *L. grayi* and *L. murrayi* were sufficiently distinct from *L. monocytogenes* to merit a separate genus for which they suggested the name "*Murraya*." The same authors also concluded that there was not sufficient difference between strains of *L. grayi* and *L. murrayi* to justify two species in the new genus and designated them as "*Murraya grayi* subsp. *grayi*" and "*Murraya grayi* subsp. *murrayi*." Support for the recognition of strains of *L. grayi* and *L. murrayi* as a genus distinct from *Listeria* comes from the results of

a more recent DNA-DNA hybridization study which showed that *L. grayi* and *L. murrayi* were, respectively, 3–29% and 1–9% related to reference strains of *L. monocytogenes, L. innocua, L. ivanovii, L. seeligeri* and *L. welshimeri* (Rocourt et al., 1982b).

Further work is required to resolve the lack of congruence between the results of the DNA hybridization studies and the results of the numerical taxonomic and chemical studies. Such studies should resolve the problems of the taxonomic status of strains now designated *L. grayi* and *L. murrayi*. Until that time they are best treated as species *incertae sedis*. For this reason, full species descriptions of each are given at the end of the *Listeria* section.

In the description of *L. monocytogenes* given in the eighth edition of the *Manual*, attention is drawn to certain strains which exhibited pronounced β-hemolysis and to others, isolated from fecal and environmental sources, which were nonhemolytic and also nonpathogenic for mice. Stuart and Welshimer (1973, 1974) noted the existence of two DNA-DNA hybridization groups among strains labeled *L. monocytogenes* but could not distinguish the two groups by any other criteria.

Seeliger (1981) proposed that nonpathogenic strains of *L. monocytogenes* which belonged to serovar 6 be recognized as a new species, *Listeria innocua* (Seeliger and Schoofs, 1979). Later, comprehensive DNA-DNA hybridization studies of a large collection of *Listeria* strains isolated from a variety of sources indicated the existence of five DNA-DNA homology groups among the strains (Rocourt et al., 1982b). One of the five genomic groups contained the type strain of *Listeria monocytogenes*, a second contained those strains designated *L. innocua*, another group contained those strains which were first described by Ivanov (1962) as exhibiting a pronounced β-hemolysis and for which the name of *L. ivanovii* (Seeliger et al., 1984) has been proposed. The two other genomic groups have been designated *L. seeligeri* and *L. welshimeri* (Rocourt and Grimont, 1983). There are at present few phenotypic characters which distinguish the five species (Rocourt et al., 1983b). It is interesting, however, that in a recent numerical taxonomic survey, Feresu and Jones, unpublished), strains of *L. monocytogenes* were grouped into seven clusters. Three of the clusters could be equated with *L. monocytogenes, L. innocua* and *L. ivanovii*. It is possible that two of the other clusters represent *L. seeligeri* and *L. welshimeri* although it is not possible to be certain because the study did not contain the same strains as those examined by Rocourt et al. (1982b). Of the remaining groups, one contained rough strains of *L. monocytogenes*, the other may represent another species. All were related at approximately 88%S. Further studies are required to detect more phenotypic features to distinguish the five species in the genus *Listeria* and to clarify their relationship to the bacteria named *L. grayi* and *L. murrayi*.

The taxonomic position of the genus *Listeria* with regard to other genera is still not resolved. Numerical taxonomic, serological, chemical and nucleic acid data (Davis et al., 1969; Stuart and Pease, 1972; Stuart and Welshimer, 1973, 1974; Jones 1975a, b; Wilkinson and Jones, 1977; Jones et al., 1979; Collins et al., 1979; Rocourt et al., 1982b; Seeliger, 1984) indicate that *Listeria* is a distinct genus not related to the coryneform bacteria. The genus may be more closely associated with the genera *Bacillus, Erysipelothrix, Lactobacillus* and *Streptococcus*. Chemical data on the composition of the cell wall peptidoglycan, cytochrome and menaquinone content, and the fatty acid composition (Schleifer and Kandler, 1972; Tadayon and Carroll, 1971; Shaw, 1974; Collins et al., 1979; Fiedler and Seger, 1983; Fiedler et al., 1984; M. D. Collins, personal communication) indicate that *Listeria* shows the greatest similarity to the genus *Brochothrix* and to the two taxa designated *L. grayi* and *L. murrayi*.

Further DNA-DNA hybridization studies, DNA-rRNA hybridization studies and studies on rRNA oligonucleotide sequences are required before the phylogenetic relationship of *Listeria* to other genera or higher taxa can be established.

Further Reading

Gray, M.L. and A.H. Killinger. 1966. *Listeria monocytogenes* and listeric infections. Bacteriol. Rev. *30*: 309–382.

Jones, D. 1975. The taxonomic position of the genus *Listeria*. *In* Woodbine (Editor). Problems of Listeriosis, Leicester University Press, Leicester, England, pp. 4–17.

Rocourt, J., F. Grimont, P.A.D. Grimont and H.P.R. Seeliger. 1982. DNA relatedness among serovars of *Listeria monocytogenes sensu lato*. Curr. Microbiol. *7:* 383–388.

Seeliger, H.P.R. 1961. Listeriosis. S. Karger, Basel.

Seeliger, H.P.R. 1984. Modern taxonomy of the *Listeria* group, relationship to its pathogenicity. Clin. Invest. Med. *7:* 217–221.

Welshimer, H.J. 1981. The genus *Listeria* and related organisms. *In* Starr, Stolp, Trüper, Balows and Schlegel (Editors), The Prokaryotes. A Handbook on Habitats, Isolation and Identification of Bacteria. Springer-Verlag, New York, pp. 1680–1687.

Differentiation and characteristics of the species of the genus **Listeria**

The differential characteristics of the species of *Listeria* and the taxa *L. grayi*, *L. murrayi* and *L. denitrificans* are listed in Table 14.12. The detailed antigenic structure of *Listeria* species and *L. grayi* and *L. murrayi* is presented in Table 14.13.

Table 14.12.
Characters differentiating the accepted species of the genus **Listeria** *and* **L. grayi**, **L. murrayi** *and* **L. denitrificans**[a,b]

Characteristics	L. monocytogenes	L. innocua	L. seeligeri	L. welshimeri	L. ivanovii	L. grayi	L. murrayi	L. denitrificans
Irregular rods with uneven staining	−	−	−	−	−	−	−	+
β-Hemolysis	+[c]	−	+	−	+[d]	−	−	−
CAMP-test (*Staphylococcus aureus*)	+[e]	−	+					−
CAMP-test (*Rhodococcus equi*)	−	−			+			
Acid production from:								
L-Arabinose	−	−			−	−	−	+
Dextrin	d	−			−	+	+	+
Galactose	d	−			d	+	+	+
Glycogen	−	−			−	−	−	+
Lactose	d	+			+	+	+	+
D-Lyxose	−	−			−	+	+	
Mannitol	−	−	−	−	−	+	+	−
Melezitose	d	d			d	−	−	−
Melibiose	−	−			−	−	−	+
α-Methyl-D-glucoside	+	+			+	+	+	−
α-Methyl-D-mannoside	+	+	−[f]	+	−			
L-Rhamnose	+	d	−	d	−	−	d	−
Sorbitol	d	−			−	−	−	−
Soluble starch	−	−			−	+	+	
Sucrose	−	d			d	−	−	+
D-Xylose	−	−	+	+	+	−	−	+
Voges-Proskauer	+	+	+	+	+	+	+	
Hydrolysis of:								
Cellulose	−	−			−	−	−	+
Hippurate	+	+			+	−	−	−
Starch	d	d			−	−	−	+
Lecithinase	d	d			+	−	−	−
Phosphatase	+	+			+	+	+	−
Reduction NO$_3$ to NO$_2$	−	−			−	−	+	+
Pathogenicity for mice	+	−	−	−	+	−	−	+?
Mol % G + C	37–39 (T_m); 38 (Bd)	36–38 (T_m); 38 (Bd)	36 (T_m)	36 (T_m)	37–38 (T_m)	41–42 (T_m); 41 (Bd)	41–42.5 (T_m); 42 (Bd)	56–58 (T_m); 57 (Bd)
Major peptidoglycan diamino acid	meso-DAP	meso-DAP	meso-DAP	meso-DAP	meso-DAP	meso-DAP	meso-DAP	L-Lysine
Major menaquinone	MK-7	MK-7			MK-7	MK-7	MK-7	MK-9

[a] Data from Chatelain and Second, 1966; Collins et al., 1983; Fiedler and Seger, 1983; Rocourt et al., 1983a, b; Rocourt and Grimont, 1983; Seeliger, 1961; Seeliger, 1981; Seeliger and Welshimer, 1974; Stuart and Welshimer, 1973, 1974; Welshimer and Meredith, 1971; Wetzler et al., 1968; Wilkinson and Jones, 1977 and unpublished data.

[b] Standard symbols: see Table 14.11; also d, 11–89% of the strains are positive.

[c] Not all strains of *L. monocytogenes* exhibit β-hemolysis—the type strain ATCC 15313 is nonhemolytic on horse, sheep and bovine blood.

[d] A very wide zone or multiple zones of hemolysis are usually exhibited by *L. ivanovii* strains.

[e] Of 30 strains listed, ATCC 15313, the type strain, did not give a positive reaction.

[f] Of 10 strains tested, 1 gave a positive reaction (Rocourt and Grimont 1983).

Table 14.13.
Serovars of the genus **Listeria** *and* **Listeria grayi** *and* **Listeria murrayi**

Paterson	Seeliger Donker-Voet	I	II	III	IV	V	VI	VII	VIII	IX	X	XI	XII	XIII	XV	XIV	H antigens
1	1/2a	I	II	(III)[a]													A B
	1/2b	I	II	(III)													A B C
2	1/2c	I	II	(III)													B D
3	3a		II	(III)	IV												A B
	3b		II	(III)	IV								(XII	XIII)			A B C
	3c		II	(III)	IV								(XII	XIII)			B D
4	4a			(III)		(V)		VII		IX							A B C
	4ab			(III)		V	VI	VII		IX	X						A B C
	4b			(III)		V	VI										A B C
	4c			(III)		V		VII									A B C
	4d			(III)		(V)	VI		VIII								A B C
	4e			(III)		V	VI		(VIII)	(IX)							A B C
	5			(III)		(V)	VI		(VIII)		X						A B C
	7 ?			(III)									XII	XIII			A B C
	6a (4f)			(III)		V	(VI)	(VII)		(IX)					XV		A B C
	6b (4g)			(III)		(V)	(VI)	(VII)		IX	X	XI					A B C
L. grayi				(III)									XII			XIV	E
L. murrayi				(III)									XII			XIV	E

[a] ( ), not always present.

List of the species of the genus **Listeria**

1. Listeria monocytogenes (Murray et al., 1926) Pirie 1940, 383.[AL] (*Bacterium monocytogenes* Murray, Webb and Swann 1926, 408.)

mo.no.cy.to′ge.nes M.L. n. *monocytum* a blood cell, monocyte; Gr. v. *gennaio* produce; M.L. adj. *monocytogenes* monocyte producing.

Morphology and general characters are as given for the genus and in Tables 14.11–14.13. β-hemolytic on blood agar with narrow zone of hemolysis around colonies, varying with species of blood. Positive CAMP reaction on sheep blood (5% v/v) agar plates with *Staphylococcus aureus* but not with *Rhodococcus equi*. All strains examined produce acid from L-rhamnose and α-methyl-D-mannoside. No acid is produced from D-xylose nor from D-mannitol. The species is antigenically varied. Strains exhibit the antigenic composition of serovars 1/2a, 1/2b, 1/2c, 3a, 3b, 3c, 4a, 4ab, 4b, 4c, 4d, 4e and serovar 7; serovar 7 has not been studied extensively. Strains exhibiting the antigenic composition of serovars 1/2b, 4c and 4d are also found in the species *L. seeligeri*.

Experimentally pathogenic for mice and guinea pigs. The Anton eye test is positive. Appears to be widely distributed in nature. Isolated from sewage, soil, silage and from feces of healthy animals and man. Also isolated from clinical conditions in man and animals. Causes meningitis, encephalitis, septicemia, endocarditis, abortion, abscesses and local purulent lesions.

The mol% G + C of the DNA is 37–39 (T_m); 38 (Bd).

Type strain: ATCC 15313.

It should be noted that the currently designated type strain, ATCC 15313, is not β-hemolytic nor does it exhibit hemolysis in the CAMP test with *Staphylococcus aureus*. Additionally, it is difficult to determine its antigenic composition. For these reasons it has been requested that strain ATCC 15313 be replaced by strain NCTC 7973 (Jones and Seeliger, 1983). The Judicial Commission of the ICSB has not yet issued an Opinion.

2. Listeria innocua (ex Seeliger and Schoofs 1979)* Seeliger 1983, 439.[VP] (Effective publication: Seeliger 1981, 492.)

in.noc′u.a. M.L. adj. *innocuus* harmless.

Morphology and general characters are as given for the genus and in Tables 14.11–14.13. Never hemolytic on 5 or 10% (v/v) sheep, rabbit, horse or human blood in liquid or solid culture media. CAMP-negative with *Staphylococcus aureus* and *Rhodococcus equi*. Most strains examined produce acid from L-rhamnose. All produce acid from α-methyl-D-mannoside. Acid is not produced from D-xylose and D-mannitol. All strains examined so far have the antigenic composition of serogroup 6 (serovars 6a and 6b) or serovar 4ab. Serovars 6a and 6b also occur in *L. welshimeri* and some strains of *L. seeligeri* have the antigenic composition of serovar 6b. Some strains of *L. monocytogenes* are antigenically serovar 4ab. Indeed, the strain used to raise antisera 4ab is a strain of *L. innocua* (Rocourt et al., 1982b).

Experimentally nonpathogenic for mice and guinea pigs when up to 10^{10} cells injected intraperitoneally. The Anton eye test is negative. One strain, Welshimer S-12, has been reported to produce a localized encephalitis in suckling mice after intracerebral inoculation. (Patočka et al., 1979). Isolated from soil, vegetation, human and animal feces and bird droppings. Has never been isolated from unequivocally pathological material from humans or animals.

The mol% G + C of the DNA is 36–38(T_m); 38 (Bd).

Type strain: SLCC 3379 (ATCC 33090, NCTC 11288).

3. Listeria welshimeri Rocourt and Grimont 1983, 867.[VP]

wel.shi′mer.i. M.L. gen. n. *welshimeri* of Welshimer, named in honor of Herbert J. Welshimer, American bacteriologist.

Morphology and general characters are as given for the genus and in Tables 14.11–14.13.

No β-hemolysis on blood agar and CAMP tests with *Staphylococcus aureus* and *Rhodococcus equi* are negative. Acid is produced from D-xylose and α-methyl-D-mannoside and may or may not be produced from L-rhamnose. The type strain does not produce acid from L-rhamnose. Acid is not produced from D-mannitol.

Strains currently assigned to this species are all of serovars 6a or 6b. The antigenic composition of the type strain is that of serovar 6b. Serovars 6a and 6b also occur in *L. innocua* and some strains of *L. seeligeri* have the antigenic composition of serovar 6b.

*** Editorial note:** this name was proposed in a paper given in 1977 but was not published until 1979. It should be considered a revived name with the original publication date of 1979.

Experimentally nonpathogenic for mice. Isolated from decaying vegetation and soil in the United States; probably widely distributed in nature.

The mol% G + C of the DNA is 36 (T_m).

Type strain: CIP 8149 (SLCC 5334, Welshimer V8).

4. Listeria seeligeri Rocourt and Grimont 1983, 869.[VP]

see′li.ger.i. M.L. gen. n. *seeligeri* of Seeliger, named in honor of Heinz P.R. Seeliger, German bacteriologist.

Morphology and general characters are as given for the genus and in Tables 14.11–14.13.

β-Hemolysis weak. The CAMP test is positive when *Staphylococcus aureus* is used but negative with *Rhodococcus equi*. Acid is produced from D-xylose. Acid is not produced from D-mannitol or L-rhamnose. The majority of strains do not produce acid from α-methyl-D-mannoside; the type strain is negative in this respect.

Strains currently assigned to this species have the antigenic composition of serovars 1/2b, 4c, 4d (which also occur in *L. monocytogenes*) and 6b (which is found in some strains of *L. innocua* and *L. welshimeri*). The type strain belongs to serovar 1/2b.

Experimentally nonpathogenic for mice. Isolated from vegetation, soil and animal feces in Europe; probably widely distributed in nature.

The mol% G + C of the DNA is 36 (T_m).

Type strain: CIP 100100 (SLCC 3954, Weiss 1120).

5. Listeria ivanovii Seeliger, Rocourt, Schrettenbrunner, Grimont and Jones 1984, 336.[VP]

i.van.ov′i.i. M.L. gen. n. *ivanovii* of Ivanov; named in honor of Ivan Ivanov, Bulgarian microbiologist.

Morphology and general characters are as given for the genus and in Tables 14.11–14.13.

Pronounced β-hemolysis with double or triple zones produced when grown on sheep or horse blood (5% v/v) agar (Fig. 14.5). A positive CAMP test is exhibited with *Rhodococcus equi* but not with *Staphylococcus aureus*. No acid is produced from D-mannitol, L-rhamnose or α-methyl-D-mannoside. All strains so far examined belong to serovar 5. This antigenic pattern has not been shown to be present in strains of any other species of the genus *Listeria*.

Experimentally pathogenic for mice. The LD_{50} for holoxenic pathogen-free mice is between 1×10^5 and 3×10^6 colony-forming units per milliliter. Pathogenic for animals, especially pregnant sheep, the species has also been isolated from healthy animals and human carriers and from the environment and, very occasionally, from clinical conditions in man.

The mol% G + C content of the DNA is 37–38 (T_m).

Type strain: SLCC 2379 (ATCC 19119).

Species Incertae Sedis

a. Listeria grayi Errebo Larsen and Seeliger 1966, 19.[AL]

gray′i. M.L. gen. n. *grayi* of Gray; named for M.L. Gray, American bacteriologist, known for his work in the field of listeriosis.

Straight rods usually 0.6–0.8 μm × 0.7–2.5 μm. The ends are rounded. Occur singly in pairs or short chains; rods are sometimes arranged at an angle to each other or lie in parallel groups along their long axis. Actively motile when grown at 20–25°C; nonmotile, or weakly so at 37°C. Capsules are not formed. Do not produce spores. Gram-positive. Not acid-fast. Do not contain metachromatic granules. Facultatively anaerobic. Grow well on Tryptose Agar (Difco) or Blood Agar Base No. 2 (Difco) with 1% (w/v) added glucose. Colonies small (1–1.5 mm in 24 h) round, smooth, butyrous, low convex with entire edge. When colonies on these media are examined with a dissecting microscope, using oblique light, small colonies (0.2 mm) give the same blue-green appearance associated with *Listeria monocytogenes*; after incubation for a longer period the colonies appear orange-red, especially around the edge. Not hemolytic. Optimum growth temperature 30–37°C. Temperature limits of growth 1–45°C. Do not survive heating at 60°C for 30 min. Catalase-positive. Oxidase-negative. Contain cytochromes a_1bdo. Organic growth factors required. The pH range of growth is 5–9, no growth at pH 9.6. Grow in presence of 10% (w/v) NaCl, 0.025% (w/v) thallous acetate, 3.75% (w/v) potassium thiocyanate, 0.04% (w/v) potassium tellurite, 0.01% (w/v) 2,3,5-triphenyltetrazolium chloride; tellurite and tetrazolium are reduced. Do not grow in the presence of 0.02% (w/v) sodium azide. Sensitive to penicillin, streptomycin, tetracycline, chloramphenicol, erythromycin, novobiocin, neomycin; resistant to sulfanilamide, polymyxin B, colistin sulfate and nalidixic acid. Do not grow on the medium of Gardner (1966). Weak growth may occur on MRS medium (de Mann et al., 1960).

Fermentative metabolism of glucose results in the production of L(+)-lactic acid and some other products. Acid but no gas is produced from esculin, amygdalin, cellobiose, dextrin, galactose, glucose, glycerol, lactose, fructose, maltose, mannitol, mannose, salicin, starch, trehalose. Acid is usually not produced from adonitol, arabinose, dulcitol, erythritol, glycogen, inositol, inulin, melibiose, melezitose, α-methyl-D-glucoside, raffinose, rhamnose, sorbitol, sucrose and xylose. Weak acid and slight reduction in litmus milk. Methyl red-positive. Voges-Proskauer-positive.

Esculin hydrolyzed. Sodium hippurate not hydrolyzed. Starch hydrolysis variable. Tween 20 hydrolyzed slowly (14 days). Tween 80 not hydrolyzed. Weak production of deoxyribonuclease, ribonuclease and phosphatase. Sulfatase not produced. Xanthine, tyrosine and chitin are not degraded. Cellulose, gelatin, casein and milk are not hydrolyzed. Indole not produced. Urea not hydrolyzed. Small amounts of H_2S (detected by lead acetate paper) produced by some strains. Nitrates not reduced to nitrites.

Serologically distinct from members of the genus *Listeria* and from *L. denitrificans*. Antigenic composition identical with that of *L. murrayi*.

The major cell wall amino acids are *meso*-diaminopimelic, glutamic acid and alanine. The cell wall amino sugars are muramic acid and glucosamine (Fiedler and Seger, 1983). Teichoic acid is present (Fiedler et al., 1984). Mycolic acids are not present. The long chain fatty acids consist predominantly of straight chain saturated, *anteiso*- and *iso*-methyl branched chain types. The major fatty acids are 14-methylhexadecanoic (*anteiso*-$C_{17:0}$) and 12-methyltetradecanoic (*anteiso*-$C_{15:0}$). Menaquinones are the sole respiratory quinones; the major menaquinone contains seven isoprene units (MK-7) (Feresu and Collins, personal communication). The mol% G + C content of the DNA is 41–42 (T_m); 41 (Bd).

Not pathogenic after intraperitoneal or intradural injection in white mice; 5×10^8 cells may be toxic to mice, strain C57/Leaden. Not pathogenic after intravenous injection into pregnant rabbits. Does not cause conjunctivitis when instilled into eyes of rabbits or guinea pigs.

Isolated from feces of chinchillas. Natural habitat not known.

Type strain: ATCC 19120 (Seeliger 332/64).

b. Listeria murrayi Welshimer and Meredith 1971, 7.[AL]

mur.ray.i. M.L. gen. n. *murrayi* of Murray; named in honor of E.G.D. Murray, co-discoverer of *L. monocytogenes*.

Straight rods usually 0.6–0.8 μm × 0.7–2.5 μm. The ends are rounded. Occur singly in pairs or short chains; rods are often arranged at an angle to each other or lie in parallel along their long axis. Actively motile when cultured at 20–25°C; nonmotile or weakly so when cultured at 37°C. Capsules not formed. Do not form spores. Gram-positive. Not acid-fast. Do not contain metachromatic granules. Facultatively anaerobic. Grow well on Tryptose Agar (Difco) or Blood Agar Base No. 2 (Difco) with 1% (w/v) added glucose. Colonies small (1–1.5 mm in 24 h), round, smooth, butyrous, low convex with entire edge. By reflected daylight colonies appear gray with a tinge of blue; by transmitted daylight the colonies appear more blue and become tinged with yellow as they become larger after further incubation. When colonies on the above media are examined with a dissecting microscope, using oblique light, small colonies (0.2 mm) give the same blue-green characteristics associated with *Listeria monocytogenes*; after incubation for 2 days, the larger colonies appear orange-red with oblique illumination (Welshimer and Meredith, 1971). When incubated at 22–25°C on nutrient agar containing added glucose, the colonies develop a distinct yellow pigment. The pigment does not develop in the absence of glucose. Not hemolytic on blood agar. Optimum growth temperature 30–37°C. Temperature limits of growth 1–45°C. Do not survive heating at 60°C for 30 min. Catalase-positive. Oxidase-negative. Contain cytochromes a_1bdo. Organic growth factors required. Grow in the presence of 10% (w/v) NaCl; 0.25% (w/v) thallous acetate; 3.75% (w/v) potassium thiocyanate; 0.04% (w/v) potassium tellurite; 0.01% (w/v) 2,3,5-triphenyltetrazolium chloride; tellurite and tetrazolium are reduced. Do not grow in the presence of 0.02% (w/v) sodium azide. Sensitive to penicillin, streptomycin, tetracycline, chloramphenicol, erythromycin, novobiocin, neomycin; resistant to sulfanilamide, polymyxin B, colistin sulfate and nalidixic acid. Do not grow on the medium of Gardner (1966).

Fermentative metabolism of glucose results in the production of L(+)-lactic acid and some other products. Acid but no gas is produced from esculin, amygdalin, cellobiose, dextrin, galactose, glucose, glycerol, lactose, fructose, maltose, mannitol, mannose, salicin, starch, trehalose. Some strains produce acid from melibiose, rhamnose and sorbitol. Weak acid production from α-methyl-D-glucoside. Acid not produced usually from adonitol, arabinose, dulcitol, erythritol, glycogen, inositol, inulin, melezitose, raffinose, sucrose, xylose. Weak acid and slight reduction in litmus milk. Methyl red-positive. Voges-Proskauer-positive.

Esculin hydrolyzed. Sodium hippurate not hydrolyzed. Starch hydrolysis variable. Tween 20 hydrolyzed slowly (14 days). Tween 80 not hydrolyzed. Weak production of deoxyribonuclease, ribonuclease and phosphatase. Sulfatase not produced. Xanthine, tyrosine and chitin are not degraded. Cellulose, gelatin, casein and milk are not hydrolyzed. Indole not produced. Urea not hydrolyzed. Some strains produce H_2S, detected by lead acetate paper. Nitrates reduced to nitrites.

Serologically distinct from members of the genus *Listeria* and from *L. denitrificans*. Antigenic composition identical with that of *L. grayi*.

The major cell wall amino acids are *meso*-diaminopimelic, glutamic acid and alanine. The cell wall amino sugars are muramic acid and glucosamine (Fiedler and Seger, 1983). Teichoic acid is present (Fiedler et al., 1984). Mycolic acids are not present. The long chain fatty acids consist predominantly of straight chain saturated, *anteiso*- and *iso*-methyl-branched chain types. The major fatty acids are 14-methylhexadecanoic (*anteiso*-$C_{17:0}$) and 12-methyltetradecanoic (*anteiso*-$C_{15:0}$). Menaquinones are the sole respiratory quinones; the major menaquinone contains seven isoprene units (MK-7) (Feresu and Collins, personal communication). The mol% G + C content of the DNA is 41–42.5 (T_m); 42 (Bd).

Does not produce conjunctivitis when instilled into the eyes of rabbits. Not pathogenic on intravenous injection into rabbits nor on intraperitoneal injection of 10^8 cells into 20 g Rockland Farm SW mice.

Isolated from decaying leaves of corn (maize); natural habitat probably soil and vegetation.

Type strain: ATCC 25401 (F9).

Seeliger et al. (1982) considered this taxon to be a nitrate-reducing biovar of *Listeria grayi* and were of the opinion that it could be designated as a subspecies of *L. grayi*.

c. *Listeria denitrificans* Prévot 1961, 512.[AL]

de.ni.tri′fi.cans. M.L. inf. *denitrificare* to denitrify; M.L. part adj. *denitrificans* denitrifying.

Irregular rods, 0.3–0.5 μm in diameter and 2–3 μm in length. Filamentous forms may develop. Gram-positive, but many cells, especially in older cultures, fail to retain the Gram stain completely. Coccoid forms which develop in older cultures always stain Gram-positive. The coccoid forms give rise to rod forms on transfer to a fresh medium. Not acid-fast. No endospores are formed. Motile by peritrichous flagella at both 25°C and 37°C. Facultatively anaerobic. Colonies on nutrient agar are 0.5–1.5 mm in diameter (24–48 h), convex, smooth, edge entire, grayish and translucent to opaque. When removed leave an impression on agar, but easily emulsified. Rough forms produce rough type colonies with a depressed center. As cultures become older (10–21 days) a

yellowish pigmentation may develop. The colonies do not appear blue or blue-green by obliquely transmitted light. Optimum growth temperature ~30°C. Temperature limits of growth 10–40°C. Do not survive heating at 60°C for 30 min. Grow in 5% but not 10% (w/v) NaCl. Grow in presence of 4.5% (w/v) potassium thiocyanate; 0.01% (w/v) potassium tellurite; 0.01% (w/v) 2,3,5-triphenyltetrazolium chloride; 0.02% (w/v) thallous acetate. Do not grow in the presence of 0.01% (w/v) sodium azide. Catalase-positive. Oxidase-negative. Acid but no gas produced from glucose, cellobiose, fructose, mannose, galactose, maltose, lactose, melibiose, sucrose, trehalose, glycerol, L-arabinose, D-xylose, amygdalin, arbutin, amidon, glycogen, starch, dextrin, β-gentobiose, D-turanose, D-lyxose, salicin, esculin. Weak acid production from melezitose. No acid produced from L-fucose, D-arabinose, sorbose, D-tagatose, ribose, L-xylose, adonitol, β-methyl-xyloside, rhamnose, erythritol, inositol, mannitol, dulcitol, sorbitol, α-methyl-D-glucoside, α-methyl-D-mannoside, n-acetylglucosamine, inulin, D-raffinose, xylitol, D-fucose, D-arabitol, L-arabitol, gluconate, 2-keto-gluconate and 5-keto-gluconate. Methyl red-positive. Voges-Proskauer-negative. Exogenous citrate not utilized. Growth requirements not determined. Sensitive, in vitro, by agar diffusion method to penicillin, streptomycin, chloramphenicol, aureomycin, terramycin, erythromycin, tetracycline, bacitracin, novobiocin, oleandomycin, kanamycin, vancomycin, colomycin, polymyxin B and nitrofurantoin. Resistant to sulfonamide, neomycin and nalidixic acid (Jones, 1975 and unpublished).

Extracellular enzymes hydrolyze DNA, RNA, cellulose and starch but not gelatin, chitin, casein, lecithin (egg yolk), xanthine, tyrosine, Tween 20 nor Tween 80; slight hydrolysis of Tweens 40 and 60 takes place after 7 days. Phosphatase, sulfatase and urease are not produced. Esculin is hydrolyzed. Sodium hippurate is not hydrolyzed or only weakly hydrolyzed (H_2SO_4 method) after 10 days. Acid produced in litmus milk. Nitrates are reduced to nitrites. H_2S-negative. Indole negative.

The cell walls contain lysine, alanine, glutamic acid and serine (Fiedler and Seger, 1983). The amino sugar content of the cell wall consists of muramic acid, glucosamine and galactosamine; the galactosamine is not a constituent of the peptidoglycan (murein) but of the polysaccharide associated with the cell wall (Fiedler and Seger, 1983). Teichoic acid is present (Fiedler et al., 1984). Mycolic acids are not present. The long chain fatty acids consist predominantly of straight chain saturated *anteiso*- and *iso*-methyl-branched chain acids. 12-Methyltetradecanoic acid (*anteiso*-$C_{15:0}$) constitutes the major fatty acid with substantial amounts of hexadecanoic (*iso*-$C_{16:0}$) and 14-methyl-hexadecanoic (*anteiso*-$C_{17:0}$) acids also present. The polar lipids are diphosphatidylglycerol, phosphatidylinositol, two phosphoglycolipids and small amounts of two unidentified phospholipids (Collins et al., 1983). Menaquinones are the sole respiratory quinones; the major component contains nine isoprene units (MK-9) (Collins et al., 1983).

Listeria denitrificans is serologically distinct from *Listeria* spp. and from *Erysipelothrix* (Welshimer and Meredith, 1971; Wilkinson and Jones, 1975). Isolated from cooked ox blood (Sohier et al., 1948), the natural habitat of the organism is not known. It is pathogenic to rats and mice when injected intraperitoneally but does not cause conjunctivitis when instilled into the eyes of rabbits and guinea pigs.

The mol% G + C of the DNA is 56 (T_m); 57 (Bd) (Stuart and Welshimer, 1973, 1974); 58 ± 0.5 (T_m) (Collins et al., 1983).

Type strain: ATCC 14870 (CIP 55134).

Taxonomic Comment

All the evidence from morphological, biochemical, serological, chemical and nucleic acid studies indicates that *L. denitrificans* is not a member of the genus *Listeria* (Chatelain and Second, 1966; Welshimer and Meredith, 1971; Stuart and Pease, 1972; Jones, 1975; Stuart and Welshimer, 1973, 1974; Wilkinson and Jones, 1975, 1977; Collins et al., 1983; Fiedler and Seger, 1983; Fiedler et al., 1984).

In the numerical phenetic study of Jones (1975a), *L. denitrificans* clustered at a similarity of approximately 80% with three coryneform strains isolated from poultry deep litter by Schefferle (1966). Similarly, a later numerical phenetic study also indicated a phenetic similarity between *L. denitrificans* and some unidentified coryneform bacteria (S. B. Feresu, personal communication). However, in neither of these studies did *L. denitrificans* form part of a named taxon.

L. denitrificans resembles members of the genus *Oerskovia* in that it is cellulolytic, contains lysine as the cell wall diamino acid, possesses major amounts of straight chain saturated and *anteiso*-methylbranched fatty acids, and contains diphosphatidylglycerol, phosphatidylinositol and two unknown phosphoglycolipids. However, *L. denitrificans* differs from *Oerskovia* in that the latter possess a high mol% G + C (approximately 71–76) and tetrahydrogenated menaquinones. On the basis of cell wall, fatty acid and menaquinone composition, together with the mol% G + C value, *L. denitrificans* appears to resemble the genus *Renibacterium*. However, *R. salmoninarum* differs from *L. denitrificans* in possessing a complex mixture of glycolipids (Collins, 1982). *Renibacterium salmoninarum* also lacks phosphatidylinositol and two unknown glycolipids detected in *L. denitrificans* (Collins et al., 1983).

Fiedler and Seger (1983) have suggested that the composition and structure of the cell wall peptidoglycan of *L. denitrificans* indicates a relationship with the "nicotianae group" of the genus *Arthrobacter*. However, the different polar lipid composition of *L. denitrificans* and the "nicotianae group" does not support this (Collins et al., 1983; Collins and Kroppenstedt, 1983).

The taxonomic assignment of *L. denitrificans* must await further study. A major drawback is that only one strain of the organism has been isolated and described (Sohier et al., 1948). Strains isolated from pork held at low temperatures and designated *L. denitrificans* (Blickstad et al., 1981) were later shown not to reduce nitrate and are no longer considered to be *L. denitrificans* (E. Blickstad, personal communication). To our knowledge, the rRNA oligonucleotide sequences of *L. denitrificans* have not been determined, nor have any DNA-rRNA hybridization studies been done with the organism and other coryneform bacteria.

Genus **Erysipelothrix** *Rosenbach 1909, 367*[AL]

DOROTHY JONES

E.ry.si.pe′lo.thrix. Gr. neut. n. *erysipelas* erysipelas; Gr. fem. n. *thrix* hair; M.L. fem. n. *Erysipelothrix* erysipelas thread

Straight, or slightly curved, slender rods with a tendency to form **long filaments**. Rods usually 0.2–0.4 μm in diameter and 0.8–2.5 μm in length with rounded ends; occur singly in short chains, in pairs at an angle to give V-forms, or in groups with no particular arrangement. **Filaments may be 60 μm or more in length. Nonmotile. Capsules not formed. Do not form spores. Gram-positive. Not acid-fast. Facultatively anaerobic.** Colonies small, usually transparent and nonpigmented. Narrow zones of α-hemolysis may occur on blood agar. **No β-hemolysis. Optimum temperature 30–37°C**; growth occurs between 5 and 42°C. Do not survive heating at 60°C for 15 min. **Catalase-negative.** Oxidase-negative. **Fermentative activity weak.** Acid but no gas produced from glucose and certain other carbohydrates. Organic growth factors are required. Cell wall contains a group B peptidoglycan based on lysine. Mycolic acids not present. Contain 2- and 3-hydroxy and nonhydroxylated long chain fatty acids. The hydroxylated fatty acids are primarily of the straight chain saturated series; the nonhydroxylated fatty acids are predominantly of the straight chain saturated and monounsaturated acid series; small amounts of *iso* and *anteiso*-methyl-branched fatty acids are also present. The mol% G + C of the DNA is 36–40 (T_m, Bd). Widely distributed in nature. Parasitic on mammals, birds and fish; some strains pathogenic for mammals and birds.

Type species: *Erysipelothrix rhusiopathiae* (Migula 1900) Buchanan 1918, 55.

Further Descriptive Information

Erysipelothrix rhusiopathiae exists in a rough and smooth form each characterized by closely associated morphological and colonial appearances. In the smooth form (S-form) the cells appear as small, straight or slightly curved Gram-positive rods as described for the genus. In the rough form (R-form) long filaments often more than 60 μm in length predominate. Both forms may be decolorized easily when Gram-stained. Both rods and filaments may appear as Gram-negative containing Gram-positive granules which give a beaded effect. In smears of blood and tissue taken from acute forms of infection, especially cases of septicemia, the organisms are of the S-form; in chronic infections with arthritis and endocarditis, the R-form is isolated frequently, and S-forms are usually present along with R-forms (Ewald, 1981). The cell morphology varies to some extent with the growth medium and conditions of incubation. It has been reported that a pH of 7.6–8.2 favors the S-form while at pH values below 7.0 the R-form predominates (Wilson and Miles, 1975). It has been claimed also that S-forms grow

better at 33°C and at 37°C the R-form is favored (Grieco and Sheldon, 1970).

S-form colonies (24–48 h) are very small, 0.3–1.5 mm in diameter, low convex, circular, transparent with smooth glistening surface and entire edge. In older cultures the colonies become slightly larger and the center becomes opaque. R-form colonies are larger, flatter and more opaque; the surface is matt, uneven and the edge irregular. R-form colonies may resemble miniature anthrax colonies (Wilson and Miles, 1975). The distinction between S-form and R-form colonies is not always sharp; intermediate forms may exist (Barber, 1939; Ewald, 1981). S-form colonies dissociate to give rise to intermediate and R-form colonies; R-forms also give rise to S-form colonies.

On blood agar α-hemolysis may be so intense that after 48 h incubation a slight clearing may be seen. Care must be taken in observing the plates because true β-hemolysis never occurs.

Erysipelothrix rhusiopathiae strains produce a characteristic "pipe cleaner" type of growth in gelatin stab cultures incubated at 22°C. At first (24 h) growth is faint and hazy and confined to an area just below the surface. After a few days the growth appears as a column extending to the bottom of the tube. The column of growth is composed of fine lateral outgrowths which in S-form organisms extend only 2 or 3 mm from the stab but may reach the sides of the tube in R-form organisms. Gelatin is not liquefied when incubation is at 22, 30 or 37°C. However, Ewald (1981) reports a small hole of liquefaction at the top of the stab after "further incubation."

E. rhusiopathiae is microaerophilic. Especially on first isolation, it grows in a band just below the surface of a soft agar culture. Whether this is due to a preference for CO_2 or reduced oxygen tension is not resolved. Laboratory cultures grow well aerobically and anaerobically.

Growth is favored by an alkaline pH . The optimum pH for growth is between pH 7.2 and 7.6; pH limits of growth have been reported as 6.8–8.2 (Karlson and Merchant, 1941) and 6.7–9.2 (Sneath et al., 1951).

Growth in nutrient agar is improved by the addition of glucose (0.2–0.5% w/v) or serum (5–10% v/v). The exact growth requirements of *E. rhusiopathiae* have not been determined; several amino acids, riboflavin and small amounts of oleic acid are required (Hutner, 1942). Ewald (1981) noted that tryptophan enhanced growth.

Acid production from carbohydrates is usually poor or inconsistent when in 1% (w/v) peptone water. Most workers recommend the addition of sterile horse serum (5–10% v/v) to the basal medium (White and Shuman, 1961; Wood, 1970; Seeliger, 1974). It is not always convenient to use serum: good results are achieved by testing for acid production in nutrient broth (Oxoid) plus the test carbohydrate (0.5–1% w/v) with phenol red as indicator. As noted by White and Shuman (1961) and Wood (1970), the fermentation pattern varies with the basal medium used. It is therefore advisable to determine the pattern of known strains in a particular medium.

The methyl red test is usually reported as negative, but Wilkinson and Jones (1977) reported all strains examined gave a weak positive reaction when incubated at 35°C for 7 days in BM broth*. Acetoin (Voges-Proskauer test) was not produced in this medium.

Fermentation of glucose results in the production of mainly L(+)-lactic acid with smaller amounts of acetic acid, formic acid, ethyl alcohol and carbon dioxide. Glucose catabolism is via the Embden-Meyerhof-Parnas pathway, although a small amount of glucose is dissimilated by the hexose monophosphate pathway (Robertson and McCullough, 1968). The evidence available indicates that the tricarboxylic acid cycle is relatively unimportant in *E. rhusiopathiae* (Robertson and Mc-Cullough, 1968). Exogenous citrate is not utilized.

The majority of *E. rhusiopathiae* strains produce H_2S but the results can vary with the medium used. The use of lead acetate paper with cultures in a liquid or solid medium does not always detect H_2S production. Tests are best carried out in triple sugar iron agar (Wood, 1970). It should be noted that an occasional old laboratory or vaccine strain does not produce detectable H_2S in triple sugar iron agar (R.L. Wood, personal communication).

Most known strains produce hyaluronidase and it has been speculated that there is a correlation between virulence and hyaluronidase production; good hyaluronidase producers usually belong to serovar 1 (Ewald, 1957; 1981). Neuraminidase is produced in differing amounts by all strains, and there appears to be a good correlation between the level of neuraminidase activity and virulence (Krasemann and Müller, 1975; Müller and Krasemann, 1976; Müller and Seidler, 1975; Nikolov and Abrashev, 1976).

E. rhusiopathiae strains may be identified serologically. Heat and acid-stable type-specific and heat labile species antigens occur. This accounts for differences observed between antisera prepared with boiled and unboiled strains. Most strains agglutinate with unabsorbed sera prepared against unboiled antigens. The type-specific antigens are polysaccharide complexes. Several serovars have been detected and different serotyping schemes proposed (see Kucsera, 1973). Dedié (1949) recognized two serovars, A and B, and proposed that all those strains which showed no reaction with A- or B-type specific antiserum be designated as a third group, N. As new or supposedly new serovars were detected within group "N" they were designated by consecutive letters of the alphabet. Unfortunately, new serovars were not always compared with strains belonging to different serovars and the serological methods used were different. To overcome this problem, Kucsera (1973) recommended the use of a uniform serological method and a uniform system for designating the serovar using arabic numbers. The recommended method for serological investigation is the double agar-gel diffusion precipitation test using autoclaved antigens and type-specific antisera.

Although capital letters of the alphabet are still used by some workers to designate the two main serovars (A and B), the numbered system introduced by Kucsera (1973) is now preferred by most investigators interested in the serotyping of *E. rhusiopathiae*. Under this system, Dedié's original serovars A and B are designated 1 and 2, respectively. All the other 20 serovars recognized to date have been found within Dedié's group N (Kucsera, 1973; Wood et al., 1978; Nørrung, 1979). Of the 22 serovars, 1 and 2 are the most common; the other 20 are relatively rare.

Cultures of serovar 2 agglutinate chicken red blood cells which lyse when complement is added to the complex (Dinter et al. 1976).

L-forms of *E. rhusiopathiae* have been described, especially by Bulgarian workers (see Ewald, 1981).

Bacteriophages active on *E. rhusiopathiae* strains have been isolated and there is some evidence that a phage typing system may prove useful. Lysogeny has been reported (see Ewald, 1981)

In vitro, most strains of *E. rhusiopathiae* are resistant to sulfonomides, colistin, gentamicin, kanamycin, neomycin, novobiocin and polymyxin; sensitive to penicillin, streptomycin, chloramphenicol, tetracycline and other antibiotics (Sneath, et al., 1951; Füzi, 1963; Wood 1965; Jones, unpublished).

The organisms grow in the presence of phenol (0.2% w/v), potassium tellurite (0.05% w/v), sodium azide (0.1% w/v), thallous acetate (0.02% w/v), 2,3,5-triphenyltetrazolium chloride (0.2% w/v) and crystal violet (0.001% w/v) (Sneath et al., 1951; Ewald, 1981).

Studies on protein gel electrophoresis patterns have been made by White and Mirikitani (1976) and investigations of amino acid decarboxylases by Nikolov and Mihailova (1975).

Neither cytochromes (M.D. Collins, personal communication) nor isoprenoid quinones have been detected in any of the strains examined (Collins et al., 1979).

The cell wall peptidoglycan diamino acid is lysine. Mann (1969) reported the presence of lysine, glycine, serine, glutamic acid and alanine while Feist (1972) found the cell wall peptidoglycan to comprise lysine, serine and alanine. However, Schleifer and Kandler (1972) reported the presence in *E. rhusiopathiae* (one strain) of a novel peptidoglycan type (B1δ) containing lysine, glycine, glutamic acid and alanine but not serine.

Feist (1972) noted the presence of a large number of sugars in the cell wall; galactose, glucose, arabinose, xylose, ribose, glucose-6-phosphate and galactose-6-phosphate. From the monosaccharide patterns, he recognized three chemovars, but there appeared to be no correlation between these and serovars.

Flossman and Erler (1972) reported the mol% G + C values of five strains of *E. rhusiopathiae* to be between 38 and 40 (T_m). White and Mirikitani (1976) found the mol% G + C values of 14 strains of *E. rhusiopathiae* to be between 36 and 38 (T_m). Stuart and Welshimer (1974) found the mol% G + C value of the type strain ATCC 19414, and strain ATCC 19416 to be 36 (Bd); the same value, 36 (T_m) was found by S.B. Feresu (personal communication).

E. rhusiopathiae has a very wide distribution in nature. It is parasitic on mammals, birds and fish. Most frequently it is found in pigs where it is the causative agent of the economically important disease swine erysipelas. Infection probably occurs by the oral route through the ingestion of contaminated material. Natural infections, with epizootics, also occur in other domestic animals (sheep, lambs, cows, horses, dogs, mice) and birds (turkeys, chickens, geese, pheasants) and also in wild and zoo animals. It has been isolated, not in association with infection, from the surfaces of fish, shellfish, fish slime and fish boxes (see Woodbine, 1950; Grieco and Sheldon, 1970; Ewald, 1981).

In man it causes a skin infection known as erysipeloid. The majority of reported infections result from direct handling of contaminated organic matter such as swine carcasses, fish and poultry. The infections are largely limited to veterinarians, butchers and fish handlers. Generally, infection is confined to the skin of the hands and lower arms where the organisms gain entry through cuts and abrasions. Only rarely does the infection become systemic, causing arthritis and endocarditis. Most fatal cases have been shown to occur among excessive users of alcohol (see Ewald, 1981).

For a nonsporing organism, *E. rhusiopathiae* is remarkably persistent in the environment. Organisms remain viable in drinking water for up to 5 days and for up to 15 days in sewage. Although it does not survive heating at 60°C for 15 min and does not grow in 10% (w/v) NaCl *E.*

---

* BM broth: peptone (Oxoid), 10.0 g; Lab Lemco (Oxoid), 8.0 g; yeast extract (Oxoid) 4.0 g; glucose, 5.0 g; Tween 80, 1 ml; K_2HPO_4, 5.0 g; $MgSO_4 \cdot 7H_2O$, 0.2 g; $MnSO_4 \cdot 5H_2O$, 0.05 g; and distilled water, 1000 ml.

rhusiopathiae has been isolated from salted or pickled bacon after several weeks and from ham 3 months after smoking. Viable organisms have been recovered from a buried carcass after 9 months. Heat and direct sunlight diminish the viability of *E. rhusiopathiae*. A low temperature, alkaline conditions and organic matter favor its survival (see Woodbine, 1950; Grieco and Sheldon, 1970; Ewald, 1981).

Contamination of soil and water occurs not only from the feces and urine of sick animals but also from the activities of asymptomatic carrier pigs (Wood, 1974). The shedding of the organisms by asymptomatic pigs into the soil of pigpens is probably the reason that *E. rhusiopathiae* may be isolated from farms on which no cases of swine erysipelas have occurred for many years (Wood and Packer, 1972). Contaminated surface water, rodents, wild birds and insects may be responsible for carrying the organisms to farms and fish factories.

Enrichment and Isolation Procedures

A number of procedures have been devised for the isolation of *E. rhusiopathiae*. Most are based on the ability of the organism to grow in the presence of various substances which are bacteriocidal or bacteriostatic for other organisms (see Wood, 1965; Ewald, 1981).

E. rhusiopathiae can be isolated easily from the blood of infected animals. A small quantity of blood is placed in a tube of semisolid nutrient agar supplemented with 0.2% (w/v) glucose or horse serum (5–10% v/v), incubated at 35°C for 24–48 h. The layer of growth which develops below the surface is then plated onto a suitable solid medium (Blood Agar Base No. 2 (Difco) plus 0.2% w/v glucose), incubated at 35°C for up to 48 h and examined for colonies.

Successful isolation of *E. rhusiopathiae* from pig or human skin requires the removal of a small piece of the entire thickness of the dermis because the organisms are located in the deeper parts of the skin. *E. rhusiopathiae* is rarely the only bacterium present in skin samples or in pieces of other tissues (spleen, kidney, liver, lung, tonsil). Isolation is best achieved by placing the skin or small pieces of tissue in 10 ml of modified ESB medium* (Wood, 1965). After overnight incubation at 35°C, 5 ml of the liquid portion is placed in a sterile tube and centrifuged for 20 min at approximately 1400 × g. The supernatant is discarded, the sediment resuspended in 1–2 ml physiological saline (0.85% w/v) and a portion plated on MBA medium† (Harrington and Hulse, 1971). After incubation at 35°C for 24–48 h the plate is examined for colonies.

Isolation from feces or contaminated soil may be achieved in much the same way. Samples (approximately 100 g) are placed in a sterile blender containing 220 ml sterile 0.1 M phosphate buffer. After mixing for 10 min, the whole is transferred to a sterile centrifuge bottle and centrifuged at a low speed for 10 min. The bottle is shaken slightly to resuspend the top 5–10 ml of the sediment and the cloudy supernatant decanted into a sterile 1-l screw-capped flask containing 200 ml double strength ESB medium (but volume of serum is not doubled, i.e. 5% v/v). After mixing, the flask is incubated at 35°C for 48 h. Samples are then plated on Packer's agar with 5% (v/v) horse serum (Packer, 1943), incubated at 35°C for 24–48 h and the plates then examined for colonies. Packer's agar is recommended for grossly contaminated specimens such as soil or feces because it is more selective for *E. rhusiopathiae* than is MBA medium (R.L. Wood personal communication).

A fluorescent antibody technique may be used to detect *E. rhusiopathiae* in tissues (Dacres and Groth, 1959; Seidler et al. 1971) and in enrichment broth cultures (Harrington et al. 1974).

Sakuma et al. (1973) used whole body autobacteriography to study localization of *E. rhusiopathiae* in the whole body of mice. Bacterial localization as demonstrated by this technique was very similar to that observed when infected mice were investigated by conventional bacteriological techniques. The inoculated organisms eventually localized in organs such as the spleen, liver and subcutaneous and intramuscular tissue. They then multiplied in these tissues and finally were found to be widely distributed as in a bacteremia.

Maintenance Procedures

The organisms may be preserved for several months by stab inoculation into screw-capped tubes of nutrient agar (pH 7.4). After overnight growth at 30°C, the tubes are tightly closed and kept at room or refrigerator temperature in the dark.

Longer term preservation (over 5 years) may be achieved by freezing in glass beads at −70°C (Feltham et al., 1978). The organisms can also be preserved by freeze drying.

Differentiation of the genus **Erysipelothrix** from other genera

Table 14.11 lists the features most useful in differentiating *E. rhusiopathiae* from other Gram-positive, nonsporing, rod-shaped bacteria. Identification of new isolates may be achieved by examination of cell morphology, growth characteristics in nutrient gelatin, maximum growth temperature, catalase and oxidase tests, production of acid from carbohydrates in a suitable medium and hydrogen sulfide production. Serological identification and the mouse protection test by specific hyperimmune antierysipelothrix serum may be necessary for certain identification of *E. rhusiopathiae* but they are not normally required. Similarly, the composition of the cell wall peptidoglycan and lipid composition are characteristic but it is not usually necessary to perform these analyses for routine identification.

The smooth form of *E. rhusiopathiae* is the most usual on first isolation. Bacteria with which it is most likely to be confused are members of the genera *Brochothrix*, *Corynebacterium*, *Lactobacillus*, *Listeria* and *Streptococcus*. The rough form of *E. rhusiopathiae* could be confused also with *Brochothrix* and with members of the genus *Kurthia*.

The novel cell wall peptidoglycan of *E. rhusiopathiae* (B1δ, Schleifer and Kandler, 1972), serves to distinguish it from all the above genera.

E. rhusiopathiae may be easily distinguished from *Kurthia* spp. by the catalase test and by their different oxygen requirements. *Kurthia* spp. are strongly catalase-positive and strictly aerobic.

The catalase test, if carried out on a suitable medium, also serves to distinguish *E. rhusiopathiae* from the catalase-positive genera, *Brochothrix*, *Corynebacterium* and *Listeria*. The presence of lysine as the cell wall diamino acid distinguishes *E. rhusiopathiae* from *Brochothrix*, *Corynebacterium* and *Listeria*, all of which contain *meso*-diaminopimelic acid (*meso*-DAP).

Brochothrix thermosphacta also differs from *E. rhusiopathiae* in inability to grow above 30°C; producing acid from far more carbohydrates, producing acetoin (Voges-Proskauer-positive); not producing H₂S; growing well on the selective medium of Gardner (1966); not exhibiting a "pipe cleaner" type of growth in nutrient gelatin; containing cytochromes and menaquinones.

Listeria spp. may be further distinguished from *E. rhusiopathiae* by growing well on the usual nutrient media; producing colonies with a marked blue-green sheen on Tryptose Agar (Difco); exhibiting motility; possessing a more marked saccharolytic activity; hydrolysis of esculin; greater tolerance of NaCl—(*L. monocytogenes* grows in nutrient broth

---

* ESB medium: Nutrient Broth No. 2 (Oxoid), 25.0 g; distilled water, 1000 ml. After sterilization at 121°C for 15 min, 50 ml horse serum; kanamycin, 400 mg; neomycin, 50 mg; vancomycin, 25 mg are added aseptically. This medium may be stored at 4°C for not more than 2 weeks.
† MBA medium: Heart Infusion Agar (Difco), 40.0 g; sodium azide, 0.4 g; distilled water, 1000 ml. After sterilization at 121°C for 15 min, 20 ml horse blood and 50 ml horse serum are added aseptically.

plus 8.5% (w/v) NaCl, *E. rhusiopathiae* does not); producing acetoin; sensitivity to neomycin (Füzi, 1963); hydrolysis of Tweens 20 and 80, not producing H₂S. *Listeria* spp. and *E. rhusiopathiae* are serologically distinct. *Listeria* spp. contain cytochromes and menaquinones.

Corynebacterium spp. may be further distinguished from *E. rhusiopathiae* by the type of colony produced on blood tellurite agar (0.04% w/v potassium tellurite); cell morphology which usually exhibits clubbing and swelling; not producing the pipe cleaner effect in nutrient gelatin at 20°C; containing corynomycolic acids (α-alkyl-β-hydroxy fatty acids) of carbon number C₂₂–C₃₆; the much higher mol% G + C content (approximately 55–66); containing cytochromes and menaquinones.

Although streptococci are also catalase-negative, careful inspection of the cell morphology should pose no problem in differentiating between members of the two genera. However, when the so-called "cell-retraction forms" of erysipelothrix are isolated, the differentiation may be difficult (Eissner and Ewald, 1973). Streptococci are, on the whole, more proteolytic and saccharolytic than *E. rhusiopathiae* and, with few exceptions, do not produce H₂S.

It is possible that if the cell wall peptidoglycan diamino acid is not determined, lactobacilli could be confused with *E. rhusiopathiae*. Both are Gram-positive, catalase-negative rods which have complex growth requirements. Neither group contains cytochromes nor menaquinones. Lactobacilli are rarely isolated from clinical sources but do occur in organic matter and rotting straw. Only rare strains are motile. However, lactobacilli are aciduric, the optimum pH is usually 5.5–5.8 or less. *E. rhusiopathiae* strains do not grow on MRS medium (de Mann et al., 1960). Lactobacilli do not produce H₂S.

The mouse protection test is considered by some clinical workers, especially veterinarians, to be the best method of identifying new isolates of *E. rhusiopathiae* with complete certainty. However, the test will detect only those strains which are virulent for mice.

Mouse Protection Test

Commercial horse antierysipelothrix serum is satisfactory for the test. It can be obtained from manufacturers specializing in veterinary products.

Three to six mice are injected subcutaneously with 0.1 ml of a 24-h broth culture of the suspected strain. The loose skin of the flank is recommended. At the same time, 0.3 ml of antiserum is injected into the opposite flank. If the organism is *E. rhusiopathiae* the mice which received culture and antiserum should survive; those receiving culture alone should die within 6 days. It is recommended that a control test using a known virulent strain of *E. rhusiopathiae* be set up in parallel as a check on the antiserum.

Taxonomic Comments

The results of the serological, biochemical and chemical studies of Erler (1972), Flossman and Erler (1972), Feist (1972) and White and

Miritikani (1976) suggest a variation in the properties of *E. rhusiopathiae* strains greater than that normally found within one species. Indeed, the variation in the gel electrophoresis protein patterns noted among strains of *Erysipelothrix* by White and Miritikani (1976) led these authors to state "that the strains' taxonomic relation might be suspect." Further studies are required to determine whether or not the genus *Erysipelothrix* contains more than one species.

The taxonomic position of the genus *Erysipelothrix* with regard to other bacterial genera is not yet resolved. The results of numerical taxonomic studies (Davis et al., 1969; Stuart and Pease, 1972; Jones, 1975; Wilkinson and Jones, 1977), cell wall peptidoglycan (Schleifer and Kandler, 1972), lipid (Tadayon and Carroll, 1971; M.D. Collins, personal communication) and DNA hybridization studies (Stuart and Welshimer, 1974) do not support the close similarity between the genera *Erysipelothrix* and *Listeria* suggested by earlier workers (Barber, 1939; Hutner, 1942).

The results of all the numerical taxonomic studies indicated that strains of *E. rhusiopathiae* formed a distinct cluster which showed the closest similarity to the streptococci. In the study of Wilkinson and Jones (1977) the closest similarity of the *Erysipelothrix* cluster was to representatives of the genus *Gemella*.

The results of recent 16S rRNA oligonucleotide studies are very interesting. They indicate that *E. rhusiopathiae* is specifically, although remotely, related to *Clostridium innocuum* (Ludwig and Stackebrandt, unpublished). Both organisms contain lysine in their cell wall, but the detailed peptidoglycan structure of *C. innocuum* has not yet been determined (Stackebrandt, personal communication). *C. innocuum* together with *C. ramosum* is a member of the RNA cluster which contains the mycoplasmas and which itself is part of the much broader clostridial group (Woese et al., 1980; Stackebrandt and Woese, 1981). The mycoplasma cluster is peripherally related to the large *Bacillus-Lactobacillus-Streptococcus* cluster. However, no specific relationship exists between *E. rhusiopathiae* and *Listeria* (Stackebrandt, personal communication). In the light of the results of the 16S rRNA studies, nucleic acid hybridization studies with *Erysipelothrix* strains and other appropriate bacteria should help clarify the taxonomic position of the genus.

Further Reading

Ewald, F.W. 1931. The genus *Erysipelothrix*. *In* Starr, Stolp, Trüper, Balows and Schlegel (Editors), The Prokaryotes. A Handbook on Habitats, Isolation and Identification of Bacteria. Springer-Verlag, New York.

Grieco, M.H. and C. Sheldon. 1970. *Erysipelothrix rhusiopathiae*. Ann. N. Y. Acad. Sci. *174*:523–532.

Kucsera, G. 1973. Proposal for standardization of the designations used for serotypes of *Erysipelothrix rhusiopathiae* (Migula) Buchanan. Int. J. Syst. Bacteriol. *23*: 184–188.

Woodbine, M. 1950. *Erysipelothrix rhusiopathiae*. Bacteriology and chemotherapy. Bacteriol. Rev. *14*:161–178.

Characteristics of the species of the genus **Erysipelothrix**

At present the genus *Erysipelothrix* contains only the type species *Erysipelothrix rhusiopathiae*.

List of the species of the genus **Erysipelothrix**

1. **Erysipelothrix rhusiopathiae** (Migula) Buchanan 1918, 55.[AL] Epit. spec. cons. Opin. 32, Jud. Comm. 1970,9. (*Bacterium rhusiopathiae* Migula 1900, 431.)

rhu.si.o.pa'.thi.ae. Gr. adj. *rhusios* red; Gr. n. *pathos*, disease; M.L. gen. n. *rhusiopathiae* of red disease.

In addition to those features given in the generic description, acid is usually produced in a suitable medium from glucose, galactose, fructose, lactose, maltose, dextrin and *N*-acetylglucosamine. Weak or delayed acid production from mannose and sometimes sucrose. Acid is not produced usually from glycerol, erythritol, arabinose, xylose, adonitol, β-methylxyloside, sorbose, rhamnose, dulcitol, inositol, mannitol, sor-

bitol, α-methyl-D-mannoside, α-methyl-D-glucoside, amygdalin, arbutin, esculin, salicin, cellobiose, melibiose, trehalose, inulin, melezitose, raffinose, amidon, glycogen, xylitol, β-gentiobiose, D-turanose, D-lyxose, D-tagatose, D-fucose or L-fucose. Weak acid or no change in litmus milk. Methyl red reaction negative or very weakly positive. Acetoin (Voges-Proskauer test) not produced. Indole not produced. Exogenous citrate not utilized. Hydrogen sulfide produced. Nitrates not reduced. Ammonia not produced from peptone.

Gelatin not liquefied but most strains exhibit characteristic pipe cleaner growth in gelatin stab cultures incubated at 22°C. Urea, sodium hippurate, esculin, starch, cellulose, casein and Tweens 20, 40, 60, not hydrolyzed. Tyrosine and xanthine not degraded. Sulfatase not produced. Deoxyribonuclease and ribonuclease not produced. Neuraminidase is produced. Most strains produce hyaluronidase. Some strains produce phosphatase and lecithinase.

On potassium tellurite (0.05% w/v) agar colonies grayish after 24 h, later (2–3 days) becoming black. Grow in presence of, but do not reduce, 2,3,5-triphenyltetrazolium chloride (0.2% w/v). Most strains do not grow in 6.5% (w/v) NaCl; none grows in 10% (w/v) NaCl. Do not grow on the medium of Gardner (1966) or on MRS medium (de Mann et al., 1960).

Twenty-two serovars can be distinguished on the basis of heat-stable somatic antigens.

Causes swine erysipelas. Sometimes pathogenic to man causing erysipeloid. Mice and pigeons are very susceptible; septicemia is produced. Rabbits are less susceptible and guinea pigs are more resistant.

The cell wall peptidoglycan is based upon lysine of the rare B1δ variation. Isoprenoid quinones absent. Cytochromes absent.

The mol% G + C of the DNA is 36–40 (T_m, Bd).

The mol% G + C of the type strain is 36 (Bd).

Type strain: ATCC 19414.

Genus **Brochothrix** *Sneath and Jones 1976, 102*[AL]

PETER H. A. SNEATH AND DOROTHY JONES

Bro.cho.thr'ix. Gr. n. *brochos* a loop; Gr. n. *thrix* a thread; M.L. fem. n. *Brochothrix* loop(ed) thread.

Regular, unbranched rods, usually 0.6–0.75 µm in diameter and 1–2 µm in length. **Occur singly, in short chains, or in long filamentous-like chains which fold into knotted masses.** In older cultures the rods give rise to coccoid forms which develop into rod forms when subcultured on to a suitable medium. **Capsules are not formed. Gram-positive,** but some cells, both rod and coccoid forms, lose the ability to retain the Gram stain. **No endospores are produced. Nonmotile. Aerobic and facultatively anaerobic.** Colonies (24–48 h) on nutrient agar are opaque, 0.75–1 mm in diameter, convex with entire margin. In older cultures (>2 days) the edge of the colony often breaks up and the center becomes raised to give a "fried-egg" appearance. **Nonpigmented.** Nonhemolytic. **Optimum temperature, 20–25°C; growth occurs within the range 0–30°C; over 30°C growth rarely occurs.** Does not survive heating at 63°C for 5 min. **Catalase and cytochromes are produced. Fermentative metabolism of glucose results in the production of L(+)-lactic acid and some other products. Methyl red-positive.** Acid but no gas is produced from a number of carbohydrates. Acetoin and acetate are the major end products of aerobic metabolism of glucose. **Voges-Proskauer-positive. Exogenous citrate is not utilized. Enzymes of the tricarboxylic acid cycle are almost totally absent.** Organic growth factors are required. The cell wall contains a directly cross-linked peptidoglycan based upon *meso*-diaminopimelic acid (*meso*-DAP). Mycolic acids are not present. The long chain fatty acid composition is predominantly of the straight chain saturated, *iso*- and *anteiso*-methyl-branched chain types. Menaquinones are the sole respiratory quinones. The mol% G + C of the DNA is 36 (T_m).

Type species: *Brochothrix thermosphacta* (McLean and Sulzbacher), Sneath and Jones 1976, 103.

Further Descriptive Information

The cellular morphology, with the characteristic loops and knots, is shown in Fig. 14.6. Two types of colony are frequently present even in young (24 h) plate cultures (see Barlow and Kitchell, 1966). The colony types can appear so different from each other that the culture may be thought to be contaminated. One type is convex, with entire edge and smooth surface, the other is flatter with irregular edge and rough surface. On restreaking, individual colonies of both types give rise to colonies which exhibit both kinds of morphology.

Colony size is increased markedly if glucose is included in the medium: on such a medium *B. thermosphacta* has a sour odor due to the production of lactic acid, acetoin, acetic acid and other fatty acids. Although *B. thermosphacta* is facultatively anaerobic, better growth is achieved in air.

There is general agreement that the temperature limits of growth are between 0 and 30°C. However, limited growth has been noted at 35°C and one strain has been reported to grow at 37 and at 45°C (see Gardner, 1981).

B. thermosphacta was originally classified in the genus *Microbacterium*, members of which are relatively heat resistant; consequently, much attention has been paid to the heat resistance of this species (see Gardner, 1981). All workers agree that the organisms do not survive heating at 63°C for 5 min. However, Gardner (1981) has suggested that it is more useful to characterize the heat resistance of an organism by its D value (decimal reduction time), defined as the time (min) for a 10-fold reduction in the population under specified conditions. The change in D with temperature is expressed as a z value defined as the change in temperature changing the D value by a factor of 10. On the basis of $D_{50°C}$ and $D_{55°C}$ values abstracted from the literature, Gardner (1981) calculated the z value to be 8°C. Therefore, he states that "... we can more accurately describe the heat resistance of *B. thermosphacta* as $D_{55°C} = 1$ min and z = 8°C. Hence, the calculated $D_{63°C} = 0.1$ min and $D_{47°C} = 10$ min." It should be noted that these values refer to skim milk and phosphate buffer and may not be directly relevant to meat, especially cured meats (Gardner, 1981).

The optimum pH for growth of *B. thermosphacta* is pH 7.0 but growth occurs within the range pH 5.0–9.0 (Brownlie, 1966).

The ability of *B. thermosphacta* to grow in the presence of NaCl has been examined by many workers (see Gardner, 1981). All strains grow in the presence of 6.5% NaCl but reports differ on the ability of the organisms to grow in the presence of higher concentrations. Many strains have been reported to tolerate concentrations of up to 10% NaCl. Whether the differences are due to differences between strains or to different methodologies is not clear (see Gardner, 1981).

Growth is inhibited by nitrite, but the degree of inhibition is related to the pH of the medium and the temperature of incubation (Brownlie, 1966). Low pH and low incubation temperatures increase the degree of inhibition. Collins-Thompson and Rodriguez-Lopez (1980) state that, unlike many lactic acid bacteria, *B. thermosphacta* does not appear to have a nitrite reductase system.

A preliminary study by Macaskie (1982) indicated that palmitic acid is inhibitory to the growth of *B. thermosphacta* in liquid culture.

B. thermosphacta does not grow on the acetate medium of Rogosa et al. (1951) and only poorly on the MRS medium of De Man et al. (1960). These media were devised, respectively, for the selective isolation and for the cultivation of lactobacilli.

Sutherland et al. (1975) reported that *B. thermosphacta* attacked tributyrin but not beef fat. However, Patterson and Gibbs (1978) found that not one of 95 isolates attacked tributyrin, an observation in agreement with the results of Davis et al. (1969). Collins-Thompson et al. (1971) demonstrated the presence of a glycerol ester hydrolase in *B. thermosphacta*. This lipase was active on tripropionin, tributyrin, tricaproin, tricaprylin and trilaurin but not tripalmitin, but the temperature optimum of the lipase was 35–37°C with little or no activity below 20°C.

B. thermosphacta possesses a high level of enzymes associated with the catabolism of glucose (Collins-Thompson et al., 1972; Grau, 1983). Fermentative metabolism of glucose always results in the production of L(+)-lactic acid. Other end products appear to depend on the conditions of growth. McLean and Sulzbacher (1953) found only L(+)-lactic acid present in quantities sufficient to detect. Davidson et al. (1968) also reported L(+)-lactic acid to be the main end product of

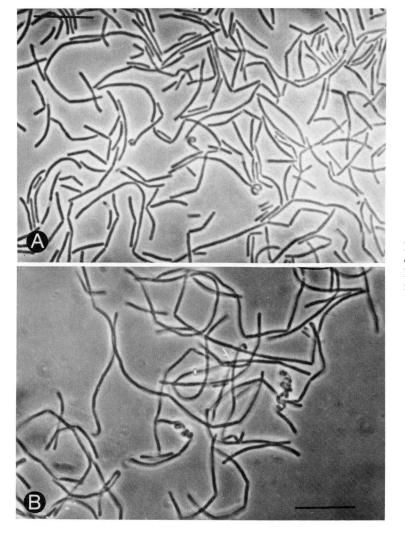

Figure 14.6.   Phase contrast photomicrograph of 24-h culture of *Brochothrix thermosphacta* ATCC 11509 grown on Blood Agar Base (Difco) showing *A*, long forms with loops and knots; *B*, more loops and knots (*bar*, 10 μm).

ofglucose fermentation but small amounts of acetic acid and propionic acids were also detected. In glucose-limited continuous culture, under anaerobic conditions, the end products of glucose metabolism have been identified as L(+)-lactic acid and ethanol (Hitchener et al., 1979). More recently, it has been reported that anaerobically *B. thermosphacta* ferments glucose to L(+)-lactate, acetate, formate and ethanol and that the ratio of these end products varies with the conditions of growth (Grau, 1983). Although McLean and Sulzbacher (1953) reported CO_2 production during fermentation of carbohydrates this has not been confirmed in any subsequent study.

The major end products of aerobic metabolism of glucose are acetoin and acetic, isobutyric and isovaleric acids (Dainty and Hibbard, 1980). The relative proportions of these end products are also affected by growth conditions. Low glucose concentrations and a near neutral pH favor fatty acid production; high glucose and a lower pH favor acetoin production. Similar results are obtained with ribose and glycerol. Acetoin and probably acetic acid are derived from the carbohydrates, and isobutyric and isovaleric acids from valine and leucine, respectively (Dainty and Hibbard, 1980).

When *B. thermosphacta* is grown in a complex medium, enzymes of the tricarboxylic acid (TCA) cycle are almost totally absent (Collins-Thompson et al., 1972). However, it has been suggested that in a defined or less complex medium the TCA cycle enzymes may be sufficiently active to provide substrates for synthesis but not to provide energy (Grau, 1979).

B. thermosphacta requires cysteine, lipoate, nicotinate, pantothenate, *p*-aminobenzoate, biotin and thiamine for aerobic growth in a glucose-mineral salts medium (Grau, 1979). The organism can also grow an-aerobically in this medium. Macaskie et al. (1981) showed that most, but not all, of the yeast extract requirement of *B. thermosphacta* can be fulfilled by thiamine.

B. thermosphacta strains contain cytochromes and are unequivocally catalase-positive when cultured on a suitable medium and incubated at 20°C (Davidson and Hartree, 1968; Davidson et al., 1968). Care must be taken when examining cultures for the presence of catalase. Production of the enzyme is dependent both on the growth medium and the temperature of incubation. Davidson et al. (1968) noted that *B. thermosphacta* strains grown on APT Medium (Difco) incubated at 20°C were always catalase-positive but that weak or negative reactions were obtained on HIA Medium (Difco) incubated at the same temperature. The same authors reported that negative results were obtained frequently if the bacteria were grown on either medium incubated at 30°C. This has been our experience, but we have found that BAB No. 2(Difco) is a satisfactory alternative medium for APT (Difco). Davidson and Hartree (1968) showed that *B. thermosphacta* contained cytochromes *aa₃b* and noted the same effects of growth medium and temperature of incubation on the quantitative cytochrome content of the organism. No satisfactory explanation can be offered for the differences in cytochrome and catalase content. It does not appear to be due to a difference in the concentration of heme compounds in the different media. In APT Medium (Difco)—a medium which favors the formation of catalase and cytochromes—the concentration of heme compounds has been reported to be too low to be detected by the sensitive hemochromogen technique (see Davidson et al., 1968). However, Davidson et al. (1968) noted that APT Medium (Difco) does contain added iron (8.0 μg/ml).

Little serological work has been carried out with *B. thermosphacta*. Wilkinson and Jones (1975) could demonstrate no serological relationships between *B. thermosphacta* and species of the genera *Listeria*, *Erysipelothrix* and *Kurthia*.

Bacteriophages active on *B. thermosphacta* have been isolated from aqueous extracts of spoiled beef (Greer, 1983). Phage plaque size and plating efficiency were reported to be increased significantly when the incubation temperature was reduced from 25 to 1°C. Fourteen distinct phage lysotypes were detected. Greer (1983) suggested that phage typing may provide a rapid method of differentiating *B. thermosphacta* strains. None of the high titer lysates of any of the *B. thermosphacta* phages was capable of lysing *Corynebacterium flavescens*, *Microbacterium lacticum*, *Lactobacillus mali*, *Lactobacillus plantarum*, *Listeria dentrificans* or *Listeria grayi*.

All the *B. thermosphacta* strains phage typed by Greer (1983) appeared to form a homogeneous group on the basis of their other phenotypic characters. However, as indicated by phage typing, the species may not be as homogeneous as currently thought. An investigation of the esterases of a number of *B. thermosphacta* strains by gel electrophoresis indicated the presence of seven groups among the strains examined. There was no association between groups based on esterase patterns and source of isolation (Gardner, personal communication). Further taxonomic studies are required to establish whether those strains of *B. thermosphacta* exhibiting esterase patterns different from the type strain represent variants within the species or new distinct species within the genus *Brochothrix*.

There is no evidence that *B. thermosphacta* is pathogenic. It is an economically important meat-spoilage organism because it grows in a wide variety of meats and meat products and produces malodorous metabolic end products which make affected meat unpalatable.

Enrichment and Isolation

Brochothrix thermosphacta is an important spoilage organism of meat and meat products stored aerobically or vacuum packed at chill temperatures (see Dainty and Hibbard, 1980; Gardner, 1981; Keddie and Jones, 1981). Consequently, almost all isolation methods described for *B. thermosphacta* are concerned with its recovery from such sources.

After its first reported isolation from pork trimmings and finished pork sausage (Sulzbacher and McLean, 1951), *B. thermosphacta* has been isolated on numerous occasions from the same sources and from a variety of animal and poultry meats or products based on them (see Gardner, 1981; Keddie and Jones, 1981). Until fairly recently, there were few reports of the isolation of *B. thermosphacta* from other sources. McLean and Sulzbacher (1953) occasionally isolated it from equipment and tables used to prepare sausages but assumed that its presence there was the result of contamination by pork trimmings since it could be isolated repeatedly from unopened barrels of such trimmings. Gardner (1966), using a selective medium, isolated bacteria which he thought resembled *B. thermosphacta* from soil and feces. Collins-Thompson et al. (1971) suggested that some lipolytic, coryneform bacteria isolated from dairy sources by Jayne-Williams and Skerman (1966) could be *B. thermosphacta* but the published description of these organisms does not support this suggestion.

However, in the last 5 years, *B. thermosphacta* has been isolated from fish fingers, smoked whiting, frozen coley, frozen cod and from a variety of foods including frozen peas, frozen runner beans, milk, cream, cottage cheese and prepacked tomato salad (see Gardner, 1981). Nickelson et al. (1980) recovered *B. thermosphacta* at all stages of the production (whole fish, scaled, headed, eviscerated and minced flesh) of minced fish flesh from six different species of finfish caught in the Gulf of Mexico. Rarely did the proportion of *B. thermosphacta* in the total bacterial flora exceed 5%. Lannelongue et al. (1982) isolated *B. thermosphacta* from stored finfish fillets packaged under an atmosphere of CO_2. Again, the numbers recovered as a proportion of the total population were very small. As noted by Gardner (1981) it is probable that *B. thermosphacta* may also occur in other situations and further work is required to establish its natural habitats.

Enrichment is not usually performed for the isolation of *B. thermosphacta*. Wolin et al. (1957) placed irradiated sliced beef in sterile Petri dishes containing 3–5 ml sterile distilled water and incubated at 2°C in a slanting position until spoilage was evident. Samples plated out on a suitable nutrient agar yielded good growth of *B. thermosphacta*. There do not appear to be any other reports of enrichment.

Brochothrix thermosphacta is often the dominant organism in prepacked meats and meat products stored at refrigerator temperature. The conditions prevailing in such storage conditions selectively favor its growth. *B. thermosphacta* grows at 1–4°C and under conditions of O_2 depletion and increased CO_2 concentration (see Gardner et al., 1967). The organism is mainly limited to the meat surface and grows at the meat/clingfilm interface (Ingram and Dainty, 1971).

Swabs of various meat surfaces are directly plated on to suitable media, e.g. glycerol nutrient agar* (Gardner, 1966) or glucose nutrient agar† (Sulzbacher and McLean, 1951; Wolin et al., 1957). Alternatively, samples of macerated meat or other materials suspended in saline (0.85%, w/v) or peptone water (0.1%, w/v) are shaken vigorously and appropriate dilutions spread on to the same media (see above) to give well separated colonies. In addition to *B. thermosphacta*, such procedures can result in the recovery of a wide variety of other bacteria, e.g. micrococci, staphylococci, lactobacilli, *Kurthia* spp. and *Pseudomonas* spp.

Selective isolation of *B. thermosphacta* is achieved by the use of the selective medium (STAA)§ of Gardner (1966). After incubation of appropriate material on the medium at 20–22°C for 2 days, almost all the colonies are those of *B. thermosphacta*. A few pseudomonads may also grow; these may be detected by their positive oxidase reaction. Members of the genera *Lactobacillus*, *Listeria*, *Erysipelothrix*, *Streptococcus* and *Bacillus* and those coryneform bacteria tested do not grow on this medium.

Maintenance Procedure

Cultures may be preserved for short periods (months rather than years) in nutrient agar (plus 0.1% (w/v) glucose) stabs in screw-capped bottles. After overnight incubation at 25–30°C, the caps should be screwed tightly and the bottles stored at room or refrigerator temperature in the dark.

Longer term preservation (over 7 years) may be achieved by freezing in glass beads at −70°C (Feltham et al., 1978). The organisms can also be preserved by lyophilization.

Differentiation of the genus **Brochothrix** from other genera

Table 14.11 lists the features most useful in differentiating the genus *Brochothrix* from other Gram-positive, nonspore-forming, rod-shaped bacteria.

Identification of new isolates may be achieved by examination of cellular morphology, maximum growth temperature, oxygen requirements, catalase production and tests such as production of acid from

---

* Glycerol nutrient agar: peptone (Oxoid), 20 g; yeast extract (Oxoid), 2 g; glycerol, 15 g; K_2HPO_4, 1 g; $MgSO_4 \cdot 7H_2O$, 1 g; agar (Oxoid No. 3), 13 g; distilled water, 1000 ml; pH 7.0. Autoclave 121°C for 15 min.

† Glucose nutrient agar: tryptone (Difco), 10 g; yeast extract (Difco), 5 g; K_2HPO_4, 5 g; NaCl, 5 g; glucose, 5 g; agar (Difco), 15 g; distilled water, 1000 ml; pH 7.0. Autoclave at 121°C for 15 min.

§ STAA agar; peptone (Oxoid), 20 g; yeast extract (Oxoid), 2 g; glycerol, 15 g; K_2HPO_4, 1 g; $MgSO_4 \cdot 7H_2O$, 1 g; agar (Oxoid No. 3), 13 g; distilled water, 1000 ml; pH accurately adjusted to 7.0. Autoclaved at 121°C for 15 min. To this molten, sterile medium are added the following solutions prepared with sterile distilled water: streptomycin sulfate (Glaxo) to a final concentration of 500 μg/ml, actidione (Upjohn) to 50 μg/ml, thallous acetate to 50 μg/ml. Mix well and dispense.

various sugars. The chemical composition of the cell wall and the lipid composition also aid the identification of *Brochothrix* but it is not usually necessary to perform these analyses for routine identification.

Morphologically, *Brochothrix* may be confused with *Kurthia* but the two genera can be distinguished by their different O₂ requirements. *Brochothrix* is facultatively anaerobic while *Kurthia* is strictly aerobic. In addition, *Brochothrix* strains produce acid from a wide variety of sugars, whereas members of the genus *Kurthia* are not saccharolytic. The presence of *meso*-diaminopimelic acid (*meso*-DAP) as the cell wall diamino acid also distinguishes *Brochothrix* from *Kurthia* (Schleifer and Kandler, 1972). The latter contains lysine in the cell wall.

The cell wall peptidoglycan composition of *B. thermosphacta* also serves to distinguish it from the genus *Erysipelothrix* which, like *Kurthia*, contains lysine as the major cell wall diamino acid. Other features which distinguish *B. thermosphacta* from *Erysipelothrix* are the inability of *B. thermosphacta* to grow at 35°C and the catalase reaction. If the test is carried out under the correct conditions, *B. thermosphacta* is always catalase-positive while *Erysipelothrix* is catalase-negative.

The cell wall peptidoglycan diamino acid of *B. thermosphacta*, *meso*-DAP, is the same as that of the genus *Corynebacterium*. However, *B. thermosphacta* cell walls contain neither arabinose nor galactose (Schleifer and Kandler, 1972), and this together with the absence of mycolic acids (see Minnikin et al., 1978), its distinctive morphology and its inability to grow at 35°C, serve to distinguish it from the genus *Corynebacterium*.

Strains of *B. thermosphacta* are most likely to be confused with some members of the genus *Lactobacillus* and with the genus *Listeria*. Its distinctive morphology and inability to grow at 35°C; presence of catalase, cytochromes and menaquinones (Davidson and Hartree, 1968; Collins et al., 1979; Collins and Jones, 1981); and inability to grow on acetate medium together with its low mol% G + C value, differentiate *B. thermosphacta* from those lactobacilli which possess a cell wall peptidoglycan containing *meso*-DAP as the diamino acid, e.g. *Lactobacillus plantarum*. The fatty acid composition of *B. thermosphacta* is also different from that of the lactobacilli (Shaw and Stead, 1970; Shaw, 1974).

B. thermosphacta shares many characters in common with members of the genus *Listeria*. Both taxa are facultatively anaerobic; produce acid from a variety of sugars; contain catalase, cytochromes and menaquinones; possess a cell wall peptidoglycan with *meso*-DAP as the diamino acid; and exhibit similar polar lipid and fatty acid profiles and similar mol% G + C values. Members of both taxa fail to grow on acetate medium. However, they may be distinguished by their morphology, growth temperature and motility and also serologically.

Colonies of *B. thermosphacta* do not show the blue-green coloration exhibited by *Listeria* species when viewed by obliquely transmitted white light. Gram-stained preparations of *B. thermosphacta* possess a distinctive morphology not found in the genus *Listeria*. The inability of *B. thermosphacta* to grow at 35°C together with its lack of motility also distinguish it from the genus *Listeria*. No serological cross-reactions have been shown between *B. thermosphacta* and species of the genus *Listeria* (Wilkinson and Jones, 1975).

Taxonomic Comments

The genus *Brochothrix* comprises those bacteria isolated by Sulzbacher and McLean (1951) and later allocated to the genus *Microbacterium* as *Microbacterium thermosphactum* by McLean and Sulzbacher (1953). The genus was tentatively placed in the family *Lactobacillaceae* by Sneath and Jones (1976) but its precise taxonomic position is not resolved.

The similarity between bacteria of the genus *Brochothrix* and the lactobacilli was noted by McLean and Sulzbacher (1953). However, at that time both the genera *Microbacterium* and *Lactobacillus* were classified, in the sixth edition of the *Manual*, in the family *Lactobacteriaceae*. Since the main distinction between the two genera was catalase

production, McLean and Sulzbacher (1953) allocated their bacteria to the genus *Microbacterium*. In the eighth edition of the *Manual*, *M. thermosphactum* (*Brochothrix thermosphacta*) is listed as *incertae sedis* because the genus *Microbacterium* is so listed.

Subsequent studies confirmed and augmented the observations of McLean and Sulzbacher (1953) that *M. thermosphactum* (*B. thermosphacta*) differed quite markedly from *M. lacticum*, the type species of the genus *Microbacterium*. These included morphological (Davidson et al., 1968; Jones, 1975), enzymic and protein profile (Robinson, 1966; Collins-Thompson et al., 1972), peptidoglycan (Schleifer, 1970; Schleifer and Kandler, 1972) and DNA base composition (Collins-Thompson et al., 1972) studies. Further, numerical taxonomic studies showed that *M. thermosphactum* strains formed a relatively homogeneous taxon (with an intragroup similarity of more than 85%) quite separate from *M. lacticum* (Davis and Newton, 1969; Davis et al., 1969; Jones, 1975; Wilkinson and Jones, 1977). On the basis of these data, Sneath and Jones (1976) considered that *M. thermosphactum* strains were sufficiently distinct from other Gram-positive bacteria to merit separate genus status and proposed that they be transferred to the genus *Brochothrix* as *B. thermosphacta*.

Sneath and Jones (1976) were aware that the taxonomic relatedness of *B. thermosphacta* to other bacteria could be controversial. The results of most numerical taxonomy studies indicated a fairly close similarity to the lactic acid bacteria and this similarity was supported by the results of enzymic and DNA-base ratio studies (Collins-Thompson et al., 1972). The presence of *meso*-DAP in the cell wall peptidoglycan was not at variance with these data because some lactobacilli also contain this diamino acid in the cell wall peptidoglycan. The presence of catalase and cytochromes in *Brochothrix* was not, in the opinion of Sneath and Jones (1976), sufficient to exclude the genus from the family *Lactobacillaceae*. However, they did note that the reported fatty acid composition of *B. thermosphacta* (Shaw and Stead, 1970) was quite distinct from that of the genus *Lactobacillus*. After careful consideration of these data, Sneath and Jones (1976) suggested that *Brochothrix* be placed in the family *Lactobacillaceae* for the present.

More recent data indicate that the genus *Brochothrix* more closely resembles the genus *Listeria* than the genus *Lactobacillus*. Both *Brochothrix* and *Listeria* possess catalase and cytochromes (Davidson and Hartree, 1968; Meyer and Jones, 1973; Jones unpublished), contain *meso*-DAP in the cell wall peptidoglycan (Schleifer and Kandler, 1972), possess menaquinones with seven isoprene units (MK-7) as the predominant isoprenoid quinone (Collins et al., 1979; Collins and Jones, 1981) and contain predominantly methyl-branched chain fatty acids (Shaw, 1974; Collins and Feresu, unpublished).

The results of studies on the degree of DNA/DNA and DNA/rRNA pairing between the genera *Brochothrix* and *Listeria* should clarify the degree of relatedness. The current evidence indicates that when rRNA oligonucleotide sequencing is done, the genus *Brochothrix*, together with *Listeria*, will be found to be part of the supercluster of taxa which embraces *Bacillus*, *Streptococcus* and *Lactobacillus* and which is, itself, a subbranch of the much broader clostridial group (Stackebrandt et al., 1980; Stackebrandt and Woese, 1981).

Further Reading

Collins-Thompson, D. L., T. Sørhaug, L. D. Witter and Z. J. Ordal. 1972. Taxonomic consideration of *Microbacterium lacticum*, *Microbacterium flavum* and *Microbacterium thermosphactum*. Int. J. Syst. Bacteriol. 22: 65–72.

Davidson, C. M., P. Mobbs and J. M. Stubbs. 1968. Some morphological and physiological properties of *Microbacterium thermosphactum*. J. Appl. Bacteriol. 31: 551–559.

Gardner, G. A. 1981. *Brochothrix thermosphacta* (*Microbacterium thermosphactum*) in the spoilage of meats: a review. *In* Roberts, Hobbs, Christian and Skovgaard (Editors), Psychrotrophic Microorganisms in Spoilage and Pathogenicity. Academic Press, London, pp. 139–173.

Keddie, R. M. and D. Jones. 1981. The genus *Brochothrix* (formerly *Microbacterium thermosphactum*, McLean and Sulzbacher) *In* Starr, Stolp, Trüper, Balows and Schlegel (Editors), The Prokaryotes. A Handbook on Habitats, Isolation and Identification of Bacteria. Springer-Verlag, Berlin, pp. 1866–1869.

Sneath, P. H. A. and D. Jones. 1976. *Brochothrix*, a new genus tentatively placed in the family *Lactobacillaceae*. Int. J. Syst. Bacteriol. 26: 102–104.

Characteristics of the species of the genus **Brochothrix**

At present, the genus *Brochothrix* contains only the type species *Brochothrix thermosphacta*.

List of the species of the genus **Brochothrix**

1. **Brochothrix thermosphacta** (McLean and Sulzbacher 1953) Sneath and Jones 1976, 103.[AL] (*Microbacterium thermosphactum* McLean and Sulzbacher, 1953, 432.)

ther′mos.phac.ta. M.L. fem. adj. *therme* Gr. n. heat; *sphactos* Gr. adj. slain; killed by heat.

In addition to the features given in the generic description, acid is produced fermentatively from arabinose, cellobiose, fructose, glucose, glycerol, lactose, maltose, mannitol, mannose, melezitose, rhamnose, salicin, sorbitol, sucrose and xylose. Weak or delayed acid production from adonitol, dulcitol, galactose, inositol and melibiose. No acid is produced from sorbose. Milk is made slightly acid, otherwise unchanged. Methyl red- and Voges-Proskauer-positive. Oxidase-negative. H₂S and indole are not produced. Sodium hippurate is not hydrolyzed.

Exogenous urea and citrate are not utilized. Nitrate is not reduced. Gelatin is not liquefied and casein is not digested. Phosphatase is produced. Deoxyribonuclease and ribonuclease activities absent or very weak; Tweens 20, 40, 60, 80, not hydrolyzed.

The cell walls contain *meso*-DAP, glutamic acid and alanine, but not arabinose or galactose. The major fatty acids are 12-methyltetradecanoic and 14-methylhexadecanoic. The major phospholipids are phosphatidylglycerol, diphosphatidylglycerol and phosphatidylethanolamine. The glycolipid fraction contains acetylated glucose and small amounts of glycosyl diglyceride. The major menaquinone contains seven isoprene units (MK-7). The mol% G + C of the DNA is 35.6–36.2 (T_m).

Type strain: ATCC 11509.

Genus **Renibacterium** Sanders and Fryer, 1980, 501[VP]*

J. E. SANDERS AND J. L. FRYER

Re.ni.bac.te′ri.um. L. n. renes the kidneys; Gr. neut. n. bakterion small rod; M.L. neut. n. *Renibacterium* kidney bacterium.

Short rods, 0.3–1.0 × 1.0–1.5 μm, often occurring in pairs and short chains sometimes observed. **Strongly Gram-positive.** Nonencapsulated. Nonmotile. Endospores are absent.

Aerobic. Slow growing bacterium; temperature for optimum growth 15–18°C; no growth at 37°C. Cysteine required for growth. Growth enhanced by addition of blood or serum to media. **No acid production from sugars. Catalase-positive.**

The diamino acid of the peptidoglycan is lysine. Cell wall sugars include galactose, rhamnose, N-acetylglucosamine and N-acetylfucosamine. **No mycolic acids are present.** The major fatty acid is 12-methyltetradecanoic acid (anteiso-C_{15}) with 14-methylhexadecanoic acid (anteiso-C_{17}) also present in signficant amounts. The major respiratory quinones are unsaturated menaquinones with nine isoprene units. **The mol% G + C of the DNA is 53 (T_m).**

Type species: *Renibacterium salmoninarum* Sanders and Fryer 1980, 501.

Further Descriptive Information

Pleomorphic forms of *Renibacterium*, evidence of Chinese letter formation and metachromatic granules were reported by Ordal and Earp (1956) and Smith (1964); however, recent studies show no evidence of these characteristics (Young and Chapman, 1978). Cells grown in culture media tend to exhibit consistent morphology while pleomorphic forms are most often observed among cells from host tissues (Fryer and Sanders, 1981).

Isolation and Enrichment Procedures

Renibacterium can be isolated from fish by streaking kidney tissue or material taken from a characteristic lesion onto either cysteine blood† or cysteine serum agar§ (Ordal and Earp, 1956; Fryer and Sanders, 1981) or KDM-2 medium¶ (Evelyn, 1977). Inoculated culture containers should be sealed to prevent loss of moisture and incubated at 15–18°C. Growth of *Renibacterium* on these media at primary

isolation is slow and depending on the amount of inoculum may require several weeks (3–5) for visible colonies to appear. Laboratory cultures adapted to either media may require 7 or more days of incubation before individual colonies appear.

These media are not selective; therefore, the presence of contaminating microorganisms may overgrow agar surfaces. Cultures should be examined periodically during incubation and contaminants aseptically removed from the agar surfaces. Addition of the antimicrobial agents cycloheximide, D-cycloserine, polymyxin B and oxolinic acid to KDM-2 medium has been reported to improve isolation of *Renibacterium* by elimination or control of many contaminating microorganisms (Austin et al., 1983).

Shaker cultures of either cysteine serum or KDM-2 broth give uniform growth throughout the medium. Initial attempts at growth in shaker cultures are often characterized by a prolonged lag period; however, subsequent transfers frequently reduce this period to less than 24 h.

Mueller Hinton medium supplemented with 0.1% cysteine has also been used for the growth of *Renibacterium* (Wolf and Dunbar, 1959; Bullock et al., 1974). However, Evelyn (1977) indicated that serum appeared to be an essential ingredient for the isolation and continued cultivation of *Renibacterium*. Paterson et al. (1979) added 10% fetal calf serum to cysteine-supplemented Mueller Hinton to improve primary isolation. A semidefined culture medium devoid of serum has been reported for the routine laboratory cultivation of *Renibacterium*; however, it is not clear if this preparation is suitable for primary isolation of the bacterium from infected fish (Embly et al., 1982).

Maintenance Procedures

Renibacterium strains can be lyophilized by common procedures used for many bacteria. Broth cultures of the bacterium distributed into 5-ml vials and frozen at −70°C maintain their viability for 6 months or longer.

---

* Technical Paper No. 6443, Oregon Agricultural Experiment Station.

† The formulation for cysteine blood agar (% w/v): tryptose, 1.0; beef extract, 0.3; NaCl, 0.5; yeast extract, 0.05; cysteine-HCl, 0.1; human blood (by vol), 20.0; and agar, 1.5.

§ The formulation for cysteine serum agar is identical to cysteine blood agar except 10% by volume calf serum is added in place of whole blood. Human and fetal calf serum can also be used.

¶ The formulation for KDM-2 agar (% w/v): peptone, 1.0; yeast extract, 0.05; cysteine-HCl, 0.1; fetal calf serum (by vol), 10–20; agar, 1.5; and pH 6.5.

Differentiation of the genus **Renibacterium** from other genera

The genus *Renibacterium* differs from *Listeria* and "*Murraya*" which have a peptidoglycan containing *meso*-diaminopimelic acid (*meso*-DAP), aspartic acid and leucine. In addition, the mol% G + C of the DNA in *Renibacterium* is 53, in contrast to values of 38–42 for *Listeria* and "*Murraya*" (Stuart and Welshimer, 1974). According to Stuart and Welshimer (1973; 1974), *Listeria denitrificans* differed from *Listeria monocytogenes*, "*Murraya murrayi*" and "*Murraya grayi*." *Listeria denitrificans* grows at 37°C, is motile, produces acids from carbohydrates, and does not require added cysteine for growth. Further, S and R colony types described for *Listeria denitrificans* have not been reported with *Renibacterium*. Jones (1975) placed *Listeria denitrificans* with the *Arthrobacter*.

Renibacterium isolates do not resemble members of the genus *Erysipelothrix*. The latter have a tendency to form long filaments, are catalase-negative and have a mol% G + C of 36.

Corynebacterium pyogenes and *Corynebacterium haemolyticum* share certain characteristics with *Renibacterium*. However, Collins and Jones (1982) and Reddy et al. (1982) propose the transfer of *Corynebacterium pyogenes* to the genus *Actinomyces* as *Actinomyces pyogenes*. *Corynebacterium haemolyticum* has also been reclassified into a new genus, *Arcanobacterium*, as *Arcanobacterium haemolyticum* (Collins et al., 1982). These workers tentatively placed this new genus within the coryneform group of bacteria. *Arcanobacterium haemolyticum* produces rapid growth with many cells arranged in V-formations, is facultatively anaerobic, catalase-negative, and produces acids from carbohydrates. Both *Actinomyces pyogenes* and *Arcanobacterium haemolyticum* have different optimum temperatures than *Renibacterium* for growth and pathogenesis.

The genus *Renibacterium* can be separated from the human and animal pathogenic corynebacteria and the genus *Caseobacter* by the presence of lysine in the cell wall and the absence of mycolic acids. The genus *Caseobacter* is further separated by a mol% G + C of 60–67 (Crombach, 1978). *Cellulomonas* and *Curtobacterium* both contain the diamino acid ornithine in their cell wall peptidoglycan and have a mol% G + C ranging between 65–72.

Numerous studies have shown overlapping characteristics between the remaining genera or groups of microorganisms included within the coryneform bacteria (Yamada and Komagata, 1972; Schleifer and Kandler, 1972; Jones, 1975; Stackebrandt and Fiedler, 1979; Suzuki et al., 1981). Among these microorganisms a cell wall peptidoglycan containing lysine occurs primarily in the *Arthrobacter* and *Brevibacterium*. DNA homology studies (Stackebrandt and Fiedler, 1979; Suzuki et al., 1981) have shown a close relationship between several species in these two genera. In summary, these bacteria have usually been isolated from the environment, are chemoorganotrophic, show a progression of morphological changes during the growth cycle and have a mol% G + C above 60. All these characteristics are distinctly different from *Renibacterium*.

List of species of the genus **Renibacterium**

1. **Renibacterium salmoninarum.** Sanders and Fryer 1980, 501.[VP]
sal.mo.ni.na'rum. M.L. *Salmoninae* subfamily of the *Salmonidae*; M.L. gen. n. *salmoninarum*, of the *Salmoninae*.

Short rods, 0.3–1.0 × 1.0–1.5 μm, often occurring in pairs and short chains sometimes observed. Strongly Gram-positive. Not acid fast. Nonmotile. Nonencapsulated. Endospores absent. Hemolysis not observed.

On cysteine serum agar and KDM-2 media, colonies are circular and convex, white to creamy yellow, and of varying sizes. On Loeffler coagulated serum, a creamy growth with a matt surface is produced and with Dorset egg medium, growth appears as a raised, smooth shiny yellow layer (Smith, 1964). Creamy yellow growth occurs only at the surface in stationary cultures of cysteine serum broth. Growth on all media is slow especially at primary isolation, often requiring several weeks for visible colonies.

Grows very slowly at 5 and 22°C; no growth at 37°C (Smith, 1964). Optimal growth occurs at 15–18°C. Maximum cell yields with laboratory cultures are obtained after 20–30 days incubation at 15–18°C and pH 6.5–7.5 (Fryer and Sanders, 1981).

Aerobic. No acid production from sugars. All strains isolated require cysteine and the majority serum or whole blood for growth. Catalase-positive. Cytochrome-oxidase negative. Does not liquefy gelatin, and proteolysis without pH change is produced in litmus milk (Smith, 1964).

The cell wall peptidoglycan contains lysine, glutamic acid, alanine and glycine. This peptidoglycan type represents a variation of A3α (Schleifer and Kandler, 1972). The structure contains a glycylalanine interpeptide bridge and the substitution of the α-carboxyl group of D-glutamic acid in position 2 of the peptide subunit with D-alanineamide (Kusser and Fiedler, 1983). Glucose, arabinose, mannose and rhamnose have been reported in cell wall hydrolysates (Sanders and Fryer, 1980); however, more recent studies have indicated the cell wall polysaccharide contains galactose, rhamnose, *N*-acetylglucosamine and *N*-acetylfucosamine (Kusser and Fiedler, 1983). The 12-methyltetradecanoic acid (*anteiso*-C_{15}) is the major fatty acid with 14-methylhexadecanoic acid (*anteiso*-C_{17}) present in significant amounts (Collins, 1982). The principal respiratory quinones are unsaturated menaquinones with nine isoprene units (Collins, 1982). Diphosphatidylglycerol and 10 other characteristic lipids are present (Collins, 1982). Mycolic acids are absent (Fryer and Sanders, 1981).

Serological tests with 10 strains showed all to be antigenically homologous (Bullock et al., 1974). Does not cross-react with Lancefield group G streptococcal antiserum.

Renibacterium salmoninarum is an obligate pathogen of salmonid fish that occurs intracellularly and produces a slowly developing chronic infection characterized by gray-white, enlarged necrotic abscesses primarily in, but not restricted to, the kidney (Wood and Yasutake, 1956; Fryer and Sanders, 1981). Infections develop over a range of water temperatures from 4–20.5°C (Sanders et al., 1978). The pathological changes in the fine structure of both the glomerulus and renal tubules of the kidney resemble those observed in mammalian glomerulonephritis and nephrotic syndrome (Young and Chapman, 1978).

Using the disk-agar diffusion method, cells were sensitive to bacitracin, chloramphenicol, cycloserine, erythromycin, terramycin, novobiocin, streptomycin, nitrofurazone and the sulfonamides. In vivo tests have indicated the sulfonamides and erythromycin give the best results; however, when these drugs are removed from the diet, the disease frequently reappears (Wolf and Dunbar, 1959).

Occurs among populations of salmonid fishes in North America, Europe, Iceland and Japan. Has only been isolated from members of the subfamily Salmoninae, the salmon, trout and char, of the family Salmonidae. The 12 fish species from which *Renibacterium salmoninarum* has been isolated are chinook salmon (*Oncorhynchus tshawytscha*), coho salmon (*Oncorhynchus kisutch*), sockeye salmon, including the landlocked kokanee, (*Oncorhynchus nerka*), cherry salmon, including the landlocked yamame, (*Oncorhynchus masou*), chum salmon (*Oncorhynchus keta*), pink salmon (*Oncorhynchus gorbuscha*), rainbow trout, including the anadromous steelhead, (*Salmo gairdneri*), cutthroat trout (*Salmo clarki*), brown trout (*Salmo trutta*), Atlantic salmon (*Salmo salar*), brook trout (*Salvelinus fontinalis*) and lake trout (*Salvelinus namaycush*).

The mol% G + C of the DNA is 53 ± 1.0 (T_m).

Type strain: ATCC 33209 (Lea-1-74).

Genus **Kurthia** Trevisan 1885, 92[AL] Nom. cons. Opin. 13 Jud. Comm. 1954, 152

RONALD M. KEDDIE AND S. SHAW

Kurth'i.a. M.L. fem. n. *Kurthia* named for H. Kurth, the German bacteriologist who described the type species.

In young cultures (12–24 h): **regular, unbranched rods** with rounded ends **occurring in chains** which are often parallel. The rods are ~0.8–1.2 μm in diameter and vary in length according to the stage of growth but are generally ~2–4 μm long; filaments may occur. **Older cultures** (3–7 days) are **usually composed of coccoid cells** formed by fragmentation of the rods but short rods may be the dominant forms in such cultures. **Do not form endospores. Gram-positive.** Not acid fast. The rods are usually motile by peritrichous flagella but nonmotile strains occur. The diamino acid in the cell wall peptidoglycan is lysine. **Strictly aerobic.** Optimum temperature ~25–30°C. Grow well on peptone-yeast extract media at neutral pH producing rhizoid colonies. Chemoorganotrophs: metabolism respiratory, never fermentative. **Do not produce acid from carbohydrates in peptone media.** Catalase-positive, oxidase-negative. Do not reduce nitrate. Gelatin not hydrolyzed. Organic growth factors required. The mol% G + C of the DNA is 36–38 T_m.

Type species: *Kurthia zopfii* (Kurth) Trevisan 1885, 92.

Further Descriptive Information

When examined by methods similar to those of Cure and Keddie (1973) but using yeast nutrient agar (Shaw and Keddie, 1983a), the long parallel chains of rods ~0.8–1.2 × 2–4 μm characteristic of late exponential phase cultures are seen. In early exponential phase (~6–12 h) long filaments, which give rise to chains of rods as growth proceeds, may occur (Fig. 14.7A). Stationary phase cultures (~3–7 days) of fresh isolates are generally composed largely or entirely of coccoid cells (Fig. 14.7B); strains maintained in artificial culture for some time often give short rods in such cultures (Shaw and Keddie, 1983a). The rods are usually motile by numerous peritrichous flagella but nonmotile strains have been described (Keddie, 1949; Gardner, 1969); the coccoid cells are nonmotile.

The cell walls contain lysine as diamino acid (Belikova et al., 1980; Shaw and Keddie, 1984). In two strains studied, the walls contain a group A peptidoglycan, type L-lysine-D-aspartic acid (Schleifer and Kandler, 1972). The major isoprenoid quinones are unsaturated menaquinones with seven isoprene units (MK-7) (six strains, Collins et al., 1979). The major fatty acid types are straight-chain saturated, anteiso- and iso-methyl-branched chain acids; the major fatty acid is 12-methyltetradecanoic (anteiso-C15) acid. The polar lipids are diphosphatidylglycerol, phosphatidylglycerol and phosphatidylethanolamine (six strains, Goodfellow et al., 1980). On yeast nutrient agar, surface colonies are usually rhizoid. When the edges of such colonies are examined under low magnification, long parallel chains of rods in loops and whorls are seen giving a characteristic "Medusa-head" appearance. Granular colonial variants may be produced. A useful diagnostic feature of the two *Kurthia* species now recognized is the so-called "bird's feather" growth on nutrient gelatin. When a yeast nutrient gelatin* slant is inoculated with a single, central streak by straight wire and is incubated in the near-vertical position, then the subsequent growth resembles a bird's feather (Fig. 14.7C; see Keddie, 1981, for early references). Growth occurs in the range ~5–35°C for *K. zopfii* and ~5–45°C or more for *K. gibsonii*. Growth occurs in 5% NaCl (Shaw and Keddie, 1983a). Glucose is not utilized as a carbon + energy source but the following substrates are utilized for growth by ~90% or more of strains tested: uridine, acetate, butyrate, succinate, fumarate, L-malate, glycerol, ethanol, L-alanine, L-serine, L-asparagine and L-aspartate; L-glutamate and L-proline are universal substrates (Shaw and Keddie, 1983a). When supplied with a suitable source of amino acids, B-vitamins are required (Shaw and Keddie, 1983b).

Although a number of strains of bacteria identified as "*Kurthia* spp," have been isolated from various clinical materials, most commonly from the feces of patients suffering from diarrhea, there is no evidence of pathogenicity in authentic members of the genus (see Keddie, 1981 for references).

K. zopfii and *K. gibsonii* are commonly isolated from meat (particularly after storage for a few days at ~16°C (Gardner, 1969)) and meat products: it is likely that the meat becomes contaminated with *Kurthia* in the abattoir (see Keddie, 1981). The other common source of *K. zopfii* and *K. gibsonii* (see Keddie 1981; Shaw and Keddie, (unpublished)) is feces of certain farm animals, especially of chickens and pigs, several hours to a few days after they have been voided. There are also reports of presumptive *Kurthia* spp. being isolated from such diverse sources as "sloughing spoilage" of ripe olives, the gut of a crab, wet-stored wood, dental plaque of beagle dogs and from air at an altitude greater than 10,000 feet (see Keddie, 1981 for references), but there is considerable doubt about the accuracy of identification of many of these isolates. However, authentic *Kurthia* spp. have been isolated occasionally from milk, soil and surface waters, presumably as a result of contamination with animal dung (see Keddie, 1981; Shaw and Keddie, unpublished). The most unusual source is that reported by Belikova et al. (1980) who isolated *Kurthia* from the stomach and intestinal contents of the Susuman mammoth. Of 13 strains referred to by these authors as *K. zopfii*, four had the characters of that species, three appeared to be *K. gibsonii* and the remainder differed from both these species, mainly in being psychrophilic.

Isolation and Enrichment Procedures

Kurthia zopfii and *K. gibsonii* may be isolated by making use of their unusual cultural properties. Plates (~20 ml) of a suitable nutrient gelatin are inoculated heavily with a single, central streak of the material to be examined, or of a suspension or macerate of solid material in sterile water. The plates are incubated at 20°C with the lids uppermost and examined daily. Liquefaction of the gelatin around the streak soon occurs but in successful cultures filamentous outgrowths of *Kurthia* appear beyond this zone in ~2–3 days. To obtain pure cultures, a small piece of nutrient gelatin containing outgrowths is streaked out on a suitable nutrient agar medium. Isolates are then examined for the characteristic properties of *Kurthia*: useful screening tests are, colony form, morphology and Gram reaction, production of "bird's feather" growth on nutrient gelatin slants and aerobic growth in glucose nutrient agar shake cultures.

The composition of the nutrient gelatin medium used is important and, in particular, the concentration and brand of gelatin used; not all brands allow the typical outgrowths. A nutrient gelatin (YNG, see Keddie, 1981) of the following composition may be used (per liter of distilled water): meat extract (Lab Lemco powder, Oxoid), 4g; peptone (Difco), 5 g; yeast extract (Difco), 2.5 g; NaCl, 5 g; gelatin (BDH), 100 g; pH 7.0. The medium is sterilized at 115°C for 30 min (for quantities up to 100 ml). The medium should be inoculated with a reference strain of *K. zopfii* (NCIB 9878) to test its ability to allow good outgrowth production. The gelatin manufactured by the BDH Chemical Co., Poole, England is satisfactory but, with some batches, a lower concentration than that stated may give more satisfactory outgrowths. Although all *Kurthia* strains tested grow well in yeast nutrient broth, YNB (YNG without gelatin), YNG prepared with some batches of gelatin has given poor growth. Dissolving the constituents of YNG in mineral base E (Owens and Keddie, 1969) to give MYNG (Shaw and Keddie, 1983a) overcomes this problem. When MYNG is autoclaved, a

---

* See Isolation and Enrichment Procedures for a suitable medium.

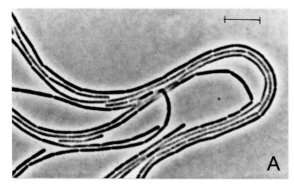
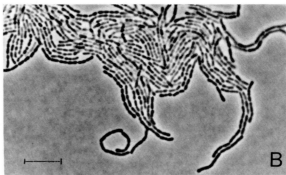
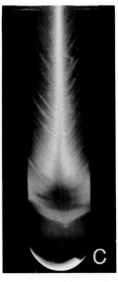

Figure 14.7. *A and B.* *Kurthia zopfii* (isolate): edge of colony on yeast nutrient agar incubated at 25°C. *A,* after 24 h, showing long filaments composed of rods. *B,* after 3 days, showing development of coccoid forms (*bar*, 10 μm). *C, Kurthia zopfii* (NC1B 9878): yeast nutrient gelatin slant showing "birds' feather" type of growth; incubated 5 days at 20°C. (Reproduced with permission from The Prokaryotes. A Handbook on Habitats, Isolation, and Identification of Bacteria, Springer-Verlag, Berlin, Heidelberg, © 1981.)

precipitate, which should be dispersed before pouring plates, is produced; it does not interfere with isolation. Overgrowth with fungi may be a problem with some materials; it may be prevented by adding nystatin to the molten YNG to a concentration of 10 units/ml before pouring plates.

A single central streak on a suitable nutrient agar often allows the detection and isolation of *Kurthia* and should be used in addition to the gelatin streak method in case rapid liquefaction of gelatin prevents the detection of *Kurthia.* YNA is inoculated with a single, central streak as described above. YNA has a composition similar to YNG but the gelatin is replaced by 12 g/liter of Bacto-Agar (Difco). Plates are incubated at 25°C and each day the edge of the streak is scanned by low power (× 100) microscope for the characteristic skein-like outgrowths of *Kurthia.* Pure cultures are obtained by picking carefully from outgrowths and streaking on YNA.

A partial enrichment of *K. gibsonii* may sometimes be achieved by inoculating YNB with some of the material being tested and incubating at 45°C for 24 h; after a second subculture in YNB incubated at 45°C, YNG and YNA plates are inoculated and incubated as described above.

If *K. zopfii* and *K. gibsonii* form a sufficiently high fraction of the population, as in some meat products, they may be enumerated and isolated by direct plating on YNA and similar media. Surface colonies are recognized by their rhizoid form and by the Medusa-head appearance of the edges of young colonies when examined at low magnifications (Gardner, 1969).

Maintenance Procedures

Kurthia spp. grow well on media such as YNA (described above) at 25°C. YNA cultures should remain viable for at least 6 months at room temperature (~20°C) provided they are not allowed to dry out. They may be preserved for long periods by lyophilization.

Procedure for Testing for Special Characters

Acid production from carbon sources is tested for by comparing cultures in 5-ml amounts of basal medium containing 1% (w/v) of membrane filter-sterilized carbon source with those on basal medium alone after 14 days incubation. The basal medium has the following composition per liter of mineral base E without nitrogen source (Min E-N, Owens and Keddie, 1969): Bacto-Yeast Extract (Difco) 1g; Bacto-Agar (Difco), 3 g; and bromothymol blue, 30 mg (Shaw and Keddie, 1983a).

Other characters may be tested for by methods described by Shaw and Keddie (1983a).

Differentiation of the genus **Kurthia** from other genera

The genus *Kurthia* may be distinguished from morphologically similar genera by the characters given in Table 14.14.

Taxonomic Comments

K. zopfii and *K. gibsonii* were shown to resemble each other closely in the numerical phenetic study of Shaw and Keddie (1983a) and, therefore, the present circumscription of the genus is based on the characters of these two species. However, the same study showed that a number of strains of Gram-positive, nonfermentative, asporogenous rods had most or all of the characters of the genus *Kurthia* as defined by Keddie and Rogosa (1974), but were excluded from the clusters containing *K. zopfii* and *K. gibsonii.* The relationship of such strains to the genus *Kurthia* remains unresolved. Similarly Belikova *et al.* (1980) described a few psychrophilic isolates which seem to have the characters

Table 14.14.

Differentiation of **Kurthia** *from morphologically similar taxa*[a]

Characteristics	Kurthia	Brochothrix	Listeria	Bacillus
Endospores	−	−	−	+
Motility	+	−	+	D
Glucose fermentation	−	+	+	D
Facultatively anaerobic	−	+	+	D
"Bird's feather" growth on nutrient gelatin slant	+	−	NA	−

[a] Symbols: −, 90% or more of strains are negative; +, 90% or more of strains are positive; D, different reactions in different taxa (species of a genus or genera of a family); and NA, not applicable.

of the genus but which do not conform to the species *K. zopfii* or *K. gibsonii*. DNA/DNA base-pairing studies of a few strains (some atypical) support the conclusion that *K. zopfii* and *K. gibsonii* are distinct species and that the psychrophilic strains mentioned are different from both species (Cherevach et al., 1983).

The "tentative" inclusion of *Kurthia* in the section on the "Coryneform Group of Bacteria" in the eighth edition of the *Manual* (Rogosa et al., 1974) created the unfortunate impression that the authors considered that *Kurthia* is "related" to the coryneform bacteria (e.g. see Ludwig et al., 1981). Some earlier numerical phenetic studies also suggested some similarity between *Kurthia* and one or more genera of coryneform bacteria (see Ludwig et al., 1981, Shaw and Keddie, 1983a).

However their G + C content (Belikova et al., 1980; Shaw and Keddie, 1984), peptidoglycan structure (Schleifer and Kandler, 1972), isoprenoid quinone composition (Collins et al., 1979) and polar lipid composition (Goodfellow et al., 1980) distinguish members of the genus *Kurthia* from the various genera of coryneform bacteria. Also, comparative analysis of the RNase T1-resistant oligonucleotides of the 16S rRNA of one *Kurthia* strain and various Gram-positive bacteria demonstrated that *Kurthia* is only remotely related to the coryneform bacteria and more closely related to the genera *Bacillus*, *Staphylococcus*, *Streptococcus* and *Lactobacillus*, and of similar taxonomic rank to these genera (Ludwig et al., 1981). Similarly, the numerical taxonomic study of Shaw and Keddie (1983a) revealed little phenetic similarity between the genus *Kurthia* and representatives of various genera of aerobic, coryneform bacteria; but a close relationship was found with some aerobic *Bacillus* spp.

Further Reading

Keddie, R.M. 1981. The Genus *Kurthia*. *In* Starr, Stolp, Trüper, Balows and Schlegel (Editors), The Prokaryotes. A Handbook on Habitats, Isolation, and Identification of Bacteria, Springer-Verlag, Berlin, pp. 1888–1893.

Ludwig, W., E. Seewaldt, K.H. Schleifer and E. Stackebrandt. 1981. The phylogenetic status of *Kurthia zopfii*. FEMS Microbiol. Lett. *10*: 193–197.

Shaw, S. and R.M. Keddie. 1983a. A numerical taxonomic study of the genus *Kurthia* with a revised description of *Kurthia zopfii* and a description of *Kurthia gibsonii* sp. nov. Syst. Appl. Microbiol. *4*: 253–276.

List of the species of the genus **Kurthia**

1. **Kurthia zopfii** (Kurth) Trevisan 1885, 92,[AL] *Nom. cons.* Opin. 13, Jud. Comm. 1954, 152. (*Bacterium zopfii* Kurth 1883, 98.)

zop′fi.i. M.L. gen. n. *zopfii* of Zopf; named for W. Zopf, a German botanist.

Description based on emended description of Shaw and Keddie (1983a).

Morphology and general characters as for generic description. Other characters are given in Tables 14.15 and 14.16.

Table 14.15.

Differential characteristics of the species of the genus **Kurthia**[a]

Characteristics	1. *K. zopfii*	2. *K. gibsonii*
Growth at 45°C	−	+
Survival at 55°C for 20 min	−	+
Acid from:		
Ethanol	+	−
Glycerol	−	+
Colonies yellow or cream	−	+
Deoxyribonuclease	−	+
Ribonuclease	+	−
Phosphatase	−	+
4-amino-*n*-butyrate used as carbon + energy source	−	+
Pantothenic acid[b] required	+	−

[a] Symbols: −, 12% or less of strains negative; and +, 88% or more of strains positive.

[b] When supplied with vitamin-free Casamino acids (Difco) and/or Casitone (Difco), biotin and thiamin (Shaw and Keddie, 1983b).

Cell wall peptidoglycan type, L-lysine-D-aspartic acid (two strains, Schleifer and Kandler, 1972).

Data on the isoprenoid quinones, fatty acids and polar lipids are available only for one strain (NCTC 404, Collins et al., 1979; Goodfellow et al., 1980); see generic description for details.

When supplied with a suitable source of amino acids require biotin + thiamine + pantothenic acid (Shaw and Keddie 1983b).

The mol% G + C of the DNA is 36–38 (T_m) (see Keddie, 1981; Shaw and Keddie, 1984).

Type strain: NCIB 9878.

Further Comments. "*Kurthia zenkeri*" (Hauser) Bergey et al. 1925, 215 is usually considered a synonym of *K zopfii*; its differentiation from the latter species was based on its inability to form arborescent growth in a gelatin stab.

2. **Kurthia gibsonii** Shaw and Keddie 1983c, 672.[VP] (Effective publication: Shaw and Keddie, 1983a, 268.)

gib′son.i.i. M.L. gen. n. *gibsonii* of Gibson; named for T. Gibson.

Description based on that of Shaw and Keddie (1983a).

Morphology and general characters as for generic description. Other characters are given in Tables 14.15 and 14.16.

Cell wall peptidoglycan contains lysine and aspartic acid (Belikova et al., 1980; Shaw and Keddie, 1984); no data on peptidoglycan type.

Data on the isoprenoid quinones, fatty acids, and polar lipids are available for four strains (Collins et al., 1979; Goodfellow et al., 1980); see generic description for details.

When supplied with a suitable source of amino acids require biotin + thiamine (Shaw and Keddie, 1983b).

The mol% G + C of the DNA is 36–38 (T_m) (Keddie, 1981; Shaw and Keddie, 1984).

Type strain: NCIB 9758.

Table 14.16.
*Other features of **Kurthia zopfii** and **Kurthia gibsonii**[a,b]*

Characteristics	1 K. zopfii	2 K. gibsonii	Characteristics	1 K. zopfii	2 K. gibsonii
Acid from:			Isobutyrate	−	−
Butan-1-ol	d	+	Succinate	+	+
Propan-1-ol	+	+	Fumarate	+	+
Pentan-1-ol	d	d	Crotonate	d	+[c]
Methanol	−	−	Tricarballylate	−	−
D-Mannitol	−	−	Aconitate	−	−
D-Sorbitol	−	−	DL-Lactate	d	+[c]
m-Inositol	−	−	L-Malate	+	+
Ethanediol	+	+	DL-2-Hydroxybutyrate	−	−
Propanediol	−	−	DL-Glycerate	d	d
2,3-Butanediol	−	−	Citrate	−	d
m-Erythritol	−	−	Glyoxylate	−	−
D-Glucose	−	−	Pyruvate	+	d
Hydrolysis of:			Oxalacetate	d	+
Aesculin	−	−	2-Oxoglutarate	d	−
Arginine	−	−	Ethanediol	d	d
Tween 20	+	+	Propanediol	−	d
Tween 40	d	d	Glycerol	+	+
Tween 60	+	+	Adonitol	d	d
Tween 80	d	−	Ethanol	+	+
Tributyrin	d	d	Propan-1-ol	d	d
Starch	−	−	Butan-1-ol	+	d
Tyrosine	−	−	Acetaldehyde	d	d
Xanthine	−	−	Benzaldehyde	d	d
Hippurate	+	+	m-Hydroxybenzoate	−	−
Cellulose	−	−	Cinnamate	d	d
Uric acid	+	+	Phenylacetate	d	d
Casein	−	−	Uric acid	−	−
Chitin	−	−	Glycine	+	d
Production of:			L-Alanine	+	+[c]
H₂S	d	−	L-Leucine	−	−
Acetoin	−	−	L-Isoleucine	−	d
Dihydroxyacetone	−	−	L-Serine	+	+[c]
Sulfatase	−	−	L-Threonine	+	d
Urease	−	−	L-Lysine	−	−
Acetamidase	−	−	L-Citrulline	d	d
Lecithinase	−	−	L-Arginine	+	d
Gluconate oxidation	−	−	L-Aspartate	+	+
Indole	−	−	L-Glutamate	+	+
Growth at:			L-Asparagine	+	+
40°C	−	+	L-Phenylalanine	−	−
50°C	−	−	L-Tyrosine	−	−
Carbon + energy source:			L-Histidine	d	d
D-Ribose	d	d	L-Tryptophan	−	d
D-Xylose	−	−	L-Proline	+	+
D-Glucose	−	−	L-Hydroxyproline	d	d
D-Mannose	−	−	DL-Homoserine	−	−
D-Fructose	−	d	L-Cysteine	−	−
Lactose	−	−	L-Methionine	−	−
D-Raffinose	−	−	Ethanolamine	−	−
Glycogen	−	−	Acetamide	−	−
D-Glucuronate	−	−	Betaine	−	−
N-Acetylglucosamine	+	d	Creatinine	−	−
Uridine	+[c]	+	Hippurate	d	−
Formate	−	d	Allantoin	d	−
Acetate	+[c]	+	L-Ascorbate	−	−
Propionate	−	d	m-Erythrytol	−	−
n-Butyrate	+	+	Methanol	−	−
n-Pentanoate	−	−	Crotonol	−	−
n-Hexanoate	d	d	Cytosine	−	−
n-Heptanoate	d	d	Xanthine	−	−
n-Octanoate	d	+			

[a] Data from Shaw and Keddie (1983a).

[b] Symbols: see Table 14.14; and d, 11–89% of strains are positive.

[c] 88% of strains positive.

1258

Genus **Caryophanon** Peshkoff 1939, 244[AL]

WILLIAM C. TRENTINI

Ca.ry.oph′ a.non. Gr. n. *caryum* nut, kernel, nucleus; Gr. adj. *phanus* bright, conspicuous; M.L. neut. n. *Caryophanon* that which has a conspicuous nucleus.

Slightly curved to straight **multicellular rods** (i.e. trichomes), 1.5–3.0 μm in diameter and 10–20 μm in length. The ends are rounded with a slight taper. Trichomes often attached in short chains. True resting stages are not known. **Gram-positive. Motile by means of peritrichous flagella. Strictly aerobic.** Preservation of true enrichment morphology is difficult during laboratory cultivation. Morphology is best preserved at 25–30°C and at pH 7.8–8.5. Chemoorganotrophic with presumed respiratory metabolism. **Acetate is the only major carbon source.** Biotin required, thiamine, stimulatory. Catalase-strongly positive. Cytochrome oxidase-negative; indole not produced. The mol% G + C of the DNA is 41–46 (T_m).

Type species: *Caryophanon latum* Peshkoff 1939, 244.

Further Descriptive Material

Trichomes of *C. latum* are about 3.0 μm in width; cells within the trichome have greater width than length. There are several growing cross-septa at various stages of closure in each cell unit. In *C. tenue*, trichomes are about 1.5 μm in width; cells within the trichome are slightly greater in length than in width. There is only one cross-septum/cell unit (Peshkoff and Marek, 1973). *C. tenue* grows noticeably slower than *C. latum*.

Cell wall peptidoglycan composition contains a molar ratio of glutamic acid:alanine:lysine:muramic acid of 2:2:1:1:. Wall material is very sensitive to egg-white lysozyme. However, septal peptidoglycan is hydrolyzed very quickly, while wall peptidoglycan is resistant (Trentini and Murray, 1975). Wall teichoic acids and *o*-acetyl groups are absent (Trentini, unpublished data).

One or two superficial wall layers containing protein are present (Trentini and Gilleland, 1974).

Numerous growth conditions which result in poor growth physiology lead to extensive lysis during which a small number of cell units are preserved as round bodies (spheroids). Under good growth conditions spheroids are able to grow into normal trichomes. The general nutrition necessary to maintain enrichment morphology consistently among many isolates is unknown. Best morphology is obtained from growth on cow dung agar containing 0.5–1% lactalbumin hydrolysate, pH 8.0, 25°C (Moran and Whitter, 1976; Smith and Trentini, 1972; Smith and Trentini, 1973). After 48-h incubation at room temperature on cow dung agar containing 0.5% lactalbumin hydrolysate, colonies of *C. latum* appeared pale yellow, about 1.5 mm in diameter, opaque, granular, convex, lobate edge and glistening in reflected light (Fig. 14.8). *C. tenue* colonies were similar in appearance, but much smaller (0.5–1.0 mm) and less irregular in edge (Fig. 14.9).

No information has been published on major metabolic pathways, genetics or antigenic structure. Fail to hydrolyze starch, gelatin and casein. Poly-β-hydroxybutyric acid is the principal carbon storage material and appears in trichomes of almost all older or malnourished cultures.

Resistant to streptomycin, nalidixic acid, several sulfa drugs and polymyxin B. *C. latum* and *C. tenue* show a similar antibiotic sensitivity pattern to 45 separate antibiotics (Trentini, unpublished data).

Found associated with cattle dung, but believed not to be a normal resident of the cow (Trentini and Machen, 1973); can be isolated from fresh dung in barn gutters of stanchioned cattle, but field samples are best taken from dung 1–2 days old.

Enrichment and Isolation

Distilled water is added to fresh cattle droppings from barn gutters of 1- to 2-day-old field samples to effect a dung "slurry" surface. Wet mounts are viewed periodically with a phase-contrast microscope. Both species, because of their size and other morphological properties, can be identified by phase microscopy. When the number of *Caryophanon* appear to have reached a peak, samples are streaked onto cow dung

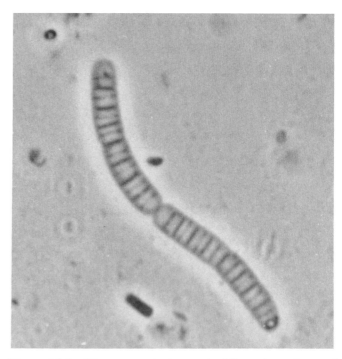

Figure 14.8. Phase contrast micrograph of *Caryophanon latum* trichome chain as seen in cattle dung enrichment culture. Septa contrast was enhanced by placing the enrichment sample onto a slide covered by a thin layer of 16% (w/v) polyvinylpyrrolidone. Note trichome width, cell shape and immature cross-walls in each cell (× 3450).

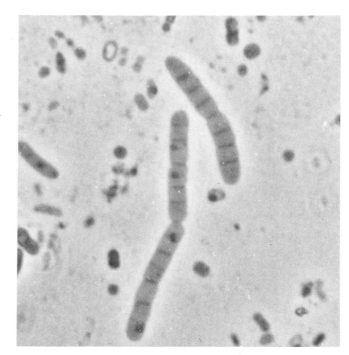

Figure 14.9. Phase contrast micrograph of *Caryophanon tenue* trichome chains as seen in cattle dung enrichment culture. Septa contrast was enhanced by placing the enrichment sample onto a slide covered by a thin layer of 16% (w/v) polyvinylpyrrolidone. Note trichome width, cell shape and absence of immature cross-walls in each cell (× 3450).

agar containing 80 μg/ml of filter-sterilized streptomycin sulfate. Incubate at room temperature for 48 h. Purify on cow dung agar without streptomycin. To date there is no method for selective species isolation (Smith and Trentini, 1972; Trentini and Machen, 1973).

Maintenance Procedures

Caryophanon isolates are readily lyophilized. Refrigerated stock cultures are kept on cow dung-lactalbumin hydrolysate agar.

Differentiation of the genus **Caryophanon** from other genera

Caryophanon is the only genus possessing the following set of characteristics: Gram-positive, peritrichously flagellated, strictly aerobic, asporogenous, multicellular large rods isolated from cattle dung.

Taxonomic Comments

The relationship between *Caryophanon* and other trichomous prokaryotes remains obscure and awaits further studies, particularly nucleic acid homologies and sequencing of rRNA oligonucleotides.

Caryophanon is a misnomer, arising from the original false interpretation of cross-walls, nuclear material and cytoplasm in stained material. The true nature of the trichome was shown by improved cytological technique (Pringsheim and Robinow, 1947).

Presently, the genus and the two species are recognized routinely only by morphological properties. However, *C. tenue* is *not* a morphovar of *C. latum*. *C. tenue* showed a different mol% G + C, a smaller genome size and only 13–30% relative DNA/DNA reassociation with *C. latum* (Adcock, Trentini and Seidler, 1976).

Further Reading

Trentini, W.C. 1978. Biology of the genus *Caryophanon*. Annu. Rev. Microbiol. *32:* 123–141.

Trentini, W.C. 1981. The genus *Caryophanon*. *In* Starr, Stolp. Trüper, Balows and Schlegel (Editors), The Prokaryotes: A Handbook of Habitats, Isolation, and Identification of Bacteria, Springer-Verlag, Berlin, pp. 1701–1707.

List of species of the genus, **Caryophanon**

1. **Caryophanon latum** Peshkoff 1939, 244[AL].

la'tum. L. neut. adj. *latum* broad.

See generic description and Table 14.17.

The mol% G + C of the DNA is 44.0–45.6

Type strain: NCIB 9533.

2. **Caryophanon tenue** (*ex* Peshkoff 1939) nom. rev.

te'nu.e. L. neut. adj. *tenue* slender.

See generic description and Table 14.17.

The mol% G + C of the DNA is 41.2–41.6.

Type strain: NCIB 9535.

Further comment. To date the distinction between the species is summated in Table 14.17. A more definitive differentiation of species awaits the development of nutritional conditions which will support predictable growth with consistent characteristic enrichment morphology. All isolates appear to have no diversity as to carbon source and are inactive to most standard biochemical tests.

Table 14.17.

Characteristics differentiating **Caryophanon latum** *from* **Caryophanon tenue**

Characteristics	1. *C. latum*	2. *C. tenue*
Morphology		
Trichrome width	2.8–3.2 μm	1.4–2.0 μm
Number-cross septa/cell unit	>1 (several)	1
DNA		
Mol% G + C (T_m)	44.0–45.6	41.2–41.6
Genome size	1100–1200 $\times 10^6$ daltons	900–1000 $\times 10^6$ daltons

SECTION 15

Irregular, Nonsporing Gram-Positive Rods

Dorothy Jones and M. D. Collins

This Section contains a diverse collection of bacterial taxa grouped together for practical purposes only. While the majority comprise irregular rods which stain Gram-positive, grow in the presence of air and do not produce endospores, not all the bacteria are irregular rods, some stain Gram-negative and some are strict anaerobes. The bacteria exhibit a very wide range of mol% G + C values (~30–78). Phylogenetic information, where available, supports the suprageneric relatedness of some of the taxa; an example is those which possess a group B type peptidoglycan and a high mol% G + C value (viz: *Agromyces, Aureobacterium, Curtobacterium, Microbacterium* and the diaminobutyric acid (DAB)-containing phytopathogens). But, for the most part, it emphasizes the diversity of the 22 taxa included in this Section. Differential properties of these are shown in Table 15.1.

The genus *Exiguobacterium* (Collins et al., 1983d), which is probably related to some of the genera in this Section, was described too recently to be included in the *Manual*. A full description of the genus is given by Gee et al. (1980) and Collins et al. (1983d). The salient characters of the genera *Renibacterium* and *Oerskovia* are included in Table 15.1 because of their similarity to some of the bacteria in this Section. Full descriptions of these two genera are to be found in Sections 14 and 17, respectively, of this Volume.

On the basis of their oxygen requirements for growth the 22 taxa described in this Section can be divided into two broad groups: (a) aerobic or facultatively anaerobic bacteria; and (b) anaerobic or aerotolerant to facultatively anaerobic bacteria, although those have been arranged somewhat differently in Table 15.1 for deterministic convenience.

Aerobic or Facultatively Anaerobic Bacteria

The genus *Corynebacterium* which includes important human and animal pathogens and saprophytes is now well defined on the basis of chemical criteria. Properties of members of this genus include a directly cross-linked peptidoglycan based upon *meso*-diaminopimelic acid (*meso*-DAP), a wall arabino-galactan polymer, short chain mycolic acids (~22–36 carbons), predominantly straight chain saturated and monosaturated long chain fatty acids (and occasionally 10-methyl-branched acids) and a mol% G + C content of ~51–65. The monospecific genus *Caseobacter* isolated from cheese also possesses these properties and may indeed be a member of the genus *Corynebacterium*. The inclusion of the genus *Corynebacterium* in this Section of the *Manual* does not imply a close phylogenetic relationship with the remaining taxa. On the basis of chemical similarity (cell wall and lipid composition) the genus *Corynebacterium* is most closely related to the genera *Mycobacterium, Nocardia* and *Rhodococcus* (the four together form the CMN group, *sensu* Barksdale, 1970). In this context it should be noted that the results of the 16S-rRNA studies of Stackebrandt and co-workers present a confusing picture. The results of Stackebrandt and

Woese (1981), as depicted in their dendrogram based on S_{AB} values, indicate that *Corynebacterium* is more closely related to *Nocardia* and *Rhodococcus* (the rhodococcus group, Figure 2 in Stackebrandt and Woese, 1981) than to the nonmycolic acid containing actinoplanes group (*Actinoplanes philippinensis, Ampulariella regularis, Dactylosporangium aurantiacum* and *Micromonospora chalcea*). However, in a later study (Stackebrandt et al., 1983b), the *Mycobacterium/Nocardia/Rhodococcus* group is depicted as being more closely related to the *Actinoplanes* group than to *Corynebacterium* (including *C. diphtheriae*, the type species). It seems highly unlikely that the common package of chemical constituents (particularly mycolic acids) characteristic of the CMN group should have emerged or evolved more than once. The apparent closer relatedness of *Mycobacterium/Nocardia/Rhodococcus* to the nonmycolic acid-containing *Actinoplanes* group (Stackebrandt et al., 1983b) is therefore in doubt. A revaluation of these conflicting results (Stackebrandt and Woese, 1981; Stackebrandt et al., 1983b) is desirable.

The DAB-containing phytopathogens are here retained under the genus *Corynebacterium* for historical reasons. There is, however, overwhelming evidence that they are distantly related to true corynebacteria. After the completion of the description of these bacteria (see Plant Pathogenic Species of *Corynebacterium*), Davis et al. (1984) proposed the genus *Clavibacter* to accommodate these organisms. The DAB-containing phytopathogens are unusual in containing a group B type peptidoglycan (i.e. cross-linkage between positions 2 and 4 of adjoining peptide subunits). The monospecific mycelium-producing genus *Agromyces* also contains a group B type peptidoglycan based upon DAB. This similarity between the DAB-containing phytopathogens and *Agromyces ramosus* is supported by lipid and nucleic acid studies (Collins, 1982; Döpfer et al., 1982). However *A. ramosus* has been isolated only from soil and is not known to be phytopathogenic.

Three of the other genera also contain the rare group B type peptidoglycan. These are *Aureobacterium, Curtobacterium* and *Microbacterium*. The cell walls of the first two taxa contain ornithine. The genus *Aureobacterium* presently contains only saprophytic bacteria isolated from a variety of habitats. The genus *Curtobacterium* contains both saprophytes and phytopathogens (i.e. *Curtobacterium flaccumfaciens* and varieties). At present there are very few reliable criteria for distinguishing between members of these two genera (Table 15.1). In contrast the genus *Microbacterium* contains L-lysine as the cell wall diamino acid. These bacteria are all saprophytes isolated from diverse sources. Although the genus *Microbacterium* is phylogenetically closely related to the genera *Agromyces, Aureobacterium, Clavibacter* and *Curtobacterium*, it is readily distinguished from these taxa (Table 15.1, and see genus *Microbacterium*).

The genus *Brevibacterium* is clearly distinguished from other *meso*-DAP containing genera by physiological and chemical criteria (Table

15.1.
Differential characteristics of the groups in Section 15[a]

Characteristics	Aerobes							
	Caseobacter	Plant pathogenic Corynebacterium species	Aureobacterium	Curtobacterium	Microbacterium	Arthrobacter	Brevibacterium	Renibacterium[c]
Cell morphology	Irregular rods, some coccoid forms	Irregular rods	Irregular rods	Irregular rods	Irregular rods, some coccoid forms	Marked rod-coccus cycle	Rod-coccus cycle	Short rods
Gram stain	+	+	+	+[d]	+	+[d]	+	+
Oxygen requirement								
Strictly aerobic	+	+	+	+	+	+	+	+
Facultatively anaerobic or microaerophilic	−	−	−	−	−	−	−	−
Strictly anaerobic	−	−	−	−	−	−	−	−
Motility	−	−	D	D	D	D	−	−
Catalase	+	+	+	+	+	+	+	+
Peptidoglycan Group[e]	A	B	B	B	B	A	A	A
Diamino acid	meso-DAP	DAB	D-Orn	D-Orn	Lys	Lys	meso-DAP	Lys
N-Glycolyl residue	+	−	+[f]	−	+	−	−	−
Wall arabino-galactan polymer	+	−	−	−	−	−	−	−
Mycolic acids	+	−	−	−	−	−	−	−
Major fatty acid types[g,h]	S, U, T	S, A, I	S, A, I	S, A, I (H)	S, A, I	S, A, I	S, A, I	A
Major menaquinones	MK-9(H₂), MK-8(H₂)	MK-10, MK-9	MK-11, MK-12	MK-9	MK-12, MK-11, MK-10	MK-9(H₂), MK-9, MK-8	MK-8(H₂), MK-7(H₂)	MK-9, MK-10
Habitat and pathogenicity	Cheese	Plant material, pathogenic for various plants	Dairy, sewage, soil, insect sources	Plants, soil, oil brine; C. flaccumfaciens is pathogenic for plants	Dairy, sewage and insect sources	Soil	Cheese, skin	Salmonid fish, causing nephrotic syndrome
Mol% G + C	65–67	67–68	67–70	68–75	69–75	59–70	60–67	53

[a] Symbols: +, 90% or more strains positive; −, 10% or less strains positive; d, 11–89% of strains positive; D, substantial proportion of species differ; and ND, not determined.

[b] May be confused with *Clostridium* if spores not observed.

[c] *Renibacterium* and *Oerskovia* are included for comparison and are dealt with in Sections 14 and 17, respectively.

[d] Often readily decolorized.

[e] Designation as Schleifer and Kandler (1972).

[f] Some strains do not possess a *N*-glycolyl residue.

[g] S, straight-chain saturated; U, monounsaturated; A, *anteiso*-methyl-branched; I, *iso*-methyl branched; C, cyclopropane ring; T, 10-methyl branched (tuberculostearic acid); 3-hydroxy FA, 3-hydroxylated long chain fatty acids; H, cyclohexyl fatty acids; and DCFA, dicarboxylic fatty acids with vicinal dimethyl branching.

[h] ( ), sometimes present.

[i] *Corynebacterium matruchotii* exhibits a characteristic whip-handle type morphology (see genus *Corynebacterium*).

[j] Usually appear Gram-negative, but sometimes Gram-variable.

[k] Usually catalase-negative, some strains give positive reaction.

[l] Usually catalase-positive, some strains give negative reaction.

[m] Occasionally some Gram-positive staining.

[n] This group B type of peptidoglycan differs from that of the high mol% G + C-containing "coryneforms" in having a seryl residue in position 1 of the peptide subunit.

[o] Relatively few species examined (Miyagawa, 1982).

Table 15.1.—continued

		Facultative anaerobes									
Characteristics	Corynebacterium[i]	Agromyces	Cellulomonas	Oerskovia[c]	Rothia	Actinomyces	Arcanobacterium	Propionibacterium[b]	Arachnia	Gardnerella
Cell morphology	Rods, clubbed forms[i]	Branched, filamentous elements	Irregular rods, some coccus forms	Extensive branching hyphae, some rod forms	Irregular rods, filamentous and coccoid forms	Rods, filaments, some branching	Short, irregular rods, coccoid forms may predominate	Irregular rods, branched forms and cocci	Rods, filaments, some branching	Irregular rods
Gram stain	+	+	+[d]	+	+	+	+	+	+	−[j]
Oxygen requirement:										
Strictly aerobic	D	d	−	−	−	−	−	−	−	−
Facultatively anaerobic or microaerophilic	D	d	+	+	+	+	+	D	+	+
Strictly anaerobic	−	−	−	−	−	−	−	D	−	−
Motility	−	−	D	d	−	−	−	−	−	−
Catalase	+	−	+	+	+	−[k]	−[k]	+[l]	−	−
Peptidoglycan Group[e]	A	B	A	A	A	A	A	A	A	A
Diamino acid	meso-DAP	DAB	L-Orn	Lys	Lys	Lys, Orn	Lys	LL-DAP, meso-DAP	LL-DAP	Lys
N-Glycolyl residue[f]	−	−	ND	−	ND	−	−	ND	ND	ND
Wall arabinogalactan polymer	+	−	−	−	−	−	−	−	−	−
Mycolic acids	+	−	−	−	−	−	−	−	−	−
Major fatty acid types[g,h]	S, U, (T)	S, A, I	S, A, I	S, A, I	S, A, I	S, U, (C)	S, U	A, I, (S)	A, (S), (I), (U)	S, U
Major menaquinones	MK-9(H₂), MK-8(H₂)	MK-11, MK-12	MK-9(H₄)	MK-9(H₄)	MK-7	MK-10(H₄)	MK-9(H₄)	MK-9(H₄)	MK-9(H₄)	—
Habitat and pathogenicity	Human, animal and other sources, some species pathogenic for man and animals	Soil	Soil and rotting vegetation	Soil, rotting vegetation, other sources including clinical material	Oral cavity; opportunistic pathogen, isolated from human clinical conditions	Mammalian sources; most are pathogenic for man and animals	Human and animal sources; pathogenic for man and animals	Dairy products, human skin also clinical material; some species pathogenic for man	Oral cavity and other human sources; can be pathogenic for man	Genital and urinary tract; believed to cause "nonspecific" bacterial vaginitis
Mol% G + C	51–65	71–76	71–76	71–75	47–53	57–69	48–52	53–68	63–65	42–44

[a] Symbols: +, 90% or more strains positive; −, 10% or less strains positive; D, substantial proportion of species differ; d, 11–89% of strains positive; and ND, not determined.

[b] May be confused with Clostridium if spores not observed.

[c] Renibacterium and Oerskovia are included for comparison and are dealt with in Sections 14 and 17, respectively.

[d] Often readily decolorized.

[e] Designation as Schleifer and Kandler (1972).

[f] Some strains do not possess a N-glycolyl residue.

[g] S, straight-chain saturated; U, monounsaturated; A, anteiso-methyl-branched; I, iso-methyl branched; C, cyclopropane ring; T, 10-methyl branched (tuberculostearic acid); 3-hydroxy FA, 3-hydroxylated long chain fatty acids; H, cyclohexyl fatty acids; and DCFA, dicarboxylic fatty acids with vicinal dimethyl branching.

[h] ( ), sometimes present.

[i] Corynebacterium matruchotii exhibits a characteristic whip-handle type morphology (see genus Corynebacterium).

[j] Usually appear Gram-negative, but sometimes Gram-variable.

[k] Usually catalase-negative, some strains give positive reaction.

[l] Usually catalase-positive, some strains give negative reaction.

[m] Occasionally some Gram-positive staining.

[n] This group B type of peptidoglycan differs from that of the high mol% G + C-containing "coryneforms" in having a seryl residue in position 1 of the peptide subunit.

[o] Relatively few species examined (Miyagawa, 1982).

Table 15.1.—*continued*

	Anaerobes					
Characteristics	*Bifidobacterium*[b]	*Eubacterium*[b]	*Acetobacterium*	*Butyrivibrio*	*Thermoanaerobacter*	*Lachnospira*
Cell morphology	Very irregular rods with branching	Irregular or regular rods	Oval shaped short rods	Curved rods, may be helical	Regular rods, older forms may be irregular, some filamentous and coccoid forms	Curved rods, filaments occur
Gram stain	+	+	+	−		−[m]
Oxygen requirement						
Strictly aerobic	−	−	−	−	−	−
Facultatively anaerobic or microaerophilic	−	−	−	−	−	−
Strictly anaerobic	+	+	+	+	+	+
Motility	−	D	+	+	+	+
Catalase	−	−	−	−	−	−
Peptidoglycan						
Group[e]	A	D	B[n]	ND	ND	ND
Diamino acid	Lys, Orn	D	Orn	ND	*meso*-DAP	ND
N-Glycolyl residue	ND	ND	ND	ND	ND	ND
Wall arabino-galactan polymer	−	−	−	−	−	−
Mycolic acids	−	−	−	−	−	−
Major fatty acid types[g,h]	S, U[o]	ND	ND	(a) A, (I), (S) (b) S, I, U, (A) (c) DCFA[g]	ND	S, U, 3-hydroxy FA
Major menaquinones		ND	ND	ND	ND	ND
Habitat and pathogenicity	Intestines, man and animals, sewage; pathogenicity doubtful, though isolated from clinical conditions	Human and animal sources; plants and soil; some species are pathogenic for man	Anaerobic fresh water and marine sediments and sewage	Rumen	Hot springs	Rumen
Mol% G + C	55–67	30–55	39–43	36–42	37–39	ND

[a] Symbols: +, 90% or more strains positive; −, 10% or less strains positive; d, 11–89% of strains positive; D, substantial proportion of species differ; and ND, not determined.

[b] May be confused with *Clostridium* if spores not observed.

[c] *Renibacterium* and *Oerskovia* are included for comparison and are dealt with in Sections 14 and 17, respectively.

[d] Often readily decolorized.

[e] Designation as Schleifer and Kandler (1972).

[f] Some strains do not possess a N-glycolyl residue.

[g] S, straight-chain saturated; U, monounsaturated; A, *anteiso*-methyl-branched; I, *iso*-methyl branched; C, cyclopropane ring; T, 10-methyl branched (tuberculostearic acid); 3-hydroxy FA, 3-hydroxylated long chain fatty acids; H, cyclohexyl fatty acids; and DCFA, dicarboxylic fatty acids with vicinal dimethyl branching.

[h] ( ), sometimes present.

[i] *Corynebacterium matruchotii* exhibits a characteristic whip-handle type morphology (see genus *Corynebacterium*).

[j] Usually appear Gram-negative, but sometimes Gram-variable.

[k] Usually catalase-negative, some strains give positive reaction.

[l] Usually catalase-positive, some strains give negative reaction.

[m] Occasionally some Gram-positive staining.

[n] This group B type of peptidoglycan differs from that of the high mol% G + C-containing "coryneforms" in having a seryl residue in position 1 of the peptide subunit.

[o] Relatively few species examined (Miyagawa, 1982).

15.1). The genus currently contains four species isolated from dairy products and skin. Two of these species, *B. casei* and *B. epidermidis* are difficult to distinguish from each other (see genus *Brevibacterium*).

Members of the strictly aerobic genus *Arthrobacter* exhibit a marked rod-coccus growth cycle. Their predominant habitat is soil. Within *Arthrobacter sensu stricto* there are two or three "groups of species" (see genus *Arthrobacter*). The taxonomic status of these "groups" is equivocal. Although the "groups" differ with respect to peptidoglycan type (A3α or A4α, Schleifer and Kandler, 1972) and lipid composition (Collins and Kroppenstedt, 1983), nucleic acid studies suggest they are closely related (Stackebrandt et al., 1983a). The results of chemical and nucleic acid studies also indicate a close relationship between true arthrobacters and some micrococci (including *Micrococcus luteus*, the type species of the genus *Micrococcus*, Stackebrandt and Woese, 1981). In this Volume the genera *Arthrobacter* and *Micrococcus* are treated separately for diagnostic purposes.

The bacteria currently designated *Arthrobacter simplex* and *A. tumescens* are phenotypically and genomically distinct from *Arthrobacter sensu stricto* (see genus *Arthrobacter*). It has been suggested that *A. simplex* be reclassified as *Nocardioides simplex* (O'Donnell et al., 1982), whereas Suzuki and Komagata (1983) proposed that *A. simplex* together with *A. tumescens* be transferred to a new genus *Pimelobacter*. Although there is no doubt that *A. simplex* is a distinct taxon the placement of *A. tumescens* in the same genus as *A. simplex* is not supported by physiological or chemical evidence. *Arthrobacter tumescens* is very unusual in containing substantial levels (~20%) of monounsaturated terminally branched long chain fatty acids, for example, iso-$C_{15:1}$ (Collins et al., 1983f). It is worth noting that a similar fatty acid composition has been detected in *Intrasporangium calvum* (M. D. Collins, unpublished).

The genus *Renibacterium* (treated in Section 14 of this Volume) displays some resemblance to the true arthrobacters in morphological, physiological and chemical properties.

Members of the genus *Cellulomonas*, which occur mainly in soil, can be distinguished by their cellulolytic activity and chemical composition. Phenotypically cellulomonads most closely resemble members of the genus *Oerskovia* (see Section 17 of this Volume). This relatedness has been confirmed by 16S-rRNA cataloging studies and indeed Stackebrandt et al. (1982) proposed the union of the two genera. The species *Listeria denitrificans* (see genus *Listeria*) and *Promicromonospora citrea* also resemble both *Cellulomonas* and *Oerskovia* in chemical composition (Collins et al., 1980, 1983j).

The monospecific genus *Rothia* is commonly found in the human oral cavity and as an opportunistic pathogen in human clinical conditions. Until recently it has been considered to closely resemble the genera *Actinomyces* and *Nocardia* (see genus *Rothia*). However, in peptidoglycan and lipid composition, *R. dentocariosa* most closely resembles certain members of the "coryneform group" of bacteria (sensu Keddie and Jones, 1981) although chemically it is a distinct genus (Collins and Shah, 1984; Table 15.1).

Anaerobic, Aerotolerant to Facultatively Anaerobic Bacteria

The taxa included in this group are phenotypically and phylogenetically diverse. The grouping is purely arbitrary.

The genus *Actinomyces* is relatively well defined and currently contains ten species all of which are associated with, or cause, disease in animals including man. *Arcanobacterium haemolyticum* (formerly *Corynebacterium haemolyticum*), also isolated from human and animal sources, possesses a similar peptidoglycan type to that found in actinomyces (type A5; Weiss, 1985). It also resembles *Actinomyces* in possessing predominantly straight chain saturated and monounsaturated long chain fatty acids. Although we believe *Arcanobacterium* is probably the closest known relative of the genus *Actinomyces*, differences between the taxa in menaquinone composition and mol% G + C values indicate that they are separate genera probably within the same family.

The genus *Bifidobacterium* is a well defined genus containing 24

species isolated from the gut of man, animals and honey bees. Historically, the genus was associated with the genus *Lactobacillus* but recent 16S-rRNA sequencing studies indicate it is quite unrelated to this genus and, in fact, is a member of the "high mol% G + C actinomycete lineage" (Stackebrandt and Woese, 1981). Bifidobacteria which are anaerobes may be distinguished from the genera *Actinomyces* and *Arcanobacterium* by their peptidoglycan composition and in lacking respiratory quinones (Fernandez et al., 1984).

The anaerobic to aerotolerant genus *Propionibacterium* contains eight species. These can be divided into two groups on the basis of habitat. One group, the "classical" propionibacteria, usually occurs in cheese and other dairy products. Members of the second group are isolated mainly from human skin and include bacteria previously designated anaerobic corynebacteria, (viz. *P. acnes*, *P. avidum*, *P. granulosum* and *P. lymphophilum*).

The monospecific genus *Arachnia* resembles the genus *Propionibacterium* in producing propionic acid as the major end product of carbohydrate metabolism. *Arachnia propionica* is found predominantly in the oral cavity of man (it may also be associated with clinical conditions) and has been shown to be serologically related to the human propionibacteria. It is also similar to propionibacteria in menaquinone composition. The report of the presence of predominantly straight chain and monounsaturated long chain fatty acids in *A. propionica* by Amdur et al. (1978) indicates a possible association with the genus *Actinomyces*. However, a re-examination of the fatty acid composition of *A. propionica* demonstrated the presence of predominantly methyl-branched fatty acids, a result more in keeping with a possible relationship with the propionibacteria (M. D. Collins, unpublished).

The anaerobic genus *Eubacterium* contains 34 named species isolated from the body cavities of man and animals, animal and plant products, and from soil. Some species may be pathogenic for man and animals. The genus is extremely heterogeneous as is evident from the wide range of mol% G + C values (~30–55) and is defined primarily on the basis of negative characters (see genus *Eubacterium*). Due to the paucity of reliable criteria the genus has served as a depository for a wide and varied collection of Gram-positive, nonsporeforming, obligately anaerobic bacteria. Relatively little is known of the phylogenetic relationships of the species currently classified in the genus and much further work is required to resolve their taxonomic status. The problem is highlighted by the uncertain taxonomic position of the new type species *E. limosum* (Cato et al., 1981; Wayne, 1982). Ribonucleic acid sequencing studies (Tanner et al., 1981; 1982) indicate *E. limosum* is phylogenetically very closely related to *Acetobacterium woodii* and *Clostridium barkeri* (Balch et al., 1977). The presence of a rare group B type peptidoglycan within both *E. limosum* and *A. woodii* (Kandler and Schoberth, 1979) supports this relationship. If this close similarity is confirmed, the taxonomic status of the type species and, therefore, the genus *Eubacterium* would be in jeopardy. Yet another example of the unsatisfactory nature of the genus is the recent inclusion of the swine pathogen *E. suis* (Wegienek and Reddy, 1982), formerly *Corynebacterium suis*. The composition of *E. suis*, peptidoglycan (type A5, N. Weiss personal communication) and long chain fatty acid (predominantly straight chain, saturated and monounsaturated (oleic) acids, M. D. Collins, unpublished), indicates a close similarity to the genus *Actinomyces*.

Members of the anaerobic genus *Acetobacterium* produce acetic acid as the end product of metabolism. Currently, two species are recognized, *A. woodii* isolated from aquatic anaerobic sediments, and *A. wieringae* isolated from sewage sludge. As noted previously, 16S-rRNA sequencing studies indicate *A. woodii* together with *Eubacterium limosum* and also *Clostridium barkeri* are related at least at the generic level (Tanner et al., 1981; 1982). The same studies indicate that these taxa constitute one of the many evolutionary lines emerging from the clostridial cluster (the so-called *Clostridium-Lactobacillus-Bacillus* phylogenetic branch, Stackebrandt and Woese, 1981). The rare group B type peptidoglycan in *A. woodii*, *C. barkeri* and *E. limosum* is different from that found in certain coryneform genera (for example, *Agromyces*, *Aureobacterium*,

Curtobacterium and others) in containing an L-seryl residue in position 1 of the peptide subunits (Kandler and Schoberth, 1979). It is interesting that *Erysipelothrix rhusiopathiae* also contains the same rare group B type peptidoglycan (Schleifer and Kandler, 1972).

The anaerobic genus *Thermoanaerobacter* contains a single species, *T. ethanolicus*, isolated from hot springs. Currently there is no information on the phylogenetic position of *T. ethanolicus* although the possession of a relatively low mol% G + C value (~37–39) indicates it is also a member of the clostridial lineage (Stackebrandt and Woese, 1981).

Included in this Section are three genera, *Butyrivibrio*, *Gardnerella* and *Lachnospira* also described in Sections 5 or 6 (Volume 1 of the *Manual*) because they stain Gram-negative or only weakly Gram-positive. The cell walls of *Butyrivibrio* and *Gardnerella* have been shown to be structurally Gram-positive and their low mol% G + C values (36–42 and 42–44, respectively) suggest a phylogenetic placement in the clostridial lineage. Members of both genera have been isolated from human and animal sources. The diversity of the fatty acid and fatty aldehyde composition of *Butyrivibrio* strains suggest that the genus is very heterogeneous. The long chain fatty acid composition indicates the presence of three groups; one group produces a novel long chain dicarboxylic fatty acid (Klein et al., 1979; Miyagawa, 1982). It should be noted that the absence of 3-hydroxy fatty acids in *Butyrivibrio* is consistent with its inclusion in the Gram-positive group. Gram-negative bacteria almost invariably possess hydroxy fatty acids in the lipopolysaccharide moiety of their cell walls.

The anaerobic genus *Lachnospira* contains a single species, *L. multiparus*, isolated from the rumen where it is important in the fermentation of pectin. Relatively little is known about these bacteria. As noted earlier, their reaction to the Gram stain is equivocal, although it has been claimed that fine structure studies indicate a Gram-positive structure (Cheng et al., 1979). However, it is evident from the long chain fatty acid composition, particularly the presence of 3-hydroxy fatty acids (Miyagawa, 1982) that the genus is Gram-negative.

In conclusion it should be stressed that the 22 genera described in detail in this Section (and the recently described genus *Exiguobacterium*) are phenotypically extremely diverse. The results of the application of molecular biological techniques (in particular 16S-rRNA cataloging) have confirmed this heterogenicity. The evidence now available indicates that the genus *Lachnospira* belongs to the Gram-negative group. The remainder belong to the Gram-positive group. With the exception of *Acetobacterium*, *Butyrivibrio*, *Gardnerella*, *Thermoanaerobacter* and the majority of the members of the genus *Eubacterium* (which belong to the *Clostridium-Lactobacillus-Bacillus* lineage), all the other genera are members of the high mol% G + C Actinomycetes phylogenetic branch (*sensu* Stackebrandt and Woese, 1981). Within this major grouping (Order or perhaps Super Order) several generic clusters are evident. These include: (a) the CMN group (contains *Caseobacter*, *Corynebacterium*, *Mycobacterium*, *Nocardia* and *Rhodococcus*), for which the name *Mycobacteriaceae* might be suitable; (b) those coryneform bacteria which contain the group B type peptidoglycan (including *Agromyces*, *Aureobacterium*, *Clavibacter*, *Curtobacterium* and *Microbacterium*), which might form the family *Microbacteriaceae*; (c) a grouping to include *Arthrobacter sensu stricto* (as defined in the section on *Arthrobacter*), and certain micrococci; (d) a grouping to include *Cellulomonas*, *Oerskovia* and possibly *Listeria denitrificans*; (e) a group centered on *Brevibacterium*; (f) the genera *Actinomyces* and *Arcanobacterium*, which might form the family *Actinomycetaceae*; (g) a group to contain both "classical" and "cutaneous" propionibacteria; (h) a group centered on the genus *Bifidobacterium*.

The higher taxonomic affiliation of the remaining taxa (viz. *Arachnia*, *Renibacterium*, *Rothia*) is presently unclear. There is no doubt that there will be major rearrangements of the majority of the taxa contained in this Section in future editions of the Manual.

Genus **Corynebacterium** *Lehmann and Neumann 1896, 350*[AL*,†]

M. D. COLLINS AND C. S. CUMMINS

Co.ry.ne.bac.ter′i.um. Gr. n. *coryne* a club; Gr. n. *bakterion* a small rod; M.L. neut. n. *Corynebacterium* a club bacterium.

Straight to slightly curved rods with tapered ends. Club-shaped forms may be observed; sometimes ellipsoidal, ovoid or rarely "whip handle" (see *C. matruchotii*). **Snapping division produces angular and palisade arrangements of cells. Gram-positive; some cells stain unevenly. Metachromatic granules formed.** Not acid-fast (Ziehl-Neelsen stain). Endospores not formed. **Nonmotile. Facultatively anaerobic although some organisms are aerobic. Catalase-positive.** Chemoorganotrophs. Most species produce acid from glucose and some other sugars in peptone media.

Cell wall peptidoglycan is based upon *meso*-diaminopimelic acid (*meso*-DAP) (variation A1γ of Schleifer and Kandler, 1972). Glycan moiety of cell wall contains acetyl residues. **Major cell wall sugars are arabinose and galactose. Short-chain mycolic acids (approximate limiting range 22–36 carbon atoms) present.** Nonhydroxylated long chain fatty acids are primarily of the straight chain saturated and monounsaturated types; some strains may also produce substantial amounts of 10-methyl-branched chain acids (e.g. 10-methyloctadecanoic acid). *Anteiso*- and *iso*-methyl-branched fatty acids are either absent or present in only trace amounts. Menaquinones (vitamin K_2) sole respiratory quinones; major components are dihydrogenated menaquinones with eight and/or nine isoprene units. Contain phosphatidylinositol and phosphatidylinositol dimannoside(s) in addition to simple phospholipids; some strains also produce trehalose dimycolates and other glycolipids. DNA base composition within the approximate limiting range of 51–63 mol% G + C (T_m, Bd, Ch).

Type species: *Corynebacterium diphtheriae* (Kruse) Lehmann and Neumann 1896, 350.

Enrichment and Isolation Procedures

Most members of the genus are nutritionally exacting, requiring one or more vitamins, amino acids, purines and pyrimidines. The addition of serum or other body fluids (5–10%) to media will generally provide all substances necessary for growth. The temperature range for the growth of most corynebacteria is from 30–37°C.

Maintenance Procedures

For short term maintenance, screw-capped bottles containing Dorset egg slants. Cultures should remain viable for 3 months or more when stored at ~20°C. Cultures grown in chopped meat medium (without glucose) for 48 h at 37°C, and then kept in the dark at room temperature, will remain viable for many months. Medium term preservation (several years) can be achieved by storing on beads at −76°C (Feltham et al., 1978). Long term preservation is by lyophilization.

---

* *AL*, denotes the inclusion of this name on the Approved Lists of Bacterial Names (1980).

† It should be noted that the phytopathogenic species that remain in the genus are dealt with in the next article (p. 1276).

Differentiation of the genus **Corynebacterium** from other genera

Facultatively anaerobic coryneform bacteria which contain meso-DAP, arabinose and galactose in the cell wall, a DNA base composition in the range 51–63 mol% G + C and corynomycolic acids (~22–36 carbon atoms) may be regarded as members of the genus *Corynebacterium*.

Table 15.2 provides the primary chemical criteria that can be used to differentiate the genus *Corynebacterium* (as defined in this chapter) from other actinomycete and coryneform taxa. In particular, thin layer chromatographic analysis of whole-organism methanolysates (method below) provides a simple and reliable means of differentiating true corynebacteria (which possess mycolic acids) from the plethora of coryneform taxa (viz. *Arthrobacter, Brevibacterium, Curtobacterium, Cellulomonas, Microbacterium*) which lack these characteristic lipids (see Goodfellow et al., 1976; Minnikin et al., 1978).

The differentiation of true corynebacteria from representatives of the genera *Mycobacterium* and *Nocardia* is relatively easy. Mycobacteria and nocardias possess relatively large mycolic acids (~60–90 and 46–60 carbon atoms, respectively). Thus chromatographic analysis of whole-organism methanolysates (see below for method) provides a simple and reliable means of differentiating these taxa from *Corynebacterium*. Mycobacteria and nocardias may also be distinguished from corynebacteria by the fact that the former two taxa possess DNA which is relatively rich in guanine plus cytosine (~62–70 mol%) and contain N-glycolyl residues in the glycan moiety of their walls (Table 15.2).

The differentiation of true corynebacteria from certain rhodococci, however, may prove very difficult. Although members of the genus *Rhodococcus* generally possess larger mycolic acids (~30–64 carbon atoms) than those of *Corynebacterium* (~22–36 carbon atoms) a clear distinction between the two taxa cannot, at present, be made by analysis of mycolic acids alone (Collins et al., 1982a). Several rhodococci possess mycolic acids which overlap in size with those of true corynebacteria (e.g. the type strains of *R. equi* and *C. hoagii*, Collins et al., 1982a). Representatives of the genus *Rhodococcus* are described as aerobic (Goodfellow and Alderson, 1977); although the vast majority of true corynebacteria are undoubtedly facultative, obligately aerobic strains do occur which cannot be assigned with confidence to one genus or the other. Other criteria which may be of value in the identification of such "difficult strains" include mol% G + C determination and peptidoglycan analysis. Representatives of the genus *Rhodococcus* generally possess a higher G + C content (~60–69 mol%) than *Corynebacterium* (~51–60 mol%), and most rhodococci examined to date contain N-glycolyl residues in the glycan moiety of their cell walls. Similarly, the presence of 10-methyloctadecanoic (tuberculostearic) acid may be of value in the assignment of "intermediate" strains. Members of the genus *Rhodococcus* possess tuberculostearic acid whereas the vast majority of true corynebacteria lack this fatty acid (see Collins et al., 1982b).

The differentiation of *Corynebacterium* and *Caseobacter* can also pose serious problems. Caseobacters are similar to corynebacteria in possessing relatively small mycolic acids (~30–36 carbon atoms, Collins et al., 1982a) and N-acetyl residues in the glycan moiety of their cell walls (Collins, unpublished). It is worth noting, however, that caseobacteria are considered obligately aerobic and possess a significantly higher guanine plus cytosine content (~65–67 mol%, Crombach, 1978)

Table 15.2.

Differential characteristics of the genus **Corynebacterium** *and other actinomycete and coryneform taxa[a,b]*

Taxon	Major peptidoglycan diamino acid	Peptidoglycan type	N-glycolyl in glycan moiety of wall	Mol% G + C	Mycolic acids	Fatty acid types[c]	Major menaquinone isoprenologue(s)	Phosphatidylinositol and phosphatidylinositol mannoside(s)[d]
Agromyces	DAB	B2γ	ND	71–76	–	S, A, I	MK-12	–
Arthrobacter	ʟ-Lysine	A3α	–	59–66	–	S, A, I	MK-9(H₂)	(PI)[e]
Brevibacterium	meso-DAP	A1γ	–	60–64	–	S, A, I	MK-8(H₂)	(PI)[f]
Caseobacter	meso-DAP	A1γ	–[g]	65–67	30–36 Carbon atoms	S, U, T	MK-9(H₂), MK-8(H₂)	+
Cellulomonas	ʟ-Ornithine	A4β	–	71–75	–	S, A, I	MK-9(H₄)	(PI)[h]
Corynebacterium	meso-DAP	A1γ	–	51–60	22 to 36 Carbon atoms	S, U (T)	MK-9(H₂), MK-8(H₂)	+
Plant pathogenic *Corynebacterium* spp.	DAB	B2γ	–	67–78	–	S, A, I	MK-9, MK-10	–
Curtobacterium	ᴅ-Ornithine	B2β	–	67–75	–	S, A, I	MK-9	–
Microbacterium	ʟ-Lysine	B1α or B1β	+	69–75	–	S, A, I	MK-11, MK-12	–
Mycobacterium	meso-DAP	A1γ	+	62–70	60–90 Carbon atoms	S, U, T	MK-9(H₂)	+
Nocardia	meso-DAP	A1γ	+	64–69	46–60 Carbon atoms	S, U, T	MK-8(H₄)	+
Rhodococcus	meso-DAP	A1γ	+	60–69	30–64 Carbon atoms	S, U, T	MK-9(H₂), MK-8(H₂)	+

[a] Data from Schleifer and Kandler, 1972; Minnikin et al., 1978; Uchida and Aida, 1977, 1979; Minnikin and Goodfellow, 1980; Keddie and Jones, 1981; Collins and Jones, 1981; Collins et al., 1982a; Collins et al., 1983c; Döpfer et al., 1982; Collins, 1982a.

[b] Standard symbols: +, 90% or more of strains are positive; –, 90% or more of strains are negative; ND, not defined.

[c] S, straight-chain saturated; A, *anteiso*-methyl-branched; I, *iso*-methyl-branched; U, monounsaturated; T, 10-methyl-branched acids; ( ) may be present.

[d] PI, phosphatidylinositol.

[e] PI present in *A. globiformis* group of species.

[f] Some strains possess trace amounts of PI.

[g] Collins, unpublished.

[h] In addition to PI, cellulomonads contain two unidentified phosphoglycolipids of similar chromatographic mobilities to phosphatidylinositol dimannosides.

than true corynebacteria. Representatives of the genus *Caseobacter* may be distinguished from most *Corynebacterium* spp. in possessing significant amounts of tuberculostearic acid (Collins et al., 1982b).

Screening for presence of mycolic acids is by whole-organism acid methanolysis (Minnikin et al., 1975). Dry bacteria (~50 mg) are mixed with dry methanol (3 ml), toluene (3 ml) and concentrated H_2SO_4 (0.1 ml) in a capped tube. The contents are mixed thoroughly and methanolysis allowed to proceed for ~12–16 h at 50°C (oven). The reaction mixture is allowed to cool to room temperature, 1.5 ml of hexane added, mixture shaken and then allowed to settle (two layers are formed). Samples from the upper layer are spotted on thin layer chromatography plates (e.g. Merck Kieselgel 60) 10 × 10 cm and the chromatograms developed (ascending chromatography) in hexane/diethyl ether (85:15 by volume). The lipids are revealed (blue spots on a yellow background) by spraying with 10% dodecamolybdophosphoric acid in ethanol and heating at 140°C for 10 min.

Components with R_f values of >0.6 are attributable to methyl esters of nonhydroxylated fatty acids whereas those with R_f values of ~0.2–0.5 correspond to methyl esters of mycolic acids. It is worth noting that mycolic acid methyl esters from mycobacteria and nocardiae have greater R_f values than those from corynebacteria (due to differences in molecular weight). In addition, mycobacteria usually produce multiple spot patterns (see Minnikin et al., 1978; Goodfellow et al., 1976 for illustrations).

Taxonomic Comments

The genus *Corynebacterium* was created essentially for the diphtheria bacillus and a few other animal pathogenic species (Lehmann and Neumann, 1896). However, over many years other nonsporing, irregularly staining, Gram-positive species, both aerobic and anaerobic, were assigned to the genus, until it came to include a very wide collection of ill-assorted organisms.

The heterogeneity of the genus *Corynebacterium* was very evident on the basis of cell wall composition alone, as was noted in the eighth edition of the *Manual* (Cummins et al., 1974). However, since then much more information has become available, and the investigation of mycolic acid structure, menaquinone content and fatty acid composition, among other things, has clarified the situation considerably.

Two major conclusions have emerged. (a) The genus *Corynebacterium*, sensu stricto, should be confined to species that are characterized by the presence of arabinogalactan and *meso*-DAP in the cell wall, contain mycolic acids of chain length between 22 and 36 carbon atoms, dihydrogenated menaquinones with eight and/or nine isoprene units, predominantly straight chain saturated and monounsaturated fatty acids, and that have a mol% G + C content of 51–63 and (b) that there is a strong chemical similarity between species at present assigned to *Corynebacterium*, *Caseobacter*, *Mycobacterium*, *Nocardia* and *Rhodococcus*.

The narrower definition of *Corynebacterium* means that many organisms traditionally assigned to it are now in other genera. Many of the plant pathogens can now be assigned to the genera *Arthrobacter*, *Curtobacterium* and *Rhodococcus*; the anaerobic or microaerophilic skin coryneforms (anaerobic diphtheroids) are in *Propionibacterium*; "*Corynebacterium pyogenes*" and "*Corynebacterium haemolyticum*" are assigned, respectively, to *Actinomyces* and *Arcanobacterium*; and some species like *C. equi* have been transferred to the genus *Rhodococcus*.

On the other hand, several new species have been added to the genus. Prominent among these is the organism formerly called *Bacterionema matruchotii*, which is included here as *Corynebacterium matruchotii* since it was found to have all the chemotaxonomic characteristics of a true *Corynebacterium* (Collins, 1982c). Similarly, there are good biological and chemical reasons for reclassifying *Microbacterium flavum* as *Corynebacterium flavescens* (Barksdale et al. 1979) and for reclassifying *Brevibacterium vitarumen* as *Corynebacterium vitarumen* (Lanéelle et al., 1980).

There is now overwhelming phenetic, chemical and genetic evidence that representatives of the genus *Corynebacterium* are quite unrelated to those coryneform taxa (such as *Arthrobacter*, *Brevibacterium*, *Cellulomonas*, *Curtobacterium*, *Microbacterium*) which lack mycolic acids (see Table 15.2 and preceding section Differentiation of the Genus . . .). There is a growing opinion, however, that *Corynebacterium* should be included in the same "suprageneric group" as certain actinomycetes (i.e. *Caseobacter*, *Mycobacterium*, *Nocardia*, *Rhodococcus*, "*aurantiaca* taxon"). Although true corynebacteria can be distinguished from these actinomycetes on the basis of numerical phenetic and chemical studies (see Table 15.2 and preceding section Differentiation of the Genus . . .) and undoubtedly warrant separate generic status, these taxa do possess many properties in common. Properties shared by the majority of these taxa include: (a) DNA which is relatively rich in G + C (~51–70 mol%); (b) a directly cross-linked peptidoglycan based upon *meso*-DAP (variation A1γ of Schleifer and Kandler, 1972); (c) a wall arabinogalactan polymer; (d) characteristic α-alkyl-β-hydroxy-long chain fatty acids (mycolic acids); (e) "cord-factors" (mycolic esters of trehalose); (f) nonhydroxylated long chain fatty acids of the straight chain saturated and monounsaturated types; the majority of these taxa also synthesize 10-methyl-branched acids; (g) phosphatidylinositol and phosphatidylinositol mannosides. However, despite this "impressive array" of common chemical criteria it is presently difficult to decide the level of this common higher grouping (i.e., family, suborder, order, etc.). It is worth noting that although the family *Mycobacteriaceae* may prove a convenient "suprageneric niche" for these mycolic acid-containing taxa, the results of the 16S-rRNA oligonucleotide cataloging studies of Stackebrandt and co-workers form a confusing picture. The results of Stackebrandt and Woese (1981), as depicted in the dendrogram based on S_{AB} values, indicate that *Corynebacterium* is more closely related to *Nocardia* and *Rhodococcus* than to the nonmycolic acid-containing actinoplanes group (*Actinoplanes philippinensis*, *Ampulariella regularis*, *Dactylosporangium aurantiacum* and *Micromonospora chalcea*). However, in a subsequent study (Stackebrandt et al., 1983), the *Mycobacterium/Nocardia/Rhodococcus* group is depicted as being more closely related to the *Actinoplanes* group than to *Corynebacterium* (including *C. diphtheriae*). It is highly unlikely, however, that the common package of chemical constituents (see above) characteristic of the CMN group should have evolved more than once. The apparent closer relatedness of *Mycobacterium/Nocardia/Rhodococcus* to the nonmycolic acid-containing *Actinoplanes* group (Stackebrandt et al., 1983) must therefore be in serious doubt.

Further Reading

Barksdale, L. 1970. *Corynebacterium diphtheriae* and its relatives. Bacteriol. Rev. *34*: 378–422

Barksdale, L. 1981. The genus *Corynebacterium*. *In* Starr, Stolp, Trüper, Balows, and Schlegel (Editors), The Prokaryotes: A Handbook on Habitats, Isolation and Identification of Bacteria, Springer-Verlag, New York, pp 1827–1837.

Collins, M.D., M. Goodfellow and D.E. Minnikin. 1982. A survey of the structures of mycolic acids of the genus *Corynebacterium* and possibly related taxa. J. Gen. Microbiol. *128*: 129–149.

Keddie, R.M. and D. Jones. 1981. Saprophytic, aerobic coryneform bacteria. *In* Starr, Stolp, Trüper, Balows and Schlegel (Editors). The Prokaryotes: A Handbook on Habitats, Isolation and Identification of Bacteria, Springer-Verlag, New York, pp 1838–1878.

Minnikin, D.E., M. Goodfellow and M.D. Collins. 1978. Lipid composition in the classification and identification of coryneform and related taxa. *In* Bousfield and Callely (Editors), Coryneform Bacteria, Academic Press, London, pp 85–160.

Stackebrandt, E. and C.R. Woese. 1981. The evolution of prokaryotes. *In* Carlile, Collins and Mosely (Editors). Molecular and Cellular Aspects of Microbial Evolution. Cambridge University Press, Cambridge, pp. 1–31.

Stackebrandt, E., B.J. Lewis and C.R. Woese. 1980. The phylogenetic structure of the coryneform group of bacteria. Zentralbl. Bakteriol. Mikrobiol. Hyg. I. Abt. Orig. *C1*: 137–149.

Stackebrandt, E., R.M. Kroppenstedt and V.J. Fowler. 1983. A phylogenetic analysis of the family *Dermatophilaceae*. J. Gen. Microbiol. *129*: 1831–1838.

Differentiation and characteristics of the species of the genus Corynebacterium

Although most of the species listed in Table 15.3 (and the List of Species) are distinct taxa, it is presently not easy to choose phenetic characters that allow their differentiation. Those characters likely to be of most value in differentiating the species are listed in Table 15.3.

Table 15.3.
Characteristics differentiating the species of the genus **Corynebacterium**[a, b]

Characteristics	1. C. diphtheriae	2. C. pseudotuberculosis	3. C. xerosis	4. C. pseudodiphtheriticum	5. C. kutscheri	6. C. minutissimum	7. C. striatum	8. C. renale	9. C. cystitidis	10. C. pilosum	11. C. mycetoides	12. C. matruchotii	13. C. flavescens	14. C. vitarumen	15. C. glutamicum	16. C. callunae	a. C. bovis	c. C. paurometabolum
Acid produced from:																		
Glucose	+	+	+	−	+	+	+	+	+	+	+	+	+	+	+	+	+	−
Arabinose	+	d	−	−	−	ND	−	−	−	−	−	−	−	ND	−	−	d	−
Xylose	−	−	−	−	−	−	−	−	+	−	−	−	−	ND	−	−	−	−
Rhamnose	+	−	−	−	−	ND	−	−	−	−	ND	−	−	ND	−	−	−	−
Fructose	+	+	+	−	+	+	+	+	+	+	ND	+	+	+	+	+	+	−
Galactose	+	+	+	−	−	ND	d[c]	−	−	−	−	−	+	+	−	−	+	−
Mannose	+	+	+[d]	−	+	d	+	+	−	+	−	+	+	+	+	+	−	−
Lactose	−	−	−	−	−	−	d	−	−	−	−	−	−	−	−	−	d	−
Maltose	+	+	−[e]	−	+	+	+	d	+	+	−	+	−	+	+	+	+	−
Sucrose	−[e]	d	+	−	+	+[d]	−[e]	−	−	−	ND	+	−	+	+	+	−	−
Trehalose	−[f]	−[e]	−[e]	−	d	−	d	d	+	+	d	−	−	+	+	+	d	−
Raffinose	d	−	−	−	−	−	−	−	−	−	−	d[e]	−	ND	−	−	−	−
Salicin	+	−	+	−	+	ND	−	−	−	−	ND	+	−	+	−	+	−	−
Dextrin	+	d	−	−	+	ND	+	+	+	+	−	+	−	ND	−	−	d	−
Starch	d	−[g]	−	−	+	−	+	−	+	+	ND	−[e]	−	−	−	−	−	−
Hydrolysis of:																		
Esculin	−	−	−	−	−	ND	−	−	−	−	−	+	ND	+	−	−	−	+
Hippurate	−	−	+	+	+	+	+	+	+	+	ND	+	−	−	+	+	+	−
Gelatin liquefaction	−[f]	d	−	−	−	d[c]	−	−	−	−	−	−	−	−	−	−	−	−
Urease	−[f]	+	−	+	+	−	−	+	+	+	−	d[c]	−	+	+	+	−	−
Phosphatase	−	−	−	−	−	+	+	−	−	−	+	−	−	−	ND	ND	+	+
Decomposition of tyrosine	−	−	−	−	−	ND	+	−	−	−	ND	ND	ND	ND	ND	ND	−	−
Pyrazinamidase	−	−	+	+	+	+	ND	+	+	+	ND	+	−	+	ND	ND	+	+
Methyl red	+[h]	+	−	−	−	−	+	−	−	−	−	−	+	+	+	+	−	−
Casein digestion	−	−	−	−	−	−	+	−	−	ND	ND	ND	ND	−	−	−	−	−
Nitrate → nitrite	+[i]	d	+	+	+	−	−	−	+	−	+	−	+	+	−	−	−	−
Tuberculostearic acid present	−	−	−	−	−	+	ND	−	ND	ND	−	−	−	−	−	−	+	+
Whip handle morphology	−	−	−	−	−	−	−	−	−	−	−	+	−	−	−	−	−	−

[a] This table lists the differential characteristics of species which are considered "legitimate" corynebacteria (i.e. those that conform to the present generic definition) and of the species *C. bovis* and *C. paurometabolum* (Addendum I).

[b] Standard symbols: see Table 15.2.

[c] Approximately 50% of strains positive.

[d] Occasional strains negative.

[e] Occasional strains positive.

[f] Except "ulcerans" type.

[g] Starch not normally fermented; although occasional strains are reported to hydrolyze starch (see Cummins et al., 1974; Yanagawa and Honda, 1978).

[h] Majority of strains (90%) are methyl red-positive.

[i] "ulcerans" type may be negative.

List of species of the genus Corynebacterium

1. **Corynebacterium diphtheriae** (Kruse 1886) Lehmann and Neumann 1896, 350.[AL] (*Bacillus diphtheriae* Kruse *in* Flügge 1886, 225.)

diph.the′ri.ae. Gr. n. *diphthera* leather, skin; M.L. fem. n. diphtheria a disease in which a leathery membrane forms in the throat; M.L. gen. n. *diphtheriae* of diphtheria.

The description given here is condensed and modified from that in the eighth edition of the *Manual*.

Straight or slightly curved rods, frequently swollen at one or both ends, 0.3–0.8 × 1.0–8.0 μm; usually stain unevenly and often contain metachromatic granules (polymetaphosphate) which stain bluish purple with methylene blue. Gram-positive, but rather easily decolorized, especially in old cultures.

Three cultural types recognized (see McLeod, 1943) designated *gravis*, *intermedius* and *mitis*. These names were originally given in accordance with the clinical severity of the cases from which the different strains were most frequently isolated. On blood agar, colonies vary considerably in size and appearance depending on type (see below) but are usually 1–3 mm at 24 h; may show a narrow band of hemolysis around and/or under colony but soluble hemolysin not produced. Colonies on blood tellurite agar (0.04% potassium tellurite) gray to black, appearance depends on type. On Loeffler's medium, growth is good: grayish to cream colored, no liquefaction. For fermentation and other reactions, see Table 15.3. It is important to note that some toxinogenic strains ferment sucrose; inability to ferment this sugar cannot be taken as proof of nontoxinogenicity.

Gravis strains show short, irregular rods, give large radially striated brittle colonies, and ferment starch; *mitis* strains show long curved irregular rods and smooth shiny rather butyrous colonies, while *intermedius* strains show long rods with marked cross-striations and give small colonies. Neither *mitis* nor *intermedius* strains ferment starch. Many strains isolated from nature have mixtures of characters which make it difficult to place them in any one type.

Most strains produce a highly lethal exotoxin which is the cause of death in diphtheria, and the toxin produced by all three cultural types is identical. However, nontoxinogenic strains occur which are typical in all other respects. Ability to produce toxin is determined by the presence of prophage carrying a specific determinant called tox+, and toxinogenicity is induced in nontoxinogenic strains by making them lysogenic for phages of this type (see Collier, 1975; Pappenheimer, 1977 for reviews on diphtheria toxin).

Complex media required for growth: for vitamin and other requirements see Koser (1968). Optimum growth temperature 30–37°C.

All strains examined conform to the genus description given above in terms of cell wall composition and type of mycolic acids present. MK-8(H$_2$) is the major menaquinone present. Polar lipids consist of predominantly diphosphatidylglycerol, phosphatidylinositol and monoacylated phosphatidylinositol dimannoside (Collins, unpublished).

The DNA base composition is 52–55 mol% G + C.

In temperate climates, the organism is generally found in the nasopharynx of man; in moist tropical areas, skin carriage is common, often associated with ulcers.

Type strain: ATCC 27010.

C. diphtheriae and related strains

It has become evident from the work of Barksdale (see Barksdale, 1981) and Maximescu (Maximescu et al., 1974), among others, that there is a group of related species or varieties which have in common the ability to produce diphtheria toxin if lysogenized by a suitable bacteriophage. These include *C. diphtheriae* (all three cultural varieties); *C. pseudotuberculosis*, which is also traditionally accorded specific rank; and strains previously designated "*C. ulcerans*" (starch-fermenting, gelatin-liquefying: see e.g. Jebb, 1948). This group also appears to differ from other members of the genus in being pyrazinamidase-negative and neuraminidase-positive (Barksdale, 1981), and in having C$_{16:0}$ and C$_{16:1}$ as the major nonhydroxylated long chain fatty acids (Collins et al., 1982b).

Strains of "*C. ulcerans*" produce an exotoxin distinct from diphtheria toxin, but resembling it pharmacologically (Vertiev, 1981a, b). *C. pseudotuberculosis* also produces a distinct toxin ("ovis" toxin) which is a phospholipase D (Carne, 1940; Carne and Onon, 1982; Goel and Singh, 1972; Barksdale et al., 1981). Strains of "*C. ulcerans*" also produce phospholipase D.

2. Corynebacterium pseudotuberculosis (Buchanan 1911) Eberson 1918, 294.^{AL} (*Bacillus pseudotuberculosis* Buchanan 1911, 238.)

pseu.do.tu.ber.cu.lo'sis. Gr. adj. *pseudes* false; M.L. fem. n. *tuberculosis* tuberculosis; M.L. gen. n. pseudotuberculosis of false tuberculosis.

Still frequently referred to in the literature as "*Corynebacterium ovis*."

In stained smears the appearance is very similar to that of *C. diphtheriae*, especially the *gravis* type; small irregular rods, 0.5–0.6 ×

1.0–3.0 μm, staining irregularly, with club forms and metachromatic granules; fimbriae (pili) present but scanty (Yanagawa and Honda, 1976).

On blood agar, yellowish white, opaque convex colonies with matt surface, about 1 mm at 24 h, often with a narrow zone of hemolysis around the colony; Lovell and Zaki (1966a, b) demonstrated a cell-free hemolysin in 9 out of 11 strains: hemolytic activity much enhanced by diffusible products from *C. equi* and *C. renale* (Fraser, 1964). On the other hand *C. pseudotuberculosis* toxin will *inhibit* the action of staphylococcal β-lysin (see Barksdale et al., 1981). On blood tellurite, colonies small, uniformly blackish, low convex with matt surface, more uniform in color than *C. diphtheriae*. Yellowish, friable growth on Loeffler serum slopes after 24 h, no liquefaction. Scanty growth in *broth* with slight pellicle and sediment; no general turbidity.

Facultatively anaerobic. Most strains produce acid from glucose, galactose, maltose and mannose; variable results reported for lactose, sucrose, xylose, dextrin, arabinose, mannitol and glycerol; starch and trehalose not fermented (Jebb, 1948, four strains). Muckle and Gyles (1982) reported that strains of caprine origin (25 strains) hydrolyzed soluble starch. Nitrate reduction variable; Knight (1969) and Biberstein and Knight (1971) reported that equine strains reduced nitrate to nitrite, while ovine strains could not. Muckle and Gyles (1982) reported that the 25 strains of caprine origin they examined failed to reduce nitrate. Many strains are reported to hydrolyze urea (Hughes and Biberstein, 1959; Muckle and Gyles, 1982). Gelatin liquefaction variable, e.g. one strain out of four positive (Jebb, 1948), eight strains out of eight positive (Hughes and Biberstein, 1959). In the eighth edition of the *Manual* it was noted that the results of fermentation and other tests quoted in the literature for *C. pseudotuberculosis* seemed to be unusually variable. This still appears to be true. Some of this variation may be due to examination of different biotypes, e.g. the differences between equine and ovine strains in nitrate reduction, quoted above. Others may be due to technical methods, e.g. the type of starch used (soluble or laundry) may determine whether the organisms hydrolyze it or not (e.g. Muckle and Gyles, 1982).

Essentially it seems that strains of *C. pseudotuberculosis* can be distinguished from other similar corynebacteria (e.g. *C. diphtheriae*, "*C. ulcerans*," *C. xerosis*) in being pyrazinamidase-negative and urease-positive and failing to hydrolyze laundry starch (Muckle and Gyles, 1982). Identification of the specific toxin will also help (see Barksdale et al., 1981). Strains of *C. pseudotuberculosis* are particularly toxic to mice.

Nutritional requirements not known in detail; growth much improved by the addition of blood or serum to medium; increased CO$_2$ concentration may facilitate primary isolation (Knight, 1969).

Cell-free filtrates of most strains are lethal to guinea pigs, mice, rabbits and sheep: the toxic material can be toxoided with formalin, and antitoxic sera produced; crude toxin causes intense cellular and fluid exudation (Jolly, 1965; Carne and Onon, 1982). All strains produce antigenically similar toxin (Doty et al., 1964). The toxin is a phospholipase D which breaks down sphingomyelin to produce ceramide phosphate and choline (see Onon, 1979; Linder and Bernheimer, 1978; Carne and Onon, 1982); however, purified toxin is not hemolytic (Linder and Bernheimer, 1978). In addition, a toxic surface lipid is produced (Carne, et al., 1956) analogous to the cord factor of *M. tuberculosis*.

Cell wall sugars are arabinose, galactose, glucose and mannose; diaminoacid of peptidoglycan is *meso*-DAP (Cummins and Harris 1956; Keddie and Cure, 1977).

Contains short chain mycolic acids. Long chain fatty acids predominantly of the straight chain saturated and monounsaturated types. Major fatty acids are hexadecanoic (C$_{16:0}$) and hexadecenoic (C$_{16:1}$) acids (Collins et al., 1982b), thus closely resembling the pattern found in *C. diphtheriae* and "*C. ulcerans*." Major menaquinone is MK-8 (H$_2$) (Collins et al., 1977). Polar lipids consist of predominantly diphosphatidylglycerol, phosphatidylinositol and monoacylated phosphatidylinositol dimannoside (Collins, unpublished).

Strains of *C. pseudotuberculosis* are sensitive to some of the bacteriophages used in typing *C. diphtheriae*, although none of the strains tested fall exactly into any of the *C. diphtheriae* lysotypes; strains of *C. diphtheriae* are sensitive to bacteriophages from *C. pseudotuberculosis* (Carne 1968; Maximescu et al., 1974). Strains of *C. pseudotuberculosis* have the ability to produce diphtheria toxin if lysogenized by suitable bacteriophages (see Maximescu et al., 1974).

May cause ulcerative lymphangitis, abscesses and other chronic purulent infections in sheep, goats, horses and other warm-blooded animals; occasional infections in man. Originally isolated from necrotic areas in the kidney of a sheep.

The G + C content of DNA is 51.8–52.5 mol% (Bouisset, et al., 1963, quoted by Hill, 1966; Pitcher, 1983).

Type strain: ATCC 19140 (NCTC 3450).

3. **Corynebacterium xerosis** (Lehmann and Neumann 1896) Lehmann and Neumann 1899, 385.[AL] (*Bacillus xerosis* Lehmann and Neumann 1896, 361.)

xe.ro'sis. Gr. fem. n. *xerosis* a parched skin. M.L. gen. n. *xerosis* of xerosis.

Irregularly staining, Gram-positive, often barred rods, with occasional granules and club forms. Small circular colonies (1 mm or less at 24 h) on plain agar, may be rough or smooth. On blood agar colonies larger than on media without blood, may be pale yellow to tan color, no hemolysis, colonies rough or smooth. Granular deposit in broth with clear supernatant. Aerobic and facultative; glucose and gluconate largely broken down by pentose phosphate pathway (Zagallo and Wang, 1967, ATCC 7084), does not liquefy serum, indole not produced, ferments glucose, galactose and sucrose, producing acid but no gas; good growth at 22 and 37°C. Does not produce urease, neuraminidase or *N*-acetylneuraminate lyase; generally reduces nitrate, although the latter property is reported to be variable (Ray and Sen, 1977; Porschen et al., 1977; Arden and Barksdale, 1976): however, ATCC 373 is nitrate positive (Arden and Barksdale, 1976).

Requires amino acids but not vitamins (Smith, 1969, three strains, including ATCC 373 and ATCC 7064).

Cell wall sugars in three strains were reported to be arabinose, galactose, and mannose (Cummins and Harris, 1956; Cummins, 1971). However, Evangelista et al. (1978) found that the major sugars of the cell wall polysaccharide in ATCC 373 were arabinose, galactose and glucose, with smaller amounts of rhamnose. Cell walls contain antigen common to other corynebacteria which have arabinose and galactose as distinctive cell wall sugars (Cummins, 1962).

Short chain mycolic acids present. Long chain fatty acids are of the straight chain saturated and monounsaturated types. Major fatty acids are hexadecanoic and octadecenoic acids ($C_{16:0}$, $C_{18:1}$, Collins et al., 1982b).

Generally nonpathogenic, but may cause endocarditis or pneumonia in severely ill patients or those receiving steroids (Porschen et al., 1977).

Originally isolated from conjunctival sac in man. Presumably inhabits skin and mucous membranes of man.

The G + C content of the DNA generally ranges from 55–59 mol% (Bouisset, et al, 1963; Marmur and Doty, 1962; Danhaive et al., 1982; Pitcher, 1983). However, Yamada and Komagata (1970) reported a G + C content of 67.3 mol% for ATCC 373 and 68.5 mol% for ATCC 7711. Pitcher (1983) also reported a G + C content of 68.0 mol% in ATCC 373.

Type strain: ATCC 373.

Further comments. As can be seen from the G + C contents reported in the literature, it is very likely that two kinds of strains are being attributed to *C. xerosis*, one with a G + C content of about 55% and the other with a much higher content of around 67%. It is assumed here that authentic *C. xerosis* strains have a G + C content of 55–59 mol%. It is unfortunate that the strain designated as type strain (ATCC 373) appears from two independent investigations to have a G + C content of 67–68% (Yamada and Komagata, 1970; Pitcher, 1983).

4. **Corynebacterium pseudodiphtheriticum** Lehmann and Neumann 1896, 361.[AL]

Often referred to as "*C. hofmanii*"; however, this name was not validly published (Buchanan et al., 1966).

pseu.do.diph.the.ri'ti.cum. Gr. adj. *pseudes* false M.L. fem. n. *diphtheria* diphtheria; M.L. adj. *diphtheriticus* diphtheritic; M.L. neut. adj. *pseudodiphtheriticum* relating to false diphtheria.

Short rather regular rods, 0.5–2.0 × 0.3–0.5 µm, which stain evenly except for a transverse medial unstained septum: club forms and metachromatic granules minimal or absent; in stained smears the organisms often lie in rows with the long axes parallel. Gram-positive, less readily decolorized than most other corynebacteria.

Good growth on all media, whether or not they contain blood or serum. On blood agar, growth white to cream colored, colonies regular and smooth, butyrous consistency; no hemolysis. Aerobic and facultatively anaerobic; although it does not appear to attack the commonly used carbohydrates it can degrade a wide variety of amides, esters, amino acids and other organic compounds (strain NCIB 10803, Grant 1973a, b; Grant and Wilson 1973); reduces nitrate, hydrolyzes urea, utilizes citrate and other Krebs cycle intermediates (Grant 1973b). Optimum growth temperature 37°C.

Major cell wall sugars are arabinose, galactose and glucose; diamino acid of peptidoglycan is *meso*-DAP (Cummins and Harris, 1956). Cell walls contain antigenic determinant common to other corynebacteria which have arabinose and galactose as major cell wall sugars (Cummins, 1962).

Short chain mycolic acids present (Welby-Guisse et al., 1970; Collins et al., 1982a).

The G + C content of DNA ranges from 54.9–56.8 mol% (Danhaive et al., 1982; Pitcher, 1983).

Found in nasopharyngeal mucosa of man. Not pathogenic, toxins not produced.

Type strain: ATCC 10700.

5. **Corynebacterium kutscheri** (Migula 1900) Bergey et al. 1925, 395.[AL] (*Bacterium kutscheri* Migula 1900, 372.)

kut'scher.i. M.L. gen. n. *kutscheri* of Kutscher, the bacteriologist who first isolated the species: frequently referred to in the literature as "*Corynebacterium murium*," a name given because of its association with rats and mice (see below).

Irregularly staining slender rods, often clubbed, sometimes with pointed ends; metachromatic granules present. Large numbers of fimbriae (pili) present (Yanagawa and Honda, 1976). Small, thin, yellowish or grayish white serrate colonies on nutrient agar. Abundant growth on Loeffler's medium. Broth turbid with sediment, no pellicle, Litmus milk, no change. Aerobic, facultative. Reduces potassium tellurite. Other properties are noted in Table 15.3. Major metabolic products are propionate and lactate, with lesser amounts of acetate, pyruvate and oxalacetate; accumulates intracellular starch from glucose-1-phosphate (Arden and Barksdale, 1976).

Pyrazinamidase-positive (Sulea et al., 1980); produces *N*-acetylneuraminate lyase but not neuraminidase (Arden et al., 1972).

All of five strains agglutinated to titer with a serum prepared against one of them (Pierce-Chase, et al., 1964).

Cell wall contains *meso*-DAP. Arabinose, galactose, mannose and rhamnose were reported as cell wall sugars by Cummins and Harris (1956) and Pitcher (1983) has reported arabinose, galactose and rhamnose as the wall sugars in ATCC 6944 and ATCC 15677; however, Bruce et al. (1969) did not find rhamnose in the wall of *C. kutscheri* ATCC 15677.

This organism appears to be a frequent, if not invariable, parasite of mice and rats, but also occurs in other small rodents, e.g. voles (Barrow, 1981). Its pathogenicity for normal animals appears to be low, but if their resistance to infection is altered, for example by cortisone or dietary deficiency, extensive pseudotuberculous lesions develop from which *C. kutscheri* can readily be isolated. However, in voles the infection occurred in the wild.

For a discussion of the question of latent infection with this organism, see Pierce-Chase, et al. (1964); Fauve, et al. (1964); also Hirst and Olds 1978a, b.

Type strain: ATCC 15677 (NCTC 11138).

Further comments. This appears to be the only organism in *Corynebacterium* (or indeed in *Mycobacterium*, *Nocardia*, *Rhodococcus* or *Caseobacter*) which has rhamnose as a cell wall sugar in addition to arabinose and galactose, and this may be a characteristic of the species. However, a larger series of strains needs to be examined. The two strains (NCTC 949, 11138) examined by Pitcher (1983) had a G + C content of 46.0–46.2 mol% as opposed to ~53 mol% for the type species, and he suggests that it may be necessary to consider placing *C. kutscheri* in a separate genus.

6. **Corynebacterium minutissimum** (*ex* Sarkany, Taplin and Blank 1962) Collins and Jones 1983, 870.[VP*]

mi.nu.tis′si.mum. L. sup. adj. *minutissimus* very small.

Surface colonies on blood agar are ~1 mm in diameter after 24 h; circular, slightly convex, shiny and moist. Short straight to slightly curved rods (1–2 × 0.3–0.6 μm); some cells arranged at an angle to give V-formations. Gram-positive although irregular staining may be observed. Metachromatic granules formed. When grown on certain rich media (e.g. 20% bovine fetal serum) colonies show a coral-red to orange fluorescence under Wood's light (365 nm).

Facultatively anaerobic. Optimum temperature 37°C. Salient cultural characteristics are given in Table 15.3. Pyrazinamidase-positive.

Cell wall peptidoglycan contains *meso*-DAP; arabinose and galactose are the wall sugars. Short chain nonhydroxylated fatty acids are primarily of the straight chain saturated, monounsaturated and 10-methyl-branched chain types. Major fatty acids are hexadecanoic ($C_{16:0}$), octadecenoic ($C_{18:1}$) and 10-methyloctadecanoic (t_{19}) acids (Collins et al., 1982b). Major menaquinones are MK-8 (H_2) and MK-9 (H_2). Polar lipid composition comprises diphosphatidylglycerol, phosphatidylinositol, phosphatidylinositol dimannoside and some unidentified glycolipids.

The mol% G + C of DNA is 56.4–58.9 (Collins and Jones, 1983; Pitcher, 1983).

Pathogenicity: May cause erythrasma of skin in man.

Type strain: NCTC 10288.

7. **Corynebacterium striatum** (Chester 1901) Eberson, 1918, 22.[AL] (*Bacterium striatum* Chester 1901, 171.)

stri.a′tum. L. part. adj. *striatus* grooved.

Pleomorphic Gram-positive rods, often club shaped, 0.25–0.5 × 2.0–3.0 μm; coccoid forms and long filaments found in old cultures; metachromatic granules present, often regularly arranged to produce a segmented effect; nonmotile; not acid-fast. Rather slow growth on agar, white, smooth, entire colonies about 1 mm in diameter at 48 h; some strains produce a yellowish green pigment soluble in the medium. On blood agar slight hemolysis around deep colonies. Clear supernatant in broth, no pellicle, finely granular white sediment. Moderate growth on Loeffler's medium, no liquefaction.

Aerobic and facultative. Ferments glucose, fructose, mannose, trehalose, dextrin, glycogen and usually also lactose, maltose and starch; about 50% of strains ferment galactose, occasional strain ferments sucrose. Nitrates not reduced to nitrites; no production of acetoin or indole. Gelatin liquefaction variable, about 50% strains positive.

The cell walls of strain ATCC 6940 (NCTC 764) contain arabinose and galactose and the G + C content is 57.6 mol% (Pitcher, 1983). ATCC 6940 has also been shown to contain mycolic acids (Goodfellow et al., 1976). MK-8 (H_2) is the major menaquinone ("*C. flavidum*"; Collins et al., 1977).

Originally isolated from human nasopharynx; also from milk of cows with mastitis. Twenty-four-hour cultures fatal to guinea pigs and mice on intramuscular injection.

Type strain: ATCC 6940.

Further comments. *C. striatum* appears to have been isolated originally from human nasal mucus and was designated *Bacterium striatum* by Chester and later *Corynebacterium flavidum* by Holland (*In* Winslow et al., 1920). When Munch-Petersen (1954) described 31 strains of a corynebacterium in milk from cows with an udder infection (a mixed infection comprising a staphylococcus, a streptococcus and corynebacterium) he noted that the corynebacterium was "apparently identical with *C. flavidum*." However, no definitive study has yet been made and it still seems quite uncertain whether descriptions of human and bovine strains refer to the same organism.

The type strain ATCC 6940 is in some places in the literature referred to as *Corynebacterium striatum* and in others as "*Corynebacterium flavidum.*"

8. **Corynebacterium renale** (Migula 1900) Ernest 1906, 89.[AL] (*Bacterium renale* Migula 1900, 504.)

re.na′le. L. neut. adj. *renale* pertaining to the kidneys.

A rather large irregularly staining bacillus 0.7 × 3.0 μm or more, often with pointed ends; fimbriae (pili) present but not numerous (Yanagawa and Honda, 1978).

Salient cultural characteristics are given in Tables 15.3 and 15.4. The organism produces urease, pyrazinamidase and hydrolyzes hippurate.

Diaminoacid of peptidoglycan is *meso*-DAP, major cell wall sugars are arabinose, galactose, glucose and mannose.

Short chain mycolic acids (~30–36 carbon atoms) present (Collins et al., 1982a); menaquinone type MK-8 (H_2) (Collins et al., 1979). Long chain fatty acids consist of predominantly straight chain saturated and monounsaturated types. Major fatty acids are hexadecanoic and octadecenoic acids (Collins et al., 1982b).

G + C content of DNA is 53–58 mol%.

Pathogenicity: Causes cystitis and pyelitis in cattle.

Type strain: ATCC 19412 (NCTC 7448).

Further discussion: relationships of **C. renale, C. pilosum** *and* **C. cystitidis**. Strains originally classified as *C. renale* were shown by Yanagawa et al. (1967) to belong to three different serological types, and these immunological differences were later found to be correlated with differences in biochemical properties, nutritional requirements and virulence for cows, among other things. Finally, an examination of

Table 15.4.
Differences between **C. renale, C. pilosum** *and* **C. cystitidis**[a–c]

Characteristics	C. renale	C. pilosum	C. cystitidis
Colony color	Yellow	Yellow	Whitish
Colony first visible at 37°C in:	24 h	24 h	48 h
Growth in broth at pH 5.4	+	−	−
Acid from:			
Xylose	−	−	+
Starch	−	+	+
Reduction of nitrate	−	+	−
Digestion of casein	+	−	−
Hydrolysis of Tween 80	−	−	+
G ± C mol% in DNA	56.7 ± 1.1	57.9 ± 1.9	53.5 ± 0.9
Percentage homology to DNA from ATCC 10848 (C. renale)	80–100	45–55	15–20
Original designation	C. renale type I	C. renale type II	C. renale type III

[a] Based on information from Yanagawa, Basri and Otsuki (1967); Honda and Yanagawa (1973), Yanagawa (1975); and Yanagawa and Honda (1978).

[b] Because strains of all three species have complex nutritional requirements, these tests were done in media containing 5% serum (see Yanagawa and Honda, 1978).

[c] Standard symbols: see Table 15.2.

---

* *VP*, denotes that this name has been validly published in the official publication, International Journal of Systematic Bacteriology.

DNA base ratios and DNA homology studies showed considerable differences, especially between the strains originally designated type III and the other two types. On this basis Yanagawa and Honda (1978) proposed two new species *C. pilosum* and *C. cystitidis*. The main properties distinguishing *C. renale*, *C. pilosum* and *C. cystitidis* are given in Table 15.4.

For references, see Yanagawa and Honda (1978).

9. **Corynebacterium cystitidis** Yanagawa and Honda, 1978, 298.[AL]

cys.ti'ti.dis. Gr. n. Bladder; M.L. n. *cystitis* cystitis; M.L. gen. n. *cystitidis* of cystitis.

The following description is condensed from Yanagawa and Honda (1978).

Gram-positive rods, straight to slightly curved, 0.5 × 2.6 μm, often occurring in angular or palisade arrangement; metachromatic granules present; numerous fimbriae (pili) visible on electron microscopy. Colonies on nutrient agar and serum agar white, entire, circular and semitranslucent, usually very small and not readily visible at 24 h. Not hemolytic on blood agar (sheep, guinea pig or rabbit blood). Slight turbidity but no pellicle in broth cultures. Grows at 41.5°C.

Aerobic and facultatively anaerobic; vitamin and amino acid requirements complex, probably include thiamin, biotin, nicotinic acid, pantothenic acid, pyridoxine, glutamic acid, valine, *iso*-leucine and tryptophane.

Salient cultural characteristics are given in Table 15.3. Hydrolyzes starch, hippurate and Tween 80, but does not reduce nitrate. Urease produced.

Diamino acid of peptidoglycan is *meso*-DAP; major cell wall sugars are arabinose, galactose and glucose. Short chain mycolic acids present.

G + C content of DNA has been reported to be 52.6–53.5 mol% (Kumazawa and Yanagawa, 1969; Honda and Yanagawa, 1973) and 70.3 mol% (Pitcher, 1983).

Pathogenicity: Causes severe hemorrhagic cystitis in cows; isolated from the prepuce in healthy bulls.

Type strain: ATCC 29593 (strain 42 Fukuya of Yanagawa and Honda, 1978).

10. **Corynebacterium pilosum** Yanagawa and Honda, 1978, 209.[AL]

pi.lo'sum. L. adj. *pilosus* having much hair; intended to mean having many fimbriae (pili).

The following description is condensed from Yanagawa and Honda (1978).

Gram-positive rods, 0.5 × 1.3 μm, occurring singly, in pairs (often at an angle) or in irregular masses; metachromatic granules present; large numbers of fimbriae (pili) visible on electron microscopy. Colonies on nutrient agar and serum agar cream to pale yellow, entire, circular, opaque, 1 mm in diameter at 24 h. No hemolysis on blood agar (sheep, guinea pig or rabbit blood); pellicle and granular sediment in broth. No growth at 41.5°C but cells will remain viable for 30 min at 56°C.

Aerobic and facultatively anaerobic; vitamin and amino acid requirements complex, probably include biotin, nicotinic acid, *p*-aminobenzoic acid, glutamic acid, valine and *iso*-leucine.

Salient cultural characteristics are given in Table 15.3. Reduces nitrate and hydrolyzes starch and hippurate, but not Tween 80. Urease produced.

Diaminoacid of peptidoglycan is *meso*-DAP; major cell wall sugars are arabinose and galactose. Short chain mycolic acids present. Menaquinone type: MK-8 (H_2).

G + C content of DNA is 57.9–60.9 mol% (Kumazawa and Yanagawa, 1969; Honda and Yanagawa, 1973; Pitcher, 1983).

Pathogenicity: Originally isolated from the urine and vagina of healthy cows; occasionally causes cystitis and pyelonephritis (Hiramune et al., 1971).

Type strain: ATCC 29592 (strain 46 Hara of Yanagawa and Honda, 1978).

11. **Corynebacterium mycetoides** (*ex* Ortali and Capocaccia 1951) Collins 1982, 399.[VP] (*Corynebacterium mycetoides* (Castellani) Ortali and Capocaccia 1956, 490).

my.cetoid'es. Gr. n. myke, fungus -oides, like M.L. adj. mycetoides, similar to fungi, referring to ulcers caused by fungi.

Straight to slightly curved rods produced (1–3 × 0.3–0.5 μm); some cells arranged at an angle to give V-formations, coccoid forms may be present. Usually stain unevenly and often contain metachromatic granules. Surface colonies on blood agar or nutrient agar are ~1 mm in diameter after 2 days; circular, convex with entire margin, shiny. Yellow pigment produced. Salient cultural characteristics are given in Table 15.3.

Cell wall peptidoglycan contains *meso*-DAP; arabinose and galactose are the wall sugars. Short chain mycolic acids (30–36 carbon atoms) present. Mycolic acids are unusual in containing significant amounts of α-alkyl branches with odd numbers of carbon atoms (i.e. $C_{15}H_{31}$, Collins et al., 1982a). Long chain nonhydroxylated fatty acids are primarily of the straight chain saturated and monounsaturated types. Major fatty acids are octadecenoic, heptadecanoic and hexadecanoic acids (Collins et al., 1982b). MK-8 (H_2) and MK-9 (H_2) predominant menaquinones. Polar lipid composition consists of diphosphatidylglycerol, phosphatidylglycerol, phosphatidylinositol, phosphatidylinositol dimannoside and some unknown glycolipids and phospholipids.

G + C content of DNA is 59 mol%.

Pathogenicity: Reported to have caused tropical ulcers in man (Castellani, 1942).

Type strain: NCTC 9864.

12. **Corynebacterium matruchotii** (Mendel 1919) Collins 1982, 365.[VP] (*Cladothrix matruchoti* Mendel 1919, 584; *Bacterionema matruchotii* Gilmour, Howell and Bibby 1961, 139.)

ma.tru.cho'ti.i. M.L. gen. n. *matruchotii*, of Matruchot; named after Professor Matruchot, a French mycologist.

Cells are pleomorphic, comprising nonseptate and septate filaments, and bacilli. Characteristic morphology is a bacillus attached to a filament ("whip handle"). Branching is frequent with aerobic and/or acid conditions. Gram-positive, nonacid-fast, and nonmotile. Metachromatic granules formed. Morphology of young microcolonies is similar under aerobic or anaerobic conditions. Microcolonies are flat, filamentous, spider-like, may have dense center, and are composed of nonseptate, septate and fragmenting filaments of varying lengths. Macrocolony appearance is variable. Aerobically incubated surface colonies are 0.5–1.5 mm in diameter, can be circular, convex, rough with entire or filamentous margin; or irregular, molar-toothed, rough, with an entire to filamentous margin at the base; or, irregular, with a low convex rough center and raised curled-up lobate margin. The three colonial forms are opaque, tough and adhere to medium. Anaerobically incubated, surface colonies are 1–2 mm diameter, filamentous, flat with filamentous edges, opaque at the center to translucent at the edge, tough and adherent to the medium.

Facultatively anaerobic; some strains aerobic. Optimum temperature 37°C. Carbohydrates are fermented to yield acid and some gas. For fermentation and other reactions, see Table 15.3. Starch hydrolyzed; urea sometimes hydrolyzed.

Vitamin and amino acid requirements are complex, and include riboflavin, thiamin, nicotinic acid, pantothenic acid and cysteine. Hemin stimulates growth under anaerobic conditions, but can be inhibitory with aerobic incubation. CO_2 is stimulatory under all growth conditions.

Diamino acid of peptidoglycan is *meso*-DAP; arabinose and galactose are cell wall sugars. Short chain mycolic acids are present (Alshamoany et al., 1977; Wada et al., 1981). The fatty acid composition is mainly straight chain saturated and monounsaturated acids. Major fatty acids are hexadecanoic and octadecenoic acids (Alshamoany et al., 1977; Collins et al., 1982b). The principal menaquinones are MK-9 (H_2) and MK-8 (H_2) (Collins and Jones, 1981). The polar lipids comprise diphosphatidylglycerol, phosphatidylglycerol, phosphatidylinositol, phosphatidylinositol dimannoside and some unidentified glycolipids (Minnikin et al., 1978).

G + C content of DNA is 55–58 mol% (Page and Krywolap, 1974).

Habitat: Found in the oral cavity of man and primates, particularly in calculus and plaque deposits on the teeth.

Type strain: ATCC 14266 (NCTC 10254).

13. **Corynebacterium flavescens** Barksdale, Lanéelle, Pollice, Asselineau, Welby and Norgard, 1979, 222.[AL] (*Microbacterium flavum*, Orla-Jensen 1919, 181).

fla.ves′cens. L. v. *flavescere* become yellow; L. part. adj. *flavescens* becoming yellow.

Gram-positive to Gram-variable pleomorphic rods often with tapered ends, and showing metachromatic granules. On Loeffler slants, growth is yellow: on BHI agar colonies growths are smooth, butyrous and cream colored. Colonies on tellurite agar are grayish black, later developing grayish white centers.

Aerobic and facultatively anaerobic. Salient cultural characteristics are given in Table 15.3. Nutritional requirements complex: *p*-amino benzoic acid, biotin, nicotinic acid, pantothenate and thiamine are required.

The G + C content of the DNA is 58.3 mol% and the DNA of the type strain was found to show 64% homology with *C. diphtheriae* PW8.

Type strain: ATCC 10340 (NCIB 8707), from cheese.

Further comments. The organism was originally described as *Microbacterium flavum* by Orla-Jensen (1919), having been isolated from dairy products. Several investigators had pointed out its similarity to strains in *Corynebacterium* (e.g. Keddie, et al., 1966; Robinson, 1966a, b; Schleifer, 1970; Bousfield 1972; Jones, 1975; Goodfellow et al., 1976) before the formal proposal for its reclassification was made by Barksdale et al. (1979).

14. **Corynebacterium vitarumen** (Bechdel, Honeywell, Dutcher and Knutsen 1928) Lanéelle, Asselineau, Welby, Norgard, Imaeda, Pollice and Barksdale 1980, 544.[VP] (*Flavobacterium vitarumen* Bechdel, Honeywell, Dutcher and Knutsen 1928, 234; *Brevibacterium vitarumen* (Bechdel, Honeywell, Dutcher and Knutsen) Breed, Murray and Smith 1957, 495.)

vi.ta.ru′men. L. n. *vita* life; L. n. *rumen* throat, gullet, rumen; M.L. n. *vitarumen* rumen-life.

Based on description of ATCC 10243 and strain 12143 from the Institute for Fermentation, Osaka, Japan (Lanéelle et al., 1980).

Pleomorphic Gram-positive rod-shaped bacteria with moderately tapered ends; may contain metachromatic granules when grown on phosphate-rich media. Lemon yellow growth on Loeffler slants, raised, smooth, butyrous yellow colonies on chocolate agar, black colonies on chocolate agar containing 0.03% tellurite.

Salient cultural characteristics are given in Table 15.3. Produces urease, nitrate reductase, and pyrazinamidase, but does not decarboxylate lysine or ornithine or produce cysteine desulfhydrase.

Meso-DAP is the diamino acid of peptidoglycan and the principal cell wall sugars are arabinose and galactose. Short chain mycolic acids are present (Lanéelle et al., 1980). MK-8 (H$_2$) is the major menaquinone (Kanzaki et al., 1974).

G + C content of DNA is 64.8 mol% (strain 12143).

Type strain: ATCC 10234 (NCIB 9291).

Further comments. This organism was originally isolated from the rumen of the cow (Bechdel et al., 1928) and called *Flavobacterium vitarumen*. It is thought to participate in the production of the vitamin B complex in the rumen (especially riboflavin: Breed et al., 1957, p. 495). It was transferred to *Corynebacterium* on the basis of biological and chemical characteristics by Lanéelle et al. (1980).

15. **Corynebacterium glutamicum** (Kinoshita, Nakayama and Akita 1958) Abe, Takayama and Kinoshita 1967, 299.[AL] (*Micrococcus glutamicus* Kinoshita, Nakayama and Akita 1958, 176.)

glu.tam′ic.um M.L. adj. *glutamicus*, of glutamic acid.

This description is taken largely from that of Abe, Takayama and Kinoshita (1967).

Short Gram-positive rods or ellipsoids 0.7–1.0 × 1.0–3.0 μm, occur-

ring singly, in pairs or in irregular masses. In lag phase cultures, quite long branching cells may be found, but in mid- to late log phase cultures the cells are very short, ellipsoidal to almost coccal.

Colonies on nutrient agar are smooth, entire, circular, dull to slightly glistening, generally pale yellow to yellow; gray to black colonies on tellurite medium; in broth, moderate turbidity with floccular sediment. For details of fermentation and other metabolic properties see Table 15.3.

Aerobic and facultatively anaerobic; optimum growth temperature 25–37°C, slight growth at 42°C. Vitamin requirements: all strains require biotin and some, in addition, need thiamine and/or *p*-amino-benzoic acid.

All strains produced large amounts of L-glutamic acid under aerobic conditions.

Cell wall contains *meso*-DAP, arabinose and galactose. Short chain mycolic acids present (Collins et al., 1982a). MK-9 (H$_2$) is the major menaquinone (Collins et al., 1977). The major nonhydroxylated long chain fatty acids are C$_{16:0}$ and C$_{18:1}$ (Collins et al., 1982b).

G + C content of DNA is 55–57.7 mol%.

Type strain: ATCC 13032.

Further comments. Because of the coccal appearance of cells in log phase cultures and the characteristic of producing large amounts of L-glutamic acid, the organism was originally labeled *Micrococcus glutamicus*. However, these strains have all the major characteristics of *Corynebacterium*, i.e. mycolic acids, arabinogalactan, and meso-DAP as the diaminoacid of peptidoglycan. Strains labeled *Corynebacterium lilium* show 100% DNA homology with *C. glutamicum* (Suzuki et al., 1981) therefore *C. lilium* is here regarded as synonymous with *C. glutamicum*.

It should be noted that Abe, Takayama, and Kinoshita (1967) described three groups of coryneform organisms which produce L-glutamic acid, with mol% G + C of 55–57 (group I), 53 (group II), and 64–65 (group III).

The organism described here as *C. glutamicum* corresponds to their group I.

16. **Corynebacterium callunae** Yamada and Komagata 1972, 412.[AL]

cal.lun′ae. M.L. n. *Calluna* generic name of heather; M.L. gen. n. *callunae* of heather.

Rather short, strongly Gram-positive rod, metachromatic granules present.

Moderate growth on nutrient agar.

Other major characteristics are listed in Table 15.3.

Cell wall contains *meso*-DAP; arabinose and galactose are the major cell wall sugars. Mycolic acids present (Collins et al., 1982a). The major fatty acids are C$_{16:0}$ and C$_{18:1}$. The major isoprenoid quinone is MK-9 (H$_2$) (Collins, et al., 1979).

The G + C content of DNA is 51.2 mol% (ATCC 15991; Yamada and Komagata, 1970).

Type strain: ATCC 15991.

Further comments. This organism is one of a number of coryneforms characterized principally by producing large amounts of glutamic acid under aerobic conditions on a scale that can be used commercially. So far only one strain (ATCC 15991; NCIB 10338; NRRL B-2244) has been examined in any detail. It appears this organism was isolated from heather, but there is no evidence that it is phytopathogenic.

Addendum I

The following species are not members of the genus *Corynebacterium*, for the reasons given below, but are retained here pending further work or to assist the reader by cross-referencing to other genera.

a. *Corynebacterium bovis* Bergey, Harrison, Breed, Hammer and Huntoon 1923, 388.[AL]

bo′vis. L. n. *bos* a cow; L. gen. n. *bovis* of a cow.

Irregular rods (0.5–0.7 × 2.5–3.0 μm), often barred and clubbed; coccobacillary forms may occur. Colonies on nutrient agar supplemented with Tween 80 (~0.1%) are white to cream, circular, entire, slightly shiny, ~1-2 mm in diameter (24 h). Aerobic. Most strains ferment glucose, fructose, maltose and glycerol. Oxidase-positive. Starch and casein not hydrolyzed. Coagulated serum not liquefied. Lipolytic for butter fat. Requires unsaturated long chain fatty acids.

Cell wall peptidoglycan based upon *meso*-DAP (variation A1γ). Arabinose and galactose are the principal wall sugars. Contains very low molecular weight mycolic acids (~22–36 carbon atoms); the side chain (α-alkyl group) has the composition of C_6H_{13} and C_8H_{17} (major component) (Collins et al., 1982a). Upon whole-organism acid methanolysis many strains produce a multiple spot pattern (see Goodfellow et al., 1976; Minnikin et al., 1978). Major nonhydroxylated fatty acids (when growing in presence of 0.01% Tween 80) are of the straight chain saturated, monounsaturated and 10-methyl-branched chain (e.g. tuberculostearic acid) types (Collins et al., 1982b). Dihydrogenated menaquinones with nine isoprene units constitute major respiratory quinones (Collins et al., 1977).

The mol% G + C of DNA ranges from 67.8–69.7 (T_m).

Found in aseptically drawn milk; commensal on the cow's udder. Causes bovine mastitis.

Type strain: ATCC 7715.

Several strains of *C. bovis* held in national culture collections do not conform to the above description and are incorrectly designated. *C. bovis* ATCC 13722 possesses a peptidoglycan based upon diaminobutyric acid (DAB) ([L-Hsr]-D-Glu-Gly-D-DAB) and a DNA base composition of 73.7 mol% G + C (Döpfer et al., 1982). DNA/DNA and DNA/rRNA homology studies (Döpfer et al., 1982) indicate that this organism may constitute the nucleus of a new taxon. *C. bovis* NCDO 1927 and 1929 produce pink colonies on glucose yeast extract agar and do not require Tween 80. These strains possess mycolic acids with chain lengths compatible with rhodococci (Collins, unpublished results).

Taxonomic comments. Corynebacterium bovis differs from all other members of the genus *Corynebacterium* in containing some very short characteristic mycolic acids (Collins et al., 1982a). The species also differs from the majority of true corynebacteria in possessing high levels of tuberculostearic acid (Collins et al., 1982b). The presence of DNA rich in guanine plus cytosine (~68 mol%) (Yamada and Komagata, (1970) also indicates this species should be excluded from the genus *Corynebacterium*.

b. *Corynebacterium hoagii* (Morse 1912, 281) Eberson 1918, 11.[AL]

hoa'gi.i. M.L. gen. n. *hoagii* of Hoag; named after Dr. Louis Hoag who first isolated the species.

C. hoagii (type strain ATCC 7005) can be distinguished from true corynebacteria by the presence of *N*-glycolyl residues within the glycan moiety of its cell wall and by its relatively high mol% G + C of ~65 mol% (Yamada and Komagata, 1970; Uchida and Aida, 1977, 1979). Numerical phenetic (Goodfellow et al., 1982), mycolic acid (Collins et al., 1982a), fatty acid (Collins et al., 1982b) and DNA homology (Suzuki et al., 1981) studies indicate *C. hoagii* and *Rhodococcus equi* represents a single species. It is considered in this volume under *R. equi*.

c. *Corynebacterium paurometabolum* Steinhaus 1941, 783.[AL]

pau.ro.me.ta'.bo.lum. Gr. adj. *paurus* little; Gr. adj. *metabolus* changeable; M.L. neut. adj. *paurometabolum* little changeable, probably meaning producing little change.

Straight to slightly curved rods (0.5–0.8 × 1.0–2.5 μm). Metachromatic granules formed. Colonies white to gray, entire, circular, dry and granular. Aerobic. No fermentation of any sugars tested. Nitrates not reduced to nitrite. Indole not formed. Gelatin not liquefied. Esculin hydrolyzed. Urease-negative. Phosphatase-positive. Originally isolated from the mycetome and ovaries of the bed bug (*Cimex lectularius*).

Type strain: ATCC 8368.

Taxonomic comments. Although the species resembles true corynebacteria in possessing a directly cross-linked peptidoglycan based upon *meso*-DAP and an arabinogalactan polymer (Cummins, 1971; Schleifer

and Kandler, 1972), it can be distinguished from them in possessing very long (68–76 carbon atoms) highly unsaturated (2–6 double bonds) mycolic acids and unsaturated menaquinones with nine isoprene units (MK-9) (Collins and Jones, 1982b). On the basis of lipid composition *C. paurometabolum* closely resembles *Mycobacterium album* and strains of the "*aurantiaca*" taxon (Goodfellow et al., 1978; Collins and Jones, 1982b).

d. *Corynebacterium pyogenes* (Glage 1903, 173) Eberson 1918, 23.[AL]

py.o'ge.nes. Gr. n. *pyum* pus. Gr. v. *gennaio* produce. M.L. adj. *pyogenes* pus-producing.

The taxonomic position of *C. pyogenes* has always been controversial. The species bears little similarity to other animal or human corynebacteria and its retention within the genus *Corynebacterium* has been questioned (see Barksdale, 1970; Slack and Gerencser, 1975; Collins et al., 1982c). On the basis of cell wall studies Cummins and Harris (1956) suggested a close relationship between *C. pyogenes* and certain streptococci. This view was supported by Barksdale et al. (1957). Recent numerical phenetic (Schofield and Schaal, 1981) and chemical (Collins et al., 1982c) studies have indicated that *C. pyogenes* is closely related to *Actinomyces bovis*. This organism has recently been reclassified in the genus *Actinomyces* as *A. pyogenes* (Reddy et al., 1982; Collins and Jones, 1982a; see genus *Actinomyces*).

e. *Corynebacterium liquefaciens* (Okabayashi and Masuo 1960, 1087). Lanéelle, Asselineau, Welby, Norgard, Imaeda, Pollice and Barksdale 1980, 544.[VP]

li.que.fa'ci.ens. L. v. *liquefacio* to liquefy; L. part. adj. *liquefaciens* liquefying.

The taxonomic position of this organism has been controversial (see genera *Arthrobacter* and *Brevibacterium*). The organism is treated under *Arthrobacter nicotianae*. The adenyl cyclase from *C. liquefaciens* has been crystallized and studied in detail (Takai et al., 1974; Umezawa et al., 1974).

Addendum II

The following species probably belong to the genus *Corynebacterium* since they all possess a cell wall peptidoglycan based upon *meso*-DAP, an arabinogalactan polymer, and relatively short chain mycolic acids (Schleifer and Kandler, 1972; Collins et al., 1982a).

Arthrobacter variabilis Muller 1961, 524.[AL] *Type strain:* ATCC 15753.

Brevibacterium ammoniagenes (Cook and Keith 1927, 318) Breed 1953, 14.[AL] *Type strain:* ATCC 6871.

Brevibacterium divaricatum Su and Yamada 1960, 74.[AL] *Type strain:* ATCC 14020.

Brevibacterium stationis (ZoBell and Upham 1944, 273) Breed 1953, 14.[AL] *Type strain:* ATCC 14403.

Corynebacterium lilium Lee and Good 1963[AL] is treated under *Corynebacterium glutamicum*.

The organism *Corynebacterium equi* is treated under *Rhodococcus* as *R. equi*.

Further Comments

The taxonomic position of *A. variabilis* is controversial. Keddie and Cure (1977) reported the presence in *A. variabilis* of free mycolic acids which had chromatographic mobilities compatible with its inclusion in the genus *Rhodococcus*. Recent structural studies by Collins et al. (1982a) however, have shown that the mycolic acids of *A. variabilis* are in fact relatively short (30–36 carbon atoms) and are more compatible with those of true corynebacteria or caseobacters. The presence of 10-methyloctadecanoic acid within *A. variabilis* (Collins et al., 1982b), however, indicates a possible relationship with caseobacters or rhodococci (although a few corynebacteria do possess this fatty acid). On the basis of 16S-rRNA oligonucleotide cataloging, *A. variabilis* displays a greater affinity to true corynebacteria than to rhodococci (Stackebrandt and Woese, 1981; Stackebrandt et al., 1980). *Arthrobacter variabilis* is also similar to true corynebacteria and caseobacters in containing *N*-acetyl residues in the glycan moiety of its cell wall peptidoglycan (Collins, unpublished). Future studies should be directed towards a comparison of *A. variabilis*, corynebacteria and caseobacters.

Brevibacterium ammoniagenes contains a DNA base composition of 53.7–54.6 mol% G + C (Yamada and Komagata, 1970) and relatively short mycolic acids (~32–36 carbon atoms, Collins et al., 1982a). *B. ammoniagenes* is quite distinct from *C. diphtheriae, C. flavescens, C. glutamicum, C. lilium, C. renale, C. vitarumen* and *C. xerosis* on the basis of DNA/DNA homology studies (Suzuki et al., 1981) and possibly warrants separate species status. Support for the "distinctiveness" of *B. ammoniagenes* comes from the report (Collins et al., 1982b) of significant amounts of 10-methyloctadecanoic acid within this species. The vast majority of true corynebacteria lack this acid.

Brevibacterium divaricatum is similar to *C. glutamicum* on the basis of DNA base composition, mycolic acid, fatty acid and menaquinone composition (Abe et al., 1967; Yamada and Komagata, 1970; Collins et al., 1979, 1982a, b). DNA/DNA homology studies (Suzuki et al., 1981) indicate that *B. divaricatum* should be reduced to synonymy with *C. glutamicum*.

Keddie and Cure (1977) reported the presence in *B. stationis* of free mycolic acids which had chromatographic mobilities compatible with its inclusion in the genus *Rhodococcus*. Structural studies by Collins et al. (1982a), however, have shown that the mycolic acids of *B. stationis* are in fact relatively short (29–36 carbon atoms) and are more compatible with those of true corynebacteria. Support for the inclusion of *B. stationis* within the genus *Corynebacterium* comes from the report of a relatively low DNA base composition (~54 mol% G + C) (Yamada and Komagata, 1970) within this species. Representatives of the genus *Rhodococcus* generally possess a higher G + C content (~60–70 mol%).

Plant pathogenic species of **Corynebacterium**

M. D. COLLINS AND J. F. BRADBURY

The plant pathogenic coryneform bacteria comprise a diverse collection of Gram-positive, nonsporeforming, rod-shaped bacteria which, over the years, have been consigned to the genus *Corynebacterium* Lehmann and Neumann. It is now generally accepted, however (see Jones, 1975; Keddie and Cure, 1978; Collins et al., 1980; Collins and Jones, 1980; Döpfer et al., 1982), if not universally (Dye and Kemp, 1977; Carlson and Vidaver, 1982), that none of them belongs to the genus *Corynebacterium sensu stricto* but for the present they are here retained under this generic name. Characters which distinguish them are shown later in Table 15.5.

In the eighth edition of *Bergey's Manual* all of the plant pathogenic coryneform bacteria were retained within the genus *Corynebacterium* (Cummins et al., 1974). Recent phenetic and chemotaxonomic studies, however, have shown that of the 14 recognized phytopathogenic "*Corynebacterium*" species (Skerman et al., 1980; Carlson and Vidaver, 1982), 7 can now be accommodated in other genera. *Corynebacterium betae, C. flaccumfaciens, C. oortii* and *C. poinsettiae* have been reclassified in the genus *Curtobacterium* (Collins and Jones, 1983), whereas *C. fascians* has been shown to be a member of the genus *Rhodococcus* (Goodfellow and Alderson, 1977; Collins et al., 1982a; see genus *Rhodococcus*). *Corynebacterum ilicis* has been reclassified in the genus *Arthrobacter*, as *A. ilicis* (Collins et al., 1981) while phenetic and chemical studies have shown *C. beticola* is, in fact, *Erwinia herbicola* (Collins and Jones, 1982).

The taxonomic position of the remaining species (viz *C. iranicum, C. insidiosum, C. michiganense, C. nebraskense, C. rathayi, C. sepedonicum, C. tritici*), however, remains equivocal (Collins and Jones, 1980; Vidaver and Starr, 1981; Carlson and Vidaver, 1982). Other than to note that none belongs to the genus *Corynebacterium* (as defined in this volume see p. 1266) little can as yet be said about their true generic locality. Thus, in the present volume the generic name *Corynebacterium* is retained for these species **only for the purpose of convenience and practicality**—although it must be stressed that **none is an authentic** *Corynebacterium* species.

General characteristics of the group

Gram-positive short rods (~0.4–0.75 × 0.8–2.5 μm). Rods may be straight to slightly curved or wedge shaped; coccoid forms may be observed. **Predominantly single cells but some V, Y and palisade arrangements usually present;** primary branching uncommon. **Non acid-fast.** Endospores are not formed. Nonmotile. Growth on most media (e.g. yeast extract-nutrient agar) is usually slow and poor, giving circular, entire, smooth, convex, semiopaque colonies; growth enhanced by addition of glucose or sucrose. **Nutritionally exacting.** Optimum temperature for growth 21–26°C; maximum between 29–35°C.

Obligately aerobic. Acid produced oxidatively in medium C of Dye and Kemp, 1977 from arabinose, fructose and sucrose, but not from rhamnose, raffinose, adonitol, sorbitol, dulcitol, α- or β-methylglucosides, esculin, dextrin, starch or glycogen. Most isolates also produce acid from glucose, xylose, galactose and glycerol. Citrate, gluconate and fumarate are used as carbon sources for growth, but not benzoate, malonate, galacturonate, oxalate or tartrate. Catalase-positive. Oxidase-, tyrosinase-, urease-, and indole-negative. Nitrate not reduced to nitrite; nitrite not reduced. Ammonia not produced from peptone. Amino acid decarboxylases not produced; phenylalanine deaminase-negative. Pectate not liquefied. H₂S produced from cysteine hydrochloride. No lipolysis of cotton seed oil.

Cell wall peptidoglycan contains diaminobutyric acid (DAB) as the dibasic amino acid. The peptidoglycan is a group B type, characterized by a cross-linkage between the α-carboxyl group of D-glutamic acid in position 2 of the peptide subunit and the C-terminal D-alanine of an adjacent subunit (Schleifer and Kandler, 1972). The peptidoglycan contains L-DAB in position 3 of the peptide subunit and D-DAB in the interpeptide bridge (Fig. 15.1).

Mycolic acids are absent. The nonhydroxylated long chain fatty acids consist of predominantly *anteiso-* and *iso-*methyl branched acids; straight chain saturated acids are present in only small amounts. 12-Methyltetradecanoic (anteiso-$C_{15:0}$), 14-methylhexadecanoic (anteiso-

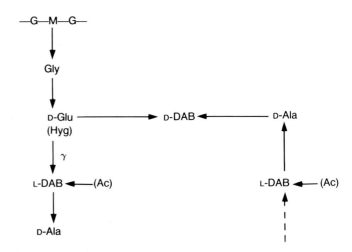

Figure 15.1. Fragment of primary structure of peptidoglycan B2γ (Schleifer and Kandler, 1972). *G*, glucosamine; *M*, muramic acid; *Gly*, glycine; *Glu*, glutamic acid; *Hyg*, hydroxyglutamic acid; *DAB*, diaminobutyric acid; *Ala*, alanine; and *Ac*, acetyl.

$C_{17:0}$) and 14-methylpentadecanoic (*iso*-$C_{16:0}$) acids are the major fatty acids (Collins and Jones, 1980; Bousfield et al., 1983). Menaquinones are the sole respiratory quinones. Unsaturated menaquinones with nine isoprene units (abbreviated MK-9) predominate in *C. michiganense, C. insidiosum, C. nebraskense* and *C. sepedonicum* whereas MK-10 constitutes the major isoprenologue in *C. iranicum, C. tritici* and *C. rathayi* (Collins and Jones, 1980; Collins 1983). Polar lipids consist of diphosphatidylglycerol, phosphatidylglycerol and some characteristic glycosyldiacylglycerols (Collins and Jones, 1980).

The mol% G + C of DNA is within the approximate range of 67–78 (T_m, ρ).

Further Descriptive Information

Cells stain unevenly, often showing deeply stained bands or granules; when grown on glycerol media, fatty inclusions which stain with Sudan black B may be observed (Cummins et al., 1974; Ramamurthi, 1959). Capsules have been reported in some strains of *C. insidiosum, C. michiganense, C. rathayi* and *C. sepedonicum* (Elliott, 1951).

Colonies may have a mucoid or nonmucoid appearance dependent upon the production of extracellular slime. Nonmucoid colonies frequently arise from mucoid forms in cultures of *C. insidiosum* (Fulkerson, 1960) and *C. michiganense* (Strider, 1969).

With the exception of the generally off-white to pale yellow colonies of *C. sepedonicum*, the remaining phytopathogenic corynebacteria are usually pigmented on complex media, in shades of yellow, orange, and pink; colors may vary between strains of a species, age of culture, medium and temperature of incubation. The pigments of *C. michiganense* have been shown to be carotenoids (Saperstein and Starr, 1954; Saperstein et al., 1954). *C. michiganense* was reported to contain lycopene and cryptoxanthin. A red mutant had only lycopene but a pink variant also apparently contained a dimethoxy carotenoid related to lycopene, possibly spirilloxanthin. Cryptoxanthin, β-carotene and a ketonic carotenoid, canthaxanthin, were reported to be present in an orange mutant of *C. michiganense*. The carotenoid content of the other plant pathogenic "corynebacteria" has not been investigated, although the dark blue-gray pigment produced by most strains of *C. insidiosum* in YDC* or YSC† medium (Starr, 1958; Dye and Kemp, 1977) has been shown to be the bipyridyl pigment, indigoidine (Starr, 1958; Kuhn et al., 1965). It is worth noting that in a detailed investigation of a large number of phytopathogenic corynebacteria, Dye and Kemp (1977) concluded that colonial pigmentation on four different media examined by them was of little value in identification. Vidaver (1980), however, is of the opinion that colonial pigmentation on NBY§ agar may be diagnostically useful (*C. nebraskense* orange; *C. michiganense* and *C. insidiosum* yellow; *C. sepedonicum* generally colorless).

Knowledge of nutritional requirements of the plant pathogenic "corynebacteria" is incomplete, but probably all require B-vitamins, and some also require amino acids. Starr (1949) reported that *C. michiganense, C. insidiosum* and *C. sepedonicum* required biotin, nicotinic acid and thiamine when grown in a glucose-mineral salt medium supplemented with vitamin-free casein hydrolysate. Lachance (1960) confirmed these requirements for *C. sepedonicum*, and later (Lachance, 1962) found L-asparagine and L-methionine to be essential, and histidine and leucine stimulatory. A single strain of *C. michiganense* has been reported to require tryptophan (Starr, 1949) whereas L-methionine is required by *C. sepedonicum* (Ikin et al., 1978). Cystine and cysteine have been found to be strongly inhibitory and some other amino acids less inhibitory for *C. sepedonicum* (Paquin and Lachance, 1970; Ikin et al., 1978). All three species have also been reported to require histidine and one or more of adenine, guanine and uracil (Ramamurthi, 1959). Vidaver (1982) reported that the growth of *C. nebraskense* and *C. michiganense* subsp. *tessellarius* in a purified mineral salts medium was greatly stimulated by the same three vitamins:

biotin, nicotinic acid and thiamine, and that these two taxa also require L-methionine. The minimal nutritional requirements of *C. iranicum, C. rathayi* and *C. tritici* are unknown.

Glucose metabolism has been shown to take place in a strain of *C. tritici* primarily via the Embden-Meyerhof-Parnas (EMP) pathway and to a lesser extent by the pentose phosphate (PP) pathway (Zagallo and Wang, 1967). In a single strain of *C. sepedonicum* the PP pathway appeared to play as important a role in glucose catabolism as the EMP pathway; pentose derived from glucose via the PP pathway was catabolized via the pentose cycle (Zagallo and Wang, 1967). There are also reports of the PP pathway in *C. insidiosum* (Zajic et al., 1956). Katznelson (1955, 1958) showed that a single isolate of *C. michiganense* did not metabolize glucose via the Entner-Doudoroff pathway.

Bacteriophages specific to *C. michiganense* are most widely known among those that attack the plant pathogenic "corynebacteria." Ercolani (1968) reported their use to detect the presence of the bacterium in tomato seed. A mutant derived from a phage with a more limited host range was isolated by Wakimoto et al. (1969). It was able to attack all their strains of *C. michiganense*, but not any other species examined. Bacteriophages have also been isolated that attack *C. insidiosum* (Cook and Katznelson, 1960), *C. nebraskense* (Shirako and Vidaver, 1981) and *C. rathayi* (Severin and Docea, 1970). The phages from *C. rathayi* did not attack *C. tritici*. Trofimets and Schneider (1969) found two morphologically distinct phage particles when examining a culture of *C. sepedonicum*. It is not known whether these particles were temperate or defective phage, since no biological activity was reported (Trofimets and Schneider, 1969).

Plasmids of moderate size (approximately 23–53 megadaltons) have been demonstrated in *C. insidiosum, C. michiganense, C. nebraskense, C. rathayi, C. sepedonicum* and *C. tritici* (Gross et al., 1979). A single strain of *C. iranicum* apparently did not contain plasmids (Gross et al., 1979). There is as yet no correlation between the presence of plasmids and bacteriocin production or pathogenicity.

Bacteriocins are produced by the majority of phytopathogenic "corynebacteria" (Echandi, 1976; Gross and Vidaver, 1979b; Carlson and Vidaver, 1982; De Boer, 1982a). Gross and Vidaver (1979b) have devised a typing scheme which may be of some value in strain and species differentiation.

Studies of the serological relationships between the various plant pathogenic corynebacteria form a confusing picture. The most comprehensive study to date is that of Lazar (1968), who investigated the serological relationships of 39 strains of "corynebacteria" from man and plants (22 phytopathogenic) using precipitin, double-gel diffusion, and immunoelectrophoretic techniques. Although a number of cross-reactions occurred, he concluded that the plant pathogenic coryneforms formed five distinct serological groups (see also Masuo and Nakagawa, 1970). Group 2 contained *C. michiganense* and *C. insidiosum*, group 3 *C. tritici* and *C. rathayi*, and group 4 *C. sepedonicum* (Lazar, 1968). *Corynebacterium nebraskense* was later found by Lazar to also belong to group 2 (cited by Schuster et al., 1975). Earlier work using tube agglutination with single strains of five species, including *C. insidiosum, C. michiganense* and *C. rathayi* (Soda and Cleverdon, 1960), and a number of strains of *C. insidiosum, C. michiganense* and *C. sepedonicum* (Rosenthal and Cox, 1953, 1954) also showed that quite strong cross-reactions can occur. Slack et al. (1979), comparing three serological methods for detecting *C. sepedonicum*, found that cross-reactions occurred with agglutination, but not with double-gel diffusion or indirect fluorescent antibody staining. In contrast, Claflin and Shepard (1977) did not observe cross-reactions when agglutination was used to detect *C. sepedonicum*. Cross-reactions have been reported to limit the usefulness of immunofluorescence methods for diagnosis of *C. sepedonicum* by De Boer and associates (DeBoer and Copeman, 1980; Crowley and De Boer, 1982; De Boer, 1982b). All strains of *C. michiganense* and *C.*

---

* YDC medium contains (g/liter): yeast extract 10.0, D-glucose 10.0, CaCO₃ 20.0 and agar 15.0.

† YSC medium contains (g/liter): yeast extract 5.0, sucrose 5.0, CaCO₃ 20.0, and agar 15.0.

§ NBY medium contains (g/liter): Difco nutrient broth 8.0, Difco yeast extract 2.0, K₂HPO₄ 2.0, KH₂PO₄ 0.5; agar 15.0; to which is added after separate sterilization 50 ml 10% glucose and 1.0 ml 1 M MgSO₄·7H₂O.

insidiosum investigated cross-reacted with *C. sepedonicum* (De Boer, 1982b), as well as nonpathogens found in potato tissue (Crowley and De Boer, 1982). Rather less serology seems to have been done with plant pathogenic species other than *C. sepedonicum*. Hale (1972) reported that 16 out of 17 strains of *C. insidiosum* agglutinated with his antiserum. Single strains of six other species of plant pathogenic coryneforms did not react with the *C. insidiosum* antiserum (Hale, 1972). *Corynebacterium michiganense* has been successfully detected in seed using immunofluorescence (Akerman et al., 1973; Trigalet and Rat, 1976).

The organisms are plant pathogens whose infections are normally systemic, or become so when the pathogen reaches the vascular region of the plant. Various symptoms may be produced. *Corynebacterium iranicum*, *C. rathayi* and *C. tritici* all produce yellow slime on leaves and inflorescences, and some distortion of these organs usually results. *Corynebacterium michiganense* subsp. *tessellarius* produces a leaf spot or mottling, and the remaining pathogens, *C. insidiosum*, *C. michiganense*, *C. nebraskense* and *C. sepedonicum*, all produce wilting as a major symptom. Chlorosis, necrosis or stunting may also be evident. Details of the hosts infected are listed under the individual species. Mechanisms of production of symptoms are not known, but most, if not all, involve metabolites (e.g. toxins, polysaccharides, enzymes) (Strobel, 1977; Daly, 1982; Vidaver, 1982 for details).

Host plants are the most usual and, in most cases, virtually the only habitat. Survival in the absence of a suitable host plant or residual materials is normally poor; survival in soil is extremely limited. All are known to survive between crops in or on seed (Richardson, 1979). *Corynebacterium sepedonicum* has been reported to survive on nonhost materials (machinery, sacks, bins) for long periods (Richardson, 1957).

Enrichment and Isolation Procedures

As with most plant pathogenic bacteria, a susceptible host plant is the preferred "enrichment medium." Such susceptible hosts can be used directly to "trap" the pathogen by planting them out in soil suspected of being infested. The method is not very sensitive or quantitative. For enrichment from plant material (perhaps rather old material, unpromising for isolations to be made directly) the procedure is as follows: carefully select the best diseased material (i.e. that showing typical symptoms without undue necrosis or signs of secondary infection) and gently rinse in tap water and then in sterile water. Place in fresh sterile water or buffer and grind or tease apart to allow the bacteria to escape from the tissue. It is usually best at this stage to allow it to stand for an hour or two. It should not, however, stand for too long or more rapidly growing saprophytes may predominate in the resulting suspension. Filter or centrifuge (preferably aseptically) and inoculate the clear filtrate into a suitable host plant. This may be done by hypodermic injection; vacuum infiltration; multiple needle puncture; atomized spray; immersion of washed, cut root tips; or other method that has been successfully used for the pathogen under test. The bacteria are grown on the plants and the appearance of symptoms awaited. Isolation should be easier from the young, growing plants with newly-formed development symptoms. Bacterial numbers should usually be high in such material.

Isolations can usually be made directly from suitable infected material on to nonselective medium. To do this, the above procedure is followed until the suspended plant tissue has stood for an hour or two. Loopfuls of the suspension are then streaked out on to nutrient medium. Suitable media for this purpose are: nutrient agar (Oxoid or Difco) with sucrose added at 5%; or Doepel's medium.* Most strains will grow well on the former, some fail.

Selective media are occasionally useful. CNS agar (Gross and Vidaver, 1979) may be used. It was developed for *C. nebraskense*, but is also useful for *C. michiganense*, *C. tritici*, and some strains of *C. rathayi*. *Corynebacterium insidiosum*, *C. iranicum* and *C. sepedonicum* do not grow on the medium.

Maintenance Procedures

For short term maintenance the plant pathogenic corynebacteria may be held at 15°C as slope cultures on yeast extract-salts agar (YSA) (Dye and Kemp, 1977), nutrient broth-yeast extract agar (NBYA) or Doepel's medium (DM*). Medium term preservation (several years) can be achieved by storing on glass beads at −76°C (Feltham et al., 1978). Long term preservation is by lyophilization (Lelliot, 1965; Vidaver, 1977), after which it is beneficial to store cultures in the dark and at reduced temperature.

Differentiation of the plant pathogenic species of **Corynebacterium** from other coryneform taxa

Table 15.5 provides the primary chemical characteristics that can be used to differentiate the plant pathogenic "corynebacteria" from possibly related actinomycete and coryneform taxa.

Taxonomic Comments

Generic Level

C. michiganense, C. insidiosum, C. iranicum, C. nebraskense, C. sepedonicum and C. tritici. The generic assignment of those plant pathogenic "corynebacteria" which possess a cell wall peptidoglycan based upon diaminobutyric acid (DAB) has always been controversial. On the basis of a recent numerical phentic study of all the phytopathogenic coryneform bacteria, Dye and Kemp (1977) concluded that all, including the DAB-containing taxa, should be retained in the genus *Corynebacterium*. A view upheld by Carlson and Vidaver (1982). This opinion is at variance, however, with all other recent phenetic (Jones, 1975; Minnikin et al., 1978; Keddie and Cure, 1978; Collins and Jones, 1980) and genetic (Starr et al., 1975; Döpfer et al., 1982) data. There is now overwhelming evidence that the genus *Corynebacterium* should be restricted to those species which possess a directly cross-linked peptidoglycan based upon *meso*-diaminopimelic acid (DAP) (variations A1γ), cell wall sugars arabinose and galactose, a DNA base composition within the approximate range of 51–63 mol% G + C, short-chain mycolic acids (~22–36 carbon atoms), straight chain saturated and monounsaturated long chain fatty acids, dihydrogenated menaquinones (MK-8 (H₂), MK-9 (H₂)), phosphatidylinositol and related phosphatidylinositol dimannoside(s). In contrast, the plant pathogens *C. iranicum*, *C. insidiosum*, *C. michiganense*, *C. nebraskense*, *C. sepedonicum* and *C. tritici* possess a cell wall peptidoglycan based upon DAB (variation B2γ), a DNA base composition within the approximate range of 68–76 mol% G + C, straight chain saturated, *iso-* and *anteiso*-methyl-branched chain fatty acids, unsaturated menaquinones (MK-9 or MK-10) but lack a cell wall arabinogalactan polymer, mycolic acids and phosphatidylinositol and related dimannosides. However, other than stress that none of the DAB-containing plant pathogenic coryneforms belong to the genus *Corynebacterium*, there is presently little agreement on where they should be reclassified. On the basis of numerical phenetic and lipid criteria the DAB-taxa closely resemble members of the genus *Curtobacterium*. Further chemical and genetical information is, however, required to ascertain whether or not the DAB-taxa constitute a new genus as suggested by Yamada and Komagata (1972) or whether they should be reclassified in the genus *Curtobacterium*.

It is worth noting that although the plant pathogenic "corynebacteria" (viz. *C. iranicum*, *C. insidiosum*, *C. michiganense*, *C. nebraskense*, *C. sepedonicum*, *C. tritici*) can be clearly distinguished from the plethora

---

* Doepel's medium contains (g/liter): glucose 10.0, casein hydrolysate (not vitamin free) 8.0, yeast extract 8.0, K₂HPO₄ 2.0, MgSO₄·7H₂O 0.2, agar 12.0 (J. Sellwood, personal communication).

Table 15.5.
Differentiation of the plant pathogenic corynebacteria from other actinomycete and coryneform taxa[a,b]

Taxon	Major peptidoglycan diamino acid	Peptidoglycan type	N-glycolyl in glycan moiety of wall	Mol% G + C	Mycolic acids	Fatty acid types[c,d]	Major menaquinone isoprenologue(s)	Polar lipids[d,e]
Plant pathogenic 'coryne-bacteria'								
C. iranicum, C. insidiosum C. michiganense, C. nebraskense C. sepedonicum, C. tritici	DAB	B2γ	–	67–78	–	S, A, I	MK-9, MK-10[f]	DPG, PG, G
C. rathayi NCPPB 797, 2980	DAB[g]	B2γ	ND	70	–	S, A, I	MK-10	DPG, PG, G
"C. rathayi" ATCC 13659	L-Lysine[h]	A3α	ND	64	–	S, A, I	MK-9 (H$_2$)	DPG, PG, PI, G
Other taxa								
Agromyces	DAB	B2γ	ND	71–76	–	S, A, I	MK-12, MK-11	DPG, PG, G
Arthrobacter	L-Lysine	A3α	–	59–66	–	S, A, I	MK-9 (H$_2$)	DPG, PG, PI, DMDG, (DGDG), (MGDG)
Brevibacterium	meso-DAP	A1γ	–	60–64	–	S, A, I	MK-8 (H$_2$)	DPG, PG, DMDG
Caseobacter	meso-DAP	A1γ	–	65–67	30–36 Carbon atoms	S, U, T	MK-9 (H$_2$), MK-8 (H$_2$)	DPG, PI, PIM
Cellulomonas	L-Ornithine	A4β	–	71–75	–	S, A, I	MK-9 (H$_4$)	DPG, PI, PGL
Corynebacterium	meso-DAP	A1γ	–	51–60	22–36 Carbon atoms	S, U, (T)	MK-9 (H$_2$), MK-8 (H$_2$)	DPG, (PG), PI, PIM, (G)
Curtobacterium	D-Ornithine	B2β	–	67–75	–	S, A, I	MK-9	DPG, PG, G
Microbacterium	L-Lysine	B1α or B1β	+	69–75	–	S, A, I	MK-11, MK-12	DPG, PG, DMDG, (MMDG), (PGL)
Rhodococcus	meso-DAP	A1γ	+	60–69	30–64 Carbon atoms	S, U, T	MK-9 (H$_2$), MK-8 (H$_2$)	DPG, PE, PI, PIM, (G)

1279

[a] Data from Schleifer and Kandler 1972; Minnikin et al., 1978; Uchida and Aida, 1977, 1979; Keddie and Jones, 1981; Collins and Jones, 1980; Collins et al., 1980, 1982a, 1983c; Döpfer et al., 1982; Collins, 1982, 1983.

[b] Standard symbols: see Table 15.2.

[c] S, straight-chain saturated; A, anteiso-methyl branched; I, iso-methyl branched; U, monounsaturated; T, 10-methyl branched acid.

[d] Those in parentheses ( ) may be present.

[e] DPG, diphosphatidylglycerol; PG, phosphatidylglycerol; PI, phosphatidylinositol; DMDG, dimannosyldiacylglycerol; DGDG, digalactosyldiacylglycerol; MGDG, monogalactosyldiacylglycerol; PIM, phosphatidylinositol mannoside; PGL, phosphoglycolipid; MMDG, monomannosyldiacylglycerol; PE, phosphatidylethanolamine.

[f] MK-9 predominates in C. michiganense, C. nebraskense, C. insidiosum and C. sepedonicum; MK-10 predominates in C. iranicum, C. tritici and C. rathayi.

[g] Keddie and Cure (1977) reported presence of L-Lysine in NCPPB 797; studies by Collins (1983) on NCPPB 797 have shown the cell wall peptidoglycan is, in fact, based upon DAB.

[h] "C. rathayi" ATCC 13659 is not a legitimate strain of C. rathayi (see Taxonomic Comments).

of coryneform genera (see Table 15.5) on the basis of its rare group B type peptidoglycan based upon DAB, the actinomycete genus *Agromyces* (Gledhill and Casida, 1969) contains an identical peptidoglycan type (Schleifer and Kandler, 1972). This similarity between the DAB-containing phytopathogens and *Agromyces* as indicated by cell wall peptidoglycan has also been demonstrated by DNA base composition and lipid studies. *Agromyces ramosus* is similar to the phytopathogenic coryneforms in possessing DNA which is relatively rich in guanine plus cytosine (~71–76 mol%), primarily *iso*- and *anteiso*-methyl-branched chain fatty acids, polar lipids comprising diphosphatidylglycerol, phosphatidylglycerol and some glycosyldiacylglycerols, and lacking mycolic acids (Collins, 1982; Döpfer et al., 1982). Although *Agromyces ramosus* is also similar to the plant pathogenic "corynebacteria" in possessing unsaturated menaquinones, the presence of unusually long menaquinones (i.e. MK-12) within the former, serves to distinguish these taxa (Collins, 1982). Thus, despite the fact that *Agromyces ramosus* forms a branching mycelium (a characteristic of the family Actinomycetaceae), chemical criteria indicate a possible relationship with the DAB-containing plant pathogens (Collins, 1982; Döpfer et al., 1982).

Most peptidoglycan types found within the Gram-positive eubacteria belong to the peptidoglycan group A (Schleifer and Kandler, 1972), defined by a cross-linkage extending from the ω-amino group of the diamino acid in position 3 of the peptide subunit to the carboxyl group of the C-terminal D-alanine in position 4 of an adjacent peptide subunit. Relatively few Gram-positive eubacteria contain peptidoglycan types of the group B, characterized by a cross-linkage between the α-carboxyl group of D-glutamic acid in position 2 of the peptide subunit and the C-terminal D-alanine of an adjacent subunit (Schleifer and Kandler, 1972). Recent DNA/RNA homology studies (Döpfer et al., 1982) have shown that the DAB-containing plant pathogenic "corynebacteria" along with other coryneform taxa containing group B type peptidoglycans (e.g. *Curtobacterium*, *Microbacterium*) are phylogenetically closely related and are members of one rRNA cistron cluster (differences between the $T_{m(e)}$ values of the homologous and heterologous reaction ($\Delta T_{m(e)}$) spread over the narrow range of ~8°C). Interestingly, the type strain of *Agromyces ramosus* (DSM 43045) is a member of the rRNA cistron similarity cluster (Döpfer et al., 1982), thereby confirming the relationship between this taxon and the plant-pathogenic "corynebacteria" as shown by cell wall and lipid studies (Collins, 1982). Döpfer et al. (1982) were faced with the problem of deciding the taxonomic rank of this phylogenetically "coherent" cluster, and suggested that all coryneform bacteria having a group B type peptidoglycan constitute a single genus, *Microbacterium*. However, while we agree that all coryneforms and actinomycetes which possess a group B type peptidoglycan are phylogenetically closely related, in our opinion this is at the suprageneric (family) level. On the basis of phenetic, cell wall peptidoglycan and lipid composition there are several recognizable groups (i.e. *Curtobacterium*, *Microbacterium*, DAB-phytopathogenic "corynebacteria") among those bacteria which possess a group B-type peptidoglycan (see Table 15.5), which we equate with genera. It therefore seems likely that the rRNA cistron cluster of Döpfer et al. (1982) should be given family status (*Microbacteriaceae*?).

Corynebacterium rathayi. The taxonomic position of *C. rathayi* is even more controversial. Jones (1975) on the basis of numerical phenetic studies, suggested that *C. rathayi* should be removed from the genus *Corynebacterium* and reclassified as an *Arthrobacter* species. This view was supported by the demonstration that the cell wall peptidoglycan of *C. rathayi* (ATCC 13659) was based upon lysine (type L-Lys-L-Thr-L-Ala₃) and not upon *meso*-DAP (variation A1γ) as is found in true corynebacteria (Schleifer and Kandler, 1972; Fiedler et al., 1973). Lysine was later reported to be present in the cell wall of a strain of *C. rathayi* (NCPPB 797) by Keddie and Cure (1977). The presence of DNA relatively rich in guanine plus cytosine (~64–70 mol%) (Starr et al., 1975) and the absence of mycolic acids (Goodfellow et al., 1976) within *C. rathayi* reinforces the case for excluding this species from the genus *Corynebacterium*.

There is some evidence of heterogeneity among strains designated *C. rathayi*. Starr et al. (1975) reported that *C. rathayi* ATCC 13659 possessed a DNA base composition of 63.8 mol% G + C, compared with 70.4 mol% G + C for strain NCPPB 797. *Corynebacterium rathayi* ATCC 13659 has been recently shown to possess dihydrogenated menaquinones with nine isoprene units (MK-9 (H₂)) and polar lipids comprising diphosphatidylglycerol, phosphatidylglycerol, phosphatidylinositol and some glycosyldiacylglycerols (Collins, 1983). The presence of MK-9 (H₂) and phosphatidylinositol within ATCC 13659 is particularly distinctive, and alongside cell wall peptidoglycan (Schleifer and Kandler, 1972; Fiedler et al., 1973) and DNA base composition (Starr et al., 1975) data, suggests this strain is a legitimate *Arthrobacter* species. As mentioned earlier, lysine was reported to be present in the cell wall of *C. rathayi* NCPPB 797 by Keddie and Cure (1977)—a result in accord with the suggestion (Jones, 1975) that this organism should be reclassified in the genus *Arthrobacter*. In the extensive numerical phenetic study of Dye and Kemp (1977) *C. rathayi* NCPPB 797 (CE9) did not show a close affinity with *C. ilicis* (*Arthrobacter ilicis* see Collins et al., 1981), but instead displayed a very high similarity to the DAB-containing plant pathogenic corynebacteria. A re-examination of the cell wall peptidoglycan of NCPPB 797 has, however, revealed DAB as the dibasic amino acid (Collins, 1983). The presence of unsaturated menaquinones with ten isoprene units (MK-10) and a polar lipid composition comprising diphosphatidylglycerol, phosphatidylglycerol and two diglycosyldiacylglycerols within NCPPB 797 (Collins, 1983) reinforces the close relationship between this strain and the other DAB-containing "corynebacteria," and suggests the earlier report (Keddie and Cure, 1977) of lysine in the wall of this strain is in error.

Species Level

There is presently little agreement as to whether or not the various plant pathogenic corynebacteria warrant separate species status. In a recent numerical phenetic study of all the plant pathogenic corynebacteria (Dye and Kemp, 1977), the species *C. michiganense*, *C. insidiosum*, *C. iranicum*, *C. nebraskense*, *C. rathayi*, *C. sepedonicum* and *C. tritici* were recovered in a single phenon (~80% similarity). Despite some physiological and biochemical differences between these taxa, Dye and Kemp (1977) concluded ". . . that they are not sufficient to justify differentiation at the subspecies level." Dye and Kemp (1977), however, did not suggest synonymy within this phenon, but considered host specificity/pathogenic ability to be an important attribute that merits taxonomic distinction. The authors, therefore, proposed that *C. insidiosum*, *C. iranicum*, *C. nebraskense*, *C. rathayi*, *C. sepedonicum* and *C. tritici* be considered as pathogenic varieties (pathovars) of *C. michiganense* (Dye and Kemp, 1977). Carlson and Vidaver (1982) reported that *C. michiganense*, *C. insidiosum*, *C. nebraskense* and *C. sepedonicum* were almost identical on the basis of polyacrylamide gel electrophoresis of cellular proteins. Carlson and Vidaver (1982) concurred with the proposal of Dye and Kemp (1977) to combine these organisms into a single species but considered they were sufficiently distinct to warrant subspecies status (i.e. *C. michiganense* subspp. *michiganense*, *nebraskense*, *insidiosum* and *sepedonicum*). The pathogens causing gummosis of wheat, *C. iranicum*, *C. rathayi* and *C. tritici*, however, differed from one another and from the other plant pathogenic "corynebacteria" in their cellular protein patterns (Carlson and Vidaver, 1982). Carlson and Vidaver (1982), therefore, suggested they should retain separate specific status. Preliminary genetic data also form a confusing picture (Starr et al., 1975; Döpfer et al., 1982). *C. michiganense* and *C. tritici* were shown to be quite separate species from *C. insidiosum* on the basis of DNA/DNA homologies by Starr et al. (1975). In the same investigation *C. insidiosum* and *C. sepedonicum* appeared to be genetically quite closely related (Starr et al., 1975). In the study of Döpfer et al. (1982), however, *C. michiganense* and *C. insidiosum* exhibited a high DNA homology (~74%), whereas, *C. insidiosum* and *C. sepedonicum* were unrelated at the species level (~50%). Taking 70% homology as the borderline for species differentiation, DNA homology values obtained for *C. michiganense* (DSM 20134), *C. nebraskense* (DSM 20400) and *C. sepedonicum* (NCPBB 378) by Döpfer et al. (1982) indicate a close relatedness (~30–50%), although not at the infrasubspecific (Dye and Kemp, 1977) or subspecific (Carlson and Vidaver, 1982) levels, **but at the species level.** This preliminary genetic information (Starr et al., 1975; Döpfer et al., 1982), when used in conjunction with biochemical, physiological, cellular protein and lipid differences (Dye and Kemp, 1977; Collins and Jones, 1980; Collins, 1983; Carlson and Vidaver, 1982), indicate that the phytopathogens, *C. michiganense, C. insidiosum, C. iranicum, C. nebraskense, C. rathayi, C. sepedonicum* and *C. tritici* are probably sufficiently distinct to justify **separate species status.**

Further Reading

Carlson, R.R. and A.K. Vidaver. 1982. Taxonomy of *Corynebacterium* plant pathogens, including a new pathogen of wheat, based on polyacrylamide gel electrophoresis of cellular proteins. Int. J. Syst. Bacteriol. *32:* 315–326.

Collins, M.D. 1983. Cell wall peptidoglycan and lipid composition of the phytopathogen *Corynebacterium rathayi* (Smith). Syst. Appl. Microbiol. *4:* 193–195.

Collins, M.D. and D. Jones. 1980. Lipids in the classification and identification of coryneform bacteria containing peptidoglycans based on 2,4-diaminobutyric acid. J. Appl. Bacteriol. *48:* 459–470.

Döpfer, H., E. Stackebrandt and F. Fiedler. 1982. Nucleic acid hybridization studies on *Microbacterium, Curtobacterium, Agromyces* and related taxa. J. Gen. Microbiol. *128:* 1697–1708.

Dye, D.W. and W.J. Kemp. 1977. A taxonomic study of plant pathogenic *Corynebacterium* species. N. Z. J. Agr. Res. *20:* 563–582.

Vidaver, A.K. 1982. The plant pathogenic corynebacteria. Annu. Rev. Microbiol. *36:* 495–517.

Vidaver, A.K. and M.P. Starr. 1981. Phytopathogenic coryneform and related bacteria. *In* Starr, Stolp, Trüper, Balows, and Schlegel (Editors). The Prokaryotes: A Handbook on Habitats, Isolation and Identification of Bacteria, Springer-Verlag, New York, pp. 1879–1887.

Differentiation and characteristics of the plant pathogenic **Corynebacterium** species

The differential characteristics of the plant pathogenic *Corynebacterium* species are indicated in Table 15.6.

Procedures for Testing Differential Characteristics
(Tables 15.5 and 15.6)

Production of Acid from Carbohydrates and Related Carbon Sources

Acid production may be tested on agar slopes of medium C ($NH_4H_2PO_4$ 0.5 g; K_2HPO_4 0.5 g; $MgSO_4 \cdot 7H_2O$ 0.2 g; NaCl 5 g; yeast extract (Difco) 1 g; bromocresol purple 0.7 ml of 1.6% (w/v) alcoholic solution; agar 12 g; H_2O 1 liter; pH 6.8) with 0.5% (w/v) carbon source (Dye and Kemp, 1977). Carbon sources are added to the autoclaved basal medium (cooled to 48°C) from filter-sterilized concentrated (5–20%) stock solutions. Cultures are examined for growth and acid production after 3, 7, 14, 21, and 28 days.

Utilization of Organic Acids

Utilization of organic acids may be examined in a modified YSA (Dye and Kemp, 1977) (yeast extract reduced to 0.08% (w/v) with bromothymol blue 0.0016% (w/v) as indicator and sodium salt of the organic acid 0.2% (w/v); pH 6.8). Cultures may be examined for growth

Table 15.6.

Characteristics differentiating the plant pathogenic species of **Corynebacterium**[a,b]

Characteristics	1. C. michiganense	2. C. insidiosum	3. C. iranicum	4. C. nebraskense	5. C. sepedonicum	6. C. tritici	7. C. rathayi
Acid production from:							
Mannose	+	+	+	+	d	+	−
Melezitose	−	−	+	−	−	−	−
Mannitol	−	−	−	−	+	+	+
Inulin	−	−	−	−	−	+	−
Utilization of:							
Acetate	+	−	−	+	+	+	−
Lactate	d	−	−	+	−	−	−
Propionate	−	−	0	w	−	−	−
Succinate	+	−	+	+	+	+	+
Gelatin hydrolysis	w	−	−	−	−	−	+
Starch hydrolysis:							
Soluble	d	−	−	d	d	−	d
Potato	+	−	−	+	+	−	+
Methyl red[c]	−	+	−	d	−	−	d
H_2S production from[d]:							
Peptone	+	−	+	d	−	+	+
$Na_2S_2O_3 \cdot 5H_2O$	+	−	d	+	−	−	−
Levan production	−	−	−	+	−	d	+
Blue pigment[e]	d	d	−	−	−	−	−
Growth on CNS[f]	+	−	−	+	−	+	d
Max NaCl tolerance (%)[g]	6	3–4	2	6–7	3	3–4	3
Major menaquinone	MK-9	MK-9	MK-10	MK-9	MK-9	MK-10	MK-10

[a] For methods see p. 1278.

[b] Standard symbols see Table 15.2; also d, different results either between cultures or when repeated; 0, no growth; w, weak reaction.

[c] Method 1 of Dye and Kemp (1977).

[d] All produced H_2S from cysteine hydrochloride (Dye and Kemp, 1977).

[e] Most strains of *C. insidiosum* produce a dark blue-gray somewhat diffusible pigment (indigoidine) on complex media (e.g., yeast extract sucrose-chalk agar and corn meal agar).

[f] CNS agar (Gross and Vidaver, 1979a).

[g] Data from Dye and Kemp (1977).

and pH changes after 3, 7, 14 and 21 days; positive reaction when there is increased alkalinity compared with the same strain inoculated on the basal medium without organic acid and shown by development of a bright blue color (pH about 7.4).

Methyl-Red Test

Shaker incubated cultures (casein hydrolysate 7 g; glucose 5 g; K_2HPO_4 5 g; H_2O 1 liter) may be tested by adding 3–4 drops methyl red (methyl red 0.1 g; ethanol (95%) 300 ml; H_2O 200 ml) after 5 days; a strong red color is indicative of a positive result (Dye and Kemp, 1977).

Gelatin Hydrolysis

Stab inoculations into nutrient gelatin (Difco), incubated at 22°C.

Starch Hydrolysis

Two media may be employed: (a) Yeast extract-nutrient agar (YNA) (yeast extract (Difco) 5 g, nutrient agar (Difco) 23 g, H_2O 1 liter) plus soluble starch (Difco) 0.2% (w/v); and (b) YNA plus potato starch (Difco) 1% (w/v). Plates are spot inoculated in duplicate. After 7–14 days (depending on amount of growth) one colony of each pair is wiped from the agar surface before flooding with dilute iodine solution. Removal of one colony allows detection of weak hydrolysis beneath but not beyond colony edge (Dye and Kemp, 1977).

Hydrogen Sulfide Production

Production of H_2S may be examined by modifying yeast extract-salts broth (YSB) ($NH_4H_2PO_4$ 0.5 g; K_2HPO_4 0.5 g; $MgSO_4 \cdot 7H_2O$ 0.2 g; NaCl 5 g; yeast extract (Difco) 5 g; H_2O 1 liter) (Dye and Kemp, 1977) with sodium thiosulfate 0.05% (w/v) or with peptone (Difco) 0.5% (w/v). H_2S production indicated by darkening of lead-acetate-impregnated paper strips suspended over shaker-incubated cultures (recorded after 3, 5, 7, and 14 days).

Levan Production

Plates of nutrient agar (Difco) supplemented with 5% sucose (w/v) are streaked with bacterial suspension. Development of domed mucoid colonies in 7 days considered to be due to production of levan (Dye and Kemp, 1977).

Growth on CNS

CNS agar (Gross and Vidaver, 1979a). Nutrient broth (Difco), 8 g; yeast extract (Difco) 2 g; K_2HPO_4 2 g; KH_2PO_4 0.5 g; glucose 5 g; $MgSO_4 \cdot 7H_2O$ 0.25 g; lithium chloride 10 g; cycloheximide (Sigma) 40 mg; polymyxin B sulfate (Sigma) 32 mg; Bravo (chlorothalonil, active ingredient, 54%; Diamond Shamrock Corp.; 1:500 dilution) 0.625 ml; agar 15 g, distilled H_2O 1 liter. Glucose (10% w/v) and $MgSO_4 \cdot 7H_2O$ (1 M) autoclaved separated and added aseptically. Nalidixic acid and polymyxin sulfate added to partially cooled autoclaved medium from freshly prepared stock solutions. Cycloheximide also added after autoclaving. The pH after autoclaving is 6.9.

Menaquinone Determination

Menaquinones are extracted from dry cells (~100 mg) with 20 ml chloroform/methanol (2:1 by volume) for 2 h using a magnetic stirrer. The cell-solvent mixture is passed through a filter funnel to remove cell debris, collected in a flask and evaporated to dryness under reduced pressure at 40°C on a rotary evaporator. The lipid extract is resuspended in a few drops of chloroform-methanol and quinones purified by thin layer chromatography using Merck Kieselgel $60F_{254}$ plastic-backed sheets (10×10 cm) with hexane/diethyl ether (85:15 by volume) as developing solvent. Qualitative/semiquantitative quinone patterns may be obtained by reverse-phase partition thin layer chromatography using Merck HPTLC RP-$18F_{254}$ plates (10×10 cm) with acetone as developing solvent (Collins, 1985). Menaquinones are revealed by observing under ultraviolet light (254 nm) or by spraying the plates with 10% dodecamolybdophosphoric acid in ethanol and heating at 80°C for 2–5 minutes. Quantitative menaquinone profiles can be obtained by reverse phase partition high performance liquid chromatography using a ODS or RP-18 column with 1-chlorobutane/methanol (1:10 by volume) as eluting solvent (1.5 ml/min). Quinones are monitored at 270 nm with an ultraviolet detector and quantitation achieved with a computer integrator.

List of plant pathogenic species of **Corynebacterium**

1. **Corynebacterium michiganense** (Smith 1910) Jensen 1934, 47.[AL] (*Bacterium michiganense* Smith, 1910, 794; *Corynebacterium michiganense* pv. *michiganense* (Smith) Dye and Kemp 1977, 578; *Corynebacterium michiganense* subsp. *michiganense* (Smith) Carlson and Vidaver 1982, 322.)

mi.chi.ga.nen'se. M.L. neut. adj. *michiganense* pertaining to Michigan (USA).

The characteristics are those of the group and as listed in Tables 15.5 and 15.6. Causes a vascular wilt, canker, leaf and fruit spot on tomato (*Lycopersicon esculentum*) and some other solanaceous plants. Reported naturally infecting *Solanum nigrum*, *S. pimpinellifolium* and *S. triflorum*. Reported to infect the following when inoculated: *Cyphomandra betacea*, *Nicotiana glutinosa*, *S. mammosum*, *S. muricatum* and *S. nigrum* var. *guineense*.

The mol% G + C of the DNA is 72.5–74.0 (ρ) (Skyring and Quadling, 1970; Starr et al., 1975) and 67.3–74.2 (T_m) (Bousfield, 1972; Cummins et al., 1974; Döpfer et al., 1982).

Type strain: NCPPB 2979.

2. **Corynebacterium insidiosum** (McCulloch 1925) Jensen 1934, 41.[AL] (*Aplanobacter insidiosum* McCulloch 1925, 497; *Corynebacterium michiganense* pv. *insidiosum* (McCulloch) Dye and Kemp 1977, 578; *Corynebacterium michiganense* subsp. *insidiosum* (McCulloch) Carlson and Vidaver 1982, 322.)

in.si.di.o'sum. L. neut. adj. *insidiosum* deceitful, insidious.

The characteristics are as described for the group and as listed in Tables 15.5 and 15.6. Causes a vascular wilt and stunting of lucerne (alfalfa) (*Medicago sativa*) and some other *Leguminosae*. Other reported natural hosts include clover (*Trifolium* spp.), *Medicago falcata*, *Melilotus alba*, *Onobrychis sativa*, and *Lotus corniculatus*. Many wild perennial species of *Medicago* are susceptible when inoculated through wounded roots.

The mol% G + C of the DNA is 71.8–73.8 (ρ) (Starr et al., 1975); 76.1–78.1 (T_m) (Yamada and Komagata 1970; Döpfer et al., 1982).

Type strain: NCPPB 1109.

3. **Corynebacterium iranicum** (ex Scharif 1961) Carlson and Vidaver 1982, 322.[VP] (*Corynebacterium iranicum* Scharif 1961, 21; *Corynebacterium michiganense* pv. *iranicum* (Scharif) Dye and Kemp 1977, 578.)

i.ran'i.cum. M.L. adj. *iranicum* pertaining to Iran.

The characteristics are those of the group and as listed in Tables 15.5 and 15.6. Causes a gumming disease of seed heads of wheat (*Triticum aestivum*). The mol% G + C of the DNA is 68.2–71.2 (ρ) (Starr et al., 1975).

Type strain: NCPPB 2253.

4. **Corynebacterium nebraskense** Vidaver and Mandel 1974, 482.[AL]

ne.bras.ken'se. M.L. neut. adj. *nebraskense* pertaining to Nebraska (USA).

The characteristics are those of the group and those listed in Tables 15.5 and 15.6. Causes a leaf spot ("freckles"), leaf blight and wilt of maize or corn (*Zea mays*).

The mol% G + C of the DNA is 73.5 (ρ) (Vidaver and Mandel, 1974) and 76.6 (T_m) (Döpfer et al., 1982).

Type strain: NCPPB 2581.

5. **Corynebacterium sepedonicum** (Spieckermann and Kotthoff 1914) Skaptason and Burkholder, 1942, 441.^{AL} (*Bacterium sepedonicum* Spieckermann and Kotthoff 1914, 674; *Corynebacterium michiganense* pv. *sepedonicum* (Spieckermann and Kotthoff) Dye and Kemp 1977, 578; *Corynebacterium michiganense* subsp. *sepedonicum* (Spieckermann and Kotthoff) Carlson and Vidaver 1982, 322.)

se.pe.don'i.cum. Gr. n. *sepedon* rottenness, decay; M.L. neut. adj. *sepedonicum* leading to decay.

The characteristics are those of the group and those listed in Tables 15.5 and 15.6. Causes a vascular wilt and tuber rot of potato (*Solanum tuberosum*) (Solanaceae), which is the natural host. By inoculation the following have been reported to show symptoms: *Athenaea* sp., *Lycopersicon esculentum*, *L. pimpinellifolium*, *Solanum antipoviczii*, *S. balleii*, *S. cardiophyllum*, *S. chacoense*, *S. citrullifolium*, *S. commersonii*, *S. corymbosum*, *S. demissum*, *S. endlicheri*, *S. fendleri*, *S. integrifolium*, *S. jujuyense*, *S. mammosum*, *S. melongena*, *S. pampasense*, *S. parodii*, *S. radicans*, *S. tequilense*, *S. thaxcalense*, *S. vavilovii*, *S. verrucosum*, and *S. warscewiczii*. Some of these species, e.g. *S. demissum*, show a high degree of resistance to the disease.

The mol% G + C of the DNA is 71.8–73.0 (ρ) (Starr et al., 1975); 69.8–74.9 (T_m) (Yamada and Komagata 1970; Döpfer et al., 1982).

Type strain: ATCC 33113.

6. **Corynebacterium tritici** (ex Hutchinson 1917) Carlson and Vidaver 1982, 324.^{VP} (*Pseudomonas tritici* Hutchinson 1917, 174; *Corynebacterium michiganense* pv. *tritici* (Hutchinson) Dye and Kemp 1977, 578.)

tri'ti.ci. M.L. gen. n. *tritici* of *Triticum*, generic name of wheat.

The characteristics are those of the group and those shown in Tables 15.5 and 15.6. Causes a gumming disease of leaves, stems and ears of wheat (*Triticum aestivum*). Weed grasses recently reported to be natural host are *Alopecurus monspeliensis*, *Lolium temulentum*, and *Phalaris minor*. *Triticum dicoccum*, *T. durum* and *T. pyramidale* have been successfully inoculated using the nematode *Anguina tritici* as vector.

The mol% G + C of the DNA is 69.6–72.8 (ρ) (Starr et al., 1975).

Type strain: ATCC 11403 (NCPPB 1857).

7. **Corynebacterium rathayi** (Smith 1913) Dowson 1942, 313.^{AL} (*Aplanobacter rathayi* Smith 1913, 926; *Corynebacterium michiganense* pv *rathayi* (Smith) Dye and Kemp, 1977, 578.)

rath'ay.i. M.L. gen. n. *rathayi* of Rathay; named for E. Rathay, Austrian plant pathologist who first isolated the organism.

The characteristics are those of the group and those given in Tables 15.5 and 15.6. Causes a gumming disease of seed heads of cocksfoot grass (*Dactylis glomerata*). *Cynodon dactylon* and *Secale cereale* have been reported as natural hosts, and by inoculating (using the nematode vector *Anguina tritici*) *Triticum aestivum*, *T. dicoccum*, *T. durum* and *T. pyramidale*.

The mol% G + C of the DNA of two strains (NCPPB 797 and ATCC 13659) is, respectively, 70.4 and 63.8 (ρ) (Starr et al., 1975). The latter strain has been shown to be an *Arthrobacter* sp. (see Taxonomic Comments).

Type strain: NCPPB 2980.

Further Comments

Carlson and Vidaver (1982) proposed a new taxon, *C. michiganense* subsp. *tessellarius*, for the causative agent of leaf spot of wheat (bacterial mosaic pathogen). Although the bacterial mosaic pathogen was reported to be virtually identical to *C. michiganense* on the basis of polyacrylamide gel electrophoresis of cellular proteins, the authors considered differences in pigmentation and morphology on NBY medium, tetrazolium salt tolerances, growth on CNS medium, and "overall" bacteriocin production were sufficient to warrant the establishment of a new subspecies of *C. michiganense*. Further studies (e.g. DNA/DNA homologies) are however necessary to establish whether or not *C. michiganense* subsp. *tessellarius* is worthy of separate specific/or subspecific status. The mol% G + C content of DNA of *C. michiganense* subsp. *tessellarius* is ~74.

Type strain: 78181 (ATCC 33566).

Editorial note. After this section was completed, a new genus *Clavibacter* Davis, Gillespie, Vidaver and Harris 1984, 113^{VP} was proposed for all the taxa in this section. The type species is *Clavibacter michiganensis*. The new genus contains *Clavibacter michiganensis* subsp. *michiganensis*, *C. michiganensis* subsp. *nebraskensis*, *C. michiganensis* subsp. *sepedonicus*, *C. michiganensis* subsp. *tessellarius*, *C. iranicus*, *C. rathayi* and *C. tritici*. In addition, Davis et al. (1984) describe a new plant pathogenic species *Clavibacter xyli* containing two subspecies, *C. xyli* subsp. *xyli*, the causative agent of ratoon stunting disease in sugar cane, and *C. xyli* subsp. *cynodontis* which causes bermuda grass stunting disease. Full descriptions of *C. xyli* subsp. *xyli* and *C. xyli* subsp. *cynodontis* are given by Davis et al. (1984). It should be noted that the epithets have been altered to the masculine form, because generic names ending in -*bacter* are masculine (Opinion 3, 1951; International Code of Nomenclature of Bacteria, 1975 p. 84).

Genus **Gardnerella** *Greenwood and Pickett 1980, 170^{VP}**

J. R. GREENWOOD AND M. J. PICKETT

Gard.ne.rel'la. M.L. dim. ending-*ella*⁻L. fem. n. *Gardnerella* named after H. L. Gardner.

Pleomorphic rods ~0.5 μm in diameter and 1.5–2.5 μm in length. Filaments do not occur. No capsules or endospores formed. **Stain Gram-negative to Gram-variable;** the cell walls are laminated. **Nonmotile.** Facultatively anaerobic. Fastidious in growth requirements. **Catalase- and oxidase-negative.** Chemoorganotrophic, having a fermentative type of metabolism. **Acid but no gas** is produced from a variety of carbohydrates including **maltose and starch.** Acetic acid is the major product of fermentation. **Hippurate is hydrolyzed. Human blood, but not sheep blood, is hemolyzed.** Found in the human genital/urinary tract. **Considered to be a major cause of bacterial "nonspecific" vaginitis.** The mol% G + C of the DNA is 42–44 (Bd).

Type species: *Gardnerella vaginalis* (Gardner and Dukes 1955) Greenwood and Pickett 1980, 170.

Further Descriptive Information

The cells appear as small pleomorphic bacilli and coccobacilli (Fig. 15.2) which stain Gram-negative to Gram-variable. They have been described as staining Gram-positive when grown on inspissated serum (Zinnemann and Turner, 1963). They do not give an acid-fast reaction. Sudanophilic inclusions and metachromatic granules occur in the cells.

The cell walls contain the amino acids *N*-acetylglucosamine, alanine, aspartic acid, glutamic acid, glycine, histidine, lysine, methionine, proline, serine, threonine and tryptophan, but diaminopimelic and

---

* Reprinted from volume 1.

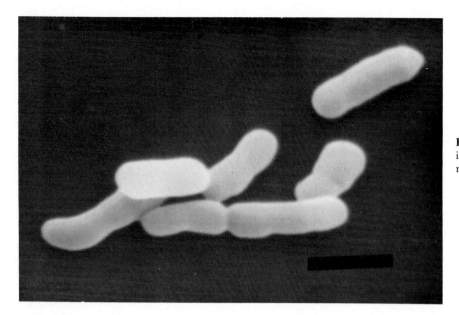

Figure 15.2. Scanning electron micrograph showing *Gardnerella vaginalis* ATCC 14018. Note pleomorphic morphology (*bar*, 1 μm).

teichoic acids have not been detected (Criswell et al., 1971). Cellular fatty acid analysis has shown laurate, myristate, stearate and oleate (Greenwood and Pickett, 1980) and palmitic, stearic and C_{18} monoenoic acids (Moss and Dunkelberg, 1969). Carbohydrate analysis has indicated 6-deoxytalose and no arabinose in the cell walls of the type strain (Vickerstaff and Cole, 1969). A lipopolysaccharide-like fraction is associated with the cell walls.

Electron microscopy of the cell walls has presented conflicting results. Reyn et al. (1966) reported a microscopic morphology, particularly of the cell walls and septa, closely resembling that of a Gram-positive organism. On the other hand, Criswell et al. (1971, 1972) reported that the fine structure of the walls was more typical of a Gram-negative organism.

Although most strains of *Gardnerella* are facultatively anaerobic, obligately anaerobic strains have been described.

No growth or only slight growth occurs on nutrient agar (Piot et al., 1980). No growth occurs on most common selective media (see Table 5.90). On Vaginalis agar* colonies are pinpoint after incubation for 24 h and are 0.4–0.5 mm in diameter after 48 h. The colonies are round, opaque and smooth. They become larger than 0.5 mm after incubation beyond 48 h, but their viability decreases rapidly.

Colonies are nonhemolytic on sheep blood agar but the majority of strains exhibit diffuse β hemolysis on human or rabbit blood (Fig. 15.3). Little or no hemolysis occurs on horse blood.

The optimum growth temperature is 35–37°C. Growth may also occur at 25 and 42°C. The optimum pH range is 6.0–6.5. No growth occurs at pH 4.0 and only slight growth occurs at pH 4.5 (Greenwood and Pickett, 1979). Obligately anaerobic strains have been described (Malone et al., 1975).

Gardnerella strains are fastidious in their nutritional requirements but do not need nicotinamide adenine dinucleotide (V factor), hemin (X factor) or coenzyme-like substances (Dunkelberg and McVeigh, 1969; Edmunds, 1962). They have been reported to require biotin, folic acid, niacin, thiamine, riboflavin and two or more purines/pyrimidines (Dunkelberg and McVeigh, 1969). Growth is improved with fermentable carbohydrates and certain peptones.

The major product of sugar fermentation is acetic acid (Moss and Dunkelberg, 1969), but some strains also produce one or more of the following organic acids: lactic, formic or succinic (Malone et al., 1975). Gas is not formed from sugar fermentation. Tables 15.7 and 15.8 (below) indicate the variety of carbohydrates that can be fermented.

A common antigen has been shown by tube agglutination and by fluorescent antiserum (Redmond and Kotcher, 1963).

All strains are susceptible to ampicillin, carbenicillin, oxacillin, penicillin and vancomycin, and all are uniformly resistant to nalidixic acid, neomycin, colistin and sulfadiazine at usual therapeutic levels. Strains differ in susceptibility to kanamycin, tetracycline, gentamicin and tobramycin (McCarthy et al., 1979; Greenwood, unpublished results). Metronidazole is effective in vivo but gives variable results in in vitro susceptibility tests (Balsdon et al., 1980).

G. vaginalis is believed by many to be the cause of bacterial "nonspecific" vaginitis. It has also occasionally been reported to cause bacteremia in postpartum women and in patients following septic abortion and transurethral resection of the prostate.

G. vaginalis is isolated from the human genital/urinary tract and appears to have worldwide distribution.

Enrichment and Isolation Procedures

Swabs of clinical material are plated on Vaginalis agar within 4–6 h of collection. One quadrant is inoculated with the swab and the plate is streaked for isolation of colonies. Inoculated plates are incubated for 48 h at 35°C in either a candle extinction jar lined with water-saturated absorbent paper or in a CO_2 incubator. *G. vaginalis* forms opaque, domed, entire colonies ~0.5 mm in diameter that are surrounded by a small zone of diffuse β hemolysis. For final purification, a single colony is streaked onto a chocolate agar plate.

Another medium for isolation is peptone-starch-dextrose medium† (Dunkelberg et al., 1970). This medium is inoculated and incubated as outlined above for Vaginalis agar. On this medium, colonies of *G. vaginalis* are 0.5–2.0 mm in diameter, dull white, convex, domed, somewhat conical in shape and entire.

Other methods for isolation of *G. vaginalis* include those described by Smith (1975), Golberg and Washington (1976), Gardner and Dukes (1955) and Mickelson et al. (1977).

---

* Vaginalis agar consists of Columbia agar base (BBL) containing 1% Proteose Peptone No. 3 (Difco). After the medium has been autoclaved and cooled to 45–50°C, human blood (5%, v/v) preserved with citrate, phosphate and dextrose is added aseptically to the medium.

† Peptone-starch-dextrose agar (g/liter): proteose peptone No. 3 (Difco), 20.0; soluble starch, 10.0; dextrose, 2.0; Na_2HPO_4, 1.0; $NaH_2PO_4 \cdot H_2O$, 1.0; agar, 15.0. A more recent formulation of this medium omits the phosphate buffer (W. E. Dunkelberg, personal communication).

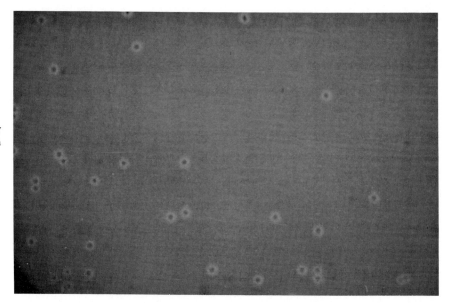

Figure 15.3. Diffuse β hemolysis produced by *Gardnerella vaginalis* colonies on V agar after 48 h incubation (×2).

Maintenance Procedures

Working stock cultures should be transferred every 48 h to ensure viability. This requirement for frequent transfer appears to be independent of the type of maintenance medium or temperature of storage.

Stock strains may be preserved indefinitely by lyophilization in either rabbit serum or 10% skim milk. Strains may also be preserved for at least 5 years by storage at −70°C in glycerol-brucella broth (brucella broth (Difco) supplemented with 15% glycerol). This is most easily accomplished by autoclaving 0.5 ml of broth in a half-dram vial. A dense suspension of organisms harvested from 24- to 48-h-old cultures grown on Vaginalis agar is prepared directly in the vial. After incubation for 20–30 min at room temperature the vial is transferred to a −70°C freezer.

Procedures for Testing Special Characters

Acid production from carbohydrates. The following basal medium is used (g/liter): proteose peptone No. 3 (Difco), 20.0; phenol red, 0.02; agar (Difco), 5.0; pH 7.3. The carbohydrate (10.0 g/liter) is added aseptically to the autoclaved, cooled basal medium from a 10% stock solution which has been sterilized by filtration. The complete medium is dispensed in 3-ml volumes into screw-capped test tubes (13 × 100 mm). The tubes are inoculated with 48-h-old growth by stabbing the medium. The cultures are incubated under an air atmosphere at 35°C for up to 5 days. Acidification is indicated by a yellow color.

Oxidase and catalase tests. For the oxidase test, growth from a 48-h-old chocolate agar plate is smeared with a platinum loop onto a filter paper strip saturated with a solution of 1% tetramethyl-*p*-phenylenediamine dihydrochloride in 0.2% ascorbic acid. For the catalase test, growth from a chocolate agar plate is placed on a glass slide with a wooden applicator stick, overlayed with a drop of 3% H_2O_2 and observed for evolution of bubbles. It is important to note that false-positive catalase reactions have been observed when growth for this test has been taken from media containing human blood (e.g. Vaginalis agar).

Hemolysis. Hemolytic activity of cultures is tested on Vaginalis agar. Blood (e.g. human, sheep, rabbit) is added to the basal medium to provide a 5% (v/v) final concentration.

Hippurate hydrolysis. This test is by the rapid method of Hwang and Ederer (1975).

Differentiation of the genus **Gardnerella** from other genera

Table 15.7 provides the primary characteristics that can be used to differentiate this genus from other morphologically or physiologically similar genera.

Taxonomic Comments

Based on either superficial growth characteristics or morphology, previous studies classified *G. vaginalis* as a member of the genus *Haemophilus* or the genus *Corynebacterium*. Other genera have also been suggested (Lapage, 1974). DNA/DNA hybridization studies by Greenwood and Pickett (1980) and by Piot et al. (1980) have shown that *G. vaginalis* is not closely related to members of the following genera: *Actinobacillus, Bifidobacterium, Branhamella, Brevibacterium, Corynebacterium, Capnocytophaga, Haemophilus, Lactobacillus, Pasteurella, Propionibacterium* and *Streptococcus*. Precise knowledge of broader phylogenetic relationships to other genera or higher taxa will depend on future studies of the ribosomal RNA, i.e. DNA/rRNA hybridization or rRNA oligonucleotide cataloging.

Because of the unusual cell wall of *Gardnerella* and the apparent lack of a genetic relationship to other genera with comparable mol% G + C values, the genus is not presently assignable to any existing family.

Further Reading

Dunkelberg, W.E. 1974. Monograph. A bibliographic review of *Corynebacterium vaginale (H. vaginalis)*, Printing Office, Fort McPherson, Georgia.

Dunkelberg, W.E. 1977. *Corynebacterium vaginale*. Sex. Trans. Dis. *4:* 69–75.

Gardner, H.L. 1980. *Haemophilus vaginalis* vaginitis after twenty-five years. Am. J. Obstet. Gynecol. *137:* 385–391.

Greenwood, J.R. and M.J. Pickett. 1979. Salient features of *Haemophilus vaginalis*, J. Clin. Microbiol. *9:* 200–204.

Greenwood, J.R. and M.J. Pickett. 1980. Transfer of *Haemophilus vaginalis* Gardner and Dukes to a new genus, *Gardnerella; G. vaginalis* (Gardner and Dukes) comb. nov. Int. J. Syst. Bacteriol. *30:* 170–178.

Pheifer, T.A., P.S. Forsyth, M.A. Durfee, H.M. Pollock and K.K. Holmes. 1978. Nonspecific vaginitis: role of *Haemophilus vaginalis* and treatment with metronidazole. N. Engl. J. Med. *298:* 1429–1434.

Piot, P., E. Van Dyck, M. Goodfellow and S. Falkow. 1980. A taxonomic study of *Gardnerella vaginalis (Haemophilus vaginalis)* Gardner and Dukes 1955. J. Gen. Microbiol. *119:* 373–396.

Table 15.7.

Differential characteristics of the genus **Gardnerella** *and other morphologically or physiologically similar genera[a]*

Characteristics	Gardnerella	Capnocytophaga	Actinobacillus	Cardiobacterium	Haemophilus
Oxidase test	−	−	−[b]	+	D
Catalase test	−	+	+	−	D
Cell morphology:					
Pleomorphic	+	−	−	+	−
Short to coccoid	−	−	+	−	+
Fusiform	−	+	−	−	−
Yellow-orange pigmented colonies	−	+	−	−	−
Nitrate reduced to nitrite	−	D	+	−	+
Indole production	−	−	−	Weak	D
Growth on:					
Blood agar, 35°C, in air	+	−	−[b]	d	−
Blood agar, 35°C, in air + 5% CO_2	+	+	+	+	−[c]
Kligler iron agar reactions:					
Slant	NG	K or A	K or A	K or NG	NG
Butt	NG	N or A	A	A or NG	NG
β-Hemolysis, 5% human blood agar	+	−	−	−	−
Acid production from:					
Lactose	d	D	D	−	D
Maltose	+	+	+	+	D
Mannitol	−	−	D	+	D
Mol% G + C of DNA	42–44	33–41	40–43	59–60	38–44

[a] Symbols: +, typically positive; D, differs among species; d, differs among strains of a genus containing only a single species; −, typically negative; A, acid; K, alkaline; N, neutral; NG, no growth.

[b] An occasional strain may be weakly positive.

[c] One species, *H. aphrophilus*, can grow on blood agar.

List of species of the genus **Gardnerella**

1. **Gardnerella vaginalis** (Gardner and Dukes 1955) Greenwood and Pickett 1980, 170.[VP] (*Haemophilus vaginalis* Gardner and Dukes 1955, 963.)

va.gi.na′lis. L. adj. *vaginalis* of the vagina.

The morphology and cultural characteristics are as described for the genus and as depicted in Figures 15.2 and 15.3.

Physiological and nutritional characteristics are as described for the genus and as listed in Tables 15.7 and 15.8.

Considered to be a major cause of bacterial "nonspecific" vaginitis. Isolated from the human genital/urinary tract.

The mol% G + C of the DNA is 42–44 (Bd).

Type strain: ATCC 14018 (strain 594 of Gardner and Dukes, 1955).

Table 15.8.

Physiological characteristics of **Gardnerella vaginalis**[a]

Test	Reaction or Result	Test	Reaction or Result
Indole production	−	Lysine dihydrolase	−
Urease	−	Ornithine decarboxylase	−
ONPG hydrolysis	d	Gluconate oxidized to 2-ketogluconate	−
Voges-Proskauer test	−	Growth on selective media:	
Methyl red test	+	Tellurite (0.01%) agar	−
Phenylalanine deaminase	−	Sodium chloride (3%) agar	−
H_2S production	−	Bile (1%) agar	−
Lipase	d	Rogosa agar	−
Hydrolysis of:		Thayer-Martin agar	−
Hippurate	+	Growth at:	
Tributyrin	−	pH 4	−
Tween 80	−	pH 8	d
Casein	d	25° C	d
Starch	+	30° C	d
Esculin	−	Acid produced from:	
Gelatin	−	Dextrose, dextrin, maltose, ribose, starch	+
H_2O_2 inhibition	+	L-Arabinose, fructose, galactose, insulin, lactose, mannose, sucrose, xylose	d
Benzidine test for cytochromes	−	Arbutin, cellobiose, glycerol, inositol, mannitol, melibiose, raffinose, rhamnose, salicin	−
Nitrite from nitrate	−		
Amino acid decarboxylases (Møller):			
Arginine decarboxylase	−		

[a] Symbols: see Table 15.2 also d, 11–89% of strains are positive; ONPG, o-nitrophenyl-β-D-galactopyranoside.

Genus **Arcanobacterium** Collins, Jones and Schofield 1983, 438^{VP} (Effective publication: Collins, Jones and Schofield 1982, 1280)

M. D. COLLINS AND C. S. CUMMINS

Ar.can.o.bac.te′ri.um. L. adj. arcanus secretive; Gr. neut. dim. n. bakterion a small rod; M.L. neut. n. Arcanobacterium secretive bacterium.

Slender, irregular, bacillary forms predominate during the first 18 h on blood agar plates; some cells exhibit V-formations. Upon extended incubation organisms become granular and segmented, and may resemble small irregular cocci. Both rods and coccoid cells are **Gram-positive, nonacid-fast** and **nonmotile.** Metachromatic granules are not formed. Endospores are not formed. **Facultatively anaerobic.** Growth considerably enhanced in an atmosphere of CO_2. Growth sparse on ordinary media but enhanced by blood or serum. **Optimum growth temperature ~37°C. Catalase reaction is generally negative;** some strains show weak catalase activity. Acid produced from glucose and some other sugars.

The cell wall peptidoglycan is based upon lysine; glutamic acid and alanine are also present. The principal menaquinones are tetrahydrogenated menaquinones with nine isoprene units (MK-9(H_4)). Mycolic acids are not present. The long chain fatty acids are primarily straight chain saturated and monounsaturated acids. The major fatty acids are hexadecanoic, octadecanoic and octadecenoic (ω9) acids.

The mol% G + C of DNA is 48.4–52 (T_m).

Type species: Arcanobacterium haemolyticum (MacLean, Liebow and Rosenberg 1946) Collins, Jones and Schofield 1983, 438.

Differentiation of the genus **Arcanobacterium** from other genera

Table 15.9 provides the primary chemical characteristics that can be used to differentiate the genus Arcanobacterium from other actinomycete and coryneform taxa.

Taxonomic Comments

The generic assignment of Arcanobacterium haemolyticum (formerly "Corynebacterium haemolyticum") has always been controversial. Arcanobacterium haemolyticum was originally isolated from nasopharyngeal and other infections among American soldiers in the Pacific, and assigned to the genus Corynebacterium by MacLean et al. (1946). However, the species bears little similarity to other animal or human corynebacteria and its placement in the genus Corynebacterium was questioned by several workers (Cummins and Harris, 1956; Barksdale et al., 1957; Rogosa et al., 1974; Jones, 1975; Schofield and Schaal, 1981; Collins et al., 1982a, b). Further, the relationship of "C. haemolyticum" to the species Corynebacterium (Actinomyces) pyogenes remained unclear. On the basis of cell wall composition, Cummins and Harris (1956) suggested that both "C. haemolyticum" and C. pyogenes were very closely related to the streptococci. This view was upheld by Barksdale et al. (1957) who suggested not only that "C. haemolyticum" and C. pyogenes should be reclassified in the genus Streptococcus, but also that "C. haemolyticum" was a mutant form of C. pyogenes. Numerical phenetic (Schofield and Schaal, 1981) and chemotaxonomic (Collins et al., 1982a) studies have, however, revealed that "C. haemolyticum" and C. pyogenes are in fact two quite distinct taxa. Corynebacterium pyogenes has now been reclassified in the genus Actinomyces (Collins and Jones, 1982; Reddy et al., 1982) whereas "C. haemolyticum" was assigned to the genus Arcanobacterium.

Isolation Procedures

We are not aware of any methods designed specifically for the isolation of Arcanobacterium haemolyticum. However, samples of infected material(s) can be emulsified under aseptic conditions, streaked out onto blood agar plates (5% horse blood agar layered on a nutrient or digest agar base) and incubated at 37°C. Growth may be enhanced by CO_2. It is worth noting that hemolysis is only partial after 24 h becoming complete (3–4 × diameter of colony) at 48 h. Suspected colonies can be picked off, purified and examined microscopically and biochemically.

Growth of A. haemolyticum in liquid media is generally poor. Growth may be considerably enhanced, however, by the addition of serum. A suitable liquid medium consists of brain heart infusion broth supplemented with 5% horse serum.

Maintenance Procedures

For short term maintenance screw-capped bottles containing slopes of brain heart infusion agar or Dorset egg. Medium term preservation (several years) can be achieved by storing on beads at −76°C (Feltham et al., 1978). Long term preservation is by lyophilization.

Presently, little can be said regarding the family or suprageneric relationships of the genus Arcanobacterium. However, it is apparent that chemical criteria do not support the view (Cummins and Harris, 1956; Barksdale et al., 1957; Cummins et al., 1974) that A. haemolyticum is closely related to streptococci. Arcanobacterium haemolyticum possesses major amounts of oleic acid ($C_{18:1}$,ω9). In contrast, members of the "lactic acid group of bacteria" (e.g. Streptococcus) generally possess cis-vaccenic acid ($C_{18:1}$,ω7). The presence of respiratory quinones and of DNA which is relatively rich in G + C (~50 mol%) within A. haemolyticum also clearly distinguishes this taxon from streptococci. Members of the genus Streptococcus generally lack respiratory quinones and contain a relatively low G + C content (~33–42 mol%). Although it would be premature to suggest any higher grouping for the genus Arcanobacterium, in terms of biochemical and chemical similarity, it probably has a greater affinity with the genus Actinomyces and other anaerobic/facultative actinomycetes than with the "Coryneform group of bacteria." Precise knowledge of the phylogenetic relationships of the genus Arcanobacterium will require a comparative study of its 16S-rRNA oligonucleotide sequences.

Further Reading

Collins, M.D., D. Jones, R.M. Kroppenstedt and K.H. Schleifer. 1982. Chemical studies as a guide to the classification of Corynebacterium pyogenes and "Corynebacterium haemolyticum." J. Gen. Microbiol. 128: 335–341.

Collins, M.D., D. Jones and G.M. Schofield. 1982. Reclassification of "Corynebacterium haemolyticum" (MacLean, Liebow and Rosenberg) in the genus Arcanobacterium gen. nov. as Arcanobacterium haemolyticum nom. rev., comb. nov. J. Gen. Microbiol. 128: 1279–1281.

Schofield, G.J. and K.P. Schaal. 1981. A numerical taxonomic study of members of the Actinomycetaceae and related taxa. J. Gen. Microbiol. 127: 237–259.

List of species of the genus **Arcanobacterium**

1. **Arcanobacterium haemolyticum** (MacLean, Liebow and Rosenberg 1946) Collins, Jones and Schofield 1983, 438.^{VP} (Effective publication: Collins, Jones and Schofield, 1982, 1280.) (Corynebacterium haemolyticum Maclean, Liebow and Rosenberg 1946, 69.)

hae.mo.ly′ti.cum. Gr. n. haema blood; Gr. adj. lyticus dissolving; M.L. neut. adj. haemolyticum blood dissolving, haemolytic.

Surface colonies on blood agar are small (~0.75 mm in diameter) after 24 h, becoming larger (1.5–2.5 mm in diameter) on extended incubation. Colonies are circular, discoid and slightly raised, and β-haemolytic. No growth apparent after 24 h at 37°C on nutrient agar, but after 72 h colonies are ~0.5 mm in diameter. Colonies are low, convex, slightly opalescent, shining and butyrous. Growth considerably enhanced by addition of serum.

On blood agar plates slender, irregular, bacillary forms predominate

Table 15.9.

Differential characteristics of the genus **Arcanobacterium** *and some other actinomycete and coryneform genera*[a, b]

Taxon	Peptidoglycan diamino acid[c]	Mol% G + C	Mycolic acids	Fatty acid types[d]	Menaquinone types
Actinomyces bovis/pyogenes group	L-Lysine	58–63	−	S, U	MK-10(H_4)
Arcanobacterium	L-Lysine	48–52	−	S, U	MK-9(H_2)
Agromyces	DAB	71–76	−	S, A, I	MK-12
Arthrobacter	L-Lysine	59–66	−	S, A, I	MK-9(H_2)
Brevibacterium	*meso*-DAP	60–64	−	S, A, I	MK-8(H_2)
Caseobacter	*meso*-DAP	65–67	30–36 Carbon atoms	S, U, T	MK-9(H_2), MK-8(H_2)
Cellulomonas	L-Ornithine	71–75	−	S, A, I	MK-9(H_4)
Corynebacterium	*meso*-DAP	51–60	22–36 Carbon atoms	S, U, (T)	MK-8(H_2), MK-9(H_2)
Curtobacterium	D-Ornithine	67–75	−	S, A, I	MK-9
Microbacterium	L-Lysine	69–75	−	S, A, I	MK-11, MK-12
Mycobacterium	*meso*-DAP	62–70	60–90 Carbon atoms	S, U, T	MK-9(H_2)
Nocardia	*meso*-DAP	64–69	46–60 Carbon atoms	S, U, T	MK-8(H_4)
Oerskovia	L-Lysine	71–76	−	S, A, I	MK-9(H_4)
Propionibacterium	*meso*-/LL-DAP	57–67	−	S, A, I, (U)	MK-9(H_4)
Rhodococcus	*meso*-DAP	60–69	30–64 Carbon atoms	S, U, T	MK-8(H_2), MK-9(H_2)

[a] Data from Schleifer and Kandler, 1972; Minnikin et al., 1978; Collins and Jones, 1981; Keddie and Jones, 1981; Collins et al., 1982a, b; Döpfer et al., 1982.

[b] Symbols: see Table 15.2.

[c] DAB, diaminobutyric acid; DAP, diaminopimelic acid.

[d] S, straight chain saturated; U, monounsaturated; A, *anteiso*-methyl-branched; I, *iso*-methyl-branched; T, 10-methyl-branched; ( ), may be present.

during the first 18 h; some cells exhibit V-formations. As growth proceeds organisms become granular and segmented so that they resemble small and irregular cocci. On Loeffler's medium they maintain a bacillary form, but become pleomorphic at 48 h, with numerous club forms. Both rods and coccoid cells are Gram-positive, nonacid-fast and nonmotile.

Facultatively anaerobic; growth enhanced in an atmosphere of CO_2. The organism will not withstand heating at 60°C for 15 min. Catalase reaction is generally negative although some strains show weak catalase activity. Cytochrome b is present; the formation of cytochrome b_1 is dependent on the presence of hemin in growth medium (Mára et al., 1973). Acid produced from glucose, dextrin, lactose, maltose, galactose and fructose. Acid not produced from arabinose, cellobiose, dulcitol, glycerol, inositol, melibiose, melezitose, rhamnose, raffinose, sorbitol, sucrose, xylose or erythritol. Acetic, lactic and, to a lesser extent, succinic acids are end products of glucose fermentation. Esculin, gelatin and casein not hydrolyzed. Starch hydrolyzed by some strains. Nitrate reduced to nitrite by most strains. Indole-negative. Extracellular DNase

is produced. Urease-negative. Also produces neuraminidase and acyl-neuraminate pyruvate-lyase (Müller, 1973). No growth in presence of sodium azide (0.005% w/v), potassium tellurite (0.01% w/v) or sodium selenite (0.01% w/v). β-Galactosidase and N-acetyl-β-glucosaminidase are produced. The organism is resistant to oxytetracycline (30 μg/disk) but sensitive to nalidixic acid (30 μg), sulfamethoxazole trimethoprim (25 μg) and amikacin (10 μg).

Pathogenicity: may cause pharyngeal infections and skin lesions in man; occasional cause of infections in farm animals.

Lysine is the dibasic amino acid of the cell wall peptidoglycan. Cell walls contain an antigenic polysaccharide in which rhamnose is a major constituent (Barksdale et al., 1957). The principal menaquinones are MK-9(H_4). Mycolic acids are not present. The major fatty acids are hexadecanoic, octadecanoic and octadecenoic ($C_{18:1}$,ω9) acids.

The mol% G + C of the DNA is 48.4–52.0 (T_m) (Julack et al., 1978; Collins et al., 1982b).

Type strain: ATCC 9345 (NCTC 8452).

Genus **Arthrobacter** *Conn and Dimmick 1947, 300*[AL]

RONALD M. KEDDIE, M. D. COLLINS AND DOROTHY JONES

Ar.thro.bac′ter. Gr. n. *arthrus* a joint; M.L. masc. n. *bacter* the masculine equivalent of the Gr. neut. n. *bactrum* a rod; M.L. masc. n. *Arthrobacter* a jointed rod.

A marked rod-coccus growth cycle occurs during growth in complex media (Fig. 15.4). Stationary phase cultures (generally 2–7 days) are composed entirely or largely of coccoid cells, ~0.6–1.0 μm in diameter. On transfer to fresh, complex medium the coccoids swell and produce one or more outgrowths, thus giving rise to the irregular rods characteristic of exponential phase cultures. Some cells are arranged at an angle to give V-formations but more complex angular arrange-

ments occur. Primary branching may occur but true mycelia (showing secondary branching) are not produced. The cells are thus variable in size but are generally ~0.6–1.2 μm in diameter. The rods become shorter as growth continues and are replaced eventually by the coccoid forms characteristic of stationary phase cultures. **Both rod and coccoid forms are Gram-positive but may be readily decolorized.** Not acid-fast. Do not form endospores. The rods are nonmotile or occasion-

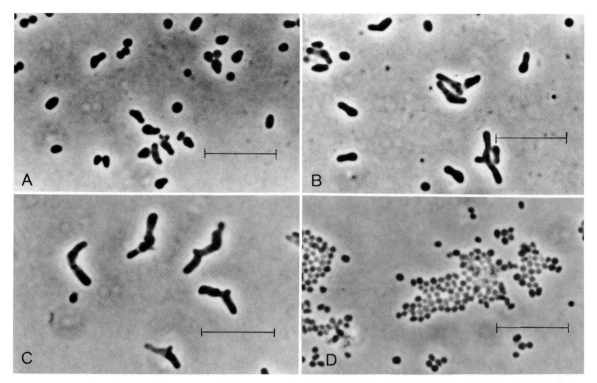

Figure 15.4. *Arthrobacter globiformis* (ATCC 8010) when grown on medium EYGA at 25°C; inoculum coccoid cells as in part *D. A*, after 6 h, showing outgrowth of rods from coccoid cells; *B*, after 12 h; *C*, after 24 h; and *D*, after 3 days (*bar*, 10 μm). (Reproduced with permission from R. M. Keddie and D. Jones. *In* Starr, Stolp, Trüper, Balows and Schlegel (Editors), *The Prokaryotes: A Handbook on Habitats, Isolation and Identification of Bacteria*, Springer-Verlag, New York, p 1848, 1981.)

ally motile. **The cell wall peptidoglycan contains lysine as the diamino acid. Obligately aerobic.** Optimum temperature 25–30°C. **Do not survive heating at 63°C for 30 min in skim milk.** Growth moderate to good on yeast extract-peptone media at near neutral pH.

Chemoorganotrophic: **metabolism respiratory, never fermentative. Little or no acid is produced from glucose and other sugars in peptone media.** Do not hydrolyze cellulose. Catalase-positive. DNase produced. Gelatin usually hydrolyzed. Most species are nonexacting or require biotin as the only organic growth factor. The mol% G + C of the DNA is ~59–70 (T_m).

Type species: *Arthrobacter globiformis* (Conn) Conn and Dimmick 1947, 301.

Further Descriptive Information

The rod-coccus growth cycle (Fig. 15.4) is a distinctive feature of the genus but also occurs in other genera such as *Brevibacterium* and in at least some members of *Rhodococcus*. The growth cycle is markedly dependent on the nutritional status of the medium (see Keddie, 1974 for references) and a suitable medium such as that described by Cure and Keddie (1973) must be used for morphological studies. In some cultures, larger coccoid cells some 2–4 times the size of the remainder may occur and may predominate, particularly when the carbon/nitrogen ratio is wide (Stephenson, 1963; Mulder et al., 1966). The large coccoid cells have in the past been referred to as "cystites," and have sometimes been considered to be specialized spore structures because initial outgrowth of slender rods from two or three places on the cell gives the appearance of germ tubes emerging from a spore. However, Stephenson (1963) considered that so-called cystites were really morphologically aberrant forms resulting from growth in conditions of nutritional stress and found no evidence to support the view that they were specialized spore structures. For more detailed accounts of morphology and morphogenesis in *Arthrobacter* see Luscombe and Gray (1971), Clark (1972), Keddie (1974) and Duxbury and Gray (1977).

On yeast-peptone agar, colonies are about 3–5 mm in diameter and are usually cream or buff in color, but in some species are some shade of yellow. A few species produce blue pigments but usually only on special media.

The cell wall peptidoglycan contains lysine as diamino acid (see Keddie and Cure, 1978 for references). The walls contain a group A peptidoglycan, i.e. one in which the cross-linkage is between positions 3 and 4 of the two peptide subunits by a number of different interpeptide bridges, depending on species (Schleifer and Kandler, 1972). However, two groups referred to as the A3α and the A4α variations (Schleifer and Kandler, 1972) occur within these numerous peptidoglycan types. In the A3α variation, found in *A. globiformis, A. citreus* (type strain) and most other *Arthrobacter* spp. the interpeptide bridge contains only monocarboxylic acids and/or glycine. In the A4α variation, on the other hand, found in *A. nicotianae*, some strains wrongly named *A. citreus* and a few other species formerly bearing the name *Brevibacterium* (see Schleifer and Kandler, 1972; Stackebrandt et al., 1983 for list), the interpeptide bridge always contains a dicarboxylic acid and in most strains also contains alanine (Schleifer and Kandler, 1972).

Two groups can also be distinguished within the genus (*Arthrobacter sensu stricto* as defined by Keddie and Jones, 1981) according to their lipid composition. Those species which, like *A. globiformis* and the type strain of *A. citreus*, have a peptidoglycan with the A3α variation, contain dihydrogenated menaquinones with nine isoprene units (MK-9(H_2)) as their major isoprenoid quinones (Yamada et al., 1976; Collins et al., 1979, 1981; Collins and Jones, 1981). In addition, all the members of this group of species which have been studied contain the polar lipids diphosphatidylglycerol, phosphatidylglycerol and phosphatidylinositol and, with the exception of the type strain of *A. citreus*, they also contain several glycolipids (Shaw and Stead, 1971; Kostiw et al., 1972; Collins et al., 1981). In *A. globiformis, A. crystallopoietes,* and *A. pascens* the glycolipids have been shown to correspond to monogalactosyldiacylglycerol, digalactosyldiacylglycerol and dimannosyldiacylglycerol

(Shaw and Stead, 1971; Kostiw et al., 1971). *A. citreus* (type strain) differs from the other species in this group in containing fewer glycolipids; it contains diphosphatidylglycerol, phosphatidylglycerol, phosphatidylinositol and a glycolipid with the chromatographic mobility of dimannosyldiacylglycerol (Collins and Kroppenstedt, 1983).

On the other hand, those species which like *A. nicotianae* have a peptidoglycan with the A4α variation always contain fully unsaturated menaquinones. Most species with this variation (*A. sulfureus* and *A. citreus* ATCC 15170 (*A. sulfureus*) excepted) have menaquinones with eight isoprene units (MK-8) as major components but with substantial amounts (~25%) of unsaturated menaquinones with nine isoprene units (MK-9) present as well (Yamada et al., 1976; Collins and Jones, 1981; Collins and Kroppenstedt, 1983). All strains with an A4α peptidoglycan that have been studied also contain the polar lipids diphosphatidylglycerol and phosphatidylglycerol but not phosphatidylinositol. With the exception of *A. sulfureus* ("*Brevibacterium sulfureum*") and *A. citreus* ATCC 15170 (*A. sulfureus*) they also contain a diglycosyldiacylglycerol with the chromatographic mobility of dimannosyldiacylglycerol (Collins and Kroppenstedt, 1983). However, *A. sulfureus* ("*Brevibacterium sulfureum*") contains MK-9 as the major menaquinone (Yamada et al., 1976; Collins et al., 1979) and *A. citreus*, ATCC 15170 (*A. sulfureus*), contains unsaturated menaquinones with comparable amounts of nine and ten isoprene units (MK-9, MK-10) (Collins and Kroppenstedt, 1983). The latter strains also differ from *A. nicotianae* and similar species in containing a diglycosyldiacylglycerol with the chromatographic mobility of digalactosyldiacylglycerol (Collins and Kroppenstedt, 1983).

All species of the *A. globiformis*/*A. citreus* group and of the *A. nicotianae* group contain similar fatty acids. All contain major amounts of iso- and anteiso-methyl-branched acids together with small amounts of straight chain saturated acids. The major fatty acid present is 12-methyltetradecanoic acid; monounsaturated acids are either absent or present only in trace amounts (Minnikin et al., 1978; Collins and Kroppenstedt, 1983; Suzuki et al., 1983a).

Most species are nonexacting or require biotin as the only organic growth factor. *A. citreus* (type strain) differs from all other species in having a complex nutrition: B-vitamins, amino acids and a siderophore are required (see Keddie, 1974).

A characteristic feature of those species which do not require vitamins or which require only biotin (i.e. probably all except *A. citreus*) is their nutritional versatility: of those whose carbon nutrition has been examined in detail, all can use a wide and varied range of substrates as sole or major carbon + energy sources (Keddie, 1974; Keddie and Jones 1981). The detailed carbon nutrition, usually of single representatives, of the following species was examined by Owens and Keddie, and Cure and Keddie (unpublished data): *A. globiformis*, *A. aurescens*, *A. ureafaciens*, *A. atrocyaneus*, *A. ramosus*, *A. crystallopoietes*, *A. histidinolovorans*, *A. nicotianae*, *A. oxydans*, *A. polychromogenes* (considered a subspecies of *A. oxydans* by Stackebrandt and Fiedler, 1979), *A. ilicis* (*Corynebacterium ilicis*) and *A. sulfureus* ("*Brevibacterium sulfureum*"), together with some 26 unnamed isolates. The following substrates were utilized by some 90% or more of the strains tested: D-xylose, D-glucose, D-mannose, D-galactose, D-fructose, cellobiose, maltose, trehalose, sucrose, raffinose, melezitose, D-gluconolactone, salicin, acetate, propionate, pentanoate, heptanoate, succinate, fumarate, DL-lactate, DL-malate, citrate, pyruvate, oxaloacetate, glycerol, mannitol, m-hydroxybenzoate, p-hydroxybenzoate, uric acid, glycine, L-α-alanine, D-α-alanine, L-isoleucine, L-threonine, L-lysine, L-arginine, L-aspartate, L-glutamate, L-phenylalanine, L-tyrosine, L-proline, L-histidine, 1,4-butanediamine, agmatine, tyramine, betaine, sarcosine and creatine. *A. citreus* (type strain) is less nutritionally versatile than the species mentioned above (Keddie, 1974). Confirmation of the nutritional versatility of most arthrobacters comes from their frequent isolation from enrichment cultures containing a variety of unusual organic compounds in otherwise mineral media (see Keddie and Jones, 1981, for examples).

Most *Arthrobacter* spp. grow in the range ~10–35°C: many strains also grow at 5°C and a few at 37°C. However, obligately psychrophilic strains of an organism considered to be a new species of *Arthrobacter* ("*A. glacialis*") were described by Moiroud and Gounot (1969): they had a growth range of ~−5 to +20°C.

The fragmentary information available on metabolic pathways in *Arthrobacter* spp. has been reviewed by Krulwich and Pelliccione (1979). In summary, those few species which have been studied fall into two groups with respect to the pathways of carbohydrate dissimilation utilized. In *A. globiformis*, *A. ureafaciens*, *A. crystallopoietes* and "*A. pyridinolis*," that used primarily is the Embden-Meyerhof-Parnas (EMP) pathway and, to a smaller extent, the hexose monophosphate (HMP) pathway. In contrast, *A. pascens* and *A. atrocyaneus* use the Entner-Doudoroff and HMP pathways (Krulwich and Pelliccione, 1979). Few data are available on bacteriophages active on *Arthrobacter* spp.: Einck et al. (1973) have summarized some of the available information.

A large number of studies has shown that soil is the usual habitat of authentic members of the genus *Arthrobacter*. Thus arthrobacters have commonly been shown to form a numerically important fraction of the indigenous flora of soils from various parts of the world and may be the most numerous bacterial group encountered in aerobic "total" plate counts (e.g. Mulder and Antheunisse, 1963; Skyring and Quadling, 1969; Holm and Jensen, 1972; Lowe and Gray, 1972; Hagedorn and Holt, 1975a). However, they become less numerous with increasing soil acidity (Lowe and Gray, 1972; Hagedorn and Holt, 1975a). Psychrophilic and psychrotrophic arthrobacters have been reported to predominate in subterranean cave silts (Gounot, 1967) and were also isolated from glacier silts (Moiroud and Gounot, 1969). They have also been isolated from sewage (Nand and Rao, 1972) and activated sludge from municipal sewage (Seiler et al., 1980). The arthrobacters reported to occur in other habitats such as fish, dairy waste-activated sludge, poultry deep litter, etc., were identified largely on the basis of their morphological features. In many such cases subsequent, more detailed, investigations have revealed that most of the isolates were not members of the genus *Arthrobacter* but belonged either to the genus *Rhodococcus* or the genus *Brevibacterium* (Keddie and Jones, 1981). It is also worth noting here that many of the species now recognized were created for single strains which possessed some unusual feature such as the ability to use a particular substrate or to produce an unusual pigment (Keddie and Jones, 1981). Therefore, in many cases, the species now recognized do not represent the most common arthrobacters in the habitats from which they were isolated. It is thus a common experience to find that new isolates from soil do not resemble the named strains used as reference cultures (Keddie et al., 1966; Skyring and Quadling, 1969; Hagedorn and Holt, 1975b).

Isolation and Enrichment Procedures

Arthrobacters have been isolated most commonly from soil and similar habitats by plating on a suitable, nonselective medium followed by picking from a large, random selection of colonies and identifying as arthrobacters those isolates which show a rod-coccus growth cycle (e.g. Skyring and Quadling, 1969; Holm and Jensen, 1972; Lowe and Gray, 1972). Mulder and Antheunisse (1963) used a method in which isolates which were coccoid in the stationary phase in a "poor" medium were screened for those which when transferred to a "rich" medium, showed outgrowth into irregular rods in the exponential phase. However, all such methods suffer from the disadvantage that they are applicable only to those habitats in which arthrobacters are among the dominant, cultivable, flora. Also, if identification of the isolates is based only or largely on the possession of a rod-coccus growth cycle, as has usually been the case, then bacteria from other genera such as *Rhodococcus* and *Brevibacterium* may be mistaken for arthrobacters.

Hagedorn and Holt (1975a) devised a medium said to be selective for arthrobacters in soil. When nine different soils were plated on the selective medium, on average 74% of the colonies that developed were identified by morphological criteria as arthrobacters. The numbers of arthrobacters estimated by using the selective medium were similar to those obtained by using a nonselective medium and were several times higher than those obtained using the poor medium of Mulder and Antheunisse (1963) (Hagedorn and Holt, 1975a).

Isolation Using Nonselective Media

Most of the nonselective media used have been based on soil extracts and give the maximum counts of soil bacteria capable of growing in aerobic conditions. It is important that the media used should contain low concentrations of carbon + energy sources in order to minimize antagonistic effects between colonies but at the same time they must supply the organic growth factors and mineral components required by a diverse, indigenous, bacterial population. Lochhead, who pioneered the use of such media, strongly advocated the use of soil extract agar without any other additions (Lochhead and Burton, 1956) while others have obtained higher counts with some soils by incorporating low concentrations of yeast extract and glucose (Jensen, 1968).

Examples of both kinds of soil extract agars are given below. The growth of fungi on plates may be suppressed by incorporating nystatin (50 µg/ml) and cycloheximide (50 µg/ml) in the medium (Williams and Davies, 1965). Other factors important in plating soil samples have been fully discussed by Jensen (1968) and the more important of these have been summarized by Keddie and Jones (1981).

Diluent. It is essential that a suitable diluent is used. Winogradsky's standard salt solution was recommended by Jensen (1968) and has the following composition (Holm and Jensen, 1972) (g/1000 ml deionized water): K_2HPO_4, 0.25; $MgSO_4$, 0.125; NaCl, 0.125; $Fe_2(SO_4)_3$, 0.0025; $MnSO_4$, 0.0025; pH 6.5–6.7.

Dilute peptone solutions (0.05–0.1%) and soil extract have also been used successfully as diluents. Owens and Keddie (1969) found that their chelated mineral salts solution, but without $(NH_4)_2SO_4$ (mineral base E–N), was a good diluent for use with coryneform bacteria (see Cure and Keddie, 1973, for the preparation of mineral base E–N).

Incubation. Plates should be incubated for a minimum of 2 weeks at 25°C.

Soil extract agar (Lochhead and Burton, 1957) has the following composition (g/1000 ml soil extract): K_2HPO_4, 0.2; agar, 15 g; final pH, 6.8. Autoclave at 121°C for 20 min. Colonies are picked into tubes of soil extract semisolid medium of the following composition (g/1000 ml soil extract): K_2HPO_4, 0.2; yeast extract, 1.0; agar, 3.0; final pH, 6.8. Autoclave at 121°C for 20 min. Soil extract may be prepared by autoclaving 1 kg of fertile garden soil in 1 liter of tapwater at 121°C for 20 min. After filtration the volume is restored to 1 liter with tap water (Lochhead and Burton, 1957).

Soil extract agar (Holm and Jensen 1972) has the following composition: soil extract, 400 ml; tap water, 600 ml; glucose, 1.0 g; peptone, 1.0 g; yeast extract, 1.0 g; K_2HPO_4, 1.0 g; agar, 20.0 g; pH 6.5–6.7. Immediately before use, a filter-sterilized solution of cycloheximide is added to the autoclaved medium to give a final concentration of 40 mg/liter. Before plates are poured a layer of sterile agar is poured in the bottom of the plates and allowed to solidify. This prevents colonies from spreading between the Petri dish and the agar. Colonies are picked onto slants of the same medium.

The Mulder and Antheunisse (1963) screening method was described fully by the authors: a brief description of the method was given by Keddie and Jones (1981).

Isolation by Selective Medium (Hagedorn and Holt, 1975a).

Plate counts are made by spreading 0.1-ml amounts of suitable dilutions over the surface of sterile medium in Petri dishes. Hagedorn and Holt used 0.5% peptone as diluent. The medium has the following composition (g/1000 ml distilled water): trypticase soy agar (BBL), 4.0; yeast extract (Difco), 2.0; NaCl, 20.0; cycloheximide, 0.1; agar, 15.0; methyl-red (Harleco), 150 µg/ml). The methyl-red is filter-sterilized and added aseptically to the cooled, autoclaved medium. The pH is adjusted to that of the soil being examined: pH values in the range 5.0–8.5 are said not to affect the selective properties of the medium. After incubation for 10 days at 25°C colonies are transferred to slants of trypticase soy agar containing 0.2% yeast extract and examined for the possession of a rod-coccus growth cycle. The medium is said to inhibit fungi, most streptomycetes, nocardias and Gram-negative bacteria as well as other Gram-positive bacteria such as bacilli and micrococci, but to have little or no effect on arthrobacters (Hagedorn and Holt, 1975a).

Identification as Arthrobacter spp. In all of the methods described above, bacteria other than arthrobacters will be isolated. Accordingly, it is necessary to apply a small number of screening tests to the isolates to identify those that are genuine arthrobacters. Demonstration of a rod-coccus cycle alone, as used by the authors of the isolation methods described, is insufficient. Obligately aerobic isolates which show the typical rod-coccus growth cycle (see Cure and Keddie, 1973 for a suitable method) and which contain lysine as cell wall diamino acid (see Bousfield et al., 1985, for suitable rapid methods) may be regarded as *Arthrobacter* spp.

Enrichment Methods

Some *Arthrobacter* spp. were originally isolated from enrichment cultures by using often unusual organic substrates as sole carbon + energy sources in mineral-based media. However, the primary purpose of such enrichment cultures was to obtain isolates capable of utilizing the substrates studied and not to isolate arthrobacters. It is not, therefore, possible to assess their efficacy in isolating particular arthrobacters. Examples of such substrates are nicotine (*A. oxydans*, Sguros, 1954, 1955; *A. nicotianae*, Giovannozzi-Sermanni, 1959); histidinol (*A. histidinolovorans*, Adams, 1954); and picolinic acid (*A. picolinophilus*, Tate and Ensign, 1974).

Maintenance Procedures

Stab cultures by loop in TSX semisolid medium (Keddie et al., 1966) or in the soil extract semisolid medium of Lochhead and Burton (1957) (see Enrichment and Isolation Procedures) will remain viable for at least 3 months at room temperature (~20°C) provided they are not allowed to dry out. Cultures may be preserved for longer periods (at least 7 years) by freezing in glass beads at −70°C (Feltham et al., 1978). For long term storage, lyophilization is suitable.

Procedures for Testing for Special Characters

The presence of a rod-coccus growth cycle may be demonstrated by using the methods described by Cure and Keddie (1973). Cell wall composition may be determined by one of the rapid methods described by Bousfield et al. (1985). Care must be taken in distinguishing lysine from ornithine. Menaquinone composition may be determined by reverse phase partition TLC or HPLC (Collins et al., 1980; 1983c). Polar lipid patterns may be obtained by two-dimensional TLC analysis of free lipid extracts (Collins et al., 1983c).

Differentiation of the genus **Arthrobacter** from other genera

Members of the genus *Arthrobacter* may be distinguished from other coryneform and morphologically similar genera which contain lysine in the cell wall, or which show a rod-coccus growth cycle, by the characters listed in Table 15.10. The differentiation from *Arthrobacter* as now defined, of some species formerly classified in the genus (Keddie, 1974), is also given in Table 15.10.

Taxonomic Comments

In the present revision, the genus has been restricted to those species which, like the type, contain lysine as the cell wall diamino acid (Fiedler et al., 1970; Yamada and Komagata, 1972b; Schleifer and Kandler, 1972; Keddie and Cure, 1977; Keddie, 1978; Keddie and Cure, 1978), i.e. *Arthrobacter sensu stricto* as defined by Keddie and Jones (1981). Thus some species formerly classified as arthrobacters (Keddie, 1974) are now excluded from the genus.

A. terregens and *A. flavescens* contain ornithine as cell wall diamino acid (Schleifer and Kandler, 1972; Keddie and Cure, 1977), and have been transferred to the proposed new genus *Aureobacterium* as *Aur. terregens* and *Aur. flavescens* (Collins et al., 1983b). While all are agreed that *A. simplex* and *A. tumescens* should be removed from *Arthrobacter*,

Table 15.10.

Characters[a] differentiating **Arthrobacter** *from similar genera which either have a rod-coccus growth cycle or which have lysine as cell wall diamino acid[b]*

Genus	Mycelium produced	Rod-coccus cycle[c]	Oxygen requirements	Acid from glucose[d]	Cell wall		Major menaquinone[f]
					Diamino acid[e]	Glycine present	
Arthrobacter	−	+	Aerobic	−	Lysine	−	MK-9 (H$_2$) **or** MK-8 and/or MK 9
Aureobacterium terregens/ Aur. flavescens[g]	−	+	Aerobic	W	Ornithine	+	MK-12 and/or MK-13
Arthrobacter simplex/[h] A. tumescens	−	+	Aerobic	−	LL-DAP	+	MK-8 (H$_4$)
Microbacterium	−	−	Equivocal	+	Lysine	+	MK-11, MK-12
Renibacterium	−	−	Aerobic	−	Lysine	+	MK-9
Oerskovia	+	−	Facultative	+	Lysine	−	MK-9 (H$_4$)
Brevibacterium	−	+	Aerobic	−	*meso*-DAP	−	MK-8 (H$_2$)
Rhodococcus	D	D	Aerobic	−	*meso*-DAP	−	MK-8 (H$_2$) **or** MK-9 (H$_2$)

[a] Data from Keddie and Cure, 1977, 1978; Keddie and Bousfield, 1980; Sanders and Fryer, 1980; Collins and Jones, 1981; Keddie and Jones, 1981; Lechevalier and Lechevalier, 1981; and Collins, 1982.
[b] Symbols: see Table 15.2; also W, weak.
[c] Similar to that in *A. globiformis*.
[d] In peptone-based media.
[e] DAP, diaminopimelic acid.
[f] MK-8, MK-9, etc., indicates the number of isoprene units in the menaquinone; (H$_2$), (H$_4$), etc., indicates the number of double bonds hydrogenated.
[g] Formerly *Arthrobacter terregens* and *A. flavescens*.
[h] It has been proposed by O'Donnell et al. (1982) that *A. simplex*, but not *A. tumescens* be transferred to the genus *Nocardioides*. Suzuki and Komagata (1983b) have proposed the new genus *Pimelobacter* to accommodate both species together with a third, *P. jensenii*.

their taxonomic position is still controversial despite the fact that both species were shown to contain LL-diaminopimelic acid (LL-DAP) as cell wall diamino acid as long ago as 1959 by Cummins and Harris. Also 16S-rRNA cataloging studies have shown that *A. simplex* is only distantly related to *A. globiformis* (Stackebrandt et al., 1980). Suzuki and Komagata (1983b) have proposed the new genus *Pimelobacter* to accommodate the LL-DAP-containing coryneform bacteria. Using DNA-DNA base-pairing methods they considered that three species could be distinguished among LL-DAP-containing coryneform bacteria previously named *Arthrobacter simplex, A. tumescens* and *"Brevibacterium lipolyticum."* The first, *Pimelobacter simplex,* contained most *A. simplex* strains and all *"B. lipolyticum"* strains, the second, *P. tumescens* contained *A. tumescens* strains while a third species, *P. jensenii*, was created for a single strain, NCIB 9770, which was originally identified as *A. simplex* by Jensen and Gundersen (1956). These authors also concluded from their DNA homology studies that *A. simplex* and *A. tumescens* were only distantly related to the nocardioform organism *Nocardioides albus*. On the other hand, O'Donnell et al. (1982) considered that *A. simplex* (but not *A. tumescens*) closely resembled *Nocardioides* spp. in chemotaxonomic (particularly lipid) characters and proposed its transfer to that genus as *Nocardioides simplex*. Collins et al. (1983b) also found a close similarity in lipid composition between *A. simplex, "B. lipolyticum,"* and *Nocardioides* spp. and considered that these taxa should be classified in the same family. These authors also reported that *A. tumescens* differed substantially in lipid composition from all other coryneform bacteria, including *A. simplex*, and should be classified separately from them, possibly in a new genus (Collins et al., 1983f). It should also be recorded that Prauser (1976) first suggested a relationship between *A. simplex* and *Nocardioides* spp. on the basis of phage susceptibility studies. It is thus clear that further studies are needed to resolve the taxonomic position of the LL-DAP-containing coryneform bacteria. The taxonomic position of *A. duodecadis* is also unresolved (Keddie and Cure, 1978); as far as we are aware only one strain, the type, has ever been described.

Most of the remaining species that conform to the present definition of *Arthrobacter* closely resemble the type species *A. globiformis* in a

large number of phenotypic (mainly nutritional) characters. For this reason an "ideal phenotype" for *A. globiformis* based on the examination of 22 named strains and unnamed isolates was included in the eighth edition of the *Manual* (Keddie, 1974). The variously named strains which contained lysine in the cell wall, which had DNA base ratios similar to that of the type strain of *A. globiformis* and which showed a high degree of conformity with the characters of the "ideal phenotype" were considered to be synonyms, and in a few cases possible synonyms of *A. globiformis* (Keddie, 1974). Thus those nomenclatural species which showed a high degree of conformity with the ideal phenotype of *A. globiformis* include the following (Keddie and Cure, 1977): *A. atrocyaneus, A. aurescens, A. crystallopoietes, A. histidinolovorans, A. nicotianae, A. oxydans, A. pascens, A. polychromogenes, A. ramosus, A. ureafaciens, "Brevibacterium sulfureum"* (*Arthrobacter sulfureus*) and *Corynebacterium ilicis* (*Arthrobacter ilicis*) i.e. the "globiformis" group of Keddie and Jones (1981). However, despite considerable phenotypic similarity (see also Skyring et al. 1971; Stackebrandt et al., 1983) subsequent work has shown much more heterogeneity among these nomenclatural species than was formerly apparent (e.g. Schleifer and Kandler, 1972) and indeed DNA-DNA base-pairing studies have now shown that most are distinct species (Stackebrandt and Fiedler, 1979; Stackebrandt et al., 1983). An exception is *A. polychromogenes* which was considered to be a subspecies of *A. oxydans* (Stackebrandt and Fiedler, 1979).

A. citreus (type strain) also has the characters of the genus and can readily be distinguished from all other species. A number of other strains named *A. citreus* are listed in culture collections but, of those which have been studied in sufficient detail, none has the characters of the type strain (ATCC 11624) (see *A. citreus* and Schleifer and Kandler, 1972; Stackebrandt et al., 1983).

As has been pointed out above, two groups of species can be distinguished within the genus according to their peptidoglycan structure and lipid composition (see Further Descriptive Information, above, for details). Those species in the *A. globiformis/A. citreus* group have peptidoglycans with the A3α variation (Schleifer and Kandler, 1972) and the major isoprenologues are dihydrogenated menaquinones with nine isoprene units (MK-9(H₂)) (Yamada et al., 1976; Collins et al., 1979, 1981; Collins and Jones, 1981). On the other hand, species in the *A. nicotianae* group have peptidoglycans with the A4α variation (Schleifer and Kandler, 1972; Stackebrandt et al., 1983) and have menaquinones with eight isoprene units (MK-8), or sometimes nine isoprene units (MK-9), as major components (Yamada et al., 1976; Collins and Jones, 1981; Collins and Kroppenstedt, 1983). For these reasons it has been suggested that the genus *Arthrobacter* be further restricted to those species which, like the type, contain peptidoglycans with the A3α variation and MK9(H₂) isoprenoid quinones (Minnikin et al., 1978; Collins and Kroppenstedt, 1983). However, Stackebrandt et al. (1983) noted that in the earlier work of Stackebrandt and Fiedler (1979) "DNA homology values found between members of the "globiformis" group" (i.e. as defined by Keddie and Jones, 1981) "did not convincingly separate *A. nicotianae* and related taxa from *A. globiformis* and related species. . . ." In attempting to resolve this situation by using 16S-rRNA cataloging studies they found a close relationship between *A. globiformis* and related species and the *A. nicotianae* group (represented by one strain). They concluded that the genus *Arthrobacter* contained two "nuclei" whose members could be distinguished from each other by the peptidoglycan type and the menaquinone composition (Stackebrandt et al., 1983). We have adopted this view in the present revision of the genus. Stackebrandt and Woese (1981) also showed that 16S-rRNA studies revealed that genuine species of the genus *Micrococcus* could not be separated from the genus *Arthrobacter* on a phylogenetic basis. While accepting that *Arthrobacter* and *Micrococcus* are very closely related phylogenetically, they are here considered as distinct genera for determinative purposes.

Further Reading

General:

Keddie R.M. and D. Jones. 1981. Aerobic saprophytic coryneform bacteria. *In* Starr, Stolp, Trüper, Balows and Schlegel (Editors), The Prokaryotes: A Handbook on Habitats, Isolation and Identification of Bacteria, Springer-Verlag, New York, pp. 1839–1878.

Cell walls:

Keddie, R.M. and G.L. Cure. 1978. Cell wall composition of coryneform bacteria. *In* Bousfield and Callely (Editors), Special Publications of the Society for General Microbiology I. Coryneform Bacteria, Academic Press, London, pp. 47–83.

Peptidoglycan:

Schleifer K.H. and O. Kandler. 1972. Peptidoglycan types of bacterial cell walls and their taxonomic implications. Bacteriol. Rev. *36:* 407–477.
Stackebrandt, E., V. J. Fowler, F. Fiedler and H. Seiler. 1983. Taxonomic studies on *Arthrobacter sulfureus* sp. nov. and reclassification of *Brevibacterium protophormiae* as *Arthrobacter protophormiae* comb. nov. Syst. Appl. Microbiol. *4:* 470–486.

DNA-DNA homology studies:

Stackebrandt, E. and F. Fiedler. 1979. DNA-DNA homology studies among *Arthrobacter* and *Brevibacterium*. Arch. Microbiol. *120:* 289–295.
Stackebrandt et al., 1983; (see under Peptidoglycan, above).

Lipids:

Minnikin, D.E., M. Goodfellow and M. D. Collins. 1978. Lipid composition in the classification and identification of coryneform and related taxa. *In* Bousfield and Callely (Editors), Special Publications of the Society for General Microbiology I. Coryneform Bacteria, Academic Press, London, pp. 85–160.
Collins, M.D. and R.M. Kroppenstedt. 1983. Lipid composition as a guide to the classification of some coryneform bacteria containing an A4α type peptidoglycan (Schleifer and Kandler). Syst. Appl. Microbiol. *4:* 95–104.

Differentiation of the species of the genus **Arthrobacter**

As explained above, species of the genus *Arthrobacter* can be divided into two groups, the *A. globiformis*/*A. citreus* group and the *A. nicotianae* group, and these have been arranged in Tables 15.11 and 15.12, respectively. The two species groups are distinguished by the peptidoglycan "variation" and by the menaquinone content but differentiation of species within the groups depends to a large extent on DNA-DNA base-pairing techniques (Fiedler and Stackebrandt, 1979; Stackebrandt et al., 1983). Apart from *A. citreus* all the remaining species now recognized share a high degree of phenetic similarity and it is not yet possible to differentiate between them by using phenetic characters. Those characters which may be of value in differentiating species in the two groups are listed in Tables 15.11 and 15.12.

List of species of the genus **Arthrobacter**

1. **Arthrobacter globiformis** (Conn 1928) Conn and Dimmick 1947, 301.*ᴬᴸ* (*Bacterium globiforme* Conn 1928, 3.)

glo.bi.for′mis. L. n. *globus* ball, globe; L. n. *forma* shape; M.L. adj. *globiformis* spherical.

Morphology and general characters as for generic description.

Other characters are given in Table 15.11 and in Further Descriptive Information.

Colonies on yeast extract-peptone media show no distinctive pigmentation. Either nutritionally nonexacting or require biotin alone. Growth occurs in a suitable mineral salts medium with an ammonium salt or nitrate as sole nitrogen source (with biotin if required) and with glucose as carbon + energy source. The widely used Conn strain ATCC 4336 (NCIB 8602), requires biotin; the type strain does not.

Common in soil.

The mol% G + C of the DNA is in the range 62.0–65.5 (T_m, Skyring and Quadling, 1970; Yamada and Komagata, 1970; Bousfield, 1972; Crombach 1972; Stackebrandt, et al., 1983) and 65.5 (Bd, type strain, Skyring and Quadling, 1970).

Type strain: ATCC 8010 (NCIB 8907).

Further comments. A. globiformis strains ATCC 4336 (NCIB 8602) and NCIB 9759 have been shown to have the same peptidoglycan type as the type strain. *A. globiformis* NCIB 8717, which, like ATCC 11822 named "*Brevibacterium helvolum*," is derived from the H. L. Jensen strain Ca3, has the peptidoglycan type L-lysine-L-serine-L-threonine-L-alanine (Schleifer and Kandler, 1972) and contains galactose, glucose, rhamnose and mannose as cell wall sugars (Keddie and Cure, 1977). DNA-DNA homology studies showed only a low degree of homology between this strain (DSM 20125) and the type strain of *A. globiformis* (27%) but a moderate degree of homology between it and *A. oxydans* (54%) and *A. polychromogenes* (47%), both of which share the same peptidoglycan type (Stackebrandt and Fiedler, 1979).

Among species which share the same peptidoglycan type, *A. globiformis* (type strain) shows 17% homology with *A. crystallopoietes*, 51% with *A. pascens* and 27% with *A. ramosus* (Stackebrandt and Fiedler, 1979).

The cytochrome composition of *A. globiformis* strain ATCC 4336 (NCIB 8602) has been shown to be $bcaa_3o$ when cells are in the logarithmic phase of growth but changes to $bcaa_3od$ in the stationary phase of growth when the cells become oxygen limited and lose their ability to retain the crystal violet-iodine complex in the Gram stain (Meyer and Jones, 1973; Jones, 1980).

2. **Arthrobacter crystallopoietes** Ensign and Rittenberg 1963, 149.*ᴬᴸ*

crys.tall.o.poie′tes. M.L. n. *crystallum*, crystal; Gr. v. *poiein*, to make, form; M.L. adj. *crystallopoietes* crystal forming.

(Description based on type strain only.)

Morphology and general characters as for generic description.

Other characters are given in Table 15.11 and in Further Descriptive Information.

Table 15.11.

Some characteristics[a] of **Arthrobacter** *spp. which have peptidoglycans of the A3α variation[b] and MK-9 (H₂)[c] as major menaquinones; the* "*A. globiformis/A. citreus group*"[d,e]

Characteristics	1. A. globiformis	2. A. crystallopoietes	3. A. pascens	4. A. ramosus	5. A. aurescens	6. A. histidinolovorans	7. A. ilicis	8. A. ureafaciens	9. A. atrocyaneus	10. A. oxydans	11. A. citreus
Peptidoglycan type[b]	Lys-Ala₃[f]	Lys-Ala	Lys-Ala₂	Lys-Ala₄	Lys-Ala-Thr-Ala	Lys-Ala-Thr-Ala	Lys-Ala-Thr-Ala	Lys-Ala-Thr-Ala	Lys-Ser-Ala₂₋₃	Lys-Ser-Thr-Ala	Lys-Thr-Ala₂
Cell wall sugars[g]	Gal, Glu	Gal, Glu	Gal, Glu	Gal, Rha, Man	Gal, (Man)	Gal, Glu	Gal, Rha, Man	Gal (Man)	Gal, Glu, (Man)	Gal, Glu	Gal
Vitamin requirement:											
None or biotin only	+	+	+	+	+	+	+[h]	+	+	+	−
Nicotine utilization	−	−	−	−	−	−	−	−	−	+	−
Starch hydrolysis	+	−	+	−	+	−	−	−	+	+	−
Motility	−	−	−	+	−	−	+	−	+	−	+

[a] Most species are represented only by their type strains and therefore the range of variation within the species is not known. For this reason data on carbon source utilization tests are not included although studies on a wider range of strains may prove them to be useful. See Further Descriptive Information for list of near-universal substrates. There are conflicting reports in the literature about the responses of the type strains of some species in some common tests such as nitrate reduction, urea hydrolysis etc. and therefore these data have been omitted. Other data are from original and/or revised descriptions of species, and from Keddie et al., 1966; Yamada and Komagata, 1972a; Cure and Keddie, Robertson and Keddie, unpublished data.

[b] Schleifer and Kandler (1972); Stackebrandt and Fiedler (1979).

[c] Collins and Jones (1981).

[d] Species are arranged according to peptidoglycan type (Stackebrandt and Fiedler, 1979). All *A. globiformis/A. citreus* group strains which have been examined contain the polar lipids diphosphatidylglycerol, phosphatidylglycerol and phosphatidylinositol (Shaw and Stead, 1971; Kostiw et al., 1971; Collins et al., 1981; 1983b). See "Further Descriptive Information" for other lipid data.

[e] Symbols: see Table 15.2.

[f] Lys, L-lysine; Ala, L-alanine; Thr, L-threonine; Ser, L-serine; Gal, galactose; Glu, glucose; Rha, rhamnose; Man, mannose; ( ), conflicting reports on occurrence.

[g] Keddie and Cure (1978).

[h] *A. ilicis* grows in a mineral salts-glucose medium only when provided with casamino acids (Robertson and Keddie, unpublished data).

Colonies on yeast extract-peptone media show no distinctive pigmentation. On 2-hydroxypyridine agar a brilliant green crystalline pigment develops in the colony mass after 3 or 4 days incubation at 30°C.

Nutritionally nonexacting: growth occurs in a suitable mineral salts medium with an ammonium salt or nitrate as sole nitrogen source and glucose as carbon + energy source.

Isolated from soil by enrichment in a mineral medium containing 2-hydroxypyridine as sole carbon + energy + nitrogen source.

The mol% G + C is in range 62.9–63.8 (Bd, Skyring and Quadling, 1970; Starr et al., 1975).

Type strain: ATCC 15481 (NCIB 9499).

Further comments. Among species which share the same peptidoglycan type as *A. crystallopoietes*, DNA-DNA homology studies show a 17% homology with *A. globiformis* (type strain), 13% with *A. pascens* and 9% with *A. ramosus* (Stackebrandt and Fiedler, 1979).

Faller and Schleifer (1981) investigated an apparent correlation between the cytochrome content and the morphology of *A. crystallopoietes*. They noted that the rods which occurred in exponential phase cultures contained large amounts of cytochrome *aa₃* but only traces of cytochrome *d*, whereas the coccoid cells in early stationary phase cultures contained relatively large amounts of cytochrome *d* with lower amounts of cytochrome *aa₃*. However, they showed that the increase in cytochrome *d* content was not correlated with morphology as such, but resulted from the oxygen limitation that occurs in late exponential/early stationary phase cultures. Oxygen limitation induces the onset of the stationary phase by reducing the growth rate which, in turn, results in the transformation of rods into coccoid cells. On the other hand,

oxygen saturation allows synthesis of cytochrome *aa₃* but inhibits synthesis of cytochrome *d*; but it also allows rapid growth to occur which, in turn, results in the rod forms characteristic of exponential phase cultures.

3. **Arthrobacter pascens** Lochhead and Burton 1953. 7.[AL]

pas′cens. L. part adj. *pascens* nourishing.

(Description based on type strain only.)

Morphology and general characters as for generic description.

Other characters are given in Table 15.11 and in Further Descriptive Information.

Colonies on yeast extract-peptone medium show no distinctive pigmentation.

Nutritionally nonexacting: growth occurs in a suitable mineral salts medium with an ammonium salt or nitrate as sole nitrogen source and glucose as carbon + energy source.

Isolated from soil.

The mol% G + C of the DNA is 63.7 (T_m, Yamada and Komagata, 1970), 64.9 (Bd, Skyring and Quadling, 1970).

Type strain: ATCC 13346 (NCIB 8910).

Further comments. Among strains which share the same peptidoglycan type *A. pascens* shows 51% homology with *A. globiformis* (type strain), 10% with *A. crystallopoietes* and 25% with *A. ramosus* (Stackebrandt and Fiedler 1979).

4. **Arthrobacter ramosus** Jensen 1960, 131.[AL]

ra.mo′sus. M.L. adj. *ramosus* branched, branching.

Morphology and general characters as for generic description. Motile by a few lateral flagella.

Table 15.12.

Characters[a] most useful in differentiating **Arthrobacter** *spp. which have peptidoglycans of the A4α variation[b] and MK-8[c] (or MK-9[d]) as major menaquinones; the "***A. nicotianae*** *group"[e,f]*

Characteristics	12. *A. nicotianae*	13. *A. protophormiae*	14. *A. uratoxydans*	15. *A. sulfureus*
Number of strains studied	6	6	2	3
Peptidoglycan type[b]	Lys-Ala-Glu[g]	Lys-Ala-Glu	Lys-Ala-Glu	Lys-Glu
Major wall sugars[h]	Gal, Glc (one strain)	ND	ND	Gal, Glc (one strain)
Hydrolysis of:				
Starch	+	−	−	−
Casein	+	d	+	−
Utilization of:				
4-Hydroxybenzoate	+	+ (5/6)[i]	−	+
Glyoxylate	d	d	+	−
L-Asparagine	+	+ (5/6)	−	+
L-Arginine	d	+	−	+
L-Histidine	d	+	−	+
D-Xylose	+	− (5/6)	−	−
D-Ribose	+	d	−	d
L-Arabinose	+	+ (5/6)	−	−
D-Galactose	+	+ (5/6)	−	d
L-Rhamnose	−	− (5/6)	+	d
2,3-Butylene glycol	+ (5/6)	−	−	d
Glycerol	+	+	−	−

[a] Adapted from Table 5 in Stackebrandt et al. (1983).

[b] Schleifer and Kandler (1972); Stackebrandt et al. (1983).

[c] Collins and Jones (1981); Collins and Kroppenstedt (1983).

[d] *A. sulfureus* ("*Brevibacterium sulfureum*") ATCC 19098 (NCIB 10355) contains MK-9 as major menaquinone (Collins and Kroppenstedt, 1983); a second strain designated *A. sulfureus* by Stackebrandt et al. (1983), ATCC 15170 (formerly named *A. citreus*), contains similar amounts of MK-9 and MK-10 as major menaquinones (Collins and Kroppenstedt, 1983).

[e] All *A. nicotianae* group strains which have been examined contain the polar lipids diphosphatidylglycerol, phosphatidylglycerol but not phosphatidylinositol (Collins and Kroppenstedt, 1983). See "Further Descriptive Information" for other lipid data. According to Stackebrandt et al. (1983) more than 90% of *A. nicotianae* group strains utilize: acetate, propionate, valerate, capronate, heptanoate, caprylate, succinate, DL-malate, citrate, DL-lactate, fumarate, D-gluconate, glycine, L-proline, L-threonine and L-aspartate; none utilizes adipate, levulinate and acetamide.

[f] Symbols: see Table 15.2; also d, 11–89% of strains are positive; ND, no data.

[g] Lys, L-lysine; Ala, L-alanine; Glu, L-glutamic acid; Gal, galactose; Glc, glucose.

[h] Keddie and Cure (1978).

[i] Fraction of strains giving indicated reaction.

Other characters are given in Table 15.11 and in Further Descriptive Information.

Colonies on yeast extract-peptone media show no distinctive pigmentation.

Nutritionally nonexacting: growth occurs in a suitable mineral salts medium with an ammonium salt or nitrate as sole nitrogen source (Keddie et al., 1966; Owens and Keddie, 1969).

Isolated from beech forest soil at depth of 30–35 cm.

Mol% G + C of the DNA is 62.2 (T_m, Yamada and Komagata, 1970), 63.1 (Bd, Skyring and Quadling, 1970).

Type strain: ATCC 13727 (NCIB 9066).

Further comments. Among species which share the same peptidoglycan type, *A. ramosus* shows 27% homology with *A. globiformis* (type strain), 15% with *A. crystallopoietes* and 30% with *A. pascens* (Stackebrandt and Fiedler, 1979).

5. Arthrobacter aurescens (ex Clark 1951) Phillips 1953, 241.[AL]

(*Arthrobacter globiforme* var. *aurescens* Clark, 1951, 180.)

au.res'cens. L. v. *auresco* to become golden; L. part. adj. *aurescens* becoming golden.

(Description based on type strain only.)

Morphology and general characters as for generic description. Other characters are given in Table 15.11 and Further Descriptive Information.

Colonies of the type strain are reported to show a yellow pigmentation on agar and other media (e.g. see Phillips, 1953) but this strain gives only pale buff colonies on yeast extract-peptone agar when incubated in the dark (Grainger, Cure and Keddie, unpublished data).

When supplied with biotin, growth occurs in a suitable mineral salts medium with an ammonium salt or nitrate as nitrogen source and glucose as carbon + energy source (Keddie et al., 1966; Owens and Keddie, 1969).

Isolated from soil.

The mol% G + C is 61.5 (T_m, Yamada and Komagata, 1970), 59.1 (Bd, Skyring and Quadling, 1970).

Type strain: ATCC 13344 (NCIB 8912).

Further comments. Among species which share the same peptidoglycan type, *A. aurescens* shows 55% homology with *A. histidinolovorans* and 35% with *A. ureafaciens* (Stackebrandt and Fiedler, 1979).

6. Arthrobacter histidinolovorans Adams 1954, 832.[AL]

hist.tid.in.ol.ov'or.ans. M.L. n. *histidinolum*, histidinol; L. *vorans*, from L. v. *voro* devour, destroy; M.L. adj. *histidinolovorans* histidinol destroying.

(Description based on type strain only.)

Morphology and general characteristics as for generic description. Other characters are given in Table 15.11 and in Further Descriptive Information.

Colonies on yeast extract-peptone media show no distinctive pigmentation.

When supplied with biotin, growth occurs in a suitable mineral salts medium with an ammonium salt as sole nitrogen source and glucose as sole carbon + energy source (Robertson and Keddie, unpublished data).

Utilizes L-histidinol as major source of carbon + energy + nitrogen when grown in mineral medium containing a low concentration of yeast extract.

Isolated from soil on mineral agar medium containing L-histidinol as carbon + energy + nitrogen source.

The mol% G + C is 61.3 (Bd, Skyring et al., 1971).

Type strain: ATCC 11442 (NCIB 9541).

Further comments. Among species which share the same peptidoglycan type, *A. histidinolovorans* shows 5% homology with *A. aurescens* and 33% with *A. ureafaciens* (Stackebrandt and Fielder, 1979).

7. Arthrobacter ilicis (Mandel, Guba and Litsky 1961) Collins, Jones and Kroppenstedt 1982, 384.[VP] (Effective publication: Collins, Jones and Kroppenstedt 1981, 321.) (*Corynebacterium ilicis* Mandel, Guba and Litsky 1961, 61.)

il.i′cis. M.L. n. *ilex*, holly; M.L. gen. n. *ilicis* of the holly.

Morphology and general characters as for generic description. Rods motile.

Other characters are given in Table 15.11 and in Further Descriptive Information.

Colonies on yeast extract-peptone media have a yellow, nondiffusible pigment.

Does not require B-vitamins: growth occurs in a mineral salts medium containing an ammonium salt and with glucose as carbon + energy source only when casamino acids are supplied (Robertson and Keddie, unpublished data).

Causes blight on foliage and twigs of American Holly (*Ilex opaca*).

The mol% G + C is 61.5 (T_m, Bousfield, 1972), 60.0 (method not stated, quoted by Schuster et al., 1968).

Type strain: ATCC 14264 (NCPPB 1228).

Further comments. Starr et al. (1975) recorded a value of 75% G + C (Bd) for *A. ilicis* (*Corynebacterium ilicis*).

8. Arthrobacter ureafaciens (Krebs and Eggleston 1939) Clark 1955, 111.[AL] (*Corynebacterium ureafaciens* Krebs and Eggleston 1939, 310.)

u.re.a.fa′ci.ens. M.L. n. *urea* urea; L. v. *facio* to make, produce: M.L. part adj. *ureafaciens* urea-producing.

(Description based on type strain only.)

Morphology and general characters as for generic description.

Other characters are given in Table 15.11 and in Further Descriptive Information.

Colonies on yeast extract-peptone media are pale gray becoming yellow especially when incubated in diffuse daylight at ~20°C.

When supplied with biotin, growth occurs in a suitable mineral salts medium with an ammonium salt as sole nitrogen source and with glucose as carbon + energy source (Robertson and Keddie, unpublished data).

Isolated from soil.

The mol% G + C is 61.7 (T_m, Yamada and Komagata, 1970), 62.9 (Bd, Skyring and Quadling, 1970).

Type strain: ATCC 7562 (NCIB 7811).

Further comments. The above description is based on that of a single strain (culture NC) isolated from a creatinine enrichment by Dubos and Miller (1937) and named "*Corynebacterium creatinovorans*" by them but the name was not validly published (Clark, 1955). Culture NC was further studied by Krebs and Eggleston (1939) who named it "*Corynebacterium ureafaciens*" because it produces urea from creatine, creatinine and uric acid.

Among species which have the same peptidoglycan type, *A. ureafaciens* shows 30% homology with *A. aurescens* and 45% homology with *A. histidinolovorans* (Stackebrandt and Fiedler 1979).

9. Arthrobacter atrocyaneus Kuhn and Starr 1960, 179.[AL]

a.tro.cy′an.e.us. M.L. adj. *ater*, black; Gr. adj. *cyaneus* dark blue; M.L. adj. *atrocyaneus*, dark blackish blue.

(Description based on type strain only.)

Morphology and general characters as for generic description. Rods motile by 1–3 lateral flagella.

Other characters are given in Table 15.11 and in Further Descriptive Information.

Colonies on yeast extract-peptone media show no distinctive pigmentation.

On peptone agar containing certain sugars or sugar alcohols an intense, nondiffusible, blue-black pigment is produced; the colonies have a metallic luster.

When supplied with biotin, growth occurs in a suitable mineral salts medium with an ammonium salt or nitrate as nitrogen source and glucose as carbon + energy source (Keddie et al., 1966; Owens and Keddie, 1969).

The only known strain was isolated as a contaminant on an agar plate.

The mol% G + C is 69.5 (T_m, Yamada and Komagata, 1970), 70.0–70.3 (Bd, Schuster et al., 1968; Skyring and Quadling, 1970).

Type strain: ATCC 13752 (NCIB 9220).

Further comments. The blue pigment of *A. atrocyaneus* was identified as indigoidine by Kuhn et al., (1965). Pigment production occurs only in peptone-based media containing a sugar or sugar alcohol e.g. glucose, galactose, xylose, maltose, lactose and sorbitol, and is best at ~26°C. In sugar agar cultures, the pigment accumulates extracellularly in granules which are usually several times as large as the bacterial cells (Kuhn and Starr, 1960).

A. atrocyaneus has a unique peptidoglycan type (Interschick et al., 1970) and a mol% G + C some 3% higher than any other Arthrobacter. In DNA-DNA base-pairing studies Stackebrandt and Fiedler (1979) found very low levels of homology with other *Arthrobacter* species.

10. Arthrobacter oxydans Sguros 1954, 21.[AL]

ox′y.dans. M.L. adj. *oxydans* oxidising.

Morphology and general characters as for generic description. (see also Sguros, 1955.)

Other characters are given in Table 15.11 and Further Descriptive Information.

Colonies on yeast extract-peptone media are either pearl-gray/white or show a yellow, nondiffusible pigment depending on strain (Sguros, 1955).

Growth on nicotine-mineral salts-yeast extract agar is abundant with production of a deep blue, diffusible pigment which turns reddish or yellow-brown in older cultures.

When supplied with biotin, growth occurs in a suitable mineral salts medium with an ammonium salt as nitrogen source and glucose as carbon + energy source (Sguros, 1954; Robertson and Keddie, unpublished data).

Isolated by enrichment in nicotine-mineral salts-yeast extract medium from cured tobacco leaves and associated air.

The mol% G + C is in the range 62.7–64.4 (T_m, Yamada and Komagata, 1970; Stackebrandt et al., 1983), 65.9 (Bd, Skyring and Quadling, 1970).

Type strain: ATCC 14358 (NCIB 9333).

Further comments. For details of the structure of the blue pigment of *A. oxydans*, "nicotine blue" see Knackmuss and Beckman (1973).

In DNA-DNA base-pairing studies, Stackebrandt and Fiedler (1979) found high degrees of homology between *A. oxydans* and *A. polychromogenes* (61 and 69%, respectively). They suggested that *A. polychromogenes* (Schippers-Lammertse, Muijsers and Klatser-Oedekerk 1963,2[AL]), type and only strain, ATCC 15216 (NCIB 10267), be regarded as a subspecies of *A. oxydans*. *A. polychromogenes* has the same peptidoglycan type as *A. oxydans* (Schleifer and Kandler, 1972) and both contain major amounts of galactose in the cell wall but, whereas the walls of *A. oxydans* also contain small to moderate amounts of glucose, those of *A. polychromogenes* do not (Keddie and Cure, 1977). A value of 65.3 mol% G + C (Bd) has been recorded for the type strain (Starr et al., 1975).

A. polychromogenes differs from *A. oxydans* in producing a blue pigment in glycerol peptone media and in other carbohydrate-peptone media and in being unable to grow on nicotine agar. The blue pigment contains a water-soluble component and a water-insoluble component identified as indigoidine. *A. polychromogenes* was shown to be lysogenic (Schippers-Lammertse et al., 1963).

11. **Arthrobacter citreus** Sacks 1954, 342.[AL]

cit′re.us. M.L. adj *citreus* lemon-colored.

(Description based on type strain only.)

Morphology and general characters as for generic description. The rods are feebly motile.

Other characters are given in Table 15.11 and Further Descriptive Information.

Colonies on yeast extract-peptone media have a yellow, nondiffusible pigment which is insoluble in ether and acetone.

Biotin, thiamine, nicotinic acid, tyrosine, methionine, cystine and a siderophore such as ferrichrome or mycobactin are required for growth; glutamic acid, while not essential, is required for maximum growth. The siderophores can be replaced by certain synthetic metal chelators (Seidman and Chan, 1969).

Owens and Keddie (unpublished data) found that the type strain used about 40 of 180 compounds tested as sole or principal sources of carbon + energy. They included a relatively wide range of carbohydrates, sugar derivatives and amino acids, together with some fatty acids, dicarboxylic acids, hydroxyacids, oxo-acids, non-nitrogenous aromatic compounds, amines and heterocyclic compounds; of those tested no simple alcohols, polyalcohols or glycols were utilized (Keddie, 1974).

Isolated from chicken feces but probably a dust or soil contaminant.

The mol% G + G is in the range 62.9–63.8 (T_m, Yamada and Komagata, 1970; Stackebrandt et al., 1983), and 65.1 (Bd, Skyring and Quadling, 1970).

Type strain: ATCC 11624 (NCIB 8915).

Further comments. A number of other strains bearing the name *A. citreus* exist in culture collections and have been widely studied; none appears to be correctly named. Thus, *A. citreus* strains ATCC 21345, 17775, 21040, 21422 and 15170 have peptidoglycans with the A4α variation (Schleifer and Kandler, 1972; Stackebrandt et al., 1983) i.e. they belong to the *A. nicotianae* group of species. Lipid data are available for ATCC 1775, 21040 and 15170 and confirm the above conclusion (Collins and Kroppenstedt, 1983). From these data and from the information provided by DNA-DNA homology studies, Stackebrandt et al. (1983) consider that *A. citreus* strains ATCC 21348, 17775, 21040 and 21422 should be redesignated as strains of *A. protophormiae* while *A. citreus* ATCC 15170 should be redesignated as a strain of *A. uratoxydans* (see *A. protophormiae* and *A. uratoxydans*).

The type strain of *A. citreus* differs from *A. globiformis* and similar species in containing fewer glycolipids (Collins and Kroppenstedt, 1983).

12. **Arthrobacter nicotianae** Giovannozzi-Sermanni 1959, 84 emend. Stackebrandt et al., 1983, 11.[AL]

ni.cot′i.an.ae. L. n. *Nicotiana*, the tobacco plant, M.L. gen. n. *nicotianae*, of the tobacco plant.

Morphology and general characters as for generic description.

Other characters are given in Table 15.12 and Further Descriptive Information.

Growth of the type strain on agar is bright lemon-yellow (Giovannozzi-Sermanni, 1959).

The type strain gives abundant growth on nicotine agar: the medium becomes blue at first and deep wine-red later (Giovanozzi-Sermanni, 1959).

Nutritionally nonexacting: growth occurs in a suitable mineral salts medium with an ammonium salt as sole nitrogen source (type strain; Robertson and Keddie, unpublished data).

Nicotine is utilized by the type strain as a sole or major carbon + energy source (Cure and Keddie, unpublished data).

The type strain was isolated from the air of tobacco warehouses, others from soil and sewage.

The mol% G + C of the DNA is in the range 60.2–63.0 (T_m, Yamada and Komagata, 1970; Stackebrandt et al., 1983).

Type strain: ATCC 15236 (NCIB 9458).

Further comments. In DNA-DNA homology studies, Stackebrandt et al. (1983) found homologies of between 81 and 93% between the type strain of *A. nicotianae* and the following strains: "*Arthrobacter nucleogenes*" ATCC 21279, *Arthrobacter* sp. NCIB 9863, *Brevibacterium* sp. AJ 1486 and *Corynebacterium liquefaciens* ATCC 14929. Accordingly, they considered that all five strains belonged to the same species, *A. nicotianae*. Data on nicotine utilization by strains other than the type are not available.

The position of strain ATCC 14929 named *Corynebacterium liquefaciens* is controversial. Formerly named *Brevibacterium liquefaciens*, it was then transferred to the genus *Corynebacterium* by Lanéelle et al. (1980), with ATCC 14929 as type strain, on the basis of a high degree of similarity in chemotaxonomic characters (peptidoglycan type, presence of arabinogalactan and shorter chain mycolic acids) and high DNA-DNA homology values with *C. diphtheriae* (PW8). On the other hand, detailed studies of the peptidoglycan structure, DNA reassociation studies (Stackebrandt et al., 1983) and the lipid composition of ATCC 14929 (NCIB 9545) indicated that it is a member of the *A. nicotianae* group of arthrobacters.

The type strain of *A. nicotianae* contains a diglycosyldiacylglycerol with the chromotographic mobility of dimannosyldiacylglycerol (Collins and Kroppenstedt, 1983).

13. **Arthrobacter protophormiae** (Lysenko 1959) Stackebrandt, Fowler, Fiedler and Seiler 1984, 270.[VP] (Effective publication; Stackebrandt et al. 1983, 482) (*Brevibacterium protophormiae* Lysenko 1959, 41.)

pro.to.phorm.′i.ae. L. n. *Protophormia*, a genus of dipteran insects, M.L. gen. n. *protophormiae* of *Protophormia*.

(Description based on those of Lysenko, 1959 and Stackebrandt et al., 1983.)

Morphology and general characters as for generic description. Nonmotile.

Other characters are given in Table 15.12 and in Further Descriptive Information.

Colonies on nutrient agar show a pale to sulfur yellow, nondiffusible pigment.

No critical work on organic growth factor requirements appears to have been done but the type strain is reported to grow on Koser's citrate medium.

The type strain was isolated from an insect *Protophormia terraenovae*, other strains from soil.

The mol% G + C of the DNA is in the range 63.2–65.9 % (T_m, see Stackebrandt et al., 1983).

Type strain: ATCC 19271 (DSM 20168).

Further comments. DNA-DNA homology studies showed a close relationship between *A. protophormiae* and four strains named *A. citreus* viz. ATCC 21348 (73%), ATCC 17775 (72%), ATCC 21040 (71%), and ATCC 21422 (75%) (Stackebrandt et al., 1983). However, the fact that all four so-called *A. citreus* strains have similar degrees of homology with *A. protophormiae* is not surprising because, according to the ATCC Catalogue (1982), all four are derived from ATCC 17775.

The type strain of *A. protophormiae* (*Brevibacterium protophormiae* CCEB 282 (ATCC 19271)) contains a diglycosyldiacylglycerol with the chromatographic mobility of dimannosyldiacylglycerol (Collins and Kroppenstedt, 1983).

14. **Arthrobacter uratoxydans** Stackebrandt, Fowler, Fiedler and Seiler 1984, 270.[VP] (Effective publication; Stackebrandt, et al., 1983, 483.)

u.ra.to′xy.dans. M.L. n. *uratum* salt of uric acid; L. part adj. *oxydans* oxidising; M.L. adj. *uratoxydans* uric acid oxidizing.

(Description based on data of Stackebrandt et al., 1983.)

Morphology and general characters as for generic description. Rods motile by peritrichous flagella or nonmotile.

Other characters are given in Table 15.12 and in Further Descriptive Information.

Colonies on nutrient agar are creamy white becoming pale yellow; water soluble, pale yellowish brown pigment produced.

No critical work on organic growth factor requirements appears to have been done but the type strain is reported to grow in Koser's citrate medium.

Uricase strongly positive.

Isolated from humus soil.

The mol% G + C of the DNA is in the range 61.2–61.5 (T_m, Stackebrandt et al., 1983).

Type strain: ATCC 21749.

Further comments. A. uratoxydans may be distinguished from *A. nicotianae* and from *A. protophormiae* by its very active oxidative decomposition of uric acid and by carbon source utilization tests listed in Table 15.12 (Stackebrandt et al., 1983).

The type strain and a second strain, ATCC 21752, are patent strains used for the production of uricase (US patent 3,767,533) and were originally described as "*Corynebacterium uratoxidans*" (Sugisaki et al., 1973) but the name was not validly published. ATCC 21749 and ATCC 21752 show 10% homology with each other (Stackebrandt et al., 1983).

15. **Arthrobacter sulfureus** Stackebrandt, Fowler, Fiedler and Seiler 1984, 270.[VP] (Effective publication; Stackebrandt et al. 1983, 484.)

sul.fu′re.us. L. adj. *sulfureus* of sulfur (i.e. sulfur colored).

(Description from those of Iizuka and Komagata, 1965 for "*Brevibacterium sulfureum*" and Stackebrandt et al., 1983.)

Morphology and general characters as for generic description. Rods motile by one to a few lateral flagella, or nonmotile.

Other characters are given in Table 15.12 and in Further Descriptive Information.

Colonies on yeast extract-peptone are dull yellow.

The type strain is nutritionally nonexacting. Growth occurs in a suitable mineral salts medium with an ammonium salt as sole nitrogen source and glucose as carbon + energy source (Robertson and Keddie, unpublished data).

The type strain was isolated from an oil brine.

The mol% G + C of the DNA is in the range 64.5–66% (T_m, Stackebrandt et al., 1983).

Type strain: ATCC 19098; (NCIB 10355).

Further comments. Other strains considered to belong to this species are ATCC 15170 (named *A. citreus*) and *Arthrobacter* sp. ATCC 21085. DNA-DNA homology studies show 62% homology between the type strain and ATCC 15170 and 65% homology between ATCC 21085 and ATCC 15170. The two strains of *A. sulfureus* examined, the type strain and ATCC 15170, differ from other *A. nicotianae* group strains in containing MK-9 as major menaquinone but with equal amounts of MK-10 in addition in ATCC 15170. These two strains also differ from other *A. nicotianae* group strains in containing a diglycosyldiacylglycerol with the chromatographic mobility of digalactosyldiacylglycerol (Collins and Kroppenstedt, 1983; see Further Descriptive Information for details).

Species Incertae Sedis

The chemotaxonomic data necessary to allow the inclusion of the following species (a–d) in the genus *Arthrobacter* are lacking.

a. *Arthrobacter mysorens* Nand and Rao 1972, 324.[AL]

(Description based on that of Nand and Rao, 1972.)

Morphology generally as for generic description; motility doubtful (sic), flagella not demonstrated. No data are available for the peptido-

glycan type, cell wall sugars or lipids of the type strain (see Further Comments).

Growth moderate on nutrient agar; colonies white becoming yellow or lemon-yellow in 3–4 days.

Obligately aerobic; weak acid production in glucose, xylose, galactose, glycerol and cellobiose in 7 days of incubation. Catalase-positive.

Growth occurs in Koser's citrate medium; ammonium salts but not nitrate are utilized as sole nitrogen source when provided with glucose as carbon + energy source.

No growth on creatine, creatinine and uric acid agar media; slight growth on nicotine agar.

Gelatin and starch hydrolyzed, lecithinase produced. Cellulose not hydrolyzed. Nitrate not reduced to nitrite; urease not produced. Indole, H_2S and acetylmethylcarbinol not produced.

Good growth in 10% NaCl broth; optimum temperature between 20–37°C, no growth at 10°C.

Glutamic acid is produced in a mineral salts medium when provided with a suitable carbohydrate; a pink water-soluble pigment is produced in the same medium.

Isolated from sewage samples in Mysore.

Type strain: NCIB 10583 (ATCC 33408).

Further comments. Stackebrandt et al. (1983) carried out a detailed study of a strain named *A. mysorens* but unfortunately used ATCC 31021, a patent strain and not the type strain. These authors considered ATCC 31021 to be a member of the *A. nicotianae* group of arthrobacters and to be distinct from those species now recognized.

b. *Arthrobacter picolinophilus* Tate and Ensign 1974, 692.[AL]

(Description based on that of Tate and Ensign, 1974, for the type strain.)

Morphology generally as for generic description; no branching observed.

Nonmotile.

Colonies on yeast-dextrose-$CaCO_3$ agar show a pink nondiffusible pigment.

Obligately aerobic; no acid from sucrose, lactose, maltose, galactose, ribose, glycerol or mannitol.

Utilizes glucose, acetate, citrate, malate or succinate as sole carbon + energy source; an ammonium salt or nitrate is utilized as sole nitrogen source.

Utilizes picolinate as sole carbon + energy source in a mineral salts medium when provided with L-methionine and thiamine.

No growth on media containing 2-hydroxypyridine, nicotine or nicotinic acid.

Cellulose, starch and gelatin not hydrolyzed. Nitrate reduced to nitrite, urease produced. Indole, H_2S and acetylmethylcarbinol not produced.

Optimum temperature 30°C; growth slow at 25 and 37°C.

Isolated from garden soil by enrichment in a medium containing mineral salts, a low concentration of yeast extract and picolinate as carbon + energy source.

The mol% G + C of the DNA is 65.2 (T_m).

Type strain: ATCC 27854.

Further comments. The type strain has been shown to contain mycolic acids with a chain length of 32–46 carbon atoms (Collins et al., 1982a) and this, together with the relatively high G + C content, suggests that *A. picolinophilus* may be a member of the genus *Rhodococcus*.

c. *Arthrobacter radiotolerans* Yoshinaka et al., 1973, 2269.[AL]

(Description based on that of Yoshinaka et al., 1973, for the type strain.)

Morphology generally as for generic description; no branching observed. Nonmotile.

Colonies on nutrient agar develop slowly, 0.5–1.0 mm in 2 weeks and 4.0–5.0 mm in 4 weeks, and show a reddish pink, nondiffusible pigment.

Obligately aerobic; growth but no acid production from glucose, fructose, xylose, arabinose, maltose and glycerol, slight growth but no acid in galactose and mannose. Sucrose, lactose sorbitol and citrate not utilized. An ammonium salt is utilized as sole nitrogen source. Catalase-positive.

Gelatin not hydrolyzed. Nitrate reduced to nitrite. Indole and acetyl methyl carbinol not produced.

Growth in 6% but not in 10% NaCl broth.

Optimum temperature 46–48°C; good growth at 55°C, no growth at 20°C.

Highly resistant to γ-irradiation.

Isolated from a radioactive hot spring after γ-irradiation of the sample.

Type strain: IAM 12072.

d. *Arthrobacter siderocapsulatus* Dubinina and Zhdanov 1975, 349.[AL]

The authors consider that the iron bacteria classified in the genus "*Siderocapsa*" and, indeed, the entire "*Siderocapsaceae*" should be considered members of the genus *Arthrobacter* (Dubinina and Zhdanov, 1975). Their evidence is almost entirely morphological and is based on their isolation, for the first time, of two strains of bacteria which they consider would previously have been called "*Siderocapsa eusphaera*" and which they now name *A. siderocapsulatus* on the basis of pure culture studies. A brief description follows.

In exponential phase on solid Pringsheim medium lacking manganese, produce long filaments which fragment to give rods in 24 h and then further divide to give cocci. Some rods swell to give "cystites" 7 μm in diameter. V-forms are stated to be present "at all times." Rods and cocci are Gram-negative; "cystites," Gram-variable. Motile by a few polar (lophotrichous) flagella. Stationary phase cells have large capsules. On solid Pringsheim medium (with manganese), colonies are brown because of deposition of manganese oxides in the capsules; on medium containing ferrous oxalate, colonies are orange because of ferric oxide deposits in the capsules. On meat peptone agar, colonies are yellowish gray and reach 10 mm in diameter.

Microaerophilic in Pringsheim semisolid medium. Acid produced from glucose, galactose and arabinose; no acid from maltose, fructose, sucrose, lactose or raffinose. Catalase-positive.

The mol% G + C of the type strain is 60.8 ± 0.5 (method not stated).

Type strain: BKM-BN 1122.

The authors also state that reference cultures of *A. globiformis, A. citreus, A simplex* and *A. tumescens* resemble *A. siderocapsulatus* in morphology and in the ability to oxidise FeS in Pringsheim medium: all but *A. tumescens* also oxidise MnSO₄.

Addendum I

Species named **Arthrobacter** *which have been shown to contain meso-diaminopimelic acid in the cell wall peptidoglycan.*

The following species (e–g) have been reported to contain a peptidoglycan based on *meso*-DAP (but this is in dispute in *A. duodecadis*) and for this reason, and usually others, are excluded from the genus.

e. *Arthrobacter duodecadis* Lochhead 1958, 170.[AL]

(Description based on type strain only.)

Morphology generally as for generic description. Nonmotile.

Growth on yeast extract-peptone medium with vitamin B_{12} added is slow and colonies show no distinctive pigmentation; inclusion of soil extract in medium results in a brownish pigmentation. Obligately aerobic; no acid production from glucose, sucrose or lactose.

Owens and Keddie (unpublished) found that the type strain utilized about 60 of 180 compounds tested as sole or principal sources of carbon + energy. They included a wide range of carbohydrates, sugar derivatives and fatty acids, a few amino acids and dicarboxylic acids, hydroxy-acids, oxo-acids, polyalcohols, non-nitrogenous aromatic compounds and amines; of those tested no simple alcohols or heterocyclic compounds were utilized (Keddie, 1974).

Vitamin B_{12} and thiamine are required for growth. Most of the usual laboratory media will not support growth unless supplemented with vitamin B_{12} and preferably also yeast extract, or with soil extract. When provided with essential vitamins and glucose as carbon + energy source, ammonium salts but not nitrate are utilized as nitrogen source (Owens and Keddie, 1969).

Gelatin hydrolyzed; starch not hydrolyzed. Nitrate is reduced to nitrite. Indole and acetylmethylcarbinol are not produced.

Growth is best at 20–30°C; grows at 10°C but not at 37°C.

The cell wall peptidoglycan is reported to have the A4γ variation, *meso*-DAP-D-aspartate₂ according to Schleifer and Kandler (1972). In conflict with this report Keddie et al. (1966) and Keddie and Cure (1977) reported that lysine is the diamino acid present in the cell wall peptidoglycan. Glucose and mannose are the major cell wall sugars (Keddie et al., 1966; Keddie and Cure, 1977).

Isolated from soil.

Mol% G + C of the DNA (CBRI 859) is 72.5 (Bd. Skyring and Quadling; 1970).

Type strain: ATCC 13347 (NCIB 9222).

f. *Arthrobacter variabilis* Muller 1961, 524.[AL]

The type strain of *A. variabilis*, ATCC 15753 (NCIB 9455) has a directly linked peptidoglycan based on *meso*-DAP and contains arabinose and galactose as major cell wall sugars (Schleifer and Kandler, 1972; Keddie and Cure, 1977). This species has also been shown to contain mycolic acids and thus it is clearly not a member of the genus *Arthrobacter*. However, its taxonomic position is controversial. Keddie and Cure (1977) considered that this species had free mycolic acids with chromatographic mobilities like those found in the genus *Rhodococcus*, whereas Minnikin et al. (1978) reported that the chain length of the mycolic acids indicated a member of the genus *Corynebacterium*. However structural studies by Collins et al. (1982) have shown that *A. variabilis* contains short chain (30–36 carbon atoms) mycolic acids like those in the genera (*Corynebacterium* and *Caseobacter*. Also 16S-rRNA cataloging studies have shown that *A. variabilis* is more closely related to the genus *Corynebacterium* than to the genus *Rhodococcus* (Stackebrandt et al., 1980). *A. variabilis*, like members of the genera *Corynebacterium* and *Caseobacter*, contains *N*-acetyl residues in the glycan moiety of the cell wall peptidoglycan (Collins, unpublished). However, *A. variabilis* resembles the genus *Caseobacter* in having a relatively high G + C content (65.5 mol%, T_m, Collins, unpublished) and in containing substantial amounts of 10-methyloctadecanoic (tuberculostearic) acid (Collins et al., 1982b).

Type strain: ATCC 15753 (NCIB 9455).

g. *Arthrobacter viscosus* Gasdorf, Benedict, Cadmus, Anderson and Jackson 1965, 150.[AL]

A. viscosus was the name given to two polysaccharide-producing bacterial strains isolated from soil by Gasdorf et al. (1965). Both strains are considered to be Gram-negative, motile, aerobic rods (ATCC catalogue 1982). The type strain, ATCC 19584 (NCIB 9729) produces from glucose large quantities of a carbohydrate polymer based on galactose, glucose and mannuronic acid. This strain (NCIB 9729) was shown to contain *meso*-DAP as cell wall diamino acid but it contained neither arabinose nor mycolic acids (Keddie and Cure, 1977). The chemotaxonomic data clearly indicate that *A. viscosus* is not a member of the genus *Arthrobacter* and are in accord with the view that it is a Gram-negative rod.

The second strain isolated by Gasdorf et al., (1965), ATCC 19583, has also been reported to contain a directly linked peptidoglycan based on *meso*-DAP and to lack arabinose in the cell wall (Schleifer and Kandler, 1972). The mol% G + C reported for ATCC 19583 is 59.4 (Skyring et al., 1971).

Type strain: ATCC 19584 (NCIB 9729).

Addendum II

Species named **Arthrobacter** *which have been shown to contain* LL-*diaminopimelic acid in the cell wall peptidoglycan.*

The following species (h. *A. simplex* and i. *A. tumescens*) contain LL-

DAP in the cell wall peptidoglycan (Cummins and Harris, 1956; Keddie and Cure, 1978) and tetrahydrogenated menaquinones with eight isoprene units (MK-8(H₄)) as their major isoprenoid quinones (Yamada et al., 1976; Collins et al., 1979, 1983f). They also have G + C contents from 7–9% higher than *A. globiformis* (Keddie and Jones, 1981) and most numerical taxonomic studies have shown that the type strains of *A. simplex* and *A. tumescens* show little resemblance to *A. globiformis* (Jones, 1978). Indeed 16S-rRNA cataloging studies have shown that *A. simplex* is only distantly related to *A. globiformis* (Stackebrandt et al., 1980). It has thus been obvious for many years that *A. simplex* and *A. tumescens* should be removed from the genus *Arthrobacter* (Keddie and Jones, 1981) but their position remains controversial (see Taxonomic Comments for discussion). Suzuki and Komagata (1983b) proposed the genus *Pimelobacter* to accommodate all the ᴌᴌ-DAP-containing coryneform species whereas O'Donnell et al., (1982) considered that *A. simplex* (but not *A. tumescens*) should be classified in the genus *Nocardioides* although it does not have a nocardioform morphology. The taxonomic position of *A. simplex* and *A. tumescens* thus remains unresolved.

See Suzuki and Komagata (1983b) for a definition of the genus *Pimelobacter*.

h. *Arthrobacter simplex* (Jensen) Lochhead 1957, 608.[AL] (*Corynebacterium simplex* Jensen 1934, 43; *Nocardioides simplex* (Jensen) O'Donnell, Goodfellow and Minnikin 1982, 327; *Pimelobacter simplex* (Jensen) Suzuki and Komagata 1983, 68).

sim'plex. L. adj. *simplex* simple.

Irregular rods in exponential phase cultures ~0.5–0.9 × 1.0–4.0 μm or more; branching is uncommon. Older cultures are composed of coccoid cells and very short rods ~0.5–0.8 × 0.7–1.5 μm. Gram-positive. Rods motile by one subpolar or a few lateral flagella or nonmotile.

Colonies on yeast extract-peptone agar show no distinctive pigmentation.

Obligately aerobic; no acid is formed from glucose and other sugars in peptone-based media.

Nutritionally nonexacting; growth occurs in a suitable mineral salts medium with an ammonium salt or nitrate as sole nitrogen source and glucose as sole carbon + energy source (Keddie et al., 1966; Owens and Keddie, 1969).

Owens and Keddie (unpublished) found that the type strain utilized about 60 of 180 compounds tested as sole or principal sources of carbon and energy (Keddie, 1974). They included a very narrow range of carbohydrates and sugar derivatives, a wide range of fatty acids, simple alcohols and amino acids, together with some hydroxy-acids, oxo-acids, amines, pyrimidines and phenol.

Gelatin hydrolyzed; cellulose and starch not hydrolyzed. Nitrate is reduced to nitrite. DNAase-positive, urease-negative (Suzuki and Komagata, 1983b).

Growth occurs at 10°C and may or may not occur at 37°C; optimum ~25–30°C. The principal amino acids in the cell wall are ᴌᴌ-DAP and glycine (Cummins and Harris, 1959, Keddie and Cure, 1977; Suzuki and Komagata, 1983b) and the major cell wall sugar is galactose (Cummins and Harris, 1959; Keddie and Cure, 1977). The peptidoglycan type is A3γ, ᴌᴌ-DAP-glycine (Schleifer and Kandler, 1972).

The major isoprenoid quinones are tetrahydrogenated menaquinones with eight isoprene units (MK-8 (H₄), (Yamada et al., 1976; Collins et al., 1979, 1983f; O'Donnell et al., 1982).

The cellular fatty acid composition is complex and consists of a mixture of nonhydroxylated and 2-hydroxylated fatty acids (O'Donnell et al., 1982; Collins et al., 1983f). The nonhydroxylated acids consist of straight chain saturated, monounsaturated, *iso*- and *anteiso*-branched and 10-methyl-branched fatty acids: tuberculostearic acid (10-methyloctadecanoic acid) is the major nonhydroxylated fatty acid (O'Donnell et al., 1982; Collins et al., 1983f). The 2-hydroxylated acids are mainly straight chain saturated, *iso*- and *anteiso*-methyl-branched fatty acids (Collins et al., 1983f).

The major polar lipids are diphosphatidylglycerol and phosphatidylglycerol (O'Donnell et al., 1982; Collins et al., 1983f) together with two closely related phospholipids which are possibly phosphatidylglycerols containing a 2-hydroxy acid (Collins et al., 1983f).

Habitat: soil.

The mol% G + C of the DNA is in the range 72.0–73.5 (T_m), Suzuki and Komagata, 1983b) and 74.0 (Bd. Skyring and Quadling, 1970).

Type strain: ATCC 6946 (NCIB 8929).

Further comments. On the basis of DNA-DNA base-pairing techniques and other studies Suzuki and Komagata (1983b) considered that two strains previously named "*Brevibacterium lipolyticum*" (IAM 1398 and IAM 1413) belonged to the species *Pimelobacter* (*Arthrobacter*) *simplex*. However, these authors also considered that one strain of *A. simplex*, (NCIB 9770) originally named *Corynebacterium simplex* by Jensen and Gundersen (1956), was sufficiently different from the others to justify the creation of a new species *Pimelobacter jensenii* to accommodate it (Suzuki and Komagata, 1983b).

i. *Arthrobacter tumescens* (Jensen) Conn and Dimmick 1947, 302.[AL] (*Corynebacterium tumescens* Jensen 1934, 45; *Pimelobacter tumescens* (Jensen) Suzuki and Komagata 1983, 70.)

tu.mes'cens L. part adj. *tumescens* swelling up.

Long, irregular rods in exponential phase cultures ~0.6–1.2 × 2.0–6.0 μm but may be much longer; the long rods show extensive primary branching (Cure and Keddie, 1973). Older cultures are composed mainly of coccoid cells ~0.6–0.9 μm in diameter. Generally nonmotile; occasional motile strains occur (Grainger and Keddie, unpublished). Gram-positive. Colonies on yeast extract peptone-agar show no distinctive pigmentation.

Obligately aerobic; no acid is formed from glucose and other sugars in peptone-based media.

Thiamin is the only organic growth factor required; when so provided, utilizes an ammonium salt or nitrate as sole nitrogen source in a suitable mineral salts medium with glucose as carbon + energy source (Keddie et al., 1966; Owens and Keddie, 1969).

A relatively wide range of organic compounds is utilized as sole or principal carbon + energy source for growth: in a study of six strains including the type strain Owens and Keddie (unpublished) found that, on average, individual strains utilized 58 out of 160 compounds tested (see Further Comments).

Gelatin hydrolyzed, cellulose not hydrolyzed. Strains differ in their ability to hydrolyze starch and to reduce nitrate to nitrite. The type strain is DNase-positive and urease-negative (Suzuki and Komagata, 1983b). Growth occurs at 10°C and may or may not occur at 37°C; optimum ~25–30°C.

The principal amino acids in the cell wall are ᴌᴌ-DAP and glycine (Cummins and Harris, 1959; Keddie and Cure, 1977; Suzuki and Komagata, 1983b) and the major cell wall sugars are galactose, glucose and mannose (Cummins and Harris, 1959; Keddie and Cure, 1977). The peptidoglycan type is A3γ, ᴌᴌ-DAP-glycine (Schleifer and Kandler, 1972).

The major isoprenoid quinones are tetrahydrogenated menaquinones with eight isoprene units (MK 8 (H₄), Yamada et al., 1976; Collins et al., 1979, 1983f; O'Donnell et al., 1982; Suzuki and Komagata, 1983b).

The cellular fatty acids (type strain) are simple, nonhydroxylated acids (O'Donnell et al., 1982; Collins et al., 1983f. Suzuki and Komagata, 1983b). They are mainly *iso*-methyl branched-chain acids: 13-methytetradecanoic acid (*iso*-C₁₅:₀) is the major component (Collins et al., 1983f). O'Donnell et al. (1982) report that 13-methylpentadecanoic acid (*iso*-15, sic) is the major component, presumably giving the same acid the wrong name in error. Substantial amounts (~20%) of monounsaturated terminally branched fatty acids (e.g. *iso*-₁₅:₁, *iso*-C₁₆:₁ etc) also occur, a feature which distinguishes this species from all other coryneform bacteria examined so far (Collins et al. 1983f). The major polar lipids (type strain) are diphosphatidylglycerol and phosphatidylglycerol (O'Donnell et al., 1982; Collins et al., 1983f) together with three unidentified phospholipids according to Collins et al. (1983f) or four phospholipids, one of which is phosphatidylinositol, according to O'Donnell et al. (1982).

Habitat: soil.

The mol% G + C of the DNA is 71.3 (T_m, Suzuki and Komagata, 1983b), 72.4 (Bd, Skyring and Quadling, 1970).

Type strain: ATCC 6947 (NCIB 8914).

Further comments. In a study of six strains (including the type strain) considered to belong to this species, Owens and Keddie (unpublished) found that individual strains generally utilized as sole or principal sources of carbon + energy, between 47 and 60 of some 180 organic compounds tested and one strain utilized 70 compounds. All six strains utilized a common core of 36 substrates which included a wide range of carbohydrates and sugar derivatives and representatives of fatty acids, dicarboxylic acids, hydroxy-acids, oxo-acids, polyalcohols, amino acids, amines and pyrimidines. The 17 characteristics which were found most useful in recognizing this species from an assemblage of more than 100 strains of coryneform bacteria, which included named strains of *Arthrobacter* and *Cellulomonas* and isolates from soil and herbage, are given in Table 15.13.

Table 15.13.

Characteristics which, when used with cell wall composition, are of greatest value in the recognition of strains of **A. tumescens**[a]

1. Vitamin requirement: thiamin only	+ (6)
2. Inorganic nitrogen utilized as sole nitrogen source[b]	+ (6)

Utilizes as carbon + energy source:

3. D-Mannose	+ (6)
4. Raffinose	+ (6)
5. α-D-Glucosamine	+ (6)
6. Acetate	+ (6)
7. Crotonate	+ (6)
8. Citrate	+ (6)
9. *Meso*-inositol	+ (6)
10. D-Alanine	+ (6)
11. Thymine and/or uracil	+ (6)
12. Suberate and/or azelate	+ (5)
13. D-Glucuronate	− (0)
14. *p*-Hydroxybenzoate and/or 3:4-dihydroxybenzoate	− (0)
15. L-Ornithine	− (0)
16. D-Phenylalanine	− (0)
17. Uric aid	− (0)

[a] Symbols: see Table 15.2; also, the figures in brackets indicate the number of strains positive for each characteristic out of six tested (Owens and Keddie, unpublished).

[b] See Owens and Keddie (1969) for method.

Genus **Brevibacterium** Breed 1953, 13[AL] emend. Collins et al. 1980, 6

DOROTHY JONES AND RONALD M. KEDDIE

Brev.i.bac.te′ri.um. L. adj. *brevis* short; Gr.dim.n. *bacterium* a small rod; M.L. neut. n. *Brevibacterium*, a short rodlet.

A marked rod-coccus cycle occurs during growth on complex media. Cells from older cultures (3–7 days) are composed largely or entirely of coccoid cells ~0.6–1.0 µm in diameter or occasionally of coccal rods. On transfer to a suitable fresh medium these forms grow out to give the irregular, slender rods characteristic of exponential phase cultures. The cells are thus variable in length but generally ~0.6–1.0 µm in diameter. Many of the cells are arranged at an angle to give V forms. Primary branching may occur but a true mycelium is not produced. **Both rod and coccoid forms are Gram-positive but some strains and older cultures decolorize readily.** Not acid-fast. No endospores are produced. Nonmotile. Optimum growth temperature is ~20–30°C or ~37°C depending on species and strain. Good growth on peptone-yeast extract (PYE) agar at near neutral pH. **Obligate aerobes.** Chemoorganotrophs. **Respiratory mode of metabolism.** Slight or no acid is produced from glucose or other carbohydrates in a peptone medium. Proteinases produced. Catalase-produced. **The cell wall peptidoglycan contains *meso*-diaminopimelic acid (DAP) as the diamino acid; the cell walls do not contain arabinose.** Mycolic acids are absent. Contain major amounts of dihydrogenated menaquinones with MK 8(H_2) or MK 8(H_2) and MK 7(H_2) predominating. The mol% G + C of the DNA is 60–67 (T_m).

Type species: *Brevibacterium linens* (Wolff) Breed 1953.

Further Descriptive Information

Although strains of *Brevibacterium* grow on a number of different nutrient media, morphology and staining reactions are best examined on EYGA medium(Cure and Keddie, 1973). As with all bacteria which exhibit a coryneform morphology, the cellular morphology and staining reactions must be examined with care. It is recommended that both young (6–24 h) and older (3–7 day) cultures be used because the cell shape (Fig. 15.5) and staining properties change during the rod-coccus growth cycle (Crombach, 1974b; Cure and Keddie, 1973; Keddie and Jones, 1981; Mulder and Antheunisse, 1963; Mulder et al., 1966; Schefferle, 1966).

Capsule formation is not a characteristic of brevibacteria, but Colwell et al. (1969) reported the production by *B. iodinum* of a capsular-like slime "which did not wash off in distilled water, sodium lauryl sulfate suspension, or acidic or alkaline rinses."

Members of the genus are nonmotile, but Colwell et al. (1969) detected "weak or paralyzed" flagella in preparations of *B. iodinum* when examined by electron microscopy. No flagella could be detected by conventional methods and the cultures were not motile (Colwell et al., 1969).

Colonies (24–48 h) on nutrient agar are opaque, small, 0.5–1 mm in diameter, convex with an entire edge and smooth, shiny surface. The growth emulsifies easily. Colonies become larger, 2–4 mm in diameter after 4–7 days of incubation.

Brevibacterium linens strains produce yellow to deep orange-red colonies when cultured on a variety of media but, for a majority of the strains, pigment production is light dependent (Crombach, 1974b; Mulder et al., 1966). Crombach (1972, 1974b) noted that one strain, B4, isolated from cheese, produced an orange pigment only if 4% (w/v) NaCl was included in the medium. This effect was not noted with other strains of *B. linens* or related organisms isolated from cheese. Jones et al. (1973) suggested that the pigment of *B. linens* is a carotenoid and that it can be recognized by the distinctive color reactions produced when the pigmented growth is treated with solutions of strong bases or glacial acetic acid (Grecz and Dack, 1961; Jones et al., 1973). Two of the most useful of these color reactions are described under Procedures for Testing for Special Characters.

Colonies of *B. casei* were described as gray-white in color by Collins et al. (1983c). However, in an earlier study of the same strains, Sharpe et al. (1976) noted the slow production of a brown, water-soluble pigment when the organisms were cultured on milk agar; no information was given on the effect of light or other factors on pigment production.

The metallic purple to blue coloration of *B. iodinum* results from the production of purple, extracellular crystals of a phenazine derivative, called iodinin, on and in the colonies and in the adjacent medium (Davis, 1939; Sneath, 1960; Colwell et al., 1969; Collins et al., 1980).

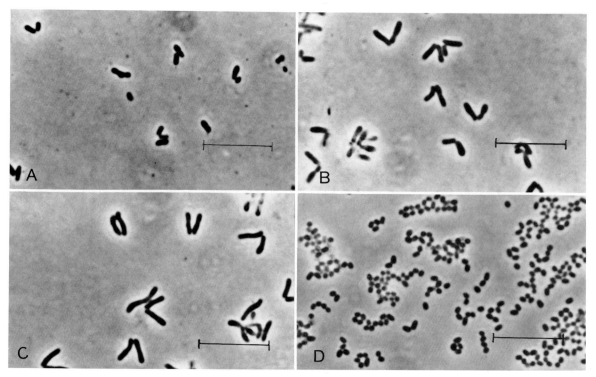

Figure 15.5. *Brevibacterium linens* (ATCC 9175) when grown on medium EYGA at 25°C; inoculum coccoid cells as in *D. A*, after 6 h, showing outgrowth of rods from coccoid cells; *B*, after 12 h; *C*, after 24 h; *D*, after 3 days (*bar*, 10 μm). (Reproduced with permission from R. M. Keddie and D. Jones. *In* Starr, Stolp, Trüper, Balows and Schlegel (Editors), The Prokaryotes: A Handbook on Habitats, Isolation and Identification of Bacteria, Springer-Verlag, New York, p. 1854, 1981).

For the chemical structure of iodinin, see Clemo and McIlwain (1938) and Clemo and Daglish (1950). The production of iodinin is not affected by light, but Sneath (1960) noted enhanced production on certain media; of those tested, a medium based on unhopped stout beerwort was found to be the most suitable. However, good production of iodinin is obtained on Blood Agar Base (Difco) incubated at 30°C for 2 days. Sneath (1960) noted more profuse iodinin production by strain NCDO 613 than by strain ATCC 9897. It should be noted that strain ATCC 9897 is derived from the original isolate, NCDO 613, of Davis (1939), as are probably all extant strains of *B. iodinum* (Colwell et al., 1969).

It has been shown that the biosynthesis of the phenazine, iodinin, and that of the much smaller amount of 2-aminophenoxazinone produced by *B. iodinum* are closely related. The source of nitrogen in both is primarily from the amide nitrogen of glutamine (Herbert et al., 1982; Römer and Herbert, 1982). Factors triggering and inhibiting the production of phenazine are discussed by Römer and Herbert (1982).

Brevibacteria usually grow poorly or not at all at 5°C but some strains show moderate growth after 2 weeks. All brevibacteria examined grow at 30°C but not all strains show optimum growth at this temperature. Most strains of *B. linens* studied show optimum growth between 20 and 25°C (Keddie and Jones, 1981) but a few strains grow best at 30°C (ATCC 19391) or 37°C (ATCC 21330) (Sharpe et al., 1977). Most *B. linens* strains tested have relatively low maximum temperatures in the range 30–33°C (G. L. Cure and R. M. Keddie, unpublished). This was also the experience of Mulder et al. (1966) but these workers reported that maximum temperatures of *B. linens* strains were slightly higher when 4% (w/v) NaCl was incorporated in the medium. However, Sharpe et al. (1977) reported that a few strains of *B. linens* grow well at 37°C (ATCC 8377, ATCC 21330) and even at 40°C (ATCC 21330) in Nutrient Broth (Oxoid). It is possible that these differences between strains of *B. linens* may be explained by the report of heterogeneity in the species when DNA-DNA hybridization studies were carried out (Fiedler et al., 1981), although there is, as yet, no evidence to support this explanation.

B. iodinum gives optimum growth at 25°C (Jones, 1975). Colwell et al. (1969) reported growth between 7 and 37°C, and Sneath (1960) reported growth between 10 and 37°C with optimum growth at 35°C, but Jones (1975) found growth to be poor or absent at 37°C.

The relatively few strains of *B. casei* and *B. epidermidis* studied grow between 22 and 40°C (Sharpe et al., 1976, 1977). Optimum growth occurs between 30 and 37°C (Pitcher and Noble, 1978; Sharpe et al., 1976, 1977) and most grow well at 40°C (Sharpe et al., 1976, 1977).

Most brevibacteria studied do not survive heating at 60°C for 30 min (Jones, 1975; Mulder et al., 1966). However, Sharpe et al. (1976, 1977, 1978) reported that most of the strains isolated from cheese and skin (and later designated *B. casei* and *B. epidermidis*, Collins et al., 1983c) survived heating at 60°C for 30 min but in considerably reduced numbers.

B. linens strains grow well at neutral pH; growth occurs between pH 6.5 and 8.5 (Mulder et al., 1966), but according to Mulder et al. (1966) the addition of 4% (w/v) NaCl to the medium allows good growth of *B. linens* strains at pH 6.0 and moderate growth at pH 5.5. *B. iodinum* (ATCC 9897) is reported to grow between pH 6.0 and 10.0 (Sneath, 1960).

All brevibacteria studied tolerate or are sometimes stimulated by the addition of NaCl to the medium. *B. linens* strains are described as salt tolerant, but there is some disagreement about the actual concentration of NaCl which will allow growth of most strains. These discrepancies may be due to the different experimental conditions and incubation times used, but they may also reflect the heterogeneity detected among strains of *B. linens* on the basis of DNA-DNA hybridization studies (Fiedler et al., 1981). When a suitable basal medium at pH 7.0 is used, it appears that all strains tested give growth in 8% (w/v) NaCl in about 1 week at 25°C, and in 12% (w/v) NaCl after about 1 month at 25°C (Crombach, 1974b; El-Erian, 1969; Mulder et al., 1966). Sharpe et al. (1977) reported that five of the seven strains (including the type strain) of *B. linens* examined by them grew in 15% (w/v) NaCl (the highest concentration tested) in a medium containing Owens and Keddie's

(1969) Mineral Base E (prepared as described by Cure and Keddie, 1973), 0.2% (w/v) yeast extract and 0.1% (w/v) glucose. The remaining strains grew in a similar medium containing 12% (w/v) NaCl. Tests were incubated at both 22 and 30°C but no details of incubation times were given. The same workers noted that all but one of the strains of brevibacteria from cheese and skin (now assigned to *B. casei* and *B. epidermidis*, Collins et al., 1983c) grew in 15% (w/v) NaCl in the medium described above (Sharpe et al., 1977). In an earlier report, Sharpe et al. (1976) noted that 12% (w/v) NaCl was stimulatory for strains of *B. casei*, while all could grow in 15% (w/v) NaCl.

Sneath (1960) reported that *B. iodinum* grew profusely on 6.5% (w/v) NaCl agar and Jones (1975) obtained good growth in a broth culture containing 10% (w/v) NaCl incubated at 30°C for 7 days.

Although few extensive comparative studies have been done, those brevibacteria tested appear to be sensitive to the usual antibiotics when tests are carried out in vitro by the agar diffusion method (Colwell et al., 1969; Jones, 1975; Sneath, 1960). Of interest is the report by Pitcher and Noble (1978) that 88% of methanethiol-producing, skin brevibacteria (*B. epidermidis*) isolated from human feet infected with *Trichophyton* spp. were resistant to penicillin G compared with only 20% of other coryneform bacteria isolated from the same source.

Information on the nitrogen and vitamin requirements of brevibacteria is limited and sometimes contradictory. Most of the orange *B. linens* strains from cheese examined by Mulder et al. (1966) required both amino acids and vitamins for growth. Only one of the 31 strains tested was able to grow on a medium containing inorganic nitrogen without added organic growth factors. Most strains required both amino acids and B-vitamins for growth but a few grew in a suitable basal medium when provided with amino acids alone. Robertson and Keddie (unpublished) found similar heterogeneity in the nutritional requirements of 18 strains of *B. linens*. Nine strains required both amino acids and vitamins while the remaining nine strains grew when supplied with amino acids only. Bousfield (1972), using an agar medium, found that *B. linens* and orange cheese coryneforms of the *B. linens* type were able to use ammonium nitrogen but presumably the agar provided organic nitrogen compounds. Smith (1969) reported that the type strain of *B. linens* (ATCC 9172) utilized ammonia as a sole source of nitrogen in a liquid medium but Mulder et al. (1966) and Robertson and Keddie (unpublished) found that this strain required both vitamins and amino acids for growth.

Sharpe et al. (1976) investigated the growth requirements of two strains of *B. casei* (CMD1 (NCDO 2048) and CMD3 (NCDO 2049)). Growth did not occur in Owens and Keddie's (1969) Mineral Base E −N (which does not contain an ammonium salt) in the presence of added ammonium salt or L-methionine but growth did occur through three transfers in the presence of ammonium salt plus yeast extract or ammonium salt plus glucose, indicating that exogenous organic growth factors were not required, and that the organisms were nutritionally nonexacting.

Colwell et al. (1969) reported that *B. iodinum* utilized ammonium phosphate as a nitrogen source and growth occurred in a basal salts solution supplemented with L-proline (0.1%), DL-alanine (0.1%) or β-alanine (0.1%). A similar medium containing L-arginine did not support growth.

The vast majority of brevibacteria examined do not produce detectable acid from glucose or other carbohydrates in a peptone water medium (Sneath, 1960; Colwell et al., 1969; Bousfield, 1972, 1978; Yamada and Komagata, 1972a, b; Crombach, 1974b; Jones, 1975; Sharpe et al., 1976, 1977, 1978; Collins et al., 1980; Keddie and Jones, 1981; Collins et al., 1983c). However, Sharpe et al. (1976) reported weak production of acid aerobically by four strains of *B. casei* when tested by the method of Hugh and Leifson (1953). When tested by the method of Latham and Legakis (1976) using gas chromatography, however, the only volatile acid produced by *B. casei* was acetic acid.

Brevibacterium linens and related organisms utilize a number of substances as carbon + energy sources. Yamada and Komagata (1972a) reported that the following organic acids (supplied as the sodium salt)

were assimilated: acetic, pyruvic, L-lactic, D-lactic, malic, succinic, fumaric, α-ketoglutaric, citric, formic, propionic, butyric, oxalic, malonic, glutaric, adipic, pimelic, glycolic, glyoxylic, gluconic, hippuric and uric when tested at a concentration of 0.5% (w/v) in a medium containing yeast extract (Difco) 0.01% (w/v); trypticase (BBL), 0.01% (w/v); glucose, 0.02% (w/v); K_2HPO_4, 0.1% (w/v); NaCl, 0.5% (w/v); phenol-red 0.0012% (w/v); agar (Difco), 2% (w/v). Assimilation of the compounds was detected by a change in the color of the indicator after 10 days. There were some differences between the strains of *B. linens* examined. These differences may be correlated with the genetic heterogeneity detected (Fiedler et al., 1981) among strains of *B. linens*, but there is no evidence to support this theory. Crombach (1974b) reported the utilization of acetate, glucose, glycerol and lactate by most strains of *B. linens*, but one strain, B3, differed in using lactose but not acetate. Schefferle (1966) found that *B. linens* (NCIB 8546) utilized acetate, citrate, gluconate, lactate, propionate and succinate and decomposed uric acid to urea. Bousfield (1972) reported the utilization of acetate, citrate, lactate and pyruvate by strains of *B. linens*. In a later study of 24 isolates, mainly from sea water and thought to be members of the *B. linens* group, and two strains of *B. linens* (a reference strain and NCMB 1322), Bousfield (1978) found that most of the strains utilized glucose, malate, fumarate, citrate, lactate, pyruvate, alanine and tyrosine as sole or major carbon sources, and over half utilized fructose, galactose, glycerol, acetate, arginine, serine and lysine. Mulder et al. (1966) noted that glucose, glycerol, lactate and acetate were good carbon sources for most *B. linens* and *B. linens*-like strains examined while sucrose and lactose were moderately good substrates but were not utilized by all strains.

There is little information on carbon source utilization by strains of *B. casei* and *B. epidermidis*. Sharpe et al. (1976, 1978) reported that all the strains of *B. casei* examined by them utilized acetate, glucose, lactate and sucrose, and one strain also utilized lactose. None utilized inulin or citrate. *B. epidermidis* strains are reported to utilize acetate and lactate (Sharpe et al., 1977, 1978).

Sneath (1960) and Colwell et al. (1969) both reported the slow utilization of citrate by strains of *B. iodinum*.

Most brevibacteria which have been studied are oxidase negative (Crombach, 1974b; Sharpe et al., 1976) but Jones (1975) found *B. linens* (ATCC 9174) to be weakly oxidase positive and Crombach (1974b) noted that two orange cheese isolates (AC 251 and AC 423) were oxidase-positive. It is of interest that strain AC 251 is a member of the second homology group (the group which did not contain the type strain, ATCC 9172) found among strains labeled *B. linens* (Fiedler et al., 1981). Strains of *B. iodinum* are strongly oxidase-positive (Colwell et al., 1969; Jones, 1975; Collins et al., 1980).

Meyer and Jones (1973) found that both logarithmic and stationary phase cultures of *B. linens* (ATCC 9174) contained cytochromes $bcaa_3o$.

All brevibacteria are proteolytic; gelatin, casein and milk are hydrolyzed by most strains (Colwell et al., 1969; Bousfield, 1972; Yamada and Komagata, 1972a, b; Crombach, 1974b; Sharpe et al., 1976, 1977, 1978 and see Keddie and Jones, 1981).

All strains of *B. linens*, *B. casei* and *B. epidermidis* examined produce methanethiol (CH_3SH) from L-methionine (Sharpe et al., 1976, 1977, 1978; Pitcher and Noble, 1978).

Brevibacteria produce extracellular deoxyribonuclease, but the amount produced may differ between species (Yamada and Komagata, 1972a; Sharpe et al., 1976, 1978; Collins et al., 1980).

Tyrosine is decomposed to some extent by all strains (Jones, 1975; Sharpe et al., 1977, 1978) but melanin production has not been noted (Colwell et al., 1969; Jones, 1975; Sharpe et al., 1977, 1978).

Starch is not hydrolyzed (Crombach, 1974b; Jones, 1975; Sharpe et al., 1976).

Urease is not produced (Colwell et al., 1969; Yamada and Komagata, 1972a, b; Jones, 1975; Sharpe et al., 1977).

Kato et al. (1984) reported the production of bacteriocins (linecins) by two strains of *B. linens* (ATCC 9175 and one other). These linecins inhibited the growth of *B. linens* ATCC 9172 (the type strain) and

ATCC 8377. Strains of *B. iodinum*, *B. casei* and *B. epidermidis* were not tested.

The cells do not contain mycolic acids (Goodfellow et al., 1976; Keddie and Cure, 1977; Sharpe et al., 1977; Collins et al., 1980).

The cell wall peptidoglycan contains *meso*-DAP as diamino acid (Fiedler et al., 1970; Schleifer and Kandler, 1972; Yamada and Komagata, 1972a; Keddie and Cure, 1977, 1978; Pitcher and Noble, 1978; Sharpe et al., 1976, 1977, 1978; Anderton and Wilkinson, 1980; Fiedler et al., 1981). The walls contain a group A peptidoglycan, i.e. one in which there is a direct cross-linkage between positions 3 and 4 of the two peptide subunits (variation A1γ, Schleifer and Kandler, 1972).

Arabinose is not present in the cell wall. Keddie and Cure (1977) reported that ribose was present in the cell wall of *B. linens*, but subsequent, more detailed, studies have failed to detect this sugar in the strains examined (Fiedler and Stackebrandt, 1978; Fiedler et al., 1981 and see Keddie and Bousfield, 1980). Fiedler et al. (1981) showed that the polysaccharide moieties of the cell walls of *B. linens* and related strains exhibited six different combinations of neutral sugars, amino sugars and sugar alcohols. Phosphate was always present. A basic pattern of glycerol, glucose, glucosamine and/or galactosamine was detected in all the strains. Two strains contained only the basic pattern of components but in some strains, including the type strain ATCC 9172, galactose was present as an additional sugar. Other strains contained ribitol in addition to glycerol and some of them also contained galactose. In yet other strains, mannitol was characteristically associated with the basic pattern and either galactose alone, or galactose and rhamnose were also present.

The studies of Fiedler et al. (1981) also demonstrated the presence of glycerol teichoic acids in the cell walls of all *B. linens* and related strains examined. In addition, some of the strains (ATCC 19391 and AC 831) contained ribitol teichoic acids, and in other strains there was evidence for the presence of unusual teichoic acids containing mannitol (strains AC 251 and AC 480) and mannitol and rhamnose (strain B3).

Anderton and Wilkinson (1980) showed that the major polysaccharide components of the cell walls of *B. iodinum* (NCTC 9742) were one or both of two carbohydrate-containing polymers. One, an unusual teichoic acid containing D-mannitol, glycerol, pyruvic acid, D-glucose and D-galactosamine was present in the walls both when the strain was grown in liquid culture (Nutrient Broth, Oxoid CM 67) with aeration or grown on Nutrient Agar (Oxoid). The second polymer, which apparently contained galactose and a 2-keto-3-deoxyaldonic acid, was present only in the cell walls from surface cultures on agar. Anderton and Wilkinson (1980) speculated that the second polymer was produced in response to phosphorus limitation of growth.

The presence of teichoic acids in the walls of brevibacteria distinguishes them from all other coryneform bacteria so far examined (Fiedler and Stackebrandt, 1978; Anderton and Wilkinson, 1980; Fiedler et al., 1981).

Sharpe et al. (1977, 1978) demonstrated galactose in the cell walls of *B. casei* and *B. epidermidis*.

All species of *Brevibacterium* contain similar fatty acids. All contain major amounts of *anteiso*- and *iso*-methyl-branched acids together with small amounts of straight chain saturated acids. The major fatty acids are 12-methyltetradecanoic (*anteiso*-C_{15}) and 14-methylhexadecanoic (*anteiso*-C_{17}) (Bowie et al., 1972; Collins et al., 1980; 1983c; Bousfield et al., 1983; Suzuki and Komagata, 1983).

All brevibacteria contain the polar lipids diphosphatidylglycerol and phosphatidylglycerol. Collins et al. (1980) reported both these lipids and a glycolipid, dimannosyldiacylglycerol in *B. linens* ATCC 9172, ATCC 9174 and NCIB 8546 and in *B. iodinum* strains NCDO 613 and NCIB 8179. However, Komura et al. (1975) also detected phosphatidylinositol in strains of *B. linens*. Collins et al. (1983a) found diphosphatidylglycerol and phosphatidylglycerol in strains of *B. casei* and *B. epidermidis*. Some of the strains examined also possessed phosphatidylinositol but others (*B. casei* NCDO 2050 and NCDO 2051) lacked this phospholipid. In addition, all strains of *B. casei* and *B. epidermidis* examined possessed a glycolipid which Collins et al. (1983a) identified

as a diglycosyldiacylglycerol. The presence of phosphatidylinositol in *B. linens*, *B. casei* and *B. epidermidis* has been shown to vary with cultural conditions (M. D. Collins, personal communication).

Menaquinones are the only isoprenoid quinones which have been detected in brevibacteria. Strains of *B. linens*, *B. iodinum* and *B. epidermidis* contain major amounts of dihydrogenated menaquinones with eight isoprene units, MK-8(H_2), although small amounts of MK-8 and MK-7(H_2) are also present (Collins et al., 1979; Collins et al., 1980; Collins et al., 1983a). In contrast, strains of *B. casei* contain comparable amounts of MK-8(H_2) and MK-7(H_2) (Collins et al., 1983a).

Recent improvements in the taxonomy of the brevibacteria have resulted in improvements in identification of isolates (Collins et al., 1980; Keddie and Jones, 1981; Sharpe et al., 1976, 1977, 1978; Pitcher and Noble, 1978; Collins et al., 1983a). Consequently, it is now recognized that, in addition to cheese, brevibacteria exist in a number of different habitats and especially those where there is a high salt concentration. The usual habitat of strains of *B. linens* is on the exterior of surface-ripened cheeses of the Limburger variety, but it occurs also on cheeses such as Brick, Camembert, Roquefort and others (Mulder et al., 1966; El-Erian, 1969). *B. linens* is believed to contribute to the surface color of Limburger and similar cheeses and also, in part, by its proteolytic activity, to the ripening of such cheeses, (Albert et al., 1944). More recently, Sharpe et al. (1977) reported the production of methanethiol from L-methionine by the seven strains of *B. linens* tested. Methanethiol is an important constituent of the aroma of Cheddar cheese and these workers suggested that the production of this compound by *B. linens* might be important also in the aroma and flavor of surface-ripened cheeses. As far as we are aware, no strain which can be allocated unequivocally to the species *B. linens* has been isolated from material other than cheese. However, ATCC 19391 which, on the basis of DNA homology studies (Fiedler et al., 1981), is a member of the species *B. linens*, is listed in the ATCC Catalog as a patent strain which produces L-lysine but its source is not given.

The strains now classified as *B. casei* (Collins et al., 1983a) were isolated from milk, cheese curd and Cheddar cheese (Sharpe et al., 1976). However, these authors were of the opinion that the natural habitat of the organisms was unlikely to be milk and suggested that their high growth temperature indicated a human or animal origin which, because of their halotolerance, was possibly the skin. Sharpe et al. (1977, 1978) noted the close similarity between their cheese isolates and methanethiol-producing coryneforms isolated from human skin by D. G. Pitcher (Sharpe et al., 1977, 1978; Pitcher and Noble, 1978). These bacteria are now classified as *B. epidermidis* (Collins et al., 1983a). Pitcher and Noble (1978) were of the opinion that *B. epidermidis* forms part of the resident flora of the human skin. This is contrary to the view of Smith (1969) who, in a study of the diphtheroids of the human skin, could find no evidence that brevibacteria formed part of the indigenous flora. However Smith (1969), who used only conventional and nutritional tests, did note that "... a precise distinction between the *Brevibacterium* and *Corynebacteriaceae* does not exist."

The only known report of the isolation of *B. iodinum* is from milk (Davis, 1939). Its natural habitat is not known.

Brevibacteria have been isolated from habitats other than dairy products and skin. Crombach (1974b) and Mulder et al. (1966) considered some orange-pigmented bacteria isolated from various sea fish to be very similar to *B. linens* in their morphological and physiological characteristics and in their DNA base ratios (Crombach, 1972; 1974a). A later study (Fiedler et al., 1981) showed that two of the fish strains examined by them (AC 470 and AC 474) were indeed members of the genus *Brevibacterium*, but on the basis of DNA homology results were members of a species distinct from *B. linens*. A group of coryneform bacteria isolated from sea water which clustered with a reference culture of *B. linens* in a numerical taxonomic study and which were also similar to *B. linens* in mol% G + C values and in cell wall composition, were considered by Bousfield (1978) to be members of the same genus.

Schefferle (1966) thought that a large number of orange coryneform bacteria isolated from poultry deep litter were closely related to *B. linens* (see also Mulder et al., 1966) but, of those subsequently examined, only two strains were grouped with reference cultures of *B. linens* in the numerical taxonomic study of Jones (1975), and few had a cell wall composition similar to that of *B. linens* (Keddie and Cure, 1977). In this context, it is interesting that Mohan (1981) isolated two methanethiol-producing coryneform bacteria which he identified as *Brevibacterium* sp. from bumble foot lesions in poultry. No other bacteria or mycoplasmas were isolated from aspirates of the lesions but the two brevibacteria proved to be nonpathogenic when injected into experimental birds. Mohan (1981) therefore concluded that the bacteria were secondary invaders. There appear to be no other published reports of such a phenomenon but Mohan (1981) quotes a personal communication from L. R. Hill that "Brevibacteria have been isolated from human lesions, normal tissue and from cattle."

Keddie and Cure (1977) suggested that three isolates from pig manure slurry were related to *B. linens*.

Brevibacteria, in particular *B. linens* and *B. casei*, have been implicated in contributing to the aroma of various cheeses by the production of methanethiol and other sulfur compounds (Manning and Robinson, 1973; Manning, 1974; Sharpe et al., 1977). Furthermore, some strains of *B. linens* have been shown to produce *S*-methylthioacetate, an important aroma component of smear-coated cheeses (Cuer et al., 1979). However, with sea fish the same compounds would contribute to spoilage and the isolation of methanethiol-producing strains designated *Arthrobacter* or corynebacteria, but not further described or identified (Cantoni et al., 1969), from canned hams suggests that brevibacteria may occur as spoilage organisms in other foods. In the case of human skin surfaces, brevibacteria probably contribute to unpleasant body odors.

Enrichment and Isolation Procedures

Brevibacteria grow well on the usual nutrient media based on peptone, yeast extract and glucose. Isolation of the organisms is usually achieved by the use of such media frequently supplemented with cheese, milk, an elevated concentration of NaCl or sea water, as appropriate, depending on the material under investigation.

Incubation should be at 20–25°C in the light. Incubation of the plates in light during the period of active growth is important because it has been noted that 50% or more of the *B. linens* group produce an orange pigment only when exposed to light (Crombach, 1974b; Mulder et al., 1966). The pigment does not develop if the plates are exposed to light only after the colonies have developed fully and growth has ceased (Mulder et al., 1966). A convenient way to overcome this problem is to incubate at 25°C in an incubator until small colonies are visible, and then to remove the plates to a bench exposed to daylight for the remainder of the incubation period.

Since the media used are not selective for brevibacteria, pigmented and unpigmented colonies are selected from plates at random and examined for the presence of a coryneform morphology. However, if identification is confined to those exhibiting a rod-coccus growth cycle, then bacteria such as *Arthrobacter* and *Rhodococcus* may be mistaken for brevibacteria. Further tests must therefore be done before isolates can be identified as brevibacteria.

Isolation from Cheese

Samples of cheese or cheese curd are homogenized in 2% (w/v) trisodium citrate in a laboratory blender then plated on tryptone soya agar containing 4% (w/v) added NaCl (El-Erian, 1969). This medium (TSAS)* is based on Tryptone Soya Broth (Oxoid). Similar products are manufactured by BBL and Difco. After incubation, part of the time at least in daylight, for 5–7 days at 20–25°C, pigmented (especially orange-pigmented colonies) and nonpigmented colonies are selected randomly, examined microscopically, and those showing a coryneform morphology are then investigated further.

A medium which has been reported to be particularly suitable for the isolation of *B. linens* from cheese is that devised by Albert et al. (1944). On this medium, which contains ripened cheese, *B. linens* is reputed to give good pigment production in 5–7 days at 21°C or at room temperature, especially when plates are incubated in an oxygen-enriched atmosphere (Keddie and Jones, 1981).

Sharpe et al. (1976) isolated brevibacteria from various dairy products by direct plating on to a medium containing 30% (w/v) skim milk and 2% (w/v) agar. All isolates were tested for the production of methanethiol from L-methionine (see Procedures for Testing for Special Characters) and then examined for a coryneform morphology.

Isolation from Skin

Swabs moistened in physiological saline are firmly rubbed over the appropriate skin area and then streaked on to TSAS agar or Blood Agar Base No. 2 (Difco). After incubation at 30°C for 5–7 days colonies are randomly selected and investigated for coryneform morphology. Further tests are then carried out.

Isolation from Fish, Fish Boxes and Other Habitats

Homogenized fish or other materials or swabs of the surfaces of fish boxes are plated out on TSAS agar or Blood Agar Base No. 2 (Difco) and treated as described above. In the case of isolation from sea water, a double-strength medium should be used. Equal volumes of sea water and cooled, molten agar can then be mixed and plated.

Enrichment Procedures

The ability of all strains of brevibacteria so far examined to produce methanethiol from L-methionine has been used as the basis of enrichment. Sharpe et al. (1976) incubated raw milk and homogenates of cheese curd and Cheddar cheese in a mineral-mix medium containing (% w/v): K_2HPO_4, 0.2; KH_2PO_4, 0.1; $CaCl_2 \cdot 2H_2O$, 0.1; $MgCl_2 \cdot 6H_2O$, 0.01; $FeCl_2 \cdot 6H_2O$, 0.001; plus DL-methionine, 0.5% (w/v). Samples are then plated on one of the usual nonselective media and randomly selected colonies further investigated.

Maintenance Procedures

The organisms may be preserved for some months by stab inoculation in Nutrient Agar (Oxoid), or other similar media, in screw-capped containers. After overnight incubation with the caps loose at 25 or 30°C, the caps should be screwed tightly to prevent evaporation. The containers may be stored at room temperature (~20°C), or better at ~4°C in the dark. Cultures may be preserved for longer periods (at least 7 years) by freezing in glass beads at ~−70°C (Feltham et al., 1978; Jones et al., 1984) or by lyophilization.

Procedures for Testing for Special Characters

Presence of rod-coccus growth cycle. This may be demonstrated by the methods described by Cure and Keddie (1973).

Cell wall composition. Satisfactory results may be obtained with one of the rapid methods described by Bousfield et al. (1985).

Screening for the presence of mycolic acids. This may be carried out by whole organism methanolysis as described by Minnikin et al. (1975).

Polar lipid patterns. These may be determined by the two dimensional TLC analysis of free lipid extracts (Collins et al., 1983a).

Color reaction for identification of B. linens (Grecz and Dack, 1961; Jones et al., 1973). The two most useful color reactions are those which result from the use of a strong base or glacial acetic acid.

---

* Tryptone Soya Agar Salt (TSAS) medium: Tryptone Soya Broth (TSB, Oxoid) to which a further 4% (w/v) NaCl and 1.2% (w/v) agar is added. Final pH 7.0. The medium is sterilized at 121°C for 15 min.

1. A small amont of growth from orange-pigmented colonies is placed on a white ceramic tile and a drop of 5 M NaOH or 5 M KOH is deposited on the growth material. A stable pink-red color which develops in ~2 min is presumptive evidence of *B. linens* or an orange-pigmented member of the second homology group detected by Fiedler et al. (1981) among strains labeled *B. linens*. If it is not intended to use the growth on the plate further, the alkali may be added directly to colonies on plates.
2. A small amount of growth from suspected colonies is removed and lightly smeared on to a disk of Whatman No. 1 filter paper moistened with glacial acetic acid. The growth is then rubbed on to the paper with a blunt-ended glass rod. A stable salmon-pink color which develops within ~1 min indicates *B. linens* or, as noted above, a member of the second DNA homology group reported by Fiedler et al. (1981).

A suitable reference strain, preferably the type strain of *B. linens* (ATCC 9172) should be used as a positive control, and it is recommended that both tests be used. In the case of small colonies which appear on isolation plates, it may be necessary to subculture onto agar slants before doing the tests.

Test for the production of methanethiol. Sharpe et al. (1978) recommend the DTNB method of Laakso (1976) because it is more rapid and, in their experience, 8 times more sensitive than the Conway diffusion method (Sharpe et al., 1978).

Organisms grown on nutrient agar slopes are suspended (to give an E_{580} of 10) in 0.05 M Tris-HCl buffer at pH 8.0; 1-ml samples are then incubated in rubber-stoppered test tubes with 12.5 mM L-methionine and 0.25 mM 5,5′-dithiobis-(2-nitrobenzoic acid) (DTNB) in a total volume of 5 ml, for 2 h at 30°C. Methanethiol production is indicated by the development of a yellow color in the test incubations. Positive tests are discernible by eye but weak methanethiol producers may be confirmed colorimetrically at E_{420} after centrifugation of the bacterial growth. Controls (a) without cell suspensions, and (b) without L-methionine should be incubated.

Differentiation of the genus **Brevibacterium** from other genera

Members of the genus *Brevibacterium* may be distinguished from other coryneform and morphologically similar genera which contain *meso*-DAP in the cell wall or which show a rod-coccus growth cycle by the characters listed in Table 15.14.

Brevibacteria are most likely to be confused with members of the genera *Arthrobacter* and *Rhodococcus*. All are strictly aerobic and exhibit a rod-coccus growth cycle. However, while brevibacteria contain *meso*-DAP as the cell wall diamino acid, arthrobacters contain lysine.

Although members of the genus *Rhodococcus* also contain *meso*-DAP as the cell wall amino acid, as do the genera *Corynebacterium*, *Nocardia* and *Mycobacterium*, members of all these genera contain arabinose as the characteristic cell wall sugar; brevibacteria do not contain arabinose. In addition, unlike the other *meso*-DAP-containing bacteria mentioned above, brevibacteria do not contain mycolic acids, do possess *anteiso*- and *iso*-methyl-branched fatty acids, and lack phosphatidyl-inositol dimannosides. The presence of teichoic acids in the wall polysaccharide of brevibacteria, distinguishes them from all other coryneform bacteria so far examined (Fiedler and Stackebrandt, 1978; Anderton and Wilkinson, 1980; Fiedler et al., 1981).

Bousfield et al. (1983) have suggested that computer-assisted interpretation of fatty acid analyses combined with morphological examination may be sufficient to identify strains as members of the genera *Brevibacterium*, *Corynebacterium*, *Arthrobacter*, *Cellulomonas*, *Oerskovia*, *Caseobacter* or *Kurthia*.

Taxonomic Comments

Many changes have taken place in the classification of the genus *Brevibacterium* since it was first proposed by Breed (1953), with *B. linens* as type species, for a number of nonsporeforming Gram-positive rods formerly classified in the genus *Bacterium*. The genus was recognized in the seventh edition of the *Manual* where it was classified with the genus *Kurthia* in the family *Brevibacteriaceae*. The genus *Brevibacterium* then contained 23 species which were described as typically short, unbranching rods that were usually nonmotile; no mention was made of a coryneform morphology.

However, later workers showed that *B. linens* had a coryneform morphology and indeed that it showed a rod-coccus growth cycle similar to that seen in *Arthrobacter globiformis* (Mulder and Antheunisse, 1963; Schefferle, 1966). Indeed, in some earlier numerical taxonomic studies, it was even proposed that *B. linens* should be reclassified as *Arthrobacter linens* (Da Silva and Holt, 1965; Davis and Newton, 1969; Bousfield, 1972). However, Fiedler et al. (1970) showed not only that the peptidoglycan type in *B. linens* was quite different from that in *Arthrobacter globiformis* and closely related species, but also that the 28 species named *Brevibacterium* that they examined were heterogeneous in peptidoglycan type.

Mainly for these reasons the genus *Brevibacterium* was listed as incertae sedis in the eighth edition of the *Manual*. Consequently, all the nomenclatural species (except *B. vitarumen*, which was omitted in error) listed in the seventh edition of the *Manual* and some additional nomenclatural species, a total of 43 in all, were listed as incertae sedis in the eighth edition of the *Manual* (Rogosa and Keddie, 1974).

Later numerical taxonomic (Seiler et al. 1977 and Jones, 1978) and chemotaxonomic (Schleifer and Kandler, 1972; Keddie and Cure, 1978; Minnikin et al., 1978) studies amply confirmed the heterogeneity of the group and also indicated that *B. linens* was a distinct taxon which

Table 15.14.

Characters differentiating **Brevibacterium** *from similar genera which either have a rod-coccus growth cycle or which have meso-DAP as cell wall diamino acid[a,b]*

Genera	Rod-coccus cycle[c]	Oxygen requirements	Acid from glucose[d]	Cell wall Diamino acid	Cell wall Arabinose present	Mycolic acid	Major menaquinone
Arthrobacter	+	Aerobic	−	Lysine	−	−	MK-9 (H₂) or MK-8 and/or MK-9
Brevibacterium	+	Aerobic	−	*meso*-DAP	−	−	MK-8 (H₂)
Caseobacter	+	Aerobic	−	*meso*-DAP	+	+	MK-9 (H₂), MK-8 (H₂)
Corynebacterium	−	Facultative[e]	+[f]	*meso*-DAP	+	+	MK-9 (H₂), MK-8 (H₂)
Rhodococcus	±	Aerobic	−	*meso*-DAP	+	+	MK-8 (H₂) or MK-9 (H₂)

[a] Data from Schleifer and Kandler, 1972; Keddie and Cure, 1977, 1978; Crombach, 1978; Keddie and Bousfield, 1980; Collins and Jones, 1981a; Keddie and Jones, 1981; Collins et al. 1983c.

[b] Symbols: see Table 15.2.

[c] Similar to that in *A. globiformis*.

[d] In peptone-based media.

[e] Some species are aerobic.

[f] A few species do not produce acid from glucose in peptone based media.

could form the basis of a redefined genus *Brevibacterium* as first suggested by Yamada and Komagata (1972b) and later by Jones (1975), Keddie and Cure (1977) and Sharpe et al. (1978). On the basis of these and further studies, Collins et al. (1980) redefined the genus *Brevibacterium* (Breed) and at the same time, reclassified the organism previously named "*Chromobacterium iodinum*" as *Brevibacterium iodinum*. Later, two groups of methanethiol-producing coryneform bacteria, one isolated from cheese and milk and the other from human skin, were respectively classified as *Brevibacterium casei* and *Brevibacterium epidermidis* by Collins et al. (1983a). The methanethiol-producing organisms from cheese and milk were first isolated and well described by Sharpe et al. (1976, 1977, 1978) and those from skin by Pitcher and Noble (1978) and Sharpe et al. (1978). All these workers emphasized the similarity between these isolates and the species *Brevibacterium linens*, but at that time the genus *Brevibacterium* was incertae sedis.

At the present time, the genus *Brevibacterium* contains the type species *B. linens* and the species *B. iodinum*, *B. casei* and *B. epidermidis*. However, the DNA-DNA homology studies of Fiedler et al. (1981) indicate that there is heterogeneity among strains presently classified as *B. linens*. Although all the strains possess the salient characters of the genus *Brevibacterium*, the results of the DNA studies indicate the presence of at least one extra species (see *Brevibacterium linens*, Further Comments). It is also likely that another as yet unnamed species of the genus has been isolated from human skin (Pitcher and Noble, 1978), and Keddie and Cure (1978) noted a somewhat heterogeneous group of coryneform bacteria which contained *meso*-DAP but not arabinose in their cell walls and which did not contain mycolic acids (see Keddie and Jones, 1981).

The results of 16S-rRNA cataloging studies (Stackebrandt et al.,

1980a; Stackebrandt and Woese, 1981) indicate that the genus *Brevibacterium* is related to the cluster formed by the coryneform taxa *Arthrobacter*, *Aureobacterium*, *Cellulomonas*, *Curtobacterium* and *Microbacterium* but, although related, it is peripheral to the group. The distinctness of the genus from the above taxa is reinforced by the results of the biochemical and nucleic acid studies of Fiedler et al. (1981).

Further Reading

Collins, M.D., J.A.E. Farrow, M. Goodfellow and D.E. Minnikin. 1983. *Brevibacterium casei* sp. nov. and *Brevibacterium epidermidis* sp. nov. Syst. Appl. Microbiol. *4*: 388–395.

Collins, M.D., D. Jones, R.M. Keddie and P.H.A. Sneath. 1980. Reclassification of *Chromobacterium iodinum* (Davis) in a redefined genus *Brevibacterium* as *Brevibacterium iodinum* nom. rev.; comb. nov. J. Gen. Microbiol. *120*: 1–10.

Crombach, W.H.J. 1974. Relationships among coryneform bacteria from soil, cheese and sea fish. Antonie van Leeuwenhoek. J. Microbiol. Serol. *40*: 347–359.

Crombach, W.H.J. 1974. Morphology and physiology of coryneform bacteria. Antonie van Leeuwenhoek. J. Microbiol. Serol. *40*: 361–376.

Fiedler, F., M.J. Schäffler and E. Stackebrandt. 1981. Biochemical and nucleic acid hybridization studies on *Brevibacterium linens* and related strains. Arch. Microbiol. *129*: 85–93.

Keddie, R.M. and D. Jones. 1981. Aerobic, saprophytic coryneform bacteria. *In* Starr, Stolp, Trüper, Balows and Schlegel (Editors), The Prokaryotes: A Handbook on Habitats, Isolation and Identification of Bacteria, Springer-Verlag, New York, pp. 1838–1878.

Mulder, E.G., A.D. Adamse, J. Antheunisse, M.H. Deinema, J.W. Woldendorp and L.P.T.M. Zevenhuizen. 1966. The relationship between *Brevibacterium linens* and bacteria of the genus *Arthrobacter*. J. Appl. Bacteriol. *29*: 44–71.

Seiler, H. 1983. Identification key for coryneform bacteria derived by numerical taxonomic studies. J. Gen. Microbiol. *129*: 1433–1471.

Differentiation of the species of the genus **Brevibacterium**

Those differential characteristics which have the most value in distinguishing the species of the genus *Brevibacterium* are listed in Table 15.15. Differentiation of *B. casei* from *B. epidermidis* relies heavily on DNA-DNA base-pairing techniques (Collins et al., 1983a).

DNA homology studies also show that *B. linens* as characterized in Table 15.15 is heterogeneous and comprises at least two species (Fiedler et al., 1981).

List of species of the genus **Brevibacterium**

1. **Brevibacterium linens** (Wolff 1910) Breed 1953, 13[AL] emend Collins et al. 1980, 7.

li'nens. L. part. adj. *linens* spreading over, smearing.

Morphology and general characteristics are as for genus description. Other characters are given in Tables 15.14 and 15.15 and in Further Descriptive Information.

On PYE media, colonies exhibit an orange to orange-red pigment. Pigment production by most strains (including the type strain ATCC 9172) is light dependent. Colonial growth of all strains gives a charac-

teristic color reaction when treated with certain acids and bases (see Further Descriptive Information and Further Comments).

Optimum growth temperature is 20–25°C; growth is generally poor or absent at 5°C and 37°C; occasional strains are reported to grow at 37°C.

When supplied with glucose in a mineral salts medium, most strains require B-vitamins and amino acids for growth; some strains require only amino acids (Mulder et al., 1966; Robertson and Keddie, unpublished) while a few require additional unknown factors (Mulder et al.,

Table 15.15.
Characters differentiating the species in the genus **Brevibacterium**[a,b]

Characteristics	1. *B. linens*	2. *B. iodinum*	3. *B. casei*	4. *B. epidermidis*
Color of colonies	Yellow-orange[c]	Gray-white[d]	Gray-white	Gray-white
Color reaction[e] with KOH	+	−	−	−
Crystals of iodinin formed	−	+	−	−
Oxidase	−	+	−	−
Survival at 60°C for 30 min	−	−	+	+
Major menaquinones	MK-8 (H_2)	MK-8 (H_2)	MK-8 (H_2), MK-7 (H_2)[f]	MK-8 (H_2)
Mol% G + C	60–64	61–63	66–67	63–64

[a] Data from Mulder et al., 1966; Colwell et al., 1969; Yamada and Komagata, 1970; Sharpe et al., 1976, 1977, 1978; Collins et al., 1980; Collins and Jones, 1981a; and Collins et al. 1983c.

[b] Symbols: see Table 15.2.

[c] When incubated in light.

[d] When cultured on suitable media, crystals of iodinin give the growth a purple coloration.

[e] Colonial growth gives a stable pink-red color with 5 M KOH (see Jones et al., 1973).

[f] MK-8 (H_2) and MK-7 (H_2) present in comparable amounts.

1966). For some strains, the amino acid requirement can be satisfied by glutamic acid (Mulder et al., 1966).

Acid is not produced from glucose and other sugars in a peptone medium. Glucose, glycerol, lactate and acetate, but not lactose, sucrose or citrate, are utilized as sole or major sources of carbon + energy (7 of 8 strains tested, Crombach, 1974b).

Gelatin is liquefied, casein hydrolyzed, hippurate hydrolyzed, extracellular DNase produced.

The mol% G + C is 60–64% T_m (Yamada and Komagata, 1970; Skyring and Quadling, 1970; Bousfield, 1972; Crombach, 1972; Collins et al., 1983a). Mol% G + C of type strain (ATCC 9172) 63.4 T_m (Yamada and Komagata, 1970); 62.5 T_m (Collins et al. 1983a).

Type strain: ATCC 9172 (NCIB 9909).

Further comments. The results of DNA-DNA homology studies have shown that strains presently designated *B. linens* in public and private collections constitute at least two distinct species (Fiedler et al., 1981). Of the strains studied by Fiedler et al. (1981), ATCC 9172 (the type strain), ATCC 19391, ATCC 9864 (almost certainly a misprint because this strain number is not that of a strain of *B. linens* in the ATCC catalogue) and strain B3 (from W. H. J. Crombach) form one homology group. This group represents the species *B. linens* because it contains the type strain. The second homology group contains strain ATCC 9175, strains AC 251, AC 252 and AC 474 (from E. G. Mulder) and strain B4 (from W. H. J. Crombach) and related to this group at a DNA homology value of 61%, strain AC 470 (from E. G. Mulder). The strains in both DNA-DNA homology groups have an identical *meso*-DAP-containing directly cross-linked peptidoglycan type which is not amidated and lack arabinogalactan in their cell walls, and on this and other grounds are members of the genus *Brevibacterium* (Fiedler et al., 1981).

The demonstration of at least two DNA-DNA homology groups is in keeping with previous evidence of heterogeneity among *B. linens* strains. Mulder et al. (1966) reported considerable differences in the nutritional requirements of different strains, and there are also reports of differences in maximum growth temperatures and tolerated salt concentrations between strains (see Further Descriptive Information). Of interest is the finding of Mulder et al. (1966) that in most *B. linens* strains examined by them (including the type strain ATCC 9172, DNA-DNA homology group 1) pigment production is light dependent, while in other strains (including strain ATCC 9175, DNA-DNA homology group 2) pigment production is not light dependent.

The results of an analysis of 15 strains of *B. linens* by an electrophoretic zymogram technique (Foissy, 1974) also showed heterogeneity within *B. linens*. On the basis of protein patterns, Foissy (1974) divided the strains into three biotypes containing ten (including strains ATCC 9174 and ATCC 9175), four and one strains, respectively. The type strain of *B. linens* was not included in the study. Strain ATCC 9175 is the only strain common to the studies of Foissy (1974) and Fiedler et al. (1981). It seems, therefore, that biotype 1 of Foissy (1974) may be equivalent to DNA homology group 2 of Fiedler et al. (1981), but it is not possible to comment on any correlation between the two other biotypes and the DNA homology groups.

Strains of both DNA-DNA homology groups give the color reaction previously deemed by Jones et al. 1973, to be characteristic of *B. linens* (Grecz and Dack, 1961; Jones et al., 1973; Sharpe et al., 1977). Members of both DNA-DNA homology groups also produce methanethiol. At the present time, homology group 2 of Fiedler et al. (1981) has not been designated by a species name. It should be noted that the numerical taxonomic study of Seiler (1983) which included representatives of both DNA-DNA homology groups and strain AC 470 did not result in the separation of the strains on physiological and biochemical characters. The strains tested did cluster as two groups (Clusters VII and VIII, Seiler, 1983) but one group (Cluster VII) contained, not only representatives of both DNA-DNA homology groups, but also four strains (labeled *Corynebacterium* sp; NCDO 2048, 2049, 2050 and 2051) which have been designated as a separate species, *B. casei* (Collins et al., 1983a).

Cuer et al. (1979) demonstrated that, in addition to methanethiol, all eight strains of *B. linens* examined produced dimethyldisulfide and 2,3,4-trithiapentane from cheeses. Four strains, ATCC 8377 and three isolates (B11, B12 and B13), also produced *S*-methylthioacetate but strains NCIB 9909 (ATCC 9172), IP 6311 (ATCC 9174), IP 6312 (ATCC 9175) and strain NIRD 1002 did not.

On the basis of the results of a study of the transport of phenylalanine, tyrosine and tryptophan by metabolizing cells of *B. linens* (strain 47), Boyaval et al. (1983) hypothesized that the transport of aromatic amino acids by *B. linens* is determined by three high-affinity permeases.

Schefferle (1966) noted the decomposition of uric acid to urea by *B. linens* strain NCIB 8546 but none of the orange pigmented strains which she considered to be closely related to *B. linens* decomposed this compound.

2. **Brevibacterium iodinum** (ex Davis 1939) Collins, Jones, Keddie and Sneath 1981, 216.[VP] (Effective publication Collins, Jones, Keddie and Sneath 1980, 7.)

i.o.di′num M.L. neut. n. *iodinum* iodine.

General characters are as for genus description. Other characters are given in Tables 15.14 and 15.15 and in Further Descriptive Information.

Characteristic extracellular purple crystals of iodinin are produced on a variety of media. Iodinin production is not influenced by light.

Cells frequently appear Gram-negative. In very young cultures (about 8 h) cells frequently stain one-half Gram-positive, the other half Gram-negative (Colwell et al., 1969).

Optimum growth temperature 25–30°C. Some growth occurs at 37°C but not at 5°C.

Nutritional requirements not determined (but see Further Descriptive Information). Acid is not produced from glucose and other sugars in a peptone medium. Citrate is utilized.

Strongly oxidase-positive (Colwell et al., 1969; Jones, 1975).

H_2S is produced from cysteine but not from sodium thiosulfate (Colwell et al., 1969; Jones, 1975). Nitrate is reduced to nitrite (Colwell et al., 1969; Jones, 1975).

Gelatin is liquefied, extracellular DNase is produced. Hydrolysis of casein is weak and hippurate is not hydrolyzed.

Isolated from milk.

The mol% G + C of the DNA is in the range 60–63 T_m (De Ley et al., 1966; Colwell et al., 1969; Collins et al., 1983a); 63 Bd (Colwell et al., 1969).

Type strain: NCDO 613.

Further comments. The species description is based on studies of strains NCDO 613, 753, ATCC 9897, 15728, 15729, NCTC 9742, NCIB 8179. It is believed that all these strains are derived from the original isolate NCDO 613 (Davis, 1939). For further information on strain histories, see Colwell et al. (1969).

3. **Brevibacterium casei** Collins, Farrow, Goodfellow and Minnikin 1984, 91.[VP] (Effective publication Collins, Farrow, Goodfellow and Minnikin 1983a, 393.)

ca′se.i. M.L. n. *caseus* cheese; M.L. gen. n. casei of cheese.

Morphology and general characters are as for genus description. Other characters are given in Tables 15.14 and 15.15 and in Further Descriptive Information.

On nutrient agar colonies are gray-white in color. On milk agar a brown, water-soluble pigment is produced (Sharpe et al., 1976).

Optimum growth temperature is between 30 and 37°C and growth occurs at 40°C; survive heating in Nutrient Broth (Oxoid) at 60°C for 30 min (Sharpe et al., 1976, 1977, 1978).

Nutritional requirements have not been fully determined, but strains CMD1 (NCDO 2048) and CMD3 (NCDO 2049) are reported not to require exogenous growth factors (Sharpe et al., 1976).

Weak acid production from glucose has been reported when tested by the method of Hugh and Leifson (1953), (Sharpe et al., 1976). All strains tested utilize glucose, sucrose, acetate and lactate as carbon and energy sources. Strain CMD3 (NCDO 2049) also utilizes lactose. None of the strains tested utilizes inulin or citrate (Sharpe et al., 1976).

Gelatin and casein are hydrolyzed and extra-cellular DNase is produced (Sharpe et al., 1976; 1977).

Isolated from Cheddar cheese, cheese curd and raw milk.

The mol% G + C of the DNA is in the range 66–67 T_m (Collins et al., 1983a).

Type strain: CMD1 (NCDO 2048).

Further comments. At present, four strains only have been studied: NCDO 2048 (CMD1), NCDO 2049 (CMD3), NCDO 2050 (C4) and NCDO 2051 (R6), (Sharpe et al., 1976, 1977, 1978; Collins et al., 1983a; Seiler, 1983).

In DNA-DNA homology studies (S_1 nuclease-treated preparations), the type strain of *B. casei* showed 22–23% homology with *B. epidermidis* strains; 19% homology with *B. linens* (type strain) and 14% homology with *B. iodinum* (Collins et al., 1983a).

4. **Brevibacterium epidermidis** Collins, Farrow, Goodfellow and Minnikin 1984, 91.[VP] (Effective publication: Collins, Farrow, Goodfellow and Minnikin 1983a, 393.)

e.pi.der′mi.dis. Gr. n. *epidermidis*, the outer skin; M.L. gen. n. *epidermidis* of the epidermis.

Morphology and general characters are as for genus description. Other characters are given in Tables 15.14 and 15.15 and in Further Descriptive Information.

On nutrient agar, colonies show no distinctive pigment but appear gray-white in color.

Optimum growth temperature is between 30 and 37°C; strain NCDO 2285 (NCTC 11083) grows well at 40°C but the type strain NCDO 2286 (NCTC 11084) shows very poor growth at this temperature (Sharpe et al., 1977). Of the strains tested, a small number of cells survive heating in Nutrient Broth (Oxoid) for 30 min at 60°C (Sharpe et al., 1977, 1978).

Acid is not produced from glucose in a peptone medium. Acetate and lactate are utilized as sole or major sources of carbon + energy (Sharpe et al., 1977).

Gelatin and casein are hydrolyzed and extracellular DNase is produced (Sharpe et al., 1977, 1978).

Isolated from human skin.

The mol% G + C of the DNA of the type strain (NCDO 2286) is 63.5 T_m and that of strain NCDO 2285 is 63.3 T_m (Collins et al., 1983a).

Type strain: D731 (NCDO 2286, NCTC 11084).

Further comments. In DNA-DNA homology studies (S_1 nuclease-treated preparations) *B. epidermidis* (strain NCDO 2285) showed 31–44% homology with *B. casei* strains; 39% homology with *B. iodinum* (type strain) and 14% homology with *B. linens* (type strain) (Collins et al., 1983a).

At present, only two of the skin isolates (NCTC 11084 (NCDO 2286) and NCTC 11083 (NCDO 2285)) studied by Sharpe et al. (1977) can be allocated with confidence to the species *B. epidermidis.* Further work is required to establish whether or not the skin isolates P 151, D 69, K 608, K 656 and K 673 (Sharpe et al., 1977) are members of this species.

At the present time it is not possible to distinguish between strains of *B. casei* and *B. epidermidis* except by menaquinone composition, mol% G + C ratio values and DNA-DNA base pairing values.

Species Incertae Sedis

The following species (a–f) are almost certainly not members of the genus *Brevibacterium* but there are insufficient data to allow them to be reclassified with confidence.

a. *Brevibacterium incertum* (Steinhaus 1941) Breed 1953, 14.[AL]

(Description based on type strain only; data based on Steinhaus, 1941; Breed, 1957; Jones, unpublished.)

Short rods about 1 μm in length with rounded ends and coccobacillary forms. Occur singly, in pairs or in chains. Gram-positive; cells from older cultures tend to lose the ability to retain the Gram stain. Not acid-fast. Endospores are not formed. Motile with one or two flagella.

On PYE media colonies are tiny with no distinctive pigmentation.

Microaerophilic. Metabolism fermentative. Acid is produced from glucose, fructose, mannose, maltose and sucrose. No acid is produced from galactose, lactose, rhamnose, mannitol, dulcitol, inositol or sorbitol (Steinhaus, 1941).

Catalase-negative, oxidase-negative. Sodium hippurate is hydrolyzed. Gelatin, casein and starch are not hydrolyzed. Cellulose is not attacked. Urease is not produced. DNase is not produced. Acetyl methyl carbinol, indole and H_2S are not produced. Nitrates are not reduced to nitrites.

Growth occurs between 25 and 37°C; no growth at 5 or 45°C.

The cell wall peptidoglycan contains lysine as the main diamino acid (peptidoglycan type L-lysine-D-glutamic acid; variation A4α, Schleifer and Kandler, 1972). Mycolic acids are not present (Minnikin et al., 1978; Collins and Kroppenstedt, 1983). Isoprenoid quinones are absent (Collins and Kroppenstedt, 1983). The fatty acids are composed of predominantly straight chain saturated and monounsaturated acids. The major fatty acids are hexadecanoic ($C_{16:0}$), hexadecenoic ($C_{16:1}$) and cis-vaccenic ($C_{18:1}$, $ω_7$) (Collins and Kroppenstedt, 1983).

Isolated from the ovaries of the lyreman cicada, *Tibicen linnei.* Habitat is not known.

Type strain: ATCC 8363; NCIB 9892.

Further comments. This bacterium is not a member of the genus *Brevibacterium.* As noted by Collins and Kroppenstedt (1983) the absence of isoprenoid quinones suggests that it may be related to one of a number of other Gram-positive taxa which lack respiratory quinones (for example, *Erysipelothrix, Gemella, Lactobacillus, Streptococcus*).

The presence of substantial amounts of monounsaturated fatty acids of the cis-vaccenic acid series (synthesized via an anaerobic pathway) also indicates a relationship to the lactic acid group of bacteria of the genera *Lactobacillus* and *Streptococcus* (see Kroppenstedt and Kutzner, 1978; Collins and Kroppenstedt, 1983).

b. *Brevibacterium acetylicum* (Levine and Soppeland 1926) Breed 1957, 502.[AL] (*Flavobacterium acetylicum* Levine and Soppeland 1926, 46.)

(Description based on type strain only; data based on Levine and Soppeland, 1926; Breed, 1957; Jones, unpublished.)

Short rods about 1 μm in length with rounded ends. Occur singly, in pairs or in chains. Gram-positive. Not acid-fast. Endospores are not formed. Motile with peritrichous flagella.

Colonies on PYE agar are flat with irregular edge and orange-yellow in color. The pigment does not diffuse into the medium. The growth does not give the color reactions typical of *B. linens* (Jones et al., 1973).

Facultatively anaerobic but better growth occurs aerobically. Acid is produced from glucose, fructose, mannose and sucrose.

Catalase-positive and oxidase-positive.

Nutritional requirements are not known.

Gelatin, casein and starch are hydrolyzed. DNase and phosphatase are produced, sulfatase is not produced. Cellulose, chitin, tyrosine, xanthine and hippurate are not attacked. Tween 20, 40, 60 and 80 are not hydrolyzed.

Urease is not produced.

Acetyl methyl carbinol is produced. Nitrates are not reduced to nitrites. Indole and H_2S are not produced.

Optimum growth temperature is about 30°C. Growth occurs at 10 and 37°C; slow growth at 5 but not 45°C.

The cell wall peptidoglycan is based on lysine (Schleifer and Kandler, 1972) (peptidoglycan type L-lysine-D-aspartic acid; variation A4α, Schleifer and Kandler, 1972). Mycolic acids are not present (Minnikin et al., 1978; Collins and Kroppenstedt, 1983). Menaquinones are the sole respiratory quinones; the principal quinone is MK-7 (Collins and Jones, 1981a; Collins and Kroppenstedt, 1983). The major polar lipids are diphosphatidylglycerol, phosphatidylglycerol and phosphatidylethanolamine. Glycolipids are not present (Collins and Kroppenstedt, 1983). The fatty acid composition is characterized by large amounts of iso-methyl-branched chain acids. The major component is 13-methyltetradecanoic acid (iso-$C_{15:0}$) and hexadecanoic acid is also present in a substantial amount (Collins and Kroppenstedt, 1983; Bousfield et al., 1983).

The strain was isolated from skimmed milk (creamery waste). Habitat is not known.

The mol% G + C of the DNA is 46.6 (T_m, Yamada and Komagata, 1970); 52 (T_m, M. D. Collins, personal communication).

Type strain: ATCC 953; NCIB 9889.

Further comments. Although clearly not a member of the genus *Brevibacterium*, the taxonomic position of *B. acetylicum* (ATCC 953) is not known.

In a numerical taxonomic study of a large number of coryneform and other Gram-positive bacteria, strain ATCC 953 did not group with any representatives of the genera *Arthrobacter, Brevibacterium, Brochothrix, Corynebacterium, Kurthia, Lactobacillus, Mycobacterium, Nocardia, Rhodococcus* nor with *Streptococcus* (Jones, 1975). The closest similarity (about 82%) was with a Gram-positive bacterium labeled "*Achromobacter liquefaciens*" ATCC 15716.

B. acetylicum has the same peptidoglycan type as *Kurthia zopfii* (Schleifer and Kandler, 1972) and, as noted by Collins and Kroppenstedt (1983), the polar lipid and menaquinone compositions of strain ATCC 953 are also identical with those found in the genus *Kurthia*. However, the ability of strain ATCC 953 to grow anaerobically and, more significantly, the mol% G + C of the DNA do not support a close relationship with the genus *Kurthia*. The peptidoglycan type of *B. acetylicum* strain ATCC 953 (A4α) is also similar to one group of arthrobacters (see genus *Arthrobacter*) but differs from these bacteria in a number of other characters (Collins and Kroppenstedt, 1983).

In morphological, general and chemical characters, *B. acetylicum* exhibits a very close similarity to the alkalophilic bacterium *Exiguobacterium aurantiacum*, the type species of the new genus *Exiguobacterium* (Collins et al., 1983d), although the ability of *B. acetylicum* ATCC 953 to grow at pH values above 9 is not known. Further studies, including DNA-DNA homology determinations, are required.

Brevibacterium acetylicum strain ATCC 954, also isolated by Levine and Soppeland (1926) from creamery waste, differs from *B. acetylicum* ATCC 953 in peptidoglycan structure. Although the peptidoglycan of strain ATCC 954 is also based on lysine, the detailed structure is different; variation A3α (Schleifer and Kandler, 1972) like that of one group of arthrobacters (see genus *Arthrobacter*). However, the mol% G + C of the DNA is reported to be 46.8 (T_m, Yamada and Komagata, 1970).

Strain ATCC 21665, also listed as *B. acetylicum*, is a patent strain used with fungi in a mixed culture method for increasing yields of fermentation products. Whether or not this strain is taxonomically related to *B. acetylicum* ATCC 953 is not known.

c. *Brevibacterium oxydans* Chatelain and Second 1966, 642.^AL
(Description based on that of Chatelain and Second, 1966.)

Rods, 1–2 μm in length occur singly or in random groups. Gram-positive. Not acid-fast. Endospores are not formed. Motile.

On nutrient agar, incubated at 30°C for 24 h, colonies are very small, round with an entire edge. After 2 to 3 days the colonies become larger and show a yellow color. The growth does not give the color reactions typical of *B. linens*.

Aerobic. Oxidative acid production from glucose, fructose, galactose, mannose, sucrose, maltose, mannitol, glycerol, salicin and dextrin (method of Veron and Chatelain, 1960). No acid is produced from inositol. Acid production from xylose, arabinose, rhamnose, lactose and sorbitol varies between strains. No growth on Simmon's citrate.

Catalase-positive, oxidase-negative. Gelatin liquefied and most strains peptonize milk. Esculin hydrolyzed. β-Galactosidase produced. H_2S produced (lead acetate paper).

Methyl red-negative. Acetyl methyl carbinol not produced. Nitrates not reduced to nitrites.

Urease-negative. Lysine decarboxylase not produced. Cellulose not attacked.

Optimum temperature about 30°C. All strains grow at 20 and most at 37°C.

Isolated from contaminated hospital material. Habitat unknown.

Mol% G + C of the DNA of the type strain is 67 (T_m, Bousfield, 1972).

Type strain: CIP 6612; NCIB 9944.

Further comments. Seiler (1983) reported that the type strain (CIP 6612) utilized caprylate, citrate, 4-hydroxybenzoate, D-gluconate and D-ribose as carbon sources.

In the numerical taxonomic study of Seiler (1983) five strains of *B. oxydans* (including the type strain) clustered in a group which contained, among others, the type strain of *Curtobacterium saperdae* (DSM 20169, ATCC 19272) and *Aureobacterium* (*Microbacterium*) *liquefaciens* (Robinson strains 15 (NCIB 11509) and 20). The mol% G + C value of 67 reported by Bousfield (1972) for *B. oxydans* (NCIB 9944) and the fatty acid studies of Bousfield et al. (1983) also indicate that further studies should be directed towards comparing the bacterium with representatives of the genera *Curtobacterium* and *Aureobacterium*.

In the study of Seiler (1983) another strain labeled *B. oxydans* (CIP 5580) did not show a close similarity to any particular taxon ("C rest," Seiler, 1983).

d. *Brevibacterium halotolerans* Delaporte and Sasson 1967, 2259.^AL
(Description based on that of Delaporte and Sasson, 1967; type strain only.)

Straight rods with rounded ends, 6–7 μm in length. Occur singly, in pairs or in chains. Gram-positive. Endospores are not formed. Motile with a few (1–9) lateral flagella.

Colonies on nutrient agar are small, somewhat flat but not of uniform depth, round and pale cream in color.

Aerobic. Metabolism oxidative. Weak and slow (10 days) acid production from sugars in media containing tryptone (0.2 and 1%, (w/v)) but strong and rapid acid production from glucose, fructose, galactose, lactose and sucrose when tested in the medium of Veron and Chatelain (1960). The end products of aerobic glucose metabolism are lactic, formic and valeric acids. Citrate is utilized as a carbon source.

Catalase-positive, oxidase-negative.

Ammonium salts do not serve as a sole source of nitrogen.

Growth occurs in the presence of 15% (w/v) NaCl.

Gelatin, casein and starch are hydrolyzed. Weak lecithinase activity is exhibited. Urease is not produced. Cellulose is not attacked. Acetyl methyl carbinol is produced. Nitrates are reduced to nitrites. Indole is not produced.

Optimum pH for growth is 7 but good growth occurs at pH 6.0 and 8.2.

Optimum temperature seems to be about 30°C.

Menaquinones are the sole isoprenoid quinones; the principal menaquinone is MK-7 (Collins et al., 1981b).

Isolated from arid soil in Morocco.

Habitat unknown.

Type strain: ATCC 25096.

Further comments. The sparse data on this bacterium make it difficult to classify. Unsaturated menaquinones with seven isoprene units are not found in members of the genus *Brevibacterium* (Collins and Jones, 1981a). Such menaquinones are present in members of the genera *Bacillus, Brochothrix, Exiguobacterium, Kurthia* and *Listeria* (Collins and Jones, 1981a; Collins et al., 1983d).

e. *Brevibacterium frigoritolerans* Delaporte and Sasson 1967, 2260.^AL
(Description based on that of Delaporte and Sasson, 1967; type strain only.)

Straight rods with rounded ends, 4–7 μm in length. Rarely occur singly or in pairs but frequently occur in long, twisted chains. Gram-positive. Endospores are not formed. Occasional cells are motile; these possess very few (1–3) lateral flagella.

Colonies on nutrient agar are small, round and cream-white in color.

Aerobic. Metabolism mainly oxidative but partially fermentative.

Weak and slow (7–10 days) acid production from sugars in media containing tryptone (0.2 and 1% (w/v)) but rapid acid production from glucose, fructose, lactose and sucrose when tested in the medium of Veron and Chatelain (1960). The end products of aerobic glucose metabolism are mainly acetic and valeric acids with small amounts of lactic and succinic acids. Citrate is utilized as a carbon source.

Catalase-positive, oxidase-negative.

Ammonium salts serve as a sole source of nitrogen.

Growth occurs in the presence of 10% (w/v) NaCl.

Gelatin is hydrolyzed. Lecithinase is produced. Cellulose, casein and starch are not hydrolyzed. Urease is not produced. Nitrates are reduced to nitrites. Acetyl methyl carbinol is not produced; indole is not produced.

Growth occurs between pH 6.0 and 8.2.

Optimum temperature for growth is about 37°C. Good growth occurs at 4°C; no growth occurs at 45°C.

Isolated from arid soil in Morocco.

Habitat unknown.

Type strain: ATCC 25097.

Further comments. Much more data are required before the taxonomic position of this bacterium can be determined. The morphological features indicate that it is not a member of one of the coryneform taxa.

f. "*Brevibacterium rufescens*" Nazina 1981, 225.

(Description based on that of Nazina, 1981.)

This bacterium was described after the publication of the Approved Lists of Bacterial Names (Skerman et al., 1980). To our knowledge, the name has not been validated by citation in the International Journal of Systematic Bacteriology.

The bacterium is described as a nonsporeforming, pleomorphic rod; motile with polar flagella. Colonies on a variety of solid nutrient media are reddish pink. Optimum growth temperature 25–30°C. Catalase-positive. Nitrates are reduced to nitrites. Cellulose, gelatin and starch are not attacked; H_2S is not produced. Under aerobic conditions, a number of carbon compounds, including methanol, are assimilated.

Mol% G + C 69 (determined chromatographically), Nazina (1981).

Isolated from oil reservoirs.

Reference strain: Nazina, strain 5057.

Further comments. There is no information on the reaction of the bacterium to the Gram strain. The bacterium was classified in the genus *Brevibacterium* "on the basis of morphological properties." No information is given on the chemical composition of the peptidoglycan or lipids. On the basis of the information presented by Nazina (1981) the bacterium is not a member of the genus *Brevibacterium*.

Addendum I

The chemical data now available indicate that the following species (g–i) should be reclassified in the genus *Corynebacterium*. However, at the present time, it is not clear whether all or some represent distinct species in the genus or whether they can be classified in one of the existing species of the genus.

g. *Brevibacterium ammoniagenes* (Cooke and Keith 1927) Breed 1953, 14.[AL] (*Bacterium ammoniagenes* Cooke and Keith 1927, 315.)

(Description based on Cooke and Keith, 1927; Breed, 1957; Yamada and Komagata, 1972a; Seiler, 1983; Cure and Keddie, unpublished; Jones, unpublished.)

In young cultures (up to 24 h) irregular rods with rounded ends ~1–4.5 μm in length and ~0.6–1.2 μm in diameter are seen. Occur singly or in pairs, some cells are arranged at an angle to give V forms. In older cultures (~7 days) the rods are shorter but V forms still occur. Gram-positive. Not acid-fast. Endospores are not formed. Nonmotile.

Colonies on PYE agar are circular, low convex with an entire edge and gray-white, pale or bright yellow in color (Cooke and Keith, 1927; Seiler, 1983). The pigment does not diffuse into the medium. The growth does not give the color reactions typical of *B. linens* (Seiler, 1983).

Facultatively anaerobic but better growth occurs aerobically. Acid production from glucose and some other carbohydrates is equivocal (see Further Comments). Some organic acids are assimilated (Yamada and Komagata, 1972a; Seiler, 1983).

Catalase-positive.

Growth occurs in the presence of 10% (w/v) NaCl (Yamada and Komagata, 1972a).

Urease is produced. Sodium hippurate and tyrosine are hydrolyzed. Nitrates are reduced to nitrites. Acetyl methyl carbinol is not produced; indole is not produced.

Gelatin, casein and starch are not hydrolyzed. Cellulose, chitin and xanthine are not attacked. DNase is not produced.

Optimum pH for growth is between 7.0 and 8.5 (Cooke and Keith, 1927).

Optimum growth temperature is about 30°C. Growth occurs at 10 and 37°C; slow growth at 5 but not at 45°C.

The cell wall peptidoglycan contains *meso*-DAP as the major diamino acid (Yamada and Komagata, 1972a; Schleifer and Kandler, 1972 but see Further Comments; Keddie and Cure, 1977 and see Keddie and Cure 1978). Mycolic acids are present (Goodfellow et al., 1976; Keddie and Cure, 1977) with carbon ranges of C_{32} to C_{36} (Minnikin et al., 1978; Collins et al., 1982a). Menaquinones are the sole respiratory quinones; the principal quinone is MK-9(H_2) (Yamada et al., 1976; Collins et al., 1979). The major polar lipids are diphosphatidylglycerol, phosphatidylglycerol, phosphatidylinositol and an unidentified phosphatidylinositol mannoside. Trace amounts of phosphatidylethanolamine are also present (Komura et al., 1975 and see Minnikin et al., 1978). The fatty acid composition consists of substantial amounts of hexadecanoic ($C_{16:0}$), octadecanoic ($C_{18:0}$), octadecenoic ($C_{18:1}$) and 10-methyloctadecanoic acids (Collins et al., 1982b; Suzuki and Komagata, 1983; Bousfield et al., 1983). The octadecenoic acid is the Δ^9 isomer synthesized by the anaerobic pathway (Suzuki et al., 1982).

B. ammoniagenes is reported to be nonpathogenic for rabbits and guinea pigs (Cooke and Keith, 1927).

Isolated from the feces of infants (Cooke and Keith, 1927) and from piggery waste (Seiler and Hennlich, 1983); probably widely distributed in putrefying materials.

Cause of diaper rash in infants (Cooke and Keith, 1927).

The mol% G + C range of the DNA is 53.7–55.8 (T_m, Yamada and Komagata, 1970; Crombach, 1972).

Type strain: ATCC 6871; NCTC 2398; NCIB 8143.

The type strain does not show yellow pigmentation.

Further comments. Reports of acid production from sugars are conflicting, presumably because of the different methods used. Cooke and Keith (1927) reported that none of "the commonly used laboratory sugars are fermented." Yamada and Komagata (1972a) reported no acid produced from a wide variety of carbohydrates by strain AJ 1444 (ATCC 6872); acid production after 7 days from fructose only by strain AJ 1443 (ATCC 6871). Schefferle (1966) reported that the type strain oxidized glucose to form acid while Cure and Keddie (unpublished), using a medium and method similar to that of Hugh and Leifson (1953) obtained fermentation of glucose by the type strain, the result to be expected from a facultative anaerobe.

The report of lysine as the main diamino acid of the peptidoglycan of *B. ammoniagenes* (type strain ATCC 6871, Schleifer and Kandler, 1972) is difficult to explain and is probably an error as other workers have detected *meso*-DAP as the main diamino acid in duplicates of this strain (Yamada and Komagata, 1972a; Keddie and Cure, 1977).

It is now generally agreed, on the basis of the results of morphological, physiological, biochemical and chemical studies, that all strains of *B. ammoniagenes* examined should be reclassified in the genus *Corynebacterium* (Abe et al., 1967; Yamada and Komagata, 1972b; Keddie and Cure, 1977, 1978; Minnikin et al., 1978; Collins et al., 1979; Suzuki et al., 1981; Collins et al., 1982a, b; Seiler, 1983). However, there is evidence that the many strains labeled *B. ammoniagenes* in public and private collections do not constitute a homogeneous taxon. In the study of Seiler (1983), *B. ammoniagenes* strains ATCC 6871, DSM 20306, ATCC 15137 and IFO 12072 showed a close phenetic similarity to each other (Cluster E II, *B. ammoniagenes*) but were less similar to *B. ammoniagenes* strains ATCC 13746 and DSM 20163 (ATCC 13745); the last two strains clustered with strains of *Corynebacterium glutamicum* (Cluster E 1, *C. glutamicum*). These results are in accord with the results of the DNA-DNA homology studies of Komatsu and Kaneko (1980) and Suzuki et al. (1981). Komatsu and Kaneko (1980), reported that *B. ammoniagenes* ATCC 13745 was closely related to *C. glutamicum* ATCC 13032, the type strain of the species. However, Suzuki et al. (1981), found that *B. ammoniagenes* strain CNF 012 (ATCC 6872, one of the original isolates of Cooke and Keith, 1927 but, unfortunately, not the type strain) gave a rather low DNA-DNA homology value with *C. glutamicum* strain CNF 016 (ATCC 13032). Furthermore, the results of Suzuki et al. (1981), indicate that *B. ammoniagenes* (CNF 012 (ATCC 6872)) is quite distinct, not only from *C. glutamicum* but also from *C. diphtheriae. C. flavescens, C. lilium, C. renale, C. vitarumen* and *C. xerosis* and possibly represents a separate species. The presence of 10-methyloctadecanoic acid in *B. ammoniagenes* strain NCIB 8143 (Collins et al., 1982b) and *B. ammoniagenes* strain CNF 096 (ATCC 6871) and strain CNF 012 (ATCC 6872) (Suzuki and Komagata, 1983) supports such a suggestion. Further studies, including DNA-DNA homology studies with the type strain of *B. ammoniagenes*, are required to resolve the taxonomic status of these organisms. However, the evidence currently available indicates that while some strains labeled *B. ammoniagenes* can probably be reduced to synonymy with *C. glutamicum*, as suggested by Abe et al. (1967), others, including the type strain, constitute a taxon distinct from *C. glutamicum*.

h. *Brevibacterium stationis* (Zobell and Upham 1944) Breed 1953, 14.[AL] (*Achromobacter stationis* Zobell and Upham 1944, 273.)

(Description based on type strain only.)

Short, ovoid rods about 0.6–1 μm in diameter with rounded ends. Occur singly or in pairs, some cells are arranged at an angle to give V forms. Gram-positive. Not acid-fast. Endospores are not formed. Nonmotile.

Colonies on PYE media are circular, low convex with an entire edge and show a yellow color. The pigment does not diffuse into the agar. The growth does not give the color reactions typical of *B. linens* (Chatelain and Second, 1966; Jones et al., 1973; Seiler, 1983).

Facultatively anaerobic but better growth occurs aerobically. Acid is produced from glucose, fructose, mannose and sucrose; acid production from trehalose and cellobiose is weak and slow after 7 days (Yamada and Komagata, 1972a). A number of organic acids are assimilated (Yamada and Komagata, 1972a; Seiler, 1983).

Catalase-positive.

Growth occurs in the presence of 10% (w/v) NaCl (Yamada and Komagata, 1972a).

Urease is produced. Phosphatase is produced. Sodium hippurate and tyrosine are hydrolyzed. Nitrates are reduced to nitrites.

Acetyl methyl carbinol is not produced. Indole is not produced. Gelatin, casein and starch are not hydrolyzed. Cellulose, chitin and xanthine are not attacked. DNase is not produced.

Optimum growth temperature between 25 and 30°C. Growth occurs at 10 and 37°C; slow growth at 5°C but no growth at 45°C.

The cell wall peptidoglycan contains *meso*-DAP as the major diamino acid (Schleifer and Kandler, 1972; Yamada and Komagata, 1972a; Keddie and Cure, 1977, 1978). The cell wall sugars are arabinose and galactose (Schleifer and Kandler, 1972); arabinose, galactose and glucose (Keddie and Cure, 1977, 1978). Mycolic acids are present (Goodfellow et al., 1976; Keddie and Cure, 1977, 1978; Minnikin et al., 1978; Collins et al., 1982a). Menaquinones are the sole respiratory quinones; MK-8(H_2) and MK-9(H_2) are present in comparable amounts (Collins et al., 1979).

The fatty acid composition (Bousfield et al., 1983) consists mainly of hexadecanoic ($C_{16:0}$), octadecenoic ($C_{18:1}$) and 10-methyloctadecanoic acids.

Isolated from a film of marine-fouling organisms; habitat is considered to be sea water.

Mol% G + C of the DNA is 53.9 (T_m, Yamada and Komagata, 1970).

Type strain: ATCC 14403.

Further comments. Different results have been obtained for the mycolic acids of *B. stationis*. Goodfellow et al. (1976) reported that the methyl esters of the total mycolic acids of *B. stationis* co-chromatographed with those of *Corynebacterium diphtheriae*. In contrast, Keddie and Cure (1977) noted that the free mycolic acids of *B. stationis* possessed a chromatographic mobility similar to those of a standard strain of *Rhodococcus erythropolis*. Subsequent mass spectral analyses of the mycolic acids of *B. stationis* (Collins et al., 1982a), however, showed that the mycolates are of the short chain type (29–36 carbon atoms) in accord with the results of Goodfellow et al. (1976).

The presence of 10-methyloctadecanoic acid does not exclude *B. stationis* from the genus *Corynebacterium* as this fatty acid also occurs in some strains of that genus (see Collins et al., 1982b), including *B. ammoniagenes*.

Numerical taxonomic studies (Chatelain and Second, 1966; Jones, 1975; Seiler, 1983) all indicate a close phenetic similarity (about 90%) between *B. stationis* (ATCC 14403) and *B. ammoniagenes* (ATCC 6871). Similarly the same strains of *B. ammoniagenes* and *B. stationis* showed a high degree of similarity in a computer-assisted analysis of fatty acid profiles of a large number of bacteria (Bousfield et al., 1983).

Further studies are required to resolve the taxonomic position of strains designated *B. ammoniagenes* and *B. stationis*, but the evidence to date indicates that the type strain of *B. ammoniagenes* (ATCC 6871) is very similar to *B. stationis* and that both taxa are members of the genus *Corynebacterium* and may even be members of the same species.

i. *Brevibacterium divaricatum* Su and Yamada 1960, 74.[AL]

Rods ~0.8–1.2 μm in diameter and up to ~4–5 μm in length. Occur singly or in pairs, some cells are arranged at an angle to give V forms. Gram-positive. Not acid-fast. Endospores are not formed. Nonmotile.

On PYE agar colonies are round, convex with an entire edge and yellowish gray in color. In areas of dense growth, the coloration appears bright yellow. The growth does not give the color reactions typical of the growth of *B. linens* (Seiler, 1983).

Facultatively anaerobic but better growth occurs aerobically. Acid is produced from glucose, fructose, mannose, sucrose and maltose. Su and Yamada (1960) also reported acid production from glycerol. A number of organic acids are assimilated (see Yamada and Komagata, 1972a).

L-glutamic acid is accumulated in large quantities when the organism is grown aerobically in the presence of carbohydrates, ammonium ion and inorganic salts (Su and Yamada, 1960).

Growth occurs in the presence of 10% (w/v) NaCl. Urease is produced. Sodium hippurate and tyrosine are hydrolyzed.

Gelatin, casein and starch are not hydrolyzed. DNase is not produced. Nitrates are not reduced to nitrites. Indole is not produced. Acetyl methyl carbinol is not produced. H_2S is not produced.

Optimum pH range for growth is between 7 and 8 (Su and Yamada, 1960).

Optimum growth temperature is about 30°C. Growth occurs at 10 and 37°C but not at 5 and 45°C.

The cell wall peptidoglycan contains *meso*-DAP as the major diamino acid (Yamada and Komagata, 1972a; Schleifer and Kandler, 1972; Keddie and Cure, 1977 and see Keddie and Cure 1978). The cell wall sugars are arabinose and galactose (Schleifer and Kandler, 1972; Keddie and Cure, 1977, and see Keddie and Cure, 1978). Mycolic acids are present (Goodfellow et al., 1976; Keddie and Cure, 1977) with carbon ranges of C_{28}–C_{36} (Minnikin et al., 1978; Collins et al., 1982a). Menaquinones are the sole respiratory quinones; the principal quinone is MK-9(H_2) (Collins et al., 1979; Collins and Jones, 1981a). The fatty acid composition consists of substantial amounts of hexadecanoic ($C_{16:0}$) and octadecenoic ($C_{18:1}$) acids; 10-methyloctadecanoic acid is not present (Collins et al., 1982b; Suzuki and Komagata, 1983; Bousfield et al., 1983).

Nonpathogenic for animals and plants (Su and Yamada, 1960). Isolated from soil.

The mol% G + C of the DNA is 54.4 (T_m, Yamada and Komagata, 1970). 51 (T_m, Bousfield, 1972).

Type strain: ATCC 14020, NCIB 9379.

Further comments. As previously suggested (Abe et al., 1967; Yamada and Komagata, 1972b; Keddie and Cure, 1977, 1978; Minnikin et al., 1978; Collins et al., 1979; Collins et al., 1982a, b), *B. divaricatum* should be reclassified in the genus *Corynebacterium*. On the basis of DNA base composition, mycolic acid, fatty acid and menaquinone composition (Abe et al., 1967; Yamada and Komagata, 1970; Collins et al., 1979, 1982a, b; Suzuki and Komagata, 1983), the bacteria are very similar to *Corynebacterium glutamicum*. In the recent numerical taxonomy study of Seiler (1983) two strains of *B. divaricatum* clustered with strains of *C. glutamicum*. Furthermore, the DNA-DNA homology studies of Suzuki et al. (1981) showed a high degree of homology (88%) between *B. divaricatum* (CNF 013 (ATCC 14020, the type strain)) and the type strain of *C. glutamicum*. These findings indicate that *B. divaricatum* should be reclassified in the species *C. glutamicum*.

Addendum II

The following species (listed alphabetically) included in the Approved Lists of Bacterial Names (Skerman et al., 1980), have been reclassified and are treated in detail in the appropriate section of the *Manual*. They are listed here only for completeness.

i. *B. albidum* Komagata and Iizuka, 1964.
ii. *B. citreum* Komagata and Iizuka, 1964.
iii. *B. fermentans* Chatelain and Second, 1966.
iv. *B. imperiale* (Steinhaus, 1941), Breed, 1953.
v. *B. liquefaciens* Okabayashi and Masuo, 1960.
vi. *B. luteum* Komagata and Iizuka, 1964.
vii. *B. lyticum* Takayama, Udagawa and Abe, 1960.
viii. *B. protophormiae* Lysenko, 1959.
ix. *B. pusillum* Iizuka and Komagata, 1965.
x. *B. saperdae* Lysenko, 1959.
xi. *B. testaceum* Komagata and Iizuka, 1964.
xii. *B. vitarumen* (Bechdel, Honeywell, Dutscher and Knutsen 1928) Breed, 1957.

Brevibacterium albidum, *B. citreum*, *B. luteum*, *B. pusillum*, *B. saperdae* and *B. testaceum* all contain ornithine as the principal cell wall diamino acid and for this and other reasons were classified in a new genus *Curtobacterium* (Yamada and Komagata, 1972b) as *C. albidum*, *C. citreum*, *C. luteum*, *C. pusillum*, *C. saperdae* and *C. testaceum*. However, because at the time the genus *Curtobacterium* was not universally recognized, all these species are listed in the Approved Lists of Bacterial Names (Skerman et al., 1980) under both the genera *Brevibacterium* and *Curtobacterium*. Evidence for heterogeneity among the members of the genus *Curtobacterium* came from cell wall peptidoglycan (Schleifer and Kandler, 1972; Uchida and Aida, 1977) and menaquinone studies (Yamada et al., 1976; Collins et al., 1979). On the basis of the results of these studies, a number of workers were of the opinion that neither *B. saperdae* nor *B. testaceum* should be classified in the genus *Curtobacterium* (Yamada et al., 1976; Uchida and Aida, 1977; Keddie and Jones, 1981). As a result of further chemical and biochemical work, Collins et al. (1983b) proposed the reclassification of *C. saperdae* and *C. testaceum* in the genus *Aureobacterium* as *A. saperdae* and *A. testaceum*, respectively. *Curtobacterium albidum*, *C. luteum* and *C. pusillum*, together with the type species *C. citreum*, remain in the genus *Curtobacterium*.

Brevibacterium fermentans and *B. lyticum* were reclassified in the genus *Cellulomonas* as strains of the species *C. cartae* (Stackebrandt et al., 1982). *Cellulomonas cartae* is dealt with as a subjective synonym of *C. cellulans*, see genus *Cellulomonas*.

It should be noted that on the basis of numerical phenetic (Goodfellow, 1971; Jones, 1975), cell wall (Schleifer and Kandler, 1972; Keddie and Cure, 1977), isoprenoid quinone (Yamada et al., 1976; Collins et al., 1979; Collins and Jones, 1981b), fatty acid (Minnikin et al., 1979; Collins and Jones, 1981b; Bousfield et al., 1983) and polar lipid (Minnikin et al., 1979; Collins and Jones, 1981b) data, *B. fermentans* and *B. lyticum* were considered to be members of the genus *Oerskovia* (Minnikin et al., 1979; Collins and Jones, 1981b). On the basis of the results of DNA-DNA homology studies (Stackebrandt et al., 1980b), comparative analysis of 16S-rRNA (Stackebrandt and Woese, 1981), together with the high similarity in physiological, biochemical and metabolic properties found between species of the genera *Cellulomonas* and *Oerskovia*, Stackebrandt et al. (1982) proposed the union of the two genera in a redefined genus *Cellulomonas*. However, in this *Manual* the genera *Cellulomonas* and *Oerskovia* are treated separately. See genus *Cellulomonas* for further information.

Brevibacterium liquefaciens (strain ATCC 14929) was reclassified by Lanéelle et al. (1980) as *Corynebacterium liquefaciens* on the basis of the results of physiological, cell wall, lipid and DNA-DNA homology studies. However, the results of Lanéelle et al. (1980) are not in agreement with those of other workers. Schleifer and Kandler (1972) reported that the peptidoglycan of *B. liquefaciens* ATCC 14929 was based upon lysine (variation A4α), a finding in keeping with the clustering of *B. liquefaciens* strain NCIB 9545 (ATCC 14929) with arthrobacters in the numerical taxonomic study of Bousfield (1972). More recent studies on *B. liquefaciens* (NCIB 9545, ATCC 14929) have shown that the lipid composition of this organism is quite different from that of *C. diphtheriae* (Collins and Kroppenstedt, 1983). Because of these discrepancies between their earlier work and that of other workers, Lanéelle and colleagues re-examined strain ATCC 14929. The results were unlike those described by Lanéelle et al. (1980) but were in accord with those of Collins and Kroppenstedt (1983). Lanéelle and colleagues, therefore, withdrew their recommendation for the assignment of *B. liquefaciens* to the genus *Corynebacterium* (Author's Correction, 1984). In the meantime, DNA homology studies have indicated that strain DSM 20579 (ATCC 14929) is a member of the species *Arthrobacter nicotianae* (Stackebrandt et al., 1983). See genus *Arthrobacter*.

Brevibacterium imperiale and *B. protophormiae* both contain lysine as the principal cell wall diamino acid but their respective peptidoglycans differ in structure and in the variation of the peptide subunit. *B. imperiale* has the peptidoglycan type [L-Hsr] D-Glu-Gly$_2$-L-Lys, variation B1β (Schleifer and Kandler, 1972) and has been transferred to the redefined genus *Microbacterium* as *M. imperiale* (Collins et al., 1983c). *B. protophormiae* has the peptidoglycan type L-Lys-Ala-Glu (Stackebrandt and Fiedler, 1979; type A4α, Schleifer and Kandler, 1972) and has been transferred to the genus *Arthrobacter* as *A. protophormiae* (Stackebrandt et al., 1983). Full descriptions of *M. imperiale* and *A. protophormiae* are given under the appropriate genera.

Brevibacterium vitarumen contains meso-DAP as the principal cell wall diamino acid (Fiedler et al., 1970; Yamada and Komagata, 1972b; Schleifer and Kandler, 1972; Keddie and Cure, 1977) but differs from *B. linens* in containing arabinose and galactose as cell wall sugars (Fiedler et al., 1970; Schleifer and Kandler, 1972; Keddie and Cure, 1977) and also by containing mycolic acids (Keddie and Cure, 1977). For these and additional reasons, *B. vitarumen* was reclassified in the genus *Corynebacterium* as *C. vitarumen* (Lanéelle et al., 1980).

Genus **Curtobacterium** *Yamada and Komagata 1972 425*[AL]

KAZUO KOMAGATA AND KEN-ICHIRO SUZUKI

Cur.to.bac.te′ri.um. L. adj. *curtus* shortened; *bacterium* Gr. neut. dim. n. a small rod; M.L. neut. n. *Curtobacterium* a short rodlet.

Small irregular rods. Pleomorphism is not distinctive. Cells become shorter to **coccoid form in older culture.** Branching is not found. Endospores not formed. Generally **motile.** Motile species show lateral flagellation. Cells multiply by **bending type of cell division.**

Gram-positive, but old cells frequently lose the Gram-positivity. Not acid-fast. Metachromatic granules not present.

Obligately aerobic. Catalase-positive. Chemoorganotrophs. Good growth on nutrient agar. **Acid is produced slowly and weakly from**

glucose, fructose and some other carbohydrates. In addition to pyruvic acid, several kinds of organic acids are assimilated. **Urease is not produced. Generally gelatin is hydrolyzed and DNase is produced.**

The cell wall peptidoglycan is based upon D-ornithine (type B2β, [L-Hsr]-D-Glu-D-Orn; Schleifer and Kandler, 1972). **Mycolic acids are not present.** The predominant fatty acids are anteiso-$C_{15:0}$ and anteiso-$C_{17:0}$, moderate amounts of iso-$C_{16:0}$ and $C_{16:0}$ are produced. Polar lipids consist of diphosphatidylglycerol, phosphatidylglycerol and several glycosyldiacylglycerols. Menaquinones are the sole respiratory quinones with MK-9 predominating. The DNA base composition ranges from 68.3–75.2 mol% G + C.

Type species: *Curtobacterium citreum* (Komagata and Iizuka 1964) Yamada and Komagata 1972, 425.

Further Descriptive Information

Curtobacterium strains grow well on nutrient agar at 25–30°C. They are obligately aerobic, and possibly possess nutritional requirements for some amino acids or vitamins.

Colonies on nutrient agar are circular, smooth, entire, raised, glistening, opaque and butyrous. The color of the colonies depends on the species as described below.

Morphologically, *Curtobacterium* strains show a bending-type cell division and often exhibit V forms (Yamada and Komagata, 1972a). Rod forms are observed in young cultures, while coccoid forms predominate in older cultures. Branching is not observed. The rod-coccus growth cycle is not as distinctive as that of *Arthrobacter*. The size of cells is 0.4–0.6 μm × 0.6–3.0 μm. *Curtobacterium* strains are Gram-positive, but sometimes they present a Gram-variable reaction in old cultures. Motility is observed in all the species of *Curtobacterium* except *C. albidum*. Motile strains have lateral flagella (Komagata and Iizuka, 1964; Iizuka and Komagata, 1965; Yamada and Komagata, 1972a).

Strains of *Curtobacterium* have the unusual peptidoglycan structure called group B type determined by Fiedler et al. (1970) and named by Schleifer and Kandler (1972). In the B type of peptidoglycan, dicarboxyl amino acid (usually D-glutamic acid) at position 2 is linked with the amino acid (D-alanine) at position 4 via an interpeptide bridge containing a D-diamino amino acid. Glutamic acid at position 2 is partly replaced by hydroxyglutamic acid. *Curtobacterium* species possess singly D-ornithine in the interpeptide bridge (see Table 15.16).

Keddie and Cure (1977) reported that the cell walls of members of the genus *Curtobacterium* contained a variety of sugars including galactose, glucose, mannose and rhamnose; not all of which are present in strains of every species. The cell wall sugars do not show a characteristic profile for the species or for the genus.

Cellular fatty acids of *Curtobacterium* are essentially anteiso isomers of $C_{15:0}$ and $C_{17:0}$ (Collins et al., 1980; Suzuki and Komagata, 1983). Iso-$C_{16:0}$ is always the third most abundant acid following the anteiso acids. Suzuki and Komagata (1983) reported that these fatty acid profiles are stable despite the culture conditions.

One species, *C. pusillum*, contains an unusual fatty acid, ω-cyclohexylundecanoic acid as its major component (Suzuki et al., 1981b). It accounts for more than 60% of the cellular fatty acids. Other than *C. pusillum*, large amounts of ω-cyclohexyl fatty acids have been found only in some thermophilic-acidophilic *Bacillus* strains.

Mycolic acids are not found in *Curtobacterium* strains (Goodfellow et al., 1976; Collins et al., 1980).

Polar lipids of *Curtobacterium* strains were studied by Komura et al. (1975) and Collins et al. (1980). *Curtobacterium* strains contain diphosphatidylglycerol, phosphatidylglycerol, and two or more unidentified glycolipids (Collins et al., 1980). Komura et al. (1975) detected diphosphatidylglycerol, phosphatidylglycerol and trace amounts of phosphatidylinositolmannosides and phosphatidylinositol in several strains of *Curtobacterium*. However, the more detailed study by Collins et al. (1980) failed to detect phosphatidylinositol and related mannosides in curtobacteria.

Curtobacterium strains possess menaquinones with nine isoprene units (MK-9) (see Table 15.16) (Yamada et al., 1976; Collins et al., 1979, 1980; Suzuki and Komagata, unpublished data).

The base composition of DNA of *Curtobacterium* determined by Yamada and Komagata (1970) ranges from 66–72 mol% G + C (T_m). Döpfer et al. (1982) reported higher values of 72–75 mol% (T_m).

DNA homology of the genus *Curtobacterium* revealed moderately high homology indices between the strains (Suzuki et al., 1981a; Döpfer et al., 1982). Döpfer et al. (1982) reported that *C. albidum* possessed the same peptidoglycan as members of the genus *Aureobacterium* and showed low DNA homology with *Aureobacterium* strains. However, Fiedler et al. (1970) and Schleifer and Kandler (1972) reported that *C. albidum* contained the same peptidoglycan as the other species of the genus *Curtobacterium*. Furthermore, *C. albidum* possesses the acetyl type of cell wall and the same menaquinone composition as the genus *Curtobacterium* (Uchida and Aida, 1979; Yamada et al., 1976). Recently it has been shown that the type strain of *C. citreum* shows a high DNA homology (75%) with the type strain of *C. albidum* (Suzuki and Komagata, unpublished data). These facts support the inclusion of *C. albidum* in the genus *Curtobacterium*.

Physiological and biochemical characteristics were studied systematically by Yamada and Komagata (1972a). Strains of *Curtobacterium* produce acid slowly from various kinds of sugars as shown later in Table 15.17. All of the *Curtobacterium* strains produce acid from glucose and fructose. Some organic acids are assimilated by strains of the genus *Curtobacterium*. They assimilate pyruvic acid and do not assimilate malonic acid, glutaric acid, adipic acid and pimelic acid.

Most of the members of the genus *Curtobacterium* have been isolated from plants. *Curtobacterium citreum*, *Curtobacterium albidum* and *Curtobacterium luteum* were isolated from rice (Komagata and Iizuka, 1964). The phytopathogenicity of these species is not known. *C. flaccumfaciens* is the only species regarded as a plant pathogen. This organism causes a vascular wilt and/or leaf spot of bean, red beet, tulips and poinsettia (see List of Species). The two strains of *Curtobacterium pusillum* were isolated from oil brine in an oil field in Japan (Iizuka and Komagata, 1965).

Enrichment and Isolation Procedures

Selective media or special methods for enrichment of the majority of curtobacteria have not been developed. The bacteria grow well on a medium based on peptone, yeast extract and glucose.

It has been reported that successful isolation of *C. flaccumfaciens* and its pathovars *C. flaccumfaciens* pathovar *betae*, *C. flaccumfaciens* pathovar *oortii* and *C. flaccumfaciens* pathovar *poinsettiae* can be achieved by the use of CNS agar medium (Gross and Vidaver, 1979). However, some strains of *C. flaccumfaciens* pathovar *poinsettiae* do not grow well on this medium (Vidaver and Starr, 1981).

Maintenance Procedures

Most of the strains of *Curtobacterium* recover well from cultures on nutrient agar slant stored at 5°C for 6 months. But, care should be taken because a few strains do not revive after 1- or 2-month storage. Lyophilization or deep freezing are recommended for long term preservation or more than 1 year.

Procedures for Testing Special Characters

Acid production from carbohydrates. The medium employed for the production of acid from carbohydrates contains 0.3% (w/v) peptone, 0.25% (w/v) NaCl, and 0.5% (w/v) carbohydrate (adjusted pH 7.2). The pH indicator is bromcresol purple (BCP). Acid production within 5 days is recorded as positive, and acid production after 7–20 days is regarded as weakly positive.

Assimilation of organic acids. The medium contains 0.5% (w/v) organic acid (sodium salt), 0.002% (w/v) glucose, 0.001% (w/v) yeast extract, 0.001% (w/v) trypticase (B.B.L.), 0.1% (w/v) K_2HPO_4, 0.5% (w/v) NaCl, 2% (w/v) agar, and 12 ppm of phenol red. The pH is adjusted to 7.0. Assimilation is determined by the color change of the indicator after 10 days of incubation.

Glycolate test. The following is the method of Uchida and Aida (1984). Ten milligrams of freeze-dried cells are hydrolyzed by 100 μl of 6 N HCl at 100°C for 2 h in a screw-capped tube. The hydrolysate is passed through a micro column (5 cm high, acetate type of Dowex 1×8). The column is then washed with 2 ml of water and 1 ml of 0.5 N HCl and the glycolate fraction is eluted by 2 ml of 0.5 N HCl. Two milliliters of DON reagent (0.02% of 2, 7-dihydroxynaphthalene dissolved in concentrated H_2SO_4) is added to 100 μl of this fraction and treated in boiling water for 10 min. The development of a clear purple red color indicates a positive glycolyl test. Quantitative determination (nmol glycolic acid by 1 mg dried cell) is carried out by colorimetry at 530 nm and compared with the standard solution of glycolic acid treated in the same manner.

Differentiation of the genus Curtobacterium *from other genera*

Members of the genus *Curtobacterium* are most likely to be confused with the genera *Aureobacterium*, *Cellulomonas* and *Microbacterium* (Table 15.16, and later in Table 15.22).

The most useful character for the differentiation of the genus *Curtobacterium* from these genera is the structure of the peptidoglycan.

Both *Curtobacterium* and *Aureobacterium* possess a group B peptidoglycan based on ornithine (Schleifer and Kandler, 1972). However, while *Curtobacterium* species possess a single D-ornithine in the interpeptide bridge, members of the genus *Aureobacterium* contain glycine in addition to the one D-ornithine residue. (D-Glu-Gly-D-Orn) (Fiedler et al., 1970; Schleifer and Kandler, 1972; Uchida and Aida, 1979). In addition, the menaquinone content of the two genera is different. *Curtobacterium* strains contain MK-9 as the predominant menaquinone while the major menaquinone isoprenologues of *Aureobacterium* are MK-11 to MK-13 (Yamada et al., 1976; Collins et al., 1979, 1980, 1983b; Tamaoka et al., 1983; Suzuki and Komagata, unpublished data). The two genera also differ in polar lipid content (Collins et al., 1980; 1983b). However, the cellular fatty acid composition is the same in both genera (Collins et al., 1980, 1983b; Suzuki and Komagata 1983). Some members of the genus *Aureobacterium* are more nutritionally exacting and require terregens factor, biotin, thiamine, pantothenic acid and L-methionine for growth (Lochhead and Burton, 1953; Lochhead, 1958).

Members of the genus *Cellulomonas* also possess ornithine as the principal amino acid in the cell wall peptidoglycan. However, the genus *Cellulomonas* contains a group A peptidoglycan based on L-ornithine while the genus *Curtobacterium* contains a group B peptidoglycan based on D-ornithine (Fiedler and Kandler, 1973). Furthermore, while curtobacteria are strictly aerobic, many strains of the genus *Cellulomonas* are facultatively anaerobic and produce acid from a wide variety of sugars (Keddie and Jones, 1981). The menaquinone composition is different: members of the genus *Cellulomonas* contain hydrogenated menaquinones MK-9 (H₄) (Minnikin et al., 1979; Yamada et al., 1976). There are also differences in the polar lipid and fatty acid composition of the genera *Cellulomonas* and *Curtobacterium* (Minnikin et al., 1979). In addition, all strains of *Cellulomonas* are cellulolytic while no curtobacteria are known to exhibit this property (Keddie and Jones, 1981).

The genus *Microbacterium* (Collins et al., 1983c) also possesses a group B peptidoglycan but the principal amino acid is lysine not ornithine (Schleifer et al., 1968; Schleifer and Kandler, 1972). *Curtobacterium* species are further distinguished from members of the genus *Microbacterium* by the possession of an acyl type cell wall and by menaquinone and polar lipid composition (Collins et al., 1983c). Fur-thermore, *Microbacterium* also differs from *Curtobacterium* in physiological and biochemical characteristics (Yamada and Komagata, 1972a).

Taxonomic Comments

The genus *Curtobacterium* was established by Yamada and Komagata (1972a) for the so-called motile brevibacteria. The genus was originally characterized by the presence of ornithine in the cell wall and weak production of acid from various carbohydrates. As defined by Yamada and Komagata (1972b), the genus contained organisms previously designated *Brevibacterium albidum*, *B. citreum*, *B. insectiphilium*, *B. luteum*, *B. pusillum*, *B. saperdae*, *B. testaceum*, *B. helvolum*, *Corynebacterium flaccumfaciens* and *C. flaccumfaciens* subsp. *aurantiacum*.

Support for the recognition of the genus *Curtobacterium* came from the results of the peptidoglycan studies of Schleifer and Kandler (1972) which grouped together most of the species listed above. However, there was also some evidence of heterogeneity within the genus. The major menaquinones in most of the *Curtobacterium* species were normal menaquinones with nine isoprene units, MK-9, whereas *B. testaceum* was shown to contain MK-11 (Yamada et al., 1976). Also, Uchida and Aida (1977) showed that the glycan moiety of the peptidoglycan of *B. testaceum* was unusual in that it contained approximately equal amounts of glycolyl and acetyl residues. There was also some discrepancy between the results obtained by different groups working with strains of *B. helvolum* (see Keddie and Jones, 1981, for detailed discussion).

For reasons such as these, only six species of *Curtobacterium* were listed in the Approved Lists of Bacterial Names (Skerman et al., 1980): *Curtobacterium albidum*, *C. citreum*, *C. luteum*, *C. pusillum*, *C. saperdae* and *C. testaceum* and these were also listed under the genus *Brevibacterium*. *Corynebacterium flaccumfaciens* was listed only under *Corynebacterium* (Skerman et al., 1980).

Recently, Collins and Jones (1983) formally proposed the transfer of *Corynebacterium flaccumfaciens* to the genus *Curtobacterium* and at the same time proposed that *Corynebacterium betae*, *C. oortii* and *C. poinsettiae* be regarded as pathovars of *Curtobacterium flaccumfaciens*.

On the basis of previous studies and further chemical investigations Collins et al. (1983b) proposed that *Curtobacterium saperdae* and *C. testaceum* be reclassified in a new genus *Aureobacterium*. Thus the genus *Curtobacterium* currently contains the species *C. albidum*, *C. citreum*, *C. flaccumfaciens*, *C. luteum* and *C. pusillum*.

The taxonomic relationship of the genus *Curtobacterium* to other Gram-positive genera is still not resolved but there is evidence that it

Table 15.16.
Chemotaxonomic characteristics of the genera **Aureobacterium** *and* **Curtobacterium**

Genera	Peptidoglycan structure[a]	Isoprenoid quinones[b]	Cell wall acyl type[c]	Cellular fatty acid[d]
Curtobacterium	[L-Hsr]D-Glu-D-Orn	MK-9	Acetyl	*Anteiso-* and *iso-*[e]
Aureobacterium	[L-Hsr]D-Glu-Gly-D-Orn (Hyg)	MK-11–MK-13	Glycolyl	*Anteiso-* and *iso-*

[a] Schleifer and Kandler (1972).

[b] Yamada et al. (1976) and Collins et al. (1979, 1980).

[c] Uchida and Aida (1979).

[d] Collins et al. (1980), Suzuki et al. (1981b), and Suzuki and Komagata (1983).

[e] *Curtobacterium pusillum* contains, in addition, large amounts of ω-cyclohexyl undecanoic acid.

is most closely related to the genera *Aureobacterium* and *Microbacterium* (Döpfer et al., 1982; Collins et al., 1983b).

Further Reading

Keddie, R.M. and D. Jones. 1981. Saprophytic, aerobic coryneform bacteria. *In* Starr, Stolp, Trüper, Balows, and Schlegel (Editors), The Prokaryotes. A

Handbook on Habitats, Isolation and Identification of Bacteria, Springer-Verlag, New York, pp. 1838–1878.

Yamada, K. and K. Komagata. 1972. Taxonomic studies on coryneform bacteria. V. Classification of coryneform bacteria. J. Gen. Appl. Microbiol. *18:* 417–431.

Differentiation and characteristics of the species of the genus **Curtobacterium**

The differential characteristics of the species of the genus *Curtobacterium* are listed in Table 15.17 and in the description of the species in the List of Species.

List of species of the genus **Curtobacterium**

Following are the characteristics essentially based on their original description. Cell morphology and some physiological characteristics in common are described in the definition of the genus. Further physiological and biochemical characteristics are shown in Table 15.17.

1. **Curtobacterium citreum** (Komagata and Iizuka 1964) Yamada and Komagata 1972b, 425.[AL] (*Brevibacterium citreum* Komagata and Iizuka 1964, 498.)

cit′re.um. L. adj. *citreum* lemon colored.

Dull yellow colonies appear on nutrient agar. Motile. Starch is hydrolyzed. Gelatin is hydrolyzed but casein is not.

Base composition of DNA: 70.5–75.2 mol% G + C (T_m).

Source: rice.

Type strain: ATCC 15828 (IAM 1614).

2. **Curtobacterium albidum** (Komagata and Iizuka 1964) Yamada and Komagata 1972b, 425.[AL] (*Brevibacterium albidum* Komagata and Iizuka 1964, 500.)

al′bi.dum. L. neut. adj. *albidum* white.

Colonies on nutrient agar present an ivory color. Nonmotile. Starch is hydrolyzed. Gelatin and casein are hydrolyzed.

Base composition of DNA: 70.0–72.0 mol% G + C (T_m).

Source: rice.

Type strain: ATCC 15831 (IAM 1631).

3. **Curtobacterium luteum** (Komagata and Iizuka 1964) Yamada and Komagata 1972b, 425.[AL] (*Brevibacterium luteum* Komagata and Iizuka 1964, 499.)

lu′te.um. L. adj. *luteum* saffron or golden yellow.

Colonies on nutrient agar are dark yellow. Motile. Starch is not hydrolyzed. Both gelatin and casein are hydrolyzed.

Base composition of DNA: 69.8–74.9 mol% G + C (T_m).

Source: rice.

Type strain: ATCC 15830 (IAM 1623).

4. **Curtobacterium pusillum** (Iizuka and Komagata 1965) Yamada and Komagata 1972b, 425.[AL] (*Brevibacterium pusillum* Iizuka and Komagata 1965, 2.)

pu′sil.lum. L. adj. *pusillum* very small.

Colonies on nutrient agar are pale yellow to grayish white. Motile.

Starch is hydrolyzed. The type strain hydrolyzes gelatin and casein. ω-Cyclohexylundecanoic acid is a major cellular fatty acid.

Base composition of DNA: 69.0–74.2 mol% G + C (T_m).

Source: oil brine.

Type strain: ATCC 19096 (IAM 1479).

5. **Curtobacterium flaccumfaciens** (Hedges 1922) Collins and Jones 1984, 270.[VP] (Effective publication: Collins and Jones 1983, 3546.) (*Bacterium flaccumfaciens* Hedges, 1922, 433; *Corynebacterium flaccumfaciens* (Hedges) Dowson 1942, 313.)

flac.cum.fa′ci.ens. L. adj. *flaccus* flabby; L. part. adj. *faciens* making; M.L. part. adj. flaccumfaciens wilt-making.

The colonies are yellow, orange or pink. Some variants may lose pigmentation, and some may produce a blue to purple water-soluble pigment. Generally motile. Thiamine, biotin and pantothenate are required for growth. Optimum growth temperature is 24–27°C and maximum temperature 35–37°C. Gelatin and pectate gel are not liquefied.

Base composition of DNA: 68.3–73.7 mol% G + C (T_m).

Type strain: NCPPB 1446.

The following pathovars are recognized in this species for practical, quarantine purposes. Their description is the same as that for the species.

(i) *Curtobacterium flaccumfaciens* pathovar *flaccumfaciens*.

This organism causes a vascular wilt of bean (*Phaseolus vulgaris*).

Reference strain: NCPPB 1446.

(ii) *Curtobacterium flaccumfaciens* pathovar *betae*.

This organism causes a vascular wilt and leaf spot of red beet (*Beta vulgaris*).

Reference strain: NCPPB 374.

(iii) *Curtobacterium flaccumfaciens* pathovar *oortii*.

This organism causes a vascular disease and leaf and bulb spot of tulips (*Tulipa* spp.).

Reference strain: ATCC 25283.

(iv) *Curtobacterium flaccumfaciens* pathovar *poinsettiae*.

This organism causes a stem canker and leaf spot of the poinsettia (*Euphorbia pulcherrima*).

Reference strain: ATCC 9682.

Table 15.17.
Physiological and biochemical characteristics of species of the genus **Curtobacterium**[a,b]

Characteristics	1. C. citreum	2. C. albidum[c]	3. C. luteum[c]	4. C. pusillum	5. C. flaccumfaciens
Acid from:					
L-Arabinose	W	−	W	W	−
Xylose	W	W	W	Wd	Wd
Rhamnose	W	W	−	Wd	Wd
Glucose	W	W	+	W	W
Fructose	W	W	+	W	W
Mannose	W	−	+	W	Wd
Galactose	W	−	−	Wd	−
Sorbose	W	−	−	−	−
Sucrose	−	−	−	W	−
Lactose	−	−	−	−	−
Maltose	W	W	−	W	Wd
Trehalose	−	−	−	−	−
Cellobiose	−	−	−	Wd	−
Raffinose	−	−	−	Wd	−
Dextrin	−	−	−	−	−
Starch	−	−	−	−	−
Inulin	−	−	−	−	−
Glycerol	−	−	−	Wd	−
Erythritol	W	−	−	Wd	−
Adonitol	−	−	−	−	−
Mannitol	−	−	−	Wd	−
Dulcitol	−	−	−	−	−
Sorbitol	−	−	−	−	−
Inositol	−	−	−	W	Wd
Arbutin	−	−	−	−	−
Esculin	−	−	−	−	−
Salicin	−	−	−	−	−
α-Methylglucoside	−	−	−	−	−
Assimilation of:					
Acetic acid	d	+	+	d	+
Pyruvic acid	+	+	+	+	+
L-Lactic acid	+	+	−	+	+
D-Lactic acid	+	+	−	+	+
Malic acid	+	−	+	−	+
Succinic acid	+	+	+	d	+
Fumaric acid	+	+	+	d	+
α-Ketoglutaric acid	+	−	−	−	−
Citric acid	d	+	−	−	+
Formic acid	−	−	−	−	−
Propionic acid	−	−	−	d	−
Butyric acid	−	−	−	−	−
Oxalic acid	−	−	−	−	−
Malonic acid	−	−	−	−	−
Glutaric acid	−	−	−	−	−
Adipic acid	−	−	−	−	−
Pimelic acid	−	−	−	−	−
Glycolic acid	+	+	+	d	d
Glyoxylic acid	+	+	−	−	−
Gluconic acid	+	+	+	−	+
Hippuric acid	−	−	−	d	−
Uric acid	−	−	−	−	−
Gelatin hydrolysis	+	+	+	d	+
Casein +hydrolysis	−	+	+	d	+
DNase	+	+	+	d	+
Urease	−	−	−	−	−
Growth in 5% +NaCl	+	+	+	+	d
Growth in 10% NaCl	−	−	−	−	

[a] Data from Yamada and Komagata, 1972b.
[b] Symbols: see Table 15.2; also d, 11–89% of strains are positive; W, weak positive (acid production between 7 and 20 days); and Wd, 11–89% of strains are positive but with weak reaction.
[c] Data for type strain only.

Genus **Caseobacter** *1978, 364*[AL]

W. H. J. CROMBACH

Ca.se.o.bac'ter. L. n. *caseus* cheese; M.L. masc. n. *bacter*, the masculine equivalent of Gr. n. *bactrum* a rod; M.L. masc. n. *Caseobacter* cheese rod.

Rods, mostly irregular, i.e. club-shaped or tapered, 0.8–1.2 × 2.0–4 µm. On fresh media the rods **become ovoid or coccoid** in shape **with the exhaustion of the medium.** Branching of the rods only rarely occurs.

After transferring the cells to a fresh medium out-growths from the coccoid and ovoid cells occur, which result in the formation of rods. The cells occur singly, in pairs with the typical V-formation, in clumps and in **palisade** formation. Capsules are not formed. **Gram-positive. Nonmotile. On yeast extract, 0.7% (w/v)-glucose, 1% (w/v) agar colonies have a dry appearance, are small (2–3 mm), circular, convex, and are gray-white, slightly pink or slightly red. Strictly aerobic.** Optimum temperature between 25 and 30°C. **Most strains require organic nitrogen compounds.**

Nearly all strains tolerate 8% NaCl (w/v) in the medium. No hydrolysis of gelatin, casein or starch. Acid is produced in Hugh-Leifson medium.

Catalase-positive. Oxidase-negative.

Meso-**diaminopimelic acid is the characteristic amino acid of the cell wall.**

Arabinose, mannose and galactose are the characteristic cell wall sugars. Low molecular weight **mycolic acids** (C_{30}–C_{36}) **are present.**

The mol% G + C of the DNA is 65–67 (T_m).

Type species: *Caseobacter polymorphus* Crombach 1978, 364.

Further Descriptive Information

Cells occur in the coccoid, ovoid or rod shape depending on the available growth factors in the medium. Most rods are short, thick and pleomorphic. Frequently, septa-like clear-cut structures can be seen in the rods. With aging the rods gradually transform into the coccoid or ovoid shape. This occurs on either a poor or a rich medium at 25°C within 3 or 7 days, respectively (for the composition of these media: see Isolation and Enrichment).

The rods can be arranged in the characteristic V formation. From electron micrographs (Crombach, 1978) it can be seen that both daughter cells are only linked indirectly by extracellular material, resulting in "snapping division."

Most strains of this genus can utilize the following compounds as carbon sources: glucose, fructose, galactose, glycerol, lactate and succinate, whereas lactose, glyoxylate and citrate cannot be utilized. Only a few strains can utilize mannitol, sorbitol, inositol, xylose, cellobiose and maltose as carbon source. Valine, tyrosine, lysine and asparagine can be utilized as sources of nitrogen. Choline and glucosamine cannot be utilized as carbon and nitrogen sources.

DNase-negative, RNase-negative, cellulase-negative, uricase-negative, indole not produced, acetylmethylcarbinol not produced.

Glucose can be oxidized nearly completely and conversion into reserve material hardly occurs. Tween 80 is only rarely hydrolyzed.

The dyes safranin, gentian violet, methylene blue and bromothymol blue, with a final concentration of 80 µg/ml in a nutritionally rich medium composed of yeast extract, 0.7% (w/v); glucose, 1% (w/v) and agar, 1.2% (w/v) (medium A), inhibit growth during 2 weeks at 30°C. Growth on this medium after 1 and 2 days at 25°C is also frequently inhibited by the following antibiotics (µg/disk): penicillin G, 5; novobiocin, 5; chloramphenicol, 50; neomycin, 30; erythromycin, 10; kanamycin, 30; streptomycin, 10; and tetracycline, 10.

The cells do not survive 30 min at 70°C in skim milk.

DNA base composition and DNA/DNA hybridization experiments (Crombach, 1972, 1974), revealed a high level of homogeneity within this taxon.

Strains of *Caseobacter* occur on the rind of different types of soft cheeses produced in some parts of Holland and Belgium (Mulder et al., 1966, El Erian, 1969). These strains may play an important role in the ripening of soft cheeses.

Enrichment and Isolation

Scrapings of the rind of Limburger and Meshanger type soft cheese are homogenated in physiological saline. Subsequently, the dilution is plated on a nutritionally poor medium containing: tryptone, 0.5% (w/v); glucose, 0.1% (w/v); yeast extract, 0.3% (w/v); NaCl, 0.1% (w/v) and agar 1% (w/v) (medium B). After incubation for 5 days at 25°C the nonorange colored colonies with a dry surface and consisting of coccoid or ovoid cells are transferred to fresh medium A. Out-growth and elongation from the coccoid and ovoid cells of *Caseobacter polymorphus* will then occur with the formation of rods. After exhaustion of the medium the rods transform again into ovoid cells and ultimately into cocci (Mulder et al., 1966).

Maintenance Procedures

Caseobacter strains can be preserved for several months on agar slants of medium A. The slants are grown for one day at 25°C followed by storage at 4°C. For longer preservation (several years) young active cells can be lyophilized in skim milk by common procedures.

Procedures for Testing Special Characters

The characteristic growth cycle can be shown clearly by incubating the cells for 2 weeks at 25°C on a poor medium of the following composition: soil extract supplied with yeast extract, 0.1% and agar, 1.2%. The cells in the coccoid stage are then transferred to fresh medium A and incubated at 15°C. At this low temperature the outgrowth from the cocci into rods occur within 2 days and, subsequently, the rods transform again into ovoids and cocci within 7 days.

The utilization of several carbon compounds as carbon sources can be tested in a basal medium consisting of yeast extract, 0.2% (w/v); $MgSO_4 \cdot 7H_2O$, 0.05% (w/v); K_2HPO_4, 0.01% (w/v); soil extract, 100 ml; tap water, 900 ml. The carbon compound under study which is autoclaved separately, is added to a final concentration of 0.5%. A 5-ml volume of medium is inoculated with 0.1 ml of a suspension of 1-day-old cells with a concentration of about 20 Eel-nephelometer units and incubated aerobically for 5 days at 25°C. Turbidity readings of 50% over the controls (the same strains cultured in the same medium but without the carbon compound) and exceeding 50 nephelometer units are considered positive.

The utilization of several amino acids as nitrogen sources can be determined in a basal medium composed of yeast extract, 0.01% (w/v); glucose, 0.5% (w/v); K_2HPO_4, 0.1% (w/v); $MgSO_4 \cdot 7H_2O$, 0.05% (w/v); soil extract, 100 ml; tap water, 900 ml. The K_2HPO_4 should be autoclaved separately to avoid precipitation. The different amino acids are added to a final concentration of 0.3% (w/v). The inoculation, growth conditions and the determination of results are the same as for the assay of carbon sources.

Differentiation of the genus **Caseobacter** from other genera

Table 15.18 shows the characteristics differentiating the genus *Caseobacter* from related taxa.

Taxonomic Comments

Originally, members of *Caseobacter* were classified as arthrobacters on account of their distinct pleomorphic morphology (Mulder and Antheunisse, 1963). However, further study (Mulder et al., 1966) revealed that the gray-white cheese coryneforms deviate from soil arthrobacters and resemble to a certain extent *Brevibacterium linens* in respect of nutritional requirements, salt tolerance and lack of production of polysaccharides. Concerning the nitrogen and vitamin requirements, the *Caseobacter* strains are intermediate between the less exacting *Arthrobacter* strains and the more exacting *Brevibacterium linens* strains.

DNA/DNA hybridization experiments (Crombach, 1974) have shown only a distant genetic relationship between representatives of *Caseobacter polymorphus* on the one hand and those of *Arthrobacter globiformis* and *Brevibacterium linens* on the other hand.

The cell wall composition of *Caseobacter polymorphus* (*meso*-diaminopimelic acid (*meso*-DAP)), arabinose and galactose suggest a relationship of that species to representatives of the genus *Rhodococcus* (Keddie and Cure, 1977). This is supported by the DNA-base composition of the two genera. The mol% G + C of *Rhodococcus* ranges from 58–69 (Mordarski et al. 1976, 1977). However, the mycolic acid profiles of *Caseobacter polymorphus* containing between 30 and 36 carbon atoms (Crombach, 1978; Keddie and Bousfield, 1980; Collins et al., 1982) suggest a closer relationship of this species with *Corynebacterium* sensu stricto with mycolic acid profiles of 22–38 carbon atoms. The mol% G + C of these two genera are not in the same range, indicating only a remote relationship (*Corynebacterium* exhibits a mol% G + C of 51–59).

DNA/DNA hybridization experiments between the representative strain of *Caseobacter* (AC 256), and some strains of *Corynebacterium* have shown only a low homology index, D, within the range of 9–30% at restricted temperature (Crombach, 1974; Suzuki et al., 1981). This low homology also supports the separate genus *Caseobacter* within the coryneform group.

The presence of a dihydrogenated menaquinone with nine isoprene units, MK-9 (H$_2$) as the major component in both *Caseobacter polymorphus*, type strain LMD AC 256 (NCDO 2097) and representatives of the genera *Arthrobacter*, *Corynebacterium* and *Rhodococcus*, indicates a certain extent of relatedness among these strains (Collins et al., 1979, 1980b).

Further comparative studies, viz. DNA/DNA and rRNA/DNA hybridization, numerical phenetic analysis, chemotaxonomy and further menaquinone analysis are required to establish the taxonomic niche of the genus *Caseobacter* within the "catch-all" group of coryneforms. Especially the determination of the extent of relatedness between the genera *Caseobacter*, *Corynebacterium* and *Rhodococcus* needs more study.

Further Comments

Within the genus *Caseobacter* a minor group of strains differs from the rest of the nonorange cheese coryneforms. These strains are strongly proteolytic, decompose starch (Crombach, 1972, 1974), lack *meso*-DAP and arabinose but contain lysine in their cell walls, and mycolic acids are not present (Sharpe et al., 1976; Keddie and Cure, 1977; Crombach, 1978). The representative strains AC 253 and EC 20 of this group of gray-white cheese strains (Crombach 1972, 1978) also have deviating mol% G + C of DNA of 60–61 and show low DNA homologies with the type strain of *Caseobacter polymorphus* (Crombach, 1974). These strains have a much more pronounced tendency than *Caseobacter polymorphus* to transform into cocci on aging.

In search of a clear-cut taxonomic allocation of the genus *Caseobacter*, these strains should not be accommodated in the genus *Caseobacter*. This enables a more distinct and tighter description of the genus *Caseobacter* than the definition initially proposed by Crombach, 1978. Further study is needed to elucidate the taxonomic position of the lysine-containing cheese coryneforms.

List of species of the genus **Caseobacter**

1. **Caseobacter polymorphus** Crombach 1978, 364.[AL]

po.ly.mor'phus. Gr. adj. multiform, organism can appear in different forms.

Morphology and description as for genus.
Type strain: LMD AC 256 (NCDO 2097).

Table 15.18.
Differential characteristics of the genus **Caseobacter** *and other related taxa*[a]

Characteristics	Caseobacter	Arthrobacter	Rhodococcus[b]	Brevibacterium[c]	Corynebacterium
Oxygen requirements	Aerobic	Aerobic	Aerobic	Aerobic	Facultative
Primary mycelium, fragmenting easily into irregular fragments	−	−	+	−	−
Rod-coccus cycle[d]	+	+	±	+	−
Casein hydrolyzed	−	+	−	+	ND
Glucosamine as carbon and nitrogen source	−	d	+[e]	ND	ND
Meso-DAP in cell wall	+	−	+	+	+
Presence of mycolic acids	+(C$_{30}$–C$_{36}$)	−	+(C$_{36}$–C$_{66}$)	−	+(C$_{22}$–C$_{38}$)
Arabinose in cell wall	+	−	+	−	+
DNase	−	+	ND	+	ND
Major menaquinones	MK-9 (H$_2$)[f]	MK-9 (H$_2$)	MK-9 (H$_2$) or MK-8 (H$_2$)	MK-8 (H$_2$)	MK-8 (H$_2$) or MK-9 (H$_2$)
Mol% G + C of DNA	65–67	59–67	58–59	60–64	51–59

[a] Symbols: see Table 15.2; also d, 11–89% of strains are positive; ±, poorly developed; and ND, data not available or incomplete.

[b] Goodfellow and Alderson, 1977; Mordarski et al., 1976, 1977.

[c] Collins, Jones, Keddie and Sneath 1980b.

[d] Keddie and Jones, 1981.

[e] With reference to the type species *Rhodococcus rhodochrous* and *Rhodococcus rubrus* only (Goodfellow and Alderson, 1977).

[f] MK-9 (H$_2$): menaquinone with one of the nine isoprene units hydrogenated.

Genus **Microbacterium** Orla-Jensen 1919, 179[AL]

M. D. Collins and Ronald M. Keddie

Mic.ro.bac.te´ri.um. Gr. adj. *micrus* small; Gr. neut. dim. n *bacterium* a small rod; M.L. neut. n. *Microbacterium* a small rodlet.

In young cultures (~12–24 h), small, slender, irregular rods ~0.4–0.8 µm in diameter by ~1.0–4.0 µm or more in length; some of the rods are arranged at an angle to each other to give **V formations.** Primary branching occurs but is uncommon; mycelia are not produced. **In older cultures (~3–7 days) the rods are shorter and a proportion may be coccoid but a marked rod-coccus growth cycle does not occur. Gram-positive** and nonacid-fast. Do not form endospores. Nonmotile or motile by one to three flagella. **The cell wall peptidoglycan contains lysine as the diamino acid. The major isoprenoid quinones are unsaturated menaquinones with 11 and 12 isoprene units (MK-11 and MK-12). Aerobic; weak anaerobic growth may occur.** Optimum temperature, ~30°C. Growth moderate on yeast extract-peptone-glucose media at neutral pH giving circular, opaque, glistening colonies, often with yellowish pigmentation.

Chemoorganotrophic: metabolism primarily respiratory but may also be fermentative (see Further Descriptive Information). **Acid is produced from glucose and from some other sugars in peptone media.** DNase produced. Catalase-positive. Nutrition complex; all require B-vitamins and some require amino acids also.

The mol% G + C of the DNA is 69–75.4 (T_m).

Type species: *Microbacterium lacticum* Orla-Jensen 1919, 179.

Further Descriptive Information

The cell wall peptidoglycan contains lysine as diamino acid (Schleifer and Kandler, 1972; Keddie and Cure, 1978). The walls contain a Group B peptidoglycan, type B1α, [L-lysine]D-glutamic acid-glycine-L-lysine (*M. lacticum* and *M. laevaniformans*), or type B1β, [L-homoserine]D-glutamic acid-glycine₂-L-lysine (*M. imperiale*, Schleifer and Kandler, 1972). The glycan moiety of the cell wall contains both glycolyl and acetyl residues (Uchida and Aida, 1977; 1979; Collins, unpublished). The nonhydroxylated long chain fatty acids are mainly *anteiso-* and *iso*-methyl-branched acids; straight chain saturated and monounsaturated acids are present only in small amounts. The major fatty acids are 12-methyltetradecanoic, 14-methylhexadecanoic and 14-methylpentadecanoic acids. The major polar lipids are diphosphatidylglycerol, phosphatidylglycerol and dimannosyldiacylglycerol; minor amounts of a monoglycosyldiacylglycerol and a phosphoglycolipid may occur (Collins et al., 1983c).

Colonies on yeast extract-peptone-glucose solid media are ~1–3 mm in diameter and are gray-white to yellow-orange depending on species. Minimum growth temperatures are 10°C or less and maximum temperatures are in the range 36–40°C. Most strains of *M. lacticum* are described as markedly heat resistant but most were isolated from laboratory pasteurized materials or from sources in which a heat treatment was used. However, Jayne-Williams and Skerman (1966) described nonthermoduric strains, otherwise indistinguishable from *M. lacticum* (only conventional tests were used), which were isolated from sources that had not been heat treated. Thermoduric strains of *M. lacticum* survive 63°C for 30 min in skim milk (Abd-el-Malek and Gibson, 1952; Jayne-Williams and Skerman, 1966) or 72°C for 15 min (Rogosa and Keddie, 1974). *M. laevaniformans* survives 63°C for 30 min; *M. imperiale* does not (Collins et al., 1983c). There are conflicting reports about the ability of *M. lacticum* to grow anaerobically (Keddie and Jones, 1981). It has variously been reported to be unable to grow in strictly anaerobic conditions (Orla-Jensen, 1919; Abd-el-Malek and Gibson, 1952) or to give weak anaerobic growth (Robinson, 1966; Jones, 1975; Keddie and Cure, 1977). Similarly, acid formation by *M. lacticum* in anaerobic conditions is described as equivocal (Jayne-Williams and Skerman, 1966) or generally positive (Robinson, 1966; Jones, 1975). *M. laevaniformans* and *M. imperiale* were reported to give weak growth in anaerobic conditions (Dias and Bhat, 1962; Jones 1975).

Some of the substances used as carbon + energy sources by *Microbacterium* spp. are listed later in Table 15.21. All the species now recognized are nutritionally exacting: B-vitamins and usually amino acids also are required (Dias and Bhat, 1964; Skerman and Jayne-Williams, 1966; Robertson and Keddie, unpublished).

M. lacticum is reported to produce L(+)-lactic acid from glucose (Rogosa and Keddie, 1974); enzymes of the Embden-Meyerhof and hexosemonophosphate pathways, and of the tricarboxylic acid cycle are reported to be present (Collins-Thompson et al., 1972).

M. lacticum has usually been isolated from milk or dairy sources after laboratory pasteurization or from dairy products that have been heat treated. This species may form a considerable part of the thermoduric bacterial flora of raw and pasteurized milk, powdered milk, cheese and dairy equipment (see Keddie and Jones, 1981, for further details and references). *M. laevaniformans* is reported to occur in raw sewage and in activated sludge (Dias and Bhat, 1962; Dias, 1963). *M. imperiale* was isolated originally from the alimentary canal of the Imperial moth *Eacles imperialis*. Bacteria resembling microbacteria have been reported to occur in fresh beef, poultry giblets and raw and pasteurized egg fluid (Kraft et al, 1966).

Enrichment and Isolation Procedures

M. lacticum is usually found in milk and dairy products, and on dairy equipment; thus the methods described for its isolation are based on those used for the examination of dairy products and equipment (see Keddie and Jones, 1981). Because most *M. lacticum* strains are thermoduric, the usual isolation procedure is to plate out laboratory pasteurized samples on a suitable, nonselective medium (Thomas et al., 1967). A common laboratory pasteurization procedure is to heat 5-ml amounts of milk, etc., in test tubes at 63°C for 30 min. Tubes are removed from the water bath and immediately cooled to 10°C in ice water. The contents are then plated on to Yeast Extract Milk Agar (YMA, Harrigan and McCance, 1976) and incubated at 30°C for up to 7 days. For rinse solutions, it is recommended that an equal volume of sterile skim milk be added to the samples before pasteurization (Thomas et al., 1967). The primary dilution of cheese and butter samples is prepared by homogenizing the material in 2% sodium citrate solution (Gillies, 1971).

M. laevaniformans can be isolated from raw sewage and more especially from activated sludge by using the enrichment procedure of Dias and Bhat (1962). A medium without added nitrogen of the following composition is used for enrichment: sucrose, 10 g; $CaSO_4 \cdot 2H_2O$, 0.1 g; $MgSO_4 \cdot 7H_2O$, 0.2 g; NaCl, 0.2 g; K_2HPO_4, 0.2 g; $Na_2MoO_4 \cdot 2H_2O$, 5 mg; $FeCl_3$, 2 mg; $CaCO_3$, 5 g; micronutrient solution, 1 ml; distilled water, 1000 ml. The micronutrient solution contains per liter of distilled water: $ZnSO_4 \cdot 7H_2O$, 11 g; $MnSO_4 \cdot H_2O$, 5 g; $CoSO_4$, 0.05 g; H_3BO_3, 0.05 g; $CuSO_4 \cdot 5H_2O$, 0.007 g (Dias and Bhat, 1962). Flasks (50 ml) containing 10-ml amounts of medium (sterilized at 121°C for 10 min) are inoculated with 1 ml of raw sewage or activated sludge and incubated at 25–28°C. After three serial passages in the above medium, the final enrichment is plated out on a similar medium solidified with agar and the plates incubated for 4 days. Large (3–4 mm), mucoid colonies are picked, purified and examined for the characters of *M. laevaniformans*. Dias and Bhat (1962) emphasize that bacteria other than *M. laevaniformans* are enriched for by this method.

We are not aware of any methods designed specifically for the isolation of *M. imperiale*. It is worth noting, however, that all microbacteria (including *M. imperiale*) grow well in air on media that contain yeast extract, peptone and milk or glucose (Keddie and Jones, 1981).

Maintenance Procedures

For short term preservation, slant cultures on yeast extract-peptone-milk agar should remain viable for a few weeks at room temperature

(~20°C). For long term preservation (several years), cultures may be preserved by lyophilization. Cultures may also be preserved for several years using the method described by Feltham et al. (1978).

Procedures for Testing for Special Characters

Morphological characteristics may be examined by using the methods described by Cure and Keddie (1973). Cell wall composition may be determined by two-dimensional paper chromatography of acid hydrolysates of cell wall preparations obtained by alkali treatment of whole cells as described by Keddie and Cure (1977). Menaquinone composition may be determined by reverse-phase partition TLC or HPLC (Collins et al., 1980b, 1983c). Polar lipid patterns may be obtained by simple two-dimensional TLC analysis of free lipid extracts (Collins et al., 1983c).

Differentiation of the genus **Microbacterium** from other genera

Members of the genus *Microbacterium* may be differentiated from other coryneform bacteria with a group B peptidoglycan (the most closely related) or which contain lysine as diamino acid in the cell wall peptidoglycan by the characters listed in Table 15.19. In addition the polar lipid pattern is diagnostic: microbacteria always contain diphosphatidylglycerol, phosphatidylglycerol and dimannosyldiacylglycerol as the major polar lipids; in addition some strains have minor amounts of a monoglycosyldiacylglycerol and a phosphoglycolipid.

Taxonomic Comments

DNA/DNA homology studies (Döpfer et al., 1982) have shown that *M. lacticum*, *M. imperiale* and *M. laevaniformans* are distinct species. Furthermore, the relatively high interspecific DNA homologies (~29–44%) exhibited by these three taxa indicate that they form a tight genus.

Members of the genus *Microbacterium* contain the unusual group B peptidoglycan (characterized by a cross-linkage between the α-carboxyl group of D-glutamic acid in position 2 of the peptide subunit and the C-terminal D-alanine of an adjacent subunit, Schleifer and Kandler, 1972). Microbacteria and other group B peptidoglycan-containing coryneform taxa (*Agromyces*, *Curtobacterium*, *Corynebacterium michiganense*, *C. insidiosum*, *C. sepedonicum*, *C. iranicum*, *C. tritici*, *C. nebraskense* etc.) share a number of features. Thus they have high G + C values (~65–75 mol%), and they contain major amounts of *anteiso*- and *iso*-methyl-branched chain fatty acids, unsaturated menaquinones, and diphosphatidylglycerol, phosphatidylglycerol and a variety of glycolipids (Collins, 1982; Collins and Jones, 1980; Collins et al., 1980a, 1983c; Döpfer et al., 1982; Keddie and Jones, 1981). DNA/rRNA homology studies (Döpfer et al., 1982) have shown that microbacteria and other group B peptidoglycan-containing coryneform bacteria are closely related phylogenetically and belong to a single rRNA cistron cluster ($T_{m(e)}$ values of DNA-rRNA heteroduplices cover a range of ~8°C). Döpfer et al. (1982) have suggested that all actinomycete and coryneform taxa with a group B peptidoglycan should be included in one genus, *Microbacterium*. While we agree that the high rRNA cistron similarities exhibited by these bacteria indicate a close phylogenetic relationship, we do not agree that all should be classified in one genus. On the basis of phenetic, cell wall peptidoglycan and lipid composition, there are several clearly recognizable groups, which we consider to represent genera, among those bacteria which have a group B peptidoglycan (e.g. *Agromyces*, *Curtobacterium*, *Microbacterium* and "diaminobutyric acid-containing plant pathogenic coryneform bacteria"). We would also consider that the rRNA cistron cluster of Döpfer et al. (1982) should be given family status.

Further Reading

Collins, M.D., D. Jones and R.M. Kroppenstedt. 1983. Reclassification of *Brevibacterium imperiale* (Steinhaus) and *Corynebacterium laevaniformans* (Dias and Bhat) in a redefined genus *Microbacterium* (Orla-Jensen), as *Microbacterium imperiale* comb. nov. and *Microbacterium laevaniformans* nom. rev.; comb. nov. Syst. Appl. Microbiol. *4*: 65–78.

Döpfer, H., E. Stackebrandt and F. Fiedler. 1982. Nucleic acid hybridization studies on *Microbacterium*, *Curtobacterium*, *Agromyces* and related taxa. J. Gen. Microbiol. *128*: 1697–1708.

Keddie, R.M. and D. Jones. 1981. Aerobic, saprophytic coryneform bacteria. *In* Starr, Stolp, Trüper, Balows, and Schlegel (Editors), The Prokaryotes. A Handbook on Habitats, Isolation, and Identification of Bacteria, Springer-Verlag, New York, pp. 1839–1878.

Table 15.19.
Characters most useful in distinguishing the genus **Microbacterium** *from similar coryneform taxa[a,b]*

Taxon	Mycelium produced	Rod-coccus cycle[c]	Oxygen requirements	Acid from glucose[d]	Peptido-glycan group[e]	Major peptidoglycan diamino acid	Acyl type of peptidoglycan	Mol% G + C	Major menaquinone(s)
Agromyces	+	−	Aerobic to microaerophilic	V	B	DAB	ND	71–76	MK-11, MK-12
Arthrobacter	−	+	Aerobic	−	A	L-Lysine	Acetyl	59–66[f]	MK-9(H₂) or MK-8 and/or MK-9
Aureobacterium	−	−	Aerobic	W	B	D-Ornithine	Glycolyl	65–75	MK-11, MK-12, MK-13
Curtobacterium	−	−	Aerobic	W	B	D-Ornithine	Acetyl	65–75	MK-9
Microbacterium	−	−	Equivocal	+	B	L-Lysine	Glycolyl	69–75	MK-11, MK-12
Renibacterium	−	−	Aerobic	−	ND	L-Lysine	ND	53–54	MK-9
DAB-containing plant pathogens[g]	−	−	Aerobic	+[h]	B	DAB	ND	67–78	MK-9, MK-10

[a] Data from Gledhill and Casida, 1969; Schleifer and Kandler, 1972; Keddie and Cure, 1977; Dye and Kemp, 1977; Uchida and Aida, 1977, 1979; Minnikin et al., 1978; Collins and Jones, 1980, 1981; Keddie and Jones, 1981; Collins, 1982; Collins et al., 1980a, 1983b, c; Döpfer et al., 1982.

[b] Symbols: see Table 15.2; also v, strain instability (*not* equivalent to "d"); DAB, diaminobutyric acid; ND, not determined; and W, weak.

[c] Marked rod-coccus cycle similar to that in *A. globiformis*.

[d] In peptone-based medium.

[e] Group A: cross-linkage between positions 3 and 4 of two peptide subunits; group B: cross-linkage between positions 2 and 4 of two peptide subunits (Schleifer and Kandler, 1972).

[f] *A. atrocyaneus* has been shown to have a relatively high G + C content (~70 mol% see Keddie and Jones, 1981).

[g] *Corynebacterium michiganense*, *C. insidiosum*, *C. iranicum*, *C. nebraskense*, *C. rathayi*, *C. sepedonicum* and *C. tritici*.

[h] Most strains positive (Dye and Kemp, 1977).

Differentiation of the species of the genus **Microbacterium**

Although DNA/DNA homology studies clearly show that *M. lacticum*, *M. laevaniformans* and *M. imperiale* are distinct species (Döpfer et al., 1982), it is not easy to choose phenetic characters that allow their differentiation. Those characters which may be of value in differentiating the species are listed in Table 15.20.

List of species of the genus **Microbacterium**

1. **Microbacterium lacticum** Orla-Jensen 1919, 179.[AL]

lac'ti.cum. L. masc.n. *lac, lactis* milk; M.L. adj. *lacticum* pertaining to milk, lactic.

Morphology and general characters as for generic description. Other characters are given in Tables 15.20 and 15.21.

Found in raw and pasteurized milk, powdered milk, cheese and on dairy equipment.

Nutrition complex: when supplied with a suitable source of amino acids most strains require thiamine and pantothenic acid and many require biotin also. (Skerman and Jayne-Williams, 1966).

The mol% G + C of the DNA is 69–74.9 (T_m, Bousfield, 1970; Yamada and Komagata, 1970; Döpfer et al., 1982; Collins et al., 1983c).

Type strain: ATCC 8180 (NCIB 8540).

2. **Microbacterium imperiale** (Steinhaus 1941) Collins, Jones and Kroppenstedt 1983i, 672.[VP] (Effective publication: Collins, Jones and Kroppenstedt 1983c.) (*Bacterium imperiale* Steinhaus 1941, 777.)

im.pe.ri.al'e. L. adj. *imperialis* imperial; from specific epithet of name of insect host (*Eacles imperiale* Dru).

Morphology and general characters as for generic description.

Other characters are given in Tables 15.20 and 15.21.

Isolated from the alimentary tract of the imperial moth, *Eacles imperialis* Dru.

Nutrition complex: amino acids and B-vitamins are required (Robertson and Keddie, unpublished).

The mol% G + C of the DNA is 69.8–75.4 (T_m, Yamada and Komagata, 1970; Döpfer et al., 1982 Collins et al., 1983c)

Type strain: ATCC 8365 (NCIB 9888.)

3. **Microbacterium laevaniformans** (Dias and Bhat 1962, 68) Collins, Jones and Kroppenstedt 1983i, 673.[VP] (Effective publication:

Table 15.20.

Characteristics differentiating the species of the genus **Microbacterium**[a,b]

Characteristics	M. lacticum	M. imperiale	M. laevaniformans
Pigmentation	Yellow-white	Red-orange	Yellow
Motility	−	+	−
Survive heating at 63°C for 30 min	+[c]	−	+
Acid produced from:			
Inulin	−	−	+
Sucrose	−	+	+
Trehalose	−	+	+
Glycerol	−	−	+
Raffinose	−	−	+
Nitrate reduced to nitrite	+	−	−
Arginine utilized	−	+	+
H₂S produced	−	−	+
Cell wall sugar			
Galactose	+	ND	+
Rhamnose	+	ND	−

[a] Data from Steinhaus (1941), Dias an Bhat (1962, 1964), Jones (1975), Keddie and Cure (1978).

[b] Symbols: see Table 15.2; also ND, not determined.

[c] Jayne-Williams and Skerman (1966) have described nonthermoduric strains.

Table 15.21.

Other characteristics of the species of the genus **Microbacterium**[a,b]

Characteristics	M. lacticum	M. imperiale	M. laevaniformans
Acid produced from:			
Glucose	+	+	+
Fructose	+	+	+
Galactose	+	+	+
Mannose	+	+	+
Maltose	+	+	+
Cellobiose	+	+	+
Inositol	−	−	−
Sorbose	−	−	−
Rhamnose	−	−	ND
Lactose	+	+	+
Hydrolysis of:			
Cellulose	−	−	−
Hippurate	−	−	−
Gelatin	−/w	−/w	+
Starch	+	−/w	+
Phosphatase	ND	+	ND
Sulfatase	ND	+	ND
Urease	−	−	−
Utilization of:			
Acetate	+	+	+
Lactate	+	+	ND
Malate	+	+	ND
Succinate	+	+	+
Formate	−	−	−
Citrate	+	+	+
Fumarate	+	+	+
Oxalate	−	−	−

[a] Data from Steinhaus (1941), Dias and Bhat (1962, 1964), Yamada and Komagata (1972a, b) and Jones (1975).

[b] Symbols: see Table 15.2; also ND, not determined; and −/w, hydrolysis is either negative, or weak and slow.

Collins, Jones and Kroppenstedt 1983c.) (*Corynebacterium laevaniformans* Dias and Bhat 1962, 68.)

lae.van.i.for'mans. L. part. adj. *laevan* levan polysaccharide *formans* forming, levan forming.

Morphology and general characters as for generic description.

Other characters are given in Tables 15.20 and 15.21.

Found in raw sewage and especially in activated sludge (Dias and Bhat, 1962).

Synthesizes a levan from sucrose or raffinose (Dias and Bhat, 1962).

Nutrition complex: requires biotin, thiamine and pantothenic acid; some strains require glutamic acid also (Dias and Bhat, 1964).

The mol% G + C of the DNA is 70–73.7 (T_m, Döpfer et al., 1982, Collins et al., 1983c).

Type strain: NCIB 9659 (ATCC 15953)

Further comments. The name "*Microbacterium liquefaciens*" (Orla-Jensen 1919, 182) was not included in the Approved Lists of Bacterial Names (Skerman et al., 1980) and, therefore, has no standing in bacterial nomenclature. The species formerly named "*Microbacterium liquefaciens*" has been reclassified in the new genus *Aureobacterium* as *Aureobacterium liquefaciens* (Collins et al., 1983i).

Genus **Aureobacterium** Collins, Jones, Keddie, Kroppenstedt and Schleifer 1983, 672[VP]
(Effective publication: Collins, Jones, Keddie, Kroppenstedt and Schleifer 1983, 244)

KAZUO KOMAGATA AND KEN-ICHIRO SUZUKI

Aur.e.o.bac.te'ri.um. L. adj. *aureus* golden; Gr. neut. dim. n. *bakterion* a small rod; M.L. neut. n. *Aureobacterium* a golden small rod.

Irregular short slender rods, 0.4–0.6 μm in diameter and 0.6–3 μm in length. Occur singly or in groups. Many of the cells are arranged at an angle to each other to give V forms. **In older cultures (3–7 days) rods become shorter but a marked rod-coccus growth cycle does not occur.** Primary branching is uncommon. **No mycelium is produced. Gram-positive but older cultures may fail to retain the Gram stain.** Metachromatic granules are not formed. **Not acid-fast. Endospores are not formed. Optimum temperature ~25–30°C.** Growth range ~10–40°C. **Nutritionally exacting.** Inhibited by 6.5% (w/v) NaCl. **Obligately aerobic. Catalase-positive. Slow and weak oxidative acid production from some carbohydrates.** Some organic acids utilized. **Noncellulolytic. Urease-negative.** The cell wall peptidoglycan is based upon D-ornithine (type B2β), [L-Hsr]-D-Glu-Gly-D-Orn; Schleifer and Kandler (1972). **Glycan moiety of the peptidoglycan of those species so far examined contains glycolyl and acetyl residues. Mycolic acids absent.** The mol% G + C of the DNA has been reported to be in the range of 65–76 (T_m, Bd).

Type species: *Aureobacterium liquefaciens* (Orla-Jensen) Collins, Jones, Keddie, Kroppenstedt and Schleifer 1983, 672.

Further Descriptive Information

The genus *Aureobacterium* was described relatively recently by Collins et al. (1983b). The description of the genus is based on a large number of publications cited by Collins et al. (1983b) as well as their own observations.

The cell wall peptidoglycan is based on D-ornithine (B2β type of Schleifer and Kandler, 1972) with a glycine residue within the interpeptide bridge. Those strains examined possess high levels of glycolate in the glycan moiety of their cell walls (Collins et al., 1983b; Uchida and Aida 1977, 1979). The presence of high levels of glycolate in the cell walls suggests that the muramic acid occurs in the N-glycolyl form rather than the more usual N-acetyl form (Table 15.16). In addition to D-ornithine and glycine the cell wall contains alanine, glutamic acid and homoserine (Collins et al., 1983b). Keddie and Cure (1977) detected the following sugars in the cell walls of those strains examined: galactose, glucose, rhamnose and 6-deoxytalose. All sugars were not present in the cell walls of all the strains.

All strains of *Aureobacterium* so far examined exhibit similar fatty acid profiles. The fatty acids are composed mainly of *anteiso*-branched acids, although *iso*-branched acids are also present in substantial amounts. Small amounts of straight acids and traces of monounsaturated acids have also been detected (Collins et al., 1980; Collins et al., 1983b). The major fatty acids in all strains examined are 12-methyltetradecanoic (*anteiso*-$C_{15:0}$) and 14-methylhexadecanoic (*anteiso*-$C_{17:0}$) with 14-methylpentadecanoic (*iso*-$C_{16:0}$) present in substantial amounts.

The polar lipid content comprises diphosphatidylglycerol, phosphatidylglycerol and a single diglycosyldiacylglycerol (Collins et al., 1980; Collins et al., 1983b). Traces of phosphatidylinositol and phosphatidylinositolmannosides reported in *Aureobacterium testaceum* by Komura et al. (1975) could not be detected by Collins et al. (1983b). Komura et al. (1975) did not investigate the possible presence of glycolipids.

All the strains investigated contain very long unsaturated menaquinones viz. MK-11, MK-12, MK-13; MK-14 has also been detected (Collins et al., 1980; Collins et al., 1983b; Yamada et al., 1976). The major menaquinone composition is not the same in all species (Collins et al., 1983b).

The DNA base ratios were reported by Collins et al. (1983b) to be between 66.9 and 69.1 mol% G + C. These data are in fairly good accord with the report of 64.4 and 68.5 mol% G + C for A. *testaceum* and A. *saperdae* (Yamada and Komagata, 1970), and 70.3 and 68.7% mol% G + C for A. *flavescens* and A. *terregens* (Skyring and Quadling, 1970; Skyring et al., 1971). They are, however, 4 to 5% lower than the mol% G + C values of 72.5, 74.4, 72.5, 72.7 and 75.6 for A. *liquefaciens*, A. *barkeri*, A. *saperdae*, A. *testaceum* and A. *terregens*, respectively, reported by Döpfer et al. (1982).

Strains of *Aureobacterium* have been isolated from a variety of sources including plants, milk, dairy products, soil, sewage and dead insects (see List of Species).

Enrichment and Isolation Procedures

There is not a great deal of information on the isolation of strains of *Aureobacterium* by special procedures. The strains are nutritionally exacting and some require terregens factor (see Lochhead, 1958; Lochhead and Burton, 1953; Keddie and Jones, 1981; and Collins et al., 1983b for pertinent references). Isolation of strains for special purposes were carried out by Dias et al. (1962) (A. *barkeri* for its pectinolytic activity) and by Ueno et al. (1983) (a "Curtobacterium" strain as a decomposer of trichothecene mycotoxin).

Maintenance Procedures

Most of the strains of *Aureobacterium* recover well from cultures on slants of suitable solid media stored at 5°C for 6 months. However, care should be exercised because a few strains do not revive after 1 or 2 months storage. Lyophilization or deep-freezing are recommended for preservation of more than 1 year.

Procedures for Testing Special Characters

Acid production from carbohydrates. The medium employed for the production of acid from carbohydrates contains 0.3% (w/v) peptone, 0.25% (w/v) NaCl and 0.5% (w/v) carbohydrate (adjusted pH 7.2) supplemented with terregens factor where appropriate. The pH indicator is bromcresol purple (BCP). Acid production within 5 days is recorded as positive, and acid production after 7–20 days is regarded as weakly positive.

Assimilation of organic acids. The medium contains 0.5% (w/v) organic acid (sodium salt), 0.002% (w/v) glucose, 0.001% (w/v) yeast extract, 0.001% (w/v) trypticase (B.B.L.), 0.1% (w/v) K_2HPO_4, 0.5% (w/v) NaCl, 2% (w/v) agar and 12 ppm of phenol-red and terregens factor where appropriate. The pH is adjusted to 7.0. Assimilation is determined by the color change of the indicator after 10 days of incubation.

Glycolate test. The following is the method of Uchida and Aida (1984). Ten milligrams of freeze-dried cells are hydrolyzed by 100 μl of 6 N HCl at 100°C for 2 h in a screw-capped tube. The hydrolysate is passed through a micro column (5 cm high, acetate type of Dowex 1×8). The column is then washed with 2 ml of water and 1 ml of 0.5 N HCl and the glycolate fraction is eluted by 2 ml of 0.5 N HCl. Two milliliters of DON reagent (0.02% of 2,7-dihydroxynaphthalene dissolved in concentrated H_2SO_4) is added to 100 μl of this fraction and treated in boiling water for 10 min. The development of a clear purple red color indicates a positive glycolyl test. Quantitative determination (1 nmol glycolic acid by 1 mg dried cell) is carried out by colorimetry at 530 nm compared with the standard solution of glycolic acid treated in the same manner.

Differentiation of the Genus **Aureobacterium** from Other Genera

The genera with which members of the genus *Aureobacterium* are likely to be confused are *Curtobacterium* and *Microbacterium*. Characters which differentiate the genera are discussed under the sections on *Curtobacterium* and *Microbacterium*.

Taxonomic Comments

The genus *Aureobacterium* was established by Collins et al. (1983b) to accommodate some bacteria previously allocated to the genera *Arthrobacter, Curtobacterium, Corynebacterium* and *Microbacterium* as *Arthrobacter flavescens, A. terregens, Curtobacterium testaceum, C. saperdae,* "*Corynebacterium barkeri*" and "*Microbacterium liquefaciens.*" The last two names are placed in quotation marks because they were not included in the Approved Lists of Bacterial Names (Skerman et al., 1980).

The genus *Aureobacterium* as presently defined exhibits the closest taxonomic relationship to bacteria which contain the unusual group B peptidoglycan, i.e. the genera *Agromyces, Curtobacterium, Microbacterium* and the diaminobutyric acid-containing plant pathogenic corynebacteria (see sections on *Curtobacterium, Microbacterium* and Döpfer et al., 1982 for detailed discussion. Indeed, there is good evidence from nucleic acid studies that all these taxa form a cluster worthy of designation as a family (Döpfer et al. 1982)).

Further Readings

Collins, M.D., D. Jones, R.M. Keddie, R.M. Kroppenstedt and K.H. Schleifer. 1983b. Classification of some coryneform bacteria in a new genus *Aureobacterium*. Syst. Appl. Microbiol. *4:* 236–252.

Döpfer, H., E. Stackebrandt and F. Fiedler. 1982. Nucleic acid hybridization studies on *Microbacterium, Curtobacterium, Agromyces* and related taxa. J. Gen. Microbiol. *128:* 1697–1708.

Keddie, R.M. and D. Jones. 1981. Saprophytic, aerobic coryneform bacteria. *In* Starr, Stolp, Trüper, Balows and Schlegel, (Editors), The Prokaryotes. A Handbook on Habitats, Isolation and Identification of Bacteria, Springer-Verlag, New York, pp. 1838–1878.

Differentiation of the species of the genus **Aureobacterium**

The characteristics of the species of the genus which enable their differentiation from each other are contained in the description of the individual species listed under List of Species.

List of species of the genus **Aureobacterium**

The description of the species listed below (and see Table 15.1) is based on the genus description and the species descriptions given by Collins et al. (1983b) which were based on the observations of a number of authors cited by Collins et al. (1983b), the observations of Collins et al. (1983b) and our own observations.

1. **Aureobacterium liquefaciens** (Orla-Jensen 1919) Collins, Jones, Keddie, Kroppenstedt and Schleifer 1983, 672.[VP] (Effective publication: Collins, Jones, Keddie, Kroppenstedt and Schleifer 1983, 246). (*Microbacterium liquefaciens* Orla-Jensen 1919, 182.)

li.que.fa'ciens. M.L. part. adj. *liquefaciens* dissolving.

On suitable solid media incubated in air, surface colonies are 1–3 mm in diameter, circular, convex, glistening, opaque with an entire margin. A bright yellow pigment is produced. Not motile. Optimum temperature is about 30°C.

Slow and weak, oxidative acid production from D-glucose, galactose and a few other sugars. Arginine, gelatin, esculin and casein are hydrolyzed. Hippurate, Tween 20, 40 and 80 are not hydrolyzed. Tellurite (0.05% w/v) is not reduced. Several organic acids are utilized. Nutritionally exacting, B-vitamins are required.

Cell wall sugar is rhamnose. Major fatty acids are *anteiso*-$C_{15:0}$ and *anteiso*-$C_{17:0}$. Major menaquinones are MK-11 and MK-12.

The mol% G + C of the DNA is 68.2–72.5 (T_m).

Usually found in milk, cheese, dairy products and on dairy equipment.

Type strain: NCIB 11509.

2. **Aureobacterium flavescens** (Lochhead 1958) Collins, Jones, Keddie, Kroppenstedt and Schleifer 1983, 672.[VP] (Effective publication: Collins, Jones, Keddie, Kroppenstedt and Schleifer 1983, 247.) (*Arthrobacter flavescens* Lochhead 1958, 170.)

fla.ves'cens. L. v. *flavesco* to become golden yellow; L. part. adj. flavescens becoming yellow.

Grows in air on a suitable medium supplemented with soil extract. On such a medium a yellow nondiffusible pigment is produced. Colonies are 0.2–0.5 mm in diameter, low, convex, circular with entire edge. Not motile. Optimum temperature is about 25°C. Growth occurs at 10 but not 37°C. Very weak and slow, oxidative acid production from glucose, fructose and maltose. Gelatin, starch and DNA hydrolyzed. H_2S produced. Nitrate reduced to nitrite. Nutritionally exacting; terregens factor, biotin and thiamine are required for growth. The terregens

factor is replaceable by other siderophores, e.g. coprogen and ferrichrome. When supplied with a suitable medium containing terregens factor, vitamins and glucose as carbon + energy source, ammonium salts or nitrate are utilized as nitrogen source.

The cell wall sugars are galactose, glucose and rhamnose. Major fatty acids are *anteiso*-$C_{15:0}$ and *anteiso*-$C_{17:0}$. Major menaquinone is MK-13 although small amounts of MK-14 and MK-12 also are present.

The mol% G + C of the DNA is 66.9–70.3 (T_m).

Found in soil.

Type strain: ATCC 13348 (NCIB 9221).

3. **Aureobacterium terregens** (Lochhead and Burton 1953) Collins, Jones, Keddie, Kroppenstedt and Schleifer 1983, 672.[VP] (Effective publication: Collins, Jones, Keddie, Kroppenstedt and Schleifer 1983, 247). (*Arthrobacter terregens* Lochhead and Burton 1953, 7).

ter're.gens. L. n. *terra* soil; L. part. adj. *egens* requiring; M.L. part. adj. *terregens* soil-requiring.

On suitable media supplemented with soil extract, growth is yellowish brown. Colonies are 1–2 mm in diameter, low, convex, circular with entire margin. Not motile. Optimum temperature is between 20–26°C. Growth occurs at 10 but not at 37°C. Starch, casein and gelatin are not hydrolyzed. Nitrate reduced to nitrite. Weak and slow oxidative acid production from glucose, fructose and a few other sugars. H_2S not produced. Nutritionally exacting; terregens factor, biotin, thiamine, pantothenic acid and L-methionine are required for growth. Most of the usual laboratory media will not support growth unless supplemented with terregens or similar factors and also yeast extract. The terregens factor may be replaced by coprogen or ferrichrome. When provided with essential growth factors and glucose as carbon + energy source, ammonium salts are utilized as a nitrogen source.

Cell wall sugars are galactose, rhamnose and 6-deoxytalose. Major fatty acids are *anteiso*-$C_{15:0}$ and *anteiso*-$C_{17:0}$. Major menaquinones are MK-12 and MK-13.

The mol% G + C of the DNA is 68.6–75.6 (T_m).

Habitat: soil.

Type strain: ATCC 13345 (NCIB 8909).

4. **Aureobacterium saperdae** (Lysenko 1959) Collins, Jones, Keddie, Kroppenstedt and Schleifer 1983, 672.[VP] (Effective publication: Collins, Jones, Keddie, Kroppenstedt and Schleifer 1983, 248). (*Brev-*

ibacterium saperdae Lysenko 1959, 41; *Curtobacterium saperdae* (Lysenko) Yamada and Komagata 1972b, 425.)

sa.per'dae. *saperdae* of *Saperda*, genus of insects; named for the source of this organism.

Colonies on solid media in air are 1–3 mm in diameter, convex, circular with entire edge, opaque, glistening and yellow in color. Motile. Optimum temperature about 28°C. Slow and weak, oxidative acid production from D-glucose and some other sugars. Arginine, esculin and starch hydrolyzed. Tween 20, 40 and 60 hydrolyzed. Hippurate and Tween 80 not hydrolyzed. Gelatin and casein not hydrolyzed. Deoxyribonuclease not produced. Tellurite (0.05% w/v) not reduced. Several organic acids utilized.

Major fatty acids are *anteiso*-$C_{15:0}$ and *anteiso* $C_{17:0}$. Major menaquinones are MK-11 and MK-12.

Mol% G + C content of the DNA is 68.5–72.5 (T_m).

Isolated from body cavity of dead insect (*Saperda caracharias*).

Type strain: ATCC 19272 (CCEB 366).

5. **Aureobacterium barkeri** (Dias, Bilimoria and Bhat 1962) Collins, Jones, Keddie, Kroppenstedt and Schleifer 1983, 672.^{VP} (Effective publication: Collins, Jones, Keddie, Kroppenstedt and Schleifer 1983, 249.) (*Corynebacterium barkeri* Dias, Bilimoria and Bhat 1962, 66.)

bar'ker.i. M.L. gen. n. *barkeri* of Barker; named for H. A. Barker, American biochemist.

Good growth on solid media in air. Colonies 1–3 mm in diameter, circular, low convex, entire edge, opaque. A yellow pigment is produced. Motile. Optimum temperature about 28°C. Slow and weak oxidative acid production from glucose and some other sugars. Esculin, starch, hippurate, Tween 20, 40, 60 hydrolyzed. Tween 80 not hydrolyzed.

Some organic acids are utilized. Tellurite (0.05% w/v) reduced. Pectinolytic. Pectin methyl esterase and pectin transeliminase produced. H_2S produced. Biotin required for growth.

Major fatty acids are *anteiso*-$C_{15:0}$ and *anteiso*-$C_{17:0}$. Major menaquinones are MK-11 and MK-12.

The mol% G + C content of the DNA is between 68.7 and 74.4 (T_m). Isolated from raw domestic sewage.

Type strain: NCIB 9658.

6. **Aureobacterium testaceum** (Komagata and Iizuka 1964) Collins, Jones, Keddie, Kroppenstedt and Schleifer 1983, 672.^{VP} (Effective publication: Collins, Jones, Keddie, Kroppenstedt and Schleifer 1983, 249.) (*Brevibacterium testaceum* Komagata and Iizuka 1964, 497; *Curtobacterium testaceum* (Komagata and Iizuka) Yamada and Komagata 1972b, 425.)

tes'ta.ceum. L. adj. *testaceum* brick-colored.

On solid media colonies are 1–3 mm in diameter, low convex, circular, entire edge, shiny. A yellow- to orange-red pigment is produced. Motile. Optimum temperature about 28°C. Slow and weak, oxidative acid production from glucose and some other sugars. A variety of organic acids utilized. Tween 20, 40, 60, 80 are hydrolyzed. Tellurite (0.05% w/v) reduced. Casein and gelatin are hydrolyzed. Esculin and starch are hydrolyzed. Deoxyribonuclease is produced.

Major fatty acids are *anteiso*-$C_{15:0}$ and *anteiso* $C_{17:0}$. The major menaquinone is MK-11.

The mol% G + C content of the DNA is 65.4–72.7 (T_m). Isolated from rice.

Type strain: ATCC 15829 (IAM 1561).

Genus *Cellulomonas* Bergey et al. 1923, 154, emend. mut. char. Clark 1952, 50^{AL}

ERKO STACKEBRANDT AND RONALD M. KEDDIE

Cel.lu.lo.mo'nas. M.L. n. *cellulosa* cellulose; Gr. n. *monas* a unit, monad; M.L. fem. n. *Cellulomonas* cellulose monad.

In young cultures, slender, irregular rods ~0.5–0.6 μm × ~2.0–4.0 μm or more which may be straight, angular or slightly curved; **some of the rods are arranged at an angle to each other giving V formations.** Occasional cells may show primary branching but **mycelia are not produced.** Cultures a week or more old usually contain mainly short rods but a small proportion of the cells may be coccoid. Do not form endospores. **Gram-positive** but the cells are **very readily decolorized.** Not acid-fast. Motile by one (usually polar or subpolar) or a few lateral flagella, or nonmotile. **The cell wall peptidoglycan contains ornithine but not glycine or homoserine; the cell wall polysaccharide does not contain galactose as a major sugar. The major isoprenoid quinones are tetrahydrogenated quinones with nine isoprene units MK-9 (H_4). Aerobic; most strains also capable of anaerobic growth.** Optimum temperature, ~30°C. Growth moderate on peptone-yeast extract media at neutral pH giving opaque, usually convex, yellow colonies.

Chemoorganotrophic: **metabolism primarily respiratory, but also fermentative; most strains produce acid from glucose both aerobically and anaerobically.** Catalase-positive.

Cellulolytic. Starch and gelatin (weakly) hydrolyzed. Nitrate reduced to nitrite, DNase produced. Require biotin and thiamine. The mol% G + C of the DNA is 71–76 (T_m).

Type species: *Cellulomonas flavigena* (Kellerman and McBeth) Bergey et al. 1923, 165.

Further Descriptive Information

The slender, irregular rods characteristic of young cultures range from ~0.4–0.8 μm in diameter but most are ~0.5–0.6 μm. The rods vary considerably in length and may appear as short filaments in late exponential phase cultures (~24 h) when examined by the methods described by Cure & Keddie (1973). They may also show primary branching but do not form true mycelia. As growth proceeds the rods

become shorter and V formations become more obvious. Cultures a week or more old are usually composed mainly of short rods but a proportion of the cells may be coccoid. When placed on fresh solid medium, growth of coccoid cells occurs by elongation from one or sometimes two parts of the cell to give rods which appear club-shaped or jointed. However, they do not show the marked rod-coccus cycle characteristic of *Arthrobacter* and *Brevibacterium*.

The cell wall contains a group A peptidoglycan with L-ornithine as diamino acid; the interpeptide bridge is D-aspartic acid in *C. flavigena* and D-glutamic acid in all other species (Fiedler and Kandler, 1973). Hydrolysates of purified cell walls contain one or more sugars depending on species but galactose is not a major component (Keddie et al., 1966; Fiedler and Kandler, 1973). The major isoprenoid quinones in the type strains of *C. biazotea*, *C. flavigena* and *C. fimi*, and in *C. gelida* NCIB 8075 ("*C. subalbus*") are tetrahydrogenated quinones with nine isoprene units [MK-9 (H_4)] (Yamada et al., 1976; Collins et al., 1979). The major fatty acids in the type strains of *C. biazotea* and *C. flavigena*, and in *C. gelida*, NCIB 8075 ("*C. subalbus*") is 12-methyltetradecanoic (*anteiso* C_{15}) acid and occurs together with other *anteiso* acids, and *iso*- and straight chain acids; a distinctive feature is the presence of 13-carbon acids and relatively high proportions of straight chain acids (Minnikin et al., 1979). The polar lipids in the three species mentioned consist of diphosphatidylglycerol, phosphatidylinositol, and two unidentified phosphoglycolipids (Minnikin et al., 1979).

Moderate growth occurs in air at 30°C on meat extract, peptone agar or yeast-extract, peptone-based agar media at near neutral pH (Keddie and Jones, 1981). Colonies on such media are opaque, usually convex, ~1–3 mm in diameter, and usually yellow, but sometimes white (Cure and Keddie, unpublished). Strains of all species grow at 10°C, some at 5°C, and the maximum temperatures are in the range 36–43°C (Cure and Keddie, unpublished). They do not survive heating at 63°C for 30 min (Keddie et al., 1966). All strains grow best aerobically and most give markedly reduced growth in anaerobic conditions; a few are aerobic or give equivocal results (Keddie, 1974; Keddie and Cure, 1977).

Of 180 organic compounds tested only between 16 and 28 were utilized as sole carbon + energy sources by the type strains of C. biazotea, C. cellasea, C. flavigena, C. gelida and C. uda, and C. gelida NCIB 8075 ("C. subalbus") (Owens and Keddie, unpublished, Keddie, 1974). Most of the substrates utilized were carbohydrates and sugar derivatives but a few organic acids and amino acids were also utilized. All of the species mentioned above utilized the following compounds as sole or major carbon + energy sources: cellulose, starch, D-xylose, L-arabinose, D-glucose, D-mannose, D-galactose, cellobiose, maltose, trehalose and sucrose (Owens and Keddie, unpublished). Data on the utilization of other carbon and energy sources by individual species are given later in Tables 15.23 and 15.24. There are some contradictions in these data, probably because of the different experimental methods used (Yamada and Komagata, 1972; Stackebrandt et al., 1982). Biotin and thiamine are the only exogenous organic growth factors required by the type strains of five of the species now recognized (data for C. fimi are not available) (Keddie et al., 1966); when provided with these vitamins, growth occurs in suitable mineral media with glucose as carbon + energy source and an ammonium salt (or nitrate for most strains) as nitrogen source (Owens and Keddie, 1969).

The main products of glucose dissimilation in resting cell suspensions are acetic acid, L-lactic acid, formic acid, succinic acid, ethanol and CO_2. The major acid produced from glucose in growing cultures is acetic acid; most strains in addition produce L-lactic acid (Stackebrandt and Kandler, 1974). The major route for glucose dissimilation is the Embden-Meyerhof-Parnas pathway; a small amount of glucose is metabolized via the hexose monophosphate pathway (Stackebrandt and Kandler, 1974; 1980a).

Braden and Thayer (1976) used a quantitative agglutination procedure to compare the type strains of C. biazotea, C. fimi, C. flavigena, C. gelida (and "C. subalbus" (C. gelida)) and C. uda together with two Cellulomonas isolates. They concluded that the named strains shared many antigens with each other but were not identical. There was little affinity between the new isolates and the named strains.

Little information is available on the habitats of Cellulomonas spp.; most of the original cultures were isolated from soil (Bergey et al., 1923) and this has generally been considered to be their major habitat. They have also been isolated from waste materials with a high cellulose content (see e.g. Kaufmann et al., 1976). Members of the genus have been the subject of many investigations on the production of single cell protein from cellulosic materials; see Han (1982) for references. Individual strains of Cellulomonas spp. have been isolated from a variety of sources, e.g. from rotting sugar cane stalks and soil (Cellulomonas sp., Han and Srinivasan (1968); from milk, probably through contamination with soil (an isolate with characters similar to "C. acidula," Whitehouse and Jackson, 1972); from "sloughing spoilage" of California ripe olives (C. flavigena, Patel and Vaughn, 1973) and from Indian sugar cane (C. uda, Stoppok et al., 1982).

The principal enzymes involved in the degradation of cellulose by Cellulomonas spp. have been widely studied; see Stoppok et al. (1982) for references.

Isolation and Enrichment Procedures

There is little published information on the isolation of Cellulomonas spp. from natural sources; such methods as have been described exploit the cellulolytic properties of members of the genus. A suitable procedure is to prepare enrichment cultures in a mineral-based medium containing a low (0.05–0.1%) concentration of yeast extract to provide the necessary organic growth factors, and filter paper as cellulose source. This is followed by plating on a similar solid medium but containing cellulose in a finely divided form. Cellulolytic bacteria produce colonies surrounded by zones of clearing. Direct plating on cellulose agar, without previous enrichment, may also be used. The methods are not selective for Cellulomonas and isolates must be screened for those with a coryneform morphology.

Suspensions or macerates of the material being examined, soil, compost, etc., may be streaked directly on to the surface of cellulose agar (Stewart and Leatherwood, 1976) of the following composition (g/100 ml of distilled water): $NaNO_3$, 0.1; K_2HPO_4, 0.1; KCl, 0.05; $MgSO_4$, 0.05; yeast extract, Difco, 0.05; agar, 1.7; ball-milled filter paper, 0.1; glucose 0.1; pH 7.0. The medium is autoclaved at 121°C for 15 min. To prepare the ball-milled filter paper, a 3% (w/v) aqueous suspension of filter paper (Whatman No. 1) is ball-milled for 3 days. Other suitable sources of finely divided cellulose may be used, e.g. microcrystalline cellulose (Avicel, FML) at a concentration of 0.1% (w/v) (Kaufmann et al., 1976). In other similar versions of the medium, the glucose is omitted. The plates are incubated at 30°C for 5–7 days; colonies showing zones of clearing are replated on the same medium until pure cultures are obtained. Isolates which have a coryneform morphology and are cellulolytic are presumptive Cellulomonas spp.

Cellulomonas enrichments may be prepared by a method similar to that described by Han and Srinivasan (1968). A liquid version of the cellulose agar described above is used but with the glucose omitted and with strips of filter paper replacing the finely divided cellulose. Other mineral bases may be used, e.g. that of Han and Srinivasan (1968); mineral base "E" described by Owens and Keddie (1969) has been used successfully (Cowan and Keddie, unpublished). The medium is dispensed in 10-ml quantities in tubes or in larger quantities in conical flasks. Before sterilization at 121°C for 20 min, a strip of filter paper (e.g. Whatman No. 1) is added to each tube, or a few strips to a flask. The tubes are inoculated with a small quantity of the material to be investigated (flasks allow a larger inoculum), and are shaken until the filter paper disintegrates. One or more serial transfers are made before a small amount of the disintegrated filter paper is macerated in sterile mineral base before plating out on cellulose agar.

Maintenance Procedures

For short term preservation (a few weeks) stab cultures (by loop) in TSX semisolid medium (Keddie et al., 1966) should remain viable for 3 months or more at room temperature (~20°C). For long term preservation (several years), cultures may be preserved by lyophilization.

Procedures for Testing for Special Characters

The absence of diaminopimelic acid (DAP) in cell walls of cellulomonads can easily be tested for by hydrolyzing 0.5 g of wet weight cells with 6 N HCl for 16 h at 120°C, followed by one-dimensional, descending paper chromatography (Schleicher and Schüll No. 2043b) of the water-soluble fraction in methanol:pyridine:10 N $HCl:H_2O$ = 80:10:2.5:27.5 (v/v) according to Rhuland et al. (1955). The presence of ornithine and aspartic acid and the absence of glycine in the cell wall can only be determined by using purified cell wall preparations. These are obtained either by alkali treatment of whole cells (Keddie and Cure, 1977) or from cells which are treated first with trichloroacetic acid and then with trypsin (Schleifer and Kandler, 1972). After HCl-hydrolysis of the cell wall preparations, amino acids are separated by two-dimensional paper chromatography (Schleifer and Kandler, 1972; Keddie and Cure, 1977) or on an amino acid analyzer (Schleifer and Kandler, 1972). Care must be taken in distinguishing ornithine from lysine.

Differentiation of the genus **Cellulomonas** from other taxa

Cellulomonads may be differentiated from Oerskovia and from other genera harboring organisms with a coryneform morphology (as defined by Keddie and Jones, 1981), and lacking both DAP as a cell wall amino acid and an Arthrobacter-type growth cycle, by the characters given in Table 15.22.

Taxonomic Comments

Most species of Cellulomonas which have been included in taxonomic studies have been represented only by single strains. It is therefore difficult to know whether or not the few phenetic differences reported

Table 15.22.
Characters differentiating **Cellulomonas** *from similar genera lacking DAP and an* **Arthrobacter**-*type growth cycle[a]*

Characteristics	Cellulomonas	Curtobacterium	Aureobacterium[b]	Oerskovia	Microbacterium	Unassigned species containing DAB[c]
Cell wall contains:						
Lysine	−	−	−	+	+	−
Ornithine	+	+	+	−	−	−
Glycine	−	+	+	−	+	+
Galactose	−[d]	(+)	ND	d	+	ND
Cellulase activity	+	−	−	d[e]	−	−
Mycelium present	−	−	−	+	−	−[f]

[a] Symbols: see Table 15.2; also DAP, diaminopimelic acid; DAB, diaminobutyric acid; (+), most strains are positive; d, 11–89% of strains are positive; and ND, data not available or incomplete.
[b] Collins et al., 1983b.
[c] This group contains "*Brevibacterium helvolum*" (some strains), "*Corynebacterium aquaticum,*" *C. insidiosum,* "*C. mediolanum,*" *C. michiganense, C. nebraskense, C. sepedonicum* and *Agromyces ramosus* (Döpfer et al., 1982).
[d] *Cellulomonas biazotea* contains minor amounts of galactose (Fiedler and Kandler, 1973).
[e] Cellulolytic activity is present in nonmotile oerskovias.
[f] *Agromyces ramosus* forms a substrate mycelium.

to occur between species (see Table 15.23) are indeed species-specific or only strain-specific. Also the different species bear a considerable phenetic resemblance to each other; for this reason only one species, *Cellulomonas flavigena,* was recognized in the eighth edition of the *Manual,* while *C. biazotea, C. cellasea, C. gelida* and *C. uda* were considered to be subjective synonyms and *C. fimi* a possible subjective synonym of *C. flavigena* (Keddie, 1974). However, in a serological study using purified cell wall preparations as antigens, although six named *Cellulomonas* species could be shown to be similar to each other, their reduction to a single species was questioned (Braden and Thayer, 1976). Furthermore, when representatives of eight species of *Cellulomonas* were subjected to DNA reassociation studies, seven genetically well defined species could be recognized (Stackebrandt and Kandler, 1979); "*C. subalbus*" NCIB 8075 was found to be identical with *C. gelida* ATCC 488 on the basis of genetic and phenotypic evidence. The Approved Lists of Bacterial Names (Skerman et al., 1980) contains 6 species: *C. flavigena, C. biazotea, C. gelida, C. uda, C. fimi* and *C. cellasea.* A seventh species, *C. cartae,* was described later and *Nocardia cellulans, Oerskovia xanthineolytica, Brevibacterium fermentans* and *B. lyticum* were considered to be subjective synonyms of *C. cartae* (Stackebrandt and Kandler, 1980b). However, the specific epithet *cellulans* (Metcalf and Brown, 1957) antedates the name *cartae* (Stackebrandt and Kandler, 1980b) which is, therefore, illegitimate. The legitimate name for this species is therefore *C. cellulans* (see Addendum) (Stackebrandt and Kandler, 1980b). Also, classification of the species now named *C. cellulans* as a cellulomonad was questioned because of differences in biochemical and morphological characters found between it and the other named species of *Cellulomonas* (Keddie and Jones, 1981). Actually, *C. cellulans* (*C. cartae*) is more closely related to members of *Oerskovia* than to any

species of *Cellulomonas* (Stackebrandt et al., 1980b, 1982) as determined by DNA homology and comparative analysis of the 16S-rRNA and supported by chemotaxonomic characters (Minnikin et al., 1979; Collins et al., 1979; Stackebrandt et al., 1982). Phylogenetically, however, representatives of *Cellulomonas* and *Oerskovia* are so closely related that a union of these two genera in a redefined genus *Cellulomonas* has been proposed recently (Stackebrandt et al., 1982) (see Addendum). However, for purposes of identification, *Cellulomonas* and *Oerskovia* are treated as two separate genera in this volume. The comparative analysis of the 16S-rRNA indicates that the genus *Cellulomonas* so defined is one of several genera forming a phylogenetically coherent group, including not only genera whose members exhibit a coryneform morphology, i.e. *Arthrobacter, Brevibacterium, Microbacterium* and *Curtobacterium,* but also genera whose members have either a simple morphology (*Micrococcus, Stomatococcus*) or a complex morphology (*Promicromonospora, Agromyces, Dermatophilus*) (Stackebrandt et al., 1980a, 1982: Döpfer et al., 1982).

Further Reading

General aspects:

Keddie and Jones (1981).

Cell wall composition:

Keddie and Cure (1978).

Peptidoglycan:

Schleifer and Kandler (1972).

Phylogeny:

Stackebrandt et al. (1982).

List of species of the genus *Cellulomonas*

1. **Cellulomonas flavigena** (Kellerman and McBeth 1912) Bergey, Harrison, Breed, Hammer and Huntoon 1923, 165.[AL] (*Bacillus flavigena* Kellerman and McBeth 1912, 488.)

fla.vi′ge.na. L. adj. *flavus* yellow; L. v. *gigno* produce; M.L. adj. *flavigena* yellow-producing.

Cell morphology as given for the genus. Colonies irregular, 0.9–3.1 mm in diameter, lemon yellow and opaque, flat, edge fimbriate.

Physiological and biochemical characters are given in Tables 15.23 and 15.24 and in the generic description.

Peptidoglycan is of the L-Orn-D-Asp type (Fiedler and Kandler, 1973).

The major cell wall sugar is rhamnose; mannose (Keddie et al., 1966; Fiedler and Kandler, 1973), and ribose (Fiedler and Kandler, 1973) are minor components.

The major fatty acid is 12-methyltetradecanoic acid (*anteiso*-$C_{15:0}$) with smaller amounts of straight chain pentadecanoic acid ($C_{15:0}$) and traces of other *iso*-, *anteiso*- and straight chain saturated acids (Minnikin et al., 1979).

The mol% G + C of the DNA is 72.7–74.8 (T_m) (Sukapure et al., 1970; Yamada and Komagata, 1970; Stackebrandt and Kandler, 1979.)

Type strain: ATCC 482.

Further comments on the relationship. Serological studies suggest a

Table 15.23.

*Characters differentiating the species of the genus **Cellulomonas**[a,b]*

Characteristics	1. C. flavigena	2. C. biazotea	3. C. fimi	4. C. gelida	5. C. uda	6. C. cellasea
Utilization of:[c]						
D-Ribose	+	−	−	−	−	−
Raffinose	−	+	−	−	−	−
Acetate	+	+	−[f]	+	+	+
L-Lactate	−	+[d]	+[d]	−	−	+
Motility	−	+	+	+	−	−[e]
Acid from:						
Lactose	−[f]	−[f]	+	d	+	−
Dextrin	+	−	+	+	+	−
Cell wall amino acid:						
Aspartic acid	+	−	−	−	−	−
Ornithine	+	+	+	+	+	+
Cell wall contains:						
Rhamnose	+	+	+	−	−	+

[a] Data from Stackebrandt and Kandler (1979) and Stackebrandt et al. (1982). In most cases the reactions are based on the investigation of a single strain only.

[b] Symbols: see Table 15.2; also d, 11–89% of strains are positive.

[c] Utilization as sole carbon source according to Stackebrandt and Kandler (1979) and Seiler et al. (1980).

[d] Negative reaction according to Yamada and Komagata (1972).

[e] Previously reported to be motile (Clark, 1953).

[f] Positive reaction according to Yamada and Komagata (1972).

Table 15.24.

*Other characteristics of the species of the genus **Cellulomonas**[a,b]*

Characteristics	1. C. flavigena	2. C. biazotea	3. C. fimi	4. C. gelida	5. C. uda	6. C. cellasea
Utilization of:[c]						
D-gluconate	+	−	−	−	−	−
Propionate	−	−	−	−	d	−
Aspartate	−	−	−	−	d	−
Capronate	−	−	d	−	−	−
Proline	−	−	−	−	−	+
Rhamnose and xylitol	−	−	d	−	−	−

[a] Data from Stackebrandt and Kandler (1979) and Stackebrandt et al. (1982). In most cases the reactions are based on the investigation of a single strain only.

[b] Symbols: see Table 15.2; also d, 11–89% of strains are positive.

[c] Utilization as sole carbon source according to Stackebrandt and Kandler (1979) and Seiler et al. (1980).

close relationship between *C. flavigena* and *C. biazotea* (Braden and Thayer, 1976). However, DNA/DNA homology values found for these two species do not support this finding (Stackebrandt and Kandler, 1979).

2. **Cellulomonas biazotea** (Kellerman, McBeth, Scales and Smith 1913) Bergey, Harrison, Breed, Hammer and Huntoon 1923, 158.[AL] (*Bacillus biazoteus* Kellerman et al. 1913, 506.)

bi.az.o'te.a. L. prefix *bi* two; Gr. *azous* without life; Fr. n. *azote* nitrogen; M.L. adj. *biazoteus* two nitrogen sources utilized (i.e. organic and inorganic).

Cell morphology as given for the genus. Colonies circular, slightly irregular, 0.6–2.8 mm in diameter, canary yellow and opaque, low convex, edge entire or erose.

Physiological and biochemical characters are given in Tables 15.23 and 15.24 and in the generic description.

Peptidoglycan is of the L-Orn-D-Glu type (Fiedler and Kandler, 1973).

The major cell wall sugar is rhamnose while galactose (Fiedler and

Kandler, 1973), mannose and 6-desoxytalose are minor components (Keddie et al., 1966; Fiedler and Kandler, 1973).

The major fatty acid is 12-methyltetradecanoic acid (*anteiso*-$C_{15:0}$) with smaller amounts of 13-methyltetradecanoic acid (*iso*-$C_{15:0}$) (Minnikin et al., 1979).

The mol% G + C of the DNA is 71.5–75.6 (T_m) (Jones and Bradley, 1971; Bousfield, 1972; Sukapure et al., 1970.)

Type strain: ATCC 486.

Further comments on the relationship. DNA homology studies reveal a specific relationship to *C. fimi* (50% homology) (Stackebrandt and Kandler, 1979), while the relationship to other *Cellulomonas* species is definitely lower (below 35% homology). The close relationship between *C. biazotea* and *C. flavigena* detected in a serological study (Braden and Thayer, 1976), is not supported by the genetic analysis.

3. **Cellulomonas fimi** (McBeth and Scales 1913) Bergey, Harrison, Breed, Hammer and Huntoon 1923, 166.[AL] (*Bacterium fimi* McBeth and Scales 1913, 30.)

fi'mi. L. n. *fimus* dung; L. gen. n. *fimi* of dung.

Cell morphology as given for the genus.

Colonies circular, 0.4–1.8 mm in diameter, cream or yellowish, opaque, glistening, edge entire.

Physiological and biochemical characters are given in Tables 15.23 and 15.24 and in the generic description.

Peptidoglycan is of the L-Orn-D-Glu type (Fiedler and Kandler, 1973).

The major cell wall sugar is rhamnose; fucose and glucose are minor components (Fiedler and Kandler, 1973).

The mol% G + C of the DNA is 71.3–72.0 (T_m) (Yamada and Komagata, 1970).

Type strain: ATCC 484.

Further comments on the relationship. C. biazotea is the only member of *Cellulomonas* to which *C. fimi* is specifically related (50% DNA homology) (Stackebrandt and Kandler, 1979).

4. **Cellulomonas gelida** (Kellerman, McBeth, Scales and Smith 1913) Bergey, Harrison, Breed, Hammer and Huntoon 1923, 162.[AL] (*Bacillus gelidus* Kellerman et al., 1913, 510.)

ge'li.da. L. adj. *gelidus* cold.

Cell morphology as given for the genus. Colonies circular, 0.8–2 mm in diameter, cream or yellowish, opaque, convex domed, edge entire or slightly erose.

Physiological and biochemical characters are given in Tables 15.23 and 15.24 and in the generic description.

Peptidoglycan is of the L-Orn-D-Glu type (Fiedler and Kandler, 1973). Glucose is present as a minor cell wall component (Keddie et al., 1966; Fiedler and Kandler, 1973). The major fatty acid is 12-methyltetradecanoic acid (*anteiso*-$C_{15:0}$) with smaller amounts of straight chain pentadecanoic acid ($C_{15:0}$) and traces of other *iso*-, *anteiso*- and straight chain saturated acids (Minnikin et al., 1979). (These studies were carried out on "*C. subalbus*" C222 (NCIB 8075) which has to be regarded as a strain of *C. gelida* (see Further Comments on the Relationship).)

The mol% G + C of the DNA is 72.4–74.4 (T_m) (Yamada and Komagata, 1970; Sukapure et al., 1970; Stackebrandt and Kandler, 1979).

Type strain: ATCC 488.

Further comments on the relationship. C. gelida is genotypically and phenotypically indistinguishable from the strain named "*C. subalbus*" (NCIB 8075) (Stackebrandt and Kandler, 1979) thus supporting the earlier proposal by Clark (1953) to reduce "*C. subalbus*" to synonymy with *C. gelida*.

DNA homology studies reveal that *C. gelida* is specifically related to *C. uda* (45% homology) (Stackebrandt and Kandler, 1979), a finding that is supported by serological studies (Braden and Thayer, 1976).

5. **Cellulomonas uda** (Kellerman, McBeth, Scales and Smith 1913) Bergey, Harrison, Breed, Hammer and Huntoon 1923, 166.[AL] (*Bacterium udum* Kellerman et al. 1913, 514.)

u'da. L. adj. *udus* moist, wet.

Cell morphology as given for the genus. Colonies punctiform or circular, 0.2–1.8 mm in diameter, off-white, opaque, smooth and shiny surface, edge entire.

Physiological and biochemical characters are given in Tables 15.23 and 15.24 and in the generic description.

Peptidoglycan is of the L-Orn-D-Glu type (Fiedler and Kandler, 1973).

Glucose (Keddie et al., 1966; Fiedler and Kandler, 1973) and mannose (Keddie et al., 1966) occur as minor cell wall sugars.

The mol% G + C of the DNA is 72 (T_m) (Yamada and Komagata, 1970).

Type strain: ATCC 491.

Further comments on the relationship. *C. gelida* is the only member of *Cellulomonas* to which *C. uda* is specifically related (45% DNA homology) (Stackebrandt and Kandler, 1979).

6. **Cellulomonas cellasea** (Kellerman, McBeth, Scales and Smith 1913) Bergey, Harrison, Breed, Hammer and Huntoon 1923, 158.[AL] (*Bacillus cellaseus* Kellerman et al. 1913, 508.)

cel.la′se.a. M.L. Adj. *cellaseus* pertaining to cellulose.

Cell morphology as given for the genus. Colonies circular, becoming irregular, 0.6–3.4 mm diameter, lemon yellow, opaque, edge entire or slightly erose.

Physiological and biochemical characters are given in Tables 15.23 and 15.24 and in the generic description.

Peptidoglycan is of the L-Orn-D-Glu type (Fiedler and Kandler, 1973).

The major cell wall sugar is rhamnose; mannose and 6-desoxytalose occur in smaller amounts (Keddie et al., 1966; Fiedler and Kandler, 1973).

The mol% G + C content of the DNA is 75.0 (T_m) (Stackebrandt and Kandler, 1979).

Type strain: ATCC 487.

Further comments on the relationship. DNA homology studies reveal no specific relationship to any other species of *Cellulomonas* (homology values below 35%) (Stackebrandt and Kandler, 1979).

Addendum

7. **Cellulomonas cellulans** (Metcalf and Brown 1957, 569) comb. nov. (*Nocardia cellulans* Metcalf and Brown 1957, 569; *Cellulomonas cartae* Stackebrandt and Kandler, 1980b, 186.)

cel.lu′lans. L. part. adj. *cellulans* cell-making.

The following description is based on the description of *C. cartae* (Stackebrandt and Kandler, 1980b) now considered a subjective synonym of *C. cellulans*.

The cell morphology differs from that of the other species of this genus in that a primary mycelium is produced which fragments later in the growth cycle (Keddie and Bousfield, 1980). After exhaustion of

the medium, the rods transform into shorter rods or even spherical cells. On peptone-yeast extract-glucose agar, colonies are circular, 0.9–5 mm in diameter, convex yellow-whitish, and glistening, edge entire.

The following are used as sole carbon sources: glucose, mannose, maltose, sucrose, D-xylose, L-arabinose, D-ribose, M-inositol, glycerol, cellobiose, lactose, gluconate, mannitol, L-lactate, acetate, pyruvate, propionate, pentanoate, capronate, heptanoate, caprylate, glyoxylate, proline, asparagine, aspartate, and histidine. Raffinose, DL-malate and D-lactate are not used as sole carbon sources.

Xanthine is hydrolyzed. Acid is produced from glucose, mannose, sucrose, D-xylose, L-arabinose, ribose, gluconate, glycerol, lactose, pyruvate and cellobiose. Starch is hydrolyzed and gelatin slowly liquefied. Catalase is produced. Nitrite is produced from nitrate. Aerobic and facultative anaerobic. Acetic acid is the main acidic intermediary product of aerobic glucose dissimilation. Resting cells ferment glucose anaerobically predominantly to CO_2, acetic acid and L-lactic acid. Ethanol and formic acid are minor end products.

Peptidoglycan is of the L-Lys-D-Ser-D-Asp type (Stackebrandt et al., 1978).

The major cell wall sugar is rhamnose, while fucose and galactose are minor components (Stackebrandt and Kandler, 1980b).

The mol% G + C of the DNA is 76.6 (T_m) (Stackebrandt et al., 1978).

Type strain: ATCC 12830.

Further comments on the relationship. DNA homology studies (Stackebrandt et al., 1980b) and the comparative analysis of the 16S-rRNA (Stackebrandt and Woese, 1981) demonstrated a very high genetic relationship between *C. cartae* and *Nocardia cellulans* (90% DNA homology, S_{AB} of 0.93). Somewhat lower DNA homology values were detected between these two species and *Brevibacterium fermentans*, *B. lyticum*, *Oerskovia xanthineolytica*, "*Arthrobacter luteus*" and "*Corynebacterium manihot*" (DNA homology values above 65%). The close genetic relationship found between these species, supported by a high similarity in chemical characters, e.g. the peptidoglycan type (Stackebrandt et al., 1980), the menaquinone composition (Collins et al., 1979; Collins and Jones, 1981) and the fatty acid composition (Minnikin et al., 1979) led to the proposal to combine *N. cellulans*, *B. fermentans*, *B. lyticum*, *O. xanthineolytica* and *C. cartae* in one species which was named *C. cartae* (Stackebrandt et al., 1982). It was accepted that differences occurred between these strains in some phenotypic characters, e.g. motility, cellulose degradation and the utilization of certain compounds (Stackebrandt et al., 1982). However, the specific epithet *cellulans* (Metcalf and Brown, 1957) antedates the epithet *cartae* (Stackebrandt and Kandler, 1980b) and, therefore, the legitimate name for this species is *Cellulomonas cellulans* (Metcalf and Brown, 1957) not *C. cartae* (Stackebrandt and Kandler, 1980b). *C. cartae* (Stackebrandt and Kandler, 1980b), *Brevibacterium fermentans* (Chatelain and Second, 1966), *B. lyticum* (Takayama et al., 1960) and *Oerskovia xanthineolytica* (Lechevalier, 1972) are considered subjective synonyms of *C. cellulans*.

Genus **Agromyces** Gledhill and Casida 1969c, 346[AL]

LESTER E. CASIDA, JR.

Ag.ro.my′ces. Gr. n. *agros* field or soil; Gr. n. *myces* fungus; M.L. n. *Agromyces* soil fungus.

Initial growth on agar as microcolonies composed of **branched, filamentous elements (1 μm or less in diameter) which subsequently undergo septation and fragmentation to yield coccoid and irregular cells. Gram-positive,** with segments of the mycelium becoming Gram-negative during fragmentation. Aerial mycelium and spores not produced; nonacid-fast, nonmotile. Pigmentation not evident during agar or broth growth. **Microaerophilic to aerobic,** poor or no growth anaerobically. CO_2 not required for growth. **Catalase-negative, benzidine-negative, oxidase-negative.** Organic nitrogen required for good growth. Sugars oxidized without gas production. **Cell walls contain 2,4-diaminobutyric acid (2,4-DAB),**

alanine, glutamic acid, glycine, and rhamnose as major components. Lysozyme-sensitive. Optimal growth at 30°C; grows well at 20°C and 37°C. Optimal pH 6.6–7.1. The mol% G + C of the DNA is 71 (Bd) or 76.7 (T_m) (Döpfer et al., 1982).

Type species: *Agromyces ramosus* Gledhill and Casida 1969c, 346.

Further Descriptive Information

A. ramosus has been isolated from a wide range of soil types (Gledhill and Casida, 1969c). In these soils it is 10- to 100-fold more numerous than the total microflora from conventional plating. It has not been isolated from some soils, apparently because special isolation techniques are required that rely on the presence of high cell numbers in the soil. On initial isolation, and for several transfers thereafter while it is adapting to laboratory cultivation, it grows very slowly and dies easily. It is morphologically unstable during this time.

The diameter of the cells depends on the growth medium used (Casida, 1977). The average cell diameter for 5 days shaken growth in a broth composed of 0.3% beef extract and 0.5% glucose is 0.65 μm (S.D. 0.07 μm). The cell diameters for growth on some other media, however, can be half or less of this value. An average cell diameter of 0.18 μm was observed for growth on Ashby's agar containing 0.2% glucose.

Although pigmentation usually is not evident under normal growth conditions, concentrated masses of cells may show a pale yellow color due to small amounts of an alcohol soluble pigment (Gledhill and Casida, 1969c). The concentration of this pigment (a carotenoid) and the rate of growth can be increased (Jones et al., 1970) by growth on fresh horse blood agar, or by growth on soil extract agar with added drops of bovine catalase or 4% MnO_2 (a substitute for catalase). The presence of other organisms can also cause increased growth and pigment formation. However, catalase added to isolation media does not increase the success of isolation (Labeda et al., 1974). Although the added catalase can extend the period of mycelial growth before onset of fragmentation (Casida, 1972), fewer of the cells initially present can initiate growth in the presence of the added catalase. It also appears that the response to the presence of other organisms is not to any extent due to their catalase content. A. ramosus kills at least some of these other types of organisms in order to obtain from them an as yet unknown stimulant for growth and pigment production (Casida, 1983). In addition, there are other catalase-positive organisms which have no effect on A. ramosus when it is grown in their presence.

The test for oxidative versus fermentative utilization of sugars is a modification of the method of Hugh and Leifson (1953). The medium contains tryptone, 0.5%; casamino acids (Difco), 0.4%; $(NH_4)_2HPO_4$, 0.07%; NaCl, 0.5%; agar, 0.2%; bromothymol blue, 0.003%; pH 7.0. Filter-sterilized carbohydrate, usually fructose, is added at 1%. Tests are considered positive if the pH value decreases by at least one unit during a 3-week incubation period.

The original description of the genus included a statement that it was weakly proteolytic. This is no longer included because *Agromyces* does not attack casein, and the results with gelatin hydrolysis can be unclear. Thus, growth may be poor in media solidified only with gelatin or in a medium in which gelatin is the only nutrient present.

Although A. ramosus is benzidine-negative, it does produce cytochromes *b*, *c*, and aa_3 but not cytochrome *o* (Jones et al., 1970). Flavoproteins are present.

Phage for A. ramosus can be isolated from soil by shaking the soil for 12 h in heart infusion broth (Difco) before diluting for phage isolation (Moore, 1974). It is not necessary to add A. ramosus to the soil for this enrichment.

A. ramosus is not antigenically related to the root nodule endophyte of *Alnus crispa* var. *mollis* Fern. (Lalonde et al., 1975), nor to the genera *Actinomyces*, *Rothia*, *Bacterionema*, or "*Ramibacterium*" (Gledhill and Casida, 1969c). It is not pathogenic for mice.

A. ramosus is readily induced into its L-phase (lacking a cell wall) when it is grown on agar media containing low levels of penicillin or glycine (Horwitz and Casida, 1975). After initial contact with these agents, the L-forms are stable in the absence of the agents. In the absence of serum in the medium, calcium stabilizes the L-form, but magnesium causes reversion to the parent bacterial form (Horwitz and Casida, 1978a). The reversion due to magnesium can be diminished or prevented by simultaneous addition of calcium. Reversion of the stable L-form also occurs during its incubation in sterile soil (Horwitz and Casida, 1978b). However, this revertant differs from the above in that it maintains its bacterial form when cultured on a medium low in NaCl, but is induced into a L-form on a medium high in NaCl.

Enrichment and Isolation Procedures

A. ramosus is isolated from soil, but its isolation is difficult to accomplish, apparently because A. ramosus has difficulty in adjusting to the conditions of laboratory cultivation. The possibility for death of the isolates decreases, however, after several transfers of the purified cultures on laboratory media. The fact that A. ramosus occurs in very high numbers in many soils is the basis of the isolation techniques. For those soils which are less numerous in A. ramosus, the isolation procedures may yield other catalase-negative bacteria, such as *Actinomyces humiferus* (Gledhill and Casida, 1969b) or *Streptococcus sanguis* (Gledhill and Casida, 1969a), instead of A. ramosus.

The isolation techniques are by dilution frequency or by plating. The dilution frequency procedure is recommended because greater numbers of isolates are obtained but neither procedure can be used for actually counting the numbers of A. ramosus in soil.

Dilution frequency procedure (Casida, 1965)

The numbers of conventional bacteria in the soil are first determined by plating or by dilution frequency. For the actual isolation of A. ramosus, a 1-g portion of soil is blended in a sterile Waring blender microhead with 100 ml of heart infusion broth adjusted to pH 7.8 with KOH, for 1 min on, 1 min off, and 1 min on, in succession. The blended soil dilution is allowed to stand for ~30 s to dissipate foam, and then a 1-ml sample is transferred to 99 ml of fresh heart infusion broth (pH 7.8) and shaken. Further 10-fold dilutions of the soil suspension are carried out in a conventional manner in heart infusion broth. The dilution is selected that corresponds to the dilution end point of the soil's population of conventional bacteria (as determined above by dilution frequency or plating). The next two greater 10-fold dilutions are also selected. For each of these dilutions, fifty 1-ml samples are transferred individually to screw-cap tubes containing 1-ml slants of 1.5% agar in water. The caps are screwed tight, and the tubes are incubated up to 4 weeks at 30°C, with periodic observations for growth. Tubes showing growth (these would be conventional bacteria) at 4 days incubation are discarded. At intervals, all tubes, including those not showing growth, are streaked on the surfaces of slants of heart infusion agar (pH 7.8) in screw-cap tubes; the caps are screwed tight, and the tubes incubated at 30°C. Addition of 1 ml of sterile heart infusion broth to these tubes and incubation for 2–3 weeks can help to stabilize the organisms so that death is less likely to occur. The growth in these tubes is streaked on plates of similar medium, and the plates are incubated in air at 30°C. The resulting small, white, opaque colonies are transferred to agar slants of similar medium for incubation in air at 30°C. The growth in these tubes is checked for catalase activity and, if found to be catalase-positive, the tubes are discarded. All catalase-negative cultures are restreaked on a similar medium for purification.

A consistent growth appearance is observed in the primary isolation tubes and on streak plates. In primary isolation tubes, initial growth occurs some time after the first week of incubation, and is evidenced by a white opaque button of cells at the butt of the slant under the liquid; the growth appears ropey on swirling the tube. Usually, turbidity is not observed. The colonies on the primary isolation streak plates are quite small (often less than 1 mm in diameter) and present a "fried egg" appearance.

Plating procedure (Labeda et al., 1974)

Nutritionally minimal media are used, and the incubation periods are shorter than with the above dilution frequency procedure. Two media can be used. For the first, plates are poured with 1.5% Noble agar (Difco), and the agar is allowed to solidify. To the surface of this agar are added 0.2 ml of an *Arthrobacter globiformis* ATCC 8010 culture filtrate and 0.1 ml of a dilution in sterile tap water of the soil that is to be plated. These are mixed and spread across the agar surface by the use of an alcohol-flamed glass spreading rod. For the second medium, the soil dilution is spread directly onto the surface of a modified Burk *Azotobacter* medium that lacks a carbon source and contains only a minimal nitrogen level. This medium contains K_2HPO_4, 0.8 g; KH_2PO_4, 0.2 g; $MgSO_4 \cdot 7 H_2O$, 0.2 g; $CaSO_4 \cdot 2 H_2O$, 0.1 g; $NaMoO_4 \cdot 2 H_2O$, 0.25 mg; $Fe(NH_4)_2(SO_4)_2 \cdot 12 H_2O$, 8.6 mg; Noble agar 15.0 g; and water, 1 liter. After a period of 7–11 days of incubation at 30°C, the plates of both media are observed for the presence of colonies having a characteristic fried-egg appearance. Detection and characterization of the colonies requires 20-fold magnification, since their size usually ranges from ~0.1–0.5 mm in diameter. The colonies are then streaked onto plates of similar medium and, after 7–11 days of incubation, catalase tests are run directly on a portion of the colonial growth on the plates. Isolates proving to be catalase-negative are subcultured to slants of either Noble agar overlaid with 0.2 ml of *A. globiformis* culture filtrate or the above-modified Burk medium without an overlay. Growth occurring in these tubes is rechecked for catalase activity and cellular morphology and physiology. Some isolates may register initially as catalase-negative but actually may be catalase-positive when greater amounts of growth are available for analysis. The partially immersed slant step is required for stabilization of the culture for continued survival when the culture filtrate plating procedure is used, but it is not a strict requirement with the Burk medium plating procedure.

The culture filtrate is prepared by inoculating heart infusion broth with *A. globiformis* ATCC 8010 and shaking for 3 days at 30°C; the pH value of the broth should rise to ~8.7 during this incubation. Most of the cells are removed by centrifugation, and then the supernatant fluid is sterilized by filtration through a 0.22 μm membrane filter. Paper chromatography of this filtrate shows only traces of residual amino acids. It does, however, contain anthranilic acid and urea (indicated by paper chromatography) as biosynthetic products of the *Arthrobacter*, but these compounds have no demonstrable effect on the success of the isolation technique, nor do the filtrate or these compounds exhibit any effect on a stock strain of *A. ramosus*.

With the plating procedure, *A. ramosus* occurs at the approximate dilution end point for the more conventional soil microorganisms, but not at the yet greater dilutions which are used in the dilution frequency procedure. The use of plating media of a very low nutritional level seems to provide a very slow growth rate for adaptation of the cells to laboratory cultivation. However, approximately 30% of the isolates can be expected to die during their adaptation to growth under laboratory conditions. Addition of catalase to the media does not prevent this. Once adapted, however, *A. ramosus* usually grows without problems on these media or on a nutritionally rich medium such as heart infusion agar.

Maintenance Procedures

Agromyces strains can be lyophilized by common procedures used for aerobes.

Differentiation of the genus **Agromyces** from other genera

The genus *Agromyces* produces Gram-positive, true-branching mycelium (no spores or aerial mycelium) that is 1 μm or less in diameter, and that subsequently fragments into diphtheroid and coccoid forms as growth progresses. It is separated from other genera with these characteristics by being catalase-, benzidine-, and oxidase-negative, although simultaneously it is microaerophilic to aerobic and has an oxidative metabolism. Its cell wall does not contain L-lysine or α,ε-diaminopimelic acid as major constituents of the wall glycopeptide, but does contain 2,4-DAB.

Taxonomic Comments

Agromyces has certain features of both the *Actinomyces* and *Nocardia* genera (Gledhill and Casida, 1969c). Therefore, it would seem that it should be placed as a separate taxon between the large groups containing these genera (the families *Actinomycetaceae* and *Nocardiaceae* in the eighth edition of the *Manual*). However, the lack of serological cross reactions with *Actinomyces*, *Rothia* and *Bacterionema*, the cell wall composition and the lysozyme sensitivity of *Agromyces* prompted Cross and Goodfellow (1973) to state that: "Until comparative studies are done with many more strains isolated from soil they are best left as a taxon without a family." Fiedler and Kandler (1973) point out that 2,4-DAB is a common component of the peptidoglycan of some coryneform bacteria.

A. ramosus (DSM43045, (ATCC 25173)) does not contain mycolic acids (Collins, 1982). The fatty acid composition is predominantly *anteiso*-methyl-branched chain acids: 12-methyltetradecanoic (*anteiso*-$C_{15:0}$); 14-methylhexadecanoic (*anteiso*-$C_{17:0}$) with minor amounts of *iso*-methyl-branched straight chain saturated and monounsaturated acids *iso*-$C_{14:0}$, *iso*-C_{15}, $C_{16:0}$, *iso*-$C_{16:0}$, *iso*-$C_{17:0}$, $C_{18:0}$ and $C_{18:1}$. Menaquinones are the only isoprenoid quinones detected, the major component being MK-12 but significant amounts of MK-13 and trace amounts of MK-11 and MK-14 also occur. The polar lipid composition comprises diphosphatidylglycerol (DPG), phosphatidylglycerol (PG) and two unidentified glycolipids.

Differentiation of the species of the genus **Agromyces**

There is only one species, *A. ramosus*. Nevertheless, the 60 isolates examined (Gledhill and Casida, 1969c) show two different carbohydrate patterns for their cell walls: 77% of the isolates, including the type strain, contain rhamnose, galactose, mannose, and xylose as the major cell wall sugars, whereas 23% demonstrate only rhamnose and glucose. The percentage of distribution among the isolates of other characteristics of the species (listed below) are also given by Gledhill and Casida (1969c).

List of species of the genus **Agromyces**

1. **Agromyces ramosus** Gledhill and Casida 1969, 346.[AL]

ra.mos'us. L. adj. *ramosus* much-branched.

Description as for the genus. In addition, mature colonies are small, opaque, entire, smooth, and convex with a pronounced dense, dark, central area adhering to the agar surface. Rough colony variants occur. Growth in broth is poor, adhering to glass surfaces and settling to the bottom of the culture vessel. Acid is produced oxidatively from glycerol, arabinose, ribose, xylose, glucose, fructose, mannose, galactose, mannitol, sucrose, maltose, cellobiose, rhamnose, raffinose, stachyose, inulin, starch, and dextrin. Glucose either is not oxidized or requires a prolonged incubation period for oxidation. Fructose and polymeric units of glucose are readily oxidized. Fructose is used in a modified Hugh-Leifson test for examining the oxidative versus fermentative capabilities of isolates. Nitrates are not reduced. Cell wall sugars consist of rhamnose, galactose, xylose, and mannose, but some strains contain only rhamnose and glucose. H_2S is produced. Indole is not produced. Methyl-red and Voges-Proskauer tests are negative. Starch and esculin are hydrolyzed. Casein and tributyrin are not hydrolyzed. Growth occurs in the presence of 4% NaCl. Sodium lactate, succinate, fumarate, pyruvate, and α-ketoglutarate are utilized for growth. Ammonia is not produced from arginine. Xanthine and tyrosine are decomposed.

Growth in most media is relatively slow, requiring 5–7 days to achieve maximal cell density.

A. ramosus has a slight ability to oxidize CO to CO_2 (Bartholomew and Alexander, 1979).

The habitat is soil.

The mol% G + C of the DNA is 71 (Bd), 76.7 (T_m).

Type strain: ATCC 25173.

Genus **Arachnia** *Pine and Georg 1969, 269*[AL]

Klaus P. Schaal

A.rach′ni.a. Gr. neut. n. *arachnion*, a cobweb; Gr. fem. n. *arachne*, a spider; M.L. fem. n. *Arachnia* referring to the cobweb-like or spider-like appearance of filamentous microcolonies.

Short diphtheroidal rods, 0.2–0.3 μm in diameter and 3.0–5.0 μm in length, which may or may not be branched, and **branching filaments**, 5–20 μm or more in length. Occasionally, cells are of uneven diameter, often with distended or clubbed ends. **Swollen spherical cells** resembling spheroplasts, up to 5.0 μm in diameter, are frequently seen, and an occasional culture may consist entirely of these coccoid forms. **Gram-positive, nonacid-fast,** and **nonmotile.** Endospores, conidia or capsules are not formed.

Facultatively anaerobic, but best growth is obtained under anaerobic conditions. Carbon dioxide is not required for either aerobic or anaerobic growth.

Microcolonies are **usually highly filamentous,** being composed of long, branched, septate and nonseptate filaments which often originate from a single, eccentric reproductive element. Mature colonies (7–14 days) measure up to 2 mm in diameter and are white to gray-white, rough or smooth and crumbly to soft, but not mucoid in texture. **No aerial filaments.** Optimum temperature 35–37°C.

Chemoorganotrophic, having a **fermentative type of metabolism.** Carbohydrates serve as fermentable substrates. **Anaerobically, fermentation of glucose** yields CO_2, **acetic acid, propionic acid** and small amounts of DL-lactic and succinic acids. **In air, glucose is fermented** to **acetic acid** and **carbon dioxide. Catalase-negative;** does not produce indole.

Cell walls contain LL-**diaminopimelic acid** (LL-DAP), glycine and **galactose,** but **no lysine, ornithine** or **arabinose.** The mol% G + C of the DNA is 63–65 (T_m).

Type species: *Arachnia propionica* Pine and Georg 1969, 269.

Further Descriptive Information

The cellular morphology of *Arachnia* is extremely variable, depending on factors such as strain variation, growth conditions and incubation time. In smears or wet mounts from mature colonies or aged broth cultures (7–14 days), short diphtheroid rods usually predominate (Fig. 15.6), but septate or nonseptate filaments may also be seen (Fig. 15.7).

The process of septation of the filaments occasionally produces chains of very short rods which may resemble certain streptococci (e.g. *Streptococcus mutans*).

Cells derived from young broth cultures (24–48 h) tend to be more filamentous, and branched structures as well as bulbous forms can easily be detected (Fig. 15.8). Wet mounts prepared from these cultures may even contain intact filamentous microcolonies (Fig. 15.9). In Gram-stained smears from clinical materials such as tissue sections or lacrimal concretions the morphology is also definitely filamentous.

The occurrence of large spherical cells has already been reported in the first detailed description of *Arachnia propionica* (*Actinomyces propionicus*) (Buchanan and Pine, 1962). These were thought to represent spheroplasts, but have been shown to resist high (20% (w/v) sucrose) as well as low (distilled water) osmotic pressures. It has been claimed (Buchanan and Pine, 1962) that the formation of such swollen coccoid elements is greatly enhanced by raffinose as the sole carbohydrate in the growth substrate. However, Slack and Gerencser (1975) have observed that these spherical cells also develop in routine media and thus may not require special growth conditions.

The peptidoglycan of *Arachnia* cell walls is composed of *N*-acetylmuramic acid, *N*-acetylglucosamine, glycine, glutamic acid, LL-DAP, and alanine (Buchanan and Pine, 1962; Pine, 1970; Schleifer and Kandler, 1972; Johnson and Cummins, 1972). Aspartic acid has been reported (Buchanan and Pine, 1962; Pine, 1963), but not confirmed by other investigators. Quantitatively, the walls of *Arachnia* contain 2 mol of glycine and 1 mol of alanine/mol of glutamic acid (Schleifer and Kandler, 1972). These findings together with preliminary results on the primary structure of the peptidoglycan suggest that L-alanine in position 1 of the peptide subunit is replaced by glycine as in peptidoglycans of group B, while a single glycine residue is responsible for the cross-linkage of two peptide subunits between LL-DAP and D-alanine, as in group B mucopeptides (Schleifer and Kandler, 1972).

Glucose and galactose have been reported to be the major cell wall sugars of arachniae (Buchanan and Pine, 1962; Pine, 1963). However,

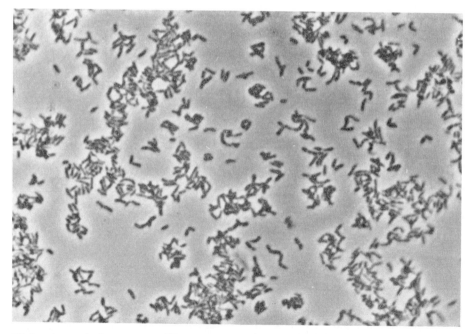

Figure 15.6. *A. propionica, serovar 1.* Wet mount in lactophenol cotton blue mounting fluid prepared from a mature colony (BHI agar, 14 days, 36°C); phase contrast micrograph (× 1200).

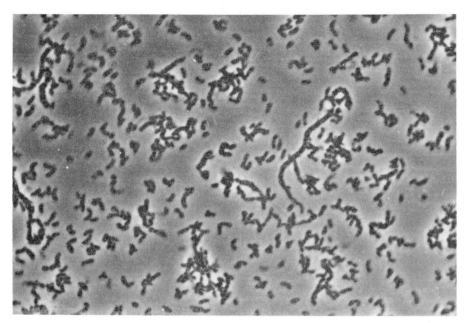

Figure 15.7. *A. propionica, serovar 1.* Wet mount in lactophenol cotton blue mounting fluid prepared from a mature colony (BHI agar, 12 days, 36°C); phase contrast micrograph (× 1600).

Figure 15.8. *A. propionica, serovar 1.* Gram-stained smear from a 24-h culture in Tarozzi broth; phase contrast micrograph (× 1200).

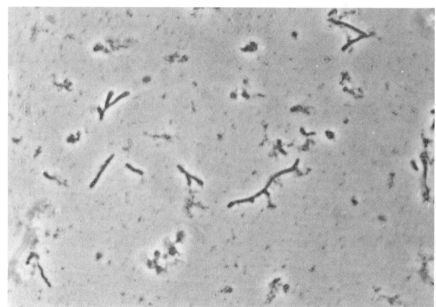

Johnson and Cummins (1972) found two distinct sugar patterns. Strains of the first group had essentially only galactose, thereby resembling *Actinomyces israelii*, whereas isolates of the second group contained both glucose and galactose together with a trace of mannose. Arabinose, 6-deoxytalose, fucose and rhamnose have not been demonstrated in *Arachnia* cell walls.

The total cellular fatty acid profile of the type strain of *Arachnia propionica* was found to be similar to that of *Actinomyces* species, comprising only even numbered, *n*-saturated and *n*-unsaturated, but no branched-chain, odd numbered, cyclopropane or hydroxy substituted fatty acids. In particular, 16:0, 18:1 and smaller amounts of 18:0 and 14:0 fatty acids were detected (Amdur et al., 1978).

The cell walls of *Arachnia* as observed by electron microscopy of ultrathin sections showed a characteristic bilayered structure (Lai and Listgarten, 1980) which is also seen in *Actinomyces* species (Duda and Slack, 1972). However, in contrast to the latter organisms, these two layers appeared as two concentric halves of similar thickness and electron density divided by an electron-lucent line. Measurements of the overall width of *Arachnia* walls range from 11.4 nm (Overman and Pine, 1963) to 32–45 nm (Lai and Listgarten, 1980). The external contour of the outer layer was described as smooth and surface fibrils (surface fuzz) were not demonstrated (Lai and Listgarten, 1980). The cytoplasmic membrane exhibits the usual trilaminar appearance.

Within the cytoplasm, the nucleoid (nuclear apparatus), ribosomes, electron-dense inclusions and mesosomes could clearly be discerned (Lai and Listgarten, 1980). Overman and Pine (1963) reported that the latter were compact and highly coiled and had no apparent contact with the cytoplasmic membrane. This feature was thought to differentiate *Arachnia* from *Actinomyces* species which, in this study, appeared to have no mesosomes or only simple ones that were connected to the cytoplasmic membrane.

Within 24 h of incubation at 36°C on brain heart infusion agar or

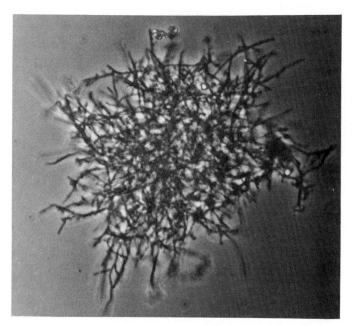

Figure 15.9. *A. propionica, serovar 1.* Wet mount in lactophenol cotton blue mounting fluid prepared from a 24-h culture in Tarozzi broth; phase contrast micrograph (× 640).

other suitable agar media, *Arachnia* usually produces filamentous microcolonies which are composed of long, irregularly branched, nonseptate threads. The filaments mostly originate from a single, rod-like propagule by apical growth at one end of the rod (Locci and Schaal, 1980). As a consequence, the original reproductive element is placed in the periphery of the young microcolony giving rise to asymmetrical, twig-like structures (Fig. 15.10). When growth continues, the typical spider- or cobweb-like microcolonies develop (Fig. 15.11) which may already show signs of septation and fragmentation of the central threads. Rarely, smooth microcolonies are formed which consist chiefly of short rods together with a few peripheral, short, branching filaments (Fig. 15.12).

As cultures age, the colonies become more variable in size and appearance. Mature macrocolonies (7–14 days) are either (a) rough, with umbonate or heaped centers, irregular edges, convoluted surfaces and a dry or crumbly texture (Figs. 15.13 and 15.14) resembling the molar-tooth or the bread-crumb type of *Actinomyces israelii*, or (b) they are smooth, convex and creamy in texture, with entire or undulate edges and dense, granular centers (Fig. 15.15). Occasionally, completely smooth, flat, circular, slightly granular forms occur (Fig. 15.16). The diameter of mature colonies grown on brain heart infusion agar for 14 days at 36°C varies from about 0.2–2.0 mm.

The first strain of *Arachnia propionica* (*Actinomyces propionicus*) described in detail (Buchanan and Pine, 1962) has been reported to produce dull orange colonies "when grown on the surface of aerobic or anaerobic tubes." However, essentially all of the subsequent isolates obtained in various microbiological laboratories since then were found to be nonpigmented, forming white to gray-white, opaque macrocolonies. Among 30 *Arachnia* strains identified at the Institute of Hygiene of the University of Cologne, West Germany, during the past 12 years (Schaal, unpublished), only two cultures were encountered which developed an orange color, especially when incubated for prolonged periods of time (3 to 6 weeks). However, these differed in several physiological characters from typical *Arachnia propionica* strains so that it remains an open question whether they are only pigmented variants of that species or represent members of a new taxon to be described in the future.

In liquid media (e.g. thioglycolate broth), the organisms primarily grow as discrete filamentous colonies of variable size (Fig. 15.9) sus-

pended in the broth or adhering to the glass surface. Depending on inoculum size and medium composition, this may lead to a granular, fluffy or pellicular appearance of the culture. With aging the clumps often dissociate in part and/or settle out. In the latter case, a felty or flocculent sediment is produced leaving a clear supernatant. Evenly turbid broth cultures are not characteristic of *Arachnia* and usually indicate contamination.

Arachniae grown on artificial media exhibit a typical life cycle which involves short rods, long branched filaments and septation and fragmentation of the filaments as successive and subsequently simultaneous morphological stages. Binary fission as a means of division has not been documented with certainty and is at least uncommon. The usual mode of division is septum formation in shorter or longer filaments. These arise from rod-like propagules by apical growth (budding) at one end of the rod (Locci and Schaal, 1980) and develop into a network of interwoven, branching filaments. With aging of the culture, the process of septation and fragmentation proceeds transforming the filamentous, spider-like microcolonies successively into smoother macrocolonies which consist predominantly of diphtheroidal rods. Intensity and rate of septation and fragmentation determine whether the mature colonies remain partially filamentous or become completely smooth. The budding process as observed by scanning electron microscopy (Locci, 1978) and immunofluorescent labeling (Locci and Schaal, 1980) takes place by active synthesis and deposition of new cell wall material at rod apices leaving the remaining cell envelope unchanged.

The minimal nutritional requirements are not known. However, growth has only been reported on complex media containing either rich biological substrates (e.g. brain heart infusion, meat extract, yeast extract) or a defined mixture of a large variety of organic and inorganic compounds (Howell and Pine, 1956; Heinrich and Korth, 1967). Thus, the anabolic capacity of *Arachnia* seems to be limited so that most of the amino acids and possibly also vitamins and other growth factors must be added to the culture medium to induce uninhibited reproduction.

Carbohydrates, sugar alcohols and similar substances are the preferred sources of carbon and energy and when utilized are fermented with production of acid but no gas (Pine and Georg, 1974). Nearly all of the strains grow well on and produce acid from fructose, glucose,

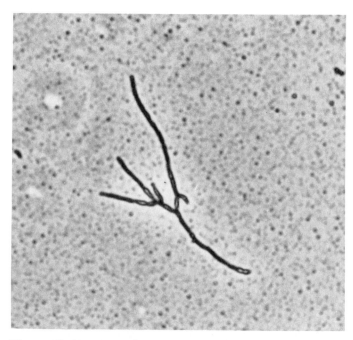

Figure 15.10. *A. propionica, serovar 1.* Young (20-h) microcolony on CC medium (slide culture); phase contrast micrograph in situ (× 1600).

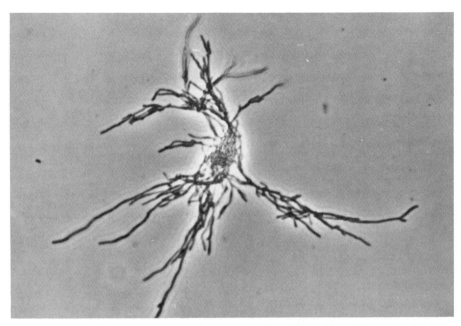

Figure 15.11. *A. propionica, serovar 1.* Microcolony (24-h) on CC medium (slide culture); phase contrast micrograph in situ (× 800).

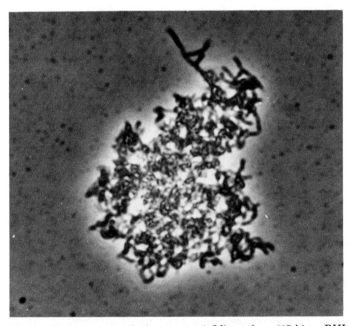

Figure 15.12. *A. propionica, serovar 2.* Microcolony (48-h) on BHI agar (slide culture); phase contrast micrograph in situ (× 1600).

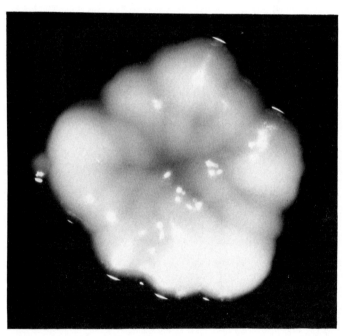

Figure 15.13. *A. propionica, serovar 1.* Molar-tooth type mature colony on BHI agar (14 days); micrograph (× 30).

maltose, mannitol, raffinose and sucrose (Buchanan and Pine, 1962; Pine and Georg, 1969, 1974; Holmberg and Nord, 1975; Schofield and Schaal, 1979a, b, 1980a, 1981). In addition, most of the isolates utilize amygdalin, dextrin, dihydroxyacetone, galactose, lactose, mannose, sorbitol, starch and trehalose (Pine and Georg, 1969, 1974; Slack and Gerencser, 1975; Schofield and Schaal, 1981). Acid production from adonitol, arabinose, glycerol, *meso*-inositol, melibiose, ribose and salicin is less common and may aid in differentiating the two serovars recognized within the species *Arachnia propionica* (Schofield and Schaal, 1980a). Cellobiose, cellulose, dulcitol, *iso*-erythritol, glycogen, inulin, melezitose, rhamnose, sorbose, xylose, α-methyl-D-glucoside and α-methyl-D-mannoside are usually not fermented (Pine and Georg, 1969, 1974; Schofield and Schaal, 1980a, 1981).

Organic compounds other than carbohydrates and sugar alcohols have rarely been studied for their ability to serve as carbon and energy sources. Buchanan and Pine (1962) and Pine and Georg (1969, 1974) reported that *Arachnia* did not ferment lactate or pyruvate nor did it grow on these acids as sole carbon sources. In contrast, Schofield and Schaal (1979a) found that lactate and pyruvate as well as alanine supported growth of the type strain and the reference strain of serovar 2 in a minimal medium to a similar extent as did glucose. Succinate is not utilized.

Arachnia propionica can grow aerobically and anaerobically without added CO_2 (Buchanan and Pine, 1962; Pine and Georg, 1969, 1974; Slack and Gerencser, 1975). It remains to be clarified, however, whether

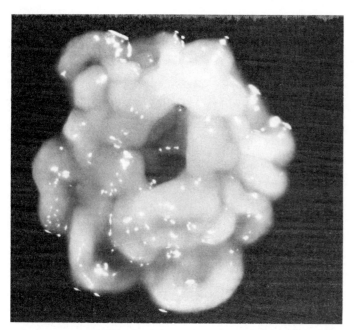

Figure 15.14. *A. propionica, serovar 1.* Bread-crumb type mature colony with central hole on BHI agar (14 days); micrograph (× 30).

Figure 15.15. *A. propionica, serovar 1.* Smooth mature colony with finely granular surface on BHI agar (14 days); micrograph (× 30).

or not CO_2 may be stimulatory under certain cultural conditions (Schofield and Schaal, 1981).

Although *Arachnia propionica* does grow quite well aerobically in liquid media (Buchanan and Pine, 1962), anaerobiosis induces growth of smaller inocula. On the surface of agar plates or slants incubated aerobically, growth is usually poor or does not occur at all. Thus, agar cultures should be incubated anaerobically (e.g. in GasPak jars) when sufficient and consistent growth yields are to be obtained. Sophisticated anaerobic techniques (e.g. roll tube method, prereduced media) are not required.

The optimum growth temperature lies between 35 and 37°C. Some strains may also grow at 45°C (Holmberg and Nord, 1975).

Anaerobically, in the presence or absence of CO_2, glucose is fermented principally to CO_2, acetic and propionic acids. In addition, small and variable amounts of DL-lactic and succinic acids are usually produced. In media containing 0.5% glucose, approximately 54.2 μmol of CO_2, 54.5 μmol of acetic acid, 113 μmol of propionic acid, 2.5 μmol of DL-lactic acid and 3.0 μmol of succinic acid were formed/100 μmol of glucose fermented (Pine and Georg, 1969). On the whole, only 16 μmol of the glucose provided in the medium were utilized under these conditions. Aerobically, the fermentation of glucose yields stoichiometric amounts of acetic acid and CO_2 and the total amount of glucose utilized increases to about 27 μmol (Buchanan and Pine, 1962).

These fermentation end products do not differ significantly, neither qualitatively nor quantitatively, from those produced by propionibacteria (Buchanan and Pine, 1962). Furthermore, extracts of *Arachnia propionica* were found to contain methylmalonyl-coenzyme A transcarboxylase together with malate dehydrogenase, fumarase, acetate kinase and coenzyme A transferase, indicating that arachniae utilize the same pathway of propionate production as do *Propionibacterium* species (Allen and Linehan, 1977). The presence of aldolase has also been demonstrated (Pine and Georg, 1969).

Arachniae have little or no proteolytic activities: casein is not hydrolyzed, meat not digested and milk not peptonized (Pine and Georg, 1969; Slack and Gerencser, 1975; Schofield and Schaal, 1981). However, some strains may acidify, reduce and/or clot milk. Gelatin may be liquefied slowly by a few isolates (Pine and Georg, 1969, 1974; Slack and Gerencser, 1975; Schofield and Schaal, 1981).

Amino acids may be deaminated or decarboxylated especially by strains belonging to serovar 1 of *Arachnia propionica*. Deamination has been observed with alanine, aspartic acid, glutamic acid and serine as substrates, decarboxylation occurred with ornithine. Arginine, leucine, lysine and methionine are usually neither deaminated nor decarboxylated (Schofield and Schaal, 1980b, 1981).

No urease activity has been detected and indole is not produced. Most strains produce H_2S when grown on triple sugar iron agar, but not on brain heart infusion agar (Pine and Georg, 1974). Esculin, hippurate, Tweens and lecithin are not hydrolyzed. Starch hydrolysis may be detected, but is usually weak. Hyaluronidase, chondroitin sulfatase and DNAase are absent. With human erythrocytes, a β-hemolytic activity may be observed. The Voges-Proskauer test is negative, the methyl red test is positive with the majority of strains (Pine

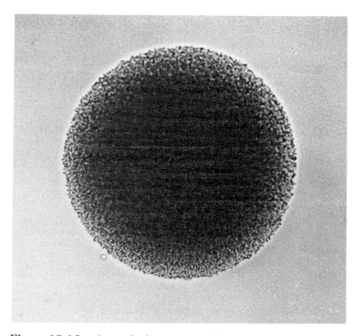

Figure 15.16. *A. propionica, serovar 2.* Mature colony on BHI agar (14 days); micrograph (× 200).

and Georg, 1974; Slack and Gerencser, 1975; Holmberg and Nord, 1975; Schofield and Schaal, 1981).

Additional enzymatic activities which can be observed using the API-zym test kit (API-Bio Merieux) are: esterase (C_4), esterase lipase (C_8), leucine arylamidase and α-glucosidase. Tests for alkaline and acid phosphatases, lipase (C_{14}), valine arylamidase, cystine arylamidase, trypsin, chymotrypsin, phosphoamidase, α-galactosidase, β-galactosidase, β-glucuronidase, β-glucosidase, N-acetyl-β-glucosaminidase, α-mannosidase and α-fucosidase were found to be negative or only weakly positive (Kilian, 1978; Schofield and Schaal, 1981).

Catalase and cytochrome oxidase are consistently absent (Buchanan and Pine, 1962; Pine and Georg, 1969, 1974; Schofield and Schaal, 1981). On the other hand, all of the *Arachnia* strains tested so far were able to reduce nitrate to nitrite. Further reduction of nitrite has not been observed (Slack and Gerencser, 1975; Schofield and Schaal, 1981).

Mutants, plasmids and bacteriocins have not been reported, and no detailed study on phages of *Arachnia* has been published. However, Overman and Pine (1963) detected dense hexagonal bodies in ultrathin sections of *Arachnia* cells which resembled heads of mature phages.

Two serovars have been recognized within the species *Arachnia propionica* using the direct modification of the fluorescent antibody technique (Gerencser and Slack, 1967; Holmberg and Forsum, 1973; Brock et al., 1973). The type strain ATCC 14157 represents serovar 1; strain "Fleischmann" isolated in this laboratory by F. Lentze (other designations: HIK* A 110, WVU† 346, ATCC 29326) and originally described as a serological and physiological variant of *Actinomyces israelii* (Lentze, 1953), is the reference strain of serovar 2.

Cross-reactions between the serovars occur (Slack and Gerencser, 1975; Holmberg and Forsum, 1973) especially when using an indirect immunofluorescence method (Schaal and Pulverer, 1973; Gatzer and Schaal, unpublished). These cross-reactions can also be demonstrated by gel diffusion (Gerencser and Slack, 1967; Gatzer and Schaal, unpublished), but can be removed by absorption. In addition, low-titered cross-reactivity has been observed with other fermentative actinomycetes, especially with *Actinomyces israelii*, serovar 1. Antisera to this organism even showed a common precipitinogen with *Arachnia propionica*, serovar 1, when the Ouchterlony technique was used (Gatzer and Schaal, unpublished). Serovar 2 antisera exhibited some cross-staining with *Actinomyces naeslundii*, serovar 1, and also with *Bacterionema* (*Corynebacterium*) *matruchotii* (Gatzer and Schaal, unpublished). Using a cell wall agglutination test, Johnson and Cummins (1972) found no relevant cross-reactivity between the two serovars of *Arachnia propionica* and between them and *Propionibacterium* species. However, with the direct or indirect immunofluorescence, low titered cross-staining may be observed between *Arachnia propionica*, serovar 1, and *Propionibacterium avidum* (Holmberg and Forsum, 1973; Gatzer and Schaal, unpublished). *Propionibacterium acnes* may also weakly stain with *Arachnia* antisera (Slack and Gerencser, 1975; Gatzer and Schaal, unpublished), whereas *Propionibacterium granulosum* obviously does not have surface antigens in common with the arachniae (Gatzer and Schaal, unpublished).

The chemical nature of the species- and serovar-specific antigens has not been determined in detail, but there is some evidence that polysaccharides are the active components (Kwapinski, 1969; Johnson and Cummins, 1972).

All of the *Arachnia* strains tested so far have proved to be highly susceptible to any of the β-lactam antibiotics. The following minimal inhibitory concentrations have been reported (Schaal et al., 1979; Schaal and Pape, 1980; Pape, 1979; Niederau et al., 1982): benzyl penicillin, 0.1–0.2 I.U./ml; propicillin, 0.1–0.2 mg/liter; azidocillin, 0.1–0.2 mg/liter; ampicillin, 0.02–0.2 mg/liter; amoxicillin, 0.2–0.78 mg/liter; ciclacillin, 1.56–3.12 mg/liter; epicillin, 0.2–0.39 mg/liter; carbenicillin, 1.56–3.12 mg/liter; ticarcillin, 1.56–3.12 mg/liter; mezlocillin, 0.39 mg/liter; azlocillin, 0.2–1.56 mg/liter; cefalothin, 0.2–0.78 mg/liter;

cefapirin, 0.2–0.39 mg/liter; cefacetrile, 0.78–1.56 mg/liter; cefradine, 0.2 mg/liter; cefazolin, 0.2–0.78 mg/liter; cefalexin, 0.39 mg/liter; cefamandole, 0.2–0.39 mg/liter; cefuroxime, 0.39–0.78 mg/liter; cefoxitin, 0.39–0.78 mg/liter; cefotaxime, 0.2–0.78 mg/liter.

Further antibacterial drugs which show a moderate to high inhibitory activity against arachniae are: tetracyclines (tetracycline-HCl, 0.2–1.56 mg/liter; doxycycline, 0.2–0.78 mg/liter; minocycline, 0.2–0.78 mg/liter), the chloramphenicols (chloramphenicol, 0.78–12.5 mg/liter; thiamphenicol, 3.12–6.25 mg/liter), the makrolids (erythromycin, 0.2–0.78 mg/liter; oleandomycin, 0.39–3.12 mg/liter; spiramycin, 0.2–1.56 mg/liter), the lincomycins (lincomycin, 0.2–0.39 mg/liter; clindamycin, 0.2–1.56 mg/liter), the rifamycins (rifampicin, 0.05–0.78 mg/liter), fusidic acid (0.2–6.25 mg/liter) and vancomycin (0.78–3.12 mg/liter). Complete resistance has only been observed against aminoglycosides (gentamicin, 25–100 mg/liter; tobramycin, 50–100 mg/liter; sisomicin, 50–100 mg/liter; amikacin, 25–100 mg/liter; spectinomycin, 12.5–50 mg/liter), nitroimidazole compounds (metronidazole, >100 mg/liter; tinidazole, 25, >100 mg/liter), peptide antibiotics (colistine, >100 I.U./ml) and antifungal drugs.

The resistance pattern to other chemical and biological inhibitors commonly used in microbiology is as follows: arachniae will grow in the presence of 5% (w/v) bile and 0.005% (w/v) sodium azide, but will usually be inhibited by 2% (w/v) sodium chloride, 0.005% (w/v) crystal violet and 0.2% (w/v) sodium taurocholate. Between one-third and one-half of the strains are resistant to 0.01% (w/v) potassium tellurite and 0.01% (w/v) sodium selenite (Schofield and Schaal, 1981).

Arachnia propionica may be pathogenic to men and laboratory animals. It causes lacrimal canaliculitis with and without conjunctivitis (Pine and Hardin, 1959; Brock et al., 1973; Jones and Robinson, 1977; Seal et al., 1981; Schaal and Pulverer, 1984) and localized or systemic human actinomycosis which predominantly involves the cervicofacial area (Brock et al., 1973; Schaal and Pulverer, 1973, 1981, 1984; Pulverer and Schaal, 1978; Schaal, 1979, 1981), but may also affect lungs, kidneys, brain and bone (Brock et al., 1973; Conrad et al., 1978; Riley and Ott, 1981). An etiologic role of arachniae in the pathogenesis of caries and periodontal disease has not been established.

After intraperitoneal injection into white mice, *Arachnia propionica* consistently produces liver and mesenteric abscesses which show no tendency to spontaneous healing (Buchanan and Pine, 1962; Georg and Coleman, 1970; Abe et al., 1978).

The normal habitat of the organism is the human oral cavity. Thus, arachniae are facultative pathogens which invade the human body endogenously and are not transmissible (an exception is "punch actinomycosis" after human bites). Dental plaque or calculus are the materials from which arachniae can be recovered most easily either by immunofluorescence or by culturing (Slack et al., 1971; Collins et al., 1973; Holmberg and Forsum, 1973; Hill et al., 1977). Occasionally, *Arachnia propionica* has also been detected in cervicovaginal secretions of women using and not using intrauterine contraceptive devices (Pine et al., 1981) and in secretions from the uninfected conjunctiva and cornea (Jones and Robinson, 1977). It has not been isolated from animals (Slack and Gerencser, 1975).

Isolation and Enrichment Procedures

The following general purpose culture media may be used successfully for isolating *Arachnia propionica* from clinical specimens: fluid thioglycolate broth, possibly supplemented with 0.1–0.2% (w/v) sterile rabbit serum; brain heart infusion broth; trypticase soy broth; brain heart infusion agar; trypticase soy agar; heart infusion agar or trypticase soy agar with 5.0% (w/v) defibrinated rabbit, sheep or horse blood; and Schaedler broth or agar (Slack and Gerencser, 1975).

However, growth is not always satisfactory in or on these media. Isolation, enrichment and subcultivation can be improved by employing the complex semisynthetic media of Pine and Watson (1959) or Hein-

---

* HIK, Institute of Hygiene, Cologne, F.R.G.
† WVU, West Virginia University, U.S.A.

rich and Korth (1967) or by using Tarozzi broth. The CC medium* of Heinrich and Korth (1967) has proved especially reliable (Schaal and Pulverer, 1981).

No detailed study on the selective isolation of *Arachnia propionica* from clinical specimens or dental materials has been reported. The medium devised by Beighton and Colman (1976) for the isolation and enumeration of oral actinomycetes was found to inhibit growth of *A. propionica* ATCC 14157 because this strain proved to be sensitive to NaF used as one of the selective principles (see genus *Actinomyces*). However, the medium described by Kornman and Loesche (1978) for the selective isolation of *Actinomyces viscosus* and *Actinomyces naeslundii* from dental plaque which utilizes metronidazole and cadmium sulfate as inhibitors might also be applicable to the arachniae.

Anaerobic growth conditions can be obtained by using a Torbal anaerobic jar with a gas mixture of N_2-H_2-CO_2 (80:10:10%) or a GasPak jar with a H_2-CO_2-generating envelope. For routine use in the diagnostic laboratory, Fortner's method† (Fortner, 1928, 1929) has proved especially useful and reliable, because cultures can be checked for growth without disturbing the anaerobic environment (Schaal and Pulverer, 1981).

All cultures are incubated at 35–37°C and should be observed at 1, 2, 4, 7 and 14 days for growth of typical micro- or macrocolonies, respectively.

Maintenance Procedures

Cultures in routine use can be maintained by weekly transfer in thioglycolate broth containing 0.2% (v/v) rabbit serum or by monthly transfer on Fortner plates with CC medium.

For longer preservation (1 year), 3–5 ml brain heart infusion broth

in screw-capped tubes are inoculated with 0.1 ml broth culture and incubated at 35–37°C for 3–5 days. As soon as good growth is obtained, the caps are tightened and the tubes placed in freezer at −70°C.

For long term storage (10 years and longer), lyophilization should be applied. Heavy suspensions in sterile skim milk or lyophilization medium§ are prepared from the sediment of a broth culture or from the growth of several agar plates. Lyophilization is performed by standard techniques (Slack and Gerencser, 1975; Schaal and Pulverer, 1981).

Procedures for Testing for Special Characters

The methods for testing for morphological, biochemical and physiological characters are the same as described for the genus *Actinomyces* (Schaal, p. 1383). It should be noted, however, that the results of some of these tests are highly dependent on the individual technique used. Thus, if other methods or modifications are employed, these may give different results.

The analysis of end products from glucose fermentation by gas-liquid chromatography can be performed according to the procedures described in the VPI Anaerobe Laboratory Manual (Holdeman et al., 1977).

Demonstration and isomer determination of DAP in whole cell hydrolysates is reliably and most rapidly achieved by a modification of the technique of Becker et al. (1964) using cellulose-coated, thin layer plates (Kutzner, 1981).

The fluorescent-antibody techniques useful for identification of *Arachnia propionica* in culture as well as its demonstration and enumeration in pus specimens or dental plaque or calculus are also the same as those described for the genus *Actinomyces* (Schaal, p. 1383).

Differentiation of the genus **Arachnia** from other genera

Arachnia strains resemble certain *Actinomyces* species, especially *Actinomyces israelii*, and members of several other filamentous or diphtheroidal genera morphologically and/or physiologically. However, they can reliably be identified and differentiated from similar organisms by the characteristics given in Table 15.25. Furthermore, immunofluorescence can be used for differentiation because cross-reactions are usually absent or weak.

Taxonomic Comments

Originally, clinical *Arachnia* isolates had been identified as *Actinomyces israelii* (Pine and Hardin, 1959) or as serological and physiological variants of that species (Lentze, 1953). Further studies showed, however, that these organisms differed markedly from *Actinomyces israelii* in that they contained LL-DAP in their walls and produced propionic acid as a major end product of glucose fermentation. There-

fore, Buchanan and Pine (1962) proposed a new species of the genus *Actinomyces*, *Actinomyces propionicus*, to take these differences into account. Later, the definition of the genus *Actinomyces* was confined to those actinomycetes which have lysine as the major dibasic amino acid in their walls and do not produce propionic acid. As a consequence, the genus *Arachnia* was created (Pine and Georg, 1969).

The phylogenetic relationships of the genus *Arachnia* to other genera are uncertain and will probably remain so until modern molecular genetic techniques such as 16S-rRNA sequencing have been applied. As yet, an intermediate position of *Arachnia* between *Actinomyces* and *Propionibacterium* is usually assumed considering the similarities in morphology and pathogenicity to the former and the physiological and biochemical resemblances to the latter genus. Because of the propionic acid produced and the presence of DAP and glycine, the question has been raised (Pine, 1970; Allen and Linehan, 1977) as to whether or not

---

* *CC medium: I. Solution of minerals and trace elements*, containing per liter of distilled water (can be stored in refrigerator for months): $MgSO_4$·$7H_2O$, 20 g; $CaCl_2$·$2H_2O$, 2 g; $FeSO_4$·$7H_2O$, 400 mg; $MnSO_4$·$2H_2O$, 15 mg; $NaMoO_4$·$2H_2O$, 15 mg; $ZnSO_4$, 4 mg; $CuSO_4$·$5H_2O$, 0.4 mg; $CoCl_2$·$4H_2O$, 0.4 mg; boric acid, 20 mg; KI, 10 mg. The solution is acidified with 10 ml 10% HCl. *II. Vitamin solution*, containing per 100 ml distilled water (should be prepared freshly): thiamine-HCl, 20 mg; pyridoxine-HCl, 20 mg; biotin, 1 mg; folic acid, 5 mg; vitamin B_{12} (1 mg/100 ml), 1 ml; p-aminobenzoic acid, 20 mg; m-inositol, 20 mg; nicotinamide, 10 mg; nicotinic acid, 10 mg; Ca-pantothenate, 20 mg. *III. Solution of amino acids and vitamins*, containing per 100 ml of distilled water (should always be prepared freshly): casein hydrolysate, 12 g; yeast extract, 12 g; L-cysteine-HCl, 500 mg; L-asparagine, 30 mg; DL-tryptophan, 20 mg; solution II, 12 ml. This solution is sterilized by Seitz filtration. *Preparation of the medium*: KH_2PO_4, 4 g, is dissolved in 250 ml distilled water and adjusted to pH 7.6 with NaOH. Then, 10 ml of solution I, 500 mg of potato starch dissolved in 70 ml boiling distilled water, about 20 g agar (depending on quality), and distilled water are added to give a final volume of 900 ml. The mixture is autoclaved at 121°C for 15 min. After cooling to 50°C, solution III is added aseptically and the final pH adjusted to 7.3. The medium is poured in 15-ml amounts into glass Petri dishes.

† *Fortner's method*: only agar media in glass Petri dishes are suitable. One-half to two-thirds of the agar surface is inoculated with the clinical material to be examined or the strain to be subcultured. The remaining agar surface is heavily seeded with a culture of *Serratia marcescens* using a spatula. The dish is placed upside down upon a glass sheet of appropriate size and fixed and sealed with plasticine to make the system air-tight. During incubation, *Serratia* progressively removes O_2 and produces CO_2. Leakage in the plasticine seal which may occur when the plasticine is too brittle or the base plate damp, can easily be detected because pigment-producing strains of *Serratia marcescens* form unpigmented growth under reduced O_2 tension, but become red when the system is not completely closed.

§ *Lyophilization medium*: one part brain heart infusion broth plus one part horse serum plus sucrose to give a final concentration of 7% (w/v) in the whole mixture.

Table 15.25.
Characteristics differentiating the genus **Arachnia** *from other genera of fermentative, filamentous or diphtheroidal organisms[a,b]*

Characteristics	Arachnia	Actinomyces	Rothia	Bacterionema[c]	Bifidobacterium	Propionibacterium	Eubacterium	Lactobacillus	Arcanobacterium[d]	Erysipelothrix
Aerobic growth[e]	d	D	+	+	−	D	−	D	+	+
Microcolonies filamentous	+	D	d	+	−	−	−	−	−	−
Catalase	−	D[f]	+	+	−	D	−	−	−	−
Nitrate reduction	+	D	+	+	−	D	D	−	d	−
Cell wall components:										
DL-DAP	−	−	−	+	−	d	D	−	−	−
LL-DAP	+	−	−	−	−	+[g]	−	−	−	−
Mycolic acid	−	−	−	+	−	−	−	−	−	−
End products from glucose fermentation:[h]										
Acetic acid	+	+	+	+	+	+	+	d	+	+
Propionic acid	+	−	−	d	−	+	D	−	−	−
iso-Butyric acid	−	−	−	−	−	−	D	−	−	−
n-Butyric acid	−	−	−	−	−	−	D	−	−	−
iso-Valeric acid	−	−	−	−	−	d	D	−	−	−
n-Valeric acid	−	−	−	−	−	−	−	−	−	−
iso-Caproic acid	−	−	−	−	−	−	−	−	−	−
n-Caproic acid	−	−	−	−	−	−	D	−	−	−
Pyruvic acid	−	−	d	d	−	−	−	−	−	−
Lactic acid	+	+	+	+	+	+	+	+	+	+
Succinic acid	+	+	d	d	D	d	d	d	d	+
Mol% G + C of the DNA	63–65[i]	58–68[i]	65–69[i]	55–57[i]	57–64[j]	59–66[i]	30–40[i]	35–53[j]	50–52[i]	36[k]

[a] Data compiled from: Collins et al., 1982; Holdeman et al., 1977; Holländer and Pohl, 1980; Schaal, 1984; Schaal and Pulverer, 1981; Schofield and Schaal, 1981; Slack and Gerencser, 1975.

[b] Symbols: +, 90% or more of strains are positive; −, 90% or more of strains are negative; d, 11–89% of strains are positive; D, different reactions in different taxa (species of a genus or genera of a family) and DAP, diaminopimelic acid.

[c] *Bacterionema matruchotii* has recently been reclassified in the genus *Corynebacterium* (Collins, 1982).

[d] The new genus *Arcanobacterium* has been created to accommodate organisms formerly named *Corynebacterium haemolyticum* (Collins et al., 1982).

[e] On the surface of suitable agar media.

[f] *Actinomyces viscosus, A. howellii* and *A. hordeovulneris* are catalase-positive.

[g] *Propionibacterium lymphophilum* does not contain DAP.

[h] In peptone-yeast extract-glucose medium (Holdeman et al., 1977).

[i] T_m.

[j] T_m or Bd.

[k] Bd.

arachniae should be more rightfully included in the genus *Propionibacterium*. However, low DNA/DNA-homologies (Johnson and Cummins, 1972), little phenetic similarities (Schaal and Schofield, 1981; Schofield and Schaal, 1981) and differences in the detailed peptidoglycan structure (Schleifer and Kandler, 1972) would primarily not support a very close relationship between arachniae and propionibacteria.

The two serovars recognized within the species *Arachnia propionica* differ from one another in several physiological characteristics and were found to form two distinct subclusters in numerical phenetic analyses (Schaal and Schofield, 1981; Schofield and Schaal, 1981). Thus, they must at least be considered biovars as well. Furthermore, the type strain (serovar 1) showed no DNA homology with the reference strain of serovar 2 (Johnson and Cummins, 1972). These findings suggest that serovar 2 strains may represent a second, independent, species (Slack and Gerencser, 1975; Schaal and Schofield, 1984; Schofield and Schaal, 1981). The taxonomic status of the rare, orange-colored strains (third species?) cannot be assessed before more such isolates have been studied.

Further Reading

Bowden, G. and J. Hardie. 1978. Oral pleomorphic (coryneform) Gram-positive rods. *In* Bousfield and Callely (Editors), Coryneform Bacteria, Academic Press, London, pp. 235–263.

Schaal, K.P. 1984. Laboratory diagnosis of actinomycete diseases. *In* Goodfellow, Mordarski, and Williams (Editors), The Biology of the Actinomycetes, Academic Press, London, pp. 425–456.

Schaal, K.P. and B.L. Beaman. 1984. Clinical significance of actinomycetes. *In* Goodfellow, Mordarski, and Williams (Editors), The Biology of the Actinomycetes, Academic Press, London, pp. 389–424.

Schaal, K.P. and G. Pulverer. 1981. The genera *Actinomyces, Agromyces, Arachnia, Bacterionema* and *Rothia. In* Starr, Stolp, Trüper, Balows, and Schlegel (Editors), The Prokaryotes. A Handbook on Habitats, Isolation, and Identification of Bacteria, Springer-Verlag, Berlin, pp. 1923–1950.

Schaal, K.P., G.M. Schofield and G. Pulverer. 1980. Taxonomy and clinical significance of Actinomycetaceae and Propionibacteriaceae. Infection 8, (Suppl) 2: 122–130.

Schofield, G.M. and K.P. Schaal. 1981. A numerical taxonomic study of members of the Actinomycetaceae and related taxa. J. Gen. Microbiol. *127*: 237–259.

Slack, J.M. and M.A. Gerencser. 1975. Actinomyces, Filamentous Bacteria. Biology and Pathogenicity, Burgess Publishing, Minneapolis.

Identification and descriptive characteristics of *Arachnia propionica*

The basic characteristics necessary for reliable identification of *Arachnia propionica* are summarized in Table 15.26. Table 15.27 indicates some of the differential features which can be used to separate the two serovars (biovars) of *A. propionica* on physiological grounds. Further descriptive details are listed in Table 15.28.

Table 15.26.

Characteristics useful for identifying **Arachnia propionica**[a,b]

Characteristics	Result	Characteristics	Result
Cells:		*meso*-Inositol	d
Filamentous	d	Lactose	d
Diphtheroidal	d	Mannitol	+
Microcolonies filamentous	+	Maltose	+
Aerobic growth[c]	d	Raffinose	+
Catalase	−	Rhamnose	−
Nitrate reduction	+	Salicin	d
Nitrite reduction	−	Sorbitol	d
Hydrolysis of:		Sucrose	+
Esculin	−	Trehalose	d
Starch	d	Xylose	−
Gelatin	d	Ammonia from:[f]	
Acid from:[d]		Arginine	−
Arabinose	−[e]	Urea (urease)	−
Cellobiose	−	Indole production	−
Glucose	+	Esterase lipase (C8)[g]	+
Glycerol	d		

[a] Data compiled from: Slack and Gerencser, 1975; Holdeman et al., 1977; Schofield and Schaal, 1981; Schaal, 1984; Schaal and Schofield, 1984.
[b] Symbols: see Table 15.25.
[c] On the surface of agar media.
[d] Methods of Schofield and Schaal, 1979b, 1980a.
[e] Strains of serovar 1 may produce acid.
[f] Method of Schofield and Schaal, 1980b.
[g] API-zym test kit.

Table 15.27.

Characteristics useful for differentiating the two serovars of **Arachnia propionica**[a,b]

Characteristics	Serovar 1	Serovar 2	Characteristics	Serovar 1	Serovar 2
Acid from:[c]			Deamination of:[d]		
Adonitol	d	−	Serine	+	−
Arabinose	d	−	Alanine	+	−
Glycerol	d	−	Decarboxylation of:[d]		
Sorbitol	+	−	Ornithine	d	−

[a] Data compiled from: Schofield and Schaal, 1980a, b, 1981.
[b] Symbols: see Table 15.25.
[c] Methods of Schofield and Schaal, 1979b, 1980a.
[d] Methods of Schofield and Schaal, 1980b.

Table 15.28.

Other characteristics of **Arachnia propionica**[a,b]

Characteristics	Result	Characteristics	Result
Cell morphology:[c]		Gram-reaction variable	−
Coccoid elements only	−	Motility	−
Cocco-bacillary elements	d	Colony morphology:[d]	
Branched rods	d	Size <2 mm	+
Rods with swollen ends	d	>2 mm	−
Square-ended ends	−	Flat	−
Round-ended ends	+	Raised	+
Large spherical bodies	d	Umbonate	d
Filaments	d	Entire margin	d
Gram-reaction positive	+	Undulate margin	d

Table 15.28— *Continued*

Characteristics	Result	Characteristics	Result
Transparent	−	Ornithine	d
Opaque	+	API enzyme tests:[h]	
White	d	Alkaline phosphatase	−
Orange	−[e]	Esterase (C4)	+
Rough	d	Lipase (C14)	−
Smooth	d	Leucine arylamidase	d
Mucoid	−	Valine arylamidase	−
Aerobic growth in surface culture:		Cystine arylamidase	−
Weak	d	Trypsin	−
Fair	d	Chymotrypsin	−
Good	−	Acid phosphatase	−
Anaerobic growth:		Phosphoamidase	−
With CO_2	+	α-Galactosidase	−
Without CO_2	+	β-Galactosidase	−
Growth at:		β-Glucuronidase	−
$36 \pm 1°C$	+	α-Glucosidase	+
45°C	d	β-Glucosidase	−
Growth on Rogosa's agar	−	N-acetyl-β-glucosaminidase	−
Cytochrome oxidase	−	α-Mannosidase	−
Voges-Proskauer test	−	α-Fucosidase	−
Methyl red test	d	β-Xylosidase	−
H_2S production:		Lysis by lysozyme + SDS (24 h)[i]	+
On BHI gar	−	Alkali produced in peptone-containing media	−
On TSI agar	d	Acid from:[i]	
Hydrolysis of:[f]		Adonitol	d
Adenine	−	Amygdalin	d
Casein	−	Cellulose	−
Guanine	−	Dextrin	d
Hippurate	−	Dulcitol	−
Hypoxanthine	−	iso-Erythritol	−
Lecithin	−	meso-Erythritol	d
Tween 20	−	Fructose	+
Tween 40	−	Galactose	d
Tween 60	−	Glycogen	−
Tween 80	−	Inulin	−
Tyrosine	−	Mannose	d
Xanthine	−	Melezitose	−
Meat digestion	−	Melibiose	d
Milk:		Ribose	d
Acid, reduction	d	Sorbose	−
Acid, clot	d	Starch	d
Peptonization	−	α-Methyl-D-glucoside	−
No change	d	α-Methyl-D-mannoside	−
Lipase (egg yolk)	d	Growth on (sole carbon source):[j]	
DNase	−	Glucose	+
Hyaluronidase	−	Glycerol	−
Chondroitin sulfatase	−	Sorbitol	d
Deamination of:[g]		Xylose	−
Alanine	d	Lactate	+
Aspartic acid	d	Pyruvate	+
Glutamic acid	d	Succinate	−
Leucine	−	Alanine	+
Lysine	−	α-Hemolysis:[k]	
Methionine	−	Human blood	−
Ornithine	−	Sheep blood	−
Serine	d	β-Hemolysis:[k]	
Decarboxylation of:[g]		Human blood	+
Aspartic acid	−	Sheep blood	−
Glutamine	−	Horse blood	d
Leucine	−	Growth in the presence of:[l]	
Lysine	−	NaCl 2% (w/v)	d
Methionine	−	NaCl 4% (w/v)	−

Characteristics	Result	Characteristics	Result
NaCl 6% (w/v)	−	Sodium selenite 0.01% (w/v)	d
Bile 5% (w/v)	+	Potassium tellurite 0.01% (w/v)	d
Bile 10% (w/v)	d	Sodium azide 0.005% (w/v)	d
Bile 20% (w/v)	−	Crystal violet 0.005% (w/v)	d
Sodium taurocholate 0.2% (w/v)	d		

[a] Data compiled from: Kilian, 1978; Holdeman et al., 1977; Holmberg and Nord, 1975; Schofield and Schaal, 1979b, 1980a, 1980b, 1981; Slack and Gerencser, 1975.

[b] Symbols: see Table 15.25; also TSI, triple sugar iron; and SDS, sodium dodecyl sulfate.

[c] When grown in thioglycolate broth.

[d] On brain heart infusion agar.

[e] The taxonomic status of strains forming dull orange colonies remains to be clarified.

[f] Methods of Schofield and Schaal, 1981.

[g] Methods of Schofield and Schaal, 1980b.

[h] Only strongly positive reactions recorded as positive.

[i] Methods of Schofield and Schaal, 1981.

[j] Methods of Schofield and Schaal, 1979a.

[k] On blood-containing BHI agar.

[l] On BHI-agar.

List of species of the genus *Arachnia*

1. **Arachnia propionica** (Buchanan and Pine 1962) Pine and Georg 1969, 269.[AL] (*Actinomyces propionicus* Buchanan and Pine 1962, 305.) pro.pi.o′ni.ca. M.L. fem. adj. *propionica* pertaining to propionic acid.

Cell and colonial morphology, ultrastructure, physiology and nutrition, and serology are as described for the genus. Differential and other descriptive characteristics are given in Tables 15.26–15.28.

The normal habitat is the human oral cavity, but the organism has also been detected in cervicovaginal secretions of healthy women and in secretions from the uninfected conjunctiva and cornea. *A. propionica* apparently does not occur in animals.

It causes human actinomycosis and lacrimal canaliculitis, and produces abscesses in laboratory mice when injected intraperitoneally.

The mol% G + C of the DNA is 63–65 (T_m).

Type strain: ATCC 14157 (ATCC 13682, WVU 471 (serovar 1)).

Reference strain for serovar 2: ATCC 29326 (WVU 346).

Genus *Rothia* Georg and Brown 1967, 68[AL]

MARY ANN GERENCSER AND GEORGE H. BOWDEN

Roth′ia, M.L. fem. n *Rothia*: named for Dr. Genevieve D. Roth, who performed basic studies with these organisms.

Coccoid, diphtheroid or **filamentous** cells (Fig. 15.17), **usually 1.0 μm in diameter**. Irregular swellings and clubbed ends up to 5.0 μm in diameter may be present. Growth may be exclusively coccoid, diphtheroid or filamentous or a mixture of these forms. No capsule formed, nonmotile, **Gram-positive**, not acid-fast.

Mature colonies (4–7 days) 2–6 mm in diameter, creamy white, smooth or rough, usually soft in texture but may be dry and crumbly or mucoid (Fig. 15.18). Optimum temperature 35–37°C. **Catalase-positive. Chemoorganotroph.** Ferments carbohydrates. **Major product of glucose fermentation is lactic acid. Does not produce propionic acid.** Cell wall peptidoglycan contains **alanine, glutamic acid** and **lysine** but no **diaminopimelic acid** (DAP). **Cell wall sugars** include **galactose, glucose** and **fructose** but not **6-deoxytalose** or **arabinose**. The mol% G + C of the DNA is 47–53. Type species: *Rothia dentocariosa* (Onisi, 1949) Georg and Brown 1967, 68.

Further Descriptive Information

Growth may consist entirely of coccoid, diphtheroidal or filamentous forms, but mixtures of these cell types are more common. Filamentous forms are seen in cultures grown on solid media while coccoid forms are more common in broth cultures. Cells in 2- to 3-day-old broth cultures may be entirely coccoid, but it is difficult to maintain a culture in the coccoid form.

The cell wall peptidoglycan contains the amino acids alanine, glutamic acid and lysine. The peptidoglycan belongs to the L-lysine-L-alanine type A3a of Schleifer and Kandler (1972).

The major cell wall sugar is galactose (Hammond, 1972) with smaller amounts of glucose, fructose, and ribose but no 6-deoxytalose (Hammond et al., 1973) or arabinose.

Dimannosyl diglyceride (Pandhi and Hammond, 1975) is a major lipid component of the cell envelopes of both coccal and filamentous cells.

A distinct capsule or slime layer is not evident in wet mounts of *Rothia* cultures, but one electron microscope study reported their presence on the cell surface (Roth et al., 1976). Neither capsules nor slime were seen in a more recent study (Lai and Listgarten, 1980). Lai and Listgarten (1980) reported that the cell wall is seen as a moderately electron-dense layer, 19–32 nm thick. Multiple cleavage planes parallel to each other and to the cell surface gave the impression of a multilayered cell wall. Markedly thickened segments of cell wall material were also seen. No periplasmic space was visible. When grown on blood agar, the contour of the wall was rough. Neither long nor short surface fuzz or other structures were present outside the cell wall. Mesosomes or intracellular membrane and electron-dense cytoplasmic inclusions were present. Nucleoids were seen in cells grown in broth but not in agar. No variation in the wall structure at the ultrastructural level was found in the various morphological types.

Young colonies (18–24 h) grown anaerobically are microscopic and usually highly filamentous, frequently resembling the "spider" colonies of *Actinomyces*. Young colonies grown aerobically are larger (±1 mm), smooth or granular and may or may not have a fringed border.

Mature colonies (4–7 days) may be smooth and convex or have rough, highly convoluted (folded) (Fig. 15.18) but glistening surfaces. Both smooth and rough colonies may be seen on the same illustration. The texture of the colonies may be creamy, dry and crumbly or mucoid.

Rothia grows well on enriched media such as brain heart infusion (BHI) and trypticase soy (TS) and on simpler medium such as nutrient broth. The best growth is obtained with aerobic incubation at 35–37°C, but some growth is obtained in an anaerobe jar with $N_2:H_2:CO_2$ (80:10:10). Little or no growth is obtained on media with inorganic

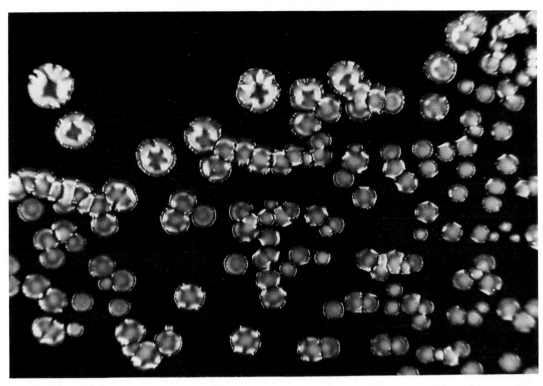

Figure 15.17. Gram-stained smear of *R. dentocariosa* taken from a colony after 48 h aerobic growth on blood agar (× 1200).

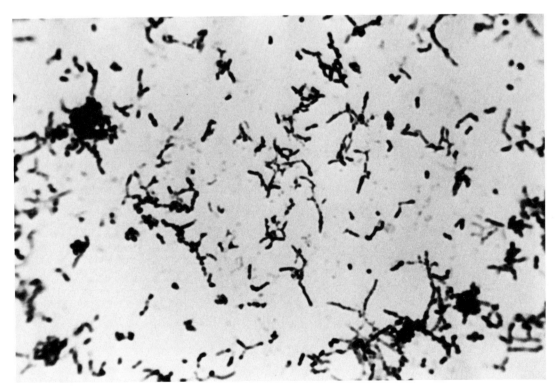

Figure 15.18. Colonies of *R. dentocariosa* after 48 h aerobic growth on blood agar (× 2.5).

nitrogen such as Czapek. Citrate cannot serve as a sole source of carbon.

Rothia has a fermentative metabolism producing mainly lactic and acetic acid from glucose. The lactate dehydrogenase (LDH) produced by *Rothia* is activated by fructose 1,6-diphosphate and inhibited by adenosine 5'-triphosphate and thus resembles the LDH of *A. viscosus* (Eisenberg et al., 1976).

Rothia is highly sensitive to most antibiotics but multiple resistant strains occasionally occur. Dzier-Zanowska et al. (1978) tested 90 strains of *Rothia* against 18 antibiotics. The strains were sensitive

(MIC 0.06–0.26 μg/ml) to cephaloridine, amoxycillin, doxycycline, erythromycin, ampicillin, chlortetracycline, cafazolin, benzylpenicillin, cefamanadole, methacycline, cephalothin, flucloxacillin, cefalaxin, cloxacillin, and carbenicillin. They were resistant to clindamycin and highly resistant to lincomycin.

Rothia is an opportunistic pathogen. It has been isolated from abdominal infection (Scharfen, 1975); from infectious endocarditis (Schafer et al., 1979; Pope et al., 1979) and from a variety of other infections (Blevins et al., 1973; Lutwick and Rockhill, 1978).

Abcess formation can be demonstrated experimentally in mice (Roth and Flannagan, 1969; Scharfen, 1975; Miksza-Zylkiewicz and Linda, 1980).

Rothia is a normal inhabitant of the human oral cavity. It is present in saliva and is isolated most frequently from supragingival dental plaque.

Isolation and Enrichment Procedures

For primary isolation and maintenance, enriched media such as BHI or TS are recommended. Incubation should be aerobic at 37°C for 2–7 days.

Three selective media have been described for the isolation of *Rothia* (Ritz, 1966, Beighton, 1976, Beighton and Colman, 1976). An extensive comparative evaluation of these media has not been done.

Maintenance Procedures

Cultures on slants or in deep agar butts of BHI agar or TS agar remain viable 3–4 months at room temperature. For long term storage, lyophilization of suspensions in milk is satisfactory.

Procedures for Testing for Special Characters

Microcolonies of *Rothia* are best demonstrated by dropping a sterile coverslip over a portion of the inoculum on a BHI agar plate and incubating the plate aerobically at 35°C for 18–24 h. Filamentous microcolonies will be seen in growth under the coverslip and smooth ones in growth outside the coverslip (Roth and Thurn, 1962).

In general, methods described for testing biochemical characteristics of *Actinomyces* species are suitable for *Rothia* except that aerobic incubation is important. The catalase test is best performed by flooding aerobic growth on slants with 3.0% H_2O_2 and observing a stream of bubbles. Anaerobic cultures must be exposed to air 15–30 min before testing.

A satisfactory fermentation base medium is meat extract 3 g, peptone 10 g, NaCl 5 g, Andrade's indicator 10 ml, distilled water 1000 ml. The Hugh and Leifson test (1953) is useful in distinguishing *Rothia* from *Nocardia* which oxidizes or does not attack glucose in this medium. Liquefaction of gelatin in deep tubes of nutrient gelatin is difficult to demonstrate and gives inconsistent results. Better results are obtained by inoculating plates of gelatin agar (BHI broth + 15 g agar + 4 g gelatin/liter). After 7 days incubation, the plates are flooded with mercuric chloride and observed for clear zones around the growth. For bulk cultures for chemotaxonomic tests, *Rothia* is best grown in trypticase soy broth or nutrient broth on a shaker.

Differentiation of the genus **Rothia** from closely related taxa

Rothia can be differentiated from *Nocardia* by its ability to ferment carbohydrates, its failure to grow with an inorganic nitrogen source and by cell wall composition.

Rothia is differentiated from *Actinomyces*, especially *A. viscosus*, by a combination of colonial and cellular morphological features and biochemical characteristics. *Rothia* does not produce succinic acid as a major product of glucose fermentation as do *Actinomyces* species and does not have ornithine in the cell wall peptidoglycan as do all species of *Actinomyces* except *A. bovis*.

Rothia may be distinguished from *Bacterionema* by the distinct cellular morphology of the latter. In addition, *Bacterionema* cell walls contain arabinose and DL-DAP and this genus produces propionic acid as an end product of glucose metabolism.

Rothia may be separated from similar genera serologically (fluorescent antibody reactions) since it has few cross-reactions. Weak reactions are sometimes seen with strains of *Actinomyces*, *Lactobacillus*, and *Bacterionema*. The major drawback to serological identification is that not all *Rothia*-like isolates react with the currently available battery of FITC-conjugated antisera.

Taxonomic Comments

The genus *Rothia* was created to accommodate organisms originally called *Nocardia* (Roth, 1957; Roth and Thurn, 1962; Davis and Freer, 1960) which are normal inhabitants of the mouth and throat. *Rothia* resembles *Actinomyces* morphologically but grows better aerobically. Stimulation of growth by CO_2 has not been reported except in a study by Schofield and Schaal (1981).

Rothia differs significantly from *Nocardia* in having a fermentative metabolism, in failure to grow on inorganic nitrogen sources and in its cell wall composition. *Nocardia* cell walls typically contain DAP and arabinose, both of which are absent in *Rothia*.

Rothia more closely resembles *Actinomyces* than *Nocardia* in morphology. It differs in being aerobic and in the major end products of glucose metabolism. The L-Lys-L-Ala₃ type cell wall has not been found in other *Actinomycetales* (Schleifer and Kandler, 1972). Sodium dodecyl sulfate (SDS) polyacrylamide gel electrophoresis shows little or no relationship between protein patterns of *Rothia* and *Actinomyces*.

At present the genus contains a single species, *R. dentocariosa*, but *Rothia*-like organisms which probably constitute additional species are being studied.

Further Reading

Bowden, G. and J. Hardie. 1978. Oral pleomorphic (coryneform) Gram-positive rods. *In* Bousfield and Callely (Editors). Coryneform Bacteria, Academic Press, London, pp. 235–263.

Schaal, K.P, and G. Pulverer. 1981. The genera *Actinomyces*, *Agromyces*, *Arachnia*, *Bacterionema*, and *Rothia*. *In* Starr, Stolp, Trüper, Balows and Schlegel (Editors), The Prokaryotes, A Handbook on Habitats, Isolation and Identification of Bacteria, Springer-Verlag, New York, pp. 1923–1950.

Schaal, K.P., G.M. Schofield and G. Pulverer. 1980. Taxonomy and clinical significance of *Actinomycetaceae* and *Propionibacteriaceae*. Infection 8, (Suppl) 2: 122–130.

Slack, J.M. and M.A. Gerencser. 1975. Actinomyces, Filamentous Bacteria. Biology and Pathogenicity. Burgess Publishing, Minneapolis.

Differentiation of the species of the genus **Rothia**

The characteristics of *Rothia dentocariosa* are given in Tables 15.29 and 15.30.

Table 15.29.
Differential characteristics of **Rothia dentocariosa**[a]

Test	Result	Test	Result
Oxygen requirements	Aerobic	Catalase	+
O-F test (Hugh and Leifson)	Fermentative	Indole	−
End products of glucose		Nitrate reduction	+
fermentation:		Nitrite reduction[b]	+
Acetic	+	Acid from:	
Propionic	−	Glucose	+
Butyric	−	Glycerol	d[c]
Isovaleric	−	Lactose	d
Caproic	−	Raffinose	−[d]
Pyruvic	v (trace)	Sucrose	+
Lactic	+	Trehalose	+
Succinic	v (trace)	Xylose	−
Cell wall	Lysine galactose		

[a] Symbols: see Table 15.25; also v, strain instability (*not* equivalent to "d").

[b] Concentration of nitrate should be 0.001%.

[c] Positive (100%) with prolonged incubation of 21 days.

[d] One source (Miksza-Zykiewicz, 1980) reports positive raffinose reactions.

Table 15.30.
Additional characteristics of **Rothia dentocariosa**[a,b]

Test	Result	Test	Result
Urease	−	Growth in presence of:	
Hydrolysis of:		NaCl 2%	+
Esculin	+	NaCl 4%	d
Starch	−	NaCl 6%	−
Gelatin	d	Bile 5%	+
Casein	−	Bile 10%	+
Hydrogen sulfide over triple	+	Bile 20%	d
sugar iron agar		Growth in presence of:	
Methyl red	v	0.01% Sodium selenite	+
Voges-Proskauer	v	0.01% Potassium tellurite	+
Acid production from:		0.005% Sodium azide	+
Adonitol	−	0.05% Crystal violet	−
Arabinose	d	0.2% Sodium taurocholate	+
Amygdalin	−	Enzyme production:	
Cellobiose	−	Alkaline phosphatase	−
Dulcitol	−	Esterase (C$_4$)	d
Dextrin	d	Esterase lipase (C$_8$)	d
Erythritol	−	Lipase (C$_{14}$)	−
Fructose	+	Leucine aminopeptidase	+
Galactose	d	Valine aminopeptidase	d
α-Methyl-d-glucoside	+	Cystine aminopeptidase	v
Glycogen	−	Trypsin	+
Inositol	−	Chymotrypsin	−
Inulin	−	Acid phosphatase	d
Maltose	+	Phosphoamidase	−
Mannitol	−	α-Galactosidase	−
Mannose	d	β-Galactosidase	−
α-Methyl-d-mannoside	−	β-Glucuronidase	−
Melezitose	d	α-Glucosidase	+
Melibiose	−	β-Glucosidase	d
Rhamnose	d	N-Acetyl β-glucosaminidase	d
Ribose	d	α-Mannosidase	−
Salicin	+	α-Mannosidase	−
Sorbitol	−	β-Xylosidase	−
Starch	−		

[a] Data from Holmberg and Hallender, 1973; Slack and Gerencser, 1975; Killian, 1978; Miksra-Zylkiewicz, 1980; Schofield and Schaal, 1981.

[b] Symbols: see Table 15.25; also v, strain instability (*not* equivalent to "d").

List of species of the genus **Rothia**

1. Rothia dentocariosa (Onishi 1949) Georg and Brown 1967, 68.[AL] (*Actinomyces dentocariosus* Onishi 1949, 282.)

den.to.car.i.o′sa. L. n. *dens, dentis* tooth; L. adj. *cariosus* decayed or decaying; M.L. fem. adj. *dentocariosa* tooth-decaying.

See Tables 15.29 and 15.30 and the generic descriptions for many features.

The cell and colonial morphology and ultrastructural features of *Rothia dentocariosa* are those described for the genus. The biochemical reactions are shown in Table 15.30. Early studies of *R. dentocariosa* indicated that this species has remarkably homogeneous biochemical reactions (Brown et al. 1969). Later studies indicated more variability in reactions of *Rothia* and *Rothia*-like isolates (Lesher et al., 1974) although separation into biovars seems to serve little purpose.

Amdur et al. (1978) found that strains labeled *N. salivae* and *R. dentocariosa* had a characteristic fatty acid profile which was different from that of similar organisms studied. The *Rothia* group was characterized by the presence of *iso-* even-numbered fatty acids and only *antiso-* odd-numbered fatty acids. Specifically *ante-iso* 15:0; *iso*-16; 16:0 and *ante*-17:0 fatty acids were present with only trace amounts of other fatty acids.

On the basis of SDS-polyacrylamide gel electrophoresis of soluble proteins, 26 *Rothia* strains could be divided into three groups (P.G. Fotos and M.A. Gerencser, unpublished data). Two of these groups contain strains of *R. dentocariosa* while the third contained strains

belonging to serovar 3 (see serotyping). This gives additional support to the concept of a second species of *Rothia*.

Serotyping

Antiserum made from killed whole cell antigens of *Rothia* can be used for serotyping, using immunofluorescence. Using this technique *Rothia* can be separated into at least three serovars. At present this serotyping scheme is not completely satisfactory. Some isolates stain well with unadsorbed FITC conjugated antiserum for both serovars 1 and 2, but fail to stain with serovar-specific (adsorbed) antiserum for either type. In addition, some isolates fail to react with any of the available antisera (Lesher et al., 1974). This suggests that additional serovars need to be recognized.

Strains belonging to serovars 1 and 2 can be placed in the species *R. dentocariosa*. Strains belonging to serovar 3 show limited reactions with serovars 1 and 2 and differ in other ways from typical *R. dentocariosa*. These strains may represent a second species.

Hammond (1970) isolated a soluble polysaccharide antigen (RPS) from the cell wall which contained fructose, glucose, galactose, and ribose, with fructose as the major antigenic determinant. This antigen was found in all strains of *Rothia* tested and may account for the cross-reactions seen between serovars.

Type strain: ATCC 17931.

Genus **Propionibacterium** *Orla-Jensen 1909, 337* [AL]

C. S. CUMMINS AND JOHN L. JOHNSON

Pro.pi.on.i.bac.te′ri.um. M.L. n. *acidum propionicum* propionic aid; Gr. dim. n. *bakterion* a small rod; M.L. neut. n. *Propionibacterium* propionic (acid) bacterium.

Pleomorphic rods, 0.5–0.8 μ in diameter × 1–5 μ in length, often diphtheroid or club-shaped with one end rounded and the other tapered or pointed: **however, cells may be coccoid, bifid or even branched.** Cells may occur singly, in pairs or short chains, in V or Y configurations, or in clumps with "Chinese character" arrangement. **Gram-positive, nonmotile, nonsporing chemoorganotrophs: fermentation products include large amounts of propionic and acetic acids,** and generally lesser amounts of *iso*-valeric, formic, succinic or lactic acids and carbon dioxide. **Anaerobic to aerotolerant, generally catalase-positive.** Growth most rapid at 30–37°C and may be white, gray, pink, red, yellow or orange in color.

G + C content of the DNA is from 53–67 mol% (T_m).

Type species: *Propionibacterium freudenreichii* van Niel 1928, 162.

Further Descriptive Information

Two principal groups of organisms from different habitats are described in the genus *Propionibacterium*.

1. Strains from cheese and dairy products. It was through an examination of such organisms that Orla-Jensen in 1909 originally described *Propionibacterium* (von Freudenreich and Orla-Jensen, 1906; Orla-Jensen, 1909) and it seems appropriate, therefore, to refer to them as "classical propionibacteria," or "dairy propionibacteria." They have also been found in other natural fermentations, e.g. silage and fermenting olives (Cancho et al., 1970, 1980; Plastourgos and Vaughn, 1957) and from soil (van Niel, 1928).

2. Strains typically found on human skin, although occurring also elsewhere, e.g. the intestine. The acne bacillus, originally described on morphological grounds as *Corynebacterium acnes*, is the exemplar of this group. These may be referred to as "acnes group strains" or "cutaneous propionibacteria" although they occur elsewhere than on the skin. They have also frequently been referred to, especially in medical and immunological literature, as "anaerobic coryneforms" or "anaerobic diphtheroids" (see e.g. *Corynebacterium parvum*, referred to below).

These two groups of strains (the "classical propionibacteria" and the "acnes group") are here described separately. The acnes group strains were formerly (up to the seventh edition of the *Manual*) described in *Corynebacterium*. They were transferred from that genus to *Propionibacterium* in the eighth edition essentially because (a) they were anaerobic, (b) they produce propionic acid as a major end product of metabolism, (c) they mostly contain L-diaminopimelic acid (L-DAP) as the diamino acid of peptidoglycan, (d) they produce C_{15} *iso*- and *anteiso*-acids as the principal fatty acids of cell lipids, and (e) they lack mycolic acids and the arabinogalactan characteristic of *Corynebacterium* sensu stricto. However, Prévot (1976) has proposed that the *acnes* group organisms, because of their pathogenic and reticulostimulatory properties, should not be transferred to *Propionibacterium* but should instead be accommodated in a separate subgenus *Coryneformis* in the family *Corynebacteriaceae*.

The choice of *P. freudenreichii* as the type species of *Propionibacterium* is perhaps rather anomalous in the light of more recent information. It seems in fact to be somewhat atypical in that (a) it has *meso*-DAP instead of the L-isomer found in other strains, (b) it has a cell wall polysaccharide containing rhamnose and (c) it is more distantly related to other species by DNA/DNA homology (Johnson and Cummins, 1972).

We have kept much of the genus and species descriptions of Moore and Holdeman as given in the eighth edition of the *Manual*, but have added to them in the light of new information, and have arranged the material somewhat differently. There is no general key, but instead there is one table (Table 15.31) for the classical propionibacteria, and another (Table 15.32) for the cutaneous propionibacteria, incorporating in each case what appears to be the most reliable tests for differentiation. In addition there is a more general table (Table 15.33) which compares important characteristics of both groups. Other reported characteristics which appear to be less important for differentiation, are collected in Table 15.34.

Morphologically, strains in both groups are irregularly staining,

Table 15.31.
Differentiation of classical species of the genus **Propionibacterium**[a]

Organism	Fermentation of sucrose and maltose	Reduction of nitrate	β-Hemolysis	Color of pigment	Isomer of DAP in cell wall
P. freudenreichii	−	d	−	Cream	meso-
P. jensenii	+	−	−	Cream	L-
P. thoenii	+	−	+	Red-brown	L-
P. acidipropionici	+	+	−	Cream to orange-yellow	L-

[a] Symbols: see Table 15.25; also DAP, diaminopimelic acid.

Table 15.32.
Principal characteristics of cutaneous species of the genus **Propionibacterium**[a, b]

Character	P. acnes Biovar I	P. acnes Biovar II	P. avidum Biovar I	P. avidum Biovar II	P. granulosum
Fermentation of:					
Glucose	+	+	+	+	+
Sorbitol	d	−	−	−	−
Sucrose	−	−	+	+	+
Maltose	−	−	+	+	+
Esculin hydrolysis	−	−	+	+	−
Biochemical tests:					
Gelatin liquefaction	+	+	+	+	−
Indole production	+	+	−	−	−
Nitrate reduction	+	+	−	−	−
β-Hemolysis of rabbit blood (5 days/37°C)	68% +	−	+	+	−[c]
Cell wall composition:					
DAP isomer	L-	L-(occasionally meso)	L-	L-(occasionally meso)	L-
Sugars in					
Cell wall	Galactose	Glucose	Galactose	Glucose	Galactose
Polysaccharide	Glucose Mannose	Mannose	Glucose Mannose	Mannose	Mannose

[a] The characteristics of *P. lymphophilum* are given in Tables 15.33 and 15.34. The results in this table are based on an examination of numerous strains in this laboratory. We have found that occasional strains of *P. granulosum* (~5%) give positive results in tests for indole production, gelatin liquefaction and nitrate reduction. A few of these anomalies may have been due to unrecognized contamination of the cultures with *P. acnes*. Kishishita et al. (1980) found all of 447 strains of *P. acnes* isolated from 30 individuals to be positive for these three tests, and all of 86 strains of *P. granulosum* to be negative.
[b] Symbols: see Table 15.25.
[c] Hoeffler (1977) reported all of 18 strains of *P. granulosum* to be β-hemolytic in rabbit blood: the reason for this discrepancy is not known.

Table 15.33.
Differential characteristics of species of the genus **Propionibacterium**[a]

Organism	Acid from Sucrose	Acid from Maltose	Esculin hydrolysis	Production of indole	Reduction of nitrate	Hydrolysis of gelatin	Hemolysis	DAP isomer in cell wall	DNA, G + C mol%
P. acnes	−	−	−	+	+	+	d	L; occasionally meso-	59 ± 1.49
P. avidum	+	+	+	−	−	+	+	L; occasionally meso-	62 ± 0.49
P. granulosum	+	+	−	−	−	−	−	L	62 ± 1.00
P. freudenreichii	−	−	+	−	d	−	−	meso-	65 ± 0.97
P. jensenii	+	+	+	−	−	−	−	L	67 ± 1.07
P. thoenii	+	+	+	−	−	−	+	L	66 ± 0.50
P. acidipropionici	+	+	+	−	+	−	−	L	67 ± 0.84
P. lymphophilum	d	+	−	−	d	−	−	No DAP: cell walls contain lysine	53 ± 0.71

[a] Symbols: see Table 15.25.

Gram-positive, nonmotile, nonsporing rods, but in general the classical propionibacteria tend to be shorter and rather thicker, although all strains may be very variable in morphology, especially in early log phase cultures. Strains of *P. acnes* in particular give long slender irregular rods in young cultures, much resembling the classical description of *C. diphtheriae mitis,* and making it easy to understand why the organism was called *Corynebacterium acnes.* In older (postlog phase) cultures all strains tend to be more coccal. Some strains of the

Table 15.34.

Additional information on fermentations and other metabolic reactions of species of the genus **Propionibacterium**[a, b]

Reaction	1. P. freuden-reichii	2. P. jensenii	3. P. thoenii	4. P. acidi-propionici	5. P. acnes	6. P. avidum	7. P. granulosum	8. P. lympho-philum
Acid produced from:								
Adonitol	d+	d+	d+	+	d+	d+	−	+
Amygdalin	−	d+	d+	−	−	−	d−	−
Arabinose	+	−	−	+	−	d+	−	−
Cellobiose	−	d−	−	+	−	−	−	−
Dulcitol	−	−	−	−	−	−	−	−
Erythritol	+	+	d+	+	d+	+	−	+
Esculin	−	−	+	d+	−	−	−	−
Esculin hydrolysis	+	+	+	+	−	+	−	−
Fructose	+	+	+	+	d+	+	+	+
Galactose	+	+	+	+	d+	+	d−	−
Glucose	+	+	+	+	sk+	+	+	+
Glycerol	+	+	+	+	d+	+	+	−
Glycogen	−	−	d+	−	−	−	−	−
Inositol	d+	d+	d+	+	d−	d+	−	d+
Inulin	−	−	−	−	−	−	−	−
Lactose	d−	d+	d−	+	−	d+	−	−
Maltose	−	d+	d+	+	−	+	d+	+
Mannitol	−	+	−	+	d−	d−	d+	−
Mannose	+	+	+	+	d+	+	+	−
Melezitose	−	d+	d+	+	−	d+	d−	−
Melibiose	d−	+	d+	d+	−	d+	d−	−
Raffinose	−	+	d+	d−	−	d+	d+	−
Rhamnose	−	−	−	+	−	−	−	−
Ribose	d+	+	+	+	d+	d+	d−	+
Salicin	−	+	d+	+	−	d+	−	−
Sorbitol	−	−	d+	+	d+	−	−	−
Sorbose	−	−	−	d+	−	−	−	−
Starch	−	−	+	+	−	−	−	d+
Starch hydrolosis	−	−	d+	−	−	−	−	d−
Sucrose	−	+	d+	+	−	+	+	d−
Trehalose	−	+	+	+	−	+	d+	−
Xylose	−	d+	d+	d+	−	d−	−	−
Gelatin	−	−	−	−	+	+	d−	d+
Milk:								
Curd	d+	d−	d−	d−	d+	+	+	−
Digestion	−	d−	−	−	d+	+	−	−
Indole-produced	−	−	−	−	d+	−	−	−
Nitrate-reduced	−	−	−	+	d+	−	−	d−
Catalase	+	d+	+	d+	d+	+	+	d+
Gas	−, 1	−, 1	−, 1	1, 2	−, 2	−, 1	−, 1	−, 1
Acetoin	−	−	−	d+	−	−	−	−
Growth in 20% bile	2, 1	d+	−, 1	4	2, 1	d+	−, 1	−, 1

[a] Symbols: +, positive reaction in 90–100% of strains, *or* pH below 5.7 in 90–100% of strains; −, negative reaction in 90–100% of strains, *or* pH 5.7 or above in 90–100% of strains; d−, reaction positive in 10–40% of strains; d+, reaction positive in 40–90% of strains. Numbers refer to amount of gas produced (gas) or amount of growth (bile) on a 1–4 scale.

[b] Reactions determined by procedures given in Holdeman et al., 1977.

classical propionibacteria are definitely capsulated in wet mount India ink preparations (e.g. *P. thoenii*, Skogen et al., 1974; *P. jensenii*, Cummins, unpublished), and it seems that a considerable number of strains of all species may produce extracellular slime not organized in the form of clear-cut individual capsules. The extracellular material is carbohydrate in nature (Skogen et al., 1974).

The cell walls of strains in the genus typically contain peptidoglycan in which L-DAP is the diamino acid, and a polysaccharide consisting of hexosamines and some combination of glucose, galactose and mannose. However, significant variations may occur in some species (or in some strains in some species) as shown in Table 15.35. A detailed analysis of peptidoglycan composition has been undertaken in only a few instances (Schleifer et al., 1968; Allsop and Work, 1963; Kandler,

personal communication), from which it appears that in strains with the L-isomer of DAP, glycine is a bridging amino acid, between DAP and the D-alanine of the adjacent chain. However, it has been suggested (Kamisango et al., 1982) that in *P. acnes*, at least, glycine, although attached to DAP, does not take part in a cross-link.

The cell lipids of both classical and cutaneous strains are characterized by the presence of large amounts of C_{15} branched-chain fatty acids (Moss et al., 1969). Mannose-containing phospholipids have been described in "*P. shermanii*" strains (Prottey and Ballou, 1968), but the mycolic acids so characteristic of the *Corynebacterium-Mycobacterium-Nocardia* group are not found.

In nutritional requirements, both groups of propionibacteria seem similar. Pantothenate is required by all strains and many also require

Table 15.35.
Cell wall composition in species of the genus **Propionibacterium**

	Amino acids in peptidoglycan	Sugars in polysaccharide[a]
Classical propionibacteria:		
P. freudenreichii	Ala, Glu, *meso*-DAP	Mannose, galactose, rhamnose
P. jensenii	Ala, Glu, Gly, L-DAP	Galactose, glucose; mannose in some
P. acidipropionici		strains
P. thoenii		
Cutaneous propionibacteria:		
P. acnes	Ala, Glu, Gly[b], L-DAP, (*meso*-DAP in	1.[c] Galactose, glucose, mannose
	a few strains)	2. Glucose, mannose
P. avidum	Ala, Glu, Gly[b], L-DAP, (*meso*-DAP in	1.[c] Galactose, glucose, mannose
	a few strains)	2. Glucose, mannose
P. granulosum	Ala, Glu, Gly, L-DAP	Galactose, mannose
P. lymphophilum	Ala, Glu, Lys	Galactose, glucose, mannose

[a] Hexosamines (glucosamine, galactosamine) also present in many cases.

[b] Gly not present in strains with *meso*-DAP.

[c] In both *P. acnes* and *P. avidum*, two serological types are present, with different cell wall sugars.

biotin; growth is generally much improved by thiamine and nicotinamide; oleate is generally also stimulatory and some strains require *p*-aminobenzoic acid (Delwiche, 1949; Ferguson and Cummins, 1978; Holland et al., 1979). Amino acid requirements are complex for the acnes group (Ferguson and Cummins, 1978) but at least some of the classical propionibacteria can grow with ammonium sulfate as nitrogen source (Wood et al., 1938). Most strains of classical propionibacteria produce vitamin B$_{12}$, and some, especially strains of *P. freudenreichii*, produce large amounts (Hettinga and Reinbold, 1972c; Janicka et al., 1976). Production of vitamin B$_{12}$ by cutaneous propionibactera has not been described.

Isolation and Maintenance Procedures

Most methods for the primary isolation of propionibacteria from dairy products have relied on yeast extract-sodium lactate media, e.g. that of Malik et al. (1968). However, more complex media such as brain-heart infusion broth or agar have often been used for the cutaneous strains. We have found that excellent growth of all propionibacteria can be obtained in a trypticase-yeast extract-glucose medium containing 0.05% of Tween 80 (Cummins and Johnson, 1981). The medium is tubed and inoculated under N$_2$ using a modified Hungate technique (Holdeman et al., 1977). The dairy propionibacteria grow best at 30–32°C and the acnes group strains at 36–37°C. In general, maximum growth is attained in 48 h.

For maintenance, chopped meat medium in stoppered tubes under N$_2$ is excellent. After 48 h at the optimum temperature for growth, cultures are kept at room temperature and will remain viable for many months. For maintenance it is better to omit glucose from the medium to avoid formation of excess aid.

Completely synthetic media for the growth of propionibacteria have been devised (Kurman, 1960; Reddy et al., 1973; Ferguson and Cummins, 1978; Holland et al. 1979).

Differentiation of the genus **Propionibacterium** from other genera

Members of the genus *Propionibacterium* need to be distinguished from other Gram-positive, nonsporing, nonmotile, mainly anaerobic organisms with a rather irregular morphology. The production of large amounts of propionic and acetic acids will generally be distinctive. However, some particular groups may cause confusion in some cases.

Clostridium

Some members of the genus *Clostridium* produce considerable amounts of propionic acid, and not all of these spore freely. This refers especially to strains of *C. haemolyticum*, *C. novyi*, *C. botulinum* types C and D, *C. propionicum*, *C. quericolum* and "*C. arcticum*." However, all of these except "*C. arcticum*" produce large amounts of hydrogen during growth (Cato, personal communication), unlike propionibacteria, which produce largely CO$_2$. Moreover, apart from the production of heat-resistant spores, members of *Clostridium* for the most part have a G + C content in the DNA of 28–30 mol%, and their cell walls contain the *meso*-isomer of DAP.

Corynebacterium

The other genus whose members are most likely to cause confusion is *Corynebacterium*. Here the range of G + C contents (53–63%) is very similar to that in the propionibacteria, and propionic acid may be an end product of metabolism. However, the corynebacteria generally grow very much better aerobically than anaerobically and are characterized by the presence of arabinogalactan and mycolic acids in the cell wall, neither of which are found in propionibacteria. On the other hand, corynebacteria do not have the large amount of C$_{15}$ branched-chain fatty acids in the membrane lipids which are found in propionibacteria.

Arachnia

Arachnia propionica is at first sight an obvious candidate for confusion, since the cell walls contain LL-DAP and propionic acid is a major end product of metabolism. However, it is catalase-negative, produces branched diphtheroid rods and filaments, and compact masses of growth in liquid media, and is in fact much more likely to be confused with *Actinomyces israelii* than with propionibacteria.

Further Reading

An exensive review of the activities of the classical propionibacteria was published in three parts by Hettinga and Reinbold (1972 a, b, c). There has been no similar comprehensive review for the cutaneous propionibacteria, but reference may be made to Johnson and Cummins (1972), Cummins and Johnson (1981) and to the various papers quoted in the present article. The taxonomy of *Propionibacterium* has been reviewed by Malik et al. (1968) and by Britz (1979).

List of species of the genus **Propionibacterium**

Section I

Classical or Dairy Propionibacteria

1. **Propionibacterium freudenreichii** van Niel 1928 162.[AL]

freu.den.reich′i.i. M.L. gen. n. *freudenreichii* of Freudenreich; named for Edouard von Freudenreich, the Swiss bacteriologist who first isolated this species.

Description based on literature descriptions, including those of Sakaguchi et al. (1941), Janoschek (1944) and Werkman and Brown (1933), and on a study of two strains of *P. freudenreichii* (Williams E1.51 and ATCC 6207) and seven strains of "*P. shermanii*" (van Niel 1.11, IAM 1714, and ATCC 8262, 9615, 9617, 13673).

Surface colonies on horse blood agar (2 days) are 0.2–0.5 mm, circular, entire, convex, semiopaque, glistening, gray to white (may become cream, tan or pink). The observations of Vedamuthu et al. (1971) suggest that they found strains of *freudenreichii* to be nonhemolytic, and this has also been our experience (Stimpson and Cummins, unpublished observations). Colonies in deep agar are lenticular, to 4 mm, white, tan or pink.

Glucose broth cultures are turbid with a smooth or granular sediment, or clear with a granular sediment and terminal pH of 4.5–4.9.

Anaerobic to aerotolerant. Rarely grows on the surface of agar incubated aerobically, grows in deep broth incubated aerobically, but more slowly than anaerobically. Catalase-positive, generally strongly so (Malik et al., 1968). Produces large amounts of free proline in peptide-containing media (Lansgrud et al., 1977, 1978).

Strains may require pantothenic acid, biotin or thiamine, but do not require *p*-aminobenzoic acid (Delwiche, 1949).

The major long chain fatty acids produced in thioglycollate cultures (Moss et al., 1969) are 12-methyltetradecanoic (about 43%) and a 17-carbon branched-chain acid (about 12%).

Peptidoglycan contains *meso*-DAP; cell wall polysaccharide has major amounts of galactose and moderate amounts of mannose and rhamnose; no glucose. The G + C content of the DNA ranges from 64–67 mol% (T_m) (Johnson and Cummins, 1972).

Type strain: ATCC 6207.

Isolated from raw milk, Swiss cheese and other dairy products; particularly associated with the flavor of Swiss cheese, probably because of proline production.

Further comments. Strains in this species differ broadly from the other classical propionibacteria in the following ways:

1. They are generally very short rods, often almost coccal in shape.
2. They are more heat resistant (Malik et al., 1968).
3. They ferment a restricted range of carbohydrates: in particular the inability to ferment sucrose and maltose seems a reliable distinction from other species.
4. Their peptidoglycan contains *meso*-DAP instead of the L-isomer and the cell wall polysaccharide contains rhamnose.

In the past, these strains have been differentiated into *P. freudenreichii* subsp. *freudenreichii* and *P. freudenreichii* subsp. *shermanii* on the basis of nitrate reduction and lactose fermentation, as follows:

	NO₃ reduction	Lactose fermentation
P. freudenreichii subsp. *freudenreichii*	+	−
P. freudenreichii subsp. *shermanii*	−	+

However, strains of both kinds have high DNA homology with each other (Johnson and Cummins, 1972) and there seems to be no justification for separation at the species level. Of the species described by Sakaguchi et al. (1941) it is possible that "*P. globosum*," "*P. orientum*" and "*P. coloratum*" were in fact other variants of *P. freudenreichii*, but no labeled strains appear to be extant and available for testing.

2. **Propionibacterium jensenii** van Niel 1928, 163.[AL]

jen.se′ni.i. M.L. gen. n. *jensenii* of Jensen; named for S. Orla-Jensen, the Danish bacteriologist who first isolated this organism

Description based on literature descriptions, including those of Werkman and Brown (1933), Sakaguchi et al. (1941) and Janoschek (1944), and on study of 13 strains including 3 of *P. jensenii* (ATCC 4867 (van Niel 24), ATCC 4868 (van Niel 29), ATCC 4869 (van Niel 1)), 2 of "*P. technicum*" (ATCC 14073 (van Niel E.6.1) and ISL 106 (van Niel 22)) 1 of "*P. raffinosaceum*" (ISL 103 (van Niel 29)), 1 of "*P. peterssonii*" (ATCC 4870 (van Niel 20)), 1 of "*P. zeae*" (ATCC 4964 (Hitchner)), and ATCC 4871. Phenotypic characteristics of these strains are similar to original descriptions of *P. jensenii*, "*P. raffinosaceum*," "*P. technicum*," "*P. peterssonii*" or "*P. zeae.*"

Surface colonies on horse blood agar (2 days) are punctiform, circular, entire, convex, glistening, semiopaque and white, cream, or pink; 1 out of 13 strains β-hemolytic. Colonies in deep agar are minute to 4 mm, lenticular and white, pink or red-brown.

Glucose broth cultures are turbid or clear with smooth, granular or ropy sediment and terminal pH of 4.4–4.9.

Anaerobic to aerotolerant or facultative. Some strains grow as well aerobically as anaerobically.

Requires pantothenate and biotin for growth: some strains require *p*-aminobenzoic acid in addition. Thiamine stimulatory but not essential (Delwiche, 1949).

In the cell wall, L-DAP is the diamino acid of peptidoglycan and the principal sugars in the polysaccharide are glucose, and galactose, usually with small amounts of mannose.

The G + C content of the DNA ranges from 65–68 mol% (T_m) (Johnson and Cummins, 1972).

Type strain: ATCC 4868.

Isolated from dairy products, silage and occasionally from infections.

3. **Propionibacterium thoenii** van Niel 1928, 164.[AL]

thoe′ni.i. M.L. gen. n. *thoenii* of Thöni; named for J. Thöni, the Swedish bacteriologist who first isolated this organism.

Description based on a study of the type strain and ATCC 4871, ("*P. rubrum*" van Niel 23) and ATCC 4872 ("*P. rubrum*" van Niel 19).

Surface colonies at 4 days are circular, entire, smooth, generally orange or red-brown. β-Hemolytic on blood agar containing human, bovine, equine, sheep, rabbit or pig blood (Vedamuthu et al., 1971; Stimpson and Cummins, unpublished).

Broth cultures show generalized turbidity with brownish red or orange-red deposit: terminal pH in glucose broth 4.7–4.9.

Less strictly anaerobic than the type species (Breed et al., 1957).

Pantothenate and biotin essential for growth: thiamine or *p*-aminobenzoic acid required by some strains, stimulatory to others (Delwiche, 1949).

Peptidoglycan has L-DAP as diamino acid: cell wall polysaccharide contains glucose and galactose. G + C content of the DNA is 66–67 mol% (T_m) (Johnson and Cummins, 1972).

Originally isolated from red spots in Emmentaler cheese (Thöni and Alleman, 1910): occurs in cheese and other dairy products.

Type strain: ATCC 4874: (van Niel 15).

Further comments. It has been recognized for some time that the strains described under the names *P. thoenii* and "*P. rubrum*" have many features in common, not least the production of an intense red or reddish brown pigment. Classically (van Niel, 1957) "*P. rubrum*" ferments raffinose and mannitol but not sorbitol, while *P. thoenii* ferments sorbitol but not raffinose or mannitol. However, strains of *P. thoenii* and "*P. rubrum*" show high DNA homology (Johnson and Cummins, 1972) and therefore it was recommended that they be combined in a single species. Apart from pigment production, they seem to be related in other way, e.g. production of β-hemolysis on blood agar. The nature of the pigment is not known.

4. **Propionibacterium acidipropionici** Orla-Jensen 1909, 337.[AL]

a.ci′di.pro.pi.on′i.ci. M.L. n. *acidum propionicum* propionic acid; M.L. gen. n. *acidipropionici* of propionic acid.

Description based on study of the type strain (ATCC 25562), ATCC 4875 ("*P. pentosaceum*", van Niel 4), ATCC 4965 ("*P. arabinosum*," Hitchner) and IAM 1725. Phenotypic characteristics of these strains are similar to original descriptions.

Surface colonies on horse blood agar (2 days anaerobic incubation) are punctiform to 1 mm, circular to slightly irregular, convex, entire or slightly scalloped, gray or white, semiopaque; usually nonhemolytic, but may show slight β-hemolysis under area of confluent growth. Colonies in deep agar are white, becoming pink after continued incubation.

Catalase reaction weak or negative (Werkman and Brown, 1933; van Niel, 1957).

Glucose broth cultures are turbid with smooth cream-colored sediment: terminal pH 4.1–4.9.

Anaerobic to aerotolerant. May grow as well in aerobic as in anaerobic conditions.

Requires pantothenate and biotin for growth: thiamine is not essential but is stimulatory (Delwiche, 1949).

In the cell wall, L-DAP is the diamino acid of peptidoglycan, and the principal sugars in the polysaccharide are glucose with galactose and/or mannose (some strain variation: see Table 5 in Johnson and Cummins, 1972).

The major long chain fatty acids produced in thioglycolate cultures (Moss et al., 1969) are 13-methyltetradecanoic (17–40%) and 12-methyltetradecanoic (12–23%).

The G + C content of the DNA is 66–68 mol% (T_m) (Johnson and Cummins, 1972).

Isolated from dairy products.

Type strain: ATCC 25562 (VPI 0399, Prévot 14 X, "*P. pentosaceum,*" from Orla-Jensen).

Further comments. The strains originally described by van Niel (1928) and Hitchner (1932) as "*P. pentosaceum*" and "*P. arabinosum*", respectively, show high DNA homology with each other and are also characterized by showing a weak or negative catalase reaction.

Strains of "*P. arabinosum*" were described as being unable to ferment xylose and rhamnose, while those of "*P. pentosaceum*" could ferment these sugars. However, as in the case of *P. freudenreichii* and "*P. shermanii*" with lactose fermentation, this is not considered sufficient to warrant speciation.

Section II

Cutaneous Propionibacteria Commonly Found on the Skin

5. **Propionibacterium acnes** (Gilchrist) Douglas and Gunter 1946, 22.[AL]

ac′nes. Gr. n. *acme* a point; incorrectly transliterated as M.L. n. *acne* acne; M.L. gen. n. *acnes* of acne.

Description based on a study of several hundred strains by Moore, Cato and Holdeman (Holdeman et al., 1977) and ourselves (Cummins and Johnson, unpublished), including the type strain ATCC 6919, and on the results of Kishishita et al. (1979) and McGinley et al. (1978).

Colonies in deep agar are lenticular, minute to 4 mm, white; colonies of some strains become tan, pink or orange in 3 weeks. Surface colonies on blood (horse or rabbit) agar (2–3 days) are punctiform to 0.5 mm, circular, entire to pulvinate, translucent to opaque, white to gray, glistening.

Glucose broth cultures are turbid or clear with a finely granular sediment which generally resuspends readily. In older cultures the sediment is often reddish. In suitable media with good growth, the final pH is 4.5–5.0.

Anaerobic to aerotolerant. Two percent of the strains tested grew on the surface of blood agar plates incubated aerobically; 30% grew slightly to well on the surface of blood agar plates incubated in a candle jar. Many strains grew in deep glucose broth without special provisions for anaerobiosis.

Lactate is converted to propionate by most strains, but often only if the initial oxidation-reduction potential of the medium is sufficiently low or if the initial growth rate is rapid. The major long chain fatty

acid produced in thioglycolate cultures (Moss et al., 1969) is 13-methyltetradecanoic acid (32–62%).

Generally catalase-positive, although cultures need to be exposed to air for a period before testing. McGinley et al. (1978) who exposed colonies on brain-heart infusion agar to air for 1 h before testing, found all of 231 strains to be positive for catalase.

A number of extracellular enzymes are produced: ribonuclease (Smith, 1969); neuraminidase (Von Nicolai et al., 1980); hyaluronidase (Ingham et al., 1979, Hoeffler, 1980); acid phosphatase (Ingham et al., 1980); lecithinase (Werner, 1967); and lipase (Smith and Willett, 1968; Ingham et al., 1981). The lipase produces fatty acids from triacylglycerols, and the acids may (a) act as tissue irritants and (b) promote the growth of *P. acnes* (e.g. oleate, see below).

Some strains produce bacteriocin-like substances inhibitory for other strains (Fujimura and Nakamura, 1978; Ko et al., 1978) and a variety of bacteriophage types exist (Prévot and Thouvenot, 1964; Pulverer et al., 1973; Jong et al., 1975; Webster and Cummins, 1978).

Nutritional requirements: all strains tested had an absolute requirement for pantothenate; thiamine, biotin and nicotinamide stimulated growth, as did the three organic acids lactate, pyruvate and α-ketoglutarate. Oleate (usually used in the form of Tween 80) is also stimulatory. Amino acid requirements are complex (for nutritional requirements see Ferguson and Cummins, 1978; Holland et al., 1979).

On the basis of cell wall composition and cell wall polysaccharide antigens, there are two types of *P. acnes* (Table 15.32). The two serological types cross-react but can be separated with absorbed sera. *P. acnes* type II also cross-reacts strongly with sera to *P. avidum* type II (Johnson and Cummins, 1972; Cummins, 1975).

Hemolytic activities: 68% of 34 type I strains were β-hemolytic on rabbit blood, while 100% of 65 type II strains were nonhemolytic (Stimpson and Cummins, unpublished). Sheep blood gives less clearcut results and few strains were hemolytic. About 30% of type I strains were hemolytic on horse blood.

Antibiotic sensitivity: most strains were susceptible to commonly used antimicrobial agents, especially penicillin G and erythromycin; they are generally resistant to aminoglycosides (Wang et al., 1977), metronidazole (Chow et al., 1975) and sulfonamides (Pochi and Strauss, 1961).

Strains of group II are often more coccoid than those of group I and are most similar to previous descriptions of "*C. parvum*" and "*C. adamsoni,*" which are synonyms of *P. acnes*.

The G + C content of the DNA of both serological groups is 57–60 mol% (T_m) (Johnson and Cummins, 1972).

Isolated from the normal skin, comedones of acne vulgaris, intestinal contents, wounds, blood, pus and soft tissue abscesses. Numbers on skin vary very widely in different persons from about 10^2–10^6/sq cm (Evans et al., 1950).

Type strain: ATCC 6919 (Zierdt et al., 1968) (Ponsonby; NCTC 737).

Further comments. Strains of *P. acnes* are common contaminants in anaerobic cultures, the source of contamination probably being scales from the skin or scalp of the person transferring cultures. This may give rise to particular problems if, for example, a strain of *P. acnes* contaminates a culture of *P. granulosum*; and such situations may have led to confusing results in the past. For similar reasons, the organism is a common contaminant of clinical specimens. However, there are well documented cases where *P. acnes* has been the only organism cultured on several occasions, under circumstances when it seemed clearly to be causing the lesions (see e.g. Yocum et al., 1982). A high percentage of the clinically normal adult human population has detectable serum antibody against *P. acnes*. Titers are higher in those with acne vulgaris (Puhvel et al., 1964; Wolberg et al., 1977). Five biovars of *P. acnes* have been proposed, based on fermentation of ribose, erythritol and sorbitol (Kishishita et al., 1979).

"*Corynebacterium parvum*." This name has been used in immunological and oncological literature to describe the organism which has been extensively used as a reticulostimulant or immunomodulator. However, an examination of 59 strains labeled "*C. parvum*" (Cummins and Johnson, 1974) showed that 90% of them were *P. acnes* by serological and physiological tests and by DNA homology, and almost all the remainder were *P. avidum*. The name is therefore regarded as a synonym of *P. acnes*.

Relationship of P. acnes *to acne vulgaris.* P. acnes was originally isolated from the lesion of acne vulgaris by Sabouraud (1897). However the idea that it was the cause of acne was shaken when Lovejoy and Hastings (1911) found it on normal skin. It is now believed that the underlying cause of acne vulgaris is endocrine (see e.g., Cunliffe, 1982) and probably associated with over-production of sebum; however, there is no doubt that the presence of large numbers of *P. acnes* in sebaceous follicles is intimately connected with the more severe manifestations of the disease. Two possible mechanisms by which inflammation may be produced in acne are (a) release of free fatty acids by bacterial lipases from triglycerides in sebaceous material and (b) local destruction of cells and attraction of phagocytes by activation of the complement system by *P. acnes* antigens (Webster et al., 1980; Webster and McArthur, 1982).

P. granulosum, which is also found in sebaceous glands, although in smaller numbers, may contribute to the process, since it can liberate free fatty acids from triglycerides and will activate complement. *P. avidum* is unlikely to contribute much since it occurs mostly in moister, less oily areas of the skin.

6. **Propionibacterium avidum** (Eggerth 1935) Moore and Holdeman 1969, 7.[AL]

a'vi.dum. L. neut. adj. *avidum* greedy, voracious.

Description based on a study of 20 strains including ATCC 25577 by Holdeman and Moore (Holdeman et al., 1977), and 23 strains by ourselves (Cummins and Johnson, unpublished).

Anaerobic or microaerophilic when freshly isolated but often grows well aerobically after a few transfers. In general, grows better and gives larger colonies than either *P. acnes* or *P. granulosum*. Surface colonies on suitable media are 0.5–1.0 mm at 2–3 days, smooth, entire, circular, white to light cream color. Generally β-hemolytic on sheep, horse or rabbit blood (19/23 strains, (Cummins and Johnson, unpublished), which agrees well with the results of Hoeffler, (1977) on sheep, human and rabbit blood. Glucose broth is generally turbid with a smooth abundant sediment which resuspends readily.

Nutritional characteristics: less exacting than either *P. acnes* or *P. granulosum* and, after several transfers, will grow in a simple medium consisting of salts, glucose and vitamins (Ferguson and Cummins, 1978). Pantothenic acid is an absolute requirement for growth.

Antigenic structure: using acid extracts containing cell wall polysaccharides, two distinct types were found corresponding to two types of sugar patterns in the wall polysaccharide: type I contains glucose, galactose and mannose, type II contains glucose and mannose only (Johnson and Cummins, 1972; Cummins, 1975) and cross-reacts to some extent with *P. acnes* type II. However Hoeffler et al. (1977) reported three types by agglutination of bacterial suspensions.

Peptidoglycan: generally, the tetrapeptide amino acids are alanine, glutamic acid, glycine and L-DAP, but a few strains of serological type II have *meso*-DAP and no glycine (thus resembling similar strains of *P. acnes* type II).

Cell wall polysaccharides: two kinds, corresponding to the two serological types (see above). Both contain glucosamine and galactosamine, but type II strains have a higher galactosamine content.

Antibiotic sensitivities: not extensively investigated but presumed to be similar to *P. acnes*. Not sensitive to lysozyme.

Extracellular enzymes: strains are almost always positive for gelatinase and deoxyribonuclease; most were negative for lecithinase, hyaluronidase and chondroitin sulfatase (Hoeffler, 1977).

Isolated principally from the moister areas of the skin: e.g. vestibule of the nose, axilla, perineum and from chronically infected areas such as sinuses. Probably not a primary pathogen.

The G + C content of the DNA is 62–63 mol% (T_m) (Johnson and Cummins, 1972).

Type strain: ATCC 25577; (VPI 0179; CIPP (Prévot) 1689B).

Further comments. Strains of *P. avidum* are found almost exclusively in the moister areas of the skin, such as the axilla and perineum: they are also the most aerotolerant of the three species. There are several points of similarity between strains of *P. acnes* type II and *P. avidum* type II. Both have cell wall polysaccharides which contain glucose and mannose, but no galactose, and there is quite strong serological cross-reaction between the two types. Also, some strains in each type have *meso*-DAP as the diamino acid of the peptidoglycan instead of the L-isomer.

P. avidum strains have been most often isolated from infected sinuses, chronically infected wounds, submaxillary abscesses, etc., but it is doubtful if they are the primary cause of those infections.

7. **Propionibacterium granulosum** (Prévot 1938) Moore and Holdeman 1970, 15.[AL]

gra.nu.lo'sum. L. neut. adj. *granulosum* full of granules.

Description based largely on a study of the reference strains and 36 other strains by Moore, Cato and Holdeman (Holdeman et al., 1977) and of 30 additional strains originally isolated by Dr. Charles Evans, University of Washington, Seattle, and examined by us (Cummins and Johnson, unpublished).

Anaerobic or microaerophilic: very poor or no growth under aerobic conditions.

Surface colonies on suitable media up to 1 mm at 2–3 days, generally white or grayish, smooth, circular, entire, usually larger and more whitish than colonies of *P. acnes*. Generally nonhemolytic on sheep, horse or rabbit blood, although reported to be β-hemolytic on rabbit blood by Hoeffler (1977). Glucose broth cultures generally turbid with a rather coarsely granular deposit: often rather viscid and difficult to centrifuge cleanly.

Nutritional characteristics: generally the same as for *P. acnes* although some strains appear to require additional unidentified factors for growth (Ferguson and Cummins, 1978).

Antigenic structure: based on acid extracts containing cell wall polysaccharide, all strains belong to a single type, with little or no cross-reaction against polysaccharides from *P. acnes* or *P. avidum* (Cummins, 1975). However, Hoeffler et al. (1977) have found at least three types by tube or slide agglutination of intact suspensions, which points to the existence of several surface antigens.

Peptidoglycan: tetrapeptide amino acids are alanine, glutamic acid, glycine and L-DAP (Johnson and Cummins, 1972). It is assumed that the diamino acid is L-DAP and that glycine is an interpeptide bridge.

Cell wall polysaccharide: sugar constituents are galactose and mannose, with glucosamine (Johnson and Cummins, 1972).

Antibiotic sensitivities: not extensively investigated but presumed to be similar to *P. acnes*. Not sensitive to lysozyme.

Extracellular enzymes: Hoeffler (1977) found that most strains had an active deoxyribonuclease and lecithinase; chondroitin sulfatase, hyaluronidase and phosphatase activity was detected in a few strains, but none produced gelatinase. *P. granulosum* has a lipase which is considerably more active than that of *P. acnes* (Greenman et al., 1981).

Isolated from the sebum-rich more oily areas of the skin but in smaller numbers than *P. acnes* (McGinley et al., 1978). *P. granulosum* is found along with *P. acnes* in acne comedones, and may play some part in the pathogenesis of acne. Probably not otherwise pathogenic.

The G + C content of the DNA is 61–63 mol% (T_m) (Johnson and Cummins, 1972).

Type strain: ATCC 25564; (VPI 0507) (isolated as a contaminant from a culture labeled *Staphylococcus aerogenes*).

Further comments. P. granulosum seems to have been first clearly separated from *P. acnes* (although those names were not used) by Brzin (1964) who recognized a group of strains that did not produce indole or reduce nitrate but could ferment sucrose and maltose. This distinction was later confirmed by Voss (1970) who showed that the indole-negative, nitrate-negative strains were serologically distinct from typical *"acnes"* strains.

Strains of *P. granulosum* show low DNA homology (about 12–15%)

to both *P. acnes* and *P. avidum*, which is about the same level of relationship as is found between *P. acnes* and the classical propionibacteria (Johnson and Cummins, 1972).

P. granulosum occurs in the same areas of skin as *P. acnes*, and is isolated along with *P. acnes* from acne comedones, although generally in smaller numbers. It is especially common in the region of the alae nasi (McGinley et al., 1978).

8. Propionibacterium lymphophilum (Torrey 1916) Johnson and Cummins 1972, 1057.[AL]

lym.pho'phi.lum. Gr. n. *lympho* lymph; Gr. neut. adj. *philum* loving. M.L. adj. *lymphophilum* lymph-loving.

Description based on study of two strains including CIPP (Prévot) 1519F (VPI 0202).

Surface colonies on horse blood in 4 days are punctiform to 0.5 mm, circular, entire, convex to pulvinate, white, glistening and smooth.

Glucose broth cultures (24 h) are turbid, becoming clear, with a ropy sediment and terminal pH of 5.4–5.7.

Anaerobic, producing no growth on agar surface incubated aerobically but growth develops in deep broth incubated aerobically.

Cell walls contain lysine, which presumably is the diamino acid of peptidoglycan, and glucose, galactose and mannose as principal cell wall sugars.

The G + C content of the DNA is 53–54 mol% (T_m) (Johnson and Cummins, 1972).

Isolated from urinary tract infections (CIPP (Prévot) 1519F, VPI 0202) and mesenteric ganglion of a monkey inoculated with "*Actinobacterium*" (CIPP (Prévot) SB, VPI 0383). Strains originally described by Torrey (1916) were from lymph glands in Hodgkin's disease.

Type strain: ATCC 27520.

Further comments. In addition to the two strains VPI 0202 and VPI 0383 (see above) only two other similar strains have been examined. These two strains were made available to us by Dr. C. Adlam of the Wellcome Research Laboratories, Kent, England. They both appear to have high DNA homology (~100%) to strain 0202 (Johnson, unpublished results) and also show cross-agglutination and precipitation with it (Adlam and Reid, 1978; Adlam, personal communication). Strain 0383 shows about 75% DNA homology to strain 0202.

None of the strains show any reaction with antisera to *P. acnes*, *P. avidum* or *P. granulosum*. These strains also differ from the other three species described in Section II in that they have lysine in place of DAP,

and the G + C content of the DNA is 53–54 mol%, which is considerably lower than that of the other described species.

In view of this and the small number of strains, the status and taxonomic position of this organism must remain somewhat provisional until more strains are collected and examined.

Further Comments on Species Common on the Skin

P. acnes, *P. granulosum* and *P. lymphophilum* are reported to have been isolated from the soils of rice paddy fields (Hayashi and Furusaka, 1979, 1980) and from preparations of green olives (Cancho et al., 1980). However, these identifications were based on biochemical tests and fermentative abilities only, without studies of cell wall composition, DNA base ratios, or serological properties.

Antigenic composition. Based on immunodiffusion tests in agar, using polysaccharides extracted with trichloroacetic acid, there are two types of *P. acnes*, two of *P. avidum* and one of *P. granulosum* (Cummins, 1975). However, tube and slide agglutination tests with bacterial suspensions (Hoeffler et al., 1977) have given somewhat different results in that all of 60 *P. acnes* strains seemed to belong to one type, while there were at least two types of *P. granulosum* and three of *P. avidum*. It appears therefore that there may be surface antigens which have not so far been investigated in detail.

Isolation of strains. Strains of *P. acnes* (both serological types) and *P. granulosum* can readily be isolated from oily areas of skin; e.g. the forehead, or between the shoulder blades. *P. avidum* is more readily isolated from moister areas, e.g. axilla or perineum. The method of Williamson and Kligman (1965) may be used, in which the skin is gently scraped into a slightly alkaline detergent solution. Tween 80 appears to be a more suitable detergent than Triton X100, which was originally recommended (Kishishita et al., 1980). Cultures are best spread in several dilutions on the surface of solid media (e.g. blood agar) and incubated anaerobically. Colonies may not be apparent for several days.

The numbers of propionibacteria isolated from the skin vary very widely from area to area and from person to person, from 10^2/sq cm to 10^6/sq cm or more (see e.g. Evans et al., 1950; Kishishita et al., 1979).

Although *P. acnes* is normally regarded as anaerobic, Evans and Mattern (1979) have reported that the organism was commonly seen in aerobic primary cultures from the skin. On subculture, the strains showed their normal anaerobic state.

Genus Eubacterium Prévot 1938, 294[AL]

W. E. C. MOORE AND LILLIAN V. HOLDEMAN MOORE

Eu.bac.te'ri.um. Gr. pref. *eu-* good-, well-, beneficial (*not* as opposed to pseudo-); Gr. neut. dim. n. *bakterion* a small rod; M.L. neut. n. *Eubacterium* beneficial bacterium.

Uniform or pleomorphic **nonsporing Gram-positive rods.** Nonmotile or motile. **Obligately anaerobic. Chemoorganotrophs,** saccharoclastic or nonsaccharoclastic. Usually **produce mixtures of organic acids from carbohydrates** or **peptone often including** large amounts of **butyric, acetic or formic acids.** Do not produce:

1. Propionic as a major acid product (see *Propionibacterium*).
2. Lactic as the sole major acid product (see *Lactobacillus*).
3. Succinic (in the presence of CO_2) and lactic acid with small amounts of acetic or formic acids (see *Actinomyces*).
4. Acetic and lactic (acetic > lactic), with or without formic, as the sole major acid products (see *Bifidobacterium*).

The mol% G + C of the DNA of the type species is 47 (T_m). The mol% G + C of the DNA of other species examined is 30–55 (T_m).

Type species: *Eubacterium limosum* (Eggerth) Prévot 1938, 295 (see Taxonomic Comments below).

Further Descriptive Information

The species and strains within species vary in sensitivity to O_2; some can be cultured only in prereduced media. Catalase usually is not

produced (trace amounts are detected in some strains); hippurate usually is not hydrolyzed. Growth usually is most rapid at 37°C and pH near 7.

Found in cavities of man and other animals, animal and plant products, infections of soft tissue, and soil. Some species may be pathogenic.

Isolation and Enrichment Procedures

Strains of *Eubacterium* species found in clinical infections can be isolated on usual complex media appropriate for culture of anaerobes. Such media include blood agar plates or prereduced media in roll tubes made with a peptone base including or supplemented with 0.5% (w/v) yeast extract, 0.2% glucose, 0.001 μl/ml vitamin K_1 and 0.005 μg/ml (w/v) hemin. Species from the rumen or feces usually are isolated in rumen fluid-glucose-cellobiose agar (RGCA), RGCA supplemented with 1% peptone, or similar medium containing hemin and volatile fatty acids. Special nutrients are required to obtain good growth of some species (e.g. arginine for *E. lentum*). Some rumen bacteria grow better in media containing 1,4-naphthoquinone than in media with menadione, vitamin K_1, or vitamin K_5 (Gomez-Alarcon et al., 1982). Although this effect was observed with a Gram-negative anaerobe (*Succinivibrio dextrinosolvens*) it also might apply to the Gram-positive anaerobes.

Maintenance Procedures

Lyophilization of cultures in the early stationary phase of growth in a medium containing no more than 0.1–0.2% fermentable carbohydrate is recommended for long term storage of most strains. For short term storage, cultures should be grown in a nutritionally adequate medium containing no, or minimal, fermentable carbohydrate and stored at room temperature (~22°C). Saccharoclastic strains should be transferred every 7–10 days, nonsaccharoclastic strains every 2–3 weeks.

Procedures and Methods for Characterization Tests

Use of heavy inoculum (~5% v/v of broth culture) of a young culture (late log or early stationary phase of growth) gives the most reliable results for biochemical reactions based upon growth of the organism in the substrate. Results of tests for constitutive enzymes with heavy inocula on dehydrated substrates may differ from results based upon growth of the organism in the substrate where adaptive enzymes are also detected. Unless otherwise cited, the characteristics listed here for the species were determined with prereduced anaerobically sterilized media by the methods described in Holdeman et al. (1977), and susceptibility to antimicrobial agents was tested by the broth-disk method of Wilkins and Thiel (1973). The basal prereduced peptone-yeast extract (PY) medium contained (per 100 ml): 0.5 g trypticase, 0.5 g peptone, 1 g yeast extract, 0.1 μl vitamin K_1, 0.5 g hemin, cysteine hydrochloride, and salts solution.

Morphology and Gram reaction. For determination of Gram reaction, a buffered Gram's stain (such as Kopeloff's modification) of cells from young cultures (log phase of growth) is recommended. Cells of many Gram-positive species stain Gram-negative in older cultures or when acid has been produced in the culture medium.

Hydrogen production. H_2 production was determined by gas chromatography of the headspace gas of cultures grown in tubes closed with rubber stoppers (Holdeman et al., 1977). Production of H_2 may not disrupt the agar in PY-glucose (PYG) agar deep cultures.

Gas production. Gas production was determined by disruption of agar in loosely covered PYG agar deep cultures. Accumulation of small (lenticular) bubbles near the colonies in the agar deep tube was recorded as "1" gas, a small split of the agar column across the tube as "2" gas, agar separated and displaced in one or more places as "3" gas and agar forced to the top of the tube as "4" gas.

Esculin hydrolysis and esculin fermentation. These are independent reactions. Hydrolytic products are detected by the appearance of a black color upon addition of ferric ammonium citrate reagent to esculin-PY broth cultures. Acid production is determined by pH measurement.

Indole. Cultures in media containing sufficient tryptophan (chopped meat medium, tryptone-yeast extract medium, etc.) and no fermentable carbohydrate are required. The culture medium should be extracted with xylene and the reagent (either Ehrlich's or Kovac's) poured down the side of the tube to layer next to the xylene. Do not shake the tube after the reagent has been added. Alternatively, a loopful of growth from a *pure culture* can be smeared on filter paper saturated with 1% *p*-dimethylaminocinnamaldehyde in 10% (v/v) HCl (Sutter et al., 1980). Development of a blue color indicates indole. This method may not detect weak indole production.

Gelatin digestion. Incubated cultures of PY medium containing 10% gelatin and 1% glucose and uninoculated tubes of the same medium are chilled until the originally liquid controls are solid (about 15 min at 4°C). Failure of the gelatin culture to solidify at 4°C indicates complete digestion (+). Liquefaction of cultures at room temperature within 30 min or in less than half the time of the control tube indicates partial liquefaction (w).

Milk proteolysis. Acid production in milk often causes protein coagulation. The curd may or may not show evidence of streaks caused by evolution of gas. Shrinking of the curd may leave a clear whey. This liquid is sometimes mistaken for digestion (proteolysis) of the milk. Digestion of curd is evidenced by dissolution of the curd (first) and increasing turbidity of the whey and often takes many days. Digestion may occur without previous curd formation and results (usually slowly) in clear liquid after extended incubation (up to 3 weeks). (*Clostridium sporogenes* is a good positive control culture for milk proteolysis.)

Acid production. Positive reactions listed for carbohydrate fermentation represent a pH below 5.5 and a decrease in pH of at least 0.5 pH units below the control PY basal medium culture. Weak (w) acid production represents a pH of 5.5–5.9 and at least 0.3 pH units below the PY culture control. A pH of 5.9 or above, or within 0.3 pH unit of the PY control culture pH, is considered negative. When in doubt, examination of the amount of growth in the sugar-containing medium vs. that in medium without sugar can help in interpretation of weak reactions; i.e. with slight pH decrease, if growth is much better in the sugar-containing medium the carbohydrate probably was fermented. For rapid-growing saccharoclastic species, the final pH is reached after incubation for 18–24 h in nutritionally adequate media.

Uninoculated xylose and arabinose, under CO_2, are often pH 5.9. For cultures in these media, a final pH of 5.4 or below is considered strong acid production.

Differentiation of the genus **Eubacterium** from other genera

The genera of anaerobic nonsporing Gram-positive rod-shaped bacteria are differentiated according to the major metabolic pathways and products of these organisms as described above. The current classification, based on assignment of species to the genus *Eubacterium* by a process of elimination from other anaerobic genera, has led to a convenient taxon that includes somewhat diverse but well-recognized species in three subgroups: those that produce butyric acid, usually in combination with other volatile fatty acids and sometimes with alcohols; those that produce various combinations of lactate, acetate and formate together with H_2 gas; and those that produce little if any detectable fermentation acid. Hydrogen gas is not produced by anaerobic species or strains of *Lactobacillus*, *Bifidobacterium* or *Actinomyces* that may also produce combinations of lactate, acetate and formate. The genus *Eubacterium* may include some species that are phylogenetically distant from each other. However, as currently defined, the genus serves an important taxonomic purpose for effective scientific communication.

All descriptions are based on strains in which no spores have been detected. Heat-resistant spores, often difficult to detect, have been found in organisms similar to some described species, especially motile or filamentous species; these strains are members of the genus *Clostridium*.

Taxonomic Comments

Eubacterium foedans (Klein) Prévot 1938 was designated the type species of *Eubacterium* by Hauduroy et al. (1953). No strains of *E. foedans* are extant and attempts to isolate representative strains from spoiled hams, the original source of *E. foedans*, were unsuccessful. Therefore, the species was not included in the Approved Lists of Bacterial Names (Skerman et al., 1980) and the species has no taxonomic standing. In keeping with the recommendation in the Introduction to the Approved Lists that action be taken to conserve genera that contained well described species but for which there was no strain to represent the type species, Cato et al. (1981) requested an opinion to designate *Eubacterium limosum* (Eggerth) Prévot 1938 as the type species of the genus. The opinion has been approved by the Judicial Commission as Opinion 57 (Judicial Commission, 1983).

Acknowledgements

We gratefully acknowledge the microbiological or technical assistance of Loretta P. Albert, Pauletta C. Atkins, Ella W. Beaver, Ruth Z.

Beyer, Maeve N. Crowgey, Ann P. Donnelly, Jackie G. Eudaly, Luba S. Fabrycky, Barbara A. Harich, Donald E. Hash, Linda K. Hoffman, Carolyn L. Hubbard, Jane L. Hungate, Daniel M. Linn, June M. McElwee, Ann C. Ridpath, Carolyn W. Salmon, Gail S. Selph, Debra B. Sinsabaugh, Sue C. Smith, Christina J. Spittle, Susan E. Stevens, Barbara C. Thompson, Dianne M. Wall, Catherine A. Waters and Bonnie M. Williams; the support assistance of Eleanor A. Johnson, Julia J. Hylton, Ruth E. McCoy, Claudine P. Saville, Phyllis V. Sparks and Margaret L. Vaught; and the secretarial assistance of Donya B. Stephens and Mary P. Harvey, whose work has contributed to our taxonomic studies since 1974.

We are especially grateful to Elizabeth P. Cato for collaborative assistance in the taxonomic studies and to Thomas O. MacAdoo, Department of Foreign Languages and Literature, Virginia Polytechnic Institute and State University, for advice concerning the derivation of specific epithets. We gratefully acknowledge the helpful comments of Victor Bokkenheuser and Ian Macdonald concerning *E. lentum* and similar organisms, and of M. P. Bryant for critical review of the manuscript.

For support of research for studies of many of these species, we appreciate the assistance of VPI & SU project 2022820 from the Commonwealth of Virginia, contracts N01-CP-33334 from the National Cancer Institute and 9-12601 from the National Aeronautics and Space Administration, and grants DE-05139 and DE-05054 from the National Institute of Dental Research.

Further Reading

Anyone with an interest in anaerobic bacteriology should find the excellent 7-volume bibliography, compiled by L. S. McClung (1982), invaluable. It is a subject index to the literature from 1940 through 1975 and is a continuation of the original publications of E. McCoy and L. S. McClung (The anaerobic bacteria and their activities in nature and disease: a subject bibliography (in 2 volumes), University of California Press, 1939; and Supplement one (literature for 1938 and 1939)).

Excellent color photographs of colonies and stains of cells of many species of eubacteria (and other anaerobes) are depicted in Mitsuoka, T. 1980. The world of intestinal bacteria—the isolation and identification of anaerobic bacteria; a color atlas of anaerobic bacteria (English title given). Sobunsha (Sobun Press) (Hisamatsu Building, 1-4-5 Enraku-cho, Kanda, Chiyoda-ku), Tokyo.

Differentiation of the species of the genus Eubacterium

Characteristics by which species in the genus can be differentiated are given in the Key to species in the genus and in Tables 15.36 and 15.37. Acid production in the key refers to both strong and weak fermentations; i.e., a species is positive if the terminal pH is below 5.9 and 0.3 pH units below the control basal medium culture. Additional characteristics are given in the text concerning each species.

Species are listed alphabetically. Reactions of saccharoclastic species are given in Table 15.36 and reactions of nonsaccharoclastic species in Table 15.37.

Colony descriptions are from blood agar plates or streak tube cultures incubated anaerobically for 2 days, unless otherwise indicated.

Reported G + C contents of the DNAs are expressed as the nearest whole number.

The drawings of each species, given in the text, are composites of the type strain and other phenotypically similar strains. Individual cultures may show somewhat less variation in cellular morphology than is depicted. Unless specified otherwise in the text, cells are nonmotile.

Key for presumptive identification of species in the genus Eubacterium

I. Butyric acid produced
  1. Glucose acid
    a. Erythritol acid
          17. *E. limosum*
    aa. Erythritol not acid
      b. Indole produced
          27. *E. saburreum*
      bb. Indole not produced
        c. Melezitose acid
           25. *E. rectale*
        cc. Melezitose not acid
          d. Mannose acid
            e. Xylose acid
              14. *E. hadrum*
            ee. Xylose not acid
              f. Melibiose acid
                g. Cells motile
                  23. *E. plexicaudatum*
                gg. Cells nonmotile
                  24. *E. ramulus*
              ff. Melibiose not acid
                g. Glycogen acid
                  5. *E. budayi**
                gg. Glycogen not acid
                  h. Esculin hydrolyzed
                    i. Maltose acid
                      j. Cellobiose acid
                  5. *E. budayi*
                20. *E. nitritogenes**
                  jj. Cellobiose not acid

---

* If gelatin hydrolyzed, see *Clostridium perfringens*.

Table 15.36.

Biochemical reactions of saccharoclastic species in the genus **Eubacterium**[a]

Characteristics	1. E. aerofaciens	2. E. alactolyticum	3. E. biforme	5. E. budayi	6. E. cellulosolvens	8. E. contortum	9. E. cylindroides	11. E. eligens	12. E. fissicatena	13. E. formicigenerans	14. E. hadrum	15. E. hallii	17. E. limosum	18. E. moniliforme	19. E. multiforme	20. E. nitritogenes	22. E. plautii	23. E. plexicaudatum	24. E. ramulus	25. E. rectale	26. E. ruminantium	27. E. saburreum	28. E. siraeum	29. E. suis	30. E. tarantellus	31. E. tenue	33. E. tortuosum	34. E. ventriosum
Cells motile	−	−	−	−	d	−	−	+	+	−	−	−	−	+	+	−	+	+	−	d	−	−	−	−	−	d	−	−
Acid produced from:																												
Amygdalin	−	−	−	−	−	−	−	−	−	−	−	−	−	−	−	−	−	±[b]	−	A−	−	−	−	−	−	−	−	−
Arabinose	−	−w	−	−w	w	−w	−	−	−	−	−	−	−	−	−	−	−	±	−w	A−	−	−	−	−	−	−	−	−A
Cellobiose	A−	−	−A	−	−	Aw	−	As	−	d	−A	−	−A	−	−	wA	−	±	A	Aw	Aw	Aw	wA	−	−	−	−w	−A
Esculin	−w	−	−w	wA	wA	A−	−w	−	−	−s	−w	−	−	−	w−	wA	−	±	A	Aw	A	d	wA	−	−	−	−w	−A
Fructose	A	A	A	A	−	Aw	A	Aw	A	A	Aw	Aw	A	Aw	Aw	Aw	A	±	A	A	w−	d	−w	−	Aw	−w	−	A
Glucose	A	A	A	A	A	A	A	A−	A	A	Aw	Aw	A	Aw	Aw	A	A	±	A	A	A	Aw	−w	A	Aw	−A	−A	Aw
Glycogen	Aw	−	−	w−	d	−	−	−	−	−	−	−	−	−	Aw	−w	A	±	−	d	−	d	−	−	−	−	−A	Aw
Lactose	A−	−	d	A−	d	A−	A−	A−	−w	A−	A−	A−	−w	A−	Aw	−w	A	±	A−	A−	−	d	w−	ws	Aw	−	−A	−A
Maltose	Aw	−A	w	w	Aw	A	−	−s	Aw	A−	−A	As	A−	−A	Aw	Aw	A	±	A−	−A	w−	d	w−	−	Aw	w−	−A	A
Mannitol	A	−A	d	A	−	d	A	−	−	−	−A	Aw	−w	−A	Aw	A	A	±	−A	−A	−	−A	−	−	Aw	−w	−	A
Mannose	A	−	−	A	−	A	−	−w	−w	−s	−A	Aw	−	−	Aw	−	A	±	−w	−A	A	d	w−	−	−	−	w−	A
Melezitose	−	−	−	−	−	−	−	−	−	−	−	−	−	−	−	−	−	−	−	−	−	−A	−	−	−	−	−	−
Melibiose	−	−	−	−	w−	A−	−w	−	−	−	Aw	Aw	−	−	Aw	−	−	±	Aw	Aw	A	d	w−	−	−	−	w−	−
Raffinose	−	−	−	−	w−	Aw	−	−	A	−A	Aw	Aw	−	−	Aw	−A	−	±	Aw	A	A−	d	−w	A	Aw	−	w−	−
Rhamnose	−	−	−	−	−	A−	−	−	d	−	Aw	−	−	−	Aw	−	−	±	−	−	A−	−A	−A	−	−	−	−	−
Ribose	−	−	−	−	−	Aw	−	−	A	−A	−A	A−	Aw	−	−	−A	−	±−	−	Aw	A	A	−A	Aw	w	w−	w−	−A
Salicin	Aw	−	−A	−w	−w	A	−w	−A	d	−A	−A	Aw	−	−	Aw	d	−	±−	d	Aw	A	d	w−	−	−	−	−	−
Sorbitol	−	−A	−	−	−	−	−	−	−A	−	−A	As	−	−	−	−	−	±−	w−	Aw	−	−	−	A	−	−	w−	−
Starch	A	−A	−A	−w	−	−w	−w	w−	−	−s	−A	−A	d	−w	−w	−w	−	±	−w	Aw	A	−w	−w	A	−	−	Aw	d
Sucrose	A	−	−A	−	Aw	A	−	A−	A	−A	Aw	−A	−	−	−	d	−	±−	−w	A	A	A	−w	A	Aw	Aw	Aw	d
Trehalose	−A	−	d	d	d	−w	−w	−	−	−s	−	−	d	−	−	d	−	±−	−w	d	A	d	−w	−	+w	d	−	d
Xylose	−	−	−	−	−	A−	−	A−	d	−A	Aw	A−	d	−w	−	−A	−	±−	Aw	Aw	A−	d	−A	−	−	−	Aw	−
Esculin hydrolyzed	+−	−	d	+	+	+	+	−+	+	−A	+−	+	+	+	+	+	+	±−	+	+	w−	+	+−	−	+	+	+	+
Starch hydrolyzed	−	−	−+	−	+	−+	−+	+	+	−	+−	−	−	−	−+	−	−	±−	w−	Aw	w−	+−	+−	A	−	−	Aw	−
Gelatin digested	−	−	−	−+	−	−	−+	−w	−	−	−	−	d	−	−+	−+	−	−+	−	−	−	−w	−w	−	+w	+	−	−
Milk reaction	−c	−	−c	w	w	−w	−w	−w	−	−	−	−	d	−w	−w	−w	−	−+	−w	Aw	w−	−w	−w	−	+w	+w	Aw	c−
Meat digested	−c	−	−c	c−	c−	−c	−w	c−	−	−	+−	c−	d	c−	c−	−+	−	−c	w−	d	A−	d	−w	A	−	dc	−c	−
Indole produced	−	−	−c	−	c−	−c	−	c−	−	−c	−c	c−	−	c−	c−	−c	−	−c	c−	−c	−	c−	−c	−	+w	+−	−c	c−
Nitrate reduced	−	−	−	+	−	−	+	−	+	−A	−	−	−	+	+	+	−	−	+	+	+	+	+	−	+	+	−	−
H₂ produced	4	4	4, 2	4	4	4	−, 1	−	4	2, 3	4	4	4	4	4	4	4	±−	4	4	−+	4	4	−	4	4	2, 4	4
Butyrate produced	−	+	+	+	+	+	+	+	+	+	+	+	+	+	+	+	+	+	+	+	+	+	+	+	+	−	+	+
Other (see footnote)	c	−	c	c	d	−	−	−	e	−	−	−	l	g	g	g	−	−	−	−	−	−	−	−	−	c	−	−

[a] Symbols: −, negative reaction for 90–100% of strains; +, positive reaction for 90–100% of strains; ±, see footnote b; A, acid (pH below 5.5); w, weak reaction (pH 5.5–5.9, sugars); d, 40–60% of strains positive or milk digested; c, curd (milk); s, growth stimulated but pH usually not lowered; numbers (hydrogen) represent abundant (4) to negative on a "−" to "4+" scale. When two reactions are given, the usual reaction is listed first.

Unless otherwise noted (in the footnotes), erythritol and inositol are not fermented and lecithinase is not produced.

[b] ±, pH 6.0–7.0; −, (for this species), pH above 7.0.
[c] Lecithinase produced.
[d] Cellulose digested.
[e] Inositol fermented.
[f] Erythritol fermented.
[g] Lecithinase may be produced.

1356

Table 15.37.
Characteristics of nonsaccharoclastic species of **Eubacterium**[a]

Characteristics	4. *E. brachy*	7. *E. combesii*	10. *E. dolichum*	16. *E. lentum*	21. *E. nodatum*	32. *E. timidum*
Carbohydrates fermented	−	−	−[b]	−	−	−
Nitrate reduced	−	−	−	+	−	−
Gelatin hydrolyzed	−	+	−w	−	−	−
Esculin hydrolyzed	−	d	−	−	−	−
Hydrogen produced	2,4	4	−	−	−	−
Products[c] (PYG cultures)	ic, iv, ib, (f, p, b, l, v)	A, B, iv, l, ib (p, f)	b (l, a)	(a, l, s)	b, a (l, s)	(a, s)

[a] Symbols: −, negative reaction; +, positive reaction; w, weak reaction; d (reactions), variable; numbers (hydrogen) represent abundant (4) to negative on a "−" to "4+" scale. When two reactions are given, the usual reaction is listed first. No acid produced from amygdalin, arabinose, cellobiose, erythritol, esculin, fructose, glucose, glycogen, inositol, lactose, maltose, mannitol, mannose, melezitose, melibiose, raffinose, rhamnose, ribose, salicin, sorbitol, starch, sucrose, trehalose, or xylose. Indole and lecithinase are not produced.

[b] Growth may be stimulated in media containing certain carbohydrates (see text).

[c] ic, isocaproate; iv, isovalerate; ib, isobutyrate; a, acetate; b, butyrate; l, lactate; s, succinate; f, formate; p, propionate; v, valerate. Capital letters represent an amount of product equal to or greater than 1 meq/100 ml of culture; small letters represent an amount of product less than 1 meq/100 ml of culture; products in parentheses are not produced uniformly.

34. *E. ventriosum**
    ii. Maltose not acid
      j. Nitrate reduced
19. *E. multiforme*
      jj. Nitrate not reduced
        k. Copious H$_2$ produced
3. *E. biforme*
        kk. Copious H$_2$ not produced
9. *E. cylindroides*
  hh. Esculin not hydrolyzed
      j. Butanol produced
        k. Lactose acid
15. *E. hallii*
        kk. Lactose not acid
18. *E. moniliforme**
      jj. Butanol not produced
3. *E. biforme*
  dd. Mannose not acid
    e. Caproic acid produced
2. *E. alactolyticum*
  ee. Caproic acid not produced
    f. Starch acid
26. *E. ruminantium*
    ff. Starch not acid
      g. Major amount of lactic acid produced
        h. Glycogen acid
22. *E. plautii*
      hh. Glycogen not acid
        i. Cellulose digested
6. *E. cellulosolvens*
        ii. Cellulose not digested, twisted filamentous cells in chains
33. *E. tortuosum*
    gg. Major amount of lactic acid not produced
24. *E. ramulus*
2. Glucose not acid
  a. Isovaleric acid produced
    b. Gelatin hydrolyzed
7. *E. combesii*
    bb. Gelatin not hydrolyzed
4. *E. brachy*
  aa. Isovaleric acid not produced
    b. Major amount of acetic acid produced
28. *E. siraeum*
    bb. Major amount of acetic acid not produced
      c. Esculin hydrolyzed

* If gelatin hydrolyzed, see *Clostridium perfringens.*

23. *E. plexicaudatum*
  cc. Esculin not hydrolyzed
    d. Cells in long chains
10. *E. dolichum*
    dd. Cells not in long chains
21. *E. nodatum*
II. Butyric acid not produced
  1. Indole produced
31. *E. tenue*
  2. Indole not produced
    a. Raffinose acid
      b. Arabinose acid
8. *E. contortum*
      bb. Arabinose not acid
6. *E. cellulosolvens**
    aa. Raffinose not acid
      b. Glycogen acid
29. *E. suis*
      bb. Glycogen not acid
        c. Inositol acid
12. *E. fissicatena*
      cc. Inositol not acid
        d. Maltose acid
          e. Glucose acid
            f. Sucrose acid
              g. Mannose acid
1. *E. aerofaciens*
              gg. Mannose not acid
6. *E. cellulosolvens*
            ff. Sucrose not acid
13. *E. formicigenerans*
          ee. Glucose not acid
28. *E. siraeum*
        dd. Maltose not acid
          e. Isovaleric acid produced
4. *E. brachy*
          ee. Isovaleric acid not produced
            f. Glucose acid
              g. Copious H_2 produced
30. *E. tarantellus*†
              gg. Copious H_2 not produced
11. *E. eligens*†
            ff. Glucose not acid
              g. Nitrate reduced
16. *E. lentum*
              gg. Nitrate not reduced
                h. Copious H_2 produced
28. *E. siraeum*
                hh. Copious H_2 not produced
                  i. Fructose acid
11. *E. eligens*
                  ii. Fructose not acid
32. *E. timidum*

1. **Eubacterium aerofaciens** (Eggerth 1935) Prévot 1938, 295.[AL] (*Bacteroides aerofaciens* Eggerth 1935, 282.)

ae.ro.fa′ci.ens. Gr. n. *aer* air, gas; L. v. *facio* make, produce; N.L. part. adj. *aerofaciens* gas producing.

Description from Eggerth (1935), Prévot et al. (1967), Moore, Cato and Holdeman (1971a), and Moore and Holdeman (1974).

Cells from PY extract-glucose broth cultures are 0.3–0.8 × 1.5–4.7 μm and occur in chains of coccoid cells and rods. Chains contain 4 to 20 elements.

Surface colonies on horse blood agar are 1–3 mm in diameter, circular,

_10 μm_ PY   PYG

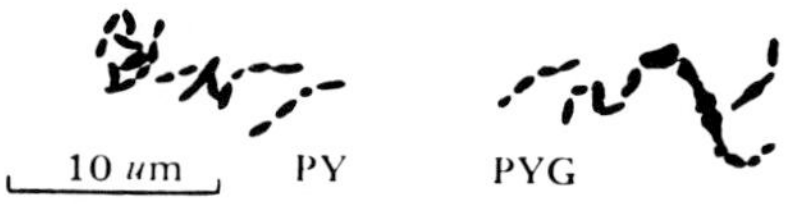

Figure 15.19. *E. aerofaciens*

entire to erose, convex, translucent or slightly opaque, smooth and glistening.

Glucose broth cultures are turbid with smooth sediment and terminal pH of 4.4–5.0.

---

* If fructose is acid, see *Clostridium ramosum*.
† If sucrose is acid or small amounts of H_2 are produced, see *Lachnospira multiparus*.

The optimum temperature for growth is 35–37°C. Most strains grow at 25–45°C. Growth and acid production are stimulated by 0.02% Tween 80 or 5% (v/v) rumen fluid. Growth may be inhibited by 20% bile. No growth occurs in broth containing 6.5% NaCl.

Gas production in PYG agar deep cultures is variable among strains.

Galactose and pectin are fermented. Dextrin usually is not fermented. Neutral red is reduced. Two of eight strains tested produce acetylmethylcarbinol. Ammonia production from peptone and arginine is variable among strains. Adonitol, dulcitol, glycerol, inulin and sorbose are not fermented. Hippurate is not hydrolyzed. No H_2S produced in SIM (sulfide-indole-motility) medium.

Products (in milliequivalents per 100 ml of culture) in PYG broth are acetic (0.5–1.8), formic (0.1–1.2), D- or DL-lactic (3.3–6.3) acids and large amounts of ethanol. Trace amounts of succinic acid sometimes are detected. Abundant H_2 is produced. Pyruvate is converted to acetate. Lactate is not utilized and threonine is not converted to propionate.

Of five strains tested, all are susceptible to chloramphenicol (12 μg/ml), clindamycin (1.6 μg/ml), erythromycin (3 μg/ml), and penicillin G (2 units/ml); one is resistant to tetracycline (6 μg/ml).

Other characteristics of the species are given in Table 15.36.

Isolated from feces and intestinal contents of man (Peach et al., 1973; Moore et al., 1978; Koornhof et al., 1979; Moore et al., 1981) and other animals. Constitutes 6–10% of the human fecal flora (Moore and Holdeman, 1974; Holdeman et al., 1976) and 2.6% of the intestinal flora of swine (Russell, 1979). Occasionally isolated from human blood cultures and various kinds of infections, including subacute bacterial endocarditis (Sans and Crowder, 1973), renal abscess fluid, and appendiceal abscess.

The mol% G + C of the DNA has not been determined.

Type strain: ATCC 25986.

Further comments. On original isolation these organisms often appear as pleomorphic cocci to short rods and may be confused with anaerobic streptococci or lactobacilli. However, the production of copious amounts of H_2 clearly differentiates it from lactobacilli and anaerobic streptococci.

Moore and Holdeman (1974) recognize three phenotypic groups of *E. aerofaciens* isolated from feces. Strains that do not ferment sucrose are designated "*E. aerofaciens* II" and those that do not ferment cellobiose and produce little or no acid in salicin are designated "*E. aerofaciens* III."

Whole cell suspensions of one strain of *E. aerofaciens* II convert deoxycholate to 12-ketolithocholate, but lysozyme treatment of cells is required for significant conversion of cholate to 7-ketodeoxycholate (Hylemon and Stellwag, 1976).

2. **Eubacterium alactolyticum** (Prévot and Taffanel 1942) Holdeman and Moore 1970, 23.[AL] (*Ramibacterium alactolyticum* Prévot and Taffanel 1942, 261.)

a.lac.to.ly′ti.cum. Gr. pref. *a* not; L. n. *lac, lactis* milk; Gr. adj. *lyticus* dissolving; N.L. neut. adj. *alactolyticum* not milk digesting.

Description from Holdeman et al. (1967) and Holdeman and Moore (1972) who studied the type strain and 37 other strains with similar characteristics.

Cells in PYG broth cultures are 0.3–0.6 × 1.6–7.5 μm in pairs resembling flying birds, clumps, and "Chinese character" arrangement.

Cell walls contain *meso*-diaminopimelic acid (meso-DAP).

Surface colonies on horse blood agar (2–3 days) are punctate to 0.5 mm, circular, entire, convex to pulvinate, smooth, and shiny. Smaller colonies may be translucent, larger colonies usually are opaque.

Glucose broth cultures usually are turbid with granular or smooth

(occasionally ropy or flocculent) sediment; a few strains produce sediment without turbidity in broth cultures. Terminal pH in glucose cultures usually is 5.0–5.6; that of a few strains is 5.8–6.0.

The optimum temperature for growth is 35–37°C. Most strains grow at 30°C; some strains grow at 25 and 45°C. Growth is stimulated by fermentable carbohydrate, is unaffected by 0.02% Tween 80 or 5% rumen fluid, and is inhibited by 20% bile.

Moderate to abundant gas is produced in glucose agar deep cultures.

Neutral red is reduced; resazurin usually is not reduced. The type strain does not ferment adonitol, dextrin, dulcitol, galactose, glycerol, inulin or sorbose and does not produce ammonia from peptone, arginine or threonine. Hippurate is not hydrolyzed; H_2S not produced in SIM medium. Acetyl methyl carbinol is not produced.

Products (milliequivalents per 100 ml of culture) in PYG broth cultures are formate (0.01–1.0), acetate (0.01–0.5), butyrate (0.1–0.5), and caproate (0.2–2.0). Small amounts of caprylate also may be produced. Abundant H_2 is produced. Pyruvate is converted to acetate and formate. Lactate is not utilized; threonine is not converted to propionate.

The type and 24 other strains tested are susceptible to chloramphenicol (12 μg/ml), clindamycin (1.6 μg/ml), erythromycin (3 μg/ml), penicillin G (2 units/ml), and tetracycline (6 μg/ml).

Other characteristics of the species are given in Table 15.36.

Isolated from dental calculus and the gingival crevice in periodontal disease (Moore et al., 1982), root canals (Kantz and Henry, 1974), and from various other kinds of infections including purulent pleurisy, jugal cellulitis, and postoperative wounds; abscesses of the brain, lung, intestinal tract, and mouth.

The mol% G + C of the DNA has not been determined.

Type strain: ATCC 23263 (CIPP DO-4).

Further comments. Although this species is rather frequent in clinical infections, it may be overlooked because of its "diphtheroid" appearance. However, its frequent occurrence in infections associated with the oral flora suggests that it may be clinically significant. In the presence of glucose and propionic acid, the organism also synthesizes valeric and heptanoic acids.

3. **Eubacterium biforme** (Eggerth 1935) Prévot 1938, 295.[AL] (*Bacteroides biformis* Eggerth 1935, 283.)

bi.for′me. L. adj. *biformis* two-formed (pertaining to cellular morphology).

The description is from Holdeman and Moore (1974).

Cells from PYG broth cultures are nonmotile, 0.6–2.7 μm in diameter × 1.6–15 μm long in pairs and chains. Definite rods and coccoid forms often occur in the same chain. Central swellings ("pelton de jardinier") are present in some cultures, particularly among cells from growth on solid media.

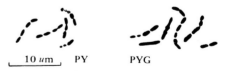

Figure 15.21. *E. biforme*

After incubation for 5 days in RGCA, subsurface colonies are 0.5–1.0 mm in diameter, white to tan, and usually lenticular and translucent, opaque or translucent with opaque centers, sometimes with diffuse edges. Occasional colonies look like "woolly balls." Surface colonies on blood agar plates incubated for 2 days are 0.5–2.0 mm in diameter, circular, entire to slightly erose, flat to low convex, white, shiny, smooth, translucent with dense centers, and nonhemolytic.

PYG-0.2% Tween 80 broth cultures have abundant growth with smooth sediment with little or no turbidity and a pH of 4.5–5.0 in 1–2 days.

The temperature for optimal growth is 37°C; some strains grow at 45°C. Most strains do not grow at 30°C. Growth of most strains is stimulated by 0.02% Tween 80 but not by rumen fluid and may be

Figure 15.20. *E. alactolyticum*

inhibited by 20% bile. Growth is markedly stimulated by fermentable carbohydrate.

Variable amounts of gas are detected in glucose agar deep cultures.

Most strains produce acid from galactose. Neutral red is reduced. A small amount of ammonia may be produced in chopped meat cultures.

Products (milliequivalents per 100 ml of culture) in PYG-0.02% Tween-80 cultures are butyric (0.4–1.2), caproic (0.01–0.07), and lactic (2.0–5.0) acids. Moderate to abundant H_2 is produced from PYG-Tween 80 cultures. Caproic acid is produced irregularly and probably is dependent upon the amount of growth and the age of the culture analyzed. Pyruvate is converted to butyrate, often with lactate and acetate, sometimes with formate and a small amount of caproate. Lactate and gluconate are not utilized.

The type strain is susceptible to 12 μg/ml of chloramphenicol, 1.6 μg/ml of clindamycin, 3 μg/ml of erythromycin, 2 units/ml of penicillin G, and 6 μg/ml of tetracycline.

Other characteristics of the species are given in Table 15.36.

Isolated from human feces at 1–3% of the flora (Moore and Holdeman, 1974; Holdeman et al., 1976).

The mol% G + C of the DNA is 32 (T_m).

Type strain: ATCC 27806.

Further comments. Definite rod-shaped cells may not appear in all cultures of each strain, but when they do, observation of cocci attached to rods is the only assurance that the culture is not a mixture. The production of caproic acid in small amounts is a most distinguishing characteristic, but sometimes is detected only on repeated culture. The morphology of the cells and the production of H_2 and of small amounts of caproic acid by *E. biforme* help distinguish this species from *E. cylindroides.*

4. **Eubacterium brachy** Holdeman, Cato, Burmeister and Moore 1980, 167.[VP]

bra'chy (bra'ky). N.L. neut. adj. *brachy* from Gr. *brachy* short; referring to the length of the cells.

The description is from Holdeman et al., 1980.

Cells from PY broth cultures are 0.4–0.8 × 1.0–3.0 μm and occur predominantly in long chains. Although some cells in a chain might be coccoid, definite rods can be seen upon close examination of the smears.

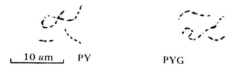

Figure 15.22. *E. brachy*

After 5 days of incubation, subsurface colonies in supplemented brain heart infusion agar (BHIA-S, Holdeman et al., 1977) enriched with 5% v/v rabbit serum and 0.02 g each of sodium formate and ammonium fumarate and 400 μg thiamine pyrophosphate/100 ml are 0.5–2.0 mm in diameter, lenticular to trifoliate, and transparent to opaque. After incubation for 2 days, surface colonies on BHIA-S streak tubes are less than 0.5 mm in diameter, circular, entire, low convex, white, shiny, and smooth. Surface colonies on chopped meat broth agar are similar to those on BHIA-S but are larger (1.0 mm in diameter). Surface growth is obtained more reliably on chopped meat broth agar than on BHIA-S tubes or anaerobic blood agar plates. The one strain (of five strains tested) that grew on anaerobically incubated blood agar plates did not lyse the rabbit red blood cells.

Broth cultures have a small amount of fine, granular to smooth sediment, often without turbidity. Broth cultures often require incubation for 2–3 days to attain maximum growth, which at best is 2+, as graded on a scale of negative to 4+. Carbohydrates are not fermented.

The temperature for optimum growth is 37°C; most strains do not grow at 30 or 45°C. Growth is inhibited by 6.5% NaCl and slightly inhibited by 2% oxgall (20% bile). Growth is neither inhibited nor

stimulated by the addition of 0.02% Tween 80, 5% (v/v) rumen fluid, or 10% (v/v) serum with thiamine pyrophosphate.

A small amount of gas is produced in glucose agar deep cultures.

Ammonia is produced in PY broth cultures.

Products (milliequivalents per 100 ml of culture) in PYG broth are acetate (0.1–0.4), *iso*-butyrate (0.1–0.4), *iso*-valerate (0.2–0.4), and *iso*-caproate (0.2–0.5), often with trace amounts of formate, propionate, butyrate or valerate. Headspace gas of 5-day-old PY broth cultures contains 2–3% H_2.

Strains tested are susceptible to chloramphenicol (12 μg/ml), clindamycin (1.6 μg/ml), erythromycin (3 μg/ml), penicillin G (2 units/ml), and tetracycline (6 μg/ml).

Other characteristics of the species are given in Table 15.37.

Isolated from subgingival samples and from supragingival tooth scrapings from persons with periodontal disease. One isolate from a human lung abscess has been reported (Rochford, 1980).

The mol% G + C of the DNA could not be determined because of the sparse amount of growth.

Type strain: ATCC 33089.

Further comments. Because of the fermentation products and the occurrence of short cells in chains, *E. brachy* resembles *Peptostreptococcus anaerobius.* It does not grow as well in usual media as does *P. anaerobius,* and definite rod-shaped cells can be seen in Gram stains. Additionally, growth of *E. brachy* is not inhibited by sodium polyanetholsulfonate.

5. **Eubacterium budayi** (Le Blaye and Guggenheim 1914) Holdeman and Moore 1970, 23.[AL] (*Bacterium budayi* Le Blaye and Guggenheim 1914, 402.)

bu.day'i. N.L. gen. n. *budayi* of Buday; named for the bacteriologist who first isolated the organism.

The description is from Prévot et al. (1967) and study of the type strain.

Cells in PYG broth cultures are 0.8–1.7 μm in diameter × 3.0–greater than 78 μm in length and are straight or slightly curved. Cells occur singly, in pairs with angular and parallel arrangements. Short forms may occur in media without a fermentable carbohydrate.

Figure 15.23. *E. budayi*

Surface colonies on McClung-Toabe egg yolk agar (EYA) are 4–5 mm, irregular, low convex, scalloped, yellowish and are surrounded by a small opaque zone in the agar. Colonies in deep agar look like snowflakes, often with rhizoids.

Glucose broth cultures are turbid with smooth sediment and pH of 5.0.

The optimum temperature for growth is 37°C. Moderately good growth occurs at 25–30°C but not at 45°C.

Growth is stimulated by 0.02% Tween 80. Moderate growth occurs in peptone-yeast extract broth; growth usually is stimulated by a fermentable carbohydrate. Growth is inhibited by 6.5% NaCl. Growth occurs in PYG with 20% bile but fermentation may be suppressed.

Abundant gas is produced in PYG agar deep cultures.

Galactose is fermented. Adonitol, dulcitol, inulin, pectin and sorbose are not fermented; little or no ammonia is produced from peptone, arginine, threonine or chopped meat medium; and H_2S is not produced in SIM medium. Slight lecithinase activity may be observed on EYA.

Products (milliequivalents per 100 ml of culture) in PYG cultures are acetic (0.4), butyric (0.8), L-(+)-lactic (1.4), and pyruvic (0.1) acids. Copious H_2 is produced.

Other characteristics of the species are given in Table 15.36.

Isolated from a cadaver by Buday and from poorly sterilized catgut, mud, and soil by Prévot et al.

The mol% G + C of the DNA has not been reported.

Type strain: ATCC 25541 (CIPP strain ECI).

Further comments. The relationship between *E. budayi* and *E. nitritogenes* is not clear.

6. **Eubacterium cellulosolvens** (Bryant et al. 1958) Holdeman and Moore 1972, 39.[AL] (*Cillobacterium cellulosolvens* Bryant, Small, Bouma and Robinson 1958, 529.)

cel.lu.lo.sol'vens. N.L. n. *cellulosum* cellulose; L. part. adj. *solvens* dissolving; N.L. part. adj. *cellulosolvens* cellulose dissolving.

The description is from Bryant et al. (1958), van Gylswyk and Hoffman (1970) and from study of one strain from van Gylswyk and Hoffman and one isolate from the intestinal tract of a hog.

Cells in PYG broth usually are motile and peritrichous and are 0.3–0.8 × 0.8–3.1 µm. They occur singly and in pairs and clumps.

Figure 15.24. *E. cellulosolvens*

Surface colonies on anaerobic streak tubes of BHIA-S are 3–5 mm, circular, entire, flat to slightly convex, translucent, and light tan to colorless. Subsurface colonies are lenticular. No growth is produced on the surface of blood agar plates incubated in anaerobic jars.

PYG-Tween 80 broth cultures are turbid with smooth sediment and have a terminal pH of 4.9–5.4.

The temperature for optimum growth is 37°C. Growth is slightly inhibited at 45°C and markedly inhibited at 30°C. Growth is stimulated by 0.02% Tween 80. Growth is inhibited by 20% bile or 6.5% NaCl and at pH 9.6. Poor to moderate growth occurs without fermentable carbohydrate.

Variable amounts of gas are detected in glucose agar deep cultures of the strains available for study.

Cellulose is digested. Pectin is fermented by some strains; xylan usually is not fermented, or only weakly so; dextrin and gum arabic are not fermented. Adonitol, dextrin, dulcitol, galactose, glycerol, inulin and sorbose are not fermented. Resazurin is not reduced. Hippurate is not hydrolyzed and acetyl methyl carbinol is not produced. Ammonia is not produced from arginine or threonine and only small amounts are produced from peptone. H₂S is not detected in SIM medium.

Products (milliequivalents per 100 ml of culture) in PYG broth are acetic (0.1–0.3), butyric (0.1–0.3) and D-(−)-lactic (3.0–5.0) acids; trace amounts of formic or succinic acids also may be detected. Little or no H₂ is detected in headspace gas of cultures of strains available for study. Pyruvate and lactate are not utilized; threonine is not converted to propionate.

Other characteristics of the species are given in Table 15.36.

Isolated from rumen contents of sheep and cows.

The mol% G + C of the DNA has not been determined.

Type strain: Bryant B348, which has been lost (M.P. Bryant, personal communication).

Further comments. The strains studied by Bryant et al. (1958) produce primarily lactic with small amounts of acetic and formic but no butyric, propionic and succinic acids or H₂. Strains isolated by van Gylswyk and Hoffman also produce small amounts of butyric and valeric acids. We detected butyric acid (0.25–0.40 meq/100 ml) in the one van Glyswyk-Hoffman strain we studied. Van Gylswyk and Hoffman suggested that the species description of Bryant et al. be amended to include organisms that produce a small amount of butyrate and valerate. In other characteristics, the strains are very much like those described by Bryant et al. Recognition of these organisms as eubacteria

rather than lactobacilli will be difficult with those strains that do not produce butyrate.

7. **Eubacterium combesii** (Prévot and Laplanche 1947) Holdeman and Moore 1970, 23.[AL] (*Cillobacterium combesi* Prévot and Laplanche 1947, 688.)

com.be'si.i. N.L. gen. n. *combesii* of Combes; named for Combes.

This description is from Prévot et al. (1967) and study of the type and 12 other strains.

Cells are motile and peritrichous. Cells in PYG broth cultures are 0.6–0.8 × 3.0–10.0 µm and occur as single cells and in pairs, short chains, and palisade arrangements.

Figure 15.25. *E. combesii*

Surface colonies of the type strain on horse blood agar are 0.5–2.0 mm, circular, entire to irregular, convex, semiopaque, whitish yellow, shiny, and smooth with fine granular or mosaic appearance when viewed by obliquely transmitted light. The type and most other strains are β-hemolytic.

Glucose broth cultures are turbid with flocculent sediment and pH of 6.4.

The optimum temperature for growth is 35–37°C. The range of temperatures at which growth occurs is variable.

Moderate to abundant gas is produced in glucose agar deep cultures.

Ammonia is produced from peptone; neutral red and resazurin are reduced. H₂S usually is produced in SIM medium.

Products (milliequivalents per 100 ml of culture) in PYG broth cultures are acetate (1.5–3.8), *iso*-butyrate (0.05–0.37), butyrate (trace to 1.8), *iso*-valerate (0.1–1.3), and lactate (0.1–0.7), usually with ethanol, propanol, *iso*-butanol, butanol and *iso*-amyl alcohols. Variable amounts of formate and small amounts of propionate may be detected. Abundant H₂ is produced. Pyruvate is converted to acetate. Lactate is not utilized and threonine is not converted to propionate.

Four clinical isolates tested are susceptible to chloramphenicol (12 µg/ml), clindamycin (1.6 µg/ml), penicillin G (2 units/ml), and tetracycline (6 µg/ml). One of these strains was resistant to erythromycin (3 µg/ml).

Culture supernatants of 1-day-old chopped meat broth cultures of two strains tested are not toxic for mice (0.5 ml intraperitoneally).

Other characteristics of the species are given in Table 15.37.

Isolated from human infections and African soil.

The mol% G + C of the DNA has not been reported.

Type strain: ATCC 25545 (CIPP A13D).

Further comments. Although no spores have been detected in the strains on which this description is based, some strains that are otherwise like *E. combesii* produce spores and have been identified by us as *Clostridium subterminale*.

8. **Eubacterium contortum** (Prévot 1947) Holdeman, Cato and Moore 1971, 306.[AL] (*Catenabacterium contortum* Prévot 1947, 414.)

con.tor'tum. L. neut. adj. *contortum* twisted.

This description is from study of the type and 10 other strains.

Cells in PYG broth cultures are ovoid, 0.4–0.9 × 1.4–2.3 µm and occur in pairs and short or long twisted chains.

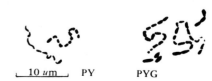

Figure 15.26. *E. contortum*

Surface colonies on horse blood agar are 0.5 mm, circular, entire to erose, low convex and translucent. There may be slight clearing of the blood under the area of heavy growth.

Glucose broth cultures usually are turbid with heavy sediment (gelatinous or granular) and a terminal pH of 4.8–5.0.

The optimum temperature for growth is 35–37°C. Strains grow well at 30 and 45°C but poorly at 25°C. Growth is not affected by 0.02% Tween 80; growth of some strains is inhibited by 20% bile.

Abundant gas is detected in PYG agar deep cultures.

Acid is produced from galactose. No acid is produced from adonitol, dextrin, dulcitol, glycerol, inulin or sorbose. Little or no ammonia is produced from peptone and no ammonia is produced from arginine or threonine. Urease and acetyl methyl carbinol are not produced. H₂S is not produced in SIM medium. Hippurate is not hydrolyzed.

Products (milliequivalents per 100 ml of culture) in PYG broth are acetic (0.5–4.6) and formic (0.1–5.3) acids, with large amounts of ethanol. Trace amounts of lactic and succinic acids may be detected.

The type and seven other strains tested are susceptible to chloramphenicol (12 μg/ml). Two of these strains are resistant to clindamycin (1.6 μg/ml), four are resistant to erythromycin (3 μg/ml), one is resistant to penicillin G (2 units/ml), and three are resistant to tetracycline (6 μg/ml).

Other characteristics of the species are given in Table 15.36.

Isolated from human feces, vaginal swab, and human clinical specimens, including blood (postkidney transplant), abdominal aortic anurism, and wound.

Sera from 36–50% of patients with Crohn's disease are reported to contain agglutinating antibodies to strains of *E. contortum* (Van de Merwe, 1981).

The mol% G + C of the DNA of one fecal strain is 45 (F. Wensinck, personal communication).

Type strain: ATCC 25540.

9. **Eubacterium cylindroides** (Rocchi 1908) Holdeman and Moore 1970, 23.[AL] (*Bacterium cylindroides* Rocchi 1908, 479.)

cy.lin.dro.i′des. Gr. n. *cylindrus* a cylinder; Gr. n. *idus* form, shape; N.L. neut. adj. *cylindroides* cylinder-shaped.

The description is from Cato et al. (1974) and study of the type and 20 other strains.

Cells in PYG broth cultures are 0.7–1.0 μm in diameter by 1.5–18.0 μm long and occur singly, in pairs and in long chains. Gram-positive cells usually are seen only in young cultures (log phase or early stationary phase of growth); cells of older cultures destain very easily.

Figure 15.27. *E. cylindroides*

Surface colonies on horse blood agar incubated for 2 days are punctate to 2 mm, circular to slightly irregular, entire to diffuse, flat to low convex, and translucent. They sometimes have a mottled appearance when viewed by obliquely transmitted light.

Glucose broth cultures are turbid with smooth or ropy sediment and terminal pH of 4.8–5.5.

Growth is most rapid at 37–45°C; most strains grow at 25°C. Glucose or fructose enhances growth. Cell density is not markedly enhanced by 0.02% Tween 80. Rumen fluid (5% final concentration) often increases the amount of acid produced. Comparatively less growth or acid production occurs in glucose media containing 20% bile.

Little or no gas is detected in glucose agar deep cultures.

The type strain ferments inulin and pectin (weakly) and produces ammonia from peptone. Neutral red is reduced. The type strain does not ferment adonitol, dextrin, dulcitol, galactose, glycerol or sorbose and does not hydrolyze hippurate or produce acetyl methyl carbinol.

Acetyl methyl carbinol is produced by 8 of 21 strains tested (Cato et al., 1974).

Products (milliequivalents per 100 ml of culture) in PYG broth cultures are butyric (0.3–1.4) and DL-lactic (1.0–3.0) acids, often with smaller amounts of acetic, formic and succinic acids. Little or no H₂ is detected in headspace gas. Pyruvate is converted to acetate, formate and butyrate. Lactate is not utilized; threonine is not converted to propionate.

The type strain is susceptible to chloramphenicol (12 μg/ml), clindamycin (1.6 μg/ml), erythromycin (3 μg/ml), and penicillin G (2 units/ml). Susceptibility to tetracycline (6 μg/ml) is variable.

Other characteristics of the species are given in Table 15.36.

Isolated from human feces.

The mol% G + C of the DNA of the type strain is 31.

Type strain: ATCC 27803.

Further comments. Because strains of *E. cylindroides* may decolorize readily, they may be confused with *Fusobacterium prausnitzii*. Strains of *E. cylindroides* are differentiated from those of *F. prausnitzii* in that all strains of *E. cylindroides* produce acid (below pH 5.5) from mannose, usually from glucose and fructose, and sometimes from salicin and sucrose (Cato et al., 1974). Strains of *F. prausnitzii* do not lower the pH in these sugar media below 5.5.

10. **Eubacterium dolichum** Moore, Johnson and Holdeman 1976, 246.[AL]

do′li.chum (do′li.kum). Gr. adj. *dolichum* long, referring to the long chains formed in broth cultures.

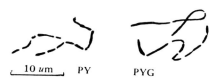

Figure 15.28. *E. dolichum*

Cells from PYG broth cultures are nonmotile, thin rods in long chains. The ends of the cells usually are slightly tapered. Cells are 0.4–0.6 × 1.6–6.0 μm.

Subsurface colonies in RGCA incubated for 5 days are 0.2–0.5 mm in diameter, lenticular and transparent to translucent. Surface colonies on BHIA-S streak tubes are 0.5–1.5 mm in diameter, circular, entire, convex, opaque to translucent, granular, dull and smooth. None of the four strains tested grew on anaerobically incubated blood agar plates. Three of four strains tested grew on EYA plates incubated anaerobically.

PY broth cultures have a stringly to ropy sediment without marked turbidity. In general, growth is poor in most broth media. Growth often is enhanced in a broth medium containing fructose, glucose, maltose, starch or sucrose, although the pH is not necessarily lowered.

Best growth is obtained at 37°C, although most strains grow at 45°C and, to a limited extent, at 30°C. Growth usually is stimulated by 0.02% Tween 80 and sometimes is inhibited by 20% bile.

No gas is produced in glucose agar deep cultures.

The type strain does not ferment adonitol, dextrin, dulcitol, galactose, glycerol, inulin or sorbose. Ammonia is not produced from peptone, arginine or chopped meat. Hippurate is not hydrolyzed. The type strain reduces neutral red and resazurin.

Products (milliequivalents per 100 ml of culture) in PYG broth are butyrate (0.2–1.7), often with small amounts of acetate and lactate. Little or no H₂ is detected in the headspace gas. Lactate and pyruvate are not utilized. Threonine is not converted to propionate.

Other characteristics of the species are given in Table 15.37.

Isolated from human feces.

The mol% G + C of the DNA could not be determined because of the sparse growth (Moore et al., 1976).

Type strain: ATCC 29143.

11. **Eubacterium eligens** Holdeman and Moore 1974, 273.[AL]

el′i.gens. L. adj. *eligens* choosy, referring to its generally poor growth without fermentable carbohydrate.

The description is from Holdeman and Moore (1974).

Cells from PYG broth cultures are straight to slightly curved rods, occurring singly or in pairs, occasionally in short chains of three to six cells, and are 0.3–0.8 × 1.9–4.9 µm. Central or eccentric swellings predominate in some strains. Twenty-four of 33 strains tested were motile and peritrichous. Cells decolorize easily in older cultures, but some Gram-positive cells usually are seen.

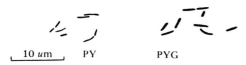

Figure 15.29. *E. eligens*

After incubation for 5 days in RGCA, subsurface colonies are 0.5–1.0 mm in diameter, white to tan, and lenticular. Surface colonies on BHIA-S are punctiform, circular, entire, transparent to translucent, white to tan, and smooth. Surface colonies on 2-day-old anaerobic BHIA-S blood agar plates are 0.5–1.0 mm in diameter, circular, entire, convex, smooth, translucent to semiopaque, shiny, white, and not hemolytic (sheep blood).

PY-fructose broth cultures are turbid with a smooth sediment and pH of 4.6–5.8 in 5 days.

The temperature for optimal growth is 37°C. Most strains grow well at 45°C; little or no growth occurs at 30°C. Tween 80 (0.02%) sometimes enhances growth. Growth usually is not affected by addition of 10–30% (v/v) rumen fluid. Bile (20%) inhibits acid production and may inhibit growth. There is limited, if any, growth in PY broth without a fermentable carbohydrate.

Little or no gas is produced in PYG agar deep cultures.

Neutral red is reduced. Adonitol, dextrin, dulcitol, galactose, glycerol, inulin, and sorbose are not fermented. Ammonia is not produced from PY broth or arginine. Acetyl methyl carbinol is not produced. H_2S is not produced in SIM medium.

Products (milliequivalents per 100 ml of culture) in PYG broth cultures are acetic (0.7–2.3), formic (0.8–3.7), lactic (0.1–0.9) acids and ethanol, occasionally with a trace of succinic acid. No H_2 is detected in headspace gas. Pyruvate and gluconate are converted to acetate, formate, lactate, and ethanol. Lactate is not utilized.

Other characteristics of the species are given in Table 15.36.

Isolated from human feces.

The mol% G + C of the DNA is 36 (T_m).

Type strain: VPI C15-48 (ATCC 27750).

Further comments. E. eligens differs from E. aerofaciens by the relatively small amount of lactate produced, extremely poor growth without fermentable carbohydrate, cellular morphology, motility of some strains, and lack of H_2 production.

12. **Eubacterium fissicatena** Taylor 1972, 462.[AL]

fiss.i.ca.te′na. L. n. *fissi* a cleft; L. n. *catena* a chain; N.L. n. *fissicatena* a broken chain.

The description is from Taylor (1972) and study of the type and seven other strains received from Dr. Taylor.

Cells in glucose broth cultures are motile, 0.3–0.8 × 1.4–8.0 µm, and occur singly and in pairs, often with parallel arrangement.

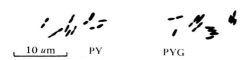

Figure 15.30. *E. fissicatena*

Surface colonies on blood agar are 0.5 mm, circular, entire to slightly scalloped, low convex, opaque, grayish white, shiny and smooth.

Glucose broth cultures are turbid with smooth sediment and terminal pH of 5.0–5.5

The optimum temperature for growth is 37°C. All strains grow at 25°C but rarely grow at 13°C. There usually is no growth at 45°C. Growth occurs at pH 9.6 and is unaffected by Tween 80. Although growth is not inhibited in PY broth containing 20% bile, acid production is decreased. Growth is inhibited by 6.5% NaCl.

Abundant gas is detected in PYG agar deep cultures.

Galactose is fermented. Neutral red is reduced. Small amounts of ammonia are produced from peptone. All strains produce H_2S from YE broth (Taylor, 1972). Some strains produce H_2S in SIM medium. Adonitol, dextrin, dulcitol, glycerol, inulin and sorbose are not fermented. Hippurate is not hydrolyzed. Acetylmethylcarbinol is not produced.

Hydroxyethylflavine is produced from riboflavine.

Products (milliequivalents per 100 ml of culture) in PYG broth are acetic (0.7–2.5) and formic (0.4–2.1) acids and large amounts of ethanol. Trace amounts of lactic and succinic acids may be detected. Carbon dioxide and abundant H_2 are produced. Pyruvate is converted to acetate, formate and ethanol. Some cultures produce methane in media without glucose. Lactate is not utilized. Threonine is not converted to propionate.

Other characteristics of the species are given in Table 15.36.

Isolated from the alimentary tract of goats.

The mol% G + C of the DNA is 45.5 (Taylor, 1972).

Type strain: NCIB 10446 (Taylor A2a).

13. **Eubacterium formicigenerans** Holdeman and Moore 1974, 274.[AL]

for.mi.ci.gen′er.ans. N.L. adj. *formicigenerans* formic acid-producing; referring to its production of large amounts of formic acid from fermentation of carbohydrates.

The description is from Holdeman and Moore (1974).

Cells are nonmotile rods, occurring in chains or pairs, and are 0.6–1.4 × 0.8–4.7 µm.

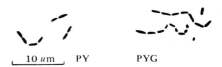

Figure 15.31. *E. formicigenerans*

After incubation for 5 days in RGCA, colonies are 0.5–1.0 mm in diameter, white to tan, circular to lenticular, and often have fuzzy edges or a woolly ball appearance. Surface colonies on BHIA-S streak tubes or blood agar plates are 0.5–3.0 mm in diameter, circular to slightly irregular, entire to slightly erose, convex to umbonate, opaque, white to tan, shiny and smooth. Slight greening is produced on sheep blood agar by the type strain, but not by six other strains tested.

PYG broth cultures have little or no turbidity, stringy or flocculent (occasionally smooth) sediment, and pH of 4.7–5.0 in 5 days. Growth in PYG generally is not affected by the addition of 0.02% Tween 80 or 10% (v/v) rumen fluid. The growth in PY broth without fermentable carbohydrate is less than in PYG broth. Growth is inhibited by 6.5% NaCl.

The temperature for optimal growth is 37°C. Most strains grow moderately well at 30 and 45°C, but usually not at 25°C.

Moderate to abundant gas is produced in glucose agar deep cultures.

Neutral red and resazurin are reduced. Galactose is fermented. Hippurate hydrolysis is variable among strains. Adonitol, dextrin, dulcitol, glycerol, inulin and sorbose are not fermented. The type strain does not produce ammonia from peptone or arginine.

Products (milliequivalents per 100 ml of culture) in PYG broth are acetic (1.2–4.0), formic (1.1–4.0), lactic (0.2–1.8) acids and large amounts of ethanol. Moderate H_2 is detected in headspace gas from cultures containing a fermentable carbohydrate. Pyruvate is converted to acetate, formate, and ethanol, usually with a trace of lactate and sometimes with a trace of succinate. Lactate and gluconate are not utilized. Threonine is not converted to propionate.

Other characteristics of the species are given in Table 15.36.

Isolated from human feces.

The mol% G + C of the DNA is 40–44 (T_m).

Type strain: ATCC 27755.

Further comments. The fermentation acids and ethanol produced and lack of fermentation of cellobiose help differentiate *E. formicigenerans* from other species with which it might be confused.

14. **Eubacterium hadrum** Moore, Johnson and Holdeman 1976, 247.[AL]

had'rum. Gr. adj. *hadrum* thick, bulky (referring to the relatively large size of the cell).

The description is from Moore et al. (1976), who studied the type and 17 other similar strains.

Cells from PYG broth cultures are nonmotile rods of uniform width with rounded ends, occurring in pairs and short chains, and are 0.7–1.0 μm in diameter and 3.0–10.0 μm long.

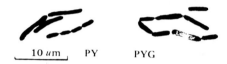

Figure 15.32. *E. hadrum*

Subsurface colonies in RGCA incubated for 5 days are woolly balls 1–2 mm in diameter. Surface colonies on BHIA-S streak tubes or BHIA-S blood agar plates incubated anaerobically are 2–3 mm in diameter, circular, entire to erose, convex, opaque to translucent and smooth. Some strains produce slight greening hemolysis on rabbit blood. The type and one other strain tested did not grow on EYA.

Glucose broth cultures have abundant growth and are turbid with smooth, sometimes ropy, sediment. After incubation for 1 day, the pH in glucose broth cultures is 4.9–5.4.

The optimum temperature for growth is 37°C; good growth also occurs at 45°C but lesser growth is obtained at 30°C. Growth is stimulated by 0.02% Tween 80. No growth occurs in medium containing 6.5% NaCl. Growth occurs in PYG with 20% bile but no acid is produced.

Abundant gas is observed in glucose agar deep cultures.

Products (milliequivalents per 100 ml of culture) from PYG broth cultures are butyric (1–6) and lactic (0.1–2) acids with little or no acetic acid; trace amounts of pyruvic and succinic acids sometimes may be detected. Abundant H_2 is produced from fermentation of carbohydrates. Pyruvate is converted to butyrate and acetate; little or no lactate is utilized.

The type strain ferments galactose and sorbose but does not ferment adonitol or glycerol. Neutral red and resazurin are reduced. Hippurate is not hydrolyzed. Ammonia is not produced from peptone or arginine.

Other characteristics of the species are given in Table 15.36.

Isolated from human feces.

The mol% G + C of the DNA is 32–33.

Type strain: VPI B2-52 (ATCC 29173).

Further comments. Although *E. hadrum* has large cells typical of those found in many species of *Clostridium*, and even though slight swellings in the cells are very occasionally observed, no typical spores are seen and cultures with swellings do not survive heating at 80°C for 10 min.

15. **Eubacterium hallii** Holdeman and Moore 1974, 275.[AL]

hall'i.i. N.L. gen. n. *hallii* of Hall (named for Ivan C. Hall, an American bacteriologist).

The description is from Holdeman and Moore (1974).

Cells from PYG broth cultures are nonmotile rods occurring singly and in pairs, occasionally in short chains. Cells are 0.8–0.8–2.4 μm in diameter × 4.7 to more than 25.0 μm in length (Fig. 15.33). Although subterminal and terminal swellings sometimes are observed, cultures do not survive heating at 80°C for 10 min.

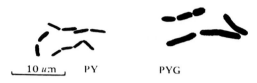

Figure 15.33. *E. hallii*

Subsurface colonies in RGCA incubated for 5 days are 0.5–1.0 mm in diameter and lenticular or woolly balls. Surface colonies in BHIA-S streak tubes or BHIA-S blood agar plates are 1–2 mm in diameter, circular to slightly irregular, entire to slightly erose, low convex, semiopaque, white to yellowish, smooth, shiny and not hemolytic (rabbit blood).

Glucose broth cultures have abundant growth and smooth or ropy sediment with little or no turbidity. The pH of 5 day-old cultures in PYG is 4.7–5.5. Moderate growth occurs in PY broth without a fermentable carbohydrate.

The temperature for optimum growth is 37°C. Most strains grow at 30°C, but less well than at 37°C. Strains usually do not grow at 25 and 45°C. Growth is inhibited by 6.5% NaCl, may be inhibited by 20% bile, but is unaffected by 0.02% Tween 80. Growth is poor in the absence of fermentable carbohydrate.

Abundant gas is produced in glucose agar deep cultures.

The type strain and five of five other strains tested produced acid from galactose and reduced neutral red. Strains do not produce acetyl methyl carbinol, ammonia, urease or H_2S in SIM medium and do not hydrolyze hippurate.

Products (milliequivalents per 100 ml of culture) from PYG broth are acetic (0.1–0.8), formic (0.1–0.8), and butyric (2.0–4.0) acids and large amounts of butanol. Trace amounts of lactic and succinic acids sometimes are detected. Abundant H_2 is produced from fermentation of carbohydrates. The type strain converts pyruvate to butyrate and acetate. Lactate and gluconate are not utilized; threonine is not converted to propionate.

The type and one other strain tested are susceptible to chloramphenicol (12 μg/ml), clindamycin (1.6 μg/ml), erythromycin (3 μg/ml), penicillin G (2 units/ml), and tetracycline (6 μg/ml).

Other characteristics of the species are given in Table 15.36.

Isolated from human feces.

The mol% G + C of the DNA of the type strain is 38 (T_m). That of one other strain tested is 36 (J. L. Johnson, personal communication).

Type strain: VPI B4-27 (ATCC 27751).

Further comments. Cultures do not survive storage at room temperature for more than a few days.

16. **Eubacterium lentum** (Eggerth 1935) Prévot 1938, 295.[AL] (*Bacteroides lentus* Eggerth 1935, 280.)

len'tum. L. neut. adj. *lentum* slow.

The description is based on that of Moore, Cato and Holdeman (1971b) who studied the type and 51 other strains.

Cells from PYG broth cultures are 0.2–0.4 × 0.2–2.0 μm and occur singly and in pairs and short chains.

Figure 15.34. *E. lentum*

Surface colonies on horse blood agar are 0.5–2.0 mm, circular, entire to erose, raised to low convex, translucent to semiopaque, dull to shiny, smooth and sometimes with mottled appearance when viewed by obliquely transmitted light.

Glucose broth cultures are moderately turbid with a small smooth (occasionally stringy) sediment; the pH is not changed.

Most strains grow at 30 and 45°C; some grow at 25°C. Arginine enhances growth.

No gas is detected in PYG agar deep cultures.

Ammonia is produced from arginine. Adonitol, dextrin, dulcitol, galactose, glycerol, inulin and sorbose are not fermented. Hydrogen sulfide is produced in the butt of TSI (triple sugar iron) slants incubated anaerobically but is not produced in SIM medium. Strains grown on agar medium containing 1% arginine decompose H_2O_2. Hippurate is not hydrolyzed.

Arginine is used as a substrate and energy is obtained by the arginine dihydrolase pathway (Sperry and Wilkins, 1976a). The type strain contains cytochromes a, b, and c and a carbon monoxide-binding pigment (Sperry and Wilkins, 1976b).

Strains degrade bile salts by oxidizing 3α- and 12α-OH groups (Macdonald et al., 1977; Macdonald et al., 1979; Hirano et al., 1980; Hirano and Masuda, 1981). The presence of each of the 3α-, 7α-, and 12α-OH oxidizing enzymes varies according to the strain studied and the E_h value of the medium (Hirano and Masuda, 1981). *E. lentum* strains also degrade corticosteroids by dehydroxylating the 21-OH group and oxidizing the 3α-OH group (Bokkenheuser et al., 1977; Bokkenheuser et al., 1979). The 3α-OH oxidizing enzyme for corticosteroids and that for bile acids appear to be the same (Bokkenheuser et al., 1979; Macdonald et al., 1979).

Products (milliequivalents per 100 ml of culture) of PYG broth cultures are acetate (0.1–0.7), often with trace amounts of lactate or succinate. Products of chopped meat (cooked meat) broth cultures are acetate (0.7–2.0), often with trace amounts of lactate or succinate. No H_2 is detected in headspace gas.

Of 12 strains tested, all are susceptible to chloramphenicol (12 μg/ml) and clindamycin (1.6 μg/ml), one is resistant to erythromycin (3 μg/ml), 4 are resistant to penicillin G (2 units/ml) and 5 are resistant to tetracycline (6 μg/ml).

Other characteristics of the species are given in Table 15.37.

Isolated from human feces (Finegold et al., 1974; Peach et al., 1974; Finegold et al., 1975; Finegold et al., 1977) and from blood, postoperative wounds, and various kinds of abscesses (brain, rectal, scrotal, pelvic).

The mol% G + C of the DNA has not been determined.

Type strain: ATCC 25559 (DMS 2243).

Further comments. Many of the seemingly inert eubacteria synthesize enzymes active upon the steroid molecule. For example, Bokkenheuser et al. (1980) isolated an organism "phenotypically related to *E. lentum*" that synthesizes 16α-steroid dehydroxylase. In another study of 37 strains of nonsaccharoclastic strains, Bokkenheuser et al. (1979) found that 70% synthesized a 21-steroid dehydroxylase and/or a 3α-hydroxysteroid dehydrogenase. Of the 24 strains that elaborated steroid enzymes, 96% metabolized arginine, reduced nitrate and formed H_2S in TSI medium. About 90% of the steroid-converting strains produced gas from H_2O_2. Strain No. 116 (Bokkenheuser et al., 1977) was the exception. It was "steroid active" but did not reduce nitrate. A few typical *E. lentum* strains that reduce nitrate and metabolize arginine do not produce H_2S or gas from H_2O_2 and are "steroid inactive." Since the syntheses of arginine-utilizing and nitrate-reducing enzymes as well as the elaboration of steroid active enzymes have not yet been correlated with DNA homology groups, the phenotypic diversity of the species has not been determined.

17. **Eubacterium limosum** (Eggerth 1935) Prévot 1938, 295.[AL] (*Bacteroides limosus* Eggerth 1935, 290; *Butyribacterium rettgeri* Barker and Haas 1944, 303.)

li.mo'sum. L. neut. adj. *limosum* full of slime, slimy.

The description is based on our examination of the type and 50 similar strains.

Cells of the type strain grown in PYG broth are nonmotile, 0.6–0.9 × 1.6–4.8 μm, often with swollen ends and bifurcations, and occur singly, and in pairs and small clumps.

10 μm   PY   PYG

Figure 15.35. *E. limosum*

Cell walls contain peptidoglycan of the group B type (Schleifer and Kandler, 1972; Guinand et al., 1969).

Surface colonies on blood agar are punctiform to 2 mm in diameter, circular, entire, convex, translucent to slightly opaque and sometimes with mottled appearance when viewed by obliquely transmitted light.

Glucose broth cultures are turbid with viscous, stringy or smooth sediment and a terminal pH of 4.5–5.2.

Growth often is better with a fermentable carbohydrate.

The optimum temperature for growth is 37°C. Most strains grow well at temperatures between 30 and 45°C but growth often is inhibited at 25°C.

Abundant gas is produced in PYG agar deep cultures.

Adonitol and pectin are fermented; some strains ferment galactose. Dextrin, dulcitol, glycerol, inulin and sorbose are not fermented. Hippurate is not hydrolyzed. Some strains produce ammonia from arginine. Catalase and urease are not produced. H_2S may be produced.

Synthesizes vitamin B_{12}.

Products (milliequivalents per 100 ml of culture) in PYG cultures are acetic (1–3), butyric (0.5–2.5) and lactic (1–3) acids, occasionally with small amounts of succinic, *iso*-butyric and *iso*-valeric acids. Carbon dioxide and copious amounts of H_2 are produced in PYG broth cultures. Products in PY broth cultures are acetate, *iso*-butyrate, butyrate (small amounts) and *iso*-valerate, sometimes with traces of propionate. Glucose spares amino acid (peptone) metabolism and the branched chain acids (*iso*-butyric and *iso*-valeric) usually are not detected in PYG cultures. Lactate and pyruvate are converted to acetate and butyrate.

Carbon monoxide can be used as the sole energy source; acetate and CO_2 are the major products (Genthner and Bryant, 1982). Other energy sources include methanol, H_2—CO_2, adonitol, arabitol, erythritol, fructose, glucose, isoleucine, lactate, mannitol, ribose and valine. Ammonia or each of several amino acids serve as the main nitrogen source. Acetate, cysteine, CO_2, calcium-D-pantothenate, and lipoic acid are required for growth on a chemically defined methanol medium. Acetate, butyrate and caproate are produced from methanol. Ammonia and branched chain fatty acids are produced from amino acids. Arabinose, galactose, galacturonic acid, gluconate, mannose, rhamnose, cellobiose, lactose, maltose, melibiose, sucrose, trehalose, melezitose, raffinose, dextrin, pectin, starch, dulcitol, salicin, xylitol, ethanol, propanol, butanol, inositol, acetate, fumarate, malate, succinate, adenine, thymine, cytosine, uracil, allantoin, urea, methylamine and esculin are not utilized as energy sources although esculin is hydrolyzed (Genthner et al., 1981). Although *E. limosum* may grow slightly when pectin is autoclaved in the medium, the organism probably is not using pectin but is growing on methanol that is hydrolyzed from the pectin during autoclaving (Rode et al., 1981).

Most strains are susceptible to chloramphenicol (12 μg/ml) and clindamycin (1.6 μg/ml). Some strains are resistant to erythromycin (3 μg/ml), penicillin G (2 units/ml), or tetracycline (6 μg/ml).

Isolated from feces of man; rumen; intestinal contents of man, rats, poultry and fish; various human and animal infections (rectal and vaginal abscesses, blood, wounds); sewage sludge; and mud.

Other characteristics of the species are given in Table 15.36.

The mol% G + C of the DNA of the type strain is 47 (T_m) (J. L. Johnson, personal communication).

Type species of the genus *Eubacterium* (Cato et al., 1981; Wayne, 1982).

Type strain: ATCC 8486 (DSM 20543; NCIB 9763).

Further comments. The 16S-rRNA sequences indicate that *E. limosum* is closely related phylogenetically to *Acetobacterium woodii* and *Clostridium barkeri* (Tanner et al., 1981). Also, *E. limosum* and *A. woodii* share a similar and uncommon (group B) murein structure (Kandler and Schoberth, 1979).

18. Eubacterium moniliforme (Repaci 1910) Holdeman and Moore 1970, 23.[AL] (*Bacillus moniliformis* Repaci 1910, 412.)

mo.ni.li.for′me. L. n. *monile, monilis* a necklace; L. n. *forma* shape; N.L. neut. adj. *moniliforme* necklace-shaped.

The description is from study of the type and 20 other strains.

Cells of young cultures are motile and peritrichous. Cells of the type strain are 0.6–0.9 μm in diameter × 1.7–9.4 μm long and occur singly and in short chains, often in palisade arrangement.

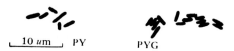

Figure 15.36. *E. moniliforme*

Surface colonies on horse blood agar (2 days) are 2–8 mm, circular with irregular edges, pulvinate or umbonate, and opaque.

Glucose broth cultures are turbid with a smooth sediment and a terminal pH of 5.0–5.6.

All strains grow at 45°C, most grow at 30°C, and some grow at 25°C. Moderate to good growth occurs in PY broth cultures. Growth is slightly greater in media with a fermentable carbohydrate. Growth is not affected by 0.02% Tween 80 or 20% bile.

Abundant gas is produced in glucose agar deep cultures.

The type strain does not produce lecithinase, but lecithinase activity is detected in some other strains. Ammonia is produced from peptone and arginine. The type strain ferments galactose, produces H_2S in SIM medium and reduces neutral red and resazurin. The type strain does not produce acetyl methyl carbinol, does not hydrolyze hippurate and does not produce acid from adonitol, dextrin, dulcitol, glycerol, inulin and sorbose.

Products (milliequivalents per 100 ml of culture) in PYG are acetic (0.5–2.5), butyric (1.3–2.3), and L-(+) (or DL)-lactic (1.0–4.9) acids and butanol, sometimes with trace amounts of formic, propionic and succinic acids. Abundant H_2 is produced. The type strain converts pyruvate principally to acetate. Lactate is not utilized and threonine is not converted to propionate.

Of nine strains tested, four were resistant to erythromycin (3 μg/ml) and one was resistant to tetracycline (6 μg/ml). All were susceptible to chloramphenicol (12 μg/ml), clindamycin (1.6 μg/ml), and penicillin G (2 units/ml).

Other characteristics of the species are given in Table 15.36.

Isolated from blood, various kinds of human clinical infections, intestinal tract, and soil.

The mol% G + C of the DNA has not been determined.

Type strain: ATCC 25546 (CIPP 2055).

Further comments. Although no spores have been detected in the strains on which this description is based, some strains, otherwise like *E. moniliforme*, produce spores and therefore belong to the genus *Clostridium*.

19. Eubacterium multiforme (Distaso 1911) Holdeman and Moore 1970, 23.[AL] (*Bacillus multiformis* Distaso 1911, 101.)

mul.ti.for′me. L. adj. *multus* much, many; L. n. *forma* shape; N.L. neut. adj. *multiforme* many shaped.

The description is based on Distaso (1911), Prévot et al. (1967), and study of the type and four other strains.

Cells are 0.6–0.8 × 0.8–8.0 μm and are motile with single or multiple subpolar flagella. One strain is not motile. Cells occur singly and in pairs, often in a palisade arrangement.

Figure 15.37. *E. multiforme*

Meso-DAP is present in the type and one other strain.

Surface colonies on blood agar incubated for 2 days are 1–2 mm, circular, erose, convex, translucent, gray-white, smooth and slightly shiny with mosaic appearance when viewed by obliquely transmitted light.

Glucose broth cultures are turbid with a smooth, granular, or flocculent sediment and pH of 5.3–5.6.

The optimum temperature for growth is 37°C. Strains grow well at temperatures between 25 and 45°C. Growth may be slightly stimulated by Tween 80 and is inhibited by 6.5% NaCl. Growth and acid production may be inhibited by 20% bile.

Abundant gas is produced in PYG agar deep cultures.

Galactose may be fermented. Adonitol, dextrin, dulcitol, inulin, pectin and sorbose are not fermented. Ammonia is produced from peptone. Hippurate may be hydrolyzed. Acetyl methyl carbinol generally is not produced. Neutral red and resazurin are reduced.

Products (milliequivalents per 100 ml of culture) in PYG cultures are acetic (1.0–3.0), butyric (1.5–4.3) and lactic (0.6–3.6) acids, sometimes with trace amounts of formic, propionic or succinic acids. Copious H_2 is produced. Pyruvate is converted to acetate, formate and butyrate. Little or no lactate is used.

Two of three strains tested were resistant to erythromycin (3 μg/ml). All three strains were susceptible to chloramphenicol (12 μg/ml), clindamycin (1.6 μg/ml), penicillin G (2 units/ml) and tetracycline (6 μg/ml).

Other characteristics of the species are given in Table 15.36.

Isolated from soil of the Ivory Coast (Africa) and a gunshot wound.

The mol% G + C of the DNA has not been determined.

Type strain: ATCC 25552 (Prévot collection 06A).

20. Eubacterium nitritogenes Prévot 1940, 355.[AL]

ni.tri.to′ge.nes. N.L. n. *nitritum* nitrite; N.L. verbal suff.-*genes* from Gr. v. *gennaio* beget, produce; N.L. adj. *nitritogenes* nitrite producing.

The description is from Prévot et al. (1967) and study of the type strain and three similar strains.

Cells of the type strain are straight rods with blunt ends and are 0.8–1.6 × 1.6–16.0 μm. Cells occur singly and in short chains and may occur in palisade arrangement. They occasionally have central swellings. In many cultures, approximately one-half of the cells stain uniformly Gram-positive, a few stain Gram-positive only at one end, and the remainder stain Gram-negative.

Figure 15.38. *E. nitritogenes*

Surface colonies on horse blood agar (2 days) are 0.5–2.0 mm, circular to slightly irregular with scalloped edge, low convex, translucent-opaque, sometimes with mottled appearance when viewed by obliquely transmitted light.

Glucose broth cultures are turbid with sediment and terminal pH of 5.1–5.3.

Growth is most rapid at a pH of 6.5–7.8. Optimum temperature is 37°C but good growth generally occurs at temperatures between 30 and

45°C; three of four strains tested grew at 25°C. Growth is not affected by 0.02% Tween 80.

Abundant gas is produced in PYG agar deep cultures.

Galactose is fermented by the type and one other strain. Adonitol, dulcitol, dextrin, inulin and sorbose are not fermented. Growth is inhibited in media containing 6.5% NaCl or 20% bile. Small amounts of ammonia are produced from PY broth. Hippurate hydrolysis is variable for the species. Lecithinase is not produced by the type and one other strain but is produced by two of the strains. Neutral red and resazurin are reduced.

Products (milliequivalents per 100 ml of culture) in PYG broth cultures are acetic (0.7–2.4), butyric (1.0–3.0), and DL-lactic (1.0–3.0) acids; small amounts of formic or succinic acids may be detected. Abundant H_2 is produced.

Other characteristics of the species are given in Table 15.36.

Isolated from soil and human infections.

The mol% G + C of the DNA has not been determined.

Type strain: ATCC 25547.

Further comments. The relationship between *E. nitritogenes* and *E. budayi* is not clear. They are exceedingly difficult, if not impossible, to differentiate by the usual phenotypic tests.

21. **Eubacterium nodatum** Holdeman, Cato, Burmeister and Moore 1980, 167.[VP]

no.da′tum. L. neut. participle *nodatum* entangled, referring to the tangled arrangement of the cells.

The description is from Holdeman et al. (1980), who studied the type and 49 other isolates.

Cells from PY broth are nonmotile and occur in clumps. Individual cells are 0.5–0.9 × 2.0–12.0 μm and appear branched, somewhat filamentous or club shaped.

Figure 15.39. *E. nodatum*

Subsurface colonies on BHIA-S enriched with 5% (v/v) rabbit serum, 0.02 g each of formate and fumarate and 400 μg thiamine pyrophosphate/100 ml, are 0.5–2.0 mm in diameter, translucent to opaque, and raspberry shaped. After incubation for 2–4 days, surface colonies on BHIA-S streak tubes or anaerobically incubated blood agar plates are less than 0.5–1.0 mm in diameter, generally circular, entire to lobate, and heaped or berry-like in appearance. There is no hemolytic action on rabbit blood cells. No growth occurs on EYA plates incubated anaerobically.

PY broth cultures are not turbid and have a small to moderate amount of flocculent, granular or bread crumb-like sediment.

The temperature for optimum growth is 37°C. Most strains grow at 30 and 45°C, but only an occasional strain grows at 25°C.

Growth is inhibited by 20% bile and 6.5% NaCl, may be stimulated by 0.02% Tween 80, and is neither inhibited nor enhanced by the addition of 5% rumen fluid, 10% serum with thiamine pyrophosphate, or 0.05% sodium polyanetholsulfonate.

Little or no gas is detected in agar deep cultures.

Neutral red is reduced. The type strain does not produce H_2S in SIM medium. Ammonia is produced from peptone and may be produced from arginine. Carbohydrates are not fermented.

Products (milliequivalents per 100 ml of culture) in PYG broth are acetate (0.04–0.36) and butyrate (0.3–1.6), often with trace amounts of formate, lactate and succinate. No H_2 is detected in headspace gas of broth cultures. Neither lactate nor pyruvate is utilized.

The 32 strains tested were susceptible to chloramphenicol (12 μg/ml), clindamycin (1.6 μg/ml), erythromycin (3 μg/ml), penicillin G (2 units/ml), and tetracycline (6 μg/ml).

Other characteristics of the species are given in Table 15.37.

Isolated from subgingival samples and from supragingival tooth scrapings from persons with periodontal disease.

The mol% G + C of the DNA is 36 to 38 (T_m).

Type strain: ATCC 33099.

Further comments. Morphologically, strains of *E. nodatum* resemble members of the genus *Actinomyces*. However, unlike species of *Actinomyces*, strains of *E. nodatum* are nonsaccharoclastic and produce butyrate, presumably from peptone.

22. **Eubacterium plautii** (Seguin) Hofstad and Aasjord 1982, 347.[VP] (*Fusobacterium plauti* Seguin 1928, 439.)

plau′tii. N.L. gen. n. *plautii* of Plaut; named for R. Plaut, the bacteriologist who first described this organism.

This description is from study of the type strain and from Hofstad and Aasjord (1982).

Cells of the type strain are straight rods with rounded ends and are 0.4–0.8 μm in diameter × 2.0–10.0 μm in length. Cells are motile and peritrichous and occur singly, and in pairs or short chains. Cells stain Gram-negative with very occasional weak Gram-positive areas.

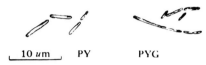

Figure 15.40. *E. plautii*

Surface colonies on horse blood agar (2 days) are 0.5 mm, circular with diffuse edges, gray-white, dull, smooth and translucent with mottled appearance when viewed by obliquely transmitted light.

Glucose broth cultures are moderately turbid with a smooth (occasionally flocculent) sediment and final pH of 5.2.

Small amounts of gas may be present in PYG agar deep cultures.

Products (milliequivalents per 100 ml of culture) in PYG broth cultures are lactic (2), butyric (0.7), and acetic (0.2) acids, often with a trace amount of succinic acid. Pyruvate is converted to acetate, butyrate and lactate. Lactate is not utilized; threonine is not converted to propionate. No H_2 is detected in headspace gas of PYG cultures.

Other characteristics of the species are given in Table 15.36.

Isolated from cultures of *Entamoeba histolytica*.

The mol% G + C of the DNA has not been reported.

Type strain: ATCC 29863.

Further comments. Although this species recently was transferred to the genus *Eubacterium* from the genus *Fusobacterium* by Hofstad and Aasjord because the cells do not have a triple-layered membrane containing the lipopolysaccharide-lipoprotein complex that is typical of Gram-negative bacteria, C.-H. Lai (personal communication) has observed that electron micrographs of cells of the type strain show a cell wall arrangement typical of Gram-negative bacteria. It appears that the taxonomic position of this species still is not resolved.

Motility is difficult to demonstrate. We have had best results from microscopic examination of cells in the water of syneresis of a PYG agar slant.

23. **Eubacterium plexicaudatum** Wilkins, Fulghum and Wilkins 1974, 408.[AL]

plex.i.cau.da′tum. L. n. *plexus* twisting or braiding, N.L. adj. *caudatus* with a tail, *plexicaudatum* N.L. adj. with a braided tail, referring to the tuft of subpolar flagella that are twisted together to form the "large flagellum" often visible by darkfield or phase-contrast microscopy.

This description is from Wilkins et al. (1974) and study of the type and 16 other strains.

Cells in PYG broth cultures are motile, have bipolar tufts of flagella, and are 0.8–1.6 × 4.0–10.0 μm and often have tapered ends. By phase-contrast or darkfield microscopy, the tufts of bipolar flagella may appear to be single polar flagella. Upon initial isolation the cells are slightly curved and may have a double curvature. After several transfers in culture media, the cells often are thinner and usually straight. Cells may decolorize easily. They may have refractile areas or swellings but do not survive heating at 70°C for 10 min or treatment with ethanol.

Figure 15.41.   *E. plexicaudatum*

Surface colonies on RGCA or BHIA-S in anaerobic streak tubes incubated for 5 days are 0.5–1 mm, circular to slightly irregular, convex, translucent, dull, smooth and white to light gray. In the lower portion of the streak tubes, the organisms often grow as a translucent film. No growth occurs on agar plates incubated in anaerobic jars.

Cultures in PYG broth incubated in O₂-free N₂ are turbid, usually with no sediment, and have a terminal pH of 6.0–6.8. The pH of PY cultures is 7.0–7.6. Optimum growth occurs at a pH near neutrality. Growth often is stimulated by 20% bile or 15% rumen fluid.

Moderate to abundant gas is produced in glucose agar deep cultures.

Growth in media containing galactose and various carbohydrates decreases the pH to between 6.0 and 7.0 (Table 15.36), but not below 6.0, in any sugar medium tested. Neutral red is reduced; resazurin is not reduced. The type strain does not produce acid from adonitol, dextrin, glycerol, inulin and sorbose. Ammonia is not produced from peptone or arginine. H₂S is not produced in SIM medium. Hippurate is not hydrolyzed.

Products (milliequivalents per 100 ml of culture) in PYG broth are butyrate (1.0–2.5), usually with acetate (0.02–0.12) and butanol. Trace amounts of pyruvate and succinate may be detected. Hydrogen is produced. Pyruvate and lactate are not utilized. Threonine is not converted to propionate.

Other characteristics of the species are given in Table 15.36.

Isolated from the ceca of mice or rats.

The mol% G + C of the DNA is 44 (T_m).

Type strain: VPI 7582 (ATCC 27514).

Further comments. Many strains of *E. plexicaudatum* are difficult to grow, and the variations obtained in carbohydrate fermentation may be due to this characteristic. The organisms are quite susceptible to oxidation and are difficult to preserve in a lyophilized or frozen state. Cultures streaked on BHIA-S streak tubes survive if stored at 37°C and transferred monthly.

24. **Eubacterium ramulus** Moore, Johnson and Holdeman 1976, 249.[AL]

ra′mu.lus. L. n. *ramulus* twig (referring to the shape of the cell).

The description is from Moore et al. (1976), who studied the type and 15 other similar strains.

Cells are regular rods with rounded ends, 0.5–0.9 μm in diameter and 1.0–5.0 μm long with filaments exceeding 25.0 μm in length. Cells usually stain boldly in young cultures and are arranged in pairs or short chains. The filaments present in some cultures appear to be either undivided cells or long chains of distinct cells. Cells in the long chains often are of unequal length and occasionally have marked swellings. Cultures having cells with swellings do not survive heating at 80°C for 10 min.

Subsurface colonies in RGCA incubated for 5 days are 1–4 mm in diameter and have the appearance of woolly balls or "balls of fuzz" or

Figure 15.42.   *E. ramulus*

are "cauliflower-like," or sometimes are of such indefinite form that there is doubt that the area picked (under × 10 magnification) truly contained a colony. Surface colonies on BHIA-S roll streak tubes are 1–4 mm in diameter, circular to slightly irregular, entire or slightly lobate, raised to low convex or umbonate, translucent and white to beige. Three of eight strains tested did not grow on the surface of anaerobically incubated blood agar or EYA plates. When there is growth, there is no reaction on EYA and no hemolytic activity on rabbit blood.

Glucose broth cultures have stringy or flocculent sediment, usually without turbidity. The pH of 1-day-old cultures in PYG is 4.8–5.3.

Best growth reliably is obtained at 37°C, although most strains grow equally well at 30 and 45°C. Strains grow not at all or poorly at 25°C. Growth is not stimulated by 0.02% Tween 80; 20% bile may inhibit growth or fermentation. Growth usually is best in media containing fermentable carbohydrate.

Abundant gas is observed in glucose agar deep cultures.

Products (milliequivalents per 100 ml of culture) in PYG cultures are acetic (0.1–1.0), formic (0.5–2.3), butyric (1.6–3.0) and lactic (0.1–0.7) acids; trace amounts of succinic acid sometimes are detected. Abundant H₂ is produced from fermentation of carbohydrates.

The type strain is susceptible to chloramphenicol (12 μg/ml), clindamycin (1.6 μg/ml), penicillin G (2 units/ml), and tetracycline (6 μg/ml) and is resistant to erythromycin (3 μg/ml).

Other characteristics of the species are given in Table 15.36.

Isolated from human feces.

The mol% G + C of the DNA is 39.

Type strain: ATCC 29099.

25. **Eubacterium rectale** (Hauduroy et al. 1937) Prévot 1938, 294.[AL] (*Bacteroides rectalis* Hauduroy, Ehringer, Urbain, Guillot and Magrou 1937, 72.)

rec.ta′le. N.L. n. *rectum* the straight bowel, rectum; N.L. neut. adj. *rectale* rectal.

The description is from Prévot et al. (1967) and Holdeman and Moore (1974) and study of the type and 22 other strains.

Cells in PYG broth cultures are 0.5–0.6 × 1.7–4.7 μm and occur singly and in short chains and small clumps. Cells may be slightly curved and may have central or terminal swellings. Some strains are motile and peritrichous.

Figure 15.43.   *E. rectale*

Surface colonies are 0.5–2.0 mm, circular to irregular, entire to scalloped, convex, translucent, smooth and shiny. They may be mottled when viewed by obliquely transmitted light. The type strain is nonhemolytic on horse blood. Some strains will not grow on the surface of blood agar plates incubated in an anaerobe jar.

Glucose broth cultures are turbid with a smooth or flocculent sediment and have a terminal pH of 4.7–5.5, usually around 5.0. Growth in prereduced PY broth is questionable or very sparse.

The optimum temperature for growth is 37°C. Most strains grow at 25–45°C. Growth is stimulated markedly by a fermentable carbohydrate. Growth usually is not stimulated by 0.02% Tween 80 and is inhibited by 20% bile.

Moderate to large amounts of gas are produced in PYG agar deep cultures.

The type strain ferments dextrin, galactose, inulin and pectin and reduces neutral red. The type strain does not ferment adonitol, dulcitol, glycerol or sorbose; does not produce acetylmethylcarbinol or catalase; does not produce ammonia from peptone, arginine or threonine; does not produce H₂S in SIM medium, does not grow in medium containing 6.5% NaCl, and does not hydrolyze hippurate.

Products (milliequivalents per 100 ml of culture) in PYG cultures are butyric (0.5–1.5), lactic (1.5–5.5), acetic (0–0.4) acids, occasionally with a trace of propionate or succinate. Copious H₂ is produced. Pyruvate is converted to acetate, butyrate and lactate. Lactate is not utilized.

Other characteristics of the species are given in Table 15.36.

Isolated from human colon and feces.

The mol% G + C of the DNA of VPI C30-7 is 30 (T_m) (J. L. Johnson, personal communication).

Type strain: ATCC 33656. VPI 0989, the type strain in the Approved Lists (Skerman et al., 1980) was deposited in ATCC (25578) but was lost in both collections. A different isolate, VPI 0990, from the same fecal sample has been deposited to represent the type strain as ATCC 33656. Because these isolates came from the same sample, we believe that they are the same strain.

Further comments. Strains of fecal bacteria with the general characteristics of *E. rectale* comprise at least five distinct groups (Moore and Holdeman, 1974). All strains ferment cellobiose, fructose, glucose, maltose and starch. They decolorize readily, and Gram-positive cells cannot be demonstrated in some strains. The rods are usually curved and motile with flagella singly, in pairs or in tufts at one or both ends of the cells.

Strains of *E. rectale,* referred to as "*E. rectale*-I" in Moore and Holdeman (1974), are slender curved rods, generally longer than the other phenotypes, frequently with large central or terminal swellings. They uniformly ferment arabinose, melezitose, melibiose, raffinose, sucrose and xylose, and produce large quantities of H₂. Strains of *Fusobacterium mortiferum* may be similar to *E. rectale* except that they have no flagella and are thicker rods showing more pleomorphism, especially when stained from growth on blood agar.

Strains designated "*E. rectale*-II" differ from "*E. rectale*-I" strains in that they fail to ferment melezitose and may or may not ferment melibiose and sucrose. Cells are generally shorter, more uniform, curved rods.

Strains of "*E. rectale*-III-H" differ from "*E. rectale*-II" strains in that they fail to ferment raffinose. Strains designated "*E. rectale*-III-F" have the same reactions as do strains of "III-H" except that they produce major amounts of formic acid and no H₂. Strains designated "*E. rectale*-IV" are similar to strains of "III-H" except that they fail to ferment xylose, may or may not ferment raffinose, and never reduce the pH in arabinose to below 5.5

Heat-resistant spores have been detected in some strains otherwise similar to *E. rectale* subgroups I, II, and III-H. These strains were called unnamed "*Clostridium* species A" (Moore and Holdeman, 1974). However, not all strains with swellings resist heating at 80°C for 10 min or treatment with absolute ethanol for 30 min.

The relationship between *E. rectale* and *Butyrivibrio fibrisolvens* is in question and has been discussed by Moore and Holdeman (1974). They report that strains of *B. fibrisolvens* stain only Gram-negative and are monotrichous and generally are more fastidious than strains of *E. rectale.*

26. Eubacterium ruminantium Bryant 1959, 140.[AL]

ru.mi.nan′ti.um. N.L. pl. n. *ruminantia* ruminants; N.L. gen. pl. n. *ruminantium* of ruminants.

The description is from Bryant (1959) who studied 20 strains and from our study of the type and one other strain.

Cells in PYG broth cultures are 0.2–0.3 × 0.8–2.5 μm and occur singly and in pairs. Cells decolorize readily.

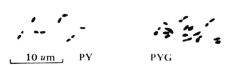

<u>10 um</u>   PY   PYG

Figure 15.44. *E. ruminantium*

Surface colonies on RGCA are entire, low convex, smooth, translucent to opaque and light buff-colored. Subsurface colonies are lenticular and do not produce gas. No growth occurs on the surface of BHIA-S with 5% blood, even when rumen fluid is added to the medium. EYA does not support growth of the type strain.

PYG broth cultures are turbid in 18 h with smooth or ropy sediment; the terminal pH (5 days) is 5.0–5.5

The optimum temperature for growth is 37°C, little or no growth occurs at 30 or 45°C, no growth occurs at 22 or 50°C. Less growth is produced when rumen fluid in the medium is replaced by 0.5% yeast extract and 1.5% trypticase. Growth is unaffected by heme (0.5 mg/100 ml), 20% bile, or 0.02% Tween 80. Growth of a test strain was inhibited by 20 μg/ml of either HgCl₂ or CuCl₂ but not by 100 μg/ml of CdCl₂ (Forsberg, 1978). No growth occurs at Na⁺ concentrations of 3.1 mM or less; best growth occurs with at least 91 mM Na⁺ (Caldwell and Hudson, 1974).

No gas is detected in glucose agar deep cultures.

The type and one other strain tested do not ferment adonitol, dextrin, dulcitol, galactose, glycerol, inulin and sorbose. Pectin is weakly fermented. Hydrogen sulfide is not produced in SIM medium. Ammonia is not produced from arginine or threonine but small quantities may be detected in peptone cultures. The type strain does not produce acetylmethylcarbinol or hydrolyze hippurate. Neutral red and resazurin are not reduced. Nine of 20 strains tested ferment xylan; none ferment gum arabic.

Products (milliequivalents per 100 ml of culture) in PYG-10% rumen fluid broth cultures are formic (0.5–2.0), acetic (0.01–2.5), butyric (0.3 to 0.6), DL-lactic (0.4–1.5), and succinic (0.05–0.15) acids. Small amounts of CO₂ are produced. Little or no H₂ is detected in headspace gas. Galacturonic acid is utilized (Tomerska and Wojciechowicz, 1973). Pyruvate and lactate are not utilized and threonine is not converted to propionate.

Exogenous ammonia is required for growth and is the preferred nitrogen source, even in very complex media (Bryant and Robinson, 1962) and amino acids (mainly alanine, valine and *iso*-leucine) are excreted into the medium during the log phase of growth (Stevenson, 1978). One or more of *n*-valerate, *iso*-valerate, 2-methyl-*n*-butyrate, or *iso*-butyrate, but not amino acids, are required as carbon sources for growth (Bryant and Robinson, 1962; Bryant and Robinson, 1963).

Other characteristics of the species are given in Table 15.36.

Isolated from bovine rumen contents where it represents up to 7.3% of the total isolates.

The mol% G + C of the DNA has not been reported.

Type strain: ATCC 17233 (Bryant GA 195).

Further comments. Bryant described two biotypes, with eight strains in biotype 1 and five in biotype 2. The other seven strains seemed to be intermediates between the two biotypes.

In differentiating human fecal isolates of *E. ruminantium* from *E. ventriosum,* Moore and Holdeman (1974) designated as *E. ruminantium* those strains that ferment cellobiose and produce no H₂; these strains did not produce acid from mannose but usually produced acid from salicin. Strains designated *E. ventriosum* either did not ferment cellobiose or they produced H₂; they often fermented mannose and did not ferment salicin.

Although the reactions of *E. ruminantium* are similar to those of *Gemmiger formicilis* (Gossling and Moore, 1975), the cells and attached "buds" of *G. formicilis* are more nearly spherical than are the cells of *E. ruminantium* (Moore and Holdeman, 1974).

27. Eubacterium saburreum (Prévot 1966) Holdeman and Moore 1970, 23.[AL] (*Catenabacterium saburreum* Prévot 1966, 171.)

sa.bur′re.um. L. n. *saburra* sand; N.L. neut. adj. *saburreum* sandy.

Nineteen strains were isolated and described but not named by Theilade and Gilmour. This description is from Theilade and Gilmour (1961), Hofstad (1967), and study of the type and 37 other strains.

Cells of the type strain in PYG broth cultures are 0.7–1.1 × 6–18 μm and occur in pairs and short chains of cells, often in parallel arrangement. Curving filaments, sometimes with swellings, often occur. Cells often stain very weakly Gram-positive or Gram-negative with a few

Gram-positive spots or areas within some cells (Kopeloff's modification of the Gram's stain). Cells from broth cultures containing fermentable carbohydrate usually stain Gram-negative.

Figure 15.45. *E. saburreum*

Cell walls contain glucose and rhamnose but no galactose.

Surface colonies on blood agar are 1–4 mm, flat, with interlaced filamentous rhizoid edges and small slightly raised granular centers which penetrate the agar.

Cultures in broth produce slight turbidity with a flocculent or granular sediment that may adhere to the tube. There is moderate growth in prereduced PY broth and heavy growth in prereduced broth containing a fermentable carbohydrate. The terminal pH in culture medium with a fermentable carbohydrate is 4.7–5.5.

Growth of most strains is stimulated by 0.02% Tween 80.

The optimum temperature for growth is 37°C. The type strain grows moderately well at 30°C and poorly at 25 and 45°C.

Abundant gas is produced in PYG agar deep cultures.

Urease is not produced by the type strain. H_2S is not produced in SIM medium. α-Methyl glucoside is fermented, sometimes only weakly.

Products (milliequivalents per 100 ml of culture) in PYG cultures are acetic (0.3–1.4), butyric (0.2–1.0), and lactic (0.1–3.5) acids, occasionally with trace amounts of formic or succinic acid. Copious H_2 and moderate amounts of CO_2 are produced.

Tested strains are susceptible to chloramphenicol (12 μg/ml), clindamycin (1.6 μg/ml), erythromycin (3 μg/ml), and penicillin G (2 units/ml). Some strains are resistant to tetracycline (6 μg/ml).

Based on double diffusion in agar gel, strains of *E. saburreum* have been classified into serotypes 1, 2, and 3 (Kondo et al., 1979). The type-specific polysaccharide antigens representative of these and other groups apparently are located on the surface of the cell and contain heptose and *o*-acetyl as major constituents (Hofstad 1972, 1975; Hoffman et al., 1974, 1976, 1980; Hofstad and Lygre, 1977; Hofstad and Skaug, 1978; Skuag and Hofstad, 1979; and Kondo et al., 1979).

Other characteristics of the species are given in Table 15.36.

Isolated from human dental plaque and gingival crevice.

The mol% G + C of the DNA has not been determined.

Type strain: VPI 11763 (ATCC 33271).

28. Eubacterium siraeum Moore, Johnson and Holdeman 1976, 250.^{AL}

si.rae'um (si.re'um). Gr. adj. *siraeum* sluggish, referring to the relative inactivity of this organism in most substrates tested.

The description is from Moore et al. (1976).

Cells are 0.5–0.6 × 1.3–3.0 μm and occur singly, in pairs or short chains, sometimes in "V" or "flying gull" arrangements. Some strains are motile and have one or two subpolar flagella.

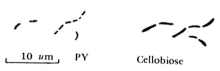

Figure 15.46. *E. siraeum*

After incubation for 5 days in RGCA, subsurface colonies are 0.5–1.0 mm in diameter, lenticular, translucent and tan or white. The larger colonies often have a dense center. Surface colonies on BHIA-S in roll streak tubes or on blood agar plates are 0.5 mm in diameter, circular, entire, low convex, smooth, shiny, and transparent to translucent. Three of eight strains tested that grew on anaerobically incubated blood agar plates did not lyse the rabbit red blood cells.

There is little growth in PY broth without fermentable carbohydrate. Cultures in PY broth with a fermentable carbohydrate are slightly turbid with abundant smooth sediment and pH of 5.3–5.8.

Most strains grow equally well at 37 and 45°C; little growth occurs at 30°C. Addition of 10–30% (v/v) rumen fluid stimulates growth of most strains. Growth is not affected by 0.02% Tween 80 and is inhibited by 20% bile.

Gas production in glucose agar deep cultures is variable.

Products (milliequivalents per 100 ml of culture) in PY-cellobiose broth are acetic acid (1.3–4.1) and large amounts of ethanol. Trace amounts of lactic, butyric or succinic acids also may be detected. Abundant H_2 (greater than 3% of the headspace gas) is produced from cellobiose, fructose or starch.

Other characteristics of the species are given in Table 15.36.

Isolated from human feces.

The mol% G + C of the DNA is 45 (T_m).

Type strain: ATCC 29066.

Further comments. Although there often is no visible turbidity in broth medium that does not contain a fermentable carbohydrate, the fact that esculin frequently is hydrolyzed when no turbidity can be seen suggests that there is slight growth in PY broth medium. Cells often can be seen in Gram stains from PY cultures, which also suggests that some cell multiplication may occur in this basal medium.

29. Eubacterium suis Wegienek and Reddy, 1982, 218.^{VP} ("*Corynebacterium suis*" Soltys and Spratling 1957, 500.)

su'is. L. n. *sus* the hog, swine; L. gen. n. *suis* of the hog.

This description is from Wegienek and Reddy (1982) and study of the type strain.

Cells in PY-maltose medium (Wegienek and Reddy, 1982) are 0.5 × 1.0–3.0 μm and occur singly and in pairs and clumps and often have a beaded appearance.

Figure 15.47. *E. suis*

The major sugar component of the cell walls is rhamnose; a small amount of glucosamine and a trace amount of mannose also are present. The major diamino acid found is lysine; small amounts of glutamic acid and alanine also are present.

Surface colonies on rabbit blood agar plates are 2.0–3.0 mm, circular, erose, raised in the center (fried egg), translucent, gray and smooth with a shiny center and dull edge. Indefinite β hemolysis may be seen on sheep blood agar plates; no hemolysis on rabbit blood agar.

Cultures in PY-maltose broth have slightly stringy to smooth sediment, no turbidity and pH of 5.8. The pH of glycogen broth cultures is 4.4–4.6.

The optimum temperature for growth is 37°C, although good growth is obtained at 30 and 43°C. Growth is not detected at room temperature after 72 h. The optimum pH for growth is between pH values of 7 and 8. Growth is stimulated by fermentable carbohydrate. Growth is not enhanced by Tween 80, hemin or menadione; 20% bile is inhibitory.

Urease is produced and hippurate is hydrolyzed. No ammonia is detected from arginine. Acetyl methyl carbinol is not produced.

Major amounts of type *b* cytochrome and minor amounts of type *c* cytochrome appear to be present in cell extracts (Wegienek and Reddy, 1982).

Products (milliequivalents per 100 ml of culture) in PY-glycogen broth are acetic (2.7–3.3) and formic (3.0) acids and ethanol, often with trace amounts of lactic and succinic acids. Lactate and pyruvate are not utilized; no propionate is produced from threonine.

The type strain is susceptible to chloramphenicol (12 μg/ml), clindamycin (1.6 μg/ml), erythromycin (3 μg/ml), penicillin G (2 units/ml), tetracycline (6 μg/ml), ampicillin (4 μg/ml), and cephalothin (6 μg/ml).

Other characteristics of the species are given in Table 15.36.

Originally isolated from cases of cystitis and pyelonephritis and cases of metritis in pregnant sows (Soltys, 1961; Soltys and Spratling, 1957). Not isolated from healthy sows, but frequently recovered from urine and semen of apparently healthy boars. Sows can be infected artificially by intrarenal injection of live organisms plus 5% saponin (Soltys, 1961). No demonstrable exotoxin is produced (Wegienek and Reddy, 1982).

The mol% G + C of the DNA is 55 (T_m).

Type strain: ATCC 33144.

30. **Eubacterium tarantellus** Udey, Young, and Sallman 1977, 407.[AL]

tar′an.tell′us. N.L. n. *tarantellus*, Ital. n., *tarantella*, a fast, whirling dance; referring to the disease symptoms of the fish from which the species was isolated.

Figure 15.48. *E. tarantellus*

This description is from Udey et al. (1977) and our studies of the type strain.

In tissue and when initially isolated, cells are long unbranched filaments. After several transfers in the laboratory, the cells are 1.3–1.6 × 10.0–17.0 μm. No motility has been detected.

Surface colonies on BHIA are 2–5 mm, flat, translucent, colorless, rhizoid, soft and slightly mucoid, with a distinct "pinwheel" appearance. A large zone of β hemolysis is present around colonies on sheep blood agar plates. Lecithinase is produced on EYA medium.

In Brewer's thioglycollate medium, the species grows as discrete clusters which are fluffy and filamentous. The clusters of growth often are surrounded by a slimy layer. After 48 h incubation, PY broth cultures have smooth to cottony sediment, no turbidity and terminal pH of 5.3–5.8.

Strains grow well at temperatures between 25 and 37°C. Growth at 15 or 45°C is marginal. Cells survive for 2 weeks at 4°C, but there is no evidence of growth at this temperature. Cultures grow within 72 h in brain heart infusion broth (BHIB) adjusted to an initial pH from 5.6–8.0 but do not grow at pH extremes outside this range. The species does, however, survive for 24 h at pH 3.4 and 8.8.

Growth occurs in Brewer's thioglycollate medium with 2% (w/v) NaCl. Cells can be recovered after 24-h incubation in concentrations of NaCl up to 10% (w/v).

Moderate to abundant gas is produced in PYG agar deep cultures.

Deoxyribonuclease is produced.

Products (milliequivalents per 100 ml of culture) in PYG broth are acetic (1.5–5.0) and formic (1.0–5.0) acids often with trace amounts of lactic or succinic acids. Abundant H_2 is detected in headspace gas.

Strains are sensitive (zones of inhibition surrounding the disk) to erythromycin, chloramphenicol, penicillin, tetracycline and novobiocin

(concentrations not given). Strains grow well in BHIB containing 100 μg/ml gentamicin, 100 μg/ml neomycin or 300 units/ml polymyxin B. Vancomycin is inhibitory at 7.5 μg/ml. Blood agar with 100 μg/ml gentamicin is a highly selective isolation medium.

The type strain is susceptible to chloramphenicol (12 μg/ml), clindamycin (1.6 μg/ml), erythromycin (3 μg/ml), penicillin G (2 units/ml), and tetracycline (6 μg/ml).

Not toxigenic for mice (intraperitoneal inoculation) or pathogenic for guinea pigs (intramuscular inoculation with $CaCl_2$). Pathogenic for channel catfish (*Ictalurus punctatus*) by intraperitoneal inoculation.

Other characteristics of the species are given in Table 15.36.

Isolated from brains of dead or moribund striped mullet (*Mugil dephalus*) in Biscayne Bay, Florida, that have evidence of a neurological disease.

The mol% G + C of the DNA has not been determined.

Type strain: ATCC 29255.

Further comments. Henley and Lewis (1976) reported isolation of two strains with characteristics similar to, but not identical with, those of *E. tarantellus* from moribund fish from the Texas coast. These strains share antigens in common with *E. tarantellus* (Udey et al., 1977).

31. **Eubacterium tenue** (Bergey et al. 1923) Holdeman and Moore 1970, 23.[AL] (*Bacteroides tenuis* Bergey, Harrison, Breed, Hammer and Huntoon 1923, 263.)

te′nu.e. L. neut. adj. *tenue* slender (originally used with *spatuliformis* to indicate forms like slender spatulas).

The description is from study of the type and three other strains.

Cells in PYG broth cultures are 0.5–0.8 × 4.9–20.0 μm and occur singly and in pairs and short chains. Individual cells often have slightly widened and blunt ends. Cells of young cultures (log phase or early stationary phase of growth) of the type strain are motile by microscopic examination but no flagella have been seen with Leifson's flagella stain. One other strain is peritrichous. Motile cells have not been seen in two strains.

Figure 15.49. *E. tenue*

Surface colonies on horse blood agar (2 days) are 4–6 mm, slightly irregular with lobate to diffuse edges, flat, translucent, gray-white, smooth and dull, with granular or mottled appearance when viewed by obliquely transmitted light. Strains are either not hemolytic or produce slight clearing of the blood beneath heavy growth.

Glucose broth cultures are turbid with sediment and terminal pH of 5.5–5.9. They have a putrid odor.

Growth is most rapid at 24–37°C. Growth is stimulated by 0.02% Tween 80 and inhibited by 6.5% NaCl.

Moderate to abundant gas is detected in PYG agar deep cultures.

Lecithinase is produced on McClung-Toabe EYA. H_2S is produced in SIM medium. Ammonia is produced from peptone. Neutral red is reduced. Hippurate is not hydrolyzed and acetylmethylcarbinol is not produced.

Products (milliequivalents per 100 ml of culture) in PYG broth are acetate (1.0–4.7) and formate (0.4–0.8), often with small amounts of propionate, *iso*-butyrate, *iso*-valerate, *iso*-caproate, ethanol, propanol, *iso*-butanol and butanol.

One human clinical isolate from blood is susceptible to chloramphenicol (12 μg/ml), clindamycin (1.6 μg/ml), erythromycin (3 μg/ml), penicillin G (2 units/ml), and tetracycline (6 μg/ml).

The type strain is nontoxic for mice (0.5 ml of 24-h chopped meat culture supernatant fluid injected intraperitoneally).

Other characteristics of the species are given in Table 15.36.

Isolated from abscess following abortion, knee synovial fluid, and blood.

The mol% G + C of the DNA has not been determined.

Type strain: ATCC 25553.

Further comments. By oligonucleotide cataloging of their 16S-rRNAs, the type strains of *E. tenue* and *Clostridium lituseburense* are highly related (Tanner et al., 1981).

32. **Eubacterium timidum** Holdeman, Cato, Burmeister and Moore, 1980, 164.[VP]

ti′mi.dum. L. neut. adj. *timidum* fearful, timid, referring to the slight or slow growth in clumps.

Figure 15.50. *E. timidum*

Cells of PYG broth cultures are regular to slightly diphtheroid rods 0.8–1.6 × 1.6–3.1 μm, often arranged in clumps. Cells from actively growing cultures are Gram-positive; they often stain Gram-negative from older cultures.

After 5 days of incubation, subsurface colonies in BHIA-S enriched with 5% (v/v) rabbit serum, 0.02 g each of sodium formate and ammonium fumarate and 400 μg thiamine pyrophosphate/100 ml are 0.5–1.0 mm in diameter, lenticular and transparent to translucent. After incubation for 2 days, surface colonies on enriched BHIA-S or BHIA-S blood streak tubes are 0.2 mm in diameter, circular, entire, low convex and transparent to translucent. Four of five strains tested did not grow on the surface of anaerobic blood agar plates, but all five strains grew on EYA.

PY broth cultures have only a small amount of smooth or slightly granular sediment and no, or only very slight, turbidity.

Strains grow equally well at temperatures between 37 and 45°C but do not grow at 30°C. Growth is equally poor in an atmosphere of 10% CO₂ in 90% N₂ or in 100% CO₂. Growth is inhibited by 20% bile and 6.5% NaCl but is neither inhibited nor enhanced by the addition of 0.02% Tween 80, 5% (v/v) rumen fluid, 10% (v/v) serum with thiamine pyrophosphate or 0.05% (w/v) sodium polyanetholsulfonate.

No gas is detected in PYG agar deep cultures.

Neutral red is reduced. Ammonia is produced from urea (five of five strains tested) but not from arginine or in PY broth cultures. Adonitol, dextrin, dulcitol, galactose, glycerol, inulin and sorbose are not fermented.

Trace amounts of acetate, succinate, or lactate may be detected from glucose broth cultures. Hydrogen is not detected in headspace gas. Pyruvate and lactate are not utilized.

Strains are susceptible to chloramphenicol (12 μg/ml), clindamycin (1.6 μg/ml), and erythromycin (3 μg/ml). Of 30 strains tested, 2 are resistant to penicillin G (2 units/ml), and 8 are resistant to tetracycline (6 μg/ml).

Other characteristics of the species are given in Table 15.37.

Isolated from subgingival samples and from supragingival tooth scrapings from persons with periodontal disease.

The mol% G + C of the DNA could not be determined because of the sparse amount of growth.

Type strain: ATCC 33093.

33. **Eubacterium tortuosum** (Debono 1912) Prévot 1938, 295.[AL] (*Bacillus tortuosus* Debono 1912, 233.)

tor.tu.o′sum. L. neut. adj. *tortuosum* full of windings.

This description from study of the type and 15 other strains.

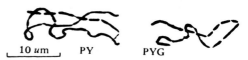

Figure 15.51. *E. tortuosum*

Cells in PYG broth cultures are 0.5–0.6 × 2.4–5.0 μm in long chains of 50 or more elements.

Surface colonies on horse blood agar (2 days) are 0.5–4.0 mm, circular, entire to erose to diffuse, convex to umbonate, translucent, gray to white and smooth to slightly rough.

Glucose broth cultures have flocculent to gelatinous (occasionally granular) sediment, no turbidity and terminal pH of 5.3–5.6.

The temperature for optimum growth is 37–41°C. Most strains grow at 30 and 45°C, some grow at 25°C. Growth is inhibited by 20% bile but not affected by 0.02% Tween 80.

Variable amounts of gas are produced in glucose agar deep cultures.

Neutral red is reduced; resazurin is not reduced. Adonitol, dextrin, dulcitol, galactose, glycerol, inulin and sorbose are not fermented; ammonia is not produced from peptone, arginine or threonine. Hippurate is not hydrolyzed. Little or no acetyl methyl carbinol is produced. Hydrogen sulfide is not produced in SIM medium.

Products (milliequivalents per 100 ml of culture) in PYG broth are acetic (0.05–0.45), butyric (0.3–0.8), DL-lactic (2.1–4.6), and succinic (0.06–0.5) acids, with trace to moderate amounts of formic acid. Moderate to abundant amounts of H₂ are detected in headspace gas. Pyruvate is converted primarily to acetate and butyrate. Lactate is not utilized. Threonine is not converted to propionate.

Other characteristics of the species are given in Table 15.36.

Isolated from turkey liver granulomas, turkey enteritis, human feces, soil, and fresh water.

The mol% G + C of the DNA has not been reported.

Type strain: ATCC 25548.

Further comments. E. tortuosum (referred to as "*Catenabacterium*," Moore and Gross, 1968) is believed to be a causative agent of turkey liver granulomas.

34. **Eubacterium ventriosum** (Tissier 1908) Prévot 1938, 295.[AL] (*Bacillus ventriosus* Tissier 1908, 204.)

ven.tri.o′sum. L. neut. adj. *ventriosum*, potbellied.

The description is from Eggerth (1935), Weinberg et al. (1937), Prévot et al. (1967) and study of the type and 20 similar strains.

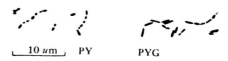

Figure 15.52. *E. ventriosum*

Cells of the type strain are 0.8–1.3 × 1.9–5.0 μm and occur singly and in pairs and short chains of pairs.

Surface colonies on blood agar are 0.5–3.0 mm, circular, entire-diffuse, convex, translucent, smooth and shiny, with slightly mottled or granular appearance when viewed by obliquely transmitted light. Colonies in agar are 1.0–2.0 mm in diameter, lenticular and translucent.

Glucose broth cultures are turbid with smooth, ropy or granular sediment and terminal pH of 4.6–5.4.

The optimum temperature for growth is 37°C. Strains grow well at 45°C and moderately well at 30°C. Some strains grow at 25°C. Growth is stimulated by a fermentable carbohydrate and by Tween 80. Growth is inhibited by 6.5% NaCl. Growth and fermentation are partially inhibited by 20% bile.

Little or no gas is produced by PYG agar deep cultures.

The type strain ferments pectin and galactose and weakly ferments dextrin. Adonitol, glycerol, inulin and sorbose are not fermented. Acetyl methyl carbinol is produced. Neutral red and resazurin are reduced. Ammonia is not produced from arginine, threonine or peptone. H_2S is not produced in SIM medium.

Products (milliequivalents per 100 ml of culture) in PYG cultures are acetic (0.1–0.5), formic (0.8–3.0), butyric (0.5–1.5), and D-(−)-lactic (0.8–2.7) acids; trace amounts of succinic acid may be detected. Little or no H_2 is detected. Pyruvate is partially converted to acetate and formate. Lactate, threonine and glucose are not utilized.

The type strain is susceptible to chloramphenicol (12 $\mu g/ml$), clindamycin (1.6 $\mu g/ml$), erythromycin (3 $\mu g/ml$), penicillin G (2 units/ml), and tetracycline (6 $\mu g/ml$).

Other characteristics of the species are given in Table 15.36.

Available strains are isolated from human feces. Isolations from dog feces, mouth abscess, neck infection, purulent pleurisy, pulmonary abscesses, and material from a bronchiectasis have been reported.

The mol% G + C of the DNA has not been determined.

Type strain: ATCC 27560.

Further comments. Holdeman et al. (1976) differentiated *E. ven-triosum* from *E. ruminatium* on the basis of H_2 production in at least small amounts by strains of *E. ventriosum*. These authors recognized two biogroups (I and II) that differed from strains of *E. ventriosum*. Characteristics of these biogroups are given in Table 15.38.

Table 15.38.
Differential reactions of **Eubacterium ventriosum** *subgroups[a]*

Substrate	Subgroup		
	E. ventriosum	*E. ventriosum* I	*E. ventriosum* II
Amygdalin	−	−A	A
Arabinose	−	−A	−
Cellobiose	−	A	w
Mannose	A−	−w	A
Melibiose	−	Aw	−

[a] Symbols: −, negative reaction; A, acid reaction, pH below 5.5; w, weak reaction, pH between 5.5 and 6.0.

Unnamed species of the genus Eubacterium

Strains of eubacteria isolated from rat cecal contents (Eysen et al., 1973) and represented by ATCC 21408, human feces (Sadzikowski et al., 1977) and baboon feces (Mott and Brinkley, 1979) reduce cholesterol to coprostanol. The baboon and rat strains were characterized by Mott et al. (1980) as nonproteolytic, nonsaccharoclastic, short, Gram-positive rods that hydrolyze esculin, do not reduce nitrate, do not produce indole, do not attack arginine and do not produce short chain fatty acids. The strains have an absolute requirement for cholesterol or other sterols.

Characteristics of 19 groups of unnamed species of eubacteria isolated from human feces are given in Moore and Holdeman (1974) and Holdeman et al. (1976).

Genus Acetobacterium Balch, Schoberth, Tanner and Wolfe 1977, 355[AL]

RALPH S. TANNER

A.ce.to.bac.te'ri.um. L. n. *acetum* vinegar; Gr. Neut. n. *bakterion* a small rod; M.L. neut. n. *Acetobacterium* vinegar rod.

Oval-shaped, short rods, about 1 μm × 2 μm. Cells frequently occur in pairs, occasionally in short chains. Endospores are absent. Gram-positive. **Motile, with one or two subterminal flagella (Fig. 15.53). Strictly anaerobic.** Optimum temperature, 30°C. Chemolithotrophic, **oxidizing hydrogen and reducing carbon dioxide to acetic acid.** Chemoorganotrophic, carrying out a **homoacetic fermentation of fructose and a limited number of other substrates. Acetic acid is the organic end product of metabolism.** Catalasenegative. The mol% G + C of the DNA is 39 (Bd) to 43 (T_m).

Type species: *Acetobacterium woodii* Balch, Schoberth, Tanner and Wolfe 1977, 355.

Further Descriptive Information

The morphology of oval-shaped, short rods occurring in pairs and the motility imparted by subterminal flagella makes *Acetobacterium* distinctive under phase microscopy. Under nonideal growth conditions cells can appear swollen and elongated.

The murein of *Acetobacterium* species contains a rare peptidoglycan type of the cross-linking group B (Kandler and Schoberth, 1979; Braun and Gottschalk, 1982). Ornithinyl residues function as the cell wall interpeptide bridges.

Acetobacterium uses the tetrahydrofolate enzyme pathway found in *Clostridium thermaceticum* and *Clostridium formicaceticum* to produce acetate from fructose or $H_2:CO_2$ (Tanner et al., 1978). Under the condition of a low HCO_3^- concentration in medium, H_2 may be produced from reduced substrates (Braun and Gottschalk, 1981). *Acetobacterium* converted methanol or the methoxyl groups of phenolic compounds to acetate and obtained energy for growth by reducing the propenoate group of compounds like ferulate or trimethoxycinnamate to the propanoate group (Bache and Pfennig, 1981). *Acetobacterium* utilizes carbon monoxide as a substrate and energy source (Sharak-Genthner, B. R., unpublished data, 1982).

Acetobacterium is probably important in anaerobic aquatic ecosystems with temperatures less than 35°C where it may produce acetate from lactate, $H_2:CO_2$, methanol or the methoxyl group of phenolic compounds like vanillate or syringate. It has not been observed in mammalian gastrointestinal systems, probably because of the higher temperature (37–41°C).

Little or no information is available on the genetics, antigenic properties or antibiotic sensitivity of *Acetobacterium*.

Enrichment and Isolation Procedures

Acetobacterium has been observed in anaerobic sediments and anaerobic digestors. Enrichment is carried out using a liquid medium and a $H_2:CO_2$ (80:20 or 67:33) atmosphere for chemolithotrophic growth. Methanogens have been inhibited in enrichment by the addition of sodium dithionite ($Na_2S_2O_4$) prepared under strict anaerobic conditions to a final concentration of 50 mg/liter of medium (Balch et al., 1977). Bromoethanesulfonic acid at a final concentration of 10–50 $\mu mol/liter$ could also be used to inhibit methanogens. *Acetobacterium* can be isolated directly from anaerobic spread plates incubated under a $H_2:CO_2$ gas phase (Braun et al., 1979). After colonies have formed, an overlay of plates with bromocresol blue, a pH indicator, in 1% agar permits the distinction of acidogenic and methanogenic colonies. *Acetobacterium* has been enriched from aquatic sediments and sewage sludge using 5 mmol/liter of a methoxylated aromatic compound (vanillate, syringate or trimethoxycinnamate) as the substrate (Bache and Pfennig, 1981). The strict anaerobic techniques helpful in isolating and culturing *Acetobacterium* have been described by Balch and Wolfe, 1976.

Maintenance Procedures

Acetobacterium strains can be lyophilized by procedures used for strict anaerobes.

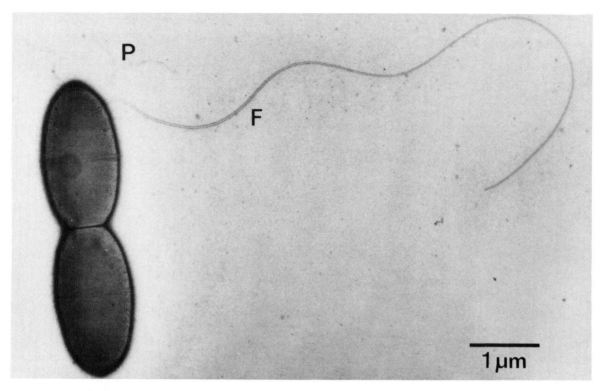

Figure 15.53. Micrograph of *Acetobacterium woodii*. (Micrograph taken by F. Mayer. Reproduced with permission from E. S. Balch, S. Schoberth, R. S. Tanner and R. S. Wolfe, 1977. International Journal of Systematic Bacteriology 27: 355–361.)

Differentiation of the genus **Acetobacterium** from other genera

The genus *Acetobacterium* is separated from the other genera of anaerobic, Gram-positive bacteria by endospore formation, end products of metabolism and the ability to grow chemolithotrophically. The members of the genus *Acetobacterium* do not form endospores, distinguishing this genus from the genus *Clostridium*. The members of the genus *Acetobacterium* do not produce the organic acid mixtures characteristic of the genera *Propionibacterium*, *Lactobacillus*, *Actinomyces* or *Bifidobacterium*. The members of the genus *Acetobacterium* differ from the members of the genus *Eubacterium* in their ability to grow by the anaerobic oxidation of hydrogen and the reduction of carbon dioxide to acetic acid, as well as in their ability to carry out only homoacetic fermentations on a limited number of substrates.

Taxonomic Comments

The original description of the genus *Acetobacterium* indicated that it was most closely related to the genus *Eubacterium*, and it was tentatively placed in the family *Propionibacteriaceae* (Balch et al., 1977).

Murein structure analysis and comparative cataloging of rRNA oligonucleotide sequences show that *Acetobacterium woodii* is closely related to *Eubacterium limosum* and *Clostridium barkeri* (Tanner et al., 1981). While further research may be required to refine the relationships, the strains of the genus *Acetobacterium*, *E. limosum* and *C. barkeri* form a group at a genus level that belongs to a much larger group basically defined by the Gram-positive, spore-forming anaerobes (Tanner et al., 1982).

Further Reading

Braun, M. and G. Gottschalk. 1982. *Acetobacterium wieringae* sp. nov., a new species producing acetic acid from molecular hydrogen and carbon dioxide. Zentralbl. Bakteriol. Mikrobiol. Hyg., Abt. I. Orig. C 3: 368–376.

Balch, W.E., S. Schoberth, R.S. Tanner and R.S. Wolfe. 1977. *Acetobacterium*, a new genus of hydrogen-oxidizing carbon dioxide-reducing, anaerobic bacteria. Int. J. Syst. Bacteriol. 27: 355–361.

Tanner, R.S., E. Stackebrandt, G.E. Fox, R. Gupta, L.J. Magrum and C.R. Woese. 1982. A phylogenetic analysis of anaerobic eubacteria capable of synthesizing acetate from carbon dioxide. Curr. Microbiol. 7: 127–132.

Differentiation of the species of the genus **Acetobacterium**

Some differential features of the species *A. woodii* and *A. wieringae* are indicated in Table 15.39.

Table 15.39.
Characteristics differentiating **Acetobacterium woodii** *and* **Acetobacterium wieringae**[a]

Characteristics	1. *A. woodii*	2. *A. wieringae*
Fermentation of		
Glucose	+	−
Glycerate	+	−
Glycerol	−	wk
Mol% G + C of DNA	39 (Bd)	43 (T_m)

[a] Symbols: see Table 15.25; also wk, weak fermentation.

List of species of the genus **Acetobacterium**

1. **Acetobacterium woodii** Balch, Schoberth, Tanner and Wolfe 1977, 355.[AL]

wood'i.i. M.L. gen. n. *woodii* of Wood, named for Harland G. Wood.

See Tables 15.39 and 15.40 and the generic description for many features.

Colonies on anaerobic agar are circular and convex. Colonies are whitish; older colonies may show a slight yellow pigmentation. Colonies grow to a diameter of 1 mm in 7–10 days with H_2:CO_2. Traces of a water-soluble yellow pigment may be excreted into the medium. Catalase-negative.

Optimal temperature is 30°C. Optimal pH is 6.6–6.8; remains viable at pH 5 for many weeks.

Obligately anaerobic. Produces homoacetic fermentations of fructose, glucose, lactate, glycerate, formate and methanol. Oxidizes hydrogen

and reduces carbon dioxide to acetic acid. Produces homoacetic fermentations of the methoxyl group of phenolic compounds such as syringate, vanillate and ferulate, converting them to gallate, protocatechuate and hydrocaffeate, respectively. Traces of succinate may be produced from organic substrates. Ethanol and higher alcohols, acetate and higher aliphatic acids, fumarate, malate, pyruvate, succinate, amino acids, disaccharides, arabinose, galactose, mannose, 2-deoxyglucose, rhamnose, ribose, xylose, mannitol, inositol, glycerol, galacturonic acid and glucuronic acid are not fermented.

Not pathogenic. Isolated from aquatic and marine anaerobic sediments.

The mol% G + C of the DNA is 39 (Bd).

Type strain: ATCC 29683.

2. **Acetobacterium wieringae** Braun and Gottschalk 1983, 438.[VP] (Effective publication, Braun and Gottschalk 1982, 368.)

wie.rin'gae. M.L. gen. n. *wieringae* of Wieringa, named for K.T. Wieringa.

See Tables 15.39 and 15.40 and the generic description for many features.

Colonies on anaerobic agar are circular and convex. Colonies are brownish. Colonies can grow up to a diameter of 3 mm with H_2:CO_2. Catalase-negative.

Optimal temperature is 30°C. Optimal pH is 7.2–7.8.

Obligately anaerobic. Produces homoacetic fermentations of fructose and lactate. Oxidizes hydrogen and reduces carbon dioxide to acetic acid. Weakly ferments melibiose, formate and glycerol. Glucose is not fermented. Methanol, ethanol, isopropanol, glycerate, fumarate, malate, succinate, citrate, amino acids, maltose, ribose, xylose, galactose, sucrose, lactose, mannitol, dulcitol, sorbitol, ribitol, hippurate and amygdalin are not fermented. This substrate spectrum may be amended by further study.

Not pathogenic. Isolated from anaerobic sewage sludge.

The mol% G + C of the DNA is 43.

Type strain: DSM 1911.

Table 15.40.
Other physiological features of **Acetobacterium woodii** *and* **Acetobacterium wieringae**[a]

Characteristics	1. *A. woodii*	2. *A. wieringae*
Chemolithotrophic growth with H_2:CO_2	+	+
Fermentation of		
Fructose	+	+
Lactate	+	+
Formate	+	+
Ribose, xylose, galactose, maltose, lactose, sucrose and mannitol	−	−
Fumarate, malate, succinate and citrate	−	−
Ethanol	−	−
Amino acids	−	−

[a] Symbols: see Table 15.45.

Genus XIII. **Lachnospira** *Bryant and Small 1956, 24*[AL] *

MARVIN P. BRYANT

Lach.no.spi'ra. Gr. n. *lachnos* woolly hair, down; L. n. *spira* a coil; M.L. fem. n. *Lachnospira* woolly (colony producing) spiral.

Curved rods, 0.4–0.6 μm wide and 2.0–4.0 μm long, with bluntly pointed ends. Weakly Gram-positive; the Gram stain may be negative except in very young cultures. Motile by monotrichous lateral to subpolar flagella. Strictly anerobic. **The colonies are filamentous, woolly in appearance.** Chemoorganotrophic, having a fermentative type of metabolism. **The products of glucose fermentation are formate, lactate, acetate, ethanol, CO_2 and H_2. Pectin is also fermented** with the same products plus **methanol** being produced. Succinate, butyrate and propionate are not produced. Detected so far only in the rumen. The mol% G + C of the DNA is not known.

Type species: *Lachnospira multiparus* Bryant and Small 1956, 24.

Further Descriptive Information

Single cells appear as curved or helical rods containing one or more turns. Some very long chains of cells which are only slightly curved and which have more bluntly rounded ends, and also filamentous cells, are often observed, especially in liquid media lacking rumen fluid.

The Gram stain may be only weakly positive or seem negative;

however, the cell wall is structurally of the Gram-positive type (Cheng et al., 1979).

Surface colonies on RGCA agar in roll tubes (see the genus *Butyrivibrio* for composition of medium and anaerobic methods) incubated at 37°C for 3 days are 2–5 mm in diameter and characteristically flat and filamentous; some spreading may occur. Deep agar colonies appear as woolly balls which penetrate the agar to some extent. In liquid rumen-fluid glucose medium growth occurs as a large flocculent sediment. The floc may be difficult to disperse; this is especially true in media in which yeast extract and Trypticase or chemically defined ingredients replace the rumen fluid.

Growth of *L. multiparus* occurs at 30–45°C but not at 22 or 50°C.

The final pH in lightly buffered glucose medium is 4.8–5.2. Abundant growth occurs at pH 6.0–7.0. The upper pH limit for growth is not known.

L. multiparus grows well in a chemically defined medium containing B-vitamins, cysteine, glucose, ammonium ions, minerals, acetate and CO_2/HCO_3^- buffer (Bryant and Robinson, 1962). Acetate is highly stimulatory, especially in defined media lacking amino acids or peptides. One strain of *L. multiparus* has been found to require only *p*-

---

* This genus is also included in volume 1, pp 661–662.

aminobenzoic acid and biotin among the B-vitamins (Emery et al., 1957). Either ammonium ions or mixtures of amino acids or peptides can serve as the nitrogen source.

Glucose, fructose, cellobiose, esculin, pectin, salicin and sucrose are fermented. Xylose fermentation is variable. Arabinose, cellulose, dextrin, galactose, glycerol, gum arabic, inositol, inulin, lactose, maltose, mannitol, trehalose and xylan are not fermented. Starch is not hydrolyzed. The organism contains pectinmethylesterase, as indicated by its production of methanol during pectin fermentation (Rode et al., 1981), but does not ferment galacturonic acid (Dehority, 1969). It presumably ferments polygalacturonic acid, since pectin is an excellent energy source. Amino acids are not fermented.

Indole and H_2S are not produced. Nitrate is not reduced. Gelatin is not liquefied. The Voges-Proskauer test differs among strains.

L. multiparus is very important in pectin fermentation in the rumen, as indicated by its greatly increased numbers in the rumen when lush legume forages such as alfalfa or ladino clover are the chief dietary components (Bryant et al., 1960). Its importance is also indicated by its ability to macerate clover and grass leaves (Cheng et al., 1979).

Enrichment and Isolation Procedures

L. multiparus can be isolated from rumen contents using nonselective procedures similar to those described for the genus *Butyrivibrio*, i.e. anaerobic roll tubes containing RGCA medium. The distinctive woolly and filamentous appearance of the colonies makes isolation of *L. multiparus* quite easy (Bryant and Small, 1956). Isolation from the rumen contents of cattle fed lush alfalfa or clover diets is especially easy because of the large numbers of *L. multiparus* present (Bryant et al., 1960).

RGCA medium can be made somewhat selective by replacing the usual carbohydrates by 0.3% pectin (Dehority, 1969); however, colonies of other pectin-fermenting bacteria such as *Bacteroides ruminicola* or *Butyrivibrio fibrisolvens* may also be quite numerous.

Maintenance Procedures

L. multiparus can be maintained on stab-inoculated RGCA slants at −70°C for 1 year or more. The organisms can also be preserved indefinitely by lyophilization.

Differentiation of the genus **Lachnospira** from other genera

See Table 15.41 in the next article on the genus *Butyrivibrio* for those characteristics useful in differentiating *Lachnospira* from other morphologically or physiologically similar genera of anaerobic bacteria. The key to the genera of the family *Bacteroidaceae* also provides useful distinguishing characteristics.

Taxonomic Comments

In volume 1 of this edition of the *Manual* the genus *Lachnospira* was included in the family *Bacteroidaceae* solely for convenience, since the cells generally stain Gram-negative. The cell wall, however, is of the

Gram-positive type. The relationship of *Lachnospira* to genera or higher taxa of Gram-positive bacteria needs to be elucidated, and nucleic acid studies such as DNA/rRNA hybridization or rRNA oligonucleotide cataloging would be quite useful.

Further Reading

Bryant, M.P., and N. Small. 1956. Characteristics of two new genera of anaerobic curved rods isolated from the rumen of cattle. J. Bacteriol. 72: 22–26.
Cheng, K.-J., D. Dinsdale and C.S. Stewart. 1979. Maceration of clover and grass leaves by *Lachnospira multiparus*. Appl. Environ. Microbiol. 38: 723–729.
Dehority, B.A. 1969. Pectin-fermenting bacteria isolated from the bovine rumen. J. Bacteriol. 99: 189–196.

List of species of the genus **Lachnospira**

1. **Lachnospira multiparus** Bryant and Small 1956, 24.[AL]
mul.ti.par′us. L. adj. *multus* much, many; L. v. suff. *parus* from L. v. *pario* to produce; M.L. adj. *multiparus* many (products) produced.
The characteristics are as described for the genus.

Occurs in the rumen of the bovine and probably other ruminants. The mol% G + C of the DNA is not known.
Type strain: ATCC 19207 (strain D32 of Bryant and Small, 1956).

Genus IV. **Butyrivibrio** Bryant and Small 1956, 18, emend. Moore, Johnson and Holdeman 1976,* 241[AL]

MARVIN P. BRYANT

Bu.ty.ri.vib′ri.o. M.L. adj. *butyricus* butyric; L. v. *vibro* to vibrate; M.L. n. *vibrio* that which vibrates; a generic name; M.L. masc. n. *Butyrivibrio* a butyric vibrio.

Curved rods, 0.3–0.8 μm × 1.0–5.0 μm, single or in chains or filaments which may or may not be helical. No resting stages known. Stain Gram-negative but the type species is structurally Gram-positive. **Motile by polar or subpolar flagella,** monotrichous or lophotrichous; some strains may be nonmotile. **Strictly anaerobic.** Chemoorganotrophic, having a **fermentative type of metabolism** with carbohydrates being the main fermentable substrates. **Glucose or maltose is fermented with butyrate as one of the important products.** Under some conditions large amounts of lactic acid and little butyric acid may be produced. Occur in the rumen of ruminants and sometimes in human, rabbit and horse feces. The mol% G + C of the DNA is 36–42 (T_m).

Type species: *Butyrivibrio fibrisolvens* Bryant and Small 1956, 19.

Further Descriptive Information

In wet mounts prepared from the water of syneresis at the base of RGCA slants,† stab-inoculated into the base and incubated overnight at 37°C, the organisms appear as curved rods with tapered and rounded ends. Cells occur as singles and in short or long chains and sometimes as filaments. Chains may or may not show a helical arrangement and pairs may be in an "S" arrangement. Motility is rapid and vibrating and often progressive, but often only a few cells in a culture show motility. Flagellation is polar or subpolar and a few nonmotile, aflagellated strains have been reported. Flagellation is monotrichous (*Butyrivibrio fibrisolvens*) or lophotrichous (*Butyrivibrio crossotus*). Cells of the latter tend to be the larger in diameter. Cell walls of *B. fibrisolvens*,

---

* This genus is also included in volume 1, pp 641–643.
† RGCA slants (g/liter): K_2HPO_4, 0.23; KH_2PO_4, 0.23; $(NH_4)_2SO_4$, 0.45; NaCl, 0.45; $MgSO_4$, 0.023; $CaCl_2$, 0.023; resazurin, 0.001; glucose, 0.5; cellobiose, 0.5; soluble starch, 0.5; rumen fluid (centrifuged), 400; cysteine·HCl, 0.25; $Na_2S·9H_2O$, 0.25; Na_2CO_3, 4.0; agar (Difco), 10.0; pH 6.7; gas phase, 100% CO_2. Prepared and used with the Hungate technique (Bryant, 1972).

while staining Gram-negative, contain teichoic acid (Sharpe et al., 1975) and have a Gram-positive fine structure, although the wall is much thinner than is usual for Gram-positive bacteria (Cheng and Costerton, 1977). The fine structure of cells of *B. crossotus* has not been determined.

The genus is fermentative and little growth occurs in media lacking in carbohydrate energy source. Formic, lactic and butyric acid are produced from glucose (*B. fibrisolvens*) or maltose (*B. crossotus*). CO_2 is a product of *B. fibrisolvens* but has not yet been determined in *B. crossotus*. *B. fibrisolvens* produces H_2 but *B. crossotus* does not. Acetic acid may be produced or used. Under some conditions of culture, lactic acid production may be increased and only a small amount of butyric acid is formed (Gill and King, 1958). Small amounts of ethanol, propionic, pyruvic and succinic acids may be formed.

Nothing is known concerning the antigenic structure of *B. crossotus*, but Margherita and Hungate (1963) found *B. fibrisolvens* strains to be antigenically very diverse. Sharpe et al. (1975) and Hewett et al. (1976) found that a number of strains contained lipoteichoic acids, but this was not found in other strains.

B. fibrisolvens is highly sensitive to Monensin and Lasalocid (0.4–2.5 µg/ml) which correlates with the Gram-positive wall structure (Chen and Wolin, 1979; Dennis et al., 1981).

Pathogenicity has not been detected.

The members of the genus are among the most numerous bacteria in the rumen of ruminants under a wide variety of dietary conditions, and are sometimes found among the predominant bacteria in human, rabbit and horse feces (Brown and Moore, 1960; Moore and Holdeman, 1974). They are highly versatile in energy sources utilized for growth and some strains ferment cellulose and many ferment starch, pectin, xylan and other polysaccharides.

Enrichment and Isolation Procedures

Butyrivibrio strains are commonly isolated by nonselective procedures from among the predominant bacteria grown as colonies in RGCA roll-tubes inoculated with high dilutions of rumen fluid (Bryant and Small, 1956) or less often from human feces (Moore et al., 1976). *B. fibrisolvens* is often isolated from high dilutions of rumen fluid using rumen fluid agar roll-tube media with finely-ground cellulose as the energy source. In this case zones of cellulose digestion around colonies are seen, although the zones are often less distinct and the colonies less numerous than those of other cellulolytic bacteria such as *Ruminococcus albus* and *R. flavefaciens* (Hungate, 1966; Shane et al., 1969). *B. fibrisolvens* is the main rumen organism fermenting the bioflavonoid rutin, and can be isolated from high dilutions of rumen fluid as colonies in RGCA roll-tube medium containing 0.2% rutin in place of other energy sources. Colonies form clear zones in the suspended rutin and a yellow precipitate of quercetin forms around some (Cheng et al., 1969; Leedle and Hespell, 1980).

Maintenance Procedures

Strains can be maintained in stabbed-slants of RGCA medium held at −70°C for periods of a year or more. They can also be preserved by lyophilization.

Differentiation of the genus **Butyrivibrio** from other taxa

Differentiation of the genus from other taxa with which it might be confused is indicated in Table 15.41.

Taxonomic Comments

The genus constitutes one of the most numerous and biochemically versatile groups of bacteria in the rumen, and there is great variation among the features of strains. For example, great variation occurs in energy sources, fermentation products (Bryant and Small, 1956; Shane et al., 1969) and nutrition (Bryant and Robinson, 1962; Roche et al., 1973). Shane et al. (1969) placed most cellulolytic strains in two groups. One group produced appreciable lactate and low levels of formate, and removed acetate from the medium during cellobiose fermentation; a second group produced acetate, more formate but little or no lactate. The latter group was more exacting in nutritional requirements (Roche et al., 1973); however, both groups varied greatly in other features.

Under somewhat acidic conditions, the rumen may contain a species similar to *B. fibrisolvens* except that it is somewhat larger and exhibits lophotrichous flagella. This group has been referred to as the "B-385-like strains" (Bryant, 1956; Bryant et al., 1961).

Moore and Holdeman (1974) observed that human fecal isolates of *Eubacterium rectale* may stain very weakly Gram-positive or Gram-negative and their differentiation from *B. fibrisolvens* is in doubt.

B. fibrisolvens is in the "Clostridium" group and closely related to *Clostridium sphenoides* and *Clostridium aminovalericum* as determined by 16S-RNA oligonucleotide catabolizing (B. Paster, R. B. Hespell and C. R. Woese, 1983, unpublished data).

Further Reading

Bryant, M.P. and N. Small. 1956. The anaerobic, monotrichous, butyric acid-producing curved rod-shaped bacteria of the rumen. J. Bacteriol. 72: 16–21.
Hungate, R.E. 1966. *The Rumen and Its Microbes*, Academic Press, New York.
Moore, W.E.C., J.L. Johnson and L.V. Holdeman. 1976. Emendation of *Bacteriodaceae* and *Butyrivibrio* and descriptions of *Desulfomonas* gen. nov. and ten new species in the genera *Desulfomonas, Butyrivibrio, Eubacterium, Clostridium* and *Ruminococcus*. Int. J. Syst. Bacteriol. 26: 238–252.
Shane, B.S., L. Gouws and A. Kistner. 1969. Cellulolytic bacteria occurring in the rumen of sheep conditioned to low-protein teff hay. J. Gen. Microbiol. 55: 445–457.

Table 15.41.
Differential characteristics of genus **Butyrivibrio** *and other morphologically similar, nonsporing anaerobic genera*[a]

Characteristics	Butyri-vibrio	Eubac-terium	Seleno-monas	Lachnospira	Succini-vibrio	Pectinatus
Curved rods	+	D	+	−	+	+
Stain Gram-positive	−	+[b]	−	+[b]	−	−
Flagellar arrangement						
Polar to subpolar	+	D	−	−	+	−
Laterally attached	−	D	+	+	−	+
Monotrichous	D	D	−	+	+	−
Tufts	D	D	+	−	−	−
Major products of glucose or maltose fermentation						
Butyrate	+	D	−	−	−	−
Succinate	−	−	v	−	+	+
Propionate	−	−	+	−	−	+

[a] Symbols: see Table 15.25.
[b] Cells of some strains may be very weakly Gram-positive or Gram-negative.

Differentiation of the species of the genus **Butyrivibrio**

Some differential characteristics of the species of *Butyrivibrio* are indicated in Table 15.42.

Table 15.42.
Characteristics differentiating **Butyrivibrio fibrisolvens** *and* **Butyrivibrio crossotus**[a]

Characteristics	1. *B. fibrisolvens*	2. *B. crossotus*
Flagella monotrichous	+	−
Flagella lophotrichous	−	+
Produce H_2 from glucose or maltose	+	−
Ferment glucose and fructose	+	Weak
Ferment sucrose, cellobiose and xylose	+	−

[a] Symbols: Table 15.25.

List of species of the genus **Butyrivibrio**

1. **Butyrivibrio fibrisolvens** Bryant and Small 1956, 19.[AL]

fi.bri.sol′vens. L. n. *fibra* fiber; L. part. adj. *solvens* dissolving; M.L. part. adj. *fibrisolvens* fiber-dissolving.

The morphology is indicated in the generic description. While strains stain uniformly Gram-negative, they have a Gram-positive ultrastructure with a very thin wall.

In RGCA roll-tubes, surface colonies are usually smooth, entire, slightly convex, translucent, light tan in color and 2–4 mm in diameter. Some strains have rough colonies that are more flat, lighter in color and have filamentous margins. Deep colonies are usually lenticular or Y-shaped but some form compound lenticular colonies.

In rumen fluid-cellulose agar, colonies of cellulolytic strains vary from lens-shaped to trianglular to compound lenticular or rhizoidal colonies. Zones of cellulose digestion around colonies vary from very narrow with slow and indistinct digestion to broad zones with rapid and complete digestion.

The type of zones formed in rumen fluid-rutin agar is indicated under Enrichment and Isolation Procedures.

Growth in liquid glucose medium varies from a uniform turbidity to a flocculant or granular sediment, some of which may adhere to the walls of tubes.

Good growth occurs at 37°C and usually at 45°C, but more slowly at 30°C. No growth occurs at 22 or 50°C.

The final pH in poorly buffered glucose medium is usually 5.0–5.6.

Most strains grow in chemically defined culture media containing glucose or cellobiose as the energy source, amino acid mixtures and ammonium salts added as nitrogen sources, minerals, B-vitamins and cysteine. Many strains grow with an ammonium salt as the nitrogen source, but amino acid mixtures are usually stimulatory. Acetate is often stimulatory to growth; propionate or branched-chain volatile acids are stimulatory to some strains but not to others (Bryant and Robinson, 1962; Bryant, 1973; Roche et al., 1973).

The energy-yielding metabolism is indicated in the generic description and in Tables 15.42 and 15.43. Some strains ferment polysaccharides such as cellulose, starch, pectin and/or hemicellulose and xylan and other materials such as the flavone glycoside, rutin.

Indole and catalase are not produced. Nitrate is usually not reduced. Hydrogen sulfide is usually not produced.

Gelatin liquefaction is variable.

The species is nonpathogenic.

Ecological information is indicated above and in the generic description.

The mol% of the DNA is 41 (absorbance ratio 245/270 nm; Van Glyswick, 1980 and by T_m, Leatherwood and Sharma (1972)).

Table 15.43.
Other characteristics of **Butyrivibrio fibrisolvens** *and* **Butyrivibrio crossotus**[a]

Characteristics	1. *B. fibrisolvens*	2. *B. crossotus*
Energy sources		
Maltose	+	−
Galactose, inulin, salicin	+	+
Lactose	+	d
Starch	d	+
Esculin, trehalose	d	−
Glycerol, inositol, mannitol	−	−
Esculin hydrolysis	d	−
Acids produced from fermentation[b]		
Butyrate, formate	+	+
Lactate, acetate	d	+

[a] Symbols: see Table 15.25.
[b] From glucose or cellobiose in *B. fibrisolvens*, and from maltose in *B. crossotus*.

Type strain: ATCC 19171 (strain D1 of Bryant and Small (1956)).

2. **Butyrivibrio crossotus** Moore, Johnson and Holdeman, 1976, 241.[AL]

cros.so′tus. Gr. adj. *crossotus*, tasseled.

The morphology is indicated in the generic description. The fine structure has not yet been elucidated.

After incubation at 37°C for 5 days in RGCA roll-tubes, subsurface colonies are 0.5–1.0 mm in diameter, lenticular and translucent to transparent. Surface colonies on brain heart infusion agar (supplemented) roll-streaks are 0.2–1.0 mm in diameter, circular, entire, convex, translucent to semiopaque and smooth (Moore et al., 1976). Only a few strains grow as surface colonies on blood agar plates incubated anaerobically, and no hemolytic activity is observed. On egg yolk agar few strains grow and no lecithinase or lipase reactions are seen.

There is poor growth in peptone-yeast extract broth unless a fermentable carbohydrate is provided. In maltose broth, cultures show abundant growth with a smooth, flocculent or ropy sediment and usually some uniform turbidity.

Optimum growth temperature is 37°C. Some strains grow at 45°C. Growth is slow at 30°C.

The energy-yielding metabolism is indicated in the generic description and in Tables 15.42 and 15.43. Materials fermented are largely limited to starch, glycogen, dextrin and maltose.

No ammonia is produced from peptone.

All strains have been isolated from human feces or rectal contents. The mol% G + C of the DNA is 36–37 (T_m).

Type strain: ATCC 29175 (T9-40A; Moore et al. (1976)).

Genus **Thermoanaerobacter** Wiegel and Ljungdahl, 1982, 384[VP] (Effective publication Wiegel and Ljungdahl 1981, 348)

JÜRGEN K. W. WIEGEL

Ther.mo.an.ae.ro.bac'ter. Gr. adj. *thermos* hot; Gr. pref. *an* not; Gr. IE. *aer* air; M.L. bacter masc. equivalent of Gr. neut. n. *bacterion* rod staff; M.L. masc. nn. *Thermoanaerobacter*, rod which grows in the absence of air at elevated temperatures.

Rods in the logarithmic growth phase 0.5–0.8 μm × 4–8 μm (Fig. 15.54a); at later growth stages pleomorphic; **chains of rods interspersed with coccoid cells**, 0.8–1.5 μm in diameter (Fig. 15.54b) and long filamentous cells (Fig. 15.54a) up to 100 μm, which sometimes divide later by constrictions. Uneven cell division (Fig. 15.54c), formation of protuberances and spheroplast-like forms are common during late logarithmic growth phases (Fig. 15.54d and e). **The organism is of the Gram-positive type** (Wiegel, 1981), although only cells during the very early logarithmic growth phase exhibit a Gram-positive staining reaction; in most instances **cells from a growing culture stain Gram-negative.** Capsules are not formed. **Resting stages** (*spores*) **are not known.** One to twelve peritrichously inserted flagella (Fig. 15.54f) causing frequently tumbling motility.

Strictly anaerobic; pH range for growth is 4.4–9.8 with a broad optimum between pH 5.8 and 8.5. The **marginal temperature data for growth are T_{max} 78°C, T_{opt} 68/69°C, T_{min} 35°C.**

Chemoorganotroph. Main fermentation products from hexoses and pentoses are ethanol and CO_2; minor products are lactate, acetate and H_2. Starch and hemicellulose hydrolyzed, but not cellulose. Yeast extract is required for growth.

Habitat: alkaline and slightly acidic hot springs in Yellowstone National Park. The mol% G + C of the DNA is 37–39 (Bd), around 32 (T_m) and 38 (chemical).

Type species: *Thermoanaerobacter ethanolicus* Wiegel and Ljungdahl 1982, 384.

Further Descriptive Information

Colony and Cell Morphology

In 2.5% agar shakes at 60°C colonies are 0.2–1 mm in diameter, lenticular and white; after 1 week of incubation, frequently colonies turn brown. No soluble pigments. In solid agar shakes, gas bubbles are formed (thus thin agar-shake-roll tubes are preferred). Surface colonies are white, smooth, round to irregular with a diameter of 2–4 mm after 40 h of incubation. Especially on agar plates, after 20–30 h incubation some colonies look more translucent than other colonies. After 50 h incubation at 60–65°C all colonies have the same appearance. On reculturing, both types of colonies yield both types, and no differences in temperature range or fermentation behavior are found (unpublished results).

The cell diameter of *Thermoanaerobacter* is variable with the growth conditions: during growth on xylose, above 1% glucose or above 1% starch, the cell diameter is up to 1.5 μm. At the extremes of its pH and temperature range, as above pH 9.0 or below 5.0 and near the T_{max} and T_{min}, the cells can be long and as thin as 0.3 μm.

Thermoanaerobacter forms coccoid cells, frequently alternating with rods in short chains. It shares this property with other extreme thermophiles as *Thermoanaerobium* (Zeikus et al., 1979), *Thermobacteroides* and *Clostridium thermohydrosulfuricum* (Wiegel, unpublished results). The protuberances and bleb formation occur mainly at temperatures above 55°C and in media containing less than 30 mM phosphate. In cultures being transferred continuously for more than 1 year, the formation of blebs and extreme long filamentous cells occurs less frequently than in subcultures of original stock culture kept at -20°C in 50% glycerol. Using phase contrast microscopy some of the bleb formation can be misjudged as (pre-) spores; however, at no stage do these formations exhibit a heat stability indicative of spores, and they lyse easily (Wiegel and Ljungdahl, 1981). Since growth conditions and requirements are very similar, cultures of *Thermoanaerobacter* easily become contaminated with the spore-forming *Clostridium thermohydrosulfuricum*, which is ubiquitous and forms very heat-stable spores (see also Taxonomic Comments). The flagella of *Thermoanaerobacter* are 150–170 Å thick and up to 50 μm in length; however, they are commonly lost during late logarithmic growth phase. Four to eight pili of the 50 Å-type are present, predominantly found at the poles.

Cell Wall and Gram-Staining Reaction

The cell wall contains the diaminopimelic acid (DAP), about 4.3% (w/w) phosphate and several unidentified neutral sugars; glycerol and ribitol have not been found in significant amounts. Lipopolysaccharides have not been detected: the limulus lysate (Sigma) reaction and the polymyxin B-LPS complex reaction (Wiegel and Quandt, 1982) gave negative results. Thus, *Thermoanaerobacter* is a positive Gram-type organism (Wiegel 1981; Wiegel and Ljungdahl, 1981). Electron photomicrographs of ultrathin sections frequently exhibit an atypical cell wall with two dense layers (Fig. 15.54i). Preparations from cells of the late logarithmic growth phase grown at 70°C exhibit weakly stained cytoplasmic membranes, and significantly thicker dense layers than preparations from cells of the early logarithmic growth phase (Wiegel and F. Mayer, unpublished results). Contrary to cells grown at 60°C or harvested in the logarithmic growth phase, these cells are difficult to break by ultrasonic or French-press treatment. The outer cell wall layer has hexagonal rotary symmetry pattern (Fig. 15.54f, *right insert*).

Antibiotic Resistance

Strain JW200 grows at 50–70°C in the presence of the following antibiotic concentrations (Wiegel and Ljungdahl, 1981; Carreira, personal communication): 10 mg/ml ampicillin; 1.5 mg/ml nalidixic acid; 1 mg/ml Novobiocin; 400 μg/ml 5-fluorodeoxyuridine; 200 μg/ml chloramphenicol, or streptomycin sulfate; 150 μg/ml (but not at 200 μg/ml) rifampicin; 25 μg/ml polymyxin B; 10 μg/ml erythromycin, tetracycline, acridine orange, or penicillin G; 20 μg/ml neomycin sulfate; 60 mM hydroxyphenylazouracil; and 200 mM L-azetidine-2-carboxylic acid. **Although JW 200 does not yield an LPS-polymyxin B complex in the presence of 100 μg/ml Polymyxin B, this concentration causes cells from the late logarithmic growth phase to lyse at 37 and 60°C.**

Enrichment and Isolation Procedures

The only identified strains are strain JW 200 and JW 201, described in the original paper (Wiegel and Ljungdahl, 1981). Without prior enrichment steps, they were obtained from serial dilutions of hot pool water of the Yellowstone National Park into agar-shake-role tubes. The method of Hungate as modified by Bryant and Robinson (1961) was employed. Strain JW 200 was obtained from a sample of the alkaline (pH 8.8) hot spring (45–50°C) in White Creek at Fire Hole

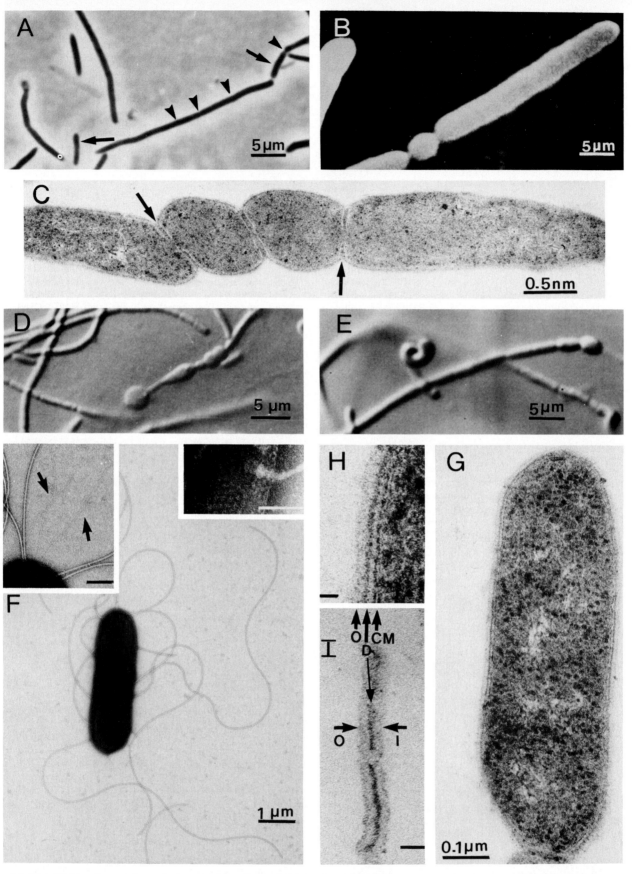

Figure 15.54. Light microscopic and electron microscopic (EM) observations of strain JW 200. *A*, normal size (*arrow*) and longer, filamentous cells which may divide to give long chains of bacteria (*arrow heads*) (light microscopy). *B*, coccoid cell between rods (scanning EM). *C*, uneven length of dividing cells; notice the uncomplete cross-wall formation (arrows) (EM). *D* and *E*, protuberances or spheroplast type cells (light microscopy, Nomarski technique). *F*, single cell with flagella; *right inset*: hexagonal pattern of the outer cell wall layer and the insertion point of a flagellum; *left inset*: Pili (*arrows*) and flagella;

bars represent 0.1 μm. G, ultrathin section showing cell envelope layers (EM). *H*, higher magnification of the cell envelope layers (*arrows*); outer cell wall layer (*O*); dense layer (*D*); cytoplasmic membrane (CM); *bar* represents 2.5 nm (EM). *I*, cell wall from a lysed cell showing three layers (*arrows*): outer cell wall layer (*O*); dense layer (*D*), and an inner cell wall layer (*I*); *bar* represents 25 nm (*EM*). (Reproduced with permission from J. Wiegel and L. G. Ljungdahl, *Archives of Microbiology 128:* 343–348, 1981.)

Lake Drive, and strain JW 201 was obtained from the outflow of the slightly acidic hot spring (pH 5.5; 55–60°C) Dragon's Mouth. No specific enrichment or isolation procedure can be recommended at the present time. Single colonies derived from agar-shake-role tubes are grown in liquid media and analyzed for ethanol production. However, Zeikus (1980) also reported high ratios of ethanol production for his strain C. thermohydrosulfuricum 39E isolated from a sample from the Yellowstone National Park.

A thermoanaerobacter-like strain has been isolated from a hot spring in New Zealand; however, no DNA/DNA homology tests have been done (Morgan, personal communication).

Metabolic Pathways and Enzymes Studies

Hexoses are mainly degraded via the Embden-Meyerhof pathway as elucidated by the fermentation analysis of specifically labeled ^{14}C-glucose. The pentose phosphate pathway seems to be concomitantly used to a significant degree (unpublished results). So far no indications of one-carbon utilization have been observed. The fermentation of glucose and starch (given as glucose equivalents) can be described as follows: glucose + 0.1 H_2O → 1.8 ethanol + 0.1 acetate + 0.1 lactate + 1.9 CO_2 + 0.2 H_2. With increasing glucose (up to 4% w/v) and starch (up to 20% w/v) the ethanol yields decrease drastically and acetate becomes a more dominant fermentation product. However, mutants have been obtained which yield high ethanol values in the presence of high (up to 20%) starch concentrations (Carreira and Ljungdahl, 1983). Contrary to C. thermohydrosulfuricum, a pH-shift from above pH 7.2 to below pH 6.9 during growth seems not to be required for the formation of a high ratio of ethanol per mole of sugar utilized (Wiegel and Ljungdahl, 1979).

Although cellulose is not degraded, various hemicelluloses are directly hydrolyzed by T. ethanolicus (Wiegel and Puls, 1983; Wiegel et al., 1983). The doubling times of the wild type on xylose, xylobiose and xylotriose are about 120 min; on xylotetraose about 140 min; and on higher oligomers and polymers the doubling time increases with the chain length up to about 5 h. The fermentation balance on xylose can be represented as follows: 1 xylose + 0.1 H_2O → 1.4 ethanol + 0.2 acetate + 0.1 lactate + 1.5 CO_2 + 0.2 H_2. Minor variations occur frequently, depending on starting pH, age of the culture and substrate concentration. At higher xylose concentration (up to 5%) or on hemicellulose, the ethanol can decrease to 1.0 mol ethanol/mol xylose (equivalent) with a corresponding increase in acetate formation. Using resting cells, ethanol is produced linearly from 1 to more than 1 h; however, at xylose concentrations above 1%, lactate becomes a major product, too. Using xylan (birch) at 4%, the major product is acetate for the first 6 h, then more and more ethanol is produced. When supplied together, pentoses such as xylose and hexoses like glucose, including its oligomers and polymers, are concomitantly metabolized. However, diauxic growth characteristics were obtained either with the monomer glucose and cellobiose or glucose polymers (Carreira et al., 1983) or when the monomer xylose and polymeric xylanes (Wiegel et al., 1983) were supplied together. Only the corresponding monomers inhibited the utilization of the oligomers and polymers. These data suggest that hexoses and pentoses are degraded via different pathways and that they are regulated separately.

Contrary to the inhibition of cellobiose or starch utilization by growing cells which is inhibited by 20 mM (0.35%) glucose (Carreira et al., 1983), the cell-associated β-1,4-glucosidase is only inhibited to 20% of the full p-nitrophenyl-β-D-glucoside-dependent activity by 2.5% (about 140 mM) glucose (Mitchell et al., 1982). The β-1,4-glucosidase exhibited a broad pH-optimum between pH 5–7 and a temperature optimum around 70°C; the K_m for cellobiose was 4.2 mM. In a crude enzyme preparation, the half-life of the activity was four and less than half a day at 45 and 65°C, respectively (Mitchell et al., 1982).

The ethanol dehydrogenase and lactate dehydrogenase from Thermoanaerobacter have been characterized. It contains at least two alcohol **dehydrogenases (A-DH) (Bryant and Ljungdahl, 1981; Bryant and Wiegel, 1983; Bryant et al., unpublished results). They are both** NADP/NADPH₂ dependent. An enzyme isolated from the wild type JW 200, grown at 55° to late logarithmic growth phase has a molecular weight of 176,000 and contains 4 subunits of identical molecular weight. Each subunit contains 4 Zn. In the oxidation of ethanol, the ethanol saturation curves are sigmoidal. However, in the presence of the activator pyruvate, the ethanol saturation curves approach Michaelis-Menton kinetics. Unlike the yeast, enzyme pyruvate can serve as a substrate. The highest oxidation rates have been observed with secondary alcohols, namely with 2-propanol, 1,2-propanediol and 2-pentanol; the rates are 16-, 2.9- and 1.8 times, respectively, faster than the oxidation of ethanol. Methanol is not oxidized. Also, acetone and 2,3-butanedione are reduced twice as fast as acetaldehyde. **The enzyme first isolated from the mutant T. ethanolicus JW 200 Fe(4) grown at 55°C to the late logarithmic growth phase contains zinc. It also has a different amino acid composition and a molecular weight of 198,000. This second enzyme exhibits the highest activity with primary alcohols and no, or very little, activity with 2-propanol and 1,2-propanediol, respectively. The rates for the primary alcohols are in the order of pentanol > butanol > propanol > ethanol > hexanol > heptanol. Both enzymes are present in the wild type and in the mutant as shown by activity stain after polyacrylamide gel electrophoresis (Bryant and Wiegel, 1983). During growth in a batch culture first the zinc-containing, secondary alcohol-DH appeared (logarithmic growth phase); but the specific activity of primary alcohol-DH peaked at the early stationary growth phase (unpublished results).**

The purified NAD-linked lactate dehydrogenase catalyzes unidirectionally the reduction of pyruvate to L(+)-lactate. Fructose-1,6-bisphosphate is a positive effector, and NADH is an inhibitor. Both effects are strongly pH-dependent (Carreira et al., 1982).

T. ethanolicus grows in mineral media containing 0.5% pyrophosphate (PP_i) and 0.2% yeast extract as sole energy and carbon sources without producing significant amounts of fermentation products (Peck et al., 1983). The purified acetate kinase catalyzes in a reversible reaction the formation of ATP and acetate from ADP plus acetylphosphate. The enzyme is a single subunit protein with molecular weight of 56,000 and a surprisingly low temperature optimum of 45°C.

T. ethanolicus contains a rubredoxin and at least two ferredoxins. The ferredoxins differ in amino acid sequence and thermostability. The predominant ferredoxin contains a single four-iron cluster and exhibits extremely high thermostability (e.g. 2 h at 80°C). Its N-terminal amino acid is methionine. The less abundant ferredoxins contain two four-iron clusters and exhibit a less pronounced thermostability (Wiegel and Ljungdahl, 1979 and unpublished results). Its N-terminal amino acid is alanine. There are indications that one of the iron-sulfur centers is actually a 3 Fe-3 S center.

Menaquinones and ubiquinones have not been observed (M. D. Collins, personal communication).

Genetic Studies on T. ethanolicus

Although indications for the presence of phages have been obtained, they have not been isolated.

Mutants of T. ethanolicus (JW 200 Fe(4) and L-large) have been isolated for commercial ethanol production. Mutants are resistant to greater than 10% ethanol and produce ethanol from starch in the presence of 20% starch (Ljungdahl et al., 1982; Carreira and Ljungdahl, 1983).

Opportunity of Industrial Application

The mutant of T. ethanolicus, JW 200 Fe(4), is regarded as having a high potential for industrial ethanol production from starch and renewable resources containing hemicellulosic material. The optimum pH range for growth and ethanol production is broad and the temperature range for growth and ethanol production is attractive (Wiegel, 1980, 1982; Zeikus, 1980). The utilization of untreated starch (e.g. ground corn), pentoses and hemicellulose is an attractive property (Wiegel and Puls, 1983). The ethanol production rates have not yet been determined. The production of ethanol by T. ethanolicus JW 200 WT, the mutant Fe(4) and mixed culture systems containing *T. ethanolicus* have been patented (Ljungdahl and Wiegel, 1981 a and b; Ljungdahl and Carreira, 1983).

Procedures for Testing Special Features

High Ratios of Ethanol Formation

The growth and maintenance of *Thermoanaerobacter* do not require extreme precautions. Oxygen has to be removed from the media for growth. This can be done by the usual procedures: boiling the medium and degassing with oxygen-free gas. In the presence of 0.2% yeast extract, the reducing agents Na_2S and/or cysteine are not required. For high yields of ethanol from carbohydrates, especially at concentrations above 0.8% hexoses (or equivalents), yeast extract and $FeSo_4$ should be added at 0.3% and 5 μg/ml media, respectively. To maintain the mutant JW 200 Fe(4) as a high-ethanol-producing derivative, it should be always kept in the presence of 5% starch and 5% (w/v) ethanol

Maintenance Procedures

Cultures kept in liquid media at 70 or 50°C have to be transferred every 4 and 7 days, respectively, to prevent cell lysis and loss in viability. For preservation of up to 3 months, cells from early logarithmic growth phases (grown between 40 and 75°C) can be stored in media at temperatures between 5 and 20°C. The mutants selected for growth on high concentrations of substrate have been successfully cultured from media containing 20% w/v ground corn as long as 18 months later. For **long term preservation (up to 10 years tested) storage at –18°C in 55-60% (v/v; final concentration glycerol) under anaerobic conditions is** recommended. Inoculations from those glycerol-containing stock cultures into prereduced media yield growing cultures in less than 20 h. *T. ethanolicus* can be lyophilized; however, survival rates are not known (H. Hippe DSM, Göttingen, personal communication).

Differentiation of the genus **Thermoanaerobacter** from other genera

The main characteristics of *Thermoanaerobacter* are its temperature range and the fermentation of various carbohydrates to ethanol and CO_2 as main products.

There are few strict anaerobic eubacteria known which grow in this temperature range. High yields of ethanol (above 1.5 mol ethanol/mol of hexose consumed) are only produced by members of the genus *Zymomonas* and the genus *Clostridium*. DNA:RNA hybridization or determination of the 16S-rRNA catalogs have not been performed. Contrary to *Thermoanaerobacter*, *Zymomonas* produces ethanol via the Entner-Doudoroff pathway. Although there are many similarities to thermophilic clostridia, especially *C. thermohydrosulfuricum*, no spores are found with *T. ethanolicus*. In all cases where spores had been observed, contamination with the ubiquitous *C. thermohydrosulfuricum* was found. In the absence of further evidence, *Thermoanaerobacter* cannot be included in the genus *Clostridium* or *Thermoanaerobium*.

Some properties which are helpful in the differentiation between various other organisms growing in the same temperature range or producing ethanol as main product are given in Table 15.44.

Taxonomic Comments

In addition to *T. ethanolicus*, *Thermoanaerobium brockii* is a non-spore-forming, extreme thermophilic anaerobe (Zeikus et al., 1979).

Both were isolated from hot springs at Yellowstone National Park. They differ morphologically and physiologically from *C. thermohydrosulfuricum*, the only well described extreme thermophilic clostridium. *C. thermohydrosulfuricum* was first isolated by Klaushofer and Parkinen from an Austrian sugar beet factory and has been shown to be ubiquitous (Wiegel et al., 1978). It is also present in the hot springs of Yellowstone National Park (Wiegel and Ljungdahl, 1979; Zeikus, 1981). However, after having the nonspore-forming bacteria for more than 3 years in culture, there appears to be more and more similarities. All three bacteria have very similar temperature ranges for growth, and all three exhibit the biphasic Arrhenius graph with the plateau at 55-62°C. *C. thermohydrosulfuricum* and *T. ethanolicus* have two sets of ferredoxins: a (4Fe—4S) ferredoxin and a (4Fe—4S)$_2$ ferredoxin, each resembling the respective compound in the other organism (unpublished results). Preliminary DNA/DNA homology studies (unpublished results) revealed DNA/DNA homologies at about 50% between each of the three thermophiles. Thus, the two new thermophiles are not nonspore-forming strains of *C. thermohydrosulfuricum*; however, the relative high homology suggests that the two might be different but nonspore-forming species of the genus *Clostridium*. Further taxonomic studies are needed.

Table 15.44.
Differential characteristics of the genus **Thermoanaerobacter**[a]

Characteristics	T. ethanolicus	Zymomonas	Thermo-anaerobium	Thermo-bacteroides	C. thermo-hydro-sulfuricum	B. stear-other-mophilus
T$_{max}$ above 75°C	+	–	+	+	+	+
Strict anaerobe	+	+	+	+	+	–
Above 1.5 mol ethanol per mol hexose utilized	+	+	–	–	d	d
Entner-Duodoroff pathway	–	+	–	–	–	–
Embden-Meyerhof pathway	+	–	+	+	+	+
Spore formation	–	–	–	–	+	+
Biphasic dependence of growth on temperature	+	–	+	+	+	+
Presence of quinones	–		+[b]	–	–	+[c]

[a] Symbols: see Table 15.25.
[b] Uncommon type of unknown quinone (Collins, personal communication).
[c] MK-7 (Collins and Jones, 1981).

List of species of the genus Thermoanaerobacter

1. **Thermoanaerobacter ethanolicus** Wiegel and Ljungdahl, 1982, 384.[VP] (Effective publication: Wiegel and Ljungdahl 1981, 348.)

e.tha.no'li.cus. M.L. masc. n. ethanol. corresponding alcohol of ethane (ethane + ol) M.L. masc. adj. *ethanolicus*, indicating the production of ethanol.

This species is the only described member of this genus; thus the description is the same as for the genus and as listed in Table 15.45.

Type strain: T. ethanolicus JW 200 (ATCC 31550; DSM 2246).

Table 15.45.
Descriptive characteristics of **Thermoanaerobacter ethanolicus**[a]

Characteristics	1. *T. ethanolicus*	Characteristics	1. *T. ethanolicus*
Rod-like cell shape	+	pH Range of growth	4.4–9.9
Occurrence of coccoid cells	+	Substrates utilized	
Cell diameter	0.3–0.8[b]	Cellulose, sorbitol, trehalose, rhamnose, fucose	−
Cell length	4–8[b]		
Motile	−	Cellobiose, sucrose, maltose, xylan, starch	+
Flagellation (peritrichous)	+		
Pili (50 Å-type)	+	Glucose, mannose, xylose, ribose, lactose	+
Gram-staining reaction	Variable[c]		
Gram-type	Positive[c]	Pyruvate	+
Cell wall type	*meso*-DAP	Esculin hydrolysis	−
Hexagonal arrays of cell envelope	+	Gelatin hydrolysis	−
Uneven cell division	+	Catalase	−
Strict anaerobic metabolism	+	Indole formation	−
Exposure to O_2 kills the organism	−	Lipase present	−
Presence of spores	−[d]	NO_2^- from NO_3^-	−
Water soluble pigments	−	Resistance to many antibiotics	+[e]
Growth temperature (°C)		Yeast extract required for growth	+
T_{min}	35	Growth on yeast extract as carbon and energy source	Weak
T_{opt}	69		
T_{max}	78	Chemoorganotroph	+
Ethanol + CO_2 as main fermentation product	+	Chemolithotroph	−

[a] Symbols: see Table 15.25.

[b] Average cell diameter is about 0.4–0.5; on xylose or high substrate concentration, cells are becoming significantly thicker. Frequently long cells up to 200 μm can be observed, which may or may not segregate into chains of cells.

[c] Normally cells stain negative, except for very young cells during early logarithmic growth phase, which stain positive. Gram-type (Wiegel, 1982) positive, LPS absent.

[d] Heat-stable spores have not been observed. Using light microscopy, frequently spore-shaped cells can be observed which, however, are not heat stable (Fig. 15.54, *d* and *e*).

[e] See text for concentrations.

Genus Actinomyces Harz 1877, 133[AL]

KLAUS P. SCHAAL

Ac.ti.no.my'ces. Gr. fem. n. *aktis, aktinos*, ray; Gr. masc. n. *mykes*, fungus; M.L. masc. n. *Actinomyces*, ray fungus referring to the radial arrangement of filaments in *Actinomyces bovis* granules.

Straight or slightly curved rods, 0.2–1.0 μm in diameter, which vary considerably in shape and size, **and slender filaments,** 1 μm or less in width and 10–50 μm or more in length, **with true branching. Short rods** (1.5–5.0 μm in length) **with or without clubbed ends** are frequently seen and **may occur singly, in pairs with diphtheroidal arrangements** (Y, V, T forms and palisades), **in short chains or in small clusters.** Longer (5.0–10.0 μm in length) branched rods are also common. Coccobacillary elements may be found occasionally. **Filaments** which may predominate in certain species **are either straight or wavy,** show varying degrees of **branching** and may have **swollen, clubbed or clavate ends. Gram-positive,** but irregular staining giving rise to a beaded or barred appearance frequently occurs. **Nonacid-fast, nonmotile,** and **nonendospore-forming.** Conidia are not produced.

Facultatively anaerobic; most species are preferentially anaerobic, some grow well aerobically. **Carbon dioxide is required for maximum growth.**

Several species produce characteristic **filamentous microcolonies** which are composed of branched, septate or nonseptate filaments with or without signs of central fragmentation. The microcolonies of the remaining species are predominantly or exclusively nonfilamentous and consist chiefly of diphtheroidal and/or branched rods.

Mature colonies (7–14 days) measure between 0.5 and 5.0 mm in diameter and **are either rough and dry to crumbly in texture or smooth and soft to mucoid,** or they show various degrees of transition between these forms. Most colonies are **white to gray-white** or **creamy white,** but *A. odontolyticus* develops a **deep red pigmentation** when grown on blood agar and *A. denticolens* forms **pink colonies** when grown anaerobically on horse blood agar. Aerial filaments are usually absent although an occasional rough isolate of *Actinomyces israelii* may produce short aerial threads under certain cultural conditions. **Optimum temperature 35–37°C** (*A. meyeri* grows equally well at 30°C).

Chemoorganotrophic, having a **fermentative type of metabolism.** Carbohydrates are fermented with the production of acid but no gas. **End products from glucose fermentation** include **formic, acetic, lactic and succinic acids,** but **not propionic acid.** Catalase-negative or -positive; nitrate reduction positive or negative; **indole not produced.**

Organic nitrogen is required for growth. Some species may show greening or complete lysis on agar media containing rabbit, sheep, horse or human red blood cells.

Characteristic **amino acids of the cell wall peptidoglycan are lysine and either aspartic acid or ornithine or none of the latter two components; diaminopimelic acid (DAP) or glycine do not occur.** Cell wall sugars may include glucose, galactose, rhamnose, 6-deoxytalose, fucose, and mannose, but not arabinose or xylose. The mol% G + C of the DNA is 57–69 (T_m).

Type species: *Actinomyces bovis* Harz 1877, 133.

Further Descriptive Information

The cellular morphology of *Actinomyces* species is usually described as being both diphtheroidal and filamentous (Slack, 1974; Slack and

Gerencser, 1975). However, members of different species as well as individual strains of a single species may vary considerably with regard to the proportions in which rods and filaments occur. Furthermore, the cellular appearance may additionally be influenced by factors such as composition of the growth medium, cultural conditions or age of cultures.

Gram-stained smears or wet mounts prepared from young broth (2–3 days) or agar (3–5 days) cultures, rough mature colonies (7–14 days) or clinical specimens often show groups or clusters of Gram-positive, intertwining filaments and branching rods which resemble small microcolonies (Fig. 15.55). The staining may be uniform (Fig. 15.55) or irregular (Fig. 15.56), sometimes producing a beaded or barred appearance (Slack and Gerencser, 1975). A granular cytoplasm that might

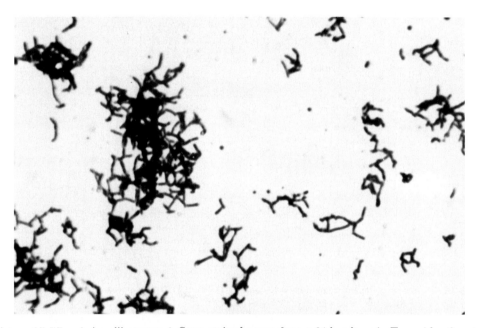

Figure 15.55. *A. israelii, serovar 1.* Gram-stained smear from a 24-h culture in Tarozzi broth; micrograph (× 1200).

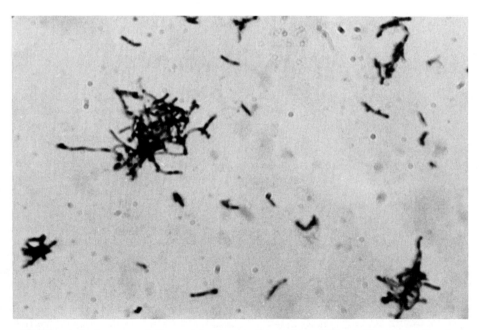

Figure 15.56. *A. bovis, serovar 2.* Gram-stained smear from a 24-h culture in Tarozzi broth; micrograph (× 1200).

account for the irregular staining properties can also be demonstrated in unstained cells observed in darkfield (Slack and Gerencser, 1975). The filaments are often wavy, sometimes straight, slender and of varying length (Figs. 15.55 and 15.56). Some may have swollen, bulbous ends where *Actinomyces hordeovulneris* occasionally develops spheroplast-like bodies. *A. israelii* and *A. hordeovulneris* predominantly occur in these branching filamentous structures (Fig. 15.55) and *A. naeslundii* frequently also produces similar microscopic pictures. Filament formation is less common in *A. viscosus*, especially after prolonged laboratory cultivation (Slack and Gerencser, 1975), and rare in *A. bovis* and *A. odontolyticus* although very young cultures or an occasional rough

strain of the latter two species (Fig. 15.56) may be definitely filamentous.

On the other hand, the cellular morphology of *Actinomyces* cultures may also be predominantly or completely diphtheroidal and even pus specimens derived from *Actinomyces* infections may only contain short or medium-sized rods. *A. viscosus* (Fig. 15.57), smooth *A. naeslundii* strains (Fig. 15.58), *A. howellii*, and some *A. israelii*, serovar 2, isolates produce comparatively long rods which often show characteristic arrangements in Y, V and T forms. Shorter rods resembling propionibacteria are frequently seen in *A. denticolens* (Fig. 15.59), *A. bovis* (Fig. 15.60), *A. odontolyticus* (Fig. 15.61) and *A. meyeri*, and may be arranged

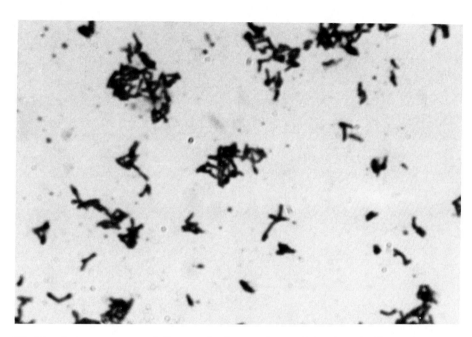

Figure 15.57. *A viscosus, serovar 2.* Gram-stained smear from a 24-h culture in Tarozzi broth; micrograph (× 1200).

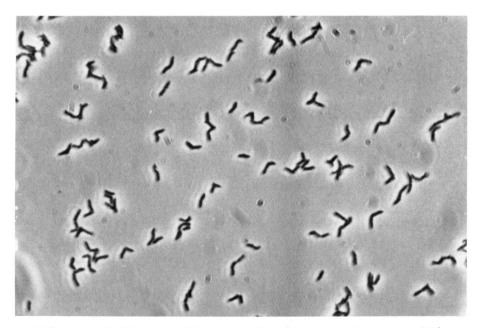

Figure 15.58. *A. naeslundii, serovar 1.* Wet mount in lactophenol cotton blue mounting fluid prepared from a mature colony (BHIA, 14 days, 36°C); phase-contrast micrograph (× 1200).

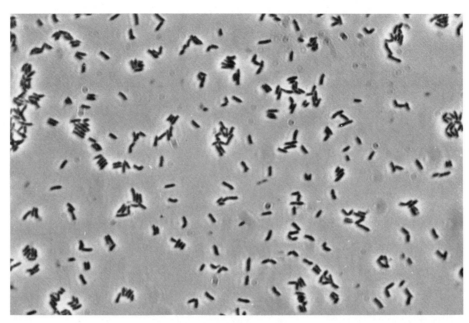

Figure 15.59. *A. denticolens.* Wet mount in lactophenol cotton blue mounting fluid prepared from a mature colony (BHIA, 10 days, 36°C); phase-contrast micrograph (× 1200).

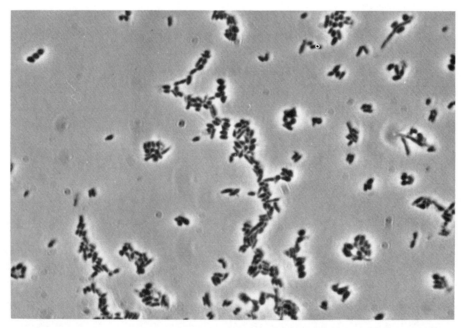

Figure 15.60. *A. bovis, serovar 1.* Wet mount in lactophenol cotton blue mounting fluid prepared from a mature colony (BHIA, 14 days, 36°C); phase-contrast micrograph (× 1200).

in palisades as well as in other diphtheroidal arrangements. Coccobacillary elements are uncommon in most of the *Actinomyces* species, but may predominate in certain *A. odontolyticus* cultures (Fig. 15.62) and in many of the *A. pyogenes* strains (Fig. 15.63) and may also be present in certain *A. meyeri* cultures.

The cell wall composition of the various *Actinomyces* species is not uniform. Originally, the species recognized in the previous edition of the *Manual* (Slack, 1974) were assigned to two different peptidoglycan types (Schleifer and Kandler, 1972). These were primarily defined on the basis of differences in their qualitative and quantitative chemical composition, but without detailed knowledge of their structures. Thus, it is not surprising that recent structural data (Schleifer and Seidl,

1985; Weiss, 1985) have necessitated a reconsideration of the number and types of peptidoglycans found among members of the genus *Actinomyces*.

The murein type described by Schleifer and Kandler (1972) for *A. bovis* was originally characterized as containing a typical L-alanine, D-glutamic acid, L-lysine, D-alanine sequence in the peptide moiety and a D-aspartic acid cross-link between L-lysine and D-alanine (Cummins and Harris, 1958; Cummins, 1965; Schleifer and Kandler, 1972). However, recent findings (Schleifer and Seidl, 1985; Weiss, 1985) indicate that this peptidoglycan is of the Lys-Lys-D-Asp type with L-lysine in position 3 of the tetrapeptide and L-lysine and D-aspartic acid in the interpeptide bridge. Those filamentous strains of *A. bovis* that were

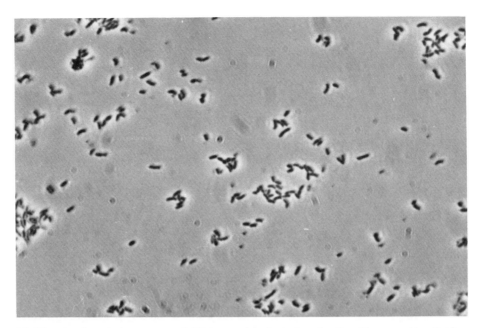

Figure 15.61. *A. odontolyticus, serovar 1.* Wet mount in lactophenol cotton blue mounting fluid prepared from a mature colony (BHIA, 14 days, 36°C); phase-contrast micrograph (× 1200).

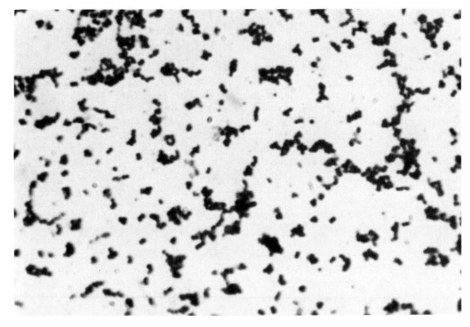

Figure 15.62. *A. odontolyticus, serovar 2.* Gram-stained smear from a 48-h culture in Tarozzi broth; micrograph (× 1200).

reported to lack aspartic acid (Pine and Boone, 1967) would thus appear to belong to a different murein type which awaits further elucidation.

Qualitatively, the cell walls of *A. pyogenes* have a composition similar to that of the *A. bovis* strains, lacking aspartic acid with L-lysine as the dibasic amino acid, the presence of alanine and D-glutamic acid, and the absence of aspartic acid (Cummins and Harris, 1956; Barksdale et al., 1957; Collins et al., 1982). Recently, this peptidoglycan has been identified as belonging to the structural type Lys-Ala-Lys-D-Glu having L-lysine in position 3 of the peptide moiety and a L-alanine, L-lysine, D-glutamic acid cross-link (Schleifer and Seidl, 1985; Weiss, 1985) and

represents the second *Actinomyces* murein type. Interestingly, the same structure was also found in walls of *Eubacterium suis* (Weiss, 1985).

All of the remaining *Actinomyces* species hitherto chemically characterized appear to belong to a third peptidoglycan type which corresponds to the second type described by Schleifer and Kandler (1972). In this type, *N*-acetylmuramic acid, *N*-acetylglucosamine, alanine, glutamic acid, lysine, and ornithine occur in molar ratios of 1:1:2:2:1:1. Such an amino acid composition has been demonstrated not only for *A. israelii, A. naeslundii, A. viscosus,* and *A. odontolyticus* (Schleifer and Kandler, 1972), but also for the newly described species *A. denticolens* (Dent and Williams, 1984a), *A. howellii* (Dent and Williams,

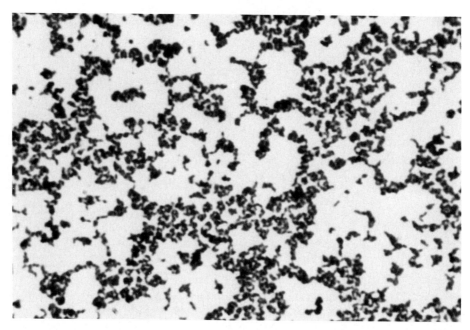

Figure 15.63. *A. pyogenes.* Gram-stained smear from a mature colony (blood agar, 7 days, 36°C); micrograph (× 1200).

1984b) and most strains of *A. hordeovulneris* (Buchanan *et al.*, 1984). Recent results on the structure of the peptidoglycans of *A. israelii, A. naeslundii, A. odontolyticus, A. viscosus* and *A. denticolens* indicate that L-ornithine is located in position 3 of the peptide subunit, and the interpeptide bridge consists of L-lysine and D-glutamic acid (Schleifer and Seidl 1984; Weiss, 1985).

The qualitative, quantitative and structural peptidoglycan data given above were predominantly taken from the articles published by Schleifer and Kandler in 1972, Schleifer and Seidl in 1985 and Weiss in 1985. Other investigators (DeWeese et al., 1968; Reed, 1972; Reed and Evans, 1973) have reported conflicting results with regard to the actual components found, the molar ratios of the components and/or the proposed murein structure. However, in respect of the more refined techniques which are now available, there appears to be no reason to doubt the recent findings described above in detail.

The cell wall sugars of members of the genus *Actinomyces* may include glucose, galactose, rhamnose, 6-deoxytalose, fucose and mannose which may be present singly or in various combinations (Pine, 1963; Slack, 1974; Slack and Gerencser, 1975). Two species, *A. israelii* and *A. denticolens*, contain only one detectable sugar component in their walls which was identified as galactose (Cummins and Harris, 1958) or rhamnose (Dent and Williams, 1984a), respectively. All of the other *Actinomyces* species exhibit more complex sugar patterns: in *A. bovis* cell walls, rhamnose, fucose and usually 6-deoxytalose (Cummins and Harris, 1958, 1959; Slack, 1974; Slack and Gerencser, 1975) were detected together with trace amounts of glucose and mannose (Cummins and Harris, 1958; Pine, 1970). *A. naeslundii* walls were reported to contain glucose, rhamnose, 6-deoxytalose, mannose and occasionally fucose (Cummins and Harris, 1958; Boone and Pine, 1968; Hammond et al., 1973; Slack, 1974; Slack and Gerencser, 1975). However, a recent re-evaluation of these data (Dent and Williams, 1984a) showed that certain *A. naeslundii* strains may lack 6-deoxytalose, and mannose was not detected at all. The sugar pattern of *A. viscosus* is similar to that of *A. naeslundii* in that it encompasses glucose, rhamnose, and 6-deoxytalose as characteristic components (Slack, 1974; Slack and Gerencser, 1975; Dent and Williams, 1984a). However, some strains may have varying amounts of galactose and traces of mannose in addition to the above components (Slack, 1974; Dent and Williams, 1984a) while

isolates from dogs may lack 6-deoxytalose (Buchanan et al., 1984). Results on the sugar composition of *A. odontolyticus* cell walls have remained contradictory: Hammond et al. (1973) as well as Slack and Gerencser (1975) claimed that glucose, galactose, and mannose, but no rhamnose or 6-deoxytalose were present in *A. odontolyticus* walls. On the other hand, Slack (1974), in his contribution to the previous edition of the *Manual*, listed deoxytalose, fucose, galactose, glucose, mannose and rhamnose as sugar constituents characteristic of this species. Recent findings of Dent and Williams (1984a) who analyzed two strains of *A. odontolyticus* (NCTC 9931, NCTC 9935) are similar to those reported by Slack in 1974 and support the view that at least certain isolates of this species may have a sugar pattern that includes glucose, mannose, rhamnose and 6-deoxytalose as principle components. The cell wall sugars of *A. pyogenes* are glucose and rhamnose and possibly traces of mannose (Cummins and Harris, 1956; Barksdale et al., 1957; Reddy et al., 1982; Collins and Jones, 1982). *A. howellii* cell wall carbohydrates include glucose and rhamnose but not 6-deoxytalose (Dent and Williams, 1984b). The walls of *A. hordeovulneris* contain glucose and galactose. Rhamnose and 6-deoxytalose are absent (Buchanan et al., 1984). Amino acid and carbohydrate composition of *A. meyeri* have not been reported.

The cellular fatty acid profiles of members of the genus *Actinomyces* were found to be comparatively simple encompassing predominantly even numbered, *n*-saturated and *n*-unsaturated, but no branched chain, odd numbered or hydroxy substituted fatty acids (Kroppenstedt and Kutzner, 1978; Amdur et al., 1978). Whole-organism extracts of *A. bovis, A. israelii, A. naeslundii, A. viscosus, A. odontolyticus,* and *A. pyogenes* may contain tetradecanoic (14:0, myristic acid), hexadecanoic (16:0, palmitic acid), octadecanoic (18:0, stearic acid) and octadecenoic (18:1, ω-9, oleic acid) acids, occasionally together with small amounts of decanoic (10:0; *A. pyogenes*), dodecanoic (12:0; *A. pyogenes, A. naeslundii*), hexadecenoic (16:1; *A. pyogenes*) and C$_{20}$ acids (Kroppenstedt and Kutzner, 1978; Amdur et al., 1978; Collins, et al., 1982; Kroppenstedt, personal communication). In *A. pyogenes*, tetradecanoic acid is a major component (Collins et al., 1982) and *A, bovis, A. naeslundii* and *A. viscosus* may contain larger amounts of octadecanoic acid (Amdur et al., 1978) while these acids constitute only a small part in the fatty acid profiles of the remaining species. Whether or not cyclopropane fatty acids do occur in *Actinomyces* species remains a controversial question: Kroppenstedt (Kroppenstedt and Kutzner, 1978, and personal communication) found significant amounts of C_{19} cyclopropane fatty acids in one strain of *A. israelii* (DSM 43 305) whereas Amdur et al. (1978) were unable to detect these acids in any of the *Actinomyces* strains examined. Fatty acid data of *A. denticolens*, *A. howellii*, *A. hordeovulneris* and *A. meyeri* have not been reported.

Polar lipids have been determined in four strains of *A. viscosus* (Yribarren et al., 1974; Pandhi and Hammond, 1978). The total lipid content of this species was found to range from 3.5–5.5% of dry weight. Pandhi and Hammond (1978) identified the polar lipids as galactosyl diglyceride, phosphatidylcholine, cardiolipin, phosphatidylglycerol and phosphatidylinositol dimannosides. Phosphatidylethanolamine and phosphatidylinositol were not detected. This is in contrast to the findings of Yribarren et al. (1974) who reported that their *A. viscosus* strain contained phosphatidylethanolamine.

The principal isoprenoid quinones identified so far in *Actinomyces* species (*A. bovis*, *A. israelii*, *A. naeslundii*, *A. viscosus*, *A. pyogenes*) are tetrahydrogenated menaquinones with 10 isoprene units (MK-10(H$_4$)), but small amounts of MK-9(H$_2$), MK-9(H$_4$) and MK-10(H$_2$) may also be present (Collins et al., 1977, 1982; Kroppenstedt, personal communication).

The cell morphology of *Actinomyces* species as observed by electron microscopy of ultrathin sections reflects the pictures obtained by light microscopy: cells usually present as bacillary forms of varying length with or without clubbed ends. Branching can readily be demonstrated (Overman and Pine, 1963; Duda and Slack, 1972; Slack and Gerencser, 1975; Lai and Listgarten, 1980). In *A. pyogenes* rounded forms are not uncommon, although the cellular morphology of this species is also predominantly rod-shaped (Reddy et al., 1982).

Septum formation can easily be seen in most of the ultrathin sections (Duda and Slack, 1972; Slack and Gerencser, 1975). The septa are not always at a right angle to the longitudinal axis of the cell and the completed cross-walls are often curved (Duda and Slack, 1972). Cells which have divided by cross-wall formation do not necessarily separate so that a straight row of rods may result (Duda and Slack, 1972).

Cross-walls dividing rod-shaped elements into two approximately equal parts were taken as an indication for the occurrence of binary fission (Slack and Gerencser, 1975). However, microorganisms which typically multiply by binary fission synthesize new cell wall material either along the whole length of the cell body or in the center part of the rod leading to conservation of the apices (Locci and Schaal, 1980). In contrast, all of the *Actinomyces* species investigated so far exclusively showed apical growth at one or both ends of the rod-shaped propagule. As evidenced by scanning electron microscopy (Locci, 1978) and immunofluorescence labeling (Locci and Schaal, 1980), new cell wall material is only formed at the tips of the cell while the interpolar part of the envelope remains unchanged. The unipolar or bipolar extension of the rods resulting from the apical synthesis of new wall material leads to elongated cells or filaments which divide by septum formation and may separate into rod-shaped elements by fragmentation (Locci, 1976; Locci and Schaal, 1980).

The mode of cell growth and division outlined above may be described more appropriately by the term "budding" than by "binary fission." Budding is obviously also responsible for the formation of side branches (Locci and Schaal, 1980). In this respect, electron microscopy of ultrathin sections provided essentially the same information (Slack and Gerencser, 1975).

The cell walls of *Actinomyces* have a bilayered structure (Duda and Slack, 1972; Lai and Listgarten, 1980). They consist of a moderately electron-dense outer layer and a thinner, more electron dense, inner layer. The thickness of the inner layer ranged from 3.6–9.0 nm depending on the species examined. In *A. bovis*, this inner layer of the cell wall was not detectable (Lai and Listgarten, 1980). Measurements of the thickness of the outer layer indicated pronounced interspecies variation ranging from 14 nm in *A. bovis* to 36 nm in *A. israelii* (Lai

and Listgarten, 1980). Analogous differences were observed for the overall width of the walls of various *Actinomyces* species although the absolute figures given by different investigators vary considerably. In the following compilation, wall thickness data are listed in this order: Overman and Pine, 1963, Duda and Slack, 1972, and Lai and Listgarten, 1980. According to these authors, the wall thickness is as follows: *A. israelii*, 29:65:35–50 nm; *A. naeslundii*, 20:45:25–35 nm; *A. viscosus*, NO: 35:23–28 nm; *A. bovis*, 10:31:14–22 nm; *A. odontolyticus*, NO: 31:22–27 nm. The cell walls of *A. pyogenes* were reported to measure from 29–30 nm in width (Reddy et al., 1982).

In most of the *Actinomyces* strains investigated, the external contour of the outer cell wall layer was described as fuzzy or shaggy (Duda and Slack, 1972; Lai and Listgarten, 1980). *A. naeslundii*, *A. bovis* and certain strains of *A. israelii* and *A. viscosus* may produce long filamentous projections radiating from the cell wall surface (Slack and Gerencser, 1975; Lai and Listgarten, 1980). These surface fibrils may play an important role in the attachment of these microbes to epithelial cells, tooth surfaces or other bacteria (Girard and Jacius, 1974; Ellen et al., 1978; McIntire et al., 1978). The detection of the "surface fuzz" is markedly influenced by the culture conditions used with broth-grown cells usually exhibiting longer appendages than blood agar-grown organisms (Lai and Listgarten, 1980). Nevertheless, *A. odontolyticus*, *A. pyogenes* and some strains of *A. israelii* and *A. viscosus* were reported to produce neither long nor short surface fibrils (Lai and Listgarten, 1980; Reddy et al., 1982).

A. odontolyticus cells show a marked increase in electron-density at cell wall surface. This increase can also be demonstrated, although to a lesser extent, in *A. israelii*, *A. naeslundii* and *A. viscosus*, but is apparently absent in *A. bovis* (Lai and Listgarten, 1980).

The cytoplasmic membrane has the usual trilaminar structure (Lai and Listgarten, 1980), but is not always readily discernible (Slack and Gerencser, 1975). A nucleoid without a membrane could be detected in most species (Duda and Slack, 1972; Lai and Listgarten, 1980). The occurrence of mesosomes has been described for all species (Overman and Pine, 1963; Duda and Slack, 1972; Lai and Listgarten, 1980; Reddy et al., 1982). However, they vary widely in number, size and type. Simple tubular structures were observed most frequently, but coiled forms were also seen (Slack and Gerencser, 1975). Most of these membranous figures appeared to be derived from the cytoplasmic membrane with which they were often found to be in continuity (Overman and Pine, 1963; Lai and Listgarten, 1980). The cytoplasm of actively growing cells may be densely packed with ribosomes, and electron-dense cytoplasmic inclusions were observed in most species (Duda and Slack, 1972; Lai and Listgarten, 1980).

The morphology of *Actinomyces* microcolonies, although often indicative of the genus, may vary considerably, depending on both species affiliation of the strain under study and media and incubation conditions employed. Nevertheless, standardization of the latter factors usually results in morphological features of the colonies which are fairly reproducible and comparatively diagnostic of the various species (Slack and Gerencser, 1975; Schaal and Pulverer, 1981). Microcolonies are observed most appropriately on transparent media (brain heart infusion agar (BHIA) or CC-medium) incubated anaerobically for 18–48 h and viewed in situ under the microscope at low magnifications in transmitted light (Lentze, 1938a; Erikson, 1940).

Under these growth conditions and observation techniques, several characteristic types of *Actinomyces* microcolonies can be discerned. *A. israelii* and *A. hordeovulneris* (Buchanan et al., 1984) produce microcolonies which are regularly filamentous, but may vary in the number and length of the filaments formed. A definitely filamentous type which is commonly referred to as "spider" colony may be considered most typical of *A. israelii* (Fig. 15.64). This microcolony consists of branching threads which often appear to originate from a single, central point with radial symmetry. The filaments are slender and moderately long with branches arising at an acute angle (Slack and Gerencser, 1975). In certain strains, the microcolonies may be smaller consisting of only

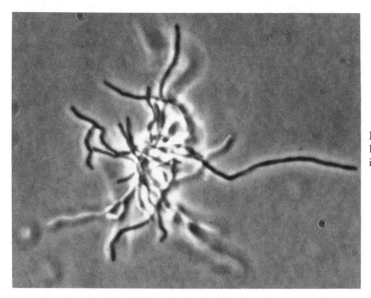

Figure 15.64. *A. israelii, serovar 2.* Spider-like microcolony on BHIA (slide culture, 26 h at 36°C); phase-contrast micrograph in situ in lactophenol cotton blue mounting fluid (× 1500).

one or two branched filaments (Fig. 15.65), or the colonies may be larger with many radial threads which cross and fragment to bacillary elements at the center (Fig. 15.66). Occasionally, small colonies with few, short projecting filaments and a fragmented center are seen (Slack and Gerencser, 1975; Schaal and Pulverer, 1981) (Fig. 15.67).

A. *naeslundii* and *A. viscosus* also produce filamentous microcolonies. In very early growth stages (8–14 h), these may resemble the spider colonies of *A. israelii* (Howell et al., 1959; Coleman et al., 1969; Locci and Schaal, 1980). However, after 18–24 h of incubation, *A. naeslundii* microcolonies are usually larger than those of *A. israellii* and have a dense center composed of diphtheroidal cells and/or partly fragmented filaments which are surrounded by long, branched and irregularly curved threads projecting in all directions (Fig. 15.68) (Slack and Gerencser, 1975; Schaal and Pulverer, 1981). Occasionally, the filamentous fringe may be very short.

In principle, the microcolonies of human and hamster isolates of *A. viscosus* resemble those of *A. naeslundii* presenting as densely centered structures with a filamentous periphery. However, the projecting threads are often shorter or even very short (Slack and Gerencser, 1975), and the colonies may be small when the organism is cultured under anaerobic conditions (Fig. 15.69). Spider-like very young microcolonies have also been reported (Howell, 1963). After serial laboratory subcultivation, *A. viscosus* tends to produce "smooth" microcolonies which are circular and have a granular surface and an entire or irregular edge (Fig. 15.70) (Slack and Gerencser, 1975). In such predominantly smooth cultures, a few filamentous colonies may be present. The heterogeneous picture resulting from this morphological dissociation is easily mistaken for a contamination of the culture.

A. *bovis*, *A. odontolyticus*, *A. pyogenes*, *A. denticolens* and *A. meyeri* all form predominantly or exclusively nonfilamentous microcolonies. These have a smooth or finely granular surface, an entire or irregular edge, are slightly raised to convex, white to colorless and soft and may show a few short projecting filaments (Figs. 15.71–15.74). In addition, they may have an optically dark center spot, and typical *A. bovis* microcolonies often have irregular, jagged edges, but do not show radiating filaments (Slack and Gerenser, 1975). Similar colonies, occasionally with a dark center, are produced by *A. denticolens* (Fig. 15.75). Nevertheless, truly filamentous strains have been reported for both *A. bovis* (Fig. 15.76) and *A. odontolyticus* (Pine et al., 1960; Georg et al., 1964; Slack and Gerencser, 1975) and may also occur in *A. denticolens* (Fig 15.77). Completely smooth, circular microcolonies with entire edges are characteristic of *A. pyogenes* (Fig. 15.73) (Reddy et al., 1982) and *A. meyeri*. Details on the morphology of *A. howellii* microcolonies have not been reported.

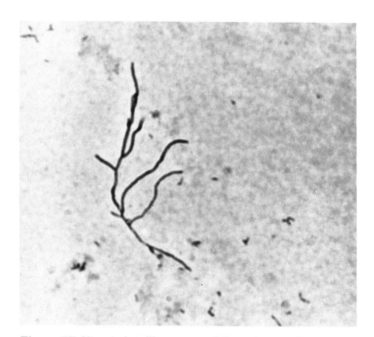

Figure 15.65. *A. israelii, serovar 2.* Microcolony on BHIA (slide culture, 24 h at 36°C); phase-contrast micrograph in situ in lactophenol cotton blue mounting fluid (× 1200).

The morphological properties of *Actinomyces* microcolonies as demonstrated by light microscopy have been confirmed and elucidated in more detail by scanning electron microscopy (Slack and Gerencser, 1975; Locci, 1976, 1978). A sharp contrast was noted between the compact nonfilamentous colonies of *A. bovis* and *A. odontolyticus* and the filamentous ones of *A. israelii*, *A. naeslundii* and *A. viscosus*.

Actinomyces colonies mature within (5–) 7–14 days of incubation on BHIA at 36°C. The colony morphology may slightly differ when growth on BHIA is compared with that on CC-medium (Heinrich and Korth, 1967) or blood agar, but the general growth characteristics may be observed on either medium. Strains which produce filamentous microcolonies usually form rough mature colonies.

Mature colonies of *A. israelii* range from 0.5–2.0 mm in diameter and have a white to gray-white color. Rough colonies, with or without central depression, are considered most typical of this species and have

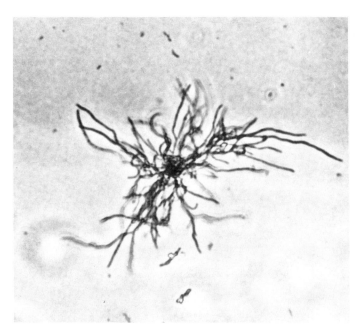

Figure 15.66. *A. israelii, serovar 1.* Microcolony on BHIA (slide culture, 48 h at 36°C); phase-contrast micrograph in situ in lactophenol cotton blue mounting fluid (× 1200).

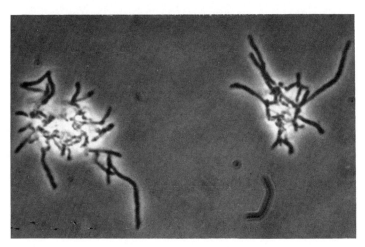

Figure 15.67. *A. israelii, serovar 2.* Microcolony on BHIA (slide culture, 24 h at 36°C); phase-contrast micrograph in situ in lactophenol cotton blue mounting fluid (× 1600).

been described as molar-tooth, bread-crumb or raspberry-like colonies (Figs. 15.78 and 15.79) (Slack and Gerencser, 1975; Schaal and Pulverer, 1981). These colonies are circular to irregular in shape with undulate, lobate or erose edges and, often, with a highly filamentous fringe (see Fig. 15.81). In addition, they are convex, pulvinate or umbonate with a convoluted or granular surface and a dry to crumbly texture. However, smooth mature colonies with entire, circular or irregular edges, a white to gray-white to creamy-white, opaque appearance and soft texture may also occur (Fig. 15.80), especially in certain strains of *A. israelii, serovar 2* (Slack and Gerencser, 1975; Schaal and Pulverer, 1981). The surface of this colony type is usually shiny, but may also be granular and matte. Rough *A. israelii* colonies may be very hard and they often adhere to the medium or come off in toto when an attempt is made to pick them. Soft and certain bread-crumb colonies, on the other hand, are friable and easy to remove from the agar surface.

Although the formation of aerial filaments is by no means characteristic of the genus, a few *A. israelii* isolates from various sources may produce numerous, short aerial hyphae (Fig. 15.81) (Erikson, 1940; Slack and Gerencser, 1975; Locci, 1978; Schaal, unpublished results).

Mature colonies of *A. hordeovulneris* are similar to those of *A. israelii*. On bovine blood agar, this species produces white, agar-adherent, molar toothed colonies with a tendency to shift to a white, conical, domed, buttery, less adherent type upon laboratory passage. Diameters are 0.5–1.0 mm after 48 h at 37°C and 2 mm after 3 days (Buchanan et al., 1984).

Mature colonies of *A. naeslundii* are predominantly smooth, circular, low convex to umbonate and entire measuring 1.0–5.0 mm in diameter (Figs. 15.82 and 15.83). Their surface structure is usually granular. Frequently, some peripheral projecting filaments can be observed under the microscope (Fig. 15.84) which often show an asymmetrical distribution. Furthermore, various degrees of roughness may be found so that certain isolates may appear in the bread-crumb or molar-tooth colony type (Fig. 15.85).

Macrocolonies of *A. viscosus* grown under reduced oxygen tension for 7 days are approximately 0.5 mm in diameter while cultures incubated aerobically with added CO_2 produce larger colonies (4–5 mm in diameter). The texture is usually soft and may be viscous, in both animal (Howell, 1963) and human isolates. The overall appearance is very similar to that described for *A. naeslundii* (Figs. 15.86 and 15.87) although hamster isolates with short, scattered aerial filaments have been reported (Slack and Gerencser, 1975).

The typical macrocolony of *A. bovis* measures 0.5–1.0 mm in diameter and is circular, with a smooth or finely granular surface and a dark center, convex, entire, white and predominantly soft (Fig. 15.88). Rough mature colonies (Fig. 15.89) may resemble the molar-tooth or bread-crumb colony type of *A. israelii* (Slack and Gerencser, 1975).

A. odontolyticus forms mature colonies which are 1.0–2.0 mm in size, circular to irregular with an entire or irregular edge, low convex to umbonate with a smooth to finely granular surface and a soft texture (Fig. 15.90). In transmitted light, they may show a dark center spot and/or dark irregular granules which decrease in size from the center to the periphery (Fig. 15.91). Rough strains may have a colony mor-

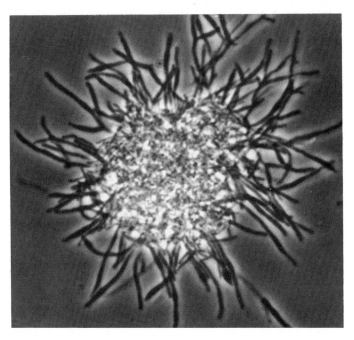

Figure 15.68. *A. naeslundii, serovar 1.* Microcolony on BHIA (slide culture, 24 h at 36°C); phase-contrast micrograph in situ in lactophenol cotton blue mounting fluid (× 1600).

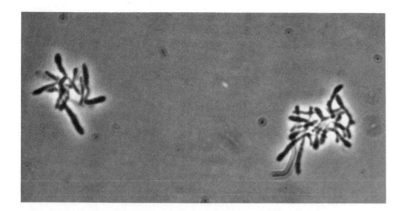

Figure 15.69. *A. viscosus, serovar 2.* Microcolony on BHIA, grown under anaerobic conditions (slide culture, 24 h at 36°C); phase-contrast micrograph in situ in lactophenol cotton blue mounting fluid (× 1600).

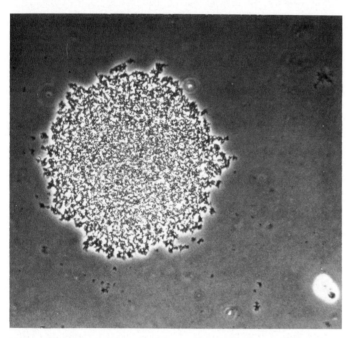

Figure 15.70. *A. viscosus, serovar 1.* Microcolony on BHIA, grown under aerobic conditions with added CO_2 (slide culture, 48 h at 36°C); phase-contrast micrograph in situ in lactophenol cotton blue mounting fluid (× 200).

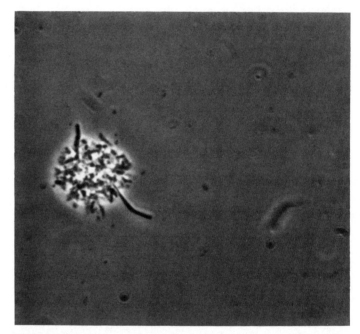

Figure 15.71. *A. bovis, serovar 1.* Microcolony on BHIA (slide culture, 24 h at 36°C); phase-contrast micrograph in situ in lactophenol cotton blue mounting fluid (× 1200).

phology similar to that of *A. israelii*. On BHIA, macrocolonies of *A. odontolyticus* are opaque and white. However, older colonies grown on blood agar develop a deep red color (Batty, 1958). This pigment may appear during anaerobic incubation in as little as 48 h, but it usually requires 5–10 days to develop or it may only be formed after the cultures have been left standing in air at room temperature following primary anaerobic incubation (Slack and Gerencser, 1975). Occasionally, the pigment formation cannot be demonstrated.

A. pyogenes grows faster than the other *Actinomyces* species. After 48–72 h, the colonies are 0.5–3.0 mm in diameter depending on the growth medium used. Their appearance is convex, opaque, white and soft with entire edges (Fig. 15.92). *A. pyogenes* is the only *Actinomyces* species which always produces β-hemolysis. After 24 h of incubation on sheep blood agar, pinpoint, β-hemolytic colonies occur whose zones of hemolysis are 2–3 times the diameter of the colony (Reddy et al., 1982).

Mature colonies of *A. denticolens* reach at least 2.0 mm in diameter and are circular, entire, convex or umbonate and smooth (Fig. 15.93). On horse blood agar, they develop a slightly pink pigmentation when

grown anaerobically, but they are white if grown aerobically (Dent and Williams, 1984a).

On horse blood agar plates after 3–5 days of anaerobic incubation at 37°C, colonies of *A. howellii* are white, smooth, shiny, translucent and convex with entire margins. Their diameter is up to 2 mm (Dent and Williams, 1984b).

Mature surface colonies of *A. meyeri* on supplemented BHI blood agar are pinpoint to 1 mm in diameter, circular, flat to convex, translucent to opaque and white with shiny, smooth surfaces and entire margins. They may be α- or nonhemolytic (Cato et al., 1984).

In liquid media (e.g. thioglycolate broth or Tarozzi broth), *Actinomyces* species either grow as discrete compact masses of variable size leaving the medium clear, or they produce a diffuse growth with varying amounts of a granular, flaky or pellicular sediment (Slack, 1974). The cotton pad-like masses which are filamentous colonies may be suspended in the broth or adhere to the glass or tissue surface (Fig. 15.94) and are commonly seen in *A. israelii* strains, but also in rough isolates of the other species. Broth cultures of *A. bovis, A. odontolyticus, A. viscosus, A. pyogenes, A. denticolens* and *A. meyeri* are often evenly turbid and, upon further incubation, may show a granular, flocculent, pellicular, stringy or smooth sediment. Many strains of *A. viscosus* and parts sooner or later exhibit signs of septation and fragmentation (Figs. 15.64, 15.66–15.68, 15.76, 15.77). Degree of filament formation as well as intensity and rate of fragmentation and cell separation vary depending on species and strain peculiarities so that it may sometimes be difficult to detect the developmental cycle. Variation in filament formation and fragmentation results in mature colonies which either remain predominantly or partly filamentous (Figs. 15.79, 15.81, 15.84) and have a rough appearance and a dry to crumbly texture or become

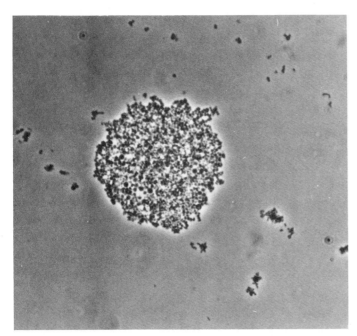

Figure 15.72. *A. odontolyticus, serovar 1.* Microcolony on BHIA (slide culture, 48 h at 36°C); phase-contrast micrograph in situ in lactophenol cotton blue mounting fluid (× 480).

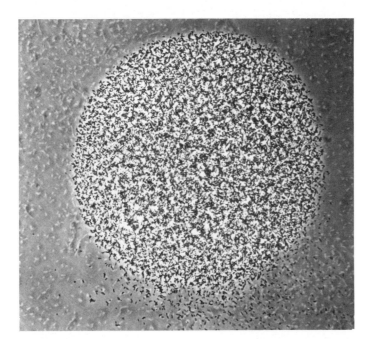

Figure 15.73. *A. pyogenes.* Microcolony on BHIA (slide culture, 24 h at 36°C); phase-contrast micrograph in situ in lactophenol cotton blue mounting fluid (× 200).

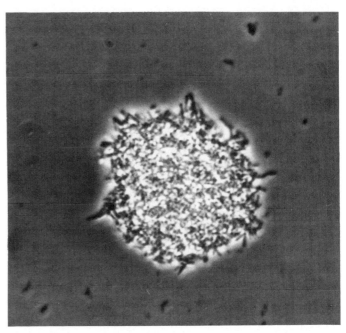

Figure 15.74. *A. denticolens.* Microcolony on BHIA (slide culture, 24 h at 36°C); phase-contrast micrograph in situ in lactophenol cotton blue mounting fluid (× 1200).

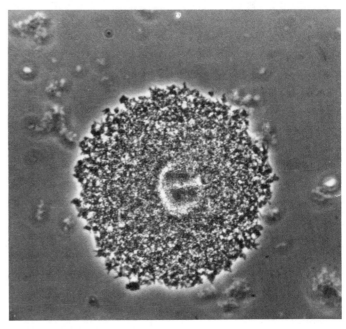

Figure 15.75. *A. denticolens.* Microcolony on BHIA (slide culture, 48 h at 36°C); phase-contrast micrograph in situ in lactophenol cotton blue mounting fluid (× 480).

a few *A. bovis* and *A. israelii* isolates produce a viscous growth which, in the case of *A. viscosus*, may give rise to a mucoid sediment (Howell, 1963).

Filamentous *Actinomyces* strains show a characteristic life cycle when cultivated on artificial media. The initial developmental step includes elongation of the original, usually rod-shaped, propagule by apical growth at one or both ends and possibly lateral budding (Fig. 15.95) (Morris, 1951; Locci and Schaal, 1980). When growth proceeds, a network of interwoven, branching filaments may form whose central

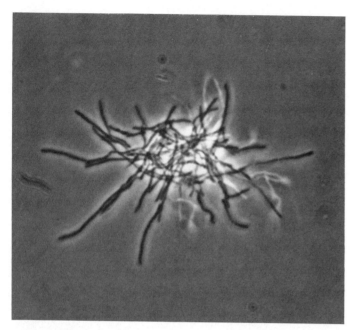

Figure 15.76. *A. bovis, serovar 2.* Microcolony on BHIA (slide culture, 24 h at 36°C); phase-contrast micrograph in situ in lactophenol cotton blue mounting fluid (× 1200).

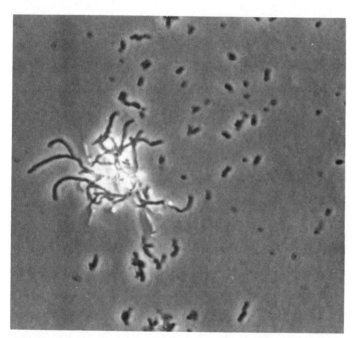

Figure 15.77. *A. denticolens.* Microcolony on BHIA (slide culture, 24 h at 36°C); phase-contrast micrograph in situ in lactophenol cotton blue mounting fluid (× 1200).

smooth and soft (Figs. 15.82, 15.83, 15.88, 15.90–15.93) being composed of bacillary forms only. In addition, various transitions between these extremes may be found.

The minimal nutritional requirements of *Actinomyces* species are not known in detail. However, good growth has only been obtained on complex media containing either rich biological ingredients (e.g. BHI, meat extract, yeast extract, serum, vitamin K₁) or a defined mixture of a large variety of organic and inorganic compounds. Studies using synthetic media were reported (Howell and Pine, 1956; Christie and

Porteus, 1962a, b, c; Keir and Porteus, 1962), but were mostly restricted to a few strains. Furthermore, several workers have developed modifications and/or simplifications of the synthetic medium of Howell and Pine (1956) (Pine and Watson, 1959; Georg et al., 1964; Heinrich and Korth, 1967) that were tested with a greater number of strains. In this laboratory, the semisynthetic medium (CC-medium) of Heinrich and Korth (1967) has proved especially useful for the primary isolation and subsequent subcultivation of *Actinomyces* species. Hundreds of strains were isolated and multipally subcultured on this medium without loss of viability. However, attempts to further simplify this medium have so far always resulted in insufficient rates of primary recovery or loss of cultures during subcultivation.

From these observations, it can be concluded that the anabolic capacity of *Actinomyces* species is limited (Slack and Gerencser, 1975). These organisms apparently require organic nitrogen (e.g. peptides and/or amino acids), a fermentable carbohydrate and possibly also vitamins and other growth factors for optimum growth. *A. meyeri* strains were found to have an absolute requirement for vitamin K₁, and their growth was greatly stimulated by 0.02% Tween 80 and by a fermentable carbohydrate (Cato et al., 1984). Serum added to the medium may also enhance growth of these organisms. The latter is also true for *A. hordeovulneris*, whose growth is greatly enhanced when the medium is supplemented with 10–20% (v/v) fetal calf serum (Buchanan et al., 1984).

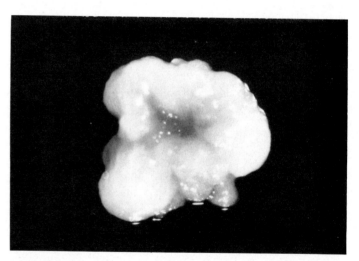

Figure 15.78. *A. israelii, serovar 1.* Mature colony on CC-medium (14 days at 36°C); micrograph (× 30).

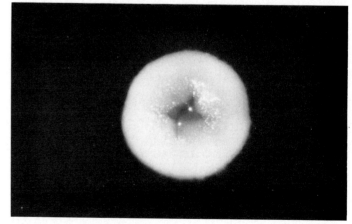

Figure 15.79. *A. israelii, serovar 1.* Mature colony on CC-medium (14 days at 36°C); micrograph (× 30).

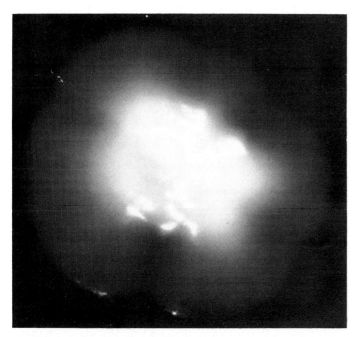

Figure 15.80. *A. israelii, serovar 2.* Mature colony on CC-medium (14 days at 36°C); micrograph (× 30).

when the medium contained a fermentable carbohydrate and hemin plus HCO_3^-. Simultaneous deletion of hemin and HCO_3^- resulted in negligible growth. Single deletions decreased the growth only to a certain extent indicating that each of these compounds could partially substitute for the other. Riboflavin, nicotinic acid, biotin and thiamin appeared to be necessary for some of the *A. pyogenes* strains although the vitamin requirements obviously vary within the species. The same is true for purines and pyrimidines (adenine, uracil) which may or may not be required. Under certain conditions, inositol may be necessary for growth, but trypticase can possibly relieve this requirement. No adverse effects were noted when lipoic acid and Tween 80 were omitted from the medium. But *A. pyogenes* definitely has a requirement for peptides because deletion of trypticase resulted in a considerable decrease in growth rate although the medium used still contained various amino acids. Upon further investigation, it was shown that peptides

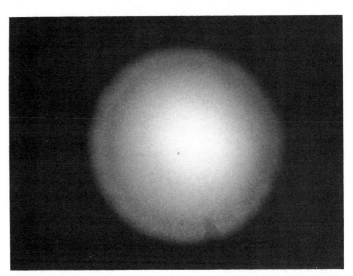

Figure 15.82. *A. naeslundii, serovar 2.* Mature colony on CC-medium (14 days at 36°C); micrograph (× 30).

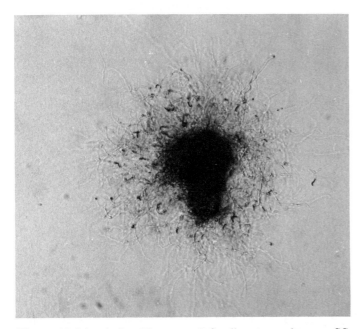

Figure 15.81. *A. israelii, serovar 2.* Small mature colony on CC-medium with short aerial hyphae (dark parts of the filaments) (14 days at 36°C); micrograph (× 200).

Purines, pyrimidines and other growth factors may be either stimulatory or inhibitory depending on the individual strain tested (Christie and Porteus, 1962c; Keir and Porteus, 1962). For example, one strain of *A. israelii* apparently required adenine and thymine to be added to the simplest medium while another strain of the same species was inhibited by one or more of the bases guanine, xanthine, and uracil (Christie and Porteus, 1962c).

The nutritional requirements of *A. pyogenes* are probably the best known of all of the *Actinomyces* species (Reddy and Cornell, 1982; Reddy et al., 1982). Members of this species only showed good growth

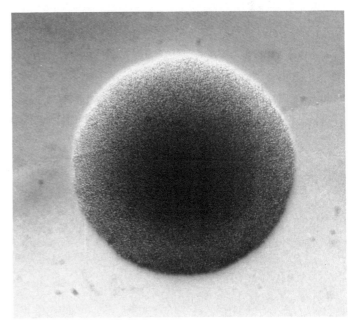

Figure 15.83. *A. naeslundii, serovar 1.* Small mature colony on CC-medium (10 days at 36°C); micrograph (× 120).

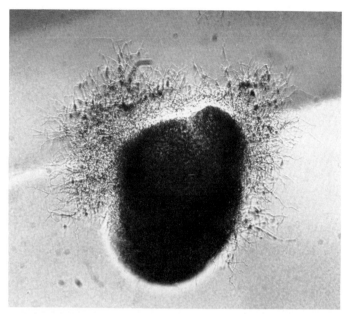

Figure 15.84. *A. naeslundii, serovar 1.* Small mature colony on CC-medium (14 days at 36°C); micrograph (× 120).

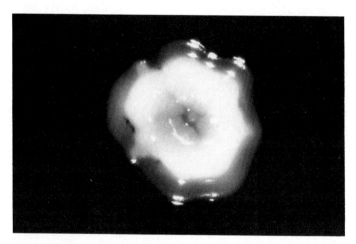

Figure 15.85. *A. naeslundii.* Molar tooth colony on CC-medium (14 days at 36°C); micrograph (× 30).

ing on the species affiliation of the isolate tested so that these characters may be used for differentiation and identification of *Actinomyces* species (see Table 15.47) although intraspecific variation in the carbohydrate fermentation patterns is not uncommon.

Organic compounds other than carbohydrates and sugar alcohols have rarely been studied for their ability to serve as sole sources of carbon and energy. Howell and Pine (1956) reported that pyruvate could support growth of certain *Actinomyces* strains. These findings were confirmed by Howell and Jordan (1963), Schofield and Schaal (1979b) and Cato et al. (1984) who found that *A. israelii, A. naeslundii, A. viscosus* and *A. meyeri* were able to grow on pyruvate as the sole carbon source. In addition, hamster strains of *A. viscosus* were reported to utilize lactate, succinate, acetate and gluconate, but not citrate and benzoate (Howell and Jordan, 1963). In contrast to *Arachnia propionica*

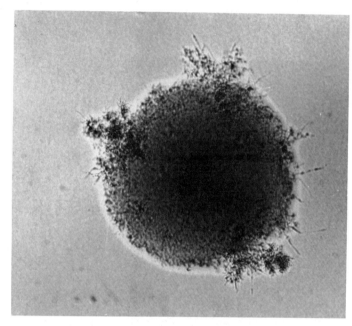

Figure 15.86. *A. viscosus, serovar 2.* Small mature colony on CC-medium (14 days at 36°C); micrograph (× 120).

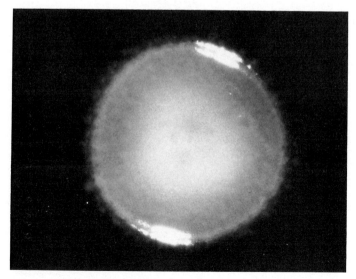

Figure 15.87. *A. viscosus, serovar 2.* Mature colony on CC-medium (14 days at 36°C); micrograph (× 30).

with a mean size of 1.5 amino acids gave the optimal growth response (Reddy et al., 1982).

Fermentable carbohydrates and similar substances are the preferred sources of carbon and energy. These compounds are usually necessary for good growth and, when utilized, are fermented with production of acid but no gas (Lentze, 1938b; Howell et al., 1959; Slack, 1974; Slack and Gerencser, 1975; Reddy et al., 1982; Dent and Williams, 1984a). Mono-, di-, tri- and polysaccharides as well as sugar alcohols and glucosides may serve as fermentable substrates. Among these, the monosaccharides glucose and fructose are used by essentially all representatives of the genus *Actinomyces* (Slack, 1974; Slack and Gerencser, 1975; Schofield and Schaal, 1979a and b, 1980a, 1981). Dextrin, galactose, *meso*-inositol, lactose, maltose and starch are fermented by many, but not all, members of the various species (Slack and Gerencser, 1975; Schofield and Schaal, 1981). However, strain variation has been observed with regard to the growth rates obtained on different sugars (Howell and Pine, 1956). Dulcitol and sorbose are usually not utilized by any *Actinomyces* strain (Slack and Gerencser, 1975; Schofield and Schaal, 1981). The fermentation of other carbohydrates varies depend-

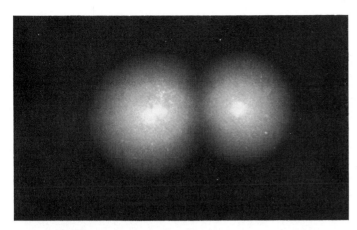

Figure 15.88. *A. bovis, serovar 1.* Mature colonies on CC-medium (14 days at 36°C); micrograph (× 30).

obtained with added CO_2 or HCO_3^- while under anaerobic conditions without CO_2, most strains will not grow at all or only produce poor growth (Howell and Pine, 1956; Schofield and Schaal, 1981). On the other hand, essentially all of the *Actinomyces* species including the aerotolerant ones will multiple well at low oxygen tension with added CO_2 so that the genus *Actinomyces* as a whole may be considered a taxon of facultative anaerobes with an obligate carbon dioxide requirement and variation in oxygen tolerance (Erickson, 1940; Howell, 1963; Slack and Gerencser, 1975). For these reasons, sophisticated anaerobic techniques such as the roll tube method or the anaerobic glove box are not required for the cultivation of fermentative actinomycetes.

The optimum growth temperature of the recognized *Actinomyces* species ranges from 30–37°C (Slack, 1974; Slack and Gerencser, 1975). Detailed studies on the temperature requirements of these organisms have rarely been performed. However, *A. pyogenes* was reported to grow between 20 and 40°C (Reddy et al., 1982) and most strains of *A. israelii, A. naeslundii* and *A. viscosus* were found to produce visible growth in the range of 28–32°C (Thompson and Lovestedt, 1951; Howell, 1963;

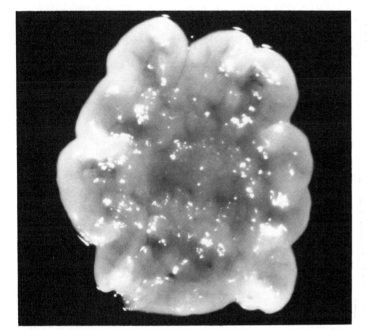

Figure 15.89. *A. bovis, serovar 2.* Bread-crumb colony on CC-medium (14 days at 36°C); micrograph (× 30).

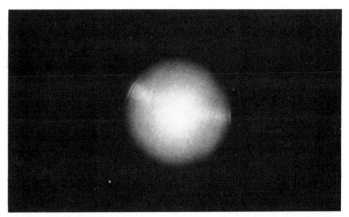

Figure 15.90. *A. odontolyticus, serovar 2.* Mature colony on CC-medium (14 days at 36°C); micrograph (× 40).

which grows well on alanine, none of the *Actinomyces* species tested (Schofield and Schaal, 1979b) could use this amino acid as sole carbon source.

Members of the genus *Actinomyces* vary considerably in respect to their oxygen tolerance. *A. viscosus, A. pyogenes* and many strains of *A. naeslundii, A. odontolyticus* and *A. denticolens* show good to moderate growth when incubated in air on the surface of suitable agar media (Howell, 1963; Slack, 1974; Schofield and Schaal, 1981; Reddy et al., 1982; Dent and Williams, 1984a). In liquid media, aerobic growth is usually better, provided that a large inoculum is used. However, aerobic growth yields are greatly increased when the incubation atmosphere contains carbon dioxide or when HCO_3^- is added to the medium. This is also true for the newly described species *A. howellii* (Dent and Williams, 1984b) and *A. hordeovulneris* (Buchanan et al., 1984). Thus, oxygen-tolerant *Actinomyces* strains may be described most appropriately as capnophilic.

Most strains of *A. bovis, A. israelii* and *A. meyeri* prefer anaerobic growth conditions although a marked strain variation in aerotolerance can be noted. Furthermore, consistently good anaerobic growth is only

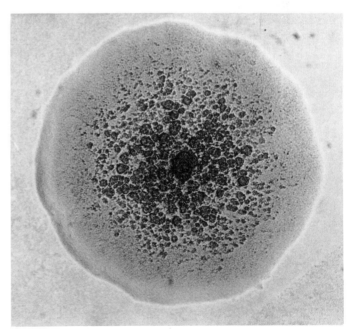

Figure 15.91. *A. odontolyticus, serovar 1.* Mature colony on CC-medium (14 days at 36°C); micrograph (× 120).

Figure 15.92. *A. pyogenes.* Mature colony on CC-medium (7 days at 36°C); micrograph (× 120).

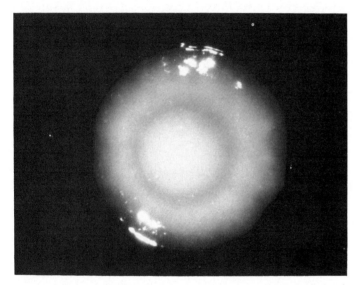

Figure 15.93. *A. denticolens.* Mature colony on CC-medium (14 days at 36°C); micrograph (× 30).

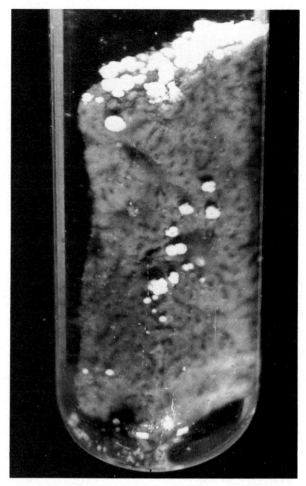

Figure 15.94. *A. israelii, serovar 1.* Mature colonies adhering to the tissue in Tarozzi broth (14 days at 36°C) micrograph (× 2.5).

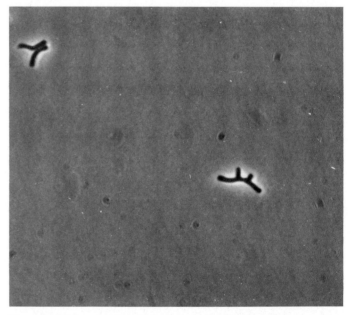

Figure 15.95. *A. odontolyticus, serovar 1.* Budding in a very young (16 h) slide culture; phase-contrast micrograph in situ in lactophenol cotton blue mounting fluid (× 1600).

Schaal, unpublished results). Growth of hamster strains of *A. viscosus* even occurred at 23°C when the organisms were incubated in air with added CO_2 or in an atmosphere of CO_2 and N_2 on BHIA (Howell, 1963). *A. naeslundii* and *A. viscosus* may also grow at 45°C (Holmberg and Nord, 1975). *A. meyeri* strains were reported to grow equally well at 30 and 37°C and nearly as well at 25°C. No growth was obtained at 45°C (Cato et al., 1984).

Anaerobically with added carbon dioxide, *Actinomyces* species ferment glucose to formic, acetic, lactic and succinic acids (Howell and Pine, 1956; Buchanan and Pine, 1965; Holdeman et al., 1977; Schofield and Schaal, 1981). The amount of succinic acid formed depends on the concentration of CO_2/HCO_3^- available in the medium. *A. meyeri* only occasionally forms small amounts of lactic acid and may also produce small quantities of pyruvic acid (Cato et al., 1984). Anaerobically without CO_2, the fermentation of *A. israelii* (ATCC 10049) was found to be homolactic (Buchanan and Pine, 1965). In air with added CO_2, acetate and carbon dioxide were the main end-products of glucose fermentation (Buchanan and Pine, 1965). Thus, the relative and absolute quantities of the end products formed may vary considerably depending on medium composition, growth conditions and strain peculiarities (Howell and Pine, 1956).

Carbon dioxide when present is utilized (Howell and Pine, 1956) and in the case of *A. pyogenes*, for each 1 mol of CO_2/HCO_3^- fixed, 1 mol of succinate, 1 mol of acetate and 1 mol of formate were formed (Reddy et al., 1982).

Energy is produced by substrate phosphorylation, but the amount of ATP formed depends on cultural conditions. The results of Buchanan and Pine (1967) indicate that 4 mol of ATP are formed per mol of glucose when *A. israelii* is incubated aerobically with added CO_2. Anaerobically, without CO_2, only 2 mol of ATP are produced.

Several enzymes involved in the initial dissimilation of exogenous carbohydrates or the energy yielding glycolytic pathway have been identified in *Actinomyces* species and some were characterized in more detail. α-Galactosidase activities were observed in *A. israelii*, *A. naeslundii* and *A. denticolens*. β-Galactosidase was found in *A. israelii*, *A. naeslundii*, *A. viscosus*, *A. pyogenes* and *A. denticolens*. *A. pyogenes* may also contain β-glucuronidase. α- and β-glucosidases were detected in *A. israelii*, *A. naeslundii* and *A. denticolens* and, in addition, the former enzyme was present in *A. viscosus* and *A. meyeri* and the latter in *A. odontolyticus*. *A. israelii* and a few *A. naeslundii* and *A. viscosus* isolates showed β-xylosidase activities. *A. odontolyticus* strains were the only fermentative actinomycetes which contained α-fucosidase. α-Mannosidase was not detected in any of the *Actinomyces* species (Kilian, 1978; Schofield and Schaal, 1981; Dent and Williams, 1984a).

The β-galactosidase of *A. viscosus* was found to have a molecular weight of 4.2×10^5, a K_m for lactose of about 6 mM and a pH optimum between pH 6.0 and 6.5 (Kiel et al., 1977). In addition, invertase activity has been demonstrated in *A. naeslundii* (Miller, 1974) and *A. viscosus* (Palenik and Miller, 1975; Kiel et al., 1977). The characteristics of the *A. viscosus* invertase were a molecular weight of 8.6×10^4, a K_m for sucrose of about 71 mM and a pH optimum between pH 5.8 and 6.3. The enzyme was noncompetitively inhibited by fructose-6-phosphate and fructose-1,6-diphosphate (Kiel and Tanzer, 1977).

Furthermore, fructose-1,6-diphosphate aldolase and phosphate acetyltransferase activities were demonstrated in cell-free extracts of *A. israelii* (Buchanan and Pine, 1967). Brown et al. (1975) purified and characterized a nicotinamide adenine dinucleotide-dependent lactate dehydrogenase from *A. viscosus*. This enzyme, which had a molecular weight of 1×10^5 and a pH optimum of pH 5.5–6.2, was under negative control by adenosine 5′-triphosphate and under positive control by fructose-1,6-diphosphate and inorganic phosphate. Malate dehydrogenase and glutamate dehydrogenase were identified in extracts of *A. viscosus*, *A. naeslundii* and *A. israelii*, serovars 1 and 2, but differed in their electrophoretic mobilities between both species and serovars (Fillery et al., 1978). 6-Phosphogluconate dehydrogenase was only detected in *A. naeslundii* and *A. viscosus* strains, but not in either serovar of *A. israelii* (Fillery et al., 1978).

Several workers have shown that *A. viscosus* is able to synthesize extracellular and cell-associated levan (Howell and Jordan, 1967; Warner and Miller, 1978). Under certain conditions, the levan formed a capsule tenaciously adhering to the cell (Warner and Miller, 1978). The enzyme levansucrase which is responsible for the production of the high-molecular-weight polymer was found to occur in the growth medium as well as associated to the cell wall (Pabst, 1977; Pabst et al., 1979). Furthermore, *A. viscosus* may exert levan-hydrolyzing activities (Miller and Somers, 1978). These are due to an extracellular and cell-associated enzyme which degrades the levan of *A. viscosus* as well as inulin and other levans. *A. naeslundii* possibly possesses an extracellular enzyme with similar activities (Miller and Somers, 1978).

Most *Actinomyces* species do not exhibit any detectable proteolytic activity. *A. bovis*, *A. israelii*, *A. naeslundii*, *A. viscosus*, *A. odontolyticus*, and *A. meyeri* do not liquefy gelatin or Loeffler's serum nor do they hydrolyze casein, peptonize milk or digest meat (Slack, 1974; Slack and Gerencser, 1975; Holdeman et al., 1977; Schofield and Schaal, 1981). However, litmus milk may be acidified, reduced and/or clotted (Slack and Gerencser, 1975). *A. pyogenes* is the only *Actinomyces* species

which has characteristic proteolytic capabilities. All strains of this taxon were reported to liquefy gelatin, hydrolyze casein and digest clots of litmus milk (Schofield and Schaal, 1981; Reddy et al., 1982).

The ability to deaminate or decarboxylate amino acids is also uncommon in the genus *Actinomyces*. Nonetheless, most strains of *A. israelii* and a few *A. naeslundii* and *A. viscosus* isolates are capable of producing ammonia from arginine. Furthermore, some *A. odontolyticus* strains may decarboxylate lysine (Schofield and Schaal, 1980b, 1981). *A. bovis*, *A. israelii*, *A. naeslundii*, *A. viscosus* and *A. odontolyticus* form hydrogen sulfide when grown on triple sugar iron agar (TSIA), although they do not blacken the indicator in the medium but only lead acetate strips suspended over the agar. In addition, certain strains of *A. bovis*, *A. naeslundii* and *A. odontolyticus* also show H_2S production over BHIA (Slack and Gerencser, 1975). Indole is not produced by any member of the genus *Actinomyces*.

With the exception of *A. naeslundii* and certain strains of *A. viscosus* and *A. meyeri*, all other *Actinomyces* species tested so far are urease-negative. All of the *A. naeslundii* and *A. viscosus* strains tested by Scharfen (1973) were found to be urease-positive. In contrast, Fillery et al. (1978) reported variation in the occurrence of this enzyme among strains of these two species. Schofield and Schaal (1981), in their numerical phenetic analysis of fermentative actinomycetes, obtained results which were again at variance with those of both previous studies. The latter authors found that the vast majority of typical *A. naeslundii* strains were urease-positive while most, but not all of the *A. viscosus* isolates were urease-negative.

Adenine, guanine, hippurate, hypoxanthine, xanthine and tyrosine are not hydrolyzed (Schofield and Schaal, 1981). With the exception of *A. pyogenes*, *A. meyeri* and many strains of *A. viscosus*, esculin hydrolysis is common to all members of the genus although strain variation may be observed, especially in *A. bovis* and *A. odontolyticus* (Slack and Gerencser, 1975; Schofield and Schaal, 1981). Starch hydrolysis is a consistent feature of *A. bovis*, but may also be found in other species (Slack and Gerencser, 1975; Schofield and Schaal, 1981). Hydrolysis of Tweens is weak or absent; lipase as observed on egg yolk-containing agar media has not been detected (Schofield and Schaal, 1981). Lecithinase activity has only been observed in a few *A. israelii* isolates (Schofield and Schaal, 1981). The Voges-Proskauer reaction is consistently negative while the methyl red test is usually positive (Slack and Gerencser, 1975; Holmberg and Nord, 1975). No chondroitin sulfatase activity has been detected and hyaluronidase was only found in a few *A. odontolyticus* strains. DNAase is absent in *A. israelii*, *A. naeslundii*, *A. viscosus* and *A. meyeri* but present in *A. bovis*, *A. pyogenes* and some strains of *A. odontolyticus* (Schofield and Schaal, 1981).

Additional enzymatic activities can be observed using the API-zym test kit (API-Bio Mérieux): All of the *Actinomyces* strains tested so far were negative for alkaline phosphatase, esterase-lipase (C_8), lipase (C_{14}), valine arylamidase (except for one *A. meyeri* strain tested), cystine arylamidase (except for one *A. meyeri* strain tested), trypsin and chymotrypsin when very weakly positive reactions which might be unspecific were excluded (Kilian, 1978; Schofield and Schaal, 1981; Dent and Williams, 1984a). Esterase (C_4) activity has been observed in some *A. israelii* and *A. viscosus* strains. Leucine arylamidase is present in most members of the genus. Strain variation was detected for acid phosphatase in *A. israelii*, *A. naeslundii*, *A. viscosus* and *A. denticolens* while the other species were consistently negative (Kilian, 1978; Schofield and Schaal, 1981; Dent and Williams, 1984a). Phosphoamidase activity is a typical character of *A. denticolens* and may also be found in *A. israelii*, *A. naeslundii* and *A. odontolyticus*. No phospholipase A activity was detected in *Actinomyces* species (Bulkacz et al., 1979).

A. viscosus, *A. howellii* (Dent and Williams, 1984b), and *A. hordeovulneris* are the only *Actinomyces* species which are catalase-positive, although the catalase activity of *A. hordeovulneris* is usually weak to moderate (Buchanan et al., 1984). In all of the other species, catalase activity is consistently absent (Slack, 1974; Slack and Gerencser, 1975; Schofield and Schaal, 1981; Reddy et al., 1982). Cytochrome oxidase activity may be found in certain strains of *A. naeslundii*, *A. viscosus* and *A. pyogenes* (Schofield and Schaal, 1981). However, the results of the benzidine test used for the determination of cytochrome oxidase do not necessarily correspond with the detection of cytochromes in cell extracts of actinomycetes by low temperature spectrophotometry (Taptykova and Kalakoutskii, 1973; Reddy et al., 1982). *A. bovis, A. viscosus, A. israelii,* serovar 1, and *A. pyogenes* apparently all contain cytochrome *b* (Taptykova and Kalakoutskii, 1973; Reddy et al., 1977). For *A. pyogenes,* it was assumed that cytochrome *b* mediates the reduction of fumarate to succinate, with reduced nicotinamide adenine dinucleotide as the electron donor (Reddy et al., 1982). *A. viscosus* may contain cytochromes *a* and *c* in addition to cytochrome *b* (Taptykova and Kalakoutskii, 1973).

Nearly all of the *A. naeslundii, A. odontolyticus* and *A. denticolens* strains are able to reduce nitrate to nitrite. Nitrate reduction is variable in *A. israelii* and *A. viscosus* and usually absent in *A. bovis, A. pyogenes, A. hordeovulneris* and *A. meyeri* (Slack, 1974; Slack and Gerencser, 1975; Schofield and Schaal, 1981; Reddy et al., 1982; Dent and Williams, 1984a; Cato et al., 1984; Buchanan et al., 1984). The ability to reduce nitrite is uncommon in the genus, but may be seen in many *A. naeslundii* and few *A. viscosus* strains (Schofield and Schaal, 1980b, 1981) although this character may often be difficult to detect (Dent and Williams, 1984a).

Some *Actinomyces* species exhibit β-hemolytic activities which may vary depending on the erythrocytes used (Slack, 1974). *A. pyogenes* consistently produces β-hemolysis presenting as large zones of clearing and decolorization on agar media containing human, sheep or horse red blood cells (Schofield and Schaal, 1981; Reddy et al., 1982). *A. bovis* usually shows β-hemolysis on human erythrocytes and may also lyse sheep and horse red blood cells. *A. odontolyticus* strains vary in their β-hemolytic activity irrespective of the erythrocytes used. Members of the other species usually do not show any signs of β-hemolysis. However, *A. odontolyticus* isolates, some *A. bovis* and *A. meyeri* strains and very few *A. naeslundii* and *A. viscosus* strains may produce an area of greening around the colonies when grown on human, sheep or rabbit blood agar (Slack, 1974; Schofield and Schaal, 1981). A few *A. hordeovulneris* strains were reported to exhibit a weak hemolytic activity on bovine blood agar while most isolates of this species were nonhemolytic (Buchanan et al., 1984).

Certain *Actinomyces* species, especially *A. naeslundii* and *A. viscosus,* were found to possess mechanisms which enable them to adhere to hard surfaces and epithelial cells, to coaggregate with each other or with other bacteria or to agglutinate erythrocytes. From the results reported so far, it appears that more than one mechanism is involved in these adherence and coaggregation functions. Thus, it was suggested that dextran and levan production by *A. naeslundii* and *A. viscosus* was related to adherence and coaggregation (Bourgeau and McBride, 1976; Miller et al., 1978). In addition, enhancement of aggregation by low pH values indicated that alterations of the cell surface potential might be involved in the aggregation process (Miller et al., 1978). Furthermore, the surface fibrils (or fimbriae) mentioned above possibly play an important role in the adherence of actinomycetes. The fibrils of *A. viscosus* T14V were found to consist of high molecular weight proteins with some carbohydrate and up to 14.3% nitrogen (Cisar and Vatter, 1979). That such proteins may be responsible for attachment and coaggregation was suggested by the finding that treatment with proteolytic enzymes or heat could impair the aggregation of *A. naeslundii* or *A. viscosus* with *Streptococcus sanguis* and *Streptococcus mitis* (Ellen and Balcerzak-Raczkowski, 1977; McIntire et al., 1978).

At least in the latter case, a lectin-like mechanism may be involved. McIntire et al. (1978) reported that the coaggregation between *A. viscosus* T14V and *Streptococcus sanguis* 34 required a protein or glycoprotein on *A. viscosus* and a carbohydrate on *S. sanguis.* The coaggregation was specifically inhibited more than 80% by lactose, β-methyl-D-galactoside and D-galactose, but not or only weakly by α-methyl-D-galactoside, melibiose, maltose, cellobiose, sucrose and a number of monosaccharides (Costello et al., 1979), and Ellen et al.

(1980) found that *A. naeslundii* and *A. viscosus* were able to agglutinate human AB and horse erythrocytes. This agglutinating property was enhanced by pretreatment of the red blood cells with neuraminidase and specifically inhibited by β-galactosidase. Furthermore, it was suggested that the actinomycetes used were able to prime erythrocytes for hemagglutination by removing sialic acid to expose more penultimate β-galactoside sites on the surface of the red blood cell. Thus, neuraminidase removal of terminal sialic acid and lectin-like binding to exposed β-galactoside-associated sites on the erythrocytes appeared to be responsible for the hemagglutinating capabilities of *A. naeslundii* and *A. viscosus.*

Only one mutant strain of *A. viscosus* has been characterized in some detail. This mutant (T14AV) appeared spontaneously in a culture of *A. viscosus* T14 which was used for studying the periodontopathic potential of this species (Hammond et al., 1976). The mutant strain T14AV differed from the parent strain (T14V) in that it proved to be avirulent under certain experimental conditions. Upon further investigation, it was shown that the morphological, chemical and antigenetic differences between the strains which could account for the impaired virulence of T14AV were quantitative rather than qualitative. Thus, the virulent parent strain produced more fimbriae and more virulence-associated antigen than the mutant strain (Cisar et al., 1978; Brecher et al., 1978). On the other hand, strain T14AV was found to form much larger amounts of extracellular heteropolysaccharide than did strain T14V (Hammond et al., 1976). These findings and corresponding animal studies indicate that the virulence of strain T14V is associated with its ability to colonize teeth by adhering strongly to tooth surfaces (Brecher et al., 1978). The lower virulence of strain T14AV might be related to its relative deficiency of fimbriae and to the presence of the heteropolysaccharide which appeared to impair the adherence functions.

Only one report deals with the isolation of a lytic phage which produced clear plaques on a human isolate of *A. viscosus* (Delisle et al., 1978). This phage (Av-1) was obtained from a sample of raw domestic sewage and belonged to Bradley's morphological group C (Bradley, 1967). It possessed a small polyhedral head measuring 40 nm in diameter and a short tail (26 nm in length). Its host range appeared to be very narrow as it only infected one human isolate of *A. viscosus* (strain MG-1), but did not show lysis or growth inhibition when plated with ten other *A. viscosus* and ten *A. naeslundii* strains. Plasmids and bacteriocins have not been reported.

Agglutination (Holm, 1930; Lentze, 1938b; Erikson, 1940; Slack et al., 1951, 1955; King and Meyer, 1957; Snyder et al., 1967; Bellack and Jordan, 1972), complement fixation (Kwapinski and Snyder, 1961; Snyder et al., 1967), cell wall agglutination (Cummins, 1962, 1970; Bowden and Fillery, 1978, Fillery et al., 1978), immunofluorescence (Slack et al., 1961, 1966; Slack and Grencser, 1966, 1970, 1975; Lambert et al., 1967; Brock and Georg, 1969a; Bellack and Jordan, 1972; Holmberg and Forsum, 1973; Schaal and Pulverer, 1973; Marucha et al., 1978; Lai and Listgarten, 1979; Buchanan et al., 1984; Cato et al., 1984; Schaal and Gatzer, 1985), immunoassay (Gillis and Thompson, 1978), immunodiffusion (King and Meyer, 1963; Snyder et al., 1967; Georg et al., 1968; Bellack and Jordan, 1972; Bowden et al., 1976; Schaal and Gatzer, 1985), immunoelectrophoresis (Holmberg et al., 1975; Bowden et al., 1976; Fillery et al., 1978), passive hemagglutination (Wicken et al., 1978) and immunoelectron microscopy (Garant et al., 1979) procedures have all been used to elucidate the antigenic structure of *Actinomyces* species, to define the serological relationships between species and serovars and to locate antigenic determinants in or on *Actinomyces* cells. Among these techniques, cell wall agglutination, immunodiffusion and fluorescent antibody tests have proved especially useful for application to the fermentative actinomycetes. Immunofluorescence has been employed most extensively and the majority of data reported below are based on direct or indirect modifications of this technique.

The species *A. bovis, A. israelii, A. naeslundii, A. viscosus, A. odontolyticus* and *A. meyeri* are all serologically heterogeneous and contain at least two serovars (Slack and Gerencser, 1975; Cato et al., 1984). Varying degrees of cross-reactions between the serovars within one species and between different species have been reported, but there is general agreement that species and type-specific antigens do exist (Bowden et al., 1976; Slack and Gerencser, 1975). Furthermore, a certain variability within some of the serovars has already been noted previously which might form the basis for the description of additional serovars or new serological groups (Schaal and Gatzer, 1985).

Serological grouping of members of the genus *Actinomyces* was first introduced by Slack and co-workers (Slack et al., 1961; Slack and Gerencser, 1966, 1970) and these serological groups could be related later on to the various *Actinomyces* species. Depending on the quality of antisera and the serological techniques used, species and/or serovar-specific antisera can be obtained either by simple dilution or by sorption procedures (Slack and Gerencser, 1975; Marucha et al., 1978; Schaal and Pulverer, 1973; Schaal and Gatzer, 1985).

A. bovis shows no or only low grade cross-reactions with other species or between the two serovars (Slack and Gerencser, 1975; Schaal and Gatzer, 1985). From the comparatively small number of strains studied so far it can possibly be concluded that serovar 1 strains usually produce smooth, nonfilamentous microcolonies (Fig. 15.71), whereas serovar 2 microcolonies are often definitely filamentous (Fig. 15.76) (Slack and Gerencser, 1975).

Within the species *A. israelii*, the serological relationships are much more complex and have not been definitely clarified. Serological heterogeneity of this taxon was already assumed by Lentze in 1938 (b). The existence of two serovars was first reported by Lambert et al. (1967) and confirmed by Blank and Georg (1968), Brock and Georg (1969a), Slack et al. (1969), Cummins (1970) and Schaal and Pulverer (1973). Cross-reactions between these serovars occur. However, Brock and Georg (1969a) found only a one-way cross-reaction in which serovar 1 antiserum reacted with serovar 2 cells, while others (Slack et al., 1969; Holmberg and Forsum, 1973; Schaal and Pulverer, 1973) obtained reciprocal cross-reactions. Cross-reactions with *A. naeslundii*, *A. viscosus*, *A. pyogenes*, *A. hordeovulneris* (Buchanan et al., 1984) and *Arachnia propionica* may be noticeable at low dilutions of the antisera, but are usually eliminated by further dilution or absorption (Slack and Gerencser, 1975; Schaal and Gatzer, 1985).

Serovar 1 strains of *A. israelii* were reported to exhibit antigen variation within the serovar (Slack and Gerencser, 1975; Bowden et al., 1976). These findings were recently confirmed by Schaal and Gatzer (1985) using an indirect fluorescent antibody test. Furthermore, these authors found two additional serological types among strains showing the general characteristics of *A. israelii*. These types showed little or no cross-reactivity with each other and with members of the two defined serovars and corresponded well with the subclusters 1 c and 1 d delineated among clinical isolates of *A. israelii* in a recent numerical phenetic analysis (Schofield and Schaal, 1981).

The second serovar of *A. naeslundii* was first reported by Bragg et al. (1972) on the basis of four strains which cross-reacted with both *A. naeslundii*, serovar 1, and *A. viscosus*, serovar 2. Further studies showed, however, that three of these strains should be considered untyped (Bragg et al., 1975) or serovar 1 variants (Gerencser, 1979) and only one strain (WVU 1523) should be accepted as serovar 2. Jordan et al. (1974) described strain N 16 (WVU 820) as serovar 3 of *A. naeslundii*. This strain and a few similar isolates appeared to be more closely related to *A. naeslundii*, serovar 2, than to serovar 1 and showed strong cross-reactions with *A. viscosus*, serovar 2. In the numerical-phenetic study of Schofield and Schaal (1981), strain WVU 820 was recovered in subgroup III of the *A. viscosus* phenon while two other strains received from the London Medical College as representatives of serovar 3 formed subgroup IV of the same phenon. The organisms tentatively designated as serovar 4 of *A. naeslundii* by Gerencser and Slack (1976) are not related to *A. viscosus*, serovar 2, and probably represent a new taxon which has provisionally been designated *Actinomyces* sp., type 963 (Gerencser, 1979).

Cross-reactions between the serovars of *A. naeslundii* do occur and may be difficult to eliminate (Schaal and Gatzer, 1985), especially when high titred antisera and an indirect immunofluorescence test are employed. In addition, low grade cross-reactions have been observed between *A. naeslundii* and *A. israelii* (Slack and Gerencser, 1975; Schaal and Gatzer, 1985).

The two serovars of *A. viscosus* which have been recognized so far (Gerencser and Slack, 1969; Slack and Gerencser, 1970) also differ in their ecology; serovar 1 strains represent the original hamster isolates of *A. viscosus* while serovar 2 isolates were primarily obtained from human sources. Cross-reactions were reported between *A. hordeovulneris* and an antiserum to serovar 1 of *A. viscosus* (ATTC 15987). Serovar 2 antisera showed no reaction with this new species (Buchanan et al., 1984).

The two serovars of *A. odontolyticus* (Slack and Gerencser, 1970) exhibit only low grade cross-reactivity between each other. However, very pronounced cross-reactions can be observed between *A. odontolyticus*, serovar 2, and *A. pyogenes* (Slack and Gerencser, 1975; Schaal and Gatzer, 1985). This cross-reactivity which resulted in indirect immunofluorescence staining of the heterologous antigens at titres which were only one dilution below the homologous reaction (Schaal and Gatzer, 1985) clearly requires further studies. Cross-reactions with other *Actinomyces* species are weak or do not occur at all.

Little is known about the antigenic structure of *A. pyogenes* and its serological relationships to other *Actinomyces* species. At present, the species appears to be serologically homogeneous and, apart from the cross-reactions with *A. odontolyticus*, there seems to be little serological affinity to other *Actinomyces* species (Schaal and Gatzer, 1985). The same is true for *A. meyeri* which showed no cross-reactions (using a monovalent FITC conjugate prepared against strain ATCC 33972) with *A. israelii*, *A. odontolyticus*, *A. naeslundii* and *A. viscosus* (Cato et al., 1984). However, several isolates that appeared to be identical to *A. meyeri* phenotypically and on the basis of their electrophoretic protein patterns (Cato et al., 1984), did not react with this antiserum, indicating that there probably exist additional serovars of this species.

Serology and antigenic structure of *A. denticolens* have not been studied. However, preliminary data obtained in this laboratory (Schaal and Gatzer, unpublished results) indicate that this species represents a separate serological entity. Cross-reactions with antisera to the other *Actinomyces* species did not occur or were only very low grade. Even with *A. naeslundii* with which *A. denticolens* shares many physiological characters, striking cross-reactions were not observed.

In principle, the various serological techniques give comparable results when applied to *Actinomyces* strains. However, the degree of cross-reactivity may vary depending on the method, the antigens and the antisera used. Cross-reactions with members of other genera have occasionally been encountered (Slack and Gerencser, 1975), but are essentially always low titred. They may be especially pronounced with cytoplasmic antigens (Kwapinski and Seeliger, 1964), but can also be demonstrated with other antigens including whole cells and fluorescent antibody procedures. Cross-reactions between *A. israelii*, serovar 1, and *Arachnia propionica* can be observed by immunofluorescence and by the Ouchterlony technique (Schaal and Gatzer, 1985). Occasionally, *A. israelii* may also cross-react with *Propionibacterium acnes* (Slack and Gerencser, 1975) and *Rothia dentocariosa* (Schaal and Gatzer, 1985). Similar cross-reactions may occur between *A. naeslundii* and *A. viscosus* on the one hand and *Arachnia* and/or *Rothia* on the other hand (Slack and Gerencser, 1975; Schaal and Gatzer, 1985). *A. bovis*, serovar 1, antisera show low-grade cross-staining with *A. meyeri* cells in indirect fluorescent antibody tests (Schaal and Gatzer, 1985). Finally, cell wall polysaccharides derived from *A. pyogenes* by formamide extraction cross-reacted with antisera to group G streptococci (Cummins et. al., 1974).

Knowledge of the number, chemical composition and cellular location of *Actinomyces* antigens is still fragmentary. Antigenic components were found in the cytoplasm (Kwapinski and Snyder, 1961; Holmberg et al., 1975), in or on the cell wall (Cummins, 1962; Bowden et al., 1976; Hammond et al., 1976; Bowden and Fillery, 1978; Fillery et al., 1978; Powell et al., 1978; Wicken et al., 1978; Wheeler and Clark, 1980) and in culture supernatants (King and Meyer, 1963; Georg et al., 1968; Lambert et al., 1967).

One group of *Actinomyces* antigens consists of cell wall-associated carbohydrates or polysaccharides which may also be released into the culture fluid (Cummins, 1962; King and Meyer, 1963; Pirtle et al., 1965; Bowden and Hardie, 1973; Bowden et al., 1976). These antigens are usually heat stable and resistant to treatment with proteases. They may carry species and serovar specificity (Cummins, 1962; Bowden and Hardie, 1973; Bowden et al., 1976). More recently, Wicken et al. (1978) characterized an amphipathic antigen of *A. viscosus* NY 1 in more detail. This antigen was found to be a fatty acid-substituted heteropolysaccharide which amounted to 1% of the cell mass. It contained 16.8% (by weight) *O*-esterified fatty acids (C_{16}, C_{18}, $C_{18:1}$) and a polysaccharide composed of mannose, glucose and galactose in a molar ratio of 1:2:3. Glycerophosphate, *N*-acetylgalactosamine, alanine and lysine were further constituents, but 6-deoxytalose and rhamnose were not identified. Evidence for the presence of lipoteichoic acids was not found.

Landfried (1972) isolated an antigen from the culture supernate of *A. israelii* ATCC 12102 which was nonmigratory in immunoelectrophoresis and was detected in several strains of both serovars of this species. Chemical analysis indicated that this antigen might be a glycan or a mixture of two or three glycans. The peptidoglycan itself may also possess antigenic properties (Reed, 1972).

A second group of *Actinomyces* antigens is composed of polypeptides or polypeptide-containing compounds (Bowden and Hardie, 1973; Bowden et al., 1976). These were found to be pronase-sensitive but trypsin resistant, charged molecules which were species-specific, but contained cross-reacting components.

The cytoplasmic antigens prepared by Kwapinski and Snyder (1961) exhibited pronounced cross-reactions. In contrast, Holmberg et al. (1975) reported that their cytoplasmic preparations gave specific and reproducible results when used for crossed immunoelectrophoresis or crossed immunoelectrofocusing. It can be assumed that at least some of the cytoplasmic antigens are proteins or polypeptides. The protein nature has been proven for the virulence-associated antigen 1 of *A. viscosus* T14V (Cisar and Vatter, 1979; Cisar et al., 1978; Wheeler and Clark, 1980) which is identical with the fimbriae produced by this organism. These fimbriae were reported (Wheeler and Clark, 1980) to consist of 95.2% protein and less than 2% carbohydrate and contained high quantities of aspartic acid, threonine, glutamic acid and alanine. A minimum molecular weight of 24,960 was calculated for this fibril protein.

Actinomyces species exhibit a moderate to high susceptibility to many of the antibacterial drugs currently in use, thereby showing little species and strain variation (Abrahams and Miller, 1946; Howell, 1953; Hanf et al., 1953; Suter and Vaughan, 1955; Hanf, 1956; Lentze, 1957, 1967; Blake, 1964; Fritsche, 1964a; Lerner, 1968, 1974; Spieckermann, 1970; Sutter and Finegold, 1976; Chow and Bednorz, 1978; Pape, 1979; Schaal et al., 1979; Schaal and Pape, 1980; Niederau et al., 1982; Buchanan et al., 1984, Cato et al., 1984). A tendency towards an increasing antibiotic resistance of the actinomycetes has not been observed. However, susceptibility results obtained in different laboratories often differ considerably despite the fact that certain reference strains were common to many studies. This indicates that methodological problems obviously play an important role in the assessment of drug sensitivity of the actinomycetes (Schaal and Pape, 1980).

The minimal inhibitory concentrations recently reported for *A. bovis*, *A. israelii*, *A. naeslundii*, *A. odontolyticus* and *A. viscosus* (Schaal et al., 1979; Pape, 1979; Schaal and Pape, 1980; Niederau et al., 1982) are summarized in Table 15.46. As can be seen from this table, the *Actinomyces* species tested are highly susceptible to essentially all of the β-lactam antibiotics (penicillins and cephalosporins). Tetracyclines, chloramphenicols, macrolides, lincomycins, rifamycins, fusidic acid and vancomycin also exhibit a moderate to high inhibitory activity.

Complete resistance has only been observed to aminoglycosides, nitroimidazole compounds, peptide antibiotics, antifungal drugs and certain antituberculotics (Table 15.46; Suter and Vaughan, 1955).

Although susceptibility data for *A. pyogenes* have rarely been reported, it appears that this species is sensitive to penicillins, cephalosporins, tetracyclines, macrolides and vancomycin. It is resistant to peptide antibiotics, but apparently slightly more sensitive to aminoglycosides than the other *Actinomyces* species. All strains of *A. meyeri* tested so far were susceptible to chloramphenicol, clindamycin, erythromycin, penicillin G and tetracycline (Cato et al., 1984). Minimal inhibitory concentrations for *A. hordeovulneris* were reported as follows: penicillin G, ≤0.25 mg/liter; chloramphenicol, 1.0 mg/liter; cotrimoxazole, 1.0 mg/liter; tetracycline, 0.5–4.0 mg/liter. The antibiotic susceptibilities of *A. denticolens* and *A. howellii* are not known.

The resistance pattern of *Actinomyces* species to other chemical and biological inhibitors is as follows: sodium chloride at a concentration of 2% (w/v) was found to be inhibitory only to *A. odontolyticus*, while many strains of *A. bovis*, *A. israelii*, *A. naeslundii*, *A. viscosus*, *A. pyogenes* and *A. meyeri* were able to grow at this concentration. Certain *A. naeslundii*, *A. viscosus* and *A. meyeri* strains also tolerated 4% (w/v) NaCl, but 6% inhibited growth of all of the *Actinomyces* strains tested. Bile at concentrations of 5 and 10% (w/v) inhibited many *A. bovis*, *A. israelii*, *A. meyeri* and all of the *A. odontolyticus* strains, whereas *A. naeslundii*, *A. viscosus* and most *A. pyogenes* isolates were resistant. Even at 20% bile, about one-half of the *A. naeslundii*, *A. viscosus* and *A. pyogenes* strains were able to grow. The results obtained with sodium taurocholate (0.2%, w/v) were at variance to these findings (see Table 15.49). Sodium selenite (0.01%, w/v) caused complete growth inhibition of the majority of the *A. israelii*, *A. odontolyticus* and *A. pyogenes* strains tested while about one-third of the *A. bovis*, *A. naeslundii* and *A. viscosus* isolates grew in the presence of this inhibitor. Potassium tellurite (0.01%, w/v) and sodium azide (0.005%, w/v) were tolerated by most of the *Actinomyces* strains, whereas crystal violet (0.005%, w/v) inhibited all of the *A. israelii*, *A. odontolyticus* and *A. pyogenes* strains and many of the *A. bovis*, *A. naeslundii* and *A. viscosus* isolates tested (Schofield and Schaal, 1981; Cato et al., 1984).

Gallagher and Cutress (1977) studied the effect of trace elements on growth and fermentation of *A. naeslundii* and *A. viscosus*, and found that sodium selenite (50 ppm) inhibited growth and acid production of both species tested. This is in accord with the results of Schofield and Schaal (1981). Similar inhibitory effects were obtained with sodium fluoride (1000 ppm), manganous chloride (500 ppm), zinc sulfate (400 ppm), strontium chloride (10,000 ppm, only *A. viscosus*), silver nitrate (30 ppm) and antimony potassium tartrate (100 ppm). Ammonium vanadate (12.5 ppm), nickelous chloride (100 ppm), copper sulfate (200 ppm), cadmium iodide (2 ppm) and barium chloride (400 ppm) showed no, or only slight, inhibition of growth and/or acid production.

Apart from *A. denticolens* and *A. howellii*, whose pathogenic potentials have not been investigated so far, all of the other typical *Actinomyces* species may cause various forms of disease in man as well as in feral, domestic and laboratory animals. Nevertheless, members of different species vary considerably with regard to virulence, types of pathological lesions induced and host specificity (Slack, 1974; Slack and Gerencser, 1975; Pulverer and Schaal, 1978; Schaal, 1979, 1981; Schaal and Pulverer, 1981; Schaal and Beaman, 1984; Buchanan et al., 1984, Cato et al., 1984).

Typical diseases and impairments caused by *Actinomyces* species in humans are: actinomycoses, ocular infections, periodontal disease, caries and intrauterine infections. Among these pathological conditions, actinomycoses are undoubtedly the most characteristic disease entities produced in man by members of the genus *Actinomyces*.

As currently recognized, actinomycoses are endogenous, subacute to chronic, granulomatous inflammatory processes that give rise to suppuration, abscess formation and development of draining sinus tracts. The etiology of actinomycotic infections is complex in that they essentially always contain microbes, so-called concomitant bacteria, in addition to the causative actinomycete (Lentze, 1948, 1953, 1969; Holm,

Table 15.46.
Susceptibility of **Actinomyces** *species to various antimicrobial drugs[a]*

Drug	Range of minimal inhibitory concentrations (in milligram/liter) observed for				
	1. *A. bovis*	2. *A. israelii*	3. *A. naeslundii*	4. *A. odontolyticus*	5. *A. viscosus*
Penicillin G[b]	0.78–1.56	0.05–0.39	0.05–0.2	0.1–0.2	0.05–0.39
Propicillin	0.39–1.56	0.05–0.78	0.05–0.2	0.1–0.2	0.05–0.78
Azidocillin	0.39	0.2–0.78	0.1–0.39	0.05–0.2	0.05–0.39
Ampicillin	0.1–0.2	0.1–0.78	0.1–0.39	0.05–0.1	0.02–0.2
Amoxicillin	0.2	0.2–1.56	0.2–0.78	0.2	0.2–0.39
Ciclacillin	0.78–6.25	0.39–3.12	0.39–3.12	0.78	0.78–6.25
Epicillin	0.2–0.39	0.2–0.78	0.2–0.39	0.2	0.2–0.39
Carbenicillin	3.12	0.78–3.12	1.56–3.12	1.56	0.78–3.12
Ticarcillin	0.39–1.56	0.39–3.12	0.39–3.12	0.78	0.78–3.12
Mezlocillin	0.78–3.12	0.39–1.56	0.39–1.56	0.39–0.78	0.2–0.78
Azlocillin	1.56–3.12	0.2–1.56	0.2–1.56	0.39–0.78	0.2–0.78
Cefalothin	0.78–1.56	0.2–0.78	0.2–0.78	0.2	0.2–0.78
Cefapirin	0.78–1.56	0.2–0.39	0.2–0.78	0.2–0.39	0.2–0.78
Cefacetrile	0.39–6.25	0.39–1.56	0.2–1.56	0.39	0.2–1.56
Cefradine	0.2–3.12	0.2–0.78	0.2–0.78	0.2	0.2–0.78
Cefazolin	1.56–3.12	0.2–0.78	0.2–0.78	0.39	0.2–0.39
Cefalexin	1.56–3.12	0.2–0.78	0.2–1.56	0.2–0.39	0.2–1.56
Cefamandole	0.2	0.2–0.78	0.2–0.78	0.2	0.2–0.78
Cefuroxime	0.2–3.12	0.2–1.56	0.2–0.78	0.2	0.2–0.78
Cefoxitin	0.78	0.2–0.78	0.2–0.78	0.2	0.2–0.78
Cefotaxime	0.2–0.78	0.05–0.39	0.05–0.39	0.2	0.1–0.78
Gentamicin	6.25–12.5	3.12–25	3.12–25	3.12–25	1.56–12.5
Tobramycin	12.5–100	3.12–50	3.12–50	25–50	3.12–50
Sisomicin	3.12	3.12–50	0.78–50	12.5	0.78–50
Amikacin	12.5–50	12.5–100	6.25–25	50	6.25–50
Spectinomycin	6.25	3.12–50	6.25–50	12.5	6.25–50
Tetracycline-HCl	0.39–3.12	0.78–3.12	0.78–3.12	0.39–1.56	0.39–3.12
Doxycycline	0.2	0.2–0.78	0.2–0.78	0.78	0.2–1.56
Minocycline	0.2–0.39	0.2–1.56	0.2–0.78	0.39	0.2–0.78
Chloramphenicol	0.39–1.56	1.56–12.5	1.56–6.25	1.56–3.12	1.56–3.12
Thiamphenicol	6.25	3.12–12.5	3.12	3.12	3.12–12.5
Erythromycin	0.2–1.56	0.1–0.78	0.1–0.78	0.2–0.78	0.1–0.78
Oleandomycin	0.2	0.2–3.12	0.2–1.56	0.2	0.2–1.56
Spiramycin	0.39	0.2–3.12	0.39–1.56	0.2	0.39–1.56
Lincomycin	0.39–0.78	0.2–0.78	0.2–1.56	0.2–0.39	0.2–1.56
Clindamycin	0.39–3.12	0.39–1.56	0.39–3.12	0.2–1.56	0.39–1.56
Rifampicin	0.05–0.2	0.1–0.39	0.1–0.39	0.05–0.39	0.02–0.39
Metronidazole	>100	25->100	50->100	≧100	25->100
Tinidazole	≧100	25->100	50->100	≧100	50->100
Fusidic acid	0.2–0.39	0.2–6.25	0.2–6.25	0.2–1.56	0.2–6.25
Vancomycin	0.39	0.39–1.56	0.39–0.78	0.78	0.39–1.56
Colistine	>100	>100	>100	>100	>100

[a] Data compiled from: Schaal et al., 1979; Schaal and Pape, 1980; Pape, 1979; Niederau et al., 1982

[b] In International Units per milliliter.

1950, 1951; Pulverer and Schaal, 1978; Schaal, 1979, 1981; Schaal and Pulverer, 1981; Schaal and Beaman, 1984).

In about 40% of human cases (Pulverer and Schaal, 1978; Schaal 1979; Schaal and Pulverer, 1981; Schaal and Beaman, 1984), the purulent discharge from actinomycotic lesions contains macroscopically visible (≤1 mm in diameter), yellowish to brownish particles which are usually referred to as "Drusen" or "sulfur granules" and which consist of a conglomerate of filamentous actinomycete microcolonies formed in vivo, various other bacteria and tissue reaction material, especially polymorphonuclear granulocytes surrounding the microbial center. Corresponding filamentous colonies are commonly found in tissue sections, but are often smaller than those from pus and show a club-shaped layer of hyaline material on the tips of peripheral hyphae (Slack and Gerencser, 1975; Schaal and Beaman, 1984). The chemical composition of these clubs has not been elucidated in detail, but from the observations reported so far, it can be concluded that the hyaline material represents a polysaccharide-protein complex containing high concentrations of various salts (Widra, 1963; Pine and Overman, 1966; Frazier and Fowler, 1967; Crawford, 1971).

Human actinomycoses most frequently affect face and neck, the cervicofacial area, but they may also be encountered in thoracic and abdominal sites (Wolff and Israel, 1891; Lentze, 1938b, 1948, 1953, 1969, 1970, 1971; Slack and Gerencser, 1975; Pulverer and Schaal, 1978; Schaal, 1979, 1981; Schaal and Pulverer, 1981; Schaal and Beaman, 1984). Dissemination with involvement of bone (Winston, 1951; Slack and Gerencser, 1975; Kannangara et al., 1981), central nervous system (Fetter et al., 1967; Stevenson and Gossman, 1968; Slack and Gerencser, 1975; Hutton and Behrens, 1979; Koshi et al., 1981), liver, spleen, kidney, testes (Slack and Gerencser, 1975), extremities (Legum et al., 1978; Schaal and Pulverer, 1984) and skin (Slack and Gerencser, 1975; Legum et al., 1978) may occur. Primary infections of the skin are very rare and result usually from human bites or fistfight traumata (punch actinomycosis) (Slack and Gerencser, 1975; Southwick and Lister, 1979). Pelvic actinomycoses in females which may involve fallopian tubes, ovaries, uterus and bladder, are often associated with the use of intrauterine contraceptive devices (IUDs), vaginal pessaries or tampons (Hart et al., 1977; Witwer et al., 1977; Barnham et al., 1978; Bhagavan and Gupta, 1978; Gupta et al., 1978; Kohoutek and Nozicka, 1978; Hager and Majmudar, 1979; Drew, 1981; Szabo et al., 1981).

The predominant causative agent of human actinomycoses is *A. israelii* (Wolff and Israel, 1891; Slack and Gerencser, 1975; Lentze, 1969; Schaal and Pulverer, 1973; Pulverer, 1974; Pulverer and Schaal, 1978, 1984; Schaal, 1981). However, other *Actinomyces* species have also been isolated from typical human actinomycotic lesions. *A. naeslundii* has been recovered comparatively frequently from cervicofacial infections (Pulverer and Schaal, 1978; Schaal, 1979, 1981, 1984) and occasionally from pulmonary (Karetzky and Garvey, 1974) and abdominal (Scharfen, 1975) suppurations. Even from an empyema of the knee joint *A. naeslundii* was repeatedly cultured as the sole pathogen (Schaal and Pulverer, 1981). Likewise, *A. viscosus* has been recovered from cervicofacial (Larsen et al., 1978; Schaal, 1979, 1981) and thoracic (Lewis and Gorbach, 1972; Mosimann et al., 1979; Thadepalli and Rao, 1979; Eng et al., 1981) actinomycoses. *A. odontolyticus* appears to be rarely involved in the etiology of invasive infections although this species has been isolated from a few cases of cervicofacial (Mitchell et al., 1977; Schaal and Pulverer, 1984) lesions and from pleural fluid and a lung abscess, respectively (Guillou et al., 1977; Baron et al., 1979). *A. meyeri* has been isolated frequently from brain abscesses and pleural fluid and less often from cervicofacial abscesses, abscesses of the hip, hand, foot, spleen and from bite wounds (Cato et al., 1984).

Actinomycete eye infections may present as conjunctivitis, lacrimal canaliculitis, dacryocystitis, keratitis or even intraocular infection (Pine et al., 1960; Ellis et al., 1961; Slack and Gerencser, 1975; Jones and Robinson, 1977; Blanksma and Slijper, 1978). *A. israelii*, *A. naeslundii*, *A. viscosus* and *A. odontolyticus* have been identified as etiologic agents of these predominantly noninvasive conditions (Slack and Gerencser, 1975; Schaal and Pulverer, 1984) and may occur singly or in various combinations.

In the complex etiology of caries and periodontal disease, *Actinomyces* species apparently constitute only one link in a long chain of cause and effect. Nevertheless, there is growing evidence that at least *A. viscosus*, *A. naeslundii* and *A. odontolyticus* play an important role in the pathogenesis of these conditions (Batty, 1958; Winford and Haberman, 1966; Socransky, 1970; Jordan and Hammond, 1972; Slack and Gerencser, 1975). *A. meyeri* has been cultured from a subgingival crevice in a patient with severe periodontitis (Cato et al., 1984). However, its etiological role in this disease remains to be definitely proven.

Apart from typical pelvic actinomycoses, *Actinomyces* species, especially *A. israelii*, are possibly also able to produce cervicitis and endometritis in women using IUDs (Bhagavan and Gupta, 1978; Luff et al., 1978).

Human diseases caused by *A. pyogenes* are clinically less characteristic than those incited by the other *Actinomyces* species. They may present as acute pharyngitis, urethritis or as a cutaneous or subcutaneous suppurative process (Barksdale et al, 1957; Collins and Jones, 1982; Reddy et al., 1982).

The best known natural actinomycete infection occurring in animals is bovine actinomycosis (lumpy jaw) which is usually also located in the cervicofacial area, but in contrast to human infections, frequently involves the bone, especially the mandible but sometimes the maxilla (Slack and Gerencser, 1975). Primary lung infections in cattle have been reported (Biever et al., 1969). The principle causative agent of bovine infections is *A. bovis* which presumably also occurs in other animals, but which has never been identified with certainty in human lesions (Bollinger, 1877; Slack and Gerencser, 1975; Schaal and Beaman, 1984). On the other hand, *A. israelii* may occasionally be isolated from bovine infections (Cummins and Harris, 1958; King and Meyer, 1957; Pine et al., 1960).

Actinomycosis of swine may involve the udder, lungs or internal organs, and bone invasion may also occur (Thompson, 1933; Grässer, 1957; Franke, 1973). *A. israelii* (Magnusson, 1928), *A. viscosus* (Georg et al., 1972) and "*A. suis*" (Grässer, 1957; Franke, 1973) have been isolated from such infections.

Canine and feline actinomycoses have been described (McGaughey et al., 1951; Georg et al., 1972; Davenport et al., 1974; Moens and Verstraeten, 1980; Hardie and Barsanti, 1982). These included infections of the soft tissue of the jaw as well as thoracic and abdominal cases. Tail, scrotum and epidural space may also be involved (Bestetti et al., 1977). Most infections encountered in dogs or cats were found to be caused by *A. viscosus* (Georg et al., 1972; Davenport et al., 1974; Bestetti et al., 1977; Moens and Verstraeten, 1980; Hardie and Barsanti, 1982). However, a new *Actinomyces* species, *A. hordeovulneris* (Buchanan et al., 1984) has recently been identified as an additional important causative agent of canine actinomycosis in California. This organism was isolated from cases of pleuritis, peritonitis, visceral abscesses, septic arthritis, and recurrent localized infections in dogs and was found to be frequently associated with injuries caused by awns of the grass genus *Hordeum* which easily penetrate skin or mucous membranes and are propelled forward through the tissue with any adjacent muscle contraction, leaving trails of inflammation and necrosis. The natural habitat of the pathogen has not been determined yet but it appears sensible to assume that it is introduced from the oral cavity or intestinal tract when mucous membranes are penetrated by the awn or the animal licks or bites a wound. Furthermore, the organisms may spread hematogenously to the necrotic focus originating from an awn injury.

Actinomycosis-like lesions have been reported from sheep, goats, horses, deer, moose, antelope and mountain sheep. However, no detailed reports are available on the *Actinomyces* species involved (Slack and Gerencser, 1975). In one case of pyogenic granulomas found in the abdomen of a mandrill, *A. israelii* was confirmed as the etiologic agent by immunofluorescence (Altman and Small, 1973).

A. pyogenes causes a variety of pyogenic disease conditions in domestic animals among which mastitis in cows and peritonitis and pleuritis in swine are most characteristic (Glage, 1903; Roberts, 1968; Reddy et al., 1982). Suppurative lesions due to this species have also been observed in sheep, rabbits and horses (Roberts, 1968).

A. viscosus has been isolated from dental plaque of hamsters with naturally occurring periodontal disease (Howell, 1963; Jordan and Keyes, 1964, 1965).

Various domestic and laboratory animals including horses, cattle, sheep, goats, pigs, dogs, rabbits, guinea pigs, rats, hamsters and mice have been used to produce experimental infections with *Actinomyces* species (Slack and Gerencser, 1975). After intraperitoneal, intravenous or subcutaneous injection, *A. bovis*, *A. israelii*, *A. naeslundii*, *A. odontolyticus* and *A. viscosus* were found to cause abscess formation which resembled to some extent naturally occurring actinomycotic lesions. However, progressive infections rarely developed from these abscesses and the infected animals usually survived. Furthermore, the virulence of *Actinomyces* species is apparently both species and strain variable and the test animals may vary considerably in their susceptibility (Wolff and Israel, 1891; Pine et al., 1960; Coleman and Georg, 1969; Georg and Coleman, 1970; Georg et al., 1972; Beaman et al., 1979). Mice and hamsters appear to be the most suitable laboratory animals for studying abscess formation by fermentative actinomycetes (Slack and Gerencser, 1975), although an animal model satisfactorily resembling natural actinomycoses has not been developed.

After oral inoculation, *A. naeslundii* and *A. viscosus* were both shown to produce periodontal disease with alveolar bone loss and/or root surface caries in hamsters (Jordan and Keyes, 1964) as well as in conventional and gnotobiotic rats (Jordan et al., 1965; Socransky et al., 1970; Llory et al., 1971; Jordan and Hammond, 1972; Crawford et al., 1978; Brecher et al., 1978; Brecher and van Houte, 1979; Burckhardt et al., 1981).

The factors of pathogenicity and/or virulence which enable the actinomycetes, especially *A. israelii* and *A. bovis*, to invade the tissue and to cause necrosis and abscess formation have not been elucidated. Only *A. pyogenes* has been shown to produce soluble toxic and hemolytic activites which are neutralized by antitoxin and which are fatal to mice and rabbits after intravenous injection (Lovell, 1944; Reddy et al., 1982).

In contrast, *A. viscosus*, *A. naeslundii* and possible *A. odontolyticus* were shown to possess a large variety of properties which apparently contribute to the formation of dental plaque and to the development of periodontal and carious lesions. Adherence and aggregation mechanisms which are involved in plaque formation have already been described. The same is true for certain surface components which may represent virulence antigens. In addition, *A. viscosus* and *A. naeslundii* were found to form a food chain with *Veillonella* (Distler et al., 1980; Distler and Kröncke, 1981), to produce chemotactic effects, a polyclonal B-cell activator and a stimulus for lysosome release in polymorphonuclear phagocytes, to induce release of mediators of inflammation and immunoglobulin production and to mark fibroblasts for immune-mediated damage (Engel et al., 1976, 1978; Taichman et al., 1978; Wicken et al., 1978, Burckhardt, 1978; Clagett et al., 1980; Mangan and Lopatin, 1981).

All of the recognized *Actinomyces* species appear to occur primarily as normal inhabitants of mucosal surfaces of man and other homoiothermic animals. Thus, the typical pathogenic species may be considered facultative or opportunistic pathogens which invade the human or animal body endogenously and are not transmissible (exceptions: "punch actinomycosis" after human bites, *A. pyogenes* infections).

The oral cavity of man and animals is apparently the principal natural habitat of members of the genus *Actinomyces*. Dental plaque or calculus and saliva are the materials from which *Actinomyces* species can consistently be recovered by either immunofluorescence or culture (Slack and Gerencser, 1975). *A. israelii*, *A. naeslundii*, *A. viscosus* and *A. odontolyticus* have all been identified in such samples from human sources (Naeslund, 1925; Slack, 1942; Lentze, 1948; Thompson and Lovestedt, 1951; Batty, 1958; Howell et al., 1959, 1962; Snyder et al., 1967; Gerencser and Slack, 1969; Socransky, 1970; Slack et al., 1971; Collins et al., 1973; Hill et al., 1977; Ellen et al., 1978; Russell and Melville, 1978). The principal habitat of *A. meyeri* appears to be human periodontal sulci (Cato et al., 1984).

In human dental calculus, the total count of fermentative actinomycetes was estimated to 1.9×10^7 particles per gram wet weight. For individual species, the counts were: *A. israelii*, 6.32×10^6; *A. naeslundii*, 5.37×10^6; *A. viscosus*, 1.75×10^6; *A. odontolyticus*, 0.1×10^6 when no signs of inflammation were observed. In patients with periodontal disease, the respective figures were: *A. israelii*, 14.01×10^6; *A. naeslundii*, 2.38×10^6, *A. viscosus*, 4.39×10^6; *A. odontolyticus*, 1.26×10^6 (Collins et al., 1973). Furthermore, *A. israelii* and *A. naeslundii* have been demonstrated in human tonsils (Emmons, 1938; Grüner, 1969; Blank and Georg, 1968; Hotchi and Schwarz, 1972) and, along with *A. viscosus*, in cervicovaginal smears from women with or without IUDs (Gupta et al., 1976, 1978; Hager et al., 1979; Pine et al., 1981; Schaal and Pulverer, 1984). Occasionally, *A. viscosus* may be recovered from the uninfected conjunctiva and/or cornea (Jones and Robinson, 1977).

None of the *Actinomyces* species has been documented with certainty as normal inhabitant of the intestinal tract. However, cases of abdominal actinomycosis and the recovery of *A. israelii* from a large proportion of diseased appendices (60.3% of the samples examined, Minsker and Moskovskaya, 1979) indicate that these organisms might at least occur as transient intestinal epiphytes. Using strong selective measures (Fritsche, 1964b), Fritsche (unpublished results) was even able to isolate *A. israelii* in low numbers from stool specimens of healthy individuals.

The occurrence of *Actinomyces* species in healthy animals is less well documented. However, by analogy to the situation in humans, it might be concluded that animal actinomycoses develop endogenously as well and that their causative agents also belong to the normal mucosal microflora of the respective animals (Slack and Gerencser, 1975). Thus,

A. viscosus has been isolated from subgingival plaque of hamsters and from cervical plaque of rats (Howell, 1963; Howell and Jordan, 1963; Jordan and Keyes, 1965; Bellack and Jordan, 1972). Similarly, *A. denticolens* and *A. howellii* have been recovered from the dental plaque of cattle (Dent and Williams, 1984a, b). *A. pyogenes* presumably also occurs as a commensal organism on the mucous surfaces of various warm-blooded animals (Reddy et al., 1982). On the other hand, Dent and Williams (1984b), in their study of the microflora of the dental plaque in cattle, were not able to identify any *Actinomyces*-like isolate from healthy cattle as *A. bovis* so that the normal habitat of this species still awaits definite clarification.

Isolation and Enrichment Procedures

The following general purpose culture media may be used for isolating *Actinomyces* species from clinical specimens: fluid thioglycolate broth, possibly supplemented with 0.1–0.2% (w/v) sterile rabbit serum; brain heart infusion broth (BHIB); trypticase soy broth; brain heart infusion agar (BHIA); trypticase soy agar; heart infusion agar or trypticase soy agar with 5% (w/v) defibrinated rabbit, sheep or horse blood; and Schaedler broth or agar (Slack and Gerencser, 1975; Schaal and Pulverer, 1981).

However, growth is not always satisfactory in or on these media. Isolation, enrichment and subcultivation of most of the species that are pathogenic to humans can be improved by employing complex semisynthetic media such as those of Pine and Watson (1959) or Heinrich and Korth (1967) or by using Tarozzi broth or cooked-meat medium. Details on the preparation of the CC-medium of Heinrich and Korth (1967) which has proved especially useful in the diagnostic laboratory are given in the chapter on *Arachnia* (this section). Best growth of *A. pyogenes* and many strains of *A. odontolyticus* is obtained on media containing defibrinated blood or serum and a fermentable carbohydrate (Slack and Gerencser, 1975; Reddy et al., 1982). Similarly, *A. hordeovulneris* requires addition of 10–20% fetal calf serum to the medium for adequate growth (Buchanan et al., 1984). Media for primary isolation and subcultivation of *A. meyeri* should be supplemented with vitamin K_1, 0.02% Tween 80 and a fermentable carbohydrate (Cato et al., 1984). *Tarozzi broth* is as follows (modification used at the Institute of Hygiene, University of Cologne, F.R.G.) I. *Nutrient broth:* beef extract, 10 g; peptone (e.g. peptone P-Oxoid), 12 g; NaCl, 3 g; K_2HPO_4, 2 g are dissolved in 1 liter of distilled water and heated to about 80°C. Then, the mixture is boiled for 20 min and the pH adjusted to 7.5 by adding NaOH. II. *Final medium:* fresh guinea pig or beef liver is cut into $2 \times 1 \times 1$-cm pieces which are washed in several changes of saline. The pieces are placed into test tubes (one piece per tube) and 8 ml of the above nutrient broth supplement with 0.1% Na thioglycolate (w/v) are added to each tube. The medium is sterilized by autoclaving at 121°C for 15 min. If it is not used immediately after cooling, it is heated at 100°C for 10 min before use. After inoculation, the tubes are sealed with sterile melted vaseline to prevent oxygen uptake.

No detailed studies on the selective isolation of *Actinomyces* species from pus specimens have been reported. However, selective principles have been used successfully for cultivating these actinomycetes from dental plaque or saliva. Beighton and Colman (1976) described a medium for the selective isolation of oral actinomycetes and Kornman and Loesche (1978) devised a selective GMC medium for *A. viscosus* and *A. naeslundii*. Although very useful in principle, these media do not necessarily allow recovery of all of the *Actinomyces* strains in question so that they should not be applied to diagnostic specimens. Beighton and Colman's medium is: I. *Basal culture medium* (BYS medium): BHIB, 3.7 g; yeast extract powder, 0.5 g; polyvinylpyrrolidone, 1.0 g; cysteine-HCl, 0.1 g; agar, 1.5 g are added to 100 ml of distilled water, autoclaved for 15 min at 10 psi, cooled to 45°C, and 5 ml of sterile horse serum are added. II. *Selective enrichment medium:* the enrichment medium (FC medium) is prepared by adding 1 ml of a sterile 25 mg/ml NaF solution and 0.5 ml of a sterile 1 mg/ml colistin sulfate solution to 100 ml of BYS medium. The inhibitor solutions are sterilized by autoclaving at 10 psi for 15 min prior to usage. *GMC medium is:* to enriched gelatin agar (Syed, 1976) as basal medium, metronidazole and cadmium sulfate ($3CdSO_4 \cdot 8H_2O$) are added to give final concentrations of 10 μm/ml and 20 μg/ml, respectively. Cadmium sulfate can be autoclaved together with the basal medium; metronidazole is added to the cooled medium after filter sterilization.

A different means of selective isolation of fermentative actinomycetes, especially of *A. israelii*, has been described by Fritsche (1964b). This method utilizes the inhibitory effect of toluene to Gram-negative bacteria and, although it may alter the colony morphology of the actinomycetes on primary isolation, was found to be very suitable for separating *A. israelii* cultures from contaminating *Enterobacteriaceae*. Fritsche's method is as follows: The specimen is suspended and dispersed in a suitable transport fluid. One milliliter of this suspension is added to 1 ml of toluene in a screw-cap tube and placed on a mechanical shaker (high speed for 20–25 min). The watery suspension at the bottom of the tube is carefully removed with a capillary pipette and added to 10 ml of transport medium. Remaining droplets of toluene on the surface of the medium are removed in a Bunsen flame. Then the medium is centrifuged and the sediment streaked onto nonselective culture media.

Occasionally, commercial disks for testing antibiotic susceptibility can aid in purifying an *Actinomyces* isolate. Depending on the contaminating organisms, disks containing metronidazole, colistin or nalidixic acid in common concentrations may be employed. From their inhibition zones, actinomycete colonies can often be picked easily and transferred to another plate.

In order to avoid loss of viability of the actinomycetes by oxygen contact, clinical specimens as well as samples for ecological studies should be inoculated onto suitable culture media as soon as possible. Alternatively, reduced transport media should be used. For clinical specimens, the Stuart medium or its commercial modifications (e.g. Port-A-Cul, BBL) has proved satisfactory (Loesche et al., 1972; Syed and Loesche, 1972). For ecological studies, the reduced transport fluid (RTF) of Syed and Loesche (1972) may be more appropriate. This medium, which can also be used for sonic or mechanical dispersal of solid materials, is as follows: I. *Stock mineral salt solution No. 1*, containing: K_2HPO_4, 0.6%. II. *Stock mineral salt solution No. 2*, containing: NaCl, 1.2%; $(NH_4)_2SO_4$, 1.2%; KH_2PO_4, 0.6%; $MgSO_4$, 0.25%. III. *Final transport medium*, containing per liter: stock solution No. 1, 75 ml; stock solution No. 2, 75 ml; 0.1 M ethylenediaminetetraacetate solution, 10 ml; 8% Na_2CO_3 solution, 5 ml; 1% dithiothreitol solution (freshly prepared), 20 ml; 0.1% resazurin solution (optional), 1 ml; distilled water, 814 ml. The medium is sterilized by membrane filtration (pore size, 0.22 μm) and dispensed into suitable screw-cap tubes. The pH should be 8 $\pm$ 0.2 without adjustment, and it decreases to 7 in 48 h in the anaerobic glove box atmosphere (85% N_2, 10% H_2, 5% CO_2).

Anaerobic growth conditions may be obtained by using a Torbal anaerobic jar with a gas mixture of N_2-H_2-CO_2 (80:10:10%) or a GasPak jar with a H_2-CO_2-generating envelope. The capnophilic members of the species *A. viscosus*, *A. naeslundii*, *A. odontolyticus*, *A. pyogenes* and *A. denticolens* may be grown in a candle jar or in a GasPak jar with a CO_2-generating envelope. For routine use in the diagnostic laboratory, Fortner's method (Fortner, 1928, 1929) has proved especially useful and reliable, because cultures can be checked for growth without disturbing the gaseous conditions, and the semi-anaerobic atmosphere of this technique allows growth of essentially all of the known *Actinomyces* species including the more aerophilic ones (Schaal and Pulverer, 1981) (for technical details see article on *Arachnia*, this Section).

All *Actinomyces* cultures are incubated at 35–37°C and should be observed at 2, 4, 7 and 14 days for growth of characteristic micro- or macrocolonies, respectively.

Maintenance Procedures

Cultures in routine use can be maintained by weekly transfer in thioglycolate broth containing 0.2% (v/v) rabbit serum or by monthly transfer on Fortner plates containing CC-medium with 0.2% (v/v) rabbit or fetal calf serum.

The procedures for longer preservation (1 year) and long term storage (10 years and longer) are the same as described for the genus *Arachnia* (this Section).

Procedures for Testing for Special Characters

The morphology of *Actinomyces* microcolonies is demonstrated most easily on BHIA or CC-medium plates that have been incubated at 36 $\pm$ 1°C for 18–24 h. The colonies may be observed directly on the medium within the plate or a small section of the agar may be removed and placed on a glass slide in order to facilitate the use of a microscope. Magnifications of $\times$ 100 to $\times$ 400 are usually satisfactory (Slack and Gerencser, 1975). If higher magnification is required for detailed study of young microcolonies (oil immersion), a slide culture technique together with a lactophenol cotton blue staining of the microbes may be more appropriate. The slide cultures are prepared as follows: Sterile glass slides are coated with a thin layer of BHIA by dipping one side of the slide briefly into the melted agar. The slides are transferred immediately to a Petri dish containing a sheet of filter paper moistened with sterile water. Inoculation is performed using a freshly drawn, thin glass filament. After 18–24 (to 48) h incubation, the agar film is air-dried. Then, 3 drops of lactophenol cotton blue mounting fluid are placed on the slide and the slide is carefully covered with a coverslip and observed under oil immersion.

Mature colonies are observed on BHIA or blood agar plates after 7 and 14 days of incubation at 36 $\pm$ 1°C using a hand lens or a dissecting microscope.

Tests for oxygen requirements are performed after Slack and Gerencser (1975): eight brain heart infusion agar slants in cotton-plugged tubes are inoculated with a standardized suspension of the test strain using a capillary pipette. The inoculum is obtained from about 3-day-old cultures, either in broth or on plates. The cells are suspended in 0.85% saline and adjusted to a density matching a MacFarland 3 standard or an OD reading of 0.5. The slants are incubated, in duplicate, under the following conditions:

1. For aerobic conditions, two slants with the original cotton plugs are placed directly in the incubator.
2. For aerobic conditions with added CO_2, the cotton plugs of two tubes are clipped off and the remaining parts are pushed into the tubes to just above the slant. Small pledgets of absorbent cotton are placed on top of the plugs, 5 drops 10% Na_2CO_3 and 5 drops 1 M KH_2PO_4 are added to each tube, and the tubes are immediately closed with rubber stoppers.
3. For anaerobic conditions with added CO_2, tubes are prepared as under 2. To the absorbent cotton, 5 drops 10% Na_2CO_3 and 5 drops pyrogallol solution (100 g pyrogallic acid in 150 ml distilled water) are added.
4. For anaerobic conditions without CO_2, to the properly prepared tubes (see 2) are added 5 drops 10% KOH and 5 drops pyrogallol solution.

Results are recorded after 3 and 7 days of incubation at 36 $\pm$ 1°C. If growth in two corresponding tubes does not seem equal, the test has to be repeated. Alternatively, the agar deep method (Slack and Gerencser, 1975) may be used.

The catalase test is best performed by flooding aerobic growth on slants with 3% H_2O_2 and observing active production of gas bubbles. Anaerobic cultures must be exposed to air for 30 min before testing. The presence of cytochrome oxidase may be demonstrated by the method of Deibel and Evans (1960).

The results of most of the other biochemical and physiological tests listed later in Tables 15.47–15.49 are considerably influenced by the media and methods used (Scharfen, 1973; Slack and Gerencser, 1975; Schofield and Schaal, 1979b). The data reported in the diagnostic and descriptive tables are chiefly based upon the techniques described by Slack and Gerencser (1975), Schofield and Schaal (1979a, b, 1980a, b, 1981), Dent and Williams (1984a, b), Buchanan et al. (1984) and Cato et al. (1984).

For nitrate and nitrite reduction tests as well as for decarboxylation and deamination reactions of amino acids, the basal medium and the procedures described by Schofield and Schaal (1980b) provide satisfactory results. The basal medium contains per liter of distilled water: Bacto peptone (Difco), 5.0 g; yeast extract (Difco), 3.0 g; glucose, 0.5 g; K_2HPO_4, 2.0 g; agar, 0.5 g; pH 6.8. After autoclaving, the respective amino acids, urea or sodium nitrite or nitrate are added from filter-sterilized solutions. The various media are pipetted in 0.15-ml volumes into the wells of U-form microtitre trays. The wells are inoculated from suspensions of the test strains prepared from 4-day cultures on BHIA in the basal medium without agar. After inoculation, the trays are sealed with plastic tape and pin-point holes are made above each well before incubation in a GasPak jar.

In order to test for nitrate and nitrite reduction, sodium salts of these compounds are added to the base to give a final concentration of 0.1% (w/v) of nitrate and 0.005% (w/v) of nitrite, respectively. After 3–4 days of incubation at $36 \pm 1°C$, production or disappearance of nitrite is assessed by adding nitrite reagents A and B (Cowan and Steel, 1975).

Decarboxylation and deamination reactions are performed with the basal medium to which the various amino acids have been added from filter-sterilized solutions to give final concentrations of 0.5% (w/v). Urease activity may be tested in the same way. One set of the triplicated amino acid tests is examined after incubation for 24 h by removing the tape from the tray and adding 1 drop of a 0.2% (w/v) solution of bromocresol purple to each well and recording acid production. After a further 2-day incubation, 1 drop of phenol red solution (0.2% (w/v)) is added to the second set and the results are compared with those after 24 h. Any distinct rise of pH would indicate that in the absence of ammonia as assessed by adding Nessler's reagent, a decarboxylation reaction had taken place. Deamination of the amino acids and urease activity are demonstrated by adding Nessler's reagent to the wells of the third replicate and to that containing urea. Alternatively, the urease test as described by Scharfen (1973) may be used.

For the Voges-Proskauer and methyl red tests the procedures of Cowan and Steel (1975) may be used with peptone yeast extract glucose medium. Hydrogen sulfide production is tested according to Slack and Gerencser (1975). The methods for testing for deoxyribonuclease, hyaluronidase, chondroitin sulfatase, hydrolysis of casein, esculin, gelatin, starch, Tweens, tyrosine, guanine, hypoxanthine, xanthine and adenine, as well as for egg-yolk reaction, indole production and lysozyme resistance, are those described by Schofield and Schaal (1981). Meat digestion is assessed by the methods of Holdeman et al. (1977) and reduction, acidification or clotting of milk by the method of Slack and Gerencser (1975). The various enzyme tests can be performed most easily by using the API-zym enzyme test kit (API-Bio Mérieux) according to the instructions of the manufacturer.

Tests for acid production from carbon compounds are especially liable to variation of the results, depending on the basal medium and the test conditions used (Slack and Gerencser, 1975). In this laboratory, the most consistent and reproducible results were obtained with a miniaturized technique described by Schofield and Schaal (1979b, 1980a). This technique is based upon the commercial Minitek system (BBL), but provides more satisfactory results when the basal medium is modified and the sugar disks are prepared in the laboratory as follows: basal medium: fluid thioglycolate medium without dextrose or indicator (BBL) plus 0.2% (w/v) yeast extract. Preparation of sugar disks: sterile blank paper disks (BBL) are impregnated with 3% filter-

sterilized sugar solutions and freeze-dried. Test procedure: the test organism is suspended in fluid thioglycolate medium + yeast extract. Two to three drops of this suspension are added to each well of a Minitek tray (BBL) containing one of the sugar disks. After 3–4 days of incubation in a GasPak jar at $36 \pm 1°C$, the plates are examined by adding 1 drop of 0.2% (w/v) bromothymol blue solution (Cowan and Steel, 1975) to each well. The appearance of a yellow color indicates a positive result. Alternatively, the media and methods described by Slack and Gerencser (1975) can be used.

Tests for utilization of carbon compounds as sole sources of carbon and energy may be performed using the medium and techniques of Schofield and Schaal (1979a). The procedures for testing for susceptibility to chemical or antibiotic inhibitors have been outlined by Schofield and Schaal (1981).

The analysis of end products from glucose fermentation is appropriately performed according to the procedures described in the VPI Anaerobe Manual (Holdeman et al., 1977). Demonstration and isomer determination of diaminopimelic acid in whole-cell hydrolysates which may be useful for separating Actinomyces species from Arachnia are easily and reliably achieved by a modification of the technique of Becker et al. (1964) using cellulose-coated thin-layer plates (Kutzner, 1981).

Rapid and reliable identification of Actinomyces species may also be obtained by immunofluorescence tests. These can be applied to culture material as well as to clinical specimens including pus, dental plaque or calculus or mucosal secretions. Both direct and indirect modifications of the fluorescent antibody technique have been applied successfully to the actinomycetes (Slack et al., 1961, 1971; Schaal and Pulverer, 1973; Slack and Gerencser, 1975). The direct fluorescent antibody procedure of Slack and Gerencser (1975) is most widely used, but its application is limited because conjugated, specific antisera are only available for a few species and serovars (Biological Reagent Section, Centers for Disease Control, Atlanta, Ga., U.S.A.). The procedure is as follows: Preparation of smears. I. Clinical material: two smears of the material are made on a clean glass slide containing two marked circles. After air-drying, they are fixed by flooding with methanol for 1 min and air-dried again. II. Cultures: Smears from a suspension of the organisms obtained by centrifugation of a broth culture or from a plate culture are made on glass slides, air-dried and gently heat-fixed. Staining procedure: (a) one drop of conjugated specific antiserum is placed on each smear and the slide incubated in a moist chamber for 30 min at room temperature; (b) excess conjugate is poured off and the slide washed in two changes of pH 7.2 buffer (FTA hemagglutination buffer, BBL) for 5 min each; (c) the preparation is counterstained in 0.5% Evans blue for 5 min; (d) excess Evans blue is removed by dipping the slide briefly into distilled water, and the slide is washed in two changes of pH 9.0 buffer for 1 min each; (e) smears are allowed to air-dry. (f) one drop of buffered glycerol mounting fluid (9 parts c.p. glycerol, 1 part pH 9.0 buffer) is placed on each smear which is subsequently covered with a cover slip. For examination with a microscope equipped for immunofluorescence work, × 54 oil-immersion or × 50 water-immersion objectives are most useful.

Alternatively, the indirect test described by Schaal and Pulverer (1973) may be used which avoids the conjugation of many antisera and which has been shown to possess at least an equal specificity and a high sensitivity (Schaal and Gatzer, 1985). However, specific unconjugated antisera are nearly as hard to obtain as conjugates since the former are also not available commercially.

Differentiation of the genus **Actinomyces** from other genera

Actinomyces species resemble a variety of other Gram-positive, filamentous or diphtheroidal genera morphologically and/or physiologically. This is especially true for Actinomyces israelii and Arachnia propionica which share many diagnostic characters. However, these

organisms can be identified reliably and differentiated from each other by a set of physiological and chemotaxonomic tests that have been summarized in Table 15.25 of the Arachnia chapter (this Section). In addition, immunofluorescence may also be used to separate members of the genus *Actinomyces* from related organisms because cross-reactions are usually absent or weak.

Taxonomic Comments

The genus *Actinomyces* was created in 1877, when Harz described the causative agent of bovine actinomycosis which he named *Actinomyces bovis*. In 1919, Breed and Conn recommended that *Actinomyces* be accepted as a *genus conservandum* with *A. bovis* as the type species and this was accepted by the Winslow Committee in 1920 (Winslow et al., 1920) thereby confirming that *Actinomyce* Meyen 1827, a name proposed for a fungus (*Tremella meteorica*), was not valid.

Originally, the description of the genus was solely based upon the morphology of actinomycete elements in the granules produced by *A. bovis* in bovine infections. Many years later, cultural, physiological and chemical characters slowly began to influence the definition of the taxon. In 1965, Pine and Georg used cellular and colonial morphology, cell wall composition, fermentation end products and certain physiological characteristics to define the family *Actinomycetaceae* and members of the genus *Actinomyces* in light of more modern approaches to taxonomy. Georg et al. (1969) modified the genus description of *Actinomyces* to include catalase-positive organisms permitting the inclusion of *A. viscosus*. At the same time, *Actinomyces propionicus* was removed from the genus and reclassified as *Arachnia propionica* because it produced propionic acid as a major fermentation end product and had diaminopimelic acid in its cell wall (Pine and Georg, 1965).

A. bovis, although validly described by Harz in 1877, was not cultured until 1890 by Mosselman and Lienaux. The principal human pathogen, *A. israelii*, was first isolated by Bujwid in 1889 and described in more detail by Wolff and Israel in 1891. For many years, it was essentially impossible to differentiate between *A. bovis* and *A. israelii* and both names were used interchangeably. However, the extensive studies of Erikson (1940) and the reports of Thompson (1950) and Pine et al. (1960) finally showed that the causative agents of human and bovine actinomycosis were two taxonomically distinct species.

In 1925, Naeslund described a filamentous bacterium occurring in the human mouth which differed from *A. israelii* and which was described as the separate species *A. naeslundii* by Thompson and Lovestedt in 1951. *A. viscosus* was first isolated from periodontal plaque in hamsters by Howell in 1963 and primarily described under the new genus designation *Odontomyces viscosus* (Howell et al., 1965). After the genus description of *Actinomyces* had been changed to include both catalase-positive and catalase-negative bacteria, *Odontomyces viscosus* was reclassified as *Actinomyces viscosus* (Georg et al., 1969). *A. odontolyticus* was first isolated from human advanced caries and described as a separate *Actinomyces* species by Batty in 1958.

Recently, five additional species were added to the list of species of the genus *Actinomyces*. In 1982, Reddy et al. and Collins and Jones reclassified *Corynebacterium pyogenes* (Glage) Eberson 1918 as *Actinomyces pyogenes*. This reclassification was based on physiological, metabolic, nutritional and biochemical characteristics and was supported by numerical phenetic and chemical data (Schofield and Schaal, 1981; Collins et al., 1982; Collins and Jones, 1982; Reddy et al., 1982). In 1984, Dent and Williams (1984a, b) described two groups of Gram-positive bacteria isolated from the dental plaque of cattle which showed some similarity to *A. naeslundii*, but could be differentiated from this species on the basis of cell wall and DNA base composition, polypeptide molecular weight distribution and a few physiological reactions. Therefore, the new species *A. denticolens* and *A. howellii* were proposed for these isolates. Also in 1984, Buchanan and co-workers performed a taxonomic study on 30 *Actinomyces* strains isolated from infections in dogs. Fifteen of these canine isolates were similar to *A. viscosus*, whereas the other 15 strains had chemotaxonomic, biochemical and serological characteristics unlike those of any previously accepted *Actinomyces* species. Thus, the new taxon *A. hordeovulneris* was proposed for these organisms. The fifth additional species recently included in the genus *Actinomyces* is *A. meyeri* (Cato et al., 1984).

Organisms conforming to the description of *A. meyeri* were first described as a "new anaerobic *Cohnistreptothrix*" species by Kurt Meyer in 1911 and were included in the genus *Actinobacterium* by Prévot in 1938. However, the type species of this genus, *Actinobacterium israelii*, was transferred to the genus *Actinomyces* Harz 1877 by Breed and Conn in 1919. This effectively invalidated the name *Actinobacterium* according to Rule 37a of the *International Code of Nomenclature of Bacteria* (Lapage et al., 1975). Thus, the genus name *Actinobacterium* was not listed in the Approved Lists of Bacterial Names (Skerman et al., 1980) and has no standing in nomenclature. On the other hand, *meyeri* organisms were recovered as separate clusters in numerical phenetic analyses (Holmberg and Nord, 1975; Schofield and Schaal, 1981) and showed considerable similarity to the accepted *Actinomyces* species on the basis of both numerical phenetic data and DNA base composition (Bouisset et al., 1968). Moreover, organisms fitting the description of *A. meyeri* have been isolated not infrequently from human pathological lesions and from healthy as well as diseased subgingival pockets (Lipton and Sonnenfeld, 1980; Rose et al., 1982; Cato et al., 1984, Schaal, unpublished results). Thus, these microbes are not uncommon and may be significant pathogens so that there is a practical need for improvements in their classification and identification. The data available so far suggest inclusion of the *meyeri* organisms in the genus *Actinomyces* as was recently proposed by Cato and co-workers (1984). However, future studies will show as to whether or not this reclassification is also of phylogenetic relevance.

Several additional *Actinomyces* species were proposed in the past. Most of these have been declared *nomina dubia* (Slack, 1974) and are not listed in the Approved Lists of Bacterial Names (Skerman et al., 1980), or they have been reclassified in other genera or reduced to objective synonyms such as *Actinomyces eriksonii* Georg, Robertstad, Brinkman and Hicklin 1965 which is phenetically more closely related to the bifidobacteria (Holmberg and Nord, 1975) and which even appears to constitute one biological variant or DNA-DNA homology group (*B. dentium*) of *Bifidobacterium adolescentis* (Scardovi et al., 1971; Mitsuoka et al., 1974).

The species *Actinomyces humiferus* was validly published by Gledhill and Casida in 1969 and has been listed in the Approved Lists of Bacterial Names (Skerman et al., 1980). However, *A. humiferus* differs from typical members of the genus *Actinomyces* in that it grows at 30°C, has a high G + C content of the DNA (73 mol%), is sensitive to lysozyme and occurs as a numerically predominant inhabitant of organically rich soils (Gledhill and Casida, 1969). Thus, its affiliation to the genus *Actinomyces* is doubtful and it will be listed as *species incertae sedis*.

A complicated taxonomic and nomenclatural problem arises with organisms named *Actinomyces suis*. This species designation was first used by Gasperini (1982), but *A. suis* Gasperini has been declared a *nomen dubium* (Slack, 1974) as it had not been published validly. Nevertheless, Grässer (1957) isolated bacteria from mastitis of swine which were again named *A. suis*. In this case, the name was validly published, but the description was inadequate and cultures are not available. Hence, the species *A. suis* was not included in the Approved Lists of Bacterial Names (Skerman et al., 1980). More recently, Franke (1973) once more described actinomycetes isolated from the udder of swine which he also called *A. suis*. The morphological and biochemical characteristics of these isolates appeared to be somewhat different from those of Grässer's strains, but they seemed to fit well within the genus *Actinomyces* (Slack and Gerencser, 1975). At present, it is impossible to decide whether *A. suis* Grässer and *A. suis* Franke are identical. On the other hand, it is obvious that certain cases of actinomycosis of the mammary gland of swine are caused by actinomycetes which cannot be included in any of the accepted *Actinomyces* species. Therefore, "*Actinomyces suis*" Franke will be listed as *species incertae sedis* at the end of this chapter because Franke's description appears to be most adequate and because there is a need for reliable recognition of typical swine isolates.

The phylogenetic relationships of the genus *Actinomyces* to other Gram-positive, filamentous or diphtheroidal genera are uncertain. Preliminary data derived from the application of the rRNA cataloging method (Stackebrandt and Woese, 1981) to *A. bovis* and *A. viscosus* indicate that these species form an independent phylogenetic group which is peripherally related to a subunit comprising *Cellulomonas*, *Micrococcus* and *Arthrobacter* species together with the plant pathogenic coryneforms and certain other genera, but very distant relationships were found to the propionibacteria and bifidobacteria. These results show clearly that there appears to be no phylogenetic basis for adhering to the family *Actinomycetaceae* as defined in the eighth edition of the *Manual* (Slack, 1974). However, additional information is required to definitely clarify the position of the genus *Actinomyces* in the phylogenetic tree.

In recent numerical taxonomic analyses (Melville, 1965; Holmberg and Hallander, 1973; Holmberg and Nord, 1975; Fillery et al., 1978; Schaal and Schofield, 1981a, b; Schofield and Schaal, 1981), most of the classical *Actinomyces* species as well as *A. pyogenes* and *A. meyeri* were recovered as well defined and clearly separated phena. However, several new and interesting observations were also made. The data of Schofield and Schaal (1981) suggest that the genus *Actinomyces* may be subdivided into three major subgroups which could be considered subgenera or even new independent genera (Schaal and Schofield, 1984; Schaal and Gatzer, 1985). The first subgroup solely contained strains labeled *A. israelii*. But these were further subdivided into four "subspecies." Two of them corresponded to the two serovars recognized so far in the species *A. israelii*. The two additional "subspecies" are serologically distinct from each other and from the classical serovars (Schaal and Gatzer, 1985). Furthermore, all of the "subspecies" within the phenon *A. israelii* may be separated from one another on physiological and biochemical grounds, although this appears to be more difficult than the differentiation between the recognized species.

The second subgroup within the genus *Actinomyces* as defined by Schofield and Schaal (1981) was composed of *A. naeslundii* and *A. viscosus* strains and two additional, hitherto unnamed phena. In contrast to previous numerical taxonomic studies (Holmberg and Nord, 1975; Fillery et al., 1978) the results of Schofield and Schaal (1981) confirm that *A. naeslundii* and *A. viscosus* constitute independent, but closely related species. In addition, the phenon *A. viscosus* could be subdivided into four subunits which corresponded to serovar 1, serovar 2 and so-called "atypical" (Coykendall and Munzenmaier, 1979) *A. viscosus* strains and to two serovar 3 isolates of *A. naeslundii*, respectively. It is interesting to note that the prototype strain of *A. naeslundii*, serovar 3 (WVU 820; WVU, West Virginia University, Morgantown, West Virginia, U.S.A.), grouped with the "atypical" *A. viscosus* and not with the other two *A. naeslundii*, serovar 3 strains. These data suggest that the *A. naeslundii/A. viscosus* group of actinomycetes is taxonomically more heterogeneous than previously thought. They also indicate that the serovars of *A. viscosus* can be distinguished phenotypically so that these serovars may be considered biovars as well, and that serovar 3 of *A. naeslundii* is apparently heterogeneous.

Using DNA hybridization techniques, Coykendall and Munzenmaier (1979) showed that *A. naeslundii* and *A. viscosus* strains from hamsters were clearly separable. With the use of the more sensitive S_1 nuclease hybridization method, typical human *A. viscosus* strains could be distinguished from typical *A. naeslundii* strains although both groups of organisms appeared to be similar. Furthermore, a group of atypical *A. naeslundii* isolates could be delineated. Thus, there seems to be considerable agreement between the results of DNA homology studies and numerical data so that the phenetic differences found apparently reflect taxonomic differences. The two additional phena recovered in the second subgroup of *Actinomyces* (Schofield and Schaal, 1981) require further investigation, but might represent new species.

The third *Actinomyces* subgroup described by Schaal and Schofield (1984) contained representatives of *A. bovis*, *A. pyogenes*, *A. odontolyticus* and a single strain of *A. meyeri*. These organisms formed distinct phena corresponding to the respective taxa, but they showed a lower phenetic similarity to the other *Actinomyces* species and some relationship to certain bifidobacteria, eubacteria, *Erysipelothrix rhusiopathiae* and *Arcanobacterium haemolyticum*. Thus, the taxonomic structure of this subgroup appears to be less well-defined and requires further investigation.

From the data available so far it seems probable that further revisions of the genus *Actinomyces* will become inevitable. These will possibly include the description of new species and a reconsideration of the genus definition which might finally lead to the division of the genus *Actinomyces* into two or more genera.

Further Reading

Bowden, G. and J. Hardie. 1978. Oral pleomorphic (coryneform) Gram-positive rods. *In* Bousfield and Callely (Editors), Coryneform Bacteria, Academic Press, London, pp. 235–263.

Schaal, K.P. 1984. Laboratory diagnosis of actinomycete diseases. *In* Goodfellow, Mordarski and Williams (Editors), Biology of the Actinomycetes, Academic Press, London, pp. 425–456.

Schaal, K.P. and B.L. Beaman. 1984. Clinical significance of actinomycetes. *In* Goodfellow, Mordarski and Williams (Editors), Biology of the Actinomycetes, Academic Press, London, pp. 389–424.

Schaal, K.P. and G. Pulverer. 1981. The genera *Actinomyces*, *Agromyces*, *Arachnia*, *Bacterionema*, and *Rothia*. *In* Starr, Stolp, Trüper, Balows and Schlegel (Editors), The Prokaryotes. A Handbook on Habitats, Isolation, and Identification of Bacteria. Springer-Verlag, Berlin, pp. 1923–1950.

Schaal, K.P. and G.M. Schofield. 1981. Current ideas on the taxonomic status of the *Actinomycetaceae*. *In* Schaal and Pulverer (Editors), Actinomycetes. Proceedings of the Fourth International Symposium on Actinomycete Biology, Cologne, September 3–7, 1979. Zentralbl. Bakteriol. Mikrobiol. Hyg., Suppl. 11, Gustav Fischer Verlag, Stuttgart, pp. 67–78.

Schaal, K.P., G.M. Schofield and G. Pulverer. 1980. Taxonomy and clinical significance of *Actinomycetaceae* and *Propionibacteriaceae*. Infection 8, *Suppl. 2:* 122–130.

Schofield, G.M. and K.P. Schaal. 1981. A numerical taxonomic study of members of the *Actinomycetaceae* and related taxa. J. Gen. Microbiol. *127:* 237–259.

Slack, J.M. and M.A. Gerencser. 1975. Actinomyces. Filamentous Bacteria. Biology and Pathogenicity. Burgess Publishing Company, Minneapolis.

Identification and descriptive characteristics of **Actinomyces** *species*

The characteristics necessary for reliable identification of the recognized *Actinomyces* species and three additional taxa that might be related to the genus *Actinomyces* are given in Table 15.47. Table 15.48 summarizes some of the differential properties which may be used to separate the serovars (biovars) of *A. israelii*, *A. naeslundii* and *A. viscosus* on physiological and biochemical grounds. Most of the important additional descriptive details are listed in Table 15.49.

Table 15.47.

Differential characteristics of the species of the genus **Actinomyces** *and related bacteria[a,b]*

Characteristics	1. A. bovis	2. A. israelii	3. A. naeslundii	4. A. odontolyticus	5. A. viscosus	6. A. pyogenes	7. A. denticolens	8. A. howellii	9. A. hordeovulneris	10. A. meyeri	11. "A. suis"	12. "A. humiferus"
Cells coccobacillary	d	−	−	d	−	+	−	−	−	d	d	d
Microcolonies filamentous	d	+	d	d	d	−	d		+	−	d	+
Aerobic growth (without CO_2)[c]	−	(d)[d]	d	d	+	+	d			−	+	+
Catalase	−	−	−	−	+	−	−	+	(+)	−	−	−
Nitrate reduction[e]	−	d	+	+	d	−	+		−	−	+	−
Nitrite reduction[e]	−	−	d	−	−	−	−			−		
Hydrolysis of[f]												
Casein	−	−	−	−	−	+				−		d
Esculin	d	+	+	d	d	−	+		+	−		+
Gelatin	−	−	−	−	−	+				−	−	d
Starch	+	d	d	d	d	d				−		+
DNAase	+	−	−	d	−	+				−		
Ammonia from[e]												
Arginine	−	+	d	−	d	−				−		−
Urea	−	−	+	−	d	−			−	d		−
Indole production	−	−	−	−	−	−				−	−	−
Valine arylamidase[g]	−	−	−	−	−	−	−			+		
Phosphoamidase	−	d	d	d	−	−	+			−		
β-Galactosidase[g]	−	+	d	−	d	d	+			−		
α-Glucosidase[g]	−	+	d	−	d	−	+			+		
N-Acetyl-β-glucosaminidase[g]	+	−	−	−	−	+	−			−		
Acid from[h]												
Amygdalin	−	+	d	−	−	−	−	−		d		
Arabinose	−	d	−	d	−	d	−	d	−	d	−	+
Cellobiose	−	+	d	−	−	d	−	−	+	−		d
Glucose	+	+	+	+	+	+	+	+	+	+	+	+
meso-Inositol	d	+	+	−	d	d	d	−	−	−	+	−
Mannitol	−	d	−	−	−	d	−		+	−	d	+
Raffinose	−	+	+	−	+	−	+	+	(+)[i]	−	+	d
Rhamnose	−	d	−	d	−	−	−	−	−	−	−	+
Trehalose	−	d	+	d	d	d	−	d	+	−	+	d
Xylose	−	+	d	d	−	d	−	d	+	+	−	d
β-Hemolysis of sheep blood	d	−	−	d	−	+				−		

[a] Data compiled from: Franke, 1973; Slack and Gerencser, 1975; Holmberg and Nord, 1975; Holdeman et al., 1977; Schofield and Schaal, 1981; Schaal, 1984; Reddy et al., 1982; Dent and Williams, 1984a, b; Cato et al., 1984; Buchanan et al., 1984.

[b] Symbols: see Table 15.25.

[c] On the surface of agar media.

[d] If positive, the reaction is usually weakly positive.

[e] Methods of Schofield and Schaal, 1980b.

[f] Methods of Schofield and Schaal, 1981.

[g] API-zym enzyme test kit (API-BioMérieux).

[h] Methods of Schofield and Schaal, 1979b, 1980a.

[i] Weakly positive.

Table 15.48.

Characteristics useful for differentiating the serovars (biovars) of certain **Actinomyces** *species[a,b]*

Characteristics	A. israelii serovars		A. naeslundii serovars			Actinomycetes intermediate between A. naeslundii and A. viscosus	A. viscosus serovars		Atypical
	1	2	1	2	3		1	2	
Catalase	−	−	−	−	−		+	+	+
Nitrate reduction[c]	+	d	+	+	+	+	d	d	+
Nitrite reduction[c]	−	−	+	+	−	+	−	−	d
Esculin hydrolysis	+	+	+	+	+	+	−	−	d
Acid from[d]									
Arabinose	+	−	−	−	−	−	−	−	d
Cellobiose	+	+	d	d	−	d	−	−	d
Glycerol	−	−	−	−	d	−	−	d	+
Ribose	+	+	d	d	+	d	−	+	+
Sorbitol	d	−	−	−	−	−	−	−	−
Ammonia from[c]									
Arginine	+	+	d	d	−	+	−	d	−
Urea	−	−	+	+	−	d	d	d	−

[a] Data compiled from: Brock and Georg, 1969b; Schofield and Schaal, 1980a, b, 1981.

[b] Symbols: see Table 15.25.

[c] Methods of Schofield and Schaal, 1980b.

[d] Methods of Schofield and Schaal, 1979b, 1980a.

Table 15.49.

Other characteristics of **Actinomyces** *species and related bacteria[a,b]*

Characteristics	1. A. bovis	2. A. israelii	3. A. naeslundii	4. A. odontolyticus	5. A. viscosus	6. A. pyogenes	7. A. denticolens	8. A. howellii	9. A. hordeovulneris	10. A. meyeri	11. "A. suis"	12. "A. humiferus"
Cell morphology[c]												
Coccoid elements	−	−	−	−	−	d	−	−	−	−	−	d
Coccobacillary elements	d	−	−	d	−	+	−	−	−	d	d	d
Irregular rods	+	+	+	+	+	d	+	+	+	d	+	+
Rods with swollen ends	d	d	d	d	d	d	d		+	d	d	d
Branched rods	d	+	+	d	d	−	d		+	−	d	d
Round-ended ends	+	+	+	d	+	+	+			d	+	+
Square-ended ends	−	−	−	−	−	−	−			d	−	−
Large spherical bodies	−	−	−	−	−	−	−		d	d	−	−
Short filaments	d	+	+	d	d	d	d	d	d	d	+	+
Long filaments	d	d	d	d	d	−	−	−	d	−	d	d
Gram-reaction positive	+	+	+	+	+	d	+	+	+	d	+	+
Gram-reaction variable	−	−	−	−	−	d	−	−	−	d	−	−
Motility	−	−	−	−	−	−	−	−	−	−	−	−
Colony morphology[d]												
Microcolonies filamentous	d	+	d	d	d	−	d		+	−	d	+
Macrocolony size												
≤2 mm	+	+	d	+	d	d	+	+	d	+	+	+
>2 mm	−	−	d	−	d	d	−	−	d	−	−	−
Flat	−	−	d	d	d	−	−	−	−	d	−	−
Raised	+	+	d	d	d	+	+	+	+	d	+	+
Umbonate	d	d	d	d	d	−	−	−	d	−	d	−
Central depression	d	d	−	−	−	−	−	−	−	−	d	−
Entire margin	d	−	d	d	d	+	+	+	−	+	d	d
Undulate margin	d	d	d	d	d	−	−	−	d	−	d	d
Filamentous margin	−	d	d	−	d	−	−	−	d	−	d	d
Transparent	−	−	d	d	−	−	−	d	−	d	−	−
Opaque	+	+	+	d	d	+	+	d	+	d	+	+
White/gray-white	d	d	d	d	d	+	d	+	+	+	+	+
Creamy white	d	d	d	−	d	−	−	−	−	−	−	−

Table 15.49—continued

Characteristics	1. A. bovis	2. A. israelii	3. A. naeslundii	4. A. odontolyticus	5. A. viscosus	6. A. pyogenes	7. A. denticolens	8. A. howellii	9. A. hordeovulneris	10. A. meyeri	11. "A. suis"	12. "A. humiferus"
Orange	−	−	−	−	−	−	−	−	−	−	−	−
Pink	−	−	−	−	−	−	d	−	−	−	−	−
Red	−	−	−	d	−	−	−	−	−	−	−	−
Rough	d	+	d	d	d	−	−	−	+	−	d	d
Smooth	d	v	d	d	d	+	+	+	−	+	d	d
Dry to crumbly	d	d	d	−	d	−	−	−	d	−	d	−
Soft	d	d	d	+	d	+	+	+	d	+	d	+
Mucoid	−	−	−	−	d	−	−	−	−	−	−	−
Aerobic growth in surface culture												
Without CO_2	−	(d)[e]	d	d	+	+	d			−	(+)[f]	+
With CO_2	d	d	+	+	+	+	+	+	+	d	+	+
Anaerobic growth												
Without CO_2	−	d	−	−	−	−	(d)			+	+	(d)
With CO_2	+	+	+	+	+	+	+	+	+	+	+	(d)
Growth at												
30°C		+	+		+	+				+		+
36 ± 1°C	+	+	+	+	+	+	+	+	+	+	+	d
45°C		−	d	−	d	−				−		−
Growth on Rogosa's agar	−	−	−	−	−							
Catalase	−	−	−	−	+	−	−	+	(+)	−	−	−
Cytochrome oxidase	−	−	d	−	d	d			−	−		−
Nitrate reduction[g]	−	d	+	+	d	−	+		−		+	−
Nitrite reduction[g]	−	−	d	−	−	−	−			−		
Voges-Proskauer test	−	−	−	−	−				−	−	−	d
Methyl red test	+	d	+	d	+					+	+	+
H₂S-production												
On BHI[h]-agar	d	−	d	d	−	−				−	−	d
On TSI[i]-agar	+	+	+	+	+	−				−		
Hydrolysis of[j]												
Adenine	−	−	−	−	−	−				−		
Casein	−	−	−	−	−	+				−		d
Esculin	d	+	+	d	d	−	+		+	−		+
Gelatin	−	−	−	−	−	+				−	−	d
Guanine	−	−	−	−	−	−				−		
Hippurate	−	−	−	−	−					−		
Hypoxanthine	−	−	−	−	−					−		
Lecithin	−	d	−	−	−	−				−		
Starch	+	d	d	d	d	d				−		+
Tween 20	−	−	−	−	−					−		
Tween 40	−	d	d	−	d	d				−		
Tween 60	−	d	d	−	d	−				−		
Tween 80	−	−	−	d	−	−				−		
Tyrosine	−	−	−	−	−	−				−		−
Xanthine	−	−	−	−	−	−				−		−
Meat digestion	−	−	−	−	−					−		
Serum liquefaction		−	−	−	−							
Milk												
Acid, reduction	d	d	d	d	d							d
Acid, clot	d	d	d	d	d	+				d		−
Peptonization	−	−	−	−	−	+				−		−
No change	d	−	d	d	−	−				d		d
Lipase (egg yolk)	−	−	−	−	−					−		
DNAase	+	−	−	d	−	+				−		
Hyaluronidase	−	−	−	d	−	−				−		
Chondroitin sulfatase	−	−	−	−	−					−		
Deamination of[g]												
Alanine	−	−	−	−	−					−		
Arginine	−	+	d	−	d	−				−		−
Aspartic acid	−	−	−	−	−					−		

Table 15.49—*continued*

Characteristics	1. A. bovis	2. A. israelii	3. A. naeslundii	4. A. odontolyticus	5. A. viscosus	6. A. pyogenes	7. A. denticolens	8. A. howellii	9. A. hordeovulneris	10. A. meyeri	11. "A. suis"	12. "A. humiferus"
Glutamic acid	−	−	−	−	−	−				−		
Leucine	−	−	−	−	−	−				−		
Lysine	−	−	−	−	−	−				−		
Methionine	−	−	−	−	−	−				−		
Ornithine	−	−	−	−	−	−				−		
Serine	−	−	−	−	−	−				−		
Urease	−	−	+	−	d	−			−	d		−
Indole production	−	−	−	−	−	−				−	−	
Decarboxylation of [g]												
Aspartic acid	−	−	−	−	−	−				−		
Glutamic acid	−	−	−	−	−	−				−		
Leucine	−	−	−	−	−	−				−		
Lysine	−	−	−	d	−	−				−		
Methionine	−	−	−	−	−	−				−		
Ornithine	−	−	−	−	−	−				−		
API enzyme tests [k]												
Alkaline phosphatase	−	−	−	−	−	−	−			−		
Esterase (C$_4$)	−	d	−	−	d	−	−			−		
Esterase-lipase (C$_8$)	−	−	−	−	−	−	−			−		
Lipase (C$_{14}$)	−	−	−	−	−	−	−			−		
Leucine arylamidase	+	d	+	d	+	d	+			+		
Valine arylamidase	−	−	−	−	−	−	−			+		
Cystine arylamidase	−	−	−	−	−	−	−			+		
Trypsin	−	−	−	−	−	−	−			−		
Chymotrypsin	−	−	−	−	−	−	−			−		
Acid phosphatase	−	d	d	−	d	−	d			−		
Phosphoamidase	−	d	d	d	−	−	+			−		
α-Galactosidase	−	d	d	−	−	−	d			−		
β-Galactosidase	−	+	d	−	d	d	+			−		
β-Glucuronidase	−	−	−	−	−	d	−					
α-Glucosidase	−	+	d	−	d	−	+			+		
β-Glucosidase	−	+	d	d	−	−	d					
N-Acetyl-β-glucosaminidase	+	−	−	−	−	+	−					
α-Mannosidase	−	−	−	−	−	−	−					
α-Fucosidase	−	−	−	d	−	−	−					
β-Xylosidase	−	+	d	−	d							
Lysis by lysozyme + SDS (24 h) [l]	+	d	+	+	d	+				+		+
Alkali produced in peptone containing media	−	−	−	−	−	−				−		−
Acid from [m]												
Adonitol	−	−	−	d	−	d				−	d	−
Amygdalin	−	+	d	−	−	−	−	−		d		
Arabinose	−	d	−	d	−	d	−	d	−	d	−	+
Cellobiose	−	+	d	−	−	d	−	−	+	−		d
Dextrin	d	+	d	d	d	d				+		+
Dulcitol	−	−	−	−	−					−	−	−
iso-Erythritol	−	−	−	−	−	d				−		
meso-Erythritol	−	−	−	−	−	d				−		
Fructose	+	+	+	+	+	+				+	+	+
Galactose	+	+	+	d	d	+				d	+	+
Glucose	+	+	+	+	+	+	+	+	+	+	+	+
Glycerol	−	−	d	d	d	−	−	−	−	d	d	d
Glycogen	+	−	d	d	d	+				d		
meso-Inositol	d	+	+	−	d	d	d	−	−	−	+	−
Inulin	−	d	d	−	d	−	d	−		−	+	−
Lactose	+	+	d	d	d	+	+	d	+	d	+	d
Maltose	d	+	+	d	+	d	+	+	+	+	+	+
Mannitol	−	d	−	−	−	−	d	−	−	−	d	+
Mannose	d	+	+	−	d	d	d	d	(+)	−	+	+

Table 15.49—*continued*

Characteristics	1. A. bovis	2. A. israelii	3. A. naeslundii	4. A. odontolyticus	5. A. viscosus	6. A. pyogenes	7. A. denticolens	8. A. howellii	9. A. hordeovulneris	10. A. meyeri	11. "A. suis"	12. "A. humiferus"
Melezitose	−	d	d	−	−	d	−			−		+
Melibiose	d	+	+	−	d	−		d	(+)	−		+
Raffinose	−	+	+	−	+	−	+	+	(+)	−	+	d
Rhamnose	−	d	−	d	−	−	−	−	−	−	−	+
Ribose	−	+	d	d	d	+	d	−	−	+	d	d
Salicin	−	+	d	d	d	−	+	−			+	d
Sorbitol	−	d	−	−	−	d	−			−	−	−
Sorbose	−	−	−	−	−					−		
Starch	+	d	d	d	d	+				+	+	+
Sucrose	+	+	+	d	+	−	+	+		+	+	+
Trehalose	−	d	+	d	d	d	−	d	+	−	+	d
Xylose	−	+	d	d	−	d	−	d	+	+	−	d
α-Methyl-D-glucoside	−	d	−	−	−	−				−		
α-Methyl-D-mannoside	−	d	d	−	−	−				−		
Growth on (sole carbon source)[n]												
Acetate					(+)							−
Alanine		−	−	−	−							
Citrate					−							−
Fumarate												+
Gluconate					(d)					−		d
Glucose		+	+	+	+							
Glycerol		−	−	d	−							
α-Ketoglutarate												+
Lactate		−	−	−	d					−		−
Oxalate												−
Propionate												−
Pyruvate	−	+	+	−	+					+		+
Sorbitol		d	−	−	−							
Succinate		−	−	−	(d)							−
Threonine										−		
Xylose		+	−	d	−							
β-Hemolysis[o]												
Human blood	+	−	−	d	−	+				−		
Sheep blood	d	−	−	d	−	+				−		
Horse blood	d	−	−	d	−	+			−	−		
Growth in the presence of[p]												
NaCl												
2% (w/v)	d	d	d	−	d	d				+		
4% (w/v)	−	−	d	−	d	−				+		d
6% (w/v)	−	−	−	−	−	−				−		
Bile												
5% (w/v)	d	d	+	−	+	d				+		−
10% (w/v)	d	d	d	−	d	d				−		−
20% (w/v)	−	−	d	−	d	d				−		−
Sodium taurocholate 0.2% (w/v)	d	d	d	−	d	d				−		
Sodium selenite 0.01% (w/v)	d	−	d	−	d	−						
Potassium tellurite 0.01% (w/v)	+	d	d	d	d	d						
Sodium azide 0.005% (w/v)	d	+	d	+	d	+				+		
Crystal violet 0.005% (w/v)	d	−	d	−	d	−				−		

[a] Data compiled from: Prévot, 1938; Thompson and Lovestedt, 1951; Batty, 1958; Howell et al., 1959; Howell, 1963; Howell and Jordan, 1963; Roberts, 1968; Gledhill and Casida, 1969; Franke, 1973; Holmberg and Hallander, 1973; Slack and Gerencser, 1975; Holmberg and Nord, 1975; Holdeman et al., 1977; Kilian, 1978; Fillery et al., 1978; Schofield and Schaal, 1979b, 1980a, b, 1981; Reddy *et al.*, 1982; Dent and Williams, 1984a, b; Cato et al., 1984; Buchanan et al., 1984.

[b] Symbols: see Table 15.25.
[c] When grown in thioglycolate broth.
[d] On brain heart infusion agar.
[e] If positive, the reaction is usually weak.
[f] Weakly positive.
[g] Methods of Schofield and Schaal, 1980b.
[h] Brain heart infusion.

[i] Triple sugar iron.
[j] Methods of Schofield and Schaal, 1981.
[k] Only strongly positive reactions recorded as positive.
[l] Methods of Schofield and Schaal, 1981.
[m] Methods of Schofield and Schaal, 1979b, 1980a.
[n] Methods of Schofield and Schaal, 1979a.
[o] On blood-containing BHIA.
[p] On BHIA.

List of species of the genus **Actinomyces**

1. Actinomyces bovis Harz 1877, 133.[AL]

bo'vis. L. masc. and fem. n. *bos*, ox, cow; L. gen. n. *bovis*, of the ox (cow).

Cellular and colonial morphology, ultrastructure, cell wall composition, nutrition and growth conditions, metabolism and metabolic pathways and drug sensitivity are as given for the genus. Physiological, biochemical and other descriptive and differential characteristics are listed in Tables 15.47 and 15.49.

The two serovars recognized so far within the species are most easily and reliably identified by direct or indirect fluorescent antibody techniques. The microcolonies of serovar 1 strains are usually smooth and nonfilamentous while serovar 2 microcolonies tend to be filamentous.

The natural habitat is not definitely known, but is presumably the oral cavity and/or the intestinal tract of cattle and possibly other animals.

A. bovis causes actinomycosis in cattle (bovine actinomycosis, lumpy jaw). The etiology of similar infections in other animals has not been clarified with certainty, but some of these diseases might also be due to this species. In contrast, *A. bovis* has never been proved to cause actinomycosis in man, nor has it been isolated from human mucosal surfaces or other human sources.

Experimental infections have been produced in hamsters, mice and a few other animal species after intraperitoneal, intravenous or subcutaneous injections of viable cell suspensions.

The mol% G + C of the DNA is (53.5 to) 57–63 (T_m).

Type strain: ATCC 13683 (WVU 116, CDC X521 (serovar 1)).

Reference strain for serovar 2: WVU 292.

2. Actinomyces israelii (Kruse 1896) Lachner-Sandoval 1898, 64.[AL]

(*Streptothrix israeli* Kruse 1896, 56.)

is.ra.e'li.i. M.L. gen. n. *of Israel*; named after Professor James Israel, one of the original describers of the organism.

Cell and colony morphology, ultrastructure, cell wall composition, nutrition and growth conditions, metabolism and metabolic pathways and drug sensitivity are as given for the genus. Physiological, biochemical and other descriptive and differential characteristics are listed in Tables 15.47 and 15.49.

The antigenic structure of *A. israelii* appears to be more complex than previously thought. The two serovars recognized so far obviously do not account for the whole range of serological variation occurring within the species. Serovar 1 strains are antigenically heterogeneous and may be separated from typical serovar 2 isolates on the basis of certain physiological differences (Table 15.48) so that they may be considered biovars as well. In addition, two new serological types were found among strains fitting the general description of the species. These serological variants differed from typical members of the taxon in a number of physiological characteristics so that it remains to be seen whether they only constitute new sero- and biovars or deserve species rank.

The normal habitat of *A. israelii* is the oral cavity of man including tonsillar crypts and dental plaque. In addition, this species may be recovered from the mucosal surfaces of the human intestinal and female genital tracts. But it remains to be clarified as to whether or not this species is indigenous to these habitats.

A. israelii is the principal causative agent of human cervicofacial, thoracic and abdominal actinomycoses. Furthermore, it may cause eye infections such as lacrimal canaliculitis, conjunctivitis or dacryocystitis and is apparently also involved etiologically in the development of cervicitis and endometritis in women using intrauterine contraceptive devices or vaginal pessaries. Occasionally, *A. israelii* has been isolated from animal infections such as actinomycosis in cattle and swine and pyogenic granulomas in a mandrill. "Sulfur granules" are produced in human and animal infections.

After intraperitoneal, intravenous or subcutaneous injection, *A. israelii* causes abscess formation in a variety of experimental animals

including hamsters, mice and rabbits. The pathological lesions thus induced resemble naturally occurring actinomycoses to a certain extent, but are usually self-limited and not progressive.

The mol% G + C of the DNA is 57–65 (T_m).

Type strain: ATCC 12102 (WVU 46, CDC X523, W855 (serovar 1)).

Reference strain for serovar 2: ATCC 29322 (WVU 307, CDC 1011).

Additional reference strains: ATCC 10048 (CDC X522 (serovar 1, but serologically different from ATCC 12102)); HIK 294 (HIK A 98/77 ("serovar 3")); HIK 47 (HIK A 563/75 ("serovar 4")) (HIK, Institute of Hygiene, University of Cologne, F.R.G.).

3. Actinomyces naeslundii Thompson and Lovestedt 1951, 175.[AL]

naes.lun'di.i. M.L. gen. n. *of Naeslund*; named after Carl Naeslund who first described this organism in some detail, but did not give it a specific epithet.

Cell and colony morphology, ultrastructure, cell wall composition, nutrition and growth conditions, metabolism and metabolic pathways, and drug sensitivity are as given for the genus. Physiological, biochemical and other descriptive and differential characteristics are listed in Tables 15.47 and 15.49.

Four serovars of *A. naeslundii* have been described. However, serovar 4 strains are now being considered a separate species (*Actinomyces* sp. type WVU 963) and the antigenic structures and taxonomic positions of serovars 2 and 3 require further investigation. This is especially true for the relationships which undoubtedly exist between the latter strains and representatives of serovar 2 of *A. viscosus*. In addition, the taxonomic status of the so-called "atypical" *A. naeslundii* isolates awaits definite clarification. Some of the physiological tests which may be used to differentiate between serovars 1 and 2 on one hand and certain serovar 3 strains on the other hand are listed in Table 15.48. Differential characteristics suitable for delineating strains intermediate between *A. naeslundii* and *A. viscosus* are also included in this table.

The normal habitat of *A. naeslundii* is the oral cavity of man including tonsillar crypts and dental plaque. *A. naeslundii* may also be detected in cervicovaginal secretions from women using and not using IUDs.

Human infections due to *A. naeslundii* most frequently present as cervicofacial, thoracic or abdominal actinomycotic lesions clinically indistinguishable from those produced by *A. israelii*. Eye infections (lacrimal canaliculitis etc.) and infections of the female genital tract as well as an empyema of the knee joint due to *A. naeslundii* have been reported. Furthermore, this species obviously plays an important role in the complex etiology and pathogenesis of caries and periodontal disease.

After parenteral inoculation, *A. naeslundii* was found to cause abscess formation in experimental animals. After oral inoculation, the organism is able to initiate periodontitis with alveolar bone loss and/or fissure lesions in hamsters and both conventional and gnotobiotic rats.

The mol% G + C of the DNA is 63–68.5 (T_m).

Type strain: ATCC 12104 (NCTC 10301, WVU 45, CDC X454 (W826); (serovar 1)).

Reference strain for serovar 2: WVU 1523 (CDC W1544).

4. Actinomyces odontolyticus Batty 1958, 455.[AL]

o.don.to.ly'ti.cus. Gr. masc. n. *odoys*, *odontos*, tooth; Gr. verb lyein, to dissolve; M.L. adj. *lyticus*, dissolving; M.L. adj. *odontolyticus*, tooth-dissolving.

Cellular morphology, ultrastructure, cell wall composition, nutrition and growth conditions, metabolism and metabolic pathways, and drug sensitivity are as given for the genus. Physiological, biochemical and other descriptive and differential characteristics are listed in Tables 15.47 and 15.49.

Macrocolonies of *A. odontolyticus* grown on BHIA are opaque and white to gray-white. After (2 to) 5–10 days of incubation on blood agar, however, the colonies usually become deep red. This red pigmentation may be seen on plates incubated anaerobically or it may only become apparent after the cultures have been left standing in air at room temperature following primary anaerobic incubation.

Two serovars of *A. odontolyticus* have been described which show only low-grade cross-reactivity between each other. However, considerable cross-reactions occur between *A. odontolyticus, serovar 2*, strains and *A. pyogenes*.

The natural habitat of *A. odontolyticus* is the oral cavity of man, especially dental plaque and calculus.

Progressive actinomycotic infections in man due to this organism have been reported, but are apparently rare. More frequently, *A. odontolyticus* is etiologically involved in the development of eye infections such as lacrimal canaliculitis and possibly also of periodontitis or caries. Primarily, it had been isolated from deep carious lesions in man.

No reports are available on the occurrence of *A. odontolyticus* in animals. However, this species was shown to produce abscess formation when injected into laboratory animals.

The mol% G + C of the DNA of the type strain is 62 (T_m).

Type strain: ATCC 17929 (NCTC 9935, WVU 867, CDC X363).

Reference strain for serovar 2: ATCC 29323 (WVU 482.).

5. **Actinomyces viscosus** (Howell, Jordan, Georg and Pine 1965) Georg, Pine and Gerencser 1969, 292.[AL] (*Odontomyces viscosus* Howell, Jordan, Georg and Pine 1965, 65.)

vis.co'sus. L. neut. n. *viscum*, birdlime; M.L. adj. *viscosus*, sticky.

Cell and colony morphology, ultrastructure, cell wall composition, nutrition and growth conditions, metabolism and metabolic pathways, and drug sensitivity are as given for the genus. Physiological, biochemical and other descriptive and differential characteristics are listed in Tables 15.47 and 15.49.

Catalase-positive.

The two serovars recognized so far within the species *A. viscosus* differ not only in their antigenic structure, but also in some physiological characteristics and in their ecology. The major physiological differences between both serovars (biovars) are listed in Table 15.48. With regard to ecology, serovar 1 strains represent the actinomycetes originally isolated from the oral cavity of hamsters while serovar 2 isolates were primarily obtained from various human sources.

Recent numerical phenetic data showed that the phenon *A. viscosus* could be subdivided into four subunits which corresponded to serovar 1, serovar 2, so-called "atypical" *A. viscosus* strains and certain serovar 3 isolates of *A. naeslundii*, respectively. Tests which are useful to delineate "atypical" *A. viscosus* isolates from "typical" ones have been included in Table 15.48.

Members of the taxon *A. viscosus* have been isolated from subgingival plaque of hamsters and cervical plaque of rats and also from human dental plaque and calculus. Thus, the oral cavity of man and other homoiothermic animals appears to be the principal natural habitat of this species. However, *A. viscosus* has also been recovered from cervicovaginal secretions of women with or without IUDs and from the uninfected conjunctiva and/or cornea of man.

Occasionally, *A. viscosus* may be isolated from cervicofacial and abdominal cases of human actinomycoses and at least in some of these infections, *A. viscosus* appears to be the primary pathogen. Furthermore, this species may be involved in the etiology of lacrimal canaliculitis and other eye infections and possibly also in the etiology of cervicitis and endometritis of women using IUDs. Among the members of the genus *Actinomyces* recognized so far *A. viscosus* is undoubtedly the species whose contribution to the etiology of caries and periodontal disease has been established most reliably.

A. viscosus has also been identified as the causative agent of animal diseases. It has been isolated from actinomycotic lesions in swine, cats and dogs and from periodontal disease occurring spontaneously in hamsters.

Experimental infections have been produced in mice and hamsters. After oral inoculation, *A. viscosus* causes periodontal disease and/or caries in hamsters as well as in conventional and gnotobiotic rats.

The mol% G + C of the DNA is 59–69.9 (T_m).

Type strain: ATCC 15987 (WVU 745, CDC X603 (A828) (serovar 1)).

Reference strain for serovar 2: ATCC 19246 (WVU 371, CDC W859).

6. **Actinomyces pyogenes** (Glage 1903) Reddy, Cornell and Fraga 1982, 427.[VP] (*Bacillus pyogenes* Glage 1903, 173; *Corynebacterium pyogenes* (Glage) Eberson 1918, 23.)

py.o'ge.nes. Gr. neut. n. *pyon*, pus; Gr. verb *gennaein*, to produce; Gr. fem. n. *gennesis*, production; M.L. adj. *pyogenes*, pus-producing.

Cellular and colonial morphology, ultrastructure, cell wall composition, nutrition and growth conditions, metabolism and metabolic pathways, and drug sensitivity are as given for the genus. Physiological, biochemical and other descriptive and differential characteristics are listed in Tables 15.47 and 15.49.

Culture filtrates are fatal to mice and rabbits after intravenous injection. The soluble hemolysin produced is active against human, guinea pig, sheep, horse and rabbit red blood cells. Both toxic and hemolytic activities of crude cell extracts are neutralized by antitoxin.

At present, *A. pyogenes* appears to be serologically homogeneous. However, pronounced cross-reactivity has been observed between this species and *A. odontolyticus, serovar 2*.

A. pyogenes presumably occurs as a commensal organism on the mucous surfaces of warm-blooded animals.

It causes a variety of pyogenic disease conditions in many species of domestic animals and in man. These include mastitis in cows and peritonitis and pleuritis in swine as well as various forms of suppurative lesions in sheep, rabbits and horses. In man, *A. pyogenes* may cause acute pharyngitis, urethritis and cutaneous or subcutaneous suppurative processes.

The mol% G + C of the DNA is 56–58 (T_m).

Type strain: ATCC 19411 (NCTC 5224).

7. **Actinomyces denticolens** Dent and Williams 1984c, 503.[VP] (Effective publication Dent and Williams 1984a, 188.)

den.ti.co'lens. L. masc. n. *dens, dentis*, tooth; L. verb *colere*, to inhabit, to dwell; L. part. praes. *colens*, dwelling; M.L. adj. *denticolens*, tooth-dwelling.

Cell morphology, cell wall composition, nutrition and growth conditions, metabolism and metabolic pathways, and drug sensitivity are as given for the genus. Physiological, biochemical and other descriptive and differential characteristics are listed in Tables 15.47 and 15.49.

Microcolonies of *A. denticolens* are usually smooth and nonfilamentous although very young microcolonies may occasionally show filament formation. Mature colonies when grown anaerobically on horse blood agar exhibit a slightly pink pigmentation, but they are white if grown anaerobically.

The patterns of polypetides as analyzed by SDS-polyacrylamide gel electrophoresis are unique, although with some intraspecies variation, and allow the species to be distinguished from both *A. naeslundii* and *A. howellii* which are similar in respect to physiology and biochemistry.

Serology and antigenic structure of *A. denticolens* have not been studied. However, preliminary data obtained with antisera to the other *Actinomyces* species indicate that *A. denticolens* represents a separate serological entity clearly distinct from other members of the genus including *A. naeslundii*.

The normal habitat of *A. denticolens* appears to be the oral cavity of cattle, especially dental plaque. Pathogenicity has not been reported.

The mol% G + C of the DNA is 65.9–67.7 (T_m).

Type strain: Sh8/4303 (NCTC 11490).

8. **Actinomyces howellii** Dent and Williams 1984b, 319.[VP]

how.el'li.i. M.L. gen. n. *of Howell*; named after Arden Howell, who studied oral actinomycetes from animals, particularly *A. viscosus*.

Cell morphology, cell wall composition, nutrition and growth conditions as well as metabolism and metabolic pathways are as given for the genus. Physiological, biochemical and other descriptive and differential characteristics are listed in Tables 15.47 and 15.49.

Catalase-positive.

Microcolonies of *A. howellii* have not been described in detail. Mature colonies when grown anaerobically on horse blood agar at 37°C are white, smooth, shiny, translucent, entire, convex and up to 2 mm in diameter.

Members of the species exhibit characteristic although not completely identical polypeptide patterns as analyzed by SDS-polyacrylamide gel electrophoresis which allow clear separation *A. howellii* from both *A. naeslundii* and *A. denticolens*.

Serology and antigenic structure of *A. howellii* have not been studied.

The normal habitat of *A. howellii* appears to be the oral cavity of cattle, especially dental plaque. Pathogenicity has not been reported.

The mol% G + C of the DNA is 65.9–67.3 (T_m).

Type strain: Sh7/4276 (NCTC 11636).

9. Actinomyces hordeovulneris Buchanan, Scott, Gerencser, Beaman, Jang and Biberstein 1984, 442.[VP]

hor.de.o.vul′ne.ris. L. neut. n. *hordeum*, barley; M.L. n. *Hordeum*, genus of grass; L. neut. n. *vulnus, vulneris*, wound, injury; M.L. gen. n. *hordeovulneris*, of (isolated from) injuries produced by *Hordeum (awns)*.

Cell morphology, cell wall composition, nutrition and growth conditions, metabolism and metabolic pathways, and drug sensitivity are as given for the genus. Growth of the organism is greatly enhanced by addition of 10–20% fetal calf serum to the culture medium. Physiological, biochemical and other descriptive and differential characteristics are listed in Tables 15.47 and 15.49.

A weak to moderate catalase activity has been observed in all of the strains tested so far.

Microcolonies of *A. hordeovulneris* grown on trypticase soy or brain heart infusion agar supplemented with 10–20% fetal calf serum are regularly filamentous and similar to those of *A. israelii*. Mature colonies are white, agar adherent and molar toothed with a tendency to shift to a white conical, domed, buttery, less adherent type upon laboratory passage.

Preliminary serological data obtained with FITC-conjugated antisera to various *Actinomyces* and *Rothia* species showed that there was some degree of cross-reactions between *A. israelii*, serovar 1 (ATCC 12102), and *A. viscosus*, serovar 1 (ATCC 15987), conjugates and *A. hordeovulneris* cells. No significant cross-reactivity was observed with antisera to *A. viscosus*, serovar 2 (ATCC 19246), *A. israelii*, serovar 2 (ATCC 29322), *A. bovis*, serovars 1 and 2, *A. naeslundii*, serovars 1, 2 and 3, *A. odontolyticus*, serovars 1 and 2, *A. suis*, *A. pyogenes*, *Rothia dentocariosa*, serovars 1, 2 and 3, and *Rothia*-like strain WVU 1556.

The normal habitat of *A. hordeovulneris* is not known. The organism has been isolated from cases of pleuritis, peritonitis, visceral abscesses, septic arthritis and recurrent localized infections in dogs. These infections often occurred secondary to tissue-migrating awns of several members of the grass genus *Hordeum* (commonly called foxtails) which are common in the western United States, especially in California. Some of the infected animals died despite ampicillin treatment although this drug was found to be effective in vitro. The unfavorable responses to ampicillin were attributed to a marked tendency of *A. hordeovulneris* to produce L-forms spontaneously with coincident uptake of calcium (Buchanan et. al., 1984). Occasionally, *A. hordeovulneris* and *A. viscosus* were found to be present in the same inflammatory process.

The mol% G + C of the type strain is 67 (Bd).

Type strain: ATCC 35275 (UCD 81-332-9).

10. Actinomyces meyeri (Prévot 1938) Cato, Moore, Nygaard and Holdeman 1984, 487.[VP] (*Actinobacterium meyeri* Prévot 1938, 303.)

mey′e.ri. M.L. gen. n. *of Meyer*; named after Kurt Meyer who described the organism as a "new anaerobic *Streptothrix* species" in 1911.

Cell morphology, nutrition and growth conditions, metabolism and metabolic pathways, and drug sensitivity are in principle as given for the genus, but a few characteristic differences should be noted:

In *A. meyeri* branching may be difficult to demonstrate. Cells are usually short, but long straight or curved filaments without branching and chains of longer rods may be seen when the growth medium lacks one or more of the components required for optimum growth. Terminal swellings are present occasionally.

Microcolonies of *A. meyeri* are smooth and nonfilamentous resembling those of *A. odontolyticus* or *A. bovis*. Mature surface colonies on supplemented BHI blood agar are pinpoint to 1 mm in diameter, circular, flat to convex, translucent to opaque, white, with a shiny, smooth surface and an entire margin. They may be α- or non-hemolytic.

Cultures in PYG broth are only slightly turbid with a smooth or stringy sediment and have a pH of 4.6–5.0 after incubation for 5 days.

Strains of *A. meyeri* have an absolute requirement for vitamin K_1 and growth is greatly stimulated by 0.02% Tween 80 and by a fermentable carbohydrate. Serum also may stimulate growth. Other descriptive and differential characteristics are listed in Tables 15.47 and 15.49.

The patterns of soluble cellular proteins as determined by polyacrylamide gel electrophoresis indicate that the strains tested so far represent a homogeneous taxon (Cato et. al., 1984).

Most of the isolates studied gave a positive fluorescent antibody reaction with a monovalent FITC-conjugate prepared against strain ATCC 33972. No cross-reactions were observed with conjugates to *A. israelii*, serovars 1 and 2, *A. odontolyticus*, serovars 1 and 2, *A. naeslundii*, serovars 1, 2 and 3, *A. viscosus* and *Actinomyces* sp. WVU 963. However, several isolates that were otherwise phenotypically identical to *A. meyeri* did not react with the antiserum to strain ATCC 33972 which indicates that there may be additional serovars of the species.

The principal natural habitat of *A. meyeri* is the human periodontal sulcus. Furthermore, the organism has been isolated frequently from brain abscesses and pleural fluid and less often from abscesses of the cervicofacial area, hips, hands, feet and spleens and from bite wounds.

The mol% G + C of the DNA of the type strain is 67 with an intraspecies variation of 64.4–67.2 (spectrophotometry/chromatography).

Type strain: ATCC 35568 (Prévot 2477 B, VPI 8617).

Species Incertae Sedis

a. "**Actinomyces suis**" Franke 1973, 123. (Not *Actinomyces suis* Gasperini 1892, 183, and presumably not *Actinomyces suis* Grässer 1957, 148.)

su′is. L. fem. n. *sus, suis*, pig, hog; L. gen. n. *suis*, of the hog.

The identity of "*A. suis*" Grässer 1957 and "*A. suis*" Franke 1973 is questionable as Grässer's description is inadequate and no cultures are available. Nonetheless, this taxon was included as *Species Incertae Sedis* because swine isolates apparently differ from all of the other *Actinomyces* species recognized so far and because there is a need for reliable identification of these organisms which can be achieved most easily following the description of Franke (1973).

The cellular morphology of "*A. suis*" Franke is predominantly diphtheroidal the rods being arranged in clusters or in V, Y or T forms. Coccoid elements may occur.

Microcolonies and macrocolonies are predominantly smooth and opaque although a few short radiating filaments may be observed.

Other descriptive and differential characteristics are given in Tables 15.47 and 15.49.

Using an agar gel precipitation procedure, all of the "*A. suis*" isolates appeared to be serologically homogeneous, but differed clearly from the other *Actinomyces* species. Only low-titred cross-reactions were observed with *A. israelii*, *A. naeslundii*, and *Arachnia propionica*.

The normal habitat of "*A. suis*" is not known. The organisms have been isolated from actinomycosis of the mammary gland of swine.

No type strain has been designated and deposited.

b. "**Actinomyces humiferus**" Gledhill and Casida 1969, 118.

hu.mi′fe.rus. L. fem. n. *humus, humi*, soil; L. verb *ferre*, to bear; M.L. adj. *humiferus*, soil-borne.

Cells are predominantly filamentous and branched and often have swollen ends. After prolonged incubation, they usually fragment into diphtheroidal or coccoid elements of varied size and shape.

The cell wall contains glucosamine, muramic acid, alanine, glutamic acid, lysine, ornithine, and aspartic acid. Rhamnose is the predominant cell wall sugar, but glucose and fucose may be present in trace amounts.

Microcolonies on agar media and initial growth in liquid media are usually filamentous. Mature colonies are small, opaque, smooth, entire, convex, with a dark central region. Rough colony variants occur occasionally. Pigmentation is not evident. In liquid media, growth is granular or flocculent forming a white sediment without turbidity.

The optimum growth temperature is approximately 30°C; poor or no growth at 37°C.

The organism does not grow on media lacking organic nitrogen. In addition, little if any growth is obtained in certain chemically defined media or media containing simple peptones.

Other descriptive and differential characteristics are listed in Tables 15.47 and 15.49.

Using the fluorescent antibody technique, no cross-reactivity was observed between "*A. humiferus*" and other *Actinomyces* or *Rothia* species. A slight cross-staining obtained with *Corynebacterium (Bacterionema) matruchotii* antiserum was considered nonspecific.

The natural habitat of "*A. humiferus*" appears to be organically rich soil from which the organism may be recovered in high numbers. Experimental infection could not be induced in mice after intraperitoneal injection of washed saline cell suspensions.

The mol% G + C of the DNA is 73 on average (density gradient). *Type strain:* ATCC 25174.

Genus **Bifidobacterium** Orla-Jensen 1924, 472[AL]

VITTORIO SCARDOVI

Bi.fi.do.bac.te′ri.um. L. adj. *bifidus* cleft, divided; Gr. dim. n. *bakterion* a small rod; M.L. neut. n. *Bifidobacterium* a cleft rodlet.

Rods of various shapes: short, regular, thin cells with pointed ends, coccoidal regular cells, long cells with slight bends or protuberances or with a large variety of branchings; pointed, slightly bifurcated **club-shaped or spatulated extremities**; single or in chains of many elements; in star-like aggregates or disposed in "V" or "palisade" arrangements. Colonies smooth, convex, entire edges, cream to white, glistening and of soft consistency. **Gram-positive, non-acid-fast; nonspore-forming, nonmotile.** Cells **often stain irregularly** with methylene blue. **Anaerobic;** some species can tolerate O_2 only in the presence of CO_2. Optimum growth temperature 37–41°C; minimum growth temperatures 25–28°C maximum 43–45°C. Optimum pH for initial growth 6.5–7.0: no growth at 4.5—5.0 or 8.0–8.5

Saccharoclastic. Acetic and lactic acid are formed primarily in the molar ratio of 3:2. CO_2 **is not produced** (except in the degradation of gluconate). Small amounts of formic acid, ethanol and succinic acid are produced. **Butyric and propionic acid are not produced. Glucose is degraded exclusively and characteristically by the fructose-6-phosphate shunt** in which fructose-6-phosphoketolase (F6PPK-EC 4.1.2.22) cleaves fructose-6-phosphate into acetylphosphate and erythrose-4-phosphate. End products are formed through the sequential action of transaldolase (EC 2.2.1.2), transketolase (EC 2.2.1.1), xylulose-5-phosphate phosphoketolase (EC 4.1.2.9) and enzymes of EMP acting on glyceraldehyde-3-phosphate. Additional acetic and formic acid may be formed through a cleavage of pyruvate.

Glucose-6-phosphate dehydrogenase (EC 1.1.1.49, NADP$^+$- or NAD$^+$-dependent) generally not determinable.

Catalase-negative except that *B. indicum* and *B. asteroides* are catalase-positive when grown in the presence of air with or without added hemin.

Ammonium is generally utilized as a source of nitrogen.

The G + C content of DNA (Bd or T_m) varies from 55–67 mol%.

The organisms occur in the intestine of man, various animals and honey bees; found also in sewage and human clinical material.

Type species: *Bifidobacterium bifidum* (Tissier) Orla-Jensen 1924, 472.

Further Descriptive Information

Morphology

The cellular morphology and its variations, as affected by different cultural conditions, have been widely investigated (see Poupard, Husain and Norris, 1973, for references). However, recent discoveries of new species from a variety of habitats have permitted a clearer picture of the morphology of the genus.

A comparison of the cell morphology of large numbers of strains

grown anaerobically (GasPak system, BBL) in stabs of trypticase-phytone-yeast extract medium (TPY) showed that some species had distinct cell shapes or arrangements which might be of help in their recognition; these traits are reported in Figs. 15.96–15.98.

Outstanding are the well known amphora-like cells of *B. bifidum* (Sundman et al., 1959) (Fig. 15.96*A*), the V or palisade arrangement of cells in *B. angulatum* (Fig. 15.96*D*), the linear groups of globular elements in *B. catenulatum* (Fig. 15.96*E*), the long chains of regular cells in *B. pullorum* (Fig. 15.97*A*), the middle-enlarged cells of *B. animalis* (Fig. 15.97*B*), the large cellular dimensions in *B. magnum* (Fig. 15.97*D*), the small cells of *B. minimum* (Fig. 15.97*F*), and the unusual starlike arrangements of cells in *B. asteroides* (Fig. 15.98*A*). The cellular shape most frequently encountered in those species not having distinct morphology (see Table 15.50) as observed in TPY stabs (see above), is depicted in Fig. 15.98*D*. Details are given under single species description.

B. asteroides (starlike clusters) and *B. indicum* (small rods or coccobacilli), the species with the most nonbifid-like morphology in the classic sense, show features common to the morphology of the other bifids only when grown in nutritionally deficient media (Scardovi and Trovatelli, 1969), which seems to be a general trend in this group of bacteria (Sundman and Björksten, 1958; Glick et al., 1960).

Cell Wall Structure

The most extensive study of cell wall murein structure of bifidobacteria has been made by O. Kandler and collaborators (reported later in Table 15.50). Closely related species can be clearly distinguished on this basis, i.e. *B. boum* from *B. thermophilum* or *B. minimum* from *B. subtile*.

On the basis of murein structure, bifidobacteria are more closely related to *Lactobacillaceae* than to *Actinomycetaceae* (Kandler and Lauer, 1974).

Lipid Cellular Composition

Some species of *Bifidobacterium* and *Lactobacillus* were studied by Exterkate et al. (1971): differences in polyglycerol phospholipids and aminoacyl phosphatidylglycerol were found to be of help in differentiating the two genera. The effects of growth conditions on the lipid and ionic composition of *B. bifidum* subsp. *pennsylvanicum* have been recently studied by Veerkamp (1977a, b).

Ultrastructure

The ultrastructure of bifidobacteria has received little attention. Overman and Pine (1963) first reported ultrastructure micrographs of *B. bifidum* subsp. *pennsylvanicum*. Recently, Zani and Severi (1982)

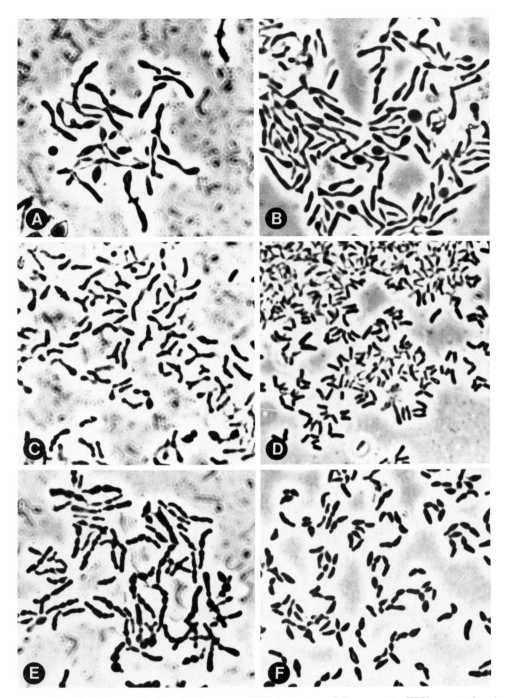

Figure 15.96. Cellular morphology in the genus *Bifidobacterium*. Cells grown in TPY agar stabs. *A, B. bifidum; B, B. longum; C, B. breve; D, B. angulatum; E, B. catenulatum;* and *F, B. globosum*. Phase-contrast photomicrographs, × 1500.

made a more extensive investigation of the ultrastructure of *B. bifidum* strain S28a of Reuter (ATCC 15696). Comparative studies on the ultrastructure of species comprising the genus have not yet been made.

Nutrition

Since its first isolation from human infants' feces (György, 1953) and its designation as *Lactobacillus bifidus* var. *pennsylvanicus* (György and Rose, 1955), this organism, the growth of which is stimulated by human milk, has been the object of numerous nutritional studies designed either to elucidate the properties of the bifidus factor(s) present in human milk, or to find a substitute for it (see reviews by

Poupard et al., 1973; Yoshioka et al., 1968; Nakamura and Tamura, 1972; Nichols et al., 1974; György et al., 1974; Yazawa et al., 1978; Beerens et al., 1980).

However, as only the György strain and a few others of ill defined taxonomy were used for these studies, they have little bearing on our knowledge of the nutritional requirements of the genus, particularly now that so many new species are known. The growth factor requirements of twelve species are reported later in Table 15.53; they form a very heterogeneous group and the vitamin requirements seem unrelated to the ecological distribution of the species.

Bifidobacteria are able to utilize ammonium salts as sole source of

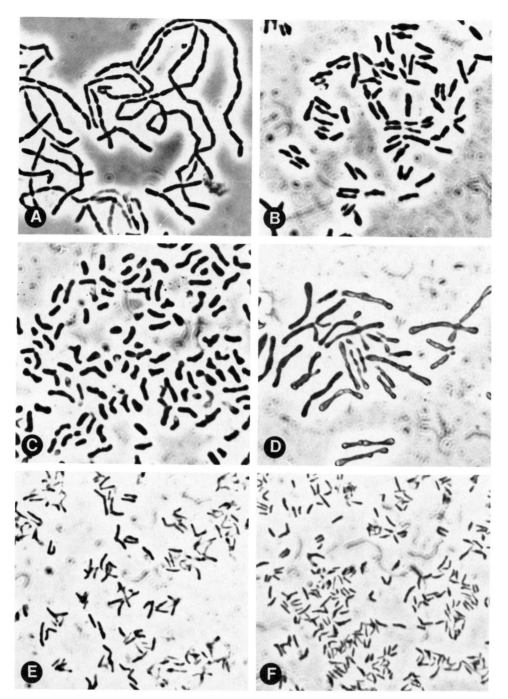

Figure 15.97. Cellular morphology in the genus *Bifidobacterium*. Cells grown in TPY agar stabs. *A, B. pullorum; B, B. animalis; C, B. cuniculi; D, B. magnum; E, B. subtile;* and *F, B. minimum.* Phase-contrast photomicrographs, × 1500.

nitrogen. This finding, reported first by Hassinen et al. (1951), is valid for most species of the genus, but *B. suis, B. magnum, B. choerinum* and *B. cuniculi* will not grow without organic nitrogen (Matteuzzi et al., 1978). (*B. choerinum* is reported under the provisory name of "*B. coirinense*" in this paper). The species which grow without organic nitrogen excrete considerable amounts of various amino acids into the medium: e.g. *B. bifidum* can produce up to 150 mg/liter threonine. Other active amino acid producers are *B. thermophilum, B. adolescentis, B. dentium, B. animalis* and *B. infantis.* The amino acids generally produced in the largest amounts are alanine, valine and aspartic acid (Matteuzzi et al., 1978).

Analog-resistant mutants were obtained from *B. thermophilum* (*B. ruminale*) showing increased production of isoleucine and valine (Matteuzzi et al., 1976; Crociani et al., 1977).

Carbohydrate Metabolism

The fermentation of hexose occurs in the genus *Bifidobacterium* through the following sequence of reactions (bifid shunt) (Scardovi and Trovatelli, 1965; De Vries et al., 1967, Veerkamp, 1969b).

$$\text{Fructose-6-P} + iP \xrightarrow{\text{F6PPK}} \text{erythrose-4-P} + \text{acetyl-P} + H_2O$$

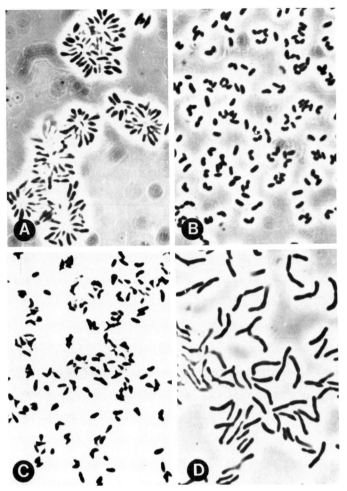

Figure 15.98. Cellular morphology in the genus *Bifidobacterium*. Cells grown in TPY agar stabs. *A, B. asteroides; B, B. indicum; C, B. coryneforme; and D, B. infantis.* Phase-contrast photomicrographs, × 1500.

Erythrose-4-P + fructose-6-P $\xrightarrow{transaldolase}$ sedoheptulose-7-P

$+$ glyceraldehyde-3-P

Sedoheptulose-7-P

$+$ glyceraldehyde-3-P $\xrightarrow{transketolase}$ ribose-5-P + xylulose-5-P

Ribose-5-P $\xrightarrow{isomerase}$ xylulose-5-P

2 xylulose-5-P + iP $\xrightarrow{xylulose\ phosphoketolase}$ 2 glyceraldehyde-3-P

$+$ 2 acetyl-P + 2H$_2$O

3 acetyl-P → 3 acetate

2 glyceraldehyde-3-P $\xrightarrow{EMP\ enzymes}$ + 2 lactate

2 hexose → 3 acetate + 2 lactate

However, the theoretical ratio of acetate 1.5:lactate 1.0 is scarcely ever found in growing cultures of bifidobacteria: phosphoroclastic cleavage of some pyruvate to formic acid and acetyl phosphate and reduction of acetyl phosphate to ethanol can often alter the fermentation balance in favor of the production of acetate and some formic acid and ethanol. Given X as the amount of formic acid produced, the general reaction

postulated was: glucose → (1.5 + 0.5 X) acetate + (1.0 − X) lactate + 0.5 X ethanol + X formate (De Vries and Stouthamer, 1968, Lauer and Kandler, 1976).

Glucose-6-phosphate dehydrogenase and aldolase are claimed to be absent or not detectable, thus ruling out the monophosphate pathway and the glycolytic system (De Vries and Stouthamer, 1967). However, low but detectable levels of these two enzymes were found in some species (Scardovi and Sgorbati, 1974; see under description of species).

The enzymes of the Leloir pathway of galactose metabolism, i.e. galactokinase (EC 2.7.1.6), hexose-1-phosphate uridylyltransferase (EC 2.7.7.12) and UDP galactose 4-epimerase (EC 5.1.3.2) **are constitutive in glucose-grown cells of *Bifidobacterium*, whereas in other microorganisms these enzymes are induced by galactose or fucose** (Lee et al., 1980).

The existence of UPD galactose pyrophosphorylase (Lee et al., 1978) suggests that, at least in species of the genus from human sources, an alternative pathway of galactose is operative (Lee et al. 1979).

The enzymatic carboxylation of phosphoenolpyruvate to oxaloacetate in some bifidobacteria from human feces and from the honey bee, has been compared with the corresponding activity in strains of *Actinomyces bovis* and *Actinomyces israelii*: in bifids this activity is independent of the phosphate acceptor and is irreversible, whereas in *Actinomyces*, it is inosine or guanosine diphosphate-dependent (Chiappini, 1966).

Urease Activity

Four hundred and fourteen strains representing 21 species of the genus were surveyed for their urease activity. The strongest ureolytic strains belong mostly to *B. suis*, in which more than 80% of the strains studied are ureolytic. Ureolytic strains were found in all species except for *B. cuniculi*. The enzyme is apparently not inducible: urea and organic nitrogen do not influence urease production. In *B. breve* and *B. longum*, i.e. "human" bifid species, less than 10% of strains are ureolytic and *B. bifidum* is only weakly ureolytic (Crociani and Matteuzzi, 1982).

Anaerobiosis

Bifidobacteria are anaerobic microorganisms: they do not develop in plates under aerobic conditions. **However, the sensitivity to oxygen is different among different strains and species,** and its intimate reasons are equivocal (De Vries and Stouthamer, 1969). While most species do not develop in slants incubated under an atmosphere of CO$_2$-enriched air (air 90%; CO$_2$ 10%) *B. globosum, B. thermophilum* and *B. suis* do so without the cells becoming catalase-positive even if hemin is added to the medium; *B. asteroides* grows under these conditions and becomes catalase-positive; *B. indicum* behaves similarly, but is catalase-positive only if hemin is added to the medium (Scardovi et al., 1969; Scardovi and Trovatelli, 1969; Matteuzzi et al., 1971).

Nitrate Reduction

Bifidobacteria are generally claimed not to reduce nitrate. However, cells grown in the presence of lysed red cells may be capable of nitrate reduction. Cytochrome *b* and cytochrome *d* are synthesized under these conditions of growth (van der Wiel-Korstanje and De Vries, 1973).

Enzymes Used for Species or Group Differentiation

Fructose-6-phosphate phosphoketolase **is the characteristic key enzyme of the** "bifid shunt." Tested with fructose-6-phosphate as substrate (see below), it is apparently absent in anaerobic Gram-positive bacteria of "pseudobifid" morphology, i.e. *Arthrobacter, Propionibacterium, Corynebacterium* and *Actinomyces* (Scardovi and Trovatelli, 1965).

Starch gel electrophoresis revealed three types of F6PPK in bifido-bacteria (see Table 15.50) (Scardovi et al., 1971a): the most anodal was detected in species found in the intestine of honey bees; the type found in species from humans (*human* type) migrated less and the least anodal type characterized the species found in animals (*animal* type). F6PPK from *B. globosum* (*animal* type) and *B. dentium* (*human* type) have been purified (Sgorbati et al., 1976). **The *animal* type has properties similar to that found in *Acetobacter xylinum*** (Schramm et al., 1958), and is **also active toward xylulose-5-P; the *human* type has different properties** (activators, pH range of activity, heat inactivation) and **cleaves only fructose-6-P** (Sgorbati et al., 1976).

These ecological groups of *Bifidobacterium* species were distinguished also on immunological basis (Sgorbati and London, 1982, see below). **Isozyme patterns could also be used to identify species.** Isozymes of transaldolase and 6-phosphogluconate dehydrogenase (6PGD in Table 15.50) were studied by starch gel electrophoresis in 1206 strains belonging to each of the 24 species (Scardovi et al., 1979a). Fourteen isozymes of transaldolase and 19 of 6PGD were identified and numbered: patterns or zymograms were obtained for each species (see Table 15.50 for relevant data): 60% of the strains could be identified on that basis. An additional 20% of the strains were assigned to species on the basis of the electrophoretic behavior of their 3-phosphoglyceraldehyde dehydrogenase (Scardovi et al., 1979a).

Antisera against eight purified transaldolases further established natural relationships among the species (Sgorbati, 1979; Sgorbati and Scardovi, 1979; Sgorbati and London, 1982). Transaldolases selected on the basis of their electrophoretic behavior (see above), were from *B. infantis*, *B. angulatum*, *B. globosum*, *B. thermophilum*, *B. suis*, *B. cuniculi*, *B. minimum* and *B. asteroides*. On the basis of microcomplement fixation data the indices of dissimilarity and the immunological distances were determined; the results, in terms of taxonomic distances, are shown diagrammatically in Fig. 15.99.

The segregation of the species into four distinct clusters which coincide so neatly with the groups which can be made on the basis of their ecological distribution, suggests that a "subdivision of the genus into four subgenera would more accurately reflect the group's natural history" (Sgorbati and London, 1982).

The various purified transaldolases cannot be distinguished on the basis of the usual parameters such as molecular weight, substrate affinity, pH range, acceptors, etc. (Scardovi, unpublished).

Plasmids

When 1461 isolates representing the 24 species of the genus were examined for the presence of plasmids, ~20% of them were found to contain plasmids. However, only four species have these elements, namely *B. longum*, the predominant bifid species in the human intestine, *B. globosum*, the most common bifid in animals, and *B. asteroides* and *B. indicum*, species found exclusively in the intestine of honey bees. *B. longum* strains have multiple-plasmid patterns (1.25–9.5 MDa); *B. globosum* strains contain one plasmid each of three classes of molecular weight (13.5, 24.5 and 46 MDa); multiple patterns were seen in *B. asteroides* (1.2–22 MDa); 60% of the plasmid-bearing *B. indicum* isolates contained one 22 MDa plasmid (Sgorbati et al., 1982). It is noteworthy that *B. infantis*, the species most closely related to *B. longum*, does not contain plasmids, although strains of both species were isolated from the same specimens. No phenotypic properties have been correlated as yet with the plasmids (Sgorbati et al., 1982).

The thirteen plasmid patterns of *B. longum* contain a few homologous structures only, whereas the fourteen patterns found in *B. asteroides* are structurally more heterogeneous (Sgorbati, unpublished).

Strains of *B. longum* liberate phage particles after UV or mitomycin c treatment, but there is no correlation with their plasmid complement (Sgorbati, unpublished).

Resistance to Antibiotics

Resistance (with respect to achievable serum levels) to kanamycin, neomycin, streptomycin, polymyxin, gentamicin, nalidixic acid and metronidazole was a general feature among 15 *Bifidobacterium* species studied (including strains from human habitats and those from honey bees). Oleandomycin, lincomycin, clindamycin, vancomycin, penicillin G, ampicillin, erythromycin, bacitracin, chloramphenicol and nitrofurantoin were strongly inhibitory; sensitivity to tetracycline was inter- and intraspecifically variable (Matteuzzi, unpublished). These results extend and substantially confirm those already reported concerning mainly *B. bifidum*, *B. longum* and *B. adolescentis* (Miller and Finegold, 1967).

TAXONOMIC DISTANCE

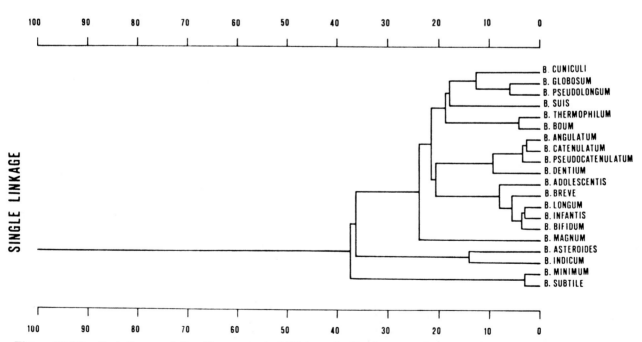

Figure 15.99. Evolutionary relationships among the bifidobacteria. The data were obtained with eight antitransaldolase sera (from Sgorbati, unpublished).

The genus *Bifidobacterium*, therefore, exhibits uniform sensitivity toward the most common antibiotics; however, sensitivity toward antibiotics such as polymyxin, neomycin, kanamycin, streptomycin, gentamicin or nalidixic acid varies greatly within species, ranging from 10–500 or more μg/ml antibiotic.

The electrophoretic patterns of cellular proteins (polyacrylamide gel electrophoresis (PAGE)) procedure of Moore et al., 1980) was used to confirm the taxonomic identification of 1094 bifidobacteria strains belonging to all known species of the genus. Excellent correlation was found between electrophoretic data and the classification of the genus presented here: on this basis two species were revived (*B. coryneforme* and *B. globosum*) and two others were proposed for previously recognized DNA homology groups (*B. minimum* and *B. subtile*) (Biavati et al., 1982).

This procedure was confirmed as one of the most useful in identifying unknown bifid isolates. In addition, a sort of *"genus band"* is clearly visible in all bifid gel electrophoresis patterns where the band migrates to the same position, with the exception of *B. boum* where it is somewhat less anodic. This appears to be a useful indication that an unknown organism belongs to the genus (Biavati et al., 1982).

Possible Pathogenicity

Most strains of *B. dentium* have been isolated from human dental caries, this species being the only bifid found in this site. (Scardovi and Crociani, 1974). Strains of *B. appendicitis* (a synonym of *B. dentium* Scardovi and Crociani, 1974) were found by Prévot in human clinical material (Prévot et al., 1967). *Actinomyces eriksonii* Georg, Robertstad, Brinkman and Hicklin, 1965, (named later *Bifidobacterium eriksonii* by Holdeman and Moore, 1970), recognized genetically as *Bifidobacterium dentium* (Scardovi et al., 1979a), was isolated primarily from human abscesses. Most of the strains studied with PAGE procedure by Biavati et al. (1982), were isolated from a variety of human clinical material and from human dental plaque. *B. longum* and *B. breve* occur occasionally in human clinical material (Biavati et al., 1982). Thus, *B. dentium* appears to have the most possible pathogenic potentiality.

Isolation Procedures

A large variety of media have been devised for isolating or enumerating bifidobacteria in natural habitats. Ingredients of substrates have been tomato juice, sheep or horse blood, human milk, liver or meat extracts, a variety of peptones etc. (for references, see Scardovi, 1981).

In order to improve selectivity, antibiotics or other ingredients have been used: kanamycin (Finegold et al., 1971); neomycin (Mata et al., 1969); paramomycin, neomycin, sodium propionate and lithium chloride (Mitsuoka et al., 1965); sorbic acid or sodium azide (Haenel and Müller-Beuthow, 1956; Haenel and Müller-Beuthow, 1957). The known bifidobacteria are resistant to certain antibiotics (see above) but the intraspecific variations are so large that the use of such antibiotics would make isolation unreliable. At present, the preference should be given to substrates that permit satisfactory growth of the largest number of bifidobacteria types presently known. The ingredients of choice are trypticase and phytone (BBL) and the formula which in our hands proved to be satisfactory for isolating the bifids from all known habitats (TPY medium) is: trypticase (BBL), 10 g; phytone (BBL), 5 g; glucose, 5 g; yeast extract (Difco), 2.5 g; Tween 80, 1 ml; cysteine hydrochloride, 0.5 g; K_2HPO_4, 2 g; $MgCl_2 \cdot 6 H_2O$, 0.5 g; $ZnSO_4 \cdot 7 H_2O$, 0.25 g; $CaCl_2$, 0.15 g; $FeCl_3$, a trace; agar, 15 g; distilled water to 1000 ml. Final pH is about 6.5 after autoclaving at 121°C for 25 min; dilutions can be made with the same liquid medium.

Petri dishes with vents are incubated at 39–40°C in anaerobic jars with palladium catalyst, filled with a gas mixture of 10% CO_2 and 90% H_2 (or GasPak system). Colonies are transferred to stabs of the same medium with 0.5% agar. After growth has recurred, stabs are kept at 3–4°C in the anaerobic jar. Transfers should be made every 2 weeks.

Procedure for Testing Special Identification Characters

Identification of a bacterial strain as *Bifidobacterium* is unreliable unless special procedures are used. Morphology can be misleading on account of unusual and unique traits shown by some species (*B. angulatum, B. asteroides, B. pullorum* etc.). Cultural and physiological characters are largely shared by other bacteria such as *Actinomyces, Corynebacterium* and *Lactobacillus*. Identification of the fermentation products by gas chromatography (Holdeman et al., 1977) may be difficult, especially for inexperienced workers, as side reactions forming substantial and variable amounts of formic and succinic acids and ethanol may occur.

The most direct and reliable characteristic assigning an organism to the genus *Bifidobacterium* is that based on the demonstration of F6PPK in cellular extracts. Proofs of the validity of this test have been furnished by (a) isolation from the bovine rumen and from sewage of anaerobic bacteria with the morphology and gross physiology of bifidobacteria but which do not possess F6PPK (Scardovi, unpublished) and (b) recognition as bifids on this basis, of bacteria of nonbifid morphology such as *B. pullorum* (Trovatelli et al., 1974) and bifids from honey bees (Scardovi and Trovatelli, 1969).

Fructose-6-Phosphate Phosphoketolase (F6PPK) Test

Reagents: 1) 0.05 M phosphate buffer pH 6.5 plus cysteine 500 mg/liter; 2) a solution containing NaF, 6 mg/ml, and K or Na iodoacetate, 10 mg/ml; 3) hydroxylamine HCl, 13.9 g/100 ml of water, freshly neutralized with NaOH to pH 6.5; 4) trichloroacetic acid (TCA), 15% (w/v) in water; 5) 4 M HCl; 6) $FeCl_3 \cdot 6H_2O$ 5% (w/v) in 0.1 M HCl; 7) fructose-6-phosphate (Na salt: 70% purity), 80 mg/ml in water.

The formation of acetyl phosphate from fructose-6-phosphate is evidenced by the reddish violet color (absorption maximum at 505 nm) formed by the ferric chelate of its hydroxamate (Lipmann and Tuttle, 1945).

Procedure

Cells harvested from 10 ml TPY broth are washed twice with buffer 1) and resuspended in 1.0 ml of the same buffer. The cells are disrupted by sonication in the cold, and 0.25 ml each of reagents 2) and 7) are added to the sonicate. After 30 min incubation at 37°C, the reaction is stopped with 1.5 ml of reagent 3). After 10 min at room temperature, 1.0 ml each of reagents 4) and 5) are added. The mixture may be left at room temperature before the final addition of 1.0 ml of the color-developing reagent 6). Invert tube for mixing. Any reddish violet color that develops immediately is taken as a positive result. A tube without fructose-6-P can serve as blank to aid the visual comparison; the color becomes more evident after standing, which allows particles to settle. Warning: avoid heating during sonication because of heat sensitivity of the enzyme.

Electrophoresis methods to detect enzymes. The starch-gel horizontal electrophoresis system of Smithies (1955) is recommended.

Transaldolase. Tris(hydroxymethyl)aminomethane (Tris) 16.3 g and citric acid monohydrate 9.0 g/liter (pH 7.0) is used as bridge buffer; dilute this 1:15 and use as gel buffer.

Hydrolyzed starch, 90 g, (Connaught Laboratories Ltd., Willondale, Ontario, Canada) is added per liter of buffer; the mixture is boiled for 5 min (keep agitated) and gas removed under reduced pressure. The liquid is poured into a plastic three-frame mold (12.0 × 37.0 × 0.9 cm are suitable to accommodate 12 samples at a time).

Samples of bacterial extracts (cells suspended in 0.05 M phosphate buffer pH 7.0, sonicated and centrifuged), 5–10 μl in 0.5 × 0.5 cm Whatman 3-mm paper cuts, are generally run for 15–20 h with a current of 15–20 mA.

The middle slab is used preferably for staining by the flooding technique. The developing solution contains (per 100 ml of distilled water): fructose-6-P (Na salt, 98% purity, Sigma), 400 mg; sodium arsenate, 370 mg; glycine, 240 mg; NAD, 13 mg; D-erythrose-4-phosphate (60–75% purity, Sigma), 16 mg; phenazine methosulfate, 2 mg; nitro blue tetrazolium (NBT, Sigma), 20 mg, and about 130–150 IU of glyceraldehyde-3-phosphate dehydrogenase.

6-Phosphogluconate dehydrogenase (6 PGD). Use trisodium citrate-2 H_2O 120 g/liter (pH 7.0 with citric acid) as bridge buffer; histidine 0.75 g plus NaCl 1.5 g/liter (pH 7.0) as gel buffer. Prepare the gel as for detecting transaldolase. The developing solution is made as follows: 0.5 M Tris-HCl buffer, pH 7.0, 10 ml; 6-phosphogluconate (trisodium salt, Sigma) 250 mg; NADP, 20 mg; NBT, 20 mg; phenazine methosulfate, 2 mg; distilled water, 90 ml.

Differentiation of the species of the genus **Bifidobacterium**

The differential characteristics of the species of *Bifidobacterium* are indicated in Table 15.50. Other characteristics are cited in Tables 15.51–15.55.

List of species of the genus **Bifidobacterium**

1. **Bifidobacterium bifidum** (Tissier 1900) Orla-Jensen 1924, 472.[AL] (*Bacillus bifidus* Tissier 1900, 86.)

bi′fi.dum. L. neut. adj. *bifidum* cleft, divided.

Although the cells are highly variable in appearance, some traits observed in cells grown in TPY agar stabs are distinct (Fig. 15.96*A*). Groupings of "amphora-like" cells are characteristic. (Sundman et al., 1959).

Distinction of two serovars have been made; serovar *a* predominates in the feces of human adults while *b* predominates in that of neonates (variants *a* and *b*, Reuter, 1963; these variants differ in sucrose, melibiose and maltose fermentation).

Type strain: Ti (Tissier) (ATCC 29521).

2. **Bifidobacterium longum** Reuter 1963, 502.[AL]

long′um. L. neut. adj. *longum* long.

In TPY agar strains of this species show mostly very elongated and relatively thin cellular elements with slightly irregular contours and rare branchings (Fig. 15.96*B*).

In TPY broth most strains develop with uniform turbidity, clearing is slow and the sediment is viscous.

Two biovars are distinguished: biovar *a*, more frequent in human adults and slowly fermenting mannose and biovar *b*, more frequent in neonates and mannose negative (Reuter, 1963). *B. longum* subsp. *animalis*, with two biovars *a* and *b*, was distinguished from *B. longum* subsp. *longum* on the basis of melezitose and mannose fermentation and ecology (Mitsuoka, 1969, 60). The biovars *a* and *b* have been recognized on DNA-homology as belonging to the species *B. animalis* Scardovi and Trovatelli, and to *B. pseudolongum* Mitsuoka respectively (Scardovi et al., 1971b; Scardovi and Trovatelli, 1974).

B. longum is apparently the only species among those usually found in human feces which possesses a large variety of plasmids (Sgorbati et al., 1982).

On DNA-homology values (50–76%), this species is most closely related to *B. infantis* (see further comments under *B. infantis*).

Type strain: ATCC 15707 (E 194 b, from feces of a human adult, Reuter, 1971).

3. **Bifidobacterium infantis** Reuter 1963, 502.[AL]

in.fant′ is. L. n. *infantis* an infant; M.L. gen. n. *infantis* of an infant.

Cellular morphology does not present specific traits so that it is similar to that of many other species of the genus (Fig. 15.98*D*).

These bifidobacteria do not ferment pentoses; found as predominant forms in breast-fed infant feces, distinguished biochemically and serologically and allotted to the species, *B. infantis* by Reuter (1963).

Bifid strains of the same ecology but fermenting xylose, were separated and distributed, on the basis of some differences in other sugars fermented, into two additional species, namely "*B. liberorum*" and "*B. lactentis*," (Reuter, 1963). DNA-homology studies indicated later that these two species were identical to *B. infantis* (Scardovi et al., 1971b).

"*Bifidobacterium parabifidum*" (Weiss and Rettger) Kandler and Lauer 1974, 40, is here considered as a synonym of *B. infantis*, because the strain ATCC 17930 (strain Timberlain isolated by Norris et al., 1950, and studied by Pine and Howell, 1956, under the label 308), displayed (a) 82% DNA homology with ATCC 27920, one of the reference strains of *B. infantis*; (b) 76% DNA homology with ATCC 15707, type strain of *B. longum*; and (c) possessed the proteins pattern of *B. infantis* (Biavati et al., 1982).

On DNA homology values (50–70%) this species is most closely related to *B. longum*.

Type strain: ATCC 15697 (S 12, from feces of human infant, Reuter, 1971).

Comments on the identification of *B. longum* and *B. infantis*. Arabinose is reported as not fermented by *B. infantis* ("*B. liberorum*" and "*B. lactentis*") while *B. longum* characteristically ferments melezitose (Reuter, 1963; Mitsuoka, 1969). Numerous strains isolated from infants' feces, which fermented arabinose or failed to ferment melezitose, were recognized by means of DNA-DNA hybridization as *B. infantis* and *B. longum*, respectively, (Scardovi, unpublished). The figures of DNA homology reported in Table 15.52 were obtained with these strains.

Isozyme patterns for transaldolase and 6PGD were determined in 63 and 126 strains allotted on DNA homology to *B. infantis* and *B. longum*, respectively: 90% of strains of *B. infantis* had the transaldolase (isozyme 5; migration 100) more anodal than that possessed by 72% of *B. longum* strains (isozyme 8, migration 90) (Scardovi et al., 1979a).

The electrophoretic patterns of soluble proteins distinguished the two species (Biavati et al., 1982).

Bifid strains isolated from the feces of the suckling calf could not be referred to either species, because they were more than 80% related on DNA homology both to *B. infantis* and *B. longum* reference strains; conversely, *B. infantis* and *B. longum* were not distinguishable when their DNA was annealed to that of the calf strains (Scardovi, unpublished). Although electrophoresis revealed that all these strains possess an "intermediate" form of transaldolase (isozyme 6, migration 96) (Scardovi et al., 1979a) and a distinct total proteins pattern (Biavati et al., 1982), no doubt the boundaries between the two species *B. longum* and *B. infantis* are tending to disappear. However 70% of the strains genetically *B. longum* carry a large array of plasmids whereas none of the strains of either *B. infantis* or "intermediates" from the calf do so, although strains of both species were isolated from the same specimens (Sgorbati et al., 1982). Pending studies on the possible taxonomic and evolutionary significance of plasmids in bifidobacteria, it is advisable to keep these two taxonomic entities separate, so diverse are they in their extrachromosomal DNA complements.

The following rule can be adopted: strains not fermenting arabinose should be retained as *B. infantis* while strains fermenting both arabinose and melezitose should be retained as *B. longum*; strains fermenting arabinose (and xylose) but not melezitose are *B. infantis* (reference strain B1269, ATCC 27920) if their transaldolase migrates more anodically (isozyme 5, migration 100) or *B. longum* (reference strain E194b, ATCC 15707) if their isozyme is less anodal (isozyme 8, migration 90); strains with an isozyme moving to an intermediate position (isozyme 6, migration 96) should be considered as "intermediates" (reference strains VT29 and VT42 from Collection of the Institute for Agricultural Microbiology, University of Bologna).

4. **Bifidobacterium breve** Reuter 1963, 502.[AL]

bre′ve. L. neut. adj. *breve* short.

The cell morphology suggests the specific epithet; the thinnest and shortest cells among bifids found in the human intestine (Fig. 15.96*C*).

Among the bifids not fermenting pentoses isolated from newborn infants, the rank of species was attributed to those fermenting mannitol and sorbitol; two biovars *a* and *b* different towards melezitose were recognized (Reuter, 1963). Serologically related strains isolated from breast-fed older infants and which did not ferment mannitol or sorbitol were referred to a separate species, namely "*B. parvulorum*" (Reuter, 1963).

Strains S50 (ATCC 15698), type strains of "*B. parvulorum*" biovar *a*, strain S17c (ATCC 15699) "*B. parvulorum*" biovar *b*, and strain S 46 (ATCC 15701) *B. breve* biovar *b*, were 88, 94 and 86%, respectively, related on DNA homology to strain S1 (ATCC 15700), type strain of *B. breve* biovar *a*; other DNA homology tests proved the genetic identity of the two Reuter's species (Scardovi et al., 1971b).

On the basis of DNA homology *B. breve* is more closely related to both *B. infantis* and *B. longum* than to any other species of the genus (40–60% homology).

One hundred and six strains of *B. breve*, including a few strains fermenting arabinose and xylose, all identified by DNA-DNA homology test, were studied for their transaldolase and 6PGD isozymes: Fifty-seven percent of the strains had unique zymograms and 25% displayed the same zymogram as *B. coryneforme*, a bifid found only in the intestine of the honey bee (Scardovi et al., 1979a).

Type strain: ATCC 15700 (S1 from feces of human infant, Reuter, 1971).

5. Bifidobacterium adolescentis Reuter 1963, 502.[AL]

a.do.les.cen'tis. L. n. *adolescens*: M.L. gen. n. *adolescentis* of an adolescent.

The cellular morphology is common to that of many other species of the genus (Fig. 15.98*D*). Reuter (1963) named as *B. adolescentis* those pentose-fermenting bifidobacteria he first found to predominate, along with *B. longum*, in the feces of human adults. Four biovars, *a*, *b*, *c* and *d* varied in fermentation of sorbitol and mannitol, and were serologically distinct (Reuter, 1963).

Among those species which are regularly found in man, *B. adolescentis* occurs most frequently in sewage (Scardovi et al. 1979a; Scardovi, unpublished). Reuter's biovars *b* and *d*, namely those which do not ferment sorbitol, cannot be distinguished phenotypically from *B. dentium* (see Table 15.51); however, transaldolase isozymes migrate differently, i.e. *B. dentium* isozyme 4 migrates 100, while that of *B. adolescentis* (number 8) migrates 87 (Scardovi et al., 1979a).

The PAGE procedure can be used alternatively (Biavati et al. 1982).

Type strain: ATCC 15703 (E194a from feces of human adult, Reuter, 1971).

6. Bifidobacterium angulatum Scardovi and Crociani 1974, 19.[AL]

an.gu.la'tum. L. Part. adj. *angulatus* with angles, angular.

Cells grown in TPY agar stabs generally and characteristically disposed in V (angular) or palisade arrangements similar to corynebacteria; rarely enlarged at the extremities; branching absent. This morphological type is unique among bifidobacteria (Fig. 15.96*D*).

Anaerobic but more sensitive to O_2 than most bifidobacteria (measured by the depth of growth in stabs). CO_2 does not affect this sensitivity but it strongly enhances anaerobic growth.

First isolated from adult human feces and then found in sewage.

Type strain: ATCC 27535 (B677, from feces of human adult, Scardovi and Crociani, 1974).

Twenty percent of strains of *B. angulatum* ferment sorbitol and can thus readily be distinguished from other species fermenting this sugar (Table 15.55). Most strains do not ferment sorbitol and so could be confused (in the case of doubtful morphology) with *B. globosum*, *B. pseudolongum* and sorbitol-negative strains of *B. pseudocatenulatum* from calf feces (nearly half of the strains from this source are inactive toward sorbitol; Scardovi et al. 1979 b). *B. angulatum* possesses the least anodal transaldolase (number 5) in respect of the named taxa.

7. Bifidobacterium catenulatum Scardovi and Crociani 1974, 18.[AL]

ca.te.nu.la'tum. M.L. adj. *catenulatum* having small chains.

Cells grown in TPY agar stabs are generally and characteristically arranged in chains of three, four or more globular elements. The distal ends of the chains are usually tapered. Distinct branchings, club-like swellings or spatula-like extremities are rare (Fig. 15.96*E*)

Anaerobic. CO_2 without effect on O_2 sensitivity or anaerobic growth. Riboflavin and pantothenate required for growth.

Although most strains of *B. catenulatum* ferment sorbitol but not mannitol similar to biovar *c* of *B. adolescentis* and *B. pseudocatenulatum* (see below), they can be distinguished from the former because they ferment melezitose and from the latter because they do not ferment starch (see Table 15.55).

Found in feces of human adult and in sewage.

The mol% G + C of the DNA is 55 (T_m). This is the lowest value found in bifidobacteria.

Type strain: ATCC 27539 (B669 from feces of human adult, Scardovi and Crociani, 1974).

Most strains of the species show DNA relatedness to *B. adolescentis* (reference DNA from strain E298b ATCC 15705, biovar *c* of Reuter) in the range 30–57% (Scardovi and Crociani, 1974): *B. catenulatum* should be considered (as also should *B. pseudocatenulatum*, see below) to be more related to *B. adolescentis* than to any other species of the genus on this basis.

8. Bifidobacterium pseudocatenulatum Scardovi, Trovatelli, Biavati and Zani 1979, 309.[AL]

pseu.do.ca.te.nu.la'tum. Gr. adj. pseudes false; L. adj *catenulatum* specific epithet; M.L. adj. *pseudocatenulatum* the false (*B.*) *catenulatum*.

Cell morphology is one of the most variable among bifidobacteria in that it shows highly diverse traits according to strain and origin; (Scardovi et al. 1979b).

Anaerobic; CO_2 is without effect upon sensitivity or anaerobic growth.

Riboflavin, pantothenate and nicotinic acid required for growth (Table 15.53).

Sorbitol is fermented by all strains found in the feces of infants and by 50% of those found in calf feces. The sorbitol fermenting strains can be distinguished from *B. adolescentis* strains because they are melezitose-negative and from *B. catenulatum* because they invariably ferment starch. The sorbitol-negative strains of this species (so far found only in calf feces) can be distinguished from *B. globosum* or *B. pseudolongum* species also found in calf feces, and from *B. angulatum*, not only on the basis of different morphology, but also by means of their transaldolase zymogram (see Table 15.50).

The mol% G + C of DNA is 57.5 (T_m) (see Table 15.50).

The DNA of this species is 46–88% related to that of *B. catenulatum* but is virtually unrelated to that of any other species of the genus (see Table 15.52; Scardovi et al. 1979b).

The structure of the interpeptide bridge of cell wall peptidoglycan is shared by *B. catenulatum* and *B. angulatum* (see Table 15.50).

Found abundantly in sewage, in the feces of breast- and bottle-fed infants and in the feces of suckling calves.

Type Strain: ATCC 27919 (B1279 from feces of human infant, Scardovi et al., 1979b).

Differentiation of *B. catenulatum* from *B. pseudocatenulatum*. There is no doubt that these two species are closely related as indicated by their DNA homology. However, the G + C content of their DNA differs by 3 mol% (see Table 15.50), the same difference which exists between *B. globosum* and *B. pseudolongum* (see below); none of the strains shown genetically to be *B. catenulatum* ferments starch or mannose, whereas strains recognized by DNA homology as *B. pseudocatenulatum* ferment these compounds; *B. catenulatum* does not require nicotinic acid for growth as does *B. pseudocatenulatum*; furthermore, *B. catenulatum* has so far never been isolated from the feces of suckling calves.

Forty-one strains of *B. catenulatum* and 120 strains of *B. pseudocatenulatum* were studied for their transaldolase and 6PGD isozymes content. The isozymes of these two species migrate very differently (see Table 15.50); the few *B. pseudocatenulatum* strains (7 out of 120) which possess a transaldolase electrophoretically identical to that of *B. catenulatum* (number 5), display a much more anodal 6PGD isozyme

Table 15.50.
Characteristics differentiating the species of the genus **Bifidobacterium**

Species	Mol% G + C of DNA[a]	Murein type[b]	Electrophoretic patterns of enzymes[c]			Immunological specificity group[f]	Cell morphology	Found in[g]
			Transaldolase	6PGD[d]	F6PPK[e]			
1. *B. bifidum*	58(Bd)	Orn(Lys)-D-Ser-D-Asp	7	7-(8)	15	B	Distinct (Fig. 15.96A)	Feces of human infant and adult; human vagina; feces of suckling calf
2. *B. longum*	58(Bd)	Orn(Lys)-Ser-Ala-Thr-Ala	(5)-6-**8**	5-(**6**)	15	B		Feces of human adult and infant; (human clinical)
3. *B. infantis*	58(Bd)	Orn(Lys)-Ser-Ala-Thr-Ala	5-(6)-(**8**)	(3)-**4**-(5)	15	B		Feces of human infant; (human vagina)
4. *B. breve*	58(Bd)	Lys-Gly	6	(5)-6-6$_a$-**7**	15	B	Distinct (Fig. 15.96C)	Feces of human infant and adult; (human vagina and clinical)
5. *B. adolescentis*	58(Bd)	Lys(Orn)-D-Asp	8	5	15	B		Feces of human adult; sewage; (rumen of cattle; feces of monkey and dog)
6. *B. angulatum*	59($_{Tm}$)	Lys(Orn)-D-Asp	5	5	15	A	Distinct (Fig. 15.96D)	Feces of human adult; sewage
7. *B. catenulatum*	55($_{Tm}$)	Lys(Orn)-Ala$_2$-Ser$_{0.2-1.0}$	5	**6**-8	15	A	Distinct (Fig. 15.96E)	Feces of human adult and infant; (sewage)
8. *B. pseudocatenulatum*	57.5($_{Tm}$)	Lys(Orn)-Ala$_2$-Ser$_{0.2-1.0}$	4-(5)	**1**-3		A		Feces of human infant and suckling calf; sewage
9. *B. dentium*	61($_{Tm}$)	Lys(Orn)-D-Asp	4	(2)[d]	15	A		Human dental caries and clinical (feces of human adult and infant; human oral cavity and vagina)
10. *B. globosum*	64($_{Tm}$)	Orn(Lys)-Ala$_{2-3}$	2	(3$_a$)-(4)-(5)-**6**-(7)	10	E		Feces of pig, suckling calf, rat, rabbit and lamb; rumen of cattle; (sewage)
11. *B. pseudolongum*	60($_{Tm}$)	Orn(Lys)-Ala$_{2-3}$	2	7	10	E		Feces of chicken, cattle, rat and mice
12. *B. cuniculi*	64($_{Tm}$)	Orn(Lys)-Ser(Ala)-Ala$_2$	1	4		C	Distinct (Fig. 15.97C)	Feces of rabbit
13. *B. choerinum*	66($_{Tm}$)	Orn(Lys)-Ser(Ala)-Ala$_2$	3	4				Feces of piglet; (sewage)
14. *B. animalis*	60($_{Tm}$)	Orn(Lys)-Ser(Ala)-Ala$_2$	5	**8**-9	10		Distinct (Fig. 15.97B)	Feces of rat, chicken, rabbit, calf, sewage
15. *B. thermophilum*	60($_{Tm}$)	Orn(Lys)-D-Glu	(7)-**8**	7-8-**9**-(9$_a$)	10	D		Feces of pig, piglet, chicken, calf, rumen of cattle; sewage
16. *B. boum*	60($_{Tm}$)	Lys-D-Ser-D-Glu	6	**8**-9-9$_a$		D		Rumen of cattle; feces of piglet
17. *B. magnum*	60($_{Tm}$)	Lys(Orn)-Ala$_2$-Ser$_{0.2-1.0}$	5	7	10	C′	Distinct (Fig. 15.97D)	Feces of rabbit
18. *B. pullorum*	67($_{Tm}$)	Lys(Orn)-D-Asp	2	Absent[d]	10		Distinct (Fig. 15.97A)	Feces of chicken
19. *B. suis*	62($_{Tm}$)	Orn(Lys)-Ser-Ala-Thr-Ala	6	5-8	10	G		Feces of piglet
20. *B. minimum*	61.5($_{Tm}$)	Lys-Ser	10	6	10	F	Distinct (Fig. 15.97F)	Sewage
21. *B. subtile*	61.5($_{Tm}$)	Lys(Orn)-D-Asp	3	2	10-15	F	Distinct (Fig. 15.97E)	Sewage

	Cell wall type	T_m (mol% G+C)	Transaldolase isozymes	6PGD isozymes	F6PPK	H-type	Total cellular proteins (electrophoresis)	Source
22. *B. coryneforme*	Lys(Orn)-D-Asp		(6)-(7)-(7_a)-**8**-(8_a)-(8_b)-(9)-(9_a)	6	16		Distinct (Fig. 15.98C)	Intestine of *Apis mellifera* L. subsp. *mellifera*
23. *B. asteroides*	Lys-Gly	59 (T_m)	6	(9)-(9_a)-(9_b)-(10)-**10_a-10_b**-(11)-(12)-(13)	16	H	Distinct (Fig. 15.98A)	Intestine of *A. mellifera* L. subspp. *mellifera, ligustica* and *caucasica;* (*A. cerana* F.)
24. *B. indicum*	Lys(Orn)-D-Asp	60 (T_m)	(6)-7-**8-9**	6-6_a-(7)-8-**(9)**-(9_a)-(9_b)	16	H	Distinct (Fig. 15.98B)	Intestine of *A. cerana* F. and *A. dorsata* F.

[a] Data based on buoyant density (Bd) were taken from Gasser and Mandel, 1968; for the other values (T_m) see under single species description.

[b] Taken from Kandler and Lauer, 1974, or kindly provided by O. Kandler, Institute of Botany, University of Münich, Germany.

[c-d] Taken from Scardovi et al., 1979a; 6PGD = 6-phosphogluconic dehydrogenase (NADP+). Numbers 1 to 10 and 1 to 13 were given to isozymes of transaldolase and 6PGD, respectively, in the order of decreasing anodic mobility (suffixed numbers indicate additional isozymes). Boldface numbers are the isozymes of the type strains. Numbers in parentheses are isozymes found in less than 10% of the strains studied. 6PGD in undetectable by spot-staining in most strains of *B. dentium* and in all strains of *B. pullorum.*

[e] Taken from Scardovi et al., 1971a; F6PPK = fructose-6-phosphate phosphoketolase. Numbers indicate the migration relative to that of *B. globosum* taken = 10. Phosphoketolases of migration 15 and those of migration 10 were ecologically distinguished as "human" and "animal" type, respectively (Sgorbati et al., 1976).

[f] Taken from Sgorbati and London, 1982.

[g] Sources are listed in the order of decreasing frequency of occurrence of the species therein. Sources in parentheses are occasional.

(Table 15.50), Scardovi et al., 1979a). Although some data suggest the existence of "intermediate" strains (Scardovi et al., 1979b) like those found between *B. longum* and *B. infantis* (see above) and, although PAGE procedure gave uncertain results (Biavati et al., 1982), it is advisable at present to maintain the rank of species for these two taxa.

9. **Bifidobacterium dentium** Scardovi and Crociani 1974, 18.[AL]

den'tium. L. mas. n. *dens* tooth; L. plural gen. n. *dentium* of teeth.

Cells grown in TPY agar show a general morphology resembling that of *B. infantis* (Fig. 15.98D) and, hence, are without any characteristic appearance.

Anaerobic: CO_2 does not affect sensitivity to O_2 or anaerobic growth.

A number of bifid strains isolated from human dental caries, feces of human adult and human vagina, phenotypically assigned to *B. adolescentis* were first recognized as forming a distinct "dentium" DNA homology group, together with strains isolated from the oral cavity (Beerens et al., 1957) and strain 3859 labeled *B. appendicitis*. The species *B. dentium* (Table 15.52) has some DNA relatedness to *B. adolescentis*, but is related less or not at all to other species of the genus. DNA of strains *Actinomyces eriksonii* ATCC 15423 and ATCC 15424 are completely homologous to that of *B. dentium* reference strain B764 (Scardovi et al. 1979a).

B. dentium requires riboflavin and pantothenate for growth, similar to *B. angulatum* and *B. catenulatum*.

Type strain: ATCC 27534 (B764, from human dental caries, Scardovi and Crociani, 1974).

Comments on the identification of *B. dentium*. *B. dentium* cannot be distinguished phenotypically from biovars b and d of *B. adolescentis* (strains not fermenting sorbitol, see Table 15.55): hence the distinction between these two species should be based either on transaldolase isozymes or on electrophoresis patterns of total cellular proteins (see under *B. adolescentis*).

10. **Bifidobacterium globosum** (ex Scardovi, Trovatelli, Crociani and Sgorbati 1969) Biavati, Scardovi and Moore 1982, 368.[VP] Considered (Rogosa, 1969) as a synonym of *B. pseudolongum* Mitsuoka; not reported in the Approved Lists (1980); its revival proposed by Biavati et al. (1982).

glo.bo'sum. from L. n. *globus* bell, sphere.

Cells grown anaerobically in TPY agar are generally short, coccoid or almost spherical to curved or tapered, arranged singly or doubly or rarely in short chains (Fig. 15.96F). This morphology, shared by *B. pseudolongum* Mitsuoka, does not change with strain or source; only cells grown in air + CO_2 have bifurcations or short cross-branchings, often with enlarged ends.

Anaerobic aerotolerant organisms. CO_2 is without effect on anaerobic growth but permits development in high O_2 tensions (slope incubated in 90% air + 10% CO_2). Aerobically grown cells do not show catalase or catalase-like activity (hemin).

Pantothenate, riboflavin, thiamin and folic acid required for anaerobic growth. Ammonia satisfies nitrogen requirements.

Initially the strains allotted to this species were isolated from the bovine rumen; they did not ferment xylose, mannose and cellobiose, and only rarely arabinose (Scardovi et al., 1969). Pentose-fermenting strains were subsequently isolated from piglets and recognized as *B. globosum* on DNA homology (Zani et al., 1974). Many other strains genetically assigned to *B. globosum* were later isolated from feces of various animals and included strains fermenting mannose, cellobiose or mannitol as well as some slowly fermenting or inactive toward fructose (Scardovi et al., 1979a).

Intraspecific DNA homologies ranged from 78–106%: relatedness to reference of *B. pseudolongum* varied in the range 69–73% (Scardovi et al., 1971b). Range 50–67% of homology was observed between *B. globosum* and *B. cuniculi* (Table 15.52).

Unlike most species of the genus, the enzyme fructose-bisphosphate aldolase and both the hexose monophosphate (HMP) dehydrogenases can be detected in cell-free extracts. Functioning of aldolase in intact cells was proved by the expected increase of acetate by degrading

Table 15.51.

Fermentative characteristics distinguishing the species of the genus **Bifidobacterium**[a]

	D-Ribose	L-Arabinose	Lactose	Cellobiose	Melezitose	Raffinose	Sorbitol	Starch	Gluconate
1. *B. bifidum*	−	−	+	−	−	−	−	−	−
2. *B. longum*	+	+	+	−	+	+	−	−	−
3. *B. infantis*	+	−	+	−	−	+	−	−	−
4. *B. breve*	+	−	+	d	d	+	d	−	−
5. *B. adolescentis*	+	+	+	+	+	+	d	+	+
6. *B. angulatum*	+	+	+	−	−	+	d	+	d
7. *B. catenulatum*	+	+	+	+	−	+	+	−	d
8. *B. pseudocatenulatum*	+	+	+	d	−	+	d	+	d
9. *B. dentium*	+	+	+	+	+	+	−	+	+
10. *B. globosum*	+	d	+	−	−	+	−	+	−
11. *B. pseudolongum*	+	+	d	d	d	+	−	+	−
12. *B. cuniculi*	−	+	−	−	−	−	−	+	−
13. *B. choerinum*	−	−	+	−	−	+	−	+	−
14. *B. animalis*	+	+	+	d	d	+	−	+	−
15. *B. thermophilum*	−	−	d	d	d	+	−	+	−
16. *B. boum*	−	−	d	−	−	+	−	+	−
17. *B. magnum*	+	+	+	−	−	+	−	−	−
18. *B. pullorum*	+	+	−	−	−	+	−	−	−
19. *B. suis*	−	+	+	−	−	+	−	−	−
20. *B. minimum*	−	−	−	−	−	−	−	+	−
21. *B. subtile*	+	−	−	−	+	+	+	+	+
22. *B. coryneforme*	+	+	−	+	−	+	−	−	+
23. *B. asteroides*	+	+	−	+	−	+	−	−	d
24. *B. indicum*	+	−	−	+	−	+	−	−	+

[a] Symbols: see Table 15.25.

glucose in the presence of iodoacetic acid (Schramm et al., 1958; Scardovi and Trovatelli, 1969). Glucose-6-phosphate dehydrogenase, undetectable in some strains, was found to be either NAD⁺ or NADP⁺ dependent (Scardovi and Sgorbati, 1974).

Zymograms of 103 strains of *B. globosum* were examined. All possessed the transaldolase isozyme 2 and more than 80% contained the 6PGD isozyme 6; four other 6PGD isozymes were found (Scardovi et al., 1979a: Table 15.50).

PAGE proteins patterns of selected strains from various sources were identical and quite distinct from those of any other species of the genus, including those of the most closely related species, *B. pseudolongum*, thus further indicating the validity of the species *B. globosum*, the revival of which was proposed by Biavati et al. (1982).

Plasmids of large molecular weights (13.5, 24.5 and 46 MDa) were found in 22% of *B. globosum* strains studied (Sgorbati et al., 1982). The mol% G + C of the DNA is 63.8 ± 0.4 (T_m) (Scardovi et al., 1971b). Originally isolated from the bovine rumen (Scardovi et al., 1969), later found in the feces of the piglet, rat, lamb, chicken, rabbit, calf, sewage and in a single specimen of feces of the human infant (Scardovi et al., 1979a).

Type strain: ATCC 25865 (RU 224 from bovine rumen, Scardovi et al., 1969).

11. Bifidobacterium pseudolongum Mitsuoka 1969, 60.[AL]

pseu.do.long'um. Gr. adj. *pseudes* false; L. adj. *longum* specific epithet; M.L. neut. adj. *pseudolongum* false (*B.*) *longum*.

Cells grown in TPY agar are morphologically identical to those of *B. globosum*.

The relations to O₂ and CO₂ have not been studied in detail. Requirements for growth factors are unknown.

Mitsuoka (1969) recognized as belonging to this species bifid strains isolated from a variety of animals; these strains fermented arabinose, xylose, starch and glycogen but slowly fermented fructose, thus differing from strains of *B. thermophilum* isolated from the same sources (Mitsuoka, 1969). Four biovars, *a, b, c* and *d* were recognized on the basis of differences in the fermentation of mannose, lactose, cellobiose and melezitose (Mitsuoka, 1969).

Fermentative characters correspond to those reported for *B. globosum* (see Tables 15.51 and 15.54).

The presence of fructose-bisphosphate aldolase or glucose-6-phosphate-dehydrogenase in cell-free extracts was not investigated.

On electrophoresis, the type strain PNC-2-9G (ATCC 25526 biovar *a*), strain 29-Sr-T representative of biovar *c* and strain Mo-2-10 representative of biovar *d*, all displayed the transaldolase isozyme 2 and the 6PGD isozyme 7, i.e. a pattern identical to that possessed by only two *B. globosum* strains of the 103 studied (Table 15.50; Scardovi et al., 1979a).

PAGE proteins patterns of some strains, including type strain ATCC 25526, were clearly distinct from those of *B. globosum* (Biavati et al., 1982).

With DNA homology, *B. pseudolongum* is related 65–70% to *B. globosum* and 10–15% to *B. cuniculi*. The Mitsuoka strain C10-45 (*B. longum* subsp. *animalis* biotype *b*, Mitsuoka 1969) is completely homologous to strain Mo-2-10 representative of biovar *d* of *B. pseudolongum* (Scardovi et al., 1971b).

Plasmids were absent in three strains studied (Sgorbati et al., 1982).

The mol% G + C of the DNA is 60.3 ± 0.45 (Scardovi et al., 1971b).

Type strain: ATCC 25526 (PNC-2-9G from feces of swine, Mitsuoka, 1969).

Comments on the differentiation between *B. globosum* and *B. pseudolongum*. Phenotypic characters cannot be used to distinguish between these two species. The interpeptide bridge of the cell wall peptidoglycan has the same structure. A similar ecological distribution is shared by the two species. Their taxonomic separation is based on (a) different content of G + C mol% of their DNA, (b) their reciprocal DNA homology and the DNA homology to other related species, (c) distinct PAGE total cellular proteins patterns, (d) differences in 6PGD zymograms and (e) differences in plasmid cellular complement. On DNA homology, *B. globosum* is much more closely related to *B. cuniculi* (50–67%) than is *B. pseudolongum* (8–16%) (Table 15.52). Differences in isozyme and plasmid content, even if biased by the small number of strains of *B. pseudolongum* investigated, are striking.

As indicated in Table 15.55, the practical distinction between the two species can be based on electrophoresis of the 6PGD dehydrogen-

Table 15.52
DNA homology relationships[a] of the species of genus **Bifidobacterium**

Competitor DNA from	Percent homology[b] to reference DNA from:[c]																						
	1. B. bifidum	2. B. longum	3. B. infantis	4. B. breve	5. B. adolescentis	6. B. angulatum	7. B. catenulatum	8. B. pseudocatenulatum	9. B. dentium	10. B. globosum	11. B. pseudolongum	12. B. cuniculi	13. B. choerinum	14. B. animalis	15. B. thermophilum	16. B. boum	17. B. magnum	18. B. pullorum	19. B. suis	20. B. minimum	21. B. subtile	23. B. asteroides	24. B. indicum
1. B. bifidum	100	2	20–28	10–20	0		25	15		0		10	27	10	5[d]	11	5	20	0	4	4	0	0
2. B. longum	40	75–101[e]	50–76[e]	12			20	22		0	75[e]	17	16	5	0[d]	23	0–29	9	0[f]	7	23	0	0
3. B. infantis	42	50–79	74–101	17–50	40		20	15		0		16–23	20–22	15[g]	5[d]	11	15[g]	13	0	4[g]	5[g]	0	0
4. B. breve	25	12	17–50	100	5–10		24	10		0		12	21	20	5[d]	18	10	18	0	10	17	0	0
5. B. adolescentis	14	22	40	5–10	70–102	20–44	20–57	30	24–49			25	20	30	18[d]	25	14	21	20	5	13		
6. B. angulatum	24	6–13	20[h]	12	20–30	76–100	20–35	20	8–20	20		17	16	11		10	5	0	25	23	8		
7. B. catenulatum	20	0–26	15[h]	10	22–57	2–37	78–101[e]	50–80[e]	3–48	5		10	32	10	20[d]	8	17	12	25	30	12		
8. B. pseudocatenulatum	20	0–13	20[h]	12		5–26	46–88	78–115	18	12		10	32	6	20[d]	5			20	30	12		
9. B. dentium	29	25	25	12	15–57		16–45	18	69–110			17	15	6	20[d]		8	0	20	5	0	13	12
10. B. globosum	35	17	25	25				5		100	75[e]	45	50	40	14[d]	14	32	16	10	35	10		
11. B. pseudolongum	7	8	17	7						65–70	100	10–15	45–50	18–25	12–20[d]	10		2	9–27				
12. B. cuniculi	32	8	17	22			10	10		50–67	8–16	94–102	9–15	4–20	15–35[d]	5		4					
13. B. choerinum	10	29	10[h]	15	20		25	25		26–57	37–62	8–20	75–120	72–100	25[d]	20	15	0	20	5	10		
14. B. animalis	22	7	29	20	25	10	10	10	4	30	34	31–26	19	72–100		21	15	10	5–12	4[i]	15[i]	0[i]	
15. B. thermophilum	38	4	7	10–22				16[i]		5[i]	0–23	15–25	15–20	23[i]	79–117	27–80[e]	11[i]	0		4[i]	15[i]	5	0[i]
16. B. boum	10	25		23			5	5				8	23	11	36–74	69–96	2		9	5	7	0[i]	0[i]
17. B. magnum	10	4	25	9	6	5	26	20	21	6		10	10	23	20[d]	12	75–106		9	20	7		
18. B. pullorum	27	39	29	33			12	12		23	0	25	37	35		13		94–107	13	10	8	13	12
19. B. suis	28	29	29	22			25	25		23		21	21	5	10[d]	7	10	10	100	10	8	5	5
20. B. minimum	5	10	6[h]	10	4	20	25	10	0	5	8	23	36	5	10[d]	0	19	8	7	100–103	16–20	3	0
21. B. subtile	5	20	5[h]	14	15	11	8	10	0	8		14	34	5	18[d]	0	0	9	11	10–31	70–100	5	4
22. B. coryneforme				6			20	5		5		20	20	0	0[d]		6	6	0	0	2	100	60
23. B. asteroides				0			25			0		11	50		0[d]				0	0	0	100	30
24. B. indicum	11	20	11	11				5		0		20	44		10[d]		17	17		0	17	23	100

[a] DNA immobilized to filter and the single-point competition procedure of Johnson and Ordal (1968) were used throughout.

[b] Data from Scardovi et al., 1970; Scardovi et al., 1971b; Scardovi and Trovatelli, 1974; Scardovi and Zani, 1974; Scardovi and Crociani, 1974; Scardovi et al., 1979b; Trovatelli et al., 1974; Scardovi, unpublished (italicized values).

[c] Species 22. B. coryneforme was not used as reference: the DNA from this species (type strain) was used as competitor. Type strains were generally used as source of reference DNA except for B. longum, B. infantis, B. catenulatum, B. pseudocatenulatum, B. thermophilum and B. boum: in these cases more than one strain were used as source of reference DNA.

[d] DNA from strain RU326 (ATCC 25866) (B. ruminale) used as reference.

[e] Groups of most closely related species (boxed). Large numbers and more than one strain used as competitor and reference respectively. Results are reported as ranges of the obtained values.

[f] DNA from B. longum strain B816 isolated from human vagina used as competitor.

[g] DNA from strain S76e Reuter (ATCC 15702) (B. liberorum) used as competitor.

[h] DNA from strain S76e Reuter (ATCC 15702) (B. liberorum) used as reference.

[i] DNA from strain RU326 (ATCC 25866) (B. ruminale) used as competitor.

Table 15.53.

Vitamin and growth factor requirement of some species of the genus **Bifidobacterium**[a, b]

Species	Riboflavin	Pantothenate	Nicotinic acid	Pyridoxine	Thiamine	Folic acid	p-Aminobenzoic acid	Biotin	Tween 80
1. *B. bifidum*[c]	+	−	+	−	−	+	+	−	+[d]
3. *B. infantis*[c]	−	+	−	−	−	−	−	−	+
5. *B. adolescentis*[c]	+	+	+	+	+	−	−	−	−
6. *B. angulatum*[c]	+	+	−	−	−	−	−	−	−
7. *B. catenulatum*[c]	+	+	−	−	−	−	−	−	−
8. *B. pseudocatenulatum*[c]	+	+	+	−	−	−	−	−	−
9. *B. dentium*[c]	+	+	−	−	−	−	−	−	−
10. *B. globosum*[c]	+	+	−	−	+	+	−	−	−
15. *B. thermophilum*[c]	+	+	−	+	−	−	−	−	−
18. *B. pullorum*[c]	−	−	+	+	+	+	+	−	+
19. *B. suis*[e]	+	−	−	−	−	−	−	−	−
23. *B. asteroides*[f]	+	+	+	+	+	−	−	+	−

[a] Data from Trovatelli and Biavati (1980).
[b] Symbols: +, required; −, not required.
[c] Ten strains, type strains included, were studied from each species.
[d] Only stimulatory.
[e] Data from Matteuzzi et al. 1971.
[f] Data from Scardovi and Trovatelli, 1969.

Table 15.54.

Additional fermentation reactions in the genus **Bifidobacterium**[a]

Species	Xylose	Mannose	Fructose	Galactose	Sucrose	Maltose	Trehalose	Melibiose	Mannitol	Inulin	Salicin
1. *B. bifidum*	−	−	+[b]	+	d[c]	−[d]	−	d	−	−	−
2. *B. longum*	d	d	+	+	+	+	−	+	−	−	−
3. *B. infantis*	d	d	+	+	+	+	−	+	−	d	−
4. *B. breve*	−	+	+	+	+	+	d	+	−	d	−
5. *B. adolescentis*	+	d	+	+	+	+	d	+	d	d	+
6. *B. angulatum*	+	−	+	+	+	+	−	+	−	+	+
7. *B. catenulatum*	+	−	+	+	+	+	d	+	d	d	+
8. *B. pseudocatenulatum*	+	+	+	+	+	+	d	+	−	−	+
9. *B. dentium*	+	+	+	+	+	+	+	+	+	−	+
10. *B. globosum*	d	−	+[e]	+	+	+	−	+	−	−[f]	−[g]
11. *B. pseudolongum*	+	+[h]	+	+	+	+	−	+	−	−	−
12. *B. cuniculi*	+	−	−	+	+	+	−	+	−	−	−
13. *B. choerinum*	−	−	−	+	+	+	−	+	−	−	−
14. *B. animalis*	+	d	+	+	+	+	d	+	−	−	+
15. *B. thermophilum*	−	−[i]	+	+	+	+	d	+	−	d	d
16. *B. boum*	−	−	+	+[j]	+	+	−	+	+	−	−
17. *B. magnum*	+	−	+	+	+	+	−	+	−	−	−
18. *B. pullorum*	+	+	+	+[h]	+	+	+	+	−	+	+
19. *B. suis*	+	d[k]	d[k]	+	+	+	−	+	−	−	−
20. *B. minimum*	−	−	+	−	+	+	−	−	−	−	−
21. *B. subtile*	−	−	+	+	+	+	d	+	−	d	d
22. *B. coryneforme*	+	−	+		+	+	−	+	−		+
23. *B. asteroides*[l]	+[m]	−[n]	+	d	+	d	−	+	−	−	+
24. *B. indicum*[l]	−	d	+	d	+	d	−	+	−	−	+

[a] Symbols: see Table 15.25.
[b] Few strains do not ferment this sugar.
[c] When positive it is fermented slowly.
[d] Some strains ferment this sugar.
[e] Some strains are negative, especially from rat and rabbit feces.
[f] Some strains from piglets are positive.
[g] Some strains can ferment weakly.
[h] Generally delayed or slight fermentation.
[i] Some strains from sewage ferment this sugar.
[j] Some strains are weak fermenters.
[k] Reported "sometimes not fermented" (Matteuzzi et al., 1971).
[l] Sugars indicated "d" gave mostly erratic results.
[m] Few strains do not ferment pentoses.
[n] Few strains ferment this sugar.

ase: the isozymes of most strains of *B. globosum* are more anodal than those of *B. pseudolongum*. Alternatively, the total proteins electrophoresis should be used. With this method strains of *B. pseudolongum* fermenting melezitose (such as Mitsuoka's biovar c) can be promptly recognized as such instead of ascertaining the isozyme of transaldolase for their distinction from the phenotypically very similar *B. animalis* (see Table 15.55).

12. **Bifidobacterium cuniculi** Scardovi, Trovatelli, Biavati and Zani 1979, 307.[AL]

cu.ni′cu.li. L. n. *cuniculus* rabbit; L. gen. n. *cuniculi* of the rabbit.

Cells grown in TPY agar stabs present a morphology very similar to that of *B. globosum* and *B. pseudolongum*. However, the short knobs or protuberances often in the center of the cells are rarely seen in other species (Fig. 15.97C).

Highly anaerobic. CO_2 has no effect on O_2 sensitivity or anaerobic growth.

Lactose and raffinose are characteristically not fermented. Fructose is regularly fermented but only in prereduced media.

Transaldolase isozyme 1 and 6PGD isozyme 4 were found in seven

Table 15.55.
Tabular key for the identification of the species of the genus **Bifidobacterium**[a]

Sorbitol	Arabinose	Raffinose	Ribose	Starch	Lactose	Cellobiose	Melezitose	Gluconate	6PGD isozymes[b]	Transaldolase isozymes[b]	Suggested species
+	+			+		+	+	+			*B. adolescentis*
					−						*B. pseudocatenulatum*
						−					*B. angulatum*
			−								*B. catenulatum*
	−				+						*B. breve*
					−						*B. subtile*
−	+	+	+	+	+	+	+	+		**4**	*B. dentium*
										8	*B. adolescentis*
								−		**5**	*B. animalis*
						−	−			**4**	*B. pseudocatenulatum*
									7	**2**	*B. pseudolongum*
									(3$_a$)-(4)-(5)-**6**-(7)	**2**	*B. globosum*
										5	*B. angulatum*
			−		+	−	+				*B. longum*
							−		(3)-**4**-(5)	**5**	*B. infantis*
									5-**(6)**	**8**	*B. longum*[c]
									7		*B. magnum*
					−	+			**6**		*B. coryneforme*
									9–13		*B. asteroides*
							−				*B. pullorum*
		−				−					*B. suis*
		−									*B. cuniculi*
	−	+	+	+	+						*B. breve*
			−		+						*B. infantis*
					−						*B. indicum*
			−							**3**	*B. choerinum*
										6	*B. boum*
										8	*B. thermophilum*
		−			+						*B. bifidum*
					−						*B. minimum*

[a] Symbols: see Table 15.25.
[b] Boldface numbers are the isozymes possessed by the type strains. Numbers in parentheses are the isozymes possessed by less than 10% of strains.
[c] Unlike *B. suis*, some strains of *B. longum* do not ferment ribose and display identical electrophoretic patterns.

strains studied by Scardovi et al., 1979. This pattern 1-4 is unique among bifidobacteria.

The DNA of the species is 50–67% related to that of *B. globosum* (reference strain RU230-ATCC 25864), but far less related to *B. pseudolongum* and other species of the genus.

The mol% G + C content of DNA is 64.1 ± 0.35 (T_m).

Found in feces of adult rabbit.

Type strain: ATCC 27916 (RA93 from feces of rabbit, Scardovi et al., 1979b).

Comments. B. cuniculi can be easily distinguished from the morphologically similar species *B. globosum, B. pseudolongum* and *B. animalis* which are also found frequently in rabbit feces, because unlike these species *B. cuniculi* does not ferment lactose, ribose or raffinose. The same fermentation pattern can be used to distinguish this species from the morphologically different *B. magnum*, another species isolated from the same source.

13. **Bifidobacterium choerinum** Scardovi, Trovatelli, Biavati and Zani 1979, 307.[AL]

choe.ri′num. M.L. adj. *choerinus* pertaining to a pig.

Cells grown in TPY agar stabs are often short or coccoid with morphology similar to that of *B. globosum*. In liquid medium cells may be elongated to 10–12 μm, bent and with rounded or spatulated ends.

Anaerobic. Effect of CO_2 not detectable.

The strains of this species which ferment raffinose but not ribose,

arabinose and sorbitol cannot be distinguished on that basis from *B. thermophilum* and *B. boum* (see Table 15.55).

Transaldolase and 6PGD isozymes are 3 and 4 respectively; this pattern is unique among bifidobacteria.

The species has distinct PAGE proteins pattern (Biavati et al., 1982). The mol% G + C of the DNA is 66.3 ± 0.15 (T_m).

The DNA of this species is related 26–57% and 37–62% to the DNA of *B. globosum* and *B. pseudolongum*, respectively. The DNA relatedness of 50% between *B. choerinum* and *B. asteroides* is unexpected.

Found in feces of piglet or, occasionally, in sewage.

Type strain: ATCC 27686 (Su806 from feces of pig, Scardovi et al., 1979b).

Comments. B. choerinum cannot be distinguished by sugar fermentation pattern from either *B. thermophilum* biovar *c* (Mitsuoka, 1969) or from strains of *B. boum* which ferment lactose but not melezitose (Table 15.51). These three species are frequently found in feces of piglets. The distinction between these species can be based on transaldolase electrophoresis: the most anodal isozyme is that of *B. choerinum* (migration 100), the least anodal that of *B. thermophilum* (isozyme 8; migration 84) and an intermediate form is that of *B. boum* (isozyme 6; migration 90). Alternatively, PAGE patterns of soluble proteins can be used.

14. **Bifidobacterium animalis** (Mitsuoka 1969) Scardovi and Trovatelli 1974, 26.[AL] (*Bifidobacterium longum* subsp. *animalis* (biotype *a*) Mitsuoka 1969, 60.)

an.i.mal'is. L. gen. n. of an animal.

Bifids isolated from feces of the calf, sheep, rat and guinea pig, phenotypically very similar to *B. longum*, but inactive toward melezitose, were referred to a subspecies of *B. longum* (*Bifidobacterium longum* subsp. *animalis* Mitsuoka 1969, 60). Two biovars *a* and *b* were distinct: biovar *a* mannose-negative and biovar *b* mannose-positive (Mitsuoka, 1969). Mitsuoka's strains R101-8 biovar *a* (ATCC 25527) and C10-45 biovar *b* were not related to *B. longum* on DNA homology but one of them (C10-45 biovar *b*) was genetically *B. pseudolongum* (Scardovi et al., 1971b). Strain R101-8 was subsequently allotted to a DNA homology group of bifids isolated from chicken, rat, rabbit and sewage and proposed as a distinct species (Scardovi and Trovatelli, 1974).

Cells grown on TPY show characteristically the central portion slightly enlarged (Fig. 15.97*B*), branchings can occur to form cross-like aggregates of four cells distally inflated.

Anerobic. CO_2 has no effect upon O_2 sensitivity or anaerobic growth.

Transaldolase and 6PGD isozymes possessed by this species (isozyme 5 and 8 or 9, respectively) give a pattern shared with only few strains of *B. catenulatum* (see Table 15.50).

PAGE proteins pattern is distinct from that of all other species (Biavati et al., 1982).

Fructose-bisphosphate aldolase and glucose-6-phosphate dehydrogenase demonstrable (Scardovi and Trovatelli, 1974).

The mol% G + C of DNA is 60.1 ± 0.3 (T_m).

DNA unrelated to that of any other species of the genus.

Found in feces of rat, chicken, rabbit, calf and in sewage.

Type strain: ATCC 25527 (R101-8 from feces of rat, Mitsuoka, 1969).

Comments. *B. animalis* is readily differentiated from other species found in animal habitats and which ferment arabinose and xylose (namely, *B. globosum*, *B. pseudolongum*, *B. cuniculi*, *B. magnum*, *B. pullorum* and *B. suis*) by lactose and salicin fermentation: if a strain ferments both sugars it can be retained as *B. animalis* (*B. pullorum* among the listed species, ferments salicin but not lactose, see Table 15.54). "Human" species which ferment pentoses but not starch, such as *B. dentium* and *B. adolescentis*, can be distinguished from *B. animalis* by the absence of gluconate fermentation in *B. animalis* (see Table 15.55).

15. Bifidobacterium thermophilum Mitsuoka 1969, 59.[AL]

ther.mo'phil.um. Gr. n. *therme* heat; Gr. adj. *philus* loving; M.L. adj. *thermophilum* heat-loving.

Mitsuoka (1969) gave the specific epithet to strains of bifids not fermenting pentoses which he isolated from the feces of swine and chicken, owing to their ability to grow at 46.5°C and to resist heating for 30 min at 60°C. Scardovi et al. (1969) named strains not fermenting pentoses or lactose which he isolated from bovine rumen, as "*B. ruminale.*" All experimental comparisons have demonstrated that these two species are identical (Scardovi et al., 1971b; Scardovi et al. 1979a; Biavati et al., 1982).

Cells grown in TPY agar stabs are long, slender, curved, arranged singly or in pairs, never in clumps; this morphology is shared by many other species of the genus.

Four biovars, *a*, *b*, *c* and *d* were distinguished by Mitsuoka (1969) according to differences in the fermentation of melezitose and lactose. Similar differences were found among strains isolated from the bovine rumen, feces of calf, sewage (Scardovi et al., 1979b) and feces of the piglet (Zani et al., 1974). A few strains genetically assigned to *B. thermophilum* which fermented arabinose and xylose, were found in sewage and in piglets (Scardovi et al., 1979b).

Both *B. globosum* and *B. thermophilum* can grow in 90% air + 10% CO_2 without the cells becoming catalase- or pseudo-catalase (hemin)-positive.

Growth factors required are riboflavin, pantothenate and pyridoxine; bases are not required; ammonia is an optimal source of nitrogen.

Fructose-bisphosphate aldolase and HMP dehydrogenases are always present in cell-free extracts (Scardovi et al., 1969). Aldolase was spot-stained after electrophoresis (Scardovi and Sgorbati, 1974).

DNA homology indicated 27–80% relatedness to *B. longum* but no relation to other species of the genus.

The mol% G + C of the DNA is 60 (T_m).

Type strain: ATCC 25525 (P2-91 from the feces of swine, Mitsuoka, 1969).

Comments. To distinguish between *B. thermophilum*, *B. boum* and *B. choerinum*, see under *B. choerinum*.

16. Bifidobacterium boum Scardovi, Trovatelli, Biavati and Zani 1979, 308.[AL]

bo'um. L. n. *bos* a cow: L. pl. gen. n. *boum* of cattle.

A few strains were isolated from the bovine rumen as morphovars of *B. thermophilum* (*B. ruminale*) (Scardovi et al., 1969); later their DNA was found to be nearly 70% related to *B. thermophilum* (Scardovi et al., 1971b). Subsequently, other strains from the bovine rumen were assigned to a DNA homology group IV, 55–75% related to *B. thermophilum* (Trovatelli and Matteuzzi, 1976).

From a large number of animal strains surveyed by DNA-DNA hybridization, 36 strains from the rumen and 5 from piglet feces were allotted to the new species *B. boum* (Scardovi et al., 1979b). Cells grown on TPY agar are in general more irregular than those of *B. thermophilum*, and vary greatly with strain. Most branched forms are seen in cells grown in air + CO_2.

Although *B. boum* does not ferment melezitose, cellobiose, trehalose or mannose, and thus has a more stable sugar fermentation pattern than *B. thermophilum*, the fermentation patterns of the two species are often the same.

Can develop in 90% air + 10% CO_2 without cells becoming catalase- or catalase-like (hemin)-positive.

Requirements for growth factors are unknown.

Fructose-bisphosphate aldolase and glucose-6-phosphate dehydrogenase present in cell-free extracts as with *B. thermophilum*.

The DNA homology relationships between *B. boum* and *B. thermophilum* have been studied extensively (Scardovi et al., 1979b).

A range of 36–74% is reported for the DNA relatedness of *B. boum* to *B. thermophilum* (Table 15.52); unrelated to any other species in the genus.

Transaldolase isozyme of *B. boum* (number 6) is clearly distinct electrophoretically from that of the majority (93%) of the strains of *B. thermophilum* (number 8).

The PAGE proteins pattern of *B. boum* is distinct and easily distinguishable from that of *B. thermophilum*. The so called "genus band" is slightly less anodal in *B. boum* than in any other species (Biavati et al., 1982).

Interpeptide bridge of the cell wall peptidoglycan is Lys-D-Ser-D-Glu; it differs from that of the closely related species *B. thermophilum* which is Orn(Lys)-D-Glu.

The mol% G + C of the DNA is 60.0 ± 0.2 (T_m).

Type strain: ATCC 27917 (RU917 from bovine rumen, Scardovi et al., 1979).

Comments. The practical distinction of *B. boum* from *B. thermophilum* and *B. choerinum* which are often found in the same ecological niches, can be achieved with transaldolase electrophoresis or with PAGE proteins electrophoresis (see *Comments* under *B. choerinum*).

17. Bifidobacterium magnum Scardovi and Zani 1974, 31.[AL]

mag'num L. adj. *magnus* large, great.

Cells grown in TPY agar are usually characteristically long and thick, with irregular contours, measuring 2×10–20 μm and occurring frequently in aggregates (Fig. 15.97*D*).

Anaerobic. CO_2 without effect on O_2 sensitivity or anaerobic growth.

Sparse growth in TPY medium not containing Tween 80, as Tween 80 is highly stimulatory (Scardovi and Zani, 1974).

The unique acidophilic species of the genus. Its original optimum pH for growth is 5.3–5.5; growth is retarded at 5.0 or at 5.9; no growth (after 2 days) at 4.2 or 7.0.

Fructose-bisphosphate aldolase present in cell-free extracts (4–5 mU/mg of proteins); NADP⁺-dependent glucose-6-phosphate dehydrogenase demonstrable. Aldolase spot-stainable on electrophoresis (Scardovi and Sgorbati, 1974).

The isozyme pattern 5-7 (transaldolase and 6PGD types resp.) is unique among bifidobacteria (Scardovi et al., 1979a).

PAGE proteins pattern clearly distinct from that of any other species of the genus (Biavati et al., 1982).

DNA unrelated to that of other species of the genus.

Found in the feces of rabbit.

The mol% G + C of the DNA is 60.0 ± 0.6 (T_m).

Type strain: ATCC 27540 (RA3 from feces of rabbit, Scardovi and Zani, 1974).

Comments. The recognition of *B. magnum* should not be based solely on the unusually large dimensions of its cells. Fermentation of lactose and starch are useful in differentiating *B. magnum* from the "animal" species *B. pullorum, B. animalis, B. globosum* and *B. pseudolongum,* (which also ferment pentoses), as *B. magnum* ferments lactose but not starch while these other species either ferment both these sugars or neither (*B. pullorum*). *B. magnum* can be distinguished from the human species *B. infantis* and *B. longum* only by virtue of the different 6PGD isozyme present (the transaldolase isozyme of *B. magnum* is shared by *B. infantis,* number 5). Alternatively, PAGE procedure can be used.

18. Bifidobacterium pullorum Trovatelli, Crociani, Pedinotti and Scardovi 1974, 197.[AL]

pul′lus. L. n. *pullus* a chicken; L. pl. gen. n. *pullorum* of chicken.

Cells grown in TPY agar stabs are slightly curved, 2–8 μm long, with tapered ends, mostly arranged in irregular chains often of great length (Fig. 15.97*A*). Cells are frequently poorly refractile and appear to be empty or vacuolized. Branchings are rare.

Anaerobic. CO_2 without effect on O_2 sensitivity or anaerobic growth.

Requires nicotinic acid, pyridoxine, thiamin, folic acid, p-aminobenzoic acid and Tween 80 for satisfactory growth.

Fructose-bisphosphate aldolase present in cell-free extracts in considerable amounts (20–30 mU/mg proteins). As with most *B. dentium* strains, neither 6PGD (NADP$^+$ or NAD$^+$ dependent) nor glucose-6-P-dehydrogenase can be detected.

Lactic and acetic acids are produced in a ratio of $1:3.5 \pm 0.2$ in TPY medium but unlike all other species of the genus, the isomeric type of lactic acid formed is DL.

The mol% G + C of the DNA is 67.4 ± 0.4 (T_m), the highest value so far found in bifidobacteria.

The DNA is not related to that of any of the other species.

Found in feces of chicken.

Type strain: ATCC 27685 (P145 from feces of chicken, Trovatelli et al., 1974).

Comments. Morphology is of help in recognizing this species. Furthermore, it can be easily distinguished from the "animal" species of the genus which also ferment arabinose, xylose and ribose because it does not ferment lactose and starch (see *Comments* under *B. magnum*). Can be distinguished from other species of the genus on the basis of the fermentative characters (Table 15.55).

19. Bifidobacterium suis Matteuzzi, Crociani, Zani and Trovatelli 1971, 393.[AL]

su′is. L. n. *sus* pig; L. sing. gen. n. *suis* of pig.

Cells grown in TPY agar stabs show a similar morphology to those of many other species of the genus (Fig. 15.98*D*).

Anaerobic. CO_2 without effect on O_2 sensitivity or anaerobic growth.

Riboflavin is the only growth factor required.

Cell-free extracts possess fructose-bisphosphate aldolase and HMP dehydrogenases. Most strains possess a constitutive urease activity (not influenced by urea or organic nitrogen sources). Fifty percent of the strains studied share with a percent of strains *B. longum* and *B. infantis* the isozyme electrophoretic pattern transaldolase 6 and 6PGD 5; the others display a pattern (6-8) which is common in *B. longum* (Scardovi et al., 1979a). PAGE proteins pattern is quite distinct from that of the other species of the genus (Biavati et al., 1982).

The mol% G + C of DNA is 62 (T_m).

Unrelated in DNA homology to any other species of the genus.

So far, found only in the feces of piglets.

Type strain: ATCC 27533 (Su859 from feces of pig, Mateuzzi et al., 1971).

Comments. *B. suis* can be distinguished readily from other bifid species commonly found in the pig, namely *B. globosum, B. pseudolongum, B. thermophilum, B. boum* and *B. choerinum,* on the basis of fermentative characters: the last three species do not ferment arabinose and xylose whereas *B. suis* does; the first two ferment starch whereas *B. suis* does not (see Tables 15.51 and 15.54; Zani et al., 1974).

20. Bifidobacterium minimum Biavati, Scardovi and Moore 1982, 368.[VP] This taxon was previously described and referred to as "minimum" DNA homology group, and consists at present only of two strains isolated from sewage (Scardovi and Trovatelli, 1974).

min′ i.mum. L. adj., least; *minimum* the least.

Cells grown in TPY agar stabs are characteristically very small (0.3 × 1.3–1.5 μm) with tapered ends; sometimes irregularly branched (Fig. 15.97*F*). This morphology resembles that *B. asteroides,* but star-like aggregates, characteristic for that species, are absent here.

Anaerobic. CO_2 without effect on O_2 sensitivity or anaerobic growth.

Sugars fermented include glucose, fructose, sucrose, maltose and starch.

The interpeptide bridge of the cell wall peptidoglycan is Lys-Ser, unique among bifidobacteria.

Cells possess the least anodal form of transaldolase among bifidobacteria (No. 10, i.e. migrating 66, whereas the most anodal isozyme, No. 1, in *B. cuniculi,* migrates 100) (Scardovi et al., 1979a).

Aldolase and glucose-6-phosphate dehydrogenase not detected.

Distinct PAGE proteins pattern.

DNA unrelated to that of any other species of the genus.

Found until now in a single specimen of waste water.

The mol% G + C of the DNA is 61.5 (T_m).

Type strain: ATCC 27538 (F392 from waste water, Scardovi and Trovatelli, 1974).

21. Bifidobacterium subtile Biavati, Scardovi and Moore 1982, 36[VP] This taxon was previously described and referred to as "subtile" DNA homology group and includes five strains isolated from sewage (Scardovi and Trovatelli, 1974).

sub′ti.le. L. adj. slender, *subtile* the slender.

Cells grown in TPY agar stabs are slender, 0.5 × 2–3 μm with rounded or tapered ends, sometimes curved; branchings are rare. This morphology (Fig. 15.97*E*) is similar to that of *B. breve,* but the cells of the latter are usually shorter and thicker, swollen and branched.

Anaerobic. CO_2 without effect on O_2 sensitivity or anaerobic growth.

Optimum temperature for growth is in the range of 34–35.5°C, being markedly lower than the range 37–41°C, valid for the other species of the genus.

Sugars fermented are similar to those fermented by *B. breve* but, unlike *B. breve,* lactose is not fermented (Table 15.55); also *B. subtile* ferments starch and gluconate whereas *B. breve* does not.

Possesses high levels (10–15 mU/mg proteins, in strain F395) of NADP$^+$-NAD$^+$-dependent glucose-6-phosphate dehydrogenase but aldolase is not measurable.

Transaldolase and 6PGD isozymes are among the most anodal in the bifidobacteria (number 3 and 2, respectively); this 3-2 pattern occurs only in this species (Scardovi et al., 1979a).

Distinct PAGE proteins pattern.

Both human and animal types of F6PPK have been detected in cell-free extracts (Scardovi and Trovatelli, 1974).

Unrelated by DNA homology to any other species of the genus.

The mol% G + C of the DNA is 65.5 (T_m).

As yet isolated only from two specimens of waste waters.

Type strain: ATCC 27537 (F395 from waste water, Scardovi and Trovatelli, 1974).

22. Bifidobacterium coryneforme (ex Scardovi and Trovatelli,

1969) Biavati, Scardovi and Moore, 1982, 368.[VP] *Bifidobacterium coryneforme* Scardovi and Trovatelli (1969) was not included in the Approved Lists (1980). On the basis of its different electrophoretic pattern of cellular proteins and the previously confirmed differential characters, its revival was proposed (Biavati et al., 1982).

co.ry.ne.for′me. Gr. n. *coryne* a club; L. n. *forma* shape, form; M.L. adj. *coryneformis* club shaped.

The strains assigned to this species had been isolated occasionally from the intestine of honey bee *Apis mellifera* subsp. *mellifera* and subsp. *caucasica*, and received from Germany (Bayern), Norway (Billingstad), England (Buckfast) and Bulgaria.

The strains grow poorly in TPY medium but profusely in MRS medium. Cells grown in MRS agar stabs are short (1.1.5 μm long) often lanceolate, single or in pairs, sometimes with short branchings or simple knobs (Fig. 15.98*C*). Radial groupings of cells are very rare and formed under extreme conditions of growth (Scardovi and Trovatelli, 1969).

Anaerobic. Does not develop in slants inoculated under 90% air + 10% CO_2. CO_2 does not influence anaerobic growth either on solid or in liquid medium.

Aldolase in cell-free extracts not detected. HMP dehydrogenases present.

Transaldolase and 6PGD isozymes pattern (6-6) is shared by some strains of *B. breve*. However zymograms of G3PD (3-phosphoglyceraldehyde dehydrogenase) are distinct (Scardovi et al., 1979a) and the PAGE proteins pattern is distinct (Biavati et al., 1982).

Unrelated by DNA homology to any other species of the genus except *B. indicum*: an interspecific DNA relatedness of 60% is reported (Scardovi et al., 1970).

Type strain: ATCC 25911 (C215 from intestine of *Apis mellifera* subsp. *caucasica* from Norway (Scardovi and Trovatelli, 1969)).

23. **Bifidobacterium asteroides** Scardovi and Trovatelli 1969, 83.[AL]

as.te.roi′des. Gr. adj. *asteroides* starlike.

Cells grown in TPY agar stabs are 2–2.5 μm long, generally in pairs with pointed ends and slightly curved; usually in radial disposition around a common mass of hold-fast material (Fig. 15.98*A*). Nutritional or CO_2 deficiencies or growth with certain sugars, may induce clavate or spatulate cells with occasional swellings, irregular or cross-like branchings in the central part of the cell body.

Colonies are circular, smooth, convex, with entire edge; consistency is such that the colony is removed by needle and can hardly be dispersed in water.

Growth in static fluid culture generally adheres to the glass walls and leaves the liquid clear.

CO_2 is required for growth in all media including stabs.

Aerobic growth occurs on slopes only if air is enriched with 10% CO_2.

Riboflavin, pantothenate, nicotinic acid, pyridoxine, thiamin and biotin are required for continued growth in a Bacto Vitamin Free Casamino Acids-containing substrate (Scardovi and Trovatelli, 1969).

Temperature relations: optimum 35–36°C; no growth at 21°C or at 42°C (after 7 days incubation).

H_2O_2 is decomposed vigorously by cells grown in 90% air + 10% CO_2. In stabs, incubated in CO_2-air, only the lower portions of the growth are catalase-negative.

Fructose-bisphosphate aldolase is not detected in cell-free extracts or in iodoacetate-poisoned cell suspensions; cell-free extracts possess both the HMP dehydrogenases NADP$^+$-dependent.

A total of 85 strains of very different geographical origins were examined electrophoretically for the content of isozymes of transaldolase and 6PGD: at least eight transaldolases and nine 6PDG were detected, each strain exhibiting only one band for each allozyme. *B. asteroides* posesses the least mobile variant of 6PGD (isozyme 13 moves 53, whereas isozyme 1, the most anodal found in *B. pseudocatenulatum*, moves 100) Scardovi et al., 1979a).

Of 224 strains of this species tested for the presence of plasmids, 74 contained a large variety of extrachromosomal elements of diverse molecular weights ranging from 1.2 to 22 MDa (Sgorbati et al., 1982); the functions coded for by these plasmids are still unknown.

DNA is about 30% homologous to DNA from *B. indicum*, another bifid found in the honey bee. Unexpectedly there is 50% homology to *B. choerinum*, a bifid from piglet, otherwise the DNA is unrelated to any other species of the genus.

Found normally in the intestine of western honey bees; occasionally found in the hind-gut of *Apis cerana*, an asiatic honey bee.

The mol% G + C of the DNA is 59 (T_m).

Type strain: ATCC 25910 (C51 from intestine of *A. mellifera* var. *ligustica* from Italy (Scardovi and Trovatelli, 1969)).

24. **Bifidobacterium indicum** Scardovi and Trovatelli 1969, 84.[AL]

in′di.cum. M.L. *neut.* adj. *indicum* from specific epithet of bee *Apis indica* F.

Cells grown in TPY agar stabs are generally short, in pairs, often in angular disposition, more or less globular, sometimes suggesting a minute morphovar of *B. globosum* (Fig. 15.98*B*): other strains have slender and longer cells; star-like clusters of cells never occur. Cells grown on gluconate are extremely small and regular in shape (Sgorbati et al., 1970).

Colonies do not have the consistency of those of *B. asteroides*.

Growth in liquid media does not adhere to the walls; the cells sediment very slowly.

Oxygen tolerance similar to that of *B. asteroides*. CO_2 is required for aerobic growth, whereas for anaerobic growth the effect of CO_2 is equivocal.

H_2O_2 is decomposed only by cells grown in 90% air + 10% CO_2 in the presence of hemin.

Cell-free extracts are fructose-bisphosphate aldolase-negative, but possess the HMP dehydrogenases: glucose-6-P dehydrogenase is both NADP$^+$- and NAD$^+$-dependent, whereas only NADP$^+$ is effective in 6-phosphogluconate dehydrogenation (Scardovi and Trovatelli, 1969).

A total of 122 strains from different sources (Malaysia, Japan, Philippines) were examined for their transaldolase and 6PGD zymograms: starch gel electrophoresis showed the presence of isozymes Nos. 4 and 7, respectively. Most strains isolated from *A. dorsata* possess a transaldolase isozyme different (number 7) from that found in strains isolated from *A. cerana* (*A. indica*) (number 9) (Scardovi et al., 1979a).

Of a total of 106 strains surveyed for the presence of plasmids, 73 were found to harbor extrachromosomal elements; 57% of these strains had a 22 MDa plasmid while 33% showed a two-banded pattern at 2.0 and 3.5 MDa (Sgorbati et al., 1982). The cellular functions coded for by these plasmids are still unknown.

Unrelated by DNA homology to any other species of the genus.

The mol% G + C of the DNA is 60 (T_m).

Type strain: ATCC 25912 (C410 from intestine of *Apis cerana* (*A. indica*) from Malaysia (Scardovi and Trovatelli, 1969)).

SECTION 16

The Mycobacteria

The mycobacteria, comprising the genus *Mycobacterium*, are aerobic, nonmotile slow-growing rod-shaped bacteria that are characteristically acid-fast. They have traditionally been considered to be rather separate, and are thus usually treated as a family, the *Mycobacteriaceae*, but there is considerable evidence that they are closely allied to the genera *Corynebacterium* and *Nocardia*.

The three genera are sometimes referred to as the CNM group, to which the genus *Rhodococcus* should be added. The reader is referred to Sections 15 and 17 for further discussion of these relationships.

The genera most easily confused with *Mycobacterium* are shown in Table 16.1 below. Further details may be found in Table 16.3, in the article on *Mycobacterium* and in Sections 15 and 17. The articles of Cross and Goodfellow (1973) and Goodfellow and Minnikin (1981) may be useful. These genera also differ in cell wall type and phospholipids (see references above and Section 17).

The property of acid-fastness, due to waxy materials in the cell walls, is particularly important for recognizing mycobacteria. These bacteria are commonly described as acid-alcohol-fast, implying that after staining they resist decolorization with acidified alcohol as well as with strong mineral acids. Some other bacteria, however, are partially acid-fast, and may be readily decolorized by alcohol although they may resist decolorization by weak acid. Artifacts may occur that retain the stain if the technique is not carefully performed. It should be noted that spores may be acid-fast, and occasionally waxy substances in tissues or other materials may cause confusion, and the degree of acid-fastness varies with technique and cultural conditions (Harrington, 1966).

Table 16.1.
Differentiation of **Mycobacterium** *from other genera*

Characteristics	*Mycobacterium*	*Nocardia*	*Rhodococcus*	*Corynebacterium*
Morphology	Rods, occasionally branched filaments; no aerial mycelium[a]	Mycelium, later fragmenting into rods and cocci; usually some aerial mycelium	Scanty mycelium, fragmenting into irregular rods and cocci; no aerial mycelium	Pleomorphic rods, often club-shaped; commonly in angular and palisade arrangement
Rate of growth: time for visible colonies	2–40 days	1–5 days	1–3 days	1–2 days
Degree of acid-fastness (not necessarily also alcohol-fastness)	Usually strongly acid-fast	Often partially acid-fast	Often partially acid-fast	Sometimes weakly acid-fast
Degree of staining in Gram stain	Weak	Usually strong	Usually strong	Strong
Reaction to penicillin	Resistant[b]	Resistant	Sensitive	Sensitive

[a] Some species may occasionally produce aerial mycelium.
[b] Some species, e.g. *M. avium* may be sensitive to penicillin.

Further Reading

Cross, T. and M. Goodfellow. 1973. Taxonomy and classification of the actinomycetes. *In* Sykes and Skinner (Editors), *Actinomycetales*: Characteristics and Practical Importance, Academic Press, London, pp. 11–112.

Goodfellow, M. and D.E. Minnikin. 1981. The genera *Nocardia* and *Rhodococcus*. *In* Starr, Stolp, Trüper, Balows and Schlegel (Editors), The Prokaryotes, Springer-Verlag, Berlin, Vol. II, pp. 2016–2027.
Harrington, B.J. 1966. A numerical taxonomic study of some corynebacteria and related organisms. J. Gen. Microbiol. *45:* 31–40.

FAMILY **MYCOBACTERIACEAE** CHESTER 1897, 63^{AL}*

Lawrence G. Wayne and George P. Kubica

My.co.bac.te.ri.a′ce.ae. M.L. neut. n. *Mycobacterium* type genus of the family; -*aceae* ending to denote a family; M.L. pl. fem. n. *Mycobacteriaceae* the *Mycobacterium* family.

Description as for genus *Mycobacterium*.

Genus **Mycobacterium** *Lehmann and Neumann 1896, 363*^{AL}

Lawrence G. Wayne and George P. Kubica

My.co.bac.te′.ri.um. Gr. n. *myces* a fungus; Gr. neut. dim. n. *bakterion* a small rod; M.L. neut. n. *Mycobacterium* a fungus rodlet.

Slightly curved or straight rods, 0.2–0.6 × 1.0–10 μm sometimes branching; filamentous or mycelium-like growth may occur but on slight disturbance usually becomes fragmented into rods or coccoid elements. Acid-alcohol-fast at some stage of growth. Not readily stainable by Gram's method but usually considered Gram-positive. Nonmotile. No endospores, conidia or capsules, no grossly visible aerial hyphae. Aerobic, although from dispersed seeding in tubed agar medium, growth of some species occurs only in the depths of the medium. One or both of two classes of catalase are produced by wild strains (Wayne and Diaz 1982), although the T-class of this enzyme can be lost in isoniazid resistant mutants.

Genus includes obligate parasites, saprophytes and intermediate forms.

Lipid content of cells and cell walls high; included are waxes having characteristic chloroform-soluble mycolic acids with long (about 60- to 90-carbon atom) branched chains (Goodfellow and Minnikin, 1980). **Diffusible pigment rare.** Colonies of some species are regularly or variably yellow or orange, or rarely pink, usually due to carotenoid pigments, the formation of which may or may not require exposure to light. The cell wall peptidoglycolipid contains *meso*-diaminopimelic acid, alanine, glutamic acid, glucosamine, muramic acid, arabinose and galactose (**wall type IV**) (Lechevalier and Lechevalier, 1970). The **G + C content of the DNA ranges from 62–70 mol%** (T_m, Bd) (Wayne and Gross, 1968; Tewfik and Bradley, 1967).

Type species: *Mycobacterium tuberculosis* (Zopf) Lehmann and Neumann 1896, 363.

Further Descriptive Information

Most strains of mycobacteria form more than one kind of colony, but colonies of some species, as *M. tuberculosis*, are regularly rough; some, as of *M. intracellulare* on primary culture from clinical specimens, are more commonly smooth. Cells of rough strains are usually compacted in curving strands; cells of smooth strains are not visibly oriented in any pattern. Some colonies, as of *M. fortuitum* and *M. xenopi*, in early growth may be mycelial, older ones exhibiting branching filamentous extensions on and into some media such as cornmeal glycerol agar; fragmentation to bacilli usually occurs in smear preparation. Aerial filamentous extensions rare, never visible without magnification (× 30–100).

Growth slow or very slow with generation times ranging (by species) from 2 to more than 20 h (David, 1973). Easily visible colonies are produced from dilute inoculum after 2 days to 8 weeks incubation at optimum temperature. Optimum temperatures vary widely according to species and range from 30°C to almost 45°C. Most species adapt readily to growth on very simple substrates, using ammonia or amino acids as nitrogen sources and glycerol as carbon source in the presence of mineral salts. Some fastidious organisms require supplements such as hemin, mycobactins or other iron transport compounds. *M. leprae* has not been cultivated outside of living cells. A growth cycle has been described for *M. avium*, consisting of an elongation phase followed by

rapid fragmentation (McCarthy and Ashbaugh, 1981), although this interpretation has been questioned on the basis of electron microscopy (Rastogi and David, 1981).

Bacteriophages lytic for many mycobacteria, including *M. tuberculosis*, have been isolated by soil enrichment techniques (Froman et al., 1954). Lysogenic (Russell et al., 1964) and plasmid-bearing (Crawford et al., 1981a) strains of mycobacteria have been found in nature. Many investigators have tried to use phage typing as a taxonomic tool, but poor interlaboratory reproducibility and the ability of the phages to cross species lines have limited its value. However, phage patterns within the species *M. tuberculosis* have been useful for epidemiologic studies (Baess, 1969; Bates and Mitchison, 1969; Jones et al., 1982). Bacteriocins have been detected in mycobacteria and applied to the classification of some species (Takeya and Tokiwa, 1972).

Genetic studies of mycobacteria have been hampered by slow growth of these organisms and difficulty in obtaining auxotrophs. Drug-resistant mutants and colonial variants occur. Although a few claims have been made for successful transformation and transductions, attempts to confirm these reports have been largely unsuccessful. Conjugation has also been reported. See review by Grange (1975). Genetic mapping of drug resistance has recently been accomplished with sequential mutagenesis technique (Woodley et al., 1981; Konickova-Radochova and Konicek, 1981).

Antigenic analyses have proven useful at several hierarchical levels. Ribosomal proteins are conserved within the genus, and immunodiffusion analyses have been used to compare members of the genus *Mycobacterium* with those of other closely allied taxa (Ridell et al., 1979). Analyses of immunodiffusion and immunoelectrophoretic patterns of whole culture filtrates and cell extracts have proven useful for defining species within the genus (Norlin, 1965; Stanford and Grange, 1974; Chaparas et al., 1978), as have quantitative comparisons of the skin hypersensitivity reactions of sensitized guinea pigs to crude culture filtrates of homologous and heterologous mycobacteria (see survey by Magnusson, 1980). Seroprecipitation studies of mycobacterial catalase has been used to establish immunologic distances between species, as an indirect measure of evolutionary divergence (Wayne and Diaz, 1982). Whole cell seroagglutination reactions, based on peptidoglycolipid surface antigens usually differentiate strains at the infrasubspecific level (Schaefer, 1967; Brennan et al., 1981) although some species appear to be comprised of single serovars (Wayne, 1971).

Mycobacteria are relatively resistant to most of the broad spectrum antibiotics, with the notable exceptions of streptomycin and rifampin. Most antimicrobial agents used to treat *M. tuberculosis* and *M. bovis* infections (e.g. isoniazid, ethambutol, *p*-aminosalicylate) are quite specific for these organisms, frequently showing little inhibitory action against other mycobacterial species.

With large enough inoculum all mycobacteria produce granulomatous lesions in experimental animals. However, only some of the species grow in host tissues and produce progressive or self-limiting disease. Individual species pathogenic for cold-blooded animals, mammals and/ or birds. Depending on species, may show predilection for internal

---

* *AL*, denotes the inclusion of this name on the Approved Lists of Bacterial Names (1980).

organs, especially the lungs (*M. tuberculosis* and many others), skin, nerves or intestinal tract. Diseases produced include tuberculosis, leprosy and other usually chronic, more or less necrotizing, limited or extensive granulomas.

Most species free-living in soil, water, but the major ecologic niche for some is the diseased tissue of warm blooded hosts.

Enrichment and Isolation Procedures

May be isolated from sputum, soil, water or other sources with mixed flora only by decontaminating with agents that destroy the rapidly growing nonacid-fast contaminants selectively. Suitable agents include dilute acids or alkali with or without quaternary ammonium compounds, followed by neutralization. Time of exposure to decontaminating agents must be carefully controlled, because resistance of mycobacteria is not absolute (Krasnow and Wayne, 1966). Decontaminated specimens are inoculated to agar or inspissated egg medium containing malachite green, which may also be supplemented with selective antimicrobial agents. Alternatively, the digested specimens may be injected into guinea pigs or mice, and cultures obtained from their internal organs 2–3 weeks later.

Additional Taxonomic Considerations

The genus *Mycobacterium* is the only one in its family. However the mycolic acid-producing bacterial genera may be so closely related to one another as to justify bringing some or all of these genera together into a single family. The similarities of mol% G + C of DNA of members of *Mycobacterium*, *Nocardia* and *Rhodococcus* support such a consolidation. If this were to occur, the senior family name, *Mycobacteriaceae* Chester 1897, would have priority. At this writing no formal proposal to effect such a consolidation has been put forward.

A natural division occurs between slowly and (relatively) rapidly growing species. For practical determinative purposes, slow growers are those that require over 7 days of incubation at optimal temperature to produce easily seen isolated colonies from highly diluted inocula. Rapid growers are seen in less than 7 days under comparable conditions. A bimodal distribution has also been reported for mol% G + C of DNA from different species, (Wayne and Gross, 1968) but this division does not correspond to the division along lines of growth rate. Tsukamura (1967a) has proposed the separation of rapid growers into a new subgenus *Mycomycobacterium*, but this division has not been adopted. The rapid and slow growers are too similar to one another in terms of

Table 16.2.

Supplementary features for distinguishing between slowly and rapidly growing mycobacteria[a]

Feature	Slow growers	Rapid growers
Growth in presence of 5% NaCl	−[b]	+[c]
Growth in presence of 0.2% picrate	−[b]	+
Iron uptake	−	+[c]
Microcolonial dye test	+	−

[a] Symbols: −, 90% or more of the strains are negative; +, 90% or more of the strains are positive.
[b] See Table 16.5 for exceptions.
[c] See Table 16.6 for exceptions.

DNA homology (Baess and Weis Bentzon, 1978; Bradley, 1973; Mordarski et al., 1980; Gross and Wayne, 1970), antigenic composition (Chaparas et al., 1978a, b; Stanford and Grange, 1974), lipid composition (Minnikin et al., 1980; Minnikin and Goodfellow, 1980a, b) and bacteriophage susceptibility (Froman et al., 1954) to justify their separation into two subgenera. Nevertheless, practical experience has demonstrated that different sets of tests are necessary for characterizing species of slow growers and of rapid growers. Therefore any determinative scheme must first permit selection of the appropriate battery of tests. Since occasional strains exhibit intermediate growth rates, supplementary features may be needed to contribute to the proper decision (Table 16.2). These tests are largely exclusionary, in that only rare exceptions are seen for the tabulated results for slow growers (i.e. *M. triviale* is the only slow grower that consistently grows on 5% NaCl, and only occasional strains of *M. simiae* grow on 0.2% picrate); a wider range of responses, which are species related, is seen with rapid growers.

Maintenance Procedures

Will remain viable on egg medium held at 5°C for many months. Survive indefinitely at −70°C or after lyophilization of a suspension. Most species readily subcultured on inspissated egg medium, Dubos oleic acid albumin or Middlebrook 7H10 or 7H11 agar, and in liquid media containing Tween 80 for dispersal and bovine serum albumin to neutralized traces of oleic acid. Laboratory-adapted strains often grow well on synthetic media containing asparagine, glycerol and mineral salts.

Differentiation of the genus **Mycobacterium** from other genera

Mycobacteria appear most closely related to members of *Nocardia*, *Rhodococcus* and *Corynebacterium*. Features useful for distinguishing *Mycobacterium* from these genera appear in Table 16.3.

Procedures for Characterizing Species

The slow growth and biochemical reactivity of mycobacteria may be due, in part, to the hydrophobic nature of the cells associated with their high lipid content, but it probably also is a reflection of the relatively low reaction rate of their RNA polymerase and their low ratio of RNA to DNA (Harshey and Ramakrishnan, 1977). Thus, although mycobacteria share metabolic pathways common to members of many other genera, the slow growth and reactivity make it necessary to employ special techniques for characterization that were developed specifically for use with members of this genus. For this reason it is essential that the diagnostic tests be performed precisely as described in the literature cited (Table 16.4), or if modifications are employed they should first be checked for concordance with results obtained with these methods.

Circumscription of Species

In the batteries of tests employed by most investigators for numerical taxonomy of mycobacteria, members of species tend to cluster together with mean matching scores of 85% or greater. The mean matching scores of members of one cluster to members of another cluster are generally less than 75% (Wayne, 1981). There is, in general, good agreement between species circumscription by numerical taxonomy and by independent criteria, such as antigenic analysis by immunodiffusion and immunoelectrophoresis, specificity of skin tests, DNA/DNA homology, and immunologic distance of selected proteins. Exceptions will be discussed in the descriptions of individual species.

Identification of Cultures

In many cases it is unnecessary to carry an identification to the level of an individual species. For example, in clinical diagnosis, it may suffice to identify an organism as belonging to a complex of similar organisms, with similar significance. In these situations, the complexities of the tests required for identification may outweigh the value received from the identification. For this reason, Tables 16.5 and 16.6 are presented as guides to a minimal test protocol which will permit identification of most slow growers (species 1–20) and rapid growers (species 27–52), respectively, at a level usually considered to be sufficient for routine practical purposes. Identification beyond these levels may require application of a larger series of tests and consideration of properties described in the narrative portions of the text. It should be recognized that intermediate growth rates are frequently seen, especially in old laboratory strains. Some organisms which have special

Table 16.3.
Differentiation between **Mycobacterium** *and allied genera[a]*

Feature	Reference	Genera			
		Mycobacterium	*Nocardia*	*Rhodococcus*	*Corynebacterium*
Mol% G + C of DNA	Goodfellow and Minnikin, 1980; Tewfik and Bradley, 1967; Wayne and Gross, 1968	62–70	60–69	59–69	51–59
Mycolic acids: number of carbons	Goodfellow and Minnikin, 1980.	60–90	46–60	34–64	22–32
Sensitivity to lysis by lysozyme	Mordarska et al., 1978.	Resistant	Variable	Sensitive	Sensitive
Acid-alcohol fastness	Goodfellow and Alderson, 1977; Cummins et al., 1974.	+[b]	–	–	–
Arylsulfatase produced	Wayne et al., 1958.	+[c]	Uncommon	–	–

[a] Symbols: see Table 16.2.

[b] Acid-alcohol-fastness is partially or completely lost at some stage of growth by a variable proportion of the cells of some species. Cells of rapid growers may be less than 10% acid-fast. Beading with metachromatic granules and banding with nonstained areas are common.

[c] Some species react very slowly.

growth requirements or have not been cultivated *in vitro* (species 21–26) are dealt with only in Table 16.7 and in the narrative text. Expanded phenetic descriptions of all species except *M. leprae* are presented in Table 16.7 (slow growers) and Table 16.8 (rapid growers).

Acknowledgments

Whenever possible, descriptions of tests and tabulations of data are based on data derived from published and unpublished cooperative studies of the International Working Group on Mycobacterial Taxonomy (IWGMT).

Further Reading

Ratledge C. and J. Stanford (Editors). 1982. The Biology of the Mycobacteria, Academic Press, London.

Kubica, G.P. and L.G. Wayne (Editors). 1984. The Mycobacteria: A Source Book. Marcel Dekker, New York.

List of the slow growing species of the genus **Mycobacterium**

1. Mycobacterium tuberculosis (Zopf) Lehmann and Neumann 1896, 363.[AL] (*Bacterium tuberculosis* Zopf 1883, 67.)

tu.ber.cu.lo′sis. L. dim. n. *tuberculum* a small swelling, tubercle; Gr. suff. *-osis* characterized by; M.L. gen. n. *tuberculosis* of tuberculosis.

Sources of data in Tables 16.5 and 16.7: Kubica, 1973; David et al., 1978; Wayne et al., 1980; Wayne and Diaz, 1982; Wayne, unpublished data; IWGMT, unpublished data.

Rods, ranging in size from 0.3–0.6 × 1–4 μm, straight or slightly curved, occurring singly and in occasional threads. Stain uniformly or irregularly, often showing banded or beaded forms. Strongly acid-fast and acid-alcohol-fast as demonstrated by Ziehl-Neelsen or fluorochrome procedures. Growth tends to be in serpentine, cordlike masses in which the bacilli show a parallel orientation. Colonies of avirulent forms are less compact.

On most solid media, colonies are rough, raised, thick, with a nodular or wrinkled surface and an irregular thin margin; may become somewhat pigmented (off-white to faint buff, or even yellow). Colonies on oleic acid albumin agar are flat, rough, corded, dry and usually nonpigmented. In liquid media lacking a dispersing agent, forms a pellicle which, with age, becomes thick and wrinkled. In Dubos' Tween albumin medium, growth is diffuse, settling if undisturbed, but readily dispersed.

Generation time in vitro under optimal conditions, 14–15 h. Optimum temperature, 37°C, some growth at 30–34°C. Optimum pH, 6.4–7.0. Growth is stimulated by incubation in air with 5–10% added CO_2 and by inclusion of glycerol to 0.5% in the medium. Bacilli grown under highly aerobic conditions die rapidly on abrupt shift to anaerobiosis; when allowed to grow and settle slowly through a self-generated oxygen gradient, they adapt a tolerance to oxygen deprivation, and exhibit synchronized growth on resuspension (Wayne and Lin, 1982).

Strain-to-strain differences in tubercle bacilli have been demonstrated by their different phage susceptibility patterns, but three major patterns are recognized (Rado et al., 1975; Bates and Mitchison, 1969).

M. tuberculosis produces tuberculosis in man, other primates, dogs and some other animals which have contact with man. Experimentally, from inoculum of 0.01 mg, it is highly pathogenic for guinea pigs and hamsters, but relatively nonpathogenic for rabbits, cats, goats, bovine animals or domestic fowls. Inocula of 0.001–1 mg are used to produce experimental disease in mice. Attenuation of virulence may occur spontaneously upon subculture in artificial media. Virulence can be maintained by selection of appropriate portions of growth on suitable media or by animal passage. Strains of *M. tuberculosis* isolated from patients from southern India may cause only localized lesions in guinea pigs and these tend to regress. These strains produce catalase and are susceptible to both hydrogen peroxide and to isoniazid (Mitchison et al., 1963).

Streptomycin, *p*-aminosalicylic acid, isoniazid, ethambutol, rifampin and some other secondary drugs are used to treat tuberculosis. Spontaneous mutants resistant to one of these drugs may replace the parent strain if treatment is improper. Resistance to isoniazid is frequently accompanied by changes in other properties, such as loss of peroxidase and catalase activity and attenuation of virulence for guinea pigs (Middlebrook and Cohn, 1953; Middlebrook, 1954).

Antigenic character: Infected animals, including man, exhibit delayed hypersensitivity to crude or purified *M. tuberculosis* culture filtrates (tuberculins) and less sensitivity to tuberculin-like preparations from other mycobacteria. Disease caused by *M. bovis* cannot be distinguished from that due to *M. tuberculosis* by use of commonly available tuberculins. Infections by other species of mycobacteria result in much lower degrees of sensitivity to tuberculin, although if "second strength" (250 tuberculin units) PPD tuberculin is used, skin tests may be interpreted erroneously as indicating infection with *M. tuberculosis*.

In cross-sensitivity studies using experimentally sensitized animals, extracts (sensitins) of *M. tuberculosis* cannot be differentiated from those of *M. bovis*, *M. microti* or *M. africanum*, although they are readily distinguished from those of other species (Magnusson, 1980). Similarly, immunodiffusion analysis of bacillary extracts shows a pattern common to *M. tuberculosis*, *M. bovis*, *M. microti* and *M. africanum* (Stanford and Grange, 1974). Monoclonal antibodies have been prepared that do appear to distinguish between *M. bovis* and *M. tuberculosis*, although the specificity of at least one such preparation appears to be at the

Table 16.4.

Sources of methods for differential tests for characterizing species of slowly growing mycobacteria and of rapidly growing mycobacteria, and for distinguishing between slow and rapid growers

Tested property	Reference(s)	Distinguish between slow and rapid growers	Species characterization of:	
			Slow growers	Rapid growers
1. Growth in presence of 5% NaCl	Wayne et al., 1974	X	X	X
2. Growth in presence of 0.2% picrate	Tsukamura, 1967	X	X	X
3. Iron uptake	Wayne and Doubek, 1968	X		X
4. Microcolonial dye test	Wayne et al., 1957	X		
Enzymes:				
5. Urease	Wayne et al., 1974; Murphy & Hawkins, 1975		X	
6. Semiquantitative catalase (>45 mm (foam)	Wayne et al., 1976		X	
7. Tween hydrolysis, 10 days	Wayne et al., 1974		X	
8. Acid phosphatase	Käppler, 1965b		X	
9. α-esterase	Käppler, 1965a		X	
10. Catalase resists 68°C	Kubica and Pool, 1960		X	
11. β-galactosidase	Wayne et al., 1974		X	
12. Pyrazinamidase in agar	Wayne, 1974		X	
13. Nitrate reduced[a]	Bönicke et al., 1962		X	X
14. Allantoinamidase[b]	Wayne et al., 1974			X
15. Benzamidase[b]	Wayne et al., 1974			X
16. Isonicotinamidase[b]	Wayne et al., 1974			X
17. Succinamidase[b]	Wayne et al., 1974			X
18. Arylsulfatase, 3 days	Wayne, 1961; Kubica & Rigdon, 1961			X
Resists inhibition by:				
19. Isoniazid, 1 μg/ml	Wayne et al., 1976		X	
20. Isoniazid, 10 μg/ml	Wayne et al., 1976		X	
21. Thiacetazone, 10 μg/ml[c]	Wayne et al., 1976		X	
22. Thiophene-2-carboxylic acid hydrazide, 1 μg/ml	Wayne et al., 1976		X	
23. Hydroxylamine HCl, 500 μg/ml	Wayne et al., 1976		X	
24. p-Nitrobenzoic acid, 500 μg/ml	Wayne et al., 1976		X	
25. MacConkey agar	Kubica & Vitvitsky, 1974			X
26. Malachite green, 0.01%	Jones & Kubica, 1963			X
27. Pyronin B, 0.01%	Jones & Kubica, 1963			X
28. Oleic acid, 250 μg/ml	Wayne et al., 1964		X	
Growth and physiology:				
29. Colonies pigmented	Wayne et al., 1974			X
30. Pigment: photochromogenic	Wayne et al., 1974		X	
31. Pigment: scotochromogenic	Wayne et al., 1974		X	
32. Niacin accumulates in medium	Wayne et al., 1976		X	
33. Growth at 25°C	Wayne et al., 1976		X	
34. Growth at 45°C	Wayne et al., 1976		X	X
34a. Growth at 52°C	Wayne et al., 1976			X
35. Acid from L-arabinose	Tsukamura, 1967a			X
36. Acid from xylose	Tsukamura, 1967a			X
37. Acid from dulcitol	Tsukamura, 1967a			X
38. Oxalate utilized	Tsukamura, 1967a			X
39. Citrate utilized	Tsukamura, 1967a			X
40. Degradation of PAS	Tsukamura, 1967a			X
41. Mannitol utilized	Silcox et al., 1981			X

[a] The nitrate reduction test is improved by incorporating the nitrate in a growth medium such as Dubos broth or Middlebrook 7H9 broth, rather than simple buffer.

[b] Technique for all amidase is as described for the hypochlorite-ammonia assay for urease cited earlier (Wayne et al., 1974).

[c] This drug is not mentioned in cited reference, but technique is as for isoniazid.

Table 16.5.

Diagnostic table of selected slowly growing mycobacteria. Data derived from references cited for each species in narrative text[a]

Characteristics	M. tuberculosis	"M. bovis complex"[b]	M. kansasii	M. marinum	M. gastri	"M. terrae complex"[b]	M. malmoense	M. gordonae	M. flavescens[c]	M. szulgai	M. simiae	M. scrofulaceum	"M. avium complex"[b]	M. xenopi
Urease	+	+	+	+	+	+	d	–	+	d	+	+	–	–
Pyrazinamidase in agar	+	–	d	+	d	d	d	d		+	d	+	+	d
Nitrate reduction	+	–	+	–	–	d	–	–	+	+	–	d	–	–
Acid phosphatase	+	+	+	+	+	+	–	d	d	+	–	–	–	–
β-Galactosidase	–	–	–	–	–	+	–	–	+	+	+	–	+	+
Catalase, >45 mm foam	d	–	+	d	–	+	–	+	+	+	+	+	d	+
Tween hydrolysis, 10 days	+	–	+	+	+	+	+	+	d	d	d	–	–	–
α-Esterase	+	+	–	–	–	d	d	+	d	d	d	d	+	+
Niacin accumulates	+	–	–	–	–	–	–	–	–	–	d	–	–	–
Photochromogenic	–	–	+	+	–	–	–	–	–	–[d]	+	–	–	–
Scotochromogenic	–	–	–	–	–	–	–	+	+	+[d]	–	+	–	d
Grows at 25°C	+	–	+	+	+	+	d	+	+	+	+	+	d	–
Grows at 45°C	–	–	–	–	–	–	–	–	d	–	–	–	d	+
Resists inhibition by:														
Picric acid, 2 mg/ml	–	–	–	–	–	+	–	–	+	–	d	–	–	–
p-Nitrobenzoic acid, 0.5 mg/ml	–	–	d	d	–	d	+	d	+	+	+	d	+	d
Hydroxylamine HCl, 0.5 mg/ml	–	–	d	+	–	+	d	d	–	–	+	d	d	–
Isoniazid, 1 μg/ml	–	–	d	d	–	+	+	d	–		+	+	+	–
Isoniazid, 10 μg/ml	–	–	d	d	–	+	–	–	–	–	d	d	d	–
Thiacetazone, 10 μg/ml	–	–	–	d	–	+	+	+	+	+	+	+	+	+
Thiophene-2-carboxylic acid hydrazide, 1 μg/ml	+	–	+	+	+	+	+	+	+	+	+	+	+	+
NaCl, 5%	–	–	–	–	–	d	–	–	+	–	–	–	–	–

[a] Symbols: +, 90% or more of strains are positive; –, 90% or more of strains are negative; d, 11–89% of strains are positive.

[b] "M. bovis complex" includes M. bovis, BCG and M. africanum. "M. terrae complex" includes M. terrae, M. nonchromogenicum and M. triviale. "M. avium complex" includes M. avium and M. intracellulare.

[c] M. flavescens is a rapid grower that can easily be confused with M. szulgai.

[d] M. szulgai may be photochromogenic when grown at 25°C.

Table 16.6.

*Diagnostic table of rapidly growing species of **Mycobacterium**, data derived from references cited for each species cited in narrative text[a]*

Characteristics	*M. chelonae* subsp. *chelonae*	*M. chelonae* subsp. *abscessus*	*M. fortuitum* subsp. *fortuitum*	*M. fortuitum* subsp. *peregrinum*	*M. chitae*	*M. senegalense*	*M. agri*	*M. smegmatis*	*M. phlei*	*M. thermoresistibile*	*M. aichiense*	*M. aurum*	*M. chubuense*	*M. duvalii*	*M. flavescens*	*M. gadium*	*M. gilvum*	*M. komossense*	*M. neoaurum*	*M. obuense*	*M. parafortuitum*	*M. rhodesiae*	*M. sphagni*	*M. tokaiense*	*M. vaccae*
Degradation of PAS	+	+	+	+	−	−	−	−	−	−	−	−	−	−	−	−	−	(67)	−	+	−	−	(33)	−	−
MacConkey without C.V. (28°C)	+	+	+	+	−	−	−	−	−	−	−	−	−	−	−	−	−	−	−	−	−	−	−	−	−
Arylsulfatase, 3 day	+	+	+	+	−	+	(24)	−	−	−	+	(40)	−	−	(33)	−	−	−	(75)	−	v	+	+	+	(20)
NH$_2$OH·HCl, 500 μg/ml[b]	+	+	+	+	−	−	−	−	−	−	−	(61)	−	−	−	−	−	−	−	−	−	−	−	−	−
Colonies pigmented	−	−	−	−	−	−	−	−	+	+	+	+	+	+	+	+	+	+	+	+	+	+	+	+	+
Colonies photochromogenic	−	−	−	−	−	−	−	−	−	−	−	−	−	−	−	−	−	−	−	−	+	−	+	+	+
Grows at 45°C	−	−	−	−	−	−	+	+	+	+	−	−	−	−	(50)	−	−	−	−	−	−	−	−	−	−
Grows at 52°C	−	−	−	−	−	−	+	−	+	+	−	−	−	−	−	−	−	−	−	−	−	−	−	−	−
Nitrate reduction	(17)	−	+	+	+	−	+	+	+	+	−	(44)	+	+	+	−	+	+	(80)	−	(67)	−	+	−	+
Iron uptake (28°C)	−	−	+	+	(43)	−	−	+	+	−	−	+	−	−	+	−	−	−	−	−	+	−	−	−	+
NaCl tolerance (28°C)	+	+	+	+	(70)	+	(24)	+	+	+	−	(22)	−	−	+	−	−	+	−	−	(36)	−	−	+	(55)
Citrate utilized (28°C)	+	−	−	−	−	+	(24)	+	+	−	+	+	(20)	−	+	+	+	+	+	+	+	+	(33)	+	+
Mannitol utilized (28°C)	−	−	−	−	−	+	−	+	+	−	+	+	−	+	(63)	+	+	+	+	+	+	+	+	+	+
Acid from arabinose	−	−	−	−	−	+	−	(67)	+	−	−	+	−	−	−	−	−	−	(80)	−	+	−	−	−	+
Acid from xylose	−	−	−	−	−	+	−	+	+	+	−	+	−	−	−	−	−	−	(80)	+	+	+	−	+	+
Acid from dulcitol	−	−	−	−	−	+	−	+	−	−	−	−	−	−	−	−	−	−	−	−	−	−	−	−	−
Malachite green, 0.01%[b]	+	+	+	+	−	−	−	+	+	−	−	−	−	−	−	−	−	+	−	−	−	−	−	+	+
Pyronin B, 0.01%[b]	+	+	+	+	+	+	−	+	+	−	−	−	−	−	−	−	−	−	−	−	−	−	−	+	+
Oxalate utilized	−	−	−	−	−	−	+	+	−	−	−	−	−	−	−	−	−	+	−	−	−	−	(88)	−	+
Allantoinamidase	−	−	+	+	−	+	+	+	−	−	−	(75)	−	−	−	−	(33)	−	+	+	−	−	−	+	+
Benzamidase	−	−	−	−	−	+	−	(55)	−	−	+	(33)	−	−	−	−	−	−	−	−	−	−	−	+	+
Isonicotinamidase	−	−	−	−	−	+	−	(67)	+	−	+	(20)	−	−	−	−	−	+	−	−	−	+	+	+	+
Succinamidase	+	+	+	+	+	−	−	+	+	−	+	−	+	+	+	−	−	+	−	−	−	+	+	+	−
Acid phosphatase	+	+	+	+	+	+	+	−	+	+	+	+	+	+	+	−	(33)	+	−	+	−	+	+	+	−
Picrate, 0.2%[b]	(17)	+	+	+	+	+	+	+	+	+	+	+	+	−	+	−	(33)	+	−	+	+	+	+	+	+

[a] Symbols: see Table 16.2; also

[b] Growth in presence of indicated substance.

Table 16.7.

Descriptive table of slowly growing mycobacteria. Data derived from references cited for each species in narrative text[a]

Characteristics	1. M. tuberculosis	2. M. microti	3. M. bovis	4. M. africanum	5. M. kansasii	6. M. marinum	7. M. gastri	8. M. nonchromogenicum	9. M. terrae	10. M. triviale	11. M. malmoense	12. M. shimoidei[c]	13. M. gordonae	14. M. asiaticum[c]	15. M. szulgai	16. M. simiae	17. M. scrofulaceum	18. M. avium	19. M. intracellulare	20. M. xenopi	21. M. ulcerans	22. M. haemophilum[c]	23. M. farcinogenes[c]	24. M. lepraemurium[c]	25. M. paratuberculosis
Growth:																									
At 22°C	−	−	−	−			−	+	+		+	−	+		+	+	+	±	+	−	−	+			
At 25°C	−	−	−	−	+	+	+	+	+	+	+	±	+		+	+	+	±	+	−	±	−	−	−	
At 42°C	−	−	−	−	±	−	−	±	−	−	−	−	−		+	−	+	+	±	+	−	−	−	−	
At 45°C	−	−	−	−	+	+	−	−	−	−	−	−	−		−	+	−	−	±	+	−	−	+	−	−
Photochromogenic	−	−	−	−	+	+	−	−	−	−	−	±	−	−	d[b]	+	−	−	−	−	−	−	−	−	−
Scotochromogenic	−	−	−	−	−	−	−	±	±	−	−	−	+	d	+[b]	−	+	±	−	+	±	−	+	−	−
Growth on glucose, sole C		−	−	+	+	±	±	−	±	−	−	−	+	−	±	−	±	±	±	+	−	−	−	−	−
Growth on pyruvate, sole C	−	−	−	±	+	+	+	−	±	−	−	−	±	−	−	±	−	−	+	±	−	+	−	−	+
Niacin accumulates	+	+	−	±	−	−	±	−	−	−	−	−	−	−	−	±	−	−	−	−	±	−	−	±	+
Special supplements required	−	−	−	−	−	−	−	−	−	−	−	−	−	−	−	−	−	−	−	−	−	+	−	(+)	+
Enzymes:																									
Urease-NH₃	+	+	+	+	+	+	+	−	+	−	±	−	−	−	±	+	+	−	−	−	−	−	±	−	−
Nicotinamidase-NH₃	+	+	+	±	+	+	+	+	±	+	+	+	−		+	±	+	+	+	+	−	+	+	+	+
Pyrazinamidase-NH₃	+	+	−	±	+	+	−	+	−	−	+	+	±		+	±	+	+	+	+	−	+		±	+
Pyrazinamidase in agar	+	+	−	±	±	+	±	−	−	−	±	−	±		+	±	+	+	+	+	−	−		−	−
Propionamidase-NH₃	−	−	−	−	−	−	+	−	−		−	−	−	−	−	−	−	−	−	−	−	−			−
β-Galactosidase	−	+	+	±	+	−	−	+	+	−	±	−	+		−	−	±	+	−	+	−	−			−
α-Esterase	+	+	+	−	+	+	−	+	±	+	±	+	±	+	±	+	±	+	+	−	±	−	+	−	−
Acid phosphatase	−	±	±	±	+	+	+	+	+	+	+	+	±	+	±	+	±	±	+	+	±	−	−	−	−
Catalase:																									
45-mm foam	−	+	+	−	+	±	−	+	+	+		−	+	+	+	+	+	+	−	−	+	−	+	−	±
Resists 68°C	+	−	+	+	+	+	−	+	+	+		+	+	+	+	+	+	±	+	+	−	−	−	−	−
"T" type produced	+	+	+	+	+	+	±	+	+	−	+	+	+	+	+	±	+	±	+	+	+	−	+	+	±
"M" type produced	−	+	−	+	+	+	+	−	+	+	−	−	±	+	+	±	+	+	+	+	−	−	−	−	−
Peroxidase	+	+	+	+	+	+	+	+	+	−	+	+	+	+	+	+	+	+	+	+	±	−	−	−	−
Tween hydrolysis (10 days)	±	+	±	−	+	+	+	+	+	+	+	+	±	+	±	−	+	±	+	+	±	−	−	±	±
Arylsulfatase (10 days)	−	±	±	±	+	+	+	+	±	±	±	−	+		+	−	+	+	+	±	±	−	−	−	−
Nitrate reduction	+	+	+	±	+	−	−	±	±	+	−	−	±		+	−	+	−	+	+	−	−	−	+	−
Nitrite reduction (7 days)	−	−	−	−	−	−	−	±	±	−	−	−	±		±	−	+	±	+	±	−	−	−	−	−
Tellurite reduction (3 days)	−	−	−	−	−	−	−	−	±	−	−	−	−		+	−	±	−	+	−	−	−	−	−	−
Tellurite reduction (9 days)	+	−	+	−	−	−	−	−	+	−	−	−	−		+	−	+	+	+	+	−	−	+	−	−

1442

Resists inhibition by:

- Oleic acid, 250 µg/ml
- Picric acid, 2 mg/ml
- p-Nitrobenzoic acid, 500 µg/ml
- Hydroxylamine·HCl, 125 µg/ml
- Hydroxylamine·HCl, 500 µg/ml
- Sodium chloride, 5%
- Toluidine blue, 300 µg/ml
- p-Amino salicyclic acid, 1 µg/ml
- Thiacetazone, 10 µg/ml
- Thiophene-2-carboxylic acid hydrazide, 1 µg/ml
- Ethambutol, 1 µg/ml
- Ethambutol, 5 µg/ml
- Isoniazid, 1 µg/ml
- Isoniazid, 10 µg/ml
- Isoniazid, 100 µg/ml

[a] −, 15% or less of strains are positive; +, 85% or more of strains are positive; ±, 51–84% of strains are positive; ∓, 16–50% of strains are positive; d, variability within a given strain.
[b] At 37°C; may exhibit photochromogenicity at 25°C.
[c] Description based on fewer than five strains.

strain rather than species level (Coates et al., 1981: see also review by Daniel and Janicki, 1978). Immunologic analysis of T-catalase extracted from *M. tuberculosis* indicates structural identity with that of *M. bovis* and *M. africanum*, and marked divergence from that of other species (Wayne and Diaz, 1979).

Further comments. Although phenotypic distinctions can be made between *M. tuberculosis* and *M. bovis*, as well as the intermediate organisms *M. microti* and *M. africanum*, numerical taxonomy places them all in a "macrocluster" distinct from other slowly growing mycobacteria. DNA homology between *M. bovis* and *M. tuberculosis* is about 100% (Baess, 1979); as discussed above, antigenic relatedness of these species also suggests that they represent a single species on an evolutionary basis. However, no formal proposal has yet been made to reduce *M. bovis*, *M. microti* and *M. africanum* to the status of subjective synonyms of *M. tuberculosis*, with possible subdivision at the subspecific or infrasubspecific level. Such a proposal will probably be forthcoming.

Type strain: ATCC 27294.

2. **Mycobacterium microti** Reed in Breed *et al.* 1957, 703.[AL]

mic.ro′ti. M.L. masc. n. *Microtus* a genus that includes the vole; M.L. gen. n. *microti* of *Microtus*.

Common name: Vole bacillus (Wells, 1937).

Sources of data in Tables 16.5 and 16.7: David et al., 1978; Wayne et al., 1980; IWGMT, unpublished data.

Rods. Primary growth on glycerol-free egg media in 28–60 days. May adapt to tolerance to glycerol. Colony morphology variable. Optimum temperature 37°C.

Cause of naturally acquired generalized tuberculosis in the vole. Local lesions produced in guinea pigs, rabbits and calves. Lose pathogenicity on repeated subculture.

Antigenic structure: See *M. tuberculosis*.

Further comments. M. microti occupies a position along a phenotypic continuum between *M. tuberculosis* and *M. bovis* and probably should be reduced to a biovar of *M. tuberculosis* (Stanford and Grange, 1974; Wayne, 1982). See discussion of *M. tuberculosis*.

Type strain: NCTC 8710.

3. **Mycobacterium bovis** Karlson and Lessel 1970, 280.[AL]

bo′vis. L. n. *bos* the ox; L. gen. n. *bovis* of the ox.

Common name: Bovine tubercle bacillus (Smith, 1896).

Sources of data in Tables 16.5 and 16.7: See sources for *M. tuberculosis*.

Short to moderately long rods. On primary isolation growth is very poor on glycerol-containing media, although repeated subculture permits adaptation to growth on such media. Furthermore, freshly isolated cultures of *M. bovis* are microaerophilic; inocula dispersed into liquid, semisolid or solid agar media grow in the medium but not on the surface, as distinguished from *M. tuberculosis* which is highly aerobic (Schmiedel and Gerloff, 1965). On repeated subculture, *M. bovis* will adapt to aerobic growth. Dilute inocula on egg media yield small, rounded, white colonies, with irregular edges and a granular surface after 21 days or more of incubation at 37°C. Colonies on transparent oleic acid albumin agar are thin, flat, generally corded; not easily emulsified in absence of a detergent.

Strains usually lose catalase on acquiring resistance to isoniazid.

Originally isolated from tubercles in cattle; generally more pathogenic for animals than is *M. tuberculosis*. Produces tuberculosis in cattle, both domestic and wild ruminants, man and other primates, carnivores including dogs and cats, swine, parrots and possibly some birds of prey. Experimentally highly pathogenic for rabbits, guinea pigs and calves; at least moderately pathogenic for hamsters and mice; slightly pathogenic for dogs, cats, horses and rats; not pathogenic for most fowl. Loss of virulence for guinea pigs and rabbits and loss of catalase activity accompany a loss of sensitivity to isoniazid as for *M. tuberculosis*. Certain strains isolated from cases of lupus and scrofuloderma in man have low pathogenicity for animals (Griffith, 1957).

Antigenic structure: See *M. tuberculosis*.

Table 16.8.

Descriptive table of rapidly growing species of the genus **Mycobacterium**[a]

Test	27a. M. chelonae subsp. chelonae	27b. M. chelonae subsp. abscessus	28. M. fortuitum	29. M. chitae	30. M. senegalense	31. M. agri	32. M. smegmatis	33. M. phlei	34. M. thermoresistibile	35. M. aichiense	36. M. aurum	37. M. chubuense	38. M. duvalii	39. M. flavescens	40. M. gadium	41. M. gilvum	42. M. komossense	43. M. neoaurum	44. M. obuense	45. M. parafortuitum	46. M. rhodesiae	47. M. sphagni	48. M. tokaiense	49. M. vaccae
1. Growth in presence of 5% NaCl	−	+	+	d	+		+	+	+	+	d	+		+			−			d		−		d
2. Growth in presence of 0.2% picrate	d	+	+	+		+	+	+	+	+	+	+		+	−	+			+	+	+	−	+	+
3. Iron uptake	−	−	+	d	−		+	+	−		+			−	−		+			+				+
4. Microcolonial dye test	+	+	+	+				+		+						d	+		−	−	+	+		
8. Acid phosphatase	d	−	+	+	−	−	−	+	+	−	d	−	−	−	−	−	+	−	−	−	+	+	−	−
13. Nitrate reduced	−	−	+	+	+	+	+	+	+	−	d	+	+	+	+	+	−	+	+	d	−	+	+	+
14. Allantoinamidase	−	−	+	−	+	+	d	−	−	−	d		−			d	−	+	+	−	−	+	+	+
15. Benzamidase	−	−	−	−	+	+	d	−	+	−	d	−	−			−	−	−	−	−	−		+	+
16. Isonicotinamidase	−	−	−	−	+	−	−	−	+	−	−	−	−			−		−	−	−	−		+	+
17. Succinamidase	+	−	−	−	+	−	+	−	+	+	−	−	−	−		−	+	−	−	−	+		+	+
18. Arylsulfatase, 3 day	+	+	−	−	+	d	−	−	−	+	d	−	−	d		−	−	−	−	−	+		+	d
23. Hydroxylamine HCl, 500 µg/ml	+	+	+	−	−	−	−	−	+	−	d	−				−	−	d	−	−	+		+	
25. MacConkey agar growth	+	+	+	−	−	−	−	−	−	−	−	−	−	−		−	−	−	−	−	−		−	−
26. Malachite green, 0.01% growth	+	+	+	−	−	−	+	+	+	+	+	+	+	−	+	+	+	−	−	−	−	+		−
27. Pyronin B, 0.01% growth	+	+	+	−		−	+	+	+	−	−							−	−	−	−			
29. Colonies pigmented	−	−	−	−	d	−	−	+	+	+	+	+	+	+	+	+	+	+	+	+	+	+	+	+
30. Colonies photochromogenic	−	−	−	−	−	−	−	−	+	+	+	+	+	+	+	+	+	+	+	+	+	+		+
34. Growth at 45°C	−	−	−	−	−	+	−	+	+	−	−	−	−	d	−	−	−	−	−	−	−			
34a. Growth at 52°C	−	−	−	−	−	−	−	+	+	−	−	−	−	−	−	−	−	−	−	−	−			
35. Acid from L-arabinose	−	−	−	−	−	−	d	+	−	−	+	−	−	d	+	−	−	d	−	+	−	+		+
36. Acid from xylose	−	−	−	−	+	−	d	+	−	−	+	−	−	d	+	−	−	d	−	+	−	+		+
37. Acid from dulcitol	−	−	−	−	+	−	+	+	−	−	+	−	−	−	−	−	−	−	−	−	−	+		
38. Oxalate utilized	+	−	+	−	+	d	+	+	+	−	+	−	−	+	−	−	+	−	−	+	+	d	−	−
39. Citrate utilized[b]	+	+	+	−	+	d	+	+	−	−	+	+	−	+	−	−	+	+	+	+	−	d	−	+
40. Degradation of PAS	+	+	+	−	−	−	−	+	+	+	+	d	+	d	−	−	d	−	+	+	−	d	−	−
41. Mannitol utilized	−	−	d	−	+	−	+	+	+	+	+	−	+	d	−	−	−	−	+	+	−	−	+	+

1444

[a] Symbols: see Table 16.5.

[b] Test 39, Citrate utilized, is listed as positive (+) for *M. fortuitum*. Recent reports of Silcox et al. (1981) and Tsukamura (1981) reveal that 100% and 82%, respectively, of strains of *M. fortuitum* are negative in this test. Reasons for this discrepancy are not known at this time. Of the tests shown in Table 16.8, ability to utilize mannitol as C-source will separate the two biovars (*fortuitum* and *peregrinum*) of the species *M. fortuitum*.

Further comments. Some differences in phage susceptibility of *M. bovis* and *M. tuberculosis* have been described (Baess, 1969).

The bacillus of Calmette-Guérin (BCG) (1908) conforms to the properties described for *M. bovis* except that it is much attenuated in pathogenicity and grows well aerobically and on glycerinated media.

As discussed above (see *M. tuberculosis*), *M. bovis* has a very close relationship to *M. tuberculosis* and probably should be reduced to a subspecies thereof.

Type strain: ATCC 19210.

4. Mycobacterium africanum Castets, Rist and Boisvert 1969, 321.[AL]

a.fri.ca′num. English n. *African* a native of Africa; M.L. gen. pl. n. *africanum* of Africans.

Sources of data in Tables 16.5 and 16.7: Kubica, 1973; David et al., 1978; Wayne and Diaz, 1979; IWGMT, unpublished data.

Rods average 3 μm. When grown on egg medium at 37°C, colonies are flat, dull and rough. Sodium pyruvate stimulates growth in egg medium. Growth homogeneous in Dubos' medium with Tween 80, and granular in Youman's medium with bovine serum. Growth in Lebek agar deeps 15 mm below surface.

Isolated from sputum of a tuberculosis patient in Senegal and a cause of human tuberculosis in tropical Africa. In guinea pigs 0.01 and 1 mg injected subcutaneously exhibit irregular pathogenicity, of lower order than *M. tuberculosis* of normal virulence. Generalized lesions seen by the 3rd month. Limited virulence on intravenous injection of 0.01 mg to rabbits (Castets et al., 1968, 1969).

David et al. (1978) reported that a large series of African strains exhibited more phenotypic heterogeneity than strains of *M. tuberculosis* and *M. bovis* from other areas, but noted that individual strains of *M. africanum* clustered with one or the other of these two species. They also noted that subclustering behavior reflected the geographic region within Africa from which they were isolated. Retention of *M. africanum* as a distinct species is probably not justified. See also discussion above, under *M. tuberculosis*.

Type strain: ATCC 25420.

5. Mycobacterium kansasii Hauduroy 1955, 73.[AL]

kan.sas′i.i. Kansas, a geographic place name. M.L. gen. n. *kansasii* of Kansas.

Common name: Group I photochromogen; yellow bacillus.

Sources of data in Tables 16.5 and 16.7: Kestle et al., 1967; Wayne et al., 1980; Wayne et al., 1981; Wayne and Diaz, 1982; Wayne, unpublished data; IWGMT, unpublished data.

Moderately long to long rods; broaden and exhibit marked crossbarring on incubation in the presence of sources of fatty acids. Dilute inocula on agar or inspissated egg media yield smooth to rough colonies after 7 or more days of incubation at 37°C. Colonies usually appear somewhat rough microscopically, but are readily emulsified in water; some strains resist emulsification. Colonies grown in dark are nonpigmented; when grown in light, or when exposed briefly to light when colonies are young, become brilliant yellow (photochromogenic). Rarely, strains produce no pigment or produce deep orange pigment when grown in dark. Most strains, if grown in a lighted incubator, form dark red crystals of β-carotene on the surface and inside of colony (Runyon, 1965).

Most strains are strongly catalase-positive; less commonly, weakly positive and inactivated at 68°C for 20 min, i.e. without M-catalase (Wayne and Diaz, 1982), and these variants appear less pathogenic for man (Wayne, 1962).

Isolated from human pulmonary lesion. Causes chronic human pulmonary disease resembling tuberculosis, although not normally considered contagious from man to man (Wolinsky, 1979a).

Subcutaneous inoculation may cause local lesions but no gross visceral lesions or death in guinea pigs. Intraperitoneal or intravenous inoculation produce self-limiting visceral lesions. Pathogenicity for guinea pigs not markedly enhanced by cortisone administration, but pretreating animals' lungs with coal dust may intensify extent of disease

(Pollak and Buhler, 1955; Tacquet et al., 1967). The usual forms cause self-limiting ulceration on intradermal inoculation of guinea pigs with 10^{-4} mg of bacilli, whereas the low catalase forms require about 100 times larger inoculum (Wayne, 1962). Intraperitoneal inoculation of hamsters usually causes death, with lesions of lymph nodes, spleen, liver and occasional invasion of capsule of the kidney; virulence enhanced by cortisone. Some deaths occur in mice inoculated intraperitoneally, but most exhibit self-limiting granulomas of liver, spleen and lymph nodes. Rats exhibit minimal lesions and chickens none (Pollak and Buhler, 1955). Rabbits, inoculated intravenously with 5 mg, develop macroscopic lesions of joints and tendon sheaths, and some gross lesions of liver and lung; rarely fatal (Engbaek et al., 1964). Intratracheal inoculation of rhesus monkey causes self-limited disease, which regresses with time (Grover et al., 1957).

Most frequently isolated from pulmonary secretions or actual tubercles of man. Occasionally associated with lesions of lungs or lymph nodes of deer, swine and cattle (Worthington and Kleeberg, 1964; Pattyn et al., 1967). Natural sources of infection unclear. Extensive soil sampling has failed to yield isolates of *M. kansasii* (Wolinsky and Rynearson, 1968), but strains have been isolated from water (Bailey et al., 1970; Engel et al., 1980; Kaustova et al., 1981).

Antigenic structure: Magnusson (1967) distinguished *M. kansasii* from 10 other mycobacterial species, including *M. marinum*, by means of dermal hypersensitivity. Although some degree of cross-reactivity does occur, dermal desensitization is effected only by homologous antigen (Worthington and Kleeberg, 1967). Only one agglutinating serovar has been established, accounting for 154 of 155 tested smooth strains whose identity has been established by biochemical methods (Hobby et al., 1967). Rough strains could not be typed because of spontaneous agglutination. Both high and low catalase varieties exhibit the same agglutinating serovar (Wayne, 1966) and this is associated with a specific phenol-soluble antigen detectable by immunodiffusion (Wayne, 1971). This antigen contains an alkali-sensitive, presumably ester-linked backbone, like corresponding antigens from *M. szulgai*, *M. gordonae*, *M. gastri*, *M. xenopi* and *M. marinum*, and distinct from alkali-resistant peptidoglycolipid surface antigens of *M. avium*, *M. intracellulare* and *M. scrofulaceum* (Brennan, 1981). No cross-reactivity was seen between the surface antigens of *M. kansasii*, *M. gastri* and *M. marinum* (Wayne et al., 1978).

Immunodiffusion analyses of cell extracts have shown four antigens that *M. kansasii* shares with no other species but *M. gastri* (Stanford and Grange, 1974). The T-catalase of *M. gastri* also shows a very close structural relatedness to that of *M. kansasii* (Wayne and Diaz, 1982).

Although clustering of phage-susceptibility patterns is seen for *M. kansasii*, some patterns also occur in *M. gastri* and *M. marinum* (Wayne et al., 1978). Thin layer chromatography of surface lipids serves to distinguish between *M. kansasii* and *M. marinum* (Szulga et al., 1966).

Type strain: ATCC 12478.

6. Mycobacterium marinum Aronson 1926, 320.[AL]

ma.ri′num. L. adj. *marinus* of the sea, marine.

Sources of data in Tables 16.5 and 16.7: Kubica, 1973; Wayne et al., 1980; Wayne et al., 1981; Wayne and Diaz, 1982; Wayne unpublished data; IWGMT, unpublished data.

Moderately long to long rods with frequent cross-barring. Colonies smooth to rough on inspissated egg medium and smooth on oleic acid-albumin agar after 7 or more days at 30°C. On primary isolation growth restricted to temperature range of 25–35°C, but may adapt to growth at 37°C. Colonies grown in the dark are nonpigmented; when grown in light or when exposed to light when colonies are young, become brilliant yellow (photochromogenic).

Isolated from diseased fish and aquariums (Aronson, 1926). In man, frequently seen in epidemic form as skin lesions resulting from abrasions incurred in swimming pools or fish tanks. Causes cutaneous granulomas ("swimming pool granuloma") in man, usually on elbow, but also found on knee, foot, finger and toe; papules or nodules, sometimes ulcerating; usually heal spontaneously over a period of months (Norden and Linell, 1951; Schaefer and Davis, 1961). Mice receiving a large inoculum intraperitoneally develop ulcerations on tail, paws and scrotum; visceral lesions and death sometimes occur; after intravenous inoculation, lesions limited to tail; foot pad inoculations lead to local swelling and some ulceration (Fenner, 1956). Guinea pigs inoculated subcutaneously or by inhalation develop no disease; intraperitoneal inoculation occasionally leads to scrotal lesions. Rats inoculated intraperitoneally develop no disease, but nodules may occur in omentum and hilar lymph nodes. Chickens develop no lesions when inoculated intraperitoneally or intravenously, but chick embryos maintained at 33°C (but not 37°C) acquire fatal infection. Rabbits develop local lesions when inoculum is applied to abraded skin sites, and may develop granuloma with caseous necrosis in scrotum after intraperitoneal or intravenous inoculation. Representatives of 50 poikilothermic species (reptiles, amphibians and fish) have been found susceptible to fatal systemic infection when maintained at 30°C (Clark and Shepard, 1963).

Antigenic structure: Magnusson (1967) distinguished *M. marinum* from 10 other mycobacterial species, including *M. kansasii*, by dermal hypersensitivity. One agglutinating serovar was originally recognized and no cross-reaction is seen with *M. kansasii* by this technique, by immunodiffusion with phenol-soluble antigen (Wayne, 1971; Wayne et al., 1978) or by immunofluorescence (Jones and Kubica, 1968). A second serovar has been proposed (Goslee et al., 1976). Castelnuovo and Morellini (1962), employing immunoelectrophoretic analysis, reported numerous precipitate lines of identity between *M. balnei*, *M. platypoecilus* and *M. marinum* and concluded that these represented a single species. Stanford and Grange (1964) detected three antigens by immunodiffusion that were shared by *M. marinum* and *M. balnei*, but by no other species. A large structural divergence between the T-catalase of *M. marinum* and those of *M. avium* and *M. kansasii* has been detected serologically (Wayne and Diaz, 1982).

See also discussion of *M. kansasii*.

Type strain: ATCC 927.

7. Mycobacterium gastri Wayne 1966, 923.[AL]

gas′tri. Gr. n. *gaster* the stomach; M.L. gen. n. *gastri* of the stomach.

Sources of data in Tables 16.5 and 16.7: See sources for *M. kansasii*.

Moderately long to long rods with cross-barring frequently seen. Colonies on inspissated egg media smooth to rough, white, and on oleic acid albumin agar, smooth or somewhat granular after 7 or more days of incubation at 37°C. Grows in temperature range 25–40°C.

Fails to produce progressive disease in the guinea pig, but usually capable of producing local ulceration at site of intradermal inoculation of 10^{-2}–10^{-3} mg of bacilli (Wayne, 1966).

Isolated from human gastric lavage specimen. Found in human gastric lavage or sputum specimens as casual residents, not considered etiologic agent of disease (Wayne, 1966; Kestle et al., 1967). Also found in soil (Wolinsky and Rynearson, 1968).

Antigenic analysis: Although closely related to *M. kansasii* in biochemical terms, this species is not agglutinated by *M. kansasii*-typing serum (Wayne, 1966) and the phenol-soluble antigen does not cross-react with this serum (Wayne, 1971). Immunodiffusion of cell extracts or culture filtrate does not permit differentiation between *M. gastri* and *M. kansasii* (Norlin et al., 1969; Stanford and Grange, 1974). Similarly, serologic analysis of the T-catalases demonstrated no significant difference between that of *M. gastri* and *M. kansasii*, but only the latter produces an M-catalase (Wayne and Diaz, 1982). *M. gastri* forms a homogeneous group by reciprocal intradermal skin testing and is distinguishable from *M. kansasii* by this technique (Magnusson, 1971).

Type strain: ATCC 15754

Further comments. The similarities of cytoplasmic antigens of *M. gastri* to those of *M. kansasii*, as well as a number of shared biochemical properties, raises the question of whether *M. gastri* represents a distinctly separate species. The distinctions based on M-catalase, nitrate reduction, pigment, drug susceptibility, surface antigens and clinical significance have tended to support separation of the species. Further studies are needed. See also discussion of *M. kansasii*.

8. Mycobacterium nonchromogenicum Tsukamura 1965, 110.[AL]

non.chrom.o.gen′i.cum. Gr. n. *chroa* color; Gr. v. *gennaio* produce; L. adj. *nonchromogenicum* not producing color.

Sources of data in Tables 16.5 and 16.7. See sources for *M. marinum*.

Moderately long to long rods. Colonies on inspissated egg media or oleic acid albumin agar smooth to rough, white to buff after 7 or more days of incubation at 37°C.

Isolated from mice injected with soil. Bacilli persist in tissues of mice without evidence of pathogenicity (Tsukamura, 1967b). Usually appears in clinical specimens only as an environmental contaminant; a very few cases are documented in which *M. nonchromogenicum* or *M. terrae* may have been considered the cause of human disease (Wolinsky, 1979a).

Antigenic structure: By dermal hypersensitivity testing a tenuous distinction can be made between *M. nonchromogenicum* and *M. terrae* (Meissner et al., 1974), but immunodiffusion analysis does not permit differentiation between these two species (Stanford and Grange, 1974). Spontaneous agglutination prevents typing by seroagglutination, but thin layer chromatography of surface lipids usually associated with agglutinating serotype among mycobacteria shows some overlap in patterns seen with these two species.

Type strain: ATCC 19530

Further comments. In numerical taxonomic analyses, two clusters corresponding to *M. nonchromogenicum* and *M. terrae* are seen, but these two clusters link to one another at a higher level than other species (Meissner et al., 1974). Positive reactions for nicotinamidase and pyrazinamidase and negative nitrate reduction provide the most definitive means of distinguishing *M. nonchromogenicum* from *M. terrae*. *M. triviale* shares most diagnostic features with both *M. terrae* and *M. nonchromogenicum*, but differs from these two species as well as all other slowly growing mycobacteria in its consistent ability to grow in the presence of 5% NaCl.

The overall similarities between *M. nonchromogenicum*, *M. terrae* and *M. triviale* raises some question of their status as separate species. Since all are generally considered clinically insignificant, it is common practice to identify strains as belonging to the "*M. terrae* complex" without further resolution into the individual species. Further taxonomic studies at the level of DNA and specific protein relatedness are required to establish whether these may be reduced to subspecies, or continue to be regarded as species.

9. Mycobacterium terrae Wayne 1966, 922.[AL]

ter′rae. L. n. *terra* earth. M.L. gen. n. *terrae* of the earth.

Sources of data in Tables 16.5 and 16.7: See sources for *M. kansasii*.

Isolated from sputum and gastric lavage specimens from humans; considered casual residents rather than pathogens. Have also been isolated from soil (Wolinsky and Rynearson, 1968).

Produces neither local nor systemic lesions after inoculation of 10^{-1} mg, intradermally, to guinea pigs.

Type strain: ATCC 15755.

Further comments. "*M. novum*" Tsukamura 1967, 163 is regarded as a synonym of *M. terrae* Wayne (Meissner *et al.*, 1974). "*M. terrae*" Tsukamura 1967b is an invalidly proposed synonym for *M. nonchromogenicum* Tsukamura 1965, and not to be confused with *M. terrae* Wayne. See also discussion of *M. nonchromogenicum*.

10. Mycobacterium triviale Kubica in Kubica, Silcox, Kilburn, Smithwick, Beam, Jones and Stottmeier 1970, 162.[AL]

tri.vi.a′le. L. n. *trivialis* that which belongs to the crossroads, i.e. common, of little importance; L. neut. adj. *triviale* of little importance.

Sources of data in Tables 16.5 and 16.7: Kestle et al., 1967; Kubica et al., 1970; Kubica, 1973; Wayne and Diaz, 1982.

From dilute inocula mature colonies do not appear on solid media for over a week. Colonies on egg medium are rough, dry, heaped up and nonchromogenic. On oleic acid agar characteristic rough R colonies are seen, which are easily confused with those of *M. tuberculosis*. Grows poorly, if at all, on cornmeal agar.

Isolated from sputum, but not considered a pathogen.

Antigenic structure: Specific bands found by immunodiffusion analysis of protoplasmic extract, and specific reaction by fluorescent antibody and agglutination tests on whole cells (Kubica et al., 1970).

Type strain: ATCC 23292.

11. **Mycobacterium malmoense** Schröder and Juhlin 1977, 241.[AL]

mal′mo.en.se. M.L. adj. *malmoensis* belonging to Malmö, Sweden, the source of the strains upon which the original description is based.

Sources of data in Tables 16.5 and 16.7: Wayne et al., 1983; IWGMT, unpublished data.

Coccoid to short rods. Growth on inspissated egg medium and oleic acid-albumin agar requires over 1 week in temperature range of 22–37°C. Colonies are smooth and nonpigmented. Growth below surface of semisolid agar medium after deep inoculation, as seen with *M. bovis*, but not with other mycobacterial species.

Isolated from sputum and biopsy specimens from patients with pulmonary disease, and considered the etiologic agent of the disease. Subcutaneous inoculation of guinea pigs with 0.01 mg of bacilli produces local, but not generalized lesions. Intravenous inoculation of chickens with 0.01 mg caused macroscopic lesions in liver and spleen in about half of the birds and some of these birds died. Intravenous inoculation of rabbits with 0.01 mg caused rare minimal lesions, containing rare viable bacilli (Schröder and Juhlin, 1977).

Antigenic structure: Seroagglutination demonstrates a single serovar distinct from that of other species. This agrees with the demonstration of a single unique thin layer chromatography pattern of their surface lipids. Dermal hypersensitivity studies also show similarity between strains of *M. malmoense* and distinction from other species (Schröder and Juhlin, 1977).

Type strain: ATCC 29571.

Further comments. Forms discrete cluster at 85% average internal match, and low linkage to other clusters in numerical taxonomy (Wayne et al., 1983).

12. **Mycobacterium shimoidei** Tsukamura 1982, 67.[VP*]

shi.moid′e.i. M.L. gen. n. *shimoidei*, of Shimoide, named for H. Shimoide, a Japanese microbiologist who first isolated a strain of this species.

Source of data in Tables 16.5 and 16.7: Wayne et al., 1981.

Moderate length rods, frequent cross-barring. On inspissated egg medium at 37°C rough nonpigmented colonies appear after 14–21 days. Optimal growth in temperature range 37–45°C; erratic growth at 28°C.

Isolated from sputum and considered to be etiologic agent of pulmonary disease.

Type strain: ATCC 27962

Further comments. The four cultures upon which this species was originally based were all isolated from the same patient in Japan (Tsukamura, 1982a). However, another culture exhibiting similar properties was isolated from a patient in Australia. Both strains differed markedly from other known species (Wayne et al., 1981) by numerical taxonomy.

13. **Mycobacterium gordonae** Bojalil, Cerbón and Trujillo 1962, 344.[AL]

gor.do′nae. L. gen. n. *gordonae* of Gordon; named after American bacteriologist, Ruth E. Gordon.

Common name: Tap water scotochromogen.

Sources of data in Tables 16.5 and 16.7: Kestle et al., 1967; Kubica, 1973; Wayne et al., 1980; Wayne et al., 1981; Wayne et al., 1971; Wayne and Diaz, 1982; Wayne, unpublished data.

Moderate to long rods. Colonies on inspissated egg medium and oleic acid albumin agar usually smooth and yellow or orange in 7 or more days of incubation at 37°C; although pigment is produced when cultures are grown in the dark, the color is often intensified by growing in continuous light. Growth optimal at 35°C.

Frequently encountered as casual resident in human sputum and gastric lavage specimens. Also found in water taps and soil. Rarely if ever implicated in disease processes (Wolinsky and Rynearson, 1968; Wolinsky, 1979a).

Antigenic structure: By comparative intradermal skin tests, *M. gordonae* may be distinguished from other mycobacteria (Runyon and Dietz, 1971; Magnusson, 1980). Similarly, immunodiffusion analysis of bacillary extracts demonstrates five precipitin lines unique to *M. gordonae* (Stanford and Grange, 1974). These bacilli were not agglutinated by sera produced against *M. scrofulaceum*, *M. avium*, *M. intracellulare*, *M. kansasii* or *M. marinum* (Hobby et al., 1967). Seven agglutinating serovars of *M. gordonae* have since been proposed (Goslee et al., 1976). Their surface lipid antigens yield characteristic thin layer chromatography patterns (Jenkins et al., 1972), and are of the alkali-labile type associated with *M. kansasii* (see discussion of *M. kansasii*) rather than the alkali-stable type seen with *M. scrofulaceum* (Brennan, 1981).

Type strain: ATCC 14470.

Further comments. For a period of time many strains of *M. gordonae* were referred to by the illegitimate name "*M. aquae*," but the type strain of "*M. aquae*" had the properties of *M. smegmatis* (Wayne, 1970). This name was then applied to strains that resembled *M. gordonae*, but yielded a different thin layer chromatography pattern (Jenkins et al., 1972). In opinion 55, the Judicial Commission placed the name "*Mycobacterium aquae*" on the list of *nomina rejicienda* (Judicial Commission, 1982).

14. **Mycobacterium asiaticum** Weiszfeiler, Karasseva and Karczag 1971, 247.[AL]

a.si.a′ti.cum. Asia, a continent, L. gen. n. *asiaticum* of Asia.

Sources of data in Tables 16.5 and 16.7: Wayne et al., 1981; Wayne and Diaz, 1982; IWGMT, unpublished data.

Coccoid rods. Dysgonic growth on inspissated egg medium after 15–21 days at 37°C. Usually photochromogenic, but occasionally fails to develop pigment after exposure to light; pigment not produced in the dark (Wayne et al., 1981).

Original strains isolated from monkeys, but one strain has been considered as cause of human lung disease (Wayne et al., 1981). Produces focal lung lesions after intravenous inoculation of mice; may kill mice within 30–60 days (Weiszfeiler et al., 1971).

Antigenic structure: Two strains of *M. asiaticum* studied by reciprocal intradermal skin testing techniques appeared homogeneous and distinct from five other species, including *M. simiae*, but were not compared to *M. gordonae* (Baess and Magnusson, 1982). Whole cells of *M. asiaticum* did not agglutinate with antiserum to a number of other mycobacteria (Wayne et al., 1981).

Type strain: ATCC 25276.

Further comments. Although *M. asiaticum* strains group in a cluster distinct from that of *M. gordonae* on numerical taxonomic analysis, only the mode of pigmentation provides a simple criterion for distinguishing between these species (Wayne et al., 1981). *M. asiaticum* has very low DNA homology to *M. simiae* but has not been tested against *M. gordonae* (Baess and Magnusson, 1982). Too few strains have been examined to provide a clear picture of distribution frequencies of diagnostic features.

15. **Mycobacterium szulgai** Marks, Jenkins and Tsukamura 1972, 211.[AL]

szul′gai. L. gen. N. *szulgai* of Szulga; named after Polish microbiologist T. Szulga.

Sources of data in Tables 16.5 and 16.7: See sources for *M. asiaticum*.

Moderately long rods, with some cross-barring. Smooth and rough colonies on inspissated egg medium in 2 weeks at 37°C. Orange pigment enhanced by growth in continuous light (Marks et al., 1972). Photochromogenic behavior more pronounced at 25°C than at 37°C (Schaefer et al., 1973).

Associated with pulmonary disease, cervical adenitis and olecranon bursitis in man (Marks et al., 1972).

---

* *VP* denotes that this name has been validly published in the official publication, International Journal of Systematic Bacteriology.

Antigenic structure: Exhibits specific agglutination with *M. szulgai* antiserum, and a unique thin layer chromatography pattern of surface lipid antigen (Marks et al., 1972; Schaefer et al., 1973); this antigen is of alkali-labile type (Brennan, 1981) (see discussion of *M. kansasii*).

Type strain: NCTC 10831

Further Comments

M. szulgai, a slow grower, may be mistaken for *M. flavescens*, nominally a rapid grower, in routine clinical diagnostic practice. Since *M. flavescens* is not a pathogen, but *M. szulgai* is, it is essential that appropriate tests be conducted to distinguish between them. In spite of superficial similarities, numerical taxonomy clearly separates these two species into different clusters (Selva-Sutter et al., 1976; Wayne et al., 1981).

16. **Mycobacterium simiae** Karassova, Weiszfeiler and Kraznay 1965, 282.[AL]

si′mi.ae. L. n. *Simia* the ape; L. gen. n. *simiae* of the ape.

Sources of data in Tables 16.5 and 16.7: Wayne et al., 1981; Wayne et al., 1983; Wayne and Diaz, 1982; IWGMT, unpublished data.

Short rods. Smooth colonies after 2–3 weeks on inspissated egg medium. Usually photochromogenic, although sometimes fail to produce pigment on exposure to light (Karassova et al., 1965).

Original strains isolated from lymph nodes of apparently healthy monkeys (Karrasova et al., 1965). Have since been implicated in a number of cases of human pulmonary disease (Wolinsky, 1979a). Organisms multiply extensively in organs of mice.

Antigenic structure: Three agglutinating serovars, labeled simiae 1, simiae 2 and avian 18 have been recognized among strains identified as *M. simiae* (Meissner and Schröder, 1975; Wayne et al., 1983). Reciprocal intradermal skin testing demonstrated a single cluster for the species, with subtle distinctions between strains of simiae serovars 1 and 2; avian serovar 18 was not included in that study (Baess and Magnusson, 1982). Immunoelectrophoretic analysis of bacillary extracts demonstrated a large number of antigens shared between simiae 1 and 2 serovars and also with *M. szulgai* (Wayne et al., 1981).

Type strain: ATCC 25275

Further comments. Numerical taxonomic analysis shows *M. simiae* to be comprised of a highly homogeneous cluster of strains of serovar 18, and a less tightly linked subcluster comprising the other two serovars (Wayne et al., 1983). Serovar 18 strains yield a negative niacin test, and the others are positive. Serovar 1 and 2 strains have been distinguished from one another by DNA homology (Baess and Magnusson, 1982), but not at a level sufficient to justify establishment of another species. Strains previously designated "*M. habana*" (Valdivia et al., 1971) are included among *M. simiae* serovar 1 (Meissner and Schröder, 1975).

17. **Mycobacterium scrofulaceum** Prissick and Masson 1956, 802.[AL]

scro.fu.la′ce.um. L. n. *scrofula* tuberculosis lymphadenitis, L. gen. n. *scrofulaceum* of scrofula.

Sources of data in Tables 16.5 and 16.7: See sources for *M. gordonae*.

Short to long rods or filaments. Colonies on oleic acid albumin agar or inspissated egg medium usually smooth and yellow to orange in 7 or more days incubation at 37°C. Occasional strains rough. Growth optimal at 35°C.

Isolated from closed lesion of cervical lymphadenitis in a child (Prissick and Masson, 1956). Most commonly encountered in human secretions. Found in pus from suppurating cervical lymph nodes (especially of children) and considered etiologic agent of the lesions (Prissick and Masson, 1957). Also found in human sputum and gastric lavage specimens, usually as a casual resident, but occasionally associated with pulmonary disease (Wolinsky, 1979a). Occasionally found in soil (Kestle et al., 1967; Wolinsky and Rynearson, 1968). Serotypes of this species have been found in swine (Schaefer, 1968).

In human disease, most commonly implicated as etiologic agent of cervical lymphadenitis in children. Does not cause extensive general-

ized disease or death in rats, hamsters or chickens; occasional lymph node involvement, are rarely localized lesions in liver or spleen. In guinea pigs inoculated subcutaneously, produces abscesses at site of inoculation and enlargement of regional nodes; intraperitoneal inoculation causes enlargement and suppuration of regional lymph nodes and, consistently, a peritonitis of various degrees and, rarely, lesions of liver or spleen, but not generalized disease or death (Penso et al., 1957; Prissick and Masson, 1957).

Antigenic structure: Reciprocal dermal skin testing establishes a discrete cluster for *M. scrofulaceum* (Magnusson, 1962, 1980; Runyon and Dietz, 1971; Wayne *et al.*, 1971; Baess and Magnusson, 1982). Distinctive patterns of cell extracts seen by immunodiffusion (Stanford and Grange, 1974) and immunoelectrophoresis (Castelnuovo and Morellini, 1962). Schaeffer (1965, 1968) established three serovars within this species by seroagglutination. These correspond to thin-layer chromatography patterns seen with surface lipid antigens (Wayne et al., 1971; Jenkins et al., 1972) which are of the alkali-stable type (Brennan, 1981). A fourth serovar, designated 44 (Goslee et al., 1976) has since been redesignated as *avium* complex serovar 27 (Wolinsky, 1979b).

Type strain: ATCC 19981.

Further comments

Mycobacterium scrofulaceum was conserved over the specific epithet *Mycobacterium marianum*, a senior subjective synonym, in Opinion 53 of the Judicial Commission, and *M. marianum* was placed on the list of *nomina rejicienda* because of its orthographic similarity to the valid epithet *M. marinum* (Judicial Commission, 1978).

See also discussion of *M. avium*.

18. **Mycobacterium avium** Chester 1901, 356.[AL] (*nom. cons.* Opin. 47, Jud. Comm. 1973.)

a′vi.um. L. n. *avis* a bird; L. gen. pl. n. *avium* of birds.

Common name: Avian tubercle bacillus.

Sources of data for Tables 16.5 and 16.7: Kestle et al., 1967; Kubica, 1973; Wayne et al., 1980; Wayne et al., 1981; Wayne and Diaz, 1982; Wayne, unpublished data.

Short to long rods, some filaments. Dilute inocula on inspissated egg or oleic acid albumin media yield usually smooth, and occasionally rough, nonpigmented colonies after 7 or more days of incubation at 37°C; on aging, colonies may become yellow. When first isolated from lesions, colonies on agar are flat and translucent, and these exhibit the characteristic pathogenicity in experimental animals described below. On repeated passage this colony type is replaced by a domed opaque type, and the latter is of greatly diminished pathogenicity (Meissner et al., 1974).

Isolated from tubercles in fowls. Widely distributed as the causal agent of tuberculosis in birds and less frequently in lesions or lymph nodes of cattle, swine and other animals. Rarely found in soil or as etiologic agent of human disease. Avian serovars 1 and 2 are widely distributed through at least three continents (Wolinsky and Rynearson, 1968), although type 3 is seen mainly in Europe (Marks et al., 1969).

Produces tuberculosis in domestic fowls and other birds; in pigs a localized disease. Experimentally in the rabbit and mouse it usually proliferates without macroscopic tubercles, producing disease of the Yersin type. Not pathogenic for guinea pig and rat. Strains of *M. avium* serotypes have been implicated in human pulmonary disease, although *M. avium*-like organisms causing human disease are usually more similar to *M. intracellulare* and fit one of that species' serovars (Hobby et al., 1967). Lesions in cattle may be caused by either *M. avium* or *M. intracellulare* serovars (Schaefer, 1968). *M. avium* serovars are highly virulent, serovar II being most consistently so, and representing the more common bird pathogen in nature (Schaefer, 1968); inocula of 0.01 mg will kill chickens. Chickens infected with 5 mg of moist bacilli, intravenously, die within 2 months, with gross lesions in spleen and microscopic lesions in lung and spleen (Engbaek et al., 1968). An inoculum of 0.01 mg will kill rabbits, but not mice. Rabbits inoculated with 5 mg moist bacilli, intravenously, usually die within 40 days with macroscopic lesions of spleen, and occasionally lungs, and microscopic lesions of spleen and lungs. Animals surviving 3 months show lesions of joints and tendon sheaths (Engbaek et al., 1964, 1968).

Antigenic structure: Schaefer (1965) established two agglutinating serovars for *M. avium*. Bacilli isolated from diseased birds almost always fall into one of these serovars, with serovar II the most frequent cause of natural bird infection. Subsequently Marks et al. (1969) showed that serovar II could be divided into two subtypes, with some cross-reaction. The serovar designations I and II have since been replaced by serovars 1, 2 and 3 (Wolinsky and Schaefer, 1973). Serovar 2 is the most consistently pathogenic in experimental infection of chickens. Occasionally a member of one of these serovars is isolated from human infection.

The seroagglutination of suspensions of *M. avium*, *M. intracellulare* and *M. scrofulaceum* is directed against a fibrillar superficial cell wall layer, which is a peptidoglycolipid (Barrow et al., 1980). Smooth strains of these species bear alkali-stable serologically active peptidoglycolipid, with a common lipopeptidyl rhamnose core, plus a variety of oligosaccharide saccharide side chains that confer specificity (Brennan, 1981). The peptidoglycolipid extracts from the various serovars among these species exhibit characteristic thin-layer chromatography patterns that correspond to their serological specificity.

In addition to common genus antigens, immunodiffusion analysis of extracts reveals at least six antigens to be shared by *M. avium*, *M. intracellulare* and *M. lepraemurium*, but not by other species (Stanford and Grange, 1974). Reciprocal intradermal testing in most cases permits discrimination between *M. avium* and *M. intracellulare* (Meissner et al., 1974; Magnusson, 1980).

Although strains of *M. avium* may be susceptible to several phages, no pattern is characteristic of any given serovar (Crawford et al., 1981b). A life cycle has been described (McCarthy and Ashbaugh, 1981).

Type strain: ATCC 25291.

Further comments. In cooperative numerical taxonomy studies based on biochemical tests, growth characteristics and drug susceptibilities, *M. avium* and *M. intracellulare* present as clusters with extensive overlap with one another at matching scores as high as 80%, and some overlap at 85% as well, whereas *M. scrofulaceum* appears as a discrete cluster with little overlap at 80% (Meissner et al., 1974; Wayne, 1982). The most definitive practically determined features for distinguishing *M. scrofulaceum* from both *M. avium* and *M. intracellulare* are pigment, urease and quantitative catalase reactions. The only known biochemical differences between *M. avium* and *M. intracellulare* are in rates of arylsulfatase and nitrite reduction and, as seen in Table 16.7, these features are not consistent. On occasion, strains of mycobacteria are encountered that yield patterns of pigment, urease and catalase that are inconsistent with *M. scrofulaceum*, *M. avium* or *M. intracellulare*, but which appear to belong to one of these species on the basis of their other properties. Hawkins (1977) proposed that these be treated as *M. avium/intracellulare/scrofulaceum* (MAIS) intermediates, until their taxonomic standing could be clarified. This proposal has since been misinterpreted and some workers refer to a "MAIS complex," which includes well-defined strains of the three component species, implying that they will bear a close taxonomic relationship to one another. There is ample evidence, based on numerical taxonomy (Meissner et al., 1974), reciprocal intradermal skin testing (Magnusson, 1980), immunodiffusion (Stanford and Grange, 1974), immunologic distance of catalases (Wayne and Diaz, 1979, 1982) and DNA homology (Baess, 1979) that *M. scrofulaceum* is a species distinct from both *M. avium* and *M. intracellulare*. These same studies provide more equivocal evidence on the relationship between *M. avium* and *M. intracellulare*, as reflected in two conflicting recommendations, presented as majority and minority opinions, on the advisability of reducing *M. intracellulare* to synonymy with *M. avium* (Meissner et al., 1974). Recent DNA/DNA hybridization studies have confirmed that *M. avium*, *M. intracellulare* and *M. scrofulaceum* are different species, and have indicated that strains belonging to serovars 4, 5, 6 and 8 appear more closely related to *M. avium* than to *M. intracellulare* (Baess, 1983). This is in agreement with results of sensitin tests on guinea pigs (Anz et al., 1970; Magnusson, 1981).

In ongoing cooperative numerical taxonomy studies, many strains that met the Hawkins (1977) criteria for MAIS intermediate status have since been found to cluster with *M. simiae*, some appear to represent minor variants of *M. avium* or *M. intracellulare* and a few strains did not cluster with any known species and require further study (Wayne et al., 1983).

19. **Mycobacterium intracellulare** (Cuttino and McCabe) Runyon 1965, 258.[AL] (*Nocardia intracellularis* Cuttino and McCabe 1949, 16; *Mycobacterium intracellularis* (sic) Runyon 1965, 258.)

in′tra.cel.lu.lar′e. L. prep. *intra* within; L. n. *cella* small room; L. adj. *intracellulare* within cell.

Common name: Battey bacillus.

Sources of data in Tables 16.5 and 16.7: See sources for *M. avium*.

Rods short to long. In new growth, transiently filamentous, eventually becoming coccobacillary. Dilute inocula on inspissated egg and oleic acid agar media yield usually smooth, rarely rough, nonpigmented colonies after 7 or more days of incubation at 37°C; on aging, colonies may become yellow.

Isolated from fatal systemic disease in a child (Cuttino and McCabe, 1949). Most frequently encountered in pulmonary secretions from people suffering from tuberculosis-like disease, and from surgical specimens from such patients. When isolated from human secretions, often represent etiologic agent of pulmonary disease, although frequently isolated as apparent casual resident (Wolinky, 1979a). Also isolated from limited disease processes in cattle and swine. Occasionally found in soil or water (Wolinsky and Rynearson, 1968; Wendt et al., 1980).

Experimentally causes limited disease in chickens and mice, much less severe than *M. avium*, and rarely fatal. In general, chicken pathogenicity corresponds to serovar, with *M. avium* serovar II more consistently pathogenic than the *M. intracellulare* serovars (Meissner et al., 1974). Forms intermediate between *M. avium* and *M. intracellulare* in terms of colonial morphology, cultural behavior and chicken pathogenicity occur. Growth at 42°C has been reported to increase pathogenicity of inoculum (Scammon et al., 1964). Lesions in guinea pigs usually limited to site of inoculation. Hamsters inoculated intratesticularly show extensive local lesions and, frequently, secondary focal lesions in liver, spleen and abdominal lymph nodes. Rabbits receiving 10^{-3} mg intravenously develop no demonstrable lesions (Feldman and Ritts, 1963) but 5 mg, although not causing death, produce rare macroscopic lesions in visceral organs, and usually moderate-to-severe lesions of joints and tendon sheaths (Engbaek et al., 1964).

Antigenic structure: At least 17 agglutinating serovars established among cultures identified as *M. intracellulare* (Wolinsky and Schaefer, 1973), labeled "avium complex" serovars 4–20. Members of these serovars are predominantly nonpathogenic for chickens and were isolated mainly from man, cattle and swine, as opposed to three distinct serovars of *M. avium*, which are virulent for chickens, and less frequently associated with mammalian disease.

See also discussion of antigenic structure of *M. avium*.

No single phage typing pattern is characteristic of this species or of individual serovars within it (Crawford et al., 1981b).

Type strain: ATCC 13950.

Further comments. Strains of "avium complex" serovar 8, once proposed as a separate species "*M. brunense*" (Kazda, 1967) and later considered to belong to *M. intracellulare* (Kubin et al., 1969) appear intermediate in virulence for experimental animals, between *M. avium* and *M. intracellulare*. Many, if not most strains of "avium complex" serovar 18 more closely resemble *M. simiae* than *M. intracellulare* (Wayne et al., 1983).

See also discussion of *M. avium*.

20. **Mycobacterium xenopi** Schwabacher 1959, 59.[AL] (*Mycobacterium xenopei* (sic) Schwabacher 1959, 59.)

xe.no′pi. *Xenopus* a genus of toad; M.L. gen. n. *xenopi* of *Xenopus*.

Sources of data in Tables 16.5 and 16.7: See sources for *M. avium*.

Long to filamentous rods. Dilute inocula on inspissated egg media yield smooth, nonpigmented colonies after 14 or more days of incubation at 37°C; on aging, most colonies become yellow. On Middlebrook 7H10 agar, uniquely characteristic colonies have compact centers, surrounded by a fringe of microscopically evident branching filaments on the agar surface. Colonies become adherent to medium by a button-like growth into the agar (Runyon, 1968). Optimal growth at 40–45°C.

Occasionally associated with chronic pulmonary disease, but more frequently isolated from human secretions without associated disease, infrequently in disease of genitourinary tract (Wolinsky, 1979a). Has been associated with waterborne nosocomial outbreaks of disease (Gross et al., 1976). First isolated from skin granulomas of the toad *Xenopus laevis* (Marks and Schwabacher, 1965). Variable response in different strains of mice inoculated intraperitoneally with 0.2–10 mg, but few animals die, and limited numbers of macroscopic lesions appear in liver, spleen, kidney or lung. Guinea pigs receiving 4 mg intramuscularly develop caseous abscess at site of inoculation; 1 mg intravenously causes no macroscopic lesions and intraperitoneally causes macroscopic nodules in omentum of some animals; 0.1 mg intracutaneously causes swelling and ulceration at site. Hens receiving 4 mg intramuscularly develop no gross lesions; 5 mg intravenously usually kill with lesions in liver and/or spleen, but 1 mg by this route induces no gross lesions. Rabbits receiving 4 mg intramuscularly develop caseous abscesses at site of inoculations; 5 mg intravenously cause few macroscopic lesions, or lesions of joints or tendon sheaths (Schwabacher, 1959; Engbaek et al., 1967).

Antigenic structure: Distinguished from other mycobacteria by reciprocal intradermal hypersensitivity reactions (Magnusson, 1980). Exhibits four species-specific antigens by immunodiffusion tests (Stanford and Grange, 1974). One agglutinating serovar has been reported (Goslee et al., 1976); the surface lipid antigen is of the alkali-labile type seen in *M. kansasii* (Brennan, 1981), not the stable type of *M. avium*.

Type strain: NCTC 10042.

Further comments. M. xenopi forms a discrete, compact cluster in numerical taxonomy. DNA homology less than 30% against *M. avium*, *M. scrofulaceum*, *M. tuberculosis*. Well separated from other species on basis of immunologic distance of T-catalase (Wayne and Diaz, 1982). Four phage-typing patterns have been described (Gunnels and Bates, 1972).

21. Mycobacterium ulcerans MacCallum, Tolhurst and Buckle in Fenner 1950, 817.[AL]

ul′ce.rans. L. part. adj. *ulcerans* making sore, causing to ulcerate.

Sources of data in Tables 16.5 and 16.7: Schröder, 1975; Boisvert, 1977.

Moderately long rods. Growth in inspissated egg medium evident after 4 weeks incubation at 30–33°C as minute, transparent domed colonies. On aging, colonies become convex to flat with irregular outline and rough surface, yellow. Rough, corded colonies on oleic acid albumin agar. Capable of growth between 30 and 33°C, but little growth at 25°C and usually none at 37°C.

Originally isolated from human skin lesion in Australia; has been isolated from ulcerative skin infections of man in Mexico, New Guinea, Malaya and Africa (Boisvert, 1977).

Causes skin ulcers in man characterized by indolent extension from areas of inconspicuous induration to involve large areas with undermining of edges. Rats and mice can be infected experimentally; guinea pigs, rabbits, fowls and lizards are resistant. Experimentally inoculated rats develop hemorrhagic necrotic lesions surrounded by zones of cellular accumulations consisting of leucocytes, lymphocytes and macrophages. There are no giant cells. The necrotic and cellular zones show large clumps of acid-fast bacilli in the extracellular spaces and in macrophages. Human lesions do not show inflammatory responses but consist of areas of lipid necrosis and tissue breakdown. Lesions develop only in cooler parts of the bodies of experimental animals; inoculation of mouse footpad consistently causes local lesions; inoculation by intranasal, intraperitoneal or intravenous route causes no visceral lesions, but after long incubation results in ulcerating lesions in hairless

peripheral parts of the body and on the scrotum (Fenner, 1956). The characteristic necrosis is associated with an extracellular heat-labile toxin of molecular weight of approximately 10^5 (Read et al., 1974).

Antigenic structure: Immunodiffusion of extracts reveals a number of antigens common to other mycobacteria, as well as five antigens that appear unique to *M. ulcerans* (Stanford and Grange, 1974). This conclusion was supported by the skin-sensitivity reactions in guinea pigs (Fenner and Leach, 1952).

Type strain: ATCC 19423.

Further comments. The name "*Mycobacterium buruli,*" which has no standing, was applied to strains isolated in Uganda (Clancey, 1964). By biochemical (Schröder, 1975) and immunologic (Stanford and Grange, 1974) criteria, strains of "*M. buruli*" are considered to belong in the species *M. ulcerans*.

22. Mycobacterium haemophilum Sompolinsky, Lagziel, Naveh and Yankilevitz 1978, 67.[AL]

hae.mo′phi.lum Gr. n. *haema* blood; Gr. adj. *philos* loving; M.L. adj. *haemophilus* blood loving.

Source of data in table 16.7: Sompolinsky et al., 1978.

Short, occasionally curved, rods, strongly acid-alcohol fast. Nonpigmented, rough to smooth colonies appear after 2–4 weeks incubation at 32°C, on inspissated egg or oleic acid-albumin media only if supplemented with 0.4% hemoglobin or 60 μM hemin, but not with $FeCl_3$ or catalase. Growth slower at 25°C and 35°C and absent at 37°C (Sompolinsky et al., 1978). Dawson and Jennis (1980) report that 15 mg/ml ferric ammonium citrate can be substituted for hemin. Strictly intracellular growth in tissue cultures of fibroblasts (Sompolinsky et al., 1979).

Isolated in Israel from subcutaneous granulomata of a patient under treatment for Hodgkins disease (Sompolinsky et al., 1978). Subsequently found in skin lesions of patients undergoing immunosuppressive treatment in Australia also (Dawson and Jennis, 1980).

Guinea pigs develop no obvious pathology after intravenous, intramuscular or subcutaneous inoculation with dense suspensions. Some mice die 2–4 weeks after inoculation; gross lesions not seen, but numerous intracellular bacilli found in monocytes and macrophages of liver, kidney and spleen. Intramuscular injections of 10^6–10^7 bacilli into frogs cause no disease when animals maintained at room temperature, but animals die within 8–20 days when maintained at 30°C, and bacilli found in liver and kidney (Sompolinsky et al., 1978).

Antigenic structure: Seroagglutination of whole bacilli specific for *M. haemophilum*, both of Israeli and Australian origin. Little if any cross-reaction with unabsorbed sera, and none using absorbed sera, with any serovar of *M. avium*/*M. intracellulare* complex, or *M. marinum*, *M. ulcerans*, *M. scrofulaceum*, *M. szulgai*, *M. asiaticum*, *M. simiae* or *M. malmoense* (Sompolinsky et al., 1979; Dawson and Jennis, 1980).

Type strain: ATCC 29548.

23. Mycobacterium farcinogenes Chamoiseau 1979, 407.[AL]

far.ci.no′ge.nes. L. v. *farcire* to stuff; L. n. *farcimen* sausage; Fr. n. *farcin* farcy or glanders; Gr. v. *gennaio* produce; L. adj. *farcinogenes* producing farcy.

Source of data in Table 16.7: Chamoiseau, 1979.

Short or long filaments, bent and branched, in clumps or tangled, lacey network, whether seen in pus from lesions or smears from culture. Do not fragment into bacillary forms. Strongly acid-alcohol fast. Rough, yellow, convoluted colonies after 15–20 days on inspissated egg medium; firmly adherent to medium and surrounded by iridescent halo.

Isolated from lesions of farcy in African bovines. On subcutaneous inoculation to guinea pigs, produces draining abscesses after 8 days, which heal slowly. Six to seven days after intraperitoneal injection to guinea pigs, abscesses seen in testes, seminal vesicles or vagina, and rare large abscesses in peritoneal walls and viscera; most animals die after prolonged infection.

Antigenic structure: No information available.

Type strain: NCTC 10955.

Further comments. A number of organisms isolated from lesions of farcy were long lumped into a group designated *Nocardia farcinica*. Chamoiseau (1973) established that the agents of farcy in African bovines belonged in the genus *Mycobacterium*, on the basis of characteristic lipids. At that time he proposed the name *M. farcinogenes*, subdividing the species into two species, designated subsp. *tchadense* and subsp. *senegalense*. Subsequently, on the basis of marked differences in growth rate, metabolic activities, lipid composition and DNA homology, Chamoiseau (1979) concluded that these represented two distinctive species, which he named *M. farcinogenes* and *M. senegalense*, respectively. *M. senegalense* is discussed among the rapid growers.

Because of discrepancies between the "type strains" of *Nocardia farcinica* held in the ATCC and the NCTC (one being a *Mycobacterium*, and the other a *Nocardia*) and uncertainty as to which is the authentic type, Tsukamura (1982b) has requested an opinion of the Judicial Commission rejecting *Nocardia farcinica* as a *nomen dubium*. Any action in this regard will have no effect on the validity of *M. farcinogenes*.

24. **Mycobacterium lepraemurium** Marchoux and Sorel 1912, 700.[AL]

lep.rae.mu′ri.um. Gr. n. *lepra* leprosy; L. n. *mus*, the mouse; M.L. gen. pl. n. *lepraemurium* of leprosy of mice.

Common name: Rat leprosy bacillus.

Source of data in Table 16.7: Saito et al., 1976.

(For many years this organism could not be cultivated in vitro, but was passed experimentally through rats, mice and hamsters. After attempts by many investigators to grow in various media and tissue cultures, Ogawa and Motomura (1970) succeeded in cultivating *M. lepraemurium* on an inspissated 1% egg yolk medium, using large inocula, confined to a concentrated area on the medium; egg white in the medium is inhibitory (Pattyn and Portaels, 1980). There is a general acceptance that these cultivated bacilli are, indeed, *M. lepraemurium* (Saito et al., 1976)).

Rods, 3–5 μm in length, with slightly rounded ends. When stained the cells often show an irregular appearance. Only the densely and uniformly stained forms appear to be infective for animals, in contrast to the "degenerate" unevenly stained forms (Rees et al., 1960). Strongly acid-fast. The bacilli from lesions are not bound together in clumps, rounded masses and palisades as in human leprosy.

Using dense compact inoculum on 1% egg yolk medium, rough, nonchromogenic colonies appear after 4–5 weeks incubation at 30–37°C. Some slow growth has also been recorded in an enriched Kirchner liquid medium; growth in this medium appears to depend on maintenance of reducing conditions (Dhople and Hanks, 1981a). Nevertheless, these organisms have been shown to produce both superoxide dismutase (Ichihara et al., 1977) and a T-type of catalase (Katoch et al., 1982a). Enzymes of the tricarboxylic acid cycle and the anaplerotic glyoxylate pathway have been demonstrated as well (Tepper and Varma, 1972; Mori et al., 1971); however, glyoxylate appears to be converted to δ-dehydroxylaeluvinate instead of malate (Mori et al., 1971). Synthesis of fatty acids also appears to follow a unique path (Kusaka, 1977).

A cause of endemic disease of rats in various parts of the world, having been found in Odessa, Berlin, London, New South Wales, Hawaii, San Francisco and elsewhere. The natural disease occurs chiefly in the skin and lymph nodes, causing induration, alopecia and eventual ulceration. Nodular diseases of the skin of other animals have been described, e.g. disease of buffalo in India, of a frog in South America, and of cats in Australia, associated with similar acid-fast bacilli.

Antigenic structure: By immunodiffusion analysis Stanford (1973) recognized three antigens previously considered to be unique to *M. avium* to be shared by *M. lepraemurium* as well. In addition, *M. lepraemurium* produced two antigens that were not shared with any other tested species. The T-catalase of *M. lepraemurium* isolated from infected mice exhibited an immunologic distance of 24 from that of *M. avium*, and occupied a unique position on the diagram of divergence of mycobacterial T-catalase between those of *M. tuberculosis* and *M. avium* (Katoch et al., 1982a). The same results were observed with

extracts of a culture of *M. lepraemurium* provided by F. Portaels (Katoch and Wayne, unpublished data).

Type strain: None specified due to difficulties in cultivation.

Further comments. Cell walls of *M. lepraemurium* contain arabinose and galactose as principal cell wall sugars, and alanine, glutamic acid and *meso*-diaminopimelic acid as mucopeptide amino acids, and a high proportion of lipid, as is typical of other, cultivable mycobacteria (Cummins et al., 1967). Characteristic mycobacterial mycolic acids have been demonstrated (Kusaka et al., 1981). DNA from *M. lepraemurium* shows a higher homology to that of *M. avium*, than to any other species tested (Imaeda et al., 1982a).

25. **Mycobacterium paratuberculosis** Bergey et al. 1923, 374.[AL]

pa.ra.tu.ber.cu.lo′sis. Gr. pref. *para* beside, related; M.L. n. *tuberculosis* tuberculosis; M.L. fem. n. *paratuberculosis* tuberculosis-like, paratuberculosis.

Common name: Johne's bacillus.

Source of data in Table 16.7: Thorel and Valette, 1976.

Plump rods, 1–2 μm in length, staining uniformly, but occasionally the longer forms show alternately stained and unstained segments.

This organism is difficult to cultivate; in primary cultures, it had originally been grown only in media containing heat killed acid-fast bacteria (Twort and Ingram, 1913). It has since been recognized that this requirement is satisfied by supplementing the medium with traces (0.03 μg/ml) of any one of a unique family of iron-binding hydroxamate compounds, called mycobactins (Snow, 1970). All mycobacteria except *M. paratuberculosis* produce mycobactins, although the side chains on the molecule vary from species to species. On mycobactin-enriched medium, growth is slow, requiring 3–4 months for primary isolation and 3–6 weeks on subculture for appearance of rough nonpigmented colonies. The absolute requirement for mycobactins may be circumvented by sequential passage in Watson-Reid medium at pH 5.5, containing mycobactin followed by subculture in the same medium without mycobactin, but kept at pH 5.5 (Morrison, 1965). This adaptation was not thought to be accompanied by acquisition of the ability to produce mycobactin. However, it has since been shown (Merkal and McCullough, 1982) that *M. paratuberculosis* can adapt to production of a mycobactin which has greater growth-stimulating effect on *M. paratuberculosis* than does a mycobactin from *M. phlei*.

Isolated from the intestinal mucosa of cattle suffering from Johne's disease, a chronic diarrhea. Apparently an obligate parasite in nature. The organisms isolated from sheep are reported to be more difficult to cultivate than are those from cattle (Dunkin and Balfour-Jones, 1935). Produces Johne's disease in cattle and sheep. Experimentally produces a similar disease in goats also. Guinea pigs, rabbits, rats and mice are not affected. Very large doses in laboratory animals produce slight nodular local lesions comparable to those produced by the nonpathogenic species *M. phlei*.

Antigenic structure: On reciprocal intradermal hypersensitivity testing, johnin, a tuberculin-like product from *M. paratuberculosis*, exhibits marked cross-reaction with avian tuberculin, but little, if any, with tuberculin derived from *M. bovis* (Thorel and Valette, 1976; Magnusson, 1980). Extracts of bacilli share more antigens with *M. avium* than with *M. tuberculosis* or *M. bovis*, as determined by immunodiffusion and immunoelectrophoresis (Tuboly, 1965).

Type strain: ATCC 19698.

Further comments. The U.S. Department of Agriculture strain 18 (corresponding to ATCC 12227) shares many antigenic properties with recent clinical isolates and is used for production of vaccine and johnin. However, this strain grows relatively rapidly, even without mycobactin (Merkal, 1979).

26. **Mycobacterium leprae** (Hansen) Lehmann and Neumann 1896, 372.[AL] (*Bacillus leprae* Hansen 1880, 32.)

lep′rae. Gr. n. *lepra* leprosy; M.L. gen. n. *leprae* of leprosy.

Common name: Leprosy bacillus or Hansen's bacillus.

Rods, 0.3–0.5 × 1.0–8.0 μm, with parallel sides and rounded ends, staining evenly or, at times, beaded. When numerous, as from lepromatous cases, generally arranged in clumps, rounded masses (globi) or in groups of bacilli side by side. Strongly acid-fast. Fisher and Barksdale (1973) report bacilli in leprosy lesions to be distinctive in losing acid-fastness on extraction with pyridine. However, clinical specimens usually contain a high proportion of nonviable bacilli, which are characterized by poor and uneven staining (McRae and Shepard, 1971). It is not clear whether the pyridine effect reflects a species characteristic, or merely the state of viability.

Despite occasional claims to the contrary, conclusive evidence of growth of *M. leprae* in culture has not yet been presented.

Causes leprosy in man. In the lepromatous form of the disease, bacilli are so abundant in the tissue as to produce stuffed-cell granulomas; in the tuberculoid and neural lesions organisms are rare. Obligate intracellular parasite. Confined largely to the skin (especially to convex and exposed surfaces), testes and to peripheral nerves. Probably do not grow in the internal organs.

Noncultivable acid-fast bacilli from human leprous tissue multiply, with an apparent generation time of 20–30 days, when inoculated into footpads of healthy mice (Shepard, 1960; Shepard and Chang, 1962). Under these conditions of in vivo cultivation in mice, the leprosy bacilli do not invade deep tissues, and their multiplication can be inhibited by diaminodiphenyl sulfone, isoniazid, rifampin, *p*-aminosalicylic acid and cycloserine. When corrected for presence of "nonviable" (i.e. non-solid-staining) bacilli, generation times as low as 10 days are seen (McRae and Shepard, 1971). Generation time in mouse footpad appears to be a consistent characteristic of a given strain (Shepard and McRae, 1971). Experimental transmission to immunosuppressed mice and rats has produced lepromatous-like model infections (Rees et al., 1967 Fieldsteel and McIntosh, 1971).

The discovery that *M. leprae* will multiply extensively in tissues of the nine-banded armadillo (*Dasypus novemcinctus*) (Kirchheimer and Storrs, 1971) led to use of this animal model to produce large quantities of bacilli which can be extracted from host liver for purification and study (Draper, 1979). Care must be taken to distinguish biological products of mycobacterial origin from those that may have been adsorbed onto the bacilli from host tissues. Leprosy bacilli from human tissues and from armadillo tissues possess an unusual *o*-phenoloxidase that is not produced by other mycobacteria (Prabhakaran et al., 1968, 1975); this activity may be obscured in very concentrated preparations due to tissue-derived inhibitors (Prabhakaran et al., 1979).

Glutamyl transpeptidase (Shelly et al., 1981) and alkaline phosphatase (Chatterjee et al., 1956) have been reported in *M. leprae* extracted from human tissues. Khanolkar (1982) has described uptake of radiolabeled glucose and protein hydrolysate products by bacilli from armadillo tissue, and Wheeler (1982) has reported the presence of enzymes of the glycolytic and hexose monophosphate-pentose phosphate pathways in comparable preparations. Armadillo-grown bacilli have also been shown to contain adenosine triphosphate (Dhople and Hanks, 1981b) and to produce *N*-acetyl-β-glucosaminidase, β-glucuronidase and acid phosphatase (Wheeler et al., 1982); in the latter study, extensive procedures were employed to distinguish between enzymes of host and of bacterial origin.

Among enzymes involved in terminal oxidation, cytochrome oxidase-linked reactions have been reported in bacilli from both human (Chat-

terjee et al., 1956) and armadillo (Ishaque et al., 1977) tissues. Superoxide dismutase has been detected in armadillo-derived leprosy bacillin; electrophoretic patterns and serologic cross-reactivity with superoxide dismutase from cultivable mycobacteria support the conclusion that superoxide dismutase found in *M. leprae* extracts is indeed derived from the bacilli, not the host (Kusunose et al., 1980, 1981; Wheeler and Gregory, 1980). The evidence for catalase is less conclusive. Wheeler and Gregory (1980) considered the catalase found to be of host origin. Katoch et al. (1982b) found no T-catalase but detected M-catalase; only 80% of the latter could be removed by precipitation with antiserum to normal armadillo liver enzyme, leaving the possibility that the remaining 20% was of mycobacterial origin. Electrophoretic studies suggest that some of the peroxidase found in *M. leprae* from armadillo liver is of mycobacterial origin as well (Wheeler & Gregory, 1980).

Mycolic acids have been identified in *M. leprae*, and these resemble those of *M. gordonae* more than those of other species tested (Young, 1980; Daffe et al., 1981; Kusaka et al., 1981). Little, if any, tuberculostearic acid is produced; in this regard, too, *M. leprae* resembles *M. gordonae* (Asselineau et al., 1981). A phenolic glycolipid resembling mycoside A of *M. kansasii*, but bearing a different trisaccharide component, has been isolated from *M. leprae* (Hunter and Brennan, 1981).

Antigenic structure: Of 12 antigens detected in *M. leprae* extracts by immunodiffusion, 6 were common to all mycobacteria and nocardiae tested, and 4 appeared specific to *M. leprae* (Stanford et al., 1975). Sera from leprosy patients contain antibodies that react with antigens from a number of mycobacterial species (Norlin et al., 1966) by immunodiffusion analysis, as well as by immunoelectrophoresis (Kronvall et al., 1976). At least 20 distinct antigenic components have been detected in *M. leprae* by immunoelectrophoresis (Closs et al., 1979), of which only 7 reacted with serum of lepromatous patients. Of 11 monoclonal antibodies against *M. leprae* extracts prepared by Gillis and Buchanan (1982), 2 reacted only against *M. leprae*; the others showed various patterns of reaction with the other 18 species tested. The greatest number shared, 8, was with *M. flavescens*, *M. gastri* and *M. gordonae*. *M. lepraemurium* was among those showing the fewest common antigens with *M. leprae*, resembling *M. intracellulare* in this regard.

Brennan and Barrow (1980) have partially purified a lipid from *M. leprae* which appears to be analogous to the C-mycosidic peptidoglycolipids responsible for specific agglutination of the serovars of the *M. avium* complex. The *M. leprae* lipid is serologically active and may be specific for this species.

Heated suspensions of the bacilli (obtained from nodules) produce a positive lepromin reactin in 75–97% of normal persons and of tuberculoid cases of leprosy, but usually produce no reaction in lepromatous individuals (Mitsuda: see Hayashi, 1932).

Type strain: Not cultivated; none designated.

Further comments. The D_{29} phage, propagated on *M. smegmatis*, can be absorbed specifically by *M. leprae* (David *et al.*, 1978a), and may cause some structural modification in the bacilli, but has not been proven to replicate in these bacilli (David et al., 1978b).

A relationship of DNA from *M. leprae* to that from a *Corynebacterium* has been proposed, in contrast to its greater antigenic and lipid similarity to mycobacteria (Imaeda et al., 1982b). This question requires further study.

List of the rapidly growing species of the genus **Mycobacterium**

27. **Mycobacterium chelonae** Bergey, Harrison, Breed, Hammer and Huntoon, 1923, 376.[AL] (Formerly *M. chelonei* (sic), see von Graevenitz and Berger, 1980, 520.)

che.lon′ae. Gr. fem. n. *chelone* a tortoise; L. gen. n. *chelonae* of a tortoise.

Organisms pleomorphic, ranging from long narrow to short thick rods (0.2–0.5 × 1–6 μm), with coccoid forms (0.5 μm) also reported. Cultures less than 5 days strongly acid-fast, but nonacid-fast forms begin to develop after 5 days. After 3–4 days incubation on most media, dilute inocula yield colonies that may be smooth, moist and shiny or

rough; usually nonchromogenic to creamy buff in color. When grown on corn meal agar, *M. chelonae* does not exhibit the extensive network of filaments observed in *M. fortuitum*. Usually capable of growth from 22–40°C, but no growth at 42°C; strains of *M. chelonae* subsp. *chelonae* often do not grow (or grow poorly) at temperatures of 37°C or higher.

As originally described, the organism produced only transient lesions in mice, hamsters, guinea pigs and rabbits. More recent cooperative studies have revealed that intravenous infection of mice with *M. chelonae* causes gross lesions visible in spleen, liver, lung and kidney. Has caused pathologic changes in the synovial tissue of knee and abscess-like lesions of gluteal region of man. Rarely isolated from sputum (with and without related disease), but have caused postoperative wound infections, including heterograft heart valve implants, mammaplasties, and corneal infections. Also found in soil.

Antigenic structure: Jenkins et al. (1971), Stanford and Beck (1969), and Stanford et al. (1972) have shown this species to be serologically and chemically distinct from closely related rapid growers (see Further Comments under *M. fortuitum*).

Evidence has been presented (Stanford and Beck, 1969; Stanford et al., 1972; Kubica et al., 1972) to show the synonymy of *M. abscessus* and *M. borstelense* with *M. chelonae*, although Jenkins et al. (1971) have shown them to have a different lipid composition. Cooperative numerical taxonomic studies showed (Kubica et al., 1972; Saito et al., 1977) *M. abscessus* to differ in some features from *M. chelonae* and *M. borstelense*; accordingly, two subspecies were recognized (Kubica et al., 1972) and this was confirmed in more recent studies (Silcox et al., 1981; Tsukamura, 1981). DNA/DNA homologies (Baess, 1982) indicate the subspecies designations to be valid.

27a. Mycobacterium chelonae subsp. chelonae Bergey et al., 1923, 376.*AL*

In addition to properties in Table 16.8, this subspecies does not grow in the presence of 1% deoxycholate, or 0.4 M nitrite, and cannot grow with nicotinamide, benzamide or nitrite as the sole sources of nitrogen.

Type strain: NCTC 946.

27b. Mycobacterium chelonae subsp. abscessus (Moore and Frerichs) Kubica, Baess, Gordon, Jenkins, Kwapinski, McDurmont, Pattyn, Saito, Silcox, Stanford, Takeya and Tsukamura 1972, 68.*AL*

In addition to characteristics listed in Table 16.8 this subspecies grows in the presence of 1% deoxycholate or 0.4 M nitrite, and can use nicotinamide and nitrite as sole sources of nitrogen.

Type strain: ATCC 19977.

28. Mycobacterium fortuitum da Costa Cruz 1938, 299.*AL*

for.tu'i.tum. Lat. neut. adj *fortuitum* casual, accidental.

Stanford and Gunthorpe (1969) presented evidence that *M. ranae*, by priority, should be the official name of the taxon *M. fortuitum*. Runyon (1972) challenged the former name as a *nomen ambiguum* and, for stability in taxonomy, requested conservation of the epithet *fortuitum*, a request acceded to in 1974 by the Judicial Commission (Opinion 51).

Rods 1–3 μm long; coccoid and short forms, even long rods are seen, with occasional beaded or swollen cells having nonacid-fast ovoid bodies at one end. Long filamentous, branching forms seen in pus. Ten to 100% of cells acid-fast after incubation 5 days at 28°C. After 2–4 days incubation on inspissated egg media, dilute inocula yield smooth, hemispheric colonies that may be butyrous, waxy, multilobate, even rosette clustered; also common are dull, waxy, rough colonies. Colonies usually off-white or cream colored, but when grown on malachite green-containing media, colonies may absorb the green dye (Hartwig et al., 1962). Young colonies (both smooth and rough) on cornmeal agar often produce an extensive network of peripheral filaments not usually seen in the closely related *M. chelonae* (Jones and Kubica, 1965).

Generalized disease rarely seen in most experimental animals, although localized kidney lesions are common in mice, guinea pigs, rabbits, monkeys, and calves; middle ear lesions lead to characteristic "spinning disease" in mice (Penso et al., 1952; Wells et al., 1955). Some strains isolated from lymph glands of cattle and systemic, nodular infection of frogs. Has been isolated from pulmonary disease, local abscesses, postoperative sternal wound infections, endocarditis, meningitis, osteomyelitis, and augmentation mammaplasties of humans. Also found in soil.

Antigenic structure: Schaefer (1967) described two serovars of *M. fortuitum*. Magnusson (1962) described dermal hypersensitivity techniques to identify the organism, while immunodiffusion and immuno-electrophoresis studies attest to the uniqueness of the species (Castel-

nuovo et al., 1960; Gimpl and Lanyi, 1965; Norlin, 1965; Stanford and Gunthorpe, 1969).

Type strain: ATCC 6841.

Further comments. The *M. fortuitum* complex, comprised of the species *M. fortuitum* and the two subspecies of *M. chelonae*, is often identified only to complex level by positive reactions in at least three of the following four tests: degradation of periodic acid-Schiff (PAS), growth on MacConkey agar without crystal violet, growth in the presence of 500 μg/ml of hydroxylamine HCl, and a positive 3-day arylsulfatase test. Differentiation to complex level is often insufficient for most clinical laboratories but, when neccessary, separation of species and subspecies may be effected as shown in Table 16.8.

It is evident from recent studies (Silcox et al., 1981; Wallace et al., 1982) that species and even subspecies within the fortuitum complex respond differently to antimicrobial drugs. At least three biovars (that may eventually become subspecies) of *M. fortuitum* have been recognized (Pattyn et al., 1974; Stanford and Grange, 1974; Grange and Stanford, 1974; Tsukamura, 1981b; Silcox et al., 1981; Wallace et al., 1982) and these subdivisions are supported by immunologic, chemical, and numerical taxonomic methods. One current biovar, *peregrinum*, was listed as a separate species in the eighth edition of the *Manual* (Runyon et al., 1974), with the proviso that it may be a subjective synonym of *M. fortuitum*. The recognition of effective therapeutic drugs for the various subspecies and biovars within the fortuitum complex doubtless will make subspecific and biovar recognition more important in the future. DNA/DNA homology studies (Baess, 1982) speak strongly for the designation of *peregrinum* as a species distinct from *M. fortuitum*.

29. Mycobacterium chitae Tsukamura 1967, 43.*AL*

chi'tae. M.L. gen. n. *chitae* coming from (of) Chita, a place.

Original description (Tsukamura, 1966e) in Japanese, but later described in English (Tsukamura, 1967c). Listed with species *incertae sedis* in eighth edition of the *Manual* (Runyon et al., 1974).

Organisms coccoid with no branching or cord formation. The strong acid-fastness in young cultures may become partially acid-fast in older cultures. After 3–5 days incubation on most media, dilute inocula produce white to cream-colored smooth, wet-looking colonies. Grows from 25–37°C, but does not grow at 45°C.

Four strains of this species originally recovered from soil samples collected by manure heaps. Suspensions of the soil were injected into chickens and homogenates of the organs yielded cultures of this rapidly growing *Mycobacterium*. Virulence studies revealed *M. chitae* to be avirulent for mice, rabbits, guinea pigs and chickens, injected with 2–10 mg of the organism. Not known to be associated with human disease.

Differential characteristics of the species are noted in Table 16.8. Other properties are found in papers by Tsukamura (1967c, 1981b) and Saito et al. (1977).

Type strain: ATCC 19627.

30. Mycobacterium senegalense (Chamoiseau 1973) Chamoiseau 1979, 407.*AL* (M. farcinogenes subsp. senegalense Chamoiseau 1973, 220.)

sen'e.gal.en'se. M.L. neut. adj. *senegalense* Coming from the West African Republic of Senegal.

Originally described as a subspecies of *M. farcinogenes* (Chamoiseau, 1973), this species was later recognized to be a totally different species.

Organisms grow as short-to-long filaments, bent and branched and assembled in clumps or a tangled, lacey network; filamentous forms do not fragment into bacillary forms. After 1–2 days of incubation at 25–37°C on most media, rough, convoluted colonies appear that are firmly attached to the medium and surrounded by an iridescent halo; usually nonchromogenic to ochre in color.

Differential characteristics are shown in Table 16.8, and a more detailed description is presented by Chamoiseau (1979). DNA/DNA homologies (Baess, 1982) attest to the uniqueness of this taxon.

M. senegalense is one of two acid-fast causative agents of farcy, a disease of the skin and superficial lymphatics in African bovines; the other, the slowly growing *M. farcinogenes*, is described elsewhere in this chapter.

Type strain: NCTC 10956.

31. **Mycobacterium agri** Tsukamura 1981, 256.*VP*

ag'ri. Gr. n. *agrus* field; L. gen. n. *agri* of a field.

Originally described by Tsukamura (1972a), the organism did not appear on the Approved List of Bacterial Names (Skerman et al., 1980) and the epithet was revived by Tsukamura (1981b).

Acid-fast rods (3–7 µm long) which frequently join to grow in long threads. Colonies rough, nonpigmented both in dark and after exposure to light. Growth occurs in less than 5 days on egg media incubated at 25–45°C; no growth at 52°C.

Original isolate recovered from alkali-treated soil sample inoculated onto egg medium and incubated at 42°C. Not known to be associated with disease.

Differential characteristics of the species are noted in Table 16.8. Other properties described by Tsukamura (1981b).

Type strain: ATCC 27406.

32. **Mycobacterium smegmatis** (Trevisan 1889) Lehmann and Neumann 1899, 403.*AL* (*Bacillus smegmatis* Trevisan 1889, 14).

smeg.ma'tis. Gr. n. *smegma* an unguent or ointment, a detergent, in M.L. sebaceous humor; M.L. gen. n. *smegmatis* of smegma.

Rods 3.0–5.0 µm long, occasionally curved with branching or Y-shaped cells seen; cells sometimes swollen with deeper staining, beaded or ovoid forms. After 5 days incubation, acid-fastness irregular (10–80%). Colonies that appear on egg media in 2–4 days are usually rough, wrinkled or coarsely folded and nonpigmented or creamy white; smooth, glistening butyrous colonies also seen, but pigmentation is rare, although it may be seen on older cultures. On oleic acid albumin agar, rough colony type appears smooth textured over a rugose but noncorded, granular colony; smooth form is domed, smooth textured, granular (Jones and Kubica, 1965). Capable of growth from 25–45°C.

Not pathogenic for mice, guinea pigs, hamsters or chickens, but positive cultures may be obtained from spleens of mice and/or guinea pigs (Durr et al., 1959). Isolated from smegma; once common in soil and water, but recent isolations from these sources have been infrequent.

Antigenic structure: One homogeneous group identified by species-specific sensitins (Magnusson, 1962) and immunodiffusion and immunologic techniques (Castelnuovo et al., 1960; Gimpl and Lanyi, 1965; Lind, 1960; Norlin, 1965).

Differential characteristics as in Table 16.8; other properties recorded by Kubica et al., 1972. A distinct taxon by DNA/DNA homology (Baess, 1982).

Type strain: ATCC 19420.

33. **Mycobacterium phlei** Lehmann and Neumann 1899, 411.*AL*

phle'i. L. neut. n. *Phleum* a grass genus, timothy; L. gen. n. *phlei* of timothy.

Short rods, 1.0–2.0 µm. Acid-fast in young cultures, but after 5–7 days staining irregular (5–100% acid-fast). Colonies on inspissated egg media usually rough, coarsely wrinkled, deep yellow to orange after 2–5 days incubation; a few cultures smooth, butyrous. On oleic acid albumin agar, rough colonies are flat, granular, loosely corded with irregular edges; smooth colonies have domed center with flat, translucent skirt and entire or irregular edges, dark granules near center (Jones and Kubica, 1965). Grows from 22–52°C.

Not pathogenic for mouse, rat, guinea pig, rabbit, chicken, frog or carp (Penso et al., 1951; Durr *et al.*, 1959).

Once widely distributed in nature, especially hay and grass, but recent environmental surveys revealed few isolates (Wolinsky and Rynearson, 1968).

Antigenic structure: One homogeneous group by immunofluorescence (Jones and Kubica, 1968), species-specific sensitins (Magnusson, 1962) and immunodiffusion and immunoelectrophoresis (Castelnuovo et al., 1960, Gimpl and Lanyi, 1965; Lind, 1960; Norlin, 1965).

Differential properties as in Table 16.8; other features described by Kubica et al., 1972. DNA/DNA homologies (Baess, 1982) confirm *phlei* as a distinct species.

Type strain: ATCC 11758.

34. **Mycobacterium thermoresistibile** Tsukamura 1966, 266.*AL*

therm'o.re.sist'i.bi.le. M.L. neut. adj. *thermoresistibile* resistant to high temperature.

Organism casually mentioned by Tsukamura (1965a, 52 in Japanese; 1966b, 266), later described in Japanese (Tsukamura 1966d, 187), and in English (Tsukamura, 1971), with more definitive features presented by Tsukamura et al. (1981). Listed with species *incertae sedis* in eighth edition of the *Manual* (Runyon et al., 1974).

May grow slowly on primary isolation and be confused with slowly growing scotochromogens, but subculture on egg-based media yields growth of smooth or rough yellow colonies that appear in 3–5 days. Lacks many properties common to rapid growers, e.g. growth on succinate, malate, fumarate, or mannose as sole C-source and negative 2-week arylsulfatase reaction.

Organism not known to be pathogenic for humans, but occasionally found in sputum.

Differential properties as described in Table 16.8, with more detailed characteristics described by Tsukamura et al. (1981) and Saito et al. (1977).

Type strain: ATCC 19527.

35. *Mycobacterium aichiense* Tsukamura, Mizuno, and Tsukamura, 1981, 274.*VP*

a.i'chi.ense. M.L. neut. adj. *aichiense* coming from Aichi prefecture, Japan.

Originally described in Japanese (Tsukamura 1973), this species did not appear on the Approved List of Bacterial Names (Skerman et al., 1980) and was revived by Tsukamura et al. (1981).

Rods less than 2 µm. Acid-fast in young cultures but may lose some of this acid-fastness on prolonged culture. Dilute inocula on egg media yield smooth yellow-orange colonies in 3–4 days or less. Grows from 25–37°C, but no growth at 45°C.

Isolated from soil and also from sputum of humans, but not associated with disease.

Differential features as in Table 16.8; other properties described by Tsukamura et al. (1981).

Type strain: ATCC 27280 (NCTC 10820).

36. **Mycobacterium aurum** Tsukamura 1966, 266.*AL*

au'rum. L. n. *aurum* gold.

Originally described in Japanese (Tsukamura and Tsukamura, 1966, 270), this was listed with species *incertae sedis* in the eighth edition of the *Manual* (Runyon et al., 1974).

Rods 1–6 µm, acid-fast. Colonies on inspissated egg media appear in less than 5 days, commonly smooth and deep yellow-orange. Grows from 25–37°C, but not at 45°C.

Recovered from soil, may occasionally be seen in sputum in humans, but not related to disease.

Differential features as in Table 16.8; other properties described by Saito et al. (1977), Tsukamura (1966b, 253) and Tsukamura et al. (1981). DNA/DNA homology studies (Baess, 1982) also attest to the uniqueness of this species.

Type strain: ATCC 23366 (NCTC 10437).

37. **Mycobacterium chubuense** Tsukamura, Mizuno and Tsukamura, 1981, 274.*VP*

chu'bu.ense. M.L. neut. adj. *chubuense* coming from soil of Chubu hospital.

Originally described in Japanese by Tsukamura (1973), the species did not appear on the Approved List of Bacterial Names (Skerman et al., 1980) and the epithet was revived by Tsukamura et al. (1981).

Organisms coccoid and acid-fast in early growth but may lose acid-fastness on prolonged incubation. Inoculation onto egg-based media yields growth of smooth yellow colonies in 3 days. Grows from 25–37°C, but not at 45°C.

Isolated from garden soil.

Differential features as in Table 16.8; other properties recorded by Tsukamura et al. (1981).

Type strain: ATCC 27278 (NCTC 10819).

38. Mycobacterium duvalii Stanford and Gunthorpe 1971, 637.[AL]

du.val'i.i. M.L. gen. n. *duvalii* of Duval; named for Professor C. W. Duval who isolated two strains.

Pleomorphic bacilli yielding bright yellow rough or smooth colonies in less than 7 days on inspissated egg media at 25–37°C, but not at 45°C. All four strains extant isolated from cases of human leprosy; not thought to be pathogenic at this time but evidence is insufficient.

Serologic specificity demonstrated by immunodiffusion (Stanford and Gunthorpe, 1971). Differential features as in Table 16.8; other properties reported by Stanford and Gunthorpe (1971) and Tsukamura et al. (1981).

Type strain: NCTC 358.

39. Mycobacterium flavescens Bojalil, Cerbón and Trujillo 1962, 344.[AL]

fla.ves'cens. L. v. *flavesco* become golden yellow; L. pres. part. *flavescens* becoming yellow.

Rod-shaped organisms. Dilute inocula on egg media usually produce soft, yellow-orange, butyrous colonies after 7–10 days at 25–37°C. Grows 25–42°C, not at 45°C. Although growth rate is intermediate, metabolic and physiologic properties are more like rapidly growing species.

Isolated from drug-treated tuberculous guinea pig; other isolations as apparent normal flora of humans suggests environmental habitat.

Serologic specificity demonstrated by immunodiffusion (Stanford and Gunthorpe, 1971). Differential features as in Table 16.8; other properties reported by Bojalil et al. (1962), Stanford and Gunthorpe (1971), Kubica et al. (1972), and Saito et al. (1977).

Type strain: ATCC 14474 (NCTC 10271).

40. Mycobacterium gadium Casal and Calero 1974, 306.[AL]

ga'di.um. A latinized name of Cadiz, a seaport town in Spain; *gadium* of Cadiz.

Short rods. Smooth yellow-orange colonies mature on egg media in less than 5 days at 25–37°C; no growth at 45°C. Older cultures become dry, more rough. First isolate recovered from known tuberculous human, but pathogenicity for man not known at this time.

Intraperitoneal injection of large numbers of organisms into mice and guinea pigs produced local lymphadenopathy in 2 weeks, with some dissemination to nearby organs, but this process resolved in 8 weeks.

Differential features as in Table 16.8; other properties reported by Casal and Calero (1974).

Type strain: ATCC 27726.

41. Mycobacterium gilvum Stanford and Gunthorpe 1971, 636.[AL]

gil'vum. L. *gilvus* pale yellow; L. gen. n. *gilvum* yellowish.

Pleomorphic bacilli producing pale yellow, smooth colonies in less than 7 days on inspissated egg media. Capable of growth at 25–37°C, but not at 45°C. Isolated both from sputum and pleural fluid; not thought to be pathogenic and never isolated more than once from same patient.

Serologic specificity demonstration by immunodiffusion (Stanford and Gunthorpe, 1971). Differential features as in Table 16.8; other properties reported by Stanford and Gunthorpe (1971) and Tsukamura et al. (1981).

Type strain: NCTC 10742.

42. Mycobacterium komossense Kazda and Müller 1979, 364.[AL]

ko.mos.sen'se. M.L. neut. adj. *komossense* belonging to Komosse sphagnum bog in south Sweden.

Short to moderately long rods, often clumped but never corded or cross-barred. Dilute inocula on both inspissated egg and oleic acid-albumin agar media yield eugonic, smooth, glistening, yellow-beige colonies in less than 7 days. Optimal growth at 31°C, but does grow at 22–37°C; no growth at 45°C.

Isolated from sphagnum vegetation of intact sphagnum bogs, southern Sweden and Atlantic coastal area of Norway; not recovered from partially cultivated moors.

Injection of rabbits, mice and guinea pigs with large inocula (1–10 mg wet weight of cells) produced neither local nor disseminated disease.

Uniqueness of species supported by antigenic analysis (immunodiffusion) and specific lipid patterns.

Differential features as in Table 16.8; other properties reported by Kazda and Müller (1979) and Tsukamura et al. (1981).

Type strain: ATCC 33013.

43. Mycobacterium neoaurum Tsukamura 1972, 229.[AL]

neo'au.rum. L. n. *aurum* gold; L. gen. n. *neoaurum* a new gold-pigmented organism.

Original description (Tsukamura, 1972) in Japanese. More detailed description in English (Tsukamura et al., 1981).

Intermediate to long rods producing golden yellow colonies in less than 5 days on inspissated egg media. Grows from 25–37°C; no growth at 45°C.

Isolated from soil; not known to be related to human disease.

Differential features as in Table 16.8; other properties described by Saito et al. (1977) and Tsukamura et al. (1981).

Type strain: ATCC 25795.

44. Mycobacterium obuense Tsukamura, Mizuno and Tsukamura, 1981, 274.[VP]

o.bu.ense'. M.L. neut. adj. *Obuense* coming from Obu, Japan.

Originally described by Tsukamura and Mizuno (1971), this species did not appear on the Approved List of Bacterial Names (Skerman et al., 1980) and was revived by Tsukamura et al. (1981).

Rods <2–6 μm. Acid-fast in young cultures but may lose some acid-fastness on prolonged culture. Dilute inocula on inspissated egg media yield smooth yellow-orange colonies in 5 days or less. Grows from 25–37°C; no growth at 45°C.

Isolated from soil and from the sputum of one patient, though association with disease not known.

Differential features as in Table 16.8; other properties described by Tsukamura and Mizuno (1971) and Tsukamura et al. (1981).

Type strain: ATCC 27023 (NCTC 10778).

45. Mycobacterium parafortuitum Tsukamura 1966, 12.[AL]

para.for.tu'i.tum. G. *para* alongside of or near; L. neut. adj. *fortuitum* casual, accidental; L. neut. adj. *parafortuitum* alongside of *fortuitum*.

Originally described in Japanese (Tsukamura, Toyama and Mizuno, 1965). Listed with species *incertae sedis* in eighth edition of the *Manual* (Runyon et al., 1974). Although the name may imply some similarity to *fortuitum*, such is not the case.

Rods 2–4 μm long. Dilute inocula on egg and agar media yield pale yellow smooth, moist colonies in less than 7 days (usually 3–4 days). Pigment increases markedly in most strains after exposure to light. Grows at 25–37°C, no growth at 45°C.

Isolated from soil. Injection of mice with 2 mg wet weight of cells caused no pathology and organisms were rapidly eliminated.

Differential features in Table 16.8; other properties described by Tsukamura et al. (1981). DNA/DNA homologies (Baess, 1982) show this species to be unique and quite different from taxonomically related *M. aurum* and *M. vaccae*.

Type strain: ATCC 19686.

46. Mycobacterium rhodesiae Tsukamura, Mizuno and Tsukamura 1981, 274.[VP]

rho.de'si.ae. M.L. gen. n. *rhodesiae* coming from Rhodesia.

Originally described by Tsukamura et al. 1971, this species did not appear on the Approved List of Bacterial Names (Skerman et al., 1980) and was revived by Tsukamura et al. (1981).

Very short rods, <2 μm, that may lose acid-fastness on prolonged culture. Dilute inocula on egg and agar media yield brilliant yellow, smooth, moist colonies in less than 5 days. Grows from 25–37°C; no growth at 45°C.

No demonstrable virulence for mice, rabbits, guinea pigs or chickens.

Isolated from sputum of patients in Rhodesia suspected to have tuberculosis; no positive evidence the species is pathogenic for humans.

Differential features as in Table 16.8; other properties described by Tsukamura et al. (1971, 1981).

Type strain: ATCC 27024 (NCTC 10779).

47. **Mycobacterium sphagni** Kazda 1980, 81.[VP]

sphag'ni. L. n. *Sphagnum* generic name of the moss of sphagnum bogs; L. gen. n. *sphagni* of sphagnum.

Short, thick often polymorphic rods; often clumped, but not corded. Dilute inocula on both egg and agar media grow eugonic, smooth, glistening, orange-yellow colonies. Grows from 22–37°C, optimal growth at 31°C in 3 days, and less than 7 days at other temperature extremes.

Isolated from sphagnum vegetation in moors of Germany and Scandinavia.

No demonstrable virulence for rabbits, guinea pigs, or mice injected with 1–10 mg wet weight of cells.

Uniqueness of taxon supported by skin testing of guinea pigs hypersensitized by injections of mycobacteria, by characteristic lipid patterns, and by immunodiffusion in agar.

Differential features as in Table 16.8; other properties described by Kazda (1980) and Tsukamura et al. (1981).

Type strain: ATCC 33027.

48. **Mycobacterium tokaiense** Tsukamura, Mizuno and Tsukamura 1981, 274.[VP]

to.kai.ense' M.L. neut. adj. *tokaiense* coming from the Tokai district of Japan.

Originally described in Japanese (Tsukamura, 1973), this species did not appear on the Approved List of Bacterial Names (Skerman et al., 1980) and was revived by Tsukamura et al. (1981).

Rods ranging from 1–7 μm long, sometimes cross-barred. Acid-fast in young cultures but may lose acid-fastness on prolonged incubation. Dilute inocula on egg media yield smooth colonies in less than 5 days. Grows from 25–37°C; no growth at 45°C.

Isolated from soil. Differential features as in Table 16.8; other properties described by Tsukamura et al. (1981).

Type strain: ATCC 27282 (NCTC 10821).

49. **Mycobacterium vaccae** Bönicke and Juhasz 1964, 133.[AL]

vac'cae. L. n. vacca a cow; L. gen. n. *vaccae* of the cow.

Short rods 1–4 μm, may be curved with rounded or thickened ends; occasionally Y-shaped cells seen. Acid-fastness irregular in old cultures. Dilute inocula on inspissated egg media yield smooth, yellow-orange, moist, shiny, butyrous, domed colonies in less than 5 days. Most strains very light sensitive, being nonpigmented if grown in complete darkness, but becoming yellow after brief (minutes) exposure to light. Occasional rough or nonpigmented colonies observed. Grow from 22–40°C; at 17 and 42°C growth restricted and pigment inhibited. No growth at 45°C.

Isolated from lacteal glands of cattle, from soil, watering ponds and wells, and even from skin lesions of cattle.

Differential features as in Table 16.8; other properties described by Bönicke and Juhasz (1964), Saito et al. (1977), and Tsukamura et al. (1981). Recent DNA/DNA homology studies (Baess, 1982) indicate this is a distinct species, thereby resolving some of the conflict in earlier international cooperative studies (Kubica et al., 1972; Saito et al., 1977).

Type strain: ATCC 15483.

Addendum

Since submission of the manuscript for this chapter, five new species of rapidly growing mycobacteria have been described. Although it was too late to incorporate these descriptions in the main text and Tables of the chapter, a brief description is presented below and readers are urged to consult the authors' original articles for detailed descriptions of the species.

50. **Mycobacterium porcinum** Tsukamura, Nemoto and Yugi, 1983, 162.[VP]

por'ci.num. L. adj. *porcinum* pertaining to swine.

Tsukamura et al. (1983) described 10 strains of a rapidly growing *Mycobacterium* that is closely similar (85%M) to *Mycobacterium fortuitum*.

On egg-based media smooth to rough, nonphotochromogenic colonies appear within 3 days; growth occurs from 28°–42°C, but not at 45°C. Rods 1.5–6 μm long, 0.5 μm wide.

The species shows positive reactions in 3-day arylsulfatase, resistance to $NH_2OH \cdot HCl$ (0.5 mg/ml) and degradation of *p*-aminosalicylate. *M. porcinum* differs from *M. fortuitum* by lacking nitrate reductase, by exhibiting a positive succinamidase activity and by utilizing benzoate as a sole source of carbon. Other differential features in original article.

All 10 strains were recovered from the lymph nodes of swine having a tuberculosis-like lymphadenitis.

When large numbers of strains of *M. fortuitum* were examined, there was some overlap of *M. porcinum* with the former taxon. Uniqueness as a species (or subspecies) must be confirmed when more strains are examined.

Type strain: ATCC 33776.

51. **Mycobacterium fallax** Lévy-Frébault, Rafidinarivo, Promé, Grandry, Boisvert and David, 1983, 336.[VP]

fal'lax. L. adj. *fallax* deceptive; in the sense that colonies resemble *M. tuberculosis.*

Cells are short rods 0.5–1.0 μm long. Growth occurs in 5 days or less on both egg- and agar-base media incubated at 30°C. Incubation at 37°C causes colonies to grow more slowly (12–21 days or more). Colonies rough, eugonic, nonpigmented and resemble *M. tuberculosis*, even to the production of cords.

Internal similarity of the 22 strains in the taxon cluster was 94.6% and percent similarity to all other species examined did not exceed 73%. Similarities to *M. tuberculosis* include colony morphology, thermolabile catalase, and positive nitrate reductase; differences include negative reactions for niacin production and β-glucosidase, and rapid growth at 30°C. Consult original article for other differences (Lévy-Frébault et al, 1983).

Most strains isolated from environmental sources; not known to be disease related.

Type strain: CIP 8139 (in Collection Nationale de Cultures de Microorganismes, Paris, France).

52. **Mycobacterium austroafricanum** Tsukamura, van der Muelen and Grabow, 1983, 460.[VP]

aus.tro.af.ri.ca'num. of South Africa.

Cells are rod-shaped, 2–6 μm long × 0.5 μm wide. Growth occurs within 3 days at 28 and 37°C, but not at 42°C. Colonies are mucoid, yellowish in dark, with pigment intensifying after exposure to light.

The species is susceptible to $NH_2OH \cdot HCl$ (250 μg/ml), isoniazid (10 μg/ml), ethambutol (5 μg/ml) and resistant to rifampin (25 μg/ml) and thiophene-2-carboxylic acid hydrazide (1 μg/ml).

M. austroafricanum is niacin negative, catalase is <45 mm foam, grows in presence of 5% NaCl, 0.1 sodium salicylate, 0.1% NaNO₂, and 0.2% picric acid. It is negative for α-esterase, β-galactosidase, acid phosphatase, Tween hydrolysis, benzamidase, isonicotinamidase, salicylamidase, allantoinase, and succinamidase. The species reduces nitrate, and is positive in 3 day arylsulfatase, β-esterase, nicotinamidase and pyrazinamidase. Other properties in original article.

All 23 strains recovered from South African waters are not known to be associated with human disease.

Type strain: ATCC 33464.

53. **Mycobacterium diernhoferi** Tsukamura, van der Muelen and Grabow, 1983, 460.[VP]

diern.ho′fer.i. of Diernhofer, who originally isolated the organisms.

Cells rod-shaped, 2–6 μm long and 0.5 μm wide. Smooth, white, nonphotochromogenic colonies appear within 3 days at 28 and 37°C, but not at 42°C.

The species is susceptible to $NH_2OH \cdot HCl$ (250 μg/ml) and ethambutol (5 μg/ml), resistant to isoniazid (10 μg/ml), rifampin (25 μg/ml) and thiophene-2-carboxylic acid hydrazide (1 μg/ml).

M. diernhoferi is niacin-negative, catalase is <45 mm foam, grows on 5% NaCl, 0.1% $NaNO_2$ and 0.2% picric acid. It is negative for α-esterase, β-galactosidase, Tween hydrolysis, benzamidase, isonicotinamidase, salicylamidase, succinamidase, and 3-day arylsulfatase. The species reduces nitrate, and is positive in acetamidase, urease, nicotinamidase, pyrazinamidase and allantoinase. Other properties in original article.

The strains were recovered from soil in a cattle field and are not known to be associated with human disease.

Type strain: ATCC 19340.

54. **Mycobacterium pulveris** Tsukamura, Mizuno and Toyama, 1983, 811.[VP]

pul′ver. is. L. gen. n. *pulvis* dust referring to the source, house dusts.

Cells are short rods or coccoid forms, less than 2 μm long and 0.5 μm wide. Growth rate intermediate between slowly and rapidly growing mycobacteria, the smooth, creamy nonphotochromogenic colonies commonly appearing in less than 8 days on egg or agar media incubated at 28–42°C; occasional strains grow at 45°C.

M. pulveris is niacin-negative, produces <45 mm foam in catalase test and gives negative reactions in tests for salicylate degradation, 3-day arylsulfatase, acetamidase, benzamidase, isonicotinamidase, salicylamidase, allantoinase, and succinamidase; it is positive in nicotinamidase, pyrazinamidase and β-esterase. The species is resistant to $NH_2OH \cdot HCl$ (125 μg/ml), ethambutol (5 μg/ml), rifampin (25 μg/ml), isoniazid (10 μg/ml), 5% NaCl, thiophene-2-carboxylic acid hydrazide (1 μg/ml), and sodium salicylate (0.5 mg/ml). Other properties in original article.

The eight strains were recovered from house dusts and are not known to be associated with human disease.

Type strain: ATCC 35154.

SECTION 17

Nocardioforms

Hubert A. Lechevalier

Table 17.1.
Differential properties of nocardioform genera[a]

Characteristics	Nocardia	Rhodococcus	Micropolyspora	Saccharopolyspora	Pseudonocardia	Oerskovia	Promicromonospora	Nocardioides	Intrasporangium
Marked fragmentation of mycelium in older cultures	d	+	−	+	+	+	+	+	d
Aerial mycelium produced	+[b]	+[c]	+	+[b]	+	−	+[b]	+	−
Conidia formed	+[b]	−	+	+[b]	+	−	−	+	−
Motile elements produced	−	−	−	−	−	+[d]	−	−	−
Strictly aerobic	+	+	+	+	+	−	+	+	+
Facultatively anaerobic	−	−	−	−	−	+	−	−	−
Cell wall type[e]	IV	IV	IV	IV	IV	VI	VI	I	I
Mycolic acids present	+[f]	+[f]	−	−	−	−	−	−	−
Phospholipid type[g]	PII	PII	PIII	PIII	PIII	PV	PV	PI	PIV
Menaquinones	MK-8(H$_4$) MK-9(H$_2$)	MK-8(H$_2$) MK-9(H$_2$)	MK-9(H$_4$) MK-9(H$_6$)	MK-9(H$_4$)	MK-9(H$_4$)	MK-9(H$_4$)	MK-9(H$_4$)	MK-8(H$_4$)	MK-8
Mol% G + C of DNA	64–72	63–72	ND	77	79	70–75	70–75	66–69	68

Symbols: +, 90% or more of strains positive; −, 10% or less of strains positive; d, 11–89% of strains positive; and ND, not determined.
[b] Lacking in some strains.
[c] Usually scanty.
[d] Some strains nonmotile (nonmotile oerskoviae (NMO)).
[e] Major constituents in cell walls of types: I, L-DAP, glycine; IV, mDAP, arabinose, galactose; and VI, lysine (with variable presence of aspartic acid and galactose).
[f] Nocardiomycolic acids.
[g] Characteristic phospholipids of types, in addition to phosphatidylinositol (which is always present); PI, phosphatidylglycerol (variable); PII, only phosphatidylethanolamine; PIII, phosphatidylcholine (with phosphatidylethanolamine, phosphatidylmethylethanolamine and phosphatidylglycerol variable, no phospholipids containing glucosamine); PIV, phospholipids containing glucosamine (with phosphatidylethanolamine and phosphatidylmethylethanolamine variable); and PV, phospholipids containing glucosamine and phosphatidylglycerol.

This last section of Volume 2 of *Bergey's Manual of Systematic Bacteriology* contains the organisms which are the most filamentous of all those discussed in this volume. They approach, and in some cases equal, the morphological complexity of the actinomycetes found in Volume 4.

The organisms grouped in Section 17 are often called "nocardioform," a term which stresses that they form a fugacious mycelium that breaks up into rod-shaped or coccoid elements. Like the term "coryneform," it is not one which can be defined satisfactorily and individual strains of nocardioforms may not exhibit this basic morphological feature. Proposed by Prauser (1967), the term "nocardioform" is intended to bring together, in an informal way, a number of organisms with a similar facies. In no way does it imply that these organisms are closely related. Thus attempts at defining nocardioforms in chemical terms are futile. Nevertheless, data from rRNA suggest that several of these genera (*Nocardia, Oerskovia, Promicromonospora*) are close phylogenetically, and are allied to *Corynebacterium, Arthrobacter* and *Mycobacterium*, and belong to a major branch of bacteria that includes *Streptomyces* and allied genera (Stackebrandt and Woese, 1981; Stackebrandt et al., 1983).

The major characteristics of the nine genera included in this section are given in Table 17.1.

All the organisms discussed in Section 17 are Gram-positive and aerobic (except for the oerskoviae which are facultative anaerobes on certain media). All the organisms of Section 17 have high mol% G + C of their DNA as do the actinomycetes found in Volume 4 of the Manual.

As far as it is known, the major respiratory quinones of all the organisms treated here are menaquinones with either eight or nine isoprene units. Minor variations in the composition of these menaquinones are observed between the genera of this section and also within the genera.

Cell wall, whole-cell sugar patterns and lipid composition furnish excellent markers for the separation of nocardioforms into genera. Most of the nocardioforms have a type IV cell wall composition (taxonomically significant constituents: meso-diaminopimelic acid (meso-DAP), arabinose, and galactose). The majority of species (the nocardiae and the rhodococci) contain mycolic acids, as do the corynebacteria (Section 15) and the mycobacteria (Section 16).

Members of the genera Nocardioides and Intrasporangium are characterized by type I cell walls (L-DAP, glycine). This wall composition is also typical of the genus Streptomyces and its satellites, descriptions of which will be found in Volume 4. The genera Oerskovia and Promicromonospora do not contain DAP but lysine in their mureins.

If one is consulting Section 17 to identify an unknown isolate, since morphology is the most rapid source of information, the characteristics listed below should furnish leads to possible generic assignments. However, it is rare, even with considerable experience, that a firm identification at the genus level can be made on the basis of morphology alone. One's attention should then turn to chemical composition. For a discussion of the methods used, see Lechevalier and Lechevalier (1976, 1980).

Morphological features	Possible generic assignment
No aerial mycelium	
Substrate mycelium breaks up into motile elements	Oerskovia (Gram+); Mycoplana (Gram−; see with Oerskovia)
Substrate mycelium breaks up into nonmotile elements	Promicromonospora (yellow to white); Rhodococcus (pinkish, salmon-colored); Mycobacterium (Section 16)
Substrate mycelium with ovoid vesicles	Intrasporangium or a culture of many other actinomycetes growing under adverse conditions.
Substrate mycelium without any special features	Actinomyces or relatives (Section 15); Nocardia, Rhodococcus, or an atypical form of any of the actinomycetes discussed in Section 17 and in Volume 4
Aerial mycelium formed	
Substrate mycelium without any special features, sterile aerial hyphae	Nocardia, Rhodococcus, or atypical forms of Micropolyspora, Saccharopolyspora, Pseudonocardia, Promicromonospora, Nocardioides or any of the actinomycetes discussed in Volume 4
Substrate mycelium breaking up into nonmotile elements; aerial hyphae without any special features	Nocardia, Rhodococcus
Both mycelia breaking up into elements of various shapes	Nocardia, Rhodococcus, Promicromonospora Pseudonocardia
Both mycelia bearing short chains of conidia	Micropolyspora, Nocardia
Aerial hyphae bearing short chains of conidia	Actinomadura or Microtetraspora (Volume 4); Nocardioides, Nocardia
Aerial mycelium with long chains of conidia or arthrospores	Saccharopolyspora, Nocardia; Actinopolyspora Nocardiopsis, Saccharothrix, Glycomyces or Streptomyces (Volume 4)
Both substrate and aerial hyphae divide into segments that may divide again into smaller elements, hyphae often present a zig-zag appearance	Nocardia, Saccharopolyspora, Pseudonocardia; Nocardiopsis or Saccharothrix (Volume 4)
Growth by budding	Pseudonocardia

Much information can be gained by the study of whole-cell hydrolysates. A one-way paper chromatogram will determine whether an unknown organism contains DAP and whether it is in the L- or the meso- form: (the D- form, which is not easily resolved from the meso-form, seems to have no taxonomic significance).

If an unknown strain contains the L-form of the DAP, there are only two possibilities in Section 17. If the descriptions of Nocardioides and Intrasporangium do not fit the unknown, one should keep in mind Streptomyces and its allies in Volume 4 and also some of the organisms in Section 15 (Arachnia, for example).

If the unknown contains the meso- form of DAP, a one-way paper chromatogram of its whole-cell sugar hydrolysate will reveal whether arabinose and galactose are present. If they are, a type IV cell wall is confirmed and a study of the mycolates will lead either to the genera Mycobacterium (Section 16), Corynebacterium, and Caseobacter (Section 15) or Nocardia and Rhodococcus in this section. Absence of mycolates suggests Micropolyspora, Saccharopolyspora, Pseudonocardia or the "mycolateless nocardiae" (this section).

Also with a cell wall of type IV are some genera treated in Volume 4 (Actinopolyspora, Saccharomonospora). The presence of meso-DAP without arabinose and galactose will also lead to actinomycetes discussed in Volume 4.

The absence of DAP would lead to an examination of the genera Oerskovia and Promicromonospora discussed in this section and to many of the organisms discussed in Section 15. Careful work in this case will require purified cell wall preparations to determine the identity of the dibasic amino acid involved.

For general reviews of the actinomycetes, see Volume II of Starr, et al., 1981 and Goodfellow, et al., 1984.

Genus **Nocardia** Trevisan 1889, 9[AL]

MICHAEL GOODFELLOW AND MARY P. LECHEVALIER

No.car′ di.a. M.L. fem. n. Nocardia, named after Edmond Nocard a French veterinarian.

Rudimentary to extensively branched vegetative hyphae, 0.5–1.2 μ in diameter, growing on the surface of, and penetrating, agar media, often fragmenting in situ or on mechanical disruption into **bacteroid, rod-shaped to coccoid, nonmotile** elements. **Aerial hyphae**, at times visible only microscopically, **almost always formed**. Short to long chains of well to poorly formed conidia may occasionally be found on the aerial hyphae and more rarely on both aerial and vegetative hyphae. No endospores, sporangia, sclerotia or synnemata formed. Nonmotile.

Gram-positive to Gram-variable. Some strains partially acid-fast at some stage of growth. **Aerobic.** Mesophilic. **Chemoorganotrophic**, having an oxidative type of metabolism. **Catalase-positive.** Cell wall contains major amounts of **meso-diaminopimelic acid**, **arabinose and galactose**. The organisms contain **diphosphatidylglycerol, phosphatidylethanolamine, phosphatidylinositol and phosphatidylinositol mannosides**, major amounts of **straight-chain, unsaturated** and **10-methyl (tuberculostearic) fatty acids, mycolic acids with 46–60 carbons and up to three double bonds**, and either tetrahydrogenated menaquinone with eight isoprene units (MK-8(H$_4$)) (most nocardial species) or **dihydrogenated menaquinone with nine isoprene units (MK-9(H$_2$))** (**N. amarae**) as the predominant isoprenologue. **The fatty acid esters released on pyrolysis gas chromatography of mycolic esters contain 12–18 carbon atoms and may be saturated or unsaturated.**

The mol% G + C of the DNA ranges from 64–72 (T_m).

Nocardiae are widely distributed and are abundant in soil. Some strains are pathogenic opportunists for man and animals.

Type species: *Nocardia asteroides* (Eppinger) Blanchard 1896, 856, Opinion 58, Judicial Commission 1985, 538.

Further Descriptive Information

Many actinomycetes with a tendency to fragment have been, and continue to be, mistakenly assigned to the genus *Nocardia* on the basis of morphology alone. This has resulted, in the past, in a genus characterized by extreme heterogeneity (Lechevalier, 1976). Following a certain number of taxonomic reassignments and formation of new genera, such as *Actinomadura*, *Nocardiopsis*, *Oerskovia*, *Rhodococcus*, *Rothia*, and *Saccharopolyspora* (see section on Differentiation from Other Genera), substantial progress has been made in refining the nocardial concept (Goodfellow and Minnikin 1981b; Goodfellow and Cross, 1984).

The genus *Nocardia* is now largely defined on the basis of cell envelope lipid and peptidoglycan composition. At present, only actinomycetes with the following characteristics are considered to be "true" nocardiae: (a) a peptidoglycan composed of *N*-acetylglucosamine, L-alanine, D-alanine, D-glutamic acid with *meso*-diaminopimelic acid as the diamino acid and muramic acid in the *N*-glycolated form rather than the *N*-acetylated form found in many other actinomycete taxa (Bordet et al., 1972; Uchida and Aida, 1977, 1979; Goodfellow and Cross, 1984); (b) a polysaccharide fraction of the wall containing arabinose and galactose (i.e. nocardiae have a wall chemotype IV and a whole cell sugar pattern type A sensu Lechevalier and Lechevalier, 1970); (c) a phospholipid pattern consisting of diphosphatidylglycerol, phosphatidylethanolamine (taxonomically significant nitrogenous phospholipid), phosphatidylinositol, and phosphatidylinositol mannosides without phosphatidylcholine or phospholipids containing glucosamine (i.e. a phospholipid pattern type PII sensu Lechevalier et al., 1977); (d) a fatty acid profile showing major amounts of straight chain, unsaturated and tuberculostearic acids (i.e. type IV fatty acid pattern sensu Lechevalier et al., 1977); (e) mycolic acids with 46–60 carbons (Lechevalier and Lechevalier, 1980; Goodfellow and Minnikin, 1981a; Goodfellow and Cross, 1984; Minnikin and O'Donnell, 1984); (f) containing tetrahydrogenated menaquinone with eight isoprene units or dihydrogenated menaquinone with nine isoprene units as the predominant isoprenologue (Yamada et al., 1976, 1977; Collins et al., 1977, 1985; Collins and Jones, 1981; Collins, 1984; Kroppenstedt, 1984).

Another group of strains which have been assigned to the genus *Nocardia* over the years is made up of actinomycetes that have a cell wall composition of chemotype IV but which lack mycolic acids (Lechevalier et al., 1971, 1977; Goodfellow and Minnikin, 1981a, 1984). Since they are thus not true nocardiae, they will be classified elsewhere in the future. However, the strains lacking mycolic acids include many of commercial and ecological importance and their descriptions are included at the end of the genus under Species Incertae Sedis for the convenience of the users of the *Manual*.

The only constant morphological feature of nocardiae is a tendency of the aerial or vegetative mycelium to fragment (Locci, 1976; Williams et al., 1976; Nesterenko et al., 1978). The consistency and composition of the growth medium can affect the growth and stability of both aerial and substrate hyphae (Williams et al., 1976). Unfortunately, fragmentation is not unique to *Nocardia* since members of the genera *Rhodococcus*, *Oerskovia* and *Nocardioides* as well as some *Streptomyces*, show this character. Other morphological features of nocardiae (also shared with other actinomycetes of different genera) include well developed conidia in *N. brevicatena* (Fig. 17.1) and less well formed spores in some *N. asteroides* strains (Fig. 17.2), in both cases borne on both vegetative and aerial hyphae. Cell pigments may be off-white, gray, yellow, orange, pink, peach, red, tan, brown or purple. Soluble pigments, when present, are usually undistinguished: brown or yellowish. Aerial mycelium, visible to the naked eye, may be lacking, sparse or very abundant; thus, colonies may have a superficially smooth appearance or, more commonly, a matte, chalky or velvety aspect. Aerial hyphae,

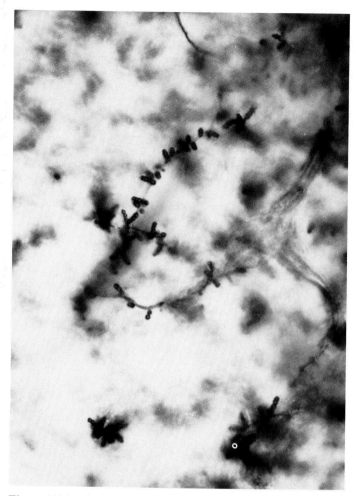

Figure 17.1.   Aerial conidial chains of *Nocardia brevicatena* (× 1250).

visible microscopically, are always present. Sections of cells show a typical Gram-positive wall, which in some strains may be quite thick (up to 0.3 μm in *N. amarae*). The thickening of the cell walls of *N. asteroides* strains has been found to be greater in stationary phase cells than in those from the exponential phase (Beaman, 1975). Mesosomes are common. L-Forms of *N. asteroides* and *N. otitidiscaviarum* are known, and there are grounds for believing that they may have a significant role in infection processes (Beaman et al., 1978; Beaman, 1980, 1984).

Most nocardiae grow readily on a variety of media containing simple nitrogen sources such as ammonium, nitrate and amino acids as well as on more complex ones containing casein, meat, soy or yeast peptones and hydrolysates. Glucose, acetate and propionate are good carbon sources. All nocardiae will grow in a temperature range of 15–37°C; many will grow at higher or lower temperatures. Doubling time, as for most actinomycetes, is appreciably longer than for other bacteria; a generation time of 5.5 h has been reported for certain *N. asteroides* and *N. brasiliensis* strains (Beadles et al., 1980). Some strains will grow to stationary phase in 3–7 days; others will grow more slowly. Colonies of nocardiae isolated from clinical sources may take up to several weeks to appear.

Autotrophs are not known among the true nocardiae (Tárnok, 1976). However, *N. autotrophica*, a strain lacking mycolic acids, was reported to grow both hetero- and autotrophically, utilizing a fixed nitrogen source in an atmosphere of CO_2, H_2, and O_2 (Takamiya and Tubaki, 1956); it can also oxidize CO to CO_2 (Bartholomew and Alexander, 1979). A *N. asteroides* strain studied at the same time was negative.

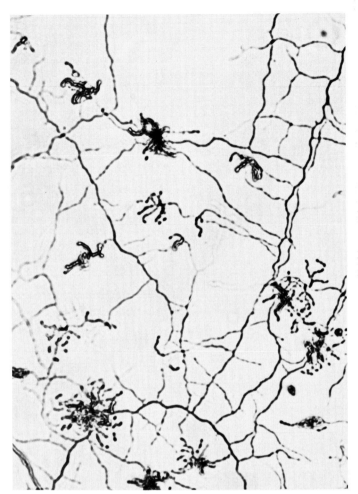

Figure 17.2. Aerial and vegetative conidial chains of *Nocardia asteroides* (× 1250).

Nocardia saturnea, another strain lacking mycolic acids, grows on silica gel plates without added carbon or nitrogen and *N. petroleophila* utilizes CO_2 while oxidizing alkanes (Hirsch, 1960). As this last strain does not have a wall chemotype IV, its taxonomic position is not clear. *Nocardia opaca* (probably a rhodococcus) had a similar capability (Aggag and Schlegel, 1973). Some nocardiae can grow on nitrogen-free media by scavenging organic substances such as ammonia, present in the air. *Nocardia hydrocarbonoxydans* (Nolof and Hirsch, 1962), which also lacks mycolic acids, can use such airborne compounds as pyridine as a source of nitrogen.

Little is known of the metabolic pathways and genetics of true nocardiae. Most studies reported on "nocardiae" were actually carried out on strains which are now classed as rhodococci (see Bradley, 1978; Brownell and Denniston, 1984) or on some of the nocardiae lacking mycolic acids (e.g. Schupp et al., 1975). Genetic recombination in *N. asteroides* has been reported (Kasweck and Little, 1982). Plasmids have been found in rhodococci (Brownell, 1978; Reh, 1981) and in *N. asteroides* (Kasweck et al., 1982). Phage (nocardiophage) have been reported for *N. asteroides* (Pulverer et al., 1974; Prauser, 1976, 1981b; Andrzejewski et al., 1978), *N. brasiliensis* (Pulverer et al., 1974), *N. carnea* (Williams et al., 1980), *N. otitidiscaviarum* and *N. vaccinii* (Prauser, 1976). In general, rhodococci and *true* nocardiae are susceptible to nocardiophages, whereas *mycolateless* nocardiae are not (Williams et al., 1980; Prauser, 1981a). Phages active on *N. mediterranei* have been found that were not active on various nocardiae tested (Thiemann et al., 1964).

Immunodiffusion analysis, using disintegrated cells as antigens, per-
mitted several clusters of nocardiae to be recognized: three groups of *N. asteroides*, one of *N. otitidiscaviarum* and one of *N. amarae* (Ridell, 1981). Delayed hypersensitivity reactions were induced by purified wall polysaccharide and ribosomal proteins from *N. asteroides* and *N. brasiliensis* (Ortiz-Ortiz et al., 1976).

Nocardiae are sensitive to antibacterial antibiotics such as aminoglycosides, tetracyclines, and to sulfonamides. Cephalosporins, penicillins and peptide antibiotics are not as active (Bach et al., 1973; Lerner and Baum, 1973; Goodfellow and Orchard, 1974). Therapy with trimethoprim-sulfamethoxazole has been used with success (González-Ochoa, 1976); minocycline (Curry, 1980), fusidic acid, and aminoglycosides such as amikacins, gentamicins and netilmicin have also been used (Schaal and Beaman, 1984). *N*-Formimidoyl thienamycin and amikacin were the most active antibiotics on *N. asteroides* among 22 β-lactam and nine aminoglycosides tested (Gutmann et al., 1983). Treatment failures with the trimethoprim-sulfamethoxazole regimen led Dewsnup and Wright (1984) to conclude that of 25 antimicrobial agents tested, doxycycline, minocycline, sulfamethoxazole or imipenem could be useful alternatives.

Actinomycetes, if pathogenic, are usually opportunists, and nocardiae are no exception to this rule. The basic types of human disease caused are: (a) pulmonary, neural and/or systemic nocardiosis, usually caused by *N. asteroides* and sometimes by *N. brasiliensis* and *N. otitidiscaviarum*; (b) actinomycotic mycetomas which are tumor-like growths of the organisms within the tissues and usually caused by *N. brasiliensis* but sometimes by *N. asteroides* and *N. otitidiscaviarum*; and (c) localized cutaneous or subcutaneous infections which usually represent primary infections by either *N. asteroides* or *N. brasiliensis* (Gordon, M. A., 1974; Gonzáles-Ochoa, 1976; Mariat and Lechevalier, 1977; Pulverer and Schaal, 1978; Schaal and Beaman, 1984). *Nocardia asteroides* strains may be quite infectious but vary in their invasiveness (Beaman, 1976; Schaal and Beaman, 1984). Using inoculation of the footpads of white mice as a test system, Ochoa (1973) concluded that *N. brasiliensis* strains were obligatory pathogens, whereas *N. asteroides* strains were opportunistic pathogens. The occurrence of nocardial infections in the United States (1972–1974) has been reviewed (Beaman et al., 1976).

Lower animals such as birds, cats, dogs, fish, goats, horses, and domestic and wild rodents are also susceptible to nocardial infection (Goodfellow and Minnikin, 1981b), *N. asteroides* being most frequently implicated; however, *N. brasiliensis* may be pathogenic in cats (Kurup et al., 1970; Gonzáles-Ochoa, 1976). *Nocardia vaccinii*, the only well documented nocardial plant pathogen, causes galls in blueberry (Demaree and Smith, 1952). "*Nocardia rhodnii*" strains are thought to be symbionts of the insect *Rhodnius prolixus* (Cross et al., 1976); the actual identity of the organisms involved remains open.

Nocardiae are widely distributed in nature, being found in soil, water, air, sewage and, as indicated above, in clinical specimens, insects and plants (Cross et al., 1976; Goodfellow and Minnikin, 1981b; Goodfellow and Williams; 1983).

Enrichment and Isolation Procedures

Once isolated, nocardiae are not difficult to grow, but their frequency of isolation may be enhanced through the use of selective techniques. *Nocardia asteroides* strains may be isolated from complex mixed communities, such as those found in soil, by plating dilutions on Diagnostic Sensitivity Test Agar (Oxoid) supplemented with the antibiotics demethylchlortetracycline at 5 µg/ml or methacycline at 10 µg/ml and antifungal antibiotics (Orchard et al., 1977). Inoculated plates are incubated for up to 21 days at 25°C. Colonies with a pink to red substrate mycelium, covered to a greater or lesser extent with white aerial hyphae, are characteristic of nocardiae. *Nocardia amarae* strains are most easily isolated from sewage treatment plant foams using Czapek's agar supplemented with (w/v) 0.4% yeast extract (Lechevalier and Lechevalier, 1974). Isolation of nocardiae from clinical specimens may be accomplished by several procedures including the use of tellurite-containing medium (Schaal, 1972), Sabouraud's dextrose agar, beefheart infusion-blood agar, and Löwenstein-Jensen medium (Schaal, 1984a). Some *N. asteroides* strains survive mycobacterial selection procedures (Gordon, M.A. 1974; Goodfellow and Minnikin, 1981b).

Maintenance Procedures

Transfer of cultures every 2–4 months onto a suitable agar medium such as Bennett's or yeast extract-glucose agars (Waksman, 1967; Gordon et al., 1974), with storage at 4°C between transfers, is satisfactory for most nocardial strains. Long-term preservation may be accomplished by lyophilization in skim milk. Shorter term storage may be effected by use of sterile mineral oil overlay of slants kept at 4°C; alternatively, stationary phase cells grown in broth culture may be quick-frozen and stored at −25°C to −70°C. Suspensions of mycelial fragments and spores may also be stored in glycerol (20%, v/v) at −20°C (Wellington and Williams, 1978).

Procedures for Testing of Special Characters

Procedures for analyses of cell wall composition, whole organism amino acids and sugars, mycolic acids and phospholipids are given in detail in Lechevalier and Lechevalier (1980). Other techniques for mycolate analysis will be found in Minnikin et al. (1975, 1980), Hecht and Causey (1976) and Schaal (1984). Detailed analytical protocols for determining menaquinone composition are also available (Yamada et al., 1976, 1977; Collins et al., 1977, 1985; Collins, 1984; Kroppenstedt, 1984).

As with all actinomycetes, the most satisfactory method of determining the micromorphology of a nocardial culture is the direct in situ observation, by means of high-dry objective (× 40–60) and a long-distance working condenser, of a Petri dish containing undisturbed colonies of the organism grown on a meager medium such as dilute Bennett's or tap-water agar. Wet mounts observed under phase, or alternatively, stained preparations, are useful for examination of tissue sections or body exudates for the characteristic *granules* often seen in nocardial infections (Gordon, M.A., 1974).

There are two principal systems for testing the physiology of nocardiae, the "Gordon" and "Goodfellow" methods. The former is summarized in Gordon et al. (1974) and Mishra et al. (1980) and the latter in Goodfellow (1971), Lacey and Goodfellow (1975), Goodfellow and Alderson (1977), Goodfellow and Pirouz (1982) and Goodfellow et al. (1982). In general, the physiological tests employed by Gordon enable one to recognize a given nocardial species on the basis of about 40 tests. Employing over 150 characters using phenetic techniques, Goodfellow and colleagues find that some of the taxa recognized by Gordon are heterogeneous. This is particularly true for *N. asteroides*, where up to five clusters have been distinguished (Orchard and Goodfellow, 1980). Data on the reactions of various nocardial species in both systems is given below.

Differentiation of the genus **Nocardia** from other genera

Nocardia is the type of the family *Nocardiaceae* (Castellani and Chalmers, 1919). The latter currently accommodates aerobic, Gram-positive, actinomycetes that have a wall chemotype IV and form rudimentary to extensive substrate mycelium which usually fragments into bacillary and coccoid elements (Goodfellow and Minnikin, 1981a, b). It is clear that the family is markedly heterogeneous (Goodfellow and Cross, 1984; Goodfellow and Minnikin, 1984) and can be divided into aggregate groups centered around the genera *Nocardia* and *Micropolyspora*. The first group is characterized by the presence of mycolic acids and major amounts of straight-chain and unsaturated fatty acids, whereas *Micropolyspora* and related taxa lack mycolic acids but contain large amounts of branched-chain *iso and anteiso*-acids (Goodfellow and Cross, 1984; Goodfellow and Minikin, 1984).

In general, nocardiae and related organisms are either amycelial (*Caseobacter, Corynebacterium, Mycobacterium*) or reproduce by mycelial fragmentation (*Nocardia, Rhodococcus*), whereas the genera in the *Micropolyspora* aggregate group show a greater differentiation of sporing structures (Locci, 1976; Locci and Sharples, 1984). However, in practical terms, morphology is of little use in distinguishing most members of the genus *Nocardia* from the taxa in the *Micropolyspora* group since among the nocardiae there is a spectrum of morphological complexity going from an undifferentiated (and often unfragmented) mycelium (e.g. *N. amarae*) to a highly developed system of short conidial chains formed on the aerial and vegetative hyphae (e.g. *N. brevicatena* and some *N. asteroides*) to long chains of well formed spores on the aerial hyphae (*N. carnea, N. vaccinii*), all morphological types found within the *Micropolyspora* aggregate. Among the latter, only saccharomonoporae are clearly differentiable from nocardiae on the basis of morphology, since the spore pattern of one to two spores on the aerial hycelium only is unique among taxa with a cell wall of chemotype IV.

The nocardioform actinomycetes i.e. those grouped around *Nocardia*, not only have many properties in common (Goodfellow et al., 1974; Goodfellow and Wayne, 1982; Goodfellow and Cross, 1984) but also form a recognizable suprageneric group (Mordarski et al., 1980; Stackebrandt and Woese, 1981; Stackebrandt et al., 1983). Nocardiae can, however, be distinguished from the other mycolic acid-containing taxa, and from strains previously assigned to the genus, using a combination of chemical and morphological properties (Tables 17.2 and 17.3). Susceptibility to the antibiotics bleomycin (2.5 μg/ml) and mitomycin (10 μg/ml) has been reported to be useful in distinguishing rhodococci from nocardiae (Tsukamura, 1981a; 1982b).

Much less attention has been paid to the second aggregate group which currently contains the genera *Actinopolyspora, Micropolyspora, Pseudonocardia, Saccharomonospora* and *Saccharopolyspora*. It is not yet clear whether all of these taxa merit generic rank or even if they are closely related. They do, however, have a number of properties in common (Lechevalier and Lechevalier, 1981; Goodfellow and Cross, 1984; Goodfellow and Minnikin, 1984) and can readily be distinguished from *Nocardia* (Table 17.4). The standing of the genus *Micropolyspora* has become a matter of dispute as the reclassification of *M. brevicatena* in the genus *Nocardia* as *N. brevicatena* (Goodfellow and Pirouz, 1982) left the genus without a type species and hence nomenclaturally invalid. McCarthy et al. (1983) requested that the name *Micropolyspora* be conserved with the type species as *M. faeni*, but more recently Kurup and Agre (1983) proposed the generic name *Faenia* for the remaining species of the genus. The resolution of these two conflicting proposals rests with the Judicial Commission.

Taxonomic Comments

Although the genus *Nocardia* is widely recognized, the name "*Proactinomyces*" is still used in some of the Russian literature to refer to this taxon. The present circumscription of *Nocardia* has been discussed under "Further Descriptive Information" (see above). The redefined genus contains only 9 of the 20 species listed under *Nocardia* in the Approved Lists (AL)*: *amarae, asteroides, brasiliensis, brevicatena, carnea, farcinica, otitidiscaviarum, transvalensis* and *vaccinii*. Using more than 50 criteria, both morphological and physiological, Gordon and her colleagues pioneered a broadly-based definition of the species *N. asteroides, N. brasiliensis, N. carnea, N. otitidiscaviarum, N. transvalensis* and *N. vaccinii* (Gordon and Mihm, 1957, 1959, 1962a, 1962b; Gordon et al., 1974, 1978; Mishra et al., 1980). Data summarized from these publications along with results on the two remaining nocardial species, *N. amarae* and *N. brevicatena* (Lechevalier, M.P., 1968, unpublished results: Lechevalier and Lechevalier, 1974; Goodfellow and Pi-

---

* *AL*, denotes the inclusion of this name on the Approved Lists of Bacterial Names (1980).

Table 17.2.

Differential characteristics of the genus **Nocardia** *and related wall chemotype IV taxa containing mycolic acids[a,b]*

Characteristics	Nocardia	Caseobacter	Corynebacterium	Mycobacterium	Rhodococcus
Morphological characters					
Substrate mycelium	+	−	−	D−	D+
Aerial mycelium, macroscopic	D	−	−	−	−
Aerial hyphae, microscopic	+	−	−	D−	D−
Conidia	D−	−	−	−	−
Entire colonies	−	+	+	D	D
Lipid characters					
Fatty acids					
Tuberculostearic acid[c]	+	+	−[d]	+[e]	+
Phospholipids					
Phosphatidylethanolamine[f]	+	ND	−	+	+
Predominant menaquinone[g]	MK-8(H$_4$) or MK-9(H$_2$)	MK-9(H$_2$)	MK-8(H$_2$) or MK-9(H$_2$)	MK-9(H$_2$)	MK-8(H$_2$) or MK-9(H$_2$)
Mycolic acids					
Overall size[h] (number of carbons)	46–60	30–36	22–38	60–90	34–64
Number of double bonds[h]	0–3	0–2	0–2	1–3	0–4
Fatty acid esters released on pyrolysis (number of carbons)[i]	12–18	14–18	8–18	22–26	12–18
Mol% G + C of DNA	64–72	60–67	51–59	62–70	59–69

[a] Data from Goodfellow and Minnikin (1981a, b, 1984) and Goodfellow and Cross (1984).

[b] Symbols: +, 90% or more of the strains are positive; −, 90% or more of the strains are negative; D, different reactions occur in different taxa (species of a genus or genera of a family); D−, uncommonly; D+, more commonly than not; and ND, not determined.

[c] Determined by gas liquid chromatography (Lechevalier et al., 1977; Collins et al., 1982b; Kroppenstedt, 1984).

[d] *Corynebacterium bovis* contains tuberculostearic acid (Lechevalier et al., 1977; Collins et al., 1982b).

[e] *Mycobacterium gordonae* lacks substantial amounts of tuberculostearic acid (Tisdall et al., 1979; Minnikin et al., 1985).

[f] Determined by thin layer chromatography and chemical analysis (Lechevalier et al., 1977, 1981; Minnikin et al., 1977).

[g] Menaquinones detected by chromatographic or physicochemical analysis (Yamada, et al., 1976, 1977; Collins et al., 1977; Collins, 1984). Abbreviations exemplified by MK-8(H$_2$), menaquinone having two of the eight isoprene units hydrogenated.

[h] Detected by mass spectrometry (Alshamaony et al., 1976; Collins et al., 1982a). In mycobacterial mycolic acids, double bonds may be converted to cyclopropane rings; methyl branches and oxygen functions may be present (Dobson et al., 1984; Minnikin et al., 1984).

[i] Esters of fatty acids detected by pyrolysis gas chromatography of mycolate esters (Lechevalier et al., 1971; Goodfellow et al., 1978; Collins et al., 1982a).

rouz, 1982), are presented later in Tables 17.5 and 17.7 ("Gordon tests"). Using numerical taxonomic methods, Goodfellow and colleagues have found that *N. amarae*, *N. brasiliensis*, *N. farcinica* and *N. otitidiscaviarum* are good taxospecies; however, these and other workers have found *N. asteroides*, as defined by Gordon, to be markedly heterogeneous (Goodfellow, 1971; Lechevalier, M.P., 1976; Tsukamura, 1977; Schaal and Reutersberg, 1978; Goodfellow and Minnikin, 1981b; Goodfellow and Cross, 1984). DNA homology determinations also have shown that many subgroupings are possible in *N. asteroides* (Bradley et al., 1978; Mordarski et al., 1978). Immunological studies also have pointed to heterogeneity in this taxon (Pier and Fichtner, 1971; Magnusson, 1976; Kurup and Scribner, 1981; Ridell, 1981). However, using more sensitive immunological techniques, Kurup et al. (1983) concluded that the seven immunotypes previously distinguished should be retained within the species *asteroides*, rather than be assigned to separate species as previously proposed by Pier and Fichtner. Additional representatives of the remaining species need to be studied for them to be defined with precision. Differentiating characters culled from Goodfellow's test results (Goodfellow tests) are presented later in Table 17.6.

The principal nomenclatural problem in the redescribed genus *Nocardia* is the status of the former type species, *N. farcinica*. The strain originally isolated by Nocard (1888) from a case of bovine farcy was, by the decision of the Judicial Commission in 1954, made the type species of the genus *Nocardia*. It is now known that Nocard's original isolate is represented by two supposedly identical, but actually very different strains, ATCC 3318 and NCTC 4524. The former contains mycolic acids characteristic of nocardiae, and the latter mycolic acids and mycosides similar to those of mycobacteria (Lanéelle et al., 1971; Lechevalier et al., 1971). A numerical phenetic study by Orchard and Goodfellow (1980) found that each of these strains fell into a different cluster along with both nocardiae and mycobacteria. Schaal and Reutersberg (1978) found that ATCC 3318 clustered with *N. asteroides* strains, but they did not examine NCTC 4524. Using both serology and physiology, Ridell (1975) noted that the latter strain was grouped with mycobacteria and ATCC 3318 with *N. asteroides*, confirming her previous results using immunodiffusion which showed that some *N. farcinica* strains were closer to the mycobacteria than to nocardiae (Ridell and Norlin, 1973). Base composition studies also showed that strains

Table 17.3.

Differential characteristics of the genus **Nocardia** *and taxa previously associated with the genus*[a,b]

Characteristics	Nocardia	Actinomadura	Nocardiopsis	Oerskovia	Rothia
Morphological characters					
Substrate mycelium					
Fragmentation	+	−	D	+	+
Motile spores	−	−	−	D	−
Aerial mycelium	D^+	D^+	+	−	−
Conidia	D	+	+	−	−
Colony characters					
Entire colonies	−	−	−	D	D
Red pink colonies	D	D	−	−	−
Metabolism of glucose	O	O	O	O/F	O/F
Wall chemotype[c]	IV	III	III	VI	VI
Whole cell sugar pattern[c]	A	B	C	Gal	NC
Lipid characters					
Fatty acids[d]					
Unsaturated	+	+	$+^e$	−	−
Tuberculostearic	+	+	+	−	−
Iso and *anteiso*	−	$+^e$	+	+	+
Phospholipid type[f]	II	I/IV	III	V	I
Predominant menaquinone[d]	MK-8(H$_4$) or MK-9(H$_2$)	MK-9(H$_4$,H$_6$,H$_8$)	MK-10(H$_2$,H$_4$,H$_6$)	MK-9(H$_4$)	MK-7
Mycolic acids[d]	+	−	−	−	−
Mol% G + C of DNA	64–72	65–77	65.7[g]	70–75	65–70

[a] Data taken from Lechevalier et al. (1977, 1981), Collins & Shah (1984), Embley et al. (1984) and Goodfellow and Cross (1984).

[b] Symbols: see Table 17.2; also NC, no characteristic sugars.

[c] See Lechevalier and Lechevalier (1980) for details of methods and explanations of wall chemotypes and sugar patterns.

[d] See footnote to Table 17.2 for references to methods.

[e] Minor components.

[f] Phospholipid types after Lechevalier et al. (1977, 1981).

[g] Data from analysis of one strain. (An, Sun-Chung, unpublished).

Table 17.4.

Differential characteristics of the genus **Nocardia** *and wall chemotype IV taxa lacking mycolic acids*[a,b]

Characteristics	Nocardia	Actinopolyspora	Micropolyspora (Faenia)	Pseudonocardia	Saccharomonospora	Saccharopolyspora
Morphological characters						
Substrate mycelium						
Fragmentation	+	D	D	D^-	−	+
Spores	D	−	+	−	−	−
Aerial mycelium						
Spores	D	+	+	+	+	+
Spores in long chains (>20 spores)	D	+	D	+	−	+
Growth characters						
Extreme halophile	−	+	−	−	−	−
Lipid characters[c]						
Fatty acids						
Unsaturated	+	−	−	D	+	+
Tuberculostearic	+	−	−	D	−	−
Iso and *anteiso*	−	+	+	+	+	+
Phospholipid type[c]	II	III	III	III	II	III
Predominant menaquinone	MK-8(H$_4$) or MK-9(H$_2$)	MK-9(H$_4$)	MK-9(H$_4$, H$_6$)	MK-8(H$_4$) or MK-9(H$_4$)	MK-9(H$_4$)	MK-9(H$_4$)
Mycolic acids	+	−	−	−	−	−
Mol% G + C of DNA	64–72	64	ND	79	74–75	77

[a] Data taken from Lechevalier et al. (1977; 1981); Lechevalier and Lechevalier (1981); Goodfellow and Cross (1984) and Goodfellow and Minnikin (1984).

[b] Symbols, see Table 17.2.

[c] See footnotes to Tables 17.2 and 17.3 for references to methods.

ATCC 3318 and NCTC 4524 were different, the mols% G + C being 68.0 and 71.6, respectively (Mordarski et al., 1978), as did phage sensitivity (Prauser, 1981a). In view of the uncertain status of *N. farcinica* (Lechevalier, H.A., 1976) the present type species has been changed to *N. asteroides*, with ATCC 19247 as type (Skerman et al., 1980); Sneath, 1982; Opinion 58, Judicial Commission, 1985) and *N. farcinica* has been retained, with ATCC 3318 as type. An appeal to the Judicial Commission to reject *N. farcinica* as a *nomen dubium* was published (Tsukamura, 1982a); however, the commission has voted to retain the *farcinica* epithet.

Five of the remaining nocardial species listed in the Approved Lists, *N. calcarea*, *N. coeliaca*, *N. corynebacteroides*, *N. globerula* and *N. restricta*, have been classified with the rhodococci (see article on *Rhodococcus*). One, *N. cellulans*, belongs with the group of oerskoviae referred to as "NMOs" (see article on *Oerskovia*) and one, *N. petroleophila*, belongs to "*incertae sedis*" as this strain lacks a wall chemotype IV and its taxonomic position is still unclear. The remaining five nocardial species in the Approved Lists, *N. autotrophica*, *N. hydrocarbonoxydans*, *N. mediterranei*, *N. orientalis* and *N. saturnea*, belong to the "mycolateless" nocardiae. Although their mol% G + C is in the same range as the "true" nocardiae (Prauser, H., unpublished) this group will ultimately be assigned to established or new wall chemotype IV genera. Many chemical data, including lack of mycolic acids, presence of major amounts of *iso-* and *anteiso-*fatty acids, and differences in phage sensitivity (Williams et al., 1980; Goodfellow and Cross, 1984; Goodfellow and Minnikin, 1984) indicate that they do not belong within the genus *Nocardia*. Other strains lacking mycolic acids that are not listed in the Approved Lists but which will be considered here because of their importance, include *N. lurida* ATCC 14930 (see *N. orientalis*), *N. rugosa* IMRU 3760 and *N. sulphurea* ATCC 27624. Differentiating characteristics of these mycolateless "nocardiae" are given later in Tables 17.8 and 17.9. Table 17.10 contains their reactions in the carbohydrate utilization tests from the International Streptomyces Project. The Gordon test reactions of the principal nocardioform taxa having a phospholipid pattern of chemotype PIII (phosphatidylcholine as diagnostic phospholipid) are given later in Table 17.11.

Two new nocardial species, *N. nova* and *N. paratuberculosis* have been proposed by Tsukamura (1982c). *Nocardia nova* is the new name for a subgroup of *N. asteroides* defined by numerical taxonomic studies. *Nocardia paratuberculosis* is a substitute name for the controversial *N. farcinica*. Until their status has been clarified, they have been included here as species *incertae sedis*. Another group, also of uncertain status, is the "*aurantiaca* taxon." Originally described by Tsukamura and Mizuno (1971) as a species of *Gordona*, a genus which is now invalid, *aurantiaca* strains have been shown to have chemical and physiological properties which clearly place them apart from rhodococci, nocardiae and mycobacteria (see article on *Rhodococcus*). They have a cell wall of chemotype IV, mycolates which are intermediate between those of *Nocardia* and *Mycobacterium* (giving rise, on pyrolysis, to C_{20} and C_{22} saturated and unsaturated fatty acid esters), as well as unusual fully saturated menaquinones having nine isoprene units (MK-9). Numerical taxonomic studies showed the strains of this group to be distinctive and possibly worthy of placement in a new genus (Goodfellow et al., 1978). Certain strains received under the name of *Mycobacterium album* Söhngen have been shown to be related to the *aurantiaca* group in immunodiffusion and lipid analyses (Horan, 1971; Ridell et al., 1985).

Further Reading

Goodfellow, M. and T. Cross. 1984. Classification. *In* Goodfellow, Mordarski and Williams (Editors), The Biology of the Actinomycetes, Academic Press, London pp. 7–164.

Goodfellow, M. and D.E. Minnikin (Editors). 1984. Chemical Methods in Bacterial Systematics, Society for Applied Bacteriology, Technical Series No. 20; Academic Press, London,

Lechevalier, M.P. 1976. The taxonomy of the genus *Nocardia*: some light at the end of the tunnel? *In* Goodfellow, Brownell and Serrano (Editors), The Biology of the Nocardiae, Academic Press, London, pp. 1–38.

Minnikin, D.E. and M. Goodfellow. 1980. Lipid composition in the classification and identification of acid-fast bacteria. *In* Goodfellow and Board (Editors), Microbiological Classification and Identification, Academic Press, London, pp. 189–256.

Minnikin, D.E. and A.G. O'Donnell. 1984. Actinomycete envelope lipid and peptidoglycan composition. *In* Goodfellow, Mordarski and Williams (Editors), The Biology of the Actinomycetes, Academic Press, London, pp. 337–388.

Williams, S.T., G.P. Sharples, J.A. Serrano, A.A. Serrano and J. Lacey. 1976. The micromorphology and fine structure of nocardioform organisms. *In* Goodfellow, Brownell and Serrano (Editors), The Biology of the Nocardiae, Academic Press, London, pp. 103–140.

Differentiation and characteristics of the species of the genus **Nocardia**

The differential characteristics of the species of the genus *Nocardia* are indicated in Tables 17.5 and 17.6. Other characteristics of the species are listed in Table 17.7.

Table 17.5.

Characteristics differentiating the species of the genus **Nocardia** *(Gordon tests)[a]*

Characteristics	1. N. asteroides[b] 2. N. farcinica	3. N. brasiliensis	4. N. otitidiscaviarum	5. N. amarae	6. N. brevicatena	7. N. carnea	8. N. vaccinii	9. N. transvalensis
Decomposition of								
Adenine	−	−	−	−	−	−	−	d
Casein	−	+	−	−	−	−	−	−
Hypoxanthine	−	+	+	−	−	−	−	+
Tyrosine	−	+	d	−	−	−	−	+
Urea	+	+	+	+	−	−	+	+
Xanthine	−	−	+	−	−	−	−	d
Acid from								
Adonitol	−	−	−	−	−	−	−	+
Arabinose	−	−	d	−	−	−	+	−
Erythritol	−	−	−	−	−	−	−	+
Galactose	d	+	−	−	−	+[c]	+[c]	+[c]
Glucose	+	+	+	+	−	+	+	+
Inositol	−	+	+	+	−	d	d	d
Maltose	−	−	−	+	−	−	d	−
Mannose	d	d	d	+	−	d	−	d
Rhamnose	d	−	−	+	−	−	d	−
Sorbitol	−	−	−	−	−	+	d	d
Decarboxylation of								
Citrate	d	+	d	−	−	d	+	+
Mucate	−	−	−	−	−	−	+	−
Production of								
Nitrate reductase	+	+	+	+	−	+	+	+
Growth at/in								
10°C	d	d	d	−	−	d	−	−
45°C	d	−	d	−	−	d	−	−
Lysozyme broth	+	+	+	−	+	+	+	+
Survival at 50°C/8 h	+	−	+	−	+	+	+	+
Pyrolytic fragments of mycolic acids	S	S	S	U	S	U[c]	U[c]	S

[a] Symbols: see Table 17.2; also d, 11–89% of the strains are positive; S, Saturated; and U, Unsaturated methyl esters of fatty acids principally released.

[b] *N. asteroides* and *N. farcinica* cannot be distinguished on the basis of Gordon tests.

[c] Based on analysis of the type strain.

Table 17.6.

Characteristics differentiating the species of the genus **Norcardia** *(Goodfellow tests)*[a]

Characteristics	1. N. aster-oides	2. N. farci-nica	3. N. brasi-liensis	4. N. otitidis-caviarium	5. N. amarae	6. N. brevica-tena	7. N. carnea	8. N. vaccinii	9. N. transva-lensis
Acid-fastness	d	d	d	d	−	d	d	d	d
Decomposition of									
Casein	−	−	+	−	−	−	−	−	−
Elastin	−	−	+	−	−	−	−	−	−
Hypoxanthine	−	−	+	+	−	−	−	−	+
Testosterone	+	+	+	−	−	+	+	−	−
Tyrosine	−	−	+	−	−	−	−	−	−
Xanthine	−	−	−	+	−	−	−	−	d
Production of									
Nitrate reductase	+	+	+	+	+	−	+	+	+
Urease	+	+	+	+	+	−	−	+	+
Resistance to lysozyme	+	+	+	+	−	+	+	+	+
Growth on sole carbon source (1%, w/v)									
Adonitol	−	−	−	−	−	+	−	−	+
L-Arabinose	−	−	−	−	−	−	+	d	−
Galactose	d	d	+	−	d	+	d	+	−
Inositol	−	−	+	+	+	+	−	+	−
Mannitol	−	−	+	+	+	+	−	+	−
Mannose	+	+	+	d	+	+	−	+	+
Melezitose	−	−	−	−	−	+	−	−	−
Rhamnose	−	+	−	−	+	+	−	+	−
Adipic acid (0.1%, w/v)	−	−	−	−	+	−	d	−	+
2,3-Butylene glycol (1%, v/v)	−	+	−						
Pimelic acid (0.1%, w/v)	−	−	−	−	+	−	−	+	−
1,2-Propylene glycol (1%, v/v)	−	+	−	−					
Sebacic acid (0.1%, w/v)	+	d	−	+	+	−	d	−	−
Predominant menaquinone	MK-8(H$_4$)	MK-8(H$_4$)	MK-8(H$_4$)	MK-8(H$_4$)	MK-9(H$_2$)	MK-8(H$_4$)	MK-8(H$_4$)	MK-8(H$_4$)	MK-8(H$_4$)

[a] Symbols: see Table 17.2; also d, 11–89% of the strains are positive.

Table 17.7.

Other characteristics of the species of the genus **Nocardia** *(Gordon tests)*[a]

Characteristics	1. N. asteroides 2. farcinica	3. N. brasi-liensis	4. N. otitidis-caviarum	5. N. amarae	6. N. brevica-tena [b]	7. N. carnea	8. N. vaccinii	9. N. transva-lensis
Acid from								
Cellobiose	−	−	−	−	−	−	−	−
Glycerol	+	+	+	+	+	+	+	+
Lactose	−	−	−	−	−	−	−	−
Mannitol	−	+	d	+	−	+	+	d
Melezitose	−	−	−	−	−	−	−	−
Melibiose	−	−	−	−	−	−	−	−
α-Methyl-D-glucoside	−	−	−	−	−	−	−	−
Raffinose	−	−	−	−	−	−	−	−
Trehalose	d	+	d	+	+	+	d	d
Xylose	−	−	−	−	−	−	d	−
Utilization of								
Benzoate	−	−	−	−	−	−	−	−
Succinate	+	+	+	+	d	+	+	+
Tartrate	−	−	−	−	−	−	−	−
Hydrolysis of								
Esculin	+	+	+	+	+	+	+	+
Potato starch	d	d	d	d	−	d	+	+
Acid fastness	d	d	d	−	weak	d	+	d

[a] Symbols: see Table 17.2; also d, 11–89% of the strains are positive.
[b] Grown at 37°C (all others at 28°C).

List of species of the genus **Nocardia**

1. **Nocardia asteroides** (Eppinger) Blanchard 1896, 856.[AL] (*Cladothrix asteroides* Eppinger 1891, 309.)

as.ter.o.i′des. Gr. adj. *asteroides* starlike.

Morphology very diverse in terms of degree of fragmentation, amount of aerial hyphae, colony color and form. The most typical strains have a salmon-colored to pinkish, or orange-tan matte growth with fringes of white to pinkish, sparse, aerial mycelium and no soluble pigment. More rarely, vegetative growth may be off-white, gray or brown and a yellowish brown soluble pigment may be formed. Entire glistening colonies are not found. Although the aerial hyphae are usually sterile or tending to break up into fragments of irregular length, some strains form typical chains of poorly-formed spores on the aerial and vegetative mycelium (Fig. 17.1). Usually weakly acid-fast.

Chemical characteristics as for true nocardiae. The predominant menaquinone is MK-8(H_4).

Physiological characteristics are given in Tables 17.5–17.7. As previously discussed (see Taxonomic Comments), *N. asteroides* cannot be distinguished chemically or physiologically by means of the Gordon tests from *N. farcinica* (as represented by its type strain ATCC 3318).

The mol% G + C of the DNA is 63–70 (T_m).

Some strains are pathogenic for man and animals but most are soil saprophytes.

An extensive review of the ecology of this taxon has been published (Orchard, 1981).

The *N. asteroides* taxon is considered to be heterogeneous by many workers (see text).

Type strain: ATCC 19247.

2. **Nocardia farcinica** Trevisan 1889, 9.[AL] *Nom. cons.* Opin. 13 Jud. Comm. 1954, 153; see Int. Code of Nom. 1958, 166 and note in *Index Bergeyana* p. 753.

far.ci′ni.ca. L.v. *farcio* stuff; L. n. *farciminum* a disease of horses; Fr. n. *farcin* farcy or glanders; M.L. fem. adj. *farcinica* relating to farcy.

See discussion under *N. asteroides*, above.

Physiological characteristics are given in Tables 17.5–17.7.

The mol% G + C of the DNA is 66–71 (T_m).

Some strains are pathogenic for man and animals.

Type strain: ATCC 3318.

3. **Nocardia brasiliensis** (Lindenberg) Pinoy 1913, 936.[AL] (*Discomyces brasiliensis* Lindenberg 1909, 279.)

bra.si.li.en′sis. M.L. adj. *brasiliensis* pertaining to Brazil, South America.

A typical strain will have pinkish or orange-tan to tan or brown vegetative growth with moderate to abundant nonfragmenting aerial hyphae usually off-white to pink-gray in color. Soluble dark pigments occur with very high frequency. Usually weakly acid-fast.

Chemical characteristics as for true nocardiae. The predominant menaquinone is MK-8(H_4).

Physiological characteristics given in Tables 17.5–17.7.

The mol% G + C of the DNA is 67–68 (T_m). Isolated from various kinds of nocardioses including mycetoma and occasionally from soil.

Type strain: ATCC 19296.

4. **Nocardia otitidiscaviarum** Snijders 1924, LXXXVII.[AL]

o.ti′ti.dis.cav.i.ar′um. M.L. n. *otitis* inflammation of the ear; M.L. n. *Cavia* (gen. plur. *caviarum*) generic name of the cavy or guinea-pig; M.L. gen. n. *otitidiscaviarum* of ear disease of guinea pigs.

Aerial hyphae usually very sparse and off-white in color, sometimes only visible microscopically, but almost always present. Cream, grayish to peach-tan to purplish vegetative hyphae. Soluble pigments variably present. Hyphae weakly acid-fast.

Chemical characteristics of true nocardiae. Predominant menaquinone MK-8(H_4).

Physiological characteristics given in Tables 17.5–17.7.

The mol% G + C of the DNA is 67–72 (T_m).

Has been isolated from soil but some strains are pathogenic for man and animals.

Type strain: ATCC 14629.

5. **Nocardia amarae** Lechevalier and Lechevalier 1974, 286.[AL]

am.ar′ae. Gr. n. amara sewage duct; M.L. gen. n. *amarae* of a sewage duct.

Colonies always tan; aerial hyphae visible only microscopically. No soluble pigments produced. Hyphae usually show banding both in the natural habitat (foam of secondary treatment sewage plants) and in vitro under phase contrast. Characteristic irregular thickenings (up to 0.3 μm) of the cell walls are visible in sections. Not acid-fast.

Chemical characteristics of true nocardiae. Like *N. vaccinii* and *N. carnea*, the mycolic acids of *N. amarae* strains have a monounsaturated α-branch. The predominant menaquinone is MK-9(H_2), and in this it differs from other true nocardiae.

Physiological characteristics are given in Tables 17.5–17.7.

The mol% G + C of the DNA is 71 (T_m) (type strain).

Isolated from foam formed on the surface of aeration tanks in activated-sludge sewage-treatment plants.

Type strain: ATCC 27808.

6. **Nocardia brevicatena** (Lechevalier, Solotorovsky and McDurmont) Goodfellow and Pirouz 1982b, 384.[VP]* (Effective publication: Goodfellow and Pirouz 1982a, 523.) (*Micropolyspora brevicatena*, Lechevalier, Solotorovsky and McDurmont 1961, 13.)

bre.vi.cat.e′na. L. adj. *brevis* short; L. n. *catena* chain; M.L. n. *brevicatena* short chain.

Colonies typically orange-tan with off-white, sparse to moderate aerial mycelium. No soluble pigments. Oval to round, **very well-formed spores** in short chains (2–7) borne on the vegetative and aerial mycelium and also at and **just above the surface of the agar.** Weakly acid-fast.

Chemical characteristics of true nocardiae. Predominant menaquinone is MK-8(H_4).

Physiological characteristics given in Tables 17.5–17.7.

The mol% G + C of the DNA is 66–68 (T_m).

Isolated from human sputum. Not pathogenic for mice.

Type strain: ATCC 15333.

7. **Nocardia carnea** (Rossi-Doria) Castellani and Chalmers 1913, 818.[AL] (*Streptothrix carnea* Rossi-Doria 1891, 415.)

car.ne′a. L. fem. adj. *carnea* fleshy.

Cream or peach-colored vegetative growth with sparse to abundant, white to pinkish aerial mycelium. Aerial hyphae are usually sterile although rudimentary "conidia" have been reported in a few strains. Rarely acid-fast. Soluble pigments are uncommon.

The physiological characteristics of *N. carnea* strains are given in Tables 17.5–17.7.

The chemical characteristics of the taxon are those of true nocardiae. Like *N. amarae* and *N. vaccinii*, the α-branches of the mycolic acids of *N. carnea* are monounsaturated. The predominant menaquinone type is MK-8(H_4).

The mol% G + C of the DNA is 64–68 (T_m).

Isolated from soil and air.

Type strain: ATCC 6847.

8. **Nocardia vaccinii** Demaree and Smith 1952, 251.[AL]

vac.cin′i.i. M.L. n. *Vaccinium* generic name of the blueberry. M.L. gen. n. *vaccinii* of *Vaccinium*.

Cream to peach-colored colonies with moderate to sparse white aerial mycelium. Conidia are rarely formed on the aerial mycelium. Partially acid-fast. Soluble pigments not usually formed.

Physiological characteristics are given in Tables 17.5–17.7.

---

* *VP*, denotes that this name has been validly published in the official publication International Journal of Systemic Bacteriology.

Chemical characteristics are those of true nocardiae. Like *N. amarae* and *N. carnea*, the α-branches of the mycolic acids of *N. vaccinii* are monounsaturated. The predominant menaquinone is MK-8(H₄).

Strains are pathogenic to blueberry plants, causing galls.

Type strain: ATCC 11092.

9. **Nocardia transvalensis** Pijper and Pullinger 1927, 155.[AL]

trans.val.en′sis. M.L. adj. *transvalensis* pertaining to the Transvaal, South Africa.

Colonies usually pale tannish cream or purplish. Moderate to heavy sterile aerial mycelium. No soluble pigments have been observed. Partially acid-fast when grown on glycerol agar.

Chemical characteristics as for true nocardiae. The predominant menaquinone is MK-8(H₄).

Physiological characteristics are given in Tables 17.5–17.7.

The mol% G + C of the DNA is 67.0 (T_m).

Isolated from mycetoma of the foot in South Africa.

Type strain: ATCC 6865.

Species Incertae Sedis

a. **Nocardia orientalis** (Pittenger and Brigham) Pridham 1970, 42.[AL] (*Streptomyces orientalis* Pittenger and Brigham 1956, 642.)

or.i.ent.al′is. L. adj. *orientalis* of the orient.

Cream, yellowish tan, peach or brown vegetative growth, aerial hyphae sparse to moderate, white to off-white to cream. Soluble pigments pale yellow-brown, light brown or greenish yellow. Long chains of smooth, squarish conidia formed by about half the cultures assigned to this taxon by Gordon et al. (1978).

Chemical characteristics of "mycolateless" nocardiae. The predominant menaquinone is MK-9(H₄).

Physiological reactions are given in Tables 17.8–17.10. Except for slightly greater resistance to lysozyme and a weak-to-negative inositol utilization, "*N. lurida*" ATCC 14930 (Grundy et al., 1957), the producer of the antibiotic, ristocetin, has the same properties as *N. orientalis*. The type strain of *N. orientalis* produces vancomycin.

The mol% G + C of the DNA of the type strain is 66.0 (T_m).

Isolated from soil, vegetable matter and clinical specimens. Pathogenicity is unknown.

Type strain: ATCC 19795 (ISP 5040).

b. **Nocardia mediterranei** (Margalith and Beretta) Thiemann, Zucco and Pelizza 1969, 148.[AL] (*Streptomyces mediterranei* Margalith and Beretta 1960, 321.)

med.i.ter.ra′ne.i. L. neut. n. *mediterraneum* interior of the land; *mediterranei*, of the interior of the land; from the Mediterranean area.

Yellowish- to pinkish tan or orange-brown vegetative mycelium; very sparse, sterile aerial mycelium visible only under the microscope on most media. When formed, the aerial mycelium en masse is white to pinkish. Soluble pigments yellowish pink, orange or orange-brown. Some strains may form chains of smooth or spiny (short spines) oblong spores.

Cell chemistry as for "mycolateless" nocardiae. Major menaquinone types are MK-9 (H₄; H₆).

Physiological reactions given in Tables 17.8–17.10.

The mol% G + C of the DNA is 67–69 (T_m).

The DNA can be isolated and restricted in the same way as that of streptomycetes (Kieser et al., 1981).

Most known strains of this taxon produce rifamycins. Isolated from soil.

Type strain: ATCC 13685 (ISP 5501).

c. **Nocardia rugosa** (ex DiMarco and Spalla 1957) nom.rev.

ru.go′sa. L. adj. *rugosa* full of wrinkles.

Off-white, yellowish to tannish, glistening, pasty vegetative growth. Aerial mycelium totally lacking even microscopically. Vegetative hyphae fragment within 24–48 h on rich media. Brownish soluble pigments sometimes formed. Not, to slightly, acid-fast.

Table 17.8.

Differential characteristics of **Nocardia** *species lacking mycolates (Gordon tests)[a]*

Characteristics	a. N. orientalis	b. N. mediter-ranei[b]	c. N. rugosa[b]	d. N. sulphurea[b]	e. N. saturnea[b]	f. N. auto-trophica	g. N. hydrocarbon-oxydans[b]
Hydrolysis of							
Adenine	−	−	−	−	−	+	−
Casein	+	+	+	+	−	−	−
Starch	+	−	−	−	−	+	+
Urea	+	+	+	−	+	+	−
Xanthine	d	−	+	−	+	d	−
Decarboxylation of							
Benzoate	d	−	+	−	+	d	−
Citrate	+	+	−	+	−	+	−
Acid from							
Adonitol	+	−	+	−	−	+	−
Arabinose	+	+	+	−	+	d	+
Cellobiose	+	+	−	−	+	d	+
Erythritol	+	−	+	−	−	+	+
Inositol	+	+	−	+	+	d	+
Lactose	+	+	−	−	−	−	+
Maltose	d	+	−	+	+	+	−
Melibiose	d	+	−	−	−	−	−
α-Methyl-D-glucoside	+	+	−	−	+	d	−
Rhamnose	d	+	+	−	−	−	+
Sorbitol	d	−	−	−	−	+	−
Lysozyme resistance	d	d	−	+	−	−	−
Growth at							
10°C	d	+	+	−	+	+	+
45°C	d	−	+	−	−	−	−
Phospholipid pattern	PII	PII	PII	PII	PIII	PIII	PIII

[a] Symbols: see table 17.2; also d, 11–89% of the strains are positive.

[b] Reactions given for the type strain.

Table 17.9.

Other characteristics of **Nocardia** *species lacking mycolates (Gordon tests)[a]*

Characteristics	a. N. orientalis	b. N. mediter- ranei[b]	c. N. rugosa[b]	d. N. sulphurea[b]	e. N. saturnea[b]	f. N. autotrophica	g. N. hydrocarbon- oxydans
Hydrolysis of							
Esculin	+	+	+	+	+	+	+
Hypoxanthine	+	+	+	+	+	+[c]	−
Tyrosine	+	+	+	+	+	+[c]	−
Production of							
Nitrate reductase	d	+	−	+	+	d	+
Decarboxylation of							
Mucate	d	−	−	−	−	−	−
Succinate	+	+	+	+	+	+	+
Tartrate	−	−	−	−	−	−	−
Acid from							
Glucose	+	+	+	+	+	+	+
Glycerol	+	+	+	+	+	+	+
Mannitol	+	+	+	+	+	+	−
Mannose	+	+	+	+	+	+	+
Raffinose	d	+	−	−	−	−	−
Trehalose	d	+	+	+	+	+	−
Xylose	+	+	+	−	+	+	+

[a] Symbols: see Table 17.2; also d, 11–89% of the strains are positive.
[b] Reactions given for the type strain.
[c] The type strain is negative for both these reactions (M.P. Lechevalier, unpublished).

Table 17.10.

Reactions of **Nocardia** *species lacking mycolates in the ISP carboydrate utilization tests[a,b]*

Characteristics	a. N. orientalis	b. N. mediter- ranei	c. N. rugosa	d. N. sulphurea	e. N. saturnea	f. N. autotrophica	g. N. hydrocarbon- oxydans
Utilization of							
Arabinose	+	+	+	−	−	+	−
Fructose	+	+	+	+	+	+	+
Glucose	+	+	+	+	+	+	+
Inositol	+	+	−	−	+	+	+
Mannitol	+	+	+	+	+	+	+
Raffinose	−	−	−	−	−	+	−
Rhamnose	d	+	+	−	−	+	−
Sucrose	+	+	−	+	+	+	+
Xylose	+	+	+	−	+	+	+

[a] Methods and data from the International Streptomyces Project (ISP) (Shirling and Gottlieb, 1968; 1972; Pridham and Lyons, 1969) and M.P. Lechevalier (unpublished); reactions given for the type species only.
[b] Symbols: see Table 17.2; also d, 11–89% of the strains are positive.

Chemical characteristics those of "mycolateless" nocardiae.
Physiological characteristics given in Tables 17.8–17.10.
The mol% G + C of the type strain is 68.9 (T_m).
The type strain produces vitamin B_{12}.
Isolated from cattle rumen.
Type strain: IMRU 3760.

d. Nocardia sulphurea sp. nov.*

sul.phu′re.a. L. adj. *sulphureus* of sulfur, referring to the yellow color of the vegetative hyphae.

Yellow to brown to blackish brown vegetative growth. White to yellowish white sterile aerial mycelium formed copiously on some media. Soluble dark pigment elaborated on most media. Very charac-teristic cracking and curling occurs in slant culture at base of tube revealing the vegetative hyphae beneath.

Cell chemistry characteristic of "mycolateless" nocardiae.
Physiological reactions given in Tables 17.8–17.10.
The mol% G + C of the DNA of the type strain is 66.8 (T_m).
Produces the antibiotic, chelocardin.
Isolated from soil.
Type strain: ATCC 27624.

e. Nocardia saturnea Hirsch 1960, 401.[AL]

sa.turn′e.a. L. n. *saturnus* Saturn, Roman god of seed-sowing M.L. adj. *saturnea* pertaining to Saturn, referring to the colonies which have a Saturnian shape.

---

* **Editorial note:** the authors of this chapter wish to revive *N. sulphurea* Oliver and Sinclair 1964, however that name was not validly published in a patent (Code, Rule 25(b)5) and, therefore, cannot be revived (Code, Rule 28a). Let it be noted that the authors have chosen the same type strain and are, in fact, reviving that name although the citation cannot mention the original authors.

Yellowish-white to bright butter-yellow vegetative mycelium. White to yellowish abundant aerial hyphae on meager media forming chains of long-rectangular fragments. No soluble pigments. Not acid-fast.

Cell chemistry typical of "mycolateless" nocardiae.

Physiological reactions given in Tables 17.8–17.10.

Isolated from air and compost.

Type strain: ATCC 15809.

f. **Nocardia autotrophica** (Takamiya and Tubaki) Hirsch 1960, 405.[AL] (*Streptomyces autotrophicus*, Takamiya and Tubaki, 1956, 59.) au.to.tro'phi.ca. G. n. *autos* self; G. part. *trophikos* nursing; M.L. fem. adj. *autotrophica* self-nourishing, referring to the ability to grow at the expense of H_2 and CO_2.

Vegetative hyphae yellow to brownish, tending to give rise to chains of chlamydospore-like cells of unequal size. Aerial mycelium white to cream, with long chains of long-oval to cylindrical spores. Pale yellow to yellow-brown soluble pigments sometimes formed late.

Cell chemistry typical of "mycolateless" nocardiae. The predominant menaquinone is MK-8(H_4).

Physiological reactions given in Tables 17.8–17.10.

Hydrogen utilized in presence of O_2 and CO_2 in a mineral medium.

The mol% G + C of the DNA of the type strain is 69.8 (T_m).

Isolated from phosphate buffer solution, aluminum hydroxide gel, vegetable matter, soil, clinical specimens. Pathogenicity is unknown.

Type strain: ATCC 19727.

g. **Nocardia hydrocarbonoxydans** Nolof and Hirsch 1962, 275.[AL] hy.dro.car.bon.ox'y.dans. G. n. *hydro* water; L. n. *carbo* ember, charcoal; Gr. adj. *oxys* sharp, acid; L. part. *dans* giving; M.L. pres. part. *hydrocarbonoxydans* oxidizing hydrocarbons.

Off-white, yellow-white, gold to brown vegetative growth, sparse to moderate white aerial mycelium. Aerial and vegetative mycelia fragment into long squarish units. No to little soluble pigment formed.

Physiological reactions given in Tables 17.8–17.10. Utilizes gaseous aliphatic hydrocarbons (C_6–C_{14}) for growth (Nolof, 1962).

Chemistry as for "mycolateless" nocardiae.

The mol% G + C of the type strain is 68.9 (T_m).

Air contaminant, isolated from a silica gel plate.

Type strain: ATCC 15104.

Species incertae sedis: not further described

Nocardia nova Tsukamura 1983, 896.[VP] (Effective publication: Tsukumura 1982c, 1115.)

"*Nocardia pseudotuberculosis*" Tsukamura 1982c, 1114.

Differentiation and characteristics of "**Nocardia**" species lacking mycolic acids

The differential characteristics of these species incertae sedis are shown in Table 17.8. Other characteristics of these species are shown in Tables 17.9 and 17.10. Table 17.11 presents a comparison of the differential characteristics of actinomycete taxa having no mycolates and a phospholipid chemotype PIII (phosphatidylcholine as diagnostic phospholipid).

Table 17.11.

Differentiating physiological characteristics of taxa lacking mycolates and containing phosphatidylcholine (phospholipid chemotype PIII) (Gordon tests)[a]

Characteristics	Nocardia autotrophica[b]	Nocardia hydrocarbonoxydans[c]	Nocardia saturnea[d]	Pseudonocardia thermophila[e]	Saccharopolyspora[f] hirsuta	Micropolyspora[g] (Faenia)
Hydrolysis of						
Adenine	+	–	–	–	+	–
Casein	–	–	–	–	+	+
Hypoxanthine	+	–	+	–	+	–
Tyrosine	+	–	+	+	+	–
Urea	+	–	+	+	+	–
Xanthine	d	–	+	–	+	+
Utilization of						
Benzoate	d	–	+	+	+	+
Citrate	+	–	–	–	+	+
Acid from						
Arabinose	d	+	+	+	–	–
Erythritol	+	+	–	–	+	+
Lactose	–	+	–	+	+	+
Rhamnose	–	+	–	+	+	–
Growth at						
45°C	–	–	–	+	+	+

[a] Symbols: see Table 17.2; also d, 11–89% of the strains are positive.

[b] Based on reactions of 31 strains.

[c] Based on reactions of the type strain IMRU 1407 (ATCC 15104).

[d] Based on reactions of type strain IMRU 1181 (ATCC 15809).

[e] Based on reactions of type strain Henssen A18 (ATCC 19285).

[f] Based on the reactions of 24 strains of *S. (Nocardia) hirsuta* (Gordon et al., 1978).

[g] Based on reactions of *Micropolyspora faeni* (*Faenia rectivirgula*) Lacey A94 (ATCC 15347; Type strain).

Genus **Rhodococcus** Zopf 1891, 28[AL]

MICHAEL GOODFELLOW

Rho.do.coc′cus. Gr. n. *rhodon* the rose; Gr. n. *coccos* a grain; M.L. neut. n. *Rhodococcus* a red coccus.

Rods to extensively branched substrate mycelium may be formed. In all strains the morphogenetic cycle is initiated with the coccus or short rod stage with different organisms showing a succession of more or less complex morphological stages by which the completion of the growth cycle is achieved. Thus, **cocci may merely germinate into short rods, form filaments with side projections, show elementary branching** or, in the most differentiated forms, produce **extensively branched hyphae.** The next generation of cocci or short rods are formed by the **fragmentation of the rods, filaments and hyphae.** Some strains produce feeble microscopically visible aerial hyphae, which may be branched, or aerial synnemata consisting of unbranched filaments that coalesce and project upwards. Rhodococci are nonmotile and form neither conidia nor endospores.

Gram-positive. Partially acid-alcohol fast at some stage of growth. **Aerobic. Chemoorganotrophic,** having an oxidative type of metabolism. **Catalase positive.** Most strains grow well on standard laboratory media at 30°C though some require thiamin. Colonies may be rough, smooth or mucoid and pigmented buff, cream, yellow, orange or red, though colorless variants do occur. Rhodococci are arylsulfatase negative, sensitive to lysozyme and are unable to degrade casein, cellulose, chitin, elastin or xylan. They are able to use a wide range of organic compounds as sole sources of carbon for energy and growth.

The cell wall peptidoglycan contains major amounts of **meso-diaminopimelic acid (meso-DAP), arabinose and galactose.** The organisms contain **diphosphatidylglycerol, phosphatidylethanolamine and phosphatidylinositol mannosides, dihydrogenated menaquinones with either eight or nine isoprene units** as the major isoprenologue, **large amounts of straight chain, unsaturated and tuberculostearic acids, and mycolic acids with 32–66 carbons and up to four double bonds. The fatty acid esters released on pyrolysis gas chromatography of mycolic esters contain 12–18 carbons. The mol% G + C of the DNA ranges from 63–72%** (T_m).

The organism is widely distributed but is particularly abundant in soil and herbivore dung. Some strains are pathogenic for man and animals.

Type species: *Rhodococcus rhodochrous* (Zopf) Tsukamura 1974a, 43.

Further Descriptive Information

Rhodococci have had a long and confused taxonomic pedigree (Cross and Goodfellow, 1973; Bradley and Bond, 1974; Bousfield and Goodfellow, 1976; Goodfellow and Minnikin, 1981a, b; Goodfellow and Wayne, 1982; Goodfellow and Cross, 1984). The epithet *rhodochrous* (Zopf, 1889) was reintroduced in 1957 by Gordon and Mihm for actinomycetes that had properties in common with both mycobacteria and nocardiae but which carried a multiplicity of generic and specific names. The taxon was provisionally assigned to the genus *Mycobacterium* but subsequent studies, based on chemical, genetic and numerical phenetic methods, showed that it merited generic rank and was heterogeneous. The genus *Rhodococcus* was subsequently resurrected and redefined for rhodochrous strains (Tsukamura, 1974a; Goodfellow and Alderson, 1977) and currently accommodates 14 species. Many of the latter were circumscribed and described in numerical phenetic surveys (Tacquet et al., 1971; Tsukamura, 1971, 1973, 1974a, 1975a, 1978; Goodfellow, 1971; Goodfellow and Alderson, 1977; Rowbotham and Cross, 1977a; Barton and Hughes, 1982; Goodfellow et al., 1974, 1982a, b; Helmke and Weyland, 1984) and some have been shown to be homogeneous on both chemical and genetic grounds (Bradley and Mordarski, 1976; Mordarski et al., 1976, 1977, 1980a, 1981; Minnikin and Goodfellow, 1980).

The genus *Rhodococcus* encompasses a wide range of morphological diversity and is defined primarily on the basis of wall envelope composition. Thus, at present, the genus should be restricted to actinomycetes that have: (a) a peptidoglycan consisting of *N*-acetylglucosamine, *N*-glycolylmuramic acid, D- and L-alanine, D-glutamic acid with meso-DAP as the diamino acid; (b) arabinose and galactose as diagnostic wall sugars {i.e. rhodococci have a wall chemotype IV and a whole-organism sugar pattern type A sensu Lechevalier and Lechevalier, 1970); (c) a phospholipid pattern consisting of diphosphatidylglycerol, phosphatidylethanolamine, phosphatidylinositol and phosphatidylinositol mannosides; (d) a fatty acid profile, containing major amounts of straight chain, unsaturated and tuberculostearic acids (i.e. a type IV fatty acid pattern sensu Lechevalier et al., 1977), mycolic acids with 32–66 carbons (Minnikin and Goodfellow, 1980, 1981), and (e) dihydrogenated menaquinones with either eight or nine isoprene units (Collins et al., 1977, 1985).

Rhodococci have no distinctive morphological features other than the ability of many strains to form hyphae that fragment into rods and cocci, but they do show considerable heterogeneity (Locci, 1976, 1981; Williams et al., 1976; Nesterenko et al., 1982; Helmke and Weyland, 1984; Locci and Sharples, 1984). Thus, *R. bronchialis, R. maris, R. rubropertinctus* and *R. terrae* are amycelial, and show a rod-coccus cycle similar to that described for *Arthrobacter* (Clark, 1979), but further morphological development is represented by *R. erythropolis, R. globerulus, R. rhodnii* and *R. rhodochrous* which exhibit elementary branching prior to fragmentation. *Rhodococcus equi* shows traces of elementary branching at early stages of growth and may represent a link between the two groups. *Rhodococcus coprophilus, R. fascians, R. marinonascens* and *R. ruber* form a third group which produces well branched substrate mycelia. The time taken to complete the developmental cycle ranges from 24 h in relatively undifferentiated forms such as *R. equi* to several days for those such as *R. coprophilus* which show the most pronounced morphological differentiation (Locci et al., 1982). However, the timing of the fragmentation process is influenced by environmental factors (Williams et al., 1976) which may act through their effects on growth rates. Rhodococci do not usually form aerial hyphae but exceptions are *R. coprophilus* and *R. ruber* which exhibit **feeble aerial hyphae and R. bronchialis strains which produce aerial synnemata (Locci and Sharples, 1984). Surfaces of colonies are covered** by a few strands to several sheets of slimy extracellular material (Williams et al., 1976). Sections of cells show a typical Gram-positive wall, lipid globules and polyphosphate granules are characteristic cell inclusions, mesosomes are common, and fibrillar materials are usually visible on the surface of negatively stained or freeze-etched cells (Williams et al., 1976; Beaman et al., 1978).

All members of the genus can be cultivated on standard nutrient media but they also grow well on very simple substrates, using ammonia, amino acids and nitrate as nitrogen sources and a host of sugars and organic acids as carbon sources. Glucose, fructose, mannose, raffinose, sucrose, acetate, butyrate and propionate are used as sole sources of carbon. Some strains require thiamin and others grow well on alkanes (Starr, 1949; Tárnok, 1976; Rowbotham and Cross, 1977a; Nesterenko et al., 1978). Oligocarbophily has been described for a rhodococcal strain labeled "*Nocardia corallina*" (Tárnok, 1976). Optimum temperatures vary between species but most strains grow between 15 and 40°C.

Genetic studies have been hampered by the slow growth of rhodococci, their tendency to clump and by their ability to form coenocytic structures. Current developments in the area have been reviewed by Brownell and Denniston (1984). Work has centered on *R. erythropolis,* an organism in which recombination was reported in 1963 by Adams and Bradley. To date, over 60 genetic traits have been used in the development of the *R. erythropolis* linkage map, and temperate phages are available to serve as cloning vectors for the development of a gene cloning system. Genetic recombination has also been demonstrated between rhodococcal strains currently labeled "*Nocardia opaca*" and "*Nocardia restricta.*" A transferable plasmid described for a strain of the former carries traits allowing for chemolithoautotrophic growth (aut⁺) which includes genes for hydrogenase, phosphoribulokinase and ribulosebisphosphate carboxylase production (Reh, 1981; Reh and Schlegel, 1981). The "N opaca" strain is able to transfer the aut⁺ trait to other organisms bearing this name as well as to *R. erythropolis*, and in the presence of CO_2 and H_2 it grows at a generation time of 7 h. Phages isolated by soil enrichment cause true lysis among *Rhodococcus* and *Nocardia* strains but not against representatives of allied taxa (Prauser and Falta, 1968; Prauser, 1976, 1981, 1984).

The catabolic potential of rhodococci not only includes the ability to assimilate proteins and carbohydrates but also unusual compounds such as alicyclic hydrocarbons, nitroaromatic compounds, polycyclic hydrocarbons, pyridine and steroids (Tárnok, 1976; Cain, 1981). Detergents and pesticides, including warfarin, are also modified (Goodfellow and Williams, 1983). Rhodococci have been implicated in the degradation of lignin-related compounds (Eggeling and Sahm, 1980, 1981; Rast et al., 1980) and humic acid (Cross et al., 1976), and have frequently been isolated from soil polluted with petroleum (Nesterenko et al., 1978). Members of the genus also produce enzymes that are exploited in the transformation of xenobiotics (Tárnok, 1976; Peczynska-Czoch and Mordarski, 1984). Indeed, many of the transformations traditionally associated with nocardiae are due to rhodococci.

Rhodococci have been examined using serological methods such as agglutination, complement fixation, immunodiffusion, immunoelectrophoresis and sensitin testing. It has been established that representatives of *Rhodococcus*, *Corynebacterium*, *Mycobacterium* and *Nocardia* have antigens in common (Cummins, 1962, 1965; Ridell, 1974, 1977; Ridell et al., 1979), and that ribosomes account for many of the cross-reactions (Ridell, 1981a). Analyses of immunodiffusion and immunoelectrophoretic patterns of whole culture filtrates and cell extracts of rhodococci have proved to be especially useful for defining species (Goodfellow et al., 1974; Ridell, 1974, 1981b, 1984; Lind and Ridell, 1976), as have quantitative comparisons of the skin hypersensitivity reactions of sensitized guinea pigs to crude culture filtrates of homologous and heterologous rhodococci (Hyman and Chaparas, 1977).

Most rhodococci are sensitive to antibacterial antibiotics such as aminoglycosides, cephalosporins, macrolides, penicillins and tetracyclines, less sensitive to sulfonamides but are resistant to most antitubercular compounds (Goodfellow and Orchard, 1974; Rowbotham and Cross, 1977a; Goodfellow et al., 1982b; Helmke and Weyland, 1984). They are also sensitive to lysozyme (Goodfellow, 1971; Mordarska et al., 1978).

Rhodococci are widely distributed in nature and have frequently been isolated from soil, fresh water and marine habitats, as well as from gut contents of blood-sucking arthropods with which they may form a mutualistic association (Cross et al., 1976; Goodfellow and Williams, 1983). *Rhodococcus coprophilus* grows on herbivore dung (Rowbotham and Cross, 1977b), *R. bronchialis* is associated with sputa of patients with cavitary pulmonary tuberculosis and bronchiectasis (Tsukamura, 1971), *R. fascians* causes leaf gall in many plants and fasciation in sweet peas (Tilford, 1936) and *R. marinonascens* has only been reported from marine sediments (Helmke and Weyland, 1984). *Rhodococcus equi* is an important equine pathogen which can infect other domestic animals, notably cattle and swine, and causes infection in human patients compromised either by immunosuppressive drug therapy and/or lymphoma (Barton and Hughes, 1980; Goodfellow et al., 1982b).

Enrichment and Isolation Procedures

Rhodococcus coprophilus has been isolated from both terrestrial and aquatic habitats by plating preheated samples (6 min at 55°C) onto M3 agar and incubating plates at 30°C for 7 days (Rowbotham and Cross, 1977b); *R. luteus* and *R. maris* from soil and the skin and intestinal contents of carp (*Cyprinus carpio*) on mineral salts agar enriched with n-alkanes and incubated at 28°C (Nesterenko et al., 1982); *R. marinonascens* from marine sediments using a number of rich media supplemented with sea water and incubated for 8–12 weeks at 18°C (Weyland, 1969, 1981), *R. erythropolis* and *R. rhodochrous* on mineral salts media supplemented with m-cresol or phenol (Gray and Thornton, 1928); and *R. bronchialis*, *R. rubropertinctus* and *R. terrae* from sputa and soil

using a decontamination method combined with a selective medium (Tsukamura, 1971). Similarly, *R. equi* has been recovered from soil, feces, lymph nodes and the intestinal contents of several animal species using a selective medium supplemented with nalidixic acid, novobiocin, cyclohexamide and potassium tellurite (Woolcock et al., 1979; Mutimer and Woolcock, 1980); a selective enrichment broth containing nalidixic acid, penicillin, cyclohexamide and potassium tellurite incubated at 30°C and used in conjunction with Tinsdale medium (Oxoid) and modified M3 medium has also been employed to good effect (Barton and Hughes, 1981). Rhodococci have also been isolated from soil using Czapek's agar (Higgins and Lechevalier, 1969), glycerol agar (Gordon and Smith, 1953) and Winogradsky's nitrite medium (Winogradsky, 1949), and from diseased sweet peas using potato dextrose agar (Tilford, 1936).

Maintenance Procedures

Long-term preservation of *Rhodococcus* strains may be achieved by lyophilization in skim milk. Suspensions of cocci and mycelial fragments can be stored in glycerol (20%, v/v) at −20°C (Wellington and Williams, 1978).

Procedures for Testing of Special Characters

Reliable identification to the generic level is best achieved using a combination of chemical and morphological techniques. Simple and accurate techniques have been introduced for the detection of lipids, wall amino acids and sugars (Goodfellow and Minnikin, 1984). Several methods can be applied to determine whether unknown organisms contain 2,6-DAP, arabinose and galactose in whole-organism hydrolysates, i.e. whether they have a wall chemotype IV characteristic of *Rhodococcus* and related taxa (Becker et al., 1964; Lechevalier, 1968; Berd, 1973; Staneck and Roberts, 1974; Richter, 1977; Lechevalier and Lechevalier, 1980). The latter also provide detailed protocols for determining wall, mycolic acid and phospholipid composition.

Fatty acid analysis is the first recommended step in the exploitation of the lipid composition of rhodococci. Indeed, since mycolic acids are restricted to some actinomycetes with a wall chemotype IV their detection (Minnikin et al., 1975, 1980; Hecht and Causey, 1976) dispenses with the need to determine whole-organism amino acid and sugar composition. The simplified procedure of Minnikin et al. (1975) facilitates the rapid analysis of mycolic acid composition. Thus, whole-organism methanolysates of rhodococci, corynebacteria and nocardiae give single spots on thin layer chromatography in contrast to the multispot pattern produced by mycolates of most mycobacteria (Minnikin et al., 1984a). Mycolic esters can be identified positively on thin-layer chromatograms as they are not removed when plates are subsequently washed with methanol-water (5:2, v/v). Mycobacterial mycolates can also be recognised as they are precipitated from ethereal solution by the addition of ethanol (Hecht and Causey, 1976); the mycolic acids from rhodococci, corynebacteria and nocardiae remain in solution but can be observed by thin layer chromatography of the supernatant.

Once mycolic acids have been detected, their esters should be isolated and characterized. Partial characterisation of mycolic acids can be achieved by pyrolysis gas chromatography of their methyl esters (Lechevalier et al., 1971; Goodfellow et al., 1978), a procedure which gives important information on the length of the fatty acid esters released. In turn, mass spectrometry of mycolic esters allows determinations of overall size, degree of unsaturation, and the size of the chain in the two position (Alshamaony et al., 1976; Collins et al., 1982a). Analysis can be taken a stage further by applying combined gas chromatography mass spectrometry of trimethylsilyl ethers of mycolic esters (Yano et al., 1978; Tomiyasu et al., 1981). This technique separates mycolic ester derivatives into their homologous components, each of which can be analyzed by mass spectrometry.

Several techniques based on gas liquid chromatography can be used to detect diagnostic fatty acids such as 10-methyloctadecanoic (tuberculostearic) acid (Lechevalier et al., 1977; Collins et al., 1982b; Bousfield et al., 1983; Kroppenstedt, 1984). In turn, menaquinone composition can be ascertained by chromatographic and physicochemical analysis

(Collins et al., 1977; Collins, 1985) and diagnostic polar lipids by applying published procedures (Lechevalier et al., 1977; Minnikin et al., 1977; O'Donnell et al., 1982). A simple small scale procedure for the sequential extraction of isoprenoid quinones and polar lipids (Minnikin et al., 1984b), and an integrated lipid and wall analysis technique may be found useful for the identification of unknown mycolic acid-containing actinomycetes (O'Donnell et al., 1984).

The micromorphology of strains should first be examined unstained on plate, slide or coverslip culture (Williams and Cross, 1971). Such preparations can also be examined by scanning electron microscopy (Williams and Davies, 1967). The morphology of undisturbed growth can be recorded conveniently on Bennett's agar (Jones, 1949) after 1, 3 and 7 days at 30°C using a long working distance objective (Rowbotham and Cross, 1977a) or in the case of R. marinonascens on yeast extract-malt extract-sea water agar at 18°C (Helmke and Weyland, 1984). Strains grown on polycarbonate membranes laid on the top of Bennett's agar can be removed at regular intervals, fixed and examined using a scanning electron microscope (Locci, 1981).

Identification to the species level is difficult but can be achieved using a combination of morphological and chemical criteria supplemented by biochemical, nutritional and physiological data. Thus, well established procedures are available to detect adenine, tyrosine and urea decomposition (Gordon et al., 1974), acid formation from sugars (Tsukamura, 1966; Goodfellow, 1971), and the ability of rhodococci to grow on sole carbon (Tsukamura, 1966; Goodfellow, 1971), nitrogen, carbon and nitrogen sources (Tsukamura, 1966), and in the presence of chemical inhibitors (Goodfellow, 1971).

There is evidence that the ability to form amidases (Bönicke, 1962; Tsukamura, 1975b), esterases (Käppler, 1965; Tsukamura, 1975b) and to grow in the presence of compounds such as ethambutol (Mizuno et al., 1966), sodium nitrite (Tsukamura and Tsukamura, 1968), sodium salicylate (Tsukamura, 1962), picric acid (Tsukamura, 1965), and rifampicin (Tsukamura, 1972) may also yield useful diagnostic data, but tests for properties such as these need to be extended to include representatives of all of the described species of Rhodococcus.

Differentiation from other closely related taxa

Rhodococci are closely related to members of the genera Caseobacter, Corynebacterium, Mycobacterium, Nocardia and the 'aurantiaca' taxon. Features useful for distinguishing Rhodococcus from these other mycolic acid-containing taxa are shown in Table 17.12. Most of these taxa have also been distinguished in comparative immunodiffusion studies (Lind and Ridell, 1976; Lind et al., 1980). It is possible to distinguish Rhodococcus from Mycobacterium and Nocardia (Ridell and Norlin, 1973), and from Corynebacterium (Ridell, 1977), using precipitinogens α and pα.

Susceptibility to bleomycin (2.5 µg/ml), 5 fluorouracil (20 µg/ml), mitomycin C (10 µg/ml) and β-galactosidase activity have been reported to be useful in distinguishing rhodococci and nocardiae (Tsukamura 1974b, 1981a, b, 1982b). There is also evidence that rhodococci are more resistant than mycobacteria to prothionamide (Ridell, 1983), and that they can be differentiated from the latter on the basis of the acid-fast stain, arylsulfatase activity, ability to use sucrose as a sole carbon source and inability to use trimethylenediamine as a simultaneous nitrogen and carbon source (Tsukamura, 1971).

Table 17.12.
Differential characteristics of the genus **Rhodococcus** *and other wall chemotype IV taxa containing mycolic acids[a,b]*

Characteristics	Rhodococcus	Caseobacter	Corynebacterium	Mycobacterium	Nocardia	'aurantiaca' taxon
Morphological characters						
Substrate mycelium	D	−	−	D	+	−
Aerial mycelium	−	−	−	D	D	−
Conidia	−	−	−	−	D	−
Lipid characters						
Fatty acids						
Tuberculostearic acid[c]	+	+	−[d]	+	+	+
Phospholipids						
Phosphatidylethanolamine[e]	+	ND	−	+	+	+
Predominant menaquinone[f]	MK-8(H$_2$) or MK-9(H$_2$)	MK-9(H$_2$)	MK-8(H$_2$) or MK-9(H$_2$)	MK-9(H$_2$)	MK-8(H$_4$) or MK-9(H$_2$)	MK-9
Mycolic acids						
Overall size (number of carbons)[g]	34–64	30–36	22–38	60–90	44–60	68–74
Number of double bonds[g]	0–4	0–2	0–2	1–2	0–3	1–5
Ester released on pyrolysis[h]	12–18	14–18	8–18	22–26	12–18	20–22
Mol% G + C of DNA	63–73	65–67	51–59	62–70	64–72	ND

[a] Data from Goodfellow and Minnikin (1981a, b, 1984), and Goodfellow and Cross (1984).

[b] Symbols: +, 90% or more of the strains are positive; −, 90% or more of the strains are negative; D, different reactions occur in different taxa (species of a genus or genera of a family); ND, not determined.

[c] Determined by gas liquid chromatography (Lechevalier et al., 1977; Collins et al., 1982b; Bousfield et al., 1983; Kroppenstedt, 1984).

[d] Corynebacterium bovis contains tuberculostearic acid (Lechevalier et al., 1977; Collins et al., 1982b).

[e] Determined by thin layer chromatography (Lechevalier et al., 1977; Minnikin et al., 1977; O'Donnell et al., 1982).

[f] Menaquinones detected by chromatographic or physiochemical analysis (Collins et al., 1977; Collins, 1985). Abbreviations exemplified by MK-8(H$_2$); menaquinone having one of the eight isoprene units hydrogenated.

[g] Detected by mass spectrometry (Alshamaony et al., 1976; Collins et al., 1982a).

[h] Esters detected by pyrolysis gas chromatography (Lechevalier et al., 1971; Goodfellow et al., 1978).

Taxonomic Comments

The application of modern taxonomic methods, notably chemotaxonomy and numerical phenetic taxonomy, led to the reintroduction of the genus *Rhodococcus* for the group of bacteria collectively known as "*Mycobacterium rhodochrous*," the '*rhodochrous*' group or the '*rhodochrous*' complex (Bousfield and Goodfellow, 1976). The genus continues to provide a niche for actinomycetes previously assigned to genera such as *Arthrobacter*, *Brevibacterium*, *Corynebacterium*, *Mycobacterium* and *Nocardia* (Collins et al., 1982a, b). Rhodococci and the other mycolic acid-containing actinomycetes constitute the nocardioform actinomycetes (Goodfellow and Cross, 1984), a group of organisms that not only have many properties in common but also form a recognizable phylogenetic entity (Mordarski et al., 1980b; Stackebrandt and Woese, 1981; Stackebrandt et al., 1983). In addition, rhodococci, mycobacteria, nocardiae and aurantiaca strains have been recovered in discrete aggregate clusters in numerical phenetic surveys by Goodfellow et al. (1982a, b) and to a lesser extent by Tsukamura et al. (1979).

Further comparative studies are needed to determine the status of the '*aurantiaca*' taxon which contains strains previously classified as "*Gordona aurantiaca*" (Tsukamura and Mizuno, 1971). Subsequent numerical taxonomic studies underlined the equivocal position of this taxon in the genus *Gordona* (Tsukamura, 1974a, 1975a) and the type strain fell outside the aggregate *Rhodococcus* cluster circumscribed by Goodfellow and Alderson (1977). In a more broadly based investigation (Goodfellow et al., 1978), *aurantiaca* strains formed a numerically well-defined taxon equivalent in rank to phena corresponding to the genera *Corynebacterium*, *Mycobacterium*, *Nocardia* and *Rhodococcus* and were shown to contain characteristic mycolic acids and unsaturated menaquinones with nine isoprene units. These findings are at variance with the view that the '*aurantiaca*' taxon be assigned to the genus *Rhodococcus* as *R. aurantiacus* (Tsukamura, 1982c). Four additional species, *R. sputi* (Tsukamura, 1978), *R. aichiense*, *R. chubuense* and *R. obuense* (Tsukamura, 1982c), also need to be studied for additional criteria before their status can be accepted. In the mean time these four taxa and *R. aurantiacus* should be considered as *species incertae sedis*.

DNA relatedness studies have shown that *R. bronchialis*, *R. coprophilus*, *R. equi*, *R. erythropolis*, *R. rhodochrous*, *R. ruber*, *R. rubropertinctus* and *R. terrae* are good species (Mordarski et al., 1976, 1977, 1980a). Rhodococcal species can, however, be divided into two well circumscribed taxa on the basis of chemical and serological data. Thus, all of the species originally assigned to the genus *Gordona* (Tsukamura, 1971) namely *R. bronchialis*, *R. rubropertinctus* and *R. terrae*, have mycolic acids with between 48 and 66 carbon atoms and major amounts of dihydrogenated menaquinones with nine isoprene units whereas the remaining species have shorter chain mycolic acids and dihydrogenated menaquinones with eight isoprene units as the major isoprenologue (see Table 17.13). The two aggregate groups can also be recognized on the basis of antibiotic sensitivity patterns (Goodfellow and Orchard, 1974), delayed skin reactions on sensitized guinea pigs and by polyacrylamide gel electrophoresis of cell extracts (Hyman and Chaparas, 1977). It is possible that further studies might underline the separation between the two aggregate taxa and thereby raise the question of whether they can be included in the same genus.

The genus *Rhodococcus* includes 5 of the 20 species listed under *Nocardia* in the Approved Lists of Bacterial Names (Skerman et al., 1980). Thus, *N. calcarea* (Metcalf and Brown, 1957) is a synonym of *N. erythropolis* (Goodfellow et al., 1982b); *N. corynebacteroides* (Serrano et al., 1972) and *N. globerula* (Gray, 1928) Waksman and Henrici 1948 are synonyms of *R. globerulus* (Goodfellow et al., 1982a); *N. restricta* (Turfitt, 1944) McClung 1974 is a synonym of *R. equi* (Goodfellow and Alderson, 1977); and *N. coeliaca* (Gray and Thornton, 1928) Waksman and Henrici 1948 has properties (Gordon et al., 1974) consistent with its inclusion in the genus *Rhodococcus*.

There is confusion over the status of *R. rubropertinctus* as defined by Goodfellow and Alderson (1977). In 1973 Tsukamura proposed the name *Gordona rubropertincta* with the type strain ATCC 14343. Since the new taxon included the hypothetical mean organism of *G. rubra* (Tsukamura, 1971) this species was considered to be a synonym of *G. rubropertincta* (Hefferan, 1904) Tsukamura 1973. The latter was renamed *R. rubropertinctus* by Tsukamura in 1974a. Goodfellow et al. (1974) recovered strain ATCC 14343 in a subcluster subsequently equated with *R. rhodochrous* (Goodfellow and Alderson, 1977), and on the basis of this reduced *R. rubropertinctus* (Hefferan) Tsukamura to a synonym of the latter species. They also classified strain NCTC 10668 as the type species of a new combination, *R. corallinus* (Goodfellow and Alderson, 1977). Additional strains conforming to Hefferan's original description were included in a further cluster which was named *R. rubropertinctus* (Hefferan) Goodfellow and Alderson with strain ATCC 14352 as the type. The type strains of *R. corallinus* and *R. rubropertinctus* (Hefferan) Goodfellow and Alderson were subsequently found to belong to a single DNA homology group (Mordarski et al., 1980a) and *R. corallinus* became a subjective synonym of *R. rubropertinctus* (Hefferan) Goodfellow and Alderson. This latter name was included in the Approved Lists of Bacterial Names (Skerman et al., 1980) with ATCC 14352 as the type strain. Tsukamura (1973, 1982b) has consistently recovered strain ATCC 14343 in a cluster equated with *R. rubropertinctus* (Hefferan 1904) Tsukamura 1978. He also believes that *R. lentifragmentus* (Kruse, 1896) Tsukamura 1978 has priority over *R. ruber* (Kruse 1896) Goodfellow and Alderson, 1980. Such tangled nomenclature matters reflect the difficulties involved in clarifying the taxonomy of very poorly classified taxa and can only be resolved by action of the Judicial Commission.

Further Reading

Bousfield, I.J. and M. Goodfellow, 1976. The "*rhodochrous*" complex and its relationships with allied taxa. *In* Goodfellow, Brownell and Serrano (Editors), The Biology of the Nocardiae. Academic Press, London, pp. 39–65.

Goodfellow, M. and T. Cross, 1984. Classification. *In* Goodfellow, Mordarski and Williams (Editors). The Biology of the Actinomycetes. Academic Press, London, pp. 7–164.

Goodfellow, M. and D.E. Minnikin (Editors) 1985. Chemical Methods in Bacterial Systematics, Society for Applied Bacteriology, Technical Series Number 20, Academic Press, London.

Differentiation and characteristics of the species of the genus **Rhodococcus**

The differential characteristics of the species of *Rhodococcus* are indicated in Table 17.13. Other properties of the species are listed in Table 17.14.

Table 17.13.

Characteristics differentiating the species of the genus **Rhodococcus**[a]

Characteristics	1. R. rhodochrous	2. R. bronchialis	3. R. coprophilus	4. R. equi	5. R. erythropolis	6. R. fascians	7. R. globerulus	8. R. luteus	9. R. marinonascens	10. R. maris	11. R. rhodnii	12. R. ruber	13. R. rubropertinctus	14. R. terrae
Morphogenetic sequence[b]	EB-R-C	R-C	H-R-C	R-C	EB-R-C	H-R-C	EB-R-C	EB-R-C	H-R-C	R-C	EB-R-C	H-R-C	R-C	R-C
Decomposition of														
Adenine	d	−	−	+	+	+	−	ND	−	ND	−	d	−	−
Tyrosine	+	−	−	−	d	+	−	−	d	−	+	+	−	−
Urea	d	+	d	+	+	+	+	+	−	d	+	+	+	+
Growth on sole carbon sources (% w/v)														
Inositol (1.0)	−	+	−	−	d	−	−	−	+	−	−	−	−	−
Maltose (1.0)	+	+	+	−	+	−	+	−	−	−	−	+	+	+
Mannitol (1.0)	+	d	−	−	+	+	d	+	−	−	+	+	+	+
Rhamnose (1.0)	−	−	−	−	−	−	−	−	−	−	−	−	−	+
Sorbitol (1.0)	+	+	−	−	+	+	d	+	d	−	+	+	+	+
p-Cresol (0.1)	+	+	d	−	d	+	−	ND	−	ND	d	+	+	+
m-Hydroxybenzoic acid (0.1)	+	−	+	d	−	+	−	ND	ND	ND	−	+	−	−
Pimelic acid (0.1)	d	+	+	−	d	−	−	ND	ND	ND	d	+	+	+
Sodium adipate (0.1)	+	+	d	−	+	−	+	−	ND	ND	+	+	+	+
Sodium benzoate (0.1)	+	+	d	−	−	−	d	−	−	−	+	+	+	+
Sodium citrate (0.1)	+	d	−	−	+	+	+	+	−	d	d	+	+	+
Sodium lactate (0.1)	+	+	d	+	+	+	−	+	−	−	d	d	−	+
Testosterone (0.1)	+	+	+	d	+	−	−	ND	−	ND	−	+	+	+
L-Tyrosine (0.1)	+	−	−	d	−	+	+	ND	ND	ND	+	+	−	+
Growth in the presence of (% w/v)														
Sodium azide (0.02)	−	+	d	d	d	−	d	ND	+	ND	d	+	+	+
Lipid characters														
Mycolic acids (number of carbons)	36–50	54–66	38–48	30–38	34–38	38–52	ND	ND	ND	ND	38–52	40–50	48–62	52–64
Predominant menaquinone	MK-8 (H₂)	MK-9 (H₂)	MK-8 (H₂)	MK-8 (H₂)	MK-8 (H₂)	MK-8 (H₂)	MK-8 (H₂)	MK-8 (H₂)	MK-8 (H₂)	MK-8 (H₂)	MK-8 (H₂)	MK-8 (H₂)	MK-9 (H₂)	MK-9 (H₂)
Mol% G + C of DNA	67–70	63–65	67–69	70–72	67–71	63–68	63–67	64	65–66	73	66	69–73	67–69	64–69

[a] Symbols: see Table 17.12; also d, 11–89% of strains are positive.

[b] EB-R-C, elementary branching-rod-coccus growth cycle; R-C, rod-coccus growth cycle; and H-R-C, hypha-rod-coccus growth cycle.

Table 17.14.
Other characteristics of the species of the genus **Rhodococcus**[a]

Characteristics	1. R. rhodochrous	2. R. bronchialis	3. R. coprophilus	4. R. equi	5. R. erythropolis	6. R. fascians	7. R. globerulus	8. R. luteus	9. R. marinonascens	10. R. maris	11. R. rhodnii	12. R. ruber	13. R. rubropertinctus	14. R. terrae
Growth on sole carbon sources (% w/v)														
Ethanol (1.0)	+	+	+	+	+	+	+	ND	−	ND	+	+	+	+
Glycerol (1.0)	+	+	−	−	+	+	+	+	d	+	ND	+	+	+
Sucrose (1.0)	+	+	d	+	+	+	+	+	−	d	+	+	+	+
Trehalose (1.0)	+	+	d	+	+	+	+	ND	−	ND	+	+	+	+
Acetamide (0.1)	d	+	−	d	+	+	d	ND	ND	ND	+	−	−	−
p-Hydroxybenzoic acid (0.1)	+	+	−	+	d	+	d	ND	−	ND	d	+	+	+
Sebacic acid (0.1)	+	+	d	−	+	+	+	ND	−	ND	d	+	+	+
Sodium fumarate (0.1)	+	+	d	+	+	+	+	+	ND	ND	d	+	+	+
Sodium gluconate (0.1)	d	d	−	−	+	+	−	ND	d	ND	+	d	+	d
Sodium malate (0.1)	+	+	−	+	+	+	+	+	d	+	+	+	+	+
Sodium pyruvate (0.1)	+	+	+	+	+	+	+	+	+	+	−	+	−	+
Sodium succinate (0.1)	+	+	−	+	+	+	+	+	−	+	+	+	+	+
Growth on sole carbon and nitrogen sources (% w/v)														
Acetamide	d	+	−	+	d	−	d	ND	ND	ND	+	d	−	−
Serine	−	−	−	−	−	−	−	ND	ND	ND	−	−	−	−
Trimethylenediamine	−	−	−	−	−	−	−	ND	ND	ND	−	−	−	−
Growth at														
10°C	+	−	d	+	+	+	+	ND	+	ND	−	ND	−	−
40°C	+	+	+	+	+	+	+	ND	−	ND	−	+	+	+
45°C	−	−	−	−	−	−	−		−		−	−	−	−
Growth in presence of (% w/v)														
Crystal violet (0.001)	+	+	+	+	d	−	+	ND	+	ND	+	+	+	+
Crystal violet (0.0001)	d	+	+	d	−	+	d	ND	d	ND	d	+	+	+
Phenol (0.1)	+	+	+	−	d	+	d	ND	+	ND	d	+	+	+
Phenyl ethanol (0.3% v/v)	+	+	d	d	d	−	+	ND	+	ND	−	+	+	+
Sodium azide (0.01)	d	+	+	d	d	−	d	ND	+	ND	+	+	+	+
Sodium chloride (5.0)	+	+	+	+	+	−	+	+	+	+	+	+	+	+
Sodium chloride (7.0)	d	+	+	d	d	−	+	+	d	+	+	+	+	+

[a] Symbols: see Table 17.12; also *d*, 11–89% of strains are positive.

List of species of the genus Rhodococcus

1. **Rhodococcus rhodochrous** (Zopf) Tsukamura 1974a, 43.[AL] (*Staphylococcus rhodochrous* Zopf 1889, 173; *Rhodococcus rubropertinctus* (Hefferan) Tsukamura 1974a, 43.*)

rho.do.ch'rous. Gr. n. *rhodon* the rose; Gr. n. *chrous*; L. adj. *rhodochrous* rose-colored.

See the generic description and Tables 17.13 and 17.14 for many of the features of *R. rhodochrous*.

Cocci germinate and give rise to branched filaments which undergo fragmentation into rods and cocci thereby completing the growth cycle. Rough, orange to red colonies are formed on glucose yeast extract agar, Sauton's agar and egg media.

Allantoinase, nicotinamidase and pyrazinamidase negative. β-Esterase weakly positive but negative for α-esterase and β-galactosidase.

Acid is produced from dextrin, ethanol, fructose, glucose, glycerol, maltose, mannitol, mannose, sorbitol, sucrose and trehalose but not from adonitol, amygdalin, D-arabinose, L-arabinose, cellobiose, dulcitol, galactose, glycogen, inositol, inulin, lactose, melezitose, raffinose, rhamnose or xylose.

Grows on inulin, *iso*-butanol, 2,3-butylene glycol, DL-norleucine, propyleneglycol, sodium octanoate and L-tyrosine as sole carbon sources but not on adonitol, amygdalin, D-arabinose, arbutin, cellobiose, dulcitol, galactose, glycogen, lactose, melezitose, raffinose, L-rhamnose, betaine HCl, D-mandelic acid, L-serine, sodium hippurate, sodium malonate or L-tryptophan.

Resistant to ethambutol (5 µg/ml), rifampicin (25 µg/ml), sodium nitrite (0.2%, w/v), sodium salicylate (0.1%, w/v) and picric acid (0.2%, w/v) but susceptible to 5-fluorouracil (5 µg/ml) and mitomycin C (5 µg/ml).

Isolated from soil.

The mol% G + C of the DNA is 67–70 (T_m).

Type strain: ATCC 13808.

2. **Rhodococcus bronchialis** (Tsukamura) Tsukamura 1974a, 43.[AL] (*Gordona bronchialis* Tsukamura 1971, 22.)

bron.chi'alis. L. adj. *bronchialis* coming from the bronchii.

See the generic description and Tables 17.13 and 17.14 for many of the features of *R. bronchialis*.

A rod-coccus life cycle is shown. Rough brownish colonies formed on

* See Taxonomic Comments.

glucose yeast extract agar, Sauton's agar and egg media. Synnemata of vertically arranged coalescing filaments are formed on the surface of colonies after 12–18 h incubation.

Acetamidase-, nicotinamidase-, pyrazinamidase- and urease-positive; allantoinase-, benzamidase-, isonicotinamidase-, malonamidase-, salicylamidase- and succinamidase-negative; α-esterase- and β-galactosidase-negative; β-esterase- and acid phosphatase-positive.

Acid is produced from glucose, inositol, maltose, mannose and trehalose but not from arabinose, galactose, raffinose, rhamnose, sorbitol or xylose.

Grows on iso-butanol, propanol and propylene glycol as sole carbon sources. Uses L-glutamate, isonicotinamide, L-methionine, nicotinamide, pyrazinamide, L-serine, succinamide and urea as sole nitrogen sources but not benzamide or nitrite. Acetamide and L-glutamate are used as sole sources of carbon and nitrogen but not benzamide or monoethanolamine.

Resistant to sodium aminosalicylate (0.2%, w/v) and picric acid (0.2%, w/v) but susceptible to 5-fluorouracil (5 μg/ml) and mitomycin C (5 μg/ml).

Isolated from sputum of patients with pulmonary disease.

The mol% G + C of the DNA is 63–65 (T_m).

Type strain: ATCC 25592.

3. **Rhodococcus coprophilus** Rowbotham and Cross 1979, 80.[AL] (Effective publication: Rowbotham and Cross 1977a, 136.)

co.pro.phi′lus. Gr. n. *copros* dung; Gr. adj. *philus* loving; M.L. adj. *coprophilus* dung loving.

See the generic description and Tables 17.13 and 17.14 for many of the features of *R. coprophilus*.

Forms a well developed primary mycelium which fragments into rods and cocci after several days incubation and aerial hyphae which may be branched. On Bennett's agar, after 2 weeks incubation at 30°C, rhizoid colonies (2 mm in diameter) are formed with a central orange papilla; growth into the agar also occurs. Pigmentation is enhanced by light. The young microcolonies on Bennett's agar are mycelial and after 24 h sparse, nonsporulating aerial hyphae are usually present. No macroscopically aerial mycelium, extracellular pigments or characteristic odors are produced. The central papilli of mature colonies are composed of complex aggregations of Gram-positive, nonacid-fast, nonmotile coccoid elements (1–1.5 μm in diameter). Cystites may occur in the mycelial fringe of the colonies on Bennett's agar. No pellicle is produced on the surface of Bennett's broth although isolated floating colonies may occur. Growth and pigmentation are reduced on media deficient in thiamin.

Urease-positive and acetamidase- and nicotinamidase-negative. Acid phosphatase-, α- and β-esterase-negative.

Grows on cetyl alcohol, D-melezitose, D-raffinose, sodium isobutyrate and sodium valerate as sole carbon sources but not on L-arabinose, galactose, D-glucosamine HCl, lactose, salicin, xylose, acetamide, benzamide, D- and L-alanine, L-asparagine, L-glycine, DL-norleucine, L-phenylalanine, L-proline, L-serine, L-tyrosine, butane-1,3-diol, butane-1,4-diol, sodium γ-aminobutyrate, sodium gluconate, or sodium phenylacetate.

Sensitive to 5-fluorouracil (20 μg/ml) and mitomycin C (5 μg/ml).

Grows on herbivore dung; it has been isolated from the dung of cows, donkeys, goats, horses, and sheep. It is common on grass and in the soil beneath grazed pastures, and is washed into streams and lakes where it can accumulate in the sediment.

The mol% G + C of the DNA is 67–69 (T_m).

Type strain: ATCC 29080.

4. **Rhodococcus equi** (Magnusson) Goodfellow and Alderson 1977.[AL] (*Corynebacterium equi* Magnusson 1923, 36; *Corynebacterium hoagii* (sic) (Morse) Eberson 1918, 11; *Nocardia restricta* (Turfitt) McClung 1974, 743.)

e′qui. L. n. *equus* horse; L. gen. nov. *equi* of the horse.

See generic description and Tables 17.13 and 17.14 for many of the features of *R. equi*. Editorial note: *Corynebacterium hoagii* is considered a synonym; for further information see genus *Corynebacterium*.

A rod-coccus life cycle is shown. Traces of elementary branching may be observed at early stages of growth. Smooth, shiny, orange to red colonies with entire margins are formed on glucose yeast extract agar. Some strains produce abundant slime which may drop onto the cover of inverted Petri dishes during incubation.

Urease-positive, acetamidase- and nicotinamidase-negative. Neither esculin nor arbutin are degraded. API-ZYM reactions: acid and alkaline phosphatase-, esterase lipase (C4)-, leucine arylamidase-, phosphoamidase-, valine arylamidase-positive, chymotrypsin-, β-glucuronidase- and α-mannosidase-negative.

Grows on amyl alcohol, butane-1,3-diol, butan-1-ol, propane-1,2-diol, propan-1-ol, sodium lactate and sodium octoate as sole carbon sources but not on erythritol, galactose, glycogen, inulin, lactose, raffinose, salicin, xylose, butane-1,4-diol, butane-2,3-diol, propan-2-ol, sodium gluconate, sodium hippurate, sodium malonate, sodium mucate, or tyrosine.

Resistant to ampicillin (20 μg/ml), erythromycin (4 μg/ml), gentamicin (8 μg/ml), lincomycin (64 μg/ml), minocycline (0.125 μg/ml), neomycin (8 μg/ml), novobiocin (4 μg/ml), penicillin (10 μg/ml), picric acid (0.2%, w/v), polymixin (256 μg/ml), rifampicin (0.25 μg/ml), streptomycin (5 μg/ml), sulfadiazine (100 μg/ml), tetracycline (50 μg/ml) and tobramycin (8 μg/ml), but susceptible to 5-fluorouracil (5 μg/ml) and mitomycin C (5 μg/ml).

Found in soil, herbivore dung and in the intestinal tract of cows, horses, sheep and pigs. Causes bronchopneumonia in foals, occasionally infects other domestic animals such as cattle and swine, and can be responsible for infection in human patients compromised by immunosuppressive drug therapy or lymphoma.

The mol% G + C of the DNA is 70–71 (T_m).

Type strain: ATCC 25729 (ATCC 6939).

5. **Rhodococcus erythropolis** (Gray and Thornton) Goodfellow and Alderson 1979, 80.[AL] (Effective publication: Goodfellow and Alderson 1977, 115) (*Mycobacterium erythropolis* Gray and Thornton 1928, 87; *Nocardia calcarea* Metcalf and Brown 1957, 568).

e.ry.thro′po.lis. Gr. adj. *erythrus* red; Gr. n. *polis* a city. M.L. n. *erythropolis* red city.

See generic description and Tables 17.13 and 17.14 for many of the features of *R. erythropolis*.

Cocci germinate to give filaments which show elementary branching. The growth cycle is completed by the appearance, through fragmentation, of cocci. Rough, orange to red colonies are formed on glucose yeast extract agar and Sauton's agar.

Acid is produced from glucose, glycerol, sorbitol, sucrose and trehalose but not from adonitol, arabinose, cellobiose, galactose, glycogen, inulin, melezitose, rhamnose or xylose.

Grows on D-glucosamine, D-salicin, D-alanine, L-asparagine, DL-norleucine, L-phenylalanine, L-proline, L-serine, acetamide, butane-1,3-diol, cetyl alcohol, propane-1,2-diol, sodium γ-aminobutyrate, sodium gluconate, sodium phenylacetate and stearic acid as sole carbon sources but not on L-arabinose, dextran, galactose, glycogen, lactose, melezitose, benzamide, butane-1,4-diol, glycine or sodium malonate.

Isolated from soil.

The mol% G + C of the DNA is 67–71 (T_m).

Type strain: ATCC 4277.

6. **Rhodococcus fascians** (Tilford) Goodfellow 1984b, 503.[VP] (Effective publication Goodfellow 1984a, 227; (*Phytomonas fascians* Tilford 1936, 394; *Corynebacterium fascians* (Tilford) Dowson 1942, 313.)

fas′ci.ans. L. part. adj. *fascians* binding together, bundling.

See generic description and Tables 17.13 and 17.14 for many of the features of *R. fascians*.

Forms branched hyphae which fragment into rods and cocci. Entire, convex, orange colonies formed on glucose yeast extract agar. Thiamin is required for growth.

Pyrazinamidase-positive, esculin and allantoin hydrolyzed.

Acid is produced from dextrin, ethanol, fructose, galactose, glucose, glycerol, mannitol, mannose, ribose, sorbitol, sucrose and trehalose but not from adonitol, amygdalin, arabinose, arbutin, cellobiose, dulcitol, glycogen, inositol, inulin, lactose, α-methyl-D-glucoside, β-methyl-D-glucoside, raffinose, rhamnose, salicin or xylose.

Grows on D- and L-alanine, sodium lactate, L-proline, L-serine and L-tyrosine as sole carbon sources, but not on benzamide, betaine, DL-norleucine, sodium hippurate, sodium malonate or L-threonine.

Causes leaf gall of many plants and fasciation of sweet peas (*Lathyrus odoratus*).

The mol% G + C of the DNA is 62.9–67.6 (T_m).

Type strain: ATCC 12974.

7. **Rhodococcus globerulus** Goodfellow, Weaver and Minnikin 1982, 741.[VP] (*Mycobacterium globerulum* Gray 1928, 265; *Nocardia globerula* (sic) (Gray) Waksman and Henrici 1948, 903; *Nocardia corynebacterioides* (sic) Serrano, Tablante, Serrano, San Blas and Imaeda 1972, 348.)

glo.be′ru.lus. M.L. dim. adj. *globerulus* globular.

See generic description and Tables 17.3 and 17.4 for many of the features of *R. globerulus*.

Cocci give rise to branched filaments which fragment into rods and cocci thereby completing the growth cycle. Entire, rough, pink to red colonies are formed on glucose yeast extract agar.

Urease positive, allantoinase and benzamidase negative.

Acid is produced from dextrin, ethanol, fructose, glucose, glycerol, maltose, mannitol, mannose, sorbitol, sucrose and trehalose but not from adonitol, amygdalin, arabinose, cellobiose, dulcitol, galactose, glycogen, inositol, lactose, melezitose, raffinose, rhamnose or xylose.

Growth on L- and D-alanine, inulin and L-serine as sole carbon sources but not on benzamide, betaine, inulin, DL-norleucine, L-proline, salicin, sodium hippurate, sodium malonate, sodium octanoate, L-threonine, L-tryptophan or L-tyrosine.

Sensitive to 5-fluorouracil (20 μg/ml) and mitomycin C (5 μg/ml). Isolated from soil.

The mol% G + C of the DNA is 63–67 (T_m).

Type strain: ATCC 25714.

8. **Rhodococcus luteus** (Söhngen) (Nesterenko, Nogina, Kasumova, Kvasnikov and Batrakov 1982, 6.[VP] (*Mycobacterium luteum* Söhngen 1913, 599.)

lu′te.us. L. adj. *luteus* golden yellow.

See generic description and Tables 17.13 and 17.14 for many of the features of *R. luteus*.

Straight or slightly curved rods give rise to shorter rods on glycerol (GA) and nutrient (NA) agars and to coccoid elements on wort agar (WA). Abundant, butyrous or mucoid, and intensely yellow growth is produced on GA and WA slants; poor to moderate, butyrous, and pale yellow to pale orange growth on NA slants; and abundant yellow growth on Lowenstein-Jensen medium. A pellicle is formed on the surface of nutrient broth.

Ammonia is formed from peptone, litmus milk is turned alkaline but p-nitrophenoloxidase is not produced. Acetyl methyl carbinol, indole, methyl red and phosphatase tests are negative.

Acid is produced from arabinose, fructose, galactose, glycerol, glucose, mannitol, mannose, sorbitol, sucrose and xylose but not from adonitol, cellobiose, dulcitol, inositol, lactose, maltose, α-methyl-D-glucoside, raffinose, rhamnose or salicin.

Growth occurs in the presence of C_9 to C_{17}, C_{19}, and C_{23} n-alkanes but not with C_8 n-alkane, ethane or methane.

Isolated from soil and from skin and intestinal tract of carp (*Cyprinus carpio*).

The mol% G + C of the DNA of the type strain is 64.1 (T_m).

Type strain: IMV 385.

9. **Rhodococcus marinonascens** Helmke and Weyland 1984, 137.[VP]

ma.ri.no.nas′cens. L. adj. *marinus*, of the sea; L. part. adj. *nascens*, born. M.L. part. adj. *marinonascens*, nascent of the the sea.

See generic description and Tables 17.13 and 17.14 for many of the properties of *R. marinonascens*.

Forms a well developed branched primary mycelium which fragments into bacillary and coccoid elements on solid media. Irregularly wrinkled,

cream-colored colonies, which are sometimes tinged with pink, are formed on yeast extract-malt extract agar after about 14 days incubation. Optimal growth is shown in media with a sea-water content of 75–100% or with an equivalent salt concentration. Very little, if any, growth occurs on media prepared with distilled water. Optimal growth occurs around 20°C, fair growth at 5°C but little, if any, is shown above 30°C.

Oxidase negative but esculin is hydrolysed.

Acid is produced from fructose, glucose, inositol, inulin and mannose but not from cellobiose, dulcitol, ethanol, galactose, glycogen, lactose, maltose, mannitol, melezitose, raffinose, rhamnose, salicin, sucrose, trehalose or xylose.

Grows on α-alanine and L-α-alanine as sole carbon sources but not on glycine, L-proline or salicylaldehyde.

Quite resistant to lysozyme but highly susceptible to penicillin, rifampicin and vancomycin.

Isolated from the uppermost layer of marine sediments from the North East Atlantic.

The mol% G + C of the DNA is 64.9–66.4 (T_m).

Type strain: 3438W (DSM 43752).

10. **Rhodococcus maris** (Harrison) Nesterenko, Nogina, Kasumova, Kvasnikov and Batrakov 1982, 11.[VP] (*Flavobacterium maris* Harrison 1929, 229.)

mar′is. L. gen. n. *maris* of the sea.

See generic description and Tables 17.13 and 17.14 for many of the features of *R. maris*.

Coccoid cells germinate into short rods which show snapping division and V-forms. Poor to moderate, butyrous, orange growth occurs on glycerol, nutrient and wort agars. Circular, raised, butyrous, glistening colonies with entire margins are formed on nutrient agar. Growth in nutrient broth is turbid.

Acetyl methyl carbinol, indole, methyl red, p-nitrophenoloxidase and phosphatase tests are negative. Hydrogen sulfide is not produced.

Acid is produced from fructose, glycerol and glucose but not from adonitol, arabinose, cellobiose, dulcitol, galactose, inositol, lactose, maltose, mannitol, α-methyl-D-glucose, raffinose, rhamnose, salicin, sorbitol, sorbose, sucrose or xylose.

Growth with C_6 to C_{17}, C_{19} and C_{23} n-alkanes but not with ethane or methane.

Isolated from soil and from skin and intestinal tract of carp (*Cyprinus carpio*).

The mol% G + C of the type strain is 73.2 (T_m).

Type strain: IMV 195.

11. **Rhodococcus rhodnii** Goodfellow and Alderson 1979, 80.[AL] (Effective publication: Goodfellow and Alderson 1977, 117.)

rhod′ni.i. M.L. masc. n. *Rhodnius* generic name of the reduvid bug; M.L. gen. n. *rhodnii* of *Rhodnius*.

See generic description and Tables 17.13 and 17.14 for many of the features of *R. rhodnii*.

Cocci germinate into rods which show limited branching. The growth cycle is completed by the appearance, through fragmentation, of short rods and cocci. Rough, red colonies are formed on glucose yeast extract agar.

Grows on sodium benzoate, sodium octanoate and L-tyrosine as sole carbon sources but not on salicin. Grows at 25–37°C.

Isolated from intestine of the reduvid bug, *Rhodnius prolixus*.

The mol% G + C of the DNA is 66 (T_m).

Type strain: KCC A-0203.

12. **Rhodococcus ruber** (Kruse) Goodfellow and Alderson 1977, 117.[AL] (*Streptothrix rubra* Kruse 1896, 63.)

rub′er. L. adj. *ruber* red.

See generic description and Tables 17.13 and 17.14 for many of the features of *R. ruber*.

Cocci give rise to multiple branched hyphae which readily fragment into rods, and cocci. Single unbranched aerial hyphae are regularly produced. Rough, pink to red colonies are formed on glucose yeast extract agar, Sauton's agar and on egg media. Single aerial hyphae, which may occasionally be branched, are formed on the surface of colonies.

Acetamidase-positive but allantoinase- and urease-negative; α- and β-esterase-, galactosidase- and acid phosphatase-negative.

Acid is formed from glucose, mannitol and sorbitol but not from inositol, mannose, rhamnose or trehalose.

Grows on butan-1-ol, iso-butanol, n-butanol, inulin, 2-methylpropan-1-ol, propan-1-ol, propane-1,2-diol, n-propanol, propylene glycol, sodium malonate, sodium octanoate and L-tyrosine but not on salicin or xylose.

Resistant to picric acid (0.2%, w/v), sodium nitrite (0.1%, v/v) and sodium salicylate (0.1%, w/v) but not to 5-fluorouracil (5 μg/ml), mitomycin C (5 μg/ml) or rifampicin (25 μg/ml).

Isolated from soil.

The mol% G + C of the DNA is 69–73 (T_m).

Type strain: KCC A-0205.

13. Rhodococcus rubropertinctus (Hefferan) Tsukamura 1974a, 43.[AL] (*Bacillus rubropertinctus* Hefferen 1904, 460; *Rhodococcus corallinus* (Bergey et al.) Goodfellow and Alderson 1977, 115.)

rub.ro.per.tinc'tus. L. adj. *ruber* red; L. pref *per* very; L. part. adj. *tinctus* dyed, coloured; M.L. adj. *rubropertinctus* heavily dyed red.

See generic description and Tables 17.13 and 17.14 for many of the features of *R. rubropertinctus.*

Rod-coccus growth cycle. Rough, orange to red colonies formed on glucose yeast extract agar, Sauton's agar and on egg media.

Allantoinase- and urease-positive; acetamidase-, benzamidase-, isonicotinamidase-, malonamidase-, nicotinamidase-, pyrazinamidase-, salicylamidase- and succinamidase-negative. β-Esterase-positive; acid phosphatase, α-esterase- and β-galactosidase-negative.

Acid is produced from glucose, mannitol, mannose and sorbitol but not from arabinose, galactose, inositol, raffinose, rhamnose or xylose.

Grows on butan-2,3-diol, 2,3-butylene glycol, butane-2,3-diol, inulin, propan-1-ol and sodium octanoate as sole carbon sources but not on propane-1,2-diol, propylene glycol, salicin or sodium malonate. Use L-glutamate, nitrate, L-serine and succinamide as sole nitrogen sources but not nitrite or urea. Glucosamine HCl and monoethanolamine are used as sole sources of carbon and nitrogen but benzamide is not.

Resistant to ethambutol (5 μg/ml), picric acid (0.2%, w/v), rifampicin (25 μg/ml), sodium-p-aminosalicylate (0.2%, w/v) and sodium nitrite (0.1%, w/v) but susceptible to 5-fluorouracil (20 μg/ml) and mitomycin C (5 μg/ml).

Isolated from soil.

The mol% G + C of the DNA is 67–69 (T_m).

Type strain: ATCC 14352.

14. Rhodococcus terrae (Tsukamura) Tsukamura 1974, 43.[AL] (*Gordona terrae* Tsukamura 1971, 22.)

ter'res. L. n. *terra* earth. M.L. gen. n. *terrae* of the earth.

See generic descriptions and Tables 17.13 and 17.14 for many of the features of *R. terrae.*

Rod-coccus growth cycle. Produces rough, pink to orange colonies on glucose yeast extract agar, Sauton's agar and on egg media.

Allantoinase-, nicotinamidase-, pyrazinamidase- and urease-positive; acetamidase-, benzamidase-, isonicotinamidase-, malonamidase- and salicylamidase-negative. β-Esterase-positive but acid phosphatase-, α-esterase- and β-galactosidase-negative.

Acid is produced from mannitol, rhamnose, sorbitol and trehalose but not from arabinose, galactose, inositol or xylose.

Grows on inulin, propanol and sodium octanoate as sole carbon sources but not on butane-2,3-diol, 1,3-butylene glycol, 1,4-butylene glycol, 2,3-butylene glycol, propane-1,2-diol or sodium malonate. Use acetamide, benzamide, isonicotinamide, nicotinamide, nitrate, pyrazinamide and urea as sole nitrogen sources but not nitrite; L-glutamate and monoethanolamine are used as sole sources of carbon and nitrogen but acetamide, benzamide and serine are not.

Resistant to ethambutol (5 μg/ml), picric acid (0.2%, w/v), rifampicin (25 μg/ml), sodium p-aminosalicylate (0.2%, w/v) and sodium nitrite

(0.1%, w/v) but not to 5-fluorouracil (40 μg/ml), mitomycin C (5 μg/ml) or sodium salicylate (0.1%, w/v).

Isolated from soil.

The mol% G + C of the DNA is 64–69 (T_m).

Type strain: ATCC 25594.

Species Incertae Sedis

15. Rhodococcus aichiensis Tsukamura 1983, 896.[VP] (Effective publication: Tsukamura 1982b, 1116.)

ai'chi.en'sis. M.L. adj. *aichiensis*, belonging to Aichi Prefecture, Japan, where the organism was isolated.

Occurs as short rods. Rough, pinkish or orange colonies formed on egg media.

Acetamidase-, nicotinamidase-, pyrazinamidase- and urease-positive, but allantoinase-, benzamidase-, isonicotinamidase-, salicylamidase- and succinamidase-negative. Reduces nitrate to nitrite, hydrolyzes Tween 80 and is acid phosphatase-positive, but is arylsulfatase, α- and β-esterase- and β-galactosidase-negative.

Grows on fructose, glucose, mannose, sucrose, trehalose, n-butanol, ethanol, iso-butanol, n-propanol, sodium acetate, sodium citrate, sodium fumarate, sodium malate, sodium pyruvate and sodium succinate as sole carbon sources, but not on arabinose, galactose, inositol, mannitol, rhamnose, sorbitol, xylose, 1,3-, 1,4-, and 2,3-butylene glycols, propylene glycol, sodium benzoate or sodium malonate. Uses acetamide, glutamate, and monoethanolamine as sole sources of carbon and nitrogen.

Resistant to picric acid (0.2, w/v) sodium nitrite (0.1%, w/v) and sodium salicylate (0.1%, w/v) but not to 5-fluorouracil (20 μg/ml), mitomycin C (5 μg/ml). Grows at 28°C, 37°C and 42°C, but not at 45°C.

Isolated from human sputum.

Type strain: E9028 (ATCC 33611).

16. Rhodococcus aurantiacus (ex Tsukamura and Mizuno 1971) Tsukamura and Yano 1985, 365.[VP]

au.ran.ti'a.cus. M.L. n. *aurantium* generic name of orange; *aurantiacus* orange colored.

Shows a rod-coccus life cycle. Rough, cream to orange colonies formed on glucose yeast extract agar and Sauton's agar. Synnemata of vertically arranged coalescing filaments are found on the surface of colonies.

Acetamidase-, allantoinase-, nicotinamidase-, pyrazinamidase- and urease-positive, but benzamidase-, isonicotinamidase-, malonamidase-, salicylamidase- and succinamidase-negative. β-Galactosidase- and urease-positive, but acid phosphatase-, arylsulfatase- and α-esterase-negative. Nitrate is not reduced. Degrades hypoxanthine, tyrosine, Tweens 20, 40, 60 and 80, but not adenine, casein or elastin.

Acid is produced from galactose, glucose, inositol, mannitol, mannose, sorbitol, trehalose and xylose, but not from rhamnose.

Grows on ethanol, fructose, galactose, glucose, glycerol, inositol, mannitol, mannose, melezitose, sorbitol, sucrose, trehalose, xylose, n-butanol, iso-butanol, 2,3-butylene glycol, propanol, propylene glycol, sodium acetate, sodium citrate, sodium fumarate, sodium malate, sodium pyruvate and sodium succinate, but not on adonitol, arabinose, inulin, lactose, raffinose or rhamnose. Uses acetamide, nicotinamide, nitrate and urea as sole nitrogen sources, but not benzamide. Acetamide, glutamate, glucosamine HCl, monoethanolamine and serine are used as sole sources of carbon and nitrogen, but not benzamide or trimethylenediamine.

Resistant to ethambutol (5 μg/ml), 5-fluorouracil (20 μg/ml), mitomycin C (10 μg/ml), picric acid (0.2%, w/v) and sodium nitrite (0.2%, w/v). Grow at 10°C, 28°C and 37°C but not at 40°C.

Isolated from sputum.

Type strain: ATCC 25938.

17. Rhodococcus chubuensis Tsukamura 1983, 896.[VP] (Effective publication: Tsukamura 1982b, 1116.)[AL]

chu'bu.en'sis. M.L. adj. *chubuensis*, belonging to Chubu Hospital where the organism was isolated.

Occurs as short rods. rough, pinkish or orange colonies formed on egg media.

Reduces nitrate to nitrite, hydrolyses Tween 80, and is α- and β-esterase- and acid-phosphatase positive, but is arylsulfatase and β-galactosidase-negative. Usually does not show any amidase activity.

Grows on fructose, glucose, mannitol, mannose, sorbitol, sucrose, trehalose, ethanol, n-propanol, sodium acetate, sodium citrate, sodium fumarate, sodium malate, sodium pyruvate and sodium succinate as **sole carbon sources, but not on arabinose, galactose, inositol, rhamnose, xylose, butanol, butylene glycol, propylene glycol, sodium benzoate** or sodium malonate. Uses acetamide and glutamate as sole sources of carbon and nitrogen, but not benzamide, glucosamine, monoethanolamine or trimethylenediamine.

Resistant to picric acid (0.2%, w/v) but sensitive to 5-fluorouracil (20 μg/ml), mitomycin C (5 μg/ml) and sodium nitrite (0.1%, w/v). Grows at 28°C and 37°C but not at 42°C.

Isolated from human sputum.

Type strain: E6324 (ATCC 33609).

18. **Rhodococcus obuensis** Tsukamura 1983, 897.[VP] (Effective publication: Tsukamura 1982b, 1117.)

obu'en'sis. M.L. adj. *obuensis*, belonging to Obu City where the organism was isolated.

Occurs as rods. Rough, pinkish or orange colonies formed on egg media.

Acetamidase-, nicotinamidase-, pyrazinamidase- and urease-positive, but allantoinase-, benzamidase-, isonicotinamidase-, salicylamidase- and succinamidase-negative. Reduces nitrate to nitrite, hydrolyses Tween 80 and is acid phosphatase positive, but is arylsulfatase, α- and β-esterase- and β-galactosidase-negative.

Grows on fructose, glucose, mannitol, sucrose, sodium acetate, sodium citrate, sodium malate, sodium pyruvate and sodium succinate as sole carbon sources, but not on arabinose, galactose, inositol, rhamnose, trehalose, xylose, *iso*-butanol, n-butanol, 1.3-, 1.4-, and 2,3-butylene glycols, ethanol, n-propanol, propylene glycol, sodium benzoate, sodium

fumarate or sodium malonate. Uses glutamate as a sole source of carbon and nitrogen but not acetamide, benzamide, glucosamine, monoethanolamine, serine or trimethylenediamine.

Resistant to picric acid (0.2%, w/v) and sodium nitrite (0.2%, w/v) but sensitive to 5-fluorouracil (20 μg/ml) and mitomycin C (5 μg/ml). Grows at 28°C and 37°C but not at 42°C.

Isolated from human sputum.

Type strain: E8179 (ATCC 33610).

19. **Rhodococcus sputi** (ex Tsukamura 1978) Tsukamura and Yano 1985, 365.[VP]

spu'ti. L. gen. *sputum* discharge from the respiratory tract.

Occurs as short rods and cocci. Rough, pink colonies formed on egg media.

Nicotinamidase-, pyrazinamidase- and urease-positive, but benzamidase-negative. Acid phosphatase-positive but negative for β-galactosidase and α-esterase.

Acid is produced from glucose, mannitol, mannose, sorbitol and trehalose, but not from arabinose, galactose, inositol, rhamnose and xylose.

Grows on ethanol, mannitol, mannose, sorbitol, sucrose, trehalose, sodium citrate and sodium fumarate as sole carbon sources, but not on galactose, inositol, rhamnose, propylene glycol, sodium benzoate or sodium malonate. Uses acetamide, nicotinamide, nitrate and succinamide as sole nitrogen sources but not benzamide, isonicotinamide, or nitrite. Acetamide and monoethanolamine used as sole sources of carbon and nitrogen but not glucosamine or serine.

Resistant to ethambutol (5 μg/ml), p-nitrobenzoic acid (0.5 mg), picric acid (0.2%, w/v) and rifampicin (25 μg/ml), but not to 5-fluorouracil (20 μg/ml), mitomycin C (5 μg/ml), sodium nitrite (0.1%, w/v) or sodium salicylate (0.1%, w/v). Grows at 28°C and 37°C but not at 42°C.

Isolated from human sputum.

Type strain: ATCC 29627.

Genus **Nocardioides** Prauser 1976, 61[AL]

HELMUT PRAUSER

No.car.di.o.i'des. M.L. fem. n. *Nocardia*, name of a genus; Gr. suff. *idea*, appearance, M.L. mas. n. *Nocardioides*, nocardia-like, referring to the similar life cycles of the two genera.

Primary mycelium shows abundantly branching hyphae growing on the surface and penetrating into agar media; they **break up into fragments** which may be irregular or rodlike or coccoid. **Aerial mycelium** consisting of irregular, sparsely and irregularly branching or unbranched hyphae which **break up into short to elongated rodlike fragments.** Both the fragments of the primary and the aerial mycelium give rise to new mycelia. No motile cells. Colonies pasty. **Gram-positive. Nonacid-fast. Catalase positive.** Strictly **aerobic.** Chemoorganotrophic. Oxidative catabolism. Grows readily on standard media. **Susceptible to specific phages.** Diagnostic amino acids of the cell wall **LL-diaminopimelic acid and glycine.** Mycolic acids lacking. Diagnostic phospholipids **phosphatidylglycerol** and **acylphosphatidylglycerol.** Phosphatidylethanolamine and other nitrogenous phospholipids lacking. 14-Methylpentadecanoic acid predominating among fatty acids. **Menaquinones primarily of MK-8(H$_4$)** type. **Mol% G + C** of DNA ranging from **66.5–68.6 (T_m).** Worldwide in **soil.**

Type species: *Nocardioides albus* Prauser 1976, 61.

Further Descriptive Information

The information given in the following four paragraphs originates from Prauser (1976, 1981a) or represents the author's unpublished results.

Fragmentation of the hyphae of the primary mycelium begins in the older parts of the colonies and hyphae (Fig. 17.3). Depending on the size of the fragments they give rise to new mycelia by extruding one, two or more hyphae. On media rich in organic nitrogen, and in submerged shaken culture, the extent of mycelia and their persistence are reduced. The irregularly shaped and branched hyphae of the aerial mycelium (Fig. 17.4) resemble those of nocardiae. Aerial hyphae break up completely into sporelike fragments (Figs. 17.5 and 17.6). They germinate by producing one or two germ tubes. Motility does not occur at any stage of the life cycle.

Hyphae and fragments show the ultrastructure typical for Gram-positive bacteria (Figs. 17.6 and 17.7). The hyphae of the primary mycelium are irregularly septated. Preceding fragmentation, additional cross-walls are formed. Branching usually takes place near to cross-walls by production of lateral outgrowths (Fig. 17.7). The process of branching can be regarded as restoration of the growth functions in the course of the individualization of the hyphal fragments (Fig. 17.7). The aerial hyphae septate and break up more regularly than those of the primary mycelium, resembling one of the types of spore formation described for streptomycetes (Fig. 17.6). The resulting sporelike fragments display a smooth surface (Fig. 17.5).

Colonies not covered by aerial mycelium have a pasty consistency. The surface of agar cultures is smooth to wrinkled and dull to bright, but usually it is faintly glistening. The aerial mycelium may totally cover the primary mycelium, may be produced only in patches or at the margins of the colonies, or may be lacking, depending on the individual strain, the medium used, and on the procedures and the duration of maintenance of strains. The sporelike fragments are rarely recognizable in situ. However, they are seen after gently pressing a

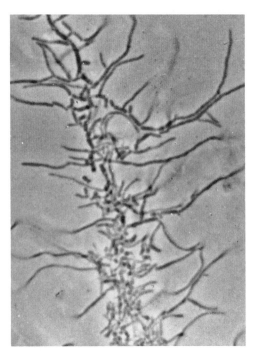

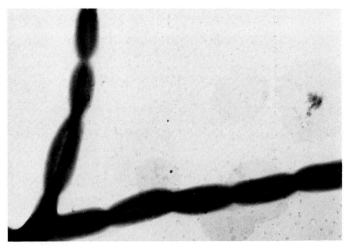

Figure 17.5. *Nocardioides albus.* Aerial hyphae fragmented into sporelike elements. Electron micrograph (× 17,000). (Reproduced with permission from H. Prauser, International Journal of Systematic Bacteriology, *26:* 58–65, 1976, © American Society for Microbiology.)

Figure 17.3. *Nocardioides albus.* Fragmentation of hyphae of the primary mycelium in situ on glycerol asparagine agar. Phase-contrast micrograph (× 1600). (Reproduced with permission from H. Prauser, International Journal of Systematic Bacteriology, *26:* 58–65, 1976, © American Society for Microbiology.)

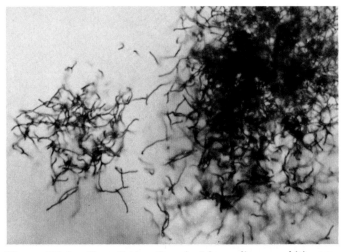

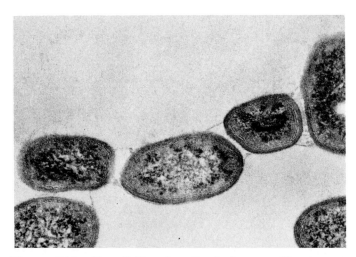

Figure 17.6. *Nocardioides albus.* Developing spore-like elements (arthrospores) still connected by the surface sheath of the aerial hypha. Electron micrograph (× 50,000). (Reproduced with permission from H. Prauser. Actinomycetes. In Schaal and Pulverer (Editors) Proceedings 4th International Symposium on Actinomycetes Biology, Zentralbl. Bakteriol. Hyg. Abt. 1 Orig. B, Supplement 11, 1981, © Fischer-Verlag, Stuttgart.)

Figure 17.4. *Nocardioides albus.* Aerial mycelium on chitin agar. Phase-contrast micrograph (× 400). (Reproduced with permission from H. Prauser, International Journal of Systematic Bacteriology, *26:* 58–65, 1976, © American Society for Microbiology.)

coverslip onto the mat of aerial mycelium followed by microscopical observation on a slide covered with a thin film of agar.

The organism grows within 1–2 days on standard media such as nutrient agar or oatmeal agar (Shirling and Gottlieb, 1966); the latter is particularly suited for the production of aerial mycelium. A variety of carbohydrates and alcohols is used as carbon source.

The large number of phages which multiply on *Nocardioides* strains do not affect other nocardioforms, coryneforms and sporoactino-mycetes (Prauser, 1976, 1984a; Prauser and Falta, 1968; Wellington and Williams, 1981) except for four phages which also propagate on strains of *Pimelobacter simplex* (*Arthrobacter simplex*) and *Pimelobacter jensenii* (Prauser 1976, 1984a, and unpublished results). On the other hand, phages which are effective against other nocardioforms, coryneforms and sporoactinomycetes do not multiply on *Nocardioides* strains (Prauser, 1974, 1976, 1984a; Wellington and Williams, 1981). Some *Streptomyces* phages cause clearing effects, i.e. phage-dependent lysis without phage propagation, on strains of the genus *Nocardioides* (Prauser, 1984a).

Diagnostic amino acids of the cell wall are LL-diaminopimelic acid and glycine (cell wall chemotype I sensu Becker et al., 1965). LL-Diaminopimelic acid of one peptide subunit is cross-linked with D-alanine of another via glycine in the interpeptide bridge (peptidoglycan type A3γ sensu Schleifer and Kandler, 1972; Prauser, 1976). The results on phospholipids are conflicting (M. P. Lechevalier et al., 1977, 1981;

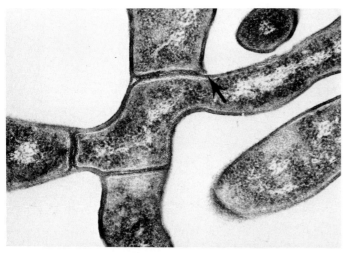

Figure 17.7. *Nocardioides albus.* Part of a hypha (vertical in the figure) of the primary mycelium with branches originating from one segment. Beginning of fragmentation of the hypha at the upper right angle (arrow). Electron micrograph (× 40,000). (Reproduced with permission from H. Prauser. Actinomycetes. In Schaal and Pulverer (Editors) Proceedings 4th International Symposium on Actinomycetes Biology, Zentralbl. Bakteriol. Hyg. Abt. 1 Orig. B, Supplement 11, 1981, © Fischer-Verlag, Stuttgart.)

O'Donnell et al., 1982; Collins et al., 1983). According to Lechevalier et al. (1977, 1981), who applied the most conclusive methods, the phospholipid composition of *Nocardioides albus* is unique among actinomycetes. In addition to phosphatidylglycerol, phosphatidylinositol and traces of phosphatidylinositol mannosides, the organism contains major amounts of acylphosphatidylglycerol. Diphosphatidylglycerol, present in nearly all actinomycetes studied so far, is lacking as well as any nitrogenous phospholipids. The predominating fatty acid seems to be 14-methylpentadecanoic acid, although there are conflicting reports on this fatty acid, and on tuberculostearic acid and its lower homologues (C_{17}, C_{18}), the saturated straight chain fatty acids (C_{15}–C_{18}), iso- and anteiso-branched fatty acids (C_{15}–C_{18}), and the straight chain unsaturated oleic acid (C_{18}) (O'Donnell et al., 1982; Collins et al., 1983; Suzuki and Komagata, 1983). Among menaquinones, tetrahydrogenated ones with eight isoprene units (MK-8(H_4)) predominate (O'Donnell et al., 1982; Collins et al., 1983; Suzuki and Komagata, 1983). Results concerning minor isoprenologues are conflicting. Mycolic acids are lacking (Prauser, 1976; O'Donnell et al., 1982). The mol% G + C of DNA

ranges from 66.5–68.6 (Prauser, 1966, 1976, and unpublished results; Suzuki and Komagata, 1983). The data given by Tille et al. (1978) are probably too high. The DNA:DNA homology between *Nocardioides albus* and *Streptomyces griseus* is very low (8%, according to Tille et al., 1978).

Pathogenicity for man, animals and plants has not been reported.

Strains have been isolated from soils of gardens, meadows, arable land, grassland, savanna and from tephra (Surtsey); the organism predominated in kaolin prepared for ceramic industry (Prauser, 1976).

Isolation Procedures

Strains can be isolated employing the usual dilution and plating techniques on several agar media appropriate for the isolation of streptomycetes and other actinomycetes. Often a dilute oatmeal agar has been used (Prauser and Bergholz, 1974: 3 g oatmeal, 0.3 g KNO_3, 0.5 g K_2HPO_4, 0.2 g $MgSO_4$, 15 g agar and 1 liter distilled water, pH = 7).

Maintenance Procedures

Strains may be maintained by serial transfers on oatmeal agar slants (once every 3 months), by lyophilization, and by storage above liquid nitrogen either as suspensions containing 5% dimethylsulfoxide or as minicultures which are likewise suitable for subsequent shipping of the strains (Prauser, 1984b).

Rapid Identification by Phage Typing

All strains of the genus *Nocardioides* hitherto examined are susceptible to at least one of a selected set of five phages: IMET 5013 (X1), IMET 5015 (X3), IMET 5017 (X5), IMET 5057 (X10), and IMET 5056 (X24) (Prauser, 1976, and unpublished results). For all procedures the complex organic medium number 79 (Prauser and Falta, 1968), may be used. Following incubation of an agar slant culture of the strain for 1–2 days the growth is suspended and one drop of the suspension is mixed with 3 ml of molten agar (0.6% agar) and spread over a basal layer (1.5% agar) in a Petri dish. One loopful of each of the five phage suspensions (<10^9 plaque-forming units/ml) is placed on the solidified agar at marked positions. Following overnight incubation at 28°C the bacterial lawn is examined for clear or turbid zones. The occurrence of one to five positive reactions identifies the strain under study with the genus *Nocardioides*. Even clearing effects—i.e. phage-dependent lysis without phage propagation—are as a rule taxon-specific (Prauser, 1981b, 1984a). Hence, differentiation between true lysis and clearing effects is not essential for correct identification. In addition, the presence of mycelia should be checked microscopically to avoid misidentification as *Pimelobacter* spp. (see above, and Table 17.13 and genus *Arthrobacter*), although none of the five phages recommended ever displayed any lytic effect against strains of the genus *Pimelobacter*.

Differentiation of genus **Nocardioides** from other taxa

Essential characteristics differentiating the genus *Nocardioides* from other more or less closely related taxa are presented in Table 17.15. The table includes, in addition to the coryneform genus *Pimelobacter* (see genus *Arthrobacter*), all LL-diaminopimelic acid-containing nocardioforms, most of the remaining nocardioforms, and the family *Streptomycetaceae*.

The data given in Table 17.15 will provide reliable recognition of *Nocardioides* strains, particularly if phage typing is included.

Taxonomic Comments

The first known representative of the genus was strain IMET 7801 isolated from kaolin (Prauser, 1966). The observed type of life cycle was termed "nocardioform," which was proposed for bacteria producing hyphae which break up completely into fragments that give rise to new mycelia (Prauser, 1967, 1978). Additional strains constituted the "IMET 7801 group" (Prauser, 1967), for which the generic designation *Nocardioides* was tentatively introduced (Prauser, 1970). The valid

publication was delayed until 1976 (Prauser) to give specialists the chance to place this widely distributed organism into an existing taxon.

Nocardioides albus and *Arthrobacter simplex* (*Pimelobacter simplex*) are closely related organisms. The two species share the peptidoglycan type A3γ (Prauser, 1976), cross-susceptibility to 4 of 26 specific phages (Prauser, 1970, 1976, 1984a, and unpublished results), similar levels (in the range around 70 mol% G + C) of DNA base composition (Prauser 1966, 1976; Suzuki and Komagata, 1983), some degree (15–20%) of DNA/DNA homology (Prauser, 1981a, and unpublished results; Suzuki and Komagata, 1983), the bending type of cell division (Prauser, 1981a), a similar pattern of fatty acids (O'Donnell et al., 1982; Collins et al., 1983) and identical predominating menaquinones (O'Donnell et al., 1982; Collins et al., 1983; Suzuki and Komagata, 1983).

On the basis of these data including the results on phospholipids, O'Donnell et al. (1982) transferred *Arthrobacter simplex* to the genus *Nocardioides* as *Nocardioides simplex* (validated in List 12, International Journal Systematic Bacteriology, *33:* 896–897, 1983). Conse-

Table 17.15.

Characteristics differentiating the genus **Nocardioides** *from other taxa[a,b]*

Characteristics	Nocardi-oides	Pimelo-bacter[c]	Intra-sporangium	Spo-richthya[d]	Arach-nia	Strepto-mycetaceae[d]	Nocar-dia	Rhodo-coccus	Myco-bacterium	Oer-skovia	Promicro-monospora
Susceptibile to *Nocar-dioides* phages[e]	+	+	−	−	−	−	−	−	−	−	−
Primary mycelium on agar media	+[f]	−	+	−	+	+	+	+	−	+	+
Hyphae of primary mycelium break up into fragments[g]	+		+		+	−	+	+	−	+	+
Aerial mycelium[h]	+	−	−	+	−	+	+	−	−	−	+
Acid-fast	−	−	−	−	−	−	p	p	+	−	−
Anaerobic growth	−	−	−	−	+	−	−	−	−	−	−
LL-Diaminopimelic acid	+	+	+	+	+	+	−	−	−	−	−

[a] For biochemical data differentiating the genera included in this table see Further Descriptive Information and the corresponding generic contributions in the present volume.

[b] Symbols: see Table 17.12; also p, partial acidfastness predominates in *Nocardia* and sometimes occurs in *Rhodococcus*; and v, strain instability (NOT equivalent to "d"). For oerskoviae anaerobic growth has been reported only on tryptic soy agar.

[c] See *Arthrobacter* in this volume 2.

[d] *Sporichthya* and the *Streptomycetaceae* will appear in volume 4.

[e] The five phages recommended are given in Rapid Identification by Phage Typing. The phages are available from the IMET Culture Collection.

[f] The primary mycelia of *Nocardioides* strains are always extensive and cannot be confused with short side projections of *Pimelobacter tumescens* and mycobacteria and the very rare cases of limited mycelia in the latter genus.

[g] Partial fragmentation (complete one in "*Nocardia italica*" (sic)) under normal growth conditions has been reported for individual strains of a few species of the genus *Streptomyces*. In some nocardiae, or even representatives of other nocardioform genera, fragmentation is not always seen on agar media in situ. Gentle pressure or submerged shaken culture will display fragmentation in nearly all cases.

[h] The data given apply at least to fresh isolates. Aerial hyphae sometimes may be seen only microscopically.

quently, the description of the genus *Nocardioides* had to be emended, i.e. to be extended in order to include these morphologically extremely different organisms. On the other hand, Suzuki and Komagata (1983), including in their comparative biochemical studies the type strains of *Nocardioides albus*, *Arthrobacter globiformis* and *Arthrobacter simplex*, established the new genus *Pimelobacter* to harbor *Pimelobacter simplex* (formerly *Arthrobacter simplex*) as the type species. This new combination has also been validated (Validation List 11, International Journal of Systematic Bacteriology, *33:* 672–674, 1983). Thus, as they are based on the same type strain, *Arthrobacter simplex*, *Nocardioides simplex* and *Pimelobacter simplex* are objective synonyms. Editorial note: the species is described in this volume under *Arthrobacter*.

In the present contribution the original description of the genus *Nocardioides* Prauser, 1976 is retained. In order to keep a classification lucid for practical application, organisms which differ markedly in morphology should not be united in one genus. Thus, a typical nocardioform organism with extended primary and aerial mycelium (*Nocardioides albus*) should not be combined with a motile, single-celled bacterium (*Pimelobacter simplex*), although they should be placed in the same family.

The precise family position of the genus *Nocardioides* is likely to remain unsolved until comparative 16S rRNA sequencing has been done. Prauser (1976), believing that the introduction of a new family might be appropriate, provisionally placed the genus in the family *Streptomycetaceae*. Relevant information is still insufficient and discordant. Lechevalier and Lechevalier (1981) regarded it as a genus "in search of a family." O'Donnell et al. (1982), Collins et al. (1983) and Suzuki and Komagata (1983) did not discuss the family placement of the genus.

Further Reading

O'Donnell et al. 1982. Lipids in the classification of *Nocardioides*: Reclassification of *Arthrobacter simplex* (Jensen) Lochhead in the genus *Nocardioides* (Prauser) emend. O'Donnell et al. as *Nocardioides simplex* comb. nov. Arch. Microbiol. *133:* 323–329.

Prauser, H. 1976. *Nocardioides*, a new genus of the order *Actinomycetales*. Int. J. Syst. Bacteriol. *26:* 58–65.

Prauser, H. 1984. Phage host ranges in the classification and identification of Gram-positive branched and related bacteria. *In* Ortiz-Ortiz et al. (Editors), Biological, Biochemical and Biomedical Aspects of Actinomycetes, Academic Press, London, pp. 617–633.

Suzuki, K.-I. and K. Komagata. 1983. *Pimelobacter* gen. nov., a new genus of coryneform bacteria with LL-diaminopimelic acid in the cell wall. J. Gen. Appl. Microbiol. *29:* 59–71.

Differentiation of the species of the genus **Nocardioides**

The differential characteristics of the genus *Nocardioides* are given in Table 17.16.

Table 17.16.

Differential characters of species of **Nocardioides**[a]

Characteristics	1. *N. albus*	2. *N. luteus*	Characteristics	1. *N. albus*	2. *N. luteus*
Yellow pigment on oatmeal agar	−	+	Acid from L-rhamnose	+	−
Well developed aerial mycelium	White	Cream	Acid from sucrose	+	−

[a] Symbols: see Table 17.12.

List of species of the genus **Nocardioides**

1. **Nocardioides albus** Prauser 1976, 61.[AL]

al' bus. M.L. adj. albus white, referring to the white aerial mycelium.

Hyphae of the primary mycelium 0.5–0.8 μm in diameter, hyphae of the aerial mycelium 0.6–1.0 μm in diameter. Primary mycelium white, whitish to faintly cream-colored on oatmeal agar, yeast extract-malt extract agar, glucose asparagine agar and other media. Aerial mycelium thin, dense and chalky. Surface of the sporelike elements smooth (Fig. 17.5). No soluble pigments, except for a reddish-brown pigment on media containing tyrosine as sole amino acid. Colonies lacking aerial mycelium are pasty, smooth to wrinkled and dull to bright. Growth optimum about 28°C, good growth at 15°C and 37°C; some strains grow at 42°C; no growth at 10°C and 50°C.

D-Glucose, L-arabinose, sucrose, D-xylose, D-mannitol, D-fructose, and L-rhamnose utilized as carbon sources; inositol and raffinose not utilized. Acid produced from D-glucose, L-arabinose, sucrose, D-xylose, D-mannitol, D-fructose and L-rhamnose, in exceptional cases from adonitol and sorbitol; no acid from inositol and raffinose. Citrate, succinate and benzoate assimilated; tartrate not assimilated. Hypoxanthine, tyrosine, esculin (most cases), casein, and starch hydrolyzed, adenine and xanthine not hydrolyzed.

Isolated from soil.

The mol% G + C of DNA 66.5–68.6 (T$_m$).

Type strain: ATCC 27980 (IMET 7807).

2. **Nocardioides luteus** Prauser 1984c647.[VP] ("*Nocardioides flavus*" Ruan and Zhang 1979, 347.)

Displays the morphological, physiological and biochemical characters of *Nocardioides albus* with a few exceptions (Table 17.16): primary mycelium yellow on oatmeal agar and other media, varying to orange-yellow in aged cultures. Aerial mycelium cream-colored if well developed, otherwise white. Most of the strains do not produce acid from L-rhamnose and sucrose.

Isolated from soil.

The mol% G + C of the DNA is 67.5 (T$_m$). DNA/DNA homology of IMET 7830 and *Nocardioides albus* IMET 7807 is 49%.

Type strain: IMET 7830.

Comment: The organism was published first as *Nocardioides* sp. IMET 7830 (Tille et al., 1978). The name "*Nocardioides luteus*," accompanied by a brief description, was used first by Prauser in a poster at the International Union of Microbiological Societies Congress, Munich 1978. The name was used in the literature (Collins et al., 1983; O'Donnell et al., 1982). Since the type strain 71–N54 of "*Nocardioides flavus*" obtained from Dr. Ruan was found to be similar to IMET 7830, synonymy was assumed.

Species Deserving More Study

a. "*Nocardioides fulvus*" Ruan and Zhang 1979, 350.

Life cycle nocardioform. Cell wall containing LL-diaminopimelic acid and glycine. Primary mycelium yellow, yellowish brown to brown; aerial mycelium cream-white or several tinges of yellow, no soluble pigment. Starch, D-mannitol, D-fructose used as carbon sources; L-arabinose, D-xylose, L-rhamnose, raffinose, mannose, lactose, inositol not tested. Acid produced from D-glucose; not produced from D-galactose, lactose, L-arabinose, D-xylose, L-rhamnose, raffinose, inositol, and mannitol. H$_2$S not produced, aminobutyric acid not fermented, D-glucose oxidized, hippuric acid and urea not decomposed. Isolated from soil samples of Beijing.

Type strain: 71-N86.

b. "*Nocardioides thermolilacinus*" Lu and Yan 1983, 224.

Life cycle nocardioform. Cell wall containing LL-diaminopimelic acid. Thermophilic, growing between 28°C and 55°C. Primary mycelium scarlet, aerial mycelium pink, soluble pigment orange-colored. Strains, T 505 and T 511, isolated at 52°C from soil samples of subtropical regions. Strains not yet studied by the present author.

Genus **Pseudonocardia** Henssen 1957, 408[AL]

AINO HENSSEN

Pseu.do.no.car′di.a Gr. adj. *pseudes* false; M.L. n. *Nocardia* a genus of the actinomycetes; M.L. fem. n. *Pseudonocadia*, the false *Nocardia*.

Substrate and aerial mycelium-bearing spores in chains, hyphae segmented, **often zig-zag shaped,** with tendency to form **apical or intercalary swellings.** Hyphal elongation by budding. Segments acting directly as spores or being secondarily divided in sporulation process. Hyphal wall with two layers, cross-walls are interspace septa, **spores without interspore pads. Gram-positive. No motile stages. Aerobic.** Growth on a variety of organic and synthetic media. **Mesophilic or thermophilic.**

Type species: *Pseudonocardia thermophila* Henssen 1957, 408.

Further Descriptive Information

Substrate and aerial hyphae varying in thickness, 0.4–1.0 μm, in swellings to 2.6 μm thick. A characteristic feature is the growth of hyphae by budding (Figs. 17.8–17.10); a constriction is produced behind the tip of the terminal segment, the tip then elongates to form a new segment, another constriction is formed near the tip and the process repeated (time-lapse photographs, Figs. 1–7, published in Henssen and Schäfer, 1971). The constrictions may be secondarily separated by septa (Fig. 17.9B), a septum may also be formed a distance behind a constriction (e.g. in *P. compacta*, Henssen et al., 1983). Side branches usually arise below a septum, more rarely from the center of a segment (Figs. 17.10 and 17.11).

Spores may be formed on substrate or aerial mycelium in three ways:

1. By successive acropetal formation (the usual way) designated as *Pseudonocardia*-type (Henssen and Schnepf, 1967),

2. By basipetal septation,
3. By irregular spore formation along senescing hyphae.

Spores are smooth walled or spiny, varying greatly in size, usually 0.5–1.0 μm wide by 1.5–3.5 μm long, but varying in length from 1.0–4.5 μm.

Cell wall constituents: *meso*-diaminopimelic acid (*meso*-DAP), arabinose and galactose: type IV (Lechevalier and Lechevalier, 1970; Henssen and Schäfer, 1971; Henssen et al., 1983), phospholipids type III (Lechevalier et al., 1981, Henssen et al., 1983) and MK-9(H$_4$) menaquinones (Minnikin and Goodfellow, 1981). No mycolic acid has been found in *P. thermophila* (Goodfellow and Minnikin, 1981) or *P. compacta* (Henssen et al., 1983).

The hyphal wall in both substrate and aerial mycelium is composed of two layers, an inner electron transparent, uniformly thick layer, and an outer electron-dense irregular layer (Figs. 17.11–17.13). Cross-walls are interspace septa (Henssen et al., 1981, 1983; cross-wall type 2 in Williams et al., 1973). Hyphal swellings may become subdivided by septa growing inwards at different angles. Spore walls are of uniform thickness, no interspore pads are formed. In spiny spores the ornamentation is formed by folds of the fibrous sheath.

Good growth on Casamino-peptone-Czapek agar and yeast starch agar. Substrate mycelium forms a compact mass, colonies yellow to ochre, pigment with a spectrum characteristic for carotenoids. Aerial mycelium white, powdery or forming a thick cover.

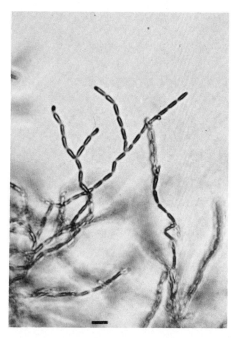

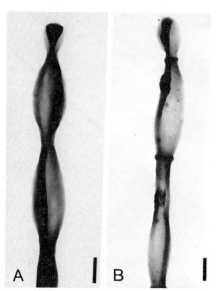

Figure 17.8. Budding, zig-zag shaped hyphae of *Pseudonocardia spinosa* strain MB SF-1. (LM, *bar*, 5 μm.) (Reproduced with permission from A. Hensson and D. Schäfer, International Journal of Systematic Bacteriology *21:* 29–34, 1971, ©American Society for Microbiology.)

Figure 17.9 *A* and *B*. Budding aerial hyphae of *Pseudonocardia thermophila* strain MB-A18, in *B*, constrictions separated by septae. (Whole mount silhouettes, TEM, *bar*, 0.5 μm.) (Reproduced with permission from A. Hensson and D. Schäfer, International Journal of Systematic Bacteriology *21:* 29–34, 1971, ©American Society for Microbiology.)

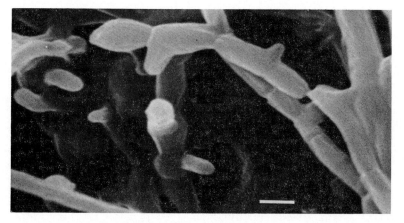

Figure 17.10. Budding substrate and aerial hyphae of *Pseudonocardia compacta* strain MB H-146. (SEM, *bar*, 1 μm.) (Reproduced with permission from A. Henssen, C. Happach-Kasan, B. Renner and G. Vobis, International Journal of Systematic Bacteriology *33:* 829–836, 1983, ©American Society for Microbiology.)

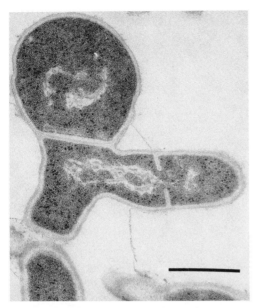

Figure 17.11. *Pseudonocardia thermophila* MB-A18, tip of aerial hyphae with apical swelling and side branch showing septum formation. (TEM, glutaraldehyde/osmium tetroxide fixation, *bar*, 0.5 μm.)

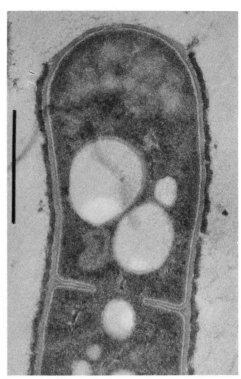

Figure 17.13. *Pseudonocardia compacta* MB H-146, tip of sporulating aerial hypha. Hyphal wall with thick inner and outer layer, crosswall formed by inwards growth of double annulus. (TEM, *bar*, 0.5 μm). (Reproduced with permission from A. Henssen, C. Happach-Kasan, B. Renner and G. Vobis, International Journal of Systematic Bacteriology *33:* 829–836, 1983, ©American Society for Microbiology.)

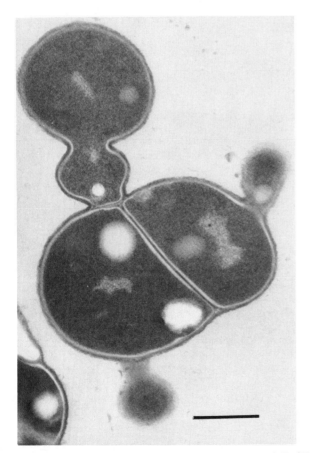

Figure 17.12. Budding cells of *Pseudonocardia spinosa* MB SF-1, the upper bud with constriction (TEM, glutaraldehyde/osmium tetraoxide fixation, *bar*, 0.5 μm.)

Ecological niches for *Pseudonocardia* are rotting plant remains. *P. thermophila* is found in self-heating hay, compost or manure heaps, the two mesophilic species occur in cultivated land.

Enrichment and Isolation

Suitable conditions for the enrichment and isolation of the thermophilic species are cellulose-containing substrates, a temperature of 50°C and a microaerophilic atmosphere. The mesophilic species are isolated on poorly nutrient media (e.g. an artificial soil agar) under aerobic conditions at 20–30°C (Henssen and Schäfer, 1971).

Maintenance Procedures

Species of *Pseudonocardia* can be lyophilized by common procedures used for aerobes; vigorously sporulating cultures should be used.

Differentiation of the genus **Pseudonocardia** from other genera

The genus *Pseudonocardia* is separated from all other genera which form aerial spores in chains by the budding process of hyphal elongation resulting in segmented hyphae. Species of *Nocardiopsis* with cell walls containing *meso*-DAP and galactose might be misinterpreted as belonging to *Pseudonocardia*. *Nocardiopsis* species are distinguished with certainty by studies of ultrathin sections. Whereas species of *Pseudonocardia* have a two-layered hyphal wall and interspace septa, species of *Nocardiopsis* have a one-layered hyphal wall and solid septa as crosswalls (Henssen et al., 1981). *Pseudonocardia* differs from the genus *Nocardia* (which also has cell wall type IV and interspace septa) in having a two-layered hyphal wall (one-layered in *Nocardia*, p. 1460), in the lack of mycolic acid and a different type of phospholipid (Lechevalier et al., 1981).

Taxonomic Comments

Precise knowledge of the phylogenetic relationships of the genus *Pseudonocardia* to other genera of *Nocardiaceae*, in which family the genus has been usually included, depend on future studies of its nucleic acids, such as the degree of similarity in rRNA oligonucleotide sequence.

Further Reading

Henssen, A. and E. Schnepf. 1967. Zur Kenntnis thermophiler Actinomyceten. Arch. Mikrobiol. *57:* 214–231.

Henssen, A. and D. Schäfer. 1971. Emended description of the genus *Pseudonocardia* Henssen and description of a new species *Pseudonocardia spinosa* Schäfer. Int. J. Syst. Bacteriol. *21:* 29–34.

Henssen, A., E. Weise, G. Vobis and B. Renner. 1981. Ultrastructure of sporogenesis in actinomycetes forming spores in chains. *In* Schaal and Pulverer (Editors), Actinomycetes. Proceedings of the Fourth International Symposium on Actinomycete Biology, Fischer-Verlag, Stuttgart, pp. 137–146.

Differentiation of the species of the genus **Pseudonocardia**

Some differential features of the species *P. thermophila*, *P. spinosa* and *P. compacta* are listed in Table 17.17.

Table 17.17.
Differential characteristics of the species of the genus **Pseudonocardia**[a]

Characteristics	1. P. thermophila	2. P. spinosa	3. P. compacta	Characteristics	1. P. thermophila	2. P. spinosa	3. P. compacta
Optimum temperature	40–50°C	20–30°C	20–30°C	Zig-zag hyphae frequent	+	+	−
Spores	Smooth	Spiny	Smooth	Hyphal swellings frequent	−	+	+
Growth on cellulose agar	+	+	−				

[a] Symbols: see Table 17.12.

List of species of the genus **Pseudonocardia**

1. **Pseudonocardia thermophila** Henssen 1957, 408.[AL]

ther.mo'phi.la. Gr. n. *thermus* heat; Gr. adj. *philus* loving; M.L. fem. adj. *thermophila* heat loving.

See Table 17.17 and generic description for some features.

Substrate hyphae septate, often zig-zag shaped, swellings usually present, becoming multidivided. Aerial hyphae often zig-zag shaped, young stages with constrictions, later on septate throughout, swellings rarely present. All three types of spores produced in substrate and aerial mycelium. Inner wall layer in hyphae and spores uniformly thin, not thickened in mature spores.

Good growth on nutrient agar and yeast agar, colonies yellow, thick cover of white aerial mycelium. Good growth on asparagine-glycerol and yeast-glucose agar, colonies yellow, aerial mycelium limited. No growth on oatmeal agar. Optimum temperature 40–50°C, growth slight at 28°C and 60°C. Isolated from fresh and rotten manure.

Type strain: ATCC 19285 (MB A-18; CBS 277.66).

2. **Pseudonocardia spinosa** Schäfer in Henssen and Schäfer 1971, 31.[AL]

spi.no'sa. L. fem. adj. *spinosa* spiny.

Substrate mycelium compact mass, hyphae irregularly branched, constricted on septate, swellings common, septate. Aerial hyphae constricted or septate. Spores in substrate and aerial mycelium formed by budding or secondary septation of hyphal segments. Inner wall layer of hyphae and spores of varying thickness.

Grows slowly. Moderate growth on asparagine-glycerol agar, colonies yellow, abundant white aerial mycelium. Moderate to good growth on oatmeal agar. Optimum temperature 20–30°C, no growth at 37°C. Isolated from soil.

Type strain: ATCC 25924 (MB SF-1; CBS 818.70).

3. **Pseudonocardia compacta** Henssen, Happach-Kasan, Renner, Vobis 1983, 834.[VP]

com.pac'ta. L. fem. adj. *compacta* compact.

Substrate mycelium septate, densely branched, swellings common, in part multidivided. Aerial mycelium compact, hyphae constricted or septate, bearing apical and intercalary swellings. Septa formed frequently at a distance behind constrictions. All three types of spores formed in substrate and aerial mycelium. Spores of varying shape and length. Inner layer of hyphal wall of varying thickness. Inner layer of spore wall thick in mature spores.

Growth moderate to good on artificial soil agar, substrate mycelium scanty, white aerial mycelium abundant. Optimum temperature 20–30°C, no growth at 35°C. Isolated from soil.

Type strain: MB H-146 (CBS 160.82; DSM 43592).

Species Incertae Sedis

Production of antibiotics has been reported for two actinomycetes included in the genus "*Pseudonocardia fastidiosa*" (Celmer et al., US Patent 4,031,206, 1977), and *Pseudonocardia azurea* Omura et al. 1983 (effective publication: Omura et al., 1979). *P. azurea* has aerial hyphae with swellings and, therefore, resembles *P. compacta*. A reexamination of the type strains of the two species revealed that the two organisms deviate in hyphal growth and fine structure and should not be included in the genus *Pseudonocardia* (Henssen et al., 1983).

Genus **Oerskovia** *Prauser, Lechevalier and Lechevalier 1970, 534; emended Lechevalier 1972, 263*[AL]

HUBERT A. LECHEVALIER AND MARY P. LECHEVALIER

Oer.sko′vi.a. M.L. fem. n. *Oerskovia*, named after Jeppe Ørskov, Danish microbiologist.

Extensively branching vegetative hyphae, about 0.5 μm in diameter, growing on the surface of and penetrating into agar media, **breaking up** into rod-shaped, **motile,** flagellate elements. Growth appears coryneform to bacteroid in smears. **No aerial mycelium** formed. **Mesophilic. Gram-positive** with part of the thallus becoming Gram-negative with age, nonacid fast, catalase-positive when grown aerobically. **Facultatively anaerobic** on trypticase-soy medium; catalase-negative when grown anaerobically. Glucose metabolized both oxidatively and fermentatively. Rods monotrichous when small and peritrichous when long. Cell wall of type VI (lysine as principal diamino acid) + major amounts of galactose; aspartic acid may be absent. Phospholipids of type PV with phospholipid fatty acids of type 1. Menaquinones of MK-9(H$_4$) type. The mol% G + C of the DNA is 70.5–75(T_m) (Sukapure et al., 1970). **Found in soil,** decaying plant materials, brewery sewage (Kaneko et al., 1969), aluminum hydroxide gels and clinical specimens, including blood samples.

Type species: *Oerskovia turbata* (Erikson) Prauser, Lechevalier and Lechevalier 1970, 534; emended Lechevalier 1972, 263.

Further Descriptive Information

Nonmotile *Oerskovia*-like strains (NMOs) can be isolated from soil, clinical specimens and other substrates. The two described species of *Oerskovia* and two types of NMOs so far isolated have the following properties in common. Most strains are yellow; some are buff-colored. A few form amorphous reddish crystals in media amended with soil extract. No soluble pigments have been observed. They have extensively branching vegetative hyphae that break up into long or very short rod-shaped elements which are either motile (*Oerskovia*) or nonmotile (NMOs). Colonies may be dull or glistening and develop dense centers with a filamentous fringe (Fig. 17.14). This fringe may be barely visible in age or on rich media. The cells stain Gram-positive

to Gram-variable and have a typical Gram-positive peptidoglycan composition with or without galactose (type VI).

Oerskoviae and NMOs produce nitrite from nitrate. They hydrolyze gelatin and starch; they utilize acetate, lactate and pyruvate but not benzoate, citrate, malate, succinate and tartrate. Acid is produced from arabinose, cellobiose, dextrin, fructose, galactose, glucose, glycerol, glycogen, α-methyl-D-glucoside, lactose, maltose, mannose, salicin, sucrose, trehalose, xylose, and β-methyl-D-xyloside. No acid is formed from adonitol, erythritol, and sorbose. DNase and β-D-galactosidase are produced but not cytochrome oxidase. No growth takes place in lysozyme (0.05%) broth. These organisms survive exposure for 8 h at 50°C and 4 h at 60°C. Neither sulfate, nitrate nor fumarate can be used as electron acceptors.

Strains of *Oerskovia* do not contain mycolic acids and have a phospholipid composition of type PV (phosphatidylglycerol, phospholipids of unknown structure containing glucosamine, phosphatidylinositol and diphosphatidylglycerol). The phospholipid fatty acids are of type 1 (principal components branched chain fatty acids of the *anteiso/iso* series), (Lechevalier et al., 1977), here specifically being *anteiso*-C$_{15}$ with lesser amounts of *anteiso*-C$_{17}$ and *iso*-C$_{16}$. Menaquinones are MK-9(H$_2$,H$_4$) (Minnikin et al., 1978).

In Table 17.18 are listed the properties that differentiate the two described species of *Oerskovia* and the two types of NMOs which have been recognized.

Enzymatic activity of oerskoviae and NMOs

Oerskoviae and NMO are prolific producers of a variety of useful enzymes. These include α-mannanases, β-(1,3)-glucanases, β-(1,6)-glucanase, β-glucosidase, proteinases, chitinases (Mann et al., 1978; Scott and Schekman, 1980), dextranase (Hayward and Sly, 1976) and keratinase (Goodfellow, 1971). Strains of these organisms degrade the walls

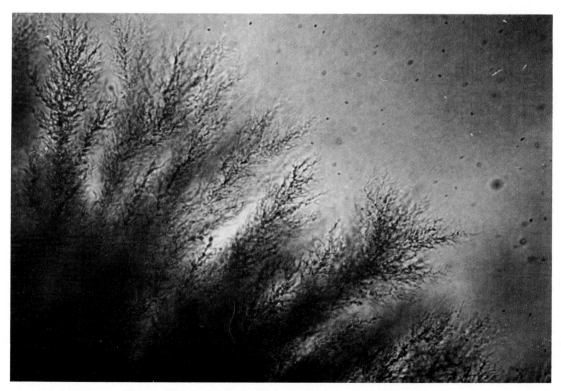

Figure 17.14. Edge of a colony of *Oerskovia turbata* (× 250).

Table 17.18.

Differential characteristics of the species of the genus **Oerskovia** *and of two types of nonmotile* **Oerskovia**-*like strains (NMOs)*[a]

Characteristics	O. turbata	O. xanthineolytica	NMOs type A	NMOs type B
Motility	+	+	−	−
Cell wall	Type VI[b] plus galactose	Type VI[b] plus galactose	Type VI[b] plus galactose (variable)	Type VI[b]
Peptidoglycan type	L-Lys-L-Thr-β-D-Asp[c] or L-Lys-L-Thr-γ-D-Glu[d]	L-Lys-D-Ser-β-D-Asp[c]		
Cytochrome a_1	+[c]	−[c]		
Growth at 42°C	−	+[d]	− (66)[e]	−
Dissimilation of				
Casein	+	+	− (66)	−
Xanthine	−	+	+	−
Hypoxanthine	−	+	+	−
Adenine	+[f]	+[g]	+[g]	+[f]
Urea	− (86)	− (88)	− (84)	+
Production of				
Cellulase	−	−	+	+
Phosphatase (24 h)	−	+	+	−
Acid from				
Melibiose	− (86)	+[d]	+ (66)	−
Raffinose	−	+[d]	+ (66)	V
Sorbitol	− (14)	+[d]	− (66)	+

[a] Symbols: see Table 17.12; also V, variable.

[b] Lechevalier, H. A. and M. P. Lechevalier, 1981.

[c] Seidl et al., 1980.

[d] Considered variable by Sottnek et al., 1977.

[e] Numbers in parentheses indicate percentages of positive or negative strains in variable tests.

[f] Transformed to hypoxanthine (M. P. Lechevalier, unpublished).

[g] Transformed to 8-hydroxyadenine (Lechevalier et al., 1982).

of both live and dead cells of yeast but there is considerable variation in lytic power from strain to strain. Oerskoviae show chemotactic activity toward yeasts and may be considered true predators of these organisms. The production of yeast-lytic enzymes (principally β-(1,3)-glucanases plus proteinase) is repressed by glucose (Mann et al., 1978; Scott and Schekman, 1980). Crude enzyme preparations from *O. xanthineolytica* are widely used for the preparation of yeast sphaeroplasts.

Oerskoviae and NMOs as pathogens

Strains of *Oerskovia* have been isolated from various clinical specimens. Sottnek et al. (1977) characterized 35 such isolates and found both species of *Oerskovia* to be represented. Reller et al. (1975) implicated *O. turbata* in a case of endocarditis after homograft replacement of an aortic valve and Cruickshank and his colleagues (1979) reported on a strain of *Oerskovia* strongly suspected of causing pyonephrosis. Their description, however, indicates that the suspected agent was an NMO. The best regimen for treatment of the infection with *O. turbata* was reported to be a prolonged course of high doses of sulfamethoxazole-trimethoprim combined with amphicillin or amoxicillin (Reller et al., 1975).

Isolation

Oerskoviae are readily isolated on a variety of media ranging from such nutrient-poor media as "tap water agar" (1.5% crude agar in tap water) to very rich media such as trypticase-soy or blood agar. They grow readily at 20–37°C usually in 24–48 h; some will grow at 42°C.

Maintenance

Good maintenance media include nutrient agar (with or without added soil extract) and trypticase soy agar. Storage for 3–4 months at 4°C between transfers has proved satisfactory. The strains may be lyophilized in skim milk for preservation.

Differentiation of genus **Oerskovia** *from other taxa*

Oerskoviae and NMOs may be differentiated from genera of the nonfilamentous bacteria on the basis of the hyphal fringe of undisturbed colonies of the former growing on nutrient agar. They may be distinguished from other nonsporing actinomycetes and their relatives by the characters given in Table 17.19.

Taxonomic position of oerskoviae

Oerskoviae are actinomycetes with many characteristics of rod-shaped bacteria and their classification has presented many difficulties. They have been placed in various genera such as *Nocardia* (Erikson, 1954), *Corynebacterium* (Sottnek et al., 1977), *Arthrobacter* (Scott and Schekman, 1980) and *Cellulomonas* (Jones and Bradley, 1964; Stackebrandt et al., 1982). Until recently, they have been considered as forming one of the genera of the *Actinomycetales* "in search of a family" (Lechevalier and Lechevalier, 1981b), and were listed under *Incertae Sedis* in the eighth edition of the *Manual*. On the basis of DNA-DNA reassociation studies supported by comparative analyses of the ribosomal 16S-RNA found in oerskoviae and cellulomonads, Stackebrandt et al. (1980, 1982) have proposed that species of *Oerskovia* be transferred to the genus *Cellulomonas*. Cellulomonads, oerskoviae and NMOs are facultative anaerobes and have the same phospholipid and menaquinone patterns. (Lechevalier, M. P., et al., 1981; Yamada et al., 1976) and are undoubtedly related phylogenetically. However, on the basis of differences in morphology (oerskoviae being hyphal and cellulomonads not), chemistry of the cell wall (oerskoviae contain lysine and cellulomonads, ornithine, in their peptidoglycan), and physiology (cellulomonads produce cellulase and oerskoviae do not), the two groups are treated here as belonging to different but closely related genera. NMOs are undoubtedly the bridging group. "*Cellulomonas cartae*" and "*Nocardia cellulans*" are both NMOs of type A (M. P. Lechevalier, unpublished) and are closely related to *O. xanthineolytica* (Stackebrandt, 1982) as witnessed by their identical peptidoglycan types.

Table 17.19.
Characters differentiating oerskoviae and nonmotile **Oerskovia***-like strains (NMOs) from other actinomycetes and related bacteria[a]*

Taxon	Branching mycelium[b]	Motility	Relation to oxygen[c]	Catalase reaction	Metabolism of glucose[d]	Cell wall type[e]	Phospholipid type[e]
Actinomyces	T	−	A, F, An	−/+	O/F	VI, V	PII
Agromyces	T	−	A, M	−	O	VII	PI
Arachnia	T	−	A, M	−	O/F	I	
Arthrobacter	N	−/+	A	+	O	I, VI	PI/PII
Bacterionema	N, T	−	A, F	+	O/F	IV	PI
Bifidobacterium	N	−	An	−	F	VIII	PI
Cellulomonas	N	−/+	A, F	−/+	O/F	VIII,	PV
Corynebacterium							
Animal spp.	N	−	A, F	+	O/F	IV	PI
Plant spp.	N	−/+	A	+	O	IV, VI, VII	PI, PII
pyogenes	N	−	A, F	−	O/F	VI	PI
Dermatophilus	T	+	A, M	+	O	III	PI
Geodermatophilus	N, T	+	A	+	O	III	PII
Intrasporangium	P	−	A	+	O	I	PIV
Mycobacterium	N, T	−	A	+	O	IV	PII
NMOs	P, T	−	A, F	−/+	O/F	VI	PV
Nocardia	P, T	−	A	+	O	IV	PII
Nocardioides	P	−	A	+	O	I	PI
Oerskovia	P, T	+	A, F	−/+	O/F	VI	PV
Promicromonospora	P	−	A	+	O	VI	PV
Rhodococcus	P, T	−	A	+	O	IV	PII
Rothia	N, T	−	A, M	+	O/F	VI	PI

[a] Symbols: see Table 17.12 and footnotes to individual columns. Where two symbols appear, the first indicates the commoner property, i.e. −/+, 11–50% positive; +/−, 51–89% positive.
[b] T, transient; N, none; P, persistent.
[c] A, aerobic; F, facultative; An, anaerobic; M, microaerophilic.
[d] O, oxidative; F, fermentative.
[e] According to Lechevalier and Lechevalier, 1981.

List of the species of the genus **Oerskovia**

1. **Oerskovia turbata** (Erikson) Prauser, Lechevalier, and Lechevalier 1970, 534.[AL] (Emended Lechevalier 1972, 263.)

tur.ba′ta. L. fem. adj. *turbata*, agitated.

Synonym: "*Nocardia turbata*" Erikson 1954, 206. *Listeria denitrificans* ATCC 14870 is a closely related species (Sottnek et al., 1977). Strains of *O. turbata* dissimilate adenine to hypoxanthine. (M. P. Lechevalier, unpublished).

Type strain: ATCC 25835.

2. **Oerskovia xanthineolytica** Lechevalier, 1972, 264.[AL]

xan.thi.neo′.ly.ti.ca from xanthine (purine); Gr. adj. lytos soluble; M.L. fem. adj. *xanthineolytica* dissolving xanthine.

Synonym: "*Arthrobacter luteus*" Kaneko, et al. 1969, 322.

Properties as given above and in Table 17.18.

As indicated in Table 17.18, xanthine and hypoxanthine are lysed by members of this species. Although no novel products of dissimilation of these compounds have been found, (xanthine is oxidized to uric acid), *O. xanthineolytica* strains transform adenine to 8-hydroxyadenine, a novel microbial product (Lechevalier, et al., 1982).

Type strain: ATCC 27402.

Comments on **Mycoplana**

The genus *Mycoplana* (Gray and Thornton, 1928) has received little subsequent study and its taxonomic position is uncertain. It was suggested by Sukapure et al. (1970) that it was related to the genus *Oerskovia* and it is included here pending further investigation. Gray and Thornton described two species, *Mycoplana dimorpha* ATCC 4279 (the type species) and *M. bullata* ATCC 4278, both of which are still extant.

The ability of these strains to form branching filaments prior to their fragmentation into motile, irregular rods has led several workers to place the genus in the order Actinomycetales (Sukapure et al., 1970; Cross and Goodfellow, 1973; Lechevalier and Lechevalier, 1981a, b). However, *Mycoplana* differs from other actinomycetes in being Gram-negative and having a cell wall of Gram-negative type, containing *meso*-diaminopimelic and numerous amino acids (Sukapure et al, 1970; Lechevalier and Lechevalier, 1981a). In the seventh edition of *Bergey's Manual* it was placed in the *Pseudomonadaceae* (Breed and Smith, 1957) and it was omitted from the eighth edition of the *Manual* (Buchanan and Gibbons, 1974). Most strains of *Mycoplana* which were tested showed nitrogenase activity (Pearson et al., 1982) and it was concluded from DNA-RNA hybridization studies that *M. dimorpha* and *M. bullata* were remote relatives of the family *Rhizobiaceae* (De Smedt and De Ley, 1977).

A description, based on the observations of Gray and Thornton (1928) and Sukapure et al. (1970) was provided by Lechevalier and Lechevalier (1981b).

Branching filaments breaking into irregular rods (0.5–0.1 μm wide × 1.25–4.5 μm long) which bear subpolar tufts of flagella. Pili may be observed on motile cells. Gram-negative, nonacid-fast. Murein contains *meso*-DAP; ribose and glycerol present in whole-cell hydrolysates. DNA G + C content, 64–69 mol%. Minimum growth temperature 24°C, maximum 42°C. No growth anaerobically. Catalase-positive. No acid from carbohydrates. No nitrite from nitrate. Gelatin liquefaction variable. Killed by 4 h at 60°C but not by 8 h at 50°C. The main difference between the two species is that *M. dimorpha* hydrolyzes starch while *M. bullata* does not.

Genus **Saccharopolyspora** *Lacey and Goodfellow 1975, 77*[AL]

J. LACEY

Sac′cha.ro.po′ly.spo.ra. M.L. n. *Saccharum* generic name of sugar cane; Gr. adj. *polus* many; Gr. n. *spora* a seed. M.L. fem. n. *Saccharopolyspora* the many spored (organism) from sugar cane.

Substrate mycelium well-developed, branched, septate 0.4–0.6 μm in diameter, **fragmenting into rod-shaped elements** about 1.0 × 0.5 μm, more often in older parts of the colony and seldom near the growing margins. **Aerial mycelium** 0.5–0.7 μm in diameter, straight or in spirals, characteristically **segmented into beadlike chains of spores,** 0.7–1.3 × 0.5–0.7 μm, **usually separated by lengths of "empty" hypha and retained in a sheath.** Gram-positive, nonacid-fast aerobic. **Colonies thin, raised or convex, slightly wrinkled, mucoid or gelatinous** in appearance with **aerial mycelium sparse, often produced in tufts** and mostly in the older parts (Fig. 17.15). **Able to utilize many organic compounds** as sole sources of carbon for energy and growth, **to degrade adenine** and other substrates; **resistant to many antibiotics** but **susceptible to lysozyme.** The mol% G + C of the DNA is 77 (T_m).

Type species: *Saccharopolyspora hirsuta* Lacey and Goodfellow 1975, 78.

Further descriptive information

Although some hyphae fragment like those of *Nocardia* species, producing long chains of cells in angular opposition (Fig. 17.16), many remain stable (Fig. 17.17). Fragmented hyphae are most abundant in the older parts of colonies but are still usually accompanied by stable hyphae. Aerial hyphae mostly arise in tufts which soon become differentiated into spore chains. These form loops and loose spirals (Fig. 17.18) but in older parts of the colony may be long and straight between tufts (Fig. 17.19). The spore chains are of indeterminate length and the spores are characteristically separated by short lengths of apparently empty hyphae giving a beadlike appearance.

The spores are round to oval, 0.7–1.3 × 0.5–0.7 μm and covered by a sheath. In *S. hirsuta,* the only species, this sheath carries tufts of long straight or curved, brittle hairs (Fig. 17.20). The morphology of the hairs is best seen on lengths of empty sheath (Fig. 17.21) or by scanning electron microscopy (Fig. 17.22). The surface of the sheath between tufts of hairs is smooth.

The cell walls contain major amounts of *meso*-diaminopimelic acid, galactose and arabinose (type IV, Becker et al., 1965). Whole-organism methanolysates yield no mycolic acids (Minnikin et al., 1975). Phos-

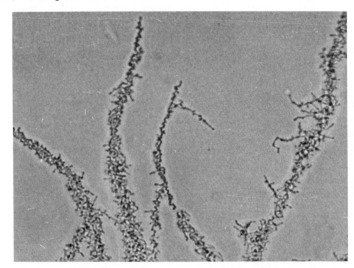

Figure 17.16. Fragmentation of substrate mycelium. Glycerol-asparagine agar, incubation 40°C. (× 500). (Reproduced with permission from J. Lacey and M. Goodfellow, Journal of General Microbiology *88:* 75–85, ©Society for General Microbiology, 1975.)

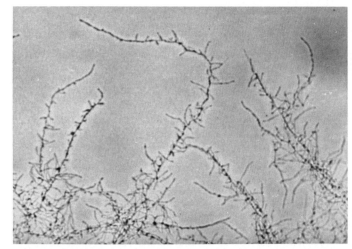

Figure 17.17. Morphology of substrate mycelium. Glycerol-asparagine agar, incubation 40°C (× 550). (Reproduced with permission from J. Lacey and M. Goodfellow, Journal of General Microbiology *88:* 75–85, ©Society for General Microbiology, 1975.)

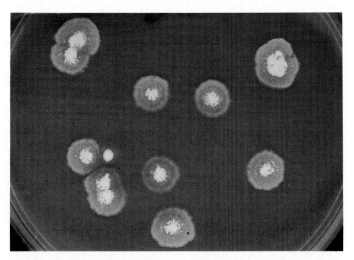

Figure 17.15. Colonies of *Saccharopolyspora hirsuta.* Half-strength nutrient agar, incubation 37°C. (× 1). (Reproduced with permission from J. Lacey and M. Goodfellow, Journal of General Microbiology *88:* 75–85, ©Society for General Microbiology, 1975.)

phatidylcholine is the diagnostic phospholipid found in cells (type III, Lechevalier et al., 1981). Also present are phosphatidylinositol and phosphatidylmethylethanolamine. *S. hirsuta* contains tetra-hydrogenated menaquinones with nine isoprene units as a major component (Collins, 1982).

In thin section, hyphae (Fig. 17.23) are bounded by a wall 22–30 nm thick. Within this, a typical unit membrane encloses granular cytoplasm with axial diffuse nuclear material. Electron-transparent vacuoles, up to 0.3 μm in diameter and resembling lipid accumulations in other nocardioform actinomycetes (Williams et al., 1976), were sometimes abundant. Also occasionally present are electron-dense granules,

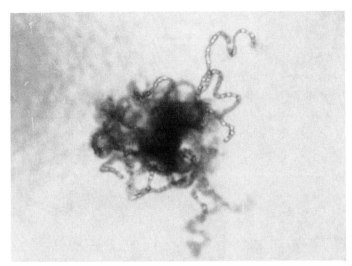

Figure 17.18. Spore chains on aerial mycelium showing tufted appearance and typical curved chains. Half-strength nutrient agar, incubation 40°C (× 800). (Reproduced with permission from J. Lacey and M. Goodfellow, Journal of General Microbiology 88: 75–85, ©Society for General Microbiology, 1975.)

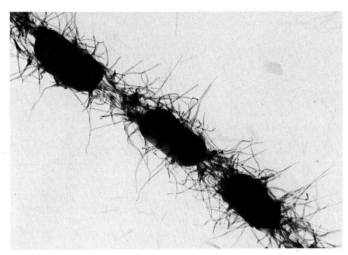

Figure 17.20. Electron micrograph of spore chain (× 18,400). (Reproduced with permission from J. Lacey and M. Goodfellow, Journal of General Microbiology 88: 75–85, ©Society for General Microbiology, 1975.)

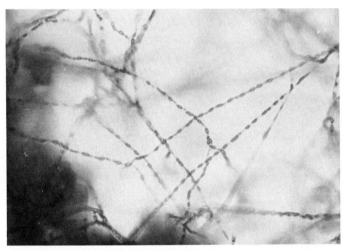

Figure 17.19. Straight spore chains from older parts of colonies. V-8 juice agar, incubation 37°C (× 650). (Reproduced with permission from J. Lacey and M. Goodfellow, Journal of General Microbiology 88: 75–85, ©Society for General Microbiology, 1975.)

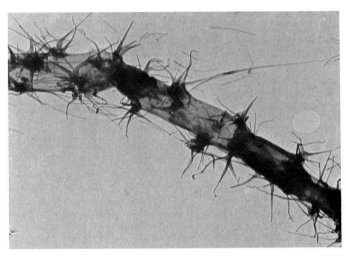

Figure 17.21. Electron micrograph of spore sheath showing tufted production of hairs (× 17,600). (Reproduced with permission from J. Lacey and M. Goodfellow, Journal of General Microbiology 88: 75–85, ©Society for General Microbiology, 1975.)

up to 0.1 μm in diameter, resembling polyphosphate or metachromatic granules. Septation occurs by double ingrowth of the wall leading to fragmentation (type II, Williams et al., 1973). This may be associated with lamellar mesosomes up to 0.25 μm in diameter.

The sheath surrounding the spores is 18–36 nm thick. It carries tufts of structureless hairs, triangular and 0.2–0.3 μm across at the base, which extend into apical filaments about 20 nm in diameter. Spore walls are thickened uniformly to 50–60 nm but their internal structure resembles that of hyphae although they lack many vacuoles (Fig. 17.24).

Colonies of *S. hirsuta* grow to about 1 cm in diameter in 7 days at 40°C with a central area of white aerial mycelium on an almost colorless substrate mycelium (Fig. 17.10). Good growth is obtained on yeast extract-malt extract agar and V-8 vegetable juice agar and a yellow soluble pigment is produced on the former.

S. hirsuta can utilize the following as sole sources of carbon for energy and growth: adonitol, cellobiose, erythritol, fructose, galactose, glucose, glycerol, inositol, lactose, maltose, mannitol, mannose, α-methyl-D-glucoside, β-methyl-D-glucoside, raffinose, rhamnose, sorbitol, sucrose, trehalose, xylose, acetate, benzoate, butyrate, citrate, fumarate, H-malate, succinate, sebacic acid and testosterone, but not arabinose, melezitose, salicin, tartrate or adipic acid. Isolates vary in their ability to utilize propionate and pyruvate. *S. hirsuta* can also degrade adenine, aesculin, casein, elastin, hypoxanthine, keratin, tyrosine, urea and xanthine but not xylan. However, it is sensitive to lysozyme. Growth occurs on agar media between 25 and 50°C with an optimum at about 37–40°C. There is no growth at 10°C. Aerial mycelium is produced only close to the optimum temperature.

Although a single strain of *S. hirsuta* was loosely associated with *Streptomyces* species in a numerical phenetic study (Williams et al., 1981, 1983), it was resistant to phages isolated from *Streptomyces* species and other wall chemotype I taxa (Wellington and Williams, 1981).

S. hirsuta is tolerant of a wide range of antibiotics (Table 17.20). However nearly all strains are susceptible to 500 μg/ml solutions of

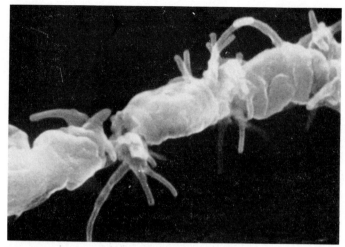

Figure 17.22.   Scanning electron micrograph of spores (× 15,000). (Reproduced with permission from J. Lacey and M. Goodfellow, Journal of General Microbiology *88:* 75–85, ©Society for General Microbiology, 1975.)

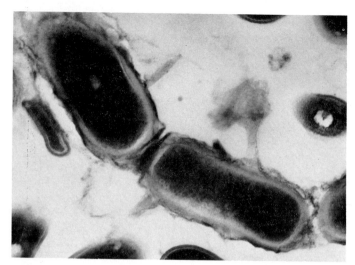

Figure 17.24.   Longitudinal section of mature spore chain showing sheath and hair bases (× 34,000). (Reproduced with permission from J. Lacey and M. Goodfellow, Journal of General Microbiology *88:* 75–85, ©Society for General Microbiology, 1975.)

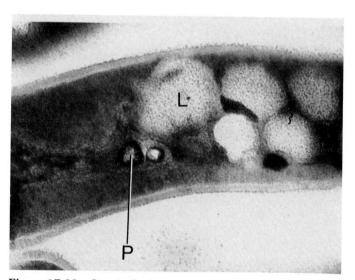

Figure 17.23.   Longitudinal section of hypha showing possible lipid accumulation (*L*) and polyphosphate granules (*P*) (× 70,000). (Reproduced with permission from J. Lacey and M. Goodfellow, Journal of General Microbiology *88:* 75–85, ©Society for General Microbiology, 1975.)

Table 17.20.
Antibiotic sensitivity of **S. hirsuta**[a]

Antibiotic	Concentration[b]	S. hirsuta
	(µg/ml)	
Gentamycin	100	−
Kanamycin	10	−
Streptomycin	100	−
Neomycin	50	−
Tobramycin	100	−
Rifampicin	50	−
Erythromycin	50	−
Fusidic acid	100	d
Minocycline	50	d
Vancomycin	50	d
Dapsone	500	+
Septrin	500	+

[a] Symbols are −, tolerant; +, susceptible; and d, 11–89% of strains are positive.
[b] Concentrations of solution used to soak filter-paper disks.

dapsone and septrin on filter paper disks and most to 50 µg minocycline and 100 µg fusidic acid/ml solution (Lacey and Goodfellow, 1975).

There is no evidence of pathogenicity but *S. hirsuta* degrades elastin, a property which in *Actinomadura* and *Nocardia* species indicates an ability to cause mycetoma.

S. hirsuta is known only from moldy sugar cane bagasse that has heated spontaneously during storage. It was found in 12% of samples originating from Puerto Rico, Trinidad, Jamaica and India but exceeded 10⁵ colony-forming units/g dry weight in only 3% (Lacey, 1974).

Isolation

S. hirsuta has been isolated only from the airborne dust from sugar cane bagasse in a small wind tunnel using an Andersen sampler (Lacey, 1974; Lacey and Goodfellow, 1975). Half-strength nutrient agar (Oxoid, CM3) containing 50 µg cycloheximide/ml medium (Gregory and Lacey, 1963) has usually been used for isolation.

Maintenance procedures

S. hirsuta can be lyophilized by usual procedures.

Differentiation of the genus **Saccharopolyspora** *from related genera*

S. hirsuta can be distinguished from related taxa chiefly by the presence of aerial mycelium, the characteristic spores, its wall chemotype IV lacking mycolic acids, carbohydrate utilization pattern, hydrolysis of adenine, casein and elastin and resistance to certain antibiotics but not to lysozyme (Table 17.21).

Taxonomic Comments

Only one species of *Saccharopolyspora* has so far been recognized but the genus clearly has affinities with other genera in the *Nocardiaceae* and also with *Nocardiopsis* and *Streptomyces* although these differ in wall chemotype.

Isolates of *S. hirsuta* have been included in two numerical taxonomic studies. In one study with *Nocardia* species (Goodfellow, Alderson and Lacey, unpublished data), 31 *S. hirsuta* isolates formed a tight cluster defined at the 92% similarity level, uniting with *Nocardia* at only the 36% similarity level regardless of whether S_{SM} or S_J coeffients were used. In another study (Williams et al., 1981), a single isolate of *S. hirsuta* was loosely associated with *Streptomyces* species while studies of rRNA:DNA pairing showed that *S. hirsuta* was quite closely related to *Streptomyces* and *Nocardia* species phylogenetically (Mordarski et al., 1980, 1981).

The morphology and fine structure of spore chains of *S. hirsuta* show similarities with those of *Nocardiopsis dassonvillei* and *Micropolyspora faeni*. All show increased wall thickness in the spores particularly in the region of cross-walls. However, *M. faeni* has chains up to only 10 spores in length. The hairy sheaths of *S. hirsuta* spores differ in their appearance from *Streptomyces* spores both in the separation of spores in the chain, in the appearance of hairs and in their brittle nature. *S. hirsuta* and *N. dassonvillei* share many nutritional and hydrolytic characters but *S. hirsuta* shows fewer similarities with *Actinomadura* species.

Further studies of actinomycetes with wall chemotype IV may lead to changes in their taxonomy and modifications in the definition of some genera. *S. hirsuta* occupies an intermediate position between *Nocardia* and *Micropolyspora*. It has never been systematically compared with *Micropolyspora* or with species of *Nocardia*, such as *N. aerocolonigenes*, *N. autotrophica*, *N. mediterranea* and *N. orientalis*, which lack mycolic acids. *Saccharopolyspora* could perhaps provide a home for some of these species or, if a close relationship were found, it could possibly be absorbed into *Micropolyspora*.

Further Reading

Lacey, J. and M. Goodfellow. 1975. A novel actinomycete from sugar-cane bagasse, *Saccharopolyspora hirsuta* gen. et sp. nov. J. Gen. Microbiol. *88:* 75–85.

Mordarski, M., A. Tkacz, M. Goodfellow, K.P. Schaal, and G. Pulverer. 1981. Ribosomal ribonucleic acid similarities in the classification of actinomycetes. Zentralbl. Bakteriol. Parasitenk. Infektionskr. Hyg., Abt. I, Suppl. *11:* 79–85.

Williams, S.T., G.P. Sharples, J.A. Serrano, A.A. Serrano, and J. Lacey. 1976. The micromorphology and fine structure of nocardioform organisms. *In* Goodfellow, Brownell and Serrano (Editors), The Biology of the Nocardiae, Academic Press, London, pp. 102–140.

Table 17.21.

Characters differentiating **Saccharopolyspora** *from related genera*[a, b]

Characteristics	Saccharo-polyspora	Micro-polyspora[c]	Saccharo-monospora	Pseudo-nocardia[d]	Nocardia	Actino-madura	Nocardiopsis
Morphological characters							
Aerial hyphae	+	+	+	+	+	+	+
Arthrospores	+	+	+	+	d	D	+
Spore surface	Hairy	Smooth	Warty	Smooth	Smooth	Smooth or Warty	Smooth
Acid fast stain	−	−	−	−	d	−	−
Chemical characters							
Wall type	IV	IV	IV	IV	IV	III	III
Mycolic acid	−	−	−	−	+		
Hydrolysis tests							
Adenine	+	−	−	−	−	−	d
Casein	+	d	+	+	D	D	d
Elastin	+	−	+	+	D	D	+
Sole carbon sources							
Cellobiose	+	d	+	+	−	D	d
Erythritol	+	+	d	+	−	D	−
Lactose	+	+	+	+	−	D	−
α-methyl-D-glucoside	+	+	d	+	−	−	d
Raffinose	+	d	+	+	−	D	−
Sorbitol	+	−	d	+	−	−	−
Benzoate	+	N/D	N/D	N/D	−	−	−
Growth in the presence of							
Lysozyme	−	d	−	+	+	D	−
Gentamycin	+	−	d	+	D	D	−
Rifampicin	+	−	−	−	+	D	d
Streptomycin	+	−	d	−	D	D	d

[a] Symbols: +, 90% or more of strains are positive; −, 90% or more of strains are negative; d, 11–89% of strains are positive; D, different reactions in different taxa (species of a genus or genera of a family); ND, not determined.

[b] References: Lacey and Goodfellow, 1975; Goodfellow and Alderson, 1977; Arden-Jones et al., 1979; Goodfellow et al., 1979; Athalye, 1981; Goodfellow and Pirouz, 1982.

[c] Results given for *M. faeni* and *M. rectivirgula* (*Faenia rectivirgula*) only.

[d] *Pseudonocardia thermophila* (type strain) only.

List of species of the genus **Saccharopolyspora**

1. **Saccharopolyspora hirsuta** Lacey and Goodfellow 1975, 78.[AL]
hir.sut'a. L. adj. shaggy, bristly, with stiff hairs. See the generic

description and Table 17.20 for features of *S. hirsuta.*
Type strain: ATCC 27875.

GENUS INCERTAE SEDIS

Genus **Micropolyspora** *Lechevalier, Solotorovsky and McDurmont 1961, 11*[AL]

J. LACEY

(*Faenia* Kurup and Agre 1983, 664[VP])

Mic.ro.po′ly.spo.ra Gr. adj. *Micros* small; Gr. adj. *polus* many; Gr. n. *spora* a seed. M.L. fem. n. *Micropolyspora*,
the small many-spored (organism)

Substrate mycelium well-developed branched septate, 0.5–0.8 μm in diameter. **Aerial mycelium** 0.8–1.2 μm in diameter, rising from the substrate mycelium. **Spores in chains** up to 20 spores long, 0.7–1.5 μm long, **on both substrate and aerial hyphae** on short, unbranched lateral or terminal sporophores. **Spore formation basipetal.** Intercalary spores occasionally found. Gram-positive, nonacid-fast, aerobic. **Colonies slow-growing, raised, with entire or filamentous margins; aerial mycelium sparse.** Thermoduric, xerotolerant. **Able to utilize a wide range of organic compounds** as sole sources of carbon for energy and growth and to degrade a number of substrates. Resistant to some antibiotics but **susceptible to lysozyme. Wall peptidoglycan containing *meso*-diaminopimelic acid (*meso*-DAP), arabinose and galactose but not mycolic acid** (wall chemotype IV).

Type species: the status of the genus and its type species is pending consideration by the Judicial Commission (see below); therefore *Micropolyspora faeni* Cross, Maciver and Lacey 1968, 354 is treated here as the type and only species until a decision is reached.

Further Descriptive Information

Characteristically, branching of the substrate hyphae is almost at right angles with chains of spores mostly on short unbranched lateral and terminal sporophores (Fig. 17.25). Aerial hyphae are usually sparse, arising from the substrate hyphae in short tufts with spore chains both lateral and terminal (Fig. 17.26). Although spore chains may be up to 20 spores long (Fig. 17.27), they are usually shorter than 5 spores. The chains are generally straight and spores form basipetally (Dorokhova et al., 1970). Occasionally intercalary spores may be observed (Fig. 17.28). Spores are round to oval, 0.7–1.5 μm long with a smooth or irregularly roughened surface in electron micrographs (Fig. 17.29).

The wall peptidoglycan contains *meso*-DAP, arabinose and galactose as diagnostic constituents (wall chemotype IV) together with glutamic acid, alanine, glucosamine and muramic acid (Becker et al., 1965). Whole organism methanolysates yield no mycolic acids (Mordarska et al., 1972) but organisms are rich in *iso*- and *anteiso*-branched chain fatty acids (Kroppenstedt and Kutzner 1976, 1978) and have polar lipid contents characterized by large amounts of phosphatidylcholine, diphosphatidylglycerol, phosphatidylglycerol, phosphatidylinositol and phosphatidylmethylethanolamine (Lechevalier et al., 1977). The major menaquinones are tetra-hydrogenated with nine isoprene units but there are usually smaller amounts of hexa- and octa-hydrogenated menaquinones also (MK-9(H$_4$, H$_6$, H$_8$); Collins et al., 1977).

Two types of hyphae have been distinguished in thin section, one having walls 19–25 nm thick and the other 11–15 nm thick (Dorokhova et al., 1970). In general, the cell structure resembles that of other actinomycetes but in the thicker-walled cells the cytoplasm is uniformly fine-grained with a large nuclear zone extending the full length of the cell. In thinner-walled cells the cytoplasm is less compact and homogeneous and the nuclear zone appears as small areas of low density. Mesosomes are also less well developed than in the thicker-walled cells. Hyphae tend to autolyze during prolonged incubation at 55°C or at room temperature.

Spore chains are surrounded by a multilayered sheath, although this is less evident on spore chains formed on the substrate mycelium than

on those formed on the aerial mycelium (Dorokhova et al., 1969; Williams et al., 1976). The spores are covered by a wall 70–100 nm thick in which two layers may be distinguished (Figs. 17.30 and 17.31), differing in thickness and electron density. Additional thickening of the cross-walls usually occurs giving characteristic interspore pads (Dorokhova et al., 1969) (Fig. 17.32). These may sometimes be observed by light microscopy of stained preparations as conspicuous nonstaining areas (Cross et al., 1968) but they may break down as the spores mature (Dorokhova et al., 1969). Plasmodesmata have also been described within these interspore pads. The protoplast is separated from the wall by a membrane and contains small, dark, densely packed ribosomes. Mesosomes are well developed and often adjoin the nuclear material. Although the spores are characteristically round or oval, spores of irregular shape are often seen in sections.

Colonies of *M. faeni*, the only species presently recognized, grow to 5 mm in 7 days at 40–50°C. The substrate mycelium may be colorless, brownish yellow or orange-yellow. Aerial mycelium is white but often sparse or absent. Good growth is obtained on yeast extract-malt extract agar or casein hydrolysate agar (Cross et al., 1968) and aerial mycelium production may be enhanced by the addition of 5% (w/v) NaCl to the medium. Spore chain length tends to be greatest on modified Umezawa's medium (Arden-Jones et al., 1979). Usually no soluble pigment is produced, but some brown pigment has been observed on V-8 agar.

M. faeni sometimes gives inconsistent results in carbon utilization tests but most authors agree that it can utilize the following as sole sources of carbon for energy and growth: amygdalin, D-arabinose, cellobiose, dextrin, erythritol, fructose, galactose, glycerol, lactose, mannitol, mannose, ribose, starch, sucrose, trehalose and xylose but not L-arabinose, dulcitol, glucosamine, melibiose, melezitose, salicin and sorbose. Isolates differ in their ability to utilize adonitol, cellobiose, glycogen, inulin, maltose, inositol, raffinose, rhamnose and sorbitol. Esculin, gelatin, guanine, hypoxanthine, ribonucleic acid, Tween 20, Tween 80, and xanthine are degraded but not adenine, allantoin, cellulose, chitin, elastin, keratin, starch, tributyrin or xylan. Degradation of arbutin, casein, deoxyribonucleic acid, hippurate, testosterone, tyrosine and urea have occasionally been demonstrated. Isolates of *M. faeni* produce catalase, galactosidase, glucosidase and phosphatase but not oxidase or prodiginins. Nitrates may be reduced to nitrites by some strains. Melanin is not produced from tyrosine nor acid from glucose but *M. faeni* is sensitive to lysozyme. Growth occurs between about 30 and 63°C with 50–55°C the optimum. The temperature range for growth may differ with strain, growth conditions and substrate (Cross et al., 1968; Arden-Jones et al., 1979; Kurup, 1981; Goodfellow and Pirouz, 1982).

Phages have been isolated from *Micropolyspora* isolates by Prauser and Momirova (1970) and by Kurup and Heinzen (1978). The latter were examined by electron microscopy and shown to have hexagonal heads 50–60 nm in diameter and long tails 7–8 × 132–145 nm. Their host range was generally restricted to *Micropolyspora* isolates but Kurup and Heinzen (1975) reported infection of some *Thermoactinomyces* cultures.

The antigenicity of *M. faeni* has received much attention since the organism was implicated as a cause of farmer's lung, the classic form of hypersensitivity pneumonitis (Pepys et al., 1963). Initially three

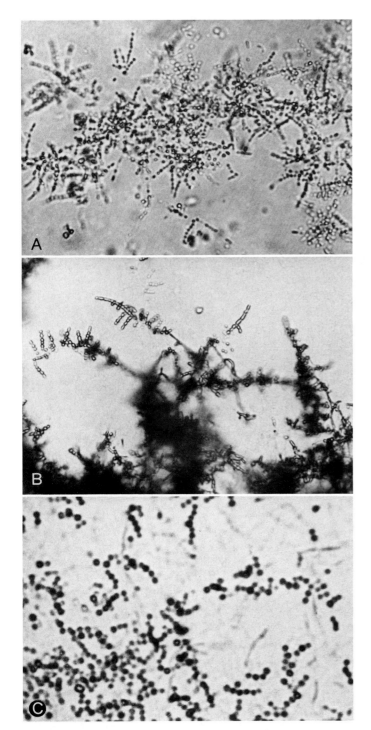

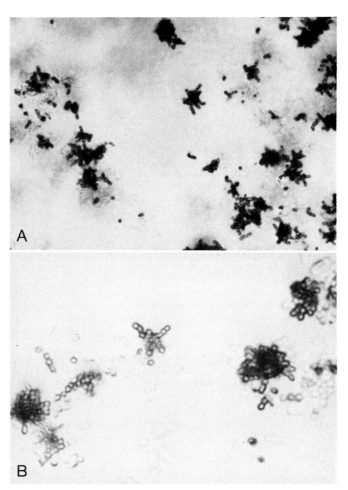

Figure 17.26. Aerial mycelium showing *A*, sparse, tufted appearance (× 390), and *B*, formation of spore chains (× 780). Half-strength nutrient agar, 55°C.

Figure 17.25. Morphology of the substrate mycelium of *Micropolyspora faeni. A*, appearance near growing margin (× 650). *B*, typical right-angle branching (× 650). *C*, spore chains in older part of colony (× 1300). Half-strength nutrient agar, 55°C.

precipitin arcs were recognized in gel diffusion and immunoelectrophoresis tests using extracts of *M. faeni* and sera from farmer's lung patients. Using gel filtration, absorption on columns of DEAE and elution with crossed immunoelectrophoresis and immunodiffusion, these three have been resolved into up to 75 individual antigenic components (Edward, 1972; Fletcher et al., 1970; Hollingdale, 1974; Kurup and Fink, 1977; Roberts et al., 1977; Walbaum et al., 1969, 1973;

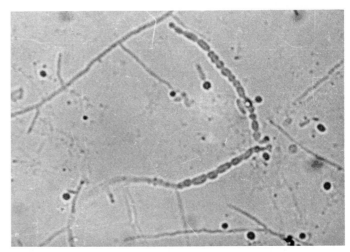

Figure 17.27. Intercalary spore formation or fragmentation of substrate mycelium in slide culture. Half-strength nutrient agar, 55°C (× 1,600). (Reproduced with permission from Cross, Lacey and Maciver, Journal of General Microbiology *50:* 351–359, © Society for General Microbiology, 1968.

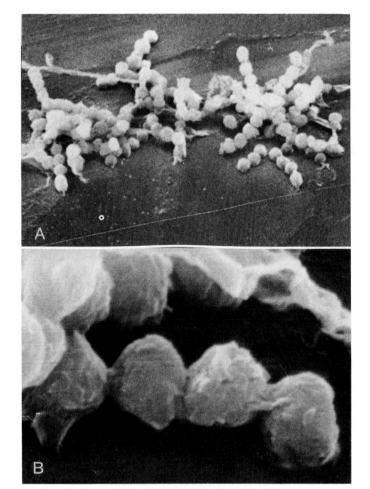

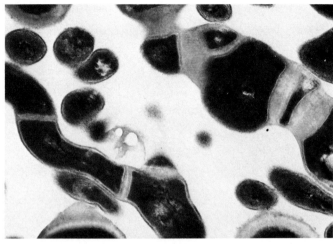

Figure 17.30. Sections of mycelium showing double septa in normal hyphae and irregularly thickened septa in enlarged hyphae or aberrant spore chains (× 25,000).

Figure 17.28. Scanning electron micrographs of *A*, sporulating hyphae (× 3000), and *B*, spores (× 13,100).

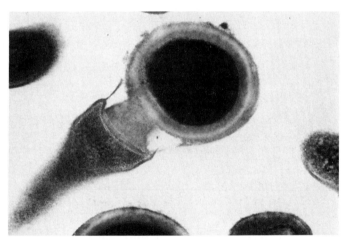

Figure 17.31. Longitudinal section through a developing spore (× 40,000).

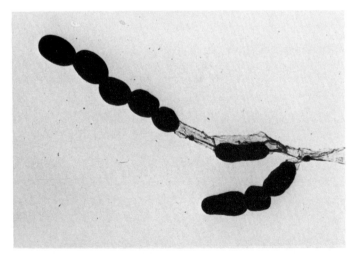

Figure 17.29. Transmission electron micrograph of spore chains (× 9600). (Reproduced with permission from Cross, Lacey and Maciver, Journal of General Microbiology *50:* 351–359, © Society for General Microbiology, 1968.

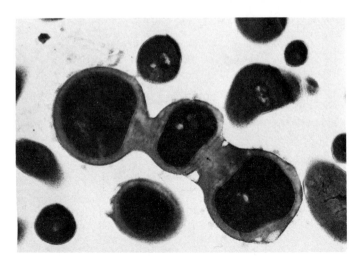

Figure 17.32. Longitudinal section through a spore chain showing interspore pads (× 25,000).

Kurup et al., 1981; Nicolet and Bannerman, 1975; Bannerman and Nicolet, 1976; Arden-Jones et al., 1979). Many of the components have been characterized and have molecular weights falling within the range $39-265 \times 10^3$ daltons. They contain protein and carbohydrate, the protein fraction yielding from 4–24 amino-acids. Sugar contents are from 2–98% and their half-lives from 8 h to >50 days. One group of heat-stable antigens consists of glycopeptides with few amino acids and a large sugar content. The remainder, some of which are very heat-labile, are protein. Enzymes have been identified in these antigenic fractions including esterases, lipases, proteases, aminopeptidases, trypsin- and chymotrypsin-like enzymes, catalases, malic dehydrogenases and peroxidases. The antigen content of "metabolic" or substrate extracts is reported to differ from that of spores and mycelium and Walbaum et al. (1969) also found that the metabolic extract was more relevant in diagnosis of farmer's lung.

M. faeni tolerates few antibiotics. It is tolerant to lincomycin (at 20 µg/ml medium), neomycin (10 µg/ml) and benzylpenicillin (20 µg/ml) but only to low concentrations of cephaloridine (2 µg/ml), demeclocyclin (2 µg/ml), gentamycin (4 µg/ml) and tobramycin (1 µg/ml). No growth occurred in the presence of rifampicin (2 µg/ml) or streptomycin (4 µg/ml) (Athalye, 1981).

M. faeni is not known to cause infections but inhalation of its spores can cause farmer's lung (extrinsic allergic alveolitis) in sensitized subjects. This disease occurs widely in Europe, the United States, Canada, and Japan, and there has been a single, unconfirmed case in Sri Lanka. In western Scotland and the Orkneys it may effect up to 8.6% of farm workers (Grant et al., 1971) while in the United States 8.4% of Wisconsin dairy farmers had precipitins to actinomycetes, 86% of these to *M. faeni* (Roberts et al., 1976). Because precipitins are found in farmer's lung patients to some of the heat-labile antigens that may be destroyed during the spontaneous heating of molding hay, it has been suggested that spores may germinate in the lungs releasing these

antigens but this is not proven. The spores may also activate complement by the alternative pathway (Edwards, 1972; Edwards et al., 1974).

M. faeni occurs widely in spontaneously heating vegetable matter. It was first isolated from moldy hay which had been baled wetter than 35% water content and which had heated to 50–65°C (Gregory et al., 1963). Subsequently it has been found in straw, cereal grain, sugar cane bagasse, cotton bales, mushroom compost, soil and in the air over pastures (Lacey, 1978). Inoculum is widespread in grass at the time of cutting and further contamination may occur in stores from previous crops. Heating of the substrate is initiated by plant cells and mesophilic fungi and bacteria. *M. faeni* commences growth at only 30–35°C and largest numbers are found in hays that heat to about 60°C after baling at about 39% water content. The change in pH from 5.5–6 to 7–8 caused by fungal proteolysis probably favors actinomycete colonization although *M. faeni* can sometimes grow in hay without pretreatment (Gregory et al., 1963; Festenstein et al., 1965) *M. faeni* may survive in moist grain stored anaerobically in silos and its spores can survive up to 20 min at 70°C.

Isolation

M. faeni has sometimes been isolated by dilution and direct plating techniques but enumeration in hay and other vegetable matter is most reliably achieved using an Andersen sampler to isolate airborne spores in a small wind tunnel or sedimentation chamber (Gregory and Lacey, 1963; Lacey and Dutkiewicz, 1976a, b). Half-strength nutrient agar (Oxoid, CM3) or half-strength tryptone soya agar (Oxoid CM131) + 0.2% casein hydrolysate (Oxoid L41) both containing 50 µg cycloheximide/ml medium have usually been used for isolation with incubation at 40–55°C.

Maintenance Procedures

M. faeni can be lyophilized by usual procedures.

Differentiation of the genus **Micropolyspora** *from related genera*

Micropolyspora faeni can be distinguished from related taxa by the presence of short chains of spores on aerial and substrate mycelium, its wall chemotype IV lacking mycolic acids, carbohydrate utilization pattern, degradation of guanine, hypoxanthine and xanthine, growth on 10% NaCl and resistance to certain antibiotics but not to lysozyme (Table 17.22). *Nocardia brevicatena* also bears short chains of spores but the walls contain mycolic acids.

Taxonomic Comments

The nomenclatural history of species of the genus *Micropolyspora* contains much controversy which is still not completely resolved. The genus was created for actinomycetes bearing short chains of spores on both aerial and substrate mycelia. The type species was designated *M. brevicatena* Lechevalier, Solotorovsky and McDurmont 1961, 13. Later, Lechevalier (1968) requested the conservation of *Micropolyspora* against "*Micropolispora*" (Shchepkina 1940). The Judicial Commission of the International Committee on Systematic Bacteriology appear to have made no ruling on this request. However, the inclusion of *Micropolyspora* in the Approved Lists (Skerman et al., 1980) indicates de facto approval.

Other species were included in *Micropolyspora* subsequently. Those in the Approved Lists include *M. faeni* Cross, Maciver and Lacey 1968, 354 (first described as "*Thermopolyspora polyspora*" Corbaz, Gregory and Lacey, 1963); *M. angiospora* (Zhukova, Tsyganov and Morozov 1968, 728); *M. rectivirgula* (Krasil'nikov and Agre 1964, 106) Prauser and Momirova 1970, 220 (first described as "*Thermopolyspora rectivirgula*," Krasil'nikov and Agre 1964); and *M. internatus* Agre, Guzeva and Dorokhova 1974, 577. Additional species excluded from the Approved Lists (1980) or never validated include "*M. caesia*" (Kalakoutskii, 1964), "*M. thermovirida*" (Kosmachev, 1964), "*M. viridinigra*" and "*M. rubrobrunea*" (Krasil'nikov, Agre and El-Registan, 1968), "*M. coerulea*" (Preobrazhenskaya et al., 1973) and "*M. fascifera*" (Prauser, 1974).

The accepted concept of the genus *Micropolyspora* has remained essentially morphological although the possession of a wall chemotype IV (Lechevalier and Lechevalier, 1970) had become incorporated into the genus definition (Cross and Goodfellow, 1973). Consequently, the genus has become heterogeneous (Goodfellow and Pirouz, 1982). Three species have been found to have wall chemotype III. "*M. rubrobrunea*" and "*M. viridinigra*" were transferred to the genus *Excellospora* Agre and Guzeva 1975, 521 while *M. angiospora* may be a species of *Actinomadura* Lechevalier and Lechevalier 1970, 400 (Lechevalier in Kurup, 1981) or *Excellospora* (Lacey, Goodfellow and Alderson, 1978). Transfer of *M. internatus* and "*M. caesia*" to *Saccharomonospora* Nonomura and Ohara 1971, 899 has been proposed but not validated (Kurup, 1981) while "*M. coerulea*" and "*M. thermovirida*" also appear typical *Saccharomonospora* spp. However, there are no isolates of "*M. thermovirida*" extant so that its position cannot be clarified. Mycolic acids have been found in "*M. fascifera*" (Prauser 1978) but so far no transfer has been proposed while the presence of mycolic acids in *M. brevicatena* together with menaquinones, phospholipids and fatty acids consistent with *Nocardia* Trevisan, 1889, 9 (Collins et al., 1977; Lechevalier, De Bièrre and Lechevalier 1977; Kroppenstedt and Kutzner, 1978) have led to its reclassification as *Nocardia brevicatena* (Lechevalier, Solorovsky and McDurmont 1961, 13) Goodfellow and Pirouz 1982, 523. Finally, there is little doubt that the two remaining species, *M. faeni* and *M. rectivirgula*, are synonymous (Dorokhova et al., 1970; Prauser and Momirova, 1970; Arden-Jones et al., 1979; Kurup, 1981) despite considerable differences between their original descriptions (Krasil'nikov and Agre, 1964; Cross, Maciver and Lacey, 1968).

Thus *Micropolyspora* is a genus without a type species and so nomenclaturally invalid under Rule 37a of the International Code of Nomenclature of Bacteria (Lapage et al., 1975). Only one species remains within the original concept of the genus, for which the epithet *rectivirgula* has priority, although *faeni* is better known, as the cause of farmer's lung and other forms of hypersensitivity pneumonitis.

Table 17.22.

Characters differentiating **Micropolyspora** *from related genera[a, b]*

Characteristics	Micro-polyspora	Saccharo-polyspora	Saccharo-monospora[c]	Pseudo-nocardia[d]	Nocardia	Actinomadura	Nocardiopsis
Morphological characters: spores or fragmentation of							
Substrate hyphae	+	v	−	−	+	−	+
Aerial hyphae	+	+	+	+	d	D	+
Spore surface	Smooth	Hairy	Warty	Smooth	Smooth	Smooth or warty	Smooth
Chemical characters							
Wall type	IV	IV	IV	IV	IV	III	III
Mycolic acid	−	−	−	−	+	−	−
Hydrolysis tests							
Adenine	−	+	−	−	−	−	d
Guanine	+	+	−	−	D	D	+
Elastin	−	+	+	+	D	D	d
Hypoxanthine	+	+	−	−	D	D	+
Xanthine	+	+	−	−	D	−	+
Sole carbon sources							
Inositol	+	+	+	+	D	D	d
Lactose	+	+	+	+	−	D	−
Maltose	−	+	+	+	D	D	+
Growth in the presence of							
10% NaCl	+	+	−	−	−	D	+
Lincomycin[e]	+	+	d	+	+	D	+
Neomycin[e]	+	+	d	+	+	D	−
Penicillin[e]	+	−	d	−	+	D	d
Lysozyme	−	−	−	+	+	−	−
Growth at 50°C	+	d	d	+	−	D	−
Growth at 60°C	+	−	d	−	−	−	−

[a] Symbols: see Table 17.21; also v, strain instability.

[b] References: Lacey and Goodfellow, 1975; Arden-Jones et al., 1979; Goodfellow et al., 1979; Athalye 1981; Goodfellow and Pirouz, 1982; McCarthy and Cross, 1984.

[c] *S. viridis* only.

[d] *P. thermophila* (type strain only).

[e] Concentrations: lincomycin, 20 µg/ml medium for *Micropolyspora*, filter paper disks soaked in 100 µg/ml solution otherwise; neomycin, 10 µg/ml medium for *Micropolyspora*, filter paper disks soaked in 50 µg/ml otherwise; 20 µg/ml medium for *Micropolyspora*, filter paper disks soaked in 10 units/ml otherwise.

Resolution of this situation has given rise to much controversy and two proposals have recently been made, one based on a strict interpretation of the International Code of Nomenclature of Bacteria (Lapage et al., 1975) and the other on the Principles on which this Code is based. Kurup and Agre (1983) propose a new genus, *Faenia* with *F. rectivirgula* Kurup and Agre 1983, 664 as the type species while McCarthy et al., (1983) argue for conservation of *Micropolyspora* with *M. faeni* Cross, Maciver and Lacey 1968, 354 as the type species.

If the proposal to conserve *M. faeni* is accepted by the Judicial Commission, the type strain would be ATCC 15347 whereas if *F. rectivirigula* is preferred it would be ATCC 33515.

Further studies of actinomycetes with wall chemotype IV could lead to changes in taxonomy and modifications in the definition of some genera. The relationships of *Micropolyspora*, *Saccharopolyspora* Lacey and Goodfellow 1975, 78, *Saccharomonospora* and species of *Nocardia* lacking mycolic acids, such as *N. autotrophica* (Takamiya and Tubaki 1956, 59) Hirsch 1961 360; *N. mediterranei* (Margalith and Beretta, 1960) Thiemann, Zucco and Pelizza 1969, 148); *N. orientalis* (Pittenger and Brigham 1956, 642) Pridham 1970, 42; and "*N. aerocolonigenes*" (Shinobu and Kawato 1960, 215) Pridham 1970, which has a wall containing galactose and mannose but not arabinose (Gordon et al.,

1978), need to be established. A possible further species of *Micropolyspora*, "*M. hordei*," has been suggested by Hill and Lacey (1983), but is not yet described.

Further Reading

Arden-Jones, M.P., A.J. McCarthy and T. Cross, 1979. Taxonomic and serological studies on *Micropolyspora faeni* and *Micropolyspora* strains bearing the specific epithet *rectivirgula*. J. Gen. Microbiol. *115*: 343–354.

Cross, T. and M. Goodfellow, 1973. Taxonomy and classification of the actinomycete. *In* Skyes and Skinner (Editors) Actinomycetales: Characteristics and Practical Importance. Academic Press, London. pp. 11–112.

Cross, T., A. Maciver and J. Lacey, 1968. The thermophilic actinomycetes in mouldy hay: *Micropolyspora faeni* sp. nov. J. Gen. Microbiol. *50*: 351–359.

Kurup, V.P. and N.S. Agre, 1983. Transfer of *Micropolyspora rectivirgula* (Krasil'nikov and Agre 1964) Lechevalier, Lechevalier, and Becker 1966 to *Faenia* gen. nov. Int. J. Syst. Bacteriol. *33*: 663–665.

Lacey, J. 1978. Ecology of actinomycetes in fodders and related substrates. Zentralbl. Bakteriol. Parasitenkd. Infektionskr. Hyg. Abt. 1. Suppl. *6*: 161–170.

Lacey, J. 1981. Airborne actinomycete spores as respiratory allergens. Zentralbl. Bakteriol. Microbiol. Hyg. Abt. 1. Suppl *11*: 243–250.

McCarthy, A.J., T. Cross, J. Lacey and M. Goodfellow, 1983. Conservation of the name *Micropolyspora* Lechevalier, Solotorovsky, and McDurmont and designation of *Micropolyspora faeni* Cross, Maciver, and Lacey as the type species of the genus: request for an opinion. Int. J. Syst. Bacteriol. *33*: 430–433.

List of species of the genus **Micropolyspora**

1. **Micropolyspora faeni** Cross, Maciver, and Lacey 1968, 354.[AL] (*Faenia rectivirgula* (Krasil'nikov and Agre 1964, 106) Kurup and Agre 1983, 664; *Micropolyspora rectivirgula* (Krasil'nikov and Agre 1964, 106) Prauser and Momirova 1970, 220; *Thermopolyspora rectivirgula* Krasil'nikov and Agre 1964, 106; *Thermopolyspora polyspora* Corbaz, Gregory and Lacey 1963, 450 non Henssen 1957, 396.)

faeni. L. n. *faenum* hay; L. gen. n. *faeni* of hay.

For features of *M. faeni* refer to the generic description and Table 17.22.

Type strain: ATCC 15347.

Corbaz et al. (1963) originally identified strains from mouldy hay as "*Thermopolyspora polyspora*" (Henssen, 1957). Subsequently Henssen examined a culture which she considered distinct from "*T. polyspora*" but no strains of the latter were available for comparison. After further study, the strains from hay which had been implicated in farmer's lung (Pepys et al., 1963) were renamed *Micropolyspora faeni* (Cross et al., 1968). Strains of *M. rectivirgula* were also first identified as a "*Thermopolyspora*" species and described as producing colorless to slightly yellow colonies with abundant yellowish aerial mycelium and straight chains of smooth spores. This contrasted with the sparse white aerial

mycelium of *M. faeni*. The status of "*T. rectivirgula*" remained in doubt until Prauser and Momirova (1970) transferred this taxon to *Micropolyspora*.

Earlier Krasil'nikov and his co-workers had questioned the separate status of "*Thermopolyspora*" and its distinction from *Micropolyspora* (Krasil'nikov, 1964; Kalakoutskii et al., 1968). However, they still did not classify *rectivirgula* isolates with *Micropolyspora*, either when they first described it (Krasil'nikov and Agre, 1964) or when they studied it in more detail (Kalakoutskii et al., 1968). However, Lechevalier et al. (1966) showed "*T. polyspora*" to have wall chemotype IV and tentatively assigned this taxon to *Micropolyspora*. Then Dorokhova et al. (1969), in using binomial *M. rectivirgula*, finally indicated Russian acceptance that it did belong to this genus but still did not propose formal transfer.

Subsequent studies by Arden-Jones et al. (1979) and Kurup (1981) have confirmed the synonymy of *M. faeni* and *M. rectivirgula* and McCarthy et al. (1983) have proposed conservation of *M. faeni* to promote stability of nomenclature and avoid confusion while Kurup and Agre (1983) have described *Faenia rectivirgula* for this taxon. The problem of an acceptable nomenclature has still to be resolved by the Judicial Commission.

Genus **Promicromonospora** Krasil'nikov, Kalakoutskii and Kirillova 1961, 107[AL]

L. V. KALAKOUTSKII, N. S. AGRE, HELMUT PRAUSER, L. I. EVTUSHENKO

Pro.mi.cro.mo.no′spo.ra. Gr. pref. *pro* before, primordial; Gr. adj. *micros* small; Gr. adj. *monos* single, solitary; Gr. fem. n. *spora* a seed. M.L. fem. n. *Promicromonospora*; the genus name was coined to reflect the combination of traits then thought to be characteristic of the actinomycete form-genera *Proactinomyces* (the tendency of the mycelium to fragment) and *Micromonospora* (the formation of single spores on the substrate mycelium).

Branching septate **hyphae** (0.5–1.0 μm in diameter) growing on the surface of and penetrating into the agar, which **break up into fragments** of various size and shape. Fragmentation finally results in nonmotile, Y- or V-shaped, rodlike, coccoid, chlamydospore-like and other spore-shaped elements. All of them may give rise to new mycelia. **Growth pasty to leathery.** Aerial hyphae in different strains may vary in abundance (sometimes discernible only microscopically). These are straight to curved, sometimes sparsely branched, usually fragmented into rodlike or elongated coccoid elements. **Gram-positive, nonacid-fast, catalase-positive. Aerobic.** Chemoorganotrophic. Glucose metabolized oxidatively, rarely also fermentatively. **Mesophilic.** Utilize a wide range of C and N sources and possess a significant spectrum of hydrolytic activities. **Susceptible to taxon-specific phages;** not susceptible to phages of various sets which attack oerskoviae and other nocardioform organisms. **Cell wall chemotype VI (lysine** as principal diagnostic amino acid); peptidoglycan type A3α (L-lysine in position 3). **Mycolic acids lacking.** No wall teichoic acids found. Among fatty acids, branching ones of the *iso-* and *anteiso-* types (*i-, a* C_{15}:O) predominate. Diagnostic phospholipids represented by phosphatidylglycerol and an unidentified glucosamine-containing phospholipids. **Menaquinones of the MK-9(H$_4$) type. Mol% G + C of the DNA in the range of 70–75 (T_m).** Mainly **found in soils.**

Type species: *Promicromonospora citrea* Krasil'nikov, Kalakoutskii and Kirillova 1961, 107.

Further Descriptive Information

Since the genus is monotypic the following more detailed information, as well as the concise genus description given above, originate only from the study of the type species *Promicromonospora citrea*, i.e. from the study of up to 30 strains isolated by various workers from different soils. Colonies develop within 1–2 d, growth is more abundant on complex peptone-containing media. Some strains respond favorably to the addition of vitamins or require yeast extract for growth on synthetic media (Evtushenko et al., 1981). Colonies usually are concave to wrinkled, but are smooth and pasty with some strains. Colors vary from yellow to white.

The length of hyphae, extent of branching and persistence of the mycelial state depend on the particular strains, the media employed, and the conditions of cultivation. Branching may be dense (Fig. 17.33A) to loose (Fig. 17.33B). Fragmentation may begin after 8 h incubation. Sometimes it is not recognizable at all in situ on and/or below the agar surface. The mycelium frequently has a "barbed wire" appearance resulting from longitudinal growth of fragments after their separation.

The process of fragmentation is more pronounced in submerged shaken cultures on complex media as compared to cultures on solid mineral salt media. Fragmentation results in the formation of coccoid, rod-like, diphtheroid and chlamydospore-like elements; sometimes enlarged cells (up to 5 μm in diameter) may occur (Fig. 17.33C). Spore-shaped elements (Fig. 17.33D) are observed mainly in solid surface cultures. By virtue of their regular shape, positioning (often terminal on short side branches), refractility and response to staining, they were regarded as spores by some authors (Krasil'nikov et al., 1961; Luedemann, 1974). Aerial hyphae are produced in all strains after 2–4 d; suitable media are oatmeal agar (Shirling and Gottlieb, 1966) and peptone-corn-extract agar (Agre, 1964). This capability tends to disappear in the course of maintaining strains by continued serial transfers. In some strains, only extremely few and short aerial hyphae are detectable after thorough microscopical examination. Other strains show a mat of aerial hyphae visible only microscopically or just detectable to the naked eye. Only a few strains produce distinct aerial mycelium. Aerial hyphae are straight to curved and not branched or only sparsely so. On observation in situ, fragments or spore-shaped elements are not visible.

More than 90% of strains were able to utilize L-arabinose, D-galactose, cellobiose, D-fructose, D-maltose, mannose, raffinose, sucrose, trehalose and D-xylose, as a sole carbon source and produce acid from them; to assimilate glycerol, acetate, fumarate, malate, malonate and succinate. The strains hydrolyzed esculin, casein, gelatin, starch; possessed catalase, DNAase, nitrate-reductase, urease activities. They were able to grow in the presence of 5% (w/v) NaCl, 7% KI, 7% KBr, but were sensitive to 7% NaCl, 0.25% phenol, 0.01% thymol, 0.0001% crystal violet. Not a single strain utilized sorbose, aconitate, benzoate,

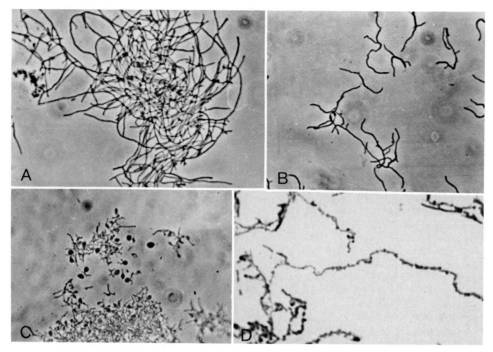

Figure 17.33. *Promicromonospora citrea. A–C,* submerged 2 d (28°C) growth in peptone-yeast extract broth. Phase contrast. *A,* strain VKM Ac 791, mycelium with well-developed hyphae (× 900). *B,* strain LL G 165, mycelium with short hyphae (× 900). *C,* strain VKM Ac 783. Cells of varying shape and size (× 500). *D,* strain VKM Ac 665, surface growth (2 d 28°C) on potato agar, crystalline violet stain. Spore-shaped cells (× 1200).

oxalate, hexadecanoic acid, paraffin, adenine, arbutin, hypoxanthine, lecithin. No growth in lysozyme broth. The following tests were variable (11–89% of strains were positive): utilization of D-arabinose, lactose, melibiose, L-rhamnose, adonitol, dulcitol, inositol, mannitol, sorbitol, citrate, formate, lactate, propionate, cellulose, Tween 20, Tween 40, Tween 60, xanthine; production of phosphatase; growth at 37°C, growth at pH 5.0.

No growth under anaerobic conditions was observed on a spectrum of complex and chemically-defined media with any of the 30 strains tested. If tryptic soy agar (Lechevalier, 1972) was employed, a very weak growth was detected with 2 strains under anaerobic conditions at 28°C. On yeast-peptone agar supplemented with 0.5% (w/v) $CaCO_3$ (Zviagintsev et al., 1980) neither of the strains tested produced clearing zones following aerobic incubation. The key enzymes of the Entner-Doudoroff pathway were not detected in type strain of *P. citrea* (Kersters and De Ley, 1968).

Cell wall chemotype (Lechevalier and Lechevalier, 1970) is VI, lysine being the diagnostic amino acid (Yamaguchi, 1965). Many strains contain galactose in the cell wall. The peptidoglycan type (Schleifer and Kandler, 1972) is A3α displaying L-Lys in position 3 and Ala₂ in the interpeptide bridge (Stackebrandt et al., 1983). Teichoic acids were not found in wall preparations (Evtushenko et al., 1984a). Branched fatty acids of the *anteiso-* and *iso-* series 12- and 13-methyltetradecanoic (*anteiso-* and *iso-*C_{15}:O) predominate in all preparations (Andreyev et al., 1983). The phospholipid type is PV, represented by phosphatidyl-glycerol and unknown glucosamine-containing phospholipids (Lechevalier et al., 1977). The phospholipid fatty acids are of type I, characterized by branched-chain fatty acids of the *anteiso-* and *iso-* series (Lechevalier et al., 1977). Tetrahydrogenated menaquinones having

nine isoprene units ($MK-9(H_4)$) are the predominating menaquinones (Collins and Jones, 1981). The mol% G + C of the DNA is in the range of 70–75 (Tsyganov et al., 1966; Yamaguchi, 1967; and unpublished results).

So far, six taxon-specific phages have been isolated from soils and used for taxonomic purposes. Most of the 40 *Promicromonospora* strains studied are susceptible, at least, to 3 of these phages (susceptibility to at least 1 phage being a rare exception). Promicromonosporae are not susceptible to any other phages, including those specific for *Oerskovia* spp. and related organisms, several nocardioforms and sporoactinomycetes. None of the *Promicromonospora* phages is effective against strains of other taxa (Prauser and Falta, 1968; Prauser, 1976, 1984).

Isolation

Most of the strains were isolated from soils and aluminium hydroxide gel antacid (Lechevalier, 1972). Among the various isolation media employed were several kinds of soil extract agar, an oatmeal agar (Prauser and Bergholz, 1974) and peptone-corn-extract agar (Agre, 1964).

Maintenance

The strains can be maintained by serial transfers (once in 3 months) on agar slants. The following media are recommended: peptone-corn extract-agar (Agre, 1964), complex organic agar number 79 of Prauser and Falta (1968) and oatmeal agar (Shirling and Gottlieb, 1966).

The strains survive lyophilization and maintenance above liquid nitrogen, applying routine procedures.

Differentiation of the genus **Promicromonospora** *from related organisms*

Promicromonospora seems to belong to a group of bacteria, which contains *Cellulomonas* (including *C. cartae*), *Oerskovia*, the so-called "nonmotile organisms" (NMOs of Lechevalier, 1972; Lechevalier and Lechevalier, 1981) as well as *Nocardia cellulans, Brevibacterium fer-* *mentans,* and "*Corynebacterium manihot*". The relationship follows from possession in common of a number of chemotaxonomic traits (mentioned above), comparison of results of 16S-rRNA cataloging (Stackebrandt et al., 1983; Stackebrandt and Schleifer, 1984), as well

Table 17.23.
Characters distinguishing **Promicromonospora** *from related organisms*[a]

Taxon	Formation of aerial hyphae	Susceptible to		Anaerobic growth on TSA[c]	Peptidoglycan type
		P phages[b]	O phages[b]		
Promicromonospora citrea	+	+	−	−	L-Lys-Ala$_2$[d]
Oerskovia turbata	−	−	+	+	L-Lys-L-Thr-D-Asp or L-Lys-L-Thr-D-Glu[e]
Oerskovia xanthineolytica	−	−	+	+	L-Lys-D-Ser-D-Asp[e]
Promicromonospora enterophila	−	−	+	+	Lys, Ala, Glu, Thr[f]
Cellulomonas cartae	−	−	+	+	L-Lys-D-Ser-D-Asp[g]
Nonmotile organisms[h]	−	−	+	+	
Cellulomonas spp.	−	−	−	−	L-Orn-D-Glu or L-Orn-D-Asp[i]

[a] Symbols: see Table 17.21.
[b] *Promicromonospora*, and *Oerskovia*, specific phages, respectively (Prauser, 1984 and unpublished results).
[c] Tryptic soy agar (Lechevalier, 1972).
[d] Stackebrandt et al. (1983) and Schleifer (1984), peptidoglycan of the type L-Lys-Ala-Glu was also reported (Evtushenko et al., 1984b).
[e] Seidl et al. (1980).
[f] Jager et al. (1983), molar ratios in cell wall hydrolysates 1:1:1.08:0.3.
[g] Stackebrandt et al. (1978).
[h] Sensu Lechevalier (1972).
[i] Fiedler and Kandler (1973).

as from results of DNA/DNA hybridization at the level of 15–23% (Prauser, unpublished).

Promicromonospora may be distinguished from these related bacteria by formation of aerial hyphae, susceptibility to taxon-specific phages, **relationship to oxygen on tryptic soy agar and peptidoglycan type (Table 17.23).**

Taxonomic Comments

Promicromonospora was regarded as an actinomycete genus "in search of a family" (Lechevalier and Lechevalier, 1981). Stackebrandt and Schleifer (1984), on the basis of comparative cataloging of 16S ribosomal RNA, proposed to place *Promicromonospora* with *Oerskovia* (which they united with *Cellulomonas* (Stackebrandt et al., 1982)), *Cellulomonas*, *Arthrobacter* and *Micrococcus* in the family *Arthrobacteraceae*. This seems to be a reasonable step towards a natural classification of coryneform and nocardioform bacteria.

Some data seem to indicate the presence of more than one species within the genus. A numerical analysis of 240 phenotypic traits in 30 strains seems to reveal three clusters, perhaps representing different species or subspecies. However, the three clusters thus revealed could not be substantiated by results of DNA/DNA hybridization. DNA/DNA homology among seven strains including the type strain of

Promicromonospora citrea as well as representatives of the three clusters was found to range from 38–71%, mainly being in the range 35–57% (Prauser, unpublished).

Jager et al. (1983) validly published *Promicromonospora enterophila* harboring actinomycetes they isolated from fecal pellets of *Chromatoiulus projectus*, a millipede. These nocardioform bacteria do not seem to possess features which distinguish the genus *Promicromonospora* (Table 17.23). *P. enterophila* resembles the NMOs which were isolated by Lechevalier (1972). Thus at present *P. enterophila* might be regarded as a species *incertae sedis*.

Further Reading

Andreev, L.V., L.I. Evtushenko, and N.S. Agre. 1983. Fatty acid composition of *Promicromonospora citrea*. Microbiologija, *52*: 58–63 (in Russian).

Krasil'nikov, N.A., L.V. Kalakoutskii and N.F. Kirillova. 1961. A new genus of ray fungi—*Promicromonospora* gen. nov. Izv. Akad. Nauk SSSR Ser. Biol., *1*: 107–112 (in Russian).

Lechevalier, H.A. and M.P. Lechevalier. 1981. Actinomycete genera "in search of a family." *In* Starr, Stolp, Trüper, Balows and Schlegel (Editors), The Prokaryotes. A Handbook on Habitats, Isolation and Identification of Bacteria. Springer-Verlag, New York, pp. 2118–2123.

Stackebrandt, E., W. Ludwig, E. Seewalt, and K.-H. Schleifer. 1983. Phylogeny of sporeforming members of the order *Actinomycetales*. Int. J. Syst. Bacteriol. *33*: 173–180.

Differentiation of the species of the genus **Promicromonospora**

1. Promicromonospora citrea Krasil'nikov, Kalakoutskii and Kirillova, 1961, 107.[AL]

ci'tre.a. M.L. adj. *citrea* lemon-yellow.

The species displays the characters of the hitherto monotypic genus. See also Further Descriptive Information.

Life cycle nocardioform, primary and aerial hyphae fragmenting.

Colors of colonies on oatmeal agar and inorganic salts-starch agar citron-yellow or white to whitish, on peptone corn extract agar citron-yellow or cream-colored; on the latter media in exceptional cases brown to orange-brown.

In some strains aerial hyphae only microscopically visible, but well-developed in others; in the latter case these are thin, white, and chalky.

Optimal temperature 28°C, growth occurs between 6 and 42°C. Physiological features described above.

Beside characteristics given in Table 17.23, the following features were found to be of value in presumptive differentiation of *P. citrea*

from related strains: utilization of malate, malonate, succinate, production of acid on rhamnose, raffinose, mannitol, absence of clearing zones on yeast-peptone agar with CaCO$_3$.

Susceptible to taxon-specific phages. Cell wall chemotype VI. Menaquinone type MK-9(H$_4$). Mol% G + C above 70 (T_m).

Mainly isolated from soil, also found on aluminium hydroxide gel antacid.

Type strain: ATCC 15908 (INMI 18; RIA 562; VKM Ac 665; KCC A 0051; IMET 7267; DSM 43110).

Species Incertae Sedis

Promicromonospora enterophila Jager, Marialigeti, Hauck and Barabas 1983, 530.[VP]

en.ter.o'phi.la. Gr. neu. pl. n. *entera* innards, guts; Gr. adj. *philo* loving; M.L. fem. adj. *enterophila* gut loving.

Type strain: HMGB B1078 (DFA 19; see Taxonomic Comments).

Genus **Intrasporangium** *Kalakoutskii, Kirillova and Krasil'nikov 1967, 79*^{AL}

L. V. KALAKOUTSKII

In.tra.spo.ran′gi.um. L. prep. *intra* within; Gr. n. *spora* a seed; Gr. n. *angeion* a vessel; M.L. neut. n. *Intrasporangium* a name coined to emphasize the possibility of intercalary formation of sporangia in mycelial filaments.

Branching mycelium, about 1.0 µm in diameter **has a tendency to break into fragments of various size and shape. Aerial mycelium never observed. Oval- and lemon-shaped vesicles** (5–15 µm in diameter) **are formed intercalary and/or at the hyphal apices.** In some of the vesicles (termed sporangia in the original description) in the older cultures one might distinguish up to several round or oval bodies (1.2–1.5 µm in diameter). These are nonmotile, but may undergo a brownian movement within the mature vesicles. When released onto fresh medium the spore-like cells will germinate giving rise to a branching mycelium. Gram-positive. Nonacid-fast. Chemoorganotrophic, having an **oxidative type of catabolism; possess catalase activity.** Aerobic. Grow best at 28–37°C, no growth at 45°C. **Prefer complex media, especially containing peptone** and **meat extract.** No growth on the majority of mineral synthetic media routinely employed for actinomycetes. The mol% G + C of the DNA is about 68.2 (T_m). **Cell wall chemotype I** (LL-diaminopimelic acid (DAP), glycine). **Peptidoglycan of the LL-DAP-Gly type. Phospholipids of the P IV type** (diagnostic of this is an unknown glucosamine-containing phospholipid). **Among cell fatty acids, straight chain saturated and unsaturated predominate. Menaquinones of MK-8 type.**

Type species: *Intrasporangium calvum* Kalakoutskii, Kirillova and Krasil'nikov 1967, 79.

Further Descriptive Information

A study of *Intrasporangium* vesicles employing electron microscopy of thin sections failed to reveal spores within them. The vesicles were said to be the result of hyphal swelling and gradual disorganization in response to environmental stress (Lechevalier and Lechevalier, 1969). The cells of *Intrasporangium*, however, were reported (Sukapure et al., 1970) to survive heating at 60°C for 4 h in aqueous suspensions.

The cell wall chemotype is type I (LL-DAP, glycine) (Prauser, 1967, personal communication; Sukapure et al., 1970). The peptidoglycan has glycine in the interpeptide bridge (Schleifer and Kandler, 1972). Phospholipids are type P IV (Lechevalier et al., 1977). Among the fatty acids, straight chain saturated and unsaturated predominate. Mycolic acids, *iso-* and *anteiso-*branched, 10-methyl-branched fatty acids not found (Kützner, 1981).

Isolation

The original strain was isolated under nonselective conditions on plates of meat-peptone agar exposed to the atmosphere of a school dining-room.

Maintenance Procedures

Survives routine procedure of lyophilization.

Differentiation of the genus **Intrasporangium** *from other genera*

The genus *Intrasporangium* can be separated from several genera of aerobic mycelium-forming actinomycetes having cell wall chemotype I and sometimes assigned to the family *Streptomycetaceae* (Pridham and Tresner, 1974) on the basis of morphology as well as phospholipid composition (Lechevalier et al., 1977), fatty acid profiles (Kützner, 1981) and susceptibility to the genus-specific phages (Wellington and Williams, 1981; Prauser and Falta, 1968). Extensive numerical taxonomic studies (Williams et al., 1983) point to a distinct and separate position of *Intrasporangium* if compared to actinomycetes of the above group.

The formation of an extensive mycelium and intramycelial vesicles, as well as fatty acid spectra (Kützner, 1981), phospholipid composition (Lechevalier et al., 1977), and menaquinone composition (Collins et al., 1984), separate *Intrasporangium* from aerobic nocardioforms and coryneforms (Collins et al., 1983; O'Donnell et al., 1982; Suzuki and Komagata, 1983) which are characterized by high mol% G + C in their DNA and LL-DAP in the cell walls.

Taxonomic Comments

The taxonomic position of *Intrasporangium* among other actinomycetes has been briefly discussed by Lechevalier and Lechevalier (1981)

as well as by Kützner (1981). The latter author suggested that *Intrasporangium* might even not belong to the whole group of *Actinomycetales*.

Clearly, a comparative study employing more isolates of *Intrasporangium* followed by examination of phylogenetic markers as well as DNA/DNA hybridization in a selected spectrum of strains is in order to understand better the relationship of *Intrasporangium* to other mycelial Gram-positive procaryotes.

Further Reading

Kalakoutskii, L.V., I.P. Kirillova and N.A. Krasil'nikov. 1967. A new genus of the Actinomycetales—*Intrasporangium* gen. nov. J. Gen. Microbiol. *48:* 79–85.
Lechevalier, H.A. and M.P. Levhevalier. 1969. Ultramicroscopic structure of *Intrasporangium calvum* (Actinomycetales). J. Bacteriol. *100:* 522–525.
Lechevalier, H.A. and M.P. Lechevalier. 1981. Actinomycete genera "in search of a family." *In* Starr, Stolp, Trüper, Balows and Schlegel (Editors), The Prokaryotes. A Handbook of Habitats, Isolation and Identification of Bacteria. Springer-Verlag, New York, pp. 2118–2123.
Sukapure, R.S., M.P. Lechevalier, H. Reber, M.L. Higgins, H.A. Lechevalier and H. Prauser. 1970. Motile nocardioid *Actinomycetales*. Appl. Microbiol. *19:* 527–533.
Williams, S.T., M. Goodfellow, G. Alderson, E.M.H. Wellington, P.H.A. Sneath and M.J. Sackin. 1983. Numerical classifications of *Streptomyces* and related genera. J. Gen. Microbiol. *129:* 1743–1813.

Differentiation of the species of the genus **Intrasporangium**

Intrasporangium calvum Kalakoutskii, Kirillova and Krasil'nikov 1967 79.^{AL}

There is only this single species recognized so far.

cal′vum L. neut. adj. bald, referring to the absence of aerial mycelium. See generic description for most of the features.

The organism grows rather slowly even on peptone-containing media. At 28°C, macroscopically visible colonies appear in 3–5 d of incubation. Colonies on meat-extract peptone agar are round, glistening and whitish, becoming creamy on aging. The colonial material is viscous, a characteristic which becomes apparent when transfers are being made

using wire loop or pipettes. On microscopical examination branching mycelium and intramycelial vesicles can be seen to extend beyond the colonial periphery (Fig. 17.34) onto and into agar.

The vesicles begin to appear following 5–6 d of cultivation on solid media, but are rare under conditions of submerged cultivation. No turbidity develops on cultivation of *I. calvum* in liquid media; growth usually occurs in a sediment form. In liquid cultures, very thin (about 0.1 µm in diameter) mycelium-like threads are occasionally seen, but their fate and origin remain obscure. Vesicles (termed sporangia in the original description) are abundant in older cultures (Figs. 17.35 and

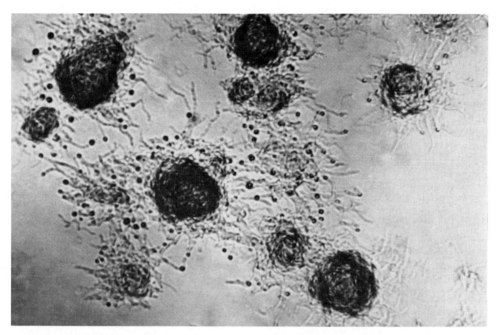

Figure 17.34. Eighteen-day-old colonies of *Intrasporangium calvum* on meat-peptone-glycerol agar (× 300). Many vesicles on mycelial branches ramifying from the colonies' edges are distinguishable. (Reproduced with permission from L. V. Kalakoutskii, I. P. Kirillova and N A. Krasil'nikov, Journal of General Microbiology *48:* 79–85, ©Society for General Microbiology 1967.)

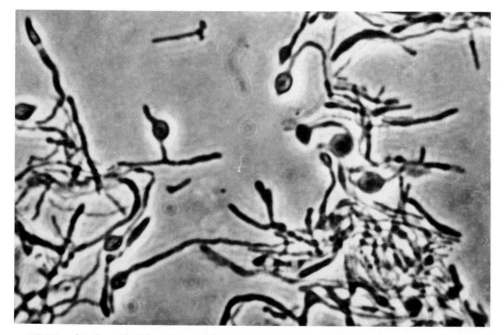

Figure 17.35. Six-day-old slide culture of *Intrasporangium calvum* on meat-peptone agar (× 4000). Phase contrast. Branching mycelium and intramycelial vesicles in both terminal and intercalary position. (Reproduced with permission from L. V. Kalakoutskii, I. P. Kirillova and N. A. Krasil'nikov, Journal of General Microbiology *48:* 79–85, ©Society for General Microbiology 1967.)

17.36). While some of them have an "empty" appearance in older cultures, in others the progression of differentiation results in formation of round to oval sporelike bodies (Fig. 17.36). These differ from the surrounding sporangial material in refractility and susceptibility to basic dyes and germinate by 1 or 2 germ tubes upon transfer to fresh media.

Temperature range for *I. calvum* grown on Bennett's agar was reported to be 10–42°C; that for another *Intrasporangium* isolate LL-12-17 (which might belong to another species) on soil-extract medium was 24–42°C (Sukapure et al., 1970).

Growth of *I. calvum* on complex media is not enhanced upon enrichment of the atmosphere by H_2 and/or CO_2. Nitrates reduced to nitrites. Physiologically not very active.

Intrasporangium calvum was reported (Williams et al., 1983) to utilize L-arginine, L-cysteine, L-methionine, L-phenylalanine and L-serine as a sole nitrogen source (0.1%, w/v) in a basal medium containing glucose and mineral salts. On ISP medium 9 the following compounds were able to support growth if added (1.0%, w/v) as sole carbon compounds: D-glucose, cellobiose, D-fructose, mannitol, L-rhamnose, salicin, trehalose, D-xylose and sodium pyruvate (0.1%, w/v). Degradation of esculin, arbutin, casein, gelatin, elastin, Tween 80 and DNA was noted on modified Bennett's agar and Bacto DNase test agar (Difco), repectively. On modified Bennett's agar growth was possible in presence of either of the following: phenylethanol (0.1%, w/v), potassium tellurite (0.01%, w/v), thallous acetate (0.001%, w/v), cephaloridine (100 μg/ml^{-1}), gentamicin (100 μg/ml^{-1}), neomycin (50 μg/ml^{-1}), tobramycin (50 μg/ml^{-1}).

No antibiotic activity was found in standard tests against Gram-positive and Gram-negative bacteria, yeasts and mycelial fungi.

Type strain: ATCC 23552 (KIP-7; VKM Ac 701; IFO 12982; DSM 43043).

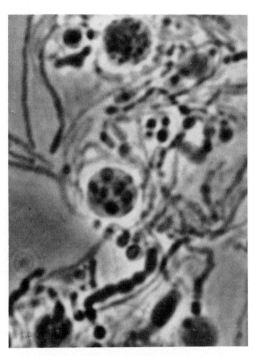

Figure 17.36. Thirty-eight-day-old culture of *Intrasporangium calvum* on meat-peptone agar (× 5000). Phase contrast. Contact preparation. Fragmented mycelium and vesicles with differentiated protoplasm are seen.

Bibliography

Aaronson, S. 1955. Biotin assay with a coccus, *Micrococcus sodonensis* nov. sp. J. Bacteriol. *69:* 67–70.

Abdelal, A.G. 1979. Arginine catabolism by microorganisms. Annu. Rev. Microbiol. *33:* 139–168.

Abd-el-Malek, Y. and T. Gibson. 1947. Studies in the bacteriology of milk. I. The streptococci of milk. J. Dairy Res. *15:* 233–248.

Abd-el-Malek, Y. and T. Gibson. 1952. Studies in the bacteriology of milk III. The corynebacteria of milk. J. Dairy Res. *19:* 153–159.

Abe, M. 1979. Immunological studies on leprosy by flourescent antibody techniques. Scientia Lepro *2:* 69–83.

Abe, P.M., J.A. Majeski and L.R. Stauffer. 1978. Histological changes observed in an *Arachnia propionica* infection of mice. J. Surg. Res. *25:* 174–179.

Abe, S., K. Takayama and S. Kinoshita. 1967. Taxonomic studies on glutamic acid-producing bacteria. J. Gen. Appl. Microbiol. *13:* 279–301.

Abo-Elnaga, I.G. and O. Kandler. 1965a. Zur Taxonomie der Gattung *Lactobacillus* Beijerinck. I. Das Subgenus *Streptobacterium* Orla-Jensen. Zentralbl. Bakteriol. II Abt. *119:* 1–36.

Abo-Elnaga, I.G. and O. Kandler. 1965b. Zur Taxonomie der Gattung *Lactobacillus* Beijerinck. II. Das Subgenus *Betabacterium* Orla-Jensen. Zentralbl. Bakteriol. II Abt. *119:* 117–129.

Abo-Elnaga, I.G. and O. Kandler. 1965c. Zur Taxonomie der Gattung *Lactobacillus* Beijerinck. III. Das Vitaminbedürfnis. Zentralbl. Bakteriol. II Abt. *119:* 661–672.

Abrahams, I. and J.K. Miller. 1946. The in vitro action of sulfonamides and penicillin on *Actinomyces*. J. Bacteriol. *51:* 145–148.

Abramson, C. 1972. Staphylococcal enzymes. *In* Cohen (Editor), The Staphylococci. Wiley-Interscience (John Wiley and Sons, Inc.), New York, London, Sydney, Toronto, pp. 187–248.

Abramson, C. and H. Friedman. 1967. Enzymatic activity of primary isolates of staphylococci in relation to antibiotic resistance and phage type. J. Infect. Dis. *117:* 242–248.

Acevedo, H.F., E.A. Campbell-Acevedo and M. Slifkin. 1980. Immunodetection of choriogonadotropin-like antigen in bacteria isolated from cancer patients. *In* Segal (Editor), Chorionic Gonadotropin, Plenum Publ. Corp., New York, pp. 435–460.

Ackermann, H.-W. 1974. La classification des bactériophages de *Bacillus* et *Clostridium*. Pathol. Biol. *22:* 909–917.

Ackermann, H.-W., A. Audurier and J. Rocourt. 1981. Morphologie de bactériophages de *Listeria monocytogenes*. Ann. Virol. (Inst. Pasteur) *132E:* 371–382.

Adams, E. 1954. The enzymic synthesis of histidine from histidinol. J. Biol. Chem. *209:* 829–846.

Adams, J.B., A.R. Archibald, J. Baddiley, H.E. Coapes and A.L. Davison. 1969. Teichoic acids possessing phosphate-sugar linkages in strains of *Lactobacillus plantarum*. Biochem. J. *113:* 191–193.

Adams, J.N. and S.G. Bradley. 1963. Recombination events in the bacterial genus *Nocardia*. Science *140:* 1392–1394.

Adams, M.E. and J.R. Postgate. 1959. A new sulphate-reducing vibrio. J. Gen. Microbiol. *20:* 252–257.

Adams, M.W.W., L.E. Mortenson and J.-S. Chen. 1981. Hydrogenase. Biochim. Biophys. Acta *594:* 105–176.

Adamse, A.D. 1980. New isolation of *Clostridium aceticum* (Wieringa). Antonie van Leeuwenhoek. J. Microbiol. Serol. *46:* 523–531.

Adcock, K.A., R.J. Seidler and W.C. Trentini. 1976. Deoxyribonucleic acid studies in the genus *Caryophanon*. Can. J. Microbiol. *22:* 1320–1327.

Adegoke, G.O. 1981. Characterization of staphylococci isolated from goats. 1. Coagulase activities and antibiotic susceptibility patterns. Zentralbl. Bakteriol. Parasitenkd. Infektionskr. Hyg. Abt. I Orig. A *249:* 431–437.

Adlam, C. and D.E. Reid. 1978. Comparative studies on the cell wall composition of some anaerobic coryneforms of varying lymphoreticular stimulatory activity. *In* Griffiths and Regamey (Editors), Developments in Biological Standardization. S. Karger, Basel, pp. 115–120.

Aftring, R.P. and B.F. Taylor. 1981. Aerobic and anaerobic catabolism of phthalic acid by a nitrate-respiring bacterium. Arch. Mikrobiol. *130:* 101–104.

Aggag, M. and H.G. Schlegel. 1973. Studies on a Gram-positive hydrogen bacterium, *Nocardia opaca* strain 1 b. Description and physiological characterization. Arch. Mikrobiol. *88:* 299–318.

Agre, N.S. 1964. A contribution to the technique of isolation and cultivation of thermophilic actinomycetes. Mikrobiologiya *33:* 913–917 (In Russian).

Agre, N.S. and L.N. Guzeva. 1975. New actinomycete genus: *Excellospora* gen. nov. Mikrobiologiya *44:* 518–523.

Agre, N.S., L.N. Guzeva and L.A. Dorokhova. 1974. A new species of the genus *Micropolyspora - Micropolyspora internatus*. Mikrobiologiya *43:* 679–685.

Ahmad, F.J. and J.H. Darrell. 1976. Significance of the isolation of *Clostridium welchii* from routine blood cultures. J. Clin. Pathol. *29:* 185–186.

Ainsworth, G.C. and P.H.A. Sneath (Editors). 1962. Microbial classification: Appendix I. Symp. Soc. Gen. Microbiol. *12:* 456–463.

Ait, N., N. Creuzet and J. Cattaneo. 1982. Properties of β-glucosidase purified from *Clostridium thermocellum*. J. Gen. Microbiol. *128:* 569–577.

Ait, N., N. Creuzet and P. Forget. 1979. Partial purification of cellulose from *Clostridium thermocellum*. J. Gen. Microbiol. *113:* 399–402.

Akagi, J.M. and V. Adams. 1973. Isolation of a bisulfite reductase activity from *Desulfotomaculum nigrificans* and its identification as the carbon monoxide-binding pigment P582. J. Bacteriol. *116:* 392–396.

Akatov, A.K., M.L. Khatenever and L.A. Devriese. 1981a. Identification of coagulase-negative staphylococci isolated from clinical sources. *In* Jeljaszewicz (Editor) Staphylococci and Staphylococcal Infections, Gustav Fischer Verlag, Stuttgart, New York, pp. 153–161.

Akatov, A.K., W. Witte and M.L. Khatenever. 1981b. Differentiation of *Staphylococcus epidermidis* and *Staphylococcus saprophyticus* strains isolated from urine. *In* Jeljaszewicz (Editor), Staphylococci and Staphyloccal Infections, Gustav Fischer Verlag, Stuttgart, New York, pp. 85–89.

Akedo, M., R. Torregrossa, C.L. Cooney and A.J. Sinskey. 1978. Production of acrylic acid by *Clostridium propionicum*. Abstr. Ann. Meeting, Am. Soc. Microbiol. *1978:* 185.

Akerman, A., D. Zutra, V. Zafrira and Y. Henis. 1973. Application of an immunofluorescent technique for detecting *Corynebacterium michiganense* and estimating its extent in tomato seed lots. Phytoparasitica *1:* 128 (Abstract).

Albert, J.O., H.F. Long and B.W. Hammer. 1944. Classification of the organisms important in dairy products. IV. *Bacterium linens*. Iowa Agric. Exp. Stn. Res. Bull. No. *328:* 234–259.

Aleman, V. and P. Handler. 1967. Dihydroorotate dehydrogenase. I. General properties. J. Biol. Chem. *242:* 4087–4096.

Aleman, V., P. Handler, G. Palmer and H. Beinert. 1968. Studies on dihydroorotate dehydrogenase by electron paramagnetic resonance spectroscopy. J. Biol. Chem. *243:* 2569–2578.

Alfano, M.C., R.E. Morhart, G. Metcalf and J.F. Drummond. 1974. Presence of collagenase from *Clostridium histolyticum* in gingival sulcal debris of a primitive population. J. Dent. Res. *53:* 142.

Ali-Cohen, C.H. 1889. Eigenbewegung bei Mikrokokken. Zentralbl. Bakteriol. Parasitenkd. Infektionskr. Hyg. Abt. I Orig. *6:* 33–36.

Alizade, M.A. and H. Simon. 1973. Zum mechanismus und zur kompartmentierung der L- und D- lactat-bildung aus L. malat bzw D-Glucose in *Leuconostoc mesenteroides*. Hoppe-Seylers Z. Physiol. Chem. *354:* 163–168.

Allcock, E.R. and D.R. Woods. 1981. Carboxymethyl cellulase and cellobiase production by *Clostridium acetobutylicum* in an industrial fermentation medium. Appl. Environ. Microbiol. *41:* 539–541.

Allen, B.T. and H.A. Wilkinson. 1969. A case of meningitis and generalized Schwartzman reaction caused by *Bacillus sphaericus*. Johns Hopkins Med. J. *125:* 8–13.

Allen, M.B. 1953. The thermophilic aerobic sporeforming bacteria. Bacteriol. Rev. *17:* 125–173.

Allen, S.H. and B.A. Linehan. 1977. Presence of transcarboxylase in *Arachnia propionica*. Int. J. Syst. Bacteriol. *27:* 291–292.

Allison, M.J. 1969. Biosynthesis of amino acids by ruminal microorganisms. J. Anim. Sci. *29:* 797–807.

Allison, M.J., M.P. Bryant and R.N. Doetsch. 1958. Volatile fatty acid growth factor for cellulolytic cocci of bovine rumen. Science *128:* 474–475.

Allison, M.J., M.P. Bryant and R.N. Doetsch. 1962a. Studies on the metabolic function of branched-chain volatile fatty acids, growth factors for ruminococci. J. Bacteriol. *83:* 523–532.

Allison, M.J., M.P. Bryant, I. Katz and M. Keeney. 1962b. Metabolic function of branched-chain volatile fatty acids, growth factors for ruminococci. II. Biosynthesis of higher branched-chain fatty acids and aldehydes. J. Bacteriol. *83:* 1084–1093.

Allison, M.J., S.E. Maloy and R.R. Matson. 1976. Inactivation of *Clostridium botulinum* toxin by ruminal microbes from cattle and sheep. Appl. Environ. Microbiol. *32:* 685–688.

Allo, M. 1980. Clostridial toxins in strangulation obstruction and antibiotic-related colitis. J. Surg. Res. *28:* 421–425.

Allsop, J. and E. Work. 1963. Cell walls of *Propionibacterium* species: Fractionation and composition. Biochem. J. *87:* 512–519.

Allsopp, A. 1969. Phylogenetic relationships of the Procaryota and the origin of the eucaryotic cell. New Phytol. *68:* 591–612.

Al-Mashat, R.R. and D.J. Taylor. 1983. Production of diarrhoea and enteric lesions in calves by the oral inoculation of pure cultures of *Clostridium sordellii.* Vet. Record *112:* 141–146.

Alpern, R.J. and V.R. Dowell, Jr. 1969. *Clostridium septicum* infections and malignancy. J. Am. Med. Assoc. *109:* 385–388.

Alpern, R.J. and V.R. Dowell, Jr. 1971. Non-histotoxic clostridial bacteremia. Am. J. Clin. Pathol. *55:* 717–722.

Alshamaony, L., M. Goodfellow and D.E. Minnikin. 1976. Free mycolic acids as criteria in the classification of *Nocardia* and the '*rhodochrous*' complex. J. Gen. Microbiol. *92:* 188–199.

Alshamaony, L., M. Goodfellow, D.E. Minnikin, G.H. Bowden and J.M. Hardie. 1977. Fatty acid and mycolic acid composition of *Bacterionema matruchotii* and related organisms. J. Gen. Microbiol. *98:* 205–213.

Al-Sheikhly, F. and R.B. Truscott. 1977. The interaction of *Clostridium perfringens* and its toxins in the production of necrotic enteritis of chickens. Avian Dis. *21:* 256–263.

Altemeier, W.A., S.A. Lewis, P.M. Schlievert, M.S. Bergdoll, H.S. Bjornson, J.L. Stareck and B.A. Crass. 1982. *Staphylococcus aureus* associated with toxic shock syndrome: phage typing and toxin capability testing. Ann. Intern. Med. *96:* 978–982.

Altman, N.H. and J.D. Small. 1973. Actinomycosis in a primate confirmed by fluorescent antibody techniques in formalin fixed tissue. Lab. Anim. Sci. *23:* 696–700.

Alvarez, E. and E. Tavel. 1885. Recherches sur le bacille de Lustgarten. Arch. Physiol. Norm. Pathol. *6:* 303–321.

Alvarez, R.J. 1982. Role of *Planococcus citreus* in the spoilage of *Penaeus* shrimp. Zentralbl. Bakteriol. Mikrobiol. Hyg. Abt. I Orig. C *3:* 503–512.

Amachi, T., S. Imamoto, H. Yoshizumi and S. Senoh. 1970. Structure and synthesis of a novel pantothenic acid derivative, the microbial growth factor from tomato juice. Tetrahedron Lett. *56:* 4871–4874.

Amador, A., J.F. Garcia, R. Estan and J. Perales. 1976. Bacteremia por *Bacillus licheniformis.* Med. Clin. *67:* 535–536.

Ambler, R.P. 1976. Amino acid sequences of prokaryotic cytochromes *c*. *In* Fasman (Editor), Handbook of Biochemistry and Molecular Biology, Proteins, Vol. 3, 3rd Ed., CRC Press, Cleveland, Ohio, pp. 292–307.

Ambler, R.P., M. Daniel, J. Hermoso, T.E. Meyer, T.G. Bartsch and M.D. Kamen. 1979. Cytochrome c_2 sequence variation among the recognized species of purple nonsulphur photosynthetic bacteria. Nature (London) *278:* 659–660.

Ambler, R.P., T.E. Meyer and M.D. Kamen. 1979. Anomalies in amino acid sequences of small cytochromes *c* and cytochromes c_1 from two species of purple photosynthetic bacteria. Nature (London) *278:* 661–662.

Ambroz, A. 1913. *Denitrobacterium thermophilum* spec. nova, ein Beitrag zur Biologie der thermophilen Bakterien. Zentralbl. Bakteriol. Parasitenkd. Infektionskr. Hyg. Abt. II *37:* 3–16.

Amdur, B.H., E.I. Szabo and S.S. Socransky. 1978. Fatty acids of gram-positive bacterial rods from human dental plaque. Arch. Oral Biol. *23:* 23–29.

Amstein, C.F. and P.A. Hartman. 1973. Differentiation of some enterococci by gas chromatography. J. Bacteriol. *113:* 38–41.

Analytab Products. 1981. STAPH-IDENT Staphylococcal System. Circular *00–41–100,* Plainview, New York, pp. 1–7.

Andersch, W., H. Bahl and G. Gottschalk. 1982. Acetone-butanol production by *Clostridium acetobutylicum* in an ammonium-linked chemostat at low pH values. Biotechnol. Lett. *4:* 29–32.

Anderson, A.W., H.C. Nordan, R.F. Cain, G. Parrish and D. Duggan. 1956. Studies on a radio-resistant micrococcus I. Isolation, morphology, cultural characteristics and resistance to gamma radiation. Food Technol. *10:* 575–582.

Anderson, R.L. and W.A. Wood. 1969. Carbohydrate metabolism in microorganisms. Annu. Rev. Microbiol. *23:* 539–578.

Anderton, W.J. and S.G. Wilkinson. 1980. Evidence for the presence of a new class of teichoic acid in the cell wall of bacterium NCTC 9742. J. Gen. Microbiol. *118:* 343–351.

Andreesen, J.R. and G. Gottschalk. 1969. The occurence of a modified Entner-Doudoroff pathway in *Clostridium aceticum.* Arch. Mikrobiol. *69:* 160–170.

Andreesen, J.R., G. Gottschalk and H.G. Schlegel. 1970. *Clostridium formicoaceticum* nov. spec. Isolation, description and distinction from *C. aceticum* and *C. thermoaceticum.* Arch. Mikrobiol. *72:* 154–174.

Andreesen, J.R., A. Schaupp, C. Neurauter, A. Brown and L.G. Ljungdahl. 1973. Fermentation of glucose, fructose, and xylose by *Clostridium thermoaceticum*: effect of metals on growth yield, enzymes, and the synthesis of acetate from CO^2. J. Bacteriol. *114:* 743–751.

Andrewes, F.W. and T.J. Horder. 1906. A study of the streptococci pathogenic for man. Lancet *2:* 708–713, 775–782, 852–855.

Andreyev, L.V., L.I. Evtushenko and N.S. Agre. 1983. Fatty acid composition of *Promicromonospora citrea.* Mikrobiologiya *52:* 58–62 (In Russian).

Andrzejewski, J., G. Müller, E. Röhrscheidt and D. Pietkiewicz. 1978. Isolation, characterization and classification of a *Nocardia asteroides* bacteriophage. Zentralbl. Bakteriol. Parasitenkd. Infektionskr. Hyg. I Abt., Suppl. *6:* 319–326.

Anz, W., D. Lauterbach, G. Meissner and I. Willers. 1970. Vergleich von Sensitintesten an Meerschweinchen mit Serotyp und Hühner-virulenz bei *M. aviumd* und *M. intracellulare*-stämmen. Zentralbl. Bakteriol. Parasitenkd. Infektionskr. Hyg. Abt. Orig. *215:* 536–549.

Aoki, H. and R.A. Slepecky. 1973. Inducement of a heat shock requirement for germination and production of increased heat resistance in *Bacillus fastidiosus* spores by manganous ions. J. Bacteriol. *114:* 137–143.

Aoyama, S. 1952. Studies on the thiamin decomposing bacterium. I. Bacteriological researches of a new thiamin decomposing bacillus, *Bacillus aneurinolyticus* Kimura et Aoyama. Acta Sch. Med. Univ. Kyoto *30:* 127–132.

API System S.A. 1979. API STAPH. Circular No. *2050,* Montalieu Vercieu (France), pp. 1–4.

Appelbaum, P.C., E.W.J. Cameron, W.S. Hutton, S.A. Chatterton and C.W. Africa. 1978. The bacteriology of chronic destructive pneumonia. S. Afr. Med. J. *53:* 541–542.

Appelbaum, P.C. and S.A. Chatterton. 1978. Susceptibility of anaerobic bacteria to ten antimicrobial agents. Antimicrob. Agents Chemother. *14:* 371–376.

Applebaum, P.C., P.S. Chaurushiya, M.R. Jacobs and A. Duffett. 1984. Evaluation of the Rapid Strep system for species identification of streptococci. J. Clin. Microbiol. *19:* 588–591.

Aragno, M. 1978. Enrichment, isolation and preliminary characterization of a thermophilic, endospore-forming hydrogen bacterium. FEMS Microbiol. Lett. *3:* 13–15.

Arbuthnott, J.P. 1981. Staphylococcal toxins: past and future. *In* Jeljaszewicz (Editor), Staphylococci and Staphylococcal Infections, Gustav Fischer Verlag, Stuttgart, New York, pp. 215–222.

Archer, G.L. 1978. Antimicrobial susceptibility and selection of resistance among *Staphylococcus epidermidis* isolates recovered from patients with infections of indwelling foreign devices. Antimicrob. Agents Chemother. *14:* 353–359.

Archer, G.L. and M.J. Tenenbaum. 1980. Antibiotic-resistant *Staphylococcus epidermidis* in patients undergoing cardiac surgery. Antimicrob. Agents Chemother. *17:* 269–272.

Archer, G.L., N. Vishiavski and H.G. Stiver. 1982. Plasmid pattern analysis of *Staphylococcus epidermidis* isolated from patients with prosthetic valve endocarditis. Infect. Immun. *35:* 627–632.

Archer, R.H., R. Chong and I.S. Maddox. 1982a. Hydrolysis of bile acid conjugates by *Clostridium bifermentans.* Eur. J. Appl. Microbiol. Biotechnol. *14:* 41–45.

Archer, R.H., I.S. Maddox and I. Chong. 1981. 7 α-Dehydroxylation of cholic acid by *Clostridium bifermentans.* Eur. J. Appl. Microbiol. Biotechnol. *12:* 46–52.

Archer, R.H., I.S. Maddox and R. Chong. 1982b. Transformation of cholic acid by *Clostridium bifermentans.* J. Appl. Bacteriol. *52:* 49–56.

Archibald, A.R. 1972. The chemistry of staphylococcal cell walls. *In* Cohen (Editor), The Staphylococcus, Wiley-Interscience (John Wiley and Sons, Inc.), New York, London, Sydney, Toronto, pp. 75–109.

Archibald, A.R. and J. Baddiley. 1966. The Teichoic Acids. Adv. Carbohyd. Chem. *21:* 323–375.

Archibald, A.R., J. Baddiley and N.L. Blumson. 1968a. The teichoic acids. Advan. Enzymol. *30:* 223–253.

Archibald, A.R., J. Baddiley, D. Button, S. Heptinstall and G.H. Stafford. 1968b. Occurrence of polymers containing *N*-acetylglucosamine 1-phosphate in bacterial cell walls. Nature (London) *219:* 855–856.

Archibald, A.R. and H.E. Coapes. 1971. The wall teichoic acids of *Lactobacillus plantarum* N.I.R.D. C 106. Biochem. J. *124:* 449–460.

Archibald, F.S. and I. Fridovich. 1981. Manganese and defenses against oxygen toxicity in *Lactobacillus plantarum.* J. Bacteriol. *145:* 442–451.

Arden, S.B. and L. Barksdale. 1976. Nitrate reductase activities in lysogenic and non-lysogenic strains of *Corynebacterium diphtheriae* and related species. Int. J. Syst. Bacteriol. *26:* 66–73.

Arden, S.B., W.-H. Chang and L. Barksdale. 1972. Distribution of neuraminidase and *N*-acetyl-neuraminate lyase activities among Corynebacteria, Mycobacteria and Nocardias. J. Bacteriol. *112:* 1206–1212.

Arden-Jones, M.P., A.J. McCarthy and T. Cross. 1979. Taxonomic and serological studies on *Micropolyspora faeni* and *Micropolyspora* strains from soil bearing the specific epithet *rectivirgula.* J. Gen. Microbiol. *115:* 343–354.

Arloing, S., Cornevin and Thomas. 1887. Le charbon symptomatique du boeuf. 2nd ed. Asselin and Houzeau, Paris, pp. 1–281.

Arnon, S.S., T.F. Midura, S.A. Clay, R.M. Wood and J. Chin. 1977. Infant botulism: epidemiological, clinical, and laboratory aspects. J. Am. Med. Assoc. *237:* 1946–1951.

Aronson, J.D. 1926. Spontaneous tuberculosis in salt water fish. J. Infect. Dis. *39:* 314–320.

Artman, M., M. Weiner and G. Frankl. 1980. Counterimmunoelectrophoresis for early detection and rapid identification of *Haemophilus influenzae* type b and *Streptococcus pneumoniae* in blood cultures. J. Clin. Microbiol. *12:* 614–616.

Arvidson, S., T. Holme and T. Wadstrom. 1970. Formation of bacteriolytic enzymes in batch and continuous culture of *Staphylococcus aureus.* J. Bacteriol. *104:* 227–233.

Asada, S., M. Takano and I. Shibasaki. 1980. Mutation induced by drying of *Escherichia coli* on a hydrophobic filter membrane. Appl. Environ. Microbiol. *40:* 274–281.

Asheshov, E.H. and A. Porthouse. 1976. The transducing abilities of different typing phages of *Staphylococcus aureus. In* Jeljaszewicz (Editor), Staphylococci and Staphylococcal Diseases, Gustav Fischer Verlag, Stuttgart, New York, pp. 307–312.

Ashwood-Smith, M.J. and E. Grant. 1976. Mutation induction in bacteria by freeze-drying. Cryobiology *13:* 206–213.

Asselineau, C., S. Clavel, F. Clement, M. Daffe, H. David, M.A. Laneelle and J.C. Prome. 1981. Constituants lipidiques de *Mycobacterium leprae* isole de tatou infecte experimentalement. Ann. Microbiol. *132A: 19–30.*

Association of Official Analytical Chemists. 1975. Changes in official methods of analysis made at the eighty-eighth annual meeting, October 14–17, 1974. J. Assoc. Off. Anal. Chem. *58:* 416–417.

Aswell, J.E., M. Ehrich, R.L. Van Tassell, C.-C. Tsai, L.V. Holdeman and T.D. Wilkins. 1979. Characterization and comparison of *Clostridium difficile* and other clostridial toxins. Microbiology (Washington, D.C.) *1979:* 272–275.

Athalye, M. 1981. Classification and isolation of actinomadurae. Ph.D. Thesis, University of Newcastle upon Tyne.

Auclair, J. and J.-P. Accolas. 1983. Use of thermophilic lactic starters in the dairy industry. Antonie van Leeuwenhoek J. Microbiol. Serol. *49:* 313–326.

Audurier, A., B. Chatelain, F. Chalons and M. Piéchaud. 1979b. Lysotypie de 823 souches de *Listeria monocytogenes* isolées en France de 1958 à 1978. Ann. Microbiol. (Inst. Pasteur) *130B:* 179–189.

Audurier, A., J. Rocourt and A.L. Courtieu. 1977. Isolement et caractérisation de bactériophages de *Listeria monocytogenes.* Ann. Microbiol. (Inst. Pasteur) *128A:* 185–198.

Audurier, A., J. Rocourt and A.L. Courtieu. 1979a. Phage typing system for *Listeria monocytogenes. In* Ivanov (Editor), Problems of Listeriosis. National Agroindustrial Union. Center for Scientific Information, Sofia, Bulgaria, pp. 108–121.

Audurier, A., A.G. Taylor, B. Carbonelle and J. McLauchlin. 1984. A phage-typing system for *Listeria monocytogenes* and its use in epidemiological studies. Clin. Invest. Med. *7:* 229–232.

Aunstrup, K., O. Andresen and H. Outtrup. 1971. Proteolytic enzymes and their production. British Pat. 1,243,784.

Aunstrup, K., H. Outtrup, O. Andresen and C. Dampmann. 1972. Proteases from alkalophilic *Bacillus* species. Proc. IV. IFS: Ferment. Technol. 299–305.

Austin, B., T.M. Embley and M. Goodfellow. 1983. Selective isolation of *Renibacterium salmoninarum.* FEMS Microbiol. Lett. *17:* 111–114.

Austrian, R. 1976. The Quellung reaction: A neglected microbiologic technique. Mt. Sinai J. Med. *43:* 699–709.

Author's Correction. 1984. Int. J. Syst. Bacteriol. *34:* 274.

Avena, R.M. and M.S. Bergdoll. 1967. Purification and some physicochemical properties of enterotoxin C, *Staphylococcus aureus* strain 361. Biochemistry *6:* 1474–1480.

Ayers, S.H. and C.S. Mudge. 1922. The streptococci of the bovine udder. J. Infect. Dis. *31:* 40–50.

Ayers, W.A. 1958. Phosphorylation of cellobiose and glucose by *Ruminococcus flavefaciens.* J. Bacteriol. *76:* 515–517.

Ayyagari, A., V.K. Pancholi, S.C. Pandhi, A. Goswami, K.C. Agarwal and Y.N. Mehra. 1981. Anaerobic bacteria in chronic suppurative otitis media. Indian J. Med. Res. *73:* 860–864.

Babcock, J.B. 1979. Tyrosine degradation in presumptive identification of *Peptostreptococcus anaerobius.* J. Clin. Microbiol. *9:* 358–361.

Bach, M.C., L.D. Sabath and M. Finland. 1973. Susceptibility of *Nocardia asteroides* to 45 antimicrobial agents *in vitro.* Antimicrob. Agents Chemother. *3:* 1–8.

Bache, R. and N. Pfennig. 1981. Selective isolation of *Acetobacterium woodii* on methoxylated aromatic acids and determination of growth yields. Arch. Microbiol. *130:* 255–261.

Back, W. 1978a. Zur Taxonomie der Gattung *Pediococcus.* Brauwissenschaft. *31:* 237–250, 312–320, 336–343.

Back, W. 1978b. Elevation of *Pediococcus cerevisiae* subsp. *dextrinicus* Coster and White to species status [*Pediococcus dextrinicus* (Coster and White) comb. nov.]. Int. J. Syst. Bacteriol. *28:* 523–527.

Back, W. 1980. Bierschädliche Bakterien. Nachweis und Kultivierung bierschädlicher Bakterien im Betriebslabor. Brauwelt *120:* 1562–1569.

Back, W. 1981. Bierschädliche Bakterien. Taxonomie der bierschädlichen Bakterien. Grampositive Arten. Monatsschr. Brau. *7:* 267–276.

Back, W. and E. Stackebrandt. 1978. DNS-DNS homologiestudien innerhalb der gattung *Pediococcus.* Arch. Microbiol. *118:* 75–79.

Backus, B.T. and L.F. Affronti. 1981. Tumor-associated bacteria capable of producing a human choriogonadotropin-like substance. Infect. Immun. *32:* 1211–1215.

Bacus, J.N. and W.L. Brown. 1981. Use of microbial cultures: meat products. Food Technol. *35:* 74–78, 83.

Baddiley, J.J., J.G. Buchanan, F.E. Hardy, R.O. Martin, U.L. Rajbhandary and A.R. Sanderson. 1961. The structure of the ribitol teichoic acid of *Staphylococcus aureus* H. Biochim. Biophys. Acta *52:* 406–407.

Bader, J., H. Günther, B. Rambeck and H. Simon. 1978. Properties of two clostridia strains acting as catalysts for the preparative stereospecific hydrogenation of 2-enoic acids and 2-alken-1-ols with hydrogen gas. Hoppe-Seylers

Z. Physiol. Chem. *359:* 19–27.

Bader, J., H. Günther, E. Schleicher, H. Simon, S. Pohl and W. Mannheim. 1980. Utilization of (E)-2-butenoate (crotonate) by *Clostridium kluyveri* and some other *Clostridium* species. Arch. Microbiol. *125:* 159–165.

Bader, J. and H. Simon. 1980. The activities of hydrogenase and formate reductase in two *Clostridium* species, their interrelationship and dependence on growth conditions. Arch. Microbiol. *127:* 279–287.

Badger, E. 1944. The structural specificity of choline for the growth of type III Pneumococcus. J. Biol. Chem. *153:* 183–191.

Baer, E.F. 1968. Proposed revision of the method for isolating coagulase positive staphylococci from foods. J. Assoc. Off. Anal. Chem. *51:* 865–866.

Baer, E.F., R.J. Gray and D.S. Orth. 1976. Methods for the isolation and enumeration of *Staphylococcus aureus. In* Speck (Editor), Compendium of Methods For the Microbiological Examination of Foods, American Public Health Association, Washington, D.C., pp. 374–386.

Baess, I. 1969. Subdivision of *M. tuberculosis* by means of bacteriophages with special reference to epidemiological studies. Acta Pathol. Microbiol. Scand. B*76:* 464–474.

Baess, I. 1979. Deoxyribonucleic acid relatedness among species of slowly-growing mycobacteria. Acta Pathol. Microbiol. Scand. B*87:* 221–226.

Baess, I. 1982. Deoxyribonucleic acid relatedness among species of rapidly-growing mycobacteria. Acta Pathol. Microbiol. Scand. B*90:* 371–375.

Baess, I. 1983. Deoxyribonucleic acid relationships between different servovars of *Mycobacterium avium, Mycobacterium intracellulare* and *Mycobacterium scrofulaceum.* Acta Pathol. Microbiol. Scand B*91:* 201–203.

Baess, I. and M. Magnusson. 1982. Classification of *Mycobacterium simiae* by means of comparative reciprocal intradermal sensitin testing on guinea-pigs and deoxyribonucleic acid hybridization. Acta Pathol. Microbiol. Scand. B*90:* 101–107.

Baess, I. and M. Weis Bentzon. 1978. Deoxyribonucleic acid hybridization between different species of mycobacteria. Acta Pathol. Microbiol. Scand. B*86:* 71–76.

Bahl, H., W. Andersch, K. Braun and G. Gottschalk. 1982b. Effect of pH and butyrate concentration on the production of acetone and butanol by *Clostridium acetobutylicum* grown in continuous culture. Eur. J. Appl. Microbiol. Biotechnol. *14:* 17–20.

Bahl, H., W. Andersch and G. Gottschalk. 1982a. Continuous production of acetone and butanol by *Clostridium acetobutylicum* in a two-stage phosphate limited chemostat. Eur. J. Appl. Microbiol. Biotechnol. *15:* 201–205.

Bailey, L. 1957. The isolation and cultural characteristics of *Streptococcus pluton* and further observations on '*Bacterium eurydice*'. J. Gen. Microbiol. *17:* 39–48.

Bailey, L. and M.D. Collins. 1982a. Taxonomic studies on *Streptococcus pluton.* J. Appl. Bacteriol. *53:* 209–214.

Bailey, L. and M.D. Collins. 1982b. Reclassification of '*Streptococcus pluton*' (White) in a new genus *Melissococcus,* as *Melissococcus pluton* nom. rev.; comb. nov. J. Appl. Bacteriol. *53:* 215–217.

Bailey, L. and M.D. Collins. 1983. Validation List No. 11. Int. J. Syst. Bacteriol. *33:* 672–674.

Bailey, L. and A.J. Gibbs. 1962. Cultural characters of *Streptococcus pluton* and its differentiation from associated enterococci. J. Gen. Microbiol. *28:* 385–391.

Bailey, R.K., S. Wyles, M. Dingley, F. Hesse and G.W. Kent. 1970. The isolation of high catalase *Mycobacterium kansasii* from tap water. Am. Rev. Resp. Dis. *101:* 430–431.

Baillie, A. and P.D. Walker. 1968. Enzymes of thermophilic aerobic spore-forming bacteria. J. Appl. Bacteriol. *31:* 114–119.

Baird, J.K., V.M.C. Longyear and D.C. Ellwood. 1973. Water insoluble and soluble glucans produced by extracellular glycosyltransferases from *Streptococcus mutans.* Microbios 8: 143–150.

Baird-Parker, A.C. 1963. A classification of micrococci and staphylococci based on physiological and biochemical tests. J. Gen. Microbiol. *30:* 409–427.

Baird-Parker, A.C. 1965. The classification of staphylococci and micrococci from world-wide sources. J. Gen. Microbiol. *38:* 363–387.

Baird-Parker, A.C. 1969. The use of Baird-Parker's medium for the isolation and enumeration of *Staphylococcus aureus. In* Shapton and Gould (Editors), Isolation Methods for Bacteriologists, Appl. Bacteriol. Tech. Series No. 3, Academic Press, London, New York, pp. 1–8.

Baird-Parker, A.C. 1970. The relationship of cell wall composition to the current classification of staphylococci and micrococci. Int. J. Syst. Bacteriol. *20:* 484–490.

Baird-Parker, A.C. 1974. Family I *Micrococcaceae* Pribram 1929. *In* Buchanan and Gibbons (Editors), Bergey's Manual of Determinative Bacteriology, 8th ed. The Williams and Wilkins Co., Baltimore, pp. 478–479.

Baker, J.A. and W.A. Hagan. 1942. Tuberculosis of the Mexican platyfish (*Platypoecilus maculatus*). J. Infect. Dis. *70:* 248–252.

Balch, W.E., G.E. Fox, L.J. Magram, C.R. Woese and R.S. Wolfe. 1979. Methanogens: Reevaluation of a unique biological group. Microbiol. Rev. *43:* 260–296.

Balch, W.E., S. Schoberth, R.S. Tanner and R.S. Wolfe. 1977. *Acetobacterium,* a new genus of hydrogen-oxidizing, carbon dioxide-reducing, anaerobic bacteria. Int. J. Syst. Bacteriol. *27:* 355–361.

Balch, W.E. and R.S. Wolfe. 1976. A new approach to the cultivation of methanogenic bacteria: 2-mercapto-ethanesulfonic acid (HS-CoM)-dependent growth of *Methanobacterium ruminantium* in a pressurized atmosphere. Appl. Environ. Microbiol. *32:* 781–791.

Balcke, J. 1884. Über faurigen Geruch des Bieres. Wschr. Brau. *1:* 257.

Balish, E., D. Cleven, J. Brown and C.E. Yale. 1977. Nose, throat, and fecal flora of beagle dogs housed in "locked" or "open" environments. Appl. Environ. Microbiol. *34:* 207–221.

Ball, L.C. and M.T. Parker. 1979. The cultural and biochemical characters of *Streptococcus milleri* strains isolated from human sources. J. Hyg. Camb. *82:* 63–78.

Balsdon, M.J., L. Pead, G.E. Taylor and R. Maskell. 1980. *Corynebacterium vaginale* and vaginitis: A controlled trial of treatment. Lancet *i:* 501–504.

Bang, S.S., L. Baumann, M.J. Woolkalis and P. Baumann. 1981. Evolutionary relationships in *Vibrio* and *Photobacterium* as determined by immunological studies of superoxide dismutase. Arch. Microbiol. *130:* 111–120.

Bannerman, E.M. and J. Nicolet. 1976. Isolation and characterization of an enzyme with esterase activity from *Micropolyspora faeni*. Appl. Environ. Microbiol. *32:* 138–144.

Banno, Y., T. Kobayashi, K. Watanabe, K. Ueno and Y. Nozawa. 1981. Two toxins (D-1 and D-2) of *Clostridium difficile* causing antibiotic-associated colitis: purification and some characterization. Biochem. Int. *2:* 629–635.

Baptist, J.N., M. Mandel and R.L. Gherna. 1978. Comparative zone electrophoresis of enzymes in the genus *Bacillus*. Int. J. Syst. Bacteriol. *28:* 229–244.

Barber, M. 1939. A comparative study of *Listerella* and *Erysipelothrix*. J. Pathol. Bacteriol. *48:* 11–23.

Barefoot, S.F. and T.R. Klaenhammer. 1983. Detection and activity of lactacin B, a bacteriocin produced by *Lactobacillus acidophilus*. Appl. Environ. Microbiol. *45:* 1808–1815.

Barile, M.F., T.C. Francis and E.A. Graykowski. 1968. *Streptococcus sanguis* in the pathogenesis of recurrent aphthous stomatitis. *In* Guze (Editor), Microbial protoplasts, spheroplasts and L-forms. Williams and Wilkins Co., Baltimore, pp. 444–456.

Barker, H.A. 1938. The fermentation of definite nitrogenous compounds by members of the genus *Clostridium*. J. Bacteriol. *36:* 322–323.

Barker, H.A. 1978. Explorations of bacterial metabolism. Annu. Rev. Biochem. *47:* 1–33.

Barker, H.A. 1981. Amino acid degradation by anaerobic bacteria. Annu. Rev. Biochem. *50:* 23–40.

Barker, H.A. and J.V. Beck. 1941. The fermentative decomposition of purines by *Clostridium acidi-urici* and *Clostridium cylindrosporum*. J. Biol. Chem. *141:* 3–27.

Barker, H.A. and J.V. Beck. 1942. *Clostridium acidi-urici* and *Clostridium cylindrosporum*, organisms fermenting uric acid and some other purines. J. Bacteriol. *43:* 291–304.

Barker, H.A. and V.Haas. 1944. *Butyribacterium*, a new genus of gram-positive, non-sporulating anaerobic bacteria of intestinal origin. J. Bacteriol. *47:* 301–305.

Barker, H.A. and S.M. Taha. 1942. *Clostridium kluyverii*, an organism concerned in the formation of caproic acid from ethyl alcohol. J. Bacteriol. *43:* 347–363.

Barksdale, L. 1970. *Corynebacterium diphtheriae* and its relatives. Bacteriol. Rev. *34:* 378–422.

Barksdale, L. 1981. The genus *Corynebacterium*. *In* Starr, Stolp, Trüper, Balows, and Schlegel (Editors). The Prokaryotes: A Handbook on Habitats, Isolation, and Identification of Bacteria. Springer-Verlag, New York.

Barksdale, L., M.-A. Lanéelle, M.C. Pollice, J. Asselineau, M. Welby and M.V. Norgard. 1979. Biological and chemical basis for the reclassification of *Microbacterium flavum* Orla-Jensen as *Corynebacterium flavescens* comb. nov. Int. J. Syst. Bacteriol. *29:* 222–233.

Barksdale, L., R. Linder, I.T. Sulea and M. Pollice. 1981. Phospholipase D activity of *Corynebacterium pseudotuberculosis* (*Corynebacterium ovis*) and *Corynebacterium ulcerans*, a distinctive marker within the genus *Corynebacterium*. J. Clin. Microbiol. *13:* 335–343.

Barksdale, W.L., K. Li, C.S. Cummins and H. Harris. 1957. The mutation of *Corynebacterium pyogenes* to *Corynebacterium haemolyticum*. J. Gen. Microbiol. *16:* 749–758.

Barkulis, S.S. 1968. Chemical and enzymatic studies on the structure and composition of group A streptococcal cell walls. *In* Caravano (Editor), Current Research on group A streptococcus. Excerpta Medica Foundation, pp. 43–50.

Barlow, J. and A.G. Kitchell. 1966. A note on the spoilage of prepacked lamb chops by *Microbacterium thermosphactum*. J. Appl. Bacteriol. *29:* 185–188.

Barnes, D.M., M.E. Bergeland and J.M. Higbee. 1975. Selected blackleg outbreaks and their relation to soil excavation. Can. Vet. J. *16:* 257–259.

Barnes, E.M. 1965. Distribution and properties of serological types of *Streptococcus faecium*, *Streptococcus durans* and related strains. J. Appl. Bacteriol. *27:* 461–470.

Barnes, E.M. and H.S. Goldberg. 1968. The relationship of bacteria within the family *Bacteroidaceae* as shown by numerical taxonomy. J. Gen. Microbiol. *51:* 313–324.

Barnes, E.M. and C.S. Impey. 1968. Psychrophilic spoilage bacteria of poultry. J. Appl. Bacteriol. *31:* 97–107.

Barnes, E.M. and C.S. Impey. 1968. Anaerobic Gram-negative non-sporing bacteria from the caeca of poultry. J. Appl. Bacteriol. *31:* 530–541.

Barnes, E.M. and C.S. Impey. 1970. The isolation and properties of the predominant anaerobic bacteria in the caeca of chickens and turkeys. Br. Poult. Sci. *11:* 467–481.

Barnes, E.M. and C.S. Impey. 1974. The occurence and properties of uric acid decomposing anaerobic bacteria in the avian caecum. J. Appl. Bacteriol. *37:* 393–409.

Barnes, E.M., C.S. Impey, B.J.H. Stevens and J.L. Peel. 1977. *Streptococcus pleomorphus* sp. nov.: an anaerobic streptococcus isolated mainly from the caeca of birds. J. Gen. Microbiol. *102:* 45–53.

Barnes, E.M., G.C. Mead, C.S. Impey and B.W. Adams. 1978. Analysis of the avian intestinal flora. *In* Lovelock and Davies (Editors), Techniques for the study of mixed populations. Academic Press, London, pp. 89–105.

Barnes, E.M., G.C. Mead, C.S. Impey and B.W. Adams. 1978. The effect of dietary bacitracin on the incidence of *Streptococcus faecalis* subspecies *liquefaciens* and related streptococci in the intestines of young chickens. Br. Poult. Sci. *19:* 713–723.

Barnes, E.M., P.W. Ross, C.D. Wilson, J.M. Hardie, P.D. Marsh, G.C. Mead, M.E. Sharpe, D.A.A. Mossel, N.P. Burman, A.W. Evans and M. Ingram. 1978. Isolation media for streptococci: Proceedings of a discussion meeting. *In* Skinner and Quesnel (Editors), Streptococci, Society for Applied Bacteriology Symposium Series No. 7, Academic Press, London, New York and San Francisco, pp. 1–49, 371–395.

Barnham, M., A.C. Burton and P. Copland. 1978. Pelvic actinomycosis associated with IUCD. Brit. Med. J. *1:* 719–720.

Baron, E.J., J.M. Angevine and W. Sundstrom. 1979. Actinomycotic pulmonary abscess in an immunosuppressed patient. Am. J. Clin. Pathol. *72:* 637–639.

Barre, P. 1978. Identification of thermobacteria and homofermentative, thermophilic, pentose utilizing lactobacilli from high temperature fermenting grape musts. J. Appl. Bacteriol. *44:* 125–129.

Barrow, P.A. 1981. *Corynebacterium kutscheri* infection in wild voles (*Microtus agrestis*). Br. Vet. J. *137:* 67–70.

Barrow, W.W., B.P. Ullom and P.J. Brennan. 1980. Peptidoglycolipid nature of the superficial cell wall sheath of smooth-colony-forming mycobacteria. J. Bacteriol. *144:* 814–822.

Bartholomew, G.W. and M. Alexander. 1979. Microbial metabolism of carbon monoxide in culture and in soil. Appl. Environ. Microbiol. *37:* 932–937.

Bartholomew, J.W. and G. Paik. 1966. Isolation and identification of obligate thermophilic sporeforming bacilli from ocean basin cores. J. Bacteriol. *92:* 635–638.

Bartlett, J.G. and S.M. Finegold. 1972. Anaerobic pleuropulmonary infections. Medicine *51:* 413–440.

Bartlett, J.G., N. Moon, T.-W. Chang, N. Taylor and A.B. Onderdonk. 1978. Role of *Clostridium difficile* in antibiotic-associated pseudomembranous colitis. Gastroenterology *75:* 778–782.

Bartlett, J.G., A.B. Onderdonk, E. Drude, C. Goldstein, M. Anderka, S. Alpert and W.M. McCormack. 1977. Quantitative bacteriology of the vaginal flora. J. Infect. Dis. *136:* 271–277.

Bartlett, J.G., H. Thadepalli, S.L. Gorbach and S.M. Finegold. 1974. Bacteriology of empyema. Lancet, March 2, *1974:* 338–340.

Barton, M.D. and K.L. Hughes. 1980. *Corynebacterium equi*: a review. Vet. Bull. *50:* 65–80.

Barton, M.D. and K.L. Hughes. 1981. Comparison of three techniques for isolation of *Rhodococcus* (*Corynebacterium*) *equi* from contaminated sources. J. Clin. Microbiol. *13:* 219–221.

Barton, M.D. and K.L. Hughes. 1982. Is *Rhodococcus equi* a soil organism? J. Reprod. Fertil., Suppl. *32:* 481–489.

Baskerville, E.N. and R. Twarog. 1972. Regulation of the tryptophan synthetic enzymes in *Clostridium butyricum*. J. Bacteriol. *112:* 304–314.

Baskerville, E.N., and R. Twarog. 1974. Regulation of a ligand-mediated association-dissociation system of anthranilate synthesis in *Clostridium butyricum*. J. Bacteriol. *117:* 1184–1194.

Batchelor, M.D. 1919. Aerobic spore-bearing bacteria in the intestinal tract of children. J. Bacteriol. *4:* 23–34.

Bates, J.H. and D.A. Mitchison. 1969. Geographic distribution of bacteriophage types of *Mycobacterium tuberculosis*. Am. Rev. Resp. Dis. *100:* 189–193.

Batty, I. 1958. *Actinomyces odontolyticus*, a new species of actinomycete regularly isolated from deep carious dentine. J. Pathol. Bacteriol. *75:* 455–459.

Bauchop, T. and R.W. Martucci. 1968. Ruminant-like digestion of the langur monkey. Science *161:* 698–700.

Bauer, A.H. and K.F. Meyer. 1926. Human intestinal carriers of tetanus spores in California. J. Infect. Dis. *38:* 295–305.

Baumann, L., S.S. Bang and P. Baumann. 1980. Study of relationship among species of *Vibrio*, *Photobacterium* and terrestrial enterobacteria by an immunological comparison of glutamine synthetase and superoxide dismutase. Curr. Microbiol. *4:* 133–138.

Bauske, R., G. Peters and G. Pulverer. 1978. Activity spectrum of micrococcal and staphylococcal phages. Zentralbl. Bakteriol. Parasitenkd. Infektionskr. Hyg. Abt. I Orig. A *241:* 24–29.

Bayer, S.S., A.W. Chow, D. Betts and L.B. Guze. 1978. Lactobacillemia - report of nine cases. Important clinical and therapeutic considerations. Am. J. Med. *64:* 808–813.

Beachey, E.H. 1975. Binding of group A streptococci to human oral mucosal cells by lipoteichoic acid. Trans. Assoc. Am. Physn. *88:* 285–292.

Beachey, E.H., J.M. Seyer and A.H. Kang. 1978. Repeating covalent structure of streptococcal M protein. Proc. Nat. Acad. Sci. *75:* 3163–3167.

Beadles, T.A., G.A. Land and D.J. Knezek. 1980. An ultrastructural comparison of the cell envelopes of selected strains of *Nocardia asteroides* and *Nocardia brasiliensis*. Mycopathologia *70:* 25–32.

Beaman, B.L. 1975. Structural and biochemical alterations of *Nocardia asteroides* cell walls during its growth cycle. J. Bacteriol. *123:* 1235–1253.

Beaman, B.L. 1976. Possible mechanisms of nocardial pathogenesis. *In* Goodfellow, Brownell and Serrano (Editors), The Biology of the Nocardiae. Academic Press, London and New York, pp. 386–417.

Beaman, B.L. 1980. Induction of L-phase variants of *Nocardia caviae* within intact murine lungs. Infect. Immun. *29:* 244–251.

Beaman, B.L. 1984. Actinomycete pathogenesis. *In* Goodfellow, Mordarski and Williams (Editors), The Biology of the Actinomycetes. Academic Press, London and New York, pp. 457–479.

Beaman, B.L., J. Burnside, B. Edwards and W. Causey. 1976. Nocardial infections in the United States, 1972–1974. J. Infect. Dis. *134:* 286–289.

Beaman, B.L., M.E. Gershwin and S. Maslan. 1979. Infectious agents in immunodeficient murine models: Pathogenicity of *Actinomyces israelii* serotype I in congenitally athymic (nude) mice. Infect. Immun. *24:* 583–585.

Beaman, B.L., J.A. Serrano and A.A. Serrano. 1978. Comparative ultrastructure within the nocardiae. Zentralbl. Bakteriol. Parasitenkd. Infektionskr. Hyg. I Abt., Suppl. *6:* 201–220.

Bechdel, S.I., H.E. Honeywell, R.A. Dutcher and M.H. Knutsen. 1928. Synthesis of vitamin B in the rumen of the cow. J. Biol. Chem. *80:* 231–238.

Becker, B., M.P. Lechevalier, R.E. Gordon and H.A. Lechevalier. 1964. Rapid differentiation between *Nocardia* and *Streptomyces* by paper chromatography of whole-cell hydrolysates. Appl. Microbiol. *12:* 421–423.

Becker, B., M.P. Lechevalier and H.A. Lechevalier. 1965. Chemical composition of cell wall preparations from strains of various form genera of aerobic actinomycetes. Appl. Microbiol. *13:* 236–243.

Beerens, H., A. Gérard and J. Guillaume. 1957. Étude de 30 souches de *Bifidobacterium bifidum* (*Lactobacillus bifidus*). Caractérization d'une variété buccale. Comparaison avec les souches d'origine fécale. Ann. Inst. Pasteur Lille *9:* 77–85.

Beerens, H., C. Romond and C. Neut. 1980. Influence of breast feeding on the bifid flora of the newborn intestine. Am. J. Clin. Nutr. *33:* 2434–2439.

Beerens, H., Y. Schaffner, J. Guillaume and M. Castel. 1963. Les bacilles anaerobies non sporules a Gram-negatif favorises par la bile. Ann. Inst. Pasteur Lille *14:* 5.

Beerens, H., S. Sugama and M. Tahon-Castel. 1965. Psychrotrophic clostridia. J. Appl. Bacteriol. *28:* 36–48.

Behrend, M. and T.B. Krouse. 1952. Postoperative bacterial synergistic cellulitis of abdominal wall: fatality following herniography. J. Am. Med. Assoc. *149:* 1122–1124.

Beighton, D. 1976. Improved medium for the recovery of *Rothia dentocariosa* from dental plaque. J. Dent. Res. *55:* 550.

Beighton, D. and G. Colman. 1976. A medium for the isolation and enumeration of oral *Actinomycetaceae* from dental plaque. J. Dent. Res. *55:* 875–878.

Beighton, D., R.R.B. Russell and H. Hayday. 1981. The isolation and characterization of *Streptococcus mutans* serotype *h* from dental plaque of monkeys (*Macaca fascicularis*). J. Gen. Microbiol. *124:* 271–279.

Beijerinck, M.W. 1893. Über die Butylalkoholgärung und das Butylferment. Verk. Kon. Akad. Wetensch. (Sect. 2) *1:* 1–51.

Beijerinck, M.W. 1901. Anhäufungsversuche mit Ureumbakterien. Ureumspaltung durch Urease und durch Katabolismus. Zentralbl. Bakteriol. Parasitenkd. Infektionskr. Hyg. Abt. 2, *7:* 33–61.

Beijerinck, M.W. 1901. Sur les ferments lactiques de l'industrie. Arch. Néer. Sci. (Sect. 2) *6:* 212–243.

Beijerinck, M.W. 1912. Die durch Bakterien aus Rohrzucker erzeugten Schleimgen Wandstaffe. Folia Mikrobiol. (Delft) *1:* 377–408.

Beijerinck, M.W. and L.E. den Dooren de Jong. 1922. On *Bacillus polymyxa*. Proc. Sect. Sci., Kon. Akad. Wetensch. Amsterdam *25:* 279–287.

Beijerinck, M.W. and D.C.J. Minkman. 1910. Bildung und Verbrauch von Stickstoffoxydul durch Bakterien. Zentralbl. Bakteriol. Parasitenkd. Infektionskr. Hyg. Abt. II *25:* 30–63.

Belikova, V.L., N.V. Cherevach, L.M. Baryshnikova and L.V. Kalakoutskii. 1980. Morphologic, physiology and biochemical characteristics of *Kurthia zopfii*. Microbiology *49:* 51–55.

Bellack, S. and H.V. Jordan. 1972. Serological identification of rodent strains of *Actinomyces viscosus* and their relationship to *Actinomyces* of human origin. Arch. Oral Biol. *17:* 175–182.

Benchetrit, L.C., L.W. Wannamaker and E.D. Gray. 1979. Immunological properties of hyaluronidases associated with temperate bacteriophages of group A streptococci. J. Exp. Med. *149:* 73–83.

Bender, J.W. and W.T. Hughes. 1980. Fatal *Staphylococcus epidermidis* sepsis following bone marrow transplantation. John Hopkins Med. J. *146:* 13–15.

Bender, R., J.R. Andreesen and G. Gottschalk. 1971. 2-keto-3-deoxygluconate, an intermediate in the fermentation of gluconate by clostridia. J. Bacteriol. *107:* 570–573.

Bennett, J.F. and E. Canale-Parola. 1965. The taxonomic status of *Lineola longa*. Arch. Mikrobiol. *52:* 197–205.

Bentley, C.M. and E.A. Dawes. 1974. The energy-yielding reactions of *Peptococcus prevotii*, their behaviour on starvation and the role and regulation of threonine dehydratase. Arch. Microbiol. *100:* 363–387.

Berd, D. 1973. Laboratory identification of clinically important aerobic actinomycetes. Appl. Microbiol. *25:* 665–681.

Berg, J.N. and W.H. Fales. 1977. Canine and feline anaerobic bacterial infections. Vet. Scope *21:* 2–8.

Berg, J.N., W.H. Fales and C.M. Scanlan. 1979. Occurrence of anaerobic bacteria in diseases of the dog and cat. Am. J. Vet. Res. *40:* 876–881.

Berg, R.W., W.E. Sandine and A.W. Anderson. 1981. Identification of a growth stimulant for *Lactobacillus sanfrancisco*. Appl. Environ. Microbiol. *42:* 786–788.

Bergan, T., K. Bøvre and B. Hovig. 1970a. Reisolation of *Micrococcus mucilaginosus* Migula 1900. Acta. Pathol. Microbiol. Scand. Sect. B, *78:* 85–97.

Bergan, T., K. Bøvre and B. Hovig. 1970b. Priority of *Micrococcus mucilaginosus* Migula 1900 over *Staphylococcus salivarius* Andrewes and Gordon 1907 with proposal of a neotype strain. Int. J. Syst. Bacteriol. *20:* 107–113.

Bergan, T. and M. Kocur. 1982. *Stomatococcus mucilaginosus* gen. nov. sp. nov. ep. rev., a member of the family *Micrococcaceae*. Int. J. Syst. Bacteriol. *32:* 374–377.

Bergdoll, M.S. 1972. The enterotoxins. *In* Cohen (Editor), The Staphylococci, Wiley-Interscience (John Wiley and Sons, Inc.), New York, London, Sydney, Toronto, pp. 301–331.

Bergdoll, M.S., B.A. Crass, R.F. Reiser, R.N. Robbins, A.C.-M. Lee, P.J. Chesney, J.P. Davis, J.M. Vergeront and P.J. Ward. 1982. An enterotoxin-like protein in *Staphylococcus aureus* strains from patients with Toxic Shock Syndrome. Ann. Intern. Med. *96:* 969–971.

Bergdoll, M.S., I.Y. Huang and E.J. Schantz. 1974. Chemistry of the staphylococcal enterotoxins. Agr. Food Chem. *22:* 9–13.

Berger, U. 1960. *Neisseria haemolysans* (Thjøtta and Bøe 1938). Untersuchungen zur Stellung im System. Z. Hyg. Infektionskr. Med. Mikrobiol. Immunol. Virol. *146:* 253–259.

Berger, U. 1961. A proposed new genus of gram-positive cocci: *Gamella*. Int. Bull. Bacteriol. Nomencl. Taxon. *11:* 17–19.

Berger, U. 1974. Pathogenicity of lactobacilli. Dtsch. Med. Wochenschr. *99:* 1200–1203.

Bergey, D.H., R.S. Breed, B.W. Hammer, F.M. Huntoon, E.G.D. Murray and F.C. Harrison. 1934. Bergey's Manual of Determinative Bacteriology. 4th ed. The Williams and Wilkins Co., Baltimore. pp. 1–664.

Bergey, D.H., F.C. Harrison, R.S. Breed, B.W. Hammer and F.M. Huntoon. 1923. Bergey's Manual of Determinative Bacteriology. 1st ed. The Williams and Wilkins Co., Baltimore.

Bergey, D.H., F.C. Harrison, R.S. Breed, B.W. Hammer and F.M. Huntoon. 1925. Bergey's Manual of Determinative Bacteriology, 2nd ed. The Williams and Wilkins Co., Baltimore.

Bergey, D.H., F.C. Harrison, R.S. Breed, B.W. Hammer and F.M. Huntoon. 1930. Bergey's Manual of Determinative Bacteriology, 3rd ed. Williams and Wilkins, Baltimore.

Berkeley, R.C.W., N.A. Logan, L.A. Shute and A.G. Capey. 1984. Identification of *Bacillus* species. *In* Bergan (Editor), Methods in Microbiology, Vol. 16, Academic Press, London, pp. 291–328.

Berkhoff, H.A. 1985. *Clostridium colinum* sp. nov., nom. rev., the causative agent of ulcerative enteritis (quail disease) in quail, chickens, and pheasants. Int. J. Syst. Bacteriol. *35:* 155–159.

Berkhoff, H.A., S.G. Campbell, H.B. Naylor and L. D. Smith. 1974. Etiology and pathogenesis of ulcerative enteritis ("quail disease"). Characterization of the causative anaerobe. Avian Dis. *18:* 195–204.

Berkhoff, H.A. and J.L. Redenbarger. 1977. Isolation and identification of anaerobes in the veterinary diagnostic laboratory. Am. J. Vet. Res. *38:* 1069–1074.

Berkowitz, R.J. and H.V. Jordan. 1975. Similarity of bacteriocins of *Streptococcus mutans* from mother and infant. Arch. Oral Biol. *20:* 725–730.

Berliner, E. 1915. Über die Schlaffsucht der Mehlmottenraupe (*Ephestia kühniella* Zell) und ihren Erreger *Bacillus thuringiensis* n. sp. Z. Angew. Entomol. *2:* 29–56.

Bernardi, G. 1969a. Chromatography of native DNA. Biochim. Biophys. Acta *174:* 423–434.

Bernardi, G. 1969b. Chromatography of nucleic acids on hydroxyapatite. II. Chromatography of denatured DNA. Biochim. Biophys. Acta *174:* 435–448.

Bernhard, K., H. Schrempf and W. Goebel. 1981. Bacteriocin and antibiotic resistance plasmids in *Bacillus cereus* and *Bacillus subtilis*. J. Bacteriol. *133:* 897–903.

Bernheimer, A.W. 1944. Parallelism in the lethal and haemolytic activity of the toxin of *Clostridium septicum*. J. Exp. Med. *80:* 309–320.

Bestetti, G., V. Bühlmann, J. Nicolet and R. Frankhauser. 1977. Paraplegia due to *Actinomyces viscosus* infection in a cat. Acta Neuropathol. *39:* 231–235.

Betz, J.V. and K.E. Anderson. 1964. Isolation and characterization of bacteriophages active on *Clostridium sporogenes*. J. Bacteriol. *87:* 408–415.

Beuchat, L.R. and R.V. Leschwich. 1968. Survival of heated *Streptococcus faecalis* as affected by phase of growth and incubation temperature. J. Appl. Bacteriol. *31:* 414–419.

Bevanger, L. and T.I. Stamnes. 1979. Group L streptococci as the cause of bacteriaemia and endocarditis. Acta Pathol. Microbiol. Scand. Sect. B *87:* 301–302.

Beveridge, T.J. 1979. Surface arrays on the wall of *Sporosarcina ureae*. J. Bacteriol. *139:* 1039–1048.

Beveridge, T.J. 1980. Cell division in *Sporosarcina ureae*. Can. J. Microbiol. *26:* 235–242.

Beveridge, T.J. 1981. Ultrastructure, chemistry and function of the bacterial wall. Int. Rev. Cytol. *72:* 229–317.

Bhagavan, B.S. and P.K. Gupta. 1978. Genital actinomycosis and intrauterine contraceptive devices. Cytopathologic diagnosis and clinical significance. Human Pathol. *9:* 567–578.

Bhandari, R.R. and T.K. Walker. 1953. *Lactobacillus frigidus* n. sp. isolated from brewery yeast. J. Gen. Microbiol. *8:* 330–332.

Biavati, B., V. Scordovi and W.E.C. Moore. 1982. Electrophoretic patterns of proteins in the genus *Bifidobacterium* and proposal of four new species. Int. J. Syst. Bacteriol. *32:* 358–373.

Biberstein, E.L., C. Brown and T. Smith. 1980. Serogroups and biotypes among beta-hemolytic streptococci of canine origin. J. Clin. Microbiol. *11:* 558–561.

Biberstein, E.L. and H.D. Knight. 1971. Two biotypes of *Corynebacterium pseudotuberculosis*. Vet. Rec. *89:* 691–692.

Biede, S.L., G.W. Reinbold and E.G. Hammond. 1976. Influence of *Lactobacillus bulgaricus* on microbiology and chemistry of Swiss cheese. J. Dairy Sci. *59:* 854–858.

Bienstock, B. 1906. *Bacillus putrificus*. Ann. Inst. Pasteur (Paris) *20:* 407–415.

Biever, L.J., G.W. Robertstad, K. van Steenbergh, E.E. Scheetz and G.F. Kennedy. 1969. Actinomycosis in a bovine lung. Am. J. Vet. Res. *30:* 1063–1066.

Bignold, L.P. and H.P.B. Harvey. 1979. Necrotizing enterocolitis associated with invasion by *Clostridium septicum* complicating cyclic neutropaenia. Aust. N. Z. J. Med. *9:* 426–429.

Birkhed, D., K-G. Rosell and K. Granath. 1979. Structure of extracellular water-soluble polysaccharides synthesized from sucrose by oral strains of *Streptococcus mutans*, *Streptococcus salivarius*, *Streptococcus sanguis* and *Actinomyces viscosus*. Arch. Oral Biol. *24:* 53–61.

Bisset, K.A. and F.W. Moore. 1950. *Jensenia*, a new genus of the *Actinomycetales*. J. Gen. Microbiol. *4:* 280.

Bissett, D.L. and R.L. Anderson. 1973. Lactose and D-galactose metabolism in *Staphylococcus aureus*: pathway of D-galactose-6-phosphate degradation. Biochem. Biophys. Res. Commun. *52:* 641–647.

Blair, D.C. and D.B. Martin. 1978. Beta hemolytic streptococcal endocarditis: predominance of non-group A organisms. Am. J. Med. Sci. *276:* 269–277.

Blair, J.E. and R.E.O. Williams. 1961. Phage typing of staphylococci. Bull. WHO *24:* 771–784.

Blake, G.C. 1964. Sensitivities of colonies and suspensions of *Actinomyces israelii* to penicillins, tetracyclines, and erythromycin. Brit. Med. J. *1:* 145–148.

Blanchard, R. 1896. Parasites végétaux à l'exclusion des bactéries. *In* Bouchard (Editor), Traité de Pathologie Générale, Vol II, G. Masson, Paris, pp. 811–932.

Blank, C.H. and L.K. Georg. 1968. The use of fluorescent antibody methods for the detection and identification of *Actinomyces* species in clinical material. J. Lab. Clin. Med. *71:* 283–293.

Blanksma, L.J. and J. Slijper. 1978. Actinomycotic dacryocystitis. Ophthalmologica *176:* 145–149.

Blaschek, H.P. and M. Solberg. 1981. Isolation of a plasmid responsible for caseinase activity in *Clostridium perfringens* ATCC 3626B. J. Bacteriol. *147:* 262–266.

Blasco, J.A. and D.C. Jordan. 1976. Nitrogen fixation in the muskeg ecosystem of the James Bay Lowlands. Can. J. Microbiol. *22:* 897–907.

Blevins, A., C. Semolic, M. Sukany and D. Armstrong. 1973. Common isolation of *Rothia dentocariosa* from clinical specimens studied in the microbiology laboratory. Abst. Annu. Meeting Am. Soc. Microbiol. *1973:* 117.

Blevins, W.T., J.J. Perry and J.B. Evans. 1969. Growth and macromolecular biosynthesis by *Micrococcus sodonensis* during the utilization of glucose and lactate. Can. J. Microbiol. *15:* 383–388.

Blickstad, E., S.O. Enfors and G. Molin. 1981 Effect of hyperbaric carbon dioxide pressure on the microbial flora of pork stored at 4 or 14°C. J. Appl. Bacteriol. *50:* 493–504.

Bliss, E.A. 1937. Studies upon minute hemolytic streptococci. 3. Serological differentiation. J. Bacteriol. *33:* 625–642.

Blobel, H., W. Schaeg and J. Brüchler. 1981. Demonstration and inactivation of staphylococcal clumping-factor. *In* Jeljaszewicz (Editor), Staphylococci and Staphylococcal Infections, Gustav Fischer Verlag, Stuttgart, New York, pp. 487–491.

Blood, R.M. 1975. Lactic acid bacteria in marinated herring. *In* Carr, Cutting and Whiting (Editors), Lactic acid bacteria in beverages and food. Academic Press, London, New York, San Francisco, pp. 195–208.

Blouse, L.E., L.N. Kolonel, C.A. Watkins, and J.M. Atherton. 1975. Efficacy of phage/typing epidemiologically related *Staphylococcus epidermidis* strains. J. Clin. Microbiol. *2:* 318–321.

Bluhm, L. and Z.J. Ordal. 1969. Effect of sublethal heat on the metabolic activity of *Staphylococcus aureus*. J. Bacteriol. *97:* 140–150.

Blumenstock, I. 1981. Isolierung und Charakterisierung Trimethylamin verwertender aerober Sporenbildner. Diplomthesis, University of Göttingen.

Blumenthal, H.J. 1972. Glucose catabolism in staphylococci. *In* Cohen (Editor), The Staphylococci, Wiley-Interscience (John Wiley and Sons, Inc.), New York, London, Sydney, and Toronto, pp. 111–135.

Bockemühl, J., H.P.R. Seeliger and R. Kathke. 1971. Acridinfarbstoffe in Selektivnährböden zur Isolierung von *Listeria monocytogenes*. Med. Microbiol. Immunol. *157:* 84–95.

Boeyé, A. and M. Aerts. 1976. Numerical taxonomy of *Bacillus* isolates from north sea sediments. Int. J. Syst. Bacteriol. *26:* 427–441.

Bogdahn, M., J.R. Andreesen and D. Kleiner. 1983. Pathways and regulation of N₂, ammonium and glutamate assimilation by *Clostridium formicoaceticum*.

Arch. Microbiol. *134:* 167–169.

Boháček, J., M. Kocur and T. Martinec. 1967. DNA base composition and taxonomy of some micrococci. J. Gen. Microbiol. *46:* 369–376.

Boháček, J., M. Kocur and T. Martinec. 1968a. Deoxyribonucleic acid base composition of some marine halophilic micrococci. J. Appl. Bacteriol. *31:* 215–219.

Boháček, J., M. Kocur and T. Martinec. 1968b. Deoxyribonucleic acid base composition of *Sporosarcina ureae*. Arch. Microbiol. *64:* 23–28.

Boháček, J., M. Kocur and T. Martinec. 1969. Deoxyribonucleic acid base composition of *Micrococcus roseus*. Antonie van Leeuwenhoek. J. Microbiol. Serol. *35:* 185–188.

Böhm, R. 1981. Sample preparation technique for the analysis of vegetative bacteria cells of the genus *Bacillus* with the laser microprobe mass analyzer (LAMMA). Fresenius Z. Anal. Chem. *308:* 258–259.

Boisvert, H. 1977. L'ulcere cutane a *Mycobacterium ulcerans* au cameroun. II. Étude bacteriologique. Bull. Soc. Pathol. Exot. *70:* 125–131.

Bojalil, L.F., J. Cerbón and A. Trujillo. 1962. Adansonian classification of mycobacteria. J. Gen. Microbiol. *28:* 333–346.

Bokkenheuser, V.D., J. Winter, P. Dehazya, O. de Leon and W.G. Kelly. 1976. Formation and metabolism of tetrahydrodeoxycorticosterone by human fecal flora. J. Steroid Biochem. *7:* 837–843.

Bokkenheuser, V.D., J. Winter, P. Dehazya and W.G. Kelly. 1977. Isolation and characterization of human fecal bacteria capable of 21-dehydroxylating corticoids. Appl. Environ. Microbiol. *34:* 571–575.

Bokkenheuser, V.D., J. Winter, S.M. Finegold, V.L. Sutter, A.E. Ritchie, W.E.C. Moore and L.V. Holdeman. 1979. New markers for *Eubacterium lentum*. Appl. Environ. Microbiol. *37:* 1001–1006.

Bokkenheuser, V.D., J. Winter, S. O'Rourke and A.E. Ritchie. 1980. Isolation and characterization of fecal bacteria capable of 16α-dehydroxylating corticoids. Appl. Environ. Microbiol. *40:* 803–808.

Bollgren, I., C.E. Nord and V. Vaclavinkova. 1981. Characterization of Coagulase-Positive and Coagulase-negative staphylococci in the periurethral flora of adult women. *In* Jeljaszewucz (Editor), Staphylococci and Staphylococcal Infections, Gustav Fischer Verlag, Stuttgart, New York, pp. 1005–1009.

Bollinger, O. 1877. Über eine neue Pilzkrankheit beim Rinde. Zentralbl. Med. Wiss. *15:* 481–485.

Bond, W.W. and M.S. Favero. 1977. *Bacillus xerothermodurans* sp. nov., a species forming endospores extremely resistant to dry heat. Int. J. Syst. Bacteriol. *27:* 157–160.

Bonde, G.J. 1975. The genus *Bacillus*. An experiment with cluster analysis. Dan. Med. Bull. *22:* 41–61.

Bonde, G.J. 1981. *Bacillus* from marine habitats: allocation to phena established by numerical techniques. *In* Berkeley and Goodfellow (Editors), The Aerobic Endospore-forming Bacteria. Academic Press, London, pp. 181–215.

Bongaerts, G.P.A. 1978. Uric acid degradation by *Bacillus fastidiosus*. Ph.D. Thesis, Nijmegen, Netherlands.

Bongaerts, G.P. and G.D. Vogels. 1976. Uric acid degradation by *Bacillus fastidiosus* strains. J. Bacteriol. *125:* 689–697.

Bönicke, R. 1962. L'identification des mycobactéries à l'aide de methodes biochimiques. Bull. Int. Union Tuberc. *32:* 13–76.

Bönicke, R. 1965. Beschreibung der neuen species *Mycobacterium borstelense* n. sp. Zentralbl. Bakteriol. Parisitenkd. Infectionskr. Hyg. Abt. I. Orig. *196:* 535–538.

Bönicke, R. and S.E. Juhansz. 1964. Beschreibung der neuen species *Mycobacterium vaccae* n. sp. Zentralbl. Bakteriol. Parasitenkd. Infectionskr. Hyg. Abt. I. Orig. *192:* 133–135.

Bönicke, R. and S.E. Juhansz. 1965. *Mycobacterium diernhoferi* n. sp., eine in der Umgebung des Rindes häufig vorkommende neue *Mycobacterium*-species. Zentralbl. Bakteriol. Parasitenkd. Infektionskr. Hyg. Abt. I. Orig. *197:* 292–294.

Bönicke, R., E. Rohrscheidt and E. Pascoe. 1962. Die Verbreitung der Nitratreduktase innerhalb der Gattung *Mycobacterium*. Sonderdr. Naturwiss. *49:* 43–44.

Boone, C.J. and L. Pine. 1968. Rapid method for characterization of actinomycetes by cell wall composition. Appl. Microbiol. *16:* 279–284.

Bordet, C., M. Karahjoli, O. Gateau and G. Michel. 1972. Cell walls of nocardiae and related actinomycetes: Identification of the genus *Nocardia* by cell wall analysis. Int. J. Syst. Bacteriol. *22:* 251–259.

Borja, C.R. and M.S. Bergdoll. 1967. Purification and partial characterization of enterotoxin C produced by *Staphylococcus aureus* strain 137. Biochemistry *6:* 1467–1473.

Bornside, G.H. and R.E. Kallio. 1956. Urea-hydrolyzing bacilli. II. Nutritional profiles. J. Bacteriol. *71:* 655–660.

Bornstein, B.T. and H.A. Barker. 1948. The nutrition of *Clostridium kluyveri*. J. Bacteriol. *55:* 223–230.

Borriello, S.P. 1980. Clostridial flora of the gastrointestinal tract in health and disease. Thesis, University of London, pp. 1–299.

Borriello, S.P. and R.J. Carman. 1983. Association of iota-like toxin and *Clostridium spiroforme* with both spontaneous and antibiotic-associated diarrhea and colitis in rabbits. J. Clin. Microbiol. *17:* 414–418.

Borriello, S.P., P. Honour, T. Turner and F. Barclay. 1983. Household pets as a potential reservoir for *Clostridium difficile* infection. J. Clin. Pathol. *36:* 84–87.

Boswell, P.A., G.F. Batstone and R.G. Mitchell. 1972. The oxidase reaction in the classification of the *Micrococcaceae*. J. Med. Microbiol. *5:* 267–269.

Bottazzi, V., M. Vescovo and F. Dellaglio. 1973. Microbiology of Grana cheese. IX. Characteristics and distribution of *Lactobacillus helveticus* biotypes in natural whey cheese starter. Sci. Tech. Latt.-casear. *24:* 23–39.

Bouisset, L., J. Breuillard and G. Michel. 1963. Etude de l'ADN chez les *Actino-mycetales*. Comparaison entre les valeure A+T/C+G et les caractères bacter-iologiques des *Corynebacterium*. Ann. Inst. Pasteur (Paris) *104:* 756–770.

Bouisset, L., J. Breuilland, G. Michel and G. Larrony. 1968. Bases nucléiques des bactéries. Application au genre *Actinobacterium*. Ann. Inst. Pasteur *118:* 1063–1081.

Bourgault, A.M. and J.E. Rosenblatt. 1979. First isolation of *Peptococcus indolicus* from a human clinical speciman. J. Clin. Microbiol. *9:* 549–550.

Bourgeau, G. and B.C. McBride. 1976. Dextran-mediated inter-bacterial aggre-gation between dextran-synthesizing streptococci and *Actinomyces viscosus*. Infect. Immun. *1:* 1228–1234.

Bourne, K.A., J.L. Beebe, Y.A. Lue and P.D. Ellner. 1978. Bacteremia due to *Bifidobacterium, Eubacterium* or *Lactobacillus*: twenty-one cases and review of the literature. Yale J. Biol. Med. *51:* 505–512.

Bousfield, I.J. 1972. A taxonomic study of some coryneform bacteria. J. Gen. Microbiol. *71:* 441–455.

Bousfield, I.J. 1978. The taxonomy of coryneform bacteria from the marine environment. *In* Bousfield and Callely (Editors), Coryneform Bacteria. Ac-ademic Press, London, pp. 217–233.

Bousfield, I.J. and M. Goodfellow. 1976. The "rhodochrous" complex and its relationships with allied taxa. *In* Goodfellow, Brownell and Serrano (Edi-tors), The Biology of the Nocardiae, Academic Press, London and New York, pp. 39–65.

Bousfield, I.J., R.M. Keddie, T.R. Dando and S. Shaw. 1985. Simple rapid methods of cell wall analysis as an aid in the identification of aerobic coryneform bacteria. *In* Goodfellow and Minnikin (Editors), Chemical Meth-ods in Bacterial Systematics. Academic Press, London, pp. 221–236.

Bousfield, I.J., G.L. Smith, T.R. Dando and G. Hobbs. 1983. Numerical analysis of total fatty acid profiles in the identification of coryneform, nocardioform and some other bacteria. J. Gen. Microbiol. *129:* 375–394.

Bouvet, A., I. van de Rijn and M. McCarty. 1981. Nutritionally variant strepto-cocci from patients with endocarditis: growth parameters in a semisynthetic medium and demonstration of a chromophore. J. Bacteriol. *146:* 1075–1082.

Bouwer-Hertzberger, S.A. and D.A.A. Mossel. 1982. Quantitative isolation and identification of *Bacillus cereus*. *In* Corry, Roberts and Skinner (Editors), Isolation and identification methods for food poisoning organisms. Academic Press, London, pp. 255–259.

Bovallius, A. and B. Zacharias. 1971. Variations in the metal content of some commercial media and their effect on microbial growth. Appl. Microbiol. *22:* 260–262.

Bovarnik, M. 1942. The function of extracellular D(-)glutamic acid polypeptide by *Bacillus subtilis*. J. Biol. Chem. *145:* 415–424.

Bøvre, K. 1980. Progress in classification and identification of *Neisseriaceae* based on genetic affinity. *In* Goodfellow and Board (Editors), Microbial classification and identification, Academic Press, London - New York, pp. 55–72.

Bowden, G.H. 1969. The components of the cell walls and extracellular slime of four strains of *Staphylococcus salivarius* isolated from human dental plaque. Arch. Oral. Biol. *14:* 685–697.

Bowden, G.H. and E.D. Fillery. 1978. Wall carbohydrate antigens of *A. israelii*. Adv. Exp. Med. Biol. *107:* 685–693.

Bowden, G.H. and J.M. Hardie. 1971. Anaerobic organisms from the human mouth. *In* Shapton and Board (Editors), Isolation of Anaerobes, Society for Applied Bacteriology Technical Series No. 5, Academic Press, London and New York, pp. 177–205.

Bowden, G.H. and J.M. Hardie. 1973. Commensal and pathogenic *Actinomyces* species in man. *In* Sykes and Skinner (Editors), Actinomycetales, Charac-teristics and Practical Importance. Academic Press, London, pp. 277–295.

Bowden, G.H. and J.M. Hardie. 1978. Oral pleomorphic (coryneform) Gram-positive rods. *In* Bousfield and Callely (Editors), Coryneform Bacteria. Academic Press, London-New York-San Francisco, pp. 235–263.

Bowden, G.H., J.M. Hardie and E.D. Fillery. 1976. Antigens from *Actinomyces* species and their value in identification. J. Dent. Res. *55* (Special Issue): A192-A204.

Bowers, E.F. and L.R. Jeffries. 1955. Optochin in the identification of *Strepto-coccus pneumoniae*. J. Clin. Pathol. (London) *8:* 58–60.

Bowie, I.S., M.R. Grigor, G.G. Dunckley, M.W. Loutit and J.S. Loutit. 1972. The DNA base composition and fatty acid constitution of some Gram-positive pleomorphic soil bacteria. Soil Biol. Biochem. *4:* 397–412.

Boyaval, P., E. Moreira and M.J. Desmazeaud. 1983. Transport of aromatic amino acids by *Brevibacterium linens*. J. Bacteriol. *155:* 1123–1129.

Boyer, E.W., M.B. Ingle and G.D. Mercer. 1973. *Bacillus alcalophilus* subsp. *halodurans* subsp. nov.: an alkaline-amylase-producing, alkalophilic orga-nism. Int. J. Syst. Bacteriol. *23:* 238–242.

Boyette, D.P. and F.L. Rights. 1952. Heretofore undescribed aerobic sporeforming bacillus in a child with meningitis. J. Am. Med. Assoc. *148:* 1223–1224.

Bracco, R.M., M.R. Krause, A.S. Rae and C.M. MacLeod. 1957. Transformation reactions between pneumococcus and three strains of streptococci. J. Exp. Med. *106:* 247.

Bracke, J.W., D.L. Cruden and A.J. Markovetz. 1978. Effect of metronidazole on

the intestinal microflora of the American cockroach, *Periplaneta americana* L. Antimicrob. Agents Chemother. *13:* 115–120.

Braden, A.R. and D.W. Thayer. 1976. Serological study of *Cellulomonas*. Int. J. Syst. Bacteriol. *26:* 123–126.

Bradley, D.E. 1967. Ultrastructure of bacteriophages and bacteriocins. Bacteriol. Rev. *31:* 230–314.

Bradley, D.E. and C.A. Dewar. 1966. The structure of phage-like objects associated with non-induced bacteriocinogenic bacteria. J. Gen. Microbiol. *45:* 399–408.

Bradley, D.E. and J.G. Franklin. 1958. Electron microscope survey of the surface configuration of spores of the genus *Bacillus*. J. Bacteriol. *76:* 618–630.

Bradley, S.G. 1973. Relationships among mycobacteria and nocardiae based upon deoxyribonucleic acid reassociation. J. Bacteriol. *113:* 645–651.

Bradley, S.G. 1978. Physiological genetics of nocardiae. Zentralbl. Bakteriol. Parasitenkd. Infektionskr. Hyg. I Abt., Suppl. *6:* 287–302.

Bradley, S.G. and J.S. Bond. 1974. Taxonomic criteria for mycobacteria and nocardiae. Adv. Appl. Microbiol. *18:* 131–190.

Bradley, S.G., L.W. Enquist and H.E. Scribner, III. 1978. Heterogeneity among deoxyribonucleotide sequences of actinomycetales. *In* Freerksen, Tárnok and Thumim (Editors), Genetics of the Actinomycetales, Gustav Fischer Verlag, Stuttgart and New York, pp. 207–224.

Bradley, S.G. and M. Mordarski. 1976. Association of polydeoxyribonucleotides of deoxyribonucleic acids from nocardioform bacteria. *In* Goodfellow, Brow-nell and Serrano (Editors), The Biology of the Nocardiae, Academic Press, London and New York, pp. 310–336.

Bragg, S., L. Georg and A. Ibrahim. 1972. Determination of a new serotype of *Actinomyces naeslundii*. Abstr. Annu. Meet. Am. Soc. Microbiol. *1972:* 38.

Bragg, S., W. Kaplan and G. Hageage. 1975. Preparation of a specific fluorescent antibody reagent for *Actinomyces naeslundii* serotype 2. Abstr. Annu. Meet. Am. Soc. Microbiol. *1975:* 86.

Branden, A.R. and D.W. Thayer. 1976. Serological study of *Cellulomonas*. Int. J. Syst. Bacteriol. *26:* 123–126.

Brandis, H. 1981. Bacteriocins with special reference to staphylococcins. *In* Jeljaszewicz (Editor), Staphylococci and Staphylococcal Infections, Gustav Fischer Verlag, Stuttgart, New York, pp. 719–729.

Brantner, H. and J. Schwager. 1979. Enzymatische Mechanismen der Onkolyse durch *Clostridium oncolyticum* M55 ATCC 13,732. Zentralbl. Bakteriol. Parasitenkd. Infektionskr. Hyg. I Abt. Orig. A *243:* 113–118.

Bratthall, D. 1969. Immunodiffusion studies on the serologic specificity of strep-tococci resembling *Streptococcus mutans*. Odont. Revy *20:* 231–243.

Bratthall, D. 1970. Demonstration of five serological groups of streptococcal strains resembling *Streptococcus mutans*. Odont. Revy *21:* 143–152.

Braun, K. and G. Gottschalk. 1981. Effect of molecular hydrogen and carbon dioxide on chemo-organotrophic growth of *Acetobacterium woodii* and *Clos-tridium aceticum*. Arch. Microbiol. *128:* 294–298.

Braun, M. and G. Gottschalk. 1982. *Acetobacterium wieringae* sp. nov., a new species producing acetic acid from molecular hydrogen and carbon dioxide. Zentralbl. Bakteriol. Mikrobiol. Hyg. Abt. I Orig. C *3:* 368–376.

Braun, M. and G. Gottschalk. 1983. *In* Validation of the publication of new names and new combinations previously effectively published outside the IJSB. Int. J. Syst. Bacteriol. *33:* 438–440.

Braun, M., F. Mayer and G. Gottschalk. 1981. *Clostridium aceticum* (Wieringa), a microorganism producing acetic acid from molecular hydrogen and carbon dioxide. Arch. Microbiol. *128:* 288–293.

Braun, M.S., S. Schoberth and G. Gottschalk. 1979. Enumeration of bacteria forming acetate from H_2 and CO_2 in anaerobic environments. Arch. Micro-biol. *120:* 201–204.

Brecher, S.M. and J. van Houte. 1979. Relationship between host age and susceptibility to oral colonization by *Actinomyces viscosus* in Sprague-Dawley rats. Infect. Immun. *26:* 1137–1145.

Brecher, S.M., J. van Houte and B.F. Hammond. 1978. Role of colonization in the virulence of *Actinomyces viscosus* strains T14-Vi and T14-Av. Infect. Immun. *22:* 603–614.

Bredemann, G. 1909. Untersuchungen über die Variation und das Stickstoffbin-dungsvermögen des *Bacillus asterosporus* A.M., ausgeführt an 27 Stämmen virschiedener Herkunft. Zentralbl. Bakteriol. Parasitenkd. Infektionskr. Hyg. Abt. II *22:* 44–89.

Bredemann, G. and H. Heigener. 1935. *In* Heigener, H. (Editor), Verwertung von Aminosäuren als gemeinsame C- und M-Quelle durch bekannte Bodenbak-terien nebst botanischer Beschreibung neu isolierter Betain- und Valin-Abbauer. Zentralbl. Bakteriol. Parasitenkd. Infektionskr. Hyg. Abt. II *93:* 81–113.

Breed, R.S. 1953. The families developed from *Bacteriaceae* Cohn with a descrip-tion of the family *Brevibacteriaceae*. Riass. Commun. VI Congr. Int. Micro-biol. Roma *1:* 10–15.

Breed, R.S. 1956. *Staphylococcus pyogenes* Rosenbach. Int. Bull. Bact. Nomencl. Taxon. *6:* 35–42.

Breed, R.S. 1957. Family IX. *Brevibacteriaceae* Breed, 1953. *In* Breed, Murray and Smith (Editors) Bergey's Manual of Determinative Bacteriology 7th Ed. Williams and Wilkins, Baltimore, pp. 490–503.

Breed, R.S. and H.J. Conn. 1919. The nomenclature of the *Actinomycetaceae*. J. Bacteriol. *4:* 585–602.

Breed, R.S., E.G.D. Murray and N.R. Smith. 1957. Bergey's Manual of Deter-minative Bacteriology, 7th ed. The Williams and Wilkins Co., Baltimore.

Brefort, G., M. Magot, H. Ionesco and M. Sebald. 1977. Characterization and transferability of *Clostridium perfringens* plasmids. Plasmid *1:* 52–66.

Brennan, P. and W. Barrow. 1980. Evidence for species-specific lipid antigens in *Mycobacterium leprae.* Int. J. Leprosy. *48:* 382–387.

Brennan, P., H. Mayer, G. Aspinall and J.N. Shin. 1981. Structures of the glycopeptidolipid antigens from serovars in the *Mycobacterium avium/Mycobacterium intracellulare/Mycobacterium scrofulaceum* serocomplex. Eur. J. Biochem. *115:* 7–15.

Brennan, P.J. 1981. Structures of the typing antigens of atypical Mycobacteria: a brief review of present knowledge. Rev. Inf. Dis. *3:* 905–913.

Brenner, D.J., G.R. Fanning, A.V. Rake and K.E. Johnson. 1969. Batch procedure for thermal elution of DNA from hydroxyapatite. Anal. Biochem. *28:* 447–459.

Brewer, D.G., S.E. Martin and Z.V. Ordal. 1977. Beneficial effects of catalase or pyruvate in a most-probable-number technique for the detecting of *Staphylococcus aureus.* Appl. Environ. Microbiol. *34:* 797–800.

Bridge, P.D. and P.H.A. Sneath. 1982. *Streptococcus gallinarum* sp. nov. and *Streptococcus oralis* sp. nov. Int. J. Syst. Bacteriol. *32:* 410–415.

Bridge, P.D. and P.H.A. Sneath. 1983. Numerical taxonomy of *Streptococcus.* J. Gen. Microbiol. *129:* 565–597.

Brimacombe, R., G. Staffer and H.G. Wittmann. 1978. Ribosome structure. Annu. Rev. Biochem. *47:* 217–249.

Brisou, J. 1958. Étude de quelques *Pseudomonadaceae.* Classification. Baillet (Editor), Bordeaux, pp. 1–214.

Britz, M.L. and R.G. Wilkinson. 1982. Leucine dissimilation to isovaleric and isocaproic acids by cell suspensions of amino acid fermenting anaerobes: the Strickland reaction revisited. Can. J. Microbiol. *28:* 291–300.

Britz, M.L. and R.G. Wilkinson. 1983. Subcellular location of enzymes involved in leucine dissimilation in *Clostridium bifermentans.* Can. J. Microbiol. *29:* 441–447.

Britz, T.J. 1979. Historical review of the taxonomy of the propionic acid bacteria. S. Afr. J. Dairy Technol. *11:* 23–27.

Brock, D.W. and L.K. Georg. 1969a. Determination and analysis of *Actinomyces israelii* serotypes by fluorescent-antibody procedures. J. Bacteriol. *97:* 581–588.

Brock, D.W. and L.K. Georg. 1969b. Characterization of *Actinomyces israelii* serotypes 1 and 2. J. Bacteriol. *97:* 589–593.

Brock, D.W., L.K. Georg, J.M. Brown and M.D. Hicklin. 1973. Actinomycosis caused by *Arachnia propionica.* Am. J. Clin. Pathol. *59:* 66–77.

Broda, P. 1979. Plasmids. W.H. Freeman Co., London - San Francisco.

Brooke, W.F. 1941. The vole acid-fast *Bacillus.* I. Experimental studies on a new type of *Mycobacterium tuberculosis.* Am. Rev. Tuberc. *43:* 806–816.

Brooker, B.E. 1976. Surface coat transformation and capsule formation by *Leuconostoc mesenteroides* NCDO 523 in the presence of sucrose. Arch. Microbiol. *111:* 99–104.

Brooker, B.E. 1977. Ultrastructural surface changes associated with dextran synthesis by *Leuconostoc mesenteroides.* J. Bacteriol. *131:* 288–292.

Brooks, B.W. and R.G.E. Murray. 1981. Nomenclature for "*Micrococcus radiodurans*" and other radiation-resistant cocci: *Deinococcaceae* fam. nov. and *Deinococcus* gen. nov., including five species. Int. J. Syst. Bacteriol. *31:* 353–360.

Brooks, B.W., R.G. Murray, J.L. Johnson, E. Stackebrandt, C.R. Woese and G.E. Fox. 1980. Red-pigmented micrococci: A basis for taxonomy. Int. J. Syst. Bacteriol. *30:* 627–646.

Brooks, J.B. 1971. Analysis by gas chromatography of *Neisseria* species. Can. J. Microbiol. *17:* 531–543.

Brooks, J.B., C.C. Alley, J.W. Weaver, V.E. Green and A.M. Harkness. 1973. Practical methods for derivatizing and analyzing bacterial metabolites with a modified automatic injector and gas chromatograph. Anal. Chem. *45:* 2083–2087.

Brooks, J.B., D.S. Kellogg, L.Thacker and E.M. Turner. 1972. Analysis by gas chromatography of hydroxyacids produced by several species of *Neisseria.* Can. J. Microbiol. *18:* 157–168.

Brooks, J.B. and W.E.C. Moore. 1969. Gas chromatographic analysis of amines and other compounds produced by several species of *Clostridium.* Can. J. Microbiol. *15:* 1433–1447.

Brooks, J.B., C.W. Moss and V.R. Dowell. 1969. Differentiation between *Clostridium sordellii* and *Clostridium bifermentans* by gas chromatography. J. Bacteriol. *100:* 528–530.

Brooks, J.B., M.J. Selin and C.C. Alley. 1976. Electron capture gas chromatography study of the acid and alcohol products of *Clostridium septicum* and *Clostridium chauvoei.* J. Clin. Microbiol. *3:* 180–185.

Broome, C.V., R.C. Moellering and B.K. Watson. 1976. Clinical significance of Lancefield groups L-T streptococci isolated from blood and cerebrospinal fluid. J. Infect. Dis. *133:* 382–392.

Brown, A.T., C.P. Christian and R.L. Eifert. 1975. Purification, characterization, and regulation of a nicotinamide adenine dinucleotide-dependent lactate dehydrogenase from *Actinomyces viscosus.* J. Bacteriol. *122:* 1126–1135.

Brown, D.W. and W.E.C. Moore. 1960. Distribution of *Butyrivibrio fibrisolvens* in nature. J. Dairy Sci. *43:* 1570–1574.

Brown, E.R. and W.B. Cherry. 1955. Specific identification of *Bacillus anthracis* by means of a variant bacteriophage. J. Infect. Dis. *96:* 34–39.

Brown, E.R., M.D. Moody, E.L. Treece and C.W. Smith. 1958. Differential diagnosis of *Bacillus cereus, Bacillus anthracis* and *Bacillus cereus* var. *mycoides.* J. Bacteriol. *75:* 499–509.

Brown, J.H. 1919. The use of blood agar for the study of streptococci. Rockefeller Institute for Medical Research Monograph No. 9. The Rockefeller Institute for Medical Research, New York.

Brown, J.M., L.K. Georg and L.C. Waters. 1969. Laboratory identification of *Rothia dentocariosa* and its occurrence in human clinical materials. Appl. Microbiol. *17:* 150–156.

Brown, R.W., O. Sandvik, R.K. Scherer and D.L. Rose. 1967. Differentiation of strains of *Staphylococcus epidermidis* isolated from bovine udders. J. Gen. Microbiol. *47:* 273–287.

Brownell, G.H. 1978. Plasmid transfer between *Nocardia erythropolis* and other nocardioform organisms. Zentralbl. Bakteriol. Parasitenkd. Infektionskr. Hyg. I Abt., Suppl. *6:* 313–317.

Brownell, G.H. and K. Denniston. 1984. Genetics of the nocardioform bacteria. *In* Goodfellow, Mordarski and Williams (Editors), The Biology of the Actinomycetes, Academic Press, London and New York, pp. 201–228.

Brownlie, L.E. 1966. Effect of some environmental factors on psychrophilic microbacteria. J. Appl. Bacteriol. *29:* 447–454.

Bruce, D.L., J.E. Bismanis and J.M. Vickerstaff. 1969. Comparative examinations of virulent *Corynebacterium kutscheri* and its presumed avirulent variant. Can. J. Microbiol. *15:* 817–818.

Brulla, W.J. and M.P. Bryant. 1980. Monensin induced changes in the major species of the rumen bacterial population. Abstr. Annu. Meet. Am. Soc. Microbiol. *1980:* 104.

Brun, Y., J. Fleurette and F. Forey. 1978. Micromethod for biochemical identification of coagulase-negative staphylococci. J. Clin. Microbiol. *8:* 503–508.

Brun, Y., J. Fleurette, F. Forey and B. Gonthier. 1981. Antibiotic susceptibility of coagulase-negative staphylococci. *In* Jeljaszewicz (Editor), Staphylococci and Staphylococcal Infections, Gustav Fischer Verlag, Stuttgart, New York, pp. 137–145.

Brundish, D.E. and J. Baddiley. 1968. Pneumococcal substance C, a ribitol teichoic acid containing choline phosphate. Biochem. J. *110:* 573–582.

Brunner, H., C.G. Gemmell, H. Huser and F.J. Fehrenbach. 1981. Chemical and biological properties of *Staphylococcus aureus* lipase. *In* Jeljaszewicz (Editor), Staphylococci and Staphylococcal Infections, Gustav Fischer Verlag, Stuttgart, New York, pp. 329–333.

Bryan, F.L. 1980. Procedures to use during outbreaks of food-borne diseases. *In* Lennette , Balows, Hausler and Truant (Editors), Manual of Clinical Microbiology, 3rd ed., American Society for Microbiology, Washington, D.C., pp. 40–51.

Bryant, M.P. 1959. Bacterial species of the rumen. Bacteriol. Rev. *23:* 125–153.

Bryant, M.P. 1972. Commentary on the Hungate technique for culture of anaerobic bacteria. Am. J. Clin. Nutr. *25:* 1324–1328.

Bryant, M.P. 1973. Nutritional requirements of the predominant rumen cellulolytic bacteria. 12th Annu. Ruminant Nutr. Conf. Fed. Proc. *32:* 1809–1813.

Bryant, M.P. 1974. Nutritional features and ecology of predominant anaerobic bacteria of the intestinal tract. Am. J. Clin. Nutr. *27:* 1313–1319.

Bryant, M.P., B.F. Barrentine, J.F. Sykes, I.M. Robinson, C.B. Shawver and L.W. Williams. 1960. Predominant bacteria in the rumen of cattle on bloat provoking ladino clover pasture. J. Dairy Sci. *43:* 1435–1444.

Bryant, M.P. and L.A. Burkey. 1953. Cultural methods and some characteristics of some of the more numerous groups of bacteria in the bovine rumen. J. Dairy Sci. *36:* 205–217.

Bryant, M.P. and L.A. Burkey. 1956. The characteristics of lactate-fermenting sporeforming anaerobes from silage. J. Bacteriol. *71:* 43–46.

Bryant, M.P. and R.N. Doetsch. 1955. Factors necessary for the growth of *Bacteroides succinogenes* in the volatile acid fraction of rumen fluid. J. Dairy Sci. *38:* 340–350.

Bryant, M.P. and I.M. Robinson. 1961a. Some nutritional requirements of the genus *Ruminococcus.* Appl. Microbiol. *9:* 91–95.

Bryant, M.P. and I.M. Robinson. 1961b. Studies on the nitrogen requirements of some ruminal cellulolytic bacteria. Appl. Microbiol. *9:* 96–103.

Bryant, M.P. and I.M. Robinson. 1961c. An improved nonselective culture medium for ruminal bacteria and its use in determining diurnal variations in numbers of bacteria in the rumen. J. Dairy Sci. *44:* 1446–1456.

Bryant, M.P. and I.M. Robinson. 1962. Some nutritional characteristics of predominant culturable ruminal bacteria. J. Bacteriol. *84:* 605–614.

Bryant, M.P. and I.M. Robinson. 1963. Apparent incorporation of ammonia and amino acid carbon during growth of selected species of ruminal bacteria. J. Dairy Sci. *46:* 150–154.

Bryant, M.P., I.M. Robinson and I.L. Lindahl. 1961. A note on the flora and fauna in the rumen of steers fed a feedlot bloat-provoking ration and the effect of penicillin. Appl. Microbiol. *9:* 511–515.

Bryant, M.P. and N. Small. 1956. The anaerobic monotrichous butyric acid-producing curved rod-shaped bacteria of the rumen. J. Bacteriol. *72:* 16–21.

Bryant, M.P. and N. Small. 1956. Characteristics of two new genera of anaerobic curved rods isolated from the rumen of cattle. J. Bacteriol. *72:* 22–26.

Bryant, M.P., N. Small, C. Bouma and I.M. Robinson. 1958. Characteristics of ruminal anaerobic cellulolytic cocci and *Cillobacterium cellulosolvens* n. sp. J. Bacteriol. *76:* 529–537.

Bryant, O.F. and L.G. Ljungdahl. 1981. Characterization of an alcohol dehydrogenase from *Thermoanaerobacter ethanolicus* active site with ethanol and secondary alcohols. Biochem. Biophys. Res. Commun. *66:* 1389–1395.

Bryant, O.F. and J. Wiegel. 1983. Comparison of alcohol dehydrogenase from wild type (JW 200) and mutant (Fe 4) strains of *Thermoanaerobacter ethanolicus.* Abstr. Annu. Meet. Am. Soc. Microbiol., K31.

Brygoo, E.R. and N. Aladame. 1953. Etude d'une espèce nouvelle anaerobic stricte du genre *Eubacterium*: *E. crispatum* n. sp. Ann. Inst. Pasteur (Paris) *84*: 640–651.

Brzin, B. 1964. Studies on the *Corynebacterium acnes*. Acta Pathol. Microbiol. Scand. *60*: 599–608.

Brzin, B. and H.P.R. Seeliger. 1975. A brief note on the CAMP phenomenon in *Listeria*. *In* Woodbine (Editor), Problems of Listeriosis. Leicester University Press, Leicester.

Buchanan, A.G. 1982. Clinical laboratory evaluation of a reverse CAMP test for presumptive identification of *Clostridium perfringens*. J. Clin. Microbiol. *16*: 761–762.

Buchanan, A.M., J.L. Scott, M.A. Gerencser, B.L. Beaman, S. Jang and E.L. Biberstein. 1984. *Actinomyces hordeovulneris* sp. nov., an agent of canine actinomycosis. Int. J. Syst. Bacteriol. *34*: 439–443.

Buchanan, B.B. and L. Pine. 1962. Characterization of a propionic acid producing actinomycete, *Actinomyces propionicus*, sp. nov. J. Gen. Microbiol. *28*: 305–323.

Buchanan, B.B. and L. Pine. 1965. Relationship of carbon dioxide to aspartic acid and glutamic acid in *Actinomyces naeslundii*. J. Bacteriol. *89*: 729–733.

Buchanan, B.B. and L. Pine. 1967. Path of glucose breakdown and cell yields of a facultative anaerobe *Actinomyces naeslundii*. J. Gen. Microbiol. *46*: 225–236.

Buchanan, R.E. 1911. Veterinary Bacteriology. W.B. Saunders Co., Philadelphia and London.

Buchanan, R.E. 1918. Studies in the nomenclature and classification of the bacteria. J. Bacteriol. *3*: 27–61.

Buchanan, R.E. and N.E. Gibbons. 1974. Bergey's Manual of Determinative Bacteriology, 8th Ed. Williams and Wilkins, Baltimore.

Buchanan, R.E., J.G. Holt and E.F. Lessel, Jr. 1966. Index Bergeyana. The Williams and Wilkins Co., Baltimore.

Bucher, E., G. Beck and K.H. Schleifer. 1980. Verkommen und Verteilung von Staphylokken und Mikrokokken in Sojaextraktions schroten. Zentralbl. Bakteriol. Mikrobiol. Hyg. Abt. I Orig. C *1*: 320–329.

Buchta, K. 1983. Lactic acid. *In* Rehm and Reed (Editors), Biotechnology, Vol. 3. Verlag Chemie, Weinheim, pp. 409–417.

Buckel, W. 1980. Analysis of the fermentation pathways of clostridia using double labeled glutamade [sic]. Arch. Microbiol. *127*: 167–169.

Buckel, W. and H.A. Barker. 1974. Two pathways of glutamate fermentation by anaerobic bacteria. J. Bacteriol. *117*: 1248–1260.

Buckwold, F.J., W.L. Albritton, A.R. Ronald, J. Lertzman and R. Henriksen. 1979. Investigations of the occurrence of gentamicin-resistant *S. aureus*. Antimicrob. Agents Chemother. *15*: 152–156.

Bujwid, O. 1889. Über die Reinkultur des *Actinomyces*. Zentralbl. Bakteriol. Hyg. *6*: 630–633.

Bulanda, M., Z. Wegrzynowicz, H. Brands, J. Jeljaszewicz, G. Pulverer and P.B. Heczko. 1981. Occurrence and characterization of staphylococcal strains showing pseudocoagulase activity. *In* Jeljaszewicz (Editor), Staphylococci and Staphylococcal Infections, Gustav Fischer Verlag, Stuttgart, New York, pp. 37–41.

Bulkacz, J., M.G. Newman, S.S. Socransky, E. Newbrun and D.F. Scott. 1979. Phospholipase A activity of micro-organisms from dental plaque. Microbios. Lett. *10*: 79–88.

Bulla, L.A., Jr. and E.S. Sharpe. 1978. Biology of *Bacillus popillae*. Adv. Appl. Microbiol. *23*: 1–18.

Bullen, C.L., P.V. Tearle and M.G. Stewart. 1977. The effect of "humanised" milks and supplemented breast feeding on the faecal flora of infants. J. Med. Microbiol. *10*: 403–413.

Bulloch, W., W.E. Bullock, S.R. Douglas, H. Henry, J. McIntoch, R.A. O'Brien, M. Robertson and C.G.L. Wolf. 1919. Report on the anaerobic infections of wounds and the bacteriological and serological problems arising therefrom. Med. Res. Comm. (Gt. Brit.) Spec. Rep. Ser. *39*: 1–182.

Bullock, G.L., H.M. Stuckey and P.K. Chen. 1974. Corynebacterial kidney disease of salmonids: growth and serological studies on the causative bacterium. Appl. Microbiol. *28*: 811–814.

Burckhardt, J.J. 1978. Rat memory T lymphocytes: *in vitro* proliferation induced by antigens of *Actinomyces viscosus*. Scand. J. Immunol. *7*: 167–172.

Burckhardt, J.J., R. Gaegauf-Zollinger, R. Schmid and B. Guggenheim. 1981. Alveolar bone loss in rats after immunization with *Actinomyces viscosus*. Infect. Immun. *31*: 971–977.

Burges, H.D. 1984. Nomenclature of *Bacillus thuringiensis* with abbreviations. Mosq. News *44*: 66–68.

Burges, H.D., K. Aizawa, H.T. Dulmage and H. de Barjac. 1982. Numbering the H-serotypes of *Bacillus thuringiensis*. J. Invert. Pathol. *40*: 419.

Burke, K.A., A.E. Brown and J. Lascelles. 1981. Membrane and cytoplasmic nitrate reductase of *Staphylococcus aureus* and application of crossed immunoelectrophoresis. J. Bacteriol. *148*: 724–727.

Burke, K.A. and J. Lascelles. 1975. Nitrate reductase system in *Staphylococcus aureus* wild type and mutants. J. Bacteriol. *123*: 308–316.

Burke, W.F. and K.O. McDonald. 1983. Naturally occurring antibiotic resistance in *Bacillus sphaericus* and *Bacillus licheniformis*. Curr. Microbiol. *9*: 69–72.

Burley, S.K. and R.G.E. Murray. 1983. Structure of the regular surface layer of *Bacillus polymyxa*. Can. J. Microbiol. *29*: 775–780.

Burri, R. and O. Ankersmit. 1906. *Bacterium clostridiiforme*. *In* P. Ankersmit (Editor), Untersuchunger über die Bakterien im Verdauungskanal des Rindes. Zentralbl. Bakteriol. Parasitenkd. Infektionskr. Hyg. Abt. I Orig. *40*: 100–118.

Butler, T. and S. Pitt. 1982. Spontaneous bacterial peritonitis due to *Clostridium tertium*. Gastroenterology *82*: 133–134.

Buxton, D. 1978a. Further studies on the mode of action of *Clostridium welchii* type-D epsilon toxin. J. Med. Microbiol. *11*: 293–298.

Buxton, D. 1978b. In-vitro effects of *Clostridium welchii* type-D epsilon toxin on guinea pig, mouse, rabbit and sheep cells. J. Med. Microbiol. *11*: 299–302.

Cagle, G.D. 1974. Fine structure and distribution of extracellular polymer surrounding selected aerobic bacteria. Can. J. Microbiol. *21*: 395–408.

Cain, R.B. 1981. Regulation of aromatic and hydroaromatic catabolic pathways in nocardioform actinomycetes. Zentralbl. Bakteriol. Mikrobiol. Hyg. I Abt. Orig., Suppl. *11*: 335–354.

Caldwell, D.R. and M.P. Bryant. 1966. Medium without rumen fluid for nonselective enumeration and isolation of rumen bacteria. Appl. Microbiol. *14*: 794–801.

Caldwell, D.R. and R.F. Hudson. 1974. Sodium, and obligate growth requirement for predominant rumen bacteria. Appl. Microbiol. *27*: 549–552.

Calmette, A. and C. Guerin. 1908. Sur quelques proprietes du bacille tubercuculeux culture sur la bile. C.R. Acad. Sci. *147*: 1456–1459.

Campbell, A. 1981. Evolutionary significance of accessory DNA elements in bacteria. Annu. Rev. Microbiol. *35*: 55–83.

Campbell, L.L., H.A. Frank and E.R. Hall. 1957. Studies on thermophilic sulfate-reducing bacteria. I. Identification of *Sporovibrio desulfuricans* as *Clostridium nigrificans*. J. Bacteriol. *73*: 516–521.

Campbell, L.L. and J.R. Postgate. 1965. Classification of the spore-forming sulfate-reducing bacteria. Bacteriol. Rev. *29*: 359–363.

Campbell, R. 1984. Plant Microbiology, Arnold, London.

Canale-Parola, E. 1970. Biology of the sugar-fermenting sarcinae. Bacteriol. Rev. *34*: 82–97.

Canale-Parola, E., R. Barasky and R.S. Wolfe. 1961. Studies on *Sarcina ventriculi*. III. Localization of cellulose. J. Bacteriol. *81*: 311–318.

Canale-Parola, E., M. Mandel and D.G. Kupfer. 1967. The classification of sarcinae. Arch. Mikrobiol. *58*: 30–34.

Canale-Parola, E. and R.S. Wolfe. 1960a. Studies on *Sarcina ventriculi*. I. Stock culture method. J. Bacteriol. *79*: 857–859.

Canale-Parola, E. and R.S. Wolfe. 1960b. Studies on *Sarcina ventriculi*. II. Nutrition. J. Bacteriol. *79*: 860–862.

Canale-Parola, E. and R.S. Wolfe. 1964. Synthesis of cellulose by *Sarcina ventriculi*. Biochem. Biophys. Acta *82*: 403–405.

Cancho, F.G., L.R. Navarro and R. de la Borbolla y Alcala. 1980. La formacion de acido propionico durante la conservacion de las aceitunas verdes de mesa. III. Microorganismos responsables. Grasas Aceites *31*: 245–250.

Cancho, F.G., M. Nosti Vega, M. Fernandez Diaz and N.J.Y. Buzcu. 1970. Especies de *Propionibacterium* relacionadas con la zapateria. Factores que influyen en su desarrollo. Microbiol. Esp. *23*: 233–252.

Candeli, A., V. Mastrandrea, G. Cenci and A. de Bartolomeo. 1978. Sensitivity to lytic agents and DNA base composition of several aerobic endospore-bearing bacilli. Zentralbl. Parasitenkd. Infektionskr. Hyg. II Abt. *133*: 250–260.

Cantoni, C., M.A. Bianchi, P. Renon and S. D'Autent. 1969. Ricerche sulla putrefazione del prosciutto crudo. Arch. Veterin. Ital. *20*: 355–370.

Carbone, D. and A. Tombolato. 1917. Sulla macerazione rustica della canapa. Staz. Sper. Agr. Ital. *50*: 563–575.

Cardenos, J. and L.E. Mortenson. 1975. Role of molybdenum in dinitrogen fixation by *Clostridium pasteurianum*. J. Bacteriol. *123*: 978–984.

Cardon, B.P. and H.A. Barker. 1946. Two new amino acid-fermenting bacteria, *Clostridium propionicum* and *Diplococcus glycinophilus*. J. Bacteriol. *52*: 629–634.

Carlile, M.J., J.F. Collins and B.E.B. Moseley (Editors). 1981. Molecular and Cellular Aspects of Microbial Evolution, Cambridge University Press, Cambridge.

Carlson, A., A. Kellner, A.W. Bernheimer and E.B. Freeman. 1957. A streptococcal enzyme that acts especially on DPN; characterization on the enzyme and its separation from streptolysin O. J. Exp. Med. *106*: 15–26.

Carlson, J.R., J.M. Sherrill, J.E. Rosenblatt and L.R. McCarthy. 1981. Penicillinase activity in three strains of *Clostridium butyricum*. Curr. Microbiol. *5*: 251–254.

Carlson, R.R. and A.K. Vidaver. 1982. Taxonomy of *Corynebacterium* plant pathogens, including a new pathogen of wheat, based on polyacrylamide gel electrophoresis of cellular proteins. Int. J. Syst. Bacteriol. *32*: 315–326.

Carlsson, J. 1965. Zooglea-forming streptococci, resembling *Streptococcus sanguis*, isolated from dental plaque in man. Odontol. Revy *16*: 348–358.

Carlsson, J. 1967a. Presence of various types of non-haemolytic streptococci in dental plaque and in other sites of the oral cavity in man. Odontol. Revy *18*: 55–74.

Carlsson, J. 1967b. A medium for isolation of *Streptococcus mutans*. Arch. Oral Biol. *12*: 1657–1658.

Carlsson, J. 1968. A numerical taxonomic study of human oral streptococci. Odontol. Revy *19*: 137–160.

Carlsson, J. 1970. Chemically defined medium for growth of *Streptococcus sanguis*. Caries Res. *4*: 297–304.

Carlsson, J. 1971. Nutritional requirements of *Streptococcus salivarius*. J. Gen. Microbiol. *67*: 69–76.

Carlsson, J., H. Grahnen, G. Jonsson and S. Wikner. 1970. Establishment of *Streptococcus sanguis* in the mouths of infants. Arch. Oral Biol. *15:* 1143–1148.

Carlsson, J., G.P.D. Granberg, G.K. Nyberg and M.-B.K. Edlund. 1979. Bacteriocidal effect of cysteine exposed to atmospheric oxygen. Appl. Environ. Microbiol. *37:* 383–390.

Carlsson, J. and C.J. Griffith. 1974. Fermentation products and bacterial yields in glucose-limited and nitrogen-limited cultures of streptococci. Arch. Oral Biol. *19:* 1105–1109.

Carlsson, J., G. Nyberg and J. Wrethén. 1978. Hydrogen peroxide and superoxide radical formation in anaerobic broth media exposed to atmospheric oxygen. Appl. Environ. Microbiol. *36:* 223–229.

Carne, H.R. 1940. The toxin of *Corynebacterium ovis.* J. Pathol. Bacteriol. *51:* 199–212.

Carne, H.R. 1968. Action of bacteriophages obtained from *Corynebacterium diphtheriae* on *C. ulcerans* and *C. ovis.* Nature (Lond.) *217:* 1066–1067.

Carne, H.R. and E.O. Onon. 1982. The exotoxin of *Corynebacterium ulcerans.* J. Hyg. *88:* 173–191.

Carne, H.R., N. Wickham and J.C. Kater. 1956. A toxic lipid from the surface of *Corynebacterium ovis.* Nature (Lond.) *178:* 701–702.

Carr, J.G. 1957. Occurrence and activity of some lactic acid bacteria from apple juice, ciders and perries. J. Inst. Brew. *63:* 436–440.

Carr, J.G. 1970. Tetrad-forming cocci in ciders. J. Appl. Bacteriol. *33:* 371–379.

Carr, J.G., C.V. Cutting and G.C. Whiting (Editors). 1975. Lactic acid bacteria in beverages and food. Proceedings of a Symposium held at Long Ashton research station University of Bristol 19–21 Sept. 1973. Academic Press, London, New York, San Francisco.

Carr, J.G. and P.A. Davies. 1970. Homofermentative lactobacilli of ciders including *Lactobacillus mali* nov. spec. J. Appl. Bacteriol. *33:* 768–774.

Carr, J.G. and P.A. Davies. 1972. The ecology and classification of strains of *Lactobacillus collinoides* nov. spec.: A bacterium commonly found in fermenting apple juice. J. Appl. Bacteriol. *35:* 463–471.

Carr, J.G., P.A. Davies, F. Dellaglio, M. Vescovo and R.A.D. Williams. 1977. The relationship between *Lactobacillus mali* from cider and *Lactobacillus yamanashiensis* from wine. J. Appl. Bacteriol. *42:* 219–228.

Carr, L.D. and W.E. Kloos. 1977. Temporal study of the staphylococci and micrococci of normal infant skin. Appl. Environ. Microbiol. *34:* 673–680.

Carreira, L.H. and L.G. Ljungdahl. 1983. High-ethanol-producing derivatives of *Thermoanaerobacter ethanolicus.* Abstr. Annu. Meet. Am. Soc. Microbiol., O4.

Carreira, L.H., L.G. Ljungdahl, O.F. Bryant, M. Szulcýnski and J. Wiegel. 1982. Control of products formation with *Thermoanaerobacter ethanolicus:* Enzymology and physiology. *In* Iteda and Beppu (Editors), Proceedings of a Conference on Genetics of Industrial Microorganisms. Tokyo, Japan. pp. 351–355.

Carreira, L.H., J. Wiegel and L.G. Ljungdahl. 1983. Production of ethanol from biopolymers by anaerobic thermophilic and extreme thermophilic bacteria. I. Regulation of carbohydrate utilization in mutants of *Thermoanaerobacter ethanolicus.* Biotech. Bioeng. Symp. *13:* 183–191.

Carroll, K.K., J.H. Cutts and E.G.D. Murray. 1968. The lipids of *Listeria monocytogenes.* Can. J. Biochem. *46:* 899–904.

Cars, O., V. Forsum and E. Hjelm. 1975. New immunofluorescence method for the identification of group A, B, C, E, F, and G streptococci. Acta Pathol. Microbiol. Scand. Ser. B *83:* 145–152.

Casal, M. and J. Rey Calero. 1974. *Mycobacterium gadium* sp. nov. A new species of rapid-growing scotochromogenic mycobacteria. Tubercle *55:*299–308.

Cashore, W.J., G. Peter, M. Lauermann, B.S. Stonestreet and W. Oh. 1981. Clostridia colonization and clostridial toxin in neonatal necrotizing enterocolitis. J. Pediatr. *98:* 308–311.

Casida, L.E., Jr. 1965. Abundant microorganism in soil. Appl. Microbiol. *13:* 327–334.

Casida, L.E., Jr. 1972. Interval scanning photomicrography of microbial cell populations. Appl. Microbiol. *23:* 190–192.

Casida, L.E., Jr. 1977. Small cells in pure cultures of *Agromyces ramosus* and in natural soil. Can. J. Microbiol. *23:* 214–216.

Casida, L.E., Jr. 1983. Interaction of *Agromyces ramosus* with other bacteria in soil. Appl. Environ. Microbiol. *46:* 881–888.

Cassity, T.R., B.J. Kolodziej and R.M. Pfister. 1978. Ultrastructure of the capsule of *B. megaterium* ATCC 19213. Microbios *21:* 153–160.

Castellani, A. 1942. *Micrococcus (Coccaceus) mycetoides,* agent étiologique de l'ulcération tropicaloide. Ann. Igiene *42:* 349.

Castellani, A. and A.J. Chalmers. 1913. Manual of Tropical Medicine, 2nd Ed. Balliere, Tindall and Cox, London, pp. 1–1719.

Castellani, A. and A.J. Chalmers. 1919. Manual of Tropical Medicine. 3rd Ed. Williams, Wood and Co., New York, pp. 1–2436.

Castelnuovo, G., A. Guandiano, M. Morellini, G. Penso and C. Rossi. 1960. Gli antigeni dei micobatteri. Rend. Ist. Super. Sanita. 23:1222–1223.

Castelnuovo, G. and M. Morellini. 1962. Gli antigeni di alcuni dei cosidetti "Micobatteri atipici" O "anonimi". Ann. Ist. Carlo Forlanini. 22:1–20.

Castets, M., H. Boisvert, F. Grumbach, M Brunel and N. Rist. 1968. Les bacilles tuberculeux de type Africain. Rev. Tuberc. Pneumol. *32:* 179–184.

Castets, M., N. Rist and H. Boisvert. 1969. La variete africaine du bacille tuberculeux humain. Med. Afr. Noire. *16:* 321–322.

Cato, E.P. 1983. Transfer of *Peptococcus parvulus* (Weinberg, Nativelle and Prévot 1937) Smith 1957 to the genus *Streptococcus: Streptococcus parvulus* (Weinberg, Nativelle and Prévot 1937) comb. nov., nom. rev., emend. Int. J. Syst. Bacteriol. *33:* 82–84.

Cato, E.P., C.S. Cummins and L. D. Smith. 1970. *Clostridium limosum* André *In* Prévot 1948, 165. Amended description and pathogenic characteristics. Int. J. Syst. Bacteriol. *20:* 305–316.

Cato, E.P., D.E. Hash, L.V. Holdeman and W.E.C. Moore. 1982a. Electrophoretic study of *Clostridium* species. J. Clin. Microbiol. *15:* 688–702.

Cato, E.P., L.V. Holdeman and W.E.C. Moore. 1981. Designation of *Eubacterium limosum* (Eggerth) Prévot as the type species of *Eubacterium.* Request for an opinion. Int. J. Syst. Bacteriol. *31:* 209–210.

Cato, E.P., L.V. Holdeman and W.E.C. Moore. 1982b. *Clostridium perenne* and *Clostridium paraperfringens:* later subjective synonyms of *Clostridium barati.* Int. J. Syst. Bacteriol. *32:* 77–81.

Cato, E.P., J.L. Johnson, D.E. Hash and L.V. Holdeman. 1983. Synonymy of *Peptococcus glycinophilus* (Cardon and Barker 1946) Douglas 1957 with *Peptostreptococcus micros* (Prévot 1933) Smith 1957 and electrophoretic differentiation of *Peptostreptococcus micros* from *Peptococcus magnus* (Prévot 1933) Holdeman and Moore 1972. Int. J. Syst. Bacteriol. *33:* 207–210.

Cato, E.P., W.E.C. Moore and L.V. Holdeman. 1968. *Clostridium oroticum* comb. nov. amended description. Int. J. Syst. Bacteriol. *18:* 9–13.

Cato, E.P., W.E.C. Moore and J.L. Johnson. 1983. Synonymy of strains of "*Lactobacillus acidophilus*" groups A2 (Johnson et al. 1980) with the type strain of *Lactobacillus crispatus* (Brygoo and Aladame 1953) Moore and Holdeman 1970. Int. J. Syst. Bacteriol. *33:* 426–428.

Cato, E.P., W.E.C. Moore, G. Nygaard and L.V. Holdeman. 1984. *Actinomyces meyeri* sp. nov., specific epithet rev. Int. J. Syst. Bacteriol. *34:* 487–489.

Cato, E.P. and W.E.C. Salmon. 1976. Transfer of *Bacteroides clostridiiformis* subsp. *clostridiiformis* (Burri and Ankersmit) Holdeman and Moore and *Bacteroides clostridiiformis* subsp. *girans* (Prévot) Holdeman and Moore to the genus *Clostridium* as *Clostridium clostridiiforme* (Burri and Ankersmit) comb. nov.: emendation of description and designation of neotype strain. Int. J. Syst. Bacteriol. *26:* 205–211.

Cato, E.P., C.W. Salmon and L.V. Holdeman. 1974. *Eubacterium cylindroides* (Rocchi) Holdeman and Moore: emended description and designation of neotype strain. Int. J. Syst. Bacteriol. *24:* 256–259.

Cavett, J.J., G.J. Dring and A.W. Knight. 1965. Bacterial spoilage of thawed frozen peas. J. Appl. Bacteriol. *28:* 241–251.

Cazzamali, P. and R. Miglierina. 1933. La batteriologia delle peritoniti acute. Arch. Ital. Chir. *34:* 573–675.

Chadwick, P. 1963. Rapid screening test for *Bacillus anthracis.* Can. J. Microbiol. *9:* 734–737.

Chamoiseau, G. 1973. *M. farcinogenes* agent causal du farcin du boeuf en Afrique. Ann. Microbiol. (Inst. Pasteur) *124*A: 214–222.

Chamoiseau, G. 1979. Etiology of farcy in African bovines: nomenclature of the causal organisms *Mycobacterium farcinogenes* Chamoiseau and *Mycobacterium senegalense* (Chamoiseau) comb. nov. Int. J. Syst. Bacteriol. *29:* 407–410.

Champion, A.B. and J.C. Rabinowitz. 1977. Ferredoxin and formyltetrahydrofolate synthetase: comparative studies with *Clostridium acidiurici, Clostridium cylindrosporum,* and newly isolated anaerobic uric acid-fermenting strains. J. Bacteriol. *132:* 1003–1020.

Chan, M., R.H. Himes and J.M. Akagi. 1971. Fatty acid composition of thermophilic, mesophilic, and psychrophilic clostridia. J. Bacteriol. *106:* 876–881.

Chaparas, S.D., T.M. Brown and I.S. Hyman. 1978a. Antigenic relationships of various mycobacterial species with *M. tuberculosis.* Am. Rev. Resp. Dis. *117:* 1091–1097.

Chaparas, S.D., T.M. Brown and I.S. Hyman. 1978b. Antigenic relationships among species of *Mycobacterium* studied by fused rocket immunoelectrophoresis. Int. J. Syst. Bacteriol. *28:* 547–560.

Chapel, T., W.J. Brown, C. Jefferies and J.A. Stewart. 1978. The microbiological flora of penile ulcerations. J. Infect. Dis. *137:* 50–56.

Chapman, G.B. 1960. Electron microscopy of cellular division in *Sarcina lutea.* J. Bacteriol. *79:* 132–136.

Chapman, G.H. 1944. The isolation of streptococci from mixed cultures. J. Bacteriol. *48:* 113–114.

Charlton, D.B., M.E. Nelson and C.H. Werkman. 1934. Physiology of *Lactobacillus fructivorans* sp. nov. isolated from spoiled salad dressing. Iowa State J. Sci. *9:* 1–11.

Charney, J.C., W.P. Fischer and C.P. Hegarty. 1951. Manganese as an essential element for sporulation in the genus *Bacillus.* J. Bacteriol. *62:* 145–148.

Chassy, B.M., E.M. Gibson and A. Fiuffrida. 1976. Evidence for extrachromosomal elements in lactobacilli. J. Bacteriol. *127:* 1576–1578.

Chassy, B.M., L.J. Lee, J.B. Hansen and K. Jagusztyn-Krynicka. 1983. Molecular cloning of *Lactobacillus casei* lactose metabolic genes. Dev. Ind. Microbiol. *24:* 97–107.

Chassy, B.M. and E. Rokow. 1981. Conjugal transfer of lactose plasmids in *Lactobacillus casei. In* Levy, Clowes and Koenig (Editors), Molecular Biology, Pathogenesis and Ecology of Bacterial Plasmids. Plenum Press, New York, p. 590.

Chassy, B.M. and J. Thompson. 1983. Regulation and characterization of the galactose-phosphoenolpyruvate-dependent phosphotransferase system in *Lactobacillus casei.* J. Bacteriol. *154:* 1204–1214.

Chatelain, R. and L. Second. 1966. Taxonomie numérique de quelques *Brevibacterium.* Ann. Inst. Pasteur *111:* 630–644.

Chatterjee, K.R., B. Dasgupta, N. Mukherjee and H.N. Ray. 1956. Some cytochemical observations in *Mycobacterium leprae.* Bull. Calcutta Sch. Trop. Med. *4:* 18

Chatton, E. and C. Pérard. 1913. Schizophytes du caecum du cobaye. I. *Oscillospira guilliermondi* n. g., n. sp. C.R. Soc. Biol. Paris *74:* 1159–1162.

Chatton, E. and C. Pérard. 1913b. Schizophytes du caecum de cobayae. II. *Metabacterium polyspora* n.g., n.sp. C.R. Soc. Biol. Paris *74:* 1232–1234.

Chen, J.-S. and D.K. Blanchard. 1978. Isolation and properties of a unidirectional H₂-oxidizing hydrogenase from the strictly anaerobic N₂-fixing bacterium *Clostridium pasteurianum*. Biochem. Biophys. Res. Comm. *84:* 1144–1150.

Chen, K.C. and A.W. Ravin. 1966. Heterospecific transformation of pneumococcus and streptococcus. J. Mol. Biol. *22:* 109.

Chen, M. and M.J. Wolin. 1979. Effect of monensin and lasalocid-sodium on the growth of methanogenic and rumen saccharolytic bacteria. Appl. Environ. Microbiol. *38:* 72–77.

Cheng, K.-J. and J.W. Costerton. 1977. Ultrastructure of *Butyrivibrio fibrisolvens*: a gram-positive bacterium? J. Bacteriol. *129:* 1506–1512.

Cheng, K.-J., D. Dinsdale and C.S. Stewart. 1979. Maceration of clover and grass leaves by *Lachnospira multiparus*. Appl. Environ. Microbiol. *38:* 723–729.

Cheng, K.-J., G.A. Jones, F.J. Simpson and M.P. Bryant. 1959. Isolation and identification of rumen bacteria capable of anaerobic rutin degradation. Can. J. Microbiol. *15:* 1365–1371.

Cheravach, L.M., T.P. Tourova and V.L. Belikova. 1983. DNA-DNA homology studies among strains of *Kurthia zopfii*. FEMS Microbiol. Lett. *19:* 243–245.

Cherniak, R. and H.M. Frederick. 1977. Capsular polysaccharide of *Clostridium perfringens* Hobbs 9. Infect. Immun. *15:* 765–771.

Chesbro, W.R. and K. Auborn. 1967. Enzymatic detection of the growth of *Staphylococcus aureus* in foods. Appl. Microbiol. *15:* 1150–1159.

Chesbro, W.R. and J.B. Evans. 1959. Factors affecting the growth of enterococci in highly alkaline media. J. Bacteriol. *78:* 858–862.

Chester, F.D. 1897. Report of the mycologist; Bacteriological work. Del. Agr. Exp. Sta. Bull. *9:* 38–145.

Chester, F.D. 1898. Report of the mycologist: Bacteriological work. Del. Agr. Expt. Sta. Bull. *10:* 47–137.

Chester, F.D. 1901. A Manual of Determinative Bacteriology. The Macmillan Co., New York.

Chiappini, M.G. 1966. Carbon dioxide fixation in some strains of the species *Bifidobacterium bifidum*, *Bifidobacterium constellatum*, *Actinomyces bovis* and *Actinomyces israelii*. Ann. Microbiol. Enzimol. *16:* 25–32.

Chiba, S. and M. Ishimoto. 1973. Ferredoxin-linked nitrate reductase from *Clostidium perfringens*. J. Biochem. *73:* 1315–1318.

Chiron, J.P., P. Mauras and F. Denis. 1977. Ultra structure des bactériophages de *Listeria monocytogenes*. C.R. Soc. Biol. (Paris) *171:* 488–492.

Cho, K.Y. and C.H. Doy. 1973. Ultrastructure of the obligately anaerobic bacteria *Clostridium kluyveri* and *Cl. acetobutylicum*. Aust. J. Biol. Sci. *26:* 547–558.

Choukévitch, J. 1911. Étude de la flore bactérienne du gros intestin du cheval. Ann. Inst. Pasteur (Paris) *25:* 245–368.

Chow, A.W. and D. Bednorz. 1978. Comparative in vitro activity of newer cephalosporins against anaerobic bacteria. Antimicrob. Agents Chemother. *14:* 668–671.

Chow, A.W., D. Bednorz and L.B. Guze. 1977. Susceptibility of obligate anaerobes to metronidazole: an extended study of 1,054 clinical isolates. *In* S.M. Finegold (Editor), Metronidazole. Excerpta Medica, Princeton, N.J. pp. 286–292.

Chow, A.W., V. Patten and L.B. Guze. 1975. Comparitive susceptability of anaerobic bacteria to minocycline, doxycycline, and tetracycline. Antimicrob. Agents Chemother. *7:* 46–49.

Chow, A.W., V. Patten and L.B. Guze. 1975. Susceptibility of anaerobic bacteria to metronidazole: Relative resistance of non-sporeforming Gram-positive bacilli. J. Infect. Dis. *131:* 182–185.

Christensen, E.A. and H. Kristensen. 1981. Radiation-resistance of microorganisms from air in clean premises. Acta Pathol. Microbiol. Scand. Sect. B *89:* 293–301.

Christensen, G.D., W.A. Simpson, A.L. Bisno and E.H. Beachey. 1982. Adherence of slime-producing strains of *Staphylococcus epidermidis* to smooth surfaces. Infect. Immun. *37:* 318–326.

Christensen, P., G. Kahlmeter, S. Jonsson and G. Kronvall. 1973. New method for the serological grouping of streptococci with specific antibodies adsorbed to Protein A-containing staphylococci. Infect. Immun. *7:* 881–885.

Christiansen, M. 1934a. Position de *M. indolicus* dans la systématique bactériénne. Acta Pathol. Microbiol. Scand. *11:* 363–366.

Christiansen, M. 1934b. Ein obligat anaerober, gasbildender, indolpositiver Mikrokokkus (*Micrococcus indolicus* n. sp.). Acta Pathol. Microbiol. Scand. Suppl. *18:* 42–63.

Christie, A.O. and J.W. Porteus. 1962a. The cultivation of a single strain of *Actinomyces israelii* in a simplified and chemically defined medium. J. Gen. Microbiol. *28:* 443–454.

Christie, A.O. and J.W. Porteus. 1962b. The growth factor requirements of the Wills strain of *Actinomyces israelii* growing in a chemically defined medium. J. Gen. Microbiol. *28:* 455–460.

Christie, A.O. and J.W. Porteus. 1962c. Growth of several strains of *Actinomyces israelii* in chemically defined media. Nature *195:* 408–409.

Christie, R., N.E. Atkins and E. Munch-Petersen. 1944. A note on a lytic phenomenon shown by group B streptococci. Aust. J. Exp. Biol. Med. Sci. *22:* 197–200.

Chung, K.T. 1976. Inhibitory effects of H₂ on growth of *Clostridium cellobioparum*. Appl. Environ. Microbiol. *31:* 342–348.

Ciccarelli, A.S., D.N. Whaley, L.M. McCroskey, D.F. Giménez, V.R. Dowell, Jr.

and C.L. Hatheway. 1977. Cultural and physiological characteristics of *Clostridium botulinum* type G and the susceptibility of certain animals to its toxin. Appl. Environ. Microbiol. *34:* 843–848.

Cisar, J.O. and A.E. Vatter. 1979. Surface fibrils (fimbriae) of *Actinomyces viscosus* T14V. Infect. Immun. *24:* 523–531.

Cisar, J.O., A.E. Vatter and F.C. McIntire. 1978. Identification of the virulence-associated antigen on the surface fibrils of *Actinomyces viscosus* T14. Infect. Immun. *19:* 312–319.

Claflin, L.E. and J.E. Shepard. 1977. An agglutination test for the seradiagnosis of *Corynebacterium sepedonicum*. Am. Potato J. *54:* 331–338.

Clagett, J., D. Engel and E. Chi. 1980. *In vitro* expression of immunoglobulin M and G subclasses by murine B lymphocytes in response to a polyclonal activator from *Actinomyces*. Infect. Immun. *29:* 234–243.

Clancey, J.K. 1964. Mycobacterial skin ulcers in Uganda: description of a new mycobacterium (*Mycobacterium buruli*). J. Pathol. Bacteriol. *88:* 175.

Clark, F.E. 1951. The generic classification of certain cellulolytic bacteria. Proc. Soil Sci. Soc. Am. *15:* 180–182.

Clark, F.E. 1953. Criteria suitable for species differentiation in *Cellulomonas* and a revision of the genus. Int. Bull. Bacteriol. Nomencl. Taxon. *3:* 179–199.

Clark, F.E. 1955. The designation of *Corynebacterium ureafaciens* Krebs and Eggleston as *Arthrobacter ureafaciens* (Krebs and Eggleston) comb. nov. Int. Bull. Bacteriol. Nomencl. Taxon. *5:* 111–113.

Clark, H.F. and C.C. Shepard. 1963. Effect of environmental temperatures on infection with *Mycobacterium marinum* (balnei) of mice and a number of poikilothermic species. J. Bacteriol. *86:* 1057–1069.

Clark, J.B. 1972. Morphogenesis in the genus *Arthrobacter*. CRC Crit. Rev. Microbiol. *1:* 521–544.

Clark, J.B. 1979. Sphere-rod transitions in *Arthrobacter*. *In* Parish (Editor), Developmental Biology of Prokaryotes, Blackwell Scientific Publications, Oxford, pp. 73–92.

Clark, J.E., S.W. Ragsdale, L.G. Ljungdahl and J. Wiegel. 1982. Levels of enzymes involved in the synthesis of acetate from CO₂ in *Clostridium thermoautotrophicum*. J. Bacteriol. *151:* 507–509.

Clarke, D.J., F.M. Fuller and J.G. Morris. 1979. The membrane adenosine triphosphatase of *Clostridium pasteurianum*. Effects of key intermediates of glycolysis on its ATP phosphohydrolase activity. FEBS Lett. *100:* 52–56.

Clarke, D.J. and J.G. Morris. 1976. Butyricin 7423: a bacteriocin produced by *Clostridium butyricum* NCIB 7423. J. Gen. Microbiol. *95:* 67–77.

Clarke, J.K. 1924. On the bacterial factor in the aetiology of dental caries. Br. J. Exp. Pathol. *5:* 141–147.

Clarke, P.A. and M.V. Tracey. 1956. The occurence of chitinase in some bacteria. J. Gen. Microbiol. *14:* 188–196.

Claus, D. 1965. Anreicherung und Direktisolierung aerober sporenbildender Bakterien. *In* Schlegel (Editor), Anreicherungskultur und Mutantenauslese. Gustav Fischer Verlag, Stuttgart, pp. 337–362.

Claus, D. 1981. The genus *Sporosarcina*. *In* Starr, Stolp, Trüper, Balows, and Schlegel (Editors), The Prokaryotes. A Handbook on Habits, Isolation, and Identification of Bacteria. Springer-Verlag, Berlin, pp. 1804–1807.

Claus, D., A.A. Chowdhury and K.V. Tilak. 1970. On cultivation and preservation of *Sarcina ventriculi*. Publ. Fac. Sci. Univ. J.E. Purkyně, Brno, *47:* 135–142.

Claus, D., F. Fahmy, H.J. Rolf and N. Tosonoglu. 1983. *Sporosarcina halophila* sp. nov., an obligate, slightly halophilic bacterium from salt marsh soils. Syst. Appl. Microbiol. *4:* 496–506.

Claus, D., F. Fahmy, H.J. Rolf and N. Tosonoglu. 1984. *In* Validation of the publication of new names and new combinations previously effectively published outside the IJSB. Int. J. Syst. Bacteriol. *34:* 270–271.

Claus, D., P. Lack and B. Neu (Editors). 1956. Catalog of strains, 3rd ed. Deutsche Sammlung von Mikroorganismen. Braunschweig, Germany.

Claus, D. and H. Wilmanns. 1974. Enrichment and selective isolation of *Sarcina maxima* Lindner. Arch. Microbiol. *96:* 201–204.

Claussen, N.H. 1903. Études sur les bactéries dites sarcines et sur les maladies quelles provoquent dans la bière. C.R. Trav. Lab. Carlsberg *6:* 64–83.

Clemo, G.R. and A.F. Daglish. 1950. The phenazine series. Part VIII. The constitution of the pigment of *Chromobacterium iodinum*. J. Chem. Soc. 1481–1485.

Clemo, G.R. and H. McIlwain. 1938. The phenazine series. Part VII. The pigment of *Chromobacterium iodinum*; the phenazine di-N-oxides. J. Chem. Soc. 479–483.

Clewell, D.B. 1981. Plasmids, drug resistance and gene transfer in the genus *Streptococcus*. Microbiol. Rev. *45:* 409–436.

Clewell, D.B. and A.E. Franke. 1974. Characterization of a plasmid determining resistance to Erythromycin, Lincomycin and Vernamycin Bα in a strain of *Streptococcus pyogenes*. Antimicrob. Agents Chemother. *3:* 534–537.

Clewell, D.B., D.R. Oliver, G.M. Dunny, A.E. Franke, Y. Yagi, J. van Houte and B.L. Brown. 1976. Plasmids in cariogenic streptococci. *In* Stiles, Loesche and O'Brien (Editors), Microbial Aspects of Dental Caries, Vol. 3. Information Retrieval Inc., Washington, D.C., pp. 713–724.

Closs, O., R.N. Mshana and M. Harboe. 1979. Antigenic analysis of *Mycobacterium leprae*. Scand. J. Immunol. *9:* 297–302.

Cmelik, S.H.W. 1980. Fermentation of 1,2-0-iso-propylidene-D-glucofuranose ("mono-acetone glucose") by anaerobic bacteria. Zentralbl. Bakteriol. Parasitenkd. Infektionskr. Hyg. Abt. Orig. A *247:* 495–501.

Coates, A.R.M., J. Hewitt, B.W. Allen, J. Ivanyi and D.A. Mitchison. 1981. Antigenic diversity of *Mycobacterium tuberculosis* and *Mycobacterium bovis* detected by means of monoclonal antibodies. The Lancet *ii*: 167–169.

Coetzee, J.N., H.C. de Klerk and T.G. Sacks. 1960. Host-range of *Lactobacillus* bacteriophages. Nature *187*: 348–349.

Cogan, T.M. 1976. The utilization of citrate by lactic acid bacteria in milk and cheese. Dairy Ind. Int. *41*: 12–16.

Cogan, T.M., M. O'Dowd and D. Mellerick. 1981. Effects of pH and sugar on acetoin production from citrate by *Leuconostoc lactis*. Appl. Environ. Microbiol. *41*: 1–8.

Cohen, J.O. 1972. Serotyping of staphylococci. *In* Cohen (Editor), The Staphylococci, Wiley-Interscience (John Wiley and Sons, Inc.), New York, pp. 419–430.

Cohen, M.L., E.S. Wong and S. Falkow. 1982. Common R-plasmids in *Staphylococcus aureus* and *Staphylococcus epidermidis* during a nosocomial *Staphylococcus aureus* outbreak. Antimicrob. Agents Chemother. *21*: 210–215.

Cohn, F. 1872. Untersuchungen über Bakterien. Beitr. Biol. Pflanz. 1875 *1* (Heft): 127–224.

Cohn, F. 1876. Untersuchunger über Bacterien. IV. Beiträge zur Biologie der Bacillen. Beitr. Biol. Pflanz. *2*: Heft 2, 249–277.

Colby, J. and L.J. Zatman. 1975. Tricarboxylic acid-cycle and related enzymes in restricted facultative methylotrophs. Biochem. J. *148*: 505–511.

Cole, R.M., G.B. Calandra, E. Huff and K.N. Nugent. 1976. Attributes of potential utility in differentiating among "Group H" streptococci or *Streptococcus sanguis*. J. Dent. Res. *55*: A142-A153.

Coleman, D.J., D. McGhie and G.M. Tebbutt. 1977. Further studies on the reliability of the bacitracin inhibition test for the presumptive identification of Lancefield group A streptococci. J. Clin. Pathol. *30*: 421–426.

Coleman, R.M. and L.K. Georg. 1969. Comparitive pathogenicity of *Actinomyces naeslundii* and *Actinomyces israelii*. Appl. Microbiol. *18*: 427–432.

Coleman, R.M., L.K. Georg and A.R. Rozzell. 1969. *Actinomyces naeslundii* as an agent of human actinomycosis. Appl. Microbiol. *18*: 420–426.

Coligan, J.E., B.A. Fraser and T.J. Kindt. 1977. A disaccharide hapten from streptococcal group C carbohydrate that cross-reacts with the Forssman glycolipid. J. Immunol. *118*: 6–11.

Coligan, J.E., W.C. Schnute and T.J. Kindt. 1975. Immunochemical and chemical studies on streptococcal group-specific carbohydrates. J. Immunol. *114*: 1654–1658.

Collee, J.G. 1965. The relationship of the hemagglutinin of *Clostridium welchii* to the neuraminidase and other soluble products of the organism. J. Pathol. Bacteriol. *90*: 13–30.

Colleen, S., B. Hovelius, A. Wieslander and P.W. Mardh. 1979. Surface properties of *Staphylococcus saprophyticus* and *Staphylococcus epidermidis* as studied by adherence tests and two polymer aqueous phase systems. Acta Pathol. Microbiol. Scand. Sect. B *87*: 321–328.

Collier, R.J. 1975. Diphtheria toxin: mode of action and structure. Bacteriol. Rev. *39*: 54–85.

Collin, B. 1913. Sur un ensemble de protistes parasites de batraciens (Note préliminaire). Arch. Zool. Exp. Gen. Notes Rev. *51*: 59–76.

Collins, F.M. and J. Lascelles. 1962. The effect of growth conditions on oxidation and dehydrogenase activity in *Staphylococcus aureus*. J. Gen. Microbiol. *29*: 531–535.

Collins, M.D. 1981. Distribution of menaquinones within members of the genus *Staphylococcus*. FEMS Microbiol. Lett. *12*: 83–85.

Collins, M.D. 1982. Lipid composition of *Renibacterium salmoninarum* (Sanders and Fryer). FEMS Microbiol. Lett. *13*: 295–297.

Collins, M.D. 1982. A note on the separation of natural mixtures of bacterial menaquinones using reverse-phase high-performance liquid chromatography. J. Appl. Bacteriol. *52*: 457–460.

Collins, M.D. 1982a. Lipid composition of *Agromyces ramosus*. FEMS Microbiol. Lett. *14*: 187–189.

Collins, M.D. 1982b. *Corynebacterium mycetoides* sp. nov., nom. rev. Zentralbl. Bakteriol. Mikrobiol. Hyg., I. Abt. Orig. C *3*: 399–400.

Collins, M.D. 1982c. Reclassification of *Bacterionema matruchotii* (Gilmour, Howell and Bibby) in the genus *Corynebacterium*, as *Corynebacterium matruchotii* comb. nov. Zentralbl. Bakteriol. Mikrobiol. Hyg. I. Abt. Orig. C *3*: 364–367.

Collins, M.D. 1983. Cell wall peptidoglycan and lipid composition of the phytopathogen *Corynebacterium rathayi* (Smith). Syst. Appl. Microbiol. *4*: 193–195.

Collins, M.D. 1984. Isoprenoid quinone analyses in bacterial classification and identification. *In* Goodfellow and Minnikin (Editors), Chemical Methods in Bacterial Systematics, Society for Applied Bacteriology, Technical Series No. 20; Academic Press, London and New York, pp. 267–287.

Collins, M.D. 1985. Isoprenoid quinone analyses in bacterial classification and identification. *In* Gottschalk (Editor), Methods in microbiology, Vol. 19, Academic Press, London.

Collins, M.D., J.A.E. Farrow, M. Goodfellow and D.E. Minnikin. 1983a. *Brevibacterium casei* sp. nov. and *Brevibacterium epidermidis* sp. nov. Syst. Appl. Microbiol. *4*: 388–395.

Collins, M.D., J.A.E. Farrow, M. Goodfellow and D.E. Minnikin. 1984. Validation of the publication of new names and new combinations previously effectively published outside the IJSB. List No. 13. Int. J. Syst. Bacteriol. *34*: 91–92.

Collins, M.D., J.A.E. Farrow, V. Katic and O. Kandler. 1984b. Taxonomic studies on streptococci of serological groups E, P, U and V: description of *Streptococcus porcinus* sp. nov. Syst. Appl. Microbiol. *5*: 402–413.

Collins, M.D., J.A.E. Farrow, V. Katic and O. Kandler. 1985. Validation List No. 17. Int. J. Syst. Bacteriol. *35*: 223–225.

Collins, M.D., J.A.E. Farrow, B.A. Phillips and O. Kandler. 1983h. Validation List No. 14. Int. J. Syst. Bacteriol. *34*: 270–271.

Collins, M.D., J.A.E. Farrow, B.A. Phillips and O. Kandler. 1983a. *Streptococcus garvieae* sp. nov. and *Streptococcus plantarum* sp. nov. J. Gen. Microbiol. *129*: 3427–3431.

Collins, M.D., S. Feresu and D. Jones. 1983j. Cell wall, DNA base composition and lipid studies on *Listeria denitrificans* (Prévot). FEMS Microbiol. Lett. *18*: 131–134.

Collins, M.D., M. Goodfellow and D.E. Minnikin. 1979. Isoprenoid quinones in the classification of coryneform and related bacteria. J. Gen. Microbiol. *110*: 127–136.

Collins, M.D., M. Goodfellow and D.E. Minnikin. 1980a. Fatty acids, isoprenoid quinones and polar lipids in the classification of *Curtobacterium* and related taxa. J. Gen. Microbiol. *118*: 29–37.

Collins, M.D., M. Goodfellow and D.E. Minnikin. 1982a. A survey of the structure of mycolic acids in *Corynebacterium* and related taxa. J.Gen. Microbiol. *128*: 129–149.

Collins, M.D., M. Goodfellow and D.E. Minnikin. 1982b. Fatty acid composition of some mycolic acid-containing coryneform bacteria. J. Gen. Microbiol. *128*: 2503–2509.

Collins, M.D., M. Goodfellow, D.E. Minnikin and G. Alderson. 1985. Menaquinone composition of mycolic acid-containing actinomycetes and some sporoactinomycetes. J. Appl. Bacteriol. *58*: 77–86.

Collins, M.D. and D. Jones. 1979. Isoprenoid quinone composition as a guide to the classification of *Sporolactobacillus* and possibly related bacteria. J. Appl. Bacteriol. *47*: 293–297.

Collins, M.D. and D. Jones. 1979. The distribution of isoprenoid quinones in streptococci of serological groups D and N. J. Gen. Microbiol. *114*: 27–33.

Collins, M.D. and D. Jones. 1980. Lipids in the classification and identification of coryneform bacteria containing peptidoglycans based on 2,4-diaminobutyric acid. J. Appl. Bacteriol. *48*: 459–470.

Collins, M.D. and D. Jones. 1981. The distribution of isoprenoid quinone structural types in bacteria and their taxonomic implications. Microbiol. Rev. *45*: 316–354.

Collins, M.D. and D. Jones. 1981b. Lipid composition of *Brevibacterium lyticum* (Takayama, Udagawa and Abe). FEMS Microbiol. Lett. *13*: 193–195.

Collins, M.D. and D. Jones. 1982a. Reclassification of *Corynebacterium pyogenes* (Glage) in the genus *Actinomyces*, as *Actinomyces pyogenes* comb. nov. J. Gen. Microbiol. *128*: 901–903.

Collins, M.D. and D. Jones. 1982b. Lipid composition of *Corynebacterium paurometabolum* (Steinhaus). FEMS Microbiol. Lett. *13*: 13–16.

Collins, M.D. and D. Jones. 1982. Taxonomic studies on *Corynebacterium beticola* (Abdou). J. Appl. Bacteriol. *52*: 229–233.

Collins, M.D. and D. Jones. 1983. Reclassification of *Corynebacterium flaccumfaciens*, *Corynebacterium betae*, *Corynebacterium oortii* and *Corynebacterium poinsettiae* in the genus *Curtobacterium*, as *Curtobacterium flaccumfaciens* comb. nov. J. Gen. Microbiol. *129*: 3545–3548.

Collins, M.D. and D. Jones. 1983. *Corynebacterium minutissimum* sp. nov., nom. rev. Int. J. Syst. Bacteriol. *33*: 870–871.

Collins, M.D. and D. Jones. 1984. Validation of the publication of new names and new combinations previously effectively published outside the IJSB. List No. 14. Int. J. Syst. Bacteriol. *34*: 270–271.

Collins, M.D., D. Jones, J.A.E. Farrow, R. Klipper-Bälz and K.H. Schleifer. 1984a. *Enterococcus avium* nom. rev., comb. nov.; *E. casseliflavus* nom. rev., comb. nov; *E. durans* nom. rev., comb. nov.; *E. gallinarum* comb. nov.; and *E. malodoratus* sp. nov. Int. J. Syst. Bacteriol. *34*: 220–223.

Collins, M.D., D. Jones, M. Goodfellow and D.E. Minnikin. 1979. Isoprenoid quinone composition as a guide to the classification of *Listeria*, *Brochothrix*, *Erysipelothrix* and *Caryophanon*. J. Gen. Microbiol. *111*: 453–457.

Collins, M.D., D. Jones, R.M. Keddie, R.M. Kroppenstedt and K.H. Schleifer. 1983b. Classification of some coryneform bacteria in a new genus *Aureobacterium*. Syst. Appl. Microbiol. *4*: 236–252.

Collins, M.D., D. Jones, R.M. Keddie, R.M. Kroppenstadt and K.H. Schleifer. 1983e. Validation of the publication of new names and new combinations previously effectively published outside the IJSB. List No. 11. Int. J. Syst. Bacteriol. *33*: 672–674.

Collins, M.D. D. Jones, R.M. Keddie and P.H.A. Sneath. 1980. Reclassification of *Chromobacterium iodinum* (Davis) in a redefined genus *Brevibacterium* (Breed) as *Brevibacterium iodinum* nom rev.; comb. nov. J. Gen. Microbiol. *120*: 1–10.

Collins, M.D., D. Jones, R.M. Keddie and P.H.A. Sneath. 1981a. Validation of the publication of new names and new combinations previously effectively published outside the IJSB. List No. 6. Int. J. Syst. Bacteriol. *31*: 215–218.

Collins, M.D., D. Jones and R.M. Kroppenstedt. 1981. Reclassification of *Corynebacterium ilicis* (Mandel, Guba and Litsky) in the genus *Arthrobacter*, as *Arthrobacter ilicis* comb. nov. Zentralbl. Bakteriol. Mikrobiol. Hyg. Abt. I. Orig. C *2*: 318–323.

Collins, M.D., D. Jones and R.M. Kroppenstedt. 1982. Validation List No. 9. Int. J. Syst. Bacteriol. *32*: 384–385.

Collins, M.D., D. Jones and R.M. Kroppenstedt. 1983a. Reclassification of *Brevibacterium imperiale* (Steinhaus) and "*Corynebacterium laevaniformans*" (Dias and Bhat) in a redefined genus *Microbacterium* (Orla-Jensen), as *Microbacterium imperiale* comb. nov. and *Microbacterium laevaniformans* nom. rev.; comb. nov. Syst. Appl. Microbiol. *4*: 65–78.

Collins, M.D., D. Jones and R.M. Kroppenstedt. 1983i. Validation of the publication of new names and new combinations previously effectively published outside the IJSB. List No. 11. Int. J. Syst. Bacteriol. 33: 672–673.

Collins, M.D., D. Jones, R.M. Kroppenstedt and K.H. Schleifer. 1982a. Chemical studies as a guide to the classification of Corynebacterium pyogenes and "Corynebacterium haemolyticum". J. Gen. Microbiol. 128: 335–341.

Collins, M.D., D. Jones and G.M. Schofield. 1982b. Reclassification of "Corynebacterium haemolyticum" (MacLean, Liebow & Rosenberg) in the genus Arcanobacterium gen. nov. as Arcanobacterium haemolyticum nom. rev., comb. nov. J. Gen. Microbiol. 128: 1279–1281.

Collins, M.D., D. Jones and G.M. Schofield. 1983. Validation of the publication of new names and new combinations previously effectively published outside the IJSB. Int. J. Syst. Bacteriol. 33: 438.

Collins, M.D., R.M. Keddie and R.M. Kroppenstedt. 1983b. Lipid composition of Arthrobacter simplex, Arthrobacter tumescens and possibly related taxa. Syst. Appl. Microbiol. 4: 18–26.

Collins, M.D. and R.M. Kroppenstedt. 1983. Lipid composition as a guide to the classification of some coryneform bacteria - containing an A4 α type peptidoglycan (Schleifer and Kandler). Syst. Appl. Microbiol. 4: 95–104.

Collins, M.D., B.M. Lund, J.A.E. Farrow and K.H. Schleifer. 1983d. Chemotaxonomic study of an alkalophilic bacterium, Exiguobacterium aurantiacum gen. nov., sp. nov. J. Gen. Microbiol. 129: 2037–2042.

Collins, M.D., G.C. Mackillop and T. Cross. 1982. Menaquinone composition of members of the genus Thermoactinomyces. FEMS Microbiol. Lett. 13: 151–153.

Collins, M.D., A.J. McCarthy and T. Cross. 1982b. New highly saturated members of the vitamin K₂ series from Thermomonospora. Zentralbl. Bakteriol. Mikrobiol. Hyg. I Abt. Orig. C 3: 358–363.

Collins, M.D., T. Pirouz, M. Goodfellow and D.E. Minnikin. 1977. Distribution of menaquinones in actinomycetes and corynebacteria. J. Gen. Microbiol. 100: 221–230.

Collins, M.D., H.N.M. Ross, B.J. Tindall and W.D. Grant. 1981b. Distribution of isoprenoid quinones in halophilic bacteria. J. Appl. Bacteriol. 50: 559–565.

Collins, M.D. and H.N. Shah. 1984. Fatty acid, menaquinone and polar lipid composition of Rothia dentocariosa. Arch. Microbiol. 137: 247–249.

Collins, M.D., H.N. Shah and D.E. Minnikin. 1980b. A note on the separation of natural mixtures of bacterial menaquinones using reverse-phase partition thin-layer chromatography. J. Appl. Bacteriol. 48: 277–282.

Collins, P.A., M.A. Gerencser and J.M. Slack. 1973. Enumeration and identification of Actinomycetaceae in human dental calculus using the fluorescent antibody technique. Arch. Oral Biol. 18: 145–153.

Collins-Thompson, D.L. and G. Rodriguez-Lopez. 1980. Influence of sodium nitrite, temperature and lactic acid bacteria on the growth of Brochothrix thermosphacta under anaerobic conditions. Can. J. Microbiol. 26: 1416–1421.

Collins-Thompson, D.L., T. Sørhaug, L.D. Witter and Z.J. Ordal. 1971. Glycerol ester hydrolase activity of Microbacterium thermosphactum. Appl. Microbiol. 21: 9–12.

Collins-Thompson, D.L., T. Sørhaug, L.D. Witter and Z.J. Ordal. 1972. Taxonomic consideration of Microbacterium lacticum, Microbacterium flavum and Microbacterium thermosphactum. Int. J. Syst. Bacteriol. 22: 65–72.

Colman, G. 1967. Aerococcus-like organisms isolated from human infections. J. Clin. Pathol. 20: 294–297.

Colman, G. 1968. The application of computers to the classification of streptococci. J. Gen. Microbiol. 50: 149–158.

Colman, G. 1976. The viridans streptococci. In De Louvois (Editor), Selected Topics in Clinical Bacteriology, Balliere Tindal, London, pp. 179–198.

Colman, G. and L.C. Ball. 1984. Identification of streptococci in a medical laboratory. J. Appl. Bacteriol. 57: 1–14.

Colman, G. and R.E.O. Williams. 1965. The cell walls of streptococci. J. Gen. Microbiol. 41: 375–387.

Colman, G. and R.E.O. Williams. 1972. Taxonomy of some human viridans streptococci. In Wannamaker and Matsen (Editors), Streptococci and Streptococcal Disease, Academic Press, London and New York, pp.281–299.

Colwell, R.R. 1973. Genetic and phenetic classification of bacteria. Adv. Appl. Microbiol. 16: 137–175.

Colwell, R.R. (Editor). 1976. The Role of Culture Collections in the Era of Molecular Biology. American Society for Microbiology, Washington, D.C.

Colwell, R.R., R.V. Citarella, I. Ryman and G.B. Chapman. 1969. Properties of Pseudomonas iodinum. Can. J. Microbiol. 15: 851–857.

Cone, D.K. 1982. A Lactobacillus sp. from diseased female rainbow trout, Salmo gairdneri Richardson, in Newfoundland, Canada. J. Fish. Dis. 5: 479–485.

Conklin, M.E. 1934. Mercurochrome as a bacteriological stain. J. Bacteriol. 27: 30–31.

Conn, H.J. 1928. A type of bacteria abundant in productive soils, but apparently lacking in certain soils of low productivity. N.Y. Agr. Exp. Sta. Geneva Bull. 138: 3–26.

Conn, H.J. and I. Dimmick. 1947. Soil bacteria similar in morphology to Mycobacterium and Corynebacterium. J. Bacteriol. 54: 291–303.

Conrad, S.E., J. Breivis and M.A. Fried. 1978. Vertebral osteomyelitis, caused by Arachnia propionica and resembling actinomycosis. J. Bone Jt. Surg. 60-A: 549–553.

Cook, A.R. 1976. Urease activity in the rumen of sheep and the isolation of ureolytic bacteria. J. Gen. Microbiol. 92: 32–48.

Cook, A.R. 1976. The elimination of urease activity in Streptococcus faecium as

evidence for plasmid-coded urease. J. Gen. Microbiol. 92: 49–58.

Cook, F.D. and H. Katznelson. 1960. Isolation of bacteriophages for the detection of Corynebacterium insidiosum, agent of bacterial wilt of alfalfa. Can. J. Microbiol. 6: 121–125.

Cooke, J.V. and H.R. Keith. 1927. A type of ureasplitting bacterium found in the human intestinal tract. J. Bacteriol. 13: 315–319.

Cooksey, R.C., F.S. Thompson and R.R. Facklam. 1979. Physiological characterization of nutritionally variant streptococci. J. Clin. Microbiol. 10: 326–330.

Cooney, J.J., H.W. Marks and A.M. Smith. 1966. Isolation and identification of canthaxanthin from Micrococcus roseus. J. Bacteriol. 92: 342–345.

Cooney, J.J. and O.C. Thierry. 1966. A defined medium for growth and pigment synthesis of Micrococcus roseus. Can. J. Microbiol. 12: 83–89.

Corbaz, R., P.H. Gregory and M.E. Lacey. 1963. Thermophilic and mesophilic actinomycetes in mouldy hay. J. Gen. Microbiol. 32: 449–455.

Cornelis, P., C. Digneffe and K. Willemot. 1982. Cloning and expression of a Bacillus coagulans amylase gene in Escherichia coli. Mol. Gen. Genet. 186: 507–511.

Costello, A.H., J.O. Cisar, P.E. Kolenbrander and O. Gabriel. 1979. Neuraminidase-dependent hemagglutination of human erythrocytes by human strains of Actinomyces viscosus and Actinomyces naeslundii. Infect. Immun. 26: 563–572.

Coster, E. and H.R. White. 1964. Further studies of the genus Pediococcus. J. Gen. Microbiol. 37: 15–31.

Costilow, R.N. 1977. Selenium requirement for the growth of Clostridium sporogenes with glycine as the oxidant in Stickland reaction systems. J. Bacteriol. 131: 366–368.

Costilow, R.N. and D. Cooper. 1978. Identity of proline dehydrogenase and Δ¹-pyrroline-5-carboxylic acid in Clostridium sporogenes. J. Bacteriol. 134: 139–146.

Courington, D.P. and T.W. Goodwin. 1955. A survey of the pigments of a number of chromogenic marine bacteria with special reference to the carotenoids. J. Bacteriol. 70: 568–571.

Cove, J.H., K.T. Holland and W.J. Cunliffe. 1981. Growth yield, phosphate and protease production by Staphylococcus epidermidis in batch and continuous culture. In Jeljaszewicz (Editor), Staphylococci and Staphylococcal Infections, Gustav Fischer Verlag, Stuttgart, New York, pp. 169–173.

Cowan, S.T. 1968. A Dictionary of Microbial Taxonomic Usage, Oliver and Boyd, Edinburgh.

Cowan, S.T. 1970. Heretical taxonomy for bacteriologists. J. Gen. Microbiol. 61: 145–154.

Cowan, S.T. 1978. In Hill (Editor), A Dictionary of Microbial Taxonomy, Cambridge University Press, Cambridge, United Kingdom.

Cowan, S.T. and K.J. Steel. 1974. Cowan and Steel's Manual for Identification of Medical Bacteria. 2nd Edition. Cambridge University Press, London.

Cox, D.J. 1978. Hydrolytic and fermentative properties of Clostridium species isolated from digesting sludge. J. Appl. Bacteriol. 45: 259–266.

Cox, R., G. Sockwell and B. Landers. 1959. Bacillus subtilis septicemia. Report of a case and review of the literature. N. Engl. J. Med. 261: 894–896.

Coykendall, A.L. 1971. Genetic heterogeneity in Streptococcus mutans. J. Bacteriol. 106: 192–196.

Coykendall, A.L. 1974. Four types of Streptococcus mutans based on their genetic, antigenic and biochemical characteristics. J. Gen. Microbiol. 83: 327–338.

Coykendall, A.L. 1977. Proposal to elevate the subspecies of Streptococcus mutans to species status, based on their molecular composition. Int. J. Syst. Bacteriol. 27: 26–30.

Coykendall, A.L. 1983. Streptococcus sobrinus nom. rev. and Streptococcus ferus nom. rev.: habitat of these and other mutans streptococci. Int. J. Syst. Bacteriol. 33: 883–885.

Coykendall, A.L., D. Bratthall, K. O'Connor and R.A. Dvarskas. 1976. Serological and genetic examination of some nontypical Streptococcus mutans strains. Infect. Immun. 14: 667–670.

Coykendall, A.L. and A.J. Munzenmaier. 1978. Deoxyribonucleic acid base sequence studies on glucan-producing and glucan-negative strains of Streptococcus mitior. Int. J. Syst. Bacteriol. 28: 511–515.

Coykendall, A.L. and A.J. Munzenmaier. 1979. Deoxyribonucleic acid hybridization among strains of Actinomyces viscosus and Actinomyces naeslundii. Int. J. Syst. Bacteriol. 29: 234–240.

Coykendall, A.L. and P.A. Specht. 1975. DNA base sequence homologies among strains of Streptococcus sanguis. J. Gen. Microbiol. 91: 92–98.

Coykendall, A.L., P.A. Specht and H.H. Samol. 1974. Streptococcus mutans in a wild, sucrose-eating rat population. Infect. Immun. 10: 216–219.

Crawford, I.P., B.P. Nichols and C. Yanofsky. 1980. Nucleotide sequence of the trpB gene in Escherichia coli and Salmonella typhimurium. J. Mol. Biol. 142: 489–502.

Crawford, J.J. 1971. Interaction of Actinomyces organisms with cationic polypeptides. I. Histochemical studies of infected human and animal tissues. Infect. Immun. 4: 632–641.

Crawford, J.M., M.A. Taubman and D.J. Smith. 1978. The natural history of periodontal bone loss in germfree and gnotobiotic rats infected with periodontopathic microorganisms. J. Periodont. Res. 13: 316–325.

Crawford, J.T., M.D. Cave and J.H. Bates. 1981a. Evidence for plasmid-mediated restriction-modification in Mycobacterium avium intracelulare (sic) J. Gen. Microbiol. 127: 333–338.

Crawford, J.T., J.K. Fitzhugh and J.H. Bates. 1981b. Phage typing of the Mycobacterium avium-intracellulare-scrofulaceum complex. Am. Rev. Resp.

Dis. *124:* 559–562.

Crisley, F.D., R. Angelotti and M.J. Foter. 1964. Multiplication of *Staphylococcus aureus* in synthetic cream fillings and pies. Publ. Health Rept. *79:* 369–376.

Criswell, B.S., J.H. Marston, W.A. Stenback, S.H. Black and H.L. Gardner. 1971. *Haemophilus vaginalis* 594, a Gram-negative organism? Can. J. Microbiol. *17:* 865–869.

Criswell, B.S., W.A. Stenback, S.H. Black and H.L. Gardner. 1972. Fine structure of *Haemophilus vaginalis.* J. Bacteriol. *109:* 930–932.

Critchley, P., J. Wood, C. Saxton and S.A. Leach. 1976. The polymerisation of dietary sugars by dental plaque. Caries Res. *1:* 112–129.

Crociani, F., O. Emaldi and D. Matteuzzi. 1977. Increase in isoleucine accumulation by α-aminobutyric acid-resistant mutants of *Bifidobacterium ruminale.* Eur. J. Appl. Microbiol. *4:* 177–179.

Crociani, F. and D. Matteuzzi. 1982. Urease activity in the genus *Bifidobacterium.* Ann. Microbiol. (Inst. Pasteur) *133 A:* 417–423.

Crombach, W.H.J. 1972. DNA base composition of soil arthrobacters and other coryneforms from cheese and sea fish. Antonie Van Leeuwenhoek J. Microbiol. Serol. *38:* 105–120.

Crombach, W.H.J. 1974a. Relationships among coryneform bacteria from soil, cheese and sea fish. Antonie van Leeuwenhoek. J. Microbiol. Serol. *40:* 347–359.

Crombach, W.H.J. 1974b. Morphology and physiology of coryneform bacteria. Antonie van Leeuwenhoek. J. Microbiol. Serol. *40:* 361–376.

Crombach, W.H.J. 1978. *Caseobacter polymorphus* gen. nov., sp. nov., a coryneform bacterium from cheese. Int. J. Syst. Bacteriol. *28:* 354–366.

Crosa, J.H., D.J. Brenner and S. Falkow. 1973. Use of a single-strand specific nuclease for analysis of bacterial and plasmid deoxyribonucleic acid homo- and heteroduplexes. J. Bacteriol. *115:* 904–911.

Cross, T. 1981. The monosporic actinomycetes. *In* Starr, Stolp, Trüper, Balows and Schlegel (Editors), The Prokaryotes, A Handbook on habitats, isolation and identification of bacteria, Springer-Verlag, Berlin, pp. 2091–2102.

Cross, T. and M. Goodfellow. 1973. Taxonomy and classification of the actinomycetes. *In* Sykes and Skinner (Editors), Actinomycetales: Characteristics and Practical Importance, Academic Press, London and New York, pp. 11–112.

Cross, T., A. Maciver and J. Lacey. 1968. The thermophilic actinomycetes in mouldy hay: *Micropolyspora faeni* sp. nov. J. Gen. Microbiol. *50:* 351–359.

Cross, T., T.J. Rowbotham, E.N. Mishustin, E.Z. Tepper, F. Antoine-Portaels, K.P. Schaal and H. Bickenbach. 1976. The ecology of nocardioform actinomycetes. *In* Goodfellow, Brownell and Serrano (Editors), The Biology of the Nocardiae. Academic Press, London and New York, pp. 337–371.

Crow, V.L. and T.D. Thomas. 1982. Arginine metabolism in lactic streptococci. J. Bacteriol. *150:* 1024–1032.

Crowle, A.J. 1962. *Corynebacterium rubrum* nov. spec. a Gram-positive non acid-fast bacterium of unusually high lipid content. Antonie Leeuwenhoek J. Microbiol. Serol. *28:* 182–192.

Crowley, C.F. and S.H. De Boer. 1982. Non-pathogenic bacteria associated with potato stems cross-react with *Corynebacterium sepedonicum* antisera in immunofluorescence. Am. Potato J. *59:* 1–8.

Cruickshank, J.G., A.H. Gawler and C. Shaldon. 1979. *Oerskovia* species: rare opportunistic pathogens. J. Med. Microbiol. *12:* 513–515.

Cuatrecasas, P., M. Wilchek and C.B. Anfinsen. 1969. The action of staphylococcal nuclease on synthetic substrates. Biochemistry *8:* 2277–2284.

Cuer, A., G. Dauphin, A. Kergomard, J.P. Dumont and J. Adda. 1979. Production of S-methylthioacetate by *Brevibacterium linens.* Appl. Environ. Microbiol. *38:* 332–334.

Cullen, G.A. 1966. The ecology of *Streptococcus uberis.* Br. Vet. J. *122:* 333–339.

Cullum, J. and H. Saedler. 1981. DNA rearrangements and evolution. *In* Carlile, Collins and Moseley (Editors), Molecular and Cellular Aspects of Microbial Evolution, Symposium No. 32 of the Society for General Microbiology, Cambridge University Press, London - New York, pp. 131–150.

Cummins, C.S. 1962. Immunochemical specificity and the location of antigens in the bacterial cell. *In* Ainsworth and Sneath (Editors), Microbial Classification, 12th Symposium of the Society for General Microbiology, Cambridge University Press, United Kingdom.

Cummins, C.S. 1962. Chemical composition and antigenic structure of cell walls of *Corynebacterium, Mycobacterium, Nocardia, Actinomyces* and *Arthrobacter.* J. Gen. Microbiol. *28:* 35–50.

Cummins, C.S. 1965. Chemical and antigenic studies on cell walls of mycobacteria, corynebacteria and nocardias. Am. Rev. Resp. Dis. *92:* 63–72.

Cummins, C.S. 1965. Ornithine in mucopeptide of Gram-positive cell walls. Nature *206:* 1272.

Cummins, C.S. 1970. *Actinomyces israelii* type 2. *In* Prauser (Editor), The Actinomycetales; The Jena International Symposium on Taxonomy. VEB Gustav Fischer Verlag, Jena, pp. 29–34.

Cummins, C.S. 1975. Identification of *Propionibacterium acnes* and related organisms by precipitin tests with trichloracetic acid extracts. J. Clin. Microbiol. *2:* 104–110.

Cummins, C.S., G. Atfield, R.J.W. Rees and R.C. Valentine. 1967. Cell wall composition in *Mycobacterium lepraemurium.* J. Gen. Microbiol. *49:* 377–384.

Cummins, C.S. and H. Harris. 1956. The chemical composition of the cell wall in some Gram-positive bacteria and its possible value as a taxonomic character. J. Gen. Microbiol. *14:* 583–600.

Cummins, C.S. and H. Harris. 1958. Studies on the cell-wall composition and taxonomy of Actinomycetales and related groups. J. Gen. Microbiol. *18:* 173–189.

Cummins, C.S. and H. Harris. 1959. Taxonomic position of *Arthrobacter.* Nature, (London) *184:* 831–832.

Cummins, C.S. and H. Harris. 1959. Cell-wall composition in strains of *Actinomyces* isolated from human and bovine lesions. J. Gen. Microbiol. *21:* ii.

Cummins, C.S. and J.L. Johnson. 1971. Taxonomy of the clostridia: wall composition and DNA homologies in *Clostridium butyricum* and other butyric acid-producing clostridia. J. Gen. Microbiol. *67:* 33–46.

Cummins, C.S. and J.L. Johnson. 1974. *Corynebacterium parvum*: a synonym for *Propionibacterium acnes*? J. Gen. Microbiol. *80:* 433–442.

Cummins, C.S. and J.L. Johnson. 1981. The genus *Propionibacterium. In* Starr, Stolp, Trüper, Balows and Schlegel (Editors), The Procaryotes: A Handbook on Habitats, Isolation and Identification of Bacteria. Springer-Verlag, New York, Berlin, pp. 1894–1902.

Cummins, C.S., R.A. Lelliott and M. Rogosa. 1974. Genus I. *Corynebacterium* Lehmann and Neumann 1896, 350. *In* Buchanan and Gibbons (Editors). Bergey's Manual of Determinative Bacteriology 8th ed. The Williams and Wilkins Co, Baltimore. pp. 602–617.

Cunliffe, W.J. 1982. Acne, hormones and treatment. Brit. Med. J. *285:* 912–913.

Cunningham, L., B.W. Catlin and M.P. deGarilhe. 1956. A deoxyribonuclease of *Micrococcus pyogenes.* J. Am. Chem. Soc. *78:* 4642–4645.

Cure, G.L. and R.M. Keddie. 1973. Methods for the morphological examination of aerobic coryneform bacteria. *In* Board and Lovelock (Editors), Sampling - Microbiological Monitoring of Environments. Academic Press, London, pp. 123–135.

Curry, J.C. and G.E. Borovian. 1976. Selective medium for distinguishing micrococci from staphylococci in the clinical laboratory. J. Clin. Microbiol. *4:* 455–457.

Curry, W.A. 1980. Human nocardiosis. A clinical review with selected case reports. Arch. Intern. Med. *140:* 818–826.

Curtis, S.N. and R.M. Krause. 1964. Antigenic relationships between groups B and G streptococci. J. Exp. Med. *120:* 629–637.

Curtiss, R. 1969. Bacterial conjugation. Annu. Rev. Microbiol. *23:* 69–136.

Cushnie, G.H., A.J. Richardson, W.J. Lawson and G.A. Sharman. 1979. Cerebrocortical necrosis in ruminants: effect of thiaminase type 1-producing *Clostridium sporogenes* in lambs. Vet. Rec. *105:* 480–482.

Cuttino, J.T. and A.M. McCabe. 1949. Pure granulomatous nocardiosis: A new fungus disease distinguished by intracellular parasitism. Am. J. Clin. Pathol. *25:* 1–34.

da Costa Cruz, J.C. 1938. *Mycobacterium fortuitum* um novo bacillo acidoresistance pathogenico para o homen. Acta Med. Rio de Janerio *1:*297–301.

Dacres, W.G. and A.H. Groth, Jr. 1959. Identification of *Erysipelothrix rhusiopathiae* with fluorescent antibody. J. Bacteriol. *78:* 298–299.

Daffé, M., M.A. Laneelle, D. Prome and C. Asselineau. 1981. Étude des lipides de *Mycobacterium gordonae* comparativement a ceux de *M. leprae* et de quelques Mycobactéries scotochromogenes. Ann. Microbiol. Paris *132:* 3–12.

Dain, J.A., A.K. Neal and H.W. Seeley. 1956. The effect of carbon dioxide on polysaccharide production by *Streptococcus bovis.* J. Bacteriol. *72:* 209–213.

Dainty, R.H. and C.M. Hibbard. 1980. Aerobic metabolism of *Brochothrix thermosphacta* growing on meat surfaces and in laboratory media. J. Appl. Bacteriol. *48:* 387–396.

Dajani, A.S. 1973. Rapid identification of beta-haemolytic streptococci by counterimmunoelectrophoresis. J. Immunol. *110:* 1702–1705.

Daly, J.M. 1982. Mechanisms of Action. *In* Durbin (Editor), Toxins in Plant Disease. Academic Press, New York, pp. 331–394.

Danahaive, P., P. Hoet and C. Cocito. 1982. Base compositions and homologies of deoxyribonucleic acids of corynebacteria isolated from human leprosy lesions and of related microorganisms. Int. J. Syst. Bacteriol. *32:* 70–76.

Daniel, T.M. and B.W. Janicki. 1978. Mycobacterial antigens: a review of their isolation, chemistry, and immunological properties. Microbiol. Rev. *42:* 84–113.

Dankert, J., W.F.A. Mensink, J.G. Aarnoudse, G.J. Meijer-Severs and H.J. Huisjes. 1979. The prevalence of anaerobic bacteria in suprapubic bladder aspirates obtained from pregnant women. Zentralbl. Bakteriol. Mikrobiol. Hyg. I Abt. Orig. A *244:* 260–267.

Darland, G. and T.D. Brock. 1971. *Bacillus acidocaldarius* sp. nov., an acidophilic, thermophilic spore-forming bacterium. J. Gen. Microbiol. *67:* 9–15.

Darling, C.L. 1975. Standardization and evaluation of the CAMP reaction for the prompt, presumptive identification of *Streptococcus agalactiae* (Lancefield group B) in clinical materials. J. Clin. Microbiol. *1:* 171–174.

Darvill, A.G., M.A. Hall, J.P. Fish and J.G. Morris. 1977. The intracellular reserve polysaccharide of *Clostridium pasteurianum.* Can. J. Microbiol. *23:* 947–953.

Dárzins, E. 1950. Tuberculose das gias (*Leptodactylus pentadactylus*). Arch. Inst. Bras. Tuberc. *9:* 29–37.

Da Silva, G.A.N. and J.G. Holt. 1965. Numerical taxonomy of certain coryneform bacteria. J. Bacteriol. *90:* 921–927.

Davenport, A.A., G.R. Carter and R.G. Schirmer. 1974. Canine actinomycosis due to *Actinomyces viscosus*: report of six cases. Vet. Med. Small Anim. Clin. *69:* 1442–1447.

David, H.L. 1973. Response of mycobacteria to ultraviolet light radiation. Am. Rev. Resp. Dis. *108:* 1175–1185.

David, H.L., S. Clavel, F. Clement, L. Meyer, P. Draper and I.D.J. Burdett. 1978a. Interaction of *Mycobacterium leprae* and mycobacteriophages D29. Ann. Microbiol. (Inst. Pasteur) *129:* 561–570.

David, H.L., F. Clement and L. Meyer. 1979. Adsorption of mycobacteriophage D29 on *Mycobacterium leprae.* Ann. Microbiol. (Inst. Pasteur) *129:* 563–566.

David, H.L., M. Jahan, A. Jumen, J. Grandy and E.H. Lehman. 1978b. Numerical taxonomy analysis of *Mycobacterium africanum*. Int. J. Syst. Bacteriol. *28:* 464–472.

Davidson, C.M. and E.F. Hartree. 1968. Cytochrome as a guide to classifying bacteria: taxonomy of *Microbacterium thermosphactum*. Nature (London). *220:* 502–504.

Davidson, C.M., P. Mobbs and J.M. Stubbs. 1968. Some morphological and physiological properties of *Microbacterium thermosphactum*. J. Appl. Bacteriol. *31:* 551–559.

Davies, M.E. 1965. Cellulolytic bacteria in some ruminants and herbivores as shown by fluorescent antibody. J. Gen. Microbiol. *39:* 139–141.

Davis, G.H.G. 1964. Notes on the phylogenetic background to *Lactobacillus* taxonomy. J. Gen. Microbiol. *34:* 177–184.

Davis, G.H.G., L. Fomin, E. Wilson and K.G. Newton. 1969. Numerical taxonomy of *Listeria*, streptococci and possibly related bacteria. J. Gen. Microbiol. *57:* 333–348.

Davis, G.H. and J.H. Freer. 1960. Studies upon an oral aerobic actinomycete. J. Gen. Microbiol. *23:* 163–178.

Davis, G.H.G. and K.G. Newton. 1969. Numerical taxonomy of some named coryneform bacteria. J. Gen. Microbiol. *56:* 195–214.

Davis, J.G. 1939. *Chromobacterium iodinum* (n. sp.). Zentralbl. Bakteriol. Parasitenkd. Infektionskr. Hyg. Abt. II. *100:* 273–276.

Davis, J.G. 1975. The microbiology of yoghurt. *In* Carr, Cutting and Whiting (Editors). Lactic acid bacteria in beverages and food. Academic Press, London, New York, San Francisco, pp. 245–263.

Davis, M.J., A.G. Gillespie, Jr., A.K. Vidaver and R.W. Harris. 1984. *Clavibacter*: a new genus containing some phytopathogenic coryneform bacteria, including *Clavibacter xyli* subsp. *xyli* sp. nov., subsp. nov. and *Clavibacter xyli* subsp. *cynodontis* subsp. nov., pathogens that cause ratoon stunting disease of sugarcane and Bermudagrass stunting disease. Int. J. Syst. Bacteriol. *34:* 107–117.

Davis, N.S., G.J. Silverman and E.B. Masurovsky. 1963. Radiation-resistant, pigmented coccus isolated from haddock tissue. J. Bacteriol. *86:* 294–298.

Dawes, E.A. and W.J. Holmes. 1958. Metabolism of *Sarcina lutea*. 1. Carbohydrate oxidation and terminal respiration. J. Bacteriol. *75:* 390–399.

Dawson, D.J. and F. Jennis. 1980. Mycobacteria with a growth requirement for ferric ammonium citrate, identified as *Mycobacterium haemophilum*. J. Clin. Microbiol. *11:* 190–192.

Dean, B.A., R.E.O. Williams, F. Hall and J. Corse. 1973. Phage typing of coagulase-negative staphylococci and micrococci. J. Hyg. Camb. *71:* 261–270.

Debard, J., F. Dubos, L. Martinet and R. Ducluzeau. 1979. Experimental reproduction of neonatal diarrhea in young gnotobiotic hares simultaneously associated with *Clostridium difficile* and other *Clostridium* strains. Infect. Immun. *24:* 7–11.

de Barjac, H. 1981a. Insect pathogens in the genus *Bacillus*. *In* Berkeley and Goodfellow (Editors), The Aerobic Endospore-forming Bacteria. Academic Press, London and New York, pp. 241–250..

de Barjac, H. 1981b. Identification of H-serotypes of *Bacillus thuringiensis*. *In* Burges (Editor), Microbial Control of Pests and Diseases 1970–1980, Academic Press, London and New York, pp. 35–43.

de Barjac, H., M. Véron and V.C. Dumanoir. 1980. Caractérisation biochimique et sérologique de souches de *Bacillus sphaericus* pathogénes ou non pour les mostiques. Ann. Microbiol. (Paris) *131B:* 191–201.

de Bary, A. 1884. Vergleichende Morphologie und Biologie der Pilze, Mycetozoen und Bacterien. Wilhelm Engelmann, Leipzig.

De Boer, S.H. 1982a. Variation among strains of *Corynebacterium sepedonicum*. Phytopathology *72:* 1001 (Abstract).

De Boer, S.H. 1982b. Cross-reaction of *Corynebacterium sepedonicum* antisera with *C. insidiosum*, *C. michiganense*, and an unidentified coryneform bacterium. Phytopathology *72:* 1474–1481.

De Boer, S.H. and R.J. Copeman. 1980. Bacterial ring rot testing with the indirect fluorescent antibody staining procedure. Am. Potato J. *57:* 457–465.

Debono, M. 1912. On some anaerobical bacteria of the normal human intestine. Zentralbl. Bakteriol. Parasitenkd. Infektionskr. Hyg. Abt. I Orig. *62:* 229–234.

De Campos, S.L., E.P.T. DeOliveira, H.G. Higashi, M.R. Lorenzetti, A.V. Diniz and D. Zappa. 1981. Ocorrencia de microrganismos do genero clostridium em material colhido de bolsa periodontal e de sulco gengival de pacientes com e sem doenca peridontal avancada, habitantes de zonas rural e urgana. Rev. Assoc. Paul. Cirurg. Dent. *35:* 422–426.

Dedié, K. 1949. Die säurelöslichen Antigene von *Erysipelothrix rhusiopathiae*. Monatsh. Veterinaermed. *4:* 7–10.

Dehority, B.A. 1963. Isolation and characterization of several cellulolytic bacteria from *in vitro* rumen fermentations. J. Dairy Sci. *46:* 217–222.

Dehority, B.A. 1969. Pectin-fermenting bacteria isolated from the bovine rumen. J. Bacteriol. *99:* 189–196.

Dehority, B.A. 1971. Carbon dioxide requirement of various species of rumen bacteria. J. Bacteriol. *105:* 70–76.

Dehority, B.A. 1977. Cellulolytic bacteria from the cecum of guinea pigs. Appl. Environ. Microbiol. *33:* 1278–1283.

Dehority, B.A., H.W. Scott and P. Kowaluk. 1967. Volatile fatty acid requirements of cellulolytic rumen bacteria. J. Bacteriol. *94:* 537–543.

Deibel, R.H. 1964. The group D streptococci. Bacteriol. Rev. *28:* 330–366.

Deibel, R.H. and J.B. Evans. 1960. Modified benzidine test for the detection of cytochrome-containing respiratory systems in microorganisms. J. Bacteriol. *79:* 356–360.

Deibel, R.H. and C.F. Niven, Jr. 1955a. Reciprocal replacement of oleic acid and

CO_2 in the nutrition of the "minute streptococci" and *Lactobacillus leichmannii*. J. Bacteriol. *70:* 134–140.

Deibel, R.H. and C.F. Niven, Jr. 1955b. The "minute" streptococci: further studies on their nutritional requirements and growth characteristics on blood agar. J. Bacteriol. *70:* 141–146.

Deibel, R.H. and C.F. Niven. 1960. Comparative study of *Gaffkya homari*, *Aerococcus viridans*, tetrad-forming cocci from meat curing brines, and the genus *Pediococcus*. J. Bacteriol. *79:* 175–180.

Deibel, R.H. and H.W. Seeley, Jr. 1974. Family II. *Streptococcaceae*. *In* Buchanan and Gibbons (Editors), Bergey's Manual of Determinative Bacteriology, 8th Ed., The Williams and Wilkins Co., Baltimore, pp. 490–515.

De Klerk, H.C. and J.N. Coetzee. 1961. Antibiosis among lactobacilli. Nature (London) *192:* 340–341.

De Klerk, H.C. and J.A. Smit. 1967. Properties of a *Lactobacillus fermenti* bacteriocin. J. Gen. Microbiol. *48:* 309–316.

De la Fuente, R., G. Suarez and K.H. Schleifer. 1985. *Staphylococcus aureus* subsp. *anaerobius* subsp. nov., the causal agent of abscess disease of sheep. Int. J. Syst. Bacteriol. *35:* 99–102.

DeLange, R.J., J.Y. Chang, J.H. Shaper and A.N. Glazer. 1976. Amino acid sequence of flagellin of *Bacillus subtilis* 168. J. Biol. Chem. *251:* 705–711.

Delaporte, B. 1963. Un phénomène singulier: des "spores mobile" chez des grandes bactéries. C.R. Acad. Sci. Paris *257:* 1414–1417.

Delaporte, B. 1964a. Etude comparée de grand spirilles formant des spores: *Sporospirillum* (*Spirillum*) *praeclarum* (Collin) n.g., *Sporospirillum gyrini* n. sp. et *Sporospirillum bisporum* n. sp. Ann. Inst. Pasteur (Paris) *107:* 246–262.

Delaporte, B. 1964b. Etude descriptive de bactéries de trés grandes dimensions. Ann. Inst. Pasteur (Paris) *107:* 845–862.

Delaporte, B. 1967. Une bactérie nouvelle de l'océan Pacifique: *Bacillus pacificus* n. sp. Acad. Sci. (Paris) Ser. D *264:* 3068–3071.

Delaporte, B. 1969. Une nouvelle espèce de *Fusosporus*: *Fusosporus minor* n. sp. C.R. Acad. Sci. Paris Ser. D Sci. Nat. *268:* 1454–1455.

Delaporte, B. 1972. Trois nouvelles espèces de *Bacillus*: *Bacillus similibadius* n.sp., *Bacillus longisporus* n.sp. et *Bacillus nitritollens* n. sp. Ann. Inst. Pasteur (Paris) *123:* 821–834.

Delaporte, B. and A. Sasson. 1967. Étude de bactéries des sols arides du Maroc: *Brevibacterium halotolerans* n. sp. et *Brevibacterium frigoritolerans* n. sp. C.R. Acad. Sci. Serie D. *264:* 2257–2260.

Delaporte, B. and A. Sasson. 1967. Étude de bactéries des sols arides du Maroc: *Bacillus maroccanus* n. sp. C.R. Acad. Sci. (Paris), Ser. D *264:* 2344–2346.

DeLey, J. 1978. Modern molecular methods in bacterial taxonomy: evaluation, application, prospects. Proc. 4th Int. Conf. Plant Pathol. Bacteriol. 347–357.

De Ley, J., H. Cattoir and A. Reynaerts. 1970. The quantitative measurement of DNA hybridization from renaturation rates. Eur. J. Biochem. *12:* 133–142.

DeLey, J. and J. DeSmedt. 1975. Improvements to the membrane filter method for DNA and RNA hybridization. Antonie Van Leeuwenhoek J. Microbiol. Serol. *41:* 287–307.

De Ley, J., I.W. Park, R. Tijtgat and J. van Ermengem. 1966. DNA homology and taxonomy of *Pseudomonas* and *Xanthomonas*. J. Gen. Microbiol. *42:* 43–56.

Delisle, A.L., R.K. Nauman and G.E. Minah. 1978. Isolation of a bacteriophage for *Actinomyces viscosus*. Infect. Immun. *20:* 303–306.

Dellaglio, F., V. Bottazzi and B. Battistotti. 1974. Caratteri e distribuzione della microflora pediococcica in alcuni formaggi italiani. Ann. Microbiol. *24:* 325–334.

Dellaglio, F., V. Bottazzi and L.D. Trovatelli. 1973. Deoxyribonucleic acid homology and base composition in some thermophilic lactobacilli. J. Gen. Microbiol. *74:* 289–297.

Dellaglio, F., V. Bottazzi and M. Vescovo. 1975. Deoxyribonucleic acid homology among *Lactobacillus* species of the subgenus *Streptobacterium* Orla-Jensen. Int. J. Syst. Bacteriol. *25:* 160–172.

Dellaglio, F., L.D. Trovatelli and P.G. Sara. 1981. DNA-DNA homology among representative strains of the genus *Pediococcus*. Zentralbl. Bakteriol. Mikrobiol. Hyg. 1 Abt. Orig. C *2:* 140–150.

Delwiche, E.A. 1949. Vitamin requirements of the genus *Propionibacterium*. J. Bacteriol. *58:* 293–398.

Delwiche, E.A. 1961. Catalase of *Pediococcus cerevisiae*. J. Bacteriol. *81:* 416–417.

De Man, J.C., M. Rogosa and M.E. Sharpe. 1960. A medium for the cultivation of lactobacilli. J. Appl. Bacteriol. *23:* 130–135.

Demaree, J.B. and N.R. Smith. 1952. *Nocardia vaccinii* n. sp. causing galls on blueberry plants. Phytopatholoty *42:* 249–252.

De Moor, C.E. 1963. Septicaemic infections in pigs, caused by haemolytic streptococci of new Lancefield groups designated R, S and T. Antonie von Leeuwenhoek J. Microbiol. Serol. *29:* 272–280.

de Moor, C.E. and E. Thal. 1968. Beta haemolytic streptococci of the Lancefield groups E, P and U: *Streptococcus infrequens*. Antonie van Leeuwenhoek J. Microbiol. Serol. *34:* 377–387.

den Dooren de Jong, L.E. 1926. Bijdrage tot de kennis van het Mineralisatieproces. Proefschrift. Niigh and van Ditmar, Rotterdam.

den Dooren de Jong, L.E. 1927. Über protaminophage Bakterien. Zentralbl. Bakteriol. Parasitenkd. Infektionskr. Hyg. Abt. II *71:* 193–232.

den Dooren de Jong, L.E. 1929. Über *Bacillus fastidiosus*. Zentralbl. Bakteriol. Parasitenkd. Infektionskr. Hyg. Abt. II *79:* 344–353.

Denhardt, D.T. 1966. A membrane-filter technique for the detection of complementary DNA. Biochem. Biophys. Res. Comm. *23:* 641–646.

Dennis, S.M., T.G. Nagaraja and E.E. Bartley. 1981. Effects of Lasolocid or

Monensin on lactate-producing or -using rumen bacteria. J. Anim. Sci. *52:* 418–426.

Dent, V.E., J.M. Hardie and G.H. Bowden. 1978. Streptococci isolated from dental plaque of animals. J. Appl. Bacteriol. *44:* 249–258.

Dent, V.E. and R.A.D. Williams. 1982. *Lactobacillus animalis* sp. nov., a new species of *Lactobacillus* from the alimentary canal of animals. Zentralbl. Bakteriol. Mikrobiol. Hyg. I Abt. Orig. C *3:* 377–386.

Dent, V.E. and R.A.D. Williams. 1983. *In* Validation of publication of new names and new combinations previously effectively published outside the IJSB. List no. 10. Int. J. Syst. Bacteriol. *33:* 438–440.

Dent, V.E. and R.A.D. Williams. 1984a. *Actinomyces denticolens* Dent & Williams sp. nov.: a new species from the dental plaque of cattle. J. Appl. Bacteriol. *56:* 183–192.

Dent, V.E. and R.A.D. Williams. 1984b. *Actinomyces howellii*, a new species from the dental plaque of dairy cattle. Int. J. Syst. Bacteriol. *34:* 316–320.

Dent, V.E. and R.A.D. Williams. 1984c. Validation List No. 16. Int. J. Syst. Bacteriol. *34:* 503–504.

Deo, Y.M., R.D. Bryant and E.J. Laishley. 1979. Differences in cellular fatty acids and lipid composition in *Clostridium pasteurianum* under nitrogen- and non-nitrogen-fixing conditions. Curr. Microbiol. *3:* 55–58.

De Rosa, M., A. Gambacorta, L. Minale and J.D. Bu'Lock. 1971. Cyclohexane fatty acids from a thermophilic bacterium. Chem. Commun. *1971:* 1334.

De Rosa, M., A. Gambacorta, L. Minale and J.D. Bu'Lock. 1973. Isoprenoids of *Bacillus acidocaldarius*. Phytochemistry *12:* 1117–1123.

De Saxe, M.J., J.A. Crees-Morris, R.R. Marples and J.F. Richardson. 1981. Evaluation of current phage-typing systems for coagulase-negative staphylococci. *In* Jeljaszewicz (Editor), Staphylococci and Staphylococcal Diseases, Gustav Fischer Verlag, Stuttgart, New York, pp. 197–204.

De Smedt, J. and J. De Ley. 1977. Intra- and inter-generic similarities of *Agrobacterium* ribosomal ribonucleic acid cistrons. Int. J. Syst. Bacteriol. *27:* 222–240.

De Stoppelaar, J.D., K.G. Konig, A.J.M. Plasschaert and J.S. van der Hoeven. 1971. Decreased cariogenicity of a mutant of *Streptococcus mutans*. Arch. Oral Biol. *16:* 971–975.

De Vries, W., S.J. Gerbrandy and A.H. Stouthamer. 1967. Carbohydrate metabolism in *Bifidobacterium bifidum*. Biochem. Biophys. Acta *136:* 415–425.

De Vries, W., W.M.C. Kapteijn, E.G. van der Beek and A.H. Stouthamer. 1970. Molar growth yields and fermentation balances of *Lactobacillus casei* 13 in batch cultures and in continuous cultures. J. Gen. Microbiol. *63:* 333–345.

De Vries, W. and A.H. Stouthamer. 1967. Pathway of glucose fermentation in relation to the taxonomy of bifidobacteria. J. Bacteriol. *93:* 574–576.

De Vries, W. and A.H. Stouthamer. 1968. Fermentation of glucose, lactose, galactose, mannitol, and xylose by bifidobacteria. J. Bacteriol. *96:* 472–478.

De Vries, W. and A.H. Stouthamer. 1969. Factors determining the degree of anaerobiosis of *Bifidobacterium* strains. Arch. Mikrobiol. *65:* 275–287.

Devriese, L.A. 1977. Isolation and identification of *Staphylococcus hyicus*. Am. J. Vet. Res. *38:* 787–792.

Devriese, L.A. 1979. Identification of clumping-factor-negative staphylococci isolated from cow's udders. Res. Vet. Sci. *27:* 313–320.

Devriese, L.A., G.N. Dutta, J.A.E. Farrow, A. van de Kerckhove and B.A. Phillips. 1983. *Streptococcus cecorum*, a new species isolated from chickens. Int. J. Syst. Bacteriol. *33:* 772–776.

Devriese, L.A., V. Hajek, P. Oeding, S.A. Meyer and K.H. Schleifer. 1978. *Staphylococcus hyicus* (Sompolinsky 1953) comb. nov. and *Staphylococcus hyicus* subsp. *chromogenes* subsp. nov. Int. J. Syst. Bacteriol. *28:* 482–490.

Devriese, L.A. and P. Oeding. 1975. Coagulase and heat-resistant nuclease producing *Staphylococcus epidermidis* strains from animals. J. Appl. Bacteriol. *39:* 197–207.

Devriese, L.A. and P. Oeding. 1976. Characteristics of *Staphylococcus aureus* strains isolated from different animal species. Res. Vet. Sci. *21:* 284–291.

Devriese, L.A., B. Pouttel, R. Kilpper-Bälz and K.H. Schleifer. 1983. *Staphylococcus gallinarum* and *Staphylococcus caprae* two new *Staphylococcus* species from animals. Int. J. Bacteriol. *33:* 480–486.

De Weese, M.S., M.A. Gerencser and J.M. Slack. 1968. Quantitative analysis of *Actinomyces* cell walls. Appl. Microbiol. *16:* 1713–1718.

Dewsnup, D.H. and D.N. Wright. 1984. *In vitro* susceptability of *Nocardia asteroides* to 25 antimicrobial agents. Antimicrob. Agents Chemother. *25:* 165–167.

Dezfulian, M. and V.R. Dowell. 1980. Cultural and physiological characteristics and antimicrobial susceptibility of *Clostridium botulinum* isolates from food-borne and infant botulism cases. J. Clin. Microbiol. *11:* 604–609.

Dhople, A.M. and J.H. Hanks. 1981a. Role of sulfhydryls in vitro growth in *M. lepraemurium*. Infect. Immun. *31:* 352–357.

Dhople, A. and J. Hanks. 1981b. Adenosine triphosphate content in *Mycobacterium leprae*. A brief communication. Int. J. Leprosy *49:* 57–59.

Dias, F.F. 1963. Studies in the bacteriology of sewage. J. Indian Inst. Sci. *45:* 36–48.

Dias, F.F. and J.V. Bhat. 1962. A new levan producing bacterium, *Corynebacterium laevaniformans* nov. spec. Antonie Van Leeuwenhoek J. Microbiol. Serol. *28:* 63–72.

Dias, F.F. and J.V. Bhat. 1964. Nutritional properties of *Corynebacterium laevaniformans*. Antonie Van Leeuwenhoek J. Microbiol. Serol. *30:* 176–184.

Dias, F.F., M.H. Bilimoria and J.V. Bhat. 1962. *Corynebacterium barkeri* nov. spec., a pectinolytic bacterium exhibiting a biotin-folic acid inter-relationship.

J. Ind. Inst. Sci. *44:* 59–67.

DiCioccio, R.A., P.J. Klock, J.J. Barlow and K.L. Matta. 1980. Rapid procedures for determination of endo-*N*-acetyl-α-D-galactosamidase in *Clostridium perfringens*, and of the substrate specificity of exo-β-D-galactosidases. Carbohydr. Res. *81:* 315–322.

Dickerson, R.E. 1980. Cytochrome *c* and the evolution of energy metabolism. Sci. Amer. *242:* 137–153.

Dickgiesser, U., N. Weiss and D. Fritsche. 1984. *Lactobacillus gasseri* as the cause of septic urinary infections. Infection *12:* 14–16.

Diekert, G.B., E.G. Graf and R.K. Thauer. 1979. Nickel requirement for carbon monoxide dehydrogenase formation in *Clostridium pasteurianum*. Arch. Microbiol. *122:* 117–120.

Diekert, G.B. and R.K. Thauer. 1978. Carbon monoxide oxidation by *Clostridium thermoaceticum* and *Clostridium formicoaceticum*. J. Bacteriol. *136:* 597–606.

Diernhofer, K. 1932. Asculinbouillon als Hilfsmittel für die Differenzierung von Euter- und Milchstreptokokken bei Massenuntersuchungen. Milchw. Forsch. *13:* 368–374.

Dillon, H.C. and L.W. Wannamaker. 1965. Physical and immunological differences among streptokinases. J. Exp. Med. *121:* 351–371.

DiMarco, A. and C. Spalla. 1957. La produzione di cobalamine da fermentazione con una nuova specie di *Nocardia*: *Nocardia rugosa*. G. Microbiol. *4:* 24–30.

Dinter, Z., H. Diderholm and G. Rockborn. 1976. Complement dependent haemolysis following haemagglutination by *Erysipelothrix rhusiopathiae*. Zentralbl. Bakteriol. Parasitenkd. Infektionskr. Hyg. Abt. I *236:* 533–535.

Dismukes, W.E., A.W. Karchmer, M.J. Buckley, W.G. Austen and M.N. Swartz. 1973. Prosthetic valve endocarditis: analysis of 38 cases. Circulation *48:* 365–377.

Distaso, A. 1911. Sur les microbes protéolytiques de la flore intestinale de l'homme et des animaux. Zentralbl. Bakteriol. Parasitenkd. Infektionskr. Hyg. Abt. I Orig. *59:* 97–103.

Distaso, A. 1912. Contribution à l'étude sur l'intoxication intestinale. Zentralbl. Bakteriol. Parasitenkd. Infektionskr. Hyg. Abt. I Orig. *62:* 433–468.

Distler, W. and A. Kröncke. 1981. Acid formation by mixed cultures of dental plaque bacteria *Actinomyces* and *Veillonella*. Arch. Oral Biol. *26:* 123–126.

Distler, W., K. Ott and A. Kröncke. 1980. Wechselwirkungen von *Streptococcus mutans*, *Actinomyces* and *Veillonella* in vitro - ein vereinfachtes Modell für den Kohlenhydratmetabolismus in der Plaque. Dtsch. Zahnaerztl. Z. *35:* 548–553.

Døbrzanski, W.T., H. Osowiecki and M.A. Jagielski. 1968. Observations on intergeneric transformation between staphylococci and streptococci. J. Gen. Microbiol. *53:* 187–196.

Dobson, G., D.E. Minnikin, S.M. Minnikin, J.H. Parlett, M. Goodfellow, M. Ridell and M. Magnusson. 1984. Systematic analysis of complex mycobacterial lipids. *In* Goodfellow and Minnikin (Editors), Chemical Methods in Bacterial Systematics, Society for Applied Bacteriology, No. 20; Academic Press, London and New York, pp. 237–265.

Docherty, A., G. Grandi, T.J. Gryczan, A.G. Shivakumar and D. Dubnau. 1981. Naturally occuring macrolide-lincosamide-streptogramin B resistance in *Bacillus licheniformis*. J. Bacteriol. *145:* 129–137.

Doi, R.H. 1977. Genetic control of sporulation. Annu. Rev. Genet. *11:* 29–48.

Döll, W. 1973. Untersuchungen über die DNase-Bildung von Clostridien. Zentralbl. Bakteriol. Parasitenkd. Infektionskr. Hyg. I. Abt. Orig. A *224:* 115–119.

Dolman, C.E. and E. Chang. 1972. Bacteriophages of *Clostridium botulinum*. Can. J. Microbiol. *18:* 67–76.

Donk, P.J. 1920. A highly resistant thermophilic organism. J. Bacteriol. *5:* 373–374.

Donker, H.J.L. 1926. Bijdrage tot de Kennis der Boterzuur-, Butylalcoholen acetonigistingen. Diss., Delft. W.E. Meinema, Delft. 1–155.

Donker-Voet, J. 1972. *Listeria monocytogenes*: Some biochemical and serological aspects. Acta Microbiol. Acad. Sci. Hung. *19:* 287–291.

Doolittle, R.F. 1981. Similar amino acid sequences: chance or common ancestry? Science (Washington) *214:* 149–159.

Doores, S. and D. Westhoff. 1981. Heat resistance of *Sporolactobacillus inulinus*. J. Food Sci. *46:* 810–812.

Doores, S. and D.C. Westhoff. 1983. Selective method for the isolation of *Sporolactobacillus* from food and environmental sources. J. Appl. Bacteriol. *54:* 273–280.

Döpfer, H., E. Stackebrandt and F. Fiedler. 1982. Nucleic acid hybridization studies on *Microbacterium*, *Curtobacterium*, *Agromyces* and related taxa. J. Gen. Microbiol. *128:* 1697–1708.

Dorn, J., J.R. Andreesen and G. Gottschalk. 1978. Fermentation of fumarate and L-malate by *Clostridium formicoaceticum*. J. Bacteriol. *133:* 26–32.

Dornbusch, K., C.-E. Nord and A. Dahlbäck. 1975. Antibiotic susceptibility of *Clostridium* species isolated from human infections. Scand. J. Infect. Dis. *7:* 127–134.

Dorokhova, L.A., N.S. Agre, L.V. Kalakoutskii and N.A. Krasil'nikov. 1969. Fine structure of sporulating hyphae and spores in a thermophilic actinomycete, *Micropolyspora rectivirgula*. J. Microsc. Biol. Cell. *8:* 845–854.

Dorokhova, L.A., N.S. Agre, L.V. Kalakoutskii and N.A. Krasil'nikov. 1970. A study on the morphology of two cultures belonging to the genus *Micropolyspora*. Mikrobiologiya *39:* 95–100.

Doty, R.B., H.W. Dunne, J.F. Hokausen and J.J. Reid. 1964. A comparison of toxins produced by various isolates of *Corynebacterium pseudotuberculosis* and the development of a diagnostic skin test for cases of lymphadenitis of sheep and goats. Am. J. Vet. Res. *25:* 1679–1685.

Douglas, H.C. 1957. Genus VI. *Peptococcus* Kluyver and van Niel 1936. *In* Breed, Murray and Smith (Editors), Bergery's Manual of Determinative Bacteriology, 7th ed. The Williams and Wilkins Co., Baltimore, pp. 474–480.

Douglas, H.C. and W.V. Cruess. 1936. A *Lactobacillus* from California wine: *Lactobacillus hilgardii*. Food Res. *1:* 113–119.

Douglas, H.C. and S.E. Gunter. 1946. The taxonomic position of *Corynebacterium acnes*. J. Bacteriol. *52:* 15–23.

Dowson, W.J. 1942. On the generic name of the gram-positive bacterial plant pathogens. Trans. Brit. Mycol. Soc. *25:* 311–314.

Doyle, M.P. 1983. Effect of carbon dioxide on toxin production by *Clostridium botulinum*. Eur. J. Appl. Microbiol. Biotechnol. *17:* 53–56.

Drapeau, G.R. 1978a. The primary structure of staphylococcal protease. Can. J. Biochem. *56:* 534–544.

Drapeau, G.R. 1978b. Role of a metalloprotease in activation of the precursor of staphylococcal protease. J. Bacteriol. *136:* 607–613.

Draper, P. 1979. Process for purification of *M. leprae* based upon report on IMMLEP. Enlarged steering committee meeting held at Geneva February 7–8, 1979. a WHO/TDR document. (Personal Communication).

Drasar, B.S., P. Goddard, S. Heaton, S. Peach and B. West. 1976. Clostridia isolated from faeces. J. Med. Microbiol. *9:* 63–71.

Drew, N.C. 1984. Genital and pelvic actinomycosis. Case report. Br. J. Obstet. Gynaecol. *88:* 776–777.

Dring, G.J., G.W. Gould and D.J. Ellar (Editors). 1985. Fundamental and applied aspects of bacterial spores. Academic Press, London.

Drow, D.L. and D.D. Manning. 1980. Indirect sandwich enzyme-linked immunosorbent assay for rapid detection of *Streptococcus pneumoniae* type 3 antigen. J. Clin. Microbiol. *11:* 641–645.

Drucker, D.B. and R.M. Green. 1978. The relative cariogenicities of *Streptococcus milleri* and other viridans group streptococci in gnotobiotic hooded rats. Arch. Oral Biol. *23:* 183–187.

Drucker, D.B. and S.M. Lee. 1981. Fatty acid fingerprints of '*Streptococcus milleri*', *Streptococcus mitis*, and related species. Int. J. Syst. Bacteriol. *31:* 219–225.

Drucker, D.B. and S.M. Lee. 1983. Possible heterogeneity of *Streptococcus milleri* determined by DNA mol % (guanine plus cytosine) measurement and physiological characterization. Microbios *38:* 151–157.

Drucker, D.B. and T.H. Melville. 1971. The classification of some oral streptococci of human or rat origin. Arch. Oral Biol. *16:* 845–853.

Dubinina, G. and A.V. Zhdanov. 1975. Recognition of the iron bacteria "Siderocapsa" as arthrobacters and description of *Arthrobacter siderocapsulatus* sp. nov. Int. J. Syst. Bacteriol. *25:* 340–350.

Dubnau, D.A. 1982. (Editor), The Molecular Biology of the Bacilli Volume 1: *Bacillus subtilis*. Academic Press, New York and London.

Dubos, R. and B.F. Miller. 1937. The production of bacterial enzymes capable of decomposing creatinine. J. Biol. Chem. *121:* 429–445.

Duboscq, O. and P. Grassé. 1930. Protistologica XXI. *Coleomitus* n.g. au lieu de *Coleonema* pour la schizophyte *C. pruvoti* Dub. et Grassé, parasite d'un *Calotermes* des Iles Loyalty. Arch. Zool. Expér. Gén. *70:* 28 (Notes et Rev.).

Duda, J.J. and J.M. Slack. 1972. Ultrastructural studies on the genus *Actinomyces*. J. Gen. Microbiol. *71:* 63–68.

Duda, V.I. and S.V. Dobritsa. 1977. Correlation between DNA nucleotide composition and physiological and cytological characteristics of anaerobic spore-forming bacteria. Mikrobiologiya *44:* 808–813.

Duncan, C.L. 1973. Time of enterotoxin formation and release during sporulation of *Clostridium perfringens* type A. J. Bacteriol. *113:* 932–936.

Duncan, C.L., E.A. Rokos, C.M. Christenson and J.I. Rood. 1978. Multiple plasmids in different toxigenic types of *Clostridium perfringens*: possible control of beta-toxin production. Microbiology Wash. D.C., *1978:* 246–248.

Duncan, C.L., D.H. Strong and M. Sebald. 1972. Sporulation and enterotoxin production by mutants of *Clostridium perfringens*. J. Bacteriol. *110:* 378–391.

Dunican, L.K. and H.W. Seeley. 1962. Starch hydrolysis by *Streptococcus equinus*. J. Bacteriol. *83:* 264–269.

Dunkelberg, W.E., Jr., I. McVeigh. 1969. Growth requirements of *Haemophilus vaginalis*. Antonie Van Leeuwenhoek J. Microbiol. Serol. *35:* 129–145.

Dunkelberg, W.E., Jr., R. Skaggs and D.S. Kellogg. 1970. Method for isolation and identification of *Corynebacterium vaginale* (*Haemophilus vaginalis*). Appl. Microbiol. *19:* 47–52.

Dunkin, G.W. and S.E.B. Balfour-Jones. 1935. Preliminary investigation of a disease of sheep possessing certain characteristics simulating Johnei disease. J. Comp. Pathol. Ther. *48:* 236–242.

Dunny, G.M., N. Birch, G. Hascall and D.B. Clewell. 1973. Isolation and characterization of plasmid deoxyribonucleic acid from *Streptococcus mutans*. J. Bacteriol. *114:* 1362–1364.

Dunny, G.M. and D.B. Clewell. 1975. Transmissible toxin (hemolysin) plasmid in *Streptococcus faecalis* and its mobilization of a noninfectious drug resistance plasmid. J. Bacteriol. *124:* 784–790.

Durham, D.R. and W.E. Kloos. 1978. Comparative study of the total cellular fatty acids of *Staphylococcus* species of human origin. Int. J. Syst. Bacteriol. *28:* 223–228.

Durr, F.E., D.W. Smith and D.P. Altman. 1959. A comparison of the virulence of various known and atypical mycobacteria for chickens, guinea pigs, hamsters and mice. Am. Rev. Resp. Dis. *80:* 876–885.

Dürre, P., W. Andersch and J.R. Andreesen. 1981. Isolation and characterization of an adenine-utilizing, anaerobic sporeformer, *Clostridium purinolyticum* sp.

nov. Int. J. Syst. Bacteriol. *31:* 184–194.

Dürre, P. and J.R. Andreesen. 1983. Purine and glycine metabolism by purinolytic clostridia. J. Bacteriol. *154:* 192–199.

Dutky, S.R. 1940. Two new spore-forming bacteria causing milky diseases of Japanese beetle larvae. J. Agric. Res. *61:* 57–68.

Dutta, G.N. and L.A. Devriese. 1981. Sensitivity and resistance to growth promoting agents in animal lactobacilli. J. Appl. Bacteriol. *51:* 283–288.

Duxbury, T. and T.R.G. Gray. 1977. A microcultural study of the growth of cystites, cocci and rods of *Arthrobacter globiformis*. J. Gen. Microbiol. *103:* 101–106.

Dye, D.W. and W.J. Kemp. 1977. A taxonomic study of plant pathogenic *Corynebacterium* species. N. Z. J. Agric. Res. *20:* 563–582.

Dzier-Zanowska, D., R. Miksza-Zythkiewica, M. Czernilawska, H. Linda and J. Browski. 1978. Sensitivity of *Rothia dentocariosa* (letter). J. Antimicrob. Chemother. *4:* 469–471.

Eady, E.A., K.T. Holland and W.J. Cuncliffe. 1981. Studies on an inhibitory strain of *Staphylococcus epidermidis*: preliminary characterization of a low molecular weight bactericidal substance. *In* Jeljaszewicz (Editor), Staphylococci and Staphylococcal Infections, Gustav Fischer Verlag, Stuttgart, New York, pp. 163–168.

Eberson, F. 1918. A bacteriologic study of the diphtheroid organisms with special reference to Hodgkin's disease. J. Infect. Dis. *23:* 1–42.

Ebisu, S., K. Kato, S. Kotani and A. Misaki. 1975. Structural differences in fructans elaborated by *Streptococcus mutans* and *Streptococcus salivarius*. J. Biochem. *78:* 879–887.

Echandi, E. 1976. Bacteriocin production by *Corynebacterium michiganense*. Phytopathology *66:* 430–432.

Ederer, G.M., M.M. Herrmann, R. Bruce, J.M. Matsen and S.S. Chapman. 1972. Rapid extraction method with pronase B for grouping beta-haemolytic streptococci. Appl. Microbiol. *23:* 285–288.

Edmunds, P.N. 1962. The biochemical, serological and haemagglutinating reactions of "*Haemophilus vaginalis*". J. Pathol. Bacteriol. *83:* 411–422.

Edson, R.S., J.E. Rosenblatt, D.T. Lee and E.A. McVey, III. 1982. Recent experience with antimicrobial susceptibility of anaerobic bacteria. Mayo Clin. Proc. *57:* 737–741.

Edwards, J. H. 1972. The isolation of antigens associated with farmer's lung. Clin. Exp. Immunol. *11:* 341–355.

Edwards, J.H., J.T. Baker and B.H. Davies. 1974. Precipitin test negative farmer's lung - activation of the alternative pathway of complement by mouldy hay dusts. Clin. Allergy *4:* 379–388.

Edwards, K.G., H.J Blumenthal, M. Kahn and M.E. Slodki. 1981. Intracellular mannitol, a product of glucose metabolism in staphylococci. J. Bacteriol. *146:* 1020–1029.

Edwards, P.R. 1934. The differentiation of hemolytic streptococci of human and animal origin by group precipitin tests. J. Bacteriol. *27:* 527–534.

Edwardsson, S. 1968. Characteristics of caries-inducing human streptococci resembling *Streptococcus mutans*. Arch. Oral Biol. *13:* 637–646.

Edwardsson, S. 1970. The caries-inducing property of variants of *Streptococcus mutans*. Odontol. Revy *21:* 153–157.

Efstathiou, J.D. and L.M. McKay. 1976. Plasmids in *Streptococcus lactis*: evidence that lactose metabolism and proteinase activity are plasmid linked. Appl. Environ. Microbiol. *32:* 38–44.

Efstratiou, A. and W.R. Maxted. 1979. Serological group of streptococci by slide agglutination. J. Clin. Pathol. *32:* 1228–1233.

Egan, F.A. 1983. Lactic acid bacteria of meat and meat products. Antonie van Leeuwenhoek J. Microbiol. Serol. *49:* 327–336.

Egerova, A.A. and Z.P. Deryugina. 1963. On the sporogenic thermophilic thiobacterium *Thiobacillus thermophilica imschenetskii* nov. sp. Mikrobiologiya *32:* 439–466. (Microbiology USSR, English edition), *32:* 376–381.

Eggeling, L. and H. Sahm. 1980. Degradation of coniferyl alcohol and other lignin-related aromatic compounds by *Nocardia* sp. DSM 1069. Arch. Mikrobiol. *126:* 141–148.

Eggeling, L. and H. Sahm. 1981. Degradation of lignin-related aromatic compounds by *Nocardia* spec. DSM 1069 and specificity of demethylation. Zentralbl. Bakteriol. Mikrobiol. Hyg. I Abt. Orig., Suppl. *11:* 361–366.

Eggerth, A.H. 1935. The Gram-positive non-spore-bearing anaerobic bacilli of human feces. J. Bacteriol. *30:* 277–290.

Ehrenberg, C.G. 1835. Dritter Beitrag zur Erkenntniss grosser Organisation in der Richtung des kleinsten Raumes. Abh. Preuss. Akad. Wiss. Phys. Kl. Berlin aus den Jahre 1833–1835, pp. 143–336.

Ehrlich, G.G., D.F. Goerlitz, J.H. Bourell, G.V. Eisen and E.M. Godsy. 1981. Liquid chromatographic procedure for fermentation product analysis in the identification of anaerobic bacteria. Appl. Environ. Microbiol. *42:* 878–885.

Einck, K.H., P.A. Pattee, J.G. Holt, C. Hagedorn, J.A. Miller and D.L. Berryhill. 1973. Isolation and characterization of a bacteriophage of *Arthrobacter globiformis*. J. Virol. *12:* 1031–1033.

Eiroma, M., J.J. Laine and H.G. Gyllenberg. 1971. DNA base composition in psychrophilic and mesophilic bacilli. Ann. Med. Exp. Biol. Fenn. *49:* 59–61.

Eisenberg, R.C. and J.B. Evans. 1963. Energy and nitrogen requirements of *Micrococcus roseus*. Can. J. Microbiol. *9:* 633–642.

Eisenberg, R.J., M. Elchisak and J. Rudd. 1976. Regulation of lactate dehydrogenase activity in *Rothia dentocariosa* by fructose 1,6-diphosphate and adenosine 5'-triphosphate. J. Bacteriol. *126:* 1344–1346.

Eissner, G. and F.W. Ewald. 1973. Rotlauf. *In* Bieling, Kathe, Köler und Mayr

(Editors), Infektionskrankheiten und ihre Erreger. Eine Sammlung von Monographien, vol. 13. Gustav Fischer Verlag, Jena.

Eklund, M.W. and F.T. Poysky. 1974. Interconversion of type C and D strains of *Clostridium botulinum* by specific bacteriophages. Appl. Microbiol. *27:* 251–258.

Eklund, M.W., F.T. Poysky, J.A. Meyers and G.A. Pelroy. 1974. Interspecies conversion of *Clostridium botulinum* type C to *Clostridium novyi* type A by bacteriophage. Science *186:* 456–458.

Eklund, M.W., F.T. Poysky, M.E. Peterson and J.A. Meyers. 1976. Relationship of bacteriophages to alpha toxin production in *Clostridium novyi* types A and B. Infect. Immun. *14:* 793–803.

Eklund, M.W., F.T. Poysky, S.M. Reed and C.A. Smith. 1971. Bacteriophage and the toxigenicity of *Clostridium botulinum* type C. Science *172:* 480–482.

Elek, S.D. 1959. *Staphylococcus pyogenes* and Its Relation to Disease. E. & S. Livingstone Ltd., Edinburgh, London.

El-Erian, A.F.M. 1969. Bacteriological studies on Limburger cheese. Thesis. Agricultural University, Wageningen, The Netherlands, Veenman and Zonen, Wageningen.

Elischerová, K. 1975. *Listeria monocytogenes* findings in raw materials of animal origin. Cesk. Epidemiol. Mikrobiol. Imunol. *24:* 47–53 (In Czechoslovakian).

El Kholy, A., L.W. Wannamaker and R.M. Krause. 1974. Simplified extraction procedure for serological grouping of beta-haemolytic streptococci. Appl. Microbiol. *28:* 836–839.

Ellen, R.P. and I.B. Balcerzak-Raczkowski. 1977. Interbacterial aggregation of *Actinomyces naeslundii* and dental plaque streptococci. J. Periodont. Res. *12:* 11–20.

Ellen, R.P., E.D. Fillery, K.H. Chan and D.A. Grove. 1980. Sialidase-enhanced lectin-like mechanism for *Actinomyces viscosus* and *Actinomyces naeslundii* hemagglutination. Infect. Immun. *27:* 335–343.

Ellen, R.P., D.L. Walker and K.H. Chan. 1978. Association of long surface appendages with adherence-related functions of the Gram-positive species *Actinomyces naeslundii*. J. Bacteriol. *134:* 1171–1175.

Eller, C., M.R. Crabill and M.P. Bryant. 1971. Anaerobic roll tube media for nonselective enumeration and isolation of bacteria in human feces. Appl. Microbiol. *22:* 522–529.

Elliott, C. 1930. Manual of Bacterial Plant Pathogens. 1st edition. Williams and Wilkins Co., Baltimore.

Elliott, C. 1951. Manual of Bacterial Plant Pathogens. 2nd edition. Chronica Botanica Company, Waltham, Massachusetts.

Elliott, J.E. and L.G. Ljungdahl. 1982. Isolation and characterization of an Fe_8-S_8 ferredoxin (ferredoxin II) from *Clostridium thermoaceticum*. J. Bacteriol. *151:* 328–333.

Elliott, S.D. 1950. The crystallization and serological differentiation of a streptococcal proteinase and its precursor. J. Exp. Med. *92:* 201–218.

Elliott, S.D. 1960. Type and group polysaccharides of group D streptococci. J. Exp. Med. *111:* 621–630.

Elliott, S.D. 1963. Teichoic acid and the group antigen of lactic streptococci (group N). Nature (London) *200:* 1184–1185.

Elliott, S.D. 1966. Streptococcal infection in young pigs. I. An immunochemical study of the causative agent (PM streptococcus). J. Hyg. *64:* 205–212.

Elliott, S.D., M. McCarty and R.C. Lancefield. 1977. Teichoic acids of group D streptococci with special reference to strains from pig meningitis (*Streptococcus suis*). J. Exp. Med. *145:* 490–499.

Ellis, P.P., S.C. Bausor and J.M. Fulmer. 1961. *Streptothrix* canaliculitis. Am. J. Ophthalmol. *52:* 36–43.

Ellwood, D.C., J.R. Hunter and V.M.C. Longyear. 1974. Growth of *Streptococcus mutans* in a chemostat. Arch. Oral Biol. *19:* 659–664.

Ellwood, D.C., P.J. Phipps and I.R. Hamilton. 1979. Effect of growth rate and glucose concentration on the activity of the phosphoenolpyruvate phosphotransferase system in *Streptococcus mutans* Ingbritt grown in continuous culture. Infect. Immun. *23:* 224–231.

Elsden, S.R. and M.G. Hilton. 1978. Volatile acid production from threonine, valine, leucine and isoleucine by clostridia. Arch. Microbiol. *117:* 165–172.

Elsden, S.R. and M.G. Hilton. 1979. Amino acid utilization patterns in clostridial taxonomy. Arch. Microbiol. *133:* 137–141.

Elsden, S.R., M.G. Hilton, K.R. Parsley and R. Self. 1980. The lipid fatty acids of proteolytic clostridia. J. Gen. Microbiol. *118:* 115–123.

Elsden, S.R., M.G. Hilton and J.M. Walker. 1976. The end products of the metabolism of aromatic amino acids by clostridia. Arch. Microbiol. *107:* 283–288.

Elwan, S.H. 1964. Some aspects of microbial stimulation in a soil treated with streptomycin. Arch. Microbiol. *47:* 277–285.

Emberger, O. 1970. A contribution to cultivation methods for the detection of aerobic sporeforming bacteria. Zentralbl. Bakteriol. Parasitenkd. Infektionskr. Hyg. Abt. II *125:* 555–565.

Embley, T.M., M. Goodfellow, D.E. Minnikin and A.G. O'Donnell. 1984. Lipid and wall amino acid composition in the classification of *Rothia dentocariosa*. Zentralbl. Bakteriol. Mikrobiol. Hyg. I. Abt. A *257:* 285–295.

Embly, T.M., M. Goodfellow and B. Austin. 1982. A semi-defined growth medium for *Renibacterium salmoninarum*. FEMS Microbiol. Lett. *14:* 299–301.

Emerson, J.E. and O.J. Weiser. 1963. Detecting cellulose digesting bacteria. J. Bacteriol. *86:* 891–892.

Emery, R.S., C.K. Smith and L.F. To. 1957. Utilization of inorganic sulfate by rumen microorganisms. II. The ability of single strains of rumen bacteria to utilize inorganic sulfate. Appl. Microbiol. *5:* 363–366.

Emmett, M. 1976. Biochemical and reversion analysis of the amino acid requirements of *Staphylococcus* auxotrophs isolated from the skin of humans and other animals. Ph. D. Dissertation, North Carolina State University.

Emmett, M. and W.E. Kloos. 1975. Amino acid requirements of staphylococci isolated from human skin. Can. J. Microbiol. *21:* 729–733.

Emmett, M. and W.E. Kloos. 1979. The nature of arginine auxotrophy in cutaneous populations of staphylococci. J. Gen. Microbiol. *110:* 305–314.

Emmons, C.W. 1938. The isolation of *Actinomyces bovis* from tonsillar granules. Public Health Rep. *53:* 1967–1975.

Endl, J., H.P. Seidl, F. Fiedler and K.H. Schleifer. 1983. Chemical structure of cell wall teichoic acids of staphylococci. Arch. Microbiol. *135:* 215–223.

Endresen, C. and A. Grov. 1976. Immunochemical analysis of an unusual cell wall polysaccharide from animal coagulase-positive staphylococci. Acta Pathol. Microbiol. Scand. Sect. B *84:* 300–308.

Endresen, C., A. Grov and P. Oeding. 1974. Immunochemical studies of three cell wall polysaccharides from animal coagulase-positive staphylococci. Acta Pathol. Microbiol. Scand. Sect. B *82:* 382–386.

Endresen, C. and P. Oeding. 1973. Purification and characterization of serologically active cell wall substances from *Planococcus* strains. Acta Pathol. Microbiol. Scand. Sect. B *81:* 571–575.

Eng, R.H.K., M.L. Corrado, D. Cleri, C. Cherubin and E.J.C. Goldstein. 1981. Infections caused by *Actinomyces viscosus*. Am. J. Clin. Pathol. *75:* 113–116.

Engbaek, H.C., A. Jespersen, D. Faber and D.W. Will. 1964. The pathology of joint disease in rabbits produced by atypical mycobacteria and *M. avium*. I. Macroscopical and bacteriological examination of organs, joints and tendon sheaths. Acta Tuberc. Pneumonol. Scand. *44:* 199–208.

Engbaek, H.C., B. Bergmann, I. Baess and D.W. Will. 1967. *M. xenopei*—A bacteriological study of *M. xenopei* including case reports of Danish patients. Acta Pathol. Microbiol. Scand. *69:* 576–594.

Engbaek, H.C., B. Bergmann, I. Baess and M. Weis Bentzon. 1968. *Mycobacterium avium*: A bacteriological and epidemiological study of *M. avium* isolated from animals and man in Denmark. Part I: Strains isolated from animals. Acta Pathol. Microbiol. Scand. *72:* 277–294.

Engel, D., H.E. Schroeder and R.C. Page. 1978. Morphological features and functional properties of human fibroblasts exposed to *Actinomyces viscosus* substances. Infect. Immun. *19:* 287–295.

Engel, D., D. van Epps and J. Clagett. 1976. *In vivo* and *in vitro* studies on possible pathogenic mechanisms of *Actinomyces viscosus*. Infect. Immun. *14:* 548–554.

Engel, H.W.B, L.G. Berwald and A.H. Havelaar. 1980. The occurance of *Mycobacterium kansasii* in tapwater. Tubercle *61:* 11.

England, D.M. and J.E. Rosenblatt. 1977. Anaerobes in human bilary tracts. J. Clin. Microbiol. *6:* 494–498.

Ensign, J.C. and S.C. Rittenberg. 1963. A crystalline pigment produced from 2-hydroxypyridine by *Arthrobacter crystallopoietes* n. sp. Arch. Mikrobiol. *47:* 137–153.

Eppinger, H. 1891. Über eine neue pathogene *Cladothrix* und eine durch sie hervorgerufene Pseudotuberculosis (Cladothrichia). Beitr. Pathol. Anat. Allg. Pathol. *9:* 287–328.

Ercolani, G.L. 1968. Effetività e misura della transmissione di *Xanthomonas vesicatoria* e di *Corynebacterium michiganense* attraverso il seme del pomodora. Ind. Conserve (Paoma) *43:* 15–22.

Erikson, D. 1940. Pathogenic anaerobic organisms of the *Actinomyces* group. Med. Res. Counc. Spec. Rep. Ser. *240:* 5–63.

Erikson, D. 1954. Factors promoting cell division in a "soft" mycelial type of Nocardia: *Nocardia turbata* n. sp. J. Gen. Microbiol. *11:* 198–208.

Erler, W. 1972. Serologisch, chemische und immunochemische Untersuchungen an Rotlaufbakterien. X. Die Differenzierung der Rotlaufbakterien nach chemischen Merkmalen. Arch. Exp. Veterinaermed. *26:* 809–816.

Ernst, W. 1906. Über pyelonephritis diphtherica bovis und die pyelonephritisbazillen. Zentralbl. Bakteriol. Parasitenkd. Infektionskr. Hyg. Abt. I. *39:* 349–558.

Errebo Larsen, H. and H.P.R. Seeliger. 1966. A mannitol fermenting *Listeria*: *Listeria grayi* sp. n. *In* Proceedings of the Third International Symposium on listeriosis. Bilthoven, The Netherlands, pp. 35–39.

Escherich, T. 1886. Die Darmbacterien des Säuglings und ihre Beziehungen zur Physiologie der Verdauung. F. Enke. Stuttgart.

Essers, L. and K. Radebold. 1980. Rapid and reliable identification of *Staphylococcus aureus* by a latex agglutination test. J. Clin. Microbiol. *12:* 641–643.

Evangelista, A.T., A. Saha, M.P. Lechevalier and G. Furness. 1978. Analysis of the cell wall constituents of *Corynebacterium genitalium*. Int. J. Syst. Bacteriol. *28:* 344–348.

Evans, A.C. 1916. The bacteria of milk freshly drawn from normal udders. J. Infect. Dis. *18:* 437–476.

Evans, C.A. and K.L. Mattern. 1978. Individual differences in the bacterial flora of the skin of the forehead: *Peptococcus saccharolyticus*. J. Invest. Dermatol. *71:* 152–153.

Evans, C.A. and K.L. Mattern. 1979. The aerobic growth of *Propionibacterium acnes* in primary cultures from skin. J. Invest. Dermatol. *72:* 103–106.

Evans, C.A., K.L. Mattern and S.L. Hallam. 1978. Isolation and identification of *Peptococcus saccharolyticus* from human skin. J. Clin. Microbiol. *7:* 261–264.

Evans, C.E., W.M. Smith, E.A. Johnson and E.R. Giblett. 1950. Bacterial flora of the normal human skin. J. Invest. Dermatol. *15:* 305–323.

Evans, J.B. 1976. Anaerobic growth of *Staphylococcus* species from human skin: effects of uracil and pyruvate. Int. J. Syst. Bacteriol. *26:* 17–21.

Evans, J.B., G.A. Ananaba, C.A. Pate and M.S. Bergdoll. 1983. Enterotoxin production by atypical *Staphylococcus aureus* from poultry. J. Appl. Bacteriol. *54:* 257–261.

Evans, J.B., L.G. Buettner and C.F. Niven, Jr. 1952. Occurrence of streptococci that give a false-positive coagulase test. J. Bacteriol. *64:* 433–434.

Evans, J.B. and W.E. Kloos. 1972. Use of shake cultures in a semisolid thioglycolate medium for differentiating staphylococci from micrococci. Appl. Microbiol. *23:* 326–331.

Evans, J.B. and C.F. Niven, Jr. 1951. Nutrition of the heterofermentative lactobacilli that cause greening of cured meat products. J. Bacteriol. *62:* 599–603.

Evelyn, T.P.T. 1977. An improved growth medium for the kidney disease bacterium and some notes on using the medium. Bull. Off. Int. Epizoot. *87:* 511–513.

Evelyn, T.P.T. and L.A. McDermott. 1961. Bacteriological studies of fresh water fish. Can. J. Microbiol. *7:* 375–382.

Evtushenko, L.I., N.A. Janushkene, G.M. Streshinskaya, I.B. Naumova and N.S. Agre. 1984a. Occurrence of teichoic acids in representatives of Actinomycetales. Dokl. Akad. Nauk SSSR *278:* 237–239.

Evtushenko, L.I., G.F. Levanova and N.S. Agre. 1984b. Nucleotide composition of DNA and amino acid composition of A 4a peptidoglycan in *Promicromonospora citrea*. Mikrobiologiya *53:* 519–520 (In Russian).

Evtushenko, L.I., D.T. Pataraya, H. Prauser and N.S. Agre. 1981. Physiological and biochemical features of *Promicromonospora citrea*. Dep. VINITI, No. 4282–81 (In Russian).

Ewald, F.W. 1957. Das Hyaluronidase - Bildungsvermögen von Rotlaufbakterien. Monatsschr. Tierheil. *9:* 333–341.

Ewald, F.W. 1981. The genus *Erysipelothrix*. *In* Starr, Stolp, Trüper, Balows and Schlegel (Editors), The Prokaryotes. A Handbook on Habitats, Isolation, and Identification of Bacteria. Springer-Verlag, New York, pp. 1688–1689.

Exterkate, F.A., B.J. Otten, H.W. Wassenberg and J.H. Veerkamp. 1971. Comparison of the phospholipid composition of *Bifidobacterium* and *Lactobacillus* strains. J. Bacteriol *106:* 824–829.

Eyssen, H.J., G.G. Parmentier, F.C. Compernolle, G. DePauw and M. Piessens-Denef. 1973. Biohydrogenation of sterols by *Eubacterium* ATCC 21408 - nova species. Eur. J. Biochem. *36:* 411–421.

Ezaki, T. and E. Yabuuchi. 1983. Deoxyribonucleic acid base composition and DNA/RNA hybridization studies among the four species of *Peptostreptococcus* Kluyver and Van Niel 1936. FEMS Microbiol. Lett. *17:* 197–200.

Ezaki, T., N. Yamamoto, K. Nonomiya, S. Suzuki and E. Yabuuchi. 1983. Transfer of *Peptococcus indolicus*, *Peptococcus asaccharolyticus*, *Peptococcus prevotii* and *Peptococcus magnus* to the genus *Peptostreptococcus* and proposal of *Peptostreptococcus tetradius* sp. nov. Int. J. Syst. Bacteriol. *33:* 683–698.

Facklam, R.R. 1973. Comparison of several laboratory media for presumptive identification of enterococci and group D streptococci. Appl. Microbiol. *26:* 138–145.

Facklam, R.R. 1974. Characteristics of *Streptococcus mutans* isolated from human dental plaque and blood. Int. J. Syst. Bacteriol. *24:* 313–319.

Facklam, R.R. 1977. Physiological differentiation of viridans streptococci. J. Clin. Microbiol. *5:* 184–201.

Facklam, R.R. 1984. The major differences in the American and British *Streptococcus* taxonomy schemes with special reference to *Streptococcus milleri*. Eur. J. Clin. Microbiol. *3:* 91–93.

Facklam, R.R. and M.D. Moody. 1970. Presumptive identification of group D streptococci: the bile-esculin test. Appl. Microbiol. *20:* 245–250

Facklam, R.R., J.F. Padula, L.G. Thacker, E.C. Wortham and B.J. Sconyers. 1974. Presumptive identification of group A, B, and D streptococci. Appl. Microbiol. *27:* 107–113.

Facklam, R.R., D.L. Rhoden and P.B. Smith. 1984. Evaluation of the Rapid Strep System for the identification of clinical isolates of *Streptococcus* species. J. Clin. Microbiol. *20:* 814–898.

Facklam, R.R. and H.W. Wilkinson. 1981. The family *Streptococcaceae* (Medical Aspects). *In* Starr, Stolp, Trüper, Balows and Schlegel (Editors), The Prokaryotes. A Handbook on Habitats, Isolation and Identification of Bacteria. Springer-Verlag, New York, pp. 1572–1597.

Fahmy, F. 1978. Untersuchungen zur Taxonomie farbstoffbildender *Bacillus*-Stämme aus Salzmarsch- und Wattböden. Ph.D. Thesis, University of Göttingen.

Fahmy, F., J. Flossdorf and D. Claus. 1985. The DNA base composition of the type strains of the genus *Bacillus*. Syst. Appl. Microbiol. *6:* 60–65.

Fahmy, F., F. Mayer and D. Claus. 1985. Endospores of *Sporosarcina halophila*: characteristics and ultrastructure. Arch. Microbiol. *140:* 338–342.

Faller, A.H., F. Götz and K.H. Schleifer. 1980. Cytochrome-patterns of staphylococci and micrococci and their taxonomic implications. Zentralbl. Bakteriol. Parasitenkd. Infektionskr. Hyg. Abt. I Orig. C *1:* 26–39.

Faller, A.H. and K.H. Schleifer. 1981. Effects of growth phase and oxygen supply on the cytochrome composition and morphology of *Arthrobacter crystallopoietes*. Curr. Microbiol. *6:* 253–258.

Faller, A.H. and K.H. Schleifer. 1981. Modified oxidase and benzidine tests for separation of staphylococci from micrococci. J. Clin. Microbiol. *13:* 1031–1035.

Farabaugh, P.J., V. Schmeissner, M. Hofer and J.H. Miller. 1978. Genetic studies of the lac repressor. VII. On the molecular nature of spontaneous hotspots in the lacI gene of *Escherichia coli*. J. Mol. Biol. *126:* 847–863.

Farrar, W.E. 1963. Serious infections due to "non-pathogenic" organisms in the genus *Bacillus*. Am. J. Med. *34:* 134–141.

Farrior, J.W. and W.E. Kloos. 1975. Amino acid and vitamin requirements of *Micrococcus* species isolated from human skin. Int. J. Syst. Bacteriol. *25:* 80–82.

Farrow, J.A.E. 1980. Lactose-hydrolyzing enzymes in *Streptococcus lactis* and *Streptococcus cremoris* and also in some other species of streptococci. J. Appl. Bacteriol. *49:* 493–503.

Farrow, J.A.E. and M.D. Collins. 1984a. DNA base composition, DNA/DNA homology and long-chain fatty acid studies on *Streptococcus thermophilus* and *Streptococcus salivarius*. J. Gen. Microbiol. *130:* 357–362.

Farrow, J.A.E. and M.D. Collins. 1984b. Validation List No. 15. Int. J. Syst. Bacteriol. *34:* 355–357.

Farrow, J.A.E. and M.D. Collins. 1984c. Taxonomic studies on streptococci of serological groups C, G and L and possibly related taxa. Syst. Appl. Microbiol. *5:* 483–493.

Farrow, J.A.E. and M.D. Collins. 1985b. Validation List No. 17. Int. J. Syst. Bacteriol. *35:* 223–225.

Farrow, J.A.E., J. Kruze, B.A. Phillips, A.J. Bramley and M.D. Collins. 1984. Taxonomic studies on *Streptococcus bovis* and *Streptococcus equinus*: description of *Streptococcus alactolyticus* sp. nov. and *Streptococcus saccharolyticus* sp. nov. Syst. Appl. Microbiol. *5:* 467–482.

Farrow, J.A.E., J. Kruze, B.A. Phillips, A.J. Bramley and M.D. Collins. 1985b. Validation List No. 17. Int. J. Syst. Bacteriol. *35:* 223–225.

Farshy, D.C. and C.W. Moss. 1970. Characterization of clostridia by gas chromatography. Differentiation of species by trimethylsilyl derivatives of whole cell hydrolysates. Appl. Microbiol. *20:* 78–84.

Fauchère, J.L., P. Berche, D. Ganeval, F. Bournerias, F. Daoulas-Lebourdelles and M. Vernon. 1977. Une septicémie à *Bacillus licheniformis*. Med. Malad. Infect. *7:* 191–195.

Fauve, R.M., C.H. Pierce-Chase and R. Dubos. 1964. Corynebacterial pseudotuberculosis in mice. II. Activation of natural and experimental latent infection. J. Exp. Med. *120:* 283–304.

Feigen, G.A., N.S. Peterson, W.W. Hofmann, G.H. Genther and W.E. Van Heyningen. 1963. The effect of impure tetanus toxin in the frequency of miniature end-plate potentials. J. Gen. Microbiol. *33:* 489–495.

Feigen, R.D., P.G. Shackelford, and J. Campbell. 1973. Assessment of the role of *Staphylococcus epidermidis* as a cause of otitis media. Pediatrics *52:* 569.

Feist, H. 1972. Serologisch, chemische und immunochemische Untersuchungen an Rotlaufbakterien. XII. Das Murein der Rotlaufbakterien. Arch. Exp. Veterinaermed. *26:* 825–834.

Feldman, W.H. and R.E. Ritts. 1963. Pathogenicity studies of group III (Battey) mycobacteria from pulmonary lesions of man. Dis. Chest *43:* 26–33.

Fellers, C.R. and R.W. Clough. 1925. Indol and skatol determination of bacterial cultures. J. Bacteriol. *10:* 105–133.

Feltham, R.K.A. 1979. A taxonomic study of the *Micrococcaceae*. J. Appl. Bacteriol. *47:* 243–254.

Feltham, R.K.A. 1979. A taxonomic study of the genus *Streptococcus*. *In* Parker (Editor), Pathogenic Streptococci, Reedbooks, Chertsey, Surrey, pp. 247–248.

Feltham, R.K.A., A.K. Power, P.A. Pell and P.H.A. Sneath. 1978. A simple method for storage of bacteria at -76°C. J. Appl. Bacteriol. *44:* 313–316.

Fenner, F. 1950. The significance of the incubation period in infectious diseases. Med. J. Aust. *2:* 813–818.

Fenner, F. 1952. Studies on *Mycobacterium ulcerans*. II. Cross reactivity in guinea pigs sensitized with *Mycobacterium ulcerans* and other mycobacteria. Aust. J. Exp. Biol. Med. Sci. *30:* 11–20.

Fenner, F. 1956. The pathogenic behavior of *Mycobacterium ulcerans* and *Mycobacterium balnei* in the mouse and the developing chick embryo. Am. Rev. Tuberc. Pulm. Dis. *73:* 650–673.

Ferguson, D.A. and C.S. Cummins. 1978. Nutritional requirements of anaerobic coryneforms. J. Bacteriol. *135:* 858–867.

Fernandez, F., M.D. Collins and M.J. Hill. 1984. Production of vitamin K by human gut bacteria. Biochem. Trans. *13:* 223–224.

Ferrieri, P., E.D. Gray and L.W. Wannamaker. 1980. Biochemical and immunological characterization of the extracellular nucleases of group B streptococci. J. Exp. Med. *151:* 56–68.

Festenstein, G.N., J. Lacey, F.A. Skinner, P.A. Jenkins and J. Pepys. 1965. Self-heating of hay and grain in Dewar flasks and the development of farmer's lung hay antigens. J. Gen. Microbiol. *41:* 389–407.

Fetter, B.F., G.K. Klintworth and W.S. Hendry. 1967. Mycoses of the Central Nervous System. Williams and Wilkins Co., Baltimore.

Fiedler, F. and O. Kandler. 1973. Die Aminosäuresequenz von 2,4-Diaminobuttersäure enthaltenden Mureinen bei verschiedenen coryneformen Bakterien und *Agromyces ramosus*. Arch. Mikrobiol. *89:* 51–66.

Fiedler, F. and O. Kandler. 1973. Die Mureintypen in der Gattung *Cellulomonas* Bergey et al. Arch. Mikrobiol. *89:* 41–50.

Fiedler, F., M.J. Schäffler and E. Stackebrandt. 1981. Biochemical and nucleic acid hybridization studies on *Brevibacterium linens* and related strains. Arch. Microbiol. *129:* 85–93.

Fiedler, F., K.H. Schleifer, B. Cziharz, E. Interschick and O. Kandler. 1970. Murein types in *Arthrobacter*, brevibacteria, corynebacteria and microbacteria. Publ. Fac. Sci. Univ. J.E. Purkyne, Brno *47:* 111–122.

Fiedler, F., K.H. Schleifer and O. Kandler. 1973. Amino acid sequence of the threonine-containing mureins of coryneform bacteria. J. Bacteriol. *133:* 8–17.

Fiedler, F. and J. Seger. 1983. The murein types of *Listeria grayi, Listeria murrayi* and *Listeria denitrificans.* Syst. Appl. Microbiol. *4:* 444–450.

Fiedler, F., J. Seger, A. Schrettenbrunner and H.P.R. Seeliger. 1984. The biochemistry of murein and cell wall teichoic acids in the genus *Listeria.* Syst. Appl. Microbiol. *5:* 360–376.

Fiedler, F. and E. Stackebrandt. 1978. Taxonomical studies on *Brevibacterium linens.* Abstracts of the XII International Congress of Microbiology, Munich, C45, p. 96.

Field, H.I., D. Buntain and J.T. Done. 1954. Studies on piglet mortality. I. Streptococcal meningitis and arthritis. Vet. Rec. *66:* 453–455.

Fieldsteel, A.H. and A.H. McIntosh. 1971. Effect of neonatal thymectomy and antithymocytic serum on susceptibility of rats to *Mycobacterium leprae* infection (35908). Proc. Soc. Exp. Biol. Med. *138:* 408–413.

Fildes, P., G.M. Richardson, B.C.J.G. Knight and G.P. Gladstone. 1936. A nutrient mixture suitable for growth of *Staphylococcus aureus.* Brit. J. Exp. Pathol. *17:* 481–484.

Filippov, V.A. 1976a. Sensitivity of some species of lactobacilli of subgenus *Betabacterium* to bacteriocins of various species. Antibiotiki *21:* 1075–1078.

Filippov, V.A. 1976b. Bacteriocin typing of lactobacilli of the *Streptobacterium* subgenus. Zh. Mikrobiol. Epidemiol. Immunobiol. *53:* 86–88.

Filippov, V.A. and E.B. Rubanenko. 1977. Differentiation of some species of lactobacilli of the *Thermobacterium* subgenus by the spectra of bacteriocin sensitivity. Zh. Mikrobiol. Epidemiol. Immunobiol. *54:* 46–50.

Fillery, E.D., G.H. Bowden and J.M. Hardie. 1978. A comparison of strains of bacteria designated *Actinomyces viscosus* and *Actinomyces naeslundii.* Caries Res. *12:* 299–312.

Finegold, S.M. 1977. Anaerobic bacteria in human disease. Academic Press, New York, pp. 1–710.

Finegold, S.M., H.R. Attebery and V.L. Sutter. 1974. Effect of diet on human fecal flora: comparison of Japanese and American diets. Am. J. Clin. Nutr. *27:* 1455–1469.

Finegold, S.M., J.G. Bartlett, A.W. Chow, D.J. Flora, S.L. Gorbach, E.J. Harder and F.P. Tally. 1975. Management of anaerobic infections. Ann. Intern. Med. *83:* 375–389.

Finegold, S.M., D.J. Flora, H.R. Attebery and V.L. Sutter. 1975. Fecal bacteriology of colonic polyp patients and control patients. Cancer Res. *35:* 3407–3417.

Finegold, S.M., P.T. Sugihara and V.L. Sutter. 1971. Use of selective media for isolation of anaerobes from humans. *In* Shapton and Board (Editors), Isolation of Anaerobes. London: Academic Press, pp. 99–108.

Finegold, S.M., V.L. Sutter and G.E. Mathisen. 1983. Normal indigenous intestinal flora. *In* Hentges (Editor), Human Intestinal Microflora in Health and Disease, Academic Press, New York, pp. 3–31.

Finegold, S.M., V.L. Sutter, P.T. Sugihara, H.A. Elder, S.M. Lehmann and R.L. Phillips. 1977. Fecal microbial flora in Seventh Day Adventist populations and control subjects. Am. J. Clin. Nutr. *30:* 1781–1792.

Finne, G. and J.R. Matches. 1974. Low-temperature-growing clostridia from marine sediments. Can. J. Microbiol. *20:* 1639–1645.

Fischer, S., H. Luczak and K.H. Schleifer. 1982. Improved methods for the detection of class II fructose-1-6-biphosphate aldolases in bacteria. FEMS Microbiol. Lett. *15:* 103–108.

Fischer, S., A. Tsugita, B. Kreutz and K.H. Schleifer. 1983. Immunochemical and protein-chemical studies of the class I fructose-1, 6-diphosphate aldolases from staphylococci. Int. J. Syst. Bacteriol. *33:* 443–450.

Fischer, U. and K.H. Schleifer. 1980. Zum Vorkommen der Gram-positiven, katalase-positiven Kokken in Rohwurst. Fleischwirt-shaft. *60:* 1046–1051.

Fisher, C.A. and L. Barksdale. 1973. Cytochemical reactions of human leprosy bacilli and mycobacteria: ultrastructural implications. J. Bacteriol. *113:* 1389–1399.

Fitz-James, P.C. and F.E. Young. 1969. Morphology of Sporulation. *In* Gould and Hurst (Editors), The Bacterial Spore, Academic Press, London and New York, pp. 39–72.

Flandrois, J.P., A. Grov, A. Ndulue, J. Fleurette and P. Oeding. 1981. Immunochemical study of *Staphylococcus aureus* type antigens. *In* Jeljaszewicz (Editor), Staphylococci and Staphylococcal Infections., Gustav Fischer Verlag, Stuttgart, New York, pp. 71–74.

Fletcher, S.M., C.J.M. Rondle and I.G. Murray. 1970. The extracellular antigens of *Micropolyspora faeni:* their significance in farmer's lung disease. J. Hyg. *68:* 401–409.

Fleurette, J. and A. Modjadedy. 1976. Attempts to combine and simplify two methods for serotyping of *Staphylococcus aureus. In* Jeljaszewicz (Editor), Staphylococci and Staphylococcal Diseases, Gustav Fischer Verlag, Stuttgart, New York, pp. 71–80.

Flossmann, K.D. and W. Erler. 1972. Serologische, chemische und immunochemische Untersuchungen an Rotlaufbakterien. XI. Isolierung und Charakterisierung von Desoxyribonukleinsäuren aus Rotlaufbakterien. Arch. Exp. Veterinaermed. *26:* 817–824.

Flügge, C. 1886. Die Mikroorganismen. F.C.W. Vogel, Leipzig.

Foglesong, M.A. and A.J. Markovetz. 1974. Morphology of bacteriophage-like particles from *Fusobacterium symbiosum.* J. Bacteriol. *119:* 325–329.

Foissy, H. 1974. Examination of *Brevibacterium linens* by electrophoretic zymo-

gram technique. J. Gen. Microbiol. *80:* 197–207.

Fontaine, F.E., W.H. Peterson, E. McCoy and M.J. Johnson. 1942. A new type of glucose fermentation by *Clostridium thermoaceticum* n. sp. J. Bacteriol. *43:* 701–715.

Food and Drug Administration. 1976. FDA Bacteriological Analytical Manual for Foods. Association of Official Analytical Chemists, Washington, D.C.

Forbes, B.A. and D.R. Schaberg. 1982. Transfer of R plasmids from *S. epidermidis* to *S. aureus:* evidence for conjugal exchange of resistance. Abstr. Annu. Meet. Am. Soc. Microbiol. *1982:* 132.

Ford, W.W. 1927. Textbook of bacteriology. Saunders, Philadelphia, pp. 1–1069.

Fornachon, J.C.M., H.C. Douglas and R.H. Vaughn. 1949. *Lactobacillus trichodes* nov. sp., a bacterium causing spoilage in appetizer and dessert wines. Hilgardia *19:* 119–132.

Forsberg, C.W. 1978. Effects of heavy metals and other trace elements on the fermentative activity of the rumen microflora and growth of functionally important rumen bacteria. Can. J. Microbiol. *24:* 298–306.

Forsgren, A. 1969. Protein A from *Staphylococcus aureus.* VIII. Production of protein A by bacterial and L-forms of *S. aureus.* Acta Pathol. Microbiol. Scand. *75:* 481–490.

Fortner, J. 1928. Ein einfaches Plattenverfahren zur Züchtung strenger Anaerobier. Zentralbl. Bakteriol. Parasitenkd. Infektionskr. Hyg. I Abt. Orig. *108:* 155–159.

Fortner, J. 1929. Zur Technik der anaeroben Züchtung. Zentralbl. Bakteriol. Parasitenkd. Infektionskr. Hyg. I. Abt. Orig. *110:* 233–256.

Foubert, E.L., Jr., and H.C. Douglas. 1948. Studies on the anaerobic micrococci. I. Taxonomic considerations. J. Bacteriol. *56:* 25–34.

Fox, E.N. 1974. M proteins of group A streptococci. Bacteriol. Rev. *38:* 57–86.

Fox, G.E., K.J. Pechmann and C.R. Woese. 1977. Comparative cataloging of 16S ribosomal ribonucleic acid: molecular approach to procaryotic systematics. Int. J. Syst. Bacteriol. *27:* 44–57.

Fox, G.E., E. Stackebrandt, R.B. Hespell, J. Gibson, J. Maniloff, T.A. Dyer, R.S. Wolfe, W.E. Balch, R.S. Tanner, L.J. Magrum, L.B. Zablen, R. Blakemore, R. Gupta, L. Bonen, B.J. Lewis, D.A. Stahl, K.R. Luehrsen, K.N. Chen and C.R. Woese. 1980. The phylogeny of prokaryotes. Science *209:* 457–463.

Frade, R. and M.T. Guenan-Siberil. 1975. Effect du pH du milieu de culture sur la teneur en phospholipedes des membranes de *Bacillus coagulans.* Biochemie *57:* 1397–1400.

Francis, J. 1943. Infection of laboratory animals with *Mycobacterium johnei.* J. Comp. Pathol. *53:* 140–150.

Franke, F. 1973. Untersuchungen zur Ätiologie der Gesäugeaktinomykose des Schweines. Zentralbl. Bakteriol. Parasitenkd. Infektionskr. Hyg. I Abt. Orig. A *223:* 111–124.

Frankland, G.C. and P.F. Frankland. 1887. Studies on some new microorganisms obtained from air. Phil. Trans. Roy. Soc. London Ser. B Biol. Sci. *178:* 257–287.

Fraser, A.G. 1978. Neuraminidase production by clostridia. J. Med. Microbiol. *11:* 269–280.

Fraser, G. 1964. The effect on animal erythrocytes of combinations of diffusible substances produced by bacteria. J. Pathol. Bacteriol. *88:* 43–53.

Frazier, P.D. and B.O. Fowler. 1967. X-ray diffraction and infrared study of the "sulphur granules" of *Actinomyces bovis.* J. Gen. Microbiol. *46:* 445–450.

Friedman, M.E. and W.L. Alm. 1962. Effect of glucose concentration in the growth medium on some metabolic activities of *Listeria monocytogenes.* J. Bacteriol. *84:* 375–376.

Friedman, S.A. and J.B. Hays. 1977. Initial characterization of hexose and hexitol phosphoenolpyruvate-dependent phosphotransferase of *Staphylococcus aureus.* J. Bacteriol. *130:* 991–999.

Friedman, S.M. 1978. Biochemistry of thermophily. Academic Press, New York.

Fritsche, D. 1964a. Untersuchungen über die Empfindlichkeit des *Actinomyces israelii* und zwei seiner häufigsten anaeroben Begleitbakterien gegen neuere Antibiotica. Z. Hyg. *150:* 50–57.

Fritsche, D. 1964b. Die Benzol- und Toluolresistenz des *Actinomyces israelii,* ein Hilfsmittel für die Strahlenpilzdiagnostik. Zentralbl. Bakteriol. Parasitenkd. Infektionskr. Hyg. I Abt. Orig. *194:* 241–244.

Frölander, F. and J. Carlsson. 1977. Bactericidal effect of anaerobic broth exposed to atmospheric oxygen tested on *Peptostreptococcus anaerobius.* J. Clin. Microbiol. *6:* 117–123.

Froman, S., D.W. Will and E. Bogen. 1954. Bacteriophage active against virulent *Mycobacterium tuberculosis* I. Isolation and activity. Am. J. Public Health *44:* 1326–1333.

Frost, W.D. and M.A. Engelbrecht. 1936. A revision of the genus *Streptococcus.* Dept. Agr. Bacteriol., Univ. Wisconsin, Madison, pp. 1–4.

Fry, R.M. 1964. Personal communication cited *In* Topley, Wilson and Miles (Editors), Topley and Wilson's principles of bacteriology and immunity. 5th Ed. Williams and Wilkins Co., Baltimore, p. 721.

Fryer, J.L. and J.E. Sanders. 1981. Bacterial kidney disease of salmonid fish. Ann. Rev. Microbiol. *35:* 273–297.

Fuchs, P.G., J. Zajdel and W.T. Dobrzanski. 1975. Possible plasmid nature of the determinant for production of the antibiotic nisin in some strains of *Streptococcus lactis.* J. Gen. Microbiol. *88:* 189–192.

Fujii, M., T. Imanaka and S. Aiba. 1982. Molecular cloning and expression of penicillinase genes from *Bacillus licheniformis* in the thermophile *Bacillus stearothermophilus.* J. Gen. Microbiol. *128:* 2997–3000.

Fujimura, S. and T. Nakamura. 1978. Purification and properties of a bacteriocin-like substance (Acnecin) of oral *Propionibacterium acnes*. Antimicrob. Agents Chemother. *14:* 893–898.

Fujita, N.K., K. Lam and A.S. Bayer. 1982. Septic arthritis due to group G *Streptococcus*. J. Am. Med. Assoc. *247:* 812–813.

Fukui, S., A. Oi, A. Obayashi and K. Kitahara. 1957. Studies on the pentose metabolism by microorganisms. I. A new type lactic acid fermentation of pentoses by lactic acid bacteria. J. Gen. Appl. Microbiol. *3:* 258–268.

Fukumoto, J. 1943. Studies on the production of bacterial amylase. 1. Isolation of bacteria secreting potent amylases and their distribution. J. Agric. Chem. Soc. Japan *19:* 487–503 (in Japanese).

Fulghum, R.S., B.P. Baldwin and P.P. Williams. 1968. Antibiotic susceptability of anaerobic ruminal bacteria. Appl. Microbiol. *16:* 301–307.

Fulkerson, J.F. 1960. Pathogenicity and stability of strains of *Corynebacterium insidiosum*. Phytopathology *50:* 377–380.

Fuller, A.T. 1938. The formamide method for the extraction of polysaccharides from haemolytic streptococci. Brit. J. Exp. Pathol. *19:* 130–139.

Fuller, R. and L.G.M. Newland. 1963. The serological grouping of three strains of *Streptococcus equinus*. J. Gen. Microbiol. *31:* 431–434.

Füzi, M. 1963. A neomycin sensitivity test for the rapid differentiation of *Listeria monocytogenes* and *Erysipelothrix rhusiopathiae*. J. Pathol. Bacteriol. *85:* 524–525.

Füzi, M. 1981. Oxygen-dependent metronidazole-resistance of *Clostridium histolyticum*. Zentralbl. Bakteriol. Mikrobiol. Hyg. I Abt. Orig. A *249:* 99–103.

Gabay, E.L., R.D. Rolfe and S.M. Finegold. 1981. Susceptibility of *Clostridium septicum* to 23 antimicrobial agents. Antimicrob. Agents Chemother. *20:* 852–853.

Gadalla, M.S.A. and J.G. Collee. 1967. The nature and properties of the haemagglutinin of *Clostridium septicum*. J. Pathol. Bacteriol. *93:* 255–274.

Gadalla, M.S.A. and J.G. Collee. 1968. The relationship of the neuraminidase of *Clostridium septicum* to the hemagglutinin and other soluble products of the organism. J. Pathol. Bacteriol. *96:* 169–185.

Galau, G.A., R.J. Britten and E.H. Davidson. 1977. Studies on nucleic acid reassociation kinetics: rate of hybridization on excess RNA with DNA, compared to the rate of DNA renaturation. Proc. Nat. Acad. Sci. USA *74:* 1020–1023.

Gale, E.F., E. Cundliffe, P.E. Reynolds, M.H. Richmond and M.J. Waring. 1981. The Molecular Basis of Antibiotic Action, 2nd ed. John Wiley and Sons, London, New York, Sydney, Toronto.

Galesloot, Th.E., F. Hassing and J. Stadhouders. 1961. Agar media voor het isoleren en tellen van aromabacterien in zuursels. Neth. Milk Dairy J. *15:* 127–150.

Gallagher, I.H.C. and T.W. Cutress. 1977. The effect of trace elements on the growth and fermentation by oral streptococci and *Actinomyces*. Arch. Oral Biol. *22:* 555–562.

Ganesan, A.T., S. Chang and J.A. Hoch. (Editors), 1982. Molecular Cloning and Gene Regulation in Bacilli. Academic Press, New York and London.

Ganesan, S., H. Kamdar, K. Jayaraman and J. Szulmajster. 1983. Cloning and expression in *Escherichia coli* of a DNA fragment from *Bacillus sphaericus* coding for biocidal activity against mosquito larvae. Mol. Gen. Genet. *189:* 181–183.

Garant, P.R., C. Moon-Il, V. Lacono and G.J. Shemaka. 1979. Immuno-electron microscopic study of antigenic surface components of *Actinomyces naeslundii* in human dental plaque. Arch. Oral Biol. *24:* 369–377.

Garcia-Martinez, D.V., A. Shinmyo, A. Madia and A.L. Demain. 1980. Studies on cellulase production by *Clostridium thermocellum*. Eur. J. Appl. Microbiol. Biotechnol. *9:* 189–197.

Gardner, G.A. 1966. A selective medium for the enumeration of *Microbacterium thermosphactum* in meat and meat products. J. Appl. Bacteriol. *29:* 455–460.

Gardner, G.A. 1969. Physiological and morphological characteristics of *Kurthia zopfii* isolated from meat products. J. Appl. Bacteriol. *32:* 371–380.

Gardner, G.A. 1981. *Brochothrix thermosphacta* (*Microbacterium thermosphactum*) in the spoilage of meats: A review. *In* Roberts, Hobbs, Christian and Skovgaard (Editors), Psychrotrophic Microorganisms in Spoilage and Pathogenicity. Academic Press, London. pp. 139–173.

Gardner, G.A., A.W. Carson and J. Patton. 1967. Bacteriology of prepacked pork with particular reference to the gas composition within the pack. J. Appl. Bacteriol. *30:* 321–333.

Gardner, H.L. and C.D. Dukes. 1955. *Haemophilus vaginalis* vaginitis. A newly defined specific infection previously classified "nonspecific" vaginitis. Am. J. Obstet. Gynecol. *69:* 962–976.

Gardner, J.M. and F.A. Troy. 1979. Chemistry and biosynthesis of the poly(gamma-D-glutamyl) capsule in *Bacillus licheniformis*. J. Biol. Chem. *254:* 6262–6269.

Gardner, P., J.R. Saffle and S.C. Schoenbaum. 1977. Management of prosthetic valve endocarditis. *In* Duma (Editor), Infections of Prosthetic Heart Valves and Vascular Grafts, University Park Press, Baltimore, pp. 123–139.

Garvie, E.I. 1960. The genus *Leuconostoc* and its nomenclature. J. Dairy Res. *27:* 283–292.

Garvie, E.I. 1967a. *Leuconostoc oenos* sp. nov. J. Gen. Microbiol. *48:* 431–438.

Garvie, E.I. 1967b. Growth factor and amino acid requirements of species of the genus *Leuconostoc* including *Leuconostoc paramesenteroides* (sp. nov.) and *Leuconostoc oenos*. J. Gen. Microbiol. *48:* 439–447.

Garvie, E.I. 1969. Lactic dehydrogenases of strains of the genus *Leuconostoc*. J. Gen. Microbiol. *58:* 85–94.

Garvie, E.I. 1974. Nomenclatural problems of the pediococci. Request for an opinion. Int. J. Syst. Bacteriol. *24:* 301–306.

Garvie, E.I. 1975. Some properties of gas-forming lactic acid bacteria and their significance in classification. *In* Carr, Cutting and Whiting (Editors), Lactic acid bacteria in beverages and food. Academic Press, London, New York, San Francisco, pp. 339–349.

Garvie, E.I. 1976. Hybridization between the deoxyribonucleic acids of some strains of heterofermentative lactic acid bacteria. Int. J. Syst. Bacteriol. *26:* 116–122.

Garvie, E.I. 1978a. *Streptococcus raffinolactis* Orla Jensen and Hansen, a group N streptococcus found in raw milk. Int. J. Syst. Bacteriol. *28:* 190–193.

Garvie, E.I. 1978b. Lactate dehydrogenases of *Streptococcus thermophilus*. J. Dairy Res. *45:* 515–518.

Garvie, E.I. 1979. Proposal of National Collection of Dairy Organisms Strain 617 as the neotype strain of *Streptococcus raffinolactis* Orla-Jensen and Hansen. Int. J. Syst. Bacteriol. *29:* 152.

Garvie, E.I. 1980. Bacterial lactate dehydrogenases. Microbiol. Rev. *44:* 106–139.

Garvie, E.I. 1981. Sub-divisions within the genus *Leuconostoc* as shown by RNA/DNA hybridization. J. Gen. Microbiol. *127:* 209–212.

Garvie, E.I. 1983. *Leuconostoc mesenteroides* subsp. *cremoris* (Knudsen and Sørensen) comb. nov. and *Leuconostoc mesenteroides* subsp. *dextranicum* (Beijerinck) comb. nov. Int. J. Syst. Bacteriol. *33:* 118–119.

Garvie, E.I. 1984. The separation of species of the genus *Leuconostoc* and the differentiation of the leuconostocs from other lactic acid bacteria. Methods Microbiol. *16:* 147–178.

Garvie, E.I. and A.J. Bramley. 1979a. *Streptococcus uberis*: an approach to classification. J. Appl. Bacteriol. *46:* 295–304.

Garvie, E.I. and A.J. Bramley. 1979b. *Streptococcus bovis* an approach to its classification and its importance as a cause of bovine mastitis. J. Appl. Microbiol. *46:* 557–566.

Garvie, E.I. and J.A.E. Farrow. 1980. The differentiation of *Leuconostoc oenos* from non-acidophilic species of *Leuconostoc* and the identification of five strains from the American Type Culture Collection. Am. J. Enol. Vitic. *31:* 154–157.

Garvie, E.I. and J.A.E. Farrow. 1981. Sub-divisions within the genus *Streptococcus* using deoxyribonucleic acid/ribosomal ribonucleic acid hybridization. Zentralbl. Bakteriol. Mikrobiol. Hyg. Abt. I Orig. C *2:* 299–310.

Garvie, E.I. and J.A.E. Farrow. 1982. *Streptococcus lactis* subsp. *cremoris* (Orla-Jensen) comb. nov. and *Streptococcus lactis* subsp. *diacetilactis* (Matuszewski et al.) nom. rev., comb. nov. Int. J. Syst. Bacteriol. *32:* 453–455.

Garvie, E.I., J.A.E. Farrow and A.J. Bramley. 1983. *Streptococcus dysgalactiae* (Diernhofer) nom. rev. Int. J. Syst. Bacteriol. *33:* 404–405.

Garvie, E.I., J.A.E. Farrow and B.A. Phillips. 1981. A taxonomic study of some strains of streptococci which grow at 10°C but not at 45°C including *Streptococcus lactis* and *Streptococcus cremoris*. Zentralbl. Bakteriol. Mikrobiol. Hyg. I Abt. Orig. C *2:* 151–165.

Garvie, E.I. and M.E. Gregory. 1961. Folinic acid requirements of strains of the genus *Pediococcus*. Nature (Lond.) *190:* 563–564.

Garvie, E.I., M.E. Gregory and L.A. Mabbitt. 1961. The effect of asparagine on the growth of a Gram-positive coccus. J. Gen. Microbiol. *24:* 25–30.

Garvie, E.I. and L.A. Mabbitt. 1967. Stimulation of the growth of *Leuconostoc oenos* by tomato juice. Arch. Microbiol. *55:* 398–407.

Garvie, E.I., V. Zezula and V.A. Hill. 1974. Guanine plus cytosine content of the deoxyribonucleic acids of the leuconostocs and some hetero-fermentative lactobacilli. Int. J. Syst. Bacteriol. *24:* 248–251.

Gasdorf, H.J., R.G. Benedict, M.C. Cadmus, R.F. Anderson and R.W. Jackson. 1965. Polymer producing species of *Arthrobacter*. J. Bacteriol. *90:* 147–150.

Gasperini, G. 1892. Ricerche morfologiche e biologiche sul genere *Actinomyces* Harz come contributo allo studio delle relative micosi. Ann. Ist. Igiene Sper. Univ. Roma *2:* 167–231.

Gasser, F. 1970. Electrophoretic characterization of lactic dehydrogenases in the genus *Lactobacillus*. J. Gen. Microbiol. *62:* 223–239.

Gasser, F. and M. Janvier. 1980. Deoxyribonucleic acid homologies of *Lactobacillus jensenii*, *Lactobacillus leichmannii*, and *Lactobacillus acidophilus*. Int. J. Syst. Bacteriol. *30:* 28–30.

Gasser, F., M. Mandel and M. Rogosa. 1970. *Lactobacillus jensenii* sp. nov., a new representative of the subgenus *Thermobacterium*. J. Gen. Microbiol. *62:* 219–222.

Gasser, F. and M. Sebald. 1966. Composition en bases nucléiques des bactéries du genre *Lactobacillus*. Ann. Inst. Pasteur (Paris) *110:* 261–275.

Gaston, L.W. and E.R. Stadtman. 1963. Fermentation of ethylene glycol by *Clostridium glycolicum* sp. n. J. Bacteriol. *85:* 356–362.

Gawehn, K. and H.U. Bergmeyer. 1974. D(-)-Lactat. *In* Bergmeyer (Editor), Methoden der enzymatischen Analyse. Verlag Chemie, Weinheim-Bergstr. pp. 1538–1541.

Gay, C.C., D.C. Blood and J.S. Wilkinson. 1975. Clinical observations of sheep with focal symmetrical encephalomalacia. Aust. Vet. J. *51:* 266–269.

Gee, J.M., B.M. Lund, G. Metcalf and J.L. Peel. 1980. Properties of a new group of alkalophilic bacteria. J. Gen. Microbiol. *117:* 9–17.

Geldreich, E.E., B.A. Kenner and P.W. Kabler. 1964. Occurrence of coliforms, fecal coliforms, and streptococci on vegetation and insects. Appl. Microbiol. *12:* 63–69.

Gemmell, C.G. and J.E. Dawson. 1982. Identification of coagulase-negative staphylococci with the API Staph system. J. Clin. Microbiol. *16:* 874–877.

Gemmell, C.G., M.S. Huhssin and A.F. McIntosh. 1981. Development of a sensitive assay for the detection of nuclease (DNAase) produced by *Staphylococcus aureus* and *Staphylococcus epidermidis*. *In* Jeljaszwicz (Editor), Staphylococci and Staphylococcal Infections, Gustav Fischer Verlag, Stuttgart, New York, pp. 363–368.

Gemmell, C.G., M. Thelestam and T. Wadstrom. 1976. Toxinogenicity of coagulase-negative staphylococci. *In* Jeljaszewicz (Editor), Staphylococci and Staphylococcal Diseases, Gustav Fischer Verlag, Stuttgart, New York, pp. 133–136.

Genthner, B.R.S. and M.P. Bryant. 1982. Growth of *Eubacterium limosum* with carbon monoxide as the energy source. Appl. Environ. Microbiol. *43:* 70–74.

Genthner, B.R.S., C.L. Davis and M.P. Bryant. 1981. Features of rumen and sewage sludge strains of *Eubacterium limosum*, a methanol- and H_2-CO_2-utilizing species. Appl. Environ. Microbiol. *42:* 12–19.

Georg, L.K. and J.M. Brown. 1967. *Rothia*, gen. nov. an aerobic genus of the family Actinomycetaceae. Int. J. Syst. Bacteriol. *17:* 79–88.

Georg, L.K., J.M. Brown, H.J. Baker and G.H. Cassell. 1972. *Actinomyces viscosus* as an agent of actinomycosis in the dog. Am. J. Vet. Res. *33:* 1457–1470.

Georg, L.K. and R.M. Coleman. 1970. Comparative pathogenicity of various *Actinomyces* species. *In* Prauser (Editor), The Actinomycetales. The Jena International Symposium on Taxonomy, VEB Gustav Fischer Verlag, Jena, G.D.R.

Georg, L.K., R.M. Coleman and J.M. Brown. 1968. Evaluation of an agar gel precipitin test for the serodiagnosis of actinomycosis. J. Immunol. *100:* 1288–1292.

Georg, L.K., L. Pine and M.A. Gerencser. 1969. *Actinomyces viscosus*, comb. nov., a catalase positive, facultative member of the genus *Actinomyces*. Int. J. Syst. Bacteriol. *19:* 291–293.

Georg, L.K., G.W. Robertstad and S.A. Brinkman. 1964. Identification of species of *Actinomyces*. J. Bacteriol. *88:* 477–490.

Georg, L.K., G.W. Robertstad, S.A. Brinkmann and M.D. Hicklin. 1965. A new pathogenic anaerobic *Actinomyces* species. J. Infect. Dis. *115:* 88–99.

Georgala, D.L. 1957. Quantitative and qualitative aspects of the skin flora of North Sea cod and the effect thereon of handling on ship and on shore. Ph.D. thesis. University of Aberdeen.

George, H.A., J.L. Johnson, W.E.C. Moore, L.V. Holdeman and J.S. Chen. 1983. Acetone, isopropanol, and butanol production by *Clostridium beijerinckii* (syn. *C. butylicum*) and *Clostridium aurantibutyricum*. Appl. Environ. Microbiol. *45:* 1160–1163.

George, W.L., B.D. Kirby, V.L. Sutter and S.M. Finegold. 1979. Antimicrobial susceptibility of *Clostridium difficile*. Microbiology (Wash., D.C.), *1979:*267–271.

George, W.L., V.L. Sutter, D. Citron and S.M. Finegold. 1979. Selective and differential medium for isolation of *Clostridium difficile*. J. Clin. Microbiol. *9:* 214–219.

George, W.L., V.L. Sutter and S.M. Finegold. 1981. Beta-lactam antimicrobials for treatment of anaerobic infections--A review of in vitro activity and therapeutic efficacy. *In* Salton and Shockman (Editors), B-lactam Antibiotics: Mode of Action, New Developments and Future Prospects, Academic Press, New York, pp. 493–550.

Gerencser, M.A. 1979. The application of fluorescent antibody techniques to the identification of *Actinomyces* and *Arachnia*. *In* Bergan and Norris (Editors), Methods in Microbiology, Vol. 13, Academic Press, London-New York-San Francisco, pp. 287–321.

Gerencser, M.A. and J.M. Slack. 1967. Isolation and characterization of *Actinomyces propionicus*. J. Bacteriol. *94:* 109–115.

Gerencser, M.A. and J.M. Slack. 1969. Identification of human strains of *Actinomyces viscosus*. Appl. Microbiol. *18:* 80–87.

Gerencser, M.A. and J.M. Slack. 1976. Serological classification of *Actinomyces* using the fluorescent antibody technique. J. Dent. Res. *55:* (Special Issue): A184-A191.

Gericke, D., F. Dietzel, W. König, I. Rüster and L. Schumacher. 1979. Further progress with oncolysis due to apathogenic clostridia. Zentralbl. Bakteriol. Parasitenkd. Infektionskr. Hyg. I. Abt. Orig. A *243:* 102–112.

Gherna, R.L. 1981. Preservation. *In* Gerhardt (Editor), Manual of Methods for General Bacteriology. Am. Soc. Microbiol., Washington, D.C., pp. 208–217.

Ghuysen, J.M. 1968. Use of bacteriolytic enzymes in determination of wall structure and their role in cell metabolism. Bacteriol. Rev. *32:* 425–464.

Gibbons, N.E. 1974. Reference collections of bacteria - the need and requirements for type and neotype strains. *In* Buchanan and Gibbons (Editors), Bergey's Manual of Determinative Bacteriology, 8th Ed., The Williams and Wilkins Co., Baltimore, pp. 14–17.

Gibbons, N.E. and R.G.E. Murray. 1978. Proposals concerning the higher taxa of bacteria. Int. J. Syst. Bacteriol. *28:* 1–6.

Gibbons, N.E., K.B. Pattee and J.G. Holt. 1981. Supplement to Index Bergeyana. The Williams and Wilkins Co., Baltimore.

Gibbons, R.J. and J. van Houte. 1973. On the formation of dental plaques. J. Periodontol. *44:* 347–360.

Gibbons, R.J. and J. van Houte. 1975. Bacterial adherence in oral microbial ecology. Ann. Rev. Microbiol. *29:* 19–44.

Gibson, T. 1935. An investigation of *Sarcina ureae*, a spore-forming, motile coccus. Arch. Mikrobiol. *6:* 73–78.

Gibson, T. 1935a. The urea-decomposing microflora of soils. I. Description and classification of the organisms. Zentralbl. Bakteriol. Parasitenkd. Infek-

Gibson, T. 1935b. The urea-decomposing microflora of soils. II. The number and types of the organisms as shown by different methods. Zentralbl. Bakteriol. Parasitenkd. Infektionskr. Hyg. Abt. II *92:* 414–424.

Gibson, T. 1974. Genus *Sporosarcina* Kluyver and Van Neil. *In* Buchanan and Gibbons (Editors), Bergey's Manual of Determinative Bacteriology, 8th ed. The Williams and Wilkins Co., Baltimore. pp. 573–574.

Gibson, T. and R.E. Gordon. 1974. Genus *Bacillus* Cohn. *In* Buchanan and Gibbons (Editors), Bergey's Manual of Determinative Bacteriology 8th Ed. The Williams and Wilkins Co., Baltimore, pp. 529–550.

Gibson, T. and L.E. Topping. 1938. Further studies of the aerobic spore-forming bacilli. Soc. Agric. Bact. Proc. Abstr. pp. 43–44.

Giesel, H. and H. Simon. 1983. Immunological relationship of enoate reductases from different clostridia and the classification of *Clostridium* species La 1. FEMS Microbiol. Lett. *19:* 43–45.

Gilbert, R.J. and J.M. Parry. 1977. Serotypes of *Bacillus cereus* from outbreaks of food poisoning and from routine foods. J. Hyg. *78:* 69–74.

Gilbert, R.J. and A.J. Taylor. 1976. *Bacillus cereus* food poisoning. *In* Skinner and Carr (Editors), Microbiology in agriculture, fisheries and food. Academic Press, London, pp. 197–213.

Gilbert, R.J., P.C.B. Turnbull, J.M. Parry and J.M. Kramer. 1981. *Bacillus cereus* and other *Bacillus* species: their part in food poisoning and other clinical infections. *In* Berkeley and Goodfellow (Editors), The Aerobic Endospore-forming Bacteria.. Academic Press, London, pp. 295–314.

Gill, J.W. and K.W. King. 1958. Nutritional characteristics of a butyrivibrio. J. Bacteriol. *75:* 666–673.

Gillert, K.-E. and Inst. Wiss. Film. 1961. Aufbau und Verhalten beweglicher Kolonien von *Bacillus circulans*. Film C838 des IWF Göttingen.

Gillespie, D. and S. Spiegelman. 1965. A quantitative assay for DNA-RNA hybrids with DNA immobilized on a membrane filter. J. Mol. Biol. *12:* 829–842.

Gillies, A.J. 1971. Significance of thermoduric organisms in Queensland cheddar cheese. Aust. J. Dairy Technol. *26:* 145–149.

Gillis, T.P. and T.M. Buchanan. 1982. Production and characterization of monoclonal antibodies to *Mycobacterium leprae*. Infect. Immun. *37:* 172–178.

Gillis, T.P. and J.J. Thompson. 1978. Quantitative fluorescent immunoassay of antibodies to, and surface antigens of, *Actinomyces viscosus*. J. Clin. Microbiol. *7:* 202–208.

Gillum, W.O., L.E. Mortenson, J.-S. Chen and R.H. Holm. 1977. Quantitative extrusions of the Fe_4S_4 cores of the active sites of ferredoxins and the hydrogenase of *Clostridium pasteurianum*. J. Am. Chem. Soc. *99:* 584–595.

Gilmour, M.N., A. Howell, Jr. and B.G. Bibby. 1961. The classification of organisms termed *Leptotrichia (Leptothrix) buccalis*. I. Review of the literature and proposed separation into *Leptotrichia buccalis* Trevisan 1879 and *Bacterionema* gen. nov., *B. matruchotii* (Mendel 1919) comb. nov. Bacteriol. Rev. *25:* 131–141.

Giménez, D.F. and A.S. Ciccarelli. 1970. Another type of *Clostridium botulinum*. Zentralbl. Bakteriol. Parasitenkd. Infektionskr. Hyg., I. Abt. Orig. *215:* 221–224.

Gimpl, F. and M. Lanyi. 1965. Use of the gel precipitation method for determining the type of mycobacteria and the clinical diagnosis. Bull. Int. Union Tuberc. *36:* 22–25.

Ginsburg, I. and N. Grossowicz. 1957. Group A hemolytic streptococci I. A chemical defined medium for growth from small inocula. Proc. Soc. Exp. Biol. Med. *96:* 108–112.

Giovannozzi-Sermanni, G. 1959. Una nuova specie di *Arthrobacter* determinante la degradazione della nicotina: *Arthrobacter nicotianae*. Il Tabacco *63:* 83–86.

Girard, A.E. 1971. A comparative study of the fatty acids of some micrococci. Can. J. Microbiol. *17:* 1503–1508.

Girard, A.E. and B.H. Jacius. 1974. Ultrastructure of *Actinomyces viscosus* and *Actinomyces naeslundii*. Arch. Oral Biol. *19:* 71–79.

Gladstone, G.P. 1937. The nutrition of *Staphylococcus aureus*: nitrogen requirements. Brit. J. Exp. Pathol. *18:* 322–333.

Glage, F. 1903. Über den *Bazillus pyogenes suis* Grips, den *Bazillus pyogenes bovis* Künnemann und den bakteriologischen Befund bei den chronischen, Abszedierenden Euterentzündungen der Milchkühe. Z. Fleisch Milchhyg. *13:* 166–175.

Glass, M. 1973. *Sarcina* species on the skin of the human forearm. Trans. St. John's Hosp. Dermatol. Soc. *59:* 56–60.

Gledhill, W.E. and L.E. Casida, Jr. 1969a. Predominant catalase-negative soil bacteria. I. Streptococcal population indigenous to soil. Appl. Microbiol. *17:* 208–213.

Gledhill, W.E. and L.E. Casida, Jr. 1969b. Predominant catalase-negative soil bacteria. II. Occurrence and characterization of *Actinomyces humiferus*, sp. N. Appl. Microbiol. *18:* 114–121.

Gledhill, W.E. and L.E. Casida, Jr. 1969c. Predominant catalase-negative soil bacteria. III. *Agromyces*, gen. n., microorganisms intermediary to *Actinomyces* and *Nocardia*. Appl. Microbiol. *18:* 340–349.

Glick, M.C., T. Sall, F. Zilliken and S. Mudd. 1960. Morphological changes of *Lactobacillus bifidus* var. *pennsylvanicus* produced by a cell-wall precursor. Biochem. Biophys. Acta *37:* 361–363.

Glinski, Z. 1972. Investigations on the properties and antigenic structure of *Streptococcus pluton*. 1. Morphological and cultural characteristics. Med. Weter. *28:* 399–405.

Goddard, P., F. Fernandez, B. West, M.J. Hill and P. Barnes. 1975. The nuclear hydrogenation of steroids by intestinal bacteria. J. Med. Microbiol. 8: 429–435.

Goel, M.C. and I.P. Singh. 1972. Purification and characterization of *Corynebacterium ovis* exotoxin. J. Comp. Pathol. 82: 345–353.

Golberg, R.L. and J.A. Washington II. 1976. Comparison of isolation of *Haemophilus vaginalis* (*Corynebacterium vaginale*) from peptone-starch-dextrose agar and Columbia colistin-nalidixic acid agar. J. Clin. Microbiol. 4: 245–247.

Goldberg, H.S. 1980. Aspects of anaerobic infections in animals. Infection 8: S131-S133.

Goldfine, H. 1982. Lipids of prokaryotes - structure and distribution. Curr. Top. Memb. Transp. 17: 1–43.

Goldman, M. and D.A. Wilson. 1977. Growth of *Sporosarcina ureae* in defined media. FEMS Microbiol. Lett. 2: 113–115.

Goldstein, E.J.C., D. Citron, H. Gonzalez, F.E. Russell and S.M. Finegold. 1979. Bacteriology of rattlesnake venom and implications for therapy. J. Infect. Dis. 140: 818–821.

Golovacheva, R.S. and G.I. Kravaik. 1978. *Sulfobacillus*, a new genus of thermophilic sporeforming bacteria. Mikrobiologiya 47: 815–822 (Microbiology, USSR, English edition, 47: 658–665.).

Golovacheva, R.S., L.G. Loginova, T.A. Salikhov, A.A. Kolesnikov and G.N. Zaitseva. 1975. A new thermophilic species, *Bacillus thermocatenulatus* nov. spec. Mikrobiologiya 44: 265–268 (Microbiology, USSR, English edition, 44: 230–233.).

Gomez-Alarcon, R.A., C. O'Dowd, J.A.Z. Leedle and M.P. Bryant. 1982. 1,4-Naphthoquinone and other nutrient requirements of *Succinivibrio dextrinosolvens*. Appl. Environ. Microbiol. 44: 346–350.

González, J.N. Jr., B.J. Brown and B.C. Carlton. 1982. Transfer of *Bacillus thuringiensis* plasmids coding for delta-endotoxin among strains of *B. thuringiensis* and *B. cereus*. Proc. Nat. Acad. Sci. USA 79: 6951–55.

González, J.N. Jr. and B.C. Carlton. 1982. Plasmid transfer in *Bacillus thuringiensis*. In Streips, Goodal, Guild and Wilson (Editors), Genetic Exchange. Marcel Dekker, New York and Basel.

González-Ochoa, A. 1976. Nocardiae and chemotherapy. In Goodfellow, Brownell and Serrano (Editors), The Biology of the Nocardiae. Academic Press, New York, pp. 429–450.

Gooder, H. and W.R. Maxted. 1961. External factors influencing structure and activities of *Streptococcus pyogenes*. Symp. Soc. Gen. Microbiol. 11: 151–173.

Goodfellow, M. 1971. Numerical taxonomy of some nocardioform bacteria. J. Gen. Microbiol. 69: 33–80.

Goodfellow, M. 1984a. Reclassification of *Corynebacterium fascians* (Tilford) Dowson in the genus *Rhodococcus*, as *Rhodococcus fascians* comb. nov. Syst. Appl. Microbiol. 5: 225–229.

Goodfellow, M. 1984b. In Validation of the publication of new names and combinations previously effectively published outside the IJSB. List No. 10. Int. J. Syst. Bacteriol. 34: 503–504.

Goodfellow, M. and G. Alderson. 1977. The actinomycete genus *Rhodococcus*: a home for the 'rhodochrous' complex. J. Gen. Microbiol. 100: 99–122.

Goodfellow, M. and G. Alderson. 1979. In Validation of the publication of new names and combinations previously effectively published outside the IJSB. List No. 2. Int. J. Syst. Bacteriol. 29: 79–80.

Goodfellow, M., G. Alderson and J. Lacey. 1979. Numerical taxonomy of *Actinomadura* and related actinomycetes. J. Gen. Microbiol. 112: 95–111.

Goodfellow, M., A.R. Beckham and M.D. Barton. 1982. Numerical classification of *Rhodococcus equi* and related actinomycetes. J. Appl. Bacteriol. 53: 199–207.

Goodfellow, M., M.D. Collins and D.E. Minnikin. 1976. Thin-layer chromatographic analysis of mycolic acid and other long-chain components in whole-organism methanolysates of coryneform and related taxa. J. Gen. Microbiol. 96: 351–358.

Goodfellow, M., M.D. Collins and D.E. Minnikin. 1980. Fatty acid and polar lipid composition in the classification of *Kurthia*. J. Appl. Bacteriol. 48: 269–276.

Goodfellow, M. and T. Cross. 1984. Classification. In Goodfellow, Mordarski and Williams (Editors), The Biology of Actinomycetes, Academic Press, London and New York, pp. 7–164.

Goodfellow, M., A. Lind, H. Mordarska, S. Pattyn and M. Tsukamura. 1974. A cooperative numerical analysis of cultures considered to belong to the 'rhodochrous' taxon. J. Gen. Microbiol. 85: 291–302.

Goodfellow, M. and D.E. Minnikin. 1980. Definition of the genus *Mycobacterium* vis a vis other allied taxa. In Kubica, Wayne and Good (Editors), 1954–1979: Twenty-Five Years of Mycobacterial Taxonomy, U.S. Department of Health, Education and Welfare, Center for Disease Control, Atlanta, pp. 115–130.

Goodfellow, M. and D.E. Minnikin. 1981a. Classification of nocardioform bacteria. Zentralbl. Bakteriol. Mikrobiol. Hyg. Suppl. 11: 7–16.

Goodfellow, M. and D.E. Minnikin. 1981b. The genera *Nocardia* and *Rhodococcus*. In Starr, Stolp, Trüper, Balows and Schlegel (Editors), The Prokaryotes: a handbook on habitiats, isolation and identification of bacteria, Springer-Verlag , Berlin, pp. 2016–2017

Goodfellow, M. and D.E. Minnikin. 1984. A critical evaluation of *Nocardia* and related taxa. In Ortiz, Bojalil and Yakoleff (Editors), Biological, biochemical and biomedical aspects, Academic Press, New York and London.

Goodfellow, M. and D.E. Minnikin (Editors). 1984. Chemical Methods in Bacterial Systematics. Society for Applied Bacteriology, Technical Series No.

20, Academic Press, London and New York.

Goodfellow, M., D.E. Minnikin, C. Todd, G. Alderson, S.M. Minnikin and M.D. Collins. 1982. Numerical and chemical classification of *Nocardia amarae*. J. Gen. Microbiol. 128: 1283–1297.

Goodfellow, M., M. Mordarski, S.T. Williams (Editors). 1984. The Biology of the Actinomycetes. Academic Press, London, New York.

Goodfellow, M. and V.A. Orchard. 1974. Antibiotic sensitivity of some nocardioform bacteria and its value as a criterion for taxonomy. J. Gen. Microbiol. 83: 375–387.

Goodfellow, M., P.A.B. Orlean, M.D. Collins, L. Alshamoany and D.E. Minnikin. 1978. Chemical and numerical taxonomy of strains received as *Gordona aurantiaca*. J. Gen. Microbiol. 109: 57–68.

Goodfellow, M. and T. Pirouz. 1982a. Numerical classification of sporoactinomycetes containing *meso*-diaminopimelic acid in the cell wall. J. Gen. Microbiol. 128: 503–527.

Goodfellow, M. and T. Pirouz. 1982b. In Validation of the publication of new names and new combinations previously effectively published outside the IJSB. List No. 9. Int. J. Syst. Bacteriol. 32: 384–385.

Goodfellow, M. and L.G. Wayne. 1982. Taxonomy and nomenclature. In Ratledge and Stanford (Editors), The Biology of the Mycobacteria, Volume 1, Academic Press, London and New York, pp. 471–521.

Goodfellow, M., C.R. Weaver and D.E. Minnikin. 1982. Numerical classification of some rhodococci, corynebacteria and related organisms. J. Gen. Microbiol. 128: 731–745.

Goodfellow, M. and S.T. Williams. 1983. Ecology of actinomycetes. Annu. Rev. Microbiol. 37: 189–216.

Goodsir, J. 1842. History of a case in which a fluid periodically ejected from the stomach contained vegetable organisms of an undescribed form. With chemical analysis of the fluid, by George Wilson. Edinburgh Med. Surg. J. 57: 430–443.

Goplerud, C.P., M.J. Ohm and R.P. Galask. 1976. Aerobic and anaerobic flora of the cervix during pregnancy and the puerperium. Am. J. Obstet. Gynecol. 126: 858–865.

Gorbach, S.L., K.B. Menda, H. Thadepalli and L. Keith. 1973. Anaerobic microflora of the cervix in healthy woman. Am. J. Obstet. Gynecol. 117: 1053–1055.

Gorbach, S.L. and H. Thadepalli. 1974. Clindamycin in pure and mixed anaerobic infections. Arch. Intern. Med. 134: 87–92.

Gorbach, S.L. and H. Thadepalli. 1975. Isolation of *Clostridium* in human infections: evaluation of 114 cases. J. Infect. Dis. 131:, Suppl: S81-S85.

Gordon, D.F. 1967. Reisolation of *Staphylococcus salivarius* from the human oral cavity. J. Bacteriol. 94: 1281–1286.

Gordon, M.A. 1974. Aerobic pathogenic *Actinomycetaceae*. In Lennette, Spaulding and Truant (Editors), Manual of clinical microbiology, American Society for Microbiology, Washington, D.C., pp. 175–188.

Gordon, R.E. 1967. The taxonomy of soil bacteria. In Gray and Parkinson (Editors), The Ecology of Soil Bacteria. An International Symposium. University of Toronto Press, Toronto, pp. 293–321.

Gordon, R.E. 1967. Some taxonomic observations on the genus *Bacillus*. In Briggs (Editor), Biological Regulation of Vectors: The Saprophytic and Aerobic Bacteria and Fungi. US Department of Health, Education and Welfare, Washington, D.C., pp. 67–82.

Gordon, R.E. 1981. One hundred and seven years of the genus *Bacillus*. In Berkeley and Goodfellow (Editors), The Aerobic Endosporeforming Bacteria. Academic Press, London, pp. 1–15.

Gordon, R.E., D.A. Barnett, J.E. Handerhan and C.H.-N. Pang. 1974. *Nocardia coeliaca*, *Nocardia autotrophica* and the nocardin strain. Int. J. Syst. Bacteriol. 24: 54–63.

Gordon, R.E., W.C. Haynes and C.H.-N. Pang. 1973. The genus *Bacillus*. Handbook No. 427. U.S. Department of Agriculture, Washington, D.C.

Gordon, R.E. and J.L. Hyde. 1982. The *Bacillus firmus-Bacillus lentus* complex and pH 7.0 variants of some alkalophilic strains. J. Gen. Microbiol. 128: 1109–1116.

Gordon, R.E., J.L. Hyde and J.A. Moore. 1977. *Bacillus firmus-Bacillus lentus*: a series or one species? Int. J. Syst. Bacteriol. 27: 256–262.

Gordon, R.E. and J.M. Mihm. 1957. A comparative study of some strains received as nocardiae. J. Bacteriol. 73: 15–27.

Gordon, R.E. and J.M. Mihm. 1959. A comparison of *Nocardia asteroides* and *Nocardia brasiliensis*. J. Gen. Microbiol. 20: 129–135.

Gordon, R.E. and J.M. Mihm. 1962a. Identification of *Nocardia caviae* (Erikson) nov. comb. Ann. N.Y. Acad. Sci. 98: 628–636.

Gordon, R.E. and J.M. Mihm. 1962b. The type species of the genus *Nocardia*. J. Gen. Microbiol. 27: 1–10.

Gordon, R.E., S.K. Mishra and D.A. Barnett. 1978. Some bits and pieces of the genus *Nocardia*: *N. carnea*, *N. vaccinii*, *N. transvalensis*, *N. orientalis* and *N. aerocolonigenes*. J. Gen. Microbiol. 109: 69–78.

Gordon, R.E. and T.K. Rynearson. 1963. Maintenance of strains of *Bacillus* species. In Martin (Editor), Culture Collections: Perspectives and Problems. University of Toronto Press, Toronto, pp. 118–128.

Gordon, R.E. and M.M. Smith. 1953. Rapidly growing acid-fast bacteria. I. Species description of *Mycobacterium phlei* Lehmann and Neumann and *Mycobacterium smegmatis* (Trevisan) Lehmann and Neumann. J. Bacteriol. 66: 41–48.

Goslee, S., J.K. Rynearson and E. Wolinsky. 1976. Additional serotypes of *M.*

scrofulaceum, M. gordonae, M. marinum and *M. xenopi* determined by agglutination. Int. J. Syst. Bacteriol. *26:* 136–142.

Gossling, J. and W.E.C. Moore. 1975. *Gemmiger formicilis,* n. gen., n. sp., an anaerobic budding bacterium from intestines. Int. J. Syst. Bacteriol. *25:* 202–207.

Gothefors, L. and I. Blenkharn. 1978. *Clostridium butyricum* and necrotizing enterocolitis. Lancet *i:* 52–53.

Gotschlich, E.C. and T.Y. Liu. 1967. Structural and immunological studies on the pneumococcal C polysaccharide. J. Biol. Chem. *242:* 463–470.

Gottheil, O. 1901. Botanische Beschreibung einiger Bodenbakterien. Zentralbl. Bakteriol. Parasitenk. Infektionskr. Hyg. Abt. II *7:* 680–691.

Gottschalk, G., J.R. Andreesen and H. Hippe. 1981. The genus *Clostridium* (nonmedical aspects). *In* Starr, Stolp, Trüper, Balows, and Schlegel (Editors), The Prokaryotes, a handbook on the habitats, isolation and identification of bacteria. Springer-Verlag, Berlin, Heidelberg, New York, pp. 1767–1803.

Gottschalk, G. and M. Braun. 1981. Revival of the name *Clostridium aceticum.* Int. J. Syst. Bacteriol. *31:* 476.

Gottwald, M., J.R. Andreesen, J. LeGall and L.J. Ljungdahl. 1975. Presence of cytochrome and menaquinone in *Clostridium formicoaceticum* and *Clostridium thermoaceticum.* J. Bacteriol. *122:* 325–328.

Götz, F., S. Ahrné and M. Lindberg. 1981. Plasmid transfer and generic recombination by protoplast fusion in staphylococci. J. Bacteriol. *145:* 74–81.

Götz, F., E.F. Elstner, B. Sedewitz and E. Lengfelder. 1980. Oxygen utilization by *Lactobacillus plantarum.* II. Superoxide and superoxide dismutation. Arch. Microbiol. *125:* 215–220.

Götz, F., S. Fischer and K.H. Schleifer. 1980. Purification and characterization of an unusually heat-stable and acid/base stable class I fructose-1, 6-biphosphate aldolase from *Staphylococcus aureus.* Eur. J. Biochem. *108:* 295–301.

Götz, F., B. Kreutz and K.H. Schleifer. 1983a. Protoplast transformation of *Staphylococcus carnosus* by plasmid DNA. Molec. Gen. Genet. *189:* 340–342.

Götz, F., E. Nürnberger and K.H. Schleifer. 1979. Distribution of class-I and classII D-fructose 1,6-biphosphate aldolases in various staphylococci, peptococci and micrococci. FEMS Microbiol. Lett. *5:* 253–257.

Götz, F. and K.H. Schleifer. 1975. Purification and properties of a fructose-1, 6-diphosphate activated L-lactate dehydrogenase from *Staphylococcus epidermidis.* Arch. Microbiol. *105:* 303–312.

Götz, F. and K.H. Schleifer. 1976. Comparative biochemistry of lactate dehydrogenases from staphylococci. *In* Jeljaszewicz (Editor), Staphylococci and Staphylococcal Diseases, Gustav Fischer Verlag, Stuttgart, New York, pp. 245–252.

Götz, F., J. Zabielski, L. Philipson and M. Lindberg. 1983b. DNA homology between the arsenate resistance plasmid pSX267 from *Staphylococcus xylosus* and the penicillinase plasmid pI258 from *Staphylococcus aureus.* Plasmid *9:* 126–137.

Gould, G.W. 1962. Microscopical observations on the emergence of cells of *Bacillus* spp. from spores under different cultural conditions. J. Appl. Bact. *25:* 35–41.

Gould, G.W. 1971. Methods for studying bacterial spores. *In* Norris and Ribbons (Editors), Methods in Microbiology, Volume 6A. Academic Press, London, pp. 327–381.

Gould, G.W. and A. Hurst. 1969. The bacterial spore. Academic Press, London.

Gounot, A.M. 1967. Role biologique des *Arthrobacter* dans les limons sonterrains. Ann. Inst. Pasteur Paris *113:* 923–945.

Graham, J.M. 1978. Inhibition of *Clostridium botulinum* type C by bacteria isolated from mud. J. Appl. Bacteriol. *45:* 205–211.

Graham, M.B. and W.A. Falkler Jr. 1979. Extractable antigen shared by *Peptostreptococcus anaerobius* strains. J. Clin. Microbiol. *9:* 507–510.

Graham, M.B., W.A. Falkler, Jr., C.A. Spiegel, S.W. Hayduk and L.G. Dober. 1982. Antibody reactive with *Peptostreptococcus micros* in the sera of patients with periodontal disease. J. Clin. Microbiol. *16:* 395–397.

Graham-Smith, G.S. 1904. A study of the virulence of the diphtheria bacilli isolated from 113 persons, and of 11 species of diphtheria-like organisms, together with measures taken to check an outbreak of diphtheria at Cambridge, 1903. J. Hyg. Camb. *4:* 258–328.

Gramoli, J. and B.J. Wilkinson. 1978. Characteristics and identification of coagulase-negative, heat-stable deoxyribonuclease-positive staphylococci. J. Gen. Microbiol. *105:* 275–285.

Grange, J.M. 1975. The genetics of mycobacteria and mycobacteriophages--A review. Tubercle *56:* 227–238.

Grange, J.M. and J.L. Stanford. 1974. Reevaluation of *Mycobacterium fortuitum* (synonym: *Mycobacterium ranae*). Int. J. Syst. Bacteriol. *24:* 320–329.

Grant, D.J.W. 1973a. Degradation and hydrolysis of esters by *Corynebacterium pseudodiphtheriticum* NCIB 10803. Microbios *8:* 35–41.

Grant, D.J.W. 1973b. Degradative versatility of *Corynebacterium pseudodiphtheriticum* NCIB 10803 which uses amides as carbon source. Antonie van Leeuwenhoek J. Microbiol. Serol. *39:* 273–279.

Grant, D.J.W. and J.V. Wilson. 1973. Degradation and hydrolysis of amides by *Corynebacterium pseudodiphtheriticum* NCIB 10803. Microbios *8:* 15–22.

Grant, J.B.W., W. Blyth, V.E. Wardrop, R.M. Gordon, J.C.G. Pearson and A. Mair. 1972. Prevalence of farmer's lung in Scotland: a pilot survey. Brit. Med. J. *1:* 530–534.

Grant, W.D. and B.J. Tindall. 1980. Enrichment and isolation procedures. *In* Gould and Corry (Editors), Microbial Growth and Survival in Extremes of Environment. Academic Press, London, pp. 27–38.

Grassé, P.P. 1924. Notes protistologiques. I. La sporulation des *Oscillospiraceae;* II. Le genre *Alysiella* Langeron 1923. Arch. Zool. Expér. Gén. *62:* (Notes et Rev.) 25–34.

Grassé, P.P. 1925. *Anisomitus denisi* n.g., n. sp. Schizophyte de l'intestine du canard domestique. Ann. Parasit. Hum. Comp. *3:* 343–348.

Grässer, R. 1957. Vergleichende Untersuchungen an Actinomyceten von Mensch, Rind und Schwein. Dissertation, Leipzig.

Grau, F.H. 1979. Nutritional requirements of *Microbacterium thermosphactum.* Appl. Environ. Microbiol. *38:* 818–820.

Grau, F.H. 1983. End products of glucose fermentation by *Brochothrix thermosphacta.* Appl. Environ. Microbiol. *45:* 84–90.

Graudal, H. 1952. Motile streptococci. Acta Pathol. Microbiol. Scand. *31:* 46–50.

Graves, M.H., J.A. Morello and F.E. Kocka. 1974. Sodium polyanethol sulfonate sensitivity of anaerobic cocci. Appl. Microbiol. *27:* 1131–1133.

Gray, M.L. 1957. A rapid method for the detection of colonies of *Listeria monocytogenes.* Zentralbl. Bakteriol. Parasitenkd. Infektionskr. Hyg. I Orig. *169:* 373–377.

Gray, M.L. and A.H. Killinger. 1966. *Listeria monocytogenes* and listeric infections. Bacteriol. Rev. *30:* 309–382.

Gray, M.L., H.J. Stafseth, F. Thorp, Jr., L.B. Sholl and W.F. Riley, Jr. 1948. A new technique for isolating listerellae from the bovine brain. J. Bacteriol. *55:* 471–476.

Gray, M.W. and W.F. Doolittle. 1982. Has the endosymbiont hypothesis been proven? Microbiol. Rev. *46:* 1–42.

Gray, P.H.H. 1928. The formation of indigotin from indol by soil bacteria. Proc. Roy. Soc. Ser. B *102:* 263–280.

Gray, P.H.H. and H.G. Thornton. 1928. Soil bacteria that decompose certain aromatic compounds. Zentralbl. Bakteriol. Parasitenkd. Infektionskr. Hyg. Abt. II *73:* 74–96.

Gray, S.J. 1980. An investigation into the origin of bacteriophage non-typable strains of *Staphylococcus aureus.* J. Med. Lab. Sci. *37:* 81–84.

Grecz, N. and G.M. Dack. 1961. Taxonomically significant color reactions of *Brevibacterium linens.* J. Bacteriol. *82:* 241–246.

Greenberg, M., J.F. Milne and A. Watt. 1970. Survey of workers exposed to dusts containing derivatives of *Bacillus subtilis.* Br. Med. J. *ii:* 629–633.

Greenman, J., K.T. Holland and W.J. Cunliffe. 1981. Effects of glucose concentration on biomass, maximum specific growth rate and extracellular enzyme production by three species of cutaneous propionibacteria grown in continuous culture. J. Gen. Microbiol. *127:* 371–376.

Greenwood, J.R. and M.J. Pickett. 1979. Salient features of *Haemophilus vaginalis.* J. Clin. Microbiol. *9:* 200–204.

Greenwood, J.R. and M.J. Pickett. 1980. Transfer of *Haemophilus vaginalis* Gardner and Dukes to a new genus, *Gardnerella: G. vaginalis* (Gardner and Dukes) comb. nov. Int. J. Syst. Bacteriol. *30:* 170–178.

Greenwood, J.R., M.J. Pickett, W.J. Martin and E.G. Mack. 1977. *Haemophilus vaginalis (Corynebacterium vaginale):* method for isolation and rapid biochemical identification. Health Lab. Sci. *14:* 102–106.

Greer, G.G. 1983. Psychrotrophic *Brochothrix thermosphacta* bacteriophages isolated from beef. Appl. Environ. Microbiol. *46:* 245–251.

Greer, S.B., W. Hsiang, G. Musil and D.D. Zinner. 1971. Viruses of cariogenic streptococci. J. Dent. Res. *50:* 1594–1604.

Gregory, P.H. and M.E. Lacey. 1963. Mycological examination of dust from mouldy hay associated with farmer's lung disease. J. Gen. Microbiol. *30:* 75–88.

Gregory, P.H., M.E. Lacey, G.N. Festenstein and F.A. Skinner. 1963. Microbial and biochemical changes during the moulding of hay. J. Gen. Microbiol. *33:* 147–174.

Gretler, A.C., P. Muccolo, J.B. Evans and C.F. Niven, Jr. 1955. Vitamin nutrition of the staphylococci with special reference to their biotin requirements. J. Bacteriol. *70:* 44–49.

Grieco, M.H. and C. Sheldon. 1970. *Erysipelothrix rhusiopathiae.* Ann. N.Y. Acad. Sci. *174:* 523–532.

Griffith, A.S. 1957. The types of tubercle bacilli in lupus and scrofuloderma. J. Hyg. *55:* 1–26.

Griffith, F. 1934. The serological classification of *Streptococcus pyogenes.* J. Hyg. (Cambridge) *34:* 542–584.

Grimme, A. 1902. Die wichtigsten Methoden der Bakterienfärbung in ihrer Wirkung auf die Membran, den Protoplasten und die Einschlüsse der Bakterienzelle. Zentrabl. Bakteriol. Parasitenk. Infectionskr. Hyg. Abt. I. Orig. *32:* 81–90.

Gross, D.C. and A.K. Vidaver. 1979a. A selective medium for the isolation of *Corynebacterium nebraskense* from soil and plant parts. Phytopathology *69:* 82–87.

Gross, D.C. and A.K. Vidaver. 1979b. Bacteriocins of phytopathogenic *Corynebacterium* species. Can. J. Microbiol. *25:* 367–374.

Gross, D.C., A.K. Vidaver and M.B. Keralis. 1979. Indigenous plasmids from phytopathogenic *Corynebacterium* species. J. Gen. Microbiol. *115:* 479–489.

Gross, W., J.E. Hawkins and R. Murphy. 1976. *M. xenopi* in clinical specimens I. Water as a source of contamination. Am. Rev. Resp. Dis. *113:* 78.

Gross, W.B. and L.D. Smith. 1971. Experimental botulism in gallinaceous birds. Avian Dis. *15:* 716–722.

Gross, W.M. and L.G. Wayne. 1970. Nucleic acid homology in the genus *Mycobacterium.* J. Bacteriol. *104:* 630–634.

Grov, A., P. Oeding, B. Myklestad and J. Aasen. 1970. Reactions of staphylococcal antigens with normal sera, γG-globulins, and γG-globulin fragments of various species origin. Acta Pathol. Microbiol. Scand. Sect. B 78: 106–111.

Grover, A.A., L.H. Schmidt and J. Rehm. 1957. The pathogenicity of various atypical mycobacteria for the rhesus monkey. Abstr. Annu. Meet. Natl. Tuberc. Assoc. p. 35.

Groves, D.J. 1979. Interspecific relationships of antibiotic resistance in Staphylococcus sp.: isolation and comparison of plasmids determining tetracycline resistance in S. aureus and S. epidermidis. Can. J. Microbiol. 25: 1468–1475.

Groves, R.D. and H.J. Welshimer. 1977. Separation of pathogenic from apathogenic Listeria monocytogenes by three in vitro reactions. J. Clin. Microbiol. 5: 559–563.

Grubb, J.A. and B.A. Dehority. 1976. Variation in colon counts of total viable anaerobic rumen bacteria as influenced by media and cultural methods. Appl. Environ. Microbiol. 31: 262–267.

Grula, E.A., S-K. Luk and Y-C. Chu. 1961. Chemically defined medium for growth of Micrococcus lysodeikticus. Can. J. Microbiol. 7: 27–32.

Grundy, W.E., A.C. Sinclair, R.J. Theriault, A.W. Goldstein, C.J. Rickher, H.B. Warren, Jr., T.J. Oliver and J.C. Sylvester. 1957. Ristocetin, microbiologic properties. Antibiot. Annu. 1956–1957, New York, pp. 687–792.

Grüner, O.N.P. 1969. Actinomyces in tonsillar tissue. Acta Pathol. Microbiol. Scand. 76: 239–244.

Gubash, S.M. 1980. Synergistic haemolysis test for presumptive identification and differentiation of Clostridium perfringens, C. bifermentans, C. sordellii, and C. paraperfringens. J. Clin. Pathol. 33: 395–399.

Guex-Holzer, S. J. Tomcsik. 1956. The isolation and chemical nature of capsular and cell-wall haptens in a Bacillus species. J. Gen. Microbiol. 14: 14–25.

Guffanti, A.A. and H.C. Eisenstein. 1983. Purification and characterization of flagella from the alkalophile Bacillus firmus RAB. J. Gen. Microbiol. 129: 3329–3242.

Guggenheim, B. 1970. Enzymatic hydrolysis and structure of water-soluble glucan produced by glucosyltransferases from a strain of Streptococcus mutans. Helv. Odontol. Acta 14: Suppl. V: 89–108.

Guggenheim, B. and H.E. Schroeder. 1967. Biochemical and morphological aspects of extracellular polysaccharides produced by cariogenic streptococci. Helv. Odontol. Acta 11: 131–152.

Guicciardi, A., M.R. Biffi, P.L. Menachim, A. Craveri, C. Scolastico, B. Rindone and R. Craveri. 1968. Novo termofilo del genere Bacillus. Ann. Microbiol. Enzymol. 18: 191–205.

Guillaume, J., H. Beerens and H. Osteux. 1956. Production de gaz carbonique et fermentation du glycocolle par Clostridium histolyticum. Ann. Inst. Pasteur 91: 721–726.

Guillot, E.P. and C.S. McCleskey. 1963. Phage susceptibility of Listeria monocytogenes. Bacteriol. Proc. 139.

Guillou, J.-P. and L. Chevrieu. 1977. Étude des lipases actives sur le glycéroltributyrate chez les bactéries anaérobies par la méthode de chromatographie gaz liquide. C. R. Acad. Sci. Paris 284: 97–99.

Guillou, J.-P., R. Durieux, A. Dublanchet and L. Chevrier. 1977. Actinomyces odontolyticus, première étude réalisée en France. C. R. Acad. Sci. Paris 285: 1561–1564.

Guinand, M., J.M. Ghuysen, K.H. Schleifer and O. Kandler. 1969. The peptidoglycan in walls of Butyribacterium rettgeri. Biochemistry 8: 200–206.

Gunn, B.A. 1976. SXT and Taxo A discs for presumptive identification of group A and B streptococci in throat cultures. J. Clin. Microbiol. 4: 192–193.

Gunnels, J. and J.H. Bates. 1972. Characterization and mycobacteriophage typing of Mycobacterium xenopi. Am. Rev. Resp. Dis. 105: 388–392.

Gunsalus, J.C. and C.F. Niven, Jr. 1942. The effects of pH on the lactic fermentation. J. Biol. Chem. 145: 131–136.

Gunther, H.L. 1959. Mode of division of pediococci. Nature (Lond.) 183: 903–904.

Gunther, H.L. and H.R. White. 1961. The cultural and physiological characters of the pediococci. J. Gen. Microbiol. 26: 185–197.

Gunther, H.L. and H.R. White. 1962a. Proposed designation of a neotype strain of Pediococcus cerevisiae Balcke. Int. Bull. Bacteriol. Nomencl. Taxon. 12: 185–187.

Gunther, H.L. and H.R. White. 1962b. Designation of the type strain of Pediococcus parvulus (Gunther and White). Int. Bull. Bacteriol. Nomencl. Taxon. 12: 189–190.

Gupta, P.K., Y.S. Erozan and J.K. Frost. 1978. Actinomycetes and the IUD: An update. Acta Cytol. 22: 281–282.

Gupta, P.K., D.H. Hollander and J.K. Frost. 1976. Actinomycetes in cervicovaginal smears: An association with IUD usage. Acta Cytol. 20: 295–297.

Guthof, O. 1955. Über eine neue serologische Gruppe alphahämolytischer Streptokokken (Serologische Gruppe O). Zentralbl. Bakteriol. Parasitenkd. Infektionskr. Hyg. Abt. 1 Orig. 164: 60–63.

Guthof, O. 1956. Ueber pathogene 'Vergrunede Streptokokken', Streptokokken Befunde bei dentogenen Abszessen und Infiltraten im Bereich der Mundhöhle. Zentralbl. Bakteriol. Parasitenkd. Infektionskr. Hyg. Abt. I 166: 533–564.

Gutmann, I. and A.W. Wahlefeld. 1974. L(+)-Lactat. Bestimmung mit Lactat-Dehydrogenase und NAD. In Bergmeyer (Editor), Methoden der enzymatischen Analyse. Verlag Chemie, Weinheim-Bergstr. pp. 1510–1514.

Gutmann, L., F.W. Goldstein, M.D. Kitzis, B. Hautefort, C. Darmon and J.F. Acar. 1983. Susceptibility of Nocardia asteroides to 46 antibiotics, including 22 β-lactams. Antimicrob. Agents Chemother. 23: 248–251.

Gyllenberg, H.G. and J.J. Laine. 1971. Numerical approach to the taxonomy of psychrophillic bacilli. Ann. Med. Exp. Biol. Fenn. 49: 62–66.

György, P. 1953. A hitherto unrecognized biochemical difference between human milk and cow's milk. Pediatrics 11: 98–108.

György, P., R.W. Jeanloz, H. von Nicolai and F. Zilliken. 1974. Undialyzable growth factors for Lactobacillus bifidus var. pennsylvanicus: protective effect of sialic acid bound to glycoproteins and oligosaccharides against bacterial degradation. Eur. J. Biochem. 43: 29–33.

György, P. and C.S. Rose. 1955. Further observations on metabolic requirements of Lactobacillus bifidis var. pennsylvanicus. J. Bacteriol. 69: 483–490.

Hachisuka, Y., I. Suzuki, K. Morikawa and S. Maeda. 1982. The effect of oxidation-reduction potential on spore germination, outgrowth, and vegetative growth of Clostridium tetani, Clostridium butyricum, and Bacillus subtilis. Microbiol. Immunol. 26: 803–811.

Haenel, H. und W. Müller-Beuthow. 1956. Vergleichende quantitative Untersuchungen über Keimzahlen in den Faeces des Menschen und einiger Wirbeltiere. Zentralbl. Bakteriol. Parasitenkd. Infektionskr. Hyg. I. Orig. 167: 123–133.

Haenel, H. and W. Müller-Beuthow. 1957. Untersuchungen über die Eignung vor Bifidusnährböden zur quantitativen Züchtung der Bifidusgruppe des Erwachsenen. Zentralbl. Bakteriol. Parasitenkd. Infektionskr. Hyg. Abt. I. Orig. 169: 196–204.

Hafiz, S., M.G. McEntegart, R.S. Morton and S.A. Waitkins. 1975. Clostridium difficile in the urogenital tract of males and females. Lancet i: 420–421.

Hafiz, S. and C.L. Oakley. 1976. Clostridium difficile; isolation and characteristics. J. Med. Microbiol. 9: 129–136.

Hagedorn, C. and J.G. Holt. 1975a. Ecology of soil arthrobacters in Clarion-Webster toposequences of Iowa. Appl. Microbiol. 29: 211–218.

Hagedorn, C. and J.G. Holt. 1975b. A nutritional and taxonomic survey of Arthrobacter soil isolates. Can. J. Microbiol. 21: 353–361.

Hager, W.D., B. Douglas, B. Majmudar, Z.M. Naib, O.J. Williams, C. Ramsey and J. Thomas. 1979. Pelvic colonization with Actinomyces in women using intrauterine contraceptive devices. Am. J. Obst. Gynecol. 135: 680–684.

Hager, W.D. and B. Majmudar. 1979. Pelvic actinomycosis in women using intrauterine contraceptive devices. Am. J. Obst. Gynecol. 133: 60–64.

Hajek, V. 1976. Staphylococcus intermedius, a new species isolated from animals. Int. J. Syst. Bacteriol. 26: 401–408.

Hajek, V. and E. Marsalek. 1971. The differentiation of pathogenic staphylococci and a suggestion for their taxonomic classification. Zentralbl. Bakteriol. Parasitenkd. Infektionskr. Hyg. I Abt. Orig. A217: 176–182.

Hajek, V. and E. Marsalek. 1976. Evaluation of classificatory criteria for staphylococci. In Jeljaszewicz (Editor), Staphylococci and Staphylococcal Diseases, Gustav Fischer Verlag, Stuttgart, New York, pp. 11–21.

Hale, C. and K. Bisset. 1958. The pattern of growth and flagellar development in motile Gram-positive cocci. J. Gen. Microbiol. 18: 688–691.

Hale, C.M. 1953. The use of phosphomolybdic acid in the demonstration of bacterial cell walls. Lab. Pract. 2: 115–116.

Hale, C.N. 1972. Rapid identification methods for Corynebacterium insidiosum (McCulloch, 1925) Jensen, 1934. N. Z. J. Agric. Res. 15: 149–154.

Hall, E.R. 1952. Investigations on the microbiology of cellulose utilization in domestic rabbits. J. Gen. Microbiol. 7: 350–357.

Hall, H.E., R. Angelotti, K.H. Lewis and M.J. Foter. 1963. Characteristics of Clostridium perfringens strains associated with food and food-borne disease. J. Bacteriol. 85: 1094–1103.

Hall, I.C. 1929. The occurrence of Bacillus sordellii in icterohemoglobinuria of cattle in Nevada. J. Infect. Dis. 45: 156–162.

Hall, I.C. 1930. Micrococcus niger, a new pigment-forming anaerobic coccus recovered from urine in a case of general arteriosclerosis. J. Bacteriol. 20: 407–415.

Hall, I.C. and K. Matsumura. 1924. Recovery of Bacillus tertius from stools of infants. J. Infect. Dis. 35: 502–504.

Hall, I.C. and E. O'Toole. 1935. Intestinal flora in newborn infants with a description of a new pathogenic anaerobe, Bacillus difficilis. Am. J. Dis. Child 49: 390–402.

Hall, I.C. and J.P. Scott. 1927. Bacillus sordellii, a cause of malignant edema in man. J. Infect. Dis. 41: 329–335.

Hall, I.C. and R.W. Whitehead. 1927. A pharmaco-bacteriologic study of African poisoned arrows. J. Infect. Dis. 41: 51–69.

Haller, E.-M. and H. Brantner. 1979. Enzymologische Untersuchungen an Clostridium oncolyticum M55 ATCC 13732. Zentralbl. Bakteriol. Parasitenkd. Infektionskr. Hyg., I. Orig. A 243: 522–527.

Halliwell, G. and M.P. Bryant. 1963. The cellulolytic activity of pure strains of bacteria from the rumen of cattle. J. Gen. Microbiol. 32: 441–448.

Hamada, S., K. Gill and H.D. Slade. 1976. Chemical and immunological properties of the type f polysaccharide antigen of Streptococcus mutans. Infect. Immun. 14: 203–211.

Hamada, S., J. Mizuno and S. Kotani. 1978. A separation of staphylococci and micrococci based on serological reactivity with antiserum specific for polyglycerolphosphate. Microbios 18: 213–221.

Hamada, S. and T. Ooshima. 1975. Production and properties of bacteriocins (mutacins) from Streptococcus mutans. Arch. Oral Biol. 20: 641–648.

Hamada, S. and H.D. Slade. 1976. Purification and immunochemical characterization of the type e polysaccharide antigen of Streptococcus mutans. Infect. Immun. 14: 68–76.

Hamada, S. and H.D. Slade. 1980a. Biology, immunology and cariogenicity of *Streptococcus mutans*. Microbiol. Rev. *44:* 331–384.

Hamada, S. and H.D. Slade. 1980b. Mechanisms of adherence of *Streptococcus mutans* to smooth surfaces in vitro. *In* Beachey (Editor), Bacterial Adherence, Chapman and Hall, London, pp. 105–135.

Hamm, A. 1912. Die puerperal Wundinfektion. Julius Springer, Berlin. 1–168.

Hammann, R. and J.C.G. Ottow. 1976. Über N₂-bindende Clostridien und Bazillen aus Böden. Zentralbl. Bakteriol. Parasitenkd. Infektionskr. Hyg. I. Abt. Orig. B *161:* 527–533.

Hammer, B.W. 1915. Bacteriological studies on the coagulation of evaporated milk. Iowa Agric. Exp. Sta. Res. Bull. *19:* 119–131.

Hammerschlag, M.R., S. Alpert, A.B. Onderdonk, P. Thurston, E. Drude, W.M. McCormack and J.G. Bartlett. 1978. Anaerobic microflora of the vagina in children. Am. J. Obstet. Gynecol. *131:* 853–856.

Hammes, W., K.H. Schleifer and O. Kandler. 1973. Mode of action of glycine on the biosynthesis of peptidoglycan. J. Bacteriol. *116:* 1029–1053.

Hammond, B.F. 1970. Isolation and serological characterization of a cell wall antigen of *Rothia dentocariosa*. J. Bacteriol. *103:* 634–640.

Hammond, B.F. 1972. Cell wall analysis of *Rothia dentocariosa*. Int. Assoc. Dent. Res. Absts. *1972:* 242.

Hammond, B.F., C.F. Steel and K. Peindl. 1973. Occurrence of 6-deoxytalose in cell walls of plaque actinomycetes. J. Dent. Res. *52:* 88.

Hammond, B.F., C.F. Steel and K.S. Peindl. 1976. Antigens and surface components associated with virulence of *Actinomyces viscosus*. J. Dent. Res. *55:* (Special Issue) A19-A25.

Hammond, R.K. and D.C. White. 1970. Carotenoid formation by *Staphylococcus aureus*. J. Bacteriol. *103:* 191–198.

Hamon, Y. and Y. Péron. 1962. Étude du pouvoir bactériocinogène dans le genre *Listeria*. I. Propriétés générales de ces bactériocines. Ann. Inst. Pasteur *103:* 876–889.

Hamon, Y. and Y. Péron. 1963. Étude du pouvoir bactériocinogène dans le genre *Listeria*. II. Individualité et classification des bactériocines en cause. Ann. Inst. Pasteur *104:* 55–65.

Han, Y.W. 1982. Nutritional requirements and growth of a *Cellulomonas* species on cellulosic substrate. J. Ferm. Technol. *60:* 99–104.

Han, Y. and V.R. Srinivasan. 1968. Isolation and characterization of a cellulose-utilizing bacterium. Appl. Microbiol. *16:* 1140–1145.

Hanáková-Bauerová, E., M. Kocur and T. Martinec. 1965. Concerning the differentiation of *Bacillus subtilis* and related species. J. Appl. Bact. *28* (3): 384–389.

Hanáková-Bauerová, E., M. Kocur and T. Martinec. 1966. Concerning the problem of the differentiation of *Bacillus cereus* and related species. Mikrobiologija (Beograd), *3:* 1–7.

Handley, P.S. and P. Carter. 1979. The occurrence of fimbriae on strains of *Streptococcus mitior*. *In* Parker (Editor), Pathogenic streptococci. Reedbooks, Chertsey, Surrey, pp. 241–242.

Handley, P.S., P.L. Carter and J. Fielding. 1984. *Streptococcus salivarius* strains carry either fibrils or fimbriae on the cell surface. J. Bacteriol. *157:* 64–72.

Hanf, U. 1956. Untersuchungen über die in vitro-Empfindlichkeit des *Actinomyces israelii* gegen Erythromycin, Magnamycin, Polymyxin B, Bacitracin, Neomycin, Tyrothricin, Xanthocillin und Suprathricin. Z. Hyg. Infektionskr. *143:* 127–129.

Hanf, U., S. Heinrich and F. Legler. 1953. Untersuchungen über die Empfindlichkeit des Erregers der Aktinomykose gegen Antibiotika (Penicillin, Streptomycin, Aureomycin, Chloromycetin, Terramycin) und Methylenblau. Arch. Hyg. *137:* 527–531.

Hanna, B. and W.J. Untereker. 1975. Endocarditis and osteomyelitis caused by *Aerococcus viridans*. Abstr. Annu. Meet. Am. Soc. Microbiol., p. 49.

Hansen, G.A. 1880. *Bacillus leprae*. Virchows Arch. *79:* 32–42.

Hansen, P.A. 1968. Type strains of *Lactobacillus* species. A report by the taxonomic subcommittee on lactobacilli and closely related organisms. American Type Culture Collection, Rockville, Maryland.

Hansen, P.A. and E.F. Lessel. 1971. *Lactobacillus casei* (Orla Jensen) comb. nov. Int. J. Syst. Bacteriol. *21:* 69–71.

Hansen, P.A. and G. Mocquot. 1970. *Lactobacillus acidophilus* (Moro) comb. nov. Int. J. Syst. Bacteriol. *20:* 325–327.

Hardegree, M.C., A.E. Palmer and N. Duffin. 1971. Tetanolysin: in vivo effects in animals. J. Infect. Dis. *123:* 51–60.

Hardie, E.M. and J.A. Barsanti. 1982. Treatment of canine actinomycosis. J. Am. Vet. Med. Assoc. *180:* 537–541.

Hardie, J.M. 1981. The microbiology of dental caries. *In* Silverstone, Johnson, Hardie and Williams (Editors), Dental Caries: Aetiology, Pathology and Prevention. The Macmillan Press Ltd., London and Badingstoke, pp. 48–69.

Hardie, J.M. and G.H. Bowden. 1974. Cell wall and serological studies on *Streptococcus mutans*. Caries Res. *8:* 301–316.

Hardie, J.M. and G.H. Bowden. 1976. Physiological classification of oral viridans streptococci. J. Dent. Res. *55:* A166-A176.

Hardie, J.M. and P.D. Marsh. 1978. Streptococci and the human oral flora. *In* Skinner and Quesnel (Editors), Streptococci. Academic Press, London, New York, San Francisco, pp. 157–206.

Hardie, J.M., R.A. Whiley and M.J. Sackin. 1982. A numerical taxonomic study of oral streptococci. *In* Holm and Christensen (Editors), Proceedings of the VIIIth International Symposium on Streptococci and Streptococcal Diseases. Reedbooks, Chertsey, pp. 59–60.

Hardman, J.K. and T.C. Stadtman. 1960. Metabolism of ω-amino acids. II. Fermentation of Δ-aminovaleric acid by *Clostridium aminovalericum*. J. Bacteriol. *79:* 549–552.

Hardy, K. 1981. Bacterial plasmids. *In* Cole and Knowles (Editors), Aspects of Microbiology Series No. 4, Thomas Nelson and Sons, Ltd., Walton-on-Thames, United Kingdom.

Hare, R. 1935. The classification of haemolytic streptococci from the nose and throat of normal human beings by means of precipitin and biochemical tests. J. Pathol. Bacteriol. *41:* 499–512.

Hare, T. and R.M. Fry. 1938. Clinical observations of the beta hemolytic streptococcal infections of dogs. Vet. Rec. *50:* 1537–1548.

Hariharan, H. and W.R. Mitchell. 1976. Observations on bacteriophages of *Clostridium botulinum* type C isolates from different sources and the role of certain phages in toxigenicity. Appl. Environ. Microbiol. *32:* 145–158.

Harmon, S.M. and D.A. Kautter. 1977. Recovery of clostridia on catalase-treated plating media. Appl. Environ. Microbiol. *33:* 762–770.

Harmsen, G.W. 1946. Onderzoekingen over de aerobe cellulose ontleding in den grond. Thesis. Wageningen, Netherlands.

Harrigan, W.F. and M.E. McCance. 1976. Laboratory methods in food and dairy microbiology, revised ed. Academic Press, London.

Harrington, B.J. 1966. A numerical taxonomic study of some corynebacteria and related organisms. J. Gen. Microbiol. *45:* 31–40.

Harrington, R., Jr., D.C. Hulse. 1971. Comparison of two plating media for the isolation of *Erysipelothrix rhusiopathiae* from enrichment broth culture. Appl. Microbiol. *22:* 141–142.

Harrington, R., Jr., R.L. Wood and D.C. Hulse. 1974. Comparison of a fluorescent antibody technique and cultural method for the detection of *Erysipelothrix rhusiopathiae* in primary broth cultures. Am. J. Vet. Res. *35:* 461–462.

Harris, J.N. and P.B. Hylemon. 1978. Partial purification and characterization of NADP-dependent 12 α-hydroxysteroid dehydrogenase from *Clostridium leptum*. Biochim. Biophys. Acta *528:* 148–157.

Harrison, A.P., Jr. and P.A. Hansen. 1950. A motile *Lactobacillus* from the cecal feces of turkeys. J. Bacteriol. *59:* 444–446.

Harrison, F.C. 1929. The discoloration of halibut. Can. J. Res. *1:* 214–239.

Harshey, R.M. and T. Ramakrishnan. 1977. Rate of ribonucleic acid chain growth in *Mycobacterium tuberculosis* H37RV. J. Bacteriol. *129:* 616–622.

Hart, B. and P.T. Hooper. 1967. Enterotoxaemia of calves due to *Clostridium welchii* type E. Aust. Vet. J. *43:* 360–363.

Hart, W.R., D. Youngadahl and R. Hnat. 1977. Full-term pregnancy after pelvic actinomycosis. J. Reprod. Med. *19:* 36–38.

Hartsell, S.E. and L.F. Rettger. 1934. A taxonomic study of "*Clostridium putrificum*" and its establishment as a definite entity *Clostridium lentoputrescens* nov. spec. J. Bacteriol. *27:* 497–514.

Hartwig, E.C., R. Cacciatore and R.P. Dunbar. 1962. *M. fortuitum*: its identification, incidence and significance in Florida. Am. Rev. Resp. Dis. *85:* 84–91.

Harvey, S. and M.J. Pickett. 1980. Comparison of adansonian analysis and deoxyribonucleic acid hybridization results in the taxonomy of *Yersinia enterocolitica*. Int. J. Syst. Bacteriol. *30:* 86–102.

Harwood, C.R. 1980. Plasmids. *In* Goodfellow and Board (Editors), Microbiological Classification and Identification, Academic Press, London - New York, pp. 27–53.

Harz, C.O. 1877. *Actinomyces bovis*, ein neuer Schimmel in den Geweben des Rindes. Jahresber. K. Zentralbl. Thierarz. Sch. München für 1877/1878, *5:* 125–140.

Hasan, S.M. and J.B. Hall. 1976. Growth of *Clostridium tertium* and *Clostridium septicum* in chemically defined media. Appl. Environ. Microbiol. *31:* 442–443.

Hasan, S.M. and J.B. Hall. 1977. Dissimilatory nitrate reduction in *Clostridium tertium*. Z. Allg. Mikrobiol. *17:* 501–506.

Hasselgren, I.L. and P. Oeding. 1972. Antigenic studies of genus *Micrococcus*. Acta Pathol. Microbiol. Scand. Sec. B *80:* 257–264.

Hassinen, J.B., G.T. Durbin, R.M. Tomarelli and F.W. Bernhart. 1951. The minimal nutritional requirements of *Lactobacillus bifidus*. J. Bacteriol. *62:* 771–777.

Hauduroy, P. 1955. Derniers aspects du monde des mycobactéries. Masson et Cie., Paris. p. 72.

Hauduroy, P., G. Ehringer, G. Guillot, J. Magrou, A.R. Prévot, Rosset and A. Urbain. 1953. Dictionnaire des bactéries pathogènes. 2nd ed. Masson et Cie, Paris.

Hauduroy, P., G. Ehringer, A. Urbain, G. Guillot and J. Magrou. 1937. Dictionnaire des bactéries pathogènes. Masson and Co., Paris.

Haukenes, G. 1967. Serological typing of *Staphylococcus aureus*. 7. Technical aspects. Acta Pathol. Microbiol. Scand. *70:* 120–128.

Hauschild, A.H.W. and L.V. Holdeman. 1974. *Clostridium celatum* sp.nov., isolated from normal human feces. Int. J. Syst. Bacteriol. *24:* 478–481.

Hawiger, J. 1968. Staphylococcal lysozyme. II. Purification of lysozyme produced by *Staphylococcus aureus* 524. Med. Dosw. Mikrobiol. *20:* 1–7.

Hawkins, J.E. 1977. Scotochromogenic mycobacteria which appear intermediate between *M. avium/intracellulare* and *M. scrofulaceum*. Am. Rev. Resp. Dis. *116:* 963–964.

Hayami, M., A. Okabe, K. Sasai, H. Hayashi and Y. Kanemasa. 1979. Presence and synthesis of cholesterol in stable staphylococcal L-forms. J. Bacteriol. *140:* 859–863.

Hayase, M., N. Mitsui, K. Tamai, S. Nakamura and S. Nishida. 1974. Isolation of *Clostridium absonum* and its cultural and biochemical properties. Infect. Immun. *9:* 15–19.

Hayashi, F. 1932. Mitsuda's skin reaction in leprosy. Int. J. Leprosy *1:* 31–38.

Hayashi, R. and H. Nakayama. 1953. Studies on the aerobic mesophilic bacteria with distinctly bulged sporangium. I. Special reference to *Bacillus thiaminolyticus*. Yamaguchi Med. Sch. Bull. *1:* 57–63.

Hayashi, S. and C. Furusaka. 1979. Studies on *Propionibacterium* isolated from paddy soils. Antonie van Leeuwenhoek. J. Microbiol. Serol. *45:* 565–574.

Hayashi, S. and C. Furusaka. 1980. Enrichment of *Propionibacterium* in paddy soils by addition of various organic substances. Antonie van Leeuwenhoek. J. Microbiol. Serol. *46:* 313–320.

Hayward, A.C. and L.I. Sly. 1976. Dextranase activity of *Oerskovia xanthineolytica*. J. Appl. Bacteriol. *40:* 355–364.

Hayward, N.J., G.M. Incledon and J.E. Spragg: 1978. Effect of firm agar on the swarming of *Proteus* and *Clostridium* species and on the colonies of clinically important bacteria. J. Med. Microbiol. *11:* 155–164.

Hayward, N.J. and A.A. Miles. 1943. Inhibition of *Proteus* in cultures from wounds. Lancet *2:* 116–117.

Headington, J.T. and B. Beyerlein. 1966. Anaerobic bacteria in routine urine culture. J. Clin. Pathol. *19:* 573–576.

Hecht, S.T. and W.A. Causey. 1976. Rapid method for the detection and identification of mycolic acids in aerobic actinomycetes and related bacteria. J. Clin. Microbiol. *4:* 284–287.

Heckley, R.J. 1978. Preservation of microorganisms. Adv. Appl. Microbiol. *24:* 1–52.

Heczko, P.B., Z. Wegrzynowicz, M. Bulanda, J. Jeljaszewicz and G. Pulverer. 1981. Taxonomic implications of the pseudocoagulase activity of staphylococci. *In* Jeljaszewicz (Editor), Staphylococci and Staphylococcal Infections, Gustav Fischer Verlag, Stuttgart, New York, pp. 43–47.

Hedges, F. 1922. A bacterial wilt of the bean caused by *Bacterium flaccumfaciens* nov. sp. Science (Washington) *55:* 433–434.

Hefferan, M. 1904. A comparative and experimental study of bacilli producing red pigment. Zentralbl. Bakteriol. Parasitenkd. Infektionskr. Hyg. Abt. II *11:* 397–404, 456–475.

Hehre, E.J. and J.M. Neill. 1946. Formation of serologically reactive dextrans by streptococci from subacute bacterial endocarditis. J. Exp. Med. *83:* 147–162.

Heidelbaugh, N.D., D.B. Rawley, E.M. Powers, C.T. Bourland and J.L. McQueen. 1973. Microbiological testing of skylab foods. Appl. Microbiol. *25:* 55–61.

Heidelberger, M. and S. Elliott. 1966. Cross-reactions of streptococcal group N teichoic acid in antipneumococcal horse sera of type VI, XIV, and XXVII. J. Bacteriol. *92:* 281–283.

Heigener, H. 1935. Verwertung von Aminosäuren als gemeinsame C- und N-Quelle durch bekannte Bodenbakterien nebst botanischer Beschreibung n Aminosäuren als gemeinsame C- und N-Quelle durch bekannte Bodenbakterien nebst botanischer Beschreibung neu isolierter Betain- und Valin-Abbauer. Zentralbl. Bakteriol. Parasitenkd. Infektionskr. Hyg. Abt. II *93:* 81–113.

Heinen, U.J. and W. Heinen. 1972. Characteristics and properties of a caldoactive bacterium producing extracellular enzymes and two related strains. Arch. Mikrobiol. *82:* 1–23.

Heinen, W., A.M. Lauwers and J.W.M. Mulders. 1982. *Bacillus flavothermus*, a newly isolated facultative thermophile. Antonie van Leeuwenhoek. J. Microbiol. Serol. *48:* 265–272.

Heinrich, S. und H. Korth. 1967. Zur Nährbodenfrage in der Routinediagnostik der Aktinomykose: Ersatz unsicherer biologischer Substrate durch ein standardisiertes Medium. *In* Heite (Editor), Krankheiten durch Aktinomyceten und verwandte Errenger. Springer-Verlag, Berlin, F.R.G., pp. 16–20.

Heller, H.H. 1922. Certain genera of the *Clostridiaceae*. Studies in pathogenic anaerobes. J. Bacteriol. *7:* 1–38.

Hellinger, E. 1944. Studies on a pink butyric acid *Clostridium*. Commemorative Vol. to Dr. Weizmann's 70th birthday- Private print Nov. 1944, pp. 37–46.

Helmke, E. and H. Weyland. 1984. *Rhodococcus marinonascens* sp. nov., an actinomycete from the sea. Int. J. Syst. Bacteriol. *34:* 127–138.

Hemme, D., P. Raibaud, R. Ducluzeau, J.V. Galpin, Ph. Sicard and J. van Heijenoort. 1980. *Lactobacillus murinus* n. sp., une nouvelle espèce de la flore dominante autochtone du tube digestif du rat et de la souris. Ann. Microbiol. (Inst. Pasteur) *131A:* 297–308.

Hemme, D., P. Raibaud, R. Ducluzeau, J.V. Galpin, Ph. Sicard and J. van Heijenoort. 1982. *In* Validation of the publication of new names and new combinations previously effectively published outside the IJSB. List No. 9. Int. J. Syst. Bacteriol. *32:* 384–385.

Hengstenberg, W., W.K. Penberthy, K.L. Hill and M.L. Morse. 1969. Phosphotransferase system of *Staphylococcus aureus:* Its requirement for the accumulation and metabolism of galactosides. J. Bacteriol. *99:* 383–388.

Henley, M.W. and D.H. Lewis. 1976. Anaerobic bacteria associated with epizootics in grey mullet (*Mugil cephalus*) and red fish (*Sciaenops ocellata*) along the Texas Gulf coast. J. Wildl. Dis. *12:* 448–453.

Henneberg. W. 1903. Kenntnis der Milchsäurebakterien der Brennereimaische, der Milch, des Bieres, der Presshefe, der Melasse, des Sauerkohls, der sauren Gurken und des Sauerteiges, sowie einige Bemerkungen über die Milchsäurebakterien des menschlichen Magens. Z. Spiritusind. *26:* Nr. 22–31; see: Zentralbl. Bakteriol. II Abt. *11:* 163.

Hennenberg, G. and R. Marwitz. 1963. Über die Struktur von Bakterienkolonien. Zentralbl. Bakteriol. Parasitenkd. Infektionskr. Hyg. I. Referate, *187:* 43–68.

Henner, D.J. and J.A. Hoch. 1982. The genetic map of *B. subtilis*. *In* Dubnau (Editor), The Molecular Biology of the Bacilli. Academic Press, New York and London, pp. 1–33.

Henrichsen, J. 1972. Bacterial surface translocation: a survey and classification. Bacteriol. Rev. *36:* 478–503.

Henriksen, S.D. and J. Henrichsen. 1975. Twitching motility and possession of polar fimbriae in spreading *Streptococcus sanguis* isolates from the human throat. Acta Pathol. Microbiol. Scand. Sect. B *83:* 133–140.

Henry, B.S. 1933. Dissociation in the genus *Brucella*. J. Infect. Dis. *52:* 374–402.

Henry, H. 1917. An investigation of the cultural reactions of certain anaerobes found in war wounds. J. Pathol. Bacteriol. *21:* 344–385.

Hensel, R. 1977. Immunologische und chemische Untersuchungen an L-Laktat-Dehydrogenasen verschiedener Milchsäurebakterien. Thesis Universität München.

Hensel, R., U. Mayr, C. Lins and O. Kandler. 1981. Amino acid sequence of a dodecapeptide from the substrate-binding region of the L-lactate dehydrogenase from *Lactobacillus curvatus*, *Lactobacillus xylosus* and *Bacillus stearothermophilus*. Hoppe-Seyler's Z. Physiol. Chem. *362:* 1031–1036.

Hensel, R., U. Mayr, K.O. Stetter and O. Kandler. 1977. Comparative studies of lactic acid dehydrogenases in lactic acid bacteria I. Purification and kinetics of the allosteric L-lactic acid dehydrogenase from *Lactobacillus casei* ssp. *casei* and *Lactobacillus curvatus*. Arch. Microbiol. *112:* 81–93.

Henssen, A. 1957. Beiträge zur Morphologie und Systematik der thermophilen Actinomyceten. Arch. Mikrobiol. *26:* 373–414.

Henssen, A., C. Happach-Kasan, B. Renner and G. Vobis. 1983. *Pseudonocardia compacta* sp. nov. Int. J. Syst. Bacteriol. *33:* 829–836.

Henssen, A. and D. Schäfer. 1971. Emended description of the genus *Pseudonocardia* Henssen and description of a new species *Pseudonocardia spinosa* Schäfer. Int. J. Syst. Bacteriol. *21:* 29–34.

Henssen, A. and E. Schnepf. 1967. Zur Kenntnis thermophiler Actinomyceten. Arch. Mikrobiol. *57:* 214–231.

Henssen, A., E. Weise, G. Vobis and B. Renner. 1981. Ultrastructure of sporogenesis in actinomycetes forming spores in chains. *In* Schaal and Pulverer (Editors). Actinomycetes. Proceedings of the IVth International Symposium on Actinomycete Biology, Fischer, Stuttgart, New York, pp. 137–146.

Herbeck, J.L. and M.P. Bryant. 1974. Nutritional features of the intestinal anaerobe *Ruminococcus bromii*. Appl. Microbiol. *28:* 1018–1022.

Herbert, R.B., J. Mann and A. Römer. 1982. Phenazine and phenoxazinone biosynthesis in *Brevibacterium iodinum*. Z. Naturforsch. *37c:* 159–164.

Heritage, A.D. and I.C. MacRae. 1977. Degradation of lindane by cell-free preparations of *Clostridium sphenoides*. Appl. Environ. Microbiol. *34:* 222–224.

Herndon, S.E. and K.F. Bott. 1969. Genetic relationship between *Sarcina ureae* and members of the genus *Bacillus*. J. Bacteriol. *97:* 6–12.

Hertlein, B.C., R. Levy and T.W. Miller, Jr. 1979. Recycling potential and selective retrieval of *Bacillus sphaericus* from soil in a mosquito habitat. J. Invertebr. Pathol. *33:* 217–221.

Hess, A., R. Hollander and W. Mannheim. 1979. Lipoquinones of some spore-forming rods, lactic acid bacteria and actinomycetes. J. Gen. Microbiol. *115:* 247–252.

Hether, N.W., P.A. Campbell, L.A. Baker and L.L. Jackson. 1983. Chemical composition and biological functions of *Listeria monocytogenes* cell wall preparations. Infect. Immun. *39:* 1114–1121.

Hether, N.W. and L.L. Jackson. 1983. Lipoteichoic acid from *Listeria monocytogenes*. J. Bacteriol. *156:* 809–817.

Hettinga, D.H. and G.W. Reinbold. 1972a. The propionic acid bacteria-a review. I. Growth. J. Milk Food Technol. *35:* 295–301.

Hettinga, D.H. and G.W. Reinbold. 1972b. The propionic acid bacteria-a review. II. Metabolism. J. Milk Food Technol. *35:* 358–372.

Hettinga, D.H. and G.W. Reinbold. 1972c. The propionic acid bacteria-a review. III. Miscellaneous metabolic activities. J. Milk Food Technol. *35:* 436–447.

Hewett, M.J., A.J. Wicken, K.W. Knox and M.E. Sharpe. 1976. Isolation of lipoteichoic acids from *Butyrivibrio fibrisolvens*. J. Gen. Microbiol. *94:* 126–130.

Hewitt, J. and J.G. Morris. 1975. Superoxide dismutase in some obligately anaerobic bacteria. FEBS Lett. *50:* 315–318.

Higgins, M.L. and M.P. Lechevalier. 1969. Poorly lytic bacteriophage from *Dactylosporangium thailandensis* (*Actinomycetales*). J. Virol. *3:* 210–216.

Higgs, T.M., F.K. Neave and A.J. Bramley. 1980. Differences in intramammary pathogenicity of four strains of *Streptococcus dysgalactiae*. J. Med. Microbiol. *13:* 393–399.

Higuchi, M., K. Endo, E. Hoshino and S. Araya. 1973. Preferential induction of rough variants in *Streptococcus mutans* by ethidium bromide. J. Dent. Res. *52:* 1070–1075.

Hill, G.B. and S. Osterhout. 1972. Experimental effects of hyperbaric oxygen on selected clostridial species. I. In-vitro studies. J. Infect. Dis. *125:* 17–25.

Hill, L.R. 1966. An index to deoxyribonucleic acid compositions of bacterial species. J. Gen. Microbiol. *44:* 419–437.

Hill, L.R. 1981. Preservation of microorganisms. *In* Norris and Richmond (Editors), Essays in Applied Microbiology. John Wiley & Sons, Chichester, pp. 2/1–2/31.

Hill, M.J. 1975. The role of colon anaerobes in the metabolism of bile acids and steroids, and its relation to colon cancer. Cancer *36:* 2387–2400.

Hill, P.E., K.W. Knox, R.G. Schamschula and M. Tabua. 1977. The identification and enumeration of *Actinomyces* from plaque of New Guinea indigenes. Caries Res. *11:* 327–335.

Hill, R.A. and J. Lacey. 1983. Factors determining the microflora of stored barley. Ann. Appl. Biol. *102:* 467–483.

Hilton, M.G., G.C. Mead and S.R. Elsden. 1975. The metabolism of pyrimidines by proteolytic clostridia. Arch. Microbiol. *102:* 145–149.

Hino, S. and P.W. Wilson. 1958. Nitrogen fixation by a facultative bacillus. J. Bacteriol. *75:* 403–408.

Hippchen, B., A. Röll and K. Poralla. 1981. Occurence in soil of thermo-acidophilic bacilli possessing w-cyclohexane fatty acids and hopanoids. Arch. Microbiol. *129:* 53–55.

Hiramune, T., S. Inui, N. Murase and R. Yanagawa. 1971. Virulence of three types of *Corynebacterium renale* in cows. Am. J. Vet. Res. *32:* 236–242.

Hirano, S. and N. Masuda. 1981. Transformation of bile acids by *Eubacterium lentum*. Appl. Environ. Microbiol. *42:* 912–915.

Hirano, S. and N. Masuda. 1982. Characterization of NADP-dependent 7β-hydroxy-steroid dehydrogenases from *Peptostreptococcus productus* and *Eubacterium aerofaciens*. Appl. Environ. Microbiol. *43:* 1057–1063.

Hirano, S., N. Masuda and N. Akimori. 1981. Gas-liquid chromatographic and mass spectrometrical studies of 3 β-hydroxy and 3-oxy substituted bile acids derived from seven 3 α-hydroxy bile acids by *Clostridium perfringens*. Acta. Med. Univ. Kagoshima *23:* 55–64.

Hirano, W., N. Masuda, T. Imamura and H. Oda. 1980. Transformation of bile acids by mixed microbial cultures from human feces and bile acid transforming activities of isolated bacterial strains. Microbiol. Immunol. *25:* 271–282.

Hirsch, P. 1960. Einige, weitere, von Luftverunreinigungen lebende Actinomyceten und ihre Klassifizierung. Arch. Mikrobiol. *35:* 391–414.

Hirsch, P. 1961. Wasserstoffaktivierung und Chemoautotrophie bei Actinomyceten. Arch. Mikrobiol. *39:* 360–373.

Hirsh, D.C., E.L. Biberstein and S.S. Jang. 1979. Obligate anaerobes in clinical veterinary practice. J. Clin. Microbiol. *10:* 188–191.

Hirst, R.G. and R.J. Olds. 1978a. *Corynebacterium kutscheri* and its alleged avirulent variant in mice. J. Hyg. *80:* 349–356.

Hirst, R.G. and R.J. Olds. 1978b. Serological and biochemical relationship between the alleged avirulent variant of *Corynebacterium kutscheri* and streptococci of group N. J. Hyg. *80:* 357–363.

Hirunmitnakorn, C. and H.J. Blumenthal. 1978. Phosphatases in *Staphylococcus* species: biosynthesis and disc-gel electrophoretic studies. Abstr. Annu. Meet. Am. Soc. Microbiol. *1978:* 146.

Hitchener, B.J., A.F. Egan and P.J. Rogers. 1979. Energetics of *Microbacterium thermosphactum* in glucose-limited continuous culture. Appl. Environ. Microbiol. *37:* 1047–1052.

Hitchener, B.J., A.F. Egan and P.J. Rogers. 1982. Characteristics of lactic acid bacteria isolated from vacuum-packaged beef. J. Appl. Bacteriol. *52:* 31–37.

Hitchener, E.R. 1932. A cultural study of the propionic acid bacteria. J. Bacteriol. *23:* 40–41.

Hitchner, E.R. and S.F. Snieszko. 1947. A study of a microorganism causing a bacterial disease of lobsters. J. Bacteriol. *54:* 48.

Hjelm, H., K. Hjelm and J. Sjöquist. 1972. Protein A from *Staphylococcus aureus*. Its isolation by affinity chromatography and its use as an immunosorbent for isolation of immunoglobulins. FEBS Lett. *28:* 73–76.

Hobbs, G. and T. Cross. 1983. Identification of endospore-forming bacteria. *In* Hurst and Gould (Editors), The Bacterial Spore, Volume 2, Academic Press, London, pp. 49–78.

Hobby, G.L., W.B. Redmond, E.H. Runyon, W.B. Schaefer, L.G. Wayne and R.H. Wichelhausen. 1967. A study on pulmonary disease associated with mycobacteria other than *Mycobacteria tuberculosis*: Identification and characterization of the mycobacteria. XVIII. A report of the Veterans Administration Armed Forces Cooperative Study. Am. Rev. Resp. Dis. *95:* 954–971.

Hoeffler, U. 1977. Enzymatic and hemolytic properties of *Propionibacterium acnes* and related bacteria. J. Clin. Microbiol. *6:* 555–558.

Hoeffler, U. 1980. Production of hyaluronidase E.C.3.2.1.36 by propionibacteria from different origins. Zentralbl. Bakteriol. Parasitenkd. Infektionskr. Hyg. I. Abt. Orig. A *245:* 1–2.

Hoeffler, U., H.L. Ko and G. Pulverer. 1977. Serotyping of *Propionibacterium acnes* and related microbial species. FEMS Microbiol. Lett. *2:* 5–9.

Hof, H., V. Sticht-Groh and K.-M. Müller. 1982. Comparative in vitro activities of niridazole and metronidazole against anaerobic and microaerophilic bacteria. Antimicrob. Agents Chemother. *22:* 332–333.

Hoffman, J., B. Lindberg and J. Lönngren. 1976. Structural studies of the polysaccharide antigen of *Eubacterium saburreum*, strain 49. Carbohydr. Res. *47:* 261–267.

Hoffman, J., B. Lindberg, N. Skaug and T. Hofstad. 1980. Structural studies of the *Eubacterium saburreum* strain O 2 antigen. Carbohydr. Res. *84:* 181–183.

Hoffman, J., B. Lindberg and S. Svensson. 1974. Structure of the polysaccharide antigen of *Eubacterium saburreum*, strain L44. Carbohydr. Res. *35:* 49–53.

Hoffman, R.E., B.J. Pincomb, M.R. Skeels and M.J. Burkhart. 1982. Type F infant botulism. Am. J. Dis. Child. *136:* 270–271.

Hofstad, T. 1967. An anaerobic oral filamentous organism possibly related to *Leptotrichia buccalis*. 1. Morphology, some physiological and serological properties. Acta Pathol. Microbiol. Scand. *69:* 543–548.

Hofstad, T. 1972. A polysaccharide antigen of an anaerobic oral filamentous microorganism (*Eubacterium saburreum*) containing heptose and O-acetyl as main constituents. Acta Pathol. Microbiol. Scand. Sect. B *80:* 609–614.

Hofstad, T. 1975. Immunochemistry of a cell wall polysaccharide isolated from *Eubacterium saburreum*, strain L49. Acta Pathol. Microbiol. Scand. Sect. B *83:* 471–476.

Hofstad, T. and P. Aasjord. 1982. *Eubacterium plautii* (Séguin 1928) comb. nov. Int. J. Syst. Bacteriol. *32:* 346–349.

Hofstad, T. and H. Lygre. 1977. Composition and antigenic properties of a surface polysaccharide isolated from *Eubacterium saburreum*, strain L452. Acta Pathol. Microbiol. Scand. Sect. B *85:* 14–17.

Hofstad, T. and N. Skaug. 1978. A polysaccharide antigen from the gram-positive organism *Eubacterium saburreum* containing dideoxyhexose as the immunodominant sugar. J. Gen. Microbiol. *106:* 227–232.

Høi Sorensen, G. 1973. *Micrococcus indolicus*. Some biochemical properties and the demonstration of six antigenically different types. Acta Vet. Scand. *14:* 301–326.

Høi Sorensen, G. 1974. Studies on the aetiology and transmission of summer mastitis. Nord. Veterinaemed. *26:* 122–132.

Høi Sorensen, G. 1975. *Peptococcus* (s. *Micrococcus*) *indolicus*. The demonstration of two varieties of hemolysis forming strains. Acta Vet. Scand. *16:* 218–225.

Høi Sorensen, G. 1976. Studies on the occurrence of *Peptococcus indolicus* and *Corynebacterium pyogenes* in apparently healthy cattle. Acta Vet. Scand. *17:* 15–24.

Holbrook, R. and J.M. Anderson. 1980. An improved selective and diagnostic medium for the isolation and enumeration of *Bacillus cereus* in foods. Can. J. Microbiol. *26:* 753–759.

Holbrook, R., J.M. Anderson and A.C. Baird-Parker. 1969. The performance of a stable version of Baird-Parker's medium for isolating *Staphylococcus aureus*. J. Appl. Bacteriol. *32:* 187–192.

Holdeman, L.V. and J.B. Brooks. 1970. Variation among strains of *Clostridium botulinum* and related clostridia. Proc. of 1st U.S.-Japan Conference on Toxic Microorganisms, 1968. U.S. Govt. Printing Office, Washington, DC. 278–286.

Holdeman, L.V., E.P. Cato, J.A. Burmeister and W.E.C. Moore. 1980. Description of *Eubacterium timidum* sp. nov., *Eubacterium brachy* sp. nov., and *Eubacterium nodatum* sp. nov. isolated from human periodontitis. Int. J. Syst. Bacteriol. *30:* 163–169.

Holdeman, L.V., E.P. Cato and W.E.C. Moore. 1967. Amended description of *Ramibacterium alactolyticum* Prévot and Taffanel with proposal of a neotype strain. Int. J. Syst. Bacteriol. *17:* 323–341.

Holdeman, L.V., E.P. Cato and W.E.C. Moore. 1971. *Clostridium ramosum* (Vuillemin) comb.nov.: emended description and proposed neotype strain. Int. J. Syst. Bacteriol. *21:* 35–39.

Holdeman, L.V., E.P. Cato and W.E.C. Moore. 1971. *Eubacterium contortum* (Prévot) comb. nov.: emendation of description and designation of the type strain. Int. J. Syst. Bacteriol. *21:* 304–306.

Holdeman, L.V., E.P. Cato and W.E.C. Moore (Editors). 1977. Anaerobe laboratory manual, 4th ed. Anaerobe Laboratory, Virginia Polytechnic Institute and State University, Blacksburg. 1–156.

Holdeman, L.V., I.J. Good and W.E.C. Moore. 1976. Human fecal flora: variation in bacterial composition within individuals and a possible effect of emotional stress. Appl. Environ. Microbiol. *31:* 359–375.

Holdeman, L.V. and W.E.C. Moore. 1970. *In* Cato, Cummins, Holdeman, Johnson, Moore, Smibert and Smith (Editors). Outline of clinical methods in anaerobic bacteriology, 2nd rev. Virginia Polytechnic Institute Anaerobe Laboratory, Blacksburg, VA pp. 57–66.

Holdeman, L.V. and W.E.C. Moore. 1970. *Eubacterium. In* Cato, Cummins, Holdeman, Johnson, Smibert and Smith (Editors). Outline of clinical methods in anaerobic bacteriology, 2nd rev. Virginia Polytechnic Institute Anaerobe Laboratory, Blacksburg, Virginia, pp. 23–30.

Holdeman, L.V. and W.E.C. Moore (Editors). 1972. Anaerobe laboratory manual. Anaerobe Laboratory, Virginia Polytechnic Institute and State University, Blacksburg, 130 pp.

Holdeman, L.V. and W.E.C. Moore. 1974. New genus, *Coprococcus*, twelve new species, and emended descriptions of four previously described species of bacteria from human feces. Int. J. Syst. Bacteriol. *24:* 260–277.

Holdeman, L.V. and W.E.C. Moore. 1974. Family *Bacteroidaceae* Pribram 1933, 10. *In* Buchanan and Gibbons (Editors), Bergey's Manual of Determinative Bacteriology, 8th ed. The Williams & Wilkins Co., Baltimore, pp. 384–418.

Holdeman, L.V., W.E.C. Moore, P.J. Churn and J.L. Johnson. 1982. *Bacteroides oris* and *Bacteroides buccae*, new species from human periodontitis and other human infections. Int. J. Syst. Bacteriol. *32:* 125–131.

Holland, D.F. 1920. *In* Winslow, C.-E.A., J. Broadhurst, R.E. Buchanan, C. Krumwiede, L.A. Rogers and G.H. Smith. 1920. The families and genera of the bacteria. Final report of the committee of the Society of American Bacteriologists on characterization and classification of bacterial types. J. Bacteriol. *5:* 191–229.

Holland, K.T., J. Greenman and W.J. Cunliffe. 1979. Growth of cutaneous propionibacteria on synthetic medium: Growth yields and exoenzyme production. J. Appl. Bacteriol. *47:* 383–394.

Hollande, A.C. 1933. La structure cytologique des *Bacillus enterothrix, camptospora* Collin et de *Bacillospora (Spirillum) praeclarum* Collin. C.R. Hebd. Sea. Acad. Sci. Ser. D Sci. Nat. *196:* 1830–1832.

Hollande, A.C. 1934. Contribution à l'étude cytologique des microbes (*Coccus, Bacillus, Vibrio, Spirillum, Spirochaeta*). Arch. Protistenk. *83:* 465–468.

Holländer, R. and S. Pohl. 1980. Deoxyribonucleic acid base composition of bacteria. Zentralbl. Bakteriol. Parasitenkd. Infektionskr. Hyg. I. Abt. Orig. A. *246:* 236–275.

Holländer, R. and S. Pohl. 1980. Deoxyribonucleic acid base composition of bacteria. Zentralbl. Bakteriol. Parasitenkd. Infektionskr. Hyg. Abt. I Orig. A *246:* 236–275.

Hollaus, F. and U. Sleytr. 1972. On the taxonomy and fine structure of some hyperthermophilic saccharolytic clostridia. Arch. Microbiol. *86:* 129–146.

Hollingdale, M.R. 1974. Antibody responses in patients with farmer's lung disease to antigens from *Micropolyspora faeni.* J. Hyg. *72:* 79–89.

Holloway, Y., M. Schaareman and J. Dankert. 1979. Identification of viridans streptococci on the Minitek Miniaturised Differentiation system. J. Clin. Pathol. *32:* 1168–1173.

Höllriegl, V., L. Lamm, J. Rowold, J. Hörig and P. Renz. 1982. Biosynthesis of Vitamin B¹². Different pathways in some aerobic and anaerobic microorganisms. Arch. Microbiol. *132:* 155–158.

Holm, E. and V. Jensen. 1972. Aerobic chemoorganotrophic bacteria of a Danish Beech forest. Oikos *23:* 248–260.

Holm, P. 1930. Comparative studies on pathogenic anaerobic *Actinomyces.* Acta Pathol. Microbiol. Scand. Suppl. *3:* 151–156.

Holm, P. 1950. Studies on the aetiology of human actinomycosis. I. The "other microbes" of actinomycosis and their importance. Acta Pathol. Microbiol. Scand. *27:* 736–751.

Holm, P. 1951. Studies on the aetiology of human actinomycosis. II. Do the "other microbes" of actinomycosis possess virulence? Acta Pathol. Microbiol. Scand. *28:* 391–406.

Holmberg, K. and U. Forsum. 1973. Identification of *Actinomyces, Arachnia, Bacterionema, Rothia,* and *Propionibacterium* species by defined immunofluorescence. Appl. Microbiol. *25:* 834–843.

Holmberg, K. and H.O. Hallander. 1973. Numerical taxonomy and laboratory identification of *Bacterionema matruchotii, Rothia dentocariosa, Actinomyces naeslundii, Actinomyces viscosus,* and some related bacteria. J. Gen. Microbiol. *76:* 43–63.

Holmberg, K. and C.-E. Nord. 1975. Numerical taxonomy and laboratory identification of *Actinomyces* and *Arachnia* and some related bacteria. J. Gen. Microbiol. *91:* 17–44.

Holmberg, K., C.-E. Nord and T. Wadström. 1975. Serological studies of *Actinomyces israelii* by crossed immunoelectrophoresis: standard antigen-antibody system for *A. israelii.* Infect. Immun. *12:* 387–397.

Holt, R.J. 1969. The classification of staphylococci from colonized ventriculoatrial shunts. J. Clin. Pathol. *22:* 475–482.

Holt, R.J. 1971. The colonization of ventriculo-atrial shunts by coagulase-negative staphylococci. *In* Finland, Marget and Bartmann (Editors), Bayer-Symposium III. Bacterial Infections: Changes in Their Causative Agents. Trends and Possible Basis, Springer-Verlag, New York, Heidelberg, Berlin, pp. 81–87.

Holt, R.J. 1972. The pathogenic role of coagulase-negative staphylococci. Br. J. Dermatol. Suppl. 8 *86:* 42–49.

Holt, S.C. and E. Canale-Parola. 1967. Fine structure of *Sarcina maxima* and *Sarcina ventriculi.* J. Bacteriol. *93:* 399–410.

Holtman, D.F. 1945. *Corynebacterium equi* in chronic pneumonia of the calf. J. Bacteriol. *49:* 159–162.

Holzapfel, W.H. 1969. Aminosauresequenz des mureins und taxonomie der gattung *Leuconostoc.* Dissertation der Technischen Hochschule, München.

Holzapfel, W.H. and E.S. Gerber. 1983. *Lactobacillus divergens* sp. nov., a new heterofermentative *Lactobacillus* species producing L(+)-lactate. Syst. Appl. Microbiol. *4:* 522–534.

Holzapfel, W.H. and E.S. Gerber. 1984. *In* Validation of the publication of new names and new combinations previously published outside the IJSB. List No. 14. Int. J. Syst. Bacteriol. *34:* 270–271.

Holzapfel, W.H. and O. Kandler. 1969. Zur Taxonomie der Gattung *Lactobacillus* Beijerinck. VI. *Lactobacillus coprophilus* subsp. *confusus* nov. subsp., eine neue Unterart der Untergattung *Betabacterium.* Zentralbl. Bakteriol. II Abt. *123:* 657–666.

Holzapfel, W.H. and E.P. van Wyk. 1982. *Lactobacillus kandleri* sp. nov., a new species of the subgenus *Betabacterium* with glycine in the peptidoglycan. Zentralbl. Bakteriol. Mikrobiol. Hyg. I Abt. Orig. C *3:* 495–502.

Holzapfel, W.H. and E.P. van Wyk. 1983. *In* Validation of the publication of new names and new combinations previously effectively published outside the IJSB. List No. 10. Int. J. Syst. Bacteriol. *33:* 438–440.

Honda, E. and R. Yanagawa. 1973. Deoxyribonucleic acid homologies among three immunological types of *Corynebacterium renale* (Migula) Ernst. Int. J. Syst. Bacteriol. *23:* 226–230.

Hontebeyrie, M. and F. Gasser. 1975. Comparative immunological relationships of two distinct sets of isofunctional dehydrogenases in the genus *Leuconostoc.* Int. J. Syst. Bacteriol. *25:* 1–6.

Hontebeyrie, M. and F. Gasser. 1977. Deoxyribonucleic acid homologies in the genus *Leuconostoc.* Int. J. Syst. Bacteriol. *27:* 9–14.

Hopkins, D.G. and J.P. Kushner. 1983. Clostridial species in pathogenesis of necrotizing enterocolitis in patients with neutropenia. Am. J. Hematol. *14:* 289–295.

Horan, A.C. 1971. Mycolic acids in the classification of nocardiae, mycobacteria and corynebacteria. Ph.D. Thesis. Rutgers University, New Brunswick, NJ, USA, pp. 124–5.

Horikoshi, K. and A. Teruhiko. 1982. Alkalophilic Microorganisms. Japan Sci-

entific Societies Press, Tokyo and Springer-Verlag, Berlin and New York.

Horwitz, A.H. and L.E. Casida, Jr. 1975. L-Phase variants of *Agromyces ramosus.* Antonie Van Leeuwenhoek J. Microbiol. Serol. *41:* 153–171.

Horwitz, A.H. and L.E. Casida, Jr. 1978a. Effects of magnesium, calcium, and serum on reversion of stable L-forms. J. Bacteriol. *136:* 565–569.

Horwitz, A.H. and L.E. Casida, Jr. 1978b. Survival and reversion of a stable L-form in soil. Can. J. Microbiol. *24:* 50–55.

Hoshino, E., F. Frölander and J. Carlsson. 1978. Oxygen and the metabolism of *Peptostreptococcus anaerobius* VPI 4330–1. Can. J. Microbiol. *107:* 235–248.

Hotchi, M. and J. Schwarz. 1972. Characterization of actinomycotic granules by architecture and staining methods. Arch. Pathol. *93:* 392–400.

Houdret, N., A. Scharfman, G. Martin and P. Roussel. 1975. Étude des neuraminidases de *Diplococcus pneumoniae* et de *Clostridium perfringens* (type 33–4A). Ann. Microbiol. (Inst. Pasteur) *126B:* 175–180.

Hovelius, B., I. Thelin and P.A. Mardh. 1979. *Staphylococcus saprophyticus* in the aetiology of nongonococcal urethritis. Br. J. Vener. Dis. *55:* 369–374.

Howell, A., Jr. 1953. *In vitro* susceptibility of *Actinomyces* to terramycin. Antibiot. Chemother. *3:* 378–381.

Howell, A., Jr. 1963. A filamentous microorganism isolated from periodontal plaque in hamsters. I. Isolation, morphology and general cultural characteristics. Sabouraudia *3:* 81–92.

Howell, A., Jr. and H.V. Jordan. 1963. A filamentous microorganism isolated from periodontal plaque in hamsters. II. Physiological and biochemical characteristics. Sabouraudia *3:* 93–105.

Howell, A. and H.V. Jordan. 1967. Production of an extracellular levan by *Odontomyces viscosus.* Arch. Oral Biol. *12:* 571–573.

Howell, A., H.V. Jordan, L.K. Georg and L. Pine. 1965. *Odontomyces viscosus,* gen. nov., spec. nov., a filamentous microorganism isolated from periodontal plaque in hamsters. Sabouraudia *4:* 65–68.

Howell, A., W.C. Murphy, F. Paul and R.M. Stephan. 1959. Oral strains of *Actinomyces.* J. Bacteriol. *78:* 82–95.

Howell, A. and L. Pine. 1956. Studies on the growth of species of *Actinomyces.* I. Cultivation in a synthetic medium with starch. J. Bacteriol. *71:* 47–53.

Howell, A., Jr., R.M. Stephan and F. Paul. 1962. Prevalence of *Actinomyces israelii, A. naeslundii, Bacterionema matruchotii,* and *Candida albicans* in selected areas of the oral cavity and saliva. J. Dent. Res. *41:* 1050–1059.

Hoyer, B.H., B.J. McCarthy and E.T. Bolton. 1964. A molecular approach in the systematics of higher organisms. Science (Washington) *144:* 959–967.

Hsu, C.Y. and G.M. Wiseman. 1971. Purification of epidermidins, new antibiotics from staphylococci. Can. J. Microbiol. *17:* 1223–1226.

Hsu, E.J. and Z.J. Ordal. 1969. Sporulation of *Clostridium thermosaccharolyticum.* Appl. Microbiol. *18:* 958–960.

Hsu, E.J. and Z.J. Ordal. 1970. Comparative metabolism of vegetative and sporulating cultures of *Clostridium thermosaccharolyticum.* J. Bacteriol. *102:* 369–376.

Hubert, J., H. Ionesco and M. Sebald. 1981. Détection de *Clostridium difficile* par isolement sur milieu minimal sélectif et par immunofluorescence. Ann. Microbiol. (Inst. Pasteur) *132A:* 149–157.

Hucker, G.J. 1948. *Micrococcaceae. In* Breed, Murray and Hitchens (Editors), Bergey's Manual of Determinative Bacteriology, 6th ed., Williams and Wilkins Co., Baltimore, pp. 235–294.

Hucker, G.J. and C.S. Pederson. 1930. Studies on the *Coccaceae* XVI. The genus *Leuconostoc.* N.Y. Agric. Exp. Sta. Tech. Bull. *167:* 3–80.

Hugh, R. and M.A. Ellis. 1968. The neotype strain for *Staphylococcus epidermidis* (Winslow and Winslow 1908) Evans 1916. Int. J. Syst. Bacteriol. *18:* 231–239.

Hugh, R. and E. Leifson. 1953. The taxonomic significance of fermentation versus oxidative metabolism of carbohydrates by various gram negative bacteria. J. Bacteriol. *66:* 24–26.

Hughes, H.P. and E.L. Biberstein. 1959. Chronic equine abscesses associated with *Corynebacterium pseudotuberculosis.* J. Am. Vet. Med. Assoc. *135:* 559–562.

Huhtanen, C.M. 1979. Bile acid inhibition of *Clostridium botulinum.* Appl. Environ. Microbiol. *38:* 216–218.

Hungate, R.E. 1944. Studies on cellulose fermentation. I. The culture and physiology of an anaerobic cellulose-digesting bacterium. J. Bacteriol. *48:* 499–513.

Hungate, R.E. 1950. The anaerobic mesophilic cellulolytic bacteria. Bacteriol. Rev. *14:* 1–49.

Hungate, R.E. 1957. Microorganisms in the rumen of cattle fed a constant ration. Can. J. Microbiol. *3:* 289–311.

Hungate, R.E. 1966. The Rumen and Its Microbes. Academic Press, New York.

Hungate, R.E. and R.J. Stack. 1982. Phenylpropanoic acid: growth factor for *Ruminococcus albus.* Appl. Environ. Microbiol. *44:* 79–83.

Hunger, W. and D. Claus. 1978. Reisolation and growth conditions of *Bacillus agar-exedens.* Antonie van Leeuwenhoek. J. Microbiol. Serol. *44:* 105–113.

Hunger, W. and D. Claus. 1981. Taxonomic studies on *Bacillus megaterium* and on agarolytic *Bacillus* strains. *In* Berkeley and Goodfellow (Editors), The Aerobic Endospore-forming Bacteria. Academic Press, London, pp. 217–239.

Hungerer, K.D. and D.J. Tipper. 1969. Cell wall polymers of *Bacillus sphaericus* 9602. I. Structure of the vegetative cell wall peptidoglycan. Biochemistry *8:* 3577–3587.

Hunter, M.I.S., D.D.M. Muir and D. Thirkell. 1973. Effect of lysozyme treatment on cell wall ultrastructure in *Sarcina flava.* J. Bacteriol. *116:* 483–487.

Hunter, R. 1973. Observations on *Listeria monocytogenes* type 5 (Iwanow) isolated in New Zealand. Med. Lab. Technol. *30:* 51–56.

Hunter, S. and P. Brennan. 1981. A novel phenolic glycolipid from *Mycobacterium leprae* possibly involved in immunogenicity and pathogenicity. J. Bacteriol. *147:* 728–735.

Hunter, S.H. 1942. Some growth requirements of *Erysipelothrix* and *Listeria*. J. Bacteriol. *43:* 629–640.

Hurst, A. 1978. Nisin: Its preservative effect and function in the growth cycle of the producer organism. *In* Skinner and Quesnel (Editors), Streptococci. Academic Press, New York, pp. 297–314.

Hurst, A. and G.W. Gould. 1984. The Bacterial Spore. Vol. 2. Academic Press, London.

Huss, V., K.H. Schleifer, E. Lindal, O. Schwan and C.J. Smyth. 1982. Peptidoglycan type, base composition of DNA, and DNA-DNA homology of *Peptococcus indolicus* and *Peptococcus asaccharolyticus*. FEMS Microbiol. Lett. *15:* 285–289.

Huss, V.A.R., H. Festl and K.H. Schleifer. 1984. Nucleic acid hybridization studies and deoxyribonucleic acid base composition of anaerobic Gram-positive cocci. Int. J. Syst. Bacteriol. *34:* 95–101.

Hutchinson, C.M. 1917. A bacterial disease of wheat in the Punjab. Mem. Dep. Agric. India Bacteriol. Ser. *1:* 169–175.

Hutchinson, M., K.I. Johnstone and D. White. 1967. Taxonomy of anaerobic thiobacilli. J. Gen. Microbiol. *47:* 17–23.

Hutton, R.M. and R.H. Behrens. 1979. *Actinomyces odontolyticus* as a cause of brain abscess. J. Infect. *1:* 195–197.

Hwang, M. and G.M. Ederer. 1975. Rapid hippurate hydrolysis method for presumptive identification of group B streptococci. J. Clin. Microbiol. *1:* 114–115.

Hylemon, P.B. and E.J. Stellwag. 1976. Bile acid biotransformation rates of selected gram-positive and gram-negative anaerobic bacteria. Biochem. Biophys. Res. Comm. *69:* 1088–1094.

Hyman, I.S. and S.D. Chaparas. 1977. A comparative study of the 'rhodochrous' complex and related taxa by delayed type skin reactions on guinea pigs and by polyacrylamide gel electrophoresis. J. Gen. Microbiol. *100:* 363–371.

Iacono, V.J., M.A. Taubman, D.J. Smith and M.J. Levine. 1975. Isolation and immunochemical characterization of the group-specific antigen of *Streptococcus mutans* 6715. Infect. Immun. *11:* 117–128.

Iandolo, J.J. and Z.J. Ordal. 1964. Germination system for endospores of *Sarcina ureae*. J. Bacteriol. *87:* 235–236.

Ichihara, K., E. Kusunose, M. Kusunose and T. Nori. 1977. Superoxide dismutase from *M. lepraemurium*. J. Biochem. *81:* 1427–1433.

Ihde, D.C. and D. Armstrong. 1973. Clinical spectrum of infection due to *Bacillus* species. Am. J. Med. *55:* 839–845.

Iida, H., K. Oguma and K. Inoue. 1974. Phage-conversion of toxigenicity in *Clostridium botulinum* types C and D. Jpn. J. Med. Sci. Biol. *27:* 101–102.

Iizuka, H. and K. Komagata. 1965. Microbiological studies on petroleum and natural gas. III. Determination of *Brevibacterium, Arthrobacter, Micrococcus, Sarcina, Alcaligenes* and *Achromobacter* isolated from oil-brines in Japan. J. Gen. Appl. Microbiol. *11:* 1–14.

Iizuka, H., H. Okazaki and N. Seto. 1969. A new sulfate-reducing bacterium isolated from antarctic. J. Gen. Appl. Microbiol. *15:* 11–18.

Ikeda, Y., H. Saito, K.J. Miura, J. Takagi and A. Aoki. 1965. DNA base composition, susceptibility to bacteriophages, and interspecific transformation as criteria for classification in the genus *Bacillus*. J. Gen. Appl. Microbiol. *11:* 181–190.

Ikin, G.J., H.J. Hope and R.A. Lachance. 1978. Des acides amines stimulateurs et inhibiteurs a la croissance de *Corynebacterium sepedonicum* (Spieck. et Knott., Skapt. et Burkh.) dans des miliex synthetiques. Can. J. Microbiol. *24:* 1087–1092.

Imaeda, T., L. Barksdale and W.F. Kirchheimer. 1982a. Deoxyribonucleic acid of *Mycobacterium lepraemurium*: its genome size, base ratio, and homology with those of other mycobacteria. Int. J. Syst. Bacteriol. *32:* 456–458.

Imaeda, T., W.F. Kirchheimer and L. Barksdale. 1982b. DNA isolated from *Mycobacterium leprae*: Genome size, base ratio, and homology with other related bacteria as determined by optical DNA-RNA reassociation. J. Bacteriol. *150:* 414–417.

Imanaka, I., T. Tanaka, H. Tsunekawa and S. Aiba. 1981. Cloning of the genes for penicillinase, *penP* and *penI*, of *Bacillus licheniformis* in some vector plasmids and their expression in *Escherichia coli, Bacillus subtilis* and *Bacillus licheniformis*. J. Bact. *147:* 776–786.

Imbriani, J.L. and R. Mankau. 1977. Ultrastructure of the nematode pathogen *Bacillus penetrans*. J. Invertebr. Pathol. *30:* 337–347.

Imhoff, D. and G. Schallehn. 1980. Charakterisierung von Bakteriophagen von *Clostridium novyi* Type A. Zentralbl. Bakteriol. Parasitenkd. Infektionskr. Hyg., I. Abt. Orig. A *247:* 101–113.

Imschenetski, A.A. 1959. Mikrobiologie der Cellulose. Akademie-Verlag, Berlin.

Ingham, E., K.T. Holland, G. Gowland and W.J. Cunliffe. 1979. Purification and partial characterization of hyaluronate lyase (E.C.4.2.2.1) from *Propionibacterium acnes*. J. Gen. Microbiol. *115:* 411–418.

Ingham, E., K.T. Holland, G. Gowland and W.J. Cunliffe. 1980. Purification and partial characterization of an acid phosphatase (E.C.3.1.3.2) produced by *Propionibacterium acnes*. J. Gen. Microbiol. *118:* 59–65.

Ingham, E., K.T. Holland, G. Gowland and W.J. Cunliffe. 1981. Partial purification and characterization of lipase (E.C.3.1.1.3) from *Propionibacterium*

acnes. J. Gen. Microbiol. *124:* 393–401.

Ingram, M. and R.H. Dainty. 1971. Changes caused by microbes in the spoilage of meats. J. Appl. Bacteriol. *34:* 21–39.

International Subcommittee on Taxonomy of Staphylococci and Micrococci. 1965. Recommendation. Int. Bull. Bacteriol. Nomencl. Taxon. *15:* 109–110.

International Committee on Systematic Bacteriology Subcommittee on the Taxonomy of *Mollicutes*. 1979. Proposal of minimal standards for descriptions of new species of the class *Mollicutes*. Int. J. Syst. Bacteriol. *29:* 172–180.

Interschick, E., F. Fiedler, K.H. Schleifer and O. Kandler. 1970. Glycine amide a constituent of the murein of *Arthrobacter atrocyaneus*. Zentralbl. Naturforsch. *25:* 714–717.

Ionesco, H. 1980. Transfert de la résistance á la tétracycline chez *Clostridium difficile*. Ann. Microbiol. (Inst. Pasteur) *131A:* 171–179.

Ionesco, H., G. Bieth, C. Dauguet and D. Bouanchaud. 1976. Isolement et identification de deux plasmides d'une souche bactériocinogéne de *Clostridium perfringens*. Ann. Microbiol. (Inst. Pasteur) *127B:* 283–294.

Ionesco, H. and A. Wolff. 1975. Sur le mode d'action de la bactériocine N₅ purifiée de *Clostridium perfringens*. C.R. Acad. Sci. Paris *281*(serie D): 2033–2036.

Isaacson, P., P.H. Jacobs, A.M.R. Mackenzie and A.W. Matthews. 1976. Pseudotumor of the lung caused by an infection with *Bacillus sphaericus*. J. Clin. Pathol. *29:* 806–811.

Isenberg, H.D., L.S. Lavine, B.G. Painter, W.H. Rubins and J.I. Berkman. 1975. Primary osteomyelitis due to an anaerobic microorganism. Am. J. Clin. Pathol. *64:* 385–388.

Isenberg, H.D., J.A. Washington II, A. Balows and A.C. Sonnenwirth. 1980. Collection, handling, and processing of specimens. *In* Lennette, Balows, Hausler and Truant (Editors), Manual of Clinical Microbiology, 3rd ed., American Society for Microbiology, Washington, D.C., pp. 52–82.

Ishaque, M., L. Kato and O.K. Skinsnes. 1977. Cytochrome-linked respiration in host grown *M. leprae* isolated from an armadillo. Int. J. Leprosy *45:* 114–119.

Ishimoto, M., M. Umeyama and S. Chiba. 1974. Alteration of fermentation products from butyrate to acetate by nitrate reduction in *Clostridium perfringens*. Z. Allg. Mikrobiol. *14:* 115–121.

Ishiwa, H. and S. Iwata. 1980. Drug resistance plasmids in *Lactobacillus fermentum*. J. Gen. Appl. Microbiol. *26:* 71–74.

Ito, H. 1977. Isolation of *Micrococcus radiodurans* occurring in radurized sawdust culture media for mushrooms. Agric. Biol. Chem. *41:* 35–41.

Ito, H., H. Watanabe, M. Takehisa and H. Iizuka. 1983. Isolation and identification of radiation-resistant cocci belonging to the genus *Deinococcus* from sewage sludges and animal feeds. Agric. Biol. Chem. *47:* 1239–1247.

Ivanov, I. 1962. Untersuchungen über die Listeriose der Schafe in Bulgarien. Monatsh. Veterinaermed. *17:* 729–736.

Ivanov, I. 1975. Establishment of non-motile strains of *Listeria monocytogenes* type 5. *In* Woodbine (Editor), Problems of Listeriosis, Leicester University Press, Leicester, pp. 18–26.

Iverson, O.J. and A. Grov. 1973. Studies on lysostaphin. Separation and characterization of three enzymes. Eur. J. Biochem. *38:* 293–300.

Iverson, W.G. and N.F. Millis. 1976. Bacteriocins of *Streptococcus bovis*. Can. J. Microbiol. *24:* 1040–1047.

Ivler, D. 1965. Comparative metabolism of virulent and avirulent staphylococci. Ann. N.Y. Acad. Sci. *128:* 62–80.

Jacob, A.E., G.J. Douglas and S.J. Hobbs. 1975. Self-transferable plasmids determining the hemolysin and bacteriocin of *Streptococcus faecalis* var. *zymogenes*. J. Bacteriol. *121:* 863–872.

Jacob, A.E. and S.J. Hobbs. 1974. Conjugal transfer of plasmid-borne multiple antibiotic resistance in *Streptococcus faecalis* var. *zymogenes*. J. Bacteriol. *117:* 360–372.

Jacobs, N.J. and P.J. Van DeMark. 1960. Comparison of the mechanism of glycerol oxidation in aerobically and anaerobically grown *Streptococcus faecalis*. J. Bacteriol. *79:* 532–538.

Jacques, N.A., L. Hardy, K.W. Knox and A.J. Wicken. 1980. Effect of Tween 80 on the morphology and physiology of *Lactobacillus salivarius* strain IV CL-37 grown in a chemostat under glucose limitation. J. Gen. Microbiol. *119:* 195–201.

Jager, J., K. Marialigeti, M. Hauck and G. Barabas. 1983. *Promicromonospora enterophila* sp. nov., a new species of monospore actinomycetes. Int. J. Syst. Bacteriol. *33:* 525–531.

Jagnow, G., K. Haider and P.-C. Ellwardt. 1977. Anaerobic dechlorination and degradation of hexachlorocyclohexane isomers by anaerobic and facultatively anaerobic bacteria. Arch. Microbiol. *115:* 285–292.

Janicka, I., M. Maliszewska and F. Pedziwilk. 1976. Utilization of lactose and production of corrinoids in selected strains of propionic acid bacteria in cheese, whey and casein media. Acta Microbiol. Pol. *25:* 205–210.

Janoschek, A. 1944. Zur Systematik der Propionsäurebakterien. Zentralbl. Bakteriol. Parasitenkd. Infektionskr. Hyg. Abt. 2. *106:* 321–337.

Jantzen, E., T. Bergan and K. Bøvre. 1974. Gas chromatography of bacterial whole cell methanolysates. VI. Fatty acid composition of strains within *Micrococcaceae*. Acta Pathol. Microbiol. Scand. Sec. B *82:* 785–798.

Jarvis, A.W. and B.D.W. Jarvis. 1981. DNA homology among lactic streptococci. Appl. Environ. Microbiol. *41:* 77–83.

Jarvis, B.D.W. 1967. Antigenic relations of cellulolytic cocci in the sheep rumen. J. Gen. Microbiol. *47:* 309–319.

Jasinska, S. 1964. Bacteriophages of lysogenic strains of *Listeria monocytogenes*.

Acta Microbiol. Pol. *13:* 29–44.

Jayne-Williams, D.J. 1976. The application of miniaturized methods for characterization of various organisms isolated from the animal gut. J. Appl. Bacteriol. *40:* 189–200.

Jayne-Williams, D.J. 1979. The bacterial flora of the rumen of healthy and bloating calves. J. Appl. Bacteriol. *47:* 271–284.

Jayne-Williams, D.J. and T.M. Skerman. 1966. Comparative studies on coryneform bacteria from milk and dairy sources. J. Appl. Bacteriol. *29:* 72–92.

Jebb, W.H.H. 1984. Starch fermenting, gelatine liquefying corynebacteria isolated from the human nose and throat. J. Pathol. Bacteriol. *60:* 403–412.

Jeffrey, C. 1977. Biological Nomenclature, 2nd Ed., Arnold, London.

Jeffries, L. 1968. Sensitivity to novobiocin and lysozyme in the classification of *Micrococcaceae.* J. Appl. Bacteriol. *31:* 436–442.

Jeffries, L. 1969. Menaquinones in the classification of *Micrococcaceae* with observations on the application of lysozyme and novobiocin sensitivity test. Int. J. Syst. Bacteriol. *19:* 183–187.

Jeffries, L., M.A. Cawthorne, M. Harris, B. Cook and A.T. Diplock. 1969. Menaquinone determination in the taxonomy of *Micrococcaceae.* J. Gen. Microbiol. *54:* 365–380.

Jelínková, J. 1982. Frequency of serotypes in group B streptococcus isolates: New type candidate. *In* Holm and Christensen (Editors), Basic Concepts of Streptococci and Streptococcal Diseases, Reedbooks Ltd., Windsor, pp. 50–51.

Jelínková, J. and V. Kubín. 1974. Proposal of a new serological group ("V") of hemolytic streptococci isolated from swine lymph nodes. Int. J. Syst. Bacteriol. *24:* 434–437.

Jeljaszewicz, J. 1972. Toxins (hemolysins). *In* Cohen (Editor), The Staphylococci. Wiley-Interscience (John Wiley and Sons, Inc.), New York, London, Sydney, Toronto, pp. 249–280.

Jeljaszewicz, J. 1973. *In* Jeljaszewicz (Editor), Staphylococci and Staphylococcal Infections. Karger, Basel.

Jellet, J.J., T.P. Forrest, I.A. Macdonald, T.J. Marrie and L.V. Holdeman. 1980. Production of indole-3-propanoic acid by 3-(p-hydroxyphenyl) propanoic acid in *Clostridium sporogenes*: a convenient thin-layer chromatography detection system. Can. J. Microbiol. *26:* 448–453.

Jenkins, P.A., J. Marks and W.B. Schaefer. 1971. Lipid chromatography and seroagglutination in the classification of rapidly growing mycobacteria. Am. Rev. Resp. Dis. *103:* 179–187.

Jenkins, P.A., J. Marks and W.B. Schaefer. 1972. Thin layer chromatography of mycobacterial lipids as an aid to classification. The scotochromogenic mycobacteria, including *Mycobacterium scrofulaceum, M. xenopi, M. aquae, M. gordonae, M. flavescens.* Tubercle *53:* 118–127.

Jensen, E.M. and H.W. Seeley. 1954. The nutrition and physiology of the genus *Pediococcus.* J. Bacteriol. *67:* 484–488.

Jensen, H.L. 1934. Studies on saprophytic mycobacteria and corynebacteria. Proc. Linn. Soc. NSW *59:* 19–61.

Jensen, H.L. and K. Gunderson. 1956. A soil bacterium decomposing organic nitro-compounds. Acta Agric. Scand. *6:* 100–114.

Jensen, J. 1963. Apocatalase of catalase-negative staphylococci. Science *141:* 45–46.

Jensen, J. and H. Kleemeyer. 1953. Die bakterielle Differential-Diagnose des Anthrax mittels eines neues spezifischen Testes ("Perlschnurtest"). Zentralbl. Bakteriol. Parasitenkd. Infektionskr. Hyg. Abt. I Orig. *159:* 494–500.

Jensen, R.A. and S.L. Stenmark. 1970. Comparitive allostery of 3-deoxy-D-arabino-heptulosonate-7-phosphate-synthetase as a molecular basis for classification. J. Bacteriol. *101:* 763–769.

Jensen, V. 1960. *Arthrobacter ramosus* spec. nov. A new *Arthrobacter* species isolated from forest soils. K. Vet. Landbohojsk. Arsskr. pp. 123–132.

Jensen, V. 1968. The plate count technique. *In* Gray and Parker (Editors), The Ecology of Soil Bacteria. Liverpool University Press, Liverpool, pp. 158–170.

John, J.F., P.K. Gramling, N.M. O'Dell. 1978. Species identification of coagulase-negative staphylococci from urinary tract isolates. J. Clin. Microbiol. *8:* 435–437.

Johne, A. and Frothingham. 1895. Ein eigenthümlicher Fall von Tuberculose beim Rind. Dtsch. Z. Tiermed. Verg. Pathol. *21:* 438–454.

Johnson, A.D. 1981. Production of biochemically different types of exfoliatin from two strains of *Staphylococcus aureus. In* Jeljasawicz (Editor), Staphylococci and Staphylococcal Infections, Gustav Fischer Verlag, Stuttgart, New York, pp. 305–309.

Johnson, D.W., J.R. Tagg and L.W. Wannamaker. 1979. Production of a bacteriocine-like substance by group-A streptococci of M-type 4 and T-pattern 4. J. Med. Microbiol. *12:* 413–427.

Johnson, E.A., A. Madia and A.L. Demain. 1981. Chemically defined minimal medium for growth of the anaerobic cellulolytic thermophile *Clostridium thermocellum.* Appl. Environ. Microbiol. *41:* 1060–1062.

Johnson, J.L. 1973. Use of nucleic-acid homologies in the taxonomy of anaerobic bacteria. Int. J. Syst. Bacteriol. *23:* 308–315.

Johnson, J.L. 1980. Classification of anaerobic bacteria. *In* Proceedings of International Symposium on Anaerobes (Tokyo, Japan, June 22, 1980), Nippon Merck-Banyu Co., Ltd., Tokyo, p. 19.

Johnson, J.L. 1981. Genetic characterization. *In* Gerhardt et al. (Editors), Manual of Methods for General Bacteriology, American Society for Microbiology, Washington, D.C., pp. 450–472.

Johnson, J.L. 1984. Nucleic acids in bacterial classification. *In* Krieg and Holt (Editors), Bergey's Manual of Systematic Bacteriology, Vol. 1. Williams and Wilkins, Baltimore, pp. 8–11.

Johnson, J.L. and D.A. Ault. 1978. Taxonomy of the *Bacteroides.* II. Correlation of phenotypic characteristics with deoxyribonucleic acid homology groupings for *Bacteroides fragilis* and other saccharolytic *Bacteroides* species. Int. J. Syst. Bacteriol. *28:* 257–265.

Johnson, J.L. and C.S. Cummins. 1972. Cell wall composition and deoxyribonucleic acid similarities among the anaerobic coryneforms, classical propionibacteria and strains of *Arachnia propionica.* J. Bacteriol. *109:* 1047–1066.

Johnson, J.L. and B.S. Francis. 1975. Taxonomy of the clostridia: ribosomal ribonucleic acid homologies among the species. J. Gen. Microbiol. *88:* 229–244.

Johnson, J.L. and E.J. Ordal. 1968. Deoxyribonucleic acid homology in bacterial taxonomy: effect of incubation temperature on reaction specificity. J. Bacteriol. *95:* 893–900.

Johnson, J.L., C.F. Phelps, C.S. Cummings and L. London. 1980. Taxonomy of the *Lactobacillus acidophilus* group. Int. J. Syst. Bacteriol. *30:* 53–68.

Johnson, W.D. 1976. Prosthetic valve endocarditis. *In* Kaye (Editor), Infectious Endocarditis, University Park Press, Baltimore, pp. 129–142.

Johnston, N.C. and H. Goldfine. 1983. Lipid composition in the classification of the butyric acid-producing clostridia. J. Gen. Microbiol. *129:* 1075–1081.

Jolly, R.D. 1965. The pathogenic action of the toxin of *Corynebacterium ovis.* J. Comp. Pathol. *75:* 417–431.

Jones, C.W. 1980. Cytochrome patterns in classification and identification including their relevance to the oxidase test. *In* Goodfellow and Board (Editors), Microbiological Classification and Identification. Symposium No. 8. Society for Applied Bacteriology. Academic Press, London, New York, pp. 127–138.

Jones, D. 1975a. A numerical taxonomic study of coryneform and related bacteria. J. Gen. Microbiol. *87:* 52–96.

Jones, D. 1975b. The taxonomic position of *Listeria. In* Woodbine (Editor), Problems of Listeriosis. Leicester University Press, Leicester, pp. 4–17.

Jones, D. 1978. Composition and differentiation of the genus *Streptococcus. In* Skinner and Quesnel (Editors), Streptococci, Society for Applied Bacteriology Symposium Series No. 7, Academic Press, London, New York and San Francisco, pp. 1–49.

Jones, D. 1978. An evaluation of the contribution of numerical taxonomy to the classification of the coryneform bacteria. *In* Bousfield and Callely (Editors), Coryneform Bacteria. Academic Press, London, pp. 13–46.

Jones, D., M.D. Collins, M. Goodfellow and D.E. Minnikin. 1979. Chemical studies in the classification of the genus *Listeria* and probably related bacteria. *In* Ivanov (Editor), Problems of Listeriosis. National Agroindustrial Union, Center for Scientific Information, Sofia, Bulgaria, pp. 17–24.

Jones, D., R.H. Deibel and C.F. Niven, Jr. 1963. Identity of *Staphylococcus epidermidis.* J. Bacteriol. *85:* 62–67.

Jones, D., R.H. Deibel and C.F. Niven, Jr. 1963. Apparent pigment production by *Streptococcus faecalis* in the presence of metal ions. J. Bacteriol. *86:* 171–172.

Jones, D., P.A. Pell and P.H.A. Sneath. 1984. Maintenance of bacteria on glass beads at -60°C to -76°C. *In* Kirsop and Snell (Editors) Maintenance of Microorganisms A Manual of Laboratory Methods. Academic Press, London, pp. 35–40.

Jones, D., M.J. Sackin and P.H.A. Sneath. 1972. A numerical taxonomic study of streptococci of serological group D. J. Gen. Microbiol. *72:* 439–450.

Jones, D. and H.P.R. Seeliger. 1983. Designation of a new type strain for *Listeria monocytogenes.* Int. J. Syst. Bacteriol. *33:* 429.

Jones, D. and P.M.F. Shattock. 1960. The location of the group antigen of group D streptococcus. J. Gen. Microbiol. *23:* 335–343.

Jones, D. and P.H.A. Sneath. 1970. Genetic transfer and bacterial taxonomy. Bacteriol. Rev. *34:* 40–81.

Jones, D., J. Watkins and S.K. Erickson. 1973. Taxonomically significant colour changes in *Brevibacterium linens* probably associated with a carotenoid-like pigment. J. Gen. Microbiol. *77:* 145–150.

Jones, D., J. Watkins and D.J. Meyer. 1970. Cytochrome composition and effect of catalase on growth of *Agromyces ramnosus.* Nature *226:* 1249–1250.

Jones, D.B. and N.M. Robinson. 1977. Anaerobic ocular infections. Tr. Am. Acad. Ophthalmol. Otolaryngol. *83:* 309–331.

Jones, D.T., A. van der Westhuizen, S. Long, E.R. Allcock, S.J. Reid and D.R. Woods. 1982. Solvent production and morphological changes in *Clostridium acetobutylicum.* Appl. Environ. Microbiol. *43:* 1434–1439.

Jones, K.L. 1949. Fresh isolates of actinomycetes in which the presence of sporogenous aerial mycelia is a fluctuating characteristic. J. Bacteriol. *57:* 141–145.

Jones, L.A. and S.G. Bradley. 1964. Phenetic classification of actinomycetes. Dev. Ind. Microbiol. *5:* 267–272.

Jones, L.E., W.A. Wirth and C.C. Farrow. 1960. Clostridial gas gangrene and septicemia complicating leukemia. South. Med. J. *53:* 863–866.

Jones, W.D., R.C. Good, N.J. Thompson and G.D. Kelly. 1982. Bacteriophage typing of *Mycobacterium tuberculosis* in the United States. Am. Rev. Resp. Dis. *125:* 640–643.

Jones, W.D. and G.P. Kubica. 1963. The differential typing of certain rapidly growing mycobacteria based on their sensitivity to various dyes. Am. Rev. Resp. Dis. *88:* 355–359.

Jones, W.D. and G.P. Kubica. 1964. The use of MacConkey's agar for differential typing of *Mycobacterium fortuitum*. Am. J. Med. Technol. *30:* 187–195.

Jones, W.D., Jr. and G.P. Kubica. 1965. Differential colonial characteristics of mycobacteria on oleic acid-albumin and modified corn meal agars. II. Investigation of rapidly growing mycobacteria. Zentralbl. Bakteriol. Parasitenkd. Infektionskr. Hyg. Abt. I Orig. *196:* 68–81.

Jones, W.D., Jr. and G.P. Kubica. 1968. Fluorescent antibody techniques with mycobacteria. III. Investigation of five serologically homogeneous groups of mycobacteria. Zentralbl. Bakteriol. Parasitenkd. Infektionskr. Hyg. Abt. I Orig. *207:* 58–62.

Jong, E.C., H.L. Ko and G. Pulverer. 1975. Studies on the bacteriophage of *Propionibacterium acnes*. Med. Microbiol. Immunol. *161:* 263–271.

Jonsson, P., O. Kinsman, O. Holmberg and T. Wadstrom. 1981. Virulence studies on coagulase-negative staphylococci in experimental infections: a preliminary report. *In* Jeljaszewicz (Editor), Staphylococci and Staphylococcal Infections, Gustav Fischer Verlag, Stuttgart, New York, pp. 661–666.

Jonsson, S., E. Clausen and J. Raa. 1983. Amino acid degradation by a *Lactobacillus plantarum* strain from fish. Syst. Appl. Microbiol. *4:* 148–154.

Jordan, D.C. and P.J. McNicol. 1979. A new nitrogen-fixing *Clostridium* species from a high Arctic ecosystem. Can J. Microbiol. *25:* 947–948.

Jordan, E.O. 1890. A report on certain species of bacteria observed in sewage. *In* Sedgewick, A report of the biological work of the Lawrence Experiment Station, including an account of methods employed and results obtained in the microscopical and bacteriological investigation of sewage and water. Report on water supply and sewerage (Part II) Rep. Mass. Bd. Publ. Hlth., pp. 821–844.

Jordan, H.V., S. Bellack, P.H. Keyes and M. Gerencser. 1974. Periodontal pathology and enamel caries in gnotobiotic rats infected with a unique serotype of *A. naeslundii*. J. Dent. Res. *53:* 73.

Jordan, H.V., R.J. Fitzgerald and H.R. Stanley. 1965. Plaque formation and periodontal pathology in gnotobiotic rats infected with an oral actinomycete. Am. J. Pathol. *47:* 1157–1167.

Jordan, H.V. and B.F. Hammond. 1972. Filamentous bacteria isolated from root surface lesions. Arch. Oral Biol. *17:* 1–12.

Jordan, H.V. and P.H. Keyes. 1964. Aerobic, gram-positive, filamentous bacteria as etiological agents of experimental periodontal disease in hamsters. Arch. Oral Biol. *9:* 401–414.

Jordan, H.V. and P.H. Keyes. 1965. Studies on the bacteriology of hamster periodontal disease. Am. J. Pathol. *46:* 843–857.

Jordan, P.A., A. Iravani, G.A. Richard and H. Baer. 1980. Urinary tract infection caused by *Staphylococcus saprophyticus*. J. Infect. Dis. *142:* 510–515.

Judical Commission. 1951. Opinion 3. The gender of generic names ending in -bacter. Int. Bull. Bacteriol. Nomencl. Taxon. *1:* 36–37.

Judicial Commission. 1958. Opinion 17. Conservation of the generic name *Staphylococcus* Rosenbach, designation of *Staphylococcus aureus* Rosenbach as the nomenclatural type of the genus *Staphylococcus* Rosenbach, and designation of a neotype culture of *Staphylococcus aureus* Rosenbach. Int. Bull. Bacteriol. Nemencl. Taxon. *8:* 153–154.

Judicial Commission. 1963. Opinion 30. Conservation of the specific epithet *faecalis* in the species name *Streptococcus faecalis* Andrewes and Horder 1906. Int. Bull. Bacteriol. Nomencl. Taxon. *13:* 167.

Judicial Commission. 1973. Opinion 47. Conservation of the specific epithet *avium* in the scientific name of the agent of avian tuberculosis. Int. J. Syst. Bacteriol. *23:* 472.

Judicial Commission. 1974. Opinion 51. Conservation of the epithet *fortuitum* in the combination *Mycobacterium fortuitum* da Costa Cruz. Int. J. Syst. Bacteriol. *24:* 552.

Judicial Commission. 1976. Opinion 52. Conservation of the generic name *Pediococcus* Claussen with the type species *Pediococcus damnosus* Claussen. Int. J. Syst. Bacteriol. *26:* 292.

Judicial Commission. 1978. Opinion 53. Rejection of the species name *Mycobacterium marianum* Penso 1953. Int. J. Syst. Bacteriol. *28:* 334.

Judicial Commission. 1979. Minute 25. Refection of the species *Peptococcus anaerobius*. Int. J. Syst. Bacteriol. *29:* 267–269.

Judicial Commission. 1982. Opinion 55. Rejection of the species name *Mycobacterium aquae* Jenkins et al., 1972. Int. J. Syst. Bacteriol. *32:* 467.

Judicial Commission. 1982. Opinion 56. Rejection of the name *Peptococcus anaerobius* (Hamm) Douglas 1957. Int. J. Syst. Bacteriol. *32:* 468.

Judicial Commission. 1983. Opinion 57. Designation of *Eubacterium limosum* (Eggerth) Prévot 1938 as the type species of *Eubacterium*. Int. J. Syst. Bacteriol. *33:* 434.

Judicial Commission. 1985 Opinion 58. Confirmation of the types in the Approved Lists as nomenclatural types including recognition of *Nocardia asteroides* (Eppinger 1891) Blanchard 1896 and *Pasteurella multocida* (Lehmann and Neumann 1899) Rosenbusch and Merchant 1939 as the respective type species of the genera *Nocardia* and *Pasteurella* and rejection of the species name *Pasteurella gallicida* (Burrill 1883) Buchanan 1925. Int. J. Syst. Bacteriol. *35:* 538.

Julák, J., M. Mára, F. Patočka, B. Potužnikova and S. Zadražil. 1978. Contribution to the taxonomy of haemolytic corynebacteria. Folia Microbiol. *23:* 229–235.

Jurtshuk, P. and J.-K. Liu. 1983. Cytochrome oxidase analyses of *Bacillus* strains: existence of oxidase-positive species. Int. J. Syst. Bacteriol. *33:* 887–891.

Justus, P.G., J.L. Martin, D.A. Goldberg, N.S. Taylor, J.G. Bartlett, R.W. Alexander and J.R. Mathias. 1982. Myoelectric effects of *Clostridium difficile*; motility-altering factors distinct from its cytotoxin and enterotoxin in rabbits. Gastroenterology *83:* 836–843.

Kadota, H., A. Uchida, Y. Sako and K. Harada. 1978. Heat-induced DNA injury in spores and vegetative cells of *Bacillus subtilis*. *In* Chambliss and Vary (Editors), Spores Volume VII. Am. Soc. Microbiol., Washington, D.C., pp. 27–30.

Kagan, B.M. 1972. L-forms. *In* Cohen (Editor), The Staphylococci. Wiley-Interscience (John Wiley and Sons, Inc.), New York, London, Sydney, Toronto, pp. 65–74.

Kagermeier, A. 1981. Taxonomie und Vorkommen von Milchsäurebakterien in Fleischprodukten. Thesis, University of München.

Kahn, C.M. 1924. Anaerobic spore-bearing bacteria of the human intestine in health and in certain diseases. J. Infect. Dis. *35:* 423–478.

Kalakoutskii, L.V. 1964. A new species of *Micropolyspora* - *Micropolyspora caesia* n. sp. Mikrobiologiya *33:* 858–862.

Kalakoutskii, L.V. and N.S. Agre. 1976. Comparative aspects of development and differentiation in actinomycetes. Bacteriol. Rev. *40:* 469–524.

Kalakoutskii, L.V., N.S. Agre and N.A. Krasil'nikov. 1968. Comparative study on some oligosporic actinomycetes. Hind. Antibiot. Bull. *10:* 254–268.

Kalakoutskii, L.V., I.P. Kirillova and N.A. Krasil'nikov. 1967. A new genus of the Actinomycetales - *Intrasporangium* gen. nov. J. Gen. Microbiol. *48:* 79–85.

Kalina, A.P. 1970. The taxonomy and nomenclature of enterococci. Int. J. Syst. Bacteriol. *20:* 185–189.

Kaltwasser, H. 1971. Studies on the physiology of *Bacillus fastidiosus*. J. Bacteriol. *107:* 780–786.

Kamisango, K., I. Saiki, Y. Tanio, H. Okomura, Y. Araki, I. Sekikawa, I. Azuma and Y. Yamamura. 1982. Structures and biological activities of peptidoglycans of *Listeria monocytogenes* and *Propionibacterium acnes*. J. Biochem. *92:* 23–33.

Kandler, O. 1981. Archaebakterien und Phylogenie der Organismen. Naturwissenschaften *68:* 183–192.

Kandler, O. 1982. Cell wall structures and their phylogenetic implications. Zentralbl. Bakteriol. Parasitenkd. Infektionskr. Hyg. Abt. Orig. C *3:* 149–160.

Kandler, O. 1983. Carbohydrate metabolism in lactic acid bacteria. Antonie van Leeuwenhoek J. Microbiol. Serol. *49:* 209–224.

Kandler, O. 1984. Current taxonomy of lactobacilli. Ind. Microbiol. *25:* 109–123.

Kandler, O. and I.G. Abo-Elnaga. 1966. Zur Taxonomy der Gattung *Lactobacillus* Beijerinck. IV. *L. corynoides* ein Synonym von *L. viridescens*. Zentralbl. Bakteriol. II Abt. *120:* 753–759.

Kandler, O., D. Claus and A. Moore. 1972. Die Aminosäuresequenz des Mureins von *Sarcina ventriculi* und *Sarcina maxima*. Arch. Mikrobiol. *82:* 140–146.

Kandler, O., H. König, J. Wiegel and D. Claus. 1983. Occurrence of poly-γ-D-glutamic acid and poly-α-L-glutamine in the genera *Xanthobacter*, *Flexithrix*, *Sporosarcina* and *Planococcus*. Syst. Appl. Microbiol. *4:* 34–41.

Kandler, O. and P. Kunath. 1983. *In* Validation of the publication of new names and new combinations previously effectively published outside the IJSB. List No. 11. Int. J. Syst. Bacteriol. *33:* 672–674.

Kandler, O. and P. Kunath. 1983. *Lactobacillus kefir* sp. nov., a component of the microflora of kefir. Syst. Appl. Microbiol. *4:* 286–294.

Kandler, O. and E. Lauer. 1974. Neuere Vorstellungen zur Taxonomie der Bifidobacterien. Zentralbl. Bakteriol. Parasitenkd. Infektionskr. Hyg. Abt. I. Orig. A *228:* 29–45.

Kandler, O., U. Schillinger and N. Weiss. 1983a. *Lactobacillus halotolerans* sp. nov., nom. rev. and *Lactobacillus minor* sp. nov., nom. rev. Syst. Appl. Microbiol. *4:* 280–285.

Kandler, O., U. Schillinger and N. Weiss. 1983b. *Lactobacillus bifermentans* sp. nov., nom. rev., an organism forming CO_2 and H_2 from lactic acid. Syst. Appl. Microbiol. *4:* 408–412.

Kandler, O., U. Schillinger and N. Weiss. 1983. *In* Validation of the publication of new names and new combinations previously effectively published outside the IJSB. List No. 11. Int. J. Syst. Bacteriol. *33:* 672–674.

Kandler, O., U. Schillinger and N. Weiss. 1983. *In* Validation of the publication of new names and new combinations previously effectively published outside the IJSB. List No. 12. Int. J. Syst. Bacteriol. *33:* 896–897.

Kandler, O. and K.H. Schleifer. 1980. Taxonomy I: Systematics of bacteria. *In* Ellenberg, Esser, Kubitzki, Schnepf and Ziegler (Editors), Progress in Botany, Fortschritte der Botanik, Vol. 42. Springer-Verlag, Berlin and Heidelberg, pp. 234–252.

Kandler, O., K.H. Schleifer and R. Dandl. 1968. Differentiation of *Streptococcus faecalis* Andrewes and Horder and *Streptococcus faecium* Orla-Jensen based on the amino acid composition of their murein. J. Bacteriol. *96:* 1934–1939.

Kandler, O. and S. Schoberth. 1979. Murein structure of *Acetobacterium woodii*. Arch. Microbiol. *120:* 181–183.

Kandler, O., K.O. Stetter and R. Köhl. 1980. *Lactobacillus reuteri* sp. nov., a new species of heterofermentative lactobacilli. Zentralbl. Bakteriol. Mikrobiol. Hyg. I Abt. Orig. C *1:* 264–269.

Kandler, O., K.O. Stetter and R. Köhl. 1982. *In* Validation of the publication of new names and new combinations previously effectively published outside the IJSB. List No. 8. Int. J. Syst. Bacteriol. *32:* 266–268.

Kaneda, T. 1977. Fatty acids of the genus *Bacillus*: an example of branched-chain preference. Bacteriol. Rev. *41:* 391–418.

Kaneko, T., K. Kitamura and Y. Yamamoto. 1969. *Arthrobacter luteus* nov. sp. isolated from brewery sewage. J. Gen. Appl. Microbiol. *15:* 317–326.

Kaneko, T., R. Nozaki and K. Aizawa. 1978. Deoxyribonucleic acid relatedness between *Bacillus anthracis*, *Bacillus cereus* and *Bacillus thuringiensis*. Microbiol. Immunol. *22:* 639–641.

Kaneuchi, C., Y. Benno and T. Mitsuoka. 1976b. *Clostridium coccoides*, a new species from the feces of mice. Int. J. Syst. Bacteriol. *26:* 482–486.

Kaneuchi, C., T. Miyazato, T. Shinjo and T. Mitsuoka. 1979. Taxonomic study of helically coiled, sporeforming anaerobes isolated from the intestines of humans and other animals: *Clostridium cocleatum* sp.nov. and *Clostridium spiroforme* sp.nov. Int. J. Syst. Bacteriol. *29:* 1–12.

Kaneuchi, C., K. Watanabe, A. Terada, Y. Benno and T. Mitsuoka. 1976a. Taxonomic study of *Bacteroides clostridiiformis* subsp. *clostridiiformis* (Burri and Ankersmit) Holdeman and Moore and of related organisms: proposal of *Clostridium clostridiiforme* (Burri and Andersmit) comb. nov. and *Clostridium symbiosum* (Stevens) comb. nov. Int. J. Syst. Bacteriol. *26:* 195–204.

Kannangara, D.W., T. Tanaka and H. Thadepalli. 1981. Spinal epidural abscess due to *Actinomyces israelii*. Neurology *31:* 202–204.

Kannenberg, E. and K. Poralla. 1980. A hapanoid from the thermo-acidophilic *Bacillus acidocaldarius* condenses membranes. Naturwissenschaften *67:* 458.

Kantz, W.E. and C.A. Henry. 1974. Isolation and classification of anaerobic bacteria from intact pulp chambers of non-vital teeth in man. Arch. Oral Biol. *19:* 91–96.

Kanzaki, T., Y. Sugiyama, K. Kitano, Y. Ashida and I. Imada. 1974. Quinones of *Brevibacterium*. Biochem. Biophys. Acta *348:* 162–165.

Kaplan, M.H. and J.E. Volanakis. 1974. Interaction of C-reactive protein complexes with the complement system. I. Consumption of human complement associated with the reaction of C-reactive protein with pneumococcal C-polysaccharide and with the choline phosphatides, lecithin and sphingomyelin. J. Immunol. *112:* 2135–2147.

Käppler, W. 1965a. Acetyl-naphthylamin-esterasen-activität von Mykobakterien. Beitr. Klin. Tuberk. Spezif. Tuberk-Forsch. *130:* 1–4.

Käppler, W. 1965b. Zur Differenzierung von Mykobakterien mit dem Phosphatase-test. Beitr. Klin. Tuberk. Spezif. Tuberk-Forsch. *130:* 223–226.

Karakawa, W.W., J.E. Wagner and J.H. Pazur. 1971. Immunochemistry of the cell-wall carbohydrate of group L hemolytic streptococci. J. Immunol. *107:* 554–562.

Karetzky, M.S. and J.W. Garvey. 1974. Empyema due to *Actinomyces naeslundii*. Chest *65:* 229–230.

Karlson, A.G. and E.F. Lessel. 1970. *Mycobacterium bovis* nom. nov. Int. J. Syst. Bacteriol. *20:* 273–282.

Karlson, A.G. and I.A. Merchant. 1941. The cultural and biochemic properties of *Erysipelothrix rhusiopathiae*. Am. J. Vet. Res. *2:* 5–10.

Kasweck, K.L. and M.L. Little. 1982. Genetic recombination in *Nocardia asteroides*. J. Bacteriol. *149:* 403–406.

Kasweck, K.L., M.L. Little and S.G. Bradley. 1982. Plasmids in mating strains of *Nocardia asteroides*. Dev. Ind. Microbiol. *23:* 279–286.

Katagiri, H., K. Kitahara and K. Fukami. 1934. The characteristics of the lactic acid bacteria isolated from moto, yeast mashes for saké manufacture. IV. Classification of the lactic acid bacteria. Bull. Agr. Chem. Soc. Jpn. *10:* 156–157.

Katayama, A., E. Ishikawa, T. Ando and T. Arai. 1978. Isolation of plasmid DNA from naturally-occurring strains of *Streptococcus mutans*. Arch. Oral Biol. *23:* 1099–1103.

Kates, M. 1978. The phytanyl ether-linked polar lipids and isoprenoid neutral lipids of extremely halophilic bacteria. Prog. Chem. Fats Other Lipids *15:* 301–342.

Katitch, R., Z. Voukitchevitch, B. Djoukitch, V. Miljkovitch and B. Tchavitch. 1964. Recherches sur la pathologénie de la gangrène gazeuse lors d'infection provoquée par *W. perfringens* A, seul ou association avec *Cl. sporogenes* et avec *Cl. sporogenes* et *Staphylococcus aureus*. Rev. Immunol. (Paris) *28:* 63–74.

Kato, F., M. Yoshimi, K. Araki, Y. Motomura, Y. Matsufune, H. Nobunaga and A. Murata. 1984. Screening of bacteriocins in amino acid or nucleic acid producing bacteria and related species. Agric. Biol. Chem. *48:* 193–200.

Katoch, V.M., L.G. Wayne and G.A. Diaz. 1982a. Characterization of catalase in tissue-derived *Mycobacterium lepraemurium*. Int. J. Syst. Bacteriol. *32:* 416–418.

Katoch, V.M., L.G. Wayne and G.A. Diaz. 1982b. Serological approaches for the characterization of catalase in tissue-derived mycobacteria. Ann. Microbiol. *138B:* 407–414.

Katznelson, H. 1950. *Bacillus pulvifaciens* (n. sp.) an organism associated with powdery scale of honeybee larvae. J. Bacteriol. *59:* 153–155.

Katznelson, H. 1955. The metabolism of phytopathogenic bacteria. I. Comparitive studies on the metabolism of representative species. J. Bacteriol. *70:* 469–475.

Katznelson, H. 1955. *Bacillus apiarius* n.sp., an aerobic spore-forming organism isolated from honeybee larvae. J. Bacteriol. *70:* 635–636.

Katznelson, H. 1958. The metabolism of phytopathogenic bacteria. II. Metabolism of carbohydrates by cell-free extracts. J. Bacteriol. *75:* 540–543.

Katznelson, H. and A.G. Lockhead. 1947. Nutritional requirements of *Bacillus alvei* and *Bacillus para-alvei*. J. Bacteriol. *53:* 83–88.

Kauffmann, F., E. Lund and B.E. Eddy. 1960. Proposals for a change in the nomenclature of *Diplococcus pneumoniae* and a comparison of the Danish and American type designations. Int. Bull. Bacteriol. Nomencl. Taxon. *10:* 31–40.

Kaufmann, A., J. Fegan, P. Doleac, C. Gainer, D. Wittech and A. Glann. 1976. Identification and characterization of a cellulolytic isolate. J. Gen. Microbiol. *94:* 405–408.

Kaustova, J., Z. Olsovsky, M. Kubin, O. Zatloukal, M. Pelikan and V. Hradil. 1981. Endemic occurrence of *Mycobacterium kansasii* in water-supply systems. J. Hyg. Epidemiol. Microbiol. Immunol. *25:* 24–30.

Kawabata, N. 1980. Studies on the sulfite reduction test for clostridia. Microbiol. Immunol. *24:* 271–279.

Kaye, D. 1976. Infecting microorganisms. *In* Kaye (Editor), Infective Endocarditis, University Park Press, Baltimore, pp. 43–54.

Kazda, J. 1967. Mykobakterien im Trinkwasser als Ursache der Parallergie gegenüber Tuberkulinen bei Tieren. III. Mitterlung: Taxonomische Studie einiger rasch wachsender Mykobakterien und Beschreibung einer neuen Art: *Mycobacterium brunense* n. sp. Zentralbl. Bakteriol. Parasitenkd. Infektionskr. Hyg. Abt. I Orig. *203:* 199–211.

Kazda, J. 1980. *Mycobacterium sphagni* sp. nov. Int. J. Syst. Bacteriol. *30:* 77–81.

Kazda, J. and K. Muller. 1979. *Mycobacterium komossense* sp. nov. Int. J. Syst. Bacteriol. *29:* 361–365.

Keddie, R.M. 1949. A study of *Bacterium zopfii* Kurth. Dissertation. Edinburgh School of Agriculture, Edinburgh, Scotland.

Keddie, R.M. 1959. The properties and classification of lactobacilli isolated from grass and silage. J. Appl. Bacteriol. *22:* 403–416.

Keddie, R.M. 1974. Genus *Cellulomonas*. *In* Buchanan and Gibbons (Editors), Bergey's Manual of Determinative Bacteriology, 8th ed. The Williams and Wilkins Co., Baltimore, pp. 629–631.

Keddie, R.M. 1974. Genus *Arthrobacter*. *In* Buchanan and Gibbons (Editors), Bergey's Manual of Determinative Bacteriology, 8th ed. The Williams and Wilkins Co., Baltimore, pp. 618–625.

Keddie, R.M. 1978. What do we mean by coryneform bacteria? *In* Bousfield and Callely (Editors), Coryneform Bacteria. Academic Press, London, pp. 1–12.

Keddie, R.M. 1981. The Genus *Kurthia*. *In* Starr, Stolp, Balows and Schlegel (Editors), The Prokaryotes. A Handbook on Habitats, Isolation and Identification of Bacteria. Springer-Verlag, Berlin. pp. 1888–1893.

Keddie, R.M. and I.J. Bousfield. 1980. Cell wall composition in the classification and identification of coryneform bacteria. *In* Goodfellow and Board (Editors), Microbial Classification and Identification. Academic Press, London, New York, San Francisco, pp. 167–188.

Keddie, R.M. and G.L. Cure. 1977. The cell wall composition and distribution of free mycolic acids in named strains of coryneform bacteria and in isolates from various natural sources. J. Appl. Bacteriol. *42:* 229–253.

Keddie, R.M. and G.L. Cure. 1978. Cell wall composition of coryneform bacteria. *In* Bousfield and Callely (Editors), Coryneform Bacteria. Academic Press, London, pp. 47–83.

Keddie, R.M. and D. Jones. 1981. Aerobic, saprophytic coryneform bacteria. *In* Starr, Stolp, Trüper, Balows and Schlegel (Editors), The Prokaryotes. A Handbook on Habitats, Isolation and Identification of Bacteria. Springer-Verlag, New York, pp. 1838–1878.

Keddie, R.M. and D. Jones. 1981. The genus *Brochothrix* (formerly *Microbacterium thermosphactum*, McLean and Sulzbacher). *In* Starr, Stolp, Trüper, Balows and Schlegel (Editors), The Prokaryotes. A Handbook on Habitats, Isolation and Identification of Bacteria. Springer-Verlag, Berlin, pp. 1866–1869.

Keddie, R.M., B.G.S. Leask and J.M. Grainger. 1966. A comparison of coryneform bacteria from soil and herbage: cell wall composition and nutrition. J. Appl. Bacteriol. *29:* 17–43.

Keddie, R.M. and M.Rogosa. 1974. Genus *Kurthia* Trevisan. *In* Buchanan and Gibbons (Editors), Bergey's Manual of Determinative Bacteriology, 8th ed. The Williams and Wilkins Co., Baltimore, pp. 631–632.

Keeler, R.F. and M.L. Gray. 1960. Antigenic and related biochemical properties of *Listeria monocytogenes*. I. Preparation and composition of cell wall material. J. Bacteriol. *80:* 683–692.

Keir, H.A. and J.W. Porteus. 1962. The amino acid requirements of a single strain of *Actinomyces israelii* growing in a chemically defined medium. J. Gen. Microbiol. *28:* 193–201.

Kellerman, F.K. and I.G. McBeth. 1912. The fermentation of cellulose. Zentralbl. Bakteriol. Parasitenkd. Infektionskr. Hyg. Abt. II *34:* 485–494.

Kellerman, F.K., I.G. McBeth, F.M. Scales and N.R. Smith. 1913. Identification and classification of cellulose-dissolving bacteria. Zentralbl. Bakteriol. Parasitenkd. Infektionskr. Hyg. Abt. II *39:* 502–522.

Kelly, K.F. and J.B. Evans. 1974. Deoxyribonucleic acid homology among strains of the lobster pathogen "*Gaffkya homari*" and *Aerococcus viridans*. J. Gen. Microbiol. *81:* 257–260.

Kelstrup, J. and R.J. Gibbons. 1969. Bacteriocins from human and rodent streptococci. Arch. Oral Biol. *14:* 251–258.

Kelstrup, J., S. Richmond, C. West and R.J. Gibbons. 1970. Fingerprinting human oral streptococci by bacteriocin production and susceptibility. Arch. Oral Biol. *15:* 1109–1116.

Kenner, B.A., H.F. Clark and P.W. Kabler. 1961. Fecal Streptococci. I. Cultivation and enumeration of streptococci in surface waters. Appl. Microbiol. *9:* 15–20.

Kerbaugh, M.A. and J.B. Evans. 1968. *Aerococcus viridans* in the hospital environment. Appl. Microbiol. *16:* 519–523.

Kerby, R. and J.G. Zeikus. 1983. Growth of *Clostridium thermoaceticum* on H_2CO_2 or CO as energy source. Curr. Microbiol. *8:* 27–30.

Kerrin, J.C. 1928. The incidence of *B. tetani* in human feces. Br. J. Exp. Pathol. *9:* 69–71.

Kersters, K. and J. De Ley. 1968. The occurrence of the Entner-Doudoroff pathway in bacteria. Antonie van Leeuwenhoek J. Microbiol. Serol. *34:* 393–408.

Kersters, K. and J. De Ley. 1975. Identification and grouping of bacteria by numerical analysis of their electrophoretic protein patterns. J. Gen. Microbiol. *87:* 333–342.

Kersters, K. and J. DeLey. 1980. Classification and identification of bacteria by electrophoresis of their proteins. *In* Goodfellow and Board (Editors), Microbiological Classification and Identification. Academic Press, London, pp. 273–297.

Kestle, G., D. Abbott and G.P. Kubica. 1967. Differential identification of mycobacteria. II. Subgroups of II and III (Runyon) with different clinical significance. Am. Rev. Resp. Dis. *95:* 1041–1052.

Khan, A.W. and W.D. Murray. 1982. Influence of *Clostridium saccharolyticum* on cellulose degradation by *Acetivibrio cellulolyticus*. J. Appl. Bacteriol. *53:* 379–383.

Khanolkar, S.R. 1982. Preliminary studies of the metabolic activity of purified suspensions of *Mycobacterium leprae*. J. Gen. Microbiol. *128:* 423–425.

Kiel, R.A. and J.M. Tanzer. 1977. Regulation of invertase of *Actinomyces viscosus*. Infect. Immun. *17:* 510–512.

Kiel, R.A., J.M. Tanzer and F.N. Woodiel. 1977. Identification, separation, and preliminary characterization of invertase and β-galactosidase in *Actinomyces viscosus*. Infect. Immun. *16:* 81–87.

Kieser, T., G. Hintermann, R. Crameri and R. Hütter. 1981. Restriction analysis of *Streptomyces* - DNA. Zentralbl. Bakteriol. I. Abt. Suppl. *11:* 561–562.

Kilian, M. 1978. Rapid identification of *Actinomycetaceae* and related bacteria. J. Clin. Microbiol. *8:* 127–133.

Kilpper, R., U. Buhl and K.H. Schleifer. 1980. Nucleic acid homology studies between *Peptococcus saccharolyticus* and various anaerobic and facultative anaerobic Gram-positive cocci. FEMS Microbiol. Lett. *8:* 205–210.

Kilpper-Bälz, R., G. Fischer and K. Schleifer. 1982. Nucleic acid hybridisation of group N and group D streptococci. Curr. Microbiol. *7:* 245–250.

Kilpper-Bälz, R. and K.H. Schleifer. 1981a. Transfer of *Peptococcus saccharolyticus* Foubert and Douglas to the genus *Staphylococcus*: *Staphylococcus saccharolyticus* (Foubert and Douglas) comb. nov. Zentralbl. Bakteriol. Mikrobiol. Hyg. Abt. I Orig. C *2:* 324–331.

Kilpper-Bälz, R. and K.H. Schleifer. 1981b. DNA-γRNA hybridization studies among staphylococci and some other gram-positive bacteria. FEMS Microbiol. Lett. *10:* 357–362.

Kilpper-Bälz, R. and K.H. Schleifer. 1984. Nucleic acid hybridization and cell wall composition studies of pyogenic streptococci. FEMS Microbiol. Lett. *24:* 355–364.

Kilpper-Bälz, R., B.L. Williams, R. Lütticken and K.H. Schleifer. 1984. Relatedness of 'Streptococcus milleri' with *Streptococcus anginosus* and *Streptococcus constellatus*. Syst. Appl. Microbiol. *5:* 494–500.

King, S. and E. Meyer. 1957. Metabolic and serologic differentiation of *Actinomyces bovis* and anaerobic diphtheroids. J. Bacteriol. *74:* 234–238.

King, S. and E. Meyer. 1963. Gel diffusion technique in antigen-antibody reactions of *Actinomyces* species and "anaerobic diphtheroids. J. Bacteriol. *85:* 186–190.

Kinoshita, S., S. Nakayama and S. Akita. 1958. Taxonomic study of glutamic acid accumulating bacteria, *Micrococcus glutamicus*, nov. sp. Bull. Agric. Chem. Soc. Japan *22:* 176–185.

Kinouchi, T., K. Takumi and T. Kawata. 1981. Characterization of two inducible bacteriophages, α1 and α2, isolated from *Clostridium botulinum* type A 190L and their deoxyribonucleic acids. Microbiol. Immunol. *25:* 915–927.

Kirchheimer, W.F. and E.E. Storrs. 1971. Attempts to establish the armadillo (Dasypus novemcinctus Linn.) as a model for the study of leprosy. (1.) Report of lepromatoid leprosy in an experimentally infected armadillo. Int. J. Leprosy *39:* 693–702.

Kiritani, K., N. Mitsui, S. Nakamura and S. Nishida. 1973. Numerical taxonomy of *Clostridium botulinum* and *Clostridium sporogenes* and their susceptibilities to induced lysins and to mitomycin C. Jap. J. Microbiol. *17:* 361–372.

Kirsop, B.E. and J.J.S. Snell. 1984. Maintenance of Microorganisms. A Manual of Laboratory Methods. Academic Press, London.

Kirsop, B.H. and L. Dolezil. 1975. Detection of lactobacilli in brewing. *In* Carr, Cutting and Whiting (Editors), Lactic acid bacteria in beverages and food. Academic Press, London, New York, San Francisco, pp. 159–164.

Kishishita, M., T. Ushijima, Y. Ozaki and Y. Ito. 1978. Biotyping of *Propionibacterium acnes* isolated from normal human facial skin. Appl. Environ. Microbiol. *38:* 585–589.

Kishishita, M., T. Ushijima, Y. Ozaki and Y. Ito. 1980. New medium for isolating propionibacteria, and its application to assay of normal flora of human facial skin. Appl. Environ. Microbiol. *40:* 1100–1105.

Kistner, A. 1960. An improved method for viable counts of bacteria of the ovine rumen which ferment carbohydrates. J. Gen. Microbiol. *23:* 565–576.

Kistner, A. and L. Gouws. 1964. Cellulolytic cocci occurring in the rumen of sheep conditioned to lucerne hay. J. Gen. Microbiol. *34:* 447–458.

Kitahara, K. 1938. Studies in lactic acid-forming bacteria in milk and milk products. J. Agr. Chem. Soc. Jpn. *14:* 1449–1465.

Kitahara, K. 1949. A new type lactic acid bacteria. Bull. Res. Inst. Food Sci. Kyoto Univ. *2:* 23–36.

Kitahara, K. 1974. Genus II. *Sporolactobacillus. In* Buchanan and Gibbons (Editors), Bergey's Manual of Determinative Bacteriology, 8th ed. The Williams and Wilkins Co., Baltimore, p. 550–554.

Kitahara, K., T. Kaneko and O. Goto. 1957. Taxonomic studies on the hiochi-bacteria, specific saprophytes of saké. I. Isolation and grouping of bacterial strains. J. Gen. Appl. Microbiol. *3:* 102–110.

Kitahara, K., T. Kaneko and O. Goto. 1957. Taxonomic studies on the hiochi-bacteria, specific saprophytes of saké. II. Identification and classification of hiochi-bacteria. J. Gen. Appl. Microbiol. *3:* 111–120.

Kitahara, K. and C.L. Lai. 1967. On the spore formation of *Sporolactobacillus inulinus*. J. Gen. Appl. Microbiol. *13:* 197–203.

Kitahara, K. and A. Nakagawa. 1958. *Pediococcus mevalorus* nov. sp. isolated from beer. J. Gen. Appl. Microbiol. *4:* 21–30.

Kitahara, K. and J. Suzuki. 1963. *Sporolactobacillus* Nov. Subgen. J. Gen. Appl. Microbiol. *9:* 59–71.

Kitahara, K. and T. Toyota. 1972. Auto-spheroplastization in *Sporolactobacillus inulinus*. J. Gen. Appl. Microbiol. *18:* 99–107.

Kitchell, A.G. and B.G. Shaw. 1975. Lactic acid bacteria in fresh and cured meat. *In* Carr, Cutting and Whiting (Editors), Lactic acid bacteria in beverages and food. Academic Press, London, New York, San Francisco, pp. 209–220.

Kitt, T. 1893. Bakterienkunde und pathologisch mikroskopie für Thierärzte und studirende de thiermedicin. Moritz Perles, Wein. *2:* 1–450.

Klaenhammer, T.R. and E.G. Kleeman. 1981. Growth characteristics, bile sensitivity, and freeze damage in colonial variants of *Lactobacillus acidophilus*. Appl. Environ. Microbiol. *41:* 1461–1467.

Klaushofer, H. and F. Hollaus. 1970. Zur Taxonomie der hoch-thermophilen, in Zuckerfabriksäften vorkommenden aeroben Sporenbildner. Z. Zuckerind. *20:* 465–470.

Klaushofer, H. and E. Parkkinen. 1965. Zur Frage der Bedeutung aerober und anaerober thermophiler Sporgenbildner als Infektionsurasache in Rübenzucker-fabriken. I. *Clostridium thermohydrosulfuricum* eine neue Art eines saccharoseabbauenden, thermophilen, schwefelwasserstoffbilden Clostridiums. Z. Zukerindustr. (Boehmen) *15:* 445–449.

Klein, E. 1884. Micro-organisms and disease. Practitioner *33:* 21–40.

Klein, E. 1899. Ein Beitrag zur Bakteriologie der Leichenverwesung. Zentralbl. Bakteriol. Parasitenkd. Infektionskr. Hyg., Abt. I. Orig. *25:* 278–284.

Klein, E. 1904. Ein neuer tierpathogenen Mikrobe - *Bacillus carnis*. Zentralbl. Bakteriol Parasitenkd. Infektionskr. Hyg., Abt. I. Orig. *35:* 450–461.

Klein, J.P. and R.M. Frank. 1973. Mise en évidence de virus dans les bactérias cariogenes de la plaque dentaire. J. Biol. Buccale *1:* 79–85.

Klein, R.A., G.P. Hazelwood, P. Kemp and R.M.C. Dawson. 1979. A new series of long chain, dicarboxylic acids with vicinal dimethyl branching found as major components of the lipids of *Butyrivibrio* spp. Biochem. J. *183:* 691–700.

Kleiner, D. 1979. Regulation of ammonium uptake and metabolism by nitrogen fixing bacteria. III. *Clostridium pasteurianum*. Arch. Microbiol. *120:* 263–270.

Klesius, P.H. and V.T. Schuhardt. 1968. Use of lysostaphin in the isolation of highly polymerized deoxyribonucleic acid and in the taxonomy of aerobic *Micrococcaceae*. J. Bacteriol. *95:* 739–743.

Kline, L. and T.F. Sugihara. 1971. Microorganisms of the San Francisco sour dough bread process. II. Isolation and characterization of undescribed bacterial species responsible for the souring activity. Appl. Microbiol. *21:* 459–465.

Kloos, W.E. 1980. Natural populations of the genus *Staphylococcus*. Annu. Rev. Microbiol. *34:* 559–592.

Kloos, W.E. 1982. Coagulase-negative staphylococci. Clin. Microbiol. Newsl. *4:* 75–79.

Kloos, W.E. and M.S. Musselwhite. 1975. Distribution and persistence of *Staphylococcus* and *Micrococcus* species and other aerobic bacteria on human skin. Appl. Microbiol. *30:* 381–395.

Kloos, W.E., M.S. Musselwhite and R.J. Zimmermann. 1976. A comparison of the distribution of *Staphylococcus* species on human and animal skin. *In* Jeljaszewicz (Editor), Staphylococci and Staphylococcal Diseases. Gustav Fischer Verlag, Stuttgart, New York, pp. 967–973.

Kloos, W.E., B.S. Orban and D.D. Walker. 1981. Plasmid composition of *Staphylococcus* species. Can. J. Microbiol. *27:* 271–278.

Kloos, W.E. and K.H. Schleifer. 1975a. Isolation and characterization of staphylococci from human skin. II. Description of four new species: *Staphylococcus warneri*, *Staphylococcus capitis*, *Staphylococcus hominis*, and *Staphylococcus simulans*. Int. J. Syst. Bacteriol. *25:* 62–79.

Kloos, W.E. and K.H. Schleifer. 1975b. Simplified scheme for routine identification of human *Staphylococcus* species. J. Clin. Microbiol. *1:* 82–88.

Kloos, W.E. and K.H. Schleifer. 1981. The genus *Staphylococcus. In* Starr, Stolp, Trüper, Balows, and Schlegel (Editors). The Prokaryotes. A handbook on habitats, isolation, and identification of bacteria. Springer Verlag, Berlin, Heidelberg, pp. 1548–1569.

Kloos, W.E. and K.H. Schleifer. 1983. *Staphylococcus auricularis* sp. nov.: an inhabitant of the human external ear. Int. J. Syst. Bacteriol. *33:* 9–14.

Kloos, W.E., K.H. Schleifer and W.C. Noble. 1976. Estimation of character parameters in coagulase-negative *Staphylococcus* species. *In* Jeljaszewicz (Editor), Staphylococci and Staphylococcal Diseases, Gustav Fischer Verlag, Stuttgart, New York, pp. 23–41.

Kloos, W.E., K.H. Schleifer and R.F. Smith. 1976. Characterization of *Staphy-

lococcus sciuri sp. nov. and its subspecies. Int. J. Syst. Bacteriol. *26:* 22–37.

Kloos, W.E. and P.B. Smith. 1980. Staphylococci. *In* Lennette, Balows, Hausler and Truant (Editors), Manual of Clinical Microbiology, 3rd ed., American Society for Microbiology, Washington, D.C., pp. 83–87.

Kloos, W.E., T.G. Tornabene and K.H. Schleifer. 1974. Isolation and characterization of micrococci from human skin, including two new species.: *Micrococcus lylae* and *Micrococcus kristinae*. Int. J. Syst. Bacteriol. *24:* 79–101.

Kloos, W.E. and J.F. Wolfshohl. 1979. Evidence for deoxyribonucleotide sequence divergence between staphylococci living on human and other primate skin. Curr. Microbiol. *3:* 167–172.

Kloos, W.E. and J.F. Wolfshohl. 1982. Identification of *Staphylococcus* species with the API STAPH-IDENT system. J. Clin. Microbiol. *16:* 509–516.

Kloos, W.E. and J.F. Wolfshohl. 1983. Deoxyribonucleotide sequence divergence between *Staphylococcus cohnii* subspecies populations living on primate skin. Curr. Microbiol. *8:* 115–121.

Kloos, W.E., J.F. Wolfshohl and V. Marquardt. 1983. Effect of penicillin therapy on the antibiotic susceptibility of *Staphylococcus* species living on human skin. Abstr. Annu. Meet. Am. Soc. Microbiol. *1983:* 13.

Kloos, W.E., R.J. Zimmerman and R.F. Smith. 1976. Preliminary studies on the characterization and distribution of *Staphylococcus* and *Micrococcus* species on animal skin. Appl. Environ. Microbiol. *31:* 53–59.

Klopotek, A., von. 1969. Über die Drehrichtung einiger Stämme von *Bacillus cereus* var. *mycoides*. Zentralbl. Bakteriol. Parasitenkd. Infektionskr. Hyg. Abt. II *123:* 683–684.

Kluyver, A.J. 1931. The chemical activities of microorganisms. University of London Press, Ltd. London.

Kluyver, A.J. and C.B. Van Neil. 1936. Prospects for a natural classification of bacteria. Zentralbl. Bakteriol. Parasitenkd. Infektionskr. Hyg. Abt. II *94:* 369–403.

Knackmuss, H.J. and W. Beckman. 1973. The structure of nicotine blue from *Arthrobacter oxidans*. Arch. Mikrobiol. *90:* 167–169.

Knaysi, G. 1951. Elements of bacterial cytology. Comstock Publishing Company, Ithaca, N.Y.

Knight, B.C.J.G. and H. Proom. 1950. A comparitive survey of the nutrition and physiology of mesophilic species in the genus *Bacillus*. J. Gen. Microbiol. *4:* 508–538.

Knight, H.D. 1969. Corynebacterial infections in the horse: problems of prevention. J. Am. Vet. Med. Assoc. *155:* 446–452.

Knisely, R.F. 1965. Differential media for the identification of *Bacillus anthracis*. J. Bacteriol. *90:* 1778–1783.

Knisely, R.F. 1966. Selective medium for *Bacillus anthracis*. J. Bacteriol. *92:* 784–786.

Knivett, V.A., J.Cullen and M.J. Jackson. 1965. Odd numbered fatty acids in *Micrococcus radiodurans*. Biochem. J. *96:* 2c-3c.

Knöll, H. 1965. Zur Biologie der Gärungssarcinen. Monatsber. Dtsch. Akad. Wiss. Berl. *7:* 475–477.

Knöll, H. and R. Horschak. 1964. Zur Ernährungs-physiologie der Gärungssarcinen. Monatsber. Dtsch. Akad. Wiss. Berl. *6:* 847–849.

Knöll, H. and R. Horschak. 1971. Zur Sporulation der Gärungssarcinen. Monatsber. Dtsch. Akad. Wiss. Berl. *13:* 222–224.

Knoop, F.C. 1979. Clindamycin-associated enterocolitis in guinea pigs: evidence for a bacterial toxin. Infect. Immun. *23:* 31–33.

Knox, K.W., L.K. Campbell, J.D. Evans and A.J. Wicken. 1980. Identification of the group G antigen of lactobacilli. J. Gen. Microbiol. *119:* 203–209.

Knox, K.W. and E.A. Hall. 1964. The relationship between the capsular and cell wall polysaccharides of strains of *Lactobacillus casei* var. *rhamnosus*. J. Gen. Microbiol. *37:* 433–438.

Knox, K.W. and A.J. Wicken. 1973. Immunological properties of teichoic acids. Bacteriol. Rev. *37:* 215–257.

Knox, K.W. and A.J. Wicken. 1976. Grouping and cross-reacting antigens of oral lactic acid bacteria. J. Dent. Res. *55:* A116-A122.

Knudsen, S. 1924. Über die Milchsäurebakterien des Sauerteiges und ihre Bedeutung für die Sauerteiggärung. Arsskr. K. Vet.-Landbohoisk.

Knudsen, S. and A. Sørensen. 1929. Beiträge zur Bakteriologie der Säureweeker. Zentralbl. Bakteriol. Parasitenkd. Infektionskr. Hyg. Abt. II *79:* 75–85.

Ko, H.L., G. Pulverer and J. Jeljaszewica. 1978. Propionicins: Bacteriocins produced by *Propionibacterium avidum*. Zentralbl. Bakteriol. Parasitenkd. Infektionskr. Hyt. Abt. I. Orig. A *241:* 325–328.

Kobatake, M., S. Tamabe and S. Hasegawa. 1973. Nouveau micrococcus radioresistant a pigment rouge, isole de feces de *Lama glama*, et son utilisation comme indicateur microbiologique de la radiosterilisation. C.R. Soc. Biol. *167:* 1506–1510.

Koch, A. 1888. Ueber Morphologie und Entwicklungsgeschichte einiger endosporer Bacterienformen. Botanische Zeitung *46:* 277–287.

Koch, A.L. 1981. Evolution of antibiotic resistance gene function. Microbiol. Rev. *45:* 335–378.

Koch, R. 1876. Die Aetiologie der Milzbrandkrankheit. Beitr. Biol. Pflanz. *2:* Heft 2, 277–310.

Kocka, F.E., E.J. Arthur and R.L. Searcy. 1974. Comparitive effects of two sulfated polyanions used in blood culture on anaerobic cocci. Am. J. Clin. Pathol. *61:* 25–27.

Kocur, M. 1974. Genus *Planococcus*. *In* Buchanan and Gibbons (Editors), Bergey's Manual of Determinative Bacteriology, 8th ed. The Williams and Wilkins Co., Baltimore, pp. 489–490.

Kocur, M., T. Bergan and N. Mortensen. 1971. DNA base composition of Grampositive cocci. J. Gen. Microbiol. *69:* 167–183.

Kocur, M. and T. Martinec. 1963. The taxonomic status of *Sporosarcina ureae* (Beijerinck) Orla-Jensen. Int. Bull. Bacteriol. Nomencl. Taxon. *13:* 201–209.

Kocur, M., Z. Páčové, W. Hodgkiss and T. Martinec. 1970. The taxonomic status of the genus Planococcus Migula 1894. Int. J. Syst. Bacteriol. *20:* 241–248.

Kocur, M. and K.H. Schleifer. 1975. Taxonomic status of *Micrococcus agilis* Ali-Cohen 1889. Int. J. Syst. Bacteriol. *25:* 294–297.

Kocur, M. and K.H. Schleifer. 1981. The genus *Planococcus*. *In* Starr, Stolp, Trüper, Ballows and Schlegel (Editors), The Prokaryotes. A Handbook on Habitats, Isolation, and Identification of Bacteria. Springer-Verlag, New York, pp. 1570–1571.

Kocur, M., K.H. Schleifer and W.E. Kloos. 1975. Taxonomic status of *Micrococcus nishinomiyaensis* Oda 1935. Int. J. Syst. Bacteriol. *25:* 290–293.

Kodaka, H., G.L. Lombard and V.R. Dowell, Jr. 1982. Gas-liquid chromatography technique for detection of hippurate hydrolysis and conversion of fumarate to succinate by microorganisms. J. Clin. Microbiol. *16:* 962–964.

Kodama, R. 1956. Studies on the nutrition of lactic acid bacteria. Part I. *Lactobacillus fructosus* nov. sp., a new species of lactic acid bacteria. J. Agr. Chem. Soc. Jpn. *30:* 705–708.

Kohoutek, M. and Z. Nozicka. 1978. Tubaraktinomykose als Komplikation der intrauterinen Antikonzeption. Zentralbl. Gynaekol. *100:* 179–182.

Kolb, E. and H. Seidel. 1960. Ein Beitrag zur Kenntnis des Stoffwechsels von *Listeria monocytogenes* (Typ 1) unter besonderer Berücksichtigung der Oxydation von Kohlenhydraten und Metaboliten des Tricarbonsäurecyclus und deren Beeinflussung durch Hemmstoffe. Zentralbl. Veterinaermed. *7:* 509–518.

Kolonaros, I.V. and A.N. Bahn. 1965. Antigenic composition of the cell wall of *Streptococcus mitis*. Arch. Oral Biol. *10:* 625–633.

Komagata, K. and H. Iizuka. 1964. New species of *Brevibacterium* isolated from rice. (Studies on the microorganisms of cereal grains, Part VII). J. Agric. Chem. Soc. Jpn. *38:* 496–502.

Komaratat, P. and M. Kates. 1975. The lipid composition of a halotolerant species of *Staphylococcus epidermidis*. Biochim. Biophys. Acta *398:* 464–484.

Komatsu, Y. and T. Kaneko. 1980. Deoxyribonucleic acid relatedness between some glutamic acid-producing bacteria. Rep. Ferment. Res. Inst. (Tsukuba) *55:* 1–5.

Komura, I., K. Yamada and K. Komagata. 1975. Taxonomic significance of phospholipid composition in aerobic gram-positive cocci. J. Gen. Appl. Microbiol. *21:* 97–107.

Komura, I., K. Yamada, S. Otsuka and K. Komagata. 1975. Taxonomic significance of phospholipids in coryneform and nocardioform bacteria. J. Gen. Appl. Microbiol. *21:* 251–261.

Kondo, I., S. Itoh and Y. Yoshizawa. 1981. Staphylococcal phages mediating the lysogenic conversion of staphylokinase. *In* Jeljaszewicz (Editor), Staphylococci and Staphylococcal Infections, Gustav Fischer Verlag, Stuttgart, New York, pp. 357–362.

Kondo, I., S. Sakurai and Y. Sarai. 1976. Staphylococcal exfoliatin A and B. *In* Jeljaszewicz (Editor), Staphylococci and Staphylococcal Diseases, Gustav Fischer Verlag, Stuttgart, New York, pp.489–498.

Kondo, W., N. Sato and T. Ito. 1979. Chemical structure of the polysaccharide antigen of *Eubacterium saburreum*, strain 02. Carbohydr. Res. *70:* 117–123.

Konickova-Radochova, M. and J. Konicek. 1981. The completion of the replication map of the *Mycobacterium phlei* chromosome. Folia Microbiol. *26:* 59–61.

Konings, W.N. and R. Otto. 1983. Energy transduction and solute transport in streptococci. Antonie van Leeuwenhoek J. Microbiol. Serol. *49:* 247–257.

Konisky, J. 1978. The bacteriocins. *In* Gunsalus (Editor), The bacteria. Vol. VI. Bacterial diversity, Academic Press, New York, pp. 71–136.

Koornhof, H.J., N.J. Richardson, D.M. Wall and W.E.C. Moore. 1979. Fecal bacteria in South African rural blacks and other population groups. Isr. J. Med. Sci. *15:* 335–340.

Koransky, J.R., S.D. Allen and V.R. Dowell, Jr. 1978. Use of ethanol for selective isolation of sporeforming microorganisms. Appl. Environ. Microbiol. *35:* 762–765.

Koransky, J.R., M.D. Stargel and V.R. Dowell, Jr. 1979. *Clostridium septicum* bacteremia: its clinical significance. Am. J. Med. *66:* 63–66.

Kornman, K.S. and W.J. Loesche. 1978. New medium for isolation of *Actinomyces viscosus* and *Actinomyces naeslundii* from dental plaque. J. Clin. Microbiol. *7:* 514–518.

Kosaric, N. and K.K. Carroll. 1971. Phospholipids of *Listeria monocytogenes*. Biochem. Biophys. Acta *239:* 428–442.

Koser, S.A. 1968. *Bacillus* and *Clostridium*. *In* Vitamin Requirements of Bacteria and Yeasts. C. Thomas, Springfield, Illinois, pp. 379–401.

Koshi, G., M.K. Lalitha, T. Samraj and K.V. Mathai. 1981. Brain abscess and other protean manifestations of actinomycosis. Am. J. Trop. Med. Hyg. *30:* 139–144.

Koshikawa, T., T.C. Beaman, H.S. Pankratz, S. Nakashio, T.R. Corner and P. Gerhardt. 1984. Resistance, germination and permeability correlates of *Bacillus megaterium* spores successively divested of integument layers. J. Bacteriol. *159:* 624–632.

Kosmachev, A.E. 1964. A new thermophilic actinomycete *Micropolyspora thermovirida* n. sp. Mikrobiologiya *33:* 267–269.

Kostiw, L.L., C.W. Boylen and B.J. Tyson. 1972. Lipid composition of growing and starving cells of *Arthrobacter crystallopoietes*. J. Bacteriol. *111:* 103–111.

Krabbenhoft, K.L., A.W. Anderson and P.R. Elliker. 1965. Ecology of *Micrococcus radiodurans*. Appl. Microbiol. *13:* 1030–1037.

Kraft, A.A., J.C. Ayres, G.S. Torrey, R.H. Salzer and G.A.N. da Silva. 1966. Coryneform bacteria in poultry, eggs and meat. J. Appl. Bacteriol. *29:* 161–166.

Krämer, J. and G. Schallehn. 1974. Enterocinwirkung auf *Clostridium perfringens* und *Clostridium septicum*. Zentralbl. Bakteriol. Parasitenkd. Infektionskr. Hyg. I. Abt. Orig. A *226:* 105–113.

Kramer, P.A. and D. Jones. 1969. Media selective for *Listeria monocytogenes*. J. Appl. Bacteriol. *32:* 381–394.

Krasemann, C. and H.E. Müller. 1975. Die Virulenz von *Erysipelothrix rhusiopathiae* - Stämmen und ihre Neuraminidase-Produktion. Zentralbl. Bakteriol. Infektionskr. Hyg. Abt. I Reihe A *231:* 206–213.

Krasil'nikov, N.A. 1941. Keys to *Actinomycetales*. Inst. Microbiol. Acad. Sci., USSR, Moscow-Leningrad. English translation by Israel Program for Scientific Translations, Jerusalem, 1966.

Krasil'nikov, N.A. 1941. Guide to Bacteria and Actinomycetes (in Russian). Akad. Nauk. SSSR. Moscow. pp. 1–830.

Krasil'nikov, N.A. 1949. Guide to the bacteria and actinomycetes. Akad. Nauk. SSSR, Moscow, pp. 1–830.

Krasil'nikov, N.A. 1959. Diagnostik der Bakterien und Actinomyceten. VEB Gustav Fischer Verlag, Jena.

Krasil'nikov, N.A. 1964. Systematic position of ray fungi among the lower organisms. Hind. Antibiot. Bull. *7:* 1–17.

Krasil'nikov, N.A. and N.S. Agre. 1964. On two new species of *Thermopolyspora*. Hind. Antibiot. Bull. *6:* 97–107.

Krasil'nikov, N.A., N.S. Agre and G.I. El-Registan. 1968. New thermophilic species of the genus *Micropolyspora*. Mikrobiologiya *37:* 1065–1072.

Krasil'nikov, N.A., L.V. Kalakoutskii and N.F. Kirillova. 1961. A new genus of ray fungi – *Promicromonospora* gen. nov. Izv. Akad. Nauk SSSR (Ser. Biol.) No. 1, 107–112 (In Russian).

Krasnow, I. and L.G. Wayne. 1966. Sputum digestion. I. The mortality rate of tubercle bacilli in various digestion systems. Am. J. Clin. Pathol. *45:* 352–355.

Krasuski, A. 1981. Urea and arginine catabolism in *Staphylococcus aureus*: I. Relationships. *In* Jeljaszewicz (Editor), Staphylococci and Staphylococcal Infections, Gustav Fischer Verlag, Stuttgart, New York, pp. 413–416.

Kraus, H. 1961. Kurze Mitteilung über das Vorkommen von Lactobazillen auf frischen Heringen. Arch. Lebensmittelhyg. *12:* 101–102.

Krause, R.M. 1958. Studies on the bacteriophages of haemolytic streptococci. II. Antigens released from the streptococcal cell wall by a phage-associated lysin. J. Exp. Med. *108:* 803–821.

Krebs, H.A. and L.V. Eggleston. 1939. Bacterial urea formation (Metabolism of *Corynebacterium ureafaciens*) Enzymologia *7:* 310–320.

Kreis, W. and C. Hession. 1973. Isolation and purification of L-methionine-α-deamino-γ-mercaptomethane lyase (L-methioninase) from *Clostridium sporogenes*. Cancer Res. *33:* 1862–1865.

Krichevsky, M.I. and L.M. Morton. 1974. Sortage and manipulation of data by computers for determinative bacteriology. Int. J. Syst. Bacteriol. *24:* 525–531.

Krieg, A. 1981. The genus *Bacillus*: Insect pathogens. *In* Starr, Stolp, Trüper, Balows and Schlegel (Editors), The Prokaryotes. A Handbook on Habitats, Isolation and Identification of Bacteria. Springer-Verlag, New York, pp. 1743–1755.

Kristensen, H. and E.A. Christensen. 1981. Radiation-resistant microorganisms isolated from textiles. Acta Pathol. Microbiol. Scand. Sect. B *89:* 303–309.

Kronvall, G. 1973. A surface component of A, C and G streptococci with non-immune reactivity for immunoglobulin G. J. Immunol. *111:* 1401–1406.

Kronvall, G., J.L. Stanford and G.P. Walsh. 1976. Studies of Mycobacterial antigens with special reference to *M. leprae*. Infect. Immunol. *13:* 1132–1138.

Kroppenstedt, R.M. 1984. Fatty acid and menaquinone analysis of actinomycetes and related organisms. *In* Goodfellow and Minnikin (Editors), Chemical methods in systematic bacteriology, Society for Applied Bacteriology, Technical Series No. 20; Academic Press, London and New York, pp. 173–199.

Kroppenstedt, R.M. and H.J. Kutzner. 1976. Biochemical markers in the taxonomy of the Actinomycetales. Experimentia *32:* 318–319.

Kroppenstedt, R.M. and H.J. Kutzner. 1978. Biochemical taxonomy of some problem actinomycetes. *In* Mordarski, Kurylowicz and Jeljaszewicz (Editors), *Nocardia* and *Streptomyces*. Proceedings of the International Symposium on *Nocardia* and *Streptomyces*, Gustav Fischer Verlag, Stuttgart-New York, pp. 125–133.

Krüger, B. and O. Meyer. 1984. Thermophilic bacilli growing with carbon monoxide. Arch. Microbiol. *139:* 402–408.

Krulwich, T.A. and N.J. Pelliccione. 1979. Catabolic pathways of coryneforms, nocardias and mycobacteria. Annu. Rev. Microbiol. *33:* 95–111.

Kruse, W. 1896. Systematik der Streptothricheen und Bakterien. *In* Flügge (Editor), Die Mikroorganismen, 3rd ed., Vogel, Leipzig, Vol. 2, 48–66, 67–96, 185–526.

Krych, V.A., J.L. Johnson and A.A. Yousten. 1980. Deoxyribonucleic acid homologies among strains of *Bacillus sphaericus*. Int. J. Syst. Bacteriol. *30:* 476–484.

Kubica, G.P. and G.L. Pool. 1960. Studies on catalase activity of acid-fast bacilli. Am. Rev. Resp. Dis. *81:* 387–391.

Kubica, G.P. 1973. Differential identification of Mycobacteria. VII key features for identification of clinically significant mycobacteria. Am. Rev. Resp. Dis. *107:* 9–21.

Kubica, G.P., I. Baess, R.E. Gordon, P.A. Jenkins, J.B.G. Kwapinski, C. McDurmont, S.R. Pattyn, H. Saito, V. Silcox, J.L. Stanford, K. Takeya and M. Tsukamura. 1972. A cooperative numerical analysis of the rapidly growing mycobacteria. J. Gen. Microbiol. *73:* 55–70.

Kubica, G.P. and A.L. Rigdon. 1961. The arylsulfatase activity of acid-fast bacilli. III. Preliminary investigation of rapidly growing acid-fast bacilli. Am. Rev. Resp. Dis. *83:* 737–740.

Kubica, G.P., V.A. Silcox, J.O. Kilburn, R.W. Smithwick, R.E. Beam, W.D. Jones Jr. and K.D. Stottmeier. 1970. Differential identification of mycobacteria. VI. *Mycobacterium triviale* sp. nov. Int. J. Syst. Bacteriol. *20:* 161–174.

Kubica, G.P. and J. Vitvitsky. 1974. Comparison of two commercial formulations of the MacConkey agar test for mycobacteria. Appl. Microbiol. *27:* 917–919.

Kubica, G.P. and L.G. Wayne. 1983. The Mycobacteria: A Source Book. Marcel Dekker, Inc. New York.

Kubin, M., E. Matuskova and J. Kazda. 1969. *Mycobacterium brunense* n. sp. identified as serotype Davis of Group III (Runyon) mycobacteria. Zentralbl. Bakteriol. Parasitenkd. Infektionskr. Hyg. Abt. I Orig. *210:* 207–211.

Kübler, O., A. Engel, H.P. Zingsheim, B. Ende, M. Hahn, W. Heisse and W. Baumeister. 1980. Structure of the HP1 layer of *Micrococcus radiodurans*. *In* Baumeister and Vogell (Editors), Electron Microscopy at Molecular Dimensions, Springer-Verlag, Berlin and Heidelberg.

Kucsera, G. 1973. Proposal for the standardization of the designations used for serotypes of *Erysipelothrix rhusiopathiae* (Migula) Buchanan. Int. J. Syst. Bacteriol. *23:* 184–188.

Kuhn, D.A. and M.P. Starr. 1960. *Arthrobacter atrocyaneus* n. sp., and its blue pigment. Arch. Mikrobiol. *36:* 175–181.

Kuhn, R., M.P. Starr, D.A. Kuhn, H. Bauer, and H.J. Knackmuss. 1965. Indigoidine and other bacterial pigments related to 3,3'-bipyridyl. Arch. Mikrobiol. *51:* 71–84.

Kumazawa, N. and R. Yanagawa. 1969. DNA base composition of the three types of *Corynebacterium renale*. Jpn. J. Vet. Res. *17:* 115–120.

Kummeneje, K. and G. Bakken. 1973. *Clostridium perfringens* enterotoxaemia in reindeer. Nord. Veterinaemed. *25:* 196–202.

Kunath, P. 1983. Die Mikroflora von Kefir. Thesis, University of München.

Kundrat, W. 1963. Zur Differenzierung aerober Sporenbildner (Genus *Bacillus* Cohn). Zentralbl. Veterinaermed. B *10:* 418–426.

Kunkee, R.E. 1967. Malo-lactate fermentation. Adv. Appl. Microbiol. *9:* 235–279.

Kuno, Y. 1951. *Bacillus thiaminolyticus*, a new thiamin decomposing bacterium. Imp. Acad. Jpn. Proc. *27:* 262–265.

Kupfer, D.G. and E. Canale-Parola. 1967. Pyruvate metabolism in *Sarcina maxima*. J. Bacteriol. *94:* 984–990.

Kupfer, D.G. and E. Canale-Parola. 1968. Fermentation of glucose by *Sarcina maxima*. J. Bacteriol. *95:* 247–248.

Kurman, I. 1960. Ein vollsynthetischer Nährboden für Propionsäurebakterien. Pathol. Microbiol. *23:* 700–711.

Kurth, H. 1883. Ueber *Bacterium zopfii*, eine neue Bacterienart. Ber. Dtsch. Bot. Ges. *1:* 97–100.

Kurtzman, C.P., M.J. Smiley, C.J. Johnson, L.B. Wickerham and G.B. Fuson. 1980. Two new and closely related heterothallic species, *Pichia amylophila* and *Pichia mississippiensis*: characterization by hybridization and deoxyribonucleic acid reassociation. Int. J. Syst. Bacteriol. *30:* 208–216.

Kurup, V.P. 1981. Taxonomic study of some members of *Micropolyspora* and *Saccharomonospora*. Microbiologica (Bologna) *4:* 249–259.

Kurup, V.P. and N.S. Agre. 1983. Transfer of *Micropolyspora rectivirgula* (Krasil'nikov and Agre, 1964) Lechevalier, Lechevalier and Becker, 1966) to *Faenia* gen. nov. Int. J. Syst. Bacteriol. *33:* 663–665.

Kurup, V.P. and J.N. Fink. 1977. Extracellular antigens of *Micropolyspora faeni* grown in synthetic medium. Infect. Immun. *15:* 608–613.

Kurup, V.P. and R.J. Heinzen. 1978. Isolation and characterization of actinophages of *Thermoactinomyces* and *Micropolyspora*. Can. J. Microbiol. *24:* 794–797.

Kurup, V.P., J.E. Piechura, E.Y. Ting and J.A. Orlowski. 1983. Immunochemical characterization of *Nocardia asteroides* antigens: support for a single species concept. Can. J. Microbiol. *29:* 425–432.

Kurup, V.P., H.S. Randhawa and N.P. Gupta. 1970. Nocardiosis: a review. Mycopathol. Mycol. Appl. *40:* 193–219.

Kurup, V.P. and G.H. Schribner. 1981. Antigenic relationship among *Nocardia asteroides* immunotypes. Microbios *31:* 25–30.

Kurup, V.P., E.Y. Ting, J.N. Fink and N.J. Calvanico. 1981. Characterization of *Micropolyspora faeni* antigens. Infect. Immun. *34:* 508–512.

Kusaka, T. 1977. Fatty acid synthesizing enzyme activity of cultured *Mycobacterium lepraemurium*. Int. J. Leprosy *45:* 132–138.

Kusaka, T., K. Kohsaka, Y. Fukunishi and H. Akimori. 1981. Isolation and identification of mycolic acids in *Mycobacterium leprae* and *Mycobacterium lepraemurium*. Int. J. Leprosy *49:* 406–416.

Kusser, W. and F. Fiedler. 1983. Murien type and polysaccharide composition of cell walls from *Renibacterium salmoninarium*. FEMS Microbiol. Lett. *20:* 391–394.

Kusunose, E., M. Kusunose, K. Ichihara and S. Izumi. 1980. Occurrence of superoxide dismutase in *Mycobacterium leprae* grown on armadillo liver. J. Gen. Appl. Microbiol. *107:* 9–21.

Kusunose, E., M. Kusunose, K. Ichihara and S. Izumi. 1981. Superoxide dismutase in cell-free extracts from *Mycobacterium leprae* grown on armadillo liver. FEMS Microbiol. Lett. *10:* 49–52.

Kützing, F.T. 1843. Phycologia Generalis. Leipzig. pp. 1–458.

Kutzner, H.J. 1981. The family Streptomycetaceae. *In* Starr, Stolp, Trüper, Balows and Schlegel (Editors), The Prokaryotes. A handbook on habitats, isolation, and identification of bacteria, Springer-Verlag, Berlin-Heidelberg-New York, pp. 2028–2090.

Kuznetsov, V.D. 1959. A new species of lactic acid bacteria. Mikrobiologiya *28:* 368–373.

Kwapinski, J.B.G. 1969. Analytical serology of Actinomycetales. *In* Kwapinski (Editor), Analytical Serology of Microorganisms. John Wiley & Sons Inc., p. 86.

Kwapinski, J.B.G. and H.P.R. Seeliger. 1964. Immunological characteristics of the Actinomycetales. A review. Zentralbl. Bakteriol. Parasitenkd. Infektionskr. Hyg. I Abt. Orig. *195:* 805–854.

Kwapinski, J.B.G. and M.L. Snyder. 1961. Antigenic structure and serological relationships of *Mycobacterium, Actinomyces, Streptococcus* and *Diplococcus.* J. Bacteriol. *82:* 632–639.

Laakso, S. 1976. The relationship between methionine uptake and demethiolation in a methionine-utilizing mutant of *Pseudomonas fluorescens* UK1. J. Gen. Microbiol. *95:* 391–394.

Labeda, D.P., C.M. Hunt and L.E. Casida, Jr. 1974. Plating isolation of various catalase-negative microorganisms from soil. Appl. Microbiol. *27:* 432–434.

Labischinski, H., G. Barnickel, W. Ronspeck, K. Roth and P. Giesbrecht. 1981. New insights into the three dimensional arrangement of the cell walls of staphylococci and other Gram-positive bacteria. *In* Jeljaszewicz (Editor), Staphyolcocci and Staphylococcal Infections, Gustav Fischer Verlag, Stuttgart, New York, pp. 427–433.

Lacey, J. 1974. Moulding of sugar-cane bagasse and its prevention. Ann. Appl. Biol. *76:* 63–76.

Lacey, J. 1978. Ecology of actinomycetes in fodders and related substances. Zentralbl. Bakteriol. Parasitenkd. Infektionskr. Hyg. Abt. I Suppl. *6:* 161–170.

Lacey, J. and J. Dutkiewicz. 1976a. Methods for examining the microflora of mouldy hay. J. Appl. Bacteriol. *41:* 13–27.

Lacey, J. and J. Dutkiewicz. 1976b. Isolation of actinomycetes and fungi using a sedimentation chamber. J. Appl. Bacteriol. *41:* 315–319.

Lacey, J. and M. Goodfellow. 1975. A novel actinomycete from sugar cane bagasse: *Saccharopolyspora hirsuta* gen. et sp. nov. J. Gen. Microbiol. *88:* 75–85.

Lacey, J., M. Goodfellow and G. Alderson. 1978. The genus *Actinomadura*, Lechevalier and Lechevalier. Zentralbl. Bakteriol. Parasitenkd. Infektionskr. Hyg. Abt. 1 Suppl. *6:* 107–117.

Lacey, R.W. 1975. Antibiotic resistance plasmids of *Staphylococcus aureus* and their clinical importance. Bacteriol. Rev. *39:* 1–32.

Lacey, R.W. and M.H. Richmond. 1974. The genetic basis of antibiotic resistance in *Staphylococcus aureus*: the importance of gene transfer in the evolution of this organism in the hospital environment. Ann. N.Y. Acad. Sci. *236:* 395–412.

Lachance, R.A. 1960. The vitamin requirements of *Corynebacterium sepedonicum* (Spieck. et Kott.) Skapt. et Burkh. Can. J. Microbiol. *6:* 171–174.

Lachance, R.A. 1962. The amino acid requirements of *Corynebacterium sepedonicum* (Spieck. & Kott.) Skapt. & Burkh. Can. J. Microbiol. *8:* 321–325.

Lachica, R.V.F. and R.H. Diebel. 1969. Detection of nuclease activity in semisolid and broth cultures. Appl. Microbiol. *18:* 174–176.

Lachica, R.V.F., C. Genigeorgis and P.D. Hoeprich. 1971. Metachromatic agar diffusion methods for detecting staphylococcal nuclease activity. Appl. Microbiol. *21:* 585–587.

Lachica, R.V.F. and P.A. Hartman. 1968. Two improved media for isolating and enumerating enterococci in certain frozen foods. J. Appl. Bacteriol. *31:* 151–156.

Lachica, R.V.F., P.D. Hoeprich and C. Genigeorgis. 1972. Metachromatic agar-diffusion microslide technique for detecting staphylococcal nuclease in foods. Appl. Microbiol. *23:* 168–169.

Lachica, R.V.F., S.S. Jang and P.D. Hoeprich. 1979. Thermonuclease seroinhibition test for distinguishing *Staphylococcus aureus* and other coagulase-positive staphylococci. J. Clin. Microbiol. *9:* 141–143.

Lachner-Sandoval, V. 1898. Über Strahlenpilze. Eine bacteriologisch-botanische Untersuchung. Inaugural-Dissertation Strassburg. Universitäts-Buchdruckerei von Carl Georgi, Bonn.

Lachowicz, T. and B. Sienienska. 1976. Production of staphylococcins in solid and fluid media by different strains of staphylococci and their selection. *In* Jeljaszewicz (Editor), Staphylococci and Staphylococcal Diseases, Gustav Fischer Verlag, Stuttgart, New York, pp. 593–597.

Lai, C.-H. and M.A. Listgarten. 1979. Immune labeling of certain strains of *Actinomyces naeslundii* and *Actinomyces viscosus* by fluorescence and electron microscopy. Infect. Immun. *25:* 1016–1028.

Lai, C.-H. and M.A. Listgarten. 1980. Comparative ultrastructure of certain *Actinomyces* species, *Arachnia, Bacterionema* and *Rothia.* J. Periodontol. *51:* 136–154.

Laine, J.J. 1971. Studies on psychrophilic bacilli of food origin. Ann. Acad. Sci. Fenn. Ser. A IV Biol. 169.

Laird, W.J., W. Aaronson, R.P. Silver, W.H. Habig and M.C. Hardegree. 1980. Plasmid-associated toxigenicity in *Clostridium tetani.* J. Infect. Dis. *142:* 623.

Lalonde, M., R. Knowles and J.A. Fortin. 1975. Demonstration of the isolation of non-infective *Alnus crispa* var. *mollis* Fern. nodule endophyte by morphological immunolabelling and whole cell composition studies. Can. J. Microbiol. *21:* 1901–1920.

Lamanna, C. 1940. The taxonomy of the genus *Bacillus.* I. Modes of spore germination. J. Bacteriol. *40:* 347–359.

Lambert, F.W., Jr., J.M. Brown and L.K. Georg. 1967. Identification of *Actinomyces israelii* and *Actinomyces naeslundii* by fluorescesnt-antibody and agar-gel diffusion techniques. J. Bacteriol. *94:* 1287–1295.

Lambert, M.A.S. and A.Y. Armfield. 1979. Differentiation of *Peptococcus* and *Peptostreptococcus* by gas-liquid chromatography of cellular fatty acid and metabolic products. J. Clin. Microbiol. *10:* 464–476.

Lambert, M.A., D.G. Hollis, C.W. Moss, R.E. Weaver and M.L. Thomas. 1971. Cellular fatty acids of nonpathogenic *Neisseria.* Can. J. Microbiol. *17:* 1491–1502.

Lambert, M.A. and C.W. Moss. 1980. Production of *p*-hydroxyhydrocinnamic acid from tyrosine by *Peptostreptococcus anaerobius.* J. Clin. Microbiol. *12:* 291–293.

Lämmler, C., G.S. Chatwal and H. Blobel. 1983. Variations in the binding of mammalian fibrinogens to streptococci of different animal origin. Med. Microbiol. Immunol. *172:* 191–196.

Lamont, I.L. and J. Mandlestam. 1984. Identification of a new sporulation locus, *spoIIIF*, in *Bacillus subtilis.* J. Gen. Microbiol. *130:* 1253–1261.

LaMont, J.L., E.B. Sonnenblick and S. Rothman. 1979. Role of clostridial toxin in the pathogenesis of clindamycin colitis in rabbits. Gastroenterology *76:* 356–361.

Lancefield, R.C. 1933. A serological differentiation of human and other groups of hemolytic streptococci. J. Exp. Med. *57:* 571–595.

Lancefield, R.C. 1934. A serological differentiation of specific types of bovine hemolytic streptococci (Group B). J. Exp. Med. *59:* 441–458.

Lancefield, R.C. 1938. Two serological types of group B haemolytic streptococci with related, but not identical, type-specific substances. J. Exp. Med. *67:* 25–39.

Lancefield, R.C. and R. Hare. 1935. The serological differentiation of pathogenic and non-pathogenic strains of hemolytic streptococci from parturient women. J. Exp. Med. *61:* 335–349.

Lancy, P., Jr., R.G.E. Murray. 1978. The envelope of *Micrococcus radiodurans*: isolation, purification and preliminary analysis of the wall layers. Can. J. Microbiol. *24:* 162–176.

Landfried, S. 1972. Isolation and characterization of an antigen from *Actinomyces israelii* ATCC 12102. Ph.D. Dissertation, West Virginia University, Morgantown, West Virginia, U.S.A.

Lanéelle, G., J. Asselineau and G. Chamoiseau. 1971. Présence de mycosides C' (formes simplifiées de mycoside C) dans les bacteries isolées de bovins atteints du farcin. FEBS Lett. *19:* 109–111.

Lanéelle, M.-A., J. Asselineau, M. Welby, M.V. Norgard, T. Imaeda, M.C. Pollice and L. Barksdale. 1980. Biological and chemical bases for the reclassification of *Brevibacterium vitarumen* (Bechdel et al.) Breed (Approved Lists, 1980) as *Corynebacterium vitarumen* (Bechdel et al.) comb. nov. and *Brevibacterium liquefaciens* Okabayashi and Musuo (Approved Lists, 1980) as *Corynebacterium liquefaciens* (Okabayashi and Musuo) comb. nov. Int. J. Syst. Bacteriol. *30:* 539–546.

Langford, E.V. 1970. Feed-borne *Clostridium chauvoei* infection in mink. Can. Vet. J. *11:* 170–172.

Langford, G.C. and P.A. Hansen. 1953. *Erysipelothrix insidiosa.* Atti Del VI. Cong. Int. Microbiol. Roma. Riassunti Communicazioni *1:* 18.

Langsrud, T., G.W. Reinhold and E.G. Hammond. 1977. Proline production by *Propionibacterium shermanii* P59. J. Dairy Sci. *60:* 16–23.

Langsrud, T., G.W. Reinhold and E.G. Hammond. 1978. Free proline production by strains of *Propionibacteria.* J. Dairy Sci. *61:* 303–308.

Langston, C.W., J. Gutierrez and C. Bouma. 1960. Motile streptococci (*Streptococcus faecium* var. *mobilis* var. N) isolated from grass silage. J. Bacteriol. *80:* 714–718.

Langston, C.W. and P.P. Williams. 1962. Reduction of nitrate by streptococci. J. Bacteriol. *84:* 603.

Langworthy, T.A. 1982. Lipids of bacteria living in extreme environments. Curr. Top. Memb. Transp. *17:* 44–77.

Langworthy, T.A., W.R. Mayberry and P.F. Smith. 1976. A sulfonolipid and novel glucosamidyl glycolipids from the extreme thermoacidophilic *Bacillus acidocaldarius.* Biochem. Biophys. Acta *431:* 550–569.

Lanigan, G.W. 1976. *Peptococcus heliotrinreducans*, sp. nov., a cytochrome-producing anaerobe which metabolizes pyrrolizidine alkaloids. J. Gen. Microbiol. *94:* 1–10.

Lanigan, G.W. and L.W. Smith. 1970. Metabolism of pyrrolizidine alkaloids in the ovine rumen. I. Formation of 7α-hydroxy-1α-methyl-8α-pyrrolizidine from heliotrine and lasiocarpine. Aust. J. Agric. Res. *21:* 493–500.

Lannelongue, M., M.O. Hanna, G. Finne, R. Nickelson, II and C. Vanderzant. 1982. Storage characteristics of finfish fillets. (*Archosargus probatocephalus*) packaged in modified gas atmospheres containing carbon dioxide. J. Food Protect. *45:* 440–444.

Lapage, S.P. 1971. Culture collections of bacteria. Biol. J. Linnean Soc. *3:* 197–210.

Lapage, S.P. 1974. *Species incertae sedis. Haemophilus vaginalis* Gardner and Dukes, 1955, 963. *In* Buchanan and Gibbons (Editors), Bergey's Manual of

Determinative Bacteriology, 8th ed. The Williams and Wilkins Co., Baltimore, pp. 368–370.

Lapage, S.P. 1975. Report of the World Federation for Culture Collections. Int. J. Syst. Bacteriol. *25:* 90–94.

Lapage, S.P., P.H.A. Sneath, E.F. Lessel, V.B.D. Skerman, H.P.R. Seeliger and W.A. Clark (Editors). 1975. International code of nomenclature of bacteria. 1975. Revision. American Society for Microbiology, Washington, D.C.

Larkin, J.M. and J.L. Stokes. 1966. Isolation of psychrophilic species of *Bacillus.* J. Bacteriol. *91:* 1667–1671.

Larkin, J.M. and J.L. Stokes. 1967. Taxonomy of psychrophilic strains of *Bacillus.* J. Bacteriol. *94:* 889–895.

Larsen, J., E.J. Bottone, S. Dikman and R. Saphir. 1978. Cervicofacial *Actinomyces viscosus* infection. J. Pediatr. *93:* 797–801.

Larsson, L., P.-A. Mardh and G. Odham. 1978. Analysis of amines and other bacterial products by head-space gas chromatography. Acta. Pathol. Microbiol. Scand. Sect. B *86:* 207–213.

Latham, M.J., B.E. Brooker, G.L. Pettipher and P.J. Harris. 1978. *Ruminococcus flavefaciens* cell coat and adhesion to cotton cellulose and to cell walls in leaves of perennial ryegrass (*Lolium perenne*). Appl. Environ. Microbiol. *35:* 156–165.

Latham, M.J. and N.J. Legakis. 1976. Cultural factors influencing the utilization or production of acetate by *Butyrivibrio fibrisolvens.* J. Gen. Microbiol. *94:* 380–388.

Latham, M.J. and M.E. Sharpe. 1971. The isolation of anaerobic organisms from the bovine rumen. *In* Isolation of Anaerobes, Shapton and Board (Editors), Soc. Appl. Bacteriol. Tech. Ser. No. 5. Academic Press, London.

Latham, M.J., M.E. Sharpe and N. Weiss. 1979. Anaerobic cocci from the bovine alimentary tract, the amino acids of their cell wall peptidoglycans and those various species of anaerobic streptococcus. J. Appl. Bacteriol. *47:* 209–221.

Lau, A.H.S., R.Z. Hawirko and C.T. Chow. 1974. Purification and properties of boticin P produced by *Clostridium botulinum.* Can. J. Microbiol. *20:* 385–390.

Laubach, C.A. 1916. Studies on aerobic spore-bearing non-pathogenic bacteria. Spore-bearing organisms in water. J. Bacteriol. *1:* 505–512.

Lauer, E., C. Helming and O. Kandler. 1980. Heterogeneity of the species *Lactobacillus acidophilus* (Moro) Hansen and Moquot as revealed by biochemical characteristics and DNA/DNA hybridization. Zentralbl. Bakteriol. Mikrobiol. Hyg. I Abt. Orig. C *1:* 150–168.

Lauer, E. and O. Kandler. 1976. Mechanismus der Variation des Verhältnisses Acetat/Lactat bei der Vergärung von Glucose durch Bifidobakterien. Arch. Microbiol. *110:* 271–277.

Lauer, E. and O. Kandler. 1980. *In* Validation of the publication of new names and new combinations previously effectively published outside the IJSB. List No. 4. Int. J. Syst. Bacteriol. *30:* 601.

Lauer, E. and O. Kandler. 1980. *Lactobacillus gasseri* sp. nov., a new species of the subgenus *Thermobacterium.* Zentralbl. Bakteriol. Mikrobiol. Hyg. I Abt. Orig. C *1:* 75–78.

Laughton, N. 1948. Canine beta hemolytic streptococci. J. Pathol. Bacteriol. *60:* 471–476.

Law, B.A. and J. Kolstad. 1983. Proteolytic systems in lactic acid bacteria. Antonie van Leeuwenhoek J. Microbiol. Serol. *49:* 225–245.

Law, B.A., E. Sezgin and M.E. Sharpe. 1976. Amino acid nutrition of some commercial cheese starters in relation to their growth in peptone-supplemented media. J. Dairy Res. *43:* 291–300.

Lazar, I. 1968. Serological relationships of corynebacteria. J. Gen. Microbiol. *52:* 77–88.

Leatherwood, J.M. 1973. Cellulose degradation by *Ruminococcus.* Fed. Proc. *32:* 1814–18–18.

Leatherwood, J.M. and M.P. Sharma. 1972. Novel anaerobic cellulolytic bacterium. J. Bacteriol. *110:* 752–753.

Le Blaye, R. and H. Guggenheim. 1914. Manuel pratique de diagnostic bactériologique et de technique appliqué à la détermination des bactéries. Vigot Frères Edition, Paris, pp. 1–444.

Lechevalier, H.A. 1968. Status of the generic names *Micropolyspora* Lechevalier et al. 1961 and *Micropolispora* Shchepkina 1940 (Actinomycetales): Request for an opinion from the Judicial Commission (ICNB) conserving the generic name *Micropolyspora* Lechevalier. Int. J. Syst. Bacteriol. *18:* 203–206.

Lechevalier, H.A. 1976. Report on the cooperative study of the generic assignment of strains labelled *Nocardia farcinica.* The Biology of the Actinomycetes. *12:* 8–16.

Lechevalier, H.A. and M.P. Lechevalier. 1969. Ultramicroscopic structure of *Intrasporangium calvum* (Actinomycetales). J. Bacteriol. *100:* 522–525.

Lechevalier, H.A. and M.P. Lechevalier. 1981a. Introduction to the order Actinomycetales. *In* Starr, Stolp, Truper, Balows and Schlegel (Editors), The Prokaryotes. A Handbook on Habitats, Isolation and Identification of Bacteria. Springer-Verlag, New York, pp. 1915–1922.

Lechevalier, H.A. and M.P. Lechevalier. 1981. Actinomycete genera "in search of a family". *In* Starr, Stolp, Trüper, Balows and Schlegel (Editors). The Prokaryotes. A Handbook on Habitats, Isolation and Identification of Bacteria. Springer-Verlag, New York, pp. 2118–2123.

Lechevalier, H.A., M.P. Lechevalier and B. Becker. 1966. Comparison of the chemical composition of cell-walls of nocardiae with that of other aerobic actinomycetes. Int. J. Syst. Bacteriol. *16:* 151–160.

Lechevalier, H.A., M. Solotorovsky and C.I. McDurmont. 1961. A new genus of *Actinomycetales: Micropolyspora* gen. nov. J. Gen. Microbiol. *26:* 11–18.

Lechevalier, M.P. 1968. Identification of aerobic actinomycetes of clinical importance. J. Lab. Clin. Med. *71:* 934–944.

Lechevalier, M.P. 1972. Description of a new species, *Oerskovia xanthineolytica,* and emendation of *Oerskovia* Prauser et al. Int. J. Syst. Bacteriol. *22:* 260–264.

Lechevalier, M.P. 1976. The taxonomy of the genus *Nocardia:* some light at the end of the tunnel? *In* Goodfellow, Brownell and Serrano (Editors), The Biology of the Nocardiae, Academic Press, New York, pp. 1–38.

Lechevalier, M.P. 1977. Lipids in bacterial taxonomy - A taxonomists view. Crit. Rev. Microbiol. *5:* 109–210.

Lechevalier, M.P., C. De Bièvre and H.A. Lechevalier. 1977. Chemotaxonomy of aerobic actinomycetes: phospholipid composition. Biochem. Syst. Ecol. *5:* 249–260.

Lechevalier, M.P., N.N. Gerber and T.A. Umbreit. 1982. Transformation of adenine to 8-hydroxyadenine by strains of *Oerskovia xanthineolytica.* Appl. Environ. Microbiol. *43:* 367–370.

Lechevalier, M.P., A.C. Horan and H.A. Lechevalier. 1971. Lipid composition in the classification of nocardiae and mycobacteria. J. Bacteriol. *105:* 313–318.

Lechevalier, M.P. and H.A. Lechevalier. 1970. Chemical composition as a criterion in the classification of aerobic actinomycetes. Int. J. Syst. Bacteriol. *20:* 435–443.

Lechevalier, M.P. and H.A. Lechevalier. 1974. *Nocardia amarae* sp. nov., an actinomycete common in foaming activated sludge. Int. J. Syst. Bacteriol. *24:* 278–288.

Lechevalier, M.P. and H.A. Lechevalier. 1976. Chemical methods as criteria for the separation of nocardiae from other actinomycetes. Biol. Actinomycetes *11:* 78–92.

Lechevalier, M.P. and H.A. Lechevalier. 1980. The chemotaxonomy of actinomycetes. *In* Dietz and Thayer (Editors), Actinomycete taxonomy, Society for Industrial Microbiology, Arlington, VA, Special Publication 6. pp. 227–291.

Lechevalier, M.P., A.E. Stern and H.A. Lechevalier. 1981. Phospholipids in the taxonomy of actinomycetes. Zentralbl. Bakteriol. Parasitenkd. Infektionskr. Hyg. I Abt. Suppl. *11:* 111–116.

Ledesma, O.V., A.P. Holgado, G. Oliver, G.S. de Giori, P. Raibaud and J.V. Galpin. 1977. A synthetic medium for comparative nutritional studies of lactobacilli. J. Appl. Bacteriol. *42:* 123–133.

Ledingham, G.A., G.A. Adams and R.Y. Stanier. 1945. Production and properties of 2,3-butanediol. I. Fermentation of wheat mashes by *Aerobacillus polymyxa.* Can. J. Res. *23:* 48–51.

Lee, C.K. and Z.J. Ordal. 1967. Regulatory effect of pyruvate on the glucose metabolism of *Clostridium thermosaccharolyticum.* J. Bacteriol. *94:* 530–536.

Lee, L. and R. Cherniak. 1974. Capsular polysaccharide of *Clostridium perfringens* Hobbs 10. Infect. Immun. *9:* 318–322.

Lee, L., A. Kimura and T. Tochikura. 1978. Presence of a single enzyme catalyzing the pyrophosphorolysis of UDP-glucose and UDP-galactose in *Bifidobacterium bifidum.* Biochem. Biophys. Acta *527:* 301–304.

Lee, L., S. Kinoshita, H. Kumagai and T. Tochikura. 1980. Galactokinase of *Bifidobacterium bifidum.* Agric. Biol. Chem. *44:* 2961–2966.

Lee, L., S. Terao, H. Kumagai and T. Tochikura. 1979. Characteristics of galactose metabolism in bifidobacteria. Abstr. Annu. Meet. Am. Soc. Microbiol. p. 349.

Lee, W.H. and R.C. Good. 1963. Amino acid synthesis (United States Patent 3,087,863). Abstr. Offic. Gaz. U.S. Patent Off. *789:* 1349.

Lee, W.H. and H. Riemann. 1970a. Correlation of toxic and non-toxic strains of *Clostridium botulinum* by DNA composition and homology. J. Gen. Microbiol. *60:* 117–123.

Lee, W.H. and H. Riemann. 1970b. The genetic relatedness of proteolytic *Clostridium botulinum* strains. J. Gen. Microbiol. *64:* 85–90.

Leedle, J.A.Z., M.P. Bryant and R.B. Hespell. 1982. Diurnal variations in bacterial numbers and fluid parameters in ruminal contents of animals fed low- or high-forage diets. Appl. Environ. Microbiol. *44:* 402–412.

Leedle, J.A.Z. and R.B. Hespell. 1980. Differential carbohydrate media and anaerobic replica plating techniques in delineating carbohydrate-utilizing subgroups in rumen bacterial populations. Appl. Environ. Microbiol. *39:* 709–719.

Legakis, N.J., H. Ioannides, S. Tzannetis, B. Golematis and J. Papavassiliou. 1981. Faecal bacterial flora in patients with colon cancer and control subjects. Zentralbl. Bakteriol. Parasitenkd. Infektionskr. Hyg. I Abt. Orig A *251:* 54–61.

Legum, L.L., K.E. Greer and St. F. Glessner. 1978. Dessiminated actinomycosis. South. Med. J. *71:* 463–465.

Lehman, T.J.A., J.J. Quinn, S.E. Siegel and J.A. Ortega. 1977. *Clostridium septicum* infection in childhood leukemia. Cancer *40:* 950–953.

Lehmann, K.B. and R. Neumann. 1896. Atlas und Grundriss der Bakteriologie und Lehrbuch der speciellen backteriologischen Diagnostik. 1st ed. J.F. Lehmann, München. pp. 1–448.

Lehmann, K.B. and R. Neumann. 1899. Lehmann's Medizin, Handatlanten X. Atlas und Grundriss der Bakteriologie und Lehrbuch der speciellen bakteriologischen Diagnostik. 2 Aufl, J.F. Lehmann, München.

Lehmann, K.B. and R.O. Neumann. 1907. Lehmann's Medizin, Handatlanten, X. Atlas und Grundriss der Bakteriologie und Lehrbuch der speciellen bakteriologischen Diagnostik, 4 Aufl. Teil 2. J.F. Lehmann, München. pp. 1–730.

Lehmann, K.B. and R.O. Neumann. 1927. Bakteriologie insbesondere Bakteriologische Diagnostik. II. Allgemeine und spezielle bakteriologie 7 Aufl. Teil 2. J.F. Lehmann, München. pp. 1–835.

Lehmer, A. and K.H. Schleifer. 1980. Untersuchunger zum pentose - und pentitolstoffwechsel bei *Staphylococcus xylosus* und *Staphylococcus saprophyticus*. Zentralbl. Bakteriol. Mikrobiol. Hyg. Abt. I Orig. C *1:* 109–123.

Leichmann, G. 1896. Über die im Brennereiprozess bei der bereitung der Kunsthefe auftretende spontane Milchsäuregärung. Zentralbl. Bakteriol. II Abt. *2:* 281–285.

Leichmann, G. 1896. Über die freiwillige Säuerung der Milch. Zentralbl. Bakteriol. II Abt. *2:* 777–780.

Leidy, J. 1850. On the existence of entophyta in healthy animals, as a natural condition. Proc. Acad. Nat. Sci. Phila. *4:* 225–229.

Leifson, E. 1960. Atlas of Bacterial Flagellation. Academic Press, New York.

Leifson, E. 1963. Determination of carbohydrate metabolism of marine bacteria. J. Bacteriol. *85:* 1183–1184.

Leininger, H.V. 1976. Equipment, media, reagents, routine tests and strains. *In* Speck (Editor), Compendium of Methods for the Microbiological Examination of Foods, American Public Health Association, Washington, D.C., pp. 10–94.

Lelliot, R.A. 1965. The preservation of plant pathogenic bacteria. J. Appl. Bacteriol. *28:* 181–193.

Le Mille, F., H. de Barjac and A. Bonnefoi. 1969a. Essai sur la classification biochemique de 97 *Bacillus* du groupe I, appartenant a 9 espéces différentes. Ann. Inst. Pasteur (Paris) *116:* 809–819.

Le Mille, F., H. de Barjac and A. Bonnefoi. 1969b. Étude sérologique de *Bacillus cereus*. Mise en évidence de divers sérotypes basés sur les antigénes flagellaires. Ann. Inst. Pasteur (Paris) *117:* 31–38.

Lennette, E.H., A. Balows, W.J. Hausler and J.P. Truant. 1980. Manual of Clinical Microbiology. Third Edition. Am. Soc. Microbiol., Washington D.C.

Lentze, F. 1938b. Zur Bakteriologie und Vakzinetherapie der Aktinomykose. Zentralbl. Bakteriol. Parasitenkd. Infektionskr. Hyg. I. Abt. Orig. *141:* 21–36.

Lentze, F. 1948. Die Aetiologie der Aktinomykose des Menschen. Dtsch. Zahnaerztl. Z. *3:* 913–919.

Lentze, F. 1953. Zur Aetiologie und mikroviologischen Diagnostik der Aktinomykose. Estratto dagli Atti del VI Congresso Internazionale di Microbiologia, Roma, *5:* Sez. XIV, 145–148.

Lentze, F. 1957. Zur antibiotischen Therapie der Aktinomykose. Fortschr. Kiefer-Gesichtschir. *3:* 306–313.

Lentze, F. 1967. Die Aktinomykose und ihre Mikrobiologie. *In* Heite (Editor), Krankheiten durch Aktinomyzeten und verwandte Errenger. Springer-Verlag, Berlin-Heidelberg-New York, pp. 1–11.

Lentze, F. 1969. Die Aktinomykose und die Nocardiosen. *In* Grumbach and Bonin (Editors), Die Infektionskrankheiten des Menschen und ihre Errerger, Vol. I., 2nd Edition. Georg Thieme Verlag, Stuttgart, pp. 954–973.

Lentze, F. 1970. Klinik, Diagnostik und Therapie der Aktinomykosen. *In* Götz und Rieth (Editors), Diagnostik und Therapie der Pilzkrankheiten und Erkenntnisse in der Biochemie der pathogenen Pilze. Grosse-Verlag, Berlin, pp. 83–92.

Lentze, F.A. 1971. Die Aktinomykosen der Lunge. *In* Barysch (Editor), Lungenmykosen. Georg Thieme Verlag, Stuttgart, pp. 43–46.

Lentze, F.A. 1938a. Die mikrobiologische Diagnostik der Aktinomykose. Müench. Med. Wochenschr. *47:* 1826–1836.

Leopold, S. 1953. Heretofore undescribed organism isolated from the genitourinary system. U.S. Armed Forces Med. J. *4:* 263–266.

Leppla, S.H. 1982. Anthrax toxin edema factor: a bacterial adenylate cyclase that increases cyclic AMP concentrations in eukaryotic cells. Proc. Nat. Acad. Sci. USA *79:* 3162–3166.

Lerche, M. and G. Reuter. 1962. Das Vorkommen aerob wachsender Grampositiver Stäbchen des Genus *Lactobacillus* Beijerinck im Darminhalt erwachsener Menschen. Zentralbl. Bakteriol. Parasitenkd. Infektionskr. Hyg. I Abt. Orig. *185:* 446–481.

Lerner, P.I. 1968. Susceptibility of *Actinomyces* to cephalosporins and lincomycin. Antimicrob. Agents Chemother. *1967:* 730–735.

Lerner, P.I. 1974. Susceptibility of pathogenic actinomycetes to antimicrobial compounds. Antimicrob. Agents Chemother. *5:* 302–309.

Lerner, P.I. and G.L. Baum. 1973. Antimicrobial susceptibility of *Nocardia* species. Antimicrob. Agents Chemother. *4:* 85–93.

Lesher, R.J., M.A. Gerencser and V.F. Gerencser. 1974. Morphological, biochemical and serological characterization of *Rothia dentocariosa*. Int. J. Syst. Bacteriol. *24:* 154–159.

Lettl, A. 1973. Large-scale production of clostridial collagenase and lecithinase C. J. Hyg. Epidemiol., Microbiol., Immunol. *17:* 385–388.

Lev, M. and C.A.E. Briggs. 1956. The gut flora of the chick. 1. The flora of newly hatched chicks. J. Appl. Bacteriol. *19:* 36–38.

Levine, M. and L. Soppeland. 1926. Bacteria in creamery wastes. Bull. Iowa State Agr. Coll. *77:* 1–72.

Lévy-Frébault, V., Rafidinarivo, J.-C. Promé, J. Gandry, H. Boisvert and H.L. David. 1983. *Mycobacterium fallax* sp. nov. Int. J. Syst. Bacteriol. *33:* 336–343.

Lewis, J.F., N. Mullins and P. Johnson. 1980. Isolation and evaluation of clostridia from clinical sources. South. Med. J. *73:* 427–432.

Lewis, N.F. 1973. Radio-resistant *Micrococcus radiophilus* sp. nov. isolated from irradiated Bombay duck (*Harpodon nehereus*). Curr. Sci. *42:* 504.

Lewis, R. and S.L. Gorbach. 1972. *Actinomyces viscosus* in man. Lancet *1:* 641–642.

Li, A.W., P.J. Krell and D.E. Mahony. 1980. Plasmid detection in a bacteriocinogenic strain of *Clostridium perfringens*. Can. J. Microbiol. *26:* 1018–1022.

Liau, D.F. and J.H. Hash. 1977. Structural analysis of the surface polysaccharide of *Staphylococcus aureus* M. J. Bacteriol. *131:* 194–200.

Libby, J.M., B.S. Jortner and T.D. Wilkins. 1982. Effects of the two toxins of *Clostridium difficile* in antibiotic-associated cecitis in hamsters. Infect. Immun. *36:* 822–829.

Liebert, F. 1909. Verslagen van de gewone vergadering der wis- en natuurkundige afdeeling. Versl. Gewone. Akad. Aust. *17:* 990–1001.

Liekweg, P.D. and L.T. Greenfield. 1977. Vascular prosthetic infections: collected experience and results of treatment. Surgery *81:* 335–342.

Liepe, H.U. 1983. Starter cultures in meat production. *In* Rehm and Reed (Editors), Biotechnology, Vol. 5, Verlag Chemie Weinheim, pp. 399–423.

Lin, A.N., A. Karasik, I.E. Salit and A.G. Fam. 1982. Group G streptococcal arthritis. J. Rheumatol. *9:* 424–427.

Lin, Y.-L. and H.P. Blaschek. 1983. Butanol production by a butanol-tolerant strain of *Clostridium acetobutylicum* in extruded corn broth. Appl. Environ. Microbiol. *45:* 966–973.

Lind, A. 1960. Serological studies of mycobacteria by means of the diffusion-ingel techniques. IV. The precipitinogenic relationships between different species of mycobacteria with special reference to *M. tuberculosis*, *M. phlei.*, *M. smegmatis* and *M. avium*. Int. Arch. Allergy *17:* 300–322.

Lind, A., O. Ouchterlony and M. Ridell. 1980. Mycobacterial antigens. *In* Meissner and Schimiedel (Editors), Infektionskrankheiten und ihre Erreger, Bd. 4. Mycobakterien und mykobakterielle Krankheiten, Fischer Verlag, Jena, pp. 275–303.

Lind, A. and M. Ridell. 1976. Serological relationships between *Nocardia*, *Mycobacterium*, *Corynebacterium* and the 'rhodochrous' taxon. *In* Goodfellow, Brownell and Serrano (Editors), The Biology of the Nocardiae, Academic Press, London and New York, pp. 220–235.

Lindberg, M. 1981. Genetic studies in *Staphylococcus aureus* using protoplasts: cell fusion and transformation. *In* Jeljaszewicz (Editor), Staphylococci and Staphylococcal Infections, Gustav Fischer Verlag, Stuttgart, New York, pp. 535–541.

Lindberg, M., J.-E. Sjöström and T. Johansson. 1972. Transformation of chromosomal and plasmid characters in *Staphylococcus aureus*. J. Bacteriol. *109:* 844–847.

Lindblow-Kull, C., A. Shrift and R.L. Gherna. 1982. Aerobic, selenium-utilizing *Bacillus* isolated from seeds of *Astralagus crotalaria*. Appl. Environ. Microbiol. *44:* 737–743.

Lindenberg, A. 1909. Un nouveau mycétome. Arch. Parasitol. *13:* 265–282.

Linder, P. 1888. Die Sarcinaorganismen der Gärungsgewerben. Dissertation. Berlin, pp. 1–59.

Linder, R. and A.W. Bernheimer. 1978. Effect on sphingomyelin-containing liposomes of phospholipase D from *Corynebacterium ovis* and the cytolysin from *Stoichactis helianthus*. Biochem. Biophys. Acta *530:* 236–246.

Lindgren, S. and G. Clevström. 1978a. Antibacterial activity of lactic acid bacteria. I. Activity of fish silage, a cereal starter and isolated organisms. Swed. J. Agric. Res. *8:* 61–66.

Lindgren, S. and G. Clevström. 1978b. Antibacterial activity of lactic acid bacteria. 2. Activity in vegetable silag-s, Indonesian fermented foods and starter cultures. Swed. J. Agric. Res. *8:* 67–73.

Lindner, P. 1887. Über ein neues in Malzmaischen vorkommendes, milchsäurebildendes. Ferment. Wschr. Brau. *4:* 437–440.

Linell, F. and A. Norden. 1952. Hundinfektioner is imhall genom ny art av-*Mycobacterium*. Nord. Med. *47:* 888–891.

Linett, P.E., R.J. Roberts and J.L. Strominger. 1974. Biosynthesis and crosslinking of the -glutamylglycine-containing peptodoglycan of vegetative cells of *Sporosarcina ureae*. J. Biol. Chem. *249:* 2497–2506.

Linzer, R., K. Gill and H.D. Slade. 1976. Chemical composition of *Streptococcus mutans* type c antigen; comparison to type a, b, and d antigens. J. Dent. Res. *55:* A109–115.

Linzer, R., H. Mukasa and H.D. Slade. 1975. Serological purification of polysaccharide antigens from *Streptococcus mutans* serotypes a and d: characterization of multiple antigenic determinants. Infect. Immun. *12:* 791–798.

Lipman, F. and L.C. Tuttle. 1945. A specific micromethod for determination of acyl-phosphates. J. Biol. Chem. *159:* 21–28.

Lipton, M. and G. Sonnenfeld. 1980. *Actinomyces meyeri* osteomyelitis: An unusual cause of chronic infection of the tibia. Clin. Orthop. Relat. Res. *148:* 169–171.

Lister, J. 1873. A further contribution to the natural history of bacteria and the germ theory of fermentative changes. Quart. J. Microsc. Sci. *13:* 380–408.

Liston, J., W. Weibe and R.R. Colwell. 1963. Quantitative approach to the study of bacterial species. J. Bacteriol. *85:* 1061–1070.

Little, J.G. and P.C. Hanawalt. 1973. Thymineless death and ultraviolet sensitivity in *Micrococcus radiodurans*. J. Bacteriol. *113:* 233–240.

Litwack, G. and A.F. Carlucci. 1958. The pigment of *Micrococcus lysodeikticus*. Nature *181:* 933–934.

Liu, C.-L. and H.D. Peck, Jr. 1981. Comparative bioenergetics of sulfate reductions in *Desulfovibrio* and *Desulfotomaculum* Spp. J. Bacteriol. *145:* 966–973.

Live, I. 1972. Staphylococci in animals: differentiation and relationship to human staphylococcosis. *In* Cohen (Editor), The Staphylococci, Wiley-Interscience (John Wiley and Sons, Inc.), New York, London, Sydney, Toronto, pp. 443–456.

Ljungdahl, L.G. and L.H. Carreira. 1983. High ethanol producing derivatives of *Thermoanaerobacter ethanolicus*. U.S. Patent 44, 385, 117.

Ljungdahl, L.G. and J.K.W. Wiegel. 1981a. Anaerobic thermophilic culture system. U.S. Patent 4, 292, 406.

Ljungdahl, L.G. and J.K.W. Wiegel. 1981b. Anaerobic thermophilic culture. U.S. Patent 4, 292, 407.

Ljungdahl, L.G. and H.G. Wood. 1969. Total synthesis of acetate from CO_2 by heterotrophic bacteria. Annu. Rev. Microbiol. *23:* 515–538.

Ljungdahl, L.G. and H.G. Wood. 1982. Acetate biosynthesis. *In* D. Dolphin (Editors), B^{12}, Vol. 2. John Wiley and Sons, New York, pp. 165–202.

Llory, H., B. Guillo and R.M. Frank. 1971. A cariogenic *Actinomyces viscosus* - a bacteriological and gnotobiotic study. Helv. Odontol. Acta *15:* 134–138.

Locci, R. 1976. Developmental micromorphology of actinomycetes. *In* Arai (Editor), Actinomycetes: The Boundary Micro-organisms, University Park Press, Baltimore, London, Tokyo, pp. 249–297.

Locci, R. 1978. Micromorphological development of *Actinomyces* and related genera. *In* Mordarski, Kurylowicz and Jeljaszewicz (Editors), *Nocardia* and *Streptomyces*, Proceedings of the International Symposium on Nocardia and Streptomyces. Gustav Fischer Verlag, Stuttgart-New York, pp. 173–180.

Locci, R. 1981. Micromorphology and development of actinomycetes. Zentralbl. Bakteriol. Mikrobiol. Hyg. Abt. I Orig. Suppl. *11:* 119–130.

Locci, R., M. Goodfellow and G. Pulverer. 1982. Micro-morphological, morphogenetic and chemical characters of rhodococci. Abstracts of the Fifth International Symposium of Actinomycete Biology, Oaxtepec, Mexico, pp. 118–119.

Locci, R. and K.P. Schaal. 1980. Apical growth in facultative anaerobic actinomycetes as determined by immunofluorescent labeling. Zentralbl. Bakteriol. Parasitenkd. Infektionskr. Hyg. I. Abt. Orig. A *246:* 112–118.

Locci, R. and G.P. Sharples. 1984. Morphology. *In* Goodfellow, Mordarski and Williams (Editors), The Biology of the Actinomycetes, Academic Press, London and New York, pp. 165–199.

Lochhead, A.G. 1958. Two new species of *Arthrobacter* requiring respectively vitamin B12 and the terregens factor. Arch. Mikrobiol. *31:* 163–170.

Lochhead, A.G. and M.O. Burton. 1953. An essential bacterial growth factor produced by microbial synthesis. Can. J. Bot. *31:* 7–22.

Lochhead, A.G. and M.O. Burton. 1956. Importance of soil extract for the enumeration and study of soil bacteria. Transactions of the 6th International Congress of Soil Science, Paris, pp. 157–161.

Lochhead, A.G. and M.O. Burton. 1957. Qualitative studies of soil micro-organisms. XIV. Specific vitamin requirements of the predominant bacterial flora. Can. J. Microbiol. *3:* 35–42.

Lockhart, W.R. and J. Liston. 1970. Methods for numerical taxonomy. American Society for Microbiology, Washington, D.C.

Loeffler, F. 1886. Experimentelle Untersuchungen über Schweine-Rotlauf. Arb. GesundhAmt. Berl. *1:* 46–55.

Loesche, W.J., R.N. Hochett and S.A. Syed. 1972. The predominant cultivated flora of tooth surface plaque removed from institutionalized subjects. Arch. Oral Biol. *17:* 1311–1325.

Loesche, W.J., K.U. Paunio, M.P. Woolfolk and R.N. Hockett. 1974. Collagenolytic activity of dental plaque associated with periodontal pathology. Infect. Immun. *9:* 329–336.

Loewe, L., N. Plummer, C.F. Niven and J.M. Sherman. 1946. *Streptococcus s.b.e.* in subacute bacterial endocarditis. J. Am. Med. Assoc. *130:* 257.

Logan, N.A. and R.C.W. Berkeley. 1981. Classification and identification of members of the genus *Bacillus*. *In* Berkeley and Goodfellow (Editors), The Aerobic Endospore-forming Bacteria. Academic Press, London, pp. 105–140.

Logan, N.A. and R.C.W. Berkeley. 1984. Identification of *Bacillus* strains using the API system. J. Gen. Microbiol. *130:* 1871–1882.

Loginova, L.G., G.I. Khraptsova, L.A. Egorova and T.J. Bagdanova. 1978. Acidophilic, obligate thermophilic bacterium *Bacillus acidocaldarius* isolated from hot springs and soil of Kunashir island. Mikrobiologiya *47:* 947–952.

Löhnis, F. 1909. Die Benennung der Milchsäurebakterien. Zentralbl. Bakteriol. Parasitenkd. Infektionskr. Abt. II *22:* 553–555.

Löhnis, F. 1911. Landwirtschaftlich-bakteriologisches Praktikum. Verlag von Gebrüder Bornträger, Berlin.

Löhr, W. and L. Rabfield. 1931. Die Bakteriologie der Wurmfortsatzentzundung und der appendikularen Peritonitis. Thieme, Stuttgart, pp. 4–83.

London, J. 1971. Detection of phylogenetic relationships between streptococci and lactobacilli by a comparative biochemical and immunological study of isofunctional malic enzymes. J. Dent. Res. *50:* 1083–1093.

London, J. 1976. The ecology and taxonomic status of the lactobacilli. Annu. Rev. Microbiol. *30:* 279–301.

London, J. and M.D. Appleman. 1962. Oxidative and glycerol metabolism of two species of enterococci. J. Bacteriol. *84:* 597–598.

London, J. and N.M. Chace. 1976. Aldolases of the lactic acid bacteria. Demonstration of immunological relationships among eight genera of Gram positive bacteria using antipediococcal aldolase serum. Arch. Microbiol. *110:* 121–128.

London, J. and N.M. Chace. 1983. Relationships among lactic acid bacteria demonstrated with glyceraldehyde-3-phosphate dehydrogenase as an evolutionary probe. Int. J. Syst. Bacteriol. *33:* 723–737.

London, J. and K. Kline. 1973. Aldolase of lactic acid bacteria: a case history in the use of an enzyme as an evolutionary marker. Bacteriol. Rev. *37:* 453–478.

London, J., E.Y. Meyer and S. Kulczyk. 1971. Comparative biochemical and immunological study of malic enzyme from two species of lactic acid bacteria: evolutionary implications. J. Bacteriol. *106:* 126–137.

Long, L. and J. Edwards. 1972. Detailed structure of a dextran from a cariogenic bacterium. Carbohydr. Res. *24:* 216–217.

Long, P.H. and E.A. Bliss. 1934. Studies upon minute hemolytic streptococci. 1. The isolation and cultural characteristics of minute B-hemolytic streptococci. J. Exp. Med. *60:* 619–631.

Long, P.H., E.A. Bliss and C.F. Walcott. 1934. Studies upon minute hemolytic streptococci. 2. The distribution of minute hemolytic streptococci in normal and diseased human beings. J. Exp. Med. *60:* 633–641.

Lorenz, R.P., P.C. Applebaum, R.M. Ward and J.J. Botti. 1982. Chorioamnionitis and possible neonatal infection associated with *Lactobacillus* species. J. Clin. Microbiol. *16:* 558–561.

Lorowitz, W.H. and M.P. Bryant. 1984. *Peptostreptococcus productus* strain that grows rapidly with CO as energy source. Appl. Environ. Microbiol. *47:* 961–964.

Love, D.N., R.F. Jones and M. Bailey. 1979. *Clostridium villosum* sp. nov. from subcutaneous abscesses in cats. Int. J. Syst. Bacteriol. *29:* 241–244.

Love, D.N., R.F. Jones, M. Bailey and R.S. Johnson. 1979. Isolation and characterization of bacteria from abscesses in the subcutis of cats. J. Med. Microbiol. *12:* 207–212.

Love, D.N., R.F. Jones, M. Bailey, R.S. Johnson and N. Gamble. 1982. Isolation and characterization of bacteria from pyothorax (empyaemia) in cats. Vet. Microbiol. *7:* 455–461.

Lovejoy, E.D. and T.W. Hastings. 1911. Isolation and growth of the acne bacillus. J. Cutaneous Dis. *29:* 80–82.

Lovell, R. 1944. Further studies on the toxin of *Corynebacterium pyogenes*. J. Pathol. Bacteriol. *56:* 525–529.

Lovell, R. and M.M. Zaki. 1966a. Studies on growth products of *Corynebacterium ovis*. I. The exotoxin and its lethal action on white mice. Res. Vet. Sci. *7:* 302–306.

Lovell, R. and M.M. Zaki. 1966b. Studies on growth production of *Corynebacterium ovis*. II. Other activities and their relationship. Res. Vet. Sci. *7:* 307–311.

Lovett, P.S. and F.E. Young. 1969. Identification of *Bacillus subtilis* NRRL B-3275 as a strain of *Bacillus pumilus*. J. Bacteriol. *100:* 658–661.

Lowbury, E.J.L. and H.A. Lilly. 1958. Contamination of operating-theatre air with *Cl. tetani*. Br. Med. J. *ii:* 1334–1336.

Lowe, W.E. and T.R.G. Gray. 1972. Ecological studies on coccoid bacteria in a pine forest. I. Classification. Soil Biol. Biochem. *4:* 459–468.

Lu, Y. and X. Yan. 1983. Studies on the classification of thermophilic actinomycetes. IV. Determination of thermophilic members of *Nocardiaceae*. Acta Microbiol. Sin. *23:* 220–228 (in Chinese).

Ludwig, W., K.H. Schleifer, G.E. Fox, E. Seewaldt and E. Stackebrandt. 1981. A phylogenetic analysis of staphylococci, *Peptococcus saccharolyticus* and *Micrococcus mucilaginosus*. J. Gen. Microbiol. *125:* 357–366.

Ludwig, W., K.H. Schleifer and E. Stackebrandt. 1984. 16S rRNA analysis of *Listeria monocytogenes* and *Brochothrix thermosphacta*. FEMS Microbiol. Lett. *25:* 199–204.

Ludwig, W., E. Seewaldt, R. Kilpper-Bälz, K.H. Schleifer, L. Magrum, C.R. Woese, G.F. Fox and E. Stackebrandt. 1985. The phylogenetic position of *Streptococcus* and *Enterococcus*. J. Gen. Microbiol. *131:* 543–551.

Ludwig, W., E. Seewaldt, K.H. Schleifer and E. Stackebrandt. 1981. The phylogenetic status of *Kurthia zopfii*. FEMS Microbiology Lett. *10:* 193–197.

Luedemann, G.M. 1974. Addendum to *Micromonosporaceae*. *In* Buchanan and Gibbons (Editors), Bergey's Manual of Determinative Bacteriology 8th Ed., The Williams and Wilkins Co., Baltimore, pp. 848–855.

Luff, R.D., P.K. Gupta, M.R. Spence and J.K. Frost. 1978. Pelvic actinomycosis and the intrauterine contraceptive device. A cyto-histomorphological study. Am. J. Clin. Pathol. *69:* 581–586.

Lund, B.M. 1967. A study of some motile group D streptococci. J. Gen. Microbiol. *49:* 67–80.

Lund, B.M. and T.F. Brocklehurst. 1978. Pectic enzymes of pigmented strains of *Clostridium*. J. Gen. Microbiol. *104:* 59–66.

Lund, B.M., T.F. Brocklehurst and G.M. Wyatt. 1981a. Characterization of strains of *Clostridium puniceum* sp.nov., a pink-pigmented pectolytic bacterium. J. Gen. Microbiol. *122:* 17–26.

Lund, B.M., T.F. Brocklehurst and G.M. Wyatt. 1981b. Validation List No. 6. Int. J. Syst. Bacteriol. *31:* 215–218.

Lund, B.M., J.M. Gee, N.R. King, R.W. Horne and J.M. Harnden. 1978. The structure of the exosporium of a pigmented *Clostridium*. J. Gen. Microbiol. *105:* 165–174.

Lund, E. 1959. Diagnosis of pneumococci by the optochin and bile tests. Acta Pathol. Microbiol. Scand. *47:* 308–315.

Luscombe, B.M. and T.R.G. Gray. 1971. Effect of varying growth rate on the morphology of *Arthrobacter*. J. Gen. Microbiol. *69:* 433–434.

Lutwick, L.I. and R.C. Rockhill. 1978. Abscess associated with *Rothia dentocariosa*. J. Clin. Microbiol. *6:* 612–613.

Lutz, A. 1886. Zur Morphologie des Mikoorganismus der Lepra. Derm. Stud. Hamburg *1:* 77–100.

Lutz, A., O. Grootten and T. Wurch. 1956. Études des charactères culturaux et biochimiques de bacilles du type *Hemophilus hemolyticus vaginalis*. Rev. Immunol. *20:* 132–138.

Lyman, F. and R.H. Fleming. 1940. Composition of sea water. J. Mar. Res. 3: 134–146.

Lynch, J.M., A. Anderson, F.R. Camacho, A.K. Winters, G.R. Hodges and W.G. Barnes. 1980. Pseudobacteremia caused by Clostridium sordellii. Arch. Intern. Med. 140: 65–68.

Lysenko, O. 1959. The occurrence of species of the genus Brevibacterium in insects. J. Insect Pathol. 1: 34–42.

Lysenko, O. 1983. Bacillus thuringiensis: evolution of a taxonomic conception. J. Invert. Pathol. 42: 295–298.

Mabeck, C.E. 1969. Significance of coagulase-negative staphylococcal bacteriura. Lancet ii: 1150–1152.

Macaskie, L.E. 1982. Inhibition of growth of Brochothrix thermosphacta by palmitic acid. J. Appl. Bacteriol. 52: 339–343.

Macaskie, L.E., R.H. Dainty and P.J.F. Henderson. 1981. The role of thiamine as a factor for the growth of Brochothrix thermosphacta. J. Appl. Bacteriol. 50: 267–273.

Macdonald, I.A., V.D. Bokkenheuser, J. Winter, A.M. McLernon and E.H. Mosbach. 1983b. Degradation of steroids in the human gut. J. Lipid Research 24: 675–700.

Macdonald, I.A. and D.M. Hutchison. 1982. Epimerization versus dehydroxylation of the 7 α-hydroxyl group of primary bile acids: competitive studies with Clostridium absonum and 7 α-dehydroxylating bacteria (Eubacterium sp.). J. Steroid Biochem. 17: 295–303.

Macdonald, I.A., D.M. Hutchison and T.P. Forrest. 1981. Formation of urso- and ursodeoxy-cholic acids from primary bile acids by Clostridium absonum. J. Lipid Res. 22: 458–466.

Macdonald, I.A., D.M. Hutchison, T.P. Forrest, V.D. Bokkenheuser, J. Winter and L.V. Holdeman. 1983a. Metabolism of primary bile acids by Clostridium perfringens. J. Steroid Biochem. 18: 97–104.

Macdonald, I.A., J.F. Jellet, D.E. Mahony and L.V. Holdeman. 1979. Bile salt 3α- and 12α-hydroxysteroid dehydrogenases from Eubacterium lentum and related organisms. Appl. Environ. Microbiol. 37: 992–1000.

Macdonald, I.A., D.E. Mahony, J.F. Jellet and C.E. Meier. 1977. DNA-dependent 3α- and 12α-hydroxysteroid dehydrogenase activities from Eubacterium lentum ATCC no. 25559. Biochim. Biophys. Acta 489: 466–476.

Macdonald, I.A. and P.D. Roach. 1981. Bile salt induction of 7α- and 7β-hydroxysteroid dehydrogenases in Clostridium absonum. Biochim. Biophys. Acta 665: 262–269.

Macdonald, I.A., B.A. White and P.B. Hylemon. 1983c. Separation of 7-α- and 7-β-hydroxysteroid dehydrogenase activities from Clostridium absonum ATCC no. 27555 and cellular response of this organism to bile acid inducers. J. Lipid Res. 24: 1119–1126.

MacDonald, R.E. and S.W. MacDonald. 1962. The physiology and natural relationships of the motile, sporeforming sarcinae. Can. J. Microbiol. 8: 795–808.

Macé, E. 1889. Traité Pratique de Bactériologie, 1st ed. J.-B. Balliéré & Sons, Paris, pp. 1–711.

Macé, E. 1897. Traité Pratique de Bactériologie. 3rd Ed. Baillière, Paris, pp. 1–1144.

MacLean, P.D., A.A. Liebow and A.A. Rosenberg. 1946. A haemolytic corynebacterium resembling Corynebacterium ovis and Corynebacterium pyogenes in man. J. Infect. Dis. 79: 69–90.

MacLennan, J.D. 1939. The nonsaccharolytic plectridial anaerobes. J. Pathol. Bacteriol. 49: 435–548.

MacLennan, J.D. 1943. Anaerobic infections of war wounds in the Middle East. Lancet 2: 94.

MacLennan, J.D. 1962. The histotoxic clostridial infections of man. Bacteriol. Rev. 26: 177–276.

Macrina, F.L., J.L. Reider, S.S. Virgil and D.J. Kopecko. 1977. Survey of extrachromosomal gene pool of Streptococcus mutans. Infect. Immun. 17: 215–226.

Madden, R.H. 1983. Isolation and characterization of Clostridium stercorarium sp. nov., cellulolytic thermophile. Int. J. Syst. Bacteriol. 33: 837–840.

Madden, R.H., M.J. Bryder and N.J. Poole. 1982. Isolation and characterization of an anaerobic, cellulolytic bacterium, Clostridium papyrosolvens sp.nov. Int. J. Syst. Bacteriol. 32: 87–91.

Magnusson, H. 1923. Spezifische infektiose Pneumonie beim Fohlen. Ein neuer Entreneger beim Pferde. Arch. Wiss. Prakt. Tierheilk. 50: 22–38.

Magnusson, H. 1928. The commonest forms of actinomycosis in domestic animals and their etiology. Acta Pathol. Microbiol. Scand. 5: 170–245.

Magnusson, M. 1962. Specificity of sensitins. III. Further studies in guinea pigs with sensitin of various species of Mycobacterium and Nocardia. Am. Rev. Resp. Dis. 86: 395–404.

Magnusson, M. 1967. Identification of species of Mycobacterium on the basis of specificity of the delayed type reaction in guinea pigs. Z. Tuberk. 127: 55–56.

Magnusson, M. 1971. A comparitive study of Mycobacterium gastri and Mycobacterium kansasii by delayed type skin reaction in guinea pigs. Am. Rev. Resp. Dis. 104: 377–384.

Magnusson, M. 1976. Sensitin tests in Nocardia taxonomy. In Goodfellow, Brownell and Serrano (Editors), The Biology of the Nocardiae, Academic Press, New York, pp. 236–265.

Magnusson, M. 1980. Classification and identification of mycobacteria on the basis of sensitin specificity. In Meissner, Schmiedel, Nelles and Pfaffenberg (Editors), Mykobakterien und Mykobakterielle Krankheiten. VEB Gustav Fisher Verlag, Jena. pp. 319–348.

Magnusson, M. 1981. Mycobacterial sensitins - where are we now? Rev. Infect. Dis. 3: 944–948.

Magot, M., J.-P. Carlier and M.R. Popoff. 1983. Identification of Clostridium butyricum and Clostridium beijerinckii by gas-liquid chromatography and sugar fermentation: correlation with DNA homologies and electrophoretic patterns. J. Gen. Microbiol. 129: 2837–2845.

Magrou, J. and A.R. Prévot. 1948. Études de systematique bactérienne. IX. Essai de classification des bactéries phytopathogènes et espèces voisines. Ann. Inst. Pasteur (Paris) 75: 99–108.

Mah, R.A., D.Y.C. Fung and S.A. Morse. 1967. Nutritional requirements of Staphylococcus aureus S-6. Appl. Microbiol. 15: 866–870.

Mahoney, D.E. and A. Li. 1978. Comparative study of ten bacteriocins of Clostridium perfringens. Antimicrob. Agents Chemother. 14: 886–892.

Mahoney, D.E., C.E. Meier, I.A. MacDonald and L.V. Holdeman. 1977. Bile salt degradation by nonfermentative clostridia. Appl. Environ. Microbiol. 34: 419–423.

Makino, T. 1981. Multimolecular form of staphylokinase. In Jeljaszewicz (Editor), Staphylococci and Staphylococcal Infections, Gustav Fischer Verlag, Stuttgart, New York, pp. 351–356.

Malette, M.F., P. Reece and E.A. Dawes. 1974. Culture of Clostridium pasteurianum in defined medium and growth as a function of sulfate concentration. Appl. Microbiol. 28: 999–1003.

Malik, A.C., G.W. Reinhold and E.R. Vedamuthu. 1968. An evaluation of the taxonomy of Propionibacterium. Can. J. Microbiol. 14: 1185–1191.

Malke, H. 1979. Conjugal transfer of plasmids determining resistance to macrolides, lincosamides and streptogramin-B type antibiotics among group A, B, D and H streptococci. FEMS Microbiol. Lett. 5: 335–335.

Mallman, W.L. and E.B. Seligmann. 1950. A comparative study of media for the detection of streptococci in water and sewage. Am. J. Public Health 40: 286–289.

Malmborg, A.S., M. Rylander and H. Selander. 1970. Primary thoracic empyema caused by Clostridium sporogenes. Scand. J. Infect. Dis. 2: 155–156.

Malone, B.H., M. Schreiber, N.J. Schneider and L.V. Holdeman. 1975. Obligately anaerobic strains of Corynebacterium vaginale (Haemophilus vaginalis). J. Clin. Microbiol. 2: 272–275.

Malveaux, F.J. and C.L. San Clemente. 1967. Elution of loosely bound acid phosphatase from Staphylococcus aureus. Appl. Microbiol. 15: 738–743.

Manachini, P.L., A. Craveri and A. Guicciardi. 1968. Compozione in basi dell'acido desossiribonuleico di forme mesofile, termofacoltative e termofile del genere Bacillus. Ann. Microbiol. Enzim. 18: 1.

Mandel, M., E.F. Guba and W. Litsky. 1961. The causal agent of bacterial blight of American holly. Bacteriol. Proc. p.61.

Mangan, D.F. and D.E. Lopatin. 1981. In vitro stimulation of immunoglobulin production from human peripheral blood lymphocytes by a soluble preparation of Actinomyces viscosus. Infect. Immun. 31: 236–244.

Mankau, R. 1975. Bacillus penetrans n. comb. causing a virulent disease of plant-parasitic nematodes. J. Invertebr. Pathol. 26: 333–339.

Mann, J.W., T.W. Jeffries and J.D. Macmillan. 1978. Production and ecological significance of yeast cell-wall degrading enzymes from Oerskovia. Appl. Environ. Microbiol. 36: 594–605.

Mann, S. 1969. Uber die Zellwandbausteine von Listeria monocytogenes und Erysipelothrix rhusiopathiae. Zentralbl. Bakteriol. Parasitenkd. Infektionskr. Hyg. Abt. I 209: 510–518.

Manning, D.J. 1974. Sulphur compounds in relation to Cheddar cheese flavour. J. Dairy Res. 41: 81–87.

Manning, D.J. and H.M. Robinson. 1973. The analysis of volatile substances associated with Cheddar cheese aroma. J. Dairy Res. 40: 63–73.

Mansson, I., R. Norberg, B. Olhagen and N.E. Bjorklund. 1971. Arthritis in pigs induced by dietary factors. Microbiologic, clinical and histologic studies. Clin. Exp. Immunol. 9: 677–693.

Manten, A. 1957. Antimicrobial susceptibility and some other properties of photochromogenic mycobacteria associated with pulmonary disease. Antonie van Leeuwenhoek J. Microbiol Serol. 23: 357–363.

Mára, M., D. Kalvodová, F. Patočka and M. Vaselska. 1973. The effect of haemin on growth and cytochrome production in Corynebacterium haemolyticum. Acta Univ. Carol. Med. 19: 197–206.

Mára, M. and C. Michalec. 1977. Chromatographic study of mycolic acid-like substances in lipids of Listeria monocytogenes. J. Chromatog. 130: 434–436.

Marchoux, E. and F. Sorel. 1912. Rescherches sur la lepre. Ann. Inst. Pasteur (Paris) 26: 675–700.

Mardh, P.A., S. Colleen, B. Hovelius. 1979. Attachment of bacteria to exfoliated cells from the urogenital tract. Invest. Urol. 16: 322–325.

Margalith, P. and G. Beretta. 1960. Rifomycin. XI. Taxonomic study on Streptomyces mediterranei nov. sp. Mycopathol. Mycol. Appl. 13: 321–330.

Margherita, S.S. and R.E. Hungate. 1963. Serological analysis of Butyrivibrio from the bovine rumen. J. Bacteriol. 86: 855–860.

Mariat, F. and H. Lechevalier. 1977. Actinomycètes aerobies pathogènes. In Dumas (Editor), Bacteriologie Médicale Flammarion, Paris, pp. 566a-566z.

Marinova, R. and Ch. Manev. 1979. A study of Listeria monocytogenes strains isolated from men and animals. In Ivanov (Editor), Problems of Listeriosis. (Proceedings of the VII International Symposium, Varna, 1977) National Agroindustrial Union, Center for Scientific Information, Sofia, Bulgaria, pp. 65–69.

Marks, J. 1964. Aspects of the epidemiology of infection of "anonymous" mycobacteria. Proc. Roy. Soc. Med. 57: 479–480.

Marks, J., P.A. Jenkins and W.B. Schaefer. 1969. Identification and incidence of a third type of *Mycobacterium avium*. Tubercle *50:* 394–395.

Marks, J., P.A. Jenkins and M. Tsukamura. 1972. *Mycobacterium szulgai* - a new pathogen. Tubercle *53:* 210–214.

Marks, J. and H. Schwabacher. 1965. Infection due to *Mycobacterium xenopei*. Brit. Med. J. *1:* 32–33.

Marmur, J. and P. Doty. 1961. Thermal renaturation of deoxyribonucleic acids. J. Mol. Biol. *3:* 585–594.

Marmur, J. and P. Doty. 1962. Determination of the base composition of deoxyribonucleic acid from its thermal denaturation temperature. J. Mol. Biol. *5:* 109–118.

Marples, M.J. 1965. The Ecology of the Human Skin. Charles C. Thomas, Springfield.

Marples, R.R., R. Hone, C.M. Motely, J.F. Richardson and J.A. Cress-Morris. 1978. Investigation of coagulase-negative staphylococci from infections in surgical patients. Zentralbl. Bakteriol. Parasitenkd. Infektionskr. Hyg. Abt. I Orig. A *241:* 140–156.

Marples, R.R. and J.F. Richardson. 1981. Characters of coagulase-negative staphylococci collected for a collaborative phage-typing study. *In* Jeljaszewicz (Editor), Staphylococci and Staphylococcal Infections, Gustav Fischer Verlag, Stuttgart, New York, pp. 175–180.

Marrie, T.J., R.S. Falkner, B.W.D. Badley, M.R. Hartlen, S.A. Comeau and H.R. Miller. 1978. Pseudomembranous colitis: isolation of two species of cytotoxic clostridia and successful treatment with vancomycin. Can. Med. Assoc. J. *119:* 1058–1060.

Marrie, T.J., E.V. Haldane, C.A. Swantee and E.A. Kerr. 1981. Susceptibility of anaerobic bacteria to nine antimicrobial agents and demonstration of decreased susceptibility of *Clostridium perfringens* to penicillin. Antimicrob. Agents Chemother. *19:* 51–55.

Marrie, T.J., C. Kwan, M.A. Noble, A. West and L. Duffield. 1982. *Staphylococcus saprophyticus* as a cause of urinary tract infections. J. Clin. Microbiol. *16:* 427–431.

Marsalek, E. and V. Hajek. 1973. The classification of pathogenis staphylococci. *In* Jeljaszewicz (Editor), Staphylococci and Staphylococcal Infections, Karger, Basel, pp. 30–37.

Marshall, B.J. and D.F. Ohye. 1966. *Bacillus macquariensis* n. sp., a psychrotrophic bacterium from sub-antarctic soil. J. Gen. Microbiol. *44:* 41–46.

Marshall, J.H. and G.J. Wilmoth. 1981. Pigments of *Staphylococcus aureus*, a series of triterpenoid carotenoids. J. Bacteriol. *147:* 900–913.

Marshall, R. and R.J. Beers. 1967. Growth of *Bacillus coagulans* in chemically defined media. J. Bacteriol. *94:* 517–521.

Marshall, R. and A.K. Kaufman. 1981. Production of deoxyribonuclease, coagulase, and hemolysins by anaerobic gram-positive cocci. J. Clin. Microbiol. *13:* 787–788.

Marshall, R., V.K. Yasui, R. Prabhala, A.K. Kaufman and I. Wallace. 1981. Growth of *Peptococcus* and *Peptostreptococcus*: effect of variations of culture media on efficiency of recovery. Appl. Environ. Microbiol. *42:* 493–496.

Marth, E.H. and B.E. Ellickson. 1959. Problems created by the presence of antibiotics in milk and milk products. J. Milk Food Technol. *22:* 266–272.

Martin, J.D. and J.O. Mundt. 1972. Enterococci in insects. Appl. Microbiol. *24:* 575–580.

Martin, P.A.W., J.R. Lohr and D.D. Dean. 1981. Transformation of *Bacillus thuringiensis* protoplasts by plasmid deoxyribonucleic acid. J. Bacteriol. *145:* 980–983.

Martin, S.M. (Editor). 1963. Culture Collections: Perspectives and Problems. Proceedings of the Specialists' Conference on Culture Collections, Ottawa, 1962. University of Toronto Press, Toronto.

Martin, S.M. and V.B.D. Skerman. 1972. World Directory of Collections of Cultures of Microorganisms. Wiley-Interscience, New York.

Martin, W.J., M. Gardner and J.A. Washington, II. 1972. In vitro antimicrobial susceptibility of anaerobic bacteria isolated from clinical specimens. Antimicrob. Agents Chemother. *1:* 148–158.

Martinovich, D., M.E. Carter, D.A. Woodhouse and I.P. McCausland. 1972. An outbreak of botulism in wild waterfowl in New Zealand. N.Z. Vet. J. *20:* 61–65.

Marucha, P.T., P.H. Keyes, C.L. Wittenberger and J. London. 1978. Rapid method for identification and enumeration of oral *Actinomyces*. Infect. Immun. *21:* 786–791.

Mashimo, P.A., S.A. Ellison and J. Slots. 1979. Microbial composition of monkey dental plaque. (*Macaca arctoides* and *Macaca fascicularis*). Scand. J. Dent. Res. *87:* 24–31.

Maskell, R. 1974. Importance of coagulase negative staphylococci as pathogens in the urinary tract. Lancet *i:* 1155–1158.

Mason, J.H. 1968. *Clostridium botulinum* type D in mud of lakes of Zululand game parks. J. S. Afr. Vet. Med. Assoc. *39:* 37–38.

Masuda, N. 1981. Deconjugation of bile salts by *Bacteroides* and *Clostridium*. Microbiol. Immunol. *25:* 1–11.

Masuda, S. and I. Kondo. 1981. Variation in reactivities of protein A extracellularly produced by mutant strains with mammalian sera. *In* Jeljaszewicz (Editor), Staphylococci and Staphylococcal Infections, Gustav Fischer Verlag, Stuttgart, New York, pp. 499–504.

Masuo, E. and T. Nakagawa. 1970. Numerical classification of bacteria. Part IV. Relationships among some corynebacteria based on serological similarity alone. Agric. Biol. Chem. *34:* 1375–1382.

Mata, L.J., C. Carrillo and E. Villatoro. 1969. Fecal microflora in healthy persons

in a preindustrial region. Appl. Microbiol. *17:* 596–602.

Matches, J.R. and J. Liston. 1974. Mesophilic clostridia in Puget Sound. Can. J. Microbiol. *20:* 1–7.

Matteuzzi, D., F. Crociani, O. Emaldi, A. Selli and R. Viviani. 1976. Isoleucine production in bifidobacteria. Eur. J. Appl. Microbiol. *2:* 185–194.

Matteuzzi, D., F. Crociani and O. Emaldi. 1978. Amino acids produced by bifidobacteria and some clostridia. Ann. Microbiol. (Inst. Pasteur), *129 B:* 175–181.

Matteuzzi, D., F. Crociani, G. Zani and L.D. Trovatelli. 1971. *Bifidobacterium suis* n. sp.: a new species of the genus *Bifidobacterium* isolated from pig feces. Allg. Mikrobiol. *11:* 387–395.

Matteuzzi, D., J. Hollaus and B. Biavati. 1978. Proposal of neotype for *Clostridium thermohydrosulfuricum* and the merging of *Clostridium tartarivorum* with *Clostridium thermosaccharolyticum*. Int. J. Syst. Bacteriol. *28:* 528–531.

Matteuzzi, D., L.D. Trovatelli, B. Biavati and G. Zani. 1977. Clostridia from Grana Cheese. J. Appl. Bacteriol. *43:* 375–382.

Matuszewski, T.E., E. Pÿenowski and J. Supinska. 1936. *Streptococcus diacetilactis*, a new species. Prace Zakl. Mikrob. Prezem. Roln. Warsz. *11:* 1–28.

Maxam, A.M. and W. Gilbert. 1977. A new method for sequencing DNA. Proc. Nat. Acad. Sci. USA *74:* 560–564.

Maximescu, P., A. Oprisan, A. Pop, and E. Potorac. 1974. Further studies on *Corynebacterium* species capable of producing diphtheria toxin (*C. diphtheriae, C. ulcerans, C. ovis*). J. Gen. Microbiol. *82:* 49–56.

Maxted, W.R. 1948. Preparation of streptococcal extracts for Lancefield grouping. Lancet *ii:* 255–256.

Maxted, W.R. 1953. The use of bacitracin for identifying group A hemolytic streptococci. J. Clin. Pathol. *6:* 224–226.

Maxted, W.R. 1957. The active agent in nascent phage lysis of streptococci. J. Gen. Microbiol. *16:* 584–595.

Maxted, W.R. 1978. Group A Streptococci: pathogenesis and immunity. *In* Skinner and Quesnel (Editors), Streptococci, Society for Applied Bacteriology Symposium Series No. 7, Academic Press, London, New York and San Francisco, pp. 107–142.

Mayr, U., R. Hensel and O. Kandler. 1980. Factors affecting the quarternary structure of allosteric L-lactate dehydrogenase from *Lactobacillus casei* and *Lactobacillus curvatus* as investigated by hybridization and ultracentrifugation. Eur. J. Biochem. *110:* 527–538.

Mayr, U., R. Hensel and O. Kandler. 1982. Subunit composition and substrate binding region of potato L-lactate dehydrogenase. Phytochemistry *21:* 627–631.

Mayrand, D. and G. Bourgeau. 1982. Production of phenyl acetic acid by anaerobes. J. Clin. Microbiol. *16:* 747–750.

Mays, T.D., L.V. Holdeman, W.E.C. Moore, M. Rogosa and J.L. Johnson. 1982. Taxonomy of the genus *Veillonella* Prévot. Int. J. Syst. Bacteriol. *32:* 28–36.

Mazanec, K., M. Kocur and T. Martinec. 1965. Electron microscopy of ultrathin sections of *Sporosarcina ureae*. J. Bacteriol. *90:* 808–816.

McBeth, I.G. and F.M. Scales. 1913. The destruction of cellulose by bacteria and filamentous fungi. U.S. Bur. Plant. Ind. *266:* 1–52.

McBryde, C.N. 1911. A bacteriological study of ham souring. U.S. Bur. Anim. Ind. *132:* 1–55.

McCarthy, A.J. and T. Cross. 1984. A taxonomic study of *Thermomonospora* and other monosporic actinomycetes. J. Gen. Microbiol. *130:* 5–25.

McCarthy, A.J., T. Cross, J. Lacey and M. Goodfellow. 1983. Conservation of the name *Micropolyspora* Lechevalier, Solotorovsky and McDurmont and designation of *Micropolyspora faeni* Cross, Maciver, and Lacey as the type species of the genus. Request for an Opinion. Int. J. Syst. Bacteriol. *33:* 430–433.

McCarthy, B.J. and E.T. Bolton. 1963. An approach to the measurement of genetic relatedness among organisms. Proc. Nat. Acad. Sci. USA *50:* 156–164.

McCarthy, C. and P. Ashbaugh. 1981. Factors that effect the cell cycle of *Mycobacterium avium*. Rev. Infect. Dis. *3:* 914–925.

McCarthy, L.R., P.A. Mickelsen and E.G. Smith. 1979. Antibiotic susceptibility of *Haemophilus vaginalis (Corynebacterium vaginale)* to 21 antibiotics. Antimicrob. Agents Chemother. *16:* 186–189.

McClung, L.S. 1935. Studies on anaerobic bacteria IV. Taxonomy of cultures of a thermophilic species causing "swells" of canned foods. J. Bacteriol. *29:* 189–202.

McClung, L.S. 1943. On the enrichment and purification of chromogenic sporeforming anaerobic bacteria. J. Bacteriol. *46:* 507–511.

McClung, L.S. 1982. The anaerobic bacteria, their activities in nature and disease. Part I, Vol. 1, The literature for 1940–1951; Part I, Vol. 2, The literature for 1952–1959; Part I, Vol. 3, The literature for 1960–1965; Part I, Vol. 4, The literature for 1966–1969; Part I, Vol. 5, The subject listings for 1940–1969; Part II, Vol. 1, The literature for 1970–1975; Part II, Vol. 2, The subject listings for 1970–1975. Marcel Dekker, Inc., New York.

McClung, L.S. and E. McCoy. 1935. Studies on anaerobic bacteria. VII. The serological relations of *Clostridium acetobutylicum, Cl. felsineum*, and *Cl. roseum*. Arch. Mikrobiol. *6:* 239–249.

McClung, L.S. and E. McCoy. 1957. Genus II *Clostridium* Prazmowski 1880. *In* Breed, Murray, and Smith (Editors), Bergey's Manual of Determinative Bacteriology, 7th ed. The Williams and Wilkins Co., Baltimore, pp. 634–693.

McClung, N.M. 1974. Genus *Nocardia* Trevisan 1889,9. *In* Buchanan and Gibbons (Editors), Bergey's Manual of Determinative Bacteriology, 8th Ed., The

Williams and Wilkins Company, Baltimore, pp. 726–746.

McCoy, E., E.B. Fred, W.H. Peterson and E.G. Hastings. 1926. A cultural study of the acetone butyl alcohol organisms. J. Infect. Dis. *39:* 457–483.

McCoy, E. and L.S. McClung. 1935. Studies on anaerobic bacteria. VI. The nature and systematic position of a new chromogenic *Clostridium.* Arch. Mikrobiol. *6:* 230–238.

McCray, A.H. 1917. Spore-forming bacteria of the apiary. J. Agric. Res. *8:* 399–420.

McCready, R.G.L., E.J. Laishley and H.R. Krouse. 1976. The use of stable sulfur isotope labelling to elucidate sulfur metabolism by *Clostridium pasteurianum.* Arch. Microbiol. *109:* 315–317.

McCulloch, L. 1925. *Aplanobacter insidiosum* n. sp. the cause of an alfalfa disease. Phytopathology 15: 496–497.

McDonald, T.J. and J.S. McDonald. 1976. Streptococci isolated from bovine intramammary infections. Am. J. Vet. Res. *37:* 377–381.

McDonald, W.C., J.C. Felkner, A. Turetsky and T.S. Matney. 1963. Similarity in base composition of deoxyribonucleates from several strains of *Bacillus cereus* and *Bacillus anthracis.* J. Bacteriol. *85:* 1071–1073.

McGaughey, C.A., J.K. Bateman and P.Z. MacKenzie. 1951. Actinomycosis in the dog. Br. Vet. J. *107:* 428–430.

McGaughey, C.A. and H.P. Chu. 1948. The egg-yolk reaction of aerobic sporing bacilli. J. Gen. Microbiol. *2:* 334–340.

McGinley, K.J., G.F. Webster and J.J. Leyden. 1978. Regional variations of cutaneous propionibacteria. Appl. Environ. Microbiol. *35:* 62–66.

McIntire, F.C., A.E. Vatter, J.B. Baros and J. Arnold. 1978. Mechanism of coaggregation between *Actinomyces viscosus* T14V and *Streptococcus sanguis* 34. Infect. Immun. *21:* 978–988.

McKay, L.L., K.A. Baldwin and P.M. Walsh. 1980. Conjugal transfer of genetic information in group N streptococci. Appl. Environ. Microbiol. *40:* 84–91.

McKay, L.L., K.A. Baldwin and E.A. Zottola. 1972. Loss of lactose metabolism in lactic streptococci. J. Bacteriol. *123:* 1090–1096.

McLean, R.A. and W.L. Sulzbacher. 1953. *Microbacterium thermosphactum,* spec. nov; a nonheat resistant bacterium from fresh pork sausages. J. Bacteriol. *65:* 428–433.

McLeod, J.W. 1943. The types Mitis, Intermedius and Gravis of *Corynebacterium diphtheriae.* Bacteriol. Rev. *7:* 1–41.

McRae, D.H. and C.C. Shepard. 1971. Relationship between the staining quality of *Mycobacterium leprae* and infectivity for mice. Infect. Immun. *3:* 116–120.

Mead, G.C. 1971. The amino acid-fermenting clostridia. J. Gen. Microbiol. *67:* 47–56.

Mead, G.C., B.W. Adams, M.G. Hilton and P.G. Lord. 1979. Isolation and characterization of uracil-degrading clostridia from soil. J. Appl. Bacteriol. *46:* 465–472.

Mead, G.C., A.M. Chamberlain and E.D. Borland. 1973. Microbial changes leading to the spoilage of hung pheasants, with special reference to the clostridia. J. Appl. Bacteriol. *36:* 270–287.

Medrek, T.F. and E.M. Barnes. 1962. The physiological and serological properties of *Streptococcus bovis* and related organisms isolated from cattle and sheep. J. Appl. Bacteriol. *25:* 169–179.

Mees, R.H. 1934. Onderzoekingen over de Biersarcina. Thesis. Technical University, Delft, Holland pp. 1–110.

Meijer-Severs, G.J., J.G. Aarnoudse, W.F.A. Mensink and J. Dankert. 1979. The presence of antibody-coated anaerobic bacteria in asymptomatic bacteriuria during pregnancy. J. Infect. Dis. *140:* 653–658.

Meissner, G. and K.H. Schröder. 1975. Relationship between *M. simiae* and *M. habana.* Am. Rev. Resp. Dis. *111:* 196–200.

Meissner, G., K.H. Schröder, G.E. Amadio, W. Anz, S. Chaparas, H.W.B. Engel, P.A. Jenkins, W. Käppler, H.H. Kleeberg, E. Kubala, M. Kubin, D. Lauterbach, A. Lind, M. Magnusson, Zd. Mikova, S.R. Pattyn, W.B. Scheaffer, J.L. Stanford, M. Tsukamura, L.G. Wayne, I. Willers and E. Wolinsky. 1974. A cooperative numerical analysis of nonscoto- and nonphotochromogenic slowly growing mycobacteria. J. Gen. Microbiol. *83:* 207–235.

Mejare, B. and S. Edwardsson. 1975. *Streptococcus milleri* (Guthof); an indigenous organism of the human oral cavity. Arch. Oral Biol. *20:* 757–762.

Melish, M.E., F.S. Chen, S. Sprouse, M. Stuckey and M.S. Murata. 1981. Epidemolytic toxin in staphylococcal infection: toxin levels and host response. *In* Jeljaszwicz (Editor), Staphylococci and Staphylococcal Infections, Gustav Fischer Verlag, Stuttgart, New York, pp. 287–298.

Melish, M.E., L.A. Glasgow, M.D. Turner and C.B. Lillibridge. 1976. The staphylococcal epidermolytic toxin. *In* Jeljaszwicz (Editor), Staphylococci and Staphylococcal Diseases, Gustav Fischer Verlag, Stuttgart, New York, pp. 473–488.

Melles, Z., I. Nikodemusz and A. Abel. 1969. Die pathogene Wirkung aerober sporen-bildender Bakterien. Zentralbl. Bakteriol. Parasitenkd. Infektionskr. Hyg. Abt. I Orig. *212:* 174–176.

Melville, T.H. 1965. A study of the overall similarity of certain actinomycetes mainly of oral origin. J. Gen. Microbiol. *40:* 309–315.

Mendel, B. 1919. *Cladothrix* et infection d'origin dentiare. C.R. Séances Soc. Biol. Filiales *82:* 583–586.

Mendez, B.S. and R.F. Gomez. 1982. Isolation of *Clostridium thermocellum* auxotrophs. Appl. Environ. Microbiol. *43:* 495–496.

Mercer, E.I., N. Modi, D.J. Clarke and J.G. Morris. 1979. The occurrence and location of squalene in *Clostridium pasteurianum.* J. Gen. Microbiol. *111:* 437–440.

Merkal, R.S. 1979. Proposal of ATCC 19698 as the neotype strain of *Mycobacterium paratuberculosis* Bergey et. al. 1923. Int. J. Syst. Bacteriol. *29:* 263–264.

Merkal, R.S. and W.G. McCullough. 1982. A new Mycobactin, mycobactin J, from *Mycobacterium paratuberculosis.* Curr. Microbiol. *7:* 333–335.

Messner, P., F. Hollaus and U.B. Sleytr. 1984. Paracrystalline cell wall surface layers of different *Bacillus stearothermophilus* strains. Int. J. Syst. Bacteriol. *34:* 202–210.

Metcalf, G. and M.E. Brown. 1957. Nitrogen fixation by new species of *Nocardia.* J. Gen. Microbiol. *17:* 567–572.

Metschnikoff, M.E. 1888. *Pasteuria ramosa,* un representant des bactéries a division longitudinale. Ann. Inst. Pasteur Paris *2:* 165–170.

Metschnikoff, E. 1888. Ueber die phagocytare Rolle der Tuberkelriesenzellen. Virchows Arch. *113:* 63–94.

Meyer, D.J. and C.W. Jones. 1973. Distribution of cytochromes in bacteria: relationship to general physiology. Int. J. Syst. Bacteriol. *23:* 459–467.

Meyer, K. 1911. Ueber eine anaerobe *Streptothrix*-Art. Zentralbl. Bakteriol. Hyg. I. Abt. Orig. *60:* 75–78.

Meyer, S.A. 1979. Nucleic acid homology within the genus *Staphylococcus.* Ph. D. Diss. Technischen Universität München.

Meyer, S.A. and K.H. Schleifer. 1978. Deoxyribonucleic acid reassociation in the classification of coagulase-positive staphylococci. Arch. Microbiol. *117:* 183–188.

Meyer, V. 1956. Die Bestimmung der Bombage-Arten bei Fischkonserven. Fischwirtschaft *8:* 212.

Meyer, V. 1962. Problems des Verderbens von Fischkonserven in Dosen. VII. Untersuchungen über die Entstehung der Aminosäuren beim Marinieren von Heringen. Veroeff. Inst. Meeresforsch. Bremerhaven *8:* 21.

Meyer, W. 1967. A proposal for subdividing the species *Staphylococcus aureus.* Int. J. Syst. Bacteriol. *17:* 387–389.

Meynell, E. and G.G. Meynell. 1964. The roles of serum and carbon dioxide in capsule formation by *Bacillus anthracis.* J. Gen. Microbiol. *34:* 153–163.

Michel, M.F. and R.M. Krause. 1967. Immunochemical studies on the group F streptococci, and the identification of a group-like carbohydrate in a type II strain with an undesignated group antigen. J. Exp. Med. *125:* 1075–1089.

Michel, M.F. and J.M.N. Willers. 1964. Immunochemistry of Group F streptococci; isolation of group specific oligosaccharides. J. Gen. Microbiol. *37:* 381–389.

Mickelsen, P.A., L.R. McCarthy and M.E. Mangum. 1977. New differential medium for the isolation of *Corynebacterium vaginale.* J. Clin. Microbiol. *5:* 488–489.

Middlebrook, G. 1954. Isoniazid resistance and catalase activity of tubercle bacilli. Am. Rev. Tuberc. *69:* 471–472.

Middlebrook, G. and M.L. Cohn. 1953. Some observations on the pathogenicity of isoniazid-resistant variants of tubercle bacilli. Science *118:* 297–299.

Midtvedt, T. and A. Norman. 1967. Bile acid transformations by microbial strains belonging to genera found in intestinal contents. Acta Pathol. Microbiol. Scand. *71:* 629–638.

Mielenz, J.R. 1983. *Bacillus stearothermophilus* contains a plasmid-borne gene for alpha-amylase. Proc. Nat. Acad. Sci. *80:* 5975–5979.

Miessner, H. and R. Berge. 1929. Die Paratuberkulose des Rindes. *In* Kolle und Wassermann (Editors), Handbuch Pathogenen Mikroorganismen. Fisher, Jena. Vol. 6. pp. 779–798.

Mieth, H. 1962. Untersuchungen über das Vorkommen von Enterokokken bei Tieren und Menschen. IV Mitteilung. Die Streptokokkenflora in den Faeces von Pferden. Zentralbl. Bakteriol. Parasitenkd. Infektionskr. Hyg. Abt. 1 Orig. *185:* 166–174.

Migula, W. 1894. Über ein neues System der Bakterien. Arb. Bakt. Inst. Karlsruhe *1:* 235–238.

Migula, W. 1900. System der Bakterien. Vol. 2. Gustav Fischer, Jena.

Mihelc, V.A., C.L. Duncan and G.H. Chambliss. 1978. Characterization of a bacteriocinogenic plasmid in *Clostridium perfringens* CW55. Antimicrob. Agents Chemother. *14:* 771–779.

Mikesell, P., B.E. Ivins, J.D. Ristroph and T.M. Dreider. 1983. Evidence for plasmid-mediated toxin production in *Bacillus anthracis.* Infect. Immun. *39:* 371–376.

Miksza-Zylkiewicz, R. 1980. Complex characteristics of *Rothia dentocariosa* strains as possible etiological factors in peridontitis. Rocz. Akad. Med. Bialymstoku. *25:* 139–162.

Miksza-Zylkiewicz, R. and H. Linda. 1980. Experimental infection in the mouse by a *Rothia dentocariosa* strain. Czas. Stomatol. *33:* 1073–1076.

Milhaud, G., J.-P. Aubert and C.B. van Niel. 1956. Etude de la glycolyse de *Zymosarcina ventriculi.* Ann. Inst. Pasteur. *91:* 363–368.

Miller, A., III, M.E. Morgan and L.M. Libbey. 1974. *Lactobacillus maltaromicus,* a new species producing a malty aroma. Int. J. Syst. Bacteriol. *24:* 346–354.

Miller, A., III, W.E. Sandine and P.R. Elliker. 1970. Deoxyribonucleic acid base composition of lactobacilli determined by thermal denaturation. J. Bacteriol. *102:* 278–280.

Miller, B.A., R.F. Reiser and M.S. Bergdoll. 1978. Detection of staphylococcal enterotoxin A, B, C, D and E in foods by radioimmunoassay, using staphylococcal cells containing protein A as immunoadsorbent. Appl. Environ. Microbiol. *36:* 421–426.

Miller, C.H. 1974. Degradation of sucrose by whole cells and plaque of *Actinomyces naeslundii.* Infect. Immun. *10:* 1280–1291.

Miller, C.H., C.J. Palenik and K.E. Stamper. 1978. Factors affecting the aggregation of *Actinomyces naeslundii* during growth and in washed cell suspensions. Infect. Immun. *21:* 1003–1009.

Miller, C.H. and P.J.B. Somers. 1978. Degradation of levan by *Actinomyces viscosus*. Infect. Immun. *22:* 266–274.

Miller, I.L. and S.J. Silverman. 1959. Glucose metabolism of *Listeria monocytogenes*. Bacteriol. Proc. p. 103.

Miller, L.G. and S.M. Finegold. 1967. Antibacterial sensitivity of *Bifidobacterium* (*Lactobacillus bifidus*). J. Bacteriol. *93:* 125–130.

Miller, T.L. and J.B. Evans. 1970. Nutritional requirements for growth of *Aerococcus viridans*. J. Gen. Microbiol. *61:* 131–135.

Miller, T.L. and M.J. Wolin. 1979. Fermentations by saccharolytic intestinal bacteria. Am. J. Clin. Nutr. *32:* 164–172.

Mills, C.K. and E.F. Lessel. 1973. Designation and description of the type strain of *Lactobacillus casei* subsp. *alactosus* Rogosa et al. Int. J. Syst. Bacteriol. *23:* 67–68.

Minett, F.C. 1932. Avian tuberculosis in cattle of Great Britian. J. Comp. Pathol. Therap. *45:* 317–330.

Minett, F.C. 1934. *Streptococcus* mastitis in cattle; bacteriology and preventive medicine. Twelfth Int. Vet. Cong. New York *2:* 511–532.

Minnikin, D.E., H. Abdolrahimzadeh and J. Baddiley. 1974. The occurence of phosphatidylethanolanmine and glycosyl diglycerides in thermophilic bacilli. J. Gen. Microbiol. *83:* 415–418.

Minnikin, D.E., L. Alshamaony and M. Goodfellow. 1975. Differentiation of *Mycobacterium*, *Nocardia* and related taxa by thin-layer chromatographic analysis of whole-organism methanolysates. J. Gen. Microbiol. *88:* 200–204.

Minnikin, D.E., M.D. Collins and M. Goodfellow. 1978. Menaquinone patterns in the classification of nocardioform and related bacteria. Zentralbl. Bakteriol. Parasitenkd. Infektionskr. Hyg. Abt. Suppl. *6:* 85–90.

Minnikin, D.E., M.D. Collins and M. Goodfellow. 1979. Fatty acid and polar lipid composition in the classification of *Cellulomonas, Oerskovia* and related taxa. J. Appl. Bacteriol. *47:* 87–95.

Minnikin, D.E., G. Dobson, M. Goodfellow, P. Draper and M. Magnusson. 1985. Quantitative comparison of the mycolic and fatty acid composition of *Mycobacterium leprae* and *Mycobacterium gordonae*. J. Gen. Microbiol. *131:* 2013–2021.

Minnikin, D.E. and M. Goodfellow. 1980a. Lipid composition in the classification and identification of acid-fast bacteria. *In* Microbiological Classification and Identification (M. Goodfellow and R.G. Board, Editors) Academic Press, London. pp. 189–256.

Minnikin, D.E. and M. Goodfellow. 1980b. Mycolic acid patterns in mycobacterial classification. *In* 1954–1979: Twenty Five Years of Mycobacterial Taxonomy (G.P. Kubica, L.G. Wayne and L.S. Good, Editors) U.S. Department of Health and Human Services, Center for Disease Control, Atlanta. pp. 159–170.

Minnikin, D.E. and M. Goodfellow. 1981. Lipids in the classification of *Bacillus* and related taxa. *In* Berkeley and Goodfellow (Editors), The Aerobic Endospore-forming Bacteria. Academic Press, London, pp. 59–90.

Minnikin, D.E. and M. Goodfellow. 1981. Lipids in the classification of actinomycetes. *In* Schaal and Pulverer (Editors). Actinomycetes. Proceedings of the IVth International Symposium on Actinomycete Biology, Fischer, Stuttgart, New York, pp. 99–109.

Minnikin, D.E., M. Goodfellow and M.D. Collins. 1978. Lipid composition in the classification and identification of coryneform and related taxa. *In* Bousfield and Callely (Editors), *Coryneform Bacteria*. Academic Press, London.

Minnikin, D.E., I.G. Hutchinson, A.B Caldicott and M. Goodfellow. 1980. Thin-layer chromatography of methanolysates of mycolic acid-containing bacteria. J. Chromatogr. *188:* 221–233.

Minnikin, D.E., S.M. Minnikin, J.M. Parlett, M. Goodfellow and M. Magnusson. 1984a. Mycolic acid patterns of some species of *Mycobacterium*. Arch. Microbiol. *139:* 225–231.

Minnikin, D.E., A.G. O'Donnell, M. Goodfellow, G. Alderson, M. Athalye, A. Schaal and J.H. Parlett. 1984b. An integrated procedure for the extraction of bacterial isoprenoid quinones and polar lipids. J. Microbiol. Methods *2:* 233–241.

Minnikin, D.E., P.V. Patel, L. Alshamaony and M. Goodfellow. 1977. Polar lipid composition in the classification of *Nocardia* and related bacteria. Int. J. Syst. Bacteriol. *27:* 104–117.

Minshew, B.H. and E.D. Rosenblum. 1970. Transduction of tetracycline resistance in *Staphylococcus epidermidis*. Bacteriol. Proc. *1970:* 22.

Minshew, B.H. and E.D. Rosenblum. 1973. Plasmid for tetracycline resistance in *Staphylococcus epidermidis*. Antimicrob. Agents Chemother. *3:* 568–574.

Minsker, O.B. and M.A. Moskovskaya. 1979. Abdominal actinomycosis: Some aspects of pathogenesis, clinical manifestation and treatment. Mykosen *22:* 393–408.

Miquel, P. 1889. Étude sur la fermentation ammoniacale et sur les ferments de l'urée. Ann. Micrographie *1:* 506–519.

Mirick, G.S., L. Thomas, E.C. Curnen and F.L. Horsfall. 1944a. Studies on a non-hemolytic streptococcus isolated from the respiratory tract of human beings. I. Biological characteristics of *Streptococcus* MG. J. Exp. Med. *80:* 391–406.

Mirick, G.S., L. Thomas, E.C. Curnen and F.L. Horsfall. 1944b. Studies on a non-hemolytic streptococcus isolated from the respiratory tract of human beings. II. Immunological characteristics of *Streptococcus* MG. J. Exp. Med. *80:* 407–430.

Mirick, G., L. Thomas, E. Curnen and F. Horsfall, Jr. 1944c. Studies on a non-haemolytic streptococcus isolated from the respiratory tract of human beings III. Immunological relationship of *Streptococcus* MG to *Streptococcus salivarius* Type I. J. Exp. Med. *80:* 431–440.

Mishra, S.K., R.E. Gordon and D.A. Barnett. 1980. Identification of nocardiae and streptomycetes of medical importance. J. Clin. Microbiol. *11:* 728–736.

Mishustin, E.N. 1950. Termofilnie mikroorganizmi v prirode i praktike. Akademi Nauk SSSR, Moskwa.

Mitchell, P.D., C.S. Hintz and R.C. Haselby. 1977. Malar mass due to *Actinomyces odontolyticus*. J. Clin. Microbiol. *5:* 658–660.

Mitchell, R.W., B. Hahn-Hagerdal, J.D. Ferchak and E.K. Pye. 1982. Characterization of β-1,4-glucosidase activity in *Thermoanaerobacter ethanolicus*. Biotech. Bioeng. Symp. *12:* 462–467.

Mitchinson, D.A., J.G. Wallace, A.L. Bhatia, J.B. Selkon, T.V. Subaiah and M.C. Lancaste. 1960. A comparison of the virulence in guinea pigs of South Indian and British tubercle bacilli. Tubercle *41:* 1–22.

Miteva, V.I., N.I. Shivarova and R.T. Grigorova. 1981. Transformation of *Bacillus thuringiensis* protoplasts by plasmid DNA. FEMS Microbiol. Lett. *12:* 253–256.

Mitsui, K., N. Mitsui, K. Kobashi and J. Hase. 1982. High-molecular-weight hemolysin of *Clostridium tetani*. Infect. Immun. *35:* 1986–1990.

Mitsuoka, T. 1969. Vergleichende Untersuchungen über die Laktobazillen aus den Faeces von Menschen, Schweinen und Hühnern. Zentralbl. Bakteriol. Parasitenkd. Infektionskr. Hyg. I Abt. Orig. *210:* 32–51.

Mitsuoka, T. 1969. Vergleichende Untersuchungen über die Bifidobakterien aus dem Berdauungstrakt von Menschen und Tieren. Zentralbl. Bakteriol. Parasitenkd. Infektionskr. Hyg. Abt. I Orig. *210:* 52–64.

Mitsuoka, T., Y. Morishita, A. Terada und K. Watanabe. 1974. *Actinomyces eriksonii* Georg, Roberstad, Brinkman und Hicklin 1965 identisch mit *Bifidobacterium adolescentis* Reuter 1963. Zentralbl. Bakteriol. Parasitenkd. Infektionskr. Hyg. I. Abt. Orig. A *226:* 257–263.

Mitsuoka, T., T. Sega and S. Yamamoro. 1965. Eine verbesserte Methodik der qualitativen und quantitativen Analyse der Darmflora von Menschen und Tieren. Zentralbl. Bakteriol. Parasitenkd. Infektionskr. Hyg. Abt. I Orig. *195:* 455–469.

Mitsuoka, T., T. Sega and S. Yamamoto. 1965. Eine verbesserte Methodik der qualitativen, und quantitativen Analyse der Darmflora von Menschen und Tieren. Zentralbl. Bakteriol. Parasitenkd. Infektionskr. Hyg. Abt. I Orig. *195:* 455–469.

Miwa, T. 1975a. Clostridia in soil of the Antarctica. Jpn. J. Med. Sci. Biol. *28:* 201–213.

Miwa, T. 1975b. Clostridia isolated from the soil in the East Coast of Lützow-Holm Bay, East Antarctica. Antarct. Rec. *53:* 89–99.

Miwa, T., I. Mochizuki, K. Watanabe, S. Kobata, H. Imamura, K. Ninomiya, K. Ueno and S. Suzuki. 1976. The susceptibility of clostridia from the antarctic soil to antibiotics. *In* Williams and Geddes (Editors), Chemotherapy, Vol. 2. Plenum Press, New York, pp. 89–93.

Miyagawa, E. 1982. Cellular fatty acid and fatty aldehyde composition of rumen bacteria. J. Gen. Appl. Microbiol. *28:* 389–408.

Miyamoto, Y., K. Takizawa, A. Matsushima, Y. Asai and S. Nakatsuka. 1978. Stepwise acquisition of multiple drug resistance by beta hemolytic streptococci and difference in resistance pattern by type. Antimicrob. Agents. Chemother. *13:* 399–404.

Miyazawa, Y. and C.A. Thomas. 1965. Composition of short segments of DNA molecules. J. Mol. Biol. *11:* 223–237.

Mizuno, S., H. Toyama and M. Tsukamura. 1966. Susceptibility of various mycobacteria to ethambutol. Differentiation between *M. avium* and *M. terrae*. Jpn. J. Bacteriol. *21:* 672–674.

Mizutami, T. and T. Mitsuoka. 1979. Effect of intestinal bacteria on incidence of liver tumors in gnotobiotic C3H/He male mice. J. Nat. Cancer Inst. *63:* 1365–1370.

Moberg, K. and E. Thal. 1954. Beta-hämolytische Streptokokken einer neuen Lancefield Gruppe. Nord Veterinaermed. *6:* 69–72.

Moeller, A. 1898. Mikroorganismen, die den Tuberkelbacillen ähnlich sind und bei Thieren eine miliare Tuberkelkrankheit verursachten. Dtsch. Med. Wochenschr. *24:* 376–379.

Moens, Y. and W. Verstraeten. 1980. Actinomycosis due to *Actinomyces viscosus* in a young dog. Vet. Rec. *106:* 344–345.

Mohan, K. 1981. *Brevibacterium* sp. from poultry. Antonie van Leeuwenhoek. J. Microbiol. Serol. *47:* 449–453.

Moir, R.J. and M.J. Masson. 1952. An illustrated scheme for the microscopic identification of the rumen-organisms of sheep. J. Pathol. Bacteriol. *64:* 343–350.

Moiroud, A. and A.M. Gounot. 1969. Sur une bactérie psychrophile obligatoire isolée de limons glaciaires. Hebd. Seances Acad. Sci. Ser. D Sci. Nat. *269:* 2150–2152.

Møller-Madsen, A.A. and H. Jensen. 1962. Transformation of *Streptococcus lactis*. *In* Contributions to the XVIth Int. Dairy Congress, Copenhagen.

Molongoski, J.J. and M.J. Klug. 1976. Characterization of anaerobic, heterotrophic bacteria isolated from fresh-water lake sediments. Appl. Environ. Microbiol. *31:* 83–90.

Momose, H., E. Yamanaka, H. Akiyama and K. Nosiro. 1974. Taxonomic study on hiochi-bacteria, with special reference to deoxyribonucleic acid base composition and chemical composition of bacterial cell wall. J. Gen. Appl.

Microbiol. *20:* 179–185.

Monot, F., J.-R. Martin, H. Petitdamage and R. Gay. 1982. Acetone and butanol production by *Clostridium acetobutylicum* in a synthetic medium. Appl. Environ. Microbiol. *44:* 1318–1324.

Montague, E.A. and K.W. Knox. 1968. Antigenic components of the cell walls of *Streptococcus salivarius.* J. Gen. Microbiol. *54:* 237–246.

Montgomery, L. and J.M. Macy. 1982. Characterization of rat cecum cellulolytic bacteria. Appl. Environ. Microbiol. *44:* 1435–1443.

Monticello, D.J. and R.N. Costilow. 1982. Interconversion of valine and leucine by *Clostridium sporogenes.* J. Bacteriol. *152:* 946–949.

Montiel, F. and H.J. Blumenthal. 1965. Factors affecting the pathways of glucose catabolism and the tricarboxylic acid cycle in *Staphylococcus aureus.* Bacteriol. Proc. *1965:* 77.

Moody, M.D., E.C. Ellis and E.L. Updyke. 1958. Staining bacterial smears with fluorescent antibody. IV. Grouping streptococci with fluorescent antibody. J. Bacteriol. *75:* 553–560.

Moore, J.A., Jr. 1974. Characterization and ecology of soil bacteriophage active against *Agromyces ramosus.* Ph.D. Thesis, The Pennsylvania State University, University Park, PA, 128 pp.

Moore, M. and B. Frerichs. 1953. An unusual acid-fast infection of the knee with subcutaneous, abcess-like lesions of the gluteal region; Report of a case with a study of the organism, *Mycobacterium abscessus,* n. sp. J. Invest. Derm. *20:* 133–169.

Moore, R.W. and H.H. Greenlee. 1975. Enterotoxaemia in chinchillas. Lab. Anim. *9:* 153–154.

Moore, W.E.C. 1966. Techniques for routine culture of fastidious anaerobes. Int. J. Syst. Bacteriol. *16:* 173–190.

Moore, W.E.C., E.P. Cato, I.J. Good and L.V. Holdeman. 1981. The effect of diet on the human fecal flora. *In* Bruce, Correa, Lipkin, Tannenbaum and Wilkins (Editors). Banbury report 7. Gastrointestinal cancer: endogenous factors. Cold Spring Harbor Laboratory, New York, pp. 11–24.

Moore, W.E.C., E.P. Cato and L.V. Holdeman. 1971a. *Eubacterium aerofaciens* (Eggerth) Prévot 1938: emendation of description and designation of the neotype strain. Int. J. Syst. Bacteriol. *21:* 307–310.

Moore, W.E.C., E.P. Cato and L.V. Holdeman. 1971b. *Eubacterium lentum* (Eggerth) Prévot 1938: emendation of description and designation of the neotype strain. Int. J. Syst. Bacteriol. *21:* 299–303.

Moore, W.E.C., E.P. Cato and L.V. Holdeman. 1972. *Ruminococcus bromii* sp. n. and emendation of the description of *Ruminococcus* Sijpestein. Int. J. Syst. Bacteriol. *22:* 78–80.

Moore, W.E.C., E.P. Cato and L.V. Holdeman. 1978. Some current concepts in intestinal bacteriology. Am. J. Clin. Nutr. *31:* S33-S42.

Moore, W.E.C. and W.B. Gross. 1968. Liver granulomas of turkeys - causative agents and mechanism of infection. Avian Dis. *12:* 417–422.

Moore, W.E.C., D.E. Hash, L.V. Holdeman and E.P. Cato. 1980. Polyacrylamide slab gel electrophoresis of soluble proteins for studies of bacterial floras. Appl. Environ. Microbiol. *39:* 900–907.

Moore, W.E.C. and L.V. Holdeman. 1969. Anaerobic diphtheroids. *In* Cato, Cummins, Holdeman, Johnson, Moore, Smibert and Smith (Editors), Outline of Clinical Methods in Anaerobic Microbiology. Virginia Polytechnic Institute Anaerobe Laboratory, Blacksburg, Va.

Moore, W.E.C. and L.V. Holdeman. 1970. *Propionibacterium, Arachnia, Actinomyces, Lactobacillus* and *Bifidobacterium. In* Cato, Cummins, Holdeman, Johnson, Moore, Smibert and Smith (Editors). Outline of Clinical Methods in Anaerobic Bacteriology, 2nd rev. ed. Virginia Polytechnic Institute Anaerobe Laboratory, Blacksburg, Virginia, pp. 15–22.

Moore, W.E.C. and L.V. Holdeman. 1972. *Fusobacterium. In* Holdeman and Moore (Editors), Anaerobe Laboratory Manual. Virginia Polytechnic Institute Anaerobe Laboratory, Blacksburg, Virginia, p. 421.

Moore, W.E.C. and L.V. Holdeman. 1972. *Lactobacillus. In* Holdeman and Moore (Editors). Anaerobe Laboratory Manual. Virginia Polytechnic Institute Anaerobe Laboratory, Blacksburg, Virginia, pp. 61–66.

Moore, W.E.C. and L.V. Holdeman. 1974. Human fecal flora: the normal flora of 20 Japanese-Hawaiians. Appl. Microbiol. *27:* 961–979.

Moore, W.E.C., L.V. Holdeman, R.M. Smibert, D.E. Hash, J.A. Burmeister and R.R. Ranney. 1982. Bacteriology of severe periodontitis in young adult humans. Infect. Immun. *38:* 1137–1148.

Moore, W.E.C., L.V. Holdeman, R.M. Smibert, I.J. Good, J.A. Burmeister, K.G. Palcanis and R.R. Ranney. 1982. Bacteriology of experimental gingivitis in young adult humans. Infect. Immun. *38:* 651–667.

Moore, W.E.C., J.L. Johnson and L.V. Holdeman. 1976. Emendation of *Bacteriodaceae* and *Butyrivibrio* and descriptions of *Desulfomonas* gen. nov. and ten new species in the genera *Desulfomonas, Butyrivibrio, Eubacterium, Clostridium* and *Ruminococcus.* Int. J. Syst. Bacteriol. *26:* 238–252.

Moran, J.W. and L.D. Witter. 1976. Effect of temperature and pH on the growth of *Caryophanon latum* colonies. Can. J. Microbiol. *22:* 1401–1403.

Mordarska, H., S. Cebrat, B. Bach and M. Goodfellow. 1978. Differentiation of nocardioform actinomycetes by lyzozyme sensitivity. J. Gen. Microbiol. *109:* 381–384.

Mordarska, H., M. Mordarski and M. Goodfellow. 1972. Chemotaxonomic characters and classification of some nocardioform bacteria. J. Gen. Microbiol. *71:* 77–86.

Mordarski, M., M. Goodfellow, P.B. Heczko, G. Peters, G. Pulverer, A. Tkacz and I. Kaszen. 1981. Deoxyribonucleic acid pairing in the classification of

the genus *Staphylococcus. In* Jeljaszewicz (Editor), Staphylococci and Staphylococcal Infections, Gustav Fischer Verlag, Stuttgart, New York, pp. 15–21.

Mordarski, M., M. Goodfellow, K. Szyba, G. Pulverer and A. Tkacz. 1977. Classification of the "rhodochrous" complex and allied taxa based upon deoxyribonucleic acid reassociation. Int. J. Syst. Bacteriol. *27:* 31–37.

Mordarski, M., M. Goodfellow, K. Szyba, A. Tkacz, G. Pulverer and K.P. Schaal. 1980a. Deoxyribonucleic acid reassociation in the classification of the genus *Rhodococcus.* Int. J. Syst. Bacteriol. *30:* 521–527.

Mordarski, M., M. Goodfellow, A. Tkacz, G. Pulverer and K.P. Schaal. 1980b. Ribosomal ribonucleic acid similarities in the classification of *Rhodococcus* and allied taxa. J. Gen. Microbiol. *118:* 313–319.

Mordarski, M., I. Kaszen, A. Tkacz, M. Goodfellow, G. Alderson, K.P. Schaal and G. Pulverer. 1981. Deoxyribonucleic acid pairing in the classification of *Rhodococcus.* Zentralbl. Bakteriol. Mikrobiol. Hyg. I. Abt. Orig., Suppl. *11:* 25–31.

Mordarski, M., K. Schaal, A. Tkacz, G. Pulverer, K. Szyba and M. Goodfellow. 1978. Deoxyribonucleic acid base composition and homology studies on *Nocardia.* Zentralbl. Bakteriol. I. Abt. Suppl. *6:* 91–97.

Mordarski, M., K. Szyba, G. Pulverer and M. Goodfellow. 1976. Deoxyribonucleic acid reassociation in the classification of the 'rhodochrous' complex and allied taxa. J. Gen. Microbiol. *94:* 235–245.

Mordarski, M., A. Tkacz, M. Goodfellow, K.P. Schaal and G. Pulverer. 1981. Ribosomal ribonucleic acid similarities in the classification of actinomycetes. Zentralbl. Bakteriol. Parasitenkd. Infektionskr. Hyg. I Abt. Suppl. *11:* 79–85.

Mori, T., K. Kohsaka and Y. Tanaka. 1971. Tricarboxylic acid cycle in *M. lepraemurium.* Int. J. Leprosy *39:* 796–812.

Morishita, T., Y. Deguchi, M. Yajima, T. Sakural and T. Yura. 1981. Multiple nutritional requirements of lactobacilli: Genetic lesions affecting amino acid biosynthetic pathways. J. Bacteriol. *148:* 64–71.

Morishita, T., T. Fudada, M. Shirota and T. Yura. 1974. Genetic basis of nutritional requirements in *Lactobacillus casei.* J. Bacteriol. *120:* 1078–1084.

Moro, E. 1900. Über die nach Gram färbbaren Bacillen des Säuglingsstuhles. Wien Klin. Wochenschr. *13:* 114–115.

Moro, E. 1900. Über den *Bacillus acidophilus* n. sp. Jahrb. Kinderheilk. *52:* 38–55.

Morris, E.J. 1955. A selective medium for *Bacillus anthracis.* J. Gen. Microbiol. *13:* 455–460.

Morris, E.O. 1951. The life cycle of *Actinomyces bovis.* J. Hyg. *49:* 46–51.

Morrison, S.J., T.G. Tornabene and W.E. Kloos. 1971. Neutral lipids in the study of the relationships of members of the family *Micrococcaceae.* J. Bacteriol. *108:* 353–358.

Morse, C.D., J.B. Brooks and D.S. Kellogg. 1977. Identification of *Neisseria* by electron capture gas-liquid chromatography of metabolites in a chemically defined growth medium. J. Clin. Microbiol. *6:* 414–481.

Morse, M.E. 1912. A study of the diphtheria group of organisms by the biometrical method. J. Infect. Dis. *11:* 253–285.

Morse, M.L. 1959. Transduction by staphylococcal bacteriophage. Proc. Natl. Acad. Sci. U.S.A. *45:* 722–727.

Morse, S.A. 1969. Regulation of staphylococcal enterotoxin B. Ph. D. Diss. University of North Carolina at Chapel Hill.

Möse, J.R., G. Fischer and C. Briefs. 1972a. Die Wirkung von *Clostridium butyricum* (Stamm M55) auf menschliches Kininogen und ihre Bedeuting für den Onkolyseprozess. Zentralbl. Bakteriol. Parasitenkd. Infektionskr. Hyg., I. Abt. Orig. A *221:* 474–491.

Möse, J.R., G. Fischer and T.B. Mobascheria. 1972b. Uber Bakterienkininasen und deren physiologische Bedeutung. 1. Mitteilung: Untersuchungen an Clostridienstammen. Zentralbl. Bakteriol. Parasitenkd. Infektionskr. Hyg., I. Abt. Orig. A *219:* 530–541.

Möse, J.R. and G. Möse. 1964. Oncolysis by clostridia. I. Activity of *Clostridium butyricum* (M-55) and other nonpathogenic clostridia against the Ehrlich carcinoma. Cancer Research *24:* 212–216.

Moseley, B.E.B. 1967. The isolation and some properties of radiation-sensitive mutants of *Micrococcus radiodurans.* J. Gen. Microbiol. *49:* 293–300.

Moseley, B.E.B. 1982. Photobiology and radiobiology of *Micrococcus (Deinococcus) radiodurans.* Photochem. Photobiol. Rev. *7:* 223–274.

Moseley, B.E.B. and D.M. Evans. 1981. Use of transformation to investigate the nuclear structure and segregation of genomes in *Micrococcus radiodurans. In* Polsinelli and Mazza (Editors), Transformation 1980, Proceedings of the Fifth European Meeting on Bacterial Transformation and Transfection, Florence, Italy, pp. 371–379. Cotswold Press Limited, Oxford.

Moseley, B.E.B. and A. Mattingly. 1971. Repair of irradiated transforming deoxyribonucleic acid in wild type and a radiation sensitive mutant of *Micrococcus radiodurans.* J. Bacteriol. *105:* 976–983.

Mosimann, J., A. Hany and F.H. Kayser. 1979. Pulmonale *Actinomyces-viscosus*-Infektion. Schweiz. Med. Wochenschr. *109:* 720–722.

Moss, C.W., V.R. Dowell, D. Farshtchi, L.J. Raines and W.B. Cherry. 1969. Cultural characteristics and fatty acid composition of propionibacteria. J. Bacteriol. *97:* 561–570.

Moss, C.W. and W.E. Dunkelberg. 1969. Volatile and cellular fatty acids of *Haemophilus vaginalis.* J. Bacteriol. *100:* 544–546.

Moss, C.W., C.L. Hatheway, M.A. Lambert and L.M. McCroskey. 1980. Production of phenylacetic and hydroxyphenylacetic acids by *Clostridium botulinum* type G. J. Clin. Microbiol. *11:* 743–745.

Moss, C.W., M.A. Lambert and D.J. Goldsmith. 1970. Production of hydrocinnamic acid by clostridia. Appl. Microbiol. 19: 375–378.

Moss, C.W., M.A. Lambert and G.L. Lombard. 1977. Cellular fatty acids of Peptococcus variabilis and Peptostreptococcus anaerobius. J. Clin. Microbiol. 5: 665–667.

Mosselman, G. and E. Lienaux. 1890. L'actinomycose et son agent infecteur. Ann. Med. Vet. 39: 409–426.

Mosser, J.L. and A. Tomasz. 1970. Choline-containing teichoic acid as a structural component of pneumococcal cell wall and its role in sensitivity to lysis by an autolytic enzyme. J. Biol. Chem. 245: 287–298.

Mott, G.E. and A.W. Brinkley. 1979. Plasmenylethanolamine growth factor for cholesterol-reducing Eubacterium. J. Bacteriol. 139: 755–760.

Mott, G.E., A.W. Brinkley and C.L. Mersinger. 1980. Biochemical characterization of cholesterol-reducing Eubacterium. Appl. Environ. Microbiol. 40: 1017–1022.

Mouches, C., J.C. Vignault, J.G. Tully, R.F. Whitcomb and J.M. Bové. 1979. Characterization of spiroplasmas by one- and two-dimensional protein analysis on polyacrylamide slab gels. Curr. Microbiol. 2: 69–74.

Movitz, J. 1976. The biosynthesis of protein A. In Jeljaszewicz (Editor), Staphylococci and Staphylococcal Diseases, Gustav Fischer Verlag, Stuttgart, New York, pp. 427–437.

Mrsvic, S., M. Ulahovic and M. Nedeljkovic. 1975. Determination of phosphatase activity in Listeria. In Woodbine (Editor), Problems of Listeriosis. Leicester University Press, Leicester, pp. 30–33.

Muckle, C.A. and C.L. Gyles. 1982. Characterization of strains of Corynebacterium pseudotuberculosis. Can. J. Comp. Med. 46: 206–208.

Mukasa, H. and H.D. Slade. 1973a. Extraction, purification, and chemical and immunological properties of the Streptococcus mutans group "a" polysaccharide cell wall antigen. Infect. Immun. 8: 190–198.

Mukasa, H. and H.D. Slade. 1973b. Structure and immunological specificity of the Streptococcus mutans group b cell wall antigen. Infect. Immun. 7: 578–585.

Mulder, E.G., A.D. Adamse, J. Antheunisse, M.A. Deinema, J.W. Woldendorp and L.P.T.M. Zevenhuizen. 1966. The relationship between Brevibacterium and bacteria of the genus Arthrobacter. J. Appl. Bacteriol. 29: 44–71.

Mulder, E.G. and J. Antheunisse. 1963. Morphologie, physiologie et écologie des Arthrobacter. Ann. Inst. Pasteur Paris 105: 46–74.

Müler, H.E. 1973. Neuraminidase und Acylneuraminat-Pyruvat-Lyase bei Corynebacterium haemolyticum und Corynebacterium pyogenes. Zentralbl. Bakteriol. Parasitenkd. Infektionskr. Hyg. Abt. I Orig. A 225: 59–65.

Müler, H.E. 1976. Neuraminidase als Pathogenitätsfaktor bei mikrobiellen Infektionen. Zentralbl. Bakteriol. Parasitenkd. Infektionskr. Hyg., I. Abt. Orig. A 235: 106–110.

Müler, H.E. and C. Kraseman. 1976. Immunität gegen Erysipelothrix rhusiopathiae - Infektion durch aktive Immunisierung mit homologer Neuraminidase. Z. Immunitaetsforsch. 151: 237–241.

Müler, H.E. and D. Seidler. 1975. Über das Vorkommen Neuraminidase - Neutralisierender Antikörper bei chronisch rotlaufkranken Schweinen. Zentralbl. Bakteriol. Parasitenkd. Infektionskr. Hyg. Abt. I Reihe A 230: 51–58.

Müller, G. 1961. Mikrobiologische Unterushungen über die "Futterverpilzung durch Selbsterhitzung" III. Mitteilung: Ausfuhrliche Beschreibung neuer Bakterien-Species. Zentralbl. Bakteriol. Parasitenkd. Infektionskr. Hyg. II Abt. 114: 520–537.

Müller, O.F. 1773. Vermium Terrestrium et Fluviatilium, seu Animalium Infusoriorum, Helminthicorum et Testaceorum, non Marionorum, Succincta Historia. 1: 1–135.

Muller, P., W. Schaeg and H. Blobel. 1980. Protein A bei Staphylococcus aureus und S. hyicus. Kurzfassung der Vorträge der Arbeitstagung der Deutschen Gesellschaft für Hygiene und Mikrobiologie, 25–26 September 1980, Mainz.

Mulligan, M.E., R.D. Rolfe, S.M. Finegold and W.L. George. 1979. Contamination of a hospital environment by Clostridium difficile. Curr. Microbiol. 3: 173–175.

Munch-Petersen, E. 1954. A corynebacterial agent which protects ruminant erythrocytes against staphylococcal β-toxin. Aust. J. Exp. Biol. Med. Sci. 32: 361–368.

Mundt, J.O. 1963a. Occurrence of enterococci in animals in a wild environment. Appl. Microbiol. 11: 136–140.

Mundt, J.O. 1963b. Occurrence of enterococci on plants in a wild environment. Appl. Microbiol. 11: 141–144.

Mundt, J.O. 1973. Litmus milk reactions as a distinguishing feature between Streptococcus faecalis of human and nonhuman origins. J. Milk Food Technol. 36: 364–367.

Mundt, J.O. 1976. Streptococci in dried and frozen foods. J. Milk Food Technol. 39: 413–416.

Mundt, J.O. 1982. The ecology of the streptococci. Microb. Ecol. 8: 355–369.

Mundt, J.O. and W.F. Graham. 1968. Streptococcus faecium var. casseliflavus nov. var. J. Bacteriol. 95: 2005–2009.

Mundt, J.O., W.F. Graham and I.E. McCarty. 1967. Spherical lactic acid-producing bacteria of southern grown raw and processed vegetables. Appl. Microbiol. 15: 1303–1308.

Mundt, J.O. and J.L. Hammer. 1968. Lactobacilli on plants. Appl. Microbiol. 16: 1326–1330.

Murphy, D.B. and J.E. Hawkins. 1975. Use of urease test disks in the identification of mycobacteria. J. Clin. Microbiol. 1: 465–468.

Murray, E.G.D., R.A. Webb and M.B.R. Swann. 1926. A disease of rabbits characterized by a large mononuclear leucocytosis, caused by a hitherto undescribed bacillus Bacterium monocytogenes (n. sp.). J. Pathol. Bacteriol. 29: 407–439.

Murray, P.A., M.J. Levine, L.A. Tabak and M.S. Reddy. 1984. Neuraminidase activity: a biochemical marker to distinguish Streptococcus mitis from Streptococcus sanguis. J. Dent. Res. 63: 111–113.

Murray, R.G.E. 1962. Fine structure and taxonomy of bacteria. In Microbial Classification, Cambridge University Press, Cambridge.

Murray, R.G.E. 1968. Microbial structure as an aid to microbial classification and taxonomy. Spisy (Faculte des Sciences de l'Universite J.E. Purkyne, Brno) 43: 249–252.

Murray, R.G.E. 1974. A place for bacteria in the living world. In Buchanan and Gibbons (Editors), Bergey's Manual of Determinative Bacteriology, 8th Ed., The Williams and Wilkins Co., Baltimore.

Murray, R.G.E. and R.H. Elder. 1949. The predominance of counterclockwise rotation during swarming of Bacillus species. J. Bacteriol. 58: 351–359.

Murray, R.G.E., M. Hall and B.G. Thompson. 1983. Cell division in Deinococcus radiodurans. Can. J. Microbiol. 29: 1412–1423.

Murray, W.D. and A.W. Khan. 1983a. Growth requirement of Clostridium saccharolyticum, an ethanologenic anaerobe. Can. J. Microbiol. 29: 348–353.

Murray, W.D. and A.W. Khan. 1983b. Ethanol production by a newly isolated anaerobe, Clostridium saccharolyticum: effects of culture medium and growth conditions. Can. J. Microbiol. 29: 342–347.

Murray, W.D., A.W. Khan and L. van den Berg. 1982. Clostridium saccharolyticum sp.nov., a saccharolytic species from sewage sludge. Int. J. Syst. Bacteriol. 32: 132–135.

Murray, W.R., A. Blackwood, K.C. Calman and C. MacKay. 1980. Faecal bile acids and clostridia in patients with breast cancer. Brit. J. Cancer 42: 856–860.

Mutimer, M.D. and J.B. Woolcock. 1980. Corynebacterium equi in cattle and pigs. Vet. Q. 2: 25–27.

Myerowitz, R.L., R.E. Gordon and J.B. Robbins. 1973. Polysaccharides of the genus Bacillus cross reactive with capsular polysaccharides of Diplococcus pneumoniae Type III, Haemophilus influenzae Type b and Neisseria meningitidis Group A, Infect Immun. 8: 896–900.

Myers, P.A. and L.J. Zatman. 1971. The metabolism of trimethylamine N-oxide by Bacillus PM6. Biochem. J. 121: 10p.

Myhre, E.B., O. Holmberg and G. Kronvall. 1979. Immunoglobulin-binding structure on bovine group G streptococci different from type III Fc receptors on human group G streptococci. Infect. Immun. 23: 1–7.

Myhre, E.B. and P. Kuusela. 1983. Binding of human fibronectin to group A, C and G streptococci. Infect. Immun. 40: 29–34.

Myrick, B.A. and P.D. Ellner. 1982. Evaluation of the latex slide agglutination test for identification of Staphylococcus aureus. J. Clin. Microbiol. 15: 275–277.

Nadaud, M. 1977. Media supplemented with nalidixic acid for the isolation of gram negative anaerobic bacteria. Zentralbl. Bakteriol. Parasitenkd. Infektionskr. Hyg. I Abt. Orig. A 239: 375–378.

Nadirova, I.M. and N.I. Aleksandrushkina. 1979. The structure of spore surfaces and morphological and physiological properties of some Bacillus megaterium strains. Izv. Akad. Nauk SSSR, Ser. Biol. (1): 88–94.

Naeslund, C. 1925. Studies of Actinomyces from the oral cavity. Acta Pathol. Microbiol. Scand. 2: 110–140.

Naguib, K. and T. Shauman. 1973. Occurrence of clostridia in white pickled domiati and gouda cheese. Zentralbl. Bakteriol. Parasitenkd. Infektionskr. Abt. 1 128(S): 84–87.

Naguib, K. and M.T. Shouman. 1972. Identification and typing of clostridia in raw milk in Egypt. J. Appl. Bacteriol. 35: 525–530.

Naidoo, J. and W.C. Noble. 1981. Transmission of plasmids between staphylococci on skin. In Jeljaszewicz (Editor), Staphylococci and Staphylococcal Infections, Gustav Fischer Verlag, Stuttgart, New York, pp. 623–625.

Nakagawa, A. and K. Kitahara. 1959. Taxonomic studies on the genus Pediococcus. J. Gen. Appl. Microbiol. 5: 95–126.

Nakagawa, A. and K. Kitahara. 1962. Pleomorphism in bacteria cells. II. Giant cell formation in Pediococcus cerevisiae induced by hop resins. J. Gen. Appl. Microbiol. 8: 142–148.

Nakamura, H. and Z. Tamura. 1972. Growth responses of Bifidobacterium bifidum to S-sulfonic acid-type pantetheine related compounds. Jpn. J. Microbiol. 16: 239–242.

Nakamura, L.K. 1981. Lactobacillus amylovorus, a new starch-hydrolyzing species from cattle waste-corn fermentations. Int. J. Syst. Bacteriol. 31: 56–63.

Nakamura, L.K. 1982. Deoxyribonucleic acid homologies of Lactobacillus amylophilus and other homofermentative species. Int. J. Syst. Bacteriol. 32: 43–47.

Nakamura, L.K. 1984a. Bacillus psychrophilus sp. nov., nom. rev. Int. J. Syst. Bacteriol. 34: 121–123.

Nakamura, L.K. 1984b. Bacillus amyloliticus sp. nov., nom. rev., Bacillus lautus sp. nov., nom. rev., Bacillus pabuli, sp. nov., sp. rev., and Bacillus validus, sp. nov., nom. rev. Int. J. Syst. Bacteriol. 34: 224–226.

Nakamura, L.K. 1984c. Bacillus polvifaciens sp. nov., nom. rev. Int. J. Syst. Bacteriol. 34: 410–413.

Nakamura, L.K. and C.D. Crowell. 1979. Lactobacillus amylophilus, a new starch-hydrolyzing species from swine waste-corn fermentation. Dev. Ind. Microbiol.

20: 531–540.

Nakamura, L.K. and C.D. Crowell. 1981. *In* Validation of the publication of new names and new combinations previously effectively published outside the IJSB. List No. 6. Int. J. Syst. Bacteriol. *31*: 215–218.

Nakamura, L.K. and J. Swezey. 1983a. Taxonomy of *Bacillus circulans* Jordan 1890: base composition and reassociation of deoxyribonucleic acid. Int. J. Syst. Bacteriol. *33*: 46–52.

Nakamura, L.K. and J. Swezey. 1983b. Deoxyribonucleic acid relatedness of *Bacillus circulans* Jordan 1890 strains. Int. J. Syst. Bacteriol. *33*: 703–708.

Nakamura, S., I. Kimura, K. Yamakawa and S. Nishida. 1983. Taxonomic relationships among *Clostridium novyi* types A and B, *Clostridium haemolyticum* and *Clostridium botulinum* type C. J. Gen. Microbiol. *129*: 1473–1479.

Nakamura, S., S. Nakashio, M. Mikawa, K. Yamakawa, S. Okumura and S. Nishida. 1982. Antimicrobial susceptibility of *Clostridium difficile* from different sources. Microbiol. Immunol. *26*: 25–30.

Nakamura, S. and S. Nishida. 1974. Reinvestigations of the relationship between sporulation, heat resistance and some biochemical properties in strains of *Clostridium perfringens*. J. Med. Microbiol. *7*: 451–457.

Nakamura, S., I. Okado, T. Abe and S. Nishida. 1979. Taxonomy of *Clostridium tetani* and related species. J. Gen. Microbiol. *113*: 29–35.

Nakamura, S., M. Sakurai, S. Nishida, T. Tatsuki, Y. Yanagase, Y. Higashi and T. Amano. 1976a. Lecithinase-negative variants of *Clostridium perfringens*; the identity of *C. plagarum* with *C. perfringens*. Can. J. Microbiol. *22*: 1497–1501.

Nakamura, S., T. Shimamura, M. Hayase and S. Nishida. 1973. Numerical taxonomy of saccharolytic clostridia, particularly *Clostridium perfringens*-like strains: descriptions of *Clostridium absonum* sp. n. and *Clostridium paraperfringens*. Int. J. Syst. Bacteriol. *23*: 419–429.

Nakamura, S., T. Shimamura, H. Hayashi and S. Nishida. 1975. Reinvestigation of the taxonomy of *Clostridium bifermentans* and *Clostridium sordellii*. J. Med. Microbiol. *8*: 299–309.

Nakamura, S., T. Shimamura and S. Nishida. 1976b. Urease-negative strains of *Clostridium sordellii*. Can. J. Microbiol. *22*: 673–676.

Nakamura, S., K. Tamai and S. Nishida. 1970. Criteria for identification of *Clostridium perfringens*. 6. *Clostridium paraperfringens* sp. nov. Med. Biol. *80*: 137–140.

Nakamura, S., K. Yamakawa, H. Hashimoto and S. Nishida. 1979. Isolation of *Clostridium absonum* from a case of gas gangrene. Microbiol. Immunol. *23*: 685–687.

Nakayama, O. 1970. Abstract. *In* Proceedings of the 45th Meeting of the Agricultural Chemical Society of Japan p. 315.

Nakayama, O. and M. Yanoshi. 1967a. Sporebearing lactic acid bacteria isolated from rhizosphere. 1. Taxonomic studies on *Bacillus laevolacticus* nov. sp. and *Bacillus racemilacticus* nov. sp. J. Gen. Appl. Microbiol. *13*: 139–153.

Nakayama, O. and M. Yanoshi. 1967b. Sporebearing lactic acid bacteria isolated from rhizosphere. II. Taxonomic studies on the catalase-negative strains. J. Gen. Appl. Microbiol. *13*: 155–165.

Nakel, M., J.M. Ghuysen and O. Kandler. 1971. Wall peptidoglycan in *Aerococcus viridans* strains 201 Evans and ATCC 11563 and in *Gaffkya homari* strain 10400. Biochemistry *10*: 2170–2175.

Nand, K. and D.V. Rao. 1972. *Arthrobacter mysorens* - a new species excreting L-glutamic acid. Zentralbl. Bakteriol. Parasitenkd. Infektionskr. Hyg. Abt. II. *127*: 324–331.

Natvig, H. 1905. Bakteriologische Verhältnesse in weiblichen Genitalsekreten. Erste Mittheilung. Studien über Streptokokken der weiblichen Genitalien in Partus und Puerperium. Arch. Gynaekol. (Berlin) *76*: 701–858.

Naumann, E., H. Hippe and G. Gottschalk. 1983. Betaine: new oxidant in the Stickland reaction and methanogenesis from betaine and L-alanine by a *Clostridium sporogenes - Methanosarcina barkeri* coculture. Appl. Environ. Microbiol. *45*: 474–483.

Naylor, H.B. and E. Burgi. 1956. Observations on abortive infections of *Micrococcus lysodeikticus* with bacteriophage. Virology *2*: 577–593.

Nazina, T.N. 1981. A facultatively anaerobic methylotrophic bacterium *Brevibacterium rufescens* comb. nov. from oil reservoirs. Mikrobiologiya *50*: 311–319.

Nazina, T.N. and E.P. Rozanova. 1978. Thermophilic sulfate-reducing bacteria from oil strata. Mikrobiologiya *47*: 142–148 (English translation pages 113–118).

Neide, E. 1904. Botanische Beschreibung einiger sporenbildenden Bakterien. Zentralbl. Bakteriol. Parasitenkd. Infektionskr. Hyg. Abt. II *12*: 337–352.

Nesterenko, O.A., E.I. Kvasnikov and S.A. Kasumova. 1978. Properties and taxonomy of some spore-forming *Nocardia*. Zentralbl. Bakteriol. Parasitenkd. Infektionskr. Hyg. Abt. I Suppl. *6*: 253–260.

Nesterenko, O.A., T.M. Nogina, S.A. Kasumova, E.I. Kvasnikov and S.G. Batrakov. 1982. *Rhodococcus luteus* nom. nov. and *Rhodococcus maris* nom. nov. Int. J. Syst. Bacteriol. *32*: 1–14.

Nesterenko, P.A., S.A. Kasumova and E.I. Kvasnikov. 1978. Microorganisms of the *Nocardia* genus and the 'rhodochrous' group in soils of the Ukranian SSR. Mikrobiologiya *47*: 866–870

Nevig, R.M., S.E. Maloy, K.D. Klaus and D.R. Kolbe. 1981. *Clostridium septicum* infection in cattle in the United States. J. Am. Vet. Med. Assoc. *179*: 479.

Ng, H. and R.H. Vaughn. 1963. *Clostridium rubrum* sp.n. and other pectinolytic clostridia from soil. J. Bacteriol. *85*: 1104–1113.

Ng, T.K., A. Ben-Bassat and J.G. Zeikus. 1981. Ethanol production by thermophilic bacteria: fermentation of cellulosic substrates by cocultures of *Clostri-*

dium thermocellum and *Clostridium thermohydrosulfuricum*. Appl. Environ. Microbiol. *41*: 1337–1343.

Ng, T.K., P.J. Weimer and J.G. Zeikus. 1977. Cellulolytic and physiological properties of *Clostridium thermocellum*. Arch. Microbiol. *114*: 1–7.

Ng, T.K. and J.G. Zeikus. 1982. Differential metabolism of cellobiose and glucose by *Clostridium thermocellum* and *Clostridium thermohydrosulfuricum*. J. Bacteriol. *150*: 1391–1399.

Nichols, J.H., A. Bezkorovainy and W. Landau. 1974. Human colostral whey M-1 glycoproteins and their *L. bifidus* var *penn.* growth promoting activities. Life Sci. *14*: 967–976.

Nickelson, R., II, G. Finne, M.O. Hanna and C. Vanderzant. 1980. Minced fish flesh from nontraditional Gulf of Mexico finfish species: bacteriology. J. Food Sci. *45*: 1321–1326.

Nicolet, J. and E.N. Bannerman. 1975. Extracellular enzymes of *Micropolyspora faeni* found in mouldy hay. Infect. Immun. *12*: 7–12.

Nicolet, J., P. Paroz and M. Krawinkler. 1980. Polyacrylamide gel electrophoresis of whole-cell proteins of porcine strains of *Haemophilus*. Int. J. Syst. Bacteriol. *30*: 69–76.

Niederau, W., W. Pape, K.P. Schaal, U. Höffler and G. Pulverer. 1982. Zur Antibiotikabehandlung der menschlichen Aktinomykosen. Dtsch. Med. Wochenschr. *107*: 1279–1283.

Nielsen, M.L. and T. Justesen. 1976. Anaerobic and aerobic bacteriological studies in biliary tract diseases. Scand. J. Gastroenterol. *11*: 437–446.

Nieves, B.M., F. Gil and F.J. Castille. 1981. Growth inhibition activity and bacteriophage and bacteriocinlike particles associated with different species of *Clostridium*. Can. J. Microbiol. *27*: 216–225.

Nikolov, P. and I. Abrashev. 1976. Comparative studies of the neuraminidase activity of *Erysipelothrix insidiosa*. Activity of virulent strains and avirulent variants of *Erysipelothrix insidiosa*. Acta Microbiol. Virol. Immunol. *3*: 28–31.

Nikolov, P. and L. Mihailova. 1975. Investigations of decarboxylases of amino acids in *Erysipelothrix insidiosa*. Acta Virol. Immunol. *1*: 78–82.

Nishida, S. and G. Nakagawara. 1964. Isolation of toxigenic strains of *Clostridium novyi* from soil. J. Bacteriol. *88*: 1636–1640.

Nishida, S. and G. Nakagawara. 1965. Relationship between toxigenicity and sporulating potency of *Clostridium novyi*. J. Bacteriol. *89*: 993–995.

Nishida, S., N. Seo and M. Nagagawa. 1969. Sporulation heat resistance and biological properties of *Clostridium perfringens*. Appl. Microbiol. *6*: 303–309.

Nisman, B. 1954. The Stickland reaction. Bacteriol. Rev. *18*: 16–42.

Niven, C.F., Jr. and J.B. Evans. 1957. *Lactobacillus viridescens* nov. spec. a heterofermentative species that produces a green discoloration of cured meat pigments. J. Bacteriol. *73*: 758–759.

Niven, C.F., Jr., Z. Kiziuta and J.C. White. 1946a. Synthesis of polysaccharide from sucrose by *Streptococcus* S.B.E. J. Bacteriol. *57*: 711–716.

Niven, C.F., Jr., K.L. Smiley and J.M. Sherman. 1941a. The production of large amounts of a polysaccharide by *Streptococcus salivarius*. J. Bacteriol. *41*: 479–484.

Niven, C.F., Jr., K.L. Smiley and J.M. Sherman. 1941b. The polysaccharides synthesized by *Streptococcus salivarius* and *Streptococcus bovis*. J. Biol. Chem. *140*: 105–109.

Niven, C.F., Jr., K.L. Smiley and J.M. Sherman. 1942. The hydrolysis of arginine by streptococci. J. Bacteriol. *43*: 651–660.

Niven, C.F., Jr., M.R. Washburn and J.M. Sherman. 1946b. Folic acid requirements of the minute streptococci. J. Bacteriol. *51*: 128.

Niven, C.F., Jr., M.R. Washburn and J.C. White. 1948. Nutrition of *Streptococcus bovis*. J. Bacteriol. *55*: 601–606.

Noble, W.C. and D.A. Somerville. 1974. Microbiology of Human Skin. W.B. Saunders, London, Philadelphia, Toronto.

Noble, W.C., R.E.O. Williams, M.P. Jevons and R.A. Shooter. 1964. Some aspects of nasal carriage of staphylococci. J. Clin. Pathol. *17*: 79–83.

Nocard, E. 1888. Note sur la maladie de boeufs de la Guadeloupe, connue sous le nom de farcin. Ann. Inst. Pasteur, Paris *2*: 293–302.

Noel, K.D. and W.J. Brill. 1980. Diversity and dynamics of indigenous *Rhizobium japonicum* populations. Appl. Environ. Microbiol. *40*: 931–938.

Nolof, G. 1962. Beitrage zur Kenntnis des Stoffwechsels von *Nocardia hydrocarbonoxydans* n. spec. Arch. Mikrobiol. *44*: 278–297.

Nolof, G. and P. Hirsch. 1962. *Nocardia hydrocarbonoxydans* n. spec: ein oligocarbophiler Actinomycete. Arch. Mikrobiol. *44*: 266–277.

Nonomura, H. 1983. *Lactobacillus yamanashiensis* subsp. *yamanashiensis* and *Lactobacillus yamanashiensis* subsp. *mali* sp. and subsp. nov., nom. rev. Int. J. Syst. Bacteriol. *33*: 406–407.

Nonomura, H. and Y. Ohara. 1971. Distribution of actinomycetes in soil X. New genus and species of monosporic actinomycetes. J. Ferment. Technol. *49*: 895–903.

Nord, C.E., S. Holta-Öje, A. Ljungh and T. Wadstrom. 1976. Characterization of coagulase-negative staphylococcal species from human infections. *In* Jeljaszewicz (Editor), Staphylococci and Staphylococcal Diseases, Gustav Fischer Verlag, Stuttgart, New York, pp. 105–111.

Nord, C.-E., T. Wadström, K. Dornbusch and B. Wretlind. 1975. Extracellular proteins in five clostridial species from human infections. Med. Microbiol. Immunol. *161*: 145–154.

Norden, A. and F. Linell. 1951. A new type of pathogenic *Mycobacterium*. Nature *168*: 826–828.

Norlin, M. 1965. Unclassified mycobacteria, a comparison between a serological and a biochemical classification method. Bull. Int. Union. Tuberc. *36:* 25–32.

Norlin, M., A. Lind and O. Ouchterlony. 1969. A serologically based taxonomic study of *M. gastri.* Z. Immunitaetsforsch. Allerg. Klin. Immunol. *137:* 241–248.

Norlin, M., R.G. Navalkar, O. Ouchterlony and A. Lind. 1966. Characterization of leprosy sera with various mycobacterial antigens using double diffusion-in-gel analysis - III. Acta Pathol. Microbiol. Scand. *67:* 555–562.

Norman, A. and R. Grubb. 1955. Hydrolysis of conjugated bile acids by clostridia and enterococci. Acta Pathol. Microbiol. Scand. *36:* 537–547.

Norris, D.O. 1963. A porcelain bead method for storing *Rhizobium.* J. Exp. Agric. *31:* 255–258.

Norris, J.R. 1981. *Sporosarcina* and *Sporolactobacillus. In* Berkeley and Goodfellow (Editors), The Aerobic, Endospore-Forming Bacteria. Academic Press, London. pp. 337–357.

Norris, J.R., R.C.W. Berkeley, N.A. Logan and A.G. O'Donnell. 1981. The Genera *Bacillus* and *Sporolactobacillus. In* Starr, Stolp, Trüper, Balows, and Schlegel (Editors), The Prokaryotes. A Handbook on Habitats, Isolation, and Identification. Springer-Verlag, Berlin. pp. 1711–1742.

Norris, J.R. and H.D. Burges. 1965. The identification of *Bacillus thuringiensis.* Entomophaga *10:* 41–47.

Norris, R.F., T. Flanders, R.M. Tomarelli and P. György. 1950. The isolation and cultivation of *Lactobacillus bifidus:* a comparison of branched and unbranched strains. J. Bacteriol. *60:* 681–696.

Nørrung, V. 1979. Two new serotypes of *Erysipelothrix rhusiopathiae.* Nord. Vet. Med. *34:* 462–465.

Novellie, L. 1968. Kaffir beer brewing; ancient art and modern industry. Wallerstein Lab. Commun. *31:* (No. 104): 17–33.

Novick, R.P. 1963. Analysis by tranduction of mutations affecting penicillinase formation in *Staphylococcus aureus.* J. Gen. Microbiol. *33:* 121–136.

Novick, R.P. 1969. Extrachromosomal inheritance in bacteria. Bacteriol. Rev. *33:* 210–235.

Novick, R.P. and D. Bouanchaud. 1971. Extrachromosomal nature of drug resistance in *S. aureus.* Ann. N.Y. Acad. Sci. *182:* 279–294.

Novick, R.P., S. Cohen, L. Yamamoto and J.A. Shapiro. 1977. Plasmids of *Staphylococcus aureus. In* Bukhari, Shapiro and Adhya (Editors), DNA Insertion Elements, Plasmids, Episomes, Cold Spring Harbor Laboratory, New York, pp. 657–662.

Novitsky, T.J. and D.J. Kushner. 1975. Influence of temperature and salt concentration on the growth of a facultatively halophilic "*Micrococcus*" sp. Can. J. Microbiol. *21:* 107–110.

Novitsky, T.J. and D.J. Kushner. 1976. *Planococcus halophilus* sp. nov. a facultatively halophilic coccus. Int. J. Syst. Bacteriol. *26:* 53–57.

Nowak, W. 1965. Gibt es eine Gelbform von *Bacillus mycoides* Flügge 1886? Naturwissenschaften *52:* 593–594.

Nowlan, S.S. and R.H. Deibel. 1967. Group Q streptococci. I. Ecology, serology, physiology, and relationship to established enterococci. J. Bacteriol. *94:* 291–296.

Nyberg, G.K. and J. Carlsson. 1981. Metabolic inhibition of *Peptostreptococcus anaerobius* decreases the bactericidal effect of hydrogen peroxide. Antimicrob. Agents Chemother. *20:* 726–730.

Nygaard, A.P. and B.D. Hall. 1963. A method for detection of RNA-DNA complexes. Biochem. Biophys. Res. Comm. *12:* 98–104.

Nygren, B., J. Haborn and P. Wahlen. 1966. Phospholipase A production in *Staphylococcus aureus.* Acta Pathol. Microbiol. Scand. *68:* 429–433.

O'Brien, R.W. and J.G. Morris. 1971. Oxygen and the growth and metabolism of *Clostridium acetobutylicum.* J. Gen. Microbiol. *68:* 307–318.

O'Brien, W.E. and L.G. Ljungdahl. 1972. Fermentation of fructose and synthesis of acetate from carbon dioxide by *Clostridium formicoaceticum.* J. Bacteriol. *109:* 626–632.

Ochi, K. and E. Freese. 1983. Effects of antibiotics on sporulation caused by the stringent response in *Bacillus subtilis.* J. Gen. Microbiol. *129:* 3709–3720.

Ochoa, A.G. 1973. Virulence of nocardiae. Can. J. Microbiol. *19:* 901–904.

Ochoa, R. and S. de Velandia. 1978. Equine grass sickness: serologic evidence of association with *Clostridium perfringens* type A enterotoxin. Am. J. Vet. Res. *39:* 1049–1051.

O'Conner, J.J., A.T. Willis and J.A. Smith. 1966. Pigmentation of *Staphylococcus aureus.* J. Pathol. Bacteriol. *92:* 585–588.

Oda, M. 1935. Bacteriological studies on water used for brewing saké (part 6). I. Bacteriological studies on "miyamizu" (8) and (9). *Micrococcus* and *Actinomyces* isolated from "miyamizu". (In Japanese) Jozogaku Zasshi *13:* 1202–1228.

O'Donnell, A.G., M. Goodfellow and D.E. Minnikin. 1982. Lipids in the classification of *Nocardioides:* reclassification of *Arthrobacter simplex* (Jensen) Lochhead in the genus *Nocardioides* (Prauser) emend. O'Donnell et al. as *Nocardioides simplex* comb. nov. Arch. Microbiol. *133:* 323–329.

O'Donnell, A.G., D.E. Minnikin and M. Goodfellow. 1984. Integrated lipid and wall analysis of actinomycetes. *In* Goodfellow and Minnikin (Editors), Chemical Methods in Bacterial Systematics, Society for Applied Bacteriology, Technical Series No. 20. Academic Press, London and New York, pp. 131–143.

O'Donnell, A.G., J.R. Norris, R.C.W. Berkeley, D. Claus, T. Kaneko, N.A. Logan and R. Nozaki. 1980. Characterization of *Bacillus subtilis, Bacillus pumilus, Bacillus licheniformis* and *Bacillus amyloliquefaciens* by pyrolysis gas-liquid

chromatography, deoxyribonucleic acid-deoxyribonucleic acid hybridization, biochemical tests, and API systems. Int. J. Syst. Bacteriol. *30:* 448–459.

Oeding, P. 1965. Antigenic properties of staphylococci. Ann. N.Y. Acad. Sci. *128:* 183–190.

Oeding, P. 1971. Serological investigation of *Planococcus* strains. Int. J. Syst. Bacteriol. *21:* 323–325.

Oeding, P. 1974. Cellular antigens of staphylococci. Ann. N.Y. Acad. Sci. *236:* 15–21.

Oeding, P. and A. Digranes. 1976. *Staphylococcus saprophyticus:* classification and infections. *In* Jeljaszewicz (Editor), Staphylococci and Staphylococcal Diseases, Gustav Fischer Verlag, Stuttgart, New York, pp. 113–117.

Oeding, P. and A. Digranes. 1977. Classification of coagulase-negative staphylococci in the diagnostic laboratory. Acta Pathol. Microbiol. Scand. *85:* 136–142.

Oeding, P. and I.L. Hasselgren. 1972. Antigenic studies of genus *Micrococcus.* 2. Double diffusion in agar gel with particular emphasis on teichoic acids. Acta Pathol. Microbiol. Scand. Sec. B *80:* 265–269.

Oeding, P., J.L. Maradon, V. Hajek and E. Marsalek. 1971. A comparison of phage patterns and antigen structure with biochemical properties of *Staphylococcus aureus* strains isolated from cattle. Acta Pathol. Microbiol. Scand. B *79:* 357–364.

Oeding, P., B. Myklested and A.L. Davison. 1967. Serological investigation on teichoic acids from the wall of *S. epidermidis* and *Micrococcus.* Acta Pathol. Microbiol. Scand. B *69:* 458–464.

O'Farrell, P. 1975. High resolution two-dimensional electrophoresis of proteins. J. Biol. Chem. *250:* 4007–4021.

Ogasawara-Fujita, N. and K. Sakahuchi. 1976. Classification of micrococci on the basis of deoxyribonucleic acid homology. J. Gen. Microbiol. *94:* 97–106.

Ogata, S. and M. Hongo. 1979. Bacteriophages of the genus *Clostridium.* Adv. Appl. Microbiol. *25:* 241–273.

Ogawa, T. and K. Motomura. 1970. Studies on murine leprosy bacillus. I. Attempt to cultivate *in vitro* the Hawaiian strain of *Mycobacterium lepraemurium.* Katasato Arch. Exp. Med. *43:* 65–80.

Oguma, K. and H. Iida. 1979. High and low toxin production by a non-toxigenic strain of *Clostridium botulinum* type C following infection with type C phages of different passage history. J. Gen. Microbiol. *112:* 203–206.

Ogura, K. 1929. Ueber Druststreptococcus, mit besonderer Berucksichtigung seiner Spezifitat. J. Jpn. Soc. Vet. Sci. *8:* 174–203.

Ohisa, N., M. Yamaguchi and N. Kurihara. 1980. Lindane degradation by cell-free extracts of *Clostridium rectum.* Arch. Microbiol. *125:* 221–225.

Okabayashi, T. and E. Masuo. 1960. Occurrence of nucleotides in the culture fluid of microorganisms. I. Screening of purine auxotrophs of *Escherichia coli.* Chem Pharm. Bull. *8:* 1084–1088.

Okabayashi, T. and E. Masuo. 1960. Occurrence of nucleotides in the culture fluids of a microorganism. II. The nucleotides in the broth of *Brevibacterium liquefaciens* nov. sp. Chem. Pharm. Bull. (Tokyo) *8:* 1089–1094.

Okada, S., Y. Suzuki and M. Kozaki. 1979. A new heterofermentative *Lactobacillus* species with *meso*-diaminopimelic acid in peptidoglycan, *Lactobacillus vaccinostercus* Kozaki and Okada sp. nov. J. Gen. Appl. Microbiol. *25:* 215–221.

Okada, S., Y. Suzuki and M. Kozaki. 1983. *In* Validation of the publication of new names and new combinations previously effectively published outside the IJSB. List No. 10. Int. J. Syst. Bacteriol. *33:* 438–440.

Okada, S., T. Toyoda, M. Kozaki and K. Kitahara. 1976. Studies on the cell wall of *Sporolactobacillus inulinus.* J. Agric. Chem. Soc. *50:* 259–263.

O'Leary, V.S. and J.H. Woychick. 1976. Utilization of lactose, glucose, and galactose by a mixed culture of *Streptococcus thermophilus* and *Lactobacillus bulgaricus* in milk treated with lactase enzyme. Appl. Environ. Microbiol. *32:* 89–94.

Oliver, T.J. and A.C. Sinclair. 1964. Antibiotic M-319. U.S. Patent 3,155,582. November 3, 1964.

Olsen, W.C., Jr., J.T. Parisi, P.A. Totten and J.W. Baldwin. 1979. Transduction of penicillinase production in *Staphylococcus epidermidis* and nature of the genetic determinant. Can. J. Microbiol. *25:* 508–511.

Omura, S., H. Tanaka, Y. Tanaka, P. Spiri-Nakagawa, R. Oiwa, Y. Takahashi, K. Matsuyama and Y. Iwai. 1979. Studies on bacterial cell wall inhibitors VII. Azureomycin A and B, new antibiotics produced by *Pseudonocardia azurea* nov. sp. J. Antibiot. *32:* 985–994.

Omura, S., H. Tanaka, Y. Tanaka, P. Spiri-Nakagawa, R. Oiwa, Y. Takahashi, K. Matsuyama and Y. Iwai. 1983. *In* Validation of the publication of new names and new combinations previously effectively published outside the IJSB. Int. J. Syst. Bacteriol. *33:* 672–674.

Onishi, H. 1972b. Salt response of amylase produced in media of different NaCl or KCl concentrations by a moderately halophilic *Micrococcus.* Can. J. Microbiol. *18:* 1617–1620.

Onishi, H. 1972a. Halophilic amylase from a moderately halophilic *Micrococcus.* J. Bacteriol. *109:* 570–574.

Onishi, H. and M. Kamekura. 1972. *Micrococcus halobius* sp. n. Int. J. Syst. Bacteriol. *22:* 233–236.

Onishi, H. and K. Sonoda. 1979. Purification and some properties of an extracellular amylase from a moderate halophile, *Micrococcus halobius.* Appl. Environ. Microbiol. *38:* 616–620.

Onishi, H. 1949. Studies on the actinomyces isolated from the deeper layer of carious dentine. J. Dent. Res. *6:* 273–282.

Onon, E.O. 1979. Purification and partial characterization of the exotoxin of

Corynebacterium ovis. Biochem. J. *177:* 181–186.

Oppenheimer, C.H. and C.E. ZoBell. 1952. The growth and viability of sixty-three species of marine bacteria as influenced by hydrostatic pressure. J. Mar. Res. *11:* 10–18.

Orchard, V.A. 1981. The ecology of *Nocardia* and related taxa. Zentralbl. Bakteriol. Mikrobiol. Hyg. Suppl. *11:* 167–180.

Orchard, V.A. and M. Goodfellow. 1980. Numerical classification of some named strains of *Nocardia asteroides* and related isolates from soil. J. Gen. Microbiol. *118:* 295–312.

Orchard, V.A., M. Goodfellow and S.T. Williams. 1977. Selective isolation and occurrence of nocardiae in soil. Soil Biol. Biochem. *9:* 233–238.

Ordal, E.J. and B.J. Earp. 1956. Cultivation and transmission of the etiological agent of kidney disease in salmonid fishes. Proc. Soc. Exp. Biol. Med. *92:* 85–88.

Oren, A. 1983. *Clostridium lortetii* sp. nov., a halophilic obligatory anaerobic bacterium producing endospores with attached gas vacuoles. Arch. Microbiol. *136:* 42–48.

Oren, A. 1984. Validation List No. 14. Int. J. Syst. Bacteriol. *34:* 270–271.

Orla-Jensen, O. 1909. Die Hauptlinien des natürlichen Bakterien-systems. Zentralbl. Bakteriol. Parasitenkd. Infektionskr. Hyg. Abt. 2, *22:* 305–346.

Orla-Jensen, S. 1916. Maelkeri-Bakteriologi. Schønberske Forlag, Copenhagen.

Orla-Jensen, S. 1919. The lactic acid bacteria. Copenhagen: Høst and Son.

Orla-Jensen, S. 1924. La classification des bactéries lactiques. Lait *4:* 468–474.

Orla-Jensen, S. 1943. Die echten Milchsäurebakterien. Ejnar Munksgaard, Copenhagen.

Orla-Jensen, S., A.D. Orla-Jensen and O. Winther. 1936. *Bacterium bifidum* und *Thermobacterium intestinale.* Zentralbl. Bakteriol. Abt. 2 *93:* 321–343.

Ortali, V. and L. Capocaccia. 1956. Una nuova specie di *Corynebacterium:* il *Corynebacterium mycetoides* (Castellani) Ortali e Capocaccia 1956. Rend. Ist. Super. Sanita. *19:* 480–491.

Ortel, S. 1968. Bakteriologische, serologische und epidemiologische Untersuchungen während einer Listeriose-Epidemie. Dtsch. Gesundheitswes. *16:* 753–759.

Ortel, S. 1978. Untersuchungen über Monocine. Zentral. Bakteriol. Parasitenkd. Infektionskr. Hyg. I. I Abt. Orig. A *242:* 72–78.

Ortel, S. 1981. Lysotypie von *Listeria monocytogenes.* Z. Gesamte Hyg. Grenzgeb. *27:* 837–840.

Ortiz-Ortiz, L., M.F. Contreras and L.F. Bojalil. 1976. *In* Goodfellow, Brownell and Serrano (Editors), The Biology of the Nocardiae, Academic Press, New York, pp. 418–428.

Ottens, H. and K.C. Winkler. 1962. Indifferent and haemolytic streptococci possessing Group-antigen F. J. Gen. Microbiol. *28:* 181–191.

Overman, J.R. and L. Pine. 1963. Electron microscopy of cytoplasmic structures in facultative and anaerobic *Actinomyces.* J. Bacteriol. *86:* 656–665.

Owens, J.D. and R.M. Keddie. 1969. The nitrogen nutrition of soil and herbage coryneform bacteria. J. Appl. Bacteriol. *32:* 338–347.

Pabst, M.J. 1977. Levan and levansucrase of *Actinomyces viscosus.* Infect. Immun. *15:* 518–526.

Pabst, M.J., J.O. Cisar and C.L. Trummel 1979. The cell wall-associated levansucrase of *Actinomyces viscosus.* Biochem. Biophys. Acta *566:* 274–282.

Packer, R.A. 1943. The use of sodium azide (NaN_3) and crystal violet in a selective medium for streptococci and *Erysipelothrix rhusiopathiae.* J. Bacteriol. *46:* 343–349.

Page, L.R. and G.N. Krywolap. 1974. Deoxyribonucleic acid base composition of *Bacterionema matruchotii.* Int. J. Syst. Bacteriol. *24:* 289–291.

Pakula, R. 1963. Can transformation be used as a criterion in taxonomy? Recent Prog. Microbiol. *8:* 617–624.

Pakula, R., E. Hulanicka and W. Walczak. 1958. Transformation reactions between streptococci, pneumococci and staphylococci. Bull. Acad. Pol. Sci. Ser. Sci. Biol. *6:* 325.

Pal, S.E. and B.G. Ray. 1982. Phosphatase activity of pathogenic staphylococci. Indian J. Med. Sci. *16:* 516–519.

Palenik, C.J. and C.H. Miller. 1975. Extracellular invertase activity from *Actinomyces viscosus.* J. Dent. Res. *54:* 186.

Pan, Y.L. and H.J. Blumenthal. 1962. Pathways of glucose catabolism of *Staphylococcus aureus.* Bacteriol. Proc. *1962:* 70.

Pandhi, P.N. and B.F. Hammond. 1975. A glycolipid from *Rothia dentocariosa.* Arch. Oral Biol. *20:* 399–401.

Pandhi, P.N. and B.F. Hammond. 1978. The polar lipids of *Actinomyces viscosus.* Arch. Oral Biol. *23:* 17–21.

Pape, W. 1979. Die Resistenz verschiedener *Actinomycetaceae* gegen neuere antimikrobielle Chemotherapeutika. Inaugural-Dissertation, Rheinische Friedrich-Wilhelms-Universität Bonn, F.R.G.

Pappenheimer, A.M. 1977. Diphtheria toxin. Annu. Rev. Biochem. *46:* 69–94.

Paquin, R. and R.A. Lachance. 1970. Sur la nutrition aminée de *Corynebacterium sepedonicum* (Spieck. et Kott.) Skapt. et Burkh. et al. résistance de la pomme de terre au flétrissement bactérien. Can. J. Microbiol. *16:* 719–726.

Parish, H.J., W.W. Hoffman and N.H. Moynihan. 1966. Mode of action of tetanus toxin on the neuromuscular junction. Am. J. Physiol. *210:* 84–90.

Parisi, J.T., H.W. Talbot and J.M. Skahan. 1978. Development of a phage typing set for *Staphylococcus epidermidis* in the United States. Zentralbl. Bakteriol. Parasitenkd. Infektionskr. Hyg. Abt. I Orig. A *241:* 60–67.

Park, S.J., V. Chong and S.Y. Lee. 1976. Bacterial and fungal species from cerebrospinal fluid in the past five years. Korean J. Pathol. *10:* 137–142.

Parker, M.T. 1978. The pattern of streptococcal disease in man. *In* Skinner and Quesnel (Editors), Streptococci, Society for Applied Bacteriology Symposium Series No. 7, Academic Press, London, New York and San Francisco, pp. 71–106.

Parker, M.T. 1983. *Streptococcus* and *Lactobacillus. In* Wilson, Miles and Parker (Editors), Topley and Wilson's Principles of Bacteriology, Virology and Immunity, Seventh Edition, Volume 2, Edward Arnold, London, pp. 173–217.

Parker, M.T. and L.C. Ball. 1976. Streptococci and aerococci associated with systemic infection in man. J. Med. Microbiol. *9:* 275–302.

Parry, J.M., P.C. Turnbull and J.R. Gibson. 1983. A colour atlas of *Bacillus* species. Wolfe Medical Publications Ltd., Ipswich.

Parsons, L.B. and W.S. Sturges. 1927. Quantitative aspects of the metabolism of anaerobes. 1. Proteolysis by *Clostridium putrefaciens* compared with that of other anaerobes. J. Bacteriol. *14:* 181–192.

Partansky, A.M. and B.S. Henry. 1935. Anaerobic bacteria capable of fermenting sulfite waste liquor. J. Bacteriol. *30:* 559–571.

Partridge, M.D, A.L. Davidson and J. Baddiley. 1973. The distribution of teichoic acids and sugar 1-phosphate polymers in walls of micrococci. J. Gen. Microbiol. *74:* 169–173.

Patel, I.B. and R.H. Vaughn. 1973. Cellulolytic bacteria associated with sloughing spoilage of california ripe olives. Appl. Microbiol. *25:* 62–69.

Paterson, J.S. 1940. The antigenic structure of organisms of the genus *Listerella.* J. Pathol. Bacteriol. *51:* 427–436.

Paterson, W.D., C. Gallant, D. Desautels and L. Marshall. 1979. Detection of bacterial kidney disease in wild salmonids in the Margaree River system and adjacent waters using an indirect fluorescent antibody technique. J. Fish Res. Board Can. *36:* 1464–1468.

Patočka, F., E. Menčiková, H.P.R. Seeliger and A. Jirásek. 1979. Neurotropic activity of a strain of *Listeria innocua* in suckling mice. Zentralbl. Bakteriol. Parasitenkd. Infektionskr. Hyg. I. Abt. Orig. A *243:* 490–498.

Pattee, P.A. and J.N. Baldwin. 1961. Transduction of resistance to chlortetracycline and novobiocin in *Staphylococcus aureus.* J. Bacteriol. *82:* 875–881.

Pattee, P.A. and D.S. Neveln. 1975. Transformation analysis of three linkage groups in *Staphylococcus aureus.* J. Bacteriol. *124:* 201–211.

Pattee, P.A., M.L. Stahl, C.J. Schroeder and J.B. Luchansky. 1982. Chromosome map of *Staphylococcus aureus.* Genet. Maps *2:* 122–125.

Pattee, P.A., N.E. Thompson, D. Haubrich and R.P. Novick. 1977. Chromosomal map locations of integrated plasmids and related elements in *Staphylococcus aureus.* Plasmid *1:* 38–51.

Patterson, F.P. and C.S. Brown. 1972. The McKee-Farrar total hip replacement. Preliminary results and complications of 368 operations performed in five general hospitals. J. Bone Jt. Surg. *54A:* 257–275.

Patterson, H., R. Irvin, J.W. Costerton and K.J. Cheng. 1975. Ultrastructure and adhesion properties of *Ruminococcus albus.* J. Bacteriol. *122:* 278–287.

Patterson, J.T. and P.A. Gibbs. 1978. Some microbiological considerations applying to the conditioning, aging and vacuum packing of lamb. J. Food Protect. *13:* 1–13.

Pattison, I.H. and P.R.J. Matthews. 1955. Type classification by Lancefield's precipitin method of human and bovine group B streptococci isolated in Britain. J. Pathol. *69:* 43–50.

Pattyn, S.R., M.T. Boveroulle, J. Mortelmans and J. Vercruysse. 1967. Mycobacteria in mammals and birds of the zoo of Antwerp. Acta Zool. Pathol. Antverpiensia *43:* 125–134.

Pattyn, S.R., M. Magnusson, J.L. Stanford and J.M. Grange. 1974. A study of *Mycobacterium fortuitum* (ranae). J. Med. Microbiol. *7:* 67–76.

Pattyn, S.R. and F. Portaels. 1980. *In vitro* cultivation and characterization of *Mycobacterium lepraemurium.* Int. J. Leprosy *48:* 7–14.

Peach, S., F. Fernandez, K. Johnson and B.S. Drasar. 1974. The non-sporing anaerobic bacteria in human faeces. J. Med. Microbiol. *7:* 213–221.

Pearson, H.W., R. Howsley and S.T. Williams. 1982. A study of nitrogenase activity in *Mycoplana* species and free-living actinomycetes. J. Gen. Microbiol. *128:* 2073–2080.

Peattie, D.A. 1979. Direct chemical method for sequencing RNA. Proc. Nat. Acad. Sci. USA *76:* 1760–1764.

Pechman, K.J., B.J. Lewis and C.R. Woese. 1976. Phylogenetic status of *Sporosarcina ureae.* Int. J. Syst. Bacteriol. *26:* 305–310.

Peck, H.D., Jr. and H. Gest. 1957. Hydrogenase of *Clostridium butylicum.* J. Bacteriol. *73:* 569–580.

Peck, H.D., Jr. and J. LeGall. 1982. Biochemistry of dissimilatory sulphate reduction. Phil Trans. R. Soc. Lond. *B298:* 443–466.

Peck, H.D., C.L. Liu, A.K. Varma, L.G. Ljungdahl, M. Szulczynski, O.F. Bryant and L.H. Carreira. 1983. The utilization of inorganic pyrophosphate, tripolyphosphate, and tetrapolyphosphate as energy source for the growth of anerobic bacteria. *In* Hollander, Laskin and Rogers (Editors), Basic Biology of New Developments in Biotechnology. Plenum Press, New York, pp. 317–348.

Peckham, M.C. 1966. An outbreak of streptococcosis (apoplectiform septicemia) in white rock chickens. Avian Dis. *10:* 413–421.

Peczynska-Czoch, W. and H. Mordarski. 1984. Transformation of xenobiotics. *In* Goodfellow, Mordarski and Williams (Editors), The Biology of the Actinomycetes, Academic Press, London and New York, pp. 287–336.

Peloux, Y.C., C. Charrel-Taranger and F. Gouin. 1976. Nouvelle affection opportuniste a bacillus: un cas de bactériéme à *Bacillus licheniformis.* Pathol. Biol. *24:* 97–98.

Pennock, C.A. and R.B. Huddy. 1967. Phosphatase reaction of coagulase-negative staphylococci and micrococci. J. Pathol. Bacteriol. 93: 685–688.

Penso, G., G. Castelnuovo, A. Guadiano, M. Princivalle, L. Vella and A. Zampieri. 1952. Studi e richerche sui micobatteri. VIII. Un nuovo bacillo tubercolare: il *Mycobacterium minetti* n. sp., Studio microbiologica e patogenetico. Rend. Ist. Super. Sanita 15: 491–548.

Penso, G., R. Noel, M. Blanc and S. Marie-Suzanne. 1957. Études et recherches sur les mycobacteries XV. Le *Mycobacterium marianum* (Penso 1953). Étude microbiologique, pathogenetique et immunologique. Rend. Acad. Naz. dei XL, Ser. IV, 8: 1–75.

Penso, G., V. Ortali, A. Guadiano, M. Princivalle, L. Vella and A. Zampieri. 1951. Studi e richerche sui micobatteri. VII. *Mycobacterium phlei* (Lehman and Neumann 1899 proparte). Rend. Ist. Super. Sanita 14: 855–908.

Pepys, J., P.A. Jenkins, G.N. Festenstein, P.H. Gregory, M.E. Lacey and F.A. Skinner. 1963. Farmer's lung: thermophilic actinomycetes as a source of "farmer's lung hay" antigen. Lancet 2: 607–611.

Perceval, A., A.J. McLean and C.V. Wellington. 1976. Emergence of gentamicin resistance in *S. aureus*. Med. J. Aust. 2: 74.

Perch, B., E. Kjems and T. Ravn. 1974. Biochemical and serological properties of *Streptococcus mutans* from various human and animal sources. Acta Pathol. Microbiol. Scand. Sect. B 82: 357–370.

Perch, B., K.B. Pedersen and J. Henrichsen. 1983. Serology of capsulated streptococci pathogenic for pigs: six new serotypes of *Streptococcus suis*. J. Clin. Microbiol. 17: 993–996.

Perea, E.J., J. Aznar, M.C. Garcia-Iglesias and M.V. Borobio. 1978. Cefoxitin sodium activity against anaerobes: effect of the inoculum size, pH variation and different culture media. J. Antimicrob. Chemother. 4 (Suppl. B): 55–60.

Perez-Diaz, J.C., M.F. Vicente and F. Baquero. 1981. Plasmids in *Listeria*. In Abstracts of VIIIth International Symposium on Problems of Listeriosis. Madrid, p. 33.

Perkins, H.R. 1963. A polymer containing glucose and aminohexuronic acid isolated from the cell walls of *Micrococcus lysodeikticus*. Biochem. J. 86: 475–483.

Perry, D. and H.D. Slade. 1964. Intraspecific and interspecific transformation in streptococci. J. Bacteriol. 88: 595–601.

Perry, J.J. and J.B. Evans. 1960. Oxidative metabolism of lactate and acetate by *Micrococcus sodonensis*. J. Bacteriol. 79: 113–118.

Perry, J.J. and J.B. Evans. 1966. Oxidation and assimilation of carbohydrates by *Micrococcus sodonensis*. J. Bacteriol. 91: 33–38.

Perry, K.D., L.G.M. Newland and C.A.E. Briggs. 1958. Group D rumen streptococci with type antigens of group N. J. Pathol. Bacteriol. 76: 589–590.

Person, D.A., P.K.W. Yu and J.A. Washington II. 1969. Characterization of *Micrococcaceae* isolated from clinical sources. Appl. Microbiol. 18: 95–97.

Peshkoff, M.A. 1939. Cytology, karyology and cycle of development of new microbes - *Caryophanon latum* and *Caryophanon tenue*. C.R. (Dokl.) Acad. Sci. URSS 25: 244–247.

Peshkoff, M.A. 1965. Variation of laboratory and freshly isolated strains of the bacterium *Caryophanon latum* Peshkoff as a possible manifestation of a mutation process. Sov. Genet. 3: 147–156.

Peshkoff, M.A. and B.I. Marek. 1973. Fine structure of *Caryophanon latum* and *Caryophanon tenue* Peshkoff. Microbiology 41: 941–945.

Peters, G., R. Locci and G. Pulverer. 1982. Adherence and growth of coagulase-negative staphylococci on surfaces of intravenous catheters. J. Inf. Dis. 146: 479–482.

Peters, G. and G. Pulverer. 1975. Bakteriophagen aus Mikrokokken. Zentralbl. Bakteriol. Parasitenkd. Infektionskr. Hyg. Abt. I Orig. A 232: 221–226.

Peters, G. and G. Pulverer. 1978. Humanmedizinische Bedeutung der Mikrokokken. Zentralbl. Bakteriol. Parasitenkd. Infektionskr. Hyg. Abt. I Ref. 256: 430–431.

Peters, G., G. Pulverer and J. Pillich. 1976. Bacteriophages of micrococci. Zentralbl. Bakteriol. Parasitenkd. Infektionskr. Hyg. Abt. I Orig. Suppl. 5: 159–163.

Peterson, E.H. 1967. The isolation of *Clostridium sporogenes* from the viscera of day-old chicks. Poult. Sci. 46: 527–528.

Petitdemange, E., F. Caillet, J. Giallo and C. Gaudin. 1984. *Clostridium cellulolyticum* sp. nov., a cellulolytic mesophilic species from decayed grass. Int. J. Syst. Bacteriol. 34: 155–159.

Petitdemange, H., H. Blusson and R. Gay. 1981. Detection of NAD(P)H-rubredoxin oxidoreductases in clostridia. Anal. Biochem. 116: 564–570.

Petitdemange, H., C. Cherrier, J.M. Bengone and R. Gay. 1977. Étude des activités NADH et NADPH-ferrédoxine oxydoréductasiques chez *Clostridium acetobutylicum*. Can. J. Microbiol. 23: 152–160.

Petre, J., R. Longin and J. Millet. 1981. Purification and properties of an endo-β-1,4-glucanase from *Clostridium thermocellum*. Biochimie 63: 629–639.

Pette, J.W. and J. van Beynum. 1943. Boekelscheurbacterien Rijkslandbauwproefstation te hoorn. Versl. Landbouwkd. Onderz. 490: 315–346.

Pettipher, G.L. and M.J. Latham. 1979a. Characteristics of enzymes produced by *Ruminococcus flavefaciens* which degrade plant cell walls. J. Gen. Microbiol. 110: 21–27.

Pettipher, G.L. and M.J. Latham. 1979b. Production of enzymes degrading plant cell walls and fermentation of cellobiose by *Ruminococcus flavefaciens* in batch and continuous culture. J. Gen. Microbiol. 110: 29–38.

Pheil, C.G. and Z.J. Ordal. 1967. Sporulation of the "thermophilic anaerobes". Appl. Microbiol. 15: 893–898.

Phillips, A.P., K.L. Martin and M.G. Broster. 1983. Differentiation between spores of *Bacillus anthracis* and *Bacillus cereus* by a quantitative immunofluorescence technique. J. Clin. Microbiol. 17: 41–47.

Phillips, B.A., M.J. Latham and M.E. Sharpe. 1975. A method for freeze drying rumen bacteria and other strict anaerobes. J. Appl. Bacteriol. 38: 319–322.

Phillips, H.C. 1953. Characterization of the soil globiforme bacteria. Iowa State Coll. J. Sci. 27: 240–241.

Phillips, I., C. Warren, J.M. Harrison, P. Sharples, P.C. Ball and M.T. Parker. 1976. Antibiotic susceptibilities of streptococci from the mouth and blood of patients treated with penicillin or lincomycin and clindamycin. J. Med. Microbiol. 9: 393–404.

Phillips, I., C. Warren, E. Taylor, R. Timewell and S. Eykyn. 1981. The antimicrobial susceptibility of anaerobic bacteria in a London teaching hospital. J. Antimicrob. Chemother. 8 (Suppl. D): 17–26.

Phillips, K.D. and P.A. Rogers. 1981. Rapid detection and presumptive identification of *Clostridium difficile* by p-cresol production on a selective medium. J. Clin. Pathol. 34: 642–644.

Phillips, W.E., Jr., R.E. King and W.E. Kloos. 1980. Isolation of *Staphylococcus hyicus* subsp. *hyicus* from a pig with septic polyarthritis. Am. J. Vet. Res. 41: 274–276.

Phillips, W.E., Jr., and W.E. Kloos. 1981. Identification of coagulase-positive *Staphylococcus intermedius* and *Staphylococcus hyicus* subsp. *hyicus* isolates from veterinary clinical specimens. J. Clin. Microbiol. 14: 671–673.

Pichinoty, F., H. de Barjac, M. Mandel and J. Asselineau. 1983. Description of *Bacillus azotoformans* sp. nov. Int. J. Syst. Bacteriol. 33: 660–662.

Pichinoty, F., H. de Barjac, M. Mandel, B. Greenway and J.-L. Garcia. 1976. Une nouvelle bactérie sporulée, dénitrifiante, mésophile: *Bacillus azotoformans* n. sp. Ann. Microbiol. (Paris) 127B: 351–361.

Pichinoty, F., M. Durand, C. Job, M. Mandel and J.-L. Garcia. 1978. Étude morphologique, physiologique et taxonomique de *Bacillus azotoformans*. Can. J. Microbiol. 24: 608–617.

Pichinoty, F. and M. Mandel. 1978. The isolation and properties of a mesophilic *Bacillus* species utilizing quinate, p-hydroxybenzoate, and phthalate as sources of carbon and energy. Curr. Microbiol. 1: 269–271.

Pichinoty, F., M. Mandel, B. Greenway and J.-L. Garcia. 1977. Isolation and properties of a denitrifying bacterium related to *Pseudomonas lemoignei*. Int. J. Syst. Bacteriol. 27: 346–348.

Pier, A.C. and R.E. Fichtner. 1971. Serologic typing of *Nocardia asteroides* by immunodiffusion. Am. Rev. Respir. Dis. 103: 398–707.

Pier, G.B. and S.H. Madin. 1976. *Streptococcus iniae* sp. nov., a beta hemolytic streptococcus isolated from an Amazon freshwater dolphin, *Inia geoffrensis*. Int. J. Syst. Bacteriol. 26: 545–553.

Pierce-Chase, C.H., R.M. Fauve and R. Dubos. 1964. Corynebacterial pseudotuberculosis in mice. I. Comparative susceptibility of mouse strains to experimental infections with *Corynebacterium kutscheri*. J. Exp. Med. 120: 267–281.

Pijper, A. and B.D. Pullinger. 1927. South African nocardioses. J. Trop. Med. Hyg. 30: 153–156.

Pillet, J. and B. Orta. 1981. Species and serotypes in coagulase-negative staphylococci. In Jeljaszweicz (Editor), Staphylococci and Staphylococcal Infections, Gustav Fischer Verlag, Stuttgart, New York, pp. 147–152.

Pillet, J., B. Orta and F. Corrieras. 1967. Serotypie des staphylocoques. Interet d'une réduction du number des souches types utilisées. Ann. Inst. Pasteur 113: 363–374.

Pine, L. 1963. Recent developments on the nature of the anaerobic actinomycetes. Ann. Soc. Belge Med. Trop. 3: 247–258.

Pine, L. 1970. Classification and phylogenetic relationship of micro-aerophilic actinomycetes. Int. J. Syst. Bacteriol. 20: 445–474.

Pine, L. and C.J. Boone. 1967. Comparative cell wall analyses of morphological forms within the genus *Actinomyces*. J. Bacteriol. 94: 875–883.

Pine, L. and L.K. Georg. 1965. The classification and phylogenetic relationships of the *Actinomycetaceae*. Int. Bull. Bacteriol. Nomencl. Taxon. 15: 143–163.

Pine, L. and L.K. Georg. 1969. Reclassification of *Actinomyces propionicus*. Int. J. Syst. Bacteriol. 19: 267–272.

Pine, L. and L.K. Georg. 1974. Genus *Arachnia* Pine and Georg 1969. In Buchanan and Gibbons (Editors), Bergey's Manual of Determinative Bacteriology, 8th Ed. The Williams and Wilkins Co., Baltimore, pp. 668–669.

Pine, L. and H. Hardin. 1959. *Actinomyces israelii*, a cause of lacrimal canaliculitis in man. J. Bacteriol. 78: 164–170.

Pine, L. and A. Howell, Jr. 1956. Comparison of physiological and biochemical characters of *Actinomyces* spp. with those of *Lactobacillus bifidis*. J. Gen. Microbiol. 15: 428–445.

Pine, L., A. Howell and S.J. Watson. 1960. Studies of the morphological, physiological and biochemical characters of *Actinomyces bovis*. J. Gen. Microbiol. 23: 403–424.

Pine, L., G. Bradley Malcolm, E.M. Curtis and J.M. Brown. 1981. Demonstration of *Actinomyces* and *Arachnia* species in cervicovaginal smears by direct staining with species-specific fluorescent-antibody conjugate. J. Clin. Microbiol. 13: 15–21.

Pine, L. and J.R. Overman. 1966. Differentiation of capsules and hyphae in clubs of bovine sulphur granules. Sabouraudia 5: 141–143.

Pine, L. and S.J. Watson. 1959. Evaluation of an isolation and maintenance medium for *Actinomyces* species and related organisms. J. Lab. Clin. Med. 54: 107–114.

Pinoy, E. 1913. Actinomycoses et mycétomes. Bull. Inst. Pasteur (Paris) *11:* 929–938.

Piot, P., E. Van Dyck, M. Goodfellow and S. Falkow. 1980. A taxonomic study of *Gardnerella vaginalis (Haemophilus vaginalis)* Gardner and Dukes 1955. J. Gen. Microbiol. *119:* 373–396.

Pirie, J.H.H. 1940. The genus *Listerella* Pirie. Science (Washington) *91:* 383.

Pirtle, E.C., P.A. Rebers and W.W. Weigel. 1965. Nitrogen-containing and carbohydrate-containing antigen from *Actinomyces bovis.* J. Bacteriol. *89:* 880–888.

Pitcher, D.G. 1983. Deoxyribonucleic acid base composition of *Corynebacterium diphtheriae* and other corynebacteria with cell wall type IV. FEMS Microbiol. Lett. *16:* 291–295.

Pitcher, D.G. and W.C. Noble. 1978. Aerobic diphtheroids of human skin. *In* Bousfield and Callely (Editors) Coryneform Bacteria. Academic Press, London, pp. 265–287.

Pittenger, R.C. and R.B. Brigham. 1956. *Streptomyces orientalis* n. sp., the source of vancomycin. Antibiot. Chemother. *6:* 642–647.

Plastourgos, S. and R.H. Vaughn. 1957. Species of *Propionibacterium* associated with Zapatera spoilage of olives. Appl. Microbiol. *5:* 267–271.

Plastridge, W.N. and S.E. Hartsell. 1937. Biochemical and serological characteristics of streptococci of bovine origin. J. Infect. Dis. *61:* 110–121.

Pochi, P.E. and J.S. Strauss. 1961. Antibiotic sensitivity of *Corynebacterium acnes (Propionibacterium acnes).* J. Invest. Dermatol. *36:* 423–429.

Pokallus, R.S., G. Gilliam and K. Young. 1976. Variations in colony morphology of *Bacillus pumilus* related to the peptones in trypticase soy agar. Abstr. Ann. Meet. Am. Soc. Microbiol. *1976:* p. 118.

Polak, J. and R.P. Novick. 1982. Closely related plasmids from *Staphylococcus aureus* and soil bacilli. Plasmid *7:* 152–162.

Pollak, A. and V.B. Buhler. 1955. The cultural characteristics and animal pathogenicity of an atypical acid-fast organism which causes human disease. Am. Rev. Tuberc. *71:* 74–87.

Poole, P.M. and G. Wilson. 1976. Infection with minute-colony-forming B-haemolytic streptococci. J. Clin. Pathol. *29:* 740–745.

Poole, P.M. and G. Wilson. 1979. Occurrence and cultural features of *Streptococcus milleri* in various body sites. J. Clin. Pathol. *32:* 764–768.

Pope, J., C. Singer, T.E. Kiehn, B.J. Lee and D. Armstrong. 1979. Infective endocarditis caused by *Rothia dentocariosa.* Ann. Int. Med. *91:* 746–747.

Pope, L. and L.J. Rode. 1969. Spore fine structure in *Clostridium cochlearium.* J. Bacteriol. *100:* 994–1001.

Pope, L., D.P. Yolton and L.J. Rode. 1967. Appendages of *Clostridium bifermentans* spores. J. Bacteriol. *94:* 1206–1215.

Poralla, K., E. Kannenberg and A. Blume. 1980. A glycolipid containing hopane isolated from the acidophilic, thermophilic *Bacillus acidocaldarius,* has a cholesterol-like function in membranes. FEBS Lett. *113:* 107–110.

Porschen, R.K., Z. Goodman and B. Rafai. 1977. Isolation of *Corynebacterium xerosis* from clinical specimens. Am. J. Clin. Pathol. *68:* 290–293.

Porschen, R.K. and S. Sonntag. 1974. Extracellular deoxyribonuclease production by anaerobic bacteria. Appl. Microbiol. *27:* 1031–1033.

Porschen, R.K. and E.H. Spaulding. 1974. Phosphatase activity of anaerobic organisms. Appl. Microbiol. *27:* 744–747.

Porter, J.R. 1976. The world view of culture collections. *In* Colwell (Editor), The Role of Culture Collections in the Era of Molecular Biology. American Society for Microbiology, Washington, D.C., pp. 62–72.

Porter, J.R., C.S. McCleskey and M. Levine. 1937. The facultatively sporulating bacteria producing gas from lactose. J. Bacteriol. *33:* 163–183.

Portnoy, D., J. Prentis and G.K. Richards. 1981. Penicillin tolerance of human isolates of group C streptococci. Antimicrob. Agents Chemother. *20:* 235–238.

Posten, S.M. and T.J. Palmer. 1977. Transduction of penicillinase production and other antibiotic-resistance markers in *Staphylococcus epidermidis.* J. Gen. Microbiol. *103:* 235–242.

Postgate, J.R. 1979. The sulfate-reducing bacteria. Cambridge University Press, Cambridge.

Postgate, J.R. and L.L. Campbell. 1963. Identification of Coleman's sulfate-reducing bacterium as a mesophilic relative of *Clostridium nigrificans.* J. Bacteriol. *86:* 274–279.

Poupard, J.A., I. Husain and R.F. Norris. 1973. Biology of the bifidobacteria. Bacteriol. Rev. *37:* 136–165.

Powell, J.F. and R.E. Strange. 1957. α, ε-Diaminopimelic acid metabolism and sporulation in *Bacillus sphaericus.* Biochem. J. *65:* 700–708.

Powell, J.T., W. Fischlschweiger and D.C. Birdsell. 1978. Modification of surface composition of *Actinomyces viscosus* T14V and T14AV. Infect. Immun. *22:* 934–944.

Poxton, I.R. and M.D. Byrne. 1981. Immunological analysis of the EDTA-soluble antigens of *Clostridium difficile* related species. J. Gen. Microbiol. *122:* 41–46.

Prabhakaran, K., E.B. Harris and W.F. Kirchheimer. 1975. O-diphenoloxidase of *Mycobacterium leprae* separated from infected armadillo tissues. Infect. Immun. *12:* 267–269.

Prabhakaran, K., E.B. Harris and W.F. Kirchheimer. 1979. Metabolic inhibitors of host-tissue origin in *Mycobacterium leprae.* Lep. India *51:* 348–357.

Prabhakaran, K., W.F. Kirchheimer and E.B. Harris. 1968. Oxidation of phenolic compounds by *Mycobacterium leprae* and inhibition of phenolase by substrate analogues and copper chelators. J. Bacteriol. *95:* 2051–2053.

Prauser, H. 1966. New and rare actinomycetes and their DNA base composition. Publ. Fac. Sci. Univ. Brno *475:* 268–270.

Prauser, H. 1967. Contributions to the taxonomy of the *Actinomycetales.* Publ. Fac. Sci. Univ. Purkyne, Brno *K40:* 196–199.

Prauser, H. 1970. Characters and genera arrangement in the *Actinomycetales. In* Prauser (Editor), The Actinomycetales, VEB Gustav Fischer, Jena, pp. 407–418.

Prauser, H. 1974. Host-phage relationships in nocardioform organisms. *In* Brownell (Editor), Proc. Intern. Conf. Biol. Nocardiae, McGowen, Augusta, pp. 84–85.

Prauser, H. 1974. *Nocardioides* Prauser, *Nocardiopsis* J. Meyer, and *Micropolyspora fascifera* Prauser - new taxa of the Actinomycetales. Actinomycetologist, Tokyo *24:* 14–15.

Prauser, H. 1976. *Nocardioides,* a new genus of the order *Actinomycetales.* Int. J. Syst. Bacteriol. *26:* 58–65.

Prauser, H. 1976. Host-phage relationships in nocardioform organisms. *In* Goodfellow, Brownell and Serrano (Editors), The Biology of the Nocardiae. Academic Press, London, pp. 266–284.

Prauser, H. 1978. Considerations on taxonomic relations among Gram-positive, branching bacteria. Zentralbl. Bakteriol. Parasitenkd. Infektionskr. Hyg. Abt. 1 Suppl. *6:* 3–12.

Prauser, H. 1981a. Nocardioform organisms: general characterization and taxonomic relationships. Zentralbl. Bakteriol. Mikrobiol. Hyg. Suppl. *11:* 17–24.

Prauser, H. 1981b. Taxon specificity of lytic actinophages that do not multiply in the cells affected. Zentralbl. Bakteriol. Mikrobiol. Hyg. Suppl. *11:* 87–92.

Prauser, H. 1984a. Phage host ranges in the classification and identification of gram-positive branched and related bacteria. *In* Ortiz-Ortiz, Bojalil and Yakoleff (Editors), Biological, Biochemical and Biomedical Aspects of Actinomycetes, Academic Press, Inc., London-New York-San Diego, pp. 617–633.

Prauser, H. 1984b. One-tube method for liquid nitrogen preservation and shipping of actinomycetes. *In* Proceedings of IV International Conference on Culture Collections, Brno, Czechoslovakia, 20–24 July, 1981, World Federation for Culture Collections, pp. 109–115.

Prauser, H. 1984c. *Nocardioides luteus* spec. nov. Z. Allg. Mikrobiol. *24:* 647–648.

Prauser, H. and M. Bergholz. 1974. Taxonomy of actinomycetes and screening for antibiotic substances. Postepy. Hig. Med. Dosw. *28:* 441–457.

Prauser, H. and R. Falta. 1968. Phagensensibilität, Zellwandzusammensetzung und Taxonomic von Actinomyceten. Z. Allg. Mikrobiol. *8:* 39–46.

Prauser, H., M.P. Lechevalier and H.A. Lechevalier. 1970. Description of *Oerskovia* gen. n. to harbor Ørskov's motile *Nocardia.* Appl. Microbiol. *19:* 534.

Prauser, H. and S. Momirova. 1970. Phagensensibilitat, Zellwand-zusammensetzung und Taxonomie einiger thermophiler Actinomyceten. Z. Allg. Mikrobiol. *10:* 219–222.

Prazmowski, A. 1880. Untersuchung uber die Entwickelungsgeschichte und Ferment-wirking einiger Bacterien-Arten. Inaug. Diss. Hugo Voigt, Leipzig, pp. 1–58.

Pregerson, B.S. 1973. The distribution and physiology of *Sporosarcina ureae.* M.S. Thesis. California State University, Northridge, California.

Preobrazhenskaya, T.P., R.S. Ukholina, N.P. Nechaeva, V.A. Filicheva, G.V. Gavrilina, M.K. Kudinova, V.N. Borisova, N.M. Patukhova, I.N. Kovsharova, V.V. Proshlyakova, O.K. Rossolimo. 1973. A new species of *Micropolyspora* and its antibiotic properties. Antibiotiki *18:* 963–970.

Prescott, J.F. 1979. Identification of some anaerobic bacteria in nonspecific anaerobic infections in animals. Can. J. Comp. Med. *43:* 194–199.

Prescott, L.M. and R.A. Altenbern. 1967. Inducible lysis in *Clostridium tetani.* J. Bacteriol. *93:* 1220–1226.

Prévot, A.R. 1924. *Diplococcus constellatus* (n. sp.) C.R. Soc. Biol. (Paris) *91:* 426–428.

Prévot, A.R. 1925. Les streptocoques anaerobies. Ann. Inst. Pasteur (Paris) *39:* 415–447.

Prévot, A.R. 1933. Études de systématique bactérienne. I. Lois générales. II. Cocci anaérobies. Ann. Sci. Nat. *15:* 23–260.

Prévot, A.R. 1938. Études de systématique bactérienne. III. Invalidité du genre *Bacteroides* Castellani et Chalmers démembrement et reclassification. Ann. Inst. Pasteur *60:* 285–307.

Prévot, A.R. 1938. Études de systématique bactérienne. IV. Critique de la conception actuelle du genre *Clostridium.* Ann. Inst. Pasteur (Paris) *61:* 72–91.

Prévot, A.R. 1940. Recherches sur la flore anaérobie de l'intestin humain: *Acuformis perennis* nov. sp. C. R. Soc. Biol. (Paris) *133:* 574–577.

Prévot, A.R. 1940. Manuel de classification et de détermination des bacteries anaérobies. Masson and Co., Paris, pp. 1–223.

Prévot, A.R. 1940. Un anaérobie strict reduisant les nitrates en nitrites *Eubacterium nitritogenes* n. sp. C. R. Soc. Biol. (Paris) *134:* 353–355.

Prévot, A.R. 1941. Sur une nouvelle espéces de streptocoque anaérobie gazogéne: *Streptococcus productus* nov. spec. C.R. Seances Soc. Biol. Fil. (France) *135:* 105–107.

Prévot, A.R. 1947. Étude de quelques bactéries anaérobies nouvelles ou mal connués. Ann. Inst. Pasteur (Paris) *73:* 409–418.

Prévot, A.R. 1948a. Manuel de classification et de détermination des bactéries anaérobies, 2nd ed. Masson and Co., Paris, pp. 1–290.

Prévot, A.R. 1948b. Étude des bactéries anaérobies d'Afrique Occidentale Francaise (Sènégal), Guinée, Côte d'Ivoire). Ann. Inst. Pasteur (Paris) *74:* 157–170.

Prévot, A.R. 1961. Trait'aae de systematique bacterienne. Dunod. Paris Vol. 2. p. 31.

Prévot, A.R. 1966. Manual for the classification and determination of the anaerobic bacteria (English translation by V. Fredette). Lea and Febiger, Philadelphia.

Prévot, A.-R. 1976. Nouvelle conception de la position des corynébactéries anaérobies. C.R. Acad. Sci., Série D. 282: 1079–1081.

Prévot, A.R., B. Besse and A.P. deFagonde. 1950. Étude d'une enzootie saisonnière à mortalite élevée des doffies de l'Herault (Chondrosoma toxostoma). Ann. Inst. Pasteur (Paris) 79: 903–905.

Prévot, A.R., F. de Cadore and H. Thouvenot. 1959. Recherches sur le pouvoir pathogéne et la pigmentation de Micrococcus niger. Ann. Inst. Pasteur (Paris) 97: 860–863.

Prévot, A.R. and J. Leplanche. 1947. Étude d'une bactérie anaérobie nouvelle de Guinée Francaise Cillobacterium combesi n. sp. Ann. Inst. Pasteur (Paris) 73: 687–688.

Prévot, A.R. and J. Taffanel. 1942. Recherches sur une nouvelle espèce anaérobie Ramibacterium alactolyticum (nov. spec.). Ann. Inst. Pasteur (Paris) 68: 259–262.

Prévot, A.-R. and H. Thouvenot. 1964. Essai de lysotype des Corynebacterium anaérobies. Ann. Inst. Past. (Paris) 101: 966–970.

Prévot, A.R., A. Turpin and P. Kaiser. 1967. Les bactéries anaérobies. Dunod, Paris, pp. 1–2188.

Prévot, A.R. and J. Zimmès-Chaverou. 1947. Étude d'une nouvelle espèce anaérobie de Côte d'Ivoire: Inflabilis mangenoti. Ann. Inst. Pasteur (Paris) 73: 602–604.

Pribram, E. 1933. Klassifikation der Schizomyceten. F. Deuticke, Leipzig.

Pridham, T.G. 1970. New names and new combinations in the order Actinomycetales Buchanan 1917. U.S. Dept. Agric. Tech. Bull. 1424: 1–55.

Pridham, T.G. 1974. Micro-organism Culture Collections: Acronyms and Abbreviations. ARS-NC-17. Agricultural Research Service, US Department of Agriculture, North Central Region, Peoria, Illinois.

Pridham, T.G. and A.J. Lyons, Jr. 1969. Progress in clarification of the taxonomic and nomenclatural status of some problem actinomycetes. Dev. Indust. Microbiol. 10: 183–221.

Pridham, T.G. and H.D. Tresner. 1974. Family VII. Streptomycetaceae Waksman and Henrici 1943, 339. In Buchanan and Gibbons (Editors), Bergey's Manual of Determinative Bacteriology, 8th Ed., Williams and Wilkins Co., Baltimore, pp. 747–842.

Priest, F.G. 1981. DNA homology in the genus Bacillus. In Berkeley and Goodfellow (Editors). The Aerobic Endospore-forming Bacteria. Academic Press, London, pp. 33–57.

Priest, F.G., M. Goodfellow and C. Todd. 1981. The genus Bacillus: a numerical analysis. In Berkeley and Goodfellow (Editors). The Aerobic Endospore-forming Bacteria. Academic Press, London, pp. 91–103.

Princewill, T.J.T. 1979. Spore antigens of Clostridium sporogenes. J. Med. Microbiol. 12: 29–41.

Princewill, T.J.T. 1980. Thiaminase activity amongst strains of Clostridium sporogenes. J. Appl. Bacteriol. 48: 249–252.

Princewill, T.J.T. and M.I. Agba. 1982. Examination of bovine faeces for the isolation and identification of Clostridium species. J. Appl. Bacteriol. 52: 97–102.

Princewill, T.J.T. and C.L. Oakley. 1972a. The deoxyribonucleases and hyaluronidases of Clostridium septicum and C. chauvoei. I. An agar plate method for testing deoxyribonuclease. Med. Lab. Tech. 29: 243–254.

Princewill, T.J.T. and C.L. Oakley. 1972b. The deoxyribonucleases and hyaluronidases of Clostridium septicum and C. chauvoei. II. An agar plate method for testing for hyaluronidase. Med. Lab. Tech. 29: 255–260.

Princewill, T.J.T. and C.L. Oakley. 1976. Deoxyribonucleases and hyaluronidases of Clostridium septicum and Clostridium chauvoei. III. Relationship between the two organisms. Med. Lab. Sci. 33: 105–118.

Pringsheim, E.G. 1950. The bacterial genus Lineola. J. Gen. Microbiol. 4: 198–209.

Pringsheim, E.G. 1967. Bakterien und Cyanophyceen. Oesterr. Bot. Z. 114: 324–340.

Pringsheim, E.G. and C.F. Robinow. 1947. Observations on two very large bacteria, Caryophanon latum Peshkoff and Lineola longa (nomen provisorium). J. Gen. Microbiol. 1: 267–278.

Prins, R.A. and A. Lankhorst. 1977. Synthesis of acetate from CO_2 in the cecum of some rodents. FEMS Microbiol. Lett. 1: 255–258.

Prissick, F.A. and A.M. Masson. 1956. Cervical lymphadentis in children caused by chromogenic mycobacteria. Can. Med. Assoc. J. 75: 798–803.

Prissick, R.H. and A.M. Masson. 1957. Yellow-pigmented pathogenic mycobacteria from cervical lymphadenitis. Can. J. Microbiol. 3: 91–100.

Proom, H. and B.C.J.G. Knight. 1950. Bacillus pantothenticus. (n.sp.) J. Gen. Microbiol. 4: 539–541.

Proom, H. and B.C.J.G. Knight. 1955. The minimal nutritional requirements of some species in the genus Bacillus. J. Gen. Microbiol. 13: 474–480.

Prottey, C. and C.E. Ballou. 1968. Diacyl myoinositol monomannoside from Propionibacterium shermanii. J. Biol. Chem. 243: 6196–6201.

Puhvel, S.M., M. Barfanti, M. Warnick and T.H. Steinberg. 1964. Study of antibody levels to Corynebacterium acnes. Arch. Dermatol. 90: 421–427.

Pulverer, G. 1974. Problems of human actinomycosis. Postepy. Hig. Med. Dosw. 28: 253–260.

Pulverer, G. and R. Halswick. 1967. Koagulase-Negative Staphylokokken (Staphylococcus albus) als Krankheitserreger. Dtsch. Med. Wochenschr. 92: 1141–1145.

Pulverer, G. and J. Jeljaszewicz. 1976. Staphylococcal microccins. In Jeljaszewicz (Editor), Staphylococci and Staphylococcus Diseases, Gustav Fischer Verlag, Stuttgart, New York, pp. 599–621.

Pulverer, G., J. Pillich and M. Haklova. 1976. Phage-typing set for the species Staphylococcus albus. In Jeljaszewicz (Editor), Staphylococci and Staphylococcal Diseases, Gustav Fischer Verlag, Stuttgart, New York, pp. 153–157.

Pulverer, G., J. Pillich and A. Klein. 1975. New bacteriophages of Staphylococcus albus. J. Inf. Dis. 132: 524–531.

Pulverer, G., J. Pillich and M. Krivankova. 1973. Differentiation of coagulase-negative staphylococci by bacteriophage-typing. In Jeljaszewicz (Editor), Staphylococci and Staphylococcal Infections, Karger, Basel, pp. 503–508.

Pulverer, G. and K.P. Schaal. 1978. Pathogenicity and medical importance of aerobic and anaerobic actinomycetes. In Mordarski, Kurylowicz and Jeljaszewicz (Editors), Nocardia and Streptomyces. Proceedings of the International Symposium on Nocardia and Streptomyces. Gustav Fischer Verlag, Stuttgart-New York, pp. 417–427.

Pulverer, G. and K.P. Schaal. 1984. Medical and microbiological problems in human actinomycoses. In Ortiz-Ortiz, Bojalil and Yakoleff (Editors), Biological, Biochemical and Biomedical Aspects of Actinomycetes. Academic Press, Orlando, London, New York, San Francisco, pp. 161–170.

Pulverer, G., H. Schütt-Gerowitt and K.P. Schaal. 1974. Bacteriophages of Nocardia. In Brownell (Editor), Proc. Intern. Conf. Biol. Nocardiae. Merida, Venezuela. McGowan, Augusta, Georgia. p. 82.

Pulverer, G., W. Songo and H.L. Ko. 1973. Bakteriophagen von Propionibacterium acnes. Zentralbl. Bakteriol. Parasitenkd. Infektionskr. Hyg. I. Abt. Orig. A. 225: 353–363.

Pulverer, G.H. and K.P. Schaal. 1978. Pathogenicity and medical importance of aerobic and anaerobic actinomycetes. Zentralbl. Bakteriol. Parasitenkd. Infektionskr. Hyg. I Abt. Suppl. 6: 417–427.

Radler, F. 1975. The metabolism of organic acids by lactic acid bacteria. In Carr, Cutting and Whiting (Editors), Lactic acid bacteria in beverages and food. Academic Press, London, New York, San Francisco, pp. 17–27.

Rado, T.A., J.H. Bates, H.W.B. Engel, E. Mankiewicz, T. Marohashi, Y. Mizugushi and L. Sula. 1975. WHO studies on bacteriophage typing of mycobacteria: subdivision of the species Mycobacterium tuberculosis. Am. Rev. Resp. Dis. 111: 459–468.

Ragsdale, S.W., J.E. Clark, L.G. Ljüngdahl, L.L. Lundie and H.L. Drake. 1983. Properties of purified carbon monoxide dehydrogenase from Clostridium thermoaceticum, a nickel, iron-sulfur protein. J. Biol. Chem. 258: 2364–2369.

Rahman, M. 1978. Free-sporing Cl. welchii in ordinary laboratory media and conditions. J. Clin. Pathol. 31: 359–360.

Raibaud, P., M. Caulet, J.V. Galpin and G. Mocquot. 1961. Studies on the bacterial flora of the alimentary tract of pigs. II. Streptococci: selective enumeration and differentiation of the dominant group. J. Appl. Bacteriol. 24: 285–306.

Rainbow, C. 1975. Beer spoilage lactic acid bacteria. In Carr, Cutting and Whiting (Editors), Lactic acid bacteria in beverages and food. Academic Press, London, New York, San Francisco, pp. 149–158.

Raj, H.D., F.L. Duryee, A.M. Deeney, C.H. Wang, A.W. Anderson and P.R. Elliker. 1960. Utilization of carbohydrates and amino acids by Micrococcus radiodurans. Can. J. Microbiol. 6: 289–298.

Ramamurthi, C.S. 1959. Comparitive studies of some Gram-positive phytopathogenic bacteria and their relationship to the corynebacteria. Mem. Cornell Univ. Exp. Sta. No. 366.

Ranftl, H. 1972. Zellwandzusammensetzung bei Bacillen und Sporosarcina. Ph. D. thesis. Technical University, Munich.

Ranftl, H. and O. Kandler. 1973. D-Aspartyl-L-alanin als Interpeptidbrücke im Murein von Bacillus pasteurii Migula. Z. Naturforsch. 28c: 4–8.

Rantz, L.A. and E. Randall. 1955. Use of autoclaved extracts for haemolytic streptococci for serological grouping. Stanford Med. Bull. 13: 290–291.

Rast, H.G., G. Engelhardt, W. Diegler and P.R. Wallhoffer. 1980. Bacterial degradation of model compounds for lignin and chlorophenol derived lignin bound residues. FEMS Microbiol. Lett. 8: 259–263.

Rastogi, N. and H.L. David. 1981. Growth and cell division of Mycobacterium avium. J. Gen. Microbiol. 126: 77–84.

Ravin, A.W. 1963. Experimental approaches to the study of bacterial phylogeny. Am. Natur. 97: 307–318.

Ravin, A.W. and J.D.H. De Sa. 1964. Genetic linkage of mutational sites affecting similar characters in pneumococcus and streptococcus. J. Bacteriol. 87: 86–96.

Ray, I. and R. Sen. 1977. Biochemical characters to differentiate strains of Corynebacterium diphtheriae, C. xerosis and C. hofmannii. Indian J. Med. Res. 65: 488–494.

Rayman, M.K., B. Aris and A. Hurst. 1981. Nisin: a possible alternative or adjunct to nitrite in the preservation of meats. Appl. Environ. Microbiol. 41: 375–380.

Razin, S. and S. Rottem. 1967. Identification of Mycoplasma and other microorganisms by polyacrylamide gel electrophoresis of cell proteins. J. Bacteriol. 94: 1807–1810.

Read, J.K., C.M. Heggie, M.W. Meyers and D.H. Connor. 1974. Cytotoxic activity of Mycobacterium ulcerans. Infect. Immun. 9: 1114–1122.

Read, S.E. and J.B. Zabriskie (Editors). 1980. Streptococcal Disease and the Immune Response. Academic Press, London, New York, Toronto, Sydney, San Francisco.

Reanney, D. 1976. Extrachromosomal elements as possible agents of adaptation and development. Bacteriol. Rev. *40:* 552–590.

Rebeyrotte, N., P. Rebeyerotte, M.J. Maviel and D. Montandon. 1979. Lipides et lipopolyosides de *Micrococcus radiodurans*. Ann. Microbiol. (Inst. Pasteur) *130:* 407–414.

Reddish, G. and L. Rettger. 1922. *Clostridium putrificum* (*B. putrificus* Bienstock) a distinct species. Abstr. Bacteriol. *6:* 9.

Reddy, C.A. and C.P. Cornell. 1982. Physiological and nutritional features of *Corynebacterium pyogenes*. J. Gen. Microbiol. *128:* 2851–2855.

Reddy, C.A., C.P. Cornell and A.M. Fraga. 1982. Transfer of *Corynebacterium pyogenes* (Glage) Eberson to the genus *Actinomyces* as *Actinomyces pyogenes* (Glage) comb. nov. Int. J. Syst. Bacteriol. *32:* 419–429.

Reddy, C.A., C.P. Cornell and M. Kao. 1977. Hemin-dependent growth stimulation and cytochrome synthesis in *Corynebacterium pyogenes*. J. Bacteriol. *130:* 965–967.

Reddy, M.S., F.D. Williams and G.W. Reinbold. 1973. Sulfonamide resistance of propionibacteria: nutrition and transport. Antimicrob. Agents Chemother. *4:* 454–458.

Redmond, D.L. and E. Kotcher. 1963. Cultural and serological studies on *Haemophilus vaginalis*. J. Gen. Microbiol. *33:* 77–87.

Reece, P., D. Toth and E.A. Dawes. 1976. Fermentation of purines and their effect on the adenylate energy change and viability of starved *Peptococcus prevotii*. J. Gen. Microbiol. *97:* 63–71.

Reed, G.B. 1957. Genus *Mycobacterium* (species affecting warm-blooded animals except those causing leprosy), *In* Breed, Murray and Smith (Editors), Bergey's Manual of Determinative Bacteriology, 7th ed. The Williams and Wilkins Co., Baltimore. pp. 695–707.

Reed, M.J. 1972. Chemical and antigenic properties of the cell wall of *Actinomyces viscosus* (strain T6). J. Dent. Res. *51:* 1193–1202.

Reed, M.J. and R.T. Evans. 1973. Cell wall peptidoglycan of *A. viscosus*. Abstr. Annu. Meet. Am. Soc. Microbiol. *73:* 182.

Rees, R.J., R.G. Valentine and P.C. Wong. 1960. Application of quantitative electron microscopy to the study of *Mycobacterium lepraemurium* and *M. leprae*. J. Gen. Microbiol. *22:* 443–457.

Rees, R.J.W., M.F.R. Waters, A.G.M. Weddell and E. Palmer. 1967. Experimental lepromatous leprosy. Nature *215:* 599–602.

Reh, M. 1981. Chemolithoautotrophy as an autonomous and transferable property of *Nocardia opaca* 1b. Zentralbl. Bakteriol. Mikrobiol. Hyg. *11:* 577–583.

Reh, M. and H.G. Schlegel. 1981. Hydrogen autotrophy as a transferable genetic character of *Nocardia opaca* 1b. J. Gen. Microbiol. *126:* 327–336.

Rehg, J.E. 1980. Cecal toxin(s) from guinea pigs with clindamycin-associated colitis, neutralized by *Clostridium sordellii* antitoxin. Infect. Immun. *27:* 387–390.

Rehg, J.E. and S.P. Pakes. 1982. Implication of *Clostridium difficile* and *Clostridium perfringens* iota toxins in experimental lincomycin-associated colitis of rabbits. Lab. Anim. Care *32:* 253–257.

Rehm, H.-J. 1983. Starter cultures for other purposes. *In* Rehm and Reed (Editors), Biotechnology Vol. 3, Verlag Chemie, Weinheim, pp. 204–208.

Reilly, S. 1980. The carbon dioxide requirements of anaerobic bacteria. J. Med. Microbiol. *13:* 573–579.

Reimer, L.G. and L.B. Reller. 1981. Growth of nutritionally variant streptococci on common laboratory and 10 commercial blood culture media. J. Clin. Microbiol. *14:* 329–332.

Reiss, H.J., J. Potel and A. Krebs. 1951a. *Granulomatosis infantiseptica*, eine durch einen spezifischen Erreger hervorgerufene fetale Sepsis. Klin. Wochenschr. *29:* 29.

Reiss, H.J., J. Potel and A. Krebs. 1951b. *Granulomatosis infantiseptica* (Eine Allgemeininfektion bei Neugeborenen und Säuglingen mit miliaren Granulomen). Z. Inn. Med. *6:* 451.

Reller, L.B. 1973. Endocarditis caused by *Bacillus subtilis*. Am. J. Clin. Pathol. *60:* 714–718.

Reller, L.B., G.L. Maddoux, M.R. Eckman and G. Pappas. 1975. Bacterial endocarditis caused by *Oerskovia turbata*. Ann. Intern. Med. *83:* 664–666.

Repaci, G. 1910. Contribution à l'étude de la flore bactérienne anaérobic des gangrènes pulmonaires. Un streptobacille anaérobie. C. R. Soc. Biol. Paris *68:* 410–412.

Reuter, G. 1963. Vergleichende Untersuchungen über die Bifidus-Flora im Säuglings- und Erwachsenenstuhl. Zentralbl. Bakteriol. Parasitenkd. Infektionskr. Hyg. Abt. I Orig. *191:* 486–507.

Reuter, G. 1970. Laktobazillen und eng verwandte Mikroorganismen in Fleisch und Fleischerzeugnissen. 2. Mitteilung: Die Charakterisierung der isolierten Laktobazillenstämme. Fleischwirtschaft *50:* 954–962.

Reuter, G. 1971. Designation of type strains for *Bifidobacterium* species. Int. J. Syst. Bacteriol. *21:* 273–275.

Reuter, G. 1975. Classification problems, ecology and some biochemical activities of lactobacilli of meat products. *In* Carr, Cutting and Whiting (Editors), Lactic acid bacteria in beverages and food. Academic Press, London, New York, San Francisco, pp. 221–229.

Reuter, G. 1983. *In* Validation of the publication of new names and new combinations previously effectively published outside the IJSB. List No. 11. Int. J. Syst. Bacteriol. *33:* 672–674.

Reuter, G. 1983. *Lactobacillus alimentarius* sp. nov., nom. rev. and *Lactobacillus farciminis* sp. nov., nom. rev. Syst. Appl. Microbiol. *4:* 277–279.

Reyn, A. 1970. Taxonomic position of *Neisseria haemolysans* (Thjøtta and Bøe 1938). Int. J. Syst. Bacteriol. *20:* 19–22.

Reyn, A., A. Birch-Andersen and U. Berger. 1970. Fine structure and taxonomic position of *Neisseria haemolysans* (Thjøtta and Bøe 1938) or *Gemella haemolysans* (Berger 1960). Acta Pathol. Microbiol. Immunol. Scand. B *78:* 375–389.

Reyn, A., A. Birch-Andersen and S.P. Lapage. 1966. An electron microscope study of thin sections of *Haemophilus vaginalis* (Gardner and Dukes) and some possibly related species. Can. J. Microbiol. *12:* 1125–1136.

Rhee, S.K. and M.Y. Pack. 1980a. Effect of environmental pH on chain length of *Lactobacillus bulgaricus*. J. Bacteriol. *144:* 865–868.

Rhodes-Roberts, M.E. 1981. The taxonomy of some nitrogen-fixing *Bacillus* species with special reference to nitrogen fixation. *In* Berkeley and Goodfellow (Editors). The Aerobic Endospore-forming Bacteria. Academic Press, London, pp. 315–335.

Rhuland, L.E., E. Work, R.F. Deham and D.S. Hoare. 1955. The behavior of the isomers of α, ϵ-diaminopimelic acid on paper chromatograms. J. Am. Chem. Soc. *77:* 4944–4946.

Richardson, G.M. 1936. The nutrition of *Staphylococcus aureus*. Necessity for uracil in anaerobic growth. Biochem. J. *30:* 2184–2190.

Richardson, L.T. 1957. Quantitative determination of viability of potato ring rot bacteria following storage, heat, and gas treatments. Can. J. Bot. *35:* 647–656.

Richardson, M.J. 1979. An annotated list of seed-borne diseases. Phytopathol. Paper *23:* 1–320.

Richmond, M.H. 1963. Purification and properties of the exopenicillinase from *Staphylococcus aureus*. Biochem. J. *88:* 452–459.

Richmond, M.H. 1972. Plasmids and extrachromosomal genetics in *Staphylococcus aureus*. *In* Cohen (Editor), The Staphylococci, Wiley-Interscience (John Wiley and Sons, Inc.), New York, London, Sydney, Toronto, pp. 159–186.

Richter, G. 1977. Routine use of thin-layer chromatography for cell wall analysis of aerobic actinomycetes, including two strains from sediments of the North Sea. Veroff. Inst. Meeresforsch. Bremerhaven. *16:* 125–138.

Ridell, M. 1974. Serological study of nocardiae and mycobacteria by using "*Mycobacterium*" *pellegrino* and *Nocardia corallina* precipitation reference systems. Int. J. Syst. Bacteriol. *24:* 64–72.

Ridell, M. 1975. Taxonomic study of *Nocardia farcinica* using serological and physiological characters. Int. J. Syst. Bacteriol. *25:* 124–132.

Ridell, M. 1977. Studies on corynebacterial precipitinogens common to mycobacteria, nocardiae and rhodochrous. Int. Arch. Allergy Appl. Immunol. *55:* 468–475.

Ridell, M. 1981a. Immunodiffusion studies of some *Nocardia* strains. J. Gen. Microbiol. *123:* 69–74.

Ridell, M. 1981b. Immunodiffusion studies of *Mycobacterium*, *Nocardia* and *Rhodococcus* for taxonomic purposes. Zentralbl. Bakteriol. Mikrobiol. Hyg. I Abt. Orig., Suppl. *11:* 235–241.

Ridell, M. 1983. Sensitivity to capreomycin and prothionamide in strains of *Mycobacterium*, *Nocardia*, *Rhodococcus* and related taxa for taxonomical purposes. Zentralbl. Bakteriol. Mikrobiol. Hyg. I Abt. A *255:* 309–316.

Ridell, M. 1984. Serotaxonomical analyses of strains referred to *Nocardia amarae* and *Rhodococcus equi*. Zentralbl. Bakteriol. Mikrobiol. Hyg. I. Abt. Orig. A *259:* 492–497.

Ridell, M., R. Baker, A. Lind and O. Ouchterlony. 1979. Immunodiffusion studies of ribosomes in classification of mycobacteria and related taxa. Int. Arch. Allergy Appl. Immunol. *59:* 162–172.

Ridell, M., M. Goodfellow and D.E. Minnikin. 1985. Immunodiffusion and lipid analyses in the classification of "*Mycobacterium album*" and the "*aurantiaca*" taxon. Zentralbl. Bakteriol. Mikrobiol. Hyg. A *259:* 1–10.

Ridell, M. and M. Norlin. 1973. Serological study of *Nocardia* by using mycobacterial precipitation reference systems. J. Bacteriol. *113:* 1–7.

Rifkind, D. and R.M. Cole. 1962. Non-beta hemolytic group M-reacting streptococci of human origin. J. Bacteriol. *84:* 163–168.

Riley, T.V. and A.K. Ott. 1981. Brain abscess due to *Arachnia propionica*. Brit. Med. J. *282:* 1035.

Rimbault, A. and G. Leluan. 1982. Composés neutres et basiques présents dans les gaz produits par *Clostridium histolyticum*, *Clostridium hastiforme*, et *Clostridium ghoni* cultivés sous vide en milieu glucosé au thioglycolate de sodium. C. R. Acad. Sci. Paris *295:* 219–221.

Rinfret, A.P. and B. LaSalle (Editors). 1975. Round table conference on the cryogenic preservation of cell cultures. National Academy of Sciences, Washington, D.C.

Rische, H., W. Meyer, H. Tschäpe, W. Voigt, D. Ziomek and R. Hummell. 1973. The taxonomy of *Staphylococcus aureus*. *In* Jeljaszewicz (Editor), Staphylococci and Staphylococcal Infections, S. Karger, Basel, pp. 24–29.

Ristroph, J.D. and B.E. Ivins. 1983. Elaboration of *B. anthracis* antigens in a new, defined culture medium. Infect. Immun. *39:* 483–486.

Ritz, H.L. 1966. Selective medium for oral nocardiae. J. Dent. Res. *45:* 411.

Ritz, H.L. and J.N. Baldwin. 1958. Induction of penicillinase production in staphylococci by bacteriophage . Bacteriol. Proc., 40.

Roberts, A.P. 1967. *Micrococcaceae* from the urinary tract in pregnancy. J. Clin. Pathol. *20:* 631–632.

Roberts, G.P., W.T. Leps, L.E. Silver and W.J. Brill. 1980. Use of two-dimensional gel electrophoresis to identify and classify *Rhizobium* strains. Appl. Environ. Microbiol. *39:* 414–422.

Roberts, R.B., A.G. Krieger, N.L. Schiller and K.C. Gross. 1979. Viridans streptococcal endocarditis: the role of various species, including pyridoxal-dependent streptococci. Rev. Infect. Dis. *1:* 955–965.

Roberts, R.C., D.P. Zais and D.A. Emanuel. 1976. The frequency of precipitins to trichloroacetic acid - extractable antigens from thermophilic actinomycetes in farmer's lung patients and asymptomatic farmers. Annu. Rev. Resp. Dis. *114:* 23–28.

Roberts, R.C., D.P. Zais, J.J. Marx and M.W. Zrenhaft. 1977. Comparative electrophoresis of the proteins and proteases in thermophilic actinomycetes. J. Lab. Clin. Med. *90:* 1075–1085.

Roberts, R.J. 1968. A numerical taxonomic study of 100 isolates of *Corynebacterium pyogenes*. J. Gen. Microbiol. *53:* 299–303.

Roberts, R.J. 1968. Biochemical reactions of *Corynebacterium pyogenes*. J. Pathol. Bacteriol. *95:* 127–130.

Roberts, T.A. 1968. Heat and radiation resistance and activation of spores of *Clostridium welchii*. J. Appl. Bacteriol. *31:* 133–144.

Roberts, T.A. and G. Hobbs. 1968. Low temperature growth characteristics of clostridia. J. Appl. Bacteriol. *31:* 75–88.

Robertson, D.C. and W.G. McCullough. 1968. Glucose catabolism of *Erysipelothrix rhusiopathiae*. J. Bacteriol. *95:* 2112–2116.

Robinow, C.F. 1951. Observations on the structure of *Bacillus* spores. J. Gen. Microbiol. *5:* 439–457.

Robinow, C.F. 1960. Morphology of bacterial spores, their development and germination. *In* Gunsalus and Stanier (Editors). The Bacteria, Volume I: Structure. Academic Press, New York, pp. 207–248.

Robinson, J.M., J.K. Hardman and G.L. Sloan. 1979. Relationship between lysostaphin endopeptidase production and cell wall composition in *Staphylococcus staphylolyticus*. J. Bacteriol. *137:* 1158–1164.

Robinson, K. 1966a. Some observations on the taxonomy of the genus *Microbacterium*. I. Cultural and physiological reactions and heat resistance. J. Appl. Bacteriol. *29:* 607–615.

Robinson, K. 1966b. Some observations on the taxonomy of the genus *Microbacterium*. II. Cell wall analysis, gel electrophoresis and serology. J. Appl. Bacteriol. *29:* 616–624.

Robinson, K. 1968. The use of cell wall analysis and gel electrophoresis for the identification of coryneform bacteria. *In* Gibbs and Shapton (Editors), Identification Methods for Microbiologists, Part B. Academic Press, London, pp. 85–92.

Robinson, R.K. 1983. Starter cultures for milk and meat processing. *In* Rehm and Reed (Editors), Biotechnology, Vol. 3, Verlag Chemie, Weinheim, pp. 191–202.

Robinson, R.W. and C.R. Spotts. 1983. The ultrastructure of sporulation in *Sporosarcina ureae*. Can. J. Microbiol. *29:* 807–814.

Rocchi, G. 1908. Lo stato attuale delle nostre cognizioni sui germi anaerobi. Bull. Sci. Méd. *8:* 457–528.

Roché, C., H. Albertyn, N.O. van Gylswyk and A. Kistner. 1973. The growth response of cellulolytic acetate-utilizing and acetate-producing butyrivibrios to volatile fatty acids and other nutrients. J. Gen. Microbiol. *78:* 253–260.

Rochford, J.C. 1980. Pleuropulmonary infection associated with *Eubacterium brachy*, a new species of *Eubacterium*. J. Clin. Microbiol. *12:* 722–723.

Rocourt, J., H-W. Ackermann, M. Martin, A. Schrettenbrunner and H.P.R. Seeliger. 1983c. Morphology of *Listeria innocua* bacteriophages. Ann. Virol. (Inst. Pasteur) *134E:* 245–250.

Rocourt J., J-M. Alonso and H.P.R. Seeliger. 1983a. Virulence comparée des cinq groupes génomiques de *Listeria monocytogenes* (*sensu lato*). Ann. Microbiol. (Inst. Pasteur) *134A:* 359–364.

Rocourt, J., F. Grimont, P.A.D. Grimont and H.P.R. Seeliger. 1982b. DNA relatedness among serovars of *Listeria monocytogenes sensu lato*. Curr. Microbiol. *7:* 383–388.

Rocourt, J. and P.A.D. Grimont. 1983. *Listeria welshimeri* sp. nov. and *Listeria seeligeri* sp. nov. Int. J. Syst. Bacteriol. *33:* 866–869.

Rocourt, J., A. Schrettenbrunner and H.P.R. Seeliger. 1982a. Isolation of bacteriophages from *Listeria monocytogenes* serovar 5 and *Listeria innocua*. Zentralbl. Bakteriol. Parasitenkd. Infektionskr. Hyg. I. Abt. Orig. A *251:* 505–511.

Rocourt, J., A. Schrettenbrunner and H.P.R. Seeliger. 1983. Différenciation biochimique des groups génomiques de *Listeria monocytogenes* (*sensu lato*). Ann. Microbiol. Inst. Past. *134A:* 65–71.

Rocourt, J., A. Schrettenbrunner and H.P.R. Seeliger. 1983b. Différenciation biochimique des groupes génomiques de *Listeria monocytogenes* (*sensu lato*). Ann. Microbiol. (Inst. Pasteur) *134A:* 65–71.

Rode, L.J., L. Pope, C. Filip and L.D. Smith. 1971. Spore appendages and taxonomy of *Clostridium sordellii*. J. Bacteriol. *108:* 1384–1389.

Rode, L.J. and L.D. Smith. 1971. Taxonomic implications of spore fine structure in *Clostridium bifermentans*. J. Bacteriol. *105:* 349–354.

Rode, L.M., B.R.S. Genthner and M.P. Bryant. 1981. Syntrophic association by cocultures of the methanol- and CO2-H2-utilizing species *Eubacterium limosum* and pectin-fermenting *Lachnospira multiparus* during growth in a pectin medium. Appl. Environ. Microbiol. *42:* 20–22.

Rogers, A.H. 1975. Bacteriocin types of *Streptococcus mutans* in human mouths. Arch. Oral. Biol. *20:* 853–858.

Rogers, A.H. 1976. Bacteriocinogeny and the properties of some bacteriocins of *Streptococcus mutans*. Arch. Oral Biol. *21:* 99–104.

Rogolsky, M., B. Wiley and L. Glasgow. 1976. Phage group II staphylococcal strains with chromosomal and extrachromosomal genes for exfoliative toxin production. Infect. Immun. *13:* 44–52.

Rogosa, M. 1971. *Peptococcaceae*, a new family to include the gram-positive, anaerobic cocci of the genera *Peptococcus*, *Peptostreptococcus*, and *Ruminococcus*. Int. J. Syst. Bacteriol. *21:* 234–237.

Rogosa, M. 1974. Genus I. *Peptococcus* Kluyver and van Niel 1936, 400. *In* Buchanan and Gibbons (Editors), Bergey's Manual of Determinative Bacteriology, 8th ed. The Williams and Wilkins Co., Baltimore, pp. 518–522.

Rogosa, M. 1974. Genus III. *Ruminococcus* Sijpesteijn 1948, 152 pp. 525–527. *In* Buchanan and Gibbons (Editors), Bergey's Manual of Determinative Bacteriology, 8th ed. The Williams and Wilkins Co., Baltimore.

Rogosa, M. 1974. Family III. *Peptococcaceae*. *In* Buchanan and Gibbons (Editors), Bergey's Manual of Determinative Bacteriology, 8th Ed., The Williams and Wilkins Co., Baltimore, pp. 517–528.

Rogosa, M., C.S. Cummins, R.A. Lelliott and R.M. Keddie. 1974. Coryneform group of bacteria. *In* Buchanan and Gibbons (Editors), Bergey's Manual of Determinative Bacteriology, 8th ed. The Williams and Wilkins Co., Baltimore, pp. 599–632.

Rogosa, M., J.G. Franklin and K.D. Perry. 1961. Correlation of the vitamin requirements with cultural and biochemical characters of *Lactobacillus* spp. J. Gen. Microbiol. *25:* 473–482.

Rogosa, M. and P.A. Hansen. 1971. Nomenclatural considerations of certain species of *Lactobacillus* Beijerinck. Int. J. Syst. Bacteriol. *21:* 177–186.

Rogosa, M. and R.M. Keddie. 1974. Genus *Microbacterium*. *In* Buchanan and Gibbons (Editors), Bergey's Manual of Determinative Bacteriology. The Williams and Wilkins Co., Baltimore, pp. 628–629.

Rogosa, M. and R.M. Keddie. 1974. Genus *Brevibacterium*. *In* Buchanan and Gibbons (Editors), Bergey's Manual of Determinative Bacteriology 8th Ed. The Williams and Wilkins Co., Baltimore, pp. 625–628.

Rogosa, M., J.A. Mitchell and R.F. Wiseman. 1951. A selective medium for isolation and enumeration of oral and fecal lactobacilli. J. Bacteriol. *62:* 132–133.

Rogosa, M. and M.E. Sharpe. 1959. An approach to the classification of the lactobacilli. J. Appl. Bacteriol. *22:* 329–340.

Rogosa, M., R.F. Wiseman, J.A. Mitchell and M. Disraely. 1953. Species differentiation of oral lactobacilli from man including descriptions of *Lactobacillus salivarius* nov. spec. and *Lactobacillus cellobiosus* nov. spec. J. Bacteriol. *65:* 681–699.

Roguinsky, M. 1971. Caractères biochemiques et serologiques *Streptococcus uberis*. Ann. Inst. Pasteur (Paris) *120:* 154–163.

Rolfe, R.D., D.J. Hentges, B.J. Campbell and J.T. Barratt. 1978. Factors related to the oxygen tolerance of anaerobic bacteria. Appl. Environ. Microbiol. *36:* 306–313.

Römer, A. and R.B. Herbert. 1982. Further observations on the source of nitrogen in phenazine biosynthesis. Z. Naturforsch. *37c:* 1070–1074.

Romero, R. and H.W. Wilkinson. 1974. Identification of group B streptococci by immunofluorescence staining. Appl. Microbiol. *28:* 199–204.

Romond, C., R. Sartory and J. Malgras. 1966. Le coefficient de Shapiro-Chargoff des streptocoques anaérobies. Ann. Inst. Pasteur (Paris) *111:* 710–718.

Rood, J.I., E.A. Maher, E.B. Somers, E. Campos and C.L. Duncan. 1978a. Isolation and characterization of multiply antibiotic-resistant *Clostridium perfringens* strains from porcine feces. Antimicrob. Agents Chemother. *13:* 871–880.

Rood, J.I., V.N. Scott and C.L. Duncan. 1978b. Identification of a transferable tetracycline resistance plasmid (p CW3) from *Clostridium perfringens*. Plasmid *1:* 563–570.

Roop, D.R., J.O. Mundt and W.S. Riggsby. 1974. Deoxyribonucleic acid hybridization studies among some strains of group D and group N streptococci. Int. J. Syst. Bacteriol. *24:* 330–337.

Rosan, B. 1973. Antigens of *Streptococcus sanguis*. Infect. Immun. *7:* 205–211.

Rosan, B. 1976. Relationship of the cell wall composition of Group H streptococci and *Streptococcus sanguis* to their serological properties. Infect. Immun. *13:* 1144–1153.

Rosan, B. and L. Argenbright. 1982. Antigenic determinant of the Lancefield Group H antigen of *Streptococcus sanguis*. Infect. Immun. *38:* 925–931.

Rose, H.D., B. Varkey and C.P. Kesavan Kutty. 1982. Thoracic actinomycosis caused by *Actinomyces meyeri*. Amer. Rev. Resp. Dis. *125:* 251–254.

Rosenbach, F.J. 1884. Micro-organismen bei den Wund-Infections-Krankheiten des Menschen. J.F. Bergmann, Weisbaden, pp. 1–122.

Rosenbach, F.J. 1909. Experimentelle Morphologische und klinische Studien über krankheitserregende Mikroorganism des Schweinrotlaufs, des Erysipeloids und der Möusesepticämie. Z. Hyg. Infektionskr. *63:* 343–371.

Rosenblum, E.D. and S. Tyrone. 1979. Deoxyribonucleic acid homologies among staphylococci: coagulase-positive reference strains. Curr. Microbiol. *2:* 171–174.

Rosenblum, E.D. and P.W. Wilson. 1949. Fixation of isotopic nitrogen by *Clostridium*. J. Bacteriol. *57:* 413–414.

Rosendal, K., J. Bang and V.T. Rosdahl. 1981. Occurrence of gentamicin-resistant *Staphylococcus* in Denmark. *In* Jeljaszewicz (Editor), Staphylococci and Staphylococcal Infections, Gustav Fischer Verlag, Stuttgart, New York, pp. 985–990.

Rosendorf, L.L. and F.H. Kayer. 1974. Transduction and plasmid deoxyribonucleic acid analysis in a multiply antibiotic-resistant strain of *Staphylococcus epidermidis*. J. Bacteriol. *120:* 679–686.

Rosenthal, S.A. and C.D. Cox. 1953. The somatic antigens of *Corynebacterium michiganense* and *Corynebacterium insidiosum*. J. Bacteriol. *65:* 532–537.

Rosenthal, S.A. and C.D. Cox. 1954. An antigenic analysis of some plant and soil corynebacteria. Phytopathology *44:* 603–604.

Ross, H.E. 1965. *Clostridium putrefaciens*: a neglected anaerobe. J. Appl. Bacteriol. *28:* 49–51.

Rossi-Doria, T. 1891. Su di alcune specie di 'Streptothrix' trovate nell'aria studate in rapporto a quelle già note a specialmente all' 'Actinomyces'. Ann. Ist. Igiene Sper. Univ. Roma *1:* 399–438.

Rosypal, S. and M. Kocur. 1963. The taxonomic significance of the oxidation of carbon compounds by different strains of *Micrococcus luteus*. Antonie van Leeuwenhoek. J. Microbiol. Serol. *29:* 313–318.

Roth, G.D. 1957. Proteolytic organisms of the carious lesion. Oral Surg. Oral Pathol. *10:* 1105–1117.

Roth, G.D. and V. Flanagan. 1969. The pathogenicity of *Rothia dentocariosa* inoculated into mice. J. Dent. Res. *49:* 957–958.

Roth, G.D., R.G. Garrison and K.S. Boyd. 1976. Ultra-structure of *Rothia dentocariosa*. J. Gen. Microbiol. *95:* 45–53.

Roth, G.D. and A.N. Thurn. 1962. Continued study of oral nocardia. J. Dent. Res. *41:* 1279–1292.

Rotimi, V.O. and B.I. Duerden. 1981. The development of the bacterial flora in normal neonates. J. Med. Microbiol. *14:* 51–62.

Rotimi, V.O. and B.I. Duerden. 1982. The bacterial flora of neonates with congenital abnormalities of the gastrointestinal tract. J. Hyg. *88:* 69–81.

Rotta, J., R.M. Krause, R.C. Lancefield, W. Everly and H. Lackland. 1971. New approaches for the laboratory recognition of M types of group A streptococci. J. Exp. Med. *134:* 1298–1315.

Roux, C. and J.-L. Bergere. 1977. Caractères taxonomiques de *Clostridium tyrobutyricum*. Ann. Microbiol. (Inst. Pasteur) *128A:* 267–276.

Rowbotham, T.J. and T. Cross. 1977a. *Rhodococcus coprophilus* sp. nov.: an aerobic nocardioform actinomycete belonging to the 'rhodochrous' complex. J. Gen. Microbiol. *100:* 123–138.

Rowbotham, T.J. and T. Cross. 1977b. *Rhodococcus coprophilus* and associated actinomycetes in fresh water and agricultural habitats. J. Gen. Microbiol. *100:* 231–240.

Rowbotham, T.J. and T. Cross. 1979. Validation List No. 2. Int. J. Syst. Bacteriol. *29:* 79–80.

Ruan, J.-S. and Y.-M. Zhang. 1979. Two new species of *Nocardioides*. Acta Microbiol. Sin. *19:* 347–352 (in Chinese).

Rubin, J., W.A. Rodgers and H.M. Taylor. 1980. Peritonitis during continuous ambulatory dialysis. Ann. Intern. Med. *92:* 7–13.

Rubin, S.J., R.W. Lyens and A.J. Murcia. 1978. Endocarditis associated with cardiac catherization due to a Gram-positive coocus designated *Micrococcus mucilaginosus* incertae sedis. J. Clin. Microbiol. *7:* 546–549.

Rúger, H.J. 1983. Differentiation of *Bacillus globisporus*, *Bacillus marinus* comb. nov., *Bacillus aminovorans*, and *Bacillus insolitus*. Int. J. Syst. Bacteriol. *33:* 157–161.

Rúger, H.J. and G. Hentzschel. 1980. Mineral salt requirements of *Bacillus globisporus* subsp. *marinus* strains. Arch. Microbiol. *126:* 83–86.

Rúger, H.J. and G. Richter. 1979. *Bacillus globisporus* subsp. *marinus* subsp. nov. Int. J. Syst. Bacteriol. *29:* 196–203.

Runyon, E.H. 1965. Pathogenic mycobacteria. Adv. Tuberc. Res. *14:* 235–287.

Runyon, E.H. 1967. *Mycobacterium intracellulare*. Am. Rev. Resp. Dis. *95:* 861–865.

Runyon, E.H. 1968. Aerial hyphae of *Mycobacterium xenopei*. J. Bacteriol. *95:* 734–735.

Runyon, E.H. 1972. Conservation of the specific epithet *fortuitum* in the name of the organism known as *Mycobacterium fortuitum* da Costa Cruz - Request for an opinion. Int. J. Syst. Bacteriol. *22:* 50–51.

Runyon, E.H. and T.M. Dietz. 1971. Skin sensitivity in guinea pigs induced by Group II mycobacteria. Am. Rev. Resp. Dis. *104:* 107–113.

Runyon, E.H., L.G. Wayne and G.P. Kubika. 1974. Genus *Mycobacterium*. *In* Buchanan and Gibbons (Editors), Bergey's Manual of Determinative Bacteriology, 8th ed. The Williams and Wilkins Co., Baltimore, pp. 682–701.

Ruoff, K.L. and L.J. Kunz. 1982. Identification of viridans streptococci isolated from clinical specimens. J. Clin. Microbiol. *15:* 920–925.

Ruoff, K.L. and L.J. Kunz. 1983. Use of the rapid STREP System for the identification of viridans streptococcal species. J. Clin. Microbiol. *18:* 1138–1140.

Rupprecht, M. and K.H. Schleifer. 1977. Comparative immunological study of catalases in the genus *Micrococcus*. Arch. Microbiol. *114:* 61–66.

Rupprecht, M. and K.H. Schleifer. 1979. A comparative immunological study of catalases from coagulase-positive staphylococci. Arch. Microbiol. *120:* 53–56.

Russell, A.D. 1982. The destruction of bacterial spores. Academic Press, London.

Russell, A.D. and T.H. Melville. 1978. A review: Bacteria in the human mouth. J. Appl. Bacteriol. *44:* 163–181.

Russell, C. and T.K. Walker. 1953. *Lactobacillus malefermentans* n. sp. isolated from beer. J. Gen. Microbiol. *8:* 160–162.

Russell, E.G. 1979. Types and distribution of anaerobic bacteria in the large intestine of pigs. Appl. Environ. Microbiol. *37:* 187–193.

Russell, R.L., W.D. Richards, L.A. Scammon and S. Froman. 1964. Isolation of a lysogenic *Mycobacterium fortuitum* from soil. Am. Rev. Resp. Dis. *89:* 287–288.

Russell, R.R.B. 1976. Classification of *Streptococcus mutans* strains by SDS gel electrophoresis. Microbios Lett. *2:* 55–59.

Rustigian, R. and A. Cipriani. 1947. The bacteriology of open wounds. J. Am. Med. Assoc. *133:* 224–230.

Ruwart, M.J., H.E. Renis, A.L. Erlandson, R.M. DeHaan, J.E. Nezamis, C. Lancaster and J.P. Davis. 1983. 16, 16-dimethyl-prostaglandin E_2 inhibits toxin release from *Clostridium difficile*. *In* Samuelsson, Paoletti and Ramwell (Editors), Advances in Prostaglandin, Thromboxane and Leukotriene Research, Vol. 12, Raven Press, New York.

Ryff, J.F. and A.M. Lee. 1946. *Clostridium* infection in range animals. J. Am. Vet. Med. Assoc. *108:* 385–388.

Sabath, L.D., C. Garner, C. Wilcox and M. Finland. 1976. Susceptability of *Staphylococcus aureus* and *Staphylococcus epidermidis* to 65 antibiotics. Antimicrob. Agents Chemother. *9:* 962–969.

Sabbaj, J., V.L. Sutter and S.M. Finegold. 1971. Comparison of selective media for isolation of presumptive group D streptococci from human feces. Appl. Microbiol. *22:* 1008–1011.

Sabouraud, R. 1897. La séborrhée grasse et la pelade. Ann. Inst. Pasteur *11:* 134–159.

Sacks, L.E. 1954. Observations on the morphogenesis of *Arthrobacter citreus*, spec. nov. J. Bacteriol. *67:* 342–345.

Sacks, L.E. and P.A. Thompson. 1977. Increased spore yields of *Clostridium perfringens* in the presence of methylxanthines. Appl. Environ. Microbiol. *34:* 189–193.

Saddler, J.N. and M.K.-H. Chan. 1982. Optimization of *Clostridium thermocellum* growth on cellulose and pretreated wood substrates. Eur. J. Appl. Microbiol. Biotechnol. *16:* 99–104.

Sadzikowski, M.R., J.F. Sperry and T.D. Wilkins. 1977. Cholesterol-reducing bacterium from human feces. Appl. Environ. Microbiol. *34:* 355–362.

Saggers, B.A. and G.T. Stewart. 1968. Lipolytic esterases in staphylococci. J. Bacteriol. *96:* 1006–1010.

Sahota, S.S., P.M. Bramley and I.S. Menzies. 1982. The fermentation of lactulose by colonic bacteria. J. Gen. Microbiol. *128:* 319–325.

Saito, H., R.E. Gordon, I. Juhlin, W. Käppler, J.B.G. Kwapinski, C. McDurmont, S.R. Pattyn, E.H. Runyon, J.L. Stanford, I. Tarnok, H. Tasaka, M. Tsukamura and J. Weisfeiler. 1977. Cooperative numerical analysis of rapidly growing mycobacteria. The second report. Int. J. Syst. Bacteriol. *27:* 75–85.

Saito, H., K. Yamaoka and K. Kiyotani. 1976. *In vitro* properties of *M. lepraemurium* strain Keishicho. Int. J. Syst. Bacteriol. *26:* 111–115.

Sakaguchi, K. 1960. Vitamins and amino-acid requirements of *Pediococcus soyae* and *Pediococcus acidilactici* Kitahara's strain. IX. Studies on the activities of bacteria in soy sauce brewing. Bull Agric. Chem. Soc. Jpn. *26:* 638–643.

Sakaguchi, K., M. Iwasaki and S. Yamada. 1941. Studies on the propionic acid fermentation. J. Agric. Chem. Soc. Jpn. I. *17:* 127–158.

Sakaguchi, K. and H. Mori. 1969. Comparative study on *Pediococcus halophilus*, *P. soyae*, *P. homari*, *P. urinae-equi* and related species. J. Gen. Appl. Microbiol. *15:* 159–167.

Sakaguchi, K. and M. Okanishi (Editors). 1980. Molecular breeding and genetics of applied microorganisms. Kokansha Ltd. Tokyo and Academic Press, New York and London.

Sakaguchi, Y. and K. Murata. 1983. Studies on the β-glucuronidase-production of clostridia. Zentralbl. Bakteriol. Mikrobiol. Hyg. I. Abt. Orig. A *254:* 118–122.

Sakuma, S., M. Sakuma, A. Okaniwa and Y. Sato. 1973. Detection of *Erysipelothrix insidiosa* in mice by whole body autobacteriography. Nat. Inst. Anim. Health Q. (Tokyo) *13:* 54–58.

Sakurai, J. and C.L. Duncan. 1978. Some properties of beta-toxin produced by *Clostridium perfringens* type C. Infect. Immun. *21:* 678–680.

Sakurai, J., Y. Fujii, M. Matsuura and K. Endo. 1981. Pharmacological effects of beta toxin of *Clostridium perfringens* type C on rats. Microbiol. Immunol. *25:* 423–432.

Salle, A.J. 1973. Fundamental principles of bacteriology. 7th edition. McGraw-Hill Book Co., New York.

Sames, T. 1898. Eine bewegliche Sarcine. Zentralbl. Bakteriol. Parasitenkd. Infektionskr. Hyg. Abt. 2, *4:* 664–669.

Samples, J.R. and H. Buettner. 1983. Ocular infection caused by a biological insecticide. J. Infect. Dis. *148:* 614.

Samsonoff, W.A., T. Hashimoto and S.F. Conti. 1970. Ultrastructural changes associated with germination and outgrowth of an appendage-bearing clostridial spore. J. Bacteriol. *101:* 1038–1045.

Sand, G. and C.O. Jensen. 1888. Die Aetiologie der Druise. Dtsch. Z. Tiermed. Verg. Pathol. *13:* 437–464.

Sanders, J.E. and J.L. Fryer. 1980. *Renibacterium salmoninarum* gen. nov., sp. nov., the causitive agent of bacterial kidney disease in salmonid fishes. Int. J. Syst. Bacteriol. *30:* 496–502.

Sanders, J.E., K.S. Pilcher and J.L. Fryer. 1978. Relation of water temperature to bacterial kidney disease in coho salmon (*Oncorhynchus kisutch*), sockeye salmon (*O. nerka*), and steelhead trout (*Salmo gairdneri*). J. Fish Res. Board Can. *35:* 8–11

Sanderson, P.J., M.W.D. Wren and A.W.F. Baldwin. 1979. Anaerobic organisms in postoperative wounds. J. Clin. Pathol. *32:* 143–147.

Sanger, F., G.G. Brownlee and B.G. Barrell. 1965. A two-dimensional fractionation procedure for radioactive nucleotides. J. Mol. Biol. *13:* 373–398.

Sanger, F., S. Nicklen and A.R. Coulson. 1977. DNA sequencing with chain-terminating inhibitors. Proc. Nat. Acad. Sci. USA *74:* 5463–5467.

Sans, M.D. and J.G. Crowder. 1973. Subacute bacterial endocarditis caused by *Eubacterium aerofaciens*: Report of a case. Am. J. Clin. Pathol. *59:* 576–580.

Saperstein, S. and M.P. Starr. 1954. The ketonic carotenoid canthazanthin isolated from a colour mutant of *Corynebacterium michiganense*. Biochem. J. *57:* 273–275.

Saperstein, S., M.P. Starr and J.A. Filfus. 1954. Alterations in carotenoid synthesis accompanying mutation in *Corynebacterium michiganense*. J. Gen. Microbiol. *10:* 85–92.

Sarkany, I., D. Taplin and H. Blank. 1962. Organism causing erythrasma. Lancet (ii): 304–305.

Sarra, P.G., M. Magri, V. Bottazzi and F. Dellaglio. 1980. Genetic heterogeneity among *Lactobacillus acidophilus* strains. Antonie van Leeuwenhoek J. Microbiol. Serol. *46:* 169–176.

Sarra, P.G., M. Magri, V. Bottazzi, F. Dellaglio and E. Bosi. 1979. Frequenza di bacilli lattici eterofermentati nelle feci di vitelli lattanti. Arch. Vet. Ital. *30:* 16–21.

Sartory, A. 1920. Champignons parasites de l'homme et des animaux. V. Arsant. Saint-Nicholas-du-Port, pp. 1–845.

Sathmary, M.N. 1958. *Bacillus subtilis* septicemia and generalised aspergillosis in patient with acute myeloblastic leukemia. N.Y. State J. Med. *58:* 1870–1876.

Sato, G., S. Miura and N. Terakado. 1972. Classification of chicken coagulase-positive staphylococci into four biological types and relation to additional characteristics including coagulase antigenic type. Jpn. J. Vet. Res. *20:* 91–110.

Saunders, K.A. and L.C. Ball. 1980. The influence of the composition of blood agar on beta haemolysis by *Streptococcus salivarius*. Med. Lab. Sci. *37:* 341–345.

Saunders, S.W. and R.B. Maxcy. 1979a. Patterns of cell division, DNA base compositions, and five structures of some radiation-resistant vegetative bacteria found in food. Appl. Environ. Microbiol. *37:* 159–168.

Saunders, S.W. and R.B. Maxcy. 1979b. Isolation of radiation-resistant bacteria without exposure to irradiation. Appl. Environ. Microbiol. *38:* 436–439.

Savage, D.C. 1977. Microbial ecology of the gastrointestinal tract. Annu. Rev. Microbiol. *31:* 107–133.

Sayre, R.M., R.L. Gherna and W.P. Wergin. 1983. Morphological and taxonomic reevaluation of *Pasteuria ramosa* Metchnikoff 1888 and "*Bacillus penetrans*" Mankau 1975. Int. J. Syst. Bacteriol. *33:* 636–649.

Sayre, R.M. and W.P. Wergin. 1977. Bacterial parasite of a plant nematode: morphology and ultrastructure. J. Bacteriol. *129:* 1091–1101.

Saz, H.J. and L.O. Krampitz. 1954. The oxidation of acetate by *Micrococcus lysodeikticus*. J. Bacteriol. *67:* 409–418.

Scammon, L., S. Froman and D.W. Will. 1964. Enhancement of virulence for chickens of Battey type mycobacteria by preincubation at 42°C. Am. Rev. Resp. Dis. *90:* 804–805.

Scardovi, V. 1981. The genus *Bifidobacterium. In* Starr, Stolp, Trüper, Balows and Schlegel (Editors), The Prokaryotes. A Handbook on Habitats, Isolation, and Identification of Bacteria. Springer, New York, pp. 1951–1961.

Scardovi, V., F. Casalicchio and N. Vincenzi. 1979. Multiple electrophoretic forms of transaldolase and 6-phosphogluconate dehydrogenase and their relationships to the taxonomy and ecology of bifidobacteria. Int. J. Syst. Bacteriol. *29:* 312–329.

Scardovi, V. and F. Crociani. 1974. *Bifidobacterium catenulatum, Bifidobacterium dentium* and *Bifidobacterium angulatum*: three new species and their deoxyribonucleic acid homology relationships. Int. J. Syst. Bacteriol. *24:* 6–20.

Scardovi, V. and B. Sgorbati. 1974. Electrophoretic types of transaldolase, transketolase, and other enzymes in bifidobacteria. Antonie van Leeuwenhoek. J. Microbiol. Serol. *40:* 427–440.

Scardovi, V., B. Sgorbati and G. Zani. 1971a. Starch gel electrophoresis of fructose-6-phosphate phosphoketolase in the genus *Bifidobacterium*. J. Bacteriol. *106:* 1036–1039.

Scardovi, V. and L.D. Trovatelli. 1965. The fructose-6-phosphate shunt as peculiar pattern of hexose degradation in the genus *Bifidobacterium*. Ann. Microbiol. *15:* 19–29.

Scardovi, V. and L.D. Trovatelli. 1969. New species of bifidobacteria from *Apis mellifica* L. and *Apis indica* F. A contribution to the taxonomy and biochemistry of the genus *Bifidobacterium*. Zentralbl. Bakteriol. Parasitenkd. Infektionskr. Hyg. Abt. II. *123:* 64–88.

Scardovi, V. and L.D. Trovatelli. 1974. *Bifidobacterium animalis* (Mitsuoka) comb. nov. and the "*minimum*" and "*subtile*" groups of new bifidobacteria found in sewage. Int. J. Syst. Bacteriol. *24:* 21–28.

Scardovi, V., L.D. Trovatelli, B. Biavati and G. Zani. 1979b. *Bifidobacterium cuniculi, Bifidobacterium choerinum, Bifidobacterium boum, and Bifidobacterium pseudocatenulatum*: four new species and their deoxyribonucleic acid homology relationships. Int. J. Syst. Bacteriol. *29:* 291–311.

Scardovi, V., L.D. Trovatelli, F. Crociani and B. Sgorbati. 1969. Bifidobacteria in bovine rumen. New species of the genus *Bifidobacterium: B. globosum* n. sp. and *B. ruminale* n. sp. Arch. Mikrobiol. *68:* 278–294.

Scardovi, V., L.D. Trovatelli, G. Zani, F. Crociani and D. Matteuzzi. 1971b. Deoxyribonucleic acid homology relationships among species of the genus *Bifidobacterium*. Int. J. Syst. Bacteriol. *29:* 291–311.

Scardovi, V. and G. Zani. 1974. *Bifidobacterium magnum* sp. nov., a large, acidophillic bifidobacterium isolated from rabbit feces. Int. J. Syst. Bacteriol. *24:* 29–34.

Scardovi, V., G. Zani and L.D. Trovatelli. 1970. Deoxyribonucleic acid homology among the species of the genus *Bifidobacterium* isolated from animals. Arch. Mikrobiol. *72:* 318–325.

Schaal, K.P. 1972. Zur mikrobiologisher Diagnostik der Nocardiose. Zentralbl. Bakteriol. Parasitenkd. Infektionskr. Hyg. I Abt. Orig. *220:* 242–246.

Schaal, K.P. 1979. Die Aktinomykosen des Menschen - Diagnose und Therapie. Dtsch. Aerztebl. *76:* 1997–2006.

Schaal, K.P. 1981. Actinomycoses. Rev. Inst. Pasteur (Lyon) *14:* 279–288.

Schaal, K.P. 1984a. Laboratory diagnosis of actinomycete diseases. *In* Goodfellow, Mordarski and Williams (Editors), The Biology of Actinomycetes, Academic Press, London and New York, pp. 425–456.

Schaal, K.P. 1984b. Identification of clinically significant actinomycetes and related bacteria using chemical techniques. *In* Goodfellow and Minnikin (Editors), Chemical Methods in Bacterial Systematics, Society for Applied Bacteriology, Technical Series No. 20, Academic Press, London and New York, pp. 359–381.

Schaal, K.P. and B.L. Beaman. 1984. Clinical significance of actinomycetes. *In* Goodfellow, Mordarski and Williams (Editors), The Biology of the Actinomycetes, Academic Press, London and New York, pp. 389–424.

Schaal, K.P. and R. Gatzer. 1985. Serological and numerical phenetic classification of clinically significant fermentative actinomycetes. *In* Arai (Editor), Filamentous Microorganisms. Biomedical Aspects. Japan Scientific Societies Press, Tokyo, Japan.

Schaal, K.P. and W. Pape. 1980. Special methodological problems in antibiotic susceptibility testing of fermentative actinomycetes. Infection *8:* (Suppl. 2): 176–182.

Schaal, K.P. and G. Pulverer. 1973. Fluoreszenzserologische Differenzierung von fakultativ anaeroben Aktinomyzeten. Zentralbl. Bakteriol. Parasitenkd. Infektionskr. Hyg. I. Abt. Orig. A *225:* 424–430.

Schaal, K.P. and G. Pulverer. 1981. The genera *Actinomyces, Agromyces, Arachnia, Bacterionema,* and *Rothia. In* Starr, Stolp, Trüper, Balows and Schlegel (Editor), The Prokaryotes. A Handbook on the Habitats, Isolation, and Identification of Bacteria. Springer Verlag, Berlin-Heidelberg-New York, pp. 1923–1950.

Schaal, K.P. and G. Pulverer. 1984. Epidemiologic, etiologic, diagnostic, and therapeutic aspects of endogenous actinomycetes infections. *In* Ortiz-Ortiz, Bojalil, and Yakoleff (Editors), Biological, Biochemical, and Biomedical Aspects of Actinomycetes. Academic Press, Orlando-New York-London, pp. 13–32.

Schaal, K.P. and H. Reutersberg. 1978. Numerical taxonomy of *Nocardia asteroides*. Zentralbl. Bakteriol. Parasitenkd. Infektionskr. Hyg. I Abt. Suppl. *6:* 53–62.

Schaal, K.P. and G.M. Schofield. 1981a. Taxonomy of Actinomycetaceae. Rev. Inst. Pasteur (Lyon) *14:* 27–39.

Schaal, K.P. and G.M. Schofield. 1981b. Current ideas on the taxonomic status of the Actinomycetaceae. *In* Schaal and Pulverer (Editors), Actinomycetes. Proceedings of the Fourth International Symposium on Actinomycete Biology, Cologne, September 3–7, 1979. Zentralbl. Bakteriol. Mikrobiol. Hyg., Suppl. 11, Gustav Fischer Verlag, Stuttgart-New York, pp. 67–78.

Schaal, K.P. and G.M. Schofield. 1984. Classification and identification of clinically significant Actinomycetaceae. *In* Ortiz-Ortiz, Bojalil and Yakoleff (Editors), Biological, Biochemical and Biomedical Aspects of Actinomycetes. Academic Press, London-New York-San Francisco, pp. 505–520.

Schaal, K.P., G.M. Schofield and G. Pulverer. 1980. Taxonomy and clinical significance of *Actinomycetaceae* and *Propionibacteriaceae*. Infection 8, Suppl. *2:* 122–130.

Schaal, K.P., H. Schutt-Gerowitt and W. Pape. 1979. Cefoxitin-Empfindlichkeit pathogener aerober und anaerober Aktinomyzeten. Infection 7, Suppl. 1: 47–51.

Schaberg, D.R., D.B. Clewell and L. Glatzer. 1982. Conjugative transfer of R-plasmids from *Staphylococcus faecalis* to *Staphylococcus aureus*. Antimicrob. Agents Chemother. *22:* 201–207.

Schachtele, C.F. 1975. Glucose transport in *Streptococcus mutans*: preparation of cytoplasmic membranes and characteristics of phosphotransferase activity. J. Dent. Res. *54:* 330–338.

Schaefer, W.B. 1965. Serological identification and classification of the atypical mycobacteria by their agglutination. Rev. Resp. Dis. *92:* 85–93.

Schaefer, W.B. 1967. Type-specificity of atypical mycobacteria in agglutination and antibody adsorption tests. Am. Rev. Resp. Dis. *96:* 1165–1168.

Schaefer, W.B. 1968. Incidence of the serotypes of *Mycobacterium avium* and atypical mycobacteria in human and animal diseases. Am. Rev. Resp. Dis. *97:* 18–23.

Schaefer, W.B. and C.L. Davis. 1961. A bacteriologic and histopathologic study of skin granuloma due to *Mycobacterium balnei*. Am. Rev. Resp. Dis. *84:* 837–844.

Schaefer, W.B., E. Wolinsky, P.A. Jenkins and J. Marks. 1973. *Mycobacterium szulgai* - a new pathogen. Serologic identification and report of five new cases. Am. Rev. Resp. Dis. *108:* 1320–1326.

Schaeffer, A.B. and M. Fulton. 1933. A simplified method of staining endospores. Science *77:* 194.

Schaefler, S. 1971. *Staphylococcus epidermidis* BV: antibiotic resistance patterns, physiological characteristics, and bacteriophage susceptability. Appl. Microbiol. *22:* 693–699.

Schaefler, S., D. Jones, W. Perry, L. Ruvinskaya, T. Baradet, E. Mayr and M.E. Wilson. 1981. Emergence of gentamicin- and methicillin-resistant *Staphylococcus aureus* strains in New York City hospitals. J. Clin. Microbiol. *13:* 754–759.

Schaeg, W. and H. Blobel. 1970. Concentration and identification of penicillinase of *Staphylococcus aureus*. Zentralbl. Bakteriol. Parasitenkd. Infektionskr. Hyg. Abt. I Orig. *214:* 62–67.

Schafer, F.J., E.J. Wing and C.W. Norden. 1979. Infectious endocarditis caused by *Rothia dentocariosa*. Ann. Int. Med. *91:* 747–748.

Schäfer, R. and A.C. Schwartz. 1976. Catabolism of purines in *Clostridium sticklandii*. Zentralbl. Bakteriol. Parasitenkd. Infektionskr. Abt. I Orig. A *235:* 165–172.

Schallehn, G., M.W. Eklund and H. Brandis. 1980. Zur Phagenkonversion von *Clostridium novyi* Type A. Zentralbl. Bakteriol. Parasitenkd. Infektionskr. Hyg. I Abt. Orig. A *247:* 95–100.

Schallehn, G. and J. Krämer. 1976. Studies on mode of action of a bacteriocin from *Clostridium septicum*. Can. J. Microbiol. *22:* 435–437.

Schallehn, G. and J. Krämer. 1981. Detection of plasmids in alpha toxin-producing and non-producing strains of *Clostridium novyi* type A. FEMS Microbiol. Lett. *11:* 313–316.

Schallehn, G. and H.E. Müller. 1973. Über die Einwirkung proteolytischer Enzyme von *Clostridium histolyticum* und *Clostridium novyi* auf menschliche Plasmaeiweisskörper in Wachstumskulturen. Zentralbl. Bakteriol. Parasitenkd. Infektionskr. Hyg. I Abt. Orig. A *224:* 102–114.

Schardinger, F. 1905. *Bacillus macerans*, ein Aceton bildender Rottebacillus. Zentralbl. Bakteriol. Parasitenkd. Infectionskr. Hyg. Abt. II *14:* 772–781.

Scharfen, J. 1973. Urease als brauchbares Kriterium bei der Klassifizierung von mikroaerophilen Aktinomyzeten. Zentralbl. Bakteriol. Parasitenkd. Infektionskr. Hyg. I. Abt. Orig. A *225:* 89–94.

Scharfen, J. 1975. Untraditional glucose fermenting actinomycetes as human pathogens. Part I: *Actinomyces naeslundii* as a cause of abdominal actinomycosis. Zentralbl. Bakteriol. Parasitenkd. Infektionskr. Hyg. I. Abt. Orig. A *232:* 308–317.

Scharfen, J. 1975. Untraditional glucose fermenting actinomycetes as human pathogens. Part II. *Rothia dentocariosa* as a cause of abdominal actinomycosis, and a pathogen for mice. Zentralbl. Bakteriol. Parasitenkd. Infektionskr. Hyg. Abt. I Orig. A *233:* 80–92.

Scharif, G. 1961. *Corynebacterium iranicum* sp. nov. on wheat (*Triticum vulgare* L.) in Iran, and a comparative study of it with *C. tritici* and *C. rathayi*. Entomol. Phytopathol. Appl. *19:* 1–24.

Schefferle, H. 1966. Coryneform bacteria in poultry deep litter. J. Appl. Bacteriol. *29:* 147–160.

Schefferle, H.E. 1966. Coryneform bacteria in poultry deep litter. J. Appl. Bacteriol. *29:* 147–160.

Schenk, A. and M. Aragno. 1979. *Bacillus schlegelii*, a new species of thermophilic, facultatively chemolithoautotrophic bacterium oxidizing molecular hydrogen. J. Gen. Microbiol. *115:* 333–341.

Schenk, A. and M. Aragno. 1981. Validation List No. 6. Int. J. Syst. Bacteriol. *31:* 215–218.

Schieblich, M. 1923. Zwei aus Futterproben isolierte, bisher noch nicht beschriebene Bazillen. Zentralbl. Bakteriol. Parasitenkd. Infektionskr. Hyg. Abt. II *58:* 204–207.

Schildkraut, C.L., J. Marmur and P. Doty. 1962. Determination of the base composition of deoxyribonucleic acid from its buoyant density in CsCl. J. Mol. Biol. *4:* 430–443.

Schiller, N.L. and R.B. Roberts. 1982. Vitamin B6 requirements of nutritionally variant *Streptococcus mitior*. J. Clin. Microbiol. *15:* 740–743.

Schink, B. 1984a. *Clostridium magnum* sp. nov., a non-autotrophic homoacetogenic bacterium. Arch. Microbiol. *137:* 250–255.

Schink, B. 1984b. Validation List No. 15. Int. J. Syst. Bacteriol. *34:* 355–357.

Schink, B., J.C. Ward and J.G. Zeikus. 1981. Microbiology of wetwood: importance of pectin degradation and *Clostridium* species in living trees. Appl. Environ. Microbiol. *42:* 526–532.

Schink, B. and J.G. Zeikus. 1980. Microbial methanol formation: a major end product of pectin metabolism. Curr. Microbiol. *4:* 387–389.

Schink, B. and J.G. Zeikus. 1982. Microbial ecology of pectin decomposition in anoxic lake sediments. J. Gen. Microbiol. *128:* 393–404.

Schink, B. and J.G. Zeikus. 1983a. *Clostridium thermosulfurogenes* sp. nov., a new thermophile that produces elemental sulphur from thiosulphate. J. Gen. Microbiol. *129:* 1149–1158.

Schink, B. and J.G. Zeikus. 1983b. Characterization of pectinolytic enzymes of *Clostridium thermosulfurogenes*. FEMS Microbiol. Lett. *17:* 295–298.

Schink, B. and J.G. Zeikus. 1983c. Validation List No. 12. Int. J. Syst. Bacteriol. *33:* 896–897.

Schippers-Lammertse, A.F., A.O. Muijsers and K.B. Klatser-Oedekerk. 1963. *Arthrobacter polychromogenes*, its pigments and a bacteriophage of this species. Antonie van Leeuwenhoek. J. Microbiol. Serol. *29:* 1–15.

Schleifer, K.H. 1970. Die Mureintypen in der Gattung *Microbacterium*. Arch. Mikrobiol. *71:* 271–282.

Schleifer, K.H. 1973. Chemical composition of staphylococcal cell walls. *In* Jeljaszewicz (Editor), Staphylococci and Staphylococcal Infections. Recent Progress. S. Karger, Basel, pp. 13–23.

Schleifer, K.H. 1981. Klassifikation von *Staphylococcus* und *Micrococcus*–ein beispiel tür die moderne Bakteriensystematik. Forum Mikrobiol. *5:* 272–278.

Schleifer, K.H. and U. Fischer. 1982. Description of a new species of the genus *Staphylococcus: Staphylococcus carnosus*. Int. J. Syst. Bacteriol. *32:* 153–156.

Schleifer, K.H., U. Geyer, R. Killper-Bälz and L.A. Devriese. 1983. Elevation of *Staphylococcus sciuri* subsp. *lentus* (Kloos et al) to species status: *Staphylococcus lentus* (Kloos et al) comb. nov. Syst. Appl. Microbiol. *4:* 382–387.

Schleifer, K.H., W.P. Hammes and O. Kandler. 1976. Effect of endogenous and exogenous factors on the primary structures of bacterial peptidoglycan. Microbiol. Physiol. *13:* 245–292.

Schleifer, K.H., A. Hartinger and F. Götz. 1978. Occurrence of D-tagatose-6-phosphate pathway of D-galactose metabolism among staphylococci. FEMS Microbiol. Lett. *3:* 9–11.

Schleifer, K.H., W. Heise and S.A. Meyer. 1979. Deoxyribonucleic acid hybridization studies among micrococci. FEMS Microbiol. Lett. *6:* 33–36.

Schleifer, K.H. and O. Kandler. 1967. Zur chemischen Zusammensetzung der Zellwand der Streptokokken. I. Die Aminosäuresequenze des Mureins von *Str. thermophilus* und *Str. faecalis*. Arch. Microbiol. *57:* 335–364.

Schleifer, K.H. and O. Kandler. 1970. Amino acid sequence of the murein of *Planococcus* and other *Micrococcaceae*. J. Bacteriol. *103:* 387–392.

Schleifer, K.H. and O. Kandler. 1972. Peptidoglycan types of bacterial cell walls and their taxonomic implications. Bacteriol. Rev. *36:* 407–477

Schleifer, K.H. and R. Kilpper-Bälz. 1984. Transfer of *Streptococcus faecalis* and *Streptococcus faecium* to the genus *Enterococcus* nom. rev. as *Enterococcus faecalis* comb. nov. and *Enterococcus faecium* comb. nov. Int. J. Syst. Bacteriol. *34:* 31–34.

Schleifer, K.H., R. Kilpper-Bälz and L.-A. Devriese. 1984a. *Staphylococcus arlettae* sp. nov., *S. equorum* sp. nov. and *S. kloosii* sp. nov.: three new coagulase-negative, novobiocin-resistant species from animals. Syst. Appl. Microbiol. *5:* 501–509.

Schleifer, K.H., R. Kilpper-Bälz and L.-A. Devriese. 1985b. Validation List No. 17. Int. J. Syst. Bacteriol. *35:* 223–225.

Schleifer, K.H., R. Kilpper-Bälz, U. Fischer, A. Faller and J. Endl. 1982. Identification of "*Micrococcus candidus*" ATCC 14852 as a strain of *Staphylococcus epidermidis* and of "*Micrococcus caseolyticus*" ATCC 13548 and *Micrococcus varians* ATCC 29750 as members of a new species, *Staphylococcus caseolyticus*. Int. J. Syst. Bacteriol. *32:* 15–20.

Schleifer, K.H., R. Kilpper-Bälz, J. Kraus and F. Gehring. 1984b. Relatedness and classification of *Streptococcus mutans* and 'mutans-like' streptococci. J. Dent. Res. *63:* 1047–1050.

Schleifer, K.H. and W.E. Kloos. 1975a. A simple test for the separation of staphylococci from micrococci. J. Clin. Microbiol. *1:* 337–338.

Schleifer, K.H. and W.E. Kloos. 1975b. Isolation and characterization of staphylococci from human skin. I. Amended descriptions of *Staphylococcus epidermidis* and *Staphylococcus saprophyticus* and descriptions of three new species: *Staphylococcus cohnii*, *Staphylococcus haemolyticus*, and *Staphylococcus xylosus*. Int. J. Syst. Bacteriol. *32:* 50–61.

Schleifer, K.H. and W.E. Kloos. 1976. Separation of staphylococci from micrococci. Zentralbl. Bakteriol. Parasitenkd. Infektionskr. Hyg. Abt. I Orig. Suppl. *5:* 3–9.

Schleifer, K.H., W.E. Kloos and M. Kocur. 1981. The genus *Micrococcus*. *In* Starr, Stolp, Trüper, Balows and Schlegel (Editors), The Prokaryotes. A Handbook on Habitats, Isolation and Identification of Bacteria. Springer-Verlag, Berlin, Heidelberg, New York, pp. 1539–1547.

Schleifer, K.H. and M. Kocur. 1973. Classification of staphylococci based on chemical and biochemical properties. Arch. Mikrobiol. *93:* 65–85.

Schleifer, K.H. and E. Krämer. 1980. Selective medium for isolating staphylococci. Zentralbl. Bakteriol. Mikrobiol. Hyg. Abt. I Orig. C *1:* 270–280.

Schleifer, K.H., J. Kraus, C. Dvorak, R. Kilpper-Bälz, M.D. Collins and W. Fischer. 1985a. Transfer of *Streptococcus lactis* and related streptococci to the genus *Lactococcus* gen. nov. Syst. Appl. Microbiol. *6:* 183–195.

Schleifer, K.H. and K. Lang. 1980. Close relationship among strains of *Micrococcus conglomeratus* and *Arthrobacter* species. FEMS Microbiol. Lett. *9:* 223–226.

Schleifer, K.H., S.A. Meyer and M. Rupprecht. 1979. Relatedness among coagulase-negative staphylococci. Deoxyribonucleic acid reassociation and comparative immunological studies. Arch. Microbiol. *122:* 93–101.

Schleifer, K.H. and E. Nimmermann. 1973. Peptidoglycan types of strains of the genus *Peptococcus*. Arch. Mikrobiol. *93:* 245–258.

Schleifer, K.H., R. Plapp and O. Kandler. 1968. Glycine as cross-linking bridge in the LL-diaminopimelic acid-containing murein of *Propionibacterium peterssonii*. FEBS Lett. *1:* 287–290.

Schleifer, K.H., R. Plapp and O. Kandler. 1968. Die Aminosäuresequenz des mureins von *Microbacterium lacticum*. Biochim. Biophys. Acta *154:* 573–582.

Schleifer, K.H., F. Schumacher-Perdreau, F. Götz, B. Popp and V. Hajek. 1976. Chemical and biochemical studies for the differentiation of coagulase-positive staphylococci. Arch. Microbiol. *110:* 263–270.

Schleifer, K.H. and H.P. Seidl. 1985. Chemical composition and structure of murein. *In* Goodfellow and Minnikin (Editors), Chemical Methods in Bacterial Systematics. Academic Press, London-New York-San Francisco, pp. 201–219.

Schleifer, K.H. and E. Stackebrandt. 1983. Molecular systematics of prokaryotes. Annu. Rev. Microbiol. *37:* 143–187.

Schleifer, K.H. and J. Steber. 1974. Chemische Untersuchungen am Phagenrezeptor von *Staphylococcus epidermidis*. Arch. Microbiol. *98:* 251–270.

Schmeidel, A. and W. Gerloff. 1965. Dreifach-differen-zeitung von Mykobakterien in der Agar-Hohen-Schict-Kultur. Prax. Pneumol. *19:* 528–536.

Schoenbaum, S.C., P. Gardner and J. Shillito. 1975. Infections of cerebrospinal fluid shunts: epidemiology, clinical manifestations, and therapy. J. Infect. Dis. *131:* 543–552.

Schofield, G.M. and K.P. Schaal. 1979a. A simple basal medium for carbon source utilisation tests with the anaerobic actinomycetes. FEMS Microbiol. Lett. *5:* 309–310.

Schofield, G.M. and K.P. Schaal. 1979b. Application of the Minitek differentiation system in the classification and identification of *Actinomycetaceae*. FEMS Microbiol. Lett. *5:* 311–313.

Schofield, G.M. and K.P. Schaal. 1980a. Carbohydrate fermentation patterns of facultatively anaerobic actinomycetes using micromethods. FEMS Microbiol. Lett. *8:* 67–69.

Schofield, G.M. and K.P. Schaal. 1980b. Rapid micromethods for detecting deamination and decarboxylation of amino acids, indole production, and reduction of nitrate and nitrite by facultatively anaerobic actinomycetes. Zentralbl. Bakteriol. Parasitenkd. Infektionskr. Hyg. I. Abt. Orig. A *247:* 383–391.

Schofield, G.M. and K.P. Schaal. 1981. A numerical taxonomic study of members of the *Actinomycetaceae* and related taxa. J. Gen. Microbiol. *127:* 237–259.

Schönheit, P., A. Brandis and R.K. Thauer. 1979. Ferredoxin degradation in growing *Clostridium pasteurianum* during periods of iron deprivation. Arch. Microbiol. *120:* 73–76.

Schottmüller, H. 1903. Die Artunterscheidung der fur den Menschen Pathogenen Streptokokken durch Blutagar. Muench. Med. Wochenschr. *50:* 849–853, 909–912.

Schötz, F. I.G. Abo-Elnaga and O. Kandler. 1965. Zur Struktur der Mesosomen bei *Lactobacillus corynoides*. Z. Naturforschg. *20b:* 790–794.

Schrader, G. and H.-P. Schau. 1978. Zur Vorkommen pathogener Clostridien in der Antarktis. Z. Gesamte Hyg. *24:* 704–708.

Schramm, M., V. Klybas and F. Racker. 1958. Phosphorolytic cleavage of fructose-6-phosphate by fructose-6-phosphate phosphoketolase from *Acetobacter xylinum*. J. Biol. Chem. *233:* 1283–1288.

Schröder, K., E. Clausen, A.M. Sandberg and J. Raa. 1980. Psychrotrophic *Lactobacillus plantarum* from fish and its ability to produce antibiotic substances. *In* Advances in Fish Science and Technology. Connell (Editor), Fishing News Books Ltd., Farnham, Surrey, England.

Schröder, K.H. 1975. Investigation into the relationship of *Mycobacterium ulcerans* to *M. buruli* and other mycobacteria. Am. Rev. Resp. Dis. *111:* 559–562.

Schröder, K.H. and I. Juhlin. 1977. *Mycobacterium malmoense* sp. nov. Int. J. Syst. Bacteriol. *27:* 241–246.

Schroeter, J. 1872. Ueber einige durch Bacterien gebildete Pigmente, *In* Cohn, Beitr. Biol. Pfl. *1:* (Heft 2) 109–126.

Schultes, L.M. and J.B. Evans. 1971. Deoxyribonucleic acid homology of *Aerococcus viridans*. Int. J. Syst. Bacteriol. *21:* 207–209.

Schultz, E.W. 1945. *Listerella* infections: a review. Stanford Med. Bull *3:* 135–151.

Schumacher-Perdreau, F., G. Peters and M. Kocur. 1981. Chemical and physiological properties of a catalase-negative *Staphylococcus aureus* strain. *In* Jeljaszewicz (Editor), Staphylococci and Staphylococcal Infections, Gustav Fischer Verlag, Stuttgart, New York, pp. 61–65.

Schumacher-Perdreau, F., G. Pulverer and K.H. Schleifer. 1979. The phage adsorption test: A simple method for differentiation between staphylococci and micrococci. J. Infect. Dis. *138:* 392–395.

Schupp, T., R. Hütter and D.A. Hopwood. 1975. Genetic recombination in *Nocardia mediterranei*. J. Bacteriol. *121:* 128–135.

Schuster, M.L., B. Hoff and W.A. Compton. 1975. Variation, maintenance, loss, and failure to recover virulence in *Corynebacterium nebraskense*. Plant Dis. Rep. *59:* 101–105.

Schuster, M.L., A.K. Vidaver and M. Mandel. 1968. A purple-pigment-producing bean wilt bacterium. *Corynebacterium flaccumfaciens* var. *violaceum*, n. var. Can. J. Microbiol. *14:* 423–427.

Schütz, H. and F. Radler. 1984. Anaerobic reduction of glycerol to propanediol-1,3 by *Lactobacillus brevis* and *Lactobacillus buchneri*. Syst. Appl. Microbiol. *5:* 169–178.

Schwabacher, H. 1959. A strain of *Mycobacterium* isolated from skin lesions of a cold blooded animal, *Xenopus laevis*, and its relation to atypical acid-fast bacilli occurring in man. J. Hyg. *57:* 57–67.

Schwan, O. 1979. Biochemical, enzymatic, and serological differentiation of *Peptococcus indolicus* (Christian) Sørensen from *Peptococcus asaccharolyticus* (Distaso) Douglas. J. Clin. Microbiol. *9:* 157–162.

Schwartz, R.M. and M.O. Dayhoff. 1978. Origins of prokaryotes, eukaryotes, mitochondria, and chloroplasts. Science (Washington) *199:* 395–403.

Schwartzel, E.H. and J.J. Cooney. 1970. Isolation and identification of echinenone from *Micrococcus roseus*. J. Bacteriol. *104:* 272–274.

Scott, H.W. and B.A. Dehority. 1965. Vitamin requirements of several cellulolytic rumen bacteria. J. Bacteriol. *89:* 1169–1175.

Scott, H.W., P. Kowaluk and B.A. Dehority. 1967. Volatile fatty acid requirements of cellulolytic rumen bacteria. J. Bacteriol. *94:* 537–543.

Scott, J., A.W. Turner and L.R. Vawter. 1935. Gas edema diseases. Twelfth Int. Vet. Congr. *2:* 168–182.

Scott, J.H. and R. Schekman. 1980. Lyticase: endogluconase and protease activities that act together in yeast cell lysis. J. Bacteriol. *142:* 414–423.

Scott, J.P. 1926. A comparative study of strains of *Clostridium chauvoei* obtained in the United States and abroad. J. Infect. Dis. *38:* 262–272.

Scott, J.P. 1928. The etiology of blackleg and methods of determining *Clostridium chauvoei* from other anaerobic organisms found in cases of blackleg. Cornell Vet. *18:* 249–271.

Scott, V.N. and C.L. Duncan. 1978. Cryptic plasmids in *Clostridium* and *C. botulinum*-like organisms. FEMS Microbiol. Lett. *4:* 55–58.

Seal, D.V., J. McGill, D. Flanagan and B. Purrier. 1981. Lacrimal canaliculitis due to *Arachnia* (*Actinomyces*) *propionica*. British J. Ophthalmol. *65:* 10–13.

Seaward, M.R.S., T. Cross and B.A. Unsworth. 1976. Viable bacterial spores recovered from an archaeological excavation. Nature (London) *261:* 407–408.

Seelemann, M. 1942. Über die Bedeutung der hämolytischen streptokokken bei einigen Infektionen des Pferdes. Dtsch. Tierarztl. Wochenschr. *50:* 8–12; 38–41.

Seeliger, H.P.R. 1961. Listeriosis. 2nd ed. S. Karger: Basel, New York.

Seeliger, H.P.R. and J. Bockemühl. 1968. Kritische Untersuchungen zur Frage einer Kapselbildung bei *Listeria monocytogenes*. Zentralbl. Bakteriol. Parasitenkd. Infektionskr. Hyg. I. Abt. Orig. *206:* 216–227.

Seeliger, H.P.R. 1972. New outlook on the epidemiology and epizoology of listeriosis. Acta Microbiol. Acad. Sci. Hung. *19:* 273–286.

Seeliger, H.P.R. 1974. Genus *Erysipelothrix*. *In* Buchanan and Gibbons (Editors), Bergey's Manual of Determinative Bacteriology, 8th ed. The Williams and Wilkins Co., Baltimore. p. 597.

Seeliger, H.P.R. 1975. Serovars of *Listeria monocytogenes* and other *Listeria* species. *In* Woodbine (Editor), Problems of Listeriosis. Leicester University Press, Leicester.

Seeliger, H.P.R. 1976. Notion actuelle sur l'épidémiologie de la listériose. Medicine et Maladies Infectieuses *6:* 6–14.

Seeliger, H.P.R. 1981. Apathogene Listerien: *Listeria innocua* sp. n. (Seeliger and Schoofs, 1977). Zentralbl. Bakteriol. Parasitenkd. Infektionskr. Hyg. I. Abt. Orig. A *249:* 487–493.

Seeliger, H.P.R. 1983. Validation of the publication of new names and new combinations previously effectively published outside the IJSB. List No. 10. Int. J. Syst. Bacteriol. *33:* 438–440.

Seeliger, H.P.R. 1984. Modern taxonomy of the *Listeria* group and relation to its pathogenicity. Clin. Invest. Med. *7:* 217–221.

Seeliger, H.P.R. and K. Höhne. 1979. Serotyping of *Listeria monocytogenes* and related species. *In* Bergan and Norris (Editors), Methods in Microbiology, Vol. 13: Academic Press, London, pp. 31–49.

Seeliger, H.P.R., J. Rocourt, A. Schrettenbrunner, P.A.D. Grimont and D. Jones. 1984. Description of *Listeria ivanovii* sp. nov. Int. J. Syst. Bacteriol. *34:* 336–337.

Seeliger, H.P.R. and M. Schoofs. 1979. Serological analysis of non-hemolyzing *Listeria*-strains belonging to a species different from *Listeria monocytogenes*. *In* Ivanov (Editor), Problems of Listeriosis, Proceedings of the VIIth International Symposium, Varna, 1977. National Agroindustrial Union, Center for Scientific Information, Sofia, Bulgaria, pp. 24–28.

Seeliger, H.P.R., A. Schrettenbrunner, G. Pongratz and H. Hof. 1982. Zur Sonderstellung stark hämolysierender Stämme der Gattung *Listeria*. Special position of strongly haemolytic strains of the genus *Listeria*. Zentralbl. Bakteriol. Parasitenkd. Infektionskr. Hyg. I. Abt. Orig. A *252:* 176–190.

Seeliger, H.P.R. and H.J. Welshimer. 1974. Genus *Listeria*. *In* Buchanan and Gibbons (Editors), Bergey's Manual of Determinative Bacteriology, 8th ed. The Williams and Wilkins Co., Baltimore. pp. 593–596.

Seguin, P. 1928. Culture du *Fusobacterium plauti* forme mobile du bacille fusiforme. C.R. Soc. Biol. (Paris) *99:* 439–442.

Seidel, G. 1962. Die aeroben Sporenbildner unter besonderer Berücksichtigung des Milzbrandbazillus. Johann Ambrosius Barth, Leipzig.

Seidl, H.P. and K.H. Schleifer. 1978. A rapid test for the serological separation of staphylococci from micrococci. Appl. Environ. Microbiol. *35:* 479–482.

Seidl, P.H., A.H. Faller, R. Loider and K.-H. Schleifer. 1980. Peptidoglycan types and cytochrome patterns of strains of *Oerskovia turbata* and *O. xanthineolytica*. Arch. Microbiol. *127:* 173–178.

Seidler, V.D., G. Trautwein and K.H. Böhm. 1971. Nachweis von *Erysipelothrix insidiosa* mit fluoreszierenden antikörpern. Zentralbl. Veterinaermed. *18:* 280–292.

Seidman, P. and E.C.S. Chan. 1969. Growth of *Arthrobacter citreus* in a chemically-defined medium and its requirement for chelating agents with schizokinen activity. J. Gen. Microbiol. *58:* V.

Seiler, H. 1983. Identification key for coryneform bacteria derived by numerical taxonomic studies. J. Gen. Microbiol. *129:* 1433–1471.

Seiler, H., R. Braatz and G. Ohmeyer. 1980. Numerical cluster analysis of the coryneform bacteria from activated sludge. Zentralbl. Bakteriol. Mikrobiol. Hyg. Abt. I. Orig. C. *1:* 357–375.

Seiler, H. and W. Hennlich. 1983. Characterization of coryneform bacteria in piggery wastes. Syst. Appl. Microbiol. *4:* 132–140.

Seiler, H., G. Ohmayer and M. Busse. 1977. Taxonomische Untersuchungen an Gram-positiven coryneformen Bakterien unter Anwendung eines EDV-Programms zur Bereshnung von Vernetzungsdiagrammen. Zentralbl. Bakteriol. Parasitenkd. Infektionskr. Hyg. I. Abt. A *238:* 475–488.

Seki, S., K. Higo and M. Ishimoto. 1982. Studies on nitrate reductase of *Clostridium perfringens*. III. Comparison of nitrate reductase from five strains of *C. perfringens*. J. Gen. Microbiol. *28:* 541–550.

Seki, T., C.K. Chung, H. Mikami and H.Y. Oshima. 1978. Deoxyribonucleic acid homology and taxonomy of the genus *Bacillus*. Int. J. Syst. Bacteriol. *28:* 182–189.

Seki, T., M. Minoda, J.-I. Yagi and Y. Oshima. 1983. Deoxyribonucleic acid reassociation between strains of *Bacillus firmus*, *Bacillus lentus* and intermediate strains. Int. J. Syst. Bacteriol. *33:* 401–403.

Seki, T., T. Oshima and Y. Oshima. 1975. Taxonomic study of *Bacillus* by deoxyribonucleic acid-deoxyribonucleic acid hybridization and interspecific transformation. Int. J. Syst. Bacteriol. *25:* 258–270.

Selva-Sutter, E.A., V.A. Silcox and H.L. David. 1976. Differential identification of *Mycobacterium szulgai* and other scotochromogenic mycobacteria. J. Clin. Microbiol. *3:* 414–420.

Serikawa, T., S. Nakamura and S. Nishida. 1977. Distribution of *Clostridium botulinum* type C in Ishikawa Prefecture, and applicability of agglutination to identification of nontoxigenic isolates of *C. botulinum* type C. Microbiol. Immunol. *21:* 127–136.

Serrano, A.M. and I.S. Schneider. 1978. New modification of Willis and Hobbs' method for identification of *Clostridium perfringens*. Appl. Environ. Microbiol. *35:* 809–810.

Serrano, J.A., R.V. Tablante, A.A. de Serrano, G.C. de San Blas and T. Imaeda. 1972. Physiological, chemical and ultrastructural characteristics of *Corynebacterium rubrum*. J. Gen. Microbiol. *70:* 339–349.

Servin-Massieu, M. 1971. Effects of freeze-drying and sporulation on microbial variation. Curr. Top. Microbiol. Immunol. *54:* 119–150.

Severin, V. and E. Docea. 1970. Contribution to the study on the biology of *Corynebacterium rathayi* (In Rumanian). Lucrările conferintei nationale de microbiologie generală si aplicată, Bucuresti, 4–7 decembrie, 1968. Microbiol. Bucuresti *1:* 541–544.

Sgorbati, B. 1979. Preliminary quantification of immunological relationships among the transaldolases of the genus *Bifidobacterium*. Antonie van Leeuwenhoek. J. Microbiol. Serol. *45:* 557–564.

Sgorbati, B., G. Lenaz and F. Casalicchio. 1976. Purification and properties of two fructose-6-phosphate phosphoketolases in *Bifidobacterium*. Antonie van Leeuwenhoek. J. Microbiol. Serol. *42:* 49–57.

Sgorbati, B. and J. London. 1982. Demonstration of phylogenetic relatedness among members of the genus *Bifidobacterium* using the enzyme transaldolase as an evolutionary marker. Int. J. Syst. Bacteriol. *32:* 37–42.

Sgorbati, B. and B. Scardovi. 1979. Immunological relationships among transaldolases in the genus *Bifidobacterium*. Antonie van Leeuwenhoek. J. Microbiol. Serol. *45:* 129–140.

Sgorbati, B., V. Scardovi and D.J. LeBlanc. 1982. Plasmids in the genus *Bifidobacterium*. J. Gen. Microbiol. *128:* 2121–2131.

Sgorbati, B., G. Zani, L.D. Trovatelli and V. Scardovi. 1970. Gluconate dissimilation by the bifid bacteria of the honey bee. Ann. Microbiol. *20:* 57–64.

Sguros, P.L. 1954. Taxonomy and nutrition of a new species of nicotinophilic bacterium. Bacteriol. Proc. *1954:* 21–22.

Sguros, P.L. 1955. Microbial transformations of the tobacco alkaloids. I. Cultural and morphological characteristics of a nicotinophile. J. Bacteriol. *69:* 28–37.

Shane, B.S., L. Gouws and A. Kistner. 1969. Cellulolytic bacteria occurring in the rumen of sheep conditioned to low-protein teff hay. J. Gen. Microbiol. *55:* 445–457.

Shanson, D.C., J.G. Kensit and R. Duke. 1976. Outbreak of hospital infection with a strain of *S. aureus* resistant to gentamicin and methicillin. Lancet *II:* 1347–1348.

Shapiro, A., D.DiLello, M.C. Loudis, D.E. Keller and S.H. Hutner. 1977. Minimal requirements in defined media for improved growth of some radio-resistant pink tetracocci. Appl. Environ. Microbiol. *33:* 1129–1123.

Sharp, R.J., K.J. Bown and A. Atkinson. 1980. Phenotypic and genotypic characterization of some thermophilic species of *Bacillus*. J. Gen. Microbiol. *117:* 201–210.

Sharpe, M.E. 1955. A serological classification of lactobacilli. J. Gen. Microbiol. *12:* 107–122.

Sharpe, M.E. 1962. Enumeration and studies of lactobacilli in food products. Dairy Sci. Abstr. *24:* 165–171.

Sharpe, M.E. 1970. Cell wall and cell membrane antigens used in the classification of lactobacilli. Int. J. Syst. Bacteriol. *20:* 509–518.

Sharpe, M.E. 1979. Identification of the lactic acid bacteria. *In* Skinner and Lovelock (Editors), Identification methods for microbiologists. Society for Applied Bacteriology Technical Series No. 14, 2nd ed., Academic Press, London, New York, Toronto, Sydney, San Francisco, pp. 233–259.

Sharpe, M.E. 1981. The genus *Lactobacillus*. *In* Starr, Stolp, Trüper, Balows and Schlegel (Editors), The Prokaryotes. A Handbook on Habitats, Isolation and Identification of Bacteria. Springer, New York, pp. 1653–1679.

Sharpe, M.E., J.H. Brock and B.A. Phillips. 1975. Glycerol teichoic acid as an antigenic determinant in a gram-negative bacterium *Butyrivibrio fibrisolvens*. J. Gen. Microbiol. *88:* 355–363.

Sharpe, M.E. and F. Dellaglio. 1977. Deoxyribonucleic acid homology in anaerobic lactobacilli and in possibly related species. Int. J. Syst. Bacteriol. *27:* 19–21.

Sharpe, M.E. and J.G. Franklin. 1962. Production of hydrogen sulphide by lactobacilli isolated from Cheddar cheese. Proc. XVI Intern. Dairy Congr., Copenhagen, 8.

Sharpe, M.E., E.I. Garvie and R.H. Tilbury. 1972. Some slime forming heterofermentative species of the genus *Lactobacillus*. Appl. Microbiol. *23:* 389–397.

Sharpe, M.E., L.R. Hill and S.P. Lapage. 1973a. Pathogenic lactobacilli. J. Med. Microbiol. *6:* 281–286.

Sharpe, M.E., M.J. Latham, E.I. Garvie, J. Zirngibl and O. Kandler. 1973b. Two new species of *Lactobacillus* isolated from the bovine rumen, *Lactobacillus ruminis* sp. nov. and *Lactobacillus vitulinus* sp. nov. J. Gen. Microbiol. *77:* 37–49.

Sharpe, M.E., B.A. Law and B.A. Phillips. 1976. Coryneform bacteria producing

methane thiol. J. Gen. Microbiol. *94:* 430–435.

Sharpe, M.E., B.A. Law, B.A. Phillips and D.G. Pitcher. 1977. Methanethiol production by coryneform bacteria: Strains from dairy and human skin sources and *Brevibacterium linens*. J. Gen. Microbiol. *101:* 345–349.

Sharpe, M.E., B.A. Law, B.A. Phillips and D.G. Pitcher. 1978. Coryneform bacteria producing methanethiol. *In* Bousfield and Callely (Editors), Coryneform Bacteria. Academic Press, London, pp. 289–300.

Shattock, P.M.F. and D.G. Smith. 1963. The location of the group D antigen in a strain of *Streptococcus faecalis* var. *liquefaciens*. J. Gen. Microbiol. *31:* iv.

Shaw, J., M. Stitt and S.T. Cowan. 1951. Staphylococci and their classification. J. Gen. Microbiol. *5:* 1010–1023.

Shaw, N. 1974. Lipid composition as a guide to the classification of bacteria. Adv. Appl. Microbiol. *17:* 63–108.

Shaw, N. 1975. Bacterial glycolipids and glycophospholipids. Adv. Microb. Physiol. *12:* 141–167.

Shaw, N. and D. Stead. 1970. A study of the lipid composition of *Microbacterium thermosphactum* as a guide to its taxonomy. J. Appl. Bacteriol. *33:* 470–473.

Shaw, N. and D. Stead. 1971. Lipid composition of some species of *Arthrobacter*. J. Bacteriol. *107:* 130–133.

Shaw, S. and R.M. Keddie. 1983a. A numerical taxonomic study of the genus *Kurthia* with a revised description of *Kurthia zopfii* and a description of *Kurthia gibsonii* sp. nov. Syst. Appl. Microbiol. *4:* 253–276.

Shaw, S. and R.M. Keddie. 1983b. The vitamin requirements of *Kurthia zopfii* and *Kurthia gibsonii*. Syst. Appl. Microbiol. *4:* 439–443.

Shaw, S. and R.M. Keddie. 1983c. Validation of the publication of new names and new combinations previously effectively published outside the IJSB. List No. 11. Int. J. Syst. Bacteriol. *33:* 672–673.

Shaw, S. and R.M. Keddie. 1984. The genus *Kurthia*: cell wall composition and DNA base content. Syst. Appl. Microbiol. *5:* 220–224.

Shchepkina, T.V. 1940. Description of endoparasites of cotton fibres. Bull. Acad. Sci. U.S.S.R. Classe Sci. Biol. *5:* 643–661.

Shehata, T.E. and E.B. Collins. 1971. Isolation and identification of psychrophilic species of *Bacillus* from milk. Appl. Microbiol. *21:* 466.

Shepard, C.C. 1960. The experimental disease that follows the injection of human leprosy bacilli into foot-pads of mice. J. Exp. Med. *112:* 445–454.

Shepard, C.C. and Y.T. Chang. 1962. Effect of several anti-leprosy drugs on multiplication of human leprosy bacilli in foot-pads of mice. Proc. Soc. Exp. Biol. Med. *109:* 636–638.

Shepard, C.C. and K.H. McRae. 1971. Hereditary characteristic that varies among isolates of *Mycobacterium leprae*. Infect. Immun. *3:* 121–126.

Sherman, J.M. 1937. The streptococci. Bacteriol. Rev. *1:* 3–97.

Sherman, J.M., C.F. Niven and K.L. Smiley. 1943. *Streptococcus salivarius* and other non-hemolytic streptococci of the human throat. J. Bacteriol. *45:* 249–263.

Sherman, J.M. and H.U. Wing. 1937. *Streptococcus durans*. J. Dairy Sci. *20:* 165–167.

Shetty, K.T., N.H. Antia and P.R. Krishnaswamy. 1980. Occurrence of γ-Glutamyl transpeptidase activity in several mycobacteria including *Mycobacterium leprae*. Int. J. Lepr. *49:* 49–56.

Shimohashi, H. and M. Mutai. 1977. Specific antigens of *Lactobacillus acidophilus*. J. Gen. Microbiol. *103:* 337–344.

Shimwell, J.L. 1948. A study of ropiness in beer. Part II. Ropiness due to tetrad forming cocci. J. Inst. Brew. *54:* 237–244.

Shinn, L.E. 1938. A cinematographic analysis of the motion of colonies of *B. alvei*. J. Bacteriol. *36:* 410–422.

Shinolu, R. and M. Kiwato. 1960. On *Streptomyces aerocolonigenes* nov. sp., forming secondary colonies on the aerial mycelia. Bot. Mag. Tokyo *73:* 212–216.

Shirako, Y. and A.K. Vidaver. 1981. Characteristics of bacteriophages Con II, Con X and Con XC infecting *Corynebacterium nebraskense*. Phytopathology *71:* 903. (Abstract)

Shirling, E.B. and D. Gottlieb. 1966. Methods for characterization of *Streptomyces* species. Int. J. Syst. Bacteriol. *16:* 313–340.

Shirling, E.B. and D. Gottlieb. 1968. Cooperative description of type cultures of *Streptomyces*. II. Species descriptions from first study. Int. J. Syst. Bacteriol. *18:* 69–189.

Shirling, E.B. and D. Gottlieb. 1972. Cooperative description of type strains of *Streptomyces* V. Additional descriptions. Int. J. Syst. Bacteriol. *22:* 265–394.

Shklair, I.L. and H.J. Keene. 1974. A biochemical scheme for the separation of the five varieties of *Streptococcus mutans*. Arch. Oral Biol. *19:* 1079–1081.

Shreeve, J.E. and E.E. Edwin. 1974. Thiaminase-producing strains of *Cl. sporogenes* associated with outbreaks of cerebrocortical necrosis. Vet. Rec. *94:* 330.

Shukla, J.S. and V.N. Iyer. 1961. Aerobic spore-forming bacilli showing the phenomenon of colony migration. Arch. Mikrobiol. *39:* 298–312.

Shulman, J.A. and A.J. Nahmias. 1972. Staphylococcal infections: clinical aspects. *In* Cohen (Editor), The Staphylococci, Wiley-Interscience (John Wiley and Sons, Inc.), New York, London, Sydney, Toronto, pp. 457–481.

Shuman, R.D. and N.A. Nord. 1974. Serologic group relationship of P and U streptococci. Cornell Vet. *64:* 376–386.

Sijpesteijn, A.K. 1948. Cellulose-decomposing bacteria from the rumen of cattle. Ph.D. Thesis, Leiden University, Eduard Ijdo N.V., Leiden.

Sijpesteijn, A.K. 1951. On *Ruminococcus flavefaciens*, a cellulose-decomposing bacterium from the rumen of sheep and cattle. J. Gen. Microbiol. *5:* 869–879.

Silcox, V.A., R.C. Good and M.M. Floyd. 1981. Identification of clinically significant *M. fortuitum* complex isolates. J. Clin. Microbiol. *14:* 686–691.

Silva, M.T., M.P. Lima, A.F. Fonseca and J.C.F. Sousa. 1973. The fine structure of *Sporosarcina ureae* as related to its taxonomic position. Submicrosc. Cytol. *5:* 7–22.

Silva, M.T., J.J. Polónia and M. Kocur. 1977. The fine structure of *Micrococcus mucilaginosus.* J. Submicrosc. Cytol. *9:* 53–66.

Silvestri, L., M. Turri, L.R. Hill and E. Gilardi. 1962. A quantitative approach to the systematics of actinomycetes based on overall similarity. Symp. Soc. Gen. Microbiol. *12:* 333–360.

Simmons, J.S. 1927. An acidfast organism isolated from a mouse *Mycobacterium muris* n. sp. J. Infect. Dis. *41:* 13–15.

Simon, M.I., S.V. Emerson, J.H. Shaper, P.D. Bernard and A.N. Glazer. 1977. Classification of *Bacillus subtilis* flagellins. J. Bacteriol. *130:* 200–204.

Simonds, J., P.A. Hansen and S. Lakshmanan. 1971. Deoxyribonucleic acid hybridization among strains of lactobacilli. J. Bacteriol. *107:* 382–384.

Simons, H. 1920. Eine saprophytische Oscillarie im Darm des Meerschweinschens. Zentralbl. Bakteriol. Parasitenk. Infektionskr. Hyg. Abt. II *50:* 356–368.

Simons, H. 1922. Saprophytische Oscillatorien des Menschen und der Tiere. Zentralbl. Bakteriol. Parasitenkd. Infektionskr. Hyg. Abt. I Orig. *88:* 501–510.

Sinell, H.J. and J. Baumgart. 1966. Selektionahrboden zur isolierung von staphylokokken aus liebensmitteln. Zentralbl. Bakteriol. Parasitenkd. Infektionskr. Hyg. Abt. I Orig. *197:* 447–461.

Skadhauge, K. and B. Perch. 1959. Studies on the relationship of some alphahemolytic streptococci of human origin to the Lancefield group M. Acta Pathol. Microbiol. Scand. *46:* 239–250.

Skalka, B. and J. Smola. 1983. Selective diagnostic medium for pathogenic *Listeria* spp.. J. Clin. Microbiol. *18:* 1432–1433.

Skalka, B., J. Smola and K. Elischerová. 1982. Routine test for in vitro differentiation of pathogenic and apathogenic *Listeria monocytogenes* strains. J. Clin. Microbiol. *15:* 503–507.

Skaptason, J.B. and W.H. Burkholder. 1942. Classification and nomenclature of the pathogen causing bacterial ring rot of potatoes. Phytopathology *32:* 439–441.

Skaug, N. and T. Hofstad. 1979. Immunochemical characterization of a polysaccharide antigen isolated from the oral microorganism *Eubacterium saburreum,* strain 02/725. Curr. Microbiol. *2:* 369–373.

Skerman, T.M. and D.J. Jayne-Williams. 1966. Nutrition of coryneform bacteria from milk and dairy sources. J. Appl. Bacteriol. *29:* 167–178.

Skerman, V.B.D. 1967. A Guide to the Identification of the Genera of Bacteria, 2nd Ed., The Williams and Wilkins Co., Baltimore.

Skerman, V.B.D., V. McGowan and P.H.A. Sneath. 1980. Approved Lists of Bacterial Names. Int. J. Syst. Bacteriol. *30:* 225–420.

Skjelkvalé, R. and C.L. Duncan. 1975. Enterotoxin formation by different toxigenic types of *Clostridium perfringens.* Infect. Immun. *11:* 563–575.

Skjelkvalé, R. and T. Uemura. 1977. Experimental diarrhoea in human volunteers following oral administration of *Clostridium perfringens.* J. Appl. Bacteriol. *43:* 281–286.

Skjelkvalé, R.L.H. and T.B. Tjaberg. 1974. Incidence of clostridia in meat products. Nord. Veterinaermed. *26:* 387–391.

Skjold, S.A. and L.W. Wannamaker. 1976. Method for phage typing group A type 49 streptococci. J. Clin. Microbiol. *4:* 232–238.

Skogen, L.O., G.W. Reinbold and E.R. Vedamuthu. 1974. Capsulation of *Propionibacterium.* J. Milk Food Technol. *37:* 314–321.

Skowronski, B. and G.A. Strobel. 1969. Cyanide resistance and cyanide utilization by a strain of *Bacillus pumilis.* Can. J. Microbiol. *15:* 93–98.

Skyring, G.W. and H.E. Jones. 1972. Guanine and cytosine contents of the deoxyribonucleic acids of some sulfate-reducing bacteria: a reassessment. J. Bacteriol. *109:* 1298–1300.

Skyring, G.W. and C. Quadling. 1969. Soil bacteria: comparisons of rhizosphere and nonrhizosphere populations. Can. J. Microbiol. *15:* 473–488.

Skyring, G.W. and C. Quadling. 1970. Soil bacteria: a principal component analysis and guanine-cytosine contents of some arthrobacter-coryneform soil isolates and of some named cultures. Can. J. Microbiol. *16:* 95–106.

Skyring, G.W., C. Quadling and J.W. Rouatt. 1971. Soil bacteria: principal component analysis of physiological descriptions of some named cultures of *Agrobacterium, Arthrobacter* and *Rhizobium.* Can. J. Microbiol. *17:* 1299–1311.

Slack, J.M. 1942. The source of infection in actinomycosis. J. Bacteriol. *43:* 193–209.

Slack, J.M. 1974. Family *Actinomycetaceae* Buchanan 1918 and genus *Actinomyces* Harz 1877. *In* Buchanan and Gibbons (Editors), Bergey's Manual of Determinative Bacteriology, Eighth edition. The Williams and Wilkins Co., Baltimore, pp. 659–667.

Slack, J.M. and M.A. Gerencser. 1966. Revision of serological grouping of *Actinomyces.* J. Bacteriol. *91:* 2107.

Slack, J.M. and M.A. Gerencser. 1970. Two new serological groups of *Actinomyces.* J. Bacteriol. *103:* 265–266.

Slack, J.M. and M.A. Gerenscer. 1975. *Actinomyces,* Filamentous Bacteria. Biology and Pathogenicity. Burgess Publishing Co., Minneapolis, Minnesota.

Slack, J.M., S. Landfried and M.A. Gerencser. 1969. Morphological, biochemical, and serological studies on 64 strains of *Actinomyces israelii.* J. Bacteriol. *97:* 873–884.

Slack, J.M., S. Landfried and M.A. Gerencser. 1971. Identification of *Actinomyces*

and related bacteria in dental calculus by the fluorescent antibody technique. J. Dent. Res. *50:* 78–82.

Slack, J.M., E.H. Ludwig, H.H. Bird and C.M. Canby. 1951. Studies with microaerophilic actinomycetes. I. The agglutination reaction. J. Bacteriol. *61:* 721–735.

Slack, J.M., D.W. Moore, Jr. and M.A. Gerencser. 1966. Use of the fluorescent antibody technique in the diagnosis of actinomycosis. W. Va. Med. J. *62:* 228–231.

Slack, J.M., R.G. Spears, W.G. Snodgrass and R.J. Kuchler. 1955. Studies with microaerophilic actinomycetes. II. Serological groups as determined by the reciprocal agglutinin adsorption technique. J. Bacteriol. *70:* 400–404.

Slack, J.M., A. Winger and D.W. Moore, Jr. 1961. Serological grouping of *Actinomyces* by means of fluorescent antibodies. J. Bacteriol. *82:* 54–65.

Slack, S.A., A. Kelman and J. Perry. 1979. Comparison of three sero-diagnostic assays for the detection of *Corynebacterium sepedonicum.* Phytopathology *69:* 186–189.

Slade, H.D. and G.D. Shockman. 1963. The protoplast membrane and the group D antigen of *Streptococcus faecalis.* Iowa State J. Sci. *38:* 83–96.

Slade, H.D. and W.C. Slamp. 1962. Cell wall composition and the grouping antigens of streptococci. J. Bacteriol. *84:* 345–351.

Sleat, R. and R.A. Mah. 1985. *Clostridium populeti* sp. nov., a cellulolytic species from a woody-biomass digestor. Int. J. Syst. Bacteriol. *35:* 160–163.

Sleat, R., R.A. Mah and R. Robinson. 1984. Isolation and characterization of an anaerobic, cellulolytic bacterium *Clostridium cellulovorans* sp. nov. Appl. Environ. Microbiol. *48:* 88–93.

Sleat, R., R.A. Mah and R. Robinson. 1985. Validation List No. 17. Int. J. Syst. Bacteriol. *35:* 223–225.

Sleesman, J.P. 1982. Preservation of phytopathogenic prokaryotes. *In* Mount and Lacy (Editors). Phytopathogenic Prokaryotes, Volume 2. Academic Press, New York, pp. 447–484.

Slepecky, R.A. 1972. Ecology of bacterial sporeformers. *In* Halvorson, Hanson and Campbell (Editors). Spores, Volume V. Am. Soc. Microbiol., Washington, D.C. pp. 297–313.

Slepecky, R.A. and E.R. Leadbetter. 1977. The diversity of spore-forming bacteria: some ecological implications. *In* Barker, Wolf, Ellar, Dring and Gould (Editors). Spore Research 1976, Volume II. Academic Press, London, pp. 869–877.

Slepecky, R.A. and E.R. Leadbetter. 1983. On the prevalence and roles of spore-forming bacteria and their spores in nature. *In* Hurst and Gould (Editors), The Bacterial Spore, Vol. 2, Academic Press, London, pp. 79–99.

Sleytr, U., H. Adam and H. Klaushofer. 1969. Die feinstruktut der zellwand und cytoplasmamembran von *Clostridium nigrificans* dergestellt mit Hilfe der Gefrierätz-und uetradünnschnett-technik. Arch. Mikrobiol. *66:* 40–58.

Sleytr, U.B. and A.M. Glauert. 1976. Ultrastructure of the cell walls of two closely related clostridia that possess different regular arrays of surface subunits. J. Bacteriol. *126:* 869–882.

Sleytr, U.B., M. Kocur, A.M. Glauert and M.J. Thornley. 1973. A study by freeze-etching of the fine structure of *Micrococcus radiodurans.* Arch. Mikrobiol. *94:* 77–87.

Sleytr, U.B. and P. Messner. 1983. Crystalline surface layers on bacteria. Annu. Rev. Microbiol. *37:* 311–339.

Sleytr, U.B., M.T. Silva, M. Kocur and N.F. Lewis. 1976. The fine structure of *Micrococcus radiophilus.* Arch. Microbiol. *107:* 313–320.

Sleytr, U.B. and K.J.I. Thorne. 1976. Chemical characterization of the regularly arranged surface layers of *Clostridium thermosaccharolyticum* and *Clostridium thermohydrosulfuricum.* J. Bacteriol. *126:* 377–383.

Sloan, G.L., J.M. Robinson and W.E. Kloos. 1982. Identification of "*Staphylococcus staphylolyticus*" NRRL B-2628 as a biovar of *Staphylococcus simulans.* Int. J. Syst. Bacteriol. *32:* 170–174.

Smarda, J. and V. Obdrzalek. 1981. Staphylococcins: incidence and some characteristics of antibiotic action. *In* Jeljaszewicz (Editor), Staphylococci and Staphylococcal Infections, Gustav Fischer Verlag, Stuttgart, New York, pp. 407–411.

Smibert, R.M. and N.R. Krieg. 1981. General characterization. *In* Gerhardt, Murray, Costilow, Nester, Wood, Krieg and Phillips (Editors). Manual of Methods for General Bacteriology. Am. Soc. Microbiol., Washington, pp. 409–443.

Smiley, K.L., C.F. Niven, Jr. and J.M. Sherman. 1943. The nutrition of *Streptococcus salivarius.* J. Bacteriol. *45:* 445–454.

Smiley, M.B. and V. Fryder. 1978. Plasmids, lactic acid production, and *N*-acetyl-D-glucosamine fermentation in *Lactobacillus helveticus* subsp. *jugurti.* Appl. Environ. Microbiol. *35:* 777–781.

Smit, J. 1930. Die Gärungssarcinen. Gustav Fischer, Jena.

Smit, J. 1933. The biology of the fermenting sarcinae. J. Pathol. Bacteriol. *36:* 455–468.

Smith, C.J., S.M. Markowitz and F.L. Macrina. 1981. Transferable tetracycline resistance in *Clostridium difficile.* Antimicrob. Agents Chemother. *19:* 997–1003.

Smith, C.W. and J.F. Metzger. 1962. Demonstration of a capsular structure on *Listeria monocytogenes.* Pathol. Microbiol. *25:* 499–506.

Smith, D.G. and P.M.F. Shattock. 1962. The serological grouping of *Streptococcus equinus.* J. Gen. Microbiol. *29:* 731–736.

Smith, D.G. and P.M.F. Shattock. 1964. The cellular location of antigens in streptococci of groups D, N and Q. J. Gen. Microbiol. *34:* 165–175.

Smith, D.L. and W.C. Trentini. 1972. Enrichment and selective isolation of *Caryophanon latum*. Can. J. Microbiol. *18:* 1197–1200.

Smith, D.L. and W.C. Trentini. 1973. On the Gram reaction of *Caryophanon latum*. Can. J. Microbiol. *19:* 757–760.

Smith, D.T., D.S. Martin, N.F. Conant, J.W. Beard, G. Taylor, H.I. Kohn and M.A. Poston. 1948. Zinsser's Textbook of Bacteriology, 9th ed. pp. 1–992.

Smith, E.F. 1910. A new tomato disease of economic importance. Science (Washington) *31:* 794–796.

Smith, E.F. 1913. A new type of bacterial disease. Science (Washington), *38:* 926.

Smith, H.W. 1959. The bacteriophages of *Clostridium perfringens*. J. Gen. Microbiol. *21:* 622–630.

Smith, I.W. 1964. The occurence and pathology of Dee disease. Freshwater and Salmon Fisheries Research , Dept. Agric. Fish. Scot. *34:* 3–12.

Smith, L.DS. 1957. Genus *Peptostreptococcus* Kluyver and van Niel. *In* Breed, Murray and Smith (Editors), Bergey's manual of determinative bacteriology, 7th ed., The Williams and Wilkins Co., Baltimore, pp. 533–541.

Smith, L.DS. 1970. *Clostridium oceanicum*, sp. n., a sporeforming anaerobe isolated from marine sediments. J. Bacteriol. *103:* 811–813.

Smith, L.DS. 1972. Factors involved in the isolation of *Clostridium perfringens*. J. Milk Food Technol. *35:* 71–76.

Smith, L.DS. 1975a. Common mesophilic anaerobes, including *Clostridium botulinum* and *Clostridium tetani*, in 21 soil specimens. Appl. Microbiol. *29:* 590–594.

Smith, L.DS. 1975b. Inhibition of *Clostridium botulinum* by strains of *Clostridium perfringens* isolated from soil. Appl. Microbiol. *30:* 319–323.

Smith, L.DS. 1975. The pathogenic anaerobic bacteria, 2nd ed. Charles C. Thomas, Springfield, IL, pp. 1–430.

Smith, L.DS. 1977. Botulism. The organism, its toxins, the disease. Charles C. Thomas, Springfield, IL, pp. 1–236.

Smith, L.DS. 1978. The occurrence of *Clostridium botulinum* and *Clostridium tetani* in the soil of the United States. Health Lab Sci. *15:* 74–80.

Smith, L.DS. 1979. Virulence factors of *Clostridium perfringens*. Rev. Infect. Dis. *1:* 254–260.

Smith, L.DS. and E.P. Cato. 1974. *Clostridium durum*, sp. nov., the predominant organism in a sediment core from the Black Sea. Can. J. Microbiol. *20:* 1393–1397.

Smith, L.DS. and V.M. Gardner. 1949. The occurrence of vegetative cells of *Clostridium perfringens* in soil. J. Bacteriol. *58:* 407–408.

Smith, L. DS. and R.L. George. 1946. The anaerobic bacterial flora of clostridial myositis. J. Bacteriol. *52:* 271–279.

Smith, L.DS. and G. Hobbs. 1974. Genus III *Clostridium* Prazmowski 1880, 23. *In* Buchanan and Gibbons (Editors), Bergey's Manual of Determinative Bacteriology, 8th Ed. The Williams and Wilkins Co., Baltimore, pp. 551–572.

Smith, L.DS. and E. King. 1962. *Clostridium innocuum*, sp. n., a spore-forming anaerobe isolated from human infections. J. Bacteriol. *83:* 938–939.

Smith, M.F., S.P. Borriello, G.S. Clayden and M.W. Casewell. 1980. Clinical and bacteriological findings in necrotising enterocolitis: a controlled study. J. Infect. *2:* 23–31.

Smith, N.R. and F.E. Clark. 1938. Motile colonies of *Bacillus alvei* and other bacteria. J. Bacteriol. *35:* 59–60.

Smith, N.R., R.E. Gordon and F.E. Clark. 1946. Aerobic mesophilic spore-forming bacteria. U.S. Dept. Agric. Misc. Pub. No. 559.

Smith, N.R., R.E. Gordon and F.E. Clark. 1952. Aerobic spore-forming bacteria. U.S. Dept. of Agric., Washington, D.C. Agric. Monogr. 16.

Smith, P.B. 1972. Bacteriophage typing of *Staphylococcus aureus*. *In* Cohen (Editor), The Staphylococci, Wiley-Interscience (John Wiley and Sons, Inc.), New York, London, Sydney, Toronto, pp. 431–441.

Smith, R.F. 1969. Role of extracellular ribonuclease in growth of *Corynebacterium acnes*. Can. J. Microbiol. *15:* 749–752.

Smith, R.F. 1969. Characterization of human cutaneous lipophilic diphtheroids. J. Gen. Microbiol. *55:* 433–443.

Smith, R.F. 1975. New medium for isolation of *Corynebacterium vaginale* from genital specimens. Health Lab. Sci. *12:* 219–224.

Smith, R.F. and N.P. Willett. 1968. Lipolytic activity of human cutaneous bacteria. J. Gen. Microbiol. *52:* 441–445.

Smith, R.M., J.T. Parisi, L. Vidal and J.H. Baldwin. 1977. Nature of the genetic determinant controlling encapsulation in *Staphylococcus aureus* Smith. Infect. Immun. *17:* 231–234.

Smith, T. 1896. Two varieties of the tubercle bacillus from mammals. Trans. Assoc. Am. Physns. *11:* 75–95.

Smith, W.R., I. Yu and R.E. Hungate. 1973. Factors affecting cellulolysis by *Ruminococcus albus*. J. Bacteriol. *114:* 729–737.

Smithies, O. 1955. Zone electrophoresis in starch gels: group variations in the serum proteins of normal human adults. Biochem. J. *61:* 629–641.

Smyth, C.J. and F.J. Fehrenbach. 1974. Isoelectric analysis of haemolysins and enzymes from streptococci of groups A, C, and G. Acta Pathol. Microbiol. Scand. Sect. B *82:* 860–870.

Sneath, P.H.A. 1960. A study of the bacterial genus *Chromobacterium*. Iowa State J. Sci. *34:* 243–500.

Sneath, P.H.A. 1962. Longevity of micro-organisms. Nature (London) *195:* 643–646.

Sneath, P.H.A. 1972. Computer taxonomy. *In* Norris and Ribbons, Methods in Microbiology, Vol. 7A, Academic Press, London - New York, pp. 29–98.

Sneath, P.H.A. 1974. Phylogeny of microorganisms. Symp. Soc. Gen. Microbiol. *24:* 1–39.

Sneath, P.H.A. 1977. The maintenance of large numbers of strains of microorganisms, and the implications for culture collections. FEMS Microbiol. Lett. *1:* 333–334.

Sneath, P.H.A. 1977. A method for testing the distinctness of clusters: a test for the disjunction of two clusters in euclidean space as measured by their overlap. J. Int. Assoc. Math. Geol. *9:* 123–143.

Sneath, P.H.A. 1978. Classification of microorganisms. *In* Norris and Richmond (Editors), Essays in Microbiology, John Wiley, Chichester, United Kingdom, pp. 9/1–9/31.

Sneath, P.H.A. 1979a. BASIC program for a significance test for clusters in UPGMA dendrograms obtained from square euclidean distances. Comput. Geosci. *5:* 127–137.

Sneath, P.H.A. 1979b. BASIC program for a significance test for two clusters in euclidean space as measured by their overlap. Comput. Geosci. *5:* 143–155.

Sneath, P.H.A. 1982. Status of nomenclatural types in the approved lists of bacterial names. Request for an opinion. Int. J. Syst. Bacteriol. *32:* 459–460.

Sneath, P.H.A., J.D. Abbott and A.C. Cunliffe. 1951. The bacteriology of erysipeloid. Brit. Med. J. *ii:* 1063–1066.

Sneath, P.H.A. and D. Jones. 1976. *Brochothrix*, a new genus tentatively placed in the family *Lactobacillaceae*. Int. J. Syst. Bacteriol. *26:* 102–104.

Sneath, P.H.A. and R.R. Sokal. 1973. Numerical taxonomy. The principles and practice of numerical classification. W.H. Freeman, San Francisco.

Snijders, E.P. 1924. Cavia-scheefkopperij, een nocardiose. Geneesk. Tijdschr. Ned. Ind. *64:* 85–87.

Snow, G.A. 1970. Mycobactins: iron-chelating growth factors from mycobacteria. Bacteriol. Rev. *34:* 99–125.

Snyder, L., M.S. Slawson, W. Bullock and R.B. Parker. 1967. Studies on oral filamentous bacteria. II. Serological relationships within the genera *Actinomyces, Nocardia, Bacterionema* and *Leptotrichia*. J. Infect. Dis. *117:* 341–345.

Snyder, M.A. 1934. A modification of Dorner spore stain. Stain Technol. *9:* 71.

Snyder, M.L. 1936. The serologic agglutination of the obligate anaerobes *Clostridium paraputrificum* (Bienstock) and *Clostridium capitovalis* (Snyder and Hall). J. Bacteriol. *32:* 401–410.

Snyder, M.L. 1940. The normal fecal flora of infants between two weeks and one year of age. J. Infect. Dis. *66:* 1–16.

Socransky, S.S. 1970. Relationship of bacteria to the etiology of peridontal disease. J. Dent. Res. *49:* 203–222.

Socransky, S.S., C. Hubersak and D. Propas. 1970. Induction of peridontal destruction in gnotobiotic rats by a human oral strain of *Actinomyces naeslundii*. Arch. Oral Biol. *15:* 993–995.

Soda, J.A. and R.C. Cleverdon. 1960. Some serological studies on the genera *Corynebacterium, Flavobacterium* and *Xanthomonas*. Antonie van Leeuwenhoek J. Microbiol. Serol. *26:* 98–102.

Söhngen, N.L. 1913. Benzin, Petroleum, Paraffinöl und Paraffin als Kohlenstoff- und Energie-quelle für Mikroben. Zentralbl. Bakteriol. Parasitenkd. Infektionskr. Hyg. Abt. 2. *37:* 595–609.

Soltys, M.A. 1961. *Corynebacterium suis* associated with a specific cystitis and pyelonephritis in pigs. J. Pathol. Bacteriol. *81:* 441–446.

Soltys, M.A. and F.R. Spratling. 1957. Infectious cystitis and pyelonephritis of pigs: a preliminary communication. Vet. Rec. *69:* 500–504.

Somerville, H.J. and M.L. Jones. 1972. DNA competition studies within the *Bacillus cereus* group of bacilli. J. Gen. Microbiol. *73:* 257–265.

Sompolinsky, D. 1950. Impetigo contagiosa suis. Maanedsskr. Dyrlaeg. *61:* 402–453.

Sompolinsky, D. 1953. De l'impetigo contagiosa suis et du *Micrococcus hyicus* n. sp. Schweiz. Arch. Tierheilkd. *95:* 302–309.

Sompolinsky, D., A. Lagziel, D. Neveh and T. Yankilevitz. 1978. *M. haemophilum* sp. nov., a new pathogen of humans. Int. J. Syst. Bacteriol. *28:* 67–75.

Sompolinsky, D., A. Lagziel and J. Rosenberg. 1979. Further studies of a new pathogenic mycobacterium (*M. haemophilum* sp. nov.). Can. J. Microbiol. *25:* 217–226.

Sonea, S. 1971. A tentative unifying view of bacteria. Rev. Can. Biol. *30:* 239–244.

Sonea, S. and M. Panisset. 1976. Pour une nouvelle bacteriologie. Rev. Can. Biol. *35:* 103–167.

Sonea, S. and M. Panisset. 1980. Introduction a la nouvelle bacteriologie. Les Presses de l'Université de Montréal et Masson, Montréal and Paris.

Sonnabend, O., W. Sonnabend, R. Heinzle, T. Sigrist, R. Dirnhofer and U. Krech. 1981. Isolation of *Clostridium botulinum* type G and identification of type G botulinal toxin in humans: report of five sudden unexpected deaths. J. Infect. Dis. *143:* 22–27.

Soriano, S. and A. Soriano. 1948. Nueva bacteria anaerobia productora de una alteracion en sordinas envasadas. Rev. Asoc. Argent. Dietol. *6:* 36–41.

Sottnek, F.O., J.M. Brown, R.E. Weaver and G.F. Carroll. 1977. Recognition of *Oerskovia* species in the clinical laboratory: Characterization of 35 isolates. Int. J. Syst. Bacteriol. *27:* 263–270.

Souhami, L., R. Feld, P.G. Tuffnell and T. Feller. 1979. *Micrococcus luteus* pneumonia: A case report and review of the literature. Med. Pediatr. Oncol. *7:* 309–314.

Southwick, G.J. and G.D. Lister. 1979. Actinomycosis of the hand: A case report. J. Hand Surg. *4:* 360–362.

Sozzi, T., F. Gnaegi, N. D'Amico and H. Hose. 1982. Difficultés de fermentation malolatique du vin dues a des bactériophages de *Leuconostoc oenos*. Rev. Suisse Vitic. Aboric. Hortic. *14:* 17–23.

Sozzi, T., J.M. Poulain and R. Maret. 1978. Etude d'un bactériophage de *Leuconostoc mesenteroides* isolé de produits laitiers. Schweiz. Milchwirtsch. Forsch. *7:* 33–40.

Sozzi, T. and M.B. Smiley. 1980. Antibiotic resistances of yogurt-starter cultures *Streptococcus thermophilus* and *Lactobacillus bulgaricus*. Appl. Environ. Microbiol. *40:* 862–865.

Sozzi, T., K. Watanabe, K. Stetter and M. Smiley. 1981. Bacteriophages of the genus *Lactobacillus*. Virology *16:* 129–135.

Speckman, R.A. and E.B. Collins. 1968. Diacetyl biosynthesis in *Streptococcus diacetilactis* and *Leuconostoc citrovorum*. J. Bacteriol. *95:* 174–180.

Speller, D.C. and R.G. Mitchell. 1973. Coagulase-negative staphylococci causing endocarditis after cardiac surgery. J. Clin. Path. *26:* 517–524.

Spengler, M.D., G.T. Rodeheaver and R.F. Edlich. 1979. Practical technique for quantitating anaerobic bacteria in tissue specimans. J. Clin. Microbiol. *10:* 331–333.

Sperry, J.F. and T.D. Wilkins. 1976a. Arginine, a growth-limiting factor for *Eubacterium lentum*. J. Bacteriol. *127:* 780–784.

Sperry, J.F. and T.D. Wilkins. 1976b. Cytochrome spectrum of an obligate anaerobe, *Eubacterium lentum*. J. Bacteriol. *125:* 905–909.

Spicher, G. and R. Schröder. 1978. Die Mikroflora des Sauerteiges. IV. Mitteilung: Untersuchungen über die Art der in "Reinzuchtsauern" anzutreffenden stäbchenförmigen Milchsäurebacterien (Genus *Lactobacillus* Beijerinck). Z. Lebensm. Unters. Forsch. *167:* 342–354.

Spieckermann, A. and P. Kotthoff. 1914. Untersuchungen über die Kartoffelpflanze und ihre krankheiten. Landhr. Jb. *46:* 659–732.

Spieckermann, Ch. 1970. Die in-vitro-Empfindlichkeit von *A. israelii, Actinobacillus actinomycetem-comitans* and *Bacteroides melaninogenicus* gegenüber Cephaloridin, Cephalothin, Gentamycin, Fusidinsäure und Lincomycin. Int. J. Clin. Pharmacol. Ther. Toxicol. *4:* 318–320.

Spizizen, J. 1958. Transformation of biochemically deficient strains of *Bacillus subtilis* by deoxyribonucleate. Proc. Nat. Acad. Sci. USA *78:* 2893–2897.

Spray, R.S. 1939. Genus II *Clostridium* Prazmowski. *In* Bergey, Breed, Murray and Hitchens (Editors), Bergey's Manual of Determinative Bacteriology, 5th ed., The Williams and Wilkins Co., Baltimore, pp. 743–790.

Spray, R.S. 1948. Genus II *Clostridium* Prazmowski. *In* Breed, Murray and Hitchens (Editors), Bergey's Manual of Determinative Bacteriology, 6th ed. The Williams and Wilkins Co., Baltimore, pp. 763–827.

Spray, R.S. and L.S. McClung. 1957. Genus II *Clostridium* Prazmowski, 1880. *In* Breed, Murray, and Smith (Editors), Bergey's Manual of Determinative Bacteriology, 7th Ed. The Williams and Wilkins Co., Baltimore, pp. 634–693.

Sriranganathan, N., R.J. Seidler, W.E. Sandine and P.R. Elliker. 1973. Cytological and deoxyribonucleic acid - deoxyribonucleic acid hybridization studies on *Lactobacillus* isolated from San Francisco sour dough. Appl. Microbiol. *25:* 461–470.

Srivastava, K.K. and I.H. Siddique. 1973. Quantitative chemical composition of peptidoglycan of *Listeria monocytogenes*. Infect. Immun. *7:* 700–703.

Stableforth, A.W. 1959. Streptococcal diseases. *In* Stableforth and Galloway (Editors), Infectious Diseases of Animals, Vol. 2, Butterworths, London, pp. 589–650.

Stack, R.J., R.E. Hungate and W.P. Opsahl. 1983. Phenylacetate stimulation of cellulose digestion by *Ruminococcus albus* 8. Appl. Environ. Microbiol. *46:* 539–544.

Stackebrandt, E. and F. Fiedler. 1979. DNA-DNA homology studies among strains of *Arthrobacter* and *Brevibacterium*. Arch. Microbiol. *120:* 289–295.

Stackebrandt, E., F. Fiedler and O. Kandler. 1978. Peptidoglycan Typ und Zusammensetzung der Zellwandpolysaccharide von *Cellulomonas cartalyticum* und einigen coryneformen Organismen. Arch. Mikrobiol. *117:* 115–118.

Stackebrandt, E., V.J. Fowler, F. Fiedler and H. Seiler. 1983. Taxonomic studies on *Arthrobacter nicotianae* and related taxa. Description of *Arthrobacter uratoxydans* sp. nov. and *Arthrobacter sulfureus* sp. nov. and reclassification of *Brevibacterium protophormiae* as *Arthrobacter protophormiae* comb. nov. Syst. Appl. Microbiol. *4:* 470–486.

Stackebrandt, E., V.J. Fowler, F. Fiedler and H. Seiler. 1984. Validation of the publication of new names and new combinations previously effectively published outside the IJSB. List No. 14. Int. J. Syst. Bacteriol. *34:* 270–271.

Stackebrandt, E., V.J. Fowler and C.R. Woese. 1983a. A phylogenetic analysis of lactobacilli, *Pediococcus pentosaceus* and *Leuconostoc mesenteroides*. Syst. Appl. Microbiol. *4:* 326–337.

Stackebrandt, E., M. Häringer and K.H. Schleifer. 1980b. Molecular genetic evidence for the transfer of *Oerskovia* species into the genus *Cellulomonas*. Arch. Microbiol. *127:* 179–185.

Stackebrandt, E. and O. Kandler. 1974. Biochemische-taxonomische Untersuchungen an der Gattung *Cellulomonas*. Zentralbl. Bakteriol. Parasitenkd. Infektionskr. Hyg. Abt. I Orig. A *228:* 128–135.

Stackebrandt, E. and O. Kandler. 1979. Taxonomy of the genus *Cellulomonas*, based on phenotypic characters and deoxyribonucleic acid-deoxyribonucleic acid homology, and proposal of seven neotype strains. Int. J. Syst. Bacteriol. *29:* 273–282.

Stackebrandt, E. and O. Kandler. 1980a. Fermentation pathway and redistribution of ^{14}C in specifically labelled glucose in *Cellulomonas*. Zentralbl. Bakteriol. Mikrobiol. Hyg. Abt. I Orig. C. *1:* 40–50.

Stackebrandt, E. and O. Kandler. 1980b. *Cellulomonas cartae* spec. nov. Int. J. Syst. Bacteriol. *30:* 186–188.

Stackebrandt, E., R.M. Kroppenstedt and V.J. Fowler. 1983b. A phylogenetic analysis of the family *Dermatophilaceae*. J. Gen. Microbiol. *129:* 1831–1838.

Stackebrandt, E., B.J. Lewis and C.R. Woese. 1980a. The phylogenetic structure of the coryneform group of bacteria. Zentralbl. Bakteriol. Mikrobiol. Hyg. Abt. II. Orig. C *1:* 137–149.

Stackebrandt, E., W. Ludwig, E. Seewaldt and K.-H. Schleifer. 1983. Phylogeny of sporeforming members of the order *Actinomycetales*. Int. J. Syst. Bacteriol. *33:* 173–180.

Stackebrandt, E., C. Scheurlein and K.H. Schleifer. 1983. Phylogenetic and biochemical studies on *Stomatococcus mucilaginosus*. Syst. Appl. Microbiol. *4:* 207–217.

Stackebrandt, E. and K.-H. Schleifer. 1984. Molecular systematics of actinomycetes and related organisms. *In* Bojalil and Ortiz-Ortiz (Editors), Actinomycetes Biology, Academic Press, Inc., New York.

Stackebrandt, E., H. Seiler and K.-H. Schleifer. 1982. Union of the genera *Cellulomonas* Bergey et al. and *Oerskovia* Prauser et al. in a redefined genus *Cellulomonas*. Zentralbl. Bakteriol. Mikrobiol. Hyg. I Abt. Orig. C *3:* 401–409.

Stackebrandt, E., B. Wittek, E. Seewaldt and K.H. Schleifer. 1982. Physiological, biochemical and phylogenetic studies on *Gemella haemolysans*. FEMS Microbiol. Lett. *13:* 361–365.

Stackebrandt, E. and C.R. Woese. 1979. A phylogenetic dissection of the family *Micrococcaceae*. Curr. Microbiol. *2:* 317–322.

Stackebrandt, E. and C.R. Woese. 1981. Towards a phylogeny of the actinomycetes and related organisms. Curr. Microbiol. *5:* 197–202.

Stackebrandt, E. and C.R. Woese. 1981. The evolution of prokaryotes. *In* Carlile, Collins and Moseley (Editors), Molecular and Cellular Aspects of Microbial Evolution, Symp. Soc. Gen. Microbiol. *32:* 1–31.

Stadtman, E.R., T.C. Stadtman, I. Pastan and L.DS. Smith. 1972. *Clostridium barkeri* sp. n. J. Bacteriol. *110:* 758–760.

Stadtman, T.C. 1954. On the metabolism of an amino acid fermenting clostridium. J. Bacteriol. *67:* 314–320.

Stadtman, T.C. 1973. Lysine metabolism by clostridia. *In* Meister (Editor), Advances in Enzymology. *38:* 413–448.

Stadtman, T.C. and L.S. McClung. 1957. *Clostridium sticklandii* nov. spec. J. Bacteriol. *73:* 218–219.

Stamm, A.M. and C.G. Cobbs. 1980. Group C streptococcal pneumonia: report of a fatal case and review of the literature. Rev. Infect. Dis. *2:* 889–898.

Staneck, J.L. and G.D. Roberts. 1974. Simplified approach to identification of aerobic actinomycetes by thin-layer chromatography. Appl. Microbiol. *28:* 226–231.

Staneck, J.L. and J.A. Washington, II. 1974. Antimicrobial susceptibilities of anaerobic bacteria: recent clinical isolates. Antimicrob. Agents Chemother. *6:* 311–315.

Stanford, J.L. 1973. An immunodiffusion analysis of *Mycobacterium lepraemurium* Marchoux and Sorel. J. Med. Microbiol. *6:* 435–439.

Stanford, J.L. and A. Beck. 1969. Bacteriological and serological studies of fast growing mycobacteria identified as *Mycobacterium friedmannii*. J. Gen. Microbiol. *58:* 99–106.

Stanford, J.L. and J.M. Grange. 1974. The meaning and structure of species as applied to mycobacteria. Tubercle *55:* 143–152.

Stanford, J.L. and W.J. Gunthorpe. 1969. Serological and bacteriological investigation of *Mycobacterium ranae (fortuitum)*. J. Bacteriol. *98:* 375–383.

Stanford, J.L. and W.J. Gunthorpe. 1971. A study of some fast-growing scotochromogenic mycobacteria including species descriptions of *Mycobacterium gilvum* (new species) and *Mycobacterium duvalii* (new species). Br. J. Exp. Pathol. *52:* 627–637.

Stanford, J.L., S.R. Pattyn, F. Portaels and W.J. Gunthorpe. 1972. Studies on *Mycobacterium chelonei*. J. Med. Microbiol. *5:* 177–182.

Stanford, J.L., G.A.W. Rook, J. Convit, T. Godal, G. Kronvall, R.J.W. Rees and H.P. Walsh. 1975. Preliminary taxonomic studies on leprosy bacillus. Br. J. Exp. Pathol. *56:* 579–585.

Stanier, R.Y. 1961. La place des bactéries dans le monde vivant. Ann. Inst. Pasteur (Paris) *101:* 297–303.

Stanier, R.Y. 1970. Some aspects of the biology of cells and their possible evolutionary significance. *In* Charles and Knight (Editors), Organization and Control in Procaryotic and Eucaryotic Cells, Cambridge University Press, Cambridge.

Stanier, R.Y. and C.B. van Niel. 1962. The concept of a bacterium. Arch. Mikrobiol. *42:* 17–35.

Stanier, R.Y., D. Wachter, D. Gasser and A.C. Wilson. 1970. Comparative immunological studies of two *Pseudomonas* enzymes. J. Bacteriol. *102:* 351–362.

Stankewich, J.P., B.J. Cosenza and A.L. Shigo. 1971. *Clostridium quercicolum* sp. n., isolated from discolored tissues in living oak trees. Antonie Van Leeuwenhoek J. Microbiol. Serol. *37:* 299–302.

Stankiewicz, R. 1962. Fibrinolytic activity of staphylococcal strains and its effect on coagulase activity. Med. Dosw. Mikrobiol. *14:* 93–100.

Stapp, C. 1920. Botanische Untersuchungen einiger neuer Bakterien-spezies, welche mit reiner Harnsäure oder Hippursäure als all-einigem organischem Nährstoff auskommen. Zentralbl. Bakteriol. Parasitenkd. Infektionskr. Hyg. Abt. II *51:* 1–71.

Stapp, C. and H. Zycha. 1931. Morphologische Untersuchungen an *Bacillus mycoides*; ein Beitrag zur Frage des Pleomorphismus der Bakterien. Arch. Mikrobiol. *2:* 493–536.

Stark, P. and J.M. Sherman. 1935. Concerning the habitat of *Streptococcus lactis*. J. Bacteriol. *30:* 639–646.

Stark, P.L. and A. Lee. 1982. Clostridia isolated from the feces of infants during the first year of life. J. Pediatrics *100:* 362–365.

Starkey, R.L. 1938. A study of spore formation and other morphological characteristics of *Vibrio desulfuricans*. Arch. Mikrobiol. *9:* 268–304.

Starr, M.P. 1949. The nutrition of phytopathogenic bacteria. III. The Grampositive phytopathogenic *Corynebacterium* species. J. Bacteriol. *57:* 253–258.

Starr, M.P. 1958. The blue pigment of *Corynebacterium insidiosum*. Arch. Mikrobiol. *30:* 325–334.

Starr, M.P., M. Mandel and N. Murata. 1975. The phytopathogenic coryneform bacteria in the light of DNA base composition and DNA-DNA segmental homology. J. Gen. Appl. Microbiol. *21:* 13–26.

Starr, M.P., R.M. Sayre and J.M. Schmidt. 1983. Assignment of ATCC 27377 to *Planctomyces staleyi* sp. nov. and conservation of *Pasteuria ramosa* Metchnikoff 1888 on the basis of type descriptive material. Request for an Opinion. Int. J. Syst. Bacteriol. *33:* 666–671.

Starr, M.P. and J.M. Schmidt. 1981. Prokaryote diversity. *In* Starr, Stolp, Trüper, Balows and Schlegel (Editors), The Prokaryotes. A Handbook on habitats, isolation and identification of bacteria. Springer-Verlag, Berlin, Heidelberg and New York.

Starr, M.P., H. Stolp, H.G. Trüper, A. Balows and H.G. Schlegel (Editors). 1981. The Prokaryotes, A Handbook on the Habitats, Isolation and Identification of Bacteria. Springer-Verlag, Berlin, Heidelberg.

Steel, K.J. 1962. The oxidase activity of staphylococci. J. Appl. Microbiol. *25:* 445–447.

Steenbergan, J.F., H.S. Kimball, D.A. Low, H.C. Shapiro and L.N. Phelps. 1977. Serological grouping of virulent and avirulent strains of the lobster pathogen *Aerococcus viridans*. J. Gen. Microbiol. *99:* 425–430.

Stefansky, W.K. 1903. Eine lepraähnliche Erkrankung der Haut und der Lymphdrüsen bei Wanderratten. Zentralbl. Bakteriol. Parasitenkd. Infektionskr. Hyg. Abt. 1, Orig. *33:* 481–487.

Steffen, E.K. and D.J. Hentges. 1981. Hydrolytic enzymes of anaerobic bacteria isolated from human infections. J. Clin. Microbiol. *14:* 153–156.

Steinhaus, E.A. 1941. A study of the bacteria associated with thirty species of insects. J. Bacteriol. *42:* 757–790.

Steinkraus, K.H. 1983. Lactic acid fermentation in the production of foods from vegetables, cereals and legumes. Antonie van Leeuwenhoek J. Microbiol. Serol. *49:* 337–348.

Stellwag, E.J. and P.B. Hylemon. 1978. Characterization of 7 α-dehydroxylase in *Clostridium leptum*. Am. J. Clin. Nutr. *31:* S243-S247.

Stellwag, E.J. and P.B. Hylemon. 1979. 7 α-dehydroxylation of cholic acid and chenodeoxycholic acid by *Clostridium leptum*. J. Lipid Res. *20:* 325–333.

Stephenson, J.R. and D.J. Wright. 1982. Clostridial bacteraemia. J. Hosp. Infect. *3:* 311–312.

Stephenson, M.P. and E.A. Dawes. 1971. Pyruvic acid and formic acid metabolism in *Sarcina ventriculi* and the role of ferredoxin. J. Gen. Microbiol. *69:* 331–343.

Sternberg, W.K. 1892. Manual of Bacteriology. W. Wood and Co., New York, pp. 1–874.

Sterne, M. and I. Batty. 1975. Pathogenic clostridia, Butterworths, London, pp. 1–144.

Sterne, M. and G.H. Warrack. 1964. The types of *Clostridium perfringens*. J. Pathol. Bacteriol. *88:* 279–283.

Stetter, H. and K.O. Stetter. 1980. *Lactobacillus bavaricus* sp. nov., a new species of the subgenus *Streptobacterium*. Zentralbl. Bakteriol. Mikrobiol. Hyg. I Abt. Orig. C *1:* 70–74.

Stetter, H. and K.O. Stetter. 1980. *In* Validation of the publication of the new names and new combinations previously effectively published outside the IJSB. List No. 4. Int. J. Syst. Bacteriol. *30:* 601.

Stetter, K.O. 1974. Production of exclusively L(+)-lactic acid-containing food by controlled fermentation. Proc. 1st Intersect. Congr. IAMS, Tokyo *2:* 164–168.

Stetter, K.O. 1977. Evidence for frequent lysogeny in lactobacilli: temperature bacteriophages within the subgenus *Streptobacterium*. J. Virol. *24:* 685–689.

Stetter, K.O. and O. Kandler. 1973. Untersuchungen zur Entstehung von DL-Milchsäure bei Lactobacillen und Charakterisierung einer Milchsäureracemase bei einigen Arten der Untergattung *Streptobacterium*. Arch. Mikrobiol. *94:* 221–247.

Stetter, K.O., H. Priess and H. Delius. 1978. *Lactobacillus casei* phage PL-1. Molecular properties and first transcription studies *in vivo* and *in vitro*. Virology *87:* 1–12.

Stevens, F.L. 1925. Plant Disease Fungi. Macmillan Co., New York, pp. 1–469.

Stevens, W.C. 1956. Taxonomic studies on the genus *Bacteroides* and similar forms Thesis, Vanderbilt University.

Stevenson, G.W. and H.H. Gossman. 1968. Dental and intracranial actinomycosis. Br. J. Surg. *55:* 830–834.

Stevenson, I.L. 1963. Some observations on the so-called 'cystites' of the genus *Arthrobacter*. Can. J. Microbiol. *9:* 467–472.

Stevenson, I.L. 1978. The production of extracellular amino acids by rumen bacteria. Can. J. Microbiol. *24:* 1236–1241.

Stewart, A.W. and M.G. Johnson. 1977. Increased numbers of heat resistant spores produced by two strains of *Clostridium difficile* bearing temperate phage s9. J. Gen. Microbiol. *193:* 43–50.

Stewart, B.J. and J.M. Leatherwood. 1976. Derepressed synthesis of cellulase by *Cellulomonas*. J. Bacteriol. *128:* 609–615.

Stewart, M. and T.J. Beveridge. 1980. Structure of the regular surface layer of *Sporosarcina ureae*. J. Bacteriol. *142:* 302–309.

Stickland, L.H. 1934. Studies in the metabolism of the strict anaerobes (genus *Clostridium*). I. The chemical reactions by which *Cl. sporogenes* obtains its energy. Biochem. J. *28:* 1746–1759.

Stieb, M. and B. Schink. 1985a. Anaerobic oxidation of fatty acids by *Clostridium bryantii* sp. nov., a sporeforming, obligately syntrophic bacterium. Arch. Microbiol. *140:* 387–390.

Stieb, M. and B. Schink. 1985b. Validation List No. 18. Int. J. Syst. Bacteriol. *35:* 375–376.

Stoppok, W., P. Rapp and F. Wagner. 1982. Formation, location, and regulation of endo-1, 4-β-glucanases and β-glucosidases from *Cellulomonas uda*. Appl. Environ. Microbiol. *44:* 44–53.

Strasters, K.C. and K.C. Winkler. 1963. Carbohydrate metabolism of *Staphylococcus aureus*. J. Gen. Microbiol. *33:* 213–229.

Strider, D.L. 1969. Bacterial canker of tomato caused by *Corynebacterium michiganense*. A literature review and bibliography. N. Carolina Agric. Exp. Sta. Tech. Bull. 193.

Stringer, J. 1980. The development of a phage-typing system for group-B streptococci. J. Med. Microbiol. *13:* 113–143.

Stringer, J. and W.R. Maxted. 1979. Phage typing of group B streptococci. Lancet *1:* 328.

Strobel, G.A. 1977. Bacterial phytotoxins. Annu. Rev. Microbiol. *31:* 205–224.

Stuart, M.R. and P.E. Pease. 1972. A numerical study on the relationships of *Listeria* and *Erysipelothrix*. J. Gen. Microbiol. *73:* 551–565.

Stuart, S.E. and H.J. Welshimer. 1973. Intrageneric relatedness of *Listeria* Pirie. Int. J. Syst. Bacteriol. *23:* 8–14.

Stuart, S.E. and H.J. Welshimer. 1974. Taxonomic re-examination of *Listeria* Pirie and transfer of *Listeria grayi* and *Listeria murrayi* to a new genus *Murraya*. Int. J. Syst. Bacteriol. *24:* 177–185.

Sturges, W.S. and E.T. Drake. 1927. A complete description of *Clostridium putrefaciens* (McBryde). J. Bacteriol. *14:* 175–179.

Su, Y. and K. Yamada. 1960. Studies on L-glutamic acid fermentation. Part I. Isolation of a L-glutamic acid producing strain and its taxonomical studies. Bull. Agr. Chem. Soc. Jpn. *24:* 69–74.

Subcommittee on Taxonomy of Staphylococci and Micrococci, International Committee on Nomenclature of Bacteria. 1965. Minutes of first meeting. Int. Bull. Bacteriol. Nomencl. Taxon. *15:* 107–108.

Subcommittee (ICBN) on Taxonomy of Staphylococci and Micrococci. 1965. Recommendations. Int. Bull. Bacteriol. Nomencl. Taxon. *15:* 109–110.

Sugar, A.M. and R.V. McCloskey. 1977. *Bacillus licheniformis* sepsis. J. Am. Med. Assoc. *238:* 1180–1181.

Sugihara, T.F. and L. Kline. 1975. Further studies on a growth medium for *Lactobacillus sanfrancisco*. J. Milk Food Technol. *38:* 667–672.

Sugisaki, Z., N. Watanabe, I. Kobayashi and N. Iguchi. 1973. Process for producing uricase. U.S. Patent 3,767,533.

Sugiyama, H. 1980. *Clostridium botulinum* neurotoxin. Microbiol. Rev. *44:* 419–448.

Sukapure, R.S., M.P. Lechevalier, H. Reber, M.L. Higgins, H.A. Lechevalier and H. Prauser. 1970. Motile nocardoid *Actinomycetales*. Appl. Microbiol. *19:* 527–533.

Sukchotiratana, M., A.H. Linton and J.P. Fletcher. 1975. Antibiotics and the oral streptococci of man. J. Appl. Bacteriol. *38:* 277–294.

Sulea, I.T., M.C. Pollice and L. Barksdale. 1980. Pyrazine carboxylamidase activity in *Corynebacterium*. Int. J. Syst. Bacteriol. *30:* 466–472.

Sulzbacher, W.L. and R.A. McLean. 1951. The bacterial flora of fresh pork sausage. Food Technol. *5:* 7–8.

Summers, R.J., D.P. Bondreaux and V.R. Srinivasan. 1979. Continuous cultivation for apparent optimization of defined media for *Cellulomonas* sp. and *Bacillus cereus*. Appl. Environ. Microbiol. *38:* 66–71.

Sumner, J.L., I.R. Perry and C.A. Reay. 1977. Microbiology of New Zealand feral venison. J. Sci. Food Agric. *28:* 829–832.

Sundman, V. and K. Björksten. 1958. The globular involution forms of the bifid bacteria. J. Gen. Microbiol. *19:* 491–496.

Sundman, V., K. Björksten and H.G. Gyllenberg. 1959. Morphology of the bifid bacteria (organisms previously incorrectly designated *Lactobacillus bifidus*) and some related genera. J. Gen. Microbiol. *21:* 371–384.

Suter, L.S. and B.F. Vaughan. 1955. The effect of antibacterial agents on the growth of *Actinomyces bovis*. Antibiot. Chemother. *10:* 557–560.

Sutherland, J.D. and I.A. Macdonald. 1982. The metabolism of primary, 7-oxo, and 7β-hydroxy bile acids by *Clostridium absonum*. J. Lipid Research *23:* 726–732.

Sutherland, J.P., J.T. Patterson, P.A. Gibbs and J.G. Murray. 1975. Some metabolic and biochemical characteristics of representative microbial isolates from vacuum-packaged beef. J. Appl. Bacteriol. *39:* 239–249.

Sutter, V.L., D.M. Citron and S.M. Finegold. 1980. Wadsworth anaerobic bacteriology manual, 3rd ed. C.V. Mosby Co., St. Louis, pp. 1–131.

Sutter, V.L. and S.M. Finegold. 1976. Susceptibility of anaerobic bacteria to 23 antimicrobial agents. Antimicrob. Agents Chemother. *10:* 736–752.

Suzuki, J. and K. Kitahara. 1964. Base composition of desoxyribonucleic acid in *Sporolactobacillus inulinus* and other lactic acid bacteria. J. Gen. Appl. Microbiol. *10:* 305–311.

Suzuki, K., T. Kaneko and K. Komagata. 1981a. Deoxyribonucleic acid homologies among coryneform bacteria. Int. J. Syst. Bacteriol. *31:* 131–138.

Suzuki, K. and K. Komagata. 1983. Taxonomic significance of cellular fatty acid composition in some coryneform bacteria. Int. J. Syst. Bacteriol. *33:* 188–200.

Suzuki, K. and K. Komagata. 1983. *Pimelobacter* gen. nov. - a new genus of coryneform bacteria with LL-diaminopimelic acid in the cell wall. J. Gen. Appl. Microbiol. *29:* 59–71.

Suzuki, K., K. Saito, A. Kawaguchi, S. Okuda and K. Komagata. 1981b. Occurrence of ω-cyclohexyl fatty acids in *Curtobacterium pusillum* strains. J. Gen. Appl. Microbiol. *27:* 261–266.

Suzuki, K.-I., A. Kawaguchi, K. Saito, S. Okuda and K. Komagata. 1982. Taxonomic significance of the position of double bonds of unsaturated fatty acids in corynebacteria. J. Gen. Appl. Microbiol. *28:* 409–416.

Suzuki, Y., T. Kishigami, K. Inoue, Y. Mizoguchi, M. Eto, M. Takagi and S. Abe. 1983. *Bacillus thermoglucosidasius* sp.nov., a new species of obligately thermophilic bacilli. Syst. Appl. Microbiol. *4:* 487–495.

Swanson, J. and M. McCarty. 1969. Electron microscopic studies on opaque colony variants of group A streptococci. J. Bacteriol. *100:* 505–511.

Sweet, D.M. and B.E.B. Moseley. 1974. Accurate repair of ultraviolet-induced damage in *Micrococcus radiodurans*. Mutat. Res. *23:* 311–318.

Swenson, J.M., C. Thornsberry, L.M. McCrosky, C.L. Hatheway and V.R. Dowell, Jr. 1980. Susceptibility of *Clostridium botulinum* to thirteen antimicrobial agents. Antimicrob. Agents Chemother. *18:* 13–19.

Swift, H.F., A.T. Wilson and R.C. Lancefield. 1943. Typing group A hemolytic streptococci by M precipitin reactions in capillary pipettes. J. Exp. Med. *78:* 127–133.

Swindle, M.M., O. Narayan, M. Luzarraga and D.L. Bobbie. 1980. Contagious streptococcal lymphadenitis in cats. J. Am. Vet. Med. Assoc. *177:* 829–830.

Switalski, L.M., A. Ljungh, C. Rydén, K. Rubin, M. Höök and T. Wadström. 1982. Binding of fibronectin to the surface of groups A, C and G streptococci isolated from human infections. Eur. J. Clin. Microbiol. *1:* 381–387.

Switalski, L.M., O. Schwan, C.J. Smyth and T. Wadström. 1978. Peptocoagulase: clotting factor produced by bovine strains of *Peptococcus indolicus*. J. Clin. Microbiol. *7:* 361–367.

Sword, C.P. 1966. Mechanisms of pathogenesis in *Listeria monocytogenes* infection. I. Influence of iron. J. Bacteriol. *92:* 536–542.

Sword, C.P. and M.J. Pickett. 1961. The isolation and characterization of bacteriophages from *Listeria monocytogenes*. J. Gen. Microbiol. *25:* 241–248.

Syed, S.A. 1976. A new medium for the detection of gelatin-hydrolyzing activity of human dental plaque flora. J. Clin. Microbiol. *3:* 200–202.

Syed, S.A. and W.J. Loesche. 1972. Survival of human dental plaque flora in various transport media. Appl. Microbiol. *24:* 638–644.

Szabo, L.G., Sz. Esztergaly and I. Dzvonyar. 1981. Genitale Aktinomykose nach Einsetzen von Intrauterinpessaren (IUD) und Präventionsmöglichkeiten. Zentralbl. Gynaekol. *103:* 115–120.

Szulga, T., P.A. Jenkins and J. Marks. 1966. Thin-layer chromatography of mycobacterial lipids as an aid to classification; *Mycobacterium kansasii*, and *Mycobacterium marinum* (*balnei*). Tubercle *47:* 130–136.

Szulmajster, J. 1979. Is sporulation a simple model for studying differentiation? Trends Biochem. Sci. *4:* 18–21.

Tabor, H.W. and M. Morrison. 1964. Electron transport in staphylococci. Properties of a particle preparation from exponential phase *Staphylococcus aureus*. Arch. Biochem. Biophys. *105:* 367–379.

Tabor, M.W., J. MacGee and J.W. Holland. 1976. Rapid determination of dipicolinic acid in the spores of *Clostridium* species by gas-liquid chromatography. Appl. Environ. Microbiol. *31:* 25–28.

Tacquet, A., A. Andrejew and V. Macquet. 1958–59. Pouvoir pathogène pour la poule et le lapin des mycobactéries aviaires resistantes a de fortes concentrations d'isoniazide. Ann. Inst. Pasteur Lille *10:* 29–36.

Tacquet, A., A. Collet, J.C. Martin, B. Devulder and C. Gernez-Rieux. 1967. Pulmonary dusting and infection of guinea pigs by *Mycobacterium kansasii*. Z. Tuberkulose Erkr. Thoraxogan *127:* 115–122.

Tacquet, A., M.T. Plancot, J. Debruyne, B. Devulder, M. Joseph and J. Losfeld. 1971. Études préliminaires sur la classification numérique des mycobactéries et des nocardias. 1) Relations taxonomiques entre *Mycobacterium rhodochrous*, *Mycobacterium pellegrino* et les genres *Mycobacterium* et *Nocardia*. Ann. Inst. Pasteur (Lille) *XXII:* 121–135.

Tadayon, R.A. and K.K. Carroll. 1971. Effects of growth condition on the fatty acid composition of *Listeria monocytogenes* and comparison with the fatty acids of *Erysipelothrix* and *Corynebacterium*. Lipids *6:* 820–825.

Tagg, J.R. and L.V. Bannister. 1979. "Fingerprinting" beta-haemolytic streptococci by their production of and sensitivity to bacteriocine-like inhibitors. J. Med. Microbiol. *12:* 397–411.

Tagg, J.R., A.S. Dajani, L.W. Wannamaker and E.D. Gray. 1973. Group A streptococcal bacteriocin. Production, purification and mode of action. J. Exp. Med. *138:* 1168–1183.

Tagg, J.R., A.S. Dajani and L.W. Wannamaker. 1975. Bacteriocin of a group B streptococcus: partial purification and characterization. Antimicrob. Agents Chemother. *7:* 764–772.

Tagg, J.R., A.S. Dajani and L.W. Wannamaker. 1976. Bacteriocins of Gram-

positive bacteria. Bacteriol. Rev. *40:* 722–756.

Taichman, N.S., B.F. Hammond, Chi-Cheng Tsai, P.C. Baehni and W.P. McArthur. 1978. Interaction of inflammatory cells and oral microorganisms. VII. *In vitro* polymorphonuclear responses to viable bacteria and to subcellular components of avirulent and virulent strains of *Actinomyces viscosus*. Infect. Immun. *21:* 594–604.

Takahashi, W., H. Yamagata, K. Yamaguchi, N. Tsukagoshi and S. Udake. 1983. Genetic transformation of *Bacillus brevis* 47, a protein secreting bacterium, by plasmid DNA. J. Bacteriol. *156:* 1130–1134.

Takai, K., C. Kurashina, Suzuki-Hovi, H. Okamoto and O. Hayaishi. 1974. Adenylate cyclase from *Brevibacterium liquefaciens*. I. Purification, crystalization and some properties. J. Biol. Chem. *249:* 1965–1972.

Takamiya, A. and K. Tubaki. 1956. A new form of *Streptomyces* capable of growing autotrophically. Arch. Mikrobiol. *25:* 58–64.

Takayama, K., K. Udagawa and S. Abe. 1960. Studies on the lytic enzyme produced by *Brevibacterium*. Part. 1. Production of the lytic substance. J. Agric. Chem. Soc. Jpn. *34:* 652–656.

Takeya, K. and H. Tokiwa. 1972. Mycobacteriocin classification of rapidly growing mycobacteria. Int. J. Syst. Bacteriol. *22:* 178–180.

Takumi, K. and T. Kawata. 1974. Isolation of a common cell wall antigen from the proteolytic strains of *Clostridium botulinum*. Jpn. J. Microbiol. *18:* 85–90.

Takumi, K., T. Kinouchi and T. Kawata. 1980. Isolation of two inducible bacteriophages from *Clostridium botulinum* type A 190L. FEMS Microbiol. Lett. *9:* 23–27.

Takumi, K., A. Takeoka and T. Kawata. 1983. Purification and characterization of a wall protein antigen from *Clostridium botulinum* type A. Infect. Immun. *39:* 1346–1353.

Tally, F.P., A.Y. Armfield, V.R. Dowell, Jr., Y.-Y. Kwok, V.L. Sutter and S.M. Finegold. 1974. Susceptibility of *Clostridium ramosum* to antimicrobial agents. Antimicrob. Agents Chemother. *5:* 589–593.

Tally, F.P., B.R. Goldin, N.V. Jacobus and S.L. Gorbach. 1977. Superoxide dismutase in anaerobic bacteria of clinical significance. Infect. Immun. *16:* 20–25.

Tally, F.P., B.R. Goldin, N. Sullivan, J. Johnston and S.L. Gorbach. 1978. Antimicrobial activity of metronidazole in anaerobic bacteria. Antimicrob. Agents Chemother. *13:* 460–465.

Tamura, G. 1956. Hiochic acid, a new growth factor for *Lactobacillus homohiochii* and *Lactobacillus heterohiochii*. J. Gen. Appl. Microbiol. *2:* 431–434.

Tanaka, Y., M. Yoh, Y. Takeda and T. Miwatani. 1979. Induction of mutation in *Escherichia coli* by freeze-drying. Appl. Environ. Microbiol. *37:* 369–372.

Tandom, S.M. and K.G. Gollakota. 1971. Chemically defined liquid medium for the growth and sporulation of *Bacillus stearothermophilus* 1518 rough variant. J. Food Sci. *36:* 542–543.

Tanner, R.S., E. Stackebrandt, G.E. Fox, R. Gupta, L.J. Magrum and C.R. Woese. 1982. A phylogenetic analysis of anaerobic eubacteria capable of synthesizing acetate from carbon dioxide. Curr. Microbiol. *7:* 127–132.

Tanner, R.S., E. Stackebrandt, G.E. Fox and C.R. Woese. 1981. A phylogenetic analysis of *Acetobacterium woodii*, *Clostridium barkeri*, *Clostridium butyricum*, *Clostridium lituseburense*, *Eubacterium limosum* and *Eubacterium tenue*. Curr. Microbiol. *5:* 35–38.

Tanner, R.S., R.S. Wolfe and L.G. Ljungdahl. 1978. Tetrahydrofolate enzyme levels in *Acetobacterium woodii* and their implication in the synthesis of acetate from CO2. J. Bacteriol. *134:* 668–670.

Tanzer, J.M. (Editor). 1981. Animal Models in Cariology. Special Supplement, Microbiology Abstracts, Information Retrieval, Washington D.C. and London.

Tanzer, J.M., M.L. Freedman, F.N. Woodiel, R.L. Eifert and L.A. Rinehimer. 1976. Association of *Streptococcus mutans* virulence with synthesis of intracellular polysaccharide. *In* Stiles, Loesche and O'Brien (Editors), Microbial Aspects of Dental Caries, Vol. 3. Information Retrieval, Washington, D.C. and London, pp. 597–616.

Taptykova, S.D. and L.V. Kalakoutskii. 1973. Low temperature cytochrome spectra of anaerobic actinomycetes. Int. J. Syst. Bacteriol. *23:* 468–471.

Tárnok, I. 1976. Metabolism in nocardiae and related bacteria. *In* Goodfellow, Brownell and Serrano (Editors), Biology of Nocardiae, Academic Press, New York, pp. 451–500.

Tarpay, M. 1978. Importance of antimicrobial susceptibility testing of *Streptococcus pneumoniae*. Antimicrob. Agents Chemother. *14:* 628–629.

Tate, R.L. and J.C. Ensign. 1974. A new species of *Arthrobacter* which degrades picolinic acid. Can. J. Microbiol. *20:* 691–694.

Taylor, M.M. 1972. *Eubacterium fissicatena* sp. nov. isolated from the alimentary tract of the goat. J. Gen. Microbiol. *71:* 457–463.

Taylor, N.S., G.M. Thorne and J.G. Bartlett. 1981. Comparison of two toxins produced by *Clostridium difficile*. Infect. Immun. *34:* 1036–1043.

Taylor, W.H., M.L. Taylor, W.E. Balch and P.S. Gilchrist. 1976. Purification and properties of dihydroorotase, a zinc-containing metalloenzyme in *Clostridium oroticum*. J. Bacteriol. *127:* 863–873.

Teather, R.M. 1982. Maintenance of laboratory strains of obligately anaerobic rumen bacteria. Appl. Environ. Microbiol. *44:* 499–501.

Telander, B. and G. Wallmark. 1975. *Micrococcus* subgroup 3 - a common cause of acute urinary tract infection in women. Lakartidningen *72:* 1967.

Tenbroeck, C. and J.H. Bauer. 1922. The tetanus bacillus as an intestinal saprophyte in man. J. Exp. Med. *36:* 261–271.

Tepper, B.S. and K.G. Varma. 1972. Metabolic activity of purified suspensions of *Mycobacterium lepraemurium*. J. Gen. Microbiol. *73:* 143–152.

Tešić, Z.P. and M.S. Todorović. 1952. Prilog ispitivani ju "Silicatnih Bakterija". Zemljiste i biljka (Belgrad) *1:* 3–18.

Tešić, Z.P. and M.S. Todorović. 1953. Sur la question de l'espece "bactéries silicatées". VI Congr. Internat., Microbiol. Rom, *3:* 149 (No. 895).

Tewfik, E.M. and S.G. Bradley. 1967. Characterization of deoxyribonucleic acids from streptomycetes and nocardiae. J. Bacteriol. *94:* 1994–2000.

Thadepalli, H., S.L. Gorbach, P.W. Broido, J. Norsen and L. Nyhus. 1973. Abdominal trauma, anaerobes and antibiotics. Surg. Gynecol. Obstet. *137:* 270–276.

Thadepalli, H. and B. Rao. 1979. *Actinomyces viscosus* infections of the chest in humans. Am. Rev. Resp. Dis. *120:* 203–206.

Thal, E. and I. Grabell. 1964. Rheumatic fever in pigs. Proc. XIV Scand. Congr. Pathol. Microbiol., Oslo, pp. 223–224.

Thalestam, M. 1976. Effects of *S. aureus* haemolysins on the plasma membrane of cultured mammalian cells. *In* Jeljaszewicz (Editor), Staphylococci and Staphylococcal Diseases, Gustav Fischer Verlag, Stuttgart, New York, pp. 679–690.

Thauer, R.K., K. Jungermann and K. Decker. 1977. Energy conservation in chemotrophic anaerobic bacteria. Bacteriol. Rev. *41:* 100–180.

Theilade, E. and M.N. Gilmor. 1961. An anaerobic oral filamentous microorganism. J. Bacteriol. *81:* 661–666.

Theodore, T.S. and A.L. Schade. 1965. Carbohydrate metabolism of iron-rich and iron-poor *Staphylococcus aureus*. J. Gen. Microbiol. *40:* 385–395.

Thiemann, J.E., C. Hengeller, A. Virgilio, O. Buelli and G. Licciardello. 1964. Rifamycin 33. Isolation of actinophages active on *Streptomyces mediterranei* and characteristics of phage-resistant strains. Appl. Microbiol. *12:* 261–268.

Thiemann, J.E., G. Zucco and G. Pelizza. 1969. A proposal for the transfer of *Streptomyces mediterranei* Margalith and Beratta 1960 to the genus *Nocardia* as *Nocardia mediterranea* (Margalith and Beretta) comb. nov. Arch. Mikrobiol. *67:* 147–155.

Thiercelin, E. 1902. Procédés faciles pour isoler l–térocoque préalable en anaérobie. C.R. Soc. Biol. Paris *54:* 1082–1083.

Thirkell, D. and E.M. Gray. 1974. Variation in the lipid fatty acid composition in purified membrane fractions from *Sarcina aurantiaca* in relation to growth phase. Antonie van Leeuwenhoek. J. Microbiol. Serol. *40:* 71–78.

Thirkell, D. and M.I.S. Hunter. 1969a. Carotenoid-glycoprotein of *Sarcina flava* membrane. J. Gen. Microbiol. *58:* 289–292.

Thirkell, D. and M.I.S. Hunter. 1969b. The polar carotenoid fraction from *Sarcina flava*. J. Gen. Microbiol. *58:* 293–299.

Thirkell, D. and R.H.C. Strang. 1967. Analysis and comparison of the carotenoids of *Sarcina flava* and *Sarcina lutea*. J. Gen. Microbiol. *49:* 53–57.

Thirkell, D. and M. Summerfield. 1977a. The effect of varying sea salt concentration in the growth medium on the chemical composition of a purified membrane fraction from *Planococcus citreus* Migula. Antonie van Leeuwenhoek. J. Microbiol. Serol. *43:* 37–42.

Thirkell, D. and M. Summerfield. 1977b. The membrane lipids of *Planococcus citreus* Migula from cells grown in the presence of three different concentrations of sea salt added to a basic medium. Antonie van Leeuwenhoek. J. Microbiol. Serol. *43:* 43–54.

Thirkell, D. and M. Summerfield. 1980. Variation in pigment production by *Planococcus citreus* Migula with cultural age and with sea salt concentration in the medium. Antonie van Leeuwenhoek. J. Microbiol. Serol. *46:* 51–57.

Thjøtta, T. and J. Bøe. 1938. *Neisseria haemolysans*. A hemolytic species of *Neisseria* Trevisan. Acta Pathol. Microbiol. Scand. Suppl. *37:* 527–531.

Thomas, P.J. 1972. Identification of some enteric bacteria which convert oleic acid to hydroxystearic acid in vitro. Gastroenterology *62:* 430–435.

Thomas, S.B., R.G. Druce, G.J. Peters and D.G. Griffiths. 1967. Incidence and significance of thermoduric bacteria in farm milk supplies: A reappraisal and review. J. Appl. Bacteriol. *30:* 265–298.

Thomas, T.D. 1976. Regulation of lactose fermentation in group N streptococci. Appl. Environ. Microbiol. *32:* 474–478.

Thomas, T.D., D.C. Ellwood and V.M.C. Longyear. 1979. Change from homo- to heterolactic fermentation by *Streptococcus lactis* resulting from glucose limitation in anaerobic chemostat cultures. J. Bacteriol. *138:* 109–117.

Thomas, T.D., B.D.W. Jarvis and N.A. Skipper. 1974. Localization of proteinase(s) near the cell surface of *Streptococcus lactis*. J. Bacteriol. *118:* 329–333.

Thompson, B.G., R. Anderson and R.G.E. Murray. 1980. Unusual polar lipids of *Micrococcus radiodurans* strain Sark. Can. J. Microbiol. *26:* 1408–1411.

Thompson, B.G. and R.G.E. Murray. 1981. Isolation and characterization of the plasma membrane and the outer membrane of *Deinococcus radiodurans* strain Sark. Can. J. Microbiol. *27:* 729–734.

Thompson, B.G. and R.G.E. Murray. 1982. The fenestrated peptidoglycan layer of *Deinococcus radiodurans*. Can. J. Microbiol. *28:* 522–525.

Thompson, B.G., R.G.E. Murray and J.F. Boyce. 1982. The association of the surface array and the outer membrane of *Deinococcus radiodurans*. Can. J. Microbiol. *28:* 1081–1088.

Thompson, L. 1933. Actinobacillosis of cattle in the United States. J. Infect. Dis. *52:* 223–229.

Thompson, L. 1950. Isolation and comparison of *Actinomyces* from human and bovine infections. Proc. Staff Meet. Mayo Clin. *25:* 81–86.

Thompson, L. and S.A. Lovestedt. 1951. An *Actinomyces*-like organism obtained from the human mouth. Proc. Staff Meet. Mayo Clin. *26:* 169–175.

Thompson, R.S. and R.E. Leadbetter. 1963. On the isolation of dipicolinic acid from endospores of *Sarcina ureae*. Arch. Mikrobiol. *45:* 27–32.

Thöni, J. and O. Alleman. 1910. Über das Vorkommen von gefärbten, makroskopischen Bakterienkolonien in Emmentalerkäsen. Zentralbl. Bakteriol. Parasitenkd. Infektionskr. Hyg. Abt. 2 *25:* 8–30.

Thore, M., L.G. Burman and S.E. Holm. 1982. *Streptococcus pneumoniae* and three species of anaerobic bacteria in experimental otitis media in guinea pigs. J. Infect. Dis. *145:* 822–828.

Thorel, M.F. and L. Valette. 1976. Etude de quelque souches de *M. paratuberculosis*: caracteres biochimiques et activite allergique. Ann. Rech. Vet. *7:* 207–213.

Thorley, C.M., M.E. Sharpe and M.P. Bryant. 1968. Modification of the rumen bacterial flora by feeding cattle ground and pelleted roughage as determined with culture media with and without rumen fluid. J. Dairy Sci. *51:* 1811–1816.

Thornley, M.J., R.W. Horne and A.M. Glauert. 1965. The fine structure of *Micrococcus radiodurans*. Arch. Mikrobiol. *51:* 267–289.

Tierno, P.M. and G. Stotzky. 1978. Serological typing of *Staphylococcus epidermidis* biotype 4. J. Infect. Dis. *137:* 514–523.

Tilbury, R.H. 1975. Occurrence and effects of lactic acid bacteria in the sugar industry. *In* Carr, Cutting and Whiting (Editors), Lactic acid bacteria in beverages and food. Academic Press, London, New York, San Francisco, pp. 177–191.

Tilford, P.E. 1936. Fasciation of sweet peas caused by *Phytomonas fascians* n. sp. J. Agr. Res. *53:* 383–394.

Tille, D., H. Prauser, K. Szyba and M. Mordarski. 1978. On the taxonomic position of *Nocardioides albus* Prauser by DNA/DNA-hybridization. Z. Allg. Mikrobiol. *18:* 459–462.

Tillet, W.S. and T. Francis, Jr. 1930. Serological reactions in pneumonia with a non-protein somatic fraction of *Pneumococcus*. J. Exp. Med. *52:* 561–571.

Timmis, K., G. Hobbs and R.C.W. Berkeley. 1974. Chitinolytic clostridia isolated from marine mud. Can. J. Microbiol. *20:* 1284–1285.

Tipper, D.J. and J.L. Strominger. 1966. Isolation of 4-0-β-N-acetylmuramyl-N-acetylglucosamine and 4-0-β-N-,6-0-diacetylmuramyl-N-acetylglucosamine and the structure of the cell wall polysaccharide of *Staphylococcus aureus*. Biochem. Biophys. Res. Commun. *22:* 48–56.

Tirunarayanan, M.O. 1968. Investigations on the enzymes and toxins of staphylococci. Study of phosphatase using p-nitrophenyl phosphate as substrate. Acta Pathol. Microbiol. Scand. *74:* 573–590.

Tirunarayanan, M.O. and H. Lundbeck. 1967. Investigations on the enzymes and toxins of staphylococci. Identification of the substrate and products formed in the "egg yolk reaction". Acta Pathol. Microbiol. Scand. *69:* 314–320.

Tisdall, P.A., G.D. Roberts and J.P. Anhalt. 1979. Identification of clinical isolates of mycobacteria with gas-liquid chromatography alone. J. Clin. Microbiol. *10:* 506–514.

Tissier, H. 1900. Recherches sur la flore intestinale normale et pathologique du nourisson. Thèse de Paris. pp. 1–253.

Tissier, H. 1908. Recherches sur la flore intestinale normale des enfants agés d'un an à cinq ans. Ann. Inst. Pasteur (Paris) *22:* 189–208.

Todd, E.W. and L.F. Hewitt. 1932. A new culture medium for the production of antigenic streptococcal haemolysin. J. Pathol. Bacteriol. *35:* 973–974.

Tomerska, H. and M. Wojciechowicz. 1973. Utilization of the intermediate products of the decomposition of pectin and of galacturonic acid by pure strains of rumen bacteria. Acta Microbiol. Pol. Ser. B *5:* 63–69.

Tomiyasu, I., S. Toriyama, I. Yano and M. Masui. 1981. Changes in molecular species composition of nocardmycolic acids in *Nocardia rubru* by the growth temperature. Chem. Phys. Lipids *28:* 41–54.

Tonomura, B., R. Malkin and J.C. Rabinowitz. 1965. Deoxyribonucleic acid base composition of clostridial species. J. Bacteriol. *89:* 1438–1439.

Tornabene, T.G., S.J. Morrison and W.E. Kloos. 1970. Aliphatic hydrocarbon contents of various members of the family *Micrococcaceae*. Lipids *5:* 929–937.

Torres-Pereira, A. 1961. Antigenic loss variation in *Staphylococcus aureus*. J. Pathol. Bacteriol. *81:* 151–156.

Torres-Pereira, A. 1962. Coagulase-negative strains of *Staphylococcus* possessing antigen 51 as agents of urinary tract infections. J. Clin. Pathol. *15:* 252–253.

Torrey, J.C. 1916. Bacteria associated with certain types of abnormal lymph glands. J. Med. Res. *34:* 65–80.

Totten, P.A., L. Vidal and J.N. Baldwin. 1981. Penicillin and tetracycline resistance plasmids in *Staphylococcus epidermidis*. Antimicrob. Agents Chemother. *20:* 359–365.

Touraille, C. and J.-L. Bergère. 1974. La germination de la spore de *Clostridium tyrobutyricum*. II. Démonstration de l'intervention de l'acétokinase et de la phosphotransacétylase par l'étude de leurs propriétés. Biochimie *56:* 404–422.

Trentini, W.C. 1978. Biology of the genus *Caryophanon*. Annu. Rev. Microbiol. *32:* 123–141.

Trentini, W.C. 1981. The genus *Caryophanon*. *In* Starr, Stolp, Trüper, Balows and Schlegel (Editors), The Prokaryotes. A Handbook on Habitats, Isolation and Identification of Bacteria. Springer-Verlag, Berlin, pp. 1701–1707.

Trentini, W.C. and H.E. Gilleland, Jr. 1974. Untrastructure of the cell envelope and septation process in *Caryophanon latum* as revealed by thin section and freeze-etching techniques. Can. J. Microbiol. *20:* 1435–1442.

Trentini, W.C. and C. Machen. 1973. Natural habitat of *Caryophanon latum*.

Can. J. Microbiol. *19:* 689–694.

Trentini, W.C. and R.G.E. Murray. 1975. Ultrastructural effects of lysozymes on the cell wall of *Caryophanon latum.* Can. J. Microbiol. *21:* 164–172.

Trevisan, V. 1885. Caratteri di alcuni nuovi generi di Batteriacee. Atti Accad. Fis-Med-Stat Milano (Ser 4) *3:* 92–107.

Trevisan, V. 1889. I generi e le specie delle Batteriacee. Zanaboni and Gabuzzi, Milano.

Trigalet, A. and B. Rat. 1976. Immunofluorescence as a tool for detecting the internally borne bacterial disease: *Corynebacterium michiganense* (E.F. Smith) Jensen and *Pseudomonas phaseolicola* (Burkholder) Dowson. Report 15th Int. Workshop Seed Pathol. 8–14 Sept. 1975. Paris, France.

Trivett, T.L. and E.A. Meyer. 1971. Citrate cycle and related metabolism of *Listeria monocytogenes.* J. Bacteriol. *107:* 770–779.

Trofimets, L.N. and Y.I Schneider. 1969. Two bacteriophages of *Corynebacterium sepedonicum* (Spieck. et Kott.) strain 51, found by electron microscope in the absence of indicator strains (in Russian). Biol. Nauki. (Mosc.) *12:* 96–100.

Troili-Petersson, G. 1903. Studien über die Mikroorganismen des schwedischen Güterkäses. Zentralbl. Bakteriol. II Abt. *11:* 120–143.

Trovatelli, L.D. and B. Biavati. 1980. Esigenze nutrizionali di alcune specie del genere *Bifidobacterium. In* Atti XVIII Congresso Nazionale della Società Italiana di Microbiologia. Fiuggi, giugno 1978. Lombardo Ed., Roma, pp. 330–333.

Trovatelli, L.D., F. Crociani, M. Pedinotti and V. Scardovi. 1974. *Bifidobacterium pullorum* sp. nov.: a new species isolated from chicken feces and a related group of bifidobacteria isolated from rabbit feces. Arch. Mikrobiol. *98:* 187–198.

Trovatelli, L.D. and D. Matteuzzi. 1976. On the presence of bifidobacteria in the rumen of calves fed different rations. Appl. Environ. Microbiol. *32:* 470–473.

Trudinger, P.A. 1970. A carbon monoxide-reacting pigment from *Desulfotomaculum nigrificans* and its possible relevance to sulfite reduction. J. Bacteriol. *104:* 158–170.

Tsai, C.-G. and G.A. Jones. 1975. Isolation and identification of rumen bacteria capable of anaerobic phloroglucinol degradation. Can. J. Microbiol. *21:* 794–801.

Tschape, H. 1973. Genetic studies on nutrient markers and their taxonomic importance. *In* Jeljaszewicz (Editor), Staphylococci and Staphylococcal Infections, Karger, Basel, pp. 57–62.

Tselenis-Kotspwilis, A.D., M.P. Koliomichalis and J.T. Papavassilou. 1982. Acute pyelonephritis caused by *Staphylococcus xylosus.* J. Clin. Microbiol. *16:* 593–594.

Tsenkovskii, L. 1878. Gel formation in sugar beet solutions. Proc. Soc. Sci. Nat. Imper. Univ. Kharkov *12:* 137–167.

Tsukamura, M. 1962. Differentiation of *Mycobacterium tuberculosis* from other mycobacteria by sodium salicylate susceptibility. Am. Rev. Resp. Dis. *86:* 81–83.

Tsukamura, M. 1965a. A thermoresistant mycobacterial strain thought to be a new species. Med. Biol. (Tokyo) *71:* 52–54.

Tsukamura, M. 1965b. A group of mycobacteria from soil resources resembling nonphotochromogens (Group 3). Med. Biol. (Tokyo) *71:* 110–113.

Tsukamura, M. 1965c. Differentiation of mycobacteria by picric acid tolerance. Am. Rev. Resp. Dis. *62:* 491–492.

Tsukamura, M. 1966a. *Mycobacterium parafortuitum*: a new species. J. Gen. Microbiol. *42:* 7–12.

Tsukamura, M. 1966b. Adansonian classification of mycobacteria. J. Gen. Microbiol. *45:* 253–273.

Tsukmura, M. 1966c. *Mycobacterium chitae*, a new species. A preliminary report (in Japanese). Med. Biol. (Tokyo) *73:* 203–205.

Tsukamura, M. 1966d. *Mycobacterium thermoresistible*, a new species. Preliminary report (in Japanese). Med. Biol. (Tokyo) *72:* 187–190.

Tsukamura, M. 1967a. Identification of mycobacteria. Tubercle *48:* 311–338.

Tsukamura, M. 1967b. Two types of slowly growing, nonphotochromogenic mycobacteria obtained from soil by the mouse passage method: *Mycobacterium terrae* and *Mycobacterium novum.* Jpn. J. Microbiol. *11:* 163–172.

Tsukamura, M. 1967c. *Mycobacterium chitae*: a new species. Jpn. J. Microbiol. *11:* 43–47.

Tsukamura, M. 1971. Differentiation between *Mycobacterium phlei* and *Mycobacterium thermoresistibile.* Am. Rev. Resp. Dis. *103:* 280–282.

Tsukamura, M. 1971. Proposal of a new genus, *Gordona*, for slightly acid-fast organisms occurring in sputa of patients with pulmonary disease and in soil. J. Gen. Microbiol. *68:* 15–26.

Tsukamura, M. 1972a. *Mycobacterium agri* Tsukamura sp. nov. A new relatively thermophilic *Mycobacterium* (in Japanese). Med. Biol. (Tokyo) *85:* 153–156.

Tsukamura, M. 1972b. A new species of rapidly growing scotochromogenic mycobacteria. *Mycobacterium neoaurum* Tsukamura sp. nov. (in Japanese). Med. Biol. (Tokyo) *85:* 229–233.

Tsukamura, M. 1972. Susceptibility of *Mycobacterium intracellulare* to rifampicin: A trial of ecological observation. Jpn. J. Microbiol. *16:* 444–446.

Tsukamura, M. 1973. A taxonomic study of some strains received as "*Mycobacterium*" rhodochrous. Description of *Gordona rhodochroa* (Zopf; Overbeck; Gordon et Mihn) Tsukamura comb. nov. Jpn. J. Microbiol. *17:* 189–197.

Tsukamura, M. 1973. New species of rapidly growing, scotochromogenic mycobacteria, *Mycobacterium chubuense* Tsukamura, *Mycobacterium aichinense*

Tsukamura, and *Mycobacterium tokaiense* Tsukamura (in Japanese). Med. Biol. (Tokyo) *86:* 13–17.

Tsukamura, M. 1974a. A further numerical taxonomic study of the rhodochrous group. Jpn. J. Microbiol. *18:* 37–44.

Tsukamura, M. 1974b. Differentiation of the *Mycobacterium rhodochrous* group from nocardiae by β-galactosidase activity. J. Gen. Microbiol. *80:* 553–555.

Tsukamura, M. 1975a. Numerical ananlysis of the relationship between *Mycobacterium*, rhodochrous group and *Nocardia* by use of hypothetical median organisms. Int. J. Syst. Bacteriol. *25:* 329–335.

Tsukamura, M. 1975b. Identification of Mycobacteria. The National Chubu Hospital, Obu, Aichi, Japan, pp. 1–75.

Tsukamura, M. 1977. Extended numerical taxonomy study of *Nocardia.* Int. J. Syst. Bacteriol. *27:* 311–323.

Tsukamura, M. 1978. Numerical classification of *Rhodococcus* (formerly *Gordona*) organisms recently isolated from sputa of patients: Description of *Rhodococcus sputi* Tsukamura sp. nov. Int. J. Syst. Bacteriol. *28:* 169–181.

Tsukamura, M. 1981a. A review of the methods of identification and differentiation of mycobacteria. Rev. Infect. Dis. *3:* 841–861.

Tsukamura, M. 1981b. Numerical analysis of rapidly growing, nonphotochromogenic mycobacteria, including *Mycobacterium agri* (Tsukamura 1972) Tsukamura sp. nov., nom. rev. Int. J. Syst. Bacteriol. *31:* 247–258.

Tsukamura, M. 1981a. Tests from susceptibility to mitomycin C as aids in differentiating the genus *Rhodococcus* from the genus *Nocardia* and for differentiating *Mycobacterium fortuitum* and *Mycobacterium chelonei* from other rapidly growing mycobacteria. Microbiol. Immunol. *25:* 1197–1199.

Tsukamura, M. 1981b. Differentiation between the genera *Mycobacterium*, *Rhodococcus* and *Nocardia* by susceptibility to 5-fluorouracil. J. Gen. Microbiol. *125:* 205–208.

Tsukamura, M. 1982a. Rejection of the name *Nocardia farcinica* Trevisan 1889 (Approved Lists 1980). Request for an opinion. Int. J. Syst. Bacteriol. *32:* 235–236.

Tsukamura, M. 1982a. *Mycobacterium shimoidei* sp. nov., nom. rev., a lung pathogen. Int. J. Syst. Bacteriol. *32:* 67–69.

Tsukamura, M. 1982b. Differentiation between the genera *Rhodococcus* and *Nocardia* and between species of the genus *Mycobacterium* by susceptibility to bleomycin. J. Gen. Microbiol. *128:* 2385–2388.

Tsukamura, M. 1982b. Rejection of the name *Nocardia farcinica* Trevisan 1889 (Approved Lists 1980) Request for an Opinion. Int. J. Syst. Bacteriol. *32:* 235–236.

Tsukamura, M. 1982c. Numerical analysis of the taxonomy of nocardiae and rhodococci. *In* Validation of the publication of new names and new combinations previously effectively published outside the IJSB. List No. 12. Int. J. Syst. Bacteriol. *33:* 896–897.

Tsukamura, M., S. Mijuno and H. Murata. 1975. Numerical taxonomy study of the taxonomic position of *Nocardia rubra* reclassified as *Gordona lentifragmenta* Tsukamura nom. nov. Int. J. Syst. Bacteriol. *25:* 377–382.

Tsukamura, M. and S. Mizuno. 1971. A new species *Gordona aurantiaca* occurring in sputa of patients with pulmonary disease. Kekkaku *46:* 93–98.

Tsukamura, M. and S. Mizuno. 1971. *Mycobacterium obuense*, a rapidly growing scotochromogenic *Mycobacterium* capable of forming a black product from p-aminosalicylate and salicylate. J. Gen. Microbiol. *68:* 129–134.

Tsukamura, M., S. Mizuno, N.F.F. Gane, A. Mills and L. King. 1971. *Mycobacterium rhodesiae* sp.nov. A new species of rapid-growing scotochromogenic mycobacteria. Jpn. J. Microbiol. *15:* 407–416.

Tsukamura, M., S. Mizuno and H. Toyama. 1983. *Mycobacterium pulveris* sp. nov., a nonphotochromogenic mycobacterium with an intermediate growth rate. Int. J. Syst. Bacteriol. *33:* 811–815.

Tsukamura, M., S. Mizuno and S. Tsukamura. 1981. Numerical analysis of rapidly growing, scotochromogenic mycobacteria, including *Mycobacterium obuense* sp. nov., nom. rev., *Mycobacterium rhodesiae* sp. nov., nom. rev., *Mycobacterium aichiense* sp. nov., nom. rev., *Mycobacterium chubuense* sp. nov., nom. rev., and *Mycobacterium tokaiense* sp. nov., nom. rev. Int. J. Syst. Bacteriol. *31:* 263–275.

Tsukamura, M., S. Mizuno, S. Tsukamura and J. Tsukamura. 1979. Comprehensive numerical classification of 369 strains of *Mycobacterium*, *Rhodococcus* and *Nocardia.* Int. J. Syst. Bacteriol. *29:* 110–129.

Tsukamura, M., H. Nemoto and H. Yugi. 1983. *Mycobacterium porcinum* sp. nov. a porcine pathogen. Int. J. Syst. Bacteriol. *33:* 162–165.

Tsukamura, M., H. Toyama and S. Mizuno. 1965. *Mycobacterium parafortuitum*, a new species. Med. Biol. (Tokyo) *70:* 232–235.

Tsukamura, M. and S. Tsukamura. 1966. *Mycobacterium aurum*, a new species. Med. Biol. (Tokyo) *72:* 270–273.

Tsukamura, M. and S. Tsukamura. 1968. Differentiation of mycobacteria by susceptibility to nitrite and propylene glycol. Am. Rev. Resp. Dis. *98:* 505–506.

Tsukamura, M., H.J. van der Meulen and W.O.K. Grabow. 1983. Numerical taxonomy of rapidly growing, scotochromogenic mycobacteria of the *Mycobacterium parafortuitum* complex: *Mycobacterium austroafricanum* sp. nov. and *Mycobacterium diernhoferi* sp. nov., nom. rev. Int. J. Syst. Bacteriol. *33:* 460–469.

Tsukamura, M. and I. Yano. 1985. *Rhodococcus sputi* sp. nov., nom. rev., and *Rhodococcus aurantiacus* sp. nov., nom. rev. Int. J. Syst. Bacteriol. *35:* 364–368.

Tsyganov, V.A., V.P. Namestnikova and N.V. Krassikova. 1966. DNA composition in various genera of the *Actinomycetales*. Mikrobiologiya *35:* 92–95 (In Russian).

Tu, K.K. and W.A. Palutke. 1976. Isolation and characterization of a catalase-negative strain of *Staphylococcus aureus*. J. Clin. Microbiol. *3:* 77–78.

Tuboly, S. 1965. Studies on the antigenic structure of mycobacteria. I. Comparison of the antigenic structure of pathogenic and saprophytic mycobacteria. Acta Microbiol. Acad. Sci. Hung *12:* 233–240.

Tunnicliff, R. 1933. Colony formation of *Diplococcus rubeoloe*. J. Infect. Dis. *52–53:* 39–53.

Turfitt, G.E. 1944. Microbiological agencies in the degradation of steriods. I. The cholesterol-decomposing organisms of soil. J. Bacteriol. *47:* 487–493.

Turnbull, P.C.B. 1981. *Bacillus cereus* toxins. Pharmacol. Ther. *13:* 453–505.

Turner, M. and D.I. Jervis. 1968. The distribution of pigmented *Bacillus* species in saltmarsh and other saline and nonsaline soils. Nova Hedwigia *56:* 293–298.

Twort, F.W. and G.L.Y. Ingram. 1913. A monograph on Johne's disease. Balliere, Tindall and Cox, London.

Tyski, S. 1981. Staphylococcal lipase: purification and some properties. *In* Jeljaszewicz (Editor), Staphylococci and Staphylococcal Infections, Gustav Fischer Verlag, Stuttgart, New York, pp. 335–342.

Uchida, K. and K. Aida. 1977. Acyl type of bacterial cell wall: its simple identification by colorimetric method. J. Gen. Appl. Microbiol. *23:* 249–260.

Uchida, K. and K. Aida. 1979. Taxonomic significance of cell-wall acyl type in *Corynebacterium-Mycobacterium-Nocardia* group by a glycolate test. J. Gen. Appl. Microbiol. *25:* 169–183.

Uchida, K. and K. Aida. 1984. An improved method for the glycolate test for simple identification of the acyl type of bacterial cell walls. J. Gen. Appl. Microbiol. *30:* 131–134.

Uchida, K. and K. Mogi. 1973. Cellular fatty acid spectra of *Sporolactobacillus* and some other *Bacillus-Lactobacillus* intermediates as a guide to their taxonomy. J. Gen. Appl. Microbiol. *19:* 129–140.

Uchida, T., L. Bonen, H.W. Schaup, B.J. Lewis, L. Zablen and C.R. Woese. 1974. The use of ribonuclease U_2 in RNA sequence determination: some corrections in the catalog of oligomers produced by ribonuclease T1 digestion of *Escherichia coli* 16s ribosomal RNA. J. Mol. Evol. *3:* 63–77.

Uchino, F. and S. Doi. 1967. Acido-thermophilic bacteria from thermal waters. Agric. Biol. Chem. *31:* 817–822.

Udey, L.R., R. Young and B. Sallman. 1977. Isolation and characterization of an anaerobic bacterium, *Eubacterium terantellus* sp. nov., associated with striped mullet (*Mugil cephalus*) mortality in Biscayne Bay, Florida. J. Fish. Res. Board Can. *34:* 402–409.

Uemura, T. and R. Skjelkvalé. 1976. An enterotoxin produced by *Clostridium perfringens* type D. Purification by affinity chromatography. Acta Pathol. Microbiol. Scand. Sect. B *84:* 414–420.

Ueno, K., H. Fujii, T. Marui, J. Takahashi, T. Sugitani, T. Ushijima and S. Suzuki. 1970. Acid phosphatase in *Clostridium perfringens*. Jpn. J. Microbiol. *14:* 171–173.

Ueno, Y., K. Nakayama, K. Ishii, F. Tashiro, Y. Minoda, T. Omori and K. Komagata. 1983. Metabolism of T-2 toxin in *Curtobacterium* sp. strain 114-2. Appl. Environ. Microbiol. *46:* 120–127.

Ullman, J.S. and B.J. McCarthy. 1973. The relationship between mismatched base pairs and the thermal stability of DNA duplexes. II. Effects of deamination of cytosine. Biochim. Biophys. Acta *294:* 416–424.

Ullmann, W.W. and J.A. Cameron. 1969. Immunochemistry of the cell walls of *Listeria monocytogenes*. J. Bacteriol. *98:* 486–493.

Umezawa, K., K. Takai, S. Tsuji, Y. Kwashina, and O. Hayaishi. 1974. Adenulate cyclase from *Brevibacterium liquefaciens*. III. *In situ* regulation of adenylate cyclase by pyruvate. Proc. Natl. Acad. Sci. *71:* 4598–4601.

Underdahl, N.R., O.D. Grace and M.J. Twiehaus. 1965. Porcine exudative epidermitis: characterization of bacterial agent. Am. J. Vet. Res. *26:* 617–624.

Ungers, G.E. and J.J. Cooney. 1968. Isolation and characterization of cartenoid pigments of *Micrococcus roseus*. J. Bacteriol. *96:* 234–241.

Upreti, G.C. and R.D. Hinsdill. 1973. Isolation and characterization of a bacteriocin from a homofermentative *Lactobacillus*. Antimicrob. Agents Chemother. *7:* 487–494.

Upreti, G.C. and R.D. Hinsdill. 1975. Production and mode of action of lactocin 27: bacteriocin from a homofermentative *Lactobacillus*. Antimicrob. Agents Chemother. *7:* 139–145.

Valdivia, J.A., R. Suarez Mendez and E. Echemendia Font. 1971. *Mycobacterium habana*: Probable nueva especie dentro dellas micobacterias no classificadas. Bol. Hig. Epidemiol. *9:* 65–73.

Valentine, R.C. 1964. Bacterial ferredoxin. Bacteriol. Rev. *28:* 497–517.

Valenton, M.J., R.F. Brubaker and H.F. Allen. 1973. *Staphylococcus epidermidis* (*albus*) endophthalmitis. Arch. Opthalmol. *89:* 94.

Valisena, S., L. Radin, P.E. Valraldo, R. Fontana and G. Satta. 1981. Purification and properties of the lytic enzymes excreted by staphylococci of different lyogroups (or species). *In* Jeljaszewicz (Editor), Staphylococci and Staphylococcal Infections, Gustav Fischer Verlag, Stuttgart, New York, pp. 397–401.

Valisena, S., P.E. Varaldo and G. Satta. 1982. Purification and characterization of three separate bacteriolytic enzymes excreted by *Staphylococcus aureus*, *Staphylococcus simulans*, and *Staphylococcus saprophyticus*. J. Bacteriol. *151:* 636–647.

Van Beynum, J. and J. W. Pette. 1935. Zuckervergärend und Laktat vergärende Buttersäurebakterien. Zentralbl. Bakteriol. Parasitenkd. Infektionskr. Hyg. Abt. II *93:* 198–212.

van Bijsterveld, O.P. and R.D. Richards. 1965. *Bacillus* infections of the cornea. Arch. Ophthalmol. *74:* 91–95.

van de Merwe, J.P. 1981. Agglutination of *Eubacterium* and *Peptostreptococcus* species as a diagnostic test for Crohn's disease. Hepato-gastroenterology. *28:* 155–156.

Van de Rijn, I. and A. Bouvet. 1984. Characterization of a pH-dependent chromophore from nutritionally variant streptococci. Infect. Immun. *43:* 28–31.

Van der Wiel-Korstanje, J.A.A. and W. De Vries. 1973. Cytochrome synthesis by *Bifidobacterium* during growth in media supplemented with blood. J. Gen. Microbiol. *75:* 417–419.

van Ermengem, E. 1896. Untersuchungen über Fälle von Fleischvergiftung mit Symptomen von Botulismus. Zentralbl. Bakteriol. Parasitenkd. Infektionskr. Hyg. I Abt. Orig. *19:* 442–444.

van Gylswyk, N.O. 1980. *Fusobacterium polysaccharolyticum* sp. nov., a gram-negative rod from the rumen that produces butyrate and ferments cellulose and starch. J. Gen. Microbiol. *116:* 157–163.

van Gylswyk, N.O. 1981. Validation List No. 7. Int. J. Syst. Bacteriol. *31:* 382–383.

van Gylswyk, N.O. and J.P.L. Hoffman. 1970. Characterization of cellulolytic cillobacteria from the rumens of sheep fed teff (*Eragrostic tef*) hay diets. J. Gen. Microbiol. *60:* 381–386.

van Gylswyk, N.O., E.J. Morris and H.J. Els. 1980. Sporulation and cell wall structure of *Clostridium polysaccharolyticum* comb. nov. (formerly *Fusobacterium polysaccharolyticum*). J. Gen. Microbiol. *121:* 491–493.

van Gylswyk, N.O., E.J. Morris and H.J. Els. 1983. Validation List No. 10. Int. J. Syst. Bacteriol. *33:* 438–440.

van Gylswyk, N.O. and E.E.G. Roché. 1970. Characteristics of ruminococcus and cellulolytic butyrivibrio species from the rumens of sheep fed differently supplemented teff (*Eragrostis tef*) hay diets. J. Gen. Microbiol. *64:* 11–17.

Van Houte, J., C. de Moor and H. Jansen. 1970. Synthesis of iodophilic polysaccharide by human oral streptococci. Arch. Oral Biol. *15:* 263–266.

Van Houte, J., H.V. Jordan and S. Bellack. 1971. Proportions of *Streptococcus sanguis*, an organism associated with subacute bacterial endocarditis, in human faeces and dental plaque. Infect. Immun. *4:* 658–659.

Van Kerken, A.E. and O. Kandler. 1966. Die Laktobazillenflora des Tilsiterkäses. Milchwissenschaft *21:* 436–440.

van Niel, C.B. 1928. The propionic acid bacteria. J.W. Boissevain & Co. Haarlem, The Netherlands.

van Niel, C.B. 1957. Genus *Propionibacterium*. *In* Breed, Murray and Smith (Editors), Bergey's Manual of Determinative Bacteriology, 7th ed. Williams and Wilkins Co., Baltimore, pp. 569–576.

Van Reenen, J.F. and M.M. Coogan. 1970. Clostridia isolated from human mouths. Arch. Oral Biol. *15:* 845–848.

van Tieghem, P. 1878. Sur la gomme du sucerie (*Leuconostoc mesenteroides*). Ann. Sci. Nat. Bot. *7:* 180–203.

Varaldo, P.E. and G. Satta. 1978. Grouping of staphylococci on the basis of their bacteriolytic-activity patterns: a new approach to the taxonomy of *Micrococcaceae*. II: Main characters of 1054 strains subdivided into "lyogroups". Int. J. Syst. Bacteriol. *28:* 148–153.

Varaldo, P.E., G. Satta and V. Hajek. 1978. Taxonomic study of coagulase-positive staphylococci: bacteriolytic activity pattern analysis. Int. J. Syst. Bacteriol. *28:* 445–448.

Varel, V.H., M.P. Bryant, L.V. Holdeman and W.E.C. Moore. 1974. Isolation of ureolytic *Peptostreptococcus productus* from feces using defined medium; failure of common urease tests. Appl. Microbiol. *28:* 594–599.

Vaughan, D.H., W.S. Riggsby and J.O. Mundt. 1979. Deoxyribonucleic acid relatedness of strains of yellow-pigmented, group D streptococci. Int. J. Syst. Bacteriol. *29:* 204–212.

Vaught, R.M. and A.S. Bleweis. 1974. Antigens of *Streptococcus mutans* II. Characterization of an antigen resembling a glycerol teichoic acid in walls of strain BHT. Infect. Immun. *9:* 60–67.

Vedamuthu, E.R., C.J. Washam and G.W. Reinbold. 1971. Isolation of inhibitory factor in raw milk whey active against propionibacteria. Appl. Microbiol. *22:* 552–556.

Vedder, A. 1934. *Bacillus alcalophilus* n. sp.; benevens enkele ervaringen met sterk alcalische voedingsbodems. Antonie van Leeuwenhoek. J. Microbiol. Serol. *1:* 141–147.

Veerkamp, J.H. 1969a. Uptake and metabolism of derivatives of 2-deoxy-2-amino-D-glucose in *Bifidobacterium bifidum* var. *pennsylvanicus*. Arch. Biochem. Biophys. *129:* 248–256.

Veerkamp, J.H. 1969b. Catabolism of glucose and derivatives of 2-deoxy-2-amino-glucose in *Bifidobacterium bifidum* var. *pennsylvanicus*. Arch. Biochem. Biophys. *129:* 257–263.

Veerkamp, J.H. 1977a. Effects of growth conditions on the lipid composition of *Bifidobacterium bifidum* subsp. *pennsylvanicum*. Antonie van Leeuwenhoek. J. Microbiol. Serol. *43:* 101–110.

Veerkamp, J.H. 1977b. Effects of growth conditions on the ion composition of *Bifidobacterium bifidum* subsp. *pennsylvanicum*. Antonie van Leeuwenhoek. J. Microbiol. Serol. *43:* 111–124.

Veillon, A. and A. Zuber. 1898. Recherches sur quelques microbes strictement anaérobies et leur role en pathologie. Arch. Med. Exp. Anat. Pathol. *10:* 517–545.

Vela, G.R. 1974. Survival of *Azotobacter* in soil. Appl. Microbiol. *28:* 77–79.

Verhoef, J., C.P.A. Van Boven and K.C. Winkler. 1971. Characters of phages from coagulase-negative staphylococci. J. Med. Microbiol. *4:* 405–424.

Verhoef, J., C.P.A. Van Boven and K.C. Winkler. 1972. Phage typing of coagulase-negative staphylococci. J. Med. Microbiol. *5:* 9–19.

Verhoeven, W. 1952. Aerobic sporeforming nitrate reducing bacteria. Ph.D. Thesis, Delft.

Veron, M. and R. Chatelain. 1960. Étude comparative de milieux complexes favorisant les fermentations glucidiques des *Pseudomonas*. Ann. Inst. Past. (Paris) *99:* 253–263.

Vertiev, Y.V., Y.V. Ezepchuk, A. Souckova, and A. Soucek. 1981a. Proof for the lack of diphtheria toxin in culture filtrates of *Corynebacterium uclerans* strain ATCC 9015. Zentralbl. Bakteriol. Mikrobiol. Hyg. I. Abt. Orig. A *249:* 520–526.

Vertiev, Y.V., Y.V. Ezepchuk, A. Souckova, and A. Soucek. 1981b. Purification and some properties of exotoxin from *Corynebacterium ulcerans* strain ATCC 9015. Zentralbl. Bakteriol. Mikrobiol. Hyg. I. Abt. Orig. A *249:* 527–537.

Vescovo, M., V. Bottazzi, P.G. Sarra and F. Dellaglio. 1981. Evidence of plasmid deoxyribonucleic acid in *Lactobacillus*. Microbiologia *4:* 413–419.

Vescovo, M., F. Dellaglio, V. Bottazzi and P.G. Sarra. 1979. Deoxyribonucleic acid homology among *Lactobacillus* species of the subgenus *Betabacterium* Orla-Jensen. Microbiologia *2:* 317–330.

Vescovo, M., L. Morelli and V. Bottazzi. 1982. Drug resistance plasmids in *Lactobacillus acidophilus* and *Lactobacillus reuteri*. Appl. Environ. Microbiol. *43:* 50–56.

Vesterberg, K. and O. Vesterberg. 1972. Studies on staphylokinase. J. Med. Microbiol. *5:* 441–450.

Vickerstaff, J.M. and B.C. Cole. 1969. Characterization of *Haemophilus vaginalis, Corynebacterium cervicis*, and related bacteria. Can. J. Microbiol. *15:* 587–594.

Vidaver, A.K. 1977. Maintenance of viability and virulence during preservation of *Corynebacterium nebraskense*. Phytopathology *67:* 825–827.

Vidaver, A.K. 1980. *Corynebacterium. In* Schaad (Editor), Laboratory guide for identification of plant pathogenic bacteria. American Phytopathological Society: St. Paul, p. 12–16.

Vidaver, A.K. 1982. The plant pathogenic corynebacteria. Annu. Rev. Microbiol. *36:* 495–517.

Vidaver, A.K. and M. Mandel. 1974. *Corynebacterium nebraskense*, a new, orange-pigmented phytopathogenic species. Int. J. Syst. Bacteriol. *24:* 482–485.

Vidaver, A.K. and M.P. Starr. 1981. Phytopathogenic coryneform and related bacteria. *In* Starr, Stolp, Trüper, Balows and Schlegel (Editors), The Prokaryotes. A handbook on habitats, isolation and identification of bacteria. Springer, New York, pp. 1879–1887.

Viljoen, J.A., E.B. Fred and W.H. Peterson. 1926. The fermentation of cellulose by thermophilic bacteria. J. Agric. Sci. Camb. *16:* 1–17.

Viscidi, R., S. Willey and J.G. Bartlett. 1981. Isolation rates and toxigenic potential of *Clostridium difficile* isolates from various patient populations. Gastroenterology *81:* 5–9.

Vogels, G.D. and C. Van der Drift. 1976. Degradation of purines and pyrimidines by microorganisms. Bacteriol. Rev. *40:* 403–468.

vonBesser, L. 1889. Über die Bakterien der normalen Luftwege. Beitr. Pathol. Anat. *6:* 331–372.

von Freudenreich, E. and S. Orla-Jensen. 1906. Über die in Emmentalerkäse stattfindende Propionsäure-gärung. Zentralbl. Bakteriol. Parasitenkd. Infektionskr. Hyg. Abt. 2 *17:* 529–546.

von Graevenitz, A. and U. Berger. 1980. A plea for linguistic accuracy. Int. J. Syst. Bacteriol. *30:* 520.

von Hugo, H., S. Schoberth, V.K. Madan and G. Gottschalk. 1972. Coenzyme specificity of dehydrogenases and fermentation of pyruvate by clostridia. Arch. Mikrobiol. *87:* 189–202.

von Nicolai, H., U. Hoeffler and F. Zilliken. 1980. Isolation, purification and properties of neuraminidase from *Propionibacterium acnes*. Zentralbl. Bakteriol. Parasitenkd. Infektionskr. Hyg. I. Abt. Orig. A *247:* 84–89.

Von Rheinbaben, K.E. and R.M. Hadlok. 1981. Rapid distinction between micrococci and staphylococci with furazolidone agars. Antonie van Leeuwenhoek. J. Microbiol. Serol. *47:* 41–51.

Vorobjeva, I.P., I.A. Khmel and I. Alföldi. 1980. Transformation of *Bacillus megaterium* protoplasts by plasmid DNA. FEMS Microbiol. Lett. *7:* 216–263.

Voss, J.G. 1970. Differentiation of two groups of *Corynebacterium acnes*. J. Bacteriol. *101:* 392–397.

Vuillemin, P. 1931. Les champignons parasites et les mycoses de l'homme. Encyclopédie Mycologique II. Paul Le Chevalier and Sons, Paris, pp. 1–290.

Waber, L.J. and H.G. Wood. 1979. Mechanism of acetate synthesis from CO_2 by *Clostridium acidiurici*. J. Bacteriol. *140:* 468–478.

Wachsman, J.T. and H.A. Barker. 1954. Characterization of an orotic acid fermenting bacterium, *Zymobacterium oroticum*, nov. gen., nov. spec. J. Bacteriol. *68:* 400–404.

Wada, H., H. Okada, H. Suginaka, I. Tomiyasu and I. Yano. 1981. Gas chromatographic and mass spectrometric analysis of molecular species of bacteri-onemamycolic acids from *Bacterionema matruchotii*. FEMS Microbiol. Lett. *11:* 187–192.

Wadström, T. 1967. Studies on extracellular proteins from *Staphylococcus aureus*. II. Separation of deoxyribonucleases by isoelectric focusing. Purification and properties of the enzymes. Biochem. Biophys. Acta *147:* 441–452.

Wadström, T., M. Kjellgren and A. Ljungh. 1976. Extracellular proteins from different *Staphylococcus* species: a preliminary study. *In* Jeljaszewicz (Editor), Staphylococci and Staphylococcal Diseases, Gustav Fischer Verlag, Stuttgart, New York, pp. 623–634.

Wadström, T., M. Thelestam and R. Möllby. 1974. Biological properties of extracellular proteins from *Staphylococcus*. Ann. N.Y. Acad. Sci. *236:* 343–361.

Wadström, T. and O. Vesterberg. 1971. Studies on endo-β-N-acetylglucosamini-dase, staphylolytic peptidase and N-acetylmuramyl-L-alanine amidase in lysostaphin and from *Staphylococcus aureus*. Acta Pathol. Microbiol. Scand. Sect. B *79:* 248–264.

Wagner, B. and G. Schallehn. 1982. Zur Koloniemorphologie von *Clostridium perfringens* type A. Zentralbl. Bakteriol. Mikrobiol. Hyg. I Orig. A *251:* 537–544.

Wagner, R. and J.R. Andreesen. 1977. Differentiation between *Clostridium aci-diurici* and *Clostridium cylindrosporum* on the basis of specific metal requirements for formate dehydrogenase formation. Arch. Microbiol. *114:* 219–224.

Wagner, R. and J.R. Andreesen. 1979. Selenium requirement for active xanthine dehydrogenase from *Clostridium acidiurici* and *Clostridium cylindrosporum*. Arch. Microbiol. *121:* 255–260.

Waites, W.M., C.E. Bayliss and N.R. Krieg. 1980. The effect of sporulation medium on spores of *Clostridium bifermentans*. J. Gen. Microbiol. *116:* 271–276.

Wakimoto, S., T. Uematsu and T. Mizukami. 1969. Bacterial canker disease of tomato in Japan. 2. Properties of bacteriophage specific for *Corynebacterium michiganense* (Smith) Jensen. Ann. Phytopathol. Soc. Jpn. *35:* 168–173.

Waksman, S.A. 1967. The Actinomycetes. Ronald Press, New York.

Waksman, S.A. and A.T. Henrici. 1948. Family II. *Actinomycetaceae. In* Breed, Murray and Hitchens (Editors), Bergey's Manual of Determinative Bacteriology, 6th ed. The Williams and Wilkins Co., Baltimore, pp. 892–928.

Walbaum, S., J. Biquet and P. Tran Van Ky. 1969. Structure antigenique de *Thermopolyspora polyspora*: repercussions pratiques sur le diagnostic du poumon du fermier. Ann. Inst. Pasteur *117:* 673–693.

Walbaum, S., T. Vaucelle and J. Biquet. 1973. Analyse de l'extrait de *Micropolyspora faeni* par immunoelectrophorese en double dimension. Localisation des activities chymotrypsique. Path. Biol. Paris *21:* 555–558.

Walden, W.C. and D.J. Hentges. 1975. Differential effects of oxygen and oxidation-reduction potential on the multiplication of three species of anaerobic intestinal bacteria. Appl. Microbiol. *30:* 781–785.

Walker, P.D. and J. Wolf. 1971. The taxonomy of *Bacillus stearothermophilus. In* Barker, Gould and Wolf (Editors), Spore Research 1971. Academic Press, London, pp. 247–262.

Walker, R.D., D.C. Richardson, M.J. Bryant and C.S. Draper. 1983. Anaerobic bacteria associated with osteomyelitis in domestic animals. J. Am. Vet. Med. Assoc. *182:* 814–816.

Wallace, R.J., Jr., J.M. Swenson, V.A. Silcox and R.C. Good. 1982. Disk diffusion testing with Polymyxin and Amikacin for differentiation of *Mycobacterium fortuitum* and *Mycobacterium chelonei*. J. Clin. Microbiol. *16:* 1003–1006.

Walter, M.R. (Editor). 1977. Life in the Precambrian. Precambrian Res. *5:* 105–219.

Walther, R., H. Hippe and G. Gottschalk. 1977. Citrate, a specific substrate for the isolation of *Clostridium sphenoides*. Appl. Environ. Microbiol. *33:* 955–962.

Wang, W.L.L., E.D. Everett, M. Johnson and E. Dean. 1977. Susceptibility of *Propionibacterium acnes* to seventeen antibiotics. Antimicrob. Agents Chemother. *11:* 171–173.

Wannamaker, L.W. 1962. Characterization of a fourth desoxyribonuclease of group A streptococci. Fed. Proc. *21:* 231.

Ward, E.S. and D.J. Ellar. 1983. Assignment of the delta-endotoxin gene of *Bacillus thuringiensis* var. *israeliensis* to a specific plasmid by curing analysis. FEBS Lett. *180:* 45–49.

Warner, T.N. and Ch.H. Miller. 1978. Cell-associated levan of *Actinomyces viscosus*. Infect. Immun. *19:* 711–719.

Warth, A.D. 1979. Exploding spores. Spore Newsletter *6:* 4–6.

Washburn, M.R., J.C. White and C.F. Niven, Jr. 1946. *Streptococcus* S.B.E.: immunological characteristics. J. Bacteriol. *51:* 723–729.

Watanuki, M. and K. Aida. 1972. Significance of quinones in the classification of bacteria. J. Gen. Appl. Microbiol. *18:* 469–472.

Watkins, J. and K.P. Sleath. 1981. Isolation and enumeration of *Listeria monocytogenes* from sewage, sewage sludge and river water. J. Appl. Bacteriol. *50:* 1–9.

Watson, B.K., R.C. Moellering, Jr. and L.J. Kunz. 1975. Identification of streptococci: use of lysozyme and *Streptomyces albus* filtrate in the preparation of extracts for Lancefield grouping. J. Clin. Microbiol. *1:* 274–278.

Watson, D.W. 1979. Characterization and pathobiological properties of group A streptococcal pyrogenic exotoxins. *In* Parker (Editor), Pathogenic streptococci, Reedbooks Ltd., Windsor, Berks, pp. 62–63.

Watt, B. 1973. The influence of carbon dioxide on the growth of obligate and facultative anaerobes on solid media. J. Med. Microbiol. 6: 307–314.

Wayne, L.G. 1959. Quantitative aspects of neutral red reactions of typical and "atypical" mycobacteria. Am. Rev. Tuberc. Pulm. Dis. 79: 526.

Wayne, L.G. 1961. Recognition of Mycobacterium fortuitum by means of a three-day phenolphthalein sulfatase test. Am. J. Clin. Pathol. 36: 185–187.

Wayne, L.G. 1962. Two varieties of Mycobacterium kansasii with different clinical significance. Am. Rev. Resp. Dis. 86: 651–656.

Wayne, L.G. 1966. Classification and identification of mycobacteria. III. Species within Group III. Am. Rev. Resp. Dis. 93: 919–928.

Wayne, L.G. 1970. On the identity of Mycobacterium gordonae Bojalil and the so-called tap water scotochromogens. Int. J. Syst. Bacteriol. 20: 149–153.

Wayne, L.G. 1971. Phenol-soluble antigens from Mycobacterium kansasii, Mycobacterium gastri and Mycobacterium marinum. Infect. Immunol. 3: 36–40.

Wayne, L.G. 1974. Simple pyrazinamidase and urease tests for routine identification of mycobacteria. Am. Rev. Resp. Dis. 109: 147–151.

Wayne, L.G. 1981. Numerical taxonomy and cooperative studies: Roles and limits. Rev. Infect. Dis. 3: 822–828.

Wayne, L.G. 1982. Actions of the Judicial Commission of the International Committee of Systematic Bacteriology on requests for opinions published between July 1979 and April 1981. Int. J. Syst. Bacteriol. 32: 464–465.

Wayne, L.G. 1982. Microbiology of tubercle bacilli. Am. Rev. Resp. Dis. (Suppl.) 125: 31–41.

Wayne, L.G., L. Andrade, S. Froman, W. Käppler, E. Kubala, G. Meissner and M. Tsukamura. 1978. A cooperative numerical analysis of Mycobacterium gastri, Mycobacterium kansasii and Mycobacterium marinum. J. Gen. Microbiol. 109: 319–327.

Wayne, L.G. and G.A. Diaz. 1979. Reciprocal immunological distances of catalase derived from strains of Mycobacterium avium, Mycobacterium tuberculosis, and closely related species. Int. J. Syst. Bacteriol. 29: 19–24.

Wayne, L.G. and G.A. Diaz. 1982. Serological, taxonomic and kinetic studies of the T and M classes of mycobacterial catalase. Int. J. Syst. Bacteriol. 32: 296–304.

Wayne, L.G., T.M. Dietz, C. Gernez-Rieux, P.A. Jenkins, W. Käppler, G.P. Kubica, J.B.G. Kwapinski, G. Meissner, S.R. Pattyn, E.H. Runyon, K.H. Schröder, V.A. Silcox, A. Tacquet, M. Tsukamura and E. Wolinsky. 1971. A cooperative numerical analysis of scotochromogenic slowly growing mycobacteria. J. Gen. Microbiol. 66: 255–271.

Wayne, L.G. and J.R. Doubek. 1968. Diagnostic key to mycobacteria encountered in clinical laboratories. Appl. Microbiol. 16: 925–931.

Wayne, L.G., J.R. Doubek and R. Russell. 1964. Classification and identification of mycobacteria. I. Tests employing tween 80 as substrate. Am. Rev. Resp. Dis. 90: 588–597.

Wayne, L.G., H.C. Engbaek, H.W.B. Engel, S. Froman, W. Gross, J. Hawkins, W. Käppler, A.G. Karlson, H.H. Kleeberg, I. Krasnow, G.P. Kubica, C. McDurmont, E.E. Nel, S.R. Pattyn, K.H. Schröder, S. Showalter, I. Tarnok, M. Tsukamura, B. Vergmann and E. Wolinsky. 1974. Highly reproducible techniques for use in systematic bacteriology in the genus Mycobacterium: Tests for pigment, urease, resistance to sodium chloride, hydrolysis of tween 80 and β-galactosidase. Int. J. Syst. Bacteriol. 24: 412–419.

Wayne, L.G., H.W.B. Engel, C. Grassi, W. Gross, J. Hawkins, P.A. Jenkins, W. Käppler, H.H. Kleeberg, I. Krasnow, E.E. Nel, S.R. Pattyn, P.A. Richards, S. Showalter, M. Slosarek, I. Szabo, I. Tarnok, M. Tsukamura, B. Vergmann and E. Wolinsky. 1976. Highly reproducible techniques for use in systematic bacteriology in the genus Mycobacterium: Tests for niacin and catalase and for resistance to isoniazid, thiophene 2-carboxylic acid hydrazide, hydroxylamine and p-nitrobenzoate. Int. J. Syst. Bacteriol. 26: 311–318.

Wayne, L.G., R.C. Good, M.I. Krichevsky, R.E. Beam, Z. Blacklock, S.D. Chaparas, D. Dawson, S. Froman, W. Gross, J. Hawkins, P.A. Jenkins, I. Juhlin, W. Käppler, H.H. Kleeberg, I. Krasnow, M.J. Lefford, E. Mankiewicz, C. McDurmont, G. Meissner, P. Morgan, E.E. Nel, S.R. Pattyn, R. Portaels, P.A. Richards, S. Rüsch, K.H. Schröder, V.A. Silcox, I. Szabo, M. Tsukamura and B. Vergmann. 1981. First report of the cooperative, open-ended study of slowly-growing Mycobacteria by the International Working Group on Mycobacterial Taxonomy. Int. J. Syst. Bacteriol. 31: 1–20.

Wayne, L.G., R.C. Good, M.I. Krichevsky, R.E. Beam, Z. Blacklock, H.L. David, D. Dawson, W. Gross, J. Hawkins, P.A. Jenkins, I. Juhlin, W. Käppler, H.H. Kleeberg, I. Krasnow, M.J. Lefford, E. Mankiewicz, C. McDurmont, E.E. Nel, F. Portaels, P.A. Richards, S. Rüsch, K.H. Schröder, V.A. Silcox, I. Szabo, M. Tsukamura, L. Vanden Breen and B. Vergmann. 1983. Second report of the cooperative, open-ended study of slowly growing Mycobacteria by the International Working Group on Mycobacterial Taxonomy. Int. J. Syst. Bacteriol. 33: 265–274.

Wayne, L.G. and W.M. Gross. 1968. Base composition of deoxyribonucleic acid isolated from mycobacteria. J. Bacteriol. 96: 1915–1919.

Wayne, L.G., W.J. Juarez and E.G. Nichols. 1958. Arylsulfatase activity of aerobic actinomycetales. J. Bacteriol. 75: 367–368.

Wayne, L.G., I. Krasnow and M. Huppert. 1957. Characterization of atypical mycobacteria and of nocardia species isolated from clinical specimens. I. Characterization by means of the microcolonial test. Am. Rev. Tuberc. Pulm. Dis. 76: 451–467.

Wayne, L.G., E.J. Krichevsky, L.L. Love, R. Johnson and M.I. Krichevsky. 1980. Taxonomic probability matrix for use with slowly growing mycobacteria. Int. J. Syst. Bacteriol. 30: 528–538.

Wayne, L.G. and K.-Y. Lin. 1982 Glyoxylate metabolism and adaptation of Mycobacterium tuberculosis to survival under anaerobic conditions. Infect. Immun. 37: 1042–1049.

Webster, G.F. and C.S. Cummins. 1978. Use of bacteriophage typing to distinguish Propionibacterium acnes type I and II. J. Clin. Microbiol. 7: 84–90.

Webster, G.F., J.J. Leyden, C.-C. Tsai, P. Baehni and W.P. McArthur. 1980. Polymorphonuclear leukocyte lysosomal release in response to Propionibacterium acnes in vitro and its enhancement by sera from inflammatory acne patients. J. Invest. Dermatol. 74: 398–401.

Webster, G.F. and W.P. McArthur. 1982. Activiation of components of the alternative pathway of complement by Propionibacterium acnes cell wall carbohydrate. J. Invest. Dermatol. 79: 137–140.

Wegienek, J. and C.A. Reddy. 1982. Taxonomic study of "Corynebacterium suis" Soltys and Spratling: proposal of Eubacterium suis (nov. rev.) comb. nov. Int. J. Syst. Bacteriol. 32: 218–228.

Weichselbaum, A. 1886. Ueber die Aetiologie der akuten Lungen-und Rippenfellentzündungen. Med. Jahrb. Wien 82: 483–554.

Weigmann, H. 1898. Ueber zwei an der Käsereifung beteiligte Bakterien. Zentralbl. Bakteriol. Parasitenkd. Infektionskr. Hyg. Abt. II 4: 820–834.

Weimer, P.J. and J.G. Zeikus. 1977. Fermentation of cellulose and cellobiose by Clostridium thermocellum in the absence and presence of Methanobacterium thermoautotrophicum. Appl. Environ. Microbiol. 33: 289–297.

Weinberg, M. and B. Ginsbourg. 1927. Données récentes sur les microbes anaérobies et leur role en pathologie. Masson et Cie, Paris, pp. 1–291.

Weinberg, M., R. Nativelle and A.R. Prévot. 1937. Les microbes anaérobies. Masson and Co., Paris, pp. 1–1186.

Weinberg, M. and P. Séguin. 1915a. Flore microbienne de la gangrène gazeuse. Le B. fallax. C.R. Seances Soc. Biol. Filiales 78: 686–689.

Weinberg, M. and P. Séguin. 1915b. Le B. oedematiens et la gangrène gazeuse. C.R. Soc. Biol. Paris 78: 507–512.

Weinberg, M. and P. Séguin. 1916. Contribution à l'étiologie de la gangrène gazeuse. C.R. Acad. Sci. Paris 163: 449–451.

Weinberg, M. and P. Séguin. 1918. La gangrène gazeuse-bactériologie, reproduction expérimentale, séreothérapie. Masson and Co., Paris, pp. 1–444.

Weinrich, A.E. and V.E. Del Bene. 1976. Beta-lactamase activity in anaerobic bacteria. Antimicrob. Agents Chemother. 10: 106–111.

Weinstein, L. and C.G. Colburn. 1950. Bacillus subtilis meningitis and bacteremia: report of a case and review of the literature on "subtilis" infections in man. Arch. Intern. Med. 86: 585–594.

Weiss, J.E. and L.F. Rettger. 1938. Taxonomic relationships of Lactobacillus bifidus, Bacillus bifidus (Tissier) and Bacteroides bifidus (Eggerth). J. Bacteriol. 35: 17–18.

Weiss, N. 1981. Cell wall structure of anaerobic cocci. Rev. Inst. Pasteur (Lyon) 14: 53–59.

Weiss, N., R. Plapp and O. Kandler. 1967. Die Aminosäuresequenz des DAP-des DAP-haltigen Mureins von Lactobacillus plantarum und Lactobacillus inulinus. Arch. Mikrobiol. 58: 313–323.

Weiss, N. and U. Schillinger. 1984. Lactobacillus sanfrancisco nov. spec., nom. rev. Syst. Appl. Microbiol. 5: 230–232.

Weiss, N. and U. Schillinger. 1984. In Validation of the publication of new names and new combinations previously effectively published outside the IJSB. List No. 16. Int. J. Syst. Bacteriol. 34: 503–504.

Weiss, N., U. Schillinger and O. Kandler. 1983a. Lactobacillus trichodes, and Lactobacillus heterohiochii, subjective synonym of Lactobacillus fructivorans. Syst. Appl. Microbiol. 4: 507–511.

Weiss, N., U. Schillinger and O. Kandler. 1983b. Lactobacillus lactis, Lactobacillus leichmannii and Lactobacillus bulgaricus, subjective synonyms of Lactobacillus delbrueckii subsp. lactis comb. nov. and Lactobacillus delbrueckii subsp. bulgaricus comb. nov. Syst. Appl. Microbiol. 4: 552–557.

Weiss, N., U. Schillinger and O. Kandler. 1984. In Validation of the publication of new names and new combinations previously published outside the IJSB. List No. 14. Int. J. Syst. Bacteriol. 34: 270–271.

Weiss, N., U. Schillinger, M. Laternser and O. Kandler. 1981. Lactobacillus sharpeae sp.nov. and Lactobacillus agilis sp. nov., two new species of homofermentative, meso-diaminopimelic acid-containing lactobacilli isolated from sewage. Zentralbl. Bakteriol. Mikrobiol. Hyg. I Abt. Orig. C 2: 242–253.

Weiss, N., U. Schillinger, M. Laternser and O. Kandler. 1982. In Validation of the publication of new names and new combinations previously effectively published outside the IJSB. List No. 8. Int. J. Syst. Bacteriol. 31: 266–268.

Weiss, N., K.H. Schleifer and O. Kandler. 1981. The peptidoglycan types of gram-positive anaerobic bacteria and their taxonomic implications. Rev. Inst. Pasteur (Lyon) 14: 3–12.

Weiszfeiler, J.G., V. Karasseva and E. Karczag. 1971. A new Mycobacterium species: Mycobacterium asiaticum n. sp. Acta Microbiol. Acad. Sci. Hung. 18: 247–252.

Weitzman, P.D.J. 1980. Citrate synthase and succinate thiokinase in classification and identification. In Goodfellow and Board (Editors), Microbiological Classification and Identification, Academic Press, London - New York, pp. 107–125.

Welborn, P.P., W.K. Hadley, E. Newbrun and D.M. Yajko. 1983. Characterization of strains of viridans streptococci by deoxyribonucleic acid hybridization and physiological tests. Int. J. Syst. Bacteriol. 33: 293–299.

Welby-Guisse, M., M.A. Lanéelle and J. Asselineau. 1970. Structure des acides corynomycoliques de *Corynebacterium hofmannii* et leur implication biogénétique. Eur. J. Biochem. *13:* 164–167.

Welch, W.H. 1981. Conditions underlying the infections of wounds. Am. J. Med. Sci. *102:* 439–465.

Weldendorp, J.W. 1963. The influence of living plants on denitrification. Meded. Landbouwhogesch. Wageningen *63:* 1–100.

Welker, N.E. and L.L. Campbell. 1967. Unrelatedness of *Bacillus amyloliquefaciens* and *Bacillus subtilis*. J. Bacteriol. *94:* 1124–1130..

Wellington, E.M. and S.T. Williams. 1978. Preservation of actinomycete inoculum in frozen glycerol. Microbios Lett. *6:* 151–157.

Wellington, E.M.H. and S.T. Williams. 1981. Host ranges of phage isolated to *Streptomyces* and other genera. Zentralbl. Bakteriol. Mikrobiol. Hyg. I Abt. Suppl. *11:* 93–98.

Wells, A.Q. 1937. Tuberculosis in wild voles. Lancet *232:* 1221.

Wells, A.Q., E. Aquis and N. Smith. 1955. *Mycobacterium fortuitum*. Am. Rev. Tuberc. *72:* 53–63.

Wells, C.L. and E. Balish. 1983. *Clostridium tetani* growth and toxin production in the intestines of germfree rats. Infect. Immun. *41:* 826–828.

Wells, C.L. and C.R. Field. 1976. Long-chain fatty acids of peptococci and peptostreptococci. J. Clin. Microbiol. *4:* 515–521.

Welshimer, H.J. 1968. Isolation of *Listeria monocytogenes* from vegetation. J. Bacteriol. *95:* 300–303.

Welshimer, H.J. 1981. The genus *Listeria* and related organisms. *In* Starr, Stolp, Trüper, Balows and Schlegel (Editors), The Prokaryotes. A Handbook on Habitats, Isolation and Identification of Bacteria. Springer, New York, pp. 1680–1687.

Welshimer, H.J. and A.L. Meredith. 1971. *Listeria murrayi* sp. n.: a nitratereducing mannitol-fermenting *Listeria*. Int. J. Syst. Bacteriol. *21:* 3–7.

Wendt, S., K. George, B. Parker, H. Gruft and J. Felkinham III. 1980. Epidemiology of infection by nontuberculosis mycobacteria. III. Isolation of potentially pathogenic mycobacteria from aerosols. Am. Rev. Resp. Dis. *122:* 259–263.

Wenner, J.J. and L.F. Rettger. 1919. A systematic study of the *Proteus* group of bacteria. J. Bacteriol. *4:* 331–353.

Werkman, C.H. and R.W. Brown. 1933. The propionic acid bacteria. II. Classification. J. Bacteriol. *26:* 393–417.

Werkman, C.H. and H.J. Weaver. 1927. Studies in the bacteriology of suphur stinker spoilage of canned sweet corn. Iowa State Coll. J. Sci. *2:* 57–67.

Werner, H. 1967. Untersuchungen über die Lipase und Lecithinase-Aktivität von Aeroben und Anaeroben Corynebacterium - und von Propionibacteriumarten. Zentralbl. Bakteriol. Parasitenkd. Infektionskr. Hyg. I Abt. Orig. *204:* 127–138.

Werner, H., M. Boehm, H. Kunstek-Santos, C. Lohner and T. Jordan. 1977. Susceptibility to erythromycin of anaerobes of the genera *Bacteroides, Fusobacterium, Sphaerophorus, Veillonella, Clostridium, Corynebacterium, Peptococcus,* and *Peptostreptococcus*. Arzneim.-Forsch. *27:* 2263–2265.

Werner, H. and G. Rintelen. 1973. Anaerobe gram-positive kokken (*Peptococcus variabilis, P. asaccharolyticus, P. prevotii* and *P. saccharolyticus*) aus pathologischem material. Zentralbl. Bakteriol. Parasitenkd. Infektionskr. Hyg. Abt. I Orig. A *223:* 496–503.

Werner, H., G. Rintelen and C. Lohner. 1973. Identifizierung und medizinische Bedeutung von 8 aus pathologischem Material isolierten *Clostridium*-Arten. Zentralbl. Bakteriol. Parasitenkd. Infektionskr. Hyg. I Abt. Orig. A *224:* 220–226.

Werner, W. 1933. Botanische Beschreibung häufinger am Buttersäureabbau beteiligter sporenbildender Bakterienspezies. Zentralbl. Bakteriol. Parasitenkd. Infektionskr. Hyg. Abt. II *87:* 446–475.

West, S.E.H. and L.V. Holdeman. 1973. Placement of the name *Peptococcus anaerobius* (Hamm) Douglas on the list of nomina rejicienda. Request for an opinion. Int. J. Syst. Bacteriol. *23:* 283–289.

Wetherell, J.R. and A.S. Bleiweis. 1975. Antigens of *Streptococcus mutans:* characterization of a polysaccharide antigen from walls of strain GS-5. Infect. Immun. *12:* 1341–1348.

Wetherell, J.R. and A.S. Bleiweis. 1978. Antigens of *Streptococcus mutans:* isolation of a serotype-specific and cross-reactive antigen from walls of strain V-100 (serotype *e*). Infect. Immun. *19:* 160–169.

Wetmur, J.G. 1976. Hybridization and renaturation kinetics of nucleic acids. Annu. Rev. Biophys. Bioeng. *5:* 337–361.

Wetmur, J.G. and N. Davidson. 1968. Kinetics of renaturation of DNA. J. Mol. Biol. *31:* 349–370.

Wetzler, T.F., N.R. Freeman, M.L.V. French, L.A. Renkowski, W.C. Eveland and O.J. Carver. 1968. Biological characterization of *Listeria monocytogenes*. Health Lab. Sci. *5:* 46–62.

Weyland, H. 1969. Actinomycetes in North Sea and Atlantic Ocean sediments. Nature *223:* 858.

Weyland, H. 1981. Distribution of actinomycetes on the sea floor. Zentralbl. Bakteriol. Mikrobiol. Hyg. I. Abt. Orig. Suppl. *11:* 185–193.

Wheeler, P.R. 1982. Metabolism of carbon sources by *Mycobacterium leprae:* a preliminary report. Ann. Microbiol. (Inst. Pasteur) *133B:* 141–146.

Wheeler, P.R., V.P. Bharadwaj and D. Gregory. 1982. N-acetyl-B-glucosaminidase, B-glucuronidase and acid phosphatase in *Mycobacterium leprae*. J. Gen. Microbiol. *128:* 1063–1071.

Wheeler, P.R. and D. Gregory. 1980. Superoxide dismutase, peroxidatic activity & catalase in *Mycobacterium leprae* purified from armadillo liver. J. Gen.

Microbiol. *121:* 457–464.

Wheeler, T.T. and W.B. Clark. 1980. Fibril-mediated adherence of *Actinomyces viscosus* to saliva-treated hydroxyapatite. Infect. Immun. *28:* 577–584.

Whiley, R.A., J.M. Hardie and P.J.H. Jackman. 1982. SDS-polyacrylamide gel electrophoresis of oral streptococci. *In* Holm and Christensen (Editors), Proceedings of the VIIIth International Symposium on Streptococci and Streptococcal Diseases, Reedbooks, Chertsey, pp. 61–62.

White, D.C. and F.E. Frerman. 1967. Extraction, characterization, and cellular localization of the lipids of *Staphylococcus aureus*. J. Bacteriol. *94:* 1854–1867.

White, D.C. and F.E. Frerman. 1968. Fatty acid composition of the complex lipids of *Staphylococcus aureus* during the formation of membrane-bound electron transport system. J. Bacteriol. *95:* 2198–2209.

White, G.F. 1906. The bacteria of the apiary, with special reference to bee diseases. Tech. Ser. Bur. Entomol. U.S. Dept. Agr. No. 14.

White, G.F. 1912. The cause of European foulbrood. U.S. Dept. Agric. Circular No. 157.

White, G.F. 1920. European foulbrood. U.S. Dept. Agric. Bull. No. 810.

White, H.R. 1963. The effect of variation in pH on the heat resistance of cultures of *Streptococcus faecalis*. J. Appl. Bacteriol. *26:* 91–99.

White, J.C. and C.F. Niven, Jr. 1946. *Streptococcus* S.B.E.: A streptococcus associated with subacute bacterial endocarditis. J. Bacteriol. *51:* 711–722.

White, P.B. 1921. The normal bacterial flora of the bee. J. Pathol. Bacteriol. *24:* 64–78.

White, P.J. 1972. The nutrition of *Bacillus megaterium* and *Bacillus cereus*. J. Gen. Microbiol. *71:* 505–514.

White, P.J. and J.K. Lotay. 1980. Minimal nutritional requirements of *Bacillus sphaericus* NCTC 9602 and 26 other strains of this species: the majority grow and sporulate with acetate as sole major source of carbon. J. Gen. Microbiol. *118:* 13–19.

White, T.G. and F.K. Mirikitani. 1976. Some biological and physical-chemical properties of *Erysipelothrix rhusiopathiae*. Cornell Vet. *66:* 152–163.

White, T.G. and R.D. Shuman. 1961. Fermentation reactions of *Erysipelothrix rhusiopathiae*. J. Bacteriol. *82:* 595–599.

Whitehouse, R.L.S. and H. Jackson. 1972. Description of *Cellulomonas acidula* isolated from milk. Antonie Van Leeuwenhoek J. Microbiol. Serol. *38:* 537–542.

Whiteley, H.R. 1952. The fermentation of purines by *Micrococcus aerogenes*. J. Bacteriol. *63:* 163–175.

Whiting, G.C. 1975. Some biochemical and flavour aspects of lactic acid bacteria in ciders and other alcoholic beverages. *In* Carr, Cutting and Whiting (Editors), Lactic acid bacteria in beverages and food. Academic Press, London, New York, San Francisco, pp. 69–85.

Whittaker, R.H. and L. Margulis. 1978. Protist classification and the kingdoms of organisms. BioSystems *10:* 3–18.

Whittenbury, R. 1964. Hydrogen peroxide formation and catalase activity in the lactic acid bacteria. J. Gen. Microbiol. *35:* 13–26.

Whittenbury, R. 1965. A study of some pediococci and their relationship to *Aerococcus viridans* and the enterococci. J. Gen. Microbiol. *40:* 97–106.

Whittenbury, R. 1965. The differentiation of *Streptococcus faecalis* and *S. faecium*. J. Gen. Microbiol. *38:* 279–287.

Whittenbury, R. 1978. Biochemical characteristics of *Streptococcus* species. *In* Skinner and Quesnel (Editors), Streptococci. Society for Applied Bacteriology Symposium Series No. 7. Academic Press, London, New York and San Francisco, pp. 51–69.

Wicken, A.J., A. Ayres, L.K. Campbell and K.W. Knox. 1983. Effect of growth conditions on production of rhamnose-containing cell wall and capsular polysaccharides by strains of *Lactobacillus casei* subsp. *rhamnosus*. J. Bacteriol. *153:* 84–92.

Wicken, A.J., K.W. Broady, J.D. Evans and K.W. Knox. 1978. New cellular and extracellular amphipathic antigen from *Actinomyces viscosus* NY1. Infect. Immun. *22:* 615–616.

Wicken, A.J. and K.W. Knox. 1975. Lipoteichoic acids: a new class of bacterial antigen. Science *187:* 1161–1167.

Widdel, F. 1980. Anaerober abbau von fettsauren und benzoesaure durch neu isolierte arten sulfate-reduzierender bakterien. Dissertation. George August University of Gottingen. Gottingen.

Widdel, F. and N. Pfennig. 1977. A new anaerobic, sporing, acetate-oxidizing, sulfate-reducing bacterium, *Desulfotomaculum* (emend.) *acetoxidans*. Arch. Microbiol. *112:* 119–122.

Widdel, F. and N. Pfennig. 1981. Sporulation and further nutritional characteristics of *Desulfotomaculum acetoxidans*. Arch. Microbiol. *129:* 401–402.

Widdowson, J.P., W.R. Maxted, D.L. Grand and A.M. Pinney. 1971. The relationship between M-antigen and opacity factor in group A streptococci. J. Gen. Microbiol. *65:* 69–80.

Wideman, P.A., V.E.L. Vargo, D. Citronbaum and S.M. Finegold. 1976. Evaluation of the sodium polyanethol sulfonate disk test for the identification of *Peptostreptococcus anaerobius*. J. Clin. Microbiol. *4:* 330–333.

Widra, A. 1963. Histochemical observations on *Actinomyces bovis* granules. Sabouraudia *2:* 264–267.

Wiegel, J. 1980. Formation of ethanol by bacteria. A pledge for the use of extreme thermophilic anaerobic bacteria in industrial ethanol fermentation processes. Experientia *36:* 1434–1446.

Wiegel, J. 1981. α-Isopropylmalate synthase as a marker for the leucine biosynthetic pathway in several clostridia and in *Bacteroides fragilis*. Arch. Microbiol. *130:* 385–390.

Wiegel, J. 1981. Distinction between the Gram reaction and the Gram type of bacteria. Int. J. Syst. Bacteriol. *31:* 88.

Wiegel, J. 1982. Ethanol from cellulose. Experientia *38:* 151–155.

Wiegel, J. 1982. *Clostridium thermoautotrophicum*: growth and sporulation in media containing C-1 compounds as substrate. Abstr. Annu. Meet. Am. Soc. Microbiol. *1982:* 112.

Wiegel, J., M. Braun and G. Gottschalk. 1981. *Clostridium thermoautotrophicum* species novum, a thermophile producing acetate from molecular hydrogen and carbon dioxide. Curr. Microbiol.*5:* 255–260.

Wiegel, J., M. Braun and G. Gottschalk. 1982. Validation List No. 9. Int. J. Syst. Bacteriol. *32:* 384–385.

Wiegel, J., L.H. Carreira, C. Mothershed and J. Puls. 1983. Production of ethanol from biopolymers by anaerobic thermophilic and extreme thermophilic bacteria. II. *Thermoanaerobacter ethanolicus* W200 and its mutants in batch culture and resting cell experiments. Biotech. Bioeng. Symp. *13:* 193–205.

Wiegel, J. and L.G. Ljungdahl. 1979. Ethanol as fermentation product of extreme thermophilic anaerobic bacteria. *In* Dellweg (Editor), Technische Mikrobiologie. Verlag Versuchs- und Lehranstalt für Spiritusfabrikation und Fermentationstechnologie im Institut für Gärungsgewerbe und Biotechnologie, Berlin, pp. 117–127.

Wiegel, J. and L.G. Ljungdahl. 1981. *Thermoanaerobacter ethanolicus* gen. nov., a new, extreme thermophilic, anaerobic bacterium. Arch. Microbiol. *128:* 343–348.

Wiegel, J. and L.G. Ljungdahl. 1982. *In* Validation of the publication of new names and new combinations previously effectively published outside the IJSB. Int. J. Syst. Bacteriol. *32:* 384–385.

Wiegel, J., L.G. Ljungdahl and J.R. Rawson. 1979. Isolation from soil and properties of the extreme thermophile *Clostridium thermohydrosulfuricum*. J. Bacteriol. *139:* 800–810.

Wiegel, J. and J. Puls. 1983. Production of ethanol from hemicelluloses of hardwoods and annual plants. *In* Strub, Chartier and Schleser (Editors), Energy from Biomass, 2nd E.C. Conference. Appl. Science Publishers, London, New York, pp. 863–867.

Wiegel, J. and L. Quandt. 1982. Determination of the Gram type using the reaction between polymyxin B and lipopolysccharides of outer cell wall of whole bacteria. J. Gen. Microbiol. *128:* 2261–2270.

Wiegel, J. and H.G. Schlegel. 1977. Leucine biosynthesis: effect of branched-chain amino acids and threonine on α-isopropylmalate synthase activity from aerobic and anaerobic microorganisms. Biochem. Syst. Ecol. *5:* 169–176.

Wiegel, J., S.-S. Yang and L.G. Ljungdahl. 1979. Properties offerredoxins and a rubredoxin from extreme thermophilic anaerobes. Abstr. Annu. Meet. Am. Soc. Microbiol. K73.

Wieringa, K.T. 1940. The formation of acetic acid from carbon dioxide and hydrogen by anaerobic spore-forming bacteria. Antonie van Leeuwenhoek J. Microbiol. Serol. *6:* 251–262.

Wieringa, K.T. 1941. *Bacillus agar-exedens*, a new species decomposing agar. Antonie van Leeuwenhoek. J. Microbiol. Serol. *7:* 121–127.

Wikström, M.B., G. Dahlén and A. Linde. 1983. Fibrinogenolytic and fibrinolytic activity in oral microorganisms. J. Microbiol. *17:* 759–767.

Wilcke, B.W., Jr., T.F. Midura and S.S. Arnon. 1980. Quantitative evidence of intestinal colonization by *Clostridium botulinum* in four cases of infant botulism. J. Infect. Dis. *141:* 419–423.

Wiley, B.B. 1972. Capsules and pseudocapsules of *Staphylococcus aureus*. *In* Cohen (Editor), The Staphylococci, Wiley-Interscience (John Wiley and Sons, Inc.), New York, London, Sydney, Toronto, pp. 41–63.

Wiley, B.B., L.A. Glasgow and M.S. Rogolsky. 1976. Studies on staphylococcal scalded skin syndrome (SSS): isolation and purification of toxin and development of a radioimmunobinding assay for antibodies to exfoliative toxin (ET). *In* Jeljasewicz (Editor), Staphylococci and Staphylococcal Diseases, Gustav Fischer Verlag, Stuttgart, New York, pp. 499–516.

Wilkins, T.D., R.S. Fulghum and J.H. Wilkins. 1974. *Eubacterium plexicaudatum* sp. nov., an anaerobic bacterium with a subpolar tuft of flagella, isolated from a mouse cecum. Int. J. Syst. Bacteriol. *24:* 408–411.

Wilkins, T.D., W.E.C. Moore, S.E.H. West and L.V. Holdeman. 1975. *Peptococcus niger* (Hall) Kluyver and van Niel 1936: emendation of description and designation of neotype strain. Int. J. Syst. Bacteriol. *25:* 47–49.

Wilkins, T.D. and T. Thiel. 1973. A modified broth-disk method for testing the antibiotic susceptibility of anaerobic bacteria. Antimicrob. Agents Chemother. *3:* 350–356.

Wilkins, T.D. and T. Thiel. 1973. Resistance of some species of *Clostridium* to clindamycin. Antimicrob. Agents Chemother. *3:* 136–137.

Wilkins, T.D. and S.E.H. West. 1976. Medium-dependent inhibition of *Peptostreptococcus anaerobius* by sodium polyanetholsulfonate in blood culture media. J. Clin. Microbiol. *3:* 393–396.

Wilkinson, B.J. and D. Jones. 1975. Some serological studies on *Listeria* and possibly related bacteria. *In* Woodbine (Editor), Problems of Listeriosis. Leicester University Press, Leicester, U.K. pp. 251–261.

Wilkinson, B.J. and D. Jones. 1977. A numerical taxonomic survey of *Listeria* and related bacteria. J. Gen. Microbiol. *98:* 399–421.

Wilkinson, B.J., S. Maxwell and S.M. Schaus. 1980. Classification and characteristics of coagulase-negative, methicillin-resistant staphylococci. J. Clin.

Microbiol. *12:* 161–166.

Wilkinson, B.J., S.P. Sisson, Y. Kim and P.K. Peterson. 1981. Chemical and biological studies of encapsulated *Staphylococcus aureus*. *In* Jeljaszewicz (Editor), Staphylococci and Staphylococcal Infections, Gustav Fischer Verlag, Stuttgart, New York, pp. 469–473.

Wilkinson, H.W. 1972. Comparison of streptococcal R antigens. Appl. Microbiol. *24:* 669–670.

Wilkinson, H.W. and R.G. Eagon. 1971. Type-specific antigens of group B type Ic streptococci. Infect. Immun. *4:* 596–604.

Willers, J.M. and G.H. Alderkamp. 1967. Loss of type antigen in a type III *Streptococcus* and identification of the determinant disaccharide of the remaining antigen. J. Gen. Microbiol. *49:* 41–51.

Willers, J.M., P.A. Deddish and H.D. Slade. 1968. Transformation of type polysaccharide antigen synthesis and hemolysin synthesis in streptococci. J. Bacteriol. *96:* 1225–1230.

Willers, J.M. and M.F. Michel. 1966. Immunochemistry of the type antigen of *Streptococcus faecalis*. J. Gen. Microbiol. *43:* 375–382.

Willers, J.M.N., H. Ottens and M.F. Michel. 1964. Immunochemical relationship between *Streptococcus* MG, FIII and *Streptococcus salivarius*. J. Gen. Microbiol. *37:* 425–431.

Williams, R.E.O. 1956. *Streptococcus salivarius* (vel. *hominis*) and its relations to Lancefield's group K. J. Pathol. Bacteriol. *72:* 15–25.

Williams, R.E.O. 1963. Healthy carriage of *Staphylococcus aureus*: its prevalence and importance. Bacteriol. Rev. *27:* 56–71.

Williams, R.E.O., A. Hirch and S.T. Cowan. 1953. *Aerococcus*, a new bacterial genus. J. Gen. Microbiol. *8:* 475–480.

Williams, S.T. and T. Cross. 1971. Isolation, purification, cultivation and preservation of actinomycetes. Methods Microbiol. *4:* 295–334.

Williams, S.T. and F.L. Davies. 1965. Use of antibiotics for selective isolation and enumeration of actinomycetes in soil. J. Gen. Microbiol. *38:* 251–261.

Williams, S.T. and F.L. Davies. 1967. Use of a scanning electron microscope for the examination of actinomycetes. J. Gen. Microbiol. *48:* 171–177.

Williams, S.T., M. Goodfellow, G. Alderson, E.M.H. Wellington, P.H.A. Sneath and M.J. Sackin. 1983. Numerical classification of *Streptomyces* and related genera. J. Gen. Microbiol. *129:* 1743–1813.

Williams, S.T., G.P. Sharples and R.M. Bradshaw. 1973. The fine structure of the Actinomycetales. Soc. Appl. Bacteriol. Symp. Ser. *2:* 113–130.

Williams, S.T., G.P. Sharples, J.A. Serrano, A.A. Serrano and J. Lacey. 1976. The micromorphology and fine structure of nocardioform organisms. *In* Goodfellow, Brownell, and Serrano (Editors), The Biology of the Nocardiae, Academic Press, London, pp. 102–140.

Williams, S.T., E.M.H. Wellington, M. Goodfellow, G. Alderson, M. Sackin and P.H.A. Sneath. 1981. The genus *Streptomyces* - a taxonomic enigma. Zentralbl. Bakteriol. Parasitenkd. Infektionskr. Hyg. I Abt. Suppl. *11:* 47–57.

Williams, S.T., E.M.H. Wellington and L.S. Tipler. 1980. The taxonomic implications of the reactions of representative *Nocardia* strains to actinophage. J. Gen. Microbiol. *119:* 173–178.

Williamson, P. and A.M. Kligman. 1965. A new method for the quantitative investigation of cutaneous bacteria. J. Invest. Dermatol. *45:* 498–503.

Willis, A.T. 1969. Clostridia of wound infection. Butterworth and Co., London.

Wilson, A.C., S.S. Carlson and T.J. White. 1977. Biochemical evolution. Annu. Rev. Biochem. *46:* 573–639.

Wilson, C.D. and G.F.H. Salt. 1978. Streptococci in animal disease. *In* Skinner and Quesnel (Editors), Streptococci. Society for Applied Bacteriology Symposium Series No. 7, Academic Press, London, New York and San Francisco, pp. 143–156.

Wilson, G.S. and A.A. Miles. 1975. Topley and Wilson's Principles of Bacteriology and Immunity. Sixth edition. Vol. I. *Erysipelothrix* Edward Arnold, London, pp. 554–558.

Wilson, K.H., M.J. Kennedy and F.R. Fekety. 1982. Use of sodium taurocholate to enhance spore recovery on a medium selective for *Clostridium difficile*. J. Clin. Microbiol. *15:* 443–446.

Wilson, P.D. 1977. Joint replacement. South. Med. J. *70:* 55s-60s.

Wilson, P.D., H.C. Amstutz, A. Czerniecki, E.A. Salvati and D.G. Mendes. 1972. Total hip replacement with fixation by acrylic cement: a preliminary study of 100 consecutive McKee-Farrar prosthetic replacements. J. Bone Joint Surg. *54A:* 207–236.

Wilson, P.D., E.A. Salvati, P. Aglietti and L.J. Kutner. 1973. The problem of infection in endoprosthetic surgery of the hip joint. Clin. Orthop. *96:* 213–221.

Winford, T.E. and S. Haberman. 1966. Isolation of aerobic Gram-positive filamentous rods from diseased gingivae. J. Dent. Res. *45:* 1159–1167.

Winogradsky, S. 1895. Recherches sur l'assimilation de l'azote libre de l'atmosphère par les microbes. Arch. Sci. Biol. St. Pétersb. *3:* 297–352.

Winogradsky, S. 1949. Microbiologie du Sol. Masson et Cie: Paris.

Winslow, C.E.A., J. Broadhurst, R.E. Buchanan, C. Krumwiede, L.A. Rogers and G.H. Smith. 1920. The families and genera of bacteria. J. Bacteriol. *5:* 191–229.

Winslow, C.-E.A. and A. Winslow. 1908. The systematic relationships of the *Coccaceae*. John Wiley and Sons, New York, pp. 1–300.

Winston, M.E. 1951. Actinomycosis of the Spine. Lancet *260:* 945.

Winter, J. and V.D. Bokkenheuser. 1978. 21-Dehydroxylation of corticoids by anaerobic bacteria isolated from human fecal flora. J. Steroid Biochem. *9:* 379–384.

Winter, J., V.D. Bokkenheuser and L. Ponticorvo. 1979. Bacterial metabolism of corticoids with particular reference to the 21-dehydroxylation. J. Biol. Chem. *254:* 2626–2629.

Winter, J., A. Cerone-McLernon, S. O'Rourke, L. Ponticorvo and V.D. Bokkenheuser. 1982. Formation of 20β-dehydrosteroids by anaerobic bacteria. J. Steroid Biochem. *17:* 661–667.

Wirtz, R. 1908. Eine einfache Art der Sporenfärbung. Zentralbl. Bakteriol Parasitenkd. Infektionskr. Hyg. Abt. I Orig. *46:* 727–728.

Witwer, M.W., M.F. Farmer, J.S. Wand and L.S. Solomon. 1977. Extensive actinomycosis associated with an intrauterine contraceptive device. Am. J. Obstet. Gynecol. *128:* 913–914.

Witz, D.F., R.W. Detroy and P.W. Wilson. 1967. Nitrogen fixation by growing cells and cell-free extracts of the *Bacillaceae.* Arch. Mikrobiol. *55:* 369–387.

Woese, C.R. 1981. Archaebacteria. Sci. Amer. *244:* 98–122.

Woese, C.R. and G.E. Fox. 1977. The concept of cellular evolution. J. Mol. Evol. *10:* 1–6.

Woese, C.R., G.E. Fox, L. Zablen, T. Uchida, L. Bonen, K. Pechman, B.J. Lewis and D. Stahl. 1975. Conservation of primary structure in 16s ribosomal RNA. Nature (London) *254:* 83–86.

Woese, C.R., L.J. Magrum and G.E. Fox. 1978. Archaebacteria. J. Mol. Evol. *11:* 245–252.

Woese, C.R., J. Maniloff and L.B. Zablen. 1980. Phylogenetic analysis of the Mycoplasmas. Proc. Nat. Acad. Sci. (USA) *77:* 494–498.

Woese, C.R., M. Sogin, D. Stahl, B.J. Lewis and L. Bonen. 1976. A comparison of the 16S ribosomal RNAs from mesophilic and thermophilic bacilli: Some modifications in the Sanger method for RNA sequencing. J. Mol. Evol. *7:* 197–213.

Wolberg, G., G.S. Duncan, C. Adlam and J.K. Whisnant. 1977. Antibody to *Corynebacterium parvum* in normal human and animal sera. Infect. Immun. *15:* 1004–1007.

Wolf, J. and R.J. Sharp. 1981. Taxonomic and related aspects of thermophiles within the genus *Bacillus. In* Berkeley and Goodfellow (Editors), The Aerobic Endospore-forming Bacteria. Academic Press, London, pp. 251–296.

Wolf, K. and C.E. Dunbar. 1959. Test of 34 therapeutic agents for control of kidney disease in trout. Trans. Am. Fish. Soc. *88:* 117–124.

Wolff, A. and H. Ionesco. 1975. Purification et caractérisation de la bactériocine N₅ de *Clostridium perfringens* BP6K-N₅ type A. Ann. Microbiol. Inst. Pasteur *126B:* 343–356.

Wolff, L. and W.F. Liljemark. 1978. Observation of beta hemolysis among three strains of *Streptococcus mutans.* Infect. Immun. *19:* 745–748.

Wolff, L.F. and J.L. Duncan. 1974. Studies on a bactericidal substance produced by group A streptococci. J. Gen. Microbiol. *81:* 413–424.

Wolff, M. 1910. Ueber eine neve Krankheit der Raupe von *Bupalus piniarius* L. Mitt. K. Wilh. Inst. Landw. Bromberg *3:* 69–92.

Wolff, M. and J. Israel. 1891. Über Reincultur des *Actinomyces* und seine Übertragbarkeit auf Thiere. Arch. Pathol. Anat. Physiol. Klin. Med. *126:* 11–59.

Wolin, E.F., J.B. Evans and C.F. Niven, Jr. 1957. The microbiology of fresh and irradiated beef. Food Res. *22:* 682–686.

Wolin, H.L. and H.B. Naylor. 1957. Basic nutritional requirements of *Micrococcus lysodeikticus* J. Bacteriol. *74:* 163–167.

Wolin, M.J. 1981. Fermentation in the rumen and human large intestine. Science *213:* 1463–1468.

Wolin, M.J., J.B. Evans and C.F. Niven, Jr. 1952. The oxidation of butyric acid by *Streptococcus mitis.* J. Bacteriol. *64:* 531–535.

Wolin, M.J., G.B. Manning and W.O. Nelson. 1959. Ammonium salts as a sole source of nitrogen for the growth of *Streptococcus bovis.* J. Bacteriol. *78:* 147–149.

Wolin, M.J. and T.L. Miller. 1980. Molybdate and sulfide inhibit H2 and increase formate production from glucose by *Ruminococcus albus.* Arch. Microbiol. *124:* 137–142.

Wolinsky, E. 1979a. Nontuberculous mycobacteria and associated diseases. Am. Rev. Resp. Dis. *119:* 107–159.

Wolinsky, E. 1979b. Emendation of proposed additional serotypes of mycobacteria determined by agglutination. Int. J. Syst. Bacteriol. *29:* 59.

Wolinsky, E. and R.K. Rynearson. 1968. Mycobacteria in soil and their relation to disease-associated strains. Am. Rev. Resp. Dis. *97:* 1032–1037.

Wolinsky, E. and R.K. Rynearson. 1973. Proposed numbering scheme for mycobacterial serotypes by agglutination. Int. J. Syst. Bacteriol. *23:* 182–183.

Wood, E.J. 1946. The isolation of *Sarcina ureae* (Beijerinck) Lönis from sea water. J. Bacteriol. *51:* 287–289.

Wood, E.M. and W.T. Yasutake. 1956. Histopathology of kidney disease in fish. Am. J. Pathol. *32:* 845–857.

Wood, H.G., A.A. Andersen and C.H. Werkman. 1938. Nutrition of the propionic acid bacteria. J. Bacteriol. *36:* 201–213.

Wood, H.G., H.L. Drake and S.I. Hu. 1982. Studies with *Clostridium thermoaceticum* and the resolution of the pathway used by acetogenic bacteria that grow on carbon monoxide or carbon dioxide and hydrogen. *In* Snell (Editor), Amino Acid Fermentations and Nucleic Acids, A Symposium. Annual Reviews Monograph, Annual Reviews Inc. Palo Alto, CA, pp. 29–56.

Wood, R.L. 1965. A selective liquid medium utilizing antibiotics for isolation of *Erysipelothrix insidiosa.* Am. J. Vet. Res. *26:* 1303–1308.

Wood, R.L. 1970. *Erysipelothrix. In* Blair, Lennette and Truant (Editors), Manual of Clinical Microbiology. American Society for Microbiology. Bethesda, Md.

Wood, R.L. 1974. Isolation of pathogenic *Erysipelothrix rhusiopathiae* from feces of apparently healthy swine. Am. J. Vet. Res. *35:* 41–43.

Wood, R.L., D.R. Haubrich and R. Harrington. 1978. Isolation of previously unreported serotypes of *Erysipelothrix rhusiopathiae* from swine. Am. J. Vet. Res. *39:* 1958–1961.

Wood, R.L. and R. Packer. 1972. Isolation of *Erysipelothrix rhusiopathiae* from soil and manure of swine-raising premises. Am. J. Vet. Res. *33:* 1611–1620.

Woodbine, M. 1950. *Erysipelothrix rhusiopathiae.* Bacteriology and chemotherapy. Bacteriol. Rev. *14:* 161–178.

Woodley, C.L., J.N. Baldwin and J. Greenberg. 1981. Nitrosoguanidine sequential mutagenesis mapping of *Mycobacterium tuberculosis* genes. J. Bacteriol. *147:* 176–180.

Woolcock, J.B., A.M.T. Farmer and M.D. Mutimer. 1979. Selective medium for *Corynebacterium equi* isolation. J. Clin. Microbiol. *9:* 640–642.

Word, N.S., A.A. Yousten and L. Howard. 1983. Regularly structured and nonregularly structured surface layers of *Bacillus sphaericus.* FEMS Microbiol. Lett. *17:* 277–282.

Work, E. 1970. The distribution of diamino acids in cell walls and its significance in bacterial taxonomy. Int. J. Syst. Bacteriol. *20:* 425–433.

Work, E. and H. Griffiths. 1968. Morphology and chemistry of cell walls of *Micrococcus radiodurans.* J. Bacteriol. *95:* 641–657.

Wort, A.J. 1975. Observations on group F streptococci from human sources. J. Med. Microbiol. *8:* 455–457.

Wort, A.J. and R.L. Ozere. 1976. Characteristics of a strain of *Clostridium carnis* causing septicaemia in a young infant. J. Clin. Pathol. *29:* 1011–1013.

Worthington, R.W., H.H. Kleeberg. 1964. Isolation of *Mycobacterium kansasii* from bovines. J. S. Afr. Vet. Med. *35:* 29–33.

Worthington, R.W. and H.H. Kleeberg. 1967. Demonstration of species-specific fractions in mycobacterial purified protein derivative (PPD) sensitins. Tubercle *48:* 211–218.

Wozny, M.A., M.P. Bryant, L.V. Holdeman and W.E.C. Moore. 1977. Urease assay and urease-producing species of anaerobes in the bovine rumen and human feces. Appl. Environ. Microbiol. *33:* 1097–1104.

Wren, M.W.D., C.P. Eldon and G.H. Dakin. 1977. Novobiocin and the differentiation of peptococci and peptostreptococci. J. Clin. Pathol. *30:* 620–622.

Yale, C.E. 1975. Etiology of pneumatosis cystoides intestinalis. Surg. Clin. N. Am. *55:* 1297–1302.

Yale, C.E. and E. Balish. 1976. The importance of clostridia in experimental intestinal strangulation. Gastroenterology *71:* 793–796.

Yamada, K. and K. Komagata. 1970. Taxonomic studies on coryneform bacteria III. DNA base composition of coryneform bacteria. J. Gen. Appl. Microbiol. *16:* 215–224.

Yamada, K. and K. Komagata. 1972a. Taxonomic studies on coryneform bacteria. IV. Morphological, cultural, biochemical and physiological characteristics. J. Gen. Appl. Microbiol. *18:* 399–416.

Yamada, K. and K. Komagata. 1972b. Taxonomic studies on coryneform bacteria. V. Classification of coryneform bacteria. J. Gen. Appl. Microbiol. *18:* 417–431.

Yamada, M., A. Hirose and M. Matsuhashi. 1975. Association of lack of cell wall teichuronic acid with formation of cell packets of *Micrococcus lysodeikticus* (*luteus*) mutants. J. Bacteriol. *123:* 678–686.

Yamada, Y., G. Inouye, Y. Tahara and K. Kondo. 1976. The menaquinone system in the classification of coryneform and nocardioform bacteria and related organisms. J. Gen. Appl. Microbiol. *22:* 203–214.

Yamada, Y., G. Inouye, Y. Tahara and K. Kondo. 1976. The menaquinone system in the classification of aerobic gram-positive cocci in the genera *Micrococcus, Staphylococcus, Planococcus* and *Sporosarcina.* J. Gen. Appl. Microbiol. *22:* 227–236.

Yamada, Y., T. Ishikawa, Y. Tahara and K. Kondo. 1977. The menaquinone system in the classification of the genus *Nocardia.* J. Gen. Appl. Microbiol. *23:* 207–216.

Yamada, Y., H. Takinami, Y. Tahara and K. Kondo. 1977. The menaquinone system in the classification of radiation-resistant micrococci. J. Gen. Appl. Microbiol. *23:* 105–108.

Yamaguchi, T. 1965. Comparison of the cell wall composition of morphologically distinct actinomycetes. J. Bacteriol. *89:* 444–453.

Yamaguchi, T. 1967. Similarity in DNA of various morphologically distinct actinomycetes. J. Gen. Appl. Microbiol. *13:* 63–71.

Yamaguchi, Y. and M. Kurokawa. 1972. Nutritional requirements of *Staphylococcus aureus.* Gumma Rep. Med. Sci. *3:* 367–371.

Yamakawa, T. and N. Ueta. 1964. Gas chromatographic studies of microbial components. I. Carbohydrate and fatty acid constitution of *Neisseria.* Jpn. J. Exp. Med. *34:* 361–374.

Yamamoto, K., I. Ohishi and G. Sakaguchi. 1979. Fluid accumulation in mouse ligated intestine inoculated with *Clostridium perfringens* enterotoxin. Appl. Environ. Microbiol. *37:* 181–186.

Yanagawa, R. 1975. A numerical taxonomic study of the strains of the three types of *Corynebacterium renale.* Can. J. Microbiol. *21:* 824–827.

Yanagawa, R., H. Basri and K. Otsuki. 1967. Three types of *Corynebacterium renale* classified by precipitin reactions in gels. Jpn. J. Vet. Res. *15:* 111–119.

Yanagawa, R. and E. Honda. 1976. Presence of pili in species of human and animal parasites and pathogens of the genus *Corynebacterium.* Infect. Immun. *12:* 1293–1295.

Yanagawa, R. and E. Honda. 1978. *Corynebacterium pilosum* and *Corynebacterium*

ۼ

ۼ

牛я.

cystitidis, two new species from cows. Int. J. Syst. Bacteriol. *28:* 209–216.

Yang, S.-S., L.G. Ljungdahl and J. Le Gall. 1977. A four-iron, four-sulfide ferredoxin with high thermostability from *Clostridium thermoaceticum*. J. Bacteriol. *130:* 1084–1090.

Yano, I., K. Kageyama, Y. Ohno, M. Masui, E. Kusunose, M. Kusunose and N. Akimori. 1978. Separation and analysis of molecular species of mycolic acids in *Nocardia* and related taxa by gas chromatography mass spectrometry. Biomed. Mass. Spectrom. *5:* 14–24.

Yazawa, K., K. Imai and Z. Tamura. 1978. Oligosaccharides and polysaccharides specifically utilizable by bifidobacteria. Chem. Pharm. Bull (Tokyo) *26:* 3306–3311.

Yoch, D.C. and R.P. Carithers. 1979. Bacterial iron-sulfur proteins. Microbiol. Rev. *43:* 384–421.

Yocum, R.C., J. McArthur, B.G. Petty, A.M. Diehl and T.R. Moench. 1982. Septic arthritis caused by *Propionibacterium acnes*. J. Am. Med. Assoc. *248:* 1740–1741.

Yokagawa, K., S. Kawata, S. Nishimura, Y. Ikeda and Y. Yoshimura. 1974. Mutanolysin, bacteriolytic agent for cariogenic streptococci: partial purification and properties. Antimicrob. Agents Chemother. *6:* 156–165.

Yokagawa, K., S. Kawata, T. Takemura and Y. Yoshimura. 1975. Purification and properties of lytic enzymes from *Streptomyces globisporus* 1829. Agric. Biol. Chem. *39:* 1533–1543.

Yokokura, T., S. Kodaira, H. Ishiwa and T. Sakurai. 1974. Lysogeny in lactobacilli. J. Gen. Microbiol. *84:* 277–284.

Yokoyama, M.T. and J.R. Carlson. 1981. Production of skatole and para-cresol by a rumen *Lactobacillus* sp. Appl. Environ. Microbiol. *41:* 71–76.

Yoshimura, K., O. Yamamoto, T. Seki and Y. Oshima. 1983. Distribution of heterogenous and homologous plasmids in *Bacillus* spp. Appl. Environ. Microbiol. *45:* 1733–1740.

Yoshinaka, T., K. Yano and H. Yamaguchi. 1973. Isolation of highly radioresistant bacterium. *Arthrobacter radiotolerans* nov. sp. Agr. Biol. Chem. *37:* 2269–2275.

Yoshioka, M., S. Yoshioka, Z. Tamura and K. Ohta. 1968. Growth responses of *Bifidobacterium bifidum* to coenzyme A, its precursors and carrot extract. Jpn. J. Microbiol. *12:* 335–342.

Young, C.L. and G.B. Chapman. 1978. Ultrastructural aspects of the causative agent and renal histopathology of bacterial kidney disease in brook trout (*Salvelinus fontinalis*). J. Fish Res. Board Can. *35:* 1234–1248.

Young, D.B. 1980. Identification of *Mycobacterium leprae*: use of wall-bound mycolic acids. J. Gen. Microbiol. *121:* 249–253.

Young, F.E., R.E. Yasbin and M.A. Courtney. 1983. Genetic engineering of bacilli. Phil. Trans. Roy. Soc. Ser. B *300:* 241–248.

Yribarren, M., E. Vilkas and J. Rozanis. 1974. Galactosyl diglyceride from *Actinomyces viscosus*. Chem. Phys. Lipids *12:* 173–175.

Yu, I. and R.E. Hungate. 1979. The extracellular cellulases of *Ruminococcus albus*. Ann. Rech. Vet. *10:* 251–254.

Yu, L. and J.N. Baldwin. 1971. Intraspecific transduction in *Staphylococcus epidermidis* and interspecific transduction between *Staphylococcus aureus* and *Staphylococcus epidermidis*. Can. J. Microbiol. *17:* 767–773.

Zabriskie, J.B. 1964a. The role of lysogeny in the production of erythrogenic toxin. *In* Uhr (Editor), The Streptococcus, Rheumatic Fever and Glomerulonephritis, Williams and Wilkins Co., Baltimore, pp. 53–70.

Zabriskie, J.B. 1964b. The role of temperate bacteriophage in the production of erythrogenic toxin by group A streptococci. J. Exp. Med. *119:* 761–779.

Zagallo, A.C. and C.H. Wang. 1967. Comparative carbohydrate catabolism in corynebacteria. J. Gen. Microbiol. *47:* 347–357.

Zajdel, M., Z. Wegrzynowicz, J. Sawecka, J. Jeljaszewicz and G. Pulverer. 1976.

Mechanism of action of staphylocoagulase. *In* Jeljaszewicz (Editor), Staphylococci and Staphylococcal Diseases, Gustav Fischer Verlag, Stuttgart, New York, pp. 549–575.

Zajic, J.E., J. DeLey, and M.P. Starr. 1956. Oxidative metabolism of *Corynebacterium insidiosum*. Bacteriol. Proc. 116 (Abstract).

Zamenhof, S. 1960. Effects of heating dry bacteria and spores on their phenotype and genotype. Proc. Nat. Acad. Sci. *46:* 101–105.

Zamenhof, S. and D. Rosenbaum-Oliver. 1968. Study of mutability of stored spores of *Bacillus subtilis*. Nature *220:* 818–819.

Zani, G., B. Biavati, F. Crociani and D. Matteuzzi. 1974. Bifidobacteria from the feces of piglets. J. Appl. Bacteriol. *37:* 537–547.

Zani, G. and A. Severi. 1982. Cellular ultrastructure and morphology in *Bifidobacterium bifidum*. Microbiologica *5:* 255–267.

Zeikus, J.G. 1980. Chemical and fuel production by anaerobic bacteria. Annu. Rev. Microbiol. *34:* 423–464.

Zeikus, J.G. 1983. Metabolism of 1-carbon compounds by chemotrophic anaerobes. Adv. Microbiol. Physiol. *24:* 222–230.

Zeikus, J.G., P.W. Hegge and M.A. Anderson. 1979. *Thermoanaerobium brockii*, gen. nov., and sp. nov., a new chemoorganotrophic, caldoactive, anaerobic bacterium. Arch. Microbiol. *122:* 41–48.

Zeikus, J.G., L.H. Lynd, T.E. Thompson, J.A. Krzycki, P.J. Weimer and P.W. Hegge. 1980. Isolation and characterization of a new, methylotrophic, acidogenic anaerobe, the Marburg strain. Curr. Microbiol. *3:* 381–386.

Zhukova, R.A., V.A. Tsyganov and V.M. Morozov. 1968. A new species of *Micropolyspora* - *Micropolyspora angiospora* (sp. nov.). Mikrobiologiya *97:* 724–728.

Zierdt, C.H. 1982. Long-term *Staphylococcus aureus* carriers state in hospital patients. J. Clin. Microbiol. *16:* 517–520.

Zierdt, C.H., C. Webster and W.S. Rude. 1968. Study of the anaerobic corynebacteria. Int. J. Syst. Bacteriol. *18:* 33–47.

Zighelboim, S. and A. Tomasz. 1981. Multiple antibiotic resistance in South African strains of *Streptococcus pneumoniae*: mechanism of resistance to beta-lactam antibiotics. Rev. Infect. Dis. *3:* 267–276.

Zimmerman, R.J. 1976. Comparative zone electrophoresis of catalase of *Staphylococcus* species isolated from mammalian skin. Can. J. Microbiol. *22:* 1691–1698.

Zimmerman, R.J. and W.E. Kloos. 1976. Comparative zone electrophoresis of esterases of *Staphylococcus* species isolated from mammalian skin. Can. J. Microbiol. *22:* 771–779.

Zinnemann, K. and G.C. Turner. 1963. The taxonomic position of "*Haemophilus vaginalis*" (*Corynebacterium vaginale*). J. Pathol. Bact. *85:* 213–219.

Zinner, D.D., J.M. Jablon, A.P. Aran and M.S. Saslaw. 1965. Experimental caries induced in animals by streptococci of human origin. Proc. Soc. Exp. Biol. Med. *118:* 766–770.

ZoBell, C.E. and C. Upham. 1944. A list of marine bacteria including descriptions of sixty new species. Bull. of the Scripps Inst. Oceanogr. Univ. Calif. Technical Series *5:* 239–292.

Zopf, W. 1883. Die Spaltpilze. Edward Trewendt, Breslau.

Zopf, W. 1885. Die Spaltpilze, third ed. Edward Trewendt. Breslau.

Zopf, W. 1889. Über das Mikrochemische Verhalten von Fettfarbstoff-haltigen. Organen. Z. Wiss. Mikrosk. *6:* 172–177.

Zuckerkandl, E. and L. Pauling. 1965. Molecules as documents of evolutionary history. J. Theoret. Biol. *8:* 357–366.

Zviagintzev, D.G., I.V. Aseyeva, I.P. Babieva and T.G. Mirchink. 1980. The methods of soil microbiology and biochemistry. Lomonosov State University, Moscow (In Russian).

Index of Scientific Names of Bacteria

Key to the fonts and symbols used in this index:

Nomenclature
Lower case, Roman:

Genera, species, and subspecies of bacteria. Every bacterial name mentioned in the *Manual* is listed in the Index. Specific epithets are listed individually and also under the genus.*

CAPITALS, ROMAN:

Names of taxa higher than genus (tribes, families, orders, classes, divisions, kingdoms).

Pagination
Roman:

Pages on which taxa are mentioned.

Boldface:

Indicates page on which the description of a taxon is given.†

---

* Infrasubspecific names, such as serovars, biovars, and pathovars, are not listed in the Index.
† A description may not necessarily be given in the *Manual* for a taxon that is considered as *incertae sedis* or that is listed in an addendum or note added in proof; however, the page on which the complete citation of such a taxon is given is indicated in boldface type.

Index of Scientific Names of Bacteria